CAMBRIDGE LIBRARY COLLECTION

Books of enduring scholarly value

Botany and Horticulture

Until the nineteenth century, the investigation of natural phenomena, plants and animals was considered either the preserve of elite scholars or a pastime for the leisured upper classes. As increasing academic rigour and systematisation was brought to the study of 'natural history', its subdisciplines were adopted into university curricula, and learned societies (such as the Royal Horticultural Society, founded in 1804) were established to support research in these areas. A related development was strong enthusiasm for exotic garden plants, which resulted in plant collecting expeditions to every corner of the globe, sometimes with tragic consequences. This series includes accounts of some of those expeditions, detailed reference works on the flora of different regions, and practical advice for amateur and professional gardeners.

The Abridgement of the Gardener's Dictionary

Trained by his father, a market gardener, Philip Miller (1691–1771) rose to become Britain's most eminent horticulturalist in the eighteenth century. Following a period as a nurseryman in Southwark, he was appointed the head gardener of the Chelsea Physic Garden by the Society of Apothecaries in 1722, upon the recommendation of Sir Hans Sloane. Under Miller's supervision, the diversity of plants at Chelsea outstripped that of all other European botanic gardens. His talent was equally reflected in his writings. Miller's most famous work, *The Gardener's Dictionary*, ran to eight editions during his lifetime, as did his celebrated abridgement, reissued here in its 1771 sixth edition. Ranging widely in coverage from agriculture to winemaking, as well as incorporating traditional gardening topics, the work reflects the progress of contemporary plant science and the breadth of knowledge acquired by one of its foremost practitioners.

The Abridgement of the Gardener's Dictionary

Containing the Best and Newest Methods of Cultivating and Improving the Kitchen, Fruit, Flower Garden, and Nursery

Philip Miller

CAMBRIDGE
UNIVERSITY PRESS

University Printing House, Cambridge, CB2 8BS, United Kingdom

Published in the United States of America by Cambridge University Press, New York

Cambridge University Press is part of the University of Cambridge.
It furthers the University's mission by disseminating knowledge in the pursuit of education, learning and research at the highest international levels of excellence.

www.cambridge.org
Information on this title: www.cambridge.org/9781108068512

This edition first published 1771
This digitally printed version 2014

ISBN 978-1-108-06851-2 Paperback

Selected books of related interest, also reissued in the
CAMBRIDGE LIBRARY COLLECTION

Amherst, Alicia: *A History of Gardening in England* (1895) [ISBN 9781108062084]

Anonymous: *The Book of Garden Management* (1871) [ISBN 9781108049399]

Blaikie, Thomas: *Diary of a Scotch Gardener at the French Court at the End of the Eighteenth Century* (1931) [ISBN 9781108055611]

Candolle, Alphonse de: *The Origin of Cultivated Plants* (1886) [ISBN 9781108038904]

Drewitt, Frederic Dawtrey: *The Romance of the Apothecaries' Garden at Chelsea* (1928) [ISBN 9781108015875]

Evelyn, John: *Sylva, Or, a Discourse of Forest Trees* (2 vols., fourth edition, 1908) [ISBN 9781108055284]

Farrer, Reginald John: *In a Yorkshire Garden* (1909) [ISBN 9781108037228]

Field, Henry: *Memoirs of the Botanic Garden at Chelsea* (1878) [ISBN 9781108037488]

Forsyth, William: *A Treatise on the Culture and Management of Fruit-Trees* (1802) [ISBN 9781108037471]

Haggard, H. Rider: *A Gardener's Year* (1905) [ISBN 9781108044455]

Hibberd, Shirley: *Rustic Adornments for Homes of Taste* (1856) [ISBN 9781108037174]

Hibberd, Shirley: *The Amateur's Flower Garden* (1871) [ISBN 9781108055345]

Hibberd, Shirley: *The Fern Garden* (1869) [ISBN 9781108037181]

Hibberd, Shirley: *The Rose Book* (1864) [ISBN 9781108045384]

Hogg, Robert: *The British Pomology* (1851) [ISBN 9781108039444]

Hogg, Robert: *The Fruit Manual* (1860) [ISBN 9781108039451]

Hooker, Joseph Dalton: *Kew Gardens* (1858) [ISBN 9781108065450]

Jackson, Benjamin Daydon: *Catalogue of Plants Cultivated in the Garden of John Gerard, in the Years 1596–1599* (1876) [ISBN 9781108037150]

Jekyll, Gertrude: *Home and Garden* (1900) [ISBN 9781108037204]

Jekyll, Gertrude: *Wood and Garden* (1899) [ISBN 9781108037198]

Johnson, George William: *A History of English Gardening, Chronological, Biographical, Literary, and Critical* (1829) [ISBN 9781108037136]

Knight, Thomas Andrew: *A Selection from the Physiological and Horticultural Papers Published in the Transactions of the Royal and Horticultural Societies* (1841) [ISBN 9781108037297]

Lindley, John: *The Theory of Horticulture* (1840) [ISBN 9781108037242]

Loudon, Jane: *Instructions in Gardening for Ladies* (1840) [ISBN 9781108055659]

Mollison, John: *The New Practical Window Gardener* (1877) [ISBN 9781108061704]

Paris, John Ayrton: *A Biographical Sketch of the Late William George Maton M.D.* (1838) [ISBN 9781108038157]

Paxton, Joseph, and Lindley, John: *Paxton's Flower Garden* (3 vols., 1850–3) [ISBN 9781108037280]

Repton, Humphry and Loudon, John Claudius: *The Landscape Gardening and Landscape Architecture of the Late Humphry Repton, Esq.* (1840) [ISBN 9781108066174]

Robinson, William: *The English Flower Garden* (1883) [ISBN 9781108037129]

Robinson, William: *The Subtropical Garden* (1871) [ISBN 9781108037112]

Robinson, William: *The Wild Garden* (1870) [ISBN 9781108037105]

Sedding, John D.: *Garden-Craft Old and New* (1891) [ISBN 9781108037143]

Veitch, James Herbert: *Hortus Veitchii* (1906) [ISBN 9781108037365]

Ward, Nathaniel: *On the Growth of Plants in Closely Glazed Cases* (1842) [ISBN 9781108061131]

For a complete list of titles in the Cambridge Library Collection please visit:
www.cambridge.org/features/CambridgeLibraryCollection/books.htm

S: Hale inv.^t et del. *J: Miller Sculp.*

What NATURE sparing gives, or half denies,
See! healthfull INDUSTRY at large supplies.

See! in BRITANNIA'S Lap profusely pours,
While heaven-born SCIENCE swells th'increasing Stores.

Ecce! ferunt Pueri Calathis Tibi Lilia plenis. VIRG.

THE

ABRIDGEMENT

OF THE

GARDENERS DICTIONARY:

CONTAINING

The beſt and neweſt Methods of CULTIVATING and IMPROVING

THE

KITCHEN, FRUIT, FLOWER GARDEN, and NURSERY;

As alſo for Performing the

Practical Parts of HUSBANDRY:

Together with

The MANAGEMENT of VINEYARDS,

AND THE

METHODS of MAKING WINE in ENGLAND.

In which likewiſe are included,

DIRECTIONS for PROPAGATING and IMPROVING,

From REAL PRACTICE and EXPERIENCE,

PASTURE LANDS and all Sorts of TIMBER TREES.

By PHILIP MILLER, F. R. S.

Gardener to the Worſhipful Company of APOTHECARIES, at their Botanic Garden at CHELSEA, and Member of the Botanic Academy at FLORENCE.

.... *Digna manet divini gloria ruris.* Virg. Georg. I. v. 168.

The SIXTH EDITION, Corrected and much Enlarged.

LONDON,

Printed for the AUTHOR;

And Sold by JOHN and FRANCIS RIVINGTON, at N° 62, in St. Paul's Church-yard; J. WHISTON, W. STRAHAN, J. HINTON, L. HAWES and W. CLARKE and R. COLLINS, W. JOHNSTON, B. WHITE, T. CASLON, S. CROWDER, T. LONGMAN, C. RIVINGTON, J. DODSLEY, T. CADELL, F. NEWBERY, T. BECKET, and T. DAVIES.

M.DCC.LXXI.

To the Moſt Noble

HUGH,

Duke and Earl of NORTHUMBERLAND,

EARL PERCY,

Baron WARKWORTH of Warkworth Caſtle,

Lord Lieutenant and Cuſtos Rotulorum of the Counties of MIDDLESEX and NORTHUMBERLAND,

Of the City and Liberty of WESTMINSTER,

And of the Town and County of NEWCASTLE upon TYNE,

VICE ADMIRAL of all AMERICA,

And of the County of NORTHUMBERLAND,

One of his MAJESTY's Moſt Honourable Privy Council,

Knight of the Moſt Noble Order of the GARTER,

And Fellow of the ROYAL SOCIETY.

MAY IT PLEASE YOUR GRACE,

YOUR Grace's kind Acceptance of the three former Editions of this Work, has emboldened me to lay this at Your Grace's Feet, as a public Acknowledgement of the many uſeful Obſervations and Inſtructions communicated to the Author for its great Improvement. If I have been ſo happy as to employ them in ſuch manner as to merit Your Grace's Approbation, I ſhall have leſs Reaſon to doubt that of the Public,

as

as the moſt ſkilful Perſons in this Branch of Science, pay the higheſt Regard to Your Grace's Judgement.

The ſeveral Improvements which Your Grace is ſo happily making upon Your various Eſtates, ſufficiently demonſtrate Your Grace's ſuperior Judgement, but more particularly in a Country almoſt deſtitute of Timber; where, if Your Grace continues planting ſo ardently as for ſome Years paſt, the whole Face of the Countries will be ſo much altered, and Your Grace's Eſtates thereby ſo much improved, as ſcarcely to be known.

That Your Grace may long live to enjoy theſe Improvements, by ſetting Examples to others, is the ſincere Wiſh of

Your GRACES

Moſt obedient humble Servant,

CHELSEA,
DEC. 15, 1770.

Philip Miller.

P R E F A C E.

THE GARDENERS DICTIONARY having already gone through ſeveral editions, it may reaſonably be ſuppoſed, the public are well acquainted with the nature of it, which renders it leſs neceſſary to enlarge on that ſubject. The author therefore thinks himſelf obliged to return his thanks for the kind reception his work has met with.

But as there may be ſome, who may think that the republiſhing it is doing them an injury, eſpecially thoſe who have purchaſed a former edition, it may not be amiſs to make ſome apology for this.

When the firſt edition was publiſhed, the art of gardening was then much leſs known than at preſent; and the number of plants has been greatly increaſed in England, therefore it became more neceſſary to enlarge on the ſubject, by adding the new plants and improvements to the former, without which it would have been deemed imperfect: for as the author's ſituation in life rendered him capable of being well informed of the progreſs made in the art, by his great correſpondence both at home and abroad, he thought it would not be unpleaſing to communicate thoſe improvements to the public: in doing which, he has been careful not to publiſh any thing imparted to him, until he was fully ſatisfied of the facts by experiments.

Others have ſuggeſted, that printing the improvements ſeparately would give ample ſatisfaction to the public; but the author had made trial of this method ſome years paſt, by publiſhing ſeveral ſheets of new articles, by way of Supplement, for which there was ſcarce any demand; ſo that the few which were ſold, would not defray the expence of paper and printing.

As the number of plants now cultivated in England, are more than double thoſe which were here when the firſt edition of this book was publiſhed, the mentioning of thoſe, together with their culture, could not well be avoided in a work of this nature, therefore the author hopes his care in inſerting them will not be cenſured.

By the title of the book, it may appear to many as a treatiſe of Gardening only; but whoever will be at the trouble of examining the contents, will find moſt of the neceſſary parts of agriculture inſerted; therefore might, with great propriety, have been intitled a Dictionary of the whole art of huſbandry. And as many of the great farming gardeners near London, have turned their thoughts on this ſubject, they have adopted a very uſeful branch of this art, which is the ſparing of their ſeeds; for where they formerly ſowed one buſhel of grain upon a certain ſpot of land, they have now contracted it to half a peck, and have far greater crops than before: and where this is applied to the conſumption of the ſeveral ſorts of Corn, it will be found a great national benefit, eſpecially in times of ſcarcity.

Throughout the whole performance, the author has principally aimed at rendering the inſtructions here given as clear and intelligible as poſſible; not only to the practitioners, but alſo to thoſe who are leſs acquainted with the art, therefore hopes for indulgence from the public, for any ſmall imperfections which may be found in the work, eſpecially as conſidering the extenſiveneſs of it.

It is amazing to ſee, in moſt of the books which have been publiſhed concerning huſbandry, that ſcarce any of the compilers have taken the leaſt notice of the common practice of ſowing eight times the quantity of Corn upon land which is neceſſary, to the great expence and detriment of the farmers, who are ſo wedded to their old cuſtoms, as not to be convinced of the error: for ſo obſtinate are they in this matter, that unleſs the whole ground be covered with the blades of Corn by the ſpring, they judge it not worth ſtanding, and in conſequence thereof frequently plough up their Wheat and winter Corn, to ſow the land with Barley, or other Lent Corn; whereas, if the former had been left ſtanding, it would have produced a better crop than any land can do where the blades are very thick, as the author has frequently obſerved. I have mentioned this to ſeveral farmers, but the anſwer has conſtantly been, that on rich ground a thin crop of roots will often produce a large crop of Corn, but on poor land it will not pay coſt which is a very great abſurdity; for how is it poſſible, that bad land can ſupply proper nouriſhment to a greater number of roots than better ground? and where this practice is obſerved, ſeldom more than three or four buſhels are reaped from one ſown; whereas, where the ſame quantity is ſown upon the ſame, or a like ſoil, and has room to grow, the produce will be at leaſt ſix or ſeven buſhels. Yet I have ſeen growing upon land not very good, and uncultivated, for more than twenty years, which land was ſown with Wheat in drills, where three gallons of ſeed were allowed to an acre, a produce of nine quarters per acre; now this is no more than an eighth part of the ſeed uſually ſown by farmers, who ſeldom reap more than one-third of this produce; by which it appears plainly, that in the common method of huſbandry, there is at leaſt eight times the quantity of ſeed ſown upon the land that is neceſſary. How great a ſaving this would be in a whole country, I leave every one to judge, eſpecially in ſcarce years, when Corn is dear; and what an expence is occaſioned by the contrary practice to the farmers, who notwithſtanding ſeem unwilling to alter their ancient cuſtoms. Theſe matters are treated of under the articles AVENA, HORDEUM, SECALE, and TRITICUM.

Nor are the common farmers better managers of their paſture lands; for on them they ſeldom are at the trouble of rooting up bad weeds, which frequently over-run them; theſe are often permitted to ſcatter their ſeeds, by which the land is ſtocked with a ſupply of weeds for ſeven years or more, though the utmoſt care be taken afterward to deſtroy them; but though there are ſome farmers who may be ſuppoſed more careful in this reſpect, yet theſe leave in their head lands, and on their banks, hedge-rows, and the ſides of ditches, a ſufficient number of weeds to ſtock their fields when the ſeeds are permitted to ſcatter: beſide, theſe paſtures have rarely a ſufficient quantity of manure allowed them, eſpecially where there is much arable land; nor is the dreſſing laid on at a proper ſeaſon; the general rule with moſt of the farmers being to carry and ſpread the dung upon their paſtures, ſoon after the crop of hay is taken off the ground; and as this is done in ſummer, the heat of the ſun draws all the moiſture from it, whereby the greater part of its goodneſs is evaporated and loſt. But as theſe points are more fully treated of in the body of this work, the author deſires the reader to refer to them.

On the article of TIMBER perhaps many may ſuppoſe, the author has been too diffuſe in his inſtructions; but if thoſe who are of that opinion will only conſider, how material an article this is to the welfare of this country, he flatters himſelf they will change their ſentiments, eſpecially when they reflect upon the great waſte that has been made of it for many years paſt, as alſo that the perſons now employed by the government to cultivate and improve it, deriving their own profits from the waſte of timber, ſeem to think, that as their predeceſſors have long practiſed it, they have a right to do the ſame; this is now carried to ſo great an extravagance, that unleſs a ſpeedy ſtop be put to it, the government will be greatly diſtreſſed for their marine. For although this practice began in the Royal Foreſts, &c. yet ſeveral of the nobility and gentry, who had very great quantities of timber growing upon their eſtates, have deſtroyed a conſiderable part of theirs alſo; therefore, from a due regard for the public, the author has treated of the beſt methods for

6 propa-

propagating and preferving timber, which he hopes may not be difpleafing to the generality of his readers.

The feveral plants here propofed for trial in the Britifh dominions in America, are fuch as there is reafon to believe will fucceed in thofe parts where the experiments are defired to be made, and confined to fuch only, as may be of utility to the public, and real advantage to the inhabitants of thofe countries: furthermore, thefe experiments are propofed to be tried upon plants which will not fucceed well in England, fo as to render their culture practicable, and therefore will not interfere with the growth or trade of this country, and the confumption of which is very great here, many of them being of very confiderable ufe in our manufactures, which cannot be carried on without them; as namely, the Safflower, Indigo, and feveral other forts ufed in dyeing, none of which will thrive in this country to advantage, with many medicinal drugs, which, if introduced into the iflands of America, will certainly thrive there as well as in their native foils. Coffee and Chocolate grow equally well there; but the former being gathered before it is ripe, ill dried, and brought over to England in fhips freighted with rum and fugars, the effluvia of thefe commodities are imbibed by the Coffee, whereby it is rendered lefs valuable: as to the latter, it was formerly cultivated by the Spaniards in the ifland of Jamaica, when they were in poffeffion of it, fo as to furnifh the inhabitants with a quantity fufficient for their own confumption; whereas the Englifh inhabitants now refident there, purchafe it of the Spaniards: thefe articles therefore require the public attention, for if the above commodities may be eafily produced in the Britifh colonies in America, they will not only fupply us with fuch as are genuine, but alfo turn the balance of trade, greatly to the advantage both of Great-Britain and her colonies.

It is alfo a great neglect of the inhabitants of the fugar iflands in America, to commit the care of their plantations to overfeers, who at beft go on in their ufual courfe, planting eight or ten fugar canes in each hill, fo that if five or fix of them grow, they will be fo clofe as to fpoil each other; for whenever thefe plants are ftinted in their growth, they are foon attacked by vermin, which fpread and multiply fo greatly, as frequently to deftroy the whole crop, or at leaft very much to damage it; and this they lay upon inclement feafons, calling it a blight, whereas it proceeds from their own covetous cuftom. A gentleman of learning, who had a confiderable eftate in Jamaica, which was bequeathed to him upon his arrival there, was determined to make trial of the horfe-hoeing hufbandry among his canes. Accordingly he fet out one acre of land in the middle of a large piece, which he caufed to be planted with canes at five feet diftance, putting but one to each hill; thefe grew to a very large fize, and when ripe were cut, as alfo an acre from the beft part of the piece in which the others were planted in the ufual manner: each of them then were boiled feparately to examine their produce, which was nearly equal in the weight of fugar; but where the plants grew fingle, the juice was boiled with a ninth part of the fuel which the other required, and he fold the fugar for fix fhillings per hundred weight more than he could get for the former. This fhews what advantages may be expected, if the poffeffors of lands were careful to make trials.

In the whole of this performance, the author has principally aimed at rendering the inftructions given, as clear and intelligible as poffible to the practitioners, as well as to thofe who are lefs acquainted with the art; in every particular he has obferved all poffible regard to truth, not having advanced any thing as fuch, but what he has been fully convinced of by his own experience: he hopes therefore for indulgence from the public, for any imperfections or omiffions which may appear in the book, fince in a work of fo great extent, it cannot be expected to be abfolutely perfect, though it is humbly hoped there will not be found in it many faults.

The Gardeners Kalendar, inferted in moft of the former editions, is in this omitted, many editions of that piece having been printed in octavo; it is prefumed therefore that few perfons who have any inclination for the innocent diverfion of gardening, are without it; and as the adding any thing to this work would have fwelled it greatly, which the author wifhes he could have ftill further fhortened; and moreover it having been obferved to him, by many of his friends, that

few

few perfons would chufe to turn over fo large a volume, to find in it the articles they may have in a portable one, the omiffion of the Kalendar was thought more advifable.

Many plants are likewife omitted in this edition, feveral of them natives of England, but rarely cultivated in our gardens; as alfo many varieties accidentally arifing from feeds, as are moft of thofe with double flowers, which, if enumerated, would have fwelled the book to an immoderate fize; however, moft of thefe varieties are cafually mentioned, to inform the reader of their refpective difference, which the author hopes will be deemed fufficient. But as the variety of fruits, as well as of efculent plants, have been moft of them, at leaft thc fine forts, greatly improved by culture, they are fully treated of under their proper genera.

On this article a long feries of obfervations has been made by the author, who for more than fifty years has applied himfelf clofely to this fubject; for as many former botanifts have enumerated a great number of varieties as fo many fpecies, the ftudy of botany was thereby rendered greatly perplexed; fome of the modern writers on this fubject, by going into the contrary extreme, have abridged the fpecies almoft as much. Indeed it muft be allowed, that afcertaining the real fpecific difference of plants, would be of great fervice to the fcience of botany; but this cannot be done otherwife, than from many years experience in their culture, efpecially by obferving the varieties which arife from the fame feeds, as alfo the difference produced by different foils and fituations, which is frequently fo great as to perplex very good judges in this matter. There are likewife many other varieties which have arifen from feeds, faved from plants which have grown near others of a different fpecies, by which means they have partaken of both; but thefe hybridine plants rarely producing any feeds afterward, the alteration goes no farther

THE

THE

ABRIDGEMENT

OF THE

GARDENERS DICTIONARY.

ABELE-Tree. See POPULUS.

ABIES. Tourn. Pinus. Lin. Gen. Plant. 956. The Fir-tree.

The CHARACTERS of this genus are,

There are male and female flowers on the same tree; the male flowers have empalements of four leaves, without petals, many stamina and naked summits. The female flowers are collected in a scaley cone, each scale covering two flowers, having neither petals or stamina, with one pointal, and are each succeeded by a winged nut. The distinguishing character of this genus, is the leaves arising singly from their base; whereas the Pines have two or more arising from the same point.

The Fir has always been separated from the Pine-trees, by all the writers on botany before Dr. Linnæus, and were generally distinguished therefrom, by their leaves being produced singly on the branches; the leaves of the Pines being produced by pairs, threes, or fives, out of sheaths which surround their base. And as this distinction is now well known among the nursery-gardeners, so it is much better to keep them separate, than to join them with the Cedar of Libanus, and Larch-tree to the Pine, as the Doctor has done, making them of one genus, especially as the culture of them is very different. For there are few of the sorts of Pines which thrive well when they are removed large, as the earth generally falls from their roots, when taken up, whereas most of the sorts of Firs may be taken up with large balls of earth to their roots, so with proper care they may be safely removed, when they are sixteen or eighteen feet high; especially if they have been digged about, and their roots frequently cut round.

The following SPECIES are now in the English gardens,

1. ABIES (*Picea*) foliis emarginatis, subtus glaucis, strobilis erectis sessilibus. *The Silver or Yew-leaved Fir*

2. ABIES (*Alba*) foliis subulatis mucronatis utrinque dispositis, strobilis pendentibus. *The Spruce or Norway Fir, sometimes called the Pitch-tree.*

3. ABIES (*Balsamea*) foliis solitariis subemarginatis, subtus linea duplici punctata. Flor. Virg. 152. *The Balm of Gilead Fir.*

4. ABIES (*Canadensis*) foliis subulatis subtus glaucis, utrinque dispositis, strobilis uncialibus laxis. *The small-coned American Spruce Fir.*

5. ABIES (*Nova Anglia*) foliis subemarginatis, subtus glaucis, utrinque dispositis, strobilis uncialibus laxis. *The white Spruce Fir of North America, called Newfoundland Spruce.*

6. ABIES (*Americana*) foliis subemarginatis, bifariam dispositis, strobilis subrotundis. *The American Hemlock Fir.*

There are some other varieties of these trees, which have been raised in England from seeds which came from North America; but as they are believed to be only accidental variations, arising from the difference of soils and situations, I shall not pretend to put them down as different species, especially as several of them have not as yet produced cones in England.

The first sort grows naturally in many parts of Germany, but the finest trees of this sort are growing upon mount Olympus, from whence I have received some of the cones, which were of an extraordinary size. The Strasburgh turpentine is drawn from this tree. The wood is white and soft, and therefore not greatly esteemed. The Balm of Gilead Fir, which is the third sort, is so near resembling this, as scarcely to be distinguished from it, after it is grown to a large size. The young trees have their leaves growing on every side their branches, by which they may then be easily known; but as the trees advance, so their leaves become ranged only on two sides of the branches, and approach nearer to the Silver Fir: the short duration of this tree in almost every soil and situation in England, has inclined many persons to believe it a distinct species; but as I have observed the same trees to alter after some years growth, so I shall suspend my judgment of this matter, until I can determine with greater certainty.

The second sort grows naturally in the low lands of Sweden, Norway, and Denmark, as also in many other parts of Europe. This is sometimes titled Abies rubra, i. e. Red Fir, which has given occasion to some persons to believe, that the red deals are cut from this sort; but we now can have no doubt of the contrary, for they are cut from the Scotch Pine, the wood of the Spruce Fir being white. The young branches of this Fir are used to make Spruce beer in Germany, and from thence had the title of Spruce Fir.

The fourth sort grows naturally in many parts of North America, from whence the cones have been brought to England. The leaves of this sort are shorter than those of the Spruce Fir, but are in shape like them, their under side being of a glaucous green colour; the cones are loose, and about an inch in length.

The fifth sort is also a native of North America, where the inhabitants make three sorts of it, by the titles of Black, White, and Red Spruce. In England, these are mostly known by the title of Newfoundland Fir, because many of their cones have been brought from thence, but the trees are found growing in most parts of North America. The Black Spruce grows commonly in swamps and bogs, and rarely rises to a great height. The White Spruce is an inhabitant of the mountains and higher lands, where it grows to a large size: and in the gardens of his Grace the late Duke of Argyle at Whitton, near Hounslow, there are some noble trees of this sort, which are not more than thirty-six years growth from seeds. By the difference in the growth of this and the Black Spruce, we may readily suppose them to be different species; but upon examining the old branches with their cones, they approach so near as to give suspicion of their being only varieties. The red sort, as it is called in America, while young, had great likeness to the Black Spruce Fir of America; but as the plants advance, there is a manifest difference in their leaves, and a much greater in their cones, which are smaller than those of the other, and are almost round: how their timber differs I know not, having never seen any of its wood.

The sixth sort is also a native of the same country; and in the northern parts of America, I am informed it grows to be a very large tree; but in England the branches spread wide every way, so that there is no appearance of the trees ever arriving to any considerable height. The leaves of this tree are short, and shaped very like those of the Yew-tree; they are ranged on two sides of the branches only, so they appear flat, like those of the Silver Fir, but are of a pale green on both sides. The cones are small, loose, and roundish. What sort of wood this tree affords I cannot say, having never seen any trees of a size fit to cut down.

From most of these Firs, the inhabitants of North America collect a clear fragrant turpentine, which they use for curing green wounds; and the physicians there make great use of it internally: and it is generally supposed, that what is now sold under the title of Balm of Gilead in England, is the turpentine of the third sort.

All the sorts of Fir are propagated by seeds; the time for sowing them is about the middle of March, when the season is mild, otherwise it had better be deferred till the end of that month, or the beginning of April. The seeds which are preserved in their cones, will keep good much longer than those which are taken out; but the cones of the Silver and Balm of Gilead Firs generally fall to pieces in the autumn, soon after the seeds are ripe; so that if they are not carefully watched, and gathered at that time, the seeds will be lost. The cones of all the sorts of Fir open with more ease than those of the Pines, and require but little trouble to get out their seeds. If they are spread on a cloth before a fire for a few hours, their scales will open and emit the seeds.

The seeds may be sown in pots or boxes filled with light fresh earth, covering them over about half an inch thick with the same earth; these should be placed to an east aspect, where they may have the sun till eleven in the morning; or if the seeds are sown in a bed of earth, it should be shaded with mats in the middle of the day: for when they are too much exposed to the sun, the surface of the ground will dry so fast (especially in dry seasons) as to hinder the seeds from vegetating; and when the plants begin to appear, if they are not screened from the sun, many of them will be soon destroyed. The seeds must be carefully guarded against mice and birds, who are very fond of them, but particularly when the plants begin to appear; for as they thrust up the cover of the seeds on their top, so the birds in pecking off these covers, will destroy the young plants; therefore the surest method is to cover them with nets until the plants have thrown off their husks, and expanded their seed-leaves, soon after which they will be out of danger.

The plants may remain in these places where the seeds were sown till the following spring, provided they are not stinted by the stiffness of the ground, or any other cause; if so, they had better be carefully transplanted into new beds about the beginning of July; but this must be done with great care, observing to raise the plants with a trowel, so as to preserve their roots as entire as possible, and to plant them again immediately, otherwise their tender fibres will soon become dry when exposed to the air at this season, and the plants thereby destroyed. The distance for planting these young plants should be four inches row from row, and about three inches asunder in the rows: for as these beds must be arched over with hoops, that the plants may be shaded with mats in the middle of the day, so the closer they are planted, there will be less trouble and expence in their covering; and as the plants are to remain in these beds no longer than the following year, so there will be room enough for their growth during that time. These young plants must be carefully weeded, for if weeds are permitted to grow among them to any size, there will be great danger of drawing the plants out of the ground with the roots of the weeds, when pulled up. If the season proves very dry, it will be of service to the plants to sprinkle them over with water once or twice a week during the hot time of the year: but this should be done with caution, for too much wet will rot the shanks of these young plants and destroy them.

The plants are very rarely hurt by frost in winter, especially those in the full ground; but such as are in pots or tubs are in more danger, if they stand upon the surface of the ground; for the frost will penetrate through the sides of the pots or tubs, and thereby may injure their roots. Therefore these should either be sunk into the ground before winter, or some old tanners bark, straw, or mulch, laid round the pots or tubs to keep out the frost.

After the young plants have remained in the seed-bed one year, they may be transplanted into beds the April following; but as these which were not transplanted in summer from the places where they were sown, may stand two years in the beds when transplanted, so they should be allowed more room than those which were removed the preceding summer. Therefore the rows may be from five to six inches distant, and the plants in the rows four inches asunder, observing to treat them in the manner before directed. When they have grown two years in these beds, they may then be transplanted into the nursery, placing them in rows at three feet distance, and in the rows a foot asunder. The best season for removing them is in April, just before they begin to shoot; though they may, and often are transplanted in autumn with success; but the other time is preferable, especially if there happen to be rain soon after, otherwise they will require watering once a week for about a month.

The smaller these trees are planted out where they are to remain, the greater will be their progress, and they will grow to a much larger size than those that are removed at a much greater age; but there are few persons who have

patience

patience to wait their growth, therefore frequently plant them at the height of fix or eight feet, at which fize they will tranfplant better than moft other evergreen trees; but thofe which are fo tall will require fupport, otherwife they will be in danger of being blown down by the wind. And if trees of fuch fizes are to be carried to a diftant place, it will be expenfive; for unlefs they have large balls of earth to their roots, there will be great hazard of their growing; and thefe will require more water than young plants: fo that upon the whole, planting of young trees, is much preferable and lefs expenfive. Therefore, where there are large plantations to be made, planting the trees very young is the moft eligible; for the expence of cleaning thefe young plantations, will not be equal to that of ftaking and fecuring tall trees: and the difference of the firft price, together with the carriage of the latter, will be very confiderable; befide, the former will in a few years outgrow the latter. I have myfelf made plantations of Firs of different ages at the fame time, upon the fame ground, and have always found that plants of two or three years old, have in ten or twelve years, been much the beft of any in the plantations; and I dare fay others, who have made the like experiment, have found the fame fuccefs.

In the choice of the plants, if they are to be purchafed from a nurfery, they fhould not be taken from good land to plant in a poor foil; therefore the better way is to procure them from ground as nearly like that into which they are to be planted, as poffible; or if it is worfe, the plants will fucceed better. Indeed, where large plantations of thefe trees are defigned, it is much the better way to make nurferies on the fame ground, where the trees fhould be raifed from feeds; for this will be a great faving of expence, and as the diftance will be fmall to remove them, fo there will be little danger of their fucceeding.

But as the wood of all the forts of Fir yet known, is much inferior to that of the Pine, fo it is not adviseable to make plantations of them for their timber, therefore they are only valuable for their beauty: fo that when they are planted for ornament, they fhould be placed fo far afunder as to admit the free air between them, otherwife the lower branches will decay, and render the trees unfightly. The great beauty of thefe trees are in their pyramidal form, and being furnifhed with lateral branches from about feven feet above the furface of the ground to the top, and thefe branches fhould be well garnifhed with leaves: to obtain which, the trees fhould not be planted nearer than eighteen or twenty feet; for when they are clofer planted, the under branches foon drop their leaves and decay; and if thefe branches are taken off, the trees never put out new ones to fupply their place. The unfkilful difpofition of thefe trees, has brought them into difrepute with many perfons; whereas, if they are properly placed, they may be made very ornamental to fine feats.

In pruning off the under branches to the defigned height, there muft be care taken not to cut off too many branches at the fame time; one tier of branches is full enough to be difplaced in a year; and if every other year this is performed, it will be fufficient; and by this gradual method of pruning, the trees will not be much retarded in their growth. The beft time for this operation is in the beginning of September.

The Silver Fir requires a deep ftrong foil, for if it is planted in a light ground it will make but little progrefs; and when it is planted in a fhallow ground, as foon as the roots meet with obftruction the trees generally decay. The largeft trees of this kind which I have feen, were growing in a deep loamy foil; thefe were upwards of ninety feet high, and were furnifhed with branches from ten feet above the ground to their tops, which being well garnifhed with leaves, made a fine appearance.

The common Spruce Fir will thrive beft on the fame land, but this will alfo do well on light ground, where the other will make little progrefs, fo is more generally planted in England: befides, it will thrive in foils and in fituations where the other will fcarce live; it is alfo of longer duration in England.

The fourth fort will fucceed beft on a moift foil, for in light dry ground it makes but little progrefs; nor does it make a good appearance where the foil is not proper for its growth.

The American Spruce Firs delight in light moift ground, where the trees grow to a large fize, and make a beautiful appearance; and if they are allowed room for their lower branches to fpread and extend, they will be garnifhed with them almoft to the ground, forming themfelves in a pyramidal figure.

The Hemlock Fir thrives beft in a ftrong loamy foil; thofe which have been planted in light dry ground, have made but little progrefs, efpecially upward, their branches taking a lateral pofition: fo that unlefs the upper fhoot is trained to a ftake to direct its upright growth, the leading fhoot will turn on one fide and become flat; but in a ftrong loam, I have feen fome of thefe trees which have naturally grown upright. As there are none of thefe trees in England which are arrived to a fize fit to cut down, fo we know little of the worth of this wood.

There are fome perfons who are fond of propagating Fir-trees from cuttings, which, if properly planted, will take root, but the plants fo raifed will never arrive to near the fize of thofe raifed from feeds: they are alfo never inclined to an upright growth, fending out lateral branches, and becoming bufhy, therefore this practice is not worthy of imitation; and unlefs for fake of the multiplying a curious fort, whofe feeds cannot be eafily procured, fhould never be attempted: nor fhould the inarching of one fort upon another be practifed for the fame reafon; for the trees fo propagated, will be of flow growth and of fhort duration.

ABROTANUM. See Artemisia.

ABROTANUM Fœmina. See Santolina.

ABSINTHIUM. See Artemisia.

ABRUS. See Glycine.

ABUTILON. See Sida.

ACACIA. See Mimosa.

ACACIA, the Common American. See Robinia.

ACAJOU, or CAJOU. See Anacardium.

ACALYPHA. Three-feeded Mercury.

There are three fpecies of this genus of plants, which are preferved in curious botanic gardens for the fake of variety; but as they have no great beauty or of any ufe, fo they are rarely propagated in other gardens, therefore are not inferted here, as the enumeration of fuch plants would fwell the work beyond the bounds intended.

ACANTHUS. Tourn. Inft. Lin. Gen. Plant. 711. Bear's breech.

The Characters of this genus are,

The empalement of the flower is two-leaved and bifid; the petal has but one lip, which is turned backward, and is divided into three at the end. The capfule has two cells, each containing one feed. It is of the ringent clafs of flowers, whofe feeds are in a capfule, and are ranged in the fecond divifion of Linnæus's fourteenth clafs.

The Species of this genus now in the Englifh gardens are,

1. Acanthus (*Mollis*) foliis finuatis inermibus. Lin. Sp. 891. *Smooth Bear's-breech with finuated obtufe leaves, or the common officinal Bear's-breech.*

2. Acanthus (*Lufitanicus*) foliis finuatis inermibus lucidis, laciniis acutis. Juffieu. *Smooth Portugal Bear's-breech, with finuated fhining leaves having acute points.*

 3. Acanthus

3. Acanthus (*Dioſcoridis*) foliis pinnatifidis ſpinoſis. Bear's-breech with acute winged leaves, having ſoft prickles.

4. Acanthus (*Spinoſus*) foliis pinnatifidis lucidis ſpinoſis. Hort. Cliff. 326. Prickly Bear's-breech.

The firſt ſort is the common Acanthus, whoſe leaves are taken for the ornaments of the Corinthian capital, and is the ſort which is uſed in medicine. It grows naturally in Italy, Sicily, and the Levant.

The ſecond ſort grows naturally in Portugal. The leaves of this are much larger than thoſe of the firſt, and are leſs jagged; the cuts of the leaves are more pointed, and the upper ſurface is lucid. This is not a variety, for the ſeeds conſtantly produce the ſame kind.

The third ſort grows naturally in Italy; the leaves of this are cut into acute ſegments, and are ſhaped like winged leaves, each ſegment terminating with a ſhort ſoft ſpine. The flower-ſtalks of this ſort riſe conſiderably higher than thoſe of the former ſort.

The fourth ſort hath large ſhining winged leaves, which are armed with ſtrong ſpines at the end of each ſegment, which renders it very troubleſome to handle the leaves, and the flower-ſtalks of this riſe as high as thoſe of the third ſort. This grows naturally in the Archipelago.

Theſe plants have all of them thick fleſhy roots, which ſtrike deep in the ground; and thoſe of the third and fourth ſorts creep in the ground to a conſiderable diſtance, that it is difficult to keep them within proper compaſs. They are alſo leſs tender than the firſt and ſecond ſorts, ſo are rarely injured by the greateſt cold in England; whereas the former are killed in ſevere winters where they are expoſed in an open ſituation, therefore require a warm ſituation and a dry ſoil. Theſe plants frequently perfect their ſeeds in England, ſo they may be propagated by ſowing them in a bed of light earth in the ſpring, where the plants generally appear in about ſix weeks after; and if they are kept clean from weeds, it is all the care they require the firſt ſummer: but in the winter, the plants of the firſt and ſecond ſorts will require ſome protection, eſpecially if the weather ſhould be very ſharp; therefore they ſhould be covered with mats, peaſe-haulm, or ſome other light covering, when the froſt is ſevere, but the covering ſhould always be removed in mild weather.

About the beginning of March, if the ſeaſon is mild, the young plants ſhould be carefully taken up, and tranſplanted in the places where they are deſigned to remain. Thoſe of the two firſt ſorts ſhould have a warm ſituation and a dry ſoil; they muſt alſo be covered in winter, if the froſt is ſevere, for a year or two, till they have obtained ſtrength enough to reſiſt the cold: but the other may be planted in the open border, where, if the ground is not too wet, they will thrive and flower very well.

But as the plants which are raiſed from ſeeds ſeldom flower till the third year, ſo few people care to wait ſo long, therefore generally propagate them by offſets from the roots; theſe are produced in great plenty by the third and fourth ſorts, which ſend them out to a conſiderable diſtance from the mother plant, ſo may be had in great abundance: but the firſt and ſecond ſorts do not increaſe near ſo faſt, therefore are leſs common in the Engliſh gardens than the other. The offſets of all theſe ſorts ſhould be taken from the old plants in March, when the danger of the hard froſt is over; for if very ſevere froſt ſhould happen ſoon after their removal, it will kill them, eſpecially thoſe of the two firſt ſorts.

When the plants have taken good root in the places where they are deſigned to remain, the only culture they will require is to keep them clean from weeds; and when they ſhoot up their flower-ſtalks, to put down ſtakes and faſten the ſtalks to them, to prevent their being broke down by the wind, for they generally grow four or five feet high, and their flowers being large, become heavy; but when the ſeeds are formed, they are commonly too weighty for the ſtalks to ſupport them. The two laſt ſorts ſhould have their offſets frequently taken off, to keep them within bounds.

ACER. The Maple-tree. In French, *Erable.*

The Characters of this genus are,

It hath hermaphrodite and male flowers on the ſame tree; the hermaphrodite flowers have an empalement of one leaf, cut into five parts: the corolla has five petals, they have five ſtamina and one pointal: the flowers are ſucceeded by two winged capſules joined at their baſe, each including a ſingle ſeed. The male flowers have the ſame characters, but have no ſtyle, ſo are not fruitful.

The Species are,

1. Acer (*Pſeudo Plataneis*) foliis quinquelobis inæqualiter ſerratis, floribus racemoſis. Lin. Sp. Pl. 1054. *The greater Maple, falſely called Sycamore.*

2. Acer (*Campeſtre*) foliis lobatis obtuſis emarginatis. Lin. Sp. Pl. 1055. *The common or leſſer Maple.*

3. Acer (*Negundo*) foliis compoſitis, floribus racemoſis. Hort. Cliff. 144. *The Aſh-leaved Maple.*

4. Acer (*Platanoides*) foliis quinquelobis acuminatis acute dentatis glabris floribus corymboſis. Lin. Flor. Suec. 303. *The Norway Maple.*

5. Acer (*Rubrum*) foliis quinquelobis ſub dentatis ſubtus glaucis pedunculis ſimpliciſſimis aggregatis. Lin. Sp. Plant. 1055. *The Scarlet-flowering Maple.*

6. Acer (*Sacchatum*) foliis quinquepartito-palmatis acuminato dentatis. Lin. Sp. Pl. 1055. *The American Sugar Maple.*

7. Acer (*Penſylvanicum*) foliis trilobis acuminatis ſerrulatis, floribus racemoſis. Lin. Sp. Pl. 1055. *The American Mountain Maple.*

8. Acer (*Monſpeſulanum*) foliis trilobis integerrimis. Prod. Leyd. 459. *The Montpelier Maple.*

9. Acer (*Creticum*) foliis trilobis ſerrulatis. *The Cretan Maple.*

The firſt ſort grows naturally in the mountains in Germany, but is now ſo common in Britain, as to be by ſome ſuppoſed to be indigenous here; for the ſeeds have been carried by the winds to a great diſtance from the trees, and the plants have riſen in great plenty without care, in all places which are fenced from cattle, in the neighbourhood of the trees; ſo that there is generally a ſupply of young plants from ſcattered ſeeds without any trouble, and this may have miſled many perſons to believe the tree is a native of this country.

This ſort grows to a tree of a large ſize; the wood is ſoft and very white, ſo is uſed by the turners, but is not eſteemed very valuable for other purpoſes. But as this tree will thrive better than moſt other ſorts near the ſea, ſo it is frequently planted to ſcreen plantations of other ſorts of trees from the ſpray of the ſea.

The ſecond ſort is very common in moſt parts of Europe, and is generally believed to be a native of this country. The wood is very hard, ſo is uſed for gun-ſtocks and ſeveral other purpoſes; but this ſort never grows to a large ſize.

The third ſort, which is commonly known by the title of Aſh-leaved Maple, is a native of North America, but is now very common in the Engliſh gardens. It is of quick growth; the trees often make ſhoots of eight or ten feet long in one year, but the wood is ſoft, and the branches of the trees are frequently ſplit off by ſtrong winds in the ſummer, when they are cloathed with leaves, if they are in an expoſed ſituation. Theſe trees abound with a ſweet ſap early in the ſpring, which, if collected, by tapping their ſtems at that ſeaſon, and boiling it, a tolerable good ſugar is produced in North America; but the ſixth ſort is that, which the inhabitants of that part of the world usually tap for that purpose. From the first sort here mentioned, Dr. Lister procured some sugar after the same manner in England; and I believe if the sap of some other species were tried, there might be a coarse sort of sugar produced, as there might also by boiling the sap of the Birch-tree.

The fourth sort grows naturally in Norway, Sweden, and other northern parts of Europe; it rises to a good height, and is well furnished with branches, which are garnished with large smooth leaves of a lucid green, which are divided in shape of a hand. These have an acrid milky juice, so are rarely eaten by insects; whereas those of the first sort are frequently eaten full of holes, which render them very unsightly; for which reason, the trees have been generally neglected of late years. This fourth sort will thrive as well near the sea as the first, so is much preferable to it.

The fifth sort is a native of North America, from whence the seeds were brought to England. This is cultivated in gardens for the beauty of its red flowers, which appear early in the spring; they are formed in roundish bunches, at the bottom of the foot-stalks of the leaves. There is a variety of this, which is commonly called Sir Charles Wager's Maple, whose flowers are produced in much larger clusters than those of the common sort, and are placed closer upon the branches; so the trees make a much better appearance than the former, though I believe it to be only a variety from it. This sort never grows to a large size in England.

The sixth sort is what the inhabitants of North America generally tap for the juice, which they boil to obtain a coarse sort of sugar, so is distinguished from the other sorts by the title of Sugar Maple. The leaves of this sort have some resemblance to those of the fourth sort, but are not so lucid, and are frequently eaten by insects like those of the first sort, therefore this tree is seldom cultivated for beauty. It grows large, and the wood may be used for the same purposes as those of the other species.

The seventh sort hath some appearance of the sixth, but the leaves are more pointed.

The eighth sort is a tree of low growth, never rising to a greater height than our Lesser Maple in its native soil. The leaves are of a thick substance, divided into three entire lobes, and are of a lucid green; they continue in beauty till late in the autumn.

The ninth sort grows naturally in the islands of the Archipelago; the leaves of the young plants of this sort are oval and entire, but as they advance their leaves become in shape like those of the Ivy; they are not of so thick consistence as those of the eighth, but are of a lucid green: and in places where the trees are well sheltered from cold, the leaves continue green most part of the year, especially while the trees are young. This sort will endure the cold of our winters in the open air.

All the sorts of Maple may be propagated by cuttings, which in dry ground should be planted in the autumn; but where the land is moist and cold, the spring season is preferable: if they are cut from the trees before the buds begin to swell, and the ground is not then fit to receive them, they may be wrapped in moss, and put in a cool place, where they may be kept a month or five weeks without injury, as I have frequently experienced; so that these cuttings will bear transporting from one country to another very well. But the trees which are raised from cuttings are not so valuable as those which are propagated by seeds, because they seldom grow so large, nor so upright.

The seeds of all the sorts of Maple should be sown in the autumn, soon after they are ripe, for if they are kept dry till the spring, they often fail, or at least lie a whole year in the ground before they vegetate. Therefore if they cannot be sown in the autumn, they should be put into sand to preserve them, and the sand and seeds sown together early in the spring upon a common bed of earth. When the plants come up they must be kept clean from weeds, and in the following autumn transplanted into the nursery, where they may grow two or three years, and then may be planted where they are to remain.

ACETOSA. Sorrel. Rumex. Lin. Gen. 407.

The Characters are,

It hath male and hermaphrodite flowers on the same plant in some of the species, and in others they are on different plants; the flowers have a three-leaved empalement, and have six stamina; the hermaphrodite flowers have a three-cornered style, and these are succeeded by a three-cornered seed.

The Sorrels and Docks are by Dr. Linnæus included in the same genus, under the title of Rumex; but as the old name of Acetosa or Sorrel is better known by physicians and in the gardens, so I have continued it under the old title.

The Species are,

1. Acetosa (*Pratensis*) floribus dioicis, foliis oblongis sagittatis. *The Common Sorrel.*

2. Acetosa (*Acetoscella*) floribus dioicis, foliis lanceolato-hastatis. *Sheeps-Sorrel.*

3. Acetosa (*Scutata*) floribus hermaphroditis, foliis cordato-hastatis. *Round-leaved Garden, or Roman Sorrel.*

4. Acetosa (*Digyna*) floribus hermaphroditis digynis. *Westmoreland Sorrel.*

5. Acetosa (*Vesicaria*) floribus hermaphroditis geminatis, valvularum alis maximis membranaceis reflexis, foliis indivisis. Hort. Cliff. *American Annual Sorrel.*

6. Acetosa (*Rosea*) floribus hermaphroditis distinctis, valvularum alis maximis membranaceis, foliis erosis. *Egyptian Sorrel with Rose-coloured bladders.*

7. Acetosa (*Clunaria*) floribus dioicis, valvulis lævibus, caule arborea, foliis subcordatis. *The Sorrel-tree.*

The first sort grows naturally in pasture lands in most parts of England, but is also cultivated in gardens for culinary uses. It is a perennial plant, so will continue many years without renewing, provided the roots are planted at a sufficient distance to allow room for digging the ground between the rows.

The second sort grows naturally upon dry banks, and on gravelly ground in most parts of England, where by its creeping roots it spreads over the land, and is often a very troublesome weed, so is rarely admitted into gardens.

The third sort is cultivated in gardens for use, and is a much better plant for the kitchen than the common Sorrel. This spreads and increases greatly by its creeping roots, so should be planted at a good distance; and in a stony soil, will do much better than in rich land.

The fourth sort grows naturally in the northern counties of England, in Wales, and Scotland; it is a low plant with creeping roots; the leaves are thick in proportion to their size, and are of a glaucous colour. It is rarely propagated in gardens.

The Annual American Sorrel is kept in some gardens for the sake of variety, but is not of any use. It grows naturally in America and Egypt.

The sixth sort grows naturally in Egypt; it is an annual plant; the bladdery covers of the seeds are of a fine Rose-colour. This is kept in gardens for variety, but is not cultivated for use.

The seventh sort grows naturally in the Canary Islands. This rises with a strong woody stalk to the height of ten or twelve feet. It is frequently kept in gardens here, but must be housed in winter, for it will not live abroad in any country where there are hard frosts in winter. This is generally propagated by cuttings, because the seeds seldom ripen well in England. If the cuttings are planted in a shady border, any time in summer, and are duly supplied with water, they

they will ſoon put out roots; then they ſhould be taken up carefully and planted in pots, for if they are permitted to remain in the border, they will ſoon grow ſo vigorous as to render their tranſplanting hazardous. When they are planted into pots, they ſhould be placed in the ſhade until they are rooted again; then they may be removed to enjoy the open air till October, when the froſts begin to be ſharp; at which time they ſhould be carried into the green-houſe, and treated in the ſame way as Myrtles, and other hardy green-houſe plants.

The common Sorrel is cultivated by ſeeds, and ſometimes by parting of the roots; but the ſeedling plants, if they are allowed room, will have larger and more ſucculent leaves than thoſe which are propagated by ſlips. The plants ſhould ſtand in rows about a foot aſunder, to give room for digging the ground between them every ſpring: and if the plants are ſix inches diſtant in the rows, they may ſtand two or three years without removing, and only require to have the ground kept clean from weeds in ſummer, and ſlightly dug in the ſpring. The beſt time to part or tranſplant the roots is in autumn, which is alſo the beſt time for ſowing of the ſeeds upon dry land.

The round-leaved or Roman Sorrel, is propagated by its creeping roots. Theſe may be tranſplanted either in ſpring or autumn, but the latter ſeaſon is the beſt for dry ground. It thrives beſt on ſtony land, for it grows naturally on rocks. This ſeldom produces good ſeeds, eſpecially when it is planted in light ground. The roots of this ſort ſhould be planted two feet aſunder each way.

The ſeeds of the annual ſorts ſhould be ſown the latter end of March, on a bed of common ground, in rows at a foot and half diſtance; and when the plants come up, they ſhould be thinned ſo as to leave them four or five inches aſunder; the ground muſt be kept clean from weeds, which is all the culture theſe plants require. In July they will flower, and their ſeeds will ripen in autumn.

ACETOSELLA. See Oxalis.

ACHILLÆA. Yarrow, Milfoil, or Noſe-bleed.

The Characters of this genus are,

The compound flowers have an oval ſcaley empalement, including many hermaphrodite florets in the diſk, and from five to ten female half florets which compoſe the ray; the ſeeds are lodged in a chaffy bed, and have no down.

The Species which are kept in the Engliſh gardens,

1. Achillæa (*Santolina*) foliis ſetaceis dentatis, denticulis ſubintegris ſubulatis reflexis. Lin. Sp. Pl. *Eaſtern Sneezwort, with a leaf like Lavender Cotton, and a large flower.*

2. Achillæa (*Tomentoſa*) foliis pinnatis hirſutis, pinnis linearibus dentatis. Lin. Sp. 897. *Woolly Yarrow with yellow flowers.*

3. Achillæa (*Abrotanifolia*) foliis pinnatis ſupra decompoſitis laciniis linearibus diſtantibus. Prod. Leyd. 175. *Tall Eaſtern Yarrow with leaves like Wormwood, and yellow flowers.*

4. Achillæa (*Clavenna*) foliis pinnatifidis planis obtuſis tomentoſis. Lin. Sp. 898. *Alpine umbelliferous Wormwood with ſilvery woolly leaves.*

5. Achillæa (*Ageratum*) foliis lanceolatis obtuſis acutè ſerratis. Lin. Sp. 898. *Sweet Maudlin.*

6. Achillæa (*Ægyptiaca*) foliis pinnatis, foliolis obtuſè lanceolatis ſerrato dentatis. Lin. Sp. 898. *Hoary Sneezwort with creſted pinnulæ.*

7. Achillæa (*Ptarmica*) foliis lanceolatis acuminatis argutè ſerratis. Lin. Sp. 898. flore peno. *Double Ptarmica, or Sneezwort.*

8. Achillæa (*Alpina*) foliis lanceolatis dentato-ſerratis, denticulatis tenuiſſimè ſerratis. Lin. Sp. 898. *Alpine white Maudlin with deep green leaves.*

The common Yarrow, and ſome other ſpecies of this genus are here omitted, as they are rarely permitted to have a place in gardens. The common ſort with white and purple flowers grow naturally in England, but the white is the moſt common, and is the ſort which has been long uſed in medicine. It grows on the ſide of foot-ways almoſt every-where, ſo may be eaſily procured.

The firſt ſort here mentioned, has large yellow flowers which ſtand upon pretty long foot-ſtalks ſingly, not in cloſe bunches, as the common ſort. It has leaves like thoſe of Lavender Cotton, which, when rubbed, emit a ſtrong oily odour. This flowers in June and July.

The ſecond ſort hath woolly leaves ſhaped like thoſe of the common ſort; the flowers are yellow, growing in cluſters at the top of the ſtalks, which ſeldom riſe more than a foot high.

The third ſort grows to the height of two feet and a half, having large umbels of yellow flowers on the top; the leaves are ſomewhat like thoſe of the common Wormwood, and are cut into long narrow ſegments. This flowers in June and July.

The fourth ſort is a native of the Alps; it is a plant of humble growth, the ſtalks ſeldom riſing higher than ſix or ſeven inches; theſe ſupport umbels of white flowers like thoſe of the common Sneezwort, which appear in April and May. The leaves are ſilvery, and ſhaped like thoſe of Wormwood, which frequently decay in the autumn or winter.

The fifth ſort was uſed in medicine, and was ſome years paſt much cultivated in the gardens, as it was frequently uſed in the kitchens: but of late years it has been almoſt totally neglected, ſo was almoſt loſt in England a few years ſince; and the markets were ſupplied with the eighth ſort, which ignorant perſons ſubſtituted in its ſtead, though the two plants are very different in appearance, and have very different flavours, and probably different qualities.

The ſixth ſort grows naturally in the Archipelago, but is hardy enough to live abroad in England, provided it is planted in a dry ſoil and a warm ſituation. It is a low plant, which puts out many heads near the roots, which are fully garniſhed with fine cut-ſilvery leaves. The ſtalks riſe from nine inches to a foot high, and are terminated by compact umbels of yellow flowers This ſort continues flowering great part of ſummer, ſo deſerves a place in gardens.

The ſeventh ſort is the common Sneezwort, of which there is a variety with double flowers that is cultivated in gardens. The common ſort, which is uſed in medicine, grows naturally in woods, and upon commons, in moſt counties in England. It creeps greatly at root, ſo that variety with double flowers ſhould be confined, otherwiſe it will ſpread to a great diſtance, and will not be handſome.

The eighth ſort grows naturally on the Alps, but is now commonly cultivated by thoſe gardeners who ſupply the markets with phyſic herbs, and is ſold for Sweet Maudlin, as is before-mentioned. This plant will riſe four feet high in good land; the leaves are long, narrow, and ſharply ſawed on their edges; they are of a dark green, and the flowers are white; the roots creep far under ground, ſo they ſhould be confined.

All the ſorts of Yarrow are eaſily propagated by ſeeds, which may be ſown either in the ſpring or autumn, upon a bed of common earth; and when the plants come up and are ſtrong enough to tranſplant, they ſhould be planted into beds in the nurſery, where they may remain till autumn, at which time they ſhould be tranſplanted to the places where they are deſigned to remain: if they are planted in a ſtony dry ſoil, they will live much longer than in rich ground.

The ſorts with creeping roots propagate themſelves ſo faſt, as to render it neceſſary to confine them, otherwiſe they

they will ſpread wide on every ſide; and the ſtalks being ſeparated to a diſtance from each other, the plants will make but an indifferent figure when the flowers are fully blown. The other ſorts whoſe roots do not creep may be propagated by ſlipping off their heads, and planting them in a ſhady border; or if in an open bed, they muſt be ſhaded with mats in the day until they have taken root, after which they will require no farther care than to keep them clean from weeds till autumn, when they ſhould be tranſplanted to the places where they are deſigned to remain.

ACHRAS. Mamme Sapota.

This is a large tree, which is propagated in the iſlands of the Weſt-Indies, but is ſuppoſed to have been tranſplanted thither from ſome other country. The leaves are nine or ten inches long, and five broad in the middle, drawing to points at both ends; they are ſmooth on their upper ſide, but have many ſlight veins running from the middle rib to the ſides: the fruit is large, oval and fleſhy, including one long oval pointed nut, which is very ſmooth, having longitudinal borders on one ſide.

As I have not ſeen any of theſe trees in the Engliſh gardens, ſo I ſhall forbear to ſay any thing more of its culture, than that if the plants can be procured they muſt be kept in the bark-bed of the ſtove, and treated in the ſame way as other exotic plants of the ſame country. I have frequently received the ſtones of this fruit from Jamaica, but they were always rotten before they arrived, for not one of them ever ſprouted; ſo that I believe, theſe ſeeds will not continue long ſound after the fruit is eaten.

ACHYRANTHES. We have no proper Engliſh name for theſe plants. One of the ſorts has been long in the gardens, and has been known by Father Boccon's title, viz. Amaranthus ſiculus ſpicatus radice perenne. This ſort grows naturally in both Indies, from whence I have ſeveral times received the ſeeds. There are three other ſpecies, whoſe ſeeds have been brought from the Cape of Good Hope, and the plants are preſerved in curious botanic gardens, but being neither uſeful or beautiful, are ſeldom kept in other gardens, therefore they are not enumerated here.

ACINOS. See THYMUS.

ACONITUM. Wolfsbane, or Monkſhood.

The CHARACTERS are,

The flower has no empalement; it has five unequal petals, the upper is hooded and inverſed; it has two forked nectariums, whoſe foot-ſtalks are recurved, and many ſmall ſtamina which incline to the petals, with five ſtyles terminated by reflexed ſtigmas. The flowers are ſucceeded by three or four capſules with one valve, containing many angular ſeeds.

The SPECIES are,

1. ACONITUM (*Lycoctonum*) foliis palmatis multifidis villoſis. Lin. Sp. Pl. 532. *Yellow Wolfsbane, or Monkſhood with hand-ſhaped leaves.*

2. ACONITUM (*Altiſſimum*) foliis palmatis nervoſis glabris. *Yellow Wolfsbane, with larger ſmooth-veined hand-ſhaped leaves.*

3. ACONITUM (*Variegatum*) foliis multifidis, laciniis ſemipartitis ſupernè latis. Hort. Cliff. 24. *Small Blue Wolfsbane, or Monkſhood with many pointed leaves.*

4. ACONITUM (*Napellus*) foliorum laciniis linearibus ſupernè latioribus linea exaratis. Hort. Cliff. 214. *Large Blue Wolfsbane or Monkſhood, whoſe under leaves are cut into many narrow ſegments, and the upper into broader.*

5. ACONITUM (*Pyramidale*) foliis palmatis multipartitis, ſpicis florum longiſſimis. *The Common Monkſhood or Blue Wolfsbane, with the longeſt ſpikes of flowers.*

6. ACONITUM (*Pyreniacum*) foliis multipartitis, laciniis linearibus incumbentibus ſquarroſis. Hort. Upſal. 152. *Yellow Wolfsbane of the Pyrenees, with leaves cut into many narrow ſegments which are rough.*

7. ACONITUM (*Anthora*) floribus pentagynis. Lin. Sp. Pl. 532. *Yellow wholeſome Wolfsbane, or Monkſhood.*

Theſe ſorts of Wolfsbane grow naturally upon the Alps, the mountains in Germany, Auſtria, and in Tartary, ſo require a cool ſhady ſituation, and a ſoil rather moiſt than dry; but not ſo wet, as to have the water ſtanding near their roots in winter: in dry ground theſe plants do not thrive or flower well, eſpecially if they are expoſed much to the ſun. They may be all of them propagated by ſowing their ſeeds in autumn, upon a north border, where they are ſcreened from the ſun. The plants will come up the following ſpring, when they muſt be kept clean from weeds during the ſummer months, and in very dry ſeaſons, if they are frequently refreſhed with water, it will greatly promote their growth; the following autumn they ſhould be tranſplanted into ſhady borders, in rows a foot aſunder, and the plants at ſix inches diſtance in the rows. In this ſituation they may remain two years, by which time they will be ſtrong enough to flower, ſo may be tranſplanted to the garden where they are deſigned to remain.

As theſe plants rarely flower in leſs time than three years from ſeeds, ſo they are generally propagated by parting of their roots; for when they are planted in a ſhady cool ſituation, the roots increaſe plentifully, eſpecially the fifth ſort; which, if not confined, will in a few years ſpread to a great diſtance. The autumn is the ſeaſon for tranſplanting and parting of their roots, and if they are planted in a loamy ſoil to a north or eaſt aſpect, they will thrive greatly.

The roots of theſe plants are thick and fleſhy, and in ſome ſorts are as large as a man's thumb; theſe put out a great number of fibres every year, which ſpread to a conſiderable diſtance every way: therefore they ſhould be allowed room, for if they have not two or three feet ſpace, they will not produce ſtrong flower-ſtalks, in which their beauty chiefly conſiſts. But the fifth ſort muſt have much more room, becauſe it ſends out offsets in great plenty to a conſiderable diſtance every way. This has been the moſt commonly cultivated in the Engliſh gardens of all the ſpecies, and the flowers are annually brought in great plenty in May to the markets for flower-pots to adorn rooms; but as it is of a very poiſonous quality, ſo it ſhould be with great caution admitted where children frequent, there having been many inſtances of its dangerous effects.

Moſt if not all the ſpecies of this genus are hurtful in a greater or leſs degree, therefore ſhould not be planted in thoſe parts of gardens where children are permitted to walk, leſt by gathering of the leaves or flowers, and putting them in their mouths, or by rubbing them about their eyes, they ſhould ſuffer by it. For the juice of the leaves will occaſion great diſorder, if only rubbed upon very tender fleſh, but if taken inwardly will kill, unleſs there is timely relief. The farina of the flowers, if accidentally blown into the eyes, will occaſion great pain and blindneſs for a time, by cauſing them to ſwell greatly, as I have myſelf experienced.

The common Monkſhood flowers in May, and is ſucceeded by the firſt and ſecond ſorts. The wholeſome Wolfsbane comes after theſe, and the other ſorts flower in Auguſt and September.

ACONITUM HYEMALE, or Winter Aconite. See HELLEBORUS.

ACORUS. The Sweet Ruſh.

This plant grows naturally in deep ſtanding waters, ſo is rarely admitted into gardens, for it will not thrive on dry land; but as the roots are uſed in medicine, ſo I would not omit the mention of it. Whoever has an inclination to propagate it, should procure some roots from the places where it grows naturally, and plant them in ditches, or close on the side of ponds, where they will thrive and increase greatly if they are not disturbed.

ACRIVIOLA. See TROPÆOLUM.

ACTÆA. Herb Christopher.

The CHARACTERS of the genus are,

The empalement of the flower is composed of four roundish concave leaves which fall off. The flower has four petals which drop off, and a great number of stamina; an oval germen with one stigma, which becomes a smooth oval berry including several roundish seeds.

The SPECIES are,

1. ACTÆA (*Spicata*) racemo ovato, fructibusque baccatis. Lin. Sp. 504. *Common Herb Christopher.*

2. ACTÆA (*Racemosa*) racemis longissimis, fructibus unicapsularibus. Lin. Sp. 504. *American Herb Christopher with the longest spikes of flowers, called Black Snake-root in America.*

3. ACTÆA (*Cimicifuga*) racemis paniculatis, fructibus quadricapsularibus. Lin. Sp. 504. *Herb Christopher with flowers disposed in panicles, and four capsules to each fruit.*

The first sort grows naturally in shady woods in some of the northern counties in England, particularly near Ingleborough-hill in Yorkshire. It is by some curious persons preserved in gardens for the sake of variety, but there is little beauty in the flowers to recommend it. This must have a shady situation and a moist soil, otherwise it will not thrive. It is propagated by seeds or parting of the roots; if by seeds, they should be sown in autumn soon after they are ripe, on a shady moist border; for if the seeds are kept out of the ground till spring, they often fail, or at least lie a year before they vegetate. The time for parting and transplanting of the roots is in autumn; they require no other culture but to keep them clear from weeds. It flowers in May, and the berries ripen in September.

The second sort is a native of North America, from whence the seeds have been brought to Europe. The fruit of this plant is frequently used in America as an antidote to poison, and to cure the bite of venomous serpents. By some persons it is used as an emetic, and is sometimes called Ipecacuana.

The roots of this sort grow large, and multiply into several heads; and when they are planted in gardens, they should be allowed three feet every way to spread, for their leaves, which are composed of many branches, will soon cover so much room. The seeds of this plant do not ripen in England, so this is propagated by parting of the roots; the best time for transplanting and parting them is in the autumn, when the leaves begin to decay. It loves a loamy soil, not too dry. If the seeds are brought over, they should be sown as soon as they arrive, in a border of loamy earth: the seeds lie a year before they grow; the seedling plants should be transplanted in the autumn.

The stalks of this sort rise five or six feet high in moist land, and sustain very long spikes of white flowers in July and August. The plants should not be often removed, for that will prevent their flowering strong.

The third sort grows naturally in Siberia, and is at present rare in England. The leaves of this sort resemble those of the Feathered Columbine; the stalks rise little more than a foot high, supporting panicles of white flowers, which appear in May. This requires a moist loamy soil and shady situation, and may be propagated as the former.

ADANSONIA. The Sour Gourd; in French, *Pain du Singe*. Monkies Bread.

The CHARACTERS are,

The empalement of the flower is cup-shaped, and cut into five parts at the top which turn backward. The flower has five roundish petals, fastened to the stamina at the base. It has a great number of stamina, which are joined and form a column at their base, but spread open above, and are crowned by prostrate summits. It has an oval germen, supporting a very long tubulous style which is variously intorted, crowned by several hairy stigmas spreading out in rays. The germen becomes a large oval woody capsule with many cells, filled with a mealy pulp, inclosing a great number of kidney-shaped seeds.

We know but one SPECIES of this genus at present,

ADANSONIA (*Bahobob.*) Juss. *The Sour Gourd, or Monkies Bread.*

This tree was first described by Prosper Alpinus, in his book of Egyptian plants; but it is now known to grow in several other countries, particularly at Senegal in Africa, where there are many trees now growing, whose stems are of much greater bulk than any other trees yet known. Mr. Adanson, who was four years in that country, to examine the natural productions of it, and is writing the natural history, measured the stems of several of these trees, which were from seventy-five to eighty feet in circumference; the greater branches of these trees, he says, are equal in size to the largest trees he had ever seen in Europe. He has not, in his account of these trees, mentioned any thing of the wood of them, or if it is used for any purposes there; but we may expect a more particular account of it, in that part of his natural history, where he is to treat of the vegetables of that country.

I have also lately received a fruit of this tree, which I was assured came from Surinam in the West-Indies, so it may probably be a native of that country. The fruit is almost as large as a man's head, the shell is woody and close, having a greenish downy coat; it is divided into ten, twelve, or fourteen cells within, which contain a good number of kidney-shaped seeds, as large as the tip of a man's little finger; these are closely surrounded with a mealy pulp of an acid taste.

The leaves of the young plants are entire, of an oblong form, about four or five inches long, and almost three broad towards the top, where they are broadest, having several veins running from the middle rib; they are of a lucid green, and stand alternately. As the plants advance in height, the leaves alter, and are divided into three parts, and afterwards into five lobes, which spread out in shape of an hand. In some of the oldest plants, I have seen leaves with seven divisions, but these are rare in the plants which are in England.

The plants rise easily from fresh seeds if they are sown in a hot-bed, and are of quick growth for two or three years, but afterwards make but little progress; the lower part of their stems then begin to swell and grow much larger than the other part, after which they do not advance much in their upright growth, but put out lateral branches, which incline to an horizontal position; the branches are covered with a light grey bark. The leaves fall off in the latter part of winter, and the young leaves do not come out till summer, so the branches are naked for near three months.

As this tree is a native of very hot countries, the plants will not thrive in the open air in England in summer, therefore they must be continually kept plunged in the bark-bed in the stove; and in warm weather the fresh air should be admitted to them every day, but in winter they must be kept warm: while the plants are in a growing state, they must be frequently refreshed with water, but when they are destitute of leaves, it must be given sparingly, for too much wet will then rot their roots. It loves a light rich loamy soil.

ADENANTHERA. Bastard Flower-fence.

The CHARACTERS are,

The empalement of the flower has five indentures; the flower has five petals, ten erect stamina having prostrate summits, whose points have globular glands, and an oblong germen supporting one style, crowned by a single stigma; the flower is succeeded by an oblong compressed pod, inclosing four or five roundish compressed seeds.

We have but one SPECIES of this genus in England, which is,

ADENANTHERA (*Pavonia*) foliis decompositis utrinque glabris. Lin. Syst. 1020. *Adenanthera with smooth decompounded leaves.*

This is a native of India, from whence the seeds have been brought to England. It grows naturally in the plains near the sea in Hiboea and Senalo, where it rises to a considerable stature; it is as large as the Tamarind-tree, spreads its branches wide on every hand, making a fine shade, so is frequently planted by the inhabitants in their gardens, and near habitations, for that purpose. The leaves of this tree are doubly winged; the flowers are small and of a yellow colour, and are disposed in a long thyrse or bunch. These are succeeded by long twisted membranaceous pods, inclosing several compressed hard seeds of a fine scarlet colour, which are lodged in the pods at a distance from each other. The inhabitants perforate these seeds, and string them for the young women, who wear them about their necks.

There is another species of this tree, which is figured and described by Rumpsius, in his History of the Amboyna plants, whose leaves are woolly on their under side, but this is not in our English gardens at present.

The sort here described requires the same treatment as the POINCIANA, and the tender kinds of ACACIA, to which articles the reader is desired to turn for the culture of it: as these agree in every part so well, as that whoever can manage one, need not fear the other thriving well with the same degree of heat and management, which renders it unnecessary to insert in this place, since it would swell the work too much.

ADIANTUM. Maidenhair.

This genus is placed in Linnæus's twenty-fourth class, intitled Cryptogamia, where he has ranged the Ferns, Maidenhairs, Polypodium, &c. with the Moss, Mushroom, and all those plants, which do not produce flowers conspicuous to the naked eye; being either concealed in their fructification, or so small as not to be perceived without the help of glasses. The first order of this class is of Ferns, &c. most of which have their flowers and seeds on the back of their leaves. There are a great number of species under this genus, which grow naturally in warm countries, but we have only two in the English gardens, viz.

1. ADIANTUM (*Capillus Veneris*) frondibus decompositis, foliis alternis, pinnis cuneiformibus lobatis pediculatis. Lin. Sp. Plant. 1096. *The officinal or true Maidenhair.*

2. ADIANTUM (*Pedatum*) frondibus pedata, foliolis pinnatis, pinnis antice gibbis incisis fructificantibus. Lin. Sp. Plant. 1095. *Canada Maidenhair.*

The first sort is the true Maidenhair, which is directed to be used in medicine; but as it does not grow naturally in England, so the Trichomanes is usually substituted for it, which is found growing wild in great plenty in several parts of England. The other is a native of the South of France, Italy, and the Levant, from whence I received the plants. It usually grows out of the joints of walls, and the fissures of rocks, so that whoever is inclinable to keep this plant in their gardens, should plant it in pots filled with gravel and lime rubbish, in which it will thrive much better than in good earth; but the pots must be sheltered under a frame in winter; otherwise the plants are often killed by the frost.

The second sort is often preserved in gardens for the sake of variety; this should be planted in pots, and treated in the same manner as the former; for although it will live through the winter in the open air in moderate seasons, yet in severe frost it is often destroyed. This sort grows naturally in Canada in such quantities, that the French send it from thence in package for other goods, and the apothecaries at Paris use it for the Maidenhair, in all their compositions in which that is ordered.

ADONIS, or FLOS ADONIS. Pheasant's Eye.

The CHARACTERS are,

The flower has a five-leaved empalement, and five or eight petals without any nectarium. It has many stamina and pointals, and the seeds are naked. It is ranged in the seventh division of Linnæus's thirteenth class.

The SPECIES are,

1. ADONIS (*Annua*) floribus octopetalis fructibus subcylindricis. Hort. Upsal. 156. *The common Adonis, or Flos Adonis, with small red flowers, of late called Red Morocco.*

2. ADONIS (*Æstivalis*) floribus pentapetalis fructibus ovatis. *The annual Adonis with pale yellow flowers.*

3. ADONIS (*Vernalis*) floribus polypetalis, fructibus obtusis, radice perenne. *Perennial Adonis with yellow flowers, by some titled Fennel-leaved black Hellebore.*

The two first sorts are annual, so perish when the seeds are ripe. If the seeds are sown in the autumn, soon after they are ripe, the plants will come up the following spring; but when the seeds are not sown till spring, they rarely come up the same year; so that when the seeds are permitted to fall on the ground, they generally succeed better than when sown by art. The first sort grows naturally in Kent, particularly by the sides of the river Medway, between Rochester and Maidstone, where it is found in great plenty in the fields which are sown with Wheat; but in the intermediate fields which are sown with spring corn, there is rarely a plant of it to be found, which shews the necessity of sowing the seeds in autumn; for those fields of spring corn, if suffered to remain undisturbed after the harvest, will abound with this plant the following year. For some years past great quantities of the flowers of this plant have been brought to London, and sold in the streets by the name of Red Morocco.

These plants will thrive best in a light soil, but may be sown in any situation, so that by sowing some in a warm situation, and others in the shade, they may be continued longer in flower. The seeds ought to be sown where the plants are to remain to flower, for they do not bear transplanting well, unless it is done when the plants are young, and therefore they should be sown in small patches in the borders of the flower-garden; and when the plants come up they should be thinned, leaving but few in each patch, which will make a better appearance than where they grow single.

The third sort hath a perennial root and an annual stalk. This grows naturally on the mountains of Bohemia, Prussia, and other parts of Germany, but has been long cultivated in gardens. It produces its flowers the latter end of March, or the beginning of April, according to the forwardness of the season; the stalks rise about a foot and a half high, and when the roots are large, and have stood unremoved for some years, they will put out a great number of stalks from each root; these are garnished with fine slender leaves, which are placed in clusters at intervals. At the top of each stalk is produced one large yellow flower, composed of an unequal number of petals, the center of which is occupied by a great number of germen, surrounded by many stamina; after the flowers drop, the germen become naked feeds, clofely adhering to the foot-ftalk, forming an obtufe fpike.

This fort is propagated by feeds, which muft be fown in the autumn foon after they are ripe, on an eaft border, where they may have only the fun in the forenoon: when the plants come up the following fpring, they muft be kept clean from weeds, and in very dry weather if they are watered, it will greatly promote their growth. The following autumn the plants fhould be carefully taken up, and planted in a nurfery-bed at four or five inches diftance, where they may remain two years to acquire ftrength, then may be tranfplanted into the pleafure-garden, where they may remain for good, becaufe thefe plants do not bear tranfplanting well when they are old.

ADOXA. Mofchatellina. Tuberous Mofchatel, or Hollow Root.

This plant grows naturally in fhady woods in feveral parts of England, fo is feldom kept in gardens; therefore all that is neceffary to be inferted of its culture is, to plant it in a fhady moift part of the garden, where it will thrive faft enough: this plant has been found growing wild in North America.

ÆSCHYNOMENE. The falfe Senfitive Plant.

The CHARACTERS are,

The flower is of the butterfly kind, having ten ftamina in two bodies; the cup is divided into two lips; the pod is erect, compreffed, and jointed.

The SPECIES are,

1. ÆSCHYNOMENE (*Afpera*) caule fcabro leguminum articulis medio fcabris. Lin. Sp. Plant. 713. *Baftard Senfitive Plant with a rough ftalk and a jointed pod.*

2. ÆSCHYNOMENE (*Americana*) caule herbaceo hifpido foliolis acuminatis, leguminum articulis femicordatis. Prod. Leyd. 384. *Baftard Senfitive Plant with a prickly ftalk, pointed leaves, and jointed pods half rounded.*

3. ÆSCHYNOMENE (*Arborea*) caule lævi arboreo leguminum articulis femicordatis glabris. Prod. Leyd. 384. *Baftard Senfitive Plant with a fmooth tree-like ftalk, and fmooth jointed pods.*

Thefe plants are natives of warm countries; the feeds of the two firft forts I have received from Africa, and thofe of the third from America, and alfo from China, and feveral parts of India.

They are generally kept in botanic gardens, but are feldom preferved in any other, as there is little beauty in their flowers, and as they are plants of no ufe; befide, they require a good ftove to preferve them in England. The firft and third forts may be preferved through the winter in a bark-bed in the ftove; but as their leaves and ftalks are fucculent, fo they fhould have but little water given to them in cold weather, for much wet at that feafon will caufe them to rot. The fecond year the plants will flower, and fometimes will perfect their feeds in England.

The fecond fort will perfect its feeds the fame year it is raifed from feeds, and if kept under a frame, or in an airy glafs-cafe, fo is generally treated here as an annual plant, tho' it may be preferved through the winter in a ftove.

Thefe plants are propagated by feeds, which fhould be fown on a hot-bed early in the fpring, and when the plants have ftrength enough to be removed, they fhould be put each into a feparate fmall pot, filled with light earth, and plunged into a frefh hot-bed to bring them forward; and as they advance in their growth, they fhould be fhifted into larger pots, but great care fhould be taken not to over-pot them, for if the pots are too large the plants will not thrive. They muft be brought forward early in the year, otherwife the fecond fort will not perfect its feeds.

ÆSCULUS. Lin. Gen. 420. The Horfe Cheftnut.

The title which Dr. Linnæus has applied to this genus, might, with greater propriety, have been given to the Cheftnut, which by that author is joined to the Beech-tree, making it only a fpecies of that genus.

The CHARACTERS are,

The empalement of the flower is flightly cut into five fegments; the flower is compofed of five unequal petals, folded at their border, and waved; it has feven ftamina; the empalement becomes a thick, roundifh, echinated capfule, opening into three cells, in one or two of which are lodged globular feeds.

We have but one SPECIES of this genus, viz.

ÆSCULUS (*Hippocaftanum*) floribus heptandriis. Hort. Upfal. 92. *The Common Horfe Cheftnut.*

The Horfe Cheftnut was brought from the northern parts of Afia about the year 1550, and was fent to Vienna about the year 1588. It was called Caftanea from the fhape of its fruit, and the title of Equini was added to it from its being a good food for horfes when ground.

This tree was in much greater efteem formerly than at prefent, for fince it is become fo very common, few perfons regard it. What has occafioned its being fo feldom planted, is the decay of the leaves early in fummer, fo that their leaves frequently begin to fall in July, and occafion a litter from that time until all the leaves are fallen; but notwithftanding this inconvenience, the tree has great merit, for it affords a noble fhade in fummer; and during the month of May, there is no tree has greater beauty, for the extremity of the branches are terminated by fine fpikes of flowers, fo that every part of the tree feems covered with them, which are finely fpotted with a Rofe colour, and thefe being intermixed with the green leaves make a noble appearance.

As this tree is quick in its growth, fo in a few years it will arrive to a fize large enough to afford a good fhade in fummer, as alfo to produce plenty of flowers. I have known trees which were raifed from nuts, in twelve or fourteen years large enough to fhade two or three chairs under the fpread of their branches, and have been covered with flowers in the feafon, fo that few trees make greater progrefs than thefe. But as their wood is of little value, fo the trees fhould not be propagated in too great plenty: a few therefore of them placed at proper diftances in parks for ornament, is as many as fhould be preferved, the wood not being fit even for burning, nor any other ufe that I know of.

Thefe trees are propagated by fowing of the nuts, the beft time for doing this is early in the fpring; but the nuts fhould be preferved in fand during the winter, otherwife they are apt to grow mouldy and rot. They may indeed be planted in autumn, but then they will be in danger of rotting if the winter fhould prove very wet, or be eat by vermin.

When the nuts fucceed and have a proper foil, the plants will fhoot near a foot the firft fummer; fo that where they grow pretty clofe together, it will be proper to tranfplant them the following autumn, when they ought to be planted in rows at three or four feet diftance, and one foot and an half afunder in the rows: in this nurfery they may remain two years, by which time they will be fit to plant where they are defigned to be continued; for the younger thefe trees are planted out, the larger they will grow. But there are many who will object to their being planted out young in parks, becaufe they will require a fence to fecure them againft the cattle; which will alfo be neceffary, whatever fize they are when planted; and if large, they muft be well ftaked to prevent their being difplaced by ftrong winds, which is another expence: fo that when we confider how much fafter a young tree will grow, than thofe which are removed at a greater age, there can be no excufe for planting large trees.

When thefe trees are tranfplanted, their roots fhould be preferved as entire as poffible, for they do not fucceed well when torn or cut; nor fhould any of the branches be fhortened, for there is fcarce any tree which will not bear amputation better than this; fo that when any branches are by accident broken, they fhould be cut off clofe to the ftem, that the wound may heal over.

There is fomething very fingular in the growth of thefe trees, which is, the whole fhoot being performed in lefs than three weeks after the buds are opened; in which time I have meafured fhoots a foot and a half long, with their leaves fully expanded.

In Turkey the nuts of this tree are ground, and mixed with the provender for their horfes, efpecially thofe which are troubled with coughs, or are broken winded, in both which diforders they are accounted very good. Deer are very fond of the fruit, and at the time of their ripening will keep much about the trees, but efpecially in ftrong winds, when the nuts are blown down, which they carefully watch, and greedily devour as they fall.

There are in fome gardens a few old trees now ftanding, which were planted fingle, at a great diftance from any other; thefe are grown to a very large fize, and their heads form a natural parabola, and when their flowers are in full beauty, there is not any tree yet known in Europe, which makes fo fine an appearance. I have meafured fome of thefe trees, whofe branches have extended more than thirty feet in diameter, and their heads have been fo clofe, as to afford a perfect fhade in the hotteft feafons. Thefe were planted in 1679, as appears by fome writings which are in the poffeffion of the perfons who have now the property of the land where they grow: fo that although they are of quick growth, yet they are not of very fhort duration.

AGAVE. Lin Gen. 390. Common American Aloe.

The CHARACTERS are,

The flower has no empalement; it is erect and fpreads open at the brim. It has fix erect ftamina, crowned by narrow fummits longer the petals; after the flower is paft, the germen becomes an oblong three-cornered feed-veffel, having three cells, which are filled with flat feeds.

Dr. Linnæus has feparated the plants of this genus from the Aloe, to which they had been joined by former botanifts, becaufe the ftamina and ftyle in thefe flowers are extended much longer than the corolla, and the corolla reft upon the germen, which in the Aloe are not fo. We may alfo mention another difference in the growth of the plants, by which they may be diftinguifhed before they flower; which is, all the plants of this genus have their center leaves clofely folding over each other, and embracing the flower-ftem which is formed in the center; fo that thefe never flower until all the leaves are expanded, to give the ftem its liberty to advance, and when the flower is paft the plants die.

The SPECIES are,

1. AGAVE (*Americana*) foliis dentato-fpinofis fcapo ramofo. Gen. Nov. 1102. *The common Great American Aloe, with a branching ftalk.*

2. AGAVE (*Virginica*) foliis dentato-fpinofis fcapo fimpliciffimo. Lin. Sp. Plant. 323. *Great American Aloe with a fimple ftalk.*

3. AGAVE (*Fœtida*) foliis integerrimis. Gen. Nov. Lin. Sp. 323. *American Aloe with ftiff whole leaves, called Piet.*

4. AGAVE (*Tuberofâ*) radice tuberofâ foliis longiffimis marginibus fpinofis. *Smaller American Aloe, with a tuberous root and very long leaves, with fpines on their edges.*

5. AGAVE (*Vivipara*) foliis reflexis, marginibus dentatis. *American Aloe with fpear fhaped reflexed leaves whofe edges are indented, called Sobolifera.*

6. AGAVE (*Karattò*) foliis longis erectis læte virentibus, marginibus fufcis minimè ferratis. *American Aloe with long deep green leaves edged with brown, and very flightly fawed. This is called in America Karattò.*

7. AGAVE (*Vera Cruz*) foliis oblongis marginibus fpinofiffimis nigricantibus. *American Aloe with oblong leaves, whofe edges are clofely befet with black fpines, commonly called broad-leaved Aloe from Vera Cruz.*

8. AGAVE (*Rigida*) foliis lineari-lanceolatis integerrimis rigidis aculeo terminatis. *Narrow-leaved Aloe from Vera Cruz.*

The firft fort here mentioned, has been long preferved in the Englifh gardens, where of late years there hath been feveral of the plants in flower. The ftems of this when the plants are vigorous, generally rife upward of twenty feet high, and branch out on every fide toward the top, fo as to form a kind of pyramid: the flender fhoots being garnifhed with greenifh yellow flowers which ftand erect, and come out in thick clufters at every joint.

When thefe plants flower they make a fine appearance, and continue a long time in beauty, if they are protected from the cold in autumn, as there will be a fucceffion of new flowers produced, for near three months, in favourable feafons. It has been generally believed, that this plant doth not flower until it is an hundred years old; but this is a great miftake, for the time of its flowering depends on the growth of the plants; fo that in hot countries where they grow faft, and expand many leaves every feafon, they will flower in a few years; but in colder climates, where their growth is flow, it will be much longer before they fhoot up their ftem. There is a variety of this fort with ftriped leaves, which is now pretty common in the Englifh gardens.

The plants of the fecond fort are fo like thofe of the firft, as not to be diftinguifhed from them but by good judges. The principal difference is, the leaves of this are narrower toward their extremity and of a paler colour: the ftems of this fort do not rife fo high as the firft, nor do they branch in the fame manner, but the flowers are collected into a clofe head at the top; they are however of the fame fhape and colour. There has been three or four plants of this fort which have lately flowered in England, one of which was in the Chelfea garden a few years paft. This fort feldom puts out fo many offsets as the common Aloe.

The feventh fort greatly refembles thefe, fo that many perfons have fuppofed it to be the fame. But the leaves of this are much thinner, the indentures on their edges abundantly clofer, and not fo deep as in either of the former; the fpines are alfo blacker. How this differs from others in flower I know not, as none of their flowers have been produced in England fo far as I know.

Thefe three forts are hardy. I have known plants of the firft fort live in the open air for fome years in mild feafons, but in fevere winters they are always killed, if not fheltered in that feafon. They are propagated by offsets, which the firft fort fends out in plenty, but the third feldom puts out any; fo thefe may be increafed by taking off fome of the larger roots, at the time when the plants are fhifted, planting them in pots filled with light fandy earth, they will fhoot out and become good plants in two years, as I have often experienced. Thefe fhould be planted in pots filled with light fandy earth, and houfed in winter with Oranges, Myrtles, &c. and during that feafon fhould have but little wet. In the fummer they muft be placed abroad in the open air, where they may remain till toward the end of October, when they fhould be houfed again. The feventh fort being a little tenderer than the other two, fhould be put into the green-houfe before them, and may ftay there a little longer in the fpring.

The

The third sort hath long narrow stiff leaves, of a pale green colour, not indented on their edges, but frequently a little waved; the side leaves spread open, but those in the center fold closely over each other, and strictly surround the bud. The plants of this sort rarely grow more than three feet high, but the flower-stem rises near twenty, and branches out much like that of the first, but more horizontally; the flowers are of the same shape, but smaller, and of a greener colour: after the flowers are past, instead of seed-vessels, young plants succeed to every flower, so that all the branches are closely beset with them. This sort never produces offsets from the root, so that it cannot be increased but when it flowers, at which time there will be plenty enough; the old plant presently after dies.

The fourth sort hath leaves somewhat like the third in shape and colour, but they are indented on the edges, and each indenture terminates in a strong thorn; the root of this sort is thick, and swells just above the surface of the ground; in other respects it agrees with the former. Dr. Linnæus supposes this sort to be the same with the third species, but whoever sees the plants, will not doubt of their being different.

The fifth sort never grows to a large size, the leaves of it are seldom more than a foot and a half long, and about two inches and a half broad at their base; these end in a slender spine, being slightly indented on their edges; they are also reflexed backward toward their extremity, and are of a dark green colour. The flower-stem rises about twelve feet high, and branches out toward the top in the same manner as the third sort; the flowers are nearly of the same size and colour as those of the third, and after they fall off, are succeeded by young plants in the same manner.

The leaves of the sixth sort are from two feet and a half to three feet long, and about three inches broad, being of a dark green colour ending in a black spine; the borders of the leaves are of a brownish red colour, and slightly serrated. These stand more erect than in the other species; but as this sort hath not flowered in England, I cannot say how it differs frow the other. The plants of it were sent me from St. Christopher's by the title of Karattò, which I suppose is given indifferently to other species of this genus, for I have frequently heard the inhabitants of America call the common great Aloe by the same name.

The eighth sort hath long narrow stiff leaves, which are entire, and are terminated by a stiff black spine. These leaves are seldom more than two feet long, and little more than an inch broad, being of a glaucous colour: the side leaves stand almost horizontally, but the center leaves are folded over each other, and inclose the flower-bud. This sort never puts out suckers from the root, nor have I seen any plants of this kind in flower, although there are many of them in the English gardens, some of which are of a considerable age.

The third, fourth, fifth, sixth, and eighth sorts, are much tenderer than the others, so cannot be preserved thro' the winter in England, unless they are placed in a warm stove; nor will they thrive if set abroad in summer, therefore they should constantly remain in the stove, observing to let them enjoy a great share of free air in warm weather. They require a light sandy earth, and should have little wet in winter; but in warm weather may be gently watered twice a week, which is as often as is necessary; for if they have much water given them it rots their roots, and then their leaves will decay and insects infest them. They should be shifted every summer into fresh earth, but must not be put into large pots, for unless their roots are confined the plants will not thrive.

AGERATUM. Lin. Gen. Plant. 842. Bastard Hemp Agrimony.

The Characters are,

The flower has a naked receptacle; it has five bristly hairs, an oblong cup almost equal, and the style is scarce any longer.

The Species are,

1. Ageratum (*Conyzoides*) foliis ovatis caule piloso. Lin. Sp. Plant. 839. *Bastard Hemp Agrimony, with oval leaves and a hairy stalk.*

2. Ageratum (*Houstonianum*) foliis oppositis petiolatis crenatis, caule hirsuto. *Bastard Hemp Agrimony, with leaves having long foot-stalks placed opposite, whose edges are bluntly indented, and a hairy stalk.*

3. Ageratum (*Altissimum*) foliis ovato cordatis rugosis floralibus alternis, caule glabro. Lin. Sp. Plant. 839. *Bastard Hemp Agrimony, with rough oval heart-shaped leaves, flower branches growing alternate, and a smooth stalk.*

The two first are annual plants; the seeds of these must be sown on a hot-bed in the spring, and when the plants are come up and are strong enough to remove, they should be transplanted into another moderate hot-bed, observing to water and shade them until they have taken root; after which time they must have a good share of air in warm weather, otherwise they will grow up very weak. In summer the plants will thrive in the open air. The seeds ripen in September and October, and when any of them scatter upon the ground, and the same earth happens to be put on a hot-bed the following spring, the plants will come up in great plenty, as they frequently do also in the open air. The first sort grows naturally in Africa, and also in the islands of America; for in tubs of earth which I received with plants from Jamaica, Barbadoes, and Antigua, I have had plenty of plants arise from the seeds which were scattered on the ground. The second sort was found growing naturally at La Vera Cruz, by the late Dr. William Houstoun, who sent the seeds to Europe, which have so well succeeded in many gardens as to become a weed in the hot-beds.

The third sort grows naturally in Carolina, but has been many years an inhabitant of the English gardens. This hath a perennial root, and an annual stalk, which dies in winter, but the roots put out fresh stalks in spring.

This sort is propagated by seeds, as also by parting of the roots; the latter method is commonly practised in England, because there are few autumns so favourable as to ripen the seeds: but the seeds are frequently brought from North America, where this plant is very common; for being light, they are easily wafted about to a great distance, where they come to maturity; so that where there are any plants growing, all the adjoining land is filled with the seeds of them.

The best time for parting and transplanting the roots of this plant is in autumn, soon after their stalks decay, that they may have good root before the drying winds in spring come on, otherwise they will not flower strong or make a good increase. The roots should be allowed three feet of room every way, for as they spread and increase very much at their roots, so when they are cramped for room the plants starve, and in dry seasons their leaves will hang as if they were dead.

AGERATUM, or MAUDLIN. See Achillæa.

AGERATUM PURPUREUM. See Erinus.

AGNUS CASTUS. See Vitex.

AGRIFOLIUM. See Ilex.

AGRIMONIA. Lin. Gen. Plant. 534. Agrimony.

The Characters are,

The empalement of the flower is indented in five parts; the flower has four or five petals, which are inserted in the empalement. In the center arises a simple style resting on the germen; it has twelve slender stamina. The empalement has two cells, which are compressed on both sides, and radiated lengthways.

The

The SPECIES are,

1. AGRIMONIA (*Eupatoria*) foliis caulinis pinnatis, foliolis unuique ferratis omnibus minutis interftinctis fructibus hifpidis. Lin. *The common Agrimony.*

2. AGRIMONIA (*Minor*) foliis pinnatis, foliolis obtufis dentatis. *The white Agrimony.*

3. AGRIMONIA (*Odorata*) altiffima, foliis pinnatis foliolis oblongis acutis ferratis. *The fweet-fcented Agrimony.*

4. AGRIMONIA (*Repens*) foliis caulinis pinnatis, ftipulis caulem obtengentibus, fpica fubrotunda feffili, fructibus hifpidis. Lin. *Eaftern Agrimony.*

5. AGRIMONIA (*Agrimonoides*) foliis caulinis ternatis fructibus glabris. Hort. Cliff. 179. *Three-leaved Agrimony with fmooth fruit.*

The firft fort grows naturally in feveral parts of England by the fides of hedges, and in woods. This is the fort which is commonly ufed in medicine, and is brought to the markets by thofe perfons who gather herbs in the fields.

The fecond fort is the fmalleft of all the fpecies; the leaves of this have not fo many pinnæ as the common fort, and the pinnæ are rounder, and the indentures on their edges blunter. The fpike of flowers is flender, and the flowers fmaller and of a dirty white colour. This fort grows naturally in Italy, from whence I received the feeds, and have conftantly found that the feeds of this when fown never vary.

The third fort grows near four feet high; the leaves of this have more pinnæ than either of the other, which are longer and narrower, ending in acute points; the ferratures of the leaves are fharper than any of the other, and when handled emit an agreeable odour.

The fourth fort is of humble growth, feldom rifing above two feet high; the pinnæ of its leaves are longer and narrower than either of the former, and the fpikes of flowers very fhort and thick. The roots of this are very thick, and fpread widely under ground, by which it multiplies fafter than either of the other.

The fifth fort greatly refembles the other in the fhape of its pinnæ (or fmaller leaves,) but there are but three upon each foot-ftalk; the flower of this hath a double empalement, the outer one being fringed. Fabius Columna, and other writers on botany, have feparated it from the Agrimony, making it a diftinct genus.

All thefe forts are hardy perennial plants, which will thrive in almoft any foil or fituation, and require no other care but to keep them clean from weeds. They may be propagated by parting their roots, which fhould be done in autumn, when their leaves begin to decay, that the plants may be well eftablifhed before the fpring. They fhould not be planted nearer than two feet, that their roots may have room to fpread. They may alfo be propagated by feed, which fhould be fown in autumn, for if they are kept out of the ground till fpring, they feldom come up the fame feafon.

AGROSTEMMA. Lin. Gen. Plant. 516. Wild Lychnis, or Campion.

The CHARACTERS are,

The flower has a thick empalement of one leaf; it has five petals which are obtufe and entire, and the capfule has one cell.

The SPECIES are,

1. AGROSTEMMA (*Githago*) hirfuta calycibus corollam æquantibus petalis integris nudis. Lin. Sp. Plant. 435. *Hairy wild Lychnis, commonly called Corn Campion or Darnel.*

2. AGROSTEMMA (*Celirofa*) tomentofa foliis ovato-lanceolatis, petalis integris coronatis. Hort. Upfal. 115. *The fingle Rofe Campion.*

3. AGROSTEMMA (*Flos Jovis*) tomentofa petalis emarginatis. Lin. Sp. Plant. 436. *The umbelliferous Mountain Campion.*

The firft fort grows naturally in the corn-fields in moft parts of England, fo is feldom admitted into gardens.

The fingle Rofe Campion has been long an inhabitant of the Englifh gardens, where, by its feeds having fcattered, it is become a kind of weed. There are three varieties of this plant, one with deep red, another with flefh-coloured, and a third with white flowers, but thefe are of fmall efteem; for the double Rofe Campion being a finer flower, has turned the others out of moft fine gardens. The fingle forts propagate faft enough by the feeds, where they are permitted to fcatter, for the plants come up better from felf-fown feeds, than when they are fown by hand, efpecially if they are not fown in autumn.

The fort with double flowers never produces any feeds, fo is only propagated by parting of the roots; the beft time for this is in autumn, after their flowers are paft; in doing of this, every head which can be flipped off with roots fhould be parted; thefe fhould be planted in a border of frefh undunged earth, at the diftance of fix inches, obferving to water them gently until they have taken root, after which they will require no more, for much wet is very injurious to them, as is alfo dung After the heads are well rooted, they fhould be planted into the borders of the flower-garden, where they will be very ornamental during the time of their flowering, which is in July and Auguft. This is a variety of the fingle fort, which was firft accidentally obtained from feeds.

The third fort grows naturally upon the Helvetian mountains. This is a low plant with woolly leaves; the flower-ftem rifes near a foot and a half high; the flowers grow in umbels on the top of the ftalk, which are of a bright red colour. This flowers in June, and the feeds ripen in Auguft or September; it fhould have a fhady fituation, and will thrive beft in a ftrong foil.

AIZOON.

This name has been by fome writers applied to the Houfe-leek, and alfo to the Aloes.

The CHARACTERS are,

It hath a permanent empalement of one leaf, which is cut into five acute fegments, with a five-cornered empalement, fupporting five ftyles which are crowned, and a fimple ftigma. The germen afterward becomes a fwelling five-cornered capfule, having five cells, in which are lodged many roundifh feeds.

This genus of plants is by Dr. Linnæus ranged in the fifth divifion of his twelfth clafs, entitled Icofandria pentagynia.

The SPECIES are,

1. AIZOON (*Canarienfe*) foliis cuneiformi-ovatis floribus feffilibus. Hort. Upfal. 127. *Sempervive or Ficoidea with oval wedge-fhaped leaves, and flowers without foot-ftalks.*

2. AIZOON (*Hifpanicum*) foliis lanceolatis floribus feffilibus. Lin. Sp. Plant. 488. *Sempervive with fpear-fhaped leaves and flowers, having no foot-ftalks.*

3. AIZOON (*Paniculatum*) foliis lanceolatis floribus paniculatis. Lin. Sp. Plant. 448. *Sempervive with fpear-fhaped leaves and flowers growing in panicles.*

As we have no Englifh name for thefe plants, fo I have adopted this of Sempervive, which hath been applied to the Aloe and Sedum, both which have been alfo titled Aizoon and Sempervivum.

The firft fort is a native of the Canary Iflands: this is an annual plant, which muft be raifed on a moderate hot-bed in the fpring, and when the plants are fit to tranfplant, they fhould be carefully taken up, and planted each into a fmall pot filled with frefh light earth, and plunged into another moderate hot-bed to bring them forward; but as the weather grows warm they muft be hardened by degrees to bear the open air, into which they fhould be removed in June, placing them in a fheltered fituation, where they

will

will flower and ripen their ſeeds in September, ſoon after which the plants will periſh.

The ſecond ſort grows naturally in Spain; this is alſo an annual plant, whoſe branches trail on the ground; the flowers have no beauty, ſo theſe plants are only preſerved by thoſe who are curious in collecting rare plants for the ſake of variety.

The third ſort grows naturally at the Cape of Good Hope, from whence the ſeeds were brought to Europe. This is alſo of humble growth, and periſhes ſoon after the ſeeds are ripe.

Theſe may be propagated in the ſame manner as the firſt, and when the plants have acquired ſtrength, they may be planted in the full ground; but they require a poor ſandy ſoil, for in rich ground they will grow very luxuriant in branches, but will not flower till late in the ſeaſon, ſo rarely perfect their ſeeds; but when they are planted in dry ſand or lime rubbiſh, they will be more productive of flowers, and leſs vigorous in the branches.

ALATERNOIDES. See PHYLICA, CLUTIA, and CEANOTHUS.

ALATERNUS Evergreen Privet.

The CHARACTERS are,

It hath male and female flowers on different plants. The male flowers have an empalement of one leaf, cut into five ſegments at the brim, and five ſmall petals; at the baſe of theſe petals are faſtened ſo many ſtamina. The female flowers have a great reſemblance to the male, but have no ſtamina; in the center is placed the germen, ſupporting a trifid ſtyle crowned by a round ſtigma. The germen afterward becomes a ſoft round berry, containing three ſeeds.

The SPECIES are,

1. ALATERNUS (*Phylica*) foliis ovatis marginibus crenatis glabris. *The common Alaternus.*

2. ALATERNUS (*Glabra*) foliis ſubcordatis ſerratis glabris. *Alaternus with ſmall heart-ſhaped leaves.*

3. ALATERNUS (*Anguſtifolia*) foliis lanceolatis profundè ſerratis glabris. *Cut leaved Alaternus.*

4. ALATERNUS (*Latifolia*) foliis ovato-lanceolatis integerrimis glabris. *Broad-leaved Alaternus.*

The varieties of theſe plants are, the firſt ſort with variegated leaves, which is commonly called Bloatched Phillyrea by the nurſery-gardeners; and the third ſort with leaves ſtriped with white, and another with yellow; theſe are known by the ſilver and gold-ſtriped Alaternus: but as theſe are accidental varieties, I have omitted placing them among the number of ſpecies.

The common diſtinction of this genus from the Phillyrea is in the poſition of their leaves, which in the plants of this are placed alternately on the branches, whereas thoſe of Phillyrea are placed by pairs oppoſite; this is obvious at all ſeaſons, but there are more eſſential differences in their characters, as will be explained under the article Phillyrea.

The firſt ſort has been long cultivated in the Engliſh gardens, but the plain ſort is now uncommon here; for the bloatched-leaved ſort has been generally cultivated in the nurſeries, and the other has been almoſt totally neglected.

The ſecond ſort was formerly in the Engliſh gardens in much greater plenty than at preſent. This was generally called Celaſtrus, or Staff-tree; the leaves of this ſort are placed at greater diſtances than thoſe of the firſt, ſo that their branches appear thinly covered with them, which may have occaſioned their being diſeſteemed.

The third ſort has been an old inhabitant in ſome gardens, but was not much propagated till of late years; the leaves of this are much longer and narrower than thoſe of either of the other ſorts, and the ſerratures on their edges are much deeper.

Theſe ſorts are by ſome ſuppoſed to be only varieties and not diſtinct ſpecies; but from many repeated trials in raiſing them from ſeeds, I can affirm they do not vary, the ſeeds conſtantly producing the ſame ſpecies as they were taken from.

All theſe ſorts are eaſily propagated by laying their branches down, as is practiſed for many other trees. The beſt time for this is in autumn, and if properly performed the layers will have made good roots by the autumn following, when they may be cut off from the old ſtock, and planted either into the nurſery, or in the places where they are deſigned to remain When they are planted in a nurſery, they ſhould not remain there longer than two or three years; for they ſhoot their roots to a great diſtance on every ſide, ſo they do not remove with ſafety after ſeveral years growth. They may be tranſplanted either in the autumn or the ſpring, but in dry land the autumn planting is beſt, whereas in moiſt ground the ſpring is to be preferred.

The plain ſorts may alſo be propagated by ſowing their berries, which they produce in great plenty; but the birds are greedy devourers of them, ſo that unleſs the berries are guarded from them, they will ſoon be gone when they begin to ripen. The plants which ariſe from ſeeds, always grow more erect than thoſe which are propagated by layers, ſo are fitter for large plantations, as they may be trained up to ſtems, and formed more like trees; whereas the layers are apt to extend their lower branches, which retards their upright growth, and renders them more like ſhrubs. They will grow to the height of eighteen or twenty feet, if their upright ſhoots are encouraged; but to keep their heads from being broken by wind or ſnow, thoſe branches which ſhoot irregular ſhould be ſhortened, which will cauſe their heads to be cloſer, and not in ſo much danger.

ALCEA. Lin. Gen. 750. The Hollyhock.

The CHARACTERS are,

The flower hath a double empalement, of which the outer is permanent, and cut into ſix parts. The inner is larger and ſlightly cut into five. In the center is placed the round germen, ſupporting a ſhort cylindrical ſtyle, with many ſtamina joined below to the pentagonal column, and ſpread open at top. The germen afterward becomes a round, depreſſed, articulated capſule, having many cells, in each of which is lodged one compreſſed kidney-ſhaped ſeed.

The SPECIES are,

1. ALCEA (*Roſea*) foliis ſinuatis anguloſis. Hort. Cliff. 348. *Hollyhock with angular ſinuated leaves.*

2. ALCEA (*Ficifolia*) foliis palmatis. Hort. Cliff. 348. *Hollyhock with hand-ſhaped leaves.*

Theſe are diſtinct ſpecies, whoſe difference in the form of their leaves always continue. The leaves of the firſt ſort are roundiſh, and cut at their extremity into angles; whereas thoſe of the ſecond are deeply cut into ſix or ſeven lobes, ſo as to reſemble a hand.

The various colour of their flowers being accidental, as alſo the double flowers being only varieties which have riſen from culture, are not by botaniſts deemed diſtinct ſpecies, ſo I have not enumerated them here; therefore ſhall only mention the various colours, which are commonly obſerved in their flowers; which are white, pale red, deep red, dark red, purple, yellow, and fleſh colour.

Although theſe varieties of double Hollyhocks are not conſtant, yet where their ſeeds are carefully ſaved from the moſt double flowers, the greateſt number of the plants will approach near to the plants from which their ſeeds were taken, both as to their colour and the fulneſs of their flowers, provided no plants with ſingle or bad-coloured flowers are permitted to grow near them. Therefore as ſoon as any ſuch appear, they ſhould be removed from the good ones, that their farina may not ſpread into the other flowers, which would cauſe them to degenerate.

Theſe

Theſe plants, although natives of warm countries, yet are hardy enough to thrive in the open air in England, and have for many years been ſome of the greateſt ornaments of our gardens toward the latter part of ſummer; but ſince they have become very common, have not been ſo much regarded as they deſerve, partly from their growing too large for ſmall gardens, and their requiring tall ſtakes to ſecure them from being broken by ſtrong winds, to which they are very liable by their tall growth. But in large gardens, where they are properly diſpoſed, they make a fine appearance; for as their ſpikes of flowers grow very tall, ſo there will be a ſucceſſion of them on the ſame ſtems more than two months; the flowers on the lower part of the ſpike appearing in July, and as their ſtalks advance, ſo new flowers are produced till near the end of September.

They are propagated by ſeeds, which, as hath been already obſerved, ſhould be carefully ſaved from thoſe plants whoſe flowers are the moſt double and of the beſt colours. If theſe are preſerved in their capſules until ſpring, the ſeeds will be better, provided they are gathered very dry, and care be taken that no damp comes to them in winter, for that will cauſe their covers to be mouldy, and thereby ſpoil the ſeeds.

The ſeeds ſhould be ſown on a bed of light earth, about the middle of April, which muſt be covered about half an inch deep with the ſame light earth; ſome perſons ſow them in ſhallow drills, and others ſcatter the ſeeds thinly over the whole bed. When they are ſown in the former method, the plants generally come up thick, ſo will require to be tranſplanted ſooner than thoſe which are ſown thinly in broad-caſt. By the firſt, the ſeeds may be more equally covered, and kept clean with leſs trouble, becauſe the ground between the drills may be hoed. When the plants have put out ſix or eight leaves, they ſhould be tranſplanted into nurſery-beds, at a foot diſtance from each other, obſerving to water them until they have taken good root; after which they will require no farther care, but to keep them clean from weeds till October, when they ſhould be tranſplanted where they are to remain.

ALCHEMILLA. Ladies Mantle.

The Characters are,

The flower hath a permanent empalement of one leaf, which is cut into eight ſegments. It hath no petals, and each flower is ſucceeded by one ſeed wrapped up in the empalement.

Dr. Linnæus ranges this genus in his fourth claſs of plants, entitled Tetrandria monogynia.

The Species are,

1. Alchemilla (*Vulgaris*) foliis lobatis ſerratis, ſegmentis involucro acuto. *The common Ladies Mantle.*

2. Alchemilla (*Hybrida*) foliis lobatis ſericeis acutè ſerratis, ſegmentis involucro ſubrotundis. *Smaller ſilvery Ladies Mantle, with lobated leaves ſharply ſerrated, and the ſegments of the involucrum cut into roundiſh ſegments.*

3. Alchemilla (*Alpina*) foliis digitatis ſerratis. Flor. Lapp. 62. *Silvery Alpine Ladies Mantle with hand-ſhaped leaves.*

4. Alchemilla (*Pentaphylla*) foliis quinatis multifidis glabris. Lin. Sp. Pl. 123. *Smooth five-leaved Ladies Mantle, cut into many ſegments.*

The firſt ſort grows naturally in moiſt meadows in ſeveral parts of England, but is not very common near London. The leaves of this ſort are uſed in medicine, and are eſteemed to be vulnerary, drying and binding, and of great force to ſtop inward bleeding.

The ſecond ſort is much ſmaller than the firſt, and the leaves are much whiter and appear ſilky; this ſort is different from the third, and always continues ſo when the plants are propagated by ſeeds, ſo that there can be no doubt of its being a diſtinct ſpecies.

The third ſort grows naturally on the mountains in Yorkſhire, Weſtmoreland, and Cumberland, generally upon moiſt boggy places. The leaves of this ſort are very white, and deeply cut into five parts like a hand.

The fourth ſort grows naturally in Sweden, Lapland, and other cold countries, ſo are only to be found in ſome few curious botanick gardens in this country. Theſe are all abiding plants, which have perennial roots and annual ſtalks, which periſh in autumn. They may be propagated by parting of their roots; the beſt time for doing this is in the autumn, that their roots may be eſtabliſhed before the drying winds of the ſpring come on. They ſhould have a moiſt ſoil and a ſhady ſituation, otherwiſe they will not thrive in the ſouthern parts of England.

ALDER-TREE. See Alnus.

ALESANDER, or ALEXANDER. See Smyrnium.

ALETRIS. Lin. Gen. 428. African Aloe.

The Characters are,

The flowers have no empalement; they have one oblong oval petal, cut into ſix ſegments at the brim, which are permanent: it hath ſix awl ſhaped ſtamina the length of the corolla, whoſe baſe are inſerted in the ſegments. Theſe are crowned by oblong erect ſummits, with an oval germen, ſupporting an awl-ſhaped ſtyle the length of the ſtamina, crowned by a trifid ſtigma. The germen afterward becomes a three-cornered capſule with three cells, filled with angular ſeeds.

The Species are,

1. Aletris (*Farinoſa*) acaulis, foliis lanceolatis membranaceis, floribus alternis. Lin. Sp. 456. *Aletris with ſpear-ſhaped membranaceous leaves, and alternate flowers.*

2. Aletris (*Hyacinthoides*) acaulis foliis lanceolatis carnoſis, floribus geminatis. Lin. Sp. 456. *Aletris without ſtalks, ſpear-ſhaped leaves which are fleſhy, and flowers by pairs, commonly called Guinea Aloe.*

3. Aletris (*Zeylanica*) foliis ſublanceolatis ſeſſilibus. *Ceylon Aloe with ſeſſile leaves.*

4. Aletris (*Fragrans*) cauleſcens, foliis lanceolatis laxis. Lin. Sp. 456. *Stalky Aloe, with looſe ſpear-ſhaped leaves embracing the ſtalks, commonly called Tree Aloe.*

The firſt ſort grows naturally in Virginia, and other parts of North America, but will not live through the winter in the open air in England, but may be placed in a frame during the winter ſeaſon, where, if it is preſerved from froſts, it will flower in June or July; but theſe make no great appearance, the flowers being of a whitiſh green colour.

The ſecond ſort is tender, ſo cannot be preſerved through the winter, unleſs the plants are removed into a ſtove; nor will they produce flowers, unleſs the ſtove be kept to a moderate degree of warmth; but when rightly managed, the plants will produce long ſpikes of white flowers in July; the borders of the petals are reflexed like ſome ſorts of Hyacinths, making a fine appearance.

The third ſort is a low plant, never riſing more than ſix inches high; the roots of this ſort creeps, and thereby multiplies very faſt; but I have never ſeen one plant flower, though in the garden at Amſterdam there were ſome ſcores of this plant in the year 1727, when I was there.

The fourth ſort becomes a ſtately plant, if they are plunged into the tan-bed in the ſtove; theſe will riſe twelve or fourteen feet high, with ſtrong ſtems, ſupporting heads compoſed of many lax leaves, and annually in the months of March or April produce fine ſpikes of white flowers, which are reflexed like thoſe of the ſecond ſort; theſe open wide in the evening, when they perfume the air of the ſtove. Theſe ſend out one or two heads or tufts toward their tops, which may be cut off, and after they have laid a week in the ſtove to heal the wounded parts, they may be planted for increaſe.

The three laſt-mentioned muſt be placed in ſtoves, where they ſhould conſtantly remain, admitting a good ſhare of air in warm weather.

ALKEKENGI. See PHYSALIS.

ALLELUJAH. See OXALIS.

ALLIUM. Garlick.

The CHARACTERS are,

The flowers are included in one common ſpatha; they are compoſed of ſix concave ſpreading petals. It hath a three-cornered capſule, opening into three parts, having three cells, filled with roundiſh ſeeds.

The SPECIES are,

1. ALLIUM (*Sativum*) caule planifolio bulbifero, radice compoſitâ, ſtaminibus tricuſpidatis. Hort. Upſal. 76. *Common or manured Garlick.*

2. ALLIUM (*Scorodophraſum*) caule planifolio bulbifero, foliis crenulatis vaginis ancipitibus ſtaminibus tricuſpidatis. Hort. Upſal. 77. *The Rocambole.*

3. ALLIUM (*Urſinum*) ſcapo nudo ſemicylindrico foliis lanceolatis petiolatis umbellâ faſtigiatâ. Lin. Sp. Plant. 300. *Broad-leaved wild Garlick or Ramſons.*

4. ALLIUM (*Ampeloproſum*) caule planifolio umbellifero umbellâ globoſâ ſtaminibus tricuſpidatis radice laterali. Lin. Sp. Pl. 294. *Great round-headed Garlick of the Holm Iſlands.*

5. ALLIUM (*Moly*) ſcapo nudo ſubcylindrico foliis lanceolatis ſeſſilibus umbellâ faſtigiatâ. Hort Upſal. 76. *The yellow Moly.*

6. ALLIUM (*Magicum*) caule planifolio umbellifero ramulo bulbifero ſtaminibus ſimplicibus. Lin. Sp. Pl. 296. *Great broad-leaved Moly with Lily flowers.*

7. ALLIUM (*Seneſcens*) ſcapo nudo ancipiti foliis linearibus ſubtus convexis lævibus umbellâ ſubrotundâ ſtaminibus ſubulatis. Hort. Upſal. 79. *Greater Mountain Garlick with leaves like Narciſſus.*

8. ALLIUM (*Anguloſum*) ſcapo nudo ancipiti foliis linearibus caniculatis ſubtus ſubangulatis umbellâ faſtigiatâ. Hort. Upſal. 79. *Garlick with a naked ſtalk, narrow hollow leaves which are angular on their lower ſide, and a compact umbel.*

9. ALLIUM (*Subhirſutum*) caule planifolio umbellifero foliis inferioribus hirſutis ſtaminibus ſubulatis. Lin. Sp. Pl. 295. *Umbelliferous Garlick with hairy under leaves, and awl-ſhaped ſtamina, commonly called Dioſcoridis Moly.*

The two firſt ſpecies are eaſily propagated by planting the cloves, or ſmall bulbs, in the ſpring, in beds about four or five inches diſtance from each other, keeping them clean from weeds. About the beginning of June, the leaves of the firſt ſort ſhould be tied in knots, to prevent their ſpindling, or running to ſeed, which will greatly enlarge the bulb. In the middle of July the leaves will begin to wither and decay, at which time they ſhould be taken out of the ground, and hanged up in a dry room, to prevent their rotting, and may be thus preſerved for winter uſe.

The roots of the ſecond ſort may remain in the ground till the leaves are decayed, when their bulbs may be taken up and dried, to be preſerved for uſe during the winter ſeaſon; but ſome of the roots may be at the ſame time planted again for the ſucceeding year; for this ſort requires to be planted in autumn, eſpecially on dry ground, otherwiſe their bulbs will not be large.

The third ſort was formerly in greater eſteem than at preſent, it being rarely cultivated in gardens, but is found wild in moiſt ſhady places in many parts of England, and may be cultivated by planting the roots in a moiſt ſhady border at almoſt any time of the year; but the beſt ſeaſon is in July, juſt as the green leaves are decaying.

The fourth ſort grows naturally in the Holm Iſlands, from whence it has been tranſplanted into ſeveral gardens, where it is preſerved more for the ſake of variety than uſe.

The fifth ſort was formerly preſerved in gardens for the ſake of its yellow flowers, but having a very ſtrong Garlick ſcent, moſt people have rooted it out of their gardens.

The ſixth ſort is alſo preſerved by many perſons in their gardens for the ſake of variety, but as this hath a very ſtrong ſcent, ſo it is not very often admitted in the flower-garden.

The ninth ſort is ſometimes permitted to a place in gardens for the ſake of variety.

They are all of them very hardy, and will thrive in almoſt any ſoil or ſituation, and are eaſily propagated, either by their roots or from ſeeds; if from the roots, the beſt time is in autumn, that they may take good root in the ground before the ſpring; which is neceſſary, in order to have them flower ſtrong the following ſummer. If they are propagated by ſeeds, they may be ſown on a border of common earth, either in autumn, ſoon after the ſeeds are ripe, or in the ſpring following, and will require no farther care but to keep them clear from weeds; in the following autumn the plants may be tranſplanted into the borders where they are to remain for good.

Theſe plants produce their flowers in May, June, and July.

ALMOND-TREE. See AMYGDALUS.

ALMOND DWARF. See PERSICA.

ALNUS. The Alder-tree.

The CHARACTERS are,

It hath male and female flowers, which are produced at remote diſtances on the ſame plant; the male flowers are digeſted into a long iuli, or katkin, which is looſe, imbricated and cylindrical. The female flowers are collected into a conical ſcaly head, and are ſucceeded by ſcaly cones.

The SPECIES are,

1. ALNUS (*Rotundifolia*) foliis obverſè ovatis rugoſis. *The common, or round-leaved Alder.*

2. ALNUS (*Longifolia*) foliis ovato lanceolatis marginibus dentatis. *The long-leaved Alder.*

The firſt ſort here mentioned is the common Alder, which is propagated in England. The ſecond ſort is very common in Auſtria and Hungary, from whence I have been furniſhed with the ſeeds. The leaves of this ſort are longer, narrower, and not ſo glutinous as thoſe of the firſt, nor are they ſo rough; they are alſo of a thinner conſiſtence.

Theſe two ſorts delight in a moiſt ſoil, where few other trees will thrive, and are a great improvement to ſuch lands; they are propagated either by layers, or planting of truncheons about three feet in length. The beſt time for this is in February, or the beginning of March; theſe ſhould be ſharpened at one end, and the ground looſened with an inſtrument before they are thruſt into it, leſt by the ſtiffneſs of the ſoil the bark ſhould be torn off, which may occaſion their miſcarriage. Theſe truncheons ſhould be thruſt into the earth at leaſt two feet, to prevent their being blown out of the ground by ſtrong winds, after they have made ſtrong ſhoots. The plantations ſhould be cleared from all ſuch weeds as grow tall, otherwiſe they will over-bear the young ſhoots; but when they have made good heads, they will keep down the weeds, and will require no farther care.

If you raiſe them by laying down the branches, it muſt be performed in October; and by the October following, they will have taken roots ſufficient to be tranſplanted out; which muſt be done by digging a hole, and looſening the earth in the place where each plant is to ſtand, planting the young tree at leaſt a foot and a half deep, cutting off the top to about nine inches above the ſurface, which will occaſion them to ſhoot out many branches.

The diſtance theſe trees ſhould be placed (if deſigned for a coppice) is ſix feet ſquare; and if the ſmall lateral ſhoots are taken off in the ſpring, it will very much ſtrengthen your upright poles, provided you leave a few ſmall ſhoots

at

at distance upon the body thereof, to detain the sap for the increase of its bulk.

These trees may be also planted on the sides of brooks (as is usual for Willows) where they will thrive exceedingly, and may be cut for poles every fifth or sixth year. This wood, is in great request with the turners, and will endure a long time under ground, or to be laid in water.

ALNUS NIGRA BACCIFERA. See FRANGULA.

ALOE.

The CHARACTERS are,

The flower is of one leaf, which is cut at the top into six parts, which spread open; in the bottom of the flower is the nectarii, and the stamina are inserted in the receptacle.

This genus of plants is by Dr. Linnæus ranged in his sixth class, which is titled Hexandria monogynia.

The SPECIES are,

1. ALOE (*Mitriformis*) floribus pedunculatis cernuis corymbosis sub-cylindricis. Lin. Sp. Plant. 319. *The Mitre-shaped Aloe.*

2. ALOE (*Barbadensis*) foliis dentatis erectis succulentibus planis maculatis, floribus luteis in thyrso dependentibus. *The common Barbadoes Aloe.*

3. ALOE (*Arborescens*) foliis amplexicaulibus reflexis, margine dentatis, floribus cylindricis caule fruticosa. *Commonly called Sword Aloe.*

4. ALOE (*Africana*) foliis latioribus amplexicaulibus, margine & dorso spinosis, floribus spicatis, caule fruticoso. *Aloe with broader leaves embracing the stalks, whose edges and back are set with spines, flowers growing in spikes, and a shrubby stalk.*

5. ALOE (*Disticha*) foliis latissimis amplexicaulibus maculatis, margine spinosis floribus umbellatis. *By some called the Sope Aloe, and by others Carolina Aloe.*

6. ALOE (*Obscura*) foliis latioribus amplexicaulibus maculatis margine spinosis floribus spicatis. *Aloe with broad spotted leaves embracing the stalks, whose edges have spines, and flowers growing in a spike.*

7. ALOE (*Plicatilis*) foliis ensiformibus] inermis ancipitibus floribus laxe spicatis caule fruticoso. *Aloe with sword-shaped smooth leaves standing two ways, the flowers growing in loose spikes, and a shrubby stalk.*

8. ALOE (*Brevifolia*) foliis amplexicaulibus utrâque spinosis, floribus spicatis. *Aloe with leaves embracing the stalks which are prickly on every side, and flowers growing in spikes.*

9. ALOE (*Variegata*) floribus pedunculatis cernuis racemosis prismaticis ore patulo æquali. Lin. Sp. Plant. 321. *Aloe with hanging branching flowers having foot-stalks, and spreading equally at the brim; commonly called Partridge-breast Aloe.*

10. ALOE (*Humilis*) foliis erectis subulatis radicatis undique inerme spinosis. Hort. Cliff. 131. *Aloe with erect awl-shaped leaves, set with soft spines on every part.*

11. ALOE (*Viscosa*) floribus sessilibus infundibuli formibus bilabiatis laciniis quinque revolutis summa erecta. Lin. Sp. Pl. 322. *Aloe with funnel-shaped flowers without foot-stalks, opening in two lips, and cut into five segments which turn backward, and are erect at the top*

12. ALOE (*Spiralis*) floribus sessilibus ovatis crenatis segmentis interioribus conniventibus. Lin. Sp. Plant. 322. *Aloe with oval crenated flowers without foot-stalks, and the interior segments scarce appearing.*

13. ALOE (*Linguiformis*) sessilis foliis lingui-formibus maculatis floribus pedunculatis cernuis. *This is commonly called Tongue-Aloe.*

14. ALOE (*Margaritifera*) floribus sessilibus bilabiatis labio superiore erecto inferiore patente. Lin. Sp. Pl. 322. *Commonly called large Pearl Aloe.*

15. ALOE (*Vera*) foliis longissimis & angustissimis marginibus spinosis, floribus spicatis. *The Succotrine Aloe.*

16. ALOE (*Glauca*) caule brevi, foliis amplexicaulibus bifariam versis spinis marginibus erectis floribus capitatis. *Aloe with a short stalk, leaves standing two ways which embrace the stalk, the spines on the edges erect, and flowers growing in a head.*

17. ALOE (*Arachnoides*) sessilis foliis brevioribus planis carnosis apice triquetris marginibus inerme spinosis. *Commonly called Cobweb Aloe.*

18. ALOE (*Herbacea*) foliis ovato-lanceolatis carnosis apice triquetris angulis inerme dentatis. Hort. Cliff. 131. *Aloe with oval spear-shaped leaves having three angles at their extremities, which are indented and set with soft spines.*

19. ALOE (*Retusa*) floribus sessilibus triquetris bilabiatis labio inferiore revoluto. Lin. Sp. Plant. 322. *Aloe with flowers divided into three parts, the under lip being turned back; commonly called Cushion Aloe.*

20. ALOE (*Verrucosa*) sessilis foliis carinatis utrâque verrucosis bifariam versis. *Low Aloe with keel-shaped leaves warted on every part, and standing two ways, commonly called Pearl-tongue Aloe.*

21. ALOE (*Carinata*) sessilis foliis carinatis verrucosis apice triquetris carnosis. *Low Aloe with fleshy, keel-shaped, spotted leaves, which are triangular at their extremities.*

22. ALOE (*Ferox*) foliis amplexicaulibus nigricantibus undique spinosis. *Aloe with dark green leaves embracing the stalks, which are beset with spines on every side; commonly, but falsly called Aloe ferox.*

23. ALOE (*Uvaria*) floribus sessilibus reflexis imbricatis prismaticis. Lin. Sp. Plant. 323. *Aloe with reflexed flowers growing close to the stalk, in form of a prism, lying over each other like tiles on a house; commonly called Iris uvaria.*

The first, third, and eighteen next following sorts, are so hardy as to be kept in a warm dry green-house in winter, and may be placed in the open air in summer, in a sheltered situation; but then the plants should not have much wet, for that will rot their stems. But with this management the plants will not grow so fast, as when they are placed in a stove, though they will be stronger, and their stems will support their heads much better; nor will their leaves be so much drawn as those which are more tenderly treated, therefore this management should be preferred.

The fifth, sixth, and seventh sorts grow from one to the height of two or three feet; the leaves of the fifth and sixth are generally a foot or more in length, being broad at their base, ending in acute points; they are armed with spines on their edges, and spread out on every side of their stems. The seventh has pliable blunt leaves without spines; the stalk will rise three feet high, and put out several heads.

The eighth sort seldom rises more than a foot high, putting out from the stalk many clustering heads; the leaves are armed with short spines on every side, but those on the under side are short. This sort does not flower so frequently as many of the other.

The ninth, tenth, eleventh, twelfth, thirteenth, fourteenth, seventeenth, eighteenth, nineteenth, twentieth, and twenty-first sorts are plants of smaller growth; these seldom rise more than a foot high, and some of them not more than half that height; their leaves are of very different shapes, some of them have tongue-shaped leaves, and others have thick succulent leaves, for the most part terminating in an awl shaped point. These sorts flower every year, and some of them flower two or three times a year.

They are all of them hardy enough to live through the winter in a good green-house, provided they have not too much water, for wet in winter will soon destroy these plants, especially when they have no warmth in cold weather. In summer they may be exposed in the open air for about three months,

months, during which time they ſhould be gently watered twice a week in dry weather. But if the autumn ſhould prove cold and wet, they ſhould be removed into ſhelter earlier than in a dry ſeaſon; for if they get too much moiſture, they are very apt to rot in the winter when they have no artificial heat.

The ſecond ſort is very common in the iſlands of America, where the plants are propagated upon the pooreſt land, for to obtain the Hepatic Aloes, which is brought to England, and is uſed chiefly for horſes, being too coarſe for medicine.

The leaves of this ſort are about four inches broad at their baſe, where they are near two inches thick, and diminiſh gradually to a point. The leaves are of a pale ſea-green colour, and when young are ſpotted with white. The flower-ſtem riſes near three feet high, and the flowers ſtand in a ſlender looſe ſpike, with very ſhort foot-ſtalks, and hang downwards. This ſort is too tender to live through the winter in our climate, in a common green-houſe, therefore it ſhould be placed in a ſtove kept to a moderate degree of warmth in that ſeaſon. I have known plants of this kind, which have had an oiled cloth tied about their roots, and hung up in a warm room more than two years, and afterwards planted in pots, which have grown very well, from whence the plant has been called Sempervivum by the inhabitants of America.

Although the third ſort will live through the winter in a good green-houſe, yet it will not flower unleſs it is placed in a moderate ſhare of warmth; for the flowers of this ſort appear in December, when they make a fine appearance.

The fourth ſort is ſomewhat like the third, but the leaves are broader, and have ſeveral ſpines on their backs towards their extremities. The flowers of this grow in a looſer ſpike, and the plants never put out any ſuckers, ſo that it is very difficult to increaſe.

The fifteenth ſort is the true Succotrine Aloe, from whence the beſt ſort of Aloe for uſe in medicine is produced. This has long narrow ſucculent leaves, which come out without any order, and form large heads. The ſtalks grow three or four feet high, and have two, three, and ſometimes four of theſe heads branching out from it: the flowers of this ſort appear generally in the winter ſeaſon. It will live through the winter in a warm green-houſe, but the plants ſo managed, will not flower ſo frequently as thoſe which have a moderate degree of warmth in winter.

The twenty-ſecond ſort riſes to the height of eight or ten feet, with a ſtrong ſtem; the leaves grow on the top, which cloſely embrace the ſtalk; theſe come out irregularly, and ſpread every way; they are near four inches broad at their baſe, and diminiſh gradually to the top, where they end in a ſpine. Theſe are near two feet long, of a dark green colour, and cloſely beſet with ſhort thick ſpines on every ſide. This ſort hath not as yet flowered in England, nor does it put out ſuckers, ſo that it is difficult to increaſe. It muſt have a warm green-houſe in winter, and very little water.

The twenty-third ſort hath very long, narrow, triangular leaves, ſhaped like thoſe of the Bullruſh; the flowers are produced in cloſe thick ſpikes, upon ſtalks near three feet high. They are of an Orange-colour, ſo that when the plants are ſtrong and produce large ſpikes, they make a fine appearance. It flowers in Auguſt and September, and will live through the winter, in a warm border, cloſe to a ſouth-aſpected wall.

The ſoil in which theſe plants thrive beſt, is one half freſh light earth from a common (and if the turf is taken with it and rotted, it is much better;) the reſt ſhould be white ſea ſand and ſifted lime rubbiſh, of each of theſe two a fourth part; mix theſe together ſix or eight months at leaſt before it is uſed, obſerving to turn it over often in this time.

The middle of July is a very proper ſeaſon to ſhift theſe plants, at which time you may take them out of the pots, and with your fingers open the roots, and ſhake out as much of the earth as poſſible, taking off all dead or mouldy roots, but do not wound or break the young freſh ones: then fill the pot about three parts full of the above-mentioned earth, putting a few ſtones in the bottom of the pot to drain off the moiſture; and after placing the roots of the plant in ſuch a manner as to prevent their interfering too much with each other, put in as much of the ſame earth, as to fill the pot almoſt to the rim, and obſerve to ſhake the plant, ſo as to let the earth in between the roots; and then with your hand ſettle it cloſe to the roots of the plant, to keep it ſteady in the pot; then water them gently, and ſet them abroad in a ſhady place, where they may remain for three weeks, giving them gentle waterings, if the weather ſhould prove hot and dry.

Toward the latter end of September, in a dry day, remove them into the houſe again, obſerving to give them as much free open air as poſſible, while the weather holds warm; but, if the nights are cool, you muſt ſhut up the glaſſes, and give them air only in the day; and, as the cold increaſes, you muſt decreaſe opening the glaſſes; but obſerve to give them gentle waterings often, till the middle of October, when you muſt abate them, according to the heat of the houſe in which they are kept. For thoſe plants which are placed in a ſtove, will require to be watered at leaſt once a week, moſt part of the winter; whereas thoſe which are kept in a green-houſe without artificial heat, ſhould not be watered oftener in winter than once a month.

The tender ſorts ſhould conſtantly remain in the ſtove, or be removed in the ſummer to an airy glaſs-caſe, where they may have free air in warm weather, but be protected from rain and cold. With this management the plants will thrive and increaſe, and ſuch of them as uſually flower may be expected to produce them in beauty at their ſeaſons.

Moſt of theſe Aloes are increaſed by offsets, which ſhould be taken from the mother plant, at the time when they are ſhifted, and muſt be planted in very ſmall pots, filled with the ſame earth as was directed for the old plants; but if, in taking the ſuckers off, you obſerve that part which joined to the mother root to be moiſt, you muſt let them lie out of the ground in a ſhady dry place for about a week, to dry before they are planted, otherwiſe they are very ſubject to rot.

After planting, let them remain in a ſhady place (as was before directed in ſhifting the old plants) for a fortnight, when you ſhould remove the tender kinds to a very moderate hot-bed, plunging the pots therein, which will greatly facilitate their taking new root; but obſerve to ſhade the glaſſes in the middle of the day, and to give them a great ſhare of air.

Toward the middle of Auguſt begin to harden theſe young plants, by taking off the glaſſes in good weather, and by raiſing them at other times with props, that the air may freely enter the bed, which is abſolutely neceſſary for their growth, and to prepare them to be removed into the houſe, which muſt be done toward the end of September, and managed as before directed for the old plants.

The African Aloes, for the moſt part, afford plenty of ſuckers, by which they are increaſed; but thoſe few that do not, may be moſt of them propagated by taking off ſome of the under leaves, laying them to dry for ten days or a fortnight, as was directed for the offsets; then plant them in the ſame ſoil as was directed for them, putting that part of the leaf which did adhere to the old plant, about an inch, or an inch and an half (according to the ſize of the leaf) into the earth, giving them a little water to ſettle the earth about them; then plunge the pots into a moderate

rate

rate hot-bed, obſerving to ſcreen them from the violence of the ſun, and give them gentle refreſhings with water once in a week or ten days. The beſt ſeaſon for this is in June, that they may puſh out heads before winter.

ALOE AMERICANA MURICATA. See AGAVE.

ALOIDES. See STRATIOTES.

ALPINIA.

The CHARACTERS are,

The flower has one ſtamina and but one ſtyle; it is cut into ſix parts at the top, and has a ſwelling tube with three ſpreading lobes.

We have but one SPECIES of this genus, viz.

ALPINIA. Royen. Prod. 12. *White branching Alpinia, with leaves like the flowering Reed.*

This plant is a native of the Weſt-Indies, from whence it has been brought into ſome of the curious gardens of Europe, where it muſt be preſerved in a warm ſtove, and the pots plunged into a hot-bed of tanners bark, otherwiſe it will not thrive in this country. The leaves decay every winter, and are puſhed out from the roots every ſpring, like the Ginger and Maranta; ſo ſhould be managed in the ſame manner as is directed for thoſe two plants, and may be propagated by parting of the roots when the leaves decay. It grows naturally in moiſt places in the Weſt-Indies.

ALSINE. Chickweed.

There are ſeveral ſpecies of this plant which are common in cultivated places, and on dunghills. Theſe being ſo well known to moſt perſons, it will be needleſs to mention the ſpecies in this place.

ALTHÆA. Marſhmallow.

The CHARACTERS are,

The flower has a double calyx; the outer is cut into nine ſegments. It hath ſeveral capſules, each containing one ſeed.

The SPECIES are,

1 ALTHÆA (*Vulgaris*) foliis ſimplicibus acuminatis acutè dentatis tomentoſis. *Common Marſhmallow.*

2. ALTHÆA (*Hirſuta*) foliis trifidis piloſo-hiſpidis. Hort. Cliff. 349. *Marſhmallow with trifid, hairy, pungent leaves.*

3. ALTHÆA (*Cannabina*) foliis inferioribus palmatis ſuperioribus digitatis caule fruticoſo. Hort. Upſal. 205. *Marſhmallow with the under leaves ſhaped like a hand, the upper leaves more divided, and a ſhrubby ſtalk.*

The firſt ſort is the common Marſhmallow, which grows naturally in moiſt places in divers parts of England, and is frequently uſed in medicine. This hath a perennial root and an annual ſtalk, which periſhes every autumn. The ſtalks of this plant grow erect to the height of four or five feet; theſe are garniſhed with leaves which are hoary and ſoft to the touch, and placed alternately on the branches; the flowers come out from the wings of the leaves, which are ſhaped like thoſe of the Mallow, but are ſmaller and of a pale colour. It may be propagated faſt enough, either by ſeeds or parting of their roots. When it is propagated by ſeeds they ſhould be ſown in the ſpring, but if by parting of the roots, the beſt time is in autumn, when the ſtalks decay. It will thrive in any ſoil or ſituation, but in moiſt places will grow larger than in dry land.

The ſecond ſort grows naturally in Spain and Portugal, from both of theſe countries I have received the ſeeds. This is a low plant, whoſe branches trail on the ground, unleſs they are ſupported by ſtakes. The leaves and ſtalks are beſet with ſtrong hairs; the flowers come out at the wings of the ſtalks, and are ſmaller than thoſe of the common ſort, having purpliſh bottoms.

If the ſeeds of this ſort are ſown in April, the plants will flower in July, and the ſeeds ripen in September. They ſhould be ſown in the places where they are to remain, for as the roots ſhoot deep into the ground, ſo unleſs the plants are removed very young, they ſeldom ſurvive tranſplanting.

The third ſort has a woody ſtem which riſes to the height of four or five feet, and puts out many ſide branches. The flowers come out from the wings of the ſtalks in the ſame manner as the other ſorts, but are not ſo large as thoſe of the common Marſhmallow; they are of a deeper red colour, and the empalement is much larger. This ſort ſeldom flowers the firſt year, unleſs the ſummer proves warm; but when the plants live through the winter they will flower early the following ſummer, ſo will produce good ſeeds. This grows naturally in Hungary and Iſtria, from both which places I have received the ſeeds.

It is propagated by ſeeds, which ſhould be ſown in the ſpring in the place where the plants are to remain; or if otherwiſe, the plants muſt be tranſplanted young, elſe they will not ſucceed. They ſhould have a ſheltered ſituation and a dry ſoil, otherwiſe they will not live thro' the winter in England. When theſe plants grow in a ſtony ſoil, or in lime rubbiſh, they will be ſtinted in their growth, but they will have leſs ſap in their branches, ſo will better endure the cold of this climate. This ſort ſeldom continues longer than two years in England, but the ſeeds ripen here in kindly ſeaſons.

ALTHÆA FRUTEX. See HIBISCUS and LAVATERA.

ALYSSOIDES. See ALYSSUM and LUNARIA.

ALYSSON ALPINUM LUTEUM. See DRABA.

ALYSSON SEGETUM. See MYAGRUM.

ALYSSON SERPILLI FOLIO. See CLYPEOLA.

ALYSSON VERONICÆ FOLIO. See DRABA.

ALYSSUM. Madwort.

The CHARACTERS are,

The flower hath four petals in form of a croſs; it hath ſix ſtamina, two of which are ſhorter than the other four. The flower is ſucceeded by an indented compreſſed pod.

The SPECIES are,

1. ALYSSUM (*Saxatile*) caulibus fruteſcentibus paniculatis foliis lanceolatis undulatis integris. Prod. Leyd. 331. *Madwort with ſhrubby ſtalks, flowers growing in panicles, and whole ſpear-ſhaped waved leaves.*

2. ALYSSUM (*Halimifolium*) foliis lanceolato-linearibus acutis integerrimis caulibus procumbentibus perennantibus. Hort. Cliff. 333. *Madwort with whole ſpear-ſhaped pointed leaves, and trailing perennial ſtalks.*

3. ALYSSUM (*Spinoſum*) ramis floreis ſenilibus ſpiniformibus nudis. Hort. Cliff. 332. *Madwort whoſe older branches have naked ſpines.*

4. ALYSSUM (*Montanum*) ramulis ſuffruticoſis diffuſis foliis punctato echinatis. Hort. Upſal. 185. *Madwort with ſhrubby diffuſed branches, and leaves having prickly punctures.*

5. ALYSSUM (*Incanum*) caule erecto foliis lanceolatis incanis integerrimis floribus corymboſis. Hort. Cliff. 332. *Hoary ſhrubby Madwort.*

6. ALYSSUM (*Clypeatum*) caule erecto herbaceo ſiliculis ſeſſilibus ovalibus compreſſo planis petalis acuminatis. Lin. Sp. Plant. 651. *Madwort with an erect herbaceous ſtalk, pods growing cloſe to the ſtalks which are oval and compreſſed, and the flower leaves pointed.*

7. ALYSSUM (*Sinuatum*) caule herbaceo foliis lanceolatis dentatis ſiliculis inflatis. Lin. Sp. Plant. 651. *Madwort with an herbaceous ſtalk, ſpear-ſhaped indented leaves, and ſwollen ſeed veſſels.*

The firſt ſort is a low perennial plant with a thick fleſhy ſtem, which ſeldom riſes more than one foot high, but divides into many ſmall branches which ſpread on the ground. The branches are garniſhed with long, ſpear-ſhaped, hoary leaves, waved on their edges, and continue thro' the year. The flowers are produced in looſe panicles at the extremity of every branch, and are of a bright yellow colour, consisting of four petals, which are placed in form of a cross: these being numerous, make a fine appearance during their continuance. They appear the latter end of April or the beginning of May, and if the season is moderate will continue three weeks in beauty.

This plant is hardy, and although brought from a more southerly climate, yet, if planted in a dry, lean, or rubbishy soil, will endure our severest winters abroad. It is increased by sowing the seeds in March in a light sandy soil, or by planting cuttings in April or May, which are very apt to take root, if kept shaded in the heat of the day, and gently refreshed with water.

The second sort seldom continues above two or three years with us, and must therefore be often sown to preserve it; or if the seeds are suffered to fall and remain upon the ground, the plants will rise without any trouble. It produces at the extremity of its branches very pretty long spikes of small white flowers. This will also grow from cuttings, if planted and managed as the former.

The third sort hath ligneous branches, which rise about two feet high; these are armed with small spines; the leaves are hoary, spear-shaped, and thinly placed on the stalks without any order. The flowers are white, cross-shaped, and grow in small clusters at the extremity of the branches.

This may be propagated in the same manner as the first sort, either by seeds or slips; and when the plants grow in rubbish or on old walls, they will last much longer, and endure the cold of our winters better than those which are in a good soil. It grows naturally in Spain, Italy, and the south of France.

The fourth sort hath trailing branches which lie on the ground; these are garnished with oblong hoary leaves, which are rough to the touch. The flowers are produced in small clusters at the extremity of the branches, which are of a dark yellow colour, and are succeeded by seed-vessels shaped like those of the third sort. This grows naturally upon rocks and ruins in Burgundy, and some other parts of France, as also about Basil. It may be propagated in the same manner as the former sorts, and when it grows in rubbish the plants will continue some years, but in rich ground they seldom live through the winter in England.

The fifth sort grows to the height of two feet, having ligneous stalks which divide into several branches toward the top, and are garnished with hoary spear-shaped leaves, which are placed alternately on the branches: at the extremity of every shoot the flowers are produced in round bunches, which are small, white, and cross-shaped. This plant grows naturally in the south of France, Spain, and Italy, chiefly on rocky or gravelly soils. It flowers in June, July, August, and September, and the seeds ripen soon after, which, if permitted to scatter, the plants will come up and require little care.

The sixth sort is a biennial plant, with oblong hoary leaves placed alternately; the flowers come out from the wings of the stalk single; these grow very close to the stalk, and are succeeded by oval compressed seed-vessels shaped like those of the Lunaria, which contain many flat seeds. It is propagated by seeds, which must be sown upon dry ground or lime rubbish, and treated like the former sorts.

The seventh sort is a low spreading plant, garnished with oblong hoary leaves which continue through the year: the flowers are produced in small clusters at the extremity of the branches; these are of a bright yellow colour, consisting of four petals placed in form of a cross. This sort grows naturally in the islands of the Archipelago, but is hardy enough to live in the open air in England in a dry soil and a warm situation. It is propagated by seeds, and seldom lasts longer than two or three years.

AMARANTHOIDES. See GOMPHRENA.

AMARANTHUS. Amaranth, or Flower-gentle.

The CHARACTERS are,

It hath male and female flowers in the same plant. The flower hath no petals, but the empalement consists of three or five leaves; this is common to both sexes. The male flowers have in some species three, and in others five slender stamina. The female flowers have three short styles. The seed-vessel has one cell, in which is lodged a single globular seed.

The SPECIES are,

1. AMARANTHUS (*Tricolor*) glomerulis triandris axillaribus subrotundis ovatis. Lin. Sp. Plant. 1403. *The Amaranthus tricolor.*

2. AMARANTHUS (*Melancholicus*) glomerulis triandris axillaribus, capitellis triandris subrotundis sessilibus foliis lanceolatis acuminatis. Lin. Sp. Plant. 1403. *Amaranthus bicolor.*

3. AMARANTHUS (*Caudatus*) racemis pentandris cylindricis pendulis longissimis. Hort. Cliff. 443. *Flower-gentle with five stamina, and very long, hanging, cylindrical spikes.*

4. AMARANTHUS (*Maximus*) racemis cylindricis pendulis, caule erecto arboreo. *The Tree Amaranthus.*

The first sort has been long cultivated in gardens for the beauty of its variegated leaves, which are of three colours, viz. green, yellow, and red; these are very elegantly mixed, and when the plants are in full vigour the leaves are large, and closely set from the bottom to the top of the stalks, and the branches form a sort of pyramid, so that there is not a more beautiful plant than this when it is in full lustre. From the leaves of this plant being party-coloured like the feathers of parrots, some botanists have separated this species from the others, and constituted a genus of it by the title of Psittacus.

The second sort hath been introduced into the English gardens much later than the former. This grows to the same height with the former, and the manner of its growth greatly resembles it; but the leaves have only two colours, which are an obscure purple and a bright crimson; these are so blended as to set off each other, and when the plants are vigorous they make a fine appearance.

The third sort grows naturally in America, from whence I received the seeds: this grows with an upright stem upward of three feet high; the leaves and stalks are of a pale green colour; the spikes of the flowers are produced from the wings of the stalks, and also in clusters at the extremity of the branches: they are very long and hang downward, being of a bright purple colour. I have measured some of these spikes which were two feet and a half long, so that many of them have reached the ground.

The fourth sort hath a strong stem, which rises to the height of seven or eight feet, sending forth many horizontal branches toward the top; these are garnished with oblong, rough green leaves. At the extremity of every shoot the cylindrical spikes of a purple colour are produced, which hang downward; but these are seldom half the length of those of the former sort, and are much thicker. This is the sort of Amaranth which is directed by the College to be used in medicine.

These two sorts must be raised upon a hot-bed, and in June they may be planted into the borders of the pleasure-garden, shading them till they have taken new root, and in dry weather they should be watered. The third sort seldom fails to ripen its seeds in the open air, but the fourth sort seldom ripens its seeds abroad when the autumn proves cold and wet; therefore one plant of this sort should be potted,

potted, to be removed to shelter early in the autumn to obtain good seeds.

The two first sorts of Amaranths must be sown on a good hot-bed in February, or the beginning of March at farthest; and in about a fortnight's time, if the bed is in good temper, the plants will rise, when you must prepare another hot-bed covered with good rich light earth, about four inches thick; when this bed is in a proper temper to receive the young plants, you should raise them up with your finger so as not to break off the tender roots, and prick them into your new hot-bed about four inches distance every way, giving them a gentle watering to settle the earth to their roots; but in doing this, be very cautious not to bear the young plants down to the ground by hasty watering, which rarely rise again, or at least so as to recover their former strength in a long time, but very often rot in the stems and die quite away.

In the middle of the day keep them screened with mats from the great heat of the sun, and give them air by raising up the glasses with a small stone; and if the glasses are wet, it will be proper to turn them every day in good weather, that they may dry; for the moisture which is occasioned by the fermentation of the dung and perspiration of the plants, is of a noxious quality, and very unkindly to plants; so that if the weather happens to prove bad that you cannot turn the glasses, it will be of great service to your plants to wipe off all the moisture two or three times a day with a woollen cloth, to prevent its dropping upon the plants. When your plants are firmly rooted and begin to grow, you must observe to give them air every day more or less as the weather is cold or hot, to prevent their drawing up too fast, which greatly weakens their stems.

In about three weeks or a month's time these plants will have grown so as to meet, and will stand in need of another hot-bed, which should be of a moderate temper, and covered with the same rich earth about six inches thick, in which they should be removed, observing to take them up with as much earth about their roots as possible, and plant them seven or eight inches distance every way, giving them some water to settle the earth about their roots; but be very careful not to water them heavily so as to bear down the plants, as was before directed, and keep them shaded in the heat of the day until they have taken fresh roots; and be sure often to refresh them gently with water, and give them air in proportion to the heat of the weather, covering the glasses with mats every night, lest the cold chill your beds and stop the growth of the plants.

In the beginning of May you must provide another hot-bed, which should be covered with a deep frame that your plants may have room to grow. Upon this hot-bed you must set as many three-penny pots as can stand within the compass of the frame; these pots must be filled with good rich earth, and the cavities between each pot filled up with any common earth to prevent the heat of the bed from evaporating, and filling the frame with noxious steams; then with a trowel or some such instrument, take up your plants from the former hot-bed, with as much earth as possible to their roots, and place each single plant in the middle of one of the pots, filling the pot up with the earth before described, and settle it close to the root of the plant with your hands; water them gently as before, and shade them in the heat of the day from the violence of the sun, by covering the glasses with mats; refresh them often with water, and give them a good quantity of air in the day-time.

In about three weeks more these plants will have grown to a considerable size and strength, so that you must now raise the glasses very much in the day-time; and when the air is soft and the sun is clouded, draw off the glasses and expose them to the open air; and repeat this as often as the weather will permit, which will harden them by degrees to be removed abroad into the places where they are to remain the whole season; but it is not adviseable to set these plants out until a week in July, observing to do it when the air is perfectly soft, and if possible, in a gentle shower of rain.

Let them at first be set near the shelter of a hedge for two or three days, where they may be screened from the violence of the sun and strong winds, to which they must be inured by degrees. These plants when grown to a good stature perspire very freely, and must be every day refreshed with water if the weather proves hot and dry, otherwise they will stint, and never produce so large leaves as those which are skilfully treated.

This is the proper management in order to have fine Amaranths, which, if rightly followed, and the kinds are good, in a favourable season will produce wonderful large fine leaves, and are the greatest ornament to a good garden for upwards of two months.

There are many more species of this genus which grow naturally in the two Indies, where some of them are cultivated as esculent plants; one of which the inhabitants title Breda, the other Cullulu; but as these plants are of no beauty, so they are rarely kept in gardens here.

AMARANTHUS CRISTATUS. See Celosia.

AMARYLLIS. Lily-daffodil.

The Characters are,

It hath an oblong compressed spatha which incloses the flower-buds; it hath six awl-shaped stamina with incumbent summits. The germen turns to an oval capsule, opening in three parts, having three cells.

The Species are,

1. Amaryllis (*Lutea*) spathâ uniflorâ, corollâ æquali, staminibus declinatis. Lin. *Commonly called Autumnal Narcissus.*

2. Amaryllis (*Atamusco*) spathâ uniflorâ, corollâ æquali, pistillo declinato. Hort. Cliff. 135. *Commonly called Atamusco Lily.*

3. Amaryllis (*Formosissima*) spathâ uniflorâ, corollâ inæquali, genitalibus declinatis. Hort. Cliff. 135. *Commonly called Jacobæa Lily.*

4. Amaryllis (*Sarniensis*) spathâ multiflorâ, corollis revolutis genitalibus strictis. Hort. Upsal. 75. *Commonly called Guernsey Lily.*

5. Amaryllis (*Belladonna*) spathâ multiflorâ, corollis campanulatis æqualibus, genitalibus declinatis. Hort. Cliff. 135. *Commonly called Belladonna Lily.*

6. Amaryllis (*Regius*) spathâ multiflorâ, corollis campanulatis marginibus reflexis genitalibus declinatis. *Commonly called Mexican Lily.*

7. Amaryllis (*Longifolia*) spathâ multiflorâ, corollis campanulatis æqualibus, scapo compresso longitudini umbellæ. Flor. Leyd. 36. *Lily Daffodil with many flowers in one cover, the petals equal, and the cover compressed the length of the umbel.*

8. Amaryllis (*Zeylanica*) spathâ multiflorâ, corollis campanulatis æqualibus, scapo tereti ancipiti. Flor. Leyd. 36. *Commonly called the Ceylon Lily.*

9. Amaryllis (*Ciliaris*) spathâ multiflorâ, foliis ciliatis. Flor. Leyd. 37. *Commonly called the African Scarlet Lily.*

10. Amaryllis (*Vernalis*) spathâ uniflorâ, corollâ æquali, staminibus erectis. *Commonly called Spring yellow Lily Narcissus.*

11. Amaryllis (*Orientalis*) spathâ multiflorâ, corollis inæqualibus foliis linguiformibus. Buttn. *Lily Daffodil with many flowers in a cover, whose petals are unequal, and leaves shaped like a tongue; or the Brunsvigia of Dr. Heister.*

The first sort is a very hardy plant, which increases very fast by offsets. The season for transplanting these roots is any

any time from May to the end of July, when their leaves are decayed, after which it will be too late to remove them; for they will begin to push out new fibres by the middle of August, if the season be moist, and many times they flower the beginning of September, so that if they are then transplanted it will spoil their flowering. This plant will grow in any soil or situation; but it will thrive best in a fresh, light, dry soil, and in an open situation, i. e. not under the dripping of trees, nor too near walls. It is commonly called by the gardeners the yellow Autumnal Narcissus, &c. and is usually sold by them with Colchicums, for autumnal ornaments to gardens; for which purpose this is a pretty plant, as it will frequently keep flowering from the middle of September to the middle of November, provided the frost is not so severe as to destroy the flowers; for although there is but one flower in each cover, yet there is a succession of flowers from the same root, especially when they are suffered to remain three or four years unremoved. The flowers seldom rise above three or four inches high; it is shaped somewhat like the flowers of the large yellow Crocus; these have their green leaves come up at the same time like the Saffron, and after the flowers are past the leaves increase all the winter.

The tenth sort is more rare in England than any of the other at present. It was formerly in several curious gardens, but as it flowers at a season when there are so many finer sorts in beauty, so it was neglected and cast out of the gardens, whereby it is almost lost in England: it grows naturally in Spain and Portugal, where it flowers early in January. This is a hardy plant, which if planted in the open borders, and covered with glasses in winter, will flower well. It should not be taken out of the ground to transplant till the end of July or the beginning of August.

The second sort is a native of Virginia and Carolina, in which countries it grows very plentifully in the fields and woods, where it makes a beautiful appearance when it is in flower, which is in the spring. The flowers of this sort are produced singly, and at their first appearance have a fine Carnation colour on their outside; but this fades away to a pale, or almost white, before the flowers decay. This plant is so hardy as to thrive in the open air in England, provided the roots are planted in a warm situation and on a dry soil; it may be propagated by offsets from the roots, which they put out pretty plentifully, especially if they are not transplanted oftener than once in three years.

The third sort, which is commonly called Jacobæa Lily, is now become pretty common in the curious gardens in England, the roots sending forth plenty of offsets, especially when they are kept in a moderate warmth in winter: for the roots of this kind will live in a good green-house, or may be preserved through the winter under a common hot-bed frame; but then they will not flower so often, nor send out so many offsets, as when they are placed in a moderate stove in winter. This sort will produce its flowers two or three times in a year, and is not regular to any season; but from March to the beginning of September, the flowers will be produced when the roots are in vigour. There is never more than one flower produced on the same stalk. These flowers are large and of a very deep red; the under petals or flower leaves are very large, and the whole flower stands nodding on one side of the stalk, making a beautiful appearance.

It is propagated by offsets, which may be taken off every year; the best time to shift and part these roots is in August, that they may take good root before winter; in doing of this, there should be care taken not to break off the fibres from their roo They should be planted in pots of a middling size, filled with light kitchen-garden earth; and if they are kept in a moderate degree of warmth, they will produce their flowers in plenty, and the roots will make great increase.

The sixth sort, which is commonly called the Mexican Lily, is not quite so hardy as the former sort, so must be placed in a warm stove; and if the pots are plunged into a hot-bed of tanners bark the roots will thrive better, and the flowers will be strong. This sort is increased by offsets, as the others of this tribe, and it flowers usually the beginning of spring, when it makes a fine appearance in the stove; the flower-stems of this sort seldom rise more than one foot and a half high, each stem supports two, three, or four flowers, rarely more than that number. The flowers are large, and of a bright copper colour, inclining to red.

The eighth sort is also tender, and must be treated in the same manner as the sixth; this is more common in the gardens in Holland than in this country; and as it is a plant which increases but slowly, so will not be very common here. This flowers usually in June and July, and sometimes the same root will flower again in autumn: for if the pots are plunged in a bed of tanners bark, the roots generally flower twice every year, but the flowers are not of long duration.

The seventh and ninth sorts are more hardy, and may be treated in the same manner as the Jacobæa Lily; these will increase pretty fast by offsets when they are properly managed, especially the ninth, which sends out many offsets, so as to fill the pots with roots, but it seldom flowers in England. The petals of the flower turn back like those of the Guernsey Lily, but are of a lighter colour, rather inclining to scarlet; the roots of this are small.

The eleventh sort is figured by Ferrarius in his Garden of Flowers, as also by Morrison in his History of Plants; but Dr. Heister has separated it from this genus, and has constituted a new genus by the title of Brunswigia, in honour of the Duke of Brunswick.

This grows naturally at the Cape of Good Hope, from whence I have received the roots, which have succeeded in the Chelsea garden.

This sort may be treated in the same manner as hath been directed for the Jacobæa Lily, with this difference only, of placing it in winter under a warm frame where there is a moderate share of warmth, for the roots of this will not endure so much cold as those, nor should they have so much water given them.

The best time to transplant these roots is about the beginning of August, when their leaves are quite decayed, and before they put out new fibres, for it will be very improper to remove them afterwards.

All these bulbous-rooted flowers delight in a loose sandy earth mixed with good kitchen-garden mould; and in the culture of them there should be but little water given them at those times when their leaves decay, and the roots are not in a growing state, for much moisture at that time will often cause them to rot; but when they are growing and putting out their flower-stems, they should be frequently refreshed with water, but not given in too great quantities at a time. The pots with the tender sorts should constantly be kept in the stove, and in summer they should have as much free air as possible; for although some of these sorts may be kept abroad in summer, yet those do not thrive so well, nor flower so constantly as those which are treated in the manner here described.

The fifth sort, which is called the Belladonna Lily, was brought to England from Portugal, where the gardens did some years ago abound with these flowers; for the roots increase very fast, especially in such countries where they live in the open air. This plant thrives so well in Italy as to need no other culture than the common Lily; and altho' it it does not flower till Auguſt, yet it commonly produces good ſeeds in that country, from which they propagate them in great plenty; but with us they require more care, otherwiſe they cannot be preſerved.

The method in which I have cultivated this plant for ſome years paſt with great ſucceſs, is as follows. I prepared a border next a ſouth-weſt aſpected wall, of about ſix feet wide, in the following manner, viz. I removed all the earth to the depth of three feet, then I put ſome very rotten dung in the bottom ſix inches thick, upon which I laid light garden mould about twenty inches deep; after making this level, I placed the roots at ſix inches diſtance every way, and then covered them over with light ſandy earth to the height of the border, whereby the upper part of the roots were five or ſix inches buried, and in the winter I covered the border all over with rotten tanners bark three inches deep, to prevent the froſt from penetrating the ground; and when the froſt was very ſevere, I laid ſome mats or ſtraw over the leaves to protect them from being killed. With this management the roots have greatly increaſed, and have conſtantly flowered every year; ſome of them have put out two or three ſtems which grew near three feet high, and produced many flowers in each umbel, which have made a fine appearance during the month of October. This plant produces its flowers in October, and the green leaves come up ſoon after, and abide all the winter and ſpring until June, at which time they decay, ſoon after which the roots ſhould be tranſplanted.

The fourth ſort is ſuppoſed to come originally from Japan, but has been many years cultivated in the gardens of Guernſey and Jerſey, in both which places they ſeem to thrive as well as if it was their native country; and from thoſe iſlands their roots are ſent annually to the curious in moſt parts of Europe, and are commonly called Guernſey Lilies.

When theſe roots come over they ſhould be planted in pots filled with freſh, light, ſandy earth, mixed with a little very rotten dung, and placed in a warm ſituation, obſerving now-and-then to refreſh the earth with water. About the middle or end of September, ſuch of the roots as are ſtrong enough to flower will begin to ſhew the bud of their flower-ſtem (which is commonly of a red colour;) therefore you ſhould remove theſe pots into a ſituation where they may have the full benefit of the ſun, and may be ſheltered from ſtrong winds; but by no means place them too near a wall, nor under glaſſes, which would draw them up weak, and render them leſs beautiful. At this ſeaſon they ſhould be gently refreſhed with water if the weather be warm and dry, but if it ſhould prove very wet they ſhould be ſcreened from it.

When the flowers begin to open, the pots ſhould be removed under ſhelter to prevent the flowers from being injured by too much wet: but they muſt not be kept too cloſe, nor placed in a ſituation too warm, which would occaſion their colour to be leſs lively and haſten their decay. The flowers of this plant will continue in beauty (if rightly managed) a full month; and though they have no ſcent, yet for the richneſs of their colour, they are juſtly eſteemed in the firſt rank of the flowery race.

After the flowers are decayed the green leaves will begin to ſhoot forth in length, and if ſheltered from ſevere cold will continue growing all the winter; but they muſt have as much free air as poſſible in mild weather, and be covered only in great rains or froſts, for which purpoſe a common hot-bed frame is the propereſt ſhelter for them; under which if they are placed the glaſſes may be taken off conſtantly every day in dry open weather, which will encourage the leaves to grow ſtrong and broad; whereas when they are placed in a green-houſe, or not expoſed to the open air, they will grow long and ſlender, and have a pale weak aſpect, whereby the roots will become weak, ſo that it ſeldom happens that they produce flowers under ſuch management.

When a perſon is poſſeſſed of a large number of theſe roots, it will be troubleſome to preſerve them in pots, therefore there ſhould be a bed prepared of the following earth, in ſome well ſheltered part of the garden, viz. Take a third part of freſh virgin earth from a paſture ground, which is light, then put near an equal part of ſea-ſand, to which you ſhould add rotten dung and ſifted lime rubbiſh, of each an equal quantity. With this earth (when well mixed and incorporated) you ſhould make your bed about two feet thick, raiſing it about four or five inches above the ſurface of the ground, if the ſituation be dry; but if the ground be wet it ſhould be raiſed eight or nine inches higher. In this bed, about the beginning of July (as was before directed) you ſhould plant the roots about ſix or eight inches aſunder each way; and in the winter when the froſt begins, you ſhould either cover the beds with a frame, or arch it over, and cover it with mats or ſtraw to prevent their leaves from being pinched with cold; but in the ſpring the covering may be entirely removed, and the bed kept conſtantly clear from weeds during the ſummer, obſerving to ſtir the ſurface of the earth now-and-then; and every year when the leaves are decayed, you ſhould ſift a little freſh earth over the beds to encourage the roots. In this bed the roots may remain until they are ſtrong enough to produce flowers, when they may be taken up and planted in pots as was before directed, or ſuffered to remain in the ſame bed to flower.

AMBROSIA.

The Characters are,

It hath male and female flowers on the ſame plant. The male flowers are of one leaf, funnel-ſhaped, and cut into five parts at the brim. The female flowers are placed under the male in the ſame ſpike: they have no petals, but an oval germen placed in the bottom of the empalement. The germen afterward becomes an oval capſule with one cell, incloſing one roundiſh ſeed.

The Species are,

1. Ambrosia (*Maritima*) foliis multifidis racemis ſolitariis piloſis. Lin. Sp. Plant. 1401. *Maritime Ambroſia.*

2. Ambrosia (*Elatior*) foliis bipinnatifidis, racemis paniculatis terminalibus glabris. Hort. Upſal. 284. *Tall unſavoury Sea Ambroſia with Mugwort leaves.*

3. Ambrosia (*Trifida*) foliis trilobis & quinquelobis ſerratis. *The largeſt Virginia Ambroſia, with an eaſtern Plane-tree leaf.*

4. Ambrosia (*Artemiſiafolia*) foliis bipinnatifidis primoribus ramulorum indiviſis integerrimis. Lin. Sp. Plant. 1401. *Greateſt unſavoury Ambroſia of Virginia, with Water Horehound leaves which are finely divided.*

5. Ambrosia (*Arboreſcens*) foliis pinnatifidis hirſutis racemis ſolitariis terminalibus, caule fruticoſo perenne. *Ambroſia with hairy winged leaves, ſingle ſpikes of flowers growing at the extremity of the branches, and a ſhrubby perennial ſtalk.*

The firſt ſort grows naturally in the eaſt, near the ſea ſhore; this riſes about two feet and a half high, ſending out many ſide branches whoſe leaves are divided into many parts, and upon being handled emit a ſtrong odour. The ſpikes of flowers are produced from the wings of the ſtalks, which are long, ſingle, and hairy. After the flowers are paſt, the female flowers are ſucceeded by hard leafy capſules having one cell, in which is included a ſingle round ſeed. This is an annual plant which ſeldom perfects its ſeeds in England, unleſs the plants are brought forward in the ſpring; therefore the ſeeds ſhould be ſown in the autumn

tumn in a warm border, and when the plants come up in the spring they should be transplanted into another warm border, but not in rich moist land, where they generally grow very luxuriant, so do not flower till late in the season, and seldom perfect their seeds. Therefore the best method to obtain good seeds is to plant some of the plants in pots filled with light earth mixed with lime rubbish, to prevent their luxuriant growth, which will cause them to flower early, whereby good seeds may be obtained.

The second sort grows naturally in the Islands of America, as also in Carolina and Virginia; from the two latter countries I have frequently received the seeds. This sort grows more than four feet high, dividing into many branches, garnished with winged leaves in shape like those of Mugwort; at the extremity of each branch the loose spikes of flowers are produced, composed of one long spike in the middle, and three or four shorter lateral spikes: these have male and female flowers ranged in the same manner as the former; the female flowers are succeeded by seeds of the same shape.

This sort will come up and thrive in the open air in England, but the plants so raised will not produce good seeds, unless the season is warm: therefore to obtain them every year, it is necessary to sow the seeds of this plant on a moderate hot-bed in March, and when the plants are come up two inches high they must be transplanted into another hot-bed, observing to water them pretty well, and shade them until they have taken new root; afterward they must have a large share of fresh air every day when the weather is warm, and frequent waterings, for they are very thirsty plants. When the plants are grown pretty strong, they must be taken up with balls of earth to their roots, and planted in May in the open borders with other hardy annual plants, among which they will make a variety. These plants will flower in August, and their seeds ripen in September.

The third sort is a native of North America, where it is a very common weed. This often grows eight or ten feet high; and if it is planted in a rich moist soil, or is often watered, it will grow much higher and spread out into many branches. The seeds of this plant, when sown in the spring, seldom come up the first year, but frequently remain in the ground until the following spring; so that when the plants do not come up, the ground must not be disturbed till after the spring following, to wait for the plants coming up. When the plants come up, some of them may be transplanted into a moist soil, allowing them at least four or five feet room every way; and if they are frequently watered in dry weather they will grow to a large size, but their branches must be supported by stakes, otherwise they are very subject to break with strong winds. These plants are only preserved by such persons as are curious in botany, for the sake of variety.

The fourth sort grows naturally in North America, from whence I have frequently received the seeds. The spikes of flowers in this sort are produced from the wings of the stalks, in which this differs from the second. This may be treated in the same manner as the second sort.

The fifth sort is a native of Peru. It grows to the height of ten or twelve feet, with a woody stem, dividing into several branches, which are garnished with hairy leaves composed of several winged lobes, and are placed alternately upon the branches; the spikes of flowers are single, hairy, and are produced at the extremity of the branches. The female flowers are succeeded by hairy capsules, each containing a single seed.

This is a perennial plant, and may be propagated by cuttings or seeds; if by the former, they should be planted in a shady border in either of the summer months. In a month or five weeks they will have good roots, therefore should be then taken up and potted for when they are left longer in the full ground they will grow very luxuriant, so will not so soon recover their removal, as those which are transplanted earlier. The plants are hardy, so may be exposed to the open air in summer; and in the winter, if they are sheltered in a common green-house with Myrtles and other hardy exotic plants, they will live several years.

The seeds of this sort seldom come up the same year, when they are sown in spring, but those which have fallen in the autumn have grown the following year, and so have those which have been sown at the same season.

AMELLUS. Starwort.

The CHARACTERS are,

The common flower-cup is round and scaly; the flower is of the compound radiated kind; the hermaphrodite flowers compose the disk, the female the rays; the hermaphrodite are tubulous and cut into five segments; the female are tongue-shaped, and divided into two or three segments; the first have five short stamina, an oval germen, and slender style with two stigmas; the female are like them, and the flower-cup contains one seed.

The SPECIES are,

1. AMELLUS (*Lychnitis*) foliis oppositis lanceolatis obtusis, pedunculis unifloris. Lin. Sp. Plant. 1276. *Star-flower with opposite spear-shaped leaves, and one flower upon each foot-stalk.*

2. AMELLUS (*Umbellatus*) foliis oppositis triplinerviis subtus tomentosis, floribus umbellatus. Amœn. Acad. 5. 407. *Star-flower with opposite hoary leaves, and flowers disposed in umbels.*

The first sort grows naturally at the Cape of Good Hope. It is a perennial plant rising about three feet high, sending out many branches on every side, so as to form a bushy plant; the branches are garnished with obtuse spear-shaped leaves placed opposite, and are terminated by single naked flower-stalks, each supporting one Violet-coloured flower, having a yellow disk, which are succeeded by oblong seeds.

This plant is easily propagated either by cuttings planted in the summer months, or by seeds sown on a moderate hot-bed in the spring, but the plants require a slight shelter in winter.

The second sort grows naturally in Jamaica, so is much tenderer than the other; this rises from two to three feet high, sending out many branches cloathed with opposite leaves, and are terminated with small flowers in umbels.

AMETHYSTEA. Lin. Gen. 32. Amethyst.

The CHARACTERS are,

The flower has one leaf, which is cut into five equal pointed segments at the brim; it hath two slender stamina which stand under the upper lip. After the flower is past, the germen becomes four naked seeds shut up in the empalement.

We know but one SPECIES of this genus, viz.

AMETHYSTEA (*Cærulea.*) Hort. Upsal. 9. *Mountain upright Amethyst.*

This plant is a native of the mountains of Siberia, from whence the seeds were sent to the Imperial Garden at Petersburgh.

It is annual, and hath an upright stalk which rises about a foot high, and toward the top puts out two or three small lateral branches, garnished with small trifid leaves sawed on their edges, and of a very dark green colour; at the extremity of the branches the flowers are produced in small umbels; these are of a fine blue colour, as are also the upper part of the branches, and the leaves immediately under the umbel; so that although the flowers are small, yet from their colour with those of the upper part of the stalks, the plants make a pretty appearance during their continuance in flower. If the seeds of this plant are sown in the autumn, or are permitted to scatter, the plants will come up early

the

the following ſpring, and theſe will flower the beginning of June; but thoſe which are ſown in the ſpring, will not flower till July.

When the plants come up they will require no other care but to keep them clean from weeds, and where they are too cloſe to thin them, for they do not thrive when tranſplanted, therefore the ſeeds ſhould be ſown where they are to remain.

AMMANNIA. Houſt. Nov. Gen.

The CHARACTERS are,

It hath a bell-ſhaped empalement, divided at the brim into four ſlender parts. The flower hath no petals; it has four ſlender ſtamina which are inſerted in the empalement. The empalement afterward becomes a round capſule with four cells, which are filled with ſmall ſeeds.

The SPECIES are,

1. AMMANNIA (*Latifolia*) foliis ſemiamplexicaulibus, caule tetragono. Hort. Cliff 344. *Ammannia with a ſquare ſtalk, and leaves embracing it half round.*

2. AMMANNIA (*Ramoſior*) foliis ſubpetiolatis caule ramoſa. Lin. Sp. Plant. 120. *Ammannia with leaves having ſhort foot-ſtalks and a branching ſtalk.*

The firſt ſort grows naturally in moiſt places in Jamaica, from whence Dr. Houſtoun ſent the ſeeds to England.

It grows about a foot high, with an upright ſquare ſtalk, and long leaves ſet in form of a triangle, whoſe baſe half ſurrounds it. They are of a pale green, and of the conſiſtence with thoſe of Purſlane; the ſtalks are alſo ſucculent, and of the ſame colour with thoſe of the plant. The flowers come out in whorles round the ſtalks at the joints, where the leaves adhere in cloſe cluſters, and are ſoon ſucceeded by round ſeed-veſſels, which are full of ſmall ſeeds.

Theſe plants muſt be raiſed on a hot-bed in the ſpring, and afterward removed to another hot-bed to bring them forward; when they have acquired ſtrength they ſhould be tranſplanted into pots filled with rich light earth, and placed under a frame, or in a glaſs-caſe or ſtove to ripen their ſeeds, for the plants are too tender to thrive in the open air in this country, unleſs the ſummer proves very warm.

The ſecond ſort grows naturally in Virginia and Carolina; this is an annual plant which riſes about a foot high, with red ſucculent ſtalks, putting out ſide branches which grow oppoſite: the flowers are produced ſingle from the wings, on the lower part of the branches. Theſe have no beauty, ſo are only preſerved in botanic gardens for the ſake of variety. This ſort will perfect its ſeeds in the open air, if the plants are raiſed on a hot-bed in the ſpring, and planted in a warm border.

AMMI. Biſhops-weed.

The CHARACTERS are,

It is an umbelliferous plant; the flowers are difform, each having five heart-ſhaped petals. They have five ſlender ſtamina, and two reflexed ſtyles crowned with obtuſe ſtigmas. The germen afterward becomes a ſmall round ſtriated fruit, compoſed of two ſeeds.

The SPECIES are,

1. AMMI (*Majus*) foliis inferioribus pinnatis lanceolatis ſerratis, ſuperioribus multifidis linearibus. Hort. Upſ. 59. *Common Biſhops-weed.*

2. AMMI (*Glaucifolium*) foliorum omnium lacinulis lanceolatis. Guett. 2. p. 433. *Biſhops-weed with all its leaves cut in ſhape of a ſpear.*

The firſt ſort is annual; of this there is a variety which is mentioned by John Bauhin as a diſtinct ſpecies, under the title of Ammi majus foliis plurimum inciſis & nonnihil criſpis; but I have frequently had this variety ariſe from the ſeeds of the former, ſo I have not enumerated it as a different ſort.

This plant is propagated by ſeeds, which ſhould be ſown in the autumn in the place where it is to remain; and in the ſpring the plants ſhould be thinned in the ſame manner as is practiſed for Carrots, leaving them four or five inches aſunder, for they will grow large and cover the ground; after this they will require no farther care but to keep them clean from weeds. In June they will flower, and their ſeeds will ripen in Auguſt, which ſhould be gathered as it ripens, otherwiſe it will ſoon ſcatter. Theſe ſeeds are uſed in medicine, ſo may be had in plenty with this management.

The ſecond ſort is a perennial plant, which is preſerved in botanic gardens for variety, but having little beauty is rarely admitted into other gardens. It may be propagated by ſeeds, which ſhould be ſown in the autumn, becauſe thoſe ſown in the ſpring ſeldom come up the ſame year. It will grow in any open ſituation, is very hardy, and thrives beſt on a moiſt ſoil.

AMMI PERENNE. See SIUM.

AMOMUM. Lin. Gen. Plant. 2. Ginger.

The CHARACTERS are,

The flower is of one leaf, divided into four parts at the brim. In the boſom of the flower is ſituated an oblong thick nectarium. Under the receptacle of the flower is placed the round germen, which afterward becomes an oval three-cornered ſeed-veſſel, opening in three parts, containing ſeveral ſeeds.

The SPECIES are,

1. AMOMUM (*Zinziber*) ſcapo nudo ſpicâ ovato. Hort. Cliff. 3. *Ginger.*

2. AMOMUM (*Zerumbet*) ſcapo nudo ſpicâ oblongâ obtusâ. Hort. Cliff. 3. *Broad-leaved wild Ginger, called Zerumbet.*

The firſt, which is the common Ginger, is cultivated for ſale in moſt of the iſlands of America, but is a native of the Eaſt-Indies, and alſo of ſome parts of the Weſt-Indies, where it is found growing naturally without culture. The dried roots of this ſort furniſh a conſiderable export from the Britiſh colonies in America. Theſe roots are of great uſe in the kitchen, as alſo in medicine; and the green roots preſerved as a ſweetmeat, are preferable to every other ſort.

The roots of this ſort are jointed, and ſpread in the ground; theſe put out many green Reed-like ſtalks in the ſpring, which riſe to the height of two feet and a half, with narrow leaves. The flower-ſtems afterwards ariſe by the ſide of theſe immediately from the root; theſe are naked, ending with an oblong ſcaly ſpike; from each of theſe ſcales is produced a ſingle blue flower, whoſe petals are but little longer than the ſquamoſe covering.

The ſecond ſort grows naturally in India; the roots of this are much larger than thoſe of the firſt, but are jointed in the ſame manner. The ſtalks grow from three to near four feet high, with oblong leaves placed alternately. The flower-ſtems ariſe immediately from the root, theſe are terminated by oblong, blunt, ſcaly heads; out of each ſcale is produced a ſingle white flower, whoſe petals extend a conſiderable length beyond the ſcaly covering.

Theſe ſorts are tender, and require a warm ſtove to preſerve them in this country. They are eaſily propagated by parting of their roots; the beſt time for doing this is in the ſpring before they put out new ſhoots, for they ſhould not be tranſplanted in ſummer when they are in full vigour; nor do they ſucceed ſo well when they are removed in autumn, becauſe they remain long after in an inactive ſtate, and during that time, if wet comes to the roots, it often cauſes them to rot. When the roots are parted they ſhould not be divided into ſmall pieces, eſpecially if they are deſigned to have flowers, for until the roots have ſpread to the ſide of the pots, they rarely put out flower-ſtems, for which reaſon they ſhould not be planted in very large pots.

The pots with theſe roots ſhould conſtantly remain plunged in the tan-bed, for if they are taken out and placed on ſhelves in the ſtove, their fibres frequently ſhrink, which often occaſions the roots to decay.

With this management theſe ſorts have multiplied greatly with me, and the common Ginger has produced roots which have weighed five or ſix ounces, but the others have been near a pound weight.

AMOMUM PLINII. See SOLANUM.

AMORIS POMUM. See LYCOPERSICON.

AMORPHA. Lin. Gen. Plant. 768. Baſtard Indigo.

The CHARACTERS are,

The flower is of the butterfly kind, having an oval concave ſtandard, but no wings or keel; this is inſerted between the two upper ſegments of the empalement. The germen afterward becomes a reflexed moon-ſhaped pod, having one cell, in which are lodged two kidney-ſhaped ſeeds.

We know but one SPECIES of this genus, viz.

AMORPHA (*Fruticoſa.*) Hort. Cliff. 353. *Baſtard Indigo.*

This ſhrub grows naturally in Carolina, where formerly the inhabitants made a coarſe ſort of Indigo from the young ſhoots, which occaſioned their giving it the title of Baſtard Indigo.

It riſes with many irregular ſtems to the height of twelve or fourteen feet, with very long winged leaves in ſhape like thoſe of the common Acacia. At the extremity of the ſame year's ſhoots the flowers are produced in long ſlender ſpikes, which are very ſmall and of a deep purple colour. After the flowers are paſt the germen turns to a ſhort pod, having two kidney-ſhaped ſeeds, but theſe do not ripen in England.

This ſhrub is become very common in all the gardens and nurſeries near London, where it is propagated as a flowering ſhrub for the ornament of the ſhrubbery. It is generally propagated by laying down of the young branches, which in one year will make good roots, and may then be taken off and planted either in the nurſery, or in the places where they are deſigned to remain. The plants muſt have a ſheltered ſituation, otherwiſe their branches will be broken by the winds. As theſe ſhoots are large and ſoft, their upper parts are frequently killed by froſt in winter; but they put out ſhoots again in plenty below the dead part, the ſpring following.

AMYGDALUS. Lin. Gen. Plant. 545. The Almond-tree.

The CHARACTERS are,

It hath a tubulous empalement of one leaf, which is cut at the brim into five obtuſe ſegments; the flower hath five oval, obtuſe, concave petals, which are inſerted in the empalement. After the flower is paſt, the germen becomes an oval, compreſſed, large fruit, with a thin, tough, hairy covering, having a longitudinal furrow; this opens and falls away, leaving an oval compreſſed nut.

The SPECIES are,

1. AMYGDALUS (*Communis*) foliis ſerraturis infimis glandulofis, floribus ſeſſilibus geminis. Hort. Cliff. 186. *Common manured Almond-tree.*

2. AMYGDALUS (*Dulcis*) foliis petiolatis marginibus crenatis, corollis calyce vix longioribus. *The tender-ſhelled Almond, commonly called Jordan Almond.*

3. AMYGDALUS (*Orientalis*) foliis lanceolatis integerrimis, argenteis perennantibus petiolo breviore. *Almond-tree with ſpear-ſhaped ſilvery leaves, which are entire, and continue great part of winter, and very ſhort foot-ſtalks.*

4. AMYGDALUS (*Nana*) foliis petiolatis ſerratis baſi attenuatis. *Dwarf Almond with ſingle flowers.*

The firſt is the common Almond, which is cultivated more for the beauty of its flowers than for its fruit. There are two varieties of this, one with ſweet, and the other bitter kernels, which ariſe from the fruit of the ſame tree.

The ſecond ſort is commonly known by the title of Jordan Almonds; the nuts of this kind are frequently brought to England. Theſe have a tender ſhell, and a large ſweet kernel. The leaves of this tree are broader, ſhorter, and grow much cloſer than thoſe of the common ſort, and their edges are crenated. The flowers are very ſmall, and of a pale colour, inclining to white.

The third ſort was found growing near Aleppo, from whence the fruit was ſent to the Duke D'Ayen in France, who raiſed ſeveral of the plants in his curious garden at St. Germains. The leaves of this tree are ſilvery, and very like thoſe of the Sea Purſlane. Theſe continue great part of the year, but the flowers are ſmall and like thoſe of the ſecond ſort.

The fourth ſort is very common in the nurſeries about London, and is uſually ſold with other flowering ſhrubs to adorn gardens. This ſort ſeldom riſes more than three feet high, ſending out many ſide branches. The roots of this are very ſubject to put out ſuckers, by which it may be increaſed in plenty; but if theſe are not annually taken away, they will ſtarve the old plants. This ſhrub flowers in April, at which time all the young ſhoots are covered with flowers, which are of a Peach-bloſſom colour, and make a fine appearance when intermixed with ſhrubs of the ſame growth.

The common Almond is cultivated in all the nurſeries, and the trees are generally planted for the beauty of its flowers. Theſe often appear in February when the ſpring is forward, but if froſt comes after the flowers are blown, they are ſoon deſtroyed, ſo that their beauty is of ſhort duration, and in thoſe ſeaſons there are few Almonds produced; whereas, when the trees do not flower till late in March they ſeldom fail to bear plenty of fruit, many of which will be very ſweet, and fit for the table when green, but they will not keep long.

They are propagated by inoculating a bud of theſe trees into a Plum, Almond, or Peach-ſtock, in the month of July. The next ſpring when the buds ſhoot, you may train them up either for ſtandards, or ſuffer them to grow for half-ſtandards, according to your own fancy. The beſt ſeaſon for tranſplanting theſe trees, if for dry ground, is in October, as ſoon as the leaves begin to decay, but for a wet ſoil February is much preferable; obſerve always to bud upon Plum-ſtocks for wet ground, and Almonds or Peaches for dry.

ALMOND, the Dwarf, with double flowers. See PERSICA.

ANACAMPSEROS. See SEDUM.

ANACARDIUM. Lin. Gen. Plant. 467. The Caſhew Nut, or Acajou.

The CHARACTERS are,

The flower is of one leaf, and cut into five parts at the top; it hath ten ſlender ſtamina. In the center is placed a round germen with an awl-ſhaped ſtyle. The germen afterward becomes a large, oval, fleſhy fruit, having a large kidney-ſhaped nut growing to its apex.

We have but one SPECIES of this genus, viz.

ANACARDIUM (*Occidentale.*) Hort. Cliff. 161. *The Caſhew, or Cajou.*

This grows to a conſiderable height in its native country, which is the Weſt-Indies, but in England the plants are with great difficulty preſerved; though by their firſt ſhoot from the ſeeds they appear ſo ſtrong and vigorous, as to promiſe a much greater progreſs than they are ever ſeen to make here.

They are eaſily raiſed from the nuts, which are annually brought from America in great plenty; each of theſe ſhould be planted in a ſmall pot filled with light ſandy earth, and plunged into a good hot-bed of tanners bark, being careful

ful to prevent their having wet till the plants come up for the nuts frequently rot with moisture. The reason for my advising the nuts to be each put in a separate pot is, because the plants seldom live when they are transplanted. If the nuts are fresh, the plants will come up in about a month after planting; and in two months more, the plants will be four or five inches high, with large leaves; and from this quick growth many persons have been deceived by supposing them hardy, and that they would continue the like progress, whereas they seldom advance much farther the same year.

The plants must be constantly kept in the stove, for they are too tender to live abroad in England in the warmest season of the year, nor will they thrive in a common green-house in summer. As these plants abound with a milky acrid juice, so they should have but little water, even in summer; and in winter, if they are sparingly watered once in a month, it will be sufficient, for their roots are tender and soon perish with moisture.

The pulpy fruit, to whose apex this nut grows, is as large as an Orange, and is full of an acid juice, which is frequently mixed in the making of punch in America. Many of these fruit have been brought to England, in casks of rum for the same purpose.

The nut is of the size and shape of a hare's kidney, but is much larger at the end which is next the fruit, than at the other. The shell contains an inflammable oil, which is very caustic; this will raise blisters on the skin, and has often been very troublesome to those who have incautiously put the nuts into their mouths to break the shell.

The milky juice of this tree will stain linen of a deep black, which cannot be washed out again; but whether this has the same property with that of the eastern Anacardium, has not yet been fully experimented, for the inspissated juice of that tree, is the best sort of lack which is used for staining of black in China and Japan.

ANACYCLUS. Is a sort of Camomile of little beauty and no use, so is seldom kept in any but gardens of botany, and not worthy to be mentioned here.

ANAGALLIS. Pimpernel.

The CHARACTERS are,

The flower hath an empalement which is cut into five sharp segments. The flower is of one leaf spread open, and cut into five parts. The germen afterward becomes a globular vessel with one cell, opening horizontally, in which are lodged several angular seeds.

The SPECIES are,

1. ANAGALLIS (*Arvensis*) foliis indivisis caule procumbente. Lin. Gen. Plant. 148. *Common Pimpernel with a red flower.*

2. ANAGALLIS (*Femina*) foliis indivisis glaucis caule procumbente flore cæruleo. *Female Pimpernel with a blue flower.*

3. ANAGALLIS (*Monelli*) foliis indivisis caule erecto. Lin. Sp. Plant. 148. *Narrow-leaved Pimpernel with a blue flower.*

1. ANAGALLIS (*Latifolia*) foliis cordatis amplexicaulibus, caulibus compressis. Lin. Sp. Plant. 149. *Broad-leaved Spanish Pimpernel with a blue flower.*

The first sort is very common in corn fields, and other cultivated places, in most parts of England. The second sort is sometimes found wild in the fields, but is less common than the first in England. There is a variety of this with a deeper blue flower, whose seeds I received from Nice, and this hath retained its colour for several years, during which time I have sown it in the Chelsea garden.

These are all annual plants, except the third sort, which arise from seeds, and, if suffered to remain till their seeds scatter, will become weeds in the place; so that they are never cultivated, except in botanic gardens for variety. The first and second sorts are directed by the College of Physicians for medicinal use.

ANAGYRIS. Stinking Bean-trefoil.

The CHARACTERS are,

The flower is of the butterfly kind; the standard is heart-shaped, and much longer than the empalement; the wings are oblong, plain, and longer than the standard, as is also the keel. The germen afterward becomes a large oblong pod, in which are lodged several kidney-shaped seeds.

We have but one SPECIES in England, viz.

1. ANAGYRIS (*Fœtida*) foliis ovatis floribus lateralibus. *Stinking Bean-trefoil with oval leaves, and flowers proceeding from the wings of the stalks.*

This sort grows wild in the south of France, as also in Spain and Italy. It is a shrub which usually rises to the height of eight or ten feet, and produces its flowers in April and May, which are of a bright yellow colour, growing in spikes, somewhat like those of the Laburnum.

It may be propagated by laying down their tender branches in the spring, observing in dry weather to supply them with water, which, if duly performed, the layers will have taken root by the following spring, when they should be cut off from the old plants a little time before they begin to put out their leaves, and planted in a warm situation; for if they are too much exposed to cold winds, they will be in danger of being destroyed in a hard winter. This method of propagating these plants, is to supply their defect in not producing ripe seeds in this country; for the plants which are produced from seeds will be much handsomer, and will rise to a much greater height.

If you propagate this plant from seeds, you should sow them in pots filled with light fresh earth, toward the end of March, and plunge the pots into a gentle hot-bed. If the seeds are good, the plants will appear in a month after they are sown; as the plants advance they should be inured to the open air, that they may be hardened before the following winter. In the autumn the pots should be placed in a hot-bed frame, to screen the plants from hard frosts; and the following spring they should be each transplanted into a separate small pot, and placed into a sheltered situation in summer, and the autumn following removed again into a frame to shelter them in winter. The second spring after the plants come up, some of them may be shaken out of the pots, and planted in a border near a south wall, where, if they are protected in winter, they may remain for good.

ANANAS, the Pine-apple.

The CHARACTERS are,

The flower consists of three oval petals; these are produced from the protuberances of the pyramidal fruit. The germen is situated below the flower, which afterwards becomes a cell, in which is lodged several angular seeds.

The VARIETIES of this are,

1. ANANAS (*Ovatis*) aculeatus, fructu ovato, carne albida. Plum. *Oval-shaped Pine-apple, with a whitish flesh.*

2. ANANAS (*Pyramidalis*) aculeatus, fructu pyramidato, carne aurea. Plum. *Pyramidal Pine-apple, with a yellowish flesh, called the Sugar-loaf Pine.*

3. ANANAS (*Glaber*) folio vix serrato. Boerh. Ind. Alt. 2. 83. *Pine-apple with smooth leaves.*

4. ANANAS (*Lucidus*) lucidè virens, folio vix serrato. Hort. Elth. *Pine-apple with shining green leaves, and scarce any spines on their edges.*

5. ANANAS (*Serrotinus*) fructu pyramidato olivæ colore, intus aureo. *Pyramidal Olive-coloured Pine-apple, with a yellow flesh.*

6. ANANAS (*Viridis*) aculeatus, fructu pyramidato ex viridi flavescente. *The green Pine-apple.*

There are fome other varieties of this fruit, which may have been obtained from feeds; and I doubt not but if the feeds were fown frequently, in the countries where they are in plenty, there may be as great variety of thefe fruits, as there are of Apples or Pears in Europe. And this I have found true by fome trials which I have made by fowing of the feeds, which have always produced a variety of forts from thofe of the fame fruit.

The firft fort is the moft common in Europe, but the fecond fort is much preferable to it, the fruit of this being larger and much better flavoured: the juice of this fort is not fo aftringent as that of the firft, fo that this fruit may be eaten in greater quantity, with lefs danger. This fort frequently produces fuckers immediately under the fruit, whereby it may be increafed much fafter than the common fort, fo that in a few years it may be the beft common fort in England.

The third fort is preferved by fome curious perfons for the fake of variety, but the fruit is not worth any thing.

The fort with very fmooth Grafs-green leaves, was raifed from feeds taken out of a rotten fruit, which came from the Weft-Indies to the late Henry Heathcote, Efq; from whom I received one plant, which hath produced large fruit: this, I am told, is what the people of America call the King Pine. I have fince raifed fome plants of this kind from feeds, which were brought me from Jamaica.

The plants are propagated by planting the crowns which grow on the fruit, or the fuckers which are produced either from the fides of the plants, or under the fruit, both which I have found to be equally good; although by fome perfons the crown is thought preferable to the fuckers, as fuppofing it will produce fruit fooner than the fuckers, which is certainly a miftake; for by conftant experience I find the fuckers (if equally ftrong) will fruit as foon, and produce as large fruit as the crowns, if not better.

The fuckers and crowns muft be laid to dry in a warm place for four or five days, or more (according to the moifture of the part which adhered to the old plant or fruit); for if they are immediately planted, they will rot. The certain rule of judging when they are fit to plant, is by obferving if the bottom is healed over and become hard; for if the fuckers are drawn off carefully from the old plants, they will have a hard fkin over the lower part, fo need not lie fo long as the crowns, or thofe whofe bottoms are moift. But whenever a crown is taken from the fruit, or the fuckers from old plants, they fhould be immediately divefted of their bottom leaves, fo high as to allow depth for their planting; fo that they may be thoroughly dry and healed in every part, left when they receive heat and moifture they fhould perifh, which often happens when this method is not obferved. If thefe fuckers or crowns are taken off late in the autumn, or during the winter, or early in the fpring, they fhould be laid in a dry place in the ftove, for a fortnight or three weeks before they are planted, but in the fummer feafon they will be fit for planting in a week at fartheft.

As to the earth in which thefe fhould be planted, if you have a rich good kitchen-garden mould, not too heavy, fo as to detain the moifture too long, nor over light and fandy, it will be very proper for them without any mixture: but where this is wanting, you fhould procure fome frefh earth from a good pafture, which fhould be mixed with about a third part of rotten neats dung, or the dung of an old Melon or Cucumber-bed, which is well confumed. Thefe fhould be mixed fix or eight months at leaft before they are ufed, but if it be a year it will be the better; and fhould be often turned, that their parts may be the better united, as alfo the clods well broken. This earth fhould not be fcreened very fine, for if you only clear it of the great ftones, it will be better for the plants than when it is made too fine. You fhould always avoid mixing any fand with the earth, unlefs it be extremely ftiff, and then it will be neceffary to have it mixed at leaft fix months or a year before it is ufed; and it muft be frequently turned, that the fand may be incorporated in the earth fo as to divide its parts: but you fhould not put more than a fixth part of fand, for too much fand is very injurious to thefe plants,

In the fummer feafon, when the weather is warm, thefe plants muft be frequently watered, but you fhould not give them large quantities at a time: you muft alfo be very careful that the moifture is not detained in the pots by the holes being ftopped, for that will foon deftroy the plants. If the feafon is warm, they fhould be watered twice a week, but in a cool feafon once a week will be often enough; and during the fummer feafon you fhould once a week water them gently all over their leaves, which will wafh the filth from off them, and thereby greatly promote the growth of the plants.

There are fome perfons who frequently fhift thefe plants from pot to pot, but this is by no means to be practifed by thofe who propofe to have large well-flavoured fruit; for unlefs the pots be filled with the roots, by the time the plants begin to fhew their fruit, they commonly produce fmall fruit, which have generally large crowns on them, therefore the plants will not require to be new potted oftener than twice in a feafon: the firft time fhould be about the end of April, when the fuckers and crowns of the former year's fruit (which remained all the winter in thofe pots in which they were firft planted) fhould be fhifted into larger pots, i. e. thofe which were in halfpenny, or three farthing pots, fhould be put into penny, or at moft three halfpenny pots, according to the fize of the plants; for you muft be very careful not to overpot them, nothing being more prejudicial to thefe plants. The fecond time for fhifting of them is in the beginning of Auguft, when you fhould fhift thofe which which are of a proper fize for fruiting the following fpring, into two-penny pots, which are full large enough for any of thefe plants. At each of thefe times of fhifting the plants, the bark-bed fhould be ftirred up, and fome new bark added, to raife the bed up to the height it was at firft made; and when the pots are plunged again into the bark bed, the plants fhould be watered gently all over their leaves, to wafh off the filth, and to fettle the earth to the roots of the plants. If the bark-bed be well ftirred, and a quantity of good frefh bark added to the bed, at this latter fhifting it will be of great fervice to the plants, for they may remain in the fame tan until the beginning of November, or fometimes later, according to the mildnefs of the feafon, and will require but little fire before that time. During the winter feafon thefe plants will not require to be watered oftener than once a week, according as you find the earth in the pots to dry; nor fhould you give them too much at each time, for it is much better to give them a little water often, than to over-water them.

You muft obferve never to fhift thofe plants which fhew their fruit into other pots, for if they are removed after the fruit appears, it will ftop the growth, and thereby caufe the fruit to be fmaller, and retard its ripening, fo that many times it will be October or November before the fruit is ripe; therefore you fhould be very careful to keep the plants in a vigorous growing ftate, from the firft appearance of the fruit, becaufe upon this depends the goodnefs and the fize of the fruit; for if they receive a check after this, the fruit is generally fmall and ill tafted.

When you have cut off the fruit from the plant whofe kind you are defirous to propagate, you fhould trim the leaves, and plunge the pots again into a moderate hot-bed, obferving to refrefh them frequently with water, which will

caufe

cause them to put out suckers in plenty; so that a person may be soon supplied with plants enough of any of the kinds, who will but observe to keep the plants in health.

There is not any thing which can happen to these plants of a more dangerous nature, than to have them attacked by small white insects, which appear at first like a white mildew, but soon after have the appearance of lice: these attack both root and leaves at the same time, and if they are not soon destroyed, will spread over a whole stove in a short time, and in a few weeks will entirely stop the growth of the plants, by sucking out the nutritious juice, so that the leaves will appear yellow and sickly, and have generally a great number of yellow transparent spots all over them. These insects, after they are fully grown, appear like bugs, and adhere so closely to the leaves, as not to be easily washed off and seem to have no local motion. They were originally brought from America upon the plants which were imported from thence, and I believe they are the same insects which have destroyed the sugar canes of late in some of the Leeward Islands; for upon some sugar canes which were sent me from Barbadoes, there were great numbers of the same insects. Since they have been in England, they have spread greatly in such stoves where there has not been more than ordinary care taken to destroy them. They have also attacked the Orange-trees in many gardens near London, and have done them incredible damage; but I do not find they will endure the cold of our climate in winter, so that they are never found on such plants as live in the open air. The only method I have yet been able to discover for destroying these insects, is by washing the leaves, branches, and stems, of such plants as they attack, frequently with water, in which there has been a strong infusion of tobacco stalks, which I find will destroy the insects, and not prejudice the plants. But this method cannot be practised on the Ananas plants, because the insects will fasten themselves so low between the leaves, that it is impossible to come at them with a sponge to wash them off; so that if all those which appear to sight are cleared off, they will soon be succeeded by a fresh supply from below, and the roots will be also equally infested at the same time. Therefore, wherever these insects appear on the plants, the safest method will be to take the plants out of the pots, and clear the earth from the roots; then prepare a large tub, which should be filled with water, in which there has been a strong infusion of tobacco stalks; into this tub you should put the plants, placing some sticks cross the tub to keep the plants immersed in water. In this water they should remain twenty-four hours; then take them out, and with a sponge wash off all the insects from the leaves and roots, and dip the plants into a tub of fair water, washing them therein, which is the most effectual way to clear them from the insects. After which you should pot them in fresh earth, and having stirred up the bark-bed, and added some new tan to give a fresh heat to the bed, the pots should be plunged again, observing to water them all over the leaves (as was before directed) and this should be repeated once a week during the summer season; for I observe these insects always multiply much faster where the plants are kept dry, than in such places where the plants are sometimes sprinkled over with water, and kept in a growing state.

As these insects are frequently brought over from America on the Ananas plants which come from thence, those persons who procure their plants from thence, should look carefully over them when they receive them, to see they have none of these insects on them; for if they have, they will soon be propagated over all the plants in the stove where these are placed; therefore, whenever they are observed, the plants should be soaked (as before directed) before they are planted into pots.

The stoves which are erected for preserving of these plants are built in different ways, according to the fancy of the contriver. Some persons build them with upright glasses in front, about four feet high, and sloping glasses over these, which rise about six feet high, so that there is just height enough for persons to walk upright on the back-side of the bark-bed. Others make but one slope of glasses, from the top of the stove down to the plate, which lies about six or eight inches above the bark pit, in the front of the stove, so that in this stove there is no walk made in the front between the bark-pit and the glasses; but the inconveniency of watering the plants, as also of coming near those plants which are placed in the front of the stove to clean them, has, in some measure, brought them into disesteem, so that few persons now build them, though the expence is much less than of the other kind of stoves; but of both these stoves, the figures and descriptions which are hereafter exhibited under the article of STOVE, will be sufficient for any person to build either of the sorts. One of these stoves, about twenty five feet long in the clear, with the pit for the tan reaching from end to end, and six feet and an half wide, will contain about an hundred plants; so that whoever is desirous to have this fruit, may easily proportion their stove to the quantity of fruit which they are willing to have.

But it will be also necessary to have a bark-pit under a deep frame, in order to raise the young plants in summer; for in this bed you should plunge the suckers, when they are taken from the old plants, as also the crowns which come from the fruit, so that this frame will be as a nursery to raise the young plants to supply the stove; but these plants should not remain in these frames longer than till the beginning of November, unless the frame is built with brick work with flues in it to warm the air, (in the manner hereafter described and figured) which are very useful, as nurseries, to keep the young plants till they are of a proper size to produce fruit; and the air in this frame may be kept either warmer or cooler than the stove, according as the plants may require, so that the stove may be every autumn filled only with bearing plants, whereby a much greater quantity of fruit may be annually produced, than can be where young and old plants must be crowded into the same stove: but where there are no inconveniencies of this kind, the young plants, about the middle or latter end of October, must be removed into the stove, and being small, may be crowded in among the larger plants; for as they will not grow much during the winter season, so they may be placed very close together. The end of March, where there is no nursery for the young plants, they must be removed out into the hot-bed again, which should be prepared a fortnight before, that the tan may have acquired a proper heat; but you should be careful that the tan be not too hot, for that might scald the fibres of the plants if they are suddenly plunged therein. Therefore if you find the bark too hot, you should not plunge the pots above two or three inches into the tan, letting them remain so until the heat of the tan is a little abated, when you should plunge the pots down to their rims in the bed. If the nights should continue cold after these plants are removed into the bed, you must carefully cover the glasses with mats, otherwise by coming out of a warm stove they may receive a sudden check, which will greatly retard their growth, which must be carefully avoided; because the sooner the plants are set growing in the spring, the more time they will have to gain strength, in order to produce large fruit the following season.

You should not plunge the pots too close together in this frame, but allow them a proper distance, that the lower part of the plants may increase in bulk, for it is on this that

that the magnitude of the fruit depends; becaule when the plants are placed too close, they draw up very tall, but do not obtain strength; so that when they are taken out of the bed, the leaves are not able to support themselves, but all the outward long leaves will fall down, leaving the smaller middle leaves naked, and this sometimes will cause them to rot in the center. You must also observe, when the sun is very warm, to raise the glasses of the hot-bed in the heat of the day with props, in order to let out the steam of the bed, and to admit fresh air; for one neglect of this kind, in a very hot day, may destroy all the plants, or at least so scald them, that they will not get over it in many months. It will be also very proper, in extreme hot weather, to shade the glasses in the middle of the day with mats, for the glasses lying so near to the leaves of the plants, will occasion a prodigious heat at such times.

There are some persons who regulate the heat of their stoves by thermometers in summer, but at that season this is unnecessary, for the outward air in hot weather is frequently greater than the Ananas heat marked on the thermometers, so that the heat of the stoves at that season will be much greater. The use of the thermometer is only in winter, during the time the fires are continued, by which it is easy to judge when to increase or diminish the fires, for at that season the stoves should not be kept to a greater warmth than five or six divisions above Ananas, nor suffered to be more than as many divisions below it. When the plants are placed into the tan for the winter season (which should be done about the middle of October) the tan-bed should be renewed, adding two-thirds of new tan to one-third of the old. If this be well mixed, and the new tan is good, the bed will maintain a proper degree of warmth till February, at which time it will be proper to stir up the bed, and add a load or two of new tan, so as to raise the bed as much as it sunk since the autumn; this will give a fresh heat to the bed, and keep the plants growing, and, as the fruit will now begin to appear, it will be absolutely necessary to keep the plants in a growing state, otherwise the fruit will not be large, for if they receive any check at this time it will greatly injure them.

In April it will be proper to stir up the tan again, and if the bed has sunk since the last stirring, it will be proper to add some fresh tan to it; this will renew the warmth of the bed, and forward the fruit. And if the tan-bed is constantly kept in a good temper, and a sufficient quantity of air admitted every day to the plants, they will succeed much better than in a cool bed kept too close.

Those plants which shew their fruit early in February, will ripen about June; some sorts are at least a month or five weeks longer in ripening their fruit than others, from the time of the appearance of the fruit: but the season in which the fruit is in greatest perfection, is from the beginning of June to the end of September; though in March, April, and October, I have frequently eaten this fruit in pretty good perfection, but then the plants have been in perfect health, otherwise they are seldom well flavoured.

The method of judging when the fruit is ripe, is by the smell, and from observation; for as the several sorts differ from each other in the colour of their fruit, that will not be any direction when to cut them; for should they remain so long as to become soft to the touch before they are cut, they become flat and dead, as they do also when they are cut long before they are eaten: therefore the surest way to have this fruit in perfection, is to cut it the same day it is eaten: but it must be cut early in the morning, before the sun has heated the fruit. otherwise it will be hot, observing to cut the stalk as long to the fruit as possible, and lay it in a cool, but dry place, preserving the stalk and crown unto it until it is eaten.

That sort with green fruit, if suffered to ripen well, is of an Olive colour; but there are some persons who cut them before they are ripe, when they are not fit to be eaten, for no other reason but to have them green; and although many persons have much recommended this sort for its excellent flavour, yet I think the sugar-loaf sort is to be preferred to it.

ANAPODOPHYLLON. See PODOPHYLLUM.

ANASTATICA. Lin. Gen. Pl. 175. Rose of Jericho.

The CHARACTERS are,

The flower hath four roundish petals placed in form of a cross. The seed-vessel in this is blunt-pointed, bordered, and crowned, and the valves open oblique to the style.

We know but one SPECIES of this genus, viz.

ANASTATICA (*Hierocuntica*) foliis obtusis spicis axillaribus brevissimis siliculis ungulatis spinosis. Lin. Sp. 895. *Rose of Jericho with obtuse leaves, short spikes of flowers at the wings of the stalks, and short prickly pods.*

This plant grows naturally on the sands near the borders of the Red-sea, and in many parts of Syria. It is a low annual plant, dividing into many irregular woody branches near the root; at each joint is placed a single, oblong, hoary leaf, and at the same places come out small single flowers of a whitish green colour, composed of four small leaves, placed in form of a cross, like the other plants of this class. These are succeeded by short wrinkled pods, having four small horns; these open into two cells, in each of which is lodged a single brown seed.

It hath had the epithet of Rosa Mariæ given to it by the monks, who have superstitiously supposed that the flowers open on the night that our Saviour was born. But the truth is, that the dry woody plant being set for some time in water, will dilate and open so as to disclose the seed-vessels and seeds. This I have seen done when the plants have been many years gathered, so that there are several curious persons who preserve them in their repositories of curiosities, for the singularity of this property.

This plant is propagated by seeds, which should be sown the beginning of March, on a moderate hot-bed in pots, in which the plants are designed to remain, for they will not bear transplanting. When the plants come up they should be thinned, leaving them about six inches distant from each other, and observe to keep them clean from weeds, this is all the care they require. If the season proves favourable, the plants will flower in August; but unless the autumn proves warm and dry, they will not ripen their seeds in England; nor could I rarely procure seeds from those plants which were raised in autumn, for if much rain happens when the plants are in flower, they never perfect any seeds.

ANCHUSA. Lin. Gen. 167.

The CHARACTERS are,

The flower is of one leaf, having a cylindrical tube; at the brim it is cut into five obtuse segments, which spread open. The germen afterward becomes four oblong blunt seeds shut up in the empalement.

The SPECIES are,

1. ANCHUSA (*Officinalis*) foliis lanceolatis hirsutis, floribus capitatis axillaribus pedunculis longissimis. *The greater Garden Bugloss.*

2. ANCHUSA (*Angustifolia*) racemis subnudis conjugatis. Prod. Leyd. 408. *Perennial wild Borage with a Carmine flower.*

3. ANCHUSA (*Undulata*) strigosa foliis linearibus dentatis pedicellis bracteâ minoribus calycibus fructiferis inflatis. Loefl. Lin. Sp. Plant. 133. *Portugal Bugloss, with a waving Viper's Bugloss leaf.*

4. ANCHUSA (*Orientalis*) ramis floribusque alternis axillaribus bracteis ovatis. Lin. Sp. Plant. 133. *Eastern Bugloss with a yellow flower.*

5. ANCHUSA

5. Anchusa (*Virginiana*) floribus sparsis caule glabro. Lin. Sp. Plant. 133. *Small yellow Alkanet of Virginia, called by the inhabitants Puccoon.*

6. Anchusa (*Sempervirens*) pedunculis diphyllis capitatis. Lin. Sp. Plant. 134. *Broad-leaved Evergreen Borage.*

7. Anchusa (*Cretica*) foliis lanceolatis verrucosis semiamplexicaulibus floribus capitatis, caule procumbente. *Warted Bugloss of Crete.*

8. Anchusa (*Tinctoria*) tomentosa foliis lanceolatis obtusis staminibus corolla brevioribus. Lin. Sp. 192. *True Alkanet.*

The first sort is the Bugloss, whose flowers are ordered to be used in medicine.

The roots of this sort seldom continue longer than two years, especially in good ground, for they are subject to rot in winter, unless when they happen to grow in rubbish, or out of an old wall, where they will live three or four years; for in such places the plants are stinted in their growth, so their branches are firmer, and not so full of juice as those which grow in better soil. The plants may be easily propagated by seeds, which may be sown either in the spring or autumn, upon a bed of light sandy earth; and when the plants are strong enough to remove, they should be planted into beds at two feet distance, observing, if the season proves dry, to water them till they have taken root, after which time they will require no farther care but to keep them clean from weeds. The plants which come up in the autumn will flower the following June, and ripen their seeds in August; but those which are sown in the spring do not often flower the same year, or if they do it is late in the season, so will not ripen their seeds. If the seeds of this plant are permitted to scatter, the plants will rise in plenty, which may be managed in the manner before directed.

The second sort grows to the height of two feet when cultivated in gardens, but in the places where it grows wild, is rarely more than a foot and a half high. The leaves of this are narrow, and less hairy than those of the first; the spikes of flowers come out double, and have no leaves about them; the flowers are small, and of a red colour. The roots will continue two years in poor land.

The third sort is a biennial plant, which perishes soon after the seeds are ripe. This grows two feet high, and sends out many lateral branches. The flowers are of a bright blue colour, and grow in an imbricated spike; and after these fall, the empalement turns to a swollen vessel inclosing the seeds.

The fourth sort is a perennial plant, with long trailing branches which lie on the ground. The flowers are yellow, and about the size of the common Bugloss, and there is a succession of these on the same plants great part of the year. This, though a native of the Levant, is hardy enough to live in the open air in England, if it hath a dry sandy soil. It may be propagated by seeds in the same manner as the first sort, and if the seeds are permitted to scatter, the plants will rise without care, and will continue several years.

The fifth sort is a native of North America, where it grows naturally in the woods, and being an early plant, generally flowers before the new leaves come out on the trees; so that in some of the woods where this plant abounds, the surface of the ground seems covered with its bright yellow flowers. It is known in that country by the title of Puccoon. It is a perennial plant, which seldom rises a foot high in good ground, but not above half that height where the soil is poor; the flowers grow in loose spikes, upon a smooth stalk. This is propagated by seeds, which, if sown in the spring, seldom grow the first year.

The sixth sort is a very hardy perennial plant, with weak trailing branches; the flowers are blue and come out between the leaves on the spike, like the fourth sort; the plants frequently grow out of the joints of old walls, in those places where any of the plants have been near; for when the seeds are permitted to scatter, there will be an abundant supply of the plants. These flower great part of the year.

The seventh sort is a low, trailing, annual plant, whose branches seldom extend more than six inches. The flowers are small, of a bright blue colour, and are collected into small bunches at the extremity of the branches. The plants perish soon after their seeds are ripe, which, if permitted to scatter, the plants will come up better than when they are sown.

The eighth sort rises near as high as the first, to which it bears great resemblance in its leaves and branches; but the flowers grow on long spikes, coming out imbricatim like the tiles on a house, in which it differs from that. It grows naturally in the Levant, but is equally hardy with the first species, and may be cultivated in the same manner.

ANCHUSA RADICE RUBRA. See Lithospermum.

ANDRACHNE. Bastard Orpine.

The Characters are,

There are male and female flowers on the same plant. The male flower hath a five-leaved empalement. The flower has five slender petals. It hath five slender stamina. The female flowers come out from the wings of the stalk near the male; these have a five-leaved empalement, but no petals; it has three slender styles which are inserted in the rudiment, and a globular capsule having three cells, in each of which are lodged two triangular obtuse seeds.

We have but one Species of this genus, viz.

Andrachne (*Telephioides*) procumbens herbacea. Lin. Sp. Plant. 1014. *Bastard Orpine with trailing branches and a white flower.*

This is a low plant, whose branches trail upon the ground. The leaves are small, of an oval shape, smooth, and of a sea-green colour. It is found wild in some parts of Italy, and in the Archipelago; and being a plant of no great beauty, it is seldom cultivated but in botanic gardens for variety. If the seeds of this plant are sown on a bed of common earth in the autumn, soon after they are ripe, the plants will come up the following spring, and produce flowers and seeds; but if it is sown in the spring, the seeds will often remain in the ground until the next year before they come up. It should be planted in pots, and sheltered from frost in the winter, and should have a light dry soil and a warm situation; this seldom continues longer than two or three years.

ANDROMEDA. Lin. Gen. Plant. 485. We have no English name for this plant.

The Characters are,

The empalement of the flower is cut into five small acute segments. The flower is of one leaf, is oval and bell-shaped, and divided into five parts at the brim, which are reflexed. It hath ten stamina. The germen afterward turns to a round pentagonal vessel having five cells, which are filled with small round seeds.

The Species are,

1. Andromeda (*Polifolia*) pedunculis aggregatis, corollis ovatis, foliis alternis lanceolatis revolutis. Lin. Sp. Plant. 393. *Andromeda with aggregate foot-stalks, oval petals, and spear-shaped leaves growing alternately.*

2. Andromeda (*Mariana*) pedunculis aggregatis corollis cylindricis foliis alternis ovatis integerrimis. Lin. Sp. Plant. 393. *Andromeda with aggregate foot-stalks, cylindrical flowers, and oval entire leaves placed alternately.*

3. Andromeda (*Paniculata*) racemis secundis nudis paniculatis, corollis subcylindricis foliis alternis oblongis crenulatis. Lin. Sp. Pl. 394. *Andromeda with naked, fruitful, loose spikes, cylindrical flowers, and oblong crenated leaves placed alternately.*

4. Andro

4. ANDROMEDA (*Arborea*) racemis fecundis nudis, corollis rotundo-ovatis. Lin. Sp. Plant. 394. *Andromeda with naked fruitful ſpikes, and oval roundiſh flowers; commonly called Sorrel-tree in Carolina.*

5. ANDROMEDA (*Caniculata*) racemis fecundis foliaceis corollis ſubcylindricis, foliis alternis lanceolatis obtuſis punctatis. Lin. Sp. Pl. 394. *Andromeda with leafy fruitful ſpikes, cylindrical flowers, and obtuſe ſpear-ſhaped leaves, with punctures placed alternately.*

The firſt ſort is a low plant, which grows naturally on bogs in the northern countries, but is with difficulty preſerved in gardens; and having little beauty, is ſeldom cultivated, except in botanic gardens. I received the ſeeds from Peterſburgh, which came up in the Chelſea garden, but did not continue more than one year.

The ſecond ſort grows naturally in North America; this is a low ſhrub, which ſends out many woody ſtalks from the root, which are garniſhed with oval leaves placed alternately; the flowers are collected in ſmall bunches: theſe are ſhaped like thoſe of the Strawberry-tree, and are of an herbaceous colour. They appear in June and July.

The third ſort is alſo a native of North America. This ſhrub grows about four feet high, ſending out ſeveral branches, which are clothed with oblong leaves placed alternately; the flowers grow in looſe ſpikes from the ends of the branches: theſe are of the pitcher-ſhape, like thoſe of the Arbutus, but are a little longer. They appear in July.

The fourth ſort grows naturally in Virginia and Carolina; in the latter it is much larger than in the former, the climate being warmer, ſo many of the trees and ſhrubs grow to a much greater height there. In Virginia this is a ſhrub growing ten or twelve feet high, but in Carolina it riſes twenty feet. The flowers grow in long naked ſpikes, coming out from the ſides of the branches, which are of an herbaceous colour, and are ranged on one ſide of the ſtalk; they are oval, and ſhaped like a pitcher.

The fifth ſort grows naturally in Siberia, and alſo in North America; it is a low ſhrub which grows on moſſy land, ſo is very difficult to keep in gardens. The leaves are ſhaped like thoſe of the Box-tree, and are of the like conſiſtence, having ſeveral ſmall punctures on them; the flowers grow in ſhort ſpikes from the extremity of the branches; theſe are produced ſingle between two leaves; they are white, and of a cylindrical pitcher-ſhape.

All the ſorts, except the fourth, are very hardy plants, which delight in moiſt ground; they increaſe by their creeping roots, which put up ſuckers at a diſtance, and may be taken off with roots, and tranſplanted where they are deſigned to remain, for they do not bear to be often removed.

The fourth ſort requires to be ſheltered from hard froſt in winter, but in the ſummer ſhould be frequently watered. It is a difficult plant to keep in gardens, as it grows naturally on boggy places, and requires a greater heat than that of this climate. It may be propagated by ſeeds, which ſhould be procured from America, where it is known by the name of Sorrel-tree.

ANDROSACE. We have no Engliſh name for this plant.

The CHARACTERS are,

The flowers grow in an umbel ſet in an involucrum; the flower is of one leaf, having an oval tube, incloſed by the empalement, and is divided into five parts. It hath five ſmall ſtamina within the tube; the empalement afterward becomes a round capſule of one cell, which is full of round ſeeds.

The SPECIES are,

1. ANDROSACE (*Maxima*) perianthiis fructuum maximis. Hort. Upſal. 36. *Common broad-leaved annual Androſace.*

2. ANDROSACE (*Septentrionalis*) foliis lanceolatis dentatis glabris perianthiis angulatis corollâ brevioribus. Flor. Suec. 160. *Spring Chickweed with heads like Androſace.*

3. ANDROSACE (*Villoſa*) foliis piloſis perianthiis hirſutis. Lin. Sp. Plant. 142. *Hairy Houſleek of the Alps with a milk white flower.*

The firſt ſort grows naturally in Auſtria and Bohemia amongſt the corn: this hath broad leaves which ſpread near the ground, from the center of theſe the foot-ſtalks ariſe, which are terminated by the umbel of flowers like thoſe of the Auricula; under the umbel of flowers is a large empalement, which is permanent; the flowers are compoſed of five ſmall white petals; theſe appear in April and May, and the ſeeds ripen in June, and the plants ſoon after periſh.

The other ſorts are much ſmaller than this, ſome of them ſeldom growing more than three inches high, and have very ſmall flowers, ſo make little appearance. They grow naturally on the Alps and Helvetian mountains, as alſo in Siberia, from whence I have received the ſeeds of three or four ſpecies. Theſe are only preſerved in botanic gardens for the ſake of variety, and all the ſorts except the firſt ſhould have a ſhady ſituation.

The ſeeds of theſe ſorts ſhould be ſown ſoon after they are ripe, otherwiſe they ſeldom grow the ſame year. Their ſeeds are ripe the end of May, which, if permitted to ſcatter, will come up, and often ſucceed better than thoſe which are ſown.

ANDROSÆMUM. See HYPERICUM.

ANDRYALA. Lin. Gen. Pl. 820. Downy Sowthiſtle.

The CHARACTERS are,

It hath a ſhort, round, hairy empalement; the flowers are compoſed of many hermaphrodite flowers, which are uniform, and are of one leaf, ſtretched out like a tongue on one ſide. The germen is ſituated at the bottom of each floret, and afterward becomes a ſingle oval ſeed, crowned with down.

The SPECIES are,

1. ANDRYALA (*Integrifolia*) foliis integris. Guett. Hort. Upſal. 240. *Downy Sowthiſtle with whole leaves.*

2. ANDRYALA (*Raguſina*) foliis lanceolatis indiviſis denticulatis, acutis tomentoſis, floribus ſolitariis. Lin. Sp. 1136. *Downy Sowthiſtle with indented, ſpear-ſhaped, woolly leaves.*

3. ANDRYALA (*Lanata*) foliis oblongo-ovatis ſubdentatis incanis, pedunculis ramoſis. Amœn. Acad. 4. p. 288. *Downy Sowthiſtle with woolly, oblong, oval leaves, and branching foot-ſtalks.*

The firſt is an annual plant, which grows naturally in the ſouth of France, Spain, and Italy, and is preſerved in botanic gardens for the ſake of variety. This grows one foot and a half high, with woolly branching ſtalks. The flowers are produced in ſmall cluſters at the top of the ſtalks, which are yellow, and like thoſe of the Sowthiſtle, ſo do not make any great appearance. It is eaſily raiſed by ſeeds, which ſhould be ſown in the ſpring, in the place where the plants are to remain, and will require no other culture but to thin them where they are too cloſe, and keep them clean from weeds. It flowers in July, and the ſeeds ripen in September.

The ſecond is a perennial plant, which grows naturally in Spain, from whence I received the ſeeds, as I have alſo from the Cape of Good Hope. The leaves of this plant are extremely white, and are much indented on their edges; the flower-ſtalks grow about a foot high, having ſmall cluſters of yellow flowers, which appear in July; the ſeeds ſometimes ripen in England, but not every year. They love a light dry ſoil, in which they will live in the open air in this country.

ANEMONE. Wind-flower.

The

The CHARACTERS are,

The flower is naked having no empalement, and consists of two or three orders of leaves or petals, which are oblong, and disposed in three series over each other. It hath many germen collected into a head, which afterward become so many seeds inclosed with a down which adhere to the foot-stalk, and forms an obtuse cone.

The SPECIES are,

1. ANEMONE (*Sylvestris*) pedunculo nudo feminibus subrotundis hirsutis. Lin. Sp. Pl. 540. *Wild Anemone with a large white flower.*

2. ANEMONE (*Nemorosa*) feminibus acutis foliolis incisis caule unifloro. Hort. Cliff. 224. *Wild or Wood Anemone with a large flower.*

3. ANEMONE (*Apennina*) feminibus acutis foliolis incisis petalis lanceolatis numerosis. Lin. Sp. Plant. 541. *Wood Anemone with a blue flower.*

4. ANEMONE (*Virginiana*) pedunculis alternis longissimis fructibus cylindricis feminibus hirsutis. Lin. Sp. Pl. 540. *Small white flowering Virginia Anemone.*

5. ANEMONE (*Coronaria*) foliis radicalibus ternato-decompositis, involucro folioso. Lin. Sp. Plant. 539. *Narrow-leaved Anemone with a single flower.*

6. ANEMONE (*Hortensis*) foliis digitatis. Lin. Sp. Plant. 540. *Broad-leaved Garden Anemone.*

The first sort grows naturally in many parts of Germany; this approaches near to our Wood Anemone, but the seeds of it are round and hairy; the flower is large and white, but having little beauty, is seldom planted in gardens.

The second sort grows wild in the woods in many parts of England, where it flowers in April and May, making a pretty appearance in those places where they are in plenty. The roots of this may be taken up when their leaves decay, and transplanted in wildernesses, where they will thrive and increase greatly, if they are not disturbed; and in the spring, before the trees are covered with leaves, they will have a very good effect, in covering of the ground and making a pleasing variety at that season.

The third sort is found growing naturally in some parts of England, but particularly at Wimbledon in Surry, in a wood near the mansion-house, in great plenty; but it is not certain that they were not originally planted there, as they are not found in any other place in that neighbourhood.

The fourth sort grows naturally in North America, from whence the seeds are frequently sent to England. This is a very hardy plant, and produces plenty of seeds in England, but having little beauty, scarce deserves a place in gardens, unless for the sake of variety.

The fifth and sixth sorts are natives of the East, from whence their roots were brought originally; but have been so greatly improved by culture, as to render them some of the chief ornaments to our gardens in the spring. The principal colours of these flowers are red, white, purple, and blue, and some are finely variegated with red, white, and purple. There are many intermediate shades of these colours; the flowers are large and very double, and, when properly managed, are extremely beautiful.

The soil in which these flowers will thrive extremely, may be composed in the following manner: Take a quantity of fresh untried earth (from a common or some other pasture land) that is of a light sandy loam, or hazel mould, observing not to take it above ten inches deep below the surface; and if the turf be taken with it the better, provided it hath time to rot thoroughly before it is used: mix this with a third part of rotten cow dung, and lay it in a heap, keeping it turned over at least once a month for eight or ten months, the better to mix it, and rot the dung and turf, and to let it have the advantages of the free air. In doing this work, be careful to rake out all great stones, and break the clods; but by no means sift or screen the earth, which I have found very hurtful to many sorts of roots.

This earth should be mixed twelve months before it is used, if possible; but if you are constrained to use it sooner, you must turn it over the oftener, to mellow and break the clods; and observe to rake out all the parts of the green sward, that are not quite rotten, before you use it, which would be prejudicial to your roots if suffered to remain. The beginning of September is a proper season to prepare the beds for planting (which, if in a wet soil, should be raised with this sort of earth six or eight inches above the surface of the ground, laying at the bottom some of the rakings of your heap to drain off the moisture; but in a dry soil, three inches above the surface will be sufficient:) this compost should be laid at least two feet and a half thick, and in the bottom there should be about four or five inches of rotten neats dung, or the rotten dung of an old Melon or Cucumber-bed, so that you must take out the former soil of the beds to make room for it.

And observe in preparing your beds, to lay them (if in a wet soil) a little round, to shoot off the water; but in a dry one, let it be nearer to a level; in wet land, where the beds are raised above the surface, it will be proper to fill up the paths between them in winter, either with rotten tan or dung, to prevent the frost from penetrating into the sides of the beds, which otherwise may destroy their roots. Your earth should be laid in the beds at least a fortnight or three weeks before you plant the roots, and a longer time would be yet better, that it may settle; and when you plant them, stir the upper part of the soil about six inches deep, with a spade; then rake it even and smooth, and with a stick draw lines each way of your bed at six inches distance, so that the whole may be in squares, that your roots may be planted regularly: then with three fingers make a hole in the center of each square, about three inches deep, laying therein a root with the eye uppermost; and when you have finished your bed, with the head of a rake draw the earth smooth, so as to cover the crown of the roots about two inches thick.

The best season for planting these roots, if for forward flowers, is about the latter end of September, and for those of a middle season any time in October; but observe to perform this work, if possible, at or near the time of some gentle showers; for if you should plant them when the ground is perfectly dry, and there should no rain fall for three weeks or a month after, the roots will be very apt to grow mouldy upon the crown; and if once they get this distemper, they seldom come to good after.

You may also reserve some of your Anemone roots till after Christmas before you plant them, lest by the severity of the winter your early planted roots should be destroyed, which does sometimes happen in very hard winters, especially in those places where they are not covered to protect them from frost: these late planted roots will flower a fortnight or three weeks after those which were planted in autumn, and many time blow equally as fair, especially if it prove a moist spring, or that care be taken to refresh them with water.

But then the increase of these roots will not be near so great as those of your first planting, provided they were not hurt in winter; and it is for this reason all those who make sale of these roots, are forward in planting; but in such gardens where these flowers are preserved with care, there is always provision made to cover them from the injuries of the weather, by arching the beds over with hoops, or frames of wood, and covering them with garden mats or cloths in frosty nights and bad weather, especially in the spring of the year, when their buds begin to appear; for otherwise, if you plant the best and most double flowers,

 the

the black frosts and cutting winds in March will often cause them to blow single, by destroying the thrum that is in the middle of the flower; and this many times hath occasioned many people who have bought the roots, to think they were cheated in the purchase of them, when it was wholly owing to their neglect of covering them, that their flowers became single.

Toward the latter end of June, the leaves of your first blown roots will begin to decay; soon after which time you must take them out of the ground, clearing them from decayed stalks, and washing them, to clean the earth from the roots; then spread them on a mat in a dry shady place till they are perfectly dried, when you may put them up in bags, and hang them out of the reach of mice, or other vermin, which will destroy many of the roots if they can come at them.

As all the fine varieties of these flowers were first obtained from seeds, so no good florist that hath garden-room, should neglect to sow them; in order to which, they should provide themselves with a quantity of good roots of the single (or what the gardeners call Poppy Anemonies) of the best colours, and such as have strong stems and large flowers, but especially such as have more leaves than common, and also other good properties; these should be planted early, that they may have strength to produce good seeds, which will be ripe in three weeks or a month's time after the flowers are past, when you must carefully gather it, otherwise it will be blown away in a short time, it being inclosed in a downy substance. You must preserve this seed till the beginning of August, when you may either sow it in pots, tubs, or a well-prepared bed of light earth: in the doing of it you must be careful not to let your seeds be in heaps, to avoid which is a thing little understood, and is what I have been informed of by the late Mr. Obadiah Lowe, gardener at Battersea, who for several years raised large quantities of these flowers from seeds. His manner was thus:

After having levelled his bed of earth, in which he intended to sow his seeds, he rubbed his seeds well between his hands with a little dry sand, in order to make them separate the better; then he sowed them as regularly as possible over the bed; but as these seeds will still adhere closely together by their down, so he made use of a strong hair brush, with which he gently swept over the whole bed, observing not to brush off the seeds; this brush will so separate the seeds, if carefully managed, as not to leave any entire lumps; then gently sift some light earth about a quarter of an inch thick over the seeds, and if it should prove hot dry weather, it will be adviseable to lay some mats hollow upon the bed in the heat of the day, and now and then give them a little water; but this must be given gently, lest by hastily watering you wash the seeds out of the ground; but be sure to uncover the bed at all times when there are gentle showers, and every night, that the seeds may have the benefit of the dews; and as the heat of the weather decreases, you may begin to uncover your bed in the day time also.

In about two months after sowing your plants will begin to appear, if the season has proved favourable, or your care in management hath not been wanting, otherwise they many times remain a whole year in the ground. The first winter after their appearing above ground, they are subject to injuries from hard frosts, or too much wet, against both of which you must equally defend them; for the frost is very apt to loosen the earth, so that the young plants are often turned out of the ground, after which a small frost will destroy them; and too much wet often rots their tender roots, so that all your former trouble may be lost in a short time for want of care in this particular; nor do I know of any thing more destructive to these tender plants, than the cold black frosts and winds of February and March, from which you must be careful to defend them, by placing a low Reed fence on the north and east sides of the bed, which may be moveable, and only fastened to a few stakes to support it for the present, and may be taken quite away as the season advances, or removed to the south and west sides of the bed, to screen it from the violence of the sun, which often impairs these plants when young.

As the spring advances, if the weather should prove dry, you must gently refresh them with water, which will greatly strengthen your roots; and when the green leaves are decayed, if your roots are not too thick to remain in the same bed another year, you must clear off all the weeds and decayed leaves from the bed, and sift a little more of the same prepared good earth, about a quarter of an inch thick over the surface, and observe to keep them clear from weeds during the summer season, and at Michaelmas repeat the same earthing; but as these roots so left in the ground will come up early in the autumn, the beds should be carefully covered in frosty weather, otherwise their leaves will be injured, whereby the roots will be weakened, if not destroyed. If your roots succeed well, many of them will flower the second year, when you may select all such as you like, by marking them with a stick; but you should not destroy any of them till after the third year, when you have seen them blow strong, at which time you will be capable to judge of their goodness; for until the roots have acquired strength, the flowers will not shew themselves to advantage.

The single (or Poppy) Anemonies, will flower most part of the winter and spring, when the seasons are favourable, if they are planted in a warm situation, at which time they make a fine appearance, therefore deserve a place in every flower-garden, especially as they require little culture. There are some fine blue colours amongst these single Anemonies, which, with the scarlets and reds, make a beautiful mixture of colours; and as these begin flowering in January or February, when the weather is cold, they will continue a long time in beauty, provided the frost is not too severe, or if they are covered with mats. The seeds of these are ripe by the middle or end of May, and must be gathered daily as it ripens, otherwise it will soon be blown away by the winds.

ANEMONOIDES. See Anemone.

ANEMONOSPERMOS. See Arctotis.

ANETHUM. Dill.

The Characters are,

It is an umbelliferous plant with many umbels, which are uniform. The flowers have five spear-shaped petals; under the flower is situated the germen, which afterward becomes two compressed seeds having borders.

We have but one Species of this genus, viz.

Anethum (*Graveolens*) fructibus compressis. Hort. Cliff. 106. *Garden or common Dill.*

This plant is propagated by sowing the seeds in autumn soon after they are ripe, for if they are kept out of the ground till spring they frequently miscarry; or if any of the plants do come up, they often decay before they have perfected their seeds. They love a light soil, and will not bear to be transplanted, but must be sown where they are to remain; for if the plants are removed, they will not produce good seeds, therefore the best way is, when the plants are come up, to hoe them out, as is practised for Onions, Carrots, &c. leaving the plants about eight or ten inches asunder every way, observing to keep them clear from weeds; when the seeds are ripe, the heads or umbels should be cut, and spread upon a cloth to dry, and then beat out for use; and if you let some seeds fall upon the ground, they will

will arise the next spring without any care, so that the trouble of sowing their seeds may be spared.

ANGELICA.

The CHARACTERS are,

It is an umbelliferous plant, the greater umbel being composed of many small ones; the empalement of the flowers are indented in five parts. The flowers of the whole umbel are uniform. The germen is situated below the flower, which afterward becomes a roundish fruit splitting into two, and composed of two seeds, which are plain on one side and convex on the other, and are bordered.

The SPECIES are,

1. ANGELICA (*Sativa*) foliorum impari lobato. Flor. Lapp. 101. *Common Garden Angelica.*

2. ANGELICA (*Archangelica*) altissima foliorum lobatis maximis serratis. *Commonly called Archangelica.*

3. ANGELICA (*Sylvestris*) foliis æqualibus ovato-lanceolatis serratis. Hort. Cliff. 97. *Greater wild Angelica.*

4. ANGELICA (*Atropurpurea*) extimo foliorum pari coadunato folio terminali petiolato. Prod. Leyd. 103. *Dark Purple Angelica of Canada.*

5. ANGELICA (*Lucida*) foliis æqualibus ovatis inciso-serratis. Hort. Cliff. 97. *Shining Angelica of Canada.*

The first sort is the common Angelica, which is cultivated in the gardens for medicinal use, as also for making a sweet-meat, which is by some greatly esteemed. This grows naturally by the side of rivers in Lapland, and other northern countries.

The second sort grows naturally in Hungary, and some parts of Germany.

The third sort grows naturally in moist meadows, and by the sides of rivers in many parts of England, so is seldom admitted into gardens.

The fourth and fifth sorts grow naturally in North America, from whence their seeds were sent to Europe, where the plants are preserved in gardens for the sake of variety; but as they are of no use, and have little beauty, so they are not admitted into many gardens. They are both very hardy plants, and may be easily propagated by seeds, which should be sown in autumn, and afterward the plants should be transplanted into a moist soil, and have a shady situation, allowing them room on every side. They grow near four feet high, and put out many shoots from the root, especially the second year from seed, when they will flower in June, and the seeds ripen in September. The roots of these sorts seldom continue longer than two or three years.

The common Angelica delights to grow in a very moist soil; the seeds of this plant should be sown soon after they are ripe, for if they are kept until the spring, seldom one seed in forty will grow. The best place for this plant is upon the sides of ditches, or pools of water, where being planted about three feet asunder, they will thrive exceedingly. The second year after sowing they will shoot up to flower, therefore if you have a mind to continue their roots, you should cut down these stems in May, which will occasion their putting out heads from the sides of the roots, whereby they may be continued for three or four years; whereas if they had been permitted to seed, their roots would perish soon after.

The gardeners near London, who have ditches of water running through their gardens, propagate great quantities of this plant, for which they have a great demand from the confectioners, who make a sweet-meat with the tender stalks of it, cut in May.

This plant is also used in medicine, as are also the seeds; therefore where it is cultivated for the seeds, there should be new plantations annually made to supply the places of those which die, for when they are permitted to seed they last but two years.

ANGUINA. See TRICOSANTHES.

ANGURIA. The Water Melon, or Citrul.

The CHARACTERS are,

It hath male and female flowers growing separate on the same plant; the flowers of both sexes are of the open bell shaped kind, of one leaf. The male flowers have three short stamina, which are joined together. The female flowers rest upon an oval germen, which afterward becomes an oblong fleshy fruit, having five cells filled with compressed seeds, which are rounded at their extremity.

We have but one SPECIES of this genus, viz.

ANGURIA (*Citrulus*) foliis multipartitis. *Water Melon, called Citrul.*

Of this there are several varieties, which differ in the form and colour of their fruit. But as these vary annually from seeds, so it is needless to enumerate them here.

The fruit is cultivated in Spain, Portugal, Italy, and most other warm countries in Europe; as also in Africa, Asia, and America, and is by the inhabitants of those countries greatly esteemed for their wholsome cooling quality; but in England the fruit is not so universally esteemed, though there are some few persons who are very fond of them.

To have this fruit good, you must first provide yourself with some seeds which should be three or four years old, for new seeds are apt to produce vigorous plants, which are seldom so fruitful as those of a moderate strength. The best sorts to cultivate in England are those with small round fruit, which come from Astracan, for those with very large fruit seldom ripen in this climate. Having provided yourself with good seed, you may sow it in the hot-bed for early Cucumbers; then you should prepare a heap of new dung the beginning of February, which should be thrown in a heap for about twelve days to heat, as is practised for early Cucumbers. When the dung is of a proper temper, the bed should be made in the same manner as for the Musk Melon, covering the dung about five inches thick with loamy earth; for the plants may be raised fit to plant out for good, in the same manner as the early Cucumbers, so the bed here mentioned is where they are to remain for good. But as these plants require much more room than either Cucumbers or common Melons, so there should be but one plant put into a three light frame; therefore a hill of the same loamy earth should be raised a foot and a half high in the middle light of each frame, into which, when the bed is of a proper temper for heat, the plants should be carefully planted, observing to water and shade them until they have taken good root.

After these plants are placed in these beds, you must be careful to admit fresh air to them, by raising of the glasses in proportion to the weather; and as their branches extend you should lead the shoots as they are produced, so as to fill each part of the frame, but not to croud each other, and be careful to keep them clear from weeds; they must also be frequently watered, but do not give it them in great quantities. In short, there is little difference to be observed in the management of these, from that of Musk Melons, but only to give them more room, earthing the beds to the same depth, and adding to the sides of the beds for the roots of the plants to run into it, and to keep the beds to a good temperature of heat; and when the fruit appears, to admit air freely to the plants, in order to set their fruit; but when the nights are cold, the glasses must be covered with mats to keep the beds warm, without which this fruit will seldom come to be good in this country.

ANIL. See INDIGOPHERA.

The whole process of making the Indigo would swell this volume beyond the intended size, so is omitted here.

ANISUM, or ANISE. See PIMPINELLA.

ANNONA. Lin. Gen. Plant. 613. The Custard Apple.

The

The CHARACTERS are,

The flower hath in some species three, and in others six petals, three large and three alternately smaller. The germen afterward becomes an oval or oblong fruit, having a scaly rind, and one cell, in which are lodged many oval smooth seeds.

The SPECIES are,

1. ANNONA (*Reticulata*) foliis lanceolatis fructibus ovatis reticulato-areolatis. Lin. Sp. Plant. 537. *The Custard Apple.*

2. ANNONA (*Muricata*) foliis ovali-lanceolatis glabris nitidis planis pomis muricatis. Hort. Cliff. 222. *The Sour Sop.*

3. ANNONA (*Africana*) foliis oblongis fructibus obtusè subsquamatis. Lin. Sp. Plant. 537. *The Sweet Sop.*

4. ANNONA (*Palustris*) foliis oblongis obtusis glabris, fructu rotundo, cortice glabro. *The Water Apple.*

5. ANNONA (*Cherimola*) foliis latissimis glabris, fructu oblongo squamato, seminibus nitidissimis. *This is called Cherimolias by the Spaniards.*

6. ANNONA (*Squamosa*) foliis ovato-lanceolatis pubescentibus fructu glabro subcæruleo. *The Sweet Apple.*

7. ANNONA (*Asiatica*) foliis lanceolatis glabris nitidis secundum nervos sulcatis. Hort. Cliff. 222. *The Purple Apple.*

8. ANNONA (*Triloba*) foliis lanceolatis fructibus trifidis. Lin. Sp. Plant. 537. *The North American Annona, called by the inhabitants Papaw.*

The first sort usually grows to the height of twenty-five feet or upwards in the West-Indies, and is well furnished with branches on every side; the leaves are oblong, pointed, and have several deep transverse ribs or veins, and are of a light green colour; the fruit is of a conical form, and as large as a tennis ball, of an Orange colour when ripe, having a soft, sweet, yellowish pulp, of the consistence of a custard, from whence this name was given to it.

The second sort does not grow so large as the first, rarely rising above twenty feet high, and not so well furnished with branches; the leaves are broader, and have a smooth surface without any furrows, and are of a shining green colour; the fruit is large, of an oval shape, irregular, and pointed at the top, being of a greenish yellow colour, and full of small knobs on the outside.

The third sort is a tree of humbler growth, seldom rising so high as twenty feet, and is well furnished with branches on every side; the leaves of this sort have an agreeable scent when rubbed; the fruit is roundish and scaly, and when ripe turns to a purple colour, and hath a sweet pulp.

The fourth sort commonly grows from thirty to forty feet in the West-Indies. This hath oblong pointed leaves, which have some slender furrows, and when rubbed have a strong scent; the fruit of this sort is seldom eaten but by the negroes.

The fifth sort is much cultivated in Peru for the fruit. This grows to be a very large tree in its natural country, and is well furnished with branches, which are garnished with leaves of a bright green colour, and much larger than those of any of the other sorts. The fruit is oblong, and is scaly on the outside, and of a dark purple colour when ripe.

The sixth and seventh sorts grow in some of the French islands, as also in Cuba in great plenty; these grow to the height of thirty feet or more; their fruit are esteemed by the inhabitants of those islands, who frequently give them to sick persons, as they reckon them very cooling and wholsome.

The eighth sort grows plentifully in the Bahama Islands, where it seldom rises to more than ten feet high, having several stems; the fruit of this sort are shaped like a Pear inverted. This is seldom eaten but by the negroes, and is the food of guanas and other animals.

This sort will thrive in the open air in England, if it is planted in a warm sheltered situation; but the plants should be trained up in pots, and sheltered in winter for two or three years, until they have acquired some strength; then they may be turned out of the pots in the spring and planted in the full ground, where they are to remain. This sort flowers in many curious gardens. The seeds of this are frequently brought to England from North America, and many plants have been lately raised in the gardens near London. The seeds of this sort are very different in shape from any of those which I have yet seen, which have been brought from the islands of the West-Indies, and the shape of the leaves are also different; this casts its leaves in autumn, whereas all the others retain their leaves until the spring, when the new leaves come out. The fruit is very different from those of the other species, two or three growing together joined at their foot-stalks.

All the sorts which are natives of the warm parts of America, are too tender to live in this country, if they are not preserved in warm stoves; they come up very easily from the seeds which are brought from America, if they are fresh; but the seeds must be sown on a good hot-bed, or in pots of light earth, and plunged into a hot-bed of tanners bark pretty early in the spring; because if the plants come up early, they will have time to get strength before the cold weather comes on in the autumn.

If these plants are kept in the bark-stove and carefully managed, they will make great progress; but in warm weather they should have plenty of fresh air admitted to them, for when the air is excluded from them too much they are apt to grow sickly; when they will soon be attacked by vermin, which will multiply and spread over the whole surface of the leaves, and cause them to decay; but if the plants are carefully managed, their leaves will continue green all the winter, and make a very good appearance in the stove at that season.

They must constantly remain in the tan-bed, otherwise they will make but little progress; for although they will live in a dry stove, yet they will not thrive, nor will their leaves appear so fine as when they are preserved in a vigorous growing state; and it is more for the beauty of their leaves, than any hopes of their producing fruit in this country, that they are preserved in stoves; for tho' there has been some of the sorts which have produced flowers in England, yet they have rarely shewn their fruit here.

ANTHEMIS. Lin. Gen. Plant. 870 Chamomile.

The CHARACTERS are,

It is a plant with a compound flower. The border or rays of the flower is composed of many female flowers, whose petals are stretched out like tongues on one side. The middle or disk of the flower is composed of many hermaphrodite florets, which are funnel-shaped, erect, and cut into five parts at the top. The germen is situated at the bottom, which afterward becomes an oblong naked seed.

The SPECIES are,

1. ANTHEMIS (*Nobilis*) foliis pinnato-compositis linearibus acutis subvillosis. Lin. Sp. Plant. 894. *Common, or noble Chamomile.*

2. ANTHEMIS (*Arvensis*) receptaculis conicis paleis setaceis seminibus coronato-marginatis. Flor. Succ. 704. *Wild Chamomile, or May Weed.*

3. ANTHEMIS (*Cotula*) receptaculis conicis paleis setaceis seminibus nudis. Lin. Sp. Plant. 894. *Stinking May Weed.*

4. ANTHEMIS (*Cota*) florum paleis rigidis pungentibus. Flor. Leyd. 172. *Chamomile with stiff pungent chaff between the flowers.*

5. AN-

5. ANTHEMIS (*Altiſſima*) erecta foliorum apicibus ſubſpinoſis. Lin. Sp. Plant. 893. *Spaniſh Chamomile with a large flower.*

6. ANTHEMIS (*Maritima*) foliis pinnatis denticulatis carnoſis caule ramoſo. Lin. Sp. Plant. 893. *Sea Chamomile.*

7. ANTHEMIS (*Tomentoſa*) foliis pinnatifidis obtuſis planis, pedunculis hirſutis, folioſis calycibus tomentoſis. Hort. Cliff. 415. *Hoary Sea Chamomile with a thick Wormwood leaf.*

8. ANTHEMIS (*Mixta*) foliis ſimplicibus dentato-laciniatis. Lin. Sp. Plant. 894. *Broad-leaved Portugal Chamomile, with a Buck's-horn leaf.*

9. ANTHEMIS (*Pyrethrum*) caulibus unifloris decumbentibus foliis pinnato-multifidis. Lin. Hort. Cliff. 414. *Pellitory of Spain.*

10. ANTHEMIS (*Tinctoria*) foliis bipinnatis ſerratis ſubtus tomentoſis caule corymboſo. Lin. Sp. Plant. 896. *Alpine Ox-eye with a white flower.*

11. ANTHEMIS (*Valentina*) caule ramoſo, foliis pubeſcentibus tripinnatis, calycibus villoſis pedunculatis. Lin. Hort. Cliff. 414. *Chamomile with a branching ſtalk, multifid hoary leaves, and hoary foot-ſtalks.*

12. ANTHEMIS (*Arabica*) caule decompoſito, calycibus ramiferis. Lin. Hort. Cliff. 413. *Chamomile with a decompounded ſtalk, and branching flower-cups.*

The firſt ſort is the common Chamomile, which grows in plenty upon commons and other waſte land. It is a trailing perennial plant, which puts out roots from the branches as they lie on the ground, whereby it ſpreads and multiplies greatly; ſo that whoever is willing to cultivate this plant, need only procure a few of the ſlips in the ſpring, and plant them a foot aſunder, that they may have room to ſpread, and they will ſoon cover the ground. The flowers of this ſort are ordered for medicinal uſe, but the market people generally ſell the double flowers, which are much larger, but not ſo ſtrong as the ſingle. The double ſort is equally hardy, and may be propagated in the ſame manner.

The ſecond ſort is the common annual weed, which grows among corn; it flowers in May, ſo is called May Weed, though ſome have applied that title improperly to the Cotula fœtida, which rarely flowers till late in June. This is the third ſort here enumerated, of which there is a variety with double flowers, which is preſerved by often planting the ſlips and cuttings; ſo that although the plant is naturally an annual, yet by this method it may be continued.

The fourth, fifth, and eighth ſorts are annual plants, which grow naturally in Spain, Portugal, Italy, and the ſouth of France; the plants are preſerved in botanic gardens for the ſake of variety, but are ſeldom allowed a place in others. They flower in July, and their ſeeds ripen in September.

The ſixth, ſeventh, ninth, and tenth ſorts are perennial plants; theſe grow naturally in Spain, Portugal, and Greece: the plants are preſerved in ſome curious gardens for the ſake of variety. They are hardy, and may be propagated either by ſeeds or ſlips; if by ſeeds, they ſhould be ſown in the ſpring upon poor land, where the plants will continue much longer than in good ground. The ſlips may be planted during any of the ſummer months, obſerving to plant them in a ſhady border, and water them until they have taken root. In the autumn they may be removed to the places where they are to remain, and will require no other care but to keep them clean from weeds. Theſe plants do not grow tall, but are buſhy, ſo ſhould be allowed room. They continue in flower from July to October, and the ſeeds ripen in autumn.

The ninth ſort is the Pellitory of Spain, the roots of which are uſed for the tooth-ach, being extremely warm; when they are applied to the part affected, they draw out the cold rheum, and are often ſerviceable. This is a perennial plant, with a long taper root like a Carrot, which grows naturally in Spain and Portugal, from whence the roots are brought to England. The branches of this trail upon the ground, and ſpread a foot or more on every ſide, and have fine winged leaves like thoſe of the common Chamomile; at the extremity of each branch is produced one large ſingle flower like Chamomile, but much larger, the rays of which are of a pure white within, but purple on their outſide. It flowers in June and July, and the ſeeds are ripe in September; but unleſs the ſeaſon is warm and dry the ſeeds do not ripen in England, for the wet falls between the ſcales, and rots the ſeeds in embryo.

The eleventh and twelfth ſorts are annual plants, which do not riſe more than two feet high; the leaves are hoary, but the twelfth ſort rarely produces good ſeeds here.

Theſe ſorts are propagated by ſeeds, and may be ſown on a bed of common earth in the ſpring; when the plants are ſtrong enough to remove, they may be tranſplanted into large open borders near ſhrubs, where they may have room to grow, for they ſpread out on every ſide, therefore require three feet diſtance from other plants; in theſe large open ſpots they will make a pretty variety from June to November, during which time they continue in flower.

ANTHERICUM. Lin. Gen. Plant. 380. Spiderwort.

The CHARACTERS are,

The flower hath no empalement, and is compoſed of ſix petals which ſpread open. It hath ſix upright ſtamina. The germen, which is ſituated in the center is three-cornered, which afterward becomes an oval ſmooth capſule, having three furrows, opening in three cells, which are filled with angular ſeeds.

The SPECIES are,

1. ANTHERICUM (*Revolutum*) foliis planis ſcapo ramoſo corollis revolutis. Lin. Sp. Plant. 310. *Aſphodel with rough compreſſed leaves, and a ſpreading ſtalk.*

2. ANTHERICUM (*Ramoſum*) foliis planis ſcapo ramoſo corollis planis. Lin. Sp. Plant. 310. *Branching Spiderwort with a ſmall flower.*

3. ANTHERICUM (*Liliago*) foliis planis ſcapo ſimpliciſſimo corollis planis piſtillo declinato. Hort. Upſal. 83. *Branching Spiderwort with a ſmall flower.*

4. ANTHERICUM (*Fruteſcens*) foliis carnoſis teretibus caule fruticoſo. Lin. Sp. Plant. 310. *Cape Spiderwort with ſtalks, and pulpy Onion leaves.*

5. ANTHERICUM (*Aloides*) foliis carnoſis ſubulatis planiuſculis. Hort. Upſal. 83. *Low Cape Spiderwort, with Aloe-ſhaped pulpy leaves.*

6. ANTHERICUM (*Aſphodeloides*) foliis carnoſis ſubulatis ſemiteretibus ſtrictis. Hort. Upſal. 83. *Low Aſphodel with awl-ſhaped ſucculent leaves.*

7. ANTHERICUM (*Annuum*) foliis carnoſis ſubulatis teretibus. Hort. Upſal. 83. *Small yellow African Spiderwort, with narrow leaves.*

8. ANTHERICUM (*Altiſſimum*) caule foliis carnoſis teretibus ſpicis florum longiſſimis laxis. Fig. Plant. pl. 39. *Low African Spiderwort with taper fleſhy leaves, and very long looſe ſpikes of flowers.*

9. ANTHERICUM (*Oſſifragum*) foliis enſiformibus filamentis lanatis. Flor. Suec. 268. *Marſh yellow Aſphodel.*

10. ANTHERICUM (*Calyculatum*) foliis enſiformibus perianthiis trilobis filamentis glabris. Flor. Suec. 269. *Marſh Alpine Spiderwort with an Iris leaf.*

The firſt ſort grows near two feet high; the ſtalk branches out on every ſide, each branch being terminated by a looſe ſpike of flowers which are white, and the petals are turned backward to their foot-ſtalk. The leaves of this ſort are flat, and the root is perennial, but the branches decay in autumn.

The

The second sort hath a perennial root, but an annual stalk which decays in autumn; the stalks of this rise about the same height as the former, sending out many lateral ones in like manner, which are terminated by loose spikes of flowers, which are white, but the petals are plain and do not turn back as in the other sort.

The third sort hath plain leaves and an unbranching stalk, in which it chiefly differs from the former. The root of this is perennial, but the stalks decay in winter.

These three sorts grow naturally in Spain, Portugal, and other warm countries; they were more common some years ago in the English gardens than at present, for the severe winter in 1740 killed most of their roots. These flower in June and July, and their seeds are ripe in September. They are propagated by seeds, which should be sown in autumn these should be sown in a bed of light sandy earth in a warm situation; when the plants come up they must be kept clean from weeds during the summer, and in autumn when their leaves decay, they should be carefully taken up and transplanted in the borders of the flower-garden, where they will last several years, if they are not killed by frost; to prevent which, some rotten tan should be laid over the roots in winter, which will always secure them.

The fourth sort has been long preserved in many gardens near London, and was formerly known among the gardeners by the title of Onion-leaved Aloe. This plant produces many ligneous branches from the root, each having a plant with long taper leaves in shape of those of the Onion, which are full of a yellow pulp very juicy. It grows naturally at the Cape of Good Hope, and requires a little shelter in the winter; but in some mild seasons I have had plants live without any cover, which were planted close to a warm wall.

The fifth and sixth sorts grow close to the ground, never rising with any stalk. The fifth hath broad, flat, pulpy leaves, resembling those of some sorts of Aloe. The leaves spread open flat on the ground, and the flowers are produced on loose spikes like the former, but they are shorter. The flowers are yellow, and appear at different seasons. This is propagated by offsets, which are put out in plenty, and must be planted in pots filled with light sandy earth, and in winter placed in the green house, and treated as other hardy succulent plants which come from the Cape of Good Hope.

The sixth sort hath long, narrow, pulpy leaves, which are almost taper, but flatted on their upper side; this sends out many offsets, by which it may be increased plentifully. It must be treated in the same manner as the former.

The seventh sort is annual: this is a low plant growing close to the ground, having pretty long succulent leaves, which are taper, but flatted on their upper side; the flowers grow in loose spikes, which are shorter than either of the other sorts. They are yellow, and are succeeded by round seed-vessels like those of the former sorts; the plants perish soon after their seeds ripen. The seeds of this sort should be sown on a warm border of light earth in April, where they are to remain, and will require no other care but to keep them clean from weeds, and to thin them where they are too close.

The eighth sort never rises to a stalk, but the leaves come out close to the ground. These are long, taper, succulent, and of a sea-green colour, growing erect; the flower-stems rise between the leaves, and are near three feet long. The plants are seldom long destitute of flowers. It must be treated in the same manner as the fourth, fifth, and sixth sorts.

The ninth and tenth sorts grow naturally on bogs in most of the northern countries; the ninth is common in many parts of England, but particularly in Lancashire, from whence it had the title of Lancashire Asphodel; it also grows upon a bog upon Putney-heath. The other grows naturally in Denmark, Sweden, and Lapland. The flower-stems rise about six inches high, being terminated by a loose spike of small yellow flowers. These plants growing naturally upon bogs, are with difficulty preserved in gardens.

ANTHOLYZA. We have no English name for this plant.

The Characters are,

It hath an imbricated sheath, which is permanent; the flower is of one leaf, and opens above with compressed jaws. The under lip is trifid and short; the middle segment turns downward. Under the flower is situated the germen, which afterward becomes a roundish three-cornered vessel having three cells, in which are lodged many triangular seeds.

The Species are,

1. Antholyza (*Ringens*) stamine unico declinato. Lin. Sp. Plant. 37. *Antholyza with one stamen declining.*

2. Antholyza (*Spicata*) foliis linearibus sulcatis floribus albis uno versu dispositis. Fig. Plant. pl. 40. *Strange Corn-flag with narrow furrowed leaves, and white flowers ranged on one side of the stalk.*

The first sort hath round, red, bulbous roots, from which arise several rough furrowed leaves, which are near a foot long, and half an inch broad; between these come out the flower-stem immediately from the root, which rise two feet high, is hairy, and hath several flowers coming out on each side. These are of one leaf, cut into six unequal parts at the top; the margins are waved and close together, wrapping up the three stamina. These flowers are red, and appear in June, and the seeds ripen in September.

The roots of the second sort are in shape and size like those of the vernal Crocus, but the outer skin is thin and white; from this arises five or six long narrow leaves which are deeply furrowed. Between these arise the flower-stem, which is a foot and a half high; the flowers come out, ranged on one side, standing erect. These have each a spatha or sheath of one leaf, divided into two, ending in points. The flower is of one leaf, having a long tube, but is divided into six unequal segments at the top which spread open. After the flower is past the germen becomes a three-cornered seed-vessel, opening in three cells, which are filled with triangular seeds.

They are propagated by offsets, which the bulbous roots send forth in pretty great plenty; or by seeds, which are sometimes perfected in Europe. These seeds should be sown soon after they are ripe; if the seeds are sown in pots of light earth, and plunged into an old bed of tan which has lost its heat, and shaded in the middle of the day in hot weather, the plants will come up the following winter, therefore they must be kept covered with glasses to screen them from cold. In summer after the leaves are decayed, the roots should be taken up and planted each into a separate small pot filled with light earth. In summer the pots may be placed in the open air in half-sun, but in winter they must be placed under a hot-bed frame, for they are not very tender; but where any damp arises, it is very apt to occasion a mouldiness upon their leaves. The roots shoot up in autumn, and the flowers begin to appear in May; the seeds ripen in August, and soon after their leaves and stalks decay. The roots may be easily transported from one country to another at the time when they are taken up. These plants are a great ornament to the green-house when they are in flower, and as they require but little culture, so deserve a place in every good garden.

ANTHOSPERMUM. Amber-tree, vulgò.

The

The CHARACTERS are,

It is male and female in different plants; the male flowers have no petals. The female flowers have the same structure as the male, but have no stamina; instead of which there is an oval germen situated in the bottom, which afterward becomes a roundish capsule having four cells, which contain several angular seeds.

We have but one SPECIES of this genus, viz.

ANTHOSPERMUM (*Æthiopicum*) mas & fœmina. Hort. Cliff. 455. *Male and female Amber-tree.*

These plants are preserved in many curious gardens, which have collections of tender plants, and is easily propagated by planting cuttings during any of the summer months in a border of light earth, which will take root in six weeks time, provided they are watered and shaded as the season may require; then they should be taken up with a ball of earth to their roots, and planted into pots filled with light sandy earth, and may be exposed to the open air until October; at which time they should be removed into the conservatory, where they should be placed as free as possible from being over-hung with other plants; and during the winter season, they must be often refreshed with water, but should not have too much given them each time.

The beauty of this shrub is in its small evergreen leaves, which grow as close as Heath; and being bruised between the fingers, emit a very fragrant odour. The plants must be frequently renewed by cuttings, for the old plants are very subject to decay, seldom continuing above three or four years.

It is but of late years there have been any of the female plants in the gardens, for all those which were formerly in the gardens were the male; which being propagated by cuttings had been continued, so that no seeds were ever produced in England till within a few years past, when I received some seeds from the Cape of Good Hope, from which I raised many plants of both sexes, and a few among them which have hermaphrodite flowers, which have produced seeds, from which many plants have been raised.

ANTHYLLIS. Lin. Gen. Plant. 773. Ladies Finger.

The CHARACTERS are,

The flower is of the butterfly kind, having a long standard reflexed on both sides beyond the empalement; the two wings are short; the keel is of the same length, and compressed. In the center is situated an oblong germen, which afterward becomes a small roundish pod inclosed by the empalement, having one or two seeds.

The SPECIES are,

1. ANTHYLLIS (*Tetraphylla*) herbacea foliis quaternopinnatis floribus lateralibus. Hort. Upf. 221. *Five-leaved Woundwort.*

2. ANTHYLLIS (*Vulneraria*) herbacea foliis pinnatis inæqualibus capitulo duplicato. Lin. Sp. Plant. 719. *Low Woundwort with a scarlet flower.*

3. ANTHYLLIS (*Rustica*) herbacea foliis pinnatis foliolis inæqualibus caulinis lineari lanceolatis, floribus capitatis simplicibus. *Rustic Woundwort, or Ladies Finger.*

4. ANTHYLLIS (*Montana*) herbacea foliis pinnatis æqualibus capitulo terminali, secundis floribus obliquatis. Lin. Sp. Plant. 719. *Purple Milk Vetch.*

5. ANTHYLLIS (*Cornicinis*) herbacea foliis pinnatis inæqualibus capitulis solitariis. Lin. Sp. Plant. 719. *Herbaceous Woundwort, with unequal winged leaves and a single head.*

6. ANTHYLLIS (*Barba Jovis*) fruticosa foliis pinnatis æqualibus floribus capitatis. Hort. Cliff. 371. *Jupiter's Beard, or Silver Bush*

7. ANTHYLLIS (*Cytisoides*) fruticosa foliis ternatis inæqualibus calycibus lanatis lateralibus. Lin. Sp. Plant. 720. *Hoary Cytisus with a longer middle leaf.*

8. ANTHYLLIS (*Erinacea*) fruticosa spinosa foliis simplicibus. Lin Sp. Plant. 720. *Prickly Broom with Duckmeat leaves, and bluish purple flowers.*

The first sort grows naturally in Spain, Italy, and Sicily. This is an annual plant with trailing branches, which spread flat on the ground; the flowers which are yellow, come out in clusters on the sides of the stalks, having large swelling empalements, and are succeeded by short pods inclosed in it. This flowers in June and July, and the seeds ripen in September. The seeds of this sort should be sown on a bed of light earth in April, where the plants are to remain, and will require no other care but to thin them to the distance of two feet, and keep them clean from weeds.

The second sort grows naturally in Spain and Portugal, from both which countries I have received the seeds; it also grows wild in Wales and the Isle of Man. This is a biennial plant, having single leaves at bottom, which are oval and hairy; but those which grow out of the stalks are winged, each being composed of two or three pair of lobes terminated by an odd one: the flowers are collected into heads at the top of the stalks; these are of a bright scarlet colour, so make a pretty appearance: it flowers in June and July, and the seeds ripen in October. When the plants of this sort grow on poor land, they will sometimes continue three years, but in gardens they seldom last longer than two.

The third sort grows naturally upon chalky grounds in many parts of England, so is rarely admitted into gardens. The heads of flowers in this species are single, whereas the other has generally double heads.

The fourth sort is a perennial plant with trailing branches, at the extremity of which the flowers are produced in heads; these are of a purple colour and globular form. It grows naturally on mountains in the south of France and Italy, and is propagated by seeds, which may be sown either in the autumn or spring; those which are sown in the autumn will rise the following spring, and more certainly grow than those which are sown in the spring, which seldom grow the same year. When the plants come up they must be kept clean from weeds, and where they are too close together they must be thinned. The following autumn they should be transplanted to the places where they are to remain, and will require no particular management afterward. This sort flowers in June and July, and the seeds ripen in October.

The fifth sort approaches near to the third, but the leaves are hoary, and the flowers are produced on the side of the branches; these are yellow, and collected into small heads. This may be propagated by seeds as the former.

The sixth sort is the Barba Jovis, or Jupiter's Beard, by many called Silver Bush, from the whiteness of its leaves. This is a shrub which often grows ten or twelve feet high; the leaves are very white and hairy; the flowers are produced at the extremity of the branches, collected into small heads; they are of a bright yellow colour, and appear in June. It is propagated either by seeds or cuttings; if by seeds, they should be sown in the autumn in pots filled with light earth, and placed under a frame in winter to protect them from frost. The following spring the plants will rise, and when they are strong enough to remove they should be each planted in a small pot filled with light earth, and placed in the shade till they have taken new root; after which they may be placed with other hardy exotic plants in a sheltered situation, where they may remain till October, when they must be removed into shelter. These plants are too tender to live in the open air here in winter, though I have had some of them live abroad two or three years, which were planted against a south-west aspected wall. It may also be be propagated by cuttings, which may be planted during any of the summer months, obferving to water and fhade them until they have taken root.

The feventh fort is a low fhrub, feldom rifing above two feet high, but fends out many flender branches, which are garnifhed with hoary leaves; they are fometimes fingle, but generally have three oval lobes, the middle being longer than the other two; the flowers are white, and come out from the fide of the branches, three or four joined together, having woolly empalements, but thefe are rarely fucceeded by feeds in England. It may be propagated by cuttings in the fame manner as the former fort, and treated as hath been directed for that.

The eighth fort grows naturally in Spain and Portugal. This is a fhrub which grows nine or ten feet high, having the appearance of one fort of Gorfe or Whin, but it hath round leaves growing fingle. It will live in the open air in mild winters, but hard froft will deftroy it. It is propagated by feeds only.

ANTIRRHINUM. Snap-dragon, or Calves-fnout.

The CHARACTERS are,

The flower is ringent, having an oblong tube divided at the top into two lips, which are clofed at the jaw. In the bottom is fituated an obtufe nectarium, which is not prominent. In the center is placed a roundifh germen, which afterward becomes a round obtufe capfule having two cells, which are full of fmall angular feeds.

To this genus Linnæus has joined the Linaria and Afarina; but as the flowers of the Linaria have fpurs to their petals, and the nectarium being very prominent, which are not fo in this genus, fo it fhould be feparated from it; efpecially as there are many fpecies of both kinds, which cannot fo well be diftinguifhed when both genera are joined in one.

The SPECIES are,

1. ANTIRRHINUM (*Minus*) foliis lanceolatis obtufis alternis caule ramofiffimo diffufo. Hort. Cliff. 324. *The leaft Field Snap-dragon.*

2. ANTIRRHINUM (*Orontium*) foliis lanceolatis petiolatis calycibus corollâ longioribus. *Greater Field Snap-dragon.*

3. ANTIRRHINUM (*Majus*) foliis lanceolatis petiolatis calycibus breviffimis racemo terminali. Vir. Cliff. 61. *Another great Snap-dragon with a longer leaf.*

4. ANTIRRHINUM (*Latifolium*) foliis lanceolatis glabris calycibus hirfutis racemo longiffimo. *Broad-leaved Snap-dragon with a large pale flower.*

5. ANTIRRHINUM (*Italicum*) foliis lineari-lanceolatis hirfutis racemo breviore. *Greater Italian long-leaved Snap-dragon, with a large fnowy flower.*

6. ANTIRRHINUM (*Siculum*) foliis linearibus floribus petiolatis axillaribus. *Sicilian Snap-dragon with a Toad-flax leaf, and a fnow-white flower.*

The two firft forts grow naturally on arable land in many parts of England, fo are feldom admitted into gardens; thefe are both annual plants, which come up from fcattered feeds. They flower in June and July, and their feeds are ripe in Auguft and September.

The third fort is not a native of England, but having been firft brought into gardens, the feeds have fcattered about in fo great plenty, that it is become very common upon walls and old buildings in many parts of England. Of this fort there are feveral varieties, which differ in the colour of their flowers, fome having red flowers with white mouths, fome with yellow mouths, others have white flowers with yellow and white mouths. There is alfo one with ftriped leaves. The laft is propagated by flips and cuttings, which readily take root any time in the fpring or fummer. The different colours of the flowers are variable from feeds.

The fourth fort grows naturally in the iflands of the Archipelago. The leaves of this are much broader, the flowers greatly larger, and the fpikes longer, than in any of the other forts. The colours of the flowers are as changeable in this fort as the former, when raifed from feeds; but as this is the moft fpecious kind, fo it better deferves propagating than the common fort, efpecially as it is equally hardy.

The fifth fort hath long narrow leaves, which are hairy; the flowers are large, and the fpike is fhorter than the former.

The fixth fort is an annual plant, which feldom grows more than a foot high: the leaves of this are very narrow and fmooth; the flowers come out from the wings of the leaves fingle, ftanding on long foot-ftalks; thefe are very white, with a dark bottom.

The third, fourth, and fifth forts are raifed from feeds, which fhould be fown in a dry foil, which is not too rich, either in April or May; and in July the plants may be planted out into large borders, where they will flower the fpring following; or they may be fown early in the fpring for flowering the fame autumn, but then they are not fo likely to endure the winter; and if the autumn prove bad, they will not perfect their feeds.

Thefe plants grow extremely well upon old walls or buildings, in which places they will endure for feveral years; whereas thofe planted in gardens feldom laft longer than two years, unlefs they are planted in a very poor foil, and the flowers often cropped, and not fuffered to feed; but any of thefe forts may be continued by planting cuttings in any of the fummer months, which will eafily take root.

Wherever thefe plants are defigned to grow on walls, or on a rocky barren foil, the feeds fhould be fown the beginning of March where they are defigned to remain; for if the plants are firft raifed in a better foil, and afterward tranfplanted into thofe places, they feldom fucceed well.

APARINE. Goofe-grafs, or Clivers.

There are three or four forts of this which are preferved in botanic gardens for variety, but are not worthy of a place in other gardens.

APHACA. Vetchling.

The CHARACTERS are,

The flower is of the butterfly kind; the ftandard being large and heart-fhaped, the wings are fhorter and obtufe; the keel is the length of the wings, and divided flightly in the middle. The germen, which is fituated in the center, afterward becomes a fhort pod, containing two or three round feeds.

We have but one SPECIES of this plant, viz.

APHACA. Lob. Icon. 70. *Yellow Vetchling.*

This plant is found wild in divers parts of England on arable land, but is feldom preferved in gardens. It is an annual plant, which perifhes foon after the feeds are perfected. The fureft method to cultivate this plant is to fow the feeds on a bed of light earth in autumn, foon after they are ripe, for if they are kept out of the ground until fpring they feldom grow; and if fome of the plants come up at that feafon, they feldom perfect their feeds fo well as thofe which were fown in autumn. Thefe feeds fhould be fown where the plants are defigned to remain, for they feldom fucceed well if they are tranfplanted. All the culture thefe plants require is to keep them clean from weeds, and to thin them where they come up too clofe, leaving them about ten inches or a foot afunder.

APIOS. See GLYCINE.

APIUM. Parfley.

The CHARACTERS are,

It is a plant with an umbelliferous flower; each flower has five ftamina. Under the flower is fituated the germen, which afterward becomes an oval channelled fruit, dividing into two parts,

parts, having two oval seeds channelled on one side, and plain on the other.

The Species are,

1. Apium (*Petroselinum*) foliolis caulinis linearibus volucellis minimis. Hort. Cliff. 108. *Common Parsley.*

2. Apium (*Crispum*) foliis radicalibus amplioribus crispis caulinis ovato-multifidis. *Curled Parsley.*

3. Apium (*Latifolium*) foliis radicalibus trifidis, serratis petiolis longissimis. *The large-rooted Parsley.*

4. Apium (*Graveolens*) foliolis caulinis cuneiformibus. Hort. Cliff. 107. *Smallage, or Water Parsley.*

5. Apium (*Dulce*) foliis erectis, petiolis longissimis foliolis quinque lobatis serratis. *Upright Celery.*

6. Apium (*Rapaceum*) foliis patulis, petiolis brevibus, foliolis quinis serratis, radice rapacea. *Celeriac, or Turnep-rooted Celery.*

The first sort is the common Parsley, which is generally cultivated for culinary use, and is what the College of Physicians have directed to be used in medicine, under the title of Petroselinum; for when Apium is prescribed, the Smallage is always intended.

The second sort may be constantly preserved, if the seeds are carefully saved from plants which are well curled, the seeds will produce the same; but there are few persons who will be at the trouble to save the seeds so carefully, as not to have some of the common sort mixed with it: therefore the only method to have it good, is to separate all those plants which have plain leaves from the curled, as soon as they are distinguishable, leaving only such as are of the right kind. It will be a very safe method for such persons who cannot well distinguish the common Parsley from the lesser Hemlock, to sow the seeds of this curled-leaved Parsley, which is easily known at first sight from Hemlock; for where the latter has been used by mistake, it has been attended with bad consequences.

The third sort is chiefly cultivated for their roots, which are now pretty commonly sold in the London markets; the leaves of this sort have much longer foot-stalks, and the subdivisions of these are not so numerous as in the common Parsley; the lobes of the leaves are much larger, and of a darker green, so that it is easily distinguished from the common sort by its leaves, but the roots are six times as large as the common Parsley can be brought to with the utmost culture. This sort was many years cultivated in Holland, before the English gardeners could be prevailed on to sow it. I brought the seeds of it from thence in 1727, and would then have persuaded some of the kitchen-gardeners to make trial of it, but they refused to accept of it, so that I cultivated it several years before it was known in the markets.

The fourth sort is commonly known by the title of Smallage. This is what the physicians intend when they prescribe Apium. This plant grows naturally by the sides of brooks and ditches in many parts of England, so is rarely cultivated in gardens.

The fifth sort is the common Celery, and the sixth sort was supposed to be a degenerate species from it; but I cannot agree to this opinion, for from many years trial I have never found it vary. The leaves of this sort are short when compared with those of the other, and spread open horizontally, and the roots grow as large as common Turneps. All the difference which I have observed to arise from culture, has been only in the size of the roots; those on rich ground, which were properly cultivated, were much larger than those on poorer land, but the leaves and outward appearance of the plants were never altered, so that I make no doubt of its being a different species.

The common Parsley should be sown early in the spring, for the seeds remain a long time in the earth, the plants seldom appearing in less than six weeks after the seeds are sown. This sort is generally sown in drills by the edges of borders in the kitchen-gardens near London, because it is much easier to keep it clear from weeds, than if the seeds are sown promiscuously on a border, and the Parsley is much sooner cut for use; but when the roots are desired for medicinal use, then the seeds must be sown thin; and when the plants are come up they should be hoed out single, as is practised for Carrots, Onions, &c. observing also to cut up the weeds: if this be observed, the roots will become fit for use by July or August.

The common Parsley is, by some skilful persons, cultivated in fields for the use of sheep, it being a sovereign remedy to preserve them from the rot, provided they are fed twice a week for two or three hours each time with this herb; but hares and rabbets are so fond of it, that they will come from a great distance to feed upon it; and in countries where these animals abound, they will destroy it, if it is not very securely fenced against them; so that whoever has a mind to have plenty of hares in their fields, by cultivating Parsley, will draw all the hares of the country to them.

The best time for sowing it in the fields is about the middle or latter end of February; the ground should be made fine, and the seeds sown pretty thick in drills drawn at about a foot asunder, that the ground may be kept hoed between the drills to destroy the weeds, which, if permitted to grow, will soon over-run the Parsley. Two bushels of seed will sow one acre of land.

The great Garden Parsley is now more known to us in England, than it was some years ago: in Holland it has been long very common in all their markets; they bring these roots in bunches, as we do young Carrots to market in summer, and the roots are much of the same size; it is called Petroseline Wortle by the Dutch, who are very fond of it for Water Souche.

It may be cultivated by sowing the seeds in good ground early in the spring; and in April, when the plants are up, cut them out with a hoe (as is practised for young Carrots) to about five or six inches square, and keep them constantly clean from weeds, and in July the roots will be fit to draw for use, and may be boiled and eaten as young Carrots; they are very palatable and wholsome, especially for those who are troubled with the gravel.

But where these plants are cut out to allow them more room, if the soil is good, the roots will grow to the size of a middling Parsnep by September; and the roots may be preserved for use all the following winter, in the same manner as Carrots.

The seeds of the two sorts of Celery should be sown at two or three different times, the better to continue it for use through the whole season without running up to seed. The first sowing should be in the beginning of March, upon a gentle hot-bed; the second may be at the end of the same month, which ought to be in an open spot of light earth, where it may enjoy the benefit of the sun; the third time of sowing should be the latter end of April or beginning of May, on a moist soil; and if exposed to the morning sun only it will be so much the better, but it should not be under the drip of trees.

The seeds which are sown on the hot-bed will come up in about three weeks or a month after sowing, when you must carefully clear it from weeds, and if the season prove dry you must frequently water it; and in about five or six weeks after it is up, the plants will be fit to transplant: you must therefore prepare some beds of moist rich earth in a warm situation, in which you should prick these young plants at about three inches square, that they may grow strong; and if the season should prove cold, the beds must be covered with mats to screen them from the morning frosts, which would retard their growth: you muſt alſo obſerve in drawing theſe plants out of the ſeed-beds, to thin them where they grow too thick, leaving the ſmall plants to get more ſtrength before they are tranſplanted, by which means one and the ſame ſeed-bed will afford three different plantings, which will accordingly ſucceed each other for uſe.

The middle of May ſome of the plants of the firſt ſowing will be fit to tranſplant for blanching, which ſhould be planted in a moiſt, rich, light ſoil; upon which this firſt planted Celery will often grow to be twenty inches long in the clean blanched parts, which upon a poor or dry ſoil ſeldom riſes to be ten inches.

The manner of tranſplanting it is as follows: After having cleared the ground of weeds, you muſt dig a trench by a line about ten inches wide, and eight or nine inches deep, looſening the earth in the bottom, and laying it level; and the earth that comes out of the trench ſhould be equally laid on each ſide the trench, to be ready to draw in again to earth the Celery as it advances in height. Theſe trenches ſhould be made at three feet diſtance from each other; then plant your plants in the middle of the trench, at about four or five inches diſtance, in one ſtrait row, having before trimmed the plants, and cut off the tops of the long leaves; and as they are planted, you muſt obſerve to cloſe the earth well to their roots with your feet, and to water them plentifully until they have taken new root. As theſe plants advance in height, you muſt obſerve to draw the earth on each ſide cloſe to them, being careful not to bury their hearts, nor ever to do it but in dry weather, otherwiſe the plants will rot.

When your plants have advanced a conſiderable height above the trenches, and all the earth which was laid on the ſides thereof hath been employed in earthing them up, you muſt then make uſe of a ſpade to dig up the earth between the trenches, which muſt alſo be made uſe of for the ſame purpoſe, continuing from time to time to earth it up until it is fit for uſe.

The firſt of your planting out will, perhaps, be fit for uſe by the end of July, and will be ſucceeded by the after plantations; and if the latter ſowings are rightly managed, there will be a ſucceſſion of it till April; but you ſhould obſerve to plant the laſt crop in a drier ſoil, to prevent its being rotted with too much wet in winter; you will do well to cover your ridges of Celery with ſome Peaſe-haulm, or ſome ſuch light covering, when the froſt is very hard, which will admit the air to the plants, for if they are covered too cloſe they will be very ſubject to rot; by this means you will preſerve your Celery till ſpring: but you muſt remember to take off the covering whenever the weather will permit, otherwiſe it will be apt to cauſe the Celery to pipe, and run to ſeed. By this method of covering the Celery, the froſt will be kept out of the ground, ſo it may be always taken up for uſe when it is wanted, which, if neglected, it cannot be taken up in hard froſt. The Celery, when fully blanched, will not continue good above three weeks or a month before it will rot or pipe; therefore, in order to continue it good, you ſhould have at leaſt ſix or ſeven different ſeaſons of planting; ſo that if it be only intended to ſupply a family, there need not be much planted at each time; but this muſt be proportioned according to the quantity required.

The other ſort of Celery, which is commonly called Celeriac, is to be managed in the ſame manner as is directed for the Italian Celery, excepting that this ſhould be planted upon the level ground, or in very ſhallow drills; for this plant ſeldom grows above eight or ten inches high, ſo requires but little earthing up; the great excellency of this being in the ſize of the root, which is often as large as ordinary Turneps. It ſhould be ſown about the end of March, or beginning of April, upon a rich border of earth, and in dry weather conſtantly watered, otherwiſe the ſeeds will not grow: when the plants are large enough to tranſplant out, they ſhould be placed eighteen inches aſunder row from row, and the plants ſix or eight inches diſtant in the rows; the ground muſt be carefully kept clean from weeds; but this ſort will require only one earthing up, which ſhould not be performed until the roots are nearly grown to their ſize: both theſe ſorts of Celery delight in a rich, light, moiſt ſoil, where they will grow to a much larger ſize, and will be ſweeter and tenderer than on a poor or dry ground.

The beſt method to ſave this ſeed, is to make choice of ſome long good roots of the upright Celery, which have not been too much blanched, and plant them out at about a foot aſunder in a moiſt ſoil early in the ſpring; and when they run up to ſeed, keep them ſupported with ſtakes, to prevent their being broken down with the wind: and in July, when the ſeed begins to be formed, if the ſeaſon ſhould prove very dry, it will be proper to give ſome water to the plants, which will greatly help their producing good ſeeds. In Auguſt theſe ſeeds will be ripe, at which time it ſhould be cut up, in a dry time, and ſpread upon cloths in the ſun to dry; then beat out the ſeeds, and preſerve it dry in bags for uſe.

APIUM MACEDONICUM. See BUBON.

APIUM ANISUM DICTUM. See PIMPINELLA.

APIUM PYRENAICUM. See CRITHMUM.

APOCYNUM. Tourn. Inſt. R. H. 91. Lin. Gen. Plant. 269. *Dogſbane.*

The CHARACTERS are,

The flower is of one leaf, cut into five acute ſegments at the top, which turn backward; in the bottom of the flower are ſituated five nectariums, which ſurround the germen: there are five ſtamina ſcarce viſible. In the center are two oval germen, which afterward become two large pointed capſules, having one cell, which is filled with compreſſed ſeeds, lying over each other like tiles on a houſe, and furniſhed with long feathery down.

The SPECIES are,

1. APOCYNUM (*Androſemi folium*) caule rectiuſculo herbaceo foliis ovatis utrinque glabris cymis terminalibus. Lin. Sp. Plant. 213. *Canada Dogſbane with greater Tutſan leaves.*

2. APOCYNUM (*Cannabinum*) caule rectiuſculo herbaceo foliis oblongis paniculis terminalibus. Lin. Sp. Plant. 213. *Greateſt Canada Dogſbane with the leaſt herbaceous flower.*

3. APOCYNUM (*Venetum*) caule rectiuſculo herbaceo foliis ovato-lanceolatis. Prod. Leyd. 411. *Venetian maritime Dogſbane, with a Willow leaf and a purple flower.*

4. APOCYNUM (*Scandens*) foliis oblongo-cordatis rigidis floribus lateralibus, caule fruticoſo volubili. *Climbing Dogſbane with a Citron leaf and ſpotted pods.*

5. APOCYNUM (*Fruteſcens*) caule erecto fruteſcente foliis lanceolato-ovalibus corollis acutis fauce villoſis. Flor. Zeyl. 114. *Dogſbane with an upright woody ſtalk, and oval pointed leaves.*

6. APOCYNUM (*Reticulatum*) caule volubili perenne foliis ovatis venoſis. Prod. Leyd. 412. *Dogſbane with a perennial twining ſtalk, and oval veined leaves.*

7. APOCYNUM (*Obliquum*) caule volubili foliis ovatis rigidis obliquis cymis lateralibus tubo floris longiſſimo. *Greater climbing Dogſbane with roundiſh leaves.*

8. APOCYNUM (*Nervoſum*) caule fruticoſo ſcandente foliis ovatis nervoſis cymis lateralibus flore luteo magno tubo longiſſimo. *Dogſbane with a climbing ſhrubby ſtalk, oval veined leaves, the flowers growing in bunches from the ſides of the ſtalks, and a large yellow flower with a very long tube.*

9. APOCYNUM (*Cordatum*) foliis oblongo-cordatis, mucronatis ſeſſilibus floribus lateralibus, caule ſcandente.

Climbing Dogsbane with oblong pointed leaves, and large yellow open flowers.

10. Apocynum (*Villosum*) foliis cordatis glabris floribus villosis lateralibus petiolis longioribus caule scandente. *Climbing Dogsbane with large, yellow, hairy flowers, and swelling angular pods, which are smooth.*

The first sort grows naturally in North America. This hath an annual stalk and a perennial root; the stalks rise about three feet, grow upright, and are garnished with smooth oval leaves, growing opposite. These, as also the stalks, abound with a milky juice, which flows out when they are broken; the flowers are collected in a kind of umbel, growing at the top of the stalks. These are white, and the nectariums in the bottom have a purplish cast; these are often succeeded by pods in England, but the seeds seldom ripen well; the plant is propagated by parting of the roots. It is hardy, so will thrive in the full ground, but the soil should be light or dry, otherwise the roots are apt to rot in winter. The best time to part the roots is in March, before they begin to put out new stalks.

The second sort is a native of the same countries; the roots of this sort creep far in the ground, so that when it is planted in a garden, it is apt to spread so much as to be troublesome. The stalks of this sort grow about two feet high, are red, and have oblong smooth leaves, set on by pairs opposite. Towards the upper part of the stalk the flowers come out from the wings of the leaves, collected in small bunches, which are of an herbaceous white colour, and very small, so make no great appearance. This is very hardy, and propagates too fast by its creeping roots. Both these sorts flower in July, and in autumn their stalks decay to the root.

The third sort grows upon a small island in the sea near Venice, but is supposed to have been originally brought from some other country. The roots of this creep pretty much, by which it is propagated, for it never produces any seeds, either in the gardens where it is cultivated, or at Venice, where it grows without care, as I have been informed by a very curious botanist, who resided many years at Venice, and constantly went to the spot several times in the season to procure the seeds, if there had been any produced; but he assured me, he never could find any pods formed on the plants. The stalks of this sort decay in autumn, and new ones are sent out from the roots in the spring. The flowers grow at the top of the stalks in small umbels, which are shaped like those of the former sorts, but are much larger, so make a pretty appearance; it flowers in July and August. The best time to remove and part the roots is in the spring, just before they begin to push out new stalks.

The fourth sort was discovered by Father Plumier in some of the French islands in America, who made a drawing of the plant. It was afterwards found by the late Mr. Robert Millar, surgeon, growing plentifully near Carthagena in New Spain. It hath twining stalks, by which it mounts to the tops of very tall trees, and stiff, oblong, heart-shaped leaves, which are smooth, of a shining green colour. The flowers are produced in small clusters from the side of the branches, and are of an herbaceous colour, so do not make any great appearance.

The fifth sort grows naturally in India, Ceylon, and upon the coast of Guinea. This plant rises with a woody stem to the height of five or six feet, garnished with oblong pointed leaves, very smooth, and of a shining green above, but pale underneath. From the wings of the leaves the flowers are produced in loose bunches. These are small, tubulous, and of a purple colour. It is a very tender plant, so must be constantly kept in a hot-house, and plunged in the tan bed, otherwise it will not thrive in England. This plant must be sparingly watered, especially in winter; and should be planted in light sandy earth.

The sixth sort grows naturally in India. This plant hath a twining stalk, by which it rises to a considerable height, and is garnished with oblong leaves, which are much veined. It is tender, so requires to be constantly preserved in the stove, otherwise it will not thrive in this country.

The seventh sort grows naturally in Jamaica. This hath a climbing stalk, by which it fastens to the neighbouring trees, and rises ten or twelve feet high. The leaves are oval, stiff, and oblique to the foot-stalk; the flowers are of a purplish colour, and have very long tubes, but spread open wide at the top. It is tender, so must constantly remain in the stove, and should have very little water.

The eighth sort hath a climbing woolly stalk, and rises to a considerable height by the support of neighbouring trees. The leaves grow by pairs opposite; the flowers come out from the wings of the leaves, each standing upon a separate long foot-stalk; they are large, and of a bright yellow colour, with very long tubes, and spread open wide at the top; they are succeeded by long compressed pods, which have borders on one side, and are filled with long channelled seeds, which are crowned with long plumes of soft down. This is propagated by seeds, which must be procured from the country where it grows naturally, for the seeds do not ripen in this country; and the plants should be treated in the same manner, as hath been before directed for the fifth sort. It flowers in August and September in England.

The ninth and tenth sorts were discovered at La Vera Cruz in New Spain, by the late Dr. William Houstoun. These plants have both climbing stalks, by which they mount to the tops of the tallest trees, where they grow naturally The ninth sort has produced flowers in England several times; but the tenth, which grows more luxuriantly than the other, never had any appearance of flowers. These are both propagated by seeds, which should be sown as the fifth sort, and the plants must be treated in the same manner afterward. The pods of all the sorts are filled with seeds, which are for the most part compressed, and lie over each other (imbricatim) like tiles on a house: these have each a long plume of a cottony down fastened to their crowns, by which, when the pods are ripe and open, the seeds are wafted by the wind to a considerable distance; so that in the countries where these plants naturally grow, they are some of the most troublesome weeds.

The down of these plants is in great esteem in France for stuffing of easy chairs, making very light quilts, which are warm and extremely light, so are very proper covering for persons afflicted with the gout; for the down is so extreme light and elastic, that it occasions no weight. This the French call Delawad, and in the southern parts of France, where some of the sorts will thrive in the open air and perfect their seeds, there are many plantations made of these plants for the sake of the down.

The other sorts which have been ranged under this genus, are now referred to the following genera, to which the reader is desired to turn for such of them as are not here enumerated, viz. Asclepias, Cynanchum, and Periploca.

APPLE-TREE. See Malus.

APPLES of LOVE. See Lycopersicon and Solanum.

MAD APPLES. See Melongena.

APRICOT, or ABRICOT; or, in Latin, Malus Armeniaca. See Armeniaca.

AQUIFOLIUM. See Ilex.

AQUILEGIA. Columbine.

The Characters are,

The flower hath no empalement, but is composed of five equal oval petals which are plain, and spread open within, and have five equal nectariums ranged alternately with the petals, each of the horns widening upward. It hath many awl-shaped stamina,

and

and five oval germen, which afterward becomes five cylindrical vessels, which are filled with oval shining seeds.

The Species are,

1. Aquilegia (*Vulgaris*) nectariis rectis petalo lanceolato brevioribus. Lin. Sp. Pl. 533. *Wild Columbine.*

2. Aquilegia (*Alpina*) nectariis rectis, petalis ovatis longioribus. *Mountain Columbine with a large flower.*

3. Aquilegia (*Inversa*) nectariis incurvis. Hort. Upsal. *Columbine with a double inverted flower.*

4. Aquilegia (*Canadensis*) nectariis rectis staminibus corollâ longioribus. Hort. Upsal. 153. *Early dwarf Canada Columbine.*

The first sort is found growing wild in the woods in some parts of England. I have frequently gathered it in the woods near Bexley in Kent, and also between Maidstone and Rochester. The flowers of this are blue, and the petals are short.

The second sort I found growing naturally near Ingleborough-hill in Yorkshire. The flowers of this are much larger than those of the Garden Columbine, and the seeds which I sowed of this in the garden at Chelsea produced the same species without the least variation.

The third is the Garden Columbine, of which there are great varieties, not only in the colour and fulness of their flower, but also in their form. In some there are no visible nectariums, but in the place of them a multiplicity of petals, so that the flowers are as double as those of the Larkspur. These are commonly called Rose Columbines; the colours of these are Chestnut, blue, red, and white, and some are finely variegated with two colours.

There are others with sharp pointed petals, which expand in form of a star; of these there are single and double flowers, of the several colours as the former. But as the sorts with variegated flowers are the greatest beauties, so those persons who are desirous to have them in perfection, should root out all those plants whose flowers are not well marked, or cut off their stems so soon as their flowers appear, leaving only the most beautiful to seed.

They are all raised by sowing the seeds, or parting the old roots; but the former method is chiefly practised, for the old roots are very apt to degenerate after they have blown two or three years, and become quite plain.

The seeds should be sown in a nursery-bed in September, for the seeds which are kept till spring seldom grow well, or at least remain in the ground a whole year. In the spring following your young plants will appear above ground, you must therefore clear them from weeds, and if the season should prove dry refresh them with water, that they may gather strength.

In the middle or latter end of May these plants will be strong enough to transplant; you must therefore prepare some beds of good fresh undunged earth, planting them therein at eight or nine inches distance every way, keeping them clear from weeds.

At Michaelmas you may remove them into the borders of your flower-garden, and the May following they will produce flowers; but if you intend to maintain their roots, you should not suffer them to seed, but crop off all their flower-stems as soon as the flowers are past.

In order to keep up a succession of good flowers, you should sow fresh seeds every year; and if you can meet with a friend at some distance, who is furnished with good flowers of this kind, it will be very advantageous to both parties, to exchange seeds once in two years, by which means they will not be so apt to degenerate into plain colours.

The Canada Columbine flowers almost a month before the other sorts, for which reason it is preserved in the gardens of the curious, though there is no very great beauty in the flowers. There is another variety of this sort with taller flower-stems, which flowers a little after the other, but doth not differ, either in the shape of its flowers or leaves from this.

The first sort is that which is directed for medicinal use in the Dispensaries, but at present is very rarely ordered.

ARABIS. Lin. Gen. Pl. 732. Bastard Tower Mustard.

The Characters are,

The flower hath four petals in form of a cross, which spread open, at the bottom of which is situated a reflexed nectarium; between these arise six upright stamina. In the center is situated a taper germen, which afterward becomes a narrow, long, compressed pod, having two valves and a thin partition, between which is lodged a row of flat seeds.

The Species are,

1. Arabis (*Thaliana*) foliis petiolatis lanceolatis integerrimis. Vir. Cliff. 64. *Bastard Tower Mustard, with whole spear-shaped leaves having foot-stalks.*

2. Arabis (*Alpina*) foliis amplexicaulibus dentatis. Hort. Cliff. 335. *Bastard Tower Mustard, with indented leaves embracing the stalks.*

3. Arabis (*Pendula*) foliis amplexicaulibus siliquis ancipitibus linearibus calycibus subpilosis. Hort. Upsal. 191. *Broad-leaved hairy Tower Mustard with hanging pods.*

4. Arabis (*Turrita*) foliis amplexicaulibus siliquis decurvis planis linearibus, calycibus subrugosis. Hort. Upsal. 192. *Bastard Tower Mustard with narrow, plain, hanging pods, and rough flower-cups.*

The first sort is a low plant, which seldom rises more than four or five inches high, branching on every side, having small white flowers growing alternately, which have each four petals in form of a cross, that are succeeded by long slender pods, filled with small round seeds. It grows naturally on sandy dry ground in many parts of England.

The second sort grows naturally in Istria, and also upon the Alps, and other mountainous countries. It is a perennial plant, which increases very fast by its creeping roots, which run obliquely near the surface of the ground, and send down roots at every joint. The leaves are whitish, and indented on their edges; the flower-stalks grow near a foot high, and are garnished with leaves placed alternately, which closely embrace the stalks; the flowers grow in loose bunches on the top; these are white, and have leaves in form of a cross, which are succeeded by long flat pods, opening lengthways.

This is a very hardy plant, so will thrive in any situation. It produces seeds in plenty, but as it multiplies so fast by its creeping roots, therefore few persons are at the trouble to sow the seeds. It flowers early in the spring.

The third sort grows naturally in Siberia. This is a perennial plant, which grows near a foot high; the leaves are broad, hairy, and indented on their edges. The flowers grow alternately in loose spikes, and are of a dirty white colour. These are succeeded by long narrow pods, which are filled with flat brown seeds. This is a biennial plant, which is very hardy, so will thrive in any situation.

The fourth sort grows naturally in Hungary, Sicily, and France, as also upon some old walls at Cambridge and Ely, but the seeds might probably come out of the gardens where they were first planted. The plants of this kind, which grow on walls or ruins, continue much longer than those which are sown in gardens, where they seldom live longer than two years; the stalks rise about a foot and a half high. Toward the top of the stalks grow long loose spikes of flowers, which are of a dirty white colour. After the flowers are past, the germen become long flat pods, which open lengthways, and have two rows of flat bordered seeds of a dark brown colour.

This sort is easily propagated by seed, which should be sown in the autumn. When the plants are strong enough to to remove, they may be tranſplanted into a ſhady border, or in rural plantations, where no other care will be neceſſary, but to prevent their being overgrown by weeds.

ARACHIS. Earth, or Ground Nut.

The CHARACTERS are,

The empalement of the flower opens in two parts. The flower is of the butterfly kind; it hath ten ſtamina, nine of which coaleſce, and the upper one ſtands off. In the center is ſituated an oblong germen, which afterward turns to an oblong pod, containing two or three oblong blunt ſeeds.

We have but one SPECIES of this plant, viz.

ARACHIS (*Hypogæa.*) Lin. Hort. Cliff. 353. *Earth, or Ground Nut.*

The native country of this plant I believe is Africa, though at preſent all the ſettlements in America abound with it; but many perſons who have reſided in that country affirm, they were originally brought by the ſlaves from Africa.

It multiplies very faſt in a warm country, but being impatient of cold, it cannot be propagated in the open air in England; therefore, whoever has an inclination to cultivate this plant, muſt plant the ſeeds in a hot-bed in the ſpring of the year, and when the weather proves warm, they may be expoſed to the open air by degrees. The branches of this plant trail upon the ground, and the flowers (which are yellow) are produced ſingle upon long footſtalks; and as ſoon as the flower begins to decay, the germen is thruſt under ground, where the pod is formed and ripened, ſo that unleſs the ground is opened they never appear: the roots of this plant are annual, but the nuts or ſeeds ſufficiently ſtock the ground in a warm country, where they are not very carefully taken up.

ARALIA. Berry-bearing Angelica.

The CHARACTERS are,

It is an umbelliferous plant with a globular umbel, having a ſmall involucrum; the flower hath five oval petals, and five ſhort ſtyles. The germen afterward turns to a roundiſh channelled berry, having five cells, each containing one oblong hard ſeed.

The SPECIES are,

1. ARALIA (*Racemoſa*) caule folioſo herbaceo lævi. Hort. Upſal. 70. *Canada Berry-bearing Angelica.*

2. ARALIA (*Nudicaulis*) caule nudo. Hort. Cliff. 113. *Berry-bearing Angelica with a naked ſtalk.*

3. ARALIA (*Spinoſa*) arboreſcens caule foliolíſque aculeata. Vir. Cliff. 26. *Angelica-tree, vulgo*

The firſt ſort is pretty common in many gardens near London, but the ſecond is at preſent more rarely met with. Both theſe plants grow naturally in North America. They are perennial plants, whoſe ſtalks decay in autumn, and new ones ariſe from their roots in the ſpring. The firſt grows about four or five feet high, and divides into many irregular branches, having ramoſe leaves placed alternately; at the wings of theſe the flower-ſtalks are produced, which are terminated by round umbels of ſmall flowers, of a whitiſh colour; theſe are ſucceeded by round channelled berries, which when ripe are black. This plant flowers in June, and the ſeeds ripen in October.

The ſecond ſort riſes to near the ſame height as the former; the leaves of this divide into two or three parts, each ending with three or five large lobes, which are ſawed on their edges. The flower-ſtalks ariſe between theſe immediately from the root, being naked, and are terminated by round umbels of flowers, in ſhape and colour like the firſt, but the berries are ſmall. This flowers toward the end of June, and the ſeeds ripen late in the autumn.

Both theſe ſorts are eaſily propagated by ſeeds, which are generally produced in plenty. Theſe ſhould be ſown in the autumn ſoon after they are ripe. When the plants appear, they muſt be kept clean from weeds during the ſummer; and in the autumn following, when their leaves decay, the roots may be taken up, and tranſplanted where they are to remain. They are very hardy plants, ſo may be planted in any ſituation; and as they grow naturally in woods, ſo they may be planted in wilderneſs quarters under trees.

The third ſort riſes with a woody ſtem to the height of eight or ten feet, dividing into ſeveral branches; theſe are garniſhed with branching leaves, which are compounded of many divaricated wings; the ribs of the leaves, as alſo the branches and ſtems of the plants, are armed with ſtrong crooked ſpines, which render the places very difficult to paſs through where they grow in plenty. The flowers of this ſort are produced in large looſe umbels at the extremity of the branches, and are of an herbaceous colour, ſo make no great figure, but the plants are preſerved in moſt of the curious gardens in England. It flowers in Auguſt, but the ſeeds do not ripen in this country.

This is propagated by ſeeds, which are eaſily procured from North America; but as they ſeldom arrive here till toward the ſpring, ſo the plants never come up the firſt year, but the following ſpring. When the plants come up they ſhould be frequently refreſhed with water, and conſtantly kept clean from weeds, and in ſummer they ſhould be inured to the open air. Theſe plants ſhould not be diſturbed the firſt ſeaſon, but as they are often injured by froſt when young, ſo in the firſt winter the plants ſhould be ſcreened from hard froſts, but in mild weather ſhould be conſtantly opened to enjoy the free air. The leaves of theſe plants fall away in the autumn. In the ſpring, before the plants begin to puſh, they ſhould be tranſplanted; a few of them ſhould be planted ſingly into ſmall pots, and the others may be planted in a bed of light earth in a warm ſituation. If thoſe which are planted in the ſmall pots are plunged in a moderate hot-bed, it will greatly forward their growth; but they muſt be early inured to bear the open air, otherwiſe they will draw up weak, and the ſpring following they may be planted where they are deſigned to remain. As theſe plants do not come out very early in the ſpring, ſo they often continue growing pretty late in the autumn, which cauſes the extreme parts of their ſhoots to be very tender, whereby they often ſuffer from the early froſts in autumn, which frequently kill the upper parts of the ſhoots; but as their woody ſtems are ſeldom injured, ſo they put out new branches below: and if in very ſevere winters the ſtems are deſtroyed, yet the roots will remain, and put out new ones the following ſummer, therefore they ſhould not be deſtroyed.

This plant may alſo be propagated by its roots, for as they ſpread far in the ground, ſo they will put out young plants at a diſtance from the ſtems, which may be taken off before they begin to ſhoot in the ſpring.

ARBOR CAMPHORIFERA. See LAURUS.

ARBOR CORAL. See ERYTHRINA.

ARBOR JUDÆ. See CERCIS.

ARBUTUS. The Strawberry-tree.

The CHARACTERS are,

The flower hath a ſmall obtuſe empalement, which is cut into five parts, upon which the germen ſits. The flower is of one leaf, ſhaped like a pitcher; at the bottom of the flower is ſituated the globular germen, which afterward becomes an oval or round berry, having five cells, which are filled with hard ſeeds.

The SPECIES are,

1. ARBUTUS (*Unedo*) foliis glabris ſerratis, baccis polyſpermis, caule erecto arboreo. *The common Strawberry-tree.*

2. ARBUTUS (*Adrachne*) foliis glabris integerrimis, baccis polyſpermis caule erecto arboreo. *The Oriental Strawberry-tree, called Adrachne.*

3. Arbutus (*Acadiensis*) caulibus procumbentibus foliis ovatis subserratis floribus sparsis baccis polyspermis. Lin. Sp. Pl. 395. *Arbutus with trailing stalks, oval leaves somewhat indented, flowers growing loosely, and many seeds.*

4. Arbutus (*Alpina*) caulibus procumbentibus foliis rugosis serratis. Flor. Lap. 161. *Arbutus with trailing stalks and rough sawed leaves.*

5. Arbutus (*Uva Ursi*) caulibus procumbentibus foliis integerrimis. Flor. Lap. 162. *Arbutus with trailing stalks and entire leaves; called Uva Ursi, or Bear-berries.*

The first sort grows naturally in Italy, Spain, and also in Ireland, and is now very common in the English gardens. Of this sort there are the following varieties, viz. one with an oblong flower and oval fruit, another with a double flower, and a third with red flowers; but these being only seminal varieties, I have not mentioned them as species.

The second sort grows naturally in the East, particularly about Magnesia, where it is so plenty as to be the principal fuel used by the inhabitants of the country. The leaves are large and oval, somewhat like those of the Bay-tree, but not quite so long; they are smooth and entire, having no serratures on their edges; the flowers are shaped like those of the common Arbutus, but grow thinly on their branches. The fruit is oval, and of the same colour and consistence with the common sort, but the seeds of this are flat, whereas those of the common sort are pointed and angular.

The common Strawberry-tree is one of the greatest ornaments in the months of October, November, and frequently great part of December, that being the season when the trees are in flower, and the fruit of the former year is ripe, for the fruit is a whole year growing to perfection; so that the fruit which is produced from the flowers of one year, do not ripen till the blossoms for the succeeding year are fully blown, so they make a goodly appearance, and at a season when most other trees are past their beauty.

The sort with double flowers is a curiosity, but as the flowers have only two rows of leaves, so they make no great appearance; nor do the trees produce fruit in any plenty, therefore the other is more preferable. The sort with red flowers makes a pretty variety, when intermixed with the other, for the outside of the flowers are of a fine red colour at their first appearance, and afterward they change to purple before they fall off. These varieties are preserved by inarching or grafting them upon the common Arbutus, for the seeds of either do not produce the same kind.

The best method to propagate the Arbutus is from seeds; therefore when the fruit is perfectly ripe, it should be gathered and mixed with dry sand, to preserve them till the middle or latter end of March, which is a proper season for sowing of them, in order to have strong plants before winter: they must be sown in pots, which should be plunged into a moderate hot-bed, which will greatly forward their vegetation, and, if they are properly managed, will grow eight or ten inches high before winter. In the summer, if the pots are plunged into an old tan-bed, it will preserve the earth in the pots from drying too fast; and if the plants are screened from the sun in the heat of the day, it will greatly forward them. The beginning of October these plants may be shaken out of the pots, and their roots carefully separated, planting them singly in small pots filled with light earth; then plunge the pots into an old bed of tanners bark under a common frame, where they should remain during the winter, observing to expose the plants to the open air at all times when the weather is favourable, but in frosty weather they must be covered. The spring following they may be plunged into the ground, in a sheltered situation, observing to water them frequently in dry weather, which will keep them growing all the summer; but it will be adviseable to screen them from frost the following winter, by covering them with mats in bad weather.

The following spring you may shake them out of the pots into the open ground in the places where they are to remain, that they may have taken good root before the winter.

These trees are tolerably hardy, and are seldom hurt, except in extreme hard winters, which many times kill the young and tender branches, but rarely destroy the trees; therefore, however dead the trees may appear after a hard winter, yet they should be suffered to remain till the succeeding summer has sufficiently demonstrated what are living and what are dead; for the winters Anno 1728-9, and 1739-40, gave us great reason to believe most of the trees of this kind were destroyed, and many people were so hasty as to dig up or cut down many of their trees; whereas all those people who had patience to let them remain, found that scarce any of them failed to come out again the next summer, and made handsome plants that season.

The very best season for transplanting of the Arbutus is in September, at which time the blossoms are beginning to appear; and at that season, if it should prove very dry and they are kept moist, they will take root very soon; but toward the beginning of November their roots should be well covered with mulch to keep out the frost.

The third sort grows naturally in Acadia, and other northern parts of America, upon swampy land, which is frequently overflowed with water; this is a low bushy shrub, with slender trailing branches, which are garnished with oval leaves, a little sawed on their edges; the flowers come out from the wings of the leaves, growing in thin loose bunches: it is with great difficulty the plants of this sort are kept alive here.

The fourth sort grows naturally on the Alps and the Helvetian mountains, also in Lapland and Siberia. This sends out from the roots many slender branches, which trail upon the ground, garnished with oblong rough leaves, of a pale green colour; the flowers are produced from the wings of the leaves upon long slender foot-stalks, and are succeeded by berries about the size of the common black Cherry, which are first green, afterward red, and when ripe they are black. This is also a very difficult plant to keep alive in gardens, for it is an inhabitant of bogs, growing among moss, where the ground is never dry.

The fifth sort grows naturally upon the mountains in Spain, and some other parts of Europe. It rises little more than a foot high, dividing into many branches, which are closely garnished with small thick leaves of an oval form; the flowers are produced in small bunches toward the extremity of the branches, which are shaped like those of the common sort, but are smaller, and are succeeded by berries of the same size with those of the former sort, which are red when ripe.

The true Adrachne is now in several gardens in England, but the leaves while young are sawed on their edges, but are after two or three years growth entire. The seeds of this must be procured from the Levant, where the trees grow in plenty. The seeds may be sown, and the plants treated in the same way as the Arbutus, but the plants are much tenderer. As the leaves of this tree are larger than those of the common Arbutus, so the trees make a fine appearance, and deserve our care to cultivate them; therefore they should be preserved in pots three or four years till they have obtained strength, and may then be planted in a warm situation and on a dry soil, for this sort will not thrive in wet ground.

ARCTIUM. Lin. Gen. 830. Burdock.

The Characters are,

The flower is composed of many florets, which are tubulous and uniform, cut into five narrow segments at the top: the germen is situated at the bottom of the tube, which afterward becomes a single pyramidal angular seed, crowned with down.

The

The Species are,

1. Arctium (*Lappa*) foliis cordatis inermibus petiolatis capitulis majoribus sparsis. *Burdock with heart-shaped leaves without prickles, having foot-stalks, and large heads growing scatteringly.*

2. Arctium (*Personata*) foliis cordatis inermibus, capitulis minoribus compactis. *Burdock with heart-shaped leaves without spines, and small heads growing close together.*

3. Arctium (*Tomentosum*) foliis cordatis inermibus, capitulis tomento-reticulatis. *Burdock with heart-shaped leaves without spines, and woolly netted heads.*

The two first sorts are common weeds, growing on the sides of roads and foot-paths in most parts of England, so are not admitted into gardens. The first is ordered for medicinal use by the College of Physicians, therefore I have inserted it here.

The leaves of the third sort are like those of the common, but are whiter on their under side; the heads are more compact, and the florets are of a bright red colour; but the greatest difference is in their heads, which in this sort are beautifully netted with a fine down all over.

As these plants are seldom admitted into gardens, so it is needless to say any thing of their culture; but where they are troublesome weeds, it may not be amiss to mention that their roots last but two years, so they may be destroyed with less trouble than such as have abiding roots; for the plants which come up from seed do not flower till the second year, and when the seeds are perfected their roots decay.

ARCTOTIS, or ANEMONOSPERMOS.

The Characters are,

The common empalement is scaly and silvery; the flower is composed of many female florets, which are ranged round the border: the germen afterward becomes a single roundish seed covered with a soft down. The middle or disk of the flower is composed of hermaphrodite florets; in the center is placed a small germen, supporting a cylindrical style with a single stigma. Those flowers are abortive.

The Species are,

1. Arctotis (*Tristis*) radiantibus vicenis tripartitis. Lin. Sp. 1306. *Anemonospermos of Africa, with hoary Dandelion leaves.*

2. Arctotis (*Calendula*) flosculis radiantibus sterilibus duodenis subintegris, foliis lyratis nigro-denticulatis. Lin. Sp. 1306. *Arctotis with narrow spear-shaped leaves which are indented on the sides.*

3. Arctotis (*Acaulis*) pedunculis radicalibus, foliis lyratis. Lin. Sp. 1306. *Arctotis without stalks, Plantain leaves, and sulphur-coloured flowers.*

4. Arctotis (*Plantaginea*) flosculis radiantibus fertilibus, foliis lanceolato-ovatis nervosis denticulatis amplexicaulibus. Lin. Sp. 1306. *Arctotis with fruitful borders or rays, and oval spear-shaped leaves which embrace the stalks.*

5. Arctotis (*Angustifolia*) flosculis radiantibus fertilibus, foliis lanceolatis integris dentatis. Lin. Sp. 1306. *Arctotis with fruitful rays, and entire spear-shaped leaves.*

6. Arctotis (*Aspera*) flosculis radiantibus fertilibus, foliis pinnato-sinuatis villosis, laciniis oblongis dentatis. Lin. Sp. 1307. *Arctotis with fruitful rays, and wing-shaped woolly leaves.*

7. Arctotis (*Alba*) flosculis radiantibus fertilibus, foliis pinnato-sinuatis subtus villosis, caulibus ramosissimis. *Arctotis with fertile rays, branching stalks, and hairy winged leaves.*

These plants are natives of the country about the Cape of Good Hope, from whence they have been brought to some curious gardens in Holland and England.

The first sort here mentioned is an annual plant, which may be sown upon a warm border of light earth in the open air in the middle of April, where they are designed to remain; and require no farther care but to thin the plants where they are too close, and keep them clean from weeds. They will flower in August, and in warm seasons will perfect seeds very well in autumn.

The second, fourth, fifth, and sixth sorts, grow to the height of four or five feet, and the seventh sometimes to six or seven, sending forth many branches, therefore will require to be frequently pruned, to keep the plants in tolerable order; for it sends forth strong rambling shoots, when their roots are not too much confined in the pots, but more so when they are duly watered.

The fifth and sixth sorts flower in May and June; these have very large beautiful flowers, especially the fifth, the rays being of a yellow or deep gold colour.

The shrubby sorts are propagated by planting cuttings in a bed of light fresh earth in any of the summer months, observing to shade them from the heat of the sun until they have taken root; then they may be planted into pots filled with the like fresh earth, setting the pots in a shady place until the plants are settled in their new earth, after which time you should expose them to the open air until the latter end of October, or later, according as you find the weather is favourable; when you must remove the pots into the green-house, where they should be placed as near the window as possible, that they may have a good quantity of free air at all times when the weather is mild; you must also frequently refresh them with water, giving them it plentifully in mild weather, otherwise their leaves and branches will hang down and wither. They will require to be shifted into other pots two or three times at least every summer, and the pots should be frequently removed, to prevent the plants from striking their roots through the holes of the pots into the ground, which they are very apt to do, and they will shoot very vigorously.

All these plants should be frequently renewed by cuttings, because the old plants are subject to decay in winter; therefore if young plants are not annually raised, the species may soon be lost.

ARGEMONE. Prickly Poppy.

The Characters are,

The flower hath five roundish petals which spread open, and are larger than the empalement; in the center is situated an oval five-cornered germen. This is attended by a great number of stamina; the germen afterward becomes an oval seed-vessel, having five angles, and so many cells, which are filled with small rough seeds.

There is but one Species of this plant known, viz.

Argemone (*Mexicana.*) Tourn. *The Prickly Poppy.*

This is an annual plant, which is very common in most parts of the West-Indies, and is by the Spaniards called Fico del Inferno, or the Devil's Fig; there is no great beauty or use of this plant amongst us, that I know of, but whoever hath a mind to cultivate it, should sow it on a bed of light earth in the spring, where it is to remain; and if it comes up too thick, the plants must be thinned out to four inches distance, where, when once it has shed its seed, there will not want a supply of plants for several years after.

ARIA THEOPHRASTI. See Cratægus.

ARISARUM. See Arum.

ARISTOLOCHIA. Birthwort.

The Characters are,

The flower is of one leaf, which is unequal; the base is swelling and globular, afterward is extended in a cylindrical tube, which spreads open at the brim, where the lower part is stretched out like a tongue. The oblong angular germen sits under the flower, which afterward turns to a large seed-vessel, differing in form, which opens in six cells, which are filled with seeds, for the most part compressed.

The

The Species are,

1. Aristolochia (*Rotunda*) foliis cordatis, subsessilibus obtusis, caule infirmo, floribus solitariis. Lin. Sp. Pl. 962. *Round-rooted Birthwort, with a black purple flower.*

2. Aristolochia (*Longa*) foliis cordatis petiolatis integerrimis obtusiusculis, caule infirmo floribus solitariis. Lin. Sp. Plant. 962. *The true long-rooted Birthwort.*

3. Aristolochia (*Clematites*) foliis cordatis caule erecto floribus axillaribus confertis. Hort. Upsal 279. *Upright or climbing Birthwort.*

4. Aristolochia (*Pistolochia*) foliis cordatis, crenulatis petiolatis, floribus solitariis. Lin. Sp. Pl. 962. *Birthwort, called Pistolochia.*

5. Aristolochia (*Sempervirens*) foliis cordato-oblongis undatis, caule infirmo, floribus solitariis. Lin. Sp. Pl. 961. *Evergreen Birthwort of Crete.*

6. Aristolochia (*Serpentaria*) foliis cordato-oblongis planis, caulibus infirmis flexuosus, teretibus floribus solitariis. Lin. Sp. Plant. 961. *The Virginia Snakeroot.*

7. Aristolochia (*Arborescens*) foliis cordato-lanceolatis caule erecto fruticoso. Lin. Sp. Pl. 960. *Virginia Birthwort with eared leaves.*

8. Aristolochia (*Indica*) foliis cordato-oblongis caule volubili pedunculis multifloris Flor. Zeyl. 323. *The Contrayerva of Jamaica.*

9. Aristolochia (*Maurorum*) hirta floribus solitariis pendulis recurvatis sublabiatis. Lin. Sp. Plant. *Long-rooted hairy Birthwort, with an oblong leaf and a large flower.*

The first and second sorts grow naturally in the south of France, in Spain, and Italy, from whence the roots are brought for medicinal use. The roots of the first sort are roundish, and grow to the size of small Turneps, and are in shape and colour like the roots of the common Cyclamen, the roots of which are frequently sold in the markets for those of the round Birthwort, which at first may have been occasioned by the supposed virtues of the roots of the Cyclamen. This sort hath three or four weak trailing branches, which lie on the ground where they are not supported, and extend to the length of two feet; the leaves are heart-haped, and rounded at their extremity; the flowers come out singly at every leaf, toward the upper part of the stalk. They are of a purplish black colour, and shaped like those of the other sorts, and are frequently succeeded by oval seed-vessels, having six cells, which are full of flat seeds.

The second sort hath long tap roots, shaped like those of Carrots; this has weak trailing branches, which extend little more than a foot; the leaves of this sort are paler, and have longer foot-stalks than the first; the flowers come out from the wings of the leaves like the other, and are of a pale purple colour: they are sometimes succeeded by oblong seed-vessels, having six cells filled with compressed seeds.

They are both propagated by seeds, which should be sown in the autumn, in pots filled with light earth, and placed under a frame to be screened from the frost. If these pots are put into a gentle hot-bed in March, it will bring up the plants much sooner than they otherwise would rise. When the plants come up they should be inured by degrees to bear the open air; in summer they must have gentle refreshings of water in dry weather, but in the autumn when their stalks begin to decay, they must have little wet: in the winter the pots must be sheltered under a frame, and in March, before the roots begin to shoot, they should be transplanted into separate small pots filled with light earth, when they may be removed into the open air, and treated in the same manner as in the former summer, and sheltered also the following winter. The next spring they may be turned out of the pots, and planted in a warm border; where, in the autumn, when their stalks are decayed, if the border is covered with old tanners bark to keep out the frost, the roots will be secured; but where this care is not taken, the roots are frequently killed by frost.

When the seeds of these plants are sown in the spring, the plants will not appear till the spring following; so that a whole season is lost, and many times they fail, therefore it should always be sown in the autumn.

The third sort grows naturally in France, Spain, Italy, and Hungary, but is preserved in some of the English gardens because it is sometimes used in medicine. This is a mischievous plant for creeping at the root, so that if once it has taken in a garden, it will be difficult to extirpate again: it will thrive in almost any soil or situation.

The fourth sort grows wild in Spain, Italy, and the south of France; but in England it is preserved for variety in botanic gardens. The plants of this sort must be planted in pots filled with light earth, and sheltered from severe cold in winter, but they should have as much free air as possible in mild weather.

The fifth sort grows naturally in Crete. The root of this sort is perennial, and sends out many trailing branches, which extend to about a foot and a half in length, with oblong heart-shaped leaves, which are evergreen. The flowers are shaped like the others of this genus, of a dark purple colour, but never produces seeds in England, so are propagated by parting of the roots: this sort is too tender to thrive in the open air in winter, so is preserved in pots, and placed under a common frame in winter, where they should have as much free air as possible in mild weather.

The sixth sort is the Snakeroot, which is greatly used in medicine; these roots are brought over from Virginia and Carolina. There are some of these plants preserved in the gardens of those who are curious, but as they are sometimes killed by frost in winter, so they are not very common in the English gardens. This sort is propagated by seeds, which should be sown in the autumn, and afterward treated in the same manner as hath been directed for the two first sorts, with which management they will produce their flowers and perfect their seeds every year.

The seventh sort grows naturally in North America, and is by some called Snakeroot, but is not near so strong as the former; the branches of this grow erect and are perennial, whereas those of the other sort decay to the root every winter: this rises about two feet high; the branches are not woody, but are strong enough to support themselves; the leaves are oblong and heart-shaped. This sort will live abroad in warm borders, with a little protection in hard frosts. It is propagated by seeds as the former, and may also be increased by parting of the roots.

The eighth sort grows naturally in Jamaica, where it is called Contrayerva; the roots are there used as such: this hath long trailing branches which climb upon the neighbouring plants, and rise to a considerable height; the flowers are produced in small clusters toward the upper part of the stalks, which are of a dark purple colour. This plant is tender, and in winter should have very little wet, therefore must be constantly kept in the stove, otherwise it will not live in England.

The ninth sort was discovered by Dr. Tournefort in the Levant. This hath some resemblance to the second sort, but the leaves are not so deeply eared at bottom, and are hairy; the flowers of this are also much larger. This may be propagated by seeds in the same manner as hath been directed for the first and second sorts, and the plants so treated will thrive very well here.

ARMENIACA. The Apricot.

The Characters are,

The flower is composed of five large roundish petals which spread open, whose base is inserted in the empalement; in the center

center is placed a round germen, attended by upward of twenty awl-shaped stamina. The germen afterward becomes a roundish pulpy fruit, having a longitudinal furrow inclosing a roundish nut, which is a little compressed on the sides.

The specific title given by Linnæus to the Apricot is, Prunus floribus subsessilibus foliis subcordatis. Sp. Pl. 474.

The Varieties are,

1. The Masculine Apricot.
2. The Orange Apricot.
3. The Algier Apricot.
4. The Roman Apricot.
5. The Turkey Apricot.
6. The Breda Apricot.
7. The Brussels Apricot.

The Masculine is the first ripe of all the Apricots; it is a small roundish fruit, of a red colour towards the sun; as it ripens, the colour fades to a greenish yellow on the other side; it has a very quick high flavour. The tree is very apt to be covered with flowers, but as they come out early in the spring, they are frequently destroyed by the cold, unless the trees are covered to protect them.

The Orange is the next ripe Apricot; this fruit is much larger than the former, and as it ripens changes to a deep yellow colour. The flesh of this is dry, and not high flavoured, it is better for tarts than for the table.

The Algier is the next in season; this is of an oval shape, a little compressed on the sides; it turns to a pale yellow or straw colour, when ripe; the flesh is dry, and not high flavoured: this, and what is by some persons called the common Apricot, are often confounded.

The Roman is the next ripe Apricot; this is a larger fruit than the former, and not compressed so much on the sides; the colour is deeper, and the flesh is not so dry as the former.

The Turkey Apricot is yet larger than either of the former, and of a globular figure; the fruit turns to a deeper colour than the former; the flesh is firmer, and of a higher flavour than either of the former.

The Breda Apricot (as it is called, from its being brought from thence into England) was originally brought from Africa: this is a large roundish fruit, changing to a deep yellow when ripe; the flesh is soft, full of juice, and of a deep Orange colour within side; the stone is rounder and larger than any of the other sorts: this is the best Apricot we have, and when ripened on a standard is preferable to all other kinds.

The Brussels is the latest ripe of all the Apricots, for when it is planted against a wall, it is generally the beginning of August before it is ripe, unless when it is planted to a full south aspect; which is what should not be practised, because the fruit is never well tasted which grows in a warm exposure. This fruit is of a middling size, rather inclining to an oval figure, red on the side next the sun, with many dark spots, and of a greenish yellow on the other side; the flesh is firm, and of an high flavour; the fruit often cracks before it is ripe.

Most people train these trees up to stems of six or seven feet high, or bud them upon stocks of that height; but this is a practice I would not recommend to the public, because the higher the heads of these trees are, the more they are exposed to the cutting winds in the spring, which too frequently destroy the blossoms; and the fruit is also more liable to be blown down in summer, especially if there should happen to be much wind at the time when the fruit is ripe, which by falling from a great height will be bruised and spoiled; therefore I prefer half standards, of about two and a half, or three feet in the stem, to those which are much taller.

These fruits are all propagated by budding them on Plum stocks, and will readily take upon almost any sort of Plum, provided the stock be free and thriving (except the Brussels kind, which is usually budded on a sort of stock, commonly called the St. Julian, which better suits this tree, as being generally planted for standards, than any other sort of Plum will.) The manner of raising the stocks and budding these trees, shall be treated of under their particular articles, to which I refer the reader, and shall proceed to their planting and management.

These trees are all (except the two last sorts) planted against walls, and should have an east or west aspect; for if they are planted full south, the great heat causes them to be meally before they are well eatable.

The borders under these walls should be six feet wide at least, and if it were more the better; but I could never advise the making of them so deep as is the general custom, for if the earth be two feet deep, or two and a half at most, it is enough.

If your ground is a wet cold loam or clay, you should raise your borders as much above the level of the surface as it will admit, laying some stones or rubbish in the bottom, to prevent the roots from running downwards; but if you plant upon a chalk or gravel, it will be better to raise the borders to a proper thickness with good loamy earth, than to sink the borders by removing the chalk or gravel; for although these are removed the whole breadth of the border, which we may allow to be eight feet, and this trench filled with good earth, yet the roots of the trees will in a few years extend this length, and then meeting with the chalk or gravel, will occasion the leaves of the trees to turn pale and fall off early in the season; the fruit will be small, dry, and ill flavoured, and the shoots of the trees will be weak. But where the borders are raised above the chalk to their full height, the roots will not strike down into the gravel or chalk, but rather extend themselves near the surface, where they will meet with better soil: and as these trees are of long duration, and old trees being not only more fruitful than young, but the fruit is also better flavoured, therefore the providing for their continuance is absolutely necessary.

The soil I would in general advise to be used for these, and all other sorts of fruit-trees, is fresh untried earth from a pasture ground, taken about ten inches deep, with the turf, and laid to rot and mellow at least twelve months before it is used; and this must be kept often turned to sweeten and imbibe the nitrous particles of the air.

Your borders being prepared, make choice of such trees as are but of one year's growth from budding; and if your soil is dry, or of a middling temper, you should prefer October as the best season for planting, especially having at that time a greater choice of trees from the nurseries, before they have been picked and drawn over by other people. The manner of preparing these trees for planting is the same in common with other fruit-trees; but do not cut off any part of the head at that time, unless there are any strong foreright shoots which will not come to the wall, which may be taken quite away.

Your trees being thus prepared, you must mark out the distances they are to stand, which in a good strong soil, or against a low wall, should be twenty feet or more; but in a moderate soil, and against taller walls, eighteen feet is a good reasonable distance; then make a hole where each tree is to stand, and place its stem about four inches from the wall, inclining the head thereto; and after having fixed the tree in the ground, nail the branches to the wall with list to prevent their being shaken. In this state the trees may remain till the middle of March, when, if the weather is good, you must unnail the branches of your trees, so as not to disturb their roots; and, being provided with a sharp knife, put your foot close to the stem of the tree, and having placed your left-hand to the bottom of the tree, to prevent its being disturbed, with your right-hand cut off the head of the tree if it has but one stem, or where it may have two or more shoots, each of them must be shortened to about four or five eyes above the bud, so that the sloping side may be toward the wall.

In the spring, if the weather proves dry, you must now-and-then give your trees a gentle refreshing with water all over their heads, which will greatly help them; and also lay some turf or other mulch round their roots, to prevent their drying during the summer season; and as new branches are produced, observe to fasten them to the wall in an horizontal position; and such shoots as are produced foreright, must be entirely displaced. This must be repeated as often as is necessary, to prevent their growing from the wall, but by no means stop any of the shoots in summer.

At Michaelmas, when the trees have done growing, you must unnail their branches, and shorten them in proportion to their strength; a vigorous branch may be left eight or nine inches long, but a weak one should not be left above five or six.

When you have shortened the shoots, be sure to nail them as horizontally as possible, for upon this it is that the future good of the tree chiefly depends.

The second summer observe, as in the first, to displace all foreright shoots as they are produced, fastening the other close to the wall horizontally, so that the middle of the tree may be kept open; and never shorten any of the shoots in summer, unless to furnish branches to fill vacant places on the wall, and never do this later than April. At Michaelmas shorten these shoots, as was directed for the first year; the strong ones may be left nine or ten inches, and the weak ones six or seven at most.

The following year's management will be nearly the same with this, but only observe, that Apricots produce their blossom buds, not only upon the last year's wood, but also upon the cursons or spurs which are produced from the two years wood; great care should therefore be had in the summer management, not to hurt or displace them; observe also to shorten your branches at the winter pruning, so as to furnish bearing wood in every part of the tree.

These few rules, well executed, together with a little observation and care, will be sufficient; and to pretend to prescribe particular directions for all the different accidents, or manner of treating fruit-trees, would be impossible.

The Brussels and Breda Apricots being for the most part planted for standards, will require very little pruning or management; only observe to take out all dead wood, or such branches as cross each other; this must be done early in autumn, or in the spring after the cold weather is past, that the part may not canker where the incision is made.

ARMERIUS. Sweet-William. See DIANTHUS.

ARNICA. Lin. Gen. Plant. 784. Leopardsbane.

The CHARACTERS are,

It hath a compound flower, the border or rays being composed of many female florets, which spread open; the disk or middle has many hermaphrodite flowers, which are tubulous, and have each five short stamina. In the hermaphrodite flowers the germen is situated below the flower, which afterward becomes a single obtuse seed, crowned with long slender down.

The SPECIES are,

1. ARNICA (*Montana*) foliis ovatis integris, caulinis geminis oppositis. Lin. Sp. Pl. 884. *Arnica with entire oval leaves, and those on the stalks growing opposite by pairs.*

2. ARNICA (*Scorpioides*) foliis alternis serratis. Hall. Helvet. 737. *Arnica with sawed leaves growing alternately.*

The first sort grows naturally upon the Alps, and also upon many of the mountains in Germany, and other cold parts of Europe, and is greatly esteemed by the Germans for its medicinal qualities, where it is prescribed by this title of Arnica.

The roots of this plant, when placed in a proper soil and situation, do greatly increase, for they send out thick fleshy roots which spread very far under the surface; these put out many oval entire leaves, from between which the flower-stems arise, which grow about a foot and a half high; the top is terminated by a single yellow flower, composed of many florets like those of Dandelion. These are succeeded by oblong seeds which are crowned with down.

This plant delights in a moist shady situation; it may be propagated by parting of the root in autumn, when the stalks begin to decay, or by the seeds sown in autumn soon after they are ripe, for those sown in the spring often fail.

The second sort grows naturally on the mountains of Bohemia, as also in Siberia. The roots of this sort are much jointed, and divide into many irregular fleshy offsets, which are variously contorted; from whence many superstitious persons have been led to imagine, that the roots would expel the poison of scorpions, and cure the wounds made by the bite of that animal. It is a very hardy plant, and is propagated in the same manner as the former.

ARTEDIA. Lin Gen. Plant. 249. We have no English name for this genus.

The CHARACTERS are,

It is an umbelliferous plant; the rays of the large umbel are difform, the flowers of the small ones in the disk are male, and the rays are hermaphrodite. These have each five slender stamina; those flowers which compose the rays, have a small germen at bottom, which afterward becomes a roundish compressed fruit with a leafy border, which splits into two, and contains two oblong seeds with scaly borders.

The SPECIES are,

1. ARTEDIA (*Squamata*) seminibus squamatis. Hort. Cliff. 89. *Artedia with squamose seeds.*

2. ARTEDIA (*Aculeata*) seminibus aculeatis. Hort. Cliff. 89. *Artedia with prickly seeds.*

The first is a native of the east, Rawvolf found it growing upon mount Libanus. It is an annual plant, whose stalks rise about two feet high, sending out a few side branches, which are garnished with narrow compound leaves resembling those of Dill; the extremity of the stalk is terminated by a large umbel of white flowers, composed of five unequal petals. These are succeeded by roundish compressed fruit, each having two seeds, whose borders are scaly.

The second sort grows upon the African shore in the Mediterranean, as also in Spain. This is also an annual plant, with an upright stalk near three feet high, and puts out many side shoots; the leaves are hairy, and greatly resemble those of the common Carrot; the stalks are terminated by umbels of large white flowers, shaped like those of the former, and are succeeded by a prickly fruit, composed of two seeds.

Both these plants decay as soon as they perfect their seeds, and many times before they are ripe, in England; for unless the seeds are sown in autumn, and the plants come up before winter, they rarely produce good seeds here. The seeds should be sown on a warm border where the plants are to remain, for they will not bear transplanting

ARTEMISIA. Mugwort.

The CHARACTERS are,

The flower is composed of hermaphrodite and female florets; the hermaphrodite flowers compose the disk or middle. In the center is placed the germen, which is accompanied by five hairy stamina. The germen afterward becomes a single naked seed, sitting upon a naked placenta.

The SPECIES are,

1. ARTEMISIA (*Vulgaris*) foliis pinnatifidis planis incisis subtus tomentosis, racemis simplicibus floribus ovatis radio quinquefloro. Lin. Sp Plant. 348. *Common Mugwort*

2. AR-

2. Artemisia (*Integrifolia*) foliis lanceolatis subtus tomentosis integerrimis dentatisque. Lin. Sp. Plant. 348. *Mugwort with spear-shaped leaves which are entire, and indented on their edges, and their under sides woolly.*

The first sort grows naturally on banks, and by the side of foot paths in most parts of England, so is rarely admitted into gardens; for the roots creep far under the surface of the ground, so that unless they are stopped, they will soon spread over a large space of ground. This flowers in June, at which time the herb is in perfection for use.

The Moxa, so famous in the eastern countries for curing the gout by burning of the part affected, is the lanugo, or down, which is on the under part of the leaves of Mugwort.

The second sort grows naturally in Siberia; it rises up with single stalks about two feet high; the flowers come out from the wings of the leaves in small loose spikes, and near the top they are often single; these are larger than those of the common sort, and are of a pale yellow colour.

This sort is as hardy as the common sort, and multiplies as fast, but is only preserved in botanic gardens for the sake of variety.

ARTICHOKE is called by the Latins Cinara.

As this plant is much better known by its English title than the Latin, I shall treat of it under this head, and refer for its characters to the Latin title of Cinara, under which the other species will be exhibited.

We have two sorts of Artichokes which are cultivated in the English gardens, which we shall distinguish here only by the names they are generally known among the gardeners.

The best sort is what the gardeners call the Globe Artichoke. This hath large heads, with broad brown scales, which turn inward; the fleshy part at the bottom of the scales is very thick, therefore is much preferred to the other, which is called the French Artichoke; the stalks of which do generally grow taller, and the heads are smaller and shaped more conical than those of the globe. The scales are narrower, of a greener colour, and frequently turned outward. The fleshy part which is eaten is not near so thick, and hath a disagreeable perfumed taste; this was almost totally rooted out of the English gardens before the hard frost in 1739-40, when the greatest part of the roots of the other sort were destroyed, so many persons were supplied the following spring with plants from Guernsey, where they cultivate only the latter sort; but since the other has been increased again, this green sort has been in most gardens rooted out, to make way for the Globe Artichoke.

The manner of propagating this plant is from slips, or suckers taken from the old roots in February or March, which, if planted in a good soil, will produce large fair heads the autumn following; but as this is a plant which few gardeners who have not been bred in the kitchen-gardens near London, understand to manage well, I shall be the more particular in my directions about it.

About the beginning of March, according to the earliness of the season, or forwardness of the old Artichoke stocks, will be the proper time for dressing them, which must be thus performed: with your spade remove all the earth from about your stock, down below the part from whence the young shoots are produced, clearing the earth from between the shoots, so as to be able to judge of the goodness of each, with their proper position upon the stock; then make choice of two of the clearest, straitest, and most promising plants that are produced from the under part of the stock, which are much preferable to the strong thick plants which generally grow upon the crown of the roots, for these have hard woody stems, so do never produce good heads, but generally are what the market gardeners call rogues, which have very little bottom, and the scales of their heads are irregularly placed; in slipping off the other shoots you must be careful not to injure the plants which you are to let remain for a crop; then with your thumb force off all the other plants and buds close to the head of the stock from whence they are produced, being very careful not to leave any of the buds, and with your spade draw the earth about the two plants which are left, and with your hands close it fast to each of them, separating them as far asunder as they can conveniently be placed without breaking them, observing to crop off the tops of the leaves which hang down with your hands; your ground being levelled between the stocks, you may sow thereon a small crop of Spinach, which will be taken off before the Artichokes will cover the ground; and toward the latter end of April, or the beginning of May, when your plants begin to shew their fruit, you must carefully look over your stocks, and draw up all young plants from them which may have been produced from the roots since their dressing, and cut off all suckers which are produced from the stems of the Artichokes, leaving only the principal head, by which means they will be larger; when your Artichokes are fit to gather, you must break or cut them down close to the surface of the ground, that your stocks may make strong fresh shoots by the middle of November, which is the season for earthing, or, as the gardeners term it, landing them up, which is thus done:

Cut off all the young shoots quite close to the surface of the ground; then dig between every stock, raising all the earth between each row of stocks into a ridge, as is done in the common method of trenching ground, in such manner as that the row of Artichokes may be exactly in the middle of each ridge; this will be sufficient to guard them against common frost; and I would here recommend it to the public as infinitely preferable to long dung, which is by the unskilful often used to cover the roots, and is the occasion of their heads being small, and almost without any bottoms to them; for there is not any thing so hurtful to these roots, as new dung being either buried near or laid about them. Observe, that although I have mentioned November as the season for earthing them, yet if the weather proves mild, it may be deferred till any time in December.

As we have experienced, that in very severe frosts these roots are sometimes destroyed, therefore it is proper to give some directions to prevent it; altho' this rarely happens in dry ground, in which we have but few instances of their being killed, except in the hard frosts of 1683, 1739-40, and 1767-8. In these winters most of the Artichokes were destroyed in England; in the second of these winters it happened from the little care which was taken of them, there having been no severe frost for so many years before which had injured them, that few people used any care to preserve them; but since that hard frost many people have run into the other extreme, of covering all their roots of Artichokes with long dung in winter, which is a very bad method, because the dung lying near the roots is very apt to rot the best plants; therefore I would advise the earthing (or as it is called by the gardeners, landing) of the Artichokes to be deferred till the latter end of November, provided the season continues mild; and towards Christmas, if there is any danger of severe frosts, to lay a quantity of long dung, Peasehaulm, tanners bark, or any other light covering over the ridges of earth, which will keep out the frost, and this being at a distance from the roots will not injure them; but this covering should be carefully taken off the beginning of February, or sooner, provided the season is mild, or at least so soon as the weather is so, otherwise the plants will be injured by its lying too long upon them.

When you have thus earthed them up, you have nothing more to do till March, by which time the plants will have grown through the ridge of the earth; therefore when the weather is proper, the roots must be dressed as was before directed.

When you have a mind to make a new plantation of Artichokes, after having digged and buried ſome very rotten dung in the ground you have allotted for that purpoſe, make choice of ſuch of your plants as were taken from your old ſtocks, which are clear, ſound, and not woody, having ſome fibres to their bottom; then with your knife cut off that knobby woody part which joined them to the ſtock; and if that cuts criſp and tender it is a ſign of its goodneſs, but if tough and ſtringy throw it away as good for nothing; then cut off the large outſide leaves of the plants intended for planting pretty low, that the middle, or heart leaves, may be above them. Your plants being thus prepared (if the weather is very dry, or the plants have been any time taken from the ſtocks, it will be convenient to ſet them upright in a tub of water for three or four hours before they are planted, which will greatly refreſh them;) you muſt then proceed to planting, which muſt be done by ranging a line acroſs the ground, in order to their being placed exactly in a row, and, with a meaſure-ſtick, plant them at two feet diſtance from each other in the rows, and if deſigned for a full crop, five feet diſtance row from row; your plants muſt be ſet about four inches deep, and the earth cloſed very faſt to their roots, obſerving, if the ſeaſon proves dry, to keep them watered two or three times a week until they are growing, after which they do not require any.

N. B. You may ſow a thin crop of Spinach upon the ground before you plant your plants, obſerving to clear it from about them after it is come up.

Theſe plants in a kindly ſeaſon, or on a moiſt ſoil, will produce the largeſt and beſt Artichokes ſome time in Auguſt and September, after all thoſe from the old ſtocks are paſt; ſo that if you intend to continue your Artichokes through the whole ſeaſon, you muſt make a new plantation every year, otherwiſe you cannot poſſibly have fruit longer than two or three months.

If any of the plants which are planted in the ſpring ſhould not fruit in autumn, you may, at the ſeaſon of earthing up your roots, tie up the leaves with a ſmall Willow twig, &c. and lay the earth up cloſe to it, ſo that the top of the plant may be above ground; and when the froſt comes on, if you will cover the top with a little ſtraw, or Peaſe-haulm, to prevent their being killed by froſt, theſe plants will produce fruit in winter, or early in the ſpring.

But in thoſe plantations where you intend to plant other things between your Artichokes, you muſt allow nine or ten feet between the rows, as is often practiſed by the kitchen-gardeners near London, who ſow the ground between with Radiſhes or Spinach, and plant two rows of Cauliflowers, at four feet diſtance row from row, and two feet and a half diſtance in the rows between them, ſo that there are always five feet allowed for the Artichokes to grow; and in May, when the Radiſhes or Spinach are taken off, they ſow a row of Cucumbers for pickling exactly between the two rows of Cauliflowers, and at three feet diſtance from each other; and between the rows of Cauliflowers and the Artichokes plant a row of Cabbages or Savoys for winter uſe, which, when the Cauliflowers are drawn off and the Artichokes gathered, will have full liberty to grow; and by this means, the ground is fully employed through the whole ſeaſon. This has long been the practice of the kitchen-gardeners near London, who pay large rents for their land, ſo are obliged to get as many crops in a year from it as poſſible.

If in the ſpring you find your ſtocks ſhoot very weak, which may have been occaſioned either by hard froſt or too much wet, you muſt then uncover them, and with your ſpade looſen and break the earth about them, raiſing a ſmall hill about the plants of each ſtock, levelling the reſt between the rows, which will greatly help them, and in three weeks or a month's time after they will be fit to ſlip.

Thoſe Artichokes which are planted in a moiſt rich ſoil will always produce the largeſt and beſt fruit, ſo that where ſuch a ſoil can be obtained, it will be proper to make a freſh plantation every ſpring, to ſucceed the old ſtocks, and ſupply the table in autumn. But the roots will not live through the winter in a very moiſt ſoil, ſo that your ſtocks which you intend ſhould remain, to ſupply the table early, and to furniſh plants, ſhould be in a drier ſituation. You ſhould always obſerve to plant theſe in an open ſpot of ground, not under the drip of trees, where they will draw up very tall, and produce ſmall inſignificant fruit.

ARTICHOKES of Jeruſalem. See Helianthus.

ARUM. Wake Robin, or Cuckow Pintle.

The Characters are,

The flower hath an oblong ſpatha; the ſpadix is ſingle, ſhaped like a club at the top, upon which the germen are ſituated. It hath no petals nor ſtamina, but many four-cornered ſummits, ſitting cloſe to the germen. There are many germen which ſurround the upper part of the ſpadix which are oval, having no ſtyles, but have bearded ſtigma: the germen afterward becomes globular berries with one cell, having round ſeeds.

The Species are,

1. Arum (*Maculatum*) acaule foliis haſtatis integerrimis ſpadice clavato. Hort. Cliff. 434. *The common Arum, or Wake Robin, with ſpotted and plain leaves.*

2. Arum (*Italicum*) acaule foliis haſtatis acutis petiolis longiſſimis ſpathâ maximâ erectâ. *Largeſt Italian Arum with white veins.*

3. Arum (*Proboſcidium*) acaule foliis haſtatis ſpathâ declinata filiformi ſubulatâ. Lin. Sp. Plant. 966. *Friars Cowl with a flower ending in a ſlender tail.*

4. Arum (*Ariſarum*) acaule foliis cordato-oblongis ſpathâ bifidâ ſpadice incurvo. Hort. Cliff. 435. *Greater broad-leaved Friars Cowl.*

5. Arum (*Tenuifolium*) acaule foliis lanceolatis ſpadice ſetaceo declinato. Hort. Cliff. 345. *Narrow-leaved Friars Cowl of Dioſcorides.*

6. Arum (*Virginicum*) acaule foliis haſtato-cordatis acutis angulis obtuſis. Hort. Cliff. 434. *Arum without ſtalk, pointed ſpear heart-ſhaped leaves with obtuſe angles.*

7. Arum (*Triphyllum*) acaule foliis ternatis. Flor. Virg. 113. *Three-leaved Arum without ſtalk.*

8. Arum (*Dracunculus*) foliis pedatis, foliolis lanceolatis integerrimis æquantibus ſpatham ſpadice longiorem. Prod. Leyd. 7. *Common Dragon.*

9. Arum (*Dracontium*) foliis pedatis, foliolis lanceolatis integerrimis ſuperantibus ſpatham ſpadice breviorem. Prod. Leyd. 7. *Smaller dwarf Arum with many leaves.*

10. Arum (*Trilobatum*) acaule foliis trilobis flore ſeſſile. Flor. Zeyl. 326. *Broad-leaved low Arum of Ceylon with a ſcarlet piſtil.*

11. Arum (*Colocaſia*) acaule foliis peltatis ovatis repandis baſi ſemibifidis. Hort. Cliff. 434. *Greateſt Egyptian Arum, vulgarly called Colocaſia.*

12. Arum (*Betifolium*) acaule foliis cordatis nervoſis floribus ſeſſilibus. *American Arum with a Beet-leaf called Scunk Weed.*

13. Arum (*Divaricatum*) acaule foliis cordatis angulatis divaricatis. Lin. Sp. Pl. 966. *Arum without ſtalk, and ſpear-ſhaped leaves.*

14. Arum (*Peregrinum*) acaule foliis cordatis obtuſis mucronatis angulis rotundatis. Hort. Cliff. 435. *Arum without ſtalk, blunt heart-ſhaped leaves which are pointed, and the angles rounded. This is commonly called Edder in America.*

15. Arum (*Eſculentum*) acaule foliis peltatis ovatis integerrimis baſi ſemibifidis. Hort. Cliff. 453. *Eatable Arum with a Water Lily leaf.*

16. Arum

16. Arum (*Sagittifolium*) acaule foliis sagittatis acuminatis nervosis. *Greatest Egyptian Arum, or Colocasia with blackish stalks.*

17. Arum (*Ceylanicum*) acaule foliis hastatis acuminatis spathâ mucronatâ revolutâ. *Dwarf broad-leaved Arum of Ceylon with a purple pistil.*

18. Arum (*Arborescens*) caulescens foliis sagittatis spathâ declinatâ clausâ. *Tree-like Arum with lance-shaped leaves, commonly called Dumb Cane.*

The first sort grows naturally in woods, and on shady banks in most parts of England, so is seldom admitted into gardens; but being a medicinal plant, it is here inserted to introduce the other species. There are two varieties of this, one with plain leaves, and the other hath leaves full of black spots; but these are only accidental varieties, which arise from the same seeds. The roots of this are ordered by the College of Physicians to be used in a powder which bears the title of the plant; but these roots are generally gathered in the spring, when the leaves are in full vigour, so that the roots shrink and soon loose their pungent quality; but those which are taken up when the leaves decay, will continue good a whole year, and retain their pungency the same as when first taken up. The not observing this, has brought the medicine into disrepute. It flowers in April, and the seeds ripen in July, when it is the best time to take up the roots.

The second sort grows naturally in Italy, Spain, and Portugal. The leaves of this sort rise a foot and a half high, are very large, running out to a point; these are finely veined with white, interspersed with black spots, which, together with the fine shining green of their surface make a pretty variety. The flowers grow near a foot high, and have very long upright spathas, which are of a pale green, inclining to white; these appear the end of April, or beginning of May; this propagates very fast by offsets from the root, and will thrive in any soil or situation. The best time to transplant them is from the time the seeds are ripe, to the end of October.

The third, fourth, and fifth sorts have been generally separated from this genus, and were distinguished by the title of Arisarum, or Friar's Cowl, from the resemblance the flower has in shape to the hoods or cowls worn by the people of that order. These are very low plants, their leaves having very short foot-stalks, and the flowers grow close to the ground. They flower in April. The time for transplanting the roots is the same as for the former.

The sixth and seventh sorts grow naturally in Virginia and Carolina; these never rise with stalks, but their leaves arise immediately from the roots, having short foot-stalks; the flowers come out between their leaves, which have short foot-stalks; they appear in May, but have little beauty, so the plants are only kept in botanic gardens for the sake of variety.

The eighth sort is the common Dragon which is used in medicine, and has been generally ranged in a separate genus from this under the title of Dracunculus.

This sort is used in medicine, so is preserved in some gardens to supply the markets: it grows naturally in most of the southern parts of Europe, from whence it was first obtained. It hath a strait stalk three or four feet high, which is spotted like the belly of a snake; at the top it spreads out into leaves, which are cut into several narrow segments almost to the bottom, and are spread open like a hand; at the top of the stalk the flower is produced, which is in shape like the common Arum, having a very long spatha of a dark purple colour standing erect, with a large pistil of the same colour, so that when it is in flower it makes no unpleasing appearance; but the flower hath so strong a scent of carrion, that few persons can endure it, for which reason it hath been banished most gardens; but was it not for this, a few of the plants might merit a place in gardens for the oddness of the flower. It is very hardy, so will grow in any soil or situation, and propagates fast by offsets from the root. There is a variety of this with variegated leaves and stalks, which is preserved in the gardens of some persons who are fond of striped-leaved plants.

The ninth sort grows naturally in moist places in Virginia and New England, but is very difficult to preserve long in a garden, especially if the soil is dry. The leaves of this sort are divided like those of the former, but are smaller, and rarely grow more than nine inches high; the flowers are like those of the common Arum.

The tenth sort grows naturally in Ceylon, and some other parts of India, so is very impatient of cold: it is a plant of humble growth; the flower rises immediately from the root, standing on a very short foot-stalk; the spatha is long, erect, and of a fine scarlet within, as is also the pistil. This plant must be placed in the tan-bed of the bark-stove, otherwise it will not thrive in England. It is propagated by offsets from the root, which come out in plenty when the plants are in health.

The eleventh, thirteenth, fourteenth, fifteenth, and sixteenth sorts, have mild roots, which are eaten by the inhabitants of all the hot countries, where they grow naturally, and some of the sorts are cultivated by the inhabitants of the Sugar Colonies as esculent plants, their roots being constantly eaten; as also are the leaves of some of the sorts, particularly the fifteenth, which they call Indian Kale: the leaves of this are boiled, and supply the want of other greens. It is esteemed a wholsome green; and in those countries where many of the common European vegetables are with difficulty produced, this proves a good succedanum. The sixteenth sort has not been many years introduced among them, for it came originally from the Spanish West-Indies, where it grows in great plenty. But these have larger roots than the fifteenth, for which reason they are preferred to it.

All these sorts are preserved in the gardens of those persons who are curious in collecting exotic plants, for the variety of their leaves, for their flowers have very little beauty, nor do they often appear in this country. The plants are propagated easily by offsets from their roots, which they put out plentifully: these must be planted in pots filled with rich earth, and plunged into a hot-bed; and if they are afterward continued in the bark-stove, they will make great progress, and their leaves will be very large.

The roots of the seventeenth sort I received from India; this annually flowers in May; the leaves grow near the ground, having short foot-stalks; the flower comes up immediately from the root upon a short foot-stalk, and is of a very deep purple colour, and smell almost like the flower of Dragon. It is tender, so must be constantly kept in the bark-stove, otherwise it will not thrive in England.

The eighteenth sort grows naturally in the Sugar Islands, and other warm parts of America, chiefly in the low grounds; the whole plant abounds with an acrid juice, so that if a leaf or part of the stalk is broken, and applied to the tip of the tongue, it occasions a very painful sensation, and occasions the salivary ducts to swell, and brings on a great defluxion of saliva; the stalks of this plant are sometimes applied to the mouths of the negroes by way of punishment. This sort is propagated by cutting off the stalks into lengths of three or four joints, which must be laid to dry six weeks or two months; for if the wounded part is not perfectly healed over before the cuttings are planted, they will rot and decay: these should be planted in small pots filled with light sandy earth, and plunged into a moderate hot-bed of tan, being careful that they have little wet, until they have made good roots; some of them may be placed in a dry stove, ftove, and others plunged into the tan bed in the bark-ftove, where they will make the greateft progrefs, and produce more flowers than the others.

ARUM ÆTHIOPICUM. See CALLA.

ARUM SCANDENS. See DRACONTIUM.

ARUNDO Lin. Gen. Plant. 76. The Reed.

The CHARACTERS are,

It is of the Grafs tribe; the flowers grow in fpikes, and are included in a chaff which opens with two valves. The petals of the flowers are bivalve, having a down at their bafe, and have three hairy ftamina; in the center is fituated an oblong germen with two flender ftyles. The germen afterward becomes an oblong pointed feed, with long down adhering to its bafe.

The CHARACTERS are,

1. ARUNDO (*Phragmitis*) calycibus quinquefloris paniculâ laxâ. Prod. Leyd. 66. *The common Marfh Reed.*

2. ARUNDO (*Donax*) calycibus trifloris paniculâ diffusâ. Prod. Leyd. 66. *The manured Reed, or Donax of Diofcorides; this is fometimes called by gardeners the Evergreen Reed.*

3. ARUNDO (*Verficolor*) Indica Laconica verficolor. Mor. Hift. 3. p. 219. *The Indian variegated Reed of Theophraftus.*

4. ARUNDO (*Bamboa*) caule arboreo foliis acuminatis fulcatis, bafi rotundioribus. *Another fpecies of Bambu.*

5. ARUNDO (*Arborea*) caule arboreo foliis utrinque acuminatis. *Reed with a tree-like ftalk, and leaves which are pointed at both ends.*

6. ARUNDO (*Orientalis*) Orientalis tenuifolia caule pleno ex quâ Turcæ calamos parant. Tourn. Cor. 39. *Eaftern Reed with a narrow leaf and a full ftalk, of which the Turks made their writing pens.*

The firft fort is fo very common by the fides of rivers and large ftanding waters in divers parts of England, that it is needlefs for me to fay any thing of its culture. This is cut in autumn, when the leaves begin to fall, and the ftems are changed brown, for making hedges in kitchen-gardens, and for many other ufes.

The fecond fort, although a native of a warm country, yet will bear our cold of moderate winters in the open ground; it dies to the furface in autumn, and rifes again the fucceeding fpring; and if kept fupplied with water in dry weather, will grow ten or twelve feet high the fame fummer. This is propagated by parting the roots early in the fpring before they begin to fhoot, and will in a year or two, if the ground be good, make very large ftools, from each of which you may have twenty or thirty large canes produced.

The ftalks of this fort are brought from Portugal and Spain, and are ufed by the weavers, as alfo to make fifhing rods.

The third fort is fuppofed to be a variety of the fecond, differing therefrom only in having variegated leaves. This plant will not grow fo large, nor will it refift the cold fo well, therefore will not live in the open air through the winter in England; fo the plants muft be kept in pots, and houfed in the autumn.

The two forts of Bambu are of great fervice to the inhabitants of India, who make moft of their common utenfils of the ftems of thefe canes, which grow to a prodigious magnitude in thofe countries.

We have plants of the fourth fort in the Englifh gardens, which are more than twenty feet high; and if the ftoves in which they are kept were high enough to admit them, they would, according to appearance, rife to twice that height. Some of thefe ftems are as large as a man's wrift, but in general as big as walking-fticks, and when dried are as fit for that purpofe as thofe which are imported. The leaves of this fort are much broader than thofe of the fifth, particularly at their bafe; thefe leaves are generally put round the tea-chefts in their package, and are faftened together fo as to form a kind of mat.

The fifth fort is more rare at prefent in Europe, though it is moft common on the coaft of Malabar.

They are both tender plants, fo will not live in this country, unlefs they are preferved in a warm ftove; and as their roots fpread very wide, fo they fhould not be confined: therefore to have them produce ftrong ftems, they muft be planted in large tubs, filled with rich earth, and plunged into the tan-bed in the bark ftove; and as they naturally grow in marfhy low places, fo they require plenty of water, efpecially when the roots have filled the tubs in which they were planted. When the tubs decay, the boards may be removed, and the plants permitted to root into the tan, which will encourage them to grow to a larger fize; but then there muft be care taken when the bed is refrefhed with new tan, to leave a fufficient quantity of old tan about the roots of the plants, for if they are too much bared, and the new tan laid near them, when that heats, it will fcorch their roots, fo that the plants are fometimes deftroyed by it.

The fixth fort is what the Turks make their writing pens withal; this grows in a valley near mount Athos, as alfo on the banks of the river Jordan, but there are none of the plants in England at prefent. This fort may be managed as the Bambu.

ARUNDO SACCHARIFERA. See SACCHARUM.

ASARINA. Tourn. Inft. R. H. 171. tab. 76. Baftard Afarum.

The CHARACTERS are,

The flower is of one leaf, of the grinning kind, divided at the top into two lips, the upper one is divided into two parts. The lower lip is flightly cut into three obtufe parts; the two lips join clofe together, fo as to form a kind of fnout. It hath four ftamina. In the center is placed a round germen, which afterward turns to a round hufk, divided into two cells, which are full of roundifh feeds.

The SPECIES are,

1. ASARINA (*Procumbens*) caule decumbente foliis oppofitis reniformibus crenatis. *Afarina, or Rock Ground-ivy.*

2. ASARINA (*Erecta*) caule erecto foliis lanceolatis amplexicaulibus paniculâ dichotomâ. *Baftard Afarum with an upright ftalk, fpear-fhaped leaves which embrace the ftalks, and fpikes of flowers coming out from the divifion of the branches.*

The firft fort is a low, trailing, annual plant; the branches extend little more than a foot each way, and are weak, fo that unlefs they are fupported, they lie upon the ground; at the wings of the leaves the flowers come out fingly on each fide the ftalk, which are fhaped like thofe of Snapdragon, but have a long tube; they are of a worn-out purple colour at the top, but below of an herbaceous colour. The feeds fhould be fown after they are ripe, or permitted to fcatter, for when they are fown in the fpring they feldom grow. The plants fhould remain where they are fown, and require no other care but to thin them where they grow too clofe. It grows naturally in Italy and the fouth of France.

The fecond fort grows naturally in North America. This plant hath upright ftalks, which grow a foot and a half high, and put out feveral fide branches; the leaves grow oppofite, and embrace the ftalks at their bafe; the flowers come out in fhort loofe fpikes from the divifions of the ftalks, which are fhaped like thofe of the former, but are lefs, and of a purple colour.

The feeds of this fort fhould be fown in the autumn, for thofe which are fown in the fpring feldom grow the fame year, but remain in the ground till the following fpring; the fecond year the plants will flower and perfect their feeds. The roots feldom laft above two or three years, fo that young plants fhould be annually raifed.

ASARUM. Afarabacca.

The

The Characters are,

The flower hath no petals, but a thick coloured empalement and twelve short stamina. At the bottom of the empalement is inclosed a thick germen, which afterward turns to a thick capsule having six cells, containing several oval seeds.

The Species are,

1. Asarum (*Europæum*) foliis reniformibus obtusis binis. Lin. Sp. Plant. 442. *Common Asarabacca.*

2. Asarum (*Canadense*) foliis reniformibus mucronatis. Lin. Sp. Plant. 442. *Canada Asarabacca.*

3. Asarum (*Virginicum*) foliis cordatis obtusis glabris petiolatis. Flor. Virg. 162. *Virginia Asarabacca, with round Pistolochia leaves marked like those of Sowbread.*

The first sort hath thick fleshy roots which are jointed; the leaves grow singly upon short foot-stalks, arising immediately from the root; the flowers grow upon very short foot-stalks close to the ground, so are hid under the leaves. They have a bell-shaped empalement, of a worn-out purple colour, which is cut into three at the top, where it turns backward.

The leaves of the second sort are much larger than those of the first, and stand on longer foot-stalks; these are pointed and hairy. The flowers are like those of the other sort, growing close to the root, but are somewhat inclining to green on their outside, in all other respects they agree.

The third sort hath smooth, blunt, heart-shaped leaves, standing on long foot-stalks; these are veined, and spotted on their upper surface like those of the autumnal Cyclamen; the flowers of this are shaped like the others, but stand on longer foot-stalks, and are of a darker purple colour.

The first of these sorts is very common, and hath been found wild in some parts of England, though but rarely; it delights in a moist shady place, and is increased by parting the roots in autumn. This is the sort which is used in medicine.

The Canada sort is tolerably hardy, and will endure our common winters in the open ground, being rarely hurt but by great frosts, or being planted in a wet soil, which often occasions the roots to rot in winter. This is propagated as the other.

The third sort will live in the open air in England, being seldom injured by frost; but if the plants are too much exposed to the sun in summer, they seldom thrive well; therefore they should be planted in a border where they may have only the morning sun, in which situation they will spread and increase.

ASCLEPIAS. Hirundinaria, or Swallow-wort.

The Characters are,

The flower hath a petal of one leaf, divided into five oval parts. In the center is situated five nectariums which encompass the parts of generation; the stamina are joined in a truncated body inclosed by five scales, and are scarce visible. It hath two oval pointed germen, which afterward become two large, oblong, swelling pods ending in a point, which open with two valves, which are filled with compressed seeds, lying over each other like tiles on a house, and are crowned with a soft down.

The Species are,

1. Asclepias (*Alba*) foliis ovatis acuminatis caule erecto umbellulis proliferis. Lin. Sp. Plant. 216. *Common Swallow-wort with a white flower.*

2. Asclepias (*Nigra*) foliis lanceolatis acutis caule supernè subvolubili. Lin. Sp. Pl. 216. *Swallow-wort with a black flower.*

3. Asclepias (*Lutea*) foliis ovatis acutis caule infirmo, umbellis simplicibus. *Narrow-leaved Swallow-wort with a yellow flower.*

4. Asclepias (*Verticillata*) foliis revolutis linearibus verticillatis caule erecto. Lin. Sp. Pl. 217. *Upright Dogsbane of Maryland, with narrow Toadflax leaves, and flowers growing in an umbel.*

5. Asclepias (*Syriaca*) foliis ovalibus subtus tomentosis caule simplicissimo umbellis nutantibus. Lin. Sp. Pl. 214. *Greater upright Syrian Dogsbane.*

6. Asclepias (*Amœna*) foliis ovatis subtus pilosiusculis caule simplici umbellis nectariisque erectis. Lin. Sp. Pl. 214. *Dogsbane with fine purple flowers and upright horns.*

7. Asclepias (*Purpurascens*) foliis ovatis subtus villosis caule simplici umbellis erectis nectariis resupinatis. Lin. Sp. Pl. 214. *Upright Dogsbane of New York, with leaves less hoary, and a worn-out, palish, purple-coloured flower.*

8. Asclepias (*Variegata*) foliis ovatis rugosis nudis caule simplici umbellis subsessilibus pedicellis tomentosis. Lin. Sp. Plant. *The old American Dogsbane, called Wisank.*

9. Asclepias (*Incarnata*) foliis lanceolatis caule supernè diviso umbellis terminalibus congestis. Lin. Sp. Pl. 215. *Smaller upright Dogsbane of Canada.*

10. Asclepias (*Decumbens*) foliis villosis caule decumbente. Lin. Sp. Pl. 216. *Hairy Orange-coloured Dogsbane of Carolina.*

11. Asclepias (*Tuberosa*) foliis alternis lanceolatis caule divaricato piloso. Lin. Sp. Pl. 217. *Hairy New England Dogsbane with a tuberose root, and an Orange-coloured flower, commonly called Orange Apocynum.*

12. Asclepias (*Glabra*) foliis lineari-lanceolatis glabris caule fruticoso umbellis lateralibus. *Upright African Dogsbane with a hairy fruit, and a narrow, smooth, Willow leaf.*

13. Asclepias (*Fruticosa*) foliis lanceolatis glabris umbellis simplicibus lateralibus caule fruticoso. *Upright African Dogsbane with a broad, smooth, Willow leaf.*

14. Asclepias (*Villosa*) foliis lanceolatis villosis acutis umbellis simplicibus erectis caule fruticoso. *Upright African Dogsbane with hairy fruit, and a broad, hairy, Willow leaf.*

15. Asclepias (*Rotundifolia*) caule erecto fruticoso, foliis subrotundis amplexicaulibus, umbellis congestis. *Upright shrubby Dogsbane with a roundish sea-green leaf.*

16. Asclepias (*Nivea*) foliis lanceolatis glabris caule simplici umbellis erectis lateralibus solitariis. Lin. Sp. Pl. 215. *American Dogsbane with longer Almond leaves.*

17. Asclepias (*Curassavica*) foliis lanceolatis petiolatis glabris caule simplici umbellis erectis solitariis. Lin. Sp. Pl. 215. *Dogsbane with a fibrous root, and scarlet petals with Saffron-coloured horns, called Bastard Ipecacuana.*

18. Asclepias (*Gigantea*) foliis amplexicaulibus oblongo-ovalibus. Flor. Zeyl. 112. *Greater, upright, broad-leaved Indian Dogsbane, and the Beid el ossar. Alp. Ægypt.* 85.

19. Asclepias (*Scandens*) foliis ovato-lanceolatis glabris caule simplici umbellis erectis terminalibus. *Upright Dogsbane with oblong pointed leaves, and white flowers growing in an umbel.*

The first sort is the common Swallow-wort of the shops. This is called Vincetoxicum and Hirundinaria, in English Swallow-wort, or tame Poison, from its supposed virtue, being accounted a mighty counter poison. The root is the only part which is used, which is composed of many strong fibres, connected at the top like those of Asparagus, from which arise many stalks, in number proportional to the size of the roots, which grow two feet high, and are very slender at the top; the leaves are placed opposite by pairs. The flowers are white, growing in umbels near the top of the stalk, from which are sent out smaller umbels. After the flower is past, the two germen become two long pointed pods, inclosing many compressed seeds, lying imbricatim, which are crowned with a soft white down. It flowers in June, and the seeds ripen in September. It grows naturally in the south of France, Spain, and Italy.

The second sort agrees with the first, in the shape of its roots, leaves and flowers, but the stalks extend to a greater length,

length, and toward their upper part twiſt round any ſticks, or other plants near it, and the flowers of this are black.

The third differs from both the other in the narrowneſs of its leaves, and weakneſs of its ſtalks; the umbels of flowers are ſingle, and of a yellow colour. There is a variety of this with broader leaves, which may have come from the ſeeds of this.

Theſe plants are generally propagated by parting their roots, eſpecially the firſt ſort, which ſeldom produces ſeeds in England. The beſt time for this is in autumn, when their ſtalks begin to decay. They ſhould not be planted nearer together than three feet, for the fibres of their roots extend to a conſiderable diſtance. They are very hardy plants, ſo will thrive in any ſituation, but love a dry ſoil.

The fourth ſort grows naturally in North America. This riſes with ſlender upright ſtalks, which are garniſhed with very narrow leaves, growing in whorles: at the top of the ſtalks grow umbels of ſmall, white, ſtarry flowers, which appear in July, but are never ſucceeded by pods in England, ſo are only propagated by parting of the roots, which ſhould be done in the ſpring before they put out new ſhoots.

The fifth ſort creeps greatly at the root; this ſends up ſtrong ſtems upward of five feet high, which have thick oval leaves placed oppoſite, and hoary on their under ſides. Towards the top of the ſtalks the umbels of flowers come out on the ſide: theſe are of a worn-out purple colour, ſmelling ſweet, and nod downward; ſometimes they are ſucceeded by large oval pods, filled with flat ſeeds, crowned by a long ſoft down. This propagates faſt enough by its creeping root, and will grow in any ſoil or ſituation. It may be tranſplanted any time after the ſtalks decay, and before the roots ſhoot in the ſpring.

The ſixth ſort hath a perennial root, which ſends up ſeveral upright ſtalks in the ſpring, about two feet high, garniſhed with oval leaves growing oppoſite; at the top of the ſtalks the umbels of flowers are produced, which are of a bright purple colour, making a pretty appearance, but are not ſucceeded by pods in England; this muſt be treated as the fourth ſort.

The ſeventh ſort grows naturally in North America. This hath a perennial root, which ſends out ſingle ſtalks near three feet high, which have oval leaves placed oppoſite; the flowers grow in erect umbels at the top, and the nectariums are declining. It is very hardy, and propagates faſt by its creeping roots, but never produces ſeeds in England.

The eighth ſort reſembles the ſeventh, but the leaves are rough, and the umbels of flowers are more compact, and come out on the ſide of the ſtalk; theſe are of an herbaceous colour: it is propagated by roots as the former ſort.

The ninth ſort came firſt from Canada, but hath ſince been found growing naturally in ſeveral other parts of America. It hath a perennial root, which puts out ſeveral upright ſtalks about two feet high, with oblong ſmooth leaves placed by pairs; at the top are produced cloſe umbels of purple flowers. It is propagated by parting of the roots, which do not increaſe very faſt, ſo that it is not very plenty in the gardens; but it is hardy enough to live abroad, if it is planted in a dry ſoil.

The tenth ſort is a native of North America, but is hardy enough to live abroad in England, if it is planted in a warm ſituation and in a dry ſoil. This hath declining ſtalks, which are a foot and a half long, and hairy; the leaves are narrow, hairy, and placed oppoſite; the umbels grow at the extremity of the branches, which are compact, and the flowers are of a bright Orange colour. It is propagated by ſeeds, which ſhould be ſown upon a hot-bed to bring up the plants. When they are of a proper ſtrength to remove, they ſhould be ſhaken out of the pots, and planted in a warm border a foot aſunder, being careful to ſhade them from the ſun until they have taken freſh root, but they muſt have very little water given them, for they are milky plants, which rot with much wet. When their ſtalks decay in autumn, ſome rotten tan ſhould be laid over the ground to keep out the froſt, which ſhould be removed in the ſpring before the plants put out new ſhoots. The ſecond ſpring the roots may be tranſplanted where they are to remain; the roots will then be ſtrong enough to flower in ſummer, and will laſt ſeveral years, eſpecially if they are covered with tan to keep out the froſt in winter, but they ſhould not be afterward removed; for when the roots are large, they will not bear tranſplanting.

The eleventh ſort is a native of the ſame countries, and is much like the former, but differs in having upright ſtalks, and the leaves growing alternate. The roots of this grow to a large ſize, ſo will not bear tranſplanting. It is propagated by ſeeds, which ſhould be treated in the manner directed for the former. Theſe flower the latter end of July, and in Auguſt. Neither of theſe plants will live long in pots, for which reaſon I have recommended their being planted in the full ground; but they ſhould have a warm ſituation.

The twelfth, thirteenth, and fourteenth ſorts grow naturally at the Cape of Good Hope. Theſe riſe with upright ſhrubby ſtems to the height of eight or ten feet, and divide into many branches. The flowers of all the ſorts are white, and grow looſely on the umbel; theſe are frequently ſucceeded by ſhort, thick, ſwelling pods, ending in a point, which are thick ſet with hairs, and are filled with compreſſed ſeeds, crowned with a ſoft down.

The thirteenth ſort differs from the twelfth, in having much broader loaves, which are of a darker green; the umbels of flowers are ſmaller, grow upon ſhorter foot-ſtalks, and the ſingle flowers are larger.

The fourteenth ſort doth not riſe ſo high as either of the former, and the branches grow at a much greater diſtance; the leaves are ſhorter, and are covered on both ſides with ſhort hairs.

Theſe are propagated by ſeeds, which may be ſown in April on a bed of light earth in the open air; and when the plants are three or four inches high, they ſhould be each planted in a ſmall pot filled with light earth, and ſhaded until they have taken new root, then they may be placed with other exotic plants in a ſheltered ſituation; in October they muſt be removed into the green-houſe, and during the winter ſhould have but little water, for as they abound with a milky juice, much wet will rot them.

Theſe three ſorts may alſo be propagated by cuttings, which, if planted in July or Auguſt, in a ſhady border, will ſoon take root, and may then be taken up and planted in pots, and managed as the ſeedling plants. The thirteenth ſort has lived in the open air in mild winters in the Chelſea garden, but in cold winters they are conſtantly deſtroyed.

The fourteenth ſort grows with an upright ſhrubby ſtalk to the height of ſix or ſeven feet, dividing toward the top into three or four branches, with ſtiff roundiſh leaves, which cloſely embrace them. Toward the upper part the flowers are produced on their ſides, growing in ſhort compact umbels. They are of an herbaceous colour, and make but little appearance; they come out chiefly in autumn and winter. This requires the ſame culture as the former ſorts.

The fifteenth ſort grows naturally in the warm parts of America. This riſes with ſingle ſtalks near two feet high, which are garniſhed with ſmooth ſpear-ſhaped leaves, ending in a point; toward the top of the ſtalk the umbels of flowers are produced from the wings of the leaves, which are white, and ſtand erect, and are ſucceeded by oblong pointed pods, filled with compreſſed ſeeds, crowned with ſoft

soft down. It flowers in June and July, and the seeds ripen in October.

This plant is tender, so must be raised in a hot-bed, and constantly remain in the stove, otherwise the plants will not thrive here.

The sixteenth sort is also a native of the warm parts of America, the roots of which have been sent to England for Ipecacuana. There have been many accounts of the bad effects of the use of these roots, as also of the poisonous quality of the plant, so that the public should be cautioned not to make use of it, and also to be careful not to let the milky juice of the plant mix with any thing which is taken inwardly.

The plant rises five or six feet high, with upright stems, and smooth oblong leaves placed opposite; toward the top of the branches the umbels of flowers come out, which stand erect; the petals of the flowers are of a scarlet colour, and the horny nectariums in the middle are of a bright Saffron colour, which make a pretty appearance; there is commonly a succession of these flowers on the same plant, from June to October. The flowers are succeeded by long taper pods filled with seeds crowned by a soft down, which ripen late in the autumn.

It is propagated by seeds, which must be sown on a hot-bed in the spring, and the plants should be treated in the same manner as is before directed for the former sort; the roots of this sort may be continued three or four years, but after the second year the plants grow naked, and do not produce so many flowers as the young plants.

The seventeenth sort rises with upright stems six or seven feet high, with thick oval leaves placed opposite. The umbels of flowers are produced from the wings of the leaves; the flowers are white, of a star figure, having five points; the pods of this sort are very large, in shape like an ox's testicles, and are filled with flat seeds, lying over each other like tiles on a house.

This plant is tender, so must be preserved constantly in the stove, and treated in the same manner a the two former sorts, and should have very little wet, especially in the winter.

The eighteenth sort approaches near to the fifteenth, from which it differs in the leaves, being broader, and the umbels of flowers terminating the stalks, whereas those of the fifteenth are produced at the wings of the leaves. This is tender, so must be managed as the fifteenth sort. It flowers in August, and the seeds ripen in October.

The nineteenth sort I received from Carthagena; this hath climbing stalks, which fasten themselves to the neighbouring plants, and rise to the height of ten or twelve feet, with spear-shaped hairy leaves, growing opposite, upon very short foot-stalks; the umbels of flowers come out from the wings of the leaves, which are very compact, and the flowers are of a sulphur colour.

This plant is tender, so must be constantly preserved in the stove, and treated in the same way as is directed for the former sorts.

ASCYRUM. Lin. Gen. Plant. 737. St. Petersword.

The CHARACTERS are,

The flower hath four oval petals, the two outer are large and placed opposite, the two inner are small. In the center is situated an oblong germen, attended by a great number of bristly stamina, which are reduced at their base to four bodies. The germen afterward becomes an oblong pointed seed-vessel, filled with small round seeds.

The SPECIES are,

1. ASCYRUM (*Crux Andreæ*) foliis ovatis caule tereti paniculâ dichotomâ. Lin. Sp. Pl. 787. *Bastard St. Johnswort of Maryland, with small yellow flowers called St. Andrew's Cross*

2. ASCYRUM (*Villosum*) foliis hirsutis caule stricto. Lin. Sp. Pl. 788. *Shrubby hairy St. Johnswort of Virginia.*

3. ASCYRUM (*Hypericoides*) foliis ovatis caule compresso. Lin. Sp. Plant. 788. *Upright shrubby Bastard St. Johnswort with a yellow flower.*

The first sort is a low plant, whose stalks seldom rise more than six inches high; these have small oval leaves, placed by pairs; the stalks are slender, and divide into two toward the top. From between the division of the branches the loose spikes of flowers are produced, which are yellow, but very small, so make no appearance; therefore the plant is scarce worthy of a place in gardens, but for the sake of variety. The root is perennial, and the plant may be propagated by laying down its branches; it loves a moist soil, and a shady situation.

The second sort grows about three feet high, with upright stalks and hairy oblong leaves; the flowers are produced at the ends of the stalks, which are of the shape and colour with common St. Johnswort, but have only four leaves. This hath a perennial root, but the stalks decay every autumn. It may be propagated by parting the roots in autumn, when the stalks decay, and should be planted in a loamy soil.

The third sort grows naturally in South Carolina. This plant rises a foot and a half high with flat stalks, which are garnished with oval smooth leaves growing opposite; the stalks are terminated by three or four yellow flowers, growing close together, which are larger than those of the common St. Johnswort, and the petals of the flowers are hollow. It may be propagated by cuttings, made of the young shoots in May, which, if planted in pots, and plunged into a very moderate hot-bed, will take root in five or six weeks, when they may be transplanted into a warm border, where they will endure the cold of our ordinary winters.

These plants have little beauty, so are seldom cultivated but in botanic gardens for the sake of variety.

ASCYRUM BALEARICUM.
ASCYRUM MAGNO FLORE. } See HYPERICUM.
ASCYRUM VULGARE.

ASH-TREE. See FRAXINUS.

ASPALATHUS. Lin. Gen. Pl. 767. African Broom.

The CHARACTERS are,

The flower is of the butterfly kind. The standard is hairy, compressed, and blunt-pointed; the wings are blunt, moon-shaped, and spread open, and are shorter than the standard; the keel is bifid, and of the same length as the wings. In the bottom is situated an oval germen, which afterward becomes an oval oblong pod, inclosing one or two kidney-shaped seeds.

The SPECIES are,

1. ASPALATHUS (*Chenopoda*) foliis confertis subulatis mucronatis hispidis floribus capitatis. Lin. Sp. Pl. 711. *Yellow African Broom, with hairy flowers collected in woolly heads, and prickly Asparagus leaves which are hairy.*

2. ASPALATHUS (*Indica*) foliis quinatis sessilibus. Lin. Sp. Pl. 712. *Indian shrubby Trefoil with single red flowers, an oblong foot-stalk, and a small pod.*

3. ASPALATHUS (*Argentea*) foliis trinis linearibus sericeis stipulis simplicibus mucronatis floribus sparsis tomentosis. Lin. Sp. Pl. 713. *Narrow-leaved, African, silvery Cytisus with a silky down, and flowers in a spike like a hare's foot.*

These plants grow naturally about the Cape of Good Hope. The first is a low shrub growing about three feet high, with slender branches, having many trifoliate leaves growing in clusters; at the ends of the branches the flowers come out, which are yellow, collected in woolly heads; these are rarely succeeded by pods in England. It is propagated by seeds, which must be obtained from the country where the plants grow naturally; it ſhould be ſown in pots, filled with light earth as ſoon as they arrive: if this happens in the autumn, the pots ſhould be plunged into an old tan-bed whoſe heat is ſpent, where they may remain till ſpring, when they ſhould be removed into a moderate hot-bed, which will bring up the plants. But when the ſeeds arrive in the ſpring, the pots in which the ſeeds are ſown ſhould be then plunged into a moderate hot-bed. Thoſe ſeeds which are ſown in the ſpring, ſeldom grow the ſame year; therefore in the autumn the pots ſhould be put into an old tan-bed, as was directed for thoſe ſown in autumn, and afterward put on a hot-bed the following ſpring. When the plants come up, and are ſtrong enough to remove, they ſhould be each planted into a ſeparate ſmall pot filled with light earth, and plunged into a moderate hot-bed, to encourage their rooting again; and ſo ſoon as they are eſtabliſhed in the pots, they ſhould by degrees be inured to the open air, into which they ſhould be removed in ſummer, placing them in a ſheltered ſituation, where they may remain till autumn, when they muſt be carried into the green-houſe, and in winter ſhould have but little water.

The ſecond ſort grows about five feet high, with ſlender branches, which are garniſhed with leaves having five lobes; the flowers come out ſingly upon long foot-ſtalks, which are of a pale red colour. This is propagated as the former, and requires the ſame treatment.

The third ſort riſes about four feet high, with a ſhrubby ſtalk dividing into ſlender branches, with ſilky white leaves, coming out by threes; the flowers are yellow, downy, and grow thinly on the branches. This is propagated as the two former, and muſt be treated in the ſame way as is directed for the firſt ſort. It flowers late in the ſummer.

ASPARAGUS. Aſparagus, Sparagus, or Sperage; corruptly called Sparrowgraſs

The CHARACTERS are,

The flower is naked, having no empalement, and is of the bell-ſhaped kind, ſpread open and reflexed at the top. Theſe are male and hermaphrodite, ſometimes in different plants, and at other times on the ſame ſtalks. The hermaphrodite flowers have a germen, which afterward becomes a round berry having three cells; in each of them is lodged one or two ſeeds. The male flowers have ſix ſtamina, but no germen or ſtyle, nor are ſucceeded by any berries.

The SPECIES are,

1. ASPARAGUS (*Hortenſis*) caule herbaceo erecto, foliis ſetaceis, ſtipulis paribus. Flor. Suec. 272. *Garden Aſparagus.*

2. ASPARAGUS (*Maritimus*) caule inermi herbaceo foliis teretibus longioribus faſciculatis. *Maritime Aſparagus with a thicker leaf.*

3. ASPARAGUS (*Acutifolius*) foliis aciformibus pungentibus caule fruticoſo inermi. Sauv. Monſ. 45. *Aſparagus with ſharp-pointed leaves.*

4. ASPARAGUS (*Albus*) aculeis ſolitariis ramis flexuoſis foliis brevioribus faſciculatis. *Prickly Aſparagus with horrid ſpines.*

5. ASPARAGUS (*Retrofractus*) aculeis ſolitariis ramis reflexis retrofractiſque, foliis faſciculatis. Lin. Sp. Pl. 313. *Narrow-leaved African Aſparagus with ſlender twigs, and many leaves growing from a point like thoſe of the Larch-tree, and ſpread in form of a ſtar.*

6. ASPARAGUS (*Aphyllus*) aphyllus ſpinis faſciculatis inæqualibus divergentibus. Hort. Cliff. 122. *Another prickly Aſparagus, with three or four ſpines riſing from the ſame point.*

7. ASPARAGUS (*Declinatus*) caule inermi ramis declinatis foliis ſetaceis. Prod. Leyd. 29. *Aſparagus with a ſmooth ſtalk, declining branches, and briſtly leaves.*

8. ASPARAGUS (*Aſiaticus*) aculeis ſolitariis caule erecto foliis faſciculatis, ramis filiformibus. Lin. Sp. Plant. 313. *Aſparagus with ſingle ſpines, an upright ſtalk, leaves growing in cluſters, and very ſlender branches.*

9. ASPARAGUS (*Capenſis*) ſpinis lateralibus terminalibuſque, ramis aggregatis foliis faſciculatis. Lin. Sp. Pl. 314. *Aſparagus with ſpines growing on the ſides and ends of the branches, which are in bunches, and leaves coming out in cluſters.*

10. ASPARAGUS (*Sarmentoſus*) foliis ſolitariis lineari lanceolatis caule flexuoſo aculeis recurvis. Flor. Zeyl. 124. *The great prickly Aſparagus of Ceylon, with buſhy ſtalks.*

The firſt ſort is the common Aſparagus, which is cultivated for the uſe of the table, and may have probably been brought by culture to the perfection it now is, from the wild ſort which grows naturally in the fens of Lincolnſhire, where the ſhoots are no larger than ſtraws; but if ſo, it muſt have been from very long culture and good management; for a friend of mine, who procured ſome ſeeds of the wild ſort, which he cultivated with great care in very rich ground, yet could not bring the roots to produce ſhoots more than half the ſize of the garden ſort, which grew on the ſame ground; but he always found the wild ſort came up a week or ten days earlier in the ſpring, and the ſhoots were exceeding ſweet.

This Aſparagus is propagated by ſeeds, in the procuring of which there ſhould be particular care to get it from a perſon of ſkill, who may be depended upon for his choice of the ſhoots, and integrity in ſupplying with the beſt ſeeds. But where a perſon is in poſſeſſion of ſome good beds of Aſparagus, it is much the beſt way to ſave it himſelf; in order to which, a ſufficient number of the faireſt buds ſhould be marked early in the ſpring, and permitted to run up for ſeeds, becauſe thoſe which run up after the ſeaſon for cutting the Aſparagus is over, are generally ſo backward, as not to ripen the ſeeds unleſs the ſummer is warm, and the autumn very favourable. In the choice of the buds to be left for ſeeds, there muſt great regard be had to their ſize and roundneſs, never leaving any that are inclinable to be flat, or that ſoon grow open headed, always chuſing the roundeſt, and ſuch as have the cloſeſt tops. But as ſeveral of theſe produce only male flowers which are barren, ſo a greater number of buds ſhould be left than might be neceſſary, if there could be a certainty of their being all fruitful, but this never happens. When the buds are left, it will be proper to thruſt a ſtake down by each, but there muſt be care had in the doing of this, not to injure the crown of the root. Theſe ſtakes will not only ſerve as marks to diſtinguiſh them from the others when they are all run up, but alſo to faſten the ſhoots to when they are advanced in height, and put out lateral branches, to prevent their being broken by winds, which frequently happens where this is not obſerved, before the other ſhoots are permitted to run up, after which there is little danger of it, becauſe they will then be ſcreened by the other ſtalks. Toward the end of September the berries will be fully ripe, when the ſtalks ſhould be cut off, and the berries ſtripped in a tub, in which they may remain three weeks or a month to ſweat, by which means the outer huſks will be rotten; then fill the tub with water, and with your hands break all the huſks, by ſqueezing them between your hands. Theſe huſks will all ſwim upon the water, but the ſeeds will ſink to the bottom; ſo that by pouring off the water gently, the huſks will be carried along with it; and by putting freſh water two or three times, and ſtirring your ſeed about, you will make it entirely clean: then ſpread your ſeed upon a mat or cloth, and expoſe it to the ſun and air in dry weather, until it is perfectly dry, when you may put it into a bag, and hang it up in a dry place till the beginning of February; at which time you muſt prepare a bed of good rich earth made very level, whereon you muſt ſow your ſeeds (but not too thick, which will cauſe your plants to be ſmall;) then tread the bed all over to bury your ſeed in the ground, and rake it over ſmooth.

In the following ſummer keep it diligently cleared from weeds, which will greatly add to the ſtrength of your plants; and toward the latter end of October, when the haulm is quite withered, you may ſpread a little rotten dung over the ſurface of the ground about an inch thick, which will preſerve the young buds from being hurt with the froſts, &c.

The ſpring following your plants will be fit to plant out for good (for I would never chuſe plants of more than one year's growth, having very often experienced them to take much better than older, and to produce finer roots:) you muſt therefore prepare your ground by trenching it well, burying therein a good quantity of rotten dung at the bottom of each trench, that it may lie at leaſt ſix inches below the ſurface of the ground; then level your whole plot very exactly, taking out all large ſtones: but this ſhould not be done long before you intend to plant your Aſparagus, in which you muſt be governed according to the nature of your ſoil or the ſeaſon; for if your ſoil is dry and the ſeaſon forward, you may plant toward the end of March; but in a very wet ſoil it is better to wait till the middle of April, which is about the ſeaſon that the plants are beginning to ſhoot. I know many people have adviſed the planting of Aſparagus at Michaelmas, but this I have experienced to be very wrong, for in two different years I was obliged to tranſplant large quantities at that ſeaſon; but I had better have thrown away the plants, for upon examination in the ſpring I found moſt of the roots were grown mouldy and decaying, and I am ſure not one in five of them ſucceeded, and thoſe which did were ſo weak as not to be worth their ſtanding.

The ſeaſon being now come for planting, you muſt with a narrow pronged dung-fork carefully fork up your roots, ſhaking them out of the earth, and ſeparating them from each other, obſerving to lay their heads even for the more convenient planting them, which muſt be performed in this manner:

Your plot of ground being levelled, you muſt begin at one ſide thereof, ranging a line very tight croſs the piece; by which you muſt throw out a trench exactly ſtrait, and about ſix inchees deep, ſo as not to turn up the dung, into which you muſt lay your roots, ſpreading them with your fingers, and placing them upright againſt the back of the trench, that the buds may ſtand forward, and be about two inches below the ſurface of the ground, and at twelve inches diſtance from each other; then with a rake draw the earth into the trench again, laying it very level, which will preſerve the roots in their right poſition; then remove your line a foot farther back, and make another trench in the like manner, laying therein your plants as before directed, and continuing the ſame diſtance row from row, only obſerving between every four rows to leave a diſtance of two feet and a half for an alley to go between the beds to cut the Aſparagus, &c.

Your plot of ground being finiſhed and levelled, you may ſow thereon a ſmall crop of Onions, which will not hurt your Aſparagus, and tread in your ſeeds, raking your ground level.

There are ſome perſons who plant the ſeeds of Aſparagus in the place where the roots are to remain, which is a very good method, if it is performed with care. The way is this: After the ground has been well trenched and dunged, they lay it level, and draw a line croſs the ground (in the ſame manner as is practiſed in planting of the young plants;) then with a dibble make holes at a foot diſtance, into each of which you muſt drop two ſeeds, for fear one ſhould miſcarry; theſe holes ſhould not be more than half an inch deep; then cover the ſeeds, by ſtriking the earth in upon it, and go on removing the line a foot back for another row; and after four rows are finiſhed, leave a ſpace for an alley between the beds, if it is deſigned to ſtand for the natural ſeaſon of cutting; but if it is to be taken up for hot-beds, there may be ſix rows planted in each bed, and the diſtance in the rows need not be more than nine inches. This ſhould be performed by the middle of February, becauſe the ſeeds lie long in the ground; but if Onions are intended to be ſown upon the ground, that may be performed a fortnight or three weeks after, provided the ground is not ſtirred ſo deep as to diſturb the Aſparagus ſeeds, in raking the Onion ſeed into the ground.

As the roots of Aſparagus always ſend forth many long fibres which run deep into the ground, ſo when the ſeeds are ſown where they are to remain, theſe roots will not be broken or injured, as thoſe muſt be which are tranſplanted; therefore they will ſhoot deeper into the ground, and make much greater progreſs, and the fibres will puſh out on every ſide, which will cauſe the crown of the root to be in the center, whereas in tranſplanting the roots are laid flat againſt the ſide of the trench.

When your Aſparagus is come up, and the Onions have raiſed their ſeed leaves upright (which will be in a month or ſix weeks after ſowing) you muſt with a ſmall hoe cut up all the weeds, and thin your crop of Onions where they may have come up in bunches: but this muſt be done carefully, and in dry weather, that the weeds may die as faſt as they are cut up, being careful not to injure the young ſhoots of Aſparagus, as alſo to cut up the Onions which grow near the ſhoots. This work muſt be repeated about three times, which, if well done, and the ſeaſon not too wet, will keep the ground clear from weeds until the Onions are fit to be pulled up, which is commonly in Auguſt, and is known when their greens fall down and begin to wither. When you have drawn off your Onions, you muſt clean your ground well from weeds, which will keep it clean till you earth the beds, which muſt be done in October, when the haulm begins to decay; for if you cut off the haulm while green, the roots will ſhoot freſh again, which will greatly weaken them. This young haulm ſhould be cut off with a knife, leaving the ſtems two or three inches above ground, which will be a guide for you to diſtinguiſh the beds from the alleys; then with a hoe clear off the weeds into the alleys, and dig up the alleys, burying the weeds in the bottom, and throw the earth upon the beds, ſo that the beds may be about five inches above the level of the alleys: then you may plant a row of Coleworts in the middle of the alleys, but do not ſow or plant any thing upon the beds, which would greatly weaken your roots; nor would I ever adviſe the planting of Beans in the alleys (as is the practice of many) for it greatly damages the two outſide rows of Aſparagus. In this manner it muſt remain till ſpring, when you muſt hoe over the beds to deſtroy all young weeds; then rake them ſmooth, and obſerve all the ſucceeding ſummer to keep them clear from weeds, and in October dig up the alleys again, as was before directed, earthing the beds, &c.

The ſecond ſpring after planting you may begin to cut ſome of your Aſparagus, though it will be much better to ſtay until the third; therefore now you muſt fork up your beds with a flat pronged fork made on purpoſe, which is commonly called an Aſparagus fork: this muſt be done before the buds ſhoot in the ſpring, and with care, leſt you fork too deep, and bruiſe the head of the root; then rake the beds over ſmooth, juſt before the buds appear above ground, which will deſtroy all young weeds, and keep your beds clean much longer than if left unraked, or done ſo ſoon as forked; and when your buds appear about four or five inches above ground, you may then cut them, but it ſhould be done ſparingly, only taking the large buds, and ſuffering the ſmall to run up to ſtrengthen the roots; for the more you cut, the greater will be the increaſe of buds, but they will be ſmaller, and the roots ſooner decay. When you cut a bud, you muſt open the ground with your knife (which ſhould be very narrow and long in the blade, and filed with teeth like a ſaw) to ſee whether any more young buds are coming up cloſe by it, which might be either broken or bruiſed in cutting the other, then with your knife ſaw it off about three inches under ground. This may appear a very troubleſome affair to people unacquainted with the practical part, but thoſe who are employed in cutting Aſparagus, will perform a great deal of this work in a ſhort time; but care in doing it, is abſolutely neceſſary to be obſerved by all who cut Aſparagus.

The manner of dreſſing your Aſparagus beds is every year the ſame as directed for the ſecond, viz. clearing them from weeds, digging the alleys in October, and forking the beds toward the end of March, &c. only obſerve every other year to lay ſome rotten dung (from a Melon or Cucumber-bed) all over your beds, burying ſome in the alleys alſo, at the time for digging them up. This will preſerve the ground in heart to maintain your roots in vigour, and by this management a plot of good Aſparagus may be continued for ten or twelve years in cutting, and will produce good buds, eſpecially if it is not cut too long each ſeaſon; for when it is not left to run up pretty early in June, the roots will be greatly weakened, ſo the buds will be ſmaller: therefore, in thoſe families where Aſparagus is required late in the ſeaſon, a few beds ſhould be ſet apart for that purpoſe, which will be much better than to injure the whole plantation, by cutting it too long.

I cannot help taking notice of a common error that has long prevailed with moſt people, which is, that of not dunging the ground for Aſparagus, believing that the dung communicates a ſtrong rank taſte to the Aſparagus; which is a great miſtake, for the ſweeteſt Aſparagus is that which grows upon the richeſt ground, and poor ground occaſions that rank taſte ſo often complained of, the ſweetneſs of Aſparagus being occaſioned by the quickneſs of its growth, which is always proportionable to the goodneſs of the ground, and the warmth of the ſeaſons: but in order to prove this, I planted two beds of Aſparagus upon ground which had dung laid a foot thick, and theſe beds were every year dunged extremely thick, and the Aſparagus produced from theſe beds was much ſweeter than any I could procure, though they were boiled together in the ſame water.

The quantity of ground neceſſary to be planted with Aſparagus, to ſupply a ſmall family, ſhould be at leaſt five or ſix rods, leſs than that will not do; for if you cannot cut one hundred at a time, it will ſcarcely be worth while; for you muſt be obliged to keep it after it is cut two or three days, to furniſh enough for one meſs; but for a larger family twelve rods of ground ſhould be planted, which, if it is a good crop, will furniſh two or three hundred each day in the height of the ſeaſon.

But as there are ſeveral people who delight in having early Aſparagus, which is become a very great trade in the kitchen-gardens near London, I ſhall give proper directions for the obtaining it any time in winter.

You muſt firſt be provided with a quantity of good roots (either of your own raiſing, or purchaſed from ſuch gardeners as plant for ſale) that have been two or three years planted out from the ſeed-bed; and having fixed upon the time you would willingly have your Aſparagus fit to cut, about ſix or ſeven weeks before, you ſhould prepare a quantity of new ſtable horſe-dung, which ſhould be thrown in a heap for ten or twelve days to ferment, mixing ſome ſea-coal aſhes with it; it ſhould be turned over to mix it well, then it will be fit for uſe. Then dig out a trench in the ground where you intend to make the bed, the width of the frames that are deſigned to cover it, and the length in proportion to the quantity you intend to have (which, if deſigned only to ſupply a ſmall family, three or four lights at a time will be ſufficient:) then lay down your dung into the trench, working it very regularly, and beat it down very tight with a fork, laying it at leaſt three feet in thickneſs or more, when the beds are made in December; then put your earth thereon about ſix inches thick, breaking the clods and laying it level; and at one end begin laying your roots againſt a little ridge of earth, raiſed about four inches high: your roots muſt be laid as cloſe as poſſible one to the other in rows, with their buds ſtanding upright; and between every row lay a ſmall quantity of fine mould, obſerving to keep the crown of the roots exactly level. When you have finiſhed laying your bed with roots, you muſt lay ſome ſtiff earth up to the roots on the outſides of the bed, which are bare, to keep them from drying, and thruſt two or three ſharp-pointed ſticks, about two feet long, down between the roots, in the middle of the bed, at a diſtance from each other. The uſe of theſe ſticks is to let you know what temper of heat your bed is in, which you may find by drawing up the ſticks, and feeling the lower part; and if, after the bed has been made a week, you find it doth not heat, you may lay a little ſtraw or litter round the ſides, or upon the top, which will greatly help it; and if you find it very hot, ſo as to endanger ſcorching of the roots, it will be adviſeable to let it remain wholly uncovered, and to thruſt a large ſtick into the dung, on each ſide of the bed, into two or three places, to make holes for the great ſteam of the bed to paſs off, which in a ſhort time will reduce the bed to a moderate heat.

After your bed has been made a fortnight, you muſt cover the crowns of the roots with fine earth, about two inches thick; and when the buds appear above ground through that earth, you muſt again lay on more earth, about three inches thick, ſo that in the whole it may be five inches above the crowns of the roots, which will be ſufficient.

Then you muſt make a band of ſtraw (or long litter) about four inches thick, which you muſt faſten round the ſides of the bed, that the upper part may be level with the ſurface of the ground: this muſt be faſtened with ſtrait ſticks about two feet long, ſharpened at the points, to run into the bed; and upon this band you muſt ſet your frames, and put your glaſſes thereon; but if, after your bed hath been made three weeks you find the heat decline, you muſt lay a good lining of freſh hot dung round the ſides of the bed, which will add a freſh heat thereto; and in bad weather, as alſo every night, keep the glaſſes covered with mats and ſtraw; but in the day-time let it be all taken off, eſpecially whenever the ſun appears, which, ſhining through the glaſſes, will give a good colour to the Aſparagus.

A bed thus made, if it works kindly, will begin to produce buds for cutting in about five weeks after it is made, and will hold about three weeks in cutting, which, if rightly planted with good roots, will produce in that time about three hundred buds in each light; ſo that, if you would continue your Aſparagus until the ſeaſon of the natural being produced, you muſt make a freſh bed every three weeks, until the beginning of March, from the ſeaſon of your firſt bed being made; for if your laſt bed is made about a week in March, it will continue till the ſeaſon of natural Aſparagus, and the laſt beds will come a fortnight ſooner to cut than thoſe made about Chriſtmas; and the buds will be larger, and better coloured, as they will then enjoy a greater ſhare of the ſun.

If you intend to follow this method of forcing early Aſparagus, you muſt keep planting every year a quantity, which you ſhall judge neceſſary (unleſs you intend to buy the roots from ſome other garden;) the quantity of roots neceſſary to plant one light, is commonly known by the meaſure of the ground where they grow; for in a good crop, where

where few roots are miſſing, one rod of ground will furniſh enough for a light; but this calculation is made from the ground planted with roots, which are deſigned to be taken up after two or three years growth for forcing, in which there are ſix rows in a bed at but ten inches diſtance, and the plants eight or nine inches aſunder in the rows; but where there is a greater ſpace between the rows, and fewer rows in a bed, then there muſt be a greater quantity of ground allotted for each light. Moſt of the kitchen-gardeners about London take up their Aſparagus roots after two years growth from planting; but where the land is not very good, it will be better to let it have three years growth, for if the roots are weak the buds of Aſparagus will be very ſmall, ſo not worth the trouble of forcing. The beſt ground for planting Aſparagus, to have large roots for hot-beds, is a low, moiſt, rich ſoil; but for thoſe that are to remain for a natural produce, a middling ſoil, neither too wet nor too dry; but a freſh ſandy loam, when well dunged, is preferable to any other.

The ſecond ſort is mentioned to grow naturally in Wales, and alſo near Briſtol, but this I have great doubts about; for thoſe who have mentioned it, ſay it does not differ from the garden kind, which is only altered by culture. But I have lately received ſpecimens of this which were gathered near Montpelier, by which I am convinced that it is a different ſpecies from that which grows in Wales, for the leaves of the wild maritime kind are taper and thick, and are thinly placed on their branches, nor do the ſtalks branch out ſo much.

This ſort is propagated by ſeeds in the ſame manner as the garden kind, but muſt have a warmer ſituation, and the roots ſhould be well covered in winter to prevent the froſt from penetrating of the ground, which will deſtroy it.

The third ſort hath white, crooked, ſhrubby ſtalks, which riſe ſix or eight feet high, but have no ſpines on them; the leaves come out in cluſters from the ſame point like thoſe of the Larch-tree; theſe are very ſhort, and end in ſharp prickles, ſo that they are troubleſome to handle. This ſort grows naturally in the ſouth of France, Spain, and Portugal: it is propagated by ſeeds as the former ſorts, but is too tender to live abroad in England, ſo the roots ſhould be planted in pots, and ſheltered in winter.

The fourth ſort hath ſhrubby ſtalks three or four feet high, with very white bark, and are armed with thorns which are ſingle, coming out juſt below each tuft of leaves. Theſe ſtalks continue ſeveral years, and put out many branches, which are garniſhed with narrow ſhort leaves. Theſe continue green all the winter, if the plants are ſcreened from ſevere froſt.

It is propagated by ſeeds as the former, which may be procured from the Mediterranean, where it grows naturally; the plants ſhould be kept in pots that they may be ſheltered in winter.

The fifth ſort grows naturally at the Cape of Good Hope. This hath very crooked irregular ſtalks which riſe eight or ten feet high, and are ſhrubby, putting out ſeveral ſide branches which are weak. Theſe have long narrow leaves, coming out in cluſters like thoſe of the Larch-tree; under each of theſe cluſters is placed a ſingle ſharp thorn. The ſtalks continue ſeveral years, and the leaves keep green all the year. It is commonly propagated by parting of the roots, becauſe the plants do not ſeed in this country; the beſt time for this is in April. The roots muſt be planted in pots, and removed into the green-houſe in autumn, for they will not live abroad in England.

The ſixth ſort grows naturally in Spain, Portugal, and Sicily, generally in rocky places. This ſends up many weak irregular ſhoots which have no leaves, but inſtead thereof are armed with ſhort ſtiff thorns, which come out four or five together from the ſame point, and ſpread from each other every way. The flowers are ſmall, of an herbaceous colour; the berries are larger than thoſe of the common ſort, and are black when ripe. This is tender, ſo muſt be treated as the third ſort.

The ſeventh ſort grows naturally at the Cape of Good Hope. This ſends up from the root ſeveral ſlender ſtalks, which put out weak branches, which decline downward; theſe are cloſely garniſhed with briſtly leaves like thoſe of garden Aſparagus, which continue green through the year. It is propagated by parting of the roots as the fifth ſort, and the plants ſhould be treated in the ſame manner.

The eighth ſort grows naturally at the Cape of Good Hope: this ſends up many weak ſhoots growing in cluſters, which are armed with ſharp ſpines both on the ſide and end of the ſhoots; the leaves come out in ſmall cluſters, which continue green all the year. This is propagated as the fifth ſort, and requires the ſame treatment.

The tenth ſort ſends out from the root many weak climbing branches which riſe five or ſix feet high, and are garniſhed with narrow ſpear-ſhaped leaves coming out ſingle; the ſhoots are armed with ſhort crooked ſpines, which are ſo cloſely ſet on that it is difficult to handle the branches. This is propagated by parting the root; but the plants muſt be placed in a moderate ſtove, otherwiſe it will not thrive in this country. It grows naturally in the iſland of Ceylon.

Theſe plants are preſerved in the gardens of the curious, where they add to the variety, being not difficult to manage, where there is conveniency to houſe them in winter. They ſhould have a place among other exotic plants.

ASPARAGUS SCANDENS. See Medeola.

ASPEN-TREE. See Populus.

ASPERUGO. Small Wild Bugloſs.

The Characters are,

The flower is of one leaf, with a ſhort cylindrical tube; it hath five ſhort ſtamina in the center, with four compreſſed germen, which afterward become four oblong ſeeds incloſed in the empalement.

We know but one Species of this genus, viz.

Asperugo (*Procumbens.*) Flor. Lapp. 76. *Small Wild Bugloſs, Great Gooſe-graſs, or German Madwort.*

This is an annual plant, which is found wild in ſome parts of England, as near Newmarket, at Boxley in Suſſex, and in Holy Iſland. It is preſerved in the botanic gardens for variety: it may be eaſily propagated by ſeeds, which ſhould be ſown in autumn; and when the plants come up, they require no other culture but to keep them clear from weeds, and in May they will flower: in June their ſeeds will be perfected.

ASPERULA. Woodroof.

This plant grows wild in ſhady woods in many parts of England, and flowers in April or May, and is ſometimes uſed in medicine; but as this grows wild in England, it is rarely admitted into gardens.

ASPHODELUS. King's Spear.

The Characters are,

The flower has no empalement; it is of one leaf, cut into ſix parts, which ſpread open; at the bottom is inſerted a globular nectarium, having ſix valves; it hath ſix ſtamina, which are inſerted in the valves of the nectarium. Between the nectarium is placed a globular germen, which afterward becomes a fleſhy ſeed-veſſel, having three cells, which are filled with triangular ſeeds.

The Species are,

1. Asphodelus (*Luteus*) caule folioſo, foliis triquetris fiſtuloſis. Hort. Cliff. 127. *Common yellow King's Spear.*

2. As-

2. Asphodelus (*Ramosus*) caule nudo ramoso foliis ensiformibus lævibus. *Male branching King's Spear with white flowers.*

3. Asphodelus (*Non Ramosus*) caule nudo simplici foliis lineari-ensiformibus. *White unbranched King's Spear.*

4. Asphodelus (*Albus*) foliis ensiformibus carinatis scapo ramoso patulo. *King's Spear with sword-like keel-shaped leaves, and a branching spreading stalk.*

5. Asphodelus (*Fistulosus*) caule nudo foliis subulatis fistulosis radice annuâ. *Annual branching Spiderwort, with a small flower and fistular leaves.*

The first sort is the yellow Asphodel, which is directed for use in medicine; this hath roots composed of many thick fleshy fibres, which are yellow, and are joined into a head at the top; from whence arise strong, round, single stalks, near three feet high, garnished on the upper part of the stalk with yellow star-shaped flowers, which appear in June, and the seeds ripen in autumn.

The second sort hath roots composed of many thick fleshy fibres, to each of which is fastened an oblong tuber, as large as small Potatoes; the leaves are long and flexible, having sharp edges; between these come out the stalks, which rise more than three feet high, sending out several side branches; the upper part of these are adorned with many white star-shaped flowers, which grow in long spikes, flowering gradually upward. They come out the beginning of June, and the seeds ripen in autumn.

The third sort hath roots like the second, but the leaves are longer and narrower; the stalks of this are single, never putting out any side branches; the flowers are of a purer white, and grow in longer spikes. This flowers at the same time with the former.

The fourth sort hath roots composed of smaller fibres than the two last, nor are the knobs at bottom half so large; the leaves are long, almost triangular, and hollow like the keel of a boat; the stalks seldom rise above two feet high, and divide into several spreading branches; these are terminated by loose spikes of white flowers, which are smaller than those of the former.

The fifth sort is an annual plant; the roots of this are composed of many fleshy fibres, which are yellow; the leaves are spread out from the crown of the root, close to the ground, in a large cluster; these are convex on their under side, but plain above; the flower-stalks rise immediately from the root, and grow about two feet high, dividing into three or four branches upward, which are adorned with white starry flowers, with purple lines on the outside. These flower in July and August, and their seeds ripen in October.

The yellow sort multiplies very fast by roots, and will soon overspread a large border if suffered to remain unremoved, or the side roots are not taken off: but the other sorts are not so productive of shoots from their sides, and are much better kept within bounds.

These sorts of Asphodel are very pretty ornaments for a flower-garden, and require very little trouble to cultivate them, so are more acceptable. They may be all propagated by seeds, which should be sown soon after they are ripe, on a border of light fresh earth: in the spring the plants will appear, and will have strength enough to be transplanted by the Michaelmas following, when they should be planted in the flower-nursery at about six inches distance every way, observing to plant them so low, as that the top of the roots may be three or four inches under the surface of the bed; and some old tan or dung should be spread over the surface of the ground, to keep out the frost: in this bed they may remain one year; when the roots have acquired strength enough to produce flowers the following year, they should, at Michaelmas, when their leaves are decayed, be carefully taken up, and transplanted into the flower garden in the middle of the borders, amongst other hardy kind of flowers, where being properly intermixed, they will make an agreeable variety, and continue a long time in flower.

The fifth sort is annual, so is only propagated by seeds; these should be sown in the autumn, in the places where they are to remain for good. If the seeds of this plant are permitted to scatter, the plants will come up without care.

ASPLENIUM, or Ceterach.

This plant is nearly allied to the Fern, and grows upon old, moist, shady walls in divers parts of England, but is rarely cultivated in gardens.

ASTER. Starwort.

The Characters are,

It hath a compound flower, composed of several female and hermaphrodite florets, which are included in one common scaly empalement; the rays of the flower are composed of female florets; the hermaphrodite florets form the disk or middle; these are funnel-shaped, and have each five short slender stamina; in the bottom is placed a germen, which afterward becomes an oblong seed crowned with down.

The Species are,

1. Aster (*Alpinus*) foliis lanceolatis hirtis, radicalibus obtusis, caule simplicissimo unifloro. Lin. Sp. Plant. 872. *Blue Mountain Starwort, with a large flower, and oblong leaves.*

2. Aster (*Amellus*) foliis lanceolatis obtusis scabris trinervis integris, pedunculis nudiusculis corymbosis squamis calycinis obtusis. Lin. Sp. Plant. 873. *Common Attic Starwort, vulgarly called Italian Starwort.*

3. Aster (*Tripolium*) foliis lanceolatis integerrimis carnosis glabris ramis inæquatis, floribus corymbosis. Lin. Sp. Plant. 872. *Sea Starwort, called Tripolium.*

4. Aster (*Linifolius*) foliis linearibus acutis integerrimis, caule corymboso ramosissimo. Hort. Cliff. 408. *Starwort with a flower of Tripolium, and a very narrow thin leaf.*

5. Aster (*Nova Angliæ*) foliis lanceolatis alternis integerrimis semiamplexicaulibus floribus terminalibus. Hort. Cliff. 408. *Tallest, hairy, New England Starwort, with large, purple, Violet flowers.*

6. Aster (*Undulatus*) foliis cordato-lanceolatis undulatis floribus racemosis adscendentibus. Hort. Cliff. 408. *Purple New England Starwort with the appearance of Golden-rod, and waved leaves.*

7. Aster (*Puniceus*) foliis semiamplexicaulibus lanceolatis serratis scabris, pedunculis alternis subunifloris calycibus discum superantibus. Hort. Cliff. 408. *Broad-leaved American Starwort with purplish stalks.*

8. Aster (*Miser*) floribus ovatis disco radiis longiore. Lin. Sp. Plant. 877. *Starwort like Heath, and the disk of the flower like Wild Melilot.*

9. Aster (*Novi Belgii*) foliis lanceolatis subserratis sessilibus caule paniculato ramulis unifloris solitariis calycibus squarrosis. Hort. Cliff. 408. *Broad-leaved umbellated Starwort of New Holland, with pale Violet flowers.*

10. Aster (*Linarifolius*) foliis lanceolato-linearibus subcarnosis integerrimis planis floribus corymbosis fastigiatis pedunculis foliolosis. Lin. Sp. Plant. 874. *Starwort with the Tripolium flower.*

11. Aster (*Concolor*) caule simplicissimo foliis ovatis sessilibus integerrimis racemo terminali. Flor. Virg 178. *Starwort with single stalks, oval entire leaves growing close to the stalks, which end in a loose spike.*

12. Aster (*Ericoides*) caule paniculato pedunculis racemosis pedicellis foliosis foliolis linearibus integerrimis. Flor. Virg. 100. *Bushy Heath like Starwort.*

13. Aster (*Cordifolia*) foliis cordatis serratis petiolatis, caule paniculato. Hort. Cliff. 408. *Broad-leaved autumnal Starwort.*

14. Aster

14. Aster (*Tenuifolius*) foliis lanceolato-linearibus medio ferratis pedunculis foliofis caule racemofo calycibus erectis. Hort. Cliff. 408. *American Starwort with leaves like Summer Cypress, and bluish white flowers growing in very long spikes.*

15. Aster (*Grandiflorus*) caule corymbofo foliis lanceolatis reflexis, floribus folitariis, calycibus patulis. Flor. Leyd. 168. *Pyramidal Virginia Starwort with rough Hyssop leaves, and leafy scales to the empalement.*

16. Aster (*Scabris*) foliis lanceolatis fcabris integris, caule ramofo pedunculis foliofis, calycibus obtufis. *Another Attic Starwort of the Alps.*

17. Aster (*Glaber*) foliis oblongo lanceolatis acutis ferratis caule ramofo floribus terminalibus calycibus linearibus erectis. *Starwort with smooth jagged leaves, growing scatteringly like those of the Peach-tree, and pale blue flowers.*

18. Aster (*Tradescanti*) foliis oblongis acutis bafi latioribus femiamplexicaulibus, caule ramofo floribus terminalibus plerumque folitariis. *Late blue shrubby Starwort of John Tradescant, commonly called Michaelmas Daisy.*

19. Aster (*Præcox*) caule erecto hirfuto foliis oblongis acutis fcabris acutè dentatis femiamplexicaulibus floribus corymbofis, calycibus hirfutis erectis. *Early Pyrenean Starwort with a large blue flower.*

20. Aster (*Altissima*) caule altiffimo hirfuto, fimpliciffimo foliis oblongis acutis bafi latioribus femiamplexicaulibus, floribus tribus feffilibus terminalibus. *Starwort with a very tall unbranched stalk, oblong pointed leaves, which are broader at the base, and half embrace the stalks, which are terminated by three flowers sitting very close.*

21. Aster (*Ramocissima*) caule ramociffimo patulo, foliis lineari-lanceolatis rigidis, floribus ferriatim pofitis pedunculis foliofis. *Starwort with a very branching spreading stalk, narrow, spear-shaped, stiff leaves, flowers placed one above another, and leafy foot-stalks.*

22. Aster (*Umbellatus*) foliis lanceolatis acutis fcabris, caule fimplici floribus umbellatis terminalibus. *Starwort with rough, pointed, spear-shaped leaves, and a single stalk, terminated by flowers growing in an umbel.*

23. Aster (*Nervosus*) foliis nervofis acutis linearibus lanceolatis, caule fimplici floribus terminalibus quafi umbellatim difpofitis. *Starwort with narrow-pointed nervous leaves, and a single stalk terminated by flowers growing almost in an umbel.*

24. Aster (*Paniculatus*) foliis inferioribus ovatis bafi femiamplexicaulibus, fuperioribus lanceolatis parvis caule paniculato, ramis unifloris pedunculis foliofis. *Starwort with the lower leaves oval, whose base half embraces the stalks, the upper leaves small and spear-shaped, a stalk terminated by a loose spike, with a single flower on each branch, and a leafy foot-stalk.*

25. Aster (*Rigidus*) floribus terminalibus folitariis foliis linearibus alternis. Flor. Virg. 98. *Starwort with single flowers at the ends of the branches, and very narrow leaves placed alternately.*

26. Aster (*Latifolius*) foliis lineari-lanceolatis acutis feffilibus caule paniculato, ramis unifloris pedunculis foliofis linearibus. *Starwort with narrow-pointed spear-shaped leaves growing close to the stalks, which end in loose spikes, and branches ending with a single flower, whose foot-stalks have narrow leaves.*

27. Aster (*Dumosus*) foliis lineari-lanceolatis glabris trinerviis floribus corymbofis terminalibus. *Starwort with smooth, narrow, spear-shaped leaves with three veins, and flowers in a corymbus which terminate the stalks.*

28. Aster (*Annuus*) foliis linearibus integerrimis caule paniculato. Hort. Cliff. 408. *New England Starwort with Toad-flax leaves, and a Chamomile flower.*

29. Aster (*Fruticosus*) foliis linearibus fafciculatis, punctatis pedunculis unifloris nudis, caule fruticofo rugofo. Hort. Cliff. 409. *Shrubby Starwort, with narrow-pointed leaves set in clusters, and naked foot-stalks with one flower.*

30. Aster (*Chinensis*) foliis ovatis angulatis dentatis, petiolatis calycibus terminalibus patentibus foliofis. Hort. Cliff. 407. *Annual Starwort with a goose-foot leaf, and a large beautiful flower, commonly called China Aster, or Queen Marguerette.*

The firft fort grows naturally upon the Alps, where it feldom rifes more than fix inches high, and when tranfplanted into a garden, not above nine or ten. It fends up a fingle ftalk from the root, at the top of which is one large blue flower, fomewhat like thofe of the Italian Starwort. The root is perennial, but muft be planted in a fhady fituation and a moift foil.

The fecond fort is the Italian Starwort, which was fome years paft more common in the gardens than at prefent; for fince the great variety of American Starworts have been introduced into England, this fort hath not been fo much cultivated, though it is by no means inferior to the beft of them, and in fome refpects preferable to moft of them; for it is not fo fubject to creep by the root as many of the American forts do, whereby they often become troublefome in fmall gardens; nor do the ftalks of thefe feldom grow more than two feet high, and are very ftrong, fo are very rarely broken by the wind. They are terminated with large flowers having blue rays, with a yellow difk. It flowers in October, and in mild feafons will often continue till the middle of November, during which time they are very ornamental plants in a garden. This fort is propagated by parting of the roots; the beft time for doing it is foon after they are out of flower, for thofe which are removed in the fpring will not flower fo ftrong the fucceeding year. Thefe roots fhould not be removed oftener than every third year, where they are expected to produce many flowers.

It grows naturally in the vallies of Italy, Sicily, and Narbonne, and is generally fuppofed to be the Amellus mentioned by Virgil in his fourth Georgic, to grow in the paftures; the leaves and ftalks being rough and bitter, the cattle feldom browfe upon it, fo that whenever there are any of thefe roots in the fields, they fend up a thick tuft of ftalks, which, being left after the Grafs is eaten bare, thefe being full of flowers make a fine appearance, and therefore might engage the poet's attention.

The third fort grows naturally in falt marfhes which are flowed by the tides, and is feldom admitted into gardens. It flowers in July and Auguft.

The fourth fort is a native of North America. It fends up many ftrong fhoots from the root every fpring, which rife between four and five feet high, with oblong leaves which half embrace the ftalk with their bafe, with a fingle flower terminating the ftalk, of a blue colour. This flowers in Auguft and September; it is eafily propagated by parting of the roots foon after the flowers are paft, and will thrive in almoft any foil or fituation.

The fifth fort fends up many ftalks from the root, which rife five feet high, with fpear-fhaped leaves which are entire, and half embrace the ftalks, and are terminated by large, purple, Violet flowers growing in a loofe panicle: it flowers in Auguft, and is very hardy, fo may be planted in any foil or fituation, and is propagated by parting the roots.

The fixth fort grows naturally in North America. This hath broad, heart-fhaped, waved leaves at the bottom; the ftalks rife between two and three feet high, upon which the flowers come out in loofe fpikes, which are of a very pale blue colour, inclining to white. This flowers in the fame feafon as the former, and may be propagated in the fame manner.

The

The ſeventh ſort ſends up ſeveral ſtrong ſtalks upward of two feet high, which are of a purple colour, with ſpear-ſhaped ſmooth leaves, whoſe baſe embraces the ſtalks half round; the flowers grow upon ſingle foot ſtalks, forming a corymbus at the top, and are of a pale blue colour; theſe appear the latter end of September. This may be propagated in the ſame way as the former.

The eighth ſort riſes with ſlender ſtalks upward of three feet high, garniſhed with very ſmall leaves; the flowers come out on ſhort foot-ſtalks on every ſide of the branches, which are ſmall, with white rays and a yellow diſk. Theſe appear in November, and often continue part of December. This may be propagated as is before directed.

The ninth ſort riſes near four feet high, having broad leaves at the bottom; the flowers are produced in a looſe kind of umbel at the top of the ſtalks, which are of a pale blue colour, and appear the latter end of Auguſt. This is hardy, and may be propagated as the former.

The tenth ſort grows three feet high; the ſtalks of this divide into a great number of branches, dividing again toward the top into ſeveral ſmaller, which are garniſhed with very narrow leaves; the flowers grow in large cluſters at the top, forming a ſort of corymbus; they are of a pale bluiſh colour, and appear the beginning of Auguſt. This is hardy, and may be propagated by parting of the roots as the former.

The eleventh ſort riſes four feet high, with a ſingle ſtalk, and oval leaves growing cloſe to the ſtalks, terminated by ſlender looſe ſpikes of pale blue flowers, which appear about Michaelmas. This is propagated as the ſorts above-mentioned.

The twelfth ſort ſends up ſlender ſtalks three feet high; theſe are garniſhed with very narrow leaves their whole length, and are terminated by ſingle flowers.

The thirteenth ſort grows about two feet high, having ſlender ſtalks, with oblong, pointed, heart-ſhaped leaves, which are ſharply ſawed on their edges, and are terminated by white flowers growing in looſe panicles. This flowers in September, and may be propagated as the former.

The fourteenth ſort ſends up ſtalks five feet high, with narrow ſpear-ſhaped leaves, and are terminated by ſpikes of ſmall white flowers, which appear the end of October. This ſort ſpreads greatly at the root, ſo is apt to over-run the borders.

The fifteenth ſort hath narrow, oblong, hairy leaves at the bottom; the ſtalks riſe three feet high, with ſmall, narrow, rough leaves which turn backward: the ſtalks are terminated by a ſingle, large, blue flower. This ſort flowers the end of October, and continues moſt part of November, when it makes a fine appearance. It doth not multiply faſt by its roots, but may be propagated in plenty by cuttings made from the young ſhoots in May, which, if planted in a bed of light earth, and ſhaded from the ſun, will take root: it is called by the gardeners Catesby's Starwort.

The ſixteenth ſort ſends up ſeveral ſtalks a foot and a half high, with rough ſpear-ſhaped leaves; theſe are terminated for the moſt part by one large blue flower, ſomewhat like thoſe of the Italian Starwort, but paler, and comes earlier to flower. It is propagated by parting of the root.

The ſeventeenth ſort riſes to the height of five feet, with branching ſtalks, and oblong ſpear-ſhaped leaves ſawed on their edges; the ſtalks are terminated by large pale blue flowers, and are in beauty in October. This is propagated by parting of the roots, as the ſorts before-mentioned.

The eighteenth ſort was brought from Virginia many years ago by Mr. John Tradeſcant, who was a great collector of rarities. It is generally known by the title of Michaelmas Daiſy, from its flowering about old Michaelmas-day. The ſtalks of this ſort are numerous, and riſe about four feet high, with oblong leaves ending in a point, whoſe baſe half embrace the ſtalks. The branches are terminated by pretty large flowers, which are of a very pale bluiſh colour, tending to white. The roots of this multiply very faſt, ſo that it propagates ſo much as often to be troubleſome; it will thrive in any ſituation.

The nineteenth ſort ſends up ſeveral ſtrong hairy ſtalks, which riſe a foot and a half high, having many oblong rough leaves ending in a point, whoſe baſe half embraces the ſtalks, and are terminated by one large blue flower, having a very hairy empalement: it flowers the latter end of July. This ſhould have a moiſt ſoil and a ſhady ſituation. It is propagated by parting of the roots.

The twentieth ſort riſes with ſtrong hairy ſtalks to the height of eight or nine feet, which are upright, unbranched, and garniſhed with oblong hairy leaves ending in a point; their baſe half ſurrounds the ſtalks, which are for the moſt part terminated by three large purple flowers inclining to red, and ſit cloſe to the top of the ſtalk, ſurrounded by a few narrow leaves: this ſort flowers in November. It is propagated by parting of the roots, and delights in a moiſt ſoil.

The twenty-firſt ſort hath ſlender purpliſh ſtalks, which riſe about three feet high, ſending out many ſide branches which ſpread horizontally, and are garniſhed with narrow ſmall ſpear-ſhaped leaves; the flowers are produced in a ſort of looſe ſpike, growing one above another on each ſide the ſtalk. Theſe are ſmall, of a pale purpliſh colour, and appear in November. It is eaſily propagated by parting of the roots.

The twenty-ſecond ſort ſends up ſtiff channelled ſtalks about two feet high, which are garniſhed with rough ſpear-ſhaped leaves ending in a point; the flowers are white, and grow in a ſort of umbel at the top of the ſtalks. It flowers the end of September, and is propagated by parting of the roots.

The twenty-third ſort hath much the appearance of the former, but the leaves are narrower, whiter on their under ſide, and have three longitudinal veins; the flowers are alſo larger and whiter. It grows about the ſame height, and flowers at the ſame time with the former.

The twenty-fourth ſort riſes four feet high; the bottom leaves are oval and half ſurround the ſtalk at their baſe, the upper leaves are ſmall and ſpear-ſhaped; the ſtalks are terminated by one large blue flower with a leafy foot ſtalk: this flowers about the end of October, and is propagated by parting of the roots.

The twenty-fifth ſort ſends up from the root ſeveral ſlender ſtalks near three feet high, with very narrow leaves, and puts out ſide branches, each being terminated by one white flower. This flowers in November, and is eaſily propagated by parting of the roots.

The twenty-ſixth ſort riſes about a foot and a half high, with very narrow ſpear-ſhaped leaves which are ſmooth; the ſtalks are terminated by one pale blue flower, having a leafy foot-ſtalk.

The twenty-ſeventh ſort grows about two feet high, with erect ſtalks, and narrow ſmooth ſpear-ſhaped leaves which come out irregularly in cluſters; the upper part of the ſtalks are garniſhed with very narrow leaves; the flowers are produced in form of a corymbus at the end of the ſtalks, which are of a pale blue colour, and appear in September. This is propagated by parting of the roots.

The twenty-eighth ſort riſes with ſlender ſtalks about three feet high, with very narrow leaves terminated by looſe panicles of flowers, whoſe rays are white and their diſks yellow. This flowers in October, and is propagated by parting of the roots.

The

The twenty-ninth sort rises with a woody stem about three feet high, sending out many side branches which are ligneous, with narrow leaves coming out in clusters from one point, like those of the Larch-tree; the flowers are produced from the side of the branches, upon long slender foot-stalks singly; these are of a pale blue colour, and appear the beginning of March: as this plant never produces seeds in Europe, so it is only propagated by cuttings, which may be performed any time in the summer. When the plants are rooted, they may be placed in the open air till the end of October, when they should be removed into shelter. This sort is at present but in few English gardens.

The thirtieth sort is a native of China, from whence the seeds were sent to France by the missionaries, where the plants were first raised in Europe. The seeds came by the title of La Reine Marguerette, or Queen of Daisies, by which title the French still call it. In 1752 I received seeds of the double flowers both red and blue, and in 1753 the seeds of the double white sort, which have retained their difference from that time without variation; yet as they are generally supposed to be only varieties, I have not inserted them as different species.

As these are annual plants, so they are only propagated by seeds, which must be sown in the spring upon a gentle hot-bed just to bring up the plants; for they should be inured to the open air as soon as possible, to prevent their being drawn up weak: when the plants are big enough to remove, they should be carefully taken up and planted in a bed of rich earth at six inches distance each way. In this bed they may remain a month or five weeks, by which time they will be strong enough to transplant into the borders of the flower-garden, where they are designed to remain for flowering; the plants should be taken up carefully, with large balls of earth to their roots, and the ground dug up and well broken with the spade, where the holes are made to receive the plants: this work should, if possible, be done when there is rain, for then the plants will soon take new root, after which time they will require no other care.

In August these plants will flower, by which time, if the ground is rich in which they are planted, they will be two feet high, and furnished with many side branches, each of which is terminated by a large radiated flower, some white, some red, and others blue. The seeds ripen the beginning of October, which should be gathered when it is perfectly dry; and in order to preserve the kinds with double flowers, there should be great care taken to save those which grow upon the side branches, which are commonly fuller of leaves than the flowers on the main stem.

ASTERISCUS. See BUPHTHALMUM.

ASTEROIDES. Bastard Starwort. See INULA.

ASTRAGALOIDES. See PHACA.

ASTRAGALUS. Wild Liquorice, Liquorice Vetch, or Milk Vetch.

The CHARACTERS are,

It hath a butterfly flower. The standard (or vexillum) is upright, blunt, and reflexed on the sides; the wings are oblong, and shorter than the standard; the keel is the same length with the wings, and bordered. At the bottom of the flower is situated a taper germen, which afterward becomes a pod having two cells, each having a row of kidney-shaped seeds.

The SPECIES are,

1. ASTRAGALUS (*Glycyphuyllos*) caulescens, prostratus leguminibus subtriquetris arcuatis foliis ovalibus pedunculo longioribus. Lin. Sp. Plant. 758. *Common, wild, perennial, trailing Milk Vetch with yellow flowers, sometimes called Wild Liquorice.*

2. ASTRAGALUS (*Hamosus*) caulescens procumbens, leguminibus subulatis recurvatis glabris. Hort. Upsal. 226. *Yellow annual Milk Vetch of Montpelier, with trailing stalks.*

3. ASTRAGALUS (*Alopecuroides*) caulescens, spicis cylindricis subsessilibus, calycibus leguminibusque lanatis. Lin. Sp. Plant. 755. *Taller Foxtail Milk Vetch of the Alps.*

4. ASTRAGALUS (*Cicer*) caulescens prostatus, leguminibus subglobosis inflatis mucronatis pilosis. Hort. Upsal. 226. *Yellow perennial Milk Vetch, with a round double pod resembling a bladder.*

5. ASTRAGALUS (*Epiglottis*) caulescens procumbens, leguminibus capitatis cordatis acutis reflexis complicatis. Lin. Sp. Plant. 759. *Larger Spanish Milk Vetch, with pods like the epiglottis, and a purple flower.*

6. ASTRAGALUS (*Montanus*) subacaulos scapis folio longioribus, floribus laxè spicatis erectis. Prod. Leyd. 392. *Milk Vetch, or Cock's-head, with large Vetch flowers of a purplish blue colour, and Goat's-thorn leaves.*

7. ASTRAGALUS (*Bæticus*) caulescens procumbens, spicis pedunculatis leguminibus prismaticis rectis triquetris apice uncinatis. Hort. Upsal. 225. *Trailing, maritime, annual Milk Vetch, with broad leaves and flowers sitting upon foot-stalks.*

8. ASTRAGALUS (*Arenarius*) subcaulescens procumbens floribus subracemosis erectis foliis tomentosis. Lin. Sp. Pl. 759. *Small, hoary, purple Milk Vetch.*

9. ASTRAGALUS (*Physodes*) acaulos scapis folia æquantibus leguminibus inflatis subglobosis nudis. Lin. Sp. Pl. 760. *Low Milk Vetch with swelling globular pods.*

10. ASTRAGALUS (*Christianus*) caulescens erectus floribus glomeratis subsessilibus ex omnibus axillis foliaceis. Lin. Sp. Plant. 755. *Greatest, hoary, upright, Eastern Milk Vetch, with flowers coming out from the bottom to the top of the stalk.*

11. ASTRAGALUS (*Ægyptiacus*) caulescens scapis folio longioribus floribus laxè spicatis erectis, leguminibus arcuatis. *Egyptian Milk Vetch, with spikes of purple flowers and incurved pods.*

12. ASTRAGALUS (*Sesameus*) caulescens diffusus capitulis subsessilibus lateralibus leguminibus erectis subulatis acumine reflexis. Hort. Cliff. 361. *Annual Milk Vetch with hairy leaves and pods, many of them growing close to the wings of the leaves.*

13. ASTRAGALUS (*Galegiformis*) caulescens strictus glaber, floribus racemosis pendulis, leguminibus triquetris utrinque mucronatis. Lin. Sp. Plant. 1066. *Tallest Eastern Milk Vetch, with a Goat's-rue leaf, and a small yellowish flower.*

14. ASTRAGALUS (*Uralensis*) acaulos scapo erecto foliis longiore leguminibus subulatis inflatis villosis erectis. Hort. Upsal. 226. *Hairy, white, unbranched Milk Vetch, with purple Violet flowers growing in spikes.*

15. ASTRAGALUS (*Carolinianus*) caulescens erectus lævis pedunculis spicatis leguminibus ovato-cylindricis stylo acuminatis. Lin. Sp. Pl. 757. *Taller upright Milk Vetch, with a yellowish green flower.*

16. ASTRAGALUS (*Canadensis*) caulescens diffusus leguminibus subcylindricis mucronatis foliolis subtus subvillosis. Lin. Sp. Plant. 757. *Canada Milk Vetch, with a yellowish green flower.*

17. ASTRAGALUS (*Pilosus*) caulescens erectus pilosus floribus spicatis leguminibus subulatis pilosis. Lin. Sp. Plant. 756. *Upright hairy Milk Vetch, with yellow flowers growing in spikes.*

18. ASTRAGALUS (*Procumbens*) incanus caulibus procumbentibus scapis folio æquantibus floribus glomeratis. *Trailing Milk Vetch with hairy glomerated pods.*

19. ASTRAGALUS (*Incanus*) caulescens incanus, leguminibus subulatis recurvatis incanis. *Hoary Milk Vetch with a crooked pod.*

20. ASTRAGALUS (*Capitatus*) caulefcens capitulis globofis, pedunculis longiffimis, foliolis emarginatis. Hort. Cliff. 360. *A moft hairy eaftern Milk Vetch, with rounder heads and purple flowers.*

21. ASTRAGALUS (*Chinenfis*) caulefcens procumbens, capitulis pedunculatis, leguminibus prifmaticis rectis triquetris apice fubulatis. *Annual trailing Milk Vetch of China.*

22. ASTRAGALUS (*Uncatus*) acaulis ex fcapus, leguminibus fubulatis hamatis folio longioribus, foliolis obcordatis. Lin. Sp. 1072. *Trailing Milk Vetch, with awl-fhaped hooked pods from Aleppo.*

The firft fort grows wild upon chalky ground in many parts of England, fo is not often admitted into gardens. The root of this is perennial, but the ftalks decay every autumn: it creeps at the root, fo that it is too apt to fpread where it is fuffered to grow. It flowers in June, and the feeds ripen in September.

The fecond fort is annual; the branches of this trail upon the ground, which are ftriated; the leaves are compofed of about eight pair of lobes, terminated by an odd one: the foot-ftalk of the flowers arife from the wings of the leaves, which is about three inches long, garnifhed toward the top with a few pale yellow flowers rifing one above another; thefe are fucceeded by oblong pods, which bend in form of a fickle. It flowers in June, and the feeds ripen in September. The feeds of this fhould be fown in April, in the place where they are to remain.

The third fort is a biennial plant. This rifes with an upright hairy ftalk near three feet high, with long winged leaves, each having eighteen or twenty pair of oval lobes terminated by an odd one. The flowers are produced in large cylindrical fpikes from the wings of the leaves, fitting very clofe to the ftalks, which are entirely covered with down, out of which the yellow flowers juft peep; thefe are fucceeded by oval pods fhut up in the woolly empalements. It flowers in June and July, and the feeds ripen in the autumn, foon after which the plants decay. The feeds of this fhould be fown on an open border, where the plants are defigned to remain, in April.

The fourth fort hath a perennial root, which fends out feveral ftriated ftalks near three feet long, which, if not fupported, proftrate themfelves toward the earth, with winged leaves placed alternately, which are compofed of about ten pair of oval fmall lobes, terminated by an odd one. The flowers arife from the wings of the leaves, upon foot-ftalks two inches long, in fmall loofe fpikes which are yellow, and fhaped like the reft of this genus, and are fucceeded by hairy, globular, fwelling pods, ending with a fharp point. It flowers in July, and the feeds ripen in the autumn. It is eafily propagated by feeds, which fhould be fown upon an open border in the fpring, in the place where they are to remain. One or two of thefe plants in a garden by way of variety, may be admitted, but they have little beauty.

The fifth fort is annual. This fends out from the root two or three hairy trailing branches, which are garnifhed with leaves compofed of ten or twelve pair of blunt lobes, terminated by an odd one: the flowers come out from the wings of the leaves upon naked foot-ftalks four or five inches long, and are gathered into a round head; thefe are fhaped like the others, but are pretty large, and of a deep purple colour, which are fucceeded by fhort pods rough on their outfides, and when opened are fhaped like a heart, ending in a fharp point, containing three or four feeds.

The feeds of this fhould be fown on an open border in April, where the plants are to remain, and treated as the other annual forts before-mentioned.

The fixth fort is a perennial plant, feldom rifing with a ftem more than three inches high, with leaves which are compofed of many pairs of narrow lobes, fet very clofe together on the midrib, terminated by an odd one. The flowers grow upon long foot-ftalks, which rife above the leaves; thefe are large and of a purple colour, growing in a loofe fpike and ftand erect, and are fucceeded by oblong crooked pods opening in two cells, filled with fquare feeds. This is propagated by feeds, which fhould be fown, and the plants treated in the fame manner as the fourth fort, but fhould have a fhady fituation and a ftronger foil.

The feventh fort is annual; it hath trailing branches near two feet long, with winged leaves, compofed of about ten pair of blunt lobes fet thinly on the midrib, terminated by an odd one: at the wing of the leaf comes out a foot-ftalk near two inches long, fuftaining four or five yellow flowers at the top, which are fucceeded by triangular brown pods fhaped like a prifm, growing erect, and opens in two cells, filled with greenifh fquare feeds. This may be treated in the fame manner as the fecond.

The eighth fort is a perennial plant, which grows in feveral parts of England, particularly in the north. It is a low plant, feldom rifing more than two or three inches high, with leaves compofed of narrow woolly lobes placed clofe on the midrib; the flowers are pretty large, of a purple colour, growing in loofe fpikes. It flowers in June, and the feeds ripen in Auguft. This fhould have a fhady fituation.

The ninth fort hath a perennial creeping root, with leaves compofed of many pair of oval lobes, terminated by an odd one; the flower-ftalks are as long as the leaves, which fupport a cylindrical fpike of yellow flowers, and are fucceeded by fwollen pods opening into two cells, containing feveral greenifh feeds. This may be propagated as the fourth fort, and muft have a fhady fituation.

The tenth fort fends up ftalks near three feet high, which are large at bottom, and gradually diminifh to the top; the leaves at bottom are very long, and diminifh upward, fo as to form a fort of pyramid; thefe are compofed of many large oval pair of lobes, which are placed thinly on the midrib, and are terminated by an odd one; the flowers come out in clufters from the wings of each leaf. Thefe are large, of a bright yellow colour, and are fucceeded by cylindrical pods opening in two cells, filled with fquare yellow feeds. It flowers in July, and in very favourable feafons will perfect feeds in England. It is propagated by feeds, which fhould be fown, and the plants afterward treated as hath been directed for the fourth fort. The third year from feed the plants will flower, and continue many years in a dry foil.

The eleventh fort is an annual plant, which rifes with upright ftalks a foot and a half high, which are thinly garnifhed with leaves, compofed of about twelve pair of oval lobes, terminated by an odd one; the foot-ftalks of the flowers arife from the wings of the leaves, and are extended beyond them; thefe are terminated by loofe fpikes of yellow flowers, which are fucceeded by fickle-fhaped pods. It flowers in July, and the feeds ripen in autumn. It may be propagated by feeds, in the fame manner as hath been before directed for the annual forts.

The twelfth fort is an annual plant, which fends out feveral weak ftalks without any order, having leaves compofed of ten or twelve pair of lobes, and fometimes terminated by an odd one; at the foot-ftalks of the leaves the flowers come out in fmall clufters fitting clofe to the fides of the ftalks, which are of a copper colour, and are fucceeded by awl-fhaped pointed pods growing erect, reflexed at their points. This is propagated by feeds in the fame manner as the other annual forts before-mentioned.

The thirteenth fort hath a perennial root, which fends out many upright ftalks more than five feet high, which are

are garnished with leaves, composed of about fourteen pair of oval lobes, terminated by an odd one; from the wings of the leaves the foot-stalks of the flowers arise, which are garnished with small yellow flowers growing in loose spikes, and are extended beyond the leaves; these are succeeded by very short triangular pods, ending in a point, which open in two cells, filled with Ash-coloured square seeds. It is propagated by seeds, which, if suffered to fall on the ground, the plant will come up and require no farther culture. The roots of this sort will abide many years.

The fourteenth sort never rises with a stalk, but sends out leaves from the root, which are composed of many blunt lobes, placed by pairs, and terminated by an odd one; the foot-stalks of the flowers arise immediately from the root, and are longer than the leaves, being terminated by spikes of blue flowers, which are succeeded by swelling awl-shaped pods that are erect and hairy, having two cells, which are filled with greenish seeds. It flowers in July, and the seeds ripen in autumn. The root is abiding, and the plant is propagated by seeds as the fourth sort, but should have an open situation.

The fifteenth sort hath a perennial root, but an annual stalk; from the root arise several upright stalks three feet high, which are garnished with leaves composed of eighteen or twenty pair of oval smooth lobes, terminated by an odd one; from the wings of the leaves arise the foot-stalks, which are terminated by spikes of greenish yellow flowers, which are succeeded by oval cylindrical pods, to which adhere the style, which extend beyond the pods in a point. This flowers in August, but unless the season is warm, the plants seldom ripen their seeds in England. It is propagated by seeds, which should be sown upon a moderate hot-bed in the spring, and when the plants come up they must be inured to the open air, into which they should be removed the end of May, and planted in a warm border where they will thrive and flower; and if the winter proves very severe, a little old tan should be laid over the roots, which will effectually preserve them.

The sixteenth sort hath a perennial root, which sends out many irregular stalks about two feet long, with leaves composed of many pair of oval lobes, which are hairy on their under side; from the wings of the leaves come out the foot-stalks, supporting spikes of greenish yellow flowers, which are succeeded by cylindrical pods ending in a point.

The seventeenth sort rises with upright stalks two feet high, which are hairy, and garnished with leaves composed of many pair of oval woolly lobes, terminated by an odd one; from the wings of the leaves arise the foot-stalks, which are terminated by close spikes of yellow flowers, which are succeeded by hairy awl-shaped pods, having two cells filled with brown seeds. It is a perennial plant, and propagated by seeds in the same manner as the fourth sort.

The eighteenth sort is a biennial plant, with many trailing stalks, which are divided into smaller branches, with leaves composed of many pair of narrow lobes, terminated by an odd one; the flowers are collected into heads which terminate the foot-stalks, and are white; the foot-stalks are about the same length as the leaves; the pods are short and triangular, and the whole plant is covered with a silvery down. The seeds of this should be sown upon an open bed of light earth where the plants are to remain, and the plants afterward treated in the manner before directed for the annual sorts.

The nineteenth sort sends up an upright stalk, seldom more than six inches high, with small winged hoary leaves; the foot-stalks arise from the wings of the leaves, supporting three or four pale flowers, which are succeeded by sickle-shaped hoary pods. This is a biennial plant, and should be treated in the same manner as the last.

The twentieth sort sends up several erect stalks, garnished with leaves, composed of several pair of lobes which are indented at the top: from the wings of the leaves come out long foot-stalks, supporting a globular head of purple flowers; these are rarely succeeded by pods in England. It flowers the end of July.

The twenty-first sort grows naturally in China, from whence I received the seeds. This plant is generally an annual, though sometimes when it does not produce seeds the first year, the plants may be preserved in shelter thro' the winter. This is propagated by seeds, which should be sown on a hot-bed early in the spring; and when the plants are come up and fit to remove, they should be each transplanted into a small pot filled with light earth, and plunged into a hot-bed to bring them forward, otherwise they seldom perfect their seeds in this country; but when the plants have spread, so as to cover the surface of the pots, they should be exposed to the open air, often refreshing them with water, which will be of service to promote their ripening their seeds.

This plant bears pretty heads of blue flowers, set upon slender foot-stalks arising from their roots.

The twenty-second sort grows naturally about Aleppo, from whence the seeds were brought to England by Dr. Russel. This is an annual plant, which sends out several trailing stalks about a foot long, garnished with narrow winged leaves whose lobes are narrow at their base, but broad at their points; from the wings of the stalk the flowers are produced upon very short foot-stalks; they are of a pale white colour, and are succeeded by long hooked pods filled with gray seeds.

If the seeds of this plant are sown upon a bed of light earth in the beginning of April, the plants will arise in three weeks or a month after, when they should be thinned and kept clean from weeds; in July they will flower, and the seeds will ripen the middle or latter end of September, soon after which the plants decay.

ASTRANTIA. Masterwort.

The CHARACTERS are,

It is a plant whose flowers grow in an umbel; the involucrum of the general umbel is composed of two large trifid leaves, and two entire, and in another species of several small leaves: the flower is composed of five petals, which are bifid; it hath five stamina. The oblong germen is situated below the receptacle, which afterward becomes an oval, blunt, channelled fruit, divided into two parts, having two oblong oval seeds inclosed in the cover.

The SPECIES are,

1. ASTRANTIA (*Major*) foliis radicalibus quinquelobatis serratis, caulinis trilobatis acutis. *Greatest Masterwort with a purplish involucrum.*

2. ASTRANTIA (*Candida*) foliis quinquelobatis lobis tripartitis. Haller. Helv. 439. *Greater Masterwort with a white involucrum.*

3. ASTRANTIA (*Minor*) foliis digitatis serratis. Lin. Sp. Plant. 255. *Smaller black Hellebore with a Sanicle leaf.*

The first sort hath many spreading leaves rising from the root, composed of five large lobes, sawed pretty deep on their edges; from between these the stalks arise near two feet high, having at each joint one leaf deeply cut into three sharp pointed lobes; at the top of the stalk is produced the umbel of flowers, at the bottom of which is situated the general involucrum, composed of two long trifid leaves, and two entire ones of the same length, which extend beyond the rays, and are of a purplish colour.

The second sort hath much the appearance of the first, but it differs from that in having five lobes to the leaves of the stalks, which are much shorter and rounder at the point than those of the other. The general involucrum of the umbel is composed of short narrow leaves, and those of the smaller umbels are shorter and white.

The third sort seldom rises a foot high; the foot-stalks of the leaves are four inches long; the leaves are divided into eight segments to the bottom, and spread out like a hand; the involucrum of the general umbel is composed of several very narrow leaves; the foot-stalks of the separate umbels are very long and slender, and toward the top often divide into three, each having a small umbel. The involucrums of these small umbels are short and white.

These plants are very hardy; they may be propagated either by sowing of their seeds or parting their roots. The seeds should be sown in the autumn soon after they are ripe, on a shady border; when the plants come up where they are too close, some of the plants should be drawn out to allow room for others to grow until Michaelmas, when they should be transplanted where they are to remain, which should always be in a moist soil and a shady situation. The distance these plants should be placed is three feet, for their roots will spread to a considerable width, if they are permitted to remain long in the same place. These plants are seldom preserved but in botanic gardens, there being no great beauty in their flowers.

ATHAMANTA. Lin. Gen. Plant. 301. Spignel.

The CHARACTERS are,

It is a plant with an umbellated flower; the involucrum of the great umbel is composed of many narrow leaves, which are shorter than the rays; those of the small ones are narrow, and equal with the rays: each flower hath five slender stamina, of the same length with the petals. The germen is situated below the receptacle, which afterward becomes an oblong channelled fruit, divided into two parts, each containing one oval channelled seed.

The SPECIES are,

1. ATHAMANTA (*Meum*) foliolis capillaribus, seminibus glabris striatis. Hort. Cliff. 93. *Spignel with Dill leaves.*

2. ATHAMANTA (*Cretensis*) foliolis linearibus planis hirsutis, petalis bipartitis, seminibus oblongis hirsutis. Lin. Mat. Med. 143. *Candy Carrot with very slender Fennel leaves.*

3. ATHAMANTA (*Sicula*) foliis inferioribus nitidis, umbellis primordialibus subsessilibus, seminibus pilosis. Hort. Upsal. 60. *The second Sicilian Carrot with a Flixweed leaf.*

The first sort is the common Spignel used in medicine. It grows naturally in Westmoreland, and by the inhabitants there is called Bald Money, or Bawd Money; by some it is called Meu. This is a perennial plant; the stalks rise a foot and a half high, and are channelled; the leaves are very ramose, and composed of many fine hair-like leaves set pretty close, and are of a deep green; the stalk is terminated by an umbel of white flowers, which are succeeded by oblong smooth seeds.

This may be propagated by parting of the roots at Michaelmas, or from seeds sown soon after they are ripe; the plants should have a shady situation and a moist soil.

The second sort is the Daucus Creticus, of which there are two sorts, whose seeds are differently used in the shops, one of which is annual; but that here mentioned is a perennial plant, whose leaves are composed of numbers of slender narrow leaves like those of Fennel, irregularly disposed. The flower-stalk rises about two feet high, sending out many branches, with the same compound capillary leaves, and at the top are terminated by compound umbels, composed of near twenty small ones; these have white flowers with five petals, which are succeeded by oblong, hairy, channelled fruit, divided into two parts, containing one oblong hairy seed.

This sort is propagated by seeds, which should be sown in autumn on an open bed of light dry ground; the following autumn the plants should be carefully taken up, and planted at about a foot distance in a bed of light sandy earth, where the roots will continue several years, and annually flower and produce ripe seeds. It flowers in June, and the seeds are ripe in September.

The third sort is a perennial plant, which sends up from the root several upright stalks near three feet high, which are terminated by compound umbels. The flowers are composed of five white petals which are not quite equal, and are succeeded by oblong woolly fruit divided into two parts, each containing one oblong channelled seed.

This may be propagated in the same manner as the former, and is equally hardy.

ATHANASIA. Goldilocks. We have no proper English title for this genus.

The CHARACTERS are,

The flower hath an imbricated oval empalement, the scales are spear-shaped: the flower is composed of many florets which are uniform; the hermaphrodite florets are funnel-shaped, and cut into five segments which are erect; they have each five hair-like stamina, with tuberose cylindrical summits, an oblong germen with a slender style, terminated by an obtuse bifid stigma: each of these florets is succeeded by an oblong seed separated by a chaffy down.

The CHARACTERS are,

1. ATHANASIA (*Dentata*) corymbis impositis, foliis inferioribus linearibus dentatis, superioribus ovatis serratis. Lin. Sp. Pl. 1181. *Athanasia with a compound corymbus, the lower leaves are linear and indented, the upper are oval and sawed.*

2. ATHANASIA (*Trifurcata*) corymbis simplicibus, foliis trilobis cuneiformibus. Lin. Sp. Plant. 1181. *Athanasia with a simple corymbus, and wedge-shaped leaves ending with three points.*

3. ATHANASIA (*Crithmifolia*) corymbis simplicibus, foliis semitrifidis linearibus. Lin. Sp. Plant. 1181. *Athanasia with a simple corymbus, and linear semitrifid leaves.*

4. ATHANASIA (*Pubescens*) corymbis simplicibus, foliis lanceolatis indivisis villosis. Amœn. Acad. 4. 329. *Athanasia with a simple corymbus, and spear-shaped, undivided, hairy leaves.*

5. ATHANASIA (*Annua*) corymbis simplicibus coarctatis, foliis pinnatifidis dentatis. Lin. Sp. 1182. *Athanasia with a simple corymbus, and wing-pointed indented leaves.*

6. ATHANASIA (*Maritima*) pedunculis unifloris sub corymbosis, foliis lanceolatis indivisis crenatis obtusis tomentosis. Lin. Sp. 1182. *Athanasia with a single flower on each foot-stalk resembling a corymbus, and obtuse, spear-shaped, woolly leaves; commonly called Sea Cudweed.*

The first sort is figured and described by Dr. Commeline in his History of Rare Plants. It grows naturally at the Cape of Good Hope; this rises about three or four feet high, sending out many side branches, garnished toward the bottom with pretty long narrow leaves, having some resemblance to those of Buckshorn Plantain; but the upper leaves are small, rigid, and end in long points, where they are indented; the flowers generally terminate the branches; they are yellow, and formed into a kind of umbel; the flowers appear early in summer, and the seeds ripen in the autumn.

The second sort grows naturally in Africa: this is a low shrubby plant about the height of the former sort, sending out several weak shoots, garnished with awl-shaped trifid leaves set close to them, of a pale silvery colour; the flowers terminate the branches in a corymbus, they are yellow, and appear in August, and are succeeded by seeds which ripen the beginning of October.

The third sort is a native of Africa: this rises with a strong shrubby stalk six or eight feet high, sending out many ligneous branches, garnished with narrow long leaves,

terminating

terminating in four or five parts like thofe of Samphire; the flowers are yellow, which terminate the branches in form of a corymbus; thefe appear in July or Auguft, and are fucceeded by feeds which ripen in October.

The fourth fort grows naturally in the fame country with the former: this hath ftrong woody ftalks four or five feet high, covered with a woolly bark, as are alfo the branches, which are clofely garnifhed with entire, fpear-fhaped, woolly leaves, and are terminated with yellow flowers, each having a long foot-ftalk; thefe are fometimes fucceeded by feeds, which ripen in the autumn.

The fifth fort is a low annual plant, a native of the fame country with the former, and feldom rifes more than a foot high, fending out two or three flender branches, garnifhed with wing-pointed leaves; the ftalks are terminated by bright yellow flowers placed in umbels; thefe appear in Auguft and September, but unlefs the feafon is very favourable, the feeds do not ripen in England.

The fixth fort grows naturally near the fea in the Mediterranean, alfo in fome places in Wales, and other parts of Britain; it rifes two or three feet high, fending out a few weak branches garnifhed with white fpear-fhaped leaves, terminated by a fingle flower upon each foot-ftalk; thefe appear in fummer, but are rarely fucceeded by good feeds in England.

Thefe plants deferve a place in thofe gardens where there are conveniency for keeping them, for they will not live through the winter in the open air in this country, fo require fhelter in winter. The four firft forts fhould be preferved in a glafs-cafe in winter, where they may enjoy the free air at all times when the weather will permit: they may be propagated either from feeds or cuttings; the feeds fhould be fown in the fpring upon a moderate hot-bed, and when the plants appear they fhould be treated hardily, giving them plenty of air; and in fummer fhould be placed abroad in the open air till October, when they fhould be houfed. The fifth kind being an annual plant, the feeds fhould be fown upon a hot-bed to bring the plants forward, but they muft not be drawn up weak. The fixth fort muft be placed under a frame in the winter.

ATRACTYLIS. Lin. Gen. Plant. 837. Diftaff Thiftle.

The CHARACTERS are,

It hath a radiated compound flower, compofed of many hermaphrodite florets, which are included in a common fcaly empalement which hath no fpines. The hermaphrodite florets compofe the rays or border, and are ftretched out on one fide like a tongue. Thofe which compofe the difk are funnel-fhaped; in thofe of the difk is fituated a fhort crowned germen, which afterward becomes a turbinated compreffed feed, crowned with a plume of down, fhut up in the empalement.

The SPECIES are,

1. ATRACTYLIS (*Cancellata*) involucris cancellatis ventricofis, linearibus dentatis calycibus ovatis, floribus flofculofis. Lin. Sp. Plant. 830. *Small Cnicus with a netted leaf, and woolly feed.*

2. ATRACTYLIS (*Humilis*) foliis dentato-finuatis, flore radiato obvallato involucro patente, caule herbaceo. Lin. Sp. Plant. 829. *Lower purple prickly Cnicus.*

3. ATRACTYLIS (*Gummifera*) flore acaule. Lin. Sp. Pl. 829. *Prickly gum-bearing Cnicus without ftalk, and a Carline Thiftle leaf.*

The firft fort is an annual plant, which feldom rifes more than eight or nine inches high, with a flender ftem garnifhed with narrow hoary leaves, having fpines on their edges; at the top of the ftalk there are two or three flender branches fent out, each being terminated by a head of flowers like thofe of the Thiftle, with an involucrum compofed of feveral narrow leaves, armed with fpines on their fide, and are longer than the head of the flowers. The empalement is curioufly netted over, and is narrow at the top, but fwelling below, containing many florets of a purplifh colour. Thefe are each fucceeded by a fingle downy feed, which in cold years never perfect here.

It is propagated by feeds, which muft be fown upon an open bed of light earth, where the plants are to remain, and the plants where they come up too clofe together fhould be thinned, leaving them three inches apart.

The fecond fort rifes with a ftalk near a foot high, with indented leaves having fmall fpines on their edges; the upper part of the ftalk is divided into two or three flender branches, each fupporting a head of purple flowers, having rays inclofed in a fcaly empalement. The roots of this will live two or three years; it flowers in June, but unlefs the fummer is warm and dry, it will not perfect feeds in England. The feeds of this fort fhould be fown where they are to remain, and will require no other culture than the former.

The third fort is what the College of Phyficians have placed among the medicinal fimples, by the title of Carline Thiftle; the root of this is perennial, and fends out many narrow leaves which are deeply finuated, and armed with fpines on their edges. Thefe lie clofe on the ground, and between them the flower is fituated without ftalk, having many florets inclofed in a prickly empalement. Thofe on the border are white, but thofe which compofe the difk of a yellowifh colour. It flowers in July, but never perfects feeds in England.

ATRAPHAXIS. Lin. Gen. Plant. 405. We have no Englifh name for this genus.

The CHARACTERS are,

The flower hath a permanent empalement. It hath two roundifh finuated petals larger than the empalement, which are permanent; it hath fix capillary ftamina; in the center is fituated a compreffed germen, which afterward becomes a roundifh compreffed feed fhut up in the empalement.

The SPECIES are,

1. ATRAPHAXIS (*Spinofa*) ramis fpinofis. Hort. Cliff. 138. *Shrubby prickly Atraphaxis of the Eaft, with a fair flower.*

2. ATRAPHAXIS (*Undulata*) inermis. Lin. Sp. Pl. 333. *African, creeping, fhrubby Atraphaxis, with leaves curled on their fides.*

The firft is a fhrub which rifes four or five feet high, fending out many weak lateral branches, which are armed with fpines, with fmall fpear-fhaped leaves of an Afh colour, which are fmooth. The flowers come out at the ends of the fhoots in clufters, compofed of two white leaves tinged with purple, and are included in a two-leaved empalement of a pale herbaceous colour; thefe appear in Auguft. The plant is propagated by cuttings, and muft be fcreened from hard froft, which commonly deftroys thofe which are planted in the open air.

The fecond fort fends out many flender branches, which trail on the ground when they are not fupported, having fmall oval leaves, about the fize of thofe of the Knot-grafs, but are waved and curled on their edges, embracing the ftalk half round at their bafe. The flowers come out from the wings of the leaves, and have much the appearance of an apetalous flower, being compofed of four herbaceous leaves, two of which are the empalement, the other two the petals. It flowers in June and July. This is a native of the country about the Cape of Good Hope. It may be eafily propagated by cuttings any time in the fummer, and in winter the plants muft be fcreened from froft.

ATRIPLEX. Orach, or Arach.

The CHARACTERS are,

It hath female and hermaphrodite flowers on the fame plant. The hermaphrodite flowers have a permanent empalement of five leaves,

leaves, with membranaceous borders. In the center is placed the orbicular germen, which afterward becomes an orbicular compressed seed, shut up in the five-cornered empalement.

The Species are

1. Atriplex (*Hortensis*) caule erecto herbaceo foliis triangularibus. Hort. Cliff. 469. *Pale green, or white Garden Orach.*

2. Atriplex (*Halimus*) caule fruticoso foliis deltoidibus integris. Hort. Cliff. 469. *Broad-leaved Orach, or shrubby Halimus, commonly called Sea Purslane-tree.*

3. Atriplex (*Portulacoides*) caule fruticoso foliis obovatis. Flor. Suec. 829. *Shrubby Sea Orach, or Halimus, called Sea Purslane, with a narrow leaf.*

There are several other species of this genus, some of which grow naturally in England, but as they are plants of no beauty, so they are rarely admitted into gardens, for which reason I shall not enumerate them here.

The first of these plants was formerly cultivated in the kitchen-gardens as a culinary herb, being used as Spinach, and is now by some persons preferred to it, though in general it i not esteemed amongst the English; but the French at present cultivate this plant for use, as the people in the northern parts of England also do.

There are three or four different varieties of this, whose difference only is in the colour of the plants; one of which is of a deep green, another of a dark purple, and a third with green leaves and purple borders. These are generally supposed to be only accidental varieties which have come from the same seeds, but in forty years which I have cultivated these sorts, I have never yet observed them to vary.

These plants are annual, so must be sown for use at Michaelmas soon after the seeds are ripe; at which time it generally succeeds better than when it is sown in the spring, and will be fit for use at least a month earlier. They require no other culture but to hoe them when they are about an inch high, to cut them down where they are too thick, leaving them about four inches asunder, and to cut down all the weeds. When your plants are grown about four inches high, it will be proper to hoe them a second time, in order to clear them from weeds; and if you observe the plants are left too close in any part, you must then cut them out. Where these plants are sown on a rich soil, and allowed a good distance, the leaves will be very large, in which the goodness of the herb consists. This must be eaten while it is young, for when the stalks become tough it is good for nothing. The first sort is ordered by the College of Physicians for medicinal use.

The second sort was formerly cultivated in gardens as a shrub, and by some persons were formed into hedges, and constantly sheared to keep them thick; but this plant is by no means fit for such purposes on many accounts, for it grows too vigorous; the shoots in one month at the growing season of the year will be two feet long, provided they have a good soil, so that a hedge of this plant cannot be kept in tolerable order, nor will it ever form a thick hedge. But a worse inconvenience attends this plant, for in very hard winters it is often destroyed.

It may be propagated by cuttings, which may be planted in any of the summer months on a shady border, they will soon take root, and be fit to transplant the Michaelmas following, when they should be planted where they are to remain.

The third sort grows wild in divers parts of England, on the sea side, from whence the plants may be procured; or it may be propagated by cuttings in the same manner as the former sort. This is a low under shrub, seldom rising above two feet and a half, or at most three feet high, but becomes very bushy. This may have a place amongst other low shrubs, and if planted on a poor gravelly soil, will abide several years, and make a pretty diversity.

ATROPA. Lin. Gen. Plant. 222. Deadly Nightshade.

The Characters are,

The flower is bell-shaped, and divided into five equal parts. It hath five stamina rising from the base of the petal. In the center is situated an oval germen, which afterward becomes a globular berry having three cells sitting on the empalement, and filled with kidney-shaped seeds.

The Species are,

1. Atropa (*Belladona*) caule herbaceo, foliis ovatis integris. Lin. Sp. Plant. 181. *Common Deadly Nightshade.*

2. Atropa (*Frutescens*) caule fruticoso. Lin. Sp. Plant. 182. *Deadly Nightshade with a shrubby stalk.*

The first sort grows wild in many parts of England, but is not very frequent near London. This plant hath a perennial root, which sends out strong herbaceous stalks of a purplish colour, which rise to the height of four or five feet, with oblong entire leaves, which toward autumn change to a purplish colour; the flowers come out between the leaves singly, upon long foot-stalks; these are large, bell-shaped, and of a dusky brown colour on their outside, but are purplish within. After the flower is past, the germen turns to a large round berry, a little flatted at the top, and is first green, but when ripe turns to a shining black, sitting close upon the empalement, and contains a purple juice of a nauseous sweet taste, and full of small kidney-shaped seeds. In some places this plant is called Dwale, but in general Deadly Nightshade, from its quality. It should not be suffered to grow in any places where children resort, for it is a strong poison; there has been several instances within a few years past of its deadly quality, by several children being killed with eating the berries.

There is also an instance of the direful effects of this plant recorded in Buchanan's History of Scotland, wherein he gives an account of the destruction of the army of Sweno, when he invaded Scotland, by mixing a quantity of the juice of these berries with the drink which the Scots by their truce were to supply them with; which so intoxicated the Danes, that the Scots fell upon them in their sleep, and killed the greatest part of them, so that there were scarcely men enough left to carry off their king.

The second sort rises with a shrubby stem to the height of six or eight feet, divided into many branches, garnished with round leaves, in shape like those of the Storax-tree. The flowers come out between the leaves upon short foot-stalks, which are shaped like those of the former, but are much less, of a dirty yellowish colour, with a few brown stripes; these are never succeeded by berries in England. It grows naturally in Spain, and is only propagated by seeds. The plants are too tender to live abroad in winter, therefore at the end of October they must be removed into the green-house, and treated as other plants from the same country. It flowers in July and August.

AVENA. Lin. Gen. Plant. 85. Oats.

The Characters are,

The flowers are collected in a loose panicle, and have a bivalvular empalement, swelling in the middle. The petal of the flower is bivalve, having a spiral beard, twisting, jointed, and reflexed. There are two oval nectariums sitting upon the upper side of the germen; they have three slender stamina. The germen afterward becomes an oblong swelling seed, having a longitudinal furrow, and closely shut up in the cover or chaff.

We have but one Species of this genus in England, which is,

Avena (*Sativa*) calycibus dispermibus seminibus lævibus. Hort. Cliff. 25. *Oats with two smooth seeds in each empalement.*

There are three varieties of these Oats cultivated in England, viz. the white, the black, and the brown or red Oat, but

but where they have been many years feparately cultivated, I have never obferved them to alter. However, as their principal difference is in the colour of the grain, I fhall not enumerate them as diftinct fpecies. There is alfo a naked Oat, which is fometimes cultivated in the diftant parts of England, but is rarely feen near London.

The white fort is the moft common about London; the black is more cultivated in the northern parts of England, and is efteemed a very hearty food for horfes; but the firft makes the whiteft meal, and is chiefly cultivated where the inhabitants live much upon Oat cakes.

Oats are a very profitable grain, and abfolutely neceffary, being the principal grain which horfes love, and are efteemed the moft wholfome food for thofe cattle, being fweet and of an opening nature; other grains being apt to bind, which is injurious to labouring horfes: but if you feed them with this grain foon after they are houfed, before they have fweat in the mow, or are otherwife dried, it is as bad on the other hand, for they are then too laxative.

This grain is a great improvement to many eftates in the north of England, Scotland, and Wales, for it will thrive on cold barren foils, which will produce no other fort of grain; it will alfo thrive on the hotteft land: in fhort, there is no foil too rich or too poor for it, too hot or too cold for it; and in wet harvefts, when other grain is fpoiled, this will receive little or no damage; the ftraw and hufks being of fo dry a nature, that if they are are houfed wet they will not heat in the mow, or become mouldy, as other grain ufually do; fo is of great advantage in the northern parts of England, and in Scotland, where their harveft is generally late, and the autumns wet.

The beft time for fowing of Oats is in February or March, according as the feafon is early or late; and fometimes I have known it fown in April upon cold land, and has been early ripe. The black and red Oats may be fown a month earlier than the white, becaufe they are hardier.

Oats are often fown on land which has the former year produced Wheat, Rye, or Barley. The common method is to plough in the ftubble about the beginning of February, and fow the Oats and harrow them in; but then they muft be harrowed the fame way as the furrows lay, for if it be done croffways the ftubble will be raifed on the furface; but this is not a good method of hufbandry, for when people have time to plough the ftubble in autumn, it will rot in winter; and then give the land another ploughing and a good harrowing juft before the Oats are fown, it will make the ground finer and better to receive the grain. Moft people allow four bufhels of Oats to an acre, but I am convinced two bufhels are more than enough; the ufual produce is about twenty-five bufhels to an acre, though I have fometimes known more than thirty-five bufhels on an acre.

Oats are alfo fown upon land when it is firft broken up, before the ground is brought to a tilth for other grain, and is frequently fown upon the fward with one ploughing; but it is much better to give the fward time to rot before the Oats are fown, for the roots of the Grafs will prevent thofe of the corn from ftriking downward.

AURANTIUM. The Orange-tree.

The CHARACTERS are,

The flower hath five oblong fpreading petals, and many ftamina, which are frequently joined in fmall feparate bodies at bottom. In the center is fituated the germen, which afterward becomes a globular flefhy fruit, compreffed at both ends, having a thick flefhy pulp, and divided into feveral cells, each containing two oval callous feeds.

The SPECIES are,

1. AURANTIUM (*Acre*) foliis ovato-lanceolatis glabris. *The common Seville Orange.*

2. AURANTIUM (*Chinenfe*) foliis lanceolatis acutis glabris. *The China Orange.*

3. AURANTIUM (*Orientale*) foliis lineari lanceolatis glabris. *Orange tree with narrow leaves, called Willow-leaved Orange, and by fome the Turkey Orange.*

4. AURANTIUM (*Documane*) foliis ovato-lanceolatis craffis lucidis, fructu maximo. *The Pampelmoes, or Shaddock.*

5. AURANTIUM (*Humile*) pumilum foliis ovatis floribus feffilibus. *The Dwarf or Nutmeg Orange.*

There are many varieties of this, as there is of moft other fruits which have arifen from culture; but thofe here enumerated may ftrictly be allowed to be diftinct fpecies. The varieties in the Englifh gardens are, 1. The yellow and white ftriped-leaved Orange. 2. The curled-leaved Orange. 3. The horned Orange. 4. The double flowering Orange. And, 5. The hermaphrodite Orange.

The China Orange is not fo hardy as the Seville, therefore muft be treated more tenderly to have good fruit in England.

The dwarf Orange is alfo tender, and the leaves are very fmall, growing in clufters. This fort, when in flower, is proper to place in a room or gallery to adorn them, the flowers being very fweet; but thefe are feldom to be found in good health, becaufe they muft be treated with more care than the common Orange and Lemon-trees; as muft alfo the Shaddock, otherwife the fruit will always drop off in winter. The Pampelmoes were brought from the Eaft to the Weft-Indies, but the inhabitants have greatly degenerated the fruit fince it has been in the Weft-Indies, by raifing the trees from feeds; the greateft part of which produce harfh four fruit, greatly inferior to the original fort; the flefh or pulp of which is red, whereas the greater part of the trees in America produce fruit with a pale yellow flefh; and by conftantly raifing thefe trees from feeds, they degenerate the fruit continually; whereas if they would bud from the good fort, they might have it in as great plenty as they pleafed.

All the forts of Orange and Lemon-trees with ftriped leaves are tender, therefore muft be placed in a warm part of the green-houfe in winter, and muft be treated with more care than the common fort, otherwife they will not thrive.

The horned Orange differs from the other forts, in the fruit dividing into parts, and the rind expanding in form of horns: this and the diftorted Orange, are preferved by fome curious perfons for variety. There is alfo a great variety of fweet Oranges both in the Eaft and Weft-Indies, fome of which are much more efteemed than thofe we now have in Europe.

If you purpofe to raife ftocks for budding of Oranges you fhould procure fome Citron feeds which were duly ripened; for the ftocks of this kind are preferable to any other, both for quicknefs of growth, as alfo that they will take buds of either Orange, Lemon, or Citron; next to thefe are the Seville Orange feeds. The beft feeds are ufually to be had from rotten fruits, which are commonly eafy to be procured in the fpring of the year; then prepare a good hot-bed of either horfe dung or tanners bark, the laft of which is much the better, if you can eafily procure it. When this bed is in a moderate temper for heat, you muft fow your feeds in pots of good rich earth, and plunge them into the hot-bed. In three weeks or a month the plants will come up, and if they are not ftinted, either for want of proper heat or moifture, they will be in fix weeks after their appearance fit to tranfplant into fingle pots: you muft therefore renew your hot-bed, and having prepared a quantity of fmall halfpenny pots, fill thefe half full of frefh loamy earth, mixed with very rotten cow dung: and then fhake out the young plants from the feed-pots with all the earth about them, that you may the better feparate

parate the plants without tearing their roots; put a ſingle plant into each of the pots, then fill them up with the ſame earth as before directed, plunging the pots into the new hot-bed, giving them a good watering to fix the earth to their roots, and ſcreen them from the ſun in the heat of the day. In this method, with due care, your plants will grow to be two feet high by the end of July, when you muſt begin to harden them by degrees, in raiſing your glaſſes very high, and when the weather is good take them quite off, but do not expoſe them to the open ſun in the heat of the day, but rather take off the glaſſes and ſhade the plants with mats, which may be taken off when the ſun declines; for the violent heat in the middle of the day would be very injurious to them in hot weather, eſpecially while young. Toward the end of September you muſt houſe them, obſerving to place them near the windows of the green-houſe, to prevent the damps from moulding their tender ſhoots. During the winter ſeaſon they may be often refreſhed with water. If the plants are plunged into a gentle hot-bed in the ſpring it will greatly forward them, but they ſhould be hardened by the beginning of June, that they may be in right order to bud in Auguſt; when you ſhould make choice of cuttings from trees that are healthy and fruitful, of whatever kind you pleaſe, obſerving that the ſhoots are round, the buds of theſe being much better and eaſier to part from the wood than ſuch as are flat. When you have budded the ſtocks, you ſhould remove them into a green-houſe to defend them from wet, turning the buds from the ſun; but let them have as much free air as poſſible, and refreſh them often with water. In a month's time after budding, you will ſee which of them has taken; you muſt then untie them, that the binding may not pinch the buds, and let them remain in the green-houſe all the winter; then in the ſpring prepare a moderate hot-bed of tanners bark, and, after having cut off the ſtocks about three inches above the buds, plunge their pots into the hot-bed, obſerving to give them air and water as the heat of the weather ſhall require, but be ſure to ſcreen them from the violence of the ſun during the heat of the day. With this management, if your buds ſhoot kindly, they will grow to the height of two feet or more by July; at which time you muſt begin to harden them before the cold weather comes on, that they may the better ſtand in the green-houſe the following winter. In the firſt winter after their ſhooting you muſt keep them warm, for by forcing them in the bark-bed they will be ſomewhat tenderer than the grown trees; however, it is very neceſſary to raiſe them to their height in one ſeaſon, that their ſtems may be ſtrait; for ſuch trees which are two or more years growing to their heading height, their ſtems are generally crooked. In the ſucceeding years their management will be the ſame as in full grown trees, which will be hereafter treated of: I ſhall therefore now proceed to treat of the management of ſuch trees as are brought over every year in cheſts from Italy, which is, indeed, by much the quicker way of furniſhing a green-houſe with large trees, for thoſe which are raiſed from ſeeds in England will not grow ſo large in their ſtems under eighteen or twenty years, as theſe will have when brought over; and although their heads are ſmall when we receive them, yet in three years, with good management, they will have large heads, and produce fruit.

In the choice of theſe trees, obſerve firſt, the difference of their ſhoots and leaves (if they have any upon them) to diſtinguiſh their different ſorts, for the Shaddock and Citrons always make much ſtronger ſhoots than the Orange; for which reaſon, the Italian gardeners, who raiſe theſe trees for ſale, generally propagate thoſe ſorts, ſo that they bring few of the Seville Orange-trees over, which are much more valuable both for their flowers and fruit; alſo prefer thoſe that have two good buds, one on each ſide of the ſtock (for many of them have but one, ſo will always have an irregular head;) the ſtraitneſs of the ſtem, freſhneſs of the branches, and plumpneſs of the bark, are neceſſary obſervations.

When you have furniſhed yourſelf with a parcel of theſe trees, you muſt prepare a moderate hot-bed of tanners bark in a forcing-frame, in length and breadth according to the number of trees to be forced; then put your trees into a tub of water upright, about half way of the ſtems, leaving the head and upper part of the ſtem out of the water, the better to draw and imbibe the moiſture. In this ſituation they may remain two or three days (according to their plumpneſs when you received them;) then take them out, and clean their roots from all filth, cutting off all broken or bruiſed roots, and all the ſmall fibres, which are quite dried by being ſo long out of the earth, and ſcrub the ſtems with a hard hair bruſh, cleaning them afterwards with a cloth; then cut off the branches about three inches from the ſtem, and having prepared a quantity of good freſh earth, mixed with very rotten neats dung, plant your trees therein, obſerving never to put them into large pots, for if they are big enough to contain their roots, it is ſufficient at firſt planting; and be ſure to put ſome potſherds and large ſtones in the bottom of each pot, to keep the holes at the bottom of the pots from being ſtopped with earth, that the water may freely paſs off, and wrap ſome haybands round their ſtems from bottom to top, to prevent the ſun from drying their bark; then plunge theſe pots into the bark-bed, watering them well to ſettle the earth to their roots, frequently repeating the ſame all over their heads and ſtems, being very careful not to over water them, eſpecially before they have made good roots; and obſerve to ſcreen the glaſſes of your hot-bed from the ſun in the heat of the day.

If your trees take to grow kindly (as there is little reaſon to doubt of, if the directions given be duly obſerved) they will have made ſtrong ſhoots by the beginning of June; at which time you ſhould ſtop their ſhoots, to obtain lateral branches to furniſh their heads; and now you muſt give them air plentifully, and begin to harden them, that in the middle of July they may be removed into the open air, in ſome warm ſituation, defended from the great heat of the ſun, and from winds, that they may be hardened before winter. About the end of September you ſhould houſe theſe plants, ſetting them at firſt in the front of the green-houſe, near the glaſſes, keeping the windows open at all times when the weather will permit; and about the latter end of October, when you bring in the Myrtles and other leſs tender trees, you muſt ſet your Oranges in the warmeſt and beſt part of the houſe, placing lower plants or trees in the front to hide their ſtems. During the winter, let your waterings be frequent, but give them not too much at a time; for now their heads are but ſmall, and therefore incapable to diſcharge too great a quantity of moiſture; and take great care to guard them from froſt.

In the ſpring, when you begin to take out ſome of your hardieſt ſorts of plants to thin your houſe, waſh and cleanſe the ſtems and leaves of your Orange-trees, taking out the upper part of the earth in the pots, filling them up again with good freſh rich earth, laying thereon a little rotten neats dung round the outſide of the pots, but do not let it lie near the ſtem of the trees; then place them at wider diſtances in the houſe, that the air may circulate round their heads, giving them air diſcretionally as the weather grows warm, but do not remove them into the open air until the middle or latter end of May, that the weather is ſettled; for many times, when they are removed out too ſoon, the mornings often proving cold, give them at leaſt a great check,

check, which will change the colour of their leaves, and many times kill the extreme weak part of the shoots. Let the situation for your Orange-trees, during the summer season, be as much defended from the sun in the heat of the day, and also from strong winds, as possible, by tall trees or hedges; both of which, if they are exposed thereto, are very hurtful to them.

As these trees advance, it will be necessary in the summer to stop strong shoots when they grow irregular, to force out lateral branches to fill the head; but do not pinch off the tops of all the shoots (as is the practice of some,) which will fill the tree with small shoots, too weak to support fruit; but endeavour to form a regular head, and obtain strong shoots, taking away weak trifling branches where they are too close.

During the summer season your Orange-trees will require frequent waterings in dry weather, especially if they are large; therefore you should endeavour to have the water as near the trees as possible, to save the trouble of carrying it, which in a large quantity of trees takes up much time. Your water should be soft, and exposed to the air, but never add dung of any sort thereto; which, although by many frequently recommended, yet has always been found destructive to these trees if much used; it being like hot liquors to human bodies, which, at first taking, seem to add vigour, yet certainly leave the body weaker after some time than before.

Your Orange-trees will require to be shifted and new potted every other year, therefore you must prepare a quantity of good earth, at least a year before you intend to use it, that it may be well mixed and perfectly rotten. The best season for this work is about the end of April, that they may have taken fresh root before they are removed out of the green-house; and when this work is performed, it will be necessary to let them remain in the house a fortnight longer than usual, to be well settled.

When you first set these trees abroad after shifting, you should place them near the shelter of hedges, and fasten their stems to strong stakes, to prevent their being disturbed by winds, which sometimes will blow fresh planted trees out of the pots, if too much exposed thereto, and thereby greatly injure their new roots.

If old Orange-trees have been ill managed, and their heads become ragged and decayed, the best method to restore them is to cut off the greatest part of their heads early in March, and prune their roots; then soak and clean their stems and branches, planting them into good earth, and setting them into a hot-bed of tanners bark, as was directed for such trees as came from abroad, managing them in the same manner: by this method they will produce new heads, and in two years time become good trees again. But if these are large trees, and have grown in tubs for several years, your best way will be to prepare a parcel of rough baskets (such as are used for basketting Evergreens, when sent to a distant place:) let these be somewhat less than the tubs you design to plant your trees into; then plant your trees herein, plunging them into the hot-bed, and about the beginning of July, when your trees have made good shoots, you may remove them into the tubs, with their baskets about them, filling the empty space with the same good earth: this will preserve your tubs from rotting in the bark, and the trees will do equally well as if planted into the tubs at first, provided you are careful in removing the baskets, not to disturb their roots; and also let them remain in the green-house a fortnight or three weeks after planting, before you set them abroad.

In the management of Orange-trees, which are in good health, the chief care should be to supply them with water duly, and not (as is sometimes practised) starve them in winter, whereby their fibres are dried, and become mouldy, to the great prejudice of the trees, nor to give them water in too great abundance; but rather let their waterings be frequent and given in moderate quantities. You must also observe, that the water has free passage to drain off, for if it be detained in the tubs or pots, it will rot the tender fibres of the trees; nor should they be placed too near each other in the green-house, but set them at such a distance, that their branches may be clear of each other, and that the air may circulate freely round their heads. In summer they should be placed where the winds are not violent, and to have the morning and evening sun, for if they are too much exposed to the mid-day sun they will not thrive. The best situation for them is near some large plantation of trees, which will break the force of the winds, and screen them from the violent heat of the sun. In such a situation they may remain until the beginning of October, or later, according as the season proves favourable; for if they are carried into the green-house early, and the autumn should prove warm, it will occasion the trees to make fresh shoots, which will be weak and tender, and so liable to perish in winter; nor should they remain so long abroad as to be injured by morning frosts.

The best compost for Orange-trees is two-thirds of fresh earth from a good pasture, which should not be too light, nor over stiff, but rather a Hazel loam; this should be taken about ten inches deep with the sward, which should be mixed with the earth to rot, and one-third part of neats dung; these should be mixed together, at least twelve months before it is used, observing to turn it over every month, to mix it well, and to rot the sward; this will also break the clods, and cause the mould to be finer.

Of late years there have been many of these trees planted against walls, and frames of glass made to fix over them in winter; and some few curious persons have planted these trees in the full ground, and have erected moveable covers to put over them in winter, which are so contrived as to be all taken away in summer: where these have been well executed, the trees have made great progress in their growth, and produced a much larger quantity of fruit than those in pots or tubs, which have ripened so well, as to be extremely good for eating. If these are planted either against walls with design of training the branches to the walls, or in borders at a small distance, so as to train them up as standards, there should be a contrivance of a fire-place or two, in proportion to the length of the wall, and flues carried the whole length of the wall, to warm the air in very cold weather, otherwise it will be very difficult to preserve the trees in very hard winters alive. The manner of making these flues is fully explained under the article of HOT WALLS. Where this contrivance is made, there will be no hazard of losing the trees, be the winter ever so severe, with a little proper care; whereas, if this is wanting, there will require great care and trouble to cover and uncover the glasses every day when there is any sun; and if the wall is not thicker than they are usually built, the frost will penetrate through the walls in severe winters; so that covering and securing the glasses of the front, will not be sufficient to preserve the trees, be it done with ever so much care: therefore the first expence of the walls will save great trouble and charge, and be the securest method.

If the ground is wet, or of a strong clay, so as to detain the moisture, the borders should be raised above the level of the ground, in proportion to the situation of the place; for where the wet lies in winter near the surface, it will greatly prejudice, if not totally destroy the trees; so that lime rubbish should be laid at least two feet thick in the bottom of the border, to drain off the wet; and the earth should be laid two and a half or three feet thick thereon, which

which will be a ſufficient depth for the roots of the trees. In theſe borders there may be a few roots of the Guernſey and Belladonna Lilies planted, or any other exotic bulbous-rooted flowers, which do not grow high, or draw too much nouriſhment from the borders; and theſe producing their flowers in autumn or winter, will make a good appearance, and thrive much better than if kept in pots.

AURICULA MURIS. Mouſe Ear.

This is a ſort of Hawkweed with ſmall hairy leaves, which are white underneath: the plant trails upon the ground, taking root at the joints, by which means it will ſoon ſpread over a large compaſs of ground.

This is very common in England; it grows chiefly on dry barren places, or upon old walls, and is too often a troubleſome weed in Graſs-plats in gardens.

AURICULA URSI. Bear's Ear, or Auricula.

To enumerate the varieties of this plant, would be almoſt endleſs and impoſſible, for every year produces vaſt quantities of new flowers, differing in ſhape, ſize, or colour of the flowers; and alſo in the leaves of theſe plants there is as great a variety, that the ſkilful floriſt is often capable of diſtinguiſhing the particular varieties thereby.

But as it ſeldom happens, that ſuch of theſe flowers as are at one time in great eſteem, continue to be regarded a few years after, (there being ſtill finer or larger flowers produced from ſeeds, which are what the floriſts chiefly ſeek after,) it would be needleſs to mention any of them; wherefore I ſhall proceed to give the characters of a good Auricula.

1. *The ſtem of the flower ſhould be lofty and ſtrong.*

2. *The foot-ſtalk of the ſingle flower ſhould be ſhort, that the umbel may be regular and cloſe.*

3. *The pipe or neck of each flower ſhould be ſhort, and the flowers large, and regularly ſpread, being no ways inclinable to cup.*

4. *That the colours are very bright, and well mixed.*

5. *That the eye of the flower be large, round, and of a good white or yellow, and that the tube or neck be not too wide.*

All the flowers of this kind, that want any of the above-mentioned properties, are now rejected by every good floriſt; for as the varieties every year increaſe from ſeeds, ſo the bad ones are turned out to make room for their betters; but in ſome people the paſſion for new flowers ſo much prevails, that, ſuppoſing the old flower greatly preferable to a new one, if it is of their own raiſing, the latter muſt take place of the old one.

In order to obtain good flowers from ſeeds, you muſt make choice of the beſt flowers you have, which ſhould be expoſed to the open air, that they may have the benefit of ſhowers, without which they ſeldom produce good ſeeds: the time of their ripening is in June and July, which you will eaſily know, by their ſeed-veſſel turning to a brown colour, and opening; you muſt therefore be careful leſt the ſeeds be ſcattered out of the veſſel, for it will not be all fit to gather at the ſame time.

The time for ſowing this ſeed is commonly in Auguſt, but if it be ſown before Chriſtmas it will be time enough.

The beſt ſoil for this ſeed is good, freſh, light, ſandy mould, mixed with very rotten neats dung, or very rotten dung from the bottom of an old hot-bed: with this you ſhould fill your pots or boxes, in which you intend to ſow your ſeeds; and having levelled the ſurface of the earth very ſmooth, ſow your ſeeds thereon, covering it very lightly with rotten Willow mould taken out of the ſtems of decayed hollow Willow-trees; then cover the box, &c. with a net or wire, to prevent the cats, fowls, &c. from ſcratching out, or burying your ſeeds too deep; for whenever this happens, the ſeeds will remain a year in the ground before the plants appear, if it ſhould grow at laſt; therefore many perſons never cover theſe ſeeds, but ſow them upon the ſurface of the earth, in the boxes uncovered with earth, for the rain to waſh them into the ground, which is often the beſt method: let theſe boxes, &c. be placed ſo as to receive only the morning ſun, during the winter ſeaſon; but in the beginning of March remove them where they may have ſcarce any ſun, for your young plants will now ſoon begin to appear, which, if expoſed to one whole day's ſun only, will be all deſtroyed.

During the ſummer ſeaſon, in dry weather, often refreſh them with water, but never give them too great quantities at once. In the July following, your plants will be large enough to tranſplant, at which time you muſt prepare a bed, or boxes, filled with the above-mentioned ſoil, in which you may plant them about three inches ſquare, and (if in beds) you muſt ſhade them every day, till they are thoroughly rooted, as alſo in very hot dry weather; but if they are in baſkets, or boxes, they may be removed to a ſhady place.

When the ſeedling Auriculas are planted in beds, there ſhould be ſome rotten neats dung laid about ten inches under the ſurface, and beaten down cloſe and ſmooth: this will prevent the worms from drawing the young plants out of the earth, which they generally do where this is not practiſed. This dung ſhould be laid about a foot thick, which will entirely prevent the worms getting through it until the plants are well eſtabliſhed in the beds; and the roots of the Auriculas will ſtrike down into the dung by the ſpring, which will make their flowers ſtronger than uſual: theſe beds ſhould be expoſed to the eaſt, and ſcreened from the ſouth ſun.

When you have taken all your plants which are come up out of the boxes or pots, level the earth gently again; for it often happens, that ſome of the ſeeds will lie in the ground two years before they appear, eſpecially if they were covered too deep when ſown, as was before obſerved.

The ſpring following many of the firſt plants will ſhew their flowers, when you may ſelect ſuch of them as have good properties, to be removed into pots of the ſame prepared earth, and preſerved until the next ſeaſon, at which time you will be capable to form a judgment of the goodneſs of the flower; but thoſe that produce plain coloured or ſmall flowers ſhould be taken out, and planted in borders in the out-parts of the garden, to make a ſhew, or gather for noſegays, &c. the others, which do not produce their flowers the ſame year, may be taken up, and tranſplanted into a freſh bed, to remain till you ſee how they will prove.

The manner of propagating theſe flowers when obtained, is from offsets or ſlips, taken from the old roots in April, when the flowers are in bloom: theſe offsets muſt be planted into ſmall pots filled with the ſame ſort of earth, as was before directed for the ſeedlings; and, during the ſummer ſeaſon, ſhould be ſet in a ſhady place, and muſt be often (but very gently) refreſhed with water, but in the autumn and winter ſhould be ſheltered from violent rains. The ſpring following theſe young plants will produce flowers, though but weak; ſoon after they are paſt flowering, you muſt put them into larger pots, and the ſecond year they will blow in perfection.

But, in order to obtain a fine bloom of theſe flowers, you muſt obſerve the following directions:

Firſt, Preſerve your plants from too much wet in winter, which often rots and ſpoils them, but let them have as much free open air as poſſible; nor ſhould they be too much expoſed to the ſun, which is apt to forward the budding for flower too ſoon; and the froſty mornings, which often happen in March, thereby deſtroy their buds, if they are not protected therefrom. To prevent which, thoſe who are

are very curious in these flowers, place their pots in the autumn under a common hot-bed frame, where, in good weather, the plants may enjoy the full air, by drawing off the glasses; and in great rains, snow, or frost, the plants may be screened by covering them. When this method is practised with judgment, the flowers will be much stronger, and the plants will increase faster than when they are exposed abroad.

Secondly, In the beginning of February, if the weather is mild, you must take off the upper part of the earth in the Auricula pots, as low as you can without disturbing their roots, and fill up the pots with fresh rich earth, which will greatly strengthen them for bloom; as also prepare your offsets for transplanting in April, by causing them to push out new roots.

Those plants which have strong single heads always produce the largest cluster of flowers; therefore curious florists pull off the offsets as soon as it can be done with safety to their growing, to encourage the mother plants to flower the stronger; they also pinch off the flowers in the autumn, where they are produced, and suffer them not to open, that the plants should not be weakened thereby.

Thirdly, You must cover your pots with mats in frosty weather, during this time of their budding for flower, lest the sharp mornings blight them and prevent their blowing.

Fourthly, When your flowerr-stems begin to advance, and the blossom buds grow turgid, you must protect them from hasty rains, which would wash off their white meally farina, and greatly deface the beauty of their flowers; but at the same time observe to keep them as much uncovered as possible, otherwise their stems will be drawn up too weak to support their flowers (which is often the case when their pots are placed under walls,) and give them gentle waterings to strengthen them; but let none of the water fall into the center of the plant, or among the leaves.

Fifthly, When your flowers begin to open, you should remove their pots upon a stage (built with rows of shelves, one above another, and covered on the top, to preserve them from wet; this should be open to the morning sun, but sheltered from the heat of the sun in the middle of the day:) in this position they will appear to much greater advantage, than when the pots stand upon the ground, for their flowers being low, their beauty is hid from us; whereas, when they are advanced upon shelves, we see them in full view: in this situation they may remain until the beauty of their flowers is past, when they must be set abroad to receive the rains, and have open free air, in order to obtain seeds, which will fail if they are kept too long under shelter. When your seed is ripe, observe to gather it when it is perfectly dry, and expose it to the sun in a window upon papers, to prevent its growing mouldy, and let it remain in the pods till the season for sowing it.

AURICULA URSI MYCONI. See VERBASCUM.

AZALEA. Lin. Gen. Plant. 195. American upright Honeysuckle.

The CHARACTERS are,

It hath a coloured empalement which is permanent, cut into five acute parts at the top. The flower is funnel-shaped, having a long naked tube, cut into five parts; the two upper segments are reflexed backward, the two sides are bent inward, and the lower one turns downward. It hath five slender stamina of unequal lengths, with a round germen, which afterward becomes a roundish capsule, having five cells filled with roundish small seeds.

The SPECIES are,

1. AZALEA (*Viscosa*) foliis margine scabris, corollis piloso-glutinosis. Lin. Sp. Plant. 151. *The American upright Honeysuckle with a white flower.*

2. AZALEA (*Nudiflora*) foliis ovatis corollis pilosis staminibus longissimis. Lin. Sp. Pl. 150. *Commonly called red American upright Honeysuckle.*

The first of these is a low shrub, rising with several stems two or three feet high. The leaves come out in clusters at the end of the shoots without order, and their edges are set with very short teeth, which are rough. The flowers come out in clusters between the leaves, at the extremity of the branches, which are white, with a mixture of dirty yellow on their outside. They have a tube an inch long, and at the top are pretty deeply cut into five segments; the two upper are reflexed, the two side ones are bent inward, and the lower is turned downward, with five stamina a little longer than the petals; the style is much longer than the stamina. These flowers have much the appearance of those of the Honeysuckle, and are as agreeably scented.

The second sort grows taller than the first, and in its native country frequently rises to the height of twelve feet, but in England is never more than half that height. This hath several stems with oblong smooth leaves. The flower-stalks arise from the division of the branches which are long and naked, supporting a cluster of red flowers; they are divided at the top into five equal segments, which spread open. The five stamina and the style are much longer than the petals, and stand erect.

These plants grow naturally in shade, and upon moist ground, in most parts of North America, so they must have a moist soil and a shady situation, otherwise they will not thrive. They can only be propagated by shoots from their roots, for they do not produce good seeds here; and if good seeds are obtained, they are difficult to raise, and will be a long time before they would flower. But when they are in a proper situation, their roots extend, and put out shoots, which may be taken off with roots, and transplanted. The autumn is the best time to remove the plants, but the ground about their roots should be covered in winter to keep out the frost; and if this is every year practised to the old plants, it will preserve them in vigour, and cause them to flower well.

AZEDARACH. See MELIA.

AZEROLE, or L'AZAROLE. See MESPILUS.

B.

BACCHARIS. Ploughman's Spikenard.

The CHARACTERS are,

The flower is compoſed of many hermaphrodite and female florets, which are included in one common, cylindrical, ſcaly empalement. The hermaphrodite florets are funnel-ſhaped, and have five ſlender ſtamina with an oval germen, which afterward becomes a ſingle ſhort ſeed crowned with a long down. The female flowers have no ſtamina, but in other reſpects are the ſame.

The SPECIES are,

1. BACCHARIS (*Ivæfolia*) foliis lanceolatis longitudinaliter dentato-ſerratis. Hort. Cliff. *African-tree Groundſel with a ſawed leaf.*

2. BACCHARIS (*Halimifolia*) foliis obverſè ovatis, ſuperne emarginato-crenatis. Lin. Hort. Cliff. *Virginia Groundſel-tree with an Orach leaf.*

The firſt ſort was brought from the Cape of Good Hope, but grows naturally in Peru, and in other parts of America. There is little beauty in the flower; it grows to the height of five or ſix feet, and is a manageable ſhrub; it may be propagated by cuttings, which ſhould be planted in a ſhady border during any of the ſummer months, or by ſeeds ſown in a common border in the ſpring of the year, which ripen well in this country, and, if permitted to ſcatter on the ground, the plants will come up the following ſpring. It is pretty hardy, and will live abroad in mild winters, if planted in a warm ſituation; but it is uſually kept in greenhouſes, and placed abroad in ſummer; it requires much water in warm weather.

The ſecond ſort is a native of Virginia and other parts of North America; it grows about ſeven or eight feet high, with a crooked ſhrubby ſtem, and flowers in October; the flowers are white, and not very beautiful; but the leaves continuing green through the year, has occaſioned this ſhrub to be admitted into many curious gardens.

This ſort may be propagated by cuttings, which ſhould be planted in April or May upon a ſhady border, and at Michaelmas they will be fit to tranſplant where they are to remain; this will live in the open air, and never is injured by the cold of our ordinary winters, but ſevere froſt will ſometimes deſtroy them.

BALAUSTIA. See PUNICA.

BALLOTE. Black Horehound.

This is a common weed, growing on the ſides of banks in moſt parts of England, ſo is ſeldom allowed a place in gardens; there are two varieties of it, one with a white, and the other a purple flower.

BALM. See MELISSA.

BALSAMINA, the female Balſamine or Balſamine. See IMPATIENS.

BALSAMITA. See TANACETUM.

BAMIA MOSCHATA. See HIBISCUS.

BANANA. See MUSA.

BANISTERIA. Houſt. MSS. Lin. Gen. 509.

The CHARACTERS are,

The flower hath five petals, which are ſhaped like thoſe of the papilionaceous tribe, but ſpread open, having in ſome ſpecies one, in others two, and in ſome ſeveral nectarious glands, with ten ſhort ſtamina. There are in ſome ſpecies three, and in others but one germen, which afterward become ſo many winged fruit, like thoſe of the Maple, each containing a ſingle ſeed.

The SPECIES are,

1. BANISTERIA (*Anguloſa*) foliis ovato-oblongis rigidis racemis terminalibus caule fruticoſo ſcandente. *Baniſteria with oblong, oval, ſtiff leaves, ſpikes of flowers terminating the branches, and a ſhrubby climbing ſtalk.*

2. BANISTERIA (*Fulgens*) foliis ovatis glabris, floribus corymboſis terminalibus, caule fruticoſo ſcandente. *Baniſteria with oval ſmooth leaves, flowers growing in a corymbus at the extremity of the branches, and a ſhrubby climbing ſtalk.*

3. BANISTERIA (*Bractiata*) foliis ovatis acuminatis floribus laxè ſpicatis, ramis diffuſis ſcandentibus. *Baniſteria with oval pointed leaves, flowers growing in looſe ſpikes, and climbing diffuſed branches.*

4. BANISTERIA (*Laurifolia*) foliis cordatis nervoſis ſubtus incanis, floribus lateralibus, caule fruticoſo ſcandente. *Baniſteria with nervous heart-ſhaped leaves, hoary on their under ſide, flowers growing from the ſide of the branches, and a ſhrubby climbing ſtalk.*

5. BANISTERIA (*Bengalenſis*) foliis ovato-oblongis acuminatis racemis lateralibus ſeminibus patentibus. Flor. Zeyl. 176. *Baniſteria with oblong, oval, pointed leaves, ſpikes of flowers growing from the ſide of the branches, and ſpreading ſeeds.*

6. BANISTERIA (*Aculeata*) foliis pinnatis foliolis ovatis ſpicis lateralibus ſeminibus erectis. *Baniſteria with winged leaves, whoſe ſmall leaves are oval, ſpikes of flowers growing from the ſide of the branches, and erect ſeeds.*

The firſt ſort grows naturally in Jamaica. This hath a woody ſtalk, which twiſts itſelf round the neighbouring trees, and raiſes itſelf to their top. It is garniſhed with leaves as large as thoſe of the Bay-tree, and of the ſame thickneſs, growing oppoſite; the flowers are produced in long branching ſpikes at the end of the branches, which are yellow, and are ſucceeded by two or three winged ſeeds like thoſe of the greater Maple.

The ſecond ſort grows naturally in Jamaica, at Campeachy, and ſeveral other parts of America. This hath ſlender winding ſtalks, which riſe five or ſix feet high, with oval ſmooth leaves; the flowers grow in a round bunch at the extremity of the branches, which are of a browniſh yellow colour, and are ſucceeded by winged ſeeds like the former, but ſmaller, and have narrower wings.

The third ſort came from Carthagena, where it naturally grows. This ſends out many branches, which divide again into others, growing without order, and become very buſhy upward, ſending out tendrils, by which they faſten themſelves to the neighbouring trees, and mount to a great height; theſe have oval ſtiff leaves, ending in a point. The flowers are produced in looſe ſpikes at the ends of the branches, which are firſt of a gold colour, and fade to a ſcarlet, and are ſucceeded by ſeeds of the ſame ſhape with the former, but are ſlender, thin, and for the moſt part ſingle.

The

The fourth ſort was ſent me from Campeachy; this hath many irregular climbing ſtalks, which faſten themſelves to the neighbouring trees, and riſe to a great height, with heart-ſhaped leaves, which are hairy on their under ſide, where they have many tranſverſe ribs. The flowers come out thinly from the ſide of the branches, which are of a pale yellow colour, and are ſucceeded by large winged ſeeds, which are double.

The fifth ſort hath ſtrong woody ſtalks, which twine about the trees that grow near it, and riſes twenty feet high, garniſhed with oblong pointed leaves like thoſe of the Bay-tree, growing by pairs oppoſite; from the wings of the leaves the flowers are produced in looſe ſpikes upon long foot-ſtalks, which are blue, and are ſucceeded by ſlender winged ſeeds, which ſpread open from each other.

The ſixth ſort hath ſtrong ligneous ſtalks, covered with an Aſh-coloured bark, and divide into many branches, garniſhed with winged leaves, compoſed of five or ſix pair of oval ſmall leaves, whitiſh on their under ſide; from the wings of the leaves are produced ſlender bunches of flowers, growing in a racemus like thoſe of the Currant-buſh, and are of a purpliſh colour; theſe are ſucceeded by broad winged ſeeds, growing erect.

Theſe plants are all of them natives of warm countries, ſo cannot be preſerved in England, unleſs they are kept in a bark-ſtove. They are propagated by ſeeds, which muſt be procured from the countries where they grow naturally Theſe ſeeds ſhould be fully ripe when gathered, and put into ſand, in which they ſhould be ſent to England, otherwiſe they will loſe their vegetative quality; for theſe ſeeds are not only in ſhape like thoſe of the Maple, but alſo are of the ſame quality, requiring to be ſown as ſoon as poſſible, when they are ripe, or preſerved in ſand till they are ſown, otherwiſe they rarely ſucceed. The ſeeds ſhould be ſown in pots, and plunged into a hot-bed of tanners bark, where the heat is very moderate, and if the plants ſhould not appear the firſt year, the pots ſhould be preſerved till the next ſpring, to ſee if the ſeeds will grow. When the plants come up, they muſt be planted in ſeparate pots, filled with light earth, and plunged into the bark-bed; after which they muſt be treated like other tender plants from the ſame countries.

BAOBOB. See ADANSONIA.

BARBA CAPRÆ. See SPIRÆA.

BARBA JOVIS. See ANTHYLLIS.

BARBAREA. See ERYSIMUM.

BARDANA. See ARCTIUM.

BARLERIA. The inhabitants of the iſland of Jamaica call it Snap-dragon.

The CHARACTERS are,

The flower is nearly of the lip kind, of one leaf, funnel-ſhaped, and divided into five parts at the top. It hath four ſlender ſtamina, two ſhort, and two longer. In the center is placed the oval germen, which afterward becomes an oblong membranaceous veſſel with two cells, which is very elaſtic, containing two roundiſh compreſſed ſeeds.

The SPECIES are,

1. BARLERIA (*Prionitis*) ſpinis axillaribus quaternis foliis integerrimis. Lin. Sp. Plant. 636. *Barleria with ſpines growing by fours from the ſide of the branches, and entire leaves.*

2. BARLERIA (*Solanifolia*) ſpinis axillaribus, foliis lanceolatis denticulatis. Lin. Sp. 887. *Barleria with ſpines on the ſide of the branches, and ſpear-ſhaped indented leaves.*

3. BARLERIA (*Buxifolia*) ſpinis axillaribus oppoſitis ſolitariis, foliis ſubrotundis integerrimis. Lin. Sp. 886. *Barleria with ſingle ſpines at the wings of the ſtalk, and roundiſh entire leaves.*

4. BARLERIA (*Coccinea*) inermis foliis ovatis denticulatis petiolatis. Lin. Sp. 888. *Barleria without ſpines, and oval indented leaves having foot-ſtalks.*

The firſt ſort riſes with a branching ſtalk four or five feet high, garniſhed with oval leaves at every joint; upon ſhort foot-ſtalks, at every diviſion of the branches, comes out a long foot-ſtalk, which divides into many ſmaller, and at each diviſion of theſe is placed a ſingle flower, of a pale colour; the upper ſegment being broad, and ſhaped like the galea, or helmet, the two lateral ſegments are narrow, and the under one is bent downward, repreſenting the under lip, and is cut into two parts. The germen becomes an oblong membranaceous capſule with two cells, each containing two compreſſed roundiſh ſeeds. This ſeed-veſſel is very elaſtic, and throws out the ſeeds with violence on their being touched when ripe.

The ſecond ſort riſes ſix or ſeven feet high, ſending out many lateral branches from the bottom upward, ſo as to form a ſort of pyramid; the leaves are oval. The flowers grow upon branching foot-ſtalks, which come out from the wings of the leaves, each ſtanding upon a ſhort ſeparate foot-ſtalk, and are ſucceeded by elaſtic ſeed-veſſels, of the ſame ſhape with the former.

The third ſort hath ſquare ſtalks four or five feet high, garniſhed with oblong entire leaves at every joint, above which the flowers come out in whorls ſurrounding the ſtalks, and under each whorl there are ſix ſharp ſpines, which are as long as the empalement of the flowers. The flowers are blue, and have more of the form of the labiated flowers than any of the other ſpecies.

The fourth ſort ſends out many ſlender ſtems from the root, which riſe ſeven or eight feet high, garniſhed with oval pointed leaves, two growing oppoſite at each joint, which are indented, and ſtand upon foot ſtalks; the flowers are produced on the ſide of the branches, which are of a ſcarlet colour, ſhaped like thoſe of the other ſorts.

The two firſt ſorts are perennial plants, which are eaſily propagated by cuttings during the ſummer months, which ſhould be planted in pots, and plunged into a moderate hot-bed, obſerving to ſhade them in the heat of the day till they have puſhed out roots; then they ſhould be each planted in a ſmall pot of light earth, and plunged into a tan-bed, and treated as other tender plants.

The roots of the other ſorts will continue three or four years, but after the ſecond year they grow too rambling, and the lower parts of the branches naked, ſo are not ſo ſightly as the young plants; therefore a ſucceſſion of theſe ſhould be preſerved, and the old ones turned out. When the ſeeds are received from abroad, they muſt be ſown upon a hot-bed in the ſpring; and when the plants are fit to remove, they muſt be each planted in a ſeparate pot, and plunged into a hot-bed of tanners bark, where they muſt conſtantly remain, and managed in the ſame manner as other tender exotics from the ſame countries, giving them water frequently in ſummer, and letting in the freſh air to them every day in warm weather, but in winter they ſhould have little water, and be kept warm. They flower in June, July, and Auguſt, and their ſeeds ripen ſoon after.

The ſecond ſort is only propagated by cuttings, which muſt be planted in pots, and plunged into a moderate hot-bed, where they will take root, and the plants ſhould be kept conſtantly in the ſtove. This ſort requires plenty of water in ſummer, and in winter they muſt be frequently refreſhed, but they muſt not have it in too great quantity.

BASELLA, or climbing Nightſhade from Malabar.

The CHARACTERS are,

The flower hath no empalement; it is ſhaped like a pitcher, cloſed toward the brim. It hath five ſtamina faſtened to the petal. The globular germen, which is ſituated in the center, ſupports three ſlender ſtyles, crowned by oblong ſtigma. The petal of the flower remains, and incloſes a roundiſh fleſhy berry, including one round ſeed.

The

The Species are,

1. Basella (*Rubra*) foliis planis, pedunculis simplicibus. Lin. Sp. 390. *Basella with red leaves and simple foot-stalks.*

2. Basella (*Alba*) foliis ovatis undatis, pedunculis simplicibus folio longioribus. Lin. Sp. 390. *White Basella with oval waved leaves, set on foot-stalks longer than the leaves.*

The first sort has thick, strong, succulent stalks and leaves, which are of a deep purple colour. The plant will climb to the height of ten or twelve feet, which will twist round stakes, provided the plants are preserved in the stove; for if they are exposed to the open air, they will not grow so large, nor will they perfect their seeds, except it be in very warm seasons; but if they are placed in the stove, they will often live till the following spring, and produce great quantities of flowers and seeds. The flowers of this plant have no great beauty, but the plant is preserved for the odd appearance of the stalks and leaves.

There is a variety of this with green stalks and leaves, and the flowers are of a whitish green colour, tipped with purple on their edges, but in all other respects the same, so is supposed to be only a seminal variation.

The second sort hath flaccid leaves, and smaller flowers and fruit, in which it essentially differs from the first.

These plants are propagated by seeds, which should be sown in a hot-bed in the spring; and when the plants are fit to remove, they should be each planted into a separate pot, and plunged into the tan-bed, where they must be treated in the same manner as other tender exotics. They may also be propagated by cuttings, which should be planted in pots, and plunged into a moderate hot-bed of tanners bark, where they will take root in a fortnight or three weeks time, when they should be treated in the same manner as the seedling plants. But as these rise so easily from seeds, so they are seldom propagated any other way, because they are plants of short duration.

These will climb to a considerable height, and send forth a great number of branches; so they should be trained up to a trellis, or fastened to the back of the stove, otherwise they will twist themselves about whatever plants stand near them, which will make a very disagreeable appearance in a stove; whereas, when they are regularly trained to a trellis, they will have a good effect.

From the berries of the first sort I have seen a beautiful colour drawn, but when used for painting did not continue very long, but changed to a pale colour; though I believe there might be a method invented, whereby this beautiful colour might be fixed, so as to become very useful; for I have been assured, that the juice of these berries has been used for staining of callicoes in India.

BASILICUM, or BASIL. See Ocymum.

BASTERIA. All-spice.

As this plant has no proper title given to it, so I have given it this in honour of my worthy friend Dr. Job Baster, F. R. S. of Zurick Zee, in Holland, who is a gentleman well skilled in botany, and has a fine garden stored with rare plants, of which he is very communicative to his friends.

The Characters are,

The flower hath a double series of narrow petals, which spread open, and turn inward at their extremity. Under the receptacle is situated an oval germen, surrounded by many short stamina, crowned by obtuse summits. The germen afterward becomes a roundish fruit, compressed at both ends, having cells, containing oblong seeds.

We have but one Species of this genus, viz.

Basteria (*Calycantha.*) *Basteria with oval leaves placed opposite, and a branching shrubby stalk; commonly called in Carolina All-spice.*

This shrub grows naturally in America; Mr. Catesby, who first introduced it into the English gardens, procured it from the continent, some hundred miles on the back of Charles Town, in Carolina.

It seldom rises more than four feet high in this country, dividing into many slender branches near the ground, with two oval leaves, placed opposite at every joint, which are entire; the flowers comes out from the wings of the leaves; they have two series of narrow thick petals, which spread open, and turn inward at the top, like those of the starry Anemone, or the Virgin's Bower: these are of a sullen purple colour, and have a disagreeable scent; they appear in May; the embryo sits beneath the flowe , and supports five stigma; this afterward appears to have five cells, but the seeds never come to perfection in this country, therefore I can only give a description of it from an imperfect rudiment. The bark of this shrub is brown, and has a very strong aromatic scent, from whence the inhabitants of Carolina gave it the title of All-spice, by which it is generally known in the nurseries near London.

This shrub will thrive in the open air in England, if it be planted in a warm situation and a dry soil: it is propagated by laying down the young branches, which will take root in one year, and may then be taken from the mother plant, and planted where they are designed to remain, for they do not bear transplanting well after they are grown to any size.

The best time for laying down the branches is in the autumn, but they should not be transplanted till the spring twelve months after, for the spring is the safest time to remove those plants. After the branches are laid down, there should be some old tanners bark laid upon the surface of the ground to keep out the frost, which should also be done every winter, while the plants are young, which will prevent the frost from penetrating to their roots, and thereby secure them.

BAUHINIA. Mountain Ebony.

The Characters are,

The flower is composed of five petals. It hath two stamina: the oblong germen sits upon the foot-stalk, which afterward becomes a long taper pod, inclosing a row of roundish compressed seeds.

The Species are,

1. Bauhinia (*Aculeata*) caule aculeato. Hort. Cliff. 156. *Bauhinia with a prickly stalk.*

2. Bauhinia (*Spicata*) caule inermi foliis cordatis lobis acutis glabris, floribus spicatis terminalibus. *Bauhinia with a smooth stalk, heart-shaped smooth leaves with pointed lobes, and spikes of flowers terminating the branches.*

3. Bauhinia (*Acuminata*) foliis ovatis lobis acuminatis semi-ovatis. Lin. Sp. Pl. 375. *Bauhinia with oval leaves, and pointed lobes which are half oval.*

4. Bauhinia (*Ungulata*) foliis oblongo-cordatis, lobis acuminatis parallellis trinerviis, siliquis planis. *Bauhinia with oblong heart-shaped leaves, with pointed parallel lobes, having three ribs, and plain flat pods.*

5. Bauhinia (*Tomentosa*) caule aculeato, foliis cordatis lobis orbiculatis subtus tomentosis. *Bauhinia with a prickly stalk, and heart-shaped leaves with round lobes, which are woolly on their under side.*

6. Bauhinia (*Purpurascens*) foliis cordatis lobis semi-orbiculatis, floribus paniculatis axillaribus. *Bauhinia with heart-shaped leaves, having roundish lobes, and flowers growing in loose spikes from the side of the branches.*

7. Bauhinia (*Sparsa*) foliis subcordatis bipartitis rotundatis caule aculeato, floribus sparsis. *Bauhinia with heart-shaped, bifid, rounded leaves, a prickly stalk, and flowers growing at a distance.*

8. BAUHINIA (*Variegata*) foliis cordatis lobis coadunatis obtusis. Lin. Sp. Plant. 375. *Bauhinia with heart-shaped leaves, and obtuse lobes which join together.*

9. BAUHINIA (*Scandens*) caule cirrhifero. Lin. Sp. Pl. 374. *Bauhinia with a stalk having tendrils.*

10. BAUHINIA (*Divaricata*) foliis ovatis lobis divaricatis. Lin. Sp. Pl. 374. *Bauhinia with oval leaves, whose lobes spread different ways, called in Jamaica Honeysuckle.*

The first sort grows plentifully in Jamaica, and the other islands in America, where it rises to the height of sixteen or eighteen feet, with a crooked stem, and divides into many irregular branches, armed with short strong spines, and garnished with compound winged leaves, each having two or three pair of lobes, ending with an odd one, which are oblique, blunt, and indented at the top. The stalks are terminated by several long spikes of yellow flowers, which are succeeded by pods about three inches long, which have borders, and contain two or three swelling seeds. The pods are glutinous, and have a strong balsamic scent, as have also the leaves when bruised. It is called in America the Indian Savin-tree, from its strong odour, somewhat resembling the common Savin.

The second sort came from Campeachy, where it grows naturally. This rises to the height of twelve or fourteen feet, with a smooth stem, dividing into many branches, garnished with heart-shaped leaves, having two smooth pointed lobes; the extremity of every branch is terminated by a long spike of yellow flowers, so that when these trees are in flower, they make a fine appearance. The pods are swelling, and about five inches long, each containing five or six roundish compressed seeds.

The third sort grows naturally in both Indies, where it rises with several pretty strong, upright, smooth stems, which send out many slender branches on every side, garnished with leaves, deeply divided into two oval lobes. The leaves come out without order, and have long foot-stalks, but are much thinner than those of the species before-mentioned. The flowers come out at the extremity of the branches, three or four in a loose bunch; the petals are red, or striped with white, others are plain upon the same branch; the stamina and style are white, and stand out beyond the petals. These flowers are succeeded by long flat pods, of a dark brown colour, each containing five or six roundish compressed seeds. The wood of this tree is very hard, and veined with black, from whence the inhabitants of America call it Mountain Ebony.

The fourth sort grows naturally at Campeachy. This rises to the height of twenty feet, with a smooth stem, which divides into many small branches, garnished with oblong heart-shaped leaves, having two pointed parallel lobes, which have each three longitudinal veins. The leaves are placed alternately on the branches, which are terminated by loose bunches of white flowers; these are succeeded by very long, narrow, compressed pods, which have eight or ten compressed roundish seeds in each.

The fifth sort seldom rises more than ten feet high, dividing into many irregular branches, which are armed with short crooked spines; the leaves grow alternate, are heart-shaped, and have two roundish lobes; they are woolly on their under side, and have short foot-stalks. The flowers grow at the extremity of the branches, two or three together; these are large, and of a dirty white colour, and are succeeded by short flat pods, each containing two or three seeds.

The sixth sort grows naturally at La Vera Cruz. It rises to the height of twenty-five or thirty feet, with many irregular stems, garnished with heart-shaped leaves, having two roundish lobes. The flowers come out in loose spikes at every joint from the wings of the leaves, with naked foot-stalks, and are of a dirty white colour. These are succeeded by oblong compressed pods, each containing three or four compressed seeds.

The seventh sort grows naturally at Carthagena in New Spain. This rises twenty feet high, with a strong upright stem, which sends out many branches, which are armed with spines growing by pairs. The leaves grow alternately, and are heart-shaped, with two rounded lobes. The flowers are large and white, coming out thinly at the end of the branches. The petals of these are near two inches long, and spread open wide; these are succeeded by long flat pods, which are narrow, each containing five or six seeds.

The eighth sort grows naturally in both Indies. It rises with a strong stem upward of twenty feet high, dividing into many strong branches, which are garnished with heart-shaped leaves, having obtuse lobes. The flowers grow in loose panicles at the extremity of the branches, which are large, and of a purplish red colour, marked with white, and have a yellow bottom. These have a very agreeable scent. The flowers are succeeded by compressed pods, about six inches long, and three quarters of an inch broad, containing three or four compressed seeds in each.

The ninth sort grows naturally in both Indies, where it rises with many slender stalks, which put out tendrils, and fasten themselves to the neighbouring trees, whereby they rise to a great height; the leaves are heart-shaped, standing upon long foot-stalks, and are deeply cut into two pointed lobes, each having three prominent ribs running longitudinally.

The tenth sort grows naturally in great plenty on the north-side of the island of Jamaica, where it is called Upright Honeysuckle. This is a low shrub, seldom rising more than five or six feet high, but divides into several branches, garnished with oval leaves, divided into two lobes, which spread from each other. The flowers grow in loose panicles at the end of the branches, which are white, and have a very agreeable scent. The flowers are succeeded by taper pods about four inches long, each containing four or five roundish compressed seeds, of a dark colour.

All these plants are natives of the warm countries, so will not thrive in England, unless they are kept in a warm stove. They are propagated by seeds, which must be procured from the countries where they grow naturally, for they do not perfect their seeds in England.

The seeds should be sown in pots, and plunged into a moderate hot-bed of tanners bark; if they are good, the plants will come up in about two months, and in a month or six weeks after will be fit to transplant, when they should be carefully shaken out of the seed-pot, so as not to tear off the roots, and each planted into a separate small pot, and plunged into the hot-bed again, being careful to shade them until they have taken fresh root; after which they should have fresh air admitted to them every day in warm weather. In the autumn they must be placed in the bark-stove, and treated in the same way as other tender exotics, giving them but little water in winter.

BAY. See LAURUS.

BEANS. See FABA.

KIDNEY or FRENCH BEANS. See PHASEOLUS or DOLICHOS.

BEAN-TREFOIL. See CYTISUS.

BEAR's-EAR. See AURICULA.

BEAR's-EAR SANICLE. See VERBASCUM.

BEAR's-FOOT. See HELLEBORUS.

BECABUNGA, or Brook-lime.

This is a sort of Veronica, or Water Speedwell, of which there are two sorts, one with a long leaf, and the other round; they are both very common in ditches and watery places, place, almoſt every-where in England. The ſecond ſort is uſed in medicine.

BEE, or GNAT-FLOWER. See ORCHIS.

BEECH-TREE. See FAGUS.

BELL-FLOWER. See CAMPANULA.

BELLADONA. See ATROPA.

BELLIS. The Daiſy.

The CHARACTERS are,

It hath a radiated diſcous flower, compoſed of many hermaphrodite flowers in the diſk, and female florets which form the border or rays, included in a common empalement. The hermaphrodite flowers have an oval germen, attended by five ſhort ſtamina; the germen afterward becomes a ſingle naked ſeed placed vertically.

The SPECIES are,

1. BELLIS (*Perennis*) ſcapo nudo unifloro. Hort. Cliff. 418. *Daiſy with a naked ſtalk, having one flower.*

2. BELLIS (*Annua*) caule ſubfolioſo. Lin. Sp. Pl. 887. *Daiſy with leaves on the lower part of the ſtalk.*

3. BELLIS (*Hortenſis*) hortenſis flore pleno majore. C. B. P. 261. *Garden Daiſy with a larger double flower.*

The firſt ſort is the common Daiſy, which grows naturally in paſture land in moſt parts of Europe, and is often a troubleſome weed in the Graſs of gardens, ſo is never cultivated.

The ſecond ſort is a low annual plant, which grows naturally on the Alps, and in the hilly parts of Italy. This ſeldom riſes more than three inches high, with an upright ſtalk, which is garniſhed with leaves on the lower part, but the upper part is naked, ſupporting a ſingle flower like that of the common Daiſy, but ſmaller.

The Garden Daiſy is generally ſuppoſed to be only a variety of the wild ſort, which was firſt obtained by culture. This may probably be true, but there has not been any inſtance of late years of the wild ſort having been altered by culture; nor have I ever obſerved the Garden Daiſy to degenerate to the wild ſort, where they have been ſome years neglected, though they have altered greatly with regard to the ſize and beauty of their flowers; therefore I ſhall not conſider them as diſtinct ſpecies, but ſhall only mention the varieties which are cultivated in the gardens.

1. The red and white Garden Daiſy with double flowers.
2. The double variegated Garden Daiſy.
3. The Childing, or Hen and Chicken Daiſy.
4. The Cockſcomb Daiſy, with red and white flowers.

The Garden Daiſies flower in April and May, when they make a pretty variety, being intermixed with plants of the ſame growth; they ſhould be planted in a ſhady border, and a loamy ſoil without dung, in which they may be preſerved without varying, provided the roots are tranſplanted and parted every autumn; which is all the culture they require, except the keeping them clear from weeds.

BELLIS MAJOR. See CHRYSANTHEMUM.

BELLONIA.

The CHARACTERS are,

The flower is wheel-ſhaped, of one leaf, with a ſhort tube, but ſpread open above, and cut into five obtuſe ſegments. It hath five ſtamina, which cloſe together. The germen is ſituated under the receptacle of the flower, which afterward becomes an oval turbinated ſeed-veſſel, ending in a point, having one cell, filled with ſmall round ſeeds.

We have but one SPECIES of this plant, viz.

BELLONIA (*Aſpera.*) Lin. Sp. Plant. 172. *Shrubby Bellonia with a rough Balm leaf.*

This plant is very common in ſeveral of the warm iſlands in America.

It hath a woody ſtem which riſes ten or twelve feet high, ſending out many lateral branches garniſhed with oval rough leaves placed oppoſite; the flowers come out from the wings of the leaves in looſe panicles, which are of the wheel ſhape, of one leaf, divided into five parts; theſe are ſucceeded by oval capſules ending in a point, which are full of ſmall round ſeeds.

It is propagated by ſeeds which ſhould be ſown in a pot, and plunged into a hot-bed of tanners bark. When the plants are come up half an inch high, they ſhould be carefully tranſplanted into pots, and plunged into the hot-bed again, obſerving to water and ſhade them until they have taken root; after which time they ſhould have air admitted to them every day when the weather is warm, and frequently watered. In autumn they muſt be plunged into the bark-ſtove, and treated in the ſame manner as other tender exotics. The ſecond year theſe plants will ſometimes flower, but they rarely produce good ſeeds in this climate; however, they may be propagated by cuttings in the ſummer months, provided they are planted in light earth on a moderate hot-bed, and carefully watered and ſhaded until they have taken root.

BELVEDERE. See CHENOPODIUM.

BENZOIN. The Benjamin-tree. See LAURUS.

BERBERIS. The Barberry, or Pipperidge Buſh.

The CHARACTERS are,

The flower hath a coloured empalement compoſed of ſix leaves, which are roundiſh and concave. It hath two coloured nectariums faſtened to the baſe of each petal, and ſix ſtamina with two ſummits faſtened on each ſide their apex. The germen is cylindrical, and afterward becomes an obtuſe, cylindrical, umbilicated berry, having a puncture, and one cell incloſing two cylindrical ſeeds.

The SPECIES are,

1. BERBERIS (*Vulgaris*) pedunculis racemoſis. Mat. Med. 290. *The common Barberry.*

2. BERBERIS (*Canadenſis*) foliis obverſè ovatis. *Canada Barberry with very broad leaves.*

3. BERBERIS (*Cretica*) pedunculis unifloris. Lin. Sp. Pl. 331. *Barberry with a ſingle flower on each foot-ſtalk.*

The firſt ſort grows naturally in the hedges in many parts of England, but is alſo cultivated in gardens for its fruit, which is pickled, and uſed for garniſhing diſhes. This ſhrub riſes with many ſtalks from the root, to the height of eight or ten feet, which have a white bark, yellow on the inſide; the ſtalks and branches are armed with ſharp thorns, which commonly grow by threes; the leaves are oval, obtuſe, and ſlightly ſawed on their edges. The flowers come out from the wings of the leaves, in ſmall ramoſe bunches like thoſe of the Currant-buſh, which are yellow; theſe are ſucceeded by oval fruit, which are firſt green, but when ripe turn to a fine red colour. The flowers appear in May, and the fruit ripens in September.

This ſort is generally propagated by ſuckers, which are put out in great plenty from the root; but ſuch plants are very ſubject to ſend out ſuckers in greater plenty than thoſe which are propagated by layers, therefore the latter method ſhould be preferred. The beſt time for laying down the branches is in the autumn, when their leaves begin to fall; the young ſhoots of the ſame year are the beſt for this purpoſe; theſe will be well rooted by the next autumn, when they may be taken off and planted where they are deſigned to remain. Where this plant is cultivated for its fruit, it ſhould be planted ſingle, (not in hedges, as was the old practice) and the ſuckers every autumn taken away, and all the groſs ſhoots pruned out: by this method the fruit will be much fairer, and in greater plenty, than upon thoſe which are ſuffered to grow wild.

The Canada ſort was more common in the Engliſh gardens ſome years paſt, than at preſent. The leaves of this are much broader, and ſhorter than thoſe of the common ſort, and the fruit is black when ripe. This may be propagated

pagated in the same way as the common sort, and is equally hardy.

The Box-leaved sort is at present very rare in England, and while young the plants are somewhat tender, so have frequently been killed by severe frost. This never rises more than three or four feet high in England, but sends out many stalks from the root, which are strongly armed with spines at every joint; the leaves are shaped like those of the narrow-leaved Box-tree; the flowers come out from between the leaves, each upon a slender foot-stalk, but these are not succeeded by fruit in England.

This sort may be propagated by laying down the branches in the same manner as the first; but when the young plants are taken off they should be planted in pots, and sheltered under a frame in winter till they have obtained strength, when they may be turned out of the pots and planted in a warm situation.

BERMUDIANA. See SISYRINCHIUM.

BERNARDIA. See CROTON.

BESLERIA.

The CHARACTERS are,

The flower is of the lip kind, and of one leaf; it hath four stamina in the tube of the flower, two of which are longer than the other, with an oval germen, which afterward becomes an oval berry with one cell, filled with small seeds.

The SPECIES are,

1. BESLERIA (*Melitifolia*) pedunculis ramosis, foliis ovatis. Lin. Sp. Plant. 619. *Besleria with branching foot-stalks and oval leaves.*

2. BESLERIA (*Lutea*) pedunculis simplicibus confertis, foliis lanceolatis. Lin. Sp. Plant. 619. *Besleria with simple foot-stalks growing in clusters, and spear-shaped leaves.*

3. BESLERIA (*Cristata*) pedunculis simplicibus solitariis, involucris pentaphyllis. Lin. Sp. Plant. 619. *Besleria with stalks growing single, and a five-leaved involucrum.*

The first sort hath a smooth, woody, jointed stalk; at each joint are placed two oval nervous leaves opposite; the flowers come out from the wings of the leaves upon short branching foot-stalks, each sustaining six or eight flowers, which stand each upon a separate smaller foot-stalk. They are of one leaf, of an anomalous figure, and quinquefid; after the flower is past, the germen becomes an oval soft berry with one cell, filled with small seeds.

The second sort rises with a ligneous stem six or seven feet high, dividing toward the top into many irregular branches, garnished with spear-shaped leaves, which have many transverse veins; the flowers come out at the wings of the leaves in large clusters, each having a separate foot-stalk: these are small, tubulous, of a pale yellow colour, and are succeeded by round soft berries, inclosing many small seeds.

The third sort hath a creeping stalk, which sends out roots at every joint, garnished with oval leaves placed opposite, which have many transverse ribs, and are sharply sawed on their edges; from the wings of the leaves come out the foot-stalks of the flowers single, each sustaining one tubulous, irregular, hairy flower, divided at the top into five obtuse parts, with a large five-leaved involucrum, deeply sawed on the border.

These plants grow naturally in the warm parts of America, so are too tender to live in this country without artificial heat. They are propagated by seeds which should be sown on a hot-bed, and when the plants are come up half an inch high, they should be each transplanted into a small pot, and plunged into a hot-bed of tanners bark, where they should have air and water in proportion to the warmth of the season. When the plants have filled these small pots with their roots, they should be shaken out of them and put into larger pots, and plunged into the hot-bed again, giving them a large share of fresh air in warm weather. In winter they must be removed into the stove, where they must be kept in a temperate warmth, and should be often, but sparingly, watered. The second year these plants will flower, and sometimes they will perfect their seeds in this country, but they must be constantly preserved in the stove.

BETA. The Beet.

The CHARACTERS are,

The flower hath a five-leaved empalement which is permanent, but no petal, and five stamina placed opposite to the leaves of the empalement. The germen is situated below the receptacle, which afterward becomes a capsule with one cell, having a single seed wrapped up in the empalement.

The SPECIES are,

1. BETA (*Maritima*) caulibus decumbentibus, foliis triangularibus petiolatis. *Beet with declining stalks, and triangular leaves having foot-stalks; commonly called Sea Beet.*

2. BETA (*Hortensis*) foliis radicalibus petiolatis, caulinis sessilibus, spicis lateralibus longissimis. *The common white Beet, or Cicla of the shops.*

3. BETA (*Vulgaris*) foliis latissimis, radice pyramidato carnoso. *Red Beet with a pyramidal root.*

The first sort grows naturally on the banks of the sea, and in salt marshes in divers parts of England. This has been supposed by many to be the same with the second species; but I have brought the seeds from the places where they grow naturally many times, and have cultivated the plants with care, but could not find any of them vary from their parent plants in their characters, so that I can make no doubt of its being a distinct species.

The second sort is cultivated in gardens for its leaves, which are frequently used in soups; the root of this sort seldom grows larger than a man's thumb; the spikes of flowers come out from the wings of the leaves, which are long, and have narrow leaves placed between the flowers. The lower leaves of the plant are thick and succulent, and their foot-stalks are broad. The varieties of this are, the white Beet, the green Beet, and the Swiss or Chard Beet. These will vary from one to the other by culture, as I have often experienced, but never alter to the first or third sort.

The third sort hath large, thick, succulent leaves, which are for the most part of a dark green, or purple colour. The roots of this are large, and of a deep red colour, on which their goodness depends, for the larger these roots grow, the tenderer they will be; and the deeper their colour, the more they are esteemed. The varieties of this are, the common red Beet, the Turnep-rooted red Beet, and the green-leaved red Beet.

The second sort, which is cultivated in gardens for its leaves, which are used in the kitchen, is commonly sown by itself, and not mixed with other crops, the beginning of March, upon an open spot of ground not too moist. When the plants have put out four leaves, the ground should be hoed as is practised for Carrots, carefully cutting up all the weeds, and also the plants where they are too near each other, leaving them at least six inches asunder. In three weeks or a month's time, the ground should be a second time hoed over to cut up the weeds, and thin the plants to a greater distance; for by this time they will be past danger, so should not be left nearer than eight or ten inches, if regard is had to the goodness of the leaves: and if it is of the Swiss kind, with broad leaves, the plants must not be nearer than a foot: in six weeks after, the ground should be hoed over a third time, which, if properly done, will destroy all the weeds; so that after this, the plants will spread and prevent the weeds from growing, therefore will want but little cleaning for a considerable time, and the

the leaves will ſoon be fit for uſe, when the outer larger leaves ſhould be firſt gathered, leaving the ſmall inner leaves to grow larger; ſo that a ſmall ſpot of ground will ſupply a moderate family, and furniſh a new ſupply of leaves the whole year, provided the plants are not permitted to run up to ſeed, for after that their leaves will not be good.

The red Beet is frequently ſown with Carrots, Parſneps, or Onions, by the kitchen-gardeners near London, who draw up their Carrots or Onions when they are young, whereby the Beets will have room to grow, when the other crops are gathered; but where the crops are not timely removed from them, it will be a better method to ſow them ſeparately. This ſort requires a deep light ſoil, for as their roots run deep in the ground, ſo in ſhallow ground they will be ſhort and ſtringy. The ſeeds ſhould be ſown in March, and muſt be treated in the ſame manner as the former ſort; but the plants ſhould not be left nearer than a foot diſtance, or in good land a foot and a half, for the leaves will cover the ground at that diſtance. The roots will be fit for uſe in the autumn, and continue good all the winter; but in the ſpring, when they begin to ſhoot, they will be hardy and ſtringy.

BETONICA. Betony.

The CHARACTERS are,

The flower is of one leaf, of the lip kind, with a cylindrical incurved tube; the upper lip is roundiſh, plain, erect, and entire; the lower lip is cut into three parts. It hath four ſtamina, two long and two ſhorter, which incline to the upper lip. The germen is quadripartite, which afterward becomes four naked oval ſeeds lodged in the empalement.

The SPECIES are,

1. BETONICA (*Officinalis*) ſpicâ interruptâ, corollarum laciniâ labii intermediâ emarginatâ. Flor. Leyd. Prod. 316. *Betony with an interrupted ſpike, and the middle ſegment of the lower lip of the flower indented at the end; Purple or Wood Betony.*

2. BETONICA (*Danica*) foliis radicalibus ovato-cordatis, caulinis lanceolatis obtuſis ſpicâ craſſiore. *Greater Daniſh Betony.*

3. BETONICA (*Alpina*) foliis triangularibus obtuſis ſpicâ breviore. *The leaſt Alpine Betony.*

4. BETONICA (*Orientalis*) ſpicâ integrâ, corollarum laciniâ labii intermediâ integerrimâ. Flor. Leyd. Prod. 316. *Eaſtern Betony with very long narrow leaves, and a thicker ſpike of flowers.*

5. BETONICA (*Incana*) foliis lanceolatis obtuſis incanis ſpicâ florum craſſiori. *Hoary Italian Betony, with a fleſh-coloured flower.*

6. BETONICA (*Annua*) verticillata calycibus ſpinoſis. Hort. Upſ. 165. *Annual Field Betony with a yellowiſh white flower.*

7. BETONICA (*Lutea*) ſpicâ baſi folioſâ. Lin. Sp. Plant. 573. *Yellow Mountain Betony.*

The firſt ſort grows naturally in woods, and on ſhady banks in moſt parts of England, ſo is ſeldom cultivated in gardens. This is the ſort which is uſed in medicine, and is greatly eſteemed as a vulnerary herb. There is a variety of this with a white flower, which I have often found growing naturally in Kent.

The ſecond ſort grows naturally in Denmark. This differs greatly from our common ſort, the lower leaves being much broader and heart-ſhaped; thoſe upon the ſtalks are ſpear-ſhaped and rounded at the end, and the ſtalks are larger, ſtand upright, and are terminated by thicker ſpikes of flowers.

The third ſort grows naturally upon the Alps, where it ſeldom riſes more than four inches high; and when cultivated in a garden, not above ſeven or eight. The leaves of this are much broader at the baſe than thoſe of the common ſort, and are very different in their ſhape, being triangular and blunt at the end. The flowers grow in very ſhort cloſe ſpikes on the top of the ſtalks.

The fourth ſort hath very long, narrow, hairy leaves, neatly crenated on their edges. The flowers grow in very cloſe thick ſpikes at the top of the ſtalks, which are larger, and of a lighter purple colour than thoſe of the common ſort.

The fifth ſort grows naturally in Italy. The leaves of this are broader, and not ſo long as thoſe of the common ſort, and are hoary; the ſtalks are ſhorter and much thicker, as are alſo the ſpikes of flowers, which are larger and of a fleſh colour.

The ſixth ſort is annual, and grows naturally on arable land in France, Italy, and Germany; and if brought into a garden, the ſeeds will ſcatter, and produce plenty of the plants without farther care.

The other ſorts are perennial plants, which may be propagated by ſeeds, or parting of their roots. They are all very hardy, but require a ſhady ſituation and moiſt ſtiff ſoil, in which they will thrive better than in rich ground.

BETONICA AQUATICA. See SCROPHULARIA.

BETONICA PAULI. See VERONICA.

BETULA. The Birch-tree.

The CHARACTERS are,

It hath male and female flowers at ſeparate diſtances on the ſame tree; the male flowers are collected in a cylindrical katkin. The flower is compoſed of three equal florets, fixed to the empalement by a ſingle ſcale, and have four ſmall ſtamina. The female flowers grow in a katkin, in the ſame manner as the male, which are heart-ſhaped. They have no viſible petals, but a ſhort oval germen. It hath no pericarpium, but the ſeeds are included in the ſcales of the katkin, which are oval and winged.

The SPECIES are,

1. BETULA (*Alba*) foliis ovatis acuminatis ſerratis. Hort. Cliff. 442. *The common Birch-tree.*

2. BETULA (*Nana*) foliis orbiculatis crenatis. Flor. Lap. 266. *Dwarf Birch with roundiſh leaves.*

3. BETULA (*Lenta*) foliis cordatis oblongis acuminatis ſerratis. Lin. Sp. Plant. 983. *Birch-tree with oblong, pointed, heart-ſhaped, ſawed leaves.*

4. BETULA (*Nigra*) foliis rhombeo-ovatis acuminatis, duplicato-ſerratis. Lin. Sp. Plant. 982. *Black Virginia Birch-tree.*

The firſt is the common Birch-tree, which is ſo well known as to need no deſcription. This is not much eſteemed for its wood, but however it may be cultivated to advantage upon barren land, where better trees will not thrive, for there is no ground ſo bad but this tree will thrive in it; for it will grow in moiſt ſpringy land, or in dry gravel or ſand, where there is little ſurface; ſo that upon ground which produced nothing but Moſs, theſe trees have ſucceeded ſo well as to be fit to cut in ten years after planting, when they have been ſold for near 10 l. per acre ſtanding, and the after produce has been conſiderably increaſed. And as many of the woods near London, which were chiefly ſtocked with theſe trees, have been of late years grubbed up, ſo the value of theſe plantations have advanced in proportion. Therefore thoſe perſons who are poſſeſſed of ſuch poor land cannot employ it better, than by planting it with theſe trees, eſpecially as the expence of doing it is not great

The beſt method to cultivate this tree, is to furniſh yourſelf with young plants from the woods where they naturally grow, and are generally found there in great plenty; but in places where there are no young plants to be procured near, they may be raiſed from ſeeds, which ſhould be carefully gathered in the autumn, as ſoon as the ſcales under which they are lodged begin to open, otherwise they will soon fall out and be lost: the seeds are small, so should not be buried deep in the ground. The autumn is the best season to sow them, and in a shady situation the plants will thrive better than when they are exposed to the full sun; for in all places where there are any large trees their seeds fall, and the plants come up well without care; so that if the young plants are not destroyed by cattle, there is generally plenty of them in all the woods where there are any of these trees. These wild plants should be carefully taken up, not to injure their roots. The ground where they are to be planted, will require no preparation; all that is necessary to be done, is to loosen the ground with a spade or mattock, in the places where the plants are to stand, making holes to receive their roots, covering them again when the plants are placed, closing the earth hard to their roots. If the plants are young, and have not much top, they will require no pruning; but where they have bushy heads, they should be shortened to prevent their being shaken and displaced by the wind. When the plants have taken root, they will require no other care but to cut down the great weeds which would over-hang the plants, being careful not to cut or injure the young trees. This need not be repeated oftener than twice in a summer the two first years, after which time the plants will be strong enough to keep down the weeds, or at least be out of danger from them.

These may be planted any time from the middle of October till the middle of March, when the ground is not frozen; but in dry land the autumn is the best season, and the spring for moist. The distance which they should be planted is four feet square, that they may soon cover the ground, and by standing close they will draw each other up; for in situations where they are much exposed, if they are not pretty close, they will not thrive so well.

If the plants take kindly to the ground, they will be fit to cut in about ten years; and afterward they may be cut every seventh or eighth year, if they are designed for the broom-makers only; but where they are intended for hoops, they should not be cut oftener than every twelfth year.

The broom-makers are constant customers for Birch, in all places within twenty miles of London, or where it is near water carriage, in other parts the hoop-benders are the purchasers; but the larger trees are often bought by the turners, and the wood is used for making ox-yokes, and other instruments of husbandry.

In some of the northern parts of Europe the wood of this tree is greatly used for making of carriages and wheels, being hard and of long duration. In France it is generally used for making wooden shoes. It makes very good fuel.

In some places these trees are tapped in the spring, and the sap drawn out to make Birch wine, which has been recommended for the stone and gravel, as is also the sap unfermented. The bark of the Birch-tree is almost incorruptible. In Sweden the houses are covered with it, where it lasts many years. It frequently happens that the wood is entirely rotten, and the bark perfectly sound and good.

The second sort grows naturally in the northern parts of Europe, and upon the Alps; this seldom rises above two or three feet high, having slender branches garnished with round leaves, but seldom produces either male or female flowers here. It is preserved in some curious gardens for the sake of variety, but is a plant of no use.

The third and fourth sorts grow naturally in North America. In Canada these trees grow to a large size, where the third sort is called Merisier. The natives of that country make canoes of the bark of these trees, which are very light, and of long duration.

Both these sorts may be propagated by seeds in the same manner as the first, and are equally hardy.

BIDENS. Tourn. Inst. R. H. 362. Water Hemp Agrimony.

The CHARACTERS are,

It hath a compound flower; the middle or disk is composed of hermaphrodite florets; these have five stamina, with an oblong germen. The female flowers which compose the border are naked; these are all succeeded by a single, angular, obtuse seed, having two or more bristles or teeth.

There are several species of this plant which are seldom admitted into gardens, some of which are common weeds in England, therefore I shall only mention those which are frequently preserved in the gardens of the curious.

The SPECIES are,

1. BIDENS (*Frondosa*) foliis pinnatis serratis seminibus erecto-constantibus calycibus frondosis corollis radiatis. Lin. Sp. Plant. 832. *Broad-leaved Canada Hemp Agrimony, with a yellow flower.*

2. BIDENS (*Nodiflora*) foliis oblongis integerrimis caule dichotomo floribus solitariis sessilibus. Lin. Sp. Plant. 832. *Hemp Agrimony with oblong entire leaves, a stalk divided into two parts, and a single flower growing close to the stalk.*

3. BIDENS (*Nivea*) foliis simplicibus serratis petiolatis, floribus globosis, pedunculis elongatis seminibus lævibus. Lin. Sp. Plant. 833. *Hemp Agrimony with single sawed leaves having foot-stalks, globular flowers with longer foot-stalks, and smooth seeds.*

4. BIDENS (*Frutescens*) foliis ovatis serratis petiolatis, caule fruticoso. Hort. Cliff. 399. *Hemp Agrimony with oval sawed leaves having foot-stalks, and a shrubby stalk.*

The first sort grows naturally in Virginia, Maryland, and Canada, where it is often a troublesome weed. It rises from three to four feet high, sending out many horizontal branches garnished with trifoliate leaves, deeply sawed on their edges; the flowers are produced at the end of the branches in small clusters, which are yellow, and succeeded by oblong square seeds, having two crooked horns, by which they fasten themselves to the clothes of those who pass near them. This is an annual plant, which decays soon after the seeds are ripe.

The second sort grows naturally in warm countries. It is an annual plant which rises near three feet high, dividing into several branches, which are garnished with oblong entire leaves; the flowers come out single at the divisions of the branches, sitting close; these are white, and succeeded by smooth seeds.

This sort must be sown upon a moderate hot-bed in the spring, and afterward treated like other hardy annual plants, planting them into the full ground the latter end of May. They will flower in June, and their seeds ripen in autumn, soon after which the plants will decay.

The third sort grows naturally in South Carolina, and also at Campeachy. This is also an annual plant, which rises three feet high; the leaves come out by pairs at each joint upon long slender foot-stalks. The flowers grow at the extremity of the branches in small globular heads, which are very white, and are succeeded by smooth seeds. This must be sown upon a hot-bed, and treated as the former. It flowers and seeds about the same time.

The fourth sort rises with a shrubby stalk to the height of six or seven feet. The flowers are produced at the end of the branches in small clusters, each standing upon a long naked foot-stalk, and are succeeded by flat seeds, having two short teeth at their extremity. This sort grows naturally in Carthagena in New Spain. It is propagated by seeds, which should be sown on a hot-bed in the spring; and the plants must be each planted into a separate small pot, and plunged into a fresh hot-bed, and treated as other tender plants from the same countries, and in autumn placed in the stove: the following summer they will flower and

produce

produce feeds, but the plants will abide fome years with proper management.

BIFOLIUM. Twyblade. See OPHRYS.

BIGNONIA. Tourn. Inft. 164. Trumpet-flower, or Scarlet Jafmine.

The CHARACTERS are,

The flower is of the ringent or grinning kind, tubulous, with long chaps, which are fwelling and bell-fhaped; it hath four ftamina fhorter than the petal, two longer than the other. In the center is an oblong germen, which afterward becomes a bi-valve pod with two cells, filled with compreffed winged feeds, lying over each other imbricatim.

The SPECIES are,

1. BIGNONIA (*Radicans*) foliis pinnatis, foliolis incifis, geniculis radicatis. Lin. Hort. Cliff. 217. *Bignonia with winged leaves which are cut on their edges, and roots coming out at their joints; commonly called Trumpet-flower.*

2. BIGNONIA (*Catalpa*) foliis fimplicibus cordatis, caule erecto, floribus diandris. Lin. Sp. Plant. 622. *Commonly called Catalpa.*

3. BIGNONIA (*Frutefcens*) foliis pinnatis, foliolis lanceolatis acutis ferratis, caule erecto, floribus paniculatis erectis. *Bignonia with winged leaves, having acute fawed lobes, an upright ftalk, and flowers growing in panicles erect.*

4. BIGNONIA (*Pubefcens*) foliis conjugatis cirrhofis foliolis cordato-lanceolatis foliis imis fimplicibus. Vir. Cliff. 59. *Bignonia with conjugated leaves, having tendrils and a fhort pod.*

5. BIGNONIA (*Unguis Cati*) foliis conjugatis, cirrho breviffimo arcuato tripartito. Lin. Sp. Pl. 623. *Bignonia with leaves by pairs, and fhort arched tendrils divided into three parts, and a very long pod.*

6. BIGNONIA (*Sempervirens*) foliis fimplicibus lanceolatis caule volubili. Lin. Sp. Plant. 623. *Bignonia with fingle fpear-fhaped leaves, and a twining ftalk, called fweet-fcented Jafmine in Carolina.*

7. BIGNONIA (*Pentaphylla*) foliis digitatis integerrimis. Hort. Cliff. 497. *Bignonia with fingered leaves, which are entire.*

8. BIGNONIA (*Paniculata*) foliis conjugatis cirrhofis, foliolis cordato-ovatis, floribus racemofo-paniculatis. Lin. Sp. Plant. 623. *Bignonia with jointed leaves having tendrils, the lobes oval and heart-fhaped, and flowers in branching panicles.*

9. BIGNONIA (*Cærulea*) foliis bipinnatis, foliolis lanceolatis integris. Lin. Sp. Plant. 625. *Trumpet-flower with double winged leaves, and blue flowers; called Baftard Guaiacum.*

10. BIGNONIA (*Leucoxylon*) foliis digitatis, foliolis integerrimis ovatis acuminatis. Lin Sp Plant. 870. *Trumpet-flower with hand-fhaped leaves, and oval entire pointed lobes.*

The firft fort grows naturally in Virginia and Canada. The fecond grows naturally in Carolina, but have both been old inhabitants in fome of the Englifh gardens, but the firft is the moft common in Europe.

Thefe plants when old have large rough ftems, which fend out many weak trailing branches, putting out roots at their joints, which faften themfelves to the trees in their natural places of growth, whereby they climb to a great height; and in Europe, where they are generally planted againft walls, they faften themfelves thereto by their roots, which ftrike into the mortar of their joints fo ftrongly, as to fupport their branches, and will rife to the height of forty or fifty feet. The branches are garnifhed with winged leaves placed oppofite, which are compofed of four pair of fmall leaves, terminated by an odd one. The flowers are produced at the ends of the fhoots of the fame year, in large bunches; they have long fwelling tubes fhaped fomewhat like a trumpet, from whence they had the appellation of Trumpet-flower; they are of an Orange colour, and appear the beginning of Auguft.

The firft fort is very hardy, fo will thrive in the open air; but as it hath trailing branches, it muft be fupported; therefore it is ufually planted againft walls or buildings where, if it has room, will rife very high, fo is very proper for covering of buildings which are unfightly.

It is propagated by feeds, but the young plants fo raifed do not flower in lefs than feven or eight years; therefore thofe which are propagated by cuttings or layers from flowering plants are moft efteemed, becaufe they will flower in two or three years after planting. The old plants alfo fend out many fuckers from their roots, which may be taken off, and tranfplanted where they are to remain, for thefe plants will not tranfplant fafely if they are old.

The third fort was brought into England by Mr. Catefby, who found it growing naturally on the back of South Carolina, at a great diftance from the Englifh fettlements, and brought the feeds to Charles Town, where the inhabitants have propagated it, and difperfed it through moft of the Englifh fettlements in North America, and is now very plenty in the Englifh gardens near London.

This fort rifes with a ftrong ftem, covered with a fmooth brown bark, dividing into many branches, which are garnifhed with very large heart-fhaped leaves, placed oppofite at every joint. The flowers are produced in large branching panicles at the end of the branches, of a dirty white colour, with a few purple fpots, and feint ftripes of yellow on their infide, and waved on their edges. The flowers are in America fucceeded by very long taper pods, filled with flat winged feeds, lying over each other like the fcales of fifh. Thefe plants, when young, are frequently injured by froft, for as they fhoot pretty late in the autumn, fo the early frofts often kill the extremity of their branches; but as the plants advance in ftrength, fo they become more hardy, and are feldom injured but in very fevere winters. It is late in the fpring before thefe trees come out, which has often caufed perfons to believe they were dead, and fome have been fo imprudent as to cut them down on that fuppofition, before the tree was fo well known.

It may be propagated by cuttings, which fhould be planted in pots in the fpring before the trees begin to pufh out their fhoots, and plunged into a moderate hot-bed. In about fix weeks thefe will have taken root, and made fhoots above, therefore fhould have air admitted to them conftantly, and hardened by degrees to bear the open air, into which they fhould be removed for the fummer, but in winter will require fome fhelter, and the fpring following planted out into a nurfery-bed, where they may ftay two years to get ftrength, and then may be removed to the place where they are to remain.

As thefe trees have very large leaves, fo they require a fheltered fituation, for where they are much expofed to ftrong winds, their leaves are often torn and rendered unfightly, and many times their branches are fplit and broken by the winds, their leaves being fo large, that the wind has great force againft them. They delight in a light moift foil, where they make great progrefs, and in a few years will produce flowers.

The fourth fort is a native of the warmer parts of America. This rifes with an upright ftem to the height of twelve or fourteen feet, fending out many branches, garnifhed at every joint by two long winged leaves placed oppofite; the fmall leaves or lobes which compofe thefe are long and fpear-fhaped, ending in a point. The flowers are produced in loofe panicles at the end of the branches, and are fhaped like thofe of the other fpecies, but fpread open more at the top. Thefe are yellow, and are fucceeded by compreffed pods about fix inches long, having

having two rows of flat winged feeds like thofe of the other fpecies.

This fort is propagated by feeds, which muft be fown on a hot-bed, and the plants afterward tranfplanted into feparate fmall pots, and plunged into a frefh hot-bed, to bring the plants forward, that they may obtain ftrength before winter; in the autumn they muft be removed into the bark-ftove, and during the winter fhould have but little water. The plants fhould conftantly remain in the bark-ftove; and be treated in the fame manner as other tender plants from thofe countries. The third year from feed they will flower, but they do not produce good feeds in England.

The fifth fort grows naturally in feveral parts of North America; this hath very flender trailing ftalks which muft be fupported, fo they require the affiftance of a wall, and to have a good afpect, for they are impatient of much cold: the branches are cloathed with oblong leaves, which remain green all the year; thefe are often fingle at bottom, but upward are placed by pairs oppofite at each joint; the flowers are produced at the wings of the leaves, which are fhaped like thofe of the Foxglove, and are yellow. This is propagated either by feeds or layers; the feeds fhould be fown on a moderate hot-bed, and the plants, when they have obtained ftrength, fhould be removed into the open air to harden them; but the firft winter they will require a little fhelter, and the following fpring may be planted where they are to remain.

The fixth fort hath flender ftalks like the former, which require the fame fupport; thefe are garnifhed with fmall oval leaves which are entire, placed oppofite by pairs at every joint; at the fame places come out the tendrils, by which they faften themfelves to the plants which grow near them; the flowers come out from the wings of the leaves, which are fhaped like thofe of the former fort, but are fmaller. This grows naturally in Carolina, and the Bahama Iflands.

The feventh fort hath very weak flender branches, which put out tendrils at the joints: at each joint there are four leaves, two on each fide; thefe are oval, pointed, and waved on their edges, of a bright green; the branches ramble very far where they have room, where it fpreads over the hedges, and during its continuance of flowering, perfumes the air to a great diftance: it is called in the country of its natural growth Yellow Jafmine. This plant is generally propagated by feeds, which are frequently brought from America, but it will take root by laying down of the branches; however this plant feldom continues long here, for the heat of the ftove is too great for it, and in a common greenhoufe it will not live through the winter. A few years paft I had two plants which lived through the winter in the full ground, and produced flowers the following fummer; but the fucceeding winter deftroyed them, though they were covered.

The tenth fort grows naturally in Jamaica. This rifes with an upright ftem near twenty feet high, fending out many lateral branches, which are covered with a white bark. The leaves come out oppofite at the joints, upon long foot-ftalks; thefe are compofed of five oval ftiff leaves, which are joined in one center at their bafe. They are of a pale green, inclining to white on their under fide; the flowers are produced at the end of the branches, four or five together, on very fhort foot-ftalks; they are narrow at bottom, but the tube enlarges upward, and at the top fpreads open wide; they are of a pale blufh colour, and fmell fweet, and are in America fucceeded by taper crooked pods about four inches long, which are filled with oval compreffed feeds, with wings of a filver colour.

This fort is a native of the warmer parts of America, therefore will not thrive in this country, but in a ftove. It is propagated by feeds, which muft be fown on a hot bed, and the plants treated in the fame manner as the fourth fort.

The ninth fort grows naturally in the Bahama Iflands. This, in the country where it grows naturally, rifes to the height of twenty feet, fending out many lateral branches, which are garnifhed with compound winged leaves, each having eleven alternate wings, with fpear-fhaped fmall lobes which grow alternate, and are entire; at the ends of the branches the flowers are produced in very loofe panicles; the foot-ftalks branching into three or four, each fuftaining a fingle blue flower, with a long fwelling tube, cut into five unequal fegments at the top, where it fpreads open. The flowers are fucceeded by oval feed-veffels, which open in two parts, and are filled with flat winged feeds.

This plant is tender, fo muft be conftantly kept in the bark-ftove, and treated in the fame manner as the fourth fort. It is propagated by feeds, which muft be obtained from the country where it grows naturally, for it doth not produce any in England.

BIHAI. See Musa.

BINDWEED. See Convolvulus.

BIRCH-TREE. See Betula.

BISCUTELLA. Lin. Gen. Pl. 724. Buckler Muftard, or Baftard Mithridate Muftard.

The Characters are,

The flower hath four petals placed in form of a crofs; it hath fix ftamina, four long and two fhort. In the center is fituated an orbicular compreffed germen, which afterward becomes a plain, compreffed, erect capfule, with two convex lobes, having two cells, terminated by the rigid ftyle, which is joined to the fide of the partition.

The Species are,

1. Biscutella (*Auriculata*) calycibus nectario utrinque gibbis, filiculis in ftylum coëuntibus. Lin. Hort. Cliff. 329. *Buckler Muftard with the cup of the nectarium fwelling on each fide, and fmall pods joined to the ftyle.*

2. Biscutella (*Didyma*) filiculis orbiculato-didymis à ftylo divergentibus. Hort. Cliff. 329. *Buckler Muftard, with a double orbicular pod diverging from the ftyle.*

3. Biscutella (*Apula*) hirfuta foliis oblongis dentatis femiamplexicaulibus floribus fpicatis ftylo breviore. *Hairy Buckler Muftard, with oblong indented leaves which half embrace the ftalk, flowers growing in fpikes, and a fhorter ftyle.*

The firft fort grows naturally in the fouth of France and Italy, where it rifes about a foot high; but in a garden generally grows two feet high, dividing into feveral branches, having oblong entire leaves a little indented, thofe on the lower part of the ftalk being broader and more obtufe than thofe of the upper. The flowers are produced at the ends of the branches in loofe panicles, of a pale yellow colour; thefe are fucceeded by double, round, compreffed feed-veffels, fwelling in the middle, where is lodged a fingle, round, flat feed.

The fecond fort grows naturally in the fouth of France, Italy, and Germany. This hath many long narrow leaves fpreading near the ground, which are deeply indented on each fide, refembling thofe of Hawkweed, and are hairy; from the center arifes the ftalk, which divides upward into many branches, having no leaves on them, and are terminated by loofe panicles of yellow flowers. Thefe are fucceeded by round compreffed feed-veffels like the former, but are fmaller, and the ftyle of the flowers bend from them.

The third fort fends out many oblong hairy leaves, which are flightly indented on their edges; from among thefe there arifes a hairy branching ftalk, which grows two feet high, and at each joint is placed one oblong indented leaf, which half embraces the ftalk at its bafe; each branch is terminated

nated by a close spike of pale yellow flowers, which are succeeded by round compressed seed-vessels like the other sorts, but the style of the flower, which is joined to them, is shorter than those of the other species.

All these sorts are annual plants, which perish soon after they have perfected their seeds. These should be sown either in spring, or in the autumn, upon a border of light earth, in an open situation, where they are to remain for good. Those which are sown in autumn, if the plants live through the winter without any protection, will flower earlier the following summer, whereby good seeds may always be obtained; whereas those which are sown in the spring, do, in bad seasons, decay before their seeds are ripe. If their seeds are permitted to scatter, there will be plenty of young plants produced without any care.

BISERRULA. Lin. Gen. Plant. 800.

The CHARACTERS are,

The flower is papilionaceous, having a large roundish standard, whose edges are reflexed. The wings are oblong, and the keel is of the same length with the wings. It hath ten stamina, nine of which are joined, and the other single. In the center is situated an oblong compressed germen, which afterward becomes a flat narrow pod, indented on both edges like the saw of the sword fish, having two cells, filled with kidney-shaped seeds.

We have but one SPECIES of this genus, viz.

BISERRULA (*Pelecinus.*) Hort. Cliff. 361. We have no English name for this plant.

This is an annual plant, which grows naturally in Italy, Sicily, Spain, and the south of France. It sends out many angular stalks which trail on the ground, and are subdivided into many branches, garnished with long winged leaves, composed of many pairs of lobes, and terminated with an odd one; toward the upper part of the branches come out the pedicle of the flowers, which sustain several small butterfly flowers, of a purplish colour, collected together, and are succeeded by plain pods, indented on both sides the whole length, containing two rows of kidney-shaped seeds.

It is propagated by seeds, which in this country should be sown in the autumn, on a bed of light earth, where the plants are to remain, for they will live in the open air very well. When the plants are come up, they will require no other care, but where they are too near they should be thinned to about a foot distance from each other. It flowers in June, and the seeds ripen in September.

BISLINGUA. See RUSCUS.

BISTORTA. Bistort, or Snakeweed.

There are three different species of this plant, which are found wild in England; but as they are seldom planted in gardens, I shall pass them over with only mentioning the common sort, which is used in medicine.

BISTORTA (*Major*) major, radice minùs intortâ. C. B. *The common great Bistort, or Snakeweed.*

This plant flowers in May, and if the season proves moist will continue to produce new spikes of flowers till August: it may be propagagated by planting the roots in a moist shady border, either in spring or autumn, which will soon furnish the gardens with plants, for it greatly increases by its creeping roots.

BIXA. Lin. Gen. Plant. 581. Anotta; by the French, *Roucou.*

The CHARACTERS are,

The flower hath a double series of petals, the outer consisting of five, which are large, the inner of the same number and shape, but narrower. It hath a great number of bristly stamina. In the center is situated an oval germen, which afterward becomes an oval heart-shaped capsule, covered with sharp bristles, opening with two valves with one cell, and filled with angular seeds.

We have but one SPECIES of this genus, viz.

BIXA (*Orellana.*) Hort. Cliff. 211. *The Anotta, or Arnotta; by the French,* Roucou.

This shrub grows naturally in the warm parts of America, where it rises with an upright stem to the height of eight or ten feet, sending out many branches at the top, forming a regular head, garnished with heart-shaped leaves ending in a point, which have long foot-stalks. The flowers are produced in loose panicles at the end of the branches; these are of a pale Peach colour, having large petals, and a great number of bristly stamina of the same colour in the center. After the flower is past, the germen becomes a heart-shaped, or rather a mitre-shaped seed-vessel, covered on the outside with bristles, opening with two valves, and filled with angular seeds, covered with a red pulp or paste, which colours the hands of those who touch it, and is collected for the use of dyers and painters.

This plant is propagated by seeds, which should be sown in a small pot, and plunged into a hot-bed of tanners bark. When the plants are about an inch high, they should be shaken out of the pot, and carefully separated, so as not to tear off their tender roots, and each planted in a small pot, and plunged into a fresh hot-bed of tanners bark, observing to shade them until they have taken new root; after which they must be treated as other tender plants from the same country, by admitting fresh air to them in proportion to the warmth of the season; when the heat of the tan declines, it should be turned up from the bottom, and, if necessary, some fresh tan added to renew the heat. If the plants are raised early in the spring, and properly managed, they will be a foot and a half high by the autumn, when they should be removed into the bark stove, and plunged into the tan-bed. During the winter, they must have but little water, and while the plants are young they must have a good share of warmth, otherwise they are very subject to cast their leaves, and frequently lose their tops, which renders them unsightly. They must be constantly kept in the bark-stove, otherwise they will not thrive well in England.

BLADDER NUT. See STAPHYLEA.

BLATTARIA. See VERBASCUM.

BLITUM. Lin. Gen. Plant. 14. Strawberry Blite.

The CHARACTERS are,

The flower hath no petals, but one bristly stamina the length of the empalement. In the center is situated an oval pointed germen; the empalement afterward becomes an oval compressed capsule, including one globular compressed seed, the size of the capsule.

The SPECIES are,

1. BLITUM (*Capitatum*) capitellis spicatis terminalibus. Hort. Upsal. 3. *Blite with spikes terminated by little heads, commonly called Strawberry Blite, or Strawberry Spinach.*

2. BLITUM (*Virgatum*) capitellis sparsis lateralibus. Hort. Upsal. 3. *Blite with small heads growing scatteringly from the sides of the stalks.*

The first sort grows naturally in Spain and Portugal, but hath been long preserved in the English gardens. This is an annual plant, which hath leaves somewhat like those of Spinach; the stalk rises two feet and a half high; the upper part of the stalk hath flowers coming out in small heads, at every joint, and is terminated by a small cluster of the same: after the flowers are past, the little heads swell to the size of Wood Strawberries, and when ripe have the same appearance, being very succulent, and full of a purple juice, which stains the hands of those who bruise them of a deep purple colour.

The second sort grows naturally in the south of France and Italy. This seldom grows more than one foot high, with smaller leaves than the first, but of the same shape; the

the flowers are produced at the wings of the leaves, almost the length of the stalk, which are small, and collected in little heads, which are shaped like those of the first, but smaller, and not so deeply coloured.

These are annual plants, which will drop their seeds if permitted, and the plants will come up in plenty the following spring: or if the seeds of either of the sorts are sown in March or April, upon a bed of common earth, in an open situation, the plants will come up, and, if they are to remain in the place where they are sown, will require no other care but to thin them out, so as to leave them eight or ten inches apart: in July the plants will begin to shew their berries, when they will make a pretty appearance: but many people transplant these plants into the borders of their flower-gardens, and others plant them in pots, to have them ready for removing to court-yards, or to place them upon low walls, among other annual flowers, to adorn those places.

When these plants are designed to be removed, they should be transplanted before they shoot up their flower-stems, for they will not bear transplanting well afterward. They will require to be duly watered in dry weather, otherwise the plants will stint, and not grow to any size; and, as the flower-stems advance, they should be supported by sticks, for if they are not, the branches will fall to the ground, when the berries are grown pretty large and weighty.

BLOODWORT. See LAPATHUM.

BOCCONIA.

The CHARACTERS are,

The flower hath four narrow petals, with a great number of very short stamina: in the center is situated a roundish germen, contracted at both ends, which afterward becomes an oval fruit, contracted at both ends, having one cell full of pulp, including a single round seed.

There is but one SPECIES of this genus at present known, viz.

BOCCONIA (*Frutescens.*) Lin. Sp. Plant. 505. *Branching Bocconia, with a woolly Cow Parsnep leaf.*

It is very common in Jamaica, and several other parts of America, where it grows to the height of ten or twelve feet, having a strait trunk as large as a man's arm, which is covered with a white smooth bark. At the top it divides into several branches, on which the leaves are placed alternately. These leaves are eight or nine inches long, and five or six broad, are deeply sinuated, sometimes almost to the mid-rib, and are of a fine glaucous colour. The whole plant abounds with a yellow juice, like the greater Celandine, which is of an acrid nature; so that it is used by the inhabitants of America, to take off warts and spots from the eyes.

It is propagated by seeds, which should be sown in a pot filled with light fresh earth early in the spring, and plunged into a hot-bed of tanners bark. When the plants are come up, they should be each transplanted into separate small pots, and plunged into the hot-bed again, observing to shade the glasses in the heat of the day until the plants have taken root, then they should have a large share of air, by raising the glasses of the hot-bed. When the plants have filled these small pots with their roots, they should be shaken out of them, and planted into pots one size larger, and plunged into the bark-stove, where they should have a good share of fresh air in warm weather. These plants must be constantly kept in the stove, being too tender to thrive in this country in any other situation. The singular beauty of this plant renders it worthy of a place in every curious collection; and it seems the Indians were very fond of it, for Hernandez tells us, the Indian kings planted it in their gardens.

BOERHAAVIA. Hogweed.

The CHARACTERS are,

The flower hath one bell-shaped petal, which is pentangular and entire. It hath in some species one, and in others two short stamina. The germen is situated below the receptacle, which afterward becomes a single oblong seed, having no cover.

The SPECIES are,

1. BOERHAAVIA (*Erecta*) caule recto. Lin. Sp. Pl. 3. *Boerhaavia with an erect stalk.*

2. BOERHAAVIA (*Diffusa*) caule diffuso. Lin. Sp. Pl. 3. *Boerhaavia with a diffused stalk.*

3. BOERHAAVIA (*Scandens*) caule scandente. Lin. Sp. Plant. 3. *Boerhaavia with a climbing stalk.*

4. BOERHAAVIA (*Coccinea*) foliis ovatis, floribus lateralibus compactis, caule hirsuto procumbente. *Boerhaavia with oval leaves, flowers coming from the wings of the leaves in close heads, and a hairy trailing stalk.*

The first sort was discovered by the late Dr. Houstoun, at La Vera Cruz, in 1731. This rises with an upright smooth stalk two feet high; at each joint it hath two oval pointed leaves growing opposite, upon foot-stalks an inch long. At the joints, which are far asunder, come out small side branches, growing erect; these, as also the large stalk, are terminated by loose panicles of flesh-coloured flowers, which are succeeded by oblong glutinous seeds.

The second sort grows naturally in Jamaica. This sends out many diffused stalks a foot and a half long, garnished with small roundish leaves at each joint. The flowers grow very scatteringly upon long branching foot-stalks from the wings of the leaves, as also at the end of the branches, which are of a pale red colour, and are succeeded by seeds like the former.

The third sort sends out several stalks from the root, which divide into many branches, and trail over whatever plants grow near them, and rise to the height of five or six feet, and are garnished with heart-shaped leaves, growing by pairs opposite at each joint upon long foot-stalks, which are of the colour and consistence of those of the greater Chickweed The flowers grow in loose umbels at the extremity of the branches, which are yellow, and are succeeded by small, oblong, viscous seeds.

The fourth sort sends out many trailing hairy stalks, which divide into smaller branches, which are garnished with oval leaves at every joint; and at the wings of the leaves come out the naked foot-stalks, sustaining a small close head of scarlet flowers, which are very fugaceous, seldom standing more than half a day before their petals drop; these are succeeded by short oblong seeds.

The first, second, and fourth sorts are annual plants, which decay in autumn, but the third sort is perennial. They are all tender plants, so will not thrive in the open air in England; they are propagated by seeds, which must be sown on a hot bed in the spring, and when the plants are fit to remove, they should be each planted in a small pot, and plunged into the hot-bed, and treated as other tender exotics. When they are grown too tall to remain under the common frame, a plant or two of each sort should be placed in the stove, the other may be turned out of the pots and planted in a warm border, where, if the season proves warm, they will perfect their seeds; but as these are subject to fail in cold seasons, so those in the stove will always ripen their seeds in autumn. The third sort may be preserved in a warm stove two or three years.

BOMBAX. Lin. Gen. Pl. 530. Silk Cotton tree.

The CHARACTERS are,

The flower is quinquefid and spreading. It hath many stamina, which are the length of the petal: in the center is situated the round germen. The empalement afterward becomes a large, oblong, turbinated capsule, having five cells, which are ligneous, containing many roundish seeds, wrapped in a soft down.

The

The SPECIES are,

1. BOMBAX (*Ceiba*) floribus polyandris, foliis quinatis. Jacq. Amer. 26. *Silk Cotton-tree with a prickly ſtalk.*

2. BOMBAX (*Pentandrum*) floribus pentandris. Jacq. Amer. 26. *Silk Cotton-tree with ſmooth ſtems.*

3. BOMBAX (*Heptophyllum*) floribus polyandris, foliis ſeptenatis. Jacq. Amer. 26. *Silk Cotton-tree with leaves cut into ſeven parts.*

The firſt and ſecond ſorts grows naturally in both Indies, where they arrive to a great magnitude, being ſome of the talleſt trees in thoſe countries; but the wood is very light, and not much valued, except for making of canoes, which is the chief uſe made of them. Their trunks are ſo large, as when hollowed, to make very large ones.

Theſe trees generally grow with very ſtrait ſtems; thoſe of the firſt ſort are armed with ſhort ſtrong ſpines, but the ſecond hath very ſmooth ſtems, which in the young plants are of a bright green, but after a few years they are covered with a gray, or Aſh-coloured bark, which turns to a brown as the trees grow older. The branches toward the top are garniſhed with leaves compoſed of five, ſeven, or nine oblong ſmooth little leaves, which are ſpear-ſhaped, and join to one center at their baſe, where they adhere to the long foot-ſtalk. The flower-buds appear at the end of the branches, and ſoon after the flowers expand, which are compoſed of five oblong purple petals, with a great number of ſtamina in the center; when theſe fall off, they are ſucceeded by oval fruit as large as a ſwan's egg, having a thick ligneous cover, which, when ripe, opens in five parts, and is full of a dark ſhort Cotton, incloſing many roundiſh ſeeds as large as ſmall Peaſe.

The third ſort was ſent me from the Spaniſh Weſt-Indies, where it grows naturally, but I do not know to what ſize; for the plants which have been raiſed here, have ſoft herbaceous ſtalks very full of joints, and do not appear as if they would become woody, for the plants of ſeveral years growth have ſoft pithy ſtems. The leaves come out on long hairy foot-ſtalks at the top of the plants; theſe have the appearance of thoſe of the Mallow-tree, but are larger, and of a thicker conſiſtence, and on their under ſide are covered with a ſhort, brown, hairy down, and are cut on their edges into five angles. Theſe plants have not as yet flowered in England, nor have I received any information what flowers they produce, but by the pods and ſeeds it appears evidently to be of this genus. The down incloſed in theſe pods is of a fine purple colour, and I have been informed that the inhabitants of the countries where the trees grow naturally, ſpin it, and work it into garments, which they wear without dyeing of any other colour.

The plants of all theſe ſorts are propagated by ſeeds, which muſt be ſown on a hot-bed in the ſpring; thoſe of the two firſt ſorts will be ſtrong enough to tranſplant in a ſhort time after they are up, when they ſhould be each planted in a ſmall pot, and plunged into a moderate hot-bed of tanners bark, being careful to ſhade them from the ſun till they have taken freſh root; after which they ſhould have a large ſhare of air admitted to them when the weather is warm, to prevent their being drawn up weak. In this bed they may remain till autumn (provided there is room for the plants under the glaſſes;) when the heat of the bed declines, the tan ſhould be ſtirred up, and freſh added to it; and if the plants have filled the pots with their roots, they ſhould be ſhifted into pots a little larger; but there muſt be care taken not to over-pot them, for nothing is more injurious to theſe plants than to be put into large pots, in which they will never thrive. In the autumn they muſt be removed into the bark-ſtove, where they muſt conſtantly remain, being too tender to thrive in this country in any other ſituation. In winter they muſt have but little wet, eſpecially if they caſt their leaves; but in the ſummer they ſhould be frequently refreſhed with water, and in warm weather muſt have plenty of freſh air admitted to them.

The plants require a large ſtove where they may have room to grow, but as they are ſeveral years old before they flower in the countries where they grow naturally, ſo there is little hopes of their producing any in England.

BONDUC. See GUILANDINA.

BONTIA. Lin. Gen. Pl. 709. Barbadoes Wild Olive.

The CHARACTERS are,

The flower is of the ringent kind, gaping at the brim; the upper lip is erect, the lower lip is trifid and turns backward. It hath four ſtamina, two of them being longer than the other. In the center is ſituated the oval germen, which afterward becomes an oval berry with one cell, including a nut of the ſame form.

The SPECIES are,

1. BONTIA (*Daphnoides.*) Lin. Sp. Pl. *Barbadoes Wild Olive.*

2. BONTIA (*Germinans*) pedunculis ſpicatis. Lin. Sp. 891. *Mangrove-tree with flowers growing in ſpikes.*

The firſt ſort is greatly cultivated in the gardens at Barbadoes for making of hedges, than which there is not a more proper plant to thrive in thoſe hot countries, it being an evergreen, and of quick growth. I have been informed, that from cuttings (planted in the rainy ſeaſon, when they have immediately taken root) there has been a complete hedge, four or five feet high, in eighteen months. In England it is preſerved in ſtoves. It may be raiſed from ſeeds, which ſhould be ſown on a hot-bed early in the ſpring, that the plants may acquire ſtrength before winter. When the plants are come up, they muſt be tranſplanted out each into a ſeparate ſmall pot, and plunged into a moderate hot-bed of tanners bark, obſerving to ſhade them until they have taken root; after which they muſt have a large ſhare of air in warm weather, and be often refreſhed with water. In winter they muſt be placed in the ſtove, where they ſhould have a moderate degree of warmth, and but little water during that ſeaſon. In ſummer they may be expoſed abroad, in very hot weather, in a ſheltered ſituation. With this management theſe plants will produce flowers and fruit in three years from ſeed. They may alſo be propagated by cuttings, which ſhould be planted in the ſpring before the plants have begun to ſhoot. Theſe muſt be put into pots, and plunged into a moderate hot-bed, obſerving to ſhade them until they have taken root, after which they muſt be treated as hath been directed for the ſeedling plants. Theſe plants being evergreen, and growing in a pyramidal form, make a pretty variety in the ſtove amongſt other exotic plants.

The ſecond ſort has, by many botanic writers, been ranged under the genus of Mangrove-tree, as it grows in ſwamps, which they alſo do. It riſes to about fourteen or ſixteen feet high, ſending out ſeveral ſmall branches, which incline downward toward the water, and as ſoon as they reach that, put out roots into the mud, whereby they propagate very faſt; theſe branches are garniſhed with leaves, placed oppoſite; they are of a thick ſubſtance like thoſe of the Bay-tree, about two inches long, and one broad, very ſmooth on their ſurface: the flowers come out in ſpikes from the upper branches, which are white, compoſed of four petals.

This plant is very impatient of cold, ſo ſhould be conſtantly preſerved in the tan-bed in the ſtove: the beſt way to procure the plants of this kind, is to have their cuttings planted in a box of earth, which, if preſerved from the ſpray or ſalt-water in their paſſage, and moderately watered, while they are in a warm climate, they will arrive ſafe.

BONUS HENRICUS. See CHENOPODIUM.

BORBONIA. Lin. Gen. Plant. 764.

The CHARACTERS are,

The flower hath five leaves, and is of the butterfly shape. The standard is obtuse and reflexed, the wings are heart-shaped and shorter than the standard: the keel hath two obtuse lunulated leaves. It hath nine stamina joined in a cylinder, and one upper standing single. In the center is situated a germen, which afterward becomes a round-pointed pod terminated with a spine, having one cell, inclosing a kidney-shaped seed.

The SPECIES are,

1. BORBONIA (*Lanceolata*) foliis lanceolatis multinerviis integerrimis. Lin. Sp. Plant. 707. *Borbonia with entire spear-shaped leaves having many nerves.*

2. BORBONIA (*Cordata*) foliis cordatis multinerviis integerrimis. Lin. Sp. Plant. 707. *Borbonia with entire heart-shaped leaves having many nerves.*

3. BORBONIA (*Trinervia*) foliis lanceolatis trinerviis integerrimis. Lin. Sp. Pl. 707. *Borbonia with entire spear-shaped leaves having three veins.*

These plants grow naturally at the Cape of Good Hope, where they rise to the height of ten or twelve feet, but in Europe they are seldom more than four or five, having slender stems dividing into several branches, garnished with stiff leaves, placed alternately; those of the first sort are narrow, long, and end in a sharp point. The flowers come out from between the leaves at the end of the branches in small clusters; these are yellow, and shaped like those of the Broom.

The second sort hath broader leaves than the first; the stalks of this are slender, covered with a white bark. The leaves embrace these at their base. The flowers are produced in small clusters at the end of the branches, which are of the same shape and colour as those of the former, but are larger.

The third sort hath stronger stalks than either of the former, which are garnished almost their whole length, as are also the branches with stiff spear-shaped leaves, having three longitudinal nerves in each. The flowers are produced at the extremity of the branches, each standing on a separate foot-stalk. They are of the same shape and colour with the former, but are larger.

As these plants do not perfect their seeds in this country, so they are with difficulty propagated here. The only method by which I have yet succeeded, hath been by laying down their young shoots; but these are commonly two years, before they put out roots fit to be separated from the old plant. In laying these down, the joint which is laid in the ground should be slit upward, as is practised in laying Carnations, and the bark of the tongue at bottom taken off. The best time to lay these down is in the beginning of September, and the shoots most proper for this purpose are those which come out immediately from the root, and of the same year's growth, not only from their situation being near the ground, and thereby better adapted for laying, but these are also more apt to put out roots than any of the upper branches.

But where good seeds can be procured, that is the more eligible method of propagating the plants, for those raised from the seeds make the straitest plants, and are quicker of growth. When good seeds are obtained, they should be sown in pots as soon as they are received, which, if it happens in the autumn, the pots should be plunged into an old bed of tanners bark under a frame, where they may remain all the winter, being careful that they have not much wet; and in the spring the pots should be plunged into a hot-bed, which will bring up the plants in five or six weeks. When these are fit to remove, they should be each planted into a separate small pot, and plunged into a moderate hot-bed, observing to shade them until they have taken fresh root. After this they must by degrees be inured to the open air, into which they should be removed in June, and placed in a sheltered situation, where they may remain till autumn, when they must be removed into the green-house, and placed where they may enjoy the air and sun; during the winter season these plants must be sparingly watered, but in summer, when they are placed abroad, they will require to be frequently refreshed, but must not have too much water given them each time.

BORRAGO. Borage.

The CHARACTERS are,

The flower is of one leaf, having a short tube, and spread wide open above. The chaps of the flower are crowned by five prominences. It hath five stamina which are joined together, and four germen situated in the center, which afterward becomes so many roundish rough seeds, inserted in the cavities of the receptacle.

The SPECIES are,

1. BORRAGO (*Officinalis*) foliis omnibus alternis, calycibus patentibus. Hort. Upsal. 34. *Borage with all the leaves growing alternate, and a spreading flower-cup.*

2. BORRAGO (*Orientalis*) calycibus tubo corollæ brevioribus, foliis cordatis. Hort. Cliff. 45. *Borage of Constantinople, with a blue reflexed flower and a swelling flower-cup.*

3. BORRAGO (*Indica*) foliis ramificationum oppositis calycinis foliolis sagittatis. Lin. Sp. Pl. 137. *Borage with opposite leaves on the branches, and spear-shaped leaves to the flower-cup.*

4. BORRAGO (*Africana*) foliis ramificationum oppositis petiolatis, calycinis foliolis ovatis acutis erectis. Hort. Cliff. 55. *African Borage with leaves on the branches placed opposite on foot-stalks, but the small leaves of the cup of the flower oval, pointed, and erect.*

The first sort is the common Borage, whose flowers are used in medicine, and the herb for cool tankards in summer. Of this there are three varieties, which generally retain their difference from seeds; one hath a white, and another a red flower: the third hath variegated leaves.

The common Borage is an annual plant, which, if permitted to scatter its seeds, the plants will come up in plenty without care; or if the seeds are sown either in spring or autumn, on a spot of open ground where the plants are designed to remain, and the ground hoed to destroy the weeds, and also to cut up the plants where they are too near each other, they will require no farther care, unless the weeds should come up again; then the ground should be a second time hoed over to destroy them, which, if well performed, and in dry weather, will clear the ground from weeds, so it will require no more cleaning till the Borage is decayed.

The second sort grows near Constantinople. This is a perennial plant, with a thick fleshy root which spreads under the surface of the ground, and is thereby propagated with great facility. It sends out many oblong heart-shaped leaves from the root, having long hairy foot-stalks; from the root arises the flower-stem, which is more than two feet high when fully grown, having at the joints a single small leaf without a foot-stalk. The upper part of the stalk branches out into several small foot-stalks, which are terminated by loose panicles of flowers; these are of a pale blue colour, and the petal is reflexed backward, so that the connected stamina and style are left naked. It flowers in March, and the seeds ripen in May.

The third and fourth sorts grow naturally in Africa; these are both annual plants, which rarely rise more than two feet high, having rough stalks; those of the fourth sort are set on by pairs, with short foot-stalks, but the leaves of the third closely embrace the stalks at their base; the

M flowers

flowers come out on short foot-stalks from the wings of the leaves, and also at the top of the stalks. Those of the fourth sort are white, and those of the third a pale blue, but neither of them make any great appearance, so are seldom cultivated, but in botanic gardens for variety.

If the seeds of these plants are sown in the autumn in pots, and plunged into a tan-bed, the plants will produce good seeds; for when they are sown in the spring, if the season is not very favourable, they do not perfect their seeds in England, unless the plants are put into a glass-case.

BOSIA.

The CHARACTERS are,

The flower hath no petals, but five stamina which are as long as the empalement; in the center is situated an oval, oblong, pointed germen, which afterward becomes a globular berry with one cell, including one pointed seed.

We have but one SPECIES of this genus, viz.

BOSIA (*Yervamora.*) Lin. Hort. Cliff. 84. *Commonly called Golden-rod-tree.*

This plant is a native of the islands of the Canaries, and it hath also been since found in some of the British islands in America; it was first brought into England from the Canaries, and has been long an inhabitant of the English gardens; but I have not as yet seen any of these plants in flower, though I have had many old plants under my care more than forty-six years: it makes a pretty strong woody shrub, growing with a stem as large as a middling person's leg; the branches come out very irregular, and make considerable shoots in summer, which should be shortened every spring. These branches retain their leaves till toward the spring, when they fall away, and new leaves are produced soon after: it may be propagated by cutttings planted in the spring, and the plants must be housed in the winter, being too tender to live through the winter in the open air in this country.

BOTRYS. See CHENOPODIUM.

BOX-TREE. See BUXUS.

BRABEJUM. African Almond, vulgò.

The CHARACTERS are,

The flower is composed of four narrow obtuse petals, which are erect; it hath four slender stamina. In the center is a small hairy germen, which afterward becomes an oval, hairy, dry berry, inclosing an oval nut.

We have but one SPECIES of this genus, viz.

BRABEJUM (*Stellatifolium.*) Hort. Cliff. *African, or Ethiopian Almond, with a silky fruit.*

This tree is a native of the country about the Cape of Good Hope.

In Europe it seldom grows above eight or nine feet high, but in its native soil it is a tree of middling growth; as it is too tender to live though the winter in the open air, so we cannot expect to see it grow to a great size.

It rises with an upright stem, which is soft, and full of pith within, and covered with a brown bark. The leaves come out all round the branches at each joint; they are indented on their edges, standing on very short foot-stalks. The flowers are produced toward the end of their shoots, which are of a pale colour, inclining to white.

This plant is with difficulty propagated by layers, which are often two years before they make roots strong enough to be taken from the plants; when the branches are laid down, it will be a good method to slit them at a joint (as is practised in laying Carnations) which will promote their taking root.

The best time to make the layers is in April, just as the plants are beginning to shoot, and the layers must always be made of the former year's shoots. As this plant is very difficult to propagate, so it is very scarce in Europe, there being but few in the Dutch gardens at present.

The plants must have a good green-house in winter, but in summer should be set abroad in a sheltered situation, where, when they arrive to a proper age, they will thrive, and annually produce flowers in the spring, so will make a pretty variety among other exotic plants in the green-house.

BRANCA URSINA. See ACANTHUS.

BRASSICA. The Cabbage.

The CHARACTERS are,

The flower is cross-shaped, having four petals, and four oval nectarious glands. It hath six stamina, which are erect, two of which are opposite, and the other four are longer. It hath a taper germen the length of the stamina, which afterward becomes a long taper pod depressed on each side, and is terminated by the apex of the intermediate partition, which divides it into two cells, filled with round seeds.

I shall first enumerate the species, which are distinct, and afterward mention the varieties, which are cultivated for the table.

The SPECIES are,

1. BRASSICA (*Sativa*) radice caulescente tereti carnosâ. Hort. Cliff. 338. *The common white Cabbage.*

2. BRASSICA (*Napiformi*) radice caulescente orbiculari carnoso, foliis sessilibus. *Turnep-rooted Cabbage.*

3. BRASSICA (*Cauliflora*) radice caulescente tereti carnosâ, floralibus multicaulis. *The Cauliflower.*

4 BRASSICA (*Maritima*) radice cauleque tenui ramoso perenni foliis alternis marginibus incisis. *Taller, shrubby, branching Sea Cabbage.*

5. BRASSICA (*Dentata*) foliis lanceolato-ovatis glabris indivisis dentatis. Hort. Upsal. 191. *Cabbage with entire, oval, spear-shaped, smooth leaves, which are indented.*

6. BRASSICA (*Perfoliata*) foliis oblongo cordatis amplexicaulibus, integerrimis. *Champaign Colewort with a Thorough-wax leaf, and a purple flower.*

7. BRASSICA (*Orientalis*) foliis cordatis semiamplexicaulibus, marginibus dentatis, siliquis tetragonis longissimis. *Eastern perfoliated Colewort, with a white flower and a quadrangular pod.*

8. BRASSICA (*Rapa*) radice caulescente tereti, foliis inferioribus petiolatis superioribus semiamplexicaulibus. *The wild Navew, or Cole Seed.*

The VARIETIES of the first sort are,

1. Brassica sabauda hyberna. Lob. Icon. *The Savoy Cabbage, commonly called Savoy.*

2. Brassica capitata rubra. C. B. P. 111. *The red Cabbage.*

3. Brassica capitata alba pyramidalis. *The sugar-loaf Cabbage.*

4. Brassica capitata alba præcox. *The early Cabbage.*

5. Brassica peregrina moschum olens. H. R. Par. *Foreign Musk Cabbage.*

6. Brassica capitata alba minor Muscovitica. H. A. *Small Russia Cabbage.*

7. Brassica capitata alba compressa. Boerh. Ind. Alt. 11. *The large-sided Cabbage.*

8. Brassica capitata viridis Sabauda. Boerh Ind. 11. *The green Savoy.*

9. Brassica fimbriata. C. B. P. 111. *The Borecole.*

10. Brassica fimbriata virescens. Boerh. Ind. 2. 12. *Green Borecole.*

11. Brassica fimbriata Siberica. Boerh. Ind. 2. 12. *Siberian Borecole, called by some Scotch Kale.*

The VARIETIES of the third sort are,

1. Brassica Italica purpurea Broccoli dicta. Juss. *Purple Broccoli.*

2. Brassica Italica alba Broccoli dicta. Juss. *White Broccoli.*

The

The fecond fort, I believe, never varies, for I have cultivated it many years, and have not found it to alter. This grows naturally on the fea-fhore near Dover. It hath a perennial branching ftalk, in which it differs from all the other fpecies. In very fevere winters, when the other forts are deftroyed, this is a neceffary plant, for the moft fevere frofts do not injure it. The flower-ftalks grow from the end of the branches, and fpread out horizontally; but thofe which arife from the center of the plants grow erect, and feldom put out branches.

The two forts of Broccoli I take to be only varieties of the Cauliflower, for although thefe may with care be kept diftinct, yet I doubt, if they were to ftand near each other for feeds, if they would not intermix; and I am rather inclined to believe this, from the various changes which I have obferved in all thefe forts, for I have frequently had Cauliflowers of a green colour, with flower-buds regularly formed at the ends of the fhoots, as thofe of Broccoli, though the colour was different; and the white Broccoli approaches fo near to the Cauliflower, as to be with difficulty diftinguifhed from it; yet when thefe are cultivated with care, and never fuffered to ftand near each other, when left to produce feeds, they may be kept very diftinct in the fame garden; for the variations of thefe plants is not occafioned from the foil, but the mixing of the farina of the flowers with each other, where they are planted near together; therefore thofe perfons who are curious to preferve the feveral varieties diftinct, fhould never fuffer the different kinds to ftand near each other for feed.

The Cauliflower has been much more improved in England than in any other part of Europe. In France they rarely have Cauliflowers till Michaelmas, and Holland is generally fupplied with them from England. In many parts of Germany there was none of them cultivated till within a few years paft, and moft parts of Europe are fupplied with feeds from hence.

The eighth fort, which is generally known by the title of Rape or Cole Seed, is much cultivated in the ifle of Ely, and fome other parts of England, for its feed, from which the Rape-oil is drawn; and it hath alfo been cultivated of late years, in other places, for feeding of cattle, to great advantage.

The Cole Seed, when cultivated for feeding of cattle, fhould be fown about the middle of June. The ground for this fhould be prepared for it in the fame manner as for Turneps. The quantity of feeds for an acre of land is from fix to eight pounds, and as the price of the feed is not great, fo it is better to allow eight pounds; for if the plants are too clofe in any part, they may be eafily thinned when the ground is hoed, which muft be performed in the fame manner as is practifed for Turneps, with this difference only, of leaving thefe much nearer together; for as they have fibrous roots and flender ftalks, fo they do not require near fo much room. Thefe plants fhould have a fecond hoeing, about five or fix weeks after the firft, which, if well performed in dry weather, will entirely deftroy the weeds, fo they will require no farther culture. Where there is not an immediate want of food, thefe plants had better be kept as a referve for hard weather, or fpring feed, when there may be a fcarcity of other green food. If the heads are cut off, and the ftalks left in the ground, they will fhoot again early in the fpring, and produce a good fecond crop in April, which may be either fed off, or permitted to run to feeds, as is the practice where this is cultivated for the feeds: but if the firft is fed down, there fhould be care taken that the cattle do not deftroy their ftems, or pull them out of the ground. As this plant is fo hardy as not to be deftroyed by froft, fo it is of great fervice in hard winters for feeding of ewes; for when the ground is fo hard frozen as that Turneps cannot be taken up, thefe plants may be cut off for a conftant fupply. This will afford late food after the Turneps are run to feed; and if it is afterward permitted to ftand for feed, one acre will produce as much as, at a moderate computation, will fell for five pounds, clear of charges.

Partridges, pheafants, turkeys, and moft other fowl, are very fond of this plant; fo that wherever it is cultivated, if there are any birds in the neighbourhood, they will conftantly lie among thefe plants.

The feeds of this plant are fown in gardens for winter and fpring fallets, this being one of the fmall fallet herbs.

The common white, red, flat, and long-fided Cabbages are chiefly cultivated for autumn and winter ufe; the feeds of thefe forts muft be fown the beginning or middle of April, in beds of good frefh earth; and when the young plants have about eight leaves, they fhould be pricked out into fhady borders, about three or four inches fquare, that they may acquire ftrength, and to prevent their growing long fhanked.

About the middle of June you muft tranfplant them out, where they are to remain for good (which in the kitchen-gardens near London is commonly between Cauliflowers, Artichokes, &c. at about two feet and a half diftance in the rows;) but if they are planted for a full crop in a clear fpot of ground, the diftance from row to row fhould be three feet and a half, and in the rows two feet and a half afunder: if the feafon fhould prove dry when they are tranfplanted out, you muft water them every other evening until they have taken frefh root; and afterwards, as the plants advance in height, you fhould draw the earth about their ftems with a hoe, which will keep the earth moift about their roots, and greatly ftrengthen the plants.

Thefe Cabbages will fome of them be fit for ufe foon after Michaelmas, and will continue until the end of February, if they are not deftroyed by bad weather; to prevent which, the gardeners near London pull up their Cabbages in November, and trench their ground up in ridges, laying their Cabbages againft their ridges as clofe as poffible on one fide, burying their ftems in the ground: in this manner they let them remain till after Chriftmas, when they cut them for the market; and although the outer part of the Cabbage be decayed (as is often the cafe in very wet or hard winters,) yet, if the Cabbages were large and hard when laid, the infide will remain found.

The Ruffian Cabbage was formerly in much greater efteem than at prefent, it being now only to be found in particular gentlemen's gardens, who cultivate it for their own ufe. This muft be fown late in the fpring of the year, and managed as thofe before directed, with this difference only, that thefe muft be fooner planted out for good, and muft have an open clear fpot of ground, and require much lefs diftance every way, for it is but a very fmall hard Cabbage. This fort will not continue long before they will break, and run up to feed.

The early and fugar-loaf Cabbages are commonly fown for fummer ufe, and are what the gardeners about London commonly call Michaelmas Cabbages. The feafon for fowing of thefe is about the end of July, or beginning of Auguft, in an open fpot of ground; and when the plants have got eight leaves, you muft prick them into beds at about three or four inches diftance every way, that the plants may grow ftrong and fhort fhanked, and toward the end of October you fhould plant them out for good: the diftance that thefe require is, three feet row from row and two feet and a half afunder in the rows. The ground muft be kept clean from weeds, and the earth drawn up about your Cabbage plants.

M 2 In

In May, if your plants were of the early kind, they will turn in their leaves for cabbaging; at which time, the gardeners near London, in order to obtain them a little sooner, tie in their leaves close with a slender Osier-twig to blanch their middle; by which means, they have them at least a fortnight sooner than they could have if they were left untied.

The early Cabbage being the first, we should chuse (if for a gentleman's use) to plant the fewer of them, and a greater quantity of the sugar-loaf kind, which comes after them; for the early kind will not supply the kitchen long, generally cabbaging apace when they begin, and as soon grow hard and burst open; but the sugar-loaf kind is longer before it comes, and is as slow in its cabbaging; and being of an hollow kind, will continue good for a long time.

Although I before have advised the planting out of Cabbages for good in October, yet the sugar-loaf kind may be planted out in February, and will succeed as well as if planted earlier; with this difference only, that they will be later before they cabbage. You should also reserve some plants of the early kind in some well-sheltered spot of ground, to supply your plantation, in case of a defect; for in mild winters many of the plants are apt to run to seed, especially when their seeds are sown too early, and in severe winters they are often destroyed.

The Savoy Cabbages are propagated for winter use, as being generally esteemed the better when pinched by the frost: these must be sown about the end of April, and treated after the manner as was directed for the common white Cabbage; with this difference, that these may be planted at a closer distance than those; two feet and a half square will be sufficient. These are always much better when planted in an open situation, which is clear from trees and hedges; for in close places they are very subject to be eaten almost up by caterpillars and other vermin, especially if the autumn prove dry.

The Borecole may be also treated in the same manner, but need not be planted above one foot asunder in the rows, and the rows two feet distance; these are never eaten till the frost hath rendered them tender, for otherwise they are tough and bitter.

The seeds of the Broccoli (of which there are several kinds, viz. the Roman or purple, the Neapolitan or white, and the black Broccoli, with some others, but the Roman is chiefly preferred to them all,) should be sown about the latter end of May, or beginning of June, and when the plants are grown to have eight leaves, transplant them into beds (as was directed for the common Cabbage;) and toward the latter end of July they will be fit to plant out for good, which should be into some well-sheltered spot of ground, but not under the drip of trees: the distance these require is about a foot and a half in the rows, and two feet row from row. The soil in which they should be planted ought to be rather light than heavy, such as are the kitchen-gardens near London: if your plants succeed well (as there will be little reason to doubt, unless the winter prove extreme hard,) they will begin to shew their small heads, which are somewhat like a Cauliflower, but of a purple colour, about the end of December, and will continue eatable till the middle of April.

The brown or black Broccoli is by many persons greatly esteemed, though it doth not deserve a place in the kitchen-garden where the Roman Broccoli can be obtained, which is much sweeter, and will continue longer in season: indeed, the brown sort is much hardier, so that it will thrive in the coldest situations, where the Roman Broccoli is sometimes destroyed in very hard winters. The brown sort should be sown in the middle of May, and managed as hath been directed for the common Cabbage, and should be planted at the same distance, which is about two feet and a half asunder. This will grow very tall, so should have the earth drawn up to their stems as they advance in height. This doth not form heads so perfect as the Roman Broccoli; the stems and hearts of the plants are the parts which are eaten.

The Roman Broccoli (if well managed) will have large heads, which appear in the center of the plants like clusters of buds. These heads should be cut before they run up to seed, with about four or five inches of the stem; the skin of these stems should be stripped off before they are boiled. After the first heads are cut off, there will be a great number of side shoots produced from the stems, which will have small heads to them, but are full as well flavoured as the large

The Naples Broccoli hath white heads very like those of the Cauliflower, and eats so like it as not to be distinguished from it.

Besides this first crop of Broccoli (which is usually sown in the end of May,) it will be proper to sow another crop the beginning of July, which will come in to supply the table the latter end of March and the beginning of April; and being very young, will be extremely tender and sweet.

In order to save good seeds of this kind of Broccoli in England, you should reserve a few of the largest heads of the first crop, which should be let remain to run up to seed, and all the under shoots should be constantly stripped off, leaving only the main stem to flower and seed. If this be duly observed, and no other sort of Cabbage permitted to seed near them, the seeds will be as good as those procured from abroad, and the sort may be preserved in perfection many years.

The Turnep-rooted Cabbage was formerly more cultivated in England than at present, for since other sorts have been introduced which are much better flavoured, this sort has been neglected. There are some persons who esteem this kind for soups, but it is generally too strong for most English palates, and is seldom good but in hard winters, which will render it tender and less strong.

At the end of June the plants should be transplanted out where they are to remain, allowing them two feet distance every way, observing to water them until they have taken root; and as their stems advance, the earth should be drawn up to them with a hoe, which will preserve a moisture about their roots, and prevent their stems from drying and growing woody, so that the plants will grow more freely; but it should not be drawn very high, for as it is the globular part of the stalk which is eaten, so that should not be covered. In winter they will be fit for use, when they should be cut off, and the stalks pulled out of the ground and thrown away, being good for nothing after the stems are cut off.

The curled Colewort or Siberian Borecole is now more generally esteemed than the former, being extreme hardy, so is never injured by cold, but is always sweeter in severe winters than in mild seasons. This may be propagated by sowing of the seeds the beginning of July; and when the plants are strong enough for transplanting, they should be planted in rows about a foot and a half asunder, and ten inches distance in the rows. These will be fit for use after Christmas, and continue good until April, so that they are very useful in a family.

The Musk Cabbage. This may be propagated in the same manner as the common Cabbage, and should be allowed the same distance: it will be fit for use in October, November, and December; but, if the winter proves hard, these will be destroyed much sooner than the common sort.

The

The common Colewort or Dorsetshire Kale, is now almost lost near London, where their markets are usually supplied with Cabbage plants instead of them. Indeed, where farmers sow Coleworts to feed their milch cattle in the spring, when there is a scarcity of herbage, the common Colewort is to be preferred, as being so very hardy that no frost will destroy it. The best method to cultivate this plant in the fields is, to sow the seeds about the beginning of July, chusing a moist season, which will bring up the plants in about ten days or a fortnight: the quantity of seed for an acre of land is nine pounds; when the plants have got five or six leaves they should be hoed, as is practised for Turneps, cutting down all the weeds from amongst the plants, and also thinning the plants where they are too thick; but they should be kept thicker than Turneps, because they are more in danger of being destroyed by the fly: this work should be performed in dry weather, that the weeds may be killed. About six weeks after the plants should have a second hoeing, which, if carefully performed in dry weather, will entirely destroy the weeds, and make the ground clean, so that they will require no farther culture: in the spring they may either be drawn up and carried out to feed the cattle, or they may be turned in to feed upon them as they stand; but the former method is to be preferred, because there will be little waste; whereas, when the cattle are turned in amongst the plants, they will tread down and destroy more than they eat, especially if they are not fenced off by hurdles.

The two last sorts of Cabbage are varieties fit for a botanic garden, but are plants of no use. They are annual plants, and perish when they have perfected their seeds.

The best method to save the seeds of all the best sorts of Cabbages is, about the end of November you should make choice of some of your best Cabbages, which you should pull up, and carry to some shed or other covered place, where you should hang them up for three or four days by their stalks, that the water may drain from between their leaves; then plant them in some border near a hedge or pale, quite down to the middle of the Cabbage, leaving only the upper part of the Cabbage above ground, observing to raise the earth about it, so that it may stand a little above the level of the ground; especially if the ground is wet, they will require to be raised pretty much above the surface.

If the winter should prove very hard, you must lay a little straw or Pease-haulm lightly upon them, to secure them from the frost, taking it off as often as the weather proves mild, lest by keeping them too close they should rot. In the spring of the year these Cabbages will shoot out strongly, and divide into a great number of small branches: you must therefore support their stems, to prevent their being broken off by the wind; and if the weather should be very hot and dry when they are in flower, you should refresh them with water once a week all over the branches, which will greatly promote their seeding, and preserve them from mildew.

When the pods begin to change brown, you will do well to cut off the extreme part of every shoot with the pods, which will strengthen your seeds; for it is generally observed, that those seeds which grow near the top of the shoots, are very subject to run to seed before they cabbage; so that by this there will be no loss, but a great advantage.

When your seeds begin to ripen, you must be particularly careful that the birds do not destroy it, for they are very fond of these seeds. The best method I know to prevent this, is to get a quantity of birdlime, and dawb over a parcel of slender twigs, which should be fastened at each end to stronger sticks, and placed near the upper part of the seed in different places, so that the birds may alight upon them, by which means they will be fastened thereto; where you must let them remain, if they cannot get off themselves: and although there should not above two or three birds be caught, yet it will sufficiently terrify the rest, that they will not come to that place again for a considerable time after, as I have experienced.

When your seeds are fully ripe, yon must cut it off; and after drying, thresh it out, and preserve it in bags for use.

But in planting of Cabbages for seed, I would advise never to plant more than one sort in a place, or near one another: as for example, never plant red and white Cabbages near each other, nor Savoy with white or red Cabbages; for I am very certain they will, by the commixture of their effluvia, produce a mixture of kinds; and it is wholly owing to this neglect, that the gardeners rarely save any good red Cabbage seed in England, but are obliged to procure fresh seeds from abroad, as supposing the soil or climate of England alters them from red to white, and of a mixed kind between both; whereas, if they should plant red Cabbages by themselves for seeds, and not suffer any other to be near them, they might continue the kind as good in England as in any other part of the world.

Cauliflowers have of late years been so far improved in England, as to exceed in goodness and magnitude what are produced in most parts of Europe, and by the skill of the gardener are continued for several months together; but the most common season for the great crop is in May, June, and July. I shall therefore begin with directions for obtaining them in this season.

Having procured a parcel of good seed, you must sow it about the 21st of August, upon an old Cucumber or Melon-bed, sifting a little earth over the seeds, about a quarter of an inch thick; and if the weather should prove extreme hot and dry, you should shade the bed with mats, to prevent the earth from drying too fast, and give it gentle waterings as you may see occasion. In about a month's time after sowing, your plants will be fit to prick out; you should therefore put some fresh earth upon your Cucumber or Melon-beds, or where these are not to be had, some beds should be made with a little new dung, which should be trodden down close, to prevent the worms from getting through it; but it should not be hot dung, which would be hurtful to the plants at this season, especially if it proves hot; into this bed you should prick your young plants at about two inches square, observing to shade and water them at first planting; but do not water them too much after they are growing, nor suffer them to receive too much rain if the season should prove wet, which would be apt to make them black shanked (as the gardeners term it, which is no less than a rottenness in their stems,) and is the destruction of the plants so affected.

In this bed they should continue till about the 30th of October, when they must be removed into the place where they are to remain during the winter season, which, for the first sowing, is commonly under bell or hand-glasses, to have early Cauliflowers, and these should be of an early kind: but in order to have a succession during the season, you should be provided with another more late kind, which should be sown four or five days after the other, and managed as was directed for them.

In order to have very early Cauliflowers, you should make choice of a good rich spot of ground that is well defended from the north, east, and west winds, with hedges, pales, or walls; but the first is to be preferred, if made with Reeds, because the winds will fall dead in these, and not reverberate as by pales or walls. This ground should be well trenched, burying therein a good quantity of rotten dung; then level your ground, and if it be naturally a wet soil, you should raise it up in beds about two feet and a half, or three feet broad, and four inches above the level of the ground; but if your ground is moderately dry, you need not raiſe it at all: then plant your plants, allowing about two feet ſix inches diſtance from glaſs to glaſs in the rows, always putting two good plants under each glaſs, which may be at about four inches from each other; and if you deſign them for a full crop, they may be three feet and a half row from row: but if you intend to make ridges for Cucumbers between the rows of Cauliflower plants, (as is generally practiſed by the gardeners near London) you muſt then make your rows eight feet aſunder; and the ground between the rows of Cauliflowers may be planted with Cabbage plants, to be drawn off for Coleworts in the ſpring.

When you have planted your plants, if the ground is very dry you ſhould give them a little water, and then ſet your glaſſes over them, which may remain cloſe down over them until they have taken root, which will be in about a week or ten days time, unleſs there ſhould be a kindly ſhower of rain; in which caſe you may ſet off the glaſſes, that the plants may receive the benefit of it; and in about ten days after planting, you ſhould be provided with a parcel of forked ſticks or bricks, with which you ſhould raiſe your glaſſes about three or four inches on the ſide toward the ſouth, that your plants may have free air: in this manner your glaſſes ſhould remain over the plants night and day, unleſs in froſty weather, when you ſhould ſet them down as cloſe as poſſible; or if the weather ſhould prove very warm, which many times happens in November, and ſometimes in December, in this caſe you ſhould keep your glaſſes off in the day time, and put them on only in the night, leſt, by keeping the glaſſes over them too much, you ſhould draw them into flower at that ſeaſon; which is many times the caſe in mild winters, eſpecially if unſkilfully managed.

Toward the latter end of February, if the weather proves mild, you ſhould prepare another good ſpot of ground to remove ſome of the plants into from under the glaſſes, which ſhould be well dunged and trenched (as before;) then ſet off your glaſſes, and, after making choice of one of the moſt promiſing plants under each glaſs, which ſhould remain for good, take away the other plant, by raiſing it up with a trowel, &c. ſo as to preſerve as much earth to the root as poſſible; but have a great regard to the plant that is to remain, not to diſturb or prejudice its roots: then plant the plants which you have taken out at the diſtance before directed, viz. If for a full crop, three feet and a half, row from row; but if for ridges of Cucumbers between them eight feet, and two feet four inches diſtance in the rows: then, with a ſmall hoe, draw the earth up to the ſtems of the plants which were left under the glaſſes, taking great care not to let the earth fall into their hearts; and ſet your glaſſes over them again, raiſing your props an inch or two higher than before, to give them more air, obſerving to take them off whenever there may be ſome gentle ſhowers, which will greatly refreſh the plants.

In a little time after, if you find your plants grow ſo faſt as to fill the glaſſes with their leaves, you ſhould then ſlightly dig about the plants, and raiſe the ground about them in a bed broad enough for the glaſſes to ſtand, about four inches high, which will give your plants a great deal of room, by raiſing the glaſſes ſo much higher when they are ſet over them; and by this means they may be kept covered until April, which otherwiſe they could not, without prejudice to the leaves of the plants: and this is a great advantage to them, for many times we have returns of ſevere froſts at the latter end of March, which prove very hurtful to theſe plants, if expoſed thereto, eſpecially after having been purſed up under glaſſes.

After you have finiſhed your beds, you may ſet your glaſſes over your plants again, obſerving to raiſe your props pretty high, eſpecially if the weather be mild, that they may have free air to ſtrengthen them; and in mild ſoft weather ſet off your glaſſes, as alſo in gentle ſhowers of rain: and now you muſt begin to harden them by degrees to endure the open air; however, it is adviſeable to let your glaſſes remain over them as long as poſſible, if the nights ſhould be froſty, which will greatly forward your plants; but be ſure do not let your glaſſes remain upon them in very hot ſun-ſhine, eſpecially if their leaves preſs againſt the ſides of the glaſſes; for I have often obſerved in ſuch caſes, that the moiſture which hath riſen from the ground, together with the perſpiration of the plants, which, by the glaſſes remaining over them, hath been detained upon the leaves of the plants, and when the ſun hath ſhone hot upon the ſides of the glaſſes, hath acquired ſuch a powerful heat from the beams thereof, as to ſcald all their larger leaves, to the no ſmall prejudice of the plants: nay, ſometimes I have ſeen large quantities of plants ſo affected therewith, as never to be worth any thing after.

If your plants have ſucceeded well, toward the end of April ſome of them will begin to fruit: you muſt therefore look over them carefully every other day, and when you ſee the flower plainly appear, you muſt break down ſome of the inner leaves over it to guard it from the ſun, which would make the flower yellow and unſightly if expoſed thereto; and when you find your flower at its full bigneſs (which you may know by its outſide parting as if it would run,) you muſt then draw it out of the ground, and not cut them off, leaving the ſtalk in the ground, as is by ſome practiſed; and if they are deſigned for preſent uſe, you may cut them out of their leaves; but if deſigned to keep, you ſhould preſerve their leaves about them, and put them into a cool place: the beſt time for pulling of them is in a morning, before the ſun hath exhaled the moiſture: for Cauliflowers pulled in the heat of the day, loſe that firmneſs which they naturally have, and become tough.

But to return to our ſecond crop (the plants being raiſed and managed as was directed for the early crop, until the end of October,) you muſt then prepare ſome beds, either to be covered with glaſs-frames, or arched over with hoops, to be covered with mats, &c. Theſe beds ſhould have ſome dung laid at the bottom, about ſix inches or a foot thick, according to the ſize of your plants; for if they are ſmall, the bed ſhould be thicker of dung to bring them forward, and ſo vice verſa: this dung ſhould be beat down cloſe with a fork, in order to prevent the worms from finding their way through it; then lay ſome good freſh earth about four or five inches thick thereon, in which you ſhould plant your plants about two inches and a half ſquare, obſerving to ſhade and water them until they have taken new root; but be ſure do not keep your coverings cloſe, for the warmth of the dung will occaſion a large damp in the bed, which, if pent in, will greatly injure the plants.

When your plants have taken root, you muſt give them as much free open air as poſſible, by keeping the glaſſes off in the day-time as much as the weather will permit; and in the night, or at ſuch times as the glaſſes require to be kept on, raiſe them up with props to let in freſh air, unleſs in froſty weather; at which time the glaſſes ſhould be covered with mats, ſtraw, Peaſe haulm, &c. but this is not to be done but in very hard froſts: you muſt alſo obſerve to guard them againſt great rain, which in winter time is very hurtful to them, but in mild weather, if the glaſſes are kept on, they ſhould be propped to admit freſh air; and if the under leaves grow yellow and decay, be ſure to pick them off; for if the weather ſhould prove very bad in winter, ſo that you ſhould be obliged to keep them cloſe covered for two or three days together, as it ſometims happens, theſe decayed leaves will render the incloſed air very noxious; and the plants perſpiring pretty much at that time, are often deſtroyed in vaſt quantities.

In the beginning of February, if the weather be mild, you muſt begin to harden your plants by degrees, that they may be prepared for tranſplantation: the ground where you intend to plant your Cauliflowers out for good (which ſhould be quite open from trees, &c. and rather moiſt than dry,) having been well dunged and dug, ſhould be ſown with Radiſhes a week or fortnight before you intend to plant out your Cauliflowers: the reaſon why I mention the ſowing of Radiſhes particularly, is this, viz. that if there are not ſome Radiſhes amongſt them, and the month of May ſhould prove hot and dry, as it ſometimes happens, the fly will ſeize your Cauliflowers, and eat their leaves full of holes, to their prejudice, and ſometimes their deſtruction; whereas, if there are Radiſhes upon the ſpot, the flies will take to them, and never meddle with the Cauliflowers ſo long as they laſt: indeed, the gardeners near London mix Spinach with their Radiſh-ſeed, and ſo have a double crop; which is an advantage where ground is dear, or where perſons are ſtraitened for room; otherwiſe it is very well to have only one crop amongſt the Cauliflowers, that the ground may be cleared in time.

Your ground being ready and the ſeaſon good, about the middle of February you may begin to plant out your Cauliflowers: the diſtance which is generally allowed by the gardeners near London (who plant other crops between their Cauliflowers to ſucceed them, as Cucumbers for pickling, and winter Cabbages) is every other row four feet and a half apart, and the intermediate rows two feet and a half, and two feet two inches diſtance in the rows; ſo that in the latter end of May or beginning of June (when the Radiſhes and Spinach are cleared off,) they put in ſeeds of Cucumbers for pickling, in the middle of the wide rows, at three feet and a half apart; and in the narrow rows plant Cabbages for winter uſe, at two feet two inches diſtance, ſo that theſe ſtand each of them exactly in the middle of the ſquare between four Cauliflower plants; and theſe, after the Cauliflowers are gone off, will have full room to grow, and the crop be hereby continued in a ſucceſſion through the whole ſeaſon.

There are many people who are very fond of watering Cauliflower plants in ſummer, but the gardeners near London have almoſt wholly laid aſide this practice, as finding a deal of trouble and charge to little purpoſe; for if the ground be ſo very dry as not to produce tolerable good Cauliflowers without water, it ſeldom happens that watering of them renders them much better; and when once they have been watered, if it is not conſtantly continued, it had been much better for them if they never had any; as alſo, if it be given them in the middle of the day, it rather helps to ſcald them: ſo that, upon the whole, if care be taken to keep the earth drawn up to their ſtems, and clear them from every thing that grows near them, that they may have free open air, you will find that they will ſucceed better without than with water where any of theſe cautions are not ſtrictly obſerved.

But in order to have a third crop of Cauliflowers, you ſhould make a ſlender hot-bed in February, in which you ſhould ſow the ſeeds, covering them a quarter of an inch thick with light mould, and covering the bed with glaſs-frames. When the plants are come up, and have gotten four or five leaves, you ſhould prepare another hot-bed to prick them into, which may be about two inches ſquare; and in the beginning of April harden them by degrees, to fit them for tranſplanting, which ſhould be done the middle of that month, at the diſtance directed for the ſecond crop, and muſt be managed accordingly: theſe (if the ſoil is moiſt where they are planted, or the ſeaſon cool and moiſt) will produce good Cauliflowers about a month after the ſecond crop is gone, whereby their ſeaſon will be greatly prolonged.

There is alſo a fourth crop of Cauliflowers, which is raiſed by ſowing the ſeed about the twenty-third of May; and being tranſplanted, as hath been before directed, will produce good Cauliflowers in a kindly ſeaſon and good ſoil after Michaelmas, and continue through October and November, and, if the ſeaſon permit, often a great part of December.

The reaſon why I fix particular days for the ſowing of this ſeed, is becauſe two or three days often make a great difference in their plants; and becauſe theſe are the days uſually fixed by the gardeners near London, who have found their crops to ſucceed beſt when ſown at thoſe times, although one day, more or leſs, will make no great odds. I have alſo, in this edition, altered the days to the new ſtyle.

BREYNIA. See Capparis.

BROMELIA. Plum. Nov. Gen. 46. tab. 8. Lin. Gen Plant. 356.

The Characters are;

The flower hath three long narrow petals, each having a nectarium joined to it above the baſe: it hath ſix ſtamina. The germen is ſituated below the receptacle, which afterward becomes an oblong capſule, divided by a partition in the middle, to which the ſeeds are fixed quite round; theſe are ſmooth and almoſt cylindrical.

Dr. Dillenius has ſuppoſed this to be the ſame with Plumier's Karatas, which miſtake he was led into by Plumier's drawings, where the flower of his Caraguata is joined to the fruit of his Karatas, and vice verſa.

The Species are,

1. Bromelia (*Nudicaulis*) foliis radicalibus dentato-ſpinoſis caulinis integerrimis. Lin. Sp Plant. 286. *Bromelia with lower leaves indented and prickly, and thoſe of the ſtalks entire.*

2. Bromelia (*Lingulata*) foliis ſerrato-ſpinoſis obtuſis, ſpicis alternis. Lin. Sp. Plant. 285. *Bromelia with ſawed prickly leaves which are obtuſe, and ſpikes of flowers growing alternate.*

The firſt ſort hath leaves very like ſome of the ſorts of Aloes, but not ſo thick and ſucculent, which are ſharply indented on their edges, where they are armed with ſtrong black ſpines; from the center of the plant ariſes the flower-ſtalk, which is near three feet high, the lower part of which is garniſhed with entire leaves, placed alternately at every joint. The upper part of the ſtalk is garniſhed with flowers, ſet in a looſe ſpike or thyrſe quite round; theſe have three narrow herbaceous petals ſitting upon the germen, and within are ſix ſlender ſtamina, with the ſtyle, which are ſhorter than the petals. Theſe are ſucceeded by oval ſeed-veſſels, having a longitudinal partition, in the center of which are faſtened cylindrical ſeeds on every ſide, which are ſmooth.

The ſecond ſort hath ſhorter leaves than the firſt, which are narrow at the baſe, increaſing in width gradually to the top, where they are broadeſt; they are ſharply ſawed on their edges, and are of a deep green colour. The flower-ſtem ariſes from the center of the plant, which divides upward into ſeveral branches; the upper part of theſe are garniſhed with ſpikes of flowers, which come out alternately from the ſide of the branches, each having a narrow entire leaf juſt below it, which are longer than the ſpike. The flowers are placed very cloſe on the ſpikes, each having three ſhort petals, ſituated upon the globular empalement; when theſe decay, the empalement turns to an oval-pointed ed feed-veffel, inclofing feeds of the fame fhape with the former.

Both thefe plants grow naturally in very warm countries. The firft fort grows alfo on the coaft of Guinea.

Thefe plants are propagated by feeds, which muft be fown in fmall pots, and plunged into a moderate hot-bed of tanners bark. If the feeds are good, the plants will appear in about five or fix weeks, and in a month or fix weeks after will be fit to tranfplant, when they fhould be carefully fhaken out of the pots, and each planted in a feparate fmall pot; then they muft be plunged again into a moderate hot-bed, obferving frequently to fprinkle them over with water, but be cautious of giving them too much, left the roots fhould be thereby rotted. During the fummer feafon the plants fhould have a moderate fhare of air, in proportion to the heat of the weather; and in autumn they muft be removed into the bark-ftove, and treated in the fame manner as the Ananas or Pine Apple, with which management they will make good progrefs.

Thefe plants make a pretty variety in the hot-houfe, fo thofe who have room, may allow a plant or two of each fort to have a place in their collection of exotic plants.

BROOM, the common. See SPARTIUM.

BROOM, the Spanifh. See SPARTIUM and GENISTA.

BROWALLIA. Lin. Gen. Plant. 691. Hort. Cliff.

The CHARACTERS are,

The flower is funnel-fhaped, of one leaf, having a cylindrical tube twice the length of the empalement; the upper part is fpread open, and divided into five parts. It hath four ftamina included in the chaps of the petal, the two upper being very fhort, and the two under broad and longer. In the center is fituated an oval germen. The empalement afterward becomes an oval obtufe veffel with one cell, opening at the top in four parts, and filled with fmall compreffed feeds.

The SPECIES are,

1. BROWALLIA (*Demiffa*) pedunculis unifloris. Hort. Upfal. 179. *Browallia with a fingle flower upon each footftalk.*

2. BROWALLIA (*Elata*) pedunculis unifloris multiflorifque. Lin. Sp. Pl. 880. *Browallia with one or many flowers on each foot-ftalk.*

The feeds of the firft fort were fent me by Mr. Robert Millar from Panama, in the year 1735. The plants are annual, fo perifh in autumn; the feeds muft be fown upon a hot-bed in the fpring, and the plants brought forward on another, otherwife they will not perfect their feeds in England. Some of thefe plants may be tranfplanted in June, into the borders of the flower-garden, where, if the feafon proves warm, they will flower and perfect feeds; but left thefe fhould fail, there fhould be a plant or two kept in the ftove to fecure feeds. They ufually grow about two feet high, and fpread out into lateral branches on every fide the ftalk, garnifhed with oval leaves which are entire, and have fhort foot-ftalks. Toward the end of the branches the flowers are produced fingly upon pretty long foot-ftalks, arifing from the wing of the leaf. Thefe have a fhort empalement of one leaf, which is cut into five parts; out of the center of the empalement the flower arifes, which is crooked and bent downward; the top of the tube is fpread open, and the brim, or open part of the flower, has fome refemblance to a lip-flower, being irregular; it is of a light blue colour, fometimes inclining to a purple or red, and often there are flowers of three colours on the fame plant. When thefe fall away, the germen in the center becomes an oval capfule of one cell, filled with fmall, brown, angular feeds. It flowers in July, Auguft, and September, and the feeds are ripe in five or fix weeks after.

The fecond fort grows naturally in Peru: the ftalk of this plant is twice the fize of that of the firft, and appears fomewhat fhrubby; the leaves upon the flower branches are fmooth: the foot-ftalks have fome with one flower, others have three, and fome five flowers, which are of a deep Violet colour, and are fucceeded by feed-veffels like thofe of the firft fort. This requires the fame culture as the former.

BRUNELLA, Self-heal. See PRUNELLA.

BRUNSFELSIA. Plum. Nov. Gen. 12. Lin. Gen. Plant. 230.

The CHARACTERS are,

The flower is of one leaf, and funnel-fhaped, having a long tube, but fpreads open at the top; it hath five ftamina the length of the tube, which are inferted in the petal. In the center is placed a fmall round germen. The empalement afterward becomes a globular berry with one cell, inclofing a great number of fmall feeds, which adhere to the fkin of the fruit.

We have but one SPECIES of this genus, viz.

BRUNSFELSIA (*Americana.*) Lin. Sp. Pl. 191. *Brunsfelfia with a white flower, and a foft Saffron-coloured fruit*

This plant rifes with a woody ftem to the height of eight or ten feet, fending out many fide branches, which are covered with a rough bark, and garnifhed with oblong oval leaves. At the extremity of the branches the flowers are produced, generally three or four together. Thefe are almoft as large as thofe of the greater Bindweed, but have very long narrow tubes, which are hairy. After the flower is paft, the empalement turns to a round foft fruit, inclofing many oval feeds, which are fituated clofe to the cover or fkin, to which they adhere.

This plant grows naturally in moft of the fugar iflands in America, but in the Englifh gardens it is at prefent very rare; it may be propagated from feeds, which fhould be fown early in the fpring in pots, and plunged into a hotbed of tanners bark. When the plants are come up and fit to remove, they fhould be tranfplanted each into a feparate fmall pot, and plunged into the hot-bed again, obferving to water and fhade the plants until they have taken root. When the plants have advanced fo high as not to be contained in the frames, they fhould be removed into the bark-ftove, where, during the fummer months, they fhould have a large fhare of free air, but in the winter they muft be kept very clofe. Thefe plants may alfo be increafed by planting their cuttings in the fpring before they fhoot, in pots filled with frefh light earth, and plunged into a hotbed of tanners bark.

BRUSCUS. See RUSCUS.

BRYONIA. Briony.

The CHARACTERS are,

It hath male and female flowers on the fame plant. The male flowers are bell-fhaped, adhering to the empalement, and cut into five fegments. They have three fhort ftamina. The female flowers fit upon the germen; the petal is the fame with thofe of the male. The germen, which is under the flower, afterward becomes a fmooth globular berry, containing oval feeds adhering to the fkin.

The SPECIES are,

1. BRYONIA (*Alba*) foliis palmatis utrinque callofo fcabris. Hort. Cliff. 453. *Rough or white Briony with red flowers.*

2. BRYONIA (*Africana*) foliis palmatis quinquepartitis utrinque lævibus, laciniis pinnatifidis. Lin. Sp. Pl. 1013. *African tuberous-rooted Briony, with indented leaves and an herbaceous flower.*

3. BRYONIA (*Cretica*) foliis palmatis fuprà callofo-punctatis. Hort. Cliff. 453. *Spotted Briony of Crete.*

4. BRYONIA (*Racemofa*) foliis trilobis fuprà callofo-punctatis, fructu racemofo ovali. *Briony with a red Olive-fhaped fruit.*

5. Bryonia (*Variegata*) foliis palmatis, laciniis lanceolatis suprà punctatis infernè lævibus, fructu ovato sparso. *American Briony with a variegated fruit.*

6. Bryonia (*Bonarienfis*) foliis palmatis quinquepartitis hirsutis, laciniis obtusis. *Briony with hairy palmated leaves divided into five parts, and obtuse segments.*

The first sort grows upon dry banks under hedges in many parts of England. The roots of this plant have been formerly by impostors brought into a human shape, and carried about the country, and shewn for Mandrakes to the common people, who were easily imposed on by their credulity, and these got good livings thereby. The method which these people practised, was to find a young thriving Broniy plant, then they opened the earth all round the plant, being careful not to disturb the lower fibres; and (being prepared with such a mould, as is used by the people who make plaister figures) they fixed the mould close to the root, fastening it with wire to keep it in its proper situation: then they filled the earth about the root, leaving it to grow to the shape of the mould, which in one summer it will do; so that if this be done in March, by September it will have the shape. The leaves of this plant are also often imposed on the people in the market for Mandrake leaves, although there is no resemblance between them, nor any agreement in quality.

The second and fourth sorts are perennial plants, but their branches decay every winter. These roots must be planted in pots filled with fresh light earth, and in winter must be placed in the green-house to protect them from frosts and great rains, which would destroy them if they were exposed thereto. In summer they may be exposed to the open air, and must be frequently refreshed with water in dry weather. These plants will flower in July, and in warm summers will perfect their seeds.

The third, fifth, and sixth sorts, are annual plants; these must be raised on a hot-bed early in the spring, and when the plants are about three inches high, they should be each transplanted into a small pot, and plunged into a hot-bed of tanners bark. When the plants are grown so large as to ramble about on the surface of the bed, and begin to entangle with other plants, they should be shifted into larger pots, and placed in the bark-stove, where their branches may be trained to the wall, or against an espalier, that they may have sun and air, which is absolutely necessary for their producing fruit. When these plants are full of fruit, they make a pretty variety in the stove amongst other exotic plants

BRYONIA NIGRA. See Tamnus.

BUBON. Lin. Gen. Plant. 312. Macedonian Parsley.

The Characters are,

It hath an umbelliferous flower; the small umbels have twenty rays. The empalement of the flower is permanent; the flower is composed of five spear-shaped petals, which turn inward; it hath five stamina. The oval germen is situated below the flower, which afterward becomes an oval, channelled, hairy fruit, dividing in two parts, each having an oval seed, plain on one side, but convex on the other.

The Species are,

1. Bubon (*Macedonicum*) foliolis rhombeo-ovatis crenatis, umbellis numerosissimus. Hort. Cliff. 95. *Macedonian Parsley.*

2. Bubon (*Rigidius*) foliolis linearibus. Hort. Cliff. 95. *Hard or rigid Ferula, with very short leaves.*

3. Bubon (*Galbanum*) foliolis rhombeis serratis glabris, umbellis paucis. Hort. Cliff. 96. *African Ferula bearing Galbanum, with a leaf and appearance of Lovage.*

4. Bubon (*Gumiferum*) foliolis glabris inferioribus rhombeis serratis, superioribus pinnatifidis tridentatis. Pr. Leyd. 100. *Galbanum-bearing African Ferula, with the Mock Chervil leaf.*

The first sort sends out many leaves from the root, the lower growing almost horizontally, spreading near the surface of the ground, which are garnished with smooth rhomb-shaped leaves, which are of a bright pale green colour, and sawed on their edges. In the center of the plant arises the flower-stem, which is little more than a foot high, dividing into many branches, each being terminated by an umbel of white flowers, which are succeeded by oblong hairy seeds.

This plant in warm countries is biennial; the plants which rise from seeds one year, produce flowers and seeds the next, and then perish: but in England they seldom flower till the third or fourth year from seed; but whenever the plant flowers, it always dies.

It is propagated by seeds, which should be sown on a bed of light sandy earth as soon as it is ripe, or in April. When the plants come up, they will require no other care but to keep them clean from weeds, till the beginning of October, when they should be carefully taken up, and planted in a warm border of dry ground; and a few of them should be put into pots, that they may be sheltered under a frame in winter; for in severe frost, those which are exposed to the open air are frequently killed, though in moderate winters they will live abroad without covering. The seeds of this plant is one of the ingredients in Venice treacle.

The second sort grows naturally in Sicily. This is a low perennial plant, having short stiff leaves which are very narrow: the flower-stalk rises a foot high, which is terminated by an umbel of small white flowers, which are succeeded by small, oblong, channelled seeds. It is propagated by seeds, and should have a dry soil and a warm situation, where the plants will continue several years. It is a plant of little beauty or use, so is only preserved for the sake of variety.

The third sort rises with an upright stalk to the height of eight or ten feet, which at bottom is woody, having a purplish bark covered with a whitish powder, which comes off when handled; the upper part of the stalk is garnished with leaves at every joint, the foot-stalks half embracing them at their base, and are set with leaves like those of Lovage, but smaller, and of a gray colour; the top of the stalk is terminated by an umbel of yellow flowers, which are succeeded by oblong channelled seeds, which have a thin membrane or wing on their border. When any part of the plant is broken, there issues out a little thin milk of a cream colour, which hath a strong scent of Galbanum.

The fourth sort, like the third, rises with a ligneous stalk about the same height, and is garnished with leaves at each joint, which branch out like the former; but the small leaves or lobes are narrow and indented, like those of Bastard Hemlock. The stalk is terminated by an umbel of small yellow flowers, which are succeeded by seeds like those of the former sort.

This is propagated by seeds, which should be sown in pots filled with light loamy earth as soon as they arrive; which, if it happens toward autumn, should be plunged into an old bed of tanners bark, where the heat is gone, and screened from the frost in winter. In the spring the plants will come up, and by the middle of April will be fit to remove, when they should be carefully shaken out of the pots, and planted each into separate small pots: then plunge the pots into the tan again, and water them to settle the earth to the roots of the plants, and shade them from the sun in the day-time until they have taken new root; after this they may be inured gradually to bear the open air, into which they should be removed in June, and placed with other exotic plants in a sheltered situation, where they may remain till autumn, when they must be removed into the green-house, and placed where they may

enjoy

enjoy as much of the ſun and air as poſſible, but defended from froſt.

Theſe plants make a pretty variety in the green-houſe in winter, and when they are placed abroad in the ſummer with other green-houſe plants they have a good effect, eſpecially when they are grown to a large ſize: in warm ſummers the plants will perfect their ſeeds in England, if they ſtand in a warm ſheltered ſituation.

The Galbanum of the ſhops is ſuppoſed to be procured from both theſe ſorts indifferently; and upon breaking of their leaves, the juice which flows out from the wound hath a ſtrong odour of the Galbanum, which is a confirmation of it.

BUCKSHORN, or HARTSHORN. See PLANTAGO.

BUDDING. See INOCULATING.

BUDDLEJA. Houſt. MSS. Lin. Gen. Plant. 131.

The CHARACTERS are,

The flower is of one leaf, bell-ſhaped, and quadrifid; it hath four ſhort ſtamina, which are placed at the diviſions of the petal. The oblong germen is ſituated in the center, which afterward becomes an oblong capſule, having two cells filled with ſmall ſeeds.

The SPECIES are,

1. BUDDLEJA (*Americana*) foliis ovatis ſerratis oppoſitis ſubtus piloſis, floribus ſpicatis racemoſis, caule fruticoſo. *Shrubby Buddleja with leaves growing by pairs, ſawed at their edges, and yellow flowers growing in ſpikes.*

2. BUDDLEJA (*Occidentalis*) foliis lanceolatis acuminatis integerrimis oppoſitis, ſpicis interruptis, caule fruticoſo ramoſo. *Buddleja with pointed ſpear-ſhaped leaves which are entire, and placed oppoſite, divided ſpikes, and a branching ſhrubby ſtalk.*

The firſt ſort grows naturally in Jamaica, and moſt of the other iſlands in America, where it riſes to the height of ten or twelve feet, with a thick woody ſtem, covered with a gray bark, and ſends out many branches toward the top, which come out oppoſite; as are alſo the leaves ſo placed, which are oval, and covered on their under ſide with a brown hairy down. At the end of the branches the flowers are produced in long cloſe ſpikes, branching out in cluſters, which are yellow, conſiſting of one leaf, cut into four ſegments; theſe are ſucceeded by oblong capſules, filled with ſmall ſeeds.

The ſecond ſort grows at Carthagena. This riſes much taller than the firſt, and divides into a great number of ſlender branches, which are covered with a ruſſet hairy bark, garniſhed with long ſpear-ſhaped leaves, ending in ſharp points: at the end of the branches are produced branching ſpikes of white flowers, growing in whorls round the ſtalks, with ſmall ſpaces between each. The leaves of this are much thinner than thoſe of the firſt ſort, and have ſcarce any down on their under ſide; the ſpikes of flowers grow more erect, ſo form a large looſe ſpike at the end of every branch.

Theſe plants grow naturally in gullies or other low ſheltered ſpots in the Weſt-Indies, their branches being too tender to reſiſt the force of ſtrong winds, ſo are rarely ſeen in open ſituations.

They may be propagated by ſeeds, which ſhould be brought over in their capſules or pods, for thoſe which are taken out before they are ſent to England ſeldom grow. They ſhould be ſown in pots as ſoon as they arrive, and very lightly covered; for as the ſeeds are very ſmall, ſo if they are buried deep in the ground they will periſh. The pots ſhould be plunged into a moderate hot-bed. If the ſeeds are freſh and good, the plants will come up in about ſix weeks; and if they grow kindly, will be large enough to tranſplant in about a month after. Then they ſhould be carefully ſeparated, and each planted into a ſeparate ſmall pot, and plunged into the hot-bed again, obſerving to ſhade them from the ſun until they have taken new root. After the plants have taken freſh roots in the pots, there ſhould be freſh air admitted to them every day, in proportion to the warmth of the ſeaſon; they muſt alſo be frequently, but moderately refreſhed with water. When the plants have filled theſe ſmall pots with their roots, it will be proper to ſhift them into pots one ſize larger, that they may have time to take good root again before the cold weather comes on. When theſe are new potted, the tan ſhould be turned over to renew the heat, and if it is wanted, ſome freſh tan muſt be added to the bed, to encourage the roots of the plants. In this bed they may remain till autumn, when they muſt be removed into the ſtove, and plunged into the tan-bed, where they muſt conſtantly remain, for they are too tender to thrive in this country if they are not ſo treated. During the winter they muſt have but little water, and ſhould be kept warm; but in ſummer they ſhould have freſh air admitted to them conſtantly when the weather is warm, and frequently ſprinkled all over with water. With this management, the plants will flower the third or fourth year from ſeeds, and continue ſo to do every year after, and will make a good appearance in the ſtove..

BUGLOSSUM. See ANCHUSA and LYCOPSIS.

BUGULA. Tourn. Inſt. R. H. 208. tab. 98. Bugle.

The CHARACTERS are,

The flower is of one leaf, of the lip kind, having an incurved cylindrical tube; the upper lip is very ſmall, erect, and bifid; the under lip is large, open, and divided into three ſegments; it hath four erect ſtamina, two of which are longer than the upper lip, and two ſhorter. In the center is ſituated the four germen, which afterward become four naked ſeeds incloſed in the empalement.

The SPECIES are,

1. BUGULA (*Reptans*) foliis caulinis ſemiamplexicaulibus, ſtolonibus reptatricibus. *Common Bugle.*

2. BUGULA (*Decumbens*) foliis oblongo-ovatis, caulibus decumbentibus, verticillis diſtantibus. *Bugle with a large leaf, and pale blue flower.*

3. BUGULA (*Pyramidalis*) foliis obtuſo dentatis, caule ſimplici. *Bugle with blunt indented leaves, and a ſingle ſtalk.*

4. BUGULA (*Genevenſis*) foliis oblongis tomentoſis, calycibus hirſutis. *Bugle with a fleſh-coloured flower.*

5. BUGULA (*Orientalis*) villoſa, foliis ovato-dentatis ſeſſilibus, floribus reſupinatis. *Hairy Eaſtern Bugle, with an inverted white flower having a purple rim.*

The firſt ſort grows naturally in woods, and ſhady moiſt places in moſt parts of England. There are two varieties of this, one with a white, and the other a pale purple flower; but theſe do not differ in any other reſpect than the colour of their flowers from the common, therefore I have only mentioned them as varieties.

The common Bugle is greatly eſteemed as a vulnerary herb, and is uſed both internally and externally; it enters as an ingredient into the vulnerary decoctions of the ſurgeons, and is commended externally, applied to ulcers. As this grows naturally wild in great plenty, ſo it is ſeldom admitted into gardens.

The ſecond ſort grows naturally on the Alps; the leaves of this are much longer than thoſe of the common Bugle, the ſtalks are weaker, and decline on every ſide, and the whorls of flowers are much ſmaller, and are ranged at a greater diſtance.

The third ſort grows naturally in France, Germany, and other countries. It grows about four or five inches high, with a ſingle ſtalk, which is garniſhed with leaves placed oppoſite. The flowers grow in whorls round the ſtalks, and toward the top form a cloſe thick ſpike, and are of a fine blue colour.

The

The fourth ſort grows naturally in many parts of Europe. This approaches near to the common Bugle, but the leaves of this are woolly, and the flower-cups are very hairy. There are two varieties of this, one with a white, and the other a red flower.

The fifth ſort was brought from the Levant by Dr. Tournefort, and is preſerved by thoſe who are curious in collecting rare plants.

This ſort requires a little protection in ſevere winters, therefore ſome of the plants ſhould be planted in pots filled with a loamy ſoil, and placed in a ſhady ſituation in ſummer; but in the winter ſhould be removed under a common frame, where they may enjoy as much free air as poſſible in mild weather.

This may be propagated by ſeeds, which ſhould be ſown ſoon after it is ripe in a pot, and placed in a ſhady ſituation till autumn, when it ſhould be removed under a frame, where it may be ſcreened from hard froſt. In the ſpring the plants will come up, which ſhould be tranſplanted into ſeparate pots as ſoon as they are ſtrong enough to remove, and in ſummer placed in the ſhade, and treated as the old plants.

All the other ſorts are hardy enough, and are eaſily multiplied by their ſide ſhoots; theſe delight in a moiſt ſhady ſituation, where they are apt to ſpread too much, eſpecially the two firſt ſorts.

BULBINE. See ANTHERICUM.

BULBOCASTANUM. See BUNIUM.

BULBOCODIUM. Tourn. Cor. 50.

The CHARACTERS are,

The flower hath no empalement, it is funnel-ſhaped, and compoſed of ſix petals, which are concave. It hath ſix ſtamina inſerted in the middle. It hath an oval, blunt, three-cornered germen, which afterward becomes a triangular pointed capſule, having three cells, which are filled with angular ſeeds

The SPECIES are,

1. BULBOCODIUM (*Alpinum*) foliis ſubulato-linearibus. Prod. Leyd. 41. *Bulbocodium with narrow awl-ſhaped leaves.*

2. BULBOCODIUM (*Vernum*) foliis lanceolatis. Prod. Leyd. 41. *Bulbocodium with ſpear-ſhaped leaves; or Spaniſh Spring Meadow Saffron.*

The firſt ſort grows naturally upon the Alps, and alſo upon Snowdon in Wales. It hath a ſmall bulbous root, which ſends out a few long narrow leaves ſomewhat like thoſe of the Saffron, but are narrower; in the middle of theſe the flower comes out, which ſtands on the top of the foot-ſtalk, growing erect, and is ſhaped like thoſe of the Crocus, but ſmaller; the foot-ſtalk riſes about three inches high, and hath four or five ſhort narrow leaves placed alternately upon it below the flower. This flowers in March, and the ſeeds are ripe in May.

The ſecond ſort grows naturally in Spain. It hath a bulbous root, ſhaped like thoſe of the Snowdrop, which ſends out three or four ſpear-ſhaped concave leaves, between which comes out the flower, ſtanding on a very ſhort foot-ſtalk; theſe, when they firſt appear, are of a pale colour, but afterward change to a white purple. It produces the flowers about the ſame time with the firſt.

Theſe plants are propagated by offſets, in the ſame manner as other bulbous-rooted flowers. The time to remove them is ſoon after their leaves decay, but the roots may be kept out of the ground two months without prejudice at that ſeaſon. They ſhould not be removed oftener than every third year, for their roots do not multiply very faſt, ſo by ſuffering them to remain, they will flower much ſtronger, and make a greater increaſe than if they are often taken up.

BUNIAS. Lin. Gen. Plant. 737.

The CHARACTERS are,

The flower hath four petals, placed in form of a croſs, joined at their baſe, and erect. It hath ſix ſtamina, two of which are oppoſite, and ſhorter than the other. In the center is ſituated an oblong germen, which afterward becomes an irregular, ſhort, oval pod with four angles, one or other of which is prominent and pointed, incloſing one or two roundiſh ſeeds.

The SPECIES are,

BUNIAS (*Orientalis*) ſiliculis ovatis gibbis verrucoſis. Lin. Sp. Plant. 670. *Bunias with oval convex pods, having protuberances; or Eaſtern Sea Kale.*

2. BUNIAS (*Erucago*) ſiliculis tetragonis angulis bicriſtatis. Lin. Sp. Pl. *Bunias with ſhort four-cornered pods, whoſe angles are doubly creſted.*

3. BUNIAS (*Cakile*) ſiliculis ovatis lævibus ancipitibus. Lin. Sp. Pl. 670. *Bunias with ſmooth oval pods, ſtanding two ways on the ſtalk.*

The firſt ſort grows naturally in the Levant. This hath a perennial root and an annual ſtalk. It ſends out many oblong leaves, which ſpread on every ſide, and are deeply jagged on their edges like thoſe of the Dandelion; from between theſe ariſe the ſtalks, which grow upwards of two feet high, ſending out branches on every ſide, which are garniſhed at each joint by one oblong ſharp-pointed leaf, eared at the baſe. The branches are terminated by long looſe ſpikes of yellow flowers, compoſed of four leaves, ſhaped like thoſe of the Cabbage; theſe are ſucceeded by ſhort, oval, rough pods, ending in a point, incloſing one round ſeed.

The ſecond ſort grows naturally in the ſouth of France and Italy. This is an annual plant, which branches on every ſide, and incline toward the ground. The branches are garniſhed with glaucous leaves, which are deeply divided into many parts, almoſt like thoſe of the Swines Creſs. The flowers are produced ſingly from the wings of the leaves; theſe are very ſmall, of a pale yellowiſh colour, which are ſucceeded by ſhort pods, creſted on each ſide, containing one or two roundiſh ſeeds.

The third ſort grows naturally about Montpelier; this is an annual plant, which ſends out many oblong leaves near the root, deeply cut on each ſide, and ſpread on the ground; between theſe ariſe two or three ſtalks, which grow a foot and a half high, ſending out ſeveral ſide branches, which are garniſhed with oblong rough leaves, indented on their edges; the upper part of the branches are deſtitute of leaves, but have flowers placed alternately on each ſide the branches, ſtanding on ſhort foot-ſtalks, which are purple; theſe are ſucceeded by oval pointed pods, containing one or two roundiſh ſeeds.

Theſe plants are all propagated by ſeed; the firſt ſort may be ſown where the plants are deſigned to remain in the beginning of April, and when the plants come up, they ſhould be thinned, leaving them two feet aſunder; after which they will require no other care but to keep them clean from weeds.

The other two ſorts muſt be ſown where they are to remain, but the beſt time is in autumn, becauſe thoſe which are ſown in the ſpring often fail, or do not come up time enough to perfect their ſeeds. Theſe require no other culture but to keep them clean from weeds, and thin the plants to one foot diſtance.

BUNIUM. Lin. Gen. Pl. 298. Pig Nut, or Earth Nut.

The CHARACTERS are,

The involucrum of the great umbel is compoſed of many ſhort narrow leaves. The proper empalement of the flower is ſcarce diſcernible. The flowers have five heart-ſhaped petals, which are equal, and turn inward; they have five ſtamina; the oblong germen is ſituated below the receptacle, which afterward becomes an oval fruit dividing in two parts, containing two oval ſeeds.

The SPECIES are,

1. BUNIUM (*Bulbocastanum*) bulbo globoso. Sauv. Monsp. 256. *Earth Nut with a globular root.*

2. BUNIUM (*Creticum*) radice turbinato. *Earth Nut with a turbinated root.*

3. BUNIUM (*Saxatile*) foliis tripartitis filiformibus linearibus. *Earth Nut with very narrow tripartite leaves.*

The first sort grows naturally in moist pastures, and in woods in many parts of England; of this there is a variety, supposed to be larger than that which grows commonly here. This hath a tuberous solid root, which lies deep in the ground. The leaves are finely cut, and lie near the ground. The stalk rises a foot and a half high, which is round, channelled, and solid; the lower part being naked, but above, where it branches out, there is one leaf placed below every branch. The flowers are white, and shaped like those of other umbelliferous plants; the seeds are small, oblong, and when ripe are channelled.

The roots of this sort are frequently dug up, and by the poorer sort of people are eaten raw, having much resemblance in taste to the Chestnut, from whence it had the title of Bulbocastanum.

The second sort was discovered by Dr. Tournefort in the island of Crete, but it grows naturally in many other parts of the Levant.

The third sort I received from the Alps. This is a very low plant, seldom rising above six inches high.

These plants delight to grow among Grass, so cannot without difficulty be made to thrive long in a garden.

BUPHTHALMUM. Lin. Gen. Pl. 876. Ox-eye.

The CHARACTERS are,

It hath a compound radiated flower, composed of hermaphrodite and female florets. The hermaphrodite flowers compose the disk, and are funnel-shaped. In the center is situated an oval compressed germen, which afterward becomes an oblong seed. The female flowers which compose the rays, are stretched out on one side like a tongue, and are indented at the top in three parts; these have no stamina, but a double-headed germen, which becomes a single compressed seed cut on each side.

The SPECIES are,

1. BUPHTHALMUM (*Helianthoides*) calycibus foliolis, foliis oppositis ovatis serratis trinerviis caule herbaceo. Hort. Upsal. 264. *Ox-eye with a leafy empalement, heart-shaped leaves having three veins, growing opposite.*

2. BUPHTHALMUM (*Grandiflorum*) alternis foliis lanceolatis subdenticulatis glabris, calycibus nudis. Hort. Cliff. 415. *Ox-eye with smooth spear-shaped leaves (indented below) and naked empalements.*

3. BUPHTHALMUM (*Salicifolium*) foliis lanceolatis subserratis villosis calycibus nudis. Hort. Cliff. 414. *Ox-eye with spear-shaped leaves, sawed below and hairy, and naked empalements.*

4. BUPHTHALMUM (*Spinosum*) calycibus acutè foliosis, ramis alternis, foliis lanceolatis amplexicaulibus integerrimis. Hort. Cliff. 414. *Ox-eye with acute leafy empalements, branches placed alternate, and entire leaves embracing the stalks.*

5. BUPHTHALMUM (*Maritimum*) calycibus obtusè foliosis pedunculatis, ramis alternis, foliis cuneiformibus. Hort. Cliff. 414. *Ox-eye with blunt leafy empalements having foot-stalks, alternate branches, and wedge-shaped leaves.*

6. BUPHTHALMUM (*Sessile*) floribus axillaribus calycibus foliosis spinis terminalibus foliis sessilibus. *Ox-eye with blunt leafy empalements sitting close to the side of the stalks, and oblong blunt leaves.*

7. BUPHTHALMUM (*Arborescens*) foliis oppositis lanceolatis petiolatis bidentatis caule fruticoso. Hort. Cliff. 415. *Ox-eye with spear-shaped leaves growing opposite, with foot-stalks having two teeth, and a shrubby stalk.*

8. BUPHTHALMUM (*Incanum*) foliis oppositis lineari-lanceolatis crassis, incanis floribus sessilibus caule fruticoso. *Ox-eye with thick, smooth, narrow, spear-shaped leaves growing opposite, and flowers having no foot-stalks.*

9. BUPHTHALMUM (*Viride*) foliis oppositis lineari-lanceolatis crassis, floribus sessilibus. *Ox-eye with thick, hoary, narrow, spear-shaped leaves placed opposite, and flowers growing close to the branches.*

The first sort grows naturally in North America; this hath a perennial root, and an annual stalk which rises six or eight feet high, garnished at each joint with two oblong heart-shaped leaves, which have three longitudinal veins, and the base on one side shorter than the other. The flowers come out at the extremity of the branches, having a leafy empalement; these are of a bright yellow colour, resembling a small Sun flower, from whence the inhabitants of America have given it that appellation. It propagates easily by parting of the roots. The best time to transplant and part the roots is toward the end of October, when the stalks begin to decay. These should be removed every other year, to prevent their spreading too far; they are very hardy, so will thrive in any situation, and are proper for large borders on the sides of rural walks, or in spaces between shrubs.

The second sort grows naturally on the Alps, as also in Austria, Italy, and the south of France. This hath also a perennial root and an annual stalk; it grows near two feet high, with slender branching stalks, garnished with oblong smooth leaves; the flowers grow at the extremity of the branches, of a bright yellow colour like those of Starwort. There are two or three varieties of this, differing in the breadth of their leaves and size of their flowers, but from the same seeds all these have been produced.

This sort may be propagated by parting of the roots, at the same time and in the same manner as is directed for the second sort.

The third sort is somewhat like the second, but the leaves are broarder and obtuse; the stalks and leaves are also hairy, in which consists their difference.

The fourth sort is an annual plant, which grows naturally in the south of France, Italy, Spain, and Sicily. The stalks rise two feet and a half high, and divide into many branches upward, and the side branches rise above the middle stalk. They are garnished with spear-shaped hairy leaves, placed alternately; the flowers are produced at the end of the branches on short foot-stalks; the empalement consists of seven long, stiff, spear-shaped leaves, ending in a sharp point. The flower sits close upon the empalement, the border or rays being composed of many female florets. The middle or disk of the flower is composed of hermaphrodite flowers, which are tubulous and funnel-shaped. They are of a bright yellow colour, and are succeeded by oblong compressed seeds.

The seeds of these should be sown in April, on open borders, where they are to remain, and will require no other care but to keep them clear of weeds, and thin them to the distance of a foot and a half, that their branches may have room to spread.

The fifth sort is a low perennial plant, with a shrubby stalk, which rarely rises a foot high, sending out many spreading branches from the stem, garnished with hairy leaves, which are very narrow at their base, but broad and rounded at their extremity; the flowers are produced at the end of the branches, which are yellow, and shaped like those of the former sorts, but the leaves of the empalement are soft and obtuse. These are rarely succeeded by seeds in England, but the plant is easily propagated by slips during the summer season, which will take root in about six weeks, when they should be carefully taken up, and each planted in a

separate

separate small pot, and placed in a shady situation till they have taken fresh root; after which they may be removed to an open situation, where they may remain till the end of October, when they must be removed into a garden-frame for the winter season, being too tender to live abroad in this country; but as they only require protection from hard frosts, so they will thrive better when they have a great share of air in mild weather, than if confined in a green-house; therefore the best method is to place them in a common frame, where they may be fully exposed in mild weather, but screened from the frost.

The seventh sort rises with several woody stems, which grow to the height of eight or ten feet, garnished with leaves very unequal in size; some are narrow and long, others are broad and obtuse; these are intermixed at the same joint, and often at the intermediate one; they are green, and placed opposite. The flowers are produced at the ends of the branches; they are of a pale yellow colour, and have scaly empalements. This sort has been long preserved in the English gardens, as it flowers all the summer.

The eighth sort grows naturally in the Bahama Islands. This seldom grows much more than three feet high. It has succulent spear-shaped leaves placed opposite; the flowers are produced at the end of the branches, which are of a bright yellow colour. I received this also from the Havannah, where it grows plentifully on the borders of the sea.

The ninth sort grows in the Bahama Islands. This sends out many slender stalks from the root, which rise near three feet high, with long, narrow, thick, succulent leaves, growing opposite, embracing the stalk at their base; the flowers are yellow, and are produced at the ends of the shoots, having very short foot-stalks.

As the three last sorts do not perfect their seeds in this country, so they are propagated by cuttings, which should be planted in July, when the plants have been for some time exposed to the open air, whereby their shoots will be hardened, and better prepared to take root than when they first come abroad. They should be planted in small pots filled with light loamy earth, and plunged into a very gentle warmth observing to shade them from the sun in the heat of the day. In about six weeks these will have taken root, when they must be gradually inured to bear the open air: and soon after they should be each planted in a separate small pot, filled with light loamy earth, and placed in the shade until they have taken fresh root; after which they may be removed to a sheltered situation, where they may remain till the middle of October, when they must be removed into the green-house. During the winter they should have but little moisture, and in very mild weather they should have fresh air admitted to them. In the summer they must be placed abroad in a sheltered situation, and treated in the same manner as other exotic plants.

BUPLEUROIDES. See Phyllis.

BUPLEURUM. Hare's-ear, or Thorough Wax.

The Characters are,

It is a plant with an umbellated flower; the rays of the principal umbel are thin; the involucrum of the great umbel is composed of many leaves, those of the small have five. The flower hath five small heart-shaped petals, which are inflexed; it hath five slender stamina. The germen is situated below the flower, which afterward becomes a roundish compressed fruit, which is channelled, dividing in two parts, containing two oblong channelled seeds.

The Species are,

1. Bupleurum (*Rotundifolium*) involucris universalibus nullis, foliis perfoliatis. Hort. Upsal. 64. *The most common or Field Thorough Wax.*

2. Bupleurum (*Angulosum*) involucellis pentaphyllis orbiculatis, universali triphyllo ovato, foliis amplexicaulibus cordato-lanceolatis. Lin. Sp. Pl. 236. *Greater narrow-leaved Thorough Wax of the Alps, with an angular leaf.*

3. Bupleurum (*Odontitis*) involucellis pentaphyllis acutis, universali triphyllo, flosculo centrali altiore, ramis divaricatis. Lin. Sp. Plant. 237. *Smaller narrow-leaved Thorough Wax, with a Hare's-ear leaf.*

4. Bupleurum (*Rigidum*) caule dichotomo subnudo, involucris minimis acutis. Lin. Sp. Plant. 238. *Hare's-ear with a stiff leaf.*

5. Bupleurum (*Tenuissimum*) umbellis simplicibus alternis pentaphyllis subtrifloris. Lin. Sp. Pl. 238. *Hare's-ear with a very narrow leaf.*

6. Bupleurum (*Fruticosum*) frutescens, foliis obovatis integerrimis. Lin. Sp. Plant. 238. *Shrubby Hartwort of Ethiopia.*

7. Bupleurum (*Difforme*) frutescens, foliis vernalibus decompositis planis incisis, æstivalibus filiformibus angulatis trifidis. Lin. Sp. Plant. 238. *Shrubby Hare's-ear, whose spring leaves are decompounded, plain, and cut, and the summer leaves are narrow, angular, and trifid.*

The first sort grows naturally upon chalky land amongst Wheat in several parts of England, so is seldom admitted into gardens. The leaves and seeds of this plant are used in medicine; the herb is esteemed good for dissolving scrophulous tumours, and is by some used for internal ailments, ruptures, and bruises from a fall. This is a biennial plant.

The second, third, fourth, and fifth sorts, are annual. The fifth sort grows naturally in several parts of England, the others are natives of the Alps and Pyrenees; these are seldom cultivated but in botanic gardens for the sake of variety. Their seeds should be sown in autumn, where the plants are designed to remain, and keep the plants kept clean from weeds, which is all the culture they require.

The sixth sort hath a woody stem, which sends out many branches on every side, so as to form a large head or bush, with oblong, oval, stiff leaves, which are very smooth, and of a sea-green colour: the ends of the branches are terminated by umbels of yellow flowers, somewhat like those of Fennel.

It is commonly known among gardeners by the title of shrubby Ethiopian Hartwort, and is now propagated in the nursery-gardens for sale. It grows four or five feet high, forming a large regular bush, and the leaves continuing green through the year render it more valuable. It is hardy, so will live in the open air, and may be intermixed with other evergreen shrubs of the same growth, where they will make an agreeable variety. It is propagated by cuttings, which should be planted in pots, and in winter sheltered under a hot-bed frame; in the spring the cuttings will put out roots, but they will not be fit to transplant till the autumn following, so the pots should be placed in a shady situation in summer. The young plants may be planted in a nursery-bed at two feet distance, for a year or two to get strength, and then transplanted where they are to remain.

The seventh sort grows naturally at the Cape of Good Hope. This rises with a shrubby stalk to the height of five or six feet, sending out some side branches, which in the spring have their lower parts garnished with leaves composed of many small plain lobes, which are finely cut like those of Coriander, and of a sea-green colour; these leaves soon fall off, and the upper part of the branches are closely covered with long Rush-like leaves, having four angles, which come out in clusters from each joint. The flowers grow in spreading umbels at the extremity of the branches, which are small, and of an herbaceous colour, and are succeeded by oblong channelled seeds.

This

This sort is propagated by cuttings, which readily take root if they are planted in April in pots, and plunged into a moderate hot-bed. When they have taken root, they should be inured to the open air by degrees, and after having obtained strength, they may be planted each into a separate pot, placing them in the shade till they have taken fresh root, when they may be placed with other exotic plants in a sheltered situation, where they may remain till the autumn, when they must be removed into the green-house, and placed with such hardy plants as require a large share of air in mild weather, and only require protection from frost.

BURNET. See POTERIUM and SANGUISORBA.

BURSA PASTORIS. Shepherds-pouch.

This is a common weed in most parts of England, which propagates so fast by seeds, as not to be easily cleared when they are permitted to shed, for there are commonly four or five generations of this plant from seeds in a year, so fast does the seed ripen, and the plants come up, therefore it cannot be too soon rooted out of a garden.

BUTOMUS. The Flowering-rush, or Water-gladiole.

The CHARACTERS are,

The flowers grow in a single umbel; they have six roundish concave petals, which are alternately smaller, and nine awl-shaped stamina, six of which surround the other. It hath six oblong pointed germen, which afterward become six oblong pointed capsules, having one cell filled with oblong seeds.

We have but one SPECIES of this genus, viz.

BUTOMUS (*Umbellatus.*) Fl. Lap. 159. *The Flowering-rush, or Water-gladiole.*

There are two varieties of this plant, one with a Rose-coloured flower, and the other with a white flower; but these are only accidental variations, therefore not to be enumerated as distinct species.

The Rose-coloured sort is pretty common in standing waters in many parts of England; the other is a variety of this, though less common with us near London. These plants may be propagated in boggy places, or by planting them in cisterns, which should be kept filled with water, that should have about a foot thickness of earth in the bottom, into which the roots should be planted, or the seed sown as soon as they are ripe; these, though common plants, yet produce very pretty flowers, and are worth propagating for variety's sake, especially if in any part of the garden there should be conveniency for an artificial bog, or where there are ponds of standing water, as is many times the case, and persons are at a loss what to plant in such places that may appear beautiful.

BUXUS. The Box-tree.

The CHARACTERS are,

It hath male and female flowers on the same plant; the male flowers have a three-leaved, and the female a four-leaved empalement. The male flowers have two, and the female three concave petals. The male flowers have four upright stamina, with a rudiment of a germen, but no style or stigma. The female flowers have roundish, blunt, three-cornered germen. The empalement afterward becomes a roundish capsule, shaped like an inverted pottage pot, opening in three cells, each having two oblong seeds

The SPECIES are,

1. BUXUS (*Arborescens*) arborescens, foliis ovatis. *Box-tree with oval leaves.*

2. BUXUS (*Angustifolius*) arborescens foliis lanceolatis. *Narrow-leaved Box.*

3. BUXUS (*Suffruticosa*) humilis foliis orbiculatis. *Dwarf or Dutch Box.*

The two sorts of Tree Box have been frequently raised from seeds, and constantly produced plants of the same kind from those the seeds were taken; and the Dwarf Box will never rise to any considerable height with any culture, nor have I ever seen this sort flower, where the plants have been encouraged to grow many years in the greatest luxuriancy. There are two or three varieties of the first sort, which are propagated in the gardens, one with yellow, and the other white-striped leaves. The other hath the tops of the leaves only marked with yellow, which is called Tipped Box.

The first and second sorts grow in great plenty upon Box-hill near Dorking in Surry, where were formerly large trees of these kinds, but of late they have been pretty much destroyed; yet there are great numbers of the trees remaining, which are of a considerable bigness.

The Tree or large Box are proper to intermix in clumps of Evergreens, &c. where they add to the variety of such plantations; these may be propagated by planting the cuttings in autumn in a shady border. When they are well rooted, they may be transplanted into nurseries till they are fit for the purposes intended. The best season for removing these trees is in October, though indeed, if care be used to take them up with a good ball of earth, they may be transplanted almost at any time, except in the middle of summer: these trees are a very great ornament to cold and barren soils, where few other things will grow.

The dwarf kind of Box is used for bordering of flower-beds or borders; for which purpose it far exceeds any other plant, it being subject to no injuries from cold or heat, and is of long duration, is very easily kept handsome, and by the firmness of its rooting keeps the mould in the borders from washing into the gravel-walks, more effectually than any plant whatever. This is increased by parting the roots, or planting the slips; but as it makes so great an increase of itself, and so easily parts, it is hardly worth while to plant the slips that have no roots.

C.

CAAPEBA. See Cissampelus.

CABBAGE. See Brassica.

CACALIANTHEMUM. See Cacalia.

CACALIA. Foreign Coltsfoot.

The Characters are,

It hath compound flowers, which are included in one common cylindrical empalement: the flowers are tubulous and funnel-shaped; they have each five short slender stamina. The germen is crowned with down, which afterward becomes a single oblong seed, crowned with long down.

The Species are,

1. Cacalia (*Alpina*) foliis reniformibus acutis denticulatis. Lin. Sp. Plant. 836. *Cacalia with kidney-shaped leaves, which are sharply indented.*

2. Cacalia (*Glabra*) foliis cutaneis acutioribus glabris. *Cacalia with smooth heart-shaped leaves sawed on their edges.*

3. Cacalia (*Suaveolens*) caule herbaceo foliis hastato-sagittatis denticulatis, petiolis superne dilatatis. Hort. Upsal. 254. *Taller American Cacalia with a triangular leaf eared at the base, and white flowers.*

4. Cacalia (*Atriplicifolia*) caule herbaceo, foliis subcordatis, dentato-sinuatis, calycibus quinquefloris. Lin. Sp. Plant. 835. *Cacalia with an herbaceous stalk, heart-shaped sinuated leaves, and five florets in each empalement.*

5. Cacalia (*Ficoides*) caule fruticoso, foliis compressis carnosis. Lin. Sp. Plant. 834. *African Tree Groundsel, with the leaf and appearance of Fig Marigold.*

6. Cacalia (*Kleniu*) caule fruticoso composito, foliis lanceolatis planis, petiolorum cicatricibus obsoletis. Lin. Sp. Pl. 834. *Cacalia with a compound shrubby stalk, plain spear-shaped leaves, and the foot-stalks leaving scars.*

7. Cacalia (*Papillaris*) caule fruticoso obvallato spinis petiolaribus truncatis. Lin. Sp. Plant. 834. *Cacalia with a shrubby stalk, guarded on every side with broken rough foot-stalks.*

8. Cacalia (*Antieuphorbium*) caule fruticoso, foliis ovato-oblongis, petiolis basi linea triplici deductis. Lin. Sp. Pl. 834. *Cacalia with a shrubby stalk, oblong oval leaves, and three lines connected to the base of the foot-stalk.*

The first sort grows naturally in Austria and the Helvetian mountains, but is frequently preserved in some curious gardens for the sake of variety. It hath a fleshy root which spreads in the ground, from which springs up many leaves, standing on single foot-stalks, shaped like those of Ground Ivy, but are of a thicker texture, of a shining green on their upper side, but white on their under side; between these arise the stalk, which is round, branching toward the top, and grows a foot and a half high; the branches are terminated by purplish flowers, growing in a sort of umbel. These are succeeded by oblong seeds, crowned with down.

The second sort hath the appearance of the first, but the leaves are heart-shaped, pointed, and sharply sawed on their edges, and on both sides very green; the stalks rise higher, and the leaves upon the stalks have much longer foot-stalks than those of the first. The flowers of this are of a deeper purple colour.

CAC

The third sort grows naturally in North America. This hath a perennial creeping root, which sends out many stalks, garnished with triangular spear-shaped leaves, sharply sawed on their edges. The stalks rise to the height of seven or eight feet, and are terminated by umbels of white flowers, which are succeeded by oblong seeds crowned with down. This plant multiplies greatly by its spreading roots, and also by the seeds, which spread to a great distance by the wind, the down which adheres to them being greatly assisting to their conveyance.

The fourth sort is a native of America. This hath a perennial root, and an annual stalk, which rises four or five feet high garnished with roundish heart-shaped leaves, greatly indented on their edges, of a sea-green colour on their under side, but darker above; the stalks are terminated by umbels of yellowish herbaceous flowers.

The first and second sorts are propagated by parting of their roots, for they seldom produce good seeds in England. The best time to transplant and part their roots is in the autumn. They require a loamy soil and a shady situation.

The third and fourth sorts propagate in great plenty both by their spreading roots, and also their seeds. The roots should be transplanted in autumn, and require a moist soil and an open situation.

The fifth sort grows naturally at the Cape of Good Hope. This rises with strong round stalks to the height of seven or eight feet, which are woody at bottom, but soft and succulent upward, sending out many irregular branches, which are garnished with thick, taper, succulent leaves, a little compressed on two sides, ending in points, and are covered with a whitish glaucous farina, which comes off when handled. At the extremity of the branches the flowers are produced in small umbels; they are white, tubulous, and cut into five parts at the top. Some of the noblemen in France have the leaves of this plant pickled; in the doing of which, they have a contrivance to preserve the white farina, with which they are covered, and thereby render them very beautiful.

This sort is easily propagated by cuttings during the summer months: these should be cut from the plants, and laid to dry a fortnight, that the wound may be healed over before they are planted. If they are planted in June or July, they will take root in the open air. I have frequently had the branches broken off by accident, and fallen on the ground, which have put out roots without any care. The plants should have a light sandy earth, and in winter be placed in an airy glass-case, where they may enjoy the sun and air in mild weather, but must be protected from frost, and have but little water; they must be treated like the Ficoides, and other succulent plants from the same country.

There is a variety of this, which is a lower plant, and is much tenderer, requiring to be placed in a stove in winter, and must not have too much water This has not produced flowers here.

The sixth sort grows naturally in the Canary Islands. This rises with a thick fleshy stem, divided at certain distances, as it were in ſo many joints; each of theſe diviſions ſwell much larger in the middle than they do at each end; toward their extremities they are garniſhed with long, narrow, ſpear-ſhaped leaves, of a glaucous colour. As theſe fall off, they leave a ſcar at the place, which always remains on the branches. The flowers are produced in large cluſters at the extremity of the branches, which are tubulous, and of a faint Carnation colour.

This plant hath been called Cabbage-tree by the gardeners, I ſuppoſe from the reſemblance which the ſtalks of it have to that of the Cabbage others have titled it Carnation-tree, from the ſhape of the leaves and colour of the flowers

It is propagated by cuttings in the ſame manner as the former ſort, and the plants require the ſame culture.

The ſeventh ſort reſembles the ſixth in its form and manner of growth, but the leaves are narrower and more ſucculent. Theſe do not fall off entire like the other, but break off at the beginning of the foot-ſtalk, which are very ſtrong and thick. This ſort hath not as yet produced any flowers in England. It is propagated in the ſame manner as the two former ſorts, from cuttings, and the plants muſt be treated in the ſame way as hath been directed for the fifth ſort, but requires to be kept drier, both in winter and ſummer. This ſort grows naturally at the Cape of Good Hope.

The eighth ſort has been long preſerved in the Engliſh gardens, and was generally titled Anti-euphorbium, ſuppoſing it to have a contrary quality to the Euphorbium. This riſes with many ſucculent ſtalks from the root as large as a man's finger, which branch out upward into many irregular ſtalks of the ſame form, but ſmaller, and are garniſhed with flat, oblong, ſucculent leaves, placed alternately round the branches; under each foot-ſtalk there are three lines or ribs, which run longitudinally through the branches joined together. This ſort very rarely flowers in Europe, but is propagated by cuttings in the ſame manner as the fifth, and is equally hardy.

CACAO. Tourn. Inſt. R. H. 660. The Chocolate-nut.

The CHARACTERS are,

The flower hath five petals, which are irregularly indented; it hath five erect ſtamina. In the center is placed the oval germen, which afterward becomes an oblong pod, ending in a point, which is divided into five cells, filled with oval, compreſſed, fleſhy ſeeds.

We know but one SPECIES of this genus, viz.

CACAO. Cluſ. Exot. The Chocolate-nut-tree.

This tree is a native of America, and is found in great plenty in ſeveral places between the Tropics, but particularly at Carracca and Carthagena, on the river Amazons, in the iſthmus of Darien, at Honduras, Guatimala, and Nicaragua. At all theſe places it grows naturally without culture; but it is cultivated in many of the iſlands which are poſſeſſed by the French and Spaniards, and was formerly planted in ſome of the iſlands which are in poſſeſſion of the Engliſh.

In the making a plantation of Chocolate-trees you muſt firſt be very careful in the choice of the ſituation and the ſoil, otherwiſe there will be ſmall hopes of ſucceſs. As to the ſituation, it ſhould be in a place where the trees may be protected from ſtrong winds, to which, if they are expoſed, they will ſoon be deſtroyed: ſo that in ſuch places where torrents of water have waſhed away the earth, ſo as to leave broad and deep furrows (which the inhabitants of thoſe iſlands call gullies,) theſe trees will thrive exceedingly. The ſoil in theſe gullies is generally rich and moiſt, which is what theſe trees require, ſo that they will make great progreſs in theſe places; but where there are not a ſufficient number of theſe gullies, choice ſhould be made of a ſituation which is well ſheltered by large trees; or, if there are not trees already grown, there ſhould be three or four rows planted round the ſpot which is deſigned for the Chocolate-trees, of ſuch ſorts which are of quickeſt growth; and within theſe rows there ſhould be ſome Plantain-trees planted at proper diſtances, which being very quick of growth, and the leaves being very large, will afford a kindly ſhelter to the young Chocolate-trees placed between them.

The Chocolate-trees which are cultivated, ſeldom grow to be more than fourteen or fifteen feet in height, nor do they ſpread their branches very wide; ſo that if the Plantain-trees are placed in rows, about twenty-four feet aſunder, there will be room enough for two rows of Chocolate-trees between each row of Plantains; and if they are planted at ten feet diſtance in the rows, it will be ſufficient room for them.

The ſoil upon which theſe trees thrive to moſt advantage, is a moiſt, rich, deep earth; for they generally ſend forth one tap-root, which runs very deep into the ground, ſo that wherever they meet with a rocky bottom near the ſurface, they ſeldom thrive, nor are of long continuance; but in a rich, deep, moiſt ſoil, they will produce fruit in pretty good plenty the third year from ſeed, and will continue fruitful for ſeveral years after.

Before the plantation is begun, the ground ſhould be well prepared by digging it deep, and clearing it from the roots of the trees and noxious plants, which, if ſuffered to remain in the ground, will ſhoot up again after the firſt rain, and greatly obſtruct the growth of the plants; ſo that it will be almoſt impoſſible to clear the ground from thoſe roots, after the Chocolate plants are come up, without greatly injuring them.

When the ground is thus prepared, the rows ſhould be marked out by a line where the nuts are to be planted, ſo as that they may be placed in a quincunx order, at equal diſtances every way; or at leaſt that the Plantain-trees between them may form a quincunx, with the two rows of Chocolate-trees, which are placed between each row of them.

In making a plantation of Chocolate-nut-trees, the nuts muſt be planted where the trees are to remain; for if the plants are tranſplanted, they ſeldom live, and thoſe which ſurvive it, will never make thriving trees; for, as I before obſerved, theſe trees have a tender tap-root, which, if broken, or any way injured, the tree commonly decays.

The nuts ſhould always be planted in a rainy ſeaſon, or at leaſt when it is cloudy weather, and ſome hopes of rain falling ſoon after. As the fruit ripens at two different ſeaſons, viz. at Michaelmas and at Chriſtmas, the plantation may be made at either of thoſe; but the chief care muſt be to chuſe ſuch nuts as are perfectly ripe and ſound, otherwiſe the whole trouble and expence will be loſt. The manner of planting the nuts is, to make three holes in the ground, within two or three inches of each other, at the place where every tree is to ſtand; and into each of theſe holes ſhould be one ſound nut planted about two inches deep, covering them gently with earth. The reaſon for putting in three nuts at every place is, becauſe they ſeldom all ſucceed; or, if moſt of them grow, the plants will not be all equally vigorous; ſo that when the plants have had one year's growth, it is very eaſy to draw up the weak unpromiſing plants, and leave the moſt vigorous; but in doing this, great care ſhould be had to the remaining plants, ſo as not to injure or diſturb their roots in drawing the other out.

It is very proper to obſerve, that the Chocolate-nuts will not retain their growing faculty long after they are taken from the trees, ſo that there is no poſſibility of tranſporting them to any great diſtance for planting; nor ſhould they

be

be kept long out of the ground, in the natural places of their growth.

When the Chocolate-trees firſt appear above ground, they are very tender, and ſubject to great injuries from the ſtrong winds, the ſcorching ſun, or great droughts, for which reaſon the planters are obliged to guard againſt all theſe enemies, firſt, by making choice of a ſheltered ſituation, or at leaſt by planting trees to form a ſhelter; and, if poſſible, to have the plantation near a river, for the conveniency of watering the plants the firſt ſeaſon, until they have made ſtrong roots, and are capable of drawing their nouriſhment from ſome depth in the earth, where they meet with moiſture. The Plantains, which will be fit to cut in about twelve months after planting, will defray the whole expence of preparing the ground, ſo that the produce of the Chocolate-trees will be neat profit; for as the Plantains produce fruit and decay, they will be ſucceeded by ſuckers, which will produce fruit in eight months after, whereby there will be a continual ſupply of food for the negroes, which will more than pay for keeping the ground wrought, and clear from weeds, until the Chocolate-trees begin to produce fruit, which is generally the third year after planting.

The planters uſually ſet the Plantain-trees two or three months before the Chocolate-nuts are ripe, that they may be large enough to afford ſhelter to the young plants when they come up. Some people plant Potatoes, others Cucumbers and Melons, or Water Melons, between the rows of Chocolate plants, which, they ſay, will prevent the weeds from riſing to injure the young plants; for as all theſe trail on the ground, they occupy the whole ſurface, and prevent the weeds from growing: but where this is practiſed, it ſhould be done with great caution, leſt, by being over covetous, you injure the young Chocolate-nuts ſo much, as that they may nev recover it.

In about ſeven or eight days after the Chocolate-nuts are planted, the young plants will begin to appear above ground, when they ſhould be carefully looked over, to ſee if any of them are attacked by inſects; in which caſe, if the inſects are not timely deſtroyed, they will ſoon devour all the young plants; or if there ſhould be any weeds produced near the plants, they ſhould be carefully cut down with a hoe; in doing which, great care ſhould be taken that the tender ſhoot nor the rind of the bark are injured. About twenty days after the plants have appeared, they will be five or ſix inches high, and have four or ſix leaves, according to the ſtrength of the plants. In ten or twelve months they will be two feet and a half high, and have fourteen or ſixteen leaves.

In two years time the plants will have grown to the height of three feet and a half, and ſometimes four feet, many of which will begin to flower; but the careful planters always pull off all theſe bloſſoms, for if they are permitted to remain to produce fruit, they will ſo much weaken the trees, that they ſeldom recover their ſtrength again ſo as to become vigorous. When theſe plants are two years and a half old, they will produce flowers again, ſome of which are often left to bear fruit; but the moſt curious planters pull off all theſe, and never leave any to produce fruit until the third year, and then but a few, in proportion to the ſtrength of the trees; by which method their trees always produce larger and better nouriſhed fruit than thoſe which are ſuffered to bear a larger quantity, and will continue much longer in vigour. The fourth year they ſuffer their trees to bear a moderate crop, but they generally pull off ſome flowers from thoſe trees which are weak, that they may recover ſtrength before they are too old.

From the time when the flowers fall off, to the maturity of the fruit, is about four months. It is eaſy to know when the fruit is ripe by the colour of the pods, which become yellow on the ſide next the ſun. In gathering of the fruit, they generally place a negroe to each row of trees, who, being furniſhed with a baſket, goes from tree to tree, and cuts off all thoſe which are ripe, leaving the others for a longer time to ripen. When the baſket is full, he carries the fruit, and lays it in a heap at one end of the plantation, where, after they have gathered the whole plantation, they cut the pods lengthways, and take out all the nuts, being careful to diveſt them of the pulp which cloſely adheres to them; then they carry them to the houſe, where they lay them in large caſks, or other veſſels of wood, raiſed above-ground, and cover them with leaves of the Indian Reed and mats, upon which they lay ſome boards, putting ſome ſtones thereon to keep them down cloſe, in order to preſs the nuts. In theſe veſſels the nuts are kept four or five days, during which time they muſt be ſtirred and turned every morning, otherwiſe they will be in danger of periſhing from the great fermentation they are uſually in. In this time they change from being white to a dark red or brown colour. Without this fermentation, they ſay the nuts will not keep, but will ſprout, if they are in a damp place, or ſhrivel and dry too much if they are expoſed to heat.

After the nuts have been thus fermented, they ſhould be taken out of the veſſels and ſpread on coarſe cloths, where they may be expoſed to the ſun and wind; but at night, or in rainy weather, they muſt be taken under ſhelter, otherwiſe the damp will ſpoil them. If the weather proves fair, three days time will be long enough to dry them, provided they are carefully turned from time to time, that they may dry equally on every ſide. When they are perfectly dry, they may be put up in boxes or ſacks, and preſerved in a dry place until they are ſhipped off or otherwiſe diſpoſed of. The freſher theſe nuts are, the more oil is contained in them; ſo that the older they are, the leſs they are eſteemed.

When the trees are full grown and vigorous, they will ſometimes produce two hundred, or two hundred and forty pods at one ſeaſon, which will make ten or twelve pounds of Chocolate when dried; ſo that it is a very profitable commodity, and can be managed with very little charge, when compared with Sugar.

The Chocolate-trees, if planted on a good ſoil and properly taken care of, will continue vigorous and fruitful twenty-five or thirty years: therefore the charge of cultivating a plantation of theſe trees, muſt be much leſs than that of Sugar; for although the ground between the rows of plants will require to be often hoed and wrought, yet the firſt working of a ground to make a new plantation of Sugar, Indigo, Caſſada, &c. is a larger expence than the after-workings are. Beſides, Sugar-canes require as much labour in their cultivation as any plant whatever.

The leaves of theſe trees being large, make a great litter upon the ground when they fall, but is not injurious, but rather of ſervice to the trees; for the ſurface of the ground being covered with them, they preſerve the moiſture in the ground, and prevent its evaporating, which is of great uſe to the young tender roots which are juſt under the ſurface; and when the leaves are rotten, they may be buried in digging the ground, and it will ſerve as good manure. Some planters let the pods, in which the Chocolate is incloſed, lie and rot in a heap (after they have taken the nuts out) which they alſo ſpread on the ground inſtead of dung.

Beſides the ordinary care of digging, hoeing, and manuring the plantations of Chocolate-trees, there is alſo another thing requiſite in order to their doing well, which is, to prune the decayed branches off, and to take away ſmall ill-placed branches wherever they are produced. But you ſhould be cautious how this work is performed, for there ſhould be no vigorous branches ſhortened, nor any large amputations made on theſe trees, becauſe they abound with a ſoft, glutinous, milky juice, which will flow out for many days whenever they are wounded, which greatly weakens the trees.

In order to cultivate this plant in Europe, by way of curioſity, it will be neceſſary to have the nuts planted into boxes of earth (in the countries where they grow) ſoon after they are ripe, becauſe, if the nuts are ſent over, they will loſe their growing quality before they arrive. Theſe boxes ſhould be placed in a ſhady ſituation, and muſt be frequently watered, in order to forward the vegetation of the nuts. In about a fortnight after the nuts are planted, the plants will appear above-ground, when they ſhould be carefully watered in dry weather, and protected from the violent heat of the ſun. When the plants are grown ſtrong enough to tranſplant, they ſhould be ſhipped and placed where they may be ſcreened from ſtrong winds, ſalt water, and the violent heat of the ſun. During their paſſage they muſt be frequently refreſhed with water; but it muſt not be given them in great quantities, leſt it rot the tender fibres of their roots; and when they come into a cool latitude, they muſt be carefully protected from the cold, when they will not require ſo frequently to be watered; for in a moderate degree of heat, if they have gentle waterings once a week it will be ſufficient.

When the plants arrive in England, they ſhould be carefully taken out of the boxes, and each tranſplanted into a ſeparate pot filled with light rich earth, and plunged into a moderate hot-bed of tanners bark. In this hot-bed the plants may remain till Michaelmas, when they muſt be removed into the bark-ſtove, and plunged into the tan, in the warmeſt part of the ſtove. During the winter ſeaſon the plants muſt be frequently refreſhed with water, but it muſt be given to them in ſmall quantities, yet in ſummer they will require a more plentiful ſhare. Theſe plants are too tender to live in the open air in this country, even in the hotteſt ſeaſon of the year, therefore muſt conſtantly remain in the bark-ſtove, obſerving in very warm weather to let in a large ſhare of freſh air to them, and in winter to keep them very warm. The leaves of theſe plants muſt be frequently waſhed to clear them from filth, which they are ſubject to contract by remaining conſtantly in the houſe; and this becomes an harbour for ſmall inſects, which will infeſt the plants and deſtroy them, if they are not timely waſhed off. If theſe rules are duly obſerved, the plants will thrive very well, and may produce flowers in this climate: but it will be very difficult to obtain fruit from them, for being of a very tender nature, they are ſubject to many accidents in a cold country.

CACHRYS. Tourn. Inſt. 325. Lin. Gen. Plant. 304.

The CHARACTERS are,

It hath an umbellated flower; the involucrum is compoſed of many narrow ſpear-ſhaped leaves: the flower hath five ſpear-ſhaped erect petals. It hath five ſingle ſtamina. The turbinated germen is ſituated under the receptacle; the empalement afterward becomes a large, oval, blunt fruit, dividing in two parts, each having one large fungous ſeed.

The SPECIES are,

1. CACHRYS (*Trifida*) foliis pinnatis, foliolis linearibus trifidis, fructu lævi. *Cachrys with very narrow, pinnated, trifid leaves, and a ſmooth fruit.*

2. CACHRYS (*Sicula*) foliis bipinnatis, foliolis linearibus ſeminibus ſubulatis hiſpidis acutis. *Cachrys with narrow, pinnated, acute leaves, and a rough fruit.*

3. CACHRYS (*Linearis*) foliis pinnatis foliolis linearibus multifidis fructu ſulcato plano. *Cachrys with very narrow, multifid, pinnated leaves, and a plain channelled fruit.*

The firſt ſort hath a thick fleſhy root which ſtrikes deep in the ground, from which ſprings out many narrow winged leaves reſembling thoſe of Giant-fennel; from between theſe ariſe a hollow fungous ſtalk about two feet high, terminated by a large umbel of yellow flowers, which are ſucceeded by oval, ſmooth, fungous fruit, dividing into two parts, each incloſing an oblong ſeed.

The ſecond ſort hath a large, firm, ſweet-ſmelling root, which ſends out ſeveral pinnated leaves like thoſe of Hog's-fennel, but ſhorter. The ſtalk is ſmooth, jointed, and riſes four or five feet high, which is terminated by large umbels of yellow flowers like thoſe of Dill.

The third ſort hath very thick roots, which ſtrike deep in the ground, ſending out very narrow pinnated leaves like thoſe of Hog's-fennel. The ſtalk riſes five or ſix feet high, and is jointed like thoſe of Fennel, terminated by large umbels of yellow flowers.

The firſt ſort grows naturally in the ſouth of France and Italy, and the ſecond in Sicily.

Theſe plants are all propagated by ſeeds, which ſhould be ſown ſoon after they are ripe, for if they are kept out of the ground until the following ſpring they often miſcarry; and when they ſucceed, they never come up until the ſpring after, ſo that by ſowing them in autumn a whole year is ſaved, and the ſeeds ſeldom miſcarry. Theſe ſeeds ſhould be ſown on a ſhady border where the plants are to remain, for the plants having long tap-roots, will not bear tranſplanting ſo well as many other kinds. The diſtance to be obſerved for the ſowing of their ſeeds ſhould be in rows three feet apart; ſo that if each kind is ſown in a drill, when the plants are come up they may be thinned, leaving two of the moſt promiſing plants of each kind to remain. Theſe plants will begin to appear early in April, when they muſt be carefully cleared from weeds; and in dry weather they ſhould be gently watered, which greatly promotes their growth the firſt year; after which time they will require no farther care but to keep them clean from weeds, and every ſpring to dig the ground carefully between them, ſo as not to injure the roots.

The plants decay to the ground every autumn, and come up again in the following ſpring; they commonly flower in the beginning of June, and their ſeeds are ripe in September: the roots ſometimes run down three or four feet deep in the earth, provided the ſoil be light, and are often as large as Parſneps: they will continue many years, and if the ſoil is moiſt and rich, they will annually produce good ſeeds; but when they grow on a dry ſoil, the flowers commonly fall away, and are not ſucceeded by ſeeds.

CACTUS. Lin. Gen. Plant. 539.

The CHARACTERS are,

The flower is compoſed of ſix petals, which reſts upon the embryo; it hath ſix long ſlender ſtamina. The oval germen, which is ſituated below the petals, afterward becomes a fleſhy fruit with one cell, filled with ſmall angular ſeeds ſurrounded with pulp.

The SPECIES are,

1. CACTUS (*Melocactus*) ſubrotundus quatuordecem angularis. Hort. Cliff. 181. *Roundiſh Cactus with fourteen angles; or the Hedge-hog Melon-thiſtle, commonly called Great Melon-thiſtle.*

2. CACTUS (*Intortus*) ſubrotundus quinquedecem angularis, angulis in ſpiram intortis, ſpinis erectis. *Roundiſh Cactus or Melon-thiſtle, with fifteen angles ſpirally twiſted, and erect ſpines.*

3. CACTUS (*Recurvus*) ſubrotundus quinquedecem angularis, ſpinis latis recurvis creberrimis. *Roundiſh Melon-thiſtle with fifteen angles, having broad recurved ſpines ſet very cloſe.*

4. CACTUS (*Mamillaris*) ſubrotundus tectus tuberculis ovatis barbatis. Hort. Cliff. 181. *Roundiſh Cactus cloſely covered*

covered with bearded tubercles; or Smaller American Melon-thistle.

5. Cactus (*Proliferus*) proliferus subrotundus, tectus tuberculis ovati, barbatis longis albidis. *Roundish prolific Cactus, with oval tubercles closely joined, having long white beards; commonly called Small Childing Melon-thistle.*

These strange plants commonly grow upon the steep sides of rocks in the warmest parts of America, where they seem to be thrust out of the apertures, having little or no earth to support them, their roots shooting down into the fissures of the rock to a considerable depth; so that it is troublesome to get the plants up, especially as they are so strongly armed with thorns, as to render it very dangerous to handle them.

The great sorts were some years since brought over to England in much greater plenty than of late, but then many of them were destroyed by the unskilfulness of those persons who had the care of them in the voyage; for, by giving them water, they generally caused them to rot before they were taken out of the ships; and some of those which have appeared to be sound, have been so replete with moisture, as to rot soon after they have been placed in the stoves; therefore whoever proposes to bring these plants from abroad, should be very careful to take up their roots as entire as possible, and to plant them in tubs filled with stones and rubbish, mixing very little earth with it, and to plant three or four plants in each tub, in proportion to their sizes, for if they are placed close together it will save room; and as they do not increase in their growth during their passage, so there need not be any room allowed them for that purpose. There should be several pretty large holes bored through the bottom of these tubs, to let the moisture pass off; and if the plants are planted in the tubs a month or more before they are put on board the ship, they will in that time have made new roots, which will be the most secure method to have them succeed; but during their continuance in the country, they should have no water given them, unless the season should prove very hot and dry; and, in that case, it should be given to them sparingly; but after they are put on board the ship, they must not have any moisture whatever: therefore it will be a good method to cover the plants with tarpaulin, to keep off the spray of the sea in bad weather, and expose them at all times to the open air when the sea is calm. By observing these directions, the plants may be brought to England in good health, provided they are brought in summer.

The third sort was brought into England by the late Dr. William Houstoun, who procured the plants from Mexico; but as they were long in their passage, and had received wet, they were decayed before they arrived in England; but from the remains of them which were left, they appeared to be the most singular of all the species yet known. This has two orders of thorns, one of which is strait, and set on at the joints in clusters, spreading out from the center each way like a star; and in the middle of each cluster is produced one broad flat thorn, near two inches in length, which stands erect, and is recurved at the point, and is of a brownish red colour. These thorns are, by the inhabitants of Mexico, set in gold or silver, and made use of for picking their teeth; and the plant is by them called Visnaga, i. e. Toothpick.

The sort with spiral ribs, as also that with white spines, I received from Antigua with the common sort; but whether these are only accidental varieties arising from the same seeds, or real different species, I cannot take upon me to determine, since in this country they are seldom propagated by seeds; nor could I observe, in the several years that I have had these plants under my care, there was the least disposition in either of them to produce fruit; when, at the same time, the common large sort produced plenty of fruit out of their caps every year, from the seeds of which I have raised some young plants; but altho' some of these have grown to a considerable size, yet none of them have as yet produced caps, therefore no fruit can be yet expected from them.

The fourth sort produces quantities of fruit annually; and as the seeds grow very readily, the plants are now very common in those gardens where there are stoves to keep them; for if the fruit is permitted to drop upon the earth of the pots, and that is not disturbed, there will be plenty of plants come up without any farther trouble; and these seedling plants may be taken up as soon as they are of a proper size to remove, and planted six or seven of them into a small halfpenny pot, where they may stand one year, by which time they will be large enough to be each planted into a separate pot, and afterward they will make great progress. This sort is much more hardy than the large kind, so may be preserved in a moderate stove, or in a warm room, but the plants will not make near the progress as those which are kept in a greater degree of heat. It will continue many years with proper care, and the plants will grow to be near a foot high; but when they are so tall, the lower part of them is not so sightly, their green being decayed, and the spines changed to a dark dirty colour, they appear as if dead, so that the upper part of these old plants only seem to have life, whereas the plants of middling side appear healthy from top to bottom. The fruit are of a fine scarlet colour, and continue fresh upon the plants through the winter, which renders them very beautiful at that season. In the spring, when the fruit shrivels and becomes dry, the seeds will be ripe, and may then be rubbed out, and sown upon the surface of the earth in small pots.

The fifth sort is but little larger than the fourth, growing nearly in the same form; but this produces a great number of young plants from the sides, by which it is increased. It produces tufts of a soft white down upon the knobs, and also between them at every joint, which makes the whole plant appear as if it was covered with fine Cotton. The flowers of this sort are produced from between the knobs, round the sides of the plants, which are in shape and colour very much like those of the fourth sort, but larger. These flowers are not succeeded by any fruit, at least all those which I have under my care have not produced any, although they have produced plenty of flowers for some years; but from the same places where the flowers have appeared, there have been young plants thrust out the following season.

All the species of this genus are plants of a singular structure, but especially the larger kinds of them, which appear like a large, fleshy, green Melon, with deep ribs, set all over with strong sharp thorns, and when the plants are cut thro' the middle, their inside is a soft, pale, green, fleshy substance very full of moisture. And I have been assured by persons of credit, who have lived in the West-Indies, that in times of great drought, the cattle repair to the barren rocks which are covered with these plants, and after having ripped up the large plants with their horns, so as to tear off the outside skin, they have greedily devoured all the fleshy moist parts of the plants, which have afforded them both meat and drink.

The fruit of all the sorts of Melon-thistles are frequently eaten by the inhabitants of the West-Indies; there is scarce any difference in the fruits of all the kinds I have yet seen, either in size, shape, colour, or taste. They are about three quarters of an inch in length, of a taper form, drawing to a point at the bottom toward the plant, but blunt at the top, where the empalement of the flower was situated.

The taste is an agreeable acid, which, in a hot country, must render the fruit more grateful.

All the larger sorts of these plants require a very good stove to preserve them through the winter in England; nor should they be exposed to the open air in summer, for although they may continue fair to outward appearance when they have been some time exposed abroad, yet they will imbibe moisture, which will cause them to rot soon after they are removed into the stove again. And this is frequently the case of those plants which are brought from abroad, which have a fair healthy appearance many times at their first arrival, but soon after decay, and this will happen very suddenly; scarce any appearance of disorder will be seen, till the whole plant is killed, which, in a few hours time, has often been the fate of the plants when they have been placed in the stove.

If these plants are plunged into a hot-bed of tanners bark in summer, it will greatly forward them in their growth; but when this is practised there should be scarce any water given to the plants, for the moisture which they will imbibe from the fermentation of the tan, will be sufficient for them, and more would cause them to rot. The best method to preserve all the large kinds, is in winter to place the pots either upon the top of the flues, or at least very near them, that they may have the warmest place of the stove, and during that season never to give them any water. The soil in which these should be planted, must be of a sandy nature, and if mixed with some dry lime rubbish it will be still better. In the bottom of the pots should be placed some stones, in order to drain off any moisture which may be in the earth; for as these plants naturally grow upon the hot, dry, burning rocks which have no earth, and, were it not for these plants, would be absolutely barren, we must imitate their natural soil as near as possible, making some allowance for the difference of the climates.

The great sorts may be propagated by seeds, which must be sown and managed as hath been directed for the smaller sort; but as the plants which are raised from seeds in England, will be some years in arriving to any considerable size, it will be much the best way to procure some plants from the West-Indies.

The two small sorts propagate so fast in England, as to render it unnecessary to send for plants of these kinds from abroad; for whoever hath a mind to be plentifully stocked with them, may be soon supplied; the fourth sort from seeds, and the fifth from the young plants which are thrust out from the side of the old.

CÆSALPINA. Plum. Nov. Gen. 9. Brasiletto.

The CHARACTERS are,

The flower hath five petals, which are situated like those of the butterfly flowers. It hath ten declining stamina which are distinct. It hath an oblong germen. The empalement afterward becomes an oblong compressed pod with one cell, inclosing three or four compressed seeds.

The SPECIES are,

1. CÆSALPINA (*Brasiliensis*) foliis duplicato-pinnatis foliolis emarginatis, floribus decandris. *Cæsalpina with doubly winged leaves, whose small leaves are indented at the end, and flowers with ten stamina; commonly called Brasiletto.*

2. CÆSALPINA (*Crista*) foliis duplicato-pinnatis foliolis ovatis integerrimis floribus pentandriis. *Cæsalpina with doubly winged leaves, whose small leaves are oval and entire, and flowers with five stamina.*

The first sort is the tree which affords the Brasiletto wood, which is much used in dyeing. It grows naturally in the warmest parts of America, from whence the wood is imported for the dyers; and the demand for it has been so great, that there are no large trees left in any of the British colonies, the biggest scarce exceeding eight inches in diameter, and fifteen feet in height. It hath very slender branches, which are armed with recurved thorns. The leaves are winged, branching out into many divisions, garnished with small oval lobes which are indented at the top. The foot-stalks of the flowers come out from the side of the branches, and are terminated by a loose pyramidal spike of white flowers, which are shaped somewhat like those of the butterfly kind.

The second sort grows naturally in the same countries with the first, but is of larger size: it sends out many weak irregular branches, armed with short, strong, upright thorns. The leaves branch out in the same manner as the first, but the lobes (or small leaves) are oval and entire. The flowers are produced in long spikes like those of the former, but are variegated with red.

These plants are propagated by seeds, which should be sown in small pots filled with light rich earth early in the spring, and plunged into a hot-bed of tanners bark. In about six weeks after the plants will begin to appear, when they must be carefully cleared from weeds, and frequently refreshed with water; in warm weather the glasses of the hot-bed should be raised in the middle of the day, to admit fresh air to the plants. When the plants are two or three inches high, they should be carefully taken out of the pots, and each transplanted into a separate small pot, and plunged into the hot-bed again, observing to water and screen them from the heat of the sun until they have taken new root; after which time the glasses of the hot-bed should be raised every day, in proportion to the heat of the weather, to admit fresh air to the plants. In this hot-bed the plants may remain till autumn, when they should be removed into the stove, and plunged into the bark-bed, where they may have room to grow. These plants being tender, should always be kept in the bark-stove, and have a moderate share of heat in the winter; and being placed among other tender exotic plants of the same country, will afford an agreeable variety.

CAINITO. See CHRYSOPHYLLUM.

CAKILE. See ROCKET and BUNIAS.

CALAMINTHA. See MELISSA.

CALCEOLUS. Ladies Slipper. See CYPRIPEDIUM.

CALENDULA. Lin. Gen. Plant. 885. Marigold.

The CHARACTERS are,

It hath a compound radiated flower, the border or rays being composed of female florets, which are stretched out on one side like a tongue. The hermaphrodite flowers, which compose the disk, are tubulous and quinquefid. The germen is situated under the petal. These flowers are barren; but the female flowers are each succeeded by one oblong incurved seed, with angular membranes.

The SPECIES are,

1. CALENDULA (*Arvensis*) foliis lineari-lanceolatis semiamplexicaulibus, seminibus echinatis. *Marigold with narrow spear-shaped leaves half embracing the stalk, and prickly seeds; or the least Marigold.*

2. CALENDULA (*Sancta*) seminibus urceolatis obovatis lævibus calycibus submuricatis. Lin. Sp. 1304. *Marigold with cup-shaped prickly seeds in the border, and those in the center bicorned.*

3. CALENDULA (*Officinalis*) seminibus cymbiformibus muricatis incurvatis omnibus. Lin. Sp. Plant. 1304. *Marigold with boat-shaped, prickly, incurved seeds; or Common Marigold.*

4. CALENDULA (*Pluvialis*) foliis lanceolatis sinuato denticulatis, caule folioso pedunculis filiformibus. Hort. Upsal. 274. *Marigold with sinuated, indented, spear-shaped leaves, and a slender stalk.*

5. CA-

5. Calendula (*Hybrida*) foliis lanceolatis dentatis caule foliofo, pedunculis fupernè incraffatis. Hort. Upfal. 274. *Marigold with indented fpear-fhaped leaves, and the upper part of the foot-ftalk fwelling.*

6. Calendula (*Graminifolia*) foliis linearibus fubintegerrimis caule fubnudo. Lin. Sp. Pl. 922. *Marigold with narrow entire leaves, and a naked ftalk.*

7. Calendula (*Fruticofo*) foliis obovatis denticulatis, caule fruticofo perenni. Prod. Leyd. 531. *Marigold with obverfe oval leaves, and a perennial fhrubby ftalk.*

The firft fort grows naturally in the fouth of France, Spain, and Italy; it rifes with a flender branching ftalk which fpreads near the ground, and is garnifhed with narrow, fpear-fhaped, hairy leaves, which half furround the ftalk at their bafe; the flowers are produced at the extremity of the branches, upon long naked foot-ftalks. Thefe are very fmall, and of a pale yellow colour; the feeds are long, narrow, and on their outfide armed with prickles. The root is annual, and perifhes foon after the feeds are ripe. If the feeds of this plant are permitted to fcatter, there will be a frefh fupply of young plants.

The third fort is the common Marigold, which is cultivated for ufe in the gardens; this is fo well known, as to require no defcription. Of this there are the following varieties; the common fingle; the double flowering; the largeft very double flower; the double with Lemon-coloured flowers; and the greater and fmaller Childing Marigold.

Thefe varieties are fuppofed to have been originally obtained from the feeds of the common Marigold; but thefe differences continue, if the feeds are properly faved; nor have I obferved the common fort approaching to either of thefe, where they have been long cultivated in the greateft plenty; but as the two Childing Marigolds, and the largeft double, are fubject to degenerate, where care is not taken in faving of their feeds, fo I conclude they are not diftinct fpecies. The beft way to preferve thefe varieties, is to pull up all thofe plants whofe flowers are lefs double, as foon as they appear, that they may not impregnate the others with their farina, and fave the feeds from the largeft and moft double flowers; and the childing fort fhould be fown by itfelf in a feparate part of the garden, and the feeds faved from the large center flowers only.

The feeds of thefe may be fown in March or April, where the plants are to remain, and will require no other culture but to keep them clean from weeds, and to thin the plants where they are too clofe, leaving them ten inches afunder, that their branches may have room to fpread.

The fourth fort grows naturally at the Cape of Good Hope. This plant is annual, and perifhes foon after the feeds are perfected.

The lower leaves are oblong, fpear-fhaped, and deeply indented on their edges. The ftalks are produced on every fide the root, which decline towards the ground, and are garnifhed with leaves from the bottom, to within two inches of the top. The upper part of the ftalk is very flender, upon which refts one flower, fhaped like thofe of the common Marigold, having a purple bottom; and the rays (or border) of the flower are of a Violet colour on their outfide, and of a pure white within; thefe open when the fun fhines, but fhut up in the evening, and remain fo in cloudy weather.

The fifth fort is a native of the Cape of Good Hope. This is alfo an annual plant, and has much the appearance of the former, but the leaves are more deeply indented on their edges; the ftalks grow about the fame length as the former: the flower is a little fmaller, and the outfide of the rays are of a fainter purple colour. The feeds of this are flat and heart-fhaped, but thofe of the former are long and narrow.

The feeds of the five firft forts fhould be fown in the fpring, in the borders of the garden where the plants are defigned to remain, for they do not bear tranfplanting well; therefore they may be treated in the fame manner, and fown at the fame time with Candy Tuft, Venus Looking-glafs, and other hardy annual plants, putting four or five feeds in each patch; if they all grow, there fhould not be more than two or three plants left, after this they require no farther care but to keep them clean from weeds. If the feeds of thefe plants are permitted to fcatter, the plants will come up the following fpring without care, and thefe will flower earlier than thofe which are fown in the fpring.

The fixth fort is alfo a native of the fame country. This is a perennial plant, which divides near the root into feveral tufted heads, which are clofely covered with long graffy leaves, coming out on every fide without order, and are for the moft part entire. From between the leaves arife naked foot-ftalks about nine inches long, fuftaining one flower at the top, which is about the fize of the common Marigold, having a purple bottom; the rays are alfo purple without, but of a pure white within. Thefe expand when the fun fhines, but always clofe in the evening, and in cloudy weather. This fort doth not often produce good feeds in Europe, but it is eafily propagated by flips taken off from the heads, in the fame manner as is practifed for Thrift. They may be planted any time in fummer in a fhady border, covering them clofe with a Melon-glafs. After they have got ftrong roots, they fhould be each planted into feparate fmall pots filled with frefh light earth, and placed in a fhady fituation till they have taken frefh root, when they may be placed in the open air in a fheltered fituation, where they may remain till autumn, and then fhould be placed in a dry airy glafs-cafe for the winter feafon, or under a common hot-bed frame; for they only require protection from froft and wet, and fhould enjoy the air at all times when the weather is mild.

The feventh fort hath been of late years introduced from the Cape of Good Hope. It hath a flender, fhrubby, perennial ftalk, which rifes to the height of feven or eight feet, but requires fupport; this fends out a great number of weak branches from the bottom to the top, which hang downward unlefs they are fupported; they are garnifhed with oval leaves, having fhort flat foot-ftalks; they are of a fhining green colour on their upper fide, but paler underneath; the flowers come out at the end of the branches on fhort naked foot-ftalks, and are in fize and colour like thofe of the fifth fort.

This is eafily propagated by cuttings, which may be planted any time in fummer in a fhady border, or otherwife fhaded with mats in the heat of the day: in five or fix weeks thefe will have taken root, when they fhould be carefully taken up, and each put into a feparate pot, and placed in the fhade till they have taken frefh root; then they may be placed with other hardy exotic plants in a fheltered fituation, where they may remain till the froft begins, when they muft be removed into the green-houfe, placing them near the windows that they may enjoy the free air, for this plant only requires protection from froft.

CALF's-SNOUT. See Antirrhinum.

CALLA. Lin. Gen. Plant. 917. Wake Robin, or Ethiopian Arum.

The Characters are,

It hath a long open fpatha of one leaf, coloured and permanent. It hath a fingle upright fpadix, to which the flowers and fruit adhere. This hath male and female flowers, intermixed toward the upper part of the club or fpadix. The male flowers confift of many

many very short stamina; the female flowers have a compressed style, resting upon an obtuse germen, which afterward becomes a globular pulpy fruit, compressed on two sides, inclosing two or three obtuse seeds.

We have but one SPECIES of this genus, viz.

CALLA (*Æthiopica*) foliis sagitato-cordatis, spathâ cucullatâ, spadice superne masculo. Hort. Cliff. 436. *Calla with arrow-headed heart-shaped leaves, a hooded spatha or sheath, and male flowers situated on the upper part of the spadix.*

This plant hath thick, fleshy, tuberous roots, which are covered with a thin brown skin, and strike down many strong fleshy fibres in the ground. The leaves have foot-stalks more than a foot long, which are green and succulent. The leaves are shaped like the point of an arrow, they are eight or nine inches in length, ending in a sharp point, which turns backward; between the leaves arise the foot-stalk of the flower, which is thick, smooth, of the same colour as the leaves, and rises above them, and is terminated by a single flower, shaped like those of the Arum, the hood or spatha being twisted at the bottom, but spreads open at the top, and is of a pure white colour. When these fade, part of those which are situated at the top of the club, are succeeded by roundish fleshy berries compressed on two sides, each containing two or three seeds.

This plant grows naturally at the Cape of Good Hope. It propagates very fast by offsets, which should be taken off the latter end of August, at which time the old leaves decay, for at this season the roots are in their most inactive state. These roots have generally a great number of offsets about them, so that unless there is a want of them, the largest only should be chosen, which should be separated from all the smaller, and each planted in a separate pot, and placed with other hardy exotic plants in the open air till autumn, when they must be removed into shelter for the winter season. This plant is so hardy as to live in the open air in mild winters without any cover, if they are planted in warm borders and have a dry soil; but with a little shelter in hard frost, they may be preserved in full growth very well.

CALLACARPA. See JOHNSONIA.

CALTHA. Lin. Gen. Plant. 623. Marsh Marigold.

The CHARACTERS are,

The flower is composed of five large oval petals, which are concave; it hath a great number of slender stamina. In the center there are several oblong compressed germen situated, which afterward become so many short pointed capsules, containing many roundish seeds.

We have but one SPECIES of this genus, viz.

CALTHA (*Palustris*) foliis orbiculatis crenatis, flore majore. *Marsh Marigold with round crenated leaves, and a larger flower.*

This plant grows upon moist boggy land in many parts of England: of this there is a variety with very double flowers, which for its beauty is preserved in gardens, and is propagated by parting of the roots in autumn. It should be planted in a moist soil and a shady situation; and as there are often such places in gardens, where few other plants will thrive, so these may be allowed to have room, and during their season of flowering will afford an agreeable variety.

CAMARA. See LANTANA.

CAMERARIA Plum. Nov. Gen. 18.

The CHARACTERS are,

The flower is of one leaf, salver-shaped, and divided at the top into five acute segments; it hath five short inflexed stamina. In the bottom of the tube are situated two roundish germen, which afterward become two long, taper, leafy capsules, filled with oblong cylindrical seeds.

The SPECIES are,

1. CAMERARIA (*Latifolia*) foliis ovatis, utrinque acutis transverse striatis. Lin. Sp. Pl. 308 *Cameraria with roundish leaves ending in points.*

2. CAMERARIA (*Angustifolia*) foliis linearibus. Lin. Sp. Pl. 210. *Cameraria with very long narrow leaves.*

The first sort was sent me from the Havanna, where it grows naturally in great plenty. This rises with a shrubby stalk to the height of ten or twelve feet, dividing into several branches, which are garnished with roundish pointed leaves placed opposite. The flowers are produced at the end of the branches in loose clusters, which have long tubes enlarging gradually upward, and at the top are cut into five segments, broad at their base, but end in sharp points: the flower is of a yellowish white colour.

The second sort hath an irregular shrubby stalk, which rises about eight feet high, sending out many branches, which are garnished with very narrow thin leaves placed opposite at each joint. The flowers are produced scatteringly at the end of the branches, which are shaped like those of the former sort, but smaller. Both these plants abound with an acrid milky juice like the Spurge. The second sort grows naturally in Jamaica.

These plants are propagated by seeds, which must be procured from the places of their growth. They may also be propagated by cuttings planted in a hot-bed during the summer months; they must have a bark-stove, for they are very tender plants, but in warm weather must have plenty of air.

CAMPANULA. Tourn. Inst. R. H. 108. tab. 38. Bell-flower.

The CHARACTERS are,

The flower is of one leaf shaped like a bell, spreading at the base. In the bottom is situated the five-cornered nectarium, which is joined to the top of the receptacle; it hath five short stamina. Below the receptacle is situated the angular germen; the empalement afterward becomes a roundish angular capsule, which in some species have three, and in others five cells, each having a hole toward the top, through which the seeds are scattered when ripe.

The SPECIES are,

1. CAMPANULA (*Pyramidalis*) foliis ovatis glabris subserratis, caule erecto paniculato, ramulis brevibus. *Bell-flower with oval smooth leaves sawed below, an upright paniculated stalk, and short branches.*

2. CAMPANULA (*Decurrens*) foliis radicalibus obovatis, caulinis lanceolato-linearibus subserratis sessilibus remotis. Lin. Sp. Pl. 164. *Peach-leaved Bell-flower.*

3. CAMPANULA (*Medium*) capsulis quinquelocularibus tectis, calycis sinubus reflexis. Vir. Cliff. 16. *Garden Bell-flower with an oblong leaf and flower, commonly called Canterbury Bell-flower.*

4. CAMPANULA (*Trachelium*) caule angulato, foliis petiolatis, calycibus ciliatis, pedunculis trifidis. Vir. Cliff. 16. *Greater and rougher Bell-flower with Nettle leaves.*

5. CAMPANULA (*Latifolia*) foliis ovato-lanceolatis, caule simplicissimo tereti, floribus solitariis pedunculatis fructibus cernuis. Vir. Cliff. 17. *Greatest Bell-flower with broadest leaves.*

6. CAMPANULA (*Rapunculus*) foliis undulatis radicalibus lanceolato-ovalibus, paniculâ coarctatâ. Hort. Upsal. 40. *Bell-flower with an esculent root, commonly called Rampion.*

7. CAMPANULA (*Glomerata*) caule angulato simplici, floribus sessilibus capitulo terminali. Vir. Cliff. 16. *Meadow Bell-flower, with flowers gathered in bunches.*

8. CAMPANULA (*Speculum*) caule ramosissimo diffuso foliis oblongis subcrenatis, calycibus solitariis corollâ longioribus, capsulis prismaticis. Hort. Ups. 41. *Upright Field Bell-flower with yellow Eye-bright leaves.*

9. CAMPANULA (*Hybrida*) caule ramofo foliis ovatis feffilibus, floribus pedunculatis terminatricibus. *Upright Field Bell-flower, or Venus Looking-glafs.*

10. CAMPANULA (*Arvenfis*) caule bafi fubramofo ftricto, foliis oblongis crenatis, calycibus aggregatis corollâ longioribus, capfulis prifmaticis. Lin. Sp. Pl. 168. *Leaft upright Field Bell-flower, or fmall Venus Looking-glafs.*

11. CAMPANULA (*Erinus*) caule dichotomo, foliis feffilibus utrinque dentatis. Hort. Cliff. 65. *Smaller annual Bell-flower with cut leaves.*

12. CAMPANULA (*Pentagonia*) caule fubdivifo ramofiffimo, foliis linearibus acuminatis. Hort. Cliff. 66. *Five-cornered Bell-flower of Thrace, with a very large flower.*

13. CAMPANULA (*Perfoliata*) caule fimplici, foliis cordatis dentatis amplexicaulibus, floribus feffilibus aggregatis. Hort. Upfal. 40. *Five-cornered perfoliate Bell-flower.*

14. CAMPANULA (*Americana*) caule ramofo, foliis linguiformibus crenulatis margine cartilagineo. Prod. Leyd. 246. *Smaller American Bell-flower with ftiff leaves, and a blue fpreading flower.*

15. CAMPANULA (*Canarienfis*) foliis haftatis dentatis oppofitis petiolatis, capfulis quinquelocularibus. Lin. Sp. Plant. 168. *Canary Bell-flower with an Orach leaf, and a tuberous root.*

There are feveral other fpecies of this genus, fome of which grow naturally in England, and others in the northern parts of Europe, which have but little beauty, therefore are feldom cultivated in gardens, fo I fhall not enumerate them here. There are alfo feveral varieties of fome of the forts here mentioned, which I fhall take notice of in their proper place; but as they are not diftinct fpecies, I have omitted them in the above lift.

The firft fort hath thick tuberous roots which are milky; this fends out ftrong, fmooth, upright ftalks, which rife four feet high, garnifhed with fmooth oblong leaves, whofe edges are a little indented. The flowers are produced from the fide of the ftalks, and are regularly fet on for more than half their length, forming a fort of pyramid; thefe are large, open, and fhaped like a bell. The moft common colour of the flowers is a light blue; but there have been fome with white flowers, which make a variety when intermixed with the blue, but the latter is moft efteemed.

This plant is cultivated to adorn halls, and to place before chimnies in the fummer when it is in flower, for which purpofe there is no plant more proper; for when the roots are ftrong they will fend out four or five ftalks, which will rife as many feet high, and are garnifhed with flowers great part of their length. When the flowers begin to open, the pots are removed into the rooms, where, being fhaded from the fun and kept from the rain, the flowers will continue long in beauty; and if the pots are every night removed into a more airy fituation, but not expofed to heavy rains, the flowers will be fairer, and continue much longer in beauty.

Thofe plants which are thus treated are feldom fit for the purpofe the following feafon, therefore a fupply of young plants fhould be annually raifed. The common method of propagating this plant, is by dividing the roots. The beft time for doing this is in September, that the offsets may have time to get ftrong roots before winter.

This method of propagating by the offsets is the quickeft, therefore generally practifed; but the plants which are raifed from feeds are always ftronger, and the ftalks will rife higher, and produce a greater number of flowers, therefore I recommend it to the practice of the curious; but in order to obtain good feeds, there fhould be fome ftrong plants placed in a warm fituation near a pale or wall in autumn; and, if the following winter fhould prove fevere, they fhould be covered either with hand-glaffes or mats, to prevent their being injured by the froft; and in the fummer, when the flowers are fully open, if the feafon fhould prove very wet, the flowers muft be fcreened from great rains, otherwife there will be no good feeds produced: the not obferving this has occafioned many to believe that the plants do not bear feeds in England, which is a great miftake, for I have raifed great numbers of the plants from the feeds of my own faving; but I have always found that the plants which have been long propagated by offsets do feldom produce feeds, which is the fame with many other plants that are propagated by flips or cuttings, which in a few years become barren.

When the feeds are obtained, they muft be fown in autumn in pots or boxes filled with light undunged earth, and placed in the open air till the froft or hard rains come on, when they fhould be placed under a hot-bed frame, where they may be fheltered from both; but in mild weather the glaffes fhould be drawn off every day, that they may enjoy the free air; with this management the plants will come up early in the fpring, and then they muft be removed out of the frame, placing them firft in a warm fituation; but when the feafon becomes warm, they fhould be removed where they may have the morning fun only. In September the leaves of the plants will begin to decay, at which time they fhould be tranfplanted; therefore there muft be one or two beds prepared, in proportion to the number of plants. Thefe beds muft be in a warm fituation, and the earth light, fandy, and without any mixture of dung, which laft is an enemy to this plant. If the fituation of the place is low, or the nature of the foil moift, the beds muft be raifed five or fix inches above the furface of the ground, and the natural foil removed a foot and a half deep, putting lime rubbifh or ftones eight or nine inches thick in the bottom of the trench, to drain off the moifture. When the beds are prepared, the plants muft be taken out of the pots or cafes very carefully, fo as not to break or bruife their roots, for they are very tender, and on being broken, the milky juice will flow out plentifully, which will greatly weaken them. Thefe fhould be planted at about fix inches diftance each way, with the head or crown of the root half an inch below the furface; if there happens a gentle fhower of rain foon after they are planted, it will be of great fervice to the plants, but as the feafon fometimes proves very dry at this time of the year, fo in that cafe it will be proper to give them a gentle watering three or four days after they are planted, and to cover the beds with mats every day, to prevent the fun from drying the earth, but thefe muft be taken off in the evening that the dew may fall on the ground. Towards the end of November the beds fhould be covered over with fome old tanners bark to keep out the froft, and where there is not conveniency of covering them with frames, they fhould be arched over with hoops, that in fevere frofts they may be covered with mats, for thefe plants, when young, are often deftroyed in winter where this care is wanting. In the fpring the coverings muft be removed, and the following fummer the plants muft be kept clean from weeds. The following autumn the furface of the ground fhould be ftirred between the plants, and fome frefh earth fpread over the beds, and in winter covered as before. In thefe beds the plants may remain two years, during which time they muft be treated in the manner before directed, by which time the roots will be ftrong enough to flower, fo in September they fhould be carefully taken up, and fome of the moft promifing planted in pots; the others may be planted into warm borders, or in a frefh bed, at a greater diftance than before, to allow them room to grow. Thofe plants which are potted fhould be fheltered in winter from great rains and hard frofts, otherwife they will be in danger of rotting, or at leaft will be fo weakened as not to flower

with

with any ftrength the following fummer; and thofe which are planted in the full ground fhould have fome old tanners bark laid round them, to prevent the froft from entering deep to the roots: with this management thefe plants may be brought to the utmoft perfection, and a conftant fucceffion of good roots raifed, which will be much preferable to thofe which are propagated by offsets.

The fecond fort grows naturally in the northern parts of Europe. Of this there are the following varieties, viz. the fingle blue, and white flower, which have been long here; and the double flower of both colours, which have not been more than forty years in England, but have been propagated in fuch plenty, as to have almoft banifhed thofe with fingle flowers from the gardens. All thefe varieties are eafily propagated by parting of their roots in autumn, every head which is then flipped off will take root; they are extreme hardy, fo will thrive in any foil or fituation, therefore are very proper furniture for the common borders of the flower-garden.

The third fort is a biennial plant, which perifhes foon after it hath ripened feeds. It grows naturally in the woods of Italy and Auftria, but is cultivated in the Englifh gardens for the beauty of its flowers. Of this fort there are the following varieties, the blue, the purple, the white, the ftriped, and double flowering.

This hath oblong, rough, hairy leaves, which are ferrated on their edges: from the center of thefe a ftiff, hairy, furrowed ftalk arifes about two feet high, fending out feveral lateral branches, which are garnifhed with long, narrow, hairy leaves, fawed on their edges: from the fetting on of thefe leaves come out the foot-ftalks of the flower; thofe which are on the lower part of the ftalk and branches, being four or five inches long, diminifhing gradually in their length upward, and thereby form a fort of pyramid. The flowers of this kind are very large, fo make a fine appearance. The feeds ripen in September, and the plants decay foon after.

It is propagated by feeds, which muft be fown in the fpring on an open bed of common earth, and when the plants are fit to remove, they fhould be tranfplanted into the flower-nurfery, in beds fix inches afunder, and the following autumn they fhould be tranfplanted into the borders of the flower-garden. As thefe plants decay the fecond year, there fhould be annually young ones raifed to fucceed them.

The fourth fort hath a perennial root, which fends up feveral ftiff hairy ftalks, having two ribs or angles. Thefe put out a few fhort fide branches, garnifhed with oblong pointed leaves, which are hairy, and deeply fawed on their edges. Toward the upper part of the ftalks the flowers come out alternately, upon fhort trifid foot-ftalks, having hairy empalements.

The varieties of this are, the deep and pale blue, the white with fingle flowers, and the fame colours with double flowers. The double forts are propagated by parting of their roots in autumn, which fhould be annually performed, otherwife the flowers are apt to degenerate to fingle. The foil fhould not be too light or rich in which they are planted, for in either of thefe they will not produce double flowers. The plants are extremely hardy, and may be planted in any fituation; thofe with fingle flowers do not merit a place in gardens.

The fifth fort grows naturally in the northern parts of England; this hath a perennial root, compofed of many flefhy fibres that abound with a milky juice, from which arife feveral ftrong, round, fingle ftalks, which never put out branches, but are garnifhed with oval fpear-fhaped leaves, flightly indented on their edges. Toward the upper part of the ftalk the flowers come out fingly upon fhort foot-ftalks. After the flowers are paft, the empalement becomes a five-cornered feed-veffel, which turns downward till the feeds are ripe, when it rifes upward again.

The varieties of this are the blue, purple, and white flowering. This fort is eafily propagated by feeds, which it furnifhes in great plenty, and if fuffered to fcatter, the plants will come up in as great plenty the following fpring, when they may be tranfplanted into the nurfery till autumn, at which time they fhould be tranfplanted where they are defigned to remain.

The fixth fort hath roundifh flefhy roots, which are eatable, and are much cultivated in France for fallets; and fome years paft it was cultivated in the Englifh gardens, but is now generally neglected. It grows naturally in feveral parts of England, but thofe roots never grow to half the fize of thofe which are cultivated. This is propagated by feeds, which fhould be fown in a fhady border, and when the plants are about an inch high, the ground fhould be hoed as is practifed for Onions, to cut up the weeds, and thin the plants to the diftance of three or four inches; and when the weeds come up again, they muft be hoed over to deftroy them; this, if well performed in dry weather, will make the ground clean for a long time, fo that being three times repeated, it will keep the plants clean till the winter, which is the feafon for eating the roots, when they may be taken up for ufe as they are wanted. Thefe will continue good till April, at which time they will fend out their ftalks, when they will become hard and unfit for ufe.

The feventh fort grows naturally upon chalky paftures in many parts of England, where the ftalks do not rife many times a foot high, and in other places it grows to double that height, which has occafioned their being taken for two diftinct plants. This hath a perennial root, which fends up feveral round hairy ftalks, which rife upward of two feet high; the bottom leaves are broad, and ftand upon long foot-ftalks. Thofe which are upon the ftalks are long, narrow, have no foot-ftalks, and are placed alternately at confiderable diftances. From the wings of the leaves towards the upper part of the ftalk, come out long naked foot-ftalks, fupporting two or three bell-fhaped flowers, clofely joined together in a head, and the main ftalk is terminated by a large clufter of the fame flowers, which are fucceeded by roundifh capfules filled with fmall feeds. This plant is eafily propagated, either by feeds or parting of the roots, and will thrive in any foil or fituation.

The eighth fort is an annual plant, which rifes with flender ftalks a foot high, branching out on every fide, garnifhed with oblong leaves a little curled on their edges; from the wings of the leaves come out the flowers, fitting clofe to the ftalks, which are of a beautiful purple, inclining to a Violet colour. In the evening they contract and fold into a pentagonal figure, from whence it is by fome called Viola Pentagonia, or five-cornered Violet. If this plant is fown in autumn, it will grow much taller, and flower a month earlier than when the feeds are fown in the fpring. The autumnal plants will flower in May, and the fpring plants in June and July.

The ninth fort is the common Venus Looking-glafs. This fort feldom rifes more than fix inches high, with a ftalk branching from the bottom upward, and garnifhed with oval leaves fitting clofe to the ftalks, from the bafe of which the branches are produced, which are terminated by flowers very like thofe of the former fort.

The tenth fort grows naturally in England, in arable land; this fends out an upright ftalk about fix or feven inches high, and near the root there are a few lateral branches, which are weak, and fpread out on every fide. Thefe are garnifhed with oval obtufe leaves, which are flightly indented on their edges, placed alternately. At the extremity of the branches the flowers are produced in clufters, which are ſmall, but of the ſame ſhape with the former, having a five-leaved empalement much longer than the petal; the ſeed-veſſel is ſhaped like thoſe of the two former, but are ſmaller.

The eleventh ſort grows naturally in the ſouth of France and Italy. This is alſo a low annual plant, which ſeldom riſes ſix inches high, but divides into many branches, which are garniſhed with ſhort oval leaves ſitting cloſe to the branches. The flowers are produced at the ends of the branches, which are ſhaped like thoſe of the other three ſorts laſt mentioned, but they are ſmall, and their colour leſs beautiful, and the leaves of the empalement are broader.

The twelfth ſort grows naturally in Thrace. This is alſo a low annual plant, which riſes little more than ſix inches high; the ſtalks divide by pairs, and frequently there ariſes a branch from the middle of the diviſions; the lower leaves are oblong and obtuſe, but thoſe which come out toward the end of the branches are much narrower and pointed. The flowers come out ſingle at the end of the branches, having a long five-leaved empalement, and are larger than thoſe of the three laſt ſorts, and of a fine blue colour.

The ninth ſort is the old Venus Looking-glaſs, which was formerly cultivated in the gardens, but ſince the eighth ſort hath been introduced, it hath almoſt ſupplanted the other; for the eighth is a much taller plant, and the flowers are larger, but their colour is leſs beautiful; however it produces a greater quantity of ſeeds, ſo is to be had in plenty, and there are few perſons that are curious enough to diſtinguiſh them.

If theſe, and the Venus Navelwort, Dwarf Lychnis, Candy Tuft, and other low annual flowers, are properly mixed in the border of the flower-garden, and ſown at two or three different ſeaſons, to have a ſucceſſion of them in flower, they will afford an agreeable variety. If theſe ſeeds are ſown in autumn, the plants will flower early in ſpring, but thoſe which are ſown in the ſpring will not flower till the middle of June; and if a third ſowing is performed about the middle of May, the plants will flower in Auguſt, but from the laſt ſowing good ſeeds muſt not be expected.

The thirteenth ſort is an annual plant, which in good ground will riſe a foot and a half, but in poor land, or where it grows wild among Corn, ſcarcely riſes to the height of ſix inches. The ſtalk is ſingle, rarely putting out any branches, unleſs near the root. The leaves are roundiſh, and embrace the ſtalk at their baſe; their edges are ſharply ſawed, and from their baſe comes out a cloſe tuft of flowers ſurrounded by the leaf, as in an empalement. The flowers are five-cornered, ſhaped like thoſe of the Venus Looking-glaſs, but are much ſmaller; theſe are produced the whole length of the ſtalk. The ſeeds are incloſed in ſhort capſules, which are ſhaped like thoſe of the former ſorts. If the ſeeds of this ſort are permitted to ſcatter the plants will come up without care; or the ſeeds may be ſown in the ſpring, in the ſame manner as thoſe of the laſt ſorts, and treated in the ſame way.

The fourteenth ſort is a native of America. This hath many rigid oblong leaves coming out from the root on every ſide, which form a ſort of head like thoſe of Houſleek, and are crenated, having a ſtrong rib running on their border longitudinally. From the center of the plant proceeds the ſtalk, which riſes about a foot high, and is thinly garniſhed with very narrow ſtiff leaves, of a ſhining green colour. From the wings of the leaves come out the foot-ſtalks of the flower, which are from two to four inches long, each being terminated by one ſpreading bell-ſhaped flower, whoſe empalement is ſhort, and cut into five acute ſegments. There is a white and a blue flower of this ſort in the gardens, and in Holland they have it with a double flower. This ſort doth not produce ſeeds in England, ſo is only propagated by offſets, which may be taken off from the old plants in Auguſt, that they may get good root before the cold weather begins; they muſt be planted in ſmall pots, and placed in the ſhade until they have taken root, then they may be placed with other hardy exotic plants, and in autumn ſome of them ſhould be removed into ſhelter; for in ſevere froſts; thoſe in the open air are often killed.

The fifteenth ſort is a native of the Canary Iſlands. This hath a thick fleſhy root, which is of an irregular form, ſometimes running downward like a Parſnep, at other times dividing into ſeveral knobs near the top, and when any part of the root is broken, there iſſues out a milky juice at the wound. From the head or crown of the root, ariſes one, two, three, or more ſtalks, in proportion to the ſize of the root; but that in the center is generally larger, and riſes higher than the others. Theſe ſtalks are very tender, round, and of a pale green; their joints are far diſtant from each other, and when the roots are ſtrong, the ſtalks will riſe ten feet high, ſending out ſeveral ſmaller ſide branches. At each joint they are garniſhed with two, three, or four ſpear-ſhaped leaves, with a ſharp pointed beard on each ſide. They are of a ſea-green, and when they firſt come out are covered ſlightly with an Aſh-coloured pounce. From the joints of the ſtalk the flowers are produced, which are of the perfect bell-ſhape, and hang downward; they are of a flame colour, marked with ſtripes of a browniſh red; the flower is divided into five parts, at the bottom of each is ſituated a nectarium, covered with a white tranſparent ſkin, much reſembling thoſe of the Crown Imperial, but are ſmaller. In the center of the flower is ſituated the ſtyle, which is longer than the ſtamina, and is crowned by a trifid ſtigma, which is reflexed. The flowers begin to open in the beginning of October, and there is often a ſucceſſion of them till March. The ſtalks decay to the root in June, and new ones ſpring up in Auguſt.

It is propagated by parting of the roots, which muſt be done with caution; for if they are broken or wounded, the milky juice will flow out plentifully, ſo that if theſe are planted before the wounds are ſkinned over, it occaſions their rotting; therefore whenever any of them are broken, they ſhould be laid in the green-houſe a few days to heal. Theſe roots muſt not be too often parted, eſpecially if they are expected to flower well; for by frequent parting, the roots are weakened. The beſt time for tranſplanting and parting of their roots is in July, ſoon after the ſtalks are decayed. The earth in which theſe ſhould be planted muſt not be rich, for that will cauſe them to be luxuriant in branches, and but thinly garniſhed with flowers. The ſoil in which they have ſucceeded beſt is a light ſandy loam, mixed with a fourth part of ſcreened lime rubbiſh; when the roots are firſt planted, the pots ſhould be placed in the ſhade, and, unleſs the ſeaſon is very dry, ſhould not be watered, for during the time they are inactive, wet is very injurious to them. About the middle of Auguſt the roots will begin to put out fibres, at which time, if the pots are placed under a hot-bed frame, and as the nights grow cool, covered with the glaſſes, but opened every day to enjoy the free air, it will greatly forward them for flowering, and increaſe their ſtrength; when the ſtalks appear, the plants muſt be now and then refreſhed with water, which muſt not be given too often, nor in great quantity. The plants thus managed, by the middle of September will have grown ſo tall, as not to be kept longer under the frame, ſo they ſhould be removed into a dry airy glaſs-caſe, where they may enjoy the free air in mild weather, but ſcreened from cold. During the winter ſeaſon they muſt be frequently refreſhed with water, and guarded from froſt; and in the ſpring, when the ſtalks begin to decay, the pots ſhould be ſet abroad in the ſhade, and not watered.

CAMPHORA. See LAURUS.

CAMPION See LYCHNIS.

CANDLE-BERRY TREE. See MYRICA.

CANDY-TUFT. See IBERIS.

CANNA. Lin. Gen. Pl. 1. Indian Cane.

The CHARACTERS are,

The flower hath one petal, which is divided into ſix parts: the three upper ſegments are erect, and broader than the lower, two of which are erect, and the other turns back and is twiſted. It hath one ſpear-ſhaped ſtamina riſing as high as the petal, having the appearance of a ſegment. Below the empalement is ſituated a roundiſh rough germen, which becomes an oblong, roundiſh, membranaceous capſule, having three longitudinal furrows crowned by the empalement, which hath three cells, filled with round ſmooth ſeeds.

The SPECIES are,

1. CANNA (*Indica*) foliis ovatis utrinque acuminatis nervoſis. Prod. Leyd. 11. *Common broad-leaved flowering Cane.*

2. CANNA (*Latifolia*) foliis ovatis obtuſis nervoſis, ſpicis florum longioribus. *Indian flowering Cane, with a pale red flower.*

3. CANNA (*Glauca*) foliis lanceolatis petiolatis enervibus. Prod. Leyd. 11. *Indian Cane with glaucous leaves, a very large flower, and the appearance of the Marſh Iris.*

4. CANNA (*Lutea*) foliis ovatis petiolatis nervoſis ſpatha floribus longiore. *Yellow Indian Reed.*

The firſt ſort grows naturally in both Indies: the inhabitants of the Britiſh Iſlands in America call it Indian Shot, from the roundneſs and hardneſs of the ſeeds.

This plant hath a thick, fleſhy, tuberous root, which divides into many irregular knobs; it ſends out many large oval leaves without any order; theſe, at their firſt appearance are twiſted like a horn, but afterwards expand and are near a foot long, and five inches broad in the middle, leſſening gradually to both ends, and terminated in a point. The ſtalks are herbaceous, riſing four feet high, and are encompaſſed by the broad leafy foot-ſtalks of the leaves; at the upper part of the ſtalk the flowers are produced in looſe ſpikes, each being at firſt covered by a leafy hood, which afterward ſtands below the flower, and turns to a brown colour. The flower is encompaſſed by a three-leaved empalement, which ſits upon a ſmall, roundiſh, rough germen, which, after the flower is fallen, ſwells to a large fruit or capſule, oblong, rough, and is crowned by the three-leaved empalement of the flower which remains. When the fruit is ripe, the capſule opens lengthways into three cells, which are filled with round, hard, black, ſhining ſeeds.

As this ſort is a native of the warmeſt parts of America, ſo it requires to be placed in a moderate ſtove in winter, where they always flower in that ſeaſon, at which time they make a fine appearance; and in the ſummer place them abroad in a ſheltered ſituation with other tender exotic plants, where they generally flower again, and produce ripe ſeeds annually.

The ſecond ſort grows naturally in Carolina, and ſome of the other northern provinces of America. The leaves of this ſort are longer than thoſe of the former, and terminate in ſharper points. The ſtalks grow taller, and the ſegments of the flower are much narrower; the colour is a pale red, ſo it makes no great appearance. The ſeeds are like thoſe of the former ſort. If the roots of this ſort are planted in warm borders and a dry ſoil, they will live through the winter, if mild, in the open air without cover, and flower well every year.

The ſeeds of the third ſort I received from Carthagena in New Spain, in the year 1733, which produced very ſtrong plants the firſt year, ſome of which flowered the ſame autumn. The roots of this are much larger than either of the former ſorts, and ſtrike down ſtrong fleſhy fibres deep in the ground. The ſtalks riſe in good earth ſeven or eight feet high. The leaves are near two feet long, narrow, ſmooth, and of a ſea-green colour. The flowers are produced in ſhort thick ſpikes at the extremity, which are large, and of a pale yellow colour; the ſegments of the petal are broad, but their ſhape like thoſe of the other ſorts. The ſeed-veſſels are larger, and much longer than thoſe of the other ſorts, but contain fewer ſeeds, which are very large.

All the ſorts are propagated by ſeeds, which ſhould be ſown on a hot-bed in the ſpring, and when the plants are fit to remove, they ſhould be tranſplanted into ſeparate ſmall pots, and plunged into a moderate hot-bed of tanners bark, obſerving to ſhade them till they have taken root; after which they ſhould have a large ſhare of free air admitted to them every day in warm weather. As theſe plants will make great progreſs in their growth, ſo they muſt be ſhifted into larger pots, and part of them plunged into the hot-bed again; the others may be placed abroad in June, with other exotic plants, in a warm ſituation. Thoſe which are placed in the hot-bed, will be ſtrong enough to flower well in the ſtove the following winter, but thoſe in the open air will not flower before the following ſummer: theſe may remain abroad till the beginning of October, when they muſt be removed into the ſtove, and treated in the ſame manner as the old plants. Theſe plants will continue many years with proper management, but as young plants always flower better than the old roots, ſo it is ſcarce worth while to continue them after they have borne good ſeeds.

The ſecond ſort, which is much hardier than either of the other, ſhould have a different treatment. The young plants of this muſt be earlier inured to the open air, where they may remain till the froſt begins; then they muſt be placed in the green-houſe, and ſhould have but little wet in winter; and the beginning of May theſe ſhould be turned out of the pots, and planted in a warm ſouth border, in a dry ſoil, where they will thrive and produce flowers annually. There is a variety of this with variegated leaves, which is preſerved in ſome gardens, and is propagated by parting of the roots; but this hath little beauty, ſo is ſcarce worth cultivating.

The ſeeds of the fourth ſort I received from the Braſils; this hath much the appearance of the firſt, but the leaves are longer, the flower-ſtalks riſe much higher, and the flowers are larger, and of a deeper crimſon colour, ſo makes a more noble appearance. This ſhould be treated as the firſt.

CANNABINA. See DATISCA.

CANNABIS. Hemp.

The CHARACTERS are,

It is male and female in different plants. The male flowers have no petals; they have five ſhort hairy ſtamina, terminated by oblong ſquare ſummits. The female flowers have no petals but a ſmall germen, which afterward becomes a globular depreſſed ſeed, incloſed in the empalement.

We have but one SPECIES of this plant, viz.

CANNABIS (*Sativa.*) Lin. Sp. Plant. 1027. *Hemp.*

This plant is propagated in the rich fenny parts of Lincolnſhire, in great quantities, for its bark, which is uſeful for cordage, cloth, &c. and the ſeeds afford an oil which is uſed in medicine.

Hemp is always ſown on a deep, moiſt, rich ſoil, ſuch as is found in Holland, Lincolnſhire, and the fens in the iſle of Ely, where it is cultivated to great advantage, as it might in many other parts of England where there is the like ſoil; but it will not thrive on clay, or ſtiff cold land: it is eſteemed very good to deſtroy weeds, which is no other way effected, but by robbing them of their nouriſhment

ment, for it will greatly impoverish the land, so that this crop must not be repeated on the same ground.

The land on which Hemp is designed to be sown should be well ploughed, and made very fine by harrowing; about the middle of April is a good season for sowing of the seed: three bushels is the usual allowance for an acre, but two is fully sufficient: in the choice of the seed, the heaviest and brightest coloured should be preferred, and particular care should be had to the kernel of the seed; so that some of them should be cracked to see if they have the germ or future plant perfect, for in some places the male plants are drawn out too soon from the female, i. e. before they have impregnated the female plants with the farina; in which case, though the seeds produced by these female plants may seem fair to the eye, yet they will not grow, as is well known to the inhabitants of Bickar, Swineshead, and Dunnington, three parishes in the fens of Lincolnshire, where Hemp is cultivated in great abundance, who have dearly bought their experience.

When the plants are come up, they should be hoed out in the same manner as is practised for Turneps, leaving them two feet apart; observe also to cut down all the weeds, which, if well performed, and in dry weather, will destroy them. This crop will require a second hoeing about six weeks after the first, in order to destroy the weeds: if this be well performed it will require no farther care, for the Hemp will soon after cover the ground, and prevent the growth of weeds.

The first season for pulling the Hemp is usually about the middle of August, when they begin to pull what they call the Fimble Hemp, which is the male plants; but it would be much the better method to defer this a fortnight or three weeks longer, until these male plants have fully shed their farina or dust, without which the seeds will prove abortive, produce nothing if sown the next year, nor will those concerned in the oil-mills give any thing for them, there being only empty husks, without any kernels to produce the oil. These male plants decay soon after they have shed their farina.

The second pulling is a little after Michaelmas, when the seeds are ripe: this is usually called Karle Hemp, it is the female plants, which were left at the time when the male were pulled. This Karle Hemp is bound in bundles of a yard compass, according to statute measure, which are laid in the sun for a few days to dry; and then it is stacked up, or housed to keep it dry, till the seed can be threshed out. An acre of Hemp on a rich soil, will produce near three quarters of seed, which, together with the unwrought Hemp, is worth from six to eight pounds.

CANNACORUS. See Canna.

CAPERS. See Capparis.

CAPNOIDES. See Fumaria.

CAPNORCHIS. See Fumaria.

CAPPARIS. Lin. Gen. Plant. 567. The Caper Bush.

The Characters are,

The flower hath four large roundish petals; it hath a great number of slender stamina. In the midst of these arise a single style longer than the stamina, with an oval germen, which afterward becomes a fleshy turbinated capsule with one cell, filled with kidney-shaped seeds.

The Species are,

1. Capparis (*Spinosa*) pedunculis solitariis unifloris stipulis spinosis, capsulis ovalibus. Lin. Sp. *Prickly Caper.*

2. Capparis (*Badducea*) pedunculis subsolitariis, foliis persistentibus ovato-oblongis nudis determinate confertis. Lin. Sp. 720. *Indian Caper, called Badducea.*

3. Capparis (*Arborescens*) foliis lanceolato-ovatis perennantibus caule arborescenti. *Tree Caper with oval spear-shaped leaves.*

4. Capparis (*Cynophalophora*) pedunculis multifloris terminalibus angulatis, foliis persistentibus ovalibus obtusis. Lin. Sp. 721. *Caper with many flowers terminating the branches.*

5. Capparis (*Racemosa*) foliis ovatis oppositis perennantibus, floribus racemosis. *Caper with opposite evergreen leaves, and flowers in bunches.*

6. Capparis (*Siliquosa*) pedunculis unifloris compressis, foliis persistentibus lanceolato-oblongis acuminatis subtus punctatis. Lin. Sp. 721. *Caper with one flower upon each foot-stalked, called Breynia by Brown.*

7. Capparis (*Breynia*) pedunculis racemosis, foliis persistentibus oblongis, pedunculis calycibusque tomentosis, floribus octandris. Jacq. Amer. tab. 103. *Caper of America with branching foot-stalks, called Breynia.*

8. Capparis (*Triflora*) foliis lanceolatis nervosis perennantibus pedunculis trifloris. *Caper with spear-shaped nervous leaves, and three flowers on each foot-stalk.*

The first is the common Caper, whose full-grown flower-bud is pickled, and brought to England annually from Italy and the Mediterranean. This is a low shrub, which generally grows out of the joints of old walls, the fissure of rocks, and amongst rubbish, in most of the warm parts of Europe: it hath woody stalks, which send out many lateral slender branches; under each of these are placed two short crooked spines, between which and the branches come out the foot-stalk of the leaves, which are single, short, and sustain a round, smooth, entire leaf; at the intermediate joints between the branches come out the flowers upon long foot-stalks; before these expand, the bud with the empalement is gathered for pickling, but those which are left expand in form of a single Rose, having five large white petals, which are roundish and concave; in the middle is placed a great number of long stamina, surrounding a style which rises above them, and is crowned with an oval germen, which afterward becomes a capsule, filled with kidney-shaped seeds.

This sort is cultivated upon old walls about Toulon, and in several parts of Italy. Mr. Ray observed it growing naturally on the walls and ruins at Rome, Sienna, and Florence.

The plants are with difficulty preserved in England, for they delight to grow in crevises of rocks, and the joints of old walls or ruins, and always thrive best in an horizontal position; so that when they are planted either in pots or the full ground, they rarely thrive, though they may be kept alive for some years. They are propagated by seeds in the warm parts of Europe, but it is very difficult to get them to grow in England. There is an old plant growing out of a wall in the gardens at Camden-House, near Kensington, which has resisted the cold for many years, and annually produces many flowers, but the young shoots of it are generally killed to the stump every winter.

The roots of these plants are annually brought from Italy, by the persons who import Orange-trees, some of which have been planted in walls, where they have lived a few years, but have not continued long.

The second sort I received from Carthagena in New Spain, near which place it grows naturally. This rises with a woody stem to the height of twelve or fourteen feet, sending out many lateral branches, covered with a russet bark, and garnished with oblong oval leaves; the flowers are produced from the side of the branches single, which are like those of the last sort.

The third sort was also sent me from Carthagena. This grows with a strong upright trunk near twenty feet high, sending out many lateral branches, which are covered with a very white bark, and are closely garnished with large, oblong, stiff leaves, of a thicker consistence than those of the common Laurel, of a splendid green; the flowers come out from the side of the branches, which are large, and the summits of the stamina are purple.

The fourth sort was sent me from the same country. This rises with a trunk about twenty feet high, sending out many long slender branches, covered with a brown bark, and garnished with leaves like those of the Bay-tree, but longer, and deeply ribbed on their under side. The flowers are produced upon long foot-stalks which terminate the branches, each sustaining several flowers, which are large, white, and are succeeded by pods two or three inches long, and the thickness of a man's little finger, which are filled with large kidney-shaped seeds.

The fifth sort was sent me from Tolu in America. This rises with a shrubby stalk to the height of eight or ten feet, sending out many ligneous branches, garnished with oval stiff leaves, which are placed opposite, upon short red foot-stalks; from the wings of the leaves are produced the foot-stalks of the flowers, which are long, slender, and branch out into many smaller, each of which sustains a small white flower, which is succeeded by an oval pod, containing many small kidney-shaped seeds.

The sixth sort rises with a shrubby stem to the height of twelve or fourteen feet, sending out many strong lateral branches, garnished with oblong oval leaves, placed alternately; the leaves are of a thicker consistence than those of the Bay-tree; at the foot-stalk of each leaf comes out a single flower, almost the whole length of the branches, which are small, and stand upon short foot-stalks; the summits of these flowers are of a purplish colour, but the stamina are white.

The seventh sort rises with a shrubby stalk to the height of ten or twelve feet, sending out slender horizontal branches on every side, which are covered with a reddish bark; the joints of these branches are far distant; at each of these come out several leaves in clusters, without order, standing upon pretty long foot-stalks, smooth on their upper side, but have many transverse ribs on their under side, which are prominent.

The eighth sort I received from Carthagena, where it grows naturally; this hath many shrubby stalks arising from the root, which send out many lateral branches on every side, closely garnished with large spear-shaped leaves, standing upon very long foot-stalks; the flowers are produced three on each foot-stalk, at the end of the branches; they are small, white, with yellowish stamina, and have a very long style, which is incurved, terminated by an oblong germen, which afterward becomes an oblong fruit.

These last seven sorts are natives of warm countries, so will not live through the winter in England without the assistance of a stove. They are propagated by seeds, which must be procured from the countries where they grow naturally. These must be sown in small pots, and plunged into a hot-bed of tanners bark. In about two months the plants will appear, provided the seeds are good, then they must have but little wet, and a good share of air in warm weather; when they are large enough to remove, they must be each transplanted into a separate small pot, and plunged into the hot-bed again, observing to shade them until they have taken fresh root, after which they should have fresh air admitted to them every day, in proportion to the warmth of the season. In autumn they must be removed into the stove, and plunged into the bark-bed, where they should constantly remain, and will require the same treatment as other tender exotic plants from the same countries, with this difference only, that they require but little water, especially during the winter, for the roots of these plants are very subject to rot with wet.

BEAN CAPER. See ZYGOPHYLLUM

CAPRARIA. Lin. Gen. Pl. 686.

The CHARACTERS are,

The flower is bell-shaped, of one leaf, divided at the top into five equal parts; it hath four stamina which are inserted in the base of the petal, and but little more than half so long, two of the under being shorter than the other. It hath a conical germen, which afterward becomes an oblong conical capsule, compressed at the point, having two cells, divided by a partition filled with roundish seeds.

We have but one SPECIES of this genus, viz.

CAPRARIA (*Biflora*) foliis alternis floribus geminis. Jacq. Amer. 15. *Capraria with alternate leaves, and the petal divided into five parts.*

This plant grows naturally in the warm parts of America, where it is often a troublesome weed in the plantations; it rises with an angular green stalk about a foot and a half high, sending out branches at every joint, which sometimes come out by pairs opposite, but often there are three at a joint standing round the stalk; the leaves are also placed round the branches by threes; these stand upon short foot-stalks, are oval, hairy, and a little indented on their edges. The flowers are produced at the wings of the leaves, coming out on each side the stalk, each foot-stalk sustaining three flowers; they are white, and succeeded by conical capsules compressed at the top, opening in two parts, and filled with small seeds.

This plant is preserved in botanic gardens for the sake of variety; but as it hath no great beauty, it is seldom admitted into other gardens.

It is propagated by seeds, which must be sown upon a hot-bed in the spring of the year, and the plants must be brought forward by planting them upon a second hot-bed; and about the middle or latter end of June they may be transplanted into a warm border, and may then be exposed to the open air, where they will perfect their seeds in autumn

CAPRIFOLIUM. See PERICLYMEN.

CAPSICUM. Lin. Gen. Plant. 225. Guinea Pepper.

The CHARACTERS are,

The flower hath but one petal, which is wheel shaped; it hath five small stamina. It hath an oval germen, which afterward becomes a soft fruit or capsule, of an indeterminate figure, having two or more cells, divided by intermediate partitions, to which adhere many compressed kidney shaped seeds.

The SPECIES are,

1. CAPSICUM (*Annuum*) caule herbaceo, fructu oblongo propendente. *Capsicum with an herbaceous stalk, and an oblong fruit hanging downward.*

2. CAPSICUM (*Cordiforme*) caule herbaceo, fructu cordiformi. *Capsicum with an herbaceous stalk, and a heart-shaped fruit.*

3. CAPSICUM (*Tetragonum*) caule herbaceo, fructu maximo anguloso obtuso. *Capsicum with an herbaceous stalk, and a large, angular, obtuse fruit; commonly called Bell Pepper.*

4. CAPSICUM (*Angulosum*) caule herbaceo, fructu cordiformi angulosa. *Capsicum with an herbaceous stalk, and an angular heart-shaped fruit.*

5. CAPSICUM (*Cerasiforme*) caule herbaceo, fructu rotundo glabro. *Capsicum with an herbaceous stalk, and a round smooth fruit.*

6. CAPSICUM (*Olivaeforme*) caule herbaceo, fructu ovato. *Capsicum with an herbaceous stalk and an oval fruit*

7. CAPSICUM (*Pyramidale*) caule fruticoso foliis lineari-lanceolatis, fructu pyramidali erecto luteo. *Capsicum with a shrubby stalk, narrow spear-shaped leaves, and yellow pyramidal fruit growing upright.*

8. CAPSICUM (*Conoide*) caule fruticoso fructu conico erecto rubro. *Capsicum with a shrubby stalk, and a conical red fruit growing erect; commonly called Hen Pepper.*

9. CAPSICUM

9. Capsicum (*Frutescens*) caule fruticoso, fructu parvo pyramidali erecto. *Capsicum with a shrubby stalk, and a small pyramidal fruit growing erect; commonly called Barberry Pepper.*

10. Capsicum (*Minimum*) caule fruticoso, fructu parvo ovato erecto. *Capsicum with a shrubby stalk, and a small oval fruit growing erect; commonly called Bird Pepper.*

The first sort is the common long-poded Capsicum, which is frequently cultivated in the gardens. Of this there is one with red, and another with yellow fruit, which only differ in the colour of the fruit, which difference is permanent.

The Varieties of yellow Capsicum are,

Capsicum fructu surrecto oblongo. Tourn. *Capsicum with oblong fruit growing erect.*

Capsicum fructu bifido. Tourn. *Capsicum with a divided fruit.*

Capsicum siliquis surrectis & oblongis brevibus. Tourn. *Capsicum with oblong and short pods growing erect.*

Capsicum fructu tereti spithameo. Tourn. *Capsicum with a taper fruit a span long.*

Of these different forms I have had both the red and yellow, but neither of them have changed their colours, though they have frequently varied in their shape.

The second sort with heart-shaped fruit is undoubtedly a different species from the first, and never alters toward it, though there are several varieties of this which arise from seeds. Of this there are red and yellow fruit, which do not alter in colour, though they produce the following varieties:

Capsicum siliqua propendente rotunda & cordiformi. Tourn. *Capsicum with round, heart-shaped, hanging pods.*

Capsicum siliqua latiore & rotundiore. Tourn. *Capsicum with a larger and a rounder pod.*

Capsicum rotundo maximo. Tourn. *Capsicum with the largest round fruit.*

Capsicum siliquis surrectis cordiformibus. Tourn. *Capsicum with upright heart-shaped pods.*

Capsicum siliquis surrectis rotundis. Tourn. *Capsicum with round upright pods.*

The third sort I have cultivated many years, and have not found it alter, nor have I seen any other but the red fruit of this. It is the only sort which is proper for pickling, the skin of the fruit being fleshy and tender, whereas those of the other sorts are thin and tough. The pods of this sort are from one inch and a half to two inches long, are very large, swelling, and wrinkled, flatted at the top, where they are angular, and sometimes stand erect, at others grow downward. When the fruit of this are designed for pickling, they should be gathered before they arrive to their full size, while their rind is tender; then they must be slit down on one side to get out the seeds, after which they should be soaked two or three days in salt and water; when they are taken out of this and drained, boiling vinegar must be poured on them, in a sufficient quantity to cover them, and closely stopped down for two months; then they should be boiled in the vinegar to make them green, but they want no addition of any sort of spice, and are the wholsomest and best pickle in the world.

The fourth sort is also a distinct species from all the other: this hath broad wrinkled leaves, the fruit is also furrowed and wrinkled, generally growing upright, and of a beautiful scarlet colour: some of the fruit will have their tops compressed like a bonnet, from whence it had the name.

The fifth sort was sent me from the Spanish West Indies: this doth not grow so tall as the other sorts but spreads near the ground. The leaves come out in clusters, which are of a shining green, and stand on long foot-stalks. The fruit is round, smooth, of a beautiful red and the size of a common Cherry.

The sixth sort I received from Barbadoes: this is like the common in its stalk and leaves, but the fruit is oval, and about the size of a French Olive.

These six sorts are annual with us, whatever they may be in their native countries, for their stalks decay soon after the fruit is ripe. They are propagated by seeds, which must be sown upon a hot-bed in the spring; and when the plants have six leaves, they should be transplanted on another hot-bed at four or five inches distance, shading them in the day-time from the sun until they have taken root; after which they must have a large share of air admitted to them in warm weather, to prevent their drawing up weak. Toward the end of May the plants must be hardened by degrees to bear the open air, and in June they should be carefully taken up, preserving as much earth about their roots as possible, and planted into borders of rich earth, observing to water them well, as also to shade them until they have taken root; after which time, they will require no other management but to keep them clean from weeds, and in very dry seasons to refresh them two or three times a week with water. These directions are for the culture of the common sorts of Capsicum, which are generally planted by way of ornament. But the plants of the third sort, which are propagated for pickling, should be planted in a rich spot of ground, in a warm situation, about a foot and a half asunder, and shaded till they have taken root, and afterward duly watered in dry weather, which will greatly promote their growth, and cause them to be more fruitful, as also enlarge the size of the fruit. By this management there may be three or four crops of fruit for pickling obtained the same year, provided the season proves not too cold

The fourth, fifth, and sixth sorts being tender, the plants should be put into pots, and placed in an old hot-bed under a deep frame, where they may have room to grow; or if they are planted in the full ground, the plants should be each covered with a bell-glass to screen them from cold. The glasses may be set off every day in warm weather, and placed over them in the evening again; and at such times as the weather is not favourable, the glasses should be raised on the contrary side to the wind, to admit the fresh air. With this care the fruit of these sorts will ripen in England, which without it rarely comes to maturity but in very warm seasons.

The four last sorts have perennial shrubby stalks, which rise four or five feet high; these are not so hardy as the other, therefore when the plants have been brought forward in the hot-bed as was directed for the common sorts, they should be planted in pots filled with rich earth, and plunged into a very moderate hot-bed, under a deep frame, where they may have room to advance; in warm weather they should have a large share of air admitted to them, but must be covered with glasses every night, or in cold weather, and frequently watered. With this management they will produce plenty of fruit in autumn, which ripen in winter; but they must be removed into the stove on the first approach of frost, and placed where they may have a temperate warmth, in which they will thrive better than in a greater heat; and the fruit will continue in beauty most part of winter, making a pretty appearance in the stove during that season.

The seeds of the seventh sort I received from Egypt: the leaves of this are much narrower than those of any other sort I have yet seen; the pods always grow erect, and are produced in great plenty, so that the plants make a good appearance for three months in the winter

The eighth sort I received from Antigua, by the title of Hen Pepper. This rises with a shrubby stalk three or four feet high, sending out many branches toward the top: the fruit is about half an inch long, shaped in form of an obtuse cone, and of a bright red, growing erect.

The

The ninth fort grows about the fame height as the eighth, but differs from it in the fhape and fize of the fruit; thofe of this fort being about the bignefs of a Barberry, and nearly of the fame fhape. This I have long cultivated, and have not obferved it to alter.

The tenth fort is commonly known by the title of Bird Pepper in America. This rifes with a fhrubby ftalk four or five feet high; the leaves are of a lucid green; the fruit grows at the divifions of the branches, ftanding erect: thefe are fmall, oval, and of a bright red; they are much more fharp and biting than thofe of the other forts. From the fruit of this fort is made the Cayan butter, or what the inhabitants of America call Pepper-pots, which they efteem as the beft of all the fpices.

CARACALLA. See Phaseolus.

CARAGANA. See Robinia.

CARDAMINDUM. See Tropæolum.

CARDAMINE. Lin. Sp. Pl. 727. In Englifh, Ladies-fmock.

The Characters are,

The flower hath four oblong petals, placed in form of a crofs; it hath fix ftamina, four of which are the length of the empalement; the other two, which are oppofite, are much longer. It hath a cylindrical germen, which afterward turns to a long compreffed cylindrical pod with two cells, opening in two valves, which twift fpirally, and caft out the feeds when ripe by their elafticity.

The Species are,

1. Cardamine (*Pratenfis*) foliis pinnatis, foliis radicalibus fubrotundis, caulinis lanceolatis. Lin. Sp. Plant. 656. *Meadow Ladies-fmock with a large purplifh flower.*

2. Cardamine (*Parviflora*) foliis pinnatis, foliolis incifis, floribus exiguis, caule erecto ramofo. *Annual Impatient Crefs with a very fmall flower.*

3. Cardamine (*Hirfuta*) foliis pinnatis, floribus tetrandis. Hort. Cliff. 336. *Ladies-fmock, or Impatient Crefs with winged leaves, and flowers with four ftamina.*

4. Cardamine (*Græca*) foliis pinnatis foliolis palmatis æqualibus petiolatis. Prod. Leyd. 345. *Sicilian Impatient Crefs with Fumitory leaves.*

5. Cardamine (*Amara*) foliis pinnatis, foliolis fubrotundis angulofis. Hall. Helv. 558. *Greater bitter Water Crefs.*

6. Cardamine (*Trifolia*) foliis ternatis obtufis, caule fubnudo. Lin. Sp. Pl. 654. *Alpine three-leaved Crefs.*

7. Cardamine (*Bellidifolia*) foliis fimplicibus ovatis integerrimis petiolis longis. Flor. Lap. 206. *Smaller Alpine Crefs with a Daify leaf.*

8. Cardamine (*Petrea*) foliis fimplicibus oblongis dentatis. Lin. Sp. Pl. 654. *Rock Crefs.*

9. Cardamine (*Chelidonia*) foliis pinnatis foliolis quinis incifis. Lin. Sp. Plant. 655. *Smooth Impatient Crefs with a Celandine leaf.*

The firft fort grows naturally in the meadows in many parts of England; it is called Cuckow-flower, and Ladies-fmock. Of this there are four varieties, viz. the fingle with purple and white flowers, which are frequently intermixed in the meadows, and the double flower of both colours. The fingle forts are feldom admitted into gardens; but as the firft fort ftands in the lift of medicinal plants, fo I have enumerated it. The young leaves of this plant have been gathered in the fpring by fome perfons, and put into fallets inftead of Crefs: it is fuppofed to be an antifcorbutic.

The two varieties with double flowers were accidentally found growing in the meadows, and were tranfplanted into gardens, where they have been propagated. Thefe deferve a place in fhady moift borders of the flower-garden, where they will thrive, and make a pretty appearance during their continuance in flower. They are propagated by parting their roots; the beft time for this is the autumn, at which time they fhould be tranfplanted. They delight in a foft loamy foil, not too ftiff, and muft have a fhady fituation.

The fixth, feventh, and ninth forts grow naturally on the Alps, and other mountainous places. Thefe are low perennial plants, which may be propagated by parting of their roots in the autumn, and require a ftrong foil and fhady fituation: they may alfo be propagated by feeds, which fhould be fown in the autumn, on a fhady border, where they will come up foon after, and are never hurt by froft.

The eighth fort is a low biennial plant, which grows naturally in feveral parts of England and Wales, and is preferved in fome gardens for the fake of variety.

The fifth fort grows naturally by the fides of rivers, and in ditches in moft parts of England, fo is not admitted into gardens.

The other forts are low annual plants, which grow naturally in feveral parts of England. Thefe have the title of Impatient Crefs, from the elafticity of their pods, which, if touched when they are ripe, fpring open, and caft out their feeds with violence to a confiderable diftance. Thefe forts when young are by the country people eaten in fallets, and have the flavour of the common Crefs, but milder.

CARDIACA. See Leonurus.

CARDINALS FLOWER. See Rapuntium.

CARDIOSPERMUM. Lin. Gen. Pl. 447. Heart Pea.

The Characters are,

The flower hath four obtufe petals, which are alternately larger, and a fmall four-leaved nectarium encompaffing the germen, with eight ftamina, three and three ftanding oppofite, the other two on each fide. The germen is three-cornered, which afterward becomes a roundifh fwollen capfule with three lobes, divided into three cells, opening at the top, each having one or two globular feeds, marked with a heart.

The Species are,

1. Cardiospermum (*Corindum*) foliis fubtus tomentofis. Lin. Sp. 526. *Heart Pea with woolly leaves; called in America Wild Parfley.*

2. Cardiospermum (*Halicacabum*) foliis lævibus. Lin. Sp. *Heart-feed with fmooth leaves.*

The firft fort rifes with a flender, channelled, climbing ftalk to the height of four or five feet, fending out many fide branches, garnifhed with leaves, upon very long foot-ftalks, coming out oppofite at the lower part of the ftalk; but upward the leaves come out on one fide, and the foot-ftalk of the flower on the other, oppofite. The foot-ftalk of the flowers are long, naked, and toward the top, divided into three fhort ones, each fuftaining a fingle flower. Immediately under thefe divifions come out a tendril, or clafper, like thofe of the Vine, but fmaller; thefe faften themfelves to whatever grows near them, and thereby are fupported. The flowers are fmall, white, and compofed of four fmall concave petals, two of which ftanding oppofite, are larger than the other; when thefe fall away, the germen becomes a large inflated bladder, having three lobes, in each of which is contained one, two, and fometimes three feeds, which are round, hard, and the fize of fmall Peas, each being marked with a black fpot in fhape of a heart.

The fecond fort differs from the firft in having taller ftalks, the leaves being firft divided into five, and again into three parts. The foot-ftalks are fhorter, and the feeds and bladders in which they are contained are much larger, and the whole plant is fmoother, in other refpects they agree.

Thefe plants grow naturally in both Indies, where they climb upon whatever fhrubs are near them, and there they rife to the height of eight or ten feet, but in England they feldom ſeldom are above half ſo high; they ſend out many ſide branches, which ſpread to a conſiderable diſtance every way.

They are annual, and periſh ſoon after they have perfected their ſeeds, and being natives of hot countries they will not thrive in England in the open air. They are propagated by ſeeds, which ſhould be ſown upon a hot-bed in the ſpring, and when the plants are two inches high they ſhould be tranſplanted into pots, then plunged into a very moderate hot-bed, where they muſt be carefully ſhaded until they have taken freſh root; after which they muſt have a large ſhare of air admitted to them every day, to prevent their being drawn up weak: then they may be removed into a glaſs-caſe, where they may have room to grow, and be ſcreened from the cold of the nights, but in warm weather they will require a large ſhare of air. With this management they will flower in July, and their ſeeds will ripen in autumn.

CARDUUS. Lin. Gen. Plant. 832. Thiſtle.

The Characters are,

It hath a compound flower, made up of many hermaphrodite florets, which are fruitful; theſe are included in one common ſcaly empalement; the florets are funnel-ſhaped, and of one leaf; each of theſe florets have five ſhort hairy ſtamina. In the center is ſituated an oval germen, crowned with down, which afterward becomes an oblong four-cornered ſeed, crowned with down, and incloſed by the prickly empalement.

The Species are,

1. Carduus (*Ptarmicifolia*) foliis integris ſubtus tomentoſis, ſpinis ramoſis lateralibus. Prod. Leyd. 113. *Low prickly Thiſtle with leaves like the Eternal-flower.*

2. Carduus (*Eriophorus*) foliis ſeſſilibus bifariam pinnatifidis laciniis alternis erectis, calycibus globoſis villoſis. Hort. Upſal. 249. *Woolly-headed Thiſtle; called by ſome Friars Cowl.*

3. Carduus (*Acarna*) foliis lanceolatis dentatis ciliatis decurrentibus, ſpinis marginalibus duplicibus. *Greater Fiſh Thiſtle.*

4. Carduus (*Marianus*) ſquamis calycinis margine apiceque ſpinoſis. Hort. Cliff. 393. *Our Ladies Thiſtle, or Milk Thiſtle.*

5. Carduus (*Cirſium*) foliis lanceolatis decurrentibus denticulis inermibus, calyce ſpinoſo. Hort. Cliff. 392. *Engliſh ſoft or gentle Thiſtle.*

6. Carduus (*Caſabonæ*) foliis lanceolatis ſeſſilibus integerrimis margine ſpinis ternatis. Hort. Cliff. 393. *The ſuppoſed true Fiſh Thiſtle of Theophraſtus.*

There are a great number of ſpecies more known than are here enumerated, ſome of which are very troubleſome weeds in the gardens and fields, therefore are better to be kept out of both, ſo I thought it needleſs to mention them here.

The firſt ſort grows naturally in Sicily. This is an annual plant, which riſes with a channelled ſtalk about two feet high, ſending out ſeveral ſide branches toward the top, garniſhed with long narrow leaves like thoſe of the Auſtrian Ptarmica, which are of a deep green above, but white on their under ſide; juſt below the foot-ſtalk of the leaf come out ſeveral unequal yellow ſpines; at the end of the branches the flowers are produced; theſe have very prickly empalements, under which are placed two long leaves; the flowers are purple, and ſhaped like thoſe of the common Thiſtle, but are ſmaller; theſe are ſucceeded by oblong ſmooth ſeeds, which have a long woolly down ſitting on their top. It is propagated by ſeeds, which ſhould be ſown on a bed of light earth in the ſpring, where the plants are to remain. The only care they will require, is to keep them clean from weeds, and thin the plants where they are too cloſe.

The ſecond ſort grows naturally in ſeveral of the midland counties of England. This is a biennial plant, which ſends out many long leaves near the ground, placed by pairs, and are joined to a winged border running on each ſide the midrib the whole length; theſe ſegments are alternately pointing upward, armed with long ſharp ſpines, ſtanding every way. The following ſpring there ariſes from the center of the plant one ſtrong channelled ſtalk four or five feet high, branching every way toward the top; each branch is terminated by a ſingle head of purple flowers, having a woolly empalement. One or two of theſe plants may be allowed a place in ſome abject part of the garden for its ſingularity.

The third ſort grows naturally in Spain and Portugal. This riſes near three feet high, branching toward the top; the leaves are long, narrow, and the edges are ſet cloſely with ſmall hairs; at every indenture of the leaves there comes out two long yellowiſh ſpines; at the end of the branches the flowers are produced from the ſide of the ſtalk, which have woolly oval empalements, cloſely armed with ſlender ſpines. The flowers are yellow, but make no great appearance. This plant may be propagated by ſeeds, in the ſame manner as the former ſort. It is called Fiſh Thiſtle, from the reſemblance which the ſpines have to the bones of fiſh.

The fourth ſort grows very common on the ſide of banks, and in waſte land in many parts of England, and is by ſome perſons blanched and dreſſed as a curious diſh. This is a biennial plant, which ſhould be ſown very thin; and when the plants are come up, ſo as to be well diſtinguiſhed, the ground ſhould be hoed, to cut down all the young weeds, and the plants left about two feet and a half diſtance; in the autumn the leaves of the plants ſhould be tied up, and the earth drawn up cloſe to blanch them; when they are properly whitened, they will be fit for uſe. This is a biennial plant, which periſhes ſoon after the ſeeds are ripe.

The fifth ſort is a perennial plant, which is by ſome cultivated for medicinal uſe, and has been ſuppoſed a remedy for ſome ſort of madneſs. This may be propagated by ſeeds, in the ſame manner as the ſecond ſort. It grows naturally in the northern parts of England, and flowers in June.

The ſixth ſort is ſuppoſed to be the true Fiſh Thiſtle of Theophraſtus. This is a biennial plant, which riſes with an upright ſtalk ſix or ſeven feet high, is garniſhed with long ſpear-ſhaped leaves, armed with triple ſpines at every indenture: on the ſide of the ſtalks the flowers come out in cluſters, which are of a purple colour, and are ſucceeded by ſmooth, oval, black ſeeds. It grows naturally in Sicily and the Levant. It is propagated by ſeeds as the ſecond ſort, which ſhould be ſown on a warm border, otherwiſe the plants will not live through the winter.

CARDUUS BENEDICTUS. See Cnicus.

CARDUUS FULLONUM. See Dipsacus.

CARICA. Lin. Gen. Plant. 1000. Papaw.

The Characters are,

It is male and female in different plants; the flowers of the male are funnel-ſhaped, and of one leaf, having a long ſlender tube, which expands at the top, where it is divided into five narrow obtuſe parts; it hath ten ſtamina, five of which are alternately longer than the other. The female flowers have a ſmall permanent empalement, indented in five parts, with five long ſpear-ſhaped petals; the oval germen afterward becomes a large, oblong, fleſhy fruit, having five longitudinal cells, which are full of ſmall oval furrowed ſeeds, incloſed in a glutinous pulp.

The Species are,

1. Carica (*Papaya*) foliorum lobis ſinuatis. Hort. Cliff 461. *Papaw with the fruit ſhaped like the Squaſh.*

2. Carica (*Proſopoſa*) foliorum lobis integris. Hort Cliff. 461. *Papaw with the lobes of the leaves entire.*

There are ſeveral varieties of the firſt ſort, which differ in the ſize and ſhape of their fruit. Plumier mentions three of of the female or fruitful Papaw, befide the male; one of which he titles Melon-fhaped, and the other fhaped like the fruit of the Gourd; and I have feen another variety in England, with a large, fmooth, pyramidal fruit: but thefe are fuppofed to be accidental varieties, which arife from the fame feeds.

The firft fort rifes with a thick, foft, herbaceous ftem to the height of eighteen or twenty feet, which is naked till within two or three feet of the top, and hath marks of the veftiges of the fallen leaves on the ftem; the leaves come out on every fide upon very long foot-ftalks; thofe which are fituated undermoft are almoft horizontal, but thofe on the top are erect: thefe leaves (in full grown plants) are very large, and divided into many parts (or lobes) which are deeply finuated, or cut into irregular divifions. The whole plant abounds with a milky acrid juice, which is efteemed good for the ringworm: the ftem of the plant, and alfo the foot-ftalks of the leaves, are hollow in the middle. The flowers of the male plant are produced from between the leaves, on the upper part of the plant, which have foot-ftalks near two feet long, at the ends of which the flowers ftand in loofe clufters, each having a feparate fhort foot-ftalk; thefe are of a pure white, and have an agreeable odour; in one of the male plants, which was raifed from feeds that came from China, one of the male flowers produced a fmall fruit, about the fize of a Catharine Pear, but was much longer, having feveral angles, and drawing to a point at each end; in it there were fome feeds, but it is a doubt if they are fruitful. The flowers of the female Papaw alfo come out between the leaves toward the upper part of the plant, upon very fhort foot-ftalks, fitting clofe to the ftem; they are large and bell-fhaped, compofed of fix petals, and are commonly yellow; when thefe fall away, the germen fwells to a large flefhy fruit, the fize of a fmall Melon, which are of different forms; fome are angular, and compreffed at both ends; others are oval and globular, and fome pyramidal; the fruit alfo abounds with the fame acrid milky juice as the plants. This fruit, when ripe, is by the inhabitants of the Caribbee Iflands eaten with pepper and fugar as Melons, but are much inferior to a common Melon in flavour in its native country, but thofe which have ripened in England were deteftable: the only ufe I have known made of this fruit was, when they were about half grown, to foak them in falt-water, to get out the milky juice, and pickle them for Mangos, for which they have been a good fubftitute.

The fecond fort was found growing in a garden at Lima, by Father Feuillée, and was the only plant he faw of that fort in his travels. This differs from the other, in having a branching ftalk, the lobes or divifions of the leaves being entire, and the fruit being fhaped like a Pear, which he fays were of different fizes; that which he defigned was about eight inches long, and three and a half thick, yellow within and without, and of a fweet flavour. The flower, he fays, was of a Rofe colour, and divided into five parts.

Thefe plants being natives of hot countries, will not thrive in England, unlefs they are preferved in a warm ftove, which fhould be of a proper height to contain the plants. When they are grown to a large fize, they make a noble appearance, with their ftrong upright ftems, which are garnifhed on every fide near the top with large fhining leaves, fpreading out near three feet all round the ftem: the flowers of the male fort come out in clufters on every fide; and the fruit of the female growing round the ftalks, between the leaves, being fo different from any thing of European production, may intitle them to the care of the curious.

They are eafily propagated by feeds, which are annually brought in plenty from the Weft-Indies, though the fruit which grows in Europe produces plenty of feeds when it ripens well. Thefe fhould be fown in a hot-bed early in the fpring: when the plants are near two inches high, they fhould be each tranfplanted into a feparate fmall pot, and plunged into a hot-bed of tanners bark, carefully fhading them from the fun till they have taken root; after which they muft be treated in the fame manner as other tender plants from the fame country; but as thefe plants have foft herbaceous ftalks, and abound with a milky juice, fo they muft not have much water, for they are frequently killed with moifture. When thefe plants are fhifted from fmall pots into larger, care muft be taken to preferve the ball of earth to their roots, for whenever their roots are left bare, they rarely furvive it.

CARLINA. Lin. Gen. Plant. 836. The Carline Thiftle.

The CHARACTERS are,

It hath a compound flower, made up of many hermaphrodite florets, which are fruitful, included in a common, fwollen, fcaly empalement. The florets are funnel-fhaped, and cut into five parts at the brim; thefe have each five fhort hairy ftamina. In the center is fituated a fhort germen, crowned with down, which afterward becomes a fingle taper feed, crowned with a branching plumofe down.

The SPECIES are,

1. CARLINA (*Vulgaris*) caule multifloro corymbofo, floribus terminalibus. Hort. Cliff. 395. *Common wild Carline Thiftle.*

2. CARLINA (*Racemofa*) floribus feffilibus, lateralibus pauciffimis. Sauv. Meth. 293. *Small wild Spanifh Carline Thiftle.*

3. CARLINA (*Acaulis*) caule unifloro flore breviore. Hort. Cliff. 395. *Low Carline Thiftle with a large white flower.*

4. CARLINA (*Lanata*) caule trifloro dichotomo intermedio feffili. Sauv. Monfp. 293. *Fifh Thiftle with a reddifh, purple, fpreading flower.*

5. CARLINA (*Corymbofa*) caule multifloro fubdivifo, floribus feffilibus. Prod. Leyd. 135. *Umbellated Fifh Thiftle of Apulia.*

The firft fort grows naturally upon fterile ground in moft parts of England, fo is rarely admitted into gardens, but the others are preferved in botanic gardens for the fake of variety. They grow naturally in the fouth of France, Spain, and Italy.

They may all be propagated by fowing of their feeds in the fpring on a bed of frefh undunged earth, where they are defigned to remain; for, as they fend forth tap-roots, they will not bear tranfplanting fo well as moft other plants. When the plants appear above ground, they fhould be carefully weeded, and, as they grow in fize, they fhould be thinned where they are too clofe, leaving them about ten inches or a foot afunder. The fecond year moft of thefe plants will flower; but, unlefs the fummer proves dry, they rarely produce good feeds in England; and fome of them decay foon after they have flowered, therefore it is pretty difficult to maintain thefe plants in this country.

CARNATION. See DIANTHUS.

CARPESIUM. Nodding Starwort.

The CHARACTERS are,

The flower hath an imbricated empalement; the hermaphrodite florets are funnel-fhaped, which compofe the difk; the female florets which compofe the border are quinquefid: the former have five fhort ftamina, and an oblong germen; thefe are fucceeded by oval naked feeds.

The SPECIES are,

1. CARPESIUM (*Cernuum*) floribus terminalibus. Lin. Sp. 1203. *Nodding Starwort.*

2. CARPESIUM (*Abrotanoides*) floribus lateralibus. Ofb. It. tab. 10. *Nodding Starwort, whofe flowers come from the fide of the ftalks.*

The

The first sort grows naturally in Italy; this is a biennial plant, with obtuse, soft, woolly leaves; the stalks rise a foot and a half high, and are terminated by large flowers, of a yellow herbaceous colour. This plant is easily propagated by seeds, which should be sown the beginning of April on a bed of light earth, and when the plants come up, they should be thinned, and kept clean from weeds; the second summer the plants will flower and produce ripe seeds, then they decay.

The second sort grows naturally in China, and is pretty scarce at present in Europe: this hath branching stalks, cloathed with broad spear-shaped leaves; the flowers grow sparingly, scattered on the sides of the stalks, and branches nodding downward.

The seeds of this should be sown upon a gentle hot-bed in the spring, and when the plants come up, they should be treated in the same manner as the Tagetes, with which they will produce their flowers and ripen their seeds.

CARPINUS. Lin. Gen. Plant. 952. The Hornbeam or Hardbeam.

The Characters are,

It hath male and female flowers, growing separate on the same plant. The male flowers are disposed in a cylindrical rope or katkin; the flowers have ten small stamina. The female flowers are disposed in the same form; these have one petal, which is shaped like a cup, cut into six parts, with two short germen. The katkin afterward grows large, and at the base of each scale is lodged an oval angular nut.

The Species are,

1. Carpinus (*Vulgaris*) squamis strobilorum planis. Hort. Cliff. 447. *Common Hornbeam.*

2. Carpinus (*Ostrya*) squamis strobilorum inflatis. Hort. Cliff. 447. *The Hop Hornbeam.*

3. Carpinus (*Orientalis*) foliis ovato-lanceolatis serratis strobilis brevibus. *Eastern Hornbeam, with a smaller leaf and shorter fruit.*

4. Carpinus (*Virginiana*) foliis lanceolatis acuminatis, strobilis longissimis. *Virginia flowering Hornbeam.*

The first sort is very common in many parts of England, but is rarely suffered to grow as a timber-tree, being generally reduced to pollards by the country people; yet where the young trees have been properly treated, they have grown to a large size. Of late years, this has been only considered as a shrub, and never cultivated but for under-wood in the country, and in the nurseries to form hedges, after the French taste; but since these sort of ornaments have been almost banished from the English gardens, there has been little demand for these trees in the nurseries.

As this tree will thrive upon cold, barren, exposed hills, and in such situations where few other sorts will grow, so it may be cultivated to great advantage by the proprietors of such land. But where these are propagated for timber, they should be raised from seeds, upon the same soil, and in the same situation where they are designed to grow. Nor should they be propagated by layers, which is the common method where they are intended for hedges or under-wood; for which those so raised will answer the purpose full as well as those raised from seeds, but the latter must always be preferred for timber-trees.

The seeds of this tree should be sown in the autumn soon after they are ripe, for if they are kept out of the ground till spring, the plants will not come up till the following year. When the plants appear, they must be kept very clean from weeds, and treated as other forest-trees; in two years time they will be fit to transplant, for the sooner all trees, which are designed for timber, are planted where they are to remain, the larger they will grow, and the wood will be firmer and more durable. If they are kept clean from weeds three or four years, it will greatly promote their growth, after which the plants will have obtained sufficient strength to keep down the weeds.

As the trees advance in their growth, so they must be thinned, which should be done with caution, cutting away the most unpromising plants gradually, so as not to let too much cold air at once to those which are left, especially on the borders of the plantation.

The timber of this tree is very tough and flexible, so might be converted to many useful purposes, when suffered to grow to a proper size; but as they have been generally treated otherwise, so the principal uses it has been applied to was for turnery-ware, for which it is an excellent wood, and also for making mill-cogs, heads of beetles, &c. It is also excellent fuel.

The Hop Hornbeam sheds its leaves in the winter, with the Elm, and other deciduous trees. It is said to grow plentifully in many parts of North America, but it is doubtful whether that is not a different sort from this. The Hop Hornbeam is of quicker growth than the common sort, but what the wood of that will be I do not know; for there are but few of the trees in England growing upon their own roots, most of them having been grafted upon the common Hornbeam, which is the usual method of propagating them in the nurseries; but the trees so raised are of short duration, for the graft generally grows much faster than the stock, so that in a few years there is a great disproportion in their size; and where they happen to stand exposed to strong winds, the graft is frequently broken from the stock, after many years growth, for which reason I would caution every person not to purchase any of these trees which have been so propagated.

The Virginia flowering Hornbeam is less common than the last, and only to be seen in curious gardens; it is equally hardy as the other, and may be increased by layers.

The Eastern Hornbeam is a tree of humble growth, rarely rising above ten or twelve feet high in this country, shooting out many horizontal irregular branches, so cannot easily be trained up to a stem. The leaves of this sort are much smaller than those of the common Hornbeam, and the branches grow closer together: it may be kept in less compass than almost any other deciduous tree. It is as hardy as any of the sorts, and may be propagated in the same manner; but at present it is rare in the English nurseries.

CARROTS. See Daucus.

CARTHAMUS. Lin. Gen. Plant. 838. Bastard Saffron, or Safflower.

The Characters are,

It hath a flower composed of several hermaphrodite florets, included in one common scaly empalement. The florets are funnel-shaped, cut into five equal segments at the top; these have five short hairy stamina. In the center is situated a short germen, which afterward becomes a single, oblong, angular seed, inclosed in the empalement.

The Species are,

1. Carthamus (*Tinctorius*) foliis ovatis integris serrato aculeatis. Hort. Cliff. 394. *Bastard Saffron of the shops, with a Saffron-coloured flower.*

2. Carthamus (*Lanatus*) caule piloso supernè lanato, foliis inferioribus pinnatifidis, summis amplexicaulibus dentatis. Hort. Upsal. 251. *Yellow Distaff Thistle.*

3. Carthamus (*Creticus*) caule glabro, foliis caulinis aculeatis, profundè dentatis, semiamplexicaulibus, capitulis ovatis. *Cnicus of Crete with a leaf and appearance of Distaff Thistle, and a whitish flower.*

4. Carthamus (*Tingitanus*) foliis radicalibus pinnatis, caulinis pinnatifidis. Lin. Sp. 1163. *Blue perennial Cnicus of Tangier.*

5. CARTHAMUS (*Cardunculatus*) foliis caulinis linearibus pinnatis longitudine plantæ. Lin. Sp. Pl. 831. *Dwarf Cnicus of mount Lupus with a blue flower.*

6. CARTHAMUS (*Cæruleus*) caule erecto piloso, foliis lanceolatis hirsutis, capite maximo. *Rougher blue Cnicus.*

7. CARTHAMUS (*Arborescens*) foliis ensiformibus sinuato-dentatis. Prod. Leyd. 136. *Stinking shrubby Cnicus of Spain.*

8. CARTHAMUS (*Corymbosus*) floribus umbellatus numerosus. *Black Chameleon.*

The first sort grows naturally in Egypt, and in some of the warm parts of Asia. It is at present cultivated in many parts of Europe, and also in the Levant, from whence great quantities of Safflower are annually imported to England, for dyeing and painting.

This is an annual plant, which rises with a stiff ligneous stalk two feet and a half, or three feet high, dividing upward into many branches, garnished with oval pointed leaves, sitting close to the branches, which are entire, and are slightly sawed on their edges, each tooth being terminated by a short spine. The flowers grow single at the extremity of each branch: the heads of flowers are large, inclosed in a scaly empalement; each scale is broad at the base, flat, and formed like a leaf of the plant, terminating in a sharp spine. The lower part of the empalement spreads open, but the scales above closely embrace the florets, which stand out near an inch above the empalement; these are of a fine Saffron colour, and this is the part which is gathered for the uses above-mentioned. If the season proves cold and moist when the plants are in flower, there will be no good seeds produced, so that there are few seasons wherein the seeds of this plant come to perfection in England

When this plant is propagated for use, the seeds should be sown in drills, drawn at two feet and a half distance from each other, in which the seeds should be scattered singly, for the plants must not be left nearer each other than a foot in the rows; but as some of the seeds will fail, so a greater quantity should be sown, as it will be easy to thin the plants at the time when the ground is hoed. If the seeds are good, the plants will appear in less than a month; and in three weeks or a month after it will be proper to hoe the ground, to destroy the weeds, and at the same time the plants should be thinned where they are too close; but at this time they should not be separated to their full distance, lest some of them should afterward fail; so that if they are now left six inches asunder, there will be room enough for the plants to grow till the next time of hoeing, when they must be thinned to the distance they are to remain for good: after this they should have a third hoeing, which, if carefully performed in dry weather, will destroy the weeds and make the ground clean, so that the plants will require no farther care till they come to flower; when, if the Safflower is intended for use, the florets should be cut off from the flowers as they come to perfection; but this must be performed when they are perfectly dry, and then they should be dried in a kiln, with a moderate fire, in the same manner as the true Saffron, which will prepare the commodity for use.

But those plants which are designed for seed, the flowers must not be gathered, for if the florets are cut off it will render the seeds abortive, though they may swell and grow to their usual size, as I have frequently experienced; yet when they are broken, there will be found nothing more than a shell without any kernel. And this frequently happens to be the case with the seeds in wet cold seasons, though in very wet years the germen will rot, and never come so forward as to form a shell.

The quantity of Safflower which is annually consumed in England is so great, as to make a very considerable article in trade, therefore might be very well worthy of the public attention. If this plant was introduced to Carolina, it might be there cultivated to as great advantage as in any part of the world, for there the seeds will constantly ripen; and, as this country is furnished with it from the Mediterranean, where there is great danger of our navigation being interrupted, so it should excite the inhabitants of our American colonies to make trial of as many of the vegetables as there is a probability of succeeding there, as can be procured, which are of real use in any of the manufactures of this country. The seeds of this plant I sent to Carolina in the year 1758, which have succeeded there; and more seeds may be easily procured from the American islands, and in one season may be multiplied in so great quantity as to furnish a whole province.

The good quality of this commodity is chiefly in the colour, which should be of a bright Saffron colour, and herein that which is cultivated in England often fails; for if there happens much rain during the time the plants are in flower, it will cause the florets to change to a dark or dirty yellow, which will also befal that which is gathered when there is any moisture remaining upon it; therefore great care must be taken not to gather it till the dew is quite dried off, nor should it be pressed together till it has been dried on the kiln. The manner of doing this being the same as for the true Saffron, I shall not mention it here.

This plant may be admitted to have a place in the borders of large gardens, where it will add to the variety, during the time of its continuance in flower, which is commonly two months or ten weeks; for if the seed be sown in the beginning of April, the first flowers will appear in the middle of July at farthest, and there will be a succession of flowers on the side branches till the end of September, or in mild warm weather till the middle of October.

When they are cultivated for this purpose, the seeds should be sown in the places where the plants are designed to remain, because they do not bear transplanting well; therefore three or four seeds should be sown in each patch, lest any of them should fail, and when the plants are grown so strong as to be out of danger, the most promising in each patch should be left, and the others pulled up, that they may not draw or injure those which are to stand.

The second sort grows naturally in the south of France, Spain, and Italy, where the women use the stalks of this plant for distaffs, from whence it had the title of Distaff Thistle.

The third sort was also discovered by Tournefort in the island of Crete. This differs from the former in having a smooth stalk; the leaves are very stiff, deeply indented, smooth, and are armed with very strong spines; the heads of flowers are oval, the florets white, and the plant grows near four feet high. This is an annual plant, which may be sown and treated in the same way as the former, and flowers about the same time.

The fourth sort hath a perennial root. This grows naturally in Spain, and was first brought to England from Tangier. It is propagated by parting of the roots. The best time for transplanting and parting them is about the beginning of March; this should have a dry soil and a warm situation, otherwise the plants are liable to be destroyed in severe winters.

The fifth sort grows naturally in the south of France, Spain, and Italy. This hath a perennial root; the stalk rises about six inches high, is channelled, hairy, and garnished with long narrow leaves, ending in several sharp spines. Each stalk is terminated by one large head of blue flowers, having a leafy empalement composed of very broad scales, each ending in a sharp spine.

This sort is difficult to propagate in England, for the roots do not put out offsets like the former, so is only to be raised from seeds, which do not come to perfection here, unless

unless the season proves warm and dry. The plants should have a dry soil and a warm situation.

The sixth sort rises with a single stalk about two feet high, which is of a purplish colour, hairy, and channelled, pretty closely garnished with broad spear-shaped leaves, which are sharply sawed on their edges, and covered with a short hairy down. The stalk is terminated by a single large head of blue flowers, having a scaly empalement. This sort may be propagated by parting of their roots, which should be performed in autumn when the leaves decay. It should have a light dry soil, in which it will endure the cold of our winters, and continue many years.

The seventh sort I received from Andulusia, where it grows naturally in great plenty. This rises with a shrubby perennial stalk to the height of eight or ten feet, which divides into many branches; these are garnished with pretty long sword-shaped leaves, which are indented, and armed with spines on their edges; they embrace the stalks with their base. The branches are terminated by large, scaly, prickly heads of yellow flowers, which come out in July, but are never succeeded by seeds in this country, so can only be propagated by side shoots, slipped from the branches in the spring, and planted in pots, and plunged into a moderate hot-bed till they have taken root; then they must be gradually hardened and removed into the open air, and when they have obtained strength they may be separated, and some of them planted in a warm dry border, where they will endure the cold of our ordinary winters; but in severe frost they are frequently destroyed, therefore a plant or two should be kept in pots, and sheltered in the winter to preserve the species.

CARUM. Lin. Gen. Plant. 327. Carui, or Caraway.

The CHARACTERS are,

It hath an umbellated flower, composed of several small umbels, which are formed as rays to the general umbel, neither of which have any involucrum; the flower hath five heart-shaped petals, and five hairy stamina. The germen is situated under the flower, which afterward becomes an oblong channelled fruit, dividing into two parts, each having an oblong furrowed seed.

The SPECIES are,

1. CARUM (*Carvi*) foliis pinnatifidis planis, umbellulis inæqualibus confertis. *Meadow Cumin, or Caraway of the shops.*

2. CARUM (*Hispanicum*) foliis capillaribus multifidis, umbellis laxis. *Caraway with capillary multifid leaves, and loose umbels.*

The first sort is the common Caraway, whose seeds are greatly used, not only in medicine, but also in the kitchen, &c. This grows naturally in some rich meadows in Lincolnshire and Yorkshire, and is sometimes found growing in the pastures near London. It is also cultivated for use in Essex, and some other counties.

It is a biennial plant, which rises from seeds one year, flowers the next, and perishes soon after the seeds are ripe. It hath a taper root like a Parsnep, but much smaller, which runs deep into the ground, and hath a strong aromatic taste, sending out many small fibres; from the root arises one or two smooth, solid, channelled stalks, about two feet high, garnished with winged leaves, having long naked foot-stalks. The stalks divide upward into several smaller branches, each of which is terminated by an umbel, composed of six or eight small separate umbels, sustaining single white flowers, with heart-shaped petals; the flowers of these small umbels are closely joined together. After the flowers are decayed, the germen becomes an oblong channelled fruit, composed of two oblong channelled seeds.

The best season for sowing the seeds of this plant is in autumn soon after they are ripe, when they will more certainly grow than those sown in the spring; and the plants which rise in the autumn generally flower the following season, so that a summer's growth is hereby saved.

The second sort grows naturally in Spain. This plant rises with a stronger stalk than the former, which seldom grows more than a foot and a half high, but is closely garnished with fine narrow leaves like those of Dill; the stalks divide upward into many branches, each being terminated by loose umbels of white flowers, which are succeeded by long furrowed seeds, having the same aromatic flavour as the common sort. This is a biennial plant, and may be treated in the same manner as the former.

CARYOPHYLLATA. See GEUM.

CARYOPHYLLUS. Lin. Gen. Plant. 594. The Clove-tree, or All-spice.

The CHARACTERS are,

It hath a double empalement; the flower is of one leaf, cut into four obtuse parts, upon which the germen is situated; the fruit hath another empalement, which is small, and slightly divided into four parts, which are permanent. The flower hath four blunt petals; it has many stamina. The germen is situated under the flower, which afterward becomes a soft berry with two cells, each containing a single kidney-shaped seed.

The SPECIES are,

1. CARYOPHYLLUS (*Aromaticus*) foliis ovato-lanceolatis oppositis, floribus terminalibus, staminibus corollâ longioribus. *The Clove-tree with oval spear-shaped leaves growing opposite, and flowers terminating the stalks, whose stamina are longer than the petals.*

2. CARYOPHYLLUS (*Pimento*) foliis lanceolatis oppositis, floribus racemosis terminalibus, & axillaribus. *The Pimento, or All-spice.*

3. CARYOPHYLLUS (*Fruticosus*) foliis lanceolatis oppositis, floribus geminatis alaribus. Brown. Hist. Jam. 248. *Clove-tree with spear-shaped leaves placed opposite, and flowers growing by pairs from the sides of the stalks.*

4. CARYOPHYLLUS (*Cotinifolia*) foliis ovatis obtusis oppositis, floribus sparsis alaribus. *Clove-tree with oval blunt leaves placed opposite, and flowers growing thinly from the sides of the branches.*

5. CARYOPHYLLUS (*Racemosus*) foliis oblongo-ovatis, emarginatis, rigidis, glabris, floribus racemosis terminalibus. *Clove-tree with oblong oval leaves, which are stiff, smooth, indented at the edges, and flowers growing in bunches terminating the stalk.*

The first sort grows naturally in the Moluccas, and the hottest parts of the world, where it rises to the height of a common Apple-tree; but the trunk generally divides at about four or five feet from the ground, into three or four large limbs, which grow erect, and are covered with a thin smooth bark, which adheres closely to the wood. These limbs divide into many small branches, which form a sort of conical figure; the leaves are like those of the Bay-tree, and are placed opposite on the branches. The flowers are produced in loose bunches at the end of the branches, which are small, white, and have a great number of stamina, which are much longer than the petals. The flowers are succeeded by oval berries, which are crowned with the empalement, divided into four parts, which spread flat on the top of the fruit, in which form they are brought to Europe, for it is the young fruit beaten from the trees before they are half grown, which are the Cloves used all over Europe.

I have not heard of any plants of this kind being in the gardens either in England or Holland, but I chose to mention it here to introduce the other.

The second sort grows naturally in Jamaica, but particularly on the north side of that island, where it is found in great plenty, and is a considerable branch of their trade.

The unripe fruit dried, being the All-ſpice ſo well known in Europe. It is now cultivated with care in many of the plantations, for the trees will thrive upon ſhallow rocky land, which is unfit for the Sugar-cane, ſo that a great advantage ariſes to the planters, from thoſe lands which would otherwiſe be of ſmall account to them.

This tree grows to the height of thirty feet or more, with a ſtrait trunk covered with a ſmooth brown bark, and divides upward into many branches which come out oppoſite; theſe are garniſhed with oblong leaves, reſembling thoſe of the Bay-tree both in form, colour, and texture, and are alſo placed oppoſite; when theſe are bruiſed or broken, they have a very fine aromatic odour, like that of the fruit. The branches grow very regular, ſo that the trees make a fine appearance, and as they retain their leaves through the year, ſo they are worthy of being propagated for ornament and ſhade about the habitations of the planters. The flowers are produced in large looſe bunches from the ſide of the branches toward their ends; the flowers are ſmall, and of an herbaceous colour, which are male and female upon diſtinct trees. The male flowers have very ſmall petals, and a great number of ſtamina in each, which are of the ſame colour with the petals, and are terminated by oval bifid ſummits; the female flowers have no ſtamina, but an oval germen ſituated below the flower, ſupporting a ſlender ſtyle, with a blunt ſtigma at the top. The germen afterward becomes a globular pulpy berry, including two kidney-ſhaped ſeeds.

When the fruit of theſe trees are deſigned for uſe, they are gathered, or beaten down from the trees, a little before they arrive to their full ſize, and are ſeparated from leaves, ſtalks, or any rubbiſh which may have accidentally mixed with them: then the fruit is expoſed every day to the ſun, ſpread on cloths for ten or twelve days to dry, but removed under cover every evening to ſcreen it from the dews; when the fruit is perfectly dry, it is packed up for exportation. If the fruit is permitted to grow to maturity, the pulp which ſurrounds the ſeeds is ſo full of moiſture and glutinous, as to ſtick to the fingers of thoſe who bruiſe them, therefore are unfit for thoſe uſes to which the dried fruit are applied.

This tree is propagated by ſeeds, which in the natural place of its growth is conveyed and ſown by birds to a great diſtance; and, it is very probable, the ſeeds paſſing through them, are rendered fitter for vegetation than thoſe which are immediately gathered from the tree; for I have received great quantities of the berries which were perfectly ripe and freſh, great part of which I ſowed in different ways, and communicated ſome of them to ſeveral other curious perſons who did the ſame, but none of them have yet ſucceeded.

The plants cannot be preſerved in England, unleſs they are placed in a ſtove during the winter ſeaſon, but they will thrive in a moderate degree of warmth; they ſhould be planted in a ſoft loamy ſoil, and in winter muſt have but little water; in the ſummer they ſhould have a large ſhare of air, and in July, if the ſeaſon proves warm, they may be placed in the open air in a warm ſheltered ſituation, but upon the approach of cold nights they muſt be removed into the ſtove again. The expoſing of theſe plants to the open air for one month only, will be of great ſervice to clean their leaves from inſects or filth, which they are ſubject to contract by remaining long in the ſtove; but if the ſeaſon ſhould prove very wet or cold, it will not be ſafe to truſt theſe plants long abroad; therefore their leaves ſhould be now and then waſhed with a ſponge to clean them, which will not only render them more ſightly, but alſo promote their growth. This plant, being an evergreen, makes a fine appearance in the ſtove at all ſeaſons of the year; and their leaves having ſuch an agreeable fragrancy when rubbed, render them as worthy of a place in the ſtove, as any other exotic plant which is preſerved for ornament.

The third ſort grows naturally in Jamaica. This riſes with a divided trunk to the height of eighteen or twenty feet, ſending out many branches which are placed oppoſite, and are covered with a gray bark: the leaves come out by pairs, which are ſhorter and rounder at their points than thoſe of the laſt ſpecies; they are alſo ſmoother, and of a firmer texture. The flowers come out from the ſide of the branches between the leaves, upon ſlender foot-ſtalks, about an inch in length, two generally ariſing from the ſame point: theſe are ſucceeded by round berries, of a brighter colour than thoſe of the former, having the empalement on their crowns. The leaves and fruit of this ſort have no aromatic flavour, ſo are not of uſe, but the characters of the flower and fruit are the ſame as in the other ſort.

The fourth ſort was ſent me from Carthagena in New Spain. This riſes with many irregular ſtems about twelve or fourteen feet high, which are covered with an Aſh-coloured bark, and divide into many branches upward: theſe are garniſhed with ſtiff oval leaves placed oppoſite. The flowers are produced from the ſide of the branches, ſometimes four, five, or ſix foot-ſtalks ariſe from the ſame point; at other times they come out ſingle, or perhaps by pairs: theſe are white, and of the ſame ſhape with thoſe of the ſecond ſort, and are ſucceeded by berries which are rounder, and for the moſt part contain but one kidney-ſhaped ſeed.

This ſort agrees with the ſecond in its general characters, but not in the virtues, for it hath none of the aromatic flavour with which that abounds; but as it retains its leaves through the year, may merit a place in the ſtove, better than many other plants which are preſerved by the curious. This is propagated by ſeeds in the ſame way as the ſecond ſort, and the plants muſt be treated in the ſame manner as thoſe.

The fifth ſort was ſent me from the iſland of Barbuda, where it riſes to the height of twenty feet: the trunk and branches are covered with a ſmooth brown bark. The branches come out oppoſite; theſe grow erect, and are garniſhed with very ſmooth lucid leaves, which are placed by pairs, and have very ſhort foot-ſtalks. The leaves vary much in their form, ſome of them are oval, others are oblong, and ſome are indented ſo deeply at their ends as to be almoſt heart-ſhaped. Their conſiſtence is much thicker than thoſe of the common Laurel, and their colour is a ſplendent green, with one deep midrib running through their middle, and many ſmall veins going from thence tranſverſly to their border. The flowers are produced in ſmall looſe bunches at the extremity of the branches, which have ſeveral narrow leaves intermixed with the bunches. The flowers are ſucceeded by berries of the ſame ſhape with thoſe of the ſecond ſort, but are larger.

This tree is propagated by ſeeds as the other ſpecies, and deſerves a place in the ſtove for the beauty of its evergreen leaves, which being of a thick conſiſtence, and of a ſhining green colour, make a fine appearance at all ſeaſons of the year; but this hath no aromatic flavour to recommend it, as hath the ſecond ſort, for which reaſon it is ſeldom noticed. I take this to be the Bay-tree, mentioned by Hughes in the Hiſtory of Barbadoes, which he deſcribes to have no flavour; for I have ſeen plants of this ſort which were brought from Barbadoes, ſo that I ſuppoſe it grows naturally there.

CASIA. See Osyris.

CASSIA. Lin. Gen. Pl. 461. Wild Senna.

The

The CHARACTERS are,

The flower hath five roundish concave petals, which spread open; it hath ten declining stamina, three of the lower are long, the three upper are shorter; the summits of the three lower are large, arched, beaked, and separated at their points. In the center is situated a long taper germen, which afterwards becomes a long pod, divided by transverse partitions, each containing one or two roundish seeds, fastened to the margin of the upper valve.

The SPECIES are,

1. CASSIA (*Occidentalis*) foliolis quinquejugatis, ovato-lanceolatis, margine scabris, exterioribus majoribus, glandulâ baseos petiolorum. Lin. Sp. Plant. 377. *Cassia with leaves composed of five pair of oval spear-shaped lobes with rough borders, the upper lobes being the largest, and a small gland at the base of the foot-stalk.*

2. CASSIA (*Frutescens*) foliolis quinquejugatis ovatis glabris, exterioribus longioribus, caule fruticoso. *Cassia with leaves composed of five pair of smooth oval lobes, the upper being the longest, and a shrubby stalk.*

3. CASSIA (*Alata*) foliolis octojugatis, ovali-oblongis, interioribus minoribus, petiolis eglandulosis stipulis patulis. Hort. Cliff. 158. *Cassia with eight pair of oblong, oval, little leaves, the inner being the least, foot-stalks without glands, and a spreading stipula.*

4. CASSIA (*Villosa*) foliolis trijugatis, oblongo-ovatis æqualibus villosis, siliquis articulatis, caule erecto arboreo. *Cassia with three pair of oblong, oval, hairy leaves, which are equal, jointed pods, and an upright woody stem.*

5. CASSIA (*Uniflora*) foliolis trijugatis, ovato-acuminatis, villosis, floribus solitariis axillaribus, siliquis erectis. *Cassia with three pair of lobes in each leaf, which are oval, pointed, and hairy, and single flowers proceeding from the sides of the stalks, with upright pods*

6. CASSIA (*Marylandica*) foliolis octojugatis ovato-oblongis, æqualibus, glandula baseos petiolorum. *Cassia with small leaves, composed of eight pair of lobes, which are oblong, smooth, and equal, and flowers proceeding from the sides of the stalks.*

7. CASSIA (*Bicapsularis*) foliolis octojugatis obovatis glabris, interioribus rotundioribus minoribus, glandulâ interjectâ globosâ. Hort. Cliff. 159. *Cassia with three pair of oval smooth leaves, the inner one being rounder and smaller, and a globular gland placed between the leaves.*

8. CASSIA (*Fistula*) foliolis quinquejugatis, ovato-lanceolatis glabris, petiolis eglandulosis Flor. Zeyl. 149. *Cassia with five pair of leaves which are oval, spear-shaped, smooth, and foot-stalks having no glands; or the purging Cassia of Alexandria.*

9. CASSIA (*Bahamensis*) foliolis sexjugatis, lanceolatis, glabris, interioribus minoribus, floribus terminatricibus. *Cassia with six pair of leaves, which are smooth and spear-shaped, the inner ones being smaller, and flowers terminating the stalk.*

10. CASSIA (*Fruticosa*) foliolis bijugatis, ovato lanceolatis, glabris, floribus terminalibus, siliquis longis teretibus, caule fruticoso. *Cassia with two pair of leaves which are oval, spear-shaped, and smooth, flowers terminating the stalks, long taper pods, and a shrubby stalk.*

11. CASSIA (*Javanica*) foliolis duodecemjugatis, oblongis, obtusis, glabris, glandulâ nullâ. Lin. Sp. Plant. 379. *Cassia with twelve pair of leaves, which are oblong, blunt, smooth, and have no glands; commonly called Horse Cassia.*

12. CASSIA (*Ligustrina*) foliolis septemjugatis, oblongo-ovatis, obtusis, floribus spicatis axillaribus, siliquis recurvis. *Cassia with seven pair of leaves which are oblong, oval, and blunt, spikes of flowers proceeding from the sides of the stalks, and recurved pods.*

13. CASSIA (*Emarginata*) foliolis trijugatis, obtusis, emarginatis, caulibus pilosis, floribus solitariis axillaribus petiolis longioribus. *Cassia with three pair of obtuse leaves indented at the top, hairy stalks, flowers growing singly from the sides of the stalks, with a long foot-stalk.*

14. CASSIA (*Biflora*) foliolis quadrijugatis oblongo-ovatis, caulibus procumbentibus, floribus axillaribus pedunculis bifloris. *Cassia with four pair of oval oblong leaves, trailing stalks, and flowers proceeding from the sides of the stalks, two growing upon each foot-stalk.*

15. CASSIA (*Arborescens*) foliolis bijugatis oblongo-ovatis, subtus villosis, floribus corymbosis, caule erecto arboreo. *Cassia with two pair of oblong oval leaves, hairy on their under side, flowers growing in round bunches, and an erect tree-like stem.*

16. CASSIA (*Flexuosa*) foliolis multijugatis, glandulâ petioli pedicellatâ, stipulis ensiformibus. Hort. Ups. 101. *Cassia with many pair of leaves, and the gland on the foot-stalk resembling an insect, and sword-shaped stipulæ.*

The first sort grows naturally in most of the islands in the West-Indies, where it is called Stinking Weed, from its unsavoury odour. This rises with a channelled stalk three or four feet high, dividing into several branches, garnished with winged leaves placed alternately; each of these is composed of five pair of lobes, which are oval, spear-shaped, sitting close to the midrib, and have rough edges. The flowers come out from the side of the stalks, two growing upon each foot-stalk, but the branches are terminated by loose spikes of flowers, which are composed of five concave yellow petals, with ten declining stamina. It hath a flat pod, having a border on each side, and is indented between each seed.

This is a biennial plant, which propagates by seeds in plenty in the countries where it grows naturally; but in England the seeds must be sown on a hot-bed in the spring, and when the plants are fit to remove, they should be each planted in a separate pot filled with light earth, and plunged into a moderate hot-bed to bring them forward; and toward the end of June, some of them may be planted into a warm border, where, if the autumn proves favourable, they will flower very well; but these will not perfect their seeds, therefore a plant or two should be put into pots that they may be removed into the stove in autumn to ripen seeds.

The second sort grows in Jamaica. This rises with a shrubby stalk five or six feet high, sending out many branches toward the top, with winged leaves composed of five pair of small oval leaves, the upper ones being longest. The flowers come out from the side of the stalks, and also terminate the branches in loose spikes; they are yellow, and shaped like those of the former, but are smaller; the pods are long, taper, and contain two rows of seed.

This plant may be preserved three or four years in the stove, and will annually flower and perfect the seeds. It is propagated by seeds, which should be sown on a hot-bed in the spring; and the plants must be treated in the same manner as the former sort, with only this difference, that these when they are too tall to remain longer under the frames on the hot-bed, must be removed into the stove.

The third sort hath an herbaceous stalk, which rises five or six feet high, with long winged leaves composed of eight or ten pair of large oval lobes, rounded at the end, where they are slightly indented. The flowers are produced in loose spikes at the top of the stalk, which are large, yellow, and of the same shape with those of the other species; the pods are long, taper, and have four borders or wings running longitudinally.

This sort seldom continues more than two years; it must be raised from seeds as the former sorts, and placed in the tan-bed in the stove, being very tender, and should have but little water in winter.

The

The fourth sort grows naturally at Campeachy. This rises with a woody stem to the height of fourteen or sixteen feet, sending out many lateral branches, garnished with winged leaves, composed of three pair of oblong, oval, hairy lobes, which are of equal size; the flowers come out in loose bunches at the end of the branches, which are of a pale straw colour and small, but shaped like the others.

This may be propagated by seeds, which must be sown upon a hot-bed, and the plants afterward treated as the former sorts.

The fifth sort is a low herbaceous plant, seldom rising a foot high; the stalk is single, and garnished with winged leaves, composed of three pair of oval pointed lobes, which are hairy; the flowers come out single from the side of the stalks, they are of a pale yellow and small; these are succeeded by narrow taper pods two inches long, which grow upright. This plant is annual; the seeds must be sown on a hot-bed, and the plants treated as the first sort.

The sixth sort grows naturally in Maryland. It hath a perennial root, composed of a great number of black fibres, and sends out several upright stalks in the spring, which rise six feet high, garnished with winged leaves, composed of nine pair of oblong smooth lobes, which are equal; toward the upper part of the stalks the flowers come out from the wings of the leaves, two or three together. The stalks decay in autumn, and rise again in the spring. The roots of this sort continue many years, and will live abroad in a warm border and a dry soil.

The seventh sort is an annual plant, which rises a foot and a half high, with an erect herbaceous stalk, which is garnished with winged leaves composed of three pair of oval lobes; the flowers come out singly from the wings of the leaves, which are small, yellow, and of the same shape with those of the other species.

This is propagated by seeds, which must be sown on a hot-bed in the spring, and the plants afterward treated in the same manner as hath been directed for the first.

The eighth sort is the tree which produces the purging Cassia, which is used in medicine. It grows naturally in Alexandria, and in both Indies, where it rises to the height of forty or fifty feet, with a larger trunk, dividing into many branches garnished with winged leaves, composed of five pair of spear-shaped lobes, which are smooth; the flowers are produced in long spikes at the end of the branches, each standing upon a pretty long foot-stalk: they are composed of five large concave petals of a deep yellow colour, and are succeeded by round pods which are from one to two feet long, with a dark brown woody shell, having a longitudinal seam on one side, and divided into many cells by transverse partitions, each containing one or two oval, smooth, compressed seeds, lodged in a sweetish black pulp, which is the part used in medicine.

This tree is propagated by seeds, which may be easily procured from the druggists, who import the pods for use, which must be sown on a hot-bed in the spring, and when the plants come up they must be treated in the same manner as the other sorts during the first summer, and in autumn they must be removed into a stove; during the winter they should have very little water, for as these trees grow naturally in dry sandy land, so moisture is a great enemy to them, but especially during that season. The plants may be exposed to the open air, in a sheltered situation in the summer, in the warmest time of the year.

The ninth sort grows naturally in the Bahama Islands. This is an annual plant, which rises with an upright stalk three feet and a half high, garnished with winged leaves composed of six pair of lobes, which are smooth, narrow, and spear-shaped, standing at wide distances: the flowers are collected into loose bunches at the top of the stalks, which are of a pale yellow, and are succeeded by long compressed pods. This must be treated as the first sort.

The tenth sort grows at La Vera Cruz in New Spain. This rises upward of twenty feet high, with several trunks covered with brown bark, which divide into many branches upward, garnished with winged leaves, composed of two pair of lobes, which in the lower leaves are oval, and those of the upper are five inches long, and two and a half broad in the middle, smooth, and of a light green. The flowers are produced in loose spikes at the extremity of the branches, which are large, and of a gold colour.

This sort is propagated by seeds, which must be sown upon a hot-bed, and the plants afterward treated in the same manner as the eighth sort.

The eleventh sort grows in great plenty in most of the islands in the West-Indies. This rises to a great magnitude, with a large trunk, dividing into many branches garnished with very long winged leaves, composed of twelve or fourteen pair of oblong blunt lobes, which are smooth, of a light green, and placed near together. The flowers come out in loose spikes at the end of the branches, which are of a pale Carnation colour, shaped like those of the other species. This is called Horse Cassia, because it is generally given to horses, and seldom taken by any persons on account of its griping quality.

It is propagated by seeds, which should be sown, and the plants afterward treated in the same manner as the eighth sort.

The twelfth sort grows at the Havannah. This hath an herbaceous stalk, which divides into many branches, rising about three feet high, garnished with winged leaves, composed of seven pair of oblong oval lobes, which are rounded at the end. The flowers come out from the side of the branches upon very long foot-stalks, and are disposed in loose spikes; these are of a pale yellow.

This is a biennial plant, which, if brought forward early in the spring, will sometimes perfect seeds the same year; but if they should fail, the plants may be kept through the winter in a stove, and good seeds may be obtained the following season.

The thirteenth sort rises with several weak shrubby stalks about two feet high, closely garnished with winged leaves, composed of three pair of lobes, very narrow at their base, enlarging to the top, where they are rounded with a little indenture at the point. The flowers come out single from the side of the branches, standing upon very long foot-stalks; they are of a bright yellow, shaped like those of the other species, and are succeeded by narrow flat pods an inch and a half long. It is propagated by seeds, which must be sown on a hot-bed, and managed as the other tender sorts; it will continue two or three years, if placed in a warm stove.

The fourteenth sort sends out from the root two or three slender stalks, which trail on the ground, garnished with winged leaves, having four pair of small roundish lobes of a pale green; at the insertion of the foot-stalks arise those of the flower, which is jointed, dividing into two shorter at the top, sustaining two small yellow flowers. This is an annual plant, whose seeds must be sown early in the spring on a hot-bed, and treated like the other kinds.

The fifteenth sort grows at La Vera Cruz in New Spain. This rises with a strong upright trunk to the height of twenty-five or thirty feet, dividing into many branches, covered with an Ash-coloured bark, garnished with winged leaves having long foot-stalks, each being composed of two pair of oblong oval lobes, which are smooth. The flowers are produced sometimes from the side of the stalks, where they are few and scattering, but the ends of the branches have large round bunches of flowers, which branch out from one center; they are of a deep yellow, inclining to an Orange colour.

The sixteenth sort grows common in all the islands of the West-Indies. It rises with a slender stalk about two feet high, sending out a few side branches upward, garnished with winged leaves, composed of many pairs of narrow pinnæ like those of the Sensitive Plant. The flowers come out upon short foot-stalks from the side of the branches, each foot-stalk sustaining two or three yellow flowers, of the same form with the other species of this genus; these are succeeded by short flat pods, containing three or four flat seeds in each.

This is an annual plant, and requires the same treatment as the first; but unless the plants are placed in a glass-case, where they may have room to grow, and be screened from the cold, they will not perfect their seeds in England.

CASSIDA. Scull-cap. See SCUTELLARIA.

CASSINE. Lin. Gen. Plant. 333. The Cassioberry-bush, or South-Sea Thea.

The CHARACTERS are,

The flower hath but one petal, which is cut into five obtuse segments; it hath five stamina, which spread from each other, and a conical germen, which afterward becomes an umbilicated berry with three cells, each containing a single seed.

The SPECIES are,

1. CASSINE (*Corymbosa*) foliis ovato-lanceolatis, serratis oppositis, floribus corymbosis axillaribus. Fig. Pl. plat. 83. fig. 1. *Cassine with oval spear-shaped leaves placed opposite, and flowers growing in round bunches from the side of the branches; or the Cassioberry-bush.*

2. CASSINE (*Paragua*) foliis lanceolatis alternis sempervirentibus, floribus axillaribus. Fig. Pl. plat. 83. fig. 2. *Cassine with evergreen spear-shaped leaves placed alternately, and flowers proceeding from the sides of the branches; Yapon, or South-Sea Thea.*

3. CASSINE (*Oppositifolia*) foliis ovatis acutis glabris, floribus axillaribus sparsis. *Cassine with oval acute-pointed leaves, and flowers coming out from the wings of the stalk; called by the gardeners Hysson Tea.*

The first sort rises with two or three stems, which send out many side branches, and become bushy; these seldom rise more than eight or nine feet high in England. The branches are garnished with oval spear-shaped leaves sawed on their edges, which grow opposite. At the end of the branches the flowers come out in roundish bunches; these are white, and divided into five parts almost to the bottom; in their center is placed the germen, attended by five stamina, which spread open near as much as the segments of the petal. After the flower is past, the germen swells to a round berry, having three cells, each containing a single seed.

This sort is now become pretty common in the nurseries near London, where it is propagated by laying down the branches, which afford shoots in plenty for that purpose from the root and lower part of the stem, so is easily increased. There are numbers of these shrubs which produce fowers in England every year, but none of them ripen their seeds.

The leaves of these plants are extremely bitter, so that if a single one is chewed, the bitterness cannot be gotten rid of in a long time.

It loves a light soil, not too dry, and should have a warm situation; for in exposed places the young shoots are frequently killed in the winter, whereby the shrubs are rendered unsightly; but where they are near the shelter of trees or walls, they are very rarely hurt.

The second sort grows naturally in Carolina, and also in some parts of Virginia, but chiefly near the sea; this, in the natural places of its growth, rises to the height of ten or twelve feet, sending out branches from the ground upward, garnished with spear-shaped leaves placed alternately, which continue green through the year. The flowers are produced in close whorls round the branches at the foot-stalks of the leaves; they are white, and of the same shape with the former.

This plant was many years preserved in several curious gardens near London, till the severe winter in 1739, when most of them were destroyed, so that there was scarce any left; but of late years there have been many of the young plants raised from seeds, which came from Carolina. If this plant can be brought to thrive well in England, and to endure the winter in the open air, it will be a fine plant to make a variety in plantations of evergreen trees. The leaves of this sort are not so bitter as those of the first, especially when green.

The inhabitants of North Carolina and Virginia, where this shrub grows in plenty, give it the title of Yapon, which I suppose to be the Indian name. The leaves are about the size and shape of those of the small-leaved Alaternus, but are somewhat shorter, and a little broader at their base; they are a little notched about their edges, and are of a thick substance and deep green colour; the flowers of this sort are produced at the joints near the foot-stalk of the leaves, but the Cassioberry-bush produces its flowers in umbels at the extremity of the shoots.

These trees are propagated by sowing their seeds (which are obtained from Carolina, where they grow in great plenty near the sea-coasts;) they should be sown in pots, for the seeds do frequently remain in the ground until the second year; therefore the pots should be placed in a shady situation, where they may remain till October when they must be removed into shelter during the winter season; and in March following put them upon a fresh hot-bed, which will forward the seeds in their vegetation.

When the plants are come up, they should by degrees be exposed to the open air, in order to inure them to our climate, placing them where they may be sheltered from cold winds; they should be sheltered during the two or three first winters under a frame, after which they may be planted abroad in a warm situation, where they will endure the cold of our common winters; but very severe frosts will kill them if they have not some protection.

In South Carolina the plant is called Cassena, or South-Sea Thea: the inhabitants of that country do not make so great use of this Tea, as those of Virginia and North Carolina; in the last of which, the white people have it in as great esteem as the Indians, and make as constant use of it.

The third sort has not been long in England. This grows naturally in South Carolina, in situations near the sea, where the frosts are not so severe as the interior parts of that country, so the plants are not so hardy as those of the other two sorts. This may be raised from seeds, when they can be procured from the country where the plants naturally grow, and if treated in the same manner as the former, will succeed very well: it may also be propagated by layers, which should be laid down in the autumn, slighting the shoots, as is commonly practised for Carnations; these in one year will have put out roots, but they should not be cut from the old plants till the spring following, when they should be each planted in a separate pot, shading them till they have taken new root; after which they should be placed with other hardy sorts of exotic plants in summer, but in the autumn must be placed in shelter, for they will not live abroad in winter.

This plant retains its leaves through the year, so makes a good appearance in winter.

CASTANEA. Tourn. Inst. R. H. 584. The Chestnut-tree.

The

The CHARACTERS are,

It hath male and female flowers on the same tree, sometimes at separate distances, and at other times near each other. The male flowers form a sort of katkin; they have no petals, but include about ten or twelve bristly stamina. The female flowers are of one leaf, divided into four parts, having no petals, but a germen fixed to the empalement, which becomes a roundish fruit armed with soft spines, including one or more nuts.

The SPECIES are,

1. CASTANEA (*Sativa*) foliis lanceolatis acuminato-serratis, subtus nudis. *The manured Chestnut.*

2. CASTANEA (*Pumila*) foliis lanceolato-ovatis acutè serratis, subtus tomentosis, amentis filiformibus nodosis. *Chestnut with oval spear-shaped leaves sharply sawed, which are woolly on their under side, and a slender knotted katkin.*

3. CASTANEA (*Sloanea*) foliis oblongo-ovatis, serratis, fructu rotundo maximo echinato. *Chestnut with oblong, oval, sawed leaves, and a very large, round, prickly fruit. This is the Sloanea of Plumier.*

The Chestnut is a tree which deserves our care, as much as any of the trees which are propagated in this country, either for its use or beauty, being one of the best sort of timber, and affording a goodly shade. It will grow to a very great size, and spread its branches finely on every side where it has room. The leaves are large, of a lucid green, and continue late in the autumn; nor are they so liable to be eaten by insects, as are those of the Oak, which of late years having frequently happened to the latter, which has rendered them very unsightly great part of summer. There is no better food for deer and many other animals than their nuts, which most of them prefer to acorns; but yet there should not be so many of these trees planted too near the habitation, because, when they are in flower, they emit a very disagreeable odour, which is very offensive to most people.

There are some varieties of this tree which have accidentally arisen from seeds, that have been supposed distinct species; but the differences are only in the size of their fruit and leaves, which have been altered and improved by culture, so that the wild and unmanured Chestnut are undoubtedly the same; for I have frequently found, that the nuts taken from the same tree, and cultivated in the same soil with equal care, have produced trees with very small fruit; and among them have been others, whose fruit have been as large as those of the parent tree, therefore they can be only esteemed as varieties. But in many countries where the trees are cultivated for their fruit, the people graft the largest and fairest fruit, upon stocks of Chestnut raised from the nut; and these grafted trees are by the French called Maronnier, but these grafted trees are unfit for timber.

There is also a Chestnut with variegated leaves, which is propagated in the nurseries by way of curiosity: this is maintained by budding, and inarching it upon common Chestnut stocks in the same manner as other fruit-trees; but these variegated trees and plants are not so much regarded at present, as the were some years past.

The third sort grows in South Carolina, from whence some of the fruit with their outward covers, were sent to his Grace the Duke of Bedford a few years past: these were as large and round as a tennis-ball, and armed all over with strong spines like a hedge-hog; these capsulæ were divided regularly in four cells, each containing one small Chestnut. At that time I compared these with Father Plumier's description and figure, which he exhibited under the title of Sloanea, and found them to agree exactly; and upon looking in the box in which they were sent, I found some of the leaves of the tree which also tallied with his description, and confirmed my former opinion.

The first of these trees was formerly in greater plenty amongst us than at present, as may be proved by the old buildings in London, which were for the most part of this timber: and in a description of London, written by Fitz-Stephens in Henry the second's time, he speaks of a very noble forest, which grew on the north part of it: Proximè, (says he) patet foresta ingens, saltus numerosi ferarum, latebræ cervorum, damarum, aprorum & taurorum sylvestrium, &c. And there are some remains of old decayed Chestnuts in Enfield-chace, not far distant from London; which plainly proves, that this tree is not so great a stranger to our climate as many people believe, and may be cultivated in England, to afford an equal profit with any of the larger timber-trees, since the wood of this tree is equal in value to the best Oak, and for many purposes far exceeding it; as particularly for making vessels for all kinds of liquor, it having a property (when once thoroughly seasoned) of maintaining its bulk constantly, and is not subject to shrink or swell as other timber is too apt to do; and I am certainly informed that all the large casks, tuns, &c. for their wines in Italy, are made of this timber: and it is for that, and many more purposes, in greater esteem among the Italians, than any other timber whatever. It is also very valuable for pipes to convey water under ground, as enduring longer than the Elm or any other wood. In Italy it is planted for coppice wood, and is very much cultivated in stools, to make stakes for their Vines, which will endure seven years, which is longer than any other stakes will do by near half the time. The usefulness of the timber, together with the beauty of the tree, renders it as well worth propagating as any tree whatever, especially in large plantations in parks.

These trees are propagated by planting the nuts in February, in beds of fresh undunged earth. The best nuts for sowing, are such as are brought from Portugal and Spain, which are commonly sold in winter for eating, provided they are not kiln-dried, which is generally the case of many of those brought from abroad, which is done to prevent their sprouting in their passage; therefore, if they cannot be procured fresh from the tree, it will be much better to use those of the growth of England, which are full as good to sow for timber or beauty, as any of the foreign nuts, though their fruit is much smaller: the nuts should be preserved, until the season for sowing, in sand, where mice or other vermin cannot come to them, otherwise they will soon destroy them: before you set them, it will be proper to put them into water to try their goodness, which is known by their ponderosity; such of them as swim upon the surface of the water should be rejected, as good for nothing; but such as sink to the bottom, you may be sure are good.

In setting these seeds or nuts, the best way is to make a drill with a hoe (as is commonly practised in setting Kidney Beans) about four inches deep, in which you should place the nuts, at about four inches distance, with their eye uppermost; then draw the earth over them with a rake, and make a second drill at about a foot distance from the former, proceeding as before, allowing three or four rows in a bed, with an alley between, three feet broad, for a conveniency of clearing the beds, &c. When you have finished your plantation, you must be careful that it is not destroyed by mice or other vermin; which is very often the case, if they are not prevented by traps, or other means.

In April these nuts will appear above ground; you must therefore observe to keep them clear from weeds, especially while young: in these beds they may remain for two years, when you should remove them into a nursery, at a wider distance. The best time for transplanting these trees, is either

either in October or the latter end of February, but October is the best season: the distance these should have in the nursery, is three feet row from row, and one foot in the rows. If these trees have a downright tap-root, it should be cut off, especially if they are intended to be removed again; this will occasion their putting out lateral roots, and render them less subject to miscarry when they are removed for good.

The time generally allowed them in the nursery, is three or four years, according to their growth, but the younger they are transplanted the better they will succeed; young trees of this sort are very apt to have crooked stems, but when they are transplanted out and have room to grow, as they increase in bulk they will grow more upright, and their stems will become strait, as I have frequently observed, where there have been great plantations.

After they have remained three or four years in the nursery, they will be fit for transplanting where they are to remain; for the younger they are planted out for good, the better they will succeed. But if they are propagated for timber, it is by much the better method to sow them in furrows (as is practised for Oaks, &c.) and let them remain unremoved; for these trees are apt to have a downright tap-root, which, being hurt by transplanting, is often a check to their upright growth, and causes them to shoot out into lateral branches, as is the case with the Oak, Walnut, &c.

If you design a large plantation of these trees for timber, after having two or three times ploughed the ground the better to destroy the roots of weeds, you should make your furrows about six feet distance from each other, in which you should lay the nuts about ten inches apart, covering them with earth about two inches deep; and when they come up, you must carefully clear them from weeds: the distance allowed between each row, is for the use of the horse-hoeing plough, which will dispatch a great deal of this work in a short time; but it should be performed with great care, so as not to injure the young plants; therefore the middle of the spaces only should be cleaned with this instrument, and a hand-hoe must be used to clean between the plants in the rows, and also on each side, where it will be unsafe for the plough to be drawn; and in hand-hoeing there must be great care taken not to cut the tender rind of the plant. But for the first two years after sowing, it will be adviseable to dig the ground each winter, because the plants will be too small to admit the hoeing plough, and in summer to hand-hoe the ground. When these have remained three or four years (if the nuts succeed well,) you will have many of these trees to remove, which should be done at the season before directed, leaving the trees about three feet distance in the rows; at which distance they may remain for three or four years more, when you should remove every other tree to make room for the remaining, which will reduce the whole plantation to six feet square; which will be distance enough for them to remain in, until they are large enough to cut for poles, when you may cut down every other of these trees (making choice of the least promising) within a foot of the ground, in order to make stools for poles, which, in eight or ten years time, will be strong enough to lop for hoops, hop-poles, &c. for which purposes they are preferable to most other trees; so that every tenth year here will be a fresh crop, which will pay the rent of the ground, and all other incumbent charges, and at the same time, a full crop of growing timber left upon the ground.

The Chinquapin, or Dwarf Virginian Chestnut, is at present very rare in England; it is very common in the woods of America, where it seldom grows above twelve or fourteen feet high, and produces great plenty of nuts, which are for the most part single, in each outer coat or capsule. This tree is very hardy, and will resist the severest of our winters in the open ground; but is very apt to decay in summer, especially if it is planted in very dry ground. The nuts of these trees, if brought from America, should be put up in sand as soon as they are ripe, and sent to England immediately, otherwise they lose their growing quality, which is the reason this tree is at present so scarce with us, for not one seed in five hundred sent over ever grew. Indeed most of the nuts which have been brought over have been kiln-dried, to preserve them from sprouting, which infallibly destroys the germen: when the nuts arrive, they should pe put into the ground as soon as possible; for if they are long kept above ground, they lose their vegetative quality. This sort of Chestnut delights in a moist soil, but if the wet continues long upon the ground in the winter, it is apt to kill the trees.

The third sort, called Sloanea, is not at present in this country, nor do I believe it will live through the winter in the open air, especially while the plants are young; therefore if any nuts of this sort should be brought to England, and the plants raised from them, it will be proper to screen them for three or four winters, till the plants have obtained strength, when they should be planted in the open air in the spring, but must have a sheltered situation.

CASTANEA EQUINA. See Esculus.

CASTOREA. See Durantia.

CATANANCHE. Lin. Gen. Plant. 824. Candia Lions Foot.

The Characters are,

The flower is composed of many hermaphrodite florets, included in one common scaly empalement, which is permanent and elegant. The florets are of one leaf, tongue-shaped, indented in five parts; they have each five short hairy stamina. The germen is situated below the flower, which afterward becomes a single oval seed, which is compressed, and crowned with bristles, inclosed in the empalement.

The Species are,

1. Catananche (*Cærulea*) squamis calycinis inferioribus ovatis. Hort. Cliff. 390. *Catanance whose under scales of the empalement are oval.*

2. Catananche (*Lutea*) squamis calycinis inferioribus lanceolatis. Hort. Cliff. 390. *Catanance whose under scales of the empalement are spear-shaped.*

The first sort sends out many long, narrow, hairy leaves, which are jagged on their edges like those of the Buckshorn Plantain, but broader; the jags are deeper, and at greater distances; these lie flat on the ground, turning their points upwards. Between the leaves come out the flower-stalks, which are in number proportionable to the size of the plant; for from an old thriving root there is frequently eight or ten, and young plants do not send out more than two or three. These stalks rise near two feet high, dividing into many small branches upward, garnished with leaves like those below, but are smaller, and have few or no jags on their edges: each of these smaller branches (or foot-stalks) are terminated with single heads of flowers, having a dry silvery, scaly empalement, in which are included three or four florets, whose petals are broad, flat, and indented at their ends; these are of a fine blue colour, having a dark spot at bottom and in each five stamina.

The second sort hath broader leaves than the first, and less jagged on their edges; from each root rise one or two stalks, which grow a foot and a half high, sending out two or three slender foot-stalks, each sustaining a single head of yellow flowers, inclosed in a dry scaly empalement, of a darker colour than those of the first.

The first of these plants is perennial, and is propagated by seeds or slips; the seeds may be sown in a bed of common earth in the ſpring; in the autumn following the plants may be tranſplanted where they are to remain. It is a pretty ornament to a garden, and is eaſily kept within bounds. Theſe plants ſhould remain unremoved, which will cauſe them to flower better, and they will produce more ſeeds. The ſeeds ripen in Auguſt.

The other ſort is an annual plant, and is propagated only by ſeeds, which ripen very well in this country. The time for ſowing them is early in March, in beds or borders of light earth where they are to remain, and will require no other care but to keep them clean from weeds, and thin the plants where they are too cloſe. Theſe flower in June, and perfect their ſeeds in Auguſt or September.

CATAPUTIA MAJOR. See RICINUS.

CATAPUTIA MINOR. See EUPHORBIA.

CATARIA. See NEPETA.

CATCH-FLY. See LYCHNIS.

CATESBÆA. Lin. Gen. Pl. 121. Hiſt. Carolin. vol. ii. p. 100. The Lily Thorn.

The CHARACTERS are,

The flower is of one leaf, funnel-ſhaped, having a very long tube, which gradually widens to the top, where it is four-cornered and ſpread open; it hath four ſtamina riſing in the neck of the tube. The roundiſh germen is ſituated under the flower, which afterward becomes an oval berry with one cell, filled with angular ſeeds.

We have but one SPECIES of this genus, viz.

CATESBÆA (*Spinoſa.*) Lin. Sp. Plant. 109. *The Lily Thorn.*

This ſhrub was diſcovered by Mr. Cateſby near Naſſau town, in the iſland of Providence, where he ſaw two of them growing, which were all he ever ſaw: from theſe he gathered the ſeeds, and brought them to England.

It riſes with a branching ſtem to the height of ten or twelve feet, covered with a pale ruſſet bark; the branches come out alternately, which are garniſhed with ſmall leaves reſembling thoſe of the Box-tree, coming out in cluſters all round the branches at certain diſtances; the flowers come out ſingle from the ſide of the branches, hanging downward; they are tubulous, and near ſix inches long, very narrow at their baſe, but widening upward toward their top, where it is divided into four parts which ſpread open, and are reflexed backward: theſe are of a dull yellow colour.

This ſhrub is propagated by ſeeds, which muſt be procured from the country where it naturally grows. If the entire fruit are brought over in ſand, the ſeeds will be better preſerved; the ſeeds muſt be ſown in ſmall pots filled with light ſandy earth, and plunged into a moderate hot-bed of tanners bark. If the ſeeds are good, the plants will appear in about ſix weeks; theſe plants make little progreſs for four or five years. If the nights ſhould prove cold, the glaſſes muſt be covered with mats every evening. As theſe plants grow ſlowly, ſo they will not require to be removed out of the ſeed-pots the firſt year, but in the autumn the pots ſhould be removed into the ſtove, and plunged into the tan-bed; in ſpring the plants ſhould be carefully taken up, and each planted in a ſeparate ſmall pot, filled with light ſandy earth, and plunged into a freſh hot-bed of tanners bark. In ſummer, when the weather is warm, they ſhould have a good ſhare of air admitted to them, but in autumn muſt be removed into the ſtove, where they ſhould conſtantly remain, and muſt be treated afterward in the ſame manner as other tender exotic plants.

CAUCALIS. Baſtard Parſley.

This is one of the umbelliferous plants with oblong ſeeds, which are a little furrowed and prickly: the petals of the flower are unequal and heart-ſhaped.

There are ſeveral ſpecies of this plant preſerved in the botanic gardens; but as there is no great beauty or uſe in any of them, I ſhall paſs them over with only obſerving that if any perſon had a mind to cultivate them, the beſt ſeaſon to ſow their ſeeds is in autumn, ſoon after they are ripe.

CEANOTHUS. Lin. Gen. Pl. 237. Euonymus. Com. Hort. New Jerſey Thea.

The CHARACTERS are,

The flower hath five roundiſh equal petals which ſpread open, and are leſs than the empalement; it hath five erect ſtamina placed oppoſite to the petals, and a three-cornered germen, which afterward becomes a dry capſule with three cells, in which are lodged three oval ſeeds.

The SPECIES are,

1. CEANOTHUS (*Americanus*) foliis trinerviis. Lin. Sp. Plant. 195. *Ceanothus with leaves having three nerves.*

2. CEANOTHUS (*Africanus*) foliis lanceolatis enerviis, ſtipulis ſubrotundis. Lin. Sp. Plant. 196. *Ceanothus with ſpear-ſhaped leaves without nerves, and roundiſh ſtipulæ.*

3. CEANOTHUS (*Arboreſcens*) foliis ovatis venoſis ſeſſilibus, floribus ſingularibus alaribus. *Ceanothus with oval leaves ſet cloſe to the branches, and ſingle flowers proceeding from the wings of the leaves; commonly called Redwood.*

The firſt ſort grows naturally in moſt parts of North America, from whence great plenty of the ſeeds have been of late years brought to Europe, by the title of New Jerſey Thea. The people of Canada uſe the root in venereal caſes.

In England this ſhrub ſeldom riſes more than four or five feet high, ſending out branches on every ſide from the ground upward. The branches are garniſhed with oval pointed leaves, having three longitudinal veins running from the foot-ſtalk to the point, which diverge in the broad part of the leaves from each other: the leaves are placed oppoſite, and are of a light green colour. At the extremity of each ſhoot the flowers are produced in cloſe thick ſpikes, which are compoſed of five ſmall leaves, and are of a clear white; and every ſhoot is terminated by one of theſe ſpikes, ſo the whole ſhrub is covered over with flowers. When the autumn proves mild, theſe ſhrubs often flower again in October. In warm ſeaſons the ſeeds will ripen pretty well in England. But this ſhrub is beſt propagated by laying down the young branches, which, in a light ſoil, will put out roots in a year's time; but theſe layers ſhould not be much watered, for as the ſhoots are tender, ſo moiſture will often occaſion their rotting, when it is given in quantities, or too often repeated; therefore the beſt method is to cover the ſurface of the ground in dry weather, all round the layers with mulch, which will preſerve a ſufficient moiſture in the ground, provided the ſeaſon is not extremely dry, in which caſe they ſhould have a little water once in eight or ten days, which will be ſufficient.

The beſt time for laying down theſe branches is in autumn, and if after this is performed, the ſurface of the ground is covered over with ſome old tan, taken from a decayed hot-bed, it will prevent the froſt from penetrating of the ground, which will ſecure them from injury, and the ſame covering will prevent the winds from drying of the ground in the ſpring, and thereby promote their putting out roots. Theſe layers, when rooted, may be taken up the following ſpring, and planted where they are to remain.

The ſecond ſort grows naturally at the Cape of Good Hope, from whence it was originally brought to Holland, where it has been long known by the title of Alaternoides, &c. and by ſome authors it is titled Ricinoides Africana arboreſcens, &c. but Dr. Linnæus, having examined the characters more exactly, has joined it to this genus.

This riſes to the height of ten or twelve feet, with a woody ſtem, covered with a rough purple bark; it ſends out many weak branches which hang downward; they are garniſhed

garnished with oblong pointed leaves, of a lucid green, which are smooth, and slightly sawed on their edges. The flowers are small, of an herbaceous colour, coming out from the side of the branches.

It may be propagated either by layers or cuttings, the latter being a very sure and expeditious method, is generally preferred The cuttings should be planted in June in a shady border; in about two months, or less, they will have taken root, when they must be taken up, and each planted in a small pot filled with light earth, placing them in the shade till they have taken fresh root. In autumn they must be housed with Myrtles, and other more hardy exotic plants, and treated in the same manner.

The third sort grows naturally in the American Islands; it rises with a shrubby stalk eighteen or twenty feet high, sending out several horizontal branches, which are garnished with oval veined leaves; the flowers come out at the wings of the leaves, with very short foot-stalks; they are of a white herbaceous colour, and are succeeded by dry capsules, shaped like those of the first sort.

This plant requires to be placed in a warm stove, otherwise it will not thrive in England; it is propagated by seeds, which must be sown upon a hot-bed in the spring, and when the plants are fit to remove, they should be each planted into a separate small pot, filled with light sandy earth, and plunged into a hot bed of tanners bark, observing to shade them till they have taken root; then they must be treated in the same manner as other tender exotic plants. In the autumn they must be placed in the bark-stove, and during the winter must be watered with great caution, for too much moisture at that season will destroy them.

CEDRUS. The Cedar-tree of Barbadoes, and Mahogony, &c.

The CHARACTERS are,

The flower is of one leaf, divided at the top into five parts; it hath five short stamina, which adhere at bottom to the germen. In the center is situated the roundish germen, which afterward becomes an oval pod, having five cells, opening from the bottom upward with five valves, having a double cover, the outer being thick and woody, the inner very thin, which immediately surrounds the seeds; these are thick at their base, but upward are flat and thin, like the wings adhering to the seeds of Firs and Pines.

As the Cedar of Libanus is by Tournefort very properly referred to the genus of Larix, and all the berry-bearing Cedars are joined to the Junipers, so I have given the title of Cedrus to this genus, as the plants were mentioned by imperfect titles by most of the authors who have treated of them; and as the first sort has been generally known by the appellation of Cedar, in the countries where it naturally grows, so the applying of the same name to those plants which agree in their essential characters with it, will join them properly together.

The SPECIES are,

1. CEDRUS (*Odorata*) foliis pinnatis, foliolis multijugatis obtusis, fructu ovali glabro. *Cedar-tree with winged leaves, composed of many pairs of small leaves (or lobes) which are obtuse, and an oval smooth fruit. This is the Barbadoes Cedar-tree.*

2. CEDRUS (*Mahogoni*) foliis pinnatis, foliolis oppositis, glabris, floribus racemosis sparsis. *Cedar with winged leaves, whose lobes are smooth and stand opposite, and flowers growing in loose bunches. This is the Mahogony-tree.*

3. CEDRUS (*Alaternifolius*) foliis alternis simplicibus, cordato-ovatis acutis, fructu pentagono mucronato. *Cedar with single leaves placed alternately, which are oval, heart-shaped, acute, and have a five-cornered pointed fruit.*

The first sort is commonly known under the title of Cedar in the British Islands of America, where this tree grows naturally, and is one of the largest trees of that country. The trunks of these trees are so large, that the inhabitants hollow them, and form them into the shape of boats and periaguas, for which purpose they are extremely well adapted; the wood being soft, it may be cut out with great facility, and being light it will carry a great weight on the water. The wood has a fragrant odour, from whence the title of Cedar has been given to it. It is often used for wainscotting of rooms, and to make chests, because vermin do not so frequently breed in it as in many other sorts of wood; this having a very bitter taste which is communicated to whatever is put into the chests, especially when the wood is fresh, for which reason it is never made into casks, because spirituous liquors will dissolve part of the resin, and thereby acquire a very bitter taste.

This tree rises with a strait stem to the height of seventy or eighty feet; while young the bark is smooth, and of an Ash colour, but as they advance the bark becomes rough, and of a darker colour. Toward the top it shoots out many side branches, which are garnished with winged leaves, composed of sixteen or eighteen pair of lobes (or small leaves) so that they are sometimes near three feet long; the lobes are broad at their base, and are near two inches long, blunt at their ends, and of a pale colour; these emit a rank odour in the summer season, so as to be very offensive. The fruit is oval, about the size of a partridge's egg, smooth, and of a very dark colour, and opens in five parts, having a five-cornered column standing in the middle, between the angles of which the winged seeds are closely placed, lapping over each other like the scales of fishes.

I have received plants of this kind from Paris, by the title of Semiruba, but whether the root of this tree is what they use in medicine under that appellation, I cannot say. The seeds of this have also been sent me from the French Islands in America, by the title of Acajou Cedre.

It is propagated by seeds, which may be easily procured from the American Islands, which must be sown upon a hot-bed in the spring, and the plants treated in the same manner as the next.

The second sort is the Mahogony, whose wood is now well known in England.

This tree is a native in the warmest parts of America, growing plentifully in the islands of Cuba, Jamaica, and Hispaniola; there are also many of them on the Bahama Islands, but I have not heard of their being found in any of the Leeward Islands. In Cuba and Jamaica there are trees of a very large size, so as to cut into planks of six feet breadth; but those on the Bahama Islands are not so large, though they are frequently four feet diameter, and rise to a great height, notwithstanding they are generally found growing upon the solid rocks, where there is scarce any earth for their nourishment. The wood which has been brought from the Bahama Islands, has usually passed under the appellation of Madeira wood, but there is no doubt of its being the same as the Mahogony.

The excellency of this wood for all domestic uses is now sufficiently known in England; and it is a matter of surprise, that the tree should not have been taken notice of by any historian or traveller to this time; the only author who has mentioned this tree is Mr. Catesby, in his Natural History of Carolina and the Bahama Islands, before whom I believe neither the tree or the wood was taken notice of by any writer on natural history, although the wood has been many years brought to England in great quantities.

The leaves of this tree are winged like those of the Ash, having commonly six or eight pair of pinnæ (or lobes) which are shorter and broader at their base than those of the Ash, where they adhere to the midrib by very short foot-stalks; these lobes are very smooth, having but one vein running through each, which is always on one side, so as to divide them unequally. We have no perfect account of the flower of this tree; those which are exhibited in Mr. Catesby's Natural History, were drawn from a withered imperfect fragment, which were the only remains of the flowers, which could be found at the time when he was there; but the fruit he has delineated very exactly, as I have had an opportunity of comparing it with some that have been brought to England. The entire fruit, before it opens, is of a brown colour; these fruit grow erect upon long foot-stalks, which closely adhere to the five-cornered column, running through the middle of the fruit, and to which the seeds are fastened, lying imbricatim like slates on a house, over each other; so that when the fruit is ripe, the outer cover divides at the bottom into five equal parts; and when these fall off, and the seeds are dispersed, the foot-stalk and the column remain some months after on the tree.

It is propagated by seeds, which may be easily procured from the Bahama Islands, from whence most of the good seeds which have come to England were brought; for most of those which have been sent from Jamaica, although brought in their pods, have not succeeded, whereas those from the Bahama Islands have grown as well as if they were immediately taken from the trees; the seeds should be sown in small pots filled with light sandy earth, and plunged into a hot-bed of tanners bark, giving them a gentle watering once a week; if the seeds are good, the plants will appear in a month or five weeks; and when the plants are two inches high, a sufficient number of small pots should be filled with light earth, and plunged into the tan-bed a day or two, that the earth may be warmed before the plants are put into the pots; then the young plants should be shaken out of the pots, and carefully separated, so as not to tear their roots, and each planted in a single pot, being careful to shade them till they have taken fresh root; after which they must be treated in the same manner as other tender plants from the same climate, being careful not to give them much water, especially in winter. If the plants are properly managed, they will make considerable progress; I have some plants now in the Chelsea-garden eight or ten feet high, which are but of six years growth from seeds.

The third sort was discovered by the late Dr. Houstoun at Campeachy, from whence he sent the seeds to England, which succeeded in several gardens; when the Doctor first observed these trees, they were destitute of leaves, but were loaded with ripe fruit; and on his second visit to the place he found the trees in full verdure, but no appearance of flowers, so he was at a loss to know what genus it belonged to; but as the fruit of this tree agrees exactly with those of the two former species, so I have ventured to join it to them.

These trees usually rise to the height of eighty feet or upward, and divide into many large branches toward the top, which are garnished with leaves, somewhat resembling those of the Witch Hazel, but are broader at their base, and cut angular at their top; these are of an Ash colour underneath, and are set on the branches without any order; the fruit of this tree is much larger than that of the Barbadoes-Cedar, being broad at the base, and diminishing gradually to the top, where it terminates in a point, being upwards of two inches long; this has also a column, or woody core, running lengthways through the fruit, to which the winged seeds adhere as in the two former; but as both their fruit are smooth on their outside, this differs from them, in having five angles running from the base upward; at each angle the fruit, when ripe, separates, and exposes the winged seeds, which are dispersed by the winds.

We have no account of the wood of this tree, whether it is ever used in buildings, or for other purposes, as there have been few persons of any curiosity in that country, the cutters of Logwood being the chief people who inhabit there, from whom there can be little known of the produce. The plants which have been raised from the seeds in England, have made great progress for the two first years, but afterward were but slow of growth; for in six years more they did not shoot so much as in the first year from the seed, when they grew more than three feet high. This may be managed in the same manner as the two foregoing sorts, and with them constantly kept in the bark-stove.

CEDAR of BERMUDAS.
CEDAR of CAROLINA.
CEDAR of LYCIA.
CEDAR of PHŒNICIA.
CEDAR of VIRGINIA.
} See JUNIPERUS.

CEDAR of JAMAICA. See THEOBROMA.

CEDAR of LIBANUS. See LARIX.

CEIBA See BOMBAX.

CELASTRUS. Lin. Gen. Pl. 239. Euonymoides Isnard. Ac. R. Sc. 1716. The Staff-tree

The CHARACTERS are,

The flower hath five oval petals, which are equal and spread open. It hath five stamina as long as the petals, and a small germen with a large receptacle, marked with ten deep channels, which afterward becomes an oval, blunt, three-cornered capsule, opening in three cells, each containing an oval smooth seed.

The SPECIES are,

1. CELASTRUS (*Bullatus*) inermis, foliis ovatis integerrimis. Lin. Sp. Pl. 196. *Smooth Staff-tree with oval entire leaves.*

2. CELASTRUS (*Scandens*) inermis, caule volubili. Lin. Sp. Pl. 196. *Smooth Staff tree with a twining stalk.*

3. CELASTRUS (*Pyracanthus*) spinis nudis, ramis teretibus, foliis acutis. Hort. Cliff. 72. *Staff tree with naked spines, taper branches, and pointed leaves.*

4. CELASTRUS (*Buxifolius*) spinis foliosis, ramis angulatis, foliis obtusis. Hort. Cliff. 73. *Staff-tree with leaves on the spines, angular branches, and obtuse leaves.*

The first sort grows naturally in Virginia, and many other parts of North America, where it rises to the height of eight or ten feet. It generally puts out two or three stems from the root, which divide upward into several branches, covered with a brown bark, and garnished with leaves near three inches long, which are placed alternately on the branches; the flowers come out on the side of the branches; these are white, made up of five oval petals, with a germen in the center attended by five stamina; when the flowers fall off, the germen swells to a three-cornered capsule, of a scarlet colour, set full of small protuberances; this opens in three parts, each containing a hard oval seed, covered with a thin red pulp.

It is propagated here by layers, which will take root in one year; the young branches only are proper for this purpose, so that where there is not any of these near the ground, the main stalks should be drawn down, and fastened with pegs to prevent their rising, and the young shoots from them should be laid. The best time for doing this is in autumn, and by that time twelvemonth they will be sufficiently rooted, when they should be cut from the old plant, and planted in a nursery for a year or two to get strength; after which they must be removed to the places where they are to remain. This shrub grows naturally in moist places, so will not thrive well in a dry soil.

The second sort sends out several ligneous stalks from the root, which are flexible, and twist themselves about whatever trees and shrubs grow near them; or when they are at a distance from such support they twine about each other, and rise to the height of twelve or fourteen feet; but when they

they faſten themſelves about trees they will grow much taller. Theſe are garniſhed with oblong leaves about three inches long, and near two broad, which are ſawed on their edges. The flowers are produced in ſmall bunches toward the end of the branches, which are of an herbaceous colour, compoſed of five roundiſh petals; theſe are ſucceeded by roundiſh three-cornered capſules, which are red, and, when ripe, ſpread open in three parts, diſcloſing the ſeeds in the ſame manner as our common Spindle-tree. This ſends out ſuckers from the root, ſo is eaſily propagated.

The third ſort is a native of Ethiopia. This riſes with an irregular ſtalk about three or four feet high, ſending out ſeveral ſide branches, covered with a brown bark, and garniſhed with leaves about two inches long, and more than half an inch broad, ſome of which are pointed, and others are obtuſe; they are ſtiff, of a lucid green, and come out irregularly from the branches; theſe continue green through the year. The flowers are produced from the ſide of the branches in looſe tufts, many of them ariſing from one point, ſtanding upon long foot-ſtalks; they are of an herbaceous white colour, compoſed of five petals, which ſpread open, and five ſpreading ſtamina, which ſurround a ſwelling germen, which afterward becomes an oval fruit, of a fine red colour, opening in three cells, containing one oblong hard ſeed, the other two cells being generally abortive.

This plant is commonly propagated by cuttings in Europe, which is more expeditious than raiſing them from ſeeds, becauſe the ſeeds rarely come up the firſt year. The cuttings may be planted any time in ſummer, but thoſe which are planted early will have more time to get ſtrength before winter. When they have taken root they muſt be expoſed to the open air, and placed in a ſheltered ſituation; when they are ſeparated, they muſt be each planted in a ſmall pot filled with good earth, then placed in the ſhade till they have taken freſh root; after which they may be placed with other exotic plants in a ſheltered ſituation till autumn, when they muſt be houſed with Myrtles, and other of the hardy green-houſe plants, and will require the ſame treatment.

The fourth ſort grows naturally at the Cape of Good Hope, from whence I received the ſeeds. This riſes with a ſlender ligneous ſtalk to the height of ten or twelve feet, covered with a light Aſh-coloured bark, and full of joints, which are armed with long ſpines, upon which grow many ſmall leaves; the branches are ſlender, and armed with the ſame ſpines at every joint, but the whole plant is ſo weak as to require ſome ſupport, without which they would fall to the ground. The leaves come out in cluſters without any order, which are ſhaped ſomewhat like thoſe of the narrow-leaved Box-tree, but are longer, and of a looſe texture; the branches are angular, and when young their bark is whitiſh.

This riſes very eaſily from ſeeds, and the plants make great progreſs, for I have raiſed them four feet high in two years from ſeeds, without any artificial heat; and ſome of the plants have lived through two winters againſt a ſouth-eaſt wall, but theſe have ſhed their leaves in winter, whereas thoſe which are removed into the green-houſe have retained their verdure through the year.

It may be propagated by cuttings, which ſhould be planted in the ſpring, and treated in the ſame manner as hath been directed for the former ſort; or if the young ſhoots are laid, they will take root in one year, and may then be tranſplanted either into pots, or againſt a good aſpected wall, where I find they will endure our ordinary winters without any protection.

CELERY, or SALARY. See Apium.

CELOSIA. Lin. Gen. Plant. 255. Amaranth.

The Characters are

The flower hath five erect ſharp-pointed petals, which are permanent, ſtiff, and ſhaped like a flower-cup. It hath a ſmall nectarium joined to the border of the germen, to which adhere the five ſtamina, which are terminated by turning ſummits. The empalement afterward becomes a globular capſule with one cell, opening horizontally, containing roundiſh ſeeds.

The Species are,

1. Celosia (*Margaritacea*) foliis ovatis ſtipulis falcatis pedunculis, angulatis ſpicis ſcarioſis. Lin. Sp. Plant. 297. *Celoſia with ſpear-ſhaped leaves, angular foot-ſtalks, and an oblong ſpike.*

2. Celosia (*Criſtata*) foliis lanceolato-ovatis recurvis ſubundatis, pedunculus angulatis ſpicis oblongis criſtatis. Lin. Sp. 297. *Creſted Amaranthus, called Cockſcomb.*

3. Celosia (*Paniculata*) foliis lanceolato-ovatis panicula diffuſa filiformi. Flor. Virg. 144. *Celoſia with oval ſpear-ſhaped leaves, and ſlender diffuſed panicles.*

4. Celosia (*Coccinea*) foliis ovatis ſtrictis inauriculatis, caule ſulcato, ſpicis multiplicibus criſtatis. Lin. Sp. 297. *Celoſia with oval leaves, a furrowed ſtalk, and many creſted ſpikes of flowers.*

Dr. Linnæus has ſeparated the plants of this genus from the Amaranths, which have been generally joined with them, becauſe thoſe have male and female flowers in the ſame plants, whereas theſe have only hermaphrodite flowers; ſo that by his ſyſtem, the other are removed to the twenty-firſt claſs, and have been before mentioned under the article Amaranthus, to which the reader is deſired to turn for the Amaranthus tricolor, &c.

The firſt ſort here mentioned grows naturally in both Indies. This riſes with an upright ſtalk about three or four feet high, garniſhed with oblong ſpear-ſhaped leaves, ending in points, of a pale colour. Toward the upper part of the ſtalk there are a few ſide branches ſent out, which ſtand erect, each of which is terminated by a ſlender ſpike of flowers, and the principal ſtalk is terminated by one which is much larger; this is two or three inches long, and about as thick as a man's middle finger, the whole ſpike being of a ſilvery colour.

The ſecond ſort is well known by its common appellation of Cockſcomb, which was given to it from the form of its creſted head of flowers, reſembling a cock's comb; of this there are many varieties, which differ in their form magnitude, and colours; but as they vary from ſeeds, ſo they are not enumerated as diſtinct ſpecies I have raiſed great varieties of theſe from ſeeds which came from China, and other countries, but have generally found them alter in a few years, notwithſtanding great care has been taken in the ſaving of their ſeeds: the principal colours of their heads are red, purple, yellow, and white; but there are ſome, whoſe heads are variegated with two or three colours. I alſo raiſed ſome from ſeeds which I received from Perſia, whoſe heads were divided like a plume of feathers, which were of a beautiful ſcarlet colour.

The third ſort grows naturally in ſome of the Sugar Iſlands: it riſes with a weak ſtalk near four feet high, which is garniſhed with oblong pointed leaves, that ſtand oppoſite at each joint, and are pretty far aſunder. The flowers come out in looſe panicles from the ſide of the ſtalks, and alſo at the end of the branches; theſe are divided into a great number of very ſlender ſpikes, which are of a pale yellow, ſhining with a gloſs like ſilk. The plants of this periſhed in the autumn, without perfecting their ſeeds.

In order to have large fine Amaranths, great care ſhould be taken in the choice of the ſeeds, for if they are not carefully collected, the whole expence and trouble of raiſing them will be loſt. When you are provided with good ſeeds, they muſt be ſown on a hot-bed (which ſhould have been prepared a few days before, that the violent heat may be abated;) about the beginning of March is a good time to ſow

ſow the ſeeds, and in leſs than a fortnight the plants will appear; but as they are tender when they firſt come up, ſo they require great care for a few days till they get ſtrength; in giving them a due proportion of air, to prevent their drawing up weak, and then to keep them from too much moiſture, for a ſmall ſhare of moiſture will cauſe their tender ſtems to rot: in ſowing of theſe ſeeds, there ſhould be care taken not to put them too cloſe, for when the plants come up in cluſters, they frequently ſpoil each other for want of room to grow. In a fortnight or three weeks time the plants will be fit to remove, when you muſt prepare another hot-bed, covered with good, rich, light earth, which ſhould be laid about four inches thick; this ſhould be made a few days, that it may have a proper temperature of heat; then raiſe up the young plants with your finger, ſo as not to break off the tender roots, and prick them into your new hot-bed about four inches diſtance every way, giving them a gentle watering to ſettle the earth to their roots: but in doing this, be very cautious not to bear the young plants down to the ground by haſty watering, which rarely riſe again, or at leaſt ſo as to recover their former ſtrength in a long time, but very often rot in the ſtems, and die quite away.

After the plants are thus planted, they muſt be ſcreened from the ſun till they have taken freſh root; but as there is generally a great ſteam ariſing from the fermentation of the dung, which condenſes to wet againſt the glaſſes, and this dropping upon the plants very frequently deſtroys them, ſo the glaſſes ſhould be frequently turned in the day-time, whenever the weather will permit; but if the weather happens to prove bad that you cannot turn your glaſſes, it will be of great ſervice to your plants to wipe off all the moiſture two or three times a day with a woollen cloth, to prevent the dropping upon the plants. When your plants are firmly rooted and begin to grow, you muſt obſerve to give them air every day (more or leſs, as the weather is cold or hot) to prevent their drawing up too faſt, which greatly weakens their ſtems.

In about three weeks or a month's time, after theſe plants will have grown ſo as to meet, they will ſtand in need of another hot-bed, which ſhould be of a moderate temper, and covered with the ſame rich earth about ſix inches thick, in which they ſhould be planted (obſerving to take them up with as much earth about their roots as poſſible) planting them ſeven or eight inches diſtance every way, giving them ſome water to ſettle the earth about their roots; but be very careful not to water them heavily, ſo as to bear down the plants, (as was before directed) and keep them ſhaded in the heat of the day, until they have taken freſh root, and be ſure to refreſh them often (but gently) with water.

In the beginning of May you muſt provide another hot-bed, which ſhould be covered with a deep frame, that your plants may have room to grow: upon this hot bed you muſt ſet as many three-penny pots as can ſtand within the compaſs of the frame; theſe pots muſt be filled with good rich earth, and the cavities between each pot filled up with any common earth, to prevent the heat of the bed from evaporating, and filling the frame with noxious ſteams; then with a trowel, or ſome ſuch inſtrument, take up your plants (from the former hot-bed) with as much earth as poſſible to the roots, and place each ſingle plant in the middle of one of the pots, filling the pot up with the earth before deſcribed, and ſettle it cloſe to the root of the plant with your hands; water them gently as before, and ſhade them in the heat of he day from the violence of the ſun.

In about three weeks more theſe plants will have grown to a conſiderable ſize and ſtrength, ſo that you muſt now raiſe the glaſſes very much in the day-time; and when the air is ſoft and the ſun is clouded, draw off the glaſſes, and expoſe them to the open air, and repeat this as often as the weather will permit, which will harden them by degrees, to be removed abroad into the places where they are to remain the whole ſeaſon: but it is not adviſeable to ſet theſe plants out until a week in July, obſerving to do it when the air is perfectly ſoft, and, if poſſible, in a gentle ſhower of rain.

Let them at firſt be ſet near the ſhelter of a hedge for two or three days, where they may be ſcreened from the violence of the ſun and ſtrong winds, to which they muſt be inured by degrees: theſe plants, when grown to a good ſtature, perſpire very freely, and muſt be every day refreſhed with water, if the weather proves hot and dry, otherwiſe they will ſtunt, and never produce their plumes ſo fine as they would do, if taken care of.

In the beginning or middle of September the Amaranths will have perfected their ſeeds, ſo that you muſt make choice of the largeſt, moſt beautiful, and leaſt branching plants of each kind for ſeed, which you ſhould remove under ſhelter, (eſpecially if the weather proves wet, or the nights froſty) that the ſeeds may be maturely ripened; and in the choice thereof, be ſure never to take any ſeeds from ſide branches, nor from the neck of the plume, but ſuch only as are produced in the middle thereof, which in many plants, perhaps, may be but a ſmall quantity; but I do aſſure you, it is thoſe only you can depend upon, to have your kinds good the ſucceeding year.

CELSIA. Lin. Gen. Pl. 675. We have no Engliſh name for it.

The CHARACTERS are,

The flower is one leaf, with a very ſhort tube, ſpread open above, and cut into five unequal parts, the two upper being ſmall, and the under larger. It hath four hairy ſtamina, which incline toward the upper ſegments of the petal. In the center is ſituated a roundiſh germen, which afterward becomes a roundiſh capſule, compreſſed at the top, ſitting upon the empalement, and hath two cells, which are filled with ſmall angular ſeeds.

There is but one SPECIES of this genus at preſent known, viz.

CELSIA (*Orientalis*) foliis duplicato-pinnatis. Hort. Cliff. 321. *Celſia with double-winged leaves.*

This plant grows naturally in Armenia, from whence Dr. Tournefort ſent the ſeeds to the royal garden at Paris. In its natural place of growth this is an annual, but in England it will rarely ripen its ſeeds, unleſs the plants come up in the autumn, and live through the winter.

It ſends out many oblong leaves, which are finely divided almoſt to the midrib on both ſides; from the center ariſes a roundiſh herbaceous ſtalk two feet high, which is garniſhed the whole length with leaves of the ſame ſhape, but diminiſhing in their ſize gradually to the top: theſe are placed alternately, and at the foot-ſtalk of each come out the flowers more than half the length of the ſtalk, which are of an iron colour on their outſide, but pale yellow within, ſpreading open like thoſe of the common Mullein, but are not ſo regular: the ſhort tube being turned downward, and the lower ſegments being larger than the upper, and the ſtamina being unequal, has occaſioned Linnæus to remove it to his ringent flowers. The ſeed-veſſel is round, compreſſed, and hath two cells filled with ſmall ſeeds. It may be ſown on a poor dry ſoil in autumn, and when the plants come up, they will require no other care but to keep them clean from weeds, and thin them if they are too cloſe; for they do not bear removing well, ſo ſhould be ſown where they are intended to remain.

I have ſometimes, when the ſeaſons have proved warm, had ripe ſeeds from plants ſown in the ſpring; but this cannot be depended on, therefore it is much better to raiſe the plants in autumn.

CELTIS. Tourn. Inst. R. H. 612. tab. 383. Lin. Gen. Pl. 1012. The Lote, or Nettle-tree.

The Species are,

It hath male and hermaphrodite flowers on the same tree: the hermaphrodite flowers are single, and situated above the male; these have no petals, but five short stamina. In the center is situated an oval germen, which afterward becomes a round berry with one cell, inclosing a roundish nut. The male flowers have their empalements divided into six parts, and have no germen or style, but in other parts are like the hermaphrodite.

The Species are,

1. Celtis (*Australis*) foliis lanceolatis acuminatis, serratis, nervosis. *Nettle-tree with spear-shaped pointed leaves, which are veined and sawed on their edges; or the Lote-tree with a black fruit.*

2. Celtis (*Occidentalis*) foliis obliquè-ovatis, serratis, acuminatis. Lin. Sp. Pl. 1044. *Nettle-tree with oblique oval leaves, which are pointed and sawed on their edges; or Lote-tree with a dark purple fruit.*

3. Celtis (*Orientalis*) foliis ovato-cordatis, denticulatis, petiolis brevibus. *Nettle-tree with oval heart-shaped leaves slightly indented, and short foot-stalks; or the Smaller Eastern Lote-tree, with smaller and thicker leaves, and a yellow fruit.*

4. Celtis (*Americana*) foliis oblongo-ovatis, obtusis, nervosis, superne glabris, subtus aureis. *Nettle-tree with oblong, oval, obtuse, nervous leaves, which are smooth on their upper surface, and of a gold colour beneath.*

The first sort grows naturally in the south of France, in Spain and Italy, where it is one of the largest trees of those countries: yet this is not so plenty in England as the second, nor do I remember to have seen but two large trees of this sort in the English gardens.

This tree rises with an upright stem to the height of forty or fifty feet, sending out many slender branches, which have a smooth dark-coloured bark, garnished with leaves placed alternately, which are near four inches long, and about two broad in the middle, ending in long sharp points, and deeply sawed on their edges, having several transverse veins, which are prominent on their under side. The flowers come out from the wings of the leaves all along the branches; they have a male and an hermaphrodite flower generally at the same place, the male flowers being situated above the others: these have no petals, but a green herbaceous empalement, so make no figure; they come out in the spring, at the same time when the leaves make their first appearance, and generally decay before their leaves have grown to half their magnitude. After their flowers are past, the germen of the hermaphrodite flowers become a round berry, about the size of a Pea, which, when ripe, is black.

The second sort grows naturally in North America: it delights in a moist rich soil, in which it becomes a very large tree. This rises with a strait stem, which in young trees is smooth, and of a dark colour, but as they advance it becomes rougher and of a lighter colour. The branches are much diffused on every side, which are garnished with oblique oval leaves, ending in points, and sawed on their edges. The flowers come out opposite to the leaves upon pretty long foot-stalks, the male flowers standing above the hermaphrodite, as in the other species; after these decay, the hermaphrodite flowers are succeeded by roundish berries, which are smaller than those of the first sort, and, when ripe, are of a dark purple colour.

Thi t ee is late in coming out in the spring, but in recompense for that it continues as long in beauty in the autumn, for it is the latest in fading of any of the deciduous trees; nor do the leaves alter their colour long before they fall, but continue in full verdure till within a few days of their dropping off. There is little beauty in the nowers or fruit of this tree, but as the branches are well clothed with leaves which are of a fine green colour, so the trees, when mixed with others in wildernesses, make a pleasing variety during the summer season. The wood of this tree being tough and pliable, is esteemed by coachmakers for the frames of their carriages.

The third sort was discovered by Dr. Tournefort in Armenia. It rises with a stem about ten or twelve feet high, dividing into many branches, which spread horizontally on every side, and have a smooth greenish bark; they are garnished with leaves about an inch and a half long, and near an inch broad, inclining to an heart-shape, but are oblique; they are of a thicker texture than those of the common sort, and are of a paler green. They are placed alternate on the branches, and have short foot-stalks. The flowers come out from the foot-stalks of the leaves, in the same manner as the former, and are succeeded by oval yellow berries, which, when fully ripe, turn of a darker colour. The wood of this tree is very white.

These trees are propagated by seeds, which should be sown soon after they are ripe, when they can be procured at that season, for these frequently come up the following spring; whereas those which are sown in the spring will not come up till a twelvemonth after, therefore it is the best way to sow them in pots or tubs, that they may be easily removed. In summer they must be constantly kept clean from seeds; if the season proves dry, they will require water two or three times a week. In autumn it will be proper to remove the pots, and place them under a hot-bed frame, to shelter them in winter from severe frost; or where there is not that conveniency, the pots should be plunged into the ground, near a wall or hedge; for as the plants, when young, are full of sap and tender, so the early frosts in autumn frequently kill the upper parts of their shoots; therefore the plants should be either covered with mats, or a little straw or Pause-haulm laid over to protect them.

In the following spring the plants should be taken out of the seed-pots, and planted in the full ground: this should be done about the middle or latter end of March, when the danger of the frost is over; therefore a bed or two should be prepared (according to the number of plants raised) in a sheltered situation, and, if possible, in a gentle loamy soil. The ground must be well trenched, and cleared from the roots of bad weeds, and when levelled, should be marked out in lines at one foot distance; then the plants should be carefully turned out of the pots and separated, so as not to tear their roots, and planted in the lines at six inches asunder, pressing the earth down close to the roots. If the ground is very dry when they are planted, and there is no appearance of rain soon, it will be proper to water the beds, to settle the ground to the roots of the plants; and after this, if the surface of the ground is covered with some old tan or rotten dung, it will keep the ground moist, and prevent the drying winds from penetrating to the roots of the plants.

The plants may remain in these nursery-beds two years, by which time they will have obtained sufficient strength to be transplanted where they are designed to remain for good, because these plants extend their roots wide every way, so that if they stand long in the nursery, their roots must be cut in removing, which will be a prejudice to their future growth.

These sorts are hardy enough to thrive in the open air in England, after they have acquired some strength; but for the two first winters after they come up from seeds, they require a little protection, especially the third sort, which is tenderer than either of the former. The young plants of

this

this sort frequently have variegated leaves, but these are more impatient of cold than the plain leaved.

The fourth sort was first discovered by Father Plumier in the French Islands of America; it was also found growing in Jamaica by Dr. Houstoun, who sent the seeds to England. This rises with a strait trunk near twenty feet high, covered with a gray bark, divided into many branches upward, which are garnished with leaves near four inches long, and two and a half broad, rounded at their extremity, of a thick texture, very smooth on their upper surface, and on their under side are of a lucid gold colour, placed alternately on the branches. The fruit is round and red, but the flowers I have not seen.

The seeds of this sort rarely come up the first year, so they should be sowed in pots, and plunged into the tan-bed in the stove, where they should remain till the plants come up. These plants must be constantly kept in the bark-stove, and treated in the same manner as other tender exotics.

CENTAUREA. Lin. Gen. Pl. 880. Greater Centaury, Knapweed, Blue Bottle, &c.

The CHARACTERS are,

It hath a compound flower, whose disk is composed of many hermaphrodite florets, and the border or rays of female florets, which are larger and looser; these are included in a common scaly empalement. The germen is situated under the petal, which afterward becomes a single seed shut up in the empalement. The female florets have a slender tube, but expands above, where it is enlarged, and cut into five unequal parts; these are barren.

The SPECIES are,

1. CENTAUREA (*Alpina*) calycibus inermibus squamis ovatis obtusis, foliis pinnatis glabris, integerrimis impari serrato. Hort. Cliff. 421. *Centaury with a globular empalement without spines, sharp pointed scales, and winged leaves white underneath.*

2. CENTAUREA (*Centaurium*) calycibus inermibus, squamis ovatis obtusis, foliis pinnatis glabris integerrimis. Hort. Cliff. 421. *Centaury with an empalement without spines, oval oblong scales, and smooth winged leaves, which are entire; or Yellow Alpine Centaury.*

3. CENTAUREA (*Glastifolia*) calycibus scariosis foliis indivisis integerrimis decurrentibus. Hort. Cliff. 421. *Centaury with a rough empalement, and undivided leaves.*

4. CENTAUREA (*Stæbe*) calycibus ciliatis oblongis, foliis pinnatifidis linearibus integerrimis. Prod. Leyd. 140. *Centaury with oblong hairy empalements, and winged pointed leaves which are very narrow and entire.*

5. CENTAUREA (*Conifera*) calycibus squamosis, foliis tomentosis, radicalibus lanceolatis, caulinis pinnatifidis caule simplici. Prod. Leyd. 142. *Centaury with a scaly empalement, woolly leaves, those near the root being spear-shaped, those on the stalk pointed, and a single stalk.*

6. CENTAUREA (*Montana*) calycibus serratis, foliis lanceolatis decurrentibus, caule simplicissimo. Hort. Cliff. 422. *Centaury with sawed empalements, spear-shaped running leaves, and a single stalk; or Greater Mountain Blue Bottle.*

7. CENTAUREA (*Angustifolia*) calycibus serratis, foliis lineari-lanceolatis decurrentibus, caule simplicissimo. *Centaury with sawed empalements, very narrow spear-shaped running leaves, and a single stalk; Narrower and longer leaved Belgic Blue Bottle.*

8. CENTAUREA (*Moschata*) calycibus inermibus, subrotundis glabris, squamis ovatis, foliis sinuatis. Hort. Cliff. 421. *Centaury with unarmed, roundish, smooth empalements, oval scales, and sinuated leaves; commonly called Sweet Sultan.*

9. CENTAUREA (*Amberboi*) calycibus inermibus, subrotundis, glabris, squamis ovatis obtusis, foliis laciniatis serratis. *Centaury with roundish, smooth, unarmed empalements, oval obtuse scales, and cut leaves which are sawed on their edges; commonly called yellow Sweet Sultan.*

10. CENTAUREA (*Cyanus*) calycibus serratis, foliis linearibus integerrimis, infimis dentatis. Hort. Cliff. 422. *Centaury with sawed empalements, very narrow entire leaves indented below; or Corn Blue Bottle.*

11. CENTAUREA (*Lippii*) calycibus inermibus, squamis mucronatis, foliis pinnatifidis obtusis decurrentibus. Lin. Sp. Pl. 910. *Centaury with unarmed empalements, having pointed scales, and wing-pointed leaves which are obtuse, running along the stalk.*

12. CENTAUREA (*Cineraria*) calycibus ciliatis terminali-sessilibus, foliis tomentosis pinnatifidis, lacinulis linearibus. Hort. Cliff. 422. *Centaury with hairy empalements closely terminating the stalks, woolly leaves with winged points, and the segments very narrow.*

13. CENTAUREA (*Ragusina*) calycibus ciliatis, foliis tomentosis pinnatifidis, foliolis obtusis ovatis integerrimis exterioribus majoribus. Hort. Cliff. 422. *Centaury with hairy empalements, woolly leaves with winged points, the small leaves oval and obtuse, the outer larger; or Silvery Knapweed of Ragusa.*

14. CENTAUREA (*Napifolia*) calycibus palmato-spinosis setis æqualibus, foliis decurrentibus sinuatis spinulosis. Prod. Leyd. 141. *Centaury with palmated spinous empalements, whose bristles are equal, and sinuated prickly leaves running along the stalks.*

15. CENTAUREA (*Rapontica*) calycibus squamosis, foliis ovato-oblongis denticulatis integris petiolatis, subtus tomentosis. Hort. Cliff. 421. *Centaury with scaly empalements, oval, oblong, indented, entire leaves, having foot-stalks, and woolly underneath.*

16. CENTAUREA (*Peregrina*) calycibus setaceo-spinosis, foliis lanceolatis petiolatis, infernè dentatis. Hort. Cliff. 423. *Centaury with bristly prickly empalements, spear-shaped leaves, with foot-stalks indented beneath.*

17. CENTAUREA (*Orientalis*) calycibus squama ciliatis, foliis pinnatifidis, pinnis lanceolatis. Lin. Sp. Plant. 913. *Centaury with hairy scales to the empalement, and wing-pointed leaves whose lobes are spear-shaped.*

18. CENTAUREA (*Eriophora*) calycibus duplicato-spinosis lanatis, foliis semidecurrentibus integris sinuatisque, caule prolifero. Hort. Upsal. 272. *Centaury whose empalement is downy, and doubly armed with spines, and running leaves with indentures terminating with spines.*

19. CENTAUREA (*Benedicta*) calycibus duplicato-spinosis lanatis involucratis, foliis semidecurrentibus denticulato-spinosis. Lin. Sp. 1296. *Centaury with a downy and doubly armed empalement, and running leaves whose indentures terminate in spines.*

There are many other species of this genus, which are preserved in botanic gardens for the sake of variety; some of which grow naturally in England, and are often troublesome weeds in the fields, so do not deserve a place in gardens, therefore I chose not to trouble the reader with mentioning their titles, but have here selected those species which have some beauty to recommend them.

The first sort grows naturally on the Apennine mountains. This hath a perennial root, which strikes many strong roots deep in the ground: the lower leaves resemble those of the Artichoke; they are green on the upper side, and hoary underneath; the stalks rise about three feet high, and are garnished with leaves of the same form and colour as those below, but are much smaller. The upper part of the stalk branches out into three or four divisions, each being terminated with a large single head of flowers, which are of a a purple colour; these are composed of many hermaphrodite flowers, which form the disk in the center, and a border of female florets which compose the rays, included in a common scaly empalement, whose scales terminate in acute points. This is usually propagated by parting of their roots: the best time for doing this is early in October, that the plants may have time to take root before the frost comes on. These roots must not be removed or parted oftener than every fourth year, if they are designed to produce strong flowers; they should be planted in a dry soil, because wet in the winter will cause them to rot. This plant is rarely injured by frost, so may be fully exposed in open borders; but each root will require to have three feet to spread, so they must not be planted too near other plants.

The second sort grows naturally upon the Alps. This hath also a perennial root, which strikes deep into the ground, sending out a great number of long winged leaves, which are smooth, entire, and of a glaucous colour; the stalks rise near four feet high, and divide upward into many branches, which are garnished with small leaves of the same form as the lower; each of these stalks are terminated by a single head of yellow flowers, of the same form with those of the first, but not more than half the size. It may be propagated by parting of their roots, in the same manner as the first, and the plants do require the same treatment.

The third sort stands in the list of medicinal plants of the College, but is very rarely used; the root is reckoned to be binding, and good for all kind of fluxes, and of great use to heal wounds. This grows naturally on the mountains in Italy and Spain; it hath a strong perennial root like the two former sorts, from which come out a great number of long winged leaves, which spread wide on every side; these are of a lucid green, and sawed on their edges; the flower-stalks are slender, but very stiff, and divide upward into many smaller, which rise six or seven feet high, having at each joint one small winged leaf, of the same form with the other: each of these stalks is terminated by a single head of purplish flowers, which are considerably longer than the empalement. It may be propagated by parting of the roots, in the same manner as the former sorts, and the plants must be treated in the same way, but should have more room to grow.

The fourth sort grows naturally in Austria. This hath a perennial root as the former, from which come out in the spring many winged leaves, which are hoary, and the segments narrow and entire; the stalks rise near three feet high, dividing into several branches, which have a winged leaf at each joint, of the same shape with the other; at the end of each stalk is one head of purple flowers, inclosed in an oblong scaly empalement, each scale being bordered with small hairs like an eye-brow. This is propagated by seeds, which may be sown in a bed of common earth in a nursery; and when the plants come up, they must be thinned and kept clean from weeds, and the following autumn the plants may be transplanted where they are designed to remain.

The fifth sort grows naturally in the south of France, and in Italy. It hath a biennial root, which doth not divide and spread as the former, but grows single, sending out in the spring several entire spear-shaped leaves, and afterward a single stalk more than a foot high, which is garnished at each joint with one divided hoary leaf, and at the top comes out a single, large, scaly head, shaped like a cone of the Pine-tree, very taper at the top, where it closely surrounds the florets, whose tops just peep out of the empalement: they are of a bright purple colour, but are not succeeded by seeds in England, so cannot be propagated, unless the seeds are procured from abroad.

The sixth sort is the common perennial Blue Bottle, which by some is titled Batchelors Button. This is so well known as to need no description. The roots of this sort creep under ground to a great distance, whereby the plant propagates too fast, and often becomes troublesome in gardens; it will grow in any soil or situation.

The seventh sort differs from the sixth, in having much longer and narrower leaves, which are not so white, the heads of flowers are also smaller; but whether this is only a variety from the other, I cannot determine, having never raised either from seeds; for these plants spread very much by their creeping roots, which renders them barren, as is frequently the case with many other creeping rooted plants, few of which produce seeds. This is equally hardy, so may be planted in any soil or situation, where many other sorts will not thrive, and during its continuance in flower will make a variety in the garden.

The eighth sort is annual, so is only propagated by seeds. This has been many years propagated in the English gardens, under the title of Sultan Flower, or Sweet Sultan. This sends up a round channelled stalk near three feet high, which divide into many branches, garnished with jagged leaves, of a pale green, smooth, and stand close to the branches; from the side of the branches come out long naked foot-stalks, each sustaining a single head of flowers shaped like those of the other species, which have a very strong odour, so as to be offensive to many people, but to others is very grateful; the flowers are in some purple, and others white, and likewise a flesh colour. There is also a variety of this with fistular flowers, and another with fringed flowers, commonly called Amberboi, or Emberboi; but these have degenerated to the common sort in a few years, although I have saved the seeds with great care, so I suppose they are only varieties. These seeds are commonly sown upon a hot-bed in the spring, to bring the plants forward, and in May they are transplanted into the borders of the flower-garden; but if the seeds are sown on a warm border in autumn, they will live through the winter; and these plants may be removed in the spring into the flower-garden, which will be stronger, and come earlier to flower than those which are raised in the spring. The seeds may also be sown in the spring on a common warm border, where the plants will rise very well, but these will be later in flowering than either of the other.

The ninth sort has been supposed to be only a variety of the former, which is a great mistake; for although there is a great similitude in their appearance, yet they are specifically different. I have cultivated this sort upward of thirty years, and have never observed the least variation in it. This is much tenderer than the former, so the seeds must be sown upon a hot-bed in the spring; and when the plants are fit to remove, they should be transplanted on a fresh hot-bed to bring them forward. When the plants have obtained strength, they must be carefully taken up, and planted in separate pots, filled with light earth, and some of them placed in the shade till they have taken root; then they may be placed with other annual plants in the pleasure-garden, where they will continue long in beauty. But as these plants, which are placed in the open air, rarely produce good seeds, so there should be two or three plants kept in a moderate hot-bed, under a deep frame, where they will come earlier to flower; and being protected from wet and cold, they will ripen their seeds every year, which is the surest method to preserve the sort.

The tenth sort is the common Blue Bottle, which grows naturally amongst the Corn in most parts of England. This stands in the list of medicinal plants; there is a distilled water of the flowers, which is esteemed good for the eyes. There are great varieties of colours in these flowers, fome of which are finely variegated: the feeds of thefe are fold by feedfmen, by the title of Bottles of all colours. Thefe are annual plants, which will rife in any common border, and require no other care but to keep them clean from weeds, and thinned where they are too clofe.

The feeds of the eleventh fort were fent me by Dr. Juffieu, from Paris, who received them from Dr. Lippi at Grand Cairo. This is an annual plant, which rifes near two feet high, fending out two or three branches toward the top; the leaves are divided into many obtufe parts, and have a border running along the ftalk; the flowers are fmall, of a bright purple, and have a fcaly empalement. If the feeds are fown in the fpring upon a border of light earth, where the plants are to remain, they will require no farther care but to keep them clean from weeds. It flowers in July, and the feeds ripen in autumn.

The twelfth fort is a perennial plant, which retains its leaves through the year. This grows naturally in Italy, on the borders of the fields. The leaves are hoary, and divided into many narrow fegments; the ftalks rife near three feet high, branching upward into many divifions, each being terminated by a head of purple flowers. This fort will live abroad in moderate winters, if it has a warm fituation and a dry foil, but in fevere winters the plants are commonly killed; fo one or two of them may be fheltered under a common frame in winter to preferve the kind. It may be eafily propagated by the young branches, which do not fhoot up to flower; if thefe are cut off, and planted in a fhady border any time in fummer, they will take root, and in autumn may be removed to warm borders, or put into pots to be fheltered in winter.

The thirteenth fort grows naturally in Mauritania, and in feveral other places on the borders of the Mediterranean fea. This feldom rifes more than three feet high in this country; it hath a perennial ftalk, that divides into many branches, which are garnifhed with many white woolly leaves, divided into many obtufe lobes that are entire; the fmall leaves, or lobes, on the exterior part of the leaf, being the largeft. The flowers are produced from the fide branches upon fhort foot-ftalks, which are of a bright yellow, and are included in a fine hairy empalement. It is propagated by planting of the young fhoots in the fame manner as the laft, and the plants require protection from hard froft.

The fourteenth fort is annual. This grows naturally in the Archipelago. It rifes with a branching ftalk about three feet high; the lower leaves are not much unlike thofe of the Turnep, being rounded at their ends, and their bafe is cut into many parts; thofe upon the ftalks and branches are nearly of the fame form, but diminifh gradually in their fize to the top; thefe have a border or wing running along the ftalk, which connect them together; the flowers are produced at the end of the branches, which have prickly empalements; the fpines come out from the border of the fcales, divided like the fingers of a hand. The flowers are of a bright purple, fo make a pretty appearance. This fort may be treated in the fame manner as the Corn Blue Bottle, by fowing the feeds in autumn, and keeping the plants clean from weeds.

The fifteenth fort grows naturally upon the Helvetian mountains It hath a perennial root, and an annual ftalk; the leaves are oblong, flightly indented on their edges, and woolly on their under fide; thefe have much refemblance of thofe of Elecampane, generally ftanding upright; the ftalks rife little more than a foot high, and are terminated by large fingle heads of purple flowers, inclofed in fcaly empalements. This, like the fifth fort, is very difficult to propagate in England, unlefs good feeds can be procured from the countries where it naturally grows.

The fixteenth fort grows naturally in Auftria and Hungary. The lower leaves of this plant fpread flat on the ground; they are foft, hairy, and end in fharp points, but toward their bafe are cut into feveral narrow fegments; the ftalks rife near three feet high, garnifhed at each joint by fpear-fhaped leaves, which are entire, and are terminated by fingle large heads of flowers, of a gold colour, inclofed in a prickly fcaly empalement. It hath a perennial root which fends out offsets; thefe may be taken from the old plants in autumn, whereby it may be eafily propagated. It is very hardy in refpect to cold, but fhould have a dry foil, the roots being very apt to rot in winter with much wet.

The feventeenth fort grows naturally in Siberia. This fends out many long winged leaves from the root, which are divided into feveral fpear-fhaped lobes; the ftalks rife near five feet high, and divide upward into many fmaller branches, which are garnifhed with leaves of the fame form with the lower, but are much fmaller, and the fegments very narrow; each of thefe is terminated by a head of yellow flowers, inclofed in a fcaly empalement; the borders of the fcales are fet with fine hairs like an eye-brow. This hath a perennial root and an annual ftalk, which, with the leaves, decay in autumn, and arife new from the root in the fpring. It may be propagated by either feed, or by parting of the roots, in the fame manner as the fifth fort.

The eighteenth fort grows naturally in Portugal. The ftalks of this rife two feet high, and are garnifhed with woolly leaves, fome of which are entire, others are finuated on their borders: the ftalks branch out, and each divifion is terminated by woolly heads of flowers, ftrongly armed with double fpines on their empalements; thefe almoft inclofe the florets: this is propagated by feeds, which in cold wet feafons feldom ripen in England.

The nineteenth fort is the Carduus Benedictus of the fhops; or Bleffed Thiftle, which is frequently ufed as an emetic. It grows naturally in Spain and in the Levant, but is hardy enough to grow any where. The plant is biennial, fo decays foon after the feeds are ripe. If thefe are permitted to fcatter, the plants will rife without trouble, and fhould be afterward planted out two feet diftance each way, keeping the ground between them clean from weeds.

CENTAURIUM MINUS. See Gentiana.

CENTINODIUM, Knot Grafs. See Polygonum.

CEPA. The Onion.

The botanic difference of the Onion from Garlick is the fwelling pipy ftalk, which is much larger in the middle than at either end.

The Varieties of the common Onion are,

The Strafburgh. This is the Cepa oblonga. C. B. P. 71.

The Spanifh Onion. This is the Cepa vulgaris, floribus & tunicis purpurafcentibus. C. B. P. 71.

The white Egyptian Onion. This is the Cepa floribus & tunicis candidis. C. B. P. 71.

All thefe vary from feeds, fo that there are feveral intermediate differences, which are not worth enumerating.

Thefe three varieties are propagated by feeds, which fhould be fown the latter end of February, or the beginning of March, on good, rich, light ground, which fhould be well dug and levelled, and cleared from the roots of all bad weeds; then the feeds fhould be fown, in a dry time, when the furface of the ground is not moift; and where they are intended for a winter crop, they muft not be fown too thick The common allowance of feed is fix pounds to one acre of land, but the generality of gardeners fow more, becaufe many of them allow for a crop to draw out, which they call cullings.

In about fix weeks after fowing the Onions will be up forward enough to hoe; at which time (chufing dry weather) you fhould, with a fmall hoe about two inches and a half a half broad, cut up lightly all the weeds from amongst the Onions; and also cut out the Onions, where they grow too close in bunches, leaving them at this first hoeing at least two or three inches apart. This, if well performed, and in a dry season, will preserve the ground clear of weeds at least a month, when you must hoe them over a second time, cutting up all the weeds as before, and also cut out the Onions to a larger distance, leaving them this time four or five inches asunder. This also, if well performed, will preserve the ground clean a month or six weeks longer, when you must hoe them over the third and last time.

Now you must carefully cut up all weeds, and single out the Onions to six inches square, by which means they will grow much larger than if left too close. This time of hoeing, if the weather proves dry, and it is well performed, will keep the ground clean until the Onions are fit to pull up; but if the weather should prove moist, and any of the weeds should take root again, you should, about a fortnight or three weeks after, go over the ground, and draw out all the large weeds with your hands; for the Onions having now begun to bulb, they should not be disturbed with a hoe.

Toward the middle of August your Onions will have arrived to their full growth, which may be known by their blades falling to the ground and shrinking; you should therefore, before their necks or blades are withered off, draw them out of the ground, cropping off the extreme part of the blade, and lay them abroad upon a dry spot of ground to dry, observing to turn them over every other day at least, to prevent their striking fresh root into the ground; which they will quickly do, especially in moist weather.

In about a fortnight's time your Onions will be dry enough to house, which must be performed in perfect dry weather; in doing of this, you must carefully rub off all the earth, and be sure to mix no faulty ones amongst them, which will in a short time decay, and spoil all those that lie near them; nor should you lay them too thick in the house, which will occasion their sweating, and thereby rot them; these should not be put in a lower room or ground floor, but in a loft or garret; and the closer they are kept from the air, the better they will keep. You should, at least once a month, look over them to see if any of them are decayed; which if you find, must be immediately taken away, otherwise they will infect all that lie near them.

But notwithstanding all the care you can possibly take in the drying and housing of your Onions, many of them will grow in the loft, especially in mild winters, which are generally moist; therefore those who would preserve them late in the season, should select a parcel of the firmest and most likely to keep from the others, and with a hot iron slightly singe their beards or roots, which will effectually prevent their sprouting; but in doing of this there must be great caution used not to scorch the pulp of the Onions, for that will cause them to perish soon after.

In order to save seeds, you must in the spring make choice of some of the firmest, largest, and best shapen Onions (in quantity proportionable to the seed you intend to save;) and having prepared a piece of good ground (which should be well dug, and laid out in beds about three feet wide,) in the beginning of March you must plant your Onions in the following manner. Having strained a line about four inches within the side of the bed, you must, with a spade, throw out an opening about six inches deep, the length of the bed, into which you should place the Onions with their roots downward, at about nine inches distance from each other, and with a rake draw the earth into the opening again to cover the bulbs; then proceed to remove the line again about a foot farther back, where you must make an opening as before, and so again till the whole is finished, by which you will have four rows in each bed; between each bed you must allow the space of two feet, for an alley to go among them to clear them from weeds, &c. In a month's time their leaves will appear above ground, and many of the roots will produce three or four stalks each; you must therefore keep them cleared from weeds, and about the beginning of June, when the heads of the flowers begin to appear upon the tops of the stalks, you must provide a parcel of stakes about four feet long, which should be driven into the ground in the rows of Onions, at about six or eight feet apart, to which you should fasten some packthread, rope yarn, or small cord, which should be run on each side of the stems of the Onions a little below their heads, to support them from breaking down with the wind and rain, for when the seeds are formed the heads will be heavy, so are very often broken down by their own weight, where they are not well secured; and if the stalks are broken before the seeds have arrived to maturity, they will not be near so good, nor keep so long as those which are perfectly ripened.

About the end of August the Onion seed will be ripe, which may be known by its changing brown, and the cells in which the seeds are contained opening; so that if it be not cut in a short time, the seeds will fall to the ground; when you cut off the heads, they should be spread abroad upon coarse cloths in the sun, observing to keep it under shelter in the night, as also in wet weather; and when the heads are quite dry you must beat out the seeds, which are easily discharged from their cells; then having cleared it from all the husks, &c. after having exposed it one day to the sun to dry, you must put it up in bags to preserve it for use.

The directions here given is for the general crop of winter Onions, but there are two other crops of this common sort of Onions, cultivated in the gardens about London to supply the market, one of which is commonly called Michaelmas Onions. These are sown in beds pretty close the beginning of August, and must be well weeded when they come up. In the spring of the year, after the winter Onions are over, they are tied up in bunches to supply the markets; but from the thinning of these they carry to market young green Onions in March, for sallets, &c.

And in the spring they sow more beds in the same manner, to draw up young for sallets, after the Michaelmas Onions are grown too large for that purpose; and where a supply of these are required, there may be three different sowings, at about three weeks distant from each other, which will be sufficient for the season.

There are also the following sorts of Onions cultivated in the kitchen-gardens.

The Shallot or Eschalottes, which is the Cepa Ascalonica. Matth. 556.

The Ciboule, or Cepa fissilis. Matth. Lugd. 1539.

The Cives, or Cepa sectilis juncifolia perennis. Mor. Hist. 2. 383.

The Welch Onion I suppose to be the same with the Ciboule, although they pass under different appellations, for I have several times received the Ciboule from abroad, which, when planted, proved to be what is generally known here by the title of Welch Onions.

The Scallion, or Escallion, is a sort of Onion which never forms any bulbs at the roots, and is chiefly used in the spring for green Onions, before the other sorts sown in July are big enough; but this sort of Onion, how much soever in use formerly, is now so scarce as to be known to few people, and is rarely to be met with. The gardeners near London substitute another sort for this which are those Onions which decay and sprout in the house; these they plant in a bed early in the ſpring, which in a ſhort time will grow large enough for uſe; when they draw them up, and after pulling off all the outer coat of the root, they tie them up in bunches, and ſell them in the market for Scallions.

The true Scallion is eaſily propagated by parting the roots, either in ſpring or autumn; but the latter ſeaſon is preferable, becauſe of their being rendered more fit for uſe in the ſpring; theſe roots ſhould be planted three or four in a hole, at about ſix inches diſtance every way, in beds or borders three feet wide, which in a ſhort time will multiply exceedingly, and will grow upon almoſt any ſoil and in any ſituation.

The Cives are a very ſmall ſort of Onion, which never produce any bulbs, and ſeldom grow above ſix inches high in the blade, which is very ſmall and ſlender, and are in round bunches like the former; this was formerly in great requeſt for ſallets in the ſpring, as being milder than the Welch Onions. Theſe are propagated by parting their roots like the former, and are alſo very hardy, and will be fit for uſe early in the ſpring.

The Welch Onions are only propagated for ſpring uſe; theſe never make any bulbs, and are therefore only fit to be uſed green for ſallets, &c. They are ſown about the end of July, in beds about three feet and a half wide, leaving alleys of two feet broad to go between the beds to clean them, and in a fortnight's time they will appear above ground, and muſt be carefully cleared from weeds; towards the middle of October their blades will die away, ſo that the whole ſpot will ſeem to be naked, which hath led many people to dig up the ground again, ſuppoſing the crop totally loſt; whereas, if they ſtand undiſturbed, they will come up again very ſtrong in January, and from that time grow very vigorouſly, reſiſting all weathers, and by March will be fit to draw for young Onions, and were ſome years paſt in the markets, more valued than any other ſort; for they are extremely green and fine, though they are much ſtronger than the common Onion in taſte, approaching nearer to Garlick, which hath occaſioned their being leſs eſteemed for the table: but as no winter, however hard, will hurt them, it is proper to have a few of them to ſupply the table, in caſe the common ſort ſhould be deſtroyed by froſts.

The roots of theſe Onions, if planted out at ſix or eight inches diſtance in March, will produce ripe ſeeds in autumn, but it will be in ſmall quantities the firſt year; therefore the ſame roots ſhould remain unremoved, which the ſecond and third year will produce many ſtems, and afford a good ſupply of ſeeds; theſe roots will abide many years good, but ſhould be tranſplanted and parted every ſecond or third year, which will cauſe them to produce ſtrong ſeeds.

CEPHALANTHUS. Lin. Gen. Plant. 105. Button-wood.

The CHARACTERS are,

It hath a number of ſmall flowers, which are collected into a ſpherical head; each particular flower hath a funnel-ſhaped empalement, divided into four parts at the top; the flower is funnel-ſhaped, of one petal, divided at the top into four parts, incloſing four ſtamina, which are inſerted in the petal. The germen is ſituated under the flower, which afterward becomes a globular hairy capſule, incloſing one or two oblong angular ſeeds; theſe are joined to an axis, and form a round head.

We have but one SPECIES of this genus, viz.

CEPHALANTHUS (*Occidentalis*) foliis oppoſitis terniſque. Flor. Virg. 15. *Button-tree with leaves growing oppoſite by threes.*

This plant grows naturally in North America, from whence the ſeeds are annually ſent to Europe, and of late years great numbers of the plants have been raiſed in the gardens of the curious.

This ſeldom riſes higher than ſix or ſeven feet in this country. The branches come out oppoſite; the leaves alſo ſtand oppoſite, ſometimes by pairs, and at other times there are three ariſing at the ſame joint; theſe are near three inches long, and one and a quarter broad, having a ſtrong vein running longitudinally through the middle: they are of a light green, and their foot-ſtalks change to a reddiſh colour next the branches; the ends of the branches are terminated by ſpherical heads, about the ſize of a marble, each of which are compoſed of many ſmall flowers, which are funnel-ſhaped, of a whitiſh yellow colour, faſtened to an axis that ſtands in the middle.

Theſe plants are propagated chiefly by ſeeds (though there has been many raiſed from cuttings;) theſe ſeeds ſhould be ſown in pots, for the greater conveniency of removing them, either into a ſhady ſituation, or where they may have ſhelter, for they generally remain a year in the ground; therefore, in ſuch caſe, the pots ſhould be placed in the ſhade the firſt ſummer, and placed the autumn following under a common frame to ſhelter them from froſt, and the ſpring following the plants will come up.

The firſt year, when the plants come up, it will be neceſſary to ſhade them in hot dry weather while they are young, at which time they are often deſtroyed by being too much expoſed; nor ſhould the watering be neglected, for as theſe plants naturally grow in moiſt ground, ſo when they are not duly watered in dry weather, the young plants will ſoon decay.

The next autumn, when the leaves begin to drop, they may be tranſplanted into nurſery-beds, which ſhould be a little defended from the cold winds; and, if the ſoil is moiſt, they will ſucceed much better than in dry ground; but where it happens otherwiſe, it will be abſolutely neceſſary to water them in dry weather, otherwiſe there will be great danger of the plants dying in the middle of ſummer, which has been the caſe in many gardens where theſe plants were raiſed.

In the nurſery-beds the plants may remain a year or two (according to the progreſs they have made, or the diſtance they were planted;) then they may be taken up in October, and tranſplanted where they are to remain for good. Although I have mentioned but one ſeaſon for tranſplanting them, yet this may alſo be performed in the ſpring, eſpecially if the ground is moiſt into which they are removed, or that the plants are duly watered if the ſpring ſhould prove dry, otherwiſe there will be more hazard of their growing when removed at this ſeaſon.

CERASTIUM. Lin. Gen. Plant. 518. Mouſe-ear, or Mouſe-ear Chickweed.

The CHARACTERS are,

The flower hath five obtuſe bifid petals; it hath ten ſlender ſtamina. In the center is ſituated an oval germen with five ſtyles. The empalement afterward becomes an oval, cylindrical, or globular capſule with one cell, containing many roundiſh ſeeds.

The SPECIES are,

1. CERASTIUM (*Repens*) foliis lanceolatis, pedunculis ramoſis, capſulis ſubrotundis. Lin. Sp. Pl. 439. *Ceraſtium with ſpear-ſhaped leaves, branching foot-ſtalks, and roundiſh capſules.*

2. CERASTIUM (*Tomentoſum*) foliis oblongis, tomentoſis, pedunculis ramoſis, capſulis globoſis. Lin. Sp. Plant. 440. *Ceraſtium with oblong woolly leaves, branching foot-ſtalks, and globular capſules.*

3. CERASTIUM (*Dichotomum*) foliis lanceolatis, caule dichotomo ramoſiſſimo, capſulis erectis. Prod. Leyd. 450. *Ceraſtium*

Cerastium with spear-shaped leaves, a forked stalk, and upright capsules.

4. Cerastium (*Pentandrum*) floribus pentandris, petalis integris. Lin. Sp. Plant. 438. *Cerastium with flowers having five stamina, and entire petals.*

5. Cerastium (*Perfoliatum*) foliis connatis. Hort. Cliff. 173. *Cerastium whose leaves are joined at their base.*

The first sort grows naturally in France and Italy, and was formerly cultivated in the English gardens under the title of Sea Pink; one of the uses made of it was to plant it as an edging, to keep up the earth of borders; but this was before the Dwarf Box was brought to England, since which all those plants which were formerly applied for this purpose have been neglected. This plant was by no means fit for this use, because its creeping branches would spread into the walks, where they put out roots into the gravel, so that unless they are frequently cut off, they cannot be kept within compass.

This sort sends out many weak stalks which trail upon the ground, and put out roots at their joints, whereby it propagates very fast; the leaves are placed by pairs opposite: these are very hoary; those next the root are much smaller than the upper; the flowers come out from the side of the stalks upon slender foot-stalks, which branch out into several smaller, each supporting a white flower, composed of five petals, which are split at the top.

The seeds of the second sort I received from Istria, where it naturally grows; this is by Parkinson titled hoary narrow-leaved Pink. The leaves of this sort are narrower than those of the former, and are much whiter; the stalks grow more erect, and the seed-vessels are rounder, in which their chief difference consists.

The third sort is annual; this grows naturally on arable land in Spain. It is allowed a place in botanic gardens for the sake of variety, but hath not much beauty; this hath branching stalks, which grow about six inches high, dividing by pairs, the flowers coming out in the middle of the divisions, which are shaped like those of Chickweed; the whole plant has a clammy moisture, which sticks to the fingers of those that handle it. If the seeds are permitted to fall, the plants will rise without care.

The fourth sort is very like the third in its whole appearance, and differs from it in having but five stamina in the flower, whereas the other hath ten.

The fifth sort was discovered by Dr. Tournefort in the Levant. It is an annual plant, which rises with an upright stalk a foot high; the lower leaves of this plant have much resemblance to those of the Lychnis, which is called Lobel's Catchfly, so that when the plants are young it is not easy to distinguish them. The stalks are garnished with leaves of the same shape, but smaller; these are placed by pairs, and embrace the stalks at their base. The flowers come out at the top of the stalks, and also from the wings of the leaves on the upper part of the stalks, which are white, and shaped like those of Chickweed.

If the seeds of this sort are sown in autumn, they will more certainly grow than those which are sown in the spring; or if the seeds are permitted to scatter, the plants will come up and live through the winter, and will require no other care but to keep them clean from weeds.

CERASUS. The Cherry-tree.

The botanical characters to this genus, according to the system of Linnæus, are the same with those of Prunus, therefore he has joined the Apricot Cherry, Laurel, and Bird Cherry together, making them only species of the same genus; but those who admit of the fruit as a character to determine the genus, must separate the Cherry from the others, because they differ greatly in the shape of their stones; but there is a more essential difference in nature between them, which is, that the Cherry will not grow upon a Plum stock by budding or grafting, nor will the Plum take upon a Cherry stock, and yet we know of no trees of the same genus which do not unite with each other by budding or grafting.

I shall first enumerate the sorts which are specifically different from each other, and then mention the varieties of these fruits which are cultivated in the English gardens, many of which seem to differ so essentially from each other, that they may be allowed as specific differences; but as I have not had an opportunity of trying the various sorts from seeds to see if they alter, so I chose to insert them only as varieties, till farther observation may better settle their boundaries.

The Species are,

1. Cerasus (*Vulgaris*) foliis ovato-lanceolatis, serratis. *The common, or Kentish Cherry.*

2. Cerasus (*Nigra*) foliis serratis lanceolatis. *Cherry-tree with spear-shaped sawed leaves; or Black Cherry.*

3. Cerasus (*Hortensis*) foliis ovato-lanceolatis, floribus confertis. *Cherry-tree with oval spear-shaped leaves, and flowers growing in clusters; commonly called the Cluster Cherry.*

4. Cerasus (*Mahaleb*) floribus corymbosis, foliis ovatis. Lin. Sp. Plant. 474. *Cherry-tree with flowers growing in round bunches, and oval leaves; the Mahaleb, or perfumed Cherry.*

5. Cerasus (*Canadensis*) foliis lanceolatis, glabris, integerrimis, subtus, cæsiis, ramis patulis. *Cherry-tree with smooth, spear-shaped, entire leaves, of a bluish green on their under side, and spreading branches.*

The first sort is the common or Kentish Cherry, which is so well known in England as to need no description. From this sort, it hath been supposed, most of the varieties which are cultivated in the English gardens have been raised; but as there are very great differences in the size and shape of their leaves, as also in the shoots of the trees, from those of this sort, so I think it very doubtful, where the boundaries of their specific differences terminate: however, I shall comply with the generality of modern botanists, in supposing the following sorts to have been produced from the seeds of this, as we have not sufficient experiments to determine otherwise.

The early May Cherry.	The Ox Heart.
The May Duke Cherry.	The Lukeward.
The Archduke Cherry.	The Carnation.
The Flemish Cherry.	The Hertfordshire Heart.
The Red Heart.	The Morello.
The White Heart.	The Bleeding Heart.
The Black Heart.	Yellow Spanish Cherry.
The Amber Heart.	

Two sorts with double flowers, one larger and fuller than the other. These are propagated for ornament.

The second sort is the Black Cherry, which is supposed to be a native of England. This grows to be a large tree fit for timber, and is frequently found growing as such in the woods. From this the only varieties which I have ever known raised by seeds, are the Black Coroun, and the small Wild Cherry; of which there are two or three varieties, which differ in the size and colour of their fruit.

The stones of this sort are generally sown for raising stocks, to graft or bud the other sorts of Cherries upon, being of quicker growth, and of longer duration than either of the other, so are very justly esteemed and preferred to them.

The wood of the fourth sort, is by the French greatly esteemed for making of cabinets, because it hath an agreeable odour. This, and the wood of the Bird Cherry, are often blended together, and pass under the appellation of Bois de Sainte Lucie; but the Bird Cherry is the true sort.

The

The fifth fort was brought from Canada, where it grows naturally. This is a low fhrub, which feldom grows more than three or four feet high, fending out many horizontal branches, which fpread on every fide, and are very fubject to fall on the ground, where they will put out roots, and thereby multiply. The young branches have a very fmooth bark, inclining toward red; the leaves are long, narrow, very fmooth, and entire, having the appearance of fome forts of Willow leaves; of a light green on their upper fide, but of a bluifh, or fea-green, on their under: the flowers come out from the fide of the branches, two, three, or four arifing at the fame joint; thefe are fhaped like thofe of the common Cherry, but are fmaller, ftanding upon long flender foot-ftalks. The fruit is like thofe of the fmall wild Cherry, but hath a bitterifh flavour.

It is eafily propagated by laying down the branches early in the fpring, which will take root by the following autumn, when they may be taken off, and either planted in a nurfery to get ftrength, or to the places where they are defigned to remain. It may alfo be propagated by fowing of the ftones in the fame manner as other Cherries.

All the forts of Cherries which are ufually cultivated in fruit-gardens, are propagated by budding or grafting the feveral kinds into ftocks of the black or wild red Cherries, which are ftrong fhooters, and of a longer duration than any of the garden kinds. The ftones of thefe two kinds are fown in a bed of light fandy earth in autumn (or are preferved in fand till fpring, and then fowed:) thefe young ftocks fhould remain in thefe nurfery-beds till the fecond autumn after fowing, at which time you fhould prepare an open fpot of good frefh earth, which fhould be well worked. In this ground, in October, you fhould plant out the young ftocks at three feet diftance row from row, and about a foot afunder in the rows, being careful in taking them up from their feed-beds, to loofen their roots well with a fpade, to prevent their breaking, as alfo to prune their roots; and if they are inclinable to root downwards, you fhould fhorten the tap-root, to caufe it to put out lateral roots; but do not prune their tops, for this is what by no means they will endure.

The fecond year after planting out, if they take to growing well, they will be fit to bud, if they are intended for dwarfs; but if they are for ftandards, they will not be tall enough till the fourth year, for they fhould be budded or grafted near fix feet from the ground, otherwife the graft will not advance much in height; fo that it will be impoffible to make a good tree from fuch as are grafted low, unlefs the graft is trained upward.

The ufual way with the nurfery-gardeners is, to bud their ftocks in fummer, and fuch of them as mifcarry they graft the fucceeding fpring, (the manner of thefe operations will be defcribed under their proper heads.) Thofe trees where the buds have taken, muft be headed off the beginning of March about fix inches above the bud; and when the bud hath fhot in fummer, if you fear its being blown out by the winds, you muft faften it up with fome bafs, or fuch foft tying, to that part of the ftock which was left above the bud. The autumn following thefe trees will be fit to remove; but if your ground is not ready to receive them, they may remain two years before they are tranfplanted; in the doing of which you muft obferve not to head them, as is by many practifed, for this very often is immediate death to them; but if they furvive it, they feldom recover this amputation in five or fix years.

If thefe trees are intended for a wall, I would advife the planting dwarfs between the ftandards; fo that while the dwarfs are filling the bottom of the walls, the ftandards will cover the tops, and will produce a great deal of fruit: but thefe, as the dwarfs arife to fill the walls, muft be cut away to make room for them; and when the dwarf trees cover the walls, the ftandards fhould be entirely taken away. But I would advife never to plant ftandard Cherries over other fruits, for there is no other fort of fruit that will profper well under the drip of Cherries.

When thefe trees are taken up from the nurfery, their roots muft be fhortened, and all the bruifed parts cut off; as alfo all the fmall fibres, which would dry, grow mouldy, and be a great prejudice to the new fibres in their coming forth; you muft alfo cut off the dead part of the ftock which was left above the bud, clofe down to the back part of it, that the ftock may be covered by the bud. If thefe trees are defigned for a wall, obferve to place the bud directly from the wall, that the back part of the ftock that was cut may be hid from fight. The foil that Cherries thrive beft in, is a frefh Hazel loam; for if the foil is a dry gravel they will not live many years, and will be perpetually blighted in the fpring.

The forts commonly planted againft walls are the Early May and May Duke, which fhould have a fouth afpected wall. The Hearts and common Duke will thrive on a weft wall; and in order to continue the Duke later in the feafon, they are frequently placed againft north and north-weft afpected walls, where they fucceed very well; and the Morello on a north wall, which laft is chiefly planted for preferving. The Hearts are all of them ill bearers, for which reafon they are feldom planted againft walls: but I am apt to believe, if they were grafted on the Bird Cherry, and managed properly, that defect might be remedied; for this ftock (as I am informed) will render Cherries very fruitful; and having the fame effect on Cherries as the Paradife ftock hath on Apples, they may be kept in lefs compafs, which is an experiment well worth the trial.

Your trees, if planted againft a wall, fhould be placed at leaft twenty or twenty-four feet afunder, with a ftandard-tree between each dwarf: this will be found a reafonable diftance, when we confider that Cherry trees will extend themfelves as far, or farther than Apricots, and many other forts of fruit.

In pruning thefe forts of fruit, you fhould never fhorten their fhoots; for the moft part of them produce their fruit buds at their extreme part, which, when fhortened, are cut off, and this often occafions the death of the fhoot: their branches fhould be therefore trained in at full length horizontally, obferving in May, where there is a vacancy in the wall, to ftop fome ftrong adjoining branches, which will occafion their putting out two or more fhoots; by which means, at that feafon of the year, you may always get a fupply of wood for covering the wall; and at the fame time fhould all foreright fhoots be difplaced by the hand, for if they are fuffered to grow till winter they will not only deprive the bearing branches of their proper fupply of nourifhment, but when they are cut out it occafions the tree to gum in that part (for Cherries bear the knife the worft of any fort of fruit-trees;) but be careful not to rub off the fides or fpurs, which are produced upon the two or three years old wood: for it is upon thefe that the greateft part of the fruit are produced, which fpurs will continue fruitful for feveral years. And it is for want of duly obferving this caution, that Cherry-trees are often feen fo unfruitful, efpecially the Morello, which the more it is cut the weaker it fhoots; and, at laft, by frequent pruning, I have known a whole wall of them deftroyed; which, if they had been fuffered to grow without any pruning, might probably have lived many years, and produced large quantities of fruit.

Cherry-trees are alfo planted for orchards in many parts of England, particularly in Kent, where there are large plantations of thefe trees. The ufual diftance allowed for their ftanding is forty feet fquare, at which fpace they are lefs fubject to blight than when they are clofer planted; and the ground may be tilled between them almoft as well as if it were entirely clear, efpecially while the trees are young; and the often ftirring the ground, provided you do not difturb their roots, will greatly help the trees; but when they are grown fo big as to overfhadow the ground, the drip of their leaves will fuffer very few things to thrive under them.

The forts beft approved of for an orchard, are the common Red or Kentifh Cherry, the Duke, and Lukeward, all which are plentiful bearers. But orchards of thefe trees are now fcarcely worth planting, except where land is very cheap; for the uncertainty of their bearing, with the trouble in gathering the fruit, together with the fmall price it commonly yields, hath occafioned the deftroying many orchards of this fruit in Kent within a few years paft.

There are fome perfons who graft the Duke, and other forts of Cherries, upon the Morello Cherry, which is but a weak fhooter, in order to check the luxuriant growth of their trees, which will fucceed for three or four years; but they are not of long duration, nor have I ever feen one tree fo grafted, which had made fhoots above fix or eight inches long; but they were clofely covered with bloffoms, fo may produce fome fruit in a fmall compafs; but thefe are experiments unfit to be carried into general ufe, and only proper to fatisfy curiofity; for is it not much better to allow the trees a greater fhare of room againft the walls, when one tree fo planted, and properly managed, will produce more fruit than twenty of thefe trees, or twice that number, when they are planted too clofe, though they are grafted upon the Black Cherry, or any other free ftock?

The Early, or May Cherry, is the firft ripe; fo one or two trees of this fort may be allowed a place in a garden, where there is room for variety. The next ripe is the May Duke, which is a larger fruit than the other, and is more valuable. After this comes the Archduke, which, if permitted to hang upon the tree till the fruit is quite ripe, is an excellent Cherry; but few perfons have patience to let them hang their full time, fo rarely have them in perfection; for thefe fhould not be gathered before Midfummer, and if they hang a fortnight longer they will be better. This is to be underftood of the fituation near London, where they ripen a fortnight earlier than in places forty miles diftant, unlefs they have a very warm fheltered fituation. When this fort is planted againft north walls, the fruit may be continued till the end of Auguft, but they muft be protected from the birds, otherwife they will deftroy them.

The Hertfordfhire Cherry, which is a fort of Heart Cherry, but a firmer and better flavoured fruit, will not ripen earlier than the end of July, or the beginning of Auguft, which makes it the more valuable, for its coming when the other forts of Cherries are gone.

The Morello Cherry, which is generally planted againft walls to a north afpect, and the fruit commonly ufed for preferving, or for tarts, yet where they are planted to a better afpect, and fuffered to hang upon the trees until they are thoroughly ripe, is a very good fruit for the table; therefore two or three of the trees of this fort fhould have a place where there is plenty of walling, upon a fouth-weft wall, where they will ripen perfectly by the middle or end of Auguft, at which time they will be an acceptable fruit.

The Carnation Cherry is alfo valuable for coming late in the feafon; this is a very firm flefhy fruit, but is not the beft bearer. This fort will ripen very well on efpaliers, and by this means the fruit may be continued longer in the feafon.

The large Spanifh Cherry is nearly allied to the Duke Cherry, from which it feems to be only a variety accidentally obtained; it ripens foon after the common Duke Cherry, and very often paffes for it.

The yellow Spanifh Cherry is of an oval fhape, and of an amber colour; this ripens late, and is a fweet Cherry, but not of a rich flavour, and being but a middling bearer, is not often admitted into curious gardens, unlefs where variety is chiefly confidered.

The Corone, or Coroun Cherry, is fomewhat like the Black Heart, but a little rounder; this is a very good bearer, and an excellent fruit, fo fhould have a place in every good fruit-garden. This ripens the middle of July.

The Lukward ripens foon after the Corone Cherry; this is a good bearer, and a very good fruit; it is of a dark colour, not fo black as the Corone, and will do well in ftandards.

The Black Cherry is feldom grafted or budded, but is generally fown for ftocks to graft the other kinds of Cherries upon; but where perfons are curious to have the beft flavoured of this fort of fruit, they fhould be propagated by grafting from fuch trees as produce the beft fruit. This fort of Cherry is frequently planted in wilderneffes, where it will grow to a large fize, and, at the time of its flowering, will make a variety, and the fruit will be food for the birds.

The double-flowering Cherry is alfo propagated for the beauty of the flowers, which are extremely fine, the flowers being as double and large as a Cinnamon Rofe; and thefe being produced in large bunches on every part of the tree, render it one of the moft beautiful trees of the fpring. Some of the flowers which are lefs double, will often produce fruit, which the very double flowers will not; but this defect is fufficiently recompenfed in the beauty of its flowers. This is propagated by budding or grafting on the Black or Wild Cherry-ftock, and the trees are very proper to intermix with the fecond growth of flowering-trees.

CERASUS RACEMOSA. See PADUS.

CERATONIA. Lin. Gen. Plant. 983. The Carob, or St. John's Bread.

The CHARACTERS are,

It is male and female in diftinct trees. The male flowers have no petals, but have five long ftamina. The female flowers have no petals, but a flefhy germen fituated within the receptacle, which afterward becomes a long, flefhy, compreffed pod, divided by tranfverfe partitions, each having one large, roundifh, compreffed feed.

We have but one SPECIES of this plant, viz.

CERATONIA (*Siliqua.*) H. L. *The Carob-tree, or St. John's Bread.*

This tree is very common in Spain, but particularly in Andalufia, and in fome parts of Italy, as alfo in the Levant, where it grows in the hedges, and produces a great quantity of long, flat, brown-coloured pods, which are thick, meally, and of a fweetifh tafte. Thefe pods are many times eaten by the poorer fort of inhabitants, when they have a fcarcity of other food, but they are apt to loofen the belly, and caufe gripings of the bowels. Thefe pods are directed by the College of Phyficians to enter fome medicinal preparations, for which purpofe they are often brought from abroad.

In England the tree is preferved by fuch as delight in exotic plants as a curiofity; the leaves always continue green, and being different in fhape from moft other plants, afford an agreeable variety, when intermixed with Oranges, Myrtles, &c. in the green-houfe.

It is propagated from feeds, which, when brought over frefh in the pods, will grow very well, if they are fown in the fpring in pots, and plunged into a moderate hot-bed In June you muft inure them to the open air by degrees, and in in July they ſhould be removed out of the hot-bed, and placed in a warm ſituation, where they may remain until the beginning of October, when they ſhould be removed into the green-houſe, placing them where they may have free air in mild weather; for they are pretty hardy, and require only to be ſheltered from hard froſts. When the plants have remained in the pots three or four years, and have gotten ſtrength, ſome of them may be turned out of the pots in the ſpring, and planted into the full ground in a warm ſituation, where they will endure the cold of our ordinary winters very well, but muſt have ſome ſhelter in very hard weather.

CERBERA. Lin. Gen. Plant. 260.

The CHARACTERS are,

The flower is of one leaf, funnel-ſhaped, ſpread open at the top, where it is divided into five large ſegments; it hath five ſtamina in the middle of the tube. In the center is ſituated a roundiſh germen, which afterward becomes a large, fleſhy, roundiſh berry, divided into two cells, each containing a ſingle, large, compreſſed nut.

The SPECIES are,

1. CERBERA (*Atroucir*) foliis ovatis. Lin. Sp. Pl. 208. *Cerbera with oval leaves.*

2. CERBERA (*Thevetia*) foliis linearibus, longiſſimis, confertis. Lin. Sp. Plant. 209. *Cerbera with very long narrow leaves, growing in cluſters.*

The firſt ſort grows naturally in the Brazils, and alſo in the Spaniſh Weſt-Indies in plenty; and there are ſome of the trees growing in the Britiſh Iſlands of America; this riſes with an irregular ſtem to the height of eight or ten feet, ſending out many crooked diffuſed branches, which toward their tops are garniſhed with thick ſucculent leaves, of a lucid green, ſmooth, and very full of a milky juice. The flowers come out in looſe bunches at the end of the branches, which are of a cream colour, having long narrow tubes, and at the top are cut into five obtuſe ſegments, which ſeem twiſted, ſo as to ſtand oblique to the tube. The wood of this tree ſtinks moſt abominably, and the kernels of the nuts are a moſt deadly poiſon, ſo that the Indians always caution their children againſt eating them, for they know of no antidote to expel this poiſon; nor will any of them uſe the wood of this tree for fuel, but they take the kernels out of the ſhells, into which they put ſmall ſtones, then bore a hole through each ſhell and ſtring them; theſe they tie about their legs to dance with, as the morris-dancers uſe bells.

The ſecond ſort grows naturally in the Spaniſh Weſt-Indies, and alſo in ſome of the French Iſlands in America, and hath lately been introduced into the Britiſh Iſlands.

This riſes with a round ſtalk about the ſame height as the former, dividing upward into many branches. Theſe, when young, are covered with a green ſmooth bark, but as they grow older, the bark becomes rough, and changes to a gray or Aſh-colour. The leaves are four or five inches long, and half an inch broad in the middle, ending in ſharp points; theſe are of a lucid green, and come out in cluſters without order, and are full of a milky juice, which flows out when they are broken. The flowers come out from the ſide of the branches upon long foot-ſtalks, each ſupporting two or three yellow flowers with long tubes, ſpreading open in the ſame manner as the former.

Theſe plants may be propagated from their nuts, which muſt be procured from the countries where they grow naturally; which ſhould be put into ſmall pots filled with light earth, and plunged into a hot-bed of tanners bark in the ſpring, and treated in the ſame manner as other tender exotic ſeeds, giving them now and then a little water to promote their vegetation. When the plants are come up about two inches high, they ſhould be tranſplanted each into a ſeparate pot, and plunged again into a hot-bed of tanners bark, obſerving to ſhade the glaſſes in the heat of the day, until the plants have taken new root. As the ſummer advances, theſe plants ſhould have air admitted to them, in proportion to the warmth of the ſeaſon; and when they have filled theſe ſmall pots with their roots, they ſhould be turned out, and tranſplanted into pots of a larger ſize. After they are new potted, they ſhould be plunged into the hot-bed again. When the plants are grown about a foot high, they ſhould have a larger ſhare of air, in order to harden them before the winter, but they ſhould not be wholly expoſed to the open air. In the winter theſe plants ſhould be placed in a warm ſtove, and during that ſeaſon they ſhould have very little water given them, eſpecially in cold weather, leſt it ſhould rot their roots. Theſe plants will not thrive well unleſs they are conſtantly kept in tan; and as they abound with milky juice, ſo they ſhould be ſparingly watered, for they are impatient of moiſture, eſpecially during the winter ſeaſon.

When by any accident the tops of theſe plants are injured, they frequently put out ſhoots from their roots, which, if carefully taken up and potted, will make good plants, ſo that they may be this way increaſed.

CERCIS. Lin. Gen. Plant. 458. The Judas-tree.

The CHARACTERS are,

The flower hath five petals, which are inſerted in the empalement, and greatly reſemble a papilionaceous flower. The ſtandard is of one roundiſh petal, and the keel is compoſed of two petals. It hath ten diſtinct ſtamina, four of which are longer than the reſt. It hath a long ſlender germen, which afterward becomes an oblong pod with an oblique point, having one cell, incloſing ſeveral roundiſh compreſſed ſeeds.

The SPECIES are,

1. CERCIS (*Siliquaſtrum*) foliis cordato-orbiculatis glabris. Hort. Cliff. 156. *Cercis with round, heart-ſhaped, ſmooth leaves; the common Judas-tree.*

2. CERCIS (*Canadenſis*) foliis cordatis pubeſcentibus. Hort Cliff. 156. *Cercis with downy heart-ſhaped leaves; commonly called Canada Arbor Judæ, or Red Bud-tree.*

The firſt ſort grows naturally in the ſouth of France, in Spain, and Italy. This riſes with an upright trunk to the height of twelve or fourteen feet, covered with a dark reddiſh bark, and divides upward into many irregular branches, garniſhed with round, heart-ſhaped, ſmooth leaves, placed irregularly on the branches, having long foot-ſtalks. The flowers come out on every ſide the branches, and many times from the ſtem of the tree in cluſters, ariſing many from the ſame point, having ſhort foot-ſtalks; they are of a very bright purple colour, ſo make a fine appearance, eſpecially when the branches are covered pretty thick with them, for they come out in the ſpring with the leaves, ſo are in full beauty before the leaves have obtained to half their ſize; theſe have an agreeable poignancy, ſo are frequently eaten in ſallets. When the flowers fall off, the germen becomes a long flat pod with one cell, containing one row of roundiſh ſeeds, which are a little compreſſed.

Theſe trees are uſually planted with other flowering-trees and ſhrubs, as ornamental to pleaſure-gardens; and for their ſingular beauty deſerve a place as well as moſt other ſorts, for when they are arrived to a good ſize they are very productive of flowers, ſo as that the branches are often cloſely covered with them; but the birds peck them off the trees, being enticed thereto by the honey liquor in the empalement. The ſingular ſhape of their leaves make a very pretty variety in the ſummer, after the flowers are paſt, and are never damaged by inſects, ſo that they are often entire,

tire, when many other trees have their leaves almoſt eaten up. This tree flowers in May, when planted in the full air, but againſt warm walls it is a fortnight or three weeks earlier.

The wood of this tree is very beautifully veined with black and green, and takes a fine poliſh, ſo may be converted to many uſes.

There are two other varieties of this tree, one with a white, and the other hath a fleſh-coloured flower, but theſe have not half the beauty of the firſt. Tournefort alſo mentions one with broader pods and pointed leaves, which I believe is only a variety of this.

The ſecond ſort grows naturally in moſt parts of North America, where it is called Red Bud, I ſuppoſe from the red flower-bud appearing in the ſpring before the leaves come out; this grows to a middling ſtature in the places where it is a native, but in England rarely riſes with a ſtem to any great height, but branches out near the root. The branches of this are weaker than thoſe of the firſt ſort; the leaves are downy, and terminate in points, whereas thoſe of the firſt are ſmooth, and round at the end where they are indented. The flowers of this are alſo ſmaller, ſo do not make ſo fine appearance as thoſe of the firſt, but the trees are equally hardy, ſo will thrive in the open air very well.

Theſe plants may be propagated by ſowing their ſeed upon a bed of light earth towards the latter end of March, or the beginning of April, (and if you put a little hot dung under the bed, it will greatly facilitate the growth of the ſeeds;) when your ſeeds are ſown, you ſhould ſift the earth over them about half an inch thick; and if the ſeaſon prove wet, it will be proper to cover the bed with mats, to preſerve it from great rains, which will burſt the ſeeds, and cauſe them to rot; theſe ſeeds will often remain till the ſpring following before they come up, ſo the ground muſt not be diſturbed till you are convinced that the plants are all come up, or their ſeeds are rotten, for ſome few may riſe the firſt year, and a greater number the ſecond.

When the plants are come up they ſhould be carefully cleared from weeds, and in very dry weather muſt be now and then refreſhed with water, which will greatly promote their growth. The winter following, if the weather is very cold, it will be proper to ſhelter the plants, by covering them either with mats or dry ſtraw, in hard froſts, but they ſhould conſtantly be opened in mild weather, otherwiſe they will grow mouldy and decay.

About the beginning of April you ſhould prepare a ſpot of good freſh ground, to tranſplant theſe out (for the beſt ſeaſon to remove them is juſt before they begin to ſhoot;) then you ſhould carefully take up the plants, being careful not to break their roots, and plant them in the freſh ground as ſoon as poſſible, becauſe if their roots are dried by the air, it will greatly prejudice them.

The diſtance theſe ſhould be planted, muſt be proportionable to the time they are to remain before they are again tranſplanted; but commonly they are planted two feet row from row, and a foot aſunder in the rows, which is full room enough for them to grow two or three years, by which time they ſhould be tranſplanted where they are deſigned to remain; for if they are too old when removed, they ſeldom ſucceed ſo well as younger plants.

When they have remained in the nurſery two or three years, they ſhould be tranſplanted in the ſpring where they are deſigned to remain, which may be in wilderneſs quarters among other flowering trees, obſerving to place them with trees of the ſame growth, ſo as that they may not be overhung, which is a great prejudice to moſt plants.

CEREFOLIUM. See CHÆREFOLIUM.

CEREUS. Par. Bat. 122. The Torch Thiſtle.

The CHARACTERS are,

The flower is compoſed of a great number of narrow pointed petals, which ſpread open like the ſun's rays. It hath a great number of declining ſtamina, which are inſerted to the baſe of the petals. The germen, which is ſituated under the empalement, afterward becomes an oblong ſucculent fruit, with a prickly ſkin, full of ſmall ſeeds incloſed in the pulp.

The SPECIES are,

1. CEREUS (*Hexagonus*) erectus, ſexangularis, longus, angulis diſtantibus. *Talleſt upright Torch Thiſtle of Surinam.*

2. CEREUS (*Tetragonus*) erectus quadrangularis, angulis compreſſis. *Upright Cereus with four compreſſed angles.*

3. CEREUS (*Lanuginoſus*) erectus octangularis, angulis obtuſis, ſupernè inermibus. *Upright Cereus with eight obtuſe angles, having no ſpines on the upper part.*

4. CEREUS (*Peruvianus*) erectus octangularis, angulis obtuſis, ſpinis robuſtioribus patulis. *Upright Cereus with eight angles which are obtuſe, and ſtrong ſpreading ſpines.*

5. CEREUS (*Repandus*) erectus novemangularis, obſoletis angulis, ſpinis lanâ brevioribus. *Upright Cereus with nine angles, and ſpines ſhorter than the down.*

6. CEREUS (*Heptagonus*) erectus octangularis, ſpinis lanâ longioribus. *Upright Cereus with eight angles, and ſpines longer than the down.*

7. CEREUS (*Royeni*) erectus novemangularis, ſpinis lanâ æqualibus. *Upright Torch Thiſtle with nine angles, and ſpines of equal length with the down.*

8. CEREUS (*Gracilis*) erectus gracilior novemangularis ſpinis brevibus, angulis obtuſis. *Slenderer upright Torch Thiſtle, having nine obtuſe angles, and ſhort ſpines.*

9. CEREUS (*Triangularis*) repens triangularis, fructu maximo rotundo, rubro, eſculento. *Creeping triangular Torch Thiſtle, with a very large, round, red, eatable fruit; commonly called in the Weſt-Indies the true Prickly Pear.*

10. CEREUS (*Compreſſis*) repens triangularis, angulis compreſſis. *Creeping triangular Torch Thiſtle, with compreſſed angles.*

11. CEREUS (*Grandifloris*) repens ſubquinquangularis. *Creeping Torch Thiſtle with five angles.*

12. CEREUS (*Flagelliformis*) repens decemangularis. *Creeping Cereus with ten angles.*

The firſt ſort has been the moſt common in the Engliſh gardens. It grows naturally in Surinam.

This riſes with an upright ſtalk, having ſix large angles, which are far aſunder, and are armed with ſharp ſpines, which come out in cluſters at certain diſtances, ariſing from a point, but ſpread open every way like a ſtar; the outer ſubſtance of the ſtem is ſoft, herbaceous, and full of juice, but in the center there is a ſtrong fibrous circle running the whole length, which ſecures the ſtem from being broke by winds. This will riſe to the height of thirty or forty feet, provided their tops are not injured, if they have room to grow; but ſome of them have grown too tall to be kept in the ſtoves, ſo have been either cut off, or the plants laid down at length in winter; but whenever the ſtems are cut or otherwiſe injured, they put out one, two, or ſometimes three ſhoots, from the angles immediately under the wounded part, and frequently one or two lower down. Theſe ſhoots, if they are not cut off, form ſo many diſtinct ſtems, and grow upright; but theſe are ſeldom ſo large as the principal ſtem, eſpecially if more than one is left at the ſame place. The flowers come out from the angles, on the ſide of the ſtem; theſe have a thick fleſhy foot-ſtalk, which is ſcaly, round, channelled, and hairy, ſupporting a ſwelling germen, upon the top of which ſits the ſcaly prickly empalement, cloſely ſurrounding the petals of the flowers, till a little time before they expand, which in moſt of the ſorts is in the evening; and their duration is very ſhort, for before the next morning they wither and decay. The flower of this ſort is compoſed of many concave petals, which, when fully expanded, are as large as thoſe of the Hollyhock; the inner petals are white, and are crenated at their extremity. The empalement is green, with ſome purple ſtripes; the middle of the flower is occupied by a great number of ſtamina, which decline in the middle, and riſe at their extremities, having roundiſh ſummits. The flowers of this kind are never ſucceeded by fruit in this country, nor do the plants often produce their flowers here, but when they do, there are generally ſeveral on the ſame plant. I have ſome years had more than a dozen upon a ſingle plant, which have all flowered within a few days of each other.

This ſort is not ſo tender as the others, ſo may be preſerved in a warm green-houſe without any artificial heat; but the plants ſhould have no water given them in winter, when they are thus ſituated; for unleſs they are placed in a ſtove, where the moiſture is ſoon evaporated, the wet will occaſion them to rot.

The ſecond ſort riſes with an upright ſtem like the firſt, but it hath only four angles, which are compreſſed, and ſtand far aſunder. This is very ſubject to put out many ſhoots from the ſides, which ſtops its upright growth, ſo that the plants rarely riſe more than four or five feet high.

The third, fourth, fifth, ſixth, ſeventh, and eighth ſorts, grow naturally in the Britiſh Iſlands of America. Theſe have the ſame form as the firſt, but differ in the ſize of their ſtems, the number of their angles, and the length of their ſpines, as is before expreſſed in their titles; but except the eighth ſort, not any of them have flowered in England as yet, though there are many of the plants which are more than fifteen feet high: the eighth ſort hath the ſmalleſt ſtem of any of the upright ſorts which I have yet ſeen; this hath nine obtuſe angles, which are armed with ſhort ſpines, placed at farther diſtances than thoſe of the other ſorts, nor are the channels between the angles near ſo deep. The flowers of this are produced from the angles, in the ſame manner as the firſt, but are ſmaller, and the empalement is of a light green, without any mixture of colour. The fruit is about the ſize and ſhape of a middling Bergamot Pear, having many ſoft ſpines on the ſkin; the outſide is a pale yellow, the inſide very white, full of pulp, having a great number of ſmall black ſeeds lodged in it. This ſort frequently flowers in July, and in warm ſeaſons will perfect its fruit, which hath very little flavour in this country.

Theſe ſorts are more impatient of cold than the firſt, ſo require a ſtove to preſerve them in winter; nor ſhould they be expoſed abroad in ſummer, but kept conſtantly in the houſe, giving them a large ſhare of air in warm weather.

The twelfth ſort grows naturally in Peru. This is not ſo tender as the other ſorts, ſo may be preſerved in a green-houſe, or under a good frame in winter, and in ſummer ſhould be expoſed to the open air, which will prevent the ſhoots from drawing weak, and thereby a greater number of flowers will be produced; but during the time they remain in the open air, they ſhould have very little water; and if the ſeaſon ſhould prove wet, the plants ſhould be ſcreened from it, otherwiſe it will cauſe them to rot the following winter. This ſort produces its flowers in great plenty in May, and ſometimes earlier, when the ſeaſon is warm.

The ninth ſort is, by the inhabitants of Barbadoes, trained up againſt their houſes for the ſake of its fruit, which is about the bigneſs of a Bergamot Pear, and of a moſt delicious flavour. This, and alſo the tenth, eleventh, and twelfth ſorts, are very tender, ſo require a very warm ſtove to preſerve them. Theſe ſhould be placed againſt the walls of the ſtove, into which they will inſinuate their roots, and extend themſelves to a great length; and with a little help in faſtening them to the wall here and there, may be led up about the cieling of the houſe, where they will appear very handſome. And the eleventh ſort, when arrived to a ſufficient ſtrength, will produce many exceeding large, beautiful, ſweet-ſcented flowers; but they are (like all the flowers of theſe kinds) of very ſhort duration, ſcarcely continuing full blown ſix hours; nor do the ſame flowers ever open again, when once cloſed: they begin to open in the evening between ſeven and eight of the clock, are fully blown by eleven, and by three or four the next morning fade, and hang down quite decayed; but, during their continuance, there is ſcarce any flower of greater beauty, or that makes a more magnificent appearance; for the calyx of the flower, when open, is near a foot diameter, the inſide of which, being of a ſplendid yellow colour, appears like the rays of a bright ſtar, the outſide of a dark brown, and the petals of the flowers being of a pure white, adds to the luſtre, and the great number of recurved ſtamina ſurrounding the ſtyle in the center of the flower, make a fine appearance; add to this the fine ſcent of the flower, which perfumes the air to a conſiderable diſtance, there is ſcarce any plant which deſerves a place in the hot-houſe ſo much as this, eſpecially as it is to be trained againſt the wall, where it will not take up room. The uſual ſeaſon of its flowering is in July; when the plants are large they will produce a great number of flowers, ſo that there will be a ſucceſſion of them for ſeveral nights, and many of them will open the ſame night. I have frequently had ſix or eight flowers open at the ſame time upon one plant, which have made a moſt magnificent appearance by candle-light, but none of them have been ſucceeded by any appearance of fruit.

The tenth ſort produces a flower little inferior to the former, as I have been informed by perſons who have ſeen them, but I never had the good fortune to have any of theſe plants which have been under my care flower; nor have I heard of more than two gardens where they have as yet flowered in England; the firſt of them was many years ſince in the Royal-gardens at Hampton-court, when there was a curious collection of exotic plants kept in good order in thoſe gardens, which have ſince been greatly neglected; the other was produced in the gardens of the Right Hon. the Marquis of Rockingham, at Wentworth-hall in Yorkſhire. Theſe are the only gardens in this country, where I have heard of this ſort having produced flowers; although there are many of theſe plants in ſeveral gardens, which are of a conſiderable age, and extend their branches to a very great diſtance.

The ninth ſort hath produced its flowers in the Phyſic-garden at Chelſea, which are not ſo beautiful as thoſe of the eleventh ſort; the petals are of a dirty white colour: theſe generally appear in October, and commonly keep open two days, and the fruit is generally eſteemed by all the inhabitants.

The twelfth ſort produces a greater number of flowers than either of the other; theſe are of a fine Pink colour, both within and without; the petals are not ſo numerous, and the tube of the flower is longer than thoſe of the other ſpecies, and, contrary to all the other ſorts, keep open three or four days, provided the weather is not too hot. During the continuance of theſe flowers, they make a fine appearance. This ſort hath very ſlender trailing branches, which require to be ſupported; but they do not extend ſo far as thoſe of the other ſort, nor are their branches jointed as thoſe are, ſo they cannot be trained ſo far againſt the walls of the houſe; but as it produces ſuch beautiful flowers, and in in so great plenty, it may be placed among the first class of exotic plants. This plant has produced fruit in the garden at Chelsea.

These are all propagated by cuttings, so that if you intend to increase the number of them, you must cut off their stems at what length you please; and the cuttings should be laid in a dry place to heal, at least ten days or a fortnight before they are planted; but if they lie three weeks it is much the better, and will be in less danger of rotting, especially those sorts which are the most succulent.

These cuttings should be planted in pots filled with the mixture of earth before directed, laying some stones in the bottom of the pots to drain off the moisture; then place the pots into a gentle hot-bed of tanners bark to facilitate their rooting, giving them once a week or ten days a gentle watering.

The best season for this work is in June, or the beginning of July, that they may have time to root well before winter; toward the middle of August you must begin to give them air by degrees, to harden them against winter, but they should not be wholly exposed to the open air or sun; at the end of September they must be removed into the stove, where they are to abide the winter, during which season you must be very careful not to let them have much water, and always observe to place the young plants, for the first winter, in a little warmer situation than the older, as being somewhat tenderer.

When you have once cut off the tops of any of these plants in order to increase them, the lower parts will put forth fresh shoots from their angles, which, when grown to be eight or nine inches long, may also be taken off to make fresh plants; and, by this means, the older plants will continually afford a supply, so that you never need cut off above one plant of a sort, which you should preserve for to multiply.

These plants being succulent, they will bear to be a long time out of the ground, therefore whoever hath a mind to get any of them from the West-Indies, need give no other instructions to their friends, but to cut them off, and let them lie two or three days to dry; then put them up in a box with dry hay, or straw, to keep them from wounding each other with their spines, and if they are two or three months on their passage, they will keep very well, provided no wet get to them.

CERINTHE. Lin. Gen. Plant. 171. Honeywort.

The CHARACTERS are,

The flower hath one petal, with a thick short tube, and at the brim is quinquefid; it hath five short stamina. In the bottom are situated four germen, two of which afterward becomes so many seeds, which are hard, smooth, plain on one side, but convex on the other, and are inclosed in the empalement.

The SPECIES are,

1. CERINTHE (*Major*) foliis ovato-oblongis, asperis, amplexicaulibus, corollis obtusiusculis, patulis. *Honeywort with oval, oblong, rough leaves embracing the stalk, and spreading blunt petals.*

2. CERINTHE (*Glaber*) foliis oblongo-ovatis, glabris, amplexicaulibus, corollis obtusiusculis, patulis. *Honeywort with a purplish red flower.*

3. CERINTHE (*Minor*) foliis amplexicaulibus, integris, fructibus geminis, corollis acutis, clausis. Lin. Sp. Pl. 137. *Smaller Honeywort.*

The first sort grows naturally in Germany and Italy. It is an annual plant, with smooth branching stalks a foot and a half long, garnished with oval, oblong, prickly leaves, which are of a sea-green, spotted with white, and embrace the stalks with their base; the flowers are produced at the end of the branches, standing between the small leaves; these are oblong, tubulous, and blunt at the top, where the tube is greatly enlarged; they are yellow, and have a mellifluous liquor in their tubes, with which the bees are much delighted; these flowers have each four embryos, or germen, but only two of them are fruitful. If the seeds are not taken as soon as they change black, they drop out of the empalement in a short time; so unless they are carefully gathered up, they will vegetate with the first moist weather.

The second sort is like the first, but the leaves are larger, and smooth, having no prickles on them. The flowers of this are of a purplish red colour, and the plants grow larger. This grows in Italy and the south of France; it is also an annual plant.

The third sort grows naturally on the Alps, and other mountainous places; it hath slenderer stalks than either of the former, which rise two feet high; the leaves embrace the stalks with their base, and are of a blue green colour The flowers are small, their upper part is deeply cut into five segments, but the mouth of the tube is closely shut up. If the seeds of this are permitted to scatter, the plants will come up in autumn, and these will grow much taller, and flower earlier than those which are sown in the spring.

The several varieties of this plant are propagated by seeds, which should be sown soon after they are ripe, for if they are kept till spring, the growing quality of some of them is often lost; the plants are hardy, and if the seeds are sown in a warm situation, they will endure the winter's cold very well without shelter; these autumnal plants are also much surer to produce ripe seeds than those which are sown in the spring, which are generally late in the season before they flower; and consequently, if the autumn should not prove very warm, their seeds would not be perfected.

CESTRUM. Lin. Gen. Pl. 231. Bastard Jasmine.

The CHARACTERS are,

The flower is funnel-shaped, of one petal, having a long cylindrical tube, which spreads open at the top, and is cut into five equal segments; it hath five slender stamina. The oval cylindrical germen is situated in the empalement, which afterward becomes an oval oblong berry with one cell, inclosing several roundish seeds.

The SPECIES are,

1. CESTRUM (*Nocturnum*) floribus pedunculatis. Hort. Cliff. 490. *Cestrum with flowers standing upon foot-stalks.*

2. CESTRUM (*Diurnum*) floribus sessilibus. Hort. Cliff. 491. *Cestrum with flowers sitting close to the branches.*

3. CESTRUM (*Nervosum*) foliis lanceolatis oppositis nervis transversalibus, pedunculis ramosis. *Cestrum with spear-shaped leaves growing opposite, having transverse veins, and branching foot-stalks to the flowers.*

4. CESTRUM (*Spicatum*) foliis ovato-lanceolatis, floribus spicatis, alaribus & terminalibus. *Cestrum with oval spear-shaped leaves, and flowers growing in spikes from the sides and tops of the branches.*

5. CESTRUM (*Confertum*) foliis oblongo-ovatis, obliquis, floribus alaribus confertis, tubo longissimo & tenuissimo. *Cestrum with oblong oval leaves, which are oblique, and flowers growing in clusters from the sides of the branches, with a very long slender tube.*

6. CESTRUM (*Venenatum*) foliis lanceolatis obliquis, floribus alaribus, pedunculis foliosis. *Cestrum with oblique spear-shaped leaves, flowers proceeding from the sides of the branches, and leafy foot-stalks.*

The first sort was many years past raised in the curious gardens of the Duchess of Beaufort at Badmington in Gloucestershire, and was from thence communicated to several gardens in England in Holland, where in the latter it passes under the title of Badmington Jasmine to this time. This grows naturally in the island of Cuba, from whence I received

ceived the feeds by the title of Dama de Noche, i. e. Lady of the Night, which appellation I fuppofe was given it from the flowers fending out a ftrong odour after the fun is fet.

It rifes with an upright ftalk about fix or feven feet high, covered with a grayifh bark, and divides upward into many flender branches, which generally incline to one fide; thefe are garnifhed with leaves placed alternate, which are near four inches long, and one and a half broad, fmooth on their upper fide, of a pale green, and on their under fide they have feveral tranfverfe veins, ftanding on fhort foot-ftalks. The flowers are produced at the wings of the leaves in fmall clufters, upon fhort foot-ftalks, each fuftaining four or five flowers, which have very fhort empalements, with long flender tubes, which are enlarged at the top, where they are cut into five parts which are reflexed; thefe are of an herbaceous colour.

The feeds of the fecond fort were fent me from the Havannah, by the title of Dama de Dio, or Lady of the Day; this rifes with an upright ftalk to the height of eight or ten feet, covered with a fmooth, light, green bark, and divides upward into many fmaller branches, garnifhed with fmooth leaves near three inches long, and one and a half broad, of a lively green colour; thefe are ranged alternately on the branches. Toward the upper part of the fhoots come out the flowers from the wings of the leaves, ftanding in clufters clofe to the branches; they are very white, fhaped like thofe of the former fort, and fmell fweet in the day-time, from whence it had the appellation of Lady of the Day.

The third fort was fent me from Carthagena in New Spain, near which place it grows naturally; this rifes with a fhrubby ftalk five or fix feet high, covered with a brown bark, and divides upward into many fmall branches, garnifhed with fpear-fhaped leaves, about four inches long, and little more than one broad; they are fmooth, of a light green, and have many horizontal veins running from the midrib to the fides, and are placed oppofite by pairs. From the wings of the leaves, toward the upper part of the branches, are produced the flowers, ftanding upon branching foot-ftalks, each fuftaining four or five flowers, whofe tubes are fwelling at their bafe, juft above the empalement, but contract upward to the mouth, where the petal is cut into five broad fegments which fpread flat; they are white, but without fcent.

The fourth fort was fent me from Carthagena with the former. This rifes with a fhrubby ftalk ten or twelve feet high, covered with a light gray bark, fending out many branches the whole length, garnifhed with oval fpear-fhaped leaves, ftanding without order; they are two inches and a half long, and one and a half broad, of a light green, with flender foot-ftalks. The flowers come out in loofe fpikes from the fide, and alfo the end of the branches, which are fhaped like thofe of the firft fort, and are of a whitifh green colour, without fcent.

The fifth fort rifes with feveral fhrubby ftalks, which grow eight or ten feet high, covered with a white fmooth bark, and fend out many irregular branches, which are garnifhed with oblong oval leaves, which at their bafe are longer on one fide, fo that the foot-ftalk is oblique; they are placed on the branches without order, and are of a pale green. The flowers come out in clufters from the fide of the branches, many of them arifing from the fame point; thefe have very flender long tubes, which are cut at the top into five acute fegments, which are erect. Thefe are of a pale yellow, and without fcent.

The fixth fort grows naturally in Jamaica. This rifes with a woody ftem eight or nine feet high, covered with a fmooth whitifh bark, fending out many branches toward the top, which are garnifhed with fpear-fhaped leaves, whofe foot-ftalks are oblique; they are three inches long, and little more than one broad, fmooth, of the confiftence with Bay leaves, and are placed alternate on the branches. From the wings of the leaves the flowers are produced; the foot-ftalks of the flowers are garnifhed with fmall leaves, ftanding between each flower in a fingular manner, the flowers rifing one above the other, and between, or oppofite to each, is one, and fometimes two leaves of the fame form with thofe on the branches. The flowers are of a pale yellow, and emit a difagreeable odour.

All thefe plants grow naturally in very hot countries, fo cannot be preferved in England without artificial heat, therefore require to be placed in a warm ftove, efpecially in the winter. The two firft are hardier than the others; thefe I have kept feveral years in a dry ftove, with a moderate fhare of heat in winter, and in the middle of fummer have fet them in the open air, in a warm fituation. With this management I have found them thrive, and produce flowers much better than when they have been placed in a greater heat; but I have often endeavoured to keep thefe plants through the winter in a green-houfe, or a glafs-cafe, without fire, but could never fucceed, for by the end of January they commonly decayed.

Thefe plants may be propagated either by feeds, or cuttings. Thofe which come from feeds are always the moft vigorous, and ftraiteft plants; but as they do not produce feeds in England, fo the other method is generally practifed, becaufe their feeds are rarely brought hither.

The beft time to plant thefe cuttings is about the end of June, by which time the fhoots will have had time to recover their ftrength, after their confinement during the winter feafon. The fhoots which come out from the lower part of the ftalks, fhould always be chofen for this purpofe. Thefe fhould be cut about four inches long, and five or fix of them may be planted in one halfpenny pot, for the cuttings of moft forts of exotic plants will fucceed better when they are planted in thefe fmall pots than they do in larger, as I have many years experienced. When the cuttings are planted, the earth muft be preffed pretty clofe to them, and then gently watered; after which the pots muft be plunged into a moderate hot-bed of tanners bark, and fhaded from the fun. With this management the cuttings will put out roots in a month or fix weeks, when they fhould be gradually expofed to the fun; and when they begin to put out fhoots, they muft have a greater fhare of frefh air admitted to them, to prevent their drawing up weak. When they have made good roots, they fhould be carefully fhaken out of the pots, and each put into a feparate fmall pot; then give them fome water, to fettle the earth to their roots, and plunge them again into the tan-bed, obferving if any of their leaves hang down, to fhade them from the fun in the middle of the day, until they have taken frefh root; after which they fhould have a large fhare of air in warm weather, to ftrengthen them before winter.

In the autumn the plants muft be removed into the bark-ftove, and plunged into the tan-bed, where they muft be treated in the fame manner as other tender exotic plants; for although the two firft forts may be treated otherwife when they have obtained ftrength, yet in the firft winter they may be managed in the fame way as the others. There muft be great care had in watering of thefe plants in winter, for they are all (except the fecond fort) very impatient of moifture, fo that they are foon killed by being over watered.

If the feeds of thefe are procured from the countries where they grow naturally, they fhould be fowed in fmall pots, and plunged into a moderate hot-bed of tanners bark, giving them now and then a little water. Sometimes the feeds will come up the fame year, but they very often lie in the ground

ground till the ſpring following; ſo that if the plants do not appear in ſix or ſeven weeks after the ſeeds are ſown, they will not come up that ſeaſon; in which caſe the pots may be plunged in the tan-bed in the ſtove between the other plants, where they will be ſhaded from the ſun, and but little water given them; in this ſituation they may remain till the following ſpring, when they ſhould be removed, and plunged into a freſh hot-bed, which will bring up the plants in a ſhort time, provided the ſeeds were good.

CETERACH. See ASPLENIUM.

CHÆROPHYLLUM. Lin. Gen. Pl. 320. Chervil.

The CHARACTERS are,

It is a plant with umbellated flowers; the principal umbel is ſpreading, and compoſed of ſeveral ſmall ones, called rays; the flowers have five heart-ſhaped inflexed petals, and five ſtamina. The germen is ſituated below the flower, ſupporting two reflexed ſtyles, which afterward becomes an oblong pointed fruit, dividing in two parts, each having one ſeed, which is convex on one ſide, and plain on the other.

The SPECIES are,

1. CHÆROPHYLLUM (*Sylveſtre*) caule ſtriato, geniculis tumidiuſculis. Flor. Suec. 2. N. 257. *Wild Chervil with ſtriated ſtalks.*

2. CHÆROPHYLLUM (*Bulboſum*) caule lævi, geniculis tumidis. Lin. Sp. Pl. 258. *Chervil with a ſmooth ſtalk, and ſwelling knots.*

3. CHÆROPHYLLUM (*Temulum*) caule ſcabro, geniculis tumidis. Lin. Sp. Pl. 258. *Wild Chervil.*

4. CHÆROPHYLLUM (*Aureum*) caule æquali, foliolis inciſis acutis. Lin. Sp. Pl. 258. *Chervil with an equal ſtalk, and leaves cut into acute ſegments.*

The firſt ſort grows naturally on the ſide of highways, and the borders of the fields in moſt parts of England, ſo is never cultivated in gardens. It is frequently called Cow Parſley, but for what reaſon I cannot ſay, becauſe there are few animals who care to eat it, except the aſs, for it is reckoned to have ſomething of the quality of Hemlock, but in a leſs degree. It is a weed which ſhould be rooted out from all paſtures early in the ſpring, for it is one of the moſt early plants in ſhooting, ſo that by the beginning of April the leaves are near two feet high.

The ſecond ſort grows naturally in Hungary and Iſtria; this plant hath a thick tuberous root, from which come forth ſeveral leaves reſembling thoſe of Wild Chervil. The ſtalks riſe ſeven or eight feet high, which are ſpotted with purple, and are garniſhed with leaves of the ſame form as thoſe below. The knots at the joints of the ſtalks ſwell out on every ſide, at which is placed one of theſe divided leaves, and the ſtalks are terminated by ſmall umbels of white flowers, which are ſucceeded by long narrow ſeeds. If the ſeeds of this plant are permitted to ſcatter, the plants will come up without any farther care, and only require to be kept clean from weeds.

The third ſort grows naturally on the ſides of foot-walks, and on the borders of woods in many parts of England, ſo is not cultivated in gardens.

The fourth ſort grows naturally in the paſtures about Geneva and in Switzerland; this hath a perennial root, the leaves are ſhaped like thoſe of the firſt, but are broader, hairy, and more divided. The ſtalks riſe three feet high, which are channelled, and are terminated by large umbels, formed of many ſmall ones, which are compoſed of white or red flowers, ſometimes both colours in the ſame umbel, having five heart-ſhaped petals which turn inward; theſe are ſucceeded by two long pointed ſeeds; the whole plant has an aromatic ſmell and taſte.

CHAMÆCERASUS. See CERASUS and LONICERA.

CHAMÆCISTUS. See HELIANTHEMUM.

CHAMÆCLEMA. See GLECHOMA.

CHAMÆCYPARISSUS. See SANTOLINA.

CHAMÆDAPHNE. See RUSCUS.

CHAMÆDRYS. See TEUCRIUM.

CHAMÆLÆA. See CNEORUM.

CHAMÆMELUM. See ANTHEMIS.

CHAMÆMESPILUS. See MESPILUS.

CHAMÆMORUS. See RUBUS.

CHAMÆNERION. See EPILOBIUM.

CHAMÆPITYS. See TEUCRIUM.

CHAMÆRHODODENDRON. See RHODODENDRON, AZALEA, and KALMIA.

CHAMÆRIPHES. See CHAMÆROPS.

CHAMÆROPS. Lin. Gen. Plant. 1084. Dwarf Palm, or Palmetto.

The CHARACTERS are,

It hath male and hermaphrodite flowers in diſtinct plants; the hermaphrodite flowers are all included in one common ſpatha; the ſpadix or club is branching; each flower hath a ſmall three-pointed empalement, and one thick upright petal, which is cut into three parts, with five compreſſed ſtamina which join at their baſe. They have three roundiſh germen, which afterward becomes ſo many round berries, having one cell, each containing a ſingle ſeed. The male flowers are like the hermaphrodite, but the ſtamina are not diſtinct, nor have they any germen.

The SPECIES are,

1. CHAMÆROPS (*Humilis*) frondibus palmatis, plicatis, ſtipitibus ſpinoſis. Hort. Cliff. 482. *Dwarf Palm with folding palmated leaves, and prickly foot-ſtalks.*

2. CHAMÆROPS (*Glabra*) foliis flabelliformibus, maximis, ſtipitibus glabris. *Dwarf Palm with very large fan-ſhaped leaves and ſmooth foot-ſtalks; commonly called ſmall Palmetto Royal.*

The firſt ſort grows naturally in Spain, particularly in Andaluſia, where, in the ſandy land, the roots ſpread and propagate ſo faſt, as to cover the ground in the ſame manner as the Fern in England. The leaves of theſe plants are tied together to make beſoms for ſweeping.

This never riſes with a ſtem, but the foot-ſtalks of the leaves riſe immediately from the head of the root, and are armed on each ſide with ſtrong ſpines; they are flat on their upper ſurface, and convex on the under; to their ends the center of the leaves are faſtened, which ſpread open like a fan, having many foldings, and at the top are deeply divided like the fingers of a hand. The borders of the leaves are finely ſawed, and have white narrow edgings; theſe leaves ſpread out on every ſide of the plant, they are from nine to eighteen inches long, and near a foot broad in their wideſt part; as the lower leaves of the plants decay, ſo their veſtiges remain, and form a ſhort ſtump above ground, in the ſame manner as our common male Fern does; from between the leaves come out the ſpadix or club, which ſuſtains the flowers; this is covered with a thin ſpatha or hood, which falls off when the bunches open and divide. As all the plants of this ſort which I have ſeen flower were male, ſo I cannot give any particular deſcription of their fructification.

Theſe plants are beſt propagated by ſeeds, which ſhould be ſown in ſmall pots, filled with light ſandy earth, and plunged into a moderate hot-bed of tanners bark. If the ſeeds are freſh, the plants will come up in ſix weeks or two months; theſe riſe with a ſingle long-pointed leaf. If the plants are not too cloſe to each other in the pots, they will not require to be tranſplanted the firſt year, therefore they ſhould remain in the tan-bed all the ſummer, but in warm weather they muſt have plenty of air admitted to them. In autumn the pots ſhould be removed into the ſtove, and, if they are plunged into the bark-bed the firſt winter, it will greatly forward the growth of the plants. The following ſpring the plants ſhould be carefully turned out of the pots,

ſo

ſo as to preſerve their roots entire, for all the ſorts of Palms have tender roots, which, if they are cut off or broken, frequently kills the plants; then they ſhould be each planted into a ſeparate ſmall pot, and plunged into a freſh hot-bed, to encourage their taking root; the following ſummer they ſhould be gradually hardened, by raiſing the glaſſes pretty high, ſo as to admit a large ſhare of air to them, but they ſhould not yet be wholly expoſed to the open air. The autumn following the plants may be placed in a dry ſtove, but as the plants advance and get ſtrength, they may be treated more hardily, and in ſummer placed in the open air in a warm ſituation, and in winter may be preſerved in a warm green houſe without artificial heat.

As the plants advance in growth, ſo they ſhould be put into larger pots; but when this is done, there muſt be great care taken that their roots are not cut or broken, nor ſhould they have pots too large. In winter they muſt have but little water, and if they are expoſed to the open air in ſummer, they will not require much, unleſs the ſeaſon proves very warm and dry, in which caſe they may be ſparingly watered two or three times a week.

The ſecond ſort grows naturally in the Weſt-Indies, where it never riſes with a tall ſtem; the foot-ſtalks of the leaves are rounder than thoſe of the former, and have no ſpines on their ſides. When the plants are old their leaves are five or ſix feet long, and upward of two broad; theſe are folded in the ſame manner as thoſe of the firſt, but the folds are broader, and the leaves are of a darker green; ſome of theſe plants have put out ſlender bunches of male flowers in England, which were too imperfect to form a deſcription.

This ſort riſes freely from ſeeds, which may be eaſily procured from the iſlands in America; theſe muſt be ſown in the ſame manner as the former, and the plants treated in the ſame way; but as they are natives of a warmer climate, ſo they ſhould be conſtantly kept in the bark-ſtove, where, if they are carefully managed, they will make good progreſs.

CHAMÆRUBUS. See Rubus.

CHAMÆSYCE. See Euphorbia.

CHEIRANTHUS. Lin. Gen. Plant. 730. Stock Gilliflower and Wall-flower.

The Characters are,

The flower hath four petals in form of a croſs; it hath ſix parallel ſtamina, two of which are between the ſwelling leaves of the empalement, the other are a little ſhorter. It hath a four-cornered priſmatic germen, which afterward becomes a long compreſſed pod with two cells, opening with two valves, filled with compreſſed ſeeds.

The Species are,

1. Cheiranthus (*Eryſimoides*) foliis lineari lanceolatis ſerratis caule erecto, ſiliquis tetragonis. *Cheiranthus with narrow, indented, ſpear-ſhaped leaves, an upright ſtalk, and four-cornered pods.*

2. Cheiranthus (*Integrifolius*) foliis lanceolatis integerrimis, caule erecto, ſiliquis tetragonis. *Cheiranthus with ſpear-ſhaped entire leaves, an upright ſtalk, and quadrangular pods.*

3. Cheiranthus (*Cheiri*) foliis lanceolatis, acutis, glabris. Hort. Cliff. 334. *Cheiranthus with ſpear-ſhaped, pointed, ſmooth leaves; or Wall-flower.*

4. Cheiranthus (*Anguſtifolius*) foliis linearibus, unguibus petalorum calyce longioribus. *Cheiranthus with narrow leaves, and the necks of the petals longer than the empalement.*

5. Cheiranthus (*Annuus*) foliis lanceolatis, ſubdentatis, obtuſis, incanis, ſiliquis cylindricis apice acutis, caule herbaceo. Lin. Sp. Plant. 662. *Cheiranthus with ſpear-ſhaped leaves ſomewhat indented, obtuſe, and hoary, cylindrical pods with acute points, and an herbaceous ſtalk; commonly called the Ten Weeks Stock.*

6. Cheiranthus (*Incanus*) foliis lanceolatis, integerrimis, obtuſis, incanis, ſiliquis apice truncatis, compreſſis, caule ſuffruticoſo. Hort. Upſal. 187. *Cheiranthus with very entire ſpear-ſhaped leaves, which are obtuſe and hoary, compreſſed pods with truncated points, and a ſhrubby ſtalk; commonly called the Queen's Stock Gilliflower.*

7. Cheiranthus (*Coccineus*) foliis lanceolatis undatis, caule erecto indiviſo. *Cheiranthus with waved ſpear-ſhaped leaves, and an upright undivided ſtalk; commonly called the Brumpton Stock Gilliflower.*

8. Cheiranthus (*Albus*) foliis lanceolatis, integerrimis, obtuſis, incanis, ramis floriferis axillaribus, caule ſuffruticoſo. *Cheiranthus with hoary, entire, ſpear-ſhaped, obtuſe leaves, flower branches proceeding from the ſides, and a ſhrubby ſtalk; the Purple, or Violet Stock Gilliflower.*

9. Cheiranthus (*Glaber*) foliis lanceolatis, acutis, petiolatis, viridibus, caule ſuffruticoſo. *Cheiranthus with ſpear-ſhaped acute leaves, which are green, and have foot-ſtalks, and a ſhrubby ſtalk; commonly called White Wall-flower.*

10. Cheiranthus (*Feneſtralis*) foliis conferto-capitatis, recurvatis, undatis. Lin. Sp. Pl. App. 1198. *Cheiranthus with leaves growing cloſe together in heads, which are turned backward, and are waved.*

11. Cheiranthus (*Littoreus*) foliis linearibus, obtuſis, incanis, integerrimis, ſiliquis acuminatis, caule ſuffruticoſo. *Cheiranthus with narrow obtuſe leaves, which are hoary, and very entire ſharp pointed pods, and a ſhrubby ſtalk; or Narrow-leaved Sea Stock Gilliflower.*

12. Cheiranthus (*Maritimus*) caule diffuſo, foliis lanceolatis, ſeſſilibus, floribus alternis. *Cheiranthus with a diffuſed ſtalk, ſpear-ſhaped leaves ſitting cloſe to the ſtalks, and flowers placed alternate; commonly called Dwarf, or Virginia Stock Gilliflower.*

13. Cheiranthus (*Chius*) foliis lanceolatis, ſubdentatis, retuſis, ſiliquis apice ſubulatis. Hort. Upſal. 187. *Cheiranthus with ſpear-ſhaped leaves indented at bottom, and pods with awl-ſhaped points.*

14. Cheiranthus (*Tricuſpidatis*) ſiliquarum apicibus tridentatis. Hort. Cliff. 335. *Cheiranthus with pods indented in three parts at the point.*

The firſt ſort grows naturally in the ſouth of France, in Spain, and Italy; this is an annual plant, which riſes a foot high, with an angular channelled ſtalk, which branches upward, garniſhed with long, narrow, green leaves, reſembling thoſe of the common Wall-flower, but are ſharply ſawed on their edges, ſitting cloſe to the ſtalks; at the extremity of the branches the flowers are produced in looſe ſpikes; theſe are yellow, having four petals ſituated in the form of a croſs, greatly reſembling thoſe of the common yellow Wall-flower, but have no ſcent, and are ſucceeded by long four-cornered pods filled with brown ſeeds.

The ſecond ſort grows naturally in Hungary and Iſtria; this is alſo an annual plant, riſing with an upright ſtalk, nearly the ſame height as the other, but doth not branch out as that doth. The leaves are broader, ſmoother, and not ſawed as thoſe of the other. The flowers come out in looſe ſpikes at the top of the ſtalks; theſe are ſmall, and of a pale yellow without ſcent, and are ſucceeded by four-cornered pods like thoſe of the former.

The third ſort grows naturally upon old walls and buildings in many parts of England, but is alſo cultivated in gardens for the fragrancy of its flowers. When theſe plants grow upon walls or buildings, they ſeldom riſe more than ſix or eight inches high, having very tough roots, and firm ſtalks; the leaves are ſhort and ſharp pointed, and the flowers are ſmall, but in gardens the plants will grow two feet high,

high, and branch out wide on every ſide; the leaves are broader, and the flowers much larger, but in ſevere winters, when theſe plants are frequently killed in the gardens, thoſe upon walls will receive no injury, though they are much more expoſed to the winds and froſts; for as theſe plants are ſtunted, and of a firmer texture, having but little juice, ſo the cold never affects them.

There is a variety of this with very double flowers, which is propagated in the gardens from ſlips, planted in the ſpring, which readily take root. There is one ſort of this with variegated leaves, which is preſerved in the gardens, but this is not quite ſo hardy as the plain.

The large, yellow, and bloody Wall-flower, are alſo ſuppoſed to be varieties of this, which have been improved by culture; and this I am inclinable to believe, becauſe I have frequently obſerved many of them degenerate to the common ſort; but although I have many years ſowed the ſeeds of the common ſort from the walls, yet I could never find them alter, except in being larger, but not any of them approached toward the other varieties. The large bloody Wall-flower will frequently riſe with double flowers from ſeeds, if they are carefully ſaved from ſuch plants as have five petals; and theſe double flowers may be propagated by ſlips, as the common ſort, but the plants ſo raiſed will not produce ſuch large ſpikes of flowers as thoſe which are propagated by ſeeds.

There is alſo another variety with double blood-coloured flowers, whoſe petals are ſhorter and more numerous, approaching nearer to the common double Wall-flower, but much larger. This is called the Old Bloody Wall-flower. It is propagated from ſlips in the ſame manner as the other double ſorts.

The fourth ſort grows naturally upon the Alps and the mountains in Italy, where it rarely riſes above ſix inches high; the leaves are very narrow, and the flowers grow in cloſe ſpikes at the end of the branches; they are of a pale yellow or brimſtone colour, and the necks of the petals are much longer than the empalement; they have but little ſcent. It was titled the Straw-coloured Wall-flower by the gardeners.

The ſorts with ſingle flowers produce ſeeds in plenty, from which the plants are raiſed, but the largeſt and deepeſt coloured flowers ſhould always be ſelected for this purpoſe, becauſe from ſeeds carefully ſaved there will be fewer of the plants degenerate. The ſeeds ſhould be ſown upon poor, or undunged ſoil, and when the plants are fit to remove, they ſhould be tranſplanted into nurſery beds, at about ſix inches diſtance each way. In the autumn they may be tranſplanted into the borders of the flower-garden where they are deſigned to remain, that the plants may get good roots before the froſt comes on. This is the method which is commonly practiſed with theſe flowers; but if the ſeeds are ſown upon poor land, where they are deſigned to remain, and not tranſplanted, they will thrive, and endure the froſt in winter, much better than thoſe which are removed; ſo that upon ruins or rubbiſh the ſeeds of theſe plants may be ſown, where they will thrive and continue much longer than in good land; and in ſuch places, if they are properly diſpoſed, they will be very ornamental, and their flowers having a ſtrong odour, will perfume the air to a conſiderable diſtance.

The fifth ſort is now generally known by the appellation of Ten Weeks Stock, but it is what was formerly titled Annual Stock Gilliflower, which of late has been applied to another ſpecies, which is biennial. This riſes with a round ſmooth ſtalk about a foot high, dividing into ſeveral branches upward, garniſhed with ſpear-ſhaped hoary leaves, which are rounded at their ends, and placed without order, of unequal ſizes; at the end of the branches the flowers are produced in looſe ſpikes, which are placed alternate; the empalement of the flower is large, erect, and ſlightly cut into ſeveral acute parts at the top; the petals are large and heart-ſhaped, ſpreading open in form of a croſs; the pods are long, cylindrical, and have a longitudinal furrow on one ſide, which opens into two cells, which are filled with flat roundiſh ſeeds, having a thin border.

Of this ſort there are the red, the purple, the white, and ſtriped, with ſingle flowers, and the ſame colours with double flowers; theſe are very great ornaments in the borders of the flower-garden in the autumn, when there is a ſcarcity of other flowers; and if the ſeeds are ſown at two or three different times, the flowers may be continued in ſucceſſion near three months.

The firſt ſowing ſhould be about the middle of February, upon a very ſlender hot-bed, juſt to bring up the plants, which muſt be guarded againſt froſt, and when they are fit to remove, they ſhould be tranſplanted into nurſery-beds, at about three or four inches diſtances. In theſe beds they may remain ſix or ſeven weeks to get ſtrength, and may then be planted into the borders of the flower-garden, where they are to remain: if theſe are tranſplanted when there is rain, they will ſoon take root, after which they will require no farther care. From theſe early plants good ſeeds may be expected, therefore ſome of the fineſt plants of each colour ſhould be preſerved, and marked for ſeeds, which, when ripe, ſhould be carefully cut before the froſt pinches, and the ſtalks tied up in ſmall bundles, and hung up in a dry room till the pods are well dried, when the ſeeds may be rubbed out, and preſerved for uſe.

The ſixth ſort is a biennial plant, though when the ſeeds are ſown early in the ſpring, the plants often flower the following autumn; but theſe plants which are ſo forward, are often killed in winter, therefore it is much better to ſow them in May, that the plants may not grow too rank the firſt ſeaſon, ſo will live through the winter, and theſe will produce large ſpikes of flowers the ſecond year.

This is commonly called the Queen's Stock Gilliflower by the gardeners, and differs greatly from the other ſorts by its branching ſtalk.

It riſes with a ſtrong ſtalk, which is almoſt ſhrubby, a foot high or more, having oblong, ſpear-ſhaped, hoary leaves, which are frequently waved on their edges; from the ſtalk is ſent out many lateral branches, which are garniſhed with the ſame ſhaped leaves, but are ſmaller; theſe ſide branches are each terminated by a looſe ſpike of flowers, each having an oblong woolly empalement, and conſiſt of four large roundiſh petals, which are indented at the end. When theſe plants grow in dry rubbiſh, they will laſt two or three years, and become ſhrubby; but thoſe with ſingle flowers are not worth preſerving after they have perfected their ſeeds.

The flowers of this ſort vary in their colour, ſome are of a pale red, others are of a bright red, and ſome are curiouſly variegated, but thoſe of the bright red are generally moſt eſteemed. There are always a great number of double flowers produced, if the ſeeds are well choſen, frequently three parts in four of the plants will be double; and as the plants divide into many branches, ſo they make a fine appearance during their continuance in flower.

The ſeventh ſort is known by the title of Brumpton Stock Gilliflower, I ſuppoſe from its having been there firſt cultivated in England. This riſes with an upright, ſtrong, undivided ſtalk to the height of two feet or more, garniſhed with long hoary leaves, which are reflexed, and waved on their edges, and at the top form a large head, out of the center of which ariſe the flower-ſtalk; when the plant is ſtrong, it is frequently a foot and a half long, putting out two or three ſhort branches toward the bottom; the flowers of this kind have longer petals than any of the other ſorts, and are formed into a pyramidal ſpike, but thoſe with ſingle

gle flowers are loosely disposed, because the flowers having but few petals do not fill the spike, as those do which are double, for these often have so many petals as to render each flower as large and full as small Roses; and when they are of a bright red, make a noble appearance, being excelled by none of the flowery tribe; but the plants of this sort produce but one spike, in which it differs from all the other kinds, and being constant in this particular, I think is sufficient to establish a distinct species. This sort is generally biennial, though many times the plants are preserved longer, but they are always stronger the first year of their flowering, than they will be after; so that the seeds are sown every spring, to continue a succession of flowering plants.

The eighth sort is the white Stock Gilliflower, which is of longer duration than either of the other sorts, I have frequently had these plants live three or four years, which have become shrubby; their stalks have been three feet high, and branched out on every side, so as to appear like shrubs; these seldom send out flower-stalks from the center of the plant, but it is the side-branches which produce the flowers, and these side-branches divide into several other, which is not common to the other sorts. There are always many double flowers rise from seeds of this sort, when they are well chosen; some years I have scarce had enough single flowers to preserve their kind.

The ninth sort is known by the title of white Wall-flower among the gardeners and florists. This rises with a greenish stalk a foot or little more high, dividing into many branches, garnished with smooth spear-shaped leaves, of a lucid green colour, of thicker consistence than those of any of the other sorts, which come out without any order; they are near three inches long, and about half an inch broad in the middle; the flowers are produced in loose spikes at the end of the branches, which are of a pure white, and have a great fragancy, especially in an evening, or in cloudy weathers; the flowers are succeeded by oblong compressed pods. like those of the other species. There is a variety of this with double flowers, which is propagated by cuttings or slips in the same manner as the double Wall flowers, and from seeds many plants have double flowers; but these plants require protection from great rains and frost in winter, so if they are planted in pots, and placed under a common frame in winter, where in mild weather they may enjoy the open free air, and covered from hard rains and frost, they may be preserved some years.

Sometimes many of the double flowers will come up from seeds, but especially from that which is titled Prussian Ten Weeks Stock. I have frequently raised more than one hundred plants in a season, without obtaining one double flower, and from the seeds of these, have the following year had more than half the plants with double flowers: but this is not to be expected often. There is a variety of this which is annual, and is commonly known by the title of Prussian Stock. This, if sown early in the spring, will ripen seeds perfectly, and the plants produce many double flowers, so it well deserves a place in the flower-garden.

The seeds of the tenth sort were sent me by Dr. Linnæus, from Upsal in Sweden. This plant rises about nine inches high, with an herbaceous swelling stalk; the leaves are produced in clusters at the top, which are very hoary, waved on their edges, have obtuse points, and set very close to the stalk; the flowers are produced in slender spikes from the side of the stalk; these are purple, but not so fragrant as many of the other sorts; the pods are woolly, and recurve backward at the end. This is an accidental variety, which has sprung up with me from seeds of the Queen's Stock Gilliflower.

All these sorts flower in May and June, at which time they are the greatest ornament to the flower-garden, therefore deserve our care to cultivate them as much as any of the flowery tribe; but in order to have many double flowers, there must be great care taken in the choice of plants for seeds, without which there can be little hopes of having these flowers in perfection. The only sure way of getting many double flowers, is to make choice of those single flowers which grow near many double ones; for I have always found those seeds which have been saved from plants growing in beds close to each other, where there happened to be many double flowers among them, have produced a much greater number of plants with double flowers than those which have been saved from plants of the same kinds, which grew single in the borders of the flower-garden; so that there should be a small bed of each kind planted, on purpose to save seeds in the flower-nursery; or if they are sown there, and the plants thinned properly when they are young, they need not be transplanted; for I have always observed the plants which have come up from scattered seeds, that have not been transplanted, endure the frost much better than those which have been removed; for as these plants send out horizontal roots, which spread near the surface of the ground, so when they are transplanted, the roots are forced downward, out of their natural direction; and if their stalks were grown tall before removal, they are generally planted low in the ground, whereby they are apt to rot, if the ground is moist, or the winter should prove wet, therefore where they can be left unremoved, there will be a better chance of their living through the winter; and as these beds need not be of great extent, so when the winter proves very severe, it will not be much trouble or expence to arch the beds over with hoops, and cover them with mats in frosty weather, by which method they may be always preserved.

The time for sowing of the seeds before mentioned, must be understood to be for the sorts which are biennial, for the annual, or Ten Weeks Stock Gilliflower, should be for the first season sown in February, as was before directed; and to succeed these, there should be another parcel sown in March; and those who are curious to continue these flowers late in the autumn, should sow a parcel of the seed the latter end of May; and if these last sown plants are upon a warm border, where they may be covered, by placing glasses before them in winter, or covering them with mats, they may be continued in flower till Christmas; and if some of the plants are potted, and put under a hot-bed frame in autumn, where they may enjoy the open air in mild weather, and screened from hard rains and frost, these plants may be kept flowering all the winter, when the winters are not very severe.

There are some who propagate the double Stock Gilliflower by slips and cuttings, which will take root when properly managed; but the plants so raised are never so strong as those which come from seeds, and their spikes of flowers are always very short, so have not half the beauty; therefore it is not worth while to practise this method, unless for those sorts which cannot be obtained with any certainty from seed.

The eleventh sort grows naturally in the south of France, in Spain and Italy, near the sea-coast. This rises near a foot high, with a ligneous stalk, dividing into many small branches, garnished with narrow hoary leaves, which are entire, and rounded at their extremity; the flowers are produced in loose spikes at the end of the branches, which are smaller than either of the sorts before mentioned, of a bright red at their first appearing, but fade to a purple before they fall off. The stalks, leaves, and the whole plant is very white, and by its woody stalks hath the appearance of a perennial plant, but it constantly perishes in autumn. The seeds of this sort should be sown in autumn upon a warm border, where the plants are designed to remain; when the plants come up they will require no farther care but to keep them clean from weeds, and thin them where they come up too close. These autumnal plants will flower early in June, so will produce good seeds; but those which are sown in the spring will flower in July and August, so that from these there cannot be any certainty of having ripe seeds; however, by sowing the seeds at two or three different seasons, there may be a succession of flowers continued for three or four months.

The twelfth sort is commonly sown in gardens, sometimes as an edging for borders, but more generally in patches between taller growing flowers: it is titled sometimes Dwarf Annual Stock Gilliflower, and by others it hath the appellation of Virginia Stock Gilliflower. This seldom rises more than six inches high, sending out many branches from the root, which spread near the ground, and grow irregular; these are garnished with spear-shaped leaves rounded at their ends, and sit close to the branches at their base; the flowers come out in loose spikes at the end of the branches, which are of a purple colour, composed of four petals in form of a cross, and are succeeded by slender pods, like those of the former sorts.

The thirteenth sort rises near two feet high, sending out many upright branches from the bottom, which are thinly garnished with spear-shaped leaves, the lower ones being a little indented; the flowers come out single, at great distances from each other, toward the upper part of the branches; these are small, of a purplish colour, and soon fall away, being succeeded by long taper pods, with awl-shaped points. This is an annual plant, which may be treated in the same manner as the last mentioned sort; but as it hath little beauty, so it is not often cultivated in gardens.

The fourteenth sort grows naturally on the sea-coasts in Italy, Spain, and Portugal. This is also an annual plant, which branches out from the root into many declining stalks; the lower leaves are about two inches long, and three quarters of an inch broad, very deeply sinuated on their edges, and hoary; those upon the stalks are of the same form, but much smaller; the flowers are produced from the side of the stalks singly, and at the top in loose spikes; the empalements of the flowers are covered with a white down, as are also the ends of the branches: the flowers are purple, composed of four leaves placed in form of a cross; the pods are about three inches long, taper, woolly, and at their ends are divided into three parts, which spread into a triangle. If the seeds are sown in autumn on a warm border, the plants will live through the winter, and these will flower early in June, so from these good seeds may be obtained.

CHELIDONIUM. Tourn. Inst. R. H. 231. tab. 116. Celandine and Horned Poppy.

The CHARACTERS are,

The flower hath four large roundish petals; in the center is situated a cylindrical germen, attended by a great number of stamina. The germen afterward becomes a cylindrical pod, with one or two cells, opening with two valves, and filled with many small seeds.

The SPECIES are,

1. CHELIDONIUM (*Majus*) pedunculis umbellatis. Lin. Gen. Plant. 505. *Celandine with an umbellated foot-stalk; or the common Celandine.*

2. CHELIDONIUM (*Laciniatum*) foliis quinque lobatis, lobis angustis acutè laciniatis. *Celandine whose leaves are composed of five narrow lobes, which are cut into many acute segments.*

3. CHELIDONIUM (*Glaucium*) pedunculis unifloris, foliis amplexicaulibus sinuatis, caule glabro. Lin. Sp. Plant. 506. *Celandine with single flowers on the foot-stalks, sinuated leaves which embrace the stalks, and a smooth stalk; or Yellow Horned Poppy.*

4. CHELIDONIUM (*Corniculatum*) pedunculis unifloris, foliis sessilibus pinnatifidis, caule hispido. Lin. Sp. Plant. 506. *Celandine with single flowers upon the foot stalks, leaves set close to the stalks, which have winged points, and a rough stalk. Hairy Glaucium, or Horned Poppy with a scarlet flower.*

5. CHELIDONIUM (*Glabrum*) pedunculis unifloris, foliis semiamplexicaulibus, dentatis glabris. *Celandine with foot-stalks having a single flower, and smooth indented leaves which half embrace the stalks. Smooth Horned Poppy with a scarlet flower.*

6. CHELIDONIUM (*Hybridum*) pedunculis unifloris, foliis pinnatifidis, linearibus, caule lævi, siliquis trivalvibus. Lin. Sp. Plant. 506. *Celandine with single flowers upon the foot-stalks, many pointed narrow leaves, and a smooth stalk. Horned Poppy with a Violet-coloured flower.*

The first sort is the common Celandine which is used in medicine, and is esteemed aperitive and cleansing, opening obstructions of the spleen and liver, and is of great use in curing the jaundice and scurvy. This grows naturally on the sides of banks, and in shady lanes in many parts of England, so is seldom cultivated in gardens; for if the seeds are permitted to scatter, the ground will be plentifully stored with plants to a considerable distance. It flowers in May, at which time the herb is in the greatest perfection for use.

The second sort is found growing in a few particular places where the seeds have been formerly sown, or the plants cast out of gardens. This is by some supposed to be only a variety of the first, but I have propagated this by seeds above forty years, and have constantly found the plants the same without variation. The leaves of this are divided into narrow long divisions, which are deeply jagged on their edges in acute segments, and the petals of the flower are cut into many parts, in which it differs from the first. If the seeds of this sort are permitted to scatter, they will fill the ground with plants. They both delight in shade. There is one with a double flower of this sort.

The third sort is known by the title of Horned Poppy; it was so called from the resemblance which the flower bears to the Poppy, and the long seed-vessels which is like a horn. It grows naturally upon the sandy and gravelly shores by the sea, in many parts of England, from whence the seeds have been brought into gardens, where it is sometimes allowed to have place for the sake of variety. This plant abounds with a yellow juice, which flows out from every part when broken. It sends out many thick gray leaves, which are deeply jagged; the stalks are strong, smooth, and jointed, which rise near two feet high, and divide into many branches. These are garnished with leaves at each joint; those on the lower part of the stalks are long, broad, and deeply jagged, but the upper leaves are entire and almost heart-shaped; they closely embrace the stalks with their base: from the bosom of the leaves come out the short foot-stalks of the flowers, each supporting one large yellow flower, composed of four broad petals, which spread open like the Garden Poppy, in the center of which are a great number of yellow stamina, surrounding a long cylindrical germen, crowned by a narrow pointed stigma, which is permanent, remaining upon the top of the horned seed-vessel, which grows nine or ten inches long, having a longitudinal furrow on one side, where it opens when ripe, and lets out the seeds. This is a biennial plant, which flowers the second year, and perishes soon after the seeds are ripe.

If the seeds of this plant are permitted to scatter, they will fill the ground near them with plants, so that it is not a pro-a proper plant for a flower-garden; but if a few of the feeds are fcattered about in rock-work, the plants will rife without trouble, and in fuch places will have a pretty effect.

The fourth fort grows naturally in Spain, Italy, and fome parts of Germany. The leaves of it are deeply jagged and hairy, of a pale green, growing clofe to the ftalks: thofe at the bottom lie on the ground, and are broader than thofe above. The ftalks are a foot and a half high, having a fingle jagged leaf placed at each joint; thefe have many divifions from the origin to the point, which is extended longer than the lower leaves. The flowers come out from the bofom of the leaves; thefe are compofed of five broad obtufe petals, which are of a dark fcarlet colour, and foon fall off. In the center of each is fituated an oblong germen, having no ftyle, but fupports a bifid ftigma; this is attended by a great number of fhort ftamina, terminated by obtufe fummits. The germen afterward becomes a long taper pod, on the apex of which the bifid ftigma remains, fitting on the middle partition, which divides the pod into two cells, which are filled with fmall feeds. It flowers in June and July, and the feeds ripen in autumn. As the flowers of this plant are of but fhort duration, fo they do not make any confiderable figure; but the foliage of the plant is very elegant, and might be introduced by way of ornament to furniture to great advantage, being very picturefque; it may alfo be wrought into patterns for filks, and painted upon porcelain, where it would have a very good effect. If the feeds of this plant are fown in the autumn, they will more certainly grow than thofe which are fown in the fpring, which frequently in dry feafons do not come up the fame year, or at leaft not before autumn; whereas thofe fown in autumn come up in the fpring, and thefe plants come early to flower, fo that good feeds may be always obtained from them. They fhould be fown where the plants are to remain, and will require no other care but to thin them where they are too clofe, and keep them clean from weeds.

The fifth fort differs from the fourth, in having broader leaves, which are not fo deeply divided; the whole plant is fmooth, and the flowers are larger, but are of the fame colour. this is alfo an annual plant, and requires the fame treatment as the laft.

The fixth fort grows naturally among the Corn in fome parts of England. This is alfo an annual plant, whofe feeds fhould be fown in autumn, for thofe which are fown in the fpring feldom fucceed. The leaves of this fort are finely jagged, and divided into narrow fegments fomewhat like thofe of Buckfhorn Plantain; they are fmooth, of a lucid green, and are commonly oppofite. The ftalks rife little more than a foot high, dividing into two or three branches upward, which are garnifhed with fmall leaves of the fame form as thofe below. The flowers are fuftained by flender foot-ftalks, which come out from the wings of the leaves; they are compofed of four obtufe petals, of a Violet colour, in the center of which is fituated a cylindrical germen, attended by a great number of ftamina; the germen afterward becomes a long cylindrical pod like thofe of the other fpecies. The flowers of this plant are very fugacious, feldom lafting above three or four hours before the petals drop off, efpecially in clear weather. If the feeds are permitted to fcatter, the plants will come up without care as the others.

CHELONE. Tourn. Act. R. S. 1706. tab. 7. fol. 2.

The CHARACTERS are,

The flower is of the ringent kind, having a fhort cylindrical tube, which is fwollen at the chaps, where it is oblong, convex above, and plain below; the mouth is almoft clofed. It hath four ftamina, the two fide ones being a little longer than the other. It hath an oval germen, which afterward becomes an oval capfule having two cells, which are filled with flat roundifh feeds having a border.

The SPECIES are,

1. CHELONE (*Glabra*) foliis lanceolatis, acuminatis, feffilibus obfoletè ferratis, radice reptatrice. *Chelone with pointed fpear-fhaped leaves fet clofe to the ftalks, with fmall ferratures on their edges, and a creeping root*; or *Chelone of Acadia with a white flower.*

2. CHELONE (*Purpurea*) foliis lanceolatis, obliquis, petiolatis, oppofitis, marginibus acute ferratis. *Chelone with oblique fpear-fhaped leaves growing oppofite on foot-ftalks, and their borders fharply fawed*; *Chelone with a purple flower.*

3. CHELONE (*Hirfuta*) caule foliifque hirfutis. Lin. Sp. Plant. 611. *Chelone with hairy ftalks and leaves.*

The firft fort grows naturally in moft parts of North America. This is called by Jofcelin, in his New England Rarities, the Humming Bird-tree. It hath a pretty thick jointed root, which creeps under ground to a confiderable diftance, fending up fmooth channelled ftalks, which rife about three or four feet high, garnifhed with two leaves at each joint, ftanding oppofite without foot-ftalks; they have fmall ferratures on their edges, which fcarcely appear. The flowers grow in a clofe fpike at the end of the ftalks; they are white, and have but one petal, which is tubular, and narrow at the bottom, but fwells upward almoft like the Foxglove flower; the upper fide is bent over and convex, but the under is flat, and flightly indented in three parts at the end. It flowers in Auguft, and when the autumn proves favourable, the feeds will ripen in England; but as the plants propagate fo faft by their creeping roots, the feeds are feldom regarded. The beft time to tranfplant the roots is in autumn, that they may be well eftablifhed in the ground before the fpring, otherwife they will not flower fo ftrong, efpecially if the feafon proves dry. They will thrive in almoft any foil or fituation, but their roots are apt to creep too far, if they are not confined, and then their ftalks ftand fo far diftant from each other, as to make but little appearance; therefore they fhould be planted in pots, which will confine their roots, fo that in each pot there will be eight or ten ftalks growing near each other, when they will make a tolerable good appearance. This plant is very hardy, fo is not injured by cold, but it muft have plenty of water in hot weather.

The fecond fort grows naturally in Virginia. The roots of this do not creep fo far as thofe of the firft, the ftalks are ftronger, and the leaves much broader, and are oblique; they are deeply fawed on their edges, and ftand upon fhort foot-ftalks: the flowers are of a bright purple colour, fo make a finer appearance. This flowers at the fame time with the firft, and is propagated by parting of the roots in the fame manner.

The third fort I received from New England, where it grows naturally: this is near to the firft fort, but the ftalks and leaves are very hairy, and the flower is of a pure white. It flowers at the fame time with the former, and requires the fame treatment.

As thefe plants flower in the autumn, when there is a fcarcity of other flowers, fo it renders them more valuable, efpecially the fecond fort, whofe flowers make a very pretty appearance when they are ftrong; and if fome of them have a fhady fituation in the fummer, they will flower later in the autumn.

CHENOPODIA MORUS. See BLITUM.

CHENOPODIUM. Tourn. Inft. R. H. 506. tab. 288. Goofe-foot, or Wild Orach.

The CHARACTERS are,

The flower hath no petal, but in the center it hath five ftamina placed oppofite to the leaves of the empalement. It hath a round germen, which afterward becomes a five-cornered fruit inclofed inclosed in the empalement, containing one roundish depressed seed.

The Species are,

1. Chenopodium (*Bonus Henricus*) foliis triangulari-sagittatis, integerrimis. Hort. Cliff. 84. *Goose-foot with arrow shaped triangular leaves, which are entire; called English Mercury, All Good, or Good Henry.*

2. Chenopodium (*Vulvaria*) foliis integerrimis rhombeo-ovatis, floribus conglomeratis. Flor. Suec. 216. *Goose-foot with entire, oval, rhomboidal leaves, and flowers growing in clusters; Stinking Orach.*

3. Chenopodium (*Stoparia*) foliis lineari-lanceolatis, planis, integerrimis. Hort. Cliff. 86. *Goose-foot with narrow spear-shaped leaves, which are plain and entire; commonly called Belvedere, or Summer Cypress.*

4. Chenopodium (*Botrys*) foliis oblongis, sinuatis, racemis nudis multifidis. Hort. Cliff. 84. *Goose-foot with oblong sinuated leaves, and naked multifid spikes of flowers; commonly called Oak of Jerusalem.*

5. Chenopodium (*Ambrosioides*) foliis lanceolatis, dentatis, racemis foliatis simplicibus. Hort. Cliff. 84. *Goose-foot with spear-shaped indented leaves, and single leafy spikes of flowers; commonly called Oak of Cappadocia.*

6. Chenopodium (*Fruticosum*) foliis lanceolatis, dentatis, caule fruticoso. *Goose-foot with spear-shaped indented leaves, and a shrubby stalk; called Shrubby Mexican Orach.*

7. Chenopodium (*Multifidum*) foliis linearibus, teretibus, carnosis, caule fruticoso. Hort. Cliff. 86. *Goose-foot with narrow, taper, fleshy leaves, and a shrubby stalk; called Stonecrop-tree, or shrubby Glasswort.*

There are many other species of this genus, some of which grow naturally on dunghills and the side of ditches in most parts of England, where they often become very troublesome weeds, for which reason I have not enumerated them here.

The first sort is found growing naturally in shady lanes in many parts of England, but it is very doubtful if the seeds have not been cast out of gardens originally, because this plant was formerly cultivated in kitchen-gardens for use; and in some of the northern counties, the people still preserve it in their gardens as an esculent herb; which in the spring season they dress in the same manner as Spinach, for which it is a substitute. But as the latter is a much better herb, so it has obtained the preference very justly, in all the countries where the culture of the kitchen-garden is understood and practised.

The second sort is very common upon dunghills and in gardens, in most parts of England: it is seldom cultivated, except in some physic-gardens; for the markets in London are supplied with it by the herbwomen, who gather it in the places where it grows wild.

The third sort is sometimes cultivated in gardens; it is a beautiful plant, which is naturally disposed to grow very close and thick, and in as regular a pyramid as if cut by art. The leaves are of a pleasant green; and were it not for that, it hath so much of the appearance of a Cypress-tree, that at some distance it might be taken for the same by good judges: the seeds should be sown in autumn, and in the spring, when the plants are come up, they may be planted into pots of good earth, and kept supplied with water in dry weather: these pots may be intermixed with other plants to adorn court-yards, &c. where they will appear handsome until their seeds begin to swell and grow heavy, which weigh down and displace the branches; at which time the pots should be removed to some abject part of the garden to perfect their seeds, which, if permitted to fall upon the ground, will come up the next spring; so that you need be at no more trouble in propagating these plants, but only to transplant them where you intend they should grow.

The fifth sort was formerly used in medicine; but altho' it still continues in the catalogue of simples annexed to the London Dispensatory, yet is very seldom used at present. This plant may be propagated by sowing the seeds in an open border of good earth in the spring, where it will perfect its seeds in autumn; which, if permitted to shed upon the ground, will arise as the former.

The fourth sort was brought from America, where the seeds are called worm seed; I suppose from some quality contained in it, which destroys worms in the body.

This is propagated by sowing the seeds in the spring, as the before-mentioned sorts, and will perfect its seed in autumn, after which the plant decays to the ground; but if the root be preserved in shelter under a common frame in winter, the stalks will rise again the following spring.

The leaves of this plant emit a very strong odour when bruised, somewhat like those of the Ambrosia, for which the plants are preserved in gardens, for the flower hath no beauty. This plant grows naturally in most parts of North America, where it is generally much used to destroy worms in children. It sends up several stalks from the root, which rise about two feet high, garnished with oblong leaves a little indented on their edges, of a light green, and placed alternately on the stalks; the flowers come out from the wings of the leaves on the upper part of the branches, in loose spikes; these appear in July, and the seeds ripen in September, which, if permitted to scatter, the plants will come up in plenty the following spring.

The seeds of all the species of this genus will succeed best if they are sown in autumn; for when they are sown in the spring, they frequently lie a whole year before the plants come up; therefore where the seeds of any of them scatter, the plants will come up much better than those which are sown by hand.

The fifth sort is annual: this also grows naturally in North America, from whence I have frequently received the seeds. It is also a native of many of the warm countries in Europe. It hath many oblong leaves at the bottom, which are deeply sinuated on both sides, somewhat like those of the Oak-tree, from whence it received the title of Oak of Jerusalem. These are purple on their under side, and when bruised emit a strong odour. Their stalks rise about eight or nine inches high, dividing into several smaller branches. The lower part of these is garnished with leaves of the same shape with those below, but are smaller. The flowers grow in naked loose spikes, divided into many parts: they are small, herbaceous, and are succeeded by small round seeds. This sort flowers in June and July, and the seeds ripen in autumn.

The seventh sort hath leaves very like those of the fourth, and have the same scent; but it hath a shrubby stalk, which rises five or six feet high, and divides into many branches. It is a native of America, and must be housed in the winter, for it will not live through the winter in England in the open air. It is easily propagated by cuttings during any of the summer months, which, if planted in a shady border and duly watered, will soon take root; and then may be planted in pots, and placed in the shade till they have taken new root, after which they may be placed with other hardy exotic plants in a sheltered situation during summer; and when the frosts come on they must be removed into the green-house, but they only require protection from hard frosts, so should have plenty of air in mild weather.

The sixth sort grows naturally on the sea-coast in Devonshire and Cornwall, but is propagated in the nurseries for sale. This sends out from the root many slender shrubby stalks which rise five or six feet high, and divide upward into smaller ligneous branches, which grow erect, and are closely garnished with small, taper, succulent leaves, like those of the lesser Housleek; these remain all the year, for which the shrub is chiefly valued. The flowers are small, and have no beauty. This is propagated by suckers or cuttings, which it sends out from the roots in plenty. It may be transplanted either in spring or autumn, and will thrive almost any where.

CHERRY-LAUREL. See PADUS.

CHERRY-TREE. See CERASUS.

CHERVIL. See SCANDIX.

CHESTNUT. See CASTANEA.

CHESTNUT, the Horse. See ESCULUS.

CHIONANTHUS. Lin. Gen. Plant. 21. The Fringe, or Snowdrop-tree.

The CHARACTERS are,

The flower is of one petal, divided into four very long narrow segments, which are erect. It hath two short stamina inserted in the tube of the petal. In the center is placed the oval germen, which afterward becomes a round berry with one cell, inclosing one hard seed.

We have but one SPECIES of this genus, viz.

CHIONANTHUS (*Virginica*) pedunculis trifidis trifloris. Lin. Sp. Plant. 8. *Snowdrop-tree, or Fringe-tree, with trifid foot-stalks supporting three flowers.*

This shrub is common in South Carolina, where it grows by the side of rivulets, and seldom is more than ten feet high: the leaves are as large as those of the Laurel, but are of a much thinner substance; the flowers come out in May, hanging in long bunches, and are of a pure white, from whence the inhabitants call it Snowdrop-tree; and, from the flowers being cut into narrow segments, they give it the name of Fringe-tree: after the flowers have fallen away the fruit appears, which becomes a black berry about the size of Sloes, having one hard seed in each.

This tree is now more common in the curious gardens in England than it was a few years since, there having been many young plants raised from the seeds which have been brought from America lately: there have also been some plants propagated by layers, though there is great uncertainty of their taking root, which they seldom do in less than two years: nor will they ever take root, unless they are well supplied with water in dry weather.

The best way to obtain good plants is from the seeds, which must be procured from America, for they never have produced any fruit in this country: the seeds should be sown in small pots filled with fresh loamy earth soon after they arrive, and should be placed under a hot-bed frame, where they may remain till the beginning of May, when they must be removed to a situation exposed to the morning sun, and screened from the sun in the middle of the day; for as these seeds lie in the ground a whole year before the plants will come up, so they should not be exposed to the sun the first summer, but the following autumn they should be removed and placed under a frame, to protect the seeds from being injured by the frost. And if the pots are plunged into a moderate hot-bed the beginning of March, it will bring up the plants much sooner than they will otherwise rise; by which means they will get more strengh the first summer, and be better able to resist the cold of the next winter: while these plants are very young, they will be in danger of suffering by severe frost; but when they have obtained strength, they will resist the greatest cold of our climate in the open air; therefore for the two or three first winters, it will be proper to keep them under shelter. In the spring before they begin to shoot, they should be shaken out of the pots, and carefully separated so as not to break off their roots, and each planted in a small pot filled with light loamy soil, and plunged into a very moderate hot-bed, just to forward the taking fresh root; then they should be gradually inured to the open air, and during the following summer the pots should be plunged into the ground, to prevent the earth from drying, in a situation where they may enjoy the morning sun, but screened from the great heat at noon. The autumn following they should be again placed under a hot bed frame to screen them from frost, but they should enjoy the free air at all times when the weather is mild. The April following the plants may be shaken out of the pots with the ball of earth to their roots, and planted where they are designed to remain.

This shrub delights in a moist, soft, loamy soil, and if it is planted in a sheltered situation, will endure the cold of our winters very well in the open air; but in dry land it is very subject to decay in warm seasons.

In the places where this shrub grows naturally, it produces great quantities of flowers, so that they seem covered with snow, which gave occasion to the inhabitants for titling it Snowdrop-tree; but in England it flowers but sparingly, and the bunches of flowers are generally produced very thinly, so that they make but little appearance.

CHIRONIA. Lin. Gen. Plant. 227.

The CHARACTERS are,

The flower hath one petal, with a roundish tube the size of the empalement, which is divided into five equal parts above: it hath five short broad stamina, which are fastened to the top of the tube. It hath an oval germen situated in the center, which afterward becomes an oval capsule with two cells, filled with small seeds.

The SPECIES are,

1. CHIRONIA (*Frutescens*) frutescens capsulifera. Lin. Sp. Plant. 190. *Shrubby Chironia bearing capsules*

2. CHIRONIA (*Baccifera*) frutescens baccifera. Lin. Sp. Plant. 190. *Shrubby berry-bearing Chironia.*

These plants grow naturally at the Cape of Good Hope.

The first sort has a fibrous root, which spreads near the surface of the ground. The stalks are round, and inclining to be ligneous, but are of a very soft texture; these grow from two to three feet high, sending out several branches which grow erect, and are garnished with succulent leaves, which are an inch or more in length, and an eighth part of an inch broad, ending in an obtuse point. At the ends of each shoot the flowers are produced, which are tubulous, and spread open at the top like those of Periwinkle; these are of a bright red colour, and when there are a large number of the flowers open on the same plant, they make a very fine appearance. In the center of the flower is placed an oval germen, upon which there is fixed a recurved style, terminated by a blunt stigma; this is surrounded by five incurved stamina, each supporting a large summit. When the flowers fall away the germen becomes an inflated capsule, which is filled with small seeds. The flowers are produced from June to autumn, and the seeds ripen in October. This plant should be placed in an airy glass-case in winter, where it may enjoy a dry air and much sun, but will not thrive in a warm stove, nor can it be well preserved in a common green-house, because a damp moist air will soon cause it to rot.

The seeds of this plant should be sown in small pots filled with light sandy earth, and plunged into a moderate hot-bed; sometimes the seeds will lie a long time in the ground, so that if the plants do not appear the same season, the pots should not be disturbed, but preserved in shelter till the following spring, and then plunged into a fresh hot-bed, which will bring up the plants in a short time if the seeds are good. When the plants are fit to remove, they should be transplanted into small halfpenny pots, four or five in each pot; then plunge the pots into a moderate hot-bed, where they must have a large share of air in warm weather, to prevent their drawing up weak; when the plants have obtained

obtained some strength, they must be gradually inured to bear the open air; but when they are exposed abroad, if there should happen much rain, the plants must be screened from it, otherwise it will cause them to rot; they must be placed in a warm sheltered situation in summer, and mixed with such other plants as require but little water; where they may remain till autumn, when they must be placed in a dry airy glass-case, and in the winter should have very little wet, but must enjoy the sun as much as possible, and in mild weather should have fresh air admitted to them, but must be protected from frost; with this management, the plants will thrive and produce flowers the second year from seed; the cuttings of this sort will take root, if properly managed.

The second sort rises with a firmer stalk than the first, which is round, jointed, and divides upward into a greater number of branches, which are garnished with short, narrow, pretty thick succulent leaves. The flowers are produced at the end of the branches in the same manner as the first, which are of a fine red colour, but not half so large as those of the first; when these fall away they are succeeded by oval pulpy berries, in which are included many small seeds. This sort continues flowering great part of summer and autumn, and in warm seasons the seeds will ripen in England.

It is propagated by seeds in the same manner as the former sort, and the plants require the same treatment.

CHONDRILLA. Lin. Gen. Plant. 815. Gum Succory.

The CHARACTERS are,

The flower is composed of many hermaphrodite florets, which are uniform, included in a cylindrical scaly empalement; these have one petal, which is stretched out on one side like a tongue; they have each five short hairy stamina. The germen is situated under the floret, which afterward becomes a single, oval, compressed seed, crowned with a single down, and inclosed in the empalement.

We have but one SPECIES of this genus, viz.

CHONDRILLA (*Juncea.*) Lin. Hort. Cliff. 383. *Gum Succory.*

This plant grows naturally in Germany, Helvetia, and France, on the borders of the fields, and is seldom preserved in gardens, because the roots are very apt to spread and become troublesome weeds; and the seeds having down on their tops, are carried by the wind to a great distance, so that the neighbouring ground is filled with the plants; the roots of this strike deep into the ground, and spread out with thick fibres on every side, each of which, when cut, or broken into many parts will shoot up a plant, so that when this plant hath obtained possession of the ground, it is very difficult to root out. The root sends out a great number of slender stalks, which at the bottom are garnished with oblong sinuated leaves, but those above are very narrow and entire. The flowers are produced from the side and top of the branches, which are like those of Lettuce, and are succeeded by seeds of the same form, crowned with down. It flowers in July, and the seeds ripen in September.

CHRISANTHEMOIDES OSTEOSPERMON. See OSTEOSPERMUM.

CHRISTMAS-FLOWER, or Black Hellebore. See HELLEBORE.

CHRISTOPHORIANA. See ACTEA.

CHRYSANTHEMUM. Tourn. Inst. R. H. 491. tab. 280. Corn Marigold.

The CHARACTERS are,

It hath a compound flower; the rays are composed of female florets, which are extended on one side like a tongue; these have an oval germen. The hermaphrodite flowers which compose the disk, are funnel-shaped, divided into five parts at the top; these have five short hairy stamina and an oval germen, which afterward becomes a single, oblong, naked seed.

The SPECIES are,

1. CHRYSANTHEMUM (*Segetum*) foliis amplexicaulibus, superne laciniatis, infernè dentato-serratis. Hort. Cliff. 416. *Corn Marigold with leaves embracing the stalks, the upper jagged, and the lower indented like a saw.*

2. CHYSANTHEMUM (*Leucanthemum*) foliis amplexicaulibus, oblongis, supernè serratis, infernè dentatis. Hort. Cliff. 416. *Corn Marigold with oblong leaves embracing the stalks, the upper ones being sawed, and the lower indented; or greater wild or Ox-eye Daisy.*

3. CHRYSANTHEMUM (*Serotinum*) foliis lanceolatis, supernè serratis, utrinque acuminatis. Hort. Cliff. 416. *Corn Marigold with spear-shaped leaves, those above being sawed, and pointed at both ends.*

4. CHRYSANTHEMUM (*Montanum*) foliis imis spathulato-lanceolatis, serratis, summis linearibus. Sauv. Monsp. 87. *Corn Marigold with lower leaves shaped like a spear-shaped spatula, and sawed, and the upper ones linear.*

5. CHRYSANTHEMUM (*Graminifolium*) foliis linearibus, subintegerrimis. Sauv. Monsp. 87. *Corn Marigold with narrow leaves, which are entire.*

6. CHRYSANTHEMUM (*Alpinum*) foliis pinnatifidis, laciniis parallelis, integris, caule unifloris. Lin. Sp. Plant. 889. *Corn Marigold with many pointed leaves, whose segments are parallel and entire, and one flower on each stalk.*

7. CHRYSANTHEMUM (*Corymbiferum*) foliis pinnatis, inciso-serratis, caule multifloro. Prod. Leyd. 174. *Corn Marigold with winged leaves with sawed segments, and many flowers upon a stalk.*

8. CHRYSANTHEMUM (*Coronarum*) foliis pinnatifidis, incisis, extrorsum latioribus. Hort. Cliff. 416. *Corn Marigold with wing-pointed cut leaves, whose exterior parts are broadest.*

9. CHRYSANTHEMUM (*Monspelienfium*) foliis imis palmatis, foliolis linearibus, pinnatifidis. Sauv. Monsp. 304. *Corn Marigold whose lower leaves are palmated, the smaller ones linear, and ending in many points.*

10. CHRYSANTHEMUM (*Frutescens*) fruticosum, foliis linearibus dentato-trifidis. Hort. Cliff. 417. *Shrubby Corn Marigold with narrow leaves having three points, and indented.*

11. CHRYSANTHEMUM (*Flosculosum*) flosculis omnibus uniformibus, hermaphroditis. Hort. Cliff. 417. *Corn Marigold, whose florets are all uniform and hermaphrodite.*

The first sort is the common Corn Marigold, which grows naturally amongst the Corn, and in the borders of the Corn fields in divers parts of England, so is rarely admitted into gardens, but we have inserted this and the next to introduce the other species.

The second sort is the greater Daisy, which stands in the list of the medicinal plants in the College Dispensatory; this grows naturally in moist pastures almost every where in this country. It rises with stalks near three feet high, which are garnished with oblong indented leaves, that embrace the stalk with their base. The stalks are each terminated by one white flower, shaped like those of the Daisy but four times as large. It flowers in June.

The third sort grows naturally in North America; the roots of this plant creep far under the surface, and send up strong stalks more than four feet high, garnished with long sawed leaves ending in points; these stalks divide upward into many smaller, each being terminated by a large, white, radiated flower; these appear the end of August and September; it multiplies very fast by its creeping roots, and will thrive in any soil or situation.

The fourth sort grows naturally upon the Alps and other mountainous places; this sends up a single stalk a foot high, garnifhed with entire leaves above, but the under leaves are fawed on their edges. The ftalk is terminated by one large white flower, fhaped like thofe of the third fort. This fort may be propagated by feeds, which, if fown on a fhady border, will come up in about fix week, and the plants when fit to remove, may be tranfplanted into a fhady border where they are to remain, and will require no other care but to keep them clean from weeds.

The fifth fort grows naturally about Montpelier; it hath a perennial root, from which fpring up many narrow Grafs-like leaves, and between them ftalks which rife a foot and a half high, garnifhed with leaves of the fame fort as thofe below. The ftalks are each terminated by one large white flower, with a yellow difk or middle. It is propagated by parting of the roots; the beft time for this is in autumn, that the plants may get good root before winter.

The feventh fort grows naturally on the Alps, or other mountainous places in Germany; this fends out upright ftalks, which are garnifhed with leaves cut into many parallel fegments, fomewhat like thofe of Buckfhorn Plantain. The ftalks rife a foot and a half high, and are each terminated by a fingle flower of the fame form with thofe of the laft; it hath a perennial root, and may be propagated in the fame manner as the other.

The eighth fort hath been many years cultivated in the gardens for the beauty of its flowers; of this there are fingle and double with white flowers, and the fame with yellow. As thefe do not differ from each other in any refpect except in the colour of their flowers, therefore they are generally efteemed but one fpecies; but this is conftant, for I have never found the feeds faved from the white, produce plants with yellow flowers, nor thofe of the yellow produce white.

There is alfo a variety of thefe colours with fiftular florets, which has accidentally rifen from feeds of the other; thefe are generally titled Quilled-leaved Chryfanthemums, but as the feeds faved from thefe degenerate to the common forts, fo they do not merit a particular denomination.

Thefe plants are always efteemed as annual, fo the feeds are ufually fown upon a flender hot-bed in the fpring, and the plants are treated in the fame manner as the African Marigold, for the culture of which we fhall refer the reader to that article; but as the plants which rife from feeds, do many of them produce fingle flowers, although the feeds are faved from the beft double flowers, therefore many perfons now propagate thefe plants from cuttings, whereby they continue the double forts only; thefe cuttings taken from the plants the beginning of September, and planted in pots will readily take root; and if they are placed under a hot-bed frame to fcreen them from the froft in winter, letting them have free air in dry weather, they will live thro' the winter, and in the fpring thefe plants may be tranfplanted into the borders of the flower-garden, where they will flower in June, and continue in fucceffion till the froft puts a ftop to them: by this method all the varieties may be continued without variation, but the plants which are propagated this way by cuttings will become barren foon, fo will not produce feeds.

The ninth fort is a perennial plant, which fends out many ftalks from the root, which divide into branches on every fide, and are garnifhed with pretty thick leaves, deeply cut into many fegments like thofe of the laft fort; thefe are of a pale green; the flowers are produced at the end of the branches, ftanding upon pretty long naked foot-ftalks; thefe are very like thofe of the common Greater Daify in fize and colour. It flowers in June, and continues till the end of September. This fort ripens feeds every year in England, by which the plant is eafily propagated. As thefe plants extend their branches pretty far on every fide, fo they fhould be allowed at leaft two feet room; therefore they are not very proper furniture for fmall gardens, where there is not room for thefe large growing plants, but in large gardens thefe may have a place for the fake of variety.

If thefe plants are planted in good dry land, or upon lime rubbifh, they will not grow fo vigorous as in good ground, fo they will endure the cold better, and continue longer; for when the leaves and branches are replete with moifture, they are very apt to rot in the winter, fo are feldom of long duration; but where the plants have grown from the joints of old walls, I have known them to continue in vigour feveral years.

The tenth fort grows naturally in the Canary Iflands, from whence it was firft brought to England, where it has been long an inhabitant in fome curious gardens. It has been frequently called by the gardeners Pellitory of Spain, from the very warm tafte which it hath, much refembling the tafte of that plant.

This rifes with a fhrubby ftalk near two feet high, dividing into many branches, which are garnifhed with pretty thick fucculent leaves, of a grayifh colour, cut into many narrow fegments, that are divided into three parts at their extremity. The flowers come out from the wings of the leaves, ftanding upon naked foot-ftalks fingly, which greatly refemble thofe of the common Chamomile; there is a fucceffion of flowers upon the fame plants great part of the year, for which it is chiefly efteemed. This plant will perfect feeds in England, when the feafons are favourable; but as the cuttings of it do take root fo eafily, if planted during any of the fummer months, fo the feeds are rarely fown.

As this plant is a native of warm countries, fo it will not live in the open air in England during the winter feafon; therefore when the cuttings have made good roots, they fhould be each planted into a feparate pot, and placed in the fhade till they have taken frefh root; then they may be removed to a fheltered fituation, where they may remain till autumn, at which time they muft be removed into the green-houfe to protect them from froft; but in mild weather they fhould have plenty of free air, and, during the winter, they fhould be frequently refrefhed with water, but it muft not be given them in too great plenty; in fummer they will require more moifture, and fhould be treated in the fame manner as other hardier kind of exotic plants.

The eleventh fort grows naturally at the Cape of Good Hope. It rifes with a fhrubby ftalk about two feet high, which divides into many flender branches upward, garnifhed with oblong leaves, much indented on their edges, each indenture terminating in a foft fpine; thefe are of a pale green, fet clofe to the branches. The flowers are produced on fhort foot-ftalks from the wings of the leaves toward the upper part of the branches; thefe are globular, and formed of a great number of hermaphrodite florets, which are tubular and even, having no rays, fo are naked, and of a deep yellow colour. The flowers appear in June, and continue in fucceffion till the froft ftops them; this may be propagated by cuttings in the fame manner as the laft, and the plants fhould be treated in the fame way.

CHRYSOBALANUS. Lin. Gen. Plant. 585. Cocoa Plum.

The CHARACTERS are,

The flower hath five petals and ten ftamina, five of which are longer than the petals, the other are fhorter. In the center is fituated an oval germen, which afterward becomes an oval flefhy berry, inclofing a nut, with five longitudinal furrows.

The

The Species are,

1. Chrysobalanus (*Icaco*) foliis ovatis, emarginatis, floribus racemosis, caule fruticoso. *Chrysobalanus with oval indented leaves, flowers growing in bunches, and a shrubby stalk; commonly called the Cocoa Plum.*

2. Chrysobalanus (*Purpurea*) foliis decompositis, foliolis ovatis integerrimis. *Chrysobalanus with decompounded leaves, whose lobes are oval and entire.*

The first sort grows naturally in the Bahama Islands, and in many other parts of America, but commonly near the sea. It rises with a shrubby stalk about eight or ten feet high, sending out several side branches, which are covered with a dark brown bark, spotted with white; these are garnished with oval stiff leaves, which are indented at their ends in form of a heart, and are placed alternately on the branches; from the wings of the leaves, and also at the division of the branches, the flowers are produced, which grow in loose bunches; these are small and white, having ten stamina in each, five of which stand out beyond the petals of the flowers, the other five are shorter, and are terminated by yellow summits. The flowers are succeeded by oval Plums about the size of Damsons; some of these are blue, some red, and others yellow; they have a sweet luscious taste. The stone of the Plum is shaped like a Pear, and hath five longitudinal ridges on it.

The seeds of the second sort were sent me from Jamaica; the stones were exactly the same shape of those of the former, but the plants have winged leaves, which are branched, each having six or seven pair of pinnæ (or lobes.) This sort hath not flowered in England, so I can give no farther account of it.

As these trees are natives of the warm parts of America, so they will not thrive in England, unless they are kept in a warm stove. They are propagated by seeds, which must be obtained from the countries where the plants naturally grow; these must be sown in the spring in small pots, and plunged into a hot-bed of tanners bark. In six weeks the plants will come up, and, if properly managed, will be fit to remove in a month's time after, when they should be carefully separated, and each planted into a separate small pot, and then plunged into the hot-bed again, observing to shade them from the sun till they have taken fresh root; after which they must have air every day in proportion to the warmth of the season, and their waterings during the summer should be frequent, but sparing. In the autumn the plants must be removed into the bark-stove, and plunged into the tan-bed, and in winter the plants must not have too much water. In summer they must have a good share of air, but the plants should be constantly treated in the same manner as other tender plants from the same countries.

CHRYSOCOMA. Lin. Gen. Pl. 845. Goldylocks.

The Characters are,

The flower is composed of many hermaphrodite florets, contained in an imbricated empalement, which are tubular, and funnel-shaped, cut into five parts at the brim; these have five short slender stamina. They have an oblong germen, which afterward becomes a single, oblong, compressed seed, crowned with hairy down.

The Species are,

1. Chrysocoma (*Lynosyris*) herbacea, foliis linearibus, glabris, calycibus laxis. Lin. Sp. Plant. 841. *Herbaceous Goldylocks with narrow smooth leaves, and loose empalements; or German Goldylocks.*

2. Chrysocoma (*Biflora*) herbaceo paniculata, foliis lanceolatis trinerviis, punctatis, nudis. Lin. Sp. Pl. 841. *Herbaceous Goldylocks with flowers growing in panicles, and spear-shaped leaves having three nerves.*

3. Chrysocoma (*Coma aurea*) fruticosa foliis linearibus dorso decurrentibus. Hort. Cliff. 397. *Shrubby Goldylocks with very narrow leaves, whose back parts run along the stalks.*

4. Chrysocoma (*Cernua*) subfruticosa, foliis linearibus subtus pilosis, floribus ante florescentiam cernuis. Hort. Cliff. 397. *Shrubby Goldylocks with very narrow leaves, which are hairy on their under side, and flowers nodding before they are blown.*

5. Chrysocoma (*Ciliata*) suffruticosa, foliis linearibus recurvis, scabris ciliatis, floribus erectis. Lin. Sp. Pl. 841. *Shrubby Goldylocks with narrow rough leaves, which are recurved and hairy, and erect flowers.*

The first sort grows naturally in Germany, and also in France and Italy; it hath a perennial root, and an annual stalk which rises about a foot and a half high, is round, stiff, and closely garnished with long, narrow, smooth leaves, which come out without any order; the upper part of the stalk divides into many slender foot-stalks, each sustaining a single head of flowers, which are composed of many hermaphrodite florets, contained in one common empalement, having very narrow scales. The flowers are of a bright yellow, and stand disposed on the top of the stalk in form of an umbel.

This plant is generally propagated by parting of the roots, that being the most expeditious method, for the seedling plants do not flower till the second or third year. The best time to remove the plants and part their roots, is soon after the stalks decay in autumn, that the plants may get fresh roots before winter. It delights in a dry loose soil, in which it will live in the open air, and propagate by its roots very fast, but in strong wet land the roots often rot in winter.

The second sort grows naturally in Siberia. This plant hath a perennial creeping root, which spreads on every side to a considerable distance, sending up many erect stalks, which are garnished with flat spear-shaped leaves, ending in points; these are rough, and have three longitudinal veins; the upper part of the stalks branch out, and form loose panicles of yellow flowers, which are larger than those of the former sort.

It propagates too fast by its creeping roots to be admitted into the flower-garden, for the roots will often extend two or three feet every way, in the compass of one year, so that they will interfere with the neighbouring flowers; but as the plants will grow in any soil or situation, so a few roots may be planted on the side of extensive rural walks round the borders of fields, where they will require no care, and their flowers will make a good appearance, and continue long in beauty.

The third sort grows naturally at the Cape of Good Hope. This rises with a ligneous stalk about two feet high, dividing into many small branches, which are garnished with narrow leaves, of a deep green, coming out on every side without order; the back part of each leaf hath a small short appendix, which runs along the stalks. The flowers are produced at the end of the branches on slender naked foot-stalks; these are of a pale yellow, and shaped like those of the former sorts, but are larger.

The most expeditious method of propagating this plant is by cuttings, which, if planted in a common border, in any of the summer months, and covered with hand-glasses, will easily take root, provided they are shaded from the sun. When these have gotten good roots, they should be carefully taken up, and each planted in a separate pot, placing them in the shade till they have taken new root; then they may be exposed with other hardy exotic plants till autumn, when they must be removed into the greenhouse during the winter season; they should enjoy a large share of free air in mild weather, for they only require

quire protection from froſt, ſo muſt not be too tenderly treated.

The fourth ſort is a native of the Cape of Good Hope; this is a much leſs plant than the former, ſeldom riſing above a foot high; it hath a ſhrubby ſtalk, branching out in the ſame manner; the leaves are ſhorter, and a little hairy; the flowers are not half ſo large, of a pale ſulphur colour, and nod on one ſide before they are blown. It is generally propagated in the ſame manner as the former, and the plants require the ſame treatment.

The fifth ſort is alſo a native of the ſame country as the two former; this hath a low ſhrubby ſtalk, which ſeldom riſes above ſix inches high, branching out on every ſide; the leaves come out on every ſide the branches, which are very narrow, ſhort, rough, and reflexed; the flowers ſtand ſingle on the top of the foot-ſtalks, which ariſe from the upper part of the branches; theſe flowers are larger than thoſe of the laſt, and ſtand erect. This plant requires the ſame treatment as the two former, and is propagated by cuttings in the ſame manner.

CHRYSOPHYLLUM. Lin. Gen. Pl. 233. Star Apple.

The Characters are,

It hath a bell-ſhaped flower cut at the brim into ten ſegments, which are alternately ſpread open, and five ſhort ſtamina, with a roundiſh germen, which afterward becomes a pulpy fruit with ten cells, in four or five of which is lodged a ſingle ſeed.

The Species are,

1 Chrysophyllum (*Cainito*) floribus racemoſis terminalibus. Lin. Syſt. 937. *The Star Apple.*

2. Chrysophyllum (*Glabrum*) floribus lateralibus. Lin. Syſt. 937. *The Damſon-tree.*

Both theſe trees grow naturally in the Weſt-Indies, where the firſt ſort is often cultivated for its fruit; but the other grows wild in thoſe parts of the iſlands, which are not cleared of trees.

The firſt riſes thirty or forty feet high, with a large trunk, covered with a brown bark, and divides into many flexible ſlender branches, which generally hang downward, garniſhed with ſpear-ſhaped leaves, whoſe under ſide are of a bright ruſſet colour. The flowers come out at the extremity of the branches, diſpoſed in oblong bunches, which are ſucceeded by fruit of the ſize of a Golden Pippin, that are very rough to the palate, and aſtringent; but if kept ſome time mellow, as is here practiſed with Medlars, they have an agreeable flavour.

The ſecond ſort never riſes to the height of the firſt, nor do the trunks grow to half the ſize of thoſe; but the branches are ſlender, and garniſhed with leaves like thoſe of the firſt. The flowers come out in cluſters from the ſide of the branches, which are ſucceeded by oval ſmooth fruit about the ſize of Olives, incloſing three or four hard compreſſed ſeeds.

The plants of both theſe ſorts are frequently preſerved in thoſe Engliſh gardens, where there are large ſtoves for keeping of exotic plants; for although there is little hopes of having fruit from them in England, yet the colour of their leaves being ſo different from thoſe of moſt other plants, makes a pleaſing variety in the ſtove; and as they retain their leaves all the year, ſo the plants are as deſerving of care as many other which are here cultivated.

They are both propagated by ſeeds, which are frequently brought from the Weſt-Indies, which ſhould be ſown in ſmall pots, four or five ſeeds in each, and the pots plunged into a hot-bed of tanners bark: if the ſeeds are freſh, the plants will come up in five or ſix weeks after, and in a few weeks more will be fit to remove; then they ſhould be ſhaken out of the pots, and their roots carefully ſeparated, and each planted in a ſmall pot; then they ſhould be plunged again into a hot-bed of tanners bark, ſhading them every day until they have taken freſh root, and afterward the plants ſhould be treated in the ſame way, as has been directed for many of the tender ſorts of Annona.

CHRYSOSPLENIUM. Lin. Gen. Plant. 493. Golden Saxifrage.

The Characters are,

The flower hath no petals, but eight or ten ſtamina, which are ſhort, erect, and ſtand oppoſite to the angles of the empalement. The germen is immerſed in the empalement, which afterward becomes a capſule with two beaks, opening with two valves, and filled with ſmall ſeeds.

The Species are,

1. Chrysosplenium (*Alternifolium*) foliis alternis. Flor. Suec. 317. *Golden Saxifrage with alternate leaves.*

2. Chrysosplenium (*Oppoſitifolium*) foliis oppoſitis. Sauv. Monſp. 128. *Golden Saxifrage with oppoſite leaves.*

Theſe two plants are found growing wild in ſeveral parts of England, but eſpecially the firſt, upon marſhy ſoils and bogs, as alſo in moſt ſhady woods, and are ſeldom propagated in gardens, where, if any perſon have curioſity to cultivate them, they muſt be planted in very moiſt ſhady places, otherwiſe they will not thrive. They flower in March and April.

CIBOULS, or CHIBOULS. See Cepa.

CICER. Lin. Gen. Pl. 783. Cicer, or Chich Peaſe.

The Characters are,

The flower is of the butterfly kind; the ſtandard is large, roundiſh, and plain; the wings are much ſhorter and obtuſe; the keel is ſhorter than the wings, and is ſharp pointed. It hath ten ſtamina, nine of them being joined, and the tenth is ſeparate. It hath an oval germen, which afterward becomes a turgid ſwelling pod of a rhomboidal figure, incloſing two roundiſh ſeeds, with a protuberance on their ſide.

We have but one Species of this genus, viz.

Cicer (*Arictinum*) foliolis ſerratis. Hort. Cliff. 370. *Chich Peaſe with ſawed leaves.* Cicer ſativum, C. B. P. 347. *Garden Chich Peaſe.*

There is a variety of this with a red ſeed, which differs from it in nothing but the colour.

It is much cultivated in Spain, being one of the ingredients in their olios, and is there called Garavance; it is alſo cultivated in France, but in England it is rarely ſown.

The plant is annual, ſhooting out ſeveral ſtalks from the root, which are about two feet long; theſe are hairy, and garniſhed with long winged leaves of a grayiſh colour, compoſed of ſeven or nine pair of ſmall roundiſh leaves (or lobes) terminated by an odd one, which are ſawed on their edges. From the ſide of the branches come out the flowers, ſometimes one, at other times two together. They are ſhaped like thoſe of Peaſe, but are much ſmaller and white, ſtanding on long foot-ſtalks, and are ſucceeded by ſhort hairy pods, including two ſeeds in each, which are the ſize of common Peaſe, but have a little knob or protuberance on one ſide.

The ſeeds of this plant may be ſown in the ſpring, in the ſame manner as Peaſe, making drills with a hoe about an inch and a half deep, in which the ſeeds ſhould be ſown at about two inches aſunder, then with a rake draw the earth into the drill to cover the ſeeds. The drills ſhould be made at three feet diſtance from each other, that there may be room for their branches to ſpread, when the plants are fully grown, as alſo to hoe the ground between them, to keep it clean from weeds, which is all the culture theſe plants require.

This plant flowers in June, and the ſeeds ripen in Auguſt; but unleſs the ſeaſon proves warm and dry, the plants decay before the ſeeds are ripe.

CICHORIUM. Lin. Gen. Plant. 825. Succory.

The

The Characters are,

The flower hath a common scaly empalement, which at first is cylindrical, but is afterward expanded; the scales are narrow, spear-shaped, and equal. The flower is composed of many hermaphrodite florets, with one petal, which is tongue-shaped, and cut into five segments. These have five short hairy stamina. The germen is situated under the petal, which afterward becomes a single seed, inclosed with a down, and shut up in the empalement.

The Species are,

1. Cichorium (*Intybus*) floribus geminis sessilibus foliis runcinatis. *Succory with two flowers sitting close to the stalks; or Wild Succory.*

2. Cichorium (*Spinosum*) caule dichotomo spinoso. Hort. Cliff. 388. *Succory with a prickly forked stalk.*

3. Cichorium (*Endivia*) floribus solitariis pedunculatis, foliis integris crenatis. Hort. Cliff. 389. *Broad-leaved Succory with entire crenated leaves.*

4. Cichorium (*Crispum*) floribus solitariis pedunculatis, foliis fimbriatis, crispis. *Curled or fringed Endive.*

The first sort grows naturally by the sides of roads and in shady lanes in many parts of England: this has been supposed to be no other way differing from the Garden Succory, but by the latter being cultivated in gardens; indeed most of the writers on botany have confounded the two sorts together, for the Garden Succory which is described in most of the old books, I take to be the broad-leaved Endive, which is the third sort here mentioned; for I have many years cultivated both sorts in the garden, without finding either of them alter. There is an essential difference between these, for the wild Succory hath a perennial creeping root, whereas the other is at most but a biennial plant; and if the seeds of the latter are sown in the spring, the plants will flower and produce seeds the same year, and perish in autumn, so that it may rather be called annual The wild Succory sends out from the roots long leaves, which are jagged to the midrib, each segment ending in a point; from between these arise the stalks, which grow from three to four feet high, garnished with leaves shaped like those at the bottom, but smaller, and embrace the stalks at their base. These branch out above into several smaller stalks, which have the same leaves, but smaller and less jagged; the flowers are produced from the side of the stalks, two at each joint, which are of a fine blue colour, and are succeeded by oblong seeds, inclosed in a down.

The second sort grows naturally on the sea-coasts in Sicily and the islands of the Archipelago. This sends out from the root many long leaves, which are indented on their edges, and spread flat on the ground; from between these arise the stalks, which have very few leaves, and those are small and entire: these stalks are divided in forks upward, from between these come out the flowers, which are of a pale blue, and are succeeded by seeds shaped like those of the common sorts; the ends of the smaller branches are terminated by star-like spines, which are very sharp. This plant is biennial with us in England, and in cold winters is frequently killed. It flowers and seeds about the same time with the former sort, and may be treated in the same way as the Endive.

The broad-leaved Succory or Endive differs from the wild sort in its duration, the root always perishing after it has ripened seeds: the leaves are broader, rounder at the top, and not laciniated on the sides as the leaves of the wild; the branches are more horizontal, and the stalks never rise so high.

There is also a variety with very long broad leaves, called in Spain Escharole, which is very tender and good, but is often injured by frosts in the autumn, so is less esteemed on that account in England.

All the sorts of Succory are esteemed aperitive and diuretic, opening obstructions of the liver, and good for the jaundice; it provokes urine, and cleanses the urinary passages of slimy humours, which may stop their passage.

The curled Endive is now much cultivated in the English gardens, being one of the principal ingredients in the sallads of autumn and winter, for which purpose it is continued as long as the season will permit. I shall therefore give directions for the managing of this plant, so as to have it in perfection during the autumn and winter months.

The first season for sowing of these seeds is in June, for those which are sown earlier in the year, generally run up to seed before they have arrived to a proper size for blanching; and it frequently happens, that the seeds sown in June in the rich grounds near London, will run to seeds the same autumn; but in situations that are colder, they are not so apt to run up, therefore there should be some seeds sown about the middle or latter end of that month. The second sowing should be about the beginning of July, and the last time in the middle of July. From these three different crops there will be a supply for the table during the whole season, for there will be plants of each sowing, very different in their growth, so that there will be three different crops of plants from the same seed-beds.

When the plants come up they must be kept clean from weeds, and in dry weather duly watered, to keep them growing till they are fit to transplant, when there should be an open spot of rich ground prepared to receive the plants, in size proportionable to the quantity intended. When the ground is well dug and levelled, if it should be very dry, it must be well watered to prepare it to receive the plants; then the plants should be drawn up from the seed-bed carefully, so as not to break their roots, drawing out all the largest plants, leaving the small ones to get more strength, which, when they have room to grow, by taking away the large ones, they will soon do. As the plants are drawn up, they should be placed with their roots even, all the same way, and every handful as they are drawn should have the tops of their leaves shortened, to make them of equal length: this will render the planting of them much easier than when the plants are promiscuously mixed, heads and tails: then the ground should be marked out in rows at one foot asunder, and the plants set ten inches distant in the rows, closing the earth well to their roots; let them be well watered, and repeat this every other evening, till the plants have taken good root, after which they must be kept clean from weeds.

When the plants of the seed-bed have been thus thinned, they should be well cleaned from weeds and watered, which will encourage the growth of the remaining plants, so that in ten days or a fortnight after there may be another thinning made of the plants, which should be transplanted in the same manner; and at about the same distance of time, the third and last drawing of plants may be transplanted.

Those plants which were the first transplanted, will be fit to blanch the latter end of August at farthest; and if they are properly managed, in three weeks or a month they will be sufficiently blanched for use, which will be as soon as these sallads are commonly required; for during the continuance of good Cos Lettuce, few persons care for Endive in their sallads, nor indeed is it so proper for warm weather. If any of the plants should put out flower-stems, they should be immediately pulled up and carried away, being good for nothing, so should not be left to incommode the neighbouring plants. As the quantity of roots necessary for the supply of a middling family is not very great, so there should not be too many plants tied up to blanch at the same time; therefore the largest should be first tied, and in a week after those of the next size; so that there may be three different times of blanching the plants on the same ſpot of ground. But as in ſome large families there is a great conſumption of this herb for ſoups, ſo the quantities of plants ſhould be proportionably greater at each time of planting and blanching. The manner of blanching is the next thing to be treated of, therefore in order to this you ſhould provide a parcel of ſmall Oſier twigs (or baſs mat) to tie up ſome of the largeſt heads to blanch, which ſhould be done in a dry afternoon, when there is neither dew nor rain to moiſten the leaves in the middle of the plants, which would occaſion their rotting ſoon after their being tied up. The manner of doing it is as follows, viz. You muſt firſt gather up all the inner leaves of the plant in a regular order, into one hand, and then take up thoſe on the outſide that are ſound, pulling off and throwing away all the rotten and decayed leaves which lie next the ground, obſerving to place the outſide leaves all round the middle ones, as near as poſſible to the natural order of their growth, ſo as not to croſs each other: then having got the whole plant cloſe up in your hand, tie it up with the twig, baſs, &c. at about two inches below the top, very cloſe; and about a week after go over the plants again, and give them another tie about the middle of the plant, to prevent the heart-leaves from burſting out on one ſide, which they are ſubject to do, as the plants grow, if not prevented this way.

In doing of this you need only tie up the largeſt plants firſt, and ſo go over the piece once a week, as the plants increaſe in their growth; by which means, you will continue the crop longer than if they were all tied up at one time; for when they are quite blanched, which will be in three weeks or a month after tying, they will not hold ſound and good above ten days or a fortnight, eſpecially if the ſeaſon proves wet: therefore it is that I would adviſe you to ſow and plant at three or four different ſeaſons, that you may have a ſupply as long as the weather will permit. But in order to this, you muſt tranſplant all the plants of the laſt ſowing under warm walls, pales, or hedges, to ſcreen the plants from froſt; and if the winter ſhould prove very ſharp, you ſhould cover them with ſome Peaſe-haulm, or ſuch other light covering, which ſhould be conſtantly taken off in mild weather. theſe borders ſhould alſo be as dry as poſſible, for theſe plants are very ſubject to rot, if planted in a moiſt ſoil in winter.

Although I before directed the tying up of the plants to blanch them, yet this is only to be underſtood for the two firſt ſowings; for after October, when the nights begin to be froſty, thoſe plants which are ſo far above ground will be liable to be much prejudiced thereby, eſpecially if they are not covered in froſty weather; therefore the beſt method for the late crop is, to take up your plants in a very dry day, and with a large flat pointed dibble plant them into the ſides of trenches of earth, which ſhould be laid very upright, ſideways, towards the ſun, with the tops of the plants only out of the ground, ſo that the haſty rains may run off, and the plants be kept dry, and ſecured from froſts.

The plants thus planted will be blanched fit for uſe in about three weeks or a month's time, after which it will not keep good long; you ſhould therefore keep planting ſome freſh ones into trenches every week or fortnight at fartheſt, that you may have a ſupply for the table; and thoſe which were laſt tranſplanted out of the ſeed-beds, ſhould be preſerved till February or March before they are planted to blanch, ſo that from this you may be ſupplied until the beginning of April, or later; for at this laſt planting into trenches, it will keep longer than in winter, the days growing longer; and the ſun advancing with more ſtrength, dries up the moiſture much ſooner than in winter, which will prevent the rotting of theſe plants.

When your Endive is blanched enough for uſe, you muſt dig it up with a ſpade, and after having cleared it from all the outſide green and decayed leaves, you ſhould waſh it well in two or three different waters to clear it the better from ſlugs and other vermin, which conſtantly ſhelter themſelves amongſt the leaves thereof, and then you may ſerve it up to the table with other ſallading.

But in order to have a ſupply of good ſeeds for the next ſeaſon, you muſt look over thoſe borders where the laſt crop was tranſplanted, before you put them into the trenches to blanch, and make choice of ſome of the largeſt, ſoundeſt, and moſt curled plants, in number according to the quantity of ſeeds required: for a ſmall family, a dozen of good plants will produce ſeeds enough; and for a large, two dozen or thirty plants.

Theſe ſhould be taken up and tranſplanted under a hedge or pale at about eighteen inches diſtance in one row, about ten inches from the hedge, &c. This work ſhould be done in the beginning of March, if the ſeaſon is mild, otherwiſe it may be deferred a fortnight longer. When the flower-ſtems begin to advance, they ſhould be ſupported with a packthread, which ſhould be faſtened to nails driven into the pale, or to the ſtakes of the hedge, and run along before the ſtems, to draw them upright cloſe to the hedge or pale, otherwiſe they will be liable to break with the ſtrong winds. Obſerve alſo to keep them clear from weeds, and about the beginning of July your ſeeds will begin to ripen; therefore, as ſoon as you find the ſeeds quite ripe, you muſt cut off the ſtalks, and expoſe them to the ſun upon a coarſe cloth to dry; and then beat out the ſeeds, which muſt be dried, and put up in bags or paper, and preſerved for uſe in ſome dry place. But I would here caution you, not to wait for all the ſeeds ripening upon the ſame plant; for if ſo, all the firſt ripe and beſt of the ſeeds will ſcatter and be loſt before the other are near ripe, ſo great a difference is there in the ſeeds of the ſame plant being ripe.

The wild Succory (of which there are ſome varieties in the colour of the flowers) is ſeldom propagated in gardens, it growing wild in unfrequented lanes and dunghills in divers parts of England, where the herbwomen gather it, and ſupply the markets for medicinal uſe.

CICUTA. Lin. Gen. Pl. 316. Water Hemlock.

The Characters are,

It is a plant with an umbellated flower; the principal umbel is compoſed of ſeveral ſmaller. The great umbel hath no involucrum, but the ſmaller have, which are compoſed of many ſhort leaves. The flowers have each five oval petals; they have five hairy ſtamina. The germen is ſituated below the flower, which afterward becomes a roundiſh channelled fruit dividing in two parts, containing two oval ſeeds, plain on one ſide and convex on the other.

The Species are,

1. Cicuta (*Viroſa*) umbellis folio oppoſitis, petiolis marginatis obtuſis. Lin. Sp. Plant. 255. *Hemlock with umbels oppoſite to the leaves, and obtuſe marginated foot-ſtalks; Water Hemlock.*

2. Cicuta (*Maculata*) foliorum ſerraturis mucronatis, petiolis membranaceis, apice bilobis. Lin. Sp. Plant. 256. *Hemlock with pointed ſerratures to the leaves, and membranaceous foot-ſtalks ending in two lobes.*

The firſt ſort grows naturally in ſtanding waters in many parts of England, ſo is never propagated; for unleſs there is a conſiderable depth of ſtanding water for the plants to root in, they will not grow.

It riſes four or five feet high, with a branching hollow ſtalk, garniſhed with winged leaves; the ſtalks are terminated by umbels of yellowiſh flowers, which are ſucceeded by ſmall channelled ſeeds like thoſe of Parſley. It flowers in June and July, and the ſeeds ripen in autumn.

The ſecond ſort grows naturally in North America. This is propagated by ſeeds, which ſhould be ſown in autumn in a ſhady border, where the plants will come up in the ſpring, and require no other care but to keep them clean.

CICUTARIA. See LIGUSTICUM.

CINARA. See CYNARA.

CINERARIA. Sea Ragwort.

The CHARACTERS are,

The flowers are compoſed of many hermaphrodite florets, which are funnel-ſhaped; theſe have each five ſtamina, with an oblong germen, crowned by two oblong ſtigmas; the border or rays are of female half florets, which are tongue-ſhaped, indented at their points, having two ſtyles; theſe are alſo fruitful.

The SPECIES are,

1. CINERARIA (*Geifolia*) pedunculis ramoſis, foliis reniformibus ſuborbiculatis ſublobatis dentatis petiolatis. Lin. Sp. 1242. *African Cineraria with roundiſh kidney-ſhaped leaves.*

2. CINERARIA (*Maritima*) floribus paniculatis, foliis pinnatifidis tomentoſis, laciniis ſinuatis, caule fruteſcente. Lin. Sp. 1244. *Sea Ragwort.*

3. CINERARIA (*Amelloides*) pedunculis unifloris, foliis ovatis oppoſitis, caule fruticoſo. Lin. Sp. 1245. *African Ragwort with one flower on each foot-ſtalk.*

4. CINERARIA (*Othonnites*) pedunculis unifloris, foliis oblongis indiviſis ſubdentatis petiolatis alternis nudis. Lin. Sp. 1244. *Shrubby African Ragwort with undivided leaves.*

The firſt, third, and fourth ſorts are natives at the Cape of Good Hope. The firſt and fourth ſorts have been ſeveral years cultivated in the Engliſh gardens under the title of Jacobæa: theſe plants are eaſily propagated by planting their cuttings in July or Auguſt, on a border of light ground, ſhading them from the ſun, and gently watering them till they have put out roots; then they ſhould be taken up and planted in pots that they may be ſheltered in winter, for they will not live abroad in that ſeaſon, though they are ſo hardy as to live under a common frame, if they are ſcreened from froſt.

The ſecond ſort grows naturally on the ſea-coaſts in ſome parts of England, therefore if this is planted in a ſheltered ſituation, it will live in the open air, and may be readily propagated by ſlips or cuttings in the ſummer ſeaſon.

The third ſort I raiſed from ſeeds which came from the Cape of Good Hope: this has greatly the appearance of an Aſter; the plant grows two or three feet high, ſends out many branches, and continues flowering great part of the year.

It is eaſily propagated either from ſeeds or cuttings, and if the plants are ſheltered under a frame in winter, they will live ſome years.

CIRCEA. Lin. Gen. Pl. 24. Enchanter's Nightſhade.

The CHARACTERS are,

The flower hath two heart-ſhaped petals; it hath two erect hairy ſtamina. The germen is ſituated under the flower; the empalement afterward becomes an oval capſule with two cells opening lengthways, each containing a ſingle oblong ſeed.

The SPECIES are,

1. CIRCEA (*Lutetiana*) caule erecto, racemis pluribus. Lin. Sp. Plant. 2. *Common Enchanter's Nightſhade.*

2. CIRCEA (*Alpina*) caule adſcendente, racemo unico. Lin. Sp. Pl. 9. *Leaſt Enchanter's Nightſhade.*

The firſt ſort grows naturally in ſhady woods and under hedges in many parts of England. It hath a creeping root, by which it multiplies greatly. The ſtalks are upright, and riſe a foot and a half high, garniſhed with heart-ſhaped leaves, placed oppoſite upon pretty long foot-ſtalks: they are of a dark green on their upper ſide, but are pale on their under. The ſtalks are terminated by looſe ſpikes of flowers, which are branched out into three or four ſmall diviſions. The flowers are ſmall and white, having but two petals, oppoſite to which are ſituated the two ſtamina. After the flowers fall away, the empalement of the flower becomes a rough capſule, incloſing two oblong ſeeds.

The ſecond ſort grows at the foot of mountains in many parts of Germany; it alſo grows naturally in a wood near the Hague, from whence I brought it to England. This ſort ſeldom riſes more than ſix inches high, with a ſlender ſtalk, garniſhed with leaves ſhaped like thoſe of the former ſort, but ſmaller, and are indented on their edges. The flowers are produced on ſingle looſe ſpikes at the top of the ſtalks, which are ſmaller than thoſe of the former ſort, but of the ſame form and colour. They both multiply exceedingly by their creeping roots, ſo are ſeldom kept in gardens, unleſs for the ſake of variety.

CIRSIUM. See CARDUUS.

CISSAMPELOS. Lin. Gen. Pl. 993.

The CHARACTERS are,

It is male and female in different plants; the male flowers have no petals, but a ſingle ſtyle riſes in the center, which extends beyond the empalement, terminated by a large ſummit, having four lobes. The female flowers have four nectariums ſtanding round the oval germen, which is hairy, and afterward becomes a ſucculent berry, incloſing a ſingle ſeed.

The SPECIES are,

1. CISSAMPELOS (*Pareira*) foliis peltatis cordatis ſubtus villoſis, floribus racemoſis alaribus. *Ciſſampelos with target heart-ſhaped leaves which are hairy, and flowers growing in bunches riſing from the ſide of the ſtalks.*

2. CISSAMPELOS (*Caapeba*) foliis cordatis, tomentoſis, floribus racemoſis, alaribus. *Ciſſampelos with woolly heart-ſhaped leaves, and flowers growing in bunches from the ſides of the ſtalks; called Velvet Leaf in America.*

Theſe plants grow naturally in the warmeſt parts of America, where they twiſt themſelves about the neighbouring ſhrubs, and riſe to the height of five or ſix feet. The firſt ſort hath round heart-ſhaped leaves, whoſe foot-ſtalks are ſet within the baſe of the leaf, reſembling an ancient target; theſe are hairy on the under ſide, and have pretty long ſlender foot-ſtalks. Toward the upper part of the ſtalks the flowers come out from the wings of the leaves; thoſe of the male plants grow in ſhort ſpikes or cluſters, and are of a pale herbaceous colour, but the female flowers are produced in long looſe racemi from the ſide of the ſtalks, and are ſucceeded by a ſingle pulpy berry incloſing a ſingle ſeed.

The ſecond ſort hath round heart-ſhaped leaves, which are extremely woolly and ſoft to the touch; theſe have their foot-ſtalks placed at the baſe between the two ears; the flowers of this come out in bunches from the ſide of the ſtalks in the ſame manner as the firſt. The ſtalks and every part of the plant is covered with a ſoft woolly down.

Theſe plants are propagated by ſeeds, which ſhould be ſown upon a hot-bed in the ſpring; and the plants muſt afterward be treated in the ſame way as other tender exotics, keeping them conſtantly in the bark-ſtove, otherwiſe they will not live in this country.

The firſt ſort is ſuppoſed to be the Pareira, whoſe root has been ſo much eſteemed as a diuretic.

CISTUS. Lin. Gen. Plant. 598. Rock-roſe.

The CHARACTERS are,

The flower hath five large roundiſh petals which ſpread open; it hath a great number of hairy ſtamina, which are ſhorter than the petals. In the center is ſituated a roundiſh germen, which afterward becomes an oval cloſe capſule, having in ſome five, and others ten cells, filled with ſmall roundiſh ſeeds.

The SPECIES are,

1. CISTUS (*Hiſpanicus*) arboreſcens foliis ovatis, ſeſſilibus, utrinque villoſis, rugoſis, floribus terminalibus. *Tree Rock-roſe*

rose with oval leaves growing close to the branches, which are hairy and rough, and flowers growing at the end of the branches.

2. Cistus (*Incanus*) arborescens foliis sessilibus, utrinque villosis, rugosis, inferioribus ovatis basi connatis, summis lanceolatis. Hort. Cliff. 205. *Tree Rock-rose with leaves set close to the branches, which are hairy and rough on each side, the under being oval and joined at their base, but the upper spear-shaped.*

3. Cistus (*Brevifolius*) arborescens, foliis ovato-lanceolatis, basi connatis, hirsutis, rugosis, pedunculis florum longioribus. *Tree Rock-rose with oval spear-shaped leaves joined at their base, which are hairy, rough, and longer foot-stalks to the flowers.*

4. Cistus (*Lusitanicus*) arborescens, foliis ovatis, obtusis, villosis, subtus nervosis rugosis, floribus amplioribus. *Tree Rock-rose with oval, obtuse, hairy leaves, which are nervous and rough on their under side, and larger flowers.*

5. Cistus (*Lineifolius*) arborescens villosus, foliis lanceolatis, viridibus, basi connatis, floribus alaribus & terminalibus sessilibus, calycibus acutis. *Hairy Tree Rock-rose with green spear-shaped leaves joined at their base, flowers proceeding from the sides and ends of the branches, sitting close to the stalks, and sharp pointed empalements.*

6. Cistus (*Arborescens*) arborescens foliis lanceolatis, suprà lævibus, petiolis basi coalitis vaginantibus. Hort. Cliff. 205. *Tree Rock-rose with spear-shaped leaves, smooth on their upper side, and their foot-stalks joining like sheaths.*

7. Cistus (*Oleafolius*) arborescens foliis oblongis, tomentosis, incanis, basi connatis, supra lævibus inferne nervosis. *Tree Rock-rose with oblong, hoary, woolly leaves, joined at their base, smooth above, but nervous on their under side.*

8. Cistus (*Salvifolius*) frutescens, ramis patulis, foliis ovatis, petiolatis, hirsutis, pedunculis nudis. *Shrubby Rock-rose with spreading branches, oval hairy leaves having foot-stalks, and the foot-stalks of the flowers naked.*

9. Cistus (*Creticus*) arborescens, foliis ovato-lanceolatis, hirsutis, marginibus undulatis, floribus terminalibus. *Tree Rock-rose with oval, spear-shaped, hairy leaves, waved on their borders, and flowers terminating the branches.*

10. Cistus (*Salicifolius*) fruticosus, foliis lineari-lanceolatis, hirsutis, sessilibus, floribus terminalibus. *Shrubby Rock-rose with narrow, spear-shaped, hairy leaves, sitting close to the branches, and flowers terminating the stalks.*

11. Cistus (*Halimifolius*) arborescens, foliis lanceolatis, suprà lævibus, petiolis basi, coalitis vaginantibus. Lin. Sp. Pl. 523. *Tree Rock-rose with spear-shaped leaves, smooth on their upper side, with foot-stalks joined at their base.*

12. Cistus (*Populifolius*) foliis oblongo-cordatis, glabris, petiolis, longioribus, caule fruticoso. *Rock-rose with oblong, heart-shaped, smooth leaves, longer foot-stalks, and a shrubby stalk.*

13. Cistus (*Monspeliensis*) arborescens, foliis lanceolatis, sessilibus, utrinque villosis, trinerviis, alis nudis. Hort. Cliff. 205. *Tree Rock-rose with spear-shaped leaves sitting close to the branches, hairy on both sides, having three nerves.*

14. Cistus (*Ladaniferus*) arborescens, foliis lineari-lanceolatis, subtus incanis, trinerviis, petalis subrotundis. *Tree Rock-rose with narrow spear-shaped leaves, hoary on their under side, having three nerves with roundish petals.*

15. Cistus (*Longifolius*) foliis lanceolatis, supernè glabris, infernè incanis, trinerviis, margine undulatis, caule fruticoso. *Rock-rose with spear-shaped leaves which are smoooth on their upper side, and hoary on their under, having three nerves, waved edges, and a shrubby stalk.*

16. Cistus (*Laurifolius*) arborescens foliis cordatis lævibus acuminatis petiolatis. Lin. Sp. Pl. 523. *Tree Rock-rose with heart-shaped pointed leaves.*

17. Cistus (*Luteus*) foliis ovatis, incanis, infernè petiolatis, supernè coalitis, caule fruticoso. *Rock-rose with oval hoary leaves, those beneath having foot-stalks, the upper ones joined at their base, and a shrubby stalk.*

18. Cistus (*Helianthemum*) foliis lineari-lanceolatis, incanis, sessilibus, floribus racemosis caule fruticoso. *Rock-rose with narrow spear-shaped leaves which are hoary, and sit close to the branches, flowers growing in clusters, and a shrubby stalk.*

Most of these plants grow naturally in the south of France, Spain, and Portugal.

The first sort hath a strong woody stem, covered with a rough bark, which rises three or four feet high, dividing into many branches, so as to form a large bushy head; these are garnished with oval hairy leaves, placed opposite, and sit close to the branches, having several smaller leaves of the same form, rising from the same joint. The flowers are produced at the end of the branches, four or five standing together, almost in the form of an umbel, but rarely more than one is open at the same time: these are composed of five large roundish petals of a purple colour, which spread open like a Rose, having a great number of stamina. These flowers are of but short duration, generally falling off the same day they expand, but there is a succession of fresh flowers every day for a considerable time. After the flowers are past, the germen swells to an oval seed-vessel, sitting in the empalement, which is hairy; these capsules have ten cells, which are full of small roundish seeds.

The second sort differs from the first in the shape of the leaves, which are longer and whiter; those on the lower part of the branches are oval, and join at their base, surrounding the stalks, but the upper leaves are spear-shaped and distinct; the flowers are larger, and of a paler purple colour.

The third sort differs from both the former, in having shorter and greener leaves, which are joined at their base, and are hairy. The foot-stalks of the flowers are much longer, and the flowers are smaller, but of a deeper purple.

The fourth sort hath much larger and rounder leaves than either of the former, which are hairy, and smooth on their upper side, but rough and full of veins on their under: the branches are white and hairy, and the flowers are very large, and of a light purple colour.

The fifth sort doth not rise so high as either of the former, but sends out branches near the root, which are hairy and erect; these are garnished with spear-shaped leaves, which are of a dark green colour, and join at their base, surrounding the stalk. At each joint comes out a very slender branch, having three pair of small leaves of the same shape with the other, terminated by a single flower, and the ends of the branches have three or four flowers sitting close without foot-stalks. The flowers are of a deep purple colour, and like those of the first.

The sixth sort rises to the height of five or six feet, with a strong woody stalk, sending out many hairy branches, which are garnished with spear-shaped leaves, smooth on their upper side, but veined on their under, having short foot-stalks, which join at their base, where they form a sort of sheath to the branch. The flowers come out at the end of the branches, which are large, of a light purple colour, resembling those of the fourth sort.

The seventh sort hath erect branches, which come out from the lower part of the stalk, and are woolly; these are garnished with oblong hoary leaves, covered with a white down, which are smooth above, but veined on the under side, joining at their base, where they surround the stalk; the flowers are produced at the end of the branches, which are of a bright purple colour, and large.

The eighth ſort hath a ſlender ſmooth ſtalk, covered with a brown bark, which never riſes more than two feet high, ſending out many horizontal weak branches, which ſpread very wide, and are garniſhed with ſmall oval leaves, which are hairy, ſtanding upon ſhort foot-ſtalks. The flowers come out at the wings of the leaves upon long naked foot-ſtalks; theſe are white, and ſomewhat ſmaller than thoſe of the other ſorts.

The ninth ſort grows naturally in the iſlands of the Archipelago, and is the plant which produces the Ladanum, as is hereafter mentioned; it riſes three or four feet high, with a woody ſtalk, ſending out many lateral branches, covered with a brown bark, garniſhed with oval, ſpear-ſhaped, hairy leaves, waved and curled on their borders; theſe in warm ſeaſons ſweat a glutinous liquid, which ſpreads on the ſurface of the leaves, is very clammy and ſweet-ſcented. The flowers come out at the end of the branches on ſhort hairy foot-ſtalks; they are of a deep purple colour, and about the ſize of a ſingle Roſe.

The tenth ſort riſes with a ſhrubby ſtalk about four feet high; the branches are very hairy, glutinous, and grow erect, and are garniſhed with long narrow leaves, ending in points, which are hairy, and of a deep green on both ſides, having a deep longitudinal furrow on their upper ſide, made by the midrib, which is prominent on the under ſide; the flowers ſtand upon long foot-ſtalks at the end of the branches, which are of a pale ſulphur colour, having a bordered empalement, which is cut into five acute parts at the top.

The eleventh ſort riſes with a ſtrong woody ſtem to the height of five or ſix feet, ſending out many erect branches, which are garniſhed with ſpear-ſhaped leaves, ending in points; theſe are thick, white on their under ſide, of a dark green above, and very glutinous in warm weather. The flowers are produced at the end of the branches upon long naked foot-ſtalks, which branch on their ſides into ſmaller foot-ſtalks, each ſuſtaining one large white flower, having a hairy empalement.

The twelfth ſort riſes with a ſmooth ſhrubby ſtalk four or five feet high, ſending out many ſlender ligneous branches, covered with a ſmooth brown bark, garniſhed with oblong heart-ſhaped leaves, which are ſmooth, and have long foot-ſtalks. The flowers are white, and are produced at the end of the branches, ſtanding upon pretty long foot-ſtalks.

The thirteenth ſort riſes with a ſlender ſhrubby ſtalk from three to four feet high, ſending out many branches from the bottom upward, which are hairy, garniſhed with ſpear-ſhaped leaves, of a very dark green colour, having three longitudinal veins in each; in warm weather they are covered with a glutinous ſweet-ſcented ſubſtance, which exudes from their pores. The flower-ſtalks which come out at the end of the branches are long, naked, and ſuſtain many white flowers, riſing above each other; their empalements are bordered, and end in ſharp points.

The fourteenth ſort riſes with a woody ſtem to the height of five or ſix feet, ſending out many ſide branches from the bottom; theſe are ſmooth, covered with a reddiſh brown bark, garniſhed with narrow ſpear-ſhaped leaves, whitiſh on their under ſide, of a dark green above, having three longitudinal veins. The flowers are produced at the end of the branches on ſhort foot-ſtalks, and are compoſed of five very large roundiſh petals, each having a large purple ſpot at their baſe. The whole plant exudes a ſweet glutinous ſubſtance in warm weather, which hath a very ſtrong balſamic ſcent, ſo as to perfume the circumambient air to a great diſtance.

There is a variety of this with white flowers, having no purple ſpots, which is in all other reſpects the ſame as this.

The fifteenth ſort riſes with a ſhrubby ſtalk to the ſame height as the laſt, ſending out many branches on every ſide, which are garniſhed with ſpear-ſhaped leaves, of a thick conſiſtence, which are ſmooth on their upper ſide, of a very dark green colour, and white on their under ſide; theſe are very clammy, eſpecially in warm weather; the flowers are produced at the end of the branches on ſhort foot-ſtalks; they are very large, white, and have a broad, dark, purple ſpot at the baſe of each petal; this differs from the former ſort in the ſhape and ſize of their leaves; thoſe of this ſort being much ſhorter and broader, and are very white on their under ſide. The ſide branches are ſhorter, and the leaves ſtand much cloſer on them; the flowers are larger, and the whole plant is much more glutinous.

The ſixteenth ſort hath a ſtiff, ſlender, woody ſtalk, which ſends out many branches; it riſes to the height of ſix or ſeven feet; the leaves are large, heart-ſhaped, thin, and of a light green colour; theſe ſit cloſe to the branches, and have many nerves; the flowers are produced at the end of the branches upon naked foot-ſtalks; they are white, and fade to a pale ſulphur colour when they fall off.

The ſeventeenth ſort hath an upright ſhrubby ſtalk, which riſes four or five feet high, ſending out many branches from the ground upward, ſo as to form a large buſh. The branches are channelled and hoary. The leaves are oval, ſtanding oppoſite; thoſe on the lower part of the branches have foot-ſtalks, but upward they coaleſce at their baſe, and ſurround the ſtalk; theſe are very white. The foot-ſtalks of the flowers, which riſe at the end of the branches, are a foot in length, naked, hairy, and put out two or four ſhorter foot-ſtalks on the ſide, each ſupporting three or four flowers, each having a ſhort foot-ſtalk. The flowers are large, of a bright yellow colour, but of ſhort duration, ſeldom continuing longer than three or four hours.

The eighteenth ſort riſes with a ſlender woody ſtalk three or four feet high, ſending out many ſlender branches, garniſhed with narrow, ſpear-ſhaped, hoary leaves, which ſit cloſe to them; from the wings of the leaves come out ſlender branches, which have two or three pair of ſmall leaves, terminated by looſe bunches of flowers, each ſtanding on a ſlender foot-ſtalk. The flowers are of a dirty ſulphur colour.

This ſort will not live abroad in the winter, therefore it is generally placed in a green-houſe.

Theſe plants are moſt of them hardy enough to live in the open air in England, unleſs in very ſevere winters, which ſometimes deſtroy them, ſo that a plant or two of each ſort may be kept in pots, and ſheltered in winter, to preſerve the kinds; the reſt may be intermixed with other ſhrubs, for where they are ſheltered by other plants, they will endure the cold much better than when they are ſcattered ſingly in the borders. Many of theſe plants will grow to the height of five or ſix feet, and will have large ſpreading heads, provided they are permitted to grow uncut; but if they are ever trimmed, it ſhould be only ſo much as to prevent their heads from growing too large for their ſtems; for whenever this happens, they are apt to fall on the ground, and appear unſightly.

Theſe ſhrubs are propagated by ſeeds, and alſo from cuttings; but the latter method is ſeldom practiſed, unleſs for thoſe ſorts which do not produce ſeeds in England; theſe are the twelfth, ſeventeenth, and eighteenth ſorts; all the others generally produce plenty of ſeeds, eſpecially thoſe plants which came from ſeeds, for thoſe which are propagated by cuttings, are very ſubject to become barren, which is alſo common to many other plants.

The ſeeds of theſe plants may be ſown in the ſpring upon a common border of light earth, where the plants will come up in five or ſix weeks, and, if they are kept clear from weeds,

weeds, and thinned where they are too close, they will grow eight or ten inches high the same year; but as these plants, when young, are liable to injury from hard frost, therefore they should be transplanted when they are about two inches high, some into small pots filled with light earth, that they may be removed into shelter in winter, and the others into a warm border, at about six inches distance each way; those which are potted must be set in a shady situation till they have taken new root, and those planted in the border must be shaded every day with mats till they are rooted; after which the latter will require no other care but to keep them clean from weeds till autumn, when they should have hoops placed over them, that they may be covered in frosty weather; those in the pots may be removed into an open situation, so soon as they have taken new root, where they may remain till the end of October; then they should be placed under a hot-bed frame to screen them from the cold in winter, but at all times, when the weather is mild, they should be fully exposed to the open air, and only covered in frosts; with this management the plants will thrive much better than when they are more tenderly treated.

In the spring following, these plants may be turned out of the pots, with all the earth preserved to their roots, and planted in the places where they are to remain (for they are bad plants to remove when grown old,) observing to give them now and then a little water until they have taken fresh root; after which time they will require no farther care, than to train them upright in the manner you would have them grow. The plants which were planted in the border also may be transplanted abroad the succeeding spring. In removing of these you should be careful to preserve as much earth about the roots as you can; and if the season should prove hot and dry, you must water and shade them until they have taken fresh root; after which they will require no other culture than was before directed.

The fourteenth and fifteenth sorts are by much the most beautiful of all these Cistuses; the flowers, which are as big as a large Rose, are of a fine white, with a deep purple spot on the bottom of each leaf. These plants also abound with a sweet glutinous liquor, which exudes through the pores of the leaves in so plentiful a manner in hot weather, that the surface of the leaves are covered therewith; from this plant Clusius thinks might be gathered great quantities of the Ladanum, which is used in medicine, in the woods in Spain, where he saw vast quantities of this shrub growing.

But it is from the ninth sort which Mons Tournefort says the Greeks in the Archipelago gather this sweet gum; in the doing of which (Bellonius says) they make use of an instrument like a rake without teeth, which they call Ergastiri; to this are tied many thongs of raw and untanned leather, which they rub gently on the bushes that produce the Ladanum, that so that liquid moisture may stick upon the thongs; after which they scrape it off with knives; this is done in the hottest time of the day, for which reason the labour of gathering this Ladanum is excessive, and almost intolerable, since they are obliged to remain on the mountains for whole days together, in the very heat of summer, or the dog days; nor is there any person almost that will undertake this labour, except the Greek monks.

CITHAREXYLON. Lin. Gen. Pl. 678. Fiddle Wood.

The CHARACTERS are,

The flower is one leaf, funnel-shaped, and divided at the top into five equal parts. It hath four stamina, which adhere to the tube, two of them being longer than the other. In the center is situated the roundish germen, which afterward becomes a capsule with two cells, each having a single seed.

The SPECIES are,

1. CITHAREXYLON (*Cinereum*) ramis angulatis, foliis ovato-lanceolatis venis candicantibus. *The common Fiddle Wood of America.*

2. CITHEREXYLON (*Album*) foliis oblongo-ovatis integris, oppositis ramis angulatis, floribus spicatis. *Fiddle Wood with oblong, oval, entire leaves growing opposite, angular branches, and flowers growing in spikes.*

The first sort grows common in most of the islands in the West-Indies, where it rises to a great height, and becomes a very large timber-tree, the wood of which is greatly esteemed for buildings, being very durable.

This rises with an upright trunk to the height of fifty or sixty feet, sending out many branches, which have several angles or ribs, running longitudinally, garnished with oval spear-shaped leaves at every joint, standing in a triangle upon short foot-stalks. The leaves are about four inches long, and one and a half broad, of a lively green colour, pretty much notched on their edges, having several deep Orange-coloured veins running from the midrib to the edges. The flowers come out from the sides, and also at the end of the branches, in loose bunches, which are succeeded by small pulpy berries, inclosing two seeds in each.

The second sort is a native of the same islands with the first. This is also a very large tree, whose timber is greatly valued in America for buildings, being very durable, and from thence I have been informed the French gave it the name of Fidelle Wood, which the English have rendered Fiddle Wood.

This tree rises with a strong upright trunk to the height of sixty feet or more, sending out many angular branches, standing opposite, garnished with oval oblong leaves, standing opposite on short foot-stalks; these are of a lucid green, and are rounded at their ends. The flowers come out in long loose spikes toward the end of the branches, which are white, and smell very sweet; these are followed by small, roundish, pulpy berries, each inclosing a single seed.

The seeds of both sorts should be sown in small pots early in the spring, and plunged into a fresh hot-bed of tanners bark, and treated in the same manner as other exotic seeds which are brought from hot countries. If the seeds are fresh, the plants will appear in five or six weeks, and in about one month more will be fit to transplant; when each should be planted in a small pot, and plunged into the hot-bed again, observing to shade them till they have taken fresh root; after which they should have a large share of air in warm weather; in autumn the plants should be removed into the bark-stove, where it will be proper to keep them the first winter till they have obtained strength, but afterward they may be kept in a dry stove in winter, and in the middle of summer they may be exposed in the open air for two months, in a warm situation, with which management the plants will make better progress than when they are more tenderly treated.

If the cuttings of this plant are planted in small pots during the summer months, and plunged into a moderate hot-bed, they will take root, and may afterward be treated in the same manner as the seedling plants.

CITRUS. Lin. Gen. Pl. 807. The Citron-tree.

The CHARACTERS are,

The flower hath five oblong thick petals; it hath ten stamina, which are not equal, and join in three bodies at their base. The oval germen in the center afterward becomes an oblong fruit, with a thick fleshy skin filled with a succulent pulp, having many cells, each containing two oval hard seeds.

The SPECIES are,

1. CITRUS (*Medica*) fructu oblongo, majori, mucronato, cortice crasso rugoso. *Citron with a larger, oblong, pointed fruit, having a thick rough rind; or Sweet Citron.*

2. CITRUS (*Tuberosa*) fructu oblongo, cortice tuberoso rugoso. *The common Citron.*

There are several other varieties of this fruit, with which the English gardens have been supplied from Genoa, where is is the great nurfery for the feveral parts of Europe for this fort, as alfo Orange and Lemon-trees.

The feveral forts of Citrons are cultivated in much the fame manner as the Orange tree, to which I fhall refer the reader to avoid repetition; but fhall only remark, that thefe are fomewhat tenderer than the Orange, and fhould therefore have a warmer fituation in winter, otherwife they are very fubject to caft their fruit. They fhould alfo continue a little longer in the houfe in the fpring, and be carried in again fooner in the autumn. And as their leaves are larger, and their fhoots ftronger than thofe of the Orange, they require a little more water in the fummer; but in the winter they fhould have little water at each time, which muft be the oftener repeated.

The common Citron is much the beft ftock to bud any of the Orange or Lemon kinds upon, it being the ftraiteft and freeft growing tree; the rind is fmoother, and the wood lefs knotty than either the Orange or Lemon, and will take either fort full as well as its own kind, which is what none of the other forts will do: and thefe ftocks, if rightly managed, will be very ftrong the fecond year after fowing, capable to receive any buds, and will have ftrength to force them out vigoroufly; whereas it often happens, when thefe buds are inoculated into weak ftocks, they frequently die, or remain till the fecond year before they put out; and thofe that fhoot the next fpring after budding, are oftentimes fo weak as hardly to be fit to remain, being incapable to make a ftrait handfome ftem, which is the great beauty of thefe trees.

CITRUL. See Pepo.

CLARY. See Sclarea.

CLAYTONIA. Gron. Flor. Virg. Lin. Gen. Pl. 253.

The Characters are,

The flower hath five oval petals, which are indented at the top, and five recurved ftamina, which are fhorter than the petals. In the center is fituated an oval germen, which afterward becomes a roundifh capfule, having three cells, opening with three elaftic valves, and filled with round feeds.

We have but one fort of this genus in the Englifh gardens, viz.

Claytonia (*Virginica*) foliis linearibus. Lin. Sp. Pl. *Claytonia with very narrow leaves.*

This plant grows naturally in Virginia. It hath a fmall tuberous root, which fends out low flender ftalks in the fpring, about three inches high, which have two or three fucculent narrow leaves about two inches long, of a deep green colour; at the top of the ftalk are four or five flowers produced, ftanding in a loofe bunch, compofed of five white petals, which fpread open, fpotted with red on their infide; after thefe fall away, the germen becomes a roundifh capfule, divided into three cells, which are filled with roundifh feeds.

It is propagated by feeds, and alfo from offsets fent out from the roots: the feeds fhould be fown on a border of light earth foon after they are ripe, for if they are kept out of the ground till fpring, the plants will not come up till the next year; whereas thofe which are fown early in the autumn, will grow the following fpring, fo that a whole year is gained. When the plants come up, they will require no other care but to keep them clean from weeds; and in the autumn, if fome old tanners bark is fpread over the furface of the ground, it will fecure the roots from being injured by froft, which, if it fhould prove very fevere, might be the cafe with young plants, but after the firft winter they will not require protection.

The beft time to tranfplant the roots is about Michaelmas, when they are inactive; but as they are fmall, fo if great care is not taken in opening the ground, the roots may be buried and loft, for they are of a dark colour, fo are not eafily diftinguifhed from the ground.

CLEMATIS. Lin. Gen. Pl. 626. Virgin's Bower.

The Characters are,

The flowers have each four loofe oblong petals, with a great number of ftamina; the fummits adhere to their fide. They have many compreffed germen, which afterward becomes fo many roundifh compreffed feeds, with the ftyle fitting on their top.

The Species are,

1. Clematis (*Recta*) foliis pinnatis, foliolis ovato-lanceolatis, integerrimis, caule erecto. Hort. Cliff. 225. *Upright white Climber.*

2. Clematis (*Integrifolia*) foliis fimplicibus, ovato-lanceolatis. Hort. Cliff. 225. *Upright blue Climber.*

3. Clematis (*Vita Alba*) foliis pinnatis, foliolis cordatis, fcandentibus. Hort. Cliff. 225 *Climber with broad entire leaves; commonly called Viorna, or Traveller's Joy.*

4. Clematis (*Canadenfis*) foliis ternatis, foliolis cordatis, acutis, dentatis, fcandentibus. *Broad-leaved Canada Climber, having three leaves.*

5. Clematis (*Flammula*) foliis inferioribus, pinnatis, laciniatis, fummis fimplicibus, integerrimis, lanceolatis. Hort. Cliff. 225. *Creeping Climber.*

6. Clematis (*Cirrhofa*) cirrhis fcandens. Hort. Cliff. 226. *Clematis with climbing tendrils.*

7. Clematis (*Viticella*) foliis compofitis decompofitifque, foliolis ovatis, integerrimis. Hort. Cliff. 225. *Single blue Virgin's Bower.*

8. Clematis (*Alpina*) foliis ternatis, ternatifque, foliolis ovatis, acutè ferratis, fcandentibus. *Clematis with trifoliate leaves, which have three oval lobes, fharply fawed, and climbing.*

9. Clematis (*Viorna*) foliis compofitis decompofitifque, foliolis quibufdam trifidis. Flor. Virg. 62. *Creeping purple Climber, with coriaceous petals to the flower.*

10 Clematis (*Orientalis*) foliis compofitis, foliolis incifis, angulatis, lobatis, cuneiformibus. Lin. Sp. Pl. 543. *Eaftern Climber with a Smallage leaf, and a reflexed, greenifh, yellow flower.*

11. Clematis (*Siberica*) foliis compofitis & decompofitis, foliolis ternatis, ferratis. Gmel. *Climber with compound and decompounded leaves, whofe fmall leaves are fawed and trifoliate.*

12. Clematis (*Crifpa*) foliis fimplicibus, ternatifque, foliolis integris trilobifve. Lin. Sp. Pl. 543. *Climber with fingle and trifoliate leaves, whofe fmall leaves are either entire, or have three lobes.*

The firft fort grows naturally in the fouth of France, in Italy, Auftria, and feveral parts of Germany. This hath a perennial root and annual ftalks, which grow upright about five feet high, garnifhed with winged leaves ftanding oppofite; thefe are compofed of three or four pair of lobes, terminated by an odd one; the flowers are produced in large loofe panicles at the top of the ftalks; thefe are compofed of four white petals, which fpread open; and the middle is occupied by a great number of ftamina, furrounding five or fix germen, which afterward become fo many compreffed feeds, each having a long tail or beard fitting on the top.

The fecond fort grows naturally in Hungary and Tartary. This is perennial, fending up many flender upright ftalks from five to fix feet high, garnifhed with fingle leaves at each joint, which ftand oppofite on very fhort foot-ftalks; they are near four inches long, and an inch and a half broad in the middle, fmooth and entire, ending in a point: the flowers come out from the upper part of the ftalks, ftanding upon very long naked foot-ftalks, which nod down, each fupporting a fingle blue flower, compofed of four narrow thick petals, which fpread open, and many hairy ftamina furrounding the germina in the center. After the flowers are paft, the

germen

germen become fo many compreffed feeds, each having a tail or beard.

The third fort grows naturally in the hedges in moft parts of England. This hath a rough climbing ftalk, fending out clafpers, by which it faftens to the neighbouring bufhes and trees, and fometimes rifes more than twenty feet high, often covering all the trees and bufhes of the hedge. This puts out many bunches of white flowers in June, which are fucceeded by flat feeds joined in a head, each having a long twifted tail fitting on the top, covered with long white hairs; in autumn, when the feeds are near ripe, they appear like beards, from whence the country people call it Old Man's Beard. The branches of this being very tough and flexible, are often ufed for tying up faggots, from whence, in fome counties, it is called Bindwith.

The fourth fort grows naturally all over North America. This is in its firft appearance very like the laft fort, but the leaves are broader, and grow by threes on the fame footftalk, whereas thofe of the former have five or feven lobes in each leaf: the flowers appear at the fame time with the former.

The fifth fort hath a climbing ftalk like the third; the lower leaves of this are winged, and deeply cut on their edges, but the upper leaves are fingle, fpear-fhaped, and entire; the flowers of this fort are white. This grows naturally in the fouth of France, and in Italy.

The fixth fort grows naturally in Spain and Portugal. This hath a climbing ftalk, which will rife to the height of ten or twelve feet, fending out branches from every joint, whereby it becomes a very thick bufhy plant; the leaves are fometimes fingle, at other times double, and frequently trifoliate, being indented on their edges. Thefe keep their verdure all the year; oppofite to the leaves come out clafpers, which faften themfelves to the neighbouring fhrubs, by which the branches are fupported, otherwife they would fall to the ground. The flowers are produced from the fide of the branches; thefe are large, of an herbaceous colour, and appear always about the end of December or beginning of January.

The feventh fort is cultivated in the nurfery-gardens for fale, and is known by the title of Virgin's Bower. There are four varieties of it which are preferved in the gardens of the curious, and have been by fome treated as fo many diftinct fpecies; but as their only differences confift either in the colour of the flowers, or the multiplicity of their petals, fo they are now only efteemed as feminal variations; but as they are diftinguifhed by the nurfery-gardeners, fo I fhall juft mention them.

1. Single blue Virgin's Bower.
2. Single purple Virgin's Bower.
3. Single red Virgin's Bower.
4. Double purple Virgin's Bower.

The ftalks of thefe plants are very flender and weak, having many joints, from whence come out fide branches, which are again divided into fmaller: if thefe are fupported, they will rife to the height of ten or twelve feet, and are garnifhed with compound winged leaves placed oppofite. Thefe branch out into many divifions, each of which hath a flender foot-ftalk, with three fmall leaves which are oval and entire; from the fame joint generally four footftalks arife, two on each fide; the two lower have three of thefe divifions, fo that they are each compofed of nine fmall leaves; but the two upper have only two oppofite leaves on each, and between thefe arife three flender foot-ftalks, each fupporting one flower. The flowers have each four petals, which are narrow at their bafe, but are broad at the top and rounded: in one they are of a dark worn-out purple, in another blue, and the third of a bright purple or red colour. The double fort, which is common in the Englifh gardens, is of the worn-out purple colour. The double flowers have no ftamina or germen, but in lieu of them there is a multiplicity of petals, which are narrow, and turn inward at the top.

The eighth fort grows naturally on the Alps, and other mountains in Italy. This hath a flender climbing ftalk, which rifes three or four feet high, fupporting itfelf by faftening to the neighbouring plants or fhrubs. The leaves of this are compofed of nine lobes or fmall leaves, three ftanding upon each foot-ftalk. The flowers come out at the joints of the ftalk in the fame manner as the common Traveller's Joy, which are white, fo make no great appearance.

The ninth fort grows naturally in Virginia and Carolina. This hath many flender ftalks, garnifhed with compound winged leaves at each joint; thefe are generally compofed of nine leaves, ftanding by threes like thofe of the eighth fort, but the fmall leaves of this are nearly of a heart-fhape. The flowers of this ftand upon fhort footftalks, which come out from the wings of the leaves, one on each fide the ftalk. They are compofed of four thick petals, which are purple on their outfide and blue within.

The tenth fort grows in the Levant. This hath weak climbing ftalks, which faften themfelves by their clafpers to any plants or fhrubs which ftand near them, and thereby rife to the height of feven or eight feet; thefe are garnifhed with compound winged leaves, confifting of nine fmall leaves (or lobes) which are angular, and fharp pointed. The flowers come out from the wings of the leaves, which are of a yellowifh green, and the petals are reflexed backward.

The eleventh fort grows naturally in Tartary. This plant hath weak climbing ftalks which require fupport; they grow from four to eight feet high; their joints are far afunder; at each of thefe come two compound winged leaves, whofe fmall leaves are placed by threes; thefe are deeply fawed on their edges, and terminate in fharp points. The flowers come out from the wings of the leaves fingle, ftanding upon long naked foot-ftalks, and are compofed of four narrow fpear-fhaped petals, which fpread open in form of a crofs; they are of a yellowifh white colour After thefe are paft, the germen become fo many compreffed feeds, each having a bearded tail.

The twelfth fort grows naturally in Carolina. This hath weak ftalks which rife near four feet high, and by their clafpers faften themfelves to the neighbouring plants, whereby they are fupported. The leaves come out oppofite at the joints: thefe are fometimes fingle, at others trifoliate, and fome of the leaves are divided into three lobes. The flowers come out fingly from the fide of the branches upon fhort foot-ftalks, with one or two pair of leaves below the flower, which are oblong and fharp-pointed. The flowers have four thick petals like thofe of the ninth fort, of a purple colour, and their inner furface is curled, with many longitudinal furrows.

The two firft forts have perennial roots, which multiply pretty faft, but their ftalks die down every autumn, and new ones arife in the fpring, in which particular they differ from the other fpecies, therefore require different management.

They are propagated either by feeds or parting of their roots; but the former being a tedious method, the latter is generally practifed. The beft feafon for parting thefe roots is in October, juft before their branches decay

They will grow almoft in any foil or fituation; the roots may be cut through their crowns with a fharp knife, obferving to preferve to every offset fome good buds or eyes; and then it matters not how fmall you divide them, for their roots increafe very faft; but if you part them very

fmall,

ſmall, you ſhould let them remain three or four years before they are again removed, that their flowers may be ſtrong, and the roots multiplied in eyes.

The plants are extreme hardy, enduring the cold of our ſevereſt winters in the open air, and are very proper ornaments for large gardens, either to be planted in large borders, or intermixed with other hardy flower-roots in quarters of flowering ſhrubs. They begin to flower about the beginning of June, and often continue to produce freſh flowers until September.

The third ſort is found wild in moſt parts of England, growing upon the ſides of banks, under hedges, and extends its trailing branches over the trees and ſhrubs that are near it. This plant in the autumn is generally covered with ſeeds, which are collected into little heads, each of which having as it were a rough plume faſtened to it, hath occaſioned the country people to give it the name of Old Man's Beard. It is titled by Lobel and Gerard, Viorna; and by Dodonæus, Vitis alba: in Engliſh it is commonly called Traveller's Joy.

The fourth and fifth ſorts have no more beauty than the third, ſo are ſeldom preſerved in gardens, unleſs for the ſake of variety. They are both as hardy as the common ſort, and may be propagated either by ſeeds or laying down their branches.

The ſixth ſort retains its leaves all the year, which renders it valuable.

This ſort doth not produce ſeeds in England, ſo it is propated by layers, and alſo from cuttings. If they are propagated by layers, the ſhoots of the ſame year only ſhould be choſen for this purpoſe, for the older branches do not put out roots in leſs than two years, whereas the tender ſhoots will make good roots in one: theſe muſt be pegged down into the ground in October, in the ſame manner as is uſually practiſed for other layers, to prevent their riſing. If the ſhoots have two inches of earth over them, it will be better than a greater depth. Theſe layers will have ſtrong roots by the following autumn, when they may be taken from the old plant, and tranſplanted where they are deſigned to remain.

All the varieties of Virgin's Bower are eaſily propagated by laying down their branches; for although the ſingle flowers do ſometimes produce ſeeds in England, yet as theſe ſeeds, when ſown, remain a whole year in the ground before they vegetate, ſo the other being the more expeditious method of increaſing theſe plants, is generally practiſed: but in order to ſucceed, theſe layers ſhould be put down at a different ſeaſon from the former; for when they are layed in the autumn, their ſhoots are become tough, ſo do rarely put out roots under two years; and after lying ſo long in the ground, not one in three of them will have made good roots, ſo that many have ſuppoſed theſe plants were difficult to propagate; but ſince they have altered the ſeaſon of doing it, they have found theſe layers have ſucceeded as well as thoſe of other plants.

The beſt time for laying down of the branches is in July, ſoon after they have made their firſt ſhoots, for it is the young branches of the ſame year which do freely take root; but as theſe are very tender, and apt to break, ſo there ſhould be great care taken in the operation: therefore thoſe branches from which theſe ſhoots are produced, ſhould be firſt brought down to the ground, and faſtened to prevent their riſing; then the young ſhoots ſhould be laid into the earth, with their tops raiſed upright, three or four inches above ground; and after the layers are placed down, if the ſurface of the ground be covered with Moſs, rotten tanners bark, or other mulch, it will prevent the ground from drying, ſo that the layers will not require watering above three or four times, which ſhould not be at leſs than five or ſix days interval; for when theſe layers have too much wet, the tender ſhoots frequently rot, or when the young fibres are newly put out, they are ſo tender as to periſh by having much wet: therefore where the method here directed is practiſed, the branches will more certainly take root than by any other yet practiſed.

As theſe plants have all of them climbing branches, ſo they ſhould be always planted where they may be ſupported, otherwiſe the branches will fall to the ground and appear unſightly; ſo that unleſs they are properly diſpoſed, inſtead of being ornaments to a garden, they will become the reverſe. When there are arbours or ſeats with trelliswork round them, theſe plants are very proper to train up againſt it; or where any walls or other fences require to be covered from the ſight, theſe plants are very proper for the purpoſe; but they are by no means proper for open borders, nor do they anſwer the expectation when they are intermixed with ſhrubs; for unleſs their branches have room to extend, they will not be productive of many flowers.

The ſort with double flowers is the moſt beautiful, ſo that ſhould be preferred to thoſe with ſingle flowers, of which a few only ſhould be planted for variety. They are all equally hardy, ſo are ſeldom injured by froſt, excepting in very ſevere winters, when ſometimes the very tender ſhoots are killed; but if theſe are cut off in the ſpring, the ſtems will put out new ſhoots.

The twelfth ſort is alſo a very hardy plant, with climbing branches, ſo may be diſpoſed in the ſame manner as the other. It is alſo propagated by layers, which will ſucceed if performed at the ſame time, and in the ſame manner as is directed for the former.

CLEOME. Lin. Gen. Plant. 740.

The Characters are,

The flower hath four petals which are inclined upward, the lower being leſs than the other; in the bottom there are three mellous glands, which are ſeparated by the empalement. It hath ſix or more incurved ſtamina fixed to their ſide, and a ſingle ſtyle ſupporting an oblong germen, which afterward becomes a long cylindrical pod, having one cell opening with two valves, and filled with roundiſh ſeeds.

The Species are,

1. Cleome (*Pentaphylla*) floribus gynandris, foliis digitatis. Hort. Cliff. 341. *Smooth, five-leaved, ſmaller Indian Baſtard Muſtard, with a fleſh-coloured flower.*

2. Cleome (*Ornithopodioides*) floribus hexandris, foliis ternatis, foliolis lanceolatis. Lin. Sp. Plant. 672. *Cleome with flowers having ſix ſtamina, trifoliate leaves, and ſpear-ſhaped lobes.*

3. Cleome (*Luſitanica*) floribus hexandris, foliis ternatis, foliolis lineari-lanceolatis, ſiliquis bivalvibus. *Cleome with flowers having ſix ſtamina, trifoliate leaves, narrow ſpear-ſhaped lobes, and pods having two valves.*

4. Cleome (*Viſcoſa*) floribus dodecandris, foliis quinatis ternatiſque. Flor. Zeyl. 241. *Cleome with flowers having twelve ſtamina, and trifoliate and quinquefoliate leaves.*

5. Cleome (*Triphylla*) floribus hexandris, foliis ternatis, foliolo intermedio majori. *Cleome with flowers having ſix ſtamina, and trifoliate leaves, whoſe middle lobe is the largeſt.*

6. Cleome (*Erucago*) floribus hexandris, foliis ſeptenis, caule ſpinoſo, ſiliquis pendulis. *Cleome with flowers having ſix ſtamina, leaves with ſeven lobes, a prickly ſtalk, and hanging pods.*

7. Cleome (*Spinoſa*) floribus hexandris, foliis quinatis ternatiſque, caule ſpinoſo. *Cleome with flowers having ſix ſtamina, leaves compoſed of five and three lobes, and a prickly ſtalk.*

8. Cleome (*Monophylla*) floribus hexandris, foliis ſimplicibus, ovato-lanceolatis. Flor. Zeyl. 243. *Cleome with*

ſix

fix ftamina to the flowers, and fingle leaves which are ovally fpear-fhaped.

The firft fort grows naturally in Afia, Africa, and America. It rifes with an herbaceous ftalk about a foot high, garnifhed with fmooth leaves, compofed of five fmall leaves or lobes, joining at their bafe to one center, and fpread out like the fingers of a hand. The leaves on the lower part of the ftalk ftand upon long foot-ftalks, which are gradually fhortened to the top of the ftalk, where they almoft join it: the flowers terminate the ftalks in loofe fpikes. Thefe have four petals of a flefh colour, which ftand erect, fpreading from each other, and below thefe are placed the ftamina and ftyle which coalefce at the bottom, and are ftretched out beyond the petals. After the flower is paft, the germen which fits upon the ftyle becomes a taper pod, about two inches long, filled with round feeds.

The fecond fort grows naturally in the Levant. This rifes with an upright ftalk about a foot high, garnifhed with leaves compofed of three fpear-fhaped lobes, ftanding upon fhort foot-ftalks; the flowers come out fingly from the fide of the ftalks, and have four red petals, which ftand in the fame form as thofe of the former fort: thefe are fucceeded by flender pods two inches long, which fwell in every divifion, where each feed is lodged, fo as to appear like joints, as thofe do of the Bird's-foot Trefoil; when the feeds are ripe the whole plant decays. If the feeds of this fort are permitted to fcatter, the plants will come up without care, and require only to be thinned and kept clean from weeds, for they will not bear tranfplanting.

The third fort grows naturally in Portugal and Spain. This rifes with an herbaceous ftalk about a foot high, fending out a few fhort fide branches, garnifhed with leaves compofed of three narrow lobes, ftanding upon fhort foot-ftalks. The flowers come out fingly from the fide of the ftalks, of a deep red colour, and are fucceeded by thick taper pods filled with round feeds. This is an annual plant, which will thrive in the open air, and requires the fame treatment as the former.

The fourth fort grows naturally in the ifland of Ceylon. This rifes a foot and a half high, fending out feveral fide branches garnifhed with leaves, fome of which have five, and others three roundifh lobes, ftanding upon fhort hairy foot-ftalks. The flowers come out fingly at the foot-ftalks of the leaves, they are of a pale yellow, and are fucceeded by taper pods between two and three inches long, ending in a point, which are full of round feeds. The whole plant fweats out a vifcous clammy juice.

The fifth fort is an annual plant which rifes two feet high, fending out many fide branches, garnifhed with leaves having one large fpear-fhaped lobe in the middle, and two very fmall ones on the fide; thefe fit clofe to the branches. The flowers come out fingly from the fide of the branches, upon long foot-ftalks: they have four large flefh-coloured petals, and fix long ftamina which ftand out beyond the petals; when the flowers fade, the germen which fits upon the ftyle becomes a taper pod four inches long, filled with round feeds.

The fixth fort grows naturally in Egypt and America. This rifes with a ftrong, thick, herbaceous ftalk, two feet and a half high, dividing into many branches, which are garnifhed with leaves compofed of feven long fpear-fhaped lobes, joining in a center at their bafe, where they fit upon a long flender foot-ftalk: juft below the foot-ftalk comes out one or two fhort, thick, yellow fpines, which are very fharp. The flowers come out fingly from the fide of the branches, forming a long loofe fpike at their extremities; this fpike hath fingle broad leaves, which half furround the ftalks at their bafe, from the bofom of which come out the foot-ftalks of the flowers, which are two inches long, each fuftaining a large flefh-coloured flower, whofe ftyle and ftamina are extended two inches beyond the petals. After the flower is paft, the germen which fits upon the ftyle becomes a thick taper pod five inches long, which hangs downward, and is filled with round feeds

The feventh fort grows naturally at the Havannah. This is alfo an annual plant, which rifes near two feet high, branching out on every fide: the lower leaves are compofed of five oblong lobes ftanding upon long foot-ftalks, but thofe on the ftalks and branches have but three lobes, and have fhort foot-ftalks: the main ftalk and alfo the branches are terminated by loofe fpikes of purple flowers, each fitting upon a flender foot-ftalk, at the bafe of which is placed a fingle oval leaf. The ftalks are armed with flender ftiff fpines, which are fituated juft under the foot-ftalks of the leaves; when the flowers fade the germen becomes a taper pod two inches long, filled with round feeds.

The eighth fort grows naturally in Ceylon; this is an annual plant, which rifes with an herbaceous ftalk a foot high, garnifhed with long, narrow, fingle leaves, ftanding alternately on the ftalks; from the wings of the leaves come out the foot-ftalks of the flower, each fuftaining a fingle yellow flower, which is fucceeded by a very flender taper pod. All thefe plants, except the fecond and third forts, are natives of very warm countries, fo will not thrive in England without artificial heat; therefore their feeds muft be fown upon a good hot-bed in the fpring, and when the plants are fit to remove, they fhould be planted in feparate fmall pots, and plunged into a frefh hot-bed, obferving to fhade them until they have taken root; after which they fhould have plenty of air in warm weather. The plants, when they are too tall to remain longer in the hot-bed, fhould be removed into an airy glafs-cafe, where they may be fcreened from cold and wet, but in warm weather may enjoy the free air. With this management the plants will flower foon after, and perfect their feeds in autumn. The fecond and third forts may be fown in the open borders of the garden where they are defigned to remain, for they do not require any artificial warmth.

CLEONIA (*Lufitanica.*) Portugal Cleonia, or Prunella odorata.

This plant grows naturally in Spain and Portugal; it is an annual, or at moft a biennial plant; for if the plants come up early in the fpring, they will produce ripe feeds the fame autumn, foon after which the plants decay; but when they come up from feeds toward autumn, they will not produce flowers till the following fummer, and as foon as the feeds are ripened the plants decay.

It is in many refpects fo like Prunella or Self-heal, that it has been by many botanifts ranged under that genus. The feeds of this fhould be fown in open borders, and the plants may be treated in the fame manner as is directed for the other fpecies of Self-heal.

CLETHRA. Gron. Fl. Virg. 43.

The CHARACTERS are,

The flower hath five oblong petals; it hath ten ftamina which are as long as the petals. In the center is fituated a roundifh germen, which afterward becomes a roundifh capfule inclofed by the empalement, having three cells, which are full of angular feeds.

We have but one SPECIES of this plant, viz.

CLETHRA (*Alnifolia.*) Gron. Fl. Virg. 43. *There is no Englifh title to this plant.*

This fhrub is a native of Virginia and Carolina, where it grows in moift places, and near the fides of rivulets, rifing to the height of eight or ten feet. The leaves are in fhape like thofe of the Alder-tree, but are longer; thefe are placed alternately upon the branches: the flowers are produced at the extremity of the branches in clofe fpikes:

they are composed of five leaves, are white, and have ten stamina in each, which are nearly of the same length with the petals.

This is hardy enough to bear the open air of England, and is one of the most beautiful shrubs at the season of its flowering; which is very little later than in its native country, being commonly in flower here by the beginning of July; and if the season is not very hot, there will be part of the spikes in beauty till the beginning of September; and as most of the branches are terminated with these spikes of flowers, so when the shrubs are strong, they make a fine appearance at that season.

This shrub will thrive best in moist land, and requires a sheltered situation, where it may be defended from strong winds, which frequently break off the branches where they are too much exposed to its violence. It is propagated by layers, but they are generally two years before they get root. They may also be propagated by suckers, which are sent out from their roots; if these are carefully taken off with fibres in the autumn, and planted into a nursery-bed, they will be strong enough in two years to transplant where they are to remain.

CLIFFORTIA. Lin. Gen. Plant. 1004.

The CHARACTERS are,

It hath male and female flowers in different plants; the male flowers hath a spreading empalement, composed of three small, oval, concave leaves, but have no petals, with a great number of hairy stamina. The female flowers have a permanent empalement composed of three leaves, sitting upon the germen; these have no petals, but the oblong germen which is situated below the empalement, supports two long, slender, feathered styles; the germen afterward becomes an oblong taper capsule with two cells crowned by the empalement, including one narrow taper seed.

The SPECIES are,

1. CLIFFORTIA (*Ilicifolia*) foliis subcordatis, dentatis. Lin. Sp. Plant. 1308. *Cliffortia with heart-shaped indented leaves.*

2. CLIFFORTIA (*Trifoliata*) foliis ternatis, intermedio tridentato. Prod. Leyd. 253. *Three-leaved Cliffortia, whose middle leaf is cut in three parts.*

3. CLIFFORTIA (*Ruscifolia*) foliis lanceolatis, integerrimis. Hort. Cliff. 463. *Cliffortia with spear-shaped leaves which are entire.*

The first sort grows naturally at the Cape of Good Hope. It rises with a weak shrubby stalk four or five feet high, sending out many diffused branches, which spread on every side, requiring some support: these are garnished with leaves which are heart-shaped at their base, but are broad at their ends, where they are sharply indented. They are very stiff, of a grayish colour, and closely embrace the stalks with their base, and are placed alternate on the branches; and from the bosom of these arise a single flower sitting close to the branch, having no foot-stalk. Before the empalement is spread open, it forms a bud, in shape and size of those of the Caper; this empalement is composed of three green leaves, which afterwards spread open, and then the numerous stamina appear standing erect.

This plant is easily propagated by cuttings, which may be planted in any of the summer months, which will soon take root, provided they are screened from the sun and duly watered; when they have taken root they may be each transplanted into a separate small pot, and placed in the shade until they have taken fresh root, after which they may be placed with other of the hardy kinds of exotic plants, in a sheltered situation till October, when they should be removed into the green-house, or placed under a common hot-bed frame, where they may be screened from the hard frost, but enjoy the free air at all times when the weather is mild. This plant has endured the cold of our ordinary winters, planted near a south-west wall without covering, but in severe winters they are always destroyed.

The second sort is a native of the same country as the first; this hath very slender ligneous stalks, which must be supported, otherwise they will fall to the ground. The branches are garnished with trifoliate leaves standing close to them; the middle lobes of these are much larger than the two side ones, and are indented in three parts. The flowers of this come out from the bosom of the leaves, having very short foot-stalks, and are shaped like those of the first, but are smaller. This sort requires the same management as the first, and is equally hardy, but must not be over watered in winter. The leaves of this sort continue green all the year.

The third sort rises with a weak shrubby stalk about four feet high, sending out lateral branches, which are covered with a whitish bark, and garnished with leaves placed in clusters without order; these are stiff, of the consistence and colour of the Butchers Broom, but are narrower, and run out to a longer point. Between these clusters of leaves the flowers come out in loose bunches; they have a great number of yellowish stamina, included in a three-leaved empalement.

This plant is tenderer than either of the former sorts, so should be placed in a good warm green-house in winter, and during that season they must have but little water. In the summer they may be exposed to the open air in a sheltered situation, but they should not remain too late abroad in the autumn; for if there should be much rain at that season, it would endanger these plants if they are exposed to it.

CLINOPODIUM. Lin. Gen. Plant. 644. Field Basil.

The CHARACTERS are,

The flower is of the lip kind, with a short tube enlarging to the mouth; the upper lip is erect, and indented at the top; the under lip is trifid, the middle segment being broad and indented. It hath four stamina under the upper lip, two of which are shorter than the other. In the center is situated the quadripartite germen, which afterward become four oval seeds shut up in the empalement.

The SPECIES are,

1. CLINOPODIUM (*Vulgare*) capitulis subrotundis, hispidis, bracteis setaceis. Lin. Sp. Plant. 587. *Field Basil with roundish prickly heads, and bristly bractea.*

2. CLINOPODIUM (*Incanum*) foliis subtus tomentosis, verticillis explanatis, bracteis lanceolatis. Lin. Sp. Plant. 588. *Field Basil with leaves which are woolly on the under side, broad plain whorls, and spear-shaped bractea.*

3. CLINOPODIUM (*Rugosum*) foliis rugosis, capitulis axillaribus, pedunculatis, explanatis, radiatis. Lin. Sp. Plant. 588. *Field Basil with rough leaves, plain heads growing on the sides of the stalks, which have foot-stalks, and are radiated.*

4. CLINOPODIUM (*Humile*) humile ramosum, foliis rugosioribus, capitulis explanatis. *Low branching Field Basil, with rougher leaves and plain heads.*

5. CLINOPODIUM (*Carolinianum*) caule erecto, non ramoso, foliis subtus villosis, verticillis paucioribus, bracteis calyce longioribus. *Field Basil with an upright unbranching stalk, leaves hairy on their under side, fewer whorls, and bractea longer than the empalement.*

6. CLINOPODIUM (*Ægyptiacum*) foliis ovatis rugosis, verticillis distantibus. *Field Basil with oval rough leaves, and the whorls of flowers standing at a great distance.*

The first sort grows naturally by the side of hedges, and in thickets, in most parts of England; this hath a perennial fibrous root, which sends up several stiff square stalks a foot and a half high, which send out a few lateral branches toward the top, garnifhed with oval hairy leaves placed by pairs; at the top of the ftalks the flowers come out in round whorls or heads; one of thefe terminate the ftalk, and there is generally another which furrounds the ftalk at the joint immediately below it. The flowers are fometimes purple, at others white. The whorls (or heads) grow very clofe, and each foot-ftalk fuftains feveral flowers. At the bafe of the empalement ftand two briftly fpines, which Linnæus terms the bractea; thefe ftand almoft horizontal under the empalement. The flower is of the labiated or lip kind. The upper lip is broad and trifid, but the under is cut into two narrow fegments; each flower is fucceeded by four naked feeds, fitting at the bottom of the empalement.

The fecond fort grows naturally in Penfylvania and Carolina; this hath a perennial root, which fends up many fquare ftalks about two feet high, which put out a few fhort fide branches toward the upper part, garnifhed with oblong oval leaves about the fize of thofe of Water Mint, ftanding by pairs clofe to the ftalk; they are hoary, and foft to the touch, and have a ftrong odour, between that of Marjoram and Bafil. The flowers grow in flat fmooth whorls round the ftalks; each ftalk hath generally three of thefe whorls, the upper which terminates the ftalk being fmaller, the two other increafing, fo that the lower is the greateft. The flowers are of a pale purple colour, and fhaped like thofe of the firft fort, but the ftamina of this ftands out beyond the petal, and the bractea at the bafe of the empalement are large, fpear-fhaped, and indented on their fides.

The third fort grows naturally in Carolina. This hath a perennial root, fending up feveral fquare ftalks, which are clofely covered with brownifh hairs; thefe rife between two and three feet high; they are garnifhed with leaves which are very unequal in their fize, thofe at the bottom, and alfo toward the top, being above three inches long, and one inch and a quarter broad, whereas thofe in other parts of the ftalk are not half fo large; they are rough on their upper fide, hairy below, fawed on their edges, and ftand oppofite by pairs: all the lower part of the ftalk, but immediately below the foot-ftalks of the flower-heads, there are three large leaves ftanding round the ftalks; between thefe arife two flender hairy foot-ftalks about three inches long, one on each fide the ftalk; thefe fuftain fmall heads of flowers, fhaped like thofe of the Scabious; they are white, fhaped like thofe of the other, but fmaller; the bractea immediately under the empalement fpread out like rays.

The fourth fort grows alfo in Carolina. This hath fome appearance of our common fort, but the ftalks do not grow more than half fo high, and divide into many long fide branches; the leaves are fmaller and rougher, and the whorls of flowers are produced half the length of the branches, whereas the common fort hath rarely more than two; the bractea at the bafe of the empalement are alfo much longer.

The fifth fort grows in Carolina. This hath a perennial root, fending up ftrait hairy ftalks which are almoft round; the joints of thefe are four or five inches afunder, at each of thefe come out two oblong leaves, hairy on their under fide, ftanding upon fhort foot-ftalks; at the bottom of thefe come out on each fide a flender branch half an inch long, having two or four fmall leaves, fhaped like the other. The flowers are produced in fmall whorls, ftanding thinly; thefe are white, and the bractea are longer than the empalement.

The fixth fort is a native of Egypt. It hath a perennial root, but annual ftalks, which grow a foot and a half high, garnifhed with oval leaves, having many tranfverfe deep furrows, and are of a dark green colour, placed oppofite, at about five or fix inches afunder. There are commonly two or four fide branches from the main ftems, produced toward the bottom; and the whorls of flowers are produced at every joint, toward the upper part of the ftalks; thefe are pretty large and hairy. The flowers are fomewhat larger than thofe of the common Field Bafil, and are of a deeper colour, ftretching a little more out of the empalement. The leaves of this have at firft fight much the fame appearance, but when they are obferved with attention, the difference is foon perceived between the two forts: but the greateft difference is in the leaves and whorls of flowers being placed at a greater diftance, and the ftalks growing fparfedly in this fpecies, nor do the plants continue fo long as thofe of the common fort.

This plant approaches near to the Clinopodium Orientale Origani folio, flore minimo. Tour. Corol. 12. But by comparing this with a fpecimen of that fort from the Paris garden, I find the leaves of that are fmoother, and placed much nearer together on the ftalks than thofe of this fort, and the flowers are fmaller.

Thefe plants may be propagated by feeds, and alfo by parting of their roots; the latter is generally practifed in England. The beft time to tranfplant and part their roots is in autumn, that they may take root before winter; thefe fhould be planted in a dry foil: they are all of them, except the third fort, hardy enough to thrive in the open air in England, and require no other care but to keep them clean from weeds, and every other year they may be tranfplanted and parted.

The third fort muft be planted in pots, and in winter fheltered under a frame, where the plants may enjoy the free air in mild weather, but fhould be fcreened from froft, otherwife it will not live in this country.

CLITORIA. Lin. Gen. Plant. 796.

The CHARACTERS are,

The flower is of the butterfly kind, having a large fpreading ftandard, which is erect; the two wings are oblong, and fhorter than the ftandard. The keel is fhorter than the wings, and is hooked It hath ten ftamina, nine of which are joined, and one ftands feparate. In the center is fituated an oblong germen, which afterward becomes a long, narrow, compreffed pod, with one cell, opening with two valves, inclofing feveral kidney-fhaped feeds.

The SPECIES are,

1. CLITORIA (*Ternatea*) foliis pinnatis. Hort. Cliff. 360. *Clitoria with winged leaves.*

2. CLITORIA (*Brafiliana*) foliis ternatis, calycibus campanulatis folitariis. Hort. Cliff. 215. *Clitoria with trifoliate leaves, and a fingle flower with a bell-fhaped empalement.*

3. CLITORIA (*Virginiana*) foliis ternatis, calycibus campanulatis geminis. Flor. Virg. 83. *Three-leaved Clitoria with two flowers joined, whofe empalements are bell-fhaped.*

4. CLITORIA (*Mariana*) foliis ternatis, calycibus cylindricis. Lin. Sp. Plant. 753. *Clitoria with trifoliate leaves, and cylindrical empalements to the flowers.*

The firft fort grows naturally in India. There is a variety of this with white flowers, and another with large blue flowers, which make a fine appearance.

This rifes with a twining herbaceous ftalk to the height of four or five feet, in the fame manner as the Kidney-bean, and requires the like fupport, for in the places where it grows naturally, it twifts itfelf about the neighbouring plants: the ftalks are garnifhed with winged leaves, compofed of two or three pair of lobes, terminated by an odd one; thefe are of a beautiful green, and are placed alternate on the ftalks; from the appendages of the leaves come out the foot-ftalks of the flowers; each of thefe is encompaffed by two very fine leaves about the middle, where they

are

are bent, ſuſtaining a very large, gaping, beautiful flower, whoſe bottom part ſeems as if growing to the top.

The ſecond ſort grows naturally in the Braſils, from whence the ſeeds were brought to Europe; this hath a twining ſtalk like the former, which riſes five or ſix feet high, and is garniſhed at each joint with one trifoliate leaf, ſtanding upon a long foot-ſtalk. The flowers come out ſingly from the foot-ſtalk of the leaves, ſtanding upon pretty long foot-ſtalks, which are encompaſſed about the middle with two ſmall oval leaves; the flowers are very large, the ſtandard being much broader than that of the firſt ſort, and the two wings are larger; the flowers are of a fine blue colour, ſo make a fine appearance.

There is one with a double flower of this, which I raiſed in the Chelſea garden ſome years paſt, from ſeeds ſent me from India, but the plants did not produce ſeeds here, and being annual the ſort was loſt. The flowers of this were very beautiful.

The ſeeds of the third ſort were ſent me from the Bahama Iſlands; this ſends out from the root two or three ſlender twining ſtalks, which riſe to the height of ſix or ſeven feet, and are garniſhed at each joint with one trifoliate leaf, whoſe lobes are oblong and pointed. At the oppoſite ſide of the ſtalk the foot-ſtalk of the flower ariſes, which is little more than an inch long, naked, and ſuſtains a ſingle flower, which is of a purple colour within, but of a greeniſh white on the outſide, not half ſo large as either of the former. Theſe flowers are each ſucceeded by long, ſlender, compreſſed pods, ending in a point, which contain one row of roundiſh kidney-ſhaped ſeeds.

The ſeeds of the fourth ſort were ſent me from Carolina, where the plants grow naturally. This riſes with a twining weak ſtalk about five feet high, and is garniſhed with trifoliate leaves like the former, whoſe lobes are narrower, and of a grayiſh colour on their under ſide; the flowers come out by pairs on the foot-ſtalks, and their empalements are cylindrical: the flowers are ſmall, and of a pale blue colour within, but of a dirty white on the outſide.

All theſe ſorts are annual with us in England, ſo that unleſs the ſeeds ripen the ſpecies are loſt; and as the two ſorts with double flowers have not formed any pods in this country, ſo far as I have been able to learn, therefore the ſeeds of theſe muſt be procured from the countries where they naturally grow.

The ſeeds of theſe plants muſt be ſown upon a good hotbed early in the ſpring, and when the plants are two inches high, they ſhould be carefully taken up, and each planted in a ſmall pot, and plunged into a hot-bed of tanners bark, obſerving to ſhade them till they have taken freſh root. After they are well rooted in the pots, they muſt have air every day in proportion to the warmth of the ſeaſon, to prevent their drawing up weak; their waterings ſhould be repeated two or three times a week, but they ſhould not have too much at each time. As theſe plants have climbing ſtalks, ſo they will ſoon grow too tall to remain under common frames, therefore they muſt then be removed into the ſtove, and plunged into the bark-bed, and afterward they muſt be treated in the ſame manner as other plants from the ſame countries.

CLUSIA. Lin. Gen. Plant. 577. Plum. Nov. Gen. 20. tab. 20. The Balſam-tree.

The CHARACTERS are,

The flower hath five large, roundiſh, concave, ſpreading petals. In the bottom is ſituated a globular nectarium, including the germen, which is pervious at the top. It hath a great number of ſingle ſtamina; the oblong oval germen is terminated by a plain ſtar-like ſtigma with ſix obtuſe indentures, which afterward becomes an oval capſule with ſix furrows and ſix cells, opening with ſix valves, which ſpread in form of a ſtar, including many angular ſeeds fixed to a column, ſurrounded with pulp.

The SPECIES are,

1. CLUSIA (*Flava*) foliis craſſis ſubrotundis nitidis, caule arboreo. *Cluſia with thick, ſhining, roundiſh leaves, and a tree-like ſtalk; commonly called Balſam-tree in America.*

2. CLUSIA (*Venoſa*) foliis venoſis. Lin. Sp. Plant. 510. *Cluſia with veined leaves.*

There are two or three varieties of the firſt ſort, which differ in the ſize and colour of their flowers and fruit; one hath a white flower and ſcarlet fruit, another hath a Roſe-flower and a greeniſh fruit, and the third hath a yellow fruit: but theſe are only ſuppoſed to be ſeminal variations.

The firſt ſort is pretty common in the Britiſh Iſlands of America, where the trees grow to the height of twenty feet, and ſhoot out many branches on every ſide, which are furniſhed with thick, round, ſucculent leaves, placed oppoſite. The flowers are produced at the ends of the branches, each having a thick ſucculent cover. After the flowers are paſt, they are ſucceeded by oval fruit. From every part of theſe trees there exudes a ſort of turpentine, which is called in the Weſt-Indies hog-gum; becauſe they ſay, that when any of the wild hogs are wounded they repair to theſe trees, and rub their wounded parts againſt the ſtems of theſe trees, till they have anointed themſelves with this turpentine, which heals their wounds.

As theſe plants are tender, ſo they muſt be conſtantly kept in the ſtove, otherwiſe they will not live through the winter in England; they muſt alſo be watered very ſparingly, eſpecially in winter; for they naturally grow in thoſe parts of the iſlands where it ſeldom rains, therefore they cannot bear much moiſture.

They may be propagated by cuttings, which muſt be laid to dry when they are cut off from the plants for a fortnight or three weeks, that the wounded part may be healed over, otherwiſe they will rot: when the cuttings are planted, the pots ſhould be plunged into a hot-bed of tanners bark, and now and then gently refreſhed with water. The beſt time for planting theſe cuttings is in July, that they may be well rooted before the cold weather comes on in autumn. In winter theſe plants may be placed upon ſtands in the dry ſtove; but if in ſummer they are plunged into the tan-bed, they will make great progreſs, and their leaves will be large, in which conſiſts the great beauty of theſe plants.

The ſecond ſort was diſcovered by the late Dr. Houſtoun at Campeachy. This hath very large, oval, ſpear-ſhaped leaves, ending in points, which are placed alternate on the branches, and have ſeven ribs, which go off from the midrib alternate, and alſo a great number of ſmall veins running horizontally between theſe ribs. The borders of the leaves are ſawed, and their under ſides are of a ſhining brown colour. The branches are covered with a woolly down, and the flowers are produced in looſe ſpikes at the end of the ſhoots; theſe are ſmaller than thoſe of the former ſort, and are of a Roſe colour. This tree riſes to the height of twenty feet; it is propagated by ſeeds. The plants are tender, ſo muſt be placed in the tan-bed of the bark-ſtove, otherwiſe they will not thrive in this country; and they muſt be treated in the ſame manner as is directed for other tender plants from the ſame countries.

CLUTIA. Lin. Gen. Plant. 1009.

The CHARACTERS are,

It is male and female in different plants. The male flowers have five heart-ſhaped petals, which are ſhorter than the empalement, and ſpread open. They have five exterior nectariums, which are ſituated in a circle at the bottom of the petals; and five interior, which are ſituated within the other. They have five ſtamina ſituated in the middle of the ſtyle. The

female

female flowers have petals like those of the male; these have five double exterior nectariums, but no interior; they have a roundish germen, which afterward becomes a globular capsule with six furrows and three cells, each containing a single seed.

The SPECIES are,

1. CLUTIA (*Alaternoides*) foliis sessilibus lanceolatis. Hort. Cliff. 500. *Clutia with spear-shaped leaves sitting close to the stalks.*

2. CLUTIA (*Pulchella*) foliis ovatis integerrimis, floribus lateralibus. Lin. Sp. Plant. 1042. *Clutia with oval entire leaves, and flowers growing from the sides of the branches.*

3. CLUTIA (*Eluteria*) foliis cordato lanceolatis. Flor. Zeyl. *Clutia with heart-formed spear-shaped leaves.*

The two first sorts are natives of Africa. The first sort with female flowers has been long an inhabitant of some curious gardens in England, but that with male flowers has been but few years here.

The first sort has also been some years in the English gardens, and was ranged in the genus of Alaternoides, but we have not the male of this sort in England at present.

The second sort rises with a shrubby stalk to the height of six or eight feet, putting out many side branches, garnished with small spear-shaped leaves placed alternate, sitting close to the branches: they are of a grayish colour and entire. The flowers come out from the joints at the sitting on of the leaves, toward the upper part of the branches; these are small and of a greenish white: they appear in June, July, and August, but being small make no great appearance.

This sort rises about the same height with the first, but hath a stronger stem; the branches are garnished with oval leaves, which are much larger than those of the first sort, having foot-stalks which are an inch long; these are of a sea-green, and entire: the flowers are like those of the first sort in shape and colour, but those on the male plants are smaller, and grow closer together than those of the female, but both are sustained upon short foot-stalks; the seeds ripen in autumn.

These plants are easily propagated by cuttings during any of the summer months, when they will soon take root, and then may be each put into a separate small pot, and placed in a sheltered situation, where they may remain until the middle of October, or later, if the weather continues mild, when they should be removed into the green-house, and placed where they may have the free air in mild weather, for they only require to be protected from frost. In summer they must be placed abroad in a sheltered situation, with other hardy exotic plants.

The third sort grows naturally in India, from whence the seeds were brought. This rises with an upright shrubby stalk, not more than three or four feet in England; but in the places where it grows naturally, it rises upward of twenty feet high, and sends out many branches at the top, so as to form a large spreading head: the branches are garnished with leaves shaped like those of the Black Poplar, which are of a lucid green, and are placed alternate, standing upon slender foot-stalks.

This plant will live through the winter in an airy glass-case without artificial heat, but in that situation they should have very little water in the winter; for the plants abound with a milky juice like the Euphorbia, so must at no season of the year have too much wet. This sort may be propagated by cuttings during the summer season; but the cuttings should be laid in a dry place for a few days, when they are taken from the old plants, that their wounded parts may dry and be healed over before they are planted. These must be planted in small pots, and plunged into a moderate hot-bed of tanners bark; and if the season is very warm, the glasses should be shaded in the heat of the day; they must be sparingly watered, for much wet will cause them to rot. When they have taken root and begin to shoot, they must have a greater share of air, anp by degrees inured to the open air, and each planted in a separate pot, and placed in the shade till they have taken fresh root, after which they may be exposed gradually to the open air. In the summer they should have free air constantly in warm weather, but they must be screened from heavy rain, and in winter placed in an airy glass-case, where they may enjoy the sun, and during that season have very little wet.

CLYPEOLA. Lin. Gen. Pl. 723. Treacle Mustard.

The CHARACTERS are,

The flower hath four oblong entire petals placed in form of a cross, and six stamina, two of which standing opposite are shorter than the other. In the center is situated a roundish compressed germen, which afterward becomes an orbicular pod, which is compressed, erect, and indented at the top, with a longitudinal fissure opening in two cells, containing round compressed seeds.

The SPECIES are,

1. CLYPEOLA (*Jonthlaspi*) siliculis unilocularibus monospermis. Hort. Cliff. 329. *Clypeola with pods having but one cell and a single seed.*

2. CLYPEOLA (*Maritima*) perennis, siliculis bilocularibus ovatis dispermis. Sauv. Monsf. 71. *Perennial Clypeola with oval pods, having two cells and two seeds.*

The first sort is a low annual plant, which seldom rises more than four inches high; the slender branches commonly lie prostrate on the ground; these are garnished with small leaves, narrow at their base, but are broader at their ends, where they are obtuse. The flowers are produced in short close spikes at the extremity of the branches, which are small, yellow, and composed of four leaves placed in form of a cross; these are succeeded by orbicular compressed seed-vessels, each having one cell, containing a single seed.

The second sort is perennial; this branches out from the root into several slender branches, garnished with very narrow hoary leaves sitting close to the branches. The flowers are produced in spikes at the end of the branches; these are small, white, and shaped like those of the other sorts, but the spikes terminate in a roundish bunch.

The first sort is a low annual plant, which grows naturally in the south of France, in Spain, and Italy; this is of a hoary white, both leaves and stalks, which is much lighter in the warm countries than in England: it is propagated by seeds, which should be sown where the plants are to remain, and will require no other culture but to thin them and keep them clean from weeds. The seeds may be sown either in the spring or autumn; those which are sown in autumn will grow much larger, and flower earlier than those which are sown in the spring, and from these there will be a greater certainty of having ripe seeds.

The second sort is a perennial plant, so should be sown upon a warm border; this grows naturally on the borders of the sea, in the south of France and Italy; but when it is cultivated in a garden, if the soil is rich and moist, the plants generally grow luxuriant in summer, and are thereby too replete with moisture, so that they are frequently killed by the frost in winter; but when they grow on a poor, dry, gravelly soil, or on a wall, their stalks will be short, ligneous, and tough, so will endure the cold of this climate, and continue several years: this is propagated by seeds, which should be sown where the plants are designed to remain; or if any of them are removed, it should be done when the plants are young, for they do not bear transplanting well when they are grown pretty large.

CNEORUM. Lin. Sp. Pl. 47. Widow-wail.

The CHARACTERS are,

The flower hath three narrow oblong petals, and three stamina which are shorter than the petals. In the center is situated an oblong three-cornered germen, which afterward becomes a globular dry berry with three lobes, having three cells, each containing one round seed.

We have but one SPECIES of this genus, viz.

CNEORUM (*Tricoccum.*) Hort. Cliff. 18. *Widow-wail. This is the Chamelæa Tricoccos of Dodonæus and Caspar Bauhin.*

It is an humble shrub, which seldom rises more than two feet and a half high in this country, but spreads out on every side with many lateral branches, so as to form a thick bush. The stems are almost as hard as those of the Box-tree, and the wood is of a pale yellow colour under the bark. The branches are garnished with leaves which are stiff, of an oval shape, about an inch and a half long, and a quarter of an inch broad, of a dark green colour, having a strong vein or rib through the middle. The flowers are produced single from the wings of the leaves, toward the extremity of the branches, which are of a pale yellow colour, composed of three petals, which spread open, and a round germen at the bottom, having a single style, which doth not rise above half the length of the stamina, which are three in number, standing erect, and are situated between the petals. After the flowers are fallen the germen becomes a fruit, composed of three seeds, joined together after the same manner as those of Tithymalus, or Spurge; these are first green, afterwards turn to a brown colour, and when ripe are black.

As this is a low evergreen shrub, so it may be very ornamental, if placed in the front of plantations of evergreen trees and shrubs; for as the branches grow pretty compact, and are well garnished with leaves, so it will hide the ground between the taller shrubs better than many other plants; and, being a durable shrub, will not want to be renewed: it rises better from scattered seeds, than if sown with care.

It is propagated by seeds, which should be sown in autumn soon after they are ripe, and then the plants will come up the following spring; whereas those which are not sown till the spring will remain a year in the ground, and often miscarry; these seeds may be sown in a bed of common earth, covering them half an inch deep, and will require no other care but to keep the plants clean from weeds the following summer; and in the autumn following, they may be transplanted where they are to remain.

CNICUS. Lin. Gen. Plant. 833. Blessed Thistle.

The CHARACTERS are,

The empalement of the flower is scaly. The flower is composed of several hermaphrodite florets, which are uniform; these are funnel-shaped, and cut at the top into five equal segments, having five short hairy stamina. In the center is situated a short germen crowned with down, which afterward becomes a single seed crowned with down, and shut up in the empalement.

The SPECIES are,

1. CNICUS (*Erisithales*) caule erecto, foliis inferioribus laciniatis, superioribus integris concavis. Hort. Cliff. 394. *Cnicus with an upright stalk, whose lower leaves are laciniated, the upper entire and concave.*

2. CNICUS (*Spinosissimus*) foliis amplexicaulibus, sinuato-pinnatis, spinosis, caule simplici, floribus sessilibus. Lin. Sp. Plant. 826. *Cnicus with winged, sinuated, prickly leaves embracing the stalk, which is single, and flowers sitting close on the top.*

3. CNICUS (*Cernuus*) foliis cordatis, petiolis crispis, spinosis, amplexicaulibus, floribus cernuis. Hort. Upsal. 251. *Cnicus with heart-shaped leaves, having curled prickly foot-stalks which embrace the stalks, and a nodding flower.*

The first sort grows naturally in the northern parts of Europe. This hath a perennial root, which sends out many long jagged leaves spreading on every side near the ground, so as to form a thick bunch; these are jagged almost to the midrib, in form of a winged leaf. The stalks are striated, smooth, and rise about five or six feet high. The leaves which grow upon the stalks, are entire and heart-shaped; they are sawed on the edges, each indenture ending in a weak spine. The stalks are terminated by large heads of flowers growing in clusters; they are of a whitish yellow colour, and inclosed in a scaly empalement. These are succeeded by small oblong seeds, crowned with a bristly down.

This sort may be propagated by seeds, or parting of the roots; the latter is commonly practised where there are any of the plants, but the seeds are more easily conveyed to a distant place. The best time to part the roots is in autumn; it delights in shade, and requires no farther care but to keep it clean from weeds.

The second sort grows naturally on the Alps, and on the mountains of Austria; this rises with an upright single stalk near four feet high, garnished with sinuated leaves, which are very prickly, and embrace the stalks with their base; the flowers are produced at the top of the stalk, surrounded by a cluster of broad prickly leaves sitting close to the stalk. This is a perennial plant, which may be propagated in the same manner as the former, and requires a moist soil and a shady situation.

The third sort grows naturally in Siberia. This hath a perennial root, composed of thick fleshy fibres. The leaves which rise immediately from the root, are near a foot long, and six inches broad in the middle, diminishing toward each end: these have scarce any foot-stalks; they are of a deep green on their upper side, but white on their under, and sharply sawed on their edges. The stalks rise more than six feet high, and are striated, of a reddish colour, garnished with heart-shaped leaves, which almost embrace the stalks with their base; each branch is terminated by one large globular head of yellowish flowers, included in a scaly empalement, each scale ending with a sharp spine. It may be propagated in the same manner as the two former sorts, but requires a moist soil and shady situation. The inhabitants of Siberia eat the tender stalks of this plant when boiled, instead of other vegetables.

COA. See HIPPOCRATEA.

COAST-MARY. See TANACETUM.

COCCIGRIA. See RHUS.

COCCOLOBA. Brown. Hist. Jam. 209. The Sea side Grape.

The CHARACTERS are,

The flower has one permanent petal, which is cut into five segments; it has a fleshy umbilical nectarium which surrounds the germen, and six, seven, or more erect spreading stamina; and an oval germen, which afterward turns to a fleshy berry, including an oval nut, which is wrapped up in the petal of the flower.

The SPECIES are,

1. COCCOLOBA (*Uvifera*) foliis cordato-subrotundis nitidis. Lin. Sp. Pl. 523. *Sea-side Grape with round, thick, heart-shaped leaves.*

2. COCCOLOBA (*Rubescens*) foliis orbiculatis pubescentibus. Lin. Sp. Pl. 523. *Sea-side Grape with orbicular hairy leaves.*

3. COCCOLOBA (*Punctata*) foliis lanceolato-ovatis. Lin. Sp. Plant. 523. *Sea-side Grape with oval spear-shaped leaves.*

4. Coc-

4. Coccoloba (*Excoriata*) foliis ovatis, ramis quasi excorticatis. Lin. Sp. Plant. 524. *Sea-ſide Grape with oval leaves, whoſe branches caſt their bark*

5. Coccoloba (*Tenuifolia*) foliis ovatis membranaceis. Amœn. Acad. 5. p. 397. *Sea-ſide Grape with oval membranaceous leaves.*

The firſt ſort grows naturally in moſt of the iſlands in the Weſt-Indies, upon the ſandy ſea-ſhores, from whence the inhabitants have given it the title of Sea ſide or Mangrove Grape; this ſends up ſeveral woody ſtalks from the root, which riſe eight or ten feet high, with a light brown ſmooth bark, garniſhed with leaves which are placed alternately; they are very thick and ſtiff, almoſt round, from five to ſeven inches diameter, of a lucid green on their upper ſide, and veined on their under, ſtanding upon ſhort foot-ſtalks. The flowers come out from the wings of the ſtalks; they are diſpoſed along the foot-ſtalk in long ſlender bunches, like thoſe of the common Currant; theſe bunches are five or ſix inches long. The flowers are white, and the petal is cut into ſix parts; theſe are ſucceeded by berries about the ſize of a common Grape, of a purpliſh red colour, incloſing a nut of the ſame ſhape.

This plant is figured by Lobel under the title of Populus novi orbis, and has been copied by Parkinſon in his Herbal, and ſeveral others.

The ſecond ſort grows naturally about Carthagena; this ſends out many ſtrong ſtalks from the root, which riſe near twenty feet high, and are covered with a ſmooth gray bark. The leaves are from ſeven to nine inches long, and from three to four broad; they are indented at the foot ſtalk like a heart, but end in a point, having ſeveral tranſverſe veins running alternately from the midrib to the border, of a thick ſtiff conſiſtence, and of a lucid green on their upper ſide. The flowers are produced in long ſlender bunches at the end of the branches; they are compoſed of five white acute-pointed petals, which ſpread open in form of a ſtar; theſe are ſucceeded by roundiſh purple fruit, ſmaller than thoſe of the former ſort, and not ſo well flavoured.

The third ſort grows naturally at La Vera Cruz in New Spain; this ſends up many ſhrubby ſtalks from the root, which riſe five or ſix feet high, having a light gray bark. The leaves are oval and heart-ſhaped, three inches long, and two and a half broad, indented at their foot-ſtalks; theſe are not ſo ſtiff as thoſe of the two former ſorts. The fruit is diſpoſed in a ſlender bunch at the end of the branches; they are ſmall, of a dark purple colour and are never eaten by the inhabitants.

The fourth ſort grows naturally in Campeachy; this ſends up many ſlender ligneous ſtalks from the root, which riſe to the height of ſeven or eight feet, covered with a gray bark, garniſhed with heart-ſhaped leaves about four inches long and three broad, whoſe foot-ſtalks are joined to the under part of the leaves like the handle of a target. The leaves are of a lucid green, and ſmooth on their upper ſide; the flowers and fruit of this ſort I have not ſeen.

The fifth ſort grows naturally in Jamaica; this ſends up many ſlender ligneous ſtalks from the root, which riſe four or five feet high, covered with a brown bark. The leaves are ſix inches long and three broad, having many ſtrong veins running from the midrib toward the border; they are of a light green colour, and are not ſo ſtiff as the former. The fruit is ſmall, of a purple colour, growing in ſlender bunches from the end of the branches.

Theſe plants riſe eaſily from ſeeds, if they are ſown in pots, and plunged into a hot-bed of tan; but as they do not produce fruit in England, the ſeeds muſt be procured from the Weſt-Indies. When the plants are come up about two or three inches high, they ſhould be each tranſplanted into a ſeparate ſmall pot, and plunged into a freſh hot bed of tan, where they muſt be ſhaded from the ſun until they have taken new root; after which they muſt be treated in the ſame manner as the Annona and other tender plants from hot countries, giving them a proper ſhare of air in warm weather, and gently refreſhing them with water; but they ſhould not have too much wet, for they do not perſpire much, their leaves being of a very cloſe contexture, eſpecially thoſe of the firſt and ſecond ſorts. In autumn theſe plants ſhould be removed into the hot-houſe, and plunged into the bark-bed, otherwiſe they will not make great progreſs; therefore they ſhould always remain in the tan-bed, giving them plenty of air in ſummer when the weather is warm.

The leaves of theſe plants continue in verdure all the year, ſo make a fine appearance in the hot-houſe in winter; but I have never ſeen the flowers of either ſort produced here.

COCHLEARIA. Lin. Gen. Plant. 720. Spoonwort, or Scurvy Graſs.

The Characters are,

The flower hath four petals placed in form of a croſs; it hath ſix ſtamina, four of which are longer than the other two. The germen is heart-ſhaped, which afterward becomes a gibbous, heart-ſhaped, compreſſed pod, faſtened to the ſtyle, having two cells, in each of which are lodged four roundiſh ſeeds.

The Species are,

1. Cochlearia (*Officinalis*) foliis radicalibus ſubrotundis, caulinis oblongis ſubſinuatis. Flor. Lapp. 256. *Common or round-leaved Scurvy Graſs.*

2. Cochlearia (*Anglica*) foliis radicalibus lanceolatis, integerrimis, caulinis ſinuatis. *Sea Scurvy Graſs.*

3. Cochlearia (*Greenlandica*) foliis reniformibus, carnoſis integerrimis. Hort. Cliff. 498. *Leaſt Welch Scurvy Graſs.*

4. Cochlearia (*Danica*) foliis haſtatis, angulatis. Flor. Suec. 196. *Daniſh, or Ivy-leaved Scurvy Graſs.*

5. Cochlearia (*Armoracia*) foliis radicalibus lanceolatis, crenatis, caulinis inciſis. Hort. Cliff. 332. *Horſe Radiſh.*

6. Cochlearia (*Glaſtifolia*) foliis caulinis cordato-ſagittatis, amplexicaulibus. Hort. Cliff. 332. *Talleſt Scurvy Graſs with a Woad leaf.*

The firſt ſort grows naturally on the ſea-ſhore in the north of England, and in Holland, but is cultivated for uſe in the gardens near London: this is an annual plant, for the ſeeds are ſown, and the plants decay within the compaſs of one year; but the ſeeds ſhould be ſown early in autumn, or permitted to ſcatter; it hath a fibrous root, from which ariſe many round ſucculent leaves, which are hollowed like a ſpoon: the ſtalks riſe from ſix inches to a foot high; theſe are brittle, and garniſhed with leaves which are oblong and ſinuated. The flowers are produced in cluſters at the end of the branches, conſiſting of four ſmall white petals, which are placed in form of a croſs; and are ſucceeded by ſhort, roundiſh, ſwelling ſeed-veſſels, having two cells, divided by a thin partition; in each of theſe is lodged four or five roundiſh ſeeds.

This is propagated in gardens for medicinal uſes, which is done by ſowing the ſeeds in July, ſoon after they are ripe, in a moiſt ſhady ſpot of ground; and when the plants are come up they ſhould be thinned, ſo as to be left at about ſix inches diſtance each way. The plants that are taken out may be tranſplanted into other ſhady borders, if you have occaſion for them; and at the ſame time all the weeds ſhould be hoed down, ſo as to clear the plants entirely from them, that they may have room to grow ſtrong. In the ſpring theſe plants will be fit for uſe; and thoſe that are ſuffered to remain will run up to ſeed in May, and perfect

their

their feeds in June. If this plant is fown in the fpring, the feeds feldom grow well, therefore the beft time is foon after they are ripe.

The Sea Scurvy Grafs is alfo ufed in medicine; but this grows in the falt-marfhes in Kent and Effex, where the falt water overflows it almoft every tide, and can rarely be made to grow in a garden, or at leaft to laft longer there than one year; but it being eafily gathered in the places before-mentioned, the markets are fupplied from thence by the herb-women, who make it their bufinefs to gather herbs.

The little Welch Scurvy Grafs is a biennial plant, and may be preferved in a garden, if planted in a ftrong foil and a fhady fituation. This plant grows plentifully in Mufcovy, as alfo in Davis's Streights.

The fourth fort is a low trailing plant, whofe ftalks grow fix inches long, and lie proftrate on the ground; the leaves are angular, and in fhape like thofe of Ivy. This is found growing naturally in fome parts of England, and is annual.

The fixth fort is a biennial plant, which ufually grows about a foot and a half high with upright ftalks, which are garnifhed with angular heart-fhaped leaves, embracing the ftalks with their bafe; the flowers are produced in loofe fpikes at the end of the branches; they are very fmall, white, and are fucceeded by fhort fwelling pods filled with round feeds. This may be propagated by feeds as the common fort, and, if fown in autumn, will more certainly fucceed than in the fpring.

The Horfe Radifh is propagated by cuttings or buds from the fides of the old roots. The beft feafon for this work is in October or February; the former for dry lands, the latter for moift; the ground fhould be trenched at leaft two fpits deep, or more, if it will allow of it. The manner of planting it is as follows: Provide yourfelf with a good quantity of offsets, which fhould have a bud upon their crowns, but it matters not how fhort they are; therefore the upper part of the roots which are taken up for ufe, fhould be cut off about two inches long with the bud to it, which is efteemed the beft for planting. Then make a trench ten inches deep, in which you fhould place the offsets at about four or five inches diftances, with the bud upward, covering them up with the mould that was taken out of the trench: then proceed to a fecond trench in like manner, and continue the fame till the whole fpot of ground is planted. After this level the furface of the ground even, obferving to keep it clear from weeds, until the plants are fo far advanced as to be ftrong enough to over-bear and keep them down. With this management the roots of the Horfe Radifh will be long and ftrait, and free from fmall lateral roots, and the fecond year after planting will be fit for ufe.

CODLIN-TREE. See MALUS.

COFFEA. Lin. Gen. Pl. 209. The Coffee-tree.

The CHARACTERS are,

The flower hath one petal, which is funnel-fhaped, having a narrow cylindrical tube, and is plain at the top, where it is indented in five parts; it hath five ftamina, which are faftened to the tube. The roundifh germen afterward becomes an oval berry, containing two hemifpherical feeds, plain on one fide, and convex on the other.

We have but one SPECIES of this genus, viz.

COFFEA (*Arabica.*) Hort. Cliff. 59. *The Coffee-tree.*

This tree is fuppofed to be a native of Arabia Felix, where it was firft cultivated for ufe, and to this day, is the country from whence the beft Coffee is brought to Europe, though the plant is now propagated in many parts of both Indies; but the produce of thofe countries being greatly inferior to that of Arabia, hath occafioned its prefent difrepute in England, fo that it is fcarce worth importing; but this might be remedied, if the Coffee planters in the Weft-Indies could be prevailed on to try a few experiments, which I fhall hereafter propofe, being founded on thofe which have been made in England, upon the berries produced here. But I fhall firft treat of the plant, with its culture in England.

This is a low tree in the native country of its growth, where it feldom rifes more than fixteen or eighteen feet high; the main ftem grows upright, and is covered with a light brown bark; the branches are produced horizontally and oppofite, which crofs each other at every joint, fo that every fide of the tree is fully garnifhed with them, and form a fort of pyramid, the leaves alfo ftand oppofite; thefe when fully grown, are about four or five inches long, and two inches broad in the middle, decreafing toward each end; the borders are waved, and the furface is of a lucid green. The flowers are produced in clufters at the bafe of the leaves, fitting clofe to the branches; thefe are tubulous, and fpread open at the top, where they are divided into five parts; they are of a pure white, and have a very grateful odour, but are of fhort duration. Thefe are fucceeded by oval berries, which are firft green; when fully grown, they turn red, and afterward change to black when fully ripe. They have a thin pulpy fkin, under which are two feeds joined, which are flat on the joined fides, with a longitudinal furrow, and convex on their outer fide.

As the Coffee-tree is an evergreen, fo it makes a beautiful appearance at every feafon in the ftove, but particularly when it is in flower, and alfo when the berries are red, which is generally in the winter, fo that they continue a long time in that ftate; therefore there is fcarce any plant that more deferves a place in the ftove than this.

It is propagated by the berries, which muft be fown foon after they are gathered from the trees, for if they are kept out of the ground any time they will not grow. I have frequently fent the berries abroad by the poft, but when they have been a fortnight in their journey they have all failed, and this has conftantly happened every where, for the berries that were fent from Holland to Paris did not grow, nor did thofe that were fent from Paris to England grow: fo that wherever thefe trees are defired, the young plants muft be fent, if it be at any diftance from the place where they grow.

The berries fhould be planted in fmall pots, and plunged into a hot-bed of tanners bark. If the bed be of a proper temperature of warmth, the plants will appear in a month or five weeks time, and in about fix weeks more will be fit to tranfplant; for as many of the berries will produce two plants, fo the fooner they are parted the better; for when they grow double till they have made large roots, they will be fo intermixed and entangled, as to render it difficult to feparate them, without tearing off their fibres, which will greatly prejudice the plants. When thefe are tranfplanted, they muft each be put into a feparate fmall pot, and plunged into the tan-bed again; which fhould be ftirred up to the bottom, and, if required, fome new tan fhould be mixed with it to renew the heat; then the plants fhould be gently watered, and the glaffes of the hot-bed muft be fhaded every day till they have taken new root; after which the plants fhould have frefh air admitted to them every day, in proportion to the warmth of the feafon: during the fummer they will require frequently to be refrefhed with water, but they muft not have it in too great plenty: for if the roots are kept too moift, they are very fubject to rot, then the leaves will foon decay and drop off, and the plants become naked; when this happens, they are feldom recovered again. The firft fign of thefe plants being difordered, is, their leaves fweating out a clammy juice, which attracts the fmall infects, which frequently infeft them in ftoves when they are not in health, and thefe cannot be deftroyed till the plants are recovered to vigour; for although the plants are ever ſo carefully waſhed and cleared from theſe inſects, yet they will be ſoon attacked by them again, if they are not recovered to health, for theſe inſects are never ſeen upon any of the plants while they are in perfect vigour; but when they are diſordered, they ſoon ſpread over all the leaves and tender parts of the plants, and multiply exceedingly; ſo that upon the firſt attack, the plants ſhould be ſhifted into freſh earth, and all poſſible care taken to recover them, without which all the waſhing and cleaning of the plants will be to little purpoſe. The diſorders attending the Coffee-trees, generally proceed from either being put into pots too large for them, nothing being of worſe conſequence than over potting them, or from the earth being too ſtiff, or over watered. If theſe errors are avoided, and the ſtove kept always in a proper temperature of heat, the plants will thrive, and produce plenty of fruit annually.

I have made trial of ſeveral compoſitions of earth for theſe plants, but have found none of them equal to that of a kitchen-garden, where the ſoil is naturally looſe, and not ſubject to bind; eſpecially if it has conſtantly been well wrought and properly dunged, this without any mixture is preferable to any other.

When the plants are tranſplanted, their roots ſhould not be too much cut or trimmed, the decayed or rotten fibres ſhould be pruned off, and thoſe which are cloſely matted to the ſide of the pots ſhould be trimmed, but not cut too near to the ſtem, for the old fibres do not put out new roots very kindly, eſpecially thoſe which are become tough, ſo that there ſhould always be a ſufficient number of young fibres left to ſupport the plants till new ones are produced.

The Coffee plants were firſt carried from Arabia to Batavia by the Dutch, and from thence they were afterward brought to Holland, where great number of the plants were raiſed from the berries which thoſe plants produced, and from theſe moſt of the gardens in Europe have been furniſhed. A great number of theſe young plants, which were raiſed at Amſterdam, were ſent to Surinam by the proprietors of that iſland, where the trees were ſoon propagated in great plenty, and from thence the plants have been diſperſed to moſt of the iſlands in the Weſt-Indies; for as the plants raiſed from the berries produce fruit in eighteen months from planting in the warm countries, ſo plantations of theſe trees may be ſoon made in any of thoſe countries, where the temperature of the air is proper for their production; but the trees will not grow in the open air in the country where there is a winter, ſo that without the tropics they cannot be expected to grow abroad.

The French have made great plantations of theſe trees in their ſettlements in the Weſt-Indies, and alſo in the iſle of Bourbon, from whence they import great quantities of Coffee annually to France, which, although greatly inferior in quality to the Arabian, yet it is conſumed, otherwiſe they would not continue that branch of commerce. In the Britiſh colonies of America there have been ſome large plantations made of Coffee-trees; and it was propoſed to the parliament ſome years paſt, to give a proper encouragement for cultivating this commodity in America, ſo as to enable the planters to underſell the importers of Coffee from Arabia; accordingly there was an abatement of the duty payable on all the Coffee which ſhould be of the growth of our colonies in America, which was at that time ſuppoſed would be a ſufficient encouragement for the planters to improve this branch of commerce: but the productions of thoſe countries being greatly inferior in quality to that of Arabia, hath almoſt ruined the project; and unleſs the planters can be prevailed on to try ſome experiments to improve its quality, there can be little hope of its becoming a valuable branch of trade; therefore I ſhall beg leave to offer my ſentiments on this article, and ſincerely wiſh what I have to propoſe may be found uſeful for the inſtruction of the Coffee planters; for as my opinion is founded upon experiment, ſo it is not mere theory or ſuppoſition.

The great fault of the Coffee which grows in America is the want of flavour, or having a diſagreeable one. The berries are much larger than thoſe which are imported from Arabia, and conſequently have not ſo much ſpirit or flavour. This may be owing to ſeveral cauſes; the firſt is that of its growing in a ſoil too moiſt, which is always known to increaſe the ſize of fruit and vegetables, but their quality is greatly diminiſhed thereby. The ſecond is from the gathering of the berries too ſoon; for I have been credibly informed, that it is the conſtant practice of the planters to gather the fruit when it is red; at which time the berries are much larger, and of greater weight, than thoſe which are permitted to ripen perfectly on the trees, which is not till they are turned black, and their outer pulp becomes dry, and the ſkins ſhrink; then the berries are much ſmaller than before, and the outer cover will eaſily ſeparate from the berry, which I have always been informed has been the complaint of the planters, that this was with great difficulty and trouble effected. A third cauſe I imagine may be in the drying of the berries when gathered, which muſt be conſtantly attended to, for they cannot be too much expoſed to the ſun and air in the day-time, but they muſt be every evening removed under cover, and carefully ſcreened from dews and rain; nor ſhould they be placed near any ſort of liquid or moiſture, for theſe berries are very ſubject to imbibe moiſture, and thereby acquire the flavour of the liquid, of whatever ſort it be; and the berries will be enlarged, but the flavour diminiſhed by it, as from many experiments I can affirm; for a bottle of rum being placed in a cloſet, in which a caniſter of Coffee-berries cloſely ſtopped was ſtanding on a ſhelf at a conſiderable diſtance, in a few days had ſo impregnated the berries, as to render them very diſagreeable; the ſame alſo hath happened by a bottle of ſpirits of wine ſtanding in the ſame cloſet with Coffee and Tea, both which were in a few days ſpoiled by it. Therefore from many experiments of this nature which I have made upon Coffee, it appears to me, that it ſhould never be brought over in ſhips freighted with rum, nor ſhould the berries be laid to dry in the houſes where the ſugars are boiled, or the rum diſtilled. I have alſo been informed by a gentleman who has a very good eſtate in Jamaica, and who has lived many years in that iſland, that the planters frequently boil the Coffee-berries before they are dried. As this information comes from a gentleman of great ſkill and veracity, I cannot doubt of the fact; and if ſo, this alone is ſufficient to ſpoil the beſt Coffee in the world; ſo that I am at a loſs to gueſs the reaſon for this practice, which, as it appears to me, can only be intended to increaſe the weight, therefore muſt be imputed to avarice, the bane of every public good.

There was ſome time paſt an imperfect account printed in the papers, of the cauſe why the American Coffee was not ſo good as that which comes from Arabia, in which it was ſuppoſed that the goodneſs of the latter proceeded from the length of time which the berries had been kept: therefore the author propoſes, that the American Coffee-berries ſhould be kept many years, which he ſays will render them equally good. This is contrary to all the experience I have had, or can learn, from thoſe who have ſeen the whole progreſs of Coffee in Arabia, with their manner of drying and packing it to ſend abroad; for two gentlemen who had lived there ſome years aſſured me, that the berries, when firſt gathered, were much better than thoſe which are kept any time. And a curious gentleman who reſided in Barbadoes two years alſo told me, that he never drank better Coffee

Coffee in any part of the world, than what he made from the fresh berries, which he gathered himself, and roasted as he had occasion for them; which is also confirmed by the trials which have been made with the berries which grow in the stoves in England, which make a better flavoured liquor than the best Arabian Coffee berries which can be procured here; therefore I wish those who are inclinable to cultivate these trees in America, would make choice of a soil rather dry than moist, in which the trees will not make so great progress as those which grow in a wet soil, nor will the produce be so great; but as the quality of the produce will be so much improved, so it will certainly be of greater advantage to them.

The next thing is to permit the berries to remain so long upon the trees till their skins are shrivelled, and will fall from the trees when they are shaken, which it is true will greatly diminish their weight, but then the commodity will be more than double the value of that which is gathered sooner; for in Arabia they always shake the berries off the trees, spreading cloths under them to receive them, and only take such as readily fall at each time.

When the berries are full ripe, they should be shaken off when the trees are perfectly dry, and spread abroad upon cloths in the sun to dry, carrying them every evening under cover, to prevent the dews from falling on them, or the rain if any should happen: and when they are perfectly dry, they should have their outer skins beaten off, then carefully packed up in cloths or bags three or four times double, and consequently kept in a dry situation. When they are shipped for England, it should be on board those vessels which have no rum, lest the Coffee should imbibe the flavour, which cannot be prevented when stowed in the same place. Some years past a Coffee ship from India had a few bags of Pepper put on board, the flavour of which was imbibed by the Coffee, and the whole cargo spoiled thereby.

As the quantity of Coffee now consumed in Britain is very much increased of late years, so it will certainly be worthy of public consideration, how far it may be necessary to encourage the growth of it in the British colonies; and certainly it deserves the attention of the inhabitants of those colonies, to improve this commodity to the utmost of their power, and not to have so much regard to the quantity as to the quality of it; for although the former may appear to have the advantage of the latter in point of profit, yet the goodness of every commodity must always claim the preference, and thereby will be found of more lasting advantage to the cultivator.

COIX. Lin. Gen. Plant. 927. Job's Tears.

The CHARACTERS are,

It hath male and female flowers on the same plant; the male flowers are disposed in a loose spike. The petal has two oval valves. These have each three hairy stamina, terminated by oblong four-cornered summits. There are a few female flowers situated at the base of the male spike in the same plant; the petal hath two oval valves. They have a small oval germen, which afterward becomes a hard, roundish, smooth seed.

We have but one SPECIES of this genus, viz.

COIX (*Lachryma Jobi*) seminibus ovatis. Hort. Cliff. 434. *Coix with oval seeds; or Job's Tears.*

This plant grows naturally in the islands of the Archipelago, and is frequently cultivated in Spain and Portugal, where the poor inhabitants grind the grain to flour in a scarcity of Corn, and make a coarse sort of bread of it.

It is an annual plant, which seldom ripens its seeds in England, unless the season proves very warm; from a thick fibrous root is sent out two or three jointed stalks, which rise two feet high, garnished with single, long, narrow leaves at each joint, resembling those of the Reed; at the base of the leaves come out the spikes of flowers, standing on short foot-stalks; these spikes are composed of male flowers only, and below them is situated one or two female flowers; the male flowers decay soon after they have shed their farina, but the germen of the female flowers swell to a large oval seed, which is hard, smooth, and of a gray colour, greatly resembling the seeds of Gromwel, from whence this plant has been by several writers titled Lithospermum.

Those who are desirous to cultivate this plant in England, may procure the seeds from Portugal; these should be sown on a moderate hot-bed in the spring to bring the plants forward, and afterward transplant them on a warm border, allowing each two feet room at least; and when the plants have taken root, they will require no farther care but to keep them clean from weeds. These will flower about Midsummer, and in warm seasons the seeds will ripen at Michaelmas.

COLCHICUM. Lin. Gen. Pl. 415. Meadow Saffron.

The CHARACTERS are,

The flower hath neither empalement or spatha; it hath one petal, rising with an angular tube from the root, and is divided at the top into six oval concave segments; it hath six stamina with four valves. The germen is situated near the root, which afterward becomes a capsule with three lobes, having a seam on the inside, dividing it into three cells, which contain several roundish rough seeds.

The SPECIES are,

1. COLCHICUM (*Autumnale*) foliis planis lanceolatis, erectis. Hort. Cliff. 140. *Colchicum with plain, erect, spear-shaped leaves; or Common Meadow Saffron.*

2. COLCHICUM (*Montanum*) foliis linearibus patentissimis. Lin. Spl. Pl. 342. *Meadow Saffron with very narrow spreading leaves.*

3. COLCHICUM (*Variegatum*) foliis undulatis patentibus. Hort. Cliff. 140. *Meadow Saffron with waved spreading leaves, and chequered flowers.*

There are several varieties of these flowers, which differ in their colour and other little accidents, which are not lasting, so must not be ranged as distinct species. But as they are cultivated in flower-gardens, so I shall beg leave to mention those varieties which I have seen cultivated. These are most of them seminal variations from the first sort.

The most common Meadow Saffron hath a purplish flower.
The Meadow Saffron with white flowers.
Meadow Saffron with striped flowers.
Broad-leaved Meadow Saffron.
Striped-leaved Meadow Saffron.
Many flowered Meadow Saffron.
Meadow Saffron with double purplish flowers.
Meadow Saffron with double white flowers.
Meadow Saffron with many white flowers.

The first sort grows naturally both in the west and north of England. I have observed it in great plenty in the meadows in Warwickshire in the beginning of September. The country-people call the flowers Naked Ladies, because they come up naked, without any leaves or cover. This hath a bulbous root, about the size and shape of the Tulip root; but not so sharp pointed at the top; the skin or cover is also of a darker colour. The flowers come out in autumn; these arise with long slender tubes from the root about four inches high, shaped like those of the Saffron, but are larger; they are of a pale purple colour, and divided into six parts at the top; the number of flowers is generally in proportion to the size of the roots, from two to seven or eight: in March the green leaves appear; these are commonly four to a full grown root; they are folded over each other below, but spread open above ground, standing cross ways; they are of a deep green, and when fully grown are five or six inches long,

long, and one broad. The ſeed-veſſel comes out from between the leaves in April, and the ſeeds ripen in May, ſoon after which the leaves decay.

The other varieties are ſuppoſed to have accidentally riſen from the ſeeds of this, ſo that thoſe who are deſirous to obtain a variety of theſe flowers, ſhould propagate them from ſeeds, by which means there may be more obtained

The ſecond ſort grows naturally on the mountains in Spain and Portugal. This hath a ſmaller root than the firſt, and a darker coat; the flowers appear in Auguſt or September; theſe are cut into ſix long narrow ſegments, of a reddiſh purple colour, having ſix yellow ſtamina. The leaves of this ſort come up ſoon after the flowers decay, and continue green all the winter like the Saffron; theſe are long, narrow, and ſpread on the ground; in June theſe decay like the firſt ſort.

The third ſort grows naturally in the Levant, but is commonly cultivated in the Engliſh gardens. It flowers at the ſame time as the firſt ſort, and the green leaves come up in the ſpring. The root of this ſpecies is ſuppoſed to be the Hermodactyl of the ſhops.

Theſe are all very pretty varieties for a flower-garden, producing their flowers in autumn, when few other plants are in beauty; and are therefore, by ſome, called Naked Ladies. The green leaves come up in the ſpring, which are extended to a great length in May, then the green leaves begin to decay; ſoon after which time is the proper ſeaſon to tranſplant their roots, for if they are ſuffered to remain in the ground till Auguſt, they will ſend forth freſh fibres, after which time it will be too late to remove them. The roots may be kept above ground till the beginning of Auguſt, at which time, if they are not planted, they will produce their flowers as they lie out of the ground, but this will greatly weaken their roots. The manner of planting their roots being the ſame as for Tulips, &c. I ſhall forbear mentioning it here, referring the reader to that article: and alſo for ſowing the ſeeds, by which means new varieties may be obtained, I ſhall refer to the article Xiphion, where will be proper directions for this work.

COLDENIA. Lin. Gen. Plant. 159.

The Characters are,

It hath a funnel-ſhaped flower of one petal, ſpreading at the top, and obtuſe; it hath four ſtamina, which are inſerted in the tube of the petal. In the center is ſituated four oval germen, each ſupporting a hairy ſtyle. The germen afterward become ſo many oval, compreſſed, rough fruit, terminated by four beaks, incloſed by the empalement.

We know but one Species of this genus, viz.

Coldenia (*Procumbens.*) Flor. Zeyl. 69. This is by Dr. Plukenet titled, Teucrii facie biſnagarica tetracoccos roſtrata. Alm. 363.

This is a native of India. It is an annual plant, whoſe branches trail on the ground; they extend about ſix inches from the root, garniſhed with ſhort leaves ſitting cloſe to the branches; theſe are deeply crenated on their edges, and have ſeveral deep veins; they are of a glaucous colour, and come out without order. The flowers are produced at the wings of the leaves, growing in ſmall cluſters; theſe have four ſtamina, and but one petal, which is funnel-ſhaped, and cut into four ſegments at the top; they are of a pale blue colour, and very ſmall; when the flower decays, the germen becomes a fruit compoſed of four cells, wrapped up in the empalement, each containing a ſingle ſeed.

This plant is propagated by ſeeds, which muſt be ſown upon a hot-bed in the ſpring; and when the plants are fit to remove, they ſhould each be put into a ſeparate ſmall pot, and plunged into a hot-bed of tanners bark, obſerving to ſhade them till they have taken freſh root, after which they ſhould have air admitted to them every day in proportion to the warmth of the ſeaſon, and gently watered two or three times a week in warm weather; but they muſt not have too much moiſture. Theſe plants muſt remain in the hot-bed, where they will flower in June, and the ſeeds ripen in September.

COLEWORTS. See Brassica.

COLOCASIA. See Arum.

COLLINSONIA. Lin. Gen. Plant. 38.

The Characters are,

The flower is funnel-ſhaped, of one petal, which is unequal, cut into five parts at the top, the upper part being ſhort and obtuſe, and two of them being reflexed; the lower lip or beard is longer, ending in many points. It hath two long briſtly ſtamina, which are erect. It hath a quadrifid obtuſe germen, with a large gland, ſupporting a briſtly ſtyle, which afterward becomes a ſingle roundiſh ſeed, ſituated in the bottom of the empalement.

We have but one Species of this plant, viz.

Collinsonia (*Canadenſis*) foliis cordatis oppoſitis. *Collinſonia with heart-ſhaped leaves growing oppoſite.*

This plant was brought from Maryland, where it grows wild, as it doth alſo in moſt other parts of North America, by the ſides of ditches, and in low moiſt grounds, where it uſually riſes to the height of four or five feet.

This hath a perennial root and an annual ſtalk, which decays in the autumn, and freſh ſhoots come out in the ſpring. The ſtalks are ſquare, and garniſhed with heart-ſhaped leaves placed oppoſite, which are ſawed on their edges. The flowers are produced at the extremity of the ſtalks in looſe ſpikes: theſe have long tubes, and are divided into five parts at the top; the flowers are of a purpliſh yellow, and the lower ſegment is terminated by long hairs. The flowers appear in July, and the ſeeds ripen in autumn.

This plant may be eaſily propagated by parting of the roots in October. Theſe roots ſhould be planted at three feet diſtance, for they require much nouriſhment, otherwiſe they will not thrive, ſo ſhould be planted in a moiſt ſoil and a ſheltered ſituation.

COLOCYNTHIS. See Cucurbita.

COLUMBINE. See Aquilegia.

COLUMNEA. Lin. Gen. Plant. 710.

The Characters are,

The flower hath one petal, of the ringent or grinning kind, with a ſwelling tube, divided above into two lips; the upper being erect and entire, the lower is divided into three parts, which ſpread open: it hath four ſtamina, two being longer than the other; theſe are incloſed in the upper lip. In the center is ſituated the roundiſh germen, which afterward becomes a globular berry with two cells, ſitting on the empalement, containing ſeveral oblong ſeeds.

We have but one Species of this genus, viz.

Columnea (*Scandens.*) Lin. Sp. Plant. 638. *Climbing Columnea with a ſcarlet flower and a white fruit.*

Plumier mentions a variety of this with a yellowiſh flower and white fruit: but this is only a ſeminal variation, ſuppoſed to have accidentally riſen from the ſeeds of the firſt.

This hath a climbing ſtalk, which faſtens itſelf to the neighbouring plants whereby it is ſupported. The leaves are oval, ſawed on the edges, and ſtand upon ſhort foot-ſtalks; theſe, and alſo the ſtalks, are very hairy; but the plants which were raiſed at Chelſea decayed the following year before they produced any flowers, ſo that I can give no deſcription of them.

Theſe plants are natives of the warmeſt parts of America, ſo are too tender to live in England, unleſs they are preſerved in the ſtove. They are propagated by ſeeds, which muſt be ſown in a good hot-bed; and when the plants come up, they muſt be treated in the ſame way as other tender exotic plants, which are kept in the bark-ſtove.

COLUTEA. Tourn. Inſt. R. H. 649. tab. 417. Bladder Sena.

The Characters are,

The flower is of the butterfly kind. The ſtandard, wings, and keel, vary in their figure in different ſpecies. It hath ten ſtamina, nine of which are joined, and the other ſtands ſeparate. In the center is ſituated an oblong germen, which afterward becomes a broad ſwollen pod with one cell, including ſeveral kidney-ſhaped ſeeds.

The Species are,

1. Colutea (*Arboreſcens*) arborea, foliolis obcordatis. Hort. Cliff. 365. *Common Bladder Sena.*

2. Colutea (*Iſtria*) foliolis ovatis, integerrimis, caule fruticoſo. *Shrubby Bladder Sena with oval leaves, which are entire.*

3. Colutea (*Orientalis*) foliolis cordatis minoribus, caule fruticoſo. *Eaſtern Bladder Sena with a blood-coloured flower, ſpotted with yellow.*

4. Colutea (*Fruteſcens*) foliolis ovato-oblongis. Hort. Cliff. 366. *Æthiopian Bladder Sena with a ſcarlet flower.*

5. Colutea (*Americana*) foliolis ovatis, emarginatis, leguminibus oblongis compreſſis acuminatis, caule arboreo. *Bladder Sena with oval leaves indented at the top, oblong, compreſſed, pointed pods, and a tree-like ſtalk.*

6. Colutea (*Herbacea*) herbacea foliis linearibus. Hort. Upſal. 266. *African annual Bladder Sena, with ſmall pointed leaves and compreſſed pods.*

7. Colutea (*Procumbens*) caulibus procumbentibus, foliolis ovato-linearibus, tomentoſis, floribus alaribus pedunculis longiſſimis. *Bladder Sena with trailing ſtalks, oval narrow leaves, which are woolly, and flowers growing from the ſides of the ſtalks with very long foot-ſtalks.*

The firſt ſort is commonly cultivated in the nurſery-gardens, as a flowering ſhrub to adorn plantations. This grows naturally in Auſtria, in the ſouth of France and Italy. It hath ſeveral woody ſtems, which grow to the height of twelve or fourteen feet, dividing into many woody branches, garniſhed with winged leaves, compoſed of four or five pair of oval leaves placed oppoſite, terminated by an odd one; theſe are indented at the top, and are of a grayiſh colour. The flowers come out from the wings of the leaves upon ſlender foot-ſtalks about two inches long, each ſuſtaining two or three flowers of the butterfly kind, whoſe ſtandard is reflexed and large. The flowers are yellow, with a dark-coloured mark on each petal; theſe are ſucceeded by inflated pods an inch and a half long, having a ſeam on the upper ſide, containing a ſingle row of kidney-ſhaped ſeeds faſtened to a placenta. There is a variety of this with reddiſh pods, which is equally common in the gardens, and is ſuppoſed to be only an accidental variety, for the plants do not differ in any other reſpect.

The ſeeds of the ſecond ſort were brought from the Levant by the Rev. Dr Pococke, late Biſhop of Oſſory. This ſeldom grows more than ſix or ſeven feet high; the branches are very ſlender, and ſpread out on every ſide; they are garniſhed with winged leaves, compoſed of nine pair of ſmall, oval, entire lobes, terminated by an odd one; the flowers ſtand upon ſlender foot-ſtalks about the ſame length of the former. The flowers alſo are like thoſe, but are of a brighter yellow. This ſort begins to flower early in May, and continues flowering till the middle of October.

The third ſort was diſcovered by Dr. Tournefort in the Levant. This hath a woody ſtem, which ſends out many branches on every ſide, which do not riſe above ſeven or eight feet high; theſe are not ſo ſtrong as thoſe of the firſt ſort, and are garniſhed with winged leaves, compoſed of five or ſix pair of ſmall heart-ſhaped lobes, terminated by an odd one. The flowers proceed from the ſide of the branches, ſtanding upon foot-ſtalks, each ſuſtaining two or three flowers, ſhaped like thoſe of the firſt ſort, but ſmaller; they are of a dark red colour, marked with yellow.

The fourth ſort grows naturally in Æthiopia, from whence the ſeeds were brought to Europe. This hath a weak ſhrubby ſtalk, which ſends out many ſide-branches growing erect, garniſhed with equal winged leaves, compoſed of ten or twelve pair of ſmall, oval, oblong, hoary lobes. The flowers are produced at the upper part of the branches from the wings of the leaves, each foot-ſtalk ſuſtaining three or four ſcarlet flowers, which are longer than thoſe of the other ſorts, and are not reflexed; theſe are ſucceeded by inflated pods, containing one row of kidney-ſhaped ſeeds.

The fifth ſort grows naturally at La Vera Cruz in New Spain. This hath a ſhrubby ſtalk, which riſes to the height of twelve or fourteen feet, ſending out many branches, garniſhed with winged leaves, compoſed of three pair of oval lobes, terminated by an odd one; theſe are indented at the top, and are of a light green. The flowers are of a bright yellow, and ſtand two or three upon each foot-ſtalk, and are ſucceeded by compreſſed pods near four inches long, which end in long points.

The ſixth ſort grows naturally at the Cape of Good Hope. This is an annual plant of little beauty, ſo is rarely cultivated but in botanic gardens for the ſake of variety. It riſes with a ſlender herbaceous ſtalk about two feet high, dividing upward into three or four branches, which are garniſhed with winged leaves, compoſed of five or ſix pair of very narrow lobes near an inch long, which are a little hoary. The flowers are ſmall, of a purpliſh colour, ſtanding three together on ſlender foot-ſtalks, which are ſucceeded by flat oval pods, each containing two or three kidney-ſhaped ſeeds.

The ſeeds of the ſeventh ſort were ſent me from the Cape of Good Hope. This plant hath many ſlender herbaceous ſtalks, which frequently trail on the ground, and are garniſhed with winged leaves, compoſed of twelve or fourteen pair of ſmall, narrow, oval lobes, terminated by an odd one; theſe, and alſo the ſtalks, are covered with a whitiſh down. The flowers are very ſmall, of a purple colour, and ſtand upon very long ſlender foot-ſtalks, each ſuſtaining three or four flowers, which are ſucceeded by compreſſed pods little more than half an inch long, which are a little bent like a ſickle, each containing a ſingle row of ſmall kidney-ſhaped ſeeds. This is a perennial plant, which, if ſheltered in the winter, will continue ſeveral years; but the branches do not extend more than a foot in length, and unleſs they are ſupported, always trail upon the ground.

The three firſt ſorts are very hardy ſhrubs, which thrive in the open air extremely well, ſo they are generally propagated for ſale in the nurſery-gardens; but the firſt ſort hath been long in England, ſo is more generally known and propagated than either of the other.

Theſe are propagated by ſowing their ſeeds any time in the ſpring in a bed of common earth; and when the plants are come up, they muſt be kept clear from weeds, and the Michaelmas following they ſhould be tranſplanted either into nurſery-rows, or in the places where they are deſigned to remain; for if they are let grow in the ſeed-bed too long, they are very ſubject to have tap-roots, which render them unfit for tranſplantation; nor ſhould theſe trees be ſuffered to remain too long in the nurſery before they are tranſplanted, for the ſame reaſon.

The firſt ſort will grow to the height of twelve or fifteen feet, and are very proper to intermix with trees of a middling growth in wilderneſs quarters, or in clumps of flowering trees.

The

The third ſort does not grow ſo tall as the common, but makes a more regular ſhrub than that. The flowers of this ſort are of a duſky red colour, ſpotted with yellow, ſo it makes a very pretty variety, and is as hardy as the common ſort, therefore may be propagated by ſeeds in the ſame manner.

The fourth ſort is tender, ſo will not live through the winters (when they are ſevere) in the open air in England; but, in mild winters, if they are planted in a dry ſoil and a warm ſituation, they will thrive very well; and thoſe plants which live abroad will flower much ſtronger, and make a finer appearance, than thoſe which are preſerved in the green houſe; for theſe plants require a large ſhare of air, otherwiſe they are apt to draw up weak, ſo ſeldom produce their flowers in plenty; therefore when any of the plants are ſheltered in winter, they muſt be placed as near the windows as poſſible, that they may have all the advantages of air; and in the ſpring, they muſt be hardened to bear the open air as ſoon as poſſible.

This ſort is propagated by ſeeds as the former; if the ſeeds are ſown early in the ſpring, upon a warm border of light earth, the plants will flower in Auguſt; and, if the autumn proves favourable, they will ripen their ſeeds very well; but there are ſome perſons who ſow the ſeeds upon a moderate hot-bed in the ſpring, whereby they bring the plants ſo forward as to flower in July, ſo that the ſeeds are always perfected from ſuch plants: when theſe plants are tranſplanted, it ſhould always be done while they are young, for they do not bear removing well when they are large.

The fifth ſort grows naturally in a warm country, ſo is too tender to thrive in the open air in England: this is propagated by ſeeds, which muſt be ſown on a hot-bed in the ſpring, and when the plants are two inches high, they ſhould be each tranſplanted into a ſeparate ſmall pot, and plunged into a hot-bed of tanners bark, obſerving to ſhade them till they have taken freſh root; after which they muſt be treated in the ſame way as other plants from the ſame climate, always keeping them in a ſtove, which ſhould be kept to a moderate temperature of heat.

The ſixth ſort is a low annual plant, which ſeldom grows more than a foot and a half in height; the flowers being ſmall, and having little beauty, it is ſeldom preſerved but in botanic gardens; the ſeeds of this ſort muſt be ſown upon a moderate hot-bed in the ſpring, and the plants muſt be planted into ſmall pots, and brought forward in another hot-bed; in July they will flower, when they may be expoſed to the open air, in a warm ſituation, where the ſeeds will ripen in September, and the plants will ſoon after decay.

The ſeventh ſort may be raiſed on a moderate hot-bed in the ſpring, and afterward expoſed to the open air in ſummer; but in winter they muſt be ſheltered under a frame, otherwiſe the froſt will deſtroy them.

COLUTEA SCORPIOIDES. See EMERUS.

COMA AUREA. See CHRYSOCOMA.

COMARUM. Lin. Gen. Pl. 563. Marſh Cinquefoil.

The CHARACTERS are,

The flower hath five oblong petals, which are inſerted in the empalement, but are much ſmaller. It hath twenty permanent ſtamina, which are inſerted into the empalement, and a great number of ſmall roundiſh germen collected into a head. The common receptacle afterward becomes a large fleſhy fruit, having many pointed ſeeds adhering to it.

We have but one SPECIES of this genus, viz.

COMARUM (*Paluſtre.*) Fl. Lapp. 214. *Red Marſh Cinquefoil.*

This plant hath creeping woody roots, which ſend out many black fibres, penetrating deep into the ground, from which ariſe many reddiſh ſtalks about two feet high, which generally incline to the ground; theſe are garniſhed at each joint with one winged leaf, compoſed of five, ſix, or ſeven lobes, which riſe above each other, the middle being the largeſt, the lower diminiſhing, and with their baſe embrace the ſtalks. The flowers are produced at the top of the ſtalks, three or four together on ſhort foot-ſtalks, which have a large ſpreading empalement, red on the upper ſide, and divided at the top into ten parts; in the center ſits the five petals, which are red, and not more than a third part the ſize of the empalement; within theſe are ſituated many germen, attended by about twenty ſtamina, terminated by dark ſummits. After the flower is paſt, the receptacle, which ſits in the bottom of the empalement, becomes a fleſhy fruit, ſomewhat like a Strawberry, but flatter, including a great number of pointed ſeeds.

As this plant grows naturally on bogs, ſo it is with difficulty preſerved in gardens, for it muſt be planted in a ſoil as near to that of its natural growth as poſſible; it is very apt to ſpread much at the root when in a proper ſituation, ſo whoever is inclinable to preſerve this plant, may remove it from the places of its growth in October, and plant it on a bog, where there will be no danger of the plants ſucceeding. There are a few of theſe plants now growing upon a bog at Hampſtead, which were planted there ſome years ago; but the neareſt place to London, where it grows wild in plenty, is in the meadows near Guildford, in Surry.

COMMELINA. Lin Gen. Plant. 58.

The CHARACTERS are,

It hath a permanent heart-ſhaped ſpatha; the flower hath ſix concave petals, three of which are ſmall and oval, the other are large, roundiſh, and coloured. It hath three nectariums, (which have been ſuppoſed to be ſtamina;) there are three awl-ſhaped ſtamina, which recline; in the center is ſituated a roundiſh germen, which afterward becomes a naked globular capſule with three furrows, having three cells, each containing two angular ſeeds.

The SPECIES are,

1. COMMELINA (*Communis*) corollis inæqualibus, foliis ovato-lanceolatis, acutis, caule procumbente, glabro. Hort. Upſal. 18. *Commelina with unequal petals, oval, ſpear-ſhaped, pointed leaves, and a ſmooth trailing ſtalk.*

2. COMMELINA (*Erecta*) corollis inæqualibus, foliis ovato-lanceolatis, caule erecto, ſcabro, ſimpliciſſimo. Hort. Upſal. 18. *Commelina with unequal petals, oval ſpear-ſhaped leaves, and a ſingle, upright, rough ſtalk.*

3. COMMELINA (*Africana*) corollis inæqualibus, foliis lanceolatis, glabris, obtuſis, caule repente. Lin. Sp. Pl. 41. *Commelina with unequal petals, ſmooth, ſpear-ſhaped, obtuſe leaves, and a creeping ſtalk.*

4. COMMELINA (*Tuberoſa*) corollis æqualibus foliis ovato-lanceolatis, ſubcilliatis. Hort. Upſal. 18. *Commelina with equal petals, and oval ſpear-ſhaped leaves, which are hairy on their under ſide.*

5. COMMELINA (*Zanonia*) corollis æqualibus, pedunculis incraſſatis, foliis lineari lanceolatis. Lin. Sp. Pl. 41. *Commelina with equal petals, thick foot-ſtalks to the flower, and narrow ſpear-ſhaped leaves.*

The firſt ſort grows naturally in the iſlands in the Weſt-Indies, and alſo in Africa; this is an annual plant, which hath ſeveral trailing ſtalks two or three feet long, which put out roots at the joints, and ſtrike into the ground; at each joint is placed one oval ſpear-ſhaped leaf, ending in a point, which embraces the ſtalk with its baſe, and hath ſeveral longitudinal veins: it is of a deep green, and ſmooth. The flowers come out from the boſom of the leaves, included in a ſpatha, which is compreſſed and ſhut up, each having two or three flowers, ſtanding upon ſhort foot-ſtalks, compoſed of two large blue petals, and four ſmaller green ones,

ones, which have been generally termed the empalement of the flower; within these are situated three nectariums, each having a slender stamina fixed on the side; these surround the germen, which afterward becomes a roundish capsule, having three cells; in each of these is lodged two angular seeds.

The second sort grows naturally in Pensylvania; this hath a perennial root, composed of many white fibres; the stalks rise two feet high, are upright, rough, herbaceous, and about the size of quills; these have a single leaf at each joint, which is shaped like those of the first sort, and embrace the stalks with its base; the flowers come out from the bosom of the leaves at the upper part of the stalk, sitting upon short foot-stalks; these are of a pale bluish colour, and are succeeded by seeds as the first sort.

The third sort grows naturally in Africa; this hath a fibrous root, which sends out many trailing stalks three or four feet high, which put out roots at every joint, and from them many more shoots are produced; so that where they have room to spread, they will cover a large surface of ground. The leaves of this sort are very like those of the first, but the flowers are larger, and of a deep yellow; the petals of this are heart-shaped, and the seed-vessels are larger.

The fourth sort grows naturally near old Vera Cruz in New Spain; this hath a thick fleshy root, composed of several tubers, somewhat like those of Ranunculus; from this arise one or two inclining stalks, which send out side branches from their lower parts, which are garnished with oval spear-shaped leaves, part of which have long foot-stalks, the others embrace the stalks with their base; they have short hairs on their under side, and toward the stalk, but are smooth above, of a deep green colour, and close every evening, or in cold weather. The flowers are produced toward the upper part of the stalks, from the bosom of the leaves, standing upon slender foot-stalks, which are composed of three blue petals, pretty large and roundish, and three smaller, which are green; the seeds are like those of the other sorts.

The fifth sort grows naturally in the West-Indies; this hath trailing stalks like the first, which are garnished with narrow grassy leaves, embracing the stalks with their base; the flowers are produced at the end of the stalks upon thick foot-stalks, three flowers generally sitting on each; they have three equal large petals, of a sky blue, and three smaller, which are green.

All the sorts are propagated by seeds; the first will thrive in the full ground, but if the seeds are sown upon a warm border of light earth in autumn, the plants will rise early in the spring, so from these good seeds may be expected, if the season proves favourable; whereas those which are sown in the spring, often lie long in the ground, so that they rarely ripen their seed. These plants have but little beauty, so that two or three of each sort is as many as most people chuse to have; therefore if the seeds are sown in autumn, where the plants are designed to remain, or the seeds permitted to scatter, the plants will require no farther care but to keep them clear from weeds.

The second sort hath a perennial root; this seldom ripens seeds in England, but the roots send out offsets, by which the plant is easily propagated; this sort will live in the full ground in winter, provided it is planted in a sheltered situation: the best time to transplant and part these roots is about the end of March.

The other sorts are tender, so their seeds must be sown on a moderate hot-bed in the spring, and when the plants are two inches high, they should be transplanted to a fresh hot-bed to bring them forward; when they have taken root, they should have a large share of fresh air admitted to them every day in warm weather, to prevent their growing weak; and in June they may be carefully taken up and transplanted on a warm border of light earth, observing to shade them till they have taken root, after which they will require no other care but to keep them clear from weeds. With this management the plant will flower and produce good seeds.

The third and fourth sorts may be continued if they are planted in pots, and in autumn placed in the bark-stove; or if the roots of the fourth sort are taken out of the ground in autumn, and kept in a warm place in winter, they may be planted again in the spring, placing them on a hot-bed to forward their shooting, and these will produce stronger plants than those which rise from seeds.

CONIUM. Lin. Gen. Plant. 299. Hemlock.

The CHARACTERS are,

It is an umbelliferous plant; the petals of the greater umbel are uniform; each flower is composed of five unequal heart-shaped petals, which turn inward; they have five stamina; the germen, which is situated under the flower, supports two reflexed styles, and afterward becomes a roundish channelled fruit, divided into two parts, containing two seeds.

The SPECIES are,

1. CONIUM (*Maculatum*) seminibus striatis. Hort. Upsf. 92. *Greater Hemlock.*

2. CONIUM (*Africanum*) seminibus aculeatis. Hort. Cliff. 92. *Hemlock with prickly seeds.*

The first sort grows naturally on the side of banks and roads in many parts of England: this is a biennial plant, which perishes after it hath ripened seeds. It hath a long taper root like a Parsnep, but much smaller. The stalk is smooth, spotted with purple, and rises from four to upwards of six feet high, branching out toward the top into several stalks, which are garnished with decompounded leaves, whose small leaves are cut at the top into three parts; these are of a lucid green, and have a disagreeable smell. The stalks are terminated by umbels of white flowers, each being composed of about ten rays (or small umbels) which have a great number of flowers spread open, each sitting upon a slender foot-stalk; the seeds are small and channelled, and like those of Anise.

The second sort grows naturally near the Cape of Good Hope in Africa. This plant rarely grows more than nine inches high; the lower leaves are divided somewhat like those of the small wild Rue, and are of a grayish colour; those upon the stalk are much narrower, but of the same colour; the stalks are terminated by umbels of white flowers, each of these large umbels being composed of three small ones; and the involucrum hath three narrow leaves, situated under the umbel.

The first sort grows wild in most parts of England, so is seldom allowed room in gardens, because it is supposed to have a poisonous quality; some physicians have affirmed, that it is so to all animals, while others have assured us, that it is eaten by the inhabitants of some parts of Italy when it is young, and is by them esteemed a great dainty. Mr. Ray mentions, that he has found the gizzard of a thrush full of Hemlock seeds, with four or five grains of Corn intermixed with it, which, in the time of harvest, that bird had neglected for Hemlock, so very fond was it of that seed, which has been reckoned pernicious: however, it is very certain, that scarce any animal will eat the green herb; for it is very common to see the Grass, and most other weeds, eat close where cattle are allowed to feed, and all the plants of Hemlock which were growing left untouched.

This plant is esteemed, by many physicians, as an excellent remedy to dissolve scirrhous tumours; and some have greatly recommended it for cancers, and most of them agree, that it may be prescribed as a good narcotic.

The

The ſecond ſort is an humble plant, and, being tender, will never become troubleſome; for, unleſs the winters are very favourable, this plant will not live in the open air in England. The ſeeds of this ſort ſhould be ſown in pots in autumn ſoon after they are ripe, and placed under a common frame in winter, where they may be expoſed to the open air at all times when the weather is mild, and only covered in bad weather. The plants will come up very early in the ſpring, and muſt then be expoſed to the open air conſtantly, when the weather will permit, otherwiſe they will draw up very weak. As theſe plants do not bear tranſplanting well, ſo they ſhould be thinned, and not more than four or five left in each pot; and, as the plants have no great beauty, ſo a few of them will be ſufficient to continue the ſort, where a variety of plants are preſerved. The other culture is only to keep them clean from weeds, and, in very dry weather, to water them.

CONOCARPODENDRON. See PROTEA.

CONOCARPOS. Lin. Gen. Plant. 213. Button-tree.

The CHARACTERS are,

The flowers are collected in a globular head, each ſtanding in a ſcaly empalement. At the bottom is ſituated a large compreſſed germen, divided into five parts at the top. The flower hath one petal, which is cut into five equal parts, and five ſlender ſtamina, which extend beyond the petal. The germen afterward becomes a ſingle ſeed, incloſed in the ſcale of the fruit, which is ſhaped like the cone of the Alder-tree.

The SPECIES are,

1. CONOCARPOS (*Erecta*) foliis lacinatis erectis. Lin. Sp. 250. *Upright Button-tree.*

2. CONOCARPOS (*Procumbens*) fruteſcens, caulibus procumbentibus. *Shrubby trailing Button-tree.*

The firſt ſort grows plentifully in moſt of the ſandy bays in all the iſlands of the Weſt-Indies. It riſes with a woody upright ſtem about ſixteen feet high, ſending out many ſide branches, which alſo grow erect, garniſhed with ſpear-ſhaped leaves, having broad ſhort foot-ſtalks, and are placed alternate on every ſide the branches. The flowers grow upon ſhort branches, which ariſe from the wings of the leaves; theſe have three or four ſmall leaves on their lower part, under the flowers; each of theſe branches are terminated by ſix or eight conical heads of flowers, which have ſome reſemblance to thoſe of Acacia, but each of theſe come out of a ſcaly covering; the flowers are ſmall, of a reddiſh colour, having five ſlender ſtamina and one ſtyle, which ſtand out farther than the petal. The flowers are ſucceeded by ſingle ſeeds, which are included in the ſcales of the conical fruit.

The ſecond ſort hath ſhort crooked branches, which divide and ſpread out on every ſide upon the ground; theſe are covered with a grayiſh bark, and their upper parts are garniſhed with oval thick leaves, a little larger than thoſe of the Dwarf Box; they have very ſhort foot-ſtalks, and are placed on every ſide the branches without order. The flowers are collected in ſmall round heads, which come out ſingle from the ſide of the branches, and in looſe ſpikes at the end; theſe are ſmall, and of an herbaceous colour; the ſcales are rough, and the cones are of a looſer texture than thoſe of the former ſort.

Both theſe ſorts are preſerved in ſome curious gardens for the ſake of variety, but they are plants of no great beauty. They are propagated from ſeeds, which muſt be obtained from the places of their natural growth. Theſe ſeeds, if they are freſh, will come up very ſoon, if they are ſown upon a good hot-bed; and if the plants are potted, and preſerved in the bark-ſtove, they will make great progreſs; but they are too tender to live in this country, unleſs they are conſtantly kept in the ſtove, and treated in the ſame manner with other exotic plants of the ſame country, obſerving, as they are natives of ſwamps, to ſupply them often with water, but in winter they muſt have it ſparingly.

CONSOLIDA MAJOR. See SYMPHYTUM.

CONSOLIDA MEDIA. See BUGULA.

CONSOLIDA MINIMA. See BELLIS.

CONSOLIDA REGALIS. See DELPHINIUM.

CONVAL LILY. See CONVALLARIA.

CONVALLARIA. Lin. Gen. Plant. 383. Lily of the Valley.

The CHARACTERS are,

The flower hath one petal, which is bell-ſhaped, ſpread open and reflexed. It hath no empalement. It hath ſix ſtamina, which are inſerted into the petal. In the center is ſituated a globular germen, which afterward becomes a globular berry with three cells, containing one roundiſh ſeed.

The SPECIES are,

1. CONVALLARIA (*Majalis*) ſcapo nudo. Flor. Lapp. 113. *White Lily of the Valley. There is a variety of this with rediſh flowers, which is preſerved in gardens.*

2. CONVALLARIA (*Latifolia*) ſcapo nudo, foliis latoribus. *Broad-leaved Lily of the Valley. There is alſo a variety of this with double variegated flowers, which is preſerved in gardens.*

3. CONVALLARIA (*Odorata*) foliis alternis, ſemiamplexicaulibus, floribus majoribus axillaribus. *Broad-leaved Solomon's Seal with a larger ſweet flower.*

4. CONVALLARIA (*Polygonatum*) foliis amplexicaulibus, caule tereti, pedunculis axillaribus multifloris. Lin. Phil. Bot. 218. *Greateſt broad-leaved Solomon's Seal.*

5. CONVALLARIA (*Stellata*) foliis alternis petiolatis, pedunculis axillaribus trifloris. *Broad-leaved Solomon's Seal with a white Hellebore leaf.*

6. CONVALLARIA (*Verticillata*) foliis verticillatis. Flor. Lapp. 114. *Convallaria with leaves growing in whorls.*

7. CONVALLARIA (*Racemoſa*) foliis ſeſſilibus, racemo terminali compoſito. Lin. Sp. Pl. 315. *Convallaria with leaves ſitting cloſe to the ſtalks, which are terminated by compound ſpikes of flowers.*

8. CONVALLARIA (*Multiflora*) foliis amplexicaulibus plurimis, racemo terminali ſimplici. Lin. Sp. Plant. 316. *Convallaria with many leaves embracing the ſtalks, which are terminated by ſingle bunches of flowers.*

9. CONVALLARIA (*Bifolia*) foliis cordatis. Flor. Lapp. 113. *Convallaria with heart-ſhaped leaves.*

The firſt ſort grows naturally in great plenty in the woods near Wooburn in Bedfordſhire, from whence the markets in London are generally ſupplied with the flowers. It is alſo cultivated in gardens for the ſweetneſs of the flowers; formerly it grew in great plenty on Hampſtead-heath, but of late years it has not been ſo common there; for ſince all the trees have been deſtroyed, the plants have not flowered there as formerly, nor have the roots increaſed.

This hath a ſlender fibrous root, which creeps under the ſurface of the ground, and thereby propagates in great plenty. The leaves come out by pairs; their foot-ſtalks, which are about three inches long, are wrapped together in one cover, and at the top divide into two parts, each ſuſtaining a ſingle leaf, one of which ariſes a little above the other; theſe leaves are from four to five inches long, and near an inch and a half broad in the middle, leſſening gradually to both ends; the foot-ſtalks of the flowers ariſe immediately from the root on one ſide the leaves; theſe are naked, about five inches long, and are adorned towards their upper parts with pendulous white flowers ranged on one ſide the ſtalk, which decline to one ſide; each flower ſtands upon a ſhort ſeparate foot-ſtalk, which are crooked. The flowers are open, of the ſhort bell-ſhaped kind; they have ſix ſtamina, which are inſerted in the petal of the flower,

flower, and are shorter than the tube, and a single style arising from the germen, which is triangular, and crowned by a three-cornered stigma: the germen afterward becomes a globular berry, of a red colour when ripe, inclosing three roundish seeds.

The second sort grows on the Alps; this has retained its difference in the garden, where it grew in the same soil and situation with the common sort, so I make no doubt of its being a distinct species. The other with a double variegated flower is supposed to be only a variety of this, therefore I have not enumerated it as a distinct species, though the flowers are much larger, and beautifully variegated with purple and white.

These plants require a loose sandy soil, and a shady situation; they are propagated by parting of their roots, which multiply in great plenty. The best time to transplant and part the roots is in autumn. They should be planted near a foot asunder, that their roots may have room to spread; for if they agree with the soil and situation, they will meet and fill the ground in one year. If these roots are planted in a rich soil they will spread and multiply greatly, but will not be so productive of flowers.

The only culture which these plants require, is to keep them clean from weeds, and to transplant and separate the roots every third or fourth year, otherwise they will be so greatly matted together, as not to have proper nourishment, so the flowers will be but small and few in number.

The third sort is the common Solomon's Seal, which is said to grow naturally wild in England, but I doubt ours is a different sort from that mentioned by Caspar Bauhin under that title; for in two places where I have found it growing, the stalks were much shorter, the leaves were broader, and their borders turned inward; and this difference continues in the garden, where it grows in the same soil and situation with the common sort of Germany.

This plant hath a fleshy white root, as large as a man's finger, which creeps in the ground, and is full of knots, from whence it had the name of Polygonatum. In the spring arises several taper stalks, which grow near two feet high, adorned with oblong oval leaves placed alternate, which embrace the stalks with their base; on the opposite side come out the foot stalks of the flowers, which are about an inch long, dividing at the top into three or four smaller, sustaining a single tubulous flower, cut into six parts at the brim: these have each six slender stamina, surrounding a single style, which arises from the germen, and is crowned by a blunt stigma; the germen afterward becomes a round berry, about the size of Ivy berries, each inclosing three seeds.

The fourth sort doth not rise so high as the third, the leaves are broader, and half embrace the stalks with their base. It hath fewer flowers on each foot-stalk, and those are much larger and smell sweet.

The fifth sort rises much higher, the leaves are broader, and embrace the stalks with their base; there are generally three flowers on each foot-stalk, which have longer and narrower tubes than either of the former. This grows naturally in the northern parts of Europe.

The sixth sort hath large fleshy roots, full of knots or joints, which (when the roots are strong) send up many stalks four feet high, garnished with oblong oval leaves near five inches long, and above two inches broad in the middle, having many deep longitudinal furrows running parallel to the midrib, somewhat like those of white Hellebore. The flowers come out on the opposite side of the stalks from the leaves, having short foot-stalks, which divide into three smaller, each sustaining one flower, with a long slender tube, more closed at the top than those of the other species, but the colour is the same.

The seventh sort rises with an upright stalk about two feet high, garnished with long broad leaves; they are four inches long, and an inch broad, smooth, and of a light green. The flowers come out on the top of the stalks in a single racemus, standing upon short foot-stalks, each supporting five or six flowers, which are smaller, and have much shorter tubes than either of the former sorts; they are of a dirty white, tipped with green, and slightly cut into six parts at the top.

The eighth sort grows naturally in most parts of North America. This rises with an upright stalk above two feet high, garnished with oblong leaves ending in sharp points, which are near five inches long, and two and a half broad, having three large longitudinal veins, with several smaller between, which join at both ends. The leaves are alternate, standing close to the stalks, and are of a light green on their upper side, but are paler on their under. The flowers are produced in branching spikes at the extremity of the stalks, each being composed of several small loose spikes of star-like flowers, of a pale yellow, which fall away without producing any seed.

All these sorts of Solomon's Seal are very hardy plants, and delight in a moist soil and a shady situation, so are very proper to plant in wilderness quarters under tall trees, where, if they are not crowded by lower shrubs, they will thrive and multiply exceedingly, and during the summer season will make an agreeable variety, the whole appearance of the plants being very singular.

They all multiply very fast by their creeping roots, especially when they are planted in a proper soil and situation. The best time to transplant and part the roots is in the autumn, soon after their stalks decay; those which are removed at that season, will grow much stronger than those which are planted in the spring; but they may be safely transplanted any time after the stalks decay, till the roots begin to shoot in the spring. As these roots greatly increase, so they should be planted at a wide distance from each other, that they may have room to spread, for they should not be removed oftener than every third or fourth year, where they are expected to grow strong, and produce a good number of stalks, in which their beauty consists. The only culture these plants require, is to dig the ground between them every spring, and keep them clean from weeds.

The ninth sort is an humble plant, which, when transplanted into gardens, seldom rises above six inches high, and where it grows naturally not much more than half so high; this hath a fibrous creeping root, which spreads and multiplies greatly in the ground, sending up many slender stalks, each having for the most part two heart-shaped leaves, one standing above the other. The stalks are terminated by loose spikes of white star-like flowers, which are succeeded by small red berries.

It grows naturally in a wood at the Hague, and in all the northern parts of Europe, and delights in a moist soil and shady situation, where it will spread and multiply in great plenty.

CONVOLVULUS. Lin. Gen. Pl. 198. Bindweed.

The Characters are,

The flower hath one large bell-shaped petal, which spreads open. It hath five short stamina, and a roundish germen. The empalement afterward becomes a roundish capsule with one, two, or three valves, containing several seeds which are convex on their outside, but on their inside angular.

The Species are,

1. Convolvulus (*Arvensis*) foliis sagitatis utrinque acutis, pedunculis unifloris. Flor. Suec. 173. *Smaller Field Bindweed; commonly called Gravel Bindweed.*

2. CONVOLVULUS (*Sepium*) foliis sagitatis posticè truncatis, pedunculis unifloris. Prod. Leyd. 427. *Larger white Bindweed, called Bearbind.*

3. CONVOLVULUS (*Scammonia*) foliis sagitatis posticè truncatis, pedunculis bifloris. Prod. Leyd. 427. *Syrian Bindweed, or Scammony.*

4. CONVOLVULUS (*Purpureus*) annuus, foliis cordatis calycibus acutis, villosis. *Purple Bindweed with a roundish heart-shaped leaf; commonly called Convolvulus major, or greater Bindweed.*

5. CONVOLVULUS (*Indicus*) foliis cordatis, acuminatis, pedunculis trifloris. *Bindweed with heart-shaped pointed leaves, and three flowers on each foot-stalk.*

6. CONVOLVULUS (*Nil*) foliis cordatis trilobis villosis, calycibus lævibus, capsulis hirsutis, pedunculis bifloris. *Blue Bindweed with an angular Ivy leaf.*

7. CONVOLVULUS (*Battatus*) foliis cordatis angulato-nervosis, caule repente tubifero. Lin. Sp. Pl. 154. *Bindweed with heart-shaped leaves, having angular nerves, and a creeping stalk bearing tubers; commonly called Spanish Potatoes.*

8. CONVOLVULUS (*Serpens*) foliis palmatis, lobis septem-sinuatis acutis, pedunculis unifloris, calycibus maximis patentibus. *Five-leaved Bindweed, with smooth indented leaves and hairy stalks.*

9. CONVOLVULUS (*Aristolochiolus*) foliis hastato-lanceolatis, auriculis rotundatis, pedunculis multifloris. *Bindweed with spear-pointed leaves having rounded ears, and many flowers on each foot-stalk.*

10. CONVOLVULUS (*Hirtus*) foliis cordatis subhastatisque villosis, caule petiolisque pilosis, pedunculis multifloris. Lin. Sp. Pl. 159. *Bindweed with heart-shaped leaves, somewhat spear-pointed and downy, with hairy stalks and foot-stalks, and many flowers on each.*

11. CONVOLVULUS (*Glaber*) foliis ovato-oblongis, glabris pedunculis unifloris, calycibus decempartitis. *Bindweed with oval, oblong, smooth leaves, and foot-stalks having a single flower, whose empalement is cut into ten parts.*

12. CONVOLVULUS (*Pentaphyllos*) hirsutissimus, foliis quinquelobatis, pedunculis longissimis bifloris. *Very hairy Bindweed with leaves having five lobes, and very long foot-stalks with two flowers.*

13. CONVOLVULUS (*Frutescens*) caule fruticoso, glabro, foliis quinquelobis, pedunculis geniculatis, unifloris, capsulis maximis. *Bindweed with a shrubby smooth stalk, leaves having five lobes, many jointed foot-stalks with one flower, and very large seed-vessels.*

14. CONVOLVULUS (*Brasiliensis*) foliis emarginatis, pedunculis trifloris. Lin. Sp. Plant. 159. *Bindweed with indented leaves, and foot-stalks having three flowers.*

15. CONVOLVULUS (*Multifloris*) foliis cordatis, glabris, pedunculis multifloris, feminibus villosis ferrugineis. *Bindweed with smooth heart-shaped leaves, foot-stalks having many flowers, and seeds covered with an iron-coloured down.*

16. CONVOLVULUS (*Canariensis*) foliis cordatis pubescentibus, caule perenni, villoso, pedunculis multifloris. Lin. Sp. Pl. 155. *Bindweed with soft, woolly, heart-shaped leaves, a hairy perennial stalk, and foot-stalks having many flowers.*

17. CONVOLVULUS (*Hederaceus*) foliis triangularibus acutis, floribus plurimis sessilibus patulis, calycibus acutis multifidis. *Bindweed with sharp-pointed triangular leaves, many spreading flowers set close to the stalk, and acute empalements ending in many points.*

18. CONVOLVULUS (*Roseus*) foliis cordatis, acuminatis, pedunculis bifloris. *Bindweed with heart-shaped pointed leaves, and foot-stalks having two flowers.*

19. CONVOLVULUS (*Repens*) foliis sagitatis posticè obtusis, caule repente, pedunculis unifloris. Lin. Sp. Plant. 158. *Bindweed with narrow-pointed leaves, which are obtuse at the foot-stalk, a creeping stalk, and one flower on each foot-stalk.*

20. CONVOLVULUS (*Betonicifolius*) foliis cordato-sagitatis, pedunculis unifloris. *Bindweed with heart-shaped arrow-pointed leaves, and foot-stalks having a single flower.*

21. CONVOLVULUS (*Siculus*) foliis cordato-ovatis, pedunculis unifloris, bracteis lanceolatis, flore sessile. Hort. Cliff. 68. *Bindweed with oval heart-shaped leaves, foot-stalks having one flower, spear-shaped bracteæ, and the flower sitting close to the stalk.*

22. CONVOLVULUS (*Elegantissimus*) foliis palmatis sericeis, pedunculis bifloris, calycibus acutis. *Bindweed with silky palmated leaves, foot-stalks having two flowers, and sharp-pointed empalements.*

23. CONVOLVULUS (*Altheoides*) foliis cordatis incisis & incanis, pedunculis bifloris, calycibus obtusis. *Bindweed with hoary heart-shaped leaves, which are jagged, foot-stalks having two flowers, and obtuse empalements.*

24. CONVOLVULUS (*Tricolor*) foliis lanceolato-ovatis glabris, caule declinato, floribus solitariis. Lin. Vir. Cliff. *Bindweed with smooth, oval, spear-shaped leaves, a declining stalk, and one flower upon each foot-stalk; commonly called Convolvulus minor.*

25. CONVOLVULUS (*Cantabrica*) villosus, foliis leneari-lanceolatis, caule erecto, pedunculis multifloris. *Hairy Bindweed with narrow spear-shaped leaves, an upright stalk, and foot-stalks having many flowers.*

26. CONVOLVULUS (*Lineatus*) foliis lanceolatis, sericeis, caule declinato, radice repente, pedunculis multifloris. *Bindweed with silky spear-shaped leaves, a declining stalk, creeping root, and foot-stalks having many flowers.*

27. CONVOLVULUS (*Cneorum*) foliis lanceolatis, obtusis sericeis pedunculis multifloris. Hort. Cliff. 68. *Bindweed with blunt, spear-shaped, silky leaves, and foot-stalks having many flowers.*

28. CONVOLVULUS (*Linarifolius*) foliis lineari-lanceolatis, acutis caule ramoso, recto, pedunculis unifloris. Hort. Cliff. 68. *Bindweed with narrow spear-shaped leaves, which are pointed, an upright branching stalk, and foot-stalks with one flower.*

29. CONVOLVULUS (*Soldanella*) foliis reniformibus, pedunculis unifloris. Hort. Cliff. 67. *Bindweed with kidney-shaped leaves, and one flower on each foot-stalk; or Sea Bindweed.*

30. CONVOLVULUS (*Turpethum*) foliis cordatis, angulatis, caule membranaceo, quadrangulari, pedunculis multifloris. Flor. Zeyl. 72. *Bindweed with angular heart-shaped leaves, a quadrangular membranaceous stalk, and foot-stalks having many flowers.*

31. CONVOLVULUS (*Jalapa*) foliis variis, pedunculis unifloris, radice tuberosâ cathartica. *Bindweed with variable leaves, foot-stalks with single flowers, and a tuberous root; or the true Jalap.*

The first sort is very common upon dry banks and in gravelly grounds in most parts of England, and is generally a sign of gravel lying near the surface. The roots of this shoot very deep into the ground, from whence some country people call it Devil's Guts: this is a troublesome weed in gardens, so should be constantly rooted out.

The second sort is also a troublesome weed in gardens, when the roots are intermixed with those of trees and shrubs, or under hedges, where the plants cannot be easily destroyed: but in an open clear spot of ground, where they are constantly hoed down for three or four months, they may be effectually destroyed; for when the stalks are broken or cut, a milky juice flows out, and thereby the roots are soon exhausted and decay: as every small piece of the root will grow, so it renders this a troublesome weed to destroy, where they are intermixed with other roots.

The

The third ſort grows naturally in Syria, where the roots of the plants are wounded, and ſhells placed under the wounds to receive the milky juice which flows out, which is inſpiſſated, and afterward put up and exported: this is what is called Scammony in the ſhops: it is a very hardy plant, and will thrive very well in the open air in England, provided it is on a dry ſoil. The roots of this are thick, run deep into the ground, and are covered with a dark bark. The branches extend themſelves on every ſide to the diſtance of ten or twelve feet; they are ſlender and trail on the ground, and are garniſhed with narrow arrow-pointed leaves. The flowers are of a pale yellow, and come out from the ſide of the branches, two ſitting upon each long foot-ſtalk; theſe are ſucceeded by roundiſh ſeed-veſſels, having three cells, filled with ſeeds ſhaped like thoſe of the former ſort, but ſmaller. If the ſeeds of this ſort are ſown in the ſpring on a border of light earth, the plants will come up, and require no other culture but to keep them clean from weeds, and thin the plants where they grow too cloſe; for as the branches extend very far, the plants ſhould not be nearer than five feet aſunder. The ſtalks decay in autumn, but the roots will abide many years.

The fourth ſort is an annual plant, which grows naturally in Aſia and America, but has been long cultivated for ornament in the Engliſh gardens, and is generally known by the title of Convolvulus major. Of this there are three or four laſting varieties; the moſt common hath a purple flower, but there is one with a white, another with a red, and one with a whitiſh blue flower, which hath white ſeeds. All theſe varieties I have cultivated many years, without obſerving them to change. If the ſeeds of theſe ſorts are ſown in the ſpring, upon a warm border where the plants are deſigned to remain, they will require no other culture but to keep them clean from weeds, and place ſome tall ſtakes down by them, for their ſtalks to twine about, otherwiſe they will ſpread on the ground, and make a bad appearance. Theſe plants, if they are properly ſupported, will riſe ten or twelve feet high in warm ſummers: they flower in June, July, and Auguſt, and will continue till the froſt kills them. Their ſeeds ripen in autumn.

The fifth ſort grows naturally in Jamaica. This ſends out long branches, which twiſt about the trees and riſe to a great height; the leaves are ſmooth, heart-ſhaped, ending in long points, and the ears at the baſe are large and rounded; they ſtand upon long ſlender foot-ſtalks. The flowers come out on the oppoſite ſide of the ſtalks, upon long foot-ſtalks, each ſuſtaining three flowers, with longer tubes than thoſe of the former, and are of a deeper blue colour. This is not ſo hardy as the former, ſo the ſeeds ſhould be ſown upon a hot-bed in the ſpring, to bring the plants forward, and toward the end of May they ſhould be planted out in warm borders, and treated in the ſame manner as the former ſort.

The ſixth ſort grows naturally in Africa and America. It is an annual plant, which riſes with a twining ſtalk eight or ten feet high, garniſhed with heart-ſhaped leaves, divided into three lobes which end in ſharp points; theſe are woolly, and ſtand upon long foot-ſtalks; the flowers come out on long foot-ſtalks, each ſuſtaining two flowers of a very deep blue colour, from whence it has been titled Anil or Indigo. This is one of the moſt beautiful flowers of the genus, and is undoubtedly a diſtinct ſpecies, though ſome have ſuppoſed it to be only a variety of the fourth ſort: the leaves of this has three deeply divided lobes, and thoſe of the fourth ſort are entire. This ſort is annual, and muſt be propagated in the ſame manner as the fifth. It flowers all the latter part of ſummer, and in good ſeaſons the ſeeds ripen very well in the open air.

The ſeventh ſort is that whoſe roots are eaten, and is generally called Spaniſh Potatoe; theſe roots are annually imported from Spain and Portugal, where they are greatly cultivated for the table, but they are too tender to thrive well in the open air in England; they are cultivated by the roots in the ſame way as the common Potatoe, but require much more room; for theſe ſend out many trailing ſtalks, which extend ſix or eight feet every way, and at their joints ſend out roots, which, in warm countries grow to be large tubers, ſo that from a ſingle root planted, forty or fifty large roots are produced. This is ſometimes propagated by way of curioſity in England, but the roots ſhould be planted on a hot-bed in the ſpring, and if the plants are kept covered in bad weather with glaſſes, they will produce flowers, and ſome ſmall roots will be produced from the joints of the ſtalks: but if they are expoſed to the open air, they ſeldom grow to be of any ſize.

The eighth ſort grows naturally at La Vera Cruz in New Spain. This riſes with a ſtrong winding ſtalk to the height of twenty feet, dividing into ſeveral ſmaller, which faſten themſelves about any of the neighbouring trees and ſhrubs; and are garniſhed with leaves in ſhape of a hand, having ſeven lobes, which are ſpear-ſhaped, and deeply cut on their borders, ending in ſharp points. The flowers are ſingle on each foot-ſtalk, which are very long. The empalement of the flower is large, ſpreading open, and is divided deeply into five parts. The flowers are large, of a purple colour, and are ſucceeded by large roundiſh ſeed-veſſels, having three cells, in each of which is lodged a ſingle ſeed.

This plant is tender, ſo the ſeeds ſhould be ſown on a hot-bed in the ſpring, and when the plants are fit to remove, they muſt be tranſplanted each into a ſeparate pot, and plunged into a moderate hot-bed, obſerving to ſhade them from the ſun till they have taken new root; then they ſhould have a large ſhare of air admitted to them every day, to prevent their drawing up weak. When the plants are grown too tall to remain in the hot-bed, they muſt be ſhifted into larger pots and placed in the bark-ſtove, where, if they are allowed room, they will riſe to a great height, and produce flowers, but it rarely produces ſeeds in England.

The ninth ſort is an annual plant; it grows naturally near Carthagena in New Spain. This riſes with a twining ſlender ſtalk ten or twelve feet high, garniſhed with arrow-pointed leaves, whoſe ears at the baſe are rounded. The flowers are produced in ſmall cluſters ſtanding on long foot-ſtalks; theſe are yellow, and are ſucceeded by three-cornered ſeed-veſſels, having three cells, in each of which are lodged two ſeeds.

This plant is annual, and too tender to thrive in the open air in England, ſo the ſeeds ſhould be ſown on a hot-bed in the ſpring, and the plants may be afterward treated in the ſame way as the eighth ſort, with which management they will flower and produce ripe ſeeds.

The ſeeds of the tenth ſort were ſent me from Jamaica. This is an annual plant, riſing with ſlender twining ſtalks eight or nine feet high, garniſhed with heart-ſhaped leaves which are downy. The flowers ſtand many together at the end of ſtrong foot-ſtalks; they are purple, and are ſucceeded by roundiſh ſeed-veſſels with three cells, containing three ſmall ſeeds.

This ſort requires the ſame treatment as the eighth, being too tender to thrive in this country in the open air.

The eleventh ſort was ſent me from the iſland of Barbuda. This is an annual plant, which riſes with twining ſtalks ſeven or eight feet high, garniſhed with oblong, oval, ſmooth leaves. The flowers come out at every joint on ſlender long foot-ſtalks, each ſupporting a large purple flower, whoſe empalement is cut almoſt to the bottom in ten parts. The ſeeds and capſule are like thoſe of the other ſpecies. This is a tender plant, ſo muſt be treated in the ſame manner as the eighth ſort.

The twelfth ſort grows naturally at Carthagena in New Spain. This is a perennial plant, which riſes with ſtrong winding ſtalks to the height of fourteen or fifteen feet, garniſhed with leaves divided into five lobes, ſtanding upon ſhort foot-ſtalks; the flowers ſtand upon long foot-ſtalks, each ſuſtaining two purple flowers. The ſtalks, leaves, and every part of the plant, is cloſely covered with pungent ſtinging hairs of a light brown colour. This ſort is tender, ſo muſt be treated in the ſame manner as the eighth.

The thirteenth ſort grows naturally about Tolu in New Spain. This hath a ligneous ſtalk covered with a purple bark, which twines about the trees, and riſes to the height of thirty feet or more, garniſhed with leaves which are deeply divided into five ſharp-pointed lobes. The flowers ſtand upon long thick foot-ſtalks, which have a knee in the middle; they are very large, of a purple colour, and are ſucceeded by round ſeed-veſſels as large as a middling Apple, divided into three cells, each containing two very large ſmooth ſeeds.

This plant is too tender to thrive in the open air in England, ſo muſt be treated in the ſame manner as the eighth ſort, but it grows too tall for the ſtoves here. I have had theſe plants upward of twenty feet high, which have ſent out many ſide branches, extending ſo wide on every ſide as to cover moſt of the neighbouring plants, ſo that I was obliged to remove them into a cooler ſituation, where they would not thrive.

The fourteenth ſort grows naturally on the ſea-ſhores in moſt of the iſlands in the Weſt-Indies, where the ſtalks trail on the ground, which are garniſhed with oval leaves, indented at the top. The flowers are large, of a purple colour, and are produced by threes on very long foot-ſtalks; theſe are ſucceeded by large oval ſeed-veſſels, with three cells, each containing a ſingle ſeed; this hath a perennial ſtalk, which ſpreads to a great diſtance, but is too tender to thrive in the open air in England, ſo muſt be treated in the ſame manner as the eighth ſort, and may be continued two or three years in a warm ſtove; but it is apt to ſpread too far for a ſmall ſtove, ſo that where there is not great room, it is not worthy of culture.

The fifteenth ſort grows naturally in Jamaica; this riſes with ſlender winding ſtalks eight or ten feet high; the leaves of theſe are ſhaped a little like thoſe of the common great white Convolvulus, but the foot-ſtalks which are pretty long, do each ſuſtain many purple flowers growing in bunches. The ſeed-veſſels of this ſort are three-cornered, and have three cells, each containing a ſingle ſeed.

The ſixteenth ſort hath been long preſerved in ſeveral curious gardens in England. It grows naturally in the Canary Iſlands; this hath a ſtrong fibrous root, from which ariſe ſeveral twining woody ſtalks, and where they have ſupport, will grow more than twenty feet high, garniſhed with oblong heart-ſhaped leaves, which are ſoft and hairy. The flowers are produced from the wings of the leaves, ſeveral ſtanding upon one foot-ſtalk; they are for the moſt part of a pale blue, but there is a variety of it with white flowers. It flowers in June, July, and Auguſt, and ſometimes ripens ſeeds here; but as the plants are eaſily propagated by layers, and alſo from cuttings, ſo the ſeeds are not ſo much regarded; nor indeed will thoſe plants which are raiſed by layers or cuttings produce ſeeds, though thoſe which come from ſeeds ſeldom fail. It may be propagated by laying down of the young ſhoots in the ſpring, which generally put out roots in three or four months; then they may be taken from the old plants, and each planted in a ſeparate pot, and placed in the ſhade till they have taken new root, after which they may be placed with other hardy green-houſe plants till autumn, when they ſhould be removed into the green-houſe, and afterward treated in the ſame way as Myrtles and other green-houſe plants. If the tender cuttings of this are planted during any of the ſummer months, and plunged into a moderate hot-bed, ſhading them from from the ſun, they will take root, and afterward ſhould be treated as the layers.

The ſeventeenth ſort is an annual plant; this riſes with a very ſlender twining ſtalk four or five feet high, garniſhed with triangular leaves which are pointed. The flowers grow in cluſters ſitting cloſe to the ſtalks, which are blue, and are ſucceeded by ſeeds like thoſe of the fourth ſort. This ſort will not ripen ſeeds in England, unleſs the plants are brought forward on a hot-bed in the ſpring, and afterward placed in a glaſs-caſe, where they may be defended from cold.

The eighteenth ſort grows naturally in Jamaica. This is one of the moſt beautiful ſpecies of this genus, the flowers being very large and of a fine Roſe colour. It riſes with a winding ſtalk ſeven or eight feet high, garniſhed with heart-ſhaped leaves ending in long points, ſitting upon very long foot-ſtalks. The flowers alſo have long foot-ſtalks, each ſupporting two flowers, whoſe empalement is divided deeply into five parts; the ſeeds of this are large, and covered with a fine down. This is an annual plant, which is too tender to thrive in the open air in this country, ſo the ſeeds ſhould be ſown on a hot-bed in the ſpring, and the plants afterward treated in the ſame manner as is directed for the eighth ſort.

The nineteenth ſort grows naturally near the ſea at Campeachy. It hath ſtrong, ſmooth, winding ſtalks, which ſend out roots at their joints, garniſhed with arrow-pointed leaves, whoſe ears (or lobes) are obtuſe; the flowers are large, of a ſulphur colour, and ſit upon very long foot-ſtalks, which proceed from the ſide of the ſtalks, each ſupporting one flower with a large ſwelling empalement; they are ſucceeded by large, ſmooth, oval capſules, having three cells, each including one large ſmooth ſeed. This is a perennial plant, whoſe ſtalks extend to a great diſtance, and put out roots at the joints, whereby it propagates in plenty, but it is too tender to thrive in England, unleſs it is preſerved in a warm ſtove, where it requires more room than can well be allowed to one plant. It muſt be treated in the ſame manner as the eighth ſort.

The twentieth ſort grows naturally in Africa; this riſes with a ſlender winding ſtalk five or ſix feet high, garniſhed with heart-ſhaped arrow-pointed leaves; the flowers ſtand on long ſlender foot-ſtalks, they are white, with purple bottoms. This may be treated in the ſame manner as the common great Convolvulus.

The twenty-firſt ſort grows naturally in Spain and Italy. It is an annual plant, which riſes about two feet high, with ſlender twining ſtalks, garniſhed with oval leaves. The flowers are ſmall, and of a bluiſh colour; each foot-ſtalk ſupporting one flower, of little beauty, ſo is not often cultivated in gardens. If the ſeeds of this ſort are permitted to ſcatter, the plants will riſe in the ſpring, and require no other culture but to keep them clean from weeds; or if the ſeeds are ſown in the ſpring where the plants are to remain, they will flower in June, and the ſeeds will ripen in Auguſt.

The twenty-ſecond ſort grows naturally in Sicily, and alſo in the iſlands of the Archipelago; this hath a perennial root, which ſends out many ſlender ſtiff ſtalks, twiſting themſelves round the neighbouring plants, and grow five or ſix feet high, garniſhed with leaves, which are divided into five or ſeven narrow lobes, and are ſoft like ſatten, ſtanding on ſhort foot-ſtalks. The flowers are produced from the ſide of the ſtalks upon long foot-ſtalks, which ſuſtain two flowers of a pale Rofe colour, with five ftripes of a deeper red. This fort creeps at the root, fo feldom produces feeds in England, but is propagated by the fhoots taken from the old plants; the beft time for parting and tranfplanting thefe plants is about the beginning of May, when they may be taken out of the green-houfe and expofed in the open air; but the young plants which are feparated from the old ones, fhould be placed under a frame, and fhaded from the fun till they have taken new root, after which they muft be gradually hardened to bear the open air; but in autumn they muft be placed in the green-houfe, and may be treated in the fame way as the Canary Convolvulus before-mentioned.

The twenty-third fort hath fome appearance of the twenty-fecond, and hath been fuppofed to be the fame fpecies by fome writers; but as I have cultivated both many years, and never have found either of them alter, fo I make no doubt of their being diftinct plants. This hath a perennial root like the former, which fends out many weak twining ftalks, rifing three or four feet high, twifting about the plants which ftand near it, or about each other, and if they have no other fupport, fall to the ground; they are garnifhed with leaves of different forms, fome are fhaped almoft like thofe of Betony, being flightly cut on their edges, others are almoft heart-fhaped, and are deeply cut on the fides, and fome are cut to the midrib; they have a fhining appearance like fatten, and are foft to the touch, ftanding on fhort foot-ftalks. The flowers are produced on the oppofite fide from the leaves, having very long foot-ftalks, each fuftaining two flowers of a pale Rofe colour, very like thofe of the former fpecies. It hath a perennial root, which fends out offsets, by which it is propagated in England, in the fame manner as the laft mentioned, and the plants muft be treated in the fame way.

The twenty-fourth fort grows naturally in Portugal, but hath been long cultivated in the flower-gardens in England for ornament; this is ufually titled Convolvulus minor, by the feedfmen and gardeners. It is an annual plant, which hath feveral thick herbaceous ftalks growing about two feet long, which do not twine like the other forts, but decline toward the ground, upon which many of the lower branches lie proftrate; they are garnifhed with fpear-fhaped leaves, which fit clofe to the branches; the foot-ftalks of the flowers come out juft above the leaves at the fame joint, and on the fame fide of the ftalks, and are about two inches long, each fuftaining one large, open, bell-fhaped flower, which in fome is of a fine blue colour, with a white bottom; in others they are pure white, and fome are beautifully variegated with both colours. The white flowers are fucceeded by white feeds, and the blue by dark coloured feeds, and this difference is pretty conftant in both; but thofe plants with variegated flowers, have frequently plain flowers of both colours intermixed with the ftriped; therefore the only method to continue the variegated fort, is to pull off the plain flowers when they appear, never fuffering any of them to remain for feed.

This fort is propagated by feeds, which fhould be fown on the borders of the flower-garden where they are defigned to remain, and when the plants come up, if the feeds grow, there fhould be but one or two left in each place. After which they will require no other culture but to keep them clean from weeds.

The twenty-fifth fort grows naturally in Italy and Sicily; this hath a perennial root, which runs deep in the ground, from which arife two or three upright branching ftalks, two or three feet high, garnifhed with narrow leaves about two inches long, which fit clofe to the ftalks; the foot-ftalks of the flower proceed from the fame place; thefe are four or five inches long, each fuftaining four or five flowers, of a pale Rofe colour, which fpread open almoft flat. It flowers in July and Auguft, and the ftalks decay in autumn; but the roots will laft feveral years, and if they are in a dry foil and warm fituation, will abide through the winters very well without covering, and may be propagated by layers.

The twenty-fixth fort hath a perennial creeping root, from which arife feveral fhort branching ftalks about four inches high, garnifhed with fpear-fhaped filky leaves; the flowers are produced on the fide, and at the top of the ftalks, in fmall clufters, fitting clofe together; they are much fmaller than thofe of the former fort, but are of a deeper Rofe colour: it feldom produces feeds in England, but the roots propagate in plenty; it may be tranfplanted either in the fpring or autumn. This is by fome fuppofed to be the fame as the laft mentioned fort, but whoever has cultivated them, can have no doubt of their being different fpecies.

The twenty-feventh fort grows naturally in Italy, Sicily, and the iflands of the Archipelago; it rifes with upright fhrubby ftalks about three feet high, clofely garnifhed with blunt, fpear-fhaped, filky leaves, which are placed on every fide the ftalks; they are near two inches long, and a quarter broad, rounding at their ends. The flowers are produced in clufters at the top of the ftalks, fitting very clofe; they are of a pale Rofe colour, and come out in June and July, but do not perfect feeds in England. This plant will live in the open air in mild winters, if it is planted in a light foil and a warm fituation, but in hard winters it is fometimes deftroyed; therefore fome of the plants fhould be kept in pots, and fheltered under a common frame in winter, where they may enjoy the free air in mild weather, and be protected from the froft, and in fummer placed abroad with other hardy exotic plants, where its fine filky leaves will make a pretty appearance. It may be propagated by laying down the branches, and alfo by cuttings.

The twenty-eighth fort grows naturally in Candia, and feveral of the iflands in the Archipelago; this hath a perennial root, which fends out feveral erect branching ftalks about two feet high, which are garnifhed with very narrow-pointed leaves, fitting clofe to the ftalks, which are hoary. The flowers come out fingly on the fide of the ftalks, fitting very clofe to them, having fcarce any foot-ftalks; thefe are of a pale bluifh colour, and fpread open almoft to the bottom. This fort is propagated in the fame manner as the twenty-fifth, and the plants require the fame treatment.

The twenty-ninth fort is ufed in medicine; this is ftiled Soldanella, and Braffica marina; it grows naturally on the fea-beaches in many parts of England, but cannot be long preferved in a garden It hath many fmall, white, ftringy roots, which fpread wide, and fend out feveral weak trailing branches, which twine about the neighbouring plants like the common Bindweed, garnifhed with kidney-fhaped leaves about the fize of thofe of the leffer Celandine, ftanding upon long foot-ftalks, and are placed alternate. The flowers are produced on the fide of the branches at each joint. Thefe are fhaped like thofe of the firft fort, and are of a reddifh purple colour; they appear in July, and are fucceeded by round capfules having three cells, each containing one black feed; every part of the plant abounds with a milky juice.

The thirtieth fort grows naturally in the ifland of Ceylon; this is a perennial plant, having thick flefhy roots which fpread far in the ground, and abound with a milky juice, which flows out when the roots are broken or wounded, and foon hardens into a refinous fubftance when expofed to the fun and air. From the root fhoots forth many twining branches, which twift about each other or the neighbouring

bouring plants like the common Bindweed. They are garnifhed with heart-fhaped leaves, which are foft to the touch, like thofe of the Marfhmallow. The flowers are produced at the joints on the fide of the ftalks, feveral ftanding together on the fame foot-ftalk; they are white, and fhaped like thofe of the common great Bindweed, and are fucceeded by round capfules having three cells, which contain two feeds in each.

The roots of this plant, which is ufed in medicine, are brought to us from India; it is titled Turpethum, or Turbith, in the fhops.

This plant is tender, fo will not thrive in the open air in England. It is propagated by feeds, which muft be fown on a hot-bed, and when the plants are fit to remove, they fhould be each planted in a feparate pot, and plunged into a hot-bed of tanners bark, and fcreened from the fun till they have taken root, and afterward muft be treated in the fame manner as hath been directed for the eighth fort.

The thirty-firft fort is the Jalap which is ufed in medicine. This grows naturally at Haleppo, in the Spanifh Weft-Indies, fituated between La Vera Cruz and Mexico. The root of this plant hath been long ufed in medicine, but it was not certainly known from what plant it was produced. The old title of this was Mechoacana nigra; but Father Plumier afferted, that it was the root of one fpecies of Marvel of Peru, from whence Tournefort was induced to conftitute a genus of that plant under the title of Jalapa. But Mr. Ray, from better information, put it among the Convolvuli, and titled it Convolvulus Americanus, Jalapium dictus. This was by the late Dr. Houftoun fully afcertained, who brought fome of the roots of this plant from the Spanifh Weft-Indies to Jamaica, where he planted them, with a defign of cultivating the plants for ufe in that ifland, where they flourifhed during his abode there: but foon after he left the country, the perfon to whofe care he committed them was fo carelefs, as to fuffer hogs to root them out of the ground and deftroy them, fo that there was no remains of them left when he returned there; nor have I heard of this plant being introduced into any of the Britifh Iflands fince.

This hath a large root of an oval form, which is full of a milky juice, from which come out many herbaceous twining ftalks, rifing eight or ten feet high, garnifhed with variable leaves, fome of them being heart-fhaped, others angular, and fome oblong and pointed; they are fmooth, and ftand upon long foot-ftalks; and from a drawing of the plant, made by a Spaniard, in the country where it grows naturally, who gave it to Dr. Houftoun, and is now in my poffeffion, the flowers are fhaped like thofe of the common Great Bindweed, each foot-ftalk fupporting one flower: but as it is only a pencil drawing, fo the colour is not expreffed, therefore I can give no further account of it.

As this plant is a native of a warm country, fo it will not thrive in England, unlefs it is preferved in a warm ftove; therefore the feeds muft be fown on a hot-bed, and the plants put into pots, and plunged into a hot-bed of tanners bark, and treated in the fame manner as the eighth fort, with this difference only, that as this hath large, flefhy, fucculent roots, fo they fhould have but little water given them, efpecially in winter, left it caufe them to rot.

CONYZA. Lin. Gen. Plant. 854. Flea-bane.

The CHARACTERS are,

It hath a compound flower, made up of many hermaphrodite florets which compofe the difk, and female half florets which form the rays; the hermaphrodite florets are funnel-fhaped, and cut into five parts at the brim, and have each five fhort hairy ftamina; in the bottom of each floret is fituated a germen. The female half florets or rays are funnel-fhaped, and cut into three parts at the top; thefe have a germen. The hermaphrodite and female florets are both fucceeded by one oblong feed, crowned with down, fitting upon a plain receptacle, and are included in the empalement.

The SPECIES are,

1. CONYZA (*Squarrofa*) foliis lanceolatis acutis, caule annuo corymbofo. Hort. Cliff. 405. *Common Greater Fleabane.*

2. CONYZA (*Bifrons*) foliis ovato oblongis, amplexicaulibus. Hort. Cliff. 405. *Pyrenean Flea-bane with a Primrofe leaf.*

3. CONYZA (*Candida*) foliis ovatis tomentofis, floribus confertis, pedunculis lateralibus terminalibufque. Hort. Cliff. 405. *Shrubby Flea-bane of Crete, with foft woolly leaves, which are very white.*

4. CONYZA (*Lutea*) foliis haftatis fcabris, caule erecto ramofo, perenni, floribus corymbofis. *Yellow tree-like Fleabane with a trifid leaf.*

The firft fort grows naturally upon dry places in feveral parts of England, fo is feldom allowed a place in gardens. This is a biennial plant, which decays foon after the feeds are ripe; it hath feveral large oblong pointed leaves growing near the ground, which are hairy; between thefe the ftalks come out, which rife more than two feet high, and divide upward into feveral branches, which are garnifhed with fmaller oblong leaves ftanding alternate; at the ends of the ftalks the flowers are produced in round bunches, which are of a dirty yellow colour; thefe are fucceeded by oblong feeds crowned with down.

The fecond fort grows naturally on the mountains in Italy, and is preferved in botanic gardens for the fake of variety: it hath a perennial root, but an annual ftalk. From a thick fibrous root arife many upright ftalks, garnifhed with oblong oval leaves, which are rough, and embrace the ftalks with their bafe; they have appendages running along the ftalk from one to the other, whereby the ftalk is winged. The upper part of the ftalks divide into many fmaller branches, garnifhed with leaves of the fame form as the other, but fmaller, ftanding alternate. The branches are terminated by yellow flowers growing in round bunches, and are fucceeded by oblong feeds crowned with down. This is propagated by feeds, which may be fown on a bed of light earth in the fpring, and when the plants come up, they fhould be thinned where they are too near, and kept clean from weeds; the following autumn they may be tranfplanted where they are defigned to remain, and require no other care but to keep them clean from weeds.

The third fort grows naturally in Crete. This hath a fhort fhrubby ftalk, which in this country feldom rifes more than fix inches high, dividing into feveral fhort branches, which are clofely garnifhed with oval woolly leaves, which are very white; from thefe branches arife the flower-ftalks, which are woolly, and about nine inches high: thefe are garnifhed with fmall, oval, white leaves, placed alternate. The flowers are produced at the fides and end of the ftalk, fometimes but one, at other times two, and fometimes three flowers ftanding on the fame foot-ftalk: they are of a dirty yellow colour, and rarely are fucceeded by feeds in this country: fo the plant is propagated here by flips, which, if taken from the old plants in June, and planted on an eaft-afpected border, and covered with hand-glaffes, will take root in fix or eight weeks; when they have taken root, they fhould be gradually expofed to the open air. In autumn thefe fhould be carefully taken up, preferving the earth to their roots; fome of them may be planted in pots, that they may be fheltered under a frame in the winter; and the others may be planted in a warm border of dry poor earth, where they will endure the cold of our ordinary winters very well, and continue many years.

The fourth ſort grows naturally in Jamaica. This is titled by Sir Hans Sloane, Virga aurea major, ſc. Herba Doria folio ſinuato hirſuto. Cat. Jam. 125. It riſes with a ſhrubby ſtalk ſeven or eight feet high, dividing into ſeveral branches, which are cloathed with rough leaves ſhaped like the point of a halbert, about four inches long. The flowers are produced in roundiſh bunches, at the extremity of the branches; theſe are yellow, and ſtand cloſe together, and are ſucceeded by oblong ſeeds crowned with down.

This plant is too tender to thrive in the open air in this country, therefore the ſeeds muſt be ſown upon a hot-bed, and when the plants are fit to remove, they muſt be each tranſplanted into a ſeparate ſmall pot, and plunged into a hot-bed, obſerving to ſcreen them from the ſun till they have taken new root; then they muſt have free air admitted to them every day, in proportion to the warmth of the ſeaſon. As the plants advance in ſtrength, ſo they muſt have a greater ſhare of air; and if the ſeaſon is warm, they may be expoſed to the open air for a few weeks in the heat of the ſummer, provided they are placed in a warm ſituation; but if the nights prove cold, or much wet ſhould fall, they muſt be removed into ſhelter. If theſe plants are placed in a moderate ſtove in winter, they will thrive better than in greater heat; and in ſummer they ſhould have a large ſhare of air. With this management I have had the plants flower well, tho' they have not perfected ſeeds here.

COPAIFERA. The Balſam of Capevi-tree.

The Characters are,

It hath a flower conſiſting of five leaves, which expands in form of a Roſe; it hath five ſhort ſtamina. The germen is fixed in the center of the flower, which afterward becomes a pod, containing one or two ſeeds, which are ſurrounded with a pulp of a yellow colour.

We have but one Species of this plant, viz.

Copaifera (*Officinalis*) folio ſubrotundo, flore rubro. *The Balſam of Capevi, with a roundiſh leaf and a red flower.*

This tree grows near a village called Ayapel, in the province of Antiochi, in the Spaniſh Weſt-Indies, which is about ten days journey from Carthagena. There are great numbers of theſe trees in the woods about this village, which grow to the height of fifty or ſixty feet. Some of theſe trees do not yield any of the balſam; thoſe which do, are diſtinguiſhed by a ridge which runs along their trunks: theſe trees are wounded in their center, and they place Calabaſh ſhells, or ſome other veſſels to the wounded part, to receive the balſam, which will all flow out in a ſhort time. One of theſe trees will yield five or ſix gallons of balſam: but though theſe trees will thrive well after being tapped, yet they never afford any more balſam.

The ſeeds of this tree were brought from the country of their growth by Mr. Robert Millar, ſurgeon, who ſowed a part of them in Jamaica, which he informed me had ſucceeded very well: ſo that there were hopes to have had theſe trees propagated in great plenty in a few years, in ſome of the Engliſh colonies; but the ſlothfulneſs of the inhabitants ſuffered them to periſh, as they have the Cinnamon tree, and ſome other uſeful plants which have been carried thither by curious perſons.

There are not at preſent any of theſe trees in Europe, that I can learn; for thoſe ſeeds which were ſent over to England, were all deſtroyed by inſects in their paſſage, ſo that not one ſucceeded in the ſeveral places where they were ſown.

CORALLODENDRON. See Erythrina.

CORCHORUS. Lin. Gen. Pl. 596. Tourn. Inſt. 259. tab. 135. Jews Mallow.

The Characters are,

The flower hath five oblong blunt petals; it hath many hairy ſtamina, which are ſhorter than the petals; in the center is ſituated an oblong furrowed germen, which afterward becomes a cylindrical pod, having five cells, which are filled with angular-pointed ſeeds.

The Species are,

1. Corchorus (*Olitorius*) capſulis oblongis, ventricoſis, foliorum infimis ſerraturis ſetaceis, reflexis. Lin. Flor. Zeyl. 213. *Common Jews Mallow.*

2. Corchorus (*Æſtuans*) foliis cordatis, ſerratis, capſulis oblongis, ventricoſis, ſulcatis. *Jews Mallow, with heart-ſhaped ſawed leaves, and oblong, ſwelling, furrowed capſules.*

3. Corchorus (*Capſularis*) capſulis ſubrotundis, depreſſis, rugoſis. Flor. Zeyl. 214. *Jews Mallow with roundiſh depreſſed capſules, which are rough.*

4. Corchorus (*Tetragonus*) foliis ovato-cordatis crenatis, capſulis tetragonis, apicibus reflexis. *Jews Mallow with a yellow flower, and fruit like a Clove.*

5. Corchorus (*Linearis*) foliis lineari-lanceolatis, ſerratis, capſulis linearibus, compreſſis, bivalvibus. *American Jews Mallow, with narrower leaves and fruit.*

The firſt ſpecies, Rauwolf ſays, is ſown in great plenty about Aleppo, as a pot-herb, the Jews boiling the leaves of this plant to eat with their meat. This he ſuppoſes to be the Olus Judaicum of Avicenna, and the Corchorum of Pliny.

This plant grows in the Eaſt and Weſt Indies, from both which places I have ſeveral times received the ſeeds. In the Eaſt-Indies the herb is uſed in the ſame manner as in the Levant, as I have been informed; but I do not hear that it is uſed by the inhabitants of America.

This is an annual plant, which riſes about two feet high, dividing into ſeveral branches, garniſhed with leaves of different ſizes and forms; ſome are ſpear-ſhaped, others are oval, and ſome almoſt heart-ſhaped; they are of a deep green, and ſlightly indented on their edges, having near their baſe two briſtly ſegments, which are reflexed. They have very long ſlender foot-ſtalks, eſpecially thoſe which grow on the lower part of the branches. The flowers ſit cloſe on the oppoſite ſide of the branches to the leaves, coming out ſingly: they are compoſed of five ſmall yellow petals, and a great number of ſtamina ſurrounding the oblong germen, which is ſituated in the center of the flower, and afterward turns to a rough ſwelling capſule two inches long, ending in a point, opening in four cells, which are filled with angular greeniſh ſeeds.

The ſecond ſort grows naturally in ſeveral iſlands of the Weſt-Indies. This is alſo an annual plant, which riſes with a ſtrong herbaceous ſtalk two feet high, divided upward into two or three branches, which are garniſhed with heart-ſhaped leaves ſawed on their edges, ſtanding upon long foot-ſtalks; and between theſe are ſeveral leaves nearly of the ſame form, ſitting cloſe to the branches. The flowers come out ſingly on the ſide of the branches as the other, which are ſhaped like them, and are ſucceeded by longer ſwelling pods, which are rough, and have four longitudinal furrows; theſe open into four parts at the top, and contain four rows of angular ſeeds.

The third ſort grows naturally in both Indies. This is alſo an annual plant, which riſes with a ſlender herbaceous ſtalk about three feet high, ſending out ſeveral weak branches, which are garniſhed at each joint by one leaf of an oblong heart-ſhape, ending in a long acute point, and are ſawed on their edges, ſtanding upon ſhort foot-ſtalks, but have no ſmall leaves between. The flowers come out ſingly on the ſide of the branches, to which they ſit very cloſe; they are ſmaller than thoſe of the former ſorts, and are ſucceeded by ſhort roundiſh ſeed-veſſels, which are rough, and flatted at the top, having ſix cells filled with ſmall angular ſeeds.

The

The fourth ſort is alſo a native of both Indies. It is an annual plant, which riſes about two feet high, dividing into ſmall branches, garniſhed with oval heart-ſhaped leaves, crenated on their edges. The flowers of this are very ſmall, of a pale yellow, and are ſucceeded by ſwelling, rough, four-cornered ſeed-veſſels, about an inch long, flatted at the top, where there are four horns, which are reflexed; ſo that theſe have ſome reſemblance in ſhape to the Clove.

The ſeeds of the fifth ſort were ſent me from Carthagena, where the plant grows naturally. This riſes about three feet high, ſending out ſeveral weak ſide branches, garniſhed with long narrow leaves which are ſawed on their edges, ſitting cloſe to the branches. The flowers are ſmall, of a pale yellow, and come out on the ſide of the branches. Theſe are ſucceeded by very narrow compreſſed pods near two inches long, opening with two valves, and filled with ſmall angular ſeeds.

All theſe plants are too tender to thrive in England in the open air, therefore their ſeeds muſt be ſown on a hot-bed in the ſpring, and when the plants are come up and fit to remove, they ſhould be tranſplanted on a freſh hot-bed to bring the plants forward. After the plants are rooted in the new hot bed, they muſt have free air admitted to them every day, for they muſt not be drawn up weak; and when they have obtained ſtrength, they ſhould be tranſplanted each into a ſeparate pot, and plunged into a hot-bed, obſerving to ſhade them from the ſun till they have taken root; in June they ſhould be gradually inured to the open air: part of them may be ſhaken out of the pots, and planted in a warm border, where, if the ſeaſon proves warm, they will flower and perfect their ſeeds; but as theſe will ſometimes fail, ſo it will be proper to put one or two plants of each ſort into pots, which ſhould be placed in a glaſs-caſe, where they may be ſcreened from bad weather, and from theſe good ſeeds may always be obtained.

CORDIA. Plum. Nov. Gen. 13. tab. 14. Sebeſten.

The CHARACTERS are,

The flower hath one funnel-ſhaped petal, whoſe tube is the length of the empalement; the top is divided into four, five, or ſix parts. It hath five ſtamina, and in the center a roundiſh pointed germen, which afterward becomes a dry berry, which is globular and pointed, faſtened to the empalement, incloſing a furrowed nut with two cells.

The SPECIES are,

1. CORDIA (*Sebeſtina*) foliis oblongo ovatis, repandis, ſcabris. Lin. Sp. Pl. 190. *Cordia with oblong, oval, rough leaves, turning backward.*

2. CORDIA (*Myxa*) foliis ſubovatis ſerrato dentatis. Lin. Hort. Cliff. 63. *The cultivated Sebeſten.*

3. CORDIA (*Macrophylla*) foliis ovatis, integerrimis. Lin. Sp. Pl. 191. *Cordia with oval entire leaves.*

The firſt ſort grows naturally in ſeveral iſlands in the Weſt-Indies, where it riſes with many ſhrubby ſtalks eight or nine feet high, which are garniſhed toward the top with oblong, oval, rough leaves, ſtanding alternate on ſhort foot-ſtalks; they are of a deep green on their upper ſide. The flowers terminate the branches, growing in large cluſters upon branching foot-ſtalks, ſome ſuſtaining one, others two, and ſome have three flowers, which are large, funnel-ſhaped, having long tubes, which ſpread open at the top, where it is divided into five obtuſe ſegments: they are of a beautiful ſcarlet, ſo make a fine appearance.

The ſecond ſort is by moſt botaniſts believed to be the Myxa of Cæſalpinus, which is the true Sebeſten of the ſhops; the fruit of which was formerly uſed in medicine, but of late years has been ſeldom brought to England. This is called Aſſyrian Plum, from the country where it naturally grows; it riſes to the height of our common Plum-trees in its native country, but is very rare in Europe at preſent.

The third ſort was diſcovered by Father Plumier in ſome of the French Iſlands in America. This ſort grows to the height of eighteen or twenty feet, in the natural places where it is found wild. It hath winged leaves, which are large, entire, and ſmooth; but it hath not yet flowered in England, ſo I can give no farther account of it.

Theſe plants, being natives of warm countries, are too tender to live through the winter in England, unleſs they are preſerved in a ſtove. They are all propagated by ſeeds, which muſt be procured from the countries of their natural growth. Theſe muſt be ſown in ſmall pots which ſhould be plunged into a good hot-bed of tanners bark in the ſpring; if the ſeeds are freſh and good, the plants will begin to appear in ſix weeks or two months after. Theſe muſt be brought forward in the hot-bed, by being treated as other tender exotic plants, obſerving frequently to water them; in ſummer they ſhould be gradually hardened, and as they obtain ſtrength will become more hardy; but, during the three or four firſt winters, it will be proper to plunge them into the tan-bed in the ſtove; but when they begin to have woody ſtems, they may be placed on ſhelves in a dry ſtove, where, if they are kept in a moderate degree of heat, they may be preſerved very well (eſpecially the ſecond ſort) which is ſomewhat hardier than the others.

Theſe plants produce very fine flowers, eſpecially the firſt ſort, which has large tufts of ſcarlet flowers produced at the extremity of the branches, after the ſame manner as the Oleander or Roſe-bay; but theſe flowers are much larger, and of a finer colour.

A ſmall piece of the wood of this tree being put on a pan of lighted coals, will ſend forth a moſt agreeable odour, which will perfume a whole houſe.

COREOPSIS. Lin. Sp. Pl. 879. Tickſeed.

The CHARACTERS are,

The common empalement of the flower is double; the diſk of the flower is compoſed of many hermaphrodite florets, which are tubular; theſe have each five hairy ſtamina; in their center is ſituated a compreſſed germen with two horns, which afterward becomes a ſingle orbicular ſeed. The border or rays is compoſed of eight female florets, which are tongue-ſhaped, indented in five parts; theſe have no ſtamina, but a germen like the other, and are abortive.

The SPECIES are,

1. COREOPSIS (*Alternifolia*) foliis lanceolatis, ſerratis, alternis, petiolatis decurrentibus. Hort. Upſ. 270. *Tickſeed with ſpear-ſhaped ſawed leaves placed alernate, are decurrent, and have foot-ſtalks.*

2. COREOPSIS (*Lanceolata*) foliis lanceolatis, integerrimis. Lin. Sp. Pl. 908. *Tickſeed with ſpear-ſhaped leaves, which are entire.*

3. COREOPSIS (*Verticillata*) foliis decompoſito-pinnatis, linearibus. Lin. Sp. Plant. 907. *Tickſeed with decompound, winged, narrow leaves.*

4. COREOPSIS (*Tripteris*) foliis ſubternatis, integerrimis. Hort. Upſal. 269. *Tickſeed with leaves growing by threes, which are entire.*

The firſt ſort grows naturally in moſt parts of North America. This hath a perennial root and an annual ſtalk, which are ſtrong, herbaceous, and riſe to the height of eight or ten feet, garniſhed with ſpear-ſhaped leaves ſawed on their edges, which are from three to four inches long, and one broad in the middle, placed alternate on every ſide the ſtalks, with a border or wing running from one to the other the whole length of the ſtalk. The flowers grow at the top of the ſtalks, forming a ſort of corymbus, each foot-ſtalk ſuſtaining one, two, or three yellow flowers, ſhaped like Sun-flowers, but much ſmaller. It is a very hardy plant,

plant, and may be propagated in plenty, by parting of the roots in autumn, when the ſtalks begin to decay. It will thrive in almoſt every ſoil and ſituation.

The ſecond ſort is an annual plant. The ſeeds of this were brought from Carolina by Mr. Cateſby, in the year 1726. It hath an upright ſtalk, garniſhed with ſmooth, narrow, ſpear-ſhaped leaves, placed oppoſite, which are entire; from the wings of the leaves come out the foot-ſtalks of the flowers, which ſtand oppoſite, and are erect; the lower part of theſe have one or two pair of very narrow leaves, but the upper is naked, and terminated by one large yellow flower, whoſe border or rays are deeply cut into ſeveral ſegments; theſe are ſucceeded by flat winged ſeeds, which, when ripe, roll up; the naked foot-ſtalks of theſe flowers are more than a foot long. This muſt be ſown upon a gentle hot-bed in the ſpring, and, when the plants are fit to tranſplant, they ſhould be each planted into a ſeparate ſmall pot, and plunged into a freſh hot-bed, to bring them forward; and in June they ſhould be inured by degrees to the open air, and afterward ſome of them may be ſhaken out of the pots, and planted in a warm border, where, if the ſeaſon is good, they will flower in the middle of July, and ripen their ſeeds the beginning of September.

The third ſort hath a perennial root, which ſends up many ſtiff angular ſtalks, which riſe upward of three feet high, garniſhed at each joint with decompound winged leaves, ſtanding oppoſite; theſe are very narrow and entire. The branches alſo come out by pairs oppoſite, as do alſo the foot-ſtalks of the flowers, which are long, ſlender, and each terminated by a ſingle flower, of a bright yellow, the rays or border being oval and entire; the diſk or middle is of a dark purple colour. This grows naturally in Maryland and Philadelphia. It is propagated by parting of the roots, in the ſame manner as the firſt ſort, and delights in a light loamy earth, and a ſunny expoſure.

The fourth ſort hath a perennial root and an annual ſtalk. This grows naturally in many parts of North America. The ſtalks of this are ſtrong, round, and ſmooth, riſing ſix or ſeven feet high, and are garniſhed at each joint with ſome trifoliate leaves, which ſtand oppoſite. The flowers are produced in bunches at the top of the ſtalks, ſtanding upon long foot-ſtalks; they are of a pale yellow, with a dark purple diſk. This ſort is propagated by parting of the roots, in the ſame manner as the firſt, but requires a better ſoil and poſition.

CORIANDRUM. Lin. Gen. Plant. 318. Coriander.

The CHARACTERS are,

It is a plant with an umbellated flower; the proper empalement is divided into five parts; the rays of the principal umbel are difform; the hermaphrodite flowers, which form the diſk, have five equal heart-ſhaped petals; they have each five ſtamina. The germen, which is ſituated under the flower, afterward becomes a ſpherical fruit, divided into two parts, each having a hemiſpherical concave ſeed.

The SPECIES are,

1. CORIANDRUM (*Sativum*) fructibus globoſis. Hort. Cliff. 100. *Great Coriander.*

2. CORIANDRUM (*Teſticulatum*) fructibus didymis. Hort. Cliff. 100. *Smaller teſticulated Coriander.*

The firſt of theſe ſpecies is the moſt common kind, which is cultivated in the European gardens and fields for the ſeeds, which are uſed in medicine. The ſecond ſort is leſs common than the firſt, and is ſeldom found but in botanic gardens in theſe parts of Europe. Theſe plants grow naturally in the ſouth of France, in Spain and Italy; but the firſt ſort has been long cultivated in the gardens and fields, though at preſent there is not near ſo much of it ſown in England as was ſome years paſt.

They are propagated by ſowing their ſeeds in the autumn, in an open ſituation, in a bed of good freſh earth, and, when the plants are come up, they ſhould be hoed out to about four inches diſtance every way, clearing them from weeds; by which management theſe plants will grow ſtrong, and produce a greater quantity of good ſeeds. The firſt ſort was formerly cultivated in the gardens as a ſallad herb.

CORINDUM. See CARDIOSPERMUM.

CORIS. Lin. Gen. Pl. 216. We have no Engliſh name for this plant.

The CHARACTERS are,

The flower hath one ringent petal, whoſe tube is the length of the empalement, ſpread open at the top, where it is divided into five oblong ſegments; it hath five briſtly ſtamina. In the center is ſituated a round germen; the empalement afterward becomes a globular capſule, having five valves, incloſing ſeveral ſmall oval ſeeds.

We have but one SPECIES of this genus, viz.

CORIS (*Monſpelienſis.*) Hort. Cliff. 68. *Blue maritime Coris.*

There are two other varieties of this plant, one with a red, and the other a white flower; but theſe are only accidental varieties, ariſing from the ſame ſeeds.

Theſe plants grow wild about Montpelier, and in moſt places in the ſouth of France; they ſeldom grow above ſix inches high, and ſpread near the ſurface of the ground like Heath; and in June, when they are full of flowers, they make a very pretty appearance.

They may be propagated by ſowing of their ſeeds on a bed of freſh earth, and when the plants are about an inch high, they ſhould be tranſplanted, ſome of them into pots, that they may be ſheltered in winter, and the others into a warm border, where they will endure the cold of our ordinary winters very well, but in ſevere froſts they are generally deſtroyed: for which reaſon, it will be proper to have ſome plants of each ſort in pots, which may be put under a common hot-bed frame in winter, where they may be covered in froſty weather, but, when it is mild, they ſhould have a great ſhare of free air. Theſe plants rarely producing ripe ſeeds in England, ſhould be increaſed from ſlips and cuttings, which will take root if planted in Auguſt on a very gentle hot-bed, and ſhaded from the ſun, and duly watered.

CORISPERMUM. Lin. Gen. Plant. 12. Tickſeed.

The CHARACTERS are,

The flower hath no empalement; it hath two compreſſed incurved petals; it hath one, two, or three ſtamina, and a compreſſed pointed germen, which afterward becomes one oval compreſſed ſeed, with an acute border.

The SPECIES are,

1. CORISPERMUM (*Hyſopifolium*) floribus lateralibus. Hort. Upſal. 2. *Tickſeed with flowers on the ſide of the ſtalks.*

2. CORISPERMUM (*Squarroſum*) ſpicis ſquarroſis. Hort. Upſal. 3. *Tickſeed with rough ſpikes.*

Theſe plants are preſerved in botanic gardens for the ſake of variety, but as they have no beauty, ſo are ſeldom cultivated in other gardens.

The firſt ſort is an annual plant, which, if ſuffered to ſcatter its ſeeds, the ground will be plentifully ſtocked with the plants, which will require no other care but to prevent the weeds from over-growing them.

The ſecond ſort will not grow but in marſhy places, where there is ſtanding water; over the ſurface of which this plant will ſoon extend, when once it is eſtabliſhed.

CORK-TREE. See QUERCUS.

CORN-FLAG. See GLADIOLUS.

CORN-MARIGOLD. See CHRYSANTHEMUM.

CORN-SALLAD. See VALERIANA.

CORNUS. Lin. Gen. Pl. 139. The Cornelian Cherry

The CHARACTERS are,

It hath many flowers, which are included in one common, four-leaved, coloured involucrum; the flowers have four plain petals, and four erect stamina which are longer than the petals; the round germen situated below the empalement, afterward becomes an oval or roundish berry, inclosing a nut with two cells, having an oblong kernel.

The SPECIES are,

1. CORNUS (*Sanguinea*) arborea, cymis nudis. It. W-goth. Lin. Sp. Pl. 117. *Female Dogwood.*

2. CORNUS (*Mas*) arborea, umbellis involucrum æquantibus. Hort. Cliff. 38. *Male Cornel, or Cornelian Cherry-tree.*

3. CORNUS (*Florida*) arborea, involucro maximo, foliolis obversè cordatis. Hort. Cliff. 38. *Male Virginia Dogwood, with flowers collected into a corymbus.*

4. CORNUS (*Fœmina*) arborea, foliis lanceolatis, acutis, nervosis, floribus corymbosis terminalibus. *Female Virginia Dogwood, with a narrower leaf.*

5. CORNUS (*Amomum*) arborea foliis oblongo ovatis, nervosis, infernè albis, floribus corymbosis terminalibus. *Wild Tartarian Dogwood, with a white fruit.*

6. CORNUS (*Suecica*) herbacea ramis binis. Fl. Lapp. 55. *Low herbaceous Dogwood, called Dwarf Honeysuckle.*

The first of these trees is very common in the hedges in most parts of England, and is seldom preserved in gardens. This tree is called Virga sanguinea, from the young shoots being of a fine red colour in winter.

The second sort was formerly very common in the English gardens, where it was propagated for its fruit, which by many people was preserved to make tarts. Of this there are two or three varieties, which differ in the colour of their fruit; but that with the red fruit is the most common in England.

As the fruit of this tree is not at present much esteemed, the nursery-men about London propagate it only as a flowering shrub, and is by some people valued for coming so early to flower; for, if the season is mild, the flowers will appear by the beginning of February; and though there is no great beauty in the flowers, yet as they are generally produced in plenty, at a season when few other flowers appear, a few plants of them may be admitted for variety. The fruit of this tree is seldom ripe before September.

The third sort is found growing naturally in all the northern parts of America. This will grow to the same height with our common female Dogberry, and make a much better appearance. It is now very common in the nerseries, where it is known by the name of Virginia Dogwood. This is well garnished with leaves, which are larger than either of the other sorts, but is not so plentiful of flowers, nor do the plants produce berries in England, though the shrubs are as hardy as the other.

The sixth sort grows upon Cheviot-hills in Northumberland, and also upon the Alps, and other mountainous places in the northern countries, but is very difficult to preserve in gardens. The only method is, to remove the plants from the places of their natural growth, with good balls of earth to their roots, and plant them in a moist shady situation, where they are not annoyed by the roots of other plants. In such a situation they may be preserved two or three years, but it rarely happens that they will continue longer.

All the sorts of Dogwood may be propagated by their seeds, which, if sown in autumn soon after they are ripe, will most of them come up the following spring; but, if the seeds are not sown in autumn, they will lie a year in the ground before the plants will appear; and, when the season proves dry, they will sometimes remain two years in the ground: therefore the place should not be disturbed where these seeds are sown under two years, if the plants should not come up sooner. When the plants are come up, they should be duly watered in dry weather, and kept clean from weeds, and the autumn following they may be removed and planted in beds in the nursery, where they may remain two years, by which time they will be fit to transplant where they are to remain for good.

They are also propagated by suckers, and laying down of the branches. Most of the sorts produce plenty of suckers, especially when they are planted on a moist soil, which may be taken off from the old plants in autumn, and planted into a nursery for a year or two, and then may be transplanted into the places where they are to remain; but those plants which are propagated by suckers, rarely have so good roots as those which are propagated by layers.

CORNUTIA. Plum. Nov. Gen. 17. Lin. Gen. Pl. 684.

The CHARACTERS are,

The flower has one petal, having a cylindrical tube, which is divided into four parts at the top; it hath four stamina, two of these are longer than the tube, the other are shorter; in the center is situated the roundish germen, which afterward becomes a globular berry, sitting upon the empalement, inclosing several kidney-shaped seeds.

We have but one SPECIES of this genus, viz.

CORNUTIA (*Pyramidata.*) Hort. Cliff. 319. *Cornutia with a blue pyramidal flower and hoary leaves.*

This plant is found in plenty in several of the islands in the West-Indies, also at Campeachy, and at La Vera Cruz. It grows to the height of ten or twelve feet, with rude branches; the leaves are placed opposite. The flowers are produced in spikes at the end of the branches, which are of a fine blue colour; these usually appear in autumn, and sometimes will remain in beauty for two months or more.

It is propagated either by seeds or cuttings. The seeds should be sown early in the spring on a hot-bed, and, when the plants are come up, they should be transplanted each into a separate halfpenny pot, and plunged into a hot-bed of tanners bark, observing to shade them until they have taken root; when the plants have filled these pots with their roots, they should be shifted into others of a larger size, and plunged into a hot-bed again, where they should be continued till October, when they must be removed into the bark-stove, and plunged into the tan, for otherwise it will be very difficult to preserve them through the winter; but a moderate share of heat will agree better with them than a very warm stove. The third year from seed these plants will flower, when they will make a very fine appearance in the stove, but they never perfect their seeds in England.

The cuttings should be planted into pots, and plunged into a bark-bed, observing to shade and water them; they will take root, and must be afterwards treated as the seedling plants.

CORONA IMPERIALIS. See FRITILLARIA.

CORONA SOLIS. See HELIANTHUS.

CORONILLA. Jointed-podded Colutea.

The CHARACTERS are,

It hath a butterfly flower with nine stamina, which are united, and one standing single, terminated by small summits; in the center is situated an oblong taper germen, which afterward becomes a taper-jointed pod, inclosing oblong seeds.

The SPECIES are,

1. CORONILLA (*Glauca*) fruticosa, foliis septenis stipulis lanceolatis. Lin. Sp. 1047. *Shrubby maritime Coronilla with a sea-green leaf.*

2. CORONILLA (*Argentea*) fruticosa foliolis undenis, extimo majore. Lin. Sp. Plant. 743. *Shrubby silvery Coronilla of Crete.*

3. CORONILLA (*Valentina*) fruticosa stipulis subrotundis. Lin. Sp. Plant. 742. *Shrubby Spanish Coronilla.*

4. CORONILLA (*Hispanica*) fruticosa enneaphylla, foliolis emarginatis, stipulis majoribus subrotundis. *Coronilla with thicker pods and seeds.*

5. CORONILLA (*Minima*) foliolis plurimis, ovatis, caule suffruticoso declinato, pedunculis longioribus. *Smallest Coronilla.*

6. CORONILLA (*Varia*) herbacea, leguminibus erectis, teretibus, torosis, numerosis, foliis glabris. Hort. Cliff. 363. *Herbaceous Coronilla with a various coloured flower.*

7. CORONILLA (*Cretica*) herbacea, leguminibus quinis, erectis, teretibus, articulatis. Prod. Leyd. 387. *Herbaceous Coronilla of Crete with a small purplish flower.*

The first sort is an humble shrub, which seldom rises more than two feet high, with a shrubby branching stalk, garnished closely with winged leaves, each being generally composed of five pair of small leaves, or lobes, terminated by an odd one; these are narrow at their base, and broad at the end, and of a sea-green colour. The flowers are produced on slender foot-stalks from the wings of the leaves, on the upper part of the branches, several standing together in a roundish bunch; they are of the butterfly, or Pea-bloom kind, of a bright yellow colour, having a very strong odour, which to some persons are agreeable, but to others the contrary.

This plant is propagated by sowing the seeds in the spring, either upon a gentle hot-bed, or on a warm border of light earth; when the plants are come up about two inches high, they should be transplanted either into pots, or into a bed of fresh earth, at about four or five inches distance every way, where they may remain until they have obtained strength enough to plant out for good, which should be either into pots filled with good fresh earth, or in a warm situated border; in which, if the winter is not too severe, they will abide very well, provided they are in a dry soil.

The second sort is a shrub of the same size with the first, from which it differs in the number of small leaves, or lobes, on each midrib, these having nine or eleven; they are of a silver colour, but the flowers and pods are the same as the former, and the plants require the same treatment.

The third sort is a shrubby plant, which will rise five feet high, but the stalks are weak, so must be supported; these are garnished with many winged leaves, composed of small oval lobes, placed along the midrib by pairs, ending in an odd one. The flowers stand upon long slender foot-stalks, which arise from the end of the branches; they are yellow, and grow together in close bunches. This sort, if sheltered from hard frosts, will flower all the winter.

This is a perennial plant, which is propagated by seeds; they may be sown on a bed of light earth in April, and, when the plants are fit to transplant, they should be planted in a warm border near a wall, a pale, or Reed hedge, observing to shade them from the sun till they have taken fresh root; after they are well rooted, they will require no other culture but to keep them clean from weeds, and to support their branches, otherwise they will fall on the ground; the next year they will flower, and if they are on a dry soil, and in a warm situation, they will continue many years.

The fourth sort is nearly like the first, but hath fewer pinnæ on each midrib; the flowers are larger, and have little scent; the pods and seeds are much larger, and the plants are not quite so hardy. It requires the same treatment as the first; but in winter the plants should be sheltered, otherwise hard frosts will destroy them.

The fifth sort is a low trailing plant with shrubby stalks which spread near the ground, and are garnished with winged leaves, composed of many pair of small leaves, placed along the midrib, terminated by an odd one; these are oval, and of a bright green: the flowers stand upon long foot-stalks in close bunches; they are yellow, and without scent. This is propagated by seeds, in the same manner as the other sorts, and requires the same treatment.

The sixth sort dies to the ground every winter, but rises again the succeeding spring; the shoots rise to the height of five or six feet, where they have support, otherwise they trail on the ground, and are garnished with winged leaves, composed of several oblong small pinnæ, which are sometimes placed by pairs, and at other times are alternate, ending in a single one; they are of a deep green. The flowers come out on long foot-stalks from the wings of the leaves, many growing together in round bunches; they are variable, from a deep to a light purple, mixed with white, and are succeeded by slender pods, from two to three inches long, standing erect. The roots of this plant creep very far under ground, by which the plant increases greatly, which, when permitted to remain unmoved for two or three years, will spread and overbear whatever plants grow near it; for which reason the roots should be confined, and should be planted at a distance from any other plants: they will grow in almost any soil and situation, but thrive best in a warm sunny exposure, in which the flowers will also be much fairer, and in greater quantities. This plant was formerly cultivated to feed cattle.

The seventh sort hath an herbaceous stalk, which rises two feet high, garnished with winged leaves, composed of six pair of small leaves placed along the midrib, which is terminated by an odd one; these are larger than those of the sixth sort, and broader at the top. The foot-stalks of the flowers come out from the sides of the stalks, but are shorter than those of the sixth sort, and sustain smaller heads of flowers, which are succeeded by taper-jointed pods near two inches long.

This is an annual plant, which grows naturally in the Archipelago, from whence Tournefort sent the seeds to the royal garden at Paris. The seeds of this sort should be sown on a bed of light earth in the spring, where the plants are designed to remain, and, when the plants come up, they should be thinned where they are too close, and afterwards kept clean from weeds, which is all the culture they will require: in June they will flower, and the seeds ripen in autumn.

CORONOPUS. See PLANTAGO.

CORTUSA. Lin. Gen. Plant. 181. Bear's-ear Sanicle.

The CHARACTERS are,

The flower hath one wheel-shaped petal, cut into five parts at the brim, having five prominent tubercles at the base; it hath five short obtuse stamina; in the center is fixed an oval germen, which afterward becomes an oval, oblong, pointed capsule, having two longitudinal furrows, and one cell, opening with two valves, and filled with small oblong seeds.

The SPECIES are,

1. CORTUSA (*Mattholi*) calycibus corollâ brevioribus. Lin. Sp. Plant. 144. *Bear's-ear Sanicle with an empalement shorter than the petal.*

2. CORTUSA (*Gmelini*) calycibus corollum excedentibus. Amœn. Acad. 2. p. 340. *Bear's-ear Sanicle with an empalement longer than the petal.*

The first sort grows naturally on the Alps, the mountains in Austria, and in Siberia. This plant sends out many oblong smooth leaves, which are a little indented on the edges, and form a sort of head like the Auricula. The foot-stalks of the flowers come out in the center of the leaves; these rise about four inches high, and support an umbel of flowers, each sitting on a slender, separate, short footstalk; they are of a flesh colour, and spread open like those of Auricula: this plant is with great difficulty kept in a garden; the only method by which I could ever preserve it, has been by planting the plants in pots, covering the earth with moss, and placing them in a shady situation, where they were duly watered in dry weather; in this place they constantly remained both summer and winter, for the cold will not destroy them; the earth for this plant should be light, and not rich. As this very rarely produces any seeds in England, the only method to propagate it is, by parting the roots in the same manner as is practised for Auriculas; the best time for this is about Michaelmas, soon after which the leaves decay.

The second sort is very like the first, but the flowers are much less, and their empalements are larger; this grows naturally in Siberia, but is with great difficulty kept in a garden.

CORYLUS. Lin. Gen. Pl. 953. The Hazel, or Nut-tree.

The CHARACTERS are,

It hath male and female flowers growing at remote distances on the same tree. The male flowers are produced in long scaly katkins, having no petals, but eight short stamina fastened to the side of the scale; the female flowers are included in the future bud, sitting close to the branches; these have no petals, but a small round germen occupies the center, which afterward becomes an oval nut, shaved at the base, and compressed at the top, ending in a point.

The SPECIES are,

1. CORYLUS (*Avellana*) stipulis ovatis obtusis. Hort. Cliff. 448. *Wild Hazel Nut.*

2. CORYLUS (*Maxima*) stipulis oblongis, obtusis, ramis erectioribus. *The Filbert.*

3. CORYLUS (*Colurna*) stipulis linearibus acutis. Hort. Cliff. 448. *Byzantine Nut.*

The first of these trees is common in many woods in England, from whence the fruit is gathered in plenty, and brought to the London markets by the country people. This is seldom planted in gardens; it delights to grow on a moist strong soil, and may be plentifully increased by suckers from the old plants, or by laying down their branches, which, in one year's time, will take sufficient root for transplanting; and these will be much handsomer and better rooted plants than suckers, and will greatly outgrow them, especially while young.

The second sort is by many supposed to be only a seminal variety from the first, which hath been improved by culture; but this is very doubtful, for I have several times propagated both from the nuts, but never have found them vary from one to the other, though they have altered in the size and colour of their fruit, from the sorts which were sown; but as the shrubs of this grow more erect than those of the other, and the stipulæ are different in their shape, so I have enumerated it as a distinct sort; of this there are the red and white Filberts, both which are so well known as to need no description.

The third sort grows naturally near Constantinople; the nuts of this are large, roundish, and in shape like those of the common Hazel, but are more than twice their size; the cups in which the nuts grow are very large, so as almost to cover the nut and is deeply cut at the brim. This sort is not common in England, but I take those large nuts, which are annually imported from Barcelona in Spain, to be of the same kind, the nuts being so like as not to be distinguished when out of their cups; and those of the Spanish sort come over naked, so I cannot with certainty say how they essentially differ.

All these sorts may be propagated by layers, or sowing their nuts in February, which, in order to preserve them good, should be kept in sand in a moist cellar, where the vermin cannot come at them to destroy them; nor should the external air be excluded from them, which would occasion their growing mouldy.

COSTUS. Lin. Gen. Plant. 3.

The CHARACTERS are,

It hath a simple spadix and spatha, with a small empalement, divided into three parts, sitting on the germen; the flower hath three concave petals, which are erect and equal, with a large oblong nectarium of one leaf, having two lips; the upper is shorter, and turns to a stamen, this is fastened to the upper lip of the nectarium; the germen is situated within the receptacle of the flower, and afterward becomes a roundish capsule with three cells, containing several triangular seeds.

We have but one SPECIES of this genus, viz.

COSTUS (*Arabicus.*) Hort. Cliff. 2. *Arabian Costus.*

This hath a fleshy jointed root like that of Ginger, which propagates under the surface as that doth, from which arise many round, taper, herbaceous stalks, garnished with oblong smooth leaves, embracing them like those of the Reed; these stalks rise two feet high, and out of the center, the club or head of flowers is produced, which is two inches long, the thickness of a man's finger, and blunt at the top, composed of several leafy scales, out of which the flowers come; these have but one thin white petal, which is of short duration, seldom continuing longer than one day before it fades, and is never succeeded by seeds in this country. The time of its flowering is uncertain, for sometimes it flowers late in the winter, and at other times it has flowered in summer, so is not constant to any season in England.

This is propagated by parting of the roots; the best time for doing this is in the spring, before the roots put out new stalks; the roots must not be divided too small because that will prevent their flowering; they should be planted in pots, and plunged into the tan-bed in the stove, where they should constantly remain, and may be treated in the same manner as the Ginger.

The roots of this plant were formerly imported from India, and were much used in medicine; but of late years they have not been regarded, the roots of Ginger being generally substituted for these.

COTINUS. See RHUS.

COTONEA MALUS. See CYDONIA.

COTONEASTER. See MESPILUS.

COTULA. Lin. Gen. Plant. 868. Mayweed.

The CHARACTERS are,

It hath a flower composed of hermaphrodite florets in the disk, and female half florets which form the rays, included in one common convex empalement. The hermaphrodite florets are tubular, cut into four unequal parts at the top, and have four small stamina, with a germen in the center, which becomes one small, oval, angular seed; the female half florets have an oval compressed germen, but have no stamina, and are succeeded by single heart-shaped seeds, plain on one side, and convex on the other.

The SPECIES are,

1. COTULA (*Anthemoides*) foliis pinnato-multifidis, corollis radio destitutis. Hort. Cliff. 417. *Yellow Chamomile with heads having no rays.*

2 COTULA (*Turbinata*) receptaculis subtus inflatis, turbinatis. Hort. Cliff. 417. *African Mayweed with an elegant empalement.*

3. COTULA (*Coronopifolia*) foliis lanceolato-linearibus, amplexicaulibus pinnatifidis. Hort. Cliff. 417. *Smaller foreign Corn Marigold without rays, having the appearance of naked Chamomile.*

The first sort grows naturally in Spain, Italy, and the Archipelago. This is an annual plant, which rises with a branching stalk half a foot high, garnished with leaves which are finely divided like those of Chamomile; the flowers

flowers are produced singly at the end of the branches, which are very like those of the naked Chamomile, but the heads rise higher in the middle like a pyramid. If the seeds of this sort are permitted to scatter, the plants will come up in the spring, and require no other care but to keep them clean from weeds, and thin them where they are too close.

The second sort grows naturally at the Cape of Good Hope; this is an annual plant, sending out many branching stalks from the root, which spread on the ground, garnished with very fine divided leaves, covered with a lanugo, or cotton; the flowers are produced singly upon long foot-stalks, arising from the side of the branches; these have a narrow border of white rays, with a pale yellow disk: this sort must be raised on a moderate hot-bed in the spring, and when the plants have obtained strength, they may be transplanted into a warm border, where they will ripen their seeds very well.

The third sort is an annual plant, which sends out trailing stalks about six inches long, garnished with succulent leaves, in shape like those of Buckshorn Plantain; the flowers grow from the divisions of the stalks upon short weak foot stalks, being destitute of rays; they are of a sulphur colour. If the seeds of this sort are sown on a warm border, where the plants are to remain, they will require no other culture but to keep them clean from weeds; the flowers of the two last sorts stand erect when they first appear, but so soon as the florets are impregnated, and their colour changes, the foot-stalks become very flaccid toward the top, and the flowers hang downward; but when the seeds are ripe, the foot-stalks become stiff, and the heads stand erect for the wind to disperse the seeds.

COTYLEDON. Lin. Gen. Plant. 512. Navelwort.

The CHARACTERS are,

The flower hath one petal, which is funnel-shaped, cut into five parts at the brim; it hath five germina, which have each a squamous concave nectarium at the base, and ten erect stamina; the germen afterward becomes so many oblong swelling capsules, opening longitudinally with one valve, and filled with small seeds.

The SPECIES are,

1. COTYLEDON (*Umbilicus*) foliis cucullatis, serrato dentatis, alternis, caule ramoso, floribus erectis. Lin. Sp. Plant. 429. *Greater Navelwort, or Umbilicus Veneris.*

2. COTYLEDON (*Spinosa*) foliis oblongis spinoso mucronatis, caule spicato. Lin. Sp. Plant. 429. *Navelwort with oblong pointed leaves, ending with a spine, and a spiked stalk.*

3. COTYLEDON (*Serrata*) foliis ovalibus, crenatis, caule spicato. Lin. Sp. Plant. 429. *Navelwort of Crete with an oblong fringed leaf.*

4. COTYLEDON (*Hemisphaerica*) foliis semiglobosis. Hort. Cliff. 176. *Navelwort with semiglobular leaves; or Navelwort of the Cape, with a semiglobular leaf.*

5. COTYLEDON (*Orbiculata*) foliis subrotundis, planis integerrimis. Hort. Cliff. 276. *Shrubby African hoary Navelwort, with roundish leaves.*

6. COTYLEDON (*Ramosissima*) caule ramosissimo, foliis rotundis, planis, marginibus purpureis. *Navelwort with a very branching stalk, and round, plain, hoary leaves, with purple edges.*

7. COTYLEDON (*Arborescens*) caule ramoso, succulento, foliis obversè ovatis, emarginatis, marginibus purpureis. *Navelwort with a branching succulent stalk, and obverse oval leaves, which are indented at the top, and have purple borders.*

8. COTYLEDON (*Ovata*) caule ramoso, succulento, foliis ovatis planis, acuminatis oppositis semiamplexicaulibus. *Navelwort with a succulent branching stalk, and oval, plain, pointed leaves, growing opposite, and half embracing the stalk.*

9. COTYLEDON (*Spuria*) caule ramoso, foliis longis, succulentissimis, supernè sulcatis infernè convexis, floribus cernuis. *Shrubby African Navelwort, with long narrow leaves, and a yellowish flower.*

10. COTYLEDON (*Laciniata*) foliis laciniatis, floribus quadrifidis. Hort. Cliff. 175. *Navelwort with cut leaves, and four-pointed flowers.*

The first sort, which is that used in medicine, grows upon old walls and buildings in divers parts of England, particularly in Shropshire and Somersetshire, in both which counties it greatly abounds upon old buildings, and on rocky places, but is not often found wild near London, nor often cultivated in gardens: this hath many round succulent leaves, whose foot-stalks are placed almost in the center, so as to resemble a target. These are alternately sawed on their edges; the upper surface of the leaves are hollowed in the middle, where the foot-stalks are joined on the lower side, so as to resemble a navel, from whence the plant was titled Navelwort. From between the leaves arise the foot-stalks of the flowers, which in some places grow near three feet high, and in others no more than six inches, their lower part being garnished with leaves, and their upper part with flowers, which stand close to the side of the branches, and grow erect; they are of a whitish yellow colour. It requires a dry rubbishy soil, and to have a shady position: it is a biennial plant, so that after it hath perfected seed the plant decays; but if the seeds are scattered on walls and old buildings as soon as it is ripe, or if the seeds are permitted to fall on such places, the plant will come up, and thrive much better than when they are sown in the ground; and when once the plants are established upon an old wall or building, they will sow their seeds, and maintain their place, better than when cultivated with more care.

The second sort grows naturally in Siberia. It is a low plant, in shape like the first, but the leaves are longer and terminate in soft spines. The flower-stalks rise eight inches high, and support four or five whitish flowers, which are cut at the brim into five parts. This sort requires a very shady situation, for if it is exposed to the sun in summer the plants will soon decay. It is propagated like the other, and requires a pretty strong soil.

The third sort grows naturally in the Levant; this hath a fibrous root, from which is produced a single, upright, succulent stalk, garnished with oblong, thick, succulent leaves, placed alternate, which are sawed on their edges. The upper part of the stalk is garnished with purplish flowers, growing in a loose spike, two or three being joined on the same foot-stalk, which is very short. It is a biennial plant, which decays soon after the seeds are ripe. If this sort is sown upon a wall, it will thrive better than in the ground, and be less liable to suffer by frost; so that where the seeds scatter themselves in such situations, the plants thrive better than when they are cultivated.

The fourth sort grows naturally at the Cape of Good Hope. This hath a thick succulent stalk, which rarely rises above a span high, dividing into many branches, garnished with thick, short, succulent leaves, which are very convex on their under side, but plain on their upper, not more than half an inch long, and a quarter broad, of a grayish colour, spotted over with small green spots, and sit close to the branches. The foot-stalks of the flower rise from the top of the branches, and are six inches long, naked, and support five or six flowers; which come out alternate from the side, sitting very close to the stalks; they are tubular, and cut into five parts at the top; they are greenish with purple tips.

The fifth sort grows naturally upon dry gravelly spots at the Cape of Good Hope. It hath a thick succulent stalk, which

which by age becomes ligneous, and rises three or four feet high, sending out crooked branches, which grow irregular, garnished with thick, fleshy, succulent leaves about two inches long, and near as wide toward the top; they are narrow at the base, and rounded at the top, of a sea-green colour, with a purple edge, which is frequently irregularly indented. The flowers grow upon thick succulent foot-stalks, which arise from the end of the branches, and are near a foot long, naked, and supporting eight or ten flowers, growing in an irregular umbel at the top; these are of a pale yellow colour, having long tubes which hang downward, cut into five parts at the brim, which turn backward, the stamina and style being longer than the tube of the flower, hanging downward.

The sixth sort is also a native of the Cape of Good Hope; this hath a short, thick, succulent stalk, which rarely rises more than a foot high, branching out on every side, so as to spread over the pots in which they are planted. These become woody by age, and are closely garnished with thick round leaves, of a grayish colour with purple borders. They are plain on their upper side, but convex on their under, very fleshy, of an herbaceous colour within, and full of moisture. This sort hath not flowered in England, so far as I can learn. It is undoubtedly a different sort from the former, although they have been supposed to be the same by some writers.

The seventh sort is somewhat like the sixth, but the stalks rise higher; the leaves are much larger, and shaped more like those of the fifth, but are spotted on their upper side with great numbers of dark green spots; they have a deep border of purple on their edges, and sit close to the branches. This hath not as yet flowered in England.

The eighth sort hath been of late years introduced from the Cape of Good Hope, where it grows naturally. This rises with a succulent stalk near three feet high, which divides into many branches, growing erect, garnished with oval succulent leaves, placed by pairs opposite; they are of a lively green, and end in points, and half embrace the stalks with their base. This sort hath not as yet produced any flowers in England.

The ninth sort grows on rocky places at the Cape of Good Hope. This hath a short, greenish, succulent stalk, which seldom rises more than a span high, dividing into several irregular branches, garnished with thick succulent leaves, four inches long, and half an inch broad, and as much in thickness, having a broad concave furrow on their upper side, running almost their whole length, and are convex on their upper side, of a bright green with a purple tip. The foot-stalks of the flowers are produced at the end of the branches, which rise a foot high having here and there an oblong pointed leaf, growing on their side. The flowers stand upon short foot stalks, which branch out from the principal stem; these are yellow, having pretty long tubes, which are cut at the top into five parts, and are reflexed backward. The flowers of this sort hang downward, and the stamina are longer than the tube of the flower; the reflexed parts of the petal are tipped with purple.

The tenth sort grows naturally in the warm parts of Africa, so is much more tender than either of the other sorts; this rises with an upright stem about a foot high, which is jointed and succulent, garnished with broad leaves, which are deeply cut on their edges; they are of a grayish colour, placed opposite, and almost embrace the stalks with their base. The foot-stalks of the flowers arise from the end of the branches, which are about six inches long, sustaining seven or eight small flowers, of a deep yellow colour, which are divided into four parts almost to the bottom. The stamina of these flowers are not longer than the short tube.

This sort requires a warm stove to preserve it through the winter in England, nor should it be exposed abroad in summer; for if it receives much wet, the stalks are very subject to rot, so that it should constantly remain either in the stove, or in summer be placed in an airy glass-case, with other tender succulent plants, where they may have free air in warm weather, and be screened from cold and wet; but in autumn they must be removed into the stove, where they should be kept in a moderate temperature of warmth. This is propagated by cuttings, which should be taken off in summer, and planted into small pots, and then plunged into a moderate hot-bed, and when they have taken root they should be removed into the stove.

The African kinds are all of them propagated by planting cuttings in any of the summer months, which should be laid in a dry place for a fortnight or more after they are taken from the plant, before they are planted; for these abound with juice, which will certainly rot the cuttings, if they are not suffered to lie out of the ground, so long as that the wounded part may heal over, and the great redundancy of sap evaporate. The soil in which these plants thrive best, is one third fresh light earth from a pasture, one third sand, and the other third part lime rubbish; these should be well mixed, and laid in a heap six or eight months before it is used, turning it over five or six times, that the parts may the better incorporate; and before it is used it will be proper to pass it through a screen, to separate the large stones, clods, &c. therefrom.

In about a month or six weeks after planting these cuttings will be rooted, when they must be inured to bear the open air by degrees, first drawing the pots out of the tan, and setting them on the top, then raise the glasses very high in the day-time; and in about three weeks after remove the pots into a green-house, and there harden them for another week; after which they may be exposed to the open air in a well defended place, observing not to set them into a place too much exposed to the sun, until they have been inured to the open air for some time.

In this place the plants may remain until the beginning of October, at which time you should remove them into the conservatory, placing them as near the windows as possible at first, letting them have as much free open air as the season will permit, by keeping the windows open whenever the weather is good: and now you must begin to abate your waterings, giving it to them sparingly; but you should not suffer their leaves to shrink for want of moisture, which is another extreme some people run into for want of a little observation.

The best method to treat these plants is, to place them in an open, airy, dry glass-case, among Ficoideses and African Housleeks, where they may enjoy as much of the sun-shine as possible, and have a free, dry, open air; for if they are placed in a common green-house amongst shrubby plants, which perspire freely, it will fill the house with a damp air, which these succulent plants are apt to imbibe; and thereby becoming too replete with moisture, often cast their leaves.

COURBARIL. See Hymenæa.

COWSLIP. See Primula.

CRAB-TREE. See Malus.

CRAMBE. Lin. Gen. Pl. 739. Sea Cabbage.

The Characters are,

The flower hath four petals, placed in form of a cross; it hath six stamina, two of which are the length of the empalement, the other four are longer. The petals have honey glands on their inside, which are longer than the stamina. It hath an oblong germen, which afterward becomes a round dry capsule with one cell, inclosing one roundish seed.

The

The Species are,

1. Crambe (*Maritima*) foliis caule glabris. Fl. Suec. 570. *Sea Cabbage with smooth stalks and leaves.*

2. Crambe (*Suecica*) foliis profundè laciniatis, caule erecto, ramoso. *Sea Cabbage with leaves deeply cut, and an upright branching stalk.*

3. Crambe (*Orientalis*) foliis scabris, caule glabro. Lin. Sp. Plant. 671. *Sea Cabbage with rough leaves, and a smooth stalk.*

4. Crambe (*Hispanica*) foliis cauleque scabris. Hort. Upsal. 193. *Sea Cabbage with rough stalks and leaves.*

The first sort sends out many broad leaves, which are jagged and furbelowed on their sides, of a grayish colour, spreading on the ground; between these arise a thick smooth foot-stalk about one foot high, which spreads out into many branches, having at each joint one leaf, of the same form of those below, but much less: these foot-stalks subdivide again into many smaller, which are garnished with white flowers, growing in a loose spike, composed of four concave petals placed in form of a cross; these are succeeded by round dry seed-vessels, about the size of large Pease, having a single seed in each. The roots of this sort creep under ground, whereby it propagates very fast.

The seeds of the second sort were sent me from Petersburgh. This hath a perennial root, which sends out many oblong smooth leaves, which are pointed, and irregularly cut on their sides into acute segments almost to the midrib; they are very smooth, and of a sea-green colour; between these arise the stalk, which grows three feet high, garnished below by oblong pointed leaves, which are acutely indented on their edges. The stalks branch out into many smaller, and subdivide again into less, which are garnished with loose spikes of white flowers like those of the first sort, which are succeeded by seeds of the same form. This differs greatly from the first, in the shape of its leaves, which are longer, ending in points, and the segments do the same, whereas those of the other are blunt, and not half so deeply cut. The stalks rise more than twice the height of the first, branch out more, and the branches grow more erect. And these differences are constant, where the plants grow in the same soil.

The third sort grows naturally in the East. This hath a perennial root, from which arise many leaves in the spring, which are alternately divided to the midrib, and these divisions are again alternately cut on their edges into many points, so that they have the appearance of winged leaves, and are of a grayish colour. The stalks rise about two feet high, and divide into many branches, which are terminated by loose panicles of small white flowers, whose petals are placed in form of a cross, which are succeeded by small round capsules, each containing a single seed.

The fourth sort is an annual plant, which grows naturally in Spain and Italy. This rises with a very branching stalk near three feet high, garnished with roundish heart-shaped leaves, indented on their edges, standing upon long foot-stalks; the branches subdivide into many slender ones, which end in long loose spikes of small white flowers, and are succeeded by small, round, dry seed-vessels, which contain a single seed in each. The leaves and stalks of this sort are rough.

The first of these species is found wild upon sea-shores in divers parts of England, but particularly in Sussex and Dorsetshire in great plenty, where the inhabitants gather it in the spring to eat, preferring it to any of the Cabbage kind, as it generally grows upon the gravelly shore, where the tide flows over it, so the inhabitants observe where the gravel is thrust up by the roots of this plant; they open the gravel, and cut the shoots before they come out, and are exposed to the open air, whereby the shoots appear as if they were blanched; and when they are cut so young, they are very tender and sweet, but if they are suffered to grow till they are green, they become tough and bitter.

This plant may be propagated in a garden, by sowing the seed soon after it is ripe in a sandy or gravelly soil, where it will thrive exceedingly, and increase greatly by its creeping roots, which will soon overspread a large spot of ground, if encouraged, but the heads will not be fit to cut until the plants have had one year's growth; and in order to have it good, the bed in which the plants grow should at Michaelmas be covered over with sand or gravel about four or five inches thick, which will allow a proper depth for the shoots to be cut before they appear above ground; and if this is repeated every autumn, in the same manner as is practised in earthing of Asparagus beds, the plants will require no other culture,

The other sorts are only preserved in curious gardens of plants for variety, but are not of any use or beauty. The perennial sort may be propagated in the same manner as the first.

CRANE's-BILL. See Geranium.

CRANIOLARIA. Lin. Gen. Pl. 670. Martynia. Houst. Gen.

The Characters are,

The flower hath a permanent empalement, composed of four short narrow leaves and a swollen head, which is cut longitudinally on the side. The flower hath one petal, which is unequal, having a very long tube, whose brim is divided into two lips. It hath four stamina, two of which are the length of the tube, and two are shorter. At the bottom of the tube is situated an oval germen, which afterward becomes an oval leathery fruit, pointed at both ends, opening with two valves, inclosing a depressed woody nut, pointed at both ends and recurved, having two or three furrows, so as to resemble a skull, opening in two parts.

The Species are,

1. Craniolaria (*Annua*) foliis cordatis, angulatis lobatis. Lin. Sp. Pl. 618. *Craniolaria with heart-shaped angular leaves.*

2. Craniolaria (*Fruticosa*) foliis lanceolatis, dentatis. Lin. Sp. Pl. 618. *Craniolaria with spear-shaped indented leaves.*

The first sort was discovered in the neighbourhood of Carthagena in New Spain. This is an annual plant, which rises with a branching stalk about two feet high; the branches come out opposite, which are hairy and viscous; the leaves also are placed opposite upon very long foot-stalks; these are of different shapes, some of them are divided into five lobes, others into three, and some are almost heart-shaped, ending in acute points; they are hairy and clammy. The flowers are produced from the side, and also at the end of the branches, standing on short foot-stalks, having an inflated sheath or cover, out of which the tube of the flower arises, which is seven or eight inches long and very slender, but at the top is divided into two lips the under being large, and divided into three broad segments, the middle being larger than the other two; the upper lip is roundish and entire; the flowers are succeeded by oblong fruit, having a thick dry skin, which opens lengthways, inclosing a hard furrowed nut, with two recurved horns. This is an annual plant, whose seeds must be sown on a hot-bed in the spring, and when the plants are fit to remove, they should be each planted in a separate small pot, and plunged into a moderate hot-bed, carefully shading them from the sun till they have taken new root; after which they should have free air admitted to them, to prevent their drawing up weak, and treated in the same manner as other tender exotic plants, being too tender to thrive in the open air in England; so that when they

they are grown too large to remain under the frames, they ſhould be removed into the bark-ſtove, and plunged into the tan-bed, where they will flower, and with good management they often perfect their ſeeds in England. But the ſeeds of this plant ſhould remain on till they drop, otherwiſe they will not grow; for the outer covers of theſe ſeeds ſplit open and drop off, like thoſe of the Almond, before the ſeeds are fully ripened.

The ſecond ſort grows naturally at the Havannah, and in ſome of the other iſlands in America. This riſes with a ſhrubby ſtalk to the height of ten or twelve feet, dividing upward into a few branches, which are garniſhed with ſpear-ſhaped leaves cut on their edges; theſe are ſoft and hairy. The flowers are produced from the ſide of the branches, growing ſeveral together on the ſame foot-ſtalk; they are ſhaped like thoſe of the Foxglove, of a greeniſh yellow colour, with brown ſpots on the inſide; the flowers have a ſwelling tube which is recurved, and the brim is ſlightly divided into five unequal ſegments.

This ſort is propagated by ſeeds, which muſt be ſown on a hot-bed in the ſpring, and when the plants are fit to remove, they ſhould be each planted into a ſeparate ſmall pot and plunged into a freſh hot-bed, where they muſt be ſhaded from the ſun till they have taken freſh root; then they muſt have air admitted to them daily. In autumn they muſt be removed into the bark-ſtove, and plunged into the tan-bed; during the winter ſeaſon the plants ſhould not have much water, and may be treated in the ſame manner as other tender plants from thoſe countries.

CRASSULA. Dillen. Hort. Elth. 114. Leſſer Orpine, or Live-ever.

The CHARACTERS are,

The flower conſiſts of five narrow petals, which are joined at their baſe, but are reflexed, and ſpread open at the brim: in the bottom of the tube are ſituated five nectaria, and five ſtamina ſituated round theſe. At the bottom of the tube are placed five oblong pointed germina; after the flower is paſt, theſe become five capſules, opening lengthways, and filled with ſmall ſeeds.

The SPECIES are,

1. CRASSULA (*Coccinea*) foliis planis cartilagineo-ciliatis, baſi connato vaginantibus. Vir. Cliff. 26. *Shrubby African Navelwort, with umbels of ſcarlet flowers.*

2. CRASSULA (*Perfoliata*) foliis lanceolato-ſubulatis ſeſſilibus connatis, canaliculatis ſubtus convexis. Hort. Cliff. 116. *Talleſt Craſſula with perfoliate leaves.*

3. CRASSULA (*Cultrata*) foliis oppoſitis, obtusè ovatis, integerrimis, hinc anguſtioribus. Hort. Cliff. 496. *Craſſula with an Orpine leaf.*

4. CRASSULA (*Ciliata*) foliis oppoſitis, oblongis, planiuſculis, diſtinctis, ciliatis. Hort. Cliff. 496. *Craſſula with leaves like Orpine placed croſſways.*

5. CRASSULA (*Scabra*) foliis oppoſitis, patentibus, ſcabris. Lin. Sp. Pl. 283. *Shrubby African Navelwort, with narrow, rough, pointed leaves, and a greeniſh flower.*

6. CRASSULA (*Nudicaulis*) foliis ſubulatis, radicatis, caule nudo. Hort. Cliff. 116. *Craſſula with a long Onion like leaf.*

7. CRASSULA (*Punctata*) caule flaccido, foliis connatis, cordatis, ſucculentibus, floribus confertis terminalibus. *Leſſer Orpine with a weak ſtalk growing through the leaves, which are heart-ſhaped and ſucculent, and flowers growing in cluſters at the end of the branches.*

8. CRASSULA (*Fruticoſa*) foliis longis, teretibus, alternis, caule fruticoſo, ramoſo. *Leſſer Orpine with long taper leaves placed alternate, and a branching ſhrubby ſtalk.*

9. CRASSULA (*Sedioides*) caule flaccido, prolifero, determinatè-folioſo, foliis patentiſſimis, imbricatis. Hort. Cliff. 496. *African Rock Houſleek, with leaves like the common ſort, ſpreading like a Roſe.*

10. CRASSULA (*Pelucida*) caule flaccido repente, foliis oppoſitis. Lin. Sp. Pl. 283. *Creeping Craſſula with the appearance of Purſlane.*

11. CRASSULA (*Portulacaria*) foliis obverſe-ovatis, oppoſitis, caule fruticoſo ſucculento. *Tree-like Craſſula with the appearance of Purſlane.*

The firſt ſort hath a round reddiſh ſtalk which is jointed, riſing about three feet high, which divides upward into many irregular branches, garniſhed with oblong plain leaves placed oppoſite, having a griſtly border ſet with ſmall ſilver hairs; they cloſely embrace the ſtalks with their baſe, and form a ſort of ſheath or cover to it. The flowers terminate the branches in cloſe umbels, ſitting very cloſe at the top of the branches; theſe are funnel-ſhaped, having pretty long tubes cut at the top into five parts, which ſpread open; they are of a fine ſcarlet colour, and ſtand erect. This is propagated by cuttings during any of the ſummer months, which ſhould be taken off three weeks before they are planted, and laid in a dry place that the wounded part may heal over; then they ſhould be each planted in a ſmall pot, and plunged into a moderate hot-bed, giving them but little water: in about ſix weeks theſe will have put out roots, when they ſhould be gradually inured to the open air, into which they ſhould be removed, placing them in a ſheltered ſituation, where they may remain till autumn; when they muſt be removed into a dry airy glaſs-caſe, where they may enjoy the ſun as much as poſſible, and be ſcreened from the wet and cold. In warm dry weather during the ſummer months, while they are abroad, theſe plants ſhould be gently watered two or three times a week, but in winter they ſhould have very little given them. Theſe plants require no artificial heat in winter, but they muſt be ſecured from froſt and wet.

The ſecond ſort will riſe with an upright ſtalk ten or twelve feet high, if it is not broken or injured, but it will require ſupport; for the ſtalks being ſlender, and the leaves very weighty, they are very ſubject to break, eſpecially if they are expoſed to the wind. The leaves of this plant are about three inches long, they are hollowed on the upper ſide, and have a convex ridge on their lower, and are placed oppoſite, ſurrounding the ſtalks with their baſe, and alternately croſs each other. They are very thick, ſucculent, and of a pale green colour, ending in acute points; at the top of the ſtalk the flowers are produced in large cluſters, which are of a whitiſh herbaceous colour, having ſhort tubes cut into five parts at the brim, which ſpread open. The ſtalk which ſuſtains the flowers is pretty thick and ſucculent, generally turning firſt downward, and then upward again, almoſt in the form of a ſyphon. This ſort is propagated by cuttings in the ſame manner as the firſt, and the plants require the ſame treatment.

The third ſort riſes with a weak ſucculent ſtalk about two feet high, ſending out many irregular branches, garniſhed with oblong, oval, thick leaves, plain on their upper ſide, but convex below, of a deep green, and their borders ſet with a few ſilvery hairs. The ſtalk which ſupports the flowers riſes from the top of the branches, and is from four to ſix inches long, putting out ſeveral ſide branches which grow erect; theſe are terminated by large cluſters of ſmall greeniſh flowers.

This is propagated by cuttings in the ſame manner as the two former, but being pretty hardy ſhould not be ſo tenderly treated; for if the cuttings of this are planted in a border of light earth they will put out roots, and may afterward be taken up and potted.

The fifth ſort hath a very weak ſucculent ſtalk, which riſes about a foot and a half high, dividing upward into ſmall branches, garniſhed with thin rough leaves, which are flat, near two inches long, and a quarter broad at their baſe, gradually narrowing to a point; they are rough, placed opposite, and embrace the stalks with their base. The flowers come out in small clusters at the end of the branches, which are small, and of an herbaceous colour, so make no figure. This may be propagated by cuttings, which may be treated in the same manner as the fourth sort.

The sixth sort never rises with a stalk, but the leaves come out close to the ground, forming a sort of head; they are taper and succulent, ending in points, and frequently put out roots: out of the center of the heads arise the flower-stalk, which grows about six inches high, and branches into two or three smaller upward, each being terminated by clusters of greenish flowers, which make no great appearance.

This is propagated by taking off the heads or side offsets, which should be laid to dry three or four days before they are planted, then they may be treated in the same manner as the other hardier sorts before mentioned.

The seventh sort hath been lately introduced from the Cape of Good Hope; this hath very slender stalks, which are reddish and full of joints; they trail upon the ground, unless they are supported, and are closely garnished with thick, succulent, heart-shaped leaves, placed opposite, which are closely joined at their base, so that the stalks run through them; they are of a grayish colour; the stalks are divided, and grow about eight or nine inches long, terminated by clusters of small white flowers, sitting very close to the top of the stalk. It is propagated by cuttings in the same manner as the other hardier sorts, and may be treated in the same way.

The eighth sort rises with a shrubby stalk four or five feet high, dividing into many branches, which at first are taper and succulent, but by age become ligneous; these are garnished with very slender, taper, succulent leaves, which are near three inches long, flaccid, and generally turning downward, especially in winter, when they are in the house, but it hath not as yet flowered here. This is equally hardy with the former sorts, and takes easily from cuttings, so may be treated in the same way.

The ninth sort is a low plant, with the appearance of Housleek, having open spreading heads, very like those of some sorts of Housleek; these grow on the ends of very slender trailing stalks, which are produced in plenty on every side the parent plant, in like manner as the Childing Marigold The flower-stalks arise from the center of these heads, which are naked, about four inches long, and are terminated by close clusters of herbaceous flowers. This plant propagates very fast by the side heads, which come out from the parent plant, and frequently puts out roots as they trail on the ground, so may be taken off and potted during any of the summer months; it is equally hardy with the former sorts, so the plants may be treated in the same way.

The tenth sort hath very slender, trailing, succulent stalks, of a reddish colour, which put out roots at the joints as they lie upon the ground. The stalks and leaves of this sort have the appearance of Purslane, and trail upon the ground like Chickweed. The flowers are produced in small clusters at the end of the branches; these are white, with a blush of purple at their brim. This sort is easily propagated by its trailing branches, and the plants require the same treatment as the other hardy sorts, but is of short duration.

The eleventh sort rises with a very thick, strong, succulent stalk, to the height of five feet, sending out branches on every side, so as to form a kind of pyramid, the lower branches being extended to a great length, and the other diminishing gradually to the top; these are of a red or purplish colour, and very succulent: they are garnished with roundish succulent leaves, very like those of Purslane, from whence the gardeners have titled it the Purslane-tree.

This sort hath not flowered in England, though it has been many years in the gardens, so that we are not sure if it is properly ranged in this genus; but from the outward appearance it seems to be nearly allied to some of the other species, on which account Dr. Dillenius has placed it here.

It is propagated with great facility by cuttings, which may be planted during any of the summer months, but these should be laid to dry for some days before they are planted, that the wounded part may be healed over, otherwise they will rot. This is somewhat tenderer than the four sorts last mentioned, so must be placed in a warm glass-case in winter, where they may enjoy the full sun, and should have very little water during that season; in summer the plants should be placed abroad in a sheltered situation, and in warm weather will require to be refreshed with water twice a week; but as the stalks are very succulent, so too much wet at any season is very hurtful to these plants.

All the hardy sorts of Crassula may be treated in the same way as the Mesembryanthemi, and other hardier kinds of succulent plants, with this difference only, not to give them so much water; but the first, second, and eleventh sorts, require to be placed in a warm dry glass-case in winter, and must not be so long exposed abroad in the summer as the other species, and should have but little water, especially in the winter.

CRATÆGUS. Tourn. Inst. R. H. 633. The Wild Service.

The CHARACTERS are,

The flower hath five roundish concave petals, and many stamina, which are inserted in the empalement. The germen is situated under the flower, which afterward becomes an oval or roundish umbilicated berry, inclosing two oblong hard seeds.

The SPECIES are,

1. CRATÆGUS (*Aria*) foliis ovatis inæqualiter serratis, subtus tomentosis. Hort. Cliff. 187. *Cratægus with a roundish sawed leaf, white on the under side; commonly called Aria Theophrasti, and in some countries the white Beam, or white Leaf-tree.*

2. CRATÆGUS (*Torminalis*) foliis cordatis septangulis, lobis infimis divaricatis. Lin. Sp. Pl. 476. *Wild, or Maple-leaved Service.*

3. CRATÆGUS (*Alpina*) foliis oblongo-ovatis serratis, utrinque virentibus. *Cratægus with an oblong sawed leaf, green on both sides.*

4. CRATÆGUS (*Virginiana*) foliis oblongo-ovatis, crenatis, subtus argenteis. *Virginia Cratægus with an Arbutus leaf*

5. CRATÆGUS (*Coccinea*) foliis ovatis repando-angulatis serratis glabris. Hort. Cliff 187. *Cratægus with oval, smooth, sawed leaves, having angles.*

6. CRATÆGUS (*Crus Galli*) foliis lanceolato-ovatis serratis glabris, ramis spinosis. Lin. Sp. Pl. 682. *Cratægus with oval spear-shaped leaves, which are sawed, and prickly branches; commonly called Cockspur Hawthorn.*

7. CRATÆGUS (*Lucida*) foliis lanceolatis serratis lucidus spinis longissimis, floribus corymbosis. *Cratægus with spear-shaped, lucid, sawed leaves, very long spines, and flowers growing in a corymbus.*

8. CRATÆGUS (*Azarolus*) foliis obtusis subtrifidis dentatis. Lin. Sp. Pl. 683. *Cratægus with obtuse, trifid, sawed leaves; commonly called L'Azarole.*

9. CRATÆGUS (*Oxyacantha*) foliis obtusis subtrifidis serratis. Hort. Cliff. 188. *Cratægus with obtuse, trifid, sawed leaves; or the common Hawthorn.*

10. CRATÆGUS (*Tomentosa*) foliis cuneiformi-ovatis serratis subangulatis, subtus villosis ramis spinosis. Lin. Sp.

 Plant.

Plant. 682. *Cratægus with oval, wedge-shaped, sawed leaves, heary on their under side, and prickly branches; or Gooseberry-leaved Hawthorn.*

The first sort grows naturally on the chalky hills in Kent, Surry, and Suffex, and in some other parts of England; it rises to the height of thirty or forty feet, with a large trunk, and divides upward into many branches; the young shoots have a brown bark, covered over with a meally down, and are garnished with oval leaves between two and three inches long, and one and a half broad in the middle, of a light green on their upper side, but very white on their under, having many prominent transverse veins running from the midrib to the border, where they are unequally sawed. The flowers are produced at the end of the branches in bunches, their foot-stalks being meally, as are also the empalements of the flowers, which are cut into five obtuse segments that are reflexed. The flowers have five short petals, which spread open, and are like those of the Pear-tree, having a great number of stamina of the same length with the petals, terminated by oval summits. The germen, which is situated below the flowers, afterward becomes an oval fruit, crowned with the empalement of the flower, having one cell, in which is inclosed three or four seeds. It flowers in May, and the fruit ripens in autumn.

This tree may be propagated by seeds, which should be sown soon after they are ripe; for if they are kept out of the ground till the spring, they remain at least one year in the ground before the plants appear. When the plants come up, they may be treated in the same manner as the Hawthorn, but they should by no means be headed or cut down: when these plants are upon a poor chalky soil they make great progress, and the wood is very white and hard, so has been often used for making cogs for mills, and many other purposes where hard tough timber is wanted.

It may also be propagated by layers in the same manner as the Lime-tree and Elm, but these should be laid in the young wood, and they are two years before they have sufficient roots to transplant.

The tree will take by grafting, or budding upon Pear-stocks very well, and Pears will take by grafting on these trees, so that there is a nearer affinity between the Cratægus and the Pear, than there is between either of these and the Mespilus; for although both these will sometimes take upon the Mespilus, yet neither of them thrive so well, or last so long when grafted or budded upon those stocks, as they do upon each other.

The second sort grows naturally in many parts of England, and is chiefly found upon strong soils; it formerly grew in great plenty in Cane-wood, near Hampstead, and lately there were some young trees growing in Bishops-wood, near the same place; this rises to the height of forty or fifty feet, with a large trunk. The young branches are covered with a purplish bark, garnished with leaves placed alternately, standing on pretty long foot-stalks, which are cut into many acute angles like those of the Maple-tree; they are near four inches long, and three broad in the middle, having several smaller indentures toward the top, of a bright green on their upper side, but a little woolly on their under. The flowers are produced in large bunches toward the end of the branches; they are white, and shaped like those of the Pear-tree, but smaller, and stand upon longer foot-stalks, and are succeeded by roundish compressed fruit which are shaped like large Haws, and ripen late in autumn, and if kept till they are soft in the same way as Medlars, they have an agreeable acid flavour. The fruit of this tree is annually sold in the London markets in autumn.

The third sort grows naturally upon mount Baldus, and on other mountainous parts of Italy; it rises with a woody trunk about twenty feet high, dividing into many branches closely garnished with oblong sawed leaves, standing alternate on very short foot-stalks; they are about three inches long, and one and a half broad, and are slightly sawed on their edges, of a deep green on both sides. The flowers are produced at the end of the branches in small bunches, which have rarely more than four or five in each; they are white, and much smaller than those of the former sorts, and are succeeded by fruit about the size of the common Haw, which is of a dark brown colour when ripe.

This sort may be propagated in the same manner as the first, but requires a strong deep soil, otherwise it will not thrive. It is very hardy in respect to cold, but at present is rare in England.

The fourth sort grows naturally in most parts of North America; this seldom rises more than five or six feet high; it hath generally many shrubby stalks arising from the same root, garnished with leaves placed alternate, standing on very short foot-stalks; they are about two inches long, and one broad, ending in a point, of a deep green on their upper side, and a little woolly on their under, of a yellowish white colour; these leaves in autumn change to purple some time before they fall off. The flowers are produced in small bunches at the end, and also from the side of the branches; they are small, white, and shaped like those of the former sorts, but the petals are narrower, and are succeeded by small fruit shaped like those of the common Haw, which turn red in autumn, and when fully ripe are of a dark brown colour.

This sort may be propagated by seeds, which should be sown in autumn, in the same manner as hath been directed for the first sort; but as these seeds are frequently brought from America, and do not arrive here till spring, so they may be buried in the ground the first summer, and taken up and sown in the autumn, as is frequently practised with Haws; and when the plants have grown one year in the place where they were sown, they may be transplanted, and treated in the same way as the other sorts.

This is generally planted among flowering shrubs of the same growth, where it will add to the variety.

The fifth sort grows naturally in North America, where it is commonly known by the title of Cockspur Hawthorn, the spines on the branches being incurved like the spur of a cock.

The sixth sort grows naturally in Virginia, and also in several other parts of North America, and has been long known in England by the title of Virginia L'Azarole: this grows to the height of twenty feet or more; the younger branches are covered with a light gray bark, but have few thorns upon them; the leaves are spear-shaped, and are very lucid, standing upon short foot-stalks; the flowers grow in bunches from the side of the branches, they are large and white, and are succeeded by oblong, large, red fruit.

The seventh sort is also a native of the same country: this rises to the height of eight or ten feet; the leaves are spear-shaped, and of a lucid green; the branches have very long thorns, but it has not produced flowers or fruit in England.

The eighth sort is the common L'Azarole, which in Italy produces a fruit frequently eaten there in winter: the trees grow pretty large, the branches grow irregular, and are garnished with leaves shaped like those of the common Hawthorn, but are much larger, and are covered with down on their under side; the flowers are pretty large and white; the fruit is large, some red, and others yellow, and when quite ripe have an agreeable acid flavour.

The

The ninth sort is our common Hawthorn, which is so well known as to need no description; but as the gardeners generally propagate the other sorts by grafting them upon this, it was necessary to mention it here.

The tenth sort is also a native of North America, but seldom rises more than eight feet high; the branches grow very irregular, and have long weak thorns closely set upon them, and are cloathed with woolly leaves shaped like those of the Gooseberry-bush; the flowers are small, and have very long foot stalks; the fruit is small, of a greenish yellow colour, having a large tuft of leaves on the top.

All these varieties may be propagated either by sowing of their seeds, which if put into the ground in the autumn, the plants will rise the spring following, or by grafting them upon the common Hawthorn, which is the method commonly practised.

CRATEVA. Lin Gen. Plant. 528. Garlick Pear.

The Characters are;

The flower hath four oval petals, which are narrow at their base, and broad at the top. It hath many bristly stamina, which are longer than the petals, and a long incurved style, upon which sits the oval germen, which afterward becomes a large, fleshy, globular fruit, with one cell, including many kidney-shaped seeds.

The Species are,

1. Crateva (*Tapia*) inermis. Flor. Zeyl. 211. *Smooth Crateva, or Garlick Pear.*

2. Crateva (*Spinosa.*) Flor. Zeyl. 212. *Prickly Crateva.*

The first sort grows naturally in both Indies, where it has a very large trunk, which rises to the height of thirty feet or upward, covered with a dark green bark, and forms a large head: the branches are garnished with trifoliate leaves standing on pretty long foot-stalks; the middle lobe, which is much larger than either of the other, is oval, about five inches long, and two and a half broad in the middle; the two side lobes are oblique, and turn at both ends toward the middle, so that their midrib is not parallel to the sides; these two end in acute points; they are smooth, of a light green on their upper side, but pale on their under. The flowers are produced at the ends of the branches, standing upon long foot-stalks; the empalements are of one leaf, which are cut into four segments almost to the bottom; the flower hath four oblong petals, which spread open and are reflexed, having many long slender stamina which are connected at their base, but spread open above, and are terminated by oblong purple summits; these surround a slender long style, upon which is situated the oval germen, which is crowned by an obtuse stigma, which afterward becomes a round fruit, about the size of an Orange, having a hard brown shell or cover, inclosing a meally pulp, filled with kidney-shaped seeds. This fruit hath a strong smell of Garlick, which is communicated to the animals that feed on it.

This is propagated by seeds, which must be sown on a hot-bed in the spring, and when the plants come up they must be treated in the same manner as hath been directed for the Annona, to which article the reader is desired to turn for the culture.

The second sort grows naturally in India, where it becomes a very large tree, sending out many long branches, garnished with trifoliate leaves, which are oblong, entire, and end in acute points; between these the branches are armed with long sharp thorns, which come out by pairs, and spread asunder; the flowers are produced in small clusters from the side of the branches, five or seven standing upon a common branching foot-stalk; these have each five acute petals which are reflexed, and many stamina which stand round a single style of the same length; the petals are green on the outside, and whitish within, and have a grateful odour. After the flower is past the germen swells to a large fruit, the size of an Orange, having a hard shell, which incloses a fleshy viscous pulp of a yellowish colour, having many oblong plain seeds situated within it; the pulp of this fruit hath an agreeable flavour when ripe, so is frequently eaten in India, where they serve up the fruit, mixed with sugar and Orange in their deserts, and is esteemed a great delicacy.

This sort is propagated by seeds, and requires the same treatment as the former.

CREPIS. Lin. Gen. Plant. 819. Bastard Hawkweed.

The Characters are,

It hath a flower composed of many hermaphrodite florets, which are included in a double empalement; these florets are of one leaf, uniform, tongue-shaped, and are indented at the top in five parts, and have each five short hairy stamina; the germen is situated in the center of the florets, which afterward becomes an oblong seed, crowned with a long feathery down, which sits upon little foot-stalks.

The Species are,

1. Crepis (*Rubra*) foliis amplexicaulibus, lyrato-runcinatis. Vir. Cliff. 79. *Hawkweed with a Dandelion leaf, and a soft red flower.*

2. Crepis (*Barbato*) foliis pinnatis angulatis, petiolatis dentatis. Prod. Leyd. 126. *Hawkweed with hairy wild Succory leaves, smelling like Castor.*

3. Crepis (*Bœtica*) involucris calyce longioribus incurvatis, foliis lanceolatis dentatis. *Greater Spanish Hawkweed, with flowers black in the middle.*

4. Crepis (*Alpina*) foliis amplexicaulibus, oblongis acuminatis inferioribus supernè, summis infernè, denticulatis. Hort. Upsal. 238. *Alpine Hawkweed with a Viper's Grass leaf.*

There are several other species of this genus, some of which grow naturally in England, and others are weeds in divers parts of Europe, so are rarely admitted into gardens, therefore I shall not enumerate them here.

The first sort grows naturally in Apulia, but is now commonly cultivated in the English gardens for ornament; it is an annual plant, which perishes after it hath ripened seeds; this hath many irregular leaves which spread on the ground, and are deeply jagged on their sides; between them arise the branching stalks, which grow a foot and a half high, garnished with oblong leaves deeply indented on their edges, embracing the stalks with their base; the stalks are each terminated by one large radiated flower of a soft red colour, composed of many half florets, which are succeeded by oblong seeds crowned with a feathery down.

The seeds of this plant should be sown in the spring, on the borders of the flower-garden where they are designed to remain, so that if six or eight seeds are sown in each patch, when the plants come up they may be reduced to three or four; and if these are kept clean from weeds, they will require no other culture. If the seeds are sown in autumn, or permitted to scatter, the plants will come up and live thro' the winter without shelter, and these will flower early in the spring.

The second sort grows naturally in the south of France, and in Italy; this is a biennial plant, and sometimes, when it is in poor ground, it will continue longer; this hath a thick tap-root which strikes deep into the ground, sending out many small fibres; the lower leaves are from four to five inches long, and about a quarter of an inch broad, having several deep jags on their edges; from the same root arise four or five stalks, which grow about nine or ten inches high; the lower part of these are garnished with leaves of the same form with those at the root, but are smaller and more jagged; the upper part of the stalks are naked, and termi-

terminated by one flower of a gold colour, inclining to copper, composed of many florets which are included in a single empalement; the flowers are succeeded by oblong narrow seeds, crowned with a feathery down; the whole plant, when bruised, emits a strong odour of Castor.

It is propagated by seeds in the same manner as the first sort, but as this continues longer, so the seeds need not be annually sown: the plants will require no other culture but to keep them clean from weeds.

The third sort is an annual plant, which grows naturally in Spain, but is now frequently propagated in the flower-gardens for ornament; this puts out leaves near the root, which are nine inches long, and almost two broad in the middle, of a light green colour, and a little jagged on their edges; the stalks rise a foot and a half high, and are garnished with leaves of the same form as those at bottom, but are smaller; the flowers are produced at the end of the branches; these have a double empalement, composed of many long very narrow leaves; the outer series are reflexed downward, and turn upward again, and are inflexed at their extremities; the flowers are composed of many florets, which spread regularly in form of rays, situated over each other like scales of fish; the bottom or middle is black, so make a pretty appearance in a garden: this plant requires the same culture as the first, and is equally hardy, so that where the seeds are permitted to scatter, the plants will come up without care.

The fourth sort grows naturally on the Alps; this is also an annual plant, which sends out many leaves near the root five inches long, and almost two broad at their base; the upper part of these are slighty indented, but their lower are entire; the stalks are strong, upright, rising two feet high, and are terminated by pale white flowers, inclosed in a strong hairy empalement; this requires the same culture as the first, and the seeds will scatter about the garden, so that if the plants are not destroyed, they will maintain themselves without any care.

CRESCENTIA. Lin. Gen. Plant. 680. Calabash-tree.

The CHARACTERS are,

The flower hath one petal, which is irregular, having a curved gibbous tube; it hath an empalement of one leaf, cut into two obtuse segments, which are concave; it hath four stamina, two of which are the length of the petal, the other are shorter, and an oval germen, which afterward becomes an oval or bottle-shaped fruit, with a hard shell, inclosing many flat heart-shaped seeds.

The SPECIES are,

1. CRESCENTIA (*Cujete*) foliis lanceolatis, utrinque attenuatis. Hort. Cliff. 327. *Calabash-tree with oblong narrow leaves, and a large oval fruit.*

2. CRESCENTIA (*Latifolia*) foliis oblongo-ovatis, fructu rotundo, cortice fragili. *Broad-leaved Calabash-tree, whose fruit hath a tender shell.*

The first sort grows naturally in Jamaica, and in all the Leeward Islands; this hath a thick trunk, covered with a hitish bark, which rises from twenty to thirty feet high, and at the top divides into many branches, forming a large regular head, garnished with leaves which come out irregularly, sometimes single, and at others many arise from the same knot; they are near six inches long, and one and a half broad in the middle, of a lucid green, and have very short foot-stalks; the flowers are produced from the side of the large branches, and sometimes from the trunk, standing upon long foot-stalks; they have but one petal, which is irregular, having an incurved tube, which is divided at the brim into two irregular segments, which turn backward; these are of a greenish yellow colour, striped and spotted with brown; they have four slender stamina of the same colour with the petal, which are of unequal lengths, two being full as long as the petal, and the other are much shorter, terminated by oblong summits, divided in the middle, which lie prostrate on the stamina. From the lower part of the tube arises a long slender foot stalk, supporting the oval germen, which afterward turns to a large fruit of different forms and size; they are spherical, sometimes they are oval, and at other times they have a contracted neck like a bottle, and are so large, as when the pulp and seeds are cleaned out, the shells will contain three pints or two quarts of liquid. These fruit or shells are covered with a thin skin of a greenish yellow when ripe, which is peeled off; and under this is a hard ligneous shell, inclosing a pale, yellowish, soft pulp, of a tart unsavoury flavour, surrounding a great number of flat heart-shaped seeds.

The shells of this fruit are cleaned of their pulp, and the outer skin taken off by the inhabitants of the islands, and are dried; then they use them for drinking-cups, some of which are tipped with silver, and to the necks they fasten handles; and some of the long small fruit are formed into the shape of spoons or ladles, and are used as such; the round ones are cut through the middle, and are used as cups for chocolate. In short, they convert these shells into many sorts of furniture, which is the principal use made of the fruit, for the pulp is seldom eaten.

The second sort seldom rises more than fifteen or twenty feet high, with an upright trunk, covered with a white smooth bark, sending out many lateral branches at the top, garnished with leaves three inches in length, and one and a quarter broad, ranged alternately on the branches, sitting upon short foot-stalks, of a deeper green than those of the first sort, and their edges are entire; the flowers come out from the side of the large branches and the trunk; they are smaller, and of a deeper yellow colour than those of the first; the fruit of this is sometimes round, at others oval, some being much larger than the other; the shells of this fruit are thin and very brittle, so are unfit for any purposes to which those of the former are employed: the wood of this tree is hard and very white, so might be useful, were it not for the plenty of other sorts which abound in many of the islands.

These trees are too tender to live abroad in England, so require a warm stove to be preserved here; they are easily propagated by seeds, which, when fully ripe, should be brought over in the fruit; for when the seeds are taken out of the pulp abroad, and sent over hither, if they are long in their passage, they will lose their growing quality before they arrive; the seeds must be sown on a good hot-bed in the spring, and when the plants are fit to remove, they should be each planted into a small half-penny pot, and plunged into a hot-bed of tanners bark, observing to shade them from the sun till they have taken fresh root; then they must be treated in the same manner as other tender plants which are natives of the same countries, keeping them in the tan-bed of the bark-stove, and should have but little water in winter; in summer they will require to be gently watered two or three times a week, according to the warmth of the season, and should have a large share of air admitted to them; these plants may also be propagated by cuttings.

CRESS, the Garden. See NASTURTIUM.

CRESS, the Indian. See TROPÆOLUM.

CRESS, the Water.
CRESS, the Winter. } See SISYMBRIUM.

CRINUM. Lin. Gen. Plant. 366. Asphodel Lily.

The CHARACTERS are,

The flower hath one petal, which is funnel-shaped, deeply cut at the top into six parts, which are reflexed; it hath six long stamina, which are inserted in the tube of the petal, spreading open;

open; the germen is situated in the bottom of the flower, which afterward becomes an oval capsule with three cells, each containing one oval seed; the flowers are included in a two-leaved sheath.

The Species are,

1. Crinum (*Africanum*) foliis sublanceolatis planis, corollis obtusis. Lin. Sp. Pl. 292. *African tuberous Crinum, with a blue umbellated flower.*

2. Crinum (*Asiaticum*) foliis carinatis. Flor. Zeyl. 127. *Crinum with keel shaped leaves.*

3. Crinum (*Americanum*) corollarum apicibus introrsum unguiculatis. Lin. Sp. Plant. 292. *Crinum with evergreen leaves, and many white flowers.*

4. Crinum (*Latifolium*) foliis carinatis, basi angustioribus, floribus profundè dissectis. *Crinum with keel-shaped leaves, which are narrower at their base, and flowers deeply cut.*

The first sort grows naturally at the Cape of Good Hope; the root of this plant is composed of many thick fleshy fibres, diverging from the same head, which strike deep into the ground; from the same head arises a cluster of leaves surrounding each other with their base, so as to form a kind of herbaceous stalk, about three inches high, from which the leaves spread only two ways, appearing flat the other two. The flower-stalk arises by the side of these leaves, which is round, hollow, and grows upward of three feet high, terminated by a large head of flowers, included in a kind of sheath, which splits into two parts, and is reflexed; the flowers stand each upon a foot-stalk about an inch long; they are tubulous, of one petal, which is cut almost to the bottom into six oblong blunt segments, which are waved on their edges; in the center is situated an oval three-cornered germen, supporting a long style, which is attended by six stamina, two of the same length, two somewhat shorter, and the two which rest upon the lower segments are the shortest; the flowers are of a bright blue colour, and grow in large bunches, so make a fine appearance; they begin to flower in September, and frequently continue in beauty till Christmas, which renders them more valuable.

It is propagated by offsets which come out from the side of the plants, which may be taken off the latter end of June, at which time they are in their greatest state of rest; when the plants should be turned out of the pots, and the earth carefully cleared away from the roots; then the fibres of the offsets should be separated from those of the old roots, and the offsets may be taken from the old plants, being careful not to break their heads; but where they adhere so closely to the old plant as not to be separated, they must be cut off with a knife, taking great care not to wound or break the roots of either of the offsets or the parent plant. When these are parted, they should be planted each into a separate pot, and placed in a shady situation, where they may enjoy the morning sun, giving them a little water twice a week if the weather proves dry; but they must not have too much wet, especially at this season, when they are almost inactive, for as the roots are fleshy and succulent, so they are apt to rot with great moisture. In about five weeks time the offsets will have put out new roots, when the pots may be removed to a more sunny situation, and then they may have a little more water, which will strengthen their flowering; but it must not be given them too liberally, for the reasons before given. In September they will put out their flower-stalks, and toward the end of that month the flowers will open, when, if the weather should be very wet or cold, they should be removed under shelter, to prevent the flowers from being injured; but they should have as much free air as possible, otherwise the flowers will be pale coloured and weak. Toward the end of October they should be removed into the green-house, and placed where they may enjoy as much free air as possible, and not be overhung by other plants; in winter they may have a little water once a week, or oftener in mild weather, but in frost they should be kept dry; they should not have any artificial warmth in winter, and must be placed in the open air in summer.

The second sort hath large bulbous roots, which send out many large fleshy fibres, which have bulbs formed at their ends; the leaves are near three feet long, hollow on their upper side, and closely fold over each other at their base; the outer leaves generally turn downward at the top; they are of a deep green: the flower-stalks arise on one side the leaves, which are thick, succulent, and hollow in the middle, and a little compressed on two sides; these grow two feet high or more, and are of the same colour with the leaves, and are terminated by large umbels of flowers, which hath a sort of sheath or cover, which splits lengthways, and is reflexed back to the stalk, where it dries and remains; the flowers have narrow tubes near four inches long, which are deeply cut into six long segments, which are reflexed back almost to the tube; in the center arises the style, attended by six long stamina, which stand out beyond the petal, and are terminated by oblong prostrate summits of a yellow colour. After the flowers are past, the germen, which is situated at the bottom of the tube, becomes a large, roundish, three-cornered capsule, having three cells, two of which are generally abortive, and the third hath one or two irregular bulbs, which if planted produce young plants.

The third sort hath broader leaves than the second, which are plain, and not hollowed on their upper side, but they are shorter and of a lighter green; these embrace each other at their base; by the side of these arise the flower-stalk, which is compressed and hollow, growing about three feet high, terminated by large umbels of white flowers like those of the former sort, but the segments of the petal are broader and not so much reflexed.

The fourth sort hath roots like those of the second; the leaves of this are narrower at their base, and are stained with purple on their under side; the flower-stalks are purple, and grow to the same height as those of the second; the flowers are in shape like them, but the tube is purple, and the segments have a purple stripe running thro' them; the stamina also are purple, which renders this more beautiful than either of the other sorts, and these differences are constant in all the plants which rise from seeds, so there can be no doubt of its being a distinct sort.

These three sorts grow naturally in both Indies, so are tender, therefore must be kept in a stove, otherwise they will not thrive in England. They are easily propagated by offsets, which the roots put out in plenty, or by the bulbs which succeed the flowers, and ripen perfectly here. These must be planted in pots, and plunged into the tan-bed in the stove, where the plants will make greater progress, and flower oftener, than when they are placed on shelves; though in the latter way they will succeed very well, provided they are kept in a good temperature of heat. The roots should be transplanted in the spring, and all the offsets taken off, otherwise they will fill the pots and starve the old plants: they must be frequently refreshed with water, but it must not be given them too plentifully, especially in winter.

CRITHMUM. Lin. Gen. Plant. Samphire.

The Characters are,

It is a plant with an umbelliferous flower; the general umbel is uniform; the flowers have five oval inflexed petals, which are almost equal, and five stamina the length of the petals; the germen

germen is situated under the flower, which afterward becomes an oval compressed fruit, dividing into two parts, each having one compressed, elliptical, furrowed seed.

The Species are,

1. Crithmum (*Maritimum*) foliolis lanceolatis carnosis Hort. Cliff. 98. *Samphire with spear-shaped fleshy leaves.*

2. Crithmum (*Pyreniacum*) foliolis lateralibus bis trifidis. Hort. Cliff. 98. *Pyrenean Parsley, with the appearance of Scerching Carret.*

The first sort grows upon the rocks by the sea side, in many parts of England. This hath a root composed of many strong fibres, which penetrate deep into the crevices of the rocks, sending up several fleshy succulent stalks which rise about two feet high, garnished with winged leaves, composed of three or five divisions, each of which hath three or five small, thick, succulent leaves, near half an inch long; the foot-stalks of the leaves embrace the stalks at their base. The flowers are produced in circular umbels at the top of the stalks; these are of a yellow colour, composed of five petals, which are near equal in size, and afterward are succeeded by seeds like those of Fennel, but are larger. This herb is pickled, and esteemed very comfortable to the stomach, and is very agreeable to the palate: it provokes urine gently, removes the obstructions of the viscera, and creates an appetite. It is gathered on the rocks, where it grows naturally; but the people who supply the markets with it seldom bring the right herb, but instead of it they bring a species of Aster, which is called Golden Samphire, of a different flavour from the true, nor has it any of its virtues.

This plant is with difficulty propagated in gardens, nor will it grow so vigorous with any culture as it does upon rocks; but if the plants are planted on a moist gravelly soil, they will thrive tolerably well, and may be preserved some years: it may be propagated either by seeds or parting of the roots.

The second sort is by Tournefort ranged in his genus of Apium. This grows naturally on the Pyrenean mountains; it is a biennial plant, which does not flower till the second year, and perishes soon after the seeds are ripe. There are two or three varieties of this plant, which differ in their outer appearance; one of these is titled by Mr. Ray, Apium montanum five petræum album. This is of humbler growth than the other; the small leaves are broader, and not so much cut on their edges, and are of a paler green: these plants are preserved in a few gardens for the sake of variety. They are propagated by seeds, which should be sown where they are designed to remain, and will require no other culture but to keep them clean from weeds, and thin them where they are too close.

CRISTA PAVONIS. See Poinciana.

CROCUS. Lin. Gen. Plant. 53. Saffron.

The Characters are,

The flower hath one petal, which is deeply cut into six oblong equal segments, and three stamina, which are shorter than the petal; the roundish germen is situated at the bottom of the tube, which afterward becomes a roundish fruit with three cells, filled with roundish seeds.

The Species are,

1. Crocus (*Sativus*) spathâ univalvi radicali, corollæ tubo longissimo. Lin. Sp. Pl. 36. *Cultivated Saffron.*

2. Crocus (*Autumnalis*) spathâ univalvi pedunculato, corollæ tubo brevissimo. *Rush-leaved autumnal Crocus, with a large purplish flower.*

3. Crocus (*Vernus*) spathâ bivalvi radicali, floribus sessilibus. *Broad-leaved spring Crocus with a variable yellow flower, commonly called Bishop's Crocus.*

4. Crocus (*Biflora*) spathâ biflorâ corollæ tubo tenuissimo. *Ordinary, spring, striped Crocus.*

There are great variety of these flowers cultivated in gardens, but as most of them are only seminal variations, so I have not enumerated them here; those which are here mentioned I think must be allowed to be specifically different, since they do not vary to each other.

The first sort is the plant which produces the Saffron, which is a well known drug. This hath a roundish bulbous root, as large as a small Nutmeg, which is a little compressed at the bottom, and is covered with a coarse, brown, netted skin; from the upper part of the root come out the flowers, which, together with the young leaves, whose tops just appear, are closely wrapped about by a thin spatha or sheath, which parts within the ground, and opens on one side. The tube of the flower is very long, arising immediately from the bulb without any foot stalk, and at the top is divided into six oval obtuse segments, which are equal, and of a purple blue colour. In the bottom of the tube is situated a roundish germen, supporting a slender style, which is not more than half the length of the petal, crowned with three oblong golden stigmas (which is the Saffron;) these spread asunder each way; the style is attended by three stamina, whose base are inserted in the tube of the petal, and rise to the height of the style, where they are terminated by arrow-pointed summits. This plant flowers in October, and the leaves keep growing all the winter, but it never produces any seeds here.

The second sort grows naturally on the Alps and Helvetian mountains. This hath a smaller bulbous root than the first, which is more compressed; the flowers appear about the same season with the former, but they rise with a short foot-stalk, having a short spatha or sheath just below the flower, which covers it before it expands; the tube of the flower is very short, the petal being divided almost to the bottom, and the segments terminate in acute points; the stamina and style are short, and the leaves of the plant are very narrow. There is a variety of this with a sky blue flower.

The third sort hath a pretty large, compressed, bulbous root, covered with a light brown netted skin, from which arise four or five leaves like those of the other vernal Crocuses, of a purplish dark colour on their lower parts; from between these come out one or two flowers, of a deep yellow colour, sitting close between the young leaves, never rising above two inches high; these have an agreeable odour: the outer segments of the petal are marked with three black streaks or stripes, running lengthways from the bottom to the top of the segment; they are narrower than the inner segments: from the double arrangement of these segments, some have called it a double flower. On the center of the tube arises a slender style, crowned by a golden stigma, which is broad, flat, and is attended by three slender stamina of the same length, terminated by yellow summits. After the flower is past the germen pushes out of the ground, and swells to a roundish three-cornered seed-vessel, which opens in three parts, and is filled with roundish brown seeds. This is one of the earliest Crocuses of the spring.

The fourth sort rises with a few very narrow leaves, which are, together with the flower-buds, closely wrapped round by a spatha or sheath, out of which arises two flowers, one of which hath a longer tube than the other, but these are very slender, and do not rise much above the spatha; there the petal enlarges, and is divided into six obtuse segments, which are of equal size; they are of a dirty white on their outside, with three or four purple stripes in each; the inside of the petal is of a purer white; the stamina and style are nearly the same as those of the former sort. This is one of the earliest sorts which flower in the spring.

The

The varieties of the autumnal Crocus are

1. The ſweet-ſmelling autumnal Crocus, whoſe flowers come before the leaves. C. B. This is our ſecond ſort.

2. The autumnal mountain Crocus. C. B. This hath a paler blue flower.

3. The many-flowering, bluiſh, autumnal Crocus. C. B. This hath many ſky blue flowers.

4. The ſmall-flowering autumnal Crocus. C. B. This hath a ſmall, deep, blue flower.

The varieties of the ſpring Crocus are,

1. Broad-leaved, purple, variegated ſpring Crocus. C. B. This hath broad leaves, and a deep blue flower ſtriped

2. Broad-leaved Crocus of the ſpring with a purple flower. C. B. This hath a plain purple flower.

3. Broad-leaved ſpring Crocus with a Violet-coloured flower. C. B. This hath a large, deep, blue flower.

4. Spring Crocus with a white flower and purple bottom. C. B.

5. Broad-leaved, white, variegated ſpring Crocus. C. B.

6. Broad-leaved ſpring Crocus, with many purple Violet flowers, ſtriped with white. C. B.

7. Broad-leaved ſpring Crocus with an Aſh-coloured flower.

8. Broad-leaved ſpring Crocus with a large yellow flower. C. B.

9. Broad-leaved ſpring Crocus with a ſmaller and paler yellow flower. C. B.

10. Broad-leaved ſpring Crocus with ſmaller yellow flowers ſtriped with black.

11. Narrow-leaved ſpring Crocus with a ſmaller brimſtone-coloured flower.

12. Narrow-leaved ſpring Crocus with a ſmall white flower.

All theſe varieties of Crocuſes are very hardy, and will increaſe exceedingly by their roots, eſpecially if they are ſuffered to remain two or three years unremoved; they will grow in almoſt any ſoil or ſituation, and are very great ornaments to a garden early in the ſpring of the year, before many other flowers appear. They are commonly planted near the edges of borders, on the ſides of walks; in doing of which, there ſhould be care taken to plant ſuch ſorts in the ſame line as flower at the ſame time, and are of an equal growth, otherwiſe the lines will ſeem imperfect. When the roots loſe their fibres and leaves, they may then be taken up, and kept dry until the beginning of September, obſerving to keep them from vermin, for the mice are very fond of them. In planting theſe roots (after having drawn a line upon the border) holes are made with a dibble, about two inches deep or more, according to the lightneſs of the ſoil, and two inches diſtance from each other, in which muſt be placed the roots, with the bud uppermoſt; then with a rake fill up the holes in ſuch a manner, as that the upper part of the root may be covered an inch or more, being careful not to leave any of the holes open; for this will entice the mice to them, which, when once they have found out, will deſtroy all your roots, if they are not prevented.

This is the way in which theſe flowers are commonly diſpoſed in gardens; but the better way is to plant them ſix or eight near each other in bunches, between ſmall ſhrubs, or on the borders of the flower-garden, where, if the varieties of theſe flowers are planted in different patches, and properly intermixed, they will make a much better appearance than when they are diſpoſed in the old method of ſtrait edgings.

The autumnal Crocuſes are not ſo great increaſers as thoſe of the ſpring, nor do they produce ſeeds in our climate, ſo that they are leſs common in the gardens, except the true Saffron, which is propagated for uſe in great plenty in many parts of England. This may be taken up every third year, as was directed for the ſpring Crocuſes, but ſhould not be kept out of the ground longer than the beginning of Auguſt, for they commonly produce their flowers in the beginning of October; ſo that if they remain too long out of the ground, they will not produce their flowers ſo ſtrong, nor in ſuch plenty, as when they are planted early.

CROTOLARIA. Lin. Gen. Plant. 771.

The CHARACTERS are,

The flower is of the butterfly kind; the ſtandard is large, heart-ſhaped, and pointed; the wings are oval, and half the length of the ſtandard; the keel is pointed, and as long as the wings; it hath ten ſtamina, which are united, and an oblong reflexed germen, that afterward becomes a ſhort turgid pod, with one cell, opening with two valves, and filled with kidney-ſhaped ſeeds.

The SPECIES are,

1. CROTOLARIA (*Verrucoſa*) foliis ſimplicibus ovatis, ſtipulis ſemicordatis, ramis tetragonis. Flor. Zeyl. 277. *Aſiatic Crotolaria, with a ſingle warted leaf and blue flower.*

2. CROTOLARIA (*Sagittalis*) foliis ſimplicibus lanceolatis piloſis, petiolis decurrentibus. *American Crotolaria with a winged ſtalk, hairy leaves, and yellow flowers diſpoſed in looſe ſpikes.*

3. CROTOLARIA (*Pedunculato*) foliis oblongo-ovatis hirſutis ſeſſilibus, ſtipulis acutis pedunculis longioribus. *Smaller, hairy, herbaceous, American Crotolaria, with an arrow-ſhaped ſtalk.*

4. CROTOLARIA (*Fruticoſa*) foliis ſimplicibus, lineari-lanceolatis hirſutis, petiolis decurrentibus, caule fruticoſo. *Shrubby hairy Crotolaria with a yellow flower, winged branches, and pointed leaves.*

5. CROTOLARIA (*Argentea*) foliis ſimplicibus lanceolatis villoſis, argenteis, ſeſſilibus, ſiliquis pendulis. *Aſiatic Crotolaria with a ſilvery hairy leaf, a yellow flower, and hanging pods diſpoſed in a ſpike.*

6. CROTOLARIA (*Perfoliata*) foliis cordato-ovatis perfoliatis. Lin. Sp. Plant. 714. *Crotolaria with a Thorough-wax leaf.*

7. CROTOLARIA (*Retuſa*) foliis ſimplicibus, oblongis cuneiformibus retuſis. Flor. Zeyl. 276. *Aſiatic Crotolaria with yellow flowers, and a ſingle heart-ſhaped leaf.*

8. CROTOLARIA (*Villoſa*) foliis ſimplicibus ovatis villoſis, petiolis ſimpliciſſimis, ramis teretibus. Hort. Cliff. 357. *African Crotolaria with a Storax-tree leaf.*

9. CROTOLARIA (*Angulata*) foliis ovatis ſeſſilibus, ramis angulatis hirſutis, floribus lateralibus ſimpliciſſimis. *Crotolaria with oval leaves ſitting cloſe to the branches, which are angular and hairy, and ſingle flowers proceeding from the ſides of the branches.*

10. CROTOLARIA (*Alba*) foliis ternis lanceolato-ovatis, caule lævi herbaceo, racemo terminali. Hort. Cliff. 499. *Crotolaria with oval ſpear-ſhaped leaves which are ternate, herbaceous ſtalks, which are terminated by a racemus of flowers; commonly called Carolina Anonis.*

The firſt ſort grows naturally in India. This is an annual plant, which hath an herbaceous four-cornered ſtalk, riſing about two feet high, dividing into three or four branches, that have four acute angles, garniſhed with oval warted leaves, of a pale green colour, ſtanding on very ſhort footſtalks. The flowers are produced in ſpikes at the end of the branches, which are of the butterfly ſhape, and of a light blue colour; theſe are ſucceeded by ſhort turgid pods, that incloſe one row of kidney-ſhaped ſeeds.

The ſecond ſort grows naturally at La Vera Cruz in New Spain. This riſes with a compreſſed winged ſtalk near three feet high, putting out ſeveral ſide branches, garniſhed with ſpear ſhaped leaves near three inches long, and one broad, covered

covered with soft hairs, sitting close to the branches alternately; from the foot-stalks of each there runs a border or leafy wing along both sides of the branches; the flowers are produced in loose spikes at the end of the branches, which are of a pale yellow colour, the standard being stretched out a considerable length beyond the wings. These are succeeded by short turgid pods, which, when ripe, are of a deep blue, having one row of small kidney-shaped seeds, which are of a greenish brown colour.

The third sort grows in South Carolina, and in several parts of America; this is an annual plant, which rises with a slender stalk a foot and a half high, dividing into three or four spreading branches, garnished with oblong oval leaves, sitting close. The upper part of the branches have two leafy borders or wings, running from one leaf to the other, but the lower part of the branches have none; the foot-stalks of the flowers arise from the side of the stalk; they are very slender, and sustain one or two pale yellow flowers at their top, which are not more than half so large as the former sort, and are succeeded by very short turgid pods, in which are inclosed three or four smooth kidney shaped seeds.

The fourth sort grows naturally in Jamaica. It rises with a shrubby taper stalk near four feet high, sending out many side branches which are very slender, ligneous, and covered with a light brown bark; they are garnished with very narrow spear-shaped leaves, which are hairy, and sit close to the branches; the younger shoots have a leafy border or wing on two sides, but the old branches have none; the flowers are produced near the end of the branches, three or four growing alternate on a loose spike; they are of a dirty yellow, and small; the pods which succeed them are about an inch long, very turgid, and of a dark blue when ripe.

The seeds of the fifth sort were brought me from the coast of Malabar. This rises with an angular stalk near four feet high, dividing upward into three or four branches, garnished with narrow spear-shaped leaves, placed alternately on very short foot-stalks, and are pretty closely covered with soft meally hairs. The flowers are produced at the end of the branches in loose spikes; they are large, of a deep yellow colour, and the style stands out beyond the standard. The flowers are succeeded by large turgid pods, containing one row of large kidney-shaped seeds.

This plant is annual in England, but by the lower part of the stalk growing woody, it appears to be of longer duration in the country where it naturally grows, though it will not live through the winter here; for if the plants are placed in a stove, the heat is too great for them, and in a green-house they are very subject to mouldiness in damp weather. I have sown the seeds of this in the full ground, where the plants have grown upward of three feet high, and have flowered very well, but no pods were formed on these; and when they have been treated tenderly, the plants have grown much larger, and produced a greater number of flowers, but these have not been succeeded by seeds. The only way which I could ever obtain any seeds, was by raising the plants in pots upon hot-beds, and the beginning of July turning them out of the pots into the full ground on a very warm border under a wall, in which situation they flowered very well, and a few pods of seeds were ripened.

The sixth sort grows naturally in South Carolina, at a great distance from the English settlements. By the description sent me with the seeds, it grows with a shrubby stalk four or five feet high; but the plants which were raised here, perished at the approach of winter, so that they only flowered, without producing any pods.

The seventh sort rises with an herbaceous stalk near three feet high, dividing into several branches, garnished with oblong leaves, which are narrow at their base, but gradually widen to the top, where they are rounded and indented in the middle in the shape of a heart; they are of a pale green, and smooth. The flowers are produced in spikes at the end of the branches; they are pretty large, and of a yellow colour. This grows naturally in the island of Ceylon, and is an annual plant, perishing soon after it perfects seeds.

The eighth sort grows naturally at the Cape of Good Hope. This rises with a shrubby stalk about five feet high, dividing into several branches, garnished with roundish leaves, sitting close; they are of a hoary green, and soft to the touch: the branches are taper and smooth; the flowers are produced at the end of the branches in loose spikes; they are about the size of those of the first sort, and of a fine blue colour.

The ninth sort grows at Campeachy; this rises with a taper upright stalk near three feet high, dividing upward into several erect branches, garnished with oval spear-shaped leaves, of a pale green colour; the flowers are produced singly from the side of the branches, which are of a bright yellow, and are succeeded by short turgid pods, having one row of kidney-shaped seeds.

The tenth sort grows naturally in Virginia and also in Carolina: from both countries I received the seeds; the plants which I first raised produced white flowers, but the seeds saved from these plants have all of them produced blue flowers. The roots of these plants are perennial, and increase annually in their size, so that in twelve or fourteen years they will produce more than twenty stalks, each of which being garnished with spikes of flowers, make a fine appearance, the flowers being large, and the spikes long, and rise near three feet high.

The plants are propagated by seeds, of which they produce plenty, especially in moist seasons; or if the plants are duly supplied with water, otherwise the seeds will not grow to half their size, nor will they grow: if the seeds are sown in the full ground about the beginning of April, the plants will appear some time in May; these should be kept clean from weeds the following summer, and when their stalks decay in the autumn, if the surface of the ground is covered with some tanners bark, it will secure the plants in winter. The spring following they should be planted where they are to remain, allowing them four feet room, that they may spread and flower.

As most of these plants are annual, so these require to be brought forward in the spring, otherwise the summers are too short for them to perfect seeds in England; so that unless the seeds are sown upon a good hot-bed in the spring, and the plants afterward carefully managed, they will not flower well here; for in general, the summers in this country are not very favourable for these tender plants. Therefore in order to have the annual sorts in perfection, there should be a low glass-case erected about five or six feet high, which should be made with glasses to open or slide down on every side, as should also the top on both sides, having sliding glasses, that the plants may have sun and air on every side: in this there should be a pit for tanners bark to make a hot-bed, in which may be placed these and other curious, tender, annual plants, where the sun will constantly shine on them, so long as he makes his appearance above the horizon; and here they may have plenty of free air admitted at all times, when the weather is warm, so may be brought to great perfection, and hereby good seeds may be annually obtained.

These plants naturally grow on sandy light soils, so they should always be planted in such, and the pots in which they are planted must not be too large, for in such they will not thrive;

thrive; so that after they have filled the small pots with their roots in which they were first planted, they should be shaken out of those, and put into penny pots, which will be large enough for all the annual kinds. The waterings of these plants should be performed with caution, for too great moisture will rot the fibres of their roots; so that in summer, if they are gently watered three or four times a week in hot weather, it will be sufficient.

CROTON. Lin. Gen. Pl. 960. Bastard Ricinus.

The CHARACTERS are,

It hath male and female flowers in the same plants; the flowers have five petals, those of the male being no larger than the leaves of the empalement, and have five nectarious glands, which are fixed to the receptacle; they have ten or fifteen stamina, which are joined at their base. The female flowers have a roundish germen, with three reflexed styles. The germen afterward becomes a roundish three-cornered capsule with three cells, each containing a single seed.

The SPECIES are,

1. CROTON (*Tinctorium*) foliis rhombeis repandis, capsulis pendulis caule herbaceo. Hort. Upsal. 290. *Bastard Ricinus, from which the Turnsol of the French is made.*

2. CROTON (*Palustre*) foliis ovato-lanceolatis, minimè serratis, caule herbaceo hirsuto, floribus alaribus pedunculis longioribus. *Marsh Bastard Ricinus, with oblong sawed leaves and a prickly fruit.*

3. CROTON (*Lobatum*) foliis inermi serratis, inferioribus quinquelobis, superioribus trilobis. Hort. Cliff. 445. *Herbaceous Bastard Ricinus, with trifid or quinquefid sawed leaves.*

4. CROTON (*Humile*) tetraphyllum, foliis lanceolatis, acuminatis, subtus cæsiis, caule herbaceo ramoso. *Dwarf Bastard Ricinus, with oblong pointed leaves, gray on their under side, and an herbaceous stalk.*

5. CROTON (*Fruticosum*) foliis lanceolatis glabris, caule fruticoso, floribus alaribus & terminalibus. *Shrubby Bastard Ricinus with a Laurel leaf, and a very large green empalement to the flower.*

6. CROTON (*Populifolium*) foliis cordatis, acuminatis, subtus tomentosis, floribus alaribus sessilibus, caule fruticoso. *Bastard Ricinus, with hairy leaves like those of Poplar.*

7. CROTON (*Rosmarinifolium*) foliis lineari-lanceolatis, glabris, subtus argenteis, caule fruticoso, floribus spicatis terminalibus. *Bastard shrubby Ricinus with narrow leaves, which are whitish on their under side; commonly called wild, or Spanish Rosemary in Jamaica.*

8. CROTON (*Salvifolium*) foliis oblongo-cordatis tomentosis, caule fruticoso ramoso, floribus spicatis terminalibus. *Shrubby American Bastard Ricinus, with a Marshmallow leaf.*

9. CROTON (*Argenteum*) foliis ovatis tomentosis, integris, serratis. Hort. Cliff. 444. *Dwarf Ricinus with roundish sawed leaves, silvery on their under side, and flowers and fruit growing in clusters.*

The first sort grows naturally in the south of France; this is an annual plant, which rises with an herbaceous branching stalk about nine inches high, garnished with irregular, or rhomboidal figured leaves, which are near two inches long, and one inch and a quarter broad, standing upon slender foot-stalks. The flowers are produced in short spikes from the side of the stalks; the upper part of the spike is composed of male flowers, having many stamina, which coalesce at the bottom; the lower hath female flowers, which have each a roundish three-cornered germen, which afterward becomes a roundish capsule with three lobes, having three cells, each including one roundish seed.

The seeds of this plant should be sown in the autumn on a border of light earth, in a warm situation, where they are designed to remain; and when the plants come up, they should be thinned where they are too close, leaving them six inches asunder; after this they will require no other care but to keep them clean from weeds. If the summer proves favourable, the plants will flower in July, and in very warm autumns they sometimes perfect their seeds in England.

The second sort was discovered by the late Dr. Houstoun at La Vera Cruz. This is also an annual plant, which grows naturally in low marshy grounds, where it hath a very different appearance from what it puts on when sown upon dry land; those of the watery places have broad flat stalks, and leaves three inches long, which are scarce a quarter of an inch broad; they are rough, and but little indented on their edges, but those plants upon dry ground have oval leaves three inches long, and upwards of two inches broad, which are sawed on their edges. The flowers are produced at the wings of the leaves in short loose spikes, having four or five herbaceous male flowers at the top of each, and three or four female flowers at bottom, which are succeeded by roundish capsules with three lobes, covered with a prickly husk, with three cells, each inclosing a single seed.

The third sort was discovered by the same gentleman, at the same place as the former; this is an annual plant, which rises with a taper herbaceous stalk a foot and a half high, dividing into several branches, garnished with smooth leaves, standing upon very long foot-stalks, which are for the most part placed opposite; the lower leaves are divided deeply into five oblong lobes, and the upper into three, which are slightly sawed on their edges, and end in acute points. The flowers are produced in loose spikes at the end of the branches, those on the upper part being male, and the lower female; they are of a whitish herbaceous colour; the female flowers are succeeded by oblong capsules, having three lobes, which open in three parts, having three cells, each containing one oblong seed.

The fourth sort grows naturally at the Havannah. This is an annual plant, which rarely grows more than nine inches high, dividing into two or three branches; the lower parts of the branches are garnished with four leaves, placed in form of a cross, two of which are three inches long, and one inch broad near their base, ending in acute points; these stand opposite, and the other two leaves between these are about two inches long, and a quarter of an inch broad; they are of a light green on their upper side, and of a gray or Ash colour on their under. The flowers are produced in long loose spikes at the top of the stalks; the upper part of these spikes have male, and the lower female flowers, of an herbaceous colour; the female flowers are succeeded by round capsules with three cells, each containing one roundish seed.

The fifth sort grows naturally in Jamaica. It rises with a shrubby stalk to the height of seven or eight feet, covered with an Ash-coloured bark, dividing into many slender branches upward, which are naked below, but toward their upper part are garnished with smooth spear-shaped leaves, about two inches and a half long, and three quarters of an inch broad, standing on pretty long foot-stalks; the flowers are produced in short spikes at the end of the branches, in the same manner as the former; they are of an herbaceous colour, and inclosed in large green empalements.

The sixth sort grows in Jamaica. This rises with a shrubby stalk seven or eight feet high, sending out many irregular branches, covered with an Ash-coloured bark, garnished with heart-shaped leaves near four inches long, and two inches broad, ending in acute points; they are of a light green on their upper side, but woolly on their under, standing on slender foot-stalks, sometimes single, and at others two or three arise from the same joint. The flowers are produced in short spikes from the side of the branches;

branches; they are of a whitish green colour, and the female flowers are succeeded by capsules having three cells, each including a single seed.

The seventh sort grows naturally in Jamaica. This rises with a shrubby stalk about six or seven feet high, sending out many side branches, which are covered with a smooth bark, of a pale yellow colour, garnished closely with narrow stiff leaves near three inches long, and about one-eighth of an inch broad, of a light green on their upper side, but the under side is the same colour as the bark. Between the branches arise a long loose spike of whitish green flowers. The whole plant hath an aromatic odour when rubbed. The upper part of the spike hath male flowers, the lower female; the seeds grow in roundish capsules having three cells, each including a single seed.

The eighth sort grows naturally in Jamaica. This rises with a shrubby stalk six or seven feet high, dividing into several branches, whose bark is covered with a yellowish down, and are garnished with long heart-shaped leaves, ending in acute points, standing on long foot-stalks, covered on both sides with a woolly down, of the same colour as the branches. The flowers are produced on long close spikes at the end of the branches; the male flowers, which are situated on the upper part of the spikes, have white flowers of one leaf, divided into five parts almost to the bottom. The female flowers, on the lower part of the spikes, have large woolly empalements, and are succeeded by round capsules with three cells, each including a single seed.

The ninth sort grows naturally at Campeachy. This is an annual plant, which rises with an herbaceous stalk two feet and a half high, dividing into several small branches, garnished with oval woolly leaves near three inches long, and two and a half broad in the middle. The flowers are produced at the extremity of the branches in short close spikes or clusters, sitting close between the leaves; they are small, of a whitish green colour; the female flowers are succeeded by small round capsules, inclosed by the woolly empalement; they have three cells, but seldom more than one of them contains seeds, which must be gathered as soon as they are ripe, otherwise they will scatter.

All these plants, except the first, are natives of warm countries, so will not thrive in England unless they are tenderly treated. They are propagated by seeds; those which are annual perfect their seeds in England, but the shrubby sorts very rarely arrive to that perfection. The seeds must be sown on a hot-bed early in the spring, and when the plants are fit to remove, they should be each transplanted into a small pot, and plunged into a moderate hot-bed of tanners bark, where they should be shaded from the sun till they have taken fresh root; then they must have air admitted to them daily, in proportion to the warmth of the season. After the plants are grown too tall to remain in the frames, they should be removed either into the stove or a glass-case, where there is a hot-bed of tanners bark, into which the pots should be plunged, and there the annual sorts will flower and perfect their seeds; but the shrubby kinds must be removed into the bark-stove in the autumn, and during the winter season they should have but little water.

As the perennial sorts retain their leaves all the year, so they make a pretty variety in winter, when they are intermixed with other plants, whose leaves are of different forms and colours from these.

CROWN IMPERIAL. See Ptellium.

CRUCIANELLA. Lin. Gen. Pl. 118. Petty Madder.

The Characters are,

The flower hath one petal, with a slender cylindrical tube longer than the empalement, cut into four parts at the brim. It hath four stamina, situated in the mouth of the tube; and hath a compressed germen, situated at the bottom of the tube, which afterward becomes two twin capsules, each containing one oblong seed.

The Species are,

1. Crucianella (*Angustifolia*) erecta, foliis senis linearibus. Hort. Upsal. 27. *Petty Madder with a narrower leaf.*

2. Crucianella (*Latifolia*) procumbens, foliis quaternis lanceolatis, floribus spicatis. Hort. Upsal. *Broad-leaved Petty Madder.*

3. Crucianella (*Maritima*) procumbens, foliis quaternis, floribus subverticillatis. Lin. Sp. Pl. 109. *Maritime Petty Madder.*

The first sort grows naturally in the south of France and Italy; it is an annual plant, which rises with several upright stalks a foot high, which have six or seven very narrow linear leaves, placed in whorls at each joint. The flowers grow in close spikes at the top and from the side of the branches; these are small, white, and not longer than the empalement, so make no great appearance.

The second sort grows in the islands of the Archipelago, and also about Montpelier. This is also an annual plant, sending out several branching stalks from the root, which lie prostrate, and are garnished with four spear-shaped leaves at each joint. The flowers are produced in long spikes at the extremity of the branches; they are very small, so make no appearance.

The third sort is like the second in the appearance of its leaves and stalks, but the flowers grow on the side of the stalks almost in whorls, and make little appearance. This grows naturally on the borders of the sea, in the south of France and Italy.

These three sorts are preserved in some gardens for the sake of variety: if the seeds are sown on a bed of light earth in the spring, where they are designed to remain, they will require no other culture but to thin them where they are too close, and keep them clean from weeds: or if the seeds are permitted to scatter, the plants will come up in the spring, and require no other treatment: they are all annual plants.

CRUCIATA. See Valantia.

CRUPINA BULGARUM. See Centaurea.

CUCUBALUS. Lin. Gen. Plant. 502. Berry-bearing Chickweed.

The Characters are,

The flower hath five petals, with tails as long as the empalement, but spread open at the top, and ten stamina, five of which are alternately inserted in the tail of the petals. In the center is situated the oblong germen, supporting three styles. The empalement afterward becomes a pointed close capsule with three cells, opening at top in five parts, and filled with many roundish seeds.

The Species are,

1. Cucubalus (*Bacciferus*) calycibus campanulatis, petalis distantibus, fructu colorato, ramis divaricatis. Lin. Sp. Pl. 414. *Climbing Berry-bearing Chickweed.*

2. Cucubalus (*Latifolius*) caulibus erectis glabris, calycibus subglobosis, staminibus corollâ longioribus. *Wild Lychnis, or white Behen of the shops; commonly called Spattling Poppy.*

3. Cucubalus (*Angustifolius*) calycibus subglobosis, caule ramoso patulo, foliis linearibus acutis. *Wild Lychnis or Spattling Poppy, with narrower pointed leaves.*

4. Cucubalus (*Beben*) calycibus subglobosis glabris reticulato-venosis, capsulis trilocularibus corollis subnudis. Flor. Suec. 360. *Swedish Lychnis with a leaf and appearance of white Behen, having a large empalement, called Gumsepungar.*

5. Cucu-

5. CUCUBALUS (*Anglicus*) caulibus procumbentibus, calycibus amplissimis nervosis, foliis lanceolatis. *English Sea Lychnis.*

6. CUCUBALUS (*Fabarius*) foliis obovatis carnosis. Prod. Leyd. 448. *Rocky maritime Lychnis with an Orpine leaf.*

7. CUCUBALUS (*Dubrensis*) floribus lateralibus decumbentibus, caule indiviso, foliis basi reflexis. Lin. Sp. Pl. 414. *Greater perennial Night-flowering Lychnis of Dover.*

8. CUCUBALUS (*Stellatus*) foliis quaternis. Hort. Upsf. 110. *Lychnis with smooth Gentian leaves, four at each joint embracing the stalk, and a large fringed flower.*

9. CUCUBALUS (*Noctiflora*) calycibus striatis acutis petalis bipartitis, caule paniculato, foliis linearibus. *Narrow-leaved, sweet-scented, Night-flowering Lychnis.*

10. CUCUBALUS (*Otites*) floribus dioicis, petalis, setaceis indivisis. Hort. Cliff. 272. *Viscous Lychnis with a mossy flower.*

11. CUCUBALUS (*Alpinus*) acaulis. Flor. Lapp. 184. *Alpine Dwarf Lychnis with a grassy flower, or Alpine Moss with a flower of Lychnis.*

12. CUCUBALUS (*Italicus*) petalis bipartitis, floribus paniculatis, staminibus longis, foliis lanceolatis acutis. *Tallest Lychnis with the appearance of Wild Campion.*

The first sort grows naturally in France, Germany, and Italy, in shady places, and is seldom kept in gardens, unless for the sake of variety; it sends out many climbing stalks, which grow four or five feet high, where they meet with support, otherwise they trail on the ground; these stalks send out side branches by pairs, opposite, at each joint; the leaves are like those of Chickweed, and are placed opposite. The flowers come out single at the end of the branches, which have large inflated empalements; they consist of five petals, which are white, and are placed at a distance from each other; these are succeeded by oval berries, which, when ripe, are black and full of juice, inclosing several flat shining seeds. This hath a perennial creeping root, whereby it is apt to multiply too fast in gardens. It delights in shade, and will thrive in almost any soil.

The second sort grows naturally in most parts of England, where it is generally called Spattling Poppy. This stands in the catalogue of medicinal plants, under the title of Behen album; the roots of it are sometimes used, and are accounted cordial, cephalic, and alexipharmic. It hath a perennial root, which strikes deep into the ground, so that they are not easily destroyed by the plough, therefore it is frequently seen growing in bunches among Corn. It is a rambling weed, so is seldom cultivated. There are two varieties of this, one with smooth, the other hath hairy leaves.

The third sort grows naturally on the Alps; this differs from the former, in having much longer and narrower leaves, and the stalks being more divided and spreading, nor do the roots creep under ground like those of the former. These differences are constant from seeds.

The fourth sort grows naturally in Sweden, and some other northern countries, where it passes for the common sort, but is certainly a distinct species, and I have been informed has been found growing naturally in England. The stalks of this are much larger, the leaves longer and more pointed; the empalement of the flower is curiously veined like net-work, of a purplish colour, whereas that of the common sort is plain. These differences are lasting, when the plants are cultivated in a garden.

The fifth sort grows naturally on the borders of the sea, in many parts of England. This is by some supposed to be the same as the second sort, from which it greatly differs; the stalks of this are weak, and trail upon the ground; the leaves are shorter, those upon the stalks are much broader, and the empalement of the flowers are netted with purple veins like the Swedish sort before mentioned.

The sixth sort was discovered by Tournefort in the Levant. This puts out many oval, thick, succulent leaves near the ground, out of the middle of which arises an upright stalk about fifteen inches high, the lower part of which is garnished with leaves, of the same form and consistence as Orpine, but smaller; they are placed opposite; the upper part of the stalk divides into two smaller, on which stand a few small herbaceous flowers at each joint. The plant is biennial, generally perishing when it has produced seeds; but unless it is sown upon dry rubbish, in a warm situation, the plants will not live through the winter in England; for when they are in good ground they grow large, and are so replete with moisture as to be affected by the first frost in the autumn; but where they have grown upon an old wall, I have known them escape, when all those were killed which grew in the ground.

The seventh sort grows naturally upon the cliffs near Dover. This hath a perennial root, from which arises single stalks about a foot high, garnished with long narrow leaves, placed opposite; the flowers are produced from the side of the stalks, each foot-stalk sustaining three flowers; the foot-stalks come out by pairs; the empalement of the flower is long and striped, the flowers are of a pale red.

The eighth sort grows naturally in Virginia and several other parts of North America This hath a perennial root, from which arise two or three slender upright stalks about three feet high, garnished with four leaves at each joint, placed in form of a cross; they are smooth, of a deep green, about an inch and a half long, and half an inch broad, terminating in acute points; the joints of the upper part of the stalk are garnished with white fringed flowers, standing single upon pretty long foot-stalks, which come out by pairs opposite.

The ninth sort grows naturally in Spain and Italy. This is a perennial plant, which rises with an upright branching stalk a foot and a half high, garnished with very narrow leaves placed opposite; the upper part of the stalk is very branching; the flowers stand upon long naked foot-stalks, each supporting three or four flowers, which have long tubes, with striped empalements; the petals are large, and deeply divided at the top; they are of a pale bluish colour. The flowers are closed all day, but when the sun leaves them they expand, and then they have a very agreeable scent. This sort may be propagated by seeds, which should be sown in the spring upon a bed of light earth; and when the plants are fit to remove, they should be planted in a nursery-bed at about four inches distance, where they may remain till autumn, when they may be planted in the borders of the flower-garden, where they are designed to remain. The following summer these will produce their flowers, and ripen their seeds in the autumn; but the roots will continue several years, provided they are not planted in rich ground, where they are very subject to rot in winter.

The tenth sort grows naturally in Austria, Silesia, and Italy. This sort is male and female in different plants; it hath a thick, fleshy, perennial root, which strikes deep into the ground, sending out many oblong leaves, narrow at their base; from between these arise the stalks, which in the male plants often grow four feet high, but those of the female plants are seldom above three; the stalks are garnished with narrow leaves placed opposite; at the joints there exsudes a viscous clammy juice, which sticks to the fingers when handled; and the small insects which settle upon those parts of the stalks, are thereby fastened so as not to get loose again. The flowers of the male plants are produced in loose spikes from the joints of the stalks in clusters; these are small, of a greenish colour, and have each ten stamina,

stamina. The female plants have three or four flowers growing upon each foot-stalk, which arise from the side of the stalk. These are succeeded by oval seed-vessels, containing many small seeds. This is propagated by seeds, which should be sown where the plants are designed to remain; for as they send out long tap-roots, so they do not bear transplanting, unless it is performed while the plants are young

The eleventh sort grows naturally on the Alps, and also upon some hills in the north of England and Wales. This is a very low plant with small leaves, which spread on the ground, and have the appearance of Moss; the flowers are small, erect, and rarely rise more than half an inch high; they are of a dirty white colour, and appear in May. This is a perennial plant, which will not thrive but in a moist soil and a shady situation.

The twelfth sort grows naturally in Italy and Sicily. This is a perennial plant, with large thick roots, sending out many long spear shaped leaves; between these arise round viscous stalks, which grow four or five feet high, garnished at each joint by two long narrow leaves, ending in acute points. The stalks branch out into many divisions; the foot-stalks of the flowers arise from each joint by pairs; each of these sustain three or four flowers of an herbaceous colour, whose petals are divided into two parts. This is propagated by seeds in the same manner as the tenth.

CUCUMIS. Lin. Gen. Pl. 969. Cucumber.

The CHARACTERS are,

It hath male and female flowers on the same plant. The flowers are bell-shaped, of one petal, which adheres to the empalement, and is cut into five oval rough segments. The male flowers have three short stamina, which are inserted in the empalement. The female flowers have no stamina, but have three small pointed filaments without summits. The germen is situated under the flower, which afterward becomes an oblong fleshy fruit with three cells, including many oval, flat, pointed seeds.

The SPECIES are,

1. CUCUMIS (*Sativus*) foliorum angulis erectis, pomis oblongis scabris. Hort. Cliff. 451. *The common Garden Cucumber.*

2. CUCUMIS (*Flexuosus*) foliorum angulis rectis, pomis longissimis scabris. *The long Turky Cucumber.*

3. CUCUMIS (*Chata*) foliis rotundo-angulatis, pomis acutangulis. Lin. Sp. Pl. 1011. *Round-leaved Egyptian Cucumber, called Chate.*

The first sort is the Cucumber which is generally cultivated for the the table, and is so well known as to need no description.

The second sort is the long Turky Cucumber, which is also pretty well known in England. The stalks and leaves of this sort are much larger than those of the common sort. The fruit is generally twice the length, and hath a smooth rind; this is undoubtedly different from the common sort. There are green and white fruit of this, and also of the common sort, which differ but little except in their colour. The white is less watery than the green, so is generally better esteemed.

The third sort here enumerated, is rarely cultivated but in botanic gardens for the sake of variety, the fruit being very indifferent, and the plants being tender require a good heat to bring them to perfection in England: these plants ramble very far, so must have much room.

The common sort is cultivated in three different seasons; the first of which is on hot-beds under garden-frames, for early fruit; the second is under bell or hand-glasses, for the middle crop; and the third is in the common ground, for a late crop, or to pickle.

I shall begin with giving directions for raising Cucumbers early, which is what most gentlemen's gardeners have an emulation to exceed each other in; and some have been at the pains and expence to have ripe fruit in every month of the year, which is rather a curiosity than real advantage; but as there are many persons who yet value themselves on their skill in raising early Cucumbers, so we may probably be censured as being deficient in what they call the essential part of gardening, should we omit the method practised for raising these fruit early in the year, therefore shall proceed to give such directions, which, if carefully attended to, will not fail of success.

Those persons who are very desirous to be early with their Cucumbers, generally sow their seeds before Christmas, but the generality of gardeners commonly put their seeds into the hot-bed about Christmas; but where persons have the conveniency of a stove for raising these plants, it is attended with less trouble than a common hot-bed, and is a much surer method, because the plants will have a much greater share of air than under frames; therefore when there is this convenience, the seeds may be sown in small pots, and plunged into the tan-bed, in the warmest part of the stove. The seeds should be at least three or four years old, but if it is more, provided it will grow, it will be the better. When the plants are up, and begin to put out their rough leaf, there should be a sufficient number of small pots filled with good earth, and plunged into the bark-bed, that the earth may be warmed to receive the plants, which should be pricked into these pots, two plants in each; but when they have taken root and are safe, one of the worst should be drawn out, being careful not to disturb the roots of that which is left. In the management of these plants there must be great care taken not to give them too much water, and it will be very proper to put the water into the stove some hours before it is used, that the cold may be taken off; but there must be caution used not to make it too warm, for that will destroy the plants; they must also be guarded from the moisture which frequently drops from the glasses of the stoves, which is very hurtful to these plants while young; then there should be a proper quantity of new dung prepared for making a hot-bed to receive them; this must be in proportion to the quantity of holes or plants intended; for a middling family six or nine lights of Cucumbers will be sufficient, and for a large family double the quantity. The dung should be new, and not too full of straw; it should be well mixed together, and thrown in a heap, mixing some sea-coal ashes with it; after it hath lain in a heap a few days, and has fermented, it should be carefully turned over and mixed, laying it up again in a heap; and if there is a great share of straw in it, there may be a necessity for turning it over a third time, after having lain a few days; this will rot the straw and mix it thoroughly with the dung, so there will be less danger of its burning when the bed is made, which should be done when the dung is in proper order. The place where the hot-bed is made should be well sheltered with Reed hedges, and the ground should be dry; then there should be a trench made in the ground, of a proper length and breadth, and a foot deep at least, into which the dung should be wheeled and carefully stirred up and mixed, so that no part of it should be left unseparated; for where there is not this care taken, the bed will settle unequally: there should also be great care taken to beat the dung down close in every part of the bed alike; when the bed is made, the frames and glasses should be put upon it to keep out the rain, but there should be no earth laid upon the dung, till two or three days after, that the steam of the dung may have time to evaporate: if there should be any danger of the bed burning, it will be proper to lay some short old dung, or some neats dung, over the top of the hot dung about two inches thick, which will keep down the heat, and

and prevent the earth from being burnt. The usual quantity of dung allowed for making of the bed at this season, is one good cart load to each light: in about three days the bed will be in a proper temperature of heat to receive the plants, at which time the dung should be covered over with dry earth about two inches thick, and in the middle of the bed it should be a foot thick. This should be laid upon the dung two or three days before the plants are removed into the bed, that the earth may be properly warmed; then the plants should be carefully shaken out of the pots, preserving all the earth to their roots, and placed on the top of the earth in the middle of the bed. Two of these planrs will be sufficient for each light, and these should be placed at about seven or eight inches asunder, but not all their roots together, as is too often practised. When the plants are thus situated in the bed, the earth which was laid a foot thick in the middle of the bed, should be drawn up round the ball which remained to the roots of the plants, into which the roots will soon strike; there should always be a magazine of good earth laid under cover to keep it dry, for the earthing of these beds, for if it is taken up wet it will occasion great damps in the bed, therefore it is quite necessary to have a sufficient quantity of earth prepared long before it is used. When the plants are thus settled, they must have proper air and water, according to the weather, being careful not to admit too much cold air, or give too much water; the glasses should also be well covered with mats every night, to keep up the warmth of the bed, and some fresh earth should be put into the bed at different times, which should be laid at some distance from the roots of the plants till it is warmed, and then should be drawn up round the heap of earth in which the plants grow: this should be raised to the full height of the former ball, that the roots of the plants may more easily strike into it; by this method of supplying the earth, the whole surface of the beds will be covered a foot deep with earth, which will be of great service to the roots of the plants; for where the earth is very shallow, the leaves of the plants will always hang in the heat of the day, unless they are shaded, and the plants will require more water to keep them alive than is proper to give them; therefore it will be found much the better way to allow a proper depth of earth to the beds: by thus gradually applying the earth it will be fresh, and much better for the roots of the plants, than that which has been long upon the bed, and has been too much moistened by the steam arising from the dung.

If the heat of the bed should decline, there should be some hot dung laid round the side of the bed to renew the heat, for if that should fail at the time that the fruit appears, they will fall off and perish, therefore this must be carefully regarded; and when the plants have put out side branches (which the gardeners call runners) they should be properly placed, and pegged down with small forked sticks to prevent their rising up to the glasses, and also from crossing and entangling with each other; so that when they are properly directed at first, there will be no necessity of twisting and tumbling the plants afterward, which is always hurtful to them.

When the earth of the bed is laid the full thickness, it will be necessary to raise the frames, otherwise the glasses will be too close to the plants; but when this is done, there must be care taken to stop the earth very close round the sides of the frame, to prevent the cold air from entering under them. The watering of the plants and admitting fresh air to them, must be diligently attended to, otherwise the plants will be soon destroyed, for a little neglect either of admitting air, or letting in too much, or by over watering, or starving the plants, will very soon destroy them past recovery.

When the fruit appears upon the plants, there will also appear many male flowers on different parts of the plant; these may at first sight be distinguished, for the female flowers have the young fruit situated under them, but the male have none, but have three stamina in their center, with their summits, which are loaded with a golden powder; this is designed to impregnate the female flowers, and when the plants are fully exposed to the open air, the soft breezes of wind convey this farina or male powder from the male to the female flowers; but in the frames, where the air is frequently too much excluded at this season, the fruit often drops off for want of it: and I have often observed, that bees which have crept into the frames when the glasses have been raised to admit the air, have supplied the want of those gentle breezes of wind, by carrying the farina of the male flowers on their hind legs into the female flowers, where a sufficient quantity of it has been left to impregnate them. These insects have taught the gardeners a method to supply the want of free air, which is so necessary for the performance of this in the natural way; this is done by carefully gathering the male flowers, at the time when this farina is fully formed, and carrying them to the female flowers, turning them down over them, and with the nail of one finger gently striking the outside of the male, so as to cause the powder on the summits to scatter into the female flowers, and this is found sufficient to impregnate them; so that by practising this method, the gardeners have now arrived at a much greater certainty than formerly to procure an early crop of Cucumbers and Melons; and by this method the florists have arrived to greater certainty of procuring new varieties of flowers from seeds, which is done by the mixing of the farina of different flowers into each other.

When the fruit of the Cucumbers are thus fairly set, if the bed is of a proper temperature of warmth, they will soon swell and become fit for use; so all that is necessary to be observed, is to water the plants properly, which should be done by sprinkling the water all over the bed, for the roots of the plants will extend themselves to the side of the beds; therefore those who are inclined to continue these plants as long as possible in vigour, should add a sufficient thickness of dung and earth all round the sides of the beds, so as to enlarge them to near double their first width; this will supply nourishment to the roots of the plants, whereby they may be continued fruitful great part of the summer; whereas, when this is not practised, the roots of the plants, when they have reached the side of the beds, are dried by the wind and sun, so that the plants languish and decay long before their time.

Those gardeners who are fond of producing early Cucumbers, generally leave two or three of their early fruit, which are situated upon the main stem of the plant near the root for seed, which, when fully ripe, they carefully save to a proper age for sowing, and by this method they find a great improvement is made of the seed; and this they always use for their early crops only, for the succeeding crops do not deserve so much care and attention.

I have here only mentioned the method of raising the young Cucumber plants in stoves, for as these conveniencies are now pretty generally made in the curious kitchen-gardens in most parts of England, so this method may be more universally practised; but in such gardens where there are no stoves, the seeds should be sown upon a well prepared hot-bed; and here it will be the best way to sow the seeds in small halfpenny pots, because these may be easily removed from one bed to another if the heat should decline; or, on the contrary, if the heat should be too great, the pots may be raised up, which will prevent the seed or the young plants from being injured thereby. When the plants are come up, as was before directed, there should be a fresh hot-bed

bed prepared, with a sufficient number of halfpenny pots plunged therein ready to receive the plants, which must be planted into them in the same manner as before directed, and the after-management of the plants must be nearly the same; but as the steam of the hot-bed frequently occasions great damps, so there must be great care to turn and wipe the glasses frequently, to prevent the condensed moisture falling on the plants, which is very destructive to them. There must also be great attention to the admitting fresh air at all proper times, as also to be careful in keeping the bed to a proper temperature of heat; for as there is a want of fire to warm the air, so that must be supplied by the heat of dung; afterward these plants must be ridged out, in the same manner as before directed.

About the middle of March, or a little later, according to the earliness of the season, you must put in your seeds for the second crop, which may be sown either under a bell-glass, or in the upper side of your early hot-bed; and when the plants are come up, they should be pricked upon another moderate hot-bed, which should be covered with bell or hand-glasses, placed as close as possible to each other; the plants should be set about two inches distance from each other, observing to shade them until they have taken root This is to be understood of such places where a great quantity of plants are required, which is constantly the case in the kitchen-gardens near London; but where it is only for the supply of a family, there may be plants enough raised on the upper side of the beds where the first crop is growing: you must raise the glasses on the opposite side from the wind, to give air to the plants every day when the weather is warm, which will greatly strengthen them; you must also water them as you shall find they require it, but this must be done sparingly while the plants are young.

About the middle of April the plants will be strong enough to ridge out, you must therefore be provided with a heap of new dung, in proportion to the quantity of holes you intend to plant, allowing one load to five or six holes. When your dung is fit for use, you must dig a trench about two feet four inches wide, and in length just as you please, or the place will allow; and if the soil be dry, it should be ten inches deep, but if wet, very little in the ground, levelling the earth in the bottom; then put in your dung, observing to stir and mix every part of it as was directed for the first hot-beds, laying it close and even.

When this is done, you must cover the ridge over with earth about four inches thick, laying the earth the same thickness round the sides, raising hills in the middle at three feet and a half asunder; then you must set the glasses upon the hills, leaving them close down about twenty-four hours, in which time the earth in the hills will be warmed sufficiently to receive the plants; then with your hand stir up the earth, making it a little hollow in the middle in form of a bason; into each of which you should plant four plants, observing to water and shade them until they have taken root; after which time you must be careful to give them air by raising the glasses on the opposite side to the wind, in proportion to the heat of the weather; but you must only raise the glasses in the middle of the day, until the plants fill the glasses, at which time you should raise the glasses with a forked stick on the south-side in height proportionable to the growth of the plants, that they may not be scorched by the sun: this also will harden and prepare the plants to endure the open air, but you should not expose them too soon thereto, for it often happens that there are morning frosts in May, which are many times destructive to these plants when exposed thereto; it is therefore the surest method to preserve them under the glasses, as long as they can be kept without prejudice to the plants; and if the glasses are raised with three bricks, they may be kept a great while without danger.

Towards the latter end of May, when the weather appears settled and warm, you should turn the plants down gently out of the glasses, but do not perform this in a very dry, hot, sunny day, but rather when there is a cloudy sky, and an appearance of rain; in doing of this raise the glasses either upon three bricks, or three forked sticks, whereby they may stand secure at about four or five inches high from the ground, that the plants may lie under them without bruising; nor should you take the glasses quite away till the latter end of June, or the beginning of July, for these will preserve the moisture much longer to their roots than if they were quite exposed to the open air; about three weeks after you have turned the plants out of the glasses, they will have made a considerable progress, especially if the weather has been favourable; at which time you should dig up the spaces of ground between the ridges, laying it up to the sides of the bed, that the roots of the plant may strike into it; then lay out the runners of the vines in exact order, and be careful in this work not to disturb the vines too much, nor to bruise or break the leaves. After this there will be no farther care needful, but only to keep them clear from weeds, and to water them as often as they shall require, which they will soon shew, by the hanging of their greater leaves. The ridges thus managed, will continue to produce large quantities of fruit from June until the latter end of August.

From these ridges people commonly preserve their Cucumbers for seed, by making choice of one or two of the fairest fruit upon each hole, situated near the root of the plant; but those persons who value themselves upon producing Cucumbers very early, commonly leave three or four Cucumbers of the first produce of their earliest crop, when the fruit is fair; and the seeds of these early fruit are generally preferred to any other for the first crop. These should remain upon the vines until the seeds are perfectly ripe; and when you gather them from the vines, it will be proper to set the fruit in a row upright against a hedge or wall, where they may remain until the outer cover begins to decay; at which time you should cut them open, and scrape out the seeds, together with the pulp, into a tub, which should be afterwards covered with a board, to prevent filth from getting amongst the pulp. In this tub it should be suffered to remain eight or ten days, observing to stir it well with a stick to the bottom every day, in order to rot the pulp, that it may be easily separated from the seeds; then pour some water into the tub, stirring it well about, which will raise the scum to the top, but the seeds will settle to the bottom; so that by two or three times pouring in water, and afterwards straining it off from the seeds, they will be perfectly cleared from the pulp; then they should be spread upon a mat, which should be exposed to the open air three or four days, until they are perfectly dry, when they may be put up in bags, and hung up in a dry place, where vermin cannot come to them, where they will keep good for several years, but are generally preferred when three or four years old, as being apt to produce less vigorous, but more fruitful plants.

I shall, in the next place, proceed to give directions for managing Cucumbers for the last crop, or what are generally called picklers.

The season for sowing these is in the beginning of June, when the weather is settled. The ground where these are commonly sown by the London gardeners, is between the wide rows of Cauliflowers, which are four feet and a half asunder. In these rows they dig up square holes at about three feet and a half distance from each other, breaking the earth well with a spade, and afterwards smoothing and hollowing it in the form of a bason with their hands; then they put eight or nine seeds into the middle of each hole,

covering

covering them over with earth about half an inch thick; if the weather is very dry, they water the holes gently in a day or two after the feeds are fown, in order to facilitate their vegetation.

In five or fix days, if the weather is good, the plants will begin to thruft their heads above ground; at which time they are careful to keep off the fparrows, which are very fond of the young tender heads of thefe plants; and, if they are not prevented, will pinch them off, and thereby deftroy the whole crop: but as it is not above a week that the plants are in this danger, it will be no great trouble to look after them during that time; for when the plants have expanded their feed leaves, the fparrows will not meddle with them.

There muft alfo be care taken to water them gently, as the feafon may require; and when the third or rough leaf of the plants begin to appear, all the weakeft plants fhould be drawn out, leaving only four of the moft promifing and beft fituated in each hole, ftirring the earth round about them with a fmall hoe to deftroy the weeds, and raife the earth about the fhanks of the plants, putting a little earth between them, preffing it gently down with the hand, that the plants may be thereby feparated from each other to a greater diftance; then they give them a little water (if the weather is dry) to fettle the earth about them, which muft be afterwards repeated as often as it fhall be neceffary, ftill being careful to keep the ground clear from weeds.

When the Cauliflowers are quite drawn off the ground from between the Cucumbers, they hoe and clean the ground, drawing the earth up round each hole of Cucumbers in form of a bafon, the better to contain the water when it is given them: then lay out the plants in exact order as they are to run, fo that they may not interfere with each other; then lay a little earth between the plants left, preffing it down gently with the hand, the better to fpread them each way, giving them a little water to fettle the earth about them, repeating it as often as the feafon fhall require. The plants thus managed will begin to produce fruit toward the latter end of July, or the beginning of Auguft, when they either gather them young for pickling, or fuffer them to grow for large fruit.

The quantity of holes neceffary for a family, is about fifty or fixty; for if there are fewer, they will not produce enough at one gathering to make it worth the trouble and expence of pickling, without keeping them too long in the houfe, for there are rarely more than two hundred fit to gather at each time from fifty holes; but this may be done twice a week during the whole feafon, which commonly lafts five weeks; fo that from fifty holes may be reafonably expected about two thoufand in the feafon, which, if they are taken fmall, will not be too many for a private family. And if fo many are not wanted, they may be left to grow to a proper fize for eating.

CUCUMIS AGRESTIS. See MOMORDICA.

CUCURBITA. Lin. Gen. Plant. 968. The Gourd.

The CHARACTERS are,

It hath male and female flowers in the fame plant. The male flowers have three ftamina, which are connected at their extremity, but are diftinct at their bafe. The female flowers have a large germen fituated under them, fupporting a conical trifid ftyle, which afterward becomes a large flefhy fruit, having three foft membranaceous cells which are diftinct, inclofing two rows of feeds which are bordered.

The SPECIES are,

1. CUCURBITA (*Lagenaria*) foliis cordatis denticulatis tomentofis, bafi fubtus biglandulofis, pomis lignofis. Lin. Sp. Plant. 1010. *Long Gourd with a foft leaf and white flower, commonly called the Long Gourd.*

2. CUCURBITA (*Pepo*) foliis lobatis, pomis lævibus. Lin. Sp. Plant. 1010. *Greater round Gourd, with a yellow flower and rough leaf; commonly called Pompion, or Pumpkin.*

3. CUCURBITA (*Verrucofa*) foliis lobatis, pomis nodofo-verrucofis. Lin. Sp. Plant. 1010. *Warted Gourd.*

4. CUCURBITA (*Melopepo*) foliis lobatis, caule erecto, pomis depreffo-nodofis. Lin. Sp. Pl. 1010. *Melopepo having a fhield-fhaped fruit; commonly called Squafh.*

The firft fort is fometimes propagated in the Englifh gardens by way of curiofity, for the fruit is very rarely eaten here; though if they are gathered when they are young, while the fkins are tender and boiled, they have an agreable flavour. In the eaftern countries thefe fruit are very commonly cultivated and fold in the markets for the table, and are a great part of the food of the common people, from June to October.

This fort doth not vary like moft of the others, but always produces the fame fhaped fruit; the plants of this extend to a great length, if the feafon proves warm and favourable, and will then produce ripe fruit, but in cold fummers the fruit feldom grows to half its ufual fize. I have meafured fome of thefe fruit when growing, which were fix feet long, and a foot and a half round; the plants were near twenty feet in length; the ftalks of this, and alfo the leaves, are covered with fine foft hairy down; the flowers are large, white, and ftand upon long foot-ftalks, being reflexed at their brim; the fruit is generally incurved and crooked, and when ripe is of a pale yellow colour. The rind of this fruit becomes hard, fo that if the feeds and pulp are taken out and the fhell dried, it will contain water; and in thofe countries where they are much cultivated, are ufed for many purpofes.

The fecond fort, which is commonly known by the title of Pumpkin, is frequently cultivated by the country people in England, who plant them upon their dunghills, where the plants run over them, and fpread to a great diftance, and when the feafons are favourable they will produce plenty of large fruit; thefe they ufually fuffer to grow to maturity, then they cut a hole on one fide, and take the feeds out of the pulp as clean as poffible, after which they fill the fhell with Apples fliced, which they mix with the pulp of the fruit, and fome add a little fugar and fpice to it; then bake it in an oven, and eat it in the fame manner as baked Apples; but this is a ftrong food, and only fit for thofe who labour hard, and can eafily digeft it.

Both thefe may be propagated by fowing their feeds in April on a hot-bed, and the plants tranfplanted on another moderate bed, where they fhould be brought up hardily, and have a great deal of air to ftrengthen them; and when they have got four or five leaves they fhould be tranfplanted into holes made upon an old dunghill, or fome fuch place, allowing them a great deal of room to run, for fome of the forts will fpread to a great diftance.

There are feveral varieties of this fruit, which differ in their form and fize; but as thefe are annually varying from feeds, fo I have omitted the mentioning them, for they feldom continue to produce the fame kinds of fruit three years together.

The third fort is very common in moft parts of America, where it is cultivated as a culinary fruit; of this fort there are alfo feveral varieties, which differ in their form and fize; fome of thefe are flat, others round; fome are fhaped like a bottle, and others are oblong, their outer cover or rind being white when ripe, and covered with large protuberances or warts. The fruit are commonly gathered when they are half grown, and boiled by the inhabitants of America to eat as a fauce with their meat.

The fourth fort is alfo very common in North America, where it is cultivated for the fame purpofes as the third. This very often grows with a ftrong, bufhy, erect ftalk, without

without putting out runners from the fide as the other forts, but frequently varies; for after it has been cultivated a few years in the fame garden, the plants will become trailing like the others, and extend their branches to as great diftance.

CUIETE. See CRESCENTIA.

CUMINOIDES. See LAGOECIA.

CUMINUM. Lin. Gen. Plant. 313. Cumin.

The CHARACTERS are,

It hath an umbelliferous flower; the involucrum is longer than the umbel. The great umbel is uniform; the flowers have five unequal petals, whofe borders are inflexed, and five fingle ftamina, with a long germen fituated under the flower, fupporting two fmall ftyles, which afterward becomes an oval ftriated fruit, compofed of two oval feeds, which are convex and furrowed on one fide, and plain on the other.

We have but one SPECIES of this plant, viz.

CUMINUM (*Cyminum.*) Lin. Mat. Med. 139. *Cumin.*

This plant is annual, perifhing foon after the feeds are ripe; it feldom rifes more than nine or ten inches high, in the warm countries where it is cultivated; but I have never feen it grow more than four in England, where I have fometimes had the plant come fo far as to flower very well, but never to produce good feeds. The leaves of this plant are divided into long narrow fegments like thofe of Fennel, but much fmaller; they are of a deep green, and generally turn backward at their extremity, the flowers grow in fmall umbels at the top of the ftalks; they are compofed of five unequal petals, which are of a pale bluifh colour, and are fucceeded by long, channelled, aromatic feeds.

This plant is propagated for fale in the ifland of Malta, where it is called Cumino aigro, i. e. Hot Cumin. But Anife, which they alfo propagate in no lefs quantity, they call Cumino dulce, i. e. Sweet Cumin. So that many of the old botanifts were miftaken, when they made two fpecies of Cumin, viz. acer and dulce.

If the feeds of this plant are fown in fmall pots, and plunged into a very moderate hot-bed to bring up the plants, and thefe after having been gradually inured to the open air, turned out of the pots, and planted in a warm border of good earth, preferving the balls of earth to their roots, the plants will flower pretty well, and by thus bringing of the plants forward in the fpring, they may perfect their feeds in warm feafons.

CUNILA. See SIDERITIS.

CUNONIA. Buttn. Cun. tab. 1.

The CHARACTERS are,

The flowers grow alternate in an imbricated fpike, each having a fpatha or fheath; they have one ringent petal with a fhort flender tube, which is dilated at the chaps and compreffed on the fides; the upper lip is arched, ftretched out beyond the alæ or wings. It hath three flender ftamina, which are fituated in the upper lip, and a flender ftyle which is fhorter than the ftamina, crowned by three cylindrical ftigmas. The germen, which is fituated below the flower, becomes an oblong capfule with three cells, filled with compreffed feeds.

We have but one SPECIES of this plant, viz.

CUNONIA (*Antholyza*) floribus feffilibus, fpathis maximis. Buttn. Cun. 211. tab. 1. *Cunonia with flowers fitting clofe to the ftalk, and very large fpathæ or fheaths.*

This plant grows naturally at the Cape of Good Hope. It hath a compreffed bulbous root fomewhat like that of the Corn Flag, covered with a brown fkin: from this arife feveral narrow fword-fhaped leaves, about nine inches long, and a quarter of an inch broad, terminating in acute points; thefe have one longitudinal midrib which is prominent, and two longitudinal veins running parallel on each fide; they are of a fea-green colour. In fpring the ftalk rifes from between the leaves, which is round, ftrong, and jointed; at each joint is fituated a fingle leaf, which almoft embraces the ftalk, which rifes near a foot and a half high, and is generally curved two oppofite ways; the upper part of the ftalk is terminated by a loofe fpike of flowers, coming out of large fpathæ or fheaths, compofed of two oblong concave leaves, terminating in acute points: thefe are at their firft appearance placed imbricatim over each other, but as the ftalks increafe in length, fo thefe are feparated; from between thefe two leaves comes out the flower, which having a flender Saffron-coloured tube near half an inch long, is then enlarged where the petal is divided, and the upper fegment is extended two inches in length, being arched over the ftamina and ftyle. This is narrow as far as to the extent of the wings, but above them is enlarged and fpread open half an inch in length, and is concave, covering the fummits and ftigmas, which are extended to that length; the two wings are alfo narrow at their bafe, but are enlarged upward in the fame manner, ending in concave obtufe points, which are compreffed together, and cover the ftamina and ftyle. This flower is of a beautiful foft fcarlet colour, fo makes a fine appearance about the latter end of April, or beginning of May, which is the feafon of its flowering. After the flowers decay, the germen becomes an oval fmooth capfule opening in three cells, which are filled with flat bordered feeds.

This plant is eafily propagated by offsets, which it fends out in great plenty, or by fowing of the feeds, which fhould be fown in pots about the middle of Auguft, and placed in a fituation where they may enjoy the morning fun; in September the pots may be removed to a warmer fituation, and in October they muft be placed under a frame, where they may be protected from froft and hard rains, but in mild weather enjoy the free air. The plants will appear in October, and continue growing all the winter, and in June their leaves will decay; then they may be taken up and planted in halfpenny pots. As this plant is a native at the Cape of Good Hope, fo it is too tender to live through the winter in England without fhelter from froft; the beft way to have this and other bulbous-rooted flowers from the fame country in perfection, is to build a frame of a proper depth, to allow room for the ftalks of their flowers to rife to their ufual height under the glaffes, and make a bed of good frefh earth two feet deep, into which the roots fhould be planted; and here they may be protected from froft, and in mild weather fully expofed to the open air, fo will thrive and flower much better than when they are placed in a green-houfe in winter. But where fuch a frame is wanting, the roots fhould be planted in pots, and fheltered in winter under a common hot-bed frame, that they may always have free air in mild weather. Thefe feedling plants muft be fheltered in the fame manner as the old roots in winter, and the third year they will flower.

CUPRESSUS. Lin. Gen. Plant. 958. The Cyprefs-tree.

The CHARACTERS are,

It hath male and female flowers on the fame plant; the male flowers are formed into oval katkins, in which the flowers are placed thinly; they have no petals nor ftamina, but have four fummits which adhere to the bottom of the fcales. The female flowers are formed in a roundifh cone, each containing eight or ten flowers; the germen is fcarce vifible, but afterward becomes a globular cone, opening in angular target-fhaped fcales, under which are fituated angular feeds.

The SPECIES are,

1 CUPRESSUS (*Sempervirens*) foliis imbricatis, ramis erectioribus. *Female or upright common Cyprefs.*

2. CUPRESSUS (*Horizontalis*) foliis imbricatis acutis, ramis horizontalibus. *Male fpreading Cyprefs.*

3. Cupressus (*Lusitanica*) foliis imbricatis, apicibus aculeatis, ramis dependentibus. *Portugal spreading Cypress with a smaller fruit.*

4. Cupressus (*Disticha*) foliis distichis patentibus. Hort. Cliff. 409. *Virginia Cypress which sheds its leaves; commonly called Deciduous Cypress.*

5. Cupressus (*Thyoides*) foliis imbricatis, frondibus ancipitibus. Lin. Sp. Pl. 1003. *Dwarf Maryland Cypress with a small blue fruit.*

6. Cupressus (*Africana*) foliis linearibus simplicibus cruciatim positis. *Cypress with narrow single leaves placed crossways.*

The first of these trees is very common in most of the old gardens in England, but at present is not so much in request as formerly, though it is not without its advantages; nor should it be entirely rejected, although some persons are of that opinion; for it is of great beauty to wildernesses, or clumps of evergreens, and where they are properly disposed they have much beauty, for they are one of the principal ornaments of the Italian villas.

The second sort is the largest growing tree, and is the most common timber in the Levant: this, if planted upon a warm, sandy, gravelly soil, will prosper wonderfully; and though the plants of this sort are not so finely shaped as the first, yet they greatly recompense for that defect, by its vigorous growth, and strength in resisting all weathers. Besides, the wood of this tree is very valuable when grown to a size fit for planks; which I am convinced it will do, in as short space as Oaks, if properly cultivated for that purpose; since there are many places in England where the soil is of a sandy or gravelly nature, so seldom produces any trees worth cultivating. Now, in such places these would thrive wonderfully, and greatly add to the pleasure of the owner, while growing, and afterward render as much profit to his successors, as perhaps the best plantation of Oaks, especially should the timber prove as good here as in the islands of the Archipelago, which I see no reason to doubt of; for we find it was so gainful a commodity in the island of Candia, that the plantations were called Dos Filiæ, the felling of one of them being reckoned a daughter's portion.

The timber of this tree is said to resist the worm, moth, and all putrefaction, and also to last many hundred years. The coffins were made of this material, in which Thucydides tells us the Athenians used to bury their heroes.

The fourth sort is a native of America, where it grows in watery places, rising to a prodigious height, and is of a wonderful bulk: I have been informed that there are trees of this kind in South Carolina which are upwards of seventy feet high, and several fathoms in circumference, which trees grow constantly in the water; therefore they may probably be of singular advantage to plant in such swampy or wet soils where few other trees will grow, especially of the resinous kind.

These trees are all propagated from seeds, which should be sown early in the spring in pots or boxes, which, if placed in a very moderate hot-bed will bring up the plants soon; or if the seeds are sown upon a moderate hot-bed, and the beds covered with mats, they will come up much sooner, and with greater certainty than when they are sown in the cold ground.

In these pots, boxes, or this bed, the young plants may remain one year, by which time they will have strength enough to be transplanted, either into boxes or a warm border; for while the plants are young they are tender, so should be covered in severe frost with mats to prevent their being injured thereby. The best season for removing them is in the beginning of April, when the drying easterly winds of March are over; and, if possible, chuse a cloudy day, when it is inclinable to rain; and in taking them out of the seed bed or pots, preserve the roots as entire as possible, and if you can, some earth to each plant; having prepared the border by carefully digging and cleansing it from all noxious weeds, you must lay it level. Then draw the lines where the plants are to be planted, at one foot asunder row from row, and the cross lines at six inches distance in the rows; the plants must be set exactly in the squares, closing the earth to their roots, and water them well to settle the earth to them; which should be repeated in dry weather twice a week, until the plants have taken fresh root.

These plants may remain in the borders two years, according to the progress they make, but if you intend to let them remain longer, you should take up every other tree in the rows, and transplant them out; for otherwise their roots will be matted together, so that it will render it difficult to transplant them, and also endanger their future growth. These plants should by no means be let stand too long in the nursery, before they are transplanted out for good; because they rarely grow when their roots are much cut, for the roots of the Cypress are apt to extend out in length, so it is one of the most difficult trees to remove when grown large; therefore most curious persons chuse to plant the young plants into small pots, when they first take them out of the border, and so train them up in pots two or three years, until they are fit to plant out where they are to stand for good; and, by this management, they are secure of all the plants; and these may be shaken out of the pots at any time of the year, except in winter, without danger, and planted with their whole ball of earth, which is likewise a great advantage. When they are planted out for good (if they are designed for timber,) they should be planted about ten or twelve feet distance every way When they are planted, you must settle the earth close to their roots as before, laying a little mulch upon the surface of the ground about their stems, to prevent the sun and wind from entering the earth to dry their fibres; and water them well to settle the ground to their roots, which must also be repeated (if the weather be dry) until they have taken root; after which time, they will require little more care than to keep them clear from weeds.

The first, which is the most common sort in England, seldom produces good seeds in this country; it is therefore the best way to have the cones brought over entire from the south parts of France or Italy, where they ripen perfectly well, and take the seeds out just before you sow them, for they will keep much better in the cones than if they are taken out. The method to get the seeds out is to expose the cones to a gentle heat, which will cause them to open, and easily emit their seeds.

The second sort grows naturally in the Levant, for what has passed under this title in many places here, is only a variety of the common sort, whose branches grow much looser, and not so upright as the first; but the cones taken from these trees and the seeds sown, have frequently produced plants of both varieties; but the spreading Cypress extends its branches much more horizontally, and the plants raised from the seeds do not vary, so that it is certainly a distinct species. This grows to be a large timber-tree in the Levant, and in Italy there are some of a considerable size.

The Virginian kind may also be propagated in as great plenty, for the cones of this may be easily procured from Carolina or Virginia, in both which places they grow in great abundance; and the seeds will rise as easily as any of the other sorts, and the plants are equally as hardy. As this tree grows in places where the water commonly covers the surface of the ground three or four feet, so it may be a very great improvement to our boggy ſoils, were they planted with them. It may alſo be propagated by cuttings, which ſhould be planted in a bed of moiſt earth in the ſpring before the trees begin to ſhoot.

The third ſort is at preſent pretty rare in the Engliſh gardens, tho' of late years there have been many plants raiſed in the nurſeries; but this ſort is not quite ſo hardy as the common Cypreſs, for the plants are frequently killed or greatly injured in ſevere winters; and in the hard froſt in 1740, there were few trees in England of this kind which were not entirely killed. There are great plenty of theſe trees growing at a place called Buſaco, near Cœmbra in Portugal, where this tree is called the Cedar of Buſaco; and there it grows to be a timber-tree, ſo that from thence the ſeeds may be eaſily procured.

But its natural place of growth is at Goa, from whence it was firſt brought to Portugal, where it has ſucceeded, and been propagated: formerly there were ſome trees of this ſort growing in the Biſhop of London's garden at Fulham, where it paſſed under the title of Cedar of Goa, by which it was ſent from thence to the Leyden garden under that name.

The fifth ſort is a native of North America, where it grows to a conſiderable height, and affords an uſeful timber to the inhabitants for many purpoſes. This ſort is extremely worth cultivating in England, for as it grows in a much colder country, there is no danger of its thriving well in the open air in England; and being an evergreen of regular growth, will add to the variety in wilderneſs quarters or other plantations of evergreen trees.

The branches of this tree are garniſhed with flat evergreen leaves, reſembling thoſe of the Arbor Vitæ; and the cones are no larger than Juniper berries, from which they are not eaſily diſtinguiſhed at a little diſtance; but upon cloſely viewing, they are eaſily diſtinguiſhed to be perfect cones, having many cells like thoſe of the common Cypreſs. If theſe trees are planted in a moiſt ſtrong ſoil they make very great progreſs, and may in ſuch ſituations become profitable for timber; but however this tree may ſucceed for timber, yet it will be a great ornament to large plantations of evergreen trees, eſpecially in ſuch places where there is naturally a proper ſoil for them; becauſe in ſuch ſituations there are not many ſorts of evergreen trees which thrive ſo well, eſpecially in cold places; and by increaſing the number of ſorts of theſe evergreens, we add to the beauty of our gardens and plantations.

The ſixth ſort grows naturally at the Cape of Good Hope, and by the accounts which I received of the ſeeds, the cones of the tree are black when ripe. The young plants which I have raiſed from ſeeds have looſe ſpreading branches, which are cloſely garniſhed with narrow ſtrait leaves, which come out oppoſite, and are alternately croſſing each other; theſe are one inch long, of a light green colour, and continue in verdure all the year. Theſe plants being young, are too tender to thrive in the open air in England as yet, but when they have obtained more ſtrength, it is very probable they may do well in warm ſituations.

CURCUMA. Lin. Gen. Plant. 6. Turmerick.

The CHARACTERS are,

The flowers have ſpathæ, which are ſingle and drop off; they have one petal with a narrow tube, which is cut at the brim into three ſegments; and an oval pointed nectarium of one leaf, inſerted in the ſinus of the largeſt ſegment; it hath five ſtamina, four of which are barren, and one is fruitful, which is ſituated within the nectarium. It hath a roundiſh germen ſituated under the flower, which afterward becomes a roundiſh capſule having three cells, which are filled with roundiſh ſeeds.

The SPECIES are,

1. CURCUMA (*Rotunda*) foliis lanceolato-ovatis, nervis lateralibus rariſſimis. Lin. Sp. Plant. 2. *Turmerick with a round root.*

2. CURCUMA (*Longa*) foliis lanceolatis nervis lateralibus numeroſiſſimis. Lin. Sp. Plant. 2. *Turmerick with a long root.*

The firſt ſort hath a fleſhy jointed root ſomewhat like that of Ginger, but rounder, which ſends up ſeveral ſpear-ſhaped oval leaves, which riſe upwards of a foot high, with one longitudinal midrib, and a few tranſveſe nerves running to the ſides; they are of a ſea-green colour; from between theſe ariſe the flower-ſtalk, ſupporting a looſe ſpike of flowers of a pale yellowiſh colour, incloſed in ſeveral different ſpathæ or ſheaths, which drop off. Theſe flowers are never ſucceeded by ſeeds in the gardens here.

The ſecond ſort hath long fleſhy roots of a deep yellow colour, which ſpread under the ſurface of the ground like thoſe of Ginger; they are about the thickneſs of a man's finger, having many round knotty circles, from which ariſe four or five large ſpear-ſhaped leaves, ſtanding upon long foot-ſtalks; they have a thick longitudinal midrib, from which a numerous quantity of veins are extended to the ſides. The flowers grow in looſe ſcaly ſpikes on the top of the foot-ſtalks, which ariſe from the larger knobs of the roots, and grow about a foot high; they are of a yellowiſh red colour, and ſhaped ſomewhat like thoſe of the Indian Reed.

Theſe plants grow naturally in India, from whence the roots are brought to Europe for uſe. They are very tender, ſo will not live in this country unleſs they are placed in a warm ſtove. They are propagated by parting of their roots; the beſt time for removing and parting theſe roots is in the ſpring, before they put out new leaves, for the leaves of theſe plants decay in autumn, and the roots remain inactive till the ſpring, when they put out freſh leaves; theſe roots ſhould be planted in pots, which ſhould be conſtantly kept plunged in a bark-bed in the ſtove. In the ſummer ſeaſon, when the plants are in a growing ſtate, they will require to be frequently refreſhed with water, but it ſhould not be given to them in large quantities; they ſhould alſo have a large ſhare of free air admitted to them in warm weather; but when the leaves are decayed they ſhould have very little wet, and muſt be kept in a warm temperature of air, otherwiſe they will periſh.

CURRANT-TREE. See RIBES.

CURURU. See PAULLINIA.

CUSTARD-APPLE. See ANNONA.

CYANUS. See CENTAUREA.

CYCLAMEN. Lin. Gen. Plant. 184. Sowbread.

The CHARACTERS are,

The flower hath one petal with a globular tube; the upper part is divided into five large ſegments which are reflexed; it hath five ſmall ſtamina ſituated within the tube of the petal; it hath a roundiſh germen ſupporting a ſlender ſtyle, which afterward becomes a globular fruit with one cell, opening in five parts at the top, incloſing many oval angular ſeeds.

The SPECIES are,

1. CYCLAMEN (*Europeum*) foliis haſtato-cordatis angulatis. *Sowbread with an Ivy leaf.*

2. CYCLAMEN (*Purpureſcens*) foliis orbiculato-cordatis, infernè purpuraſcentibus. *Round-leaved Sowbread, with a purple under ſide.*

3. CYCLAMEN (*Perſicum*) foliis cordatis ſerratis. *This is the heart-ſhaped Spring, or Perſian Cyclamen.*

4. CYCLAMEN (*Vernale*) foliis cordatis anguloſis integris. *Winter and Spring flowering Sowbread, with a large angular leaf, and a white flower, purple at the bottom; called Perſian Cyclamen.*

5. CYCLAMEN (*Orbiculatum*) radice inæquali, foliis orbiculatis. *Sowbread with a root the ſize of a Cheſtnut.*

6. CY-

6. CYCLAMEN (*Cœum*) foliis orbiculatis planis, pediculis brevibus floribus minoribus. *Winter Cyclamen with orbicular leaves, red on their under side, and a purplish flower; or the Cœum of the herbalists.*

The first sort is the most common in the English gardens; this grows naturally in Austria, Italy, and other parts of Europe, so will thrive in the open air in England, and is never hurt by the frost; it hath a large, orbicular, compressed root, from which arise a great number of angular heart-shaped leaves upon single foot stalks, which are six or seven inches long, marked with an angular circle of black in their middle: the flowers appear before the leaves, rising immediately from the root; they appear in August and September, and soon after the leaves come out, and continue growing all the winter and spring till May, when they begin to decay; after the flowers are fallen, the foot-stalks twist up like a screw, inclosing the germen in the center, and lay down close to the surface of the ground between the leaves, which serve as a protection to the seed; this germen becomes a round fleshy seed-vessel with one cell, inclosing several angular seeds, which ripen in June, and should be sown in August. There are two varieties of this, one with a white, and the other with a purplish flower, which appear at the same time.

The second sort flowers in autumn; this is at present rare in England; the leaves of this sort are large, orbicular, heart-shaped at their base, and of a purple colour on their under side; the leaves and flowers of this come up from the root at the same time; the flowers are of a purplish colour, and their bottoms are of a deep red; it flowers late in the autumn, and requires protection from the frost in winter.

The third sort hath stiff heart-shaped leaves which are sawed on their edges, and have strong fleshy foot-stalks of a purple colour; the flowers rise with single foot-stalks from the root; these are pure white, with a bright purple bottom; the petal is divided into nine segments to the bottom, which are twisted and reflexed backward like the other sorts; this flowers in March and April, and the seeds ripen in August.

The fourth sort is commonly called the Persian Cyclamen; this hath large, angular, heart-shaped leaves, which are veined and marbled with white on their upper side, and stand upon pretty long foot-stalks; the flowers are large, of a pale purple colour, with a bright red or purple bottom; these appear in March and April, and the seeds ripen in August.

The fifth sort hath a small irregular root, not larger than a Nutmeg; the leaves are orbicular and small; the flowers are of a flesh colour, small, and have purple bottoms; they appear in spring, but rarely produce seeds in England.

The sixth sort is not so tender as the four last-mentioned, so may be planted in warm borders, where, if they are covered in hard frost, they will thrive and flower very well; it hath plain orbicular leaves, which have shorter and weaker foot-stalks than either of the other; their under sides are very red in the beginning of the winter, but that colour goes off in the spring; their upper sides are smooth, of a lucid green, and spread open flat, whereas the other sorts are hollowed and reflexed at their base; the flowers are of a very bright purple colour, and appear in the middle of winter, at a time when there are a few other flowers, which renders the plants more valuable; the seeds of this sort ripen in the end of June.

There are some other varieties of this plant, which chiefly differ in the colour of their flowers, particularly among the Persian kind, of which there is one with an entire white flower, which smells very sweet; but as these are accidental variations, I have not enumerated them, those which are here mentioned being undoubtedly distinct species.

All the sorts are propagated by seeds, which should be sown soon after they are ripe in boxes or pots, and covered about half an inch deep, placing them where they may have only the morning sun till the beginning of September, when they may be removed to a warmer exposure. Those of the first sort may be plunged into the ground, close to a south wall, a pale or Reed-hedge in October, where, if it should be very severe frost, it will be proper to cover them either with mats or Pease-haulm, but in common winters they will not require any. The pots or tubs in which the Persian kinds are sown, should then be placed under a common hot-bed frame, where they may be protected from frost and hard rains, but in mild weather the glasses may be taken off every day to admit fresh air to them; those of the Persian kinds will come up early in the spring, and continue green till June, when they will begin to decay; then they should be removed to an east aspect, where they will have only the morning sun; in which situation they may remain till the middle of August, during which time they should have very little water, as the roots are then in an inactive state, when much wet will rot them; in the beginning of October there should be some fresh earth spread over the tubs or pots, and then they should be removed again into shelter in the same manner as before; and the following summer they must be managed also in the same way till their leaves decay, when they should be carefully taken up, and those of the first sort planted in a warm border at four or five inches distance; but the other sorts must be planted in pots or tubs, to be sheltered in winter.

The third, fourth, and fifth sorts, are more impatient of cold and wet than the three other; these must constantly be preserved in pots, and sheltered in winter either under common hot-bed frames, or in an airy glass-case, where they may enjoy as much free open air as possible in mild weather; for if they are crouded under other plants, and are kept too close, they are very subject to mould and rot; nor should they have much water in winter, which is also very injurious to them; but whenever they want water, it should be given them sparingly. In summer these plants may be exposed to the open air, when their green leaves will decay, at which time you should remove them to a place where they may have the morning sun until eleven o'clock; but during the time that the roots are destitute of leaves, they should have very little water given them, because at that season they are not capable of discharging the moisture; when their leaves are decayed, it is the proper season to transplant the roots, or to fresh earth them; and as the autumn comes on that the heat decreases, they may be removed into places more exposed to the sun, where they may remain until October before they need be sheltered.

Toward Christmas, if the roots are in good health, the sixth sort will begin to flower, and continue producing fresh flowers until the middle of February; these will be succeeded by the Persian sorts, which continue till May; but if you intend to have good seeds, the pots of these sorts should be placed so as to receive a great share of fresh air; for if their flowers are drawn up weak, they seldom produce good seeds. The seeds are ripe about July, when they should be immediately sown in pots or cases of good light undunged earth; which should be sheltered in winter under a frame, and exposed in summer in the same manner as is directed for the older roots. When they are two years old they should be taken up, and each root planted in a small separate pot, in which they may remain two years; then they should be removed into pots of a size larger, to give them room, and in about four or five years time they will begin to flower, when they must be put into larger pots.

Several of these sorts have been planted under warm walls in the full ground, where, in mild winters they have succeeded very well, but in severe frost they have been destroyed; therefore, whenever these roots are planted in an open border, there should be common hot-bed frames placed over them in winter, or some other covering, that they may be screened from frost; and when they are thus managed, the plants will produce more flowers, and those will be much fairer than what are produced from the roots in the pots, and from these there may always be good seeds expected. Therefore such persons who are curious in flowers, should have a border framed over on purpose for these, and the Guernsey and Belladonna Lilies, with some other of the curious bulbous-rooted flowers; in which borders there may be many of these curious flowers cultivated to more advantage than any other method now practised.

CYDONIA. Tourn. Inst. R. H. 632. The Quince-tree.

The CHARACTERS are,

The flower is composed of five large, roundish, concave petals, which are inserted in the permanent empalement. The germen is situated under the flower; it hath five slender styles with twenty stamina. The germen afterward becomes a pyramidal or roundish fruit, which is fleshy and divided into five cells, in which are lodged several hard kernels or seeds.

The SPECIES are,

1. CYDONIA (*Oblonga*) foliis oblongo-ovatis subtus tomentosis, pomis oblongis basi productis. *Quince-tree with oblong oval leaves, woolly on their under side, and an oblong fruit lengthened at their base; commonly called the Pear Quince.*

2. CYDONIA (*Maliforma*) foliis ovatis, subtus tomentosis, pomis rotundioribus. *Quince-tree with oval leaves, woolly on their under side, and a rounder fruit; commonly called the Apple Quince.*

3. CYDONIA (*Lusitanica*) foliis obversè-ovatis subtus tomentosis. *Quince-tree with obverse oval leaves, woolly on their under side; commonly called the Portugal Quince.*

The Portugal Quince is the most valuable, the pulp of it turning to a fine purple when stewed or baked, and becomes much softer and less austere than the others, so is much better for making of marmalade.

All the sorts are easily propagated either by layers, suckers, or cuttings, which must be planted in a moist soil. Those raised from suckers are seldom so well rooted as those which are obtained from cuttings or layers, and are subject to produce suckers again in greater plenty, which is not so proper for fruit-bearing trees. The cuttings should be planted early in the autumn on a moist border. The second year after they should be removed into a nursery at three feet distance row from row, and one foot asunder in the rows, where they must be managed as was directed for APPLES. In two years time these trees will be fit to transplant, where they are to remain for good, which should be either by the side of a ditch, river, or some other moist place, where they will produce a greater plenty, and much larger fruit than in a dry soil; though those in the dry soil will be better tasted, and earlier ripe. The trees require very little pruning; the chief thing to be observed is, to keep their stems clear from suckers, and cut off such branches as cross each other; likewise all upright luxuriant shoots from the middle of the tree should be taken entirely out, that the head may not be too much crowded with wood, which is of ill consequence to all sorts of fruit-trees. If they are propagated by budding, or grafting upon stocks raised by cuttings, to multiply the best sorts, the trees so raised will bear fruit much sooner and be more fruitful, than those which come from suckers or layers.

Quince stocks are also in great esteem for to graft and bud Pears on, which on a moist soil will greatly improve some sorts, especially those designed for walls and espaliers: for the trees upon these stocks do not shoot so vigorously as those upon free stocks, and therefore may be kept in less compass, and are sooner disposed to bear fruit: but hard winter fruits do not succeed so well upon these stocks, their fruit being very subject to crack, and are commonly stony, especially all the breaking Pears, but more especially if they are planted in dry ground; therefore these stocks are only proper for the melting Pears, and for a moist soil. The best stocks are those which are raised from cuttings or layers.

As the Pear will take upon the Quince by grafting or budding, and so vice versa, we may conclude there is a near alliance between them; but as neither of these will take upon the Apple, nor that upon either of these, so we should separate them under different genera, as will be farther mentioned under the article MALUS.

CYNANCHUM. Lin. Gen. Pl. 268. Bastard Dogsbane.

The CHARACTERS are,

The flower hath one petal, which is spread open, plain, and divided into five parts; the nectarium, which is situated in the center of the flower, is erect, cylindrical, and the length of the petal. It hath five stamina, which are parallel to the nectarium, and an oblong bifid germen; the empalement becomes a capsule with two oblong pointed pods, which open longitudinally, and are filled with seeds lying over each other imbricatim, crowned with long down.

The SPECIES are,

1. CYNANCHUM (*Acutum*) caule volubili herbaceo, foliis cordato-oblongis glabris. Hort. Cliff. 79. *Montpelier Periploca with acute-pointed leaves; commonly called Montpelier Scammony.*

2. CYNANCHUM (*Monspeliacum*) caule volubili herbaceo, foliis reniformi-cordatis acutis. Hort. Cliff. 79. *Montpelier Periploca with rounder leaves; or round-leaved Montpelier Scammony.*

3. CYNANCHUM (*Hirtum*) caule volubili infernè suberoso fisso, foliis cordatis acuminatis. Hort. Cliff. 79. *Carolina Periploca with a smaller starry flower.*

4. CYNANCHUM (*Suberosum*) caule volubili fruticoso, infernè suberoso fisso, foliis ovato-cordatis. Hort. Cliff. 79. *Periploca with a climbing stalk, a Citron leaf, and a large fruit.*

5. CYNANCHUM (*Erectum*) caule erecto divaricato, foliis cordatis glabris. Hort. Cliff. 79. *Upright Dogsbane with a roundish leaf.*

6. CYNANCHUM (*Asperum*) caule volubili fruticoso, foliis cordatis acutis asperis, floribus lateralibus. *Climbing Dogsbane with heart-shaped rough leaves, and large, yellow, spreading flowers.*

The first and second sorts grow naturally at Montpelier; they have perennial creeping roots, but annual stalks, which twist themselves like Hops round whatever plants are near them, and rise to the height of eight or ten feet, the first of these has oblong, heart-shaped, smooth leaves, ending in acute points, placed by pairs on long foot-stalks; the flowers come out in small bunches from the wings; they are of a dirty white colour, divided into five acute segments, which spread open in form of a star. They appear in June and July, but are not succeeded by any seed-vessels in England, which may be occasioned by their roots creeping so far under ground.

The second sort differs from the first in the shape of its leaves, which are broader and rounder at their base. The roots of this sort are very thick, running deep into the ground, so that where this plant hath gotten possession of the ground it is not easily extirpated. Both these plants abound with a milky juice like the Spurge, which issues out wherever they are broken; and this milky juice when concreted, has been frequently sold for Scammony.

These plants propagate too fast by their creeping roots, when they are admitted into gardens, so few people care to preserve them: the roots may be transplanted any time after their stalks decay.

The third sort grows naturally in Carolina; this is a perennial plant with twining hairy stalks, which, if supported, will rise six or seven feet high; the lower part of the stalks are covered with a thick fungous bark, somewhat like cork, which is full of fissures; the stalks are slender, and garnished at each joint with two oblong, heart shaped, pointed leaves, standing on long hairy foot-stalks. The flowers are produced in small bunches at the wings of the leaves, which are star-shaped and green when they first appear, and afterward fade to a worn-out purple colour.

This plant will live in the open air in England, if it is planted in a dry soil and warm situation. It may be propagated by laying down some of the young shoots about Midsummer, which, if they are now and then refreshed with water, will put out roots, so may be transplanted in autumn where they are designed to remain. The roots of this plant should be covered in winter with some rotten tan to keep out the frost, otherwise in severe winters they are liable to be destroyed.

The fourth sort grows naturally in Jamaica. This rises with a twining stalk to the height of twenty feet or upward, provided it hath support; the lower part of the stalks are covered with a thick fungous bark, full of fissures, which gape open; the leaves are oblong and smooth, and placed by pairs, standing upon long foot-stalks: the flowers are produced from the wings of the leaves in small bunches; they are star-shaped, and of a yellowish green colour, but are not succeeded by pods in England.

This is tender, so will not thrive in this country unless it is placed in a warm stove, and requires the same treatment as other tender plants from the same country. It is propagated by laying down of the young shoots, which in three or four months will put out roots, and may then be transplanted into pots, and plunged into the tan-bed in the bark-stove, where the plants should continue all the year.

The fifth sort grows naturally in Syria. This is a perennial plant, which rises with slender upright stalks about three feet high, which are garnished with broad, smooth, heart-shaped leaves, ending in points, placed opposite: the flowers come out from the wings of the leaves in small bunches, standing on branching foot-stalks; these are small and white, greatly resembling those of the common white Asclepias, or Swallow-wort, and are succeeded by oblong taper pods, filled with flat seeds, crowned with down, but these rarely ripen in this country.

It is propagated by parting of the root; the best time for doing of this or transplanting of the roots is in the spring, before they shoot: it requires a warm situation, otherwise it will not live abroad in England.

The sixth sort grows naturally at La Vera Cruz in New Spain; this hath a shrubby twining stalk, which twists about any prop that is near it, and rises to the height of twenty feet or upward; the stalks are very slender, and are armed with small stinging hairs, and garnished with broad heart shaped leaves, which end in acute points, placed by pairs at each joint, and have slender foot-stalks; they are covered with rough hairs on their under side; the flowers are produced in small clusters, sitting close to the side of the stalks; they are pretty large, yellow, and star-shaped, spreading open to the bottom, and are succeeded by long swelling pods, filled with flat seeds, lying imbricatim, which are crowned with long down.

This sort is tender, so requires the same treatment as the fourth, and is propagated the same way.

CYNARA. Lin. Gen. Plant. 835. Artichoke.

The CHARACTERS are,

It hath a compound flower, made up of many hermaphrodite florets, included in one common scaly empalement. The florets are tubulous, equal, and uniform, divided at the top into five narrow segments. They have five short hairy stamina. At the bottom of each is situated an oval germen, which afterward becomes a single, oblong, compressed, four-cornered seed, crowned with long hairy down.

The SPECIES are,

1. CYNARA (*Scolymus*) foliis subspinosis, pinnatis indivisisque, calycinis squamis ovatis. Lin. Sp. Pl. 827. *The green or French Artichoke.*

2. CYNARA (*Hortensis*) foliis pinnatis inermibus, calycinis squamis obtusis emarginatis. *The Globe Artichoke.*

3. CYNARA (*Cardunculus*) foliis spinosis, omnibus pinnatifidis, calycinis squamis ovatis. Lin. Sp. Pl. 827. *The Cardoon.*

4. CYNARA (*Humilis*) foliis spinosis, pinnatifidis, subtus tomentosis, calycibus squamis subulatis. Lin. Sp. Pl. 828. *Wild Artichoke of Spain.*

The first sort is commonly known here by the title of French Artichoke, being the sort which is most commonly cultivated in France. The leaves of this sort are terminated by short spines, the head is oval, and the scales do not turn inward at the top like those of the Globe Artichoke; they are also of a green colour; the bottoms of these are not near so thick of flesh as those of the Globe, and have a perfumed taste, which to many persons is very disagreeable, so it is very seldom cultivated in the gardens near London, where the Globe or Red Artichoke is the only sort in esteem. The leaves of this are not prickly; the head is globular, a little compressed at the top; the scales lie close over each other, and their ends turn inward, so as to closely cover the middle.

The culture of these having been fully treated under the article ARTICHOKE, the reader is desired to turn to that, to avoid repetition.

The Chardoon, or Cardoon, is propagated in the kitchen-gardens to supply the markets; this is annually raised from seeds, which should be sown upon a bed of light earth in March, and when the plants come up, they should be thinned where they are too close; and if the plants are wanted, those which are drawn out may be transplanted into a bed at about three or four inches distance, where they should remain till they are fit to transplant out for good. In June they must be transplanted out on a moist rich spot of ground, at about four feet asunder every way; the ground should be well dug before they are planted, and the plants should be well watered until they have taken new root; after which the ground must be kept clean from weeds, to encourage the growth of the plants; and as they advance in height, there should be some earth drawn up about each; and when they are fully grown, their leaves should be closely tied up with a hay-band, and the earth drawn up in hills about each plant almost to their tops, being careful to keep the earth from falling between the leaves, which may occasion the rotting of the plants. This earth should be smoothed over the surface that the wet may run off, and not fall into the center of the plants. In about five or six weeks after the plants have been thus earthed, they will be blanched enough for use; so that if a succession of them are wanted for the table, there should be but few plants earthed up at the same time, but every week or ten days there may be a part of them earthed, in proportion to the quantity desired.

Toward the middle or latter end of November, if the frost should be severe, it will be proper to cover the tops of those plants which remain with Pease-haulm, or straw, to prevent the frost from penetrating to the tender leaves, which

which frequently pinches them where there is not some covering, but this should be taken off again in mild weather; if this care is taken, the plants may be preserved for use all the winter.

The fourth sort grows naturally in Spain, and also on the African shore, and is preserved in gardens for the sake of variety; this is very like the third sort, but the stems of the leaves are much smaller, and do not grow more than half so high. The heads of this have some resemblance to those of the French Artichoke, but have no meat, or fleshy substance in their bottoms; this may be planted in the same manner as the third sort, at about three or four feet apart, and will require no other treatment than the keeping them clean from weeds; the second year they will flower, and, if the season proves dry, they will ripen their seeds.

CYNOGLOSSUM. Lin. Gen. Pl. 168. Hounds Tongue.

The CHARACTERS are,

It hath a funnel-shaped flower of one leaf, with a long tube. It hath five short stamina in the chaps of the petal. At the bottom of the tube are situated four germen; the empalement afterward becomes four capsules, inclosing four oval seeds.

The SPECIES are,

1. CYNOGLOSSUM (*Officinale*) staminibus corolla brevioribus, foliis lato-lanceolatis tomentosis sessilibus. Lin. Sp. Plant. 134. *Common Greater Hounds Tongue.*

2. CYNOGLOSSUM (*Appeninum*) staminibus corollam æquantibus. Hort. Upf. 33. *Greatest Mountain Hounds Tongue.*

3. CYNOGLOSSUM (*Creticum*) foliis oblongis tomentosis, amplexicaulibus, caule ramoso, spicis florum longissimis sparsis. *Broad-leaved stinking Hounds Tongue of Crete.*

4. CYNOGLOSSUM (*Cherifolium*) corollis calyce duplo longioribus foliis lanceolatis. Prod. Leyd. 406. *Hounds Tongue of Crete, with a narrow silvery leaf.*

5. CYNOGLOSSUM (*Virginianum*) caule ramoso hirsuto, foliis lanceolatis asperis, floribus sparsis. *Virginia Hounds Tongue with a very small white flower.*

6. CYNOGLOSSUM (*Lusitanicum*) caule erecto ramoso, foliis lanceolatis scabris sessilibus, spicis florum longissimis. *Taller Portugal Navelwort with a Hounds Tongue leaf.*

7. CYNOGLOSSUM (*Linifolium*) foliis lineari-lanceolatis glabris. Hort. Cliff. 47. *Portugal Navelwort with a Flax leaf; commonly called Venus Navelwort.*

8. CYNOGLOSSUM (*Omphalodes*) repens, foliis radicalibus cordatis. Hort. Cliff. 47. *Low Spring Navelwort with a Comfry leaf.*

The first sort grows naturally by the side of hedges and foot-ways in most parts of England, so is seldom admitted into gardens; the roots of this sort are used in medicine, which are gathered by the herb-folks in the fields. The leaves of this plant have a strong odour like that of mice in a trap.

The second sort grows naturally on the Apennine mountains; the leaves of this sort are much larger, the petal of the flower is shorter, and the plants grow taller than those of the first, and come earlier to flower in the spring; this is equally hardy with the common sort, and where the seeds are permitted to scatter, there will be plenty of the plants arise without care.

The third sort grows naturally in Andalusia; this hath a tall branching stalk, garnished with oblong woolly leaves, which embrace the stalk with their base. The flowers are produced in loose spikes, which come out from the side of the stalk, and are from six to eight inches long, and are thinly placed on one side; they are blue, striped with red. The seeds ripen in autumn, soon after which the root decays.

The fourth sort grows naturally in Spain, and also in the island of Crete. This rises with an upright stalk little more than a foot high, garnished with long, narrow, silvery leaves, having no foot-stalks. The flowers are produced from the side, and at the top of the stalks, which are but thinly dispersed on the side, but at the top of the stalk are in small clusters; they are of a deep purple colour, and much longer than the empalement; these are succeeded by four broad buckler-shaped seeds, which are rough.

The fifth sort grows naturally in Virginia, and in other northern parts of America; this rises with an upright branching stalk near four feet high. The stalks and leaves are covered with rough hairs; the branches are spread out on every side, and are but thinly garnished with leaves, from three to near four inches in length, and little more than one inch broad in the middle, gradually lessening to both ends; they have very short foot-stalks, and are placed alternate; the flowers grow scatteringly toward the end of the branches; these are small, white, and are succeeded by four small seeds, which ripen in autumn.

The sixth sort grows naturally in Portugal, but the seventh sort hath been long cultivated in the gardens for ornament by the title of Venus Navelwort, but of late years that has been lost in England; and the sixth sort is now generally sown in the gardens, the seeds of which are sold by the seedsmen under that title; this is a much larger plant than the other, so makes a finer appearance. The leaves of the sixth sort are broad at their base, and are gradually narrowed to the end; they are slightly covered with hairs. The stalks grow nine or ten inches high, and divide into many branches, each being terminated by a long loose spike of white flowers, standing on separate foot-stalks, and are succeeded by four umbilicated seeds, from whence it had the title of Navelwort.

The seventh sort seldom rises more than five or six inches high; the stalks do not branch near so much as those of the sixth. The leaves are very narrow and long, of a grayish colour, and smooth. The flowers grow in short loose panicles at the end of the branches; these are white, but smaller than those of the other sort, and are succeeded by seeds of the same form.

These are both annual plants, which have been commonly sown in gardens with other low annual flowers, to adorn the borders of the flower-garden; but these should be sown in autumn, for those which are sown in the spring often fail, especially in dry seasons; and the autumnal plants always grow much larger than those which arise from the spring sowing, and come to flower earlier in the year. The seeds should be sown where the plants are designed to remain, for they do not bear transplanting, unless it is performed while they are young. The plants require no other culture, but to be thinned where they are too close, and kept clean from weeds.

The eighth sort is a low perennial plant, which grows naturally in the woods of Spain and Portugal, where it usually flowers about Christmas. It hath trailing branches, which put out roots from their joints, whereby it propagates very fast. The leaves are heart-shaped, of a bright green colour, and stand upon long slender foot-stalks. The flowers grow in loose panicles, which arise from the divisions of the stalk; they are shaped like those of Borage, but are smaller, and of a lively blue colour; they appear in March and April, and in a cool shady situation continue great part of May, but are rarely succeeded by seeds; but the plants propagate themselves so fast by their trailing branches, as to render the cultivation of them by seeds unnecessary. It delights in a moist cool situation.

CYPRIPEDIUM. Lin. Gen. Pl. 906. Ladies Slipper.

The CHARACTERS are,

It hath a simple spadix. The germen sits under the flower, which is covered with a sheath. The flowers have four or five narrow spear-shaped petals, which expand. The nectarium, which is situated between the petals, is swollen and hollow, in shape of a shoe, or slipper. It hath two short stamina. Below the flower is fixed a slender contorted germen, which afterward becomes an oval blunt capsule with three corners, having three valves, and one cell, filled with small seeds.

The Species are,

1. Cypripedium (*Calceolus*) radicibus fibrosis, foliis ovato-lanceolatis caulinis. Act. Upsal. 1740. *Our Ladies Slipper.*

2. Cypripedium (*Bulbosum*) scapo unifloro, foliis oblongis glabris, petalis angustis acuminatis. *Yellow Ladies Slipper.*

3. Cypripedium (*Hirsutum*) foliis oblongo-ovatis venosis hirsutis, flore maximo. *Ladies Slipper with a larger flower.*

The first sort grows naturally in some shady woods in the north of England. I found it in the park of Borough-hall, in Lancashire, the seat of the late Robert Fenwick, Esq; It hath a root composed of many fleshy fibres, from which arise two, three, or more stalks, in proportion to the strength of the root; these grow nine or ten inches high, garnished with oval spear-shaped leaves, having a few longitudinal veins; in the bosom of one of the upper leaves is inclosed the flower-bud, which is supported by a slender foot-stalk, which generally turns a little on one side. The flower hath four dark purple petals, placed in form of a cross, which spread wide open. In the center is situated the large hollow nectarium, almost as large as a bird's egg, shaped like a wooden shoe, of a pale yellowish colour, with a few brown streaks; the opening is covered with two ears; the upper one is tender, white, and spotted with purple; the lower is thick, and of an herbaceous colour.

The second sort grows naturally in Virginia, and other parts of North America. This hath longer and smoother leaves than the former. The two side petals of the flower are long, narrow, and terminate in acute points, and are wreathed, or undulated on their sides. The nectarium is oblong, and narrower than in the first sort, and is yellow, spotted with brownish red. The stalks rise near a foot and a half high.

The third sort grows naturally in America, where the inhabitants call it Moccasin flower; this rises a foot and a half high. The leaves are of an oblong oval form, and are deeply veined. The flower is large, of a reddish brown colour, marked with a few purple veins.

All these sorts are with difficulty preserved in gardens; they must be planted in a loamy soil, and in a situation where they may have the morning sun only. They must be procured from the places where they naturally grow, for they cannot be easily propagated in gardens. The roots should be seldom removed, for transplanting them prevents their flowering.

CYSTICAPNOS. See Fumaria.

CYTISO GENISTA. See Spartium.

CYTISUS. Lin. Gen. Pl. 785. Base Tree Trefoil.

The Characters are,

It hath a butterfly flower. The standard of the flower is rising, oval, and reflexed on the sides. The wings are obtuse, erect, and the length of the standard. The keel is bellied and acute. It hath ten stamina, nine joined, and one standing separate, and an oblong germen, which afterward becomes an oblong blunt pod, filled with kidney-shaped flat seeds.

The Species are,

1. Cytisus (*Laburnum*) foliis oblongo-ovatis, racemis brevioribus pendulis, caule arboreo. *Broad-leaved Cytisus of the Alps, with pendulous bunches of flowers; commonly called Laburnum.*

2. Cytisus (*Alpinus*) foliis ovato-lanceolatis, racemis longioribus pendulis, caule fruticoso. *Narrow-leaved Cytisus of the Alps, with longer pendulous bunches of flowers; commonly called long-spiked Laburnum.*

3. Cytisus (*Nigricans*) racemis simplicibus erectis, foliolis ovato-oblongis. Hort. Cliff. 354. *Blackish smooth Cytisus.*

4. Cytisus (*Hirsutis*) villosus, foliolis cuneiformibus perennantibus, caulibus ramosissimis, racemis terminalibus. *Ever-green hoary Cytisus of the Canary Islands.*

5. Cytisus (*Sessilis*) racemis erectis, calycibus bracteâ triplici unctis, foliis floralibus sessilibus. Lin. Sp. Pl. 739. *Smooth Cytisus with roundish leaves, and very short foot-stalks; commonly called by gardeners Cytisus secundus Clusii.*

6. Cytisus (*Græcus*) foliis simplicibus lanceolato-linearibus ramis angulatis. Lin. Sp. 1043. *Cytisus with single narrow leaves and angular branches.*

7. Cytisus (*Argenteus*) floribus subsessilibus, foliis tomentosis, caulibus herbaceis. Lin Sp. Pl. 740. *Low silvery Cytisus with narrow leaves.*

8. Cytisus (*Supinus*) floribus capitatis, ramis decumbentibus. Prod. Leyd. 376. *Low Cytisus with the under side of the leaves, and pods covered with a soft down.*

9. Cytisus (*Hirsutus*) pedunculis simplicibus lateralibus, calycibus hirsutis trifidis ventricoso oblongis. Hort. Upsal. 211. *Italian Cytisus with hairy leaves.*

10. Cytisus (*Austriacus*) floribus capitatis, foliolis ovato-oblongis, caule fruticoso. *Cytisus with flowers collected in heads, oblong oval leaves, and a shrubby stalk; commonly called Tartarian Cytisus.*

11. Cytisus (*Æthiopicus*) racemis lateralibus strictis, ramis angulatis, foliolis cuneiformibus. Lin. Sp. Pl. 740. *Ethiopian Cytisus with smaller, roundish, hoary leaves, and a small yellow flower.*

12. Cytisus (*Cayan*) racemis axillaribus erectis, foliolis sublanceolatis tomentosis, intermedio longius petiolato. Flor. Zeyl. 357. *Tree Cytisus with eatable fruit; commonly called Pigeon Pea in America.*

The first sort is the common broad-leaved Laburnum, which was formerly in greater plenty in the English gardens than at present; for since the second sort hath been introduced, it hath almost turned this out; the spikes of flowers being much longer, so they make a finer appearance when they are in flower, which has occasioned their being more generally cultivated; but the first grows to be the largest tree, and the wood of it is very hard, of a fine colour, and will polish very well; it approaches near to green Ebony. In England there are few of these trees which have been suffered to stand long enough to arrive to any considerable size, for as they have been only considered as an ornamental tree, so the frequent alterations which most of the gardens in England have undergone, have occasioned their being rooted out wherever they were growing; but in some of the old gardens in Scotland, where they have been permitted to stand, there are large trees of this kind, which are fit to cut down for the use of the timber. They grow very fast, and are extremely hardy, so may be well worth propagating upon poor shallow soils, and in exposed situations. His Grace the Duke of Queensberry sowed a great quantity of the seeds of this tree upon the side of the downs, at his seat near Amesbury in Wiltshire, where the situation was very much exposed, and the soil so shallow as that few trees would grow there; yet in this place the young trees were twelve feet high in four years growth, so became a shelter to the other plantations, for which purpose they were designed; but the hares and rabbits are great enemies to these trees, by barking them in winter, so that where these trees are cultivated, they should be fenced from these animals.

Both the sorts are easily propagated by seeds, which the trees produce in great plenty; if these are sown upon a common

mon bed of earth in March, the plants will appear by the middle or end of April, and will require no other care but to be kept clean from weeds during the following fummer; and if the plants are too clofe together, they may be tranf planted the autumn following, either into a nurfery where they may grow a year or two to get ftrength, or into the places where they are defigned to remain; but where peo ple would cultivate them for their wood, it will be the beft way to fow the feeds upon the fpot where they are intended to grow, becaufe thefe trees fend out long, thick, flefhy roots to a great diftance, which will penetrate gravel or rocks, and if the roots are cut or broken, it greatly retards their growth; therefore when they are not fown upon the in tended fpot, they fhould be tranfplanted thither young, otherwife they will not grow to near the fize; though where they are defigned for ornament, the removing the plants twice will ftop their growth, and caufe them to be more productive of flowers; but all trees intended for timber are much better fown on the ground where they are to remain, than if they are tranfplanted ı

If the feeds of thefe trees are permitted to fcatter in winter, the plants will rife in great plenty the following fpring, fo that a few trees will foon fupply any perfon with a fufficient number of the plants

If the firft fort comes to be confidered as a ufeful wood, which there is no reafon to doubt it may be, it may be planted in large clumps in parks, where they will be very ornamental; and I am certain, from long experience, that this tree will thrive upon many foils, and in fuch fituations as few other trees will make any progrefs; the objection to fencing is the fame here as for any other trees, for wherever plantations are made, if they are not well fecured from animals, they will not anfwer the defign of the planters.

The fecond fort differs from the firft, in having narrower leaves, longer bunches of flowers, and the trees do not grow fo large and ftrong; this difference I find is conftant from feed. There is another fort mentioned by Tournefort, with fhorter bunches of flowers than either of thefe, one tree of which kind I thought I had found in a garden; the bunches of the flowers upon this tree were clofe and almoft round, but I fowed the feeds of it, and the plants proved to be only the common fort.

The third fort grows naturally in Auftria, Italy, and Spain, and at prefent is pretty rare in the Englifh gardens; it was formerly in fome of the curious gardens here, but had been long loft till a few years ago, when I procured the feeds from abroad.

This fhrub feldom rifes more than three or four feet high in England; it naturally puts out many lateral branches near the ground, which fpread out on every fide, fo as to form a low fhrubby bufh, and is with difficulty raifed to a ftem: the branches are very flender, and their ends are frequently killed if the winter is fevere; thefe are garnifhed with oblong oval leaves, growing by threes on each footftalk; they are equal in fize, and of a dark green colour; the branches grow erect, and are terminated by fpikes of yellow flowers about four or five inches long, ftanding upright; and as all the branches are thus terminated, fo when the fhrubs are in flower they make a fine appearance; it flowers in July, after moft of the other forts are paft. This is propagated by feeds, which fhould be fown on a bed of light earth in March. In the beginning of May the plants will appear, when they muft be carefully weeded. In autumn the bed fhould be arched over with hoops, that in frofty weather the plants may be covered with mats, to prevent their tender fhoots from being killed; for as thefe young plants are apt to continue growing later in the autumn than thofe which are become woody, fo they are much more fufceptible of cold; therefore where there is not fome care taken to cover them, if the winter fhould prove fevere, many of them may be entirely deftroyed. The fpring following, after the danger of hard froft is over, the plants fhould be carefully taken up, and planted out at the diftance of one foot row from row, and fix inches afunder in the rows; this fhould be in a fheltered fituation: as thefe plants do not fhoot till late in the fpring fo they need not be tranfplanted before the beginning of April; and if the feafon fhould then prove warm and dry, it will be proper to give the plants fome water to fettle the earth to their roots. After they have taken new root, they will require no farther care but to keep them conftantly clean from weeds: in this nurfery the plants may remain two years, by which time they will have acquired ftrength enough to be tranfplanted where they are to remain.

The fourth fort grows naturally in the Canary Iflands. This is too tender to live through the winter in the open air here, but requires the fhelter of a green-houfe, and will thrive in fuch places where Myrtles and Amomum Plinii do well. It is a very bufhy fhrub, which rifes with rough pliable ftalks to the height of eight or ten feet, fending out many flender hairy branches, which are very clofely garnifhed with fmall wedge-fhaped leaves; placed by threes on each foot-ftalk: they are of a dark green, and very hairy; the branches are terminated by clofe bunches of bright yellow flowers, which are frequently fucceeded by fhort hairy feed-pods, which ripen in Auguft.

It is propagated by feeds, which fhould be fown upon a very temperate hot-bed in March, which will bring up the plants in a fhort time; then they may be tranfplanted, each into a fmall halfpenny pot, filled with light earth, and plunged into a moderate hot-bed, juft to forward their taking frefh root; after which they fhould be inured gradually to the open air, and the middle or latter end of May they fhould be placed abroad in a fheltered fituation, and afterward treated in the fame way as other hardy kinds of greenhoufe plants.

The fifth fort grows naturally in the fouth of France, in Spain and Italy, but has been long cultivated in the nurfery-gardens as an ornamental flowering fhrub, by the title of Cytifus fecundus Clufii, which is a great miftake, for the ninth fpecies here enumerated is the fecond of Clufius. This rifes with a woody ftalk, putting out many branches, covered with a brownifh bark, and garnifhed by obverfe, oval, fmall leaves, growing by threes on very fhort foot-ftalks. The flowers are produced in clofe fhort fpikes at the end of the branches; they are of a bright yellow colour, and appear in June; thefe are fucceeded by fhort broad pods, which contain one row of kidney-fhaped feeds, which ripen in Auguft. Thefe fhrubs will rife to the height of feven or eight feet, and become very bufhy; they are very hardy, fo will thrive in any fituation, and upon almoft any foil, which is not too wet.

The fixth fort hath a foft fhrubby ftalk, dividing into many branches, which grow erect, and rife to the height of eight or ten feet; the ftalks and leaves of this are very hairy; the leaves are oval, growing three upon each foot ftalk; the flowers come out from the fide of the branches in fhort bunches; they are of a pale yellow, and are fucceeded by long, narrow, hairy pods, with one row of kidney-fhaped feeds.

The feventh fort grows naturally in the fouth of France and in Italy. This is a low perennial plant, which puts out feveral weak ftalks from the root, which fpread on the ground, and are from fix to eight inches long, garnifhed with fmall filvery leaves growing by threes; the flowers are produced at the end of the branches, two or three growing together upon fhort foot-ftalks; they are of a pale yellow colour, but unlefs the feafon proves very warm, they do not

not produce feeds in England It is propagated by feeds, which fhould be fown in the fpring in a warm border, where the plants are to remain; for as they have commonly one downright root, fo they feldom live if they are tranfplanted. The plants require no other culture but to keep them clean from weeds, and the fecond year they will flower.

The eighth fort grows naturally in Sicily, Italy, and Spain; this is a perennial plant, from whofe downright root proceed feveral weak branches, which trail upon the ground, and extend to the length of eight or ten inches, garnifhed with oblong leaves, placed by threes upon pretty long foot ftalks; they are hoary on their under fide, but fmooth above; the flowers are collected in heads at the ends of the ftalks, having a clufter of leaves under them; they are of a deep yellow colour, and in warm feafons they are fucceeded by flat woolly pods, containing one row of fmall kidney-fhaped feeds. This plant is propagated by feeds, which fhould be fown where they are to remain, and treated in the fame manner as the feventh fort.

The ninth fort grows naturally in Italy; this hath two branching ftalks, fending out many fide branches, which are ftreaked or furrowed, and garnifhed with oval hairy leaves; the flowers are produced in fhort fpikes at the ends of the branches, which are of a pale yellow, and are fucceeded by hairy pods, which ripen in autumn. This may be propagated in the fame manner as the other forts.

The tenth fort grows naturally in Auftria; this is a weakly fhrub, rifing four or five feet high, with a ftrong downright root, fending out tough ligneous branches, which extend a foot and a half in length, covered with a green bark, garnifhed with very fmall Trefoil leaves; the flowers are produced at the extremity of the branches in clufters; they are but fmall, and of a deep yellow, inclining to an Orange colour, and in very warm feafons are fucceeded by fhort pods, containing three or four kidney-fhaped feeds in each. This fort is propagated in the fame manner as the feventh.

The eleventh fort came from Siberia; this is a low fhrub, which feldom rifes more than three feet high in England, fending out fide branches, garnifhed by oval fmooth leaves, having pretty long foot-ftalks; the flowers are produced in clufters or heads at the end of the branches; they are of a bright yellow, and appear the end of March, or the beginning of April, but are feldom fucceeded by pods in England. This is propagated by feeds as the other forts, but requires a cool fituation; for the plants are fubject to fhoot upon the firft mild weather in February, and fo are often cut down by the frofts in March, which fometimes kill the fhoots down to the old wood.

The twelfth fort grows naturally in the iflands of America; it rifes with a fhrubby ftalk eight or ten feet high, garnifhed by fpear-fhaped woolly leaves, placed by threes on each pedicle; that in the middle having a diftinct foot ftalk, the two fide lobes growing clofe to the principal foot-ftalk. The flowers come out from the fide of the branches, fometimes fingly, at other times in clufters; they are of a deep yellow colour, and about the fize of thofe of the common Laburnum, and are fucceeded by hairy pods about three inches long, which are fickle-fhaped, ending with a long acute point; the feeds are roundifh, a little inclined to a kidney fhape, and are efteemed an excellent food for pigeons in America, from whence it had the title of Pigeon Pea.

This plant grows only in very warm countries, fo cannot be preferved in England, unlefs it is placed in a warm ftove. It rifes eafily from feeds in a hot-bed, and will grow three or four feet high the firft year, provided they have a proper heat, and the fecond year they will produce flowers and feeds. The plants muft be placed in the bark-bed in the ftove, and treated in the fame manner as other tender plants from the fame countries: they fhould have but little water in winter, and in the fummer fhould have a large fhare of free air admitted to them in warm weather.

D.

DAFFODIL. See NARCISSUS.

DAISIES. See BELLIS.

DALECHAMPIA. Lin. Gen. Plant. 1022.

The CHARACTERS are,

It hath male and female flowers on the fame plant; the male flowers have no petals, but many ftamina. The female flowers have a roundifh three-cornered germen, which afterward becomes a round three-cornered capfule, having three cells, each containing one roundifh feed.

We have but one SPECIES of this genus, viz.

DALECHAMPIA (*Scandens*) foliis trilobis glabris, floribus axillaribus caule volubili. *Dalechampia with fmooth trifoliate leaves, flowers growing from the fides of the branches, and a twining ftalk.*

This plant grows naturally in Jamaica. From the root compofed of many fibres arife feveral weak twining ftalks, which faften themfelves to the neighbouring plants, and mount up to a confiderable height; they are garnifhed at each joint by one trifoliate leaf, or more properly by a leaf divided into three lobes, (for thefe are joined together at their bafe) which are fmooth; the two fide lobes are oblique to the midrib. The flowers are produced from the fide of the ftalks, three or four growing upon each foot-ftalk; fome of thefe are male, and others female; they are of an herbaceous colour, and fmall, fo make no appearance; they have each a double involucrum, made up of two orders of leaves, which are narrow, and armed with fmall briftly hairs, which fting the hands of thofe who unwarily touch them; the flowers are fucceeded by roundifh capfules, having three prominent lobes which are fmooth, each inclofing a fingle feed.

This plant is propagated by ſeeds, which muſt be ſown early in the ſpring on a hot bed; and when the plants are three inches high, they ſhould be carefully tranſplanted, each into a ſeparate ſmall pot, and plunged into a hot-bed of tanners bark, being careful to ſcreen them from the ſun until they have taken new root; after which they ſhould have a great ſhare of freſh air in warm weather. When the plants have grown ſo large as to fill theſe pots with their roots, they ſhould be removed into larger, and placed in the bark-bed in the ſtove, where they muſt be ſupported either with ſtakes or a trellis, round which they will twine, and riſe to the height of eight or ten feet.

The plants muſt be kept conſtantly in the ſtove, for they are too tender to bear the open air in this country, even in the ſummer ſeaſon; therefore they ſhould be placed with tender Convolvuluſes, and other twining plants, near the back of the ſtove, where a ſupport ſhould be made for them; in ſummer they will flower, and in warm ſeaſons will perfect their ſeeds in this country. Theſe plants do not continue longer than one year, ſo that young plants ſhould be raiſed annually to preſerve the kind.

DAMASONIUM. Star-headed Water Plantain.

The CHARACTERS are,

It hath a flower compoſed of three petals, included in a three-leaved empalement, with a ſtar-ſhaped fruit, which is full of oblong ſeeds.

The SPECIES are,

1. DAMASONIUM (*Aliſma*) ſtellatum. Lugd. *Star-headed Water Plantain.*

2. DAMASONIUM (*Flava*) Americanum maximum, plantaginis folio, flore flaveſcente, fructu globoſo. Plum. *Greateſt American Water Plantain, with a Plantain leaf, a yellowiſh flower, and a globular fruit.*

The firſt of theſe plants is a native of England; it grows commonly in ſtanding waters, which are not very deep. It is ſometimes uſed in medicine, but never cultivated in gardens, ſo muſt be gathered for uſe in the places of its growth.

The ſecond ſort grows in Jamaica, Barbadoes, and ſeveral other places in the warm parts of America, where it is generally found in ſtagnating waters, and other ſwampy places; ſo that it would be difficult to preſerve this plant in England, for it will not live in the open air, and requires a bog to make it thrive; but as it is a plant of no great beauty or uſe, it is not worth the trouble of cultivating in this country.

DANDELION. See LEONTODON.

DAPHNE. Lin. Gen. Plant. 436. Spurge Laurel, or Mezereon.

The CHARACTERS are,

The flower hath no empalement; it is of one petal, cut into four parts at the top; it hath eight ſhort ſtamina. The oval germen is ſituated at the bottom of the tube, which afterward becomes a roundiſh berry with one cell, incloſing one roundiſh fleſhy ſeed.

The SPECIES are,

1. DAPHNE (*Laureola*) racemis axillaribus, foliis lanceolatis glabris. Lin. Sp. Pl. 357. *Male Laureola; commonly called Spurge Laurel.*

2. DAPHNE (*Mezeron*) floribus ſeſſilibus ternis caulinis, foliis lanceolatis deciduis. Lin. Sp. Pl. 357. *Thymelæa with a deciduous Bay leaf; commonly called Mezereon.*

3. DAPHNE (*Thymelea*) floribus ſeſſilibus axillaribus, foliis lanceolatis, caulibus ſimpliciſſimis. Lin. Sp. Pl. 356. *Thymelæa with ſmooth Milkwort leaves.*

4. DAPHNE (*Tarton-raire*) floribus ſeſſilibus aggregatis axillaribus, foliis ovatis utrinque pubeſcentibus nervoſis. Lin. Sp. Pl. 356. *Thymelæa with ſoft, white, ſattiny leaves; commonly called Tarton-raire.*

5. DAPHNE (*Alpina*) floribus ſeſſilibus aggregatis lateralibus, foliis lanceolatis obtuſiuſculis ſubtus tomentoſis. Lin. Sp. Pl. 356. *Alpine Chamelæa with obtuſe leaves, hoary on their under ſide.*

6. DAPHNE (*Cneorum*) floribus congeſtis terminalibus ſeſſilibus, foliis lanceolatis nudis. Lin. Sp. Pl. 357. *The Cneorum.* Matth. Hiſt. 46.

7. DAPHNE (*Gnidium*) paniculâ terminali, foliis lineari-lanceolatis acuminatis. Lin. Sp. Pl. 357. *Thymelæa with Flax leaves.*

8. DAPHNE (*Squarroſa*) floribus terminalibus pedunculatis, foliis ſparſis linearibus patentibus mucronatis. Lin. Sp. Pl. 358. *Thymelæa with a woolly head, and many ſmall pointed leaves.*

The firſt ſort grows in the woods in many parts of England, and is commonly known by the title of Spurge Laurel; of late years there are poor people who get the young plants out of the woods, and carry them about London to ſell in the ſpring. This is a low evergreen ſhrub, which riſes with ſeveral ſtalks from the root to the height of three feet, which are garniſhed with thick ſpear-ſhaped leaves, ſitting pretty cloſe to the branches, and are of a lucid green; between theſe, toward the upper part of the ſtalks, come out the flowers in ſmall cluſters, of a yellowiſh green colour, and appear ſoon after Chriſtmas, if the ſeaſon is not very ſevere; theſe are ſucceeded by oval berries, which are green till June, when they ripen and turn black, ſoon after which they fall off. The whole plant is of a hot cauſtic taſte, burning and inflaming the mouth and throat. The leaves continue green all the year, which renders the plants ornamental in winter; and as they will thrive under tall trees, ſo are very proper to fill up the ſpaces in plantations.

The ſecond ſort grows naturally in Germany, and of late there hath been a diſcovery made of its growing in England, in ſome woods near Andover, from whence a great number of plants have been taken. This is a very ornamental ſhrub in gardens; the flowers come out very early in the ſpring, before others make their appearance. There are two diſtinct ſorts of this, one with a white flower, which is ſucceeded by yellow berries, the other with Peach-coloured flowers and red berries. Theſe are by ſome ſuppoſed to be accidental varieties, ariſing from the ſame ſeeds; but I have ſeveral times raiſed theſe plants from ſeeds, and always found the plants come up the ſame, as thoſe from which the ſeeds were taken, ſo they may be called different ſpecies. There is a variety of the Peach-coloured Mezereon, with flowers of a much deeper colour than the common, but theſe I have always found to vary in their colours when raiſed from ſeeds.

This ſhrub grows to the height of five or ſix feet, with a ſtrong woody ſtalk, putting out many woody branches, ſo as to form a regular head; the flowers come out very early in the ſpring, before the leaves appear, growing in cluſters all round the ſhoots of the former year; there are commonly three flowers produced from each knot or joint, ſtanding on the ſame ſhort foot-ſtalk, which have ſhort ſwelling tubes divided into four parts at the top, which ſpread open; they have a very fragrant odour, ſo that where there are plenty of the ſhrubs growing together, they perfume the air to a conſiderable diſtance round them. After the flowers are paſt, the leaves come out, which are ſmooth, ſpear-ſhaped, and placed without order; they are about two inches long, and three quarters broad in the middle, gradually leſſening to both ends; the flowers are ſucceeded by oval berries, which ripen in June; thoſe of the Peach coloured flowers are red, and thoſe of the white yellow.

This is propagated by ſeeds, which ſhould be ſown on a border expoſed to the eaſt, ſoon after the berries are ripe; for if they are not ſown till the following ſpring, they often miſcarry,

or

or at leaſt remain a year in the ground before the plants appear; whereas thoſe which are ſown in Auguſt, will many of them grow the following ſpring. When the plants come up, they will require no other care but to keep them clean from weeds, and may continue in the ſeed bed two ſummers, eſpecially if they do not make great progreſs the firſt year; then at Michaelmas, when the leaves are ſhedding, they ſhould be carefully taken up ſo as not to break or tear their roots, and planted in a nurſery at about ſixteen inches row from row, and eight inches aſunder in the row. In this nurſery they may remain two years, by which time they will be fit to remove to the places where they are deſigned to remain for good; the beſt time to tranſplant theſe ſhrubs is in autumn, for as theſe plants begin to vegetate very early in the ſpring, ſo it is not proper to tranſplant them at that ſeaſon. The plants grow beſt in a light earth which is dry, for in cold wet land they become moſſy, and make little progreſs.

Although the berries of this tree are ſo very acrid, as to burn the mouth and throat of thoſe who may incautiouſly taſte them, yet the birds greedily devour them as ſoon as they begin to ripen; ſo that unleſs the ſhrubs are covered with nets to preſerve the berries, they will all be deſtroyed before they are fit to gather.

The third ſort grows naturally in Spain, Italy, and the ſouth of France, where it riſes to the height of three or four feet, with a ſingle ſtalk covered with a light coloured bark; the flowers come out in cluſters on the ſides of the ſtalks, which are of an herbaceous colour, ſo make but little appearance; they appear early in the ſpring, and are ſucceeded by ſmall berries, which are yellowiſh when ripe.

The fourth ſort grows naturally in the ſouth of France; this is a low ſhrubby plant, which ſends out ſeveral weak ſtalks from the root about a foot long, which ſpread about irregularly; they ſeldom become woody in England, but are tough and ſtringy, covered with a light bark; the leaves are ſmall, of an oval form, and are very ſoft, white, and ſhining like ſattin; theſe ſit pretty cloſe to the ſtalks; between theſe the flowers come out in thick cluſters from the ſide of the ſtalks; they are white, and are ſucceeded by roundiſh berries, having one hard ſeed. This flowers here in June, but doth not produce ripe ſeeds.

The fifth ſort grows in the mountains near Geneva, and in other parts of Italy, where it riſes about three feet high; the flowers of this come out in cluſters from the ſide of the branches early in the ſpring. The leaves are ſpear-ſhaped, ending in blunt points, and are hoary on their under ſide. The flowers are ſucceeded by ſmall roundiſh berries, which turn red when ripe.

The ſixth ſort grows naturally on the Alps, as alſo upon the mountains of Verona. This is a very humble ſhrub, ſeldom growing more than one foot high, with ligneous ſtalks, garniſhed with narrow ſpear-ſhaped leaves, which are placed round the ſtalks without order; the branches are terminated by ſmall cluſters of purple flowers, which ſtand erect, having no foot-ſtalks; the tubes of theſe flowers are longer and narrower than thoſe of the Mezereon, and the mouth is cut into four acute parts, which are erect. The flowers emit a pleaſant odour; they appear early in the ſpring, but do not produce ſeeds here.

The ſeventh ſort grows naturally about Montpelier. This hath a ſhrubby ſtalk about two feet high, dividing into many ſmaller branches, cloſely garniſhed with narrow ſpear-ſhaped leaves, ending in acute points; the ends of the branches are terminated by panicles of flowers, which are much ſmaller than thoſe of the Mezereon, having ſwelling tubes, which are contracted at the mouth.

The eighth ſort grows naturally at the Cape of Good Hope. This ſhrub riſes to the height of five or ſix feet, dividing into ſeveral branches, which grow erect, and are covered with a white bark, and cloſely garniſhed with ſmall narrow leaves, which come out on every ſide of the branches without order. The tops of the branches are terminated by woolly heads, out of which the flowers come in ſmall cluſters; they are white, having oblong tubes, which are divided into four obtuſe ſegments at the mouth, which ſpread open.

The third, fourth, and ſeventh ſorts are hardy, ſo will live through the winters in England in the open air, provided they are in a dry ſoil. The fifth and ſixth ſorts are as hardy as the common Mezereon, ſo are not in danger of being hurt by froſt in England; but theſe are very difficult to keep in gardens, becauſe neither of them will bear to be tranſplanted. I have ſeveral times raiſed the plants from ſeeds, which have ſucceeded well in the places where they were ſown; but whenever they were removed, they certainly died, though performed at different ſeaſons, and with the greateſt care, and the ſame has happened to every other perſon who has raiſed any of theſe plants; and ſome of my correſpondents have aſſured me, they have frequently attempted to remove theſe plants from their natural places of growth into their gardens, and have choſen plants of all ſizes, from the youngeſt ſeedlings to the oldeſt plants, yet have never ſucceeded in it. Therefore thoſe who are deſirous to have theſe plants in their gardens, muſt procure their ſeeds, which ſhould be immediately ſown where they are deſigned to remain. The third, fourth, and ſeventh ſorts ſhould be on a warm dry border, where, if there is a foundation of lime rubbiſh or chalk under the ſurface of the ground, the plants will thrive better and continue much longer than in good ground; all the culture they require is to keep the place clean from weeds, for the leſs the ground is ſtirred near their roots, the better the plants will thrive; for they naturally grow on poor ſhallow land, and out of crevices of rocks, ſo the nearer the ſoil approaches to this, the more likely the plants will be to ſucceed.

The fifth and ſixth ſorts may have a cooler ſituation; if theſe are ſown where they may have only the morning ſun, they will thrive better than in a warmer place, and the ground near the roots of theſe ſhould not be diſturbed; therefore in the choice of the ſituation there ſhould be regard had to this, not to ſow them near other plants, which may require tranſplanting, or to have the ground dug or looſened. The ſeeds of theſe plants ſhould be put up in wet Moſs, if they are to be ſent to any conſiderable diſtance, and ſhould be ſown as ſoon as they arrive.

The eighth ſort grows naturally at the Cape of Good Hope, ſo will not live abroad through the winter in England, but requires a good green-houſe to preſerve it. This plant is very difficult to keep or propagate in gardens.

DATISCA. Lin. Gen. Pl. 1003. Baſtard Hemp.

The CHARACTERS are,

They have male and female flowers in different plants; the male flowers have a five-leaved empalement, but no petals, with ten ſummits. The female flowers have no petals, but empalements like the male, with an oblong pervious germen, and three ſtyles, an oblong triangular capſule with three valves, filled with ſmall ſeeds adhering to the three ſides of the capſule.

The SPECIES are,

1. DATISCA (*Cannabina*) caule lævi. Lin. Sp. Pl. 1037. *Datiſca with a ſmooth ſtalk.*

2. DATISCA (*Hirta*) caule hirſuto. Lin. Sp. Pl. 1037. *Datiſca with a rough ſtalk.*

The firſt ſort grows naturally in Crete. It hath a perennial root, from which ariſe ſeveral herbaceous ſtalks five or ſix feet high, garniſhed with winged leaves placed alternately, compoſed of three pair of lobes, terminated by an odd one; they are deeply ſawed on their edges, and of a

 light

light green colour. The flowers come out in long loose spikes from the side of the stalks at the wings of the leaves, but having no petals make but a poor appearance. The summits of the male flowers being pretty long, and of a bright yellow colour, are the only visible parts of the flowers to be discerned at any distance.

The flowers on the female plants are succeeded by oblong three-cornered capsules, filled with small seeds, which adhere to the three valves.

This sort may be propagated by parting the roots, which should be done in autumn, when the stalks decay (which is the best time to transplant the roots) but they must not be parted too small; they may be planted in any open beds, where they are not under the drip of trees, and will require no other culture but to keep them clean from weeds.

The second sort grows naturally in Canada, and other parts of North America. This differs from the other, in having hairy stalks, which grow taller; the leaves are larger, and do not stand so near each other upon the stalks. It is equally hardy with the first sort, and may be propagated in the same manner, but should have a more shady situation and a moister soil.

DATURA. Lin. Gen. Plant. 218. Thorn Apple.

The CHARACTERS are,

The flower hath one funnel-shaped petal with a long cylindrical tube, which is pentangular, each angle being pointed; it hath five stamina, and an oval germen, which afterward becomes an oval capsule with four cells, which are filled with kidney-shaped seeds adhering to the partition.

The SPECIES are,

1. DATURA (*Stramonium*) fructu rotundo erecto, pericarpio spinoso. *Thorn Apple with a round prickly fruit, and a single white flower.*
2. DATURA (*Tatula*) fructu ovali erecto, pericarpio spinoso. *Thorn Apple with an oblong prickly fruit, and a Violet-coloured flower.*
3. DATURA (*Metel*) pericarpiis spinosis nutantibus, foliis cordatis subintegris pubescentibus. Hort. Cliff. 55. *Datura with a globular nodding fruit, having a prickly cover, and heart-shaped entire leaves.*
4. DATURA (*Ferox*) pericarpiis spinosis ovatis, spinis supremis maximis convergentibus. Amœn. Acad. 3. p. 403. *Datura with an oval erect fruit, whose upper spines are largest, and converge together.*
5. DATURA ((*Inoxia*) pericarpiis spinosis inoxiis ovatis propendentibus, foliis cordatis pubescentibus. *Datura with an oval hanging fruit, whose cover is beset with harmless spines, and hoary leaves.*
6. DATURA (*Fastuosa*) pericarpiis tuberculosis nutantibus globosis, foliis lævibus. Lin. Sp. 256. *Datura with a globular nodding fruit, whose cover is beset with tubercles, and has smooth leaves.*

The sorts here enumerated are undoubtedly distinct species, though some have supposed part of them to be only seminal variations; but these never vary from one sort to another, all the difference which any of them have ever shewn, has been in the double flowers becoming single; for which reason I have only mentioned one with double flowers, which sometimes changes to single, at other times they are with double and treble tubes, stretched out beyond each other like those Primroses called Hose in Hose. There is also a double flower of the third sort, which is much esteemed by the curious; but this frequently degenerates to single, so is not to be mentioned as a distinct species.

The first sort here enumerated is the most common sort of Thorn Apple, and was probably first introduced from Italy or Spain, where it naturally grows; but it is now become so common about London, and near other great towns in England, as to appear like a native plant; for there are few gardens or dunghills without this plant in summer, though it is only near such places where the plants may have been cultivated in gardens; for wherever any of these plants are permitted to seed, they will furnish a supply of the plants for some years to come, as they produce a vast quantity of seeds, some of which will lie years in the ground, and when they are turned up to the air will vegetate.

This sort seldom grows much more than two feet high, dividing into many strong, irregular, hollow branches, garnished with large, smooth, angular leaves, which, when handled, emit a fœtid odour. The flowers come out first from the divisions of the branches, and afterward near their extremities; they have long swelling tubes, which are dilated at the top into large pentagonal brims, each angle ending in a long point or ligula, and are succeeded by large roundish seed-vessels, covered with large thorns, and are divided by four furrows, to which adhere the partitions, which separate the four cells, filled with black kidney-shaped seeds. The seeds ripen in autumn, which, if permitted to scatter, will fill the ground about them with plants the following years.

The second sort grows naturally in many parts of America, the islands of the West-Indies, and also in some of the northern parts of America. It rises with a purple strong stalk to the height of five or six feet, dividing into many strong branches, garnished with leaves shaped somewhat like those of the former sort, but larger, and have a greater number of angles and laciniæ on their edges; the flowers have longer and narrower tubes, of a purple colour; the fruit is also much longer, and these differences are permanent. This is equally hardy with the former, and if the seeds are permitted to scatter, the plants will become troublesome weeds.

The third sort hath a strong woolly stem, which rises three feet high; the leaves of this sort are woolly and almost entire, having only two or three slight indentures on their edges: the flowers have very long tubes, which spread out very broad at the brim, and is divided into ten obtuse angles; they are of a pure white above, but the tubes have a tincture of green within. They are succeeded by roundish fruit, closely covered with long soft thorns, and are divided into four cells as the other, but the seeds of this are of a light brown colour when ripe.

This plant is not so hardy as the others, so the seeds must be sown upon a gentle hot-bed in the spring, and the plants must be afterward treated in the same manner as the Marvel of Peru, and other of the hardier kinds of annual plants, so may be transplanted into the full ground the latter end of May. They will flower in July, and the seeds will ripen in autumn.

The fourth sort is of humbler growth, seldom rising more than a foot and a half high, spreading out into many branches, garnished with leaves somewhat like those of the first sort, but smaller, and stand upon longer foot-stalks; the flowers are like those of the first sort, but smaller; the fruit is round, and armed with very strong sharp thorns. The seeds of this are black when ripe.

This sort is too tender to be sown in the full goound in England, so the plants should be raised on a hot-bed, and afterward transplanted into borders as the former sort.

The fifth sort grows naturally in North America; this resembles the first sort while young, but afterward greatly differs from it; the stalks of this rise five feet high, are very smooth, of a light lucid green, as are also the leaves; the flowers are longer, and have a greater resemblance to those of the second sort. These differences have always continued for above thirty years, where it has been cultivated in the gardens. It is equally hardy with the first,

firſt, ſo will propagate itſelf in plenty where it remains to ripen ſeeds.

The ſixth ſort grows naturally in Egypt, and alſo in India; this riſes with a fine poliſhed purple ſtalk four feet high. dividing into ſeveral branches, garniſhed with large, ſmooth, ſinuated leaves, ſtanding upon pretty long footſtalks. The flowers are produced at the diviſions of the branches, and have large ſwelling tubes which expand very broad at the top, where their brims turn backward. The flowers are of a beautiful purple on their outſide, and a fattery white within; ſome of theſe are ſingle, others have two or three flowers ſtanding one within another. They have an agreeable odour at firſt, but if long ſmelt to become leſs agreeable, and are narcotic. If theſe plants are brought forward upon a hot bed in the ſpring, and in June planted out on a warm border of rich earth and covered with glaſſes, they will flower very finely in July and Auguſt, otherwiſe the ſeeds ſeldom ripen well in England. The fruit of this ſort is round, and grows nodding downward; the ſeed-veſſel is thick and fleſhy, as are alſo the intermediate partitions which divide the cells. The outſide of the fruit is covered with blunt protuberances, and the ſeeds are of a bright brown colour when ripe.

DAUCUS. Lin. Gen. Plant. 296. The Carrot.

The CHARACTERS are,

It hath an umbelliferous flower. The flowers are all hermaphrodite, and have five heart ſhaped petals which turn inward; they have each five hairy ſtamina. The germen ſits under the flower, which afterward becomes a roundiſh ſtriated fruit, covered with ſtinging briſtly hairs.

The SPECIES are,

1. DAUCUS (*Sylveſtris*) ſeminibus hiſpidis, radice tenuiore fervido. *Common wild Carrot.*

2. DAUCUS (*Carota*) ſeminibus hiſpidis, radice carnoſo eſculento. *Manured Carrot with an Orange-coloured root; commonly called Orange Carrot.*

3. DAUCUS (*Gingidium*) radiis involucri planis, laciniis recurvis. Prod. Leyd. 97. *Shining Maritime Carrot.*

4. DAUCUS (*Viſnaga*) ſeminibus nudis. Hort. Cliff. 89. *Gingidium with an oblong umbel.*

The firſt ſort is the common wild Carrot, which grows by the ſide of fields, and in paſture grounds in many parts of England. The plants of this ſort do not differ greatly in appearance from the Garden Carrot, which has led ſome perſons into an opinion of their being the ſame plant; but thoſe who have attempted to cultivate the wild ſort, are fully convinced of their being diſtinct. The ſeeds of this ſort are uſed in medicine, and are eſteemed good to bring away gravel; it is an excellent diuretic, but inſtead of theſe ſeeds, the ſhops are uſually ſupplied with old ſeeds of the Garden Carrot; when they have loſt their vegetative quality. the ſeedſmen then ſell them to druggiſts for medicinal uſe, when they cannot vend them to gardeners; but certainly all ſeeds which are too old to grow, can have but little medicinal virtue remaining in them.

There is ſome varieties of the Garden Carrot differing in the colour of their roots, which variations may be continued, where there is proper care taken not to mix the different ſorts together, when they are left for ſeed; but the Orange Carrot is generally eſteemed in London, ſo the yellow, the purple, and the white Carrots, are ſeldom cultivated.

The ſecond ſort is commonly cultivated in gardens for the kitchen; and the different varieties of it are in ſome places eſteemed, though in London the Orange Carrot is preferred to all the other.

Carrots are propagated at three different ſeaſons, or ſometimes oftener, where people are fond of having them young. The firſt ſeaſon for ſowing the ſeeds is ſoon after Chriſtmas, if the weather is open, which ſhould be in warm borders, near walls, pales, or hedges, but not immediately cloſe thereto; but a border of Lettuce, or other young ſallad herbs, of about a foot wide, ſhould be next the wall, &c. for if the Carrots are ſown cloſe to the wall, &c. they would draw up weak, without growing to have tolerable roots.

Carrots delight in a deep, warm, light, ſandy ſoil, which ſhould be dug two ſpades deep, that the roots may the better run down; for if they meet with any obſtruction they are very apt to grow forked, and ſhoot out lateral roots, but eſpecially where the ground is too much dunged the ſame year that the ſeeds are ſown, which will alſo occaſion their being worm-eaten; but where there may be a neceſſity for dunging it the ſame year as the Carrots are ſown, the dung ſhould be well rotted, and thinly ſpread over the ground; and in the digging of it into the ground great care ſhould be taken to diſperſe it through the different parts, and not to bury it in heaps. Where the ground is inclinable to bind, there cannot be too much care taken to break and divide the parts; therefore in digging the land for Carrots, there ſhould never be large ſpits taken, but they muſt be very thin and the clods well broken; which, if not attended to by the maſter, is ſeldom properly performed by workmen, who are too apt to hurry over their work, if they are not well obſerved.

The ground when dug ſhould be laid level and even, otherwiſe when the ſeeds are ſown and the ground is raked over, part of the ſeeds will be buried too deep, and others will be in danger of being drawn up into heaps; ſo the plants will come up in bunches, and other parts of the ground be naked, which ſhould always be carefully avoided.

As theſe ſeeds have a great quantity of ſmall forked hairs upon their borders, by which they cloſely adhere, ſo they are difficult to ſow even, not to come up in patches; therefore they ſhould be well rubbed with both hands, to ſeparate them before they are ſown; a calm day ſhould be choſen to ſow them, for if the wind blows it will be impoſſible to ſow the ſeeds equal, for as they are very light they will be blown into heaps. When the ſeed is ſown, the ground ſhould be trodden pretty cloſe to bury them, then rake the ground level.

When the plants are come up and have got four leaves, the ground ſhould be hoed with a ſmall hoe about three inches wide, cutting down all young weeds, and ſeparating the plants to three or four inches diſtance each way, that they may get ſtrength; and in about three weeks or a month after, when the weeds begin to grow again, the ground ſhould be hoed over a ſecond time, when there ſhould be care taken not to leave two Carrots cloſe to each other, and alſo to ſeparate them to a greater diſtance, cutting down all weeds, and ſlightly ſtirring the ſurface of the ground in every place, the better to prevent young weeds from ſpringing, as alſo to facilitate the growth of the Carrots.

In about five or ſix weeks after you muſt hoe them a third time, to clear them from weeds as before; and now the Carrots ſhould be ſeparated to the diſtance they are to remain, which muſt be proportioned to the ſize they are intended to grow: if they are to be drawn while young, four or five inches aſunder will be ſufficient; but if they are to grow large before they are pulled up, they ſhould be left eight or ten inches diſtant every way.

The ſecond ſeaſon for ſowing theſe ſeeds is in February, on warm beds ſituated near the ſhelter of a wall, pale, or hedge; but thoſe which are intended for the open large quarters, ſhould not be ſown before the beginning of March, nor ſhould you ſow any later than the end of the ſame month; for thoſe which are ſown in April or May will run up to feed before

before their roots have any bulk, especially if the weather should prove hot and dry.

In July you may sow again for an autumnal crop, and at the end of August you may sow some to stand through the winter; by which method you will have early Carrots in March, before the spring sowing will be fit to draw; but these are seldom so well tasted, and are often very tough and sticky. However, as young Carrots are generally expected early in the spring, so most people sow some at this season; but these should be sown on warm borders and upon dry land, otherwise they are seldom good. Many people mix other sorts of seeds with these, as Leek, Onion, Parsnep, Radish, &c. especially in the kitchen gardens near London; but this method is not good, for if there is a full crop of any one of these plants, there can be no room for any thing else amongst them, so that what is got by the one is lost by another.

The covetousness of some gardeners will not permit them to cut out their Carrots to a proper distance when they hoe them, so that by leaving them close they draw each other up weak; and if part of them are drawn up while young, those which are left never recover their strength afterward so perfectly, as to grow near the size of those which are properly thinned at their first hoeing.

This root has been long cultivated in gardens for the table, but has not till of late years been cultivated in the fields for cattle, nor has it been practised as yet, but in few parts in England; it is therefore greatly to be wished that the culture of this root was extended to every part of England, where the soil is proper for the purpose; for there is scarce any root yet known which more deserves it, being a very hearty good food for most sorts of animals. One acre of Carrots, if well planted, will fatten a greater number of sheep or bullocks than three acres of Turneps, and the flesh of these animals will be firmer and better tasted. Horses are extremely fond of these roots, and for hogs there is not any better food. I have also known these roots cultivated for feeding of deer in parks, which has proved of excellent use in hard winters, when there has been a scarcity of other food, at which time great numbers of deer have perished in some parks for want; and those which have escaped, have been so much reduced as not to recover their flesh till very late the following summer; whereas those fed with Carrots have been kept in good condition all the winter, and upon the growth of the Grass in the spring, have been fat early in the season, which is an advantage, especially where the Grass is generally backward in its growth.

There is also an advantage in the cultivation of this root beyond that of the Turnep, because the crop is not so liable to fail; for as the Carrots are sown in the spring, the plants generally come up well; and unless the months of June and July prove very bad, there is no danger of the crop succeeding, whereas Turneps are frequently destroyed by the flies at their first coming up; and in dry autumns they are attacked by caterpillars, which in a short time devour whole fields, but Carrots are not liable to such accidents: therefore every farmer who has a stock of cattle or sheep, should always have a supply of these roots, if he has land proper for the purpose, which must be light, and of a proper depth to admit of the roots running down.

In preparing of the land for Carrots in the open field, if it has not been in tillage before, it should be ploughed early in autumn, and then ploughed across again before winter, laying it up in ridges to mellow by the frost: and if the ground is poor, there should be some rotten dung spread over it in winter, which should be ploughed in about the end of January; then in March the ground should be ploughed again to receive the seeds; in the doing of which, some farmers have two ploughs, one following the other in the same furrow, so that the ground is loosened a foot and a half deep or more; others have men with spades following the plough in the furrows, turning up a spit of earth from the bottom, which they lay upon the top, levelling it smooth, and breaking the clods; the latter method is attended with a little more expence, but is much to be preferred to the first; because in this way the clods are more broken, and the surface of the ground is laid much evener.

If the land has been in tillage before, it will require but three ploughings; the first just before winter, when it should be laid in high ridges for the reasons before given; the second cross ploughing should be in January, after which, if it is well harrowed to break the clods, it will be of great service; the last time must be in March to receive the seeds; this should be performed in the manner before mentioned: after this third ploughing, if there remains great clods of earth unbroken, it will be proper to harrow it well before the seeds are sown. One pound and a half of seed will be sufficient for an acre of land, but as they are apt to adhere together, so it renders them more difficult to sow than most other seeds; therefore some mix a quantity of dry sand with their seeds, rubbing them well together, so as to separate the Carrot seeds from each other, which is a good method. After the seeds are sown, they must be gently harrowed in to bury them; and when the plants come up they should be hoed in the manner before directed, with this caution, to leave the plants at a greater distance.

But in order to preserve Carrots for use all the winter and spring, they should, about the beginning of November, when the green leaves are decayed, be digged up, and laid in sand in a dry place, where the frost cannot come to them, taking them out from time to time as there is occasion for them.

The third sort grows naturally about Montpelier; this hath smoother stalks than the common Carrot, the segments of the leaves are broader, and of a lucid green; the umbels of flowers are larger, and not so regular.

The fourth sort is an annual plant, which grows naturally in Spain and Italy; it rises with an upright channelled stalk three feet high, garnished with smooth leaves, which are divided into many fine segments like those of Fennel; the stalks branch out upward, and each branch is terminated by a large umbel composed of many small ones; the involucrum is shorter than the umbel, and each of the leaves which compose it is trifid: the foot-stalks which sustain the small umbels (or rays) are long and stiff; these are by the Spaniards used for picking their teeth, from whence the plant had the title of Visnaga, or Pick-tooth. The seeds of this plant should be sown in autumn, for those which are sown in the spring frequently fail, or at least remain in the ground till the following year before they grow; the plants require no other culture but to keep them clean from weeds, and thin them where they are too close.

DAUCUS CRETICUS. See ATHAMANTA.

D'AYENA. Monier.

This genus of plants receives its title from Monseigneur Le Duc D'Ayen, who is a great lover and promoter of the science of botany, and has a noble garden at St. Germains, which is well stored with rare plants from many different parts of the world.

The CHARACTERS are,

The flowers arise from the wings of the stalk; they have a five-leaved empalement, and five heart-shaped petals. It hath five stamina inserted in a short nectarium, and a five-cornered germen situated at the bottom of the nectarium, which afterward becomes

becomes a roundish five-cornered capsule, having five cells, each having one kidney-shaped seed.

We have but one SPECIES of this genus, viz.

D'AYENA (*Pusilla*) foliis cordatis, glabris. Lin. Sp. *D'Ayena with heart-shaped smooth leaves.*

This plant grows naturally in Peru; it hath a weak ligneous stalk a foot high, which divides into several slender horizontal branches, garnished with oblong heart-shaped leaves, which are slightly indented on their edges, standing upon pretty long foot-stalks; they are of a lucid green, and end in acute points. At the base of each foot-stalk from the side of the branches come out the flowers, two, three, or four, arising from the same point, each standing upon a separate slender foot-stalk, and have much resemblance to malvaceous flowers; they have a five cornered germen at the bottom of the nectarium, which afterward becomes a roundish five-cornered capsule having five cells, in each of these is lodged one kidney-shaped seed. The flowers are tubulous, and spread open at the top; they are purple, and continue in succession on the same plants from July to the winter.

It is propagated by seeds, which should be sown upon a moderate hot-bed early in the spring; and when the plants have four leaves, they should be transplanted on a fresh hot-bed to bring them forward; part of them may be planted in small pots, and the others may be planted on the bed: those in the pots should be plunged into a hot-bed of tanners bark, and shaded till they have taken new root, then they must have free air admitted to them every day in proportion to the warmth of the season. These plants should continue all the summer in the hot-bed, where they must have a good share of air, for those which are exposed to the open air will not thrive; and if they are too much drawn, they do not flower well, nor will they perfect their seeds unless they are brought forward in the spring, and sheltered in summer.

DAY LILY See HEMEROCALIS.

DELPHINIUM. Lin. Gen. Plant. 602. Larkspur, or Larksheel.

The CHARACTERS are,

The flower is composed of five unequal petals; the upper petal is extended at the hinder part, into a tubular obtuse tail. It has a bifid nectarium situated in the center of the petals, and many small stamina, with three oval germen, which afterward become so many capsules joined together.

The SPECIES are,

1. DELPHINIUM (*Consolida*) nectariis monophyllis, caule subdiviso. Hort. Cliff. 217. *Corn Larkspur.*

2. DELPHINIUM (*Ambiguum*) nectariis monophyllis, caule ramoso, foliis multifidis linearibus. *Garden Larkspur with a larger, single, blue flower; commonly called branching Larkspur.*

3. DELPHINIUM (*Ajacis*) nectariis monophyllis, caule simplici. Hort. Cliff. 213. *Upright or unbranching Larkspur.*

4. DELPHINIUM (*Peregrinum*) nectariis diphyllis, capsulis solitariis, foliis multipartitis obtusis. Hort. Cliff. 213. *Broad-leaved Larkspur with a small flower.*

5. DELPHINIUM (*Elatum*) nectariis diphyllis, labellis bifidis, apice barbatis, foliis incisis, caule erecto. Hort. Upsal. 151. *Perennial, hairy, Mountain Larkspur, with a Monkshood leaf; commonly called Bee Larkspur.*

6. DELPHINIUM (*Grandiflorum*) nectariis diphyllis, labellis integris, floribus subsolitariis, foliis compositis lineari-multipartitis. Hort. Upf. 150. *Dwarf, narrow-leaved, perennial Larkspur, with an azure flower.*

7. DELPHINIUM (*Americanum*) nectariis diphyllis, labellis integris, floribus spicatis, foliis palmatis multifidis glabris. *Siberian Larkspur.*

8. DELPHINIUM (*Staphisagria*) nectariis diphyllis, foliis palmatis lobis integris. Hort. Cliff. 213. *Larkspur with a Plane-tree leaf, called Stavesacre.*

The several varieties of the Garden Larkspur are not here enumerated, but as the gardeners distinguish the Garden Larkspurs into those which are branched, and such as have upright stalks, which difference is permanent, so I shall just mention the varieties of both sorts. And first of the branched Larkspur, there are of the following colours with single and double flowers.

Blue, purple, white, flesh, Ash, and Rose colours; and some have flowers beautifully variegated with two or three of these different colours.

The upright or unbranched Larkspur, produces a greater variety of colours than the branched, and the flowers are larger and fuller than those; but the principal colours run nearly the same with those of the other, tho' many of the colours are deeper, and there are more different shades of these colours in the flowers of this sort.

The first sort grows naturally amongst the Wheat in Cambridgeshire, and some other parts of England, where the flowers are of two colours, viz. blue and white.

The branching Larkspur, which is the second sort, comes later to flower than the upright, and has a very branching stalk; the branches come out horizontally from the side of the stalks, but afterward turn that part on which the spike of flowers grow upward, so as to make an angle; the leaves are long and finely divided; the flowers are placed thinner in the spikes than those of the upright sort; they are large, and some of them very double, and of various colours.

The third sort hath upright stalks, which scarce put out any branches; the spikes of flowers grow erect, and the flowers are placed very close together, so that they make a fine appearance. These plants flower in July and August, and are very great ornaments to the borders of the flower-garden.

The plants are annual, so are every year propagated by seeds, which should be sown in the autumn in the places where the plants are designed to remain; but to continue a succession of these flowers, there should be some seeds sown in autumn and some in the spring. Where the seeds are sown on the borders of the flower-garden for ornament, it should be in small patches in the middle of the borders at proper distances; in each of these patches may be scattered a few seeds, covering them over about a quarter of an inch with earth; in the spring the plants may be thinned, leaving three or four of the upright sorts in each patch to stand for flowering, but of the branching sort not more than two, because they require room; after this the plants will require no farther care but to keep them clean from weeds, and when they begin to flower should be supported, to prevent their being broken by wind, especially if they are not in a sheltered situation. If the seeds are well chosen, there will be very few ordinary flowers among them; and if there are seeds of the different coloured flowers sown in each patch, they will make a pleasing variety: but the upright sort should never be mixed in the same patches with the branching, because they do not flower at the same time.

In order to preserve the two sorts fine without degenerating to single or bad colours, there should be a bed of each sort sown in autumn, in some separate part of the garden, where the plants should be properly thinned and kept clean from weeds, till they begin to shew their flowers, when they should be carefully looked over every other day to pull out all those plants, whose flowers are not very double or of good colours; for if these are permitted to stand among the others till their farina has impregnated them, it will certainly cause them to degenerate; so that those persons who are contented with only marking their good flowers for seed, and

and suffer the others to stand for seed among them will always find themselves disappointed: therefore those who propose to have their flowers in perfection, should never gather the seeds of such as grew in the borders of the flower-garden, because there it will be almost impossible to preserve them so true as when they are in beds at a distance from all other kinds.

When the seed vessels turn brown they must be carefully watched, to gather them before they open and discharge the seeds; so that those which are situated on the lower part of the stalk will open long before those on the upper part are ripe, for which reason the pods should from time to time be gathered as they ripen.

The fourth sort grows naturally in Sicily and Spain. This hath a very branching stalk, which rises about two feet high; the lower leaves are divided into many broad obtuse segments, but those which are upon the stalks are generally single; the flowers grow scatteringly toward the upper part of the branches, they are small and of a deep blue colour; these are succeeded by very small seed-vessels, which are sometimes single, and at others double, and very rarely three together, as in the common sorts. This is an annual plant, whose seeds should be sown in autumn, and the plants treated as the common sort; it hath little beauty, and is only kept in some gardens for the sake of variety.

The fifth sort hath a perennial root, which sends out several upright stalks in the spring, rising five or six feet high, garnished with leaves which are divided into many broad segments in form of a spreading hand; these segments are cut at their extremities into several acute points; they are hairy, and stand upon long foot-stalks; the flowers terminate the stalks, growing in long spikes; they are of a light blue, covered toward their hinder part with a meally down.

The sixth sort grows naturally in America. This hath a perennial root, which puts out two or three branching stalks, which rise two feet and a half high, garnished at each joint with leaves composed of many narrow segments, which terminate with several acute points; they are smooth, and of a light green colour; the flowers come out toward the upper part of the stalks singly, each standing upon a long naked foot-stalk; they are large, and of a fine azure colour.

The seventh sort grows naturally in Portugal. This is a perennial plant, which rises with many strong branching stalks seven or eight feet high; the upper part of the stalks are of a fine purple colour, and are garnished with hand-shaped leaves, which are divided into four or five broad lobes, ending with many acute points; they are smooth, and stand upon long foot-stalks; the flowers terminate the stalks, growing in long spikes; they are of a fine blue colour, with a large bearded nectarium, having two lips, of a dark colour, resembling at a small distance the body of a bee, from whence some have titled this, and also the fifth sort, Bee Larkspur.

All these perennial Larkspurs are propagated by seeds, which, if sown in autumn, will more certainly succeed than those which are sown in the spring; when the plants come up they should be kept clean from weeds, and where they are too close together, part of them should be drawn out to allow room for others to grow till the following autumn, when they must be planted where they are to remain; the following summer they will flower, and the roots will continue many years growing in magnitude, so will produce a greater number of flower stalks.

The eighth sort is an annual plant, which grows naturally in the Levant, as also in Calabria; this rises with a strong hairy stalk three feet high, garnished with hand-shaped hairy leaves, composed of five or seven oblong lobes, which have frequently one or two acute indentures on their sides; the flowers form a loose spike at the upper part of the stalk, each standing on a long foot-stalk; they are of a pale blue or purple colour, and have a two-leaved nectarium. This is propagated by seeds, which should be sown in autumn, for those sown in the spring never grow the same year. The seeds should be sown where the plants are to remain, and require no other treatment than the common Larkspur.

DENS CANIS. See ERYTHRONIUM.

DENS LEONIS. See LEONTODON.

DENTARIA. Lin. Gen Plant. 726. Toothwort.

The CHARACTERS are,

The flower hath four obtuse petals placed in form of a cross, and six stamina, four of which are as long as the empalement; the other two are shorter. In the center is situated an oblong germen, which afterward becomes a long taper pod with two cells, divided by an intermediate partition, opening with two valves, including many roundish seeds.

The SPECIES are,

1. DENTARIA (*Pentaphyllo*) foliis summis digitatis. Lin. Sp. Plant. 912. *Five-leaved Toothwort with sawed leaves.*

2. DENTARIA (*Bulbifera*) foliis inferioribus pinnatis, summis simplicibus. Hort. Cliff. 335. *Seven-leaved bulb-bearing Toothwort.*

3. DENTARIA (*Enneaphylla*) foliis ternis ternatis. Lin. Sp. Plant. 653. *Three-leaved Toothwort.*

The first sort rises with a strong stalk a foot and a half high, garnished with a leaf at each joint, composed of five lobes, which are four inches long, and near two broad, deeply sawed on their edges; they are smooth, and stand on long foot-stalks; the flowers grow in loose spikes at the top of the stalks; they are small, of a blush colour, and are succeeded by long taper pods filled with small roundish seeds. It grows in the shady woods in the south of France and Italy.

The second sort rises with slender stalks about a foot high; the leaves at the bottom have seven lobes, those a little above five, others but three, and at the upper part of the stalk they are single: the flowers grow in clusters at the top of the stalk; these have four obtuse purple petals, and are succeeded by taper pods filled with roundish seeds.

The third sort rises with an upright stalk a foot high; the leaves are composed of nine lobes, three growing together, so that one leaf has three times three; the flowers grow in small bunches on the top of the stalks, and are succeeded by small taper pods filled with roundish seeds.

These plants grow on the mountains in Italy, and in the woods of Austria. The second sort is found wild in some parts of England, but particularly near Harefield in Hertfordshire. This produces bulbs on the side of the stalks, where the leaves are set on, which, if planted, will grow and produce plants. The plants are propagated by seeds, or parting their roots; the seeds should be sown soon after they are ripe, in a light soil and a shady situation: in the spring the plants may be taken up where they grow too close, and transplanted out in the like soil and situation; where, after they have taken root, they will require no farther care but to keep them clean from weeds: the second year they will produce flowers, and sometimes perfect their seeds.

The best time to transplant the roots is in October, when they should be planted in a moist soil and a shady situation; for they will not live in a dry soil, or if they are too much exposed to the sun.

DIANTHERA. Lin. Gen. Plant. 37.

The

The CHARACTERS are,

The flower is of the grinning kind with a short tube; it hath two stamina, one of which hath a twin summit, the other is a little taller, and an oblong germen. The empalement afterward becomes a capsule with two cells, which open with an elasticity, casting a single flat seed out of each cell.

We have but one SPECIES of this genus, viz.

DIANTHERA (*Americana*) spicis solitariis alternis. Lin. Syst. *Dianthera with a single spike.*

This plant grows naturally in Virginia, and other parts of North America. It is a low herbaceous plant with a perennial root, which sends out upright stalks a foot high, garnished with long narrow leaves of an aromatic odour, standing close to the stalks; from the side of the stalks the foot-stalks of the flowers are produced, sustaining small spikes of flowers.

This plant is very difficult to preserve in this country; for although it is hardy enough to live in the open air in England, yet it is very subject to rot in winter; and if it is placed under shelter, it is apt to draw up weak, and soon after decay, so that at present the plants are rare in this country.

It may be propagated by seeds, which must be sown upon a gentle hot-bed, and when the plants are fit to remove they must be planted in pots, and plunged into a fresh hot-bed; they must be kept in the dry stove in winter, and treated like other tender plants from the same country.

DIANTHUS. Lin. Gen. Plant. 500. Clove Gillyflower, Carnation and Pink.

The CHARACTERS are,

The flower hath a cylindrical empalement of one leaf, scaly below, with five petals, whose tails are as long as the empalement, but their upper part is broad, plain, and spread open. It hath ten stamina. In the center is situated an oval germen, which afterward becomes a cylindrical capsule with one cell, opening in four parts at the top, filled with compressed angular seeds.

The SPECIES are,

1. DIANTHUS (*Deltoides*) floribus solitariis, squamis calycinis, lanceolatis binis, corollis crenatis. Hort. Cliff. 164. *Common narrow-leaved wild Pink.*

2. DIANTHUS (*Virgineus*) caule subunifloro, corollis crenatis, squamis calycinis brevissimis, foliis subulatis. Lin. Sp. Pl. 412. *English small creeping, or Maiden Pink; commonly called the Matted Pink by seedsmen.*

3. DIANTHUS (*Glaucus*) floribus subsolitariis, squamis calycinis lanceolatis quaternis, corollis crenatis. Hort. Cliff. 164. *Branching Pink with a white flower having a purple circle; commonly called Mountain Pink.*

4. DIANTHUS (*Plumarius*) floribus solitariis, squamis calycinis subovatis brevissimis, corollis multifidis fauce pubescentibus. Lin. Sp. Plant. 411. *Single wild Pink with a small, pale, reddish flower.*

5. DIANTHUS (*Caryophyllus*) floribus solitariis, squamis calycinis subovatis brevissimis, corollis crenatis. Hort. Cliff. 164. *Single Garden Carnation with a large flower.*

6. DIANTHUS (*Armeria*) floribus aggregatis fasciculatis, squamis calycinis lanceolatis villosis tubum æquantibus. Hort. Cliff. 165. *Bearded wild Pink; called Deptford Pink.*

7. DIANTHUS (*Barbatus*) floribus aggregatis fasciculatis, squamis calycinis linearibus, foliis lanceolatis. *Broad-leaved Garden Sweet William.*

8. DIANTHUS (*Prolifer*) floribus aggregatis capitatis, squamis calycinis ovatis obtusis muticis, tubum superantibus. Lin. Sp. Plant. 410. *Wild childing Sweet William.*

9. DIANTHUS (*Ferrugineus*) floribus aggregatis capitatis, squamis calycinis lanceolatis aristatis, corollis crenatis. *Italian umbellated Mountain Pink, with flowers varying from yellow to an iron colour in the same cluster.*

10. DIANTAUS (*Chinensis*) floribus solitariis, squamis calycinis subulatis patulis, tubum æquantibus, corollis crenatis. Hort. Cliff. 164. *The China Pink.*

11. DIANTHUS (*Arenarius*) caulibus unifloris squamis calycinis ovatis, corollis multifidis, foliis linearibus. Flor. Suec. 318. *Dwarf wild Pink with one flower.*

12. DIANTHUS (*Alpinus*) caule unifloro, corollis crenatis, squamis calycinis exterioribus tubum æquantibus, foliis linearibus obtusis. Lin. Sp. Pl. 412. *Dwarf broad-leaved Pink.*

13. DIANTHUS (*Superbus*) floribus paniculatis, squamis calycinis brevibus acuminatis, corollis multifido-capillaribus, caule erecto. Amœn. Acad. 4. *The Superb Pink.*

14. DIANTHUS (*Diminutus*) floribus solitariis, squamis calycinis octonis florum superantibus. Lin. Sp. Plant. 587. *The least wild Pink.*

The first sort hath a short ligneous stalk, from which come out several tufted heads closely garnished with long narrow leaves, whose base lie over each other embracing the stalks; between these arise the flower-stalks, which grow about nine inches high, garnished at every joint by narrow grassy leaves placed opposite. The stalks are terminated by a single flower of a pale red colour. This is rarely admitted into gardens, the flower having little beauty.

The second sort is a low trailing plant, whose stalks lie on the ground; they grow very close together, and are garnished with short, narrow, grassy leaves, of a deep green colour; the stalks are terminated by small red flowers each standing upon a separate foot-stalk. This sort grows naturally in several parts of England, so is not now often cultivated in gardens; but formerly the seeds were sown to make edgings for the borders of the flower-garden by the title of Matted Pink, by which the seeds were sold in the shops.

The third sort grows naturally upon Chidder rocks in Somersetshire, and some other parts of England; this was formerly cultivated in the gardens by the title of Mountain Pink. It hath a resemblance of the second sort, but the leaves are shorter and of a grayish colour; the stalks grow taller and branch more; the flowers are larger, of a white colour with a purple circle at the bottom, like that sort of Pink called Pheasant's Eye. As the flowers of this sort have no scent, the plants are seldom kept in gardens.

The fourth sort grows naturally in several parts of England, frequently upon old walls; it is a small single Pink of a pale red colour, so is not cultivated in gardens.

The fifth sort is a small single Carnation, which has been long cast out of all the gardens; from some one of this sort it is supposed, the fine flowers now cultivated in the gardens had their original.

The sixth sort grows naturally in several parts of England, and particularly in a meadow near Deptford in Kent, from whence it had the title of Deptford Pink. This is of the kind called Sweet William; the flowers of these grow in close clusters at the end of the branches; they are red, and have long bearded empalements. I have never observed this to vary.

The seventh sort is the common Sweet William which has been long cultivated in the gardens for ornament, of which there are now great varieties which differ in the form and colour of their flowers, as also in the size and shape of their leaves; those which have narrow leaves were formerly titled Sweet Johns by the gardeners, and those with broad leaves were called Sweet Williams; there are some of both these sorts with double flowers, which are very ornamental plants in gardens.

The eighth sort grows naturally in the south of France, in Spain and Italy; this is an annual plant, which rises with an upright stalk about a foot high, garnished with

 narrow

narrow grassy leaves, and is terminated by a small head of pale red flowers, which are included in one common scaly empalement. They have little beauty, so the plants are seldom kept in gardens.

The ninth sort is a perennial plant, which rises with an upright stalk a foot and a half high, having long narrow leaves placed opposite at each joint, which embrace the stalk with their base; they are of a deep green colour, stiff, and end in acute points. The flowers grow in close clusters at the top of the stalks, having stiff bearded empalements; they are yellow and iron-coloured intermixed on the same stalk, and frequently there are of both colours in the same head. The roots will continue two years, and produce flowers and seeds; but the young plants of the first year always produce the strongest flowers.

The tenth sort came originally from China, so it is titled the China Pink; the flowers of this have no scent, but there are a great variety of lively colours among them, and of late years there has been great improvements made in the double flowers of this sort, some of which are as full of petals as the double Pink, and their colours are very rich. The plants seldom grow more than eight or nine inches high, branching out on every side; the branches grow erect, and are terminated each by a single flower. They are commonly raised every year from seeds, but the roots will continue two years in dry ground.

The eleventh sort is found growing naturally upon old walls and buildings in many parts of England; this is a single small Pink of a sweet odour, but of a pale colour, so makes no appearance; and since the great improvement which has been made in these flowers by culture, this hath been entirely neglected.

The twelfth sort grows naturally on the Alps; this hath broad, short, blunt leaves; the stalks seldom rise more than four inches high, each being terminated by a single flower of a pale red colour. It is sometimes preserved in botanic gardens for the sake of variety, but is rarely admitted into other gardens.

The thirteenth sort grows naturally in some parts of Germany, and in Denmark; the leaves of this sort are short, but narrow like those of Sweet William; the stalk rises more than one foot high, and is terminated by a single flower, having five large petals of a pale red colour, which are cut into many segments quite round. The roots of this seldom live more than two years, but they are in their greatest beauty the first year of their flowering, therefore young plants should be annually raised from seeds.

The fourteenth sort is a very diminutive plant, with short narrow leaves; the stalk rises little more than six inches high, and is terminated by a single flower of a pale red colour, so is seldom cultivated unless in botanic gardens for the sake of variety.

The sorts here enumerated are such as the botanists allow to be distinct species; and all the varieties of fine flowers, which are now cultivated in the gardens of the curious, are supposed to be only accidental variations which have been produced by culture; the number of these are greatly increased annually in many different parts of Europe, so that as new varieties are obtained, the old flowers are rejected.

The plants of this genus may be properly enough divided into three sections. The first to include all the variety of Pinks, the second all the Carnations, and the third those of the Sweet William; for although these agree so nearly in their principal characters, as to be included under the same genus by the botanists, yet they never vary from one to the other, though they frequently change and vary in the colour of their flowers.

I shall proceed therefore to treat of these under their different sections: and first I shall begin with the Pink, of which there are a great variety now cultivated in the gardens; the principal of which are, the Damask Pink, the White Shock, the Pheasant Eye with double and single flowers, varying greatly in their size and colour, the Cob Pink, Dobson's Pink, and Bat's Pink. The old Man's Head, and Painted Lady Pink, rather belong to the Carnation.

The Damask Pink is the first which flowers of the double sorts; this hath a short stalk; the flower is not very large, and not so double as many others; the colour is of a pale purple inclining to red, but is very sweet.

The next which flowers is the White Shock, which was so called from the whiteness of its flowers, and the borders of the petals being much jagged and fringed; the scent of this is not so agreeable as that of some others.

Then comes all the different kinds of Pheasants Eye, of which there are frequently new varieties raised, which are either titled from the persons who raised them, or the places where they were raised; some of these have very large double flowers, but those which burst their pods are not so generally esteemed.

The Cob Pink comes after these to flower; the stalks of this are much taller than those of any of the former; the flowers are very double, and of a bright red colour; these have the most agreeable odour of all the sorts, so merit a place in every good garden. The time of the Pinks flowering is from the latter end of May to the middle of July, and frequently that sort of Pheasant Eye, which is called Bat's Pink, will flower again in autumn.

The old Man's Head Pink, and the Painted Lady, do not flower till July, coming at the same season with the Carnation, to which they are more allied than the Pink. The first, when it is in its proper colours, is purple and white, striped and spotted, but this frequently is of one plain colour, which is purple; this sort will continue flowering till the frost in autumn puts a stop to it, and the flowers having an agreeable scent renders them valuable. The Painted Lady is chiefly admired for the liveliness of its colour, for it is not so sweet, or of so long continuance as the other.

The common Pinks are propagated either by seeds, which is the way to obtain new varieties, or by making layers of them as is practised for Carnations, or by planting slips, which, if carefully managed, will take root very well.

If they are propagated by seeds, there should be care taken in the choice of them, and only the seeds of the best sorts saved, where persons are curious to have the finest flowers. These seeds may be sown in the spring, and the plants afterward treated in the same manner as is hereafter directed for the Carnation; with this difference only, that as the Pinks are less tender, so they may be more hardily treated. Those which are propagated by layers must be also managed as the Carnation, for which there are full instructions hereafter given. The old Man's Head and Painted Lady Pinks are commonly propagated this way, but most of the other sorts are propagated from slips.

The best time to plant the slips of Pinks is about the end of July, when, if there should happen rain, it will be of great service to them; but if the weather be dry they will require to be constantly watered every other day until they have taken root; these should be planted in a shady border, and the ground should be dug well and all the clods broken, and if no rain falls it should be well soaked with water a few hours before the slips are planted; then the slips should be taken from the plants, and all their lower leaves stripped off, and planted as soon as possible after, for if they are suffered to lie long after they are taken from the plants, they will wither and spoil; these need not be planted at a

greater

greater diſtance than three inches ſquare, and the ground muſt be cloſed very hard about them; then they muſt be well watered, and this muſt be repeated as often as it is found neceſſary, till the cuttings have taken root; after which they will require no other care but to keep them clean from weeds till autumn, when they ſhould be tranſplanted to the borders of the flower-garden where they are to remain: there are ſome who plant the ſlips of Pinks later in the ſeaſon than is here directed, but theſe plants are never ſo ſtrong, nor flower ſo well as thoſe which are early planted.

We ſhall next proceed to the culture of the Carnation; theſe the floriſts diſtinguiſh into four claſſes.

The firſt they call Flakes; theſe are of two colours only, and their ſtripes are large, going quite thro' the petals.

The ſecond are called Bizarrs; theſe have flowers marked or variegated with three or four different colours, in irregular ſpots and ſtripes.

The third are called Piquettes; the flowers of theſe have always a white ground, and are ſpotted (or pounced, as they call it) with ſcarlet, red, purple, or other colours.

The fourth are called Painted Ladies; theſe have their petals of a red or purple colour on their upper ſide, and are white underneath.

Of each of theſe claſſes there are numerous varieties, but particularly of the Piquettes, which ſome years ago were chiefly in eſteem with the floriſts, but of late years the Flakes have been in greater requeſt than any of the other kinds. To enumerate the varieties of the principal flowers in any one of theſe claſſes, would be needleſs, ſince every country produces new flowers almoſt every year; ſo that thoſe flowers, which, at their firſt raiſing were greatly valued, are in two or three years become ſo common as to be of little worth, eſpecially if they are defective in any one property. Therefore (where flowers are ſo liable to mutability, either from the fancy of the owner, or that better kinds are yearly produced from ſeeds, which, with good floriſts, always take place of older, which are turned out of the garden to make room for them) it would be but ſuperfluous in this place to give a liſt of their names, which are generally borrowed either from the names and titles of noblemen, or from the perſons names or places of abode who raiſed them.

Theſe flowers are propagated either from ſeeds (by which new flowers are obtained) or by layers, for the increaſe of thoſe ſorts which are worthy maintaining; but I ſhall firſt lay down the method of propagating them from ſeed, which is thus:

Having obtained ſome good ſeeds, either of their own ſaving, or from a friend that you can confide in, about the middle of April, prepare ſome pots or boxes (according to the quantity of ſeed you have to ſow;) theſe ſhould be filled with freſh light earth mixed with rotten neats dung, which ſhould be well incorporated together; then ſow your ſeeds thereon (but not too thick,) covering it about a quarter of an inch with the ſame light earth, placing the pots or caſes ſo as to receive the morning ſun only till eleven of the clock, obſerving alſo to refreſh the earth with water as often as it may require; in a month or five weeks the plants will come up, and if kept clear from weeds and duly watered, will be fit to tranſplant about the latter end of June; at which time ſhould be prepared ſome beds (of the ſame ſort of earth as was directed to ſow them in) in an open airy ſituation, in which they ſhould be planted at the diſtance of three inches ſquare, obſerving to water and ſhade them till they have taken new root, then they muſt be kept clear from weeds; in theſe beds they may remain till the end of Auguſt, by which time they will have grown ſo large as almoſt to meet each other; then prepare ſome more beds of the like good earth (in quantity proportionable to the flowers raiſed,) in which they ſhould be planted at ſix inches diſtance each way, and not above four rows in each bed, for the more conveniently laying ſuch of them as may prove worthy preſerving, for in theſe beds they ſhould remain to flower.

The alleys between theſe beds ſhould be two feet wide, that perſons may paſs between the beds to weed and clean them. If the ſeaſon ſhould prove very dry at this time, they ſhould not be tranſplanted till there is ſome rain, ſo that it may happen to be the middle, or latter end of September ſome years, before there is wet enough to moiſten the ground for this purpoſe; but if there is time enough for the plants to get good root before the froſt comes on, it will be ſufficient. If the winter ſhould prove ſevere the beds ſhould be arched over with hoops that they may be covered with mats, otherwiſe many of the plants may be deſtroyed, for the good flowers are not ſo hardy as the ordinary ones of this genus. There will be no other culture wanting to theſe, but to keep them clean from weeds, and when they ſhoot up their ſtalks to flower, they muſt be ſupported by ſticks to prevent their breaking. When the flowers begin to blow, they muſt be looked over to ſee which of them proffer to make good flowers; as ſoon as that can be diſcovered, all the layers upon them ſhould be laid; thoſe which are well marked, and blow whole without breaking their pods, ſhould be reſerved to plant in borders, to furniſh ſeeds; and thoſe which burſt their pods, and ſeem to have good properties, ſhould be planted in pots, to try what their flowers will be when managed according to art; and it is not till the ſecond year of flowering, that any perſon can pronounce what the value of a flower will be; but in order to be well acquainted with what the floriſts call good properties, I ſhall here ſet them down.

1. The ſtem of the flower ſhould be ſtrong, able to ſupport the weight of the flower without nodding down.

2. The petals of the flower ſhould be long, broad, and ſtiff, and pretty eaſy to expand, or (as the floriſts term them) ſhould be free flowerers.

3. The middle pod of the flower ſhould not advance too high above the petals in the other part of the flower.

4. The colours ſhould be bright, and equally marked all over the flower.

5. The flower ſhould be very full of leaves, ſo as to render it when blown very thick and high in the middle, and the outſide perfectly round.

Having made choice of ſuch flowers as promiſe well, theſe ſhould be marked ſeparately for pots, and the round whole blowing flowers for borders; all ſingle flowers, or ſuch as are ill-coloured and not worth preſerving, ſhould be drawn out, that the good flowers may have the more air and room to grow ſtrong; when the layers of the good flowers have taken root (which will be ſome time in Auguſt,) they ſhould be taken off and planted out; thoſe that blow large in pots, and the other in borders (as hath been already directed.)

Of late years the whole-blowing flowers have been much more eſteemed than thoſe large flowers which burſt their pods; but eſpecially thoſe round flowers which have broad ſtripes of beautiful colours, and round Roſe leaves, of which kinds there have been a great variety introduced from France within theſe few years; but as theſe French flowers are extremely apt to degenerate to plain colours, and being much tenderer than thoſe which are brought up in England, there are not ſuch great prices given for the plants now as was a few years paſt: from the preſent taſte of theſe whole-blowing flowers, many of the old varieties which had been turned out of the gardens of the floriſts have been received

 again;

again; and large prices have been paid of late for ſuch flowers as ſome years ago were ſold for one ſhilling a dozen, or leſs, which is a ſtrong proof of the variableneſs of the fancies of the floriſts.

I ſhall next proceed to give ſome directions for propagating theſe by layers, and the neceſſary care to be taken of them to have large and fair flowers.

The beſt ſeaſon for laying theſe flowers is in July, as ſoon as the ſhoots are ſtrong enough for that purpoſe: it is performed in the following manner: after having ſtripped off the leaves from the lower part of the ſhoot intended to be laid, make choice of a ſtrong joint about the middle of the ſhoot (not too near the heart of the ſhoot, nor in the hard part next the old plant;) then with a penknife make a ſlit in the middle of the ſhoot, from the under ſide of the joint upward half the way or more, to the joint next above it, according to the diſtance of the joints; then with your knife cut the tops of the leaves, and alſo cut off the ſwelling part of the joint where the ſlit is made, ſo that the part ſlit may be ſhaped like a tongue; (when that outward ſkin is pared off, which, if left on, would prevent their puſhing out roots;) having looſened the earth round the plant, and if need be, raiſed it with freſh mould, that it may be level with the ſhoot intended to be laid, leſt by the ground being too low, by forcing down the ſhoot it ſhould be ſplit off; then make a hollow place in the earth, juſt where the ſhoot is to come, and bend the ſhoot gently into the earth, obſerving to keep the top as upright as poſſible, that the ſlit may be open: and being provided with forked ſticks for that purpoſe, thruſt one of them into the ground, ſo that the forked part may take hold of the layer, to keep it down in its proper place; then gently cover the ſhank of the layer with the ſame ſort of earth, giving it a gentle watering to ſettle the earth about it, obſerving to repeat the ſame as often as is neceſſary to promote their rooting. In about five or ſix weeks after this, the layers will have taken root ſufficient to be tranſplanted; then thoſe which are intended for pots ſhould be each planted in a ſeparate ſmall pot, placing them in the ſhade until they have taken freſh root; after which they may be removed into a more open ſituation, where they may remain till the middle of November (if the weather continues ſo long good,) when the pots ſhould be put under a common frame, where they may enjoy the open air at all times when the air is mild, but ſcreened from hard rains, ſnow and froſt.

Where there is conveniency, the layers, which are intended for the common borders, may be planted upon a bed at about three inches aſunder each way, and in winter covered with a frame, or elſe arched over with hoops, and in bad weather covered with mats, which will ſecure them till ſpring, when they may be taken up with balls of earth, and planted where they are deſigned to flower.

Thoſe layers which were planted in ſmall pots in the autumn, ſhould in the ſpring be turned out of thoſe pots, preſerving the earth to their roots, paring off the outſide with the matted roots, and put into the pots they are deſigned to remain in for good. The beſt compoſt for theſe flowers is as follows:

Make choice of ſome good upland paſture, or a common that is of an hazel earth, or light ſandy loam; dig from the ſurface of this about eight inches deep, taking all the turf with it; let this be laid in a heap to rot and mellow for one year, turning it once a month that it may ſweeten; then mix about a third part of rotten neats dung, or for want of that ſome rotten dung from a Cucumber or a Melon-bed; let this be well mixed together, and if you can get it time enough before hand, let them lie mixed ſix or eight months before it is uſed, turning it ſeveral times, the better to incorporate their parts. But as the layers which are made from ſuch roots as have been forced to flower the ſame year, do ſeldom ſucceed ſo well the next, it will be a good method to plant two or three layers of each of the beſt kinds in a bed of freſh earth not over dunged; which plants ſhould only be ſuffered to ſhew their flowers, that their colours may be known to be perfect in their kind; and when ſatisfied in that particular, the flowers ſhould be cut off the ſtems, and not ſuffered to ſpend the roots in blowing, by which means the layers will be ſtrengthened. And from ſome of the beſt plants of theſe, the layers ſhould be taken for the next year's blowing, always obſerving to have a ſucceſſion of them yearly, by which means every year a fine bloom of theſe flowers may be expected, ſuppoſing the ſeaſon favourable. When the plants which are intended to flower, are put into the larger pots in the ſpring, they ſhould be placed in a ſituation where they may be defended from the north wind, obſerving to give them gentle waterings as the ſeaſon may require.

Here they may remain till the middle or latter end of April, when a ſtage of boards ſhould be made to ſet the pots upon, which ſhould be ſo ordered as to have little ciſterns of water round each poſt, to prevent the inſects from getting to the flowers in their bloom; which, if they are ſuffered to do, they will deſtroy all the flowers in a ſhort time: the chief and moſt miſchievous inſect in this caſe is the earwig, which will gnaw off all the lower parts of the petals of the flowers (which are very ſweet,) and thereby cauſe the whole flower to fall to pieces; but ſince the making one of theſe ſtages is ſomewhat expenſive, and not very eaſy to be underſtood by ſuch as have not ſeen them, I ſhall deſcribe a very ſimple one, which I have uſed for ſeveral years, which anſwers the purpoſe full as well as the beſt and moſt expenſive one can do. Firſt, prepare ſome common flat pans about two feet over, and three inches deep, place theſe two and two oppoſite to each other, at about two feet diſtance, and at every eight feet in length. In each of theſe turn a flower-pot upſide down, then lay a piece of flat timber, about two feet and a half long, and three inches thick, croſs from pot to pot, the whole length of the ſtage; then lay the planks lengthways upon theſe timbers, which will hold two rows of the ſize for theſe pots which are proper for the Carnations; and when you have ſet the pots upon the ſtage, fill the flat pans with water, always obſerving as it decreaſes in the pans, to repleniſh it, which will effectually guard your flowers againſt inſects, for they do not care to ſwim over water; ſo that if by this, or any other contrivance, the paſſage from the ground to the ſtage on which the pots are placed, is defended by a ſurface of water four or five inches broad, and as much in depth, will effectually prevent theſe vermin from getting to the flowers.

This ſtage ſhould be placed in a ſituation open to the ſouth eaſt, but defended from the weſt winds, to which theſe ſtages muſt not be expoſed, leſt the pots ſhould be blown down by the violence of that wind, which is often very troubleſome at the ſeaſon when theſe flowers blow; indeed they ſhould be defended by trees at ſome diſtance, from the winds of every point; but theſe trees ſhould not be too near the ſtage, nor by any means place them near walls, or tall buildings, for in ſuch ſituations the ſtems of the flowers will draw up too weak. About the middle of April the layers will begin to ſhoot up for flower, therefore there ſhould be provided ſome deal ſticks, about four feet and a half long, which ſhould be thicker toward the bottom, and planed off taper at the top; theſe ſticks ſhould be carefully ſtuck into the pots as near as poſſible to the plant, without injuring it; then with a ſlender piece of baſs mat, faſten the ſtalk of the flower to the ſtick to prevent its being broken; this muſt be often repeated as the ſtalk advances in height, and all the ſide ſtalks muſt be pulled off as they are produced, never letting more than two ſtalks remain upon one root, nor above one, if they are intended to blow exceeding large. Toward the beginning of June the flowers will moſt of them have attained their height, and their pods will begin to ſwell, and about the end ſome of the earlieſt begin to open on one ſide; therefore the pods muſt be opened in two other places, at equal angles; this muſt be done as ſoon as the pod breaks, otherwiſe the flower will run out on one ſide, and be in a ſhort time paſt recovering, ſo as to make a compleat flower. In a few days after the flowers begin to open, they muſt be covered with glaſſes which are made for that purpoſe, in the following manner:

Upon the top of the glaſs, exactly in the center, is a tin collar or ſocket, about three-fourths of an inch ſquare, for the flower-ſtick to come through; to this ſocket are foldered eight ſlips of lead at equal diſtances, which are about ſix inches and a half long, and ſpread open at the bottom about four inches aſunder; into theſe ſlips of lead are faſtened ſlips of glaſs, cut according to the diſtances of the lead, which, when they are fixed in, are bordered round the bottom with another ſlip of lead quite round, ſo that the glaſs hath eight angles, with the ſocket in the middle, and ſpread open at the bottom about eleven inches wide.

When the flowers are open enough to be covered with theſe glaſſes, a hole muſt be made with an awl through the flower-ſtick, exactly to the height of the under part of the pod, through which ſhould be put a piece of ſmall wire about ſix inches long, making a ring at one end of the wire to contain the pod, into which ring ſhould the ſtem of the flower be fixed, taking off all the tyings of baſs; and the ſtem of the flower muſt be placed ſo far from the ſtick, as may give convenient room for the flower to expand without preſſing againſt it; to which diſtance it may be fixed, turning the wire ſo as not to draw back through the hole; then make another hole through the ſtick, at a convenient diſtance above the flower, through which ſhould be put a piece of wire an inch and a half long, to ſupport the glaſſes from ſliding down upon the flowers, obſerving that the glaſſes are not placed ſo high as to admit the ſun and rain under them to the flowers, nor ſo low as to ſcorch their leaves with the heat. At this time alſo, or a few days after, ſhould be cut ſome ſtiff paper, cards, or ſome ſuch thing, into collars about four inches over, and exactly round, cutting a hole in the middle of it about three-fourths of an inch diameter, for the bottom of the flower to be let through; then place theſe collars about them, to ſupport the petals of the flower from hanging down; this collar ſhould be placed within-ſide the calyx of the flower, and ſhould be ſupported thereby. If, as the flowers blow, one ſide comes out faſter than the other, the pots ſhould be turned about, to ſhift the other ſide toward the ſun; and, if the weather proves very hot, the glaſſes ſhould be ſhaded in the heat of the day with Cabbage leaves, &c. to prevent their being ſcorched, or forced out too ſoon; and, when the middle pod begins to riſe, the calyx muſt be pulled out with a pair of nippers made for that purpoſe: but this ſhould not be done too ſoon, leſt the middle part of the flower ſhould advance too high above the ſides, which will greatly diminiſh the beauty of it. And when the flowers are fully blown, if they are cut off, a freſh collar of ſtiff paper ſhould be put on, which ſhould be cut exactly to the ſize of the flower, that it may ſupport the petals to their full width, but not to be ſeen wider than the flower in any part: when this is put on, the wideſt leaves ſhould be ſpread out, to form the outſide of the flower, which although they ſhould happen to be in the middle (as is often the caſe,) yet by removing the other leaves they may be drawn down, and ſo the next longeſt leaves upon them again, that the whole flower may appear equally globular without any hollow parts. In the doing of this, ſome floriſts are ſo curious as to render an indifferent flower very handſome; and on this depends, in a great meaſure, the ſkill of the artiſt to produce large fine flowers.

The directions here given are chiefly for the management of thoſe large Carnations which require the greateſt ſkill of the floriſts to have them in perfection; but of late years theſe have not been ſo much in eſteem as formerly, and thoſe flowers which do not break their pods have now the preference. Theſe are generally planted in pots, and treated in the ſame way as the large flowers, but do not require ſo much trouble to blow them: all that is neceſſary to be done for theſe, is to faſten their ſtems up to flower-ſticks to prevent their being broken, and to take off the pods which proceed from the ſide of the ſtalks, leaving only the top bud to flower, if they are intended to be large and fair; and when the flowers begin to open, if they are ſcreened from the ſun in the heat of the day, and alſo from wet, they will continue much longer in beauty.

The layers which are planted in the full ground, generally produce ſeeds better than thoſe in pots; therefore whoever propoſes to raiſe a ſupply of new flowers from ſeeds, muſt always obſerve to ſave the beſt of their ſeedling flowers for this purpoſe; for it is well known, that after any of theſe flowers have been a few years propagated by layers, they become barren, and do not ſeed; which is alſo the caſe with many other plants which are propagated by ſlips, layers, or cuttings; ſo that the young plants which have been newly obtained from ſeeds, are always the moſt productive of ſeeds.

I ſhall next proceed to the culture of that ſpecies, which is commonly known by the title of Sweet Willam: of this there are a great variety of different colours which are ſingle, and three with double flowers; ſome of theſe have narrow leaves, which were formerly titled Sweet Johns, but of late that diſtinction has not been made, becauſe they are found to vary when raiſed from ſeeds.

Some of the ſingle flowers have very rich colours, which frequently vary in thoſe of the ſame bunch; there are others with fine variegated flowers, and others whoſe middles are of a ſoft red, bordered with white, which are called Painted Ladies; but where perſons are deſirous to preſerve any of theſe varieties in perfection, the beſt flowers of each ſhould be particularly marked, and no other permitted to ſtand near them, leſt their farina ſhould impregnate them, which would cauſe them to vary.

That which is called the Painted Lady Sweet William, is a very beautiful variety; the ſtalks of this do not riſe ſo high as moſt of the other; the bunches of flowers are larger, and produced more in the form of an umbel, the flowers ſtanding equal in height, make a better appearance: of this variety there have been ſome lately raiſed with double flowers, having many beautiful ſtripes, but I have found them apt to degenerate: there are others whoſe ſtalks riſe three feet high, and the flowers are of a very deep red or ſcarlet colour. Theſe all flower at the ſame time with the Carnations, which renders them leſs valuable, becauſe they have no ſcent.

The ſingle kinds of theſe flowers are generally propagated by ſeeds, which muſt be ſown the beginning of April in a bed of light earth, and in June they will be fit to tranſplant out, at which time muſt be prepared ſome beds ready for them; they ſhould be planted ſix inches diſtance every way: in theſe beds they may remain till Michaelmas, at which time they may be tranſplanted into the borders of the pleaſure garden. Theſe will flower the next year in June, and perfect their ſeeds in Auguſt, which you ſhould ſave from the beſt coloured flowers for a ſupply.

The

The three ſorts with double flowers, are: 1. The broad-leaved ſort, which hath very double flowers, of a deep purple colour inclining to blue, which burſts in pods, ſo that it is not ſo much eſteemed as the others, and therefore has been leſs regarded, and is now almoſt totally baniſhed the gardens of the curious. 2. The Double Roſe Sweet William, whoſe flowers are of a fine deep Roſe-colour, and ſmell ſweet; this is much valued for the beauty and ſweetneſs of its flowers; the empalement (or pods) of theſe flowers never burſt, ſo the flowers remain with their petals fully expanded, and do not hang down looſely as thoſe of the former. 3. The Mule, or Fairchild's Sweet William; it hath narrower leaves than either of the former, and is of that variety called Sweet John: this was ſaid to have been produced from ſeeds of a Carnation, which had been impregnated by the farina of the Sweet William; the flowers of this are of a brighter red colour than either of the former; their bunches are not quite ſo large, but their flowers have an agreeable ſcent.

The double kinds are propagated by layers, as the Carnations, or by ſlips as Pinks; they love a middling ſoil, not too light, nor too heavy or ſtiff, nor too much dunged, which very often occaſions their rotting: theſe continue flowering for a long time, and are extremely beautiful: but they are very ſubject to canker and rot away, eſpecially if planted in a ſoil over wet or too light. Theſe flowers when planted in pots, are very proper to adorn court-yards at the time they are in flower.

The China Pink is generally eſteemed an annual plant, becauſe the plants which are raiſed from ſeeds flower and produce ripe ſeeds the ſame ſeaſon, ſo their roots are not often preſerved; but where they are planted in a dry ſoil, they will continue two years, and the ſecond year will produce a greater number of flowers than the firſt. There are a great variety of very rich colours in theſe flowers, which annually vary when raiſed from ſeeds. The double flowers of this are moſt eſteemed, though the colours of the ſingle are more diſtinct and beautiful; for the multiplicity of petals in the double flowers, in a great meaſure, hide the deep ſhades which are toward the lower part of the petals.

Theſe plants are propagated by ſeeds, which ſhould be ſown upon a gentle hot-bed about the beginning of April; this moderate heat is only intended to forward the vegetation of the ſeeds, therefore when the plants come up, they muſt have a large ſhare of air admitted to them, to prevent their drawing up weak; and as ſoon as the weather will permit, they muſt be expoſed to the open air; in about a month after, the plants will be fit to remove, when they ſhould be carefully taken up with good roots, and planted in a bed of rich earth at about three inches aſunder, being careful to ſhade them from the ſun till they have taken new root. The farther care is to keep them clean from weeds till the end of May, at which time they may be tranſplanted to the places where they are deſigned to remain for flowering, when they may be taken up with large balls of earth to their roots, ſo as ſcarcely to feel their removal, eſpecially if it happens to rain at that time.

As theſe plants do not grow large, ſo when they are planted ſingly in the borders of the flower-garden, they do not make ſo fine an appearance, as where they are planted by themſelves in beds; or if they are planted in ſmall clumps of ſix or eight roots in each, where the flowers being of different colours ſet off each other to advantage.

Thoſe who are curious in theſe flowers, take particular care in ſaving their ſeeds, for they never permit any ſingle flowers to ſtand among their double, but pull them up as ſoon as they ſhew their flowers, and alſo draw out all thoſe which are not of lively good colours; where this is obſerved, the flowers may be kept in great perfection; but where perſons have truſty friends, who live at ſome diſtance, with whom they can exchange ſeeds once in two or three years, it is much better ſo to do than to continue ſowing ſeeds in the ſame place many years in ſucceſſion, and this holds true in moſt ſorts of ſeeds: but the great difficulty is to meet with an honeſt perſon of equal ſkill, who will be as careful in the choice of his plants for ſeed as if he was to ſow them himſelf.

DIAPENSIA. See Sanicula.

DICTAMNUS. Lin. Gen. Plant. 468.

The Characters are,

The flower hath five petals which are unequal, and ten ſtamina which are as long as the petals. In the center is ſituated a five-cornered germen, which afterward becomes a capſule with five cells joined together, incloſing ſeveral roundiſh, hard, ſhining ſeeds.

We have but one Species of this plant, viz.

Dictamnus (*Albus.*) Hort. Cliff. 161. *White Dittany; commonly called Fraxinella.*

There are three varieties of this plant, one with a pale red flower ſtriped with purple, another with a white flower, and one with ſhorter ſpikes of flowers; but as I have obſerved them to vary when proprogated by ſeeds, ſo I eſteem them only ſeminal varieties.

This is a very ornamental plant for gardens, and as it requires very little culture, ſo deſerves a place in all good gardens. It hath a perennial root, which ſtrikes deep into the ground; the head annually increaſes in ſize; it ſends up many ſtalks in proportion to its bigneſs, which riſe from two to three feet high, garniſhed with winged leaves placed alternate; they are compoſed of three or four pair of oblong lobes, terminated by an odd one, which are ſmooth and ſtiff, ſitting cloſe to the midrib; the lobes placed on each ſide the midrib are oblique, but thoſe which terminate the leaf have their ſides equal. The flowers are produced in a long, pyramidal, looſe ſpike or thyrſe, on the top of the ſtalk, which is nine or ten inches long; the flowers of one ſort is white, and the other of a pale red, marked with red or purple ſtripes. The whole plant when gently rubbed, emits an odour like that of Lemon peel, but when bruiſed has ſomething of a balſamic ſcent.

Theſe plants are propagated by ſeeds, which, if ſown in the autumn ſoon after they are ripe, the plants will appear the following April; but when they are kept out of the ground till the ſpring, the ſeeds ſeldom ſucceed; or if they do grow, it is the following ſpring before the plants appear, ſo that a whole year is loſt. When the plants come up, they muſt be conſtantly kept clean from weeds; and in the autumn, when their leaves decay, the roots ſhould be carefully taken up, and planted in beds at ſix inches diſtance every way; theſe beds may be four feet broad, and the paths between them two, that there may be room enough to paſs between the beds to weed them. In theſe beds the plants may ſtand two years, during which time they muſt be conſtantly kept clean from weeds; and if they thrive well, they will be ſtrong enough to flower; ſo in the autumn they ſhould be carefully taken up, and planted in the middle of the borders of the flower-garden, where they will continue thirty or forty years, producing more ſtems of flowers in proportion to the ſize of the roots. All the culture theſe require is to be kept clean from weeds, and the ground about them dug every winter.

DICTAMNUS CRETICUS. See Origanum.

DIERVILLA. Tourn. Act. R. Par. 1706.

The Characters are,

The flower is of one leaf, cut into five parts at the top; it hath five ſtamina, which are equal with the petal; at the bottom of the flower is ſituated an oval germen, which afterward becomes a pyramidal berry, divided into four cells, which contain small round seeds.

We have but one SPECIES of this plant, viz.

DIERVILLA (*Lonicera*) Acadiensis fruticosa, flore luteo. Act. R. Par. 1706. *Shrubby Diervilla of Acadia with a yellow flower.*

This plant grows naturally in the northern parts of America; it hath woody roots which spread far in the ground, and put out shoots at a distance from the principal stalk, whereby it multiplies greatly: the stalks are ligneous, and seldom rise more than three feet high, and are garnished with oblong heart-shaped leaves, ending in acute points, which are very slightly sawed on their edges, and are placed opposite, sitting close to the stalks; the upper part of the stalk is garnished with flowers, which come out from the side, and also at the top of the stalks, two or three together, sustained upon short foot-stalks: they are of a pale yellow, and being small make no great appearance.

It is easily propagated by suckers, which it sends out in plenty, and loves a moist soil and shady situation, where the cold will never injure it.

DIGITALIS. Lin. Gen. Plant. 676. Foxglove.

The CHARACTERS are,

The flower is bell-shaped, of one petal, with a large open tube, whose brim is slightly divided into five parts; it hath four stamina, which are inserted in the base of the petal, two being longer than the other; the germen afterward swells to an oval capsule, having two cells, inclosing many small angular seeds.

The SPECIES are,

1. DIGITALIS (*Purpura*) calycinis foliolis ovatis acutis, corollis obtusis, labio superiore integro. Hort. Upsal. 178. *Common or purple Foxglove with a rough leaf.*

2. DIGITALIS (*Thapsi*) foliis decurrentibus. Lin. Sp. Pl. 867. *Lesser Spanish Foxglove with running leaves.*

3. DIGITALIS (*Lutea*) calycinis foliolis acutis, corollis obtusis, labio superiore integro, foliis lanceolatis obtusis. *Lesser yellow Foxglove with a small flower.*

4. DIGITALIS (*Grandiflora*) foliolis calycinis linearibus, corollis acutis, labio superiore integro, foliis lanceolatis. *Yellow Foxglove with a larger yellow flower.*

5. DIGITALIS (*Ferruginea*) calycinis foliolis ovatis obtusis, corollæ labio inferiore longitudine floris. Lin. Sp. Plant. 622. *Narrow-leaved Foxglove with an iron coloured flower.*

6. DIGITALIS (*Canariensis*) calycinis foliolis lanceolatis, corollis bilabiatis acutis, caule fruticoso. Lin. Sp. Pl. 622. *Shrubby Canary Foxglove like Bearsbreech, with a golden flower.*

The first sort grows naturally by the side of hedges and in shady woods in most parts of England, so is rarely cultivated in gardens; it is a biennial plant, which the first year produces a great tuft of oblong rough leaves; the second year it shoots up a strong herbaceous stalk, which rises from three to five feet high, garnished with leaves of the same form as the lower, but they gradually lessen upward; the flowers grow in a long loose thyrse, standing only on one side of the stalk; they are large, tubulous, and shaped like a thimble, of a purple colour, with several white spots on the under lip; these are succeeded by oval capsules with two cells, which are filled with dark brown seeds: whoever has a mind to cultivate it, should sow the seeds in autumn, for those which are sown in the spring seldom succeed.

There is a variety of this with a white flower, which is found growing naturally in some parts of England, which differs from this only in the colour of the flower; but this difference is permanent.

The second sort grows naturally in Spain; this seldom rises much more than a foot high; the lower leaves are ten inches long, and three broad in the middle; they are soft, woolly, and roughly veined on their under side; the stalks are garnished with leaves of the same shape, but smaller; the upper part of the stalk hath a short thyrse of purple flowers, like those of the common sort, but smaller, and the segments of the petal are acute: this plant retains its difference when cultivated in gardens.

The third sort hath very long obtuse leaves near the root; the stalk is small, and rises from two to three feet high, the lower part being pretty closely garnished with smooth leaves, about three inches long and one broad, ending in obtuse points: the upper part of the stalk, for ten inches in length, is adorned with small yellow flowers, which are closely ranged on one side, having a few very small acute leaves placed between them, which are situated on the opposite side of the stalk; the upper lip of the flower is entire, and the petal is obtuse.

The fourth sort hath long, smooth, veined leaves at the bottom; the stalk is strong, and rises two feet and a half high, garnished with leaves which are five inches long, one and a half broad, ending in acute points; these have many longitudinal veins, and are slightly sawed on their edges; the upper part of the stalk is adorned with small yellow flower, the brim having acute points, and the upper lip entire.

The fifth sort hath narrow smooth leaves, which are entire; the stalk rises five or six feet high; the lower part of the stalk is garnished with very narrow small leaves; the flowers terminate the stalk, growing in a long spike, with very few leaves between them, and those very small; the empalement is divided into four obtuse parts, the lower lip extending much longer than the upper; the flowers are of an iron colour; there is a variety of this with broader leaves.

The sixth sort grows naturally in the Canary Islands; this plant hath a shrubby stalk, which rises to the height of five or six feet, dividing into several branches, garnished with rough spear-shaped leaves near five inches long, and one and a half broad in the middle, with a few short serratures on their edges; they are placed alternately on the branches; each of the branches is terminated by a loose spike of flowers, about six inches in length; the empalement is cut into five acute segments almost to the bottom; the upper lip is long and entire; this is arched, and immediately under it the stamina and style are situated; the lower lip is obtuse, and indented at the top; there are two acute segments on the side, which compose the chaps of the flower; two of the stamina are longer than the other; in the bottom of the flower is situated the germen, supporting a slender style, crowned by an oval stigma; the germen afterward becomes an oval capsule, filled with small angular seeds.

This plant begins to flower in May, and there is generally a succession of flowers on the same plant till the winter puts a stop to them, which renders the plant more valuable; it is propagated by seeds, which should be sown in pots in the autumn soon after the seeds are ripe; these pots should be plunged into an old bed of tanners bark, whose heat is gone, and in mild weather the glasses should be drawn off to admit the air, but in hard rains and frost they must be kept on to protect the seeds from both, which frequently destroys them; in the spring the plants will come up, when they should enjoy the free air in mild weather, but must be protected from the cold: as soon as these are large enough to transplant, they should be each put into a separate small pot, and placed under the frame till they have taken new root, then they should be gradually inured to the open air. During the summer season the plants should remain abroad in a sheltered situation, but in the autumn they must be placed in a green house, for they will not live abroad in winter; they must not be kept too warm and

and close in the house, for they only want protection from the frost.

All these sorts should be sown in the autumn, for if the seeds are sown in the spring they commonly fail, or at least lie a whole year in the ground before they vegetate. The plants are most of them biennial (except the seventh sort) and perish soon after their seeds are ripe.

DIOSCOREA. Plum. Nov. Gen. 9. tab. 26. Lin. Gen. Plant. 995.

The CHARACTERS are,

It hath male and female flowers in different plants; the male flowers have a perianthium cut into six parts, and have no petals, but have six short hairy stamina; the female flowers have the same perianthium; they have no petals, but have a three-cornered germen; the perianthium afterward becomes a triangular capsule with three cells, opening with three valves, containing two compressed bordered seeds in each.

The SPECIES are,

1. DIOSCOREA (*Sativa*) foliis cordatis alternis, caule lævi. Hort. Cliff. 459. *Climbing Dioscorea with black Briony leaves, and fruit growing in long bunches.*

2. DIOSCOREA (*Hastata*) foliis hastato-cordatis, caule lævi, racemis longissimis. *Climbing Dioscorea with a spear-pointed leaf, and fruit growing in long bunches.*

3. DIOSCOREA (*Villosa*) foliis cordatis rotundis, alternis, caule volubili lævi. *Climbing Dioscorea with a roundish pointed leaf, and fruit growing in long bunches.*

4. DIOSCOREA (*Bulbifera*) foliis cordatis, caule alato bulbifera. Flor. Zeyl. 360. *Dioscorea with heart-shaped leaves, and a winged stalk bearing bulbs; commonly called Yam.*

5. DIOSCOREA (*Oppositifolia*) foliis cordatis alternis oppositisque, caule lævi. Lin. Sp. Plant. 1033. *Dioscorea with heart-shaped leaves growing alternate and opposite, and a smooth stalk.*

The first sort grows naturally in most of the islands in the West-Indies; this hath slender climbing stalks, which fix themselves to any support near them, and rise to the height of eighteen or twenty feet, garnished with heart-shaped leaves, ending in acute points, with five longitudinal veins, which arise from the foot-stalk, and diverge towards the sides, but meet again at the point of the leaves; they stand upon pretty long foot-stalks, from the base of which arise the branching spikes of flowers, which are small, and have no beauty; the female flowers are succeeded by three-cornered oblong capsules, having three cells, each containing two compressed seeds.

The second sort differs from the first in the shape of its leaves, these having two round ears at their base, but the middle extends to an acute point, like that of an halbert. The bunches of flowers are longer, and they are looser placed than those of the former sort.

The third sort hath broad, round, heart-shaped leaves, ending in acute points, having many longitudinal veins, which arise from the foot-stalk, and diverge to the side, but afterward join at the point of the leaf; the flowers come out on long loose strings, standing on short foot-stalks; the female flowers are succeeded by three-cornered oblong capsules with three cells, having compressed bordered seeds.

The fourth sort hath triangular winged stalks, which trail upon the ground, and extend to a great length; these frequently put out roots from their joints as they lie upon the ground, whereby the plants are multiplied. The roots of this plant are eaten in many parts of both Indies, where the plants are much cultivated.

The fifth sort grows naturally in Virginia, and in other parts of North America; this hath a smooth stalk, which climbs on the neighbouring plants, and rise five or six feet high, garnished with heart-shaped leaves, which are placed sometimes alternate, and at others they are opposite, and have several longitudinal veins; the flowers come out from the side of the stalk in the same manner as the other sorts, but have no beauty.

These plants may be propagated by laying their branches into the ground, which in about three months will put out roots, and may then be taken from the old plants, and planted into separate pots, which should be plunged into the tan-bed in the stove; during the winter these plants should have but little water given them, but in summer, when they are growing vigorously, they require more; in warm weather they should have a large share of air. When the seeds of these plants are brought to England, they should be immediately sown in pots, and plunged into a hot-bed, where, if the seeds are fresh, the plants will come up in two months; but sometimes they remain in the ground till the following spring before the plants appear; therefore, when the plants do not come up the first season, the pots should be screened from the frost in the winter, and put into a new hot-bed in the spring, which will bring up the plants.

The fourth sort is much cultivated by the inhabitants of the islands in America, and is of great use to them for feeding of their negroes; and the white people make puddings of their roots, when ground to a sort of flour. This plant is supposed to have been brought from the East to the West-Indies, for it has not been discovered to grow wild in any part of America; but in the island of Ceylon, and on the coast of Malabar, it grows in the woods, and there are in those places a great variety of sorts.

This plant is propagated by cutting of the root into pieces, observing to preserve an eye or bud to each, as is practised in planting of Potatoes; each of these being planted will grow, and produce three or four large roots: in America they are commonly six or eight months in the ground before the roots are taken up for use; the roots are roasted or boiled, and eaten by the inhabitants, and sometimes are made into bread.

This plant will not thrive in the open air in the warmest time of the year, so must constantly be kept in the bark-stove.

DIOSMA. Lin. Gen. Pl. 241. African Spiræa, vulgò.

The CHARACTERS are,

The flower hath five obtuse petals, which are as long as the empalement, and five stamina, with a five-pointed hollow nectarium sitting on the germen, which afterward becomes a fruit composed of five compressed capsules, joined together, each inclosing one smooth oblong seed.

The SPECIES are,

1. DIOSMA (*Oppositifolia*) foliis subulatis acutis oppositis. Hort. Cliff. 71. *African Spiræa with leaves placed in the form of a cross.*

2. DIOSMA (*Hirsuta*) foliis linearibus hirsutis. Hort. Cliff. 71. *Diosma with narrow hairy leaves.*

3. DIOSMA (*Rubra*) foliis linearibus acutis glabris, subtus bifarium punctatis. Lin. Sp. Plant. 198. *Diosma with smooth, narrow, acute leaves, which are spotted on their under side.*

4. DIOSMA (*Ericoides*) foliis lineari-lanceolatis subtus convexis, bifariam imbricatis. Lin. Sp. Plant. 198. *African Spiræa with leaves like the berry-bearing Heath.*

The first sort rises to the height of three or four feet; the branches are slender, and produced from the stem very irregularly; the leaves are placed crossways; the flowers are produced at the end of the branches between the leaves: these plants continue a long time in flower, and make a fine appearance when they are intermixed with other exotics in the open air.

The second sort makes a very handsome shrub, growing to the height of five or six feet: the stalks are of a fine coral colour; the leaves come out alternately on every side of the branches,

branches, which are narrow-pointed and hairy: the flowers are produced in small clusters at the end of the shoots, which are small and white; these are succeeded by starry seed-vessels, having five corners like those of the starry Anise; each of these corners is a cell, having one smooth, shining, oblong, black seed: these seed-vessels abound with a resin, which affords a grateful scent, as doth also the whole plant.

The third sort rises from two to three feet high, forming a bushy head; the leaves are smooth, narrow, and acute-pointed, having two or three spots on their under side; the flowers are small, of a blush colour, and come out at the end of the shoots; but this sort rarely produces seeds in England.

The fourth sort is of humbler growth than either of the former, seldom rising above two feet high, and spreads out into many branches; the leaves of this sort are smooth, and resemble those of the Heath; and the plant from thence had the name of Erica Æthiopica, &c. given to it by Dr Plukenet: the flowers of this kind are produced in clusters at the end of the branches, like those of the second sort, but are smaller, and the bunches not so large.

All these plants are propagated by cuttings, which may be planted during any of the summer months in pots, and plunged into a moderate hot-bed, where they should be shaded in the day-time from the sun, and frequently refreshed with water; in about two months the cuttings will have taken root, when they should be each transplanted into a small pot, and placed in a shady situation until they have taken fresh root, when they may be placed among other exotic plants in a sheltered situation: the plants may remain abroad until the beginning of October or later, if the season continue favourable, for they only require to be sheltered from frost; so that in a dry airy green-house they may be preserved very well in winter, and in summer they may be exposed to the open air with other green-house plants.

DIOSPYROS. Lin. Gen. Pl. 1027. The Indian Date Plumb.

The CHARACTERS are,

It hath hermaphrodite and male flowers on separate plants; the hermaphrodite flowers have a large, obtuse, permanent empalement, divided into four parts; the flower hath one pitcher-shaped petal, cut at the brim into four segments, and eight short bristly stamina firmly joined to the empalement; in the center is situated a roundish germen, which afterward becomes a large globular berry with cells, each including one oblong, compressed, hard seed; the male flowers have the same calyx and flower, with eight short stamina, but have no germen.

The SPECIES are,

1. DIOSPYROS (*Lotus*) foliorum paginis discoloribus. Lin. Sp. Plant. 1057. *Diospyros with the surface of the leaves of two colours, or the Indian Date Plumb.*

2. DIOSPYROS (*Virginiana*) foliorum paginis concoloribus. Lin. Sp. Plant. 1057. *The Pishamin or Persimon, and by some Pitchumon Plumb.*

The first sort is supposed to be a native of Africa, and was transplanted from thence into several parts of Italy, and also the south of France. The fruit of this tree is by some supposed to be the Lotus, which Ulysses and his companions were inchanted with. This is a tree of middling growth in the warm parts of Europe, where it rises upward of thirty feet high; in the botanic garden at Padua there is one very old tree, which has been described by some of the former botanists, under the title of Guaiacum Patavinum. This tree produces plenty of fruit every year, from the seeds of which many plants have been raised.

The second sort is a native of America, but particularly in Virginia and Carolina; the seeds of this sort are frequently brought to England, where the trees are now become pretty common in nurseries about London. This rises to the height of twelve or fourteen feet, but generally divides into many irregular trunks near the ground, so that it is very rare to see a handsome tree of this sort: it produces plenty of fruit in England, but they never come to perfection here: in America the inhabitants preserve the fruit until it be rotten (as is practised by Medlars in England) when they are esteemed a pleasant fruit.

Both sorts are propagated by seeds, which will come up very well in the open ground; but if they are sown upon a moderate hot-bed, the plants will come up much sooner, and make a greater progress; but in this case the seeds should be sown in pots or boxes of earth, and plunged into the hot-bed, because the plants will not bear transplanting till autumn, when the leaves fall off; so that when the plants are up and have made some progress, they may be inured by degrees to the open air; and in June they may be wholly exposed, and may remain abroad until November, when it will be proper to set the pots under a hot-bed frame to protect them from hard frost, which, while they are very young, may kill the tops of the plants; but they must have as much free air as possible in mild weather: the following spring, before the plants begin to shoot, they should be transplanted into a nursery in a warm situation, where they may be trained up for two years, and then removed to the places where they are designed to remain. These are both hardy enough to resist the greatest cold of this country, after the plants have acquired strength.

DIPSACUS. Lin. Gen. Plant. 107. The Teazel.

The CHARACTERS are,

It hath many florets collected in one common perianthium; they have but one petal, which is tubular, cut into four parts at the top. They have four hairy stamina. The germen is situated below the flower, which afterward becomes a column-shaped seed, inclosed in the common conical fruit, which is divided by long prickly partitions.

The SPECIES are,

1. DIPSACUS (*Sylvestris*) foliis sessilibus serratis, aristis fructibus erectis. *Wild Teazel.*

2. DIPSACUS (*Fullonum*) foliis connatis, aristis fructibus recurvis. *Cultivated Teazel.*

3. DIPSACUS (*Laciniatus*) foliis connatis sinuatis. Lin. Sp. Plant. 97. *Teazel with a laciniated leaf.*

4. DIPSACUS (*Pilosus*) foliis petiolatis appendiculatis. Hort. Upsal. 25. *Wild Teazel with a smaller head, or smaller Shepherd's Rod.*

The first of these plants is very common upon dry banks in most parts of England, and is seldom cultivated in gardens, unless for the sake of variety.

The fourth sort grows naturally in many places near London, and is rarely admitted into gardens.

The third sort grows naturally in Alsace, and is kept in botanic gardens for the sake of variety; this differs from the wild Teazel, in having the leaves deeply cut and jagged.

But it is the second sort only which is cultivated for use, which is called Carduus Fullorum, or Fullonum, being of singular use in raising the knap upon woollen cloth; for which purpose, there are great quantities of this plant cultivated in the west country.

This plant is propagated by sowing the seeds in March, upon a soil that has been well prepared. About one peck of seed will sow an acre, for the plants should have room to grow, otherwise the heads will not be so large, nor in so great quantity. When the plants are come up, they must be hoed in the same manner as is practised for Turneps, cutting down all the weeds, and singling out the plants to about eight inches distance; and as the plants advance, and the weeds begin to grow again, they must be hoed a fecond time, cutting out the plants to a wider diftance; for they fhould be, at laft, left at leaft a foot afunder, and fhould be kept clear from weeds, efpecially the firft fummer, for when the plants have fpread fo as to cover the ground, the weeds will not fo readily grow between them. The fecond year after fowing the plants will fhoot up heads, which will be fit to cut about the beginning of Auguft; at which time they fhould be cut, and tied up in bunches, fetting them in the fun, if the weather be fair; but if not, they muft be fet in rooms to dry them. The common produce is about an hundred and fixty bundles or ftaves upon an acre, which they fell for one fhilling a ftave. Some people fow Caraway and other feeds amongft their Teazels, but this is not a good method, for the one fpoils the other, nor can the weeds be fo well cleaned away from the Teazels.

DIRCA. Lin. Gen. 1078.

The CHARACTERS are,

The flower is tubulous, has no empalement; it has eight ftamina, which are longer than the tube. The flower is fucceeded by a berry with one cell, including a fingle feed.

We have but one SPECIES of this genus, viz.

DIRCA (*Paluftris.*) Lin Gen. Nov. 1078. The French call it Bois de Plomb, i. e. Leadwood: the Englifh in America call it Leatherwood, from its lightnefs.

This is a low fhrub in this country; it feldom rifes more than four or five feet high, and is very nearly allied to the Mezereon; the flowers come out early in the fpring before the leaves; they are fmall, tubulous, and of a light herbaceous colour, fo make very little appearance. The leaves are oval, fmooth, and of a pale green; they fall off in the autumn; the fhoots are jointed like knees.

It grows naturally in moift places in North America, but is at prefent pretty rare in the Englifh gardens. It may be propagated by layers, but they are commonly two years before they put out roots. It fhould have a moift foil and a fhady fituation.

DITTANY, the white. See DICTAMNUS.

DOCK. See RUMEX.

DODARTIA. Lin. Gen. Plant. 698.

The CHARACTERS are,

The empalement is cut into five parts at the top; the flower hath one petal; the upper lip is twice as long as the lower. It hath four ftamina, two of which are fhorter than the other. In the center is fituated a round germen, which afterward becomes a globular capfule with two cells, filled with fmall feeds.

The SPECIES are,

1. DODARTIA (*Orientalis*) foliis linearibus integerrimis glabris. Lin. Sp. Plant. 633. *Dodartia with very narrow, fmooth, entire leaves.*

2. DODARTIA (*Linaria*) foliis radicalibus oblongo-ovatis, ferratis, caulinis linearibus integerrimis floribus fpicatis terminalibus. *Dodartia with oblong, oval, fawed leaves at the bottom, thofe on the ftalk narrow and entire, and flowers growing in fpikes at the end of the ftalks.*

The firft fort was difcovered by Dr. Tournefort near Mount Ararat in Armenia. It hath a perennial root, which creeps far in the ground, and fends out ftalks at a great diftance from the parent plant; the ftalks are firm, a little compreffed, and grow a foot and a half high, fending out feveral fide branches, garnifhed with long, flefhy, narrow leaves, placed oppofite; thofe on the lower part of the ftalk are fhorter and broader than thofe above, and have two or three fharp indentures on their edges. At the joints the flowers come out fingly on each fide the ftalk, fitting clofe to it; they are tubulous, but divide into two lips; the upper lip is hollow like a fpoon, and is divided into two parts; the lower lip is divided into three parts, the middle being the narroweft. The flower is of a deep purple colour, but is rarely fucceeded by feeds in England. It propagates very faft by its creeping roots, fo that when it is once eftablifhed in a garden, it will multiply faft enough.

The fecond fort is a biennial, or at leaft a triennial plant, which perifhes foon after the feeds are ripe; this fends out feveral oblong leaves from the root, which are narrow at their bafe, but increafes in width upward, and are rounded at the end: between thefe arife the ftalks which grow a foot high, the upper parts being garnifhed with leaves of the fame form as the lower leaves, but are much fmaller; the upper leaves are very narrow and entire. The flowers grow in fpikes at the top of the ftalks; they are very fmall and white, but are fhaped like thofe of the former fort.

This is propagated by feeds, which fhould be fown in autumn foon after they are ripe, upon a border of light earth, where they are defigned to remain. When the plants appear the following fpring, they muft be thinned, and kept clean from weeds, which is all the culture they require; the fecond or third year they will flower and feed, after which the plants decay; when the feeds are fown in the fpring, the plants never come up the fame year.

DOG's TOOTH. See ERYTHRONIUM.

DOG-WOOD. See CORNUS.

DOLICHOS. Kidney-bean.

The CHARACTERS are,

The flower is of the butterfly kind having a large round vexillum, which is reflexed, and is diftinguifhed from Phafeolus, by the keel of the flower not being fpiral.

The SPECIES are,

1. DOLICHOS (*Lablab*) volubilis leguminibus ovato-aciniformibus, feminibus ovatis hilo arcuato verfus alteram extremitatem. Prod. Leyd. 368. *Dilochos with a winding ftalk, oval bill-fhaped pods, and oval feeds.*

2. DOLICHOS (*Urens*) volubilis leguminibus ramofis herbis tranfverfim lamellatis, feminibus hilo cinctis. Jacq. Amer. 27. *Dilochos with a winding ftalk; commonly called Cowitch.*

There are many other fpecies of this genus, as there are alfo of the Phafeolus, which are rarely cultivated in the Englifh gardens, therefore to enumerate them would only fwell the work to little purpofe, as they are not cultivated for ufe.

The firft fort here enumerated is frequently cultivated in warm countries for the fupply of the table, though there are many fpecies which are more valuable for that purpofe; but in England it would be ridiculous, becaufe the plants muft be raifed on a hot-bed, otherwife they will not flower or feed.

The fecond fort is fometimes cultivated in gardens, for the pods are covered clofely with ftinging hairs, called Cowitch. The feeds of this fort muft be fown on a hot-bed, and the plants muft be kept in the ftove through the winter, and treated tenderly; the fecond year they will flower and produce pods.

DORIA. See SOLIDAGO and OTHONNA.

DORONICUM. Lin. Gen. Pl. 862. Leopard's Bane.

The CHARACTERS are,

It hath a flower compofed of feveral hermaphrodite florets, which form the difk, and female florets which compofe the rays; thefe are included in one common empalement, which hath a double feries of leaves as long as the rays. In the bottom of the hermaphrodite florets is fituated the germen, which afterward becomes a fingle, oval, compreffed feed, crowned with hairy down. The female florets are formed like a tongue, and compofe the border; thefe have a germen, crowned by two reflexed ftigmas, but have no ftamina; the germen becomes a fingle furrowed feed without down.

The

The Species are,

1. Doronicum (*Pardalianches*) foliis cordatis obtusis, radicalibus petiolatis, caulinis amplexicaulibus. Lin. Mat. Med. 394. *Leopard's Bane with obtuse heart-shaped leaves, those from the root having foot-stalks, and those above embracing the stalks.*

2. Doronicum (*Plantaginum*) foliis ovatis acutis, subdentatis, ramis alternis. Hort. Cliff. 411. *Leopard's Bane with oval pointed leaves indented at bottom, and alternate branches.*

3. Doronicum (*Bellidiastrum*) caule nudo simplicissimo unifloro. Hort Cliff. 500. *Leopard's Bane with a naked single stalk, having one flower.*

The first sort grows naturally in Hungary, and upon the Helvetian mountains, but is frequently preserved in the English gardens. It hath thick fleshy roots, which divide into many knots or knees sending out strong fleshy fibres, which penetrate deep into the ground; from these arise in the spring a cluster of heart shaped leaves, which are hairy, and stand upon foot-stalks; between these arise the flower-stalks which are channelled and hairy, near three feet high, putting out one or two smaller stalks from the side, which grow erect, and are garnished with one or two heart-shaped leaves closely embracing the stalks with their base; each stalk is terminated by one large yellow flower, composed of about twenty-four rays or female florets, long, plain, and indented in three parts at the top. In the center is situated a great number of hermaphrodite florets, which compose the disk; these are tubulous, and slightly cut at the top in five parts.

This plant multiplies very fast by its spreading roots, and if the seeds are permitted to scatter, they will produce plants wherever they happen to fall, so that it becomes a weed where it is once established; it loves a moist soil and a shady situation.

The second sort hath oval leaves, ending in acute points; these are indented on their edges toward their base, but their upper parts are entire; the stalks rise about two feet high, each is terminated by a large yellow flower like those of the former sort; the stalks of this sort have two or three leaves, which are placed alternately, and their base sits close to the stalks; these are not so hairy as those of the former sort. This grows naturally in Portugal, Spain, and Italy, but is equally hardy with the first, and multiplies in as great plenty; the root is perennial.

The third sort grows naturally on the Alps and the Pyrenean mountains; it hath a perennial root; the leaves are like those of the lesser Daisy, but longer, and not so broad. The flower grows upon a naked foot-stalk, which is near a foot long; the roots seldom send out more than one stalk; the rays of the flower are white, and very like those of the common Daisy; the disk of the flower is yellow, which is composed of hermaphrodite flowers.

This plant is preserved in botanic gardens for the sake of variety, but the flowers make little better appearance than those of the common Field Daisy, only they stand upon much taller foot stalks. It must have a shady situation and a moist soil, otherwise it will not thrive in this country; it is propagated by parting of the roots, for the seeds do not ripen well in England.

The roots of the first sort have been sometimes used in medicine; some have commended it as an expeller of the poison of scorpions, but others reckon it to be a poison, and affirm that it will destroy wolves and dogs.

DORSTENIA. Plum. Nov. Gen. 29. tab. 8. Lin. Gen. Plant. 147. Contrayerva.

The Characters are,

It hath one common involucrum situated vertically, upon which sit many small flowers which have no petals, but have four short stamina. In the center is situated a roundish germen, which afterward becomes a single seed, inclosed in the common fleshy receptacle.

The Species are,

1. Dorstenia (*Contrayerva*) foliis serratis, floribus quadrangulis acaulis, pinnatifidis-palmatis. Lin. Sp. *Dwarf Dorstenia with hand-shaped leaves.*

2. Dorstenia (*Houstoni*) foliis cordato-angulatis acutis, placentâ quadrangulari dentato. *Contrayerva with angular, heart-shaped, acute leaves, and a quadrangular placenta which is indented.*

3. Dorstenia (*Drakena*) foliis palmato-angulatis, angulis acutis, placentâ oblongâ tetragono *Contrayerva with angular handed leaves, whose angles are very acute, and an oblong four-cornered placenta.*

The first of these plants was discovered by my late ingenious friend Dr. William Houston, near Old La Vera Cruz in New Spain. The second was found by the same gentleman, on the rocky grounds about Campeachy. The third sort was found in great plenty in the island of Tobago, by Mr. Robert Millar, Surgeon. But the roots of all these species are indifferently brought over, and used in medicine, and for dyeing.

The first sort sends out several leaves from the root, which are about four inches long, and as much in breadth, deeply laciniated into five or seven obtuse parts, standing upon long foot-stalks; they are smooth, of a deep green. The stalk which supports the placenta arises from the root, is four inches high, upon which the fleshy placenta is vertically placed; this is of a square form about an inch long, and three quarters broad. Upon the upper surface of this the small flowers are closely situated, the fleshy part becoming an involucrum to them; these are very small, and scarce conspicuous at a distance, being of an herbaceous colour.

The second sort sends out several angular heart shaped leaves from the root, which have foot-stalks eight or nine inches in length, and very slender; the leaves are about three inches and a half long, and almost four broad at their base, the two ears having two or three angles, which are acute, and the middle of the leaves are extended and end in acute points like an halbert; they are smooth and of a lucid green; the foot-stalk which sustains the placenta is nine inches long, and about half an inch square; the upper surface is closely set with small flowers like the first.

The third sort sends out leaves of different forms; some of the lower leaves are heart-shaped, having a few indentures on their edges, ending in acute points, but the larger leaves are deeply cut like the fingers on a hand, into six or seven acute segments. They are five inches long, and six broad in the middle, of a deep green, and stand upon long foot-stalks. The placenta is very thick and fleshy, an inch and a half long, and three quarters broad having four acute corners; these have a number of small flowers placed on the upper surface like the other species.

These plants are at present very rare in Europe, nor was it known what the plant was, whose roots were imported, and had been long used in medicine in England, until the late Dr Houstoun informed us; for although Father Plumier had discovered one species of this plant, and given the name of Dorstenia to the genus, yet he seems not to have known, that the Contrayerva was the root of that plant.

It will be difficult to obtain these plants, because the seeds are seldom to be found good; nor will they grow, if they are kept long out of the ground; so that the only sure method to obtain them is, to have the roots taken up at the time when their leaves begin to decay, and planted very close in boxes of earth, which may be brought very safe to England, provided they are preserved from salt-water, and are not over watered with fresh in their passage.

When the plants arrive, they ſhould be each tranſplanted into a ſeparate pot, and plunged into the bark-ſtove, which ſhould be kept of a moderate heat; they may be increaſed by parting their roots in the ſpring, before the plants put out their leaves.

DORYCNIUM. See Lotus.

DOUGLASSIA. See Volkameria.

DRABA. Dillen. Gen. Lin. Gen. Pl. 714.

The Characters are,

The flower hath four petals in form of a croſs, and ſix ſtamina, four of which are as long as the empalement, the other two are much ſhorter. In the center is ſituated a bifid germen, which afterward becomes an oblong, oval, entire pod, with two cells, ſeparated by the ſwelling ſtyle, which is oblique. The valves are parallel to the middle, and opens oblique, each cell containing a ſingle ſeed.

The Species are,

1. Draba (*Alpina*) ſcapo nudo ſimplici, foliis lanceolatis integerrimis. Fl. Lap. 255. *Yellow hairy Alpine Madwort.*

2. Draba (*Verna*) ſcapis nudis, foliis ſubſerratis. Lin. Syſt. *Draba with naked ſtalks and cut leaves.*

3. Draba (*Pyrenaica*) ſcapo nudo, foliis cuneiformibus trifidis. Lin Syſt. *Leaſt perennial Madwort of the Pyrenees, with trifid leaves.*

4. Draba (*Muralis*) caule ramoſo, foliis cordatis dentatis amplexicaulibus Prod. Leyd. 33. *Draba with a branching ſtalk, and heart-ſhaped indented leaves embracing the ſtalks.*

5. Draba (*Polygonifolia*) caule ramoſo, foliis ovatis ſeſſilibus dentatis. Lin. Sp. Pl. 643. *Draba with a branching ſtalk, and oval indented leaves growing cloſe to the branches.*

6. Draba (*Incana*) foliis caulinis numeroſis incanis, ſiliculis obliquis. Flor. Suec. 526. *Draba with many hoary leaves on the ſtalks, and oblique pods.*

The firſt ſort grows naturally on the Alps, and other mountainous parts of Europe. It is a very low plant, which divides into ſmall heads like ſome ſorts of Houſeleek, and from thence it was titled Sedum Alpinum, &c. or Alpine Houſeleek. The leaves are ſhort, narrow, and hairy; from each of theſe heads come out a naked flower-ſtalk an inch and a half high, terminated by a looſe ſpike of yellow flowers, having four obtuſe petals placed in form of a croſs; when theſe fade, they are ſucceeded by heart-ſhaped pods, which are compreſſed, and incloſe three or four roundiſh ſeeds.

This plant is eaſily propagated by parting of the heads; the beſt time for doing this is in autumn, becauſe it ſhoots up to flower very early in the ſpring. It ſhould have a moiſt ſoil and a ſhady ſituation.

The ſecond ſort is an annual plant, which grows naturally upon walls and dry banks in many parts of England, ſo is never cultivated in gardens. This flowers in April, and the ſeeds ripen in May.

The third ſort grows naturally on the Alps, and other mountainous parts of Europe. This is a low perennial plant, which ſeldom riſes more than two feet high; it has a ſhrubby ſtalk, which divides into many ſmall heads like the firſt ſort. The leaves are ſmall, ſome of them are winged, having five ſhort narrow lobes placed on a midrib, others have but three. The flowers come out in cluſters, ſitting cloſe to the leaves. They are of a bright purple colour, and appear early in the ſpring. This is a perennial plant, which may be propagated by parting of the heads in the ſame manner as the firſt, and requires the ſame treatment.

The fourth ſort grows naturally in ſhady woods in many parts of Europe, and is but ſeldom kept in gardens. unleſs for the ſake of variety It is an annual plant, riſing with an upright branching ſtalk, garniſhed with heart-ſhaped indented leaves, which embrace the ſtalks with their baſe. The ſtalks are terminated by looſe ſpikes of white flowers, which appear the beginning of May; in June the ſeeds ripen, and the plants ſoon after decay.

The fifth ſort is an annual plant, which grows in ſhady woods in the northern parts of Europe. This is like the former ſort, but the leaves are larger, rounder, and do not embrace the ſtalks; they are alſo hairy, and the flowers are yellow. If the ſeeds of this are permitted to ſcatter, the plants will maintain themſelves if they have a ſhady ſituation.

The ſixth ſort riſes with an upright ſtalk about a foot high, the lower part being very cloſely garniſhed by oblong hoary leaves, which are indented on their edges. The upper part of the ſtalk puts out two or three ſhort branches, which are almoſt naked of leaves. The flowers come looſely out at the top of the ſtalk; they are compoſed of four ſmall white petals placed in form of a croſs, which are ſucceeded by oblong pods which are twiſted, containing three or four roundiſh compreſſed ſeeds.

DRACO ARBOR. See Palma.

DRACO HERBA, or Tarragon. See Abrotanum.

DRACOCEPHALUM. Lin. Gen. Pl. 648. Dragon's Head.

The Characters are,

The flower hath one ringent petal, with large, oblong, inflated chaps. The upper lip is obtuſe and arched, the under lip is trifid. It hath four ſtamina, two being ſhorter than the other, and a four-parted germen, which afterward becomes four oval oblong ſeeds, incloſed in the empalement.

The Species are,

1. Dracocephalum (*Virginianum*) foliis lanceolatis ſerratis, floribus ſpicatis. Lin. Sp. *American Dragon's Head.*

2. Dracocephalum (*Canarienſe*) floribus ſpicatis, foliis compoſitis. Lin. Hort. Cliff. *Three-leaved American Balm, having a ſtrong ſmell; commonly called Balm of Gilead.*

3. Dracocephalum (*Moldavica*) floribus verticillatis bracteis lanceolatis, ſerraturis capillaceis. Lin. Hort. Upſ. 166. *Moldavian Balm with a Betony leaf and blue flower.*

4. Dracocephalum (*Ocymifolia*) floribus verticillatis bracteis ſerrato ciliatis orbiculatis. Lin. Hort. Upſal. *Leſſer eaſtern Moldavian Balm, with a Willow leaf and a bluiſh flower.*

5. Dracocephalum (*Caneſcens*) floribus verticillatis, bracteis oblongis, ſerraturis ſpinoſis, foliis ſubtomentoſis. Hort. Upſal. 166. *Eaſtern Moldavian Balm, with a Betony leaf and a large blue flower.*

6. Dracocephalum (*Nutans*) floribus verticillatis, bracteis oblongis ovatis integerrimis, corollis majuſculis nutantibus. Hort. Upſal. 167. *Moldavian Balm with a Betony leaf, and larger blue pendulous flowers.*

7. Dracocephalum (*Thymiflorum*) floribus verticillatis, bracteis oblongis integerrimis, corollis vix calyce majoribus. Hort. Upſal. 167. *Moldavian Balm with a Betony leaf and very ſmall blue flowers.*

8. Dracocephalum (*Grandiflora*) floribus verticillatis foliis ovatis inciſo-crenatis, bracteis lanceolatis integerrimis. Lin. Sp. Pl. 595. *Dragon's Head with flowers growing in whorls, and oval leaves which are cut and crenated, and ſpear-ſhaped bracteæ, which are entire.*

The firſt ſort is a native of North America. This ſort riſes with an upright four-cornered ſtalk near three feet high, garniſhed with narrow ſpear-ſhaped leaves ſitting cloſe to the ſtalk, ſawed on their edges, and are placed oppoſite at each joint. The flowers grow in long ſpikes on the top of the ſtalks; they are of a purple colour. This is a perennial plant, which will live in the open air, but requires a moiſt ſoil, or ſhould be duly watered in dry weather. It is propagated by parting the roots in autumn.

The second sort is a native of the Canary Islands. It is usually called by the gardeners Balm of Gilead, from the strong resinous scent which the leaves emit on being rubbed. This is a perennial plant, which rises with several square stalks to the height of three feet or more, becoming ligneous at their lower parts, garnished with compound leaves at each joint, which are placed opposite; they have three or five lobes, which are oblong, pointed, and sawed on their edges. The flowers terminate the stalks in short thick spikes; they are of a pale blue colour, and are succeeded by small angular seeds. This plant continues producing flowers most part of summer; it is usually kept in green-houses, but, in mild winters the plants will live abroad, if they are planted in warm borders; and those plants which are kept in pots will thrive much better when they are sheltered under a frame, than if placed in a green-house, where the plants are apt to draw up weak, for they should have as much free air as possible in mild weather, and only require to be sheltered from severe frost. This may be propagated by seeds, which, if sown in autumn, will more certainly grow than those which are sown in the spring; if the seeds are sown in the full ground, it should be in a warm border. It may also be propagated by cuttings, which, if planted in a shady border any time in the summer, will very soon take root, and furnish plenty of rooted plants.

The third sort is a native of Moldavia; it is an annual plant, which rises with branching stalks two feet high, garnished with oblong leaves placed opposite, which are deeply sawed on their edges. The flowers come out in whorls round the stalks at every joint; they are blue, and are succeeded by seeds which ripen in September. The seeds should be sown in small patches in the spring, upon the borders where they are to remain, and when the plants come up they should be thinned where they grow too near together, and kept clean from weeds, which is the only culture they require. Of this there is a variety with white flowers, which is pretty common in the gardens.

The fourth sort was discovered by Dr. Tournefort in the Archipelago. This rises with upright stalks about a foot high, which seldom put out branches, they are garnished with very long, narrow, entire leaves, placed opposite at each joint, where the flowers come out in whorls almost the whole length of the stalks; they are of a pale blue, but as they are very small, so make no great appearance.

The fifth sort was also discovered by Dr. Tournefort in the Levant; this hath hoary square stalks, which rise a foot and a half high, putting out two or three side branches, garnished with hoary leaves a little indented on their edges; they are placed opposite at their joints just under the whorls of flowers, which sit close to the stalk; the flowers are larger than those of the other species, and are of a fine blue colour, which between the hoary leaves of the plants make a pretty appearance. It flowers and seeds about the same time with the former sorts. There is a variety of this with white flowers, the seeds of which generally produce the same coloured flowers.

The sixth sort grows naturally in Siberia. This is an annual plant, with many square weak stalks a foot long; those are at the bottom garnished with oval spear-shaped leaves, crenated on their edges. The upper part of the stalks have smaller leaves which sit close to the joints, from whence come out the flowers in whorls, of a deep blue colour, and hang downward; these appear at the same time with the former, and the seeds ripen in autumn.

The seventh sort grows also in Siberia. It hath square stalks, which rise a foot and a half high; the lower leaves are very like those of Betony, and stand upon very long foot-stalks. The upper leaves are small, and sit closer to the stalks. The flowers come out in whorls at every joint; these are very small, and of a pale purple colour, so make little appearance.

The eighth sort grows naturally in Siberia. This is an annual plant with a square stalk, sending out two or three small side branches from the lower part. The leaves are oval, and deeply crenated on their edges. The flowers are large, of a blue colour, and come out in whorls round the stalks, having two spear-shaped, entire, small leaves (called bractea) immediately under them. This sort flowers and seeds at the same time with the former.

All these sorts are propagated by seeds, which may be sown either in the spring or autumn, in the places where the plants are to remain, and will require no other treatment than the third sort.

DRACONTIUM. Lin. Gen. Plant. 916. Dragon.

The Characters are,

It hath a close cylindrical spadix, and a boat-shaped sheath. The flowers have no empalement, but have five oval concave petals. They have an oval germen, which afterward becomes a roundish berry, inclosing several seeds.

The Species are,

1. Dracontium (*Pertusum*) foliis pertusis, caule scandente. Lin. Sp. Pl. 968. *Dragon with leaves having holes, and a climbing stalk.*

2. Dracontium (*Polyphyllum*) scapo brevissimo, petiolo radicato, lacero, foliolis tripartitis, laciniis pinnatifidis. Hort. Cliff. 434. *Many-leaved Arum, with a rough purple stalk.*

3. Dracontium (*Spinosum*) foliis sagittatis, pedunculis petiolisque aculeatis. Flor. Zeyl. 328. *Dragon with arrow-pointed leaves, whose foot-stalks have spines.*

The first sort grows naturally in most of the islands in the West-Indies. This hath trailing stalks, which put out roots at every joint, that fasten to the trunks of trees, walls, or any support which is near them, and thereby rise to the height of twenty-five or thirty feet. The leaves are placed alternately upon long foot-stalks; they are four or five inches long, and two and a half broad, and have several oblong holes in each, which on the first view appear as if eaten by insects, but they are natural to the leaves. The flowers are produced at the top of the stalk, which always swells to a much larger size in that part immediately under the flower than in any other; these are covered with an oblong spatha (or hood) of a whitish green colour, which opens longitudinally on one side, and shews the pistil, which is closely covered with flowers, of a pale yellow, inclining to white. When this plant begins to flower, it seldom advances farther in height, so that these seldom are more than seven or eight feet high; but the leaves are much larger on these, than those of the plants which ramble much farther.

It is propagated easily by cuttings, which, if planted in pots filled with poor sandy earth, and plunged into a hot-bed, will soon put out roots, if they had none before, but there are few of the joints which are destitute of roots; the plants are tender, so will not live in the open air in England; therefore the pots should be placed near the walls in the hot-house, against which the plants will climb, and fasten their roots into the wall, and thereby support themselves. They should have but little water given them in the winter, but in warm weather it must be given them frequently; in the summer the free air should be admitted to them in plenty. The plants have no particular season of flowering, for they sometimes flower in autumn, and at other times in the spring, but they do not ripen their seeds in England.

The second sort grows naturally in several of the islands of America. This hath a large, knobbed irregular root, covered

covered with a rugged brown ſkin. The ſtalk riſes about a foot high, is naked to the top, where it is garniſhed with a tuft of leaves, which are divided into many parts. The ſtalk is ſmooth, of a purple colour, but is full of ſharp protuberances of different colours, which ſhine like the body of a ſerpent. The ſpadix (or ſtalk) of the flower riſes immediately from the root, and is ſeldom more than three inches high, having an oblong ſwelling hood at the top, which opens lengthways, ſhewing the ſhort, thick, pointed piſtil within, upon which the flowers are cloſely ranged.

This ſort is tender, ſo requires a warm ſtove to preſerve it in England. The roots muſt be planted in pots, and plunged into the tan bed in the ſtove, where they ſhould conſtantly remain; in the winter they muſt be watered very ſparingly, but in warm weather when the plants are in vigour, they muſt be often refreſhed, but it ſhould not be given them in too great quantities.

The third ſort grows naturally in the iſland of Ceylon, and in ſeveral parts of India; this hath an oblong thick root, full of joints, from which ariſe ſeveral leaves ſhaped like thoſe of the common Arum, but their foot-ſtalks are covered with rough protuberances. The ſtalk which ſupports the flower is ſhort, and ſet with the like protuberances, and at the top is a hood, or ſpatha, about four inches long, which open longitudinally, and expoſes the piſtil which is ſet with flowers. This is a tender plant, and requires the ſame treatment as the former ſort.

Theſe plants are preſerved in the gardens of the curious in England and Holland, more for the ſake of variety than their beauty, for except the firſt ſort, there is not any of them which makes much appearance; that indeed may be ſuffered to have a place againſt the wall of the ſtove, over which it will ſpread, and cover the nakedneſs of the wall, and the leaves remaining all the year, which is ſo remarkably perforated, make a ſingular appearance.

DRACUNCULUS PRATENSIS See ACHILLÆA.

DRAGON. See DRACONTIUM.

DROSERA, or Sun-dew, or Ros Solis.

We have three ſpecies of this genus, which grow naturally in ſeveral parts of England upon bogs, but are with difficulty preſerved in gardens, therefore deſerve no mention in this work.

DULCAMARA. See SOLANUM.

DURANTIA. Lin. Gen. Plant. 704.

The CHARACTERS are,

The flower is of the ringent kind, with one petal, which is erect and concave, and hath an empalement divided into five equal ſegments. It hath four ſhort ſtamina, the two middle being a little ſhorter than the other. The germen, which is ſituated under the flower, afterward becomes a globular berry, having one cell incloſing four angular ſeeds.

The SPECIES are,

1. DURANTIA (*Plumeiri*) ſpinoſa. Lin. Sp. Plant. 637. *Prickly Durantia*

2. DURANTIA (*Racemoſa*) inermis. Lin. Sp. Plant. 637. *Durantia without thorns.*

The firſt ſort hath many trailing branches, which are armed with hooked thorns at every joint, and are garniſhed with oblong leaves, which are placed without order; the flowers come out from the ſide of the ſtalks in pretty long bunches, like thoſe of the common Currant; they are of a pale bluiſh colour, and are ſucceeded by brown berries not unlike thoſe of the Hawthorn; theſe have one cell, and incloſe four angular ſeeds.

The ſecond ſort hath a branching woolly ſtalk, which riſes ſeven or eight feet high; the branches are garniſhed with oval ſpear ſhaped leaves ſawed on their edges, of a lucid green colour, and ſtand oppoſite. The flowers are produced in long bunches at the end of the branches; theſe are blue, and are ſucceeded by pretty large, round, yellow berries, which contain four angular ſeeds.

Theſe plants are natives of warm countries, ſo they require a ſtove to preſerve them in winter. They are propagated by ſeeds, which ſhould be ſown in ſmall pots, and plunged into a hot-bed of tanners bark; and when the plants are fit to remove, they muſt be planted each into a ſeparate ſmall pot and plunged into a hot-bed again, obſerving to ſhade them till they have taken new root, then they muſt be treated in the ſame manner as other plants from the ſame country.

They may alſo be propagated by cuttings, which may be planted in any of the ſummer months; but theſe ſhould be plunged into a moderate hot-bed, and ſhaded from the ſun till they have taken root, then they may be treated in the ſame manner as the ſeedling plant.

DWARF-TREES were formerly in much greater requeſt than they are at preſent; for though they have ſome advantages to recommend them, yet the diſadvantages attending them greatly over-balance; and ſince the introducing of eſpaliers into the Engliſh gardens, Dwarf-trees have been in little eſteem for the following reaſons:

1. The figure of a Dwarf-tree is very often ſo much ſtudied, that in order to render the ſhape beautiful, little care is taken to procure fruit, which is the principal deſign in planting theſe trees.

2. The branches being ſpread horizontally near the ſurface of the ground, render it very difficult to dig or clean the ground between them.

3. Their taking up too much room in a garden (eſpecially when they are grown to a conſiderable ſize,) for nothing can be ſown or planted between them.

It is alſo very difficult to get to the middle of theſe Dwarf-trees in the ſummer, when their leaves and fruit are on the branches, without beating off ſome of the fruit, and breaking the young ſhoots; whereas the trees on an eſpalier can at all times be come at on each ſide, to tie up the new ſhoots, or to diſplace all vigorous ones, which if left on, would rob the trees of their nouriſhment.

Add to this, the fruit-buds of many ſorts of Pears and Apples are produced at the end of the former year's ſhoot, which muſt be ſhortened in order to keep the Dwarfs to their proper figure, ſo that the fruit-buds are cut off, and a greater number of branches are obtained than can be permitted to ſtand; ſo that all thoſe ſorts of fruit-trees, whoſe branches require to be trained at full length, are very improper to train up as Dwarfs.

Theſe evils being entirely remedied by training the trees to an eſpalier, hath juſtly gained them the preference.

E.

EBENUS. Lin. Gen. Nov. Ebony.

The CHARACTERS are,

The empalement of the flower is ſlenderly indented and hairy; the flower is of the butterfly kind. It hath ten ſtamina. In the bottom is ſituated an oblong germen, which afterward becomes an oblong ſwelling pod, incloſing one kidney-ſhaped ſeed.

This is diſtinguiſhed from Trifolium by the bracteæ, which are ſituated between the flowers on the ſpikes.

We have but one SPECIES of this genus, viz.

EBENUS (*Cretica.*) Lin. Sp. Plant. 764. *Ebony.*

This plant grows naturally in Crete, and in ſome of the iſlands of the Archipelago; it riſes with a ſhrubby ſtalk three or four feet high, which put out ſeveral ſide branches, garniſhed with hoary leaves at each joint, compoſed of five narrow ſpear-ſhaped lobes, which join at their tails to the foot-ſtalk, and ſpread out like the fingers of a hand: the branches are terminated by thick ſpikes of large purple flowers, which are of the butterfly or Pea-bloom kind.

This is propagated by ſeeds, which ſhould be ſown in the autumn, for thoſe which are ſown in the ſpring often fail; they ſhould be ſown in pots, and placed under a frame in the winter, where they may be protected from froſt; in the ſpring the plants will come up. When theſe have acquired ſtrength enough to be removed, they ſhould be each planted in a ſeparate ſmall pot, and ſhaded till they have taken new root; then they ſhould be placed in a ſheltered ſituation, where they may remain till autumn, when they muſt be removed into ſhelter, for theſe plants will not live in the open air through the winter; nor ſhould they be too tenderly treated: I have found them ſucceed beſt when placed in an airy glaſs-caſe without fire in the winter, where they will have more ſun and air than in a green-houſe: during the winter ſeaſon the plants muſt be ſparingly watered, but in the ſummer they will require to be often refreſhed.

EBULUS. See SAMBUCUS.

ECHINOMELOCACTUS. See CACTUS.

ECHINOPHORA. Lin. Gen. Plant. 292. Prickly Parſnep.

The CHARACTERS are,

It hath an umbellated flower; the flowers have five unequal petals; they have each five ſtamina, terminated by roundiſh ſummits. Under the perianthium is ſituated an oblong germen within the empalement, which afterward turns to two ſeeds, which are incloſed in the empalement.

The SPECIES are,

1. ECHINOPHORA (*Spinoſa*) foliolis ſubulato-ſpinoſis integerrimis. Lin. Sp. Plant. 344. *Prickly-headed Parſnep.*

2. ECHINOPHORA (*Tenuifolia*) foliolis inciſis inermibus. Lin. Sp. Plant. 344. *Prickly-headed Parſnep with narrow leaves.*

Theſe plants grow naturally on the borders of the Mediterranean ſea; they are preſerved in the gardens of botany for the ſake of variety: they have both perennial roots, which creep in the ground; the firſt have branching ſtalks, growing five or ſix inches high, which are garniſhed with ſhort thick leaves, that terminate in two or three ſharp thorns, which are placed oppoſite: the flowers grow in an umbel, ſitting upon a naked foot-ſtalk, which ariſes from the ſide of the ſtalk; the flowers are white, and under the umbel is ſituated an involucrum, compoſed of ſeveral leaves, which terminate in ſharp ſpines.

The ſecond ſort riſes near a foot and an half high; from the principal ſtalk are ſent out two ſide branches at each joint, placed oppoſite; the lower part is garniſhed with leaves, which are finely divided like thoſe of the Carrot; the flowers grow in ſmall umbels at the extremity of the branches, having a ſhort prickly involucrum. The ſeeds of theſe plants rarely ripen in England.

Theſe plants are propagated by their creeping roots in England; the beſt time to tranſplant them is the beginning of March, a little before they ſhoot: the roots ſhould be planted in a gravelly or ſandy ſoil, and in a warm ſituation, or otherwiſe they ſhould be covered in the winter to prevent the froſt from deſtroying them.

ECHINOPS. Lin. Gen. Plant. 829. Globe Thiſtle.

The CHARACTERS are,

The flower hath one funnel-ſhaped petal, included in an imbricated empalement, divided at the top into five parts; it hath five ſhort hairy ſtamina, terminated by cylindrical ſummits. In the bottom of the tube is ſituated an oblong germen, which afterward becomes an oblong oval ſeed narrowed at the baſe, but obtuſe and hairy at the top.

The SPECIES are,

1. ECHINOPS (*Sphærocephalus*) caliculis unifloris, caule multifloro, foliis ſpinoſis ſupra nudis. Lin. Sp. Plant. 814. *Greater Globe Thiſtle.*

2. ECHINOPS (*Strigoſus*) calycibus unifloris, caule unicapitato. Lin. Sp. Plant. 815. *Smaller Globe Thiſtle.*

3. ECHINOPS (*Uniflorus*) calycibus faſciculatis unifloris, lateralibus ſterilibus, foliis pinnatifidis ſupra ſtrigoſis. Lin. Sp. Plant. 815. *Smaller annual Globe Thiſtle, with a large head.*

4. ECHINOPS (*Græcus*) caule unicapitato, foliis ſpinoſis, omnibus pinnatifidis villoſis, radice reptatrice. *Greek Globe Thiſtle, whoſe leaves are divided into narrow ſegments and are woolly, with a ſmaller blue head.*

The firſt is the common Globe Thiſtle. This grows naturally in Italy and Spain; it hath a perennial root, from which ariſe many ſtalks that grow to the height of four or five feet, garniſhed with long jagged leaves, which are divided into many ſegments almoſt to the midrib, the jags ending in ſpines; they are of a dark green on their upper ſide, but woolly on their under; the flowers are collected in globular heads, ſeveral of theſe grow upon each ſtalk; the common ſort hath blue flowers, but there is a variety of it with white.

This plant is eaſily propagated by ſeeds, which, if permitted to ſcatter, the plants will come up in plenty, ſo a few of them may be tranſplanted to the places where they are deſigned to remain to flower; they require no other culture but to keep them clean from weeds: the ſecond year they

they will flower and produce seeds, and the roots will often continue two or three years after.

The second sort grows in the south of France and in Italy; this hath a perennial root, which sends up several stalks that rise two feet high, garnished with leaves which are cut into many fine segments to the midrib, which are set with prickles, and are white on their under side: the stalks are each terminated by a globular head of flowers, which are smaller than those of the first, and of a deeper blue; there is also a variety of this with white flowers; this is propagated in the same way as the first.

The third sort grows naturally in Spain and Portugal: this is an annual plant, which rises with a stiff white stalk two or three feet high, garnished with divided leaves, ending in many points which have spines; their upper side is green, and covered with brown hairs, their under side white and woolly; the stalk is terminated by one large head of pale blue flowers; and if the season proves warm and dry the seeds will ripen in autumn, but in wet cold years they rarely ripen here.

These seeds should be sown in the spring upon a border of light earth, where the plants are to remain; they require no other management but to thin them where they are too close.

The fourth sort grows naturally in Greece; this hath a perennial creeping root, by which it multiplies fast enough; the stalks rise about two feet high, and are closely garnished with leaves which are shorter and much finer divided than either of the former sorts; these are hoary, and armed on every side with sharp thorns; the stalks are terminated by one globular head of flowers, which in some are blue and in others white: in warm seasons the seeds will ripen well in England, but it is easily propagated by its creeping roots; it loves a dry soil and a warm situation.

ECHIUM. Lin. Gen. Pl. 157. Viper's Buglofs.

The CHARACTERS are,

The flower hath one petal, with a short tube, and an erect broad brim cut into five irregular parts, and hath naked chaps; it hath five awl-shaped stamina; in the bottom are situated four germen, which afterward become so many roundish pointed seeds, inclosed in the rough empalement.

The SPECIES are,

1. ECHIUM (*Anglicum*) caule simplici erecto, foliis caulinis lanceolatis hispidis, floribus spicatis lateralibus, staminibus corolla æquantibus. *Common Viper's Buglofs.*

2. ECHIUM (*Vulgare*) caule simplici erecto, foliis caulinis lanceolato-linearibus hispidis, floribus lateralibus spicatis sessilibus, staminibus corollâ longioribus. *This is the English Lycopsis.*

3. ECHIUM (*Italicum*) corollis vix calycem excedentibus, margine villosis. Hort. Upsal. 35. *Great rough Viper's Buglofs, with a white flower.*

4 ECHIUM (*Lusitanicum*) foliis radicalibus lanceolatis amplissimis, caulinis linearibus hirsutis, corollis stamine longioribus. *Portugal Viper's Buglofs, with a large leaf.*

5. ECHIUM (*Creticum*) calycibus fructescentibus distantibus, caule procumbente. Lin. Hort. Upsal. 35. *Broad-leaved Viper's Buglofs of Candia, with a red flower.*

6. ECHIUM (*Angustifolium*) caule ramoso, aspero, foliis calloso verrucosis, staminibus corollâ longioribus. *Narrow-leaved Viper's Buglofs of Candia, with a red flower.*

7. ECHIUM (*Fruticosum*) caule fruticoso. Hort. Cliff. 43. *Shrubby African Viper's Buglofs, with hairy leaves.*

The first sort grows naturally in Germany and Austria: this and our common Viper's Buglofs, which is the second, have been confounded by most of the writers on botany, who have supposed they were the same plant, whereas they are very different; for the leaves of this are shorter, and much broader than those of the second; the spikes of flowers are much longer, and the stamina of the flowers are in this equal in length with the petal; whereas those of the second stand out much beyond it, which is an essential difference.

The second sort grows naturally upon chalky lands in most parts of England: this is what Lobel titles Lycopsis Anglica, and has been generally taken for the common Echium.

The third sort grows naturally in the south of France, and in Italy; this rises with an upright stalk, which is very hairy, as are also the leaves; the flowers are produced in short spikes on the side of the branches; they are small, and scarce appear above the empalements; some plants have white flowers, and others are purplish; the empalements of the flowers are very hairy, and cut into acute segments.

The fourth sort grows naturally in Portugal and Spain; the lower leaves of this are more than a foot long, and two inches broad in the middle, gradually lessening to both ends; these are covered with soft hairs. The stalks grow two feet high; the flowers are in short spikes coming from the side of the stalks; the petals of these are longer than the stamina.

The fifth sort grows naturally in Crete; this hath trailing hairy stalks, which grow about a foot long, and put out several side branches, garnished with hairy spear-shaped leaves sitting close to the stalks. The flowers come out on slender spikes upon long foot-stalks, which come from the wings of the leaves; they are large, of a reddish purple colour, which turns to a fine blue when they are dried; these stand at a distance from each other on the spike; it is an annual plant, which flowers in July and decays in autumn.

The sixth sort hath branching stalks, which grow a foot and a half long, declining toward the ground; they are covered with stinging hairs, which are warted; the leaves are four inches long, and not more than half an inch broad; the flowers grow in loose spikes from the side of the stalks, and also at the end of the branches; they are of reddish purple colour, but not so large as those of the former sort, and the stamina of these are longer than the petals: this is also an annual plant, which grows naturally in Crete.

These are all of them biennial plants, except the fifth and sixth sorts, which are annual, and are the most beautiful of all the kinds: the seeds of these must be sown every year in the places where they are designed to remain, and the plants require no other culture but to keep them clean from weeds, and thin them where they grow too close. In July they flower, and their seeds ripen in five or six weeks after. The seeds of the other sorts being sown in the spring, will the second summer after produce flowers and seeds, after which they seldom continue; they all delight in a rubbishy gravelly soil, and will grow upon the tops of old walls or buildings, where, when once they have established themselves, they will drop their seeds, and thereby maintain a succession of plants without any care, and on these places they appear very beautiful.

The seventh sort grows naturally at the Cape of Good Hope: this rises with a shrubby stalk two or three feet high, dividing upward into several branches, which are garnished with oval leaves placed alternate, whose base sits close to the stalk; they are hairy, and of a light green colour; the flowers are produced singly between the leaves at the end of the branches; they are of a purple colour, and in shape much like those of the fifth sort.

It is propagated by seeds, which should be sown in pots soon after they are ripe, and may be exposed to the open air till the beginning of October, when the pots should be placed under a frame to guard them from frost; but in mild weather they should have the free air, to prevent the

seeds

ſeeds from vegetating till the winter is paſt; for if the plants come up at that ſeaſon, their ſtems will be weak and full of juice, and very liable to rot with damps; therefore it is much better if the plants do not come up till toward March, which is the uſual time of their appearing, when the ſeeds are not forced by warmth. When the plants are fit to remove, they ſhould be each planted into a ſmall pot, and placed under a frame to forward their putting out new roots, then they ſhould be gradually inured to bear the open air, and the latter end of May they ſhould be placed abroad in a ſheltered ſituation, where they may remain till the beginning of October, at which time they muſt be removed into an airy glaſs-caſe, where they may enjoy the ſun, and have free air in mild weather. During the winter ſeaſon theſe plants muſt be ſparingly watered, for as their ſtems are ſucculent, ſo much moiſture will cauſe them to rot. In the ſummer they ſhould be ſet abroad in a ſheltered ſituation, and treated in the ſame manner as other plants from the ſame country.

EDERA QUINQUEFOLIA. See VITIS.

EDGINGS. The beſt and moſt durable plant for Edgings in a garden is Box, which, if well planted and rightly managed, will continue in beauty ſeveral years: the beſt ſeaſon for planting this, is either in the autumn or very early in the ſpring; for if you plant it late, and the ſeaſon ſhould prove hot and dry, it will be very ſubject to miſcarry, unleſs great care be taken to ſupply it with water; the beſt ſort for this purpoſe is the dwarf Dutch Box.

EHRETIA. Trew. Tab. 25. Baſtard Cherry-tree.

The CHARACTERS are,

The flower has one bell-ſhaped petal, cut into ſix ſegments, which are reflexed, and ſix ſtamina which are longer than the petal; it has a roundiſh germen, which afterward becomes a ſucculent berry with two cells, including two ſtones, each having two kernels.

The SPECIES are,

1. EHRETIA (*Tinifolia*) foliis alternis oblongis acuminatis, ſpicâ florum ſparſâ petalis reflexis albis. Trew. Tab. 25. *It is called Baſtard Cherry-tree in the Weſt-Indies.*

2. EHRETIA (*Bourreria*) foliis ovatis integerrimis lævibus, floribus ſubcorymboſis, calycibus glabris. Lin. Sp. Plant. 275. *Baſtard Cherry with entire ſmooth leaves, and flowers in a corymbus.*

The firſt ſort hath a ſtrong woody trunk, which in the Weſt-Indies grows to the ſize of a middling Pear-tree, covered with a gray furrowed bark, divided into many branches, garniſhed with oblong, acute-pointed, ſmooth leaves, five or ſix inches long, and two inches and a half broad, of a dark green colour on their upper ſide, placed alternate, having ſhort foot-ſtalks: the flowers are produced in panicles at the end of the branches; they are ſmall, white, and of one petal, having a bell-ſhaped tube, but cut into ſix ſegments to the middle, which are reflexed; theſe are ſucceeded by ſmall, oval, ſucculent berries, containing one or two ſtones; it grows in moſt of the iſlands in the Weſt-Indies.

This is uſually propagated by ſeeds when they can be obtained, which ſhould be ſown in pots and plunged into a hot-bed of tan; when the plants come up they may be treated in the ſame way as the MALPIGHIA, to which article the reader is deſired to turn to avoid repetition; it may alſo be increaſed by layers. This plant has produced its flowers ſeveral times in the Chelſea garden, but it has not perfected its fruit here.

The title of this genus was given to it by the learned Dr. Trew of Nuremberg, in honour of Mr. George Denis Ehret, a curious botaniſt, who ſent a drawing of the plant to the Doctor, taken from one of the plants that flowered in the Chelſea garden, from which he has publiſhed a curious print in his twenty-fifth table.

The ſecond ſort came from Surinam: this hath an upright woody ſtem covered with a brown bark, ſending out branches regularly toward the top, which are garniſhed with oval ſmooth leaves placed alternate; theſe are ſix inches long and two broad, ending with blunt oval points. As this tree has not produced flowers in England, ſo I can give no farther account of it.

This plant may be propagated by ſeeds as the former ſort, and the plants may be treated in the ſame manner as is directed for that; for both of them are too tender to live through the winter in England, unleſs they are placed in a ſtove.

ELÆAGNUS. Lin. Gen. Plant. 148. Oleaſter, or wild Olive.

The CHARACTERS are,

The flower hath a bell-ſhaped empalement of one leaf, which is quadrifid, rough on the outſide, but coloured within; it hath no petals, but four ſhort ſtamina which are inſerted in the diviſions of the empalement: at the bottom is ſituated a roundiſh germen, which afterward becomes an obtuſe oval fruit, with a puncture at the top, incloſing one obtuſe nut.

The SPECIES are,

1. ELÆAGNUS (*Spinoſa*) aculeatus, foliis lanceolatis. *Prickly wild Olive with ſpear-ſhaped leaves; or Eaſtern broad-leaved wild Olive with a large fruit.*

2. ELÆAGNUS (*Inermis*) inermis, foliis lineari-lanceolatis. *Wild Olive without thorns, and narrow ſpear-ſhaped leaves.*

3. ELÆAGNUS (*Latifolia*) foliis ovatis. Prod. Leyd. 250. *Wild Olive with oval leaves.*

The firſt and ſecond ſorts Dr. Tournefort found growing naturally in the Levant; the firſt I take to be the common ſort, which grows naturally in Bohemia, of which I ſaw ſome trees growing in the curious garden of the late Dr. Boerhaave, near Leyden in Holland. The leaves of this ſort are not more than two inches long, and about three quarters of an inch broad in the middle; they are of a ſilver colour, placed alternate; at the foot-ſtalk of every leaf there comes out a pretty long ſharp thorn, which are alternately longer: the flowers are ſmall, the inſide of the empalement is yellow, and they have a ſtrong ſcent when fully open.

The ſecond ſort hath no thorns on the branches; the leaves are more than three inches long, and half an inch broad, and have a ſhining appearance like ſattin. The flowers come out at the foot-ſtalks of the leaves, ſometimes ſingly, at other times two, and frequently three at the ſame place; the outſide of the empalement is ſilvery and ſtudded, the inſide of a pale yellow, having a very ſtrong ſcent; this flowers in July, and ſometimes the flowers are ſucceeded by fruit. This is the ſort which is moſt commonly preſerved in the Engliſh gardens.

Theſe plants may be propagated by laying down the young ſhoots in autumn, which will take root in one year, when they may be cut off from the old trees, and either tranſplanted into a nurſery for two or three years to be trained up, or into the places where they are to remain. The beſt ſeaſon for tranſplanting of theſe trees is in the beginning of March, or early in the autumn, provided the roots are mulched, to protect them from ſevere froſt in winter; they ſhould be placed where they may be ſcreened from ſtrong winds, for they grow very freely, and are very ſubject to be ſplit down by the wind if they are too much expoſed.

The third ſort grows naturally in Ceylon, and in ſome other parts of India. This is pretty rare at preſent in the Engliſh gardens, but ſome years paſt there were ſeveral pretty large plants of it growing in the garden at Hampton-court: this in England riſes with a woody ſtem to the height of eight or nine feet, dividing into many crooked branches, garnifhed with oval filvery leaves, which have feveral irregular fpots of a dark colour on their furface; they are placed alternately on the branches, and continue all the year. The flowers I have not feen, though fome of the trees at Hampton-court produced flowers, but I was not fo lucky as to fee them.

This fort requires a warm ftove to preferve it in this country, for it is too tender to live in the open air, excepting for a fhort time in the warmeft part of fummer.

ELATERIUM. See MOMORDICA.

ELATINE. See LINARIA.

ELECAMPANE. See INULA.

ELEPHANTOPUS. Lin. Gen. Plant. 827. Elephant's Foot.

The CHARACTERS are,

The flowers are collected in one common involucrum; each contains four or five florets which are tubulous, and divided into five equal parts; they have five very fhort hairy ftamina: in the bottom is fituated an oval germen, which afterward becomes a fingle compreffed feed crowned with briftles, and fitting on a naked placenta.

The SPECIES are,

1. ELEPHANTOPUS (*Scaber*) foliis oblongis fcabris. Hort. Cliff. 390. *Elephantopus with oblong rough leaves.*

2. ELEPHANTOPUS (*Tomentofis*) foliis ovatis tomentofis. Gron. Virg. 90. *Elephantopus with oval woolly leaves.*

The firft fort grows naturally in both the Indies; this fends out many oblong rough leaves, which fpread near the ground; between thefe, in the fpring, arifes a branching ftalk little more than a foot high; the fide branches are fhort, and are generally terminated by two heads of flowers, each ftanding upon a fhort foot-ftalk; the heads contain a great number of hermaphrodite florets, included in a common involucrum, compofed of four oval leaves, ending in acute points; the florets are of a pale purple colour, but it rarely produces feeds in England.

The fecond fort grows naturally in South Carolina; the plants of this have frequently come up in the earth, which has been fent from thence with other plants; this hath feveral oval woolly leaves growing from the root, which have many tranfverfe nerves running from the midrib to the fides; they fpread flat on the ground, and between thefe arife a ftiff ftalk about a foot high, which divides into many branches, each being terminated by two flowers, which are compofed of feveral florets inclofed in a four-leaved involucrum; two of thefe leaves are alternately larger than the other: the involucrum is longer than the florets, fo they do but juft appear within the two larger leaves; the flowers make no appearance; they appear in July, but the feeds never ripen in this country.

The firft fort hath a perennial root, but an annual ftalk. If this is planted in pots, and fheltered in the winter from froft, it may be preferved feveral years, and the plants will annually flower; but the fecond fort feldom continues longer than two years.

They are propagated by feeds, which fhould be fown on a hot-bed in the fpring; and when the plants are come up they muft be tranfplanted into pots, and plunged into a hot-bed of tanners bark, obferving to water and fhade them until they have taken root; then they fhould have a large fhare of frefh air in warm weather, and muft be frequently refrefhed with water.

ELICHRYSUM. See GNAPHALIUM.

ELLISIA. Brown. Hift. Jam. 262.

The CHARACTERS are,

The flower has a fmall cylindrical empalement of one leaf, indented in five parts at the brim; it has one tubulous petal, whofe brim is cut into five fegments, which fpread open; it has four ftamina, two of which are longer than the other, and a roundifh germen, which becomes a roundifh berry crowned by the empalement, inclofing eight nuts, which have two cells, with one angular feed in each.

We have but one SPECIES of this genus, viz.

ELLISIA.

This grows naturally in Jamaica, and fome of the other iflands in the Weft-Indies; it has a fhrubby ftalk, which divides into many flender, angular, ligneous branches, garnifhed with oval leaves which ftand oppofite, fome of which are obtufe, and others end in acute points; they are fawed on their edges toward the top, and are of a light green colour when they are in the ftove; but in fummer when they are expofed to the open air, they change black, and contiuue fo till after they have been fome time in the ftove again. The flowers (according to Dr. Brown's figure) are difpofed in loofe fpikes, which come out from the wings, and alfo at the end of the branches; they are white and of the ringent kind: the branches are armed at each joint with two flender erect thorns, which are fituated oppofite immediately above the leaves, and are unequal, one being longer than the other.

This plant is propagated by cuttings, which put out roots with great facility in any of the fummer months; if they are planted in pots and plunged into a moderate hot-bed of tan, and fhaded from the fun, they will foon take root; when they begin to fhoot they fhould have free air admitted to them, and gradually hardened; then they fhould be carefully feparated, and each planted in a fmall pot, placing them under cover till they have taken new root; when they may be removed into a warm fheltered fituation, where they may remain till autumn; then they muft be removed into a ftove, and during the winter feafon fhould have a temperate heat, in which they will thrive beft; for when they are too tenderly treated their fhoots are weak, and fubject to be attacked by vermin, nor will the plants live through the winter in a green-houfe. In fummer they fhould be placed in the open air in a warm fheltered fituation; with this management the plants have fucceeded beft.

ELM. See ULMUS.

EMERUS. Tourn Inft. R. H. 650. Scorpion Sena, vulgo.

The CHARACTERS are,

The flower hath an empalement of one leaf, divided into five parts; the flower is of the butterfly kind; the ftandard is narrow, and fhorter than the wings, over which it is arched; the wings are large and concave; the keel is heart-fhaped and reflexed; it hath ten ftamina; in the empalement is fituated an oblong flender germen, which afterward becomes a taper cylindrical pod, fwelling in thofe parts where the feeds are lodged.

The SPECIES are,

1. EMERUS (*Major*) caule fruticofo, pedunculis longioribus. *Scorpion Sena with a fhrubby ftalk, and longer foot-ftalks to the flowers.*

2. EMERUS (*Minor*) foliolis obcordatis, pedunculis brevioribus, caule fruticofo. *Scorpion Sena with long heart-fhaped leaves, fhorter foot-ftalks to the flowers, and a fhrubby ftalk.*

3. EMERUS (*Herbacea*) caule erecto, herbaceo, foliolis multijugatis, floribus fingularibus, filiquis longiffimis erectis. *Scorpion Sena with an erect herbaceous ftalk, the leaves compofed of many pair of lobes, fingle flowers proceeding from the fides of the ftalks, and very long erect pods.*

The firft of thefe fhrubs is very common in all the nurferies near London; this rifes with weak fhrubby ftalks to the height of eight or nine feet, dividing into many flender branches, garnifhed with winged leaves, compofed of three pair of lobes, terminated by an odd one. The flowers come out upon long foot-ftalks from the fide of the branches,

branches, two or three of these foot-ſtalks arising from the ſame point, each ſuſtaining two, three, or four yellow butterfly flowers, which are ſucceeded by long ſlender pods, ſwelling in thoſe parts where the ſeeds are lodged; theſe ſhrubs continue long in flower, eſpecially in cool ſeaſons, and frequently flower again in autumn, which renders them valuable.

The ſecond ſort riſes with many ſhrubby ſtalks like the firſt, but not more than half the height; this hath larger leaves, which are of an oblong heart-ſhape. The flowers are rather larger than thoſe of the firſt, and ſtand upon ſhorter foot-ſtalks; theſe differences hold in the plants which are raiſed from ſeeds, therefore I think they may be allowed to ſtand as diſtinct ſpecies, though there is a great likeneſs at firſt ſight in them.

Theſe ſhrubs are eaſily propagated by laying down their tender branches, which will take root in one year, and may then be tranſplanted into a nurſery, and managed in the ſame manner as other flowering ſhrubs.

The third ſort grows naturally in the Weſt-Indies, where Plumier firſt diſcovered it in the French ſettlements, but has ſince been found growing in plenty at La Vera Cruz in New Spain, by the late Dr. Houſtoun. This plant is annual; it riſes with a round herbaceous ſtalk three feet high, garniſhed at each joint with one long winged leaf, compoſed of about twenty pair of lobes, terminated by an odd one; the flowers come out ſingly from the ſide of the ſtalk, immediately above the foot-ſtalks of the leaves, ſtanding upon ſlender foot-ſtalks; they are larger than thoſe of either of the former ſorts, and are of a pale yellow colour; theſe are ſucceeded by ſlender compreſſed pods, which are more than ſix inches long, having a border on each ſide, and ſwelling where each ſeed is lodged.

The ſeeds of this plant muſt be ſown upon a hot-bed in the ſpring, and when the plants are fit to remove, they ſhould be each planted into a ſeparate ſmall pot, and plunged into a moderate hot-bed of tanners bark, ſhading them from the ſun until they have taken new root; then they muſt be treated in the ſame manner as other exotic plants from thoſe warm countries. If the plants are brought forward in the ſpring, and kept under a deep frame in a tan-bed, or plunged into the bark-bed in the ſtove, when they are grown too tall to remain under common frames, they will ripen ſeeds in England, for thoſe ſeeds which I received did not arrive here till May, and yet thoſe plants flowered well in Auguſt; but the autumn proving cold prevented their perfecting ſeeds, and thoſe ſeeds which I reſerved till the the next year did not grow.

EMPETRUM. Lin. Gen. Plant. 977. Black-berried Heath.

The Characters are,

It hath male and female flowers on different plants; the male flowers have a three-pointed empalement, and three petals which are narrow at their baſe, and three long hanging ſtamina; the female flowers have the ſame empalement and petals as the male; in the center is ſituated a depreſſed germen, with nine reflexed ſpreading ſtigmas; the germen afterward becomes a depreſſed round berry of one cell, incloſing nine ſeeds placed circularly.

We have but one Species of this genus, viz.

Empetrum (*Nigrum*) procumbens. Hort. Cliff. 470. *Trailing Berry-bearing Heath, Crow Berries, or Crake Berries.*

This little ſhrub grows wild upon the mountains of Staffordſhire, Derbyſhire, and Yorkſhire, and is ſeldom propagated in gardens, unleſs for the ſake of variety; the plants ſhould be procured from the places where they grow naturally, for the ſeeds remain a year in the ground before they vegetate, and afterward are very ſlow in their growth, ſo they are not worth the trouble of cultivating from ſeeds. If the plants are planted on a moiſt boggy ſoil in autumn, they will get roots in the winter, and will require no farther care than to clear them from weeds. As theſe low ſhrubs commonly grow upon the tops of wild mountains, where the ſoil is generally peaty and full of bogs, ſo the heath-cocks feed much upon the berries of this plant; and wherever there is a plenty of theſe low ſhrubs, there are commonly many of theſe fowls to be found.

ENULA CAMPANA. See Inula.

EPHEDRA. Lin. Gen. Plant. 1007. Shrubby Horſetail, vulgo

The Characters are,

It hath male and female flowers in different plants; the male flowers are collected in ſcaly katkins, under each ſcale is a ſingle flower; they have no petal, but ſeven ſtamina, which are joined in form of a column. The female flowers have a perianthium, compoſed of five ſeries of leaves, which alternately lie over the diviſions of the lower range; they have no petals, but two oval germen, which afterward turn to oval berries, each having two ſeeds.

We know but one Species of this genus, viz.

Ephedra (*Diſtachia*) pedunculis oppoſitis, amentis geminis. Hort. Cliff. 465. *Shrubby Horſe-tail with oppoſite foot-ſtalks, and twin katkins.*

This is a low ſhrubby plant, which grows naturally upon the rocks by the ſea in the ſouth of France, in Spain, and Italy; it is alſo preſerved in ſeveral gardens for the ſake of variety, but has little beauty. This hath a low ſhrubby ſtalk, which puts out a few ſhort branches, riſing about two feet high, which have many protuberant joints, at which come out ſeveral narrow ruſhy leaves like thoſe of the Horſe-tail, which continue green all the year, but the plants rarely flower in gardens.

It may be propagated by offsets, which the plants ſend forth in plenty, for the roots creep under ground, and put up ſuckers, which may be taken off to tranſplant in the ſpring; they love a pretty moiſt ſtrong ſoil, and will endure the cold of our ordinary winters very well in the open air. Some of theſe plants were formerly preſerved in pots, and were houſed in winter; but by later experience, they are found to thrive better in the full ground.

EPHEMERUM. See Tradescantia.

EPIGÆA. Lin. Gen. Plant. 486. Trailing Arbutus.

The Characters are,

The flower hath a double empalement; the outer is compoſed of three, and the inner of one leaf, divided into five parts; the flower is of the ſalver ſhape; it hath ten ſlender ſtamina the length of the tube. In the center is ſituated a globular germen, which afterward becomes a depreſſed five-cornered fruit, with five cells, containing ſeveral ſeeds.

We have but one Species of this genus, viz.

Epigæa (*Repens.*) Lin. Gen. Plant. 486. *Trailing Arbutus.*

This plant grows naturally in North America; it is a low plant, with a trailing ſhrubby ſtalk, which puts out roots at the joints, and when in a proper ſoil and ſituation multiplies pretty faſt; the ſtalks are garniſhed with oblong rough leaves, which are waved on their edges; the flowers are produced at the end of the branches in looſe bunches; theſe are white, and divided at the top into five acute ſegments, which ſpread open in form of a ſtar.

The plants are eaſily propagated by their trailing ſtalks, which put out roots at the joints, ſo may be cut off from the old plant, and placed in a ſhady ſituation and a moiſt ſoil; the beſt time for this is in autumn, that the plants may be well rooted before the ſpring. If the winter ſhould prove ſevere, it will be proper to lay a few dried leaves, or ſome ſuch light covering over them, which will prevent their being injured by froſt; and after they are well rooted,

 they

they will require no other care but to keep them clean from weeds.

EPILOBIUM. Lin. Gen. Plant 426. Willow Herb, or French Willow.

The CHARACTERS are,

The empalement of the flower is four-pointed; the flower hath four petals, and eight stamina, which are alternately shorter; under the flower is situated a long cylindrical germen, which afterward becomes a cylindrical furrowed capsule with five cells, filled with oblong seeds, crowned with down.

The SPECIES are,

1. EPILOBIUM (*Angustifolium*) foliis lanceolatis integerrimis. Lin. Hort. Cliff. 157. *Common broad-leaved Willow Herb, or French Willow.*

2. EPILOBIUM (*Hirsutum*) foliis oppositis lanceolatis serratis. Lin. Hort. Cliff. 145. *Hoary Willow Herb with a large flower; commonly called Codlins and Cream.*

There are several other species of this genus, some of which grow naturally in shady woods and moist places in most parts of England, where they are often very troublesome weeds, therefore are seldom admitted into gardens, so I shall not trouble the reader with their distinctions.

The first sort here mentioned, was formerly planted in gardens for the beauty of its flowers; but as it usually spreads far by the creeping roots, whereby it over-runs all the neighbouring plants, it has been cast out of most gardens; but it deserves to have room in some low moist places, or in great shade, where it will make a good appearance when it is in flower, and these flowers are very proper to cut for basons to adorn chimnies in the summer season. This usually grows about four feet high, with slender stiff branches, which are beset with leaves, resembling those of the Willow, from whence it had the name of Willow Herb, or French Willow. On the upper part of the stalks the flowers are produced in long spikes, which are of a fine Peach colour, and if the season is not very hot they will continue near a month in beauty: this sort is found wild in divers part of England; it is a great creeper at the root, so may be easily propagated.

There is a variety of this with white flowers, which is planted in gardens, but differs from it only in the colour of the flower; however some persons are fond of propagating these varieties, for which reason I mentioned it here.

The second sort is found wild by the side of ditches and rivers in many parts of England; this plant grows about three feet high, and produces its flowers on the top of the stalks, but these are much less beautiful than those of the first; and the plant being a great rambler at the root, is never admitted into gardens. The leaves of this plant being rubbed, emit a scent like scalded Apples, from whence some have given the name of Codlins and Cream to this plant.

EPIMEDIUM. Lin. Gen. Plant. 138. Barrenwort.

The CHARACTERS are,

The flower hath a three-leaved empalement, which falls off; it hath four obtuse petals and four nectariums, which are cup-shaped, and as large as the petals, with four stamina, and an oblong germen, which afterward becomes an oblong pod with one cell, inclosing many oblong seeds.

We have but one SPECIES of this genus, viz.

EPIMEDIUM (*Alpinum.*) *Alpine Barrenwort.*

This plant hath a creeping root, from which arise many stalks about nine inches high, divided at the top into three, each of which is again divided into three smaller; upon each of these stands a stiff heart-shaped leaf, ending in a point, of a pale green on the upper side, but gray on the under. A little below the first division of the stalk comes out the foot-stalk of the flower, which is near six inches long, dividing into smaller, each of these sustaining three flowers: these are composed of four leaves placed in form of a cross; they are of a reddish colour, with yellow stripes on the border. In the center of the flower arises the style, situated upon the germen, which afterward turns to a slender pod, containing many oblong seeds, which seldom ripen with us. The roots, if planted in a good border, should be every year reduced, so as to keep them within bounds, otherwise it will spread its roots and interfere with the neighbouring plants.

EQUISETUM. Horse-tail.

There are several species of this plant, which are found in England on the sides of ditches, or in shady woods; but as they are plants which are never cultivated in gardens, I shall pass them over.

ERANTHEMUM. See ADONIS.

ERICA. Lin. Gen. Plant. 435. Heath.

The CHARACTERS are,

The flower hath a coloured empalement of four leaves, and one swelling petal, which is quadrifid, with eight stamina fixed to the receptacle; in the bottom is situated the germen, which afterward becomes a round capsule having four cells, which are filled with small seeds.

The SPECIES are,

1. ERICA (*Vulgaris*) antheris bicornibus inclusis, corollis inæqualibus, campanulatis mediocribus foliis oppositis sagittatis. Lin. Sp. Pl. 352. *Common smooth Heath.*

2. ERICA (*Herbacea*) antheris bifidis simplicibus exsertis, corollis campanulatis longioribus, foliis quinis linearibus patentibus. Lin. Sp. Pl. 355. *Pine-leaved Heath with many flowers.*

3. ERICA (*Cinerea*) antheris bicornibus inclusis, corollis ovatis racemosis, foliis ternis glabris linearibus. Lin. Sp. Plant. 352. *Dwarf Heath with an Ash-coloured bark, and Strawberry-tree flower.*

4. ERICA (*Ciliaris*) antheris simplicibus inclusis, corollis ovatis irregularibus, floribus terno racemosis, foliis ternis ciliatis. Lœfl. Epist. 2. p. 9. Lin Sp. Pl. 354. *Heath with single summits, oval irregular petals, triple branching flowers, and hairy leaves placed by threes.*

5. ERICA (*Africana*) antheris bifidis exsertis, corollis globosis mediocribus, pedunculis triphyllis foliis quaternis. Lin. Sp. Pl. 355. *Shrubby African Heath.*

The four first sorts grow wild upon barren uncultivated places in divers parts of England; but notwithstanding their commonness, yet they deserve a place in small quarters of humble flowering shrubs, where, by the beauty and long continuance of their flowers, together with the diversity of their leaves, they make an agreeable variety.

These are seldom propagated in gardens, and so not to be had from the nurseries, but may be taken up with a ball of earth to their roots, from the natural places of their growth in autumn, and may be transplanted into the garden. The soil where they are planted should not be dunged, nor should they have any other culture than clearing them from weeds; for the less the ground is dug, the better these will thrive.

The fifth sort grows naturally at the Cape of Good Hope, and in Portugal: it hath a shrubby stalk, which rises four or five feet high, spreading into many branches, garnished with narrow smooth leaves: the flowers come out at the end of the shoots, they are of a bright purple colour, but are not succeeded by seeds in England.

This sort is preserved in some curious gardens, but is very difficult to increase; it requires protection from hard frost, but in mild winters will live abroad in a warm situation.

ERICA BACCIFERA. See EMPETRUM.

ERIGERON. Lin. Gen. Plant. 855. Groundsel.

The

The CHARACTERS are,

It hath a compound flower, composed of many hermaphrodite florets which form the disk, and female half florets which make the rays, contained in one oblong scaly empalement. The hermaphrodite florets are funnel-shaped, and have five stamina, and a small germen crowned with down, which afterward becomes a small oblong seed crowned with long down, sitting on a naked receptacle.

The SPECIES are,

1. ERIGERON (*Viscosum*) pedunculis unifloris lateralibus calycibus squarrosis. Hort. Upsal. 258. *Male Fleabane of Theophrastus, and greater Fleabane of Dioscorides.*

2. ERIGERON (*Acre*) pedunculis alternis, unifloris. Hort. Cliff. 407. *Blue acrid Fleabane.*

3. ERIGERON (*Bonariense*) foliis basi revolutis. Lin. Sp. Pl. 863. *Purplish Groundsel of Buenos Ayres, with under leaves like Hartshorn Plantain.*

4. ERIGERON (*Canadense*) caule floribusque paniculatis. Hort. Cliff. 407. *Annual Virginia Golden Rod.*

5. ERIGERON (*Alpinum*) caule subbifloro, calyce subhirsuto. Lin. Sp. Pl. 864. *Blue Alpine Fleabane.*

The first sort grows naturally in the south of France, and in Italy; it hath a perennial root, from which arise several upright stalks near three feet high, garnished with oblong, oval, hairy leaves, sitting close to the stalk; they are placed alternate; these in warm weather sweat out a clammy juice; the flowers are produced single upon pretty long foot-stalks, some arising from the side of the stalk, and others terminate it; they are yellow, and have an agreeable odour; they flower in July, and the seeds ripen in autumn.

This plant is propagated by seeds, which, if sown in autumn, will more certainly succeed than those which are sown in the spring; it delights in a dry soil and a sunny exposure. The second year the plants will flower and perfect their seeds, but the roots will continue some years, and annually produce their flowers and seeds.

The other four sorts are preserved in botanic gardens for the sake of variety, but are seldom admitted into gardens for pleasure. The fifth sort is a perennial plant, which grows naturally on the Alps, and may be propagated by seeds in the same manner as the first sort, but should have a shady situation and a moist soil.

The others are annual plants, which, if once admitted into a garden, and suffered to scatter their seeds, will become very troublesome weeds there.

ERINUS. Lin. Gen. Plant. 689.

The CHARACTERS are,

The flower hath a five-leaved empalement, and one tubulous petal, which is of the ringent kind, cut into five equal segments, and is short and reflexed; it hath four stamina situated within the tube, two of which are a little longer than the other; in the bottom of the tube is situated the oval germen, which afterward becomes an oval capsule with two cells, filled with small seeds.

The SPECIES are,

1. ERINUS (*Alpinus*) floribus racemosis. Lin. Sp. Plant. 630. *Erinus with branching flowers.*

2. ERINUS (*Tomentosus*) tomentosus, caulibus procumbentibus, floribus sessilibus axillaribus. *Woolly Erinus with trailing stalks, and flowers sitting close to their sides.*

The first sort grows naturally upon the Alps and Helvetian mountains; it is a very low plant, whose leaves lie close to the ground, growing in close tufts; they are about half an inch long, and one-eighth of an inch broad, sawed on their edges; between these arise the flower-stalk, which is scarce two inches high, supporting a loose bunch of purple flowers, which stand erect.

It is propagated by parting of the roots; the best time for this is in autumn; it must have a shady situation, and a loamy soil without dung, for in rich earth these plants are very subject to rot.

The second sort was sent me by the late Dr. Houstoun from La Vera Cruz, where he found it growing naturally; this sends out several trailing stalks about six inches long, which are closely garnished with small oval leaves placed on every side; they are very white and woolly, and at the joints, just above the leaves, come out the flowers, sitting very close to the stalks; they are white, and are succeeded by round capsules, having two cells filled with small seeds; this plant has great resemblance at a distance to the Sea Cudweed.

This sort is annual, so is propagated by seeds, which should be sown in pots, and plunged into a moderate hot-bed, where sometimes the plants will come up in five or six weeks, and at other times the seeds do not vegetate till the following spring; this happens frequently, when the seeds have been kept long after they were gathered. When the plants are fit to remove, they should be each planted in a separate small pot, and plunged into a hot-bed of tanners bark; when they have taken new root, they should be treated in the same way as other plants from warm countries, by admitting proper air to them at all times when the weather is warm, and frequently refreshing them with water; if the plants are brought forward early in the spring, they will perfect their seeds, otherwise the winter will come on before the seeds ripen.

ERIOCEPHALUS. Dill. Hort. Elth. 110. Lin. Gen. Plant. 890.

The CHARACTERS are,

It hath a radiated flower, composed of female half florets which form the rays, and hermaphrodite florets which form the disk; the hermaphrodite flowers are funnel-shaped and cut into five parts at the brim, and have five stamina, with a naked germen; the female florets have their petals stretched out on one side like a tongue, divided at the end into three small lobes; they have no stamina, but an oval naked germen; these have one naked seed, sitting on the naked plain receptacle.

We have but one SPECIES of this genus, viz.

ERIOCEPHALUS (*Africanus.*) Lin. Sp. Plant. 926. We have no proper title for this in English.

This plant hath a shrubby stalk, which rises from six to eight feet high, putting out many side branches, closely garnished with hairy leaves, which come out in clusters, and are divided into three or five parts, which spread open like a hand; they have a strong smell when bruised, approaching to that of the Lavender Cotton, but not quite so rank; the flowers are produced in small clusters at the extremity of the branches, standing erect, and are tubulous; the female florets which compose the rays form a hollow, in the middle of which the hermaphrodite flowers are situated which form the disk; but the plants seldom flower in this country.

It is propagated by cuttings, which may be planted any time from May to the middle of August, that there may be time for them to get good root before the winter; they should be shaded from the sun till they have taken root; then they should be removed into the open air, and placed in a sheltered situation, where they may remain till October, when they must be removed into shelter, where they may be secured from frost: in the summer, when the plants are placed in the open air, they will require to be frequently refreshed with water in hot weather.

These plants retain their leaves all the year, so they add to the variety of exotics in the winter season.

ERUCA. Tourn. Inst. R. H. 226. tab. 111. Rocket.

The CHARACTERS are,

The flower has a four-leaved empalement; it hath four oblong petals, placed in form of a cross; it hath six stamina, four of which are a little longer than the empalement, the other two are shorter; it hath an oblong taper germen, which afterward becomes a taper-corner'd pod with two cells, filled with roundish seeds.

The Species are,

1. Eruca (*Sativa*) foliis pinnato-laciniatis, laciniis exterioribus majoribus. *Greater Garden annual Rocket, with a white striped flower.*

2. Eruca (*Bellidifolia*) foliis lanceolatis, pinnato-dentatis, caule nudo simplici. *Rocket with a Daisy leaf.*

3. Eruca (*Perennis*) foliis pinnatis glabris, caule ramoso, floribus terminalibus. *Narrow-leaved perennial Rocket with a yellow flower.*

4. Eruca (*Aspera*) foliis dentato-pinnatifidis hirsutis, caule hispido, siliquis lævibus. *Greater wild Saffron-coloured Rocket, with a rough stalk.*

5. Eruca (*Tanacetifolia*) foliis pinnatis, foliolis lanceolatis pinnatifidis. Prod. Leyd. 342. *Rocket with a Tansy leaf.*

6. Eruca (*Viminea*) foliis sinuato pinnatis, sessilibus, caule ramoso. *Sicilian Rocket with a Shepherd's Purse leaf.*

The first sort is an annual plant, which was formerly much cultivated in the gardens as a sallad herb, but at present is little known by the gardeners, for it has been long rejected on the account of its strong ungrateful smell. It stands in the list of medicinal plants, but is seldom used, though it is reckoned a provocative and a good diuretic: when it is propagatd for sallads the seeds should be sown in drills, in the same manner as is practised for other small sallad herbs; for it must be eaten young, otherwise it will be too strong for most palates. The winter and spring seasons are the times when this herb is usually eaten, for when it is sown in the summer the plants soon run up to seed, and are then too rank for use.

The second sort grows naturally in the south of France and Italy, where it is often eaten as a sallad herb. This hath many spear-shaped leaves arising from the root; the stalks are single, naked, and rise about a foot high; the flowers grow in loose bunches on the top of the stalks, which are succeeded by pods having two cells, filled with small round seeds; this is an annual plant, which may be propagated by seeds in the same manner as the former.

The third sort grows naturally about Paris, and in many other parts of Europe; the leaves of this are narrow, and regularly divided in form of a winged leaf, the stalks branch out upward, and are terminated by loose spikes of yellow flowers. This hath a perennial root and an annual stalk.

The fourth sort grows naturally upon old walls and buildings in many parts of England, where it continues flowering all the summer, but is rarely admitted into gardens. It is sometimes used in medicine, for which reason I have here mentioned it.

The fifth sort grows naturally about Turin. This hath fine divided leaves, somewhat like those of Tansy, but are of a hoary green colour; the stalks rise a foot and a half high, garnished with leaves of the same form, but gradually diminish in their size upward; the flowers are produced in clusters at the top of the stalks; they are small, and of a pale yellow colour, and are succeeded by slender taper pods, which contain two rows of small round seeds.

The sixth sort grows naturally in Italy and Spain; it is an annual plant, with many oblong leaves, which are smooth, and regularly sinuated on their sides in form of a winged leaf, of a light green, having a hot biting taste the stalks rise a foot high, they are strong, and divide into several branches, garnished with a single leaf at each joint, shaped like those below, but smaller. The flowers are produced in loose clusters at the end of the branches; they are white, and near as large as those of the Garden Rocket, and are succeeded by taper pods, containing two rows of round seeds.

These plants are preserved in botanic gardens for the sake of variety, so are here mentioned, that those who are inclined to cultivate them as such, may do it by sowing their seeds in an open situation. When the plants come up, they will require no other culture but to thin them, and keep them clear from weeds. They flower in June and July, and their seeds ripen in August and September.

ERUCAGO. See Bunias.

ERVUM. Lin. Gen. Pl. 784. Bitter Vetch.

The Characters are,

The flower is of the butterfly kind, having a large, roundish, plain standard, two obtuse wings half the length of the standard, and a shorter keel which is pointed, with ten stamina, nine joined, and one standing separate, terminated by single summits; and an oblong germen, which afterward becomes an oblong taper pod, jointed between each seed.

The Species are,

1. Ervum (*Ervilia*) germinibus undato plicatis. Hort. Upsal. 224. *The true Bitter Vetch.*

2. Ervum (*Lens*) seminibus compressis convexis. Lin. Sp. Pl. 738. *Common Lentils.*

3. Ervum (*Monanthos*) pedunculis unifloris. Lin. Sp. Pl. 738. *Lentil with one flower upon each foot-stalk.*

4. Ervum (*Tetraspermum*) pedunculis subbifloris, seminibus globosis quaternis. Flor. Suec. 606. *Corn Vetch with single smooth pods.*

5. Ervum (*Hirsutum*) pedunculis multifloris, seminibus globosis binis. Lin. Sp. Pl. 738. *Corn Vetch having many hairy pods.*

The first sort grows naturally in Italy in Spain: it is an annual plant, which rises with angular weak stalks a foot and a half high, garnished at each joint with one winged leaf, composed of fourteen or fifteen pair of lobes, very like those of the Vetch, but narrower; the flowers come out from the side of the stalks on foot-stalks, each sustaining two pale coloured flowers, which are succeeded by short pods, a little compressed, containing three or four round seeds; the pods swell at the place where each seed is lodged, so that it is called a jointed pod by many.

The second sort is the common Lentil, which is cultivated in many parts of England, either as fodder for cattle, or for the seeds, which are frequently used for meagre soups. This is also an annual plant. It rises with weak stalks a foot and a half high, garnished with winged leaves at each joint, composed of several pair of narrow lobes, terminated by a tendril or clasper, which fastens to any neighbouring plant, and is thereby supported; the flowers come out three or four together upon short foot-stalks, from the side of the branches; they are small, of a pale purple colour, and are succeeded by short flat pods, containing two or three seeds, which are flat, round, and a little convex in the middle. The seeds of this plant are commonly sown in March, where the land is dry, but in moist ground the best time is in April. The usual quantity of seed allowed for an acre of land, is from one bushel and a half to two bushels. If these are sown in drills in the same manner as Pease, they will succeed better than when they are sown in broad-cast: the drills should be a foot and a half asunder, to allow room for the Dutch hoe to clean the ground between them; for if the weeds are permitted to grow among them, they will get above the Lentils and starve them.

There is another sort of Lentil which has been cultivated of late years in England, by the title of French Lentil. This is the Lens major of Caspar Bauhin, and is undoubtedly a different species from the common, being twice the size, both in plant and seed, and constantly produces the same fame from feeds, though they do not differ much in their characters; but this is much better worth cultivating than the other. This pulfe is frequently called Tills in many parts of England.

The third fort is very like the common Lentil, but differs from it, in having but one flower on each foot-ftalk, whereas the other has three or four, but in other refpects is the fame, fo may be cultivated in the fame manner.

The fourth and fifth forts are fmall annual Vetches, which grow naturally among the Wheat and Rye in many parts of England, fo are not admitted into gardens; they are only mentioned here as weeds, which may be eafily rooted out of the fields, if they are cut up when they begin to flower, and not permitted to ripen their feeds; for as they have annual roots, fo if they do not fcatter their feeds, they may be foon deftroyed.

ERVUM ORIENTALE. See Sophora.

ERYNGIUM. Lin. Gen. Plant. 287. Sea Holly, or Eryngo.

The Characters are,

It hath many fmall flowers fitting upon one common conical receptacle; the flowers have a five-leaved erect empalement fitting upon the germen, and form a roundifh general umbel, which is uniform; each flower has five oblong petals, and five ftamina, ftanding above the flowers. Under the empalement is fituated a prickly germen, which afterward becomes an oval fruit divided in two parts, each having one oblong taper feed.

The Species are,

1. Eryngium (*Maritimum*) foliis radicalibus fubrotundis plicatis fpinofis, capitulis pedunculatis. Hort. Cliff. 87. *Common Sea Holly, or Eryngo.*

2. Eryngium (*Campeftre*) foliis radicalibus pinnatis tripartitis. Hort. Cliff. 87. *Common wild Eryngo.*

3. Eryngium (*Planum*) foliis radicalibus ovalibus planis crenatis, capitulis pedunculatis. Hort. Cliff. 87. *Broad-leaved plain Eryngo.*

4. Eryngium (*Amethyftinum*) foliis radicalibus digitato-multifidis. Lin. Sp. Pl. 233. *Purple, Violet-coloured, Mountain Eryngo.*

5. Eryngium (*Pallefcens*) foliis radicalibus rotundato-multifidis, capitulis pedunculatis. *Alpine Eryngo with a large pale-coloured head.*

6. Eryngium (*Orientale*) foliis radicalibus pinnatis, ferrato-fpinofis, foliolis trifidis. *Oriental Eryngo with trifid leaves.*

7. Eryngium (*Aquaticum*) foliis gladiatis ferrato-fpinofis, floralibus indivifis. Lin. Sp. Pl. 232. *American Sea Holly with leaves like the Aloe, lightly fawed; commonly called Rattlefnake Weed in America.*

8. Eryngium (*Pufillum*) foliis radicalibus oblongis incifis, caule dichotomo, capitulis feffilibus. Hort. Cliff. 87. *Leffer plain Eryngo.*

9. Eryngium (*Alpinum*) foliis radicalibus cordatis oblongis, caulinis pinnatifidis, capitulo fubcylindrico. Lin. Sp. Pl. 233. *Blue Alpine Eryngo with heads like the Teazel.*

10. Eryginum (*Fœtidum*) foliis gladiatis ferrato-fpinofis, floralibus multifidis. Lin. Sp. Plant. 232. *Stinking Eryngo, having narrow fawed leaves; commonly called Fever-weed.*

The firft of thefe fpecies grow in great plenty on the gravelly fhores in divers parts of England, the roots of which are candied, and fent to London for medicinal ufe, and is the true Eryngo. This hath creeping roots, which fpread far in the ground; the leaves are roundifh, ftiff, and of a gray colour, fet with fharp fpines on the edges. The ftalks rife a foot high; they are fmooth, garnifhed at each joint with leaves of the fame form as the lower, but fmaller, and embrace the ftalks with their bafe; at the end of the branches come out the flowers in roundifh prickly heads; under each is fituated a range of narrow, ftiff, prickly leaves, fpreading like the rays of a ftar, the flowers are of a pale blue colour.

This fort will grow in a garden, if the roots are planted in a gravelly foil, and produce their flowers annually; but the roots will not grow near fo large and flefhy as thofe which grow on the fea-fhore, where they are flowed by every tide with falt-water. The beft time to tranfplant the roots is in autumn, when their leaves decay; the young roots are much better to remove than the old. If they are kept clean from weeds, it is all the culture they will require.

The fecond fort grows naturally in feveral parts of England, where it is a very troublefome weed; for the roots fpread greatly in the ground, fo are not eafily deftroyed by the plough, therefore it is not admitted into gardens.

The third fort makes a very pretty appearance when it is in flower, efpecially that with blue ftalks and flowers, for there is a variety of this with white flowers and ftalks, though not fo common. As this doth not fpread at the root, but keeps within bounds, fo it deferves a place in the pleafure-garden. It is propagated by feeds, which, if fown in the autumn, will more certainly fucceed than when it is fown in the fpring, for the latter commonly remains in the ground a year before they vegetate; and if the feeds are fown where the plants are to remain, they will flower ftronger than thofe which are tranfplanted; for as they have long downright roots, fo they are frequently broken in taking out of the ground, which greatly weakens the plants. The culture they require is to thin them where they are too near, keep them clean from weeds, and dig the ground about them every fpring before they fhoot.

The fourth fort grows naturally upon the mountains of Syria, and alfo upon the Apennines. The lower leaves of this fort are divided like the fingers of a hand, into five or fix fegments, which are very much cut at their extremities, and have fmall fpines; the ftalk rifes about two feet high, garnifhed with fmaller and more divided leaves; the upper part of the ftalk, and alfo the head of flowers, are of the fineft amethyft colour, fo that they make a very fine appearance. This is propagated by feeds, in the fame manner as the former fort.

The bottom leaves of the fifth fort are very much divided, and the extremity of the fegments form an oval or circle, ending in fpines; they are of a whitifh gray in the middle, and green on the borders. The ftalks rife about two feet high, garnifhed at the joints with fmaller leaves, which are finely cut: the flowers terminate the ftalk; they are of a light blue colour, and grow in larger heads than either of the former forts. This grows naturally on the Alps; it is a perennial plant, and may be propagated by feeds in the fame manner as the former.

The fixth fort was difcovered by Dr. Tournefort in the Levant. That hath a perennial root; the lower leaves are regularly divided into feven or nine parts to the midrib; thefe fegments are fawed on their edges, and end in fharp thorns. The ftalks rife a foot and a half high, fending out fide branches, garnifhed with ftiff leaves, which are divided into narrower fegments than the lower, terminated by three points. The flowers terminate the ftalks, fitting clofe among the leaves, and are of a fine blue, as are alfo the leaves on the upper part of the ftalks, fo they make a pretty appearance. It is propagated in the fame manner as the three former forts, and the plants require the fame treatment.

The feventh fort grows naturally in Virginia and Carolina, where it is titled Rattlefnake Weed, from its virtues of curing the bite of that venomous reptile. This hath a perennial root, from which arife feveral long leaves, fawed on their

their edges; the leaves are difpofed round the root, after the fame form of the Alpe or Yucca; they are of a grav colour, near a foot long, and one inch and a half broad, ftiff, ending in fpines. The ftalk is ftrong, growing two feet and a half high, dividing upward into feveral foot-ftalks, each being terminated by an oval head of flowers fhaped like thofe of the former forts; they are white, with a little caft of pale blue.

This fort is propagated by feeds, which, if fown in pots and plunged into a moderate hot-bed, the plants will come up much fooner than thofe which are fown in the full ground, whereby the plants will be much ftronger before the winter. When the plants are fit to remove, they fhould be each planted in a feparate fmall pot; and if they are plunged into a moderate hot-bed, it will forward their taking root; then they muft be gradually inured to bear the open air, into which they may be removed toward the latter end of May, and placed among other hardy exotic plants. When the plants have filled thefe pots with their roots, fome of them may be fhaken out, and planted in warm borders; the others may be put into larger pots, and in the autumn placed under a common frame, where they may be expofed to the free air in mild weather, but fheltered from fevere froft: the following fpring thefe may be turned out of the pots, and planted in a warm fituation, where they will endure the cold of our ordinary winters very well; and if in fevere froft they are covered with ftraw, Peafe-haulm, or any fuch light covering, it will fecure them from injury.

The eighth fort grows naturally in Spain and Italy; this puts out oblong plain leaves from the root, which are cut on their edges; the ftalks rife a foot high, branch out into many divifions, regularly by pairs; at each of thefe divifions is fituated a fmall head of flowers, fitting very clofe between the branches. Thefe have no great beauty, fo the plants are feldom cultivated in gardens, except for the fake of variety.

The ninth fort grows naturally on the mountains of Helvetia and Italy. The root is perennial, the lower leaves are oblong, heart fhaped, and plain; the ftalks rife from two to three feet high, branching out on their fide, garnifhed with ftiff leaves, which are deeply divided, ending in many points with fharp fpines; the flowers terminate the ftalks; they are collected into conical heads, and are of a light blue colour, as are alfo the upper part of the ftalks. It is propagated by feeds in the fame manner as the other forts.

The tenth fort grows naturally in the Weft-Indies, where it is much ufed in medicine, being accounted of great fervice in the cure of fevers, from whence it hath the appellation of Feverweed in thofe countries. The roots of this plant are compofed of many fmall fibres, which fpread near the furface; the lower leaves are fix or feven inches long, narrow at their bafe, and enlarge upward to an inch in breadth near the top, where they are rounded off on one fide like a fcymitar; they are finely fawed on their edges, and are of a light green colour; the ftalk rifes about a foot high, fpreading out into many branches, garnifhed with fmall leaves, which end in many points; the flowers are produced in fmall heads which fit clofe at every divifion of the ftalks, and alfo at the end of the branches; they are of a dull white colour, fo make little appearance.

As this plant is biennial, and grows naturally in hot countries, it will not thrive in England but in a warm ftove. It is propagated by feeds, which muft be fown on a hot-bed; and when the plants are fit to remove, they fhould be each planted into a fmall pot, and plunged into the bark-bed, and afterward treated like other tender plants from the fame country; the fecond year they will produce flowers, and frequently ripen feeds, foon after which they commonly decay.

ERYSIMUM. Lin. Gen. Pl. 729. Hedge Muftard.

The CHARACTERS are,

The flower hath four petals placed in form of a crofs and two nectareous glands fituated between the ftamina. It hath fix ftamina, four of which are the length of the empalement, the other two are fhorter. It hath a narrow four-cornered germen, which afterward becomes a long, narrow, four-cornered pod with two cells, filled with fmall round feeds.

The SPECIES are,

1. ERYSIMUM (*Officinale*) filiquis fpicæ adpreffis. Hort. Cliff. 337. *Common Hedge Muftard.*

2. ERYSIMUM (*Barbarea*) foliis lyratis extimo fubrotundo. Flor. Suec. 557. *Winter Crefs with a Rocket leaf and yellow flower.*

3. ERYSIMUM (*Vernum*) foliis radicalibus lyratis, caulinis pinnato-finuatis, floribus laxe fpicatis. *Smaller early Winter Crefs with a fmooth Rocket leaf.*

4. ERYSIMUM (*Orientale*) foliis radicalibus ovatis integerrimis, petiolis decurrentibus, caulinis oblongis dentatis feffilibus. *Oriental Sifymbrium with the appearance of Water Crefs, and a Plantain leaf.*

5. ERYSIMUM (*Alliaria*) foliis cordatis. Hort. Cliff. 338. *Dames Violet fmelling like Garlick; commonly called Alliaria, Sauce alone, or Jack by the Hedge.*

6. ERYSIMUM (*Cheiranthoides*) foliis integris lanceolatis. Flor Lapp. 263. *Cilliflower with a Dames Violet Leaf.*

The firft fort is ufed in medicine; it grows naturally on the fide of foot-paths, and upon old walls in moft parts of England, fo is rarely cultivated in gardens, where, if it was once admitted, it would foon become a troublefome weed.

The fecond and third forts alfo grow naturally on the banks in many parts of England; thefe were formerly eaten in winter fallads, before the Englifh gardens were furnifhed with better plants; fince when they have been rejected, for they have a rank fmell, and are difagreeable to the palate.

The fourth fort is not a native of this country, but it has propagated by the fcattered feeds in fo plentiful a manner, in thofe gardens where it has been fown, as to become a troublefome weed. The lower leaves of this fort are entire, and of an oblong form; the upper leaves are oblong and indented, in which this differs from the preceding.

The fifth fort grows naturally on the fides of banks in many parts of England, fo is not fuffered to have a place in gardens. This was formerly eaten as a fallad herb by the poorer fort of people, who gave it the title of Sauce alone. It hath a rank fmell and tafte of Garlick, is very biting and hot on the palate, it is frequently prefcribed in medicine.

The fixth fort is fometimes found growing naturally upon old walls in fome parts of England, particularly at Cambridge and Ely, at both which places I have obferved it. This hath pretty long, hairy, foft leaves at the top; the ftalks rife a foot high, their upper part being garnifhed with fmall, greenifh, white flowers in loofe fpikes, which are fucceeded by long compreffed pods, hanging downward. The roots will abide feveral years, if they have a dry lean foil, or grow upon a wall, for in rich land they foon decay.

ERYTHRINA. Lin. Gen. Pl. 762. Coral-tree.

The CHARACTERS are,

The flower is of the butterfly kind, compofed of five petals; the ftandard is fpear-fhaped, long, and rifes upward; the two wings are fcarce longer than the empalement. The keel is compofed of two petals, which are no longer than the wings. It hath ten ftamina, which are joined below, and unequal in their length, with an awl-fhaped germen, which afterward becomes a long fwelling swelling pod, ending in an acute point, having one cell, filled with kidney-shaped seeds.

The SPECIES are,

1. ERYTHRINA (*Herbacea*) foliis ternatis, caule simplicissimo inermi. Hort. Cliff. 354. *Low Coral-tree with a very long spike of flowers and thick root; commonly called the Carolina Coral-tree.*

2. ERYTHRINA (*Corollodendron*) inermis, foliis ternatis, caule arboreo. *Smooth American Coral-tree.*

3. ERYTHRINA (*Spinosa*) foliis ternatis, caule arboreo aculeato. Hort. Cliff. 354. *Prickly, three-leaved, American Coral-tree, with a very red flower.*

4. ERYTHRINA (*Picta*) foliis ternatis acutis, caule fruticoso aculeato. *Smaller, three-leaved, American Coral-tree, with blacker spines and seeds.*

5. ERYTHRINA (*Americana*) foliis ternatis acutis, caule arboreo aculeato, floribus spicatis longissimis. *Three leaved American Coral-tree, with acute-pointed leaves and scarlet seeds.*

6. ERYTHRINA (*Inermis*) foliis ternatis acutis, caule fruticoso inermi, corollis longioribus clausis. *Coral-tree without spines, having a longer closer flower.*

The first sort grows naturally in South Carolina, from whence Mr. Catesby first sent the seeds to England in the year 1724. This hath a very large knotty root, which seldom rises more than a foot high, from which come out fresh shoots every spring, which grow two feet high, their lower part being garnished with trifoliate leaves, of a deep green colour, which are shaped like the point of an arrow; the upper part of the stalks are terminated by a long spike of scarlet flowers, composed of five petals, the upper one being much longer than the other, so that at a small distance the flowers appear to have but one petal. The pods are five or six inches long, swelling in every part where the seeds are lodged, opening in one cell, containing five or six kidney-shaped scarlet seeds.

The second sort hath a thick, irregular, woody stem, which rises about ten or twelve feet high, sending out many strong branches covered with a brown bark, garnished with trifoliate leaves standing upon long foot-stalks, the middle lobe which terminates the leaf being much longer than the other two; they are heart-shaped, smooth, and of a deep green colour; the flowers come out at the end of the branches in short, thick, close spikes, of a deep scarlet colour, and make a fine appearance. These commonly are in beauty in May and June in this country, but are not succeeded by pods here. The flowers seldom appear till the leaves drop, so that the branches are often naked at the time when the flowers are out.

The third sort chiefly differs from the second, in having its trunk, branches, and the foot-stalks of the leaves armed with short crooked spines, the leaves and flowers being very like those of the second sort.

The fourth sort hath shrubby stalks, which divide into branches, armed in every part with strong, crooked, black spines; the leaves are smaller than those of the two last sorts, and have a nearer resemblance to those of the first; the foot-stalks and midrib of the leaves are armed with the same sort of spines; the flowers are of a paler scarlet, and grow in looser spikes. The seeds are as large as those of the second sort, but are of a dark purple colour.

The seeds of the fifth sort were first sent me from La Vera Cruz, where the plants grow naturally; and since I have received seeds of the same sort from the Cape of Good Hope, so that it is a native of both countries. These are not half so large as those of the second or third sorts, and are of a bright scarlet colour; the leaves are also much smaller, and have long acute points; the branches are very closely armed with crooked greenish spines, as are also the ribs and foot-stalks of the leaves. The flowers grow in very long close spikes, and are of a beautiful scarlet colour.

The sixth sort grows in Jamaica, and some of the other islands in America. The pods of this sort are longer, and not more than half so thick as those of the second sort; the seeds are of a bright scarlet colour; the leaves are small and acute-pointed, the stalks are smooth and without spines; this doth not grow very large, but shoots out into branches at a little distance from the ground, which grow erect, so form a bushy shrub. The flowers come out at the end of the branches in short spikes; the standard of the flower is long, and the sides turn down over the wings, which are also longer than those of the other species, and the whole flower is more closed.

These plants when they produce their flowers, are some of the greatest ornaments to the stoves, for their flowers are produced in large spikes, and are of a beautiful scarlet, so they make a fine appearance, but they do not often flower in any of the northern parts of Europe; yet in the countries where they naturally grow, they produce flowers in great plenty every year, so that it is very common there to see most of their branches terminated by large spikes of flowers, when they have no leaves upon them.

These plants are best propagated by seeds, when they can be procured from the countries where they naturally grow. The seeds should be sown in small pots, and plunged into a moderate hot-bed, where, if they are good, the plants will come up in a month or five weeks; when they are two inches high, they should be carefully shaken out of the pots, and each planted in a separate small pot, and plunged into a moderate hot-bed of tanners bark, where they must be shaded from the sun till they have taken new root; then they should have a large share of air admitted to them when the weather is warm: as the plants increase in strength, so they must have a larger share of air. In the autumn the plants should be removed into the stove, and for the two or three first winters while the plants are young, they will require more warmth than when they have acquired more strength. During the time the leaves continue in vigour, the plants will require to have water two or three times a week; but when they are destitute of leaves, it must be sparingly given, for moisture then is very hurtful to them. As the plants grow in strength, so they may be more hardily treated, and by managing them differently, there will be a greater chance of getting them to flower. The third sort is frequently planted in the gardens near Lisbon, where they annually flower and ripen their seeds.

These plants may also be propagated by cuttings, which, if planted in pots during the summer months, and plunged into a hot-bed, will take root, but the seedling plants are best.

ERYTHRONIUM. Lin. Gen. Pl. 375. Dog's Tooth, or Dog's Tooth Violet.

The CHARACTERS are,

The flower is bell-shaped, composed of six oblong petals, which spread open to the base. It hath six stamina joined to the style. In the center is situated an oblong three-cornered germen, which afterward becomes an oblong obtuse capsule with three cells, filled with flat seeds.

The SPECIES are,

1. ERYTHRONIUM (*Dens Canis*) foliis ovatis. *Erythronium with oval leaves; or Dog's Tooth Violet with a broader and rounder leaf, and a purple red flower.*

2. ERYTHRONIUM (*Longifolium*) foliis lanceolatis. *Dog's Tooth Violet with a longer and narrower leaf, and a purplish white flower.*

The first sort sends out two oval leaves, which are joined at their base, three inches long, and one and a half broad in the middle; these at first embrace each other, inclosing the

the flower, but afterward they ſpread flat upon the ground; they are ſpotted with purple and white ſpots all over their ſurface; between theſe riſes a ſingle naked ſtalk about four inches high, of a purple colour; this ſuſtains one flower, compoſed of ſix ſpear-ſhaped petals, which in this are purple, but in ſome they are white; the flower hangs downward, and the petals reflex and ſpread open to their baſe. In the center is ſituated the oblong three-cornered germen, ſupporting a ſingle ſtyle which is longer than the ſtamina, and crowned by a triple ſtigma. The plant flowers early in April. The root of this plant is white, oblong, and fleſhy, and ſhaped like a tooth, from whence it had the title of Dog's Tooth.

The ſecond ſort differs from the firſt in the ſhape of its leaves, which are longer and narrower, and the flowers are a little larger, but not ſo well coloured. They grow naturally in Hungary, and in ſome parts of Italy.

They are propagated by offsets from their roots. They love a ſhady ſituation and a gentle loamy ſoil, but ſhould not be too often removed. They may be tranſplanted any time after the beginning of June, when their leaves will be quite decayed, till the middle of September; but the roots ſhould not be kept very long out of the ground, for if they ſhrink it will often cauſe them to rot. The roots of theſe flowers ſhould not be planted ſcattering in the borders of the flower-garden, but in patches near each other, where they will make a good appearance.

ESPALIERS are either formed of rows of trees planted about a whole garden or plantation, or in hedges, ſo as to incloſe quarters or ſeparate parts of a garden, which are trained up flat like a hedge, for the defence of tender plants againſt the violence of wind and weather. See HEDGES.

The moſt commonly received notion of Eſpaliers are hedges of fruit-trees, which are trained up regularly to a lattice of wood work, formed either of Aſh-poles or ſquare long timbers cut out of Fir, &c. and it is of this ſort of Eſpalier that I ſhall treat in this place.

Eſpaliers of fruit-trees are commonly planted to ſurround the quarters of a kitchen-garden, for which purpoſe they are of admirable uſe and beauty; for by laying out the walks of this garden regularly, which are bounded on each ſide by theſe hedges, when they are handſomely managed, they have a wonderful effect in ſheltering the kitchen plants in the quarters, and alſo ſcreening them from the ſight of perſons in the walks; ſo that a kitchen-garden well laid out in this manner and properly managed, will be equal to the fineſt regular parterre for beauty.

The trees chiefly planted for Eſpaliers are Apples, Pears, and ſome Plumbs, but the two former are moſtly uſed: ſome plant Eſpaliers of Apples grafted upon Paradiſe ſtocks, but theſe being of humble growth and ſhort duration, are not ſo proper for this purpoſe, unleſs for very ſmall gardens; therefore I ſhould rather adviſe the having them upon Crab ſtocks, or (if in ſmaller gardens where the trees cannot be allowed to grow ſo high) upon what the gardeners call the Dutch ſtock; which will cauſe them to bear ſooner, and prevent their growing too luxuriant, and theſe will continue many years in vigour.

In chooſing the trees for an Eſpalier, the ſeveral ſorts which are nearly of the ſame growth, ſhould be planted in the ſame line, that the Eſpalier may be the more regular, and of an equal height, which greatly adds to their beauty; for if they are planted with trees which ſhoot unequally, it will be impoſſible to have the Eſpalier regular: beſides, the diſtance of the trees muſt be in proportion to their growth; for ſome trees, viz. thoſe of a larger growth, ſhould be planted thirty or thirty-five feet aſunder; whereas thoſe of ſmaller growth, need not be above twenty-five feet diſtance.

The width of the walks and borders between theſe Eſpaliers ſhould (in a large garden) be fourteen or ſixteen feet at leaſt; and if the trees are deſigned to be carried up pretty high, the diſtance ſhould be greater, that each ſide may receive the advantage of the ſun and air, which is abſolutely neceſſary, if you would have the fruit well taſted. And if your ground is ſo ſituated, that you are at full liberty which way to make the Eſpaliers, I would adviſe the placing the lines from the eaſt a little inclining to the ſouth, and toward the weſt a little inclining to the north, that the ſun may ſhine between the rows in the morning and evening when it is low; for in the middle of the day when the ſun is advanced far above the horizon, it will ſhine over the tops of the Eſpaliers, and reach the ſurface of the earth about the roots, which is a matter of more conſequence than many people are aware of.

The ſort of Apples proper for Eſpaliers are the Golden Pippin, Nonpareil, Rennette Griſe, Aromatic Pippin, Holland Pippin, French Pippin, Wheeler's Ruſſet, Pile's Ruſſet, with ſome others. The ſeaſon for planting, and the method of pruning and training theſe trees may be ſeen under the articles of APPLES and PRUNING.

The ſorts of Pears proper for an Eſpalier, are chiefly the ſummer and autumn fruits, for ſome of the winter Pears ſeldom ſucceed in an Eſpalier. Theſe trees, if deſigned for a ſtrong moiſt ſoil, ſhould be upon Quince ſtocks, but if for a dry ſoil upon free ſtocks. Their diſtance of planting muſt alſo be regulated by the growth of the trees, which are more unequal in Pears than Apples, and ſhould therefore be more carefully examined before they are planted. As for thoſe Pears upon free ſtocks, the diſtance ſhould never be leſs than thirty feet for moderate growing trees; but for vigorous ſhooters forty feet is little enough, eſpecially if the ſoil be ſtrong, in which caſe they ſhould be planted at a greater diſtance. The particular ſorts of Pears I would recommend for an Eſpalier, are the Jargonelle, Blanquette, Summer Boncretien, Hamden's Burgamot, Autumn Burgamot, L'Ambrette, Gros Rouſſelet, Chaumonteile, Beurre du Roy, Creſſane, Holland Bergamot, and La Chaſſery; always remembering, that thoſe Pears which are of the melting kind, will do better in the Eſpalier than the breaking Pears, which ſeldom ripen ſo well on an Eſpalier; as alſo that many ſorts of Pears will ripen well on an Eſpalier in a warm ſoil and ſituation, which require a wall in other places. As to the method of planting, ſee the article PEAR, and for the pruning and managing, ſee PRUNING.

I ſhall now give directions for making the Eſpalier, to which the trees are to be trained; but this ſhould not be done until the third year after the trees are planted, for while they are young it will be ſufficient to drive a few ſtakes into the ground on each ſide of the trees, to which the branches ſhould be faſtened in an horizontal poſition, in order to train them properly for the Eſpalier; which ſtakes may be placed nearer, or at a farther diſtance, according as the ſhoots produced may require, and theſe will be ſufficient for the three firſt years; for ſhould you frame the Eſpalier the firſt year the trees are planted, the Eſpalier would rot before the trees will cover it. The cheapeſt method to make theſe Eſpaliers is with Aſh-poles of two ſorts; one of the largeſt ſize, which contains thirteen poles in a bundle, and the other ſize thoſe of half an hundred. The firſt or largeſt ſize poles, ſhould be cut about ſeven feet and a half long; theſe are intended for upright ſtakes, and muſt be ſharpened at the largeſt end, that they may with more eaſe be driven into the ground; theſe ſhould be placed at a foot diſtance from each other in a direct line, and of an equal height, about ſix feet above ground; then a row of ſtrait ſlender poles ſhould be nailed upon the top of the upright ſtakes, which will keep them exactly even, and continue

tinue to crofs the ftakes with the fmaller poles at about nine inches diftance row from row, from the top to within a foot of the bottom of the ftakes. Thefe rows of poles fhould be faftened with wire to the ftakes, which, if painted or oiled, will laft a long time.

When the Efpalier is thus framed, the branches of the trees fhould be faftened thereto, either with fmall Ofier twigs rope yarn, or fome fuch binding, training them in a horizontal pofition, and at equal diftances, being carefu not to crofs any of the branches, nor to lay them in too thick. The diftance which fhould be allowed for the branches of Pears and Apples, muft be proportioned to the fize of their fruit; fuch of thofe whofe fruit is large, as the Summer Boncretien, Monfieur John, and Beurre du Roy Pears, and the Rennet Grife, Holland Pippin, French Pippin, and other large Apples, fhould have their branches fix or eight inches diftance at leaft; and to thofe of leffer growth four or five inches will be fufficient. But for farther directions I fhall refer to the articles of the feveral fruits, as alfo that of PRUNING, where thefe particulars will be fufficiently explained.

But befides this fort of Efpalier made with Afh-poles, there is another fort that is by many people preferred, which is framed with fquare timbers cut to a proper fize. Thefe, though they appear more fightly when well fixed and painted, are not of longer duration than one of the former, provided it is well made, and the poles are ftrong which are fet upright; nor will they anfwer the purpofe better, though they are vaftly more expenfive; for the great beauty of Efpaliers confifts in the regular training the branches of the trees, which, efpecially in fummer, when the leaves are on, will entirely hide from the fide the frame of the Efpalier; therefore all expence in erecting thefe is needlefs, farther than making provifion to keep the branches of the trees in good order.

EVERGREEN THORN. See MESPILUS.

EVERLASTING PEA. See LATHYRUS.

EUGENIA. Mich. Nov. Gen. 108. We have no Englifh title for this genus.

The CHARACTERS are,

The empalement of the flower is of one leaf, cut into four fegments; the fower has four oblong petals, and many ftamina which are inferted in th empalement, terminated by fmall fummits, with a turbinated germen fituated under the flower, fupporting a fingle ftyle, which afterward becomes a Plumb-fhaped fruit, inclofing a fmooth nut.

The SPECIES are,

1. EUGENIA (*Malaccenfis*) foliis integerrimis, pedunculis racemofis lateralibus. Flor. Zeyl. 187. *Eugenia with entire leaves, and branching foot-ftalks of flowers.*

2. EUGENIA (*Jamboo*) foliis integerrimis, pedunculis racemofis terminalibus. Flor. Zeyl. 188. *Eugenia with entire leaves, and branching foot-ftalks of flowers terminating the branches.*

The firft fort rifes with a woody ftem from twenty to thirty feet high, in the countries where it grows in the full ground, fending out many branches toward the top, garnifhed with oblong entire leaves placed oppofite; the flowers are produced on the fide of the branches each footftalk fupporting three or four, others only a fingle flower, which are fucceeded by a Plumb-fhaped fruit inclofing one nut.

The fecond fort rifes to the fame height as the firft, and has much the fame appearance, but the leaves are longer and narrower, and the flowers terminate the branches; the fruit is fmaller and rounder, and not efteemed.

Thefe plants grow naturally in the Eaft-Indies, from whence the Portuguefe brought fome of the plants or feeds to Madeira, where there are fome trees now growing; but they are too tender to thrive in the open air in England, therefore require a ftove; where, if they are plunged into the bark-bed while young, they may be preferved, and as they advance in their growth they may be treated more hardily.

EUONYMUS. Lin. Gen. Plant 240. The Spindle-tree or Prickwood.

The CHARACTERS are,

The flower hath four or five oval petals which fpread open, and four or five fhort ftamina joined at their bafe to the germen. In the center is fituated a large oval germen, which afterward becomes a fucculent four or five-cornered coloured capfule, having as many cells as angles, each containing one oval feed.

The SPECIES are,

1. EUONYMUS (*Vulgaris*) foliis lanceolatis, floribus tetandriis, fructu tetragono. *The common Spindle-tree.*

2. EUONYMUS (*Latifolius*) foliis ovato-lanceolatis, floribus pentandriis, fructu pentagono, pedunculis longiffimis. *Broad-leaved Spindle-tree*

3. EUONYMUS (*Americanus*) floribus omnibus quinquefidis. Lin. Sp. Plant. 197. *Virginian Evergreen Spindle, with rough warted feed-veffels.*

4. EUONYMUS (*Pinnatus*) foliis pinnatis, fructu racemofo trigono. *Spindle-tree with an unbranching ftalk, a winged leaf, and a round fruit having three feeds.*

The firft fort grows naturally in England, and is very common in hedges. This, when growing in hedges, is feldom feen of any confiderable fize, but rather appears like a fhrub; but if planted fingle, and trained up like other trees, will have a ftrong woody ftem, and rife more than twenty feet high, dividing into many branches, garnifhed with fpear-fhaped leaves three inches long, and one inch and a quarter broad; they are entire, of a deep green colour, and are placed oppofite. The flowers come out in fmall bunches from the fide of the branches, upon flender foot-ftalks; they are compofed of four whitifh petals, which expand in form of a crofs. The empalement is divided into four parts. The flowers have four ftamina, and the fruit is four-cornered, and opens into four cells The fruit ripens in October, at which time the feed-veffels fpread open and expofe the feeds, which are of a beautiful red colour, fo that when the branches are well ftored with them the trees make a good appearance at that feafon. The wood of this tree is ufed by the mufical inftrument-makers for toothing of organs and virginal keys; the branches are cut into tooth-picks, for making of fkewers, and fpindles are made of this wood, from whence the tree was titled Spindle-tree; but in fome counties it is called Dog-wood.

The fecond fort grows naturally in Auftria and Hungary; this rifes with a ftronger ftem than the firft, and grows to a larger fize. The leaves are oval and fpear-fhaped, about four inches long, and two inches broad in the middle, of a light green colour; they are placed oppofite on the branches, with fhort foot ftalks. The flowers come out from the fide of the branches upon very long flender footftalks; thefe branch out into a loofe bunch, fo that the flowers ftand upon feparate foot-ftalks. The flowers have five petals, which at firft are white, but afterward change to a purple colour: the empalement of the flower is divided into five parts. It hath five ftamina, and the fruit is fivecornered; the fruit is alfo much larger than that of the common fort.

The third fort grows naturally in Virginia, Carolina, and other parts of North America; it rifes with a fhrubby ftalk eight or ten feet high, dividing into many branches, which come out oppofite at every joint, garnifhed with fpear-fhaped leaves two inches long, and three quarters of an inch broad; they are placed oppofite, and continue all the year. The flowers are produced at the end of the

 branches,

branches, and alſo from the ſides, in ſmall cluſters, which are ſucceeded by round capſules, which are cloſely armed with rough protuberances.

As this is an evergreen ſhrub, ſo it merits a place in curious gardens, and particularly in all plantations of evergreen trees and ſhrubs; there is a variety of this with variegated leaves, which is preſerved in the nurſery-gardens.

The fourth ſort grows naturally in Jamaica, and ſome of the other iſlands in the Weſt-Indies; this riſes with an upright woody ſtalk ten or twelve feet high; it divides into two or three ſhort branches, garniſhed with winged leaves, compoſed of ſix or ſeven pair of ſmall leaves (or lobes;) theſe leaves come out without order, ſtanding upon long foot-ſtalks. The flowers come out in cluſters from the ſide of the branches toward their end, and are ſucceeded by roundiſh capſules, having a thick brown cover which open in three cells, each containing a ſingle hard ſeed.

The two firſt ſorts may be propagated either by ſeeds or layers; if by ſeeds, they ſhould be ſown in autumn ſoon after they are ripe, then the plants will come up the ſpring following; but if the ſeeds are not ſown till the ſpring, the plants will not appear till the year after, whereby a whole year is loſt. If they are ſown upon a ſhady border, they will ſucceed better than when they are more expoſed to the ſun. When the plants come up they will require no other care but to keep them clean from weeds till the following autumn, when, as ſoon as their leaves decay, the plants ſhould be taken up and tranſplanted into a nurſery, in rows two feet diſtant, and the plants one foot aſunder in the rows; in this place they may remain two years, and then they may be removed to the places where they are to remain.

They may be alſo propagated by laying down the young ſhoots in the autumn, which will take root in one year, and may then be planted in a nurſery to remain a year or two, then may be planted where they are to remain.

The laſt ſort is too tender to live in this country without the aſſiſtance of an hot-houſe, therefore when any of the plants are brought over, they ſhould be planted in pots and plunged into the tan-bed, and afterward treated in the ſame way as other plants from hot countries.

EUPATORIOPHALACRON. See VERBESINA.

EUPATORIUM. Lin. Gen. Plant. 842. Hemp Agrimony.

The CHARACTERS are,

It hath a compound flower, compoſed of hermaphrodite florets, included in one common ſcaly empalement, whoſe ſcales are narrow, erect, and unequal. The florets have each five ſhort ſtamina. In the bottom is ſituated a ſmall germen, which afterward becomes an oblong ſeed crowned with down, ſitting in the empalement.

The SPECIES are,

1. EUPATORIUM (*Cannabinum*) foliis digitatis. Hort. Cliff. 396. *Common Hemp Agrimony.*

2. EUPATORIUM (*Maculatum*) foliis lanceolato-ovatis, ferratis, petiolatis, caule erecto. Hort. Cliff. 396. *New England Hemp Agrimony with Nettle leaves, purpliſh flowers, and ſpotted ſtalks.*

3. EUPATORIUM (*Purpureum*) foliis ſubverticillatis, lanceolatis ferratis petiolatis rugoſis. Lin. Sp. Plant. 838. *Canada Hemp Agrimony, with a long rough leaf and a purpliſh ſtalk.*

4. EUPATORIUM (*Scandens*) caule volubili, foliis cordatis dentatis acutis. Hort. Cliff. 396. *Climbing American Hemp Agrimony, with a ſpear-like ſharp-pointed leaf.*

5. EUPATORIUM (*Rotundifolium*) foliis ſeſſilibus diſtinctis ſubrotundo-cordatis. Lin. Sp. Pl. 837. *American Hemp Agrimony with round leaves, having no foot-ſtalk.*

6. EUPATORIUM (*Odoratum*) foliis ovatis, obtuſè ferratis petiolatis trinerviis, calycibus ſimplicibus. Lin. Sp. Pl. 839. *American Hemp Agrimony, with a Tree Germander leaf and a white flower.*

7. EUPATORIUM (*Perfoliatum*) foliis connatis tomentoſis. Hort. Cliff. 396. *Virginian perfoliate Hemp Agrimony, with long Sage-like leaves cloſely ſurrounding the ſtalks.*

8. EUPATORIUM (*Moriſolium*) foliis cordatis ferratis caule erecto arboreo. *Eupatorium with heart ſhaped ſawed leaves, and an upright tree-like ſtalk.*

9. EUPATORIUM (*Hyſſopifolium*) foliis lanceolato-linearibus trinerviis integerrimis. Lin. Sp. Pl. 836. *Virginian Hemp Agrimony, with a narrow leaf and white flowers.*

10. EUPATORIUM (*Houſtonis*) foliis cordatis acuminatis, caule volubili, floribus ſpicatis racemoſis. *Eupatorium with heart-ſhaped pointed leaves, a twining ſtalk, and branching ſpiked flowers.*

11. EUPATORIUM (*Cæleſtinum*) foliis cordato-ovatis, obtuſè ferratis petiolis, calycibus multifloris. Lin. Sp. Pl. 838. *Hemp Agrimony, with a Wood Sage leaf and a blue flower.*

This laſt ſort grows naturally in Carolina, from whence the late Dr. Dale ſent me the ſeeds; the plants flowered very finely the year after they were raiſed, but never have flowered ſince, for the roots creep greatly in the ground, but never ſend up any ſtalks.

The firſt ſort grows naturally by the ſide of rivers and ditches in moſt parts of England, and is the only ſpecies of this genus, which is known to grow naturally in Europe; this is eſteemed as a very good vulnerary herb, ſo ſtands in the liſt of medicinal plants. It is ſeldom admitted into gardens, becauſe, wherever it is ſuffered to ſeed, the ground will be well ſtored with the plants to a great diſtance.

The ſecond ſort grows naturally in ſeveral parts of North America; this hath a perennial root, but an annual ſtalk, which riſes two feet and a half high; it is purple, and has many dark ſpots upon it. The leaves are rough, oval, and ſpear-ſhaped, having foot-ſtalks; they are placed by threes round the ſtalk toward the bottom, but upward by pairs oppoſite at each joint. The ſtalks are terminated by cluſters of purple flowers, growing in a ſort of corymbus.

The third ſort grows naturally in North America; this riſes with an upright ſtalk ſix or ſeven feet high, garniſhed with long, narrow, ſpear-ſhaped leaves at each joint; theſe are deeply ſawed on their edges, and the midrib is oblique to the foot-ſtalks; they are placed by fours round the ſtalk in whorls, and are of a dark green colour. The ſtalks are terminated by bunches of purple flowers like the laſt. This hath a perennial root and an annual ſtalk.

The fourth ſort grows naturally in Virginia and Carolina; this hath a perennial root, which ſends out many twining ſtalks in the ſpring, which twiſt about the neighbouring ſupport, and riſe to the height of five or ſix feet, garniſhed at each joint with two heart-ſhaped leaves, which are indented on their edges, and terminate in acute points; and at each joint there are two ſmall ſide branches come out, which are terminated by cluſters of white flowers, ſo that the ſtalks ſeem covered with them moſt part of their length; but as theſe come pretty late in the ſeaſon, ſo unleſs the ſummers prove warm, the plants do not flower in England.

There is another of theſe plants with purple flowers, ſtanding upon longer foot-ſtalks, which was ſent me from Campeachy, but the ſtalks and leaves are like thoſe of this ſort, ſo that I doubt whether it be a diſtinct ſpecies.

The fifth ſort grows naturally in New England and Virginia, from both which countries I have received the ſeeds; it hath a perennial root and an annual ſtalk, which riſes a foot high, with joints pretty near each other, garniſhed with roundish heart-shaped leaves sitting close to the stalks; they are sawed on their edges, and of a light green colour. The flowers are produced in small loose panicles at the top of the stalks, which are white, and have two small green leaves immediately under the flowers.

The sixth sort rises with upright stalks three feet high, garnished with oval sawed leaves at each joint placed opposite, and have very short foot-stalks; from the sides of the stalks at every joint is produced two slender branches, which stand erect; these, and the principal stalks also, are terminated by clusters of white flowers, which appear in August and September, and the stalks decay in winter, but the root is perennial. This grows naturally in Pensylvania, and other parts of America.

The second sort grows naturally in Virginia and Philadelphia; this hath a perennial root and an annual stalk, which rises three feet high, hairy, and garnished with rough leaves at each joint, which are from three to four inches long, and about an inch broad at their base, gradually lessening to a very acute point; the two leaves are joined at their base, so the stalks seem to grow through them. The upper part of the stalk divides into several slender foot-stalks, each sustaining a close cluster of white flowers. These come out in July, and in warm seasons the seeds will sometimes ripen in England.

The eighth sort was sent me by the late Dr Houstoun from La Vera Cruz, where he found it growing naturally; this hath a thick woody stalk, which rises twelve or fourteen feet high, sending out many channelled branches, covered with a brown bark, garnished with regular heart-shaped leaves as large as those of the Mulberry-tree; they are of a light green colour, and sawed on their edges, placed opposite upon foot-stalks near two inches long; the upper part of the branches are terminated by four or five pair of foot-stalks, which come out opposite from the joints, and the top is terminated by an odd one; these sustain branching panicles of white flowers, which together form a long, loose, pyramidal thyrse, and make a fine appearance, for there are no leaves intermixed with the flowers, but so far as the spike reaches the stalks are naked. This sort has flowered in the Chelsea garden, but did not produce seeds.

The ninth sort rises with an upright round stalk to the height of three feet, sending out several branches toward the top, which come out regularly by pairs; they are garnished with leaves placed opposite, which are two inches and a half long, and about one third of an inch broad, having three longitudinal veins; they are of a light green colour, and entire. The flowers stand upon long foot-stalks at the end of the branches, some sustaining one, some two, and others three or four flowers; they are white, and appear late in autumn. This grows naturally in Carolina.

The tenth sort was sent me from Jamaica by the late Dr. Houstoun; this hath slender twining stalks, which fasten themselves to any neighbouring support, and rise eight or ten feet high, sending out small branches by pairs at most of the upper joints. The leaves on the lower part of the stalk are heart-shaped, ending in acute points; the upper leaves are almost triangular, they are smooth, and of a lucid green; the upper part of the stalks have long branching spikes of white flowers, which are small, and sit close to the foot-stalks.

The eleventh sort grows naturally in Carolina; this hath a creeping root which spreads and multiplies very fast. The stalks rise about two feet high; they are garnished with oval, heart-shaped, sawed leaves. The flowers are produced at the top of the stalks in a sort of corymbus; they are of a fine blue colour, but the roots spread so much as to cause barrenness of flowers after the first year.

All these sorts may be propagated by seeds; several of them ripen their seeds in England: these should be sown in autumn as soon as they are ripe, for then the plants will come up the following spring; but if they are kept out of the ground till spring, the plants will not come up till the year after; and those seeds which are procured from America should be sown as soon as they arrive, for though they may not grow the first year, yet there will be a greater certainty of their succeeding than when they are kept longer out of the ground.

The second, third, fifth, sixth, seventh, and eleventh sorts, are hardy plants, so the seeds of these may be sown in the full ground, but there must be care taken in the sowing to keep the sorts separate; for as the seeds of these plants have a light down adhering to them, they are easily displaced by the least wind, so that the best way will be to sow them in drills; but these should be but shallow, for if the seeds are buried too deep they will not grow. The bed in which these are sown should not be too much exposed to the sun, but rather have an east aspect where the morning sun only reaches it; the ground should be kept pretty moist, for as these plants generally grow in moist shady situations in their native countries, they will succeed better when they have a soil and situation somewhat like that.

When the young plants come up they must be kept clean from weeds, and where they are too close, some of them should be drawn out to give room for the others to grow; and if these are wanted they must be planted in another bed, where, if they are shaded and watered, they will soon take root; after which they will require no farther care but to keep them clean from weeds till the following autumn, when they may be transplanted to the places where they are to remain. As the roots of some of the sorts spread out to a considerable distance, they should not be allowed less than three feet from any other plants, and some of the largest growing should be allowed four feet. If the soil in which they are planted is a soft gentle loam, they will thrive much better, and flower stronger than in light dry ground, in which, if they are not duly watered in dry summers, their leaves will shrink, and the stalks will not grow to half their usual height.

All these sorts have perennial roots, by which they may be propagated; for as some of them do not perfect their seeds in England, so the other is the only way of increasing the plants here; those which have creeping roots send out offsets in great plenty, so are easily propagated, and the others may be taken up, or the heads taken off from them every other year, in doing of which there should be care taken not to cut or injure the old plants too much. The best time to remove these plants is in autumn as soon as they have done growing, that they may get fresh roots before the frost comes on; but if that should happen soon after their removal, if the surface of the ground is covered with tan or dried leaves to keep out the frost, it will effectually secure them; and if this is done to the old plants in very severe winters, it will always preserve them.

The fourth sort sends out many weak twining stalks which require support, so there should be some stakes fixed down by their roots in the spring when they begin to shoot, to which the young stalks should be led and fastened, and afterward they will naturally twine round them, and rise four or five feet high; if they are supplied with water, in warm seasons they will produce plenty of white flowers in August. This sort is sometimes killed in very severe winters, if they are not covered; but if when the stalks decay in the autumn, the ground about them should be covered with some old tanners bark, it will effectually secure the roots.

EUPHORBIA. Lin. Gen. Plant. 536. Spurge.

The

The CHARACTERS are,

The flower hath four or five thick truncated petals, and twelve or more stamina which are inserted in the receptacle. In the center is situated a three-cornered germen, supporting three bifid styles, which afterward becomes a roundish capsule with three cells, each containing one roundish seed.

The SPECIES are,

1. EUPHORBIA (*Antiquorum*) aculeata triangularis subnuda articulata, ramis patentibus. Lin. Hort. Cliff. 196. *Prickly triangular-pointed Euphorbia, with spreading branches; commonly called the true Euphorbium of the ancients.*

2. EUPHORBIA (*Canariensis*) aculeata nuda subquinquangularis, aculeis geminatis. Hort. Cliff. 196. *Canary Euphorbium with four or five angles, which have twin spines.*

3. EUPHORBIA (*Trigonum*) aculeata nuda triangularis articulata, ramis erectis. *Prickly Euphorbium, having three and four angles with compressed branches.*

4. EUPHORBIA (*Officinarum*) aculeata nuda multangularis, aculeis geminatis. Lin. Hort. Cliff. 196. *Torch-shaped Euphorbium, with thick stalks armed with strong twin spines.*

5. EUPHORBIA (*Neriifolia*) aculeata seminuda, angulis oblique tuberculatis. Lin. Hort. Cliff. 196. *Angular Euphorbium with broad Oleander leaves.*

6. EUPHORBIA (*Heptagona*) aculeata nuda, septem-angularis, spinis solitariis subulatis floriferis. Lin. Hort. Cliff. 196. *Euphorbium with seven angles, and very long single spines bearing fruit at their tops.*

7. EUPHORBIA (*Caput Medusæ*) inermis tecta tuberculis imbricatis, foliolo lineari instructis. Lin. Hort. Cliff. 197 *African Euphorbium with a thick scaly stalk, and branches disposed like Medusa's Head.*

8. EUPHORBIA (*Mamillaris*) aculeata nuda, angulis tuberosis spinis interstinctis. Lin. Sp. Plant. 451. *Euphorbium with many angles, and long spines growing out from between the knots.*

9. EUPHORBIA (*Cereiformis*) aculeata nuda, multangularis, spinis solitariis subulatis. Prod. Leyd. 195. *Euphorbium with the appearance of Torch Thistle, and a slender stalk.*

10. EUPHORBIA (*Fructus Pini*) inermis imbricata tuberculis foliolo lineari instructis. Hort. Cliff. 197. *African Euphorbium with the appearance of Pine fruit; commonly called Little Medusa's Head.*

11. EUPHORBIA (*Patula*) inermis, ramis patulis simplicibus teretibus foliolis linearibus instructis. *Euphorbia without spines, and single spreading branches which are taper, terminated with very narrow leaves.*

12. EUPHORBIA (*Procumbens*) inermis ramis teretibus procumbentibus tuberculis quadragonis. *Euphorbia without spines, with trailing branches and quadrangular tubercles.*

13. EUPHORBIA (*Inermis*) inermis, ramis plurimis procumbentibus, squamosis, foliolis deciduis. *Euphorbia without spines, having many trailing branches which are scaly, and deciduous leaves.*

14. EUPHORBIA (*Tiruiaculii*) inermis fruticosa subnuda filiformis erecta, ramis patulis determinate confertis. Lin. Hort. Cliff. 197. *Indian shrubby Spurge.*

15. EUPHORBIA (*Viminalis*) inermis fruticosa nuda filiformis volubilis, cicatricibus oppositis. Hort. Cliff. 197. *Indian Spurge, with slender twining stalks entirely without leaves.*

16. EUPHORBIA (*Mauritanica*) inermis fruticosa seminuda filiformis flaccida, foliis alternis. Lin. Hort. Cliff. 197. *Mauritanian Spurge without leaves.*

17. EUPHORBIA (*Cotinifolia*) foliis oppositis subcordatis petiolatis emarginatis integerrimis, caule fruticoso. Lin. Sp. Plant. 453. *Tree American Spurge, with a Venice Sumach leaf.*

18. EUPHORBIA (*Lathyris*) umbellâ quadrifidâ, dichotomâ, foliis oppositis integerrimis. Lin Sp. Pl. 457. *Broad-leaved Spurge, called Cataputia minor.*

19. EUPHORBIA (*Myrsinites*) umbellâ suboctifidâ, bifida, involucellis subovatis, foliis spathulatio patentibus carnosis mucrotanis margine seabris. Lin. Sp. Plant. 461. *Broad-leaved Myrtle Spurge.*

20. EUPHORBIA (*Dendroides*) umbellâ multifidâ, dichotomâ, involucellis subcordatis, primariis triphyllis, caule arboreo. Lin. Sp. Plant. 462. *Myrtle-leaved Tree Spurge.*

21. EUPHORBIA (*Amygdaloides*) umbellâ multifidâ, dichotomâ, involucellis perfoliatis orbiculatis, foliis obtusis. Lin. Sp. Pl. 463. *Common Wood Spurge.*

22. EUPHORBIA (*Palustris*) umbellâ multifidâ subtrifidâ, bifidâ, involucellis ovatis, foliis lanceolatis, ramis sterilibus. Lin. Sp. Pl. 462. *Shrubby Marsh Spurge.*

23. EUPHORBIA (*Orientalis*) umbellâ quinquefidâ, quadrifidâ, dichotomâ, involucellis subrotundis acutis, foliis lanceolatis. Lin. Sp. Pl. 460. *Eastern Spurge with a Willow leaf, a purple stalk, and large flower.*

24. EUPHORBIA (*Characias*) umbellâ quinquefidâ, trifidâ, dichotomâ, involucellis ovatis, foliis lanceolatis, capsulis lanatis. Lin. Sp. Plant. 460. *Tree Spurge with a red stalk, a St. John's-wort leaf, and bearded capsule.*

25. EUPHORBIA (*Hyberna*) umbellâ sextifidâ, dichotomâ, involucellis ovalibus, foliis integerrimis, ramis nullis capsulis verrucosis. Lin. Sp. Plant. 462. *Irish Spurge, called Machingboy.*

26. EUPHORBIA (*Apios*) umbellâ quinquefidâ, bifidâ, involucellis obcordatis. Lin. Sp. Plant. 457. *Spurge with a tuberous Pear-shaped root.*

27. EUPHORBIA (*Aleppica*) umbellâ quinquefidâ, dichotomâ, involucellis ovato-lanceolatis mucronatis, foliis inferioribus setaceis. Lin. Sp. Pl. 458. *Cypress Spurge.*

28. EUPHORBIA (*Cretica*) umbellâ multifidâ, bifidâ, involucellis orbiculatis, foliis lineari-lanceolatis villosis. *Cretan Wood Spurge, with narrow, hairy, and hoary leaves.*

29. EUPHORBIA (*Sylvatica*) umbellâ multifidâ, dichotomâ, involucellis perfoliatis, subcordatis, foliis lanceolatis integerrimis. Lin. Sp. Plant. 463. *Wood Spurge with a moon-shaped flower.*

30. EUPHORBIA (*Heterophylla*) inermis foliis serratis petiolatis difformibus ovatis lanceolatis panduriformibus. Lin. Sp. Pl. 453. *Spurge from Curassoa, with variable leaves like Willow and Orach, and a green stalk.*

31. EUPHORBIA (*Hypericifolia*) dichotomâ, foliis serratis ovali-oblongis glabris, corymbis terminalibus, ramis divaricatis. Lin. Sp. Pl. 454. *Upright acrid Spurge, with smooth Pellitory leaves, and flowers growing in clusters from the joints of the stalk.*

32. EUPHORBIA (*Ocymoides*) inermis, herbacea, ramosa, foliis subcordatis integerrimis petiolatis floribus solitariis. Lin. Sp. Pl. 453. *Upright, annual, branching Spurge of America, with leaves like Small Basil.*

The first sort has been generally taken for the true Euphorbium of the ancients, and as such hath been directed for medicinal use; but it is from the second sort that the drug now imported under that title in England is taken. Dr. Linnæus supposes the fourth to be the sort which should be used, though as they are all nearly of the same quality, so it may be indifferent which of them that drug is taken from, which is the inspissated juice of the plant.

The first sort hath a triangular, compressed, succulent stalk, which is jointed, and rises to the height of seven or eight feet, sending out many irregular twisting branches, for the most part three-cornered, but have sometimes only two, and at others four angles; they are compressed, succulent, and spread out on every side the stalk; at the extremity of the branches there are a few short roundish leaves, which soon fall off; and near these come out now and then a few flowers, which have five thick whitish petals, with a large three-cornered germen in the center, which soon drop

drop off without having any feeds. It grows naturally in India.

The fecond fort grows naturally in the Canary Iflands, from whence I have been credibly informed the Euphorbium which is imported in England is now brought, and is the infpiffated juice of this plant. In its native country it grows to the height of twenty feet or more, but in England it is rarely feen more than fix or eight; nor is it of any advantage to have them fo tall here, becaufe they fend out many branches which are large and fucculent, fo render the plants too heavy to be eafily removed. It hath a very thick fucculent ftalk, with four or five large angles or corners, clofely armed with black crooked fpines, which come out by pairs at every indenture; the ftalks fend out large fucculent branches of the fame form, which turn their ends upwards, fo that when the plants are well grown they have fome refemblance to a branched chandelier; thefe are clofely armed with black fpines like the ftalks; at the end of the branches come out the flowers, which are fhaped like thofe of the firft fort.

The third fort hath a naked, three-cornered, compreffed ftalk, fending out a great number of erect branches, which join up to the main ftalk, and are generally three-cornered, but fome vary to four; they are jointed and armed with fhort crooked fpines, but have no leaves, nor do the plants produce flowers here. This grows naturally in India.

The fourth fort puts out many ftalks juft above the furface of the ground, which are thick, fucculent, and taper, having eight or ten angles while they are young, but as they grow old they lofe their angles, and become round; the branches grow diftorted and irregular, firft horizontal, and afterward turn upward; they are armed with fmall crooked fpines on their angles, and on the upper part of the branches come out the flowers, which are fmall, and of a greenifh white, fhaped like thofe of the fecond fort. This grows naturally in India.

The fifth fort rifes with a ftrong upright ftalk five or fix feet high, which hath irregular angles, and protuberances which are oblique to the angles; the lower part of the ftalk is naked, the upper part is branching, and the branches are armed with crooked fpines; at every protuberance they are garnifhed with oblong leaves of a lucid green, which are very fmooth, entire, and rounded at their ends; thefe fall off, and the plants remain naked for fome months, and then the flowers come out, which fit clofe to the branches, of a greenifh white colour; the leaves come out in the autumn, and fall off in the fpring.

The fixth fort rifes with a roundifh upright, fucculent ftalk about three feet high, putting out feveral branches on the fide of the fame form, which have feven angles or furrows, armed with long, fingle, black thorns; at the end of which come out fmall flowers of the fame form with thofe of the other forts, and are fometimes fucceeded by fmall fruit.

The feventh fort hath thick, roundifh, fucculent ftalks, which are fcaly, fending out many branches from their fides of the fame form, which are twifted, and run one over another fo as to appear like a parcel of ferpents coming out from the ftalks, from whence it had the appellation of Medufa's Head. The ends of the branches are garnifhed with narrow, thick, fucculent leaves, which foon drop off, and round the upper part of the branches the flowers come out, which are white, and of the fame form with thofe of the other fpecies, but larger, and are frequently fucceeded by round fmooth capfules with three cells, each including a fingle roundifh feed.

The eighth fort hath roundifh ftalks which fwell out like a belly in the middle, and have knobbed angles, between which come out long fpines which are ftrait; thefe ftalks rife two feet high, and put out a few branches on their fide of the fame form; the flowers are produced at the end of the branches, fitting clofe upon the angles; they are fmall, of a yellowifh green colour, and fhaped like thofe of the other fpecies.

The ninth fort hath ftalks and branches very like thofe of the fourth, but much flenderer; the fpines of this are fingle, and thofe of the other double, and the ends of the branches are clofely garnifhed with flowers on every angle, in which it differs from the fourth fort.

The tenth fort hath a thick fhort ftalk, which feldom rifes more than eight or ten inches high, from which come out a great number of trailing branches which are flender, and grow about a foot in length: thefe intermix with each other like thofe of the feventh fort, but they are much fmaller, and do not grow near fo long, but has the fame appearance, from whence it is called Little Medufa's Head; the ends of thefe branches are befet with narrow leaves, between which the flowers come out which are white, and fhaped like thofe of the other fpecies.

The eleventh fort rifes with a taper ftalk fix or feven inches high, fending out from the top a few taper branches, which fpread out on every fide; thefe are not fcaly like thofe of the laft fort, but taper, and garnifhed at their ends with feveral fmall narrow leaves which drop off.

The twelfth fort hath a fhort thick ftalk, which never rifes three inches high, fo that the branches fpread on the furface of the ground; thefe feldom grow more than fix inches long, and their fcales fwell into a fort of protuberances which are fquare; they have no leaves, and very rarely produce flowers in England.

The thirteenth fort is very like the feventh, but the ftalks never rife more than a foot or fifteen inches high, fo that the branches fpread out near the ground; thefe are much fhorter than thofe of the feventh, but have the fame appearance, and are garnifhed with narrow leaves at their end, which fall off as the branches are extended in length: this produces a great number of fmall white flowers at the end of the branches, which are fhaped like thofe of the other fpecies, and are frequently fucceeded by fmooth round capfules with three cells, including one or two roundifh feeds which ripen here.

Thefe forts have been by moft of the modern botanifts ranged under the title of Euphorbium, and have been diftinguifhed from the Tithymali, more from the ftructure and outward appearance of the plants, than any real difference in their characters; but as the number of fpecies of thofe commonly called Spurge was very great, fo many of the writers were willing to feparate the Euphorbia from that genus, to leffen the number of fpecies.

Thefe plants are preferved in many curious gardens more for the oddnefs of their ftructure than any real beauty; but being fo extremely different in their form from almoft any plants of European production, many curious perfons have been induced to preferve the feveral forts in their gardens.

They are all of them full of a milky acrid juice, which flows out on their being wounded in any part; this juice will blifter the flefh, if it happens to lie upon any tender part for a fhort time, and will burn linen almoft as bad as aqua fortis, therefore the plants fhould be handled with great caution: nor fhould the ends of their branches be ever bruifed or injured, for if they are it frequently occafions their rotting down to the next joint, and fometimes will deftroy the whole plant if thofe injured branches are not cut off in time; fo that whenever the branches appear to have been injured, the fooner they are cut from the plants

the

the less danger there will be of their suffering from it; nor should any of the branches be cut between the joints for the same reason.

The greatest part of these milky succulent plants grow naturally upon barren rocky places, or in dry sandy soils, where few other plants will thrive; therefore they should never be planted in rich earth here, nor suffered to receive much wet, which will cause them to rot. The best mixture of earth for these plants is, about a fourth part of screened lime rubbish, a fourth part of sea sand, and half of light fresh earth from a common; these should be mixed well together, and frequently turned over before it is used, that the parts may be incorporated, and the compost sweetened by being exposed to the air.

All the sorts are easily propagated by cuttings, which should be taken from the old plants in June; they must be cut at a joint, otherwise they will rot: when these cuttings are taken off, the milky juice of the old plants will flow out in plenty; therefore there should be some dry earth or sand applied upon the wounded part, which will harden and stop the sap from flowing out; and the wounded part of the cuttings should also be rubbed in sand or dry earth for the same purpose; then the cuttings should be laid in a dry part of the stove, for ten days or a fortnight; and some of those whose branches are large and very succulent, may lie a month or more before they are planted, that their wounds may be healed and hardened, otherwise they will rot. When the cuttings are planted, they should be each put into a small halfpenny pot, laying stones or rubbish in the bottom, and filling the pots with the mixture before directed; then plunge the pots into a moderate hotbed, and if the weather is very hot, the glasses of the hotbed should be shaded in the middle of the day, and the cuttings should be gently watered once a week, according as the earth may dry: in about six weeks the cuttings will have put out roots, so if the bed is not very warm, the plants may continue here, provided they have free air admitted to them every day, otherwise it will be better to remove them into the stove, where they may be hardened before the winter; for if they are too much drawn in summer they are very apt to decay in winter, unless they are very carefully managed. During the summer season these plants should be gently watered once a week, according to the warmth of the season; but in winter they must not be watered oftener than once in a fortnight or three weeks, which should be given more sparingly at that season, especially if the stove is not warm; the first sort will require more warmth in the winter than any of the other, as also less water. This, if well managed, will grow seven or eight feet high, but the plants must constantly remain in the stove, giving them a large share of air in warm weather, and in winter the stove should be kept in a temperate degree of warmth.

The sixth sort is at present the most rare in England, for most of the plants have been destroyed by placing them in stoves, where, by the heat, they have in one day turned black, and rotted immediately after. This sort may be placed in winter in a dry airy glass-case with other succulent plants, where they may have free air in mild weather, and be protected from frost; in summer the plants may be exposed in the open air in a warm situation, but should be screened from too much wet; with this treatment they will thrive much better than if they were more tenderly nursed.

The seventh, eighth, tenth, eleventh, twelfth, and thirteenth sorts, are also pretty hardy, so will live in a good glass-case in winter without fire, provided the frost is kept entirely out, and in summer they may be placed abroad in a warm situation: as these are very succulent plants they should not have too much wet, therefore if the summer should prove very moist, it will be very proper to place them under some shelter, where they may enjoy the free air, but may be screened from the rain.

The seventh sort will require to be supported, otherwise the weight of the branches will draw them into the pots; and, by training of the stems up to the stakes, they will grow four or five feet high, and a great number of side branches will be produced; these being very succulent and heavy, are very apt to draw down the stem if it hath not support.

The following sorts have been by all the writers on botany placed under the title of Tithymalus; but the fourteenth and fifteenth sorts should, according to their own distinction, have been placed in the genus of Euphorbia, because they are as destitute of leaves as most of the species which they have there placed.

The fourteenth sort rises with a taper succulent stalk to the height of eighteen or twenty feet, sending out many branches of the same form, which subdivide into many smaller, which are jointed; they are smooth, and of a deep green colour, with a few small leaves at their extremities, which soon fall off. As the plants grow older, so their stalks become stronger and less succulent, especially toward the bottom, when they turn to a brown colour, and become a little woody. The branches grow diffused and intermix with each other, so form a sort of bush toward the top.

The fifteenth sort sends out a great number of slender taper stalks of a dark green colour, which are smooth, and twist about each other, or any neighbouring support, whereby they will rise to the height of ten or twelve feet, putting out smaller branches upward, which also twine and intermix with the other stalks; they are naked, having no leaves.

The sixteenth sort sends out many taper succulent stalks from the root, which rise about four feet high; they are slender and weak, so require support to prevent their falling to the ground; these have a light green bark; their lower parts are naked, but their upper parts are garnished with oblong leaves, which are smooth, entire, and placed alternate on every side the stalks: the flowers are produced in small clusters at the end of the branches, they are of a yellowish green colour, and are sometimes succeeded by smooth round fruit, but the seeds rarely ripen in England. This sort grows naturally on the African shore in the Mediterranean.

The seventeenth sort grows naturally in most of the islands in the West-Indies, and also upon the continent there. This hath an upright stalk, which rises to the height of six or seven feet, covered with a light brown bark; it divides upward into many branches, garnished with roundish leaves indented at their ends; they are smooth and of a beautiful green, but fall away in winter, so that in the spring they are almost naked; the flowers come out from the end of the branches, which are yellow and small, soon falling away without having any fruit succeed them here.

These sorts are propagated by cuttings in the same manner as the Euphorbia, and the plants must be treated in the same way as hath been directed for them.

The fourteenth, fifteenth, and seventeenth sorts are tender, so require a stove, and must have the same treatment as the tender kinds of Euphorbiums, but the sixteenth sort will live in a common green-house in winter, and may be exposed abroad in the summer.

The eighteenth sort stands in the list of medicinal plants, but is rarely used in England at present; this is a biennial plant, which perishes after the seeds are ripe. It grows naturally in Italy and the south of France, and where it is allowed

allowed to ſcatter its ſeeds in a garden, becomes a weed here. This riſes with an upright ſucculent ſtalk from three to four feet high, garniſhed with oblong ſmooth leaves placed oppoſite, ſitting cloſe to the ſtalks; the upper part of the ſtalk divides by pairs into ſmaller branches, and from the fork between theſe diviſions come out the umbels of flowers, each fork having one; that which is ſituated in the firſt diviſion being the largeſt, and thoſe in the upper are the ſmalleſt. The flowers are of a greeniſh yellow colour, the fruit follows ſoon after, which is divided into three lobes, has three cells, each containing one roundiſh ſeed, which is caſt out to a diſtance by the elaſticity of the capſule. This ſort will propagate itſelf faſt enough when it is once introduced into gardens, ſo requires no care but to keep it clean from weeds.

The nineteenth ſort grows naturally in the ſouth of France, in Spain, and Italy. This ſends out many trailing branches from the root, which grow about a foot long, lying upon the ground, cloſely garniſhed with thick, flat, ſucculent leaves placed alternate, ſitting cloſe to the ſtalks: the flowers are produced in large umbels at the end of the branches; the involucrum of the principal umbel is compoſed of ſeveral oval-pointed leaves, but thoſe of the ſmall umbels have only two heart-ſhaped concave leaves, whoſe borders are rough; the flowers are yellow, and are ſucceeded by three ſeeds, incloſed in a roundiſh capſule with three cells. This plant will continue two or three years upon a dry warm ſoil, and will ripen ſeeds annually, which, if permitted to ſcatter, the plants will come up, and require no other care but to keep them clean from weeds.

The twentieth ſort grows naturally in Crete, and in ſeveral iſlands of the Archipelago; this hath a great reſemblance to the common Wood Spurge, but the leaves are narrower, longer, and are hoary. It is eaſily propagated by cuttings during any of the ſummer months, and requires a little protection from the froſt in hard winters.

The twenty-firſt ſort grows naturally in the woods in many parts of England; it riſes with a ſhrubby ſtalk three feet high; the flowers are produced in umbels ſitting cloſe to the ſtalks, ſo form a long ſpike; the empalements are of a greeniſh yellow, and the petals black, ſo they make an odd appearance. It flowers in May, and the ſeeds ripen in July. If the ſeeds of this are ſown under ſhelter in the autumn, the plants will riſe the following ſpring, and require no culture.

The twenty-ſecond ſort ſtands in the liſt of medicinal plants by the title of Eſula major, but at preſent is ſeldom uſed: it grows naturally in France and Germany upon marſhy places, where it riſes three or four feet high. It hath a perennial root, by which it may be propagated better than by ſeeds, which ſeldom grow unleſs they are ſown ſoon after they are ripe.

The twenty-third ſort was diſcovered in the Levant by Dr. Tournefort. This hath a perennial root, from which ariſe ſucculent ſtalks three feet high, covered with a purple bark, garniſhed with oblong ſmooth leaves ſhaped like thoſe of Willow, of a dark green colour. The upper part of the ſtalks divide, and in the fork is ſituated an umbel of flowers of a greeniſh yellow colour, which are ſucceeded by round capſules with three cells, each containing a ſingle ſeed. It flowers in June, and the ſeeds are ripe in Auguſt; this may be propagated by parting of the roots, or by ſowing of the ſeeds in autumn. The plant is hardy ſo will endure the greateſt cold of this country if it is planted in a dry ſoil.

The twenty-fourth ſort grows naturally in Sicily, and on the borders of the Mediterranean ſea; this riſes with ſeveral ſhrubby ſtalks to the height of five or ſix feet, garniſhed with oblong, ſmooth, blunt leaves, which are placed alternate. The flowers are yellow, and grow in ſmall umbels from the diviſion of their branches, and are ſucceeded by roundiſh capſules which are rough, having three cells like the other ſpecies. This is eaſily propagated by cuttings during any of the ſummer months, and requires protection from the froſt in winter.

The twenty-fifth ſort grows naturally in Ireland. This hath thick fibrous roots, which ſend out ſeveral erect ſtalks about a foot long, garniſhed with oblong leaves placed alternate. The flowers are yellow, and are produced in ſmall umbels at the top of the ſtalks, and are ſucceeded by rough warted capſules with three cells. This may be propagated by the roots, which ſhould be planted in a ſhady ſituation and a moiſt ſoil.

The twenty-ſixth ſort grows naturally in the Levant; this hath a knobbed Pear-ſhaped root, from which ariſe two or three ſtalks about a foot high, garniſhed with oblong hairy leaves placed alternate. The flowers are produced in ſmall umbels from the diviſions of the ſtalk; they are ſmall, of a greeniſh yellow colour, and are ſeldom ſucceeded by ſeeds here. It may be propagated by offsets, ſent out from the main root; theſe may be taken off in autumn and planted in a ſhady ſituation, where they will thrive better than in the full ſun.

The twenty-ſeventh ſort grows naturally at Aleppo, and in other parts of the Levant; this hath a perennial creeping root, by which it multiples very faſt where it is once eſtabliſhed. The ſtalks of this riſe a foot and a half high; the lower leaves are narrow, ſtiff, and briſtly, but thoſe on the upper part of the ſtalk are ſhaped live the narrow-leaved Myrtle. The flowers are yellow, and are produced in large umbels from the diviſions of the ſtalk, but are rarely ſucceeded by ſeeds in this country. The roots of this ſhould be confined in pots, for when they are planted in the full ground they creep about to a great diſtance.

The twenty-eighth ſort grows naturally in many parts of the Levant, and alſo in Spain and Portugal. The ſeeds of this were ſent me from Portugal by Robert More, Eſq; who found the plants growing there naturally. It riſes with a purple ſhrubby ſtalk three feet high, garniſhed with narrow, ſpear-ſhaped, hairy leaves, ſet cloſely on the ſtalk alternately; the upper part of the ſtalk is terminated by umbels of flowers, which form a ſort of ſpike. The greater umbels are multifid, but the ſmall ones are bifid. The involucrum of the flowers are yellow, and the petals of the flowers black. The young plants which have been raiſed from ſeeds are generally fruitful, but the old ones and thoſe raiſed by cuttings are barren: this may be propagated by cuttings, and will live abroad if planted in a dry rubbiſhy ſoil and a warm ſituation, otherwiſe they are frequently killed by ſevere froſt.

The twenty-ninth ſort grows naturally in the ſouth of France, in Spain and Italy; it is a biennial plant, from whoſe root ariſe two or three ſtalks, which grow two or three feet high, garniſhed with ſpear-ſhaped leaves which are entire. The umbels of flowers ariſe from the diviſion of the branches; the involucrums are heart-ſhaped, and ſurround the pedicle at their baſe. The flowers are yellow, and appear in June. The ſeeds ripen in Auguſt, which, if permitted to ſcatter, the plants will come up and require no care.

The thirtieth ſort grows naturally at La Vera Cruz. This is an annual plant, which riſes from two to three feet high. The leaves are ſometimes narrow and entire, at other times oval, and divided in the middle almoſt to the midrib, in ſhape of a fiddle; they alſo vary in their colour, ſome being inclinable to purple, others of a light green, ſawed on their edges, and ſtand upon ſhort foot-ſtalks. The flowers are produced in ſmall umbels at the end of the branches, of of a greenish white, and are succeeded by small round capsules with three cells.

The thirty-first sort grows naturally in most of the islands in the West-Indies; it is an annual plant, which rises with a branching stalk about two feet high, garnished with oblong, oval, smooth leaves, sawed on their edges. The flowers grow in small umbels at the foot-stalks of the leaves, gathered into close bunches; they are white, and are succeeded by small round capsules, inclosing three seeds.

The seeds of the thirty-second sort were sent me from La Vera Cruz, by the late Dr. Houstoun; this is also an annual plant, which rises with an upright stalk about a foot high, dividing into a great number of branches which spread very wide, garnished with roundish heart-shaped leaves, which are entire, standing upon pretty long foot-stalks. The flowers come out singly from the divisions of the stalk; they are small, and of an herbaceous colour, and are succeeded by small round capsules, containing three seeds.

The last three sorts are annual; the seeds of these must be sown upon a hot-bed in the spring, and when the plants are fit to remove, they should be each planted in a small pot filled with light earth, and plunged into the hot-bed again, and must afterward be treated in the same manner as other tender annual plants from warm countries.

EUPHRASIA. Eyebright.

This is a medicinal plant, which grows naturally in sterile fields and commons in most parts of England, always among Grass, Heath, Furz, or some other cover, and will not grow when these are cleared from about it, nor will the seeds grow when they are sown in a garden; for which reason I shall not trouble the reader with a description, or any farther account of it, than that the herb-women supply the markets with it in plenty from the fields.

F.

FABA. Tourn. Inst. R. H. 391. tab. 212. The Bean.

The Characters are,

The flower is of the butterfly kind. The standard is large, oval, and indented at the end; it hath two oblong erect wings, which inclose the keel, being much longer. The keel is short, swelling, and closely covers the parts of generation; the nine stamina are in three parts, and one stands separate. At the bottom is situated an oblong compressed germen, which afterward becomes a long, compressed, leathery pod, having one cell, filled with compressed kidney-shaped seeds.

There are several varieties of the Garden Bean, which are known and distinguished by the gardeners, but do not essentially differ from each other, so I shall not enumerate them as distinct species, but shall not join these to the Horse Bean as some have done, who have supposed them to be but one species; for, from having cultivated them more than fifty years, without finding the Garden Bean degenerate to the Horse Bean, or the latter improving to the former, I conclude they are distinct.

There are a great variety of the Garden Beans now cultivated in the kitchen-gardens in England, which differ in size and shape; some of them producing their pods much earlier in the year than others, for which they are greatly esteemed by the gardeners, whose profit arises from their early crops of most esculent plants.

I shall begin with the Garden Bean, called by the botanists Faba major, to distinguish it from the Horse Bean, which they have titled Faba minor seu equina, and shall only mention the names of each variety, by which they are known among the gardeners, placing them according to their time of ripening for the table.

The Mazagan Bean is the first and best sort of early Beans at present known; these are brought from a settlement of the Portuguese of the same name on the coast of Africa, just without the streights of Gibraltar; the seeds of this sort are much smaller than those of the Horse Bean, and as the Portuguese are but slovenly gardeners, there is commonly a great number of bad seeds among them. If this sort is sown in October under a warm hedge, pale, or wall, and carefully earthed up when the plants are advanced, they will be fit for the table by the beginning of May: the stems of this sort are very slender, therefore should be supported by strings close to the hedge or pale, to preserve them from the morning frosts, which are sometimes severe in the spring, and retards their growth; these Beans bear plentifully, but they ripen nearly together, so that there are seldom more than two gatherings from the same plants; if the seeds of this sort are saved two years in England, the Beans will become much larger, and not ripen so soon, which is called a degeneracy.

The next sort is the early Portugal Bean, which appears to be the Mazagan sort saved in Portugal, for it is very like those which are the first year saved in England; this is the most common sort used by the gardeners for their first crop, but they are not near so well tasted as the Mazagan; therefore when the Mazagan Bean can be procured, no person of skill would plant the other.

The next is the small Spanish Bean; this will come in soon after the Portugal sort, and is a sweeter Bean, therefore should be preferred to it.

Then comes the broad Spanish, which is a little later than the other, but comes in before the common sorts, and is a good bearer, therefore is frequently planted.

The Sandwich Bean comes soon after the Spanish, and is almost as large as the Windsor Bean, but being hardier, is commonly sown a month sooner; this is a plentiful bearer.

The Toker Bean, as it is generally called, comes about the same time with the Sandwich, and is a great bearer, therefore is now much planted.

The white and black Blossom Beans are also by some persons much esteemed; the Beans of the former are, when boiled, almost as green as Pease, and being a tolerable sweet Bean, renders it more valuable; these sorts are very apt to degenerate, if their seeds are not saved with great care.

The

The Windsor Bean is allowed to be the beſt of all the ſorts for the table, except the Mazagan; when theſe are planted on a good ſoil, and are allowed ſufficient room, their ſeeds will be very large, and in great plenty, and when they are gathered young are the beſt taſted of moſt of the ſorts; but theſe ſhould be carefully ſaved, by pulling out ſuch of the plants as are not perfectly right, and afterward by ſorting out all the good from the bad Beans when they are out of the pods.

This ſort of Bean is ſeldom planted before Chriſtmas, becauſe it will not bear the froſt ſo well as many of the other ſorts, ſo it is generally planted for the great crop, to come in June and July.

All the early Beans are generally planted on warm borders, near walls, pales, and hedges; and thoſe which are deſigned to come firſt, are uſually planted in a ſingle row pretty cloſe to the fence: but here I cannot help taking notice of a very bad cuſtom, which too generally prevails in gentlemen's kitchen-gardens, which is that of planting Beans cloſe to the garden walls, on the beſt aſpects, immediately before the fruit-trees, which certainly is a greater prejudice to the trees than the value of the Beans, or any other early crop, therefore this practice ſhould be every where diſcouraged; for it is much better to run ſome Reed-hedges acroſs the quarters of the kitchen-garden, where early Beans and Peaſe may be planted, in which places they may with more conveniency be covered in ſevere froſt; and to theſe hedges the Beans may be cloſely faſtened, as they advance in their growth, which, if practiſed againſt their walls where good fruit-trees are planted, will greatly prejudice them.

But to return to the culture of the Beans. Thoſe which are planted early in October, will come up by the beginning or middle of November; and as ſoon as they are two inches above ground, the earth ſhould be carefully drawn up with a hoe to their ſtems; and this muſt be two or three times repeated, as the Beans advance in height, which will protect their ſtems from the froſt. If the winter ſhould prove ſevere, it will be very proper to cover the Beans with Peaſe-haulm, Fern, or ſome other light covering, which will ſecure them from the injury of froſt; but this covering muſt be conſtantly taken off in mild weather, otherwiſe the Beans will draw up tall and weak, ſo come to little; and if the ſurface of the border is covered with tanners bark, it will prevent the froſt from penetrating of the ground to the roots of the Beans, and be of great ſervice to protect them from the injury which they might otherwiſe receive.

In the ſpring, when the Beans are advanced to be a foot high, they ſhould be faſtened up to the hedge with packthread, ſo as to draw them as cloſe as poſſible, which will ſecure them from being injured by the morning froſts, which are often ſo ſevere in April, as to lay thoſe Beans flat on the ground which are not thus guarded; at this time all ſuckers which come out from the roots ſhould be very carefully taken off, for theſe will retard the growth of the Bean, and prevent their coming early. When the bloſſoms begin to open toward the bottom of the ſtalks, the top of the ſtems ſhould be pinched off, which will cauſe thoſe firſt pods to ſtand, and thereby bring them forward. If theſe rules are obſerved, and the ground kept clean from weeds or other plants, there will be little danger of their failing.

But leſt this firſt crop ſhould be deſtroyed by froſt, it will be abſolutely neceſſary to plant more about three weeks after the firſt, and ſo to repeat planting more every three weeks, or a month, till February; but thoſe which are planted towards the end of November, or the beginning of December, may be planted on ſloping banks at a diſtance from the hedges; for if the weather ſhould prove mild, theſe will not appear above ground before Chriſtmas, therefore will not be in ſo much danger as the firſt and ſecond planting, eſpecially if the ſurface of the ground is covered with tan to keep the froſt out of the ground. The ſame directions which are before given, will be ſufficient for the management of theſe, but only it muſt be obſerved, that the larger Beans ſhould be planted at a greater diſtance than the ſmall ones, as alſo that thoſe which are firſt planted muſt be put cloſer together, to allow for ſome miſcarrying; therefore where a ſingle row is planted, the Beans may be put two inches aſunder, and thoſe of the third and fourth planting may be allowed three inches; and when they are planted in rows acroſs a bank, the rows ſhould be three feet aſunder; but the Windſor Beans ſhould have a foot more ſpace between the rows, and the Beans in the rows ſhould be planted five or ſix inches aſunder. This diſtance may, by ſome perſons, be thought too great; but from many years experience I can affirm, that the ſame ſpace of ground will produce a greater quantity of Beans, when planted at this diſtance, than if double the quantity of ſeeds are put on it. In the management of theſe later crops of Beans, the principal care ſhould be to keep them clear from weeds, and other plants, which would draw away their nouriſhment; to keep earthing them up, and, when they are in bloſſom, to pinch off their tops, which, if ſuffered to grow, will draw the nouriſhment from the lower bloſſoms, which will prevent the pods from ſetting, and ſo only the upper parts of the ſtems will be fruitful; and another thing ſhould be obſerved in planting of the ſucceeding crops, which is, to make choice of moiſt ſtrong land for the later crops, for if they are planted on dry ground, they rarely come to much, unleſs the ſummer proves wet.

Theſe after crops ſhould be planted at about a fortnight diſtance from February to the middle of May, after which time it is generally too late to plant, unleſs the land is very ſtrong and moiſt; for in warm, dry, light land, all the late crops of Beans are generally attacked by the black inſects, which cover all the upper part of the ſtems, and ſoon cauſe them to decay.

When the ſeeds of theſe Beans are deſigned to be ſaved, a ſufficient number of rows ſhould be ſet apart for that purpoſe, according to the quantity deſired; theſe ſhould be managed in the ſame way as thoſe which are deſigned for the table, but none of the Beans ſhould be gathered; though there are ſome covetous perſons who will gather all the firſt ripe for the table, and are contented to ſave the after crop for ſeed, but theſe are never ſo large and fair as the firſt; ſo that if theſe are for ſale, they will not bring near the price as the other, therefore what is gained to the table, is loſt in the value of the ſeed; but thoſe who are deſirous to preſerve the ſeveral varieties as pure as poſſible, ſhould never ſuffer two of the varieties to grow for ſeeds in the ſame place, for by their farina mixing with each other, they will not continue ſo pure, but be apt to vary; but in order to keep the early kinds perfect, thoſe which come the earlieſt ſhould be ſaved for ſeeds, which is what few people chuſe to do, becauſe they are then the moſt valuable.

When the ſeed is ripe, the ſtalks ſhould be pulled up, and ſet upright againſt a hedge to dry, obſerving to turn them every third day, that they may dry equally; then they may be threſhed out and cleaned for uſe, or otherwiſe ſtacked up in a barn, till there is more leiſure for threſhing them out; and afterward the ſeed ſhould be drawn over, to take out all thoſe that are not fair, preſerving the beſt for uſe or ſale, and the ordinary Beans will feed cattle.

It is a very good method to change the ſeeds of all ſorts of Beans, and not to ſow and ſave the ſeeds long in the ſame ground, for they do not ſucceed ſo well; therefore if the land is ſtrong where they are to be planted, it will be the beſt way to procure the ſeeds from a lighter ground, and ſo vice versa; for by this method the crops will be larger, the Beans fairer, and not so liable to degenerate.

Having given directions for the culture of the Garden Bean, I shall next proceed to that of the Horse Bean, which is cultivated in the fields: there are two or three varieties of these Beans, which differ in their size and colour, but that which is now in the greatest esteem is called the Tick Bean; this doth not grow so high as the other, is a more plentiful bearer, and succeeds better on light land than the common Horse Bean, so is preferred to it.

The Horse Bean delights in a strong moist soil and an open exposure, for they never thrive well on dry warm land, or in small inclosures, where they are very subject to blight, and are frequently attacked by a black insect, which the farmers call the black dolphin; these insects are often in such quantities as to cover the stems of the Beans entirely, especially all the upper part of them; and whenever this happens, the Beans seldom come to good; but in the open fields, where the soil is strong and the plants have room, this rarely happens.

These Beans are usually sown on land which is fresh broken up, because they are of use to break and pulverize the ground, as also to destroy weeds; so that the land is rendered much better for Corn after a crop of Beans, than it would have been before, especially if they are sown and managed according to the new husbandry, with a drill plough, and the horse-hoe, used to stir the ground between the rows of Beans, which will prevent the growth of weeds, and pulverize the ground, whereby a much greater crop of Beans may, with more certainty, be expected, and the land will be better prepared for whatever crop it is designed for after.

The season for sowing of Beans is from the middle of February to the end of March, according to the nature of the soil; the strongest wet land should always be last sown; the usual quantity of Beans, sown on an acre of land, is about three bushels, but this is double the quantity that need be sown, especially according to the new husbandry; but I shall first set down the practice according to the old husbandry, and then give directions for their management according to the new. The method of sowing is after the plough, in the bottom of the furrows, but then the furrows should not be more than five, or at most six inches deep. If the land is new broken up, it is usual to plough it early in autumn, and let it lie in ridges till after Christmas; then plough it in small furrows, and lay the ground smooth; these two ploughings will break the ground fine enough for Beans, and the third ploughing is to sow the Beans when the furrows should be made shallow, as was before mentioned.

Most people set their Beans too close, for, as some lay the Beans in the furrows after the plough, and others lay them before the plough and plough them in, so by both methods the Beans are set as close as the furrows are made, which is much too near; for when they are on strong good land, they generally are drawn up to a very great height, and are not so apt to pod as when they have more room, and are of a lower growth; therefore I am convinced by some late trials, that the better way is to make the furrows three feet asunder, or more, wh ch will cause them to branch out into many stalks, and bear in greater plenty than when they are closer; by this method less than half the quantity of Beans will be sufficient for an acre of land, and the air being admitted between the rows, the Beans will ripen much earlier and more equally than in the common way.

What has been mentioned must be understood as relating to the old husbandry; but where Beans are planted according to the new, the ground should be four times ploughed before the Beans are set, which will break the clods, and render it much better for planting; then with a drill-plough, to which a hopper is fixed for setting of the Beans, the drills should be made at three feet asunder, and the spring of the hopper set so as to scatter the Beans at three inches distance in the drills. By this method, less than one bushel of seed will plant an acre of land. When the Beans are up, if the ground is stirred between the rows with a horse-plough, it will destroy all the young weeds; and when the Beans are advanced about three or four inches high, the ground should be again ploughed between the rows, and the earth laid up to the Beans; and if a third ploughing, at about five or six weeks after is given, the ground will be kept clean from weeds, and the Beans will stalk out, and produce a much greater crop than in the common way.

When the Beans are ripe, they are reaped with a hook, as is usually practised for Pease: and after having lain a few days on the ground they are turned, and this must be repeated several times, until they are dry enough to stack; but the best method is to tie them in small bundles, and set them upright, for then they will not be in so much danger to suffer by wet as when they lie on the ground, and they will be more handy to carry and stack than if they are loose. The common produce is from twenty to twenty-five bushels on an acre of land.

The Beans should lie in the mow to sweat, before they are threshed out; for as the haulm is very large and succulent, so it is very apt to give and grow moist; but there is no danger of the Beans receiving damage, if they are stacked tolerably dry, because the pods will preserve the Beans from injury, and they will be much easier to thresh after they have sweat in the mow than before; and after they have once sweated, and are dry again, they never after give.

By the new husbandry, the produce has exceeded the old by more than ten bushels on an acre; for if the Beans which are cultivated in the common method are observed when they are in pod, it will be found that more than half way of their stems have no Beans on them; for by standing close, they are drawn up very tall, so the tops of the stalks only produce, and all the lower part is naked; whereas in the new method they bear almost to the ground, and as the joints of the stems are shorter, so the Beans grow closer together.

FABA ÆGYPTIACA. See Arum Ægyptiacum.

FABA CRASSA. See Anacampseros, or Sedum.

FABAGO. See Zygophyllum.

FAGARA. Brown. Hist. Jam. tab. 5. fig. 1. Ironwood.

The Characters are,

It hath male and hermaphrodite flowers on different plants; the male flowers have a small empalement, but have no petals, and six stamina, terminated by round summits; these are barren: the hermaphrodite flowers have a larger concave permanent empalement, and four spreading petals with four stamina, crowned with oval summits, and an oval germen, supporting a slender style, terminated by an obtuse stigma; the germen afterward becomes a globular capsule with two cells, containing two seeds.

The Species are,

1. Fagara (*Pterota*) foliolis emarginatis. Amœn. Acad. 5. p. 393. *Fagara whose small leaves or lobes are indented at the top.*

2. Fagara (*Fragodes*) articulis pinnarum subtus aculeatus. Jacq. Amer. 13. *Fagara with spines at the joints under the leaves.*

The first sort grows naturally in all the warm parts of America. The late Dr. Houstoun sent me several specimens of it from Campeachy; by the flowers upon these I am fully convinced, that there are male and hermaphrodite flowers on separate trees: they rise with woody stems more than

than twenty feet high, fending out branches great part of their length, garnifhed with fmall winged leaves, each having three or five lobes; the flowers are produced from the fide of the branches, which are of an herbaceous colour, ftanding four or five together upon fhort foot-ftalks.

The fecond fort I have placed here according to Dr. Linnæus, but I am not fure he is right in ranging it here; for although I have fome large plants of this fort in the Chelfea garden, yet none of them have flowered, but by the external face of the plants they feem to agree.

They are eafily propagated by feeds, which muft be procured from the countries where they naturally grow; thefe muft be fown on a good hot-bed, when the plants are up fit to tranfplant, they fhould be each planted in a fmall pot, filled with light earth, and plunged into a hot-bed of tanners bark, and fhould always be kept in a tan-ftove.

FAGONIA. Tourn. Inft. R. H. 265. tab. 141. Lin. Gen. Plant. 475.

The CHARACTERS are,

The flower hath five heart-fhaped petals, which fpread open, and are narrow at their bafe, where they are inferted in the empalement; it hath ten erect ftamina. In the center is fituated a five-cornered germen, which afterward becomes a roundifh capfule, having five lobes, ending in a point, and five cells, each having a fingle roundifh feed.

The SPECIES are,

1. FAGONIA (*Erecta*) fpinofa, foliolis lanceolatis planis lævibus. Hort. Upfal. 103. *Thorny Trefoil of Candia.*

2. FAGONIA (*Hifpanica*) inermis. Lin. Sp. Plant. 386. *Spanifh Fagonia without thorns.*

3. FAGONIA (*Arabica*) fpinofa, foliolis linearibus convexis. Lin. Sp. Pl. 386. *Arabian Fagonia armed with very long fpines.*

The firft fort is a native of the ifland of Candia, and has been defcribed by fome botanifts under the title of Trifolium fpinofum Creticum, which occafioned my giving it the Englifh name of Thorny Trefoil of Crete; though there is no other affinity between this and the Trefoil, than that of this having three leaves or lobes on the fame foot-ftalk.

It is a low plant, which fpreads its branches clofe to the ground, extending to the length of a foot or more every way, garnifhed with fmall trifoliate leaves placed oppofite; at each joint, immediately below the leaves, come out two pair of fpines, one on each fide the ftalk; and at the fame places come out a fingle blue flower, ftanding upon a fhort foot-ftalk, compofed of five fpear-fhaped petals, which are narrow at their bafe, where they are inferted into the empalement; after thefe fall away the germen turns to a roundifh five-lobed capfule, ending in an acute point, having five cells, each containing one roundifh feed: it flowers in July and Auguft, but unlefs the feafon proves warm the feeds do not ripen in England.

The fecond fort grows naturally in Spain; this differs from the firft in being fmooth, the branches of this having no thorns; and the plant will live two years, whereas the firft is annual.

The third fort was difcovered by the late Dr. Shaw in Arabia; this is a low plant with a fhrubby ftalk, from which come out feveral weak branches armed with long thorns; the leaves of this are thick, narrow, and convex on their lower fide; the flowers come out in the fame manner as in the firft fort.

Thefe plants are propagated by feeds, which fhould be fown upon a border of frefh light earth, where the plants are defigned to remain; when the plants come up they may be thinned out to the diftance of ten inches or a foot, and if they are kept lean from weeds, they will require no other care.

The firft and fecond forts are annual plants, which do not perfect their feeds in England, unlefs the feafons prove very warm, therefore the beft way is to fow their feeds upon a warm border in the autumn, and in frofty weather fhelter the plants with mats, or fome other covering, to fecure them; or if they are fown in pots, and placed under a frame in the winter, and the following fpring fhaken out of the pots, and planted in a warm border, they will come early to flower, and thereby ripe feeds may be more certainly obtained.

The other fort may be treated in the fame way, for as the plants feldom flower the firft year from feeds till the autumn, fo the plants fhould be either kept in pots, or fheltered under a frame in winter, or placed in a warm border, where they may be fheltered with mats, or fome other covering, to preferve them from the froft, and the following fummer it will flower and produce ripe feeds.

FAGOPYRUM. See HELXINE.

FAGUS. Tourn Inft. R. H. 584. tab. 351. The Beech-tree.

The CHARACTERS are,

It hath male and female flowers on the fame tree; the male flowers are collected into globular heads, and have no petals, but have feveral ftamina included in an empalement of one leaf; the female flowers have a one-leaved empalement cut into four parts, but have no petals; the germen is fixed to the empalement, which afterward becomes a roundifh capfule, armed with foft fpines, opening in three cells, each containing a triangular nut.

We have but one SPECIES of this genus, viz.

FAGUS (*Sylvatica.*) Dod. Pempt. 832. *The Beech-tree.*

There are fome who fuppofe there are two diftinct fpecies of this tree; one they call the Mountain Beech, which they fay is a whiter wood than the other, and they diftinguifh it by the title of Wild Beech; but it is certain, that this difference in the colour of the wood arifes from the difference of the foils in which they grow, for I have not feen any fpecific difference in the trees. There have been feeds of a Beech-tree brought from North America, by the title of Broad leaved Beech; but the plants which were raifed from them, proved to be the common fort, fo that we know of no other variety, excepting that with ftriped leaves, which is accidental.

This tree is propagated by fowing the maft; the feafon for which is any time from October to February, only obferving to fecure the feeds from vermin when early fowed, which, if carefully done, the fooner they are fown the better, after they are fully ripe: a fmall fpot of ground will be fufficient for raifing a great number of thefe trees from feed, for if the plants come up very thick, the ftrongeft of them fhould be drawn out the autumn following, that thofe left may have room to grow; fo that a feed-bed carefully managed, will afford a three years draught of young plants, which fhould be planted in a nurfery, and, if defigned for timber-trees, at three feet diftance row from row, and eighteen inches afunder in the rows.

But if they are defigned for hedges (to which the tree is very well adapted) the diftance need not be fo great; two feet row from row, and one foot in the rows, will be fufficient. In this nurfery they may remain two or three years, obferving to dig up the ground between the roots at leaft once a year, that their tender roots may the better extend themfelves each way: but be careful not to cut or bruife their roots, which is injurious to all young trees; and never dig the ground in fummer, when the earth is hot and dry, which, by letting in the rays of the fun to the roots, is often the deftruction of young trees.

This tree will grow to a confiderable ftature, though the foil be ftony and barren, as alfo upon the declivity of hills, and chalky mountains, where they will refift the winds better

better than moſt other trees; but then the nurſeries for the young plants ought to be made upon the ſame ſoil, for if they are raiſed in good ground and a warm expoſure, and afterwards tranſplanted into a bleak barren ſituation, they ſeldom thrive, which holds true in moſt other trees; therefore I would adviſe the nurſery to be made upon the ſame ſoil where the plantation is intended, and to annually draw out plants to extend the plantation.

The tree is very proper to form large hedges to ſurround plantations or wilderneſs quarters, and may be kept in a regular figure, if ſheared twice a year, eſpecially if they ſhoot ſtrong; in which caſe, if they are neglected but a ſeaſon or two, it will be difficult to reduce them again. The ſhade of this tree is very injurious to moſt ſorts of plants which grow near it, but is generally believed to be very ſalubrious to human bodies.

The timber is of great uſe to turners for making trenchers, diſhes, trays, buckets; and likewiſe to the joiner for ſtools, bedſteads, coffins, and is eſteemed the beſt wood for firing, &c. The maſt is very good to fat ſwine and deer, and alſo affords a ſweet oil.

It delights in a chalky or ſtony ground, where it generally grows very faſt; the bark of the trees in ſuch land is clear and ſmooth, and although the timber is not ſo valuable as that of many other trees, yet as it will thrive on ſuch ſoils, and in ſuch ſituations where few better trees will grow, the planting of them ſhould be encouraged, eſpecialy as the trees afford an agreeable ſhade, and the leaves make a fine appearance in ſummer, and continue green as long in autumn as any of the deciduous trees; therefore in parks, and other plantations for pleaſure, this tree deſerves to be cultivated among thoſe of the firſt claſs, eſpecially where the ſoil is adapted to it.

FEATHERFEW, or FEAVERFEW. See MATRICARIA.

FENNEL. See FOENICULUM.

FENNEL-FLOWER. See NIGELLA.

FERRARIA. Burm.

The CHARACTERS are,

The flower is incloſed in a double ſpatha or ſheath; out of each ſheath is produced one flower; the flower has ſix oblong acute-pointed petals, three of which are alternately larger than the other; theſe ſeem to be joined at their tails; their borders are fringed, turned backward, and ſilky; it has three ſtamina which riſe above the petals, and a roundiſh three-cornered germen ſituated under the flower, which afterward turns to an oblong, ſmooth, three-cornered capſule, crowned by the decayed petals of the flower with three cells, wrapped in the permanent ſheath of the flower; in each cell is lodged many ſmall round ſeeds.

The SPECIES are,

1. FERRARIA (*Undulata*) foliis lanceolatis. Burm. *Ferraria with ſpear-ſhaped leaves.*

2. FERRARIA (*Enſiformi*) foliis enſiformibus. Burm. *Ferraria with ſword-ſhaped leaves.*

Theſe plants grow naturally at the Cape of Good Hope; the root of the firſt ſort was ſent me by Dr. Job Baſter, F. R. S. of Zirkzee, who had received it from the Cape of Good Hope; the root is tuberous, roundiſh, and compreſſed, ſhaped like the root of Indian Corn Flag, but larger; in the center of the upper ſide of the root there is a hollow like a navel, from whence the flower-ſtalk comes out; the outer cover is of a light brown colour; the ſtalk riſes a foot and a half high, as thick as a man's little finger, garniſhed with leaves the whole length, which are placed alternate, and embrace the ſtalks with their baſe; they are ſmooth, a little keel-ſhaped, and of a light green colour: the upper part of the ſtalk divides into two or three branches, garniſhed with leaves of the ſame ſhape with thoſe below, but much ſhorter; each of the branches is terminated by a ſpatha or ſheath, which is at firſt of the ſame colour with the leaves, but when the flower fades, this withers and dies, but remains upon the ſtalk; theſe ſheaths are double one within the other. The flower riſes out of the top upon a very ſhort foot-ſtalk; it has ſix oblong petals, three of which are alternately larger than the other, finely fringed on their edges, and are reflexed at their points; they are of a whitiſh green on their outſide, but of a tawny Violet colour within, having many ſilky hairs on their ſurface. In the center of the flower is ſituated an erect ſtyle, which ſits upon the germen, and is crowned with three bifid ſtigmas, which terminate in hairs. On one ſide the ſtyle is ſituated three ſtamina, which divide toward the top, and are terminated by roundiſh twin ſummits. The germen, which is ſituated under the flower, afterward becomes an oblong ſmooth capſule with three cells, filled with roundiſh ſeeds.

There is a great ſingularity in the root of this plant, which is, that it vegetates but every other year, and ſometimes every third year; the intermediate time it remains unactive, though very ſound and good.

The ſecond ſort is figured by Dr. Burman, but I have not ſeen it as yet in England; it differs from the former in having longer leaves, which are ſword-ſhaped and furrowed like thoſe of Byzantine Corn Flag. The ſtalk riſes about the ſame height, but does not divide ſo much; the flowers are ſmaller, and the petals are not ſo much fringed.

Theſe plants are propagated by ſeeds, when they can be procured from the country where they naturally grow; theſe ſhould be ſown in pots as ſoon as they arrive, and require the ſame treatment as the Ixia. The roots alſo ſend out offsets, but it is ſparingly in England, and theſe are ſome years before they arrive to a ſize for flowering, which occaſions the ſcarcity of theſe plants in Europe. The roots alſo require the ſame culture with the Ixia and Watſonia, for they are too tender to live through the winter in the open air here, ſo muſt be planted in pots filled with light loamy earth, and ſheltered under a frame in winter in the ſame manner as hath been before directed for WATSONIA and IXIA, under which articles the reader will find proper directions for their culture and management; but thoſe years when theſe roots do not vegetate, they ſhould have very little water; and in the ſummer the pots ſhould be placed where they may have only the morning ſun, for if they are expoſed to the mid-day ſun, the earth will dry too much, ſo may require water often, which frequently rots the roots when they are in an inactive ſtate.

The mice are very fond of theſe roots, therefore they ſhould be protected from thoſe vermin, otherwiſe they will deſtroy them.

FERRUM EQUINUM. See HIPPOCREPIS.

FERULA. Lin. Gen. Plant. 305. Fennel Giant.

The CHARACTERS are,

It hath an umbellated flower; the principal umbel is compoſed of ſeveral ſmaller, called rays; the involucrum is compoſed of ſeveral narrow leaves, which fall off; the principal umbel is uniform. The flowers have five oblong petals and five ſtamina of the ſame length; under the flower is ſituated a turbinated germen, which afterward becomes an elliptical, compreſſed, plain fruit, dividing in two parts, each having a large, elliptical, plain ſeed.

The SPECIES are,

1. FERULA (*Communis*) foliolis linearibus longiſſimis ſimplicibus. Hort. Cliff. 95. *Pliny's Female Fennel Giant.*

2. FERULA (*Galbanifera*) foliolis multipartitis, laciniis linearibus planis. Hort. Cliff. 95. *Galbanum-bearing Fennel Giant of Lobel.*

3. FERULA

3. Ferula (*Tingitana*) foliolis laciniatis, lacinulis tridentatis inæqualibus. Hort. Cliff. 95. *Broad-leaved shining Fennel Giant from Tangier.*

4. Ferula (*Ferulago*) foliis pinnatifidis, pinnis linearibus planis trifidis. Hort. Cliff. 95. *Fennel Giant with a broader leaf.*

5. Ferula (*Orientalis*) foliorum pinnis basi nudis, foliolis setaceis. Hort. Cliff. 95. *Eastern Fennel Giant, with the leaf and appearance of Cachrys.*

6. Ferula (*Meoides*) foliorum pinnis utrinque basi acutis, foliolis setaceis. Hort. Cliff. 95. *Eastern Laserwort, with a Spignel leaf and yellow flower of Tournefort.*

7. Ferula (*Nodiflora*) foliolis appendiculatis, umbellis subsessilibus. Lin. Sp. Pl. 247. *Libanotis with a Fennel Giant leaf and seed.*

8. Ferula (*Glauca*) foliis pinnatifidis, pinnis linearibus planis brevioribus. *Fennel Giant with a narrower leaf of Tournefort.*

The first of these plant is pretty common in the English gardens: this, if plante in a good soil, will grow to the height of ten or twelve eet, and divide into many branches: the lower leaves of this sort spread more than two feet every way, which are subdivided into many small ones, garnished with very long, narrow, small leaves, of a lucid green. From the center of the plant comes out the flower-stalk, which, when the plants are strong, will be near as large as a common broomstick, and will rise ten or twelve feet high, with many joints; if the stalks are cut, there issues from the vessels a fœtid yellowish liquor, which will concrete on the surface of the wound. The stalks are terminated by large umbels of yellow flowers, which come out the latter end of June, or in the beginning of July, which are succeeded by oval compressed seeds, having three lines running longitudinally on each side. These ripen in September, and the stalks decay soon after.

The second sort doth not grow quite so large as the first, but the stalks of this will rise seven or eight feet high; the lower leaves are large, and greatly divided; the small leaves are flat, and not so long as those of the former, and are of a lucid green colour; the umbels of flowers are smaller, and the seeds are less.

The third sort hath large spreading leaves near the root, which are divided and subdivided into many parts; the small leaves of this are much broader than in any of the other sorts, and are divided at their end into three unequal segments. The stalks are strong, and rise to the height of eight or ten feet, terminated by large umbels of yellow flowers, which are succeeded by large, oval, compressed seeds, like those of the first sort. This flowers and ripens its seeds about the same time as the former sort: it grows naturally in Spain and Barbary.

The fourth sort grows to much the same height as the second; the smaller lobes or divisions of the leaves are broader than those of the others (excepting the third) but they are longer than those, and of a darker green colour, ending in three points. The umbels of flowers are large; the flowers are yellow, and are succeeded by oval compressed seeds like those of the other species. This grows naturally in Sicily.

The fifth sort is of much humbler growth than either of the former; the stalks of this seldom rise much more than three feet high; the lower leaves branch into many narrow bristly divisions; the umbel of flowers is small, when compared with the others, as are also the seeds. It grows naturally in the Levant.

The sixth sort hath very branching leaves; the foot-stalks are angular and channelled; this sends out at every joint two side branches opposite; those toward the bottom are nine or ten inches long, and the others are diminished gradually to the top, garnished with very fine leaves like those of Spignel, which stand round the stalks in shape of whorls; the flower-stalks grow three feet high, with a pretty large umbel of yellow flowers at the top, which are succeeded by oval flat seeds. It grows naturally in the Levant.

The seventh sort rises about three feet high; the leaves of this sort are much divided, and the small leaves on the divisions are very narrow and entire; the umbels of flowers are small, and are situated close to the stalks between the leaves at the joints; these are like those of the other sorts. It grows naturally in Istria and Carniola.

The eighth sort is like the second, but the small leaves are much broader; the large leaves are not so long, and the stalk grows taller, but in other respects agrees with that. It grows naturally in Spain.

All these sorts have perennial roots, which will continue several years, and run deep in the ground: the stalks are annual, and decay soon after they have perfected their seeds. As these plants spread very wide, so they should have each four or five feet room; nor should they stand near to other plants, for their roots will rob whatever plants grow near them of their nourishment.

They are all propagated by seeds, which should be sown in the autumn; for if they are kept out of the ground till the spring, they frequently fail, and those which succeed remain a year in the ground, so that much time is lost. The seeds may be sown in drills, by which method the ground may be easier kept clean; the drills must not be nearer than a foot, and the seeds may be scattered two or three inches asunder in the drills. When the plants come up, if they should be too close together, they should be thinned, to allow them room to grow, for they will not be strong enough to remove till they have had two years growth; then in the autumn, so soon as their leaves decay, the roots should be taken up with great care, so as not to cut or injure the tap or downright root, and planted in the places where they are designed to remain, for after this transplanting they should not be removed. They delight in a soft, gentle, loamy soil, not too wet, and are very rarely injured by the hardest frost.

FICOIDES. See Mesembryanthemum.

FICUS. Lin. Gen. Pl. 1032. The Fig-tree.

The Characters are,

It hath male and female flowers, which are included within the skin of the fruit, so do not appear unless the covering is opened; the male flowers are but few in number, and are situated in the upper part of the fruit; the female flowers are numerous, situated in the lower part. The male flowers sit each upon a separate foot-stalk; they have no petals, but three bristly stamina; the female flowers sit upon distinct foot-stalks; they have no petals, but a germen, which afterward becomes a large seed, sitting in the empalement.

The Species are,

1. Ficus (*Carica*) foliis palmatis. Hort. Cliff. 471. *Fig-tree with hand-shaped leaves; or the common Fig-tree.*

2. Ficus (*Sycomorus*) foliis cordatis subrotundis integerrimis. Hort. Cliff. 471. *Fig-tree with a Mulberry leaf, bearing fruit on the body or stem; commonly called Sycamore.*

3. Ficus (*Religiosa*) foliis cordatis integerrimis acuminatis. Hort. Cliff. 471. *Malabar Fig with a long-pointed leaf, and small, double, round fruit.*

4. Ficus (*Benghalensis*) foliis ovatis integerrimis obtusis, caule inferne radicato. Hort. Cliff. 471. *Bengal Fig with a roundish leaf and orbicular fruit.*

5. Ficus (*Indica*) foliis lanceolatis petiolatis, pedunculis aggregatis, ramis radicantibus. Lin. Sp. Pl. 1060. *Indian Fig of Theophrastus.*

6. Ficus (*Maxima*) foliis lanceolatis integerrimis. Hort. Cliff. 471. *The largest Indian Fig, with an oblong leaf, send-*

ing

ing out roots from the tops of the branches, and a small, spherical, blood-coloured fruit.

7. Ficus (*Racemoso*) foliis ovatis acutis integerrimis, caule arboreo, fructu racemoso. Lin. Sp. Plant. 1060 *Fig-tree with oval, entire, acute leaves, a tree-like stalk, and fruit growing in bunches.*

8. Ficus (*Pumila*) foliis ovatis acutis integerrimis, caule repente. Lin. Sp. Pl. 1060. *Trailing wild Fig-tree, having single leaves.*

9. Ficus (*Nymphæcefolia*) foliis ovato-cordatis integerrimis glabris. *Fig-tree with oval, heart-shaped, entire, smooth leaves, vulgarly called Fig-tree with a Water Lily leaf.*

10. Ficus (*Citrifolia*) foliis oblongo-cordatis acuminatis, petiolis longissimis. *Fig-tree with a Citron leaf, and small purple fruit.*

The first sort, which is the only Fig whose fruit is valuable, is cultivated in most parts of Europe; of this there are great varieties in the warm countries. In England we had not more than four or five sorts till within a few years past, for as the generality of the English were not lovers of this fruit, so there were few who troubled themselves with the culture of it. But some years past I had a large collection of these trees sent me from Venice, by my honoured friend the Chevalier Rathgeb, all which I planted and preserved to taste of their fruits; several of them proved excellent, these I have preserved and propagated, and those whose fruit were inferior have been neglected. As the variety of them is very great, so I shall here mention only such of them as are the best worth cultivating, placing them in the order of their ripening.

1. The Brown and Chestnut-coloured Ischia Fig, is the largest fruit of any I have yet seen; it is of a globular figure, with a large eye, pinched in near the foot-stalk, of a brown or Chestnut colour on the outside, and a purple within; the grains are large, the pulp sweet and high flavoured: this sort very often bursts open when it ripens, which is the latter end of July, or the beginning of August. I have had this fruit ripen well on standards, in a warm soil. If this sort is planted against hot walls, two plentiful crops of fruit may be annually ripened; for against a south-east wall, many of the second crop do annually ripen without art.

2. The Black Genoa Fig. This is a long fruit, which swells pretty large at the top where it is obtuse, but is very slender toward the stalk; the skin is of a dark purple colour, almost black, and hath a purple farina over it like that on some Plumbs; the inside is of a bright red, and the flesh is very high flavoured. It ripens early in August.

3. The small, white, early Fig. This hath a roundish fruit, a little flatted at the crown, with a very short foot-stalk; the skin, when fully ripe, is of a pale yellowish white colour is thin, the inside white, and the flesh very sweet, but not high flavoured. This ripens in August.

4. The large white Genoa Fig. This is a large globular fruit, a little lengthened toward the stalk: the skin is thin, of a yellowish colour when fully ripe, and red within. It is a good fruit, but the trees are not good bearers.

5. The Black Ischia Fig. This is a short fruit, of a middling size, a little flatted at the crown; the skin is black when ripe, and the inside is of a deep red; the flesh is very high flavoured, and the trees produce a good crop of fruit, but the birds are great devourers of them, if they are not protected from them. This ripens in August.

6. The Malta Fig. This is a small brown fruit, much compressed at the top, and greatly pinched toward the foo stalk; the skin is of a pale brown colour, as is also the in side; the flesh is very sweet, and well flavoured. If this sort is permitted to hang upon the trees till the fruit is shrivelled, it becomes a fine sweetmeat.

7. The Murrey, or Brown Naples Fig. This is a pretty large globular fruit, of a light brown colour on the outside, with some faint marks of a dirty white; the inside is nearly of the same colour, the grains are pretty large, and the flesh is well flavoured. It ripens the latter end of August, but is a bad bearer.

8. The Green Ischia Fig. This is an oblong fruit, almost globular at the crown; the skin is thin, of a green colour, but when it is fully ripe, it is stained through by the pulp to a brownish cast; the inside is purple, and will stain linen or paper; the flesh is well flavoured, especially in warm seasons. It ripens toward the end of August.

9. The Madonna Fig, commonly called here the Brunswick or Hanover Fig, is a long pyramidal fruit of a large size; the skin is brown; the flesh is of a lighter brown colour, coarse, and hath little flavour. This ripens the end of August and the beginning of September; the leaves of this sort are much more divided than of most other.

10. The common Blue, or Purple Fig, is so well known as to need no description.

11. The long brown Naples Fig. The leaves of this tree are deeply divided. This fruit is long, somewhat compressed at the crown. The foot-stalks are pretty long; the skin of a dark brown when fully ripe, the flesh inclining to red; the grains are large, and the flesh well flavoured. It ripens in September.

12. The Gentile Fig. This is a middle-sized oval fruit; the skin, when ripe, is yellow, the flesh also inclines to the same colour; the grains are large, and the flesh is well flavoured, but it ripens very late, and the trees are bad bearers, so that it is not propagated much in England.

The first, second, third, ninth, and tenth sorts, will ripen their fruits on standards, where they are in a warm situation; but the others require the assistance of walls exposed to good aspects, otherwise their fruit will not ripen well in England.

Fig-trees generally thrive in all soils and in every situation, but they produce a greater quantity of fruit upon a strong loamy soil than on dry ground; for if the season proves dry in May and June, those trees which grow upon very warm dry ground, are very subject to cast their fruit; therefore, whenever this happens, such trees should be well watered and mulched, which will prevent the fruit from dropping off; but the fruit upon these trees are better flavoured than any of those which grow upon cold moist land. I have always observed those Fig-trees to bear the greatest quantity of well-flavoured fruit, which were growing upon chalky land, where there has been a foot or more of a gentle loamy soil on the top. They also love a free open air, for although they will shoot and thrive very well in close places, yet they seldom produce any fruit in such situations; and all those which are planted in small gardens in London, will be well furnished with leaves, but I have never seen any fruit upon them which have grown to maturity.

These trees are always planted as standards in all warm countries, but in England they are generally planted against walls, there being but few standard Fig trees at present in the English gardens; however, since some of the sorts are found to ripen their fruit well upon standards, and the crop of Figs is often greater upon them than upon those trees against walls, it is worthy of our care to plant them either in standards or espaliers; the latter, I think, will succeed best in England, if they were managed as in Germany, where they untie the Fig-trees from the espalier, and lay them down, covering them in winter with straw or litter, which prevents their shoots being injured by the frost; and this covering is taken away gradually in the spring, and not wholly removed until all the danger of frost is over, by which they generally have a very great crop of Figs; whereas

whereas in England, where the trees grow in warm situations, if the ſpring proves warm, the young Figs are puſhed out early, and the cold, which frequently returns in May, cauſes the greateſt part of the fruit to drop off; ſo that our crop of Figs is generally more uncertain than moſt other ſorts of fruit; and it frequently happens, that trees which are planted againſt north and eaſt aſpected walls, produce a greater quantity of fruit in England than thoſe which are planted againſt ſouth and ſouth-eaſt aſpects, which muſt happen from the latter putting out their fruit ſo much earlier in the ſpring than the former; and if there happen cold froſty nights after the Figs are come out (which is frequently the caſe in this country) the forwardeſt of the Figs are generally ſo injured as to drop off from the trees ſoon after. In Italy, and the other warm countries, this firſt crop of Figs is little regarded, being few in number; for it is the ſecond crop of Figs which are produced from the ſhoots of the ſame year, which is their principal crop; but theſe rarely ripen in England, nor are there above three or four ſorts which ever ripen their ſecond crop, let the ſummer prove ever ſo good, unleſs they have hot walls, therefore it is the firſt crop which we muſt attend to in England; ſo that when theſe trees are growing againſt the beſt aſpected walls, it will be a good method to looſen them from the wall in autumn; and after having diveſted the branches of all the latter fruit, to lay the branches down from the wall, faſtening them together in ſmall bundles, ſo that they may be tied to ſtakes, to keep them from lying upon the ground, the damp whereof, when covered in froſty weather, might cauſe them to grow mouldy; this will ſecure the branches from being broken by the wind. When they are thus managed in autumn, if the winter ſhould prove very ſevere, the branches may be eaſily covered with Peaſe-haulm, ſtraw, or any other light covering, which will guard the tender fruit-bearing branches from the injury of froſt; and when the weather is mild, the covering muſt be removed, otherwiſe the Figs will come out too early; for the intention of this management is, to keep them as backward as poſſible; then in the ſpring, when the Figs begin to puſh out, the trees may be faſtened to the wall again. By this management I have ſeen very great crops of Figs produced in two or three places.

I have alſo ſeen great crops of Figs in ſome particular gardens, after very ſharp winters, when they have, in general, failed in other places, by covering up the trees with Reeds made into pannels, and fixed up againſt the walls.

In the pruning of Fig-trees, the branches muſt never be ſhortened, becauſe the fruit are all produced at the upper part of the ſhoots of the former year, ſo, if theſe are cut off, there can be no fruit expected, beſide the branches are very apt to die after the knife; ſo that when the branches are too cloſe together, the beſt way is to cut out all the naked branches quite to the bottom, leaving thoſe which are beſt furniſhed with lateral branches at a proper diſtance from each other, which ſhould not be nearer than a foot; when they are well furniſhed with lateral branches, if they are laid four or five inches farther aſunder it will be better.

The beſt time for pruning of Fig-trees is in autumn, becauſe at that time the branches are not ſo full of ſap, ſo they will not bleed ſo much as when they are pruned in the ſpring; and at this ſeaſon the branches ſhould be diveſted of all the autumnal Figs, and if the bud at the extremity of the ſhoots are rubbed off with the finger, it will cauſe them to put out a great crop of fruit in the ſpring; the ſooner this is done, when the leaves begin to fall off, the better will the young ſhoots reſiſt the cold of the winter. There are ſome ſeaſons ſo cold and moiſt, that the young ſhoots of the Fig-trees will not harden, but are ſoft, and full of juice; when this happens, there is little hope of a crop of Figs the ſucceeding year, for the firſt froſt in autumn will kill the upper part of theſe ſhoots. Whenever this happens, it is the beſt way to cut off all the decayed part of the ſhoots, which will prevent the infection from ſpreading to the lower part of the branches.

Thoſe trees which are laid down from the eſpaliers, ſhould not be faſtened up again till the end of March, or the beginning of April, for the reaſons before given, and thoſe againſt walls may remain ſome time longer; when the large ſhoots of theſe are nailed up, if the ſmall lateral branches are thruſt behind them to keep them cloſe to the wall, it will ſecure the young Figs from being injured by the morning froſts, and when this danger is over, they may be brought forward to their natural poſition again; during the ſummer ſeaſon theſe trees will require no other pruning, but to ſtop the ſhoots in the ſpring, where lateral branches are wanting; and as the branches are often blown down by wind, therefore, whenever this happens, they ſhould be immediately faſtened up again, otherwiſe they will be in danger of breaking, for the leaves of theſe trees being very large and ſtiff, the wind has great power on them, ſo that where the branches are not well ſecured they are frequently torn down.

Thoſe trees which are planted againſt eſpaliers, which are not laid down as before directed, may be protected from the injury of froſt in winter, by placing Reeds on each ſide the eſpalier, which may be taken down every day, and put up again at night; but this need not be practiſed in warm weather, but only at ſuch times as there are cold winds and froſty mornings; and although there is ſome trouble and expence attending this management, yet the plentiful crop of Figs which may be this way obtained, will ſufficiently recompenſe for both: the beſt way of making this covering is, to faſten the Reeds with rope-yarn, in ſuch a manner, as that it may be rolled up like a mat, ſo that the whole may with great facility be put up or taken down; and if theſe Reeds are carefully rolled up after the ſeaſon for uſing them is over, and put up in a dry ſhed, they will laſt ſeveral years.

I am aware, that what I have here advanced in relation to the pruning and dreſſing of Fig-trees, will be condemned by great numbers of people, who will not give themſelves time to conſider and examine the reaſons upon which I have founded this practice, nor to make one ſingle experiment to try the truth of it, as being vaſtly different from the general practice of moſt gardeners, who always imagine, that Fig-trees ſhould never have any pruning, or, at leaſt, that they ſhould always be ſuffered to grow very rude from the wall to ſome diſtance. By this management I have often ſeen great quantities of fruit, I cannot deny, but then this has been only after mild winters; for it is very certain, that in ſharp froſts few of theſe outſide ſhoots eſcape being greatly injured where they are not covered; whereas it rarely happens, that thoſe ſhoots which are cloſely faſtened to the wall in autumn, or laid down and covered, ſuffer the leaſt damage, and the fruit are alway produced a fortnight ſooner upon theſe branches than they are upon thoſe which grow from the wall; but although the trees which are ſuffered to grow rude from the walls, may produce a good quantity of fruit for a year or two, yet afterward the trees will only bear at the ends of the ſhoots, which will then be ſo far from the wall, as to receive little benefit from it; nor can the trees be reduced again to any regularity, without cutting away the greateſt number of their branches, by which a year or two will be loſt before they will come to bear again.

The ſeaſon alſo for pruning, which I have laid down, being vaſtly different from the common practice and opinion of moſt gardeners, may alſo be objected to; but I am ſure, if any one will but make trial of it, I doubt not his experience will confirm what I have here advanced; for as one great injury to this tree proceeds from the too great effusion of fap at the wounded parts, fo by this autumn pruning this is prevented; for at that feafon all the parts of European trees which caft their leaves, are lefs replete with moifture than at any other time of the year; for by the long continuance of the fummer's heat, the juices of plants having been exhaufted in the nourifhment and augmentation of wood, leaves, fruits, &c. and alfo great quantities being evaporated by perfpiration, the root not being able to fend up a fupply equivalent to this great confumption, the branches muft contain a much lefs quantity of fap in autumn than in fpring, when it has had feveral months fupply from the root, which though but fmall in proportion to what is fent up when the heat is greater, yet there being little or no wafte, either by perfpiration or augmentation, there muft be a greater quantity contained in the branches; which alfo is eafily to be obferved, by breaking or cutting off a vigorous branch of a Fig-tree at both feafons (the fap being milky, may be readily difcerned;) when that cut in autumn fhall be found to ftop its bleeding in one day's time or lefs, whereas that cut in the fpring will often flow a week or more, and the wound will be proportionably longer before it heals.

Of late years there has been fome of thefe trees planted againft fire-walls, which have fucceeded very well where they have been properly managed, but where they have been kept too clofe, and drawn by glaffes, they have not produced much fruit; therefore whenever this is practifed, the heat fhould not be too great, nor the glaffes or other covering kept too clofe, but at all times when the weather is favourable, a good fhare of free air fhould be admitted; and if the trees are young, that their roots are not extended beyond the reach of the covering, they muft be frequently watered when they begin to fhew fruit, otherwife it will drop off; but old trees, whofe roots are extended to a great diftance, will only require to have their branches now and then fprinkled over with water. If thefe trees are properly managed, the firft crop of fruit will be greater than upon thofe which are expofed to the open air, and will ripen fix weeks or two months earlier, and a plentiful fecond crop may alfo be obtained, which will ripen early in September, and fometimes in Auguft, which is about the feafon of their ripening in the warmer parts of Europe; but the fires fhould not be ufed to thefe trees till the beginning of February, becaufe when they are forced too early, the weather is frequently too cold to admit a fufficient quantity of frefh air to fet the fruit, but the covers fhould be put over the trees two months before, to prevent the fhoots from being injured by the froft.

Fig-trees are propagated in England, either by the fuckers, which are fent out from their roots, and by layers made by laying down of their branches, which in one year will put out roots fufficient to be removed, or by planting of cuttings, which, if properly managed, will take root; the firft of thefe is a bad method, becaufe all thofe trees which are raifed from fuckers, are very fubject to fend out great quantities of fuckers again from the roots. Thofe plants which are propagated by layers are the beft, provided the layers are made from the branches of fruitful trees, for thofe which are made from the fuckers or fhoots produced from old ftools, have very foft branches full of fap, fo are in danger of fuffering by the froft, and thefe will fhoot greatly into wood, but will not be very fruitful; for when the trees have acquired a vicious habit while young, it is feldom they are ever brought to be fruitful afterward; therefore the fhoots that are laid down fhould be fuch as are woody, compact, and well ripened, not young fhoots full of fap, whofe veffels are large and open. The beft time for laying down of the branches is in autumn; if the winter fhould prove very fevere, if they are covered with fome old tan, or any other mulch, to keep the froft from penetrating of the ground, it will be of great fervice to them; by the autumn following thefe will be fufficiently rooted for removing, when they fhould be cut off from the old plants, and tranfplanted where they are to remain. As thefe plants do not bear tranfplanting well when they are large, fo it is the better way to plant them at firft in the places where they are to remain, and after they are planted, the furface of the ground about their roots fhould be covered over with mulch to keep out the froft, if the winter fhould prove very fevere; and if the branches are covered with Reeds, Peafe-haulm, ftraw, or fome other light covering, it will prevent their tender ends being killed by the froft, which frequently happens where this care is wanting.

If fruitful branches of thefe trees are cut off and planted in pots or tubs filled with good earth, and plunged into a good hot-bed of tanners bark in the ftove, they will put out fruit early in the fpring, which will ripen the middle of May.

We fhall now return to the other forts of Figs which grow naturally in warm countries, but are preferved in the gardens of thofe who are curious in collecting of rare exotic plants, for thefe do not bear eatable fruit in their native foil, but their leaves being large and beautiful, the plants make a pleafing variety in the ftove.

The fecond fort grows naturally in the Levant, where it becomes a large tree, dividing into many branches, garnifhed with leaves fhaped like thofe of the Mulberry, fo affords a friendly fhade in thofe hot countries. The fruit is produced from the trunk and large branches of the tree, and not on the fmaller fhoots as in moft other trees; the fhape is like the common Fig, but is little efteemed. This is called the Sycamore, or Pharaoh's Fig tree.

The third fort grows naturally in India, where the trees are facred, fo that none dare deftroy them; it is called by fome the Indian God-tree. It rifes with a woody ftem to a great height, fending out many flender branches, which are garnifhed with fmooth heart-fhaped leaves, fomewhat like thofe of our Black Poplar, ending in a long tail or point; they are entire, fmooth, and of a light green, having pretty long foot-ftalks. The fruit comes out on the branches, which are fmall, round, and of no value.

The fourth fort rifes with many ftalks, which grow to the height of thirty or forty feet, dividing into a great number of branches, which fend out roots from their under fide, many of which reach to the ground; fo that in fuch places where thefe trees grow naturally, their roots and branches are fo interwoven with each other as to render the places impaffable. In India the Banyans trail the branches of thefe trees into irregular archades, and fet up their pagods under them, thefe being the places of their devotion. In America, where thefe trees are equally plenty, they form fuch thickets as neither man or beaft can pafs through. The leaves of this fort are of a thick fubftance, fmooth and oval, fix inches long, and four broad, with obtufe ends. The fruit is the fize of a marble, round, but of no ufe.

The fifth fort grows naturally in both Indies; this rifes with a woody ftalk to the height of thirty feet, fending out many branches, garnifhed with oblong leaves, ftanding upon long foot-ftalks; they are about fix inches long and two broad, ending in an obtufe point, of a dark green, fmooth on their upper fide, but of a light green and veined on their under fide. The fruit is fmall, of no value. The branches of thefe trees fend out roots from the lower fide, which fometimes reach the ground

The fixth fort grows naturally in the Weft-Indies, where it rifes to the height of thirty or forty feet, fending out many flender branches, which put out roots in the fame manner as the former. The leaves of this are eight or nine inches long, and two broad, ending in points. The fruit is fmall, round, of a blood colour when ripe, but is not eatable.

The feventh fort grows naturally in India, where it rifes to the height of twenty-five feet, divides into many branches, garnifhed with oval-pointed fmooth leaves, of a lucid green. The fruit is fmall, and grows in clufters from the fide of the branches, but are not eatable.

The eighth fort grows naturally in India; this is a low trailing fhrub, whofe ftalks put out roots at their joints, which ftrike into the ground, fo is propagated plentifully where it naturally grows. The leaves are two inches and a half long, and two broad, ending in points; they are of a lucid green; the fruit is fmall, and not eatable.

The ninth fort rifes with a ftrong, upright, woody ftalk twenty feet high, fending out feveral fide branches, garnifhed with large, oval, ftiff leaves, about fourteen inches long, and a foot broad, rounded at their ends, and indented at the foot-ftalk; the upper fide of the leaves are of a lucid green, the under fide is of a gray or fea-green colour, of a thick fubftance, and very fmooth; this grows naturally in India.

The tenth fort grows naturally in the Weft-Indies, where it rifes twenty feet high, fending out many fide branches which are covered with a white bark, garnifhed with oblong heart-fhaped leaves, ending in acute points; of a lucid green on their upper fide, but of a pale green on their under, ftanding upon very long foot-ftalks. The fruit comes out from the fide of the branches toward their ends; they are about the fize of large gray Peafe, of a deep purple colour, fitting clofe to the branches, but are not eatable.

The fecond fort, I believe, is not in England at prefent. I raifed two or three of thefe plants from feeds in the year 1736, which were deftroyed by the fevere froft in 1740, fince which time I have not been able to procure any of the feeds. The other forts are preferved in feveral curious gardens; they are eafily propagated by cuttings during the fummer feafon When the cuttings are taken from the plants, they fhould be laid in a dry fhady place for two or three days that the wounds may be healed over, otherwife they are apt to rot; for all thefe plants abound with a milky juice, which flows out whenever they are wounded, for which reafon the cuttings fhould have their wounded part healed over and hardened before they are planted; after which they fhould be planted in pots filled with fandy light earth, and plunged into a moderate hot-bed, where they fhould be fhaded from the fun, and two or three times a week gently refrefhed with water if the feafon is warm, but they muft not have too much moifture, for that will infallibly deftroy them. When the cuttings have taken root fufficient to tranfplant, they fhould be each planted into a feparate fmall pot filled with light undunged earth, and plunged into the hot-bed again, being careful to fhade them until they have taken frefh root; then they fhould have a large fhare of free air admitted to them at all times when the weather is favourable, to prevent their drawing up weak, and to give them ftrength before the cold comes on. In autumn the pots fhould be removed into the ftove, and plunged into the tan-bed, where they fhould conftantly remain, and muft be treated in the fame manner as other tender plants from the fame countries; for although two or three forts may be treated in a hardier manner, yet they will not make much progrefs.

FICUS INDICA. See OPUNTIA.

FILBERT. See CORYLUS.

FILIPENDULA. See SPIRÆA.

FIR-TREE. See ABIES.

FLOS AFRICANUS. See TAGETES.

FLOS PASSIONIS. See PASSIFLORA.

FLOS SOLIS. See HELIANTHUS.

FLOS TRINITATIS. See VIOLA.

FLOWER. A Flower is a natural production which precedes the fruit, that includes the grain or feed. Though a Flower is a thing fo well known, yet the definition of this part of a plant is as various almoft as the authors who define it. Jungius defines it to be the more tender part of a plant, remarkable for its colour or form, or both, cohering with the fruit. Yet this author himfelf confeffes, that this definition is too narrow; for fome of thofe bodies which he allows to be Flowers, are remote from the fruit.

A Flower may be thus defined, viz. it contains the organs of generation of both fexes adhering to a common placenta, together with their common coverings; or of either fex feparately, with its proper coverings, if they have any.

FŒNICULUM. Tourn. Inft. R. H. 311. tab. 164. Fennel.

The CHARACTERS are,

It hath an umbellated flower; the great umbel is compofed of many fmaller, which have no involucrum; the flowers have five incurved petals, and five ftamina; the germen is fituated under the flower, which afterward turns to an oblong fruit, deeply channelled, dividing into two parts, each containing a fingle feed.

The SPECIES are,

1. FOENICULUM (*Vulgare*) foliis decompofitis, foliolis brevioribus multifidis, femine breviore. *Common Fennel.*

2. FOENICULUM (*Dulce*) foliis decompofitis, foliolis longioribus, femine longiori. *Sweet Fennel having a larger white feed.*

3. FOENICULUM (*Azorium*) humilius, radice caulefcente carnofo, feminibus recurvis, radice annuâ. *Sweet Azorian Fennel, called Finochio.*

The firft fort is the common Fennel which is cultivated in the gardens, and has fown itfelf in many places, where it has been introduced in fuch plenty, as to appear as if it were a native in England: but it is no where found at a great diftance from gardens, fo has been undoubtedly brought from abroad. There are two varieties of this, one with light green, the other with very dark leaves, but thofe I believe are only varieties which arife from the fame feeds.

The common Fennel is fo well known as to need no defcription. It has a ftrong flefhy root which penetrates deep into the ground, and will continue feveral years. It flowers in July, and the feeds ripen in autumn. The beft time to fow the feeds, is foon after they are ripe; the plants will come up in the fpring, and require no other care but to keep them clean from weeds; it will grow in any foil or fituation. The leaves, feeds, and roots of this are ufed in medicine; the root is one of the five opening roots, and the feed one of the greater carminative feeds. There is a fimple water made from the leaves, and a diftilled oil from the feed.

The fweet Fennel has been by many fuppofed only a variety of the common fort, but I have cultivated it in the fame ground with that, where it has retained its differences. The leaves of this are very long and flender, growing more fparfedly, and do not end in fo many points as thofe of the common fort; the ftalks do not rife fo high; the feeds are longer, narrower, and of a lighter colour. Thefe feeds are generally imported from Germany to Italy, and are by fome preferred to thofe of the common fort for ufe, being much fweeter.

This may be propagated in the ſame manner as the former ſort, being very hardy, but the roots are not of ſo long duration.

The third ſort is ſuppoſed to have been originally brought from the Azorian Iſlands; it has been long cultivated in Italy as a ſallad herb, under the title of Finochio; and there are ſome few gardens in England, where it is now cultivated in ſmall quantities, for there are not many Engliſh palates which reliſh it, nor is it eaſy to be furniſhed with good ſeeds; thoſe which are annually brought from Italy ſeldom prove good, and it is difficult to ſave it in England, becauſe the winter frequently kills thoſe plants which are left for ſeeds; and when any good plants of the early ſowing are left for ſeeds, they do not ripen unleſs the autumn proves very favourable.

This ſort hath very ſhort ſtalks, which ſwell juſt above the ſurface of the ground, to four or five inches in breadth, and almoſt two thick, being fleſhy and tender: this is the part which is eaten when blanched with oil, vinegar, and pepper, as a cold ſallad. When theſe plants are permitted to run for ſeeds, the ſtalks do not riſe more than a foot and a half high, having a large ſpreading umbel ſtanding on the top. The ſeeds of this ſort are narrow, crooked, and of a bright yellow colour; they have a very ſtrong ſmell like Aniſeed, and are very ſweet to the taſte.

The manner of cultivating this plant is as follows: The firſt care is to procure ſome good ſeeds from ſome perſon who has been careful in the choice of the plants, otherwiſe there will be little hope of having it good: then make choice of a good ſpot of light rich earth, not dry nor very wet, for in either extreme this plant will not thrive. The firſt crop may be ſown about a fortnight in March, which, if it ſucceeds, will be fit for uſe in July; and ſo by ſowing at ſeveral times, there may be a ſupply for the table till froſt puts a ſtop to it. After having well dug and levelled the ground ſmooth, the ſeed ſhould be ſown in a ſhallow drill by a line, ſcattering them pretty thin, for the plants muſt be left ſix inches aſunder in the rows; but however, ſome of the ſeeds may fail, therefore they ſhould be ſcattered two inches diſtance, then cover the ſeeds about half an inch thick with earth, laying it ſmooth. Theſe drills ſhould be made eighteen inches aſunder, or more, that there may be room to clean the ground, as alſo to earth up the plants when they are full grown. When the plants come up, which will be in about three weeks or a month after ſowing, according to the ſeaſon, all the weeds between them muſt be cut up, and where the plants are too cloſe, they ſhould be thinned to about four inches diſtance; and as they advance, and the weeds ſpring again, they ſhould from time to time be hoed; and at the laſt time of thinning them, they ſhould be left ſeven or eight inches aſunder at leaſt. If the kind be good, the ſtems of the plants will increaſe to a conſiderable bulk juſt above the ſurface of the ground; which part ſhould be earthed up in the manner of Celery, to blanch, about a fortnight or three weeks before it is uſed, which will cauſe it to be very tender and criſp.

The ſecond crop ſhould be ſown about three weeks after the firſt, and ſo continued every three weeks or a month till the end of July, after which time it will be too late for the plants to come to any perfection. But the ſeeds which are ſown in May and June, ſhould have a moiſter ſoil than that which you ſowed the firſt; as alſo what is ſown the latter part of July, ſhould be on a drier ſoil and in a warmer ſituation; becauſe this crop will not be fit for uſe till late in autumn, and therefore will be ſubject to injuries from too much wet or cold, if on a moiſt ſoil. If the ſeaſon ſhould prove dry the plants muſt be watered, otherwiſe they will run up to ſeed before they are of any ſize; therefore there ſhould be a channel made where every row of plants grow, to detain the water which is poured on them, to prevent its running off. In the autumn, if there ſhould happen ſharp froſts, it will be very proper to cover the plants with ſome Peaſe-haulm, or other light covering, to prevent their being pinched, by which method they may be continued for uſe till the middle of winter.

FOENUM BURGUNDIACUM. See MEDICA SATIVA.

FOENUM GRÆCUM. See TRIGONELLA.

FRAGARIA. Lin. Gen. Pl. 558. The Strawberry.

The CHARACTERS are,

The flower hath five roundiſh petals, which are inſerted in the empalement, and twenty ſtamina, with a great number of germen collected into a head, which afterward becomes a large, ſoft, pulpy fruit, having many ſmall angular ſeeds in the empalement.

The SPECIES are,

1. FRAGARIA (*Veſca*) foliis ovatis ſerratis, calycibus, brevibus, fructu parvo. *The common, or Wood Strawberry.*

2. FRAGARIA (*Virginiana*) foliis oblongo-ovatis ſerratis, infernè incanis, calycibus longioribus, fructu ſubrotundo. *Virginia Strawberry with a ſcarlet fruit; commonly called the Scarlet Strawberry.*

3. FRAGARIA (*Muricata*) foliis ovato-lanceolatis rugoſis, fructu ovato. *Strawberry with fruit as large as a ſmall Plumb; commonly called Hautboy Strawberry.*

4. FRAGARIA (*Chiloenſis*) foliis ovatis carnoſis hirſutis, fructu maximo. *Strawberry of Chili with a large fruit, and hairy fleſhy leaves; called Frutilla in America.*

There are ſeveral other varieties of this fruit which are now cultivated in England, but theſe, I think, may be allowed to be diſtinct ſpecies, for they never alter from one to the other by any cultivation, tho' the fruit is frequently improved, ſo as to be of a larger ſize thereby; I ſhall next mention the varieties of Strawberries, which are at preſent to be found in the Engliſh gardens, under the ſeveral ſpecies to which they naturally belong.

The firſt ſort is the common Wood Strawberry, which grows naturally in the woods in many parts of England, and is ſo well known as to need no deſcription; of this there are three varieties, 1. The common ſort with red fruit. 2. The white Wood Strawberry, which ripens a little later in the ſeaſon, and is by many perſons preferred to it for its quick flavour; but as it ſeldom produces ſo large crops as the red ſort, ſo it is not very generally cultivated. 3. The green Strawberry, by ſome called the Pine Apple Strawberry, from its rich flavour. The fruit of this is green, with a faint ſhade of red when ripe; it is very firm, hath a very high flavour, and is a late ripe fruit; but unleſs it is planted in a moiſt loamy ſoil it is a very bad bearer, but in ſuch land where it does ſucceed, it merits cultivation as much as any of the ſorts. There is alſo a variety of this which has been raiſed from ſeeds by the Right Hon. Lord Willoughby of Parham, which continues to produce fruit from the firſt ſeaſon of Strawberries, till prevented by froſt; the fruit has a higher flavour than the Wood Strawberry, ſo deſerves encouragement.

The Scarlet Strawberry is the ſort which is firſt ripe, for which reaſon it merits eſteem, had it nothing elſe to recommend it; but the fruit is ſo good, as by many perſons of good taſte to be preferred to all the other ſorts; this was brought from Virginia, where it grows naturally in the woods, and is ſo different from the Wood Strawberry in leaf, flower, and fruit, that there need to be no doubt of their being diſtinct ſpecies.

There is a variety of this which hath been of late years introduced from the northern parts of America, which has the appearance of a diſtinct ſpecies. The leaves of this are rounder, and not ſo deeply veined; the crenatures on their

their edges are broader and more obtuſe. The leaves which compoſe the empalement are much longer, and are hairy, the fruit is alſo much larger; but in other reſpects it approaches near to the Scarlet Strawberry, ſo I have choſen to join it to that, rather than make a diſtinct ſpecies of it; this I have been informed grows naturally in Louiſiana.

The Hautboy Strawberry, which the French call Capitons, came originally from America, and is very different from the other ſorts in leaf, flower, and fruit, as that no one can doubt of their being different ſpecies; there is an improvement of this ſort, which is commonly called the Globe Hautboy. The fruit of this is larger, and of a globular form, but this difference has certainly ariſen from culture; for where theſe have been neglected a year or two, they have degenerated to the common Hautboy again; where the ground is proper for this plant, and their culture is well managed, the plants will produce great plenty of fruit, which will be large and well flavoured, and by ſome perſons are preferred to all the other ſorts.

The Chili Strawberry was brought to Europe by Monſ. Frazier, an engineer, who was ſent to America by the late king of France; in the year 1727 I brought a parcel of the plants to England, which were communicated to me by Mr. George Clifford of Amſterdam, who had large beds of this ſort growing in his curious gardens at Hartecamp. The leaves of this ſort are hairy, oval, and of a much thicker ſubſtance than thoſe of any ſort yet known, and ſtand upon very long hairy foot-ſtalks; the runners from the plants are very large, hairy, and extend to a great length, putting out plants at ſeveral diſtances. The foot-ſtalks which ſuſtain the flowers are very ſtrong; the leaves of the empalement are long and hairy. The flowers are large, and often deformed, ſo are the fruit alſo when cultivated in very ſtrong land, in which the plants produce plenty, which are firm and very well flavoured, but as it is a bad bearer in moſt places where it has been cultivated, ſo in general it has been neglected.

Strawberries love a gentle hazelly loam, in which they will thrive and bear greater plenty of fruit than in a light rich ſoil. The ground ſhould alſo be moiſt, for if it is very dry, all the watering which is given to the plants in warm dry ſeaſons will not be ſufficient to procure plenty of fruit; nor ſhould the ground be much dunged, for that will cauſe the plants to put out many runners and grow luxuriant, and will render them leſs fruitful.

The beſt time to remove theſe plants is in September, or the beginning of October, that they may get new roots before the hard froſt ſets in, which looſens the ground, ſo that if the roots of the plants are not pretty well eſtabliſhed, the plants frequently are turned out of the ground after froſt by the firſt thaw; therefore the ſooner they are planted when the autumnal rains begin, the better will their roots be eſtabliſhed; and ſometimes thoſe which are well rooted, will produce a few fruit the firſt year; there are ſome who tranſplant their plants in the ſpring, but where this is done they muſt be duly ſupplied with water in dry weather, otherwiſe they will not ſucceed.

The ground in which theſe are planted ſhould be thoroughly cleaned from the roots of Couch, and all other bad weeds, for as the Strawberry plants are to remain three years before they are taken up, ſo if any of the roots of thoſe bad weeds are left in the ground, they will have time to multiply ſo greatly as to fill the ground, and overbear the Strawberry plants. The uſual method is to lay the ground out into beds of four feet broad, with paths two feet, or two feet and a half broad between each; theſe paths are neceſſary for the convenience of gathering the fruit, and for weeding and dreſſing of the beds: after the beds are marked out, there ſhould be four lines drawn in each at a foot diſtance, which will leave ſix inches ſpace on each ſide, between the outſide rows and the paths; then the plants ſhould be planted at a foot diſtance from each other in the rows, in a quincunx order, being careful to cloſe the ground to the roots of the plants when they are planted, and if there ſhould not happen rain ſoon after, the plants ſhould be well watered to ſettle the earth to their roots.

The diſtance here mentioned for the plants to be placed, muſt be underſtood for the Wood Strawberries only, for as the other ſorts grow much larger, their diſtances muſt be proportioned to their ſeveral growths; therefore the Scarlets and Hautboys ſhould have but three rows of plants in each bed, which ſhould be at fifteen inches diſtance, and the plants in the rows ſhould be allowed the ſame ſpace from each other; and the Chili Strawberry muſt have but two rows of plants in each bed, which ſhould alſo be two feet apart in the rows, for as theſe grow very ſtrong, if they have not room to ſpread, they will not be very fruitful.

In the chuſing proper plants of any of the ſorts depends the whole ſucceſs, for if they are promiſcuouſly taken from beds without care, great part of the plants will become barren; theſe are generally called blind, which is when there are plenty of ſmall flowers but no fruit produced; if theſe flowers are well examined, they will be found to want the female organs of generation, moſt of them abounding with ſtamina, but they have few, if any ſtyles, ſo that it frequently happens among theſe barren plants that ſome of them will have a part of an imperfect fruit formed, which will ſometimes ripen; this barrenneſs is not peculiar to Strawberries, but is general to all thoſe plants which have creeping roots or ſtalks; and the more they increaſe from either, the ſooner they become barren, and this in ſome degree runs through the whole vegetable kingdom; for trees and ſhrubs which are propagated by cuttings are generally barren of ſeeds in two generations, that is, when they are propagated by cuttings, which were taken from plants raiſed by cuttings; this I have conſtantly found to hold in great numbers of plants; and in fruit-trees it often happens, that thoſe ſorts which have been long propagated by grafts and buds, have no kernels. But to return to the choice of the Strawberry plants; theſe ſhould never be taken from old neglected beds, where the plants have been ſuffered to ſpread or run into a multitude of ſuckers nor from any plants which are not very fruitful; and thoſe offſets which ſtand neareſt to the old plants, ſhould always be preferred to thoſe which are produced from the trailing ſtalks at a farther diſtance; the Wood Strawberry is beſt when the plants are taken freſh from the woods, provided they are taken from fruitful plants, becauſe they are not ſo liable to ramble and ſpread, as thoſe which are taken from plants which have been long cultivated in gardens; therefore thoſe who are curious in cultivating of this fruit, ſhould be very careful in the choice of their plants.

When the plants have taken new root, the next care is, if the winter ſhould prove ſevere, to lay ſome old tanners bark over the ſurface of the bed between the plants to keep out the froſt; this care is abſolutely neceſſary to the Chili Strawberry, which is frequently killed in hard winters, where they are expoſed without any covering; therefore where tanners bark cannot be eaſily procured, ſaw-duſt or ſea-coal aſhes may be uſed; or in want of theſe, if decayed leaves of trees, or the branches of evergreen trees with their leaves upon them, are laid over the beds to prevent the froſt from penetrating deep into the ground, it will ſecure the plants from injury.

The following ſummer the plants ſhould be conſtantly kept clean from weeds, and all the runners ſhould be pulled off as faſt as they are produced; if this is conſtantly practiſed, the plants will become very ſtrong by the following

lowing autumn, whereas when this is neglected (as is too frequently seen) and all the runners permitted to stand during the summer season, and then pulled off in autumn, the plants will not be half so strong as those where that care has been taken; therefore there will not be near the same quantity of fruit upon them the following spring, nor will the fruit be near so large and fair; where proper care is taken of the plants the first summer, there is generally a plentiful crop of fruit the second spring; whereas when this is neglected, the crop will be thin and the fruit small.

As this fruit is very common, there are but few persons who cultivate it with proper attention, therefore I shall give some directions for the doing of it, which, if carefully practised, will be attended with success.

The old plants of Strawberries are those which produce the fruit, for the suckers never produce any till they have grown a full year; therefore it appears how necessary it is to divest the old plants of them, for wherever they are suffered to remain, they rob the fruitful plants of their nourishment in proportion to their number; for each of these suckers send out a quantity of roots, which interfere, and are so closely matted together as to draw away the greatest part of the nourishment from the old roots, whereby they are greatly weakened; and these suckers also render each other very weak, so from hence the cause of barrenness arises; for I have known where the old plants have been constantly kept clear from suckers, they have continued very fruitful three years without being transplanted; however it is the best way to have a succession of beds, that after two years standing they may be taken up, because by that time they will generally have exhausted the ground of those vegetable salts necessary for the nourishment of that species of plants; for it is always observed, that Strawberries planted on fresh land are the most fruitful.

The next thing to be observed is, in autumn to divest the plants of any strings or runners which may have been produced, and also of all the decayed leaves, and the beds cleared from weeds; then the paths should be dug up, and the weeds buried which were taken from the beds, and some earth laid over the surface of the beds between the plants, this will strengthen and prepare them for the following spring; and if after this, there is some old tanners bark laid over the surface of the ground between the plants, it will be of great service to them. In the spring, after the danger of hard frost is over, the ground between the plants in the beds should be forked, with a narrow three-pronged fork, to loosen the ground and break the clods; and in this operation the tan which was laid over the surface of the ground in autumn will be buried, which will be a good dressing to the Strawberries, especially in strong land; then about the end of March, or the beginning of April, if the surface of the bed is covered with Moss, it will keep the ground moist, and prevent the drying winds from penetrating the ground, and thereby secure a good crop of fruit, and the Moss will preserve the fruit clean; when heavy rains fall after the fruit is full grown, there will be no dirt washed over them, which frequently happens where this is not practised, so the fruit must be washed before it is fit for the table, which greatly diminishes its flavour.

The soil in which the Chili Strawberry is found to succeed best, is a very strong brick earth, approaching near to clay; in this soil I have seen them produce a tolerable good crop, and the fruit has been extremely well flavoured; and where care has been taken to pull off the runners as they are produced from the old plants, they have been as fruitful as the common Hautboy; this I mention from two or three experiments which have been made by my direction, and not from theory.

There are some persons who are so fond of Strawberries, as to be at any expence to obtain them early in the year, and to continue them as late in the season as possible; therefore should I omit to give some directions for these purposes, they would suppose the book very defective; therefore I shall mention the practice of some few, who have succeeded best in the management of these fruits. I shall begin with directions for obtaining of this fruit early in the spring.

When there are any hot walls erected in gardens for the producing early fruit, it is very common to see Strawberries planted in the borders, that the fire which is applied for ripening of the fruit against the wall, may serve the purpose of bringing forward the Strawberries; but where this is practised, the Strawberry plants should be annually renewed, and all the earth of the borders should also be taken out at least two feet deep, and fresh earth brought in, which will be equally good for the wall trees; but as was before observed, that the old plants of Strawberries only, are those which produce the fruit, there should be a sufficient number of plants kept in pots to supply the border annually, and the same must be done if they are to be raised on a common hot-bed, or in stoves; therefore I shall begin with giving directions for the raising and preparing the plants for those purposes.

The sort which is the most proper for forcing early, is the Scarlet Strawberry, for the Hautboy grows too large for this purpose, and the Wood is too backward. In the choice of the plants there should be especial care taken to have them from the most fruitful plants, and those which grow immediately to the old plants; these should be taken off in the spring, and each planted in a separate pot filled with loamy soil, and placed in the shade till they have taken root, after which they may be removed to a shady situation, where they may remain till the middle of November, when the pots should be plunged into the ground up to their rims, to prevent the frost from penetrating through the side of the pots; if these are placed near a wall, pale, or hedge, exposed to an east aspect, they will succeed better than in a warmer situation, because they will not be forced too forward.

Those which are designed for the borders near a hot-wall, may then be turned out of the pots, and planted into the borders, that they may have time to get fresh rooting before the fires are made to heat the walls; when these are planted they may be placed pretty close to each other, for as they are designed to remain there no longer than till they have ripened their fruit, they will not require much room, because their roots will find sufficient nourishment below, and also from the earth which is filled into the spaces between the balls of earth about their roots; for it is of consequence to get as much fruit as possible in a small space, where there is an expence to force them early. If the fires are lighted about Christmas, the Strawberries in these borders will be ripe in March; or if the season should prove very cold, it may be the middle of April before they are fit for the table.

In the management of the plants there must be care taken to supply them with water when they begin to shew their flowers, otherwise they will fall off without producing any fruit; and in mild weather there should be fresh air admitted to them every day; but as fruit-trees against the wall must be so treated, the same management will agree with the Strawberries.

If the Strawberries are intended to be forced in a stove, where there are Pine Apples, and no room to plunge them in the tan-bed, then the plants ſhould be tranſplanted into larger pots in September, that they may be well rooted before they are removed into the ſtove, which ſhould not be till December; but if they are placed under a frame the beginning of November, where they may be ſcreened from the froſt, it will prepare the plants the better for forcing; and thoſe who are deſirous to have them very early, make a hot-bed under frames, upon which they place their plants the latter end of October, which will bring them forward to flower, and then they remove the plants into the ſtove; when theſe plants are removed into the ſtove, they ſhould be placed near to the glaſſes, that they may enjoy as much of the ſun as poſſible, for when they are placed backward the plants will draw up weak, and the flowers will drop without producing fruit. The earth in the pots will dry pretty faſt when they ſtand dry upon the pavement of the hot-houſe, therefore the plants muſt be duly watered, but it muſt be done with diſcretion, and not too much given to them at one time, which will be equally hurtful to them; if theſe plants are properly managed, they will produce ripe fruit in February, which is as early as moſt people will chuſe to eat them. When the fruit is all gathered from the plants, they ſhould be turned out of the ſtove, for as they will be of no farther ſervice, they ſhould not remain to take up the room; nor ſhould thoſe plants which are planted in the borders near the hot-walls, be left there after their fruit is gathered, but immediately taken up, that they may rob the fruit-trees of their nouriſhment as little as poſſible.

Where there is no conveniency of ſtoves or hot-walls for this purpoſe, the fruit may be ripened upon common hot-beds; and though they may not be quite ſo early as with the other advantages, yet I have ſeen great crops of the fruit ripe in April, which were upon common hot-beds under frames, and done with a ſmall expence in the following manner.

The plants were prepared in pots after the manner before directed, which were placed in a warm ſituation in the beginning of October, and about Chriſtmas the hot-bed was made, in the ſame manner as for Cucumbers, but not ſo ſtrong, and as ſoon as the firſt violent ſteam of the dung was over, ſome old rotten dung laid over the hot-bed to keep down the heat, or where it can be eaſily procured, neats dung is preferable for this purpoſe; then the pots ſhould be placed upon the bed as cloſe together as poſſible, filling up the interſtices between the pots with earth; the plants muſt have air admitted to them every day, and if the heat of the bed is too great, the pots ſhould be raiſed up to prevent their roots being ſcorched, and when the bed is too cold, the ſides of it ſhould be lined with ſome hot dung; this firſt bed will bring the plants to flower by the middle or latter end of February, by which time the heat of the bed will be ſpent, therefore another hot-bed ſhould be prepared to receive the plants, which need not be ſo ſtrong as the firſt; but upon the hot dung ſhould be laid ſome neats dung about two inches thick, which ſhould be equally ſpread and ſmoothed; this will prevent the heat of the bed from injuring the roots of the plants; upon this ſhould be laid two inches of a loamy ſoil; when this has laid two days to warm, the plants ſhould be taken out of the firſt hot-bed, and turned carefully out of the pots, preſerving all the earth to their roots, and placed cloſe together upon this new hot-bed, filling up the vacuities between the balls with loamy earth; the roots of the plants will ſoon ſtrike out into this freſh earth, which will ſtrengthen their flowers, and cauſe their fruit to ſet in plenty; and if proper care is taken to admit freſh air to the plants, and ſupply them properly with water, they will have plenty of ripe fruit in April, which will be full two months before their natural ſeaſon.

The methods practiſed to retard this fruit is firſt by planting them in the coldeſt part of the garden, where they may be as much in ſhade as poſſible, and the ſoil ſhould be ſtrong and cold; when there are ſuch places in a garden, the fruit will be near a month later than in a warm ſituation; the next is to cut off all the flowers when they firſt appear, and if the ſeaſon proves dry to water them plentifully, which will cauſe them to put out a freſh crop of flowers; and if they are ſupplied with water, there will be a large crop of fruit, but theſe are not ſo well flavoured as thoſe which ripen in their natural ſeaſon. But this new Alpine Strawberry will naturally ſupply the table great part of the ſummer and autumn, and the fruit will be well flavoured.

FRANGULA. Tourn. Inſt. R. H. 612. tab. 383. Berry-bearing Alder.

The CHARACTERS are,

The flower hath one petal cut into five ſegments. It hath five ſtamina, which are the length of the petal; in the center is ſituated a globular germen, which afterward becomes a round berry, incloſing two plain roundiſh ſeeds.

The SPECIES are,

1. FRANGULA (*Alnus*) foliis ovato-lanceolatis glabris. *Black Berry-bearing Alder.*
2. FRANGULA (*Latifolia*) foliis lanceolatis rugoſis. *Berry-bearing Alder with a larger and rougher leaf.*
3. FRANGULA (*Rotundifolia*) foliis ovatis nervoſis. *Low Mountain rocky Berry-bearing Alder, with a round leaf.*

The firſt ſort grows naturally in the woods in many parts of England, ſo is ſeldom planted in gardens; it riſes with a woody ſtem to the height of ten or twelve feet, ſending out many irregular branches, covered with a dark bark, garniſhed with oval ſpear-ſhaped leaves about two inches long, and one broad, having ſeveral tranſverſe veins from the midrib to the ſides. The flowers are produced in cluſters at the end of the former year's ſhoots, and alſo upon the firſt and ſecond joints of the ſame year's ſhoot, each ſtanding upon a ſhort ſeparate foot-ſtalk; they are ſmall, of an herbaceous colour, and are ſucceeded by ſmall round berries, which turn firſt red, but are black when ripe. The flowers appear in June, and the berries ripen in September; this ſtands in the Diſpenſatory as a medicinal plant, but is ſeldom uſed.

The ſecond ſort hath larger and rougher leaves than the firſt. It grows naturally on the Alps, and other mountainous parts of Europe, and is preſerved in ſome gardens for the ſake of variety.

The third ſort is of humble growth, ſeldom riſing above two feet high; this grows on the Pyrenean mountains, and is ſeldom preſerved, unleſs in botanic gardens for variety; it may be increaſed by laying down the branches, but muſt have a ſtrong ſoil.

Theſe ſhrubs are eaſily propagated by ſeeds, which ſhould be ſown as ſoon as they are ripe, and then the plants will come up the ſpring following; but if they are kept out of the ground till ſpring, the plants will not come up till the ſecond year. When the plants come up they muſt be kept clean from weeds till autumn, then they may be taken up and planted in a nurſery, in rows two feet aſunder, and at one foot diſtance in the rows; in this nurſery they may remain two years, and may then be planted where they are to remain; they may alſo be propagated by layers and cuttings, but the ſeedling plants are beſt.

FRAXINELLA. See DICTAMNUS.

FRAXINUS. Lin. Gen. Plant. 1026. The Aſh-tree.

The CHARACTERS are,

It hath hermaphrodite and female flowers on the ſame tree, and ſometimes on different trees. The hermaphrodite flowers have no petals, but a ſmall empalement, including two erect ſtamina

In the center is situated an oval compressed germen, which afterward becomes a compressed bordered fruit, shaped like a bird's tongue, having one cell, inclosing a seed of the same form. The female flowers are the same, but have no stamina.

The SPECIES are,

1. FRAXINUS (*Excelsior*) foliolis serratis, floribus apetalis. Lin. Sp. Pl. 1057. *The common Ash.*

2. FRAXINUS (*Rotundifolia*) foliolis ovato-lanceolatis serratis, floribus coloratis. *Ash-tree with a rounder leaf; commonly called Manna Ash.*

3. FRAXINUS (*Ornus*) foliolis serratis, floribus corollatis. Lin. Sp. Pl. 1057. *Dwarf Ash of Theophrastus, with smaller and narrower leaves.*

4. FRAXINUS (*Paniculata*) foliolis lanceolatis glabris, floribus paniculatis terminatricibus. *The flowering Ash.*

5. FRAXINUS (*Nova Anglia*) foliolis integerrimis, petiolis teretibus. Flor. Virg. 122. *New England Ash, with long acute points to the wings of the leaves.*

6. FRAXINUS (*Caroliniana*) foliolis lanceolatis, minimè serratis, petiolis teretibus pubescentibus. *Carolina Ash with a broad fruit.*

The first sort is the common Ash-tree, which grows naturally in most parts of England, and is so well known as to need no description. The leaves of this sort have generally five pair of lobes, terminated by an odd one; they are of a very dark green, and their edges are slightly sawed. The flowers are produced in loose spikes from the side of the branches, which are succeeded by flat seeds, which ripen in autumn; there is a variety of this with variegated leaves, which is preserved in some gardens.

The second sort grows naturally in Calabria, and is generally supposed to be the tree from whence the manna is collected, which is an exsudation from the leaves of the tree. The shoots of this tree are much shorter, and the joints closer together, than those of the first sort; the small leaves are shorter, and deeper sawed on their edges, and are of a lighter green. The flowers come out from the side of the branches, which are of a purple colour, and appear in the spring before the leaves come out. This tree is of humble growth, seldom rising more than fifteen or sixteen feet high in England.

The third sort is a low tree, which rises about the same height as the second; the leaves of this sort are much smaller and narrower than these of the first, but are sawed on their edges, and are of the same dark colour. The flowers of this sort have petals, which are wanting in the common Ash.

The fourth sort was raised by the late Dr. Uvedale at Enfield, from seeds which were brought from Italy by Dr. William Sherard, where the trees grow naturally; and was supposed to be a different sort from that mentioned by Dr. Morrisson, in his Præludia Botanica, but by comparing them together they appear to be the same.

The leaves of this sort have but three or four pair of lobes (or small leaves) which are short, broad, and smooth, of a lucid green, irregularly sawed on the edges. The flowers grow in loose panicles at the end of the branches; these are most of them male, having two stamina in each, but no germen or style; they are of a white herbaceous colour, and appear in May. As this sort very rarely produces seeds in England, so it is propagated by grafting or budding it upon the common Ash.

The fifth sort was raised from seeds, which were sent from New England in the year 1724, by Mr. Moore. The leaves of this tree have but three, or at most but four pair of lobes, which are placed far distant from each other, and are terminated by an odd one, which runs out into a very long point; they are of a light green and entire, having no serratures on their edges: this tree shoots into strong irregular branches, but doth not grow to a large size in the trunk. It is propagated by grafting it upon the common Ash.

The sixth sort was raised from seeds, which were sent from Carolina in the year 1724, by Mr. Catesby. The leaves of this sort have seldom more than three pair of lobes, the lower being the least, and the upper are the largest, of a light green colour, and slightly sawed on their edges; the foot-stalk, or rather the midrib of the leaves is taper, and has short downy hairs; the seeds are broader than those of the common Ash, and are of a very light colour. As this sort hath not yet produced seeds in England, it is propagated by grafting it upon the common Ash.

These trees are now propagated in plenty in the nurseries for sale, as there has been of late years a great demand for all the hardy sorts of trees and shrubs which will live in the open air; but all those trees which are grafted upon the common Ash, are not so valuable as those which are raised from seeds, because the stocks generally grow much faster than the grafts; whereby the lower part of the trunk, so far as the stock rises, will often be twice the size of the upper; and if the trees stand much exposed to the winds, the grafts are frequently broken off to the stock after they are grown to a large size, which is a great disappointment to a person after having waited several years, to see their trees suddenly destroyed. Beside, if the wood of either of the sorts is valuable, it can be of little use when the trees are so raised.

The fourth sort is generally planted for ornament, the flowers making a fine appearance when they are in beauty, every branch being terminated by a large loose panicle; so that when the trees are large, and covered with flowers, the are distinguishable at a great distance.

All the other sorts serve to make a variety in plantations, but have little beauty to recommend them; and as their wood seems to be greatly inferior to that of the common Ash, so there should be few of these planted, because they will only fill up the space where better trees might grow.

The common Ash propagates itself in plenty by the seeds which scatter in the autumn, so that where the seeds happen to fall in places where cattle do not come, there will be plenty of the plants come up in the spring; but where any person is desirous to raise a quantity of the trees, the seeds should be sown as soon as they are ripe, and then the plants will come up the following spring; but if the seeds are kept out of the ground till spring, the plants will not come up till the year after, which is the same with all the sorts of Ash; so that when any of their seeds are brought from abroad, as they seldom arrive here before the spring, the plants must not be expected to appear till the next year; therefore the ground should be kept clean all the summer where they are sown, and not disturbed, lest the seeds should be turned out of the ground, or buried too deep to grow; for many persons are too impatient to wait a year for the growth of seeds, so that if they do not come up the first year, they dig up the ground, and thereby destroy the seeds.

When the plants come up, they must be kept clean from weeds during the summer; and if they make good progress in the seed-bed, they will be fit to transplant by the following autumn, therefore there should be some ground prepared to receive them; and as soon as their leaves begin to fall, they may be transplanted. In the taking of them up, there should be care taken not to break or tear off their roots; to prevent which, they should be taken up with a spade, and not drawn up, as is frequently practised; for as many of the plants which rise first from seeds, will

out-

outstrip the others in their growth, so it is frequently practised to draw up the largest plants, and leave the smaller to grow a year longer before they are transplanted; and to avoid hurting those which are left, the others are drawn out by hand, and thereby many of their roots are torn off or broken; therefore it is much the better way to take all up, little or big together, and transplant them out, placing the large ones together in rows, and the smaller by themselves. The rows should be three feet asunder, and the plants a foot and a half distance in the rows; in this nursery they may remain two years, by which time they will be strong enough to plant where they are to remain; for the younger they are planted out, the larger they will grow; so that where they are designed for use, they should be planted very young, and the ground where the plants are raised, should not be better than that where they are designed to grow; for when any plants are raised in good land, and afterward transplanted into worse, they very rarely thrive; so that it is much the best method to make the nursery upon a part of the same land, where the trees are designed to be planted, and then a sufficient number of the trees may be left standing upon the ground, and these will outstrip those which are removed, and will grow to a larger size.

Where people live in the neighbourhood of Ash-trees, they may supply themselves with plenty of self sown plants, provided cattle are not suffered to graze on the land, for if cattle can come to them they will eat off the young plants, and not suffer them to grow; but where the seeds fall in hedges, and are protected by bushes, the plants will come up and thrive; in these hedges the trees frequently are permitted to grow till they have destroyed the hedge, for there is scarce any tree so hurtful to all kinds of vegetables as the Ash, which robs every plant of its nourishment within the reach of its roots, therefore should never be suffered to grow in hedge-rows; for they not only kill the hedge, but impoverish Corn, or whatever is sown near them. Nor should any Ash-trees be permitted to grow near pasture-ground, for if any of the cows eat of the leaves or shoots of the Ash, all the butter which is made of their milk will be rank, and of little or no value; which is always the quality of the butter which is made about Guildford, Godalming, and some other parts of Surry, where there are Ash-trees growing about all their pastures, so that it is very rare to meet with any butter in those places which is fit to eat; but in all the good dairy counties, they never suffer an Ash-tree to grow.

If a wood of these trees is rightly managed, it will turn greatly to the advantage of its owner; for by the underwood, which will be fit to cut every eight or ten years, there will be a continual income more than sufficient to pay the rent of the ground, and all other charges; and still there will be a stock preserved for timber, which in a few years will be worth forty or fifty shillings per tree.

This timber is of excellent use to the wheelwright and cartwright for ploughs, axle-trees, wheel-rings, harrows, bulls, oars, blocks for pullies, and many other purposes.

The best season for felling of these trees is from November to February; for if it be done either too early in autumn, or too late in the spring, the timber will be subject to be infested with worms, and other insects; but for lopping of pollards, the spring is preferable for all soft woods.

FRITILLARIA. Lin. Gen. Plant. 372. Fritillary, or chequered Tulip and Crown Imperial.

The Characters are,

The flower hath no empalement; it hath six oblong bell-shaped petals; in the hollow, at the base of each petal, is situated a nectarium; the flower hath six stamina standing near the style. In the center is situated an oblong three-cornered germen, which afterward becomes an oblong capsule with three lobes, having three cells, which are filled with flat seeds ranged in a double order.

The Species are,

1. Fritillaria (*Milagris*) foliis linearibus alternis, floribus terminalibus. *Early, purple, variegated, chequered Tulip.*

2. Fritillaria (*Aquitanica*) foliis infimis oppositis. Hort. Cliff. 81. *Aquitain chequered Tulip, with an obscure yellow flower.*

3. Fritillaria (*Nigra*) floribus adscendentibus. *Fritillary with flowers growing above each other; or Black chequered Tulip.*

4. Fritillaria (*Lutea*) foliis lanceolatis, caule uniflora maximo. *Largest yellow Italian Fritillary.*

5. Fritillaria (*Umbellata*) floribus umbellatis. *Fritillary with flowers growing in umbels.*

6. Fritillaria (*Persica*) racemo nudiusculo, foliis obliquis. Hort. Upsal. 82. *The Persian Lily.*

7. Fritillaria (*Racemosa*) floribus racemosis. *Branching Fritillary; or smaller Persian Lily.*

8. Fritillaria (*Imperialis*) racemo comoso inferne nudo, foliis integerrimis. Lin. Hort. Upsal. 82. *Crown Imperial.*

9. Fritillaria (*Regia*) racemo comoso infernè nudo, foliis crenatis. Lin. Sp. Plant. 303. *Royal Crown, with a crenated Lily leaf.*

The first sort grows naturally in Italy, and other warm parts of Europe; and from the seeds of this there have been great varieties raised in the gardens of the florists, which differ in the size and colour of their flowers; and as there are frequently new varieties produced, so it would be to little purpose to enumerate those which are at present in the English and Dutch gardens.

The first hath a round compressed root, in shape like that of Corn Flag, but is of a yellowish white colour; the stalk rises about fifteen inches high, having three or four narrow long leaves placed alternately; the top is divided into two slender foot-stalks, which turn downward, each sustaining one bell-shaped inverted flower, composed of six petals, which are chequered with purple and white like a chess-board; in the center is situated a germen, supporting one style, crowned by a trifid stigma. At the bottom of each petal there is a cavity, in which is situated a nectarium, filled with a sweet liquor; after the flower is fallen, the germen swells to a pretty large, three-cornered, blunt capsule, then the foot-stalk is turned and stands erect; when the seeds are ripe, the capsule opens in three parts and lets out the flat seeds, which were ranged in a double order.

The second sort grows naturally in France; the leaves of this are broader, and of a deeper green than the former; the lower leaves are placed opposite, but those above are alternate; the stalk rises a foot and a half high, is terminated by two flowers of an obscure yellow colour, which spread more at the brim than those of the first sort, but are turned downward in the same manner. This flowers three weeks after the first. This grows naturally in some parts of England.

The third sort seldom rises more than a foot high; the leaves are narrow like those of the first, but are shorter; each stalk is terminated by three or four flowers, which arise above each other, of a very dark purple, chequered with yellowish spots.

The fourth sort rises about a foot high; the stalk is garnished with spear-shaped leaves, of a Grass green colour; these are sometimes placed opposite, but generally alternate; the stalk is terminated by one large bell-shaped flower of a yellowish colour, chequered with light purple.

The

The fifth sort rises a foot and a half high; the stalk is garnished with shorter and broader leaves than the first sort, of a grayish colour; the flowers are produced round the stalks like those of the Crown Imperial, of a dark purple colour, chequered with a yellowish green.

The sixth sort is commonly called the Persian Lily, and is supposed to grow naturally in Persia; the root of this sort is round and large; the stalk rises three feet high; the lower part of it is closely garnished with long twisted leaves of a gray colour, standing on every side of the stalks; the flowers grow in a loose spike at the top of the stalk, forming a pyramid; they are shaped like those of the other species, but are much shorter, and spread wider at their brims, and are not bent downward like those. They are of a dark purple colour, but are seldom succeeded by seeds in England, so are only propagated by offsets.

The seventh sort has a much shorter stalk than the last, but are garnished with leaves like those, only they are smaller; the stalks branch out at the top into several small foot-stalks, each sustaining one dark coloured flower. This is commonly called the small Persian Lily, from its resemblance to the former sort.

These plants are propagated either by seeds, or offsets from the old roots; by the first of which methods new varieties may be obtained, as also a larger stock of roots in three years than can be obtained in twenty or thirty years in the latter method: I shall therefore first treat of their propagation by seeds.

Having provided yourself with some good seeds, saved from the fairest flowers, you must procure some shallow pans or boxes, which must have holes in their bottoms to let out the moisture; these should be filled with light fresh earth, laying a few potsheards over the holes, to prevent the earth from stopping them; then, having laid the earth very level in the boxes, &c you must sow the seeds thereon pretty close, covering it with fine sifted earth a quarter of an inch thick. The time for sowing the seed is about the beginning of August, for if it be kept much longer out of the ground it will not grow. The farther management of the seeds, being the same as for the seeds of Tulips and Hyacinths, need not be repeated here.

When the seedling plants shew their flowers, which is generally the third year from sowing, you should put down a mark to the roots of all such as produce fair flowers, that at the time of taking them out of the ground (which ought to be soon after their green leaves are decayed) they may be selected for to plant in the borders of the parterre-garden, where, being intermixed with other flowers of different seasons, they will make a good appearance.

When a stock of good flowers are obtained, they may be preserved and increased in the same manner as other bulbous-rooted flowers, which is by offsets sent out from their roots, which should be taken off every other year from the finest sorts; but the ordinary flowers may remain three years undisturbed, in which time they will have multiplied so much, as that each root will have formed a cluster; so that if they are left longer together, the roots will be small, and the flowers very weak; therefore if these are taken up every other year, the roots will be the stronger. These roots may be treated in the same manner as Tulips, and other bulbous-rooted flowers, with this difference only, that the roots of this will not bear to be kept out of the ground so long; therefore if there should be a necessity for keeping them out of the ground for a longer time, it will be best to put the roots into sand to prevent their shrinking.

The eighth sort is the Crown Imperial, which is now very common in the English gardens. This grows naturally in Persia, from whence it was first brought to Constantinople, and about the year 1570 was introduced to these parts of Europe; of this there are a great variety of sorts now preserved in the gardens of florists, but as they have been produced accidentally from seeds, so they may be included as one species; however, for the satisfaction of the curious, I shall here mention the varieties which have come to my knowledge.

1. The common Crown Imperial; this is of a dirty red colour.
2. The yellow Crown Imperial; this is of a bright yellow.
3. The bright red Crown Imperial, called Fusai.
4. The pale yellow Crown Imperial.
5. The yellow striped Crown Imperial.
6. The large flowering Crown Imperial.
7. The broad leaved late red Crown Imperial.
8. The double and triple crowned Crown Imperial.
9. The double red Crown Imperial.
10. The double yellow Crown Imperial.
11. The silver striped leaved Crown Imperial.
12. The yellow striped leaved Crown Imperial.

There are some few other varieties which are mentioned in the catalogues of the Dutch florists, but their differences are so minute, that they are not distinguishable, so I shall pass them over.

The Crown Imperial hath a large, round, scaly root, of a yellow colour, and a strong odour of a fox; the stalk rises to the height of four feet or upward; it is strong, succulent, and garnished two thirds of the length on every side, with long, narrow, smooth leaves, ending in points; the upper part of the stalk is naked, a foot or more in length; then the flowers come out all round the stalk upon short foot-stalks, which turn downward, each sustaining one large, spreading, bell-shaped flower, composed of six spear-shaped petals; at the base of each petal is a pretty large cavity, in which is situated a large white nectarium, filled with a mellous liquor. In the center of the flower is fixed a three-cornered oblong germen, upon which rests the single style, which is the length of the petals, crowned by a spreading obtuse stigma; round the style there are six awl shaped stamina, which are shorter than the style, terminated by oblong four-cornered summits. These flowers hang downward; above and among them arises a spreading tuft of green leaves, which are erect; from between and below these come out the foot-stalks of the flowers; when the flowers decay, the germen swells to a large hexagonal capsule, shaped like a water-mill, having six cells, which are filled with flat seeds, and the capsule turns erect.

The sort with yellow flowers, that with large flowers, and those with double flowers, are the most valuable; but that which hath two or three whorls of flowers above each other, makes the finest appearance; though this seldom produces its flowers after the same manner the first year after removing, but the second and third year the stalks will be taller, and frequently have three tier of flowers one above the other, which is called the triple crown.

As this is one of the earliest tall flowers of the spring, so it makes a fine appearance in the middle of large borders, at a season when such flowers are much wanted to decorate the pleasure-garden; but the rank fox-like odour which they emit, is too strong for most people, so hath rendered these flowers less valuable than they would have been, for there is something very pleasing in the sight of them at a distance.

These may be propagated by seeds or offsets from the root, but the first is too tedious for most of the English florists, because the plants so raised are six or seven years before they flower; but the Dutch and Flemish gardeners, who have more patience, frequently raise them from seeds, whereby they

they get some new varieties, which rewards their labour. The method of propagating these flowers from seeds, being nearly the same as for the TULIP, the reader is desired to turn to that article, where there are full directions for performing it.

The common method of propagating them here, is by offsets sent out from the old roots, which will produce strong flowers the second year after they are taken from the roots; but in order to have plenty of these, the roots should not be transplanted oftener than every third year, by which time each root will have put out several offsets, some of which will be large enough to flower the following year, so may be planted in the borders of the flower-garden where they are to remain, and the smaller roots may be planted in a nursery-bed to grow a year or two according to their size; therefore they should be sorted, and the smallest roots planted in a bed together, which should remain there two years, and the larger by themselves to stand one year, by which time they will have acquired strength enough to flower, then may be removed into the pleasure-garden.

The time for taking up of these roots is in the beginning of July, when their stalks will be decayed; they may be kept out of the ground two months, but they should be laid single in a dry shady room, but not in heaps, or in a moist place, which will cause them to grow mouldy and rot. The offsets should be first planted, for as these are small, they will be apt to shrink if they are kept long out of the ground.

FRITILLARIA CRASSA. See STAPELIA.

FRUTEX PAVONIUS. See POINCIANA.

FUCHSIA. Plum. Nov. Gen. 14.

The CHARACTERS are,

The flower hath one petal, with a closed tube, slightly cut into eight parts at the brim; it hath four stamina the length of the tube. The oval germen is situated under the flower, which afterward becomes a succulent berry with four cells, containing several small oval seeds.

We know but one SPECIES of this genus, viz.

FUCHSIA (*Triphylla*) Lin. Sp. Plant. 1191. *Three-leaved Fuchsia, with a scarlet flower.*

This plant is a native in the warmest parts of America; it was discovered by Father Plumier, in some of the French islands in America, and was since found by the late Dr. William Houstoun at Carthagena in New Spain.

It is propagated by seeds, which must be sown in pots, and plunged into a hot-bed of tanners bark, and treated in the same way as other seeds from warm countries. In about a month or six weeks after the seeds are sown, the plants will begin to appear; and when they are about two inches high, they should be shaken out of the pots, and separated carefully; planting each into a small pot, and plunged again into a hot-bed of tanners bark, screening them from the sun until they have taken new root; then they must have fresh air admitted to them, in proportion to the warmth of the season. As the season advances and becomes warm, the glasses of the hot-bed should be raised higher, to admit a greater share of air to the plants; and when the plants are grown so tall as to reach the glasses, they should be removed into the bark-stove, and plunged into the tan bed. In winter these plants require to be kept very warm, and at that season they must not have much water; but in summer it must be often repeated, and should have much air in warm weather.

FUMARIA. Lin. Gen. Plant. 760. Fumatory.

The CHARACTERS are,

The flower is of the ringent kind, approaching near to the butterfly flowers. The upper lip is plain, obtuse, and reflexed; the nectarium at the base is obtuse, and little prominent. The base is keel-shaped; the nectarium at the base is less prominent. The chaps of the flower is four-cornered, and perpendicularly bifid. In the center is situated an oblong germen, which afterward becomes a short pod with one cell, including roundish seeds.

The SPECIES are,

1. FUMARIA (*Officinalis*) pericarpiis monospermis racemosis, caule diffuso. Lin. Sp. Plant. 700. *Common Fumatory with a purple flower.*

2. FUMARIA (*Spicata*) pericarpiis monospermis spicatis, caule erecto, foliolis filiformibus. Sauv. Monsp. 263. *Lesser narrow-leaved Fumatory.*

3. FUMARIA (*Capnoides*) siliquis linearibus tetregonis, caulibus diffusis acutangulis. Lin. Sp. Plant. 700. *Evergreen Fumatory with a white flower.*

4. FUMARIA (*Tingitana*) siliquis teretibus, caulibus diffusis, angulis obtusis. *Yellow Tangier Fumatory.*

5. FUMARIA (*Claviculata*) siliquis linearibus, foliis cirrhiferis. Lin. Sp. Plant. 701. *Fumatory with tendrils.*

6. FUMARIA (*Capreolata*) pericarpiis monospermis racemosis, foliis scandentibus subcirrhosis. Lin. Sp. Pl. 701. *Greater climbing Fumatory with a paler flower.*

7. FUMARIA (*Bulbosa*) caule simplici, bracteis longitudine florum. Lin. Sp. Plant. 699. *Greater bulbous Fumatory with a hollow root.*

8. FUMARIA (*Solida*) caule simplici, bracteis brevioribus multifidis, radice solidâ. *Greater bulbous Fumatory with a solid root.*

9. FUMARIA (*Cucullaria*) scapo nudo. Hort. Cliff. 351. *Tuberous insipid Fumatory.*

10. FUMARIA (*Vesicaria*) siliquis globosis inflatis. Hort. Upsal. 207. *Climbing African Cysticapnos.*

11. FUMARIA (*Enneaphylla*) foliis triternatis, foliolis cordatis. Lin. Sp. Plant. 700. *Nine-leaved Rock Fumatory of Spain.*

12. FUMARIA (*Sempervirens*) siliquis linearibus paniculatis, caule erecto. Hort. Upsal. 207. *Bastard Fumatory.*

The first sort is the common Fumatory which is used in medicine, which grows naturally on arable land in most parts of England; it is a low annual plant, and flowers in April, May, and June; and very often from plants which rise late in the summer, there will be a second crop in autumn. The juice of this plant is greatly commended for bilious cholics. It is never cultivated in gardens.

The second sort grows naturally in the south of France, Spain, and Portugal, but is preserved in botanic gardens for the sake of variety. It is an annual plant, which rises from the scattered seeds better than when it is sown with care; the stalks of this grow more erect, the leaves are very finely divided, and the flowers grow in a close spike; they are of a deep red colour, and flower about the same time as the common sort.

The third sort grows naturally on the borders of the Mediterranean sea; it was first brought to England from Tangier. This is a perennial plant, which sends out from the root many branching stalks, which rise about six or eight inches high, growing in tufts or branches; the leaves are very much divided; the stalks are angular, and the flowers grow in loose panicles upon naked foot-stalks, which come out from the divisions of the branches; they are of a whitish yellow colour, and there is a succession of them most part of the year.

The fourth sort hath an appearance very like the third, and by some it is supposed to be only a variety of that but is undoubtedly a distinct species; for I have cultivated both more than thirty years, and never yet found either of them to vary. The stalks of this sort have blunt angles whereas those of the third are acute; they are of a purplish colour, and the flowers grow in looser panicles, each having a longer foot-stalk than those of the other; they are of a bright yellow colour, and there is a fucceffion of them great part of the year.

Thefe two forts continue green all the year, and except in very fevere froft are always in flower, which make a pretty appearance; they grow beft on walls or rocks, and are very proper for the joints of grottos, or any rock-work; where, if a few plants are planted, or the feeds fcattered, they will multiply faft enough from their fcattering feeds, which are caft out of the pods by the elaftic fpring of the valves when ripe, to a confiderable diftance; and as the plants will require no care to cultivate them, they fhould not be wanting in gardens.

The fifth fort grows in ftony and fandy places in fome parts of England; it is an annual plant with trailing ftalks, fending out clafpers from the leaves, which faften to any of the neighbouring plants. It flowers in May and June, but is never cultivated in gardens.

The fixth fort is an annual plant with many trailing ftalks, which grow about a foot long, fending out a few fhort tendrils, whereby they faften to any neighbouring fupport; the flowers come out from the fide of the ftalks in loofe bunches; they are of a whitifh herbaceous colour, with a purple fpot on the upper lip. This flowers in May and June. It grows in France and Italy, on ftony places in the fhade.

The feventh fort grows naturally in the fouth of France and Italy, and was fome years paft preferved in the Englifh gardens by way of ornament, but is now rarely to be found here; it was titled Radix cava, or hollow root, from its having a pretty large tuberous root hollowed in the middle. The ftalk of this fort rifes about fix inches high, and does not divide, but is garnifhed toward the bottom with one ramous leaf, fomewhat like the common Fumatory, but the lobes are broader; the flowers grow in a fpike at the top of the ftalk; they are of a pale purplifh colour, and appear in May. This plant delights in the fhade, and is multiplied by offsets, for it rarely ripens feeds in England.

The eighth fort is pretty common in many of the old gardens in England; it grows naturally in the fouth of France, in Germany and Italy. It hath a pretty large, round, folid root, of a yellowifh colour, from which come out branching leaves like thofe of the laft fort, but the lobes are longer; the flowers grow in fpikes on the top of the ftalks; they are of a purple colour, and come out early in April. The ftalks of this fort are fingle, and rife about four or five inches high.

There is a variety of this with green flowers, which is mentioned in moft of the books; but all the plants of this fort which I have yet feen are only abortive, having no real flower, only a green bractea, which has been generally taken for the flowers.

The ninth fort grows naturally in North America; this hath a fcaly root about the fize of a large Hazel nut, from which comes out three or four leaves upon flender foot-ftalks, divided into three parts; each of thefe parts is compofed of many fmaller divifions, which have narrow lobes, divided into three parts almoft to the bottom; the flower-ftalk is naked, and eight or nine inches long, terminated by four or five flowers, growing in a loofe fpike; thefe have two petals, which are reflexed backward, and form a fort of fork toward the foot-ftalk, and at their bafe are two horned nectariums, which ftand horizontal. The flowers are of a dirty white colour, and appear in May, but rarely produce feeds here.

This is propagated by offsets from the root; it loves a fhady fituation and a light foil; the beft time to tranfplant the roots is in autumn, when the leaves are decayed, for it fhoots pretty early in the fpring, therefore it would not be fafe to remove them at that feafon.

The tenth fort grows naturally at the Cape of Good Hope; this is an annual plant with trailing ftalks, which are two or three feet long, dividing into many fmaller, which are garnifhed with fine branching leaves, fhaped like thofe of the common Fumatory, but end with tendrils, which clafp to any neighbouring plants, and thereby the ftalks are fupported; the flowers are produced in loofe panicles which proceed from the fide of the ftalks; they are of a whitifh yellow colour, and are fucceeded by globular fwollen pods, in which are contained a row of fmall fhining feeds.

This is propagated by feeds, which fhould be fown upon a moderate hot-bed in the fpring, and when the plants are fit to remove, they muft be each planted in a fmall pot, filled with light earth, and plunged again into the hot-bed, and fhaded from the fun till they have taken new root; then they fhould have a large fhare of air admitted to them, at all times in mild weather; as foon as the feafon is favourable, they fhould be inured to bear the open air, into which they may be removed the beginning of June, when they may be fhaken out of the pots, preferving all the earth to their roots, and planted in a warm border, where their ftalks fhould be fupported with fticks to prevent their trailing on the ground; in July the plants will flower, and continue a fucceffion of flowers till the froft deftroys the plants; the feeds ripen in autumn.

The eleventh fort grows naturally upon old walls or rocky places in Spain and Italy; this hath weak trailing ftalks, which are much divided, garnifhed with fmall leaves divided into three parts, each of which hath three heart-fhaped lobes; the flowers are produced in fmall loofe panicles from the fide of the ftalks; they are of a greenifh white, and appear moft of the fummer months. It is an abiding plant, which propagates itfelf by the feeds that fcatter, and thrives beft in a fhady fituation, and on old walls or buildings.

The twelfth fort is an annual plant with an upright ftalk, which grows a foot and a half high, fending out feveral branches upward, garnifhed with fmooth branching leaves of a pale colour, divided like the common fort, but the lobes are larger and more obtufe; the flowers are produced in loofe panicles from the fides of the ftalks, and at the extremity of the branches, of a pale purple colour, with yellow chaps (or lips;) thefe are fucceeded by taper narrow pods, which contain many fmall fhining black feeds. This flowers during moft of the fummer months, and the feeds ripen in July, Auguft, and September. If the feeds of this plant are permitted to fcatter, the plants will come up without any trouble, and require no other care but to thin them where they are too clofe, and keep them clean from weeds.

The fifth, fixth, feventh, and eighth forts are propagated by offsets, as other bulbous-rooted flowers; thefe are pretty ornaments to borders in a fmall flower-garden early in the fpring. They are extreme hardy; they love a light fandy foil, and fhould be fuffered to remain three years undifturbed, in which time they will produce many offsets. The beft feafon for tranfplanting them is from May to Auguft, when the leaves die off; for if they are taken up when their leaves are frefh, it will greatly weaken their roots.

FURZ. See GENISTA.

G.

GALANTHUS. Lin. Gen Plant. 362. The Snow-drop.

The CHARACTERS are,

The flower has three oblong concave petals, which spread open; in the bottom is situated the three-leaved nectarium, which is cylindrical; under the flower is situated the oval germen, attended by six stamina, which are gathered together. The germen afterward becomes an oval capsule, opening in three cells, filled with roundish seeds.

We have but one SPECIES of this genus, viz.

GALANTHUS (*Nivalis.*) Lin. Hort. Cliff. 134. *The common Snow-drop.*

There is a variety of this with double flowers.

This is valued for its early appearance in the spring, for the flowers usually blow in February when the ground is often covered with snow. The single sort comes out the first, and though the flowers are but small, yet when the roots are in bunches they make a very pretty appearance; therefore these roots should not be planted single, as is sometimes practised by way of edging to borders; for when they are so dispofed, they make very little appearance. But when there are twenty or more roots growing in a close bunch, the flowers have a very good effect; and as these flowers thrive well under trees or hedges, they are very proper to plant on the sides of wood-walks, and in wilderness quarters, where, if they are suffered to remain undisturbed, the roots will multiply exceedingly. The roots may be taken up the latter end of June, when their leaves decay, and may be kept out of the ground till the end of August; but they must not be removed oftener than every third or fourth year: these plants are got scarce in the gardens near London.

GALE. See MYRICA.

GALEGA. Lin. Gen. Plant. 770. Goat's-rue.

The CHARACTERS are,

The flower is of the butterfly kind; the standard is oval, large, and reflexed; the wings are near the length of the standard; the keel is erect, oblong, and compressed; it has ten stamina, which join above their middle. In the center is situated a narrow cylindrical germen, which afterward becomes a long pointed pod, inclosing several oblong kidney-shaped seeds.

The SPECIES are,

1. GALEGA (*Officinalis*) foliolis lanceolato-linearibus, siliquis tenuioribus. *Common Goat's-rue with blue flowers.*

2. GALEGA (*Africana*) foliolis lanceolatis obtusis, floribus spicatis longioribus, siliquis crassioribus. *African Goat's-rue with larger flowers and thicker pods.*

3. GALEGA (*Frutescens*) foliis ovatis, floribus paniculatis alaribus, caule fruticoso scandente. *American Goat's-rue with roundish leaves, scarlet flowers, and shrubby climbing stalks.*

The first sort grows naturally in Italy and Spain, but is propagated in the English gardens for medicinal use. It hath a perennial root, from which arise many channelled hollow stalks, from two to three feet high, garnished with winged leaves, composed of six or seven pair of narrow spear-shaped lobes, terminated by an odd one: the flowers terminate the stalk, growing in spikes, of a pale blue colour, and are succeeded by taper pods, with one row of kidney-shaped seeds.

There is a variety of this with white flowers, and another with variegated flowers, which have accidentally been produced from seeds, so are not constant, therefore are only mentioned here.

The second sort grows naturally in Africa; this differs from the former in having larger leaves, which are composed of eight or ten pair of lobes, broader and blunter at their ends than those of the common sort; the flowers are larger, and the spikes are longer; the seed-pods are also much thicker than those of the common sort, but in other respects are very like it.

These plants are propagated by seeds, which may be sown either in the spring or autumn, in an open situation; when the plants come up, and are strong enough to remove, a spot of ground should be prepared, in size proportionable to the quantity of plants designed, which should be well dug, and cleared from the roots of all noxious weeds; then the plants should be carefully taken up and planted in rows at a foot and a half distance every way, observing to water them till they have taken new root; after which they will require no farther care but to keep them clean from weeds; if their stalks are cut down before the seeds are formed every year, the roots will continue the longer, especially if they grow on a light dry soil. The seeds of these will grow wherever they are permitted to scatter, so that plenty of the plants will come up without any care, and these may be transplanted and managed in the same manner as is before directed.

The third sort was discovered by the late curious botanist Dr. William Houstoun at Campeachy. This plant is propagated by seeds, which must be sown on a hot-bed early in the spring; when the plants come up and are fit to transplant, they must be each put into a separate small pot, and plunged into a hot-bed of tanners bark, shading them from the sun till they have taken new root; then they must be treated as hath been directed for other tender plants, which are kept in the bark-stove. With this management they will flower, and in warm seasons will perfect their seeds, but the plants may be preserved through one winter in the bark-stove.

GALENIA. Lin. Gen. Plant. 443.

The CHARACTERS are,

The flower hath no petals, but eight hairy stamina the length of the empalement. In the center is situated a roundish germen. The empalement afterward becomes a roundish capsule with two cells, containing two oblong angular seeds.

We have but one SPECIES of this genus, viz.

GALENIA (*Africana.*) Hort. Cliff. 150. *Shrubby Galenia.*

This plant grows naturally at the Cape of Good Hope, and in other parts of Africa; it rises with a shrubby stalk four feet high, sending out many weak branches, garnished with very narrow leaves placed irregularly, of a light green, with

with a furrow running longitudinally through the middle; the flowers are produced in loose panicles from the side and at the end of the branches; they are very small, and have no petals like the Chenopodium.

This plant must be placed in the green-house in winter, with other hardy exotic plants, where it may have a large share of air in mild weather, for it only requires to be protected from frost. In the summer it may be exposed in the open air, with other plants of the same country. It may be propagated by cuttings, which, if planted during any of the summer months, and watered frequently, will take root in about five or six weeks, and may then be treated as is directed for the old plants.

GALEOPSIS. Lin. Gen. Plant. 637. Stinking Dead Nettle.

The Characters are,

The flower is of the lip kind; the chaps are broad; from the base to the under lip, it is on both sides sharply indented; the upper lip is concave, and sawed at the top; the under lip is trifid, the middle segment being the largest. It hath four stamina, two being shorter than the other. In the center is situated a quadrifid germen, which afterward become four naked seeds, sitting in the rigid empalement.

The Species are,

1. Galeopsis (*Ladanum*) internodiis caulinis æqualibus, verticillis omnibus remotis. Lin. Sp. Plant. 579. *Red narrow-leaved Field Ironwort.*

2. Galeopsis (*Tetrahit*) internodiis supernè incrassatis, verticillis summis subcontiguis. Lin. Sp. Plant. 579. *Common Dead Nettle with a Hemp leaf.*

3. Galeopsis (*Speciosa*) corollâ flavâ, labio inferiore maculato. Flor. Lapp. 193. *Prickly Hemp Dead Nettle, with a beautiful yellow flower and purple lips.*

4. Galeopsis (*Galeobdolon*) verticillis sexfloris, involucro tetraphyllo. Lin. Sp. Plant. 780. *Stinking Dead Nettle with a yellow flower.*

These are all of them annual plants, except the fourth sort; they grow naturally in England. The first is found upon arable land in many places; the second grows upon dunghills, and by the side of paths, in most parts of England. The third sort grows chiefly in the northern counties, but I have accidentally found it growing wild in Essex, within ten miles of London. These plants are seldom cultivated in gardens, for if their seeds are permitted to scatter, the plants will come up as weeds wherever they are allowed a place.

The fourth is a perennial plant with a creeping root, which grows in the woods and under hedges in most parts of England.

GALEOPSIS FRUTESCENS. See Prasium.

GALIUM. Lin. Gen. Plant. 117. Ladies Bedstraw, or Cheese-rennet.

The Characters are,

The flower hath one petal, divided into four parts, and four awl-shaped stamina. It hath a twin germen situated under the flower, which afterward become two dry berries, joined together, each inclosing a large kidney-shaped seed.

The Species are,

1. Galium (*Verum*) foliis octonis linearibus sulcatis, ramis floriferis brevibus. Hort. Cliff. 34. *Yellow Ladies Bedstraw.*

2. Galium (*Mollugo*) foliis octonis ovatis linearibus subserratis patentissimis mucronatis, caule flaccido, ramis patentibus. Lin. Sp. Plant. 107. *Branching broad-leaved Mountain Mollugo.*

3. Galium (*Purpureum*) foliis verticillatis lineari-setaceis, pedunculis folio longioribus. Hort. Cliff. 34. *Narrow-leaved, Mountain, Ladies Bedstraw, with a black purple flower.*

4. Galium (*Glaucum*) foliis verticillatis linearibus, pedunculis dichotomis, summo caule floriferis. Prod. Leyd. 256. *Rock Ladies Bedstraw with a gray leaf.*

5. Galium (*Rubrum*) foliis verticillatis linearibus, pedunculis brevissimis. Hort. Cliff. 34. *Red Ladies Bedstraw.*

6. Galium (*Boreale*) foliis quaternis lanceolatis trinerviis glabris, caule erecto, seminibus hispidis. Flor. Lapp. 60. *Smooth Meadow Madder with an acute leaf.*

7. Galium (*Album*) foliis quaternis obovatis inæqualibus, caulibus diffusis. Flor. Suec. 119. *White Marsh Ladies Bedstraw.*

The first of these plants (which is the sort commonly used in medicine) is very common in moist meadows and in pasture-grounds in several parts of England. The other varieties are preserved in curious botanic gardens, but as they are plants of very little beauty, and are subject to spread very far, and over-run whatever plants grow near them they are seldom cultivated in other gardens.

These sorts may any of them be propagated by parting their roots, which spread and increase very fast, either in the spring or autumn, and will grow almost in any soil or situation, especially the first sort; all the other sorts, except the last, require a drier soil, but will all grow in any situation.

GARCINIA. Lin. Gen. Plant. 526. The Mangosteen.

The Characters are,

The flower hath four roundish petals, which are larger than the empalement. It hath sixteen stamina, formed into a cylinder. In the center is situated an oval germen, crowned by a buckler-shaped plain stigma, divided into eight parts, which afterward becomes a thick globular berry with one cell, including eight hairy fleshy seeds, which are convex and angular.

We have but one Species of this genus, viz.

Garcinia (*Mangostena.*) Hort. Cliff. 182. *The Mangostan, or Mangosteen.*

This tree grows naturally in the Molucca Islands, and also in the inland parts of New Spain, from whence I received perfect specimens, which were sent me by Mr. Robert Millar, who gathered them near Tolu, but did not know the tree. It rises with an upright stem twenty feet high, sending out many branches placed opposite, which stand oblique to each other, not at right angles; the bark of the branches is smooth, of a gray colour, that on the tender shoots is green. The leaves are spear shaped and entire, of a lucid green on their upper side, and of an Olive colour on their under, having a prominent midrib, with several small veins running from that to both sides of the leaf. The flower is like that of the single Rose, composed of four roundish petals, which are thick at their base, but thinner toward their ends, of a dark red colour. The fruit which succeeds the flower is round, the size of a middling Orange; the top is covered by a cap, which was the stigma on the top of the style, and remains to the top of the fruit, which is indented in six or seven obtuse rays. The shell of the fruit is like that of the Pomegranate, but softer, thicker, and fuller of juice; the inside of the fruit is of a Rose colour, divided into several parts by thin partitions, as in Oranges, in which the seeds are lodged, surrounded by a soft juicy pulp of a delicious flavour; it is esteemed one of the richest fruits in the world: the trees naturally growing in the form of pyrabolas, whose branches being well garnished with large shining green leaves, make an elegant appearance, and afford a kindly shade in hot countries.

There are but few of the seeds in this fruit which come to perfection (the greatest part of them are abortive) most of those which have been brought to Europe have failed; therefore the surest way to obtain the plants, is to sow their seeds in tubs of earth in the country, and when the plants

plants have obtained sufficient strength they may be brought to Europe; but there should be great care taken in their passage, to screen them from the spray of the sea, as also not to give them much water, especially when they arrive in a cool or temperate climate, for these plants are very impatient of wet. When the plants arrive in Europe they should be carefully transplanted, each into a separate pot, and plunged into the tan-bed, observing to shade them from the sun till they have taken new root; then they must be treated in the same manner as other tender plants from hot countries.

GARDENS are frequently distinguished into Flower-gardens, Fruit-gardens, and Kitchen-gardens; the first being designed for pleasure and ornament, so should be placed in the most conspicuous parts, i. e. next to, or just against, the back front of the house; the two latter being principally intended for use and service, are placed less in sight.

Though the Fruit and Kitchen-gardens are often mentioned as two distinct Gardens, and have by the French gardeners, as also by some of our own countrymen, been contrived as such, yet they are now usually in one; and with good reason, since they both require a good soil and exposure, and to be placed out of the view of the house. As the Kitchen-garden should be inclosed with walls, that no person may have access to it, who have no business in it, for the sake of preserving the product, so these walls answer the purposes of both.

In the choice of a place to plant a Garden, the situation and exposure of the ground are the most essential points to be regarded; since, if a failure be made in that point, all the care and expence will in a manner be lost.

The second thing to be considered in chusing a spot for a Garden, is a good earth or soil.

It is scarce possible to make a fine Garden in a bad soil; there are indeed ways to meliorate ground, but they are very expensive; and sometimes, when the expence has been bestowed of laying good earth over the whole surface, the whole Garden has been ruined, when the roots of the trees have come to reach their natural bottom.

The quality of good ground is neither to be stony or hard to work; neither too dry nor too moist; nor too sandy and light, nor too strong and clayey, which is the worst of all for Gardens.

The third requisite is water. The want of this is one of the greatest inconveniencies that can attend a Garden, and will bring a certain mortality upon whatever is planted in it, especially in the greater droughts that often happen in a hot and dry summer.

GARIDELLA. Tourn. Inst. R. H. 655. tab. 43. Lin. Gen. Plant. 507.

The CHARACTERS are,

The flower hath no petals, but five oblong equal nectariums occupy their place, which are bilabiate. It hath eight or ten awl-shaped stamina, which are shorter than the empalement. In the center is situated three compressed germina, which become three oblong compressed capsules with two valves, inclosing several small seeds.

We have but one SPECIES of this genus, viz.

GARIDELLA (*Nigellastrum.*) Hort. Cliff. 170. *Garidella with very narrow divided leaves.*

This plant is very near akin to the Nigella, or Fennel-flower, to which genus it was placed by the writers on botany before Dr. Tournefort, and was by him separated from it, as differing in the form of the flower.

It grows wild in Candia, and on mount Baldus in Italy, as also in Provence, where it was discovered by Dr. Garidel, who sent the seeds to Dr. Tournefort, for the royal garden at Paris.

This is an annual plant, which rises with an upright stalk, dividing into three or four slender branches, garnished at their joints with very narrow leaves like those of Fennel. The stalks are terminated by one small flower of a whitish colour, which is succeeded by three capsules, each containing two or three small seeds. It flowers in June and July, and the seeds ripen in September. It is propagated by seeds, which should be sown in autumn on a bed of light fresh earth, where the plants are designed to remain; when the plants come up they must be carefully cleared from weeds, and where they are too close they must be thinned, leaving them about four or five inches apart; which is all the culture the plants require, and if the seeds are permitted to scatter, the plants will come up without any farther care.

GENISTA. Lin. Gen. Plant. 766. Broom.

The CHARACTERS are,

The flower is of the butterfly kind; the standard is long and wholly reflexed; the wings are a little shorter, and loose; the keel is erect, and longer than the standard. It hath ten stamina joined, which are situated in the keel. In the center is an oblong germen, which afterward becomes a roundish turgid pod with one cell, opening with two valves, inclosing kidney-shaped seeds.

The SPECIES are,

1. GENISTA (*Sagittalis*) ramis ancipitibus articulatis, foliis ovato-lanceolatis. Hort. Cliff. 355. *Dwarf Ancre-shaped Broom.*

2. GENISTA (*Florida*) foliis lanceolatis, ramis striatis teretibus erectis. Hort. Cliff. 355. *Narrow-leaved Dyer's Broom.*

3. GENISTA (*Tinctoria*) foliis ovato-lanceolatis, ramis striatis teretibus. *Common Dyer's Broom, or Wood-waxen.*

4. GENISTA (*Lusitanica*) foliis inferioribus cuneformibus, ramis floriferis linearibus, floribus majoribus erectioribus. *Greater Portugal Dyer's Broom, called Piurna by the Portugueze.*

The first sort grows naturally in France, Italy, and Germany. It sends out several stalks from the root, which spread flat on the ground, divide into many flat branches which are jointed, and their two sides are edged like a broad sword; they are herbaceous, but perennial. At each of the joints is placed one small spear-shaped leaf, without any foot-stalk. The flowers are produced in close spikes at the end of the branches; they are of the Pea-bloom kind, of a dirty yellow colour, and are succeeded by short hairy pods, which contain three or four kidney-shaped seeds.

This sort is propagated by seeds, which, if sown in the autumn, the plants will come up the following spring; but when they are sown in the spring, the plants rarely come up the same year: when the plants come up, they will require no other culture but to keep them clean from weeds, and thin them where they are too close; at Michaelmas they may be transplanted where they are designed to remain. The plants are very hardy, and will live several years.

The second sort rises with ligneous stalks two or three feet high, garnished with small spear-shaped leaves placed alternate, and are terminated by several spikes of yellow flowers of the Pea-bloom kind, but small, which are succeeded by short pods, black when ripe, and contain four or five kidney-shaped seeds.

The third sort grows naturally in England. It hath shrubby stalks three feet high, garnished with spear-shaped leaves, which are broader than those of the former; the branches which come out from the side of the stalks do not grow so upright as those of the second, but are terminated by loose spikes of yellow flowers, which are succeeded by

pods

pods like those of the second sort. The branches of the plant are used by dyers to give a yellow colour, from whence it is called Dyer's Broom, Green-wood, Wood-waxen, or Dyer's-weed.

The fourth sort grows naturally in Portugal and Spain. This rises with shrubby channelled stalks four feet high, sending out several branches which grow erect; the lower leaves are wedge-shaped, very narrow at their base; those which are higher on the flower branches are narrow, and equal at both ends; the flowers are produced in pretty long spikes at the end of the branches, which are larger than those of the other sorts, and of a paler yellow colour; these are succeeded by pods like the former sorts.

This sort is a little tender, and in very severe frosts is sometimes killed in England, where the plants are not protected.

All these sorts of Brooms are propagated by seeds, which, if sown in autumn, will succeed much better than if sown in the spring, and a year will be thereby saved: as these plants send out long, stringy, tough roots, which run deep into the ground, they do not bear transplanting well, especially if they are not removed young; therefore the best way is to sow a few seeds in those places where the plants are designed to remain, and to pull up all except the most promising plants, as soon as they are past danger; after this the plants will require no other culture but to keep them clean from weeds; but where this cannot be practised, the seeds may be sown thin upon a bed of light earth, and when the plants come up they may remain till the following autumn, when they should be carefully taken up and transplanted where they are designed to remain.

GENISTA SPINOSA. The Furz, Whins, or Gorse. See Ulex.

GENTIANA. Lin. Gen. Plant. 285. Gentian, or Fellwort.

The Characters are,

It hath a permanent empalement to the flower, cut into five acute segments. The flower hath one petal, which is tubulous. It hath five awl-shaped stamina. In the center is situated an oblong cylindrical germen, which afterward becomes an oblong taper-pointed capsule with one cell, containing many small seeds fastened by the valves to the capsule.

The Species are,

1. Gentiana (*Lutea*) corollis quinquefidis rotatis verticillatis, calycibus spathaceis. Hall. Helv. 479. *Greater yellow Gentian.*

2. Gentiana (*Pneumonanthe*) corollis quinquefidis campanulatis oppositis pedunculatis, foliis linearibus. Lin. Sp. Plant. 228. *Greater, narrow-leaved, autumnal Gentian.*

3. Gentiana (*Asclepioides*) corollis quinquefidis campanulatis oppositis sessilibus, foliis amplexicaulibus. Lin. Sp. Plant. 227. *Gentian with a Swallow-wort leaf.*

4. Gentiana (*Acaulis*) corollâ quinquefidâ campanulatâ, caulem excedente. Lin. Sp. Plant. 228. *Broad-leaved Alpine Gentian, with a large flower; commonly called Gentianella.*

5. Gentiana (*Nivalis*) corollis quinquefidis infundibuliformibus, ramis unifloris alternis. Lin. Sp. Plant. 229. *Annual Gentian with lesser Centaury leaves.*

6. Gentiana (*Cruciata*) corollis quadrifidis imberbibus verticillatis sessilibus. Lin. Sp. Plant. 231. *Crosswort Gentian.*

7. Gentiana (*Ciliata*) corollis quadrifidis margine ciliatis. Lin. Sp. Plant. 231. *Blue Gentian with hairy brims.*

8. Gentiana (*Utriculosa*) corollis quinquefidis hypocrateriformibus, calycibus inflatis plicatis. Lin. Sp. Plant. 229. *Gentian with a bellied tube.*

9. Gentiana (*Centaureum*) corollis quinquefidis infundibuliformibus, caule dichotomo. Lin. Sp. Pl. 229. *Lesser Centaury.*

10. Gentiana (*Perfoliata*) corollis octofidis, foliis perfoliatis. Lin. Sp. Plant. 232. *Yellow perfoliate Centaury.*

11. Gentiana (*Spicata*) corollis quinfidis infundibuliformibus, floribus alternis sessilibus. Lin. Sp. Plant. 230. *Lesser Centaury with a white spiked flower.*

12. Gentiana (*Exaltata*) corollis quinquefidis infundibuliformibus, pedunculis longissimis. *Lesser maritime Centaury with a large blue flower.*

The first sort is the common Gentian of the shops, whose root is one of the principal ingredient in bitters.

This hath a large thick root of a yellowish brown colour, and a very better taste; the lower leaves are of an oblong oval shape, a little pointed at the end, stiff, of a yellowish green, and have five large veins on the back of each. The stalk rises four or five feet high, garnished with leaves, growing by pairs at each joint, almost embracing the stalk at their base; they are of the same form with the lower, but diminish gradually in their size to the top. The flowers come out in whorls at the joints on the upper part of the stalks, standing on short foot-stalks, whose origin is from the wings of the leaves; they are of a pale yellow, have one petal, divided almost to the bottom, and an oblong cylindrical germen, which afterward swells to an oblong taper capsule, bifid at the point, and opens in two cells, filled with small flat seeds.

It grows naturally in the pastures in Switzerland, and in the mountainous parts of Germany, from whence the roots are brought to England for medicinal use.

A few years ago there was a mixture of Henbane roots brought over with Gentian, which was unhappily used, and occasioned great disorders in the persons to whom it was administered.

This plant delights in a light loamy soil and a shady situation, where it will thrive much better than in an open exposure. It is propagated by seeds, which should be sown in pots soon after it is ripe, for if it is kept till the spring it will not succeed; the pots should be placed in a shady situation. In the spring the plants will appear, when they must be duly watered in dry weather; the following autumn they should be carefully shaken out of the pots, so as not to break or injure their roots, and planted in a shady border of loamy earth, at six inches distance each way, observing to let the top of the roots be a little below the surface of the ground, then press the earth close to the roots; after this they will require no farther care but to keep them constantly clean from weeds. In this border the plants may stand two years, by which time they will be fit to transplant where they are designed to remain; therefore in autumn, so soon as their leaves decay, they may be removed; but as the roots of these plants run deep into the ground like Carrots, there must be great care taken in digging them up not to cut or break them. After the plants are well fixed in their places, they require no other culture but to dig the ground about them early in the spring, before they begin to shoot. The roots of this plant will continue many years, but the stalks decay every autumn; the roots seldom flower oftener than every third year, but when they flower strong they make a fine appearance; and as these delight in a shady moist ground, where few ornamental plants will thrive, so they should not be wanting in good gardens.

The second sort grows naturally in moist pastures in many parts of England, but particularly in the north; it rises with an upright stalk about a foot high, garnished with smooth leaves placed opposite, without foot-stalks. The flowers are produced on the top of the stalk three or four in number, standing upon foot-stalks alternately above each other; they are large, bell-shaped, and divided into five points at the brim, of a deep blue colour, so make a fine appearance; these come out the latter end of July in the warm parts of England, but in the north they are full a month later.

It may be propagated by seeds in the same manner as the first sort, and the plants may be treated in the same way: but as this sort doth not shoot its roots deep into the ground, it may be transplanted with less hazard; however, if these are removed with a ball of earth to their roots, they will not feel their removal so much as when the earth is all taken from them. This should be planted in a strong, moist, loamy soil, in which the plants will thrive and flower annually, but in a warm dry soil they will not thrive or flower.

The third sort grows naturally upon the Helvetian mountains; this rises with an upright stalk a foot high, garnished with smooth leaves which embrace the stalk, but end in acute points; they are of a fine green, and are diminished in their size as they are nearer the top; and have five longitudinal veins which join at both ends, but diverge from each other in the middle. The flowers come out by pairs opposite from the bottoms of the leaves, standing on short foot-stalks; they are pretty large, bell-shaped, and of a fine blue colour, so make a fine appearance when they are open. This sort flowers in June and July.

It may be propagated by seeds in the same manner as the first sort, and the plants treated in the same way, but they must have a moist loamy soil, otherwise they will not thrive. It may also be propagated by offsets, which may be divided from the roots; these should be taken off in autumn, which is the best season for removing all these sorts of plants, but these should not be removed or parted oftener than every third year, where they are expected to produce strong flowers.

The fourth sort grows naturally on the Alps and Helvetian mountains; this is commonly known by the title of Gentianella. It is a low plant, the stalks seldom growing more than three or four inches high, garnished with smooth leaves placed opposite, which sit close to the stalk. The flower grows erect on the top of the stalk, so stand quite above it; but sometimes when the plants are strong, there will be two or three at the end of each stalk; they are large, bell shaped, and of a deep azure blue. It usually flowers in May, but sometimes the plants flower again in autumn.

This is commonly propagated by parting of the roots, in the same manner as is before directed for the third sort, but the plants must not be often transplanted or parted; they should have a soft loamy soil and a shady situation, where the plants will thrive and flower well every year.

It may also be propagated by seeds, which, in a good soil, the plants will produce in plenty; these should be sown in autumn, in the same manner as is before directed for the first sort, and if the plants are planted in a good soil, they will be strong enough to flower the second year after they are come up, and these seedling plants will flower much stronger than those which are propagated by offsets.

The fifth and eighth sorts are low annual plants, which grow naturally upon the Alps and other mountainous places in Europe, and are very rarely cultivated in gardens. The fifth sort seldom rises more than two inches high, branching out from the root into several slender stalks, garnished with very small leaves placed by pairs; each stalk is terminated by one smaller blue flower standing erect. The eighth sort grows about four inches high, with a single upright stalk of a purple colour. The leaves at the roots are oval, but those upon the stalk are narrow, and stand opposite. The stalk is terminated by one blue flower, with a large bellied empalement, which is plaited, and the petal of the flower rises but little above the empalement, so does not make much appearance. After the top flower decays, there are frequently two smaller flowers which come out from the side of the stalk at the two upper joints.

As these plants usually grow upon moist spongy ground, it is very difficult to cultivate them in gardens, for unless they have a soil approaching near to that in which they naturally grow they will not thrive; the only method to obtain them is, either to sow them in pots, or upon a moist boggy ground in autumn, but it must be in the shade; and when the plants come up they may be thinned, and the surface of the ground about them covered with Moss. which should be constantly kept moist; with this management I have seen the plants thrive and flower very well.

The sixth sort is a perennial plant, which grows naturally upon the Apennines and the Helvetian mountains; this rises with an upright stalk about six inches high, garnished with smooth spear-shaped leaves about two inches long, and one broad in the middle, sitting close to the stalk; they are placed opposite, and each pair of leaves cross one another, from whence it is called Crosswort Gentian. The flowers are produced in whorls round the upper part of the stalks, sitting very close to them; at the top there is a large cluster growing in the same form, these are of a light blue colour. This may be propagated by seeds or offsets, in the same manner as the third and fourth sorts, and the plants must be treated in the same way.

The seventh sort grows naturally upon the Alps, and other mountainous parts of Europe; this is a low perennial plant, whose stalks are very slender, rarely rising more than three or four inches high, garnished with small, narrow, acute-pointed leaves, placed by pairs; each stalk is terminated by one large blue flower, which is hairy on the inside at the brim. This flowers in July and August, and may be propagated in the same manner as the third and fourth sorts.

The ninth sort is the Lesser Centaury of the shops, which grows naturally upon dry pastures in most parts of England, where it rises in height proportionable to the goodness of the soil, for in good land it is frequently a foot high, but in poor soils not more than three or four inches. It is an annual plant, with upright branching stalks, garnished with small leaves placed by pairs. The flowers grow in form of an umbel at the top, and are of a bright purple colour; they come out in July, and the seeds ripen in autumn. This plant cannot be cultivated in gardens.

The tenth sort grows naturally upon chalky grounds in many parts of England. It is an annual plant, rising with an upright stalk a foot high, garnished with oval-pointed leaves placed opposite, whose base surrounds the stalk; they are of a gray colour; the stalks and leaves are very smooth. The flowers grow in form of an umbel on the top of the stalk, of a bright yellow colour, cut into eight parts at the top. These appear in July, and the seeds ripen in autumn.

The eleventh sort is an annual plant, which grows naturally in the south of France and Italy; this rises with an upright stalk about a foot high, sending out several branches toward the top, garnished by small leaves placed opposite. The flowers are produced from the side, and at the top of the stalk, in form of loose irregular umbels; they are white, and about the size of those of the common Centaury.

The twelfth sort grows naturally in the West-Indies, where it was discovered by Father Plumier, and the late Dr. Houstoun found it growing in plenty at La Vera Cruz, in low moist places where the water stagnates, but at a remote distance from the sea. This rises with an upright stalk near two feet high, garnished with oblong, smooth, acute-pointed leaves placed opposite; the upper part of the ſtalk divides into ſix or ſeven long naked foot-ſtalks, each ſuſtaining one large blue flower, divided into five ſegments at the brim. The flowers are ſucceeded by oblong capſules with one cell, filled with ſmall ſeeds.

This is propagated by ſeeds, which muſt be ſown on a hot-bed, and the plants afterward treated in the ſame manner as other tender annual plants from the ſame country, being too tender to thrive in the open air in England. If the ſeeds of this plant are ſown in autumn, in pots placed in the tan-bed of the ſtove, they will ſucceed better than when they are ſown in the ſpring, and the plants will flower early, ſo good ſeeds may be obtained.

GENTIANELLA. See GENTIANA.

GERANIUM. Lin. Gen. Pl. 346. Crane's-bill.

The CHARACTERS are,

The flowers have oval or heart-ſhaped petals, ſpread open, which are in ſome ſpecies equal, and in others the upper two are much larger than the three lower. It hath ten ſtamina, which are alternately longer. In the bottom of the flower is ſituated a five-cornered germen, which is permanent. The flower is ſucceeded by five ſeeds, each being wrapped up in the huſk of the beak, where they are twiſted together at the point, ſo as to form the reſemblance of a ſtork's beak.

The SPECIES are,

1. GERANIUM (*Pratenſe*) pedunculis bifloris, peltatis multipartitis rugoſis, pinnato-laciniatis, acutis. Hort. Cliff. 344. *Crane's-bill with a Crowfoot leaf, and large blue flowers.*

2. GERANIUM (*Macrorrhizum*) pedunculis bifloris, calycibus inflatis, piſtillo longiſſimo. Hort. Cliff. 343. *Long-rooted ſweet-ſmelling Crane's-bill, with a Crowfoot leaf.*

3. GERANIUM (*Sanguineum*) pedunculis unifloris, foliis quinquepartitis trifidis orbiculatis. Lin. Sp. Plant. 685. *Bloody Crane's-bill with a large flower.*

4. GERANIUM (*Lancaſtrenſe*) pedunculis unifloris, foliis quinquepartitis laciniis obtuſis brevibus, caulibus decumbentibus. *Bloody Crane's-bill with a variegated flower.*

5. GERANIUM (*Nodoſum*) pedunculis bifloris, foliis caulinis trilobis integris ſerratis, ſummis ſubſeſſilibus. Hort. Cliff. 343. *Knotty Crane's-bill.*

6. GERANIUM (*Phæum*) pedunculis bifloris, foliiſque alternis, calycibus ſubariſtatis, caule erecto. Lin. Sp. Plant. 681. *Brown Crane's-bill with reflexed petals, and leaves not ſpotted.*

7. GERANIUM (*Fuſcum*) pedunculis bifloris, foliis quinquelobatis inciſis, petalis reflexis. *Brown Crane's-bill with plain petals, and ſpotted leaves.*

8. GERANIUM (*Striatum*) pedunculis bifloris, foliis caulinis trilobis obtuſè-crenatis, inferne hirſutis. *Roman Crane's-bill with ſtriped flowers.*

9. GERANIUM (*Sylvaticum*) pedunculis bifloris, foliis peltatis inciſo-ſerratis, caule erecto. Flor. Lapp. 266. *Mountain Crane's-bill with a Crowfoot leaf.*

10. GERANIUM (*Orientale*) pedunculis bifloris, foliiſque oppoſitis, petalis integris, calycibus brevioribus. *Oriental Dove's-foot Crane's-bill, with an Aſphodel root and large flowers.*

11. GERANIUM (*Perenne*) pedunculis bifloris, foliiſque oppoſitis, caule erecto ramoſo, petalis bifidis. *Greateſt, perennial, Dove's-foot Crane's-bill of the Pyrennes.*

12. GERANIUM (*Alpinum*) pedunculis longiſſimis multifloris, calycibus ariſtatis, foliis bipinnatis. *Alpine Crane's-bill with a Coriander leaf, a long root, and a large purple flower.*

13. GERANIUM (*Argenteum*) pedunculis bifloris, petalis emarginatis, foliis peltatis ſeptempartitis trifidis, tomentoſo-ſericeis. Lin. Syſt. *Silvery Alpine Crane's-bill.*

14. GERANIUM (*Maculatum*) pedunculis bifloris, caule dichotomo erecto, foliis quinquepartitis inciſis ſummis ſeſſilibus. Flor. Virg. 78. *American ſpotted Crane's-bill, with obſolete blue flowers.*

15. GERANIUM (*Bohemicum*) pedunculis bifloris, petalis emarginatis arillis hirtis coteledonibus trifidis, medio truncatis. Lin. Syſt. *Leſſer annual Crane's-bill of Bohemia, with a purple Violet flower.*

16. GERANIUM (*Sibericum*) pedunculis ſubunifloris, foliis quinquepartitis acutis, foliolis pinnatifidis. Lin. Sp. Plant. 683. *Crane's-bill with one flower on a foot-ſtalk, leaves divided into five acute parts, and the ſmaller leaves wing-pointed.*

17. GERANIUM (*Moſchatum*) pedunculis multifloris, calycibus pentaphyllis, floribus pentandris foliis pinnatis inciſis obtuſis. Hort. Cliff. 334. *Muſked Crane's-bill, frequently called Muſcovy.*

18. GERANIUM (*Gruinum*) pedunculis bifloris, calycibus pentaphyllis, floribus pentandris; foliis pinnato-inciſis crenatis. *Broad-leaved annual Crane's-bill, with a blue flower and a very long beak.*

19. GERANIUM (*Ciconium*) pedunculis multifloris, calycibus pentaphyllis, floribus pentandris, foliis pinnatis acutis ſinuatis. Lin. Sp. Pl. 680. *Crane's-bill with a Hemlock leaf, and very long beaks to the ſeed.*

20. GERANIUM (*Viſcoſum*) pedunculis multifloris, calycibus pentaphyllis, floribus pentandris, foliis bipinnatis multifidis, caule erecto viſcoſo. *Erect viſcous Crane's-bill with a Hemlock leaf, and very long beaks to the ſeed.*

21. GERANIUM (*Cuculatum*) calycibus monophyllis, foliis cuculatis dentatis. Hort. Cliff. 345. *African Tree Crane's-bill with a round Marſhmallow leaf, and the ſcent of Carline Thiſtle.*

22. GERANIUM (*Anguloſum*) calycibus monophyllis, foliis cuculatis anguloſis, acute dentatis. *African Tree Crane's-bill, with an angular Marſhmallow leaf, and large purple flowers.*

23. GERANIUM (*Zonale*) calycibus monophyllis, foliis cordato-orbiculatis inciſis zonâ notatis. Hort. Upſal. 196. *African Tree Crane's-bill, with an hairy Ladies Mantle leaf, and red flowers.*

24. GERANIUM (*Inquinans*) calycibus monophyllis, foliis orbiculato-reniformibus tomentoſis crenatis integriuſculis. Hort. Upſal. 195. *African Tree Crane's-bill with a plain ſhining Mallow leaf, and an elegant ſcarlet flower; commonly called Scarlet Geranium.*

25. GERANIUM (*Capitatum*) calycibus monophyllis, foliis lobatis undatis villoſis. Hort. Upſ. 196. *African ſhrubby Crane's-bill with a jagged, ſweet-ſmelling, Mallow leaf; commonly called Roſe-ſcented Geranium.*

26. GERANIUM (*Vitifolium*) calycibus monophyllis, foliis adſcendentibus lobatis pubeſcentibus. Hort. Upſal. 196. *African ſhrubby Crane's-bill, with a jagged Mallow leaf ſmelling like Balm, and a purpliſh coloured flower.*

27. GERANIUM (*Papilionaceum*) calycibus monophyllis, corollis papilionaceis, vexillo dipetalo maximo, foliis angulatis. Hort. Cliff. 345. *African Tree Crane's-bill with a pointed Mallow leaf, and the under petals of the flower ſcarce diſcernible.*

28. GERANIUM (*Acetoſum*) calycibus monophyllis, foliis glabris ſubovatis carnoſis crenatis. Hort. Cliff. 345. *African ſhrubby Crane's-bill with a thick glaucous leaf, and an acid taſte like Sorrel.*

29. GERANIUM (*Carnoſum*) calycibus monophyllis, petalis linearibus, caule carnoſo nodoſo, foliis duplicato-pinnatifidis. Lin. Vir. 67. *African ſhrubby Crane's-bill with a leaf like the Alcea, the petals of the flower white and narrow, and a fleſhy jointed ſtalk.*

30. GE-

30. GERANIUM (*Gibbosum*) calycibus monophyllis, caule carnoso gibboso, foliis subpinnatis. Lin. Sp. Pl. 677. *African Crane's-bill smelling sweet in the night, with knotty tuberous stalks, and leaves like Columbine.*

31. GERANIUM (*Fulgidum*) calycibus monophyllis, foliis tripartitis incisis intermedia majore, pedunculis umbelliferis geminis, caule carnosa. Lin. Vir. 67. *African Crane's-bill with a Vervain Mallow leaf, and a deep scarlet flower.*

32. GERANIUM (*Peltatum*) calycibus monophyllis, foliis quinquelobis integerrimis glabris peltatis. Hort. Cliff. 345. *African Crane's-bill with the under leaves like Asarabacca, and the upper leaves like Stavesacre, shining, spotted, and tasting like Sorrel.*

33. GERANIUM (*Alchimilloides*) calycibus monophyllis, foliis orbiculatis palmatis incisis pilosis, caule herbaceo. Lin. Vir. 67. *African Crane's-bill with a hairy Ladies Mantle leaf, and whitish flowers.*

34 GERANIUM (*Odoratissimum*) calycibus monophyllis, caule carnoso brevissimo, ramis longis, foliis cordatis. Hort. Cliff. 345. *African Crane's-bill with a thick, soft, sweet-smelling Mallow leaf, and a small white flower composed of five leaves.*

35. GERANIUM (*Triste*) calycibus monophyllis, foliis bipinnatis multifidis villosis, radice subrotundo. *American tuberous-rooted Crane's-bill, with a dark flower smelling sweet in the night.*

36. GERANIUM (*Myrrhifolium*) calycibus monophyllis, foliis duplicato pinnatifidis, radice subrotundo. *Tuberous-rooted African Crane's-bill with an Anemony leaf, and a pale flesh-coloured flower.*

37. GERANIUM (*Pastinacæfolium*) calycibus monophyllis, foliis decompositis villosis pinnatifidis, acutis pedunculis longissimis. *Night smelling Crane's-bill with a tuberous root, broad, woolly, heary Carrot leaves, and a pale yellowish flower.*

38. GERANIUM (*Villosum*) calycibus monophyllis, foliis pinnatifidis villosis, laciniis linearibus. *Night smelling Ethiopian Crane's-bill with a tubercus root, and narrow Cicely leaves.*

39. GERANIUM (*Lobatum*) calycibus monophyllis, tubis longissimo subsessilibus, radice subrotundâ, foliis lobatis. Prod. Leyd. 352. *Night, sweet-smelling, African Crane's-bill, with a hairy Vine leaf, and a tuberous root.*

40. GERANIUM (*Coriandrifolium*) calycibus monophyllis, corollis papilionaceis, vexillo dipetalo, foliis bipinnatis. *Smaller African Crane's-bill with a Coriander leaf, and a flesh-coloured flower.*

41. GERANIUM (*Uvæcrispefolium*) calycibus monophyllis, foliis palmatis subrotundis crenatis, caulibus filiformibus procumbentibus. *African Crane's-bill with a Gooseberry leaf, and small reddish flowers.*

The first sort grows naturally in moist meadows in many parts of England, but is frequently planted in gardens for the beauty of its large blue flowers; of this there is a variety with white flowers. and another with variegated flowers, but these are apt to degenerate to the common sort if they are raised from seeds, but by parting of their roots they may be continued. It hath a perennial root which sends up many stalks three feet high, garnished with target-shaped leaves divided into six or seven parts, cut into several acute segments after the manner of winged leaves, ending in many points. The flowers are produced at the top of the stalks, each foot-stalk sustaining two flowers, whose petals are large and equal; they are of a fine blue colour, and appear in May and June.

The second sort grows naturally in Germany and Switzerland; it hath a thick, fleshy, perennial root, from which arise several branching stalks one foot high, garnished with leaves divided into five lobes, which are again divided into many short segments, crenated on their edges. The flowers are produced at the end of the branches; each short foot-stalk sustains two flowers, so it may have the title given it by Linnæus, though at the first appearance it seems as a many-flowered foot-stalk, and strictly is so, because the naked foot-stalk sustains the whole bunch. The petals are pretty large, equal, of a fine bright purple colour; the stamina and style are much longer than the petals; the whole plant, when rubbed, emits an agreeable odour. This may be propagated and treated in the same manner as the first.

The third sort grows naturally in many parts of England; this hath pretty thick, fleshy, fibrous roots, from which arise many stalks, garnished with leaves divided into five parts or lobes, which are again divided almost to the midrib. The flowers stand upon long hairy foot-stalks, which come out from the side of the stalk, each sustaining one flower, composed of five broad regular petals, which are of a deep purple colour. There are two varieties mentioned of this sort as distinct species, one whose stalk grows more erect, and the other hath leaves much deeper divided, but the plants which I have raised from seeds of these do not come up the same as the parent plants, so they are only seminal varieties.

It hath a perennial root, which may be parted in autumn, and thereby propagated; or it may be propagated by seeds, and the plants treated in the same manner as the first.

The fourth sort hath been supposed by some to be only a variety of the third, but is undoubtedly a distinct species. The stalks of this plant are shorter than those of the third, and spread flat on the ground; the leaves are much less, and not so deeply divided, the flowers much smaller and of a pale colour, marked with purple; it grows naturally in Lancashire and Westmoreland, where I saw it in plenty. This may be propagated and treated in the same manner as the others.

The fifth sort is a perennial plant, of smaller growth than either of the former. It rises with branching stalks about six inches high, garnished with leaves having three pretty broad lobes, crenated on their edges: those on the lower part of the stalks are placed opposite, upon pretty long foot-stalks, but the upper leaves sit close to the stalks and are single. The flowers are produced at the end of the stalks, standing together upon two short foot-stalks; they are of a dirty purple colour. It grows naturally in France. This sort may be propagated and treated in the same manner as the first.

The sixth sort grows naturally on the Alps and Helvetian mountains, and is found in some places in the north of England: it hath a perennial root, from which arise several stalks a foot high, garnished with leaves which are divided into five or six lobes, laciniated on their edges; those which grow near the root have long foot-stalks but those on the upper part of the stalk sit close; the stalk branches out at the top into three or four divisions, each being terminated by two or three foot-stalks, sustaining two flowers of a dark purple colour, with erect petals. This may be propagated by seeds or parting of the roots, in the same manner as the first sort.

The seventh sort is very like the sixth, but the leaves are larger, the lobes shorter, broader, and not so much cut, and are marked with black; the stalks rise higher, the flowers are larger, and the petals are reflexed. This may be propagated and treated in the same manner as the first sort. It grows naturally on the Alps.

The eighth sort hath a perennial root, which sends up many branching stalks a foot and a half high, garnished with light green leaves; these on the lower part of the stalk have five lobes, and stand upon long foot-stalks; but those on the upper part have but three, sitting closer to the stalks, and

and are sharply indented on their edges; the flowers stand upon long slender foot-stalks, each sustaining two flowers composed of five obtuse petals, which are deeply indented at the top; they are of a dull white, with many purple stripes running longitunally through them. This sort is very hardy, so may be propagated by dividing of the roots, or from seeds, in the same manner as the first sort.

The ninth sort grows plentifully in the meadows in Lancashire and Westmoreland; it hath a perennial root, which sends out three or four upright stalks about nine inches high, garnished with leaves having five lobes, which are sawed on their edges, and placed opposite; those on the lower part having pretty long foot-stalks, but those on the upper part sit closer to the stalks. The flowers terminate the stalks, standing upon short foot-stalks, each sustaining two pretty large blue flowers, with entire petals. This may be propagated and treated in the same manner as the first sort.

The tenth sort was discovered by Dr. Tournefort in the Levant; it hath a perennial root, from which arise a few weak stalks about nine inches long, garnished with leaves divided into five lobes, which are indented at the top; and placed opposite on the stalks. The flowers stand on pretty long foot-stalks, which come out singly from the joints of the stalks, each sustaining two purplish flowers with entire petals, having very short empalements. It may be propagated either from seeds or by parting of their roots, in the same manner as the first sort, but the plants require a drier soil and a warmer situation.

The eleventh sort grows naturally on the Pyrenean mountains; it hath a perennial root, sending out many branching stalks a foot and a half high, garnished with round leaves, divided into many obtuse segments at the top. The flowers are produced by pairs upon short foot-stalks, which come out at the divisions on the sides, and at the top of the stalks; they are in some of a pale purple colour, and in others white. The petals of the flowers are bifid like those of the common Dove's-foot Crane's-bill, to which the whole plant bears some resemblance, but the stalks are erect, the leaves and flowers much larger, and the root is perennial; this will propagate itself fast enough by its scattered seeds where it has once got possession, and will thrive in any soil or situation.

The twelfth sort grows naturally upon the Alps. It hath a perennial root, which runs very deep into the ground. The lower leaves of the plant have very long foot-stalks, and are doubly winged and smooth. The stalks rise a foot and a half high, garnished with leaves of the same form as the lower, but smaller, and stand opposite. The flowers are purple, many growing together upon very long foot-stalks; it hath awns to the segments of the empalement. This seldom produces seeds in England. The plant is hardy, and lives in the open air, but as the roots put out no offsets, and seldom perfect seeds here, so we have not been able to propagate it in plenty.

The thirteenth sort grows naturally on the Alps; this hath a very thick perennial root, and roundish silvery leaves divided into many parts, standing upon pretty long foot-stalks. The flower-stalks rise about four or five inches high, garnished with one or two small leaves like those below, which sit close to the stalk. The stalks are terminated by two pretty large pale flowers, whose petals are entire. It flowers in June, but rarely ripens seed here; it may be propagated by parting of the roots in the same manner as the first, and must have a shady situation.

The fourteenth sort grows naturally in North America; this hath a perennial root, sending out several stalks one foot high, which divide by pairs, and from the middle of the divisions come out the foot-stalks of the flowers, which are pretty long and naked; each sustaining two pale purple flowers with entire petals. The leaves are divided into five parts, which are cut on their edges, and are placed opposite. It flowers in June, and frequently ripens seeds, from which the plant may be propagated.

The fifteenth sort grows naturally in Bohemia; this is an annual plant, which sends out many stalks, dividing into several parts, garnished with leaves divided into five lobes, crenated on their edges, standing upon long foot-stalks, for the most part opposite. The flowers stand by pairs upon pretty long slender foot-stalks, which come out from the side of the stalk; they are of a fine blue colour, and are succeeded by seeds whose capsules and beaks are black. If the seeds are permitted to scatter, there will be a supply of plants, which want no other care but to keep them clean from weeds.

The sixteenth sort grows naturally in Siberia. It hath a perennial root; the leaves are divided into five acute lobes, which are cut into many sharp wing-like segments on their edges, and are placed opposite upon long slender foot-stalks. The foot-stalks of the flower come out from the wings of the stalk; they are long, slender, each sustaining one pale purplish flower. It ripens seeds very well, so may be easily propagated, and will grow on any soil or situation.

The seventeenth sort is an annual plant, which is sometimes found growing naturally in England, but is frequently preserved in gardens for the musky odour of the leaves; which in dry weather is very strong. The leaves of this are irregularly winged; the lobes grow alternate, and are cut into many obtuse segments on their edges. The stalks branch into many divisions, and frequently decline to the ground. The flowers are produced in umbels upon long foot stalks, which arise from the wings of the stalks; they are small, blue, and have but five stamina in each; their empalements are composed of five leaves. If the seeds are permitted to scatter, there will be a supply of plants without care, which will require no other culture but to keep them clean from weeds.

The eighteenth sort grows naturally in Crete; this is an annual plant with very broad leaves, which are cut on their sides regularly in form of winged leaves, and are crenated on their borders. The flowers are produced on pretty long foot-stalks, which come out from the wings of the stalk; they are composed of five entire blue petals, and are succeeded by the largest and longest beaks of any species of this genus yet known. It ripens seeds very well, and if they are permitted to scatter, the plants will come up without care; or they may be sown in the spring, where they are designed to remain.

The nineteenth sort grows naturally in Germany and Italy. It is an annual plant, with several prostrate stalks a foot long, garnished with winged leaves, cut in several acute parts placed opposite. The flowers come out from the wings of the stalk upon pretty long foot-stalks, some sustaining many flowers, others have no more than two; they are of a pale blue colour, and are succeeded by very long beaks, but not so long or large as those of the former sort. The seeds of this and the former sort are frequently used for hygrometers, to shew the moisture of the air; if the seeds of this are permitted to scatter, the plants will come up and thrive without any other care than to keep them clear from weeds.

The twentieth sort is an annual plant, which hath upright stalks two feet high, garnished with double winged leaves, ending in many points; the whole plant is viscous. The flowers are produced on long naked foot-stalks, standing many together upon each; they are of a pale blue colour, and have but five stamina; their empalements are composed of five leaves, which end with awns. This requires no other culture than the two former sorts.

There are several other sorts of annual Geraniums, some of which grow naturally in England, and are troublesome weeds in a garden; others grow naturally in France, Spain, Italy, and Germany, and are preserved in botanic gardens for the sake of variety; but as they are plants of little beauty, they are rarely admitted into other gardens, therefore I shall not trouble the reader with an enumeration of the species, which would swell this article too much.

The twenty-first sort grows naturally at the Cape of Good Hope; it rises with a shrubby stalk eight or ten feet high, sending out several irregular branches garnished with roundish leaves, whose sides are erect, so form a sort of hood by a hollow cavity made in the leaf. The base of the leaves is cut in form of a heart-shaped leaf, and from the foot-stalk run many nerves arising from a point, but diverge toward the sides; the borders of the leaves are sharply indented; those on the lower part of the branches have long foot-stalks; but those on the upper part have shorter, and stand opposite. The flowers are produced in large panicles on the top of the branches; their empalements are of one leaf, deeply cut into five segments, and closely covered with hairs. The petals are large, entire, and of a blue colour. The flowers are succeeded by seeds with short hairy beaks.

The twenty-second sort has some appearance of the twenty-first, but the leaves are of a thicker substance, divided into many acute angles, and have purple edges which are acutely indented. The stalks and leaves are very hairy. The branches are not so irregular as those of the former, nor are the bunches of the flowers near so large; these differences are permanent in the plants which are raised from seeds, so it is undoubtedly a distinct species.

The twenty-third sort comes from the Cape of Good Hope, but is one of the oldest, and the most common sort in the English gardens; it rises with a shrubby stalk five or six feet high, and divides into a great number of irregular branches, which are garnished with roundish heart-shaped leaves, indented on their edges with several obtuse segments, which are cut in short teeth at their brims; these have a purple circle or mark like a horse-shoe through the leaf, going from one side of the base to the other, corresponding with the border of the leaf; the leaves, when gently rubbed, have a scent like scalded Apples. The flowers are produced in pretty close bunches, standing upon long foot-stalks, which come out from the wings of the stalk toward the end of the branches; they are of a reddish purple colour, and continue in succession great part of summer; there are three or four varieties of this, one with fine variegated leaves, one with crimson, and another with Pink-coloured flowers, which have been accidentally raised from seeds.

The twenty-fourth sort grows naturally at the Cape of Good Hope; this rises with a soft shrubby stem to the height of eight or ten feet, sending out several branches, which are generally erect, garnished with roundish kidney-shaped leaves of a thick substance and a lucid green, standing on pretty long foot-stalks; they are covered with soft hairs on their under side. The flowers grow in loose bunches upon long stiff foot-stalks, which come out from the wings of the stalk; they are of a bright scarlet colour, so make a fine appearance, and there is a succession of these flowers during all the summer months.

The twenty-fifth sort grows naturally at the Cape of Good Hope; this rises with a shrubby stalk four or five feet high, dividing into several weak irregular branches, garnished with leaves divided into three unequal lobes, which are hairy and waved on their edges, placed alternate on the branches; and stand upon hairy foot-stalks. The flowers grow in close roundish heads on the top of the foot-stalks, forming a sort of corymbus; they are of a purplish blue colour, and continue in succession great part of the summer. The leaves of this sort, when rubbed, have an odour like dried Roses, from whence many have given it the title of Rose Geranium.

The twenty-sixth sort is a native of the Cape of Good Hope; this rises with an upright shrubby stalk seven or eight feet high, sending out many pretty strong branches, garnished with leaves shaped somewhat like those of the Vine; those on the lower part stand upon long foot-stalks, but the upper have short ones; when the leaves of this are rubbed they have a scent of Balm. The flowers grow in compact clusters on the top of long naked foot-stalks, which come out from the wings of the stalk; they are small, and of a pale blue colour; so make no great figure.

The twenty-seventh sort rises with an upright shrubby stalk seven or eight feet high, sending out several side branches; garnished with large, angular, rough leaves, standing upon long foot-stalks. The flowers are produced in large panicles at the end of the branches, which are shaped somewhat like a Butterfly flower; the two upper petals, which are pretty large, turn upward like a standard in the leguminous flowers; these are finely variegated, but the three under petals are so small as not to appear at a little distance; these are reflexed downward, so are screened from sight, unless the flowers are viewed near. This sort flowers in May, at which time time the plants make a fine appearance, but they are not succeeded by any more afterward, as most of the other sorts are.

The twenty-eighth sort is from the same country; this rises with a shrubby stalk six or seven feet high, sending out several side branches, garnished with oblong, oval, fleshy, smooth leaves, of a gray colour, crenated on their edges, and have an acid flavour like Sorrel. The flowers stand upon pretty long foot-stalks, which arise from the wings of the stalk, each sustaining three or four flowers, whose petals are narrow, and unequal in size; they are of a pale bluish colour, with some stripes of a light red; these continue in succession most part of the summer. There is a variety of this with scarlet flowers, which is said to have been raised from the seeds of this sort. The leaves of it are larger, and seem to be an intermediate species between this and the twenty-fourth sort, for the flowers are larger than those of the twenty-eighth sort, and are of a pale scarlet colour.

The twenty-ninth sort hath a thick, fleshy, knotted stalk, which rises about two feet high, sending out a few slender fleshy branches, garnished thinly with double winged leaves, which, on the lower part of the stalk stand upon foot-stalks, but those above sit close to the branches. The flowers are produced in small clusters at the end of the branches; these have five narrow white petals which make no appearance.

The thirtieth sort hath a round fleshy stalk with swelling knots at the joints, which rise about three feet high, sending out several irregular branches, which are smooth, thinly garnished with smooth, fleshy, winged leaves, ending in obtuse points, of a gray colour, and stand upon short foot-stalks. The flowers stand four or five upon each foot-stalk, which arises from the wings of the stalk, and are of a dark purple colour. The petals are broader than those of the former sort, and the flowers have a very agreeable scent in the evening, after the sun has left them some time.

The thirty-first sort hath a fleshy stalk, which seldom rises a foot high, and puts out very few branches, which are garnished with smooth, light, green leaves, divided into three lobes, the middle segment being much larger than the others. The flowers stand upon short foot-stalks, each sustaining two or three flowers on the top, which are of a very deep scarlet colour, and have unequal petals. The leaves of this sort fall off, so that the stalks are frequently destitute of them for three or four months in the summer, and appear as if they were dead, but in autumn they put out fresh leaves again.

The thirty-second sort hath many long, weak, shrubby stalks, which require support to prevent their falling on the ground, garnished with fleshy leaves, divided into five oblong lobes, which are entire, and have slender foot-stalks, which are fastened to the middle of the leaf like the handle of a target. The leaves have a circular purple mark in their middle, and have an acid flavour. The flowers are produced upon pretty long foot-stalks, which come out from the wings of the stalk, each foot-stalk sustaining four or five purple flowers, composed of five unequal petals.

The thirty-third sort sends out several herbaceous stalks a foot and a half long, which trail upon the ground if they are not supported, and are garnished with roundish hand-shaped leaves which are hairy, and cut into many parts. The flowers are of a pale blush colour, standing several together upon very long foot-stalks; there is a succession of these during all the summer months, and the seeds ripen accordingly about a month after the flowers are fallen.

The thirty-fourth sort hath a very short fleshy stalk, which divides near the ground into several heads, each garnished with many leaves, which arise on separate foot-stalks from the heads; they are heart-shaped, soft, and downy, and have a strong scent like Aniseed; from these heads come out several slender stalks, which lie prostrate on the ground, garnished with rounder leaves than those near the root, but are of the same texture and have the like odour. The flowers are produced from the side of these stalks, three, four, or five, standing together upon slender foot-stalks; they are very small and white, so make little appearance.

The thirty-fifth sort hath a thick, roundish, tuberous root, from which arise several hairy leaves, which are finely divided, almost like those of the garden Carrot; these spread near the ground, and between them come out the stalks, which rise about a foot high, garnished with two or three leaves of the same sort with those below, but are smaller, and sit closer to the stalks; from these arise two or three naked foot-stalks, which are terminated by a truss of yellowish flowers, marked with dark purple spots, which smell very sweet after the sun hath left them; these are frequently succeeded by seeds, which ripen in autumn. It is known by the title of Geranium noctu olens, or Night-scented Crane's-bill.

The thirty-sixth sort hath a knobbed tuberous root like the last, from which come out several pretty large leaves, composed of many lobes, set along the midrib in the form of a winged leaf, which are narrow at the base, but are very much enlarged at their ends, where they are rounded and cut into many acute points; the stalks which sustain the flowers arise immediately from the root, and sometimes have one or two small leaves toward the bottom, where they often divide into two naked foot-stalks, each being terminated by a truss of pale red flowers, which smell sweet at night.

The thirty-seventh sort hath oblong tuberous roots, from which come out several decompound winged leaves, ending in many acute points; the segments of these leaves are broader than those of the thirty-fifth sort, with leaves very hairy. The stalks rise a foot and a half high, garnished with a single leaf at the two lower joints, these are singly winged; the lobes are narrow, standing at a wider distance, and the segments are more acute than those of the lower leaves; at the two lower knots or joints arise two naked foot-stalks, each being terminated by a truss of yellowish flowers, which have long tubes, and smell sweet in the evening when the sun has left them.

The thirty-eighth sort hath a tuberous root like the former, from which spring out many hairy leaves, which are finely divided like those of the Pulsatilla, which are very hoary, and rise immediately from the root. The foot-stalk of the flower is naked, and rises from the root; this grows about nine inches high, and is terminated by a loose truss of flowers, which are of a very dark purple colour, and smell sweet in the evening.

The thirty-ninth sort hath fleshy tuberous roots like those of the former sorts, from which come out three or four broad hairy leaves, divided on their borders into several lobes in form of a Vine leaf, which spread flat on the ground, crenated on their edges, standing upon short foot-stalks. The foot-stalks of the flowers are naked, and arise immediately from the root, and grow about a foot high, terminated by a truss of dark purple flowers with long tubes, which have a very agreeable odour in the evening.

The fortieth sort is an annual plant, which grows naturally at the Cape of Good Hope; it rises with herbaceous branching stalks a foot high, garnished with doubly winged leaves at each joint; the lower leaves stand upon long foot-stalks, but those on the upper part sit close to the stalks. The flowers stand upon naked foot-stalks, which proceed from the side of the stalks on the opposite side to the leaves; they grow three or four together upon short separate foot-stalks, and are shaped somewhat like a papilionaceous flower; the two upper petals, which are large, form a kind of standard, the other three petals are narrow and reflexed downward; they are of a pale flesh colour. The seeds ripen in autumn, soon after which the plants decay.

The forty-first sort is a biennial plant, which grows naturally at the Cape of Good Hope; this sends out a great number of very slender trailing stalks which lie prostrate on the ground, and extend a foot and a half in length, garnished with small, roundish, hand-shaped leaves, crenated on their edges. The flowers are very small, red, and sit upon short slender foot-stalks, which come out at every joint from the side of the stalks; sometimes they are single, and at other times there are two or three flowers upon a foot-stalk. They continue in succession all the summer, and the seeds ripen in about five weeks after the flowers decay.

All the sorts of African Crane's-bill may be propagated by seed, which should be sown upon a gentle hot-bed toward the middle of March, which will bring up the plants in a month or five weeks; afterward they should be gradually hardened to bear the open air, so as not to draw them up weak; when the plants are fit to remove, they should be put into separate small pots, and placed under a common frame, where they should be shaded from the sun till they have taken new root. In the middle of May they should be removed to a sheltered situation with other exotic plants. If these plants are brought forward in the spring, most of the sorts will flower the same summer, and the plants will be very strong before the winter, so will make a better appearance in the green-house.

All the shrubby African Geraniums, from the twenty-first to the thirty-second inclusive, and also the forty-first, are commonly propagated by cuttings, which, if planted in a shady border in June or July, will take good root in five or six weeks, and may then be taken up and planted into separate pots, placing them in the shade till they have taken

taken new root; after which they may be removed into a sheltered situation, and treated in the same manner as the seedling plants. The twenty-ninth, thirtieth, thirty-first, and thirty-second sorts, have more succulent stalks than either of the other, so the cuttings of these sorts should be planted in pots, filled with light kitchen-garden earth, and plunged into a very moderate hot bed, where they should be shaded from the sun in the heat of the day, and should have but little water, for these are very apt to rot with much moisture; when these are well rooted they may be separated and planted in pots, and placed in the shade till they have taken new root, then they may be removed into a sheltered situation, where they may remain till autumn. These four sorts should be sparingly watered at all times, but especially in winter, for they are apt to take mouldiness with moisture, so will thrive much better in an airy glass-case, where they may have more sun and air than in a green-house. All the other shrubby sorts are proper furniture for the green-house, where they will only require protection from frost, but should have a large share of free air when the weather is mild. These plants should be hardened in the spring gradually, and in the middle of May they may be taken out of the green-house, and at first placed in the shelter of hedges, where they may remain a fortnight or three weeks to harden, then should be removed into a situation where they may be defended from strong winds, and enjoy the morning sun till eleven o'clock, where they will thrive better than in a warmer situation.

The compost in which I have always found these plants thrive best (where there has not been a conveniency of getting some good kitchen-garden earth) was fresh Hazel loam from a pasture mixed with a fifth part of rotten dung; if the earth is inclinable to bind, then a mixture of rotten tan is preferable to dung, but if it is light and warm, then a mixture of neat's dung is best; this compost should be mixed three or four months before it is used, and should be turned over three or four times, that the parts may be well mixed and incorporated; but where a quantity of good kitchen-garden earth can be had, which has been well worked and is clean from the roots of bad weeds, there will need no composition, for in that they will thrive full as well as in any mixture which can be made for them, especially if the earth has laid in a heap for some time, and has been two or three times turned over to break the clods, and make it fine.

The thirty-third sort hath herbaceous stalks, so is best propagated by seeds, which the plants produce in great plenty; but the cuttings of this will take root as freely as either of the other, but the seedling plants are preferable to those propagated by cuttings; and where the seeds of this and many other of the African sorts are permitted to scatter, there will be a supply of young plants come up the spring following, provided the seeds are not buried too deep in the ground.

The thirty-fourth sort may be propagated by seeds, or from heads slipped off from the short fleshy stalk; these heads should have their lower leaves stripped off, that the stalk which is to be planted may be clear of them; then they may be planted each into a small pot, but if the heads are small, there may be two or three put into one small pot; then they should be plunged into a very moderate hot-bed, and shaded from the sun, which will forward their putting out roots; after this they must be hardened gradually, and removed into the open air, where they may remain till autumn, when they must be removed into shelter for the winter season.

The thirty-fifth, thirty-sixth, thirty-seventh, thirty-eighth, and thirty-ninth sorts, are generally propagated by parting of their roots; the best time for doing this is in August, that the young roots may be established before the cold comes on. Every tuber of these roots will grow, provided they have a bud or eye to them; these may be planted in the same sort of earth as was before directed, and if the pots are plunged into an old tan-bed, under a good frame, in winter, the plants will thrive better than in a green-house; the glasses of the frame may be drawn off every day in mild weather, whereby the plants will enjoy the free air; and if in hard frost, they are well covered to prevent the cold penetrating to the plants, it is all the shelter they will require; in this situation they should have but little wet in winter, therefore the glasses should be kept over them in heavy rains to keep them dry; in mild weather the glasses may be raised on the upper side to admit fresh air to the plants, which will give them greater slope to carry off the wet. With this management the roots will thrive and flower very strong every year. These sorts may also be propagated by seeds.

The fortieth sort is an annual plant, and is only propagated by seeds, which should be sown upon a gentle hot-bed in the spring to bring the plants forward, otherwise, if the season should not prove very warm, the plants will not perfect their seeds in this country. When the plants are come up, and grown strong enough to remove, they should be each planted into a separate small pot, and plunged into a moderate hot-bed again, shading them till they have taken new root; then they must be gradually hardened to bear the open air, into which they should be removed in June, and shifted into larger pots; the plants will flower in July, and the seeds ripen in autumn, and soon after the plants will decay.

GERMANDER. See TEUCRIUM.

GEROPOGON. Goat's-beard.

The CHARACTERS are,

The flower has a keel-shaped empalement, is single, and longer than the corolla; it is composed of several hermaphrodite florets, which are imbricated; these have each five short stamina, terminated by cylindrical summits, with an oblong germen, and a slender style supporting two very narrow stigma; the seeds are included in the empalement, and are crowned by five-bearded spreading rays.

The SPECIES are,

1. GEROPOGON (*Glabrum*) foliis glabris. Lin. Sp. 1109. *Grass-leaved smooth Goat's-beard.*

2. GEROPOGON (*Hirsutum*) foliis pilosis. Lin. Sp. 1109. *Goat's-beard with thin hairy leaves; commonly called Red Goat's-beard.*

The first sort grows naturally in Italy. This has an erect stalk more than a foot high, garnished with smooth Grass-like leaves: the stalk divides toward the top into two or three branches, each being terminated by one flesh-coloured flower.

The second sort grows naturally in Sicily; this rises with an erect stalk a foot high, which is garnished with narrow hairy leaves; the stalks seldom branch out, but are terminated by flowers, composed of four or five hermaphrodite florets, which are succeeded by so many bearded seeds.

These plants are annual, and require the same treatment as the TRAGOPOGON, to which article the reader is desired to turn for the culture.

GESNERA. Plum. Nov. Gen. 27. tab. 9.

The CHARACTERS are,

The flower hath one petal, which is tubulous, and first bent inward, and afterward out again like a bugle horn; the brim is divided into five obtuse segments; it hath four stamina which are shorter than the petal; the germen which sits under the petal, after-

afterward becomes a roundish capsule with two cells, filled with small seeds, which are fixed on each side the partition.

The SPECIES are,

1. GESNERA (*Tomentosa*) foliis lanceolatis crenatis hirsutis, pedunculis lateralibus longissimis corymbiferis. *Gesnera with a large, woolly, Fox-glove leaf.*

2. GESNERA (*Humilis*) foliis lanceolatis serratis sessilibus, pedunculis ramosis multifloris. Lin. Sp. Plant. 612. *Low Gesnera with a yellowish flower.*

The first sort grows naturally in Jamaica; this rises with a shrubby stalk six or seven feet high, and divides into two or three irregular branches, covered with a russet wool, garnished with hairy leaves which are seven or eight inches long, and two and a half broad in the middle, with a russet woolly midrib; the edges are crenated, and they have short foot-stalks; towards the end of the branches come out the foot-stalks of the flowers at the joints, arising from the wings of the stalk, which are naked, branching at the top into many smaller foot-stalks, each sustaining a single flower with a short crooked tube, indented at the top in five obtuse segments, of an obsolete purple colour. These are succeeded by roundish capsules sitting close in the empalement, the divisions of which arise above the capsule, which is divided into two cells, filled with small seeds. It flowers in July and August, but hath not ripened seeds in England.

The second sort is a plant of humbler growth; this seldom rises more than three feet high; the leaves are much smaller, sawed on their edges, and sit close to the stalk; the flowers stand upon branching foot-stalks, each sustaining many yellowish flowers, which are deeper cut at their brims than those of the first sort. It grows naturally at Carthagena in New Spain.

These plants are propagated by seeds, which should be sown in pots, and plunged into a hot-bed of tanners bark as soon as they arrive in England, for they sometimes lie long in the ground; those which I have sown in autumn, came up the following spring; therefore when they happen to arrive here at that season, the pots in which the seeds are sown, should be plunged into the tan-bed in the stove, and during the winter the earth should be now and then gently watered to prevent its drying too much, but it must not be too moist. In the spring the pots should be removed out of the stove, and plunged into a fresh hot-bed, which will bring up the plants soon after. When these are fit to remove, they should be each planted into a separate pot, and plunged into a good hot-bed of tan, observing to shade them till they have taken new root; then they must be treated in the same way as other tender plants from the same countries.

In autumn they must be plunged into the tan-bed in the stove, where they must constantly remain, for they will not thrive out of the tan-bed. In the summer they should have free air admitted to them, at all times when the weather is warm. As the plants advance in growth, they will require larger pots, but there must be care taken not to over-pot them, for they will not thrive in large pots.

GEUM. Lin. Gen. Pl. 561. Avens, or Herb-Bennet.

The CHARACTERS are,

The flower has five roundish petals, which are narrow at their base, where they are inserted in the empalement, and a great number of awl-shaped stamina, which are the length of the empalement, into which they are inserted. In the center of the flower is situated a great number of germen, collected into a head, which afterward become so many flat, rough, hairy seeds, with the style which is bent like a knee adhering to them.

The SPECIES are,

1. GEUM (*Urbanum*) floribus erectis, fructu globoso, aristis uncinatis nudis, foliis lyratis. Hort. Cliff. 195. *Common Avens, or Herb Bennet.*

2. GEUM (*Rivale*) floribus nutantibus, fructu oblongo, aristis plumosis. Hort. Cliff. 195. *Aquatic Herb Bennet with a nodding flower.*

3. GEUM (*Pyrenaicum*) floribus nutantibus, fructu globoso, aristis nudis, foliis lyratis, foliolis rotundioribus. *Pyrenean Avens with a very large and rounder leaf, and a nodding flower.*

4. GEUM (*Montanum*) flore erecto solitario, fructu oblongo, aristis plumosis. Lin. Sp. Pl. 501. *Mountain Avens with a large yellow flower.*

5. GEUM (*Alpinum*) flore solitario erecto, fructu globoso, aristis tenuioribus nudis. *Smaller Alpine Avens.*

6. GEUM (*Virginianum*) floribus erectis, fructu globoso, aristis uncinatis nudis, foliis ternatis. Hort. Cliff. 195. *Virginia Avens with a smaller white flower, and a scentless root.*

The first sort grows plentifully by the side of hedges, and in woods in most parts of England, so is rarely admitted into gardens. This stands in the list of medicinal plants, the root is the only part used; it is esteemed cephalic and alexipharmic, and is manifestly of a binding nature, so is useful in all fluxes, &c.

The second sort grows naturally in moist meadows in the northern parts of England. This is of humbler growth than the first; the lower leaves have two pair of small lobes at bottom, and three large ones at the top, that which terminates being the largest. The leaves upon the stalks are composed of three acute lobes, which sit close to the stalk; the flowers are of a purplish colour, and nod on one side; they appear in May, and the seeds ripen in July.

The third sort grows upon the Alps, and also on the mountains in the north; this hath some resemblance to the second sort, but the leaves are much larger and rounder, and are indented on their edges; the flowers are larger and of a gold colour. This flowers about the same time as the second.

The fourth sort grows naturally upon the Alps; this hath leaves much larger than either of the other species; the lower leaves are composed of three or four pair of small irregular pinnæ set along the midrib, which is terminated by one very broad roundish lobe, crenated on the edge. The flowers are large, of a bright yellow colour, standing single on the top of the stalk, which seldom rises more than five or six inches high. It flowers in May and June.

The fifth sort grows naturally on the Alps; it is a very low plant; the flower-stalks are about three inches long, and bend on one side; they are each terminated by one bright yellow flower, about the size of those of the common sort. This flowers about the same time as the former.

The sixth sort grows naturally in North America; the stalks of this sort rise two feet and a half high, and branch out at the top into small foot-stalks, each being terminated by a small white flower; the leaves of this sort are trifoliate, and the root has no scent. These are all very hardy plants, which require a shady situation, but will thrive in any soil; they may be easily propagated by seeds, which should be sown in autumn, for when they are sown in the spring they do not grow the same year.

GILLIFLOWER, or JULY-FLOWER. See DIANTHUS.

GILLIFLOWER, or STOCK-GILLIFLOWER. See CHEIRANTHUS.

GILLIFLOWER, the Queen's, or Dame's Violet. See HESPERIS.

GINGER. See AMOMUM.

GINGIDIUM. See ARTEDIA.

GLADIOLUS. Lin. Gen. Plant. 55. Cornflag.

The CHARACTERS are,

The flowers are included in sheaths; the petal of the flower is cut into six parts, and form a short incurved tube with their base;

base; they have three awl-shaped stamina, which are inserted into every other petal. The germen is situated below the flower, which afterward becomes an oblong, swelling, three-cornered capsule with three cells, filled with roundish seeds.

The SPECIES are,

1. GLADIOLUS (*Communis*) foliis ensiformibus, floribus distantibus. Lin. Sp. Plant. 36. *Cornflag with flowers disposed on one side the stalk.*

2. GLADIOLUS (*Italicus*) foliis ensiformibus, utrinque floribus. *Cornflag with flowers on each side the stalk.*

3. GLADIOLUS (*Byzantinus*) foliis ensiformibus, spathis maximis. *Greater Cornflag of Byzantium.*

4. GLADIOLUS (*Indicus*) foliis ensiformibus, floribus maximis incarnatis. *Greatest Indian Cornflag.*

5. GLADIOLUS (*Angustifolius*) foliis linearibus, floribus distantibus, corollarum tubo limbis longiore. Lin. Sp. Plant. 37. *Cornflag with very narrow leaves, flowers standing at a distance from each other, and the tube longer than the margins of the petal.*

6. GLADIOLUS (*Tristis*) foliis linearibus sulcatis, caule biflorâ, tubo longissimo, segmentis æqualibus. *Cornflag with very narrow channelled leaves, and a stalk bearing two flowers, with a very long tube and equal segments.*

The first sort grows naturally in arable land in most of the warm countries in Europe, and was formerly cultivated in the English gardens, where the roots have multiplied so greatly as to become a most troublesome weed, and is very difficult to eradicate; this hath a round, compressed, tuberous root, which is of a yellowish colour, and covered with a brown furrowed skin like those of the large yellow vernal Crocus; from the root arise two flat sword-shaped leaves, which embrace each other at their base; and between these arise the flower-stalk, which grows near two feet high, having one or two narrow leaves embracing it like a sheath; the stalks are terminated by five or six purple flowers, standing above each other at some distance, and are ranged on one side of the stalk; each of these has a spatha (or sheath) which covers the flower-bud before it expands, but splits open lengthways when the flowers blow, and shrivel up to a dry skin, remaining about the seed-vessel till the seeds are ripe. The flower hath one petal, which is cut almost to the bottom in six parts, so as to appear like a flower of six petals; the three upper segments stand near together, and rise like a labiated flower; the under one turns downward, and the two side segments form the chaps of the flower, and spread open at the top, but are curved downward at the bottom. They are of a purplish red colour: it requires no care, for when it is once planted in a garden it will multiply too fast, so as to become a troublesome weed.

There is a variety of this with white flowers, and another with flesh-coloured flowers, which have accidentally risen from seeds, so are not different species.

The second sort differs from the first, in having the flowers ranged on both sides the stalk, but in other respects it is very like to that. Of this there is a variety with white flowers, but these are not so common in the English gardens as the former.

The third sort hath larger roots than either of the former, but are of the same form; the leaves are also much broader and longer, the veins or channels of the leaves are deeper; the flower-stalks rise higher; the flowers are much larger, and of a deeper red colour than those of the former sorts, and the sheaths are longer. This plant makes a fine appearance when in flower, so is worthy of a place in every good garden. This is propagated by offsets, which are sent off from the roots in the same manner as Tulips, in great plenty. The roots may be taken out of the ground in the end of July, when their stalks decay, and may be kept out of the ground till the latter end of September, or the beginning of October, at which time they should be planted in the borders of the flower-garden; they will thrive in any situation, and being intermixed with other flowers of the same growth, they will add to the variety.

The fourth sort grows naturally at the Cape of Good Hope. This has been many years cultivated in the English gardens, but very rarely flowers here; for in near thirty years that I have cultivated this sort, I have never seen it but once in flower, though I have kept it in all situations, and planted it in various soils. The roots of this sort are broader and flatter than those of any of the other sorts, and are covered with a netted skin; the leaves come out in the same manner, embracing each other as the former sorts; they are longer, smoother, and of a brighter green than any of the others; these begin to appear in September, and continue growing in size till after Christmas, and begin to decay in March; by the latter end of May are quite withered, when the roots may be taken up, and kept out of the ground till August; the time of its flowering is in January. The flowers of this sort are placed on each side the stalk, and sit close to it, like the grains of the flat Barley; the sheaths between the flowers are not so long as those of the other sorts, and form a kind of scaly covering to them. The flowers are of a pale red colour without, but the three lower segments are yellowish within, with a few stripes of red. The flowers do not all open at the same time, but the lower ones decay before those on the upper part of the spike are in beauty; however, they make a fine appearance at a season when all flowers are valuable.

This sort propagates by offsets very fast; these should be planted in borders, covered with glasses to be protected in winter from frost, but they do not require any artificial warmth. I have always found that those plants which were hardily treated, grew much stronger than those which were placed in a moderate degree of warmth; so that where there is a conveniency of covering a warm border with glasses in the winter, if these roots are planted in the full ground, where they may be protected from the frost, there will be a greater probability of their flowering than in any other method of culture.

The fifth sort grows naturally at the Cape of Good Hope. This hath a round, smooth, bulbous root, which is covered with a thin dark-coloured skin, from which come out in autumn two or three very narrow grassy leaves, folded over each other at their base, but open flat above; these rise near two feet high. In the spring arises a single stalk from between the leaves, about two feet long, which always bends on one side; toward the upper part of this come out two or three flowers, ranged on one side of the stalk, standing upright, each having a narrow spatha or hood, with long slender tubes, which swell large upward, and are divided into six almost equal parts. The flower is of a dusky flesh colour, and each segment of the petal has a rhomboidal mark of a dark red, or purple colour; afterward the tube of the flower opens, and the deep division of the petal is seen, and the three stamina, with their summits, appear, attended by the style with its trifid stigma, arising from the germen. This plant requires protection from the frost in winter, therefore the bulbs should be planted in pots, or in a border where the roots may be kept from frost in winter; or, where there is not such conveniency, they may be put under a hot-bed frame during that season, where they may have air in mild weather, and be screened from the frost; in such situations I have had them thrive and flower very well.

This is propagated by offsets from the root, in the same manner as the last, and also by seeds, which are frequently perfected in England, which should be sown the latter end

 of

of Auguſt in pots, and placed in a ſhady ſituation till the middle or end of September; then the pots ſhould be removed where they may have the ſun great part of the day, and in October they muſt be placed under a hot-bed frame, where they may be protected from froſts and great rains, but enjoy the free air in mild weather. In May, when the danger of froſt is over, the pots ſhould be removed to a ſheltered ſituation, where they may have the morning ſun till noon. Toward the latter end of June the leaves of theſe plants will decay, then the roots ſhould be taken up, and may be mixed with ſand, and kept in a dry room till the end of Auguſt, when they ſhould be planted again, and the roots being ſmall, four or five may be planted into each pot, and treated in the ſame way as the old roots.

The ſixth ſort is alſo a native of the Cape of Good Hope; the root of this ſort is oval, not compreſſed as thoſe of the other. The leaves are very long and narrow, having two deep furrows running the whole length, the middle one riſing very prominent, ſo as to have the appearance of a four-cornered leaf. There are ſeldom more than two of theſe leaves ariſing from one root; the ſtalk is ſlender and round, and riſes about two feet high; the top is garniſhed with two flowers, which are placed about two inches and a half aſunder on the ſame ſide of the ſtalk, each having a ſhort ſpatha or ſheath, embracing the germen and the baſe of the tube, which is long, narrow, and recurved, but enlarges greatly before it is divided. The upper part of the flower is cut into ſix equal ſegments, which end in acute points, of a purpliſh colour, which form a ſtripe through the middle of each ſegment. The petal is of a cream colour, and fades to a ſulphur colour before it decays. This may be propagated by offsets from the root, or by ſeeds, in the ſame manner as the fifth ſort, and the plants require the ſame treatment.

GLASTENBURY THORN. See MESPILUS.

GLAUCIUM. See CHELIDONIUM.

GLAUX. Sea Chickweed, or Milkwort, and by ſome Black Saltwort; it is a low, trailing, perennial plant, with leaves ſomewhat like Chickweed, but of a thicker conſiſtence, which ſit cloſe to the ſtalks. The flowers come out from the boſom of the leaves; they are white, and like thoſe of Chickweed. This is ſeldom cultivated in gardens, ſo I ſhall not trouble the reader with a farther account of it. It grows upon the ſea-ſhores in moſt parts of England.

GLECHOMA. Ground Ivy, Gill go by the Ground, Ale Hoof, or Turn Hoof.

This plant grows naturally under hedges, and upon the ſides of banks in moſt parts of England, ſo is rarely cultivated in gardens, for which reaſon I ſhall paſs over it, with barely mentioning it here.

GLEDITSIA. Lin. Gen. Plant. 1025. Honey Locuſt, or three-thorned Acacia.

The CHARACTERS are,

It hath male and hermaphrodite flowers on the ſame plant, and female flowers in different plants. The male have a three-leaved empalement, and three roundiſh petals, with a turbinated nectarium; they have ſix ſtamina, terminated by compreſſed ſummits. The hermaphrodite flowers have quadrifid empalements, with four petals and ſix ſtamina, and have a germen, ſtyle, and pod: like the female, which are ſituated on different trees; theſe have a five-leaved empalement, and five petals with two nectariums, and a broad germen longer than the petals, which afterward becomes a large flat pod, with ſeveral tranſverſe partitions, having a pulp in each diviſion, ſurrounding one hard roundiſh ſeed.

The SPECIES are,

1. GLEDITSIA (*Triacanthus*) ſpinis ternatis, foliis pinnatis, ſiliquis latis longiſſimis caule arboreo. *Gleditſia with three thorns, winged leaves, very long broad pods, and a tree-like ſtalk; commonly called the three-thorned Acacia.*

2. GLEDITSIA (*Inermis*) ſpinis paucioribus, foliis pinnatis, ſiliquis ovalibus monoſpermis. *Gleditſia with a few ſpines, winged leaves, and oval pods, containing one ſeed.*

Theſe trees grow naturally in moſt parts of North America, where the firſt is known by the title of Honey Locuſt; this has been many years cultivated in the Engliſh gardens, and is known among the gardeners by the title of three-thorned Acacia. It riſes with an erect trunk to the height of thirty or forty feet, armed with long ſpines, which have two or three ſmaller coming out from their ſide; theſe are frequently produced in cluſters at the knots, and are ſometimes three or four inches long. The branches of this tree are alſo armed with the ſame ſort of ſpines, and are garniſhed with winged leaves, compoſed of ten pair of ſmall leaves, which ſit cloſe to the midrib, of a lucid green. The flowers come out from the ſide of the young branches in form of katkins; they are of an herbaceous colour, ſo make no figure. The hermaphrodite flowers are ſucceeded by pods near a foot and a half long, and two inches broad, divided into many cells by tranſverſe partitions, each containing one ſmooth, hard, oblong ſeed, ſurrounded by a ſweet pulp.

The ſecond ſort hath much the appearance of the firſt, but hath fewer ſpines, which are very ſhort. The leaves are ſmaller, and the pods are oval, containing but one ſeed; this was diſcovered by the late Mr. Cateſby in Carolina, from whence he ſent the ſeeds to England by the title of Water Acacia, by which it is known in the gardens.

Theſe trees are propagated by ſeeds: thoſe of the firſt ſort are annually ſent to England in plenty, by the title of Locuſt or Honey Locuſt, to diſtinguiſh it from the falſe Acacia, which is frequently called Locuſt tree in America: the ſeeds may be ſown upon a bed of light earth in the ſpring, burying them half an inch deep, and if the ſpring ſhould prove dry, they muſt be frequently watered, otherwiſe the plants will not come up the firſt year, for I have ſometimes had the ſeeds remain two years in the ground before the plants have come up; therefore thoſe who are deſirous to ſave time, ſhould ſow the ſeeds as ſoon as they arrive, and plunge the pots into a moderate hot-bed, obſerving to water them frequently, by this method moſt of the plants will come up the ſame ſeaſon: they ſhould be gradually inured to bear the open air, for if they are continued in the hot-bed they will draw up weak. In autumn the pots ſhould be placed under a hot-bed frame to protect them from froſt, for theſe young plants generally keep growing late in the ſummer, ſo the upper part of their ſhoots is tender, and the early froſts of the autumn often kill the ends of them if they are not protected, and this frequently occaſions great part of the ſhoots decaying in winter; for which reaſon thoſe plants in the full ground, ſhould be covered with mats in the autumn on the firſt appearance of froſt, for a ſmall froſt in autumn will do more miſchief to theſe young ſhoots which are full of ſap, than ſevere froſt when the ſhoots are hardened.

The following ſpring the plants may be tranſplanted into nurſery-beds, at a foot diſtance row from row, and ſix inches aſunder in the rows, but this ſhould not be performed till April, after the danger of hard froſt is over; for as the plants do not put out their leaves till very late, ſo there will be no hazard in removing them any time before May. If the ſeaſon ſhould prove dry they muſt be watered, and if the ſurface of the beds is covered with Moſs or mulch, to prevent the earth from drying, it will be of great ſervice to the plants. In theſe beds the plants may remain two years, by which time, if the plants thrive well, they will be be fit to tranſplant to the places where they are to remain, for they do not bear removing when large; the beſt ſeaſon for tranſplanting of theſe trees is late in the ſpring; they thrive beſt in a deep loamy ſoil, for in ſtrong ſhallow ground they become moſſy, and never grow large; they ſhould alſo have a ſheltered ſituation, for when they are much expoſed to winds, their branches are frequently broken in the ſummer ſeaſon, when they are fully cloathed with leaves.

GLOBULARIA. Lin. Gen. Plant. 106. Blue Daiſy.

The CHARACTERS are,

It hath a flower compoſed of many florets, which are included in one common ſcaly empalement; each floret is tubulous, and cut into four parts at the top. They have four ſtamina. terminated by diſtinct ſummits; in the bottom of the tube is ſituated an oval germen, which afterward becomes an oval ſeed, ſitting in the common empalement.

The SPECIES are,

1. GLOBULARIA (*Vulgaris*) caule herbaceo, foliis radicalibus tridentatis, caulinis lanceolatis. Flor. Suec. 109. *Common Globularia.*

2. GLOBULARIA (*Nudicaulis*) caule nudo, foliis integerrimis lanceolatis. Lin. Sp. Pl. 97. *Pyrenean Globularia, with an oblong leaf and naked ſtalk.*

3. GLOBULARIA (*Alypum*) caule fruticoſo, foliis lanceolatis tridentatis integriſque. Prod. Leyd. 190. *Globularia with a ſhrubby ſtalk, and ſpear-ſhaped leaves ending in three points.*

4. GLOBULARIA (*Spinoſa*) foliis radicalibus crenato-aculeatis, caulinis integerrimis mucronatis. Lin. Sp. Pl. 96. *Prickly Globularia.*

5. GLOBULARIA (*Cordifolia*) caule ſubnudo, foliis cuneiformibus tricuſpidatis, intermedio minimo. Lin. Sp. Pl. 96. *Smalleſt Alpine Globularia, with a wild Marjoram leaf.*

6. GLOBULARIA (*Orientalis*) caule ſubnudo, capitulis alternis ſeſſilibus, foliis lanceolato-ovatis integris. Lin. Sp. Plant. 97. *Eaſtern Globularia with flowers ſcattered along the ſtalks.*

The firſt of theſe plants grows plentifully about Montpelier, as alſo at the foot of the mountains Jura and Saleva, and in many other parts of Italy and Germany. It hath leaves very like thoſe of the Daiſy, but are thicker and ſmoother. The flowers grow on long foot-ſtalks about ſix inches high, and are of a globular form, compoſed of ſeveral florets, which are included in one common ſcaly empalement; they are of a fine blue colour, and are ſucceeded by ſeeds which ſit in the empalement, and ripen in autumn.

The ſecond ſort grows plentifully in the woods near the convent of the Carthuſians, and on the Pyrenean mountains; this is much larger than the former, and the foot-ſtalk is quite naked. The leaves are narrower, and much longer.

The firſt and ſecond ſorts may be propagated by parting of their roots after the manner of Daiſies. The beſt ſeaſon for this is in September, that they may take new root before the froſty weather comes on. They ſhould be planted in a ſhady ſituation, and have a loamy ſoil, in which they will thrive much better than in light ground and an open ſituation; but the plants ſhould not be removed oftener than every other year, if they are required to flower ſtrong.

The third ſort grows about Montpelier in France, and in Valentia, and ſeveral other parts of Spain. This has a hard woody ſtem which riſes about two feet high, dividing into many woody branches, beſet with leaves like thoſe of the Myrtle-tree. On the top of the branches the flowers are produced, which are of a blue colour, and globe-ſhaped; this plant is very difficult to propagate in England, where it does not produce ſeeds; the cuttings, when carefully managed, will ſometimes put out roots, but with great difficulty. In ſummer the plants may be expoſed with other hardy exotics, and in winter they ſhould be placed under a hot-bed frame, where they may enjoy the free air in mild weather, but ſhould be ſcreened from hard froſt, which will deſtroy them if they are expoſed thereto, tho' in mild winters they will live in the open air.

The fourth ſort was found in the mountains of Granada, by Dr. Albinus; this plant is of low growth, and may be propagated as the firſt; as may alſo the fifth ſort, which is the leaſt of all the kinds, and the moſt hardy, therefore ſhould have a ſhady ſituation and a cool moiſt ſoil.

The ſixth ſort was found by Dr. Tournefort in the Levant; this is tender, and ſhould be ſheltered from the froſt in winter under a frame, but in ſummer it ſhould be expoſed with other hardy exotic plants; it will require to be frequently watered in dry weather, and is propagated by ſeeds, or by parting of the roots, as was directed for the former ſorts.

GLORIOSA. Lin. Gen. Plant. 374. The Superb Lily.

The CHARACTERS are,

The flower hath no empalement; it hath ſix ſpear-ſhaped petals, which are waved and reflexed to the bottom. It hath ſix ſtamina, which ſpread each way, terminated by proſtrate ſummits. In the center is ſituated a globular germen, which afterward becomes an oval thin capſule with three cells, filled with globular ſeeds diſpoſed in a double range.

The SPECIES are,

1. GLORIOSA (*Superba*) foliis longioribus capreolis terminalibus. *Superb Lily with longer leaves ending with claſpers.*

2. GLORIOSA (*Cærulea*) foliis ovato-lanceolatis acutis. *Superb Lily with oval, ſpear-ſhaped, acute leaves.*

The firſt ſort grows naturally on the coaſt of Malabar, and alſo in Ceylon; this hath oblong fleſhy roots of a whitiſh colour, and a nauſeous bitter taſte, from which ariſes a round weak ſtalk, which requires ſupport to prevent its trailing on the ground. The ſtalks grow to the height of eight or ten feet, garniſhed with leaves placed alternate; they are ſmooth, about eight inches long, and one and a half broad at their baſe, growing narrower till within two inches of the end, which runs out in a narrow point, ending with a tendril or claſper, by which it faſtens to the neighbouring plants for ſupport. At the upper end of the ſtalk the flowers are produced, ſtanding upon ſlender foot-ſtalks; they are compoſed of ſix oblong petals, ending with acute points, which, on their firſt opening are of a yellowiſh herbaceous colour, ſtanding at firſt erect, but when fully opened, hang downward as the Crown Imperial and Fritillary; the petals turn quite back, and change to a very beautiful red flame colour, their acute points meeting at the foot-ſtalk; theſe petals are finely waved on their edges. The ſix ſtamina ſpread out every way almoſt horizontal, and are terminated by proſtrate ſummits. In the center of the flower is ſituated a roundiſh germen, ſupporting an inclining ſtyle, crowned by a triple ſtigma. It flowers in July, and often perfects ſeeds in this country. The ſtalks decay in autumn, and the roots remain inactive all the winter. The roots and every part of this plant is very poiſonous, ſo ſhould not be put in the way of children.

The ſeeds of the ſecond ſort were ſent me by Monſ. Richard, gardener to the French King at Trianon, which were brought from Senegal by Monſ. Adanſon, who diſcovered this plant growing there naturally; this is ſaid to have a blue flower, but the plants which are in the Chelſea garden have not yet flowered. It hath a climbing ſtalk, which is garniſhed with ſmooth leaves about two inches long and two broad, ending in acute points, with ſhort tendrils

 or

or claſpers. The ſtalks as yet have not grown more than four feet high here. The leaves have a ſtrong diſagreeable ſcent on being handled, ſo as to be troubleſome to the head if too near, or long ſmelt to.

Theſe plants are propagated by their roots; thoſe of the firſt ſort creep and multiply pretty faſt. Theſe roots may be taken out of the ground when their ſtalks are decayed, and preſerved in ſand during the winter ſeaſon, but they muſt be kept in the ſtove or a warm room, where they can receive no injury from the cold, and in the ſpring they muſt be planted in pots filled with light earth, and plunged into the tan-bed in the ſtove; but others chuſe to let the roots continue in the ground all the winter, keeping the pots always in the tan-bed; where this is practiſed, the plants ſhould have very little water in the winter, for as they are then in an inactive ſtate, ſo moiſture at that time frequently rots them.

Toward the latter end of March, or the beginning of April, their ſtalks will appear, when there ſhould be ſome tall ſticks put down by them to ſupport them, otherwiſe they will trail over the neighbouring plants, and faſten to them by the tendrils, which are at the ends of the leaves. The ſtalks will riſe ten or twelve feet high if the roots are ſtrong; ſome of them will produce two or three flowers, which come out from the wings of the ſtalk near the top; theſe flowers make a fine appearance in the ſtove during their continuance, which is ſeldom more than ten days or a fortnight. In the ſummer when the plants are growing, they will require frequently to be watered, but they muſt not have it in too large quantities. Theſe roots which are not taken out of the pots in winter, ſhould be tranſplanted and parted the beginning of March, before they put out new fibres or ſtalks, for they muſt not be removed when they are in a growing ſtate; the pots in which theſe roots are planted, ſhould not be too large, for unleſs they are confined, they will not put out ſtrong ſtalks; the largeſt roots may be planted in twopenny pots, but the ſmall ones will require only pots of about five or ſix inches over at the top.

GLYCINE. Lin. Gen. Plant. 797. Knobbed rooted Liquorice Vetch.

The CHARACTERS are,

The flower is of the butterfly kind. The ſtandard is deflexed on the ſides, and indented at the point. The wings bend backward. The keel is ſickle-ſhaped, turning upward with its point to the ſtandard. The empalement hath two lips. It hath ten ſtamina, nine of which are joined in one body, and the other ſtands ſingle. In the center is ſituated an oblong germen, which afterward becomes an oblong pod with two cells, incloſing kidney-ſhaped ſeeds.

The SPECIES are,

1. GLYCINE (*Apios*) foliis pinnatis ovato-lanceolatis. Hort. Upſal. *Glycine with oval, ſpear-ſhaped, winged leaves.* Apios Americana. Cornut. 200.

2. GLYCINE (*Fruteſcens*) foliis impari-pinnatis caule perenni. Hort. Cliff. 361. *Glycine with winged leaves and a perennial ſtalk.*

3. GLYCINE (*Abrus*) foliis abruto-pinnatis conjugatis, pinnis numeroſis obtuſis. Flor. Zeyl. 284. *This is commonly called wild Liquorice in the Weſt-Indies.*

4. GLYCINE (*Comoſa*) foliis ternatis hirſutis, racemis lateralibus. Lin. Sp. Pl. 754. *Glycine with hairy trifoliate leaves, and flowers growing in long bunches from the ſides of the ſtalks.*

5. GLYCINE (*Tomentoſa*) foliis ternatis tomentoſis, racemis axillaribus breviſſimis, leguminibus diſpermis. Lin. Sp. Pl. 754. *Glycine with woolly trifoliate leaves, and very ſhort ſpikes of flowers proceeding from the ſides of the ſtalks, with pods containing two ſeeds.*

The firſt ſort grow naturally in Virginia; it hath roots compoſed of ſeveral knobs or tubers, which hang to each other by ſmall ſtrings; from theſe come out in the ſpring ſlender twining ſtalks, which riſe to the height of eight or ten feet, garniſhed with winged leaves, compoſed of three pair of oval ſpear-ſhaped lobes, terminated by an odd one. The flowers come out in ſhort ſpikes from the ſide of the ſtalks; they are of a Pea-bloſſom kind, of a dirty freſh colour, having little ſcent. The ſtalks decay in autumn, but the roots continue; this is propagated by parting of the roots; each of the tubers being ſeparated from the principal root will grow; the beſt time for this is about the end of March, or the beginning of April, before they put out ſhoots. The roots ſhould be planted in a warm ſituation, and in hard froſt covered to protect them, otherwiſe they will not live abroad in this country; where they have been planted againſt a ſouth wall, they have thriven and flowered extremely well, which they ſeldom do in any other ſituation; and thoſe roots which are planted in pots rarely flower, nor do their ſtalks riſe near ſo high as thoſe which are planted in the full ground; ſome ignorant perſons call this the Twickenham Climber.

The ſecond ſort was brought from Carolina, but has been ſince obſerved in Virginia, and ſome other parts of North America; this has woody ſtalks, which twiſt themſelves together, and alſo twine round any trees that grow near, and will riſe to the height of fifteen feet or more. The leaves are winged, and in ſhape ſomewhat like the Aſh-tree, but have a greater number of pinnæ. The flowers are produced from the wings of the leaves, which are of a purple colour, and are ſucceeded by long cylindrical pods, ſhaped like thoſe of the Scarlet Kidney Bean, containing ſeveral kidney-ſhaped ſeeds; but theſe are never perfect in England.

This climbing ſhrub is raiſed for ſale in ſeveral nurſeries near London, where it is known by the name of Carolina Kidney Bean-tree. It is propagated by laying down the young branches in October, which will be rooted by that time twelvemonth, and may then be tranſplanted either in a nurſery for a year to get ſtrength, or to the place where they are to remain for good, which ſhould be in a warm light ſoil and a ſheltered ſituation, where they will endure the cold of our ordinary winters very well; and if their roots are covered with ſtraw, Fern, Peaſe-haulm, or any other light covering, there will be no danger of their being deſtroyed by great froſt.

The third ſort grows naturally in both Indies, and alſo in Egypt. This is a perennial plant, with ſlender twining ſtalks, which twiſt round any neighbouring ſupport, and riſe to the height of eight or ten feet, garniſhed with winged leaves, compoſed of ſixteen pair of oblong blunt lobes, ſet cloſe together, which have the taſte of Liquorice, from whence the inhabitants of the Weſt-Indies have given it the name of Wild Liquorice, and uſe the herb for the ſame purpoſe of Liquorice as uſed in Europe. The flowers are produced from the ſide of the ſtalks in ſhort ſpikes or bunches; they are of a pale purple colour, and ſhaped like thoſe of the Kidney Bean; theſe are ſucceeded by ſhort pods, each containing three or four hard round ſeeds of a ſcarlet colour, with a black ſpot or eye on the ſide, which is faſtened to the pod. The ſeeds are frequently brought to England from the Weſt-Indies, and are wrought into various forms, with ſhells and other hard ſeeds.

It is propagated by ſeeds, which muſt be ſown upon a good hot-bed in the ſpring; but as they are very hard, ſo unleſs they are ſoaked in water a day or two before they are ſown, they frequently lie in the ground a whole year before they vegetate; but when ſoaked, the plants will appear in three weeks after the ſeeds are ſown, if they are good,

good, and the bed in a proper temperature of heat. When the plants are two inches high, they ſhould be each tranſplanted into a ſeparate pot, and plunged into a hot-bed of tanners bark, where they ſhould be ſhaded from the ſun till they have taken new root; after which they muſt be treated in the ſame manner as other tender plants from the ſame countries, always keeping them in the bark-ſtove, for they are too tender to thrive in any other ſituation in England.

There are two other varieties of this plant, one with a white, and the other a yellow ſeed, but the plants do not differ from the other in leaf or ſtalk; but as theſe have not as yet flowered in England, ſo I do not know how their flowers may differ.

The fourth ſort hath a perennial root and an annual ſtalk. This riſes from two to three feet high, with ſlender herbaceous ſtalks, garniſhed with trifoliate hairy leaves ſitting cloſe to them; the ſmall leaves or lobes are of the oval ſpear-ſhape, ending in acute points. The flowers come out from the ſide of the ſtalks, upon foot-ſtalks about two inches long; the ſpike of flowers is about the ſame length, and is recurved; they are of the Pea-bloſſom kind, ſitting cloſe together, but ſmall, and of a fine blue colour.

This ſort grows naturally in North America, and is hardy enough to live in the open air in England. It may be propagated by ſeeds, or parting of the roots; the former is the beſt method, where good ſeeds can be obtained; theſe ſhould be ſown on a bed of light earth in the ſpring, and if the ſeaſon ſhould prove dry, they muſt be frequently refreſhed with water, otherwiſe they will remain a long time in the ground before they vegetate; in the autumn when their ſtalks are decayed, if ſome rotten tanners bark is ſpread over the ſurface of the ground, it will protect the roots from being injured by the froſt. In the ſpring the roots ſhould be tranſplanted to the places where they are deſigned to remain, which muſt be in a warm ſheltered ſituation and a light ſoil, where they will thrive and produce flowers annually. If this is propagated by parting of the roots, it ſhould be done in the ſpring before the roots begin to ſhoot, which is the beſt ſeaſon for tranſplanting of the plants.

The fifth ſort is a perennial plant, with a climbing ſtalk which riſes ſix or ſeven feet high, garniſhed with woolly trifoliate leaves: the flowers come out in ſhort bunches from the ſide of the ſtalks; they are ſmall, of a yellowiſh colour, and are ſucceeded by ſhort pods, which contain two roundiſh ſeeds in each. It grows naturally in Virginia, and alſo in India. This is propagated by ſeeds, but the plants require protection in winter.

GLYCYRRHIZA. Lin. Gen. Plant. 788. Liquorice.

The CHARACTERS are,

The flower hath a permanent empalement of one leaf, divided into two lips. The flower is of the butterfly kind, having a long erect ſtandard, with oblong wings, and a two-leaved keel which is acute. It hath ten ſtamina, nine joined and one ſtanding ſingle. In the bottom is ſituated a ſhort germen; which afterward becomes an oblong, or oval compreſſed pod with one cell, including two or three kidney-ſhaped ſeeds.

The SPECIES are;

1. GLYCYRRHIZA (*Glabra*) leguminibus glabris. Hort. Cliff. 490. *Liquorice with ſmooth pods. This is the common Liquorice.*

2. GLYCYRRHIZA (*Echinata*) leguminibus echinatis. Prod. Leyd. 386. *Liquorice with prickly pods.*

3. GLYCYRRHIZA (*Hirſuta*) leguminibus hirſutis. Prod. Leyd. 386. *Eaſtern Liquorice with hairy pods.*

The firſt ſort is that which is commonly cultivated in England for medicine; the other two kinds are preſerved in curious botanic gardens for variety but their roots are not ſo full of juice as the firſt, nor is the juice ſo ſweet, though the ſecond ſort ſeems to be that which Dioſcorides has deſcribed and recommended; but I ſuppoſe the goodneſs of the firſt, has occaſioned its being ſo generally cultivated in England.

The roots of this run very deep into the ground, and creep to a conſiderable diſtance, eſpecially where they are permitted to ſtand long unremoved; from them ariſe ſtrong herbaceous ſtalks four or five feet high, garniſhed with winged leaves compoſed of four or five pair of oval lobes, terminated by an odd one; the flowers come out in ſpikes from the upper part of the ſtalks, ſtanding erect; they are of a pale blue colour, and are ſucceeded by ſhort compreſſed pods, each containing two or three kidney-ſhaped ſeeds, which rarely ripen in England.

This plant delights in a light ſandy ſoil, which ſhould be three feet deep at leaſt, for the goodneſs of Liquorice conſiſts in the length of the roots. The greateſt quantity of Liquorice which is propagated in England, is about Pontefract in Yorkſhire, and Godalming in Surry, though of late years there have been great quantities cultivated in the gardens near London: the ground intended for Liquorice ſhould be well dug and dunged the year before, that the dung may be perfectly rotted and mixed with the earth, otherwiſe it will be apt to ſtop the roots from running down; and before it is planted, the ground ſhould be dug three ſpade deep with two ſhovelings; when it is thus well prepared, a ſufficient quantity of freſh plants taken from the ſides or heads of the old roots ſhould be provided, each of them ſhould have a good bud or eye, otherwiſe they are ſubject to miſcarry; theſe plants ſhould be about eight or ten inches long, and perfectly ſound.

The beſt ſeaſon for planting them is toward the end of February, or the beginning of March, which muſt be done in the following manner; viz. Firſt ſtrain a line croſs the ground, beginning at one end of the piece, then with a long dibble made on purpoſe, put in the ſhoot, ſo that the whole plant may be ſet ſtrait in the ground, with the head about an inch under the ſurface, about a foot aſunder or more in the rows, and two feet diſtance row from row; and after having finiſhed the whole ſpot of ground, it may be ſown with a thin crop of Onions, which do not root deep into the ground, nor ſpread much above, ſo will do the Liquorice no damage the firſt year; for the Liquorice will not ſhoot very high the firſt ſeaſon, and the hoeing of the Onions will alſo keep the ground clear from weeds; but in doing this, care ſhould be taken not to cut off the ſhoots of the Liquorice plants as they appear above ground, which will greatly injure them, and to cut up all the Onions which grow near the heads of the Liquorice; after the Onions are full grown and pulled up, the ground ſhould be cleared from weeds; and in October, when the ſhoots of the Liquorice are decayed, a ſmall quantity of very rotten dung ſpread upon the ſurface of the ground will prevent the weeds from growing during the winter, and the rain will waſh the dung into the ground, which will greatly improve the plants.

In the beginning of March following, the ground between the rows of Liquorice ſhould be ſlightly dug, burying the remaining part of the dung; but in doing this care muſt be taken not to cut the roots. This ſtirring the ground will not only preſerve it clean from weeds a long time, but alſo greatly ſtrengthen the plants.

The diſtance I have allowed for planting theſe plants will, I doubt not, by ſome, be thought too great; but in anſwer to that, I would only obſerve, that as the largeneſs of the roots is the chief advantage to the planter, the only method to obtain it is by giving them room; beſides this will give a greater liberty to ſtir and dreſs the ground, which is of

great

great ſervice to Liquorice; and if the plantation deſigned were to be of an extraordinary bigneſs, I would adviſe the rows to be made at leaſt three feet diſtant, whereby it will be eaſy to ſtir the ground with a hoeing plough, which will greatly leſſen the expence of labour.

Theſe plants ſhould remain three years from the time of planting, when they will be fit to take up for uſe; which ſhould not be done until the ſtalks are perfectly decayed, for when it is taken up too ſoon, it is ſubject to ſhrink greatly and loſe of its weight.

The ground near London being rich, increaſes the bulk of the root very faſt; but when it is taken up it appears of a very dark colour, and not near ſo ſightly as that which grows upon a ſandy ſoil in an open country.

The ſecond ſort grows naturally in ſome parts of Italy, and the Levant; the ſtalks and leaves of this are very like thoſe of the firſt, but the flowers are produced in ſhorter ſpikes, and the pods which ſucceed them are very ſhort, broad at their baſe, ending in acute points, and are armed with ſharp prickles. This flowers about the ſame time as the firſt, and in warm ſeaſons will perfect ſeeds in England.

The third ſort grows naturally in the Levant. This hath much the appearance of the other two ſpecies, but the pods of it are hairy, and longer than thoſe of the other. Both theſe ſorts may be propagated in the ſame manner as the firſt, or from ſeeds, which may be ſown in the ſpring on a bed of light earth; but as neither of theſe are uſed, ſo they are ſeldom propagated, unleſs for the ſake of variety.

GNAPHALIUM. Lin. Gen. Plant. 850. Goldylocks, or Eternal Flower.

The CHARACTERS are,

It hath a compound flower, made up of hermaphrodite florets and female half florets, included in one ſhining ſcaly empalement; the hermaphrodite florets are tubulous, and cut into five parts at the brim; theſe have five ſhort hairy ſtamina. In the center is ſituated a germen, which afterward becomes a ſingle ſeed, which in ſome ſpecies is crowned with a hairy down, and in others a feathery down. The female florets have no ſtamina, but a crowned germen ſupporting a ſlender ſtyle. Theſe are in ſome ſpecies fruitful, and in others they are barren.

The SPECIES are,

1. GNAPHALIUM (*Stœchas*) tomentoſum, foliis caulinis linearibus, caule fruticoſo, corymbo compoſito. *Caſſidony, or narrow-leaved Goldylocks.*

2. GNAPHALIUM (*Anguſtiſſimum*) foliis linearibus, caule fruticoſo ramoſo, corymbo compoſito. Hort. Cliff. 401. *Goldylocks with a branching ſhrubby ſtalk, very narrow leaves, and a compound corymbus of flowers.*

3. GNAPHALIUM (*Uniflorum*) foliis alternis, acutè dentatis, ſubtus villoſis, pedunculis longiſſimis unifloris. *Goldylocks with alternate leaves ſharply indented, hairy on their under ſide, with very long foot-ſtalks ſuſtaining one flower.*

4. GNAPHALIUM (*Luteo-album*) foliis ſemiamplexicaulibus enſiformibus, repandis obtuſis, utrinque pubeſcentibus, floribus conglomeratis. Prod. Leyd. 149. *Broad-leaved wild Goldylocks, with heads growing in cluſters.*

5. GNAPHALIUM (*Aquaticum*) caule ramoſo diffuſo, floribus confertis terminalibus. Flor. Lapp. 300. *Goldylocks with a diffuſed branching ſtalk, and flowers in cluſters at the top.*

6. GNAPHALIUM (*Sylvaticum*) caule ſimpliciſſimo, floribus ſparſis. Flor. Lapp. 298. *Goldylocks with a ſingle ſtalk, and flowers growing ſcatteringly.*

7. GNAPHALIUM (*Dioicum*) caule ſimpliciſſimo corymbo ſimplici terminali, ſarmentis procumbentibus. Hort. Cliff. 400. *Goldylocks with a ſingle ſtalk terminated by a ſingle corymbus, and trailing branches.*

8. GNAPHALIUM (*Orientale*) foliis confertis anguſto-lanceolatis, caule fruticoſo, corymbo compoſito. Hort. Cliff. 402. *Eaſtern Goldylocks, called Immortal Flower.*

9. GNAPHALIUM (*Igneſcens*) foliis lineari-lanceolatis obtuſis, utrinque tomentoſis, corymbo compoſito terminali. *German Goldylocks with a reddiſh gold-coloured empalement.*

10. GNAPHALIUM (*Margaritaceum*) foliis lineari-lanceolatis acuminatis, alternis, caule ſupernè ramoſo corymbis faſtigiatis. Hort. Cliff. 401. *Broad-leaved American Goldylocks.*

11. GNAPHALIUM (*Fœtidum*) foliis amplexicaulibus, ovatis nervoſis, utrinque lanuginoſis. Lin. Sp. Plant. 850. *Goldylocks with oval nervous leaves embracing the ſtalks, which are downy on both ſides.*

12. GNAPHALIUM (*Argenteum*) foliis amplexicaulibus integerrimis acutis, ſubtus tomentoſis, caule ramoſo. Hort. Cliff. 402. *Moſt ſtinking African Goldylocks with a very large leaf, and a ſilvery empalement to the flower.*

13. GNAPHALIUM (*Undulatum*) foliis decurrentibus acutis, undatis, ſubtus tomentoſis, caule ramoſo. Hort. Cliff. 402. *Stinking Goldylocks, with an acute leaf and winged ſtalk.*

14. GNAPHALIUM (*Cymoſum*) foliis lanceolatis ſemiamplexicaulibus, caule infernè ramoſo terminali. Hort. Cliff. 401. *Goldylocks with ſpear-ſhaped leaves embracing the ſtalks, whoſe under branches are terminated with flowers.*

15. GNAPHALIUM (*Rutilans*) foliis lineari-lanceolatis, caule infernè ramoſo, corymbo compoſito terminali. Hort. Cliff. 401. *Goldylocks with narrow ſpear-ſhaped leaves, the under part of the ſtalk branching, and a compound corymbus terminating the branches.*

16. GNAPHALIUM (*Sanguineum*) foliis linearibus tomentoſis, integerrimis ſeſſilibus, corymbis alternis conglobatis, floribus globoſis. Prod. Leyd. 149. *Goldylocks with a ſoft red flower.*

17. GNAPHALIUM (*Fruticoſum*) fruticoſum, foliis infernè lanceolatis caulinis lineari-lanceolatis, utrinque tomentoſis, corymbo compoſito terminali. *Shrubby African Goldylocks, with longer and narrower leaves which are hoary.*

18. GNAPHALIUM (*Odoratiſſimum*) foliis decurrentibus obtuſis infernè villoſis, corymbis conglobatis terminalibus. *Goldylocks with obtuſe running leaves, hoary on their under ſide, and a cluſtered corymbus of flowers terminating the ſtalks.*

The firſt ſort hath a ſhrubby ſtalk about three feet high, dividing into many ſlender branches garniſhed with obtuſe leaves; thoſe upon the flower-ſtalks are very narrow, ending in acute points; the whole plant is very woolly; the flowers terminate the ſtalks in a compound corymbus; their empalements are of a ſilvery colour at firſt, and very neat, but afterward turn to a yellowiſh ſulphur colour. If theſe are gathered before the flowers are much open, the heads will continue in beauty many years, eſpecially if they are kept from the air and duſt. This is generally ſuppoſed to be the true golden Caſſidony of the ſhops, but the ſecond ſort is uſually ſubſtituted for it in England.

It is propagated by ſlips or cuttings, which may be planted in June or July in a bed of light earth, and covered with glaſſes, or ſhaded with mats, they will put out roots in five or ſix weeks; when they are well rooted they ſhould be taken up and planted in pots, and placed in a ſhady ſituation till they have taken new root; then they may be removed to an open ſituation, and placed among other hardy exotics till about the middle or end of October; at which time they ſhould be placed under a common frame, where they may be protected from froſt, but in mild weather they ſhould be expoſed to the open air. With this management, in winter, the plants will be much ſtronger than thoſe which are kept in the green-houſe, where they generally draw too weak; for they are ſo hardy, as in very mild winters

winters to live abroad in warm borders near walls with little shelter.

The second sort hath a shrubby stalk, which divides into many slender branches, covered with a white bark; these form a bushy under shrub near three feet high, garnished with very narrow leaves, hoary on their under side, but green on their upper, placed without order on every side the stalks; the flowers are produced in a compound corymbus at the end of the branches; their heads are small, and are of a yellow colour when fully blown. This grows naturally in France and Germany, and is hardy enough to live in the open air in England. It is propagated by slips or cuttings, in the same way as the former, and in the autumn they may be transplanted into the place where they are designed to remain.

The third sort is an annual plant, which grows naturally in Italy and Sicily; this hath an herbaceous stalk, little more than a foot high, garnished with acute indented leaves, which are hoary on their under side; the flowers stand upon long foot-stalks, which rise far above the branches, each sustaining one small whitish flower. It is propagated by seeds, which would be sown upon a bed of light earth where the plants are designed to remain; and when the plants come up, they should be thinned where they are too close, and kept clean from weeds, which is all the culture they require.

The fourth sort is an annual plant, with woolly stalks about eight inches high, garnished with oblong leaves which embrace them with their base; the flowers grow in close clusters at the top and from the side of their stalks, which are included in dry silvery empalements.

There is another species of this with narrower leaves, not quite so woolly; the stalks rise higher, and are more branched; the flowers grow in close bunches on the top of the stalks, and are of a pale yellow colour.

Both these sorts will come up better from the scattered seeds than when they are sown by art; but if the seeds are sown, it must be soon after they are ripe, otherwise they will not succeed. The plants require no other care but to be kept clean from weeds, and to be thinned where they are too close.

The fifth sort is an annual plant, which grows naturally in many parts of England, on places which are covered with water in the winter; this is a low branching plant, with silvery leaves and dark heads of flowers, but being of no use is not cultivated in gardens.

The sixth sort is also an annual plant with narrow leaves, which are hoary on their under side; the stalks grow erect about a foot high, and at every joint is produced a short spike of white flowers. This is found growing naturally in some parts of England, so is not often admitted into gardens. If the seeds of this sort are permitted to scatter, the plants will come up in the spring with greater certainty than if sown, and they will require no culture.

The seventh sort grows naturally in the northern parts of England, upon the tops of hills and mountains, where the shoots which are sent out from every side of the plant put out roots, whereby it is propagated in great plenty: the leaves of this grow close to the ground; they are narrow at their base, but rounded at the end where they are broad; they are not an inch long, and hoary on their under side; the stalks are single, and rise about four inches high, terminated by a corymbus of flowers which is single.

There are two varieties of this, one with a purple and the other a variegated flower, which continue their difference in the gardens. They are easily propagated by offsets, which should be planted in the autumn in a shady situation, where they will require no other care but to keep them clean from weeds. This plant is called Pes Cati, or Cats-foot.

The eighth sort is supposed to have been brought first from India to Portugal, where it has been long propagated for the beauty of its golden heads of flowers, which, if gathered before they are too open, will continue in beauty several years; so that in the winter season they ornament their churches with these flowers, and many of them are annually brought to England, and sold for ornaments to the ladies. These plants have a short shrubby stalk, seldom rising more than three or four inches high, putting out many short branches; the leaves are narrow and woolly on both sides, and come out without order; the flower-stalks grow eight or ten inches high, garnished all the way with narrow hoary leaves, terminated by a compound corymbus of bright yellow flowers, with large heads. This is propagated by slipping off the branches during any of the summer months, and after stripping off the lower leaves, planted in a bed of light earth, covering them with hand-glasses, which must be shaded every day when the sun is warm; when these are rooted, they should be planted in pots, and treated in the same manner as hath been directed for the first sort. These plants in mild winters will live abroad in a very warm border with little shelter, and the hardier they are treated, the greater number of flowers they will produce; for when they are grown weak by keeping in a green-house they never flower so strong.

The ninth sort hath very woolly stalks and leaves, which are much longer than those of the eighth; the stalks rise a foot high, sending out a few side branches, terminated by a compound corymbus of flowers, whose heads are small, and of a gold colour, changing a little red as they fade. This is propagated by slips in the same manner as the last mentioned, but the plants will live in the open air if they are planted on a dry soil.

The tenth sort grows naturally in North America, but has been long in the English gardens. This hath a creeping root, which spreads far in the ground, so as to become a troublesome weed very often, unless it is kept within bounds; the stalks of this are woolly, rising a foot and a half high, garnished with long, narrow, woolly leaves, ending in acute points, placed alternate; the upper part of the stalk branches into two or three divisions, each being terminated by a close corymbus of flowers, with pretty large silvery empalements, which, if gathered and properly dried, will retain their beauty several years. This sort will thrive in almost any soil or situation, and is easily propagated by its creeping roots.

The eleventh sort grows naturally at the Cape of Good Hope. It is an annual plant, which sends out many oblong blunt leaves near the root; the stalks rise a foot and a half high, garnished with leaves placed alternate, which are broad at their base, where they embrace the stalks, but end in acute points; they are woolly, and, when handled, emit a very rank odour; the stalks are terminated by a corymbus of flowers, in large silvery empalements, which will retain their beauty several years.

The twelfth sort grows naturally at the Cape of Good Hope, and is an annual plant, very like the former sort, but the leaves are of a yellowish green on the upper side, and woolly on their under; the stalks, branch, and the heads of flowers, are of a bright yellow colour, and these differences are permanent.

Both these plants are propagated by seeds, which, if sown in the autumn on a warm border, will more certainly succeed than when they are sown in the spring; or if the seeds are permitted to scatter, the plants will come up without care, and may be transplanted while they are young, to the places where they are designed to remain; when the plants have taken root, they will require no other care but to keep them clean from weeds. They flower in July, and the seeds ripen in autumn.

The thirteenth sort grows in Africa, and also in North America, from both these countries I have received the seeds. It is an annual plant, with oblong leaves at the bottom, which are a little waved, and hoary on their under side. The stalks rise about a foot high, garnished with acute-pointed leaves; from their base runs a border or wing along the stalk; the whole plant has a disagreeable odour. The flowers grow in a corymbus on the top of of the stalks; they are white, and appear in July. The seeds ripen in the autumn, which, if permitted to scatter, the plants will come up without care, as the two former sorts.

The fourteenth sort rises with a shrubby stalk three or four feet high, sending out many branches at bottom, garnished with narrow spear-shaped leaves, which half embrace the stalks with their base; they are of a dark green on their upper side, but hoary on their under; the stalks are terminated by a compound corymbus of yellow flowers, whose heads are small: these continue in succession great part of the summer, but are rarely succeeded by seeds in England It is easily propagated by cuttings in any of the summer months, which may be planted in a shady border. These will take root in a month or five weeks, and may then be taken up and planted in pots, placing them in a shady situation till they have taken fresh root; after which they may be removed to a sheltered situation, and placed with other hardy green-house plants till autumn, when they must be carried into the green house, where, during the winter season, they should have as much free air as possible in mild weather, for they only require protection from frost, so they should be treated in the same manner as other hardy green-house plants.

The fifteenth sort grows naturally at the Cape of Good Hope; this rises with a slender shrubby stalk, which sends out many lateral branches, garnished with very narrow leaves, which are hoary on their under side. The flowers are produced in a compound corymbus at the end of the branches; they are at their first appearance of a pale red colour, but afterward change to a gold colour; the empalements of this sort are small, and dry like other species of this genus. This sort is propagated by cuttings in the same manner as the former, and the plants require the same treatment.

The sixteenth sort grows naturally in Istria. This is a perennial plant, whose under leaves spread on the ground, which are obtuse, and woolly on their under side; the stalks rise about six inches high; the leaves upon these are narrow, ending in acute points; at each of the joints on the upper part of the stalks, is produced a compact corymbus of flowers, which are placed alternate on each side; the stalk is terminated by a larger corymbus sitting very close; these are of a fine soft red colour, so make a pretty appearance in the month of June, when they are in beauty.

This sort is propagated by offsets from the roots, in the same manner as the seventh sort, but this doth not produce them in plenty, so is very uncommon in the English gardens at present: it requires a drier soil than the seventh, and a warmer situation, but not too much exposed to the mid-day sun, so should be planted to an east aspect, for it is rarely injured by the cold in England.

The seventeenth sort grows naturally at the Cape of Good Hope, but has been long preserved in many curious gardens in Europe. Dr. Linnæus supposes this and the eighth to be but one species, which he might easily be inclined to do, by seeing the dried specimens only; for these seem only to differ in colour of their flowers or heads, which he never admits as a specific difference. But those who have observed the growing plants, cannot doubt of their being two distinct species; for the stalks of this rise three or four feet high, sending out several long irregular branches, which are terminated by a compound corymbus of flowers; whereas the eighth sort never rises with a stalk, but the branches sit close to the ground, and never divide: the branches of this sort are garnished with leaves which are much longer than those of the other, and the heads of the flowers are of a bright silver colour. This is propagated by cuttings, which should be planted in the same manner as hath been directed for the eighth sort, and the plants should also be treated the same way.

The eighteenth sort was raised from seeds in the Chelsea garden, which came from the Cape of Good Hope. The lower leaves of this are oblong and blunt; the stalks are shrubby, and divide into many irregular branches, which rise about three feet high, garnished with oblong blunt-pointed leaves, hoary on their under side, but of a dark green above; from the base of the leaves run a border along the stalk like a wing, of the same consistence with the leaves, which Dr. Linnæus calls a running leaf. The stalks are terminated by a compound corymbus of flowers, which are very closely joined together, of a bright gold colour, but the flowers are small, and change to a darker colour as they fade; there is a succession of these flowers most part of the summer, and the early flowers are frequently succeeded by seeds in England. This sort may be propagated by slips or cuttings in the same manner as the former, and the plants may be treated accordingly.

GNAPHALODES. See MICROPUS.

GOMPHRENA. Lin. Gen. Plant. 279.

The CHARACTERS are,

The flower hath a large coloured empalement, which is permanent. The petal is erect, and cut into five parts at the brim, which spread open; it hath five stamina scarcely discernible, situated in the brim of the nectarium. In the center is situated an oval-pointed germen, which germen afterward becomes one large roundish seed, inclosed in a thin crusted capsule with one cell.

The SPECIES are,

1. GOMPHRENA (*Globosa*) caule erecto, foliis ovato-lanceolatis capitulis solitariis, pedunculis diphyllis. Hort. Cliff. 86. *Globe Amaranthus with large purple heads.*

2. GOMPHRENA (*Serrata*) caule erecto spicâ interruptâ. Prod. Leyd. 419. *Gomphrena with an erect stalk, and an interrupted spike of flowers.*

3. GOMPHRENA (*Perennis*) foliis lanceolatis, capitulis diphyllis, flosculis perianthio proprio distinctis. Lin. Sp. Pl. 224. *Perennial Globe Amaranthus, with radiated straw-coloured flowers.*

The first sort grows naturally in India, but it has been many years cultivated in all the curious gardens in Europe; it is an annual plant, which rises with an upright branching stalk about two feet high, garnished with spear-shaped leaves placed opposite. The branches also come out opposite; the foot-stalks of the flowers, which are long and naked, have two short leaves close under each head of flowers. The heads at their first appearance are globular, but as they increase in size become oval; they are composed of dry scaly leaves, placed imbricatim like the scales of fish; under each of these is situated a tubulous flower, which just peeps out of the covering, but are not much regarded by the generality of people, for it is the scaly empalement which covers them that is so beautiful as to attract the eye, and these, if gathered before they are too much faded, will retain their beauty several years. After the flowers are past, the germen, which is situated in the bottom of each, becomes a large oval seed, inclosed in a chaffy covering, which ripens late in autumn, and the plants decay soon after.

There are two varieties of this sort, one with fine bright purple heads, the other hath white or silvery heads, which never alter, so that they are permanent varieties, though in other respects they do not differ; there is also one with mixed colours, but whether this arose accidentally from the seeds of either of the former I cannot determine, for this variety continues from seeds, and the other two I have cultivated more than thirty years, and have never found either of them vary.

The second sort hath much slenderer stalks than the first, it grows taller. The leaves are smaller, but of the same shape. The flowers grow in spikes at the end of the branches, which are broken, or divided into three or four parts, with spaces between them. They are small, and of a pale purple colour.

The third sort hath slender upright stalks, garnished with hairy spear-shaped leaves placed opposite, which sit close to the stalks; these are terminated by small heads of flowers, which spread open from each other, so as that their empalements appear distinct; they are of a pale straw-colour, so make no great appearance. The seeds, sometimes, will ripen in England, but the plants will live two or three years, if they are preserved in a stove.

The two sorts which are first mentioned, one with a purple, and the other with silver-coloured heads, are very ornamental plants in gardens, and are now very commonly cultivated in the English gardens as choice annuals. In Portugal, and other warm countries, they are cultivated to adorn their churches in the winter, for if these are gathered when they are in perfection, and dried in the shade, they will retain their beauty a long time, especially if they are not too much exposed to the air: these plants are propagated by seeds, which should be sown in a good hot-bed the beginning of March; but if the seeds are not taken out of their chaffy covering, it will be proper to soak them in water for twelve hours before they are sown, which will greatly facilitate their growing. When the plants are come up half an inch high, they should be transplanted on a fresh hot-bed, at about three inches distance, observing to shade them till they have taken root; then they should have fresh air admitted to them every day, in proportion to the warmth of the season. In about a month or five weeks the plants will have grown so large as to nearly meet, so they will require more room, otherwise they will draw up weak; then a fresh hot-bed should be prepared, into which there should be a sufficient number of three-farthing pots plunged, filled with light rich earth, and when the bed is in a proper temperature of warmth, the plants should be carefully taken up with balls of earth to their roots, and each planted into a separate pot, observing to shade them till they have taken new root, and afterward they must be treated in the same manner as other tender exotic plants. When the plants have filled these pots with their roots, they should be shaken out of them, and their roots on the outside of the ball of earth must be carefully pared off; then they should be put into larger pots, and where there is a conveniency of a deep frame, to plunge the pots into another gentle hot-bed, it will bring the plants early to flower, and cause them to grow much larger than those which are placed abroad. In July the plants should be inured gradually to bear the open air, into which some of them may be removed about the middle of that month, and intermixed with other annual plants to adorn the pleasure-garden; but it will be proper to keep a plant or two of each sort in shelter for seeds, because when the autumn proves cold or wet, those plants which are exposed abroad seldom produce good seeds.

The perennial climbing sort must be removed into the bark-stove, and plunged into the back side of the tan bed, where it may have room to grow in height; the plants will live three or four years, if they are preserved in a stove.

These sorts are not very ornamental, so they are seldom cultivated but in botanic gardens for the sake of variety.

GOOSEBERRY. See GROSSULARIA.

GORTERIA. A kind of Anemonospermus.

The CHARACTERS are,

The empalement is scaly, ending in bristly spines; the flower is composed of hermaphrodite florets in the disk, and female in the border; the hermaphrodite florets are funnel-shaped, having five short stamina, terminated by cylindrical summits, with a hairy germen, supporting a slender style, crowned by a bifid stigma: the germen afterward becomes one roundish hairy seed. The female florets are barren.

The SPECIES are,

1. GORTERIA (*Ringens*) scapis unifloris, foliis lanceolatis pinnatifidis, caule depresso. Amœn. Acad. 6. p. 86. *Gorteria with one flower on each foot-stalk, spear wing-pointed leaves, and a depressed stalk.*

2. GORTERIA (*Fruticosa*) foliis laciniatis integris dentato-spinosis subtus tomentosis, caule fruticoso. Lin. Sp. 1284. *Gorteria with entire spear-shaped leaves, having spiny indentures, are woolly on their under sides, and a shrubby stalk.* This was under the title of CARTHAMUS in the former edition.

These plants grow naturally near the Cape of Good Hope. The first is a low spreading plant, whose branches, though ligneous, yet trail upon the ground; these are garnished with narrow leaves, which are green on their upper side, but are white on their under side, and are generally divided at their extremities into three or five parts. The foot-stalks of the flowers, which arise from each branch, are naked, near six inches long, each terminating with one large Orange-coloured flower, which appear in May, but are rarely succeeded by seeds in England.

This plant is easily propagated by cuttings, which, if planted during any of the summer months in a shady border, will readily put out roots; these afterwards should be planted each into a separate small pot, and must be treated after the manner directed for ARCTOTIS, to which the reader is desired to turn for instructions.

The second sort rises with a shrubby stalk about three feet high, sending out a few weakly branches, garnished with oblong leaves, smooth on their upper surface, but are woolly on their under, indented on their sides, each terminating with a weak spine. The flowers are produced at the extremity of the branches, having leafy empalements, ending with spines.

This is propagated by planting of the small heads, stripped from the stalks, in June or July, putting them into small pots, filled with light loamy earth, plunging the pots into an old hot-bed, covering them with glasses, which should be shaded every day from the sun, and refreshing the heads frequently, but not give them too much water. When the heads have put out roots, they should be each planted in a separate small pot, and placed in the shade until they have taken root; after this they may be treated as the ARCTOTIS.

GORZ. See ULEX.

GOSSYPIUM. Lin. Gen. Plant. 755. Cotton.

The CHARACTERS are,

The flower has a double empalement. It hath five plain heart-shaped petals, which join at their base, and a great number of stamina, which are joined at bottom in a column, and are inserted into the petals. It hath a round germen, supporting four styles, joined in the column. The germen afterward becomes a roundish capsule, ending in a point, having four cells, which are filled with oval seeds, wrapped up in down.

The Species are,

1. Gossypium (*Herbaceum*) foliis quinquelobis, caule herbaceo. Hort. Upſal. 203. *Cotton with leaves having five lobes, and an herbaceous ſtalk.*

2. Gossypium (*Barbadenſe*) foliis trilobis integerrimis. Hort. Upſal. 204. *Cotton with entire leaves, having three lobes.*

3. Gossypium (*Arboreum*) foliis palmatis, lobis lanceolatis, caule fruticoſo. Lin. Sp. Plant. 693. *Cotton with hand-ſhaped leaves, having five ſpear-ſhaped lobes, and a ſhrubby ſtalk.*

4. Gossypium (*Hirſutum*) foliis trilobis & quinquelobis acutis, caule ramoſo hirſuto. *Finest American Cotton with a green ſeed.*

The firſt ſort is the common Levant Cotton, which is cultivated in ſeveral iſlands of the Archipelago, as alſo in Malta, Sicily, and the kingdom of Naples; it is ſown in tilled ground in the ſpring of the year, and is ripe in about four months after, when it is cut down in harveſt as Corn in England, the plants always periſhing ſoon after the ſeeds are ripe: this plant grows about two feet high, with an herbaceous ſtalk, garniſhed with ſmooth leaves divided into five lobes. The ſtalks ſend out a few weak branches upward, which are garniſhed with leaves of the ſame form, but are ſmaller. The flowers are produced at the extremity of the branches; theſe have two large empalements, the outer is cut into three parts, and the inner into five. The petals of the flower are of a pale yellow colour, inclining to white; theſe are ſucceeded by oval capſules, which open in four parts, having four cells, which are filled with ſeeds, wrapped up in a down, which is the Cotton.

The ſecond ſort grows naturally in ſeveral iſlands of the Weſt-Indies; this riſes with a ſhrubby ſmooth ſtalk four or five feet high, ſending out a few ſide branches, which are garniſhed with ſmooth leaves, divided into three lobes. The flowers are produced at the end of the branches, which are ſhaped like thoſe of the former ſort, but are larger, and of a deeper yellow colour. The pods are larger, and the ſeeds are black.

The third ſort hath a perennial ſhrubby ſtalk, which riſes ſix or eight feet high, and divides into many branches, which are ſmooth, and garniſhed with hand-ſhaped leaves, having four or five lobes. The flowers are produced at the end of the branches; theſe are larger than thoſe of the two former ſorts, and are of a deep yellow colour. The pods of this ſort are larger than thoſe of the former.

The fourth ſort is a native of the Eaſt and Weſt-Indies; this is an annual plant, which periſhes ſoon after the ſeeds are ripe. It riſes to the height of three feet or more, and ſends out many lateral branches where they are allowed room to grow; theſe are hairy, and garniſhed with leaves, having in ſome three, and others five acute-pointed hairy lobes. The flowers are produced from the ſide, and at the ends of the branches, which are large, of a dirty ſulphur colour, each petal having a large purple ſpot at the baſe; the flowers are ſucceeded by oval pods, which open in four cells, filled with oblong green ſeeds, wrapped up in a ſoft down. The ſtaple of this is much finer than either of the other ſpecies, therefore it is well worth the attention of the inhabitants of the Britiſh colonies in America to cultivate and improve this ſort, ſince it will ſucceed in Carolina, where it has been cultivated for ſome years; and might be a commodity worthy of encouragement by the public, could they contrive a proper gin to ſeparate the Cotton from the ſeeds, to which this ſort adheres much cloſer than any of the other ſorts, the Cotton from this ſhrub being preferable to any other yet known.

All theſe ſorts are tender plants, therefore will not thrive in the open air in England, but they are frequently ſown in curious gardens for variety; the firſt and fourth ſorts will produce ripe ſeeds in England, if their ſeeds are ſown early in the ſpring, upon a good hot-bed, and the plants afterwards planted each into ſeparate pots, and plunged into a hot-bed of tanners bark to bring them forward; when they are grown too tall to remain under the frames, they ſhould be removed into the tan-bed in the ſtove, and ſhifted into larger pots, if their roots have filled the other; with this management their flowers will appear in July, and towards the end of September the ſeeds will ripen, and the pods will be as large as thoſe produced in the Eaſt and Weſt-Indies; but if the plants are not brought forward early in the ſpring, it will be late in the ſummer before the flowers will appear, and there will be no hopes of the pods coming to perfection.

The ſhrub Cotton will riſe from the ſeeds very eaſily, if they are ſown upon a good hot-bed; and when they are ſown early in the ſpring, and brought forward in the ſame manner as hath been directed for the former ſorts, the plants will grow to be five or ſix feet high the ſame ſummer; but it is difficult to preſerve the plants through the winter, unleſs they are hardened gradually in Auguſt during the continuance of the warm weather; for when they are forced on in ſummer, they will be ſo tender as to render them incapable of reſiſting the leaſt injury. The plants of this ſort muſt be placed in the bark-ſtove in autumn, and kept in the firſt claſs of heat, otherwiſe they will not live through the winter in England.

GRAFTING is the taking a ſhoot from one tree, and inſerting it into another, in ſuch a manner, as that both may unite cloſely, and become one tree; this is called, by the ancient writers in huſbandry and gardening, inciſion, to diſtinguiſh it from inoculation or budding, which they call inſerere oculos.

The uſe of Grafting is to propagate any curious ſorts of fruits, ſo as to be certain of the kinds, which cannot be done by any other method; for as all the good fruits have been accidentally obtained from ſeeds, ſo the ſeeds of theſe, when ſown, will many of them degenerate, and produce ſuch fruit as are not worth cultivating; but when ſhoots are taken from ſuch trees as produce good fruit, theſe will never alter from their kind, whatever be the ſtock or tree on which they are grafted; for though the grafts receive their nouriſhment from the ſtocks, yet they are never altered by them, but continue to produce the ſame kind of fruit as the tree from which they were taken; the only alteration is, that when the ſtocks on which they are grafted do not grow ſo faſt, nor afford a proper ſupply of nouriſhment to the grafts, they will not make near ſo great progreſs as they otherwiſe would do, nor will the fruit they produce be ſo fair, and ſometimes not ſo well flavoured.

The ſhoots to be ingrafted are termed cions or grafts; in the choice of theſe the following directions ſhould be carefully obſerved. 1ſt, That they are ſhoots of the former year, for when they are older they never ſucceed well. 2dly, Always to take them from healthy fruitful trees, for if the trees are ſickly from whence they are taken, the grafts very often partake ſo much of the diſtemper, as rarely to get the better of it, at leaſt for ſome years; and when they are taken from young luxuriant trees, whoſe veſſels are generally large, they will continue to produce luxuriant ſhoots, and are ſeldom ſo fruitful as thoſe which are taken from fruitful trees, whoſe ſhoots are more compact, and the joints cloſer together, at leaſt it will be a great number of years before theſe luxuriant grafts begin to produce fruit, if they are managed with the greateſt ſkill. 3dly, You ſhould prefer thoſe grafts which are taken from the lateral or horizontal branches, to thoſe from the ſtrong perpendicular ſhoots, for the reaſons before given.

Theſe

Theſe grafts or cions ſhould be cut off from the trees before their buds begin to ſwell, which is generally three weeks or a month before the ſeaſon for Grafting; therefore, when they are cut off, they ſhould be laid in the ground with the cut-end downwards, burying them half their length, and covering their tops with dry litter, to prevent their drying; if a ſmall joint of the former year's wood is cut off with the cion, it will preſerve it the better, and when they are grafted this may be cut off; for at the ſame time the cions muſt be cut to a proper length before they are inſerted to the ſtocks, but till then the ſhoots ſhould remain their full length, as they were taken from the tree; if theſe cions are to be carried to a conſiderable diſtance, it will be proper to put their cut-ends into a lump of clay, and to wrap them up in Moſs, which will preſerve them freſh for a month, or longer.

In the choice of young ſtocks for Grafting, you ſhould always prefer ſuch as have been raiſed from the ſeed, and that have been once or twice tranſplanted. Next to theſe are thoſe ſtocks which have been raiſed from cuttings or layers, but thoſe which are ſuckers, taken from roots of other trees, ſhould always be rejected, for theſe are never ſo well rooted as the others, and conſtantly put out a great number of ſuckers from their roots, whereby the borders and walks of the garden will be always peſtered with them during the ſummer ſeaſon, which is not only unſightly, but they alſo take off part of the nouriſhment from the trees.

If theſe ſtocks have been allowed a proper diſtance in the nurſery where they have grown, the wood will be better ripened, and more compact than thoſe which have grown cloſe, and have been thereby drawn up to a greater height; the wood of which will be ſoft, and their veſſels large, ſo that the cions grafted into them will ſhoot very ſtrong, but they will be leſs diſpoſed to produce fruit than the other; and when trees acquire an ill habit at firſt, it will be very difficult to reclaim them afterwards.

Having directed the choice of cions and ſtocks, we come next to the operation, in order to which you muſt be provided with the following tools.

1. A neat ſmall hand-ſaw to cut off the heads of large ſtocks.

2. A good ſtrong knife with a thick back, to make clefts in the ſtocks.

3. A ſharp penknife to cut the grafts.

4. A Grafting chiſſel and a ſmall mallet.

5. Baſs-ſtrings, or woollen yarn, to tie the grafts with, and ſuch other inſtruments and materials as you ſhall find neceſſary, according to the manner of Grafting you are to perform.

6. A quantity of clay, which ſhould be prepared a month before it is uſed, and kept turned and mixed like mortar every other day, which is to be made after the following manner:

Get a quantity of ſtrong fat loam (in proportion to the quantity of trees intended to be grafted) then take ſome new ſtonehorſe dung, and break it in amongſt the loam, and if you cut a little ſtraw or hay very ſmall, and mix amongſt it, the loam will hold together the better; and if there be a quantity of ſalt added, it will prevent the clay from dividing in dry weather; theſe muſt be well ſtirred together, putting water to them after the manner of making mortar; it ſhould be hollowed like a diſh, and filled with water, and kept every other day ſtirred; but it ought to be remembered, that it ſhould not be expoſed to the froſt, or drying winds, and that the oftener it is ſtirred and wrought the better.

Of late years ſome perſons have made uſe of another compoſition for Grafting, which they have found to anſwer the intention of keeping out the air, better than the clay before preſcribed. This is compoſed of turpentine, bees-wax, and roſin, melted together, which when of a proper conſiſtence, may be put on the ſtock round the graft, in the ſame manner as the clay is uſually applied, and though it be not above a quarter of an inch thick, yet will keep out the air more effectually than the clay; and as cold will harden this, there is no danger of its being hurt by froſt, which is very apt to cauſe the clay to cleave, and ſometimes fall off; and when the heat of ſummer comes on, this mixture will melt, and fall off without any trouble. In uſing of this, there ſhould be a tin, or copper pot, with conveniency under it to keep a very gentle fire with ſmall-coal, otherwiſe the cold will ſoon condenſe the mixture; but you muſt be careful not to apply it too hot, leſt you injure the graft. A perſon who is a little accuſtomed to this compoſition, will apply it very faſt, and it is much eaſier for him than clay, eſpecially if the ſeaſon ſhould prove cold.

There are ſeveral ways of Grafting, the principal of which are four:

1. Grafting in the rind, called alſo Shoulder-grafting, which is only proper for large trees; by ſome called Crown-grafting, becauſe the grafts are ſet in form of a circle, or crown, and is generally performed about the latter end of March, or the beginning of April.

2. Cleft-grafting, which is called Stock or Slit-grafting; this is proper for trees or ſtocks of a leſſer ſize, from an inch, to two inches or more diameter; this Grafting is to be performed in the month of March, and ſupplies the failure of the eſcutcheon way, which is practiſed in June, July, and Auguſt.

3. Whip-grafting, which is alſo called Tongue-grafting; this is proper for ſmall ſtocks of an inch, half an inch, or leſs, diameter; this is the moſt effectual way of any, and that which is moſt in uſe.

4. Grafting by approach, or ablactation; this is to be performed when the ſtock you would graft on, and the tree from which you take your graft, ſtand ſo near together, that they may be joined; this is to be performed in the month of April, and is alſo called inarching; it is chiefly uſed for Jaſmines, Oranges, and other tender exotic trees.

The manner of performing the ſeveral ways of Grafting being generally known, need not be here mentioned.

The next thing which is neceſſary to be known by thoſe who would practiſe this art, is, what trees will take and thrive by being grafted upon each other; in this article there have been no ſure directions given by any of the writers on this ſubject, for there will be found great miſtakes in moſt of the books which have treated on this ſubject; but as it would ſwell this article too great, if all the ſorts of trees were to be here enumerated, which will take upon each other by Grafting, I ſhall only mention ſuch general directions, as, if attended to, will be ſufficient to inſtruct perſons ſo as they may ſucceed.

All ſuch trees as are of the ſame genus, i. e. which agree in their flower and fruit, will take upon each other; for inſtance, all the Nut-bearing trees may be ſafely grafted on each other, as may all the Plumb-bearing trees, under which head I reckon not only the ſeveral ſorts of Plumbs, but alſo the Almond, Peach, Nectarine, Apricot, &c. which agree exactly in their general characters, by which they are diſtinguiſhed from all other trees; but as many of theſe are very ſubject to emit large quantities of gum from thoſe parts of the trees where they are deeply cut and wounded, ſo the tender trees of this kind, viz. Peaches and Nectarines, which are moſt ſubject to this, it is found to be the ſureſt method to bud or inoculate theſe ſorts of fruits, for which ſee INOCULATION.

Secondly, all ſuch trees as bear cones will do well upon each other, though they may differ in one being evergreen,

 and

and the other ſhedding its leaves in winter, as is obſervable in the Cedar of Lebanus, and the Larch-tree, which are found to ſucceed upon each other very well; but theſe muſt be grafted by approach, for they abound with a great quantity of reſin, which is apt to evaporate from the graft, if ſeparated from the trees before it is joined with the ſtock, whereby they are often deſtroyed; as alſo the Laurel on the Cherry, or the Cherry on the Laurel. All the maſt-bearing trees, which alſo take upon each other, and thoſe which have a tender ſoft wood, will do well if grafted in the common way, but thoſe that are of a more firm contexture, and are ſlow growers, ſhould be grafted by approach.

By ſtrictly obſerving this rule we ſhall ſeldom miſcarry, provided the operation be righly performed, and at a proper ſeaſon, unleſs the weather ſhould prove very bad, as it ſometimes happens, whereby whole quarters of fruit-trees miſcarry; it is by this method that many kinds of exotic trees are not only propagated, but alſo rendered hardy enough to endure the cold of our climate in the open air; for, being grafted upon ſtocks of the ſame ſort which are hardy, the grafts are rendered more capable to endure the cold, as hath been experienced in moſt of our valuable fruits now in England, which were formerly tranſplanted hither from more ſoutherly climates, and were at firſt too impatient of our cold to ſucceed well abroad, but have been by budding or Grafting upon more hardy trees, rendered capable of reſiſting our ſevereſt cold.

GRAMEN., Tourn Inſt. R. A. 516. Graſs.

To enumerate all the ſpecies of Graſs which are found growing naturally in England, would ſwell this article greatly beyond the deſign of the work; therefore I ſhall only mention a few ſpecies, which are either uſed in medicine, or cultivated as a pabulum for cattle; for there is ſcarce a paſture in this country, where at leaſt twenty different ſpecies are not to be found intermixed, and in moſt of them more than twice that number.

1. Gramen (*Repens*) ſpicâ triticeâ repens vulgare, caninum dictum. Raii Syn. 2. p. 247. *Common creeping Graſs with a ſpike like Wheat, called Dog or Couch Graſs.*

2. Gramen (*Perenne*) loliaceum, anguſtiore folio & ſpica. C. B. P. *Darnel with a chaffy ſpike; commonly called Ray, or Rye Graſs.*

3. Gramen (*Anguſtifolium*) pratenſe, paniculatum majus, anguſtiore folio. C. B. P. 2. *Meadow Graſs with larger panicles and a narrower leaf.*

4. Gramen (*Pratenſis*) pratenſe paniculatum majus, latiore folio. C. B. P. 2. *Meadow Graſs with a larger panicle and broader leaf.*

5. Gramen (*Flaveſcens*) avenacium pratenſe elatius, paniculâ flaveſcente, locuſtis parvis. Raii Syn. 407. *Taller Meadow Oat Graſs, with a yellowiſh panicle and ſmall huſks.*

6. Gramen (*Murinum*) ſecalinum. Ger. Emac. lib. i. cap. 22. n. 4. *Tall Meadow Rye Graſs.*

7. Gramen (*Maxima*) tremulum maximum. C. B. P. 2. *Greateſt quaking Graſs, or Cowquakes.*

The firſt ſort of Graſs is that which is directed for uſe in medicine; the roots of this are what are chiefly uſed, and are accounted aperitive and diuretic, opening obſtructions of the reins and bladder, provoking urine, and are of ſervice againſt the gravel and ſtone. The juice of the leaves and ſtalks was greatly eſteemed by Dr Boerhaave, who generally preſcribed this in all caſes where he ſuppoſed there were any obſtructions in the bile conduit.

This hath a creeping root, which ſpreads far in the ground, and is a very troubleſome weed in gardens and arable land; for every ſmall piece of the root will grow and multiply exceedingly, ſo it is very difficult to extirpate where it once gets poſſeſſion in gardens; the common method of deſtroying it is, to fork out the roots as often as the blades appear above ground; where this is two or three times carefully repeated, it may be totally rooted out; but when the ſurface of the ground is very full of the roots of this Graſs, the ſhorteſt way of deſtroying it is to trench the ground two ſpits and a ſhovelling deep, turning all the couch into the bottom, where it will rot, and never ſhoot up; but this can only be practiſed, where there is ſufficient depth of ſoil, for in ſhallow ground the roots cannot be buried ſo deep, as to lie below the depth which they naturally ſhoot.

Where the roots of this Graſs get poſſeſſion in arable fields, it is very difficult to root out again; the uſual method is by laying the land fallow in ſummer, and frequently harrowing it well over to draw out the roots; where this is carefully practiſed, the ground may be ſo well cleaned in one ſummer, as that the roots cannot much injure the crop which may be ſown upon it; but ſuch land ſhould be cropped with ſuch things as require the horſe-hoeing culture, for where the land can be frequently ſtirred and harrowed afterward, it will be of great ſervice in cleaning it from the roots of this Graſs and other bad weeds. The blade of this Graſs is ſo rough, that cattle will not feed upon it.

The ſecond ſort is frequently cultivated, eſpecially in ſtrong cold land, upon which this Graſs will ſucceed better than any other ſpecies, and is an earlier feed in the ſpring; but this is a very coarſe Graſs, and unleſs it is cut very early for hay, it becomes hard and wiery in the ſtalks, ſo that few cattle care to eat it: this ſpecies has few leaves, running all to ſtalk, ſo is uſually called bents, and in ſome counties bennet; when this Graſs is fed, it will be proper to mow off the bents in the beginning of June, otherwiſe they will dry, and have the appearance of a ſtubble field all the latter part of the ſummer; ſo that it will not only be very diſagreeable to the ſight, but alſo troubleſome to the cattle that feed on it, by tickling their noſtrils; ſo that the want of better paſture only, will force them to eat of the young Graſs which ſprings up between theſe bents, for thoſe they will not touch; therefore thoſe who ſuppoſe that theſe are eaten in ſcarcity of feed by the cattle, are greatly miſtaken, for I have many years cloſely attended to this, and have always found theſe bents remaining on the ground untouched, till the froſt, rain, and winds deſtroy it in winter; and by permitting theſe to ſtand, the after growth of the Graſs is greatly retarded, and the beautiful verdure is loſt for three or four months; ſo that it is good huſbandry to mow them before they grow too dry, and rake them off the ground; if theſe are then made into hay, it will ſerve for cart-horſes feed in winter, and will pay the expence of mowing it.

There is another ſpecies of this Graſs called Red Darnel, which is of a worſe nature than the firſt, the ſtalks growing hard much ſooner, and having narrower leaves: this is very common in moſt paſture grounds, for as it comes early to flower, ſo the ſeeds are generally ripe before the hay is cut, and from the falling ſeeds the ground is ſupplied with plenty of this ſort; therefore thoſe who are deſirous to keep their paſtures as clear from this Graſs as poſſible, ſhould always mow it before the ſeeds are ripe.

This Graſs is uſually ſown with Clover, upon ſuch lands as are deſigned to be ploughed again in a few years, and the common method is to ſow it with ſpring Corn; but from many repeated trials I have always found, that by ſowing theſe ſeeds in Auguſt, where there has happened a few ſhowers to bring up the Graſs, that the crop has anſwered much better than any which has been ſown in the common way; ſo that I am convinced of that being the beſt ſeaſon for ſowing theſe Graſſes, though it will be very difficult to perſuade thoſe perſons to alter their practice, who have been long wedded to old cuſtoms. The quantity of ſeed, which

which I allow to an acre is about two bushels, and eight pounds of the common white Clover, which, together, will make as good plants upon the ground as can be desired; but this is not to be practised upon such lands where the beauty of the verdure is principally regarded, therefore is fit for those who have only profit in view.

The third and fourth sorts are the two best species of Grass for pastures, so that if the seeds of these were carefully collected and sown separately, without any other mixture of Grass seeds, they would not only afford a greater quantity of feed on the same space of land, but the Grass would also be better, the hay sweeter, and the verdure more lasting than of any other sort; but there requires some attention to the saving of these seeds pure without mixture. I have tried to save the seeds of several species of Grass separately, in order to determine their qualities, but have found it very difficult to keep them distinct in gardens where the seeds of other sorts of Grass have been scattered; the only method in which I could succeed was by sowing each species in a distinct pot, and when the plants come up, to weed out all the other kind of Grass which came up in the pots; by this means I preserved a great variety of the grassy tribe several years, but not having ground enough to propagate the most useful species in any quantity, I was obliged to abandon the pursuit; but I must recommend this to persons of leisure and skill, who have a sufficient quantity of land for the purpose, to carry this project into execution, which may be of singular benefit to the public; for we have an instance of the advantage which the inhabitants of the Netherlands have made, by saving the seeds of the White Clover, or Honeysuckle Trefoil, which is a plant common to most of the English pastures, yet no person in this country ever gave themselves the trouble, till within a few years past, to collect the seeds from the fields for sowing, but have purchased vast quantities of it annually, at a considerable price from Flanders, where the peasants have been so industrious, as to collect the seeds and sow great quantities of land with it, with a view of sale to this country only.

The fifth and sixth sorts are also very good Grasses for pastures, and have perennial roots, so are the next best sorts for sowing to those before mentioned, which, in my opinion, deserve the preference to all the other; but as it will be difficult to save a sufficient quantity of seeds of those alone, to supply the demand which may be for their seeds, so these two species may be admitted in aid of the other, as they are very leafy kinds of Grass, and their stalks do not become stiff and harsh like many other species, but with proper care may be made very fine, and if duly rolled, their roots will mat and form a very close sward, therefore these should be included in the number of sown Grasses.

The seventh sort is mentioned for the sake of variety, and not for use; this hath an annual stalk, which sends up many broad hairy leaves, between which arise slender stiff stalks from a foot to near two feet high, dividing upward into a large loose panicle, garnished with heart-shaped small spikes, each having about seventeen small floscules or florets, which have a single seed succeeding them; the heads hang by slender long foot-stalks, which are moved by every wind, so that they generally appear shaking, from whence it had the title of Quaking Grass. There are four species of this Grass, two of them grow naturally in England, which coming to head in May, occasioned the following English proverb, "May, come she early come "she late, makes the cow quake." The large sort here mentioned grows naturally in the south of France and Italy, and is only preserved in some English gardens for the sake of variety

The land upon which Grass seed is intended to be sown should be well ploughed, and cleared from the roots of noxious weeds, especially if the Grass is to remain for pasture, such as Couch Grass, Fern, Rushes, Heath, Gorse, Broom, Rest-harrow, &c. which, if left in the ground, will soon get the better of the Grass, and over-run the land. Therefore in such places where either of these weeds abound, it will be a good method to plough up the surface in April, and let it lie some time to dry; then harrow the roots into small heaps, and burn them. The ashes so produced, when spread on the land, will be a good manure for it. But where Couch Grass, Fern, or Rest-harrow is in plenty, whose roots run far under ground, the land must be ploughed two or three times pretty deep in dry weather, and the roots carefully harrowed off after each ploughing, which is the most sure method to destroy them. Where the land is very low, and of a stiff clayey nature, which holds w ter in winter, it will be of singular service to make some under-ground drains to carry off the wet, which, if detained too long on the ground, will render the Grass sour.

Before the seed is sown, the surface of the ground should be made fine and level, otherwise the seed will be buried unequal. When the seed is sown, it must be gently harrowed in, and the ground rolled with a wooden roller, which will make the surface even, and prevent the seeds being blown in patches. When the Grass comes up, if there should be any bare spots where the seed has not grown, they may be sown again, and the ground rolled, which will fix the seeds; and the first kindly showers will bring up the Grass, and make it very thick.

The proper management of pasture land, is certainly the least understood of any part of agriculture; the farmers never have attended to this, being more inclined to the plough; though the profits attending it have not of late years been so great, as to encourage them in that part of husbandry; but these people never think of laying down land for pasture, to continue longer than three years, at the end of which time they plough it up again, to sow it with grain.

Clover Grass. See Trifolium.
Saint Foin. See Onobrychis.
La Lucern. See Medica.
Nonesuch. See Melilotus.
Trefoil. See Trifolium.
Spurry. See Spergula.

GRANADILLA. See PASSIFLORA.

GRAPES. See VITIS.

GRATIOLA. Lin. Gen. Plant. 27. Hedge Hyssop.

The CHARACTERS are,

The flower hath one petal of the grining kind, cut at the top into four small segments. It hath five awl-shaped stamina; the other two are longer, and adhere to the tube of the petal. In the center is situated a conical germen, which afterward becomes an oval capsule ending in a point, having two cells, which are filled with small seeds.

The SPECIES are,

1. GRATIOLA (*Officinalis*) floribus pedunculatis, foliis lanceolatis serratis. Lin. Mat. Med. 18. *Hedge Hyssop with flowers standing on foot-stalks, and spear-shaped sawed leaves.*

2. GRATIOLA (*Virginiana*) foliis lanceolatis obtusis sub dentatis. Flor. Virg. 6. *Hedge Hyssop with obtuse indented leaves.*

3. GRATIOLA (*Peruviana*) floribus subsessilibus. Lin. Sp. Plant 17. *Hedge Hyssop with flowers sitting close to the branches.*

The first sort grows naturally on the Alps, and other mountainous parts of Europe. It hath a thick, fleshy, fibrous creeping root, which propagates very much when planted in a proper soil or situation, from which arise several

veral upright ſquare ſtalks near a foot high, garniſhed with narrow ſpear-ſhaped leaves placed oppoſite; the flowers are produced on the ſide of the ſtalks at each joint; they are ſhaped like thoſe of the Foxglove, but are ſmall, and of a pale yellowiſh colour.

It is eaſily propagated by parting of the roots; the beſt time to do this is in the autumn, when the ſtalks decay; the plants ſhould have a moiſt ſoil and a ſhady ſituation, in which they will thrive exceedingly; but in dry ground they often decay in ſummer, unleſs they are plentifully watered.

The ſecond ſort grows naturally in North America, where it riſes more than a foot high, but in England I have not ſeen it more than eight inches; the leaves are blunt and indented at their extremities; the flowers are white, and come out from the ſide of the ſtalks like thoſe of the other, but are not ſucceeded by ſeeds here. It may be propagated in the ſame manner as the firſt ſort, and requires the ſame treatment.

The ſeeds of the third ſort were ſent me from Carthagena, where it was found growing naturally in places where there had been ſtanding waters, which were then dried up; this plant grew about nine inches high, with a weak ſtalk, and the leaves placed oppoſite; they were about three quarters of an inch long, and half an inch broad, ſawed on their edges; the flowers came out ſingle on each ſide the ſtalk; they were white, and much ſmaller than thoſe of the firſt ſort, but were not ſucceeded by ſeeds, ſo the plant was loſt here.

GRAVEL and Graſs are naturally ornaments to a country-ſeat, and the glory of the Engliſh gardens, in which we excel moſt other nations. Indeed, moſt people who have had the deſigning of gardens, have too much conſidered Gravel as an ornament, ſo have made too many walks in gardens, and thoſe have been much too broad, for the ſight of Gravel is not very pleaſing, ſo it ought only to be conſidered as uſeful, a dry walk quite round, or to each part of a garden being abſolutely neceſſary; but theſe, in gardens of great extent, need not be more than nine or ten feet wide, and in ſmall ones five or ſix.

There are different ſorts of Gravel, but for thoſe who can conveniently have it, I approve of that Gravel on Black-heath, as preferable to moſt that we have in England; it conſiſting of ſmooth even pebbles, which, when mixed with a due quantity of loam, will bind exceeding cloſe, and look very beautiful, and continue handſome longer than any other ſort of Gravel which I have yet ſeen.

There are many kinds of Gravel which do not bind, and thereby cauſe a continual trouble of rolling to little or no purpoſe; as for ſuch,

If the Gravel be looſe or ſandy, you ſhould take one load of ſtrong loam to two or three of Gravel, and ſo caſt them well together, and turn this mixture over three or four times, that they may be well blended together; if this is done in proper proportion, it will bind well, and not ſtick to the feet in wet weather.

There are many different opinions about the choice of Gravel; ſome are for having the Gravel as white as poſſible, and in order to make the walks more ſo, they roll them well with ſtone rollers, which are often hewn by the maſons that they may add a whiteneſs to the walks; but this renders it very troubleſome to the eyes, by reflecting the rays of light too ſtrongly; therefore this ſhould ever be avoided, and ſuch Gravel as will lie ſmooth, and reflect the leaſt, ſhould be preferred.

Some are apt to lay Gravel-walks too round, but this is likewiſe an error, becauſe they are not ſo good to walk upon, and, beſides, it makes them look narrow; one inch riſe is enough in a crown for a walk of five feet, and it will be ſufficient if a walk be ten feet wide, that it lies two inches higher in the middle than it does on each ſide.

For the depth of Gravel-walks, ſix or eight inches may do well enough, but a foot thickneſs will be ſufficient for any; but then there ſhould always be a depth of rubbiſh laid under the Gravel, eſpecially if the ground is wet, in which caſe there cannot be too much care taken to fill the bottom of the walks with large ſtones, flints, brick rubbiſh, chalk, or any other materials which can be beſt procured, which will drain off the moiſture from the Gravel, and prevent its being poachy in wet weather; but as it may be difficult in ſome places to procure a ſufficient quantity of theſe materials to lay in the bottom of the walks, ſo there may be a bed of Heath or Furze, which ever can be procured at the leaſt expence, laid under the Gravel to keep it dry: if either of theſe are uſed green, they will lie a long time, as they will be covered from air, and theſe will prevent the Gravel from getting down into the clay and will always keep the Gravel dry; where there is not this precaution in the firſt laying of Gravel upon clay, the water being detained by the clay, will cauſe the Gravel to be poachy whenever there is much rain.

In the making of Gravel-walks, there muſt be great regard had to the level of the ground, ſo as to lay the walks with eaſy deſcents toward the low parts of the ground, that the wet may be drained off eaſily; for when this is omitted, the water will lie upon the walks a conſiderable time after hard rains, which will render them unfit for uſe, eſpecially where the ground is naturally wet or ſtrong; but where the ground is level, and there are no declivities to carry off the wet, it will be proper to have ſink-ſtones laid by the ſides of the walks, at convenient diſtances to let off the wet; and where the ground is naturally dry, that the water will ſoon ſoak away, the drains from the ſink-ſtones may be contrived ſo as to convey the water into ſeſſpools, from which the water will ſoak away in a ſhort time; but in wet land there ſhould be under-ground drains to convey the wet off, either into ponds, ditches, or the neareſt place to receive it, for where this is not well provided for, the walks will never be ſo handſome or uſeful.

The month of March is the propereſt time for laying Gravel; it is not prudent to do it ſooner, or to lay walks in any of the winter months before that time.

If conſtant rolling them after the rains and froſt will not effectually kill the weeds and Moſs, you ſhould turn the walks in March, and lay them down at the ſame time.

In order to deſtroy worms that ſpoil the beauty of Gravel, or Graſs-walks, ſome recommend the watering them well with water in which Walnut-tree leaves have been ſteeped, and made very bitter, eſpecially thoſe places moſt annoyed with them; and this, they ſay, as ſoon as it reaches them, will make them come out haſtily, ſo that they may be gathered up and deſtroyed; but if, in the firſt laying of the walks, there is a good bed of lime rubbiſh laid in the bottom, it is the moſt effectual method to keep out the worms, for they do not care to harbour near lime.

GREEN-HOUSE, or Conſervatory.

As of late years there have been great numbers of curious exotic plants introduced into the Engliſh gardens, ſo there has been an increaſe of conſervatories to preſerve them; and not only a greater ſkill in the management and ordering of theſe plants has increaſed therewith, but alſo a greater knowledge of the ſtructure and contrivance of theſe places, ſo as to render them both uſeful and ornamental, hath been acquired. Therefore I have not only given the beſt inſtructions for this I was capable of, but alſo a deſign of one in the manner I would chuſe to erect it, upon the annexed copper plate.

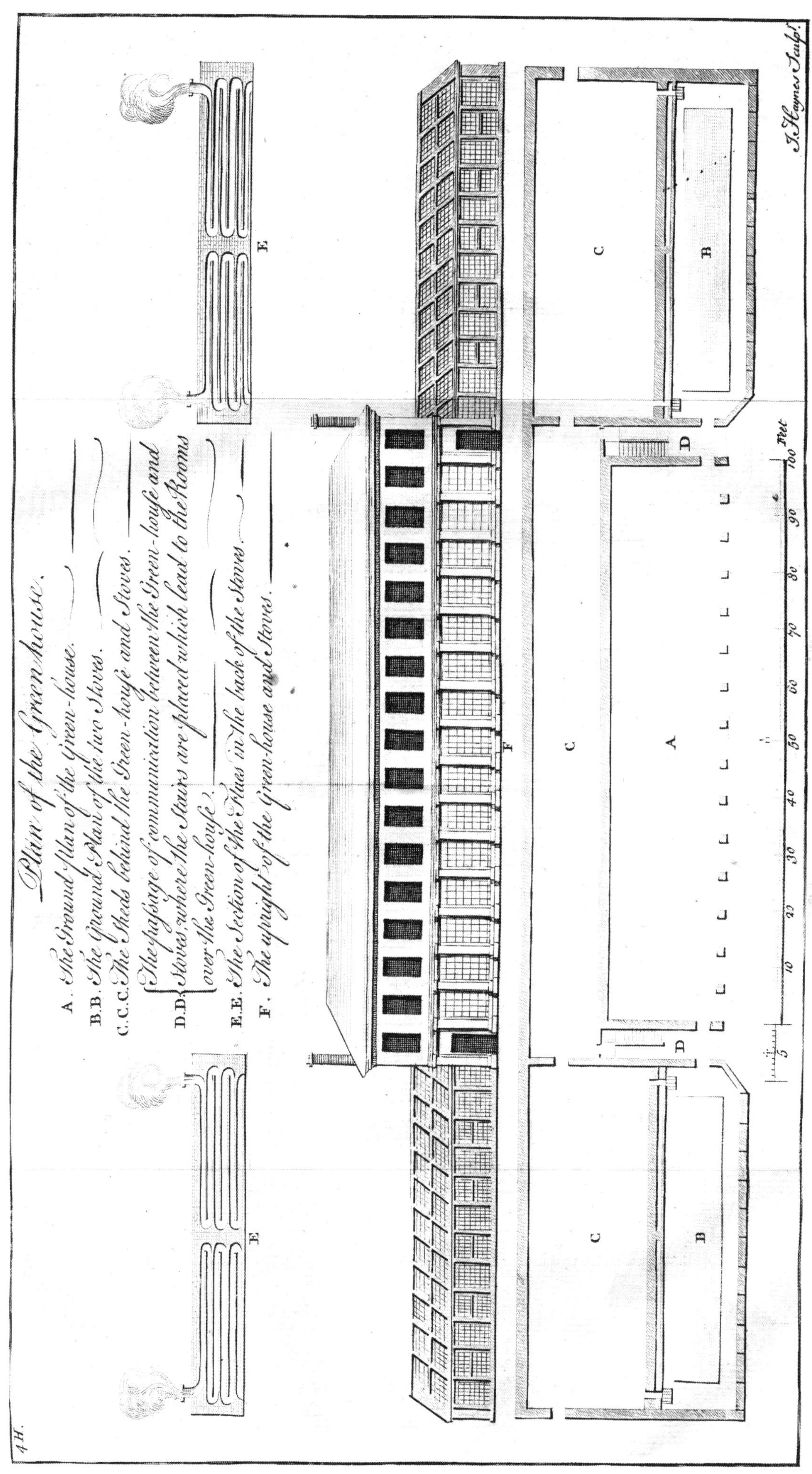

The material originally positioned here is too large for reproduction in this reissue. A PDF can be downloaded from the web address given on page iv of this book, by clicking on 'Resources Available'.

As to the length of these houses, that must be proportioned to the number of plants they are to contain, or the fancy of the owner; but their depth should never be greater than their height in the clear, which in small or middling houses may be sixteen or eighteen feet, but for large ones, from twenty to twenty-four feet is a good proportion; for if the Green-house is long and too narrow, it will have a bad appearance both within and without, nor will it contain so many plants, if proper room be allowed for passing in front, and on the backside of the stands on which the plants are placed; and, on the other hand, if the depth of the Green-house is more than twenty-four feet, there must be more rows of plants placed to fill the house, than can with conveniency be reached in watering and cleaning; nor are houses of too great depth so proper for keeping of plants, as those of a moderate size.

The windows in front should extend from about one foot and a half above the pavement, to within the same distance of the cieling, which will admit of a cornice round the building over the heads of the windows. As it is necessary to have these windows so long, it will be impossible to make them in proportion as to their breadth; for if in the largest buildings the sashes are more than seven, or seven feet and a half broad, they will be so heavy and troublesome to move up and down, as to render it very difficult for one person to perform it; besides, their weight will occasion their soon decaying. There is also another inconvenience in having the windows too broad, which is, that of fixing proper shutters to them, in such a manner, as that they may fall back close to the piers, so as not to be incommodious, or when open, to obstruct any part of the rays of light from reaching the plants. The piers between these windows should be as narrow as possible to support the building, for which reason I should chuse to have them of stone, or of hard well-burnt bricks, for if they are built with fine rubbed brick, those are generally so soft that the piers will require to be made thicker, or the building will be too weak to support the weight, especially if there are rooms over the Green-house; which is what I would always advise, as being of great use to keep the frost out in very hard winters. If these piers are made of stone, I would advise them to be two feet and a half in front, and sloped off backward to about eighteen inches broad, whereby the rays of the sun will not be taken off, or obstructed by the corners of the piers, which it would be if they were square; but if they are built with bricks, it will be proper to make them three feet in front, otherwise they will be too slender; these I would also advise to be sloped off, in the manner directed for the stone.

At the back of the Green-house there may be erected a house for tools, and many other purposes, which will be extremely useful to prevent the frost from entering the house that way, so that the wall between these need not be more than three bricks, or three and a half in thickness; whereas, were it quite exposed behind, it should be at least four bricks in thickness; by having a shed behind, if you are willing to make a handsome building, and to have a noble room over the Green-house, you may make part of the room over the tool-house, and carry up the stair-case in the back, so as not to be seen in the Green-house, and hereby a room twenty-five or thirty feet in width may be contrived, and of a proportionable length; and under this stair-case there may be a private door into the Green-house, at which the gardener may enter in hard frosty weather, when it will not be safe to open any of the glasses in front. The floor of the Green-house, which should be laid either with Bremen squares, Purbeck stone, or bread tiles, according to the fancy of the owner, must be raised two feet above the surface of the ground whereon the house is placed, which, in dry ground, will be sufficient; but if the situation is moist and springy, and thereby subject to damps, it should be raised at least three feet above the surface; and if the whole is arched with low brick arches under the floor, it will be of great service in preventing the damps rising in winter, which are often very hurtful to the plants, especially in thaws, when the air is often too cold to be admitted into the house to take off the damps. Under the floor, about two feet from the front, I would advise a flue of about twelve inches in width, and two feet deep, to be carried the whole length of the house, which may be returned along the back part, and be carried up into proper funnels adjoining to the tool-house, by which the smoke may pass off. The fire-place may be contrived at the ends of the house, and the doors at which the fuel is put in, as also the ash-grate, may be contrived to open into the tool-house, so that it may be hid from the sight, and be in the dry, and the fuel may be laid in the same place, whereby it will always be ready for use.

I suppose many people will be surprised to see me direct the making of flues under a Green-house, which has been disused so long, and by most people thought of ill consequence, as indeed they have often proved, when under the direction of unskilful managers, who have thought it necessary, whenever the weather was cold, to make fires therein; but however injurious flues may have been under such management, yet, when skilfully used they are of very great service; for though perhaps it may happen, that there will be no necessity to make any fires in them for two or three years together, when the winters prove mild, yet in very hard winters they will be extremely useful to keep out the frost, which cannot be effected any other way but with great trouble and difficulty.

Withinside of the windows, in front of the Green-house, you should have good strong shutters, which should be made with hinges to fold back, that they may lie quite close to the piers, that the rays of the sun may not be obstructed thereby. These shutters need not to be above an inch and a half thick, or little more, when wrought, which, if made to join close, will be sufficient to keep out our common frost; for when the weather is so cold as to endanger the freezing in the house, it is but making a fire in your flues, which will effectually prevent it, and without this conveniency it will be very difficult to effect; for where persons are obliged to nail mats before their windows, or to stuff the hollow space between the shutters and the glass with straw, this is commonly suffered to remain till the frost goes away; which, if it should continue very long, the keeping the Green-house closely shut up will prove very injurious to the plants; and as it frequently happens that we have an hour or two of sun-shine in the middle of the day in continued frosts, which is of great service to plants when they can enjoy the rays thereof through the glasses, so, when there is nothing more to do than to open the shutters, which may be performed in a very short time, and as soon shut again when the sun is clouded, the plants may have the benefit thereof whenever it appears; whereas, where there is so much trouble to uncover, and as much to cover again, it would take up the whole time in uncovering and shutting them up, and thereby the advantage of the sun's influence would be lost. Besides, where there is so much trouble required to keep out the frost, it will be a great chance if it be not neglected by the gardener; for if he be not as fond of preserving his plants, and as much in love with them as his master, this labour will be thought too great by him; and if he takes the pains to cover the glasses up with mats, &c. he will not care to take them away again until the weather alters, so that the plants will

will be ſhut up cloſe during the whole continuance of the froſt.

The back part and ends of the houſe withinſide ſhould be either laid over with ſtucco, and painted white, or plaſtered with good mortar, and white-waſhed, for otherwiſe the air in ſevere froſt will penetrate through the walls, eſpecially when the froſt is attended with a ſtrong wind, which is often the caſe in moſt ſevere winters.

Where Green-houſes are built in ſuch places as will not admit of rooms over them, they ſhould be contrived after the manner of ſtoves, with upright glaſſes in front, and ſloping glaſſes over theſe toward the back. If the building is well executed with proper flues in it, plants may be better preſerved therein, and have more air and ſun than in the moſt extenſive buildings of the other ſort.

In the Green-houſe there ſhould be moveable treſſels, which may be carried out and into the houſe occaſionally, upon which ſhould be fixed rows of planks for the pots, or tubs of plants to ſtand in regular rows one above another, whereby the heads of the plants may be ſo ſituated as not to interfere with each other. The loweſt row of plants, which ſhould be the forwardeſt towards the windows, ſhould be placed about four feet therefrom, that there may be a convenient breadth left next the glaſſes to walk in front; and at the backſide of the houſe there ſhould be allowed a ſpace of at leaſt four or five feet, for the convenience of watering and paſſing behind the plants. The plants ſhould never be crouded too cloſe to each other, but room left for the air to paſs freely between them.

To avoid the inconvenience which attends the placing of plants of very different natures in the ſame houſe, it will be very proper to have two wings added to the main Green-houſe, which, if ſituated in the manner expreſſed in the annexed plan, will greatly add to the beauty of the building, and alſo collect a greater ſhare of heat. In this plan the Green-houſe is placed exactly fronting the ſouth, and one of the wings faces the ſouth-ſouth-eaſt, and the other the ſouth-ſouth-weſt; ſo that from the time of the ſun's firſt appearance upon any part of the building, until it goes off at night, it is conſtantly reflected from one part to the other, and the cold winds are alſo better kept off from the front of the main Green-houſe hereby; and in the area of this place may be contrived a place where many of the moſt tender exotic plants, which will bear to be expoſed in the ſummer ſeaſon to be ſet abroad; and in the ſpring, before the weather will permit to ſet out the plants, the beds and borders of this area may be full of Anemonies, Ranunculuſes, early Tulips, &c. which will be paſt flowering, and the roots fit to take out of the ground by the time the plants can be carried out, which will render this place very agreeable during the ſpring ſeaſon, when the flowers are blown: and here a perſon may walk and divert himſelf in a fine day, when, perhaps, the air in moſt other parts of the garden will be too cold for thoſe who are not much uſed thereto.

In the center of this area may be contrived a ſmall baſon for water, which will be very convenient for watering of the plants, and add much to the beauty of the place; beſides, the water being thus ſituated, will be ſoftened by the heat which will be reflected from the glaſſes upon it, whereby it will be rendered much better than raw cold water for theſe tender plants.

The two wings of the building ſhould be contrived ſo as to maintain plants of different degrees of tenderneſs, which may be effected by the ſituation and manner of conducting the flues, a particular account of which will be exhibited under the article of STOVES. But I would here obſerve, that the wing facing the ſouth-ſouth-eaſt ſhould always be preferred as the warmeſt ſtove, its ſituation being ſuch, as that the ſun upon its firſt appearance in the morning, ſhines directly upon the glaſſes, which is of great ſervice in warming the air of the houſe, and adding life to the plants, after having been ſhut up during the long nights in the winter ſeaſon. Theſe wings, being in the draught annexed, allowed ſixty feet in length, may be divided in the middle by partitions of glaſs, with glaſs doors to paſs from one to the other. To each of theſe there ſhould be a fire-place, with flues carried up againſt the back wall, through which the ſmoke ſhould be made to paſs, as many times the length of the houſe, as the height will admit of their number; for the longer the ſmoke is paſſing before it is vented, the more heat will be given to the houſe, with a leſs quantity of fuel, which is an article worth conſideration, eſpecially where fuel is dear. By this contrivance you may keep ſuch plants as require the ſame degree of heat in one part of the houſe, and thoſe which will thrive in a much leſs warmth in the other part, but this will be more fully explained under the article of STOVES.

In building theſe wings, if there are not ſheds running behind them their whole length, the walls ſhould not be leſs than three bricks thick; and if they are more it will be better, becauſe where the walls are thin and expoſed to the open air, the cold will penetrate them, and when the fires are made, the heat will come out through the walls, ſo that it will require a larger quantity of fuel to maintain a proper temperature of warmth in the houſe. The back part of theſe houſes having ſloping roofs, which are covered either with tiles or ſlates, ſhould alſo be lined with Reeds, &c. under the covering, which will keep out the cold air, and ſave a great expence of fuel; for the cloſer and better theſe houſes are built, and the cloſer the glaſſes of the ſlope and front ſhut, the leſs fuel will be required to warm the houſes; ſo that the firſt expence in building theſe houſes properly will be the cheapeſt, when the after expence of fires is taken into conſideration.

The ſloping glaſſes of theſe houſes ſhould be made to ſlide and take off, ſo that they may be drawn down more or leſs, in warm weather, to admit air to the plants; and the upright glaſſes in front may be ſo contrived, as that every other may open as doors upon hinges, and the alternate glaſſes may be divided into two; the upper part of each ſhould be contrived ſo as to be drawn down like ſaſhes, ſo that either of theſe may be uſed to admit air in a greater or leſs proportion, according as there may be occaſion.

GREWIA. Lin. Gen. Plant. 914.

The CHARACTERS are,

The flower has a thick coloured empalement. It hath five petals, which are indented at their baſe, where is ſituated a ſcaly nectarium to each. It hath many briſtly ſtamina. In the center is ſituated the roundiſh germen, which is lengthened in form of a column, afterward becomes a four-cornered berry with four cells, each incloſing one globular ſeed.

The SPECIES are,

1. GREWIA (*Occidentalis*) foliis ſubovatis crenatis. *Grewia with oval crenated leaves.*

2. GREWIA (*Africana*) foliis ovato-lanceolatis ſerratis. *Grewia with oval ſpear-ſhaped leaves, which are ſawed.*

The firſt ſort has been long preſerved in many curious gardens both in England and Holland, and is figured by Dr. Plukenet by the title of Ulmifolia arbor Africana baccifera, floribus purpureis. It grows naturally at the Cape of Good Hope, from whence I have received the ſeeds.

This will grow to the height of ten or twelve feet; the ſtem and branches are very like thoſe of the ſmall-leaved Elm, the bark being ſmooth, and of the ſame colour as that when young; the leaves are alſo very like thoſe of the Elm,

Elm, and fall off in autumn; the flowers are produced singly along the young branches from the wings of the leaves, which are of a bright purple colour.

This may be propagated by cuttings or layers; if by cuttings, they should be taken off and planted in March, before the buds begin to swell, for they do not succeed so well after; these cuttings should be planted in small pots filled with loamy earth, and the pots plunged into a moderate hot-bed of tanners bark, and shaded from the sun in the middle of the day; these will take good root in about four months, and may then be gradually inured to bear the open air, into which they should be removed, and placed in a sheltered situation, where they may remain till autumn, when they must be removed into the green-house; the best time to lay down the layers of this plant is in the spring, before the buds come out, and these will be rooted by the same time the following year, when they may be cut off from the old plants, and planted each into a separate pot filled with soft loamy earth.

The best time to remove or transplant this plant is either in the spring, just before the buds begin to swell, or in autumn when the leaves begin to drop; for in summer when the plants are in full leaf, it will not be so proper to disturb them.

In winter these plants should be placed in the green-house, for they are too tender to live abroad in England; but they should have as much free air as possible in mild weather, as they only require to be protected from frost, and after the leaves are fallen, they will require very moderate watering, but in summer they should have it more constantly in dry weather.

The seeds of the second sort were sent me by Monf. Richard, gardener to the King of France at Marseilles, which were brought from Senegal in Africa, by Monf. Adanson; it rises in this country with a shrubby stalk five or six feet high, sending out many lateral branches, which are covered with a brown hairy bark, and garnished with oval spear-shaped leaves, having several transverse veins from the midrib to the sides, where they are sawed; they are placed alternately on the branches; the plants are young, so have not as yet flowered in England, therefore I can give no further account of them.

This sort is tender, so will not thrive in England, unless it is placed in a warm bark-stove, for the plants which have been placed on shelves in a dry stove have made little progress; therefore the only method to have them succeed, is to place them in the bark-bed in the tan-stove, where the plants have grown very well. In summer these plants require a good share of free air to be admitted to them, and should have water three or four times a week in warm weather; but in winter they must be sparingly watered, and require to be kept warm.

GRONOVIA. Martyn. Cent. 4. Lin. Gen. Plant. 248.

The CHARACTERS are,

The flower hath a permanent coloured empalement. It hath five small petals which are fixed to the cuts of the empalement, and five hairy stamina which are inserted into the empalement. The germen is situated under the flower, which afterward becomes a roundish coloured fruit with one cell, inclosing one large roundish seed.

We know but one SPECIES of this genus, viz.

GRONOVIA (*Scandens.*) Hort. Cliff. 74. Houst. *Climbing Burry Gronovia.*

This plant was discovered by the late Dr. Houstoun at La Vera Cruz. It is an annual plant, which sends forth many trailing branches like those of the Cucumber, which are closely set with broad green leaves, in shape like those of the Vine, which are covered with small spines on both sides, which sting like the Nettle; the branches have many tendrils or claspers, by which they fasten themselves to whatever plants they grow near, and will rise to the height of six or eight feet; the flowers are small, and of a greenish yellow colour, so make no great appearance.

This being a very tender plant, must be raised on a hotbed early in the spring, and afterward placed in the bark-stove, and treated in the same way as the Momordica, with which management it will produce ripe seeds; but this having neither use or beauty, is rarely cultivated but in botanic gardens for the sake of variety.

GROSSULARIA. Raii Meth. Plant. 145. Gooseberry.

The CHARACTERS are,

The flower has a permanent empalement, cut into five parts at the top, which is concave and coloured. It hath five small, obtuse, erect petals, which rise from the border of the empalement; and five awl-shaped stamina, which are inserted into the empalement. The germen is situated below the flower, and afterward becomes a globular berry having a navel, with one cell, which is filled with roundish compressed seeds included in a pulp.

This and the common Currant agreeing with each other in their characters, are by the botanists joined in the same genus; but I have chosen to treat of them separately, for the benefit of those who have not any knowledge in botany.

The SPECIES are,

1. GROSSULARIA (*Reclinata*) ramis reclinatis aculeatis, pedunculis triphyllis. *Prickly Gooseberry with a dark purplish fruit.*

2. GROSSULARIA (*Hirsuta*) ramis aculeatis, baccis hirsutis. *Gooseberry with prickly branches and hairy berries.*

3. GROSSULARIA (*Uva Crispa*) ramis aculeatis, erectis, baccis glabris. *Gooseberry with erect prickly branches, and smooth berries.*

4. GROSSULARIA (*Oxyacanthoides*) ramis undique aculeatis. *Gooseberry whose branches are armed on all sides with spines.*

5. GROSSULARIA (*Cynobati*) aculeis subaxillaribus, baccis aculeatis racemosis. *Gooseberry with spines on the lower part of the branches, and prickly berries growing in clusters.*

The sorts which are here enumerated are supposed to be distinct species, but there are several other varieties which have been obtained from seeds, and are propagated for sale in the nurseries; most of these are titled from the persons who raised them, as Lamb's Gooseberry, Hunt's Gooseberry, Edwards's Gooseberry, &c. and as there are frequently new varieties obtained, so it is needless to enumerate them here, therefore I shall proceed to their culture.

These are propagated either by suckers taken from the old plants, or by cuttings; the latter of which I prefer to the former, because those plants which are produced from suckers are always more disposed to shoot out a greater number of suckers from their roots, than such as are raised from cuttings, which generally form much better roots.

The best season for planting these cuttings is in autumn, just before their leaves begin to fall; observing always to take the handsomest shoots, and from such branches as generally produce the greatest quantity of fruit; for if you take those which are produced from the stem of the old plants (which are commonly very luxuriant,) they will not be near so fruitful as those taken from bearing branches: these cuttings should be about six or eight inches long, and must be planted in a border of light earth, exposed to the morning sun, about three inches deep; in the summer, when they have put out shoots, those near the bottom should be rubbed off, leaving only the uppermost or strongest, which should be trained upright to form a regular stem. In October following these plants will be fit to remove; at which time should be prepared an open spot of fresh earth, which should be well dug, and cleansed from noxious weeds, roots, &c. and being levelled the plants ſhould be taken up, and their roots trimmed; then plant them at three feet diſtance row from row, and one foot aſunder in the rows. In this place they may remain one year, being careful to keep them clear from all lateral ſhoots which are produced below the head of the plant, ſo that they may have clear ſtems about a foot in height above the ſurface of the earth, which will be tall enough: as the branches are generally produced very irregular, ſo ſuch of them as croſs each other, or where they are too cloſe, ſhould be cut out, whereby the head of the plant will be open, and capable of admitting the air freely into the middle, which is of great uſe to the fruit.

After theſe plants have remained in this nurſery one year, they will be fit to tranſplant to the places where they are deſigned to remain; for they ſhould not grow in the nurſeries too large, becauſe when their roots become woody, there is a hazard in removing of the plants. The ſoil in which theſe plants thrive to the greateſt advantage, is a rich light earth, though they will do very well upon middling ſoils, which are not too ſtrong or moiſt, and in moſt ſituations; but where the fruit is cultivated to have it in the greateſt perfection, they ſhould never be planted in the ſhade of other trees, but muſt have a free open expoſure. The diſtance they ought to be planted is eight feet row from row, and ſix feet aſunder in the rows, where they are planted for a full crop. The beſt ſeaſon for tranſplanting them is in October, when their leaves begin to decay; obſerving, as was before directed, to prune their roots, and trim off all lateral ſhoots, or ſuch as croſs each other, ſhortening all long branches, ſo as to make the head regular.

In pruning of theſe ſhrubs moſt people make uſe of garden ſhears, obſerving only to cut the head round, as is practiſed for ever-greens, &c. whereby the branches become ſo much crowded, that what fruit is produced never grows to half the ſize it would do, were the branches thinned and pruned according to art; which ſhould always be done with a pruning knife, ſhortening the ſtrong ſhoots to about ten inches, and cutting out all thoſe which grow irregular, thinning the fruit-bearing branches where they are too cloſe, obſerving always to cut behind a leaf bud. With this management the fruit will be near twice as large as thoſe which are produced upon ſuch buſhes as are not thus pruned, and the ſhrubs will continue in vigour much longer; but the ground ſhould be dug at leaſt once a year, and every other year a very little rotten dung ſhould be dug into the ground, which will greatly improve the fruit.

It is a common practice with the gardeners near London, who have great quantities of theſe buſhes in order to ſupply the markets, to prune them ſoon after Michaelmas, and then to dig up the ground between the rows, and plant it with Coleworts for ſpring uſe, whereby their ground is employed all the winter, without prejudicing the Gooſeberries; and in hard winters theſe Coleworts often eſcape, when thoſe which are planted in an open expoſure are deſtroyed; and theſe are generally pulled up for uſe in February or March, ſo that the ground is clear before the Gooſeberries come out in the ſpring: it is a piece of huſbandry well worth practiſing where ground is dear, or where perſons are confined for room.

GROVES are the greateſt ornaments to a garden, nor can a garden of any extent be complete which has not one or more of theſe. In ſmall gardens there is ſcarce room to admit of Groves, yet in theſe there ſhould be a few trees diſpoſed in imitation of a Grove.

Theſe Groves are not only great ornaments to gardens, but are alſo the greateſt relief againſt the violent heats of the ſun, affording ſhade to walk under in the hotteſt part of the day, when the other parts of the garden are uſeleſs; ſo that every garden is defective which has not ſhade.

Groves are of two ſorts, viz. open and cloſe Groves; open Groves are ſuch as have large ſhady trees, which ſtand at ſuch diſtances, as that their branches may approach ſo near each other, as to prevent the rays of the ſun from penetrating through them; but as ſuch trees are a long time in growing to a proper ſize for affording a ſhade, ſo where new Groves are planted the trees muſt be placed cloſer together, in order to have ſhade as ſoon as poſſible; but in planting of theſe Groves, it is much the beſt way to diſpoſe all the trees irregularly, which will give them a greater magnificence, and alſo form a ſhade ſooner, than when the trees are planted in lines; for when the ſun ſhines between the rows of trees, as it muſt do ſome part of the day in ſummer, the walks between them will be expoſed to the heat at ſuch times, until the branches of theſe trees meet; whereas in the irregular plantations, the trees intervene, and obſtruct the direct rays of the ſun.

When a perſon who is to lay out a garden is ſo happy as to meet with large full grown trees upon the ſpot, they ſhould remain inviolate if poſſible, for it will be better to put up with many inconveniencies than to deſtroy them; ſo that nothing but that of offending the habitation, by being ſo near as to occaſion great damps or obſtructing fine views, ſhould tempt the cutting of them down.

Moſt of the Groves which have been planted either in England, or in thoſe celebrated gardens of France, are compoſed only of a few regular lines of trees; but theſe do not appear ſo grand as thoſe which have been made in woods, where the trees have grown accidentally, and at irregular diſtances; where they have large ſpreading heads, and are left at ſuch diſtance as to permit the Graſs to grow under them, then they afford the greateſt pleaſure; for nothing is more noble than fine ſpreading trees, with large ſtems, growing through Graſs, eſpecially if the Graſs is well kept, and has a good verdure; beſides, moſt of theſe planted Groves have generally a gravel-walk, made in a ſtrait line between them, which greatly offends the ſight of perſons who have true taſte; therefore, whenever a gravel-walk is abſolutely neceſſary to be carried through theſe Groves, it will be much better to twiſt it about, according as the trees naturally ſtand, than to attempt regularity; but dry walks under large trees are not ſo uſeful as in open places, becauſe the dropping of the trees will render theſe walks uſeleſs after rain for a conſiderable time.

Cloſe Groves have frequently large trees ſtanding in them, but the ground is filled under theſe with ſhrubs, or under wood; ſo that the walks which are made in them are private, and ſkreened from winds, whereby they are rendered agreeable for walking, at ſuch times when the air is too violent or cold for walking in the more expoſed parts of the garden.

Theſe are often contrived ſo as to bound the open Groves, and frequently to hide the walls or other incloſures of the garden; and when they are properly laid out, with dry walks winding through them, and on the ſides of theſe ſweet-ſmelling ſhrubs and flowers irregularly planted, they have a charming effect; for here a perſon may walk in private, ſheltered from the inclemency of cold or violent winds, and enjoy the greateſt ſweets of the vegetable kingdom: therefore where it can be admitted, if they are continued round the whole incloſure of the garden, there will be a much greater extent of walk; and theſe ſhrubs will appear the beſt boundary, where there are not fine proſpects to be gained.

Theſe cloſe Groves are by the French termed Boſquets, from the Italian word Boſquetto, which ſignifies a little wood, and in moſt of the French gardens there are many of them

them planted; but theſe are reduced to regular figures, as ovals, triangles, ſquares, and ſtars, which have neither the beauty or uſe which thoſe have that are made irregularly, and whoſe walks are not ſhut up on each ſide by hedges, as thoſe in France generally are, which prevents the eye from ſeeing the tall trees growing in the quarters; and theſe want the fragrancy of the ſhrubs and flowers, which are the great delight of theſe private walks; add to this the keeping of the hedges in good order, which is attended with a great expence, which is the capital thing to be conſidered in the making of gardens.

GUAIABARA. See Coccolobos.

GUAJACANA. See Diospyros.

GUAJACUM. Plum. Nov. Gen. 39. tab. 17. Lignum Vitæ, or Pockwood.

The Characters are,

The flower hath a concave empalement. It hath five oblong, oval, concave petals, which are inſerted in the empalement, and ten erect ſtamina inſerted in the empalement. The germen, which is oval and pointed, afterward becomes a berry, which is roundiſh, with an oblique point and deeply furrowed, incloſing one oval hard ſeed.

The Species are,

1. Guajacum (*Officinale*) foliolis bijugatis obtuſis. Lin. Sp. Plant. 381. *Guajacum with obtuſe lobes placed by pairs.*

2. Guajacum (*Sanctum*) foliolis multijugatis obtuſis. Lin. Sp. Plant. 382. *Guajacum with many pair of obtuſe lobes.*

3. Guajacum (*Apum*) foliolis multijugatis obtuſis. Lin. Sp. Plant. 382. *Guajacum with many pair of blunt-pointed leaves.*

The firſt ſort is the common Lignum Vitæ, or Guajacum, which is uſed in medicine, which grows naturally in moſt of the iſlands in the Weſt-Indies, where it becomes a very large tree, having a hard, brittle, browniſh bark, not very thick; the wood is firm, ſolid, and ponderous, very reſinous, of a blackiſh yellow colour in the middle, and of a hot aromatic taſte; the ſmaller branches have an Aſh-coloured bark, and are garniſhed with leaves divided by pairs, each pair having two pair of ſmall, oval, blunt pinnæ, of a ſtiff conſiſtence, and a lucid green; the flowers are produced in cluſters at the end of the branches, compoſed of five oval concave petals, of a fine blue colour; in the center of theſe is fixed a ſtyle with an oval germen, crowned by a ſlender ſtigma; and round this is ſituated a great number of ſtamina, which are as long as the ſtyle, terminated by ſickle ſhaped ſummits. Dr. Linnæus ſuppoſes the flowers to have but ten ſtamina, whereas they certainly have twenty, ſo it ſhould be ranged in his twelfth claſs of plants; nor is it the empalement, but the germen which becomes the fruit.

The bark and wood of this tree are much of the ſame nature, only the wood is accounted hotter: they are uſed in diet drinks, to purify and cleanſe the blood, and to cauſe ſweating; they are eſteemed good for the gout and dropſy, the king's evil, and particularly for the French pox. The gum or reſin, which is black, ſhining, and brittle, and when powdered of a greeniſh white colour, of an aromatic ſmell, and poignant taſte, is ſomewhat cathartic, and a good purge in rheumatic caſes, to the quantity of two ſcruples mixed with the yolk of an egg, and given in a convenient vehicle.

The wood of this tree is ſo hard as to break the tools in felling them, ſo they are ſeldom cut down for fire-wood, being difficult to burn; but it is of great uſe to the ſugar planters, for making of wheels and cogs for the ſugar-mills, &c. It is alſo frequently brought to Europe, and wrought into bowls and other utenſils.

This tree can only be propagated by ſeeds, which muſt be procured from the countries where it naturally grows; they ſhould be freſh, otherwiſe they will not grow; they ſhould be ſown in pots, and plunged into a good hot-bed: if the ſeeds are good, and the bed in which they are plunged is of a proper temperature of heat, the plants will appear in ſix or eight weeks after, and in ſix weeks or two months more will be of ſtrength enough for tranſplanting; they ſhould be carefully taken out of the ſeed-pots, ſo as to preſerve their roots as entire as poſſible, and each planted in a ſeparate ſmall pot, and plunged into a new hot-bed of tanners bark, where they muſt be ſhaded from the ſun till they have taken freſh root; then they muſt be treated in the ſame manner as other tender exotic plants from warm countries, admitting a large ſhare of free air to them when the weather is warm. While the plants are young, they may be kept during the ſummer ſeaſon in a hot-bed of tanners bark under a frame; but in the autumn they muſt be removed into the bark-ſtove, and plunged into the hot-bed of tan, where they ſhould conſtantly remain, and muſt be treated in the ſame manner as other tender plants, being careful not to give them too much water in winter: in ſummer they ſhould have a large ſhare of free air admitted to them every day. With this treatment the plants will thrive very well, but they are plants of ſlow growth in their own country, ſo cannot be expected to make great progreſs in Europe.

The ſecond ſort hath many ſmall leaves placed along the midrib by pairs, which are rounded and obtuſe at their ends, but narrow at their baſe: they are of the ſame conſiſtence with thoſe of the former ſort, but of a darker green colour; the flowers are produced in looſe bunches toward the end of the branches, which are of a fine blue colour; their petals are fringed on their edges. This is called in ſome of the iſlands Baſtard Lignum Vitæ; I received it from Antigua by that title. It requires the ſame treatment as the firſt ſort, and is propagated by ſeeds in the ſame way.

I have alſo received ſpecimens from the iſland of Burbuda of one, which ſeems different from either of thoſe before-mentioned: the branches have the ſame appearance with thoſe of the firſt ſort, but the leaves are larger, indented at their extremities, and are placed all round the branches, on very ſhort foot-ſtalks; the flowers were broken off, ſo I cannot determine the difference between them, but by all appearance they ſeem to be of the ſame genus.

The third ſort has been long an inhabitant in ſome of the curious gardens in England and Holland, but ſeldom produces flowers here. This grows naturally at the Cape of Good Hope; however, Dr. Linnæus has removed it from the Acacia, where it had been placed, and has added it to this genus; and as I have not yet ſeen the flowers, ſo I do not know if it is right placed. The plants retain their leaves all the year, and will live in a good green-houſe in winter, but in ſummer muſt be placed abroad with other green-houſe plants. It is of ſlow growth, and is with difficulty propagated by layers.

GUAJAVA. See Psidium.

GUANABANUS. See Annona.

GUAZUMA. See Theobroma.

GUIDONIA. See Samyda.

GUILANDINA. Lin. Gen. Plant. 464. The Nickar-tree.

The Characters are,

The empalement of the flower is bell-ſhaped, cut at the rim into five equal parts; the flower has five concave ſpear-ſhaped petals which are equal, inſerted into the empalement, and ten awl-ſhaped ſtamina inſerted in the empalement. In the center is ſituated an oblong germen, which afterward becomes a rhomboid pod, with a convex ſuture on the upper ſide, having one cell, including oval hard ſeeds which are ſeparated by partitions.

The Species are,

1. Guilandina (*Bonduc*) aculeata, foliis bipinnatis, foliolis aculeis solitariis. Lin. Sp. Plant. 545. *Prickly Gualandinia with doubly winged leaves, whose lobes are set with spines; called yellow Nickar.*

2. Guilandina (*Bonducella*) aculeata, pinnis oblongo-ovatis foliolis aculeis geminis. Lin. Sp. Plant. 545. *Prickly Guilandina with oval oblong leaves; called gray Nickar.*

3. Guilandina (*Glabra*) inermis foliis bipinnatis, foliolis ovatis acutis alternis. *Smooth Guilandina with doubly-winged leaves.*

4. Guilandina (*Moringa*) inermis, foliis subpinnatis, foliolis inferioribus ternatis. Flor. Zeyl. 155. *Smooth Guilandina with winged leaves, whose under small leaves are trifoliate; called Morunga.*

5. Guilandina (*Dioica*) inermis foliis bipinnatis, basi apiceque simpliciter pinnatis. Lin. Sp. Plant. 545. *Guilandina with smooth branches, doubly winged leaves, but those at the top and bottom are singly winged; called Canada Nickar-tree.*

The first and second sorts grow naturally in most of the islands of the West Indies, where they twine their stalks about any neighbouring support, and rise to the height of twelve or fourteen feet. The leaves of the first sort are near a foot and a half long, composed of six or seven pair of pinnæ or wings, each of which has as many pair of lobes, or small leaves set along the midrib; these are oval and entire; the foot-stalk or principal midrib of the leaf is armed with short crooked thorns, which are placed irregularly; the stalks are also closely armed with the like thorns, which are larger. The stalks at first grow erect, but as they advance twine about the neighbouring trees or shrubs, being too weak to stand without support: the flowers come out in long spikes from the wings of the stalk, composed of five concave yellow petals, which are equal; in the center is situated the oblong germen, surrounded by ten stamina. After the flower is past the germen becomes a pod about three inches long and two broad, closely armed with slender spines, opening with two valves, each inclosing two hard seeds about the size of children's marbles, of a yellowish colour.

The second sort differs from the first in having much smaller leaves, which are set closer together; and below each pair of lobes are situated two short, stiff, crooked spines, which are placed opposite; the flowers are of a deeper yellow colour than those of the first sort, and the seeds are of an Ash colour.

The third sort was discovered by the late Dr. Houstoun at Campeachy, from whence he sent the dried samples to England, but there was no fruit on the trees at the time when he was there; but he mentions that this sort had an upright stem, which was of a large size, dividing into many branches, garnished with smooth double-winged leaves; the wings come out opposite, each leaf being composed of four pair, but the lobes are placed alternate upon the middle rib; they are oval, but end in a point, and are of a light green colour.

The fourth sort grows naturally in the island of Ceylon, and in several places on the Malabar coast. This in its native country rises to the height of twenty-five or thirty feet, with a strong stem, covered with a smooth bark, which in the young branches is green, but on the older it is of an Ash colour; the root grows knobbed and very thick. This, when young, is scraped and used by the inhabitants as Horse-radish in Europe, having much the same sharp taste; the branches are garnished with decompound winged leaves; those which are situated at the base have but three leaves, but above the leaves are branched out into several divisions, which are again divided into smaller, which have five or six pair of oval lobes, terminated by an odd one; they are of a light green, and a little hoary on their under side. The flowers are produced in loose bunches from the side of the branches, composed of an unequal number of petals, from five to ten; they have ten short stamina surrounding the germen, which afterwards turns to a long taper pod, including several angular seeds, covered with a thin membrane. These have a flavour like the root.

These four sorts are natives of warm countries, so will not live through the winter in England, unless they are placed in a warm stove, and the pots plunged into the tan-bed. They are propagated by seeds, but those of the two first sort are so hard, that unless they are soaked two or three days in water before they are put into the ground, or placed under the pots in the tan-bed to soften their covers, they will remain years in the ground without vegetating; when the plants come up they will be fit to transplant in a short time, when they should be each transplanted into a small pot, and plunged into a moderate hot-bed of tanners bark, shading them till they have taken fresh root; then they must be treated in the same manner as other tender exotic plants, giving them a large share of air in warm weather, and but little water; and when the plants have advanced to be too tall to remain in the frames, they must be removed into the bark-stove and plunged into the hot-bed, where they will make great progress, provided they have not too much water, especially during the winter season, for these plants are very impatient of moisture in cold weather.

The fourth sort requires the same treatment as those before-mentioned, but the seeds will grow without being steeped in water, and the plants are with difficulty shifted from one pot to another, for their roots are large, fleshy, and have but few fibres; so that unless great care is taken, all the earth will fall away from them, which often causes their stalks to decay almost to the root, and sometimes occasions the loss of the plants. This plant must be sparingly watered at all times, but particularly in cold weather, when moisture will cause them to rot in a short time.

The fifth sort grows naturally in Canada, from whence the plants were brought to Paris, where it has been some years cultivated, and a few years past it was brought to England. This in the country where it naturally grows, rises with an erect stem to the height of thirty feet or more, dividing into many branches, which are covered with a bluish Ash-coloured bark very smooth, garnished with large decompounded winged leaves which are of the oval shape, very smooth and entire, but are ranged alternate on the midrib; these fall off in the autumn, and new ones come out late in the spring.

There are male and female of this sort in different plants; as these have not as yet flowered in any of the English gardens, so I can give no farther account of it, nor of the fruit, having never seen either of them. This sort lives abroad in the open air, and is never hurt by the frost. It is only propagated by cutting off some of the roots and planting them in pots, which should be plunged into a gentle hot-bed, which will cause them to shoot upward, so may be taken from the old root and multiplied. It requires a light soil, not too moist.

GUNDELIA. Tourn. Cor. 51. tab. 486.

The Characters are,

It hath an uniform tubulous flower, composed of many hermaphrodite florets, which have but one petal slightly cut into five parts: they have five short hairy stamina. The oval germen is situated at the bottom of the flower, which afterward becomes a roundish single seed inclosed in the common receptacle, which is conical, and the seeds are separated by a chaffy down.

We have but one SPECIES of this genus, viz.

GUNDELIA (*Tournefortia.*) Lin. Sp. Plant. 814.

There is no English title to this plant, but there is a variety of it mentioned by Tournefort, which is supposed to arise from the same seed, as it was found growing promiscuously: which is,

GUNDELIA (*Glabra*) Orientalis, acanthi aculeati foliis, floribus intensè purpureis, capite araneosâ lanugine obsito. Tourn. Cor. 51. *Eastern Gundelia with prickly Bearsbreech leaves, deep purple flowers, and a head covered with a down like a cobweb.*

This plant was discovered by Dr. Gundelscheimer, in company with Tournefort, near Baibout in Armenia, but has since been found growing naturally in several places in the Levant. The stalks of this plant seldom rise more than two feet high; the under leaves are long, narrow, and sawed on their edges, their teeth ending in a spine; the other leaves are broader, which are irregularly slashed to the midrib, and armed at the points with sharp prickles; the stalks divide upward into several branches, which are armed with leaves of the same form, but are narrower, and each is terminated by a conical head of flowers, resembling those of Fuller's Thistle, being surrounded at the base by a circle of long, narrow, prickly leaves: these heads are composed of many hermaphrodite florets, which are shut up in the scales, each having a germen with five stamina surrounding it; but the seeds do not all ripen perfectly in each head, in the natural places of its growth. It has perfected seeds in the Chelsea garden.

These plants are propagated by seed, which should be sown the beginning of March in a warm border of fresh earth, in the places where the plants are designed to remain: for these plants have tap-roots which run very deep in the ground, so do not bear transplanting well. When the plants come up they should be carefully cleared from weeds, and as they grow large they should be thinned, leaving the plants which are designed to remain about two feet asunder, that they may have room to spread. After this there is no other culture required, but to keep them clear from weeds; in two years from seeds the plants will produce their flowers, which will make a fine appearance amongst other hardy plants in the pleasure-garden: the roots will continue several years in a dry soil.

GYPSOPHYLA. Lin. Gen. Plant. 498. We have no English title for this genus.

The CHARACTERS are,

The flower hath a permanent bell-shaped empalement, cut into five parts at the top. It hath five oval blunt petals, and ten awl-shaped stamina. In the center is situated a globular germen, which afterward becomes a globular capsule with one cell, opening with five valves, filled with small roundish seeds.

The SPECIES are,

1. GYPSOPHYLA (*Aggregata*) foliis mucronatis recurvatis, floribus aggregatis. Lin. Sp. Pl. 406. *Gypsophyla with pointed recurved leaves, and flowers gathered in a head.*

2. GYPSOPHYLA (*Fastigiata*) foliis lanceolato-linearibus, obsoletè triquetris lævibus obtusis secundis. Lin. Sp. Pl. 407. *Gypsophyla with narrow spear-shaped leaves, having three blunt angles, and smooth obtuse leaves in clusters.*

3. GYPSOPHYLA (*Prostrata*) foliis lanceolatis lævibus, caulibus diffusis, pistillis corollâ campanulatâ longioribus. Lin. Sp. Plant. App. 1195. *Gypsophyla with smooth spear-shaped leaves, diffused stalks, and the pointal longer than the petal, which is bell-shaped.*

4. GYPSOPHYLA (*Perfoliata*) foliis ovato-lanceolatis, semiamplexicaulibus. Lin. Sp. Plant. 408. *Gypsophola with oval spear-shaped leaves, half embracing the stalks.*

5. GYPSOPHYLA (*Paniculata*) foliis lanceolatis scabris, corollis revolutis. Lin. Sp. Plant. 407. *Gypsophyla with rough spear-shaped leaves, and the petals of the flowers recurved.*

The first sort grows naturally in the south of France, in Spain, and Italy, upon the mountains. This hath a perennial root, from which arise many narrow leaves ending in acute points, which are recurved; the stalks rise about a foot high, garnished with narrower leaves placed opposite, and at some of the joints there are smaller leaves growing from the stalks in clusters; the upper part of the stalk divides into smaller branches, each being terminated by a close bunch of small white flowers. These appear in July, and are succeeded by small oval capsules filled with small seeds.

The second sort is somewhat like the first, but the leaves are much narrower, and almost three-cornered; they are placed in clusters, which come out from the side of the stalk; the bunches of flowers are smaller, and not so closely joined. This hath a perennial root, and grows naturally upon the Helvetian mountains.

The third sort hath a perennial root, from which arise smooth spear-shaped leaves in clusters; the stalks are near a foot long, but lie prostrate upon the ground; the flowers have a purplish cast, and the stamina are much longer than the petals of the flowers. This flowers in June and July, and the seeds ripen in autumn.

The fourth sort grows naturally in the Levant, and also in Spain. It hath a strong, fleshy, fibrous root, which strikes deep in the ground, sending up several thick fleshy stalks, which rise two or three feet high, garnished with oval spear-shaped leaves which half embrace the stalks with their base; the upper part of the stalk divides into many smaller branches, which terminate with loose bunches of small white flowers, which make but little appearance.

The fifth sort grows naturally in Siberia and Tartary. This hath a perennial root, from which arise many branching stalks a foot and a half high, garnished with narrow smooth-pointed leaves shaped like those of Gilliflower; at the top of the stalks are produced loose clusters of very small white flowers, which appear at the same time with the former sorts, and the seeds ripen in autumn.

These plants have no great beauty, so are rarely cultivated but in botanic gardens for the sake of variety.

They are propagated by seeds, which should be sown in a bed of light earth, and when the plants are fit to remove they may be transplanted into the places where they are designed to remain, and will require no other culture but to keep them clean from weeds; for the roots will continue several years, and annually produce flowers and seeds.

H.

HÆMANTHUS. Tourn. Inst. R. H. 657. tab. 433. Blood-flower.

The CHARACTERS are,

The flower has a permanent empalement of six leaves, shaped like an umbel. It hath one erect petal, cut into six parts, and six awl-shaped stamina which are inserted in the petal. The germen is situated under the flower, which afterward becomes a roundish berry with three cells, each containing one triangular seed.

The SPECIES are,

1. HÆMANTHUS (*Coccineus*) foliis linguiformibus planis. Prod. Leyd. 42. *Blood-flower with plain tongue-shaped leaves.*

2. HÆMANTHUS (*Carinatus*) foliis longioribus carinatis. *Blood-flower with longer keel-shaped leaves.*

3. HÆMANTHUS (*Puniceus*) foliis lanceolato ovatis undulatis erectis. Hort. Cliff. 127. *Blood-flower with spear-shaped waved leaves.*

The first sort has been many years in several curious gardens in Europe. This hath a large bulbous root, from which in the autumn comes out two broad flat leaves of a fleshy consistence, shaped like a tongue, which turn backward on each side, and spread flat on the ground, so have a singular appearance all the winter; in the spring these decay, so that from May to the beginning of August they are destitute of leaves: this produces its flowers always in the autumn, just before the new leaves come out. The stalk rises a foot or more in height, supporting a cluster of bright red tubulous flowers, inclosed in a common leafy empalement, with one petal cut into six parts, each having six long stamina standing out beyond the petal, and in the center appears the germen sitting under the flower, supporting a single style, crowned with a stigma. The germen never ripens to a seed in England, but decays with the flower, and then the green leaves grow and spread on the ground.

The second sort hath a large bulbous root like the first, which sends out three or four leaves, which grow a foot long or more; these are not flat like those of the other, but are hollowed like the keel of a boat, and stand more erect than those of the former sort, but are not quite so broad; the flowers of this are like those of the first, but the stalk is taller, and the flowers are of a paler red; this is certainly a different species from the other, and their differences are permanent.

The third sort hath roots composed of many thick fleshy tubers, which join at the top, where they form a head, out of which arises a fleshy spotted stalk like that of the Dragon, which spreads out at the top into several spear-shaped leaves, waved on their edges. The stalks grow about a foot high; the leaves are six or eight inches long, and two broad in the middle; from the side of the stalk near the ground, breaks out a strong fleshy foot-stalk, about six or eight inches long, sustaining at the top a large cluster of flowers, included in one common empalement or covering, which is permanent; the flowers are shaped like those of the other sorts, but are of a yellowish red colour. These appear in May, June, or July, and are succeeded by berries, which are of a beautiful red colour when ripe.

The two first sorts do not propagate very fast in Europe, their roots seldom putting out many offsets; the gardens in Holland have been supplied with roots from the Cape of Good Hope, where they naturally grow and produce seeds; they are too tender to thrive in this country in winter, if planted in the full ground and exposed to the open air, therefore the roots are generally planted in pots, and in winter placed in a green-house, where by their large leaves spreading upon the pots, they make a pretty appearance, but with this treatment the roots seldom flower here: the only way to have the flowers in perfection, is to prepare a bed of good earth in a bricked pit, where they may be covered with glasses, and in hard frosts with mats and straw; the earth in the bed should be two feet deep, and the frame should rise two feet above the surface, to allow height for the flower-stems to grow. The roots should be planted nine or ten inches asunder, and in winter, if they are protected from frost, and not suffered to have too much wet, but in mild weather exposed to the open air, the roots will flower every year, and the flowers will be much stronger than with any other management.

The third sort is a native of the Cape of Good Hope; this may be propagated by parting of the roots in the spring, before the plants put out new stalks, which is also a right time to shift and new pot them; but as the roots do not multiply very fast in offsets, so the best way is to propagate them from seeds, which they ripen plentifully in England; these should be sown soon after they are ripe in pots, and kept in the stove all the winter; if these pots are plunged into the tan-bed in the bark-stove, in the vacancies between the plants, the seeds will be sooner prepared to vegetate in the spring, when the pots may be taken out of the stove and plunged into a moderate hot-bed, which will bring up the plants in a little time; soon after they are up, they must have air admitted to them every day in mild weather, to prevent their drawing up weak; and when they are fit to remove, they may be each planted in a separate small pot filled with light earth, and plunged into the hot-bed again to promote their taking new root; then they must be gradually hardened, and afterward may be removed into the dry stove, where they should constantly remain, otherwise the plants will not thrive and flower in this country. In the winter season they must not have too much wet, for as their roots are fleshy and succulent, so they are apt to rot with moisture. In the summer they must have a large share of air in warm weather, and require to be frequently watered, especially during the time of their flowering.

HÆMATOXYLUM. Lin. Gen. Plant. 417. Blood-wood, Logwood, or Campeachy Wood.

The CHARACTERS are,

The flower hath a permanent empalement, cut into five oval segments. It hath five oval petals, and ten awl-shaped stamina which

which are longer than the petals. In the center is situated an oblong oval germen, which afterward becomes a compressed obtuse capsule with one cell, opening with two valves, containing two or three oblong, flat, kidney-shaped seeds.

We have but one SPECIES of this genus, viz.

HÆMATOXYLUM (*Campechianum.*) Hort. Cliff. 160. *Logwood or Campeachy Wood.*

This tree grows naturally in the Bay of Campeachy, at Honduras, and other parts of the Spanish West-Indies, where it rises from sixteen to twenty-four feet high. The stems are generally crooked, and very deformed; they are seldom thicker than a man's thigh. The branches which come out on every side are crooked, irregular, and armed with strong thorns, garnished with winged leaves, composed of three or four pair of obtuse lobes, indented at the top. The flowers come in a racemus from the wings of the leaves, standing erect; they are of a pale yellowish colour, with a purple empalement, and are succeeded by flat oblong pods, each containing two or three kidney-shaped seeds.

The wood of this tree is brought to Europe, where it is used for dyeing purples, and for the finest blacks, so is a valuable commodity; but the Spaniards, who claim a right to the possession of those places where it naturally grows, are for excluding all other countries from cutting of the wood, which has occasioned many disputes with their neighbours, but particularly with the English: this it is to be hoped will soon be over, as there are some of the planters in Jamaica, and the other islands in America, belonging to the crown of Great-Britain, who have propagated this tree in so great plenty, as to have hopes of supplying the demand for this wood in Britain in a very few years; for the trees grow so fast there, as to be fit for use in ten or twelve years from seed; and as they produce great plenty of seeds in the British colonies, so those seeds scattering about, the plants come up in all the neighbouring lands, therefore will soon be like an indigenous plant of the country.

This plant is preserved in some curious gardens in England for the sake of variety. The seeds are frequently brought from America, which, if fresh, readily grow when sown upon a good hot-bed; and if the plants are kept in a moderate hot bed, they will grow to be upward of a foot high the same year; and, while the plants are young, they are generally well furnished with leaves, but afterward they make but little progress, and are frequently but thinly clothed with leaves. The plants are very tender, so should be constantly kept in the bark-stove, where, if they are duly watered, and the stove kept in a good degree of heat, the plants may be preserved very well. There are some of these plants now in England which are upward of six feet high, and as thriving as those in their native soil.

HALESIA.

The CHARACTERS are,

The flower has a permanent empalement of one leaf; the flower is bell-shaped, divided into four parts at the brim; it hath from twelve to sixteen stamina, terminated by oblong summits; the germen is situated below the flower, supporting a slender style, crowned by a simple stigma; the germen afterward becomes an oblong nut, having four angles with two cells, inclosing a single seed in each.

The SPECIES are,

1. HALESIA (*Tetraptera*) foliis lanceolato-ovatis petiolis glandulosis. Lin. Sp. 636. *Halesia with oval spear-shaped leaves and glandulous foot-stalks.*

2. HALESIA (*Diptera*) foliis ovatis, petiolis lævibus. Lin. Sp. 636. *Halesia with oval leaves and smooth foot-stalks.*

Both these trees grow naturally in South Carolina, but are hardy enough to thrive in the open air in England. The first sort often rises with two or three stems from one root, which rise near twenty feet high, sending out many branches, which are garnished with oval spear-shaped leaves, sawed on their edges; the flowers are produced from the side of the branches in clusters.

The second sort has a great resemblance to the first, but the leaves are oval and their foot-stalks are smooth; the fruit has but two angles.

They are both propagated by seeds, when they can be procured from the countries where they naturally grow; but as the seeds remain a year in the ground, so the plants are generally propagated by layers, which, if properly laid, and in dry weather are supplied with water, will put out roots sufficient for transplanting in one year, and toward the end of March they may be planted where they are to remain.

HALICACABUM. See PHYSALIS.

HALICACABUS PEREGRINA. See CARDIOSPERMUM.

HALIMUS. See ATRIPLEX.

HALLERIA. Lin. Gen. Pl. 679. African Fly Honeysuckle.

The CHARACTERS are,

The flower hath a permanent empalement of one leaf. It hath one petal of the grining kind, whose chaps are swollen and inflexed. It hath four stamina, which are bristly, two being longer than the other. In the bottom of the tube is situated an oval germen, which afterward becomes a roundish berry with two cells, each containing one hard seed.

We have but one SPECIES of this genus, viz.

HALLERIA (*Lucida.*) Hort. Cliff. 323. This plant has its title from Dr. Haller, who was professor of botany at Gottingen in Germany. *African Fly Honeysuckle.*

The English name which I have here added, has been given to this plant by some gardeners, who observed that the shape of the flower had some resemblance to that of the Upright, or Fly Honeysuckle, and, for want of an English name, gave this to it.

This plant grows to the height of six or eight feet, with a woody stem, which is well furnished with branches, garnished with oval sawed leaves placed opposite, which continue green through the year; the flowers come out singly, and are of a red colour, but being intermixed with the leaves, are not seen unless they are looked after, for they grow scatteringly on the branches; these come out in June, and the seeds ripen in September: the plants make a variety in the green-house during the winter season.

It may be propagated by cuttings, which, if planted in pots in the spring, and plunged into a gentle hot-bed, will soon take root; these plants must be exposed in summer, and will require plenty of water in that season; in the winter they must be housed with Myrtles, and other hardy exotic plants, which require a large share of air in mild weather.

HAMAMELIS. Lin. Gen. Pl. 155. The Witch Hazel.

The CHARACTERS are,

It is male and female in different plants; the male flowers have a four-leaved empalement, four narrow petals, which are reflexed, and four narrow stamina which are shorter than the petals. The female flowers have a four-leaved involucrum, in which are four flowers, with a four-leaved coloured empalement; they have four narrow petals, and four nectariums adhering to the petals. In the center is situated an oval hairy germen, which afterward becomes an oval capsule sitting in the involucrum, having two cells, each containing one hard, oblong, smooth seed.

We have but one SPECIES of this genus, viz.

HAMAMELIS (*Virginiana.*) Flor. Vir. 129. *The Witch Hazel.*

This plant grows naturally in North America, from whence the seeds have been brought to Europe, and many of the plants have been raised in the English gardens, where they are propagated for sale by the nursery gardeners. It hath a woody stem from two to three feet high, sending out many slender branches, garnished with oval leaves, indented on their edges, having great resemblance to those of the Hazel; they fall away in autumn, and when the plants are destitute of leaves, the flowers come out in clusters from the joints of the branches; these sometimes appear the latter end of October, and often not till December, but are not succeeded by seeds in this country.

As the flowers of this shrub make very little appearance, so it is only preserved in the gardens of the curious, more for the sake of variety than its beauty.

It is propagated by laying down the young branches in autumn, which will take root in one year, and may then be taken from the old plants, and planted where they are to remain. The seeds of this plant always remain a whole year in the ground, so they should be sown in pots, which may be plunged into the ground in a shady part of the garden, where they may remain all the summer, and will require no other care but to keep the pots clean from weeds, and in very dry weather to water them now and then; in autumn the pots may be removed to a warmer situation, and plunged into the ground under a warm hedge, and if the winter should prove very severe, they should have some light covering thrown over them, which will secure the seeds from being destroyed. In the spring the plants will come up, and as the season advances the pots may be removed where they may have the morning sun till eleven o'clock. In the autumn they should be transplanted, either into small pots or in a nursery-bed, where in one, or at most two years time, they will be strong enough to plant where they are designed to remain; they love a moist soil and a shady situation.

HARMALA. See Peganum.

HAWTHORN. See Mespilus.

HAZEL. See Corylus.

HEDERA. Lin. Gen. Plant. 249. The Ivy-tree.

The Characters are,

The flowers are disposed in form of a corymbus. The empalement is cut into five parts, and sit upon the germen. The flower hath five oblong petals, and five awl-shaped stamina. The germen, which is situated below the flower, afterward becomes a globular berry with one cell, inclosing four or five large seeds, convex on one side, and angular on the other.

The Species are,

1. Hedera (*Helix*) foliis ovatis lobatisque. Flor. Lapp. 91. *Ivy with oval and lobed leaves; common Ivy.*

2. Hedera (*Quinquefolia*) foliis quinatis, ovatis, serratis. Hort. Cliff. 74. *Ivy with leaves composed of five lobes, which are sawed; commonly called Virginia Creeper.*

The first sort grows naturally in most parts of England, and where it meets with any neighbouring support, the stalks will fasten to it, and rise to a very great height, sending out roots on every side, which get into the joints of walls, or the bark of trees, and thereby are supported; or if there is no support near, the stalks trail upon the ground, and take root all their length, so that they closely cover the surface, and are difficult to eradicate; for where any small parts of the stalks are left, they will soon spread and multiply. While these are fixed to any support or trail upon the ground, their stalks are slender and flexible; but when they have reached to the top of their support, they shorten and become woody, forming themselves into large bushy heads, and their leaves are larger, more of an oval shape, and not divided into lobes like the lower leaves, so that it hath a different appearance, which has occasioned some to take them for distinct species.

There are two varieties of this, one with silver striped leaves, and the other with yellowish leaves on the top of the branches; these are preserved in some gardens for the sake of variety.

These plants are easily propagated by their trailing branches, which send forth roots their whole length; which branches being cut off and planted, will grow in almost any soil or situation, and may be trained up to stems, or suffered to remain as climbers, to cover walls, pales, &c.

They may also be propagated by seeds, which should be sown soon after they are ripe, which is in the beginning of April; if these are kept moist and shaded, they will grow the same spring, otherwise they will remain a year in the ground, therefore few persons trouble themselves to propagate the plants in this way, the other being much more expeditious.

While the stalks of this plant trail either on the ground or upon walls, or other support, they seldom produce flowers, which has occasioned its being called sterile, or barren Ivy; but when the branches get above their support, or grow from it, they produce flowers at the end of every shoot: these appear in September, and are succeeded by berries, which turn black before they are ripe, and are formed into round bunches, which are called corymbi, and from these the epithet of corymbus, so frequently used by botanists, is taken.

There is another species of Ivy, which grows naturally about Constantinople, and other parts of the Levant, with yellow berries, titled Hedera Poetica by Caspar Bauhin; but as I have not seen this plant, I cannot give a farther account of it. Dr. Linnæus supposes it to be only a variety, though he has not seen the plant; but Tournefort, who gathered it in the Levant, puts it down as a different sort.

The second sort grows naturally in all the northern parts of America; it was first brought to Europe from Canada, and has been long cultivated in the English gardens chiefly to plant against walls, or buildings to cover them, which these plants will do in a short time, for they will shoot almost twenty feet in one year, and will mount up to the top of the highest building; but as the leaves fall off in autumn, the plants make but an indifferent appearance in winter; and as it is late before they come out in the spring, they are not much esteemed, unless it is for such situations where better things will not thrive; for this plant will thrive in the midst of London, and is not injured by smoke, or the closeness of the air, so is very proper for such situations. The stalks of the plants put out roots, which fasten themselves into the joints of the walls, whereby they are supported.

It may be propagated by cuttings, which, if planted in autumn on a shady border, will take root, and by the following autumn will be fit to plant where they are designed to remain.

HEDERA TERRESTRIS. See Glechoma.

HEDGES. Hedges are either planted to make fences round inclosures, or to part off and divide the several parts of a garden; when they are designed for outward fences, they are planted either with Hawthorn, Crabs, or Black Thorn, which is the Sloe; but those Hedges which are planted in gardens, either to surround wilderness quarters, or to screen the other parts of a garden from sight, are planted with various sorts of plants, according to the fancy of the owner, some preferring evergreen Hedges, in which case the Holly is best, next the Yew, then Laurel, Lauruſtinus, Phillyrea, &c. Others, who make choice of the deciduous plants, prefer the Beach and Hornbeam, Engliſh Elm, or the Alder, to any other; I ſhall firſt treat of thoſe Hedges which are planted for outſide fences, and afterward briefly touch on the other

Theſe Hedges are moſt commonly made of Quick, yet it will be proper, before planting, to conſider the nature of the land, and what ſort of plants will thrive beſt in that ſoil, whether it be clay, gravel, ſand, &c. As for the ſize, the ſets ought to be about the bigneſs of a gooſe quill, and cut within about four or five inches of the ground; they ſhould be freſh taken up, ſtrait, ſmooth, and well rooted. Thoſe plants which are raiſed in the nurſery are to be preferred to all others, and if raiſed on a ſpot near the place, it will be beſt.

Secondly, If the Hedge has a ditch, it ſhould be made ſix feet wide at top, and one foot and a half at bottom, and three feet deep, that each ſide may have a proper ſlope; for when the banks are made too upright, they are very ſubject to fall down after every froſt or hard rain; beſides, if the ditches are made narrower, they are ſoon choked up in autumn by the falling leaves and the growth of weeds, nor are they a ſufficient fence to the Hedge againſt cattle, where they are narrower.

Thirdly, If the bank be without a ditch, the ſets ſhould be ſet in two rows almoſt perpendicular, at the diſtance of a foot from each other, in the quincunx order, ſo that in effect they will be but ſix inches aſunder.

The uſual method of planting Quick Hedges is to lay the plants ſloping on the ſide of the bank, in two rows, one above the other, which is by no means right; for the wet cannot get to the roots of the Quick when they are planted in ſuch poſition, for the ſlope of the bank throws it off: therefore I recommend it as the beſt method to plant the ſets upright upon the top of the bank, where they will be farther from the reach of the cattle; and if they are duly cleared from weeds, they will grow as much in one year, as thoſe which are laid on the ſide of the bank will do in three. But if there are not two ditches, one on each ſide the bank to face againſt cattle, there ſhould be a dead Hedge made within for that purpoſe; otherwiſe where cattle are admitted into the field, they will browſe upon the young ſhoots of the Quick, and ſpoil the Hedge.

In making of theſe dead Hedges, there ſhould be ſtakes driven into the earth at about two feet and a half diſtance, ſo low as to reach the firm ground.

Oak ſtakes are the beſt, and Black Thorn the next; then let the ſmall buſhes be laid at bottom, but not too thick, for that will cauſe the buſhes to rot; but the upper part of the Hedge ſhould be laid with long buſhes to bind the ſtakes in with, by interweaving them.

And, in order to render the Hedge yet ſtronger, you may edder it, (as it is called,) i. e. bind the top of the ſtakes in with ſome ſmall poles on each ſide, and when the eddering is finiſhed, drive the ſtakes a-new, becauſe the waving of the Hedge and eddering is apt to looſen the ſtakes.

When a Hedge is of about eight or nine years growth, it will be proper to plaſh it; the beſt time for this work is either in October or February.

In plaſhing Quicks there are two extremes to be avoided; the firſt is, laying it too low and too thick; becauſe it makes the ſap run all into the ſhoots, and leaves the plaſhes without nouriſhment, which, with the thickneſs of the Hedge, kills them.

Secondly, It muſt not be laid too high, becauſe this draws all the ſap into the plaſhes, and ſo cauſes but ſmall ſhoots at the bottom, and makes the Hedge ſo thin, that it will neither hinder the cattle from going through, nor from cropping of it.

If the ſtems are very old, cut them quite down, and ſecure them with good dead Hedges on both ſides, till the young ſhoots are got up tall enough to plaſh, and plant new ſets in the void ſpaces.

If you would have a good Hedge or fence, you ſhould new lay it once in fourteen or fifteen years, and conſtantly root out Elder, Travellers Joy (which ſome call Bull bine,) Briony, &c. and do not leave high ſtandards or pollards in it, nor any dead wood is to be left in the bottom of the Hedges, for that will choke the Quick; but if there be a gap, the dead Hedge ſhould be made at a diſtance.

The Crab is alſo frequently planted for Hedges, and if the plants are raiſed from the kernels of the ſmall wild Crab, they are much to be preferred to thoſe which are raiſed from the kernels of all ſorts of Apples, without diſtinction, becauſe the plants of the true ſmall Crab never ſhoot ſo ſtrong as thoſe of the Apples, ſo may be better kept within the proper compaſs of a Hedge; and as they have generally more thorns upon them, they are better guarded againſt cattle, &c. than the other; beſides, the plants of the Crab will grow more equal than thoſe which are raiſed from the kernels of various kinds of Apples, for theſe always produce a variety of plants, which differ from each other in their manner of growth, as much as in the ſize and flavour of their fruits, ſo that Hedges made of theſe will not appear ſo neat, nor can be ſo well managed as the other.

The Black Thorn, or Sloe, is alſo frequently planted for Hedges, and is a ſtrong durable plant for that purpoſe, eſpecially as it is ſo ſtrongly armed with thorns that cattle ſeldom care to browſe upon it; but where this is planted, the beſt way is to raiſe the plants from the ſtones of the fruit; for all thoſe which are taken from the roots of old trees ſpawn, and put out ſuckers in ſuch plenty from their roots as to ſpread over, and fill the neighbouring ground to a conſiderable diſtance on each ſide of the Hedge; and this plenty of ſuckers drawing away the nouriſhment from the old plants of the Hedge, they never grow ſo well as where there are few or no ſuckers produced, which thoſe plants which are propagated from the ſtones ſend not forth, or at leaſt but ſparingly, therefore may with little trouble be kept clear of them. The beſt method of raiſing theſe Hedges is, to ſow the ſtones in the place where the Hedge is intended (where it can be conveniently done,) for then the plants will make a much greater progreſs than thoſe which are tranſplanted; but the objection to this method will ariſe from the difficulty of ſecuring the young plants from the cattle; but this can have little force, when it muſt be conſidered, that if the Hedge is planted, it muſt be fenced for ſome years, to prevent the cattle from deſtroying it; therefore the ſame fence will do for it when ſown, nor will this require a fence much longer than the other. For the plants which ſtand unremoved, will make a better fence in ſeven years than that which is planted, though the plants ſhould be of three or four years growth when planted. The ſtones of this fruit ſhould be ſown early in January if the weather will permit, but when they are kept out of the ground longer it will be proper to mix them with ſand, and keep them in a cool place. The buſhes of the Black Thorn are by much the beſt of any for making of dead Hedges, being of longer duration, and having many thorns, neither the cattle nor the Hedge-breakers will care to meddle with them; theſe buſhes are alſo the beſt to be uſed for under ground drains, for the draining of land, for they will remain ſound a long time when the air is excluded from them.

The Holly is ſometimes planted for Hedges, and is a very durable ſtrong fence; but where it is expoſed there will be great difficulty to prevent its being deſtroyed, otherwiſe it is by far the most beautiful plant; and being an evergreen, will afford much better shelter to cattle in winter than any other sort of Hedge, and the leaves being armed with thorns, the cattle will not care to browse upon it. Another objection to this plant is the slow growth, so that Hedges planted with this plant require to be fenced a much longer time than most others. This is a reason which must be admitted; but in such grounds as lie contiguous to, or in sight of gentlemen's houses, these sort of Hedges will have an exceeding good effect, especially when they are well kept, as they will appear beautiful at all seasons of the year; and in the spring of the year, when the sharp winds render it unpleasant to walk abroad in exposed places, these Hedges will afford good shelter, and will appear beautiful at all seasons of the year; they will also effectually keep off the cold winds, if they are kept close and thick. The surest method of raising these Hedges is, by sowing the berries in the places where they are to stand; but these berries should be buried in the ground one year before they are sown, by which method they will be prepared to grow the following spring The way of doing this is, to gather the berries about Christmas (which is the time they are usually ripe,) and put them into larger sized flower-pots, mixing some sand with them; then dig holes in the ground, into which the pots must be sunk, covering them over with earth about a foot thick; in this place they may remain till the following October, when they should be taken up, and sown in the place where the Hedge is intended. The ground for this Hedge should be well trenched, and cleared from the roots of all bad weeds, bushes, trees, &c. Then two drills should be made at about a foot distance from each other, and about two inches deep, into which the seeds should be scattered pretty close, lest some should fail: for it is better to have too many plants come up, than to want. The reason of my advising two drills is, that the Hedge may be thick to the bottom, which in a single row rarely happens, especially if there is not great care taken of them in the beginning. When the plants come up they must be carefully weeded, for if the weeds are permitted to grow among them, they will soon destroy them, or weaken them so much that they will not recover their strength in a long time.

When these Holly Hedges are designed to be kept very neat, they should be sheared twice a year, in May and August; but if they are only designed as fences, they need not be sheared oftener than once a year, which should be about the latter end of June or the beginning of July; and if this is well performed, the Hedges may be kept very beautiful.

When a Hedge of Holly is intended to be made by plants, the ground should be well trenched, as was before advised for the seeds, and (unless the ground be very wet) the plants should be set in October, but in wet ground March is preferable. The plants should not be taken from a better soil than that in which they are to be planted; for when it so happens, they are much longer before they recover this change than those are which are taken from a leaner soil. If the plants have been before removed two or three times, they will have better roots, and will be in less danger of miscarrying, because they may be removed with balls of earth to their roots. When the frost comes on, if some mulch be laid upon the ground near the roots of the plants, it will prevent the tender fibres, which may then have been put out, from being destroyed by the cold. I would never advise the planting of Hedges with Holly plants, of about four or five years growth from the berries; for when the plants are older, if they take to grow, they seldom make so close a fence at the bottom as young ones; and if the plants have been twice before transplanted, they will more certainly grow.

In the old method of laying out gardens, it was a general practice to surround the wilderness quarters and other parts of the garden with Hedges of evergreen or deciduous trees, which were then esteemed ornamental; but since a better taste has been introduced, these have rarely been admitted. For if the expence of keeping Hedges in good order, together with the litter occasioned whenever they are sheared be considered, it will be found to greatly exceed any pleasure arising from them; therefore as these are not likely to take place again in the English gardens, it is needless to give farther directions for planting and the after-management of them, especially as the instructions here given for the raising Holly Hedges may, with a little variation, serve for any other sort.

HEDYPNOIS. See HYOSERIS.

HEDYSARUM. Lin. Gen. Plant. 793. French Honeysuckle.

The CHARACTERS are,

The flower hath a permanent empalement of one leaf. It is of the butterfly kind; the wings are oblong and narrow; the keel is compressed, and convex at the base. It hath nine stamina joined, and one standing separate. In the center is situated a long narrow germen, which afterward becomes a jointed compressed pod, each joint being roundish, and incloses a single kidney-shaped seed.

The SPECIES are,

1. HEDYSARUM (*Coronarium*) foliis pinnatis, leguminibus articulatis aculeatis, nudis, rectis, caule diffuso. Hort. Cliff. 365. *French Honeysuckle with winged leaves, naked, prickly, jointed pods, and a diffused stalk.*

2. HEDYSARUM (*Spinosissimum*) foliis pinnatis, leguminibus articulatis aculeatis, tomentosis, caule diffuso. Hort. Upsal. 231. *French Honeysuckle with winged leaves, jointed, prickly, woolly pods, and a diffused stalk.*

3. HEDYSARUM (*Canadense*) foliis simplicibus ternatisque, floribus racemosis. Hort. Cliff. 232. *French Honeysuckle with single and trifoliate leaves, and flowers in bunches; called French Honeysuckle of Canada.*

4. HEDYSARUM (*Flexuosum*) foliis pinnatis, leguminibus articulatis, aculeatis, flexuosis, caule diffuso. Lin. Sp. Plant. 750. *French Honeysuckle with winged leaves, jointed prickly pods which are waved, and a diffused stalk.*

5. HEDYSARUM (*Diphyllum*) foliis binatis petiolatis, floralibus sessilibus. Flor. Zeyl. 291. *French Honeysuckle with two leaves upon a foot-stalk, and those upon the flower-stalks sitting close.*

6. HEDYSARUM (*Purpureum*) foliis ternatis, foliolis obovatis, floribus paniculatis terminalibus, leguminibus intortis. *French Honeysuckle with trifoliate oval leaves, flowers growing in panicles at the ends of the stalks, and intorted pods.*

7. HEDYSARUM (*Canescens*) foliis ternatis subtus nervosis, caule glabro fruticoso decumbente floribus spicatis terminalibus. *Three-leaved French Honeysuckle with veins on their under side, a smooth, shrubby, declining stalk, with flowers growing in spikes at the ends.*

8. HEDYSARUM (*Sericeum*) foliis ternatis, foliolis ovatis subtus sericeis, floribus spicatis alaribus terminalibusque. *Three-leaved French Honeysuckle with oval leaves, sattiny on their under side, and flowers in spikes from the side and at the end of the stalks.*

9. HEDYSARUM (*Villosum*) foliis ternatis, caulibus diffusis villosis, floribus spicatis terminalibus, calycibus villosissimis. *Three-leaved French Honeysuckle with diffused stalks which are hairy, flowers growing in spikes at the ends of the branches, and very hairy empalements.*

10. HEDYSARUM (*Procumbens*) foliis ternatis, caulibus procumbentibus racemosis, floribus laxe spicatis terminalibus, leguminibus contortis. *Three-leaved French Honeysuckle with branching trailing stalks, flowers growing in loose spikes at the ends of the branches, and twisted pods.*

11. HEDYSARUM (*Intortum*) foliis ternatis, foliolis obcordatis, caule erecto triangulo villoso, racemis terminalibus, leguminibus articulatis incurvis. *French Honeysuckle with trifoliate leaves, whose lobes are heart-shaped, a triangular, upright, hairy stalk, flowers growing in long bunches at the ends of the branches, and jointed incurved pods.*

12. HEDYSARUM (*Glabrum*) foliis ternatis obcordatis, caule paniculato, leguminibus monospermis glabris. *French Honeysuckle with trifoliate heart-shaped leaves, a paniculated stalk, and smooth pods containing one seed.*

13. HEDYSARUM (*Scandens*) foliis ternatis, foliolis obverse-ovatis, caule volubili, spicâ longissimâ reflexâ. *Three-leaved French Honeysuckle with obverse oval lobes, a twining stalk, and a very long reflexed spike of flowers.*

14. HEDYSARUM (*Repens*) foliis ternatis obcordatis, caulibus procumbentibus villosis, pedunculis unifloris. *Three-leaved French Honeysuckle with oval heart-shaped leaves, trailing hairy stalks, and foot-stalks with a single flower.*

15. HEDYSARUM (*Maculatum*) foliis simplicibus ovatis obtusis. Hort. Cliff. 449. *French Honeysuckle with oval, obtuse, single leaves.*

16. HEDYSARUM (*Frutescens*) foliis ternatis ovato-lanceolatis, subtus villosis, caule frutescente villoso. *Trifoliate French Honeysuckle with oval spear-shaped leaves, hairy on their under side, and a shrubby hairy stalk.*

17. HEDYSARUM (*Pedunculatum*) foliis ternatis, foliolo intermedio pediculo longiore, racemis alaribus erectis longissimis. *French Honeysuckle with trifoliate leaves, the middle lobe standing on a longer foot-stalk, and very long bunches of flowers coming from the sides of the stalks.*

18. HEDYSARUM (*Alhagi*) foliis simplicibus lanceolatis obtusis, caule fruticoso spinoso. Lin. Sp. Pl. 745. *French Honeysuckle with single, spear-shaped, obtuse leaves, and a prickly shrubby stalk; or the Alhagi of the Moors.*

19. HEDYSARUM (*Triquetrum*) foliis simplicibus, cordato-oblongis integerrimis glabris. *French Honeysuckle with single, oblong, heart-shaped leaves, which are smooth and entire.*

The first sort has been long cultivated in the English gardens for ornament. It grows naturally in Italy; there are two varieties of this, one with a bright red, and the other a white flower, which very rarely vary from one to the other; but as there is no other difference but in the colour of their flowers, so they are allowed to be the same species.

It is a biennial plant, which decays after the seeds are ripe. This sends up several hollow smooth stalks, which branch out, and rise from two to three feet high, garnished with winged leaves, composed of five or six pair of oval lobes, terminated by an odd one; from their base comes out foot-stalks, which are five or six inches long, sustaining spikes of beautiful red flowers, which are succeeded by compressed prickly-jointed pods; in each of the joints is lodged one kidney-shaped seed. This sort flowers in June and July, and the seeds ripen in September. The white is only a variety of this, and as such is sometimes preserved in gardens.

They are propagated by sowing their seeds in April in a bed of light fresh earth, and when the plants are large enough to remove, they should be transplanted into other beds in an open situation, at about six or eight inches distance from each other, leaving a path between every four rows, to go between them to hoe, and clear them from weeds. In these beds they may remain until Michaelmas, when they may be transplanted into the large borders of a parterre or pleasure-garden, allowing them at least three feet distance from other plants, amongst which they should be interspersed, to continue the succession of flowers; where they will make a fine appearance when blown, especially, the red sort, which produces very beautiful flowers.

As these plants decay after they have perfected their seeds, so there should annually be a fresh supply of plants raised. They are very proper ornaments for large borders, or to fill up vacancies among shrubs, but they grow too large for small borders.

The second sort is an annual plant, which grows naturally in Spain and Portugal. The leaves of this are narrow and oblong, four or five pair being placed along the mid-rib, with an odd one at the end; the stalks are terminated by small spikes of purple flowers, which are succeeded by small rough pods, shaped like those of th former sort. This plant is preserved in botanic gardens or the sake of variety; it is propagated by seeds, which should be sown the beginning of April, in the place where the plants are to remain, and will require no other culture but to thin them where they are too near, and keep them clean from weeds.

The third sort hath a perennial root, which will abide many years if planted in a dry soil. This is propagated by sowing the seeds in the manner directed for the former; but when the plants are come up two inches high, they should be transplanted where they are to remain for good; but if they are not too thick in the seed-bed, they may be suffered to remain there until the following autumn, at which time they should be carefully taken up and transplanted into the borders where they are designed to stand; for their roots generally run down very deep, so that it is not safe to remove them after they are large.

The fourth sort is an annual plant, which grows naturally in the Levant. This hath some resemblance of the first, but is much smaller; the stalks rise near a foot high, garnished with winged leaves, composed of two or three pair of oval lobes, terminated by an odd one; the flowers come out in spikes at the top of the stalks, which are of a pale red, intermixed with a little blue. This is propagated in the same way as the second sort, and is equally hardy.

The fifth sort grows naturally in both Indies. This is an annual plant, with a long tap root which runs deep in the ground, sending out one or two stalks, which rise about nine inches high, the lower part being garnished with oval leaves by pairs on each foot-stalk, but the upper part of the stalk, where the flowers come out, is garnished with small leaves ending in acute points, sitting close to the stalks; at each of these is situated a single, small, yellow flower, inclosed by the two leaves, and are succeeded by oblong pods, containing one kidney-shaped seed.

The sixth sort grows naturally at La Vera Cruz, and also in Jamaica It is an annual plant, which rises with a shrubby stalk upward of four feet high, dividing into several branches, garnished with oblong, oval, trifoliate leaves, standing upon pretty long foot-stalks, the middle lobe standing an inch beyond the other two; the branches are terminated by long loose panicles of purple flowers, which are succeeded by narrow, jointed, twisted pods. This flowers in July, and the seeds ripen in autumn.

The two last mentioned are tender plants, so their seeds must be sown in the spring upon a hot-bed; and when the plants are fit to remove, they should be each planted in a separate small pot, and plunged into a hot-bed, shading them from the sun till they have taken new root; then they must be treated in the same way as other tender plants

 from

from hot countries, always keeping them in the ftove or glafs cafe, otherwife they will not flower or produce feeds in England.

The feventh fort grows naturally in Jamaica. This is a fhrubby plant, which rifes about five feet high, and divides into feveral branches, garnifhed with trifoliate leaves which are oval, the middle lobe being much larger than the other two; the ftalks are terminated by long fpikes of fmall purple flowers, which are fucceeded by narrow pods, ftrait on one fide, but jointed on the other.

The eighth fort grows at La Vera Cruz. This rifes with a fhrubby ftalk fix or feven feet high, dividing into feveral branches, garnifhed with trifoliate oval leaves, filky and white on their under fide, but of a pale green on their upper; the flowers come out in long narrow fpikes from the wings, and at the end of the branches, fitting clofe to the ftalks; they are fmall, of a bright purple colour, and are fucceeded by flat, fmooth, jointed pods, about one inch long, each joint having one kidney-fhaped feed.

The two laft forts will continue two or three years, if the plants are placed in the bark-ftove. They are propagated by feeds, which muft be fown upon a hot-bed, and the plants treated in the fame manner as thofe juft before mentioned.

The ninth fort is an annual plant, which grows naturally at La Vera Cruz. This feldom rifes more than eight or nine inches high, fending out feveral branches from the root, which are diffufed and hairy, clofely garnifhed with fmall, oval, trifoliate leaves, a little hoary. The flowers grow in clofe fhort fpikes; they are purple, and have very hairy empalements.

The tenth fort grows naturally in Jamaica. This hath ligneous trailing ftalks a foot and a half long, fending out feveral branches on each fide, garnifhed with fmall, round, trifoliate leaves, of a pale green colour; the flowers are produced in very loofe fpikes at the end of the branches; they are of a pale purplifh colour, and are fucceeded by narrow twifted pods which are jointed, containing a fingle, fmall, compreffed feed.

The two laft forts being annual, require the fame treatment as the fifth and fixth forts before mentioned, with which they will flower and ripen their feeds in this country.

The eleventh fort grows in Jamaica. It is a fhrubby plant, which rifes with triangular ftalks five or fix feet high, dividing into feveral branches, garnifhed with heart-fhaped trifoliate leaves ending in acute points; the flowers are produced in very long fpikes at the end of the branches, which are of a pale purple colour, and are fucceeded by narrow jointed pods, which are varioufly twifted; the feeds are fmall and compreffed.

The twelfth fort is annual, it grows at Campeachy. This hath a paniculated ftalk, which rifes about two feet high, garnifhed with heart-fhaped trifoliate leaves; the upper part of the ftalk branches out into panicles of flowers, which are of a pale purple colour, and are fucceeded by lunulated compreffed pods, ftanding oblique to the ftalk, each containing one compreffed kidney-fhaped feed. This fort is propagated by feeds, and requires the fame treatment as the fifth and fixth forts.

The thirteenth fort grows at La Vera Cruz. This hath a twining ftalk, which twifts round the trees and fhrubs that grows near it, and climbs to the height of ten or twelve feet, garnifhed with obverfe, oval, trifoliate leaves, ftanding upon pretty long foot-ftalks; the flowers are produced in very long fpikes, which are reflexed; they are of a dark purple colour, and fit clofe to the ftalk. This is an abiding plant, which requires a ftove to preferve it in this country, fo the plants fhould be treated in the fame manner as the feventh and eighth forts.

The fourteenth fort is an annual plant, which grows naturally in both Indies. It hath trailing branches near a foot long, garnifhed with round trifoliate leaves, a little indented at the top, very like in fhape to thofe of the Strawberry Trefoil; the ftalks and under fide of the leaves are hairy; the flowers are produced toward the end of the branches, fometimes fingle, and at other times two at a joint: they are of a purple colour and fmall; thefe are fucceeded by pods about an inch long, which are ftrait on one fide and jointed on the other.

The fifteenth fort is a low annual plant, having flender ftalks near a foot long, their lower part being garnifhed with fingle oval leaves, ftanding upon flender foot-ftalks; their upper is adorned with flowers, which come out by pairs above each other to the end of the ftalk; they are but fmall, and of a reddifh yellow colour, and are fucceeded by jointed, narrow, fickle-fhaped pods, which fit clofe to the ftalk. The two laft mentioned are annual plants, which require the fame culture as the fifth and fixth forts.

The fixteenth fort grows in South Carolina. This hath a perennial root, from which arife two or three fhrubby hairy ftalks two feet high, branching on every fide near the top, garnifhed with oval, fpear-fhaped, trifoliate leaves, which are hairy on their under fide, and ftand upon fhort foot-ftalks; the flowers are produced at the end of the branches in fhort fpikes; they are of a purplifh yellow colour and fmall; the ftalks of this fort decay in autumn, and new ones arife in the fpring. It is propagated by feeds, which fhould be fown upon a hot-bed in the fpring, and when the plants are fit to remove, they fhould be planted in feparate fmall pots, and plunged into a moderate hot-bed, obferving to fhade them till they have taken new root; then they fhould have a large fhare of air admitted to them in warm weather; in fummer they muft be expofed to the open air, but in the autumn they muft be placed under a frame to fcreen them from froft; the following fpring fome of thefe plants may be fhaken out of the pots, and planted in a warm border, where, if the fummer proves warm, they will flower; but thefe feldom perfect their feeds, therefore two or three plants fhould be put into larger pots, and plunged into a moderate hot-bed, which will bring them early into flower; fo that if the glaffes are kept over them in bad weather, thefe will ripen their feeds in autumn, and the roots will continue fome years if they are fcreened from froft in winter.

The feventeenth fort alfo grows in South Carolina. This hath a perennial root and an annual ftalk, which grows erect about two feet high, garnifhed with long trifoliate leaves, rounded at their bafe where they are broadeft, and narrowed all the way to a point; they are near three inches and a half long, and half an inch broad at their bafe, of a light green and fmooth; the two fide lobes fit pretty clofe to the ftalk, but the middle one fits upon a foot-ftalk an inch long; the flowers are produced in long fpikes from the wings of the ftalk, growing erect; the lower part of the fpike is but thinly fet with flowers, but on the upper part they are difpofed very clofe; thefe are fmall, and of a bright yellow colour, fitting very clofe to the ftalks, and are fucceeded by jointed pods ftrait on one fide.

This plant is propagated by feeds, and requires the fame treatment as the laft mentioned, with which it will flower and produce ripe feeds.

The eighteenth fort grows naturally in Syria, where it is one of the beauties of the country. It rifes with fhrubby ftalks about three feet high, which branch out on every fide, garnifhed with fingle fmooth leaves, fhaped like thofe of the broad-leaved Knot-grafs, of a pale green, and ftand

on

on ſhort foot-ſtalks; under theſe leaves come out thorns, which are near an inch long, of a reddiſh brown colour; the flowers come out from the ſide of the branches in ſmall cluſters; they are of a purple colour in the middle, and reddiſh about the rims; theſe are ſucceeded by pods, which are ſtrait on one ſide, and jointed on the other, bending a little in ſhape of a ſickle. This plant is at preſent pretty rare in the Engliſh gardens; it is propagated by ſeeds, which will frequently lie a year in the ground before they vegetate, therefore ſhould be ſown in pots filled with light earth, and plunged into a moderate hot-bed; and if the plants do not appear by the beginning of June, the pots ſhould be taken out of the bed, and placed where they may have only the morning ſun, keeping them clean from weeds; and in the autumn they ſhould be plunged into an old bed of tanners bark under a frame, where they may be ſcreened from the froſt and hard rains in the winter; in ſpring they ſhould be plunged into a freſh hot-bed, which will bring up the plants: when theſe are fit to remove they ſhould be each planted into a ſeparate ſmall pot, and plunged into a very moderate hot-bed, ſhading them from the ſun till they have taken new root; then they ſhould be gradually hardened to bear the open air, into which they ſhould be removed in June, placing them in a ſheltered ſituation, where they may remain till the autumn, when, if they are plunged into an old tan-bed under a frame, where in mild weather they may enjoy the free air, and be protected from froſt, they will ſucceed better than if placed in a green-houſe, or tenderly treated. I have ſeen this plant growing in the full ground in a very warm border, where by covering it in froſty weather it had endured two winters, but a ſevere froſt happening the third winter entirely deſtroyed it.

From this ſhrub the Perſian manna is collected, which is an exſudation of the nutritious juice of the plant. This drug is chiefly gathered about Tauris, a town in Perſia, where the ſhrub grows plentifully. Sir George Wheeler found it growing in Tinos, and ſuppoſed it was an undeſcribed plant. Tournefort found it in plenty in many of the plains in Armenia and Georgia, and made a particular genus of it under the title of Alhagi.

The nineteenth ſort grows naturally in India, from whence the ſeeds have been lately brought to Europe, and ſeveral plants have been raiſed in the Engliſh gardens; theſe have leaves ſo like thoſe of the Orange-tree, as ſcarcely to be diſtinguiſhed while young; but as there are not any plants here of a large ſize, ſo I can give no further account of it at preſent.

HEDYSARUM Zeylanicum majus & minus. See ÆSCHYNOMENE.

HELENIUM. Lin. Gen. Plant. 863. Baſtard Sunflower.

The CHARACTERS are,

It hath a flower compoſed of ſeveral hermaphrodite florets, which form the diſk, and female half florets which compoſe the rays. The hermaphrodite florets are tubulous; theſe have each five ſhort hairy ſtamina, and an oblong germen which afterward becomes an angular ſingle ſeed, crowned by a ſmall five-pointed empalement. The female half florets in the border are ſtretched out on one ſide like a tongue to form the ray; theſe are cut into five ſegments at their points; they have no ſtamina, but an oblong germen which turns to a ſingle ſeed, like thoſe of the hermaphrodite flowers; theſe are all included in one common ſingle empalement.

The SPECIES are,

1. HELENIUM (*Autumnale*) foliis lineari-lanceolatis integerrimis glabris, pedunculis nudis unifloris. *Helenium with ſpear-ſhaped narrow leaves, which are ſmooth, entire, and naked foot-ſtalks with ſingle flowers.*

2. HELENIUM (*Latifolium*) foliis lanceolatis acutis ſerratis, pedunculis brevioribus, calycibus multifidis. *Helenium with pointed, ſpear-ſhaped, ſawed leaves, ſhorter foot-ſtalks, and a many-pointed empalement.*

Theſe plants riſe to the height of ſeven or eight feet in good ground; the roots when large ſend up a great number of ſtalks, which branch toward the top; thoſe of the firſt ſort are garniſhed with ſmooth leaves, which are three inches and a half long, and half an inch broad in the middle, with entire edges fitting cloſe to the ſtalks, and from their baſe is extended a leafy border along the ſtalk, ſo as to form what was generally termed a winged ſtalk, but Linnæus calls it a running leaf; the upper part of the ſtalk divides, and from each diviſion ariſes a naked foot-ſtalk, about three inches long, ſuſtaining one yellow flower at the top, ſhaped like a Sun-flower, but much ſmaller, having long rays, which are jagged pretty deep into four or five ſegments.

The ſecond ſort hath the appearance of the firſt, but the leaves are ſhorter and broader, ending in acute points, and are ſharply ſawed on their edges. The flowers ſtand upon ſhorter foot-ſtalks, growing cloſer together.

Theſe plants are natives of North America, where they grow wild in great plenty. They may be propagated by ſeeds, or by parting their roots; the latter is generally practiſed in this country.

The beſt ſeaſon to tranſplant and part the old roots is in October, when their leaves are paſt, or the beginning of March juſt before they begin to ſhoot; but if the ſpring ſhould prove dry they muſt be duly watered, otherwiſe they will not produce many flowers the ſame year: theſe plants ſhould not be removed oftener than every other year, if they are expected to flower ſtrong; they delight in a ſoil rather moiſt than dry, provided it be not too ſtrong, or hold the wet in winter.

HELENIUM. See INULA.

HELIANTHEMUM. Tourn. Inſt. R. H. 248. tab. 128. Dwarf Ciſtus, or Sun-flower.

The CHARACTERS are,

The flower has a three-leaved empalement. It hath five roundiſh petals, with a great number of erect ſtamina. In the center is ſituated an oval germen, which afterward becomes a roundiſh or oval capſule of one cell, opening in three parts, filled with ſmall roundiſh ſeeds.

The SPECIES are,

1. HELIANTHEMUM (*Chamæciſtus*) caulibus procumbentibus ſuffruticoſis, foliis oblongis ſubpiloſis, ſtipulis lanceolatis. *Dwarf Ciſtus with trailing ſhrubby ſtalks, oblong hairy leaves, and ſpear-ſhaped ſtipulæ.*

2. HELIANTHEMUM (*Germanicum*) caulibus procumbentibus ſuffruticoſis ramoſiſſimis, ſpicis florum longioribus. *Dwarf Ciſtus with trailing ſhrubby ſtalks full of branches, and longer ſpikes of flowers.*

3. HELIANTHEMUM (*Piloſum*) caulibus ſuffruticoſis piloſis, foliis lanceolatis obtuſis, ſpicis reflexis. *Dwarf Ciſtus with hairy ſhrubby ſtalks, blunt ſpear-ſhaped leaves, and reflexed ſpikes of flowers.*

4. HELIANTHEMUM (*Apenninum*) incanum caulibus ſuffruticoſis erectis, foliis lanceolatis hirſutis. *Hoary Dwarf Ciſtus with erect ſhrubby ſtalks, and hairy ſpear-ſhaped leaves.*

5. HELIANTHEMUM (*Umbellatum*) caule procumbente non ramoſo, foliis linearibus incanis oppoſitis. *Dwarf Ciſtus with an unbranched trailing ſtalk, and narrow hoary leaves placed oppoſite.*

6. HELIANTHEMUM (*Fumara*) caule ſuffruticoſo procumbente, foliis linearibus alternis, floribus auriculatis. *Dwarf Ciſtus with a ſhrubby trailing ſtalk, very narrow leaves placed alternate, and auriculated flowers.*

7. HELI-

7. Helianthemum (*Sampſuchefolium*) caule ſuffruticoſo procumbente, foliis lanceolatis oppoſitis, pedunculis longioribus, calycibus hirſutis. *Dwarf Ciſtus with a ſhrubby trailing ſtalk, ſpear-ſhaped leaves placed oppoſite, longer foot-ſtalks to the flowers, and hairy empalements.*

8. Helianthemum (*Serpillifolium*) caule fruticoſo procumbente, foliis linearibus oppoſitis, floribus umbellatis. *Dwarf Ciſtus with a ſhrubby ſtalk, very narrow leaves placed oppoſite, and flowers growing in an umbel.*

9. Helianthemum (*Ciſtifolium*) caulibus procumbentibus ſuffruticoſis glabris, foliis ovato-lanceolatis oppoſitis, pedunculis longioribus. *Dwarf Ciſtus with ſhrubby trailing ſtalks which are ſmooth, oval ſpear-ſhaped leaves placed oppoſite, and longer foot-ſtalks to the flowers.*

10. Helianthemum (*Tuberaria*) caule lignoſo perenne foliis radicalibus ovatis trinerviis tomentoſis caulinis glabris lanceolatis alternis. *Perennial Dwarf Ciſtus with a woody ſtalk, whoſe lower leaves have three veins, are oval, woolly, and thoſe on the ſtalks ſmooth, ſpear-ſhaped, and placed alternate.*

11. Helianthemum (*Polifolium*) caulibus ſeſſilibus ſuffruticoſis, foliis lanceolatis oppoſitis tomentoſis caule florali racemoſo. *Dwarf Ciſtus with very ſhort ſhrubby ſtalks, woolly ſpear-ſhaped leaves placed oppoſite, and a branching flower-ſtalk.*

12. Helianthemum (*Nummularium*) caule ſuffruticoſo procumbente, foliis ovatis nervoſis, ſubtus incanis. *Dwarf Ciſtus with a ſhrubby trailing ſtalk, and oval veined leaves, white on their under ſide.*

13. Helianthemum (*Lavendulæfolium*) caule ſuffruticoſo, foliis lineari-lanceolatis oppoſitis ſubtus tomentoſis. *Dwarf Ciſtus with a ſhrubby ſtalk, and narrow ſpear-ſhaped leaves placed oppoſite, which are woolly on their under ſide.*

14. Helianthemum (*Hirtum*) caule ſuffruticoſo erecto, foliis linearibus margine revolutis ſubtus incanis. *Dwarf Ciſtus with a ſhrubby erect ſtalk, and narrow leaves reflexed on their edges, with their under ſide hoary.*

15. Helianthemum (*Surrejanum*) caulibus ſuffruticoſis procumbentibus, foliis oblongo-ovatis ſubhirſutis, petalis acuminatis reflexis. *Dwarf Ciſtus with trailing ſhrubby ſtalks, oblong, oval, hairy leaves, and acute-pointed reflexed petals to the flowers.*

16. Helianthemum (*Luſitanicum*) caule ſuffruticoſo erecto, foliis lanceolatis incanis glabris caule florali ramoſo. *Dwarf Ciſtus with a ſhrubby upright ſtalk, hoary ſpear-ſhaped leaves which are ſmooth, and branching flower-ſtalks.*

17. Helianthemum (*Roſeum*) caule ſuffruticoſo, foliis oblongo-ovatis, oppoſitis, ſummis linearibus alternis. *Dwarf Ciſtus with a ſhrubby ſtalk, oblong oval leaves placed oppoſite, thoſe toward the top being narrow and alternate.*

18. Helianthemum (*Guttatum*) caule herbaceo hirſute, foliis lanceolato-linearibus piloſis, pedunculis longioribus. *Dwarf Ciſtus with an herbaceous ſtalk which is hairy, narrow, ſpear-ſhaped, hairy leaves, and longer foot-ſtalks to the flowers.*

19. Helianthemum (*Fugacium*) caule herbaceo, foliis ſubovatis piloſis, flore fugaci. *Dwarf Ciſtus with an herbaceous ſtalk, hairy oval leaves, and a fugacious flower.*

20. Helianthemum (*Ledifolium*) caule herbaceo erecto, foliis lanceolatis oppoſitis, floribus ſolitariis, capſulis maximis. *Dwarf Ciſtus with an erect herbaceous ſtalk, ſpear-ſhaped leaves placed oppoſite, flowers growing ſingly, and very large capſules.*

21. Helianthemum (*Salicifolium*) caule herbaceo ramoſo, foliis oblongo-ovatis oppoſitis, ſummis alternis, floribus ſolitariis. *Dwarf Ciſtus with a branching herbaceous ſtalk, oblong oval leaves placed oppoſite, thoſe toward the top growing alternate, and ſolitary flowers.*

22. Helianthemum (*Faſciculatum*) foliis faſciculatis. Royen. *Dwarf Ciſtus with leaves growing in bunches.*

23. Helianthemum (*Fruticoſum*) caule fruticoſo ſucculento, foliis ovatis carnoſis, floribus racemoſis. *Dwarf Ciſtus with a ſhrubby ſucculent ſtalk, oval fleſhy leaves, and branching flowers.*

24. Helianthemum (*Marifolium*) caule herbaceo procumbente, foliis ovatis tomentoſis ſeſſilibus. *Dwarf Ciſtus with an herbaceous trailing ſtalk, and oval woolly leaves ſitting cloſe to the branches.*

The firſt ſort grows naturally on the chalky hills and banks in many parts of England; the ſtalks of this plant are ligneous and ſlender, trailing upon the ground, extending themſelves near a foot each way, garniſhed with ſmall oblong leaves, of a dark green on their upper ſide, but of a grayiſh colour on their under. The flowers are produced at the end of the ſtalks in looſe ſpikes; they are compoſed of five deep yellow petals, which ſpread open in the day, but ſhut cloſe in the evening.

The ſecond ſort grows naturally in Germany; the ſtalks of this are much larger, and extend farther than thoſe of the firſt; the leaves are longer and hoary: there are three acuminated ſtipula at each of the lower joints, which are erect. The ſpikes of flowers are much longer than thoſe of the former, and the flowers are white and larger.

The third ſort grows naturally in the ſouth of France, in Italy and Germany. The ſtalks of this grow more erect than either of the former, and are ligneous. The joints are farther aſunder; the leaves are longer and hairy; the ſpikes of flowers are generally reflexed; they are white, and the ſize of thoſe of the ſecond; the ſtipula of this are very narrow.

The fourth ſort grows naturally on the Apennine mountains; the ſtalks of this are more erect than thoſe of the third. The leaves are not ſo long, the ſtipula are very ſmall, and the whole plant is very hoary. The flowers are white, and the ſpikes are ſhorter and more compact than either of the former.

The fifth ſort grows naturally in the ſouth of France, in Spain, and Iſtria; this hath low trailing ſtalks, which are ligneous, but ſeldom branch, and are not more than four or five inches long. The leaves are narrow and hoary, and grow in ſmall cluſters at the end of the ſtalks; this ſort ſeldom continues longer than two years.

The ſixth ſort hath trailing ſhrubby ſtalks, which extend a foot in length, garniſhed with very narrow ſmooth leaves placed alternate; theſe have no ſtipula at their baſe. The flowers are placed thinly toward the end of the branches, they are yellow and auriculated; this grows in the ſouth of France and Italy.

The ſeventh ſort hath very long, trailing, ligneous ſtalks, garniſhed with ſpear-ſhaped leaves placed oppoſite, which are very hairy, and gray on their under ſide, having at their baſe three long narrow ſtipula. The ſpikes of flowers are near a foot in length, but grow thinly; they are large, and of a deep yellow colour, with very hairy empalements; this grows naturally in the ſouth of France and Spain.

The eighth ſort hath very ſhrubby crooked ſtalks, covered with a purpliſh brown bark like the common Heath. The branches are ſlender, garniſhed with narrow ſtiff leaves like thoſe of Thyme, which ſtand oppoſite, having no ſtipula at their baſe. The flowers are produced on naked foot-ſtalks, which terminate the branches in a ſort of umbel; they are of a pale yellow colour, and ſmaller than thoſe of the common ſort; this grows naturally on the ſands near Fontainebleau in France.

The ninth ſort grows naturally in Germany; this ſends out from a ligneous root a great number of trailing ſtalks, which extend more than a foot each way; they are ſmooth, with a dark brown bark, garniſhed with oval, ſpear-ſhaped,

ſmooth

ſmooth leaves, placed oppoſite, having at their baſe three ſpear-ſhaped ſtipula. The flowers are large, yellow, and grow in ſhort cluſters at the end of the branches.

The tenth ſort grows naturally in Spain; this hath a ſhort, thick, woody ſtalk, from which come out ſeveral ſide branches, garniſhed with oval woolly leaves, having three longitudinal veins. The flower-ſtalk which ariſes from the main ſtem, grows about nine inches high, having two or three narrow leaves placed alternate. The flowers are produced on pretty long pedicles toward the top of the ſtalk, and have very ſmooth empalements.

The eleventh ſort was ſent from Verona, where it grows naturally; this hath a low ſhrubby ſtalk, from which come out a few ſhort branches, garniſhed with ſmall, woolly, ſpear-ſhaped leaves, placed oppoſite. The flower-ſtalk riſes about ſix inches high; it branches toward the top, where the flowers are produced on pretty long foot-ſtalks; they are white, and ſmaller than thoſe of the common ſort.

The twelfth ſort hath long ſhrubby ſtalks, which trail on the ground, and divide into many branches, garniſhed with oval veined leaves, of a light green on their upper ſide, but grayiſh below, with three narrow erect ſtipula at their baſe. The flowers are pretty large, white, and grow in cluſters at the end of the branches.

The thirteenth ſort hath ſhrubby ſtalks, which grow pretty upright, garniſhed with narrow ſpear-ſhaped leaves placed oppoſite, woolly on their under ſide, with three very narrow ſtipula growing at their baſe. The flowers are white, growing in long ſpikes at the end of the branches; this grows naturally in the ſouth of France.

The fourteenth ſort hath an erect ſhrubby ſtalk, which ſends out many ſide branches, whoſe joints are pretty cloſe, garniſhed with very narrow leaves, placed oppoſite, whoſe borders are reflexed; their upper ſide is of a lucid green, and their under ſide hoary. The flowers are pretty large, white, and grow in ſmall cluſters at the end of the branches; this grows naturally in Spain.

The fifteenth ſort was found by Mr. Edmund du Bois, near Croydon in Surry, and was at firſt only ſuppoſed to be an accidental variety of the common ſort, but the ſeeds of this always produce the ſame. This is very like the common ſort, but the leaves are hairy. The petals of the flowers are ſtar-pointed, and ſmaller than thoſe of the common ſort.

The ſixteenth ſort hath ſhrubby ſtalks, which riſe a foot and a half high, ſending out branches the whole length, garniſhed with ſmall, ſpear-ſhaped, ſmooth, ſilvery leaves, placed oppoſite. The flower-ſtalks branch, and the flowers, which are white, are produced in ſhort ſpikes at the end of the branches.

The ſeventeenth ſort was found growing naturally by the late Dr. William Sherrard near Smyrna; this hath ſhrubby ſtalks, garniſhed with oblong oval leaves placed oppoſite, but thoſe toward the top are narrow, and placed alternate. The flowers are produced at the end of the branches in long looſe ſpikes; they are of a Roſe colour, and the ſize of thoſe of the common ſort.

The eighteenth ſort is annual; this grows naturally in France, Spain, Italy, and in Jerſey, where the late Dr. William Sherrard found it; this hath a branching herbaceous ſtalk, which riſes four or five inches high, garniſhed with narrow ſpear-ſhaped leaves placed oppoſite, which are covered with hairs; thoſe on the upper part of the ſtalks are placed alternate, and are narrower. The flowers are produced in looſe ſpikes at the end of the branches, ſtanding upon long foot-ſtalks; they are ſmall, and compoſed of five yellow petals, with a dark purple ſpot at the baſe of each; theſe flowers are very fugacious, for they open early in the morning, and their petals drop off in a few hours after, ſo that by ten of the clock the flowers are all fallen.

The nineteenth ſort grows naturally upon mount Baldus; this is an annual plant, which ſends out many herbaceous ſtalks from the root, garniſhed with oval leaves, which are hairy. The flowers are produced in looſe ſpikes at the end of the branches; they are of a pale yellow colour, and very fugacious, ſeldom laſting two hours before the petals fall off; there is another variety of this which grows about Verona, with upright ſtalks.

The twentieth ſort grows naturally in the ſouth of France and Italy, and was found by the late Dr. William Sherrard growing near Smyrna, who ſent the ſeeds to England and Holland by a new title, ſuppoſing it to be a different plant; but when it was cultivated here, it proved to be the ſame with that growing in the ſouth of France, for this plant puts on different appearances, according to the ſoil and ſituation where it grows; where the plants ſtand ſingle, and are not injured by weeds, they will riſe near a foot and a half high, the leaves will be two inches and a half long, and near half an inch broad in the middle; but in a poor ſoil, or where the plants ſtand too cloſe, or are injured by weeds, or neighbouring plants, they do not riſe more than half that height; the leaves are much narrower, and the ſeed-veſſels not half ſo large, ſo that any perſon finding theſe plants in two different ſituations, may be deceived, and take them for different ſpecies; but when they are cultivated in a garden in the ſame ſoil and ſituation, they do not differ in any particular. This is an annual plant, which periſhes ſoon after the ſeeds are ripe.

The twenty-firſt ſort is an annual plant, which grows naturally in Spain and Portugal; this hath branching ſtalks which riſe a foot high, garniſhed with oval oblong leaves, placed oppoſite on the lower part of the ſtalk, but on the upper part they are alternate and narrow, a ſingle leaf being placed between each flower, which occaſions the title of Solitary Flowers, for they grow in looſe ſpikes at the end of the branches, in the ſame manner as the other ſpecies.

The twenty-ſecond ſort was ſent me by Dr. Adrian Van Royen, who received the ſeeds from the Cape of Good Hope. This riſes with a ſhrubby ſtalk about nine inches high, garniſhed with very narrow fine leaves, growing in cluſters; the flowers come out from the ſide and at the end of the branches, ſtanding upon ſlender foot-ſtalks; they are of a pale ſtraw colour, and are very fugacious, ſeldom continuing longer than two hours before the petals fall off. This ſeldom continues longer than two years.

The twenty-third ſort was ſent me by the late Dr. William Houſtoun from Campeachy, where he found it growing naturally. This hath a ſucculent ſtalk, which riſes near three feet high, garniſhed with oval, fleſhy, ſucculent leaves like thoſe of Purſlain, which are placed alternate; the flower-ſtalks ariſe from the main ſtem, which are naked, and near two feet high, branching out on each ſide in ſmaller foot-ſtalks, which are again branched into many ſmaller, each ſuſtaining a ſmall Roſe-coloured flower.

The twenty-fourth ſort grows naturally about Kendal in Weſtmoreland, and in ſome parts of Lancaſhire, upon rocky ſituations. This hath trailing herbaceous ſtalks, which ſeldom extend more than three or four inches, garniſhed with oval leaves, which are very woolly, and ſit cloſe to the branches; the flowers are produced at the upper part of the branches; they are white and ſmall, ſo make no great appearance.

All the perennial ſorts of dwarf Ciſtus (except the twenty-third) are hardy, ſo will thrive in the open air in England; they are propagated by ſeeds, which may be ſown in the places where the plants are to remain, and will require no other

other care but to keep them clean from weeds, and thin them where they are too close, always observing to leave those sorts at a farther distance, whose stalks trail on the ground, and grow to the greatest length. These plants will continue several years, especially in a poor dry soil, but in rich ground or moist situations they seldom last long; but as they ripen seeds in plenty, so they may be easily repaired. They all flower about the same time as the common sort, and their seeds ripen in the autumn.

The annual sorts may be propagated with as great facility, for if their seeds are sown upon a bed of common earth in April, the plants will come up in May, and require no other culture but to thin them where they are too close, and keep them clear from weeds. These will flower in July, and the seeds ripen in the autumn. The twenty-second sort will thrive in the full ground in the same manner as the other, but unless the summer proves favourable, the seeds will not ripen: the roots have stood through the winter when the season has proved mild, without any shelter, and have flowered the following summer.

The twenty-third sort grows naturally in the warm parts of America. Father Plumier discovered it first in some of the French islands, and Dr. Houstoun found it growing plentifully about Campeachy. This sort is propagated by seeds, which should be sown upon a hot-bed in the spring; and when the plants are fit to remove, they should be each planted in a small pot, filled with light, sandy, undunged earth, and plunged into a moderate hot-bed of tanners bark, and treated in the same manner as other tender plants from the same country; in the autumn they must be placed in a warm stove, and the second year the plants will flower, but they have not as yet produced seeds in England.

The twenty-fourth sort requires a shady situation, otherwise it will not thrive here.

HELIANTHUS. Lin. Gen. Plant. 877. Sun-flower.

The CHARACTERS are,

It hath a compound radiated flower, the border or rays being composed of female half florets which are barren, and the disk of hermaphrodite florets which are fruitful, contained in one common scaly empalement. The hermaphrodite florets are cylindrical, cut at the brim into five acute segments; these have five stamina. The germen, which is situated at the bottom of the tube, afterwards becomes an oblong, blunt, four-cornered seed. The female half florets which compose the border are not fruitful.

The SPECIES are,

1. HELIANTHUS (*Annuus*) foliis omnibus cordatis, nervis ponè basin unitis, extrorsum denudatis. Lin. Sp. Pl. 904. *Sun-flower whose leaves are all heart shaped, veins uniting behind at the base, but toward the border naked; commonly called annual Sun-flower.*

2. HELIANTHUS (*Multiflorus*) foliis inferioribus cordatis, nervis ponè basin unitis denudatis, superioribus ovatis. Lin. Sp. Plant. 905. *Sun-flower whose under leaves are heart-shaped, veins united at their base, and the upper leaves oval; commonly called perennial Sun-flower.*

3. HELIANTHUS (*Tuberosus*) foliis ovato-cordatis, nervis intra folium unitis. Lin. Sp. Plant. 905. *Sun-flower with oval heart-shaped leaves, whose nerves unite in the leaf; commonly called Jerusalem Artichoke.*

4. HELIANTHUS (*Strumosus*) radice fussi formi. Hort. Cliff. 420. *Sun-flower with a spindle-shaped root.*

5. HELIANTHUS (*Gigantius*) foliis lanceolatis scabris caule stricto, infernè glabro. Lin. Sp. Plant. 905. *Sun-flower with rough spear shaped leaves, a slender stalk, smooth toward the bottom.*

6. HELIANTHUS (*Divaricatus*) foliis oppositis sessilibus ovato-oblongis trinerviis, paniculâ dichotomâ. Lin. Sp. Plant. 906. *Sun-flower with oblong, opposite, oval leaves, having three veins, and sitting close to the stalk, and a dichotomous panicle.*

7. HELIANTHUS (*Trachelifolius*) foliis lanceolatis oppositis, supernè scabris, infernè trinerviis, caule dichotomo romoso. *Sun-flower with spear-shaped leaves placed opposite, whose upper surface is rough, the under leaves having three veins, and a divided stalk.*

8. HELIANTHUS (*Ramosissimus*) caule ramosissimo, foliis lanceolatis scabris, inferioribus oppositis, summis alternis petiolat , calycibus foliosis. *Sun-flower with a very branching stalk, rough spear-shaped leaves placed opposite at bottom, but alternate toward the top, having foot-stalks, and leafy empalements.*

9. HELIANTHUS (*Atrorubens*) foliis ovatis crenatis trinerviis scabris, squamis calycinis erectis longitudine disci. Flor. Virg. 103. *Sun-flower with oval, rough, crenated leaves, having three nerves, the scales of the empalement being erect, and as long as the disk of the flower.*

10. HELIANTHUS (*Decapetalus*) caule infernè lævi foliis lanceolato-cordatis, radiis decapetalis. Lin. Sp. Plant. 905. *Sun-flower with a stalk smooth on the upper side, heart spear-shaped leaves, and ten petals in the rays.*

All these species of Sun-flowers are natives of America, from whence we are often supplied with new kinds; and it is very remarkable, that there is not a single species of this genus that is European; so that before America was discovered, we were wholly unacquainted with these plants. But although they are not originally of our own growth, yet are they become so familiar with our climate, as to thrive and increase full as well as if they were in their native country (some of the very late flowering kinds excepted, which require a longer summer than we generally enjoy to bring them to perfection;) and many of them are now so plentiful in England, that persons unacquainted with the history of these plants, would imagine them at least to have been inhabitants of this island many hundred years; particularly the Jerusalem Artichoke, which, though it doth not produce seeds in our climate, yet doth so multiply by its knobbed roots, that, when once well fixed in a garden, it is not easily to be rooted out.

The first sort is annual, and so well known as to require no description. There are single and double flowers of two different colours, one of a deep yellow, and the other of a sulphur colour; but these vary, so are not worthy to be mentioned as different. They are easily propagated by seeds, which should be sown in March upon a bed of common earth; and when the plants come up, they must be thinned where they are too close, and kept clean from weeds; when the plants are grown six inches high, they may be taken up with balls of earth to their roots, and planted into the large borders of the pleasure-garden, observing to water them till they have taken new root, after which they will require no other care but to keep them clear from weeds.

In July the great flowers upon the tops of the stems will appear, amongst which, the best and most double flowers of each kind should be preserved for seeds; for those which flower later upon the side branches are neither so fair, nor do they perfect their seeds so well, as those which first appear: when the flowers are quite faded, and the seeds are formed, they should be carefully guarded from the sparrows, which will otherwise devour most of the good seeds: about the beginning of October, when the seeds are ripe, the heads should be cut off, with a small part of the stem, and hung up in a dry airy place for about a month, by which time the seeds will be perfectly dry and hard; when they may be easily rubbed out, and put up in bags or papers, to preserve them from vermin until the season for sowing them.

The seeds of this sort of Sun-flower are excellent food for domestic poultry, therefore where a quantity of it can be be saved, it will be of great use where there are quantities of these fowls.

The other perennial sorts rarely produce seeds in England, but most of them increase very fast at their roots, especially the creeping rooted kinds, which spread too far for small gardens. The second sort, which is the most common in the English gardens, is the largest and most valuable flower, and is a very proper furniture for large borders in great gardens, as also for bosquets of large growing plants, or to intermix in small quarters with shrubs, or in walks under trees, where few other plants will thrive; it is also a great ornament to gardens within the city, where it grows in defiance of the smoke better than most other plants; and for its long continuance in flower, deserves a place in most gardens, for the sake of its flowers for basons, &c. to adorn halls and chimneys in a season when we are at a loss for other flowers. It begins flowering in July, and continues until October. The sort with single flowers is now little valued, since that with double flowers is become common.

The third, fourth, fifth, sixth, and seventh sorts, may also have a place in some large borders of the garden for the variety of their flowers, which, though not so fair as those of the common sort, yet will add to the diversity; and as many of them are late flowerers, so by encouraging these plants we may continue the succession of flowers longer.

These sorts are all of them very hardy, and will grow in almost any soil or situation; they are propagated by parting their roots into small heads, which in one year's time will spread and increase greatly. The best season for this work is in the middle of October, soon after the flowers are past, or very early in the spring, that they may be well rooted before the droughts come on, otherwise their flowers will be few in number, and not near so fair.

The Jerusalem Artichoke is propagated in many gardens for the roots, which are by some people esteemed; but they are watery and flashy, and very subject to trouble the belly by their windy quality, which hath brought them almost into disuse.

These are propagated by planting the smaller roots, or the larger ones cut into pieces, observing to preserve a bud to each separate piece, either in the spring or autumn, allowing them a good distance, for their roots will greatly multiply; the autumn following, when their stems decay, the roots may be taken up for use. These should be planted in some remote corner of the garden, for they are very unsightly while growing, and their roots are apt to over-run whatever grows near them, nor can they be easily destroyed when they are once well fixed in a garden.

The other species which have been ranged under this genus by Tournefort and others, are now removed to the following genera, under which titles they may be found.

Corona Solis. See { Coreopsis. Helenium. Rudbeckia. Silphium. }

HELICTERES. Lin. Gen. Plant. 913. Screw-tree.

The Characters are,

The flower has a coriaceous empalement of one leaf. It hath five oblong equal petals, which are fixed to the empalement, and ten short stamina at the base of the germen, with five nectariums surrounding the germen, which have the appearance of petals. The style supports the germen at the top, which afterward turns to a twisted spiral fruit with one cell, inclosing many kidney-shaped seeds.

The Species are,

1. Helicteres (*Isora*) foliis cordato ovatis serratis, subtus tomentosis, fructu tereti contorto. *Helicteres with heart-shaped leaves which are sawed, woolly on their under side, and a taper twisted fruit.*

2. Helicteres (*Brevior*) foliis cordatis acuminatis serratis, subtus tomentosis, fructu brevi contorto. *Helicteres with heart-shaped, pointed, sawed leaves, woolly on their under side, and a short twisted fruit.*

3. Helicteres (*Arborescens*) caule arboreo villoso, foliis cordatis crenatis nervosis subtus tomentosis fructu ovato contorto villosissimo. *Helicteres with a tree-like hairy stalk, heart-shaped, nervous, crenated leaves, woolly on their under side, and an oval, twisted, very hairy fruit.*

The first sort grows naturally in the Bahama Islands. This rises with a shrubby stalk five or six feet high, sending out several lateral branches, covered with a soft yellowish down, and garnished with heart-shaped leaves four inches long, and two inches and a half broad, sawed on their edges, woolly on their under side, standing on long foot-stalks; at the upper part of the branches the flowers come out opposite to the leaves upon slender foot stalks, which are jointed; these are composed of five oblong white petals, and in the center arises the style, which is three inches long, and curved; upon the top of which is situated the germen; crowned by an acute stigma. The germen afterward turns to a taper fruit two inches and a half long, composed of five capsules, which are closely twisted over each other like a screw; these are hairy, and have each one cell, containing several kidney-shaped seeds.

The second sort grows naturally in Jamaica. This rises with a shrubby stalk nine or ten feet high, sending out many lateral branches, covered with a smooth brown bark, garnished with heart-shaped sawed leaves, which end in acute points, woolly on their under side; the flowers are produced on the side of the branches on shorter foot stalks than the former: they are composed of five petals; the style in the center is strait, upright, and not half so long as in the other; the fruit is thicker, not an inch long, but twisted in the same manner.

The third sort rises with a strong woody stalk twelve or fourteen feet high, sending out many ligneous branches, closely covered with hairy down, and garnished with large heart-shaped leaves, which are crenated on their edges, having large veins running from the midrib to the sides; the leaves are of a yellowish green, and a little woolly on their under side: the flowers are produced from the side of the branches, and are of a yellowish white colour, larger than those of the other sorts. The style is near four inches long, curved like that of the first sort; the fruit is oval, about one inch long, very thick at the bottom, and closely covered with hairy down. This sort grows naturally at Carthagena.

These plants are propagated by seeds, which must be sown upon a hot-bed in the spring; and when the plants are strong enough to remove, they should be each planted in a separate small pot, and plunged into a moderate hot-bed of tan, observing to shade them from the sun till they have taken new root; then they should be treated in the same way as other tender plants from hot countries, raising the glasses every day in proportion to the weather, that the plants may enjoy fresh air, which will strengthen them, and prevent their drawing up weak. In the summer the plants may remain under the frames, if there is a sufficient height for them to grow; but in autumn they must be plunged into the tan-bed in the stove, where they should always remain, being careful to shift them into larger pots when they require it, and not to give them too much wet in winter; in summer they should have a large share of air in warm weather, and require to be often refreshed with water: the second year from the seeds these plants have often flowered in the Chelsea garden, and the seeds have some years ripened there, but the plants will live several years with proper management.

HELIOCARPOS. Lin. Gen. Plant. 533.

The CHARACTERS are,

The flower hath one petal which is cut into five segments. It hath an empalement of one leaf, which is cut into five parts. In the center is situated a roundish germen, attended by several stamina, which afterward becomes an oval compressed capsule, about three lines long and two broad, with a tranverse partition dividing it in two cells, each containing a single roundish seed ending in a point; the borders of the capsule are set with hairs, resembling rays.

We have but one SPECIES of this genus, viz.

HELIOCARPOS (*Americana.*) Hort. Cliff. 211. tab. 16.

This plant was discovered by the late Dr. Houstoun, growing naturally about Old La Vera Cruz in New Spain. It rises with a thick, soft, woody stalk, from fifteen to eighteen feet high, sending out several lateral branches toward the top, garnished with heart-shaped leaves full of veins, sawed on their edges and ending in acute points; the flowers are produced at the end of the shoots, in branching clusters; they are of a yellowish green, and are succeeded by flat compressed seed-vessels of an oval shape, whose borders are closely set with threads, representing rays, of a brownish colour when ripe; these capsules are divided into two cells by an intermediate partition, in each of these is lodged a single roundish seed, ending in a point.

The plant is propagated by seeds, which must be sown upon a hot-bed in the spring; and when the plants are fit to remove, they should be each planted in a separate small pot, and plunged into a hot-bed, treating them in the same way as other tender plants. While the plants are young, they require to be plunged in the tan-bed, but after they have acquired strength, they will thrive in the dry stove: in winter they should have but little water, and must be kept warm; but in summer they should have plenty of fresh air in mild weather, and must be frequently refreshed with water. With this management the plants will flower the third year, and produce good seeds, but may be preserved several years with proper care.

I have sowed seeds of this plant which had been kept ten years, and came up as well as if it had been saved the former year; though from the appearance of the seeds, it seems as unlike to grow after the first year as any which I know.

HELIOTROPIUM. Lin. Gen. Plant. 164. Turnsole.

The CHARACTERS are,

The empalement of the flower is of one leaf, cut into five segments at the brim. The flower hath one petal with a tube the length of the empalement, cut into five parts, which are alternately larger than the other; the chaps of the tube is closed, and hath five prominent scales, joined in form of a star. It hath five short stamina and four germen at the bottom of the tube, which afterward becomes so many seeds, sitting in the empalement.

The SPECIES are,

1. HELIOTROPIUM (*Europæum*) foliis ovatis integerrimis tomentosis rugosis, spicis conjugatis. Hort. Upsal. 33. *Heliotrope with oval, entire, woolly, rough leaves, and conjugated spikes.*

2. HELIOTROPIUM (*Indicum*) foliis cordato-ovatis acutis scabriusculis, spicis solitariis, fructibus bifidis. Flor. Zeyl. 70. *Heliotrope with heart-shaped oval leaves which are pointed, rough single spikes of flowers, and bifid seeds.*

3. HELIOTROPIUM (*Horminifolium*) foliis ovato lanceolato-acuminatis rugosis, spicis solitariis gracilioribus alaribus & terminalibus. *Heliotrope with spear-shaped oval leaves which end in acute points, and a rough, slender, single spike of flowers, proceeding from the sides and tops of the stalks.*

4. HELIOTROPIUM (*Capitatum*) foliis oblongo-ovatis integerrimis glabris subtus incanis, floribus capitatis alaribus, caule arborescente. *Heliotrope with oblong, oval, entire, smooth leaves, which are hoary on their under side, flowers growing in heads from the wings of the stalks, and a tree-like stalk.*

5. HELIOTROPIUM (*Canariense*) foliis ovatis crenatis oppositis, floribus capitatis alaribus dichotomis, caule arborescente. *Heliotrope with oval crenated leaves placed opposite, flowers growing in heads from the wings of the stalks, which diverge, and a tree-like stalk.*

6. HELIOTROPIUM (*Peruvianum*) foliis ovato-lanceolatis, spicis plurimis confertis, caule fruticoso. *Heliotrope with oval spear-shaped leaves, many spikes of flowers growing in clusters, and a shrubby stalk.*

7. HELIOTROPIUM (*Curassavicum*) foliis lanceolato linearibus glabris aveniis, spicis conjugatis. Hort. Cliff. 45. *Heliotrope with narrow, spear-shaped, smooth leaves without veins, and conjugated spikes of flowers.*

8. HELIOTROPIUM (*Gnaphalodes*) foliis lineari-lanceolatis obtusis tomentosis, floribus alaribus sessilibus, caule arboreo. *Heliotrope with narrow, obtuse, spear-shaped, woolly leaves, flowers sitting close to the side of the branches, and a tree-like stalk.*

9. HELIOTROPIUM (*Fruticosum*) foliis lanceolatis sessilibus, spicis solitariis alaribus & terminalibus, caule fruticoso. *Heliotrope with spear-shaped leaves sitting close to the branches, single spikes of flowers proceeding from the sides and tops of the stalk, which is shrubby.*

10. HELIOTROPIUM (*Procumbens*) caule procumbente, foliis ovatis tomentosis integerrimis, spicis solitariis terminalibus. *Heliotrope with a trailing stalk, oval, woolly, entire leaves, and single spikes of flowers terminating the branches.*

11. HELIOTROPIUM (*Americanum*) foliis oblongo-ovatis-tomentosis, spicis conjugatis terminalibus, caule fruticoso. *Heliotrope with oblong, oval, woolly leaves, and double spikes of flowers terminating the stalk, which is shrubby.*

The first sort grows naturally in the south of France, in Spain, Italy, and most of the warmer countries in Europe. It is an annual plant, which succeeds better from seeds which scatter in the autumn, or if sown at that season than in the spring, for when they are sown in the spring, they seldom come up the same year; but if the plant is once obtained, and the seeds suffered to shed, it will maintain itself without any trouble, requiring no other culture but to keep it clean from weeds, and thin the plants where they are too close.

This rises about seven or eight inches high, dividing into two or three branches, garnished with oval rough leaves, of a light green, standing upon pretty long foot-stalks alternately; the flowers are produced at the end of the branches in double spikes joined at the bottom, which are about an inch and a half long, turning backward like a scorpion's tail. The flowers are white, and appear in June and July; the seeds ripen in the autumn, soon after which the plant decays.

The second sort grows naturally in the West-Indies. This is annual; the stalk rises two feet high, branching out toward the top: the leaves are rough and hairy, standing upon pretty long foot-stalks; the flowers are produced toward the end of the branches in single spikes, which are six inches long, turning backward at the top like the other species. The flowers are blue, and appear in July and August; the seeds ripen in September and October.

The third sort grows naturally in the West Indies. This is a smaller plant than the former, seldom growing above a foot high; the leaves are one inch and a half long, and about half an inch broad; the spikes of flowers are very slender, and not more than two inches long; the flowers are small, and of a light blue colour. They appear at the same time with the former, and the seeds ripen in autumn.

The

The seeds of these two sorts must be sown on a hot-bed in the spring, and when the plants come up they must be transplanted on another hot bed to bring them forward, treating them in the same way as the Balsamine, and other tender annual plants; in June they may be taken up with balls of earth, and planted in the borders of the flower-garden, where they will flower and produce ripe seeds.

The fourth sort rises with a shrubby stalk six or seven feet high; the young branches are closely covered with a white down; the leaves on those are very hoary and entire, but those on the older branches are greener, and notched on their edges; at each joint of the stalks come out two short branches opposite, garnished with small hoary leaves placed opposite: these when bruised emit a strong odour, which to some persons is very disagreeable, but others are pleased with it. These plants rarely flower in England The flowers are white, collected in roundish heads, which turn backward, and sit close to the branches; the leaves continue all the year, for which the plants are preserved in green-houses, to add to the variety in winter.

The fifth sort grows naturally in the Canary Islands. This rises with a woody stalk three or four feet high, dividing into many branches, garnished with oval hairy leaves, notched on their edges, growing opposite upon long foot-stalks, of an Ash colour on their under side; the flowers are produced from the side of the branches on pretty long foot-stalk, each sustaining four short roundish spikes or heads of flowers which divide by pairs, and spread from each other The flowers are white, and appear in June and July, but are not succeeded by seeds in England. The leaves of this plant when bruised, emit an agreeable odour, for which it is by some persons much esteemed; the gardeners have given it the title of Madam Maintenon.

The two next sorts are too tender to live through the winter in the open air in this country, so must be kept in a green-house during that season, but they only require to be screened from frost, so may be placed with Myrtles, and the other hardy green-house plants, where they may have a large share of air in mild weather, and be treated in the same way; they are easily propagated by cuttings during any of the summer months, which, if planted in a shady border, and duly supplied with water, will take root in five or six weeks, then they may be potted and placed in a shady situation till they have taken new root, after which they may be treated as the old plants.

The sixth sort grows naturally in Peru, from whence the seeds were sent by the younger Jussieu to the royal garden at Paris, where the plants produced flowers and seeds; and from the curious garden of Duke D'Ayen at St. Germains, I was supplied with some of the seeds, which have succeeded in the Chelsea garden, where the plants have flowered and perfected their seeds for some years.

This rises with a shrubby stalk three or four feet high, dividing into many branches, garnished with oval, spear-shaped, hairy, rough leaves, set on without order on short foot-stalks. The flowers are produced at the end of the branches in short reflexed spikes, growing in clusters. The foot-stalks divide into two or three, and these divide again into less, each sustaining a spike of bluish flowers, which have a strong sweet odour. The plants continue in flower great part of the year, and those flowers which come out in summer are succeeded by ripe seeds in autumn.

It may be propagated either by seeds or cuttings. The seeds should be sown upon a moderate hot-bed in the spring, and when the plants are fit to remove they should be transplanted into small pots, and plunged into a hot-bed, where they should be shaded till they have taken new root; then they should be inured to the open air by degrees, into which they should be removed in summer, placing them in a sheltered situation; and in autumn they must be housed with other exotic plants in a good green-house, where they will flower great part of winter, so will make a good appearance among the Orange-trees, and other green-house plants, with whose culture this plant will thrive. If the cuttings of this plant are put into pots during any of the summer months, and plunged into a moderate hot-bed, they will take root very freely, but these do not make so good plants as those raised from seeds.

The seventh sort grows naturally on the sea-shore in the West-Indies. This is an annual plant, whose branches trail upon the ground, and grow a foot long, garnished with narrow grayish leaves, which are smooth The flowers are produced in double spikes from the side of their branches; they are white and small, so make no great appearance. It is propagated by seeds, and requires the same treatment as the second and third sorts.

The eighth sort rises with an upright woody stalk six or seven feet high, with a hoary bark full of marks where the leaves have grown; the upper part of the stalk divides into two or three strong woody branches, which grow erect, and are very closely garnished with long, narrow, woolly leaves, which stand on every side the branches without order. The flowers come out from the side of the stalks, to which they sit close; they are short and reflexed, like those of the other species: the flowers are purple, sitting in very woolly empalements, which are divided into five parts; the whole plant is very white and woolly, like the Sea Cudweed, so makes an odd appearance. This is propagated by seeds, which must be procured from the places where it naturally grows, and should be sown in a tub of earth in the country; for when the dried seeds come over they seldom grow, and if they do it is not before the second year; so that if the seeds are sown as soon as they are ripe in a tub of earth to preserve them, when they arrive in England, the tub should be plunged into a hot-bed of tanners bark, which will bring up the plants; and when these are fit to remove they should be each planted in a separate small pot, filled with earth, composed of sand and light undunged earth, with a little lime rubbish well mixed together, then plunged into a hot-bed of tanners bark, and shaded until they have taken new root, after which they must be treated as other tender exotic plants, always keeping them in the tan-bed in the stove, giving them but little water, especially during the winter season.

The ninth sort is a native of the West-Indies, where it grows plentifully on the sea-shore; it rises with an upright shrubby stalk a foot and a half high, garnished with small spear-shaped leaves scarce one inch long, and one-third of an inch broad in the middle, ending in acute points, sitting close to the stalk; they are hoary on their under side, but smooth above. The flowers are produced in single slender spikes, which come out from the side and at the top of the stalks; they are but little recurved, especially those on the side, but those at top are more bent; they are white, so make but little appearance.

The tenth sort was sent me from Carthagena in New Spain, where it grows naturally on the sandy shores. This is an annual plant with trailing stalks, which grow six or seven inches long, garnished with small oval leaves, which are woolly and entire. The flowers are produced at the end of the branches in single short spikes, which are reflexed; they are small and white, so make little appearance.

The eleventh sort was sent me by the late Dr. Houstoun from La Vera Cruz, where he found it growing in plenty. This rises with a shrubby stalk three feet high, dividing into slender branches, which are closely garnished with oblong, oval, woolly leaves, placed without order. The flowers are produced at the end of the branches in double

 spikes,

ſpikes, which are ſlender, ſhort, and ſtrait, not recurved as the other ſpecies. The flowers are ſmall, white, and the plant is perennial.

Theſe three laſt mentioned are propagated by ſeeds, but the difficulty of getting them freſh from America, and the uncertainty of their growing, unleſs they are ſown abroad, and brought over in earth, has rendered them rare in Europe; and as they are plants of little beauty, ſo few perſons have taken the trouble to procure them; beſides, as they require a ſtove to preſerve them in this country, and muſt have a peculiar ſoil and management like the eighth ſort, ſo, unleſs for the ſake of variety in botanic gardens, they are not worth cultivating here.

HELLEBORE. See HELLEBORUS.

HELLEBORINE. See SERAPIAS and LIMADORUM.

HELLEBOROIDES HYEMALIS. See HELLEBORUS.

HELLEBORO RANUNCULUS. See TROLLIUS.

HELLEBORUS. Lin. Gen. Plant. 622. Black Hellebore, or Chriſtmas Flower.

The CHARACTERS are,

The flower hath no empalement; it hath five large, roundiſh, permanent petals, and many ſmall nectaria placed circularly. It hath a great number of ſtamina, and ſeveral compreſſed germen, which afterward turn to compreſſed capſules with two keels, the lower being ſhort, and the upper convex, filled with round ſeeds adhering to the ſeam.

The SPECIES are,

1. HELLEBORUS (*Fœtidus*) caule multifloro folioſo, foliis pedatis. Lin. Sp. Plant. 588. *Hellebore with many flowers on a ſtalk, which are intermixed with leaves, and ramoſe leaves ſitting on the foot-ſtalk; Bears Foot, or Setterwort.*

2. HELLEBORUS (*Viridis*) caule multifloro folioſo, foliis digitatis. Lin. Sp. Plant. 558. *Hellebore with many flowers on a ſtalk, which are intermixed with leaves, and fingered leaves.*

3. HELLEBORUS (*Niger*) ſcapo ſub-unifloro ſub-nudo, foliis pedatis. Hort. Upſal. 157. *Hellebore with one flower on a ſtalk which is naked, and hand-ſhaped leaves ſitting on the foot-ſtalk; true Black Hellebore, or Chriſtmas Roſe.*

4. HELLEBORUS (*Trifolius*) caule multifloro, foliis ternatis integerrimis. *Hellebore with many flowers on a ſtalk, and leaves compoſed of three entire lobes.*

5. HELLEBORUS (*Hyemalis*) flore folio inſidente. Hort. Cliff. 227. *Hellebore with the flower ſitting upon the leaf, or Winter Aconite.*

6. HELLEBORUS (*Latifolius*) caule multifloro folioſo, foliis digitatis ſerratis amplioribus. *Hellebore with many flowers upon a ſtalk intermixed with leaves, and large fingered leaves which are ſawed.*

The firſt ſort grows naturally in woods in ſeveral parts of England, but particularly in Suſſex, where I have ſeen it in great plenty. This hath a jointed herbaceous ſtalk, which riſes two feet high, dividing into two or three heads, garniſhed with leaves, compoſed of eight or nine long narrow lobes, which join at their baſe. Theſe are ſawed on their edges, and end in acute points; thoſe on the lower part of the ſtalk are much larger than the upper, which are ſmall and narrow. The flower-ſtalk ariſes from the center of the plant, dividing into many branches, each ſuſtaining ſeveral ſmaller foot-ſtalks, with one entire ſpear-ſhaped leaf upon each, and one large greeniſh flower at the top with purpliſh rims; theſe appear in winter, and the ſeeds ripen in the ſpring, which, if permitted to ſcatter, the plants will riſe without care, and may be tranſplanted into woods, or in wilderneſs quarters, where they will grow in great ſhade, and make a good appearance at a ſeaſon when there are but few plants in beauty.

The ſecond ſort grows naturally at Ditton near Cambridge, and in the woods near Stoken Church in Oxfordſhire. The ſtalks of this ſort grow more upright than thoſe of the firſt, and do not branch ſo much. The leaves are compoſed of nine long lobes, which unite to the foot-ſtalk at their baſe, and are ſharply ſawed on their edges; they are of a lighter green than thoſe of the firſt ſort The flowers are produced at the top of the ſtalk; they are compoſed of five oval green petals, with a great number of ſtamina ſurrounding the germen, ſituated in the middle; theſe appear the beginning of February, and the ſeeds ripen the end of May, which, if ſown ſoon after they are ripe, the plants will come up early the following ſpring, and when they have obtained ſtrength may be planted in ſhady places under trees, where they will thrive and flower very well. The leaves of this ſort decay in autumn, and new ones ariſe from the roots in the ſpring, but the firſt ſort is always green.

The third ſort is ſuppoſed to be the Hellebore of the ancients; this grows naturally on the Alps and Apennine mountains, and alſo in the Archipelago. The root of this ſort is compoſed of many thick fleſhy fibres, which ſpread far into the ground, from which ariſe the flowers upon naked footſtalks immediately from the root, each ſupporting one large white flower, compoſed of five roundiſh petals, with a great number of ſtamina in the middle. The leaves of this are compoſed of ſeven or eight thick, fleſhy, obtuſe lobes, which are ſlightly ſawed on their edges, and unite with the foot-ſtalk at their baſe; this plant flowers in winter, from whence the title of Chriſtmas Roſe was applied to it. This is propagated by parting of the roots in autumn; for the ſeeds ſeldom ripen well in England.

The fourth ſort is like the ſecond, but differs from it in having trifoliate leaves, which are broader. This flowers early in winter, and the ſtalks riſe higher than either of the former ſorts, but is at preſent rare in England.

The fifth ſort is the common Winter Aconite, which is ſo well known as to need no deſcription. It flowers very early in the ſpring, which renders it worthy of a place in all curious gardens, eſpecially as it requires but little room. This is propagated by offſets, which the roots ſend out in plenty. Theſe roots may be taken up and tranſplanted any time after their leaves decay, which is generally by the beginning of June till October, when they will begin to put out new fibres; but as the roots are ſmall, and nearly of the colour of the ground, ſo if care is not taken to ſearch them, many of the roots will be left in the ground. Theſe roots ſhould be planted in ſmall cluſters, otherwiſe they will not make a good appearance, for ſingle flowers ſcattered about the borders of theſe ſmall kinds are ſcarce ſeen at a diſtance; but when theſe and the Snowdrops are alternately planted in bunches, they will have a good effect, as they flower at the ſame time, and are much of a ſize.

The ſixth ſort is like the firſt, but the lobes of the leaves are broader, and the ſtalks grow taller. This grows naturally in Iſtria and Dalmatia, from whence I received ſome of the ſeeds; it has been ſuppoſed to be only a ſeminal variety of the firſt, and as ſuch I ſowed the ſeeds, but the plants had a very great appearance, and the firſt winter proving ſevere they were all deſtroyed, ſo that it is not ſo hardy as our common ſort, and depending on their being ſo, occaſioned the loſs of the plants.

HELLEBORUS ALBUS. See VERATRUM.

HELMET FLOWER, or MONK's HOOD. See ACONITUM.

HEMEROCALLIS. Lin. Gen. Plant. 391. Lily Aſphodel, or Day Lily.

The CHARACTERS are,

The flower has no empalement; in ſome ſpecies the flower is of one petal cut into ſix parts, in others it hath ſix petals. There are ſix awl-ſhaped declining ſtamina ſurrounding the ſtyle. The roundiſh furrowed germen is ſituated in the middle, which afterward

ward

ward becomes an oval three-cornered capsule with three lobes, opening with two valves, filled with roundish seeds.

The Species are,

1. Hemerocallis (*Flava*) scapo ramoso, corollis monopetalis. Hort. Upsal. 88. *Day Lily with a branching stalk, and the flower of one petal.*

2. Hemerocallis (*Minor*) scapo compresso, corollis monopetalis campanulatis. *Day Lily with a compressed stalk, and a bell-shaped flower of one petal.*

3 Hemerocallis (*Fulva*) scapo ramoso, corollis monopetalis staminibus longioribus. *Day Lily with a branching stalk, flowers of one petal, and longer stamina.*

4. Hemerocallis (*Liliastrum*) scapo simplici, corollis hexapetalis campanulatis. Hort. Cliff. 128. *Day Lily with an unbranched single stalk, and bell-shaped flowers with six petals; or St. Bruno's Lily.*

The first sort grows naturally in Hungary, Dalmatia, and Istria, but has long been an inhabitant in the English gardens. This hath strong fibrous roots, to which hang knobs or tubers like those of the Asphodel, from which come out keel-shaped leaves a foot and a half long, with a rigid midrib, the two sides drawing inward, so as to form a sort of gutter on the upper side. The flower-stalks are naked, and rise a foot and a half high, having two or three longitudinal furrows; at the top they divide into three or four short foot-stalks, each sustaining one pretty large yellow flower shaped like a Lily, having but one petal, with a short tube, spreading open at the brim, where it is divided into six parts: these have an agreeable scent, from whence some have given it the title of yellow Tuberose. It flowers in June, and the seeds ripen in August: this plant is easily propagated by offsets, which the roots send out in plenty, which may be taken off in autumn, that being the best season for transplanting of the roots, and planted in any situation, for they are extremely hardy, and will require no other culture but to keep them clean from weeds, and to allow them room that their roots may spread: they may also be propagated by seeds, which, if sown in autumn, the plants will come up the following spring, and these will flower in two years; but if the seeds are not sown till spring, the plants will not come up till the year after.

The second sort grows naturally in Siberia. This hath roots like those of the former sort, but are smaller. The leaves are not near so long, nor more than half the breadth of the former, and of a dark green colour. The flower-stalk rises a foot high, is naked and compressed, but has no furrows; at the top is produced two or three yellow flowers, which are nearer the bell-shape than those of the other species, and stand on shorter foot-stalks. It is propagated by offsets from the root, or by seeds, in the same manner as the former, but the roots do not increase so fast; it should have a moist soil and a shady situation, where it will thrive much better than in dry ground.

The third sort is a much larger plant than either of the former, the roots spread and increase much more, therefore is not proper furniture for small gardens. The leaves are near three feet long, hollowed like those of the former, turning back toward the top. The flower-stalks are as thick as a man's finger, and rise near four feet high; they are naked, without joints, and branching at the top, where are several large copper-coloured flowers, shaped like those of the red Lily, and as large. The stamina of this sort are longer than those of the other, and their summits are charged with a copper-coloured farina, which sheds on being touched; or if a person smells to the flowers it will fly off and spread over the face, dyeing it all over of a copper colour, which is a trick often played by some unlucky people to the ignorant. These flowers never continue longer than one day, but there is a succession of flowers on the same plants for a fortnight or three weeks. The roots of this sort propagate too fast for those gardens where there is but little room. It will grow on any soil or situation; the best time to transplant the roots is in autumn.

The Savoy Spiderwort, or, as the French call it, St. Bruno's Lily, is a plant of humbler growth than either of the former: there are two varieties of this, one is titled Liliastrum Alpinum majus, and the other Liliastrum Alpinum minus, by Tournefort: the first of these rises with a flower-stalk more than a foot and a half high; the flowers are much larger, and there is a greater number upon each stalk than of the second; but as there is no other essential difference between them, I have not put them down as different species; but the first is by much the finer plant, though not common in England. The leaves of this sort are somewhat like those of the Spiderwort, are pretty firm, and grow upright; the flower-stalks grow about a foot high, and have several white flowers at the top, shaped like those of the Lily, which hang on one side, and have an agreeable scent; these are but of short duration, seldom continuing in beauty above three or four days; but when the plants are strong they will produce eight or ten flowers upon each stalk, so they make a good appearance while they last.

These sorts are usually propagated by parting the roots; autumn is the best season for doing this work, as it also is for transplanting of the roots; for when they are removed in the spring they seldom flower the same year, or if they do, it is but weakly. These plants should not be transplanted oftener than every third year, when the roots may be parted to make an increase of the plants, but they should not be divided too small, for if they are it will be two years before they flower: they delight in a light loamy soil, and in an open exposure, so must not be planted under the drip of trees; but if they are planted to an east aspect, where they may be protected from the sun in the heat of the day, they will continue in beauty longer than when they are more exposed.

HEMIONITIS. Moonfern.

This is a plant which is seldom propagated in gardens, therefore I shall not trouble the reader with any account of it more than this, That whoever hath a mind to cultivate any of the sorts, must procure the plants from the countries where they naturally grow. There are two sorts which are natives of the warmer parts of Europe, but in America there is a great number of very different kinds; these must be planted in pots filled with loamy undunged earth, and such of them as are natives of hot countries must be placed in the stove; the others may be sheltered under a common frame in winter, and during the summer they must be frequently watered, but in winter they will require but little. In summer they should also have plenty of free air admitted to them; with this management the plants will thrive.

HEPATICA. Boerh. Ind. Plant. Hepatica, or Noble Liverwort.

The Characters are,

The flower hath a three-leaved empalement. It hath six petals, which are oval, with a great number of slender stamina shorter than the petals, and several germen collected into a head, which afterward turn to acuminated seeds sitting round the styles.

The Species are,

1. Hepatica (*Nobilis*) trifolia, cœruleo flore. Clus. *The single blue Hepatica, or Noble Liverwort.*

2. Hepatica (*Plena*) trifolia, flore cœruleo pleno. Clus. *The double blue Hepatica, or Noble Liverwort.*

3. Hepa-

3. HEPATICA (*Alba*) trifolia, flore albo fimplici. Boerh. Ind. *The fingle white Hepatica, or Noble Liver-wort.*

4. HEPATICA (*Vulgaris*) trifolia, rubro flore. Cluf. *Single red Hepatica, or Noble Liverwort.*

5. HEPATICA (*Rubra*) trifolia, flore rubro pleno. Boer. Ind. *Double red, or Peach-coloured Hepatica.*

Thefe plants are fome of the greateft beauties of the fpring; their flowers are produced in February and March in great plenty, before the green leaves appear, and make a very beautiful figure in the borders of the pleafure-garden, efpecially the double forts, which commonly continue a fortnight longer in flower than the fingle, and the flowers are much fairer. I have feen the double white kind often mentioned in books, but could never fee it growing, tho' I do not know but fuch a flower might be obtained from feeds of the fingle white, or blue kinds. I have fometimes known the double blue fort produce fome flowers in autumn which were inclining to white, and thereby fome people have been deceived, who have procured the roots at that feafon, and planted them in their gardens, but the fpring following their flowers were blue as before; and this is what frequently happens, when the autumn is fo mild as to caufe them to flower.

The fingle forts produce feeds every year, whereby they are eafily propagated, and alfo new flowers may be that way obtained. The beft feafon for fowing of the feeds is in the beginning of Auguft, either in pots or boxes of light earth, which fhould be placed fo as to have only the morning fun until October, when they fhould be removed into the full fun, to remain during the winter feafon; but in March, when the young plants will begin to appear, they muft be removed again to a fhady fituation, and in dry weather fhould be frequently watered; about the beginning of Auguft they will be fit to be tranfplanted, at which time you fhould prepare a border, facing the eaft, of good frefh loamy earth, into which you fhould remove the plants, placing them at about fix inches diftance each way, clofing the earth pretty faft to their roots, to prevent the worms from drawing them out of the ground, which they are very apt to do at that feafon; the fpring following they will begin to fhew their flowers, but it will be three years before they flower ftrong; till then you cannot judge of their goodnefs, when, if you find any double flowers, or any of a different colour from the common forts, they fhould be taken up and tranfplanted into the borders of the flower-garden, where they fhould continue at leaft two years before they are taken up and parted; for it is remarkable in this plant, that where they are often removed and parted, they are very fubject to die; whereas, when they are permitted to remain undifturbed for many years, they will thrive exceedingly, and become very large roots.

The double flowers, which never produce feeds, are propagated by parting their roots, which fhould be done in March, at the time when they are in flower; but you fhould be careful not to feparate them into very fmall heads, nor fhould they be parted oftener than every third or fourth year, if you intend to have them thrive, for the reafon before given. They delight in a loamy foil and in an eaftern pofition, where they may have only the morning fun.

HEPATORIUM. See EUPATORIUM.

HEPTAPHYLLUM. See POTENTILLA.

HERACLEUM. See SPHONDYLIUM and PANAX.

HERBA GERARDI. See ANGELICA SYLVESTRIS MINOR.

HERBA PARIS. See PARIS.

HERMANNIA. Tourn. Inft. R. H. 656. tab. 432.

The CHARACTERS are,

The flower hath a pitcher-fhaped permanent empalement. It hath five petals, which twift againft the fun within the tubulous empalement, but fpread open above. It hath five broad ftamina joined in one body. In the center is fituated a roundifh five-cornered germen, which afterward becomes a five-cornered roundifh capfule, with five cells opening at the top, inclofing many feeds.

The SPECIES are,

1. HERMANNIA (*Alnifolia*) foliis cuneiformibus plicatis, crenato emarginatis. Hort. Cliff. 342. *Hermannia with wedge-fhaped folded leaves, which are crenated and indented.*

2. HERMANNIA (*Groffulariæfolia*) foliis obovatis acute incifis, pedunculis bifloris. Prod. Leyd. 347. *Hermannia with oval leaves acutely cut, and foot-ftalks having two flowers.*

3. HERMANNIA (*Althææfolia*) foliis obovatis plicatis crenatis tomentofis. Hort. Cliff. 343. *Hermannia with oval, folded, woolly leaves, which are crenated.*

4. HERMANNIA (*Hyffopifolia*) foliis lanceolatis obtufis ferratis. Hort. Cliff. 342. *Hermannia with obtufe fpear-fhaped leaves which are fawed.*

5. HERMANNIA (*Trifoliata*) foliis oblongo-ovatis crenatis tomentofis flore mutabili. *Hermannia with oblong, oval, crenated, woolly leaves, and a changeable flower.*

6. HERMANNIA (*Pinnata*) foliis pinnatifidis linearibus. Hort. Cliff. 342. *Hermannia with narrow leaves ending in many points.*

7. HERMANNIA (*Lavendulæfolia*) foliis lanceolatis obtufis integerrimis. Hort. Cliff. 342. *Hermannia with obtufe fpear-fhaped leaves which are entire.*

The firft fort rifes with a fhrubby ftalk fix or feven feet high, dividing into many irregular branches, covered with a brown bark, garnifhed with wedge-fhaped leaves, which are narrow at their bafe, but broad and round at the top. The flowers are produced in fhort fpikes on the upper part of the branches, of a deep yellow colour, but fmall; thefe appear in April and May, and are fucceeded by feeds in Auguft.

The fecond fort is a fhrub of lower ftature than the firft, but fends out a great number of branches, garnifhed with fmaller leaves than thofe of the former, which are rough, and fit clofe to the branches. The flowers are produced in fhort clofe fpikes at the end of every fhoot, fo that the whole fhrub feems covered with flowers; they are of a bright yellow, and appear toward the end of April, but are not fucceeded by feeds in England.

The third fort is a plant of humbler growth than the former, feldom rifing more than two feet and a half high; the ftem is not fo woody, and the branches are foft and flender, garnifhed with oval woolly leaves which are plaited, and crenated on the edges; the flowers are produced in loofe panicles at the end of the branches; they are larger than thofe of the other fpecies, and have very hairy empalements. This fort flowers in June and July.

The fourth fort has been longer in the European gardens than either of the other. This rifes with a fhrubby upright ftalk to the height of feven or eight feet, fending out many ligneous branches from the fide, which alfo grow more erect than any of the other; thefe are cloathed with obtufe fpear-fhaped leaves, fawed on the edges toward the end; the flowers come out in fmall bunches from the fide of the ftalk: they are of a pale ftraw colour, and appear in May and June; thefe are frequently fucceeded by feeds, which ripen the latter end of Auguft.

The fifth fort feldom rifes more than two feet high, with a foft ligneous ftalk, fending out flender irregular branches, garnifhed with oblong, oval, woolly leaves, ftanding upon pretty long foot-ftalks; the flowers are produced in loofe fpikes at the end of the branches; thefe are at their firft appear-

appearance of a gold colour, but after they have been some days open they change to yellow. This flowers in June and July.

The sixth sort rises with a shrubby stalk near three feet high, sending out many slender branches covered with a reddish bark, garnished with narrow wing-pointed leaves; the flowers come out from the side of the branches in small clusters; they are small, and of a deep yellow colour.

The seventh sort hath shrubby branching stalks, which are very bushy, but seldom rise more than a foot and a half high; the branches are very slender, garnished with hairy, pale, green leaves, of different sizes; they are entire, and sit pretty close to the branches; the flowers come out from the side of the stalk singly; they are small, and of a deep yellow colour. This sort flowers most part of summer.

All the species of this genus yet known, are natives of the country about the Cape of Good Hope.

The plants are generally propagated by cuttings, which may be planted in any of the summer months, on a shady border, observing to water them until they are well rooted, which will be in about six weeks after planting; then they should be taken up, preserving a ball of earth to their roots, and planted into pots, placing them in a shady situation until they have taken fresh root; after which they may be exposed to the open air, with Myrtles, Geraniums, &c. until the middle or latter end of October, when they must be removed into the green-house, observing to place them in the coolest part of the house, where they may have as much free air as possible; for if they are too much drawn in the house, they will appear sickly, and seldom produce many flowers; whereas, when they are only sheltered from the frost, and have a great share of free air, they will appear strong, healthy, and produce large quantities of flowers: they must be frequently watered, and will require to be new potted at least twice every year, otherwise their roots will be so matted as to prevent their growth.

HERMODACTYLUS. The Hermodactyl, commonly called Snakes-head Iris.

This genus is by Dr. Linnæus joined to Iris, the characters of the flower agreeing pretty well with those of that genus, from which Tournefort has separated it by the difference of the root. As this plant requires a particular treatment, I have continued it under Tournefort's title.

The CHARACTERS are,

It hath a Lily-shaped flower, consisting of one leaf, shaped exactly like an Iris, but has a tuberous root divided into two or three digs, like oblong bulbs.

We have but one SPECIES of this plant, viz.

HERMODACTYLUS (*Quadrangularis*) folio quadrangulo. C. B. P. This is also called Iris tuberosa Belgarum. *Tuberous Iris of the Dutch.*

This plant is easily propagated by its tubers, which should be taken off soon after the green leaves decay, which is the proper season for transplanting the roots; but they should not be kept long out of the ground lest they shrink, which will cause them to rot when they are planted. They should have a loamy soil, not too strong, and must be planted to an east aspect, where they will flower very well. These roots should not be removed oftener than once in three years. The distance at which these roots should be planted is six inches square, and four inches deep in the ground. These produce their flowers in May, and their seeds are ripe in August; but as they multiply pretty fast by their roots, few people are at the trouble of raising them from seeds; but those who have an inclination so to do, must treat them in the manner directed for the bulbous Irises.

The roots of this plant are very apt to run deep into the ground, and then they seldom produce flowers; and many times they shoot so deep as to be lost, especially where the soil is very light; therefore, to prevent this, it will be proper to lay a thickness of rubbish under the border where these are planted, to hinder them from getting down; This should always be practised in light ground, but in strong land there will be no occasion to make use of this precaution, because they do not shoot downward so freely in that.

HERNANDIA. Plum. Nov. Gen. 8. tab. 40. Jack-in-a-box, vulgò.

The CHARACTERS are,

It hath male and female flowers on the same plant; the male flowers have a partial involucrum, composed of four oval small leaves, which inclose three flowers; each of these has a proper bell-shaped empalement of one leaf; the petal is funnel-shaped, cut into six parts at the brim; it hath three short stamina inserted in the empalement. The female flowers are shaped like the male, but want stamina; they have a roundish germen. The empalement afterward becomes a large, swollen, oblong fruit, perforated at each end, inclosing one hard globular nut.

We have but one SPECIES of this genus, viz.

HERNANDIA (*Sonora.*) Hort. Cliff. 485. tab. 13. *Commonly called in the West-Indies Jack-in-a-box.*

This plant is very common in Jamaica, Barbadoes, St. Christopher's, and many other islands in the West-Indies, where it is known by the name of Jack-in-a-box. The fruit of this plant when ripe is perforated, and the nut in the inside becomes hard; so that when the wind blows through the fruit it makes a whistling noise, which may be heard at a distance, from whence I suppose the inhabitants gave this name to the plant. It grows in the gullies, where there are rills of water.

In Europe this plant is preserved in curious gardens, with other tender exotic plants. It is propagated by sowing the seeds on a hot-bed in the spring, and when the plants are two inches high they should be transplanted each into a separate pot, and plunged into the hot-bed again, observing to water and shade them until they have taken root; after which time they must have air admitted to them, in proportion to the warmth of the air, or the heat of the bed in which they are placed. As the plants advance, they should be removed into larger pots; but in doing this care should be taken not to break the roots, as also to preserve a good ball of earth to them; the plants must be screened from the sun until they have taken new root. The best time to shift these plants is in July, that they may be well rooted before the cold approaches; they must be constantly kept in the bark-stove; in winter they should have a moderate share of heat, and in the summer they must have plenty of air in hot weather. With this management the plants will grow to the height of sixteen feet or more, and the leaves being very large, will make a beautiful appearance in the stove. It hath not as yet flowered in England.

HERNIARIA. Tourn. Inst. R. H. 507. tab. 283. Rupturewort.

The CHARACTERS are,

The flower hath a coloured empalement of one leaf. It hath five small stamina, situated in the divisions of the empalement, and five others which are barren, placed alternately between them. In the center is an oval germen, which afterward turns to a small capsule inclosed in the empalement, having one oval-pointed seed.

The SPECIES are,

1. HERNIARIA (*Glabra*) glabra herbacea. J. B. 3. 378. *Smooth Rupturewort.*

2. HERNIARIA (*Hirsuta*) hirsuta herbacea. J. B. 3. 379. *Rough or hairy Rupturewort.*

3. HER-

3. Herniaria (*Alsines folia*) alsines folia. Tourn. Inst. 507. *Rupturewort with a Chickweed leaf.*

4. Herniaria (*Fruticosa*) caulibus fruticosis floribus quadrifidis. Lin. Syst. *Rupturewort with shrubby stalks, and four-pointed flowers.*

The two first sorts grow naturally in England, but not very common; they are low trailing plants, their branches lying on the ground; they have leaves like the smaller Chickweed, the first is smooth, and those of the second are hairy; the flowers come out in clusters from the side of the stalks at the joints; they are small, of a yellowish green, so make no appearance.

The third sort is an annual plant, which grows naturally in France and Italy, and also in Cornwal. This doth not spread so much as either of the other sorts, but the flowers and leaves are somewhat like the first, but larger.

The fourth sort hath shrubby stalks, which trail upon the ground, garnished with small hairy leaves like the second sort; the flowers are also very like, but are four-cornered.

These plants are seldom cultivated but in botanic gardens for the sake of variety. The three first are annual plants, which do not continue longer than one year, so should be permitted to shed their seeds, whereby they are better preserved than if sown with art. The fourth sort is an abiding plant, which may be propagated by cuttings; but as they are plants of no beauty, they are rarely preserved in gardens.

HESPERIS. Tourn. Inst. R. H. 222. tab. 108. Dame's Violet, Rocket, or Queen's Gilliflower.

The Characters are,

The flower is composed of four oblong petals in form of a cross. It hath six awl-shaped stamina, two of which are much shorter than the other. It hath a honey gland situated between two short stamina, and a four-cornered germen the length of the stamina, but no style. The germen afterward becomes a plain, long, compressed pod with two cells, divided by an intermediate partition, inclosing many oval compressed seeds.

The Species are,

1. Hesperis (*Matronalis*) caule simplici erecto, foliis lanceolatis denticulatis, petalis emarginatis. *Dame's Violet with a single erect stalk, spear-shaped indented leaves, and the petals of the flower indented at the top.*

2. Hesperis (*Alba*) caule simplici erecto, foliis ovato-lanceolatis integerrimis, petalis integris. *Dame's Violet with a single upright stalk, oval, spear-shaped, entire leaves, and the petals of the flower entire.*

3. Hesperis (*Inodora*) caule simplici erecto, foliis lanceolatis acutis serratis, petalis mucrone emarginatis. *Dame's Violet with a single upright stalk, spear-shaped, acute, sawed leaves, and the tips of the petals indented.*

4. Hesperis (*Tristis*) caule hispido ramoso patente. Hort. Upsal. 187. *Dame's Violet with a prickly, branching, spreading stalk.*

5. Hesperis (*Siberica*) caule erecto, ramoso, hirsuto, foliis oblongo-cordatis, acutis sessilibus denticulatis, petalis integris. *Dame's Violet with a hairy, erect, branching stalk, oblong, heart-shaped, pointed, indented leaves, sitting close to the stalk, and the petals of the flower entire.*

6. Hesperis (*Exigua*) caule ramosissimo diffuso, foliis lineari-lanceolatis dentatis, siliquis apice truncatis. *Dame's Violet with a very branching diffused stalk, narrow, spear-shaped, indented leaves, and the points of the pods shaped like a truncheon.*

7. Hesperis (*Dentata*) foliis dentato-pinnatifidis, caule lævi. Lin. Sp. Pl. 664. *Dame's Violet with wing-pointed indented leaves, and a smooth stalk.*

8. Hesperis (*Africana*) caule ramosissimo diffuso, foliis lanceolatis serratis scabris, siliquis sessilibus. Lin. Sp. Pl. 663. *Dame's Violet with very branching diffused stalks, spear-shaped, rough, sawed leaves, and the pods sitting close to the stalks.*

9. Hesperis (*Verna*) caule erecto ramoso, foliis cordatis amplexicaulibus, serratis villosis. Lin. Sp. Plant. 664. *Dame's Violet with an erect branching stalk, and hairy, sawed, heart-shaped leaves embracing the stalk.*

The first sort grows naturally in Italy. This was formerly in greater plenty in the English gardens than at present, having been long neglected because the flowers were single, and made but little appearance; however as the flowers have a very grateful scent, so the plant is worthy of a place in every good garden. This rises with an upright stalk two feet and a half high, garnished with spear-shaped leaves which sit close to the stalk; they are slightly indented on their edges, and end in acute points: the flowers are produced in a loose thyrse on the top of the stalks, composed of four roundish petals, indented at their points, of a deep purple colour, and smell very sweet, especially in the evening or in cloudy weather. It flowers in June, and the seeds ripen the latter end of August. It is a biennial plant, so that young plants should be raised every year, to supply the place of those which decay: if the seeds are permitted to scatter, the plants will come up without trouble in the spring; and if the seeds are sown, it should be in the autumn; because those which are sown in the spring often fail if the season proves dry, or will remain a long time in the ground before they vegetate.

There is a variety of this with double flowers in some of the gardens in France, but that which we have in England is a variety of the third sort with unsavoury flowers.

The second sort has been generally supposed only a variety of the first, differing in the colour of the flower, but is certainly a distinct species; the leaves of this are not so long, but much broader than those of the first, and their borders are entire; the flowers are not quite so large, nor do they form so good spikes; they are white, and have not so fine a scent as the first. This is also a biennial plant, requiring the same treatment as the first.

The third sort grows naturally in Hungary and Austria. This rises with an upright stalk two feet high, garnished with spear-shaped leaves, ending in acute points, sharply indented on their edges, of a dark green colour, and sit close to the stalks; the flowers grow in loose spikes on the top of the stalks; in some they are white, in others purple, and sometimes both colours striped in the same flower; these have no odour, so are not deserving of a place in gardens, but may be propagated in the same manner as the two former.

From this sort the double white and purple Rockets have been accidentally obtained, which are much esteemed for the beauty of their flowers; and if they had the agreeable odour of the garden Rocket, they would be some of the best furniture for the borders of the flower-garden, but they are without scent; however for the beauty of their flowers, they are by some greatly esteemed.

These plants with single flowers rarely survive the second year; nor will those with double flowers continue much longer, therefore young plants should be annually raised to supply the place of the old ones, otherwise there will soon be a want of them, which is what few persons are careful enough to observe, but thinking the roots to be perennial, trust to their putting out offsets, and finding them decay, are apt to think their soil very improper for them, and are at a loss to account for their decaying; whereas, when the plants have flowered, they have finished their period, and seldom continue to flower a second time from the same root; though in poor land they will often put out a few weak offsets, which may flower again, but seldom so strong as the principal roots; therefore thofe who are defirous to propagate thefe plants, fhould do it in the following manner:

There fhould be fome ftrong roots of each fort kept apart for this purpofe, which fhould have their flower-ftalks cut down foon after they come out, which fhould be planted in a gentle loamy foil to an eaft expofure, where they may have only the morning fun, and covered with hand or bell-glaffes, which fhould be put over them, after the cuttings have been well watered, and clofely fhut down, drawing the earth round the rim of the glaffes to exclude the air; the glaffes fhould be fhaded with mats every day when the fun is hot. With this management the cuttings will put out roots in five or fix weeks, and will begin to fhoot above; then the glaffes fhould be gently raifed on one fide to admit the air to them, and fo gradually harden the plants to the open air, to prevent their drawing up weak. When thefe have made good roots, they fhould be carefully removed, and planted in an eaft border at about eight or nine inches afunder, obferving to fhade and water them until they have taken new root; after which they will require no other care, but to keep them clean from weeds till the autumn, when they may be tranfplanted into the borders of the pleafure-garden, where they are defigned to flower.

The roots, whofe ftalks are thus cut down, will put out many heads or offsets, which may be divided when they have proper ftrength, and treated as the former; by this method a fupply of plants for the flower-garden may be always obtained.

Thefe plants are very fubject to canker and rot when they are planted in a light rich foil, but in ftrong ground I have feen them thrive and flower in the utmoft perfection, where the ftems of flowers have been as large, and the flowers as fair as the fineft double Stock-gilliflowers. Their feafon of flowering is in June; I have frequently raifed young plants from the ftalks after the flowers have decayed, by cutting them in lengths, and planting them in the manner before directed; but thefe feldom make fo good plants as the young cuttings, nor are they fo certain to grow, therefore the other method is to be preferred.

The fourth fort grows naturally in Hungary. This is much cultivated in the gardens abroad, for the great fragrancy of its flowers, which in the evening is fo ftrong as to perfume the air to a great diftance, efpecially where there are any number of the plants. The ladies in Germany are very fond of this plant, and during the feafon of their flowering have the pots placed in their apartments every evening, that they may enjoy the fragrancy of their flowers; for they have but little beauty, being fmaller than thofe of the garden Rocket, and of a pale colour, but the fcent of their flowers is much preferable to them; though in the day-time, if the weather is clear, they have very little odour; but when the fun leaves them, their fragrancy is expanded to a great diftance. To this fpecies it is fuppofed that the title of Dame's Violet was firft applied.

This fort is very rarely feen in the Englifh gardens: I fuppofe it has been neglected, becaufe the flowers make no appearance. It is a biennial plant, like the garden Rocket, which is propagated by feeds in the fame manner; but the plants are not quite fo hardy, and are very fubject to rot in winter, efpecially on a moift foil or in rich land, where they are apt to grow very rank, fo are foon injured by wet and cold in the winter; therefore the plants of this fort fhould be planted in a dry poor foil and a warm fituation; and if fome of them are planted in pots to be placed under a common frame in winter, where they may be fheltered from hard rains and froft, and enjoy the free air at all times when the weather is mild, it will be a fure way to preferve them.

The leaves of this fort are much larger, of a paler green than thofe of the garden Rocket; the ftalks are clofely fet with briftly hairs; the flowers grow in loofe panicles at the top of the ftalk, and appear about the fame time with the garden Rocket.

The feeds of the fifth fort were fent me from Germany without any title, nor any account of the country from whence it came; but as it was fent with the feeds of fome Siberian plants, I fuppofe this came from the fame country. This is a biennial plant, which rifes with a ftrong branching ftalk three feet high, which is hairy, garnifhed with oblong heart-fhaped leaves, ending in acute points, fitting clofe to the ftalk, and are flightly indented on their edges; the upper part of the ftalk divides into many branches, garnifhed with fmall leaves of the fame fhape with thofe below, and are terminated with loofe panicles of fingle, large, purple flowers, of great fragrancy.

The fixth fort grows naturally in the warm parts of Europe. This is annual; the ftalks rife about eight or nine inches high, branching out greatly on every fide in a confufed order, garnifhed with fmall, narrow, indented leaves, and are terminated by clufters of fmall yellow flowers which make no appearance.

The feventh fort grows naturally in Sicily. This is an annual plant, which feldom rifes more than fix inches high: the ftalk branches toward the top into three or four fmaller, which are terminated by fmall white flowers.

The eighth fort grows naturally in Africa. This is an annual plant with a very branching ftalk, which rifes about nine inches high, garnifhed with rough fpear-fhaped leaves fawed on their edges, terminated by loofe panicles of fmall purple flowers, which appear in June and July, and are fucceeded by long pods fitting clofe to the ftalks, filled with fmall feeds, which ripen in September.

Thefe three forts are rarely cultivated, except in botanic gardens for the fake of variety. If the feeds of thefe are permitted to fcatter, the plants will come up without care, and only require to be kept clean from weeds; or they may be fown either in the fpring or autumn where they are to ftand, for they do not bear tranfplanting well.

The ninth fort is an annual plant, which grows naturally in the fouth of France. This fends out feveral heart-fhaped hairy leaves from the root, which fpread on the ground: the ftalk rifes nine inches high, branching toward the top, garnifhed with leaves of the fame fhape, which embrace the ftalks with their bafe; the flowers are produced in loofe panicles at the end of the branches, of a lively purple colour. If thefe feeds are fown in the autumn, they fucceed much better than in the fpring.

HEUCHERA. Lin. Gen. Plant. 283. Sanicle.

The CHARACTERS are,

The flower is compofed of five narrow petals, which are inferted in the border of the empalement. It hath five ftamina which are much longer than the empalement, and a roundifh bifid germen, which afterward turns to an oval-pointed capfule with two horns, which are reflexed, having two cells filled with very fmall feeds.

We have but one SPECIES of this genus, viz.

HEUCHERA (*Americana.*) Hort. Cliff. 82. *American Sanicle with a dirty purple flower.*

This plant grows naturally in Virginia. It hath a perennial root, which fends out many heart-fhaped oval leaves, indented into four or five lobes, crenated on their edges, of a lucid green, and fmooth; from between thefe come out the foot-ftalks of the flower, which are naked, and rife a foot high, dividing at the top into a loofe panicle, fuftaining many fmall hairy flowers, of an obfolete purple colour. This flowers in May, and the feeds ripen in Auguft.

It is propagated by parting of the roots in autumn, and ſhould be planted in a ſhady ſituation; there is little beauty in this plant, but it is preſerved for the ſake of variety.

HIBISCUS. Lin. Gen. Plant. 756. Syrian Mallow.

The CHARACTERS are,

The flower hath a double permanent empalement; the outer hath eight or ten narrow leaves; the inner is of one leaf, cut at the brim into five acute points. It hath five heart-ſhaped petals, which join at their baſe into one; and many ſtamina, which are joined in form of a column, but expand toward the top. It has a round germen, which afterward turns to a capſule with five cells, opening in five parts, incloſing kidney-ſhaped ſeeds.

The SPECIES are,

1. HIBISCUS (*Syriacus*) foliis ovato-lanceolatis, ſupernè inciſo ſerratis, caule arboreo. Hort. Cliff. 350. *Hibiſcus with oval ſpear-ſhaped leaves, whoſe upper parts are cut and ſawed, and a tree-like ſtalk; commonly called Althæa frutex.*

2. HIBISCUS (*Chinenſis*) foliis cordato quinquangularis obſoletè ſerratis, caule arboreo. Hort. Upſal. 205. *Hibiſcus with heart-ſhaped leaves, having five angles which are bluntly ſawed, and a tree-like ſtalk; commonly called China Roſe.*

3. HIBISCUS (*Abelmoſchus*) foliis ſubpeltato cordatis ſeptemangularibus, ſerratis hiſpidis. Hort. Cliff. 349. *Hibiſcus with heart-ſhaped target leaves, having ſeven angles, which are ſawed, and ſet with prickly hairs; commonly called Muſk.*

4. HIBISCUS (*Manihot*) foliis palmato digitatis quinquepartitis, laciniis ſupernè dentatis, caule lævi. *Hibiſcus with fingered leaves, which are divided into five parts, indented toward the top, and a ſmooth ſtalk.*

5. HIBISCUS (*Tomentoſus*) foliis cordatis angulatis ſerratis tomentoſis, caule arboreo. *Hibiſcus with angular, heart-ſhaped, ſawed, woolly leaves, and a tree-like ſtalk.*

6 HIBISCUS (*Tiliaceus*) foliis cordatis integerrimis. Flor. Zeyl. 258. *Hibiſcus with entire heart-ſhaped leaves.*

7. HIBISCUS (*Javanica*) foliis ovatis acuminatis ſerratis glabris, caule arboreo. Flor. Zeyl. 260. *Hibiſcus with oval, pointed, ſawed, ſmooth leaves, and a tree-like ſtalk.*

8. HIBISCUS (*Vitifolius*) foliis ſerratis, inferioribus ovatis indiviſis, ſuperioribus quinquepartitis, caule aculeato. *Hibiſcus with ſawed leaves, the lower oval and undivided, the upper divided into five parts, and a prickly ſtalk.*

9. HIBISCUS (*Sabdariffa*) foliis ſerratis, inferioribus cordatis, mediis tripartitis, ſummis quinquepartitis, caule aculeato. *Hibiſcus with ſawed leaves, the lower ones being heart-ſhaped, the middle divided into three parts, the upper into five, and a prickly ſtalk.*

10. HIBISCUS (*Goſſypifolius*) foliis quinquelobatis ſerratis, caule glabro. *Hibiſcus with ſawed leaves divided into five lobes, and a ſmooth ſtalk.*

11. HIBISCUS (*Ficulneus*) foliis quinquefido-palmatis, caule aculeato. Hort. Cliff 498. *Hibiſcus with hand-ſhaped five-pointed leaves, and a prickly ſtalk.*

12. HIBISCUS (*Surratenſis*) foliis quinquepartitis, lobis ovato lanceolatis hirſutis crenatis, caule ſpinoſiſſimo. *Hibiſcus with leaves divided into five lobes, which are oval, ſpear-ſhaped, hairy, and crenated, and a very prickly ſtalk.*

13. HIBISCUS (*Cordifolius*) foliis cordatis hirſutis crenatis, floribus lateralibus, caule arboreo ramoſo. *Hibiſcus with heart-ſhaped, hairy, crenated leaves, flowers growing from the ſides of the branches, and a tree-like branching ſtalk.*

14. HIBISCUS (*Bahamenſis*) foliis oblongo-cordatis glabris, denticulatis, ſubtus incanis, floribus ampliſſimis. *Hibiſcus with oblong, heart-ſhaped, ſmooth, indented leaves, hoary on their under ſide, and very large flowers.*

15. HIBISCUS (*Ficifolius*) foliis quinquepartito pedatis, calycibus inferioribus latere rumpentibus Lin. Sp. Plant. 696. *Hibiſcus with leaves like a foot, divided into five parts, and the lower empalement torn ſideways.*

16. HIBISCUS (*Pentacarpos*) foliis inferioribus cordatis angulatis, ſuperioribus ſubhaſtatis, floribus ſubnutantibus, piſtillo cernuo. Lin. Sp. Plant. 697. *Hibiſcus with lower leaves heart-ſhaped and angular, the upper ones ſomewhat ſpear-ſhaped, nodding flowers, and a recurved piſtil.*

17. HIBISCUS (*Populneus*) foliis ovatis acuminatis ſerratis, caule ſimpliciſſimo, petiolis floriferis. Hort. Upſal. 205. *Hibiſcus with oval, pointed, ſawed leaves, a ſingle ſtalk, and flowering foot-ſtalks.*

18. HIBISCUS (*Paluſtris*) caule herbaceo ſimpliciſſimo, foliis ovatis ſubtrilobis, ſubtus tomentoſis, floribus axillaribus. Lin. Sp. Plant. 693. *Hibiſcus with a ſingle herbaceous ſtalk, oval leaves having three lobes, and woolly on their under ſide.*

19. HIBISCUS (*Trionum*) foliis tripartitis inciſis, calycibus inflatis. Hort. Upſal. 206. *Hibiſcus with tripartite cut leaves, and a bladdery empalement.*

20. HIBISCUS (*Africanus*) foliis tripartitis dentatis, lobis anguſtioribus caule hirſuto calycibus inflatis. *Hibiſcus with tripartite indented leaves, having narrower lobes, a hairy ſtalk, and bladdery empalements.*

21. HIBISCUS (*Hiſpidus*) foliis inferioribus trilobis, ſummis quinquepartitis obtuſis crenatis calycibus inflatis, caule hiſpido. *Hibiſcus with under leaves having three lobes, the upper ones being cut into five obtuſe ſegments, which are crenated, ſwollen empalements, and a prickly ſtalk.*

22. HIBISCUS (*Malvaviſcus*) foliis cordatis crenatis, angulis lateralibus extimis parvis, caule arboreo. Hort. Cliff. 349. *Hibiſcus with heart-ſhaped crenated leaves, whoſe outward lateral angles are ſmall, and a tree-like ſtalk.*

The firſt ſort is commonly called Althæa frutex by nurſery-gardeners, who propagate the ſhrubs for ſale; of this there are four or five varieties, which differ in the colour of their flowers; the moſt common hath pale purple flowers with dark bottoms, another hath bright purple flowers with black bottoms, a third hath white flowers with purple bottoms, a fourth variegated flowers with dark bottoms; there are alſo two with variegated leaves, which are by ſome much eſteemed.

Theſe ſhrubs grow naturly in Syria; they are great ornaments in the autumn ſeaſon to a garden. They riſe with ſhrubby ſtalks to the height of eight or ten feet, ſending out many woody branches, covered with a ſmooth gray bark, garniſhed with oval ſpear-ſhaped leaves, whoſe upper parts are frequently divided into three lobes. The flowers come out from the wings of the ſtalks at every joint of the ſame year's ſhoot; they are large, and ſhaped like thoſe of the Mallow, having five large roundiſh petals, which join at their baſe, which ſpread open at the top in ſhape of an open bell; theſe appear in Auguſt, and, if the ſeaſon is not too warm, there will be a ſucceſſion of flowers part of September, and are ſucceeded by ſhort capſules with five cells, filled with kidney-ſhaped ſeeds; but unleſs the ſeaſon proves warm, they will not ripen in this country.

Theſe plants are generally propagated by laying down the young branches in the ſpring, which in one year will take root, ſo may be cut off from the old plants, and planted in a nurſery, at three feet diſtance row from row, and one foot aſunder in the rows; where they may remain two or three years to get ſtrength, and then may be tranſplanted to the places where they are to remain. But when good ſeeds can be procured, the plants which are raiſed from them will grow larger and more upright than thoſe which are propagated by layers. The ſeveral varieties may be propagated by grafting upon each other, which is the common

common method of propagating the forts with ftriped leaves.

The fecond fort grows naturally in India, from whence the French firft carried the feeds to their fettlements in the Weft-Indies, and the inhabitants of the Britifh colonies have been fupplied with the feeds from them, fo have given it the title of Martinico Rofe; of this there are the double and fingle flowering; the feeds of the double frequently produce plants with fingle flowers, but the feeds of the fingle feldom vary to the double. The flowers of thefe plants alter in their colour; at their firft opening they are white, then they change to a blufh Rofe colour, and as they decay turn to a purple. In the Weft-Indies all thefe alterations happen the fame day, as I fuppofe the flowers in thofe hot countries are not of longer duration; but in England, where the flowers laft longer in beauty, the changes are not fo fudden.

This plant hath a foft fpongy ftem, which, by age, becomes ligneous and pithy. It rifes to the height of twelve or fourteen feet, fending out branches toward the top, which are hairy, garnifhed with heart-fhaped leaves, cut into five acute angles on their borders, and are flightly fawed on their edges, of a lucid green on their upper fide, but pale below. The flowers are produced from the wings of the leaves; the fingle one is compofed of five large petals, which fpread open, and are firft white, but afterward change in the manner before mentioned; thefe are fucceeded by fhort, thick, blunt capfules, which are very hairy, having five cells, which contain many fmall kidney-fhaped feeds, having a fine plume of fibrous down adhering to them.

It is propagated by feeds, which muft be fown upon a hot-bed in the fpring, and when the plants are fit to remove, they fhould be each planted in a feparate fmall pot, and plunged into a moderate hot-bed, where they muft be fhaded till they have taken new root; then they muft be treated as other plants from warm countries, but not too tenderly, for thefe require a large fhare of air in warm weather, otherwife they will draw up very weak; they will bear the open air in fummer, in a warm fheltered fituation, and will live through the winter in a very good green-houfe, provided they have not too much wet, but the plants thus hardily treated, will not make fo great progrefs, nor flower fo well as with a little additional warmth in winter; and if they are too tenderly managed, they will draw up weak, fo will be lefs likely to flower. This fort flowers at different times of the year, as in its native country.

The third fort grows naturally in the Weft-Indies, where it is commonly known by the title of Mufk; the French cultivate great quantities of thefe plants in their American iflands, the feeds of which are annually fent to France, fo that they certainly have fome way of rendering it ufeful, as it feems to be a confiderable branch of their trade. This rifes with an herbaceous ftalk about three or four feet high, fending out two or three fide branches, garnifhed with large leaves cut into fix or feven angles, which are acute, fawed on their edges, have long foot-ftalks, and are placed alternately. The ftalks and leaves of this are very hairy. The flowers come out from the wings of the leaves upon pretty long foot-ftalks, which ftand erect; they are large, of a fulphur colour, with dark purple bottoms, and are fucceeded by pyramidal five-cornered capfules, which open in five cells, filled with large kidney-fhaped feeds of a very mufky odour.

This fort feldom lives more than two years in England, but in its native country continues much longer. It is propagated by feeds, which, if fown on a good hot-bed in the fpring, and the plants afterward planted in pots, and plunged into a frefh hot-bed, treating them afterward in the fame way as the Amaranthus, to bring them forward, they will flower in July, and their feeds will ripen in the autumn.

The fourth fort grows naturally in both the Indies; this rifes with an herbaceous fmooth ftalk three or four feet high, garnifhed with leaves divided into five fegments almoft to the bottom, which are indented at their extremities, having long foot-ftalks. The flowers are produced from the wings of the leaves toward the top, ftanding on fhort foot-ftalks; they are compofed of five large fulphur-coloured petals, which, when open, fpread five inches wide, and have a dark purple bottom, with a column of ftamina and ftyles rifing in the center; they are fucceeded by large, pyramidal, five-cornered feed-veffels, opening in five cells, which are filled with pretty large kidney-fhaped feeds, which have little fmell or tafte.

It is propagated by feeds in the fame manner as the former fort, and if fo managed will produce flowers and perfect feeds the fame feafon, but the plants may be continued through the winter in a moderate warmth.

The fifth fort grows naturally in the Weft-Indies, where it rifes with a woody ftalk feven or eight feet high, fending out many fide branches toward the top, covered with a whitifh bark, garnifhed with angular heart-fhaped leaves, which are woolly. The flowers are produced from the wings of the leaves upon long foot-ftalks; they are compofed of five roundifh petals, which are joined at their bafe, of a yellow colour, turning to red as they decay; thefe are fucceeded by large, obtufe, five-cornered, hairy feed-veffels, which open in five cells, filled with large kidney-fhaped feeds.

This is propagated by feeds, which muft be fown upon a hot-bed in the fpring, and the plants afterward treated in the fame way as the two laft mentioned, during the firft fummer, but in autumn they muft be plunged into the tan-bed in the ftove, where they fhould conftantly remain, and be treated in the fame way as other tender plants from the fame country; the fecond year the plants will flower, but they have not as yet perfected feeds in England.

The fixth fort grows naturally in both Indies; this rifes with a woody pithy ftem eight or ten feet high, dividing into feveral branches toward the top, which are covered with a woolly down, garnifhed with round heart-fhaped leaves, ending in acute points; they are of a lucid green on their upper fide, and hoary on their under, full of large veins, and are placed alternately on the ftalks. The flowers are produced at the end of the branches in loofe fpikes; they are of a whitifh yellow colour, and are fucceeded by fhort acuminated capfules, opening in five cells, filled with large kidney-fhaped feeds.

This fort is propagated in the fame way, and the plants require the fame treatment as the fifth; they flower the fecond year, provided they are brought forward, otherwife they will not flower before the third or fourth feafon, but they will bear the open air in fummer in a warm fituation, though they will not make great progrefs there.

The feventh fort grows naturally on the coaft of Malabar; this rifes with a woody ftalk twelve or fourteen feet high, dividing into many fmall branches toward the top, garnifhed with oval fawed leaves, ending in acute points, of a lucid green above, but pale on their under fide, placed without order. The flowers come out from the fide of the branches at the wings of the leaves, on pretty long foot-ftalks; they are compofed of many oblong roundifh petals of a red colour, which expand like the Rofe, the flowers being as large when fully blown as the common red Rofe, and as double. This is a perennial plant, which is propagated by cuttings; the plants muft conftantly be kept in the tan-bed in the ſtove, giving them a large ſhare of air in warm weather, and but little water in winter. The flowers of theſe plants are uſed by the ladies in India to colour their hair and eyebrows black, and by men to black their ſhoes. It is from this called Shoe-flower.

The eighth ſort is an annual plant, which riſes with an upright ſtalk ſeven or eight feet high; the lower leaves are oval, ſerrated, and entire, but the upper leaves are divided almoſt to the foot-ſtalk, into five ſpear-ſhaped ſegments like the finger of a hand, ſtanding on very long foot-ſtalks, having thorns at their baſe. The flowers come out from the wings of the ſtalks; they are large, of a pale ſulphur colour, with a purple bottom, and are ſucceeded by oval, acuminated, prickly capſules, which open in five cells, filled with large kidney-ſhaped ſeeds.

This ſort is propagated by ſeeds, which muſt be ſown upon a hot-bed, and the plants treated in the ſame way as the third ſort.

The ninth ſort is near of kin to the eighth, but the ſtalks do not grow ſo tall; the lower leaves are heart-ſhaped and entire, the middle leaves are divided into three, and the upper into five ſegments, almoſt to the foot-ſtalks; they are ſawed on their edges, and the ſtalk is prickly. The flowers come out from the wings of the ſtalks; they are of a very pale ſulphur colour, with dark bottoms, but are not ſo large as thoſe of the laſt.

This is propagated by ſeeds in the ſame way as the eighth, and the plants require the ſame treatment.

The bark of both theſe plants is full of ſtrong fibres, which I have been informed the inhabitants of the Malabar coaſt prepare, and make into a ſtrong cordage; and by what I have obſerved it may be wrought into fine ſtrong thread of any ſize, if properly manufactured.

The tenth ſort grows naturally in the Weſt-Indies, where the inhabitants uſe the green pods to add an acid taſte to their viands: there are two varieties of this, one with a light green, and the other a deep red pod, which always maintain their difference, but as there is no other difference but that of the colour of their pods, they do not deſerve ſeparate titles. This riſes with an herbaceous ſtem about four feet high, ſending out a few lateral branches, garniſhed with ſmooth leaves, divided into three or five lobes. The flowers come out from the ſide of the branches, they are of a dirty white, with dark purple bottoms, and are ſucceeded by obtuſe ſeed-veſſels divided into five cells, which are filled with kidney-ſhaped ſeeds.

This ſort is propagated in the ſame way as the third, and will flower and perfect ſeeds the ſame year, ſo is ſeldom preſerved longer in England.

The eleventh ſort is a native of Ceylon; this riſes with an herbaceous ſtalk, which is prickly, from two to three feet high, dividing upward into ſmall branches, garniſhed with hand-ſhaped leaves, divided into five parts. The flowers come out from the wings of the leaves; they are ſmall, white, with purple bottoms, and are ſucceeded by ſhort obtuſe capſules with five cells, filled with kidney-ſhaped ſeeds.

This plant is annual, ſo muſt be treated in the ſame way as the third.

The twelfth ſort is alſo annual with us; this riſes with an herbaceous ſtalk three feet high, which hath prickly hairs, and divides into branches upwards, garniſhed with hand-ſhaped leaves, divided into five ſpear-ſhaped lobes, ending in acute points, crenated on their edges, ſtanding upon very long foot-ſtalks; the flowers come out from the wings of the leaves; they are very like thoſe of the third, and the plants muſt be treated in the ſame way.

The thirteenth ſort grows naturally in the iſland of Cuba. This riſes with a woody ſtalk twelve or fourteen feet high, ſending out many lateral branches, garniſhed with hairy, heart-ſhaped leaves, crenated on their edges; the flowers come out ſingle from the wings of the leaves; they are of a very bright yellow colour, but not ſo large as either of the former ſorts, and are ſucceeded by ſhort capſules ending in acute points, divided into five cells, which are filled with kidney-ſhaped ſeeds. This plant requires the ſame treatment as the fifth, and other tender kinds.

The fourteenth ſort hath a perennial root, but an annual ſtalk. This riſes with ſeveral ſtalks from the root, which grow four feet high, garniſhed with oblong, heart-ſhaped, ſmooth leaves, ending in acute points, of a light green on their upper ſide, but hoary on their under, and are ſlightly indented on their edges; the flowers are produced on the top of the ſtalks; they are very large, of a light purple colour with dark bottoms, and are ſucceeded by ſhort capſules, divided into five cells, filled with kidney-ſhaped ſeeds.

This is propagated by ſeeds, which muſt be ſown on a moderate hot-bed in the ſpring; and when the plants are fit to remove, they ſhould be each planted in a ſeparate pot, and plunged into a hot-bed, treating them in the ſame way as the other tender ſorts, but allowing them a greater ſhare of air in warm weather; thoſe plants which flowered in the Chelſea garden, were plunged into a tan-bed whoſe heat was declining, under a deep frame, where they produced plenty of flowers, but they came too late to ripen ſeeds. The ſtalks decay in the autumn, but if the pots are ſheltered from froſt in winter under a hot bed frame, they will continue ſeveral years, and put out new ſtalks in the ſpring.

The fifteenth ſort is very common in the Weſt Indies, where the inhabitants cultivate it for the pods or ſeed-veſſels, which they gather green to put into their ſoups; theſe having a ſoft viſcous juice, add a thickneſs to their ſoups, and renders them very palatable; it is called Ocra. It riſes with a ſoft herbaceous ſtalk three or four feet high, dividing upward into many branches, garniſhed with hand-ſhaped leaves, divided into five parts; the flowers are produced from the wings of the ſtalk; they are of a pale ſulphur colour, with dark purple bottoms, but are ſmaller than either of the other ſorts, and of very ſhort duration, opening in the morning with the riſing ſun, but fade long before noon in warm weather. Theſe are ſucceeded by capſules of very different forms, in the different varieties; in ſome the capſules are not thicker than a man's finger, and five or ſix inches long; in others they are very thick, and not more than two or three inches long; in ſome plants they grow erect, in others they are rather reclined; and theſe varieties are conſtant, for I have many years cultivated theſe plants, and have not found them vary.

This ſort is propagated by ſeeds in the ſame way as the third, and the plants require the ſame treatment, for they are too tender to thrive in the open air in this country; I have often tranſplanted the plants into warm borders, after they have acquired proper ſtrength, and have ſometimes in very warm ſeaſons had them thrive for a ſhort time, but the firſt cold bad weather their leaves have all dropped off, and then they have decayed gradually, ſo that they have but rarely flowered, and have never in the beſt ſeaſons perfected their ſeeds in the open air; therefore thoſe who are inclined to cultivate theſe plants, muſt conſtantly ſhelter them in bad weather.

The ſixteenth ſort grows naturally near Venice, in moiſt land. This hath a perennial root and an annual ſtalk, which riſes three or four feet high; the lower leaves are angular and heart-ſhaped, but the upper are ſpear-ſhaped, and ſlightly indented on their edges; the flowers are produced from the wings of the leaves upon long foot-ſtalks; they

they are of a purple colour, with a dark bottom, and are ſucceeded by five-cornered compreſſed capſules, filled with kidney-ſhaped ſeeds.

This ſort is propagated by ſeeds, which muſt be ſown on a hot-bed, and the plants ſhould be treated in the ſame way as the fourteenth ſort, otherwiſe they will not flower; for although the roots will live in the full ground here, yet the ſummers are ſeldom warm enough to bring them to flower.

The ſeventeenth ſort grows naturally in North America. This hath a perennial root and an annual ſtalk; the roots of this ſort will live in the full ground, but unleſs the ſummer is warm, the flowers ſeldom open. It riſes with ſingle ſtalks from the root two or three feet high; the leaves are oval and ſawed, the flowers are large and purple.

The eighteenth ſort grows naturally in North America, in moiſt ground. This hath a perennial root and an annual ſtalk like the former, which is herbaceous and never branches; the leaves are oval, with three lobes, which are not deeply divided: they are of a bright green on their upper ſide, but of an Aſh colour on their under; the flowers are produced from the wings of the ſtalk; they are large, of a bright purple colour. This ſort, like the former, ſeldom flowers in the open air here, unleſs the ſummer proves very warm; but the roots will live in the full ground, if they are planted in a ſheltered ſituation.

The nineteenth ſort is an annual plant, which grows naturally in ſome parts of Italy, and has been long cultivated in the Engliſh gardens by the title of Venice Mallow. Gerard and Parkinſon title it Alcea Veneta, and Fios Hora, or Flower of an Hour, from the ſhort duration of its flowers, which in hot weather continue but few hours open; however, there is a ſucceſſion of flowers which open daily for a conſiderable time, ſo that a few of theſe plants may be allowed a place in every curious garden.

It riſes with a branching ſtalk a foot and a half high; the leaves are divided into three lobes, which are deeply jagged almoſt to the midrib; theſe jags are oppoſite, and the ſegments are obtuſe; the flowers come out at the joints of the ſtalks upon pretty long foot-ſtalks, having a double empalement, the outer being compoſed of ten long narrow leaves, which join at their baſe; the inner is one thin leaf, ſwollen like a bladder, cut into five acute ſegments at the top, having many longitudinal purple ribs, and is hairy; both theſe are permanent, and incloſe the capſule after the flower is paſt. The flower is compoſed of five obtuſe petals, which ſpread open at the top, the lower part forming an open bell-ſhaped flower; theſe are of a ſulphur colour, with dark purple bottoms; the ſtamina and apices are joined in a column; after the flower is paſt, the germen turns to a blunt capſule, opening in five cells, which are filled with ſmall kidney-ſhaped ſeeds. This ſort is propagated by ſeeds, which ſhould be ſown where the plants are deſigned to remain. Theſe require no other culture but to keep them clean from weeds, and thin them where they are too cloſe; if the ſeeds are permitted to ſcatter, the plants will come up full as well as when ſown, ſo that it will maintain its ſituation, unleſs it is weeded out.

The twentieth ſort grows naturally at the Cape of Good Hope, this is alſo an annual plant, which reſembles the former, but the ſtalks grow more erect, are of a purpliſh colour, and very hairy; the leaves are compoſed of three narrow lobes, which are divided almoſt to the foot-ſtalk, the middle lobe ſtretching out more than twice the length of the two ſide ones; the flowers are larger, and their colour deeper than thoſe of the other.

The twenty-firſt ſort grows naturally at the Cape of Good Hope. This is an annual plant, having at firſt ſight ſome reſemblance of the ſorts before mentioned; but it has ſtrong, hairy, branching ſtalks, garniſhed with much broader leaves than either of the former, the lower being divided into three, and the upper into five obtuſe lobes, which are crenated on their edges; the flowers are large, but of a paler colour than thoſe of the other.

Theſe three are as hardy as the nineteenth ſort, ſo may be treated in the ſame way.

The twenty-ſecond ſort grows naturally at Campeachy; this differs ſo eſſentially from the other ſpecies in its fructification, as to deſerve another title, for all the other have dry capſules with five cells, including many kidney-ſhaped ſeeds; but this hath a ſoft viſcous berry, with a hard ſhell incloſed, containing five roundiſh ſeeds: it riſes with a ſhrubby ſtalk ten or twelve feet high, dividing into many branches, garniſhed with ſmooth, heart-ſhaped, angular leaves, which are crenated on their edges; the flowers come out from the wings of the ſtalks ſingly, ſtanding on ſhort foot-ſtalks; they are compoſed of five oblong petals, which are twiſted together and never expand; they are of a fine ſcarlet colour when blown, incloſing a hard ſhell, which opens in five cells, each containing a ſingle roundiſh ſeed.

This ſort is generally propagated here by cuttings, becauſe the ſeeds do not often ripen in England; if the cuttings are planted in pots, and plunged into a gentle hot-bed, keeping the air from them, they will ſoon take root, and ſhould be gradually inured to bear the open air. The plants require a moderate ſtove to preſerve them through the winter, and if they are kept in warmth in ſummer, they will flower, and ſometimes ripen fruit, though they may be placed abroad in a ſheltered ſituation for two or three months in ſummer, but the plants ſo treated ſeldom flower.

HIERACIUM. Lin. Gen. Plant. 818. Hawkweed.

The Characters are,

It hath a flower compoſed of many hermaphrodite florets, included in one common ſcaly empalement; the florets are equal and uniform; they have one petal ſhaped like a tongue, indented in five parts at the point; they have each five ſhort hairy ſtamina. At the bottom of the petal is ſituated the germen, which afterward becomes a ſhort four-cornered ſeed, crowned with down, ſitting in the empalement.

There are a great number of ſpecies of this genus, many of which grow naturally as weeds in England, therefore I ſhall only ſelect thoſe which are beſt worth cultivating, for to enumerate all the ſpecies would ſwell this work greatly beyond its intended bulk.

The Species are,

1. Hieracium (*Aurantiacum*) foliis integris caule ſubnudo ſimpliciſſimo piloſo corymbiſero. Hort. Cliff. 388. *Hawkweed with entire leaves, a ſingle hairy ſtalk, terminated by a corymbus of dark red flowers, called Grim the Collier.*

2. Hieracium (*Cerinthoides*) foliis radicalibus obovatis denticulatis, caulinis oblongis ſemiamplexicaulibus. Prod. Leyd. 124. *Hawkweed with oval indented leaves at the roots, thoſe on the ſtalks oblong, and half embracing them.*

3. Hieracium (*Blattaroides*) foliis lanceolatis amplexicaulibus dentatis floribus ſolitariis, calycibus laxis. Hort. Cliff. 387. *Hawkweed with ſpear-ſhaped indented leaves embracing the ſtalks, flowers growing ſingly, having looſe empalements.*

4. Hieracium (*Umbellatum*) foliis linearibus ſubdentatis ſparſis, floribus ſubumbellatis. Flor. Lapp. 287. *Hawkweed with linear indented leaves placed thinly, and flowers almoſt in an umbel.*

5. Hieracium (*Amplexicaule*) foliis amplexicaulibus cordatis ſubdentatis, pedunculis unifloris hirſutis, caule ramoſo. Hort. Cliff. 387. *Hawkweed with heart-ſhaped indented leaves embracing the ſtalks, hairy foot-ſtalks bearing one flower, and a branching ſtalk.*

The

The first sort grows naturally in Syria, but has been long an inhabitant of the English gardens, where it is cultivated for its fine purple flowers. This was formerly known by the title of Grim the Collier, which was formerly given it from the dark colour of the empalement.

This hath a perennial fibrous root, which sends out many oblong, oval, entire leaves, between which arises a single stalk near a foot high, which is naked toward the bottom, and terminated by a corymbus of gold-coloured flowers, which appear in succession from the beginning of June to September, and are succeeded by seeds crowned with down, which ripen in August and September. This sort sends out many offsets from the root, by which it is easily propagated.

It may also be propagated by seeds, which should be sown in an east aspected border in March; and when the plants are strong enough to remove, they should be transplanted to a shady border of undunged earth, at six inches distance, observing to water them if the weather should prove dry, till they have taken new root; after which, if they are kept clean from weeds, they will require no other culture: in the autumn they should be transplanted where they are designed to remain, the following summer they will flower and produce ripe seeds; the roots will continue many years if they are not planted in a rich or moist soil, which frequently occasions their rotting in winter.

The second sort hath very hairy, oval, indented leaves; the stalks rise about the same height as the former, branching into several divisions, which are each terminated by a large, single, yellow flower, of the shape of the former. This flowers in June, and the seeds ripen in August; it grows naturally in some places in the north of England; it is a perennial plant, and may be propagated in the same manner as the former sort.

The third sort naturally grows on the Pyrenees; this hath a perennial root, which sends up several erect stalks, garnished with spear-shaped leaves which are indented, and embrace the stalks with their base; the flowers are produced from the wings of the stalks upon short foot-stalks, each sustaining one large yellow flower. This flowers in June, but seldom perfects seeds here, so is propagated by parting of the roots in autumn; it will thrive in any situation.

The fourth sort grows also on the Pyrenean mountains. It is a perennial plant, whose lower leaves are oval, indented, and of a grayish colour; those on the stalks are smaller, but of the same shape and colour, and half embrace the stalks with their base; the stalks rise a foot high, branching into several divisions, each being terminated by one small flower. This is propagated by seeds as the first sort.

The fifth sort rises with a branching stalk a foot and a half high, garnished with heart-shaped leaves, which are indented at their base, where they embrace the stalks; each division of the branches terminate in a hairy foot-stalk, sustaining one large yellow flower, which appears in May, and the seeds ripen in July. This is a perennial plant, which is propagated by seeds as the first sort, and requires the same treatment.

HIPPOCASTANUM. See ÆSCULUS.

HIPPOCRATEA. Lin. Gen. Plant. 1098.

The CHARACTERS are,

It hath a large spreading empalement of one leaf, cut into five segments; the flower hath one oval petal, which is entire: it hath six slender short stamina, and an oval germen situated below the petal, which afterwards becomes an oblong capsule, winged at the top, inclosing a single seed.

We have but one SPECIES of this genus, viz.

HIPPOCRATEA (*Volubilis*) fructu trigemino subrotundo, caule volubili. *Hippocratea with a triple roundish fruit, and a twining stalk.*

The seeds of this plant were brought from Campeachy, and several of the plants were raised in England, which continued two years in several gardens, but none of them lived to flower; they grow to the height of eight or ten feet, twining round stakes, but their stalks are very slender, and decayed at the bottom, probably from their having too much wet.

It is a very tender plant, so must be constantly kept in the bark-bed in the stove, and should have but little wet in winter.

HIPPOCREPIS. Lin. Gen. Pl. 791. Horseshoe Vetch.

The CHARACTERS are,

The empalement of the flower is permanent, divided into five parts; the flower is of the butterfly kind; it hath ten stamina, nine joined and one separate, and an oblong narrow germen, sitting on an awl-shaped style, which afterward becomes a long, plain, compressed pod, cut into many parts from the under seam to the upper, each part forming a roundish sinus with obtuse three-cornered joints connected to the upper seam, each joint being shaped like a horseshoe, inclosing a single seed.

The SPECIES are,

HIPPOCREPIS (*Unisiliquosa*) leguminibus sessilibus solitariis. Hort. Cliff. 264. *Horseshoe Vetch with single pods sitting close to the stalk.*

2. HIPPOCREPIS (*Comosa*) leguminibus pedunculatis confertis, margine exteriore repandis. Prod. Leyd. 384. *Horseshoe Vetch with pods growing in clusters upon foot-stalks, whose outward border is turned inward.*

3. HIPPOCREPIS (*Multisiliquosa*) leguminibus pedunculatis confertis, margine altero lobatis. Hort. Cliff. 364. *Horseshoe Vetch with pods growing in clusters upon foot-stalks, one border of which has lobes.*

The first sort grows naturally in Italy and Spain; this is an annual plant, which sends from the root several trailing stalks a foot long, that divide upward into smaller branches, garnished with winged leaves, composed of four or five pair of small lobes, terminated by an odd one; these are obtuse, and indented at their ends; from the wings of the stalk come out single flowers of the butterfly kind, which are yellow, and are succeeded by single pods sitting close to the stalk, which are about two inches long, and a third of an inch broad, bending inward like a sickle, and divided into many joints, shaped like a horseshoe: this flowers in June and July, and the seeds ripen in the autumn, soon after which the plants decay.

The second sort is found growing naturally in some parts of England upon chalky hills; this is a smaller plant than the former, and hath a perennial root, sending out several trailing stalks about ten inches long, which are garnished with narrow winged leaves; the flowers grow in clusters on the top of long foot-stalks, which are succeeded by shorter pods, twisted inward in roundish curves, but have joints shaped like those of the former sort.

The third sort grows naturally in the south of France, Germany, and Italy. This is an annual plant with trailing stalks, greatly resembling the first, but the flowers are produced in clusters on the top of pretty long foot-stalks; they are shaped like those of the other sorts, and the pods are jointed in like manner but the joints are fixed to the opposite border. These [illegible]nts flower in June and July, and the seeds ripen in August and September.

These are propagated by seeds, which should be sown in the autumn, where the plants are designed to remain; and when the plants come up, they must be kept clean from weeds, and thinned where they are too close, which is all the culture they require. The two annual sorts will decay in the autumn, after they have perfected their seeds, but the roots of the other will continue two or three years, provided they are not in too good ground.

HIPPOLAPATHUM. See RUMEX.

HIPPOMANE. Lin. Gen. Pl. 1099. The Manchineel.

The CHARACTERS are,

It hath male and female flowers in the same spike; the male flowers come out in small clusters, from a short cup-shaped empalement, and have no petals; the female flowers have no petal, but an oval germen, which afterward becomes a roundish fruit with a fleshy cover, inclosing a rough hard shell with several cells, each inclosing one seed.

The SPECIES are,

1. HIPPOMANE (*Mancinella*) foliis ovatis serratis. Hort. Cliff. 484. *Hippomane with oval sawed leaves.*

2. HIPPOMANE (*Biglandulosa*) foliis ovato-oblongis serratis, basi glandulosis. Lin. Sp. Plant. 1191. *Hippomane with oval oblong leaves which are sawed, and have glands at their base.*

3. HIPPOMANE (*Spinosa*) foliis subovatis dentato-spinosis. Lin. Sp. Plant. 1191. *Hippomane with oval leaves, which have prickly indentures.*

The first sort grows naturally in most of the islands in the West-Indies; it is a very large tree in its native soil, almost equalling the Oak in size; the wood is much esteemed for making of cabinets, book-cases, &c. being very durable, and taking a fine polish; it is also said, that the worms will not eat it: but as the trees abound with a milky caustic juice, so before they are felled they make fires round their trunks to burn out their juice, otherwise those who fell them would be in danger of losing their sight, by the juice flying in their eyes; for wherever this falls on the skin, it will raise blisters; and if it comes upon linen it will immediately turn it black, and on being washed will come in holes: it is also dangerous working of the wood after it is sawn out, for if any of the saw-dust happens to get into the workmens eyes, it causes inflammations, and the loss of sight for some time; to prevent which, they generally cover their faces with fine lawn during the time they are working the wood.

This tree hath a smooth brownish bark, the trunk divides upward into many branches, garnished with oblong leaves, ending in acute points, slightly sawed on their edges, and are of a lucid green, standing on short foot-stalks. The flowers come out in short spikes at the end of the branches, being of both sexes in the same spike, but having no petals they make but little appearance; these are succeeded by fruit about the size and of the same shape as the Golden Pippin, turning of a yellow colour when ripe, which has often tempted strangers to eat of them to their cost, for they inflame the mouth and throat to a great degree, causing violent pains in the throat and stomach, which is dangerous, unless remedies are timely applied.

The inhabitants of America believe it is dangerous to sit or lie under these trees, and affirm, that the rain or dew, which falls from the leaves, will raise blisters; but it is very certain, that unless the leaves are broken, and the juice of them mix with the rain, it will do no injury

The second sort grows naturally at Carthagena in New Spain, and the third at Campeachy. The second sort grows to as large a size as the first; the leaves of this are much longer than those of the first, and have two small glandules growing at their base; they are sawed on their edges, and are of a lucid green.

The third sort is of humbler growth, seldom rising more than twenty feet high: the leaves of this greatly resemble those of the common Holly, and are set with sharp prickles at the end of each indenture; they are of a lucid green, and continue all the year.

These plants are preserved in some of the curious gardens in Europe, where they can never be expected to rise to any great height, for they are too tender to live in these northern countries but in stoves; they rise easily from seeds, provided they are good. The seed must be sown upon a good hot-bed, and when the plants come up they should be each planted in a small separate pot, and plunged into a good bed of tanners bark, treating them in the same way as other tender plants, but they must not have much wet, for these plants abound with an acrid milky juice, and it is certain that such plants are soon killed by much moisture; these plants must be removed into the stove, and plunged into the tan-bed in autumn, where they should constantly remain, giving them very little water in winter; and in summer, when the weather is warm, they should have a good share of air admitted to them, and gently refreshed with water.

HIPPOPHAE. Lin. Gen. Plant 980. Sea Buckthorn.

The CHARACTERS are,

It is male and female in different plants: the male flowers have an empalement of one leaf, cut into two parts; they have no petals, but have four short stamina. The female flowers have no petals, but a one-leaved empalement, which is oval and bifid at the brim; these have no stamina, but in the center is situated a small roundish germen, which afterward turns to a globular berry with one cell, inclosing one roundish seed.

The SPECIES are,

1. HIPPOPHAE (*Rhamnoides*) foliis lanceolatis. Lin. Sp. Pl. 1023. *Hippophae with spear-shaped leaves, or Sea Buckthorn with a Willow leaf.*

2. HIPPOPHAE (*Canadensis*) foliis ovatis. Lin. Sp. Pl. 1024. *Hippophae with oval leaves, called Canada Sea Buckthorn.*

The first sort grows naturally on the sea-banks in Lincolnshire, and also on the sand-banks between Sandwich and Deal in Kent; there are two varieties of this, one with yellow, and the other with red fruit; but it is the first only which I have observed growing naturally in England, the other I saw growing on the sand-banks in Holland.

These rise with shrubby stalks eight or ten feet high, sending out many irregular branches, which have a brown bark silvered over, garnished with very narrow spear-shaped leaves, of a dark green on their upper side, but hoary on their under, having a prominent midrib; the two borders of the leaves are reflexed like the Rosemary; the flowers come out from the side of the younger branches, to which they sit very close; the male flowers growing in small clusters, but the female come out singly; these make but little appearance. They appear in July, and the berries on the female plants are ripe in autumn.

This sort is easily propagated by suckers from the root, for the roots spread wide, and send up a great number of shoots, so as to form a thicket; if these are taken off in autumn, and transplanted into a nursery, they will be fit to transplant after one year's growth to the places where they are to remain; as there is little beauty in this plant, so one or two of them may be allowed a place in a plantation of shrubs for the sake of variety.

The second sort grows naturally in North America; this hath much the appearance of the former sort, but the leaves differ in their shape, these being much shorter and broader, and are not so white on their under side; this may be easily propagated by suckers or layers.

HIPPOSELINUM. See SMYRNIUM.

HIRUNDINARIA. See ASCLEPIAS.

HOLCUS. Lin. Gen. Plant. 1015. Indian Millet, or Corn.

The CHARACTERS are,

It hath male and hermaphrodite flowers on the same plant; the male flowers are small, and have a twisted bivalve chaff, ending with an acute beard; they have neither petals, or any proper empalement, but have three hairy stamina. The hermaphrodite

phrodite

phrodite flowers are single, sitting in a stiff bivalva chaff; they have three hairy stamina with a roundish germen, which afterward becomes an oval single seed, wrapped up in the chaff.

The SPECIES are,

1. HOLCUS (*Sorgum*) glumis villosis, seminibus aristatis. Hort. Upsal. 301. *Holcus with hairy chaff, and bearded seeds.*

2. HOLCUS (*Saccharatus*) glumis glabris, seminibus muticis. Lin. Sp. Pl. 1047. *Holcus with smooth husks, and seeds without awns.*

There are several other of the grassy tribe which belong to this genus, but as they are not cultivated for use, so I shall not enumerate them here.

The two sorts here mentioned grow naturally in India, where their grain is often used to feed poultry, and the seeds of these are frequently sent to Europe for the same purpose, but the summers are seldom warm enough to ripen the seeds in the open air in England. The stalks of these plants rise five or six feet high, with strong reedy stalks, like those of the Maize, or Turkey Wheat. The leaves are long and broad, having a deep furrow through the center of the leaf, where the midrib is depressed on the upper surface, and is very prominent below. The leaves are two feet and a half long, and three inches broad in the middle, embracing the stalks with their base. The flowers come out in large panicles at the top of the stalks, resembling, at first appearance, the male spikes of the Turkey Wheat; these are succeeded by roundish seeds, which are wrapped round with the chaff.

These plants are propagated in a few gardens for the sake of variety, but as they are late in ripening their grain here, so they are not worth cultivating for use. The seeds should be sown on a warm border the beginning of April, and when the plants come up, they should be thinned to the distance of a foot asunder in the rows, and the rows should be three feet distance; the culture after this, is to keep the ground clean from weeds, and draw the earth up with a hoe to the stems of the plants; if the season proves warm, their panicles will appear in July, and the grain will ripen in September, but in bad seasons their grain will not ripen here.

HOLLOW ROOT. See FUMARIA.

HOLLY. See ILEX.

HOLLYHOCKS. See ALCEA.

HONEYSUCKLE. See PERICLYMENUM.

HOPS. See LUPULUS.

HORDEUM. Lin. Gen. Plant. 94. Barley.

The CHARACTERS are,

It hath a partial involucrum, which contain three flowers; the petal of the flower opens with two valves; the flower hath three hairy stamina shorter than the petal, and an oval turned germen, which afterward becomes an oblong bellied seed, pointed at both ends, having a longitudinal furrow, surrounded by the petal of the flower, which does not fall off.

The SPECIES are,

1. HORDEUM (*Vulgare*) flosculis omnibus hermaphroditis aristatis ordinibus duobus erectioribus. Lin. Sp. Pl. 84. *Barley with all the flowers hermaphrodite, and two orders of beards, which are erect; or Spring Barley.*

2. HORDEUM (*Zeocriton*) flosculis lateralibus masculis muticis, seminibus angularibus imbricatis. Hort. Ups. 23. *Barley with male flowers on the side, without awns, and angular seeds placed over each other; or common long-eared Barley.*

3. HORDEUM (*Distichon*) flosculis lateralibus masculis muticis, seminibus angularibus patentibus corticatis. Hort. Upsal. 23. *Barley with male flowers on the side, without awns, and angular spreading seeds with husks; commonly called Sprat, or Battledore Barley.*

4. HORDEUM (*Hexastichon*) flosculis omnibus hermaphroditis aristatis, seminibus sexfariam æqualiter positis. Hort. Upsal. 23. *Barley with all the flowers hermaphrodite and bearded, and six rows of seeds equally ranged; called Winter, or Square Barley, Bear Barley, or Big.*

The first sort is the common Spring Barley, which is principally cultivated in England; of this the farmers make two sorts, viz. the common and Rath-ripe Barley, which are the same; for the Rath-ripe has only been an alteration occasioned by being long cultivated upon warm gravelly lands. The seeds of this, when sown in cold or strong land, will, the first year, ripen near a fortnight earlier than the seeds taken from strong land, therefore the farmers in the vales generally purchase their seed Barley from the warm land, for if saved in the vales two or three years, it will become full as late in ripening as the common Barley of their own product, and the farmers on the warm land are also obliged to procure their seed Barley from the strong land, otherwise their grain would degenerate in bulk and fulness, which by thus changing is prevented. This sort of Barley is easily distinguished by the two orders of beards or awns, which stand erect; the chaff is also thinner than that of the two last species, so is esteemed better for malting.

The second sort is the long-eared Barley, which is cultivated in many parts of England, and is an exceeding good sort; but some farmers object to this sort, because they say the ears being long and heavy, it is more apt to lodge; this hath the grains regularly ranged in a double row, lying over each other like tiles on a house, or the scales of fish. The husk or chaff of this Barley is also very thin, so is much esteemed for malting.

The third sort is usually called Sprat Barley; this hath shorter and broader ears than either of the other sorts; the awns or beards are longer, and the grains are placed closer together, the awns being long, the birds cannot so easily get out the grains; this seldom grows so tall as the other species; the straw is shorter and coarser, so not very good fodder for cattle.

The fourth sort is rarely cultivated in the southern parts of England, but in the northern counties, and in Scotland is generally sown, being much hardier than the other species, so will bear the cold; this hath its grains disposed in six rows; the grain is large and plump, but is not so good for malting, which is the reason of its not being cultivated in the southern parts of England, where the other sorts which are much better for that purpose do thrive so well.

All the sorts of Barley are sown in the spring of the year, in a dry time; in some very dry light land it is sown early in March, but in strong clayey soils it is not sown till April, and sometimes not until the beginning of May; but when it is sown so late, if the season doth not prove very favourable, it is very late in autumn before it is fit to mow, unless it be the Rath-ripe sort, which is often ripe in nine weeks from the time of sowing.

Some people sow Barley upon land where Wheat grew the former year; but when this is practised, the ground should be ploughed the beginning of October in a dry time, laying it in small ridges, that the frost may mellow it the better, and this will improve the land greatly; and if this can be ploughed again in January, or the beginning of February, it will break and prepare the ground better; in March the ground should be ploughed again deeper, and laid even where it is not very wet; but in strong wet lands the ground should be laid in round lands, and the furrows made deep to receive the wet. When this is finished, the common method is to sow the Barley-seed with a broad-cast at two sowings; the first being harrowed in once, the second is harrowed barrowed until the seed is buried; the common allowance of seed is four bushels to an acre.

This is the quantity of grain usually sown by the farmers, but if they could be prevailed on to alter this practice, they would soon find their account in it, for if a third of that quantity is sown there will be a much greater produce, and the Corn will be less liable to lodge, as I have many times experienced; for when Corn or any other vegetable stands very close, the stalks are drawn up weak, so are incapable to resist the force of winds, or bear up under heavy rains; but when they are at a proper distance, their stalks will be more than twice the size of the other, so are seldom laid. I have frequently observed in fields where there has been a foot-path through their middle, that the Corn which has stood thin on each side the path hath stood upright, when all the rest on both sides has been laid flat on the ground and whoever will observe these roots of Corn near the paths, will find them tiller out (i.e. have a greater number of stalks) to more than four times the quantity of the other parts of the field. I have seen experiments made by sowing Barley in rows across divers parts of the same field, and the grains sowed thin in the rows, so that the roots were three or four inches asunder in the rows, and the rows a foot distance: the intermediate spaces of the same field were at the same time sown broad cast in the usual way; the success was this, the roots which stood thin in the rows tillered out from ten or twelve, to upward of thirty stalks on each root, the stalks were stronger, the ears longer, and the grains larger than any of those sown in the common way; and when those parts of the field where the Corn was sown in the usual way has been lodged, these parts sown thin have supported their upright position against wind and rain, though the rows have been made not only lengthways, but cross the lands in several positions, so that there could be no alteration in regard to the goodness of the land, or the situation of the Corn; therefore where such experiments have been frequently made, and always attended with equal success, there can be no room to doubt which of the two methods is more eligible; since if the crops were only supposed to be equal in both, the saving two thirds of the Corn sown is a very great advantage, and deserves a national consideration, as such a saving in scarce times might be a very great benefit to the public. This saving of seed Corn must be understood to regard such as is sown broad cast; for if it is sown in drills, an eighth part of the seeds usually sown will be sufficient for an acre of land, and the produce will be greater; for all sorts of Corn is naturally inclined to send out several stalks from each root, which they rarely fail to do where the roots are at a proper distance and have room; nor do the stalks grow in this case near so tall, but are much stronger than when they are near together, when they rarely have more than two or three stalks, whereas those roots which have proper room seldom have less than ten or twelve. I have had eighty-six stalks upon one root of Barley, which were strong, produced longer ears, and the grain was better filled than any which I ever saw grow in the common method of husbandry, and the land upon which this grew was not very rich; but I have frequently observed on the sides of hot beds in the kitchen-gardens, where Barley straw has been used for covering the beds, that some of the grains left in the ears have dropped out and grown, the roots have produced from thirty to sixty stalks each, and those have been three or four times larger than the stalks ever arrive at in the common way: but to this I know it will be objected, that although upon rich land in a garden, these roots of Corn may probably have so many stalks, yet in poor land they will not have such produce; therefore unless there is a greater quantity of seeds sown, their crop will not be worth standing, which is one of the greatest fallacies that can be imagined; for to suppose that poor land can nourish more than twice the number of roots in the same space as rich land, is such an absurdity as one could hardly suppose any person of common understanding guilty of; and yet so it is, for the general practice is to allow a greater quantity of seed to poor land than for richer ground, not considering, that where the roots stand so close, they will deprive each other of the nourishment, so starve themselves, which is always the case where the roots stand close, which any person may at first sight observe in any part of the fields where the Corn happens to scatter when they are sowing it; or in places, where, by harrowing, the seed is drawn in heaps, those patches with starve, and never grow to a third part of the size as the other parts of the same field; and yet common as this is, it is little noticed by farmers, otherwise they surely would not continue their old custom of sowing I have made many experiments for several years in the poorest land, and have always found that all crops which are sown or planted at a greater distance than usual, have succeeded best upon such land; and I am convinced, if the farmers could be prevailed on to quit their prejudices, and make trial of this method of sowing their Corn thin, they would soon see the advantage of this husbandry.

The noblemen and gentlemen in France are very busy in setting examples of this husbandry in most of their provinces, being convinced by many trials of its great utility, and it were to be wished the same was done in England.

After the Barley is sown, and harrowed in, the ground should be rolled after the first shower of rain, to break the clods and lay the earth smooth, which will render it better to mow the Barley, and also cause the earth to lie closer to the roots of the corn, which will be of great service to it in dry weather.

Where Barley is sown upon new broken up land, the usual method is to plough up the land in March, and let it lie fallow until June, at which time it is ploughed again, and sown with Turneps, which are eaten by sheep in winter, by whose dung the land is greatly improved, and then in March following the ground is ploughed again, and sown with Barley as before.

There are many people who sow Clover with their Barley, and some have sown the Lucern with Barley; but neither of these methods is to be commended, for where there is a good crop of Barley, the Clover or Lucern must be so weak as not to pay for standing; so that the better way is to sow the Barley alone without any other crop among it, and then the land will be at liberty for any other crop when the Barley is taken off the ground; but this practice of sowing Clover, Rye Grass, and other Grass-seeds, with Corn, has been so long and universally established among farmers, that there is little hope of prevailing with those people to alter a custom which has been handed down to them from their predecessors, although there should be many examples produced to shew the absurdity of this practice.

When the Barley has been up three weeks or a month, it will be a very good method to roll it over with a weighty roller, which will press the earth close to the roots of the Corn, and thereby prevent the sun and air from penetrating the ground, which will be of singular service in dry seasons; and this rolling of it before it stalks, will cause it to till out into a greater number of stalks; so that if the plants should be thin, this will cause them to spread so as to fill the ground, and likewise strengthen the stalks.

The time for cutting of Barley is, when the red colour of the ears is off, and the straw turns yellow, and the ears begin to hang down: in the north of England they always reap their Barley, and make it up in sheaves, as is practised here for Wheat, by which method they do not lose near so much Corn, and it is also more handy to stack; but this

 method

method cannot ſo well be practiſed where there are many weeds amongſt the Corn, which is too frequently the caſe in the rich lands near London, eſpecially in moiſt ſeaſons; therefore when this is the caſe, the Barley muſt lie on the ſwarth till all the weeds are dead; but as it is apt to ſprout in wet weather, it muſt be ſhook up, and turned every fair day after rain to prevent it. When it is carried it ſhould be thoroughly dry, otherwiſe if it be ſtacked wet it will turn muſty; or if too green, it is ſubject to burn the mow. The common produce of Barley is two and a half, or three quarters on an acre, but I have ſometimes known eight or ten quarters on an acre.

HORMINUM. Tourn. Inſt. R. H. 178. Clary.

The CHARACTERS are,

The empalement of the flower is permanent, of one leaf, having two lips; the upper ending in three acute points, the under ending in two. The flower has one petal, divided into two lips; the upper is concave, and incurved with a ſlight indenture at the point, the lower is broader and more indented. It hath two ſhort ſtamina ſituated in the tube of the flower. In the bottom of the tube are four roundiſh germen, which afterward become four ſeeds lodged in the empalement.

The SPECIES are,

1. HORMINUM (*Verbenaceum*) foliis ſinuatis obtuſis crenatis, calycibus acutis. *Clary with obtuſe ſinuated leaves which are crenated, and the empalement ending in acute points; called Wild Clary.*

2. HORMINUM (*Lyrata*) foliis pinnato-ſinuatis rugoſis, calycibus corollâ longioribus. *Clary with wing-ſhaped ſinuated leaves which are rough, and the empalements longer than the petal of the flower; or Oak-leaved Clary.*

3. HORMINUM (*Verticillatum*) foliis cordatis crenato-dentatis, verticillis ſubnudis, ſtylo corollarum labio inferiore incumbente. *Clary with heart-ſhaped, crenated, indented leaves, naked whorls, and the ſtyle lying upon the under lip of the petal.*

4. HORMINUM (*Napifolium*) foliis radicalibus pinnato-inciſis, caulinis cordatis crenatis, ſummis ſemiamplexicaulibus. *Clary with lower leaves cut and winged, thoſe on the ſtalks heart-ſhaped and crenated, and thoſe on the top half embracing the ſtalks.*

5. HORMINUM (*Sativum*) foliis obtuſis crenatis, bracteis ſummis ſterilibus majoribus corollatis. *Clary with obtuſe crenated leaves, and the bracteæ on the top of the ſtalks large, coloured, and barren; Clary with a purple Violet top.*

The firſt ſort grows naturally on ſandy and gravelly ground in many parts of England, ſo is rarely cultivated in gardens; but as it has been long uſed in medicine, I have enumerated it to introduce the other ſpecies.

This is ſometimes called Oculus Chriſti, from the ſuppoſed virtues of its ſeeds in clearing of the ſight, which it does by its viſcous covering; for when any thing happens to fall into the eye, if one of the ſeeds is put in at one corner, and the eye-lid kept cloſe over it, moving the ſeed gently along the eye, whatever happens to be there will ſtick to the ſeed, and ſo be brought out. The virtues of this are ſuppoſed to be the ſame as the garden Clary, but not quite ſo powerful.

The ſecond ſort grows naturally in the South of France and Italy. This is by ſome ſuppoſed to be a variety of the firſt; but the leaves of this are regularly ſinuated on both ſides, in form of a winged leaf; the ſtalks riſe about the ſame height with the former, but all the leaves upon the ſtalks are ſinuated in the ſame manner as the lower; the flowers are ſmaller than thoſe of the firſt, but grow in whorled ſpikes like them. It is a perennial plant, very hardy, and will propagate in plenty by its ſcattering ſeeds.

The third ſort is a perennial plant, which grows naturally in Auſtria and Bohemia. This ſends out from the root a great number of heart-ſhaped leaves, which are ſawed on their edges and deeply veined, ſtanding upon pretty long foot-ſtalks which are hairy; the ſtalks are ſquare, and grow two feet and a half high, garniſhed with two heart-ſhaped leaves at each joint, whoſe baſe half embrace the ſtalks, which are garniſhed with whorls of ſmall blue flowers, not much unlike thoſe of the common ſort, but larger; the ſpikes are more than a foot long, and toward the top the whorls are nearer together.

The fourth ſort grows naturally in the ſouth of France, and in Italy. This is alſo a perennial plant, which has ſome reſemblance of the third, but the lower leaves of this are cut at their baſe to the midrib, into two or three pair of ears or lobes, which are ſmall, and at a diſtance from each other: the leaves are not ſawed, but bluntly indented; the ſtalks of this are ſlenderer, and do not grow ſo tall as thoſe of the third, nor are the ſpikes of flowers ſo long.

Both theſe ſorts may be eaſily propagated by ſeeds, which, if ſown in the ſpring on an open ſpot of ground, the plants will come up, and require no other care but to keep them clean from weeds, and allow them room to grow; for they ſhould not be nearer than two feet apart, as they grow very large, and will laſt ſeveral years.

The fifth ſort is an annual plant, which grows naturally in Spain; of this there are three varieties which are conſtant, one with purple tops, another with red tops, and a third with green tops. As they differ in nothing but the colour of their bracteæ on the top of the ſtalks, ſo I have not put them down as different ſpecies, though from more than thirty years cultivating them, I have not known them alter from one to the other.

Theſe plants have obtuſe crenated leaves ſhaped like thoſe of the common red Sage; the ſtalks are ſquare and grow erect, about a foot and a half high; their lower parts are garniſhed at each joint, with two oppoſite leaves of the ſame ſhape, but gradually diminiſhing in ſize toward the top: the ſtalks are garniſhed upward with whorls of ſmall flowers, and are terminated by cluſters of ſmall leaves, which in one are red, in another blue, in a third green, which make a pretty appearance, and are preſerved in gardens for ornament.

The ſeeds of theſe are ſown in the ſpring, in the places where they are deſigned to remain, and require no other care but to keep them clean from weeds, and thin them where they come up too cloſe.

Garden Clary. See Sclarea.

HORNBEAM. See CARPINUS.

HORSE CHESTNUT. See ÆSCULUS.

HORSE-DUNG is of great uſe in gardens, firſt to make hot-beds for the raiſing of all ſorts of early garden crops, as Cucumbers, Melons, Aſparagus, Salleting, &c. for which purpoſe no other ſort of dung will do ſo well, this fermenting the ſtrongeſt; and, if mixed with litter and ſea-coal aſhes in a due proportion, will continue its heat much longer than any other ſort of Dung whatſoever; and afterwards when rotted, becomes an excellent manure for moſt ſorts of lands, more eſpecially for ſuch as are of a cold nature; and for ſtiff clayey lands, when mixed with ſea-coal aſhes and the cleanſing of London ſtreets, it will cauſe the parts to ſeparate much ſooner than any other compoſt will do; ſo that where it can be obtained in plenty, I would always recommend the uſe of it for ſuch lands.

HOT-BEDS are of general uſe in theſe northern parts of Europe, without which we could not enjoy ſo many of the products of warmer climes as we do now; nor could we have the tables furniſhed with the ſeveral products of the garden, during the winter and ſpring months, as they are at preſent in moſt parts of England, better than in any other country in Europe; for although we cannot boast of the clemency of our climate, yet the London markets and noblemen's tables are better furnished with all sorts of esculent plants much earlier in the season, and in greater quantities, than any of our neighbours, which is owing to our skill in Hot-beds.

The ordinary Hot-beds which are commonly used in the kitchen gardens, are made with new horse-dung, in the following manner:

1st, There is a quantity of new horse-dung from the stable (in which there should be part of the litter or straw, which is commonly used in the stable, but not in too great proportion to the dung,) the quantity of this mixture must be according to the length of the bed intended, which, if early in the year, should not be less than one good load for each light; this dung should be thrown up in a heap, mixing therewith some sea-coal ashes, which will be of service to continue the heat of the dung; it should remain six or seven days in this heap, then it should be turned over, and the parts well mixed together, and cast into a heap again, where it may continue five or six days longer, by which time it will have acquired a due heat; then in some well sheltered part of the garden, a trench should be dug out in length and width proportionable to the frames intended for it; if the ground is dry, about a foot, or a foot and a half deep; but if wet, not above six inches; then the dung should be wheeled into the opening, and every part of it stirred with a fork, to lay it exactly even and smooth thro' every part of the bed; as also to lay the bottom of the heap (which has commonly less litter) upon the surface of the bed; this will prevent the steam from rising so plentifully as it would otherwise do. To prevent this, and the heat from rising so violently as to burn the roots of whatever plants are put into the ground, it will be a very good way to spread a layer of neats-dung all over the surface of the horse-dung, which will prevent the mould from burning: if the bed is intended for Cucumbers or Melons, the earth should not be laid all over the bed at first, only a hill of earth should be first laid in the middle of each light, on which the plants should be planted, and the remaining space should be filled up from time to time, as the roots of the plants spread, but this is fully explained under those two articles. But if the Hot-bed is intended for other plants, then after the bed is well prepared, it should be left two or three days for the steam to pass off before the earth is laid upon the dung.

In the making of these Hot-beds, it must be carefully observed to settle the dung close with a fork; and if it be pretty full of long litter, it should be equally trod down close in every part, otherwise it will be subject to heat too violently, and consequently the heat will be much sooner spent, which is one of the greatest dangers these sort of beds are liable to. During the first week or ten days after the bed is made, the glasses should be but slightly covered in the night, and in the day-time they should be raised to let out the steam, which is subject to rise very copiously while the dung is fresh; but as the heat abates, so the covering should be increased.

But although the Hot-bed I have described is what the kitchen-gardeners commonly use, yet those made with tanners bark are much preferable, especially for all tender exotic plants or fruits, which require an even degree of warmth to be continued for several months, which is what cannot be effected by horse-dung only. The manner of making these beds is as follows:

There must be a trench dug in the earth about three feet deep, if the ground be dry; but if wet, it must not be above six inches deep at most, and must be raised in proportion above ground, so as to admit of the tan being laid three feet thick. The length must be proportioned to the frames intended to cover it, but that should never be less than eleven or twelve feet, and the width not less than six, which is but a sufficient body to continue the heat. This trench should be bricked up round the sides to the abovementioned height of three feet, and should be filled with fresh tanners bark (i. e. such as the tanners have lately drawn out of their vats, after they have used it for tanning leather,) which should be laid in a round heap for a week or ten days before it is put into the trench, that the moisture may the better drain out of it, which, if detained in too great a quantity, will prevent its fermentation; then put it into the trench, and gently beat it down equally with a dung-fork; but it must not be trodden, which would also prevent its heating, by settling it too close; then you must put on the frame over the bed, covering it with the glasses, and in about a fortnight it will begin to heat; at which time may be plunged pots of plants, or seeds into it, observing not to tread down the bark in doing of it.

HOTTONIA. Boerh. Ind. alt. 1. p. 207. Water Violet.

The Characters are,

The flower has one petal, cut above into five oblong oval segments; it hath five short awl-shaped stamina, standing on the tube of the petal. In the center is situated a globular germen, which afterward becomes a capsule of the same form, with one cell filled with globular seeds, sitting upon the empalement.

We know but one Species of this genus, viz.

Hottonia (*Palustris.*) Boerh. Ind. alt. 1. p. 207. *Water Milfoil, or Water Violet, with a naked stalk.*

This plant grows naturally in standing waters in many parts of England; the leaves which are for the most part immersed in the water, are finely winged and flat, like most of the sea plants, and at the bottom have long fibrous roots, which strike into the mud: the flower-stalks rise five or six inches above the water; they are naked, and toward the top have two or three whorls of purple flowers, terminated by a small cluster of the same. These flowers have the appearance of those of the Stock-gilliflower, so make a pretty appearance on the surface of the water.

It may be propagated in deep standing waters, by procuring its seeds when they are ripe, from the places of their natural growth; which should be immediately dropped into the water where they are designed to grow, and the spring following they will appear; and if they are not disturbed, they will soon propagate themselves in great plenty.

HOEING is necessary and beneficial to plants, for two things: first, for destroying of weeds; 2dly, because it disposes the ground the better to imbibe the night dews, keeps in a constant freshness, and adds a vigour to the plants and trees, whose fruit by that means becomes better conditioned than otherwise they would be.

HUMULUS. See Lupulus.

HURA. Lin. Gen. Plant. 965. Sand Box-tree.

The Characters are,

It hath male and female flowers on the same plant. The male flowers have no petal, but a column of stamina, which are joined at bottom into a cylinder. The female flowers have a swelling empalement of one leaf, with one tubulous petal; the roundish germen is situated in the bottom of the empalement, which afterward becomes an orbicular ligneous fruit, depressed at top and bottom, having twelve deep furrows with so many cells, which open at the top with an elasticity, each containing one round flat seed.

We have but one Species of this genus, viz.

Hura (*Crepitans.*) Hort. Cliff. 486. *Commonly called in the West-Indies Sand Box-tree.*

This grows naturally in the Spanish West Indies, from whence it has been introduced into the British colonies of America, where some of the plants are preserved for their shade, and by some for curiosity. It rises with a soft ligneous stem to the height of twenty-four feet, dividing into many branches, which abound with a milky juice, and have scars on their bark, where the leaves have fallen off. The branches are garnished with heart-shaped leaves; those which are the biggest are eleven inches long, and nine inches broad in the middle, indented on their edges, having a prominent midrib, with several transverse veins from that to the sides, which are alternate. The male flowers come out from between the leaves, upon foot-stalks which are three inches long; these are formed into a close spike or column, lying over each other like the scales of fish. The female flowers are situated at a distance from the male; these have a swelling cylindrical empalement, out of which arises the petal of the flower, which hath a long funnel-shaped tube, spreading at the top, where it is divided into twelve parts, which are reflexed. After the flower is past the germen swells and becomes a round, compressed, ligneous capsule, having twelve deep furrows, each being a distinct cell, containing one large, round, compressed seed; when the pods are ripe they burst with an elasticity, and throw out their seeds to a considerable distance.

It is propagated by seeds, which should be sown in pots, and plunged into a hot-bed of tanners bark. If the seeds are fresh, the plants will appear in about five weeks after the seeds are sown. As the plants will advance very fast, where due care is taken of them, so they should have a large share of fresh air admitted to them in warm weather, otherwise they will draw up too weak. When the plants are about two inches high, they should be transplanted each into a separate small pot, and plunged again into the hot-bed of tanners bark, and shaded from the sun until they have taken new root, after which they must have free air admitted to them, by raising the glasses in proportion to the warmth of the season. In this hot-bed they should remain till Michaelmas, provided the plants have room to grow, without touching of the glasses, at which time they must be removed into the bark-stove, and plunged in the warmest part thereof: during the winter season they must be sparingly watered, for as the plants have succulent stalks abounding with a milky juice, so much moisture will rot them. In summer they must have a large share of fresh air in warm weather, but should not be removed into the open air, for they are too tender to live abroad in the warmest part of the year in this country.

As this plant has ample leaves, which are of a beautiful green colour, it makes an agreeable variety among other tender exotic plants in the stove; for where they are kept warm and duly refreshed with water, they retain their leaves all the year in verdure.

HYACINTHUS. Tourn. Inst. R. H. 344. tab. 180. Hyacinth.

The CHARACTERS are,

The flower has no empalement. It has one bell-shaped petal, whose rim is cut into six parts, and three nectariums on the point of the germen, with six short awl-shaped stamina. In the center is situated a roundish three-cornered germen, having three furrows, which afterward becomes a roundish three-cornered capsule, having three cells, which contain roundish seeds.

The SPECIES are,

1. HYACINTHUS (*Non scriptus*) corollis campanulatis sexpartitis apice revolutis. Hort. Cliff. 125. *Hyacinth with bell-shaped petals divided into six parts, which are reflexed at their tops; English Hyacinth, or Hare Bells.*

2. HYACINTHUS (*Serotinus*) corollarum exterioribus petalis distinctis, interioribus coadunatis. Lin. Sp. Pl. 317. *Hyacinth whose exterior part of the flower hath distinct petals but the interior are joined.*

3. HYACINTHUS (*Campanulata*) corollis campanulatis sexpartitis, floribus utrinque dispositis. *Hyacinth with bell-shaped petals, which are divided into six parts, and flowers ranged on each side of the stalk.*

4. HYACINTHUS (*Cernuus*) corollis campanulatis sexpartitus racemo cernuo. Lin. Sp. Plant. 317. *Hyacinth with bell-shaped petals, which are divided into six parts, and a nodding branch of flowers.*

5. HYACINTHUS (*Amethystinus*) corollis campanulatis semisexfidis basi cylindricis. Hort. Upsal. 85. *Hyacinth with bell-shaped petals, cut half way into six parts, and a cylindrical base.*

6. HYACINTHUS (*Orientalis*) corollis infundibuliformibus semisexfidis basi ventricosis. Hort. Upsal. 85. *Hyacinth with funnel shaped petals cut half into six parts, and swelling at the base; or early, white, eastern Hyacinth.*

The sorts here mentioned are all of them distinct species, of which there are great varieties, especially of the sixth, which have been cultivated with so much art, as to render some of them the most valuable flowers in the spring; in Holland the gardens abound with them, especially at Haerlem, where the florists have raised so many varieties as to amount to some hundreds; and some of their flowers are so large, double, and finely coloured, as that their roots are valued at twenty or thirty pounds sterling each root. To enumerate these varieties here, would swell this work to very little purpose, as every year produces new kinds.

The first sort grows naturally in woods, and near hedges, in lands which have lately been woods, in many parts of England, so is seldom admitted into gardens; but the poor people, who make it their business to gather the wild flowers of the fields and woods for nosegays, &c. bring great quantities of these in the spring to London, and sell them about the streets.

There is a variety of this with white flowers, which is kept in some gardens, and only differs in the colour of their flowers from the other.

The second sort is preserved in some gardens for the sake of variety; but as it hath as little beauty as the first, sorts seldom allowed a place in the flower-garden. The flowers of this are narrower than those of the first sort, and seem as if their petals were divided to the bottom, three of the outer segments being separated from the other, standing at a small distance from the three interiors, but they are all joined at their base; the flowers are of a light blue colour, but they fade to a worn-out purple.

The third sort grows naturally in Spain and Italy. This hath blue flowers of the open spread bell-shape, which are divided into six segments almost to the bottom, and are disposed on every side the stalk.

The fourth sort seems to be a variety of the first, the flowers being ranged for the most part upon one side of the stalk, and the top of the bunch is always bent on one side. The flowers are of a blush Peach colour, and appear about the same time as the first.

The fifth sort grows naturally in Spain. This hath a smaller flower than either of the former sorts, and comes earlier in the season. The petal is cut into six parts half the length, and is reflexed at the brim; the lower part is cylindrical, a little swelling at the base, and is of a deeper blue than either of the former. This was formerly called by the gardeners the Coventry blue Hyacinth.

The sixth sort is the eastern Hyacinth, of which we formerly had very few varieties in the English gardens, but the single and double white and blue flowering, but from the seeds of these there has been many others raised in England; but the gardeners in Holland have, within the last fifty years,

years, raised so many fine varieties, as to render the former sorts of no value.

Those who are desirous to preserve any of the old sorts, need not be at much trouble about it, for their roots propagate in great plenty in any soil or situation, and will require no other care but to take up their roots every other year, soon after their leaves decay, and plant them again in autumn; for if they are permitted to remain longer in the ground, their roots will have multiplied to so great a degree, as to render their flowers very small and weak.

All the different sorts of Hyacinths are propagated by seeds or offsets from the old bulbs; the former method has been but little practised in England till very lately, but in Holland and Flanders it hath been followed for many years, whereby they have obtained so great a variety of most beautiful flowers. Few florists in England think it worth their trouble to wait four or five years for the flower of a plant, which when produced perhaps might not deserve to be preserved; but they do not consider, that it is only the loss of the four or five first years after sowing; for if they continue sowing every year after they begin, there will be a succession of flowers annually, which will constantly produce some sorts different from what they are before possessed of; and new flowers being always the most valuable to skilful florists (provided they have good properties to recommend them,) it will always be a sufficient recompence for the trouble and loss of time.

The method of raising these flowers from seeds is as follows: first, to be provided with some good seed (which should be saved from either semi-double, or such single flowers as are large, and have good properties:) secondly, one or more shallow boxes should be provided, according to the quantity of seeds to be sown, which must be filled with fresh, light, sandy soil, laying the surface very level; then the seeds should be sown thereon as equally as possible, covering it about half an inch thick with the same light earth: the time for this work is about the middle of August. These boxes should be placed where they may enjoy the morning sun only until the latter end of September, at which time they should be removed into a warmer situation, and towards the end of October they should be placed under a common hot-bed frame, where they may remain during the winter and spring months, to be protected from hard frosts; but they should be exposed to the open air when the weather is mild, by taking off the glasses. In February or March the young plants will begin to appear above ground, at which time they must be carefully screened from frosts, otherwise they will be destroyed when they are so young; but they must not be covered at that season except in the night, or in very bad weather; for when the plants are come up, if they are too close covered, they will draw up tall and slender, and thereby prevent the growth of their roots. At the end of March, if the weather proves good, they may be removed out of the frame, placing them in a warm situation; and if the season proves dry they should now and then have a little water, and kept very clear from weeds, which would soon overspread the tender plants and destroy them.

Towards the latter end of April or the beginning of May, these boxes should be removed into a cooler situation; for the heat of the sun at that season would be too great for these tender plants, causing their blades to decay much sooner than they would, if they were screened from its violence. In this shady situation they should remain during the heat of summer, observing to keep them constantly clear from weeds; but you must not place them under the dripping of trees, &c. nor should you give them any water after their blades are decayed, for that would infallibly rot the roots. About the latter end of August you should sift a little light rich earth over the surface of the boxes, an then remove them again into a warmer situation, and trea them during the winter, spring, and summer months, as was before directed; and the second year, about the middle of August, should be prepared a bed of light, rich, sandy soil, in proportion to the quantity of seedling roots, the surface of which should be very even; then take out the earth from the boxes in which the plants were raised into a sieve, in order to get out all the roots, which by this time (if they have grown well) will be about the size of small Pease: these roots should be placed upon the bed at about three inches asunder, observing to set the bottom part of their roots downwards; then they should be covered over two inches thick with the same light earth; but as it will be impossible to get all the small roots out of the earth in the boxes, the earth should be spread upon another bed equally, and covered over with light earth, by which method none will be lost be they ever so small.

These beds must be arched over with hoops, and in very hard frosty weather must be covered with mats, &c. to protect them from frost; and in the spring when the green leaves are above ground, if the weather should be very dry, they should have a little water sparingly, for nothing is more injurious to these bulbs than too great quantities of moisture. During the summer season the beds must be kept clear from weeds, but after the blades are decayed, should not have any water: in autumn the surface of the bed should be stirred with a very short hand-fork, being exceeding careful not to thrust it so deep as to touch the roots, which, if hurt, are very subject to perish soon after. Then a little fresh, light, rich earth should be sifted over the bed about an inch thick, or somewhat more; in winter the bed should be covered again (as before.) In this bed the roots may continue two years; the third summer when the leaves are decayed, the roots should be carefully taken up, and may be kept out of the ground till August, when they should be planted into new beds prepared as before, at the distance of six inches asunder; in these beds the roots may remain till they flower, during which time they should be treated as before, with this difference only, that instead of covering them with mats in the winter, the surface of the ground should be covered with tanners bark.

When their flowers begin to shew themselves, those which have good properties should be marked, by thrusting a small stick down by each root; which root, at the time for taking them up, should be selected from the rest, and planted by themselves, though I would by no means advise the rejecting any of the other roots until they have blown two years, before which their worth cannot be ascertained. When their roots are taken up they should be laid into the earth again in a horizontal position, leaving the green leaves hanging downwards from the roots, whereby the great moisture contained in their very succulent leaves and flower-stalks will be exhaled, and prevented from entering the roots, which, when suffered to return into them, is very often the cause of their rotting. In this position the roots should remain until the leaves are quite dried off, when they must be taken up, and after being cleared from all manner of filth which would be hurtful to them, they must be laid up in boxes, where they may be preserved dry until September, which is the proper season for planting them again.

I shall now proceed to the culture of such Hyacinths as have either been obtained from Holland, or have been produced from seeds in England. The want of skill in this particular has occasioned the ill success most people have had with them here, which has occasioned their being so much neglected, supposing their roots to degenerate after they have flowered in England; which is a great mistake,

 for

for were the roots managed with the fame art as is practifed in Holland, I am fully convinced they would thrive full as well as there; for, from fome hundreds of roots which I have received from Holland at different times, I have had a very great increafe of their roots, which were as large, and produced as many flowers upon their ftems, as the fame forts do in moft parts of Holland.

The foil in which thefe flowers fucceed beft, is a light, fandy, frefh, rich, and loamy, which may be compofed after the following manner: Take half frefh earth from a common, or pafture-land, which is of a fandy loam; this fhould not be taken above eight or nine inches deep at moft; and if taken with the turf, or green fward with it, it will ftill be better, provided it has time to rot; to this fhould be added a fourth part of fea-fand, and the other fourth part of rotten cow-dung; thefe fhould be well mixed together, and caft into a heap, where it may remain until it is wanted, but it fhould be turned over once every month. If this compoft be made two years before it is ufed, it will be much the better, but if ufed fooner it fhould be oftener turned, that the parts may the better unite.

This foil fhould be laid two feet deep in the beds which are defigned for Hyacinths, and a little rotten cow-dung or tanners bark may be laid at the bottom, which will be within reach of the fibres, but fhould by no means touch the bulb. If the foil be very wet where thefe beds are made, they fhould be raifed ten or twelve inches above the furface, but if it be dry they need not be raifed above three or four.

The beft feafon for planting thefe roots is toward the middle or latter end of September, according to the earlinefs or latenefs of the feafon, or the weather which then happens; but I would advife never to plant them when the ground is extremely dry, unlefs there is a profpect of fome rain foon after; for if the weather fhould continue dry for a confiderable time after, the roots will receive a mouldinefs, which will certainly deftroy them.

Thefe beds will require no farther care until the froft comes on, at which time they fhould have fome rotten tan fpread over the bed about four inches thick; and if the alleys on each fide of the bed are filled up, either with rotten tan, dung, or fand, it will prevent the froft from penetrating the ground to the roots, and fecure them from being deftroyed; but when the winters prove very fevere, it will alfo be proper to have fome Peafe-haulm, or fuch like covering laid over them, which will keep out the froft better than mats. But this covering fhould be taken off whenever the weather is mild, and only continued on in very hard frofts; for where the beds are covered with tan or fea-coal afhes, no common froft can penetrate through, fo the other coverings are ufelefs, except in very fevere froft. In February, when the leaves begin to appear, the beds muft be arched over with hoops, that they may be covered either with mats, canvas, or fome other light covering, to prevent the froft from injuring the buds as they arife above ground; but thefe coverings muft be conftantly taken off every day when the weather is mild, otherwife the flower-ftems will be drawn up weak, and the foot-ftalks of the flowers will be flender, and fo rendered incapable of fupporting the bells, which is a great difadvantage to the flowers. When thefe hoops are fixed over the beds, the rotten tan fhould be taken off; in the doing of which, great care fhould be taken not to bruife or injure the leaves of the Hyacinths which are then coming up.

When the ftems of the flowers are advanced to their height before the flowers are expanded, fhort fticks fhould be placed by each root, to which, with a wire formed into a hoop, the ftem of the flowers fhould be faftened, to fupport them from falling, otherwife, when the bells are fully expanded, their weight will incline them to the ground.

During their feafon of flowering they fhould be covered in the heat of the day from the fun, and alfo from all heavy rains; but they fhould be permitted to receive gentle fhowers, as alfo the morning and evening fun; but if the nights are frofty, they muft be conftantly defended therefrom. With this management the flowers may be continued in beauty at leaft three weeks or a month, and fometimes more, according to their ftrength or the favourablenefs of the feafon.

When their flowers are quite decayed, and the tops of their leaves begin to change their colour, the roots fhould be lifted with a narrow fpade, or fome other handy inftrument: in the doing of this, the inftrument muft be carefully thruft down by the fide of the root, fo as not to bruife or injure it, as alfo to put it below the bottom of the root; then by the forcing of this inftrument on one fid , the fibres of the root are raifed and feparated from the ground. The defign of this is to prevent their receiving any more nourifhment from the ground, for by imbibing too much moifture at this feafon, the roots frequently rot after they are taken up; about a fortnight after this operation the roots fhould be entirely taken out of the ground, and then the earth of the beds fhould be raifed into a fharp ridge, laying the roots into it in a horizontal pofition, with their leaves hanging out, by which means a great part of the moifture contained in their thick fucculent ftalks and leaves will evaporate, which, if permitted to return back to the roots, would caufe them to rot and decay after they are taken up, which has been the general defect of moft of the Hyacinths in England.

In this pofition the roots fhould remain until the green leaves are entirely decayed, which perhaps may be in three weeks time. This is what the Dutch gardeners term the ripening of their roots, becaufe by this method the roots become firm, and the outer cover is fmooth, and of a bright purple colour; whereas thofe roots which are permitted to remain undifturbed, till the leaves and ftalks are quite decayed, will be large, fpongy, and their outer coats will be of a pale colour, for the ftems of many of thefe flowers are very large, and contain a great quantity of moifture, which, if fuffered to return into the roots, will infallibly caufe many of them to perifh. After they are fo ripened they may be taken out of the ground, and wiped clean with a foft woollen cloth, taking off all the decayed parts of the leaves and fibres, putting them into open boxes where they may lie fingly, and be expofed to the air; but they muft be preferved carefully from moifture, nor fhould they be fuffered to remain where the fun may fhine upon them; in this manner they may be preferved out of the ground until September, which is the feafon for planting them again, at which time you muft feparate all the ftrong flowering roots, planting them in beds by themfelves, that they may make an equal appearance in their flowers; but the offsets and fmaller roots fhould be planted in another feparate bed for one year, in which time they will acquire ftrength, and by the fucceeding year will be as ftrong as the older roots.

There are fome perfons who let their Hyacinth roots remain two or three years unremoved, by which they have a much greater increafe of roots than when they are annually taken up; but the roots by this great increafe are frequently degenerated, fo as to produce fingle flowers; therefore I fhould advife the taking up of the roots every year, which is the moft certain method to preferve them in their greateft perfection, though the increafe may not be fo great; thofe roots which are annually removed will be rounder and firmer, than fuch as ftand two years unremoved.

For the other sorts of Hyacinth, see MUSCARI and ORNITHOGALUM.

HYACINTHUS TUBEROSUS. See CRINUM.

HYDRANGEA. Gron. Flor. Vir. 50.

The CHARACTERS are,

The flower hath a small permanent empalement of one leaf, and five roundish petals which are equal. It hath ten stamina, which are alternately longer than the petal. Under the flower is situated a roundish germen, which afterward turns to a roundish capsule, crowned with two horned stigmas, divided transversely into two cells, filled with small angular seeds.

We have but one SPECIES of this genus, viz.

HYDRANGEA (*Arborescens.*) Gron. Flor. Virg. 50.

This plant grows naturally in North America, from whence it has been brought within a few years past to Europe, and is preserved in gardens for the sake of variety more than its beauty. It hath a spreading fibrous root, from which is sent up many soft, pithy, ligneous stalks, which rise about three feet high, garnished at each joint with two oblong heart-shaped leaves placed opposite; the leaves are three inches long and two broad at their base, sawed on their edges, and have many veins running from the middle upward to their borders; the flowers are produced at at the top of the stalks in form of a corymbus, they are white, composed of five petals, with ten stamina surrounding the style.

This is easily propagated by parting of the roots; the best time for this is the latter end of October, which is also the best time to transplant them: the plants should have a moist soil, for they grow naturally in marshy places; they require no other culture but to keep them clean from weeds, and dig the ground between them every winter. The roots are perennial, and, if in very severe frosts the stalks are killed, they will put out new ones the following spring.

HYDROCOTYLE. Water Navelwort.

This plant grows in great plenty in moist places in most parts of England, and is never cultivated for use, so I shall pass it over with only naming it.

HYDROLAPATHUM. See RUMEX.

HYDROPHYLLON. Lin. Gen. Plant. 187. Water Leaf.

The CHARACTERS are,

The flower has a permanent empalement of one leaf, cut into five segments. It hath one bell-shaped petal divided into five parts; under each of these segments is fixed a nectarium, which is situated about the middle. It hath five stamina which are longer than the petal, and an oval-pointed germen, which afterward becomes a globular capsule with one cell, inclosing one large round seed.

We have but one SPECIES of this genus, viz.

HYDROPHYLLON (*Virginianum*) Morini. Joncq. Hort. *Water Leaf of Morinus.*

This plant grows naturally in Canada, and many other parts of America, on moist spongy ground. The root is composed of many strong fleshy fibres, from which arise many leaves with foot-stalks five or six inches long, jagged into three, five, or seven lobes, almost to the midrib, indented on their edges, and have several veins running from the midrib to the sides. The flowers rise with foot-stalks from the roots, having one or two small leaves of the same shape with the lower; the flowers are produced in loose clusters hanging downward, they are of a dirty white and bell-shaped, so make no great figure.

This plant is very hardy in respect to cold, but requires a moist rich soil: for if it is planted in dry ground it will not live, unless it is constantly watered in dry weather. It may be propagated by parting of the roots, which should be done in autumn, that the plants may be well rooted before spring, otherwise they will require a great deal of water.

HYMENÆA. Lin. Gen. Plant. 1100. Locust-tree.

The CHARACTERS are,

The outer involucrum of the flower is divided into two parts; the inner is of one leaf indented in five parts; the flower hath five equal petals. It hath ten short declining stamina. In the center is situated an oblong germen, which afterward becomes a large oblong pod, with a thick ligneous shell, divided into several partitions transversly, in each of which is lodged one compressed large seed, surrounded with a farinaceous pulp.

We have but one SPECIES of this plant, viz.

HYMENÆA (*Courbaril.*) Hort. Cliff. 484. *Commonly called Locust-tree in America.*

This is a very large spreading tree in the West-Indies, where it grows in great plenty; the stem is covered with a russet bark, which divides into many spreading branches, garnished with smooth stiff leaves, which stand by pairs, their base joining at the foot-stalk, to which they stand oblique, the two outer sides being rounded, and their inside strait, so that they resemble a pair of sheep-shears. The flowers are produced in loose spikes at the end of the branches, some of the short ligneous foot-stalks supporting two, and others three flowers, which are composed of five yellow petals striped with purple; the stamina are much longer than the petals, of a purplish colour; the flowers are succeeded by thick, fleshy, brown pods, shaped like those of the garden Bean, but much larger, of a purplish brown colour, and of a ligneous consistence, with a large suture on both edges; these contain three or four roundish compressed seeds, divided by transverse partitions.

The wood of this tree is esteemed a good timber in the West-Indies, and it yields a fine clear resin, which is called Gum Anime in the shops, which makes an excellent varnish.

It is easily raised from the seeds if they are fresh, which should be sown in pots, and plunged into a hot-bed of tanners bark: there should be but one seed put into each pot, or if there is more, when the plants appear, they should be all drawn out but one soon after they come up, before their roots entangle, when it will be hazardous doing it; for if great care is not taken, the plants intended to be left may be drawn out with the other. As the roots of this plant are but slender, so they are very difficult to transplant; for unless a ball of earth is preserved to them, they seldom survive their removal, therefore they must be seldom transplanted from one pot to another. The plants must constantly remain in the tan-bed in the stove, and should be treated in the same way with other tender plants of the same country, giving but little water to them, especially in the winter. When these plants first appear, they make considerable progress for two or three months, after which time they are at a stand perhaps a whole year without shooting, being in their growth very like the Anacardium, or Cashew Nut, so is very difficult to preserve long in this country.

HYOSCYAMUS. Tourn. Inst. R. H. 117. tab. 42. Henbane.

The CHARACTERS are,

The flower has a cylindrical empalement of one leaf, which is permanent. It hath one funnel-shaped petal, cut into five obtuse parts, with five inclined stamina. In the center is situated a roundish germen, which afterward becomes an obtuse capsule, divided into two cells by an intermediate partition, opening with a lid at the top, to let out the many small seeds which adhere to the partition.

The

The SPECIES are,

1. HYOSCYAMUS (*Niger*) foliis amplexicaulibus. Hort. Cliff. 56. *Henbane with leaves embracing the stalks; or common black Henbane.*

2. HYOSCYMUS (*Major*) foliis petiolatis, floribus pedunculatis terminalibus. *Henbane with leaves having foot-stalks, and flowers with foot-stalks terminating the branches.*

3. HYOSCYAMUS (*Albus*) foliis petiolatis, floribus sessilibus. Hort. Upsal. 56. *Henbane with leaves having foot-stalks, and flowers sitting close to the branches.*

4. HYOSCYAMUS (*Minor*) foliis lanceolatis infernè pinnato-incisis summis integerrimis. *Henbane with spear-shaped leaves, the lower being cut into regular segments and the upper ones are entire.*

5. HYOSCYAMUS (*Aureus*) foliis petiolatis acutè dentatis, pistillo corollâ longiore. *Henbane with acute indented leaves standing on foot-stalks, and a pointal longer than the petal of the flower.*

6. HYOSCYAMUS (*Pusillus*) foliis lanceolatis subdentatis, calycibus spinosis. Hort. Upsal. 44. *Henbane with spear-shaped leaves somewhat indented, and a prickly empalement.*

The first of these sorts is the common black Henbane, which grows wild in England upon the sides of banks and old dunghills almost every where. It is a biennial plant, with long fleshy roots, which strike deep into the ground, sending out several large soft leaves, which are deeply slashed on their edges; the following spring the stalks come out, which rise about two feet high, garnished with leaves of the same shape, but smaller, which embrace the stalks with their base; the upper part of the stalk is garnished with flowers standing on one side in a double row, sitting close to the stalks alternately; they are of a dark purplish colour with a black bottom, and are succeeded by roundish capsules sitting within the empalement; these open with a lid at the top, and have two cells filled with small irregular seeds. This is a very poisonous plant, and should be rooted out in all places where children are suffered to come; for in the year 1729, there were three children poisoned with eating the seeds of this plant near Tottenham Court; two of which slept two days and two nights before they could be awakened, and were with difficulty recovered; but the third being older and stronger, escaped better.

The roots of this plant are used for anodyne necklaces to hang about children's necks, being cut to pieces and strung like beads, to prevent fits and cause an easy breeding of their teeth, but they are very dangerous to use inwardly. Some years past there was a mixture of these roots brought over with Gentian, and used as such, which was attended with very bad effects, as hath been mentioned under the article of GENTIAN, so I shall not repeat it here.

The second sort grows naturally in the islands of the Archipelago. This hath rounded leaves, which are obtusely sinuated on their borders, and stand upon long foot-stalks; the stalks branch more than those of the first, and the flowers grow in clusters at the end of the branches, standing upon short foot-stalks; they are of a pale yellow colour, with very dark purple bottoms.

The third sort is much like the second, but the flowers are in closer bunches, sitting very close on the ends of the branches; they are of a greenish yellow colour, with green bottoms. It grows naturally in the warm parts of Europe, and is the sort whose seeds should be used in medicine, as the white Henbane of the shops.

The fourth sort grows naturally in Syria; this rises with a branching stalk two feet high, garnished with long spear-shaped leaves sitting close to the stalk; the lower leaves are regularly cut on both sides into acute segments, but the upper leaves are entire; the flowers grow at the end of the stalks in close bunches; they are of a worn-out red colour, and shaped like those of the common sort.

All these are biennial plants, which perish soon after they have perfected their seeds. They flower in June and July, and their seeds ripen in the autumn, which, if permitted to scatter, will produce plenty of the plants the following spring; or if the seeds are sown at that season, they will succeed much better than in the spring; for when they are sown in spring the plants seldom come up the same year. They are all hardy, and require no other culture but to keep them clean from weeds, and thin the plants where they are too close. The fourth sort should have a warm situation and a dry soil, in which it will stand much better through the winter than in rich ground.

The fifth sort grows naturally in Candia. This is a perennial plant, with weak hairy stalks which require support; the leaves are roundish, hairy, and acutely indented on their edges, standing upon pretty long foot-stalks; the flowers come out at each joint of the stalk; they are large, of a bright yellow, with a dark purple bottom; the style of this sort is much longer than the petal. It flowers most part of summer, but seldom ripen seeds in England. This sort will continue several years, if they are kept in pots and sheltered in winter, for they will not live in the open air; it only requires to be protected from frost: therefore if these plants are placed under a common hot-bed frame in winter, where they may enjoy as much free air as possible in mild weather, they will thrive better than when they are more tenderly treated. It may be easily propagated by cuttings, which, if planted in a shady border, and covered with hand-glasses in any of the summer months, they will make root in a month or six weeks, and may be afterward planted in pots, and treated like the old plants.

HYPECOUM. Tourn. Inst. R. H. 230. tab. 115.

The CHARACTERS are,

The flower hath four petals, and a two leaved empalement. It hath four stamina situated between the petals. In the center is placed an oblong cylindrical germen, which afterward becomes a long, compressed, jointed pod, which is incurved, with one roundish compressed seed in each joint.

The SPECIES are,

1. HYPECOUM (*Procumbens*) siliquis arcuatis compressis articulatis. Hort. Upsal. 31. *Hypecoum with compressed jointed pods bent inward.*

2. HYPECOUM (*Pendulum*) siliquis cernuis teretibus cylindricis. Hort. Upsal. 31. *Hypecoum with taper, cylindrical, nodding pods.*

3. HYPECOUM (*Erectum*) siliquis erectis teretibus torulosis. Hort. Upsal. 32. *Hypecoum with taper, erect, wreathed pods.*

The first sort hath many wing-pointed leaves of a grayish colour, which spread near the ground, and slender branching stalks which lie prostrate on the ground; they are naked below, but at the top are garnished with two or three small leaves of the same shape and colour with those below; from between these leaves come out the foot-stalks of the flower, each sustaining one yellow flower with four petals, and a pointal stretched out beyond the petals, which afterward turns to a jointed compressed pod about three inches long, which bends inward like a bow, having one roundish compressed seed in each joint.

The second sort hath slender stalks which stand more erect; the segments of the leaves are longer and much narrower than those of the first; the flowers are smaller, and come out at the division of the branches, which are succeeded by narrow taper pods hanging downward.

The third sort grows in the east. This hath much the appearance of the second sort in leaf and flower, but the pods grow erect, and are wreathed and twisted about.

These

These plants are all of them annual, so their seeds should be sown in the autumn, on a bed of fresh earth where they are to remain, for they seldom succeed when they are transplanted. When the plants are come up they should be carefully cleared from weeds, and where they are too close they must be thinned, leaving them about or six eight inches apart: after this they will require no other culture but to keep them constantly clear from weeds.

The juice of this plant is of a yellow colour, resembling that of Celandine, and is affirmed by some eminent physicians to have the same effect as opium.

HYPERICUM. Tourn. Inst. R. H. 254. tab. 131. St. Johnswort.

The CHARACTERS are,

The flower has a permanent empalement, divided into five oval concave segments; it hath five oblong oval petals, and a great number of hairy stamina, joined at their base in five distinct bodies. It hath in the center a roundish germen, supporting one, three, or five styles. The germen afterward becomes a roundish capsule, having the same number of cells as there are styles in the flower, which are filled with oblong seeds.

The SPECIES are,

1. HYPERICUM (*Perfoliatum*) floribus trigynis, caule ancipiti, foliis obtusis pellucido-punctatis. Hort. Cliff. 380. *St. Johnswort with three styles to the flower, and obtuse leaves having pellucid punctures; or common St. Johnswort.*

2. HYPERICUM (*Quadrangulum*) floribus trigynis, caule quadrato herbaceo. Hort. Cliff. 380. *St. Johnswort with three styles to the flowers, and a square herbaceous stalk; or St. Johnswort with a square stalk, commonly called St. Peterswort.*

3. HYPERICUM (*Hircinum*) floribus trigynis, staminibus corolla longioribus, caule fruticoso ancipiti. Hort. Cliff. 331. *St. Johnswort with three styles to the flower, stamina longer than the petals, and a shrubby stalk looking two ways; stinking shrubby St. Johnswort.*

4. HYPERICUM (*Canariense*) floribus trigynis, calycibus obtusis, staminibus corollâ longioribus caule fruticoso. *St. Johnswort with three styles to the flower, obtuse empalements, stamina longer than the petals, and a shrubby stalk; shrubby St. Johnswort from the Canaries.*

5. HYPERICUM (*Olympicum*) floribus trigynis, calycibus acutis, staminibus corollâ brevioribus, caule fruticoso. Hort. Cliff. 380. *St. Johnswort with three styles to the flower, acute empalements, stamina shorter than the petals, and a shrubby stalk.*

6. HYPERICUM (*Inodorum*) floribus trigynis, calycibus obtusis, staminibus corollâ longioribus, capsulis, coloratis, caule fruticoso. *St. Johnswort with three styles to the flower, obtuse empalements, stamina longer than the petals, coloured seed-vessels, and a shrubby stalk.*

7. HYPERICUM (*Ascyron*) floribus pentagynis, caule tetragono herbaceo simplici, foliis lævibus integerrimis. Hort. Upsal. 236. *St. Johnswort with five styles to the flower, a square, single, herbaceous stalk, and smooth entire leaves.*

8. HYPERICUM (*Balearicum*) floribus pentagynis, caule fruticoso, foliis ramisque cicatrisatis. Lin. Sp. Plant. 783. *St. Johnswort with five styles to the flower, a shrubby stalk, and scarified leaves and branches.*

9. HYPERICUM (*Androsæmum*) floribus trigynis, fructu baccato, caule fruticoso ancipiti. Hort. Upsal. 237. *St. Johnswort with three styles to the flower, a fleshy seed-vessel, and a shrubby stalk looking two ways; common Tutsan, or Park Leaves.*

10. HYPERICUM (*Bartramium*) floribus pentagynis calycibus obtusis, staminibus corollâ æquantibus, caule erecto herbaceo. *St. Johnswort with five styles to the flower, obtuse empalements, stamina equalling the petals, and an erect herbaceous stalk.*

11. HYPERICUM (*Monogynum*) floribus monogynis, staminibus corollâ longioribus, calycibus coloratis, caule fruticoso. *St. Johnswort with one style to the flowers, stamina longer than the petals, coloured empalements, and a shrubby stalk.*

The first and second sorts are both very common plants, growing in the fields in most parts of England; the first is used in medicine: these are rarely admitted into gardens, but I mention them in order to introduce the other, which are more deserving.

The first sort hath a perennial root, from which are several round stalks a foot and a half high, dividing into many small branches, garnished at each joint with two small oblong leaves, standing opposite without foot-stalks; the branches also come out opposite. The leaves have many pellucid spots in them, which appear like so many holes when held up against the light. The flowers are numerous on the top of the branches, standing on slender foot-stalks; they are composed of five oval petals, of a yellow colour, with a great number of stamina, not quite so long as the petals, terminated by roundish summits. In the center is situated a roundish germen supporting three styles, crowned by single stigmas. The germen afterward becomes an oblong angular capsule with three cells, filled with small brown seeds. The leaves and flowers of this are used in medicine; it is esteemed an excellent vulnerary plant, and of great service in wounds, bruises, and contusions; there is a compound oil made from this plant, which is of great use in the foregoing accidents.

The second sort hath square stalks, which rise about the same height with the first, but do not branch so much. The leaves are shorter and broader, and have no pellucid spots. The flowers sit upon short foot-stalks at the end of the branches, which are shaped like those of the other.

The third sort grows naturally in Sicily, Spain, and Portugal. This rises with shrubby stalks about three feet high, sending out small branches at each joint opposite, garnished with oblong oval leaves placed by pairs, sitting close to the stalks, which have a rank scent like a he-goat. The flowers are produced in clusters at the end of the branches; they are composed of five oval yellow petals, with a great number of stamina which are longer than the petals, and three styles which are longer than the stamina. The germen which supports these, afterward becomes an oval capsule with three cells, filled with small seeds.

The fourth sort grows naturally in the Canary Islands, so was formerly preserved in green-houses during the winter season, but is found to be hardy enough to resist the greatest cold of this country, and is now cultivated in the nurseries as a flowering shrub; this rises with a shrubby stalk six or seven feet high, dividing into branches upward, garnished with oblong leaves set by pairs close to the branches, which have a rank scent like the former. The flowers are produced at the end of the stalks in clusters, are very like those of the former sort, having a great number of stamina which are longer than the petals.

These two sorts are propagated by suckers, which are plentifully sent forth from the old plant. They should be planted in a light dry soil, in which they will endure the severe cold of our climate very well.

The fifth sort grows naturally on mount Olympus, where it was discovered by Sir George Wheeler, who sent the seeds to the Oxford garden. This rises with many upright ligneous stalks about a foot high, garnished with small spear-shaped leaves sitting close to the stalks opposite. The flowers are produced at the top of the stalks three or four together, composed of five oblong petals, of a bright yellow colour, with a great number of stamina, which are of unequal lengths, some being longer and others shorter than the

the petals, terminated by ſmall roundiſh ſummits. In the center is ſituated an oval germen, ſupporting three ſlender ſtyles which are longer than the ſtamina. The germen afterward becomes an oval capſule, with three cells filled with ſmall ſeeds.

This plant is uſually propagated by parting of the roots, becauſe the ſeeds do not always ripen in this country; the beſt time for doing of this is in September, that the plants may have time to get root before winter: it will live in the open air, if it is planted in a warm ſituation and a dry ſoil, but a plant or two ſhould be kept in pots, to be ſheltered under a frame in winter, leſt in very ſevere froſt thoſe in the open air ſhould be deſtroyed. If this is propagated by ſeeds, they ſhould be ſown ſoon after they are ripe, in pots filled with light earth, and placed under a frame in the winter to ſhelter them from froſt, and in the ſpring the plants will appear; when theſe are fit to remove, ſome of them may be planted in a warm border, and others into pots, and treeated in the ſame way as the old plants.

The ſixth ſort riſes with a ſhrubby ſtalk ſeven or eight feet high, with a reddiſh bark, and divides into ſmall branches, garniſhed with oval heart-ſhaped leaves, ſitting cloſe to the ſtalks oppoſite. The flowers are produced at the end of the ſtalks in cluſters; they are ſmaller than thoſe of the third ſort, and have obtuſe empalements. The ſtamina are longer than the petals, and of a deeper colour. The flowers are ſucceeded by conical capſules of a purpliſh red colour, having three cells filled with ſmall ſeeds. This is now propagated in the nurſeries as a flowering ſhrub, and may be treated in the ſame way as the third and fourth ſorts.

The ſeventh ſort was firſt brought from Conſtantinople, but has been long very common in the Engliſh gardens, for the roots ſpread and increaſe very faſt, where it is permitted to ſtand long unremoved. The ſtalks of this are ſlender, and incline downward; they are garniſhed with oval, ſpear-ſhaped, ſmooth leaves, placed by pairs. The flowers are produced at the end of the ſtalks; they are very large, and of a bright yellow colour, with a great number of ſtamina, which ſtand out beyond the petals; there are five ſtyles in each flower, which are of the ſame length with the ſtamina.

This plant is eaſily propagated by parting of the root in October, that the plants may be well eſtabliſhed before the drought of ſpring. As this will grow under trees, ſo it is a very proper plant to place under ſhrubs and trees to cover the ground, where they will make a good appearance during their ſeaſon of flowering.

The eighth ſort grows naturally in the iſland of Minorca, from whence the ſeeds were ſent to England by Mr. Salvador, an apothecary at Barcelona, in the year 1718. This riſes with a ſlender ſhrubby ſtalk about two feet high, ſending out ſeveral weak branches, with a reddiſh coloured bark, marked where the leaves have fallen off with a cicatrice. The leaves are ſmall, oval, and waved on their edges, having ſeveral ſmall protuberances on their ſurface, and ſit cloſe to the ſtalks, half embracing them with their baſe. The flowers are produced at the top of the ſtalks; they are large, of a bright yellow colour, with a great number of ſtamina which are a little ſhorter than the petals; the flowers have five ſtyles, and are ſucceeded by pyramidal capſules with five cells, which have a ſtrong ſmell of turpentine, and filled with ſmall brown ſeeds. This plant has a ſucceſſion of flowers great part of the year, which renders it valuable; it is too tender to live through the winter in the open air in England, but requires no artificial heat; if the plants are placed in a dry airy glaſſcaſe in winter, where they may be protected from froſt, and enjoy a good ſhare of freſh air in mild weather, they will thrive better than in a warmer ſituation, but in a damp air their ſhoots ſoon grow mouldy and decay; nor ſhould the plants have much water during the winter, but in ſummer they ſhould be expoſed in the open air, and in warm weather they ſhould be frequently watered. This is propagated by cuttings, which ſhould be planted in June, in pots filled with light earth, and plunged into a gentle hotbed, and covered with a hand-glaſs. Theſe will put out roots in ſix or ſeven weeks, when they ſhould be carefully taken up, and each planted into a ſeparate ſmall pot, placing them in the ſhade till they have taken new root; then they may be removed to a ſheltered ſituation, where they may remain till the froſt comes, when they ſhould be removed into ſhelter.

If theſe are propagated by ſeeds, they ſhould be ſown in autumn, in the ſame way as is before directed for the fifth ſort, and the plants treated in the ſame manner as thoſe raiſed from cuttings.

The ninth ſort is the common Tutſan, or Park Leaves, which is ſometimes uſed in medicine. It grows naturally in woods in ſeveral parts of England, ſo is not often admitted into gardens. This hath a ſhrubby ſtalk, which riſes two feet high; the ſtalks are garniſhed with oval heart-ſhaped leaves ſitting cloſe to them with their baſe, which are placed oppoſite. The flowers are produced in ſmall cluſters at the end of the ſtalk; they are yellow, but ſmaller than either of the ſorts before mentioned, and have many long ſtamina which ſtand out beyond the flower; it has three ſtyles. The germen afterward turns to a roundiſh fruit, covered with a moiſt pulp, which, when ripe, is black. The capſule has three cells, containing ſmall ſeeds. It hath a perennial root, and may be propagated by parting it in autumn; it loves ſhade and a ſtrong ſoil.

The tenth ſort grows naturally in North America; this riſes with an upright herbaceous ſtalk three feet high, garniſhed with oblong leaves placed oppoſite, which half embrace the ſtalk with their baſe. At the end of each ſtalk is produced one pretty large yellow flower, with an obtuſe empalement, having many ſtamina, which are equal in length with the petals, and five ſtyles which are ſo cloſely joined as to appear but one. The ſtigmas are reflexed, which denote their number; it is propagated by parting of the roots; the beſt time for this is in autumn; it ſhould have a light ſoil and an open ſituation.

The eleventh ſort grows naturally in China, from whence the ſeeds were brought to the Right Hon. his Grace the Duke of Northumberland, and the plants were raiſed in his Grace's curious garden at Stanwick, and by his Grace's generoſity the Chelſea garden was furniſhed with this plant.

The root of this is compoſed of many ligneous fibres, which ſtrike deep in the ground, from which ariſe ſeveral ſhrubby ſtalks near two feet high, covered with a purpliſh bark, garniſhed with ſtiff ſmooth leaves about two inches long, and a quarter of an inch broad, placed oppoſite, ſitting cloſe to the ſtalk, of a lucid green on their upper ſide, and gray on their under, having many tranſverſe veins running from the midrib to the border. The flowers are produced at the top of the ſtalks, growing in ſmall cluſters, each ſtanding upon a ſhort diſtinct foot-ſtalk; they have an empalement of one leaf, divided into five obtuſe ſegments almoſt to the bottom, which is of a deep purple colour. The flower is compoſed of five large obtuſe petals of a bright yellow colour, which are concave, and in the center is ſituated an oval germen ſupporting a ſingle ſtyle, crowned by five ſlender ſtigmas, which bend on one ſide; the ſtyle is attended by a great number of ſtamina which are longer than the petals, and are terminated by roundiſh ſummits.

This plant continues in flower great part of the year, which renders it the more valuable, and if it is planted in a very warm ſituation, it will live in the open air; but thoſe plants which ſtand abroad will not flower in winter, as thoſe do which are removed into ſhelter in autumn.

It may be propagated by ſlips from the root, or by laying down of the branches; if by ſlips they ſhould be planted in the ſpring on a moderate hot-bed, which will forward their putting out new roots; the layers ſhould alſo be laid down at the ſame time, which will have taken root by autumn, when they may be tranſplanted into pots, and ſheltered under a frame in winter, and in the ſpring part of theſe may be planted in a warm border, and the others planted in pots to be ſcreened in winter, leſt thoſe in the open air ſhould be killed.

HYPERICUM FRUTEX. See SPIRÆA.

HYSSOPUS. Tourn. R. H. 200. tab. 95. Hyſſop.

The CHARACTERS are,

The empalement of the flower is cylindrical and permanent. The flower is of one petal, of the grinning kind, with a narrow cylindrical tube; the chaps are inclining; the upper lip is ſhort, plain, roundiſh, erect, and indented at the top. It hath four ſtamina, which ſtand apart; two of them are longer than the petal, the other two are ſhorter, and four germen, which afterward becomes ſo many oval ſeeds ſitting in the empalement.

The SPECIES are,

1. HYSSOPUS (*Officinalis*) ſpicis fecundis Hort. Cliff. 304. *Hyſſop with fruitful ſpikes; or the common Hyſſop.*

2. HYSSOPUS (*Rubra*) ſpicis brevioribus, verticillis compactis. *Hyſſop with ſhorter ſpikes, and whorls more compact; Hyſſop with a red flower.*

3. HYSSOPUS (*Nepetoides*) caule acuto quadrangulo. Hort. Upſal. 163. *Hyſſop with an acute ſquare ſtalk.*

4. HYSSOPUS (*Lophantus*) corollis tranſverſalibus, ſtaminibus inferioribus corollâ brevioribus. Hort. Upſ. 162. *Hyſſop with tranſverſe petals, and the lower ſtamina ſhorter than the petal.*

The firſt ſort, which is the only one cultivated for uſe, hath a perennial root; the ſtalks are firſt ſquare, but afterward become round, garniſhed with ſmall ſpear-ſhaped leaves placed oppoſite, with ſeven or eight very narrow erect leaves (or bractæa) riſing from the ſame joint. The upper part of the ſtalk is garniſhed with whorls of flowers in a ſpike. There are four ſtamina in each flower, which ſpread at a diſtance from each other; the two upper are the ſhorteſt, which are ſituated on each ſide the upper lip; the two longer ſtand cloſe to the two ſide ſegments; they are terminated by twin ſummits. At the bottom of the tube are ſituated four naked germen, which afterward becomes four oblong black ſeeds, ſitting in the empalement; it grows naturally in the Levant. There is a variety of this with white flowers, but doth not differ from the blue in any other particular.

The ſecond ſort doth not grow ſo tall as the firſt; the ſtalks branch more, and the ſpikes of flowers are much ſhorter than thoſe of the firſt. The whorls are cloſer together, and have long narrow leaves ſituated under each; the flowers are of a fine red colour. This ſort is not quite ſo hardy as the common, for in 1739 the plants were all deſtroyed by the cold; this is certainly a diſtinct ſpecies.

Theſe ſorts of Hyſſop are propagated either by ſeeds or ſlips; if by the ſeeds, they muſt be ſown in March upon a bed of light ſandy earth, and when the plants come up, they ſhould be tranſplanted out to the places where they are to remain, placing them at leaſt a foot aſunder each way; but if they are deſigned to abide in thoſe places for a long time, two feet diſtance will be ſmall enough, for they grow pretty large, eſpecially if they are not frequently cut to keep them within compaſs; they thrive beſt upon a poor dry ſoil, in which ſituation they will endure the cold of our climate better than when they are planted on rich ground. If they are propagated by ſlips, they ſhould be planted either in ſpring or autumn in a bed of light earth, where they will take root in about two months, after which they may be tranſplanted where they are to continue, managing them as was before directed for the ſeedling plants.

The third ſort grows naturally in North America; this hath a perennial root, and riſes with an upright ſquare ſtalk four or five feet high, garniſhed with oblique heart-ſhaped leaves ſawed on their edges, ending in acute points; they are placed by pairs on ſhort foot-ſtalks. The flowers grow in cloſe thick ſpikes four or five inches long at the top of the ſtalks: there are two varieties of this, one with pale yellow, and the other has purple flowers; the ſeeds of both ſorts never vary to each other.

The fourth ſort grows naturally in Siberia; this is a perennial plant with a ſtrong fibrous root, ſending out many ſquare ſtalks, garniſhed with oblong leaves placed oppoſite. The flowers are produced at each joint toward the upper part of the ſtalks in ſmall cluſters, ariſing from the baſe of the leaves. The tube of the petal is longer than the empalement; the lips of the flower are oblique to it, being ſituated horizontally. The two upper ſtamina and the ſtyle ſtand out beyond the petal, but the other are ſhorter. The flowers are blue.

Both theſe ſorts are very hardy, and may be eaſily propagated by ſeeds, which ſhould be ſown in autumn, for thoſe ſown in the ſpring do often lie a year in the ground before they vegetate; when the plants come up, they muſt be kept clean from weeds, and thinned where they are too cloſe. The following autumn they ſhould be tranſplanted where they are to remain, for the roots will abide ſome years.

J.

JACEA. See CENTAUREA.

JACOBÆA. See SENECIO and OTHONNA.

JACQUINIA. Lin. Gen. 254.

The CHARACTERS are,

The empalement of the flower is composed of five roundish concave leaves, and is permanent. The flower has one bell-shaped petal, which is bellied, cut into ten segments. It hath five awl-shaped stamina arising from the receptacle, terminated by halbert-shaped summits, and an oval germen supporting a style the length of the stamina, crowned by a headed stigma. The germen afterward becomes a roundish berry with one cell, containing one seed.

The SPECIES are,

1. JACQUINIA (*Ruscifolia*) foliis lanceolatis acuminatis. Jacq. Amer. 15. *Jacquinia with spear-shaped acute-pointed leaves.*

2. JACQUINIA (*Armillaris*) foliis obtusis cum acumine. Jacq. Amer. 15. *Jacquinia with blunt leaves, ending in acute points.*

3. JACQUINIA (*Linearis*) foliis linearibus acuminatis. Jacq. Amer. 15. *Jacquinia with linear sharp-pointed leaves.*

The first sort grows naturally in the island of Cuba; it rises with a shrubby stalk a foot high, garnished with stiff leaves like those of Butchers Broom; the flowers, according to Father Plumier's figure, are produced between the leaves, on the upper part of the branches; but having never seen the flowers, I can give no further account of the plant.

The second sort grows naturally at Carthagena, and also in several islands of the West-Indies. This rises with a shrubby stalk five feet high, garnished with oblong blunt leaves; the flowers are produced in long bunches; they are white, and have the scent of Jasmine, which they retain after they are decayed, so are worn by the ladies of the country.

The third sort grows on the borders of the sea in the island of Dominica. This is of humble growth, seldom rising two feet high, so is a plant of little notice.

These being natives of warm countries, must be preserved in stoves in England, and should be treated as other plants of the same countries.

JALAPA. See MIRABILIS.

JASIONE. Lin. Gen. Plant. 896. Rampions with scabious heads. This plant grows naturally on sterile ground in most parts of England, and is rarely admitted into gardens.

JASMINOIDES. See CESTRUM and LYCIUM.

JASMINUM. Tourn. Inst. R. H. 597. tab. 368. Lin. Gen. Plant. 17. The Jasmine, or Jessamine-tree.

The CHARACTERS are,

The flower hath a tubulous empalement, which is permanent, cut into five segments at the brim; the flower is of one petal, cut into five segments at the top, which spread open. It hath two short stamina, situated within the tube of the petal. In the center is situated a roundish germen, which afterward turns to an oval berry, with a soft skin inclosing two seeds, which are flat on those sides which join, and convex on the other.

The SPECIES are,

1. JASMINUM (*Officinale*) foliis oppositis pinnatis, foliolis acuminatis. *Jasmine with winged leaves placed opposite, whose lobes end in acute points; or the common white Jasmine.*

2. JASMINUM (*Humile*) foliis alternis ternatis pinnatisque, ramis angulatis. Hort. Upsal. 5. *Jasmine with trifoliate and winged leaves placed alternate, and angular branches; or the Italian yellow Jasmine.*

3. JASMINUM (*Fruticans*) foliis alternis ternatis simplicibusque, ramis angulatis. Hort. Cliff. 5. *Jasmine with trifoliate and single leaves placed alternate, and angular branches; or the common yellow Jasmine.*

4. JASMINUM (*Grandiflorum*) foliis oppositis pinnatis, foliolis brevioribus obtusis. *Jasmine with winged leaves placed opposite, whose lobes are shorter and obtuse; or Catalonian Jasmine.*

5. JASMINUM (*Odoratissimum*) foliis alternis ternatis, foliolis ovatis, ramis teretibus. *Jasmine with trifoliate leaves placed alternate, whose lobes are oval, and taper branches; or the yellow Indian Jasmine.*

6. JASMINUM (*Azoricum*) foliis oppositis ternatis, foliolis cordato-acuminatis. *Jasmine with trifoliate leaves placed opposite, whose lobes are heart-shaped and pointed; the Azoriam Jasmine, commonly called the Ivy-leaved Jasmine.*

7. JASMINUM (*Capense*) foliis lanceolatis oppositis integerrimis, floribus pedunculis unifloris. *Jasmine with spear-shaped leaves placed opposite, and one flower upon each foot-stalk, having three stamina.*

The first sort is the common white Jasmine, which is a plant so generally known as to need no description. This grows naturally at Malabar, and in several parts of India, yet has been long inured to our climate, so as to thrive and flower extremely well, but never produces any fruit in England. It is easily propagated by laying down the branches, which will take root in one year, and may then be cut from the old plant, and planted where they are designed to remain: it may also be propagated by cuttings, which should be planted early in the autumn, and if the winter should prove severe, the surface of the ground between them should be covered with tan, sea-coal ashes, or saw-dust, which will prevent the frost from penetrating deep into the ground, and thereby preserve them.

When these plants are removed, they should be planted where they are designed to be continued, which should be either against some wall, pale, or other fence, where the flexible branches may be supported. These plants should be permitted to grow rude in the summer, otherwise there will be no flowers; but after the summer is past the luxuriant shoots should be pruned off, and the others must be nailed to the support.

There are two varieties of this with variegated leaves, one with white, and the other yellow stripes, but the latter is the most common: these are propagated by budding them on on the plain Jasmine, and it often happens, that when the buds do not take, yet they will communicate their gilded miasma to the plants; so that in a short time after, many of the branches both above and below the places where the buds have been inserted, have been thoroughly tinctured.

The two striped sorts should be planted in a warm situation, especially the white striped, for they are much more tender than the plain, and are very subject to be destroyed by great frosts, if they are exposed thereto; therefore the white striped should be planted to a south or south-west aspect, and in very severe winters their branches should be covered with mats or straw, to prevent their being killed: the yellow striped is not so tender, so may be planted against walls to east or west aspects; but these plants with variegated leaves are not so much in esteem as formerly.

The second sort is frequently called Italian Jasmine by the gardeners, the plants being annually brought from thence by those who come over with Orange-trees; these are generally grafted upon the common yellow Jasmine stocks, so that if the graft decays the plants are of no value. This sort is somewhat tenderer than the common, yet will endure the cold of our ordinary winters, if it is planted in a warm situation. The flowers of this kind are generally larger than those of the common yellow sort, but have very little scent, and are not produced so early in the season. It may be propagated by laying down the tender branches, as was directed for the common white sort; or by budding or inarching it upon the common yellow Jasmine, the latter of which is preferable, by making the plants hardier than those which are obtained from layers.

The third sort was formerly more cultivated in the gardens than at present, for as the flowers have no scent, so few persons regard them. This hath weak angular branches, which require support, and will rise to the height of eight or ten feet, if planted against a wall or pale; but the plants do often produce a great number of suckers from their roots, whereby they become troublesome in the borders of the pleasure-garden; it is easily propagated by suckers or layers.

The fourth sort grows naturally in India, and also in the island of Tobago, where the woods are full of it; this hath much stronger branches than the common white sort; the leaves are winged, and are composed of three pair of short obtuse lobes, terminated by an odd one, ending in obtuse points; these lobes are placed closer than those of the common Jasmine, and are of a lighter green; the flowers come out from the wings of the stalks, standing on long foot-stalks, each sustaining three or four flowers, which are of a blush red on their outside, but white within; the tube of the flower is longer; the segments are obtuse, twisted at the mouth of the tube, and are of a much thicker texture than those of the common sort, so that there is no doubt of its being a distinct species; and the reason for Dr. Linnæus's supposing it to be only a variety was a mistake; for as these plants are generally grafted upon stocks of the common Jasmine, so there are always shoots coming out from the stocks below, which, if permitted to stand, will produce flowers; and these do often starve and kill the grafts, so that there will be only the common sort left; and this has been the case with some plants which he examined, therefore supposed the difference of the other sort was wholly owing to culture; whereas, if he had only observed the difference of their leaves, he would have certainly made two distinct species of them, as he has since done.

This plant is propagated by budding or inarching it upon the common white Jasmine, on which it takes very well, and is rendered hardier than those which are upon their own stocks. But as those of this kind are brought over from Italy every spring in so great plenty, they are seldom raised here: I shall therefore proceed to the management of such plants as are usually brought into England, which are generally tied up in small bunches, containing four plants; their roots are wrapped about with Moss to preserve them from drying, which, if it happen that the ship has a long passage, will often occasion them to push out strong shoots from their roots, which must always be taken off before they are planted, otherwise they will exhaust the whole nourishment of the plant, and destroy the graft.

In the making choice of these plants, you should carefully observe if their grafts are alive, and in good health; for if they are brown and shrunk they will not push out, so that there will be only the stock left, which is of the common sort.

When you receive these plants, you must clear the roots of the Moss, and all decayed branches should be taken off; then place their roots into a pot or tub of water, which should be set in the green-house, or some other room, where it may be screened from the cold; in this situation they may continue two days, after which you must prune off all the dry roots, and cut down the branches within four inches of the place where they are grafted, and plant them into pots filled with fresh light earth; then plunge the pots into a moderate hot-bed of tanners bark, observing to water and shade them, as the heat of the season may require. In about three weeks or a month's time they will begin to shoot, when you must carefully rub off all such as are produced from the stock below the graft; they must now have a great share of air to strengthen them, and by degrees they must be hardened to endure the open air, into which they should be removed in June, placing them in a warm situation the first summer; for if they are too much exposed to the winds, they will make but indifferent progress, being rendered tender by the hot-bed. If the summer proves warm, and the trees have succeeded well, they will produce some flowers in the autumn following, though they will be few in number, and not near so strong as they will be the succeeding years, when the trees are stronger and have better roots.

These plants are preserved in green-houses with Oranges, Myrtles, &c. in the winter season, and require the same treatment: but notwithstanding most people preserve these plants in green-houses, yet they will endure the cold of our ordinary winters in the open air, if planted against a warm wall, and covered with mats in frosty weather, in which situation they will produce ten times as many flowers in one season as those kept in pots, and the flowers will likewise be much larger; but they should not be planted abroad till they have some strength, so that it will be necessary to keep them in pots two or three years; and when they are planted against the wall, which should be in May, that they may take good root in the ground before the succeeding winter, you must turn them out of the pots, preserving the earth to their roots, and nail up their shoots to the wall, shortening such of them as are very long, that they may push out new shoots below to furnish the wall, continuing to nail up all the shoots as they are produced. In the middle, or toward the latter end of July, they will begin to flower, and continue to produce new flowers until the frost prevents them.

Toward the middle of November, if the nights are frosty, you must begin to cover your trees with mats, which should be nailed over them pretty close; but this should be done when the trees are perfectly dry, otherwise the wet being lodged upon the branches, will often cause a mouldiness upon them, and the air being excluded therefrom, will rot them in a short time: it will also be very necessary to take off these mats when the weather will permit, to prevent this mouldiness, and only keep them close covered in frosty weather;

weather; if a little mulch is laid upon the furface of the ground about their roots, and fome bands of hay faftened about their ftems, to guard them from the froft in very fevere weather, it will preferve them; in the fpring, as the weather is warmer, fo by degrees the covering fhould be taken off; but they fhould not be expofed too foon to the open air, for the morning froft and dry eafterly winds, which often reign in March, do frequently pinch thefe plants if they are too early expofed. When the covering is taken off the trees fhould be pruned, and cut out all decayed branches, fhortening the ftrong fhoots to about two feet long, which will caufe them to fhoot ftrong, and produce many flowers.

The fifth fort grows naturally in India; this rifes with an upright woody ftalk eight or ten feet high, covered with a brown bark, fending out feveral ftrong branches which want no fupport, garnifhed with trifoliate leaves of a lucid green, which are placed alternate on the branches; they are oval and entire, continuing green all the year: the flowers are produced at the end of the fhoots in bunches of a bright yellow, and have a moft grateful odour. They come out in July, Auguft, September, and October, and fometimes continue to the end of November; they are frequently fucceeded by oblong oval berries, which turn black when ripe, and have each two feeds.

This fort of Jafmine is propagated by laying down the tender branches; the fhoots fhould be laid down in March, and if they are flit at the joint, as is practifed in laying of Carnations, it will promote their rooting: when the weather is dry thefe layers muft be frequently watered, which, if carefully attended to, the plants will be rooted by the fucceeding fpring, fit to be tranfplanted, when they muft be planted in pots filled with light earth, and managed as was before directed for the feedling plants.

This fort is frequently propagated by inarching the young fhoots into ftocks of the common yellow Jafmine, but the plants fo raifed do not grow fo ftrong as thofe which are upon their own ftock; befides, the common yellow Jafmine is very apt to fend out a great number of fuckers from the root, which renders the plants unfightly; and if thefe fuckers are not conftantly taken off as they are produced, they will rob the plants of their nourifhment. The cuttings of this plant will alfo take root, if they are planted in pots in the fpring, and plunged into a moderate hot-bed, covering them clofe with hand-glaffes; when thefe are well rooted, they may be tranfplanted into feparate pots, and treated as the layers.

The fixth fort grows naturally in the Azores; this hath long flender branches which require fupport, which may be trained twenty feet high, garnifhed with trifoliate leaves, whofe lobes are large and heart fhaped, of a lucid green, placed oppofite on the branches. The flowers are produced at the end of the branches in loofe bunches, which are cut into five fegments fpreading open; they are of a clear white, and have a very agreeable fcent. This the gardeners call the Ivy-leaved Jafmine.

This Jafmine is alfo pretty hardy, and requires no more fhelter than only from hard frofts; and I am apt to think, if this fort was planted againft a warm wall, and managed as hath been directed for the Catalonian Jafmine, it would fucceed very well; for I remember to have feen fome plants of this kind growing againft a wall in the gardens in Hampton-court, where they had endured the winter, and were in a more flourifhing ftate than ever. I faw any of the kind in pots, and produced a greater quantity of flowers. Thefe plants are propagated in the fame manner as the yellow Indian Jafmine, and both require the fame treatment as the Catalonian.

The feventh fort was brought from the Cape of Good Hope, by Capt. Hutchinfon of the Godolphin, who difcovered it growing naturally a few miles up the land from the fea, being drawn to it by the great fragrancy of its flowers, which he fmelt at fome diftance from the plant, which was then in full flower; and after having viewed the plant, and remarked the place of its growth, he returned thither the following day with proper help, and a tub to put it in, and caufed it to be carefully taken up, and planted in the tub with fome of the earth on the fpot, and conveyed it on board his fhip, where it continued flowering great part of the voyage to England, where it arrived in good health; and for fome years continued flowering, in the curious garden of Richard Warner, Efq; at Woodford in Effex, who was fo obliging as to favour me with branches of this curious plant in flower, to embelifh one of the numbers of my figures of plants, where it is reprefented in the 180th plate.

This plant feems not to have been known to any of the botanifts, for I have not met with any figure or defcription of it in any of the books; there is one fort which is figured in the Malabar garden, and alfo in Burman's plants of Ceylon, which approaches near this; it is titled Nandi ervatum major. Hort. Mal. But it differs from this in having longer and narrower leaves, the tube of the flower is larger, and the fegments do not fpread fo open as this; but it is furprifing that this plant fhould be unknown to the people at the Cape of Good Hope, for there was not one plant of it in their curious garden, nor could the captain fee any other plant of it but that which he brought away.

The ftem of this plant is large and woody, fending out many branches, which are firft green, but afterward the bark becomes gray, and is fmooth; the branches come out by pairs, and have fhort joints; the leaves, which are of a thick confiftence, are alfo fet by pairs clofe to the branches; they are five inches long, and two inches and a half broad in the middle, leffening to both ends, terminating in a point, of a lucid green, having feveral tranfverfe veins from the midrib to the border, which are entire. The flowers are produced at the end of the branches, fitting clofe to the leaves, one upon each foot-ftalk; they have a tubulous empalement, with five corners or angles cut deep at the brim into five long narrow fegments, ending in very acute points: the flower hath but one petal, for although it is cut into many deep fegments at the top, yet thefe are all joined in one tube below; fome of thefe flowers are much more double than others, having three or four orders of petals. In thofe flowers which I have examined, where they have been the moft double, there has not been more than one ftamen; but in fome that were lefs double I have found two, and in others three ftamina, fo that it is impoffible to determine the clafs to which this plant belongs, by them, nor indeed can any clafs of plants be diftinguifhed by double flowers, whofe parts of generation vary according as the flowers are more or lefs full of petals; nor can the genus be determined (as fome have ignorantly pretended) by the imperfect germen below the flower, which, when firft formed, viewed with glaffes which magnify greatly, feemed as if compofed of a capfule containing many feeds (which may be the cafe of many berries with one feed, when examined in the like manner by perfons who are fond of their own imaginations:) but as I have fince received feeds of the fingle plants of this fort of Jafmine from Ceylon, which are berries divided into two feeds like the Coffee and Azorian Jafmine, which have grown in the Chelfea garden; and as thefe plants have produced flowers, each having three ftamina, fo I have applied the title to this plant. It is eafily propagated by cuttings, which, if taken from the young branches, and planted in pots filled with a loamy foil, ſoil, and plunged into a moderate hot-bed, covering them cloſe down with hand-glaſſes, will ſoon put out roots: then they may be tranſplanted each into a ſeparate ſmall pot, filled with the like loamy earth, and plunged again into the hot-bed to forward their putting out new roots. When theſe young plants have obtained ſtrength, they may be treated hardily; for ſome plants I have kept under a common frame, where the pots have been plunged into an old bed of tanners bark which had no heat, others I have kept through the winter in a glaſs-caſe without any artificial heat, both which have ſucceeded, and the plants have flowered very well, and they have been more vigorous than thoſe which were treated tenderly. This flower, when fully blown, is as large as a middling Damaſk Roſe, having a very agreeable odour; on the firſt approach it is ſomething like that of the Orange flower, but when more cloſely ſmelt to, has the odour of the common double white Naciſſus. The ſeaſon of this plant flowering in England, is in July and Auguſt, but in its native country it is ſuppoſed to flower great part of the year; for Capt. Hutchinſon, who brought the plant over, ſaid there was a ſucceſſion of flowers on it till the ſhip arrived in a cold climate, which put a ſtop to its growth.

JASMINUM ARABICUM. See Coffee and Nyctanthes.

JASMINUM ILICIS FOLIO. See Lantana.

JASMINE, the Perſian. See Syringa.

JATROPHA. Lin. Gen. Plant. 961. Caſſada, or Caſſava.

The Characters are,

It hath male and female flowers in the ſame plant; the male flowers are ſalver-ſhaped, of one petal, whoſe brim is cut into five roundiſh ſegments, which ſpread open; they have ten awl-ſhaped ſtamina, five being alternately ſhorter than the other. The female flowers, which are ſituated in the ſame umbel, have five petals ſpread open like a Roſe. In the center is a roundiſh germen with three deep furrows, which afterward becomes a roundiſh capſule with three cells, each containing one ſeed.

The Species are,

1. Jatropha (*Manihot*) foliis palmatis, lobis lanceolatis integerrimis lævibus. Lin. Sp. Plant. 1007. *Jatropha with hand-ſhaped leaves, whoſe lobes are ſpear-ſhaped, entire, and ſmooth; or Caſſava of John Bauhin.*

2. Jatropha (*Quinquelobata*) foliis quinquelobatis, lobis acuminatis, acutè dentatis lævibus, caule fruticoſo. *Jatropha with leaves compoſed of five ſmooth lobes ending in points, which are ſharply indented on their edges, and a ſhrubby ſtalk.*

3. Jatropha (*Urens*) aculeata, foliis quinquelobatis acute inciſis, caule herbaceo. *Prickly Jatropha with leaves having five lobes which are ſharply cut on their edges, and an herbaceous ſtalk.*

4. Jatropha (*Herbacea*) aculeata, foliis trilobis caule herbaceo. Lin. Sp. Pl. 1007. *Prickly Jatropha with leaves having three lobes, and an herbaceous ſtalk.*

5. Jatropha (*Vitifolia*) foliis palmatis dentatis aculeatis. Hort. Cliff. 445. *Jatropha with hand-ſhaped, indented, prickly leaves.*

6. Jatropha (*Multifidis*) foliis multipartitis lævibus, ſtipulis ſetaceis multifidis. Hort. Cliff. 445. *Jatropha with leaves divided into many parts, and briſtly ſtipulæ with many points; commonly called French Phyſic Nut in America.*

7. Jatropha (*Curcas*) foliis cordatis angulatis. Hort. Cliff. 445. *Jatropha with angular heart-ſhaped leaves; commonly called Phyſic Nut in America.*

8. Jatropha (*Staphiſagrifolia*) foliis quinquepartitis, lobis ovatis integris, ſetis glandulofis ramoſis. Flor. Leyd. Prod. 202. *Jatropha with leaves divided into five parts, the lobes whereof are oval and entire, and branching briſtles ariſing from the glands; commonly called Belly-ach weed in America.*

The firſt ſort here mentioned is the common Caſſada, or Caſſava, which is cultivated for food in the warm parts of America, where, after the juice is expreſſed out of the root (which has a poiſonous quality) it is ground into a kind of flour, and made up in cakes or puddings, and is eſteemed a wholeſome food.

This riſes with a ſhrubby ſtalk ſix or ſeven feet high, garniſhed with ſmooth leaves, compoſed of ſeven lobes, which are joined at their baſe in one center, where they are narrow, but increaſe in their breadth till within an inch and a half of the top, where they diminiſh to an acute point. The flowers are produced in umbels at the top of the ſtalks; theſe are ſome male and others female, compoſed of five roundiſh petals which ſpread open; the male flowers have ten ſtamina joined together in a column, and the female flowers have a roundiſh germen with three furrows in the center, ſupporting three ſtyles crowned by ſingle ſtigmas. The germen afterward turns to a roundiſh capſule with three lobes, each having a diſtinct cell, containing one ſeed.

The ſecond ſort was diſcovered by the late Dr. Houſtoun, at the Havannah: this riſes with an upright ſtalk ten or twelve feet high, which is firſt green and herbaceous, but afterward becomes ligneous, ſending out a few branches at the top, which are garniſhed with ſmooth leaves compoſed of five oval lobes, which end in acute points; the edges are indented in ſeveral irregular acute points. The flowers are produced in an umbel at the extremity of the ſtalks, of an herbaceous white colour, and are male and female in the ſame umbel, as the other ſort; the capſule is ſmooth, and has three cells, each including a ſingle ſeed.

The third ſort was diſcovered by the late Dr. Houſtoun, growing naturally about the town of La Vera Cruz: this hath a very thick fleſhy root, in ſhape like the white Spaniſh Radiſh; the ſtalk riſes from one to two feet high, it is taper, herbaceous, and branching, cloſely armed on every ſide with long white ſpines, which are pungent and ſtinging; the leaves are divided into five lobes, which are deeply jagged and waved on their edges; all the veins of the leaves are cloſely armed with ſtinging ſpines, ſo that it is dangerous handling them. At the end of the branches the flowers are produced in umbels; they are white, and have empalements cloſely armed with the ſame ſpines as the ſtalks and leaves: there are male and female flowers in the ſame umbel, the female flowers are ſucceeded by tricapſular veſſels, containing three ſeeds.

The fourth ſort riſes with an herbaceous ſtalk about a foot high, dividing into two or three branches, garniſhed with leaves ſtanding alternate upon long foot-ſtalks, compoſed of three oblong lobes which are ſlightly ſinuated on their edges, ending in acute points; the whole plant is cloſely armed with ſtinging ſpines. The flowers are ſmall, and grow in an umbel at the end of the branches, of a dirty white colour, male and female in the ſame umbel: the female flowers are ſucceeded by oval capſules with three lobes, which are covered with the ſame ſpines as the plant, and have three cells, each containing a ſingle ſeed. This plant is annual.

The fifth ſort grows naturally at Carthagena in New Spain; this hath a thick, ſwelling, fleſhy root, from which ariſes an herbaceous ſtalk as big as a man's thumb, which is four or five feet high, divided into ſeveral branches, cloſely armed with long brown ſpines; the leaves are deeply cut into five lobes, which are jagged deeply on their ſides, and the nerves are armed with ſtinging ſpines; the the flowers are produced in umbels at the top of the branches,

branches, standing upon long naked foot-stalks, of a pure white colour, and are male and female in the same umbel: the male flowers appear first, which are composed of five petals, forming a short tube at the bottom, and the stamina arise the length of the tube, joined in a column: the petals spread open flat above, and the stamina fills the mouth of the tube, shutting it up: the female flowers are smaller, but of the same shape, having no stamina, but an oval three-cornered germen, which afterward becomes a capsule with three lobes, each having a distinct cell, with one seed inclosed.

The sixth sort is now very common in most of the islands in the West-Indies, but was introduced from the continent first into the French islands, and from thence it was brought into the British islands, where it is titled French Physic Nut, to distinguish it from the following sort, which is called Physic Nut from its purging quality.

This rises with a soft thick stem eight or ten feet high, dividing into several branches covered with a grayish bark. The leaves come out on strong foot-stalks; they are divided into nine or ten lobes in form of a hand, which are joined at their base, with many jagged points on their borders standing opposite. The upper side of the leaves are of a lucid green, but their under side gray, and a little cottony. The flowers come out upon long foot-stalks from the end of the branches, formed into an umbel, in which there are male and female flowers, as in the other species; these umbels are large, and the flowers being of a bright scarlet make a fine appearance; the leaves being also very remarkable for their beauty, has occasioned the plant being cultivated for ornament in most of the islands of the West-Indies.

The seventh sort grows naturally in all the islands of the West-Indies; this rises with a strong stalk twelve or fourteen feet high, divided into several branches, garnished with angular heart-shaped leaves, which end in acute points. The flowers come out in umbels at the end of the branches; they are male and female, of an herbaceous colour, so make but little appearance; the female flowers are succeeded by oblong oval capsules, with three cells, each containing one oblong black seed.

The seeds of the two last sorts have been used as a purgative by the inhabitants of the West-Indies, but they operate so violently that now they are seldom used; three or four of these nuts have worked upward and downward near forty times, on a person who was ignorant of their effects; but it is affirmed, that this purgative quality is contained in a thin film, situated in the center of the nut, which, if taken out, the nuts are harmless, and may be eaten with safety.

The eighth sort grows naturally in all the islands of the West Indies, where it is sometimes called wild Cassada, or Cassava, and in others Belly ach Weed, the leaves of this plant being accounted a good remedy for the dry belly-ach. This plant rises with a soft herbaceous stalk to the height of three or four feet, covered with a purple bark, the joints having branching bristly hairs, rising in small bunches, not only upon the principal stalk, but also on the branches, and the foot-stalks of the leaves. The stalk divides upward into two or three branches, garnished with leaves standing on very long foot-stalks, which are divided into five oval entire lobes, ending in acute points. The flowers are produced at the end of the branches, upon slender naked foot-stalks, in small umbels; they are of a dark purple colour, having male and female flowers in the same umbel; the female flowers are succeeded by oblong, smooth, tricapsular vessels, in each of the cells is lodged one seed.

All these plants are natives of the warm parts of America, so are too tender to thrive in the open air in England. The first sort is cultivated in the West-Indies for food, where it is propagated by cutting the stalks into lengths of seven or eight inches, which, when planted, put out roots; the method of doing this having been mentioned in various books, I shall not repeat it here.

The other sorts are easily propagated by seeds, which should be sown on a good hot-bed in the spring, and when the plants are fit to remove they should be each transplanted into a small pot, and plunged into a fresh hot-bed of tanners bark, carefully shading them till they have taken fresh root, after which they must be treated in the same manner as other tender plants from hot countries, admitting fresh air to them daily, in proportion to the warmth of the season; but as many of the sorts have succulent stalks, some of which have a milky juice, they should have but little water given them, for they are soon destroyed by wet.

The fourth sort is an annual plant, so if the seeds are sown early in the spring, and the plants are brought forward, they will perfect their seeds the same year; but the other sorts are perennial, so do not flower till the second or third year, therefore the plants should be plunged into the tan-bed in the stove, where they should constantly remain, giving them a large share of air in warm weather, but in winter they must be tenderly treated, and then must have very little water. With this management the plants will continue several years, and produce their flowers, and frequently perfect their seeds in England.

IBERIS. Dillen. Nov. Gen. 6. Sciatica Cress.

The CHARACTERS are,

The flower hath an empalement of four oval leaves, which fall away. It hath four unequal petals. It hath six awl-shaped erect stamina, the two on the sides being shorter than the rest. In the center of the tube is situated a round compressed germen, which afterward becomes a roundish compressed vessel, having two cells, each containing one oval seed.

The SPECIES are,

1. IBERIS (*Semperflorens*) frutescens, foliis cuneiformibus obtusis integerremis. Lin. Hort. Cliff. 330. *Shrubby Sciatica Cress with entire, wedge-shaped, blunt leaves; commonly called Tree Candy Tuft.*

2. IBERIS (*Sempervirens*) frutescens, foliis linearibus acutis integerrimis. Lin. Hort. Cliff. 330. *Shrubby Sciatica Cress, with narrow, pointed, whole leaves; commonly called perennial Candy Tuft.*

3. IBERIS (*Umbellata*) foliis lanceolatis acuminatis, inferioribus serratis, superioribus integerrimis. Lin. Hort. Cliff. 330. *Sciatica Cress with spear-shaped pointed leaves, the under ones being sawed, but the upper entire; commonly called Candy Tuft.*

4. IBERIS (*Odorata*) foliis linearibus superne dilatatis serratis. Flor. Leyd. 330. *Sciatica Cress with narrow leaves dilated at their top, and sawed.*

5. IBERIS (*Nudicaulis*) foliis sinuatis, caule nudo simplici. Lin. Hort. Cliff. 328. *Sciatica Cress with sinuated leaves and a single naked stalk; or Rock Cress.*

6. IBERIS (*Amara*) foliis lanceolatis acutis subdentatis, floribus racemosis. Lin. Hort. Upsal. 184. *Sciatica Cress with acute, spear-shaped, indented leaves, and flowers growing in bunches.*

7. IBERIS (*Rotundifolia*) foliis subrotundis crenatis. Royen. Lin. Sp. Plant. 649. *Iberis with roundish crenated leaves.*

8. IBERIS (*Linifolia*) frutescens, foliis linearibus acutis, corymbis hemisphæricis. *Shrubby Sciatica Cress with narrow acute leaves, and hemispherical bunches of flowers.*

The first sort here mentioned is a low shrubby plant, which seldom rises above a foot and a half high, having many slender diffused branches, which fall to the ground if they are

are not ſupported. Theſe are well furniſhed with wedge-ſhaped leaves toward their extremity, which continue green all the year; in autumn the flowers are produced at the end of the ſhoots, which are white, and grow in an umbel. Theſe continue long in beauty, and are ſucceeded by others, ſo that the plants are rarely deſtitute of flowers for near eight months, from the end of October to the beginning of June.

This plant is ſomewhat tender, therefore is generally preſerved in green-houſes in winter. But although it is commonly ſo treated, yet in moderate winters this plant will live in the open air, if it is planted in a warm ſituation and on a dry ſoil; and if, in very hard froſt, they are covered either with mats, Reeds, ſtraw, or Peaſe-haulm, they may be preſerved very well, and the plants which grow in the full ground will thrive better, and produce a greater number of flowers than thoſe which are kept in pots; but the ſoil in which theſe are planted ſhould not be over rich, nor too wet, for in either of theſe they will grow too vigorous in ſummer, ſo will be in greater danger of ſuffering by the froſt in winter; but when they grow on a gravelly ſoil, or among lime rubbiſh, their ſhoots will be ſhort, ſtrong, and not ſo replete with moiſture, ſo will better reſiſt the cold.

This plant is propagated by cuttings, which, if planted during any of the ſummer months, and ſhaded from the ſun, will be rooted in two months, and may afterward be either planted in pots, or into the borders where they are deſigned to ſtand.

There is a variety of this with variegated leaves, which is preſerved in ſome of the gardens where perſons delight in theſe ſtriped-leaved plants; but this is not ſo hardy as the plain ſort, therefore muſt be treated more tenderly in winter.

The ſecond ſort is a plant of humbler growth than the firſt; this ſeldom riſes more than ſix inches high, nor do the branches grow woody, but are rather herbaceous; the leaves of this plant continue green through the year, and the flowers are of as long duration as thoſe of the firſt ſort, which renders it valuable. This rarely produces ſeeds in England, but it is propagated by ſlips, which in ſummer eaſily take root, and the plants may be treated in the ſame manner as hath been directed for the firſt ſort, but is hardier.

The third ſort is a low annual plant, the ſeeds of which were formerly ſown to make edgings for borders in the pleaſure-garden, for which purpoſe all the low annual flowers are very improper, becauſe they do not anſwer the intent, which is to prevent the earth of the borders falling into the walks, which theſe plants never can do; and although they make a pretty appearance during their continuance in flower, which is ſeldom more than a fortnight or three weeks, yet after their flowers are paſt they become very unſightly; therefore all theſe ſorts of flowers ſhould be ſown in ſmall patches in the borders of the flower-garden, where, if they are properly mixed with other flowers, they will have a very good effect, and by ſowing of them at three or four different ſeaſons, there may be a ſucceſſion of them continued in flower till autumn.

There are two different varieties of this third ſort, one with red, and the other hath white flowers; but the white is not common in the gardens, but the ſeeds of the ſixth ſort is generally ſold for it, and is ſeldom diſtinguiſhed but by thoſe who are ſkilled in botany; the ſeeds ſhould be ſown thin in patches, and when the plants are grown pretty ſtrong they ſhould be thinned, leaving but ſix or eight in each patch to flower; and by thus treating them, they will put out ſide branches, and flower much ſtronger, and continue longer in beauty than when they are left cloſer together; they will require no other culture but to keep them clean from weeds.

The fourth ſort ſeldom grows ſo large as the third, and the flowers are much ſmaller, but have an agreeable odour. It grows naturally in Helvetia, and is preſerved in botanic gardens for variety. It is annual, and requires the ſame treatment as the third.

The fifth ſort grows on ſandy and rocky places in ſeveral parts of England, ſo is rarely admitted into gardens. The leaves of this are ſmall, and cut to the midrib into many jags; theſe ſpread on the ground, and between them ariſe a naked foot-ſtalk two or three inches long, ſuſtaining ſmall umbels of white flowers. This is an annual plant, whoſe ſeeds ſhould be ſown in autumn, where the plants are deſigned to remain, and require no other care but to keep them clean from weeds.

The ſixth ſort is very like the third, but differs in the ſhape of the leaves. The flowers of this are white, ſo may be ſown to make a variety with the red. It requires the ſame treatment.

The ſeventh ſort grow naturally on the Alps; this is a perennial plant. The lower leaves which riſe from the root are round, fleſhy, and crenated on their edges. The ſtalk riſes four or five inches high, garniſhed with ſmall oblong leaves, which half embrace the ſtalks with their baſe. The flowers terminate the ſtalk in a round compact umbel; they are of a purple colour.

It is propagated by ſeeds, which ſhould be ſown on a ſhady border in autumn, and when the plants are ſtrong enough to remove, they ſhould be tranſplanted on a ſhady border where they are deſigned to remain, and will require no other care but to keep them clean from weeds.

The eighth ſort grows naturally in Spain and Portugal; this hath a great reſemblance of the ſecond, but the ſtalks do not ſpread ſo much; they grow erect, about ſeven or eight inches high, are ligneous and perennial. The leaves are very narrow, and ſeldom more than an inch long, ſtanding thinly upon the ſtalks, having no foot-ſtalks. The flowers grow in hemiſpherical umbels on the top of the ſtalks, and are of a purple colour.

This ſort may be propagated by cuttings, which ſhould be treated in the ſame way as is before directed for the firſt ſort, and ſome of the plants may be planted on a warm border in a dry ſoil, where they will endure the cold of our ordinary winters very well; but it will be proper to have two or three plants in pots, which may be ſheltered under a frame in winter to preſerve the kind, if, by ſevere froſt, thoſe in the open air ſhould be deſtroyed.

IBISCUS. See Hibiscus.

ICACO. See Chrysobalanus.

ILEX. Lin. Gen. Plant. 158. The Holly-tree.

The Characters are,

They have male, female, and hermaphrodite flowers on different plants, and often on the ſame tree. The male flowers have a ſmall permanent empalement of one leaf, indented in four parts; they have but one petal cut into four ſegments; they have four awl-ſhaped ſtamina. The female flowers have their empalements and petals the ſame as the male, but have no ſtamina; in their center is placed the roundiſh germen, which afterward becomes a roundiſh berry with four cells, each containing a ſingle hard ſeed.

The Species are,

1. Ilex (*Aquifolium*) foliis oblongo-ovatis, undulatis, ſpinis acutis. *Holly-tree with oblong oval leaves which are waved, and have acute ſpines; the common Holly.*

2. Ilex (*Echinata*) foliis ovatis, undulatis, marginibus aculeatis, paginis ſuperne ſpinoſis. *Holly-tree with oval waved*

waved leaves, whose borders are armed with strong thorns, and their upper surface prickly; commonly called Hedge-hog Holly.

3. Ilex (*Caroliniana*) foliis ovato-lanceolatis serratis. Hort. Cliff. 40. *Holly with oval, spear-shaped, sawed leaves; commonly called Dahoon Holly.*

There are several varieties of the common Holly with variegated leaves, which are propagated by the nursery-gardeners for sale, and some years past were in very great esteem, but at present are but little regarded, the old taste of filling gardens with shorn evergreens being pretty well abolished; however, in the disposition of clumps, or other plantations of evergreen trees and shrubs, a few of the most lively colours may be admitted, which will have a good effect in the winter season, if they are properly disposed. As the different variegations of the leaves of Hollies, are by the nursery-gardeners distinguished by different titles, so I shall here mention the most beautiful of them by the names they are generally known:

Painted Lady Holly, British Holly, Bradley's best Holly, Phyllis, or Cream Holly, Milkmaid Holly, Pritchet's best Holly, Gold-edged Hedge-hog Holly, Cheyney's Holly, Glory of the West Holly, Broaderick's Holly, Partridge's Holly, Herefordshire white Holly, Blind's Cream Holly, Longstaff's Holly, Eales's Holly, Silver-edged Hedge-hog Holly.

All these varieties are propagated, by budding or grafting them upon stocks of the common green Holly: there is also a variety of the common Holly with smooth leaves, but this is frequently found intermixed with the prickly-leaved on the same tree, and often on the same branch there are both sorts of leaves.

The common Holly grows naturally in woods and forests in many parts of England, where it rises from twenty to thirty feet high, and sometimes more, but their ordinary height is not above twenty-five feet: the stem by age becomes large, and is covered with a grayish smooth bark; and those trees which are not lopped or browsed by cattle, are commonly furnished with branches the greatest part of their length, so form a sort of cone; the branches are garnished with oblong, oval leaves, of a lucid green on their upper surface, but are pale on their under, having a strong midrib: the edges are indented and waved, with sharp thorns terminating each of the points, so that some of the thorns are raised upward, and others are bent downward, and being very stiff renders them troublesome to handle. The leaves are placed alternate on every side of the branches, and from the base of their foot-stalks come out the flowers in clusters, standing on very short foot-stalks; each of these sustain five, six, or more flowers. In some plants I have observed the flowers were wholly male, and produced no berries; in others I have observed female and hermaphrodite flowers, but upon some large old trees growing on Windsor forest, I have observed all three upon the same trees. The flowers are of a dirty white, and appear in May; these are succeeded by roundish berries, which turn to a beautiful red about Michaelmas, but continue on the trees, if they are not destroyed, till after Christmas, before they fall away.

The second sort grows naturally in Canada. The leaves of this sort are not so long as those of the common Holly, and their edges are armed with stronger thorns standing closer together; the upper surface of the leaves is set very close with short prickles, from whence the gardeners have given it the title of Hedge-hog Holly. This sort is usually propagated in the nurseries by budding or grafting it upon the common Holly, but I have raised it from the berries, and found the plants to be the same as those from whence the seeds were taken, so make no doubt of its being a distinct species.

There are two varieties of this with variegated leaves, one of which is yellow, and the other white. There is also a variety of the common Holly with yellow berries, which is also accidental, and is generally found on those plants which have variegated leaves, and but seldom on plain Hollies.

The common Holly is a very beautiful tree in winter, therefore deserves a place in all plantations of evergreen trees and shrubs, where its shining leaves and red berries make a fine variety; and if a few of the best variegated kinds are properly intermixed, they will enliven the scene. The Holly was also formerly planted for hedges, and is a very proper plant for that purpose.

It is propagated by seeds, which never come up the first year, but lie in the ground as the Haws do; therefore the berries should be buried in the ground one year, and then taken up and sown at Michaelmas, upon a bed exposed only to the morning sun; the following spring the plants will appear, which must be kept clean from weeds; and if the spring should prove dry, it will be of great service to the plants if they are watered once a week; but they must not have it oftener, nor in too great quantity, for too much moisture is very injurious to these plants when young.

In this seed-bed the plants may remain two years, and then should be transplanted in the autumn, into beds at about six inches asunder, where they may stand two years longer; during which time they must be constantly kept clean from weeds, and if the plants have thriven well, they will be strong enough to transplant where they are designed to remain; for when they are transplanted at that age, there will be less danger of their failing, and they will grow to a larger size than those which are removed when they are much larger; but if the ground is not ready to receive them at that time, they should be transplanted into a nursery in rows at two feet distance, and one foot asunder in the rows, in which place the plants may remain two years longer; and if they are designed to be grafted or budded with any of the variegated kinds, that should be performed after the plants have grown one year in this nursery; but the plants so budded or grafted should continue two years after in the nursery, that they may make good shoots before they are removed; though the plain ones should not stand longer than two years in the nursery, because when they are older they do not transplant so well. The best time for removing Hollies is in the autumn, especially in dry land; but where the soil is cold and moist they may be transplanted with great safety in the spring, if the plants are not too old, or have not stood long unremoved, for if they have, it is great odds of their growing when removed.

The Dahoon Holly grows naturally in Carolina, of which there are two sorts, one with spear-shaped, the other with linear leaves. This rises with an upright branching stem to the height of eighteen or twenty feet; the bark of the old stems is of a brown colour, but that of the branches or younger stalks is green and smooth, garnished with spear-shaped leaves, which are more than four inches long, and one and a quarter broad in the broadest part, of a light green and thick consistence; the upper part of the leaves are sawed on their edges, each serrature ending in a small sharp spine. The flowers come out in thick clusters from the side of the stalks; they are white, and shaped like those of the common Holly, but are smaller; these are succeeded by small, roundish, red berries, in its native country, which make a fine appearance in winter, but they have not as yet perfected fruit in England so far as I can learn.

Dr. Linnæus ſuppoſes this plant and the evergreen Caſſine to be the ſame, but they are undoubtedly diſtinct plants; he may probably have been led into this miſtake, by receiving ſeeds of both ſorts mixed together from America, which I have more than once done; but whoever ſees the two plants growing, cannot doubt of their being different.

This ſort is tender while young, ſo requires protection in the winter till the plants are grown ſtrong and woody, when they may be planted in the full ground in a warm ſituation, where they will endure the cold of our ordinary winters pretty well: but in ſevere froſt they ſhould be protected, otherwiſe the cold will deſtroy them.

This is propagated from ſeeds, in like manner as the common ſort; the ſeeds of it will lie as long in the ground, ſo the berries ſhould be buried in the ground a year, and then taken up and ſown in pots filled with light earth, and placed under a frame in winter; in the ſpring the pots ſhould be plunged into a hot bed, which will bring up the plants. Theſe ſhould be preſerved in the pots while young, and ſheltered in winter under a common frame until they have obtained ſtrength, when they may be turned out of the pots and planted in the full ground, in a warm ſituation.

IMPATIENS. Rivin. Ord. 4. Lin. Gen. Plant. 899. Female Balſamine.

The CHARACTERS are,

The flower has a two-leaved ſmall empalement. It hath five petals which are unequal, and ſhaped like a lip-flower. It hath a nectarium in the bottom of the flower, ſhaped like a hood or cowl, which is oblique to the mouth, riſing on the out-ſide, whoſe baſe enas in a tail or ſpur, and five ſhort ſtamina which are incurved. In the bottom is ſituated an oval ſharp-pointed germen, which afterward becomes a capſule with one cell, opening with an elaſticity in five valves which twiſt ſpirally, and contain ſeveral roundiſh ſeeds fixed to a column.

The SPECIES are,

1. IMPATIENS (*Noli tangere*) pedunculis multifloris ſolitariis, foliis ovatis, geniculis caulinis tumentibus. Flor. Suec. 722. *Impatiens with foot-ſtalks ſuſtaining many ſingle flowers, oval leaves, and ſtalks having ſwelling joints; Yellow Balſamine, or Touch me not.*

2. IMPATIENS (*Balſamina*) pedunculis unifloris aggregatis, foliis lanceolatis, nectariis floris brevioribus. Hort. Upſal. 276. *Impatiens with foot-ſtalks ſuſtaining ſingle flowers, which ariſe in cluſters, ſpear-ſhaped leaves, and nectariums which are ſhorter than the flower; the Female Balſamine.*

3. IMPATIENS (*Triflora*) pedunculis trifloris ſolitariis, foliis anguſto-lanceolatis. Flor. Zeyl. 315. *Impatiens with three flowers on a foot-ſtalk, and narrow ſpear-ſhaped leaves; upright or Female Balſamine of Ceylon, with a narrow Peach leaf.*

The firſt ſort grows naturally in ſeveral parts of Weſtmoreland and Yorkſhire, but is frequently introduced into gardens by way of curioſity. It is an annual plant, which riſes about two or three feet high, with an upright ſucculent ſtalk, whoſe joints are ſwollen, garniſhed with oval ſmooth leaves, which ſtand alternate on every ſide the ſtalk. The flowers come out from the wings of the ſtalks upon long ſlender foot-ſtalks, which branch into ſeveral other ſmaller, each ſuſtaining one yellow flower, compoſed of five petals, which in front are ſhaped like the lip or grinning flowers, but at their baſe have a long tail like the flowers of Indian Creſs; theſe are ſucceeded by taper pods, which, when ripe, burſt open upon being touched, and twiſt ſpirally like a ſcrew, caſting out the ſeeds with great elaſticity. If the ſeeds of this plant are permitted to ſcatter, they generally ſucceed better than when they are ſown; for unleſs they are ſown in the autumn ſoon after they are ripe, they very rarely grow. The plants require no care but to keep them clean from weeds, and thin them where they are too cloſe. This delights in a ſhady ſituation and a moiſt ſoil. There is a variety of this with red flowers, which came from North America, which only differs in the colour of the flowers, and growing much taller, it is equally hardy.

The ſecond ſort is the Female Balſamine, of which there are ſeveral varieties; the common ſort has been long an inhabitant in the Engliſh gardens; of this there is the white, the red, and ſtriped flowered, and likewiſe the ſingle and double flowering. The common ſingle ſort is ſo hardy as to riſe in the full ground, and where the ſeeds ſcatter the plants will come up the following ſpring, but theſe do not come to flower ſo early as thoſe which are raiſed upon a hot-bed; however they generally are ſtronger plants, and continue much longer in the autumn in flower than the others, ſo are an ornament to the garden at ſuch times when there is a greater ſcarcity of flowers.

There are two varieties, if not diſtinct ſpecies, with double variegated flowers; one of them grows naturally in the Eaſt, and the other in the Weſt-Indies; that which comes from the Eaſt-Indies by the title of Immortal Eagle Flower, is a moſt beautiful plant; the flowers are twice the ſize of thoſe of the common ſort, and are very double; they are in ſome ſcarlet and white variegated, and purple and white in others, and the plants producing plenty of the flowers render them very valuable: if the ſeeds of theſe are carefully ſaved, the kinds may always be preſerved. I have raiſed ſome plants from foreign ſeeds, whoſe flowers were ſo very double as to loſe their male parts, ſo did not produce any ſeeds.

The ſeeds of theſe plants ſhould be ſown on a moderate hot-bed in the ſpring, and when the plants are come up about an inch high, they ſhould be tranſplanted on another moderate hot-bed at about four inches diſtance, obſerving to ſhade them from the ſun till they have taken new root; after which they ſhould have a large ſhare of free air, to prevent their drawing up tall and weak: they will require to be often refreſhed with water, but it ſhould not be given to them in too great plenty; for as their ſtems are very ſucculent, ſo they are apt to rot with much moiſture. When the plants are grown ſo large as to touch each other, they ſhould be carefully taken up with balls of earth to their roots, and each planted into a ſeparate pot filled with light rich earth, and plunged into a very moderate hot-bed, covered with a deep frame, to admit the plants to grow, ſhading them from the ſun until they have taken freſh root; then they ſhould have a large ſhare of air admitted to them, and by degrees hardened, ſo as to bear the open air, into which part of the plants may be removed in July, placing them in a warm ſheltered ſituation, where, if the ſeaſon proves favourable, they will flower and make a fine appearance; but it will be proper to keep part of the plants either in a glaſs-caſe or a deep frame, in order to get good ſeeds, becauſe thoſe in the open air will not ripen their ſeeds unleſs the ſummer proves very warm; and the plants in ſhelter muſt have a good ſhare of free air every day, otherwiſe they will grow pale and ſickly; nor ſhould they have too much of the ſun in the middle of the day, in very hot weather, for that occaſions their leaves hanging, and their requiring water, which is often very hurtful; therefore if the glaſſes are ſhaded in the middle of the day for three or four hours, the plants will flower better, and continue longer in beauty than when they are expoſed to the great heat. Thoſe who are curious to preſerve theſe plants in perfection, pull off all the ſingle and plain coloured flowers from the plants which they preſerve for ſeeds, leaving only thoſe flowers which are double and of good colours; where this

 is

is carefully done, they may be continued without the leaſt degeneracy.

The ſort which grows in the Weſt-Indies is there called Cockſpur. This hath ſingle or ſemi-double flowers, which are as large as the laſt mentioned ſort, but I never ſaw any of them more than half double, and only with white and red ſtripes: the plants are apt to grow to a very large ſize before they produce any flowers, ſo that it is late in the autumn before they begin to flower, and ſometimes in bad ſeaſons they will ſcarce have any flowers, and but rarely ripen their ſeeds here, ſo that few perſons care to cultivate this ſort, eſpecially if they can have the other.

The third ſort here mentioned grows naturally in Ceylon, and in many parts of India. This hath very narrow ſpear-ſhaped leaves, which are ſawed on their edges; the foot-ſtalks ſuſtain each three flowers, which are ſmaller than thoſe of the common ſort, ſo are not worthy of a place in gardens, except for the ſake of variety. This is a tender plant, and requires the ſame treatment as the Immortal Eagle Flower.

IMPERATORIA. Lin. Gen. Plant. 321. Maſterwort.

The Characters are,

It hath an umbellated flower; the principal umbel is plain, and has no involucrum, but the ſmall ones have. The principal umbel is uniform; the flowers have five heart-ſhaped inflexed petals. They have five hairy ſtamina. The germen is ſituated under the petals, which afterward becomes a roundiſh compreſſed fruit divided in two parts, containing two oval bordered ſeeds.

We have but one Species of this plant, viz.

Imperatoria (*Oſtruthium.*) Hort. Cliff. 103. *Maſterwort.*

This plant grows naturally on the Auſtrian and Styrian Alps, and other mountainous places in Italy; the root is as thick as a man's thumb, running obliquely in the ground; it is fleſhy, aromatic, and has a ſtrong acrid taſte, biting the tongue and mouth like Pellitory of Spain; the leaves ariſe immediately from the root; they have long foot-ſtalks dividing into three very ſhort ones at the top, each ſuſtaining a trilobate leaf, indented on the border; the foot-ſtalks are deeply channelled, and when broken emit a rank odour. The flower-ſtalks riſe about two feet high, dividing into two or three branches, each being terminated by a pretty large umbel of white flowers, whoſe petals are ſplit; theſe are ſucceeded by oval compreſſed ſeeds, ſomewhat like thoſe of Dill, but larger.

This plant is cultivated in gardens to ſupply the markets. It may be propagated either by ſeeds, or by parting the roots; if it is propagated by ſeeds, they ſhould be ſown in autumn ſoon after they are ripe, on a bed or border, in a ſhady ſituation. In the ſpring the plants will appear, when they ſhould be carefully weeded. Toward the beginning of May, if the plants come up too cloſe together, they ſhould be thinned, leaving them about ſix inches aſunder, planting thoſe which are drawn up into another bad about the ſame diſtance apart every way, being careful to water them duly, if the ſeaſon ſhould prove dry, until they have taken root; after which time theſe plants (as alſo thoſe remaining in the ſeed-beds) will require no other culture but to keep them clear from weeds till the following autumn, when the plants ſhould be tranſplanted where they are deſigned to remain, which ſhould be in a rich moiſt ſoil and a ſhady ſituation, for they delight in ſhade and moiſture; ſo that where theſe are wanting, the plants will require a conſtant ſupply of water in dry weather. The diſtance which theſe plants ſhould be placed muſt not be leſs than two feet every way, for where they like their ſituation they will ſpread and increaſe much. When theſe plants are rooted, they will require no other culture but to keep them clean from weeds; but in the ſpring before they ſhoot, the ground ſhould be every year gently dug between the plants; in doing of which, great care ſhould be had not to cut or bruiſe their roots. Theſe plants, with this management, will continue ſeveral years, and will produce ſeeds in plenty.

If theſe plants are propagated by offsets, their roots ſhould be parted at Michaelmas, and planted in a ſhady ſituation, at the ſame diſtance as hath been directed for the ſeedling plants.

The roots of this plant are uſed in medicine, and are greatly recommended for their virtue in contagious diſtempers or the bites of venomous creatures; they are alexipharmic and ſudorific; by ſome they are recommended for cholics and aſthmas, for the cramp, and all cold diſeaſes of the nerves.

INARCHING is a method of grafting, which is commonly called grafting by approach. This method of grafting is uſed, when the ſtock intended to graft on, and the tree from which the graft is to be taken, ſtand ſo near (or can be brought ſo near) that they may be joined together. The method of performing it is as follows: Take the branch you would Inarch, and having fitted it to that part of the ſtock where you intend to join it, pare away the rind and wood on one ſide about three inches in length. After the ſame manner cut the ſtock or branch in the place where the graft is to be united, ſo that the rind of both may join equally together, that the ſap may meet; then cut a little tongue upwards in the graft, and make a ſlit in the ſtock downward to admit it; ſo that when they are joined the tongue will prevent their ſlipping, and the graft will more cloſely unite with the ſtock. Having thus placed them exactly together, they muſt be tied with ſome baſs, or other ſoft bandage; then cover the place with grafting clay, to prevent the air from entering to dry the wound, or the wet from getting in to rot the ſtock: you ſhould alſo fix a ſtake into the ground to which that part of the ſtock, as alſo the graft, ſhould be faſtened, to prevent the wind from breaking them aſunder, which is often the caſe when this precaution is not obſerved.

In this manner they are to remain about four months, in which time they will be ſufficiently united, and the graft may then be cut from the mother tree, obſerving to ſlope it off cloſe to the ſtock; and if at this time you cover the joined parts with freſh grafting clay, it will be of great ſervice to the graft.

INDIGOPHERA. Lin. Gen. Plant. Indigo.

The Characters are,

The empalement of the flower is of one leaf, cut into five ſegments; the flower is of the butterfly kind, having a ſpreading, indented, reflexed ſtandard: the wings are oblong, with their under borders ſpreading. The keel is obtuſe; it hath ten ſtamina, formed like a cylinder, whoſe points aſcend, and are terminated by roundiſh ſummits; it hath a cylindrical germen ſupporting a ſhort ſtyle, crowned by an obtuſe ſtigma. The germen afterward becomes a taper pod, incloſing kidney-ſhaped ſeeds.

The Species are,

1. Indigophera (*Tinctoria*) leguminibus arcuatis incanis, racemis folio brevioribus. Flor. Zeyl. 273. *Indigo with hoary arched pods, whoſe bunches of flowers are ſhorter than the leaves.*

2. Indigophera (*Suffruticoſa*) leguminibus arcuatis incanis, caule fruticoſa. *Indigo with a ſhrubby ſtalk, and hoary arched pods.*

3. Indigophera (*Caroliniana*) leguminibus teretibus, folioſis quinis, radice perenne. *Indigo with taper pods, five lobes to the leaves, and a perennial root.*

Indi-

4. INDIGOPHERA (*Indica*) leguminibus pendulis lanatis compressis, foliis pinnatis. *Indigo with woolly, compressed, hanging pods, and winged leaves.*

5. INDIGOPHERA (*Glabra*) leguminibus glabris teretibus, foliolis trifoliatis. *Indigo with smooth taper pods, and trefoil leaves.*

The first and fifth sorts are annual plants in England; the seeds of these should be sown early in the spring of the year, and when the plants come up near two inches high, they should be transplanted into small pots filled with good fresh earth, and the pots must be plunged into a hot-bed of tanners bark: as the plants increase in their size, so they must have larger pots, but by no means should they have pots too large; and in warm weather they should have a large share of fresh air, and in cold weather they require to be screened from it; with this management the plants will flower and produce ripe seeds, when the plants will soon after decay.

The second sort is a perennial plant, which rises four or five feet high, sending out many lateral branches which grow erect; this may be propagated by seeds, which should be sown on a hot-bed, and treated in the same manner as the first and fifth sorts; and if the plants are afterward plunged into the tan-bed in the stove, they may be preserved two or three years, and will produce flowers, and perfect seeds the second year, but rarely ripen seeds the first year.

The third sort grows naturally in South Carolina, where the inhabitants made great use of it for two or three years; but as the stalks and leaves were thin, so it did not produce quantity enough of the dye to answer the expence of cultivating, therefore was soon after neglected.

The fourth sort is supposed to be promiscuously used in making Indigo in India, but this is not cultivated in America for this purpose; whether it makes a better or worse sort of Indigo I cannot affirm, but the Indigo merchants give a greater price for it. This sort may be propagated for curiosity in the same manner as the others.

INGA. See MIMOSA.

INOCULATING, or Budding. This is commonly practised upon all sorts of stone fruit in particular, such as Peaches, Nectarines, Cherries, Plumbs, &c. as also upon Oranges and Jasmines, and is preferable to any sort of grafting for most sorts of tender fruit. The method of performing it is as follows: You must be provided with a sharp penknife, having a flat haft (the use of which is to raise the bark of the stock to admit the bud) and some sound bass mat, which should be soaked in water to increase its strength, and make it more pliable; then having taken off the cuttings of the trees you would propagate, you should chuse a smooth part of the stock about five or six inches above the surface of the ground, if designed for dwarfs, but if for standards they should be budded six feet above ground; then with your knife make an horizontal cut cross the rind of the stock, and from the middle of that cut make a slit downwards about two inches in length, so that it may be in the form of a T; but you must be careful not to cut too deep, lest you wound the stock: then having cut off the leaf from the bud, leaving the foot-stalk remaining, you should make a cross cut about half an inch below the eye, and with your knife slit off the bud, with part of the wood to it, in form of an escutcheon: this done, you must with your knife pull off that part of the wood which was taken with the bud, observing whether the eye of the bud be left to it or not (for all those buds which lose their eyes in stripping, should be thrown away, being good for nothing:) then having gently raised the bark of the stock where the cross incision was made, with the flat haft of your penknife cleave the bark from the wood, and thrust the bud therein, observing to place it smooth between the rind and the wood of the stock, cutting off any part of the rind belonging to the bud, which may be too long for the slit made in the stock; and so having exactly fitted the bud to the stock, you must tie them closely round with bass mat, beginning at the under part of the slit, and so proceed to the top, taking care that you do not bind round the eye of the bud, which should be left open.

When your buds have been inoculated three weeks or a month, you will see which of them have taken; those of them which appear shrivelled and black being dead, but those which remain fresh and plump you may depend are joined; at this time you should loosen the bandage, which, if not done in time, will pinch the stock, and greatly injure, if not destroy, the bud.

The March following you must cut off the stock about three inches above the bud, sloping it that the wet may pass off, and not enter the stock; to this part of the stock left above the bud, it is very proper to fasten the shoot which the bud makes in summer, to secure it from being blown out; but this part of the stock must continue on no longer than one year, after which it must be cut off close above the bud, that the stock may be covered thereby.

The time for Inoculating is from the middle of June until the middle of August, according to the forwardness of the season, and the particular sorts of trees to be increased; which may easily be known by trying the buds, whether they will come off well from the wood. But the most general rule is, when you observe the buds formed at the extremity of the same year's shoots, which is a sign of their having finished their spring growth.

INTYBUS. See CICHORIUM.

INULA. Lin. Gen. Plant. 860. Elecampane.

The CHARACTERS are,

It hath a radiated compound flower, with an imbricated empalement. The disk or middle of the flower is composed of hermaphrodite florets; the border or ray of female half florets, stretched out like a tongue. The hermaphrodite florets are funnel-shaped, and cut into five parts at the top; these have five short slender stamina, which coalesce at the top; they have one long germen crowned with down. The female half florets have narrow, entire, tongue-shaped petals, no stamina, but a long crowned germen with a hairy style. The germen in both flowers become a single, narrow, four-cornered seed, crowned with a down, sitting on a naked receptacle.

The SPECIES are,

1. INULA (*Helenium*) foliis amplexicaulibus ovatis, rugosis, subtus tomentosis, calycum squamis ovatis. Amœn. Acad. 1. p. 410. *Elecampane with oval rough leaves, woolly on their under side, and the scales of the empalement oval; called Elecampane.*

2. INULA (*Odora*) foliis amplexicaulibus dentatis hirsutis, radicalibus ovatis, caulinis lanceolatis amplexicaulibus, caule paucifloro. Lin. Sp. Pl. 881. *Inula with hairy indented leaves, those at the bottom oval, but the upper are spear-shaped, embracing the stalks, with few flowers.*

3. INULA (*Salicina*) foliis sessilibus lanceolatis recurvis serrato ciliatis, floribus solitariis, ramis angulatis. Amœn. Acad. 1. p. 410. *Inula with spear-shaped recurved leaves, which are sawed, hairy, flowers growing singly, and angular branches.*

4. INULA (*Germanica*) foliis lanceolatis recurvis, subdentatis, scabris, floribus subfasciculatis. Lin. Sp. Pl. 883. *Inula with spear-shaped recurved leaves, which are rough and indented, and flowers growing in clusters.*

5. INULA (*Crithmoides*) foliis linearibus carnosis tricuspidatis. Lin. Sp. Pl. 883. *Inula with narrow fleshy leaves ending in three points; called Golden Samphire.*

6. Inula (*Montana*) foliis lanceolatis hirsutis integerrimis, caule unifloro. Lin. Sp. Plant. 884. *Inula with soft, hairy, spear-shaped, entire leaves, and one flower on a stalk.*

7. Inula (*Oculus Christi*) foliis oblongis, integris, hirsutis, caule piloso, corymboso, floribus confertis. Lin. Sp. Plant. 881. *Inula with oblong, entire, rough leaves, a hairy stalk divided by a corymbus, and flowers growing in clusters.*

8. Inula (*Britannica*) foliis lanceolatis, serratis, subamplexicaulibus, subtus villosis, caule ramoso erecto. Lin. Sp. Pl. 882. *Inula with spear-shaped sawed leaves embracing the stalk, hairy on their under side, and an erect branching stalk.*

9. Inula (*Hirta*) foliis lanceolatis semiamplexicaulibus hispidis, caulibus subunifloris teretibus. Lin. Sp. Pl. 883. *Inula with spear-shaped prickly leaves half embracing the stalk, which is taper, and has but one flower.*

10. Inula (*Bifrons*) foliis oblongis hirsutis, semiamplexicaulibus, caule ramoso piloso, squamis calycinis lanceolatis. *Inula with oval hairy leaves half embracing the stalk, which is branching and hairy, and spear-shaped scales to the empalement.*

11. Inula (*Squarosa*) foliis ovato-lanceolatis denticulatis sessilibus, caule piloso erecto corymboso. *Inula with oval, spear-shaped, indented leaves sitting close to the stalk, and an upright hairy stalk, terminating in a corymbus.*

12. Inula (*Canariensis*) foliis linearibus carnosis tricuspidatis, caule fruticoso. *Inula with narrow, fleshy, three-pointed leaves, and a shrubby stalk.*

13. Inula (*Mariana*) caule erecto hispido, foliis lanceolatis asperis, floribus alaribus solitariis sessilibus, terminalibus umbellatis. *Inula with an erect prickly stalk, spear-shaped rough leaves, flowers proceeding singly from the sides of the stalks, sitting close, and terminating in an umbel.*

The first sort grows naturally in several parts of England, but it is cultivated in gardens for the roots, which are used in medicine, and are accounted carminative, sudorific, and alexipharmic, of great service in shortness of breath, coughs, stuffing of the lungs, and infectious distempers.

It hath a perennial root, which is thick, branching, and of a strong odour. The lower leaves are eight or nine inches long, and four broad in the middle, rough on their upper side, but downy on their under. The stalks rise about four feet high, and divide toward the top into several smaller branches, garnished with oblong oval leaves indented on their edges, ending in acute points. The flowers terminate the stalks, each branch ending with one large, yellow, radiated flower, sitting in a scaly empalement, whose scales are oval, and placed like scales on fishes over each other. The flowers are succeeded by narrow four-cornered seeds crowned with down.

This sort may be propagated by seeds, which should be sown in autumn soon after they are ripe, for if they are kept till the spring they seldom grow; but where the seeds are permitted to scatter, the plants will come up the following spring without any care, and may be either transplanted the following autumn, or if they are designed to remain they should be hoed out to the distance of ten inches, or a foot each way, and constantly kept clean from weeds; these roots will be fit for use the second year.

But most people propagate this by offsets, which, if taken from the old roots, with a bud or eye to each, will take root very easily; the best time for this is in the autumn, as soon as the leaves begin to decay; these should be planted in rows about a foot asunder, and nine or ten inches distance in the rows; the spring following the ground must be kept clean from weeds, and if in autumn it is slightly dug, it will promote the growth of the roots, which will be fit for use after two years growth; the young roots are preferable to those which are old and stringy. It loves a gentle loamy soil, not too dry.

The second sort hath a perennial root, from which arise several stalks about two feet high. The leaves at bottom are oval, indented, and hairy; those above embrace the stalks with their base. The stalks are divided into several branches, garnished with a few scattering yellow flowers. The root has a very sweet odour when broken.

The third sort hath a perennial root, from which arise many spear-shaped, smooth, recurved leaves. The stalks are angular, and rise near two feet high, branching at the top into several foot-stalks, each sustaining one yellow radiated flower.

The fourth sort rises with an upright stalk between three and four feet high, with spear-shaped leaves which turn backward, indented on their edges, and rough on their upper side. The flowers are collected in close bunches on the upper part of the stalks; they are small and yellow. It grows on the Alps, and other mountainous parts of Europe.

The fifth sort grows naturally on the sea-coasts in many parts of England. This rises with an upright stalk a foot and a half high, garnished with fleshy succulent leaves, which come out in clusters. The flowers are yellow, and come out at the top of the stalks in small umbels. The young branches of this plant are frequently sold in the London markets for Samphire, but this is a great abuse, because this plant has none of the warm aromatic taste of the true Samphire.

The sixth sort grows naturally in Germany. This rises with upright stalks a foot and a half high, garnished with spear-shaped leaves, covered with soft hairs. The stalks do each support one large yellow flower.

The seventh sort has a perennial root. This grows naturally in Hungary; the leaves are oblong and hairy; the stalks branch at the top in form of a corymbus. The flowers are small, yellow, and in close clusters.

The eighth sort grows naturally in Austria, Bohemia, and other parts of Germany. This hath a perennial root; the stalk rises two feet high, garnished with spear shaped woolly leaves, which closely embrace the stalks with their base. The upper part of the stalk divides into two or three erect branches or foot-stalks, each sustaining one pretty large, deep, yellow flower.

The ninth sort grows naturally in the south of France, Spain, and Italy. This hath a perennial root, from whence arise several stalks about one foot high; the lower leaves are spear-shaped and prickly, the upper embrace the stalks which divide into several branches, each being terminated by one yellow flower.

The tenth sort hath a perennial root; the stalk rises about a foot high, dividing into many branches, garnished by oval hairy leaves, which half embrace the stalks with their base; each of the branches are terminated by one large yellow flower, whose empalement is composed of oval scales.

The eleventh sort grows naturally in Hungary. This rises with single upright stalks two feet high, garnished with oval spear-shaped leaves slightly indented on their edges, which sit close to the stalks, and divide in form of a corymbus at the top. The flowers are pretty large, and of a pale yellow colour.

The twelfth sort grows naturally in the Canary Islands. This hath several shrubby stalks near four feet high, which divide into smaller branches, garnished with clusters of narrow fleshy leaves, divided into three parts at their points. The flowers come out on the top of the stalks; they are small, and of a pale yellow colour.

The second, third, fourth, sixth, seventh, eighth, ninth, tenth, and eleventh sorts, are abiding plants, which will

thrive

thrive and flower in the open air in England; they may be all propagated by parting of their roots. The beſt time for doing of this is in autumn, when the plants ſhould be removed; theſe may be intermixed with other flowering plants in the borders of large gardens, where they will make an agreeable variety during their continuance in flower. As theſe roots multiply pretty faſt, they ſhould be allowed room to ſpread, therefore ſhould not be planted nearer than two feet from other plants; and if they are removed every third year, it will be often enough, provided the ground between them be dug every winter, and in ſummer, if they are kept clean from weeds, they will require no other care.

The twelfth ſort will not live abroad in the open air in England during the winter ſeaſon, ſo muſt be removed into ſhelter in autumn, but ſhould have as much free air as poſſible at all times, when the weather is mild, otherwiſe it is apt to draw up weak. This is eaſily propagated by cuttings any time in ſummer, which, if planted in a ſhady border, will take root in a ſhort time.

The thirteenth ſort was ſent me from Maryland, where it grows naturally. This riſes with a ſtrong ſtalk about a foot and a half high, cloſely ſet with prickly hairs, and garniſhed with rough ſpear-ſhaped leaves; toward the upper part of the ſtalk there are ſingle flowers coming from the wings at each joint, and the ſtalk is terminated by a cluſter of ſmall yellow flowers, diſpoſed in form of an umbel.

JOHNSONIA. Dale.

The CHARACTERS are,

The flower hath an empalement of one leaf, cut at the brim into ſhort ſegments. It hath one tubulous petal, divided into four parts at the brim, and four ſlender ſtamina which are longer than the petal. In the center is ſituated a roundiſh germen, which afterward becomes a ſmooth globular berry, incloſing four hard oblong ſeeds.

We have but one SPECIES of this genus, viz.

JOHNSONIA (*Americana*) floribus verticillatis ſeſſilibus, foliis ovato-lanceolatis oppoſitis, caule fruticoſo. Dale. *Shrubby Johnſonia with oval ſpear-ſhaped leaves ſitting cloſe to the ſtalks, placed oppoſite, and flowers growing in whorls.*

This ſhrub grows plentifully in the woods near Charles-Town in South Carolina. It riſes from four to ſix feet high, ſending out many branches from the ſide, which are woolly when young, like thoſe of the Wayfaring-tree, garniſhed with oval ſpear-ſhaped leaves placed oppoſite, ſtanding on pretty long foot-ſtalks. The flowers come out in whorls round the ſtalks, ſitting very cloſe to the branches; they are ſmall, tubulous, cut into four obtuſe ſegments at the top, which expand, and are of a deep purple colour; theſe are ſucceeded by ſoft ſucculent berries, which turn firſt to a bright red colour, but afterward change to a deep purple when ripe, and incloſe four hard oblong ſeeds.

The ſeeds of this plant were ſent by Mr. Cateſby from Carolina in 1724, and many of the plants were then raiſed in ſeveral curious gardens in England; moſt, if not all of them, were afterward planted in the open air, where they flouriſhed very well for ſeveral years, and ſeveral of the plants produced flowers for ſome years, but were killed in the ſevere froſt 1740; but the young plants which were raiſed from Dr. Dale's ſeeds the year before, which were only ſheltered under a frame, were ſaved.

This plant riſes eaſily from ſeeds, if they are ſown in a moderate hot-bed; the beſt way is to ſow the ſeeds in pots, and plunge them into a tan-bed, and when the plants come up, and have obtained ſome ſtrength, they ſhould be gradually inured to the open air, into which they ſhould be removed in June, and placed in a ſheltered ſituation, where they may remain till autumn. Theſe young plants ſhould be placed under a frame before the early froſt comes on, for a ſmall froſt in autumn will kill the tender part of their ſhoots, which often cauſes their ſtalks to decay moſt part of their length before the ſpring. During the winter ſeaſon they ſhould be ſcreened from froſt, but in mild weather they muſt enjoy the free air. The following ſpring, juſt before the plants ſhoot, they ſhould be carefully turned out of the pots, ſo as not to break their roots; and part of them may be planted in ſmall pots, and the others into a nurſery-bed in a warm ſituation, at about four or five inches aſunder; thoſe in the pots ſhould be plunged into a moderate hot-bed, which will forward their taking freſh root, but afterward muſt be hardened to bear the open air as before; theſe plants in pots ſhould be ſheltered under a frame in winter three or four years till they have obtained ſtrength, then they may be turned out of the pots, and planted in a warm ſituation, where they will live in the open air in common winters, but in ſevere froſt they are in danger of being killed if they are not ſheltered.

Thoſe plants in the beds ſhould alſo be covered with mats or ſtraw in froſty weather, and after they have obtained ſtrength, they may be tranſplanted into a warm ſituation, as the other.

JONQUIL. See NARCISSUS.

IPOMOEA. Lin. Gen. Pl. 199. Quamoclit, or Scarlet Convolvulus.

The CHARACTERS are,

The flower hath a ſmall permanent empalement, cut into five parts at the top. The petal is funnel-ſhaped, having a long cylindrical tube, whoſe brim is five-pointed, ſpreading open flat. It hath five awl-ſhaped ſtamina. In the bottom of the tube is ſituated a round germen, which afterward becomes a roundiſh capſule with three cells, incloſing three oblong ſeeds.

The SPECIES are,

1. IPOMOEA (*Quamoclit*) foliis pinnatifidis linearibus, floribus ſolitariis. Hort. Cliff. 60. *Ipomoea with very narrow many-pointed leaves, and ſolitary flowers.*

2. IPOMOEA (*Coccinea*) foliis cordatis acuminatis, baſi angulatis, pedunculis multifloris. Hort. Upſal. 39. *Ipomoea with heart-ſhaped pointed leaves, angular at the baſe, and many flowers on a ſtalk; commonly called Scarlet Convolvulus.*

3. IPOMOEA (*Solanifolia*) foliis cordatis acutis integerrimis, floribus ſolitariis. Prod. Leyd. 430. *Ipomoea with acute, heart-ſhaped, entire leaves, and ſolitary flowers.*

4. IPOMOEA (*Violacea*) foliis cordatis integerrimis, floribus confertis, corollis indiviſis. Sauv. Monſp. 114. *Ipomoea with heart-ſhaped entire leaves, flowers growing in cluſters, with undivided petals.*

5. IPOMOEA (*Tuberoſa*) foliis palmatis, lobis ſeptenis lanceolatis integerrimis. Hort. Upſal. 39. *Ipomoea with hand-ſhaped leaves, compoſed of ſeven ſpear-ſhaped entire lobes; called Spaniſh Arbour Vine.*

6. IPOMOEA (*Triloba*) foliis trilobis cordatis, pedunculis trifloris. Lin. Sp. Plant. 161. *Ipomoea with heart-ſhaped leaves having three lobes, and three flowers on a foot-ſtalk.*

7. IPOMOEA (*Hepaticæfolia*) foliis palmatis, floribus aggregatis. Flor. Zeyl. 79. *Ipomoea with hand-ſhaped leaves, and flowers growing in cluſters; called Tyger's Foot.*

8. IPOMOEA (*Digitata*) foliis digitatis glabris, foliolis ſeſſilibus, caule lævi. Lin. Sp. Plant. 162. *Ipomoea with ſmooth hand-ſhaped leaves, whoſe lobes ſit cloſe, and a ſmooth ſtalk.*

The firſt ſort grows naturally in both Indies; in the Weſt-Indies it is called Sweet William, and by ſome Indian Pink. It riſes with a twining ſtalk ſeven or eight feet high, ſending out many ſlender twining branches, which twiſt about any neighbouring plants for ſupport; the leaves are winged, being compoſed of ſeveral pair of very fine narrow lobes, not thicker than fine ſowing thread; they are about an inch long, of a deep green, and ſometimes are by pairs oppofite, and at others they are alternate; the flowers come out fingly from the fide of the ftalks, ftanding upon flender foot-ftalks about one inch long; they are funnel-fhaped, having a tube an inch long, which is narrow at bottom, but gradually widens to the top, and fpreads open flat, with five corners or angles; they are of a moft beautiful fcarlet colour, fo make a fine appearance. This is an annual plant in England, but whether it is fo in its native place I cannot tell; for as the feeds fall to the ground, fo there may be a fucceffion of young plants come up among the old ones, which, if not carefully obferved, may occafion the plants to be thought perennial.

This is a tender plant, fo will not thrive in the open air in England. It is propagated by feeds, which fhould be fown on a hot-bed in the fpring, and as the plants will foon appear, they fhould be each tranfplanted into a fmall pot, before they twine about each other, for then it will be difficult to difengage them without breaking their tops. When they are potted they fhould be plunged into a new hot-bed, and fticks placed down by each plant for their ftalks to twine about; after they have taken new root, they fhould have a good fhare of air in warm weather, to prevent their drawing up weak; and when they are advanced too high to remain under the frame, they fhould be removed into the tan-bed in the ftove, where they fhould have fupport, for their branches will extend to a confiderable diftance. In this place they will begin to flower in June, and there will be a fucceffion of flowers till the end of September, and the feeds will ripen very well in autumn.

The fecond fort grows naturally in Carolina, and the Bahama Iflands. This is alfo an annual plant in England, but not fo tender as the former. It hath a twining ftalk, which rifes five or fix feet high, garnifhed with heart-fhaped leaves, ending in acute points, divided into angles at their bafe; the flowers come out from the fide of the branches, upon flender foot-ftalks, which fupport three or four flowers of the fame form and fize as the former, but not fo deep coloured. There is a variety of this with Orange-coloured flowers, but they do not differ in any other refpect. If the feeds of this fort are fown on a warm border of light earth in the fpring, the plants will come up, and in favourable feafons will flower and produce good feeds; but moft people raife the plants on a very gentle hot-bed, and tranfplant them afterwards into the borders, by which method they are brought forward, fo will perfect their feeds earlier.

The third fort is like the fecond, but the leaves have no angles, and the flowers are of a Rofe colour, each foot-ftalk fuftaining one flower. This may be treated in the fame manner as the fecond.

The fourth fort grows naturally in the Weft-Indies, where it twines about any neighbouring fupport, and rifes ten or twelve feet high, garnifhed with large, heart-fhaped, entire leaves; the flowers are blue, and come out from the fide of the branches upon flender foot-ftalks in clufters; their brims are not angular as in the former fpecies, but entire. This fort is propagated by feeds, which fhould be fown on a hot bed in the fpring, and the plants afterward treated in the fame way as is before directed for the firft fort, for it is too tender to thrive in the open air here.

The fifth fort is cultivated in moft of the iflands in the Weft-Indies, but is fuppofed to have been introduced there from the Spanifh main. The plants rife to a very great height, and fend out many branches, fo are planted to cover arbours for fhade in the iflands, from whence it had the appellation of Spanifh Arbour Vine. The ftalks of this plant are covered with a purple bark; they twine about any neighbouring fupport, fending out many fide branches, fo that one plant will cover an arbour of fifty feet long. The leaves are divided into feven lobes almoft to the bottom; the flowers come out from the fide of the ftalks; they are large, funnel-fhaped, of a bright yellow colour, and fmell very fweet; thefe are fucceeded by large roundifh capfules with three cells, containing one large feed in each, which is of a dark colour.

This is a perennial plant, but too tender to thrive in the open air in England; the feeds of this muft be fown upon a hot-bed in the fpring, and when the plants come up they muft be tranfplanted into feparate pots, and plunged into a frefh hot-bed; but as they will foon grow too tall to ftand under a frame, they fhould be removed into the bark-ftove, where they fhould conftantly remain. As thefe plants extend their fhoots to a very great length, they require a tall ftove, where they may have room to grow, without which they will never produce any flowers. I have had thefe plants feveral years, but have only feen one flower produced from them; for they grow fo very large before they begin to have flowers, as that few of the ftoves in England have height enough for their growth.

The fixth fort grows naturally in moft of the iflands in the Weft-Indies; this hath a twining ftalk, which rifes ten or twelve feet high, garnifhed with leaves divided into three lobes, which are heart-fhaped; the foot-ftalks arife from the fide of the ftalks, each fuftaining three purple flowers. This is alfo tender, fo the plants muft be raifed on a hot-bed in the fpring, and afterwards planted in feparate pots, plunging them into another hot-bed, where they may remain till they reach the glaffes; then they fhould be removed into a glafs-cafe where they may have room, and be fcreened from the cold, but fhould have a large fhare of free air admitted to them in warm weather; with this treatment the plants will flower and produce ripe feeds.

The feventh fort grows naturally in India; this rifes with a twining hairy ftalk four or five feet high, garnifhed with hairy hand-fhaped leaves, divided to the bottom into feveral lobes; the flowers come out in clufters, inclofed in a five-cornered involucrum; they are of a purplifh colour, but fmall, and open only in the evening, fo make no figure. This is propagated by feeds, and requires the fame treatment as the fixth fort.

The eighth fort grows naturally in the Weft-Indies; this hath a fmooth twining ftalk, which rifes four or five feet high, garnifhed with hand-fhaped leaves, having three lobes, which fit clofe to the ftalks; the flowers come out from the fide of the ftalks upon fhort foot-ftalks, which fuftain two or three purple flowers, which are fucceeded by round tricapfular feed-veffels; in each cell there is one brown feed.

This fort requires the fame treatment as the two former, with which it will produce flowers, and perfect its feeds in England.

IRIS. Tourn. Inft. R. H. 358. tab. 186, 187, 188. Flower-de-luce.

The CHARACTERS are,

The flowers are inclofed in fpathæ (or fheaths) which are permanent; the flowers are divided into fix parts; the three outer are oblong, obtufe, and reflexed; the three inner are erect, and end in acute points; they have three awl-fhaped ftamina, which lie upon the reflexed petals. Under the flower is fituated an oblong germen, which afterward becomes an oblong angular capfule with three cells, filled with large feeds.

The SPECIES are,

1. IRIS (*Pfeudoacorus*) corollis imberbibus, petalis interioribus ftigmate minoribus, foliis enfiformibus. Hort. Cliff. 19. *Iris with an unbearded flower, the inner petals fmaller than the ftigma, and fword-fhaped leaves; or yellow Marfh Flower-de-luce.*

2. IRIS (*Squallens*) corollis barbatis, caule foliis longiore multifloro. Hort. Cliff. 18. *Iris with bearded flowers, the*

the stalks longer than the leaves, with many flowers; or common German Flower-de-luce.

3. Iris (*Aphylla*) corollis barbatis, scapo nudo longitudine foliorum multifloro. Prod. Leyd. 17. *Iris with a bearded flower, and a naked stalk the length of the leaves, with many flowers.*

4. Iris (*Variegata*) corollis barbatis, caule subfolioso longitudine foliorum multifloro. Prod. Leyd. 17. *Iris with a bearded flower, and a leafy stalk the length of the leaves, with many flowers.*

5. Iris (*Susiana*) corollis barbatis, caule foliis longiore unifloro. Hort. Cliff. 18. *Iris with a bearded flower, and a stalk longer than the leaves, having one flower; commonly called Chalcedonian Iris.*

6. Iris (*Biflora*) corollis herbatis, caule foliis breviore multifloro. Hort. Upsal. 17. *Iris with a bearded flower, and a stalk shorter than the leaves, with many flowers.*

7. Iris (*Pumila*) corollis barbatis, caule foliis breviore unifloro. Lin. Sp. Pl. 38. *Iris with a bearded flower, and a stalk shorter than the leaves, with one flower.*

8. Iris (*Germanica*) corollis barbatis, caule foliis longiore multifloro. *Iris with a bearded flower, a stalk longer than the leaves, with many flowers; called greater Dalmatian Iris.*

9. Iris (*Orientalis*) corollis barbatis, germinibus trigonis, foliis ensiformibus longissimis caule foliis longiore bifloro. Pluk. 154. *Iris with a bearded flower, a three-cornered germen, very long sword-shaped leaves, and a stalk longer than the leaves, with two flowers.*

10. Iris (*Graminea*) corollis imberbibus, germinibus sexangularibus, caule ancipiti, foliis linearibus. Hort. Cliff. 19. *Iris with flowers having no beards, a six-cornered germen, a stalk alike on both sides, and narrow leaves.*

11. Iris (*Maritima*) corollis imberbibus, caule foliis breviore trifloro, foliis lineari-ensiformibus. *Iris whose flowers are not bearded, the stalk shorter than the leaves, with three flowers, and narrow sword-shaped leaves.*

12. Iris (*Angustifolia*) corollis imberbibus, caule foliis æqualibus multifloro. *Iris whose flowers have no beards, the stalks equal in length with the leaves, with many flowers.*

13. Iris (*Bicolor*) corollis imberbibus, caule foliis longiore multifloro, germinibus sexangularibus, foliis linearibus. *Iris whose flowers have no beards, the stalks longer than the leaves, with many flowers, a six-cornered germen, and very narrow leaves.*

14. Iris (*Spuria*) corollis imberbibus, germinibus sexangularibus, caule tereti, foliis sublinearibus. Hort. Cliff. 19. *Iris whose flowers have no beards, with a six-cornered germen, a taper stalk, and very narrow leaves.*

15. Iris (*Sativa*) corollis imberbibus, caule unifloro foliis brevioribus, radice fibrosâ. Flor. Virg. 10. *Iris with an unbearded flower, a stalk shorter than the leaves, with one flower, and a fibrous root.*

16. Iris (*Picta*) corollis imberbibus, germinibus subtrigonis, caule tereti, foliis ensiformibus. Lin. Sp. Pl. 39. *Iris with an unbearded flower, a three-cornered germen, a taper stalk, and sword-shaped leaves.*

17. Iris (*Verna*) corollis imberbibus petalis internis longitudine stigmatis, foliis ensiformibus. Hort. Cliff. 19. *Iris with an unbearded flower, the inner petals as long as the stigma, and sword-shaped leaves; called Stinking Gladwyn.*

18. Iris (*Versicolor*) corollis imberbibus, germinibus trigonis, caule tereti, foliis linearibus. Lin. Hort. Cliff. 19. *Iris with an unbearded flower, a three-cornered germen, a taper stalk, and narrow leaves.*

19. Iris (*Tuberosa*) corollis imberbibus, foliis tetragonis. Vir. Cliff. 6. *Iris with an unbearded flower and four-cornered leaves; called Hermodactyle.*

The first sort grows naturally in ditches and standing waters in most parts of England; this is titled in the Pharmacopeia, Acorus adulterinus, or Pseudo acorus, i. e. Bastard Acorus. The roots of this are pretty thick, feshy, and spread near the surface of the ground; the leaves are sword-shaped, very long, of a deep green colour, and no so stiff as those of the garden Iris; the stalks rise three feet high, toward the top of which grow three or four flowers, one above another, shaped like the ordinary Flower-de-luce, but the three inner petals are shorter than the stigma.

This sort is not cultivated in gardens, but grows wild in standing waters; but being an officinal plant, it is here mentioned to introduce the other.

The second sort grows naturally in Germany; the roots of this are very thick, feshy, and divided into joints, spreading just under the surface of the ground; the leaves arise in clusters embracing each other at their base, but spread asunder upward; they are a foot and a half long, and two inches broad, having sharp edges, ending in points like swords; the stalks between these, which are a little longer than the leaves, divide into three branches, each of which produce two or three flowers one above another at distances, inclosed in sheaths; they have three large Violet-coloured petals, which turn backward, called falls; these have beards near an inch long on their midrib toward their base, and have a short arched petal which cover the beard, with three broad erect petals of the same colour, called standards; the stamina lie upon the reflexed petals. Under each flower is situated an oblong germen, which turns to a large three-cornered capsule with three cells, filled with large compressed seeds.

There is a variety of this with blue standards and purple falls, which is titled, Iris hortensis latifolia, by Caspar Bauhin; and one with pale purple standards, another with white, and a third with a smaller flower; but these are accidental varieties, which have come from seeds.

The third sort has broader leaves than the second; the flower-stalks have no leaves, but are equal in length with the leaves; they have three or four large, bright, purple flowers, which stand above each other, with purplish sheaths or hoods; the three bending petals or falls are striped with white from the base to the end of the beard; the flowers are succeeded by large, blunt, triangular capsules with three cells, filled with compressed seeds.

The fourth sort grows naturally in Hungary; the leaves of this are like those of the second sort, but are of a darker green; the stalks rise as tall as the leaves, and toward the bottom are garnished with one leaf at each joint, whose base embrace the stalks; the upper part is naked, and branches into three, each having two or three flowers above each other; the three upright petals or standards are yellow, and the bending petals or falls are variegated with purple stripes.

The fifth sort grows naturally near Constantinople, and in other parts of the east. The leaves of this sort are not so broad as those of the second, and are of a grayish colour; the stalks rise two feet and a half high, supporting one very large flower; the three upright petals are almost as broad as a hand, but very thin, striped with black and white; the three bending petals or falls are of a darker colour, from whence some gardeners have called it the Mourning Iris.

The sixth sort hath broad leaves like those of the second sort, but shorter; the stalks rise nine or ten inches high, branching into two or three at the top, each sustaining two deep purple flowers.

The seventh sort hath narrower and shorter leaves than the former; the stalks are shorter than the leaves, and support one flower on the top of a light purple colour. There are two or three varieties of this, which differ in the colour of their flowers.

The eighth fort hath the largeft leaves of any of this genus, of a grayifh colour and fpread white, embracing each other at the bafe, where they are purplifh. The ftalks rife four feet high, and divide into feveral branches, each fupporting three or four flowers above each other at diftances, covered with a thin fheath; the three bending petals or falls, of a faint purple inclining to blue, with purple veins running lengthways; the beard is yellow, and the three erect petals or ftandards are of a bright blue with fome faint purple ftripes; the flowers have an agreeable fcent.

The feeds of the ninth fort were brought from Carolina, by the Right Rev. Dr. Pococke, late Bifhop of Offory, who found the plants growing there naturally.

This plant hath a thick flefhy root, divided into many knots or tubers, which fpread and multiply in the ground, putting out many ftrong, thick, flefhy fibres. From thefe roots arife clufters of flat fword-fhaped leaves, of a deep green colour, which are more than three feet long, and little more than one inch broad in the broadeft part, ending in points; between thefe arife the flower-ftalks, which grow four feet high, having very long fpathæ or fheaths at each of the upper joints, which include the flowers. Thefe ftalks generally fuftain two flowers, which are divided into nine leaves; three of thefe ftand erect, which are white, and fix turn down, and are joined together at their bafe, the lower fpreading out in a broad, obtufe, reflexed fall, having a beard, which is of a bright yellow colour; the upper fegment is arched over the lower, fo as to form a fort of lip, which is reflexed backward; under thefe is fituated an oblong three-cornered germen, which afterward becomes an oblong, fwollen, three-cornered feed-veffel, ending in a long point, which opens into three longitudinal cells, in which the feeds are ranged; it is very hardy, and thrives in the open air without any protection. The roots propagate very faft, when they are in a light moift foil, fo that it will foon be common in England without waiting for plants from feeds.

The tenth fort grows naturally in Auftria; this hath narrow, flat, Grafs-like leaves about a foot long, of a light green colour; between thefe arife the ftalks about fix inches high, having two narrow green leaves, which are much longer than the ftalks; thefe ftalks fuftain two or three flowers, which are fmaller than any of the former fpecies; the petals have no beards, but have a broad yellow line adorned with purple ftripes; the three falls are of a light purple colour ftriped with blue, and have a convex ridge running longitudinally, the other are of a reddifh purple variegated with violet; the flowers have a fcent like frefh Plumbs.

The eleventh fort grows naturally near the fea, in the fouth of France and in Italy. This hath narrow fword-fhaped leaves more than a foot long, of a deep green colour; the ftalks do not rife fo tall as the leaves; they fuftain at the top two or three flowers which ftand near together, of a bright purple colour with very deep falls, the three ftandards are blue; the bending petals have no beards, but inftead of that white broad ftripes through the middle.

The twelfth fort hath narrower leaves than the former, but of the fame deep green colour; the ftalks do not rife higher than the leaves, and fupport two or three flowers, which have long permanent empalements, ftanding erect, which cover the feed-veffel till the feeds are ripe; the flowers are fmaller, and of a paler colour than thofe of the eleventh fort.

The thirteenth fort has very narrow, long, Grafs-like leaves, of a light green; the ftalks rife two feet and a half high, fuftaining three or four flowers above each other, which have blue falls, and purple ftandards ftriped with pale blue lines.

The fourteenth fort grows naturally in Germany; the leaves are like thofe of the eleventh fort, which, when broken, have a difagreeable fcent; but this is accidental, and not common to all the plants; the ftalks of this are taper, and rife a little above the leaves, fuftaining three or four flowers one above another, which have light blue ftandards, and purple variegated falls without beards; inftead of which, they have a broad white line in the middle; thefe are fucceeded by fhort thick capfules, which have fcarce any angles, opening in three cells, which are filled with angular feeds.

The fifteenth fort grows naturally in North America; this hath tufted fibrous roots, from which arife many narrow fword-fhaped leaves; from between thefe come out the ftalks, which are fhorter than the leaves, fupporting one purple flower with blue ftandards.

The fixteenth fort grows alfo in North America; this hath broader fword-fhaped leaves than the former, of a light green colour; the ftalks rife a little above the leaves, and fupport two or three flowers one above another; the ftandards are of a light blue, the falls are purple variegated, with a broad white line inftead of a beard through the middle.

The feventeenth fort grows naturally in moift places in many parts of England, fo is feldom admitted into gardens. This hath thick, tufted, fibrous roots; the leaves are of a Grafs-green colour, fword-fhaped, and when broken emit a ftrong odour, not much unlike that of hot roaft beef at the firft fcent, but if fmelt too clofe becomes difagreeable. It is generally called Stinking Gladwyn in England; the ftalks rife about the fame height with the leaves, fupporting two fmall flowers of a purple colour, variegated.

The eighteenth fort grows naturally in Auftria and Bohemia; this hath narrow fword-fhaped leaves near a foot and a half long, of a dark green colour; the flower-ftalks rife above the leaves, and fupport two or three flowers with light blue ftandards, and deep blue falls, with a broad ftripe of white inftead of the beard.

All thefe forts are generally propagated by parting of their roots, which do moft of them multiply faft enough. The beft time to remove and part the roots is in autumn, that they may get good root before the fpring, otherwife they will not flower ftrong the following fummer. All thofe forts which fpread much at their roots, fhould be tranfplanted every other year, to keep them within bounds, otherwife they will fpread fo much as to become troublefome, efpecially if they are planted near other flowers; indeed the large growing kinds are moft of them too fpreading for the flower-garden, fo are only fit to fill up the fpaces between trees and fhrubs in large plantations, where they will have a good effect during the time of their flowering.

The fifth, fixth, feventh, tenth, eleventh, and thirteenth forts, grow in lefs compafs, fo may be admitted into large borders, or in clumps of flowers in the pleafure-garden, where they will add to the variety. The fifth fort fhould have a warmer fituation, being a little tender, but all the other forts will grow in almoft any foil or fituation, and may be propagated by feeds, which fhould be fown foon after they are ripe, then the plants will come up the following fpring; but if the feeds are fown in the fpring, they will lie a year in the ground before they vegetate; but as moft of the forts are fo eafily propagated by their roots, few people care to wait for feedling plants, unlefs of thofe forts which are fcarce.

The nineteenth fort grows naturally in the iflands of the Archipelago; this hath a tuberous knobbed root, from which arife

arife five or fix long, narrow, four-cornered leaves, between which arife the ftalk, which fupports one flower, fhaped like thofe of the Iris, but fmall, and of a dark purple colour. This is propagated by the roots, which fend out offfets, which may be taken up, and tranfplanted when their leaves decay, but fhould not be kept too long out of the ground. If thefe are planted in a deep loofe foil, the roots will run down, and be loft in a few years where they are not difturbed, fo they fhould be annurally tranfplanted, and have a fhallow foil; they are hardy in refpect to cold, and require no farther care but to keep them clean from weeds.

IRIS Bulbofa. } See XIPHIUM.
IRIS Perfica. }

ISATIS. Tourn. Inft. R. H. 211. tab. 100. Woad.

The CHARACTERS are,

The empalement of the flower falls away; the flower hath four oblong petals, placed in form of a crofs. It hath fix ftamina, four of which are as long as the petals, the other two are fhorter, with an oblong compreffed germen, which becomes an oblong compreffed pod with one cell, opening with two valves, inclofing one oval compreffed feed in the center.

The SPECIES are,

1. ISATIS (*Tinctoria*) foliis radicalibus oblongo-ovatis obtufis integerrimis, caulinis fagittatis filiculis oblongis. *Woad with oblong, oval, blunt, entire leaves at bottom, but thofe on the ftalks arrow-pointed, and oblong pods; or cultivated Woad.*

2. ISATIS (*Lufitanica*) foliis radicalibus lanceolatis integerrimis, caulinis fagittatis, filiculis anguftioribus. *Woad with fpear-fhaped entire lower leaves, thofe on the ftalks arrow-fhaped, and narrower pods; wild narrow-leaved Woad.*

3. ISATIS (*Armena*) foliis radicalibus crenatis, caulinis fagittatis, filiculis anguftioribus villofis. *Woad with crenated lower leaves, thofe on the ftalks arrow-pointed, and narrow hairy pods; fmaller Portugal Woad.*

The firft fort is cultivated in feveral parts of England for the purpofes of dyeing, being ufed as a foundation for many of the dark colours.

This is a commodity well worth propagating in all places where the land is fuitable for it, which muft be a pretty ftrong foil, but not too moift.

The plant is biennial, in which it differs from moft of the other forts, which are annual. The lower leaves of this are of an oblong oval figure, and pretty thick confiftence, and end in obtufe roundifh points; they are entire on their edges, and of a lucid green. The ftalks rife four feet high, dividing into feveral branches, garnifhed with arrow-fhaped leaves fitting clofe to the ftalks; the branches are terminated by fmall yellow flowers, in very clofe clufters, which are compofed of four fmall petals, placed in form of a crofs, which are fucceeded by pods fhaped like a bird's tongue, which, when ripe, turn black, and open with two valves, having one cell, in which is fituated a fingle feed.

The fecond fort has been fuppofed to be the fame fpecies as the firft, and only differing by culture; but I have propagated both forts more than forty years, and have not found either of them alter.

The third fort grows naturally in Portugal: it is an annual plant, whofe lower leaves are narrow and crenated, but thofe on the ftalks are arrow-pointed; they are of a pale green, and much thinner than thofe of the other forts. The flowers are fmall, of a fulphur colour, and the feedveffels are narrow and hairy.

The two laft forts are not cultivated for ufe, fo are only preferved in botanic gardens for the fake of variety; they are propagated by feeds as the firft fort.

The firft fort which is propagated for ufe, is fown upon frefh land which is in good heart, for which the cultivators of Woad pay a large rent; they generally chufe to have their land fituated near great towns, where there is plenty of dreffing, but they never ftay long on the fame fpot, for the beft ground will not admit of being fown with Woad more than twice, and if it is oftener repeated, the crop feldom pays the charges of culture, &c.

Thofe who cultivate this commodity have gangs of people who have been bred to this employment, fo that whole families travel about from place to place, wherever their principal fixes on land for the purpofe; but thefe people go on in one track, juft as their predeceffors taught them, nor have their principals deviated much from the practice of their anceftors, fo that there is a large field for improvement, if any of the cultivators of Woad were perfons of genius, and could be prevailed on to introduce the garden culture fo far as it may be adapted to this plant; this I know from experience, having made numbers of trials on the culture of this plant, therefore I fhall infert them here, for the benefit of thofe who may have ingenuity enough to ftrike out of the old beaten track.

As the goodnefs of Woad confifts in the fize and fatnefs or thicknefs of the leaves, the only method to obtain this, is by fowing the feed upon ground at a proper feafon, and allow the plants proper room to grow, as alfo to keep them clean from weeds, which, if permitted to grow, will rob the plants of their nourifhment. The method practifed by fome of the moft fkilful kitchen-gardeners in the culture of Spinach, would be a great improvement to this plant, for fome of them have improved the round-leaved Spinach fo much by culture, as to have the leaves more than fix times the fize they were formerly, and their fatnefs has been in the fame proportion, though fown upon the fame land as formerly, which has been effected by thinning of the plants when young, and keeping the ground conftantly clean from weeds. But to return to the culture of Woad.

After having made choice of a proper fpot of land, which fhould not be too light and fandy, nor over ftiff and moift, but rather a gentle hazel loam, whofe parts will eafily feparate, the next is to plough this up juft before winter, laying it in narrow high ridges, that the froft may penetrate through the ridges to mellow and foften the clods; then in the fpring plough it again crofsway, laying it again in narrow ridges; after it has lain fome time in this manner, and the weeds begin to grow, it fhould be well harrowed to deftroy them; this fhould be twice repeated while the weeds are young, and if there are any roots of large perennial weeds, they muft be harrowed out, and carried off the ground. In June the ground fhould be a third time ploughed, when the furrows fhould be narrow, and the ground ftirred as deep as the plough will go, that the parts may be as well feparated as poffible, and when the weeds appear again, the ground fhould be well harrowed to deftroy them. Toward the end of July, or the beginning of Auguft, it fhould be ploughed the laft time, when the land fhould be laid fmooth, and when there is a profpect of fhowers the ground muft be harrowed to receive the feeds, which fhould be fown either in rows with the drill-plough, or in broad-caft after the common method; but it will be proper to fteep the feeds one night in water before they are fown, which will prepare them for vegetation: if the feeds are fown in drills, they will be covered by an inftrument fixed to the plough for that purpofe, but thofe which are fown broad-caft in the common way muft be well harrowed in. If the feeds are good, and the feafon favourable, the plants will appear in a fortnight, and in a month or five weeks after will be fit to hoe; for the fooner this is performed when the plants are diftinguifhable, the better they will thrive, and the weeds being then young will be foon deftroyed. The method of hoeing thefe plants is the fame as for Turneps,

 with

with this difference only, that these plants need not be thinned so much, for at the first hoeing, if they are separated to the distance of four inches, and at the last to six inches, it will be space enough for the growth of the plants; if this is carefully performed, and in dry weather, most of the weeds will be destroyed; but as some of them may escape in this operation, and young weeds will arise, so the ground should be a second time hoed in the beginning of October, always chusing a dry time for this work; at this second operation, the plants should be singled out to the distance they are to remain. After this, if carefully performed, the ground will be clean from weeds till the spring, when young weeds will come up, therefore about the middle of March will be a good time to hoe the ground again, for while the weeds are young, it may be performed in less than half the time it would require if the weeds were permitted to grow large, and the sun and wind will much sooner kill them; this hoeing will also stir the surface of the ground, and greatly promote the growth of the plants; if this is performed in dry weather, the ground will be clean till the first crop of Woad is gathered, after which it must be again well cleaned; if this is carefully repeated after the gathering each crop, the land will always lie clean, and the plants will thrive the better. The expence of the first hoeing will be about six shillings per acre, and for the after hoeings half that price will be sufficient, provided they are performed when the weeds are young, for if they are suffered to grow large, it will require more labour, nor can it be so will performed; therefore it is not only the best husbandry to do this work soon, but it will be found the cheapest method; for the same number of men will hoe a field of ten acres three times when it is done in time, as is required to hoe it twice only, because the weeds will have longer time to grow between the operations.

If the land, in which this seed is sown, should have been in culture before for other crops, so not in good heart, it will require dressing before it is sown, in which case rotten stable-dung is preferable to any other; but this should not be laid on till the last ploughing, just before the seeds are sown, and not spread till the land is ploughed, that the sun may not exhale the goodness of it, which in summer is soon lost when spread on the ground. The quantity should not be less than twenty loads to each acre, which will keep the ground in heart till the crop of Woad is spent.

The time for gathering of the crop is according to the season, but it should be performed as soon as the leaves are fully grown, while they are perfectly green, for when they begin to change pale, great part of their goodness is over, for the quantity will be less, and the quality greatly diminished.

If the land is good, and the crop well husbanded, it will produce three or four gatherings, but the two first are the best; these are commonly mixed together in the manufacturing of it, but the after crops are always kept separate, for if these are mixed with the other, the whole will be of little value. The two first crops will sell from twenty-five to thirty pounds a ton, but the latter will not bring more than seven or eight pounds, and sometimes not so much. An acre of land will produce a ton of Woad, and in good seasons near a ton and a half.

When the planters intend to save the seeds, they cut three crops of the leaves, and then let the plants stand till the next year for seed; but if only one crop is cut, and that only of the outer leaves, letting all the middle leaves stand to nourish the stalks, the plants will grow stronger, and produce a much greater quantity of seeds.

These seeds are often kept two years, but it is always best to sow new seeds when they can be obtained. The seeds ripen in August; when the pods turn to a dark colour the seeds should be gathered; it is best done by reaping the stalks in the same manner as Wheat, spreading the stalks in rows upon the ground, and in four or five days the seeds will be fit to thresh out, provided the weather is dry; for if it lies long, the pods will open and let out the seeds.

There are some of the Woad planters, who feed down the leaves in winter with sheep, which is a very bad method, for all plants which are to remain for a future crop, should never be eaten by cattle, for that greatly weakens the plants, therefore those who eat down their Wheat in winter with sheep are equally blameable.

ISOPYRUM. Lin. Gen. Plant. 621.

The CHARACTERS are,

The flower has no empalement. It hath five oval petals, and five short tubulous nectariums situated within the petals. It hath a great number of short hairy stamina, and several oval germen, which afterward become so many recurved capsules with one cell, filled with small seeds.

The SPECIES are,

1. ISOPYRUM (*Fumarioides*) stipulis subulatis, petalis acutis. Hort. Upsal. 157. *Isopyrum with awl-shaped stipulæ, and acute petals.*

2. ISOPYRUM (*Thalictroides*) stipulis ovatis, petalis obtusis. Lin. Sp. Plant. 557. *Isopyrum with oval stipulæ, and obtuse petals.*

3. ISOPYRUM (*Aquilegioides*) stipulis obsoletis. Lin. Sp. Plant. 557. *Isopyrum with obsolete stipulæ.*

The first sort grows naturally in Siberia; this is an annual plant, which seldom rises more than three or four inches high. The leaves are shaped like those of Fumitory. The stalk is naked to the top, where there is a circle of leaves just under the flowers. The flowers are small, of an herbaceous colour on their outside, but yellow within, having five acute petals, and as many honey glands, with a great number of stamina which are shorter than the petals, and several reflexed moon-shaped germen. The flowers are succeeded by many recurved seed-vessels with one cell, filled with small shining black seeds.

The seeds of this plant should be sown in a shady border soon after they are ripe, for when they are kept long out of the ground, they seldom grow the first year; therefore, when the seeds are permitted to scatter, they succeed very well, and require no other care but to keep them clean from weeds: as there is no great beauty in this plant, so a small patch or two of them in a shady part of the garden, by way of variety, will be sufficient.

The second and third sorts grow naturally about Verona. The second sort hath leaves very like those of the smallest meadow Rue. The stalks rise four or five inches high, supporting a few small white flowers, with obtuse petals; these are succeeded by several recurved seed-vessels, containing many small seeds.

The third sort hath leaves like the second, but a little larger, and of a greener colour. The stalks rise about six inches high, supporting two or three small white flowers, shaped like those of the second sort, which are succeeded by recurved seed-vessels, filled with small seeds.

Both these plants delight in a moist shady situation; they are propagated by seeds in the same way as the first sort, but these will live two or three years.

ISORA. See HELICTERES.

ITEA. Lin. Gen. Plant. 243.

The CHARACTERS are,

The empalement of the flower is small and permanent, ending in five acute points. The flower has five petals, which are inserted in the empalement. It hath five awl-shaped stamina inserted in the empalement, and an oval germen, which afterward becomes a long oval capsule, with the style at the top, having one cell, filled with small seeds.

We have but one SPECIES of this genus, viz.

ITEA (*Virginica.*) Flor. Virg. 143.

This shrub grows in several parts of North America, by the sides of rivers, and in other moist land, where it rises to the height of eight or ten feet, sending out many branches from the ground upward, garnished with spear-shaped leaves, placed alternately, and slightly sawed on their edges, of a light green. At the extremity of the same year's shoots, are produced fine spikes of white flowers, three or four inches long, standing erect; when these shrubs are in vigour, they will be entirely covered with these spikes of flowers, so that they make a fine appearance at their season of flowering, which is in July.

This shrub thrives well in the open air in England, the cold never injuring it; but upon dry gravelly ground it is very apt to die in the summer, in a dry season. It is propagated by layers, but these should be made of young shoots of the same year, for the old branches do not put out roots very kindly. The shoots should be laid down in the autumn, and will be rooted in one year.

IVA. Lin. Gen. 1059.

The CHARACTERS are,

It hath male and female flowers in the same plant; the flowers have a permanent empalement, including several convex florets; the male florets have one funnel-shaped petal, indented in five parts at the brim; they have five bristly stamina, terminated by erect summits; the female florets have neither petal or stamina, but have an oblong germen, supporting two styles, crowned by acute stigmas; the empalement becomes a capsule, including one naked seed.

The SPECIES are,

1. IVA (*Annua*) foliis lanceolato-ovatis, caule herbaceo. Hort. Upsal. 285. *Iva with oval spear-shaped leaves, and an herbaceous stalk.*

2. IVA (*Frutescens*) foliis lanceolatis, caule fruticoso. Amœn. Acad. 3. p. 25. *Iva with spear-shaped leaves and a shrubby stalk; commonly called Jesuits Bark-tree.*

The first sort is an annual plant, which grows naturally in many parts of the West-Indies, where it rises from two to three feet high, branching from the main stem, garnished with oval spear-shaped leaves, sawed on their edges; the stalks are terminated by small clusters of pale blue flowers, which appear in July or August.

The second sort has been long in England, under the title of Jesuits Bark-tree. This hath weak shrubby stalks, which rise seven or eight feet high, garnished with spear-shaped sawed leaves; and in warm seasons, the branches are terminated by small clusters of pale purple flowers.

The first sort is propagated by seeds, which should be sown upon a hot-bed in the spring, to bring the plants forward, treating them afterward as the Marvel of Peru, with which management they will flower and produce seeds.

The second sort was formerly preserved in green-houses in winter, supposing it too tender to live abroad in England; but lately it has been planted abroad in warm situations, where it endures the cold of our ordinary winters very well, and is seldom killed by frost.

This is propagated by the nursery-gardeners for sale. It may be propagated by planting of cuttings during the summer season, or by laying of the tender branches in the spring, which will put out roots in five or six months.

JUDAICA ARBOR. See CERCIS.

JUGLANS. Lin. Gen. Plant. 950. Walnut.

The CHARACTERS are,

It hath male and female flowers at distances on the same tree. The male flowers are disposed in an oblong cylindrical katkin; each scale has one flower, with one petal, divided into six equal parts; in the center is situated many short stamina. The female flowers grow in small clusters, sitting close to the branches; these have short, erect, four-pointed empalements, sitting on the germen, and an acute erect petal, divided into four parts. Under the empalement sits a large oval germen, which afterward becomes a large oval dry berry, with one cell, inclosing a large oval nut with netted furrows.

The SPECIES are,

1. JUGLANS (*Regia*) foliolis ovalibus glabris subserratis subæqualibus. Hort. Cliff. 449. *Walnut with oval lobes, which are smooth, sawed, and equal.*

2. JUGLANS (*Nigra*) foliolis lanceolatis acute serratis, intermediis majoribus. *Black Virginia Walnut.*

3. JUGLANS (*Oblonga*) foliolis cordato-lanceolatis inferne nervosis, pediculis foliorum pubescentibus. *Black Virginia Walnut, with an oblong fruit very deeply furrowed.*

4. JUGLANS (*Alba*) foliolis lanceolatis serratis, exterioribus latioribus. Lin. Sp. Plant. 927. *White Virginia Walnut; or Hickery-nut.*

5. JUGLANS (*Glabra*) foliolis cuneiformibus serratis, exterioribus majoribus. *White Walnut with a smaller fruit, and a smooth bark.*

6. JUGLANS (*Ovata*) foliolis lanceolatis serratis glabris subæqualibus. *White Walnut with an oval compressed fruit, a sweet kernel, and a scaly bark; commonly called Shagbark in America.*

There are several varieties of the common Walnut, which are distinguished by the following titles; the large Walnut, the thin-shelled Walnut, the French Walnut, the late ripe Walnut, and the double Walnut; but these do all of them vary when raised by seeds, so that the nuts from the same tree will produce plants, whose fruit will differ, therefore there can be no dependence upon the trees which are raised from nuts, till they have produced fruit; so that those persons who plant the trees for their fruit, should make choice of them in the nurseries when they have their fruit upon them, otherwise they may be deceived by having such as they would not chuse.

The second sort is commonly called Black Virginia Walnut; this grows to a large size in North America. The leaves of this sort are composed of five or six pair of spear-shaped lobes, which end in acute points, and are sawed on their edges: the lower pair of lobes are the least, the other gradually increase in their size to the top, where the pair at the top, and the single lobe which terminate the leaf, are smaller; these leaves, when bruised, emit a strong aromatic flavour, as do also the outer cover of the nuts, which are rough, and rounder than those of the common Walnut. The shell of the nut is very hard, thick, and the kernel small, but very sweet.

The third sort grows naturally in North America, where the trees grow to a large size. The leaves of this sort are composed of seven or eight pair of long heart-shaped lobes, broad at their base, where they are divided into two round ears, but terminate in acute points; they are rougher, and of a deeper green than those of the second sort, and have nothing of the aromatic scent which they have. The fruit is very long; the shell is deeply furrowed, and hard; the kernel is small, but well flavoured.

The fourth sort is very common in most parts of North-America, where it is called Hickery-nut. The leaves of this sort are composed of two or three pair of oblong lobes, terminated by an odd one; these are of a light green, sawed on their edges: the lower pair of lobes are the smallest, and the upper the largest. The fruit is shaped like the common Walnut, but the shell is not furrowed, and is of a light colour.

The fifth sort is not so large as the fourth. The leaves are composed of two pair of lobes, terminated by an odd one; these are narrow at their base, but broad and rounded at their ends: they are sawed on their edges, and of a light green,

green. The nuts are ſmall, and have a very hard ſmooth ſhell.

The ſixth ſort grows naturally in North America, where it riſes to a middling ſtature. The leaves of this ſort are compoſed of three pair of ſmooth ſpear-ſhaped lobes, of a dark green colour, ſawed on their edges, ending in acute points. The fruit is oval, the ſhell white, hard, and ſmooth; the kernel ſmall, but very ſweet. The young ſhoots of the tree are covered with a very ſmooth browniſh bark, but the ſtems and older branches have a rough ſcaly bark, from whence it had the appellation of Shagbark in America.

The common Walnut is propagated in many parts of England for the fruit, and formerly the trees were propagated for their wood, which was in very great eſteem, till the quantity of Mahogany, and other uſeful woods which have been of late years imported into England, have almoſt baniſhed the uſe of Walnut.

The trees are propagated by planting of their nuts, which, as was before obſerved, ſeldom produce the ſame ſort of fruit as was ſown; ſo that the only way to have the deſired ſort, is to ſow the nuts of the beſt kinds; and if this is done in a nurſery, the trees ſhould be tranſplanted out when they have had three or four years growth, to the place where they are deſigned to remain, for theſe trees do not bear tranſplanting when they are of a large ſize; therefore there may be a good number of the trees planted, which need not be put at more than ſix feet apart, which will be diſtance enough for them to grow till they produce fruit, when thoſe whoſe fruit are of the deſired kind may remain, and the others cut up, to allow them room to grow; by this method a ſufficient number of the trees may be generally found among them to remain, which will thrive and flouriſh greatly when they have room; but as many people do not care to wait ſo long for the fruit, ſo the next beſt method is to make choice of ſome young trees in the nurſeries, when they have their fruit upon them; but though theſe trees will grow and bear fruit, yet they will never be ſo large or ſo long lived, as thoſe which are planted young.

All the ſorts of Walnuts which are propagated for timber, ſhould be ſown in the places where they are to remain; for the roots of the trees always incline downward, which being ſtopped or broken, prevent their aſpiring upward, ſo that they afterwards divaricate into branches, and become low ſpreading trees; but ſuch as are propagated for fruit, are greatly mended by tranſplanting; for hereby they are rendered more fruitful, and their fruit are generally larger and fairer; it being a common obſervation, that downright roots greatly encourage the luxuriant growth of timber in all ſorts of trees, but ſuch trees as have their roots ſpreading near the ſurface of the ground are always the moſt fruitful.

The nuts ſhould be preſerved in their outer covers in dry ſand until February, when they ſhould be planted in lines, at the diſtance you intend them to remain; but in the rows they may be placed pretty cloſe, for fear the nuts ſhould miſcarry; and the young trees, where they are too thick, may be removed, after they have grown two or three years, leaving the remainder at the diſtance they are to ſtand.

In tranſplanting theſe trees, you ſhould always obſerve never to prune either their roots or large branches, both which are very injurious to them; nor ſhould you be too buſy in lopping or pruning the branches of theſe trees while growing, for it often cauſes them to decay; but when there is a neceſſity of cutting any of their branches off, it ſhould be done early in September, that the wound may heal over before the cold increaſes; and the branches ſhould always be cut off quite cloſe to the trunk, otherwiſe the ſtump which is left will decay, and rot the body of the tree.

The beſt ſeaſon for tranſplanting theſe trees is as ſoon as the leaves begin to decay, at which time if they are carefully taken up, and their branches preſerved entire, there will be little danger of their ſucceeding, although they are eight or ten years old; though, as was before obſerved, theſe trees will not grow ſo large, or continue ſo long, as thoſe which are removed young.

This tree delights in a firm, rich, loamy ſoil, or ſuch as is inclinable to chalk or marl, and will thrive very well in ſtony ground, and on chalky hills, as may be ſeen by thoſe large plantations near Leatherhead, Godſtone, and Carſhalton in Surry, where are great numbers of theſe trees planted upon the downs, which annually produce large quantities of fruit, to the great advantage of their owners; one of which I have been told, farms the fruit of his trees, to thoſe who ſupply the markets, for 30 l. per annum.

The diſtance theſe trees ſhould be placed, ought not to be leſs than forty feet, eſpecially if regard be had to their fruit; though when they are only deſigned for timber, if they ſtand near, it promotes their upright growth. The black Virginia Walnut is much more inclinable to grow upright than the common ſort, and the wood being generally of a more beautiful grain, renders it preferable to that, and better worth cultivating. I have ſeen ſome of this wood which hath been beautifully veined with black and white, which, when poliſhed, has appeared at a diſtance like veined marble. This wood was ſome years paſt greatly eſteemed by the cabinet-makers for inlaying for tables, and cabinets, and is a durable wood for thoſe purpoſes, being leſs infected with inſects of any kind, than moſt others of Engliſh growth (which may proceed from its extraordinary bitterneſs;) but it is not proper for buildings of ſtrength, it being of a brittle nature, and exceeding ſubject to break very ſhort, though it commonly gives notice thereof, by its cracking ſome time before it breaks.

The general opinion, that the beating down this fruit improves the trees, I do not believe, ſince in the doing of this the younger branches are generally broken and deſtroyed; but as it would be exceeding troubleſome to gather it by hand, ſo in beating it off great care ſhould be taken that it be not done with violence, for the reaſon before aſſigned. In order to preſerve the fruit, it ſhould remain upon the trees till it is thoroughly ripe, and drops from the trees, then laid in heaps for two or three days, when their huſks will eaſily part from the ſhells; then dried well in the ſun, and laid up in a dry place, where mice or other vermin cannot come to them, where they will remain good four or five months; but there are ſome perſons who put their Walnuts into an oven gently heated, where they let them remain four or five hours to dry, and then put them up in oil jars, or any other cloſe veſſel, mixing them with dry ſand, by which method they will keep good ſix months. The putting of them in the oven is to dry the germ, and prevent their ſprouting; but if the oven be too hot it will cauſe them to ſhrink, therefore great care muſt be had to that.

All the other ſorts are propagated in the ſame way as the common Walnut, but as few of the ſorts produce fruit in England, ſo their nuts muſt be procured from North America, which ſhould be gathered when fully ripe, and put up in dry ſand, to preſerve them in their paſſage to England; when they arrive here, the ſooner they are planted, the greater chance there will be of their ſucceeding; when the plants come up, they ſhould be kept clean from weeds; and if they ſhoot late in the autumn, and their tops are full of ſap, they ſhould be covered with mats, to prevent the early froft from pinching their tender fhoots, which often caufes them to die down a confiderable length before fpring; but if they are fcreened from thefe early frofts, the fhoots will become firmer and better able to refift the cold. Some of thefe forts are tender while young, fo require a little care the two firft winters, but afterward will be hardy enough to refift the greateft cold of this country.

The black Virginia Walnut is full as hardy as the common fort; there are fome large trees of this kind in the Chelfea garden, which have produced great quantities of fruit upward of forty years; the nuts have generally ripened fo well there as to grow, but their kernels are fmall, fo are of little value.

Thefe trees all require the fame culture as the common Walnut, but they grow beft in a foft loamy foil not too dry, and where there is a depth of foil for their roots to run down. The Hickery when young is very tough and pliable, fo the fticks of it are much efteemed; but the wood when grown large is very brittle, fo not of any great ufe. The black Virginia Walnut is the moft valuable wood of all the forts; fome of the trees are beautifully veined, and will take a good polifh, but others have very little beauty, which is the cafe of many other forts of wood.

JUJUBE. See ZIZIPHUS.

JULIANS, or ROCKETS. See HESPERIS.

JULY FLOWER. See DIANTHUS.

JUNCUS. Tourn. Inft. R.H. 246. tab. 127. Rufh.

The CHARACTERS are,

It hath a chaff opening with two valves, and an empalement with fix oblong pointed leaves; the flower hath no petals, but the coloured empalement is by fome taken for petals. It hath fix fhort hairy ftamina, and a three-cornered germen, which afterward becomes a three-cornered capfule with one cell, opening with three valves, inclofing roundifh feeds.

The SPECIES are,

1. JUNCUS (*Acutus*) culmo fubnudo tereti mucronato, paniculâ terminali, involucro diphyllo fpinofo. Lin. Sp. Pl. 325. *Prickly large Sea Rufh.*

2. JUNCUS (*Conglomeratus*) culmo nudo, apice membranaceo incurvo, paniculâ laterali. Lin. Sp. Pl. 326. *Common hard Rufh.*

3. JUNCUS (*Diffufus*) culmo nudo ftricto, paniculâ laterali. Flor. Leyd. 44. *Larger common foft Rufh, with a fpreading panicle.*

4. JUNCUS (*Filiformis*) culmo nudo ftricto, capitulo laterali. Prod. Leyd. 44. *Soft Rufh, with a more compact panicle.*

There are many other fpecies of this genus which grow naturally in England, and are very troublefome weeds in many paftures, fo are not worthy of being enumerated here; for thofe here mentioned is only to point out a method of deftroying them.

The firft and fecond forts grow on the fea-fhores, where they are frequently watered by the falt water. Thefe two forts are planted with great care on the banks of the fea in Holland, in order to prevent the water from wafhing away the earth; which being very loofe, would be in danger of removing every tide, if it were not for the roots of thefe Rufhes, which faften themfelves very deep in the ground, and mat themfelves near the furface, fo as to hold the earth clofely together. Therefore whenever the roots of thefe Rufhes are deftroyed, the inhabitants immediately repair them to prevent farther damage. In the fummer time, when the Rufhes are fully grown, the inhabitants cut them, and tie them up into bundles, which are dried, and afterward carried into the larger towns and cities, where they are wrought into bafkets, and feveral other ufeful things which are frequently fent into England. Thefe forts do not grow fo ftrong in England as they do on the Maefe, and fome other places in Holland, where I have feen them upward of four feet high.

The third and fourth forts grow on moift, ftrong, uncultivated lands in moft parts of England, and confume the herbage where they are fuffered to remain. The beft method of deftroying thefe Rufhes is, to fork them up clean by the roots in July, and after having let them lie a fortnight or three weeks to dry, lay them in heaps, and burn them gently, and the afhes which thefe afford will be tolerable manure for the land; but in order to prevent their growing again, and to make the pafture good, the land fhould be drained, otherwife there will be no deftroying thefe Rufhes entirely; but after it is well drained, if the roots are annually drawn up, and the ground kept duly rolled, they may be fubdued.

JUNIPERUS. Tourn. Inft. R. H. 588. tab. 361. Juniper.

The CHARACTERS are,

It hath male and female flowers in different plants, and fometimes at feparate diftances on the fame plant. The male flowers have a conical katkin; the flowers are placed by threes oppofite, and terminated by a fingle one; the fcales are broad, lying on each other, and fixed to the column by a very fhort foot-ftalk. The flower has no petal, but three ftamina joined in one body below. The female flowers have a fmall three-pointed empalement fitting upon the germen; they have three ftiff, acute, permanent petals; the germen fitting below the empalement, afterward becomes a roundifh berry, inclofing three ftony feeds which are oblong, angular on one fide, but convex on the other.

The SPECIES are,

1. JUNIPERUS (*Communis*) foliis ternis patentibus mucronatis. Lin. Sp. Pl. 1040. *The common Englifh Juniper.*

2. JUNIPERUS (*Suecica*) foliis ternis patentibus, longioribus acutioribufque, ramis erectioribus. *The Tree, or Swedifh Juniper.*

3. JUNIPERUS (*Virginiana*) foliis ternis omnibus patentibus. *Cedar of Virginia, or red Cedar.*

4. JUNIPERUS (*Caroliniana*) foliis ternis bafi adnatis, junioribus imbricatis, fenioribus patulis. Hort. Cliff. 464. *Commonly called Carolina Cedar.*

5. JUNIPERUS (*Bermudiana*) foliis inferioribus ternis patentibus, fuperioribus quadrifariàm imbricatis. *Commonly called Cedar of Bermudas.*

6. JUNIPERUS (*Thurifera*) foliis quadrifariàm imbricatis acutis. *Great Juniper with blue berries.*

7. JUNIPERUS (*Phœnicea*) foliis inferioribus ternis brevioribus patentibus, fuperioribus imbricatis acutis. *Greater Juniper, or Cedar with a Cyprefs leaf and yellowifh fruit.*

8. JUNIPERUS (*Lycia*) foliis undique imbricatis ovatis obtufis. Flor. Leyd. 90. *Middle Juniper, or Cedar with a Cyprefs leaf, and larger berries.*

9. JUNIPERUS (*Barbadenfis*) foliis omnibus quadrifariàm imbricatis. *Greateft Juniper with a Cyprefs leaf; commonly called Jamaica berry-bearing Cedar.*

10. JUNIPERUS (*Sabina*) foliis oppofitis erectis decurrentibus, ramis patulis. *Common Savin.*

11. JUNIPERUS (*Lufitanica*) foliis oppofitis patulis, decurrentibus, ramis erectioribus. *Upright berry-bearing Savin.*

12. JUNIPERUS (*Oxycedrus*) foliis undique imbricatis obtufis, ramis teretibus. *Greater Juniper with a brownifh berry.*

13. JUNIPERUS (*Hifpanica*) foliis quadrifariàm imbricatis acutis. Prod. Leyd. 90. *Taller Spanifh Cedar, with a very large black fruit.*

The firft fort grows naturally upon chalky lands in many parts of England. This is a low fhrub, feldom rifing more than three feet high, fending out many fpreading branches, covered with a brown bark, garnifhed with narrow awl-fhaped leaves, ending in acute points, placed by threes round

round the branches, which are of a grayiſh colour, and continue through the year; the male flowers ſometimes are ſituated at diſtances on the ſame plant with the female, at other times they are upon diſtinct plants: the female flowers are ſucceeded by roundiſh berries which are firſt green, but when ripe are of a dark purple colour. The berries ripen in the autumn.

The wood, the berries, and the gum are uſed in medicine; the gum is titled ſandaracha.

The ſecond ſort is known in the gardens by the title of Swediſh Juniper; this is by many ſuppoſed to be only a variety of the firſt, but is undoubtedly a diſtinct ſpecies, for I have many years raiſed both ſorts from the ſeeds, and have never found them alter. This riſes to the height of ten or twelve feet, the branches grow more erect, the leaves are narrower, and end in more acute points; they are placed farther aſunder on the branches, and the berries are larger. It grows naturally in Sweden, Denmark, and Norway.

The third ſort grows naturally in moſt parts of North America, where it is called red Cedar, to diſtinguiſh it from a ſort of Cypreſs, which is there called white Cedar. Of this there are two, if not three varieties, beſides the ſpecies here enumerated; one of which has leaves in every part like thoſe of the Savin, and upon being rubbed emit a very ſtrong ungrateful odour, and is commonly diſtinguiſhed in America by the title of Savin-tree. There is another with leaves very like thoſe of Cypreſs, but as theſe generally ariſe from the ſame ſeeds when they are ſent from America, ſo they are only ſeminal variations.

The lower leaves of the fourth ſort are like thoſe of the Swediſh Juniper, but the upper leaves are like thoſe of the Cypreſs; and this difference is conſtant, when the ſeeds are carefully gathered from the ſame tree; but as moſt of thoſe people who ſend over theſe ſeeds are not very careful to diſtinguiſh the difference, ſo it often happens that the ſeeds of two or three ſorts are mixed together, which has given occaſion to people to imagine them but one; but all the leaves of the third are like thoſe of the Juniper, ſo the gardeners call that the red Virginia Cedar, and this Carolina Cedar, though they grow naturally in Virginia.

The fifth ſort is the Bermudas Cedar, whoſe wood has a very ſtrong odour; it was formerly in great eſteem for wainſcotting of rooms, and alſo for furniture, but the odour being too powerful for many perſons has rendered it leſs valuable, ſo at preſent there is not much of it imported into England. Theſe plants, while young, have acute-pointed leaves, which ſpread open, and are placed by threes round the branches; but as the trees advance, ſo their leaves alter, and the branches are four-cornered; the leaves are very ſhort, and placed by fours round the branches, lying over each other like the ſcales of fiſh; the berries are produced toward the end of the branches, theſe are of a dark red colour, inclining to purple. As there are few of theſe trees of any great ſize in England, ſo I have not had an opportunity of examining their flowers, therefore do not know if they are on different plants; for although I have received very fine ſpecimens from Bermudas, yet they were all with fruit on them almoſt fully grown, and not one with flowers: as theſe trees are commonly deſtroyed in England whenever there happens a ſevere winter, we have little hopes of ſeeing them in flower here.

The ſixth ſort grows naturally in Iſtria. This hath ſpreading branches growing thinly, garniſhed with awl-ſhaped acute-pointed leaves placed by threes, of a dark green, and not very cloſe to each other; they grow horizontally, pointing outward; the berries are much larger than thoſe of the common Juniper, and are blue when ripe.

The ſeventh ſort grows naturally in Portugal. This ſort grows with its branches in a pyramidal form, the lower ones are garniſhed with ſhort, acute-pointed, grayiſh leaves, placed by threes, pointing outward; but thoſe on the upper branches are of a dark green, lying over each other like the ſcales of fiſh, ending in acute points. The male flowers are produced at the extremity of the branches; they are ſituated in a looſe, ſcaly, conical katkin, ſtanding upon a ſhort foot-ſtalk erect; the fruit is produced ſometimes upon the ſame tree, at diſtances from the flowers, and at other times they are upon ſeparate trees; the berries of this are of a pale yellow when ripe, and about the ſize of thoſe of the common Juniper.

The eighth ſort grows naturally in Spain and Italy. The branches of this ſort grow erect, and are covered with a brown bark; the leaves are ſmall, obtuſe, and lie over each other like the ſcales of fiſh; the male flowers grow at the extremity of the branches in a conical katkin, and the fruit grows ſingle from the ſide of the branches below the katkins on the ſame branch; the berries are large, oval, and when ripe are brown.

The ninth ſort grows naturally in Jamaica, and alſo in the other iſlands in the Weſt-Indies, where it riſes to be one of the largeſt timber trees in thoſe countries; the wood is frequently fetched from thence by the inhabitants of North America, for building of ſhips. This is generally confounded with the Bermudas Cedar, and taken for the ſame, but the ſpecimens of it which were ſent by the late Dr. Houſtoun, prove them to be different trees; for the branches of this ſpread wide, the leaves are extremely ſmall, and are every where lying imbricatim over each other; the bark is rugged, and ſplits off in ſtrings, and is of a very dark colour; the berries are ſmaller than thoſe of the Bermudas Cedar, and are of a light brown colour when ripe.

The tenth ſort is the common Savin; this grows naturally in Italy, Spain, and the Levant, upon the mountains where it is cold. It ſends out its branches horizontally, ſo ſeldom riſes more than three or four feet high; the branches are garniſhed with very ſhort acute-pointed leaves placed oppoſite, and their ends point upward. This ſort very rarely produces either flower or ſeed when it is tranſplanted into gardens; the berries are ſmaller than thoſe of the common Juniper, but of the ſame colour, and a little compreſſed; the whole plant has a very rank ſtrong odour when touched. The leaves of this are much uſed by the farriers for horſes when they have worms, and Mr. Ray commends the juice of it mixed with milk, and ſweetened with ſugar, as an excellent medicine for children who are troubled with worms.

The eleventh ſort has by many been ſuppoſed to be only an accidental variety of the former, but there is a manifeſt difference between them; for the branches of this grow more erect, the leaves are ſhorter, and end in acute points which ſpread outward. This will riſe to the height of eight or ten feet, and produces great quantities of berries. I have propagated this ſort from ſeeds, but have never found it vary. It has been diſtinguiſhed by moſt of the old botaniſts by the title of berry-bearing Savin. It grows naturally on the Alps.

The twelfth ſort grows naturally in Spain, Portugal, and the ſouth of France, where it riſes ten or twelve feet high, ſending out ſmall taper branches without angles the whole length of the ſtem, garniſhed with ſmall obtuſe leaves, lying over each other like the ſcales of fiſh; the male flowers are ſituated at the end of the branches in conical ſcaly katkins, and the berries grow below from the ſide of the ſame branches. Theſe are larger than thoſe of the common Juniper, and when ripe are brown.

The thirteenth sort grows naturally in Spain and Portugal, where it rises from twenty-five to thirty feet high, sending out many branches which form a pyramid, garnished with acute-pointed leaves which lie over each other four ways, so as to make the branches four-cornered; the berries of this sort are very large, and black when ripe.

These plants are all propagated by seeds, which should be sown as soon as they are ripe, if they can be procured; for when they are kept out of the ground till spring, they will not come up until the second year. The hardy sorts may be sown on a border exposed to the east, sifting some earth over them about half an inch thick; toward the middle or latter end of April, some of the plants will appear above ground, though perhaps part of them may lie till the spring following before they come up, therefore the ground should not be disturbed till after that time; for as these plants which come up the first season, will not make great progress while they are young, so they will not require moving till after two summers growth. The second autumn after sowing, some beds must be prepared to transplant them into, which should be of fresh undunged soil, well dug and cleansed from all noxious weeds and roots; then in the beginning of October, which is the proper season for removing these plants, they should be raised up with a trowel, preserving as much earth as possible to their roots, and planted into the beds about five or six inches asunder each way, giving them some water to settle the earth to their roots; and if it should prove very dry weather, this should be repeated two or three times. As some of the seeds may yet remain in the ground, so the beds should not be disturbed too much in taking up the plants; for I have known a bed sown with these berries, which has supplied plants for three years drawing, some of the berries having lain two years in the ground before they sprouted; therefore the surface of the beds should be kept level, and constantly clean from weeds.

The plants may remain two years in the beds after planting, observing to keep them clean from weeds; in the spring the ground should be gently stirred between them, that their roots may with greater ease strike into it; after they have stood so long they should be transplanted, either into a nursery, at the distance of three feet row from row, and eighteen inches asunder in the rows, or into the places where they are to remain for good. The best season to transplant them (as I have before observed) is the beginning of October.

In order to have these trees aspire in height, their under branches should be taken off, especially where they are inclined to grow out strong; but they must not be kept too closely pruned, which would retard their growth; for all these evergreen trees do more or less abound with a resinous juice, which in hot weather is very apt to flow out from such places as are wounded; so that it will not be advifeable to take off too many branches at once, which would make so many wounds, from which their sap in hot weather would flow in such plenty, as to render the trees weak and unhealthy.

The other sorts are also propagated by seeds, which must be procured from the countries where they grow naturally, and sown as was directed for the other Junipers. When the plants come up they must be carefully weeded, and in dry weather should be refreshed with water, which will greatly forward their growth; the autumn following they should have a little rotten tan laid between them to keep out the frost. In this bed the plants may remain till they have had two years growth, then they should be transplanted into other beds, and managed as was directed before for the other sorts.

In these beds they may remain two years, observing to keep them clear from weeds; and in winter lay a little fresh mulch upon the surface of the ground round their roots, which will prevent the frost from penetrating to them, and effectually preserve them; for while the plants are young, they are liable to be injured by very hard frosts; but when they have attained a greater strength, they will resist the severest of our cold.

After two years they should be either removed into a nursery (as was directed for the common Juniper) or transplanted where they are designed to remain, observing always to take them up carefully, otherwise they are subject to fail upon transplanting; as also to mulch the ground, and water them as was before directed, until they have taken root; after which they will require no farther care, than only to keep the ground clear about their roots, and to prune up their side branches to make them aspire in height.

The timber of these trees is of excellent use in America, for building of vessels, wainscotting houses, and for making many sorts of utensils, it abounding with a bitter resin, which prevents its being destroyed by vermin, but it is very brittle, and so not proper for stubborn uses; but however, by increasing the number of our timber trees, we shall find many advantages, besides the pleasure which their variety affords; for we may hereby have trees of very different kinds, which are adapted to grow in various soils and situations, whereby we shall never want proper trees for all the different soils in England, if proper care be taken in their choice; which would be a great improvement to many parts of this kingdom which now lie unplanted, because the owner perhaps finds, that neither Oaks nor Elms will thrive there, so concludes that no other sort will, which is a very great mistake.

The Bermudas Cedar being a native of that island, and also of the Bahama Islands, is much tenderer than either of the former sorts, so is not likely to thrive well in this country; for although many of these plants have lived several years in the open air in England, yet whenever a severe winter happens, it either kills them, or so much defaces them, as that they do not recover their verdure in a year or two after.

The timber of this tree is of a brown colour, and very sweet; it is commonly known in England by the name of Cedar Wood, though there are divers sorts of wood called by that name, which come from very different trees, especially in the West-Indies, where there are several trees of vastly different appearance and genera, which have that appellation; it is this wood which is used for pencils, as also to wainscot rooms, and make stair-cases. In America they build ships with this wood, for they say, the worms do not eat the bottoms of the vessels built with this wood, as they do those built with Oak; so that the vessels built with Cedar are much preferable, especially for the use of the West-India seas; but they are not fit for ships of war, the wood being so brittle as to split to pieces with a cannon ball.

The Jamaica Juniper is more impatient of cold than the Bermudas, so will not live through the winter in the open air in England, so the plants must be preserved in pots, and housed in the winter, especially while young; this is propagated by seeds in the same way as the Bermudas Cedar, but if the pots are plunged into a moderate hot-bed the second spring after the seeds are sown, it will bring up the plants sooner, and they will have more time to get strength before winter.

The common Savin should not be neglected, because it is so very hardy as never to be injured by the severest frost; and as this spreads its branches near the ground, so if the plants are placed on the borders of woods, they will have a good effect in winter, by screening the nakedness of the ground from sight.

Most of these plants may be propagated by cuttings, if they are planted in the autumn, and the tender sorts screened in winter with a common frame; but the plants so raised, will not be so good as those which come from seeds.

JUSSIÆA. Lin. Gen. Plant. 478.

The CHARACTERS are,

It hath a small permanent empalement divided into five parts, sitting upon the germen. The flower has five roundish petals, and ten short slender stamina. The oblong germen afterward becomes a thick oblong capsule, crowned by the empalement, which opens lengthways, and is filled with small seeds.

The SPECIES are,

1. JUSSIÆA (*Suffruticosa*) erecta villosa, floribus tetrapetalis, octandriis pedunculatis. Lin. Sp. Pl. 388. *Upright hairy Jussiæa with flowers standing upon long foot-stalks, having four petals and eight stamina.*

2. JUSSIÆA (*Pubescens*) villosa, caule erecto ramoso, floribus pentapetalis, decandriis sessilibus. *Hairy Jussiæa with an erect branching stalk, flowers having five petals, and ten stamina which sit close to the stalk.*

3. JUSSIÆA (*Erecta*) erecta glabra, floribus tetrapetalis, octandriis sessilibus. Flor. Zeyl. 170. *Smooth upright Jussiæa with four petals, and eight stamina to the flowers, which sit close to the stalk.*

4. JUSSIÆA (*Onagra*) caule erecta ramoso glabro, floribus tetrapetalis, octandriis sessilibus, foliis lanceolatis. *Jussiæa with an upright, branching, smooth stalk, flowers having four petals, eight stamina sitting close to the stalk, and spear-shaped leaves.*

5. JUSSIÆA (*Hirsuta*) caule erecto simplici hirsuto, foliis lanceolatis, floribus pentapetalis decandriis sessilibus. *Jussiæa with a single, upright, hairy stalk, spear-shaped leaves, flowers which have five petals, and ten stamina sitting close to the stalk.*

The first sort grows naturally at Campeachy. This rises with a shrubby stalk three feet high, sending out several side branches, garnished with oblong hairy leaves placed alternate. The flowers come out from the side of the stalks singly, upon short foot-stalks, having four small yellow petals, with eight stamina sitting upon the germen, which afterward becomes an oblong seed-vessel, crowned by the four-leaved empalement, including many small seeds.

The second sort grows naturally in Jamaica. This rises with a hairy branching stalk two feet high, garnished with narrow spear-shaped leaves placed alternate. The flowers come out toward the end of the branches singly from the wings of the leaves, sitting close to the stalk; they are composed of five pretty large yellow petals, and ten stamina, which sit upon a long germen, which afterward becomes the seed-vessel, crowned by the empalement, filled with small seeds.

The third sort grows naturally in Jamaica. This rises with a smooth erect stalk three feet high, garnished with long, narrow, smooth, spear-shaped leaves. The flowers are large and yellow, sitting close to the stalk, and are succeeded by long seed-vessels, shaped like those of the other sorts.

The fourth sort grows near Carthagena. This hath a branching smooth stalk three feet high, garnished with spear-shaped leaves, standing upon short foot-stalks. The flowers are small, yellow, and are composed of four petals and eight stamina; these sit very close to the stalk, and are succeeded by seed-vessels, shaped like those of the former sorts.

The fifth sort was sent me from La Vera Cruz. This rises with single, upright, red, hairy, channelled stalks three feet high. The leaves are spear-shaped, and placed alternate, standing nearer together than in any of the other sorts. The flowers come out from the wings of the leaves on the upper part of the stalk; they are composed of five large yellow petals, and ten stamina sitting close to the stalks, and are succeeded by seed-vessels, which are one inch long, and shaped like those of the former sorts.

The first, second, and fourth sorts are annual plants, at least they are so in England; for if the plants are raised early in the spring they will flower in July, and ripen their seeds the beginning of October, and the plants soon after decay.

The third and fifth sorts will continue through the winter in the bark-stove; but these must be such plants as do not flower and seed the first year, for after they have perfected seeds, the following summer the plants decay.

All these sorts are propagated by seeds, which should be sown early in the spring on a moderate hot-bed. When the plants come up and are fit to remove, they should be each planted into a small separate pot, and plunged into a hot-bed of tanners bark, where they should be shaded from the sun till they have taken new root, after which they should have free air admitted to them every day in proportion to the warmth of the season. When the roots of the plants have filled these small pots, the plants should be removed into others a size larger; and if the plants are too tall to stand under the frames of the hot-bed, they should be removed into the bark-stove, where they may remain to flower and perfect their seeds: for when the plants rise early in the spring, and are brought forward in hot-beds, all the sorts will flower and perfect their seeds the same year.

JUSTICIA. Houst. Nov. Gen. Lin. Gen. Plant. 26.

The CHARACTERS are,

The flower hath one petal, which is divided into two lips almost to the bottom, which are entire. The upper lip is raised archways, and the under is reflexed. It hath two awl-shaped stamina situated under the upper lip, with an oblong germen, which afterward becomes an oblong capsule with two cells, which open with an elasticity, and cast out the roundish seeds.

The SPECIES are,

1. JUSTICIA (*Scorpioides*) foliis oblongo-ovatis hirsutis, sessilibus, floribus spicatis alaribus, caule fruticoso. *Justicia with oblong, oval, hairy leaves, sitting close to the stalks, and flowers growing in spikes proceeding from the side of the stalks, which are shrubby.*

2. JUSTICIA (*Sexangularis*) caule erecto ramoso hexangulari, foliis ovatis oppositis, bracteis cuneiformibus confertis. *Justicia with an erect branching stalk, having six angles, oval leaves placed opposite, and wedge-shaped small leaves (or bracteæ) growing in clusters.*

3. JUSTICIA (*Fruticosa*) foliis ovato-lanceolatis, pediculatis, hirsutis, bracteis cordatis acuminatis, caule fruticoso. *Justicia with oval spear-shaped leaves growing on foot-stalks, heart-shaped acute-pointed bractea, and a shrubby stalk.*

4. JUSTICIA (*Adhatoda*) arborea, foliis lanceolato-ovatis, bracteis ovatis persistentibus, corollarum galeâ concavâ. Flor. Zeyl. 16. *Tree Justicia with oval spear-shaped leaves, oval permanent bracteæ, and a concave helmet to the flower; commonly called Malabar Nut.*

5. JUSTICIA (*Hyssopifolia*) fruticosa, foliis lanceolatis integerrimis, pedunculis trifloris ancipitibus, bracteis calyce brevioribus. Lin. Sp. Plant. 15. *Shrubby Justicia with entire spear-shaped leaves, foot-stalks having three flowers looking different ways, and a bracteæ shorter than the empalement; commonly called Snap-tree.*

6. JUSTICIA (*Spinosa*) spinosa, foliis oblongo-ovatis emarginatis, caule fruticoso ramoso. *Prickly Justicia with oblong oval leaves indented at their edges, and a shrubby branching stalk.*

7. Justicia (*Arborea*) arborea, foliis lanceolato-ovatis sessilibus, subtus tomentosis, floribus spicatis congestis terminalibus. *Tree Justicia with spear-shaped oval leaves, woolly on their under side, sitting close to the stalks, and spikes of flowers growing in clusters at the ends of the branches.*

8. Justicia (*Ecbolium*) arborea, foliis lanceolato-ovatis bracteis ovatis deciduis mucronatis, corollarum galeâ reflexâ. Flor. Zeyl. 17. *Tree Justicia with spear-shaped oval leaves, oval-pointed bracteæ which fall off, and a reflexed helmet to the flowers.*

The first sort grows naturally at La Vera Cruz. This rises with a shrubby brittle stalk five or six feet high, sending out many branches, garnished with oblong, oval, hairy leaves placed opposite; from the wings of the leaves come out the spikes of flowers, which are reflexed like a scorpion's tail. The flowers are large, of a carmine colour, ranged on one side of the spikes; these are succeeded by short pods about half an inch long.

The second sort is a native of the same country. This is an annual plant, with an upright stalk with six angles, which rises from two to three feet high, dividing into many branches, garnished with oval leaves placed opposite. At each joint come out clusters of small wedge-shaped leaves, which are termed bracteæ. The flowers are produced in small spikes at the end of the branches, sitting very close among the leaves; they are of a beautiful carmine colour, of one petal, which has two lips. The upper lip is arched, bending over the lower, which is also a little reflexed, but both are entire. The flowers are succeeded by short wedge-shaped capsules, opening lengthways, inclosing two small oval seeds.

The third sort grows naturally at Campeachy. This rises with a hairy shrubby stalk four or five feet high, dividing into several branches, garnished with oval, spear-shaped, hairy leaves, standing upon foot-stalks, placed opposite. The flowers come out in loose clusters from the wings of the stalks toward the end of the branches; they are of a pale red colour, shaped like those of the former sort.

These plants are propagated by seeds, which should be sown early in the spring in small pots, and plunged into a moderate hot-bed of tanners bark. The seeds of these plants frequently lie a year in the ground, so that the pots must not be disturbed if the plants do not come up the same year; but in the winter should be kept in the stove, and the spring following plunged into a fresh hot-bed, which will bring up the plants if the seeds were good.

When the plants are about two inches high they should be carefully taken up, and each transplanted into a separate small pot, and plunged into the hot-bed again, being careful to water and shade them until they have taken new root; after which time they should have air admitted to them every day, in proportion to the warmth of the season, and duly watered in hot weather.

As the plants advance in their growth, they should be shifted into larger pots; and as they are too tender to endure the open air in this country, so they should always remain in the hot-bed, being careful to let them have a due proportion of air in hot weather; and the annual sort should be brought forward as fast as possible in the spring, that the plants may flower early, otherwise they will not produce good seeds in England.

The first and third sorts should remain in the hot-bed during the summer season (provided there be room under the glasses, without being scorched;) but at Michaelmas they should be removed into the stove, and plunged into the bark-bed, where they must remain during the winter season. The following summer these plants will flower, and abide several years, but they rarely produce good seeds in Europe.

The fourth sort grows naturally in the island of Ceylon, but has been long in the English gardens, where it is commonly known by the title of Malabar Nut. This, though a native of so warm a country, is hardy enough to live in a good green-house in England, without any artificial heat. It rises here with a strong woody stalk to the height of twelve or fourteen feet, sending out many spreading branches, garnished with spear-shaped oval leaves six inches long and three inches broad, placed opposite. The flowers are produced on short spikes at the end of the branches, which are white, with some dark spots, but are not succeeded by any seeds in England.

This sort may be propagated by cuttings, which, if planted in pots in June and July, and plunged into a very moderate hot-bed, will take root, but they must be screened from the sun; and if the external air is excluded from them, they will succeed better than when it is admitted to them. It may also be propagated by laying down the young branches, which will take root in one year, and then should be put each into a separate pot, and placed in the shade till they have taken new root; then they may be removed to a sheltered situation during the summer, and in the autumn they must be housed, and treated in the same way as the Orange trees, with only this difference, that these require more water.

The fifth sort grows naturally in India. This rises with a shrubby stalk from three to four feet high, sending out branches on every side from the bottom, so as to form a kind of pyramid, garnished with spear-shaped entire leaves near two inches long, and one third of an inch broad; they are smooth, stiff, and of a deep green, standing opposite. At the base of the foot-stalks come out clusters of smaller leaves, of the same shape and texture. The flowers come out upon short foot-stalks from the side of the branches, each foot-stalk supporting one, two, or three white flowers, with long empalements, which are succeeded by oblong seed-vessels, which, when ripe, cast out their seeds with an elasticity, from whence it had the title of Snap-tree.

This is propagated by cuttings during any of the summer months, which should be planted in pots, and plunged into a hot-bed which has lost its great heat, and shaded from the sun. In about two months the cuttings will have taken root, then they must be gradually inured to bear the open air, into which they should be removed, placing them in a sheltered situation, where they may stay till autumn; but if they get root pretty early in the summer, it will be proper to separate them each into a single small pot, setting them in the shade till they have taken new root, after which they may be placed as before directed; but when it is late in the season before they take root, it will be better to let them remain in the same pots till the following spring. In winter these plants must be placed in a warm green-house, or in a moderate warm stove, for they are impatient of cold and damp, nor will they thrive in too much warmth; they will often require water in winter, but during that season it must be given them moderately; in summer they must be removed into the open air, but should have a warm sheltered situation, and in warm weather they must have plenty of water. This plant flowers at different seasons, but never produces fruit here.

The sixth sort grows naturally in Jamaica. This rises with many shrubby slender stalks about five feet high, sending out branches on every side, which are covered with a whitish bark, garnished with small, oblong, oval leave, coming out two on each side the stalk opposite, and under the leaves are placed at every joint two sharp thorns like those of the Berberry; the flowers come out singly from the wings

wings of the leaves; they are ſmall, and of a pale red colour, and ſhaped like thoſe of the other ſorts.

The ſeventh ſort grows naturally at Campeachy. This riſes with a ſtrong woody ſtem twenty feet high, dividing into many crooked irregular branches, covered with a light brown bark, garniſhed with ſpear-ſhaped oval leaves, near four inches long and two broad, which are covered with a ſoft down on their under ſide. The flowers grow in ſpikes from the end of the branches, three, four, or five of theſe ſpikes ariſing from the ſame point, the middle ſpike being near three inches long, and the others about half that length. The flowers are ſmall, white, and ſhaped like thoſe of the other ſpecies.

The eighth ſort grows naturally at Malabar and in Ceylon. This riſes with a ſtrong woody ſtem ten or twelve feet high, dividing into many branches, garniſhed with ſpear-ſhaped oval leaves five inches long, and two and a half broad, of a lucid green, placed oppoſite. The flowers grow in very long ſpikes from the end of the branches; they are of a greeniſh colour, with a ſhade of blue; the helmet of the flower is reflexed.

Theſe three ſorts are propagated by ſeeds in the ſame manner as the three firſt, and the plants muſt be treated in the ſame way, eſpecially while they are young, but afterward the eighth ſort may be more hardily treated. This ſort may alſo be propagated by cuttings, in the ſame manner as the fifth ſort; and when the plants are two or three years old they will thrive in a moderate degree of warmth in winter, and in the ſummer they may be placed abroad for two months in the warmeſt ſeaſon of the year, but they ſhould have a warm ſheltered ſituation; and when the nights begin to grow cold, they muſt be removed into the ſtove, but they muſt have free air admitted to them at all times when the weather is warm. The other two ſorts ſhould conſtantly remain in the bark-ſtove, and require the ſame treatment as other tender plants from the warmeſt countries.

IXIA. Lin. Gen. Plant. 54.

The CHARACTERS are,

It hath oblong permanent ſpathæ (or ſheaths) which incloſe the germen; the flower has ſix petals which are equal, and three awl-ſhaped ſtamina. It hath an oval three-cornered germen ſituated below the flower, with a ſingle ſtyle; the germen afterward becomes an oval capſule with three cells, filled with roundiſh ſeeds.

The SPECIES are,

1. IXIA (*Chinenſis*) foliis enſiformibus, floribus remotis. Hort. Upſal. 16. *Ixia with ſword-ſhaped leaves, and flowers ſtanding diſtant.*

2. IXIA (*Africana*) floribus capitatis, ſpathis laceris. Lin. Sp. Pl. 36. *Ixia with flowers growing in heads, and ragged ſheaths.*

3. IXIA (*Scillaris*) foliis gladiolatis, nervoſis, hirſutis, floribus ſpicatis terminalibus. Icon. tab. 155. fig. 1. *Ixia with ſword ſhaped, hairy, veined leaves, and flowers growing in ſpikes at the ends of the ſtalks.*

4. IXIA (*Polyſtachia*) foliis lineari-gladeolatis, floribus alaribus & terminalibus. Icon. tab. 155. fig. 2. *Ixia with narrow ſword-ſhaped leaves, and flowers proceeding from the ſides and tops of the ſtalk.*

5. IXIA (*Crocata*) foliis gladiolatis glabris, floribus corymboſus terminalibus. Icon. tab. 156. *Ixia with ſmooth ſpear-ſhaped leaves, and flowers growing in a corymbus terminating the ſtalk.*

6. IXIA (*Bulbifera*) foliis lineari-gladiolatis, floribus alternis, caule bulbifera. Lin. Syſt. *Ixia with narrow ſword-ſhaped leaves, flowers placed alternate, and ſtalks bearing bulbs at the joints.*

7. IXIA (*Sparſa*) foliis gladiolatis, floribus diſtantibus. *Ixia with ſpear-ſhaped leaves, and flowers growing diſtant.*

8. IXIA (*Flexuoſa*) foliis lineari-gladiolatis, floribus ſpicatis ſeſſilibus terminalibus. *Ixia with narrow ſword-ſhaped leaves, and ſeſſile flowers growing in ſpikes at the top of the ſtalk.*

The firſt ſort grows naturally in India, and alſo at the Cape of Good Hope; the ſtalks riſe to the height of three or four feet. It hath a pretty thick fleſhy root, divided in knots or joints, of a yellowiſh colour, ſending out many fibres; the ſtalk is pretty thick, ſmooth, and jointed, garniſhed with ſword-ſhaped leaves a foot long, and one inch broad, with ſeveral longitudinal furrows embracing the ſtalks with their baſe, ending in acute points; the upper part of the ſtalk divaricates into two ſmaller, with a foot-ſtalk ariſing between them, which ſupports one flower; the ſmaller branches divaricate again in the ſame manner into foot-ſtalks, which are two inches long, each ſuſtaining one flower. At each of theſe joints is a permanent ſpatha or ſheath embracing the ſtalk, ending in an acute point; the flowers are compoſed of ſix equal petals, of a yellow colour within, variegated with dark red ſpots; the outſide is of an Orange colour. Theſe appear in July and Auguſt, and in warm ſeaſons are ſucceeded by ſeeds.

This ſort may be propagated either by ſeeds or parting of the roots: if by ſeeds, they ſhould be ſown in pots ſoon after they are ripe, and plunged into an old hot-bed under a frame, to ſcreen them from froſt; and in the ſpring the plants will come up, when they ſhould be inured to the open air by degrees, for in ſummer they muſt be wholly expoſed thereto. The following autumn they muſt be ſeparated; ſome of the plants may be planted in a warm border, where they will abide through the common winters very well, but in ſevere froſts they are often killed, unleſs they are covered with tan, or other covering to keep out the froſt, therefore a few of the plants may be kept in pots, and ſheltered under a frame in winter.

The ſtalks and leaves of this plant decay to the root in autumn, ſo that if the ſurface of the ground about the roots is covered two or three inches thick with tan, it will ſecure them from the danger of froſt; and in the ſpring, before the roots ſhoot, will be the beſt time to remove and part them; but this ſhould not be done oftener than every third year, for when they are often parted they will be weak, and will not flower ſo well.

The ſecond ſort grows naturally at the Cape of Good Hope. This is a low plant, which riſes three or four inches high; the leaves are narrow and veined; the flowers are ſmall, growing in a downy head on the top of the ſtalk, but they make little appearance, ſo are only kept for the ſake of variety.

The third ſort I raiſed from ſeeds, which were ſent me from the Cape of Good Hope. This hath a bulbous root a little compreſſed, covered with a red ſkin, from which ariſe five or ſix ſword-ſhaped leaves about three or four inches long, hairy, with ſeveral longitudinal furrows; theſe embrace each other at their baſe: between theſe come out the flower-ſtalk, which riſes ſix or eight inches high, naked to the top, and terminated by a cluſter of flowers, each having a ſpatha or hood, which dries and is permanent; the flowers are of a deep blue colour, and are ſucceeded by roundiſh three-cornered ſeed-veſſels with three cells, filled with roundiſh ſeeds.

The fourth ſort was raiſed from ſeeds in the Chelſea garden, which came with thoſe of the former. This hath a ſmall, round, bulbous root, from which ariſe four or five narrow, long, ſword-ſhaped leaves ſix or ſeven inches long; between theſe come out a very ſlender round ſtalk about

 ten

ten inches long, from the ſide of which there comes out one or two cluſters of flowers, ſtanding upon ſhort foot-ſtalks, and at the top of the ſtalk the flowers grow in a looſe ſpike; they are of a pure white, and ſhaped like thoſe of the other ſpecies.

The ſeeds of the fifth ſort were ſent me from the Cape of Good Hope. This has an oval bulbous root, which is a little compreſſed, from which come out three or four narrow, thin, ſword-ſhaped leaves near a foot long; the flower-ſtalk riſes a little above the leaves; it is very ſlender, naked, and terminated by a round cluſter of flowers, compoſed of ſix pretty large, oblong, concave petals, of a deep yellow colour, each having a large black ſpot at the baſe. This flowers early in May, and the ſeeds ripen the latter end of June.

The ſixth ſort hath narrow ſword-ſhaped leaves about ſix or ſeven inches long. The ſtalk riſes near a foot and a half high, garniſhed with one leaf at each of the lower joints, of the ſame ſhape with the other, but ſmaller; theſe embrace the ſtalk with their baſe, and ſtand erect; the upper part of the ſtalk is adorned with flowers compoſed of ſix oblong, oval, whitiſh petals, having a blue ſtripe on their outſide, which are placed alternate on the ſtalk, which is bent at each joint where the flowers ſtand; the flowers have three ſhort ſtamina, which are joined at their baſe, terminated by long, flat, erect ſummits; the germen is ſituated under the flower, ſupporting a long ſlender ſtyle, crowned by a trifid ſtigma; the germen afterward becomes a roundiſh capſule with three cells, filled with roundiſh ſmall ſeeds. The ſtalks at each of the lower joints thruſt out ſmall bulbs, which, if planted, will grow and produce flowers.

The ſeventh ſort hath ſhorter and broader leaves than the former. The ſtalk is ſlender and furrowed; each of the lower joints is garniſhed with one leaf of the ſame ſhape, embracing the ſtalk with their baſe; the flowers come out toward the top of the ſtalk at two or three inches diſtance, each ſtalk ſupporting two or three ſulphur-coloured flowers, which are each compoſed of ſix ſpear-ſhaped petals an inch and a half long, equal in their ſize and regular in poſition; they have a ſhort permanent empalement, cut into two long and two ſhorter acute ſegments; theſe are ſucceeded by round capſules with three cells, filled with round ſeeds.

The eighth ſort hath very ſmall, round, bulbous roots, from which ariſe three or four long, ſlender, ſmooth, Graſs-like leaves, of a dark green colour; between theſe come out the ſtalk, which is very ſlender, riſing a foot and a half high; at the top the flowers are collected in a ſpike ſitting cloſe to the ſtalk, each having a thin, dry, permanent ſpatha or ſheath, which covers the capſule after the flower is fallen. The flowers are of a pure white, and ſhaped like thoſe of the other ſpecies, but are ſmaller; they are ſucceeded by ſmall round ſeed-veſſels with three cells, each containing two or three round ſeeds.

All the ſorts multiply very faſt by offſets, ſo that when once obtained there will be no occaſion to raiſe them from ſeeds; for the roots put out offſets in great plenty, moſt of which will flower the following ſeaſon, whereas thoſe from ſeeds are three or four years before they flower. Theſe plants will not live through the winter in the full ground in England, ſo ſhould be planted in pots, and placed under a frame in winter, where they may be protected from froſt, but in mild weather ſhould enjoy the free air; but they muſt be guarded from mice, who are very fond of theſe roots, and, if not prevented, will devour them. If a frame is made for theſe in the ſame manner as is directed for the African Gladioluſes, and other bulbous roots which require no artificial heat, they will thrive and flower much better than when they are planted in pots.

K.

KALI. See SALSOLA.

KALMIA. Lin. Gen. Plant. 482.

The CHARACTERS are,

The flower has a ſmall permanent empalement, and one petal cut into five ſegments, which ſpread open. It hath ten ſtamina which are the length of the petal, and decline in the middle. In the center is ſituated a roundiſh germen, which afterward becomes an oval globular capſule with five cells, filled with very ſmall ſeeds.

The SPECIES are,

1. KALMIA (*Latifolia*) foliis ovatis, corymbis terminalibus. *Kalmia with oval leaves, and flowers growing in bunches terminating the branches.*

2. KALMIA (*Anguſtifolia*) foliis lanceolatis corymbis lateralibus. Lin. Gen. Nov. 1079. *Kalmia with ſpear-ſhaped leaves, and flowers growing in round bunches on the ſides of the ſtalk.*

The firſt ſort grows in ſeveral parts of North America, where it riſes from ſix to twelve feet high, dividing into many ligneous branches, covered with a dark gray bark; they are generally crooked and irregular, but are cloſely garniſhed with ſtiff leaves about three inches long and one broad, of a lucid green, ſtanding upon ſlender foot-ſtalks; the flowers grow in looſe bunches at the end of the branches, upon long foot-ſtalks; they are of one petal, with a ſhort tube, which ſpreads open at the top, where it is cut into five angles; the flowers are of a bright red colour when they firſt open, but afterward fade to a bluſh or Peach-bloom colour; theſe are ſucceeded by roundiſh compreſſed ſeed-veſſels crowned by the permanent ſtyle, divided into five cells, which are filled with ſmall roundiſh ſeeds.

The leaves of this elegant plant are ſuppoſed to have a noxious quality, deſtroying ſheep and oxen when they feed upon them, yet the deer eat them with impunity.

The ſecond ſort grows naturally in Penſylvania, where it riſes four or five feet high, but in England I have not ſeen any which were more than half that height.

The leaves of this sort are about two inches long, and half an inch broad in the middle; they are stiff, of a lucid green, and stand opposite; sometimes they are by pairs at each joint, and at others there are four, two on each side, standing upon very short foot-stalks; the flowers come out in clusters on every side the stalks; they are of a beautiful red colour, and shaped like those of the first sort, but smaller; they are succeeded by short, roundish, compressed capsules with five cells, crowned by the permanent style, and filled with very small seeds. This shrub, in its native country, continues flowering most part of summer.

Both these sorts multiply by their creeping roots in their native soil, and where they have stood unremoved a considerable time, they put out suckers in pretty great plenty; and as these plants which come from suckers, are much more likely to produce others than those which are raised from seeds, and will flower much sooner, so the plants should not be removed, but encouraged to spread their roots, whereby they may be propagated; they love a moist, light, boggy soil, in which they will thrive and flower.

KARATAS. The Penguin or wild Ananas.

The CHARACTERS are,

It hath a tubulous bell-shaped flower, divided into three parts at the mouth, from whose calyx arises the pointal, fixed like a nail in the hinder part of the flower, which afterward becomes a fleshy almost conical fruit, divided by membranes into three cells, which are full of oblong seeds.

We have but one SPECIES of this genus, viz.

KARATAS (*Penguin*) foliis altissimis, angustissimis & aculeatis. Plum. Nov. Gen. *The wild Ananas or Penguin.*

Father Plumier has made a great mistake in the figure and description of the characters of this plant, and the Caraguata; for he has joined the flower of the Caraguata to the fruit of the Karatas, and vice versâ; and this has led many persons into mistakes, who have joined the Bromelia and Ananas to this, making them of the same genus.

This plant is very common in the West-Indies, where the juice of its fruit is often put into punch, being of a sharp acid flavour. There is also a wine made of the juice of this fruit, which is very strong, but it will not keep good long, so is only for present use. This wine is very intoxicating, and heats the blood, therefore should be drank very sparingly.

In England this plant is preserved as a curiosity, for the fruit seldom arrives to any degree of perfection for use in this country, though it often produces fruit in England, which has ripened pretty well; but if it were to ripen as thoroughly here as in its native country, it will be little valued on account of its great austerity, which will often take the skin off from the mouths and throats of those people who eat it incautiously.

This plant is propagated by seeds, for though there are often suckers sent forth from the old plants, yet they come out from between the leaves, and are so long, slender, and ill-shapen, that if they are planted they seldom make regular plants. These seeds should be sown early in the spring in small pots, and plunged into a hot-bed of tanners bark, where the plants will come up in six weeks. When the plants are strong enough to transplant, they should be carefully taken up, each planted into a separate pot, and plunged into the hot-bed again; when the plants have taken new root, they should have air and water in proportion to the warmth of the season. In this bed the plants may remain till Michaelmas, then they should be removed into the stove and plunged into the bark-bed, where they should be treated in the same manner as the Ananas.

The leaves of this plant are strongly armed with crooked spines, which renders it very troublesome to shift or handle them; for the spines catch hold of whatever approaches them by their crooked form, being some bent one way, and others the reverse, so that they catch both ways, and tear the skin or clothes of the persons who handle them, where there is not the greatest care taken to avoid them.

The fruit of this plant is produced in clusters, growing upon a stalk about three feet high, with a tuft of leaves growing on the top, so at first sight has the appearance of a Pine Apple; but when closer viewed, they will be found to be a cluster of oblong fruit, each being about the size of a finger.

KETMIA. See HIBISCUS.

KIGGELARIA. Lin. Gen. Plant. 1001.

The CHARACTERS are,

The male and female flowers are situated on different trees; the male flowers have five concave petals, which are shaped like a pitcher, each having a honey gland fastened to their base, and have ten small stamina. The female flowers have petals like the male, but no stamina. In the center is situated a roundish germen, which afterward becomes a rough globular fruit with a thick cover, having one cell, filled with angular seeds.

We have but one SPECIES of this genus, viz.

KIGGELARIA (*Africana.*) Hort. Cliff. 462. fol. 29.

This plant grows naturally at the Cape of Good Hope, where it rises to be a tree of middling stature; but as it will not live in the open air here, they cannot be expected to grow to a great magnitude in England. There were plants of it in the Chelsea garden upward of ten feet high, with strong woody stems and pretty large heads, which were killed in the winter 1768; the leaves were about three inches long and one broad, of a light green colour, sawed on their edges, and stand upon short foot-stalks alternately. The flowers come out in clusters from the side of the branches, hanging downward; they are of an herbaceous white colour. The male flowers fall away soon after their farina is shed, but the female flowers are succeeded by globular fruit about the size of common red Cherries, with a rough cover, of a thick consistence, opening in four parts at the top, to each of which adhere many small angular seeds. These fruits have grown to their full size in the Chelsea garden, but the seeds never came to maturity.

The plants are not very common in Europe, being very difficult to propagate; for when any of the young branches are laid down, very few of them take root, and those which do are two years before they put out roots; nor do the cuttings succeed much better, for very few of them will take root, when planted with the utmost care: the best time to plant the cuttings is in May, just before the plants begin to shoot; these should be planted in pots, and plunged into a very moderate hot-bed, covering them close with a glass, to exclude the air from them, and shade them every day from the sun; they should have very little water after their first planting. Those which do grow should be planted into separate small pots, and shaded till they have taken fresh root, then may be exposed to the air in a sheltered situation till autumn, when they must be removed into the green-house, and treated in the same manner as Orange-trees, and other green-house plants.

KITCHEN-GARDEN. A good Kitchen-garden is almost as necessary to a country-seat, as a kitchen to the house; for without one, there is no way of being supplied with a great part of necessary food, the markets in the country being but poorly furnished with esculent herbs, and those only upon the market days, which are seldom oftener than once a week; so that unless a person has a garden of his own, there will be no such thing as procuring them fresh, in which their goodness consists; nor can any variety of these be had in the country markets, therefore whoever proposes to reside in the country should be careful to make choice of a proper spot of ground for this purpose; and the fooner that is made and planted, the produce of it will be earlier in perfection; for fruit-trees and Afparagus require three years to grow, before any produce can be expected from them; fo that the later the garden is made, the longer it will be before a fupply of thefe things can be had for the table. And although the ufefulnefs of this garden is acknowledged by almoft every one, yet there are few who are careful to make a proper choice of foil and fituation for fuch a garden: the modern tafte, which is perhaps carried to as extravagant lengths, in laying open and throwing every obftruction down, as the former cuftom of inclofing within walls was ridiculous; fo that now one frequently fees the Kitchen-garden removed to a very great diftance from the houfe and offices, which is attended with great inconveniencies; and often fituated on a very bad foil, fometimes too moift, and at others without water, fo that there is a great expence in building walls and making the garden, where there can be little hopes of fuccefs.

Nor will a Kitchen-garden be well attended to, when it is fo fituated as to be out of fight of the poffeffor, efpecially if the gardener has not a love and value for it, or if it lies at a great diftance from the manfion-houfe, or the other parts of the garden, a great part of the labourer's time will be loft in going from one part to the other: therefore, before the general plan of the pleafure-garden is fettled, a proper piece of ground fhould be chofen for this purpofe, and the plan fo adapted, as that the Kitchen-garden may not become offenfive to the fight, which may be effected by proper plantations of fhrubs to hide the walls; and through thefe fhrubs may be contrived fome winding walks, which will have as good an effect as thofe which are now commonly made in gardens for pleafure only. In the choice of the fituation, if it does not fhut out any material profpect, there can be no objection to the placing it at a reafonable diftance from the houfe or offices; for as particular things may be wanted for the kitchen, which were not thought of at the time when directions were given to the gardener what to bring in, fo if the garden is fituated at a great diftance from the houfe, it will be found very inconvenient to fend thither as often as things are wanting: therefore it fhould be contrived as near the ftables as poffible, for the conveniency of carrying the dung thither; which, if at a great diftance, will add to the expence of the garden.

As to the figure of the ground, that is of no great moment, fince in the diftribution of the quarters all irregularities may be hid; though if you are at full liberty, an exact fquare or an oblong is preferable to any other figure.

The great thing to be confidered is, to make choice of a good foil, not too wet nor over dry, but of a middling quality; nor fhould it be too ftrong or ftubborn, but of a pliable nature, and eafy to work; and if the place where you intend to make the Kitchen-garden fhould not be level, but high in one part and low in another, I would by no means advife the levelling it; for by this fituation you will have an advantage which could not be obtained on a perfect level, which is, the having one part of dry ground for early crops, and the low part for late crops, whereby the kitchen may be the better fupplied throughout the feafon with the various forts of herbs, roots, &c. And in very dry feafons, when in the upper part of the garden the crops will greatly fuffer with drought, then the lower part will fucceed, and fo vice verfâ; but I would by no means direct the chufing a very low moift fpot of ground for this purpofe, for although in fuch foils garden herbs are commonly more vigorous and large in the fummer feafon, yet they are feldom fo well tafted or wholefome as thofe which grow upon a moderate foil; and efpecially fince in this garden your choice fruits fhould be planted, it would be wrong to make choice of a very wet foil for this purpofe.

This garden fhould be fully expofed to the fun, and by no means overfhadowed with trees, buildings, &c. which are very injurious to fruit trees; but if it be defended from the north wind by a diftant plantation, it will greatly preferve the early crops in the fpring. But thefe plantations fhould not be too near nor very large, for I have generally found where Kitchen-gardens are placed near woods or large plantations, they have been much more troubled with blights in the fpring than thofe which have been more expofed.

The quantity of ground neceffary for a Kitchen-garden, muft be proportioned to the largenefs of the family, or the quantity of herbs defired: for a fmall family, one acre and a half of ground may be fufficient; but for a large family there fhould not be lefs than four or five acres; becaufe, when the ground is regularly laid out, and planted with efpaliers of fruit-trees, this quantity will be found little enough, notwithftanding what fome perfons have faid on this head.

This ground muft be walled round, and if it can be conveniently contrived, fo as to plant both fides of the walls which have good afpects, it will be a great addition to the quantity of wall fruit: and thofe flips of ground which are without fide of the walls, will be very ufeful for planting of Goofeberries, Currants, Strawberries, and fome forts of kitchen plants, fo that they may be rendered equally ufeful with any of the quarters within the walls; but thefe flips fhould not be too narrow, left the hedge, pale, or plantation of fhrubs which inclofe them, fhould fhade the borders where the fruit-trees ftand: the leaft width of thefe flips fhould be twenty-five or thirty feet, but if they are double that it will be yet better, the flips will be more ufeful, and the fruit-trees will have a larger fcope of good ground for their roots to run. The walls fhould be built about ten or twelve feet high, which will be fufficient height for moft forts of fruit.

The foil of this garden fhould be at leaft two feet deep, but if deeper it will be ftill better, otherwife there will not be depth enough for many forts of efculent roots, as Carrots, Parfneps, Beets, &c. which run down pretty deep in the ground, and moft other forts of efculent plants delight in a deep foil; and many plants, whofe roots appear fhort, yet if their fibres, by which they receive their nourifhment, are traced, they will be found to extend to a confiderable depth in the ground; fo that when thefe are ftopped by meeting with gravel, chalk, clay, &c. the plants will foon fhew it, by their colour and ftinted growth.

In the diftribution of this garden, next the fouth and other good afpected walls, the borders fhould be at leaft eight or ten feet broad, whereby the roots of the fruit-trees will have greater liberty than in fuch places where the borders are not above three or four feet wide; and upon thefe borders you may fow many forts of early crops, if expofed to the fouth; and upon thofe expofed to the north, you may have fome late crops; but I would by no means advife the planting any fort of deep rooting plants too near the fruit-trees, efpecially Peafe and Beans; though for the advantage of the walls, to preferve them in winter, and to bring them forward in the fpring, the gardeners in general are too apt to make ufe of thofe borders, which are near the beft afpected walls, to the great prejudice of their fruit-trees; but for thefe purpofes it is much better to have fome Reed hedges fixed in fome of the warmeft quarters, clofe to which you fhould fow and plant early Peafe, Beans, &c. where they will thrive as well as if planted near a wall, and hereby the fruit-trees will be entirely freed from fuch troublefome plants.

The

The walks of this garden ſhould be alſo proportioned to the ſize of the ground, which in a ſmall garden ſhould be ſix feet, but in a large one the middle walks ſhould be ten or twelve; on each ſide of the walk ſhould be allowed a border four or five feet wide between the eſpalier and the walk, whereby the diſtance between the two eſpaliers will be greater, and theſe borders being kept conſtantly worked and manured, will be of great advantage to the roots of the trees; in theſe borders may be ſown ſome ſmall Sallad, or any other herbs, which do not continue long or root deep, ſo that the ground will not be loſt.

The breadth of theſe middle walks, which I have here aſſigned them, may by many perſons be thought too great; but my reaſon for this is to allow proper room between the eſpaliers, that they may not ſhade each other, or their roots interfere and rob each other of their nouriſhment; but where the walks are not required of this breadth, it is only enlarging of the borders on each ſide, and ſo reducing the walks to the breadth deſired.

The walks of theſe gardens ſhould not be gravelled, for as there will be conſtantly an occaſion to wheel manure, water, &c. upon them, they would ſoon be defaced, and rendered unſightly; nor ſhould they be laid with turf, for in green walks, when they are wheeled upon or much trodden, the turf is ſoon deſtroyed, and thoſe places where they are much uſed, become very unſightly alſo, therefore the beſt walks for a Kitchen-garden are thoſe which are laid with a binding ſand; but where the ſoil is ſtrong and apt to detain the wet, there ſhould be ſome narrow underground drains made by the ſide of the walks to carry off the wet, otherwiſe there will be no uſing of the walks in bad weather; and where the ground is wet, ſome lime rubbiſh, flints, chalk, or any ſuch material as can be procured with the leaſt expence, ſhould be laid at the bottom of them; and if neither of theſe can be had, a bed of Heath or Furze ſhould be laid, and the coat of ſand laid over it, by which the ſand will be kept drier, and the walks will be found and good in all ſeaſons. Theſe ſand-walks are by much the eaſieſt kept of any, for when either weeds or Moſs begin to grow, it is but ſcuffling them over with a Dutch hoe in dry weather, and raking them over a day or two after, and they will be as clean as when firſt laid; or if the walks are covered with the duſt taken from the great roads, it will bind and make a firm walk.

The beſt figure for the quarters to be diſpoſed, is a ſquare or an oblong, where the ground is adapted to ſuch a figure; otherwiſe they may be triangular, or of any other ſhape, which will beſt ſuit the ground.

When the garden is laid out in the ſhape intended, if the ſoil is ſtrong, and ſubject to detain moiſture, or is naturally wet, there ſhould always be under-ground drains made, to carry off the wet from every quarter of the garden, for otherwiſe moſt ſorts of kitchen plants will ſuffer greatly by moiſture in winter; and if the roots of the fruit-trees get into the wet, they will never produce good fruit, ſo that there cannot be too much care taken to let off all ſuperfluous moiſture from the Kitchen-garden.

In one of theſe quarters, which is ſituated neareſt to the ſtables, and beſt defended from the cold winds, or if either of the ſlips without the garden wall, which is well expoſed to the ſun, lies convenient, and is of a proper width, that ſhould be preferred for a place to make hot-beds for early Cucumbers, Melons, &c. The reaſons for my giving the preference to one of theſe ſlips, is, firſt, there will be no dirt or litter carried over the walks of the Kitchen-garden in winter and ſpring, when the weather is generally wet, ſo that the walks will not be rendered unſightly; ſecondly, the view of the hot-beds will be excluded from ſight; and laſtly, the convenience of carrying the dung into theſe ſlips, for by making of a gate in the hedge or pale, wide enough for a ſmall cart to enter, it may be done with much leſs trouble than that of borrowing it through the garden; and where there can be a ſlip long enough to contain a ſufficient number of beds for two or three years, it will be of great uſe, becauſe by the ſhifting of the beds annually, they will ſucceed much better than when they are continued for a number of years on the ſame ſpot of ground; and as it will be abſolutely neceſſary to fence this Melon ground round with a Reed hedge, it may be ſo contrived as to move away in pannels, and then that hedge, which was on the upper ſide the firſt year, being carried down to a proper diſtance below that which was the lower hedge, and which may remain, there will be no occaſion to remove more than one of the croſs hedges in a year; therefore, I am perſuaded, whoever will make trial of this method, will find it the moſt eligible.

The moſt important points of general culture conſiſt in well digging, keeping clean, and manuring the ſoil, and giving proper diſtance to the trees and plants, according to their different growths (which is conſtantly exhibited in their ſeveral articles in this book.) The dunghills ſhould alſo be kept always clear from weeds, for it will be to little purpoſe to keep the garden clean, if this is not obſerved; for if the ſeeds of weeds are ſuffered to ſcatter upon the dung, they will be brought into the garden, whereby there will be a conſtant ſupply of weeds yearly introduced, to the no ſmall damage of your plants, and a perpetual labour occaſioned to extirpate them again. Another thing which is abſolutely neceſſary to be obſerved, is, to carry off all the refuſe leaves of Cabbages, the ſtalks of Beans, and haulm of Peaſe, as ſoon as they have done bearing, for the ill ſcent which moſt people complain of in Kitchen gardens, is wholly occaſioned by theſe things being ſuffered to rot upon the ground; therefore when the Cabbages are cut, all the leaves ſhould be carried out of the garden while they are freſh, at which time they may be very uſeful for feeding of hogs, or other animals, and this will always keep the garden neat, and free from ill ſcents. As for all other neceſſary directions, they will be found in the articles of the ſeveral ſorts of kitchen plants, which renders it needleſs to be repeated in this place.

KLEINIA. See CACALIA.

KNAUTIA. Lin. Gen. Plant. 109.

The CHARACTERS are,

It hath ſeveral floſcular flowers incloſed in one common cylindrical empalement, which have their petals ranged ſo as to appear like a regular flower, but each ſeparate floſcule is irregular; in the bottom of each floret is ſituated the germen, attended by four ſtamina, which germen afterward changes to a ſingle, oblong, naked ſeed.

We have but one SPECIES of this genus, viz.

KNAUTIA. Lin. Hort. Cliff. This plant is very near akin to the Scabious, under which genus it has been ranged by ſeveral botaniſts; but the appearance of the flower at firſt ſight being like a Lychnis, Dr. Boerhaave ſeparated it from the Scabious, and gave it the title of Lychni Scabioſa, which being a compound name, Dr. Linnæus has altered to this of Knautia.

This is an annual plant; the ſeeds of it were brought from the Archipelago, where it is a native; but when it is allowed to ſcatter its ſeeds in a garden, it will propagate itſelf in as great plenty as if it were a native of England; and theſe autumnal plants, which ariſe from the ſcattered ſeeds, will grow much ſtronger than thoſe which are ſown in the ſpring. All the culture this plant requires is, to keep it clear from weeds, for it will thrive on almoſt any ſoil or in any ſituation.

KŒMPFERIA. Lin. Gen. Plant. 7.

The CHARACTERS are,

The flower hath a single spatha or sheath of one leaf; it hath one petal having an oblong slender tube, cut into six parts at the brim; three of them are alternately spear-shaped, the other are oval and divided into two segments; it hath but one stamina, which is membranaceous, terminated by a linear summit, which scarcely emerges out of the petal: it hath a round germen supporting a single style, crowned by an obtuse stigma, which afterward becomes a roundish capsule with three cells, filled with seeds.

We have but one SPECIES of this genus, viz.

KOEMPFERIA (*Galanga*) foliis ovatis sessilibus. Flor. Zeyl. 8. *The Wanhom, or Galangale.*

This plant grows naturally in the East-Indies, where the roots are greatly used in medicine as sudorific and carminative, and have greatly the scent of Ginger when first taken out of the ground, being divided into several fleshy tubers joined, about four inches long; the leaves are oval, growing close to the root without foot-stalks; from between these leaves the flowers are produced singly, having no foot-stalk; the flowers are white, with a bright purple bottom, but are not succeeded by seeds in England.

This plant is easily propagated by parting of the roots in March, before they put out any leaves; but as it is a native of a hot country, so the plants must be constantly preserved in the tan-bed in the stove, for if they are placed on shelves in a dry stove, the roots will decay in winter, so they should be treated in the same manner as Ginger.

L.

LABLAB. See PHASEOLUS and DILOCHOS.

LABRUM VENERIS. See DIPSACUS.

LABURNUM. See CYTISUS.

LACRYMA JOBI. See COIX.

LACTUCA. Tourn. Inst. R. H. 473. tab. 267. Lettuce.

The CHARACTERS are,

The flowers are composed of several hermaphrodite florets, inclosed in a scaly oblong empalement. The florets have one petal, which is stretched out on one side like a tongue; these have five short hairy stamina. The germen afterward becomes one oblong pointed seed, crowned with a single down.

It would be to little purpose to mention in this place the several sorts of Lettuce that are to be found in botanic writers, many of which are plants of no use, and are never cultivated but in botanic gardens for variety, and some of them are found wild in many parts of England. I shall therefore pass over those, and only mention here the several sorts which are commonly cultivated in the kitchen-garden for use: 1. Common, or garden Lettuce. 2. Cabbage Lettuce. 3. Cilicia Lettuce. 4. Dutch brown Lettuce. 5. Aleppo Lettuce. 6. Imperial Lettuce. 7. Green capuchin Lettuce. 8. Versailles, or upright white Cos Lettuce. 9. Black Cos. 10. Red Cos. 11. Red capuchin Lettuce. 12. Roman Lettuce. 13. Prince Lettuce. 14. Royal Lettuce. 15. Egyptian Cos Lettuce.

The first of these sorts is commonly sown for cutting very young, to mix with other small Sallad herbs, and is only different from the second sort, in being a degeneracy therefrom, or otherwise the second is an improvement by frequent cultivation from the first; for if the seeds are saved from such plants of the second sort as did not cabbage closely, the plants produced from that seed will all degenerate to the first sort, which is by the gardeners called Lapped Lettuce, to distinguish it from the other, which they call Cabbage Lettuce. The seeds of the first, which are commonly saved from any of the plants, without having regard to their goodness, are generally sold at a very cheap rate (especially in dry seasons, when these plants always produce the greatest quantity of seeds,) though sometimes this seed is sold in the seed-shops, and by persons who make a trade of selling seeds, for the Cabbage Lettuce, which is often the occasion of peoples being disappointed in their crop; so that this sort should never be cultivated but to be cut up very young, for which purpose this is the only good sort, and may be sown any time of the year, observing only in hot weather to sow it on shady borders, and in the spring and autumn upon warm borders, but in winter it should be sown under glasses, otherwise it is subject to be destroyed by severe frosts.

The Cabbage Lettuce may also be sown at different times of the year, in order to have a continuation of it through the whole season. The first crop is generally sown in February, which should be upon an open warm spot of ground, and when the plants are come up, they should be thinned out to the distance of ten inches each way, which may be done by hoeing them out, as is practised for Turneps, Carrots, Onions, &c. provided you have no occasion for the superfluous plants, otherwise they may be drawn up, and transplanted into another spot of good ground at the same distance, which, if done before the plants are too large, they will succeed very well, though they will not be so large as those which were left upon the spot where they were sown, but they will come somewhat later, which will be of service where people do not continue sowing every fortnight.

You must also observe in sowing the succeeding crops, as the season advances, to chuse a shady moist situation, but not under the drip of trees, otherwise, in the heat of summer, they will run up to seed before they cabbage. In the middle of August you should sow the last crop, which is to stand over winter; the seeds should be sown thin upon a good light soil, in a warm situation; and when the plants are come up, they must be hoed out so as to stand singly, and cut down all the weeds to clear them. In the beginning of October they should be transplanted into warm borders, where, if the winter is not very severe, they will stand very well; but in order to be sure of a crop, it will be adviseable to plant a few upon a bed pretty close, where they may be arched over with hoops, and in severe frosts they should be covered with mats and straw, or Pease haulm, to secure them from being destroyed; and in the spring of the year, they may be transplanted out into a warm rich soil, at the distance before-mentioned; but still those which grew under the wall, if they escaped the winter, and were suffered to remain, will cabbage sooner than those which are removed; but you must observe not to place them too close to the pale or wall, which would occasion their growing up tall, and prevent their being large or hard.

In order to save good seeds of this kind, you should look over your Lettuces when they are in perfection, and such of them as are very hard, and grow low, should have sticks thrust into the ground, by the sides of as many of them as you intend for seed, to mark them from the rest; and you should carefully pull up all the rest from amongst them as soon as they begin to run up, if any happen to be left, lest when they are run up to flower, they should, by intermixing their farina with the flowers of the good ones, degenerate the seeds.

The Cilicia, imperial, royal, black, white, and red Cos Lettuces may be sown at the following times: the first season is the beginning of February, upon a gentle hot-bed covered with a frame: the second is the latter end of February, or the beginning of March, upon a warm border of light soil, in an open situation, i. e. not over-shadowed with trees; when the plants come up on the hot-bed, they should have a great share of fresh air admitted to them, to prevent their drawing up weak, and when they have four or six leaves, they should be transplanted upon another hot-bed to bring them forward, but this bed may be arched over with hoops and covered with mats. When the plants are strong enough to plant out for good, they should be set at sixteen inches distance each way. Those which were sown on the warm border, should also be transplanted into another spot of ground, at the same distance as the former, observing, if the season is dry, to water them till they have taken root; after which they must be carefully kept clean from weeds, which is the only culture they will require, except the black Cos Lettuce, which should be tied up when they are full grown (in the manner as was directed for the blanching of Endive) to whiten their inner leaves, and render them crisp, otherwise they are seldom good for much, rarely cabbaging without this assistance.

When the Lettuces are in perfection, they should be looked over, and as many of the best of them as you intend for seed, should be marked (in the same manner as was before directed for the common Cabbage Lettuce) being very careful not to suffer any ordinary ones to seed amongst or near them, as was before observed, which would prove more injurious to these sorts than to the common, as being more inclinable to degenerate with us, if they are not carefully preserved.

But to continue these sorts of Lettuce through the season, the seeds must be sown in April, May, and June, observing (as was before directed) to sow the late crops in a shady situation, otherwise they will run up to seed before they grow to any size; in the middle or latter end of September there should be some seeds sown of these sorts, to abide the winter, which plants should be transplanted either under glasses, or in a bed, which should be arched over with hoops, in order to be covered in the winter, otherwise in hard winters they are often destroyed; but these plants should have as much free air as possible when the weather is mild, only covering them in hard rains, or frosty weather, for if they are kept too closely covered in winter, they will be subject to mouldiness, which soon rots them.

In the spring these plants should be planted out into a rich light soil, allowing them at least eighteen inches distance each way, for if they are planted too close, they are very subject to grow tall, but seldom cabbage well; and from this crop, if they succeed well, it will be proper to save the seeds; though some plants should also be marked of that crop sown in the spring, because sometimes it happens, that the first may fail by a wet season, when the plants are full in flower, and the second crop may succeed, by having a more favourable season afterward; and if they should both succeed, there will be no harm in that, since the seeds will grow very well when two years old, and if well saved, at three, but this will not always happen.

The most valuable of all the sorts of Lettuce in England are the Egyptian green Cos, and the white Cos, the Cilicia and red Cos. Though some people are very fond of the Royal and Imperial Lettuces, but they seldom sell so well in the London markets as the other, nor are so generally esteemed. Indeed of late years, since the white Cos has been commonly cultivated, it has obtained the preference of all the other sorts, until the Egyptian green and the red Cos was introduced, which are so much sweeter and tenderer than the white Cos, that they are by all good judges esteemed the best sorts of Lettuce yet known. These will endure the cold of our ordinary winters full as well as the white Cos; but at the season of their cabbaging, if there happens to be much wet, they are very subject to rot.

The brown Dutch and green capuchin Lettuces are very hardy, and may be sown at the same seasons as was directed for the common Cabbage Lettuce, and are very proper to plant under a wall or hedge, to stand the winter, where many times these will live, when most of the other sorts are destroyed, therefore they will prove very acceptable, at a time when few other sorts are to be had; they will also endure more heat and drought than most other sorts of Lettuce, which renders them very proper for late sowing; for it often happens, in very hot weather, that the other sorts of Lettuce will run up to seed in a few days after they are cabbaged, whereas these will abide near three weeks in good order, especially if care be taken to cut the forwardest first, leaving those that are not so hard cabbaged to the last. In saving of these seeds, the same care should be taken to preserve only such as are very large and well cabbaged, otherwise the seeds will degenerate, and be good for little.

If these sorts of Lettuce are planted upon a moderate hot-bed in autumn, and covered with a good frame, they may be cabbaged so well as to be fit for use in February and March, and may be continued till those in the open air are fit for use.

In saving seeds of all these sorts of Lettuce, you should observe never to let two sorts stand near each other, for by their farina mixing, they will both vary from their original, and partake of each other; there should also be a stake fixed down by the side of each, to which the stem should be fastened, to prevent their being broken, or blown out of the ground by wind, to which the Cilicia, Cos, and the other large growing Lettuces, are very subject when they are in flower. When the seeds begin to ripen, such branches of the large growing Lettuces as ripen first should be cut, and not wait to have the seed of the whole plant ripe together, which never happens; but, on the contrary, some branches will be ripe a fortnight or three weeks before others, and when they are cut they must be spread upon a coarse cloth in a dry place, that the seeds may dry; after which they should be beat or rubbed out, and dried again, and then carefully hanged up where mice and other vermin cannot come at them, for if they do, they will soon eat them up.

LACTUCA AGNINI. See VALERIANELLA.

LADY's SLIPPER. See CYPRIPEDUM.

 LADY's

LADY's SMOCK. See CARDAMINE.

LAGŒCIA. Baſtard Cumin.

The CHARACTERS are,

It hath many flowers collected into a head, in one common empalement, compoſed of eight indented leaves. The flower conſiſts of five horned petals, at the bottom of each flower is ſituated the germen, attended by five ſtamina; the germen afterward changes to an oval ſeed, crowned with the empalement.

We have but one SPECIES of this plant, viz.

LAGOECIA (*Cuminoides.*) Lin. Hort. Cliff. Baſtard, or Wild Cumin.

We have no other Engliſh name for this plant, nor is this a very proper one, but as it has been titled by ſome of the ancient botaniſts Cuminum ſylveſtre, ſo it may be ſtiled wild, or Baſtard Cumin in Engliſh.

This is an annual plant, which grows about a foot high. The leaves reſemble thoſe of the Honewort. The flowers, which are of a greeniſh yellow colour, are collected in ſpherical heads at the extremity of the ſhoots; but there being little beauty in the plant, it is rarely cultivated, except in botanic gardens. It grows plentifully about Aix in Provence, as alſo in moſt of the iſlands of the Archipelago. It is annual, and periſhes ſoon after the ſeeds are ripe. The ſeeds of this plant ſhould be ſown in autumn ſoon after they are ripe, or if they are permitted to ſcatter the plants will come up, and require no other care but to clear them from weeds. When the ſeeds are ſown in the ſpring, they commonly remain in the ground a year before they grow, and ſometimes I have known them lie two or three years in the ground, ſo that if the plants do not come up the firſt year, they ſhould not be diſturbed.

LAGOPUS. See TRIFOLIUM.

LAMIUM. Tourn. Inſt. R. H. 183. tab. 89. Dead Nettle, or Archangel.

The CHARACTERS are,

The flower hath a permanent empalement, which is cut into five equal ſegments at the top, which end in beards. The flower is of the lip kind, with one petal, ſwollen at the chaps and compreſſed; the upper lip is arched, obtuſe, and entire; the under is heart-ſhaped, and indented at the end. It hath four awl-ſhaped ſtamina, two of which are longer than the other. It hath a four-cornered germen, which afterward becomes four three-cornered ſeeds, ſitting in the open empalement.

The SPECIES are,

1. LAMIUM (*Purpureum*) foliis cordatis obtuſis petiolatis. Hort. Cliff. 341. *Purple ſtinking Archangel, or Dead Nettle.*

2. LAMIUM (*Album*) foliis cordatis acuminatis ſerratis petiolatis. Hort. Cliff. 314. *White Archangel, or Dead Nettle.*

3. LAMIUM (*Garganicum*) foliis cordatis crenatis villoſis, labio floris ſuperiore crenato. *Hoary dead Nettle with a purpliſh flower, whoſe upper lip is crenated.*

4. LAMIUM (*Moſchatum*) foliis cordatis obtuſis glabris, floralibus ſeſſilibus, calycibus profundè inciſis. *Eaſtern dead Nettle, ſometimes ſweet ſcented, and ſometimes ſtinking, with a large flower.*

5. LAMIUM (*Meliſſifolium*) foliis cordatis nervoſis ſerratis, petiolis longioribus, caule erecto. *Mountain dead Nettle with a Balm leaf.*

There are other ſpecies of this genus, as alſo ſome varieties of it, but as they are many of them weeds, ſo there are few who care to admit them into their gardens.

The firſt ſort grows naturally in moſt parts of England, under hedges and by the ſide of highways; it is alſo a troubleſome weed in gardens, but as it ſtands in moſt of the Diſpenſaries as a medicinal plant, I have choſen to inſert it. This is an annual plant, whoſe ſtalks ſeldom riſe more than ſix inches high; the under leaves are heart-ſhaped, blunt, and ſtand upon pretty long foot-ſtalks, but the upper leaves ſit nearer to the ſtalks; the flowers come out in whorls on the upper part of the ſtalk; they are of a pale purple colour, and are ſucceeded by four naked ſeeds ſitting in the empalement; after the ſeeds are ripe the plant decays. It flowers from the middle of March, when the autumnal ſelf-ſown plants begin, which are ſucceeded by others, which continue in ſucceſſion all the ſummer.

The ſecond ſort is commonly called Archangel; this is alſo uſed in medicine, for which I have enumerated it here. The roots of this ſort are perennial, and creep much in the ground, ſo are difficult to extirpate, eſpecially where the happen to grow under buſhes or hedges, for the roots in termix with thoſe of the buſhes, and every ſmall piece o them will grow and ſpread. The ſtalks of this riſe much higher than thoſe of the laſt; the flowers are larger, white, and grow in whorls round the ſtalks; theſe continue in ſucceſſion moſt part of the ſummer.

The third ſort grows naturally upon the mountains in Italy; this hath a perennial creeping root, from which ariſe many thick ſquare ſtalks a foot high, garniſhed with heart-ſhaped leaves which are hairy, placed oppoſite, ſtanding upon pretty long foot-ſtalks; the flowers come out in whorls at the joints of the ſtalk; they are large, of a pale purpliſh colour, and continue in ſucceſſion moſt part of the ſummer; the flowers are ſucceeded by ſeeds, which ripen about ſix weeks after. This may be propagated by ſeeds, but as the roots ſpread greatly in the ground, ſo when once it is obtained, it will propagate faſt enough without care.

The fourth ſort grows naturally in the Archipelago; this is an annual plant, which, if permitted to ſcatter its ſeeds, the plants will come up in the autumn, and thrive better than when ſown by hand. The plants during the winter make a pretty appearance, for the leaves are marked with white ſpots, ſomewhat like thoſe of the autumnal Cyclamen; the ſtalks riſe eight or nine inches high, garniſhed with ſmooth heart-ſhaped leaves placed oppoſite; theſe in dry weather have a muſky ſcent, but in wet weather are fœtid; the flowers are white, ſtanding in whorls round the ſtalks. They appear in April, and the ſeeds ripen in June, then the plants decay; this requires no culture, but to keep the plants clear from weeds.

The fifth ſort grows naturally in Portugal; this hath a perennial root; the ſtalk riſes a foot and a half high; they are ſtrong, ſquare, and grow erect; the leaves are large, heart-ſhaped, and much veined; they are deeply ſawed on their edges, and are placed oppoſite. The flowers come out in whorls round the ſtalks at every joint; they are very large, of a deep purple colour; thoſe on the lower part of the ſtalks appear the beginning of May, which are ſucceeded by others above, ſo that there is a continuance of flowers almoſt two months on the ſame ſtalks. This plant very rarely produces good ſeeds in England, nor do the roots propagate very faſt, ſo that it is not common in England.

The beſt time to remove and part theſe roots is in October, but they muſt not be tranſplanted oftener than every third year if they are required to flower ſtrongly, for the great beauty of this plant conſiſts in the number of ſtalks, which are always proportional to the ſize of the roots; for ſmall ones will put one or two ſtalks only, whereas the large ones will have eight or ten. The roots are hardy, and will thrive beſt in a ſoft loamy ſoil.

LAMPSANA. See LAPSANA.

LAND. Its improvement.

1. By Incloſing.

Incloſing of Lands, and dividing the ſame into ſeveral fields, for paſture or tillage, is one of the principal ways of improvement; firſt, by aſcertaining to every man his juſt property, and thereby preventing an infinity of treſpaſſes and injuries, that Lands in common are ſubject unto, beſide the

the disadvantage of being obliged to keep the same seasons with the other people who have Land in the same field; so that the sowing, fallowing, and tilling the ground, must be equally performed by all the landholders; and when there happens a slothful negligent person, who has Land intermixed with others, it is one of the greatest nuisances imaginable. Secondly, where Land is properly inclosed, and the hedge-rows planted with timber-trees, &c. it preserves the Land warm, and defends and shelters it from the violent cold nipping winds, which, in severe winters, destroy much of the Corn, pulse, or whatever grows on the champaign grounds. And where it is laid down for pasture, it yields much more Grass than the open fields, and the Grass will begin to grow much sooner in the spring. The hedges and trees also afford shelter for the cattle from the cold winds in winter, and shade for them in the great heats of summer. These hedges also afford the diligent husbandman plenty of fuel, as also plough-boot, cart-boot, &c. And where they are carefully planted and preserved, furnish him with mast for his swine; or where the hedge-rows are planted with fruit-trees, there will be a supply of fruit for cyder, perry, &c. which in most parts of England are of no small advantage to the husbandman.

By this method of inclosing, there is also much more employment for the poor, and is therefore a good remedy against beggary; for in those open countries, where there are great downs, commons, heaths, and wastes, there is nothing but poverty and idleness to be seen amongst the generality of their inhabitants.

In inclosing of Land, regard should be had to the nature of the soil, and what it is intended for, because Corn-land should not be divided into small fields; for besides the loss of ground in hedges, &c. the Corn doth seldom thrive so well in small inclosures, as in more open Land, especially where the trees are large in the hedge-rows. The Grass also in pasture is not so sweet near hedges, or under the drip of trees, as in an open exposure; so that where the inclosures are made too small, or the Land over-planted with trees, the herbage will not be near so good, nor in so great plenty, as in larger fields; therefore, before a person begins to inclose, he should well consider how he may do it to the greatest advantage: as for instance, it is always necessary to have some smaller inclosures near the habitation, for the shelter of cattle, and the conveniency of shifting them from one field to another, as the season of the year may require; and hereby the habitation, barns, stables, and out-houses, will be better defended from strong winds, which often do great damage to those that are exposed to their fury. These small inclosures may be of several dimensions, some of them three, four, six, or eight acres in extent; but the larger divisions for Corn should not contain less than twenty or thirty acres or more, according to the size of the farm.

The usual method of inclosing Land is, with a ditch and bank set with Quick. But in marsh Land, where there is plenty of water, they content themselves with only a ditch, by the sides of which they usually plant Sallows or Poplars, which being quick of growth, in a few years afford shade to the cattle; and when they are lopped, produce a considerable profit to their owners. In some counties the divisions of their Lands is by dry walls made of flat stones, laid regularly one upon another, and laying the top course of stones in clay, to keep them together, the weight of which secures the under ones. But in some parts of Sussex and Hampshire they often lay the foundation of their banks with flat stones, which is of a considerable breadth at bottom; upon which they raise the bank of earth, and plant the hedge on the top, which in a few years makes a strong durable fence, especially if they are planted with Holly, as some of those in Sussex are.

I shall now mention the most proper plants for making of fences for the different soils and situations, so as to answer the expectation of the planter: and first, the white Thorn is esteemed the best for fencing, and will grow upon almost any soil and in any situation, but it succeeds best on a Hazle loam.

The next to the white is the black Thorn, which, though not so generally esteemed as the white, yet it will make an excellent fence, where proper care is taken in the planting and after-management of it; and the loppings of this hedge make much the best bushes, and are of longer duration for dead hedges than any other sort, and are very proper to mend gaps in fences. These hedges will be better, if the plants are raised from the stones of the fruit, which should be sown on the spot where they are to remain, than where the plants are taken from a nursery.

The Crab will also make a strong durable fence; this may be raised by sowing the kernels in the place where the hedge is designed, but then there should be great care taken of the plants while they are young, to keep them clear from weeds, as also to guard them from cattle. When these stocks have obtained strength, some of them may be grafted with Apples for cyder, where the fence is not exposed to a public road; but the grafts should not be nearer than thirty-five or forty feet, lest they spoil the hedge, by their heads overgrowing and dripping on it.

The Holly is also an excellent plant for ever-green hedges, and would claim the preference to either of the former, were it not for the slowness of its growth while young, and the difficulty of transplanting the plants when grown to a moderate size. This will grow best in cold stony lands, where, if once it takes well, the hedges may be rendered so close and thick, as to keep out all sorts of animals, and will grow to a great height, and is of long duration.

The Alder will also make a good hedge, when planted on a moist soil, or on the side of rivers, or large ditches; and will preserve the bank from being washed away, where there are running streams; for they spread pretty much at bottom, and send forth suckers from their roots in great plenty.

Of late years the Furz has been propagated for hedges in several parts of England, and indeed will make a good fence on poor, sandy, or gravelly soils, where few other plants will grow. The best method of raising these hedges is, to sow the seed about the latter end of March, or the beginning of April, in the place where the hedge is designed; for the plants will not bear to be transplanted, unless it be done while they are young, and then there is great hazard of their taking.

Elder is sometimes planted for hedges, being very quick of growth; so that if sticks or truncheons about four or five feet long, be thrust into a bank slopewise each way, so as to cross each other, and thereby form a sort of chequer work, it will make a fence for shelter in one year. But as this is a vigorous growing plant, it will never form a close fence; and the young shoots being very soft and pithy, are soon broken by cattle, or boys in their sport.

There are some other plants which have been recommended for fences, but those here enumerated are the only useful sorts for such purposes; wherefore I shall pass over the others, as not worthy of the care of the husbandman. And as to the farther directions for planting and preserving of hedges, with instructions for plashing or laying them, the reader is desired to t rn to the articles of FENCES and HEDGES, where there are particular directions for these works exhibited, which I shall not here repeat.

The draining of Land is also another great improvement to it; for though meadows and pastures, which are capable of of being overflowed, produce a greater quantity of herbage than dry Land, yet where the wet lies too long upon the ground, the Grafs will be four and extremely coarfe; and where there is not care taken in time to drain this Land, it will produce little Grafs, and foon be over-run with Rufhes and Flags, fo as to be of fmall value.

The beft method for draining of thefe Lands is, to cut feveral drains acrofs the Land, in thofe places where the water is fubject to lodge; and from thefe crofs drains to make a convenient number of other drains, to carry off the water to either ponds or rivers in the lower parts of the Land. Thefe drains need not be made very large, unlefs the ground be very low, and fo fituated as not to be near any river to which the water may be conveyed; in which cafe there fhould be large ditches dug at proper diftances, in the loweft part of the ground, to contain the water, and the earth which comes out of the ditches fhould be equally fpread on the Land, to raife the furface. But where the water can be conveniently carried off, the beft method is to make under-ground drains at proper diftances, which may empty themfelves into large ditches, which are defigned to carry off the water. Thefe fort of drains are the moft convenient, and as they are hid from the fight, do not incommode the Land, nor is there any ground loft where thefe are made.

The ufual method of making thefe drains is to dig trenches, and fill the bottoms with ftones, bricks, Rufhes, or bufhes, which are covered over with the earth, which was dug out of the trenches; but this is not the beft method, becaufe the water has not a free paffage through thefe drains, fo that whenever there is a flood, they are often ftopped by the foil which the water frequently brings down with it. The beft method I have yet found to make thefe drains is to dig trenches to a proper depth for carrying off the water, which for the principal drains fhould be three feet wide at their top, and floped down two feet in depth, where there fhould be a fmall bank or fhoulder left on each fide, upon which the crofs ftakes or bearers fhould be laid, and below thefe banks there fhould be an open drain left, at leaft one foot deep, and nine or ten inches wide, that there may be room for the water to pafs through: thefe larger drains fhould be at convenient diftances, and fmaller drains of about fix or feven inches wide, and the hollow under the bufhes eight or nine inches deep, fhould be cut acrofs the ground, which fhould difcharge the water into thefe larger drains. The number and fituation of them muft be in proportion to the wetnefs of the Land, and the depth of earth above the bufhes muft alfo be proportioned to the intended ufe of the Land; for if it is arable Land to be ploughed, it muft not be fhallower than a foot or fourteen inches, that there may be fufficient depth for the plough, without difturbing the bufhes, but for Pafture-land nine inches will be full enough; for when the bufhes lie too deep in ftrong Land, they will have little effect, the ground above will bind fo hard as to detain the wet on the furface. When the drains are dug, there fhould be prepared a quantity of good Brufh-wood; the larger fticks fhould be cut to lengths of about fixteen or eighteen inches, which fhould be laid acrofs upon the two fide-banks of the drain, at about four inches diftance; then cover thefe fticks with the fmaller Brufh-wood, Furz, Broom, Heath, or any other kind of Brufh, laying it lengthwife pretty clofe; on the top of thefe may be laid Rufhes, Flags, &c. and then the earth laid on to cover the whole. Thefe fort of drains will continue good for a great number of years, and are never liable to the inconveniencies of the other, for the water will find an eafy paffage through them under the bufhes; and where there is plenty of Brufh-wood, they are made at an eafy expence; but in places where wood is fcarce, it would be chargeable to make them: however, in this cafe, it would be a great advantage to thefe Lands to plant a fufficient number of cuttings of Willow, or the black Poplar, on fome of the moift places, which would furnifh Brufh-wood for thefe purpofes in four or five years; and as the expence of planting thefe cuttings is trifling, there cannot be a greater advantage to an eftate which wants draining than to practife this method, which is in every perfon's power, fince there is little expence attending it.

In countries where there is plenty of ftone, that is the beft material for making thefe under-ground drains; for when thefe are properly made, they will never want repairing.

The beft time of the year for making thefe drains is about Michaelmas, before the heavy rains of autumn fall, becaufe at this feafon the Land is ufually dry, fo that the drains may be dug to a proper depth; for when the ground is wet, it will be very difficult to dig to any depth, becaufe the water will drain in wherever there is an opening in the ground.

As the draining of cold wet Lands is a great improvement to them, fo the floating or watering of dry loofe Land is not a lefs advantage to them. This may be eafily effected where there are rivers, or refervoirs of water, which are fituated above the level of the ground defigned to be floated, by under-ground drains (made after the manner of thofe before directed for draining of Land) through which the water may be conveyed at proper feafons, and let out on the ground: in order to this, there muft be good fluices made at the heads of the drains, fo that the water may never get out, but as fuch times as is required; for if this be not taken care of, the water, inftead of improving the Land, will greatly damage it.

The time for drowning of Land is ufually from November to the end of April; but though this is the general practice, yet I cannot approve of it for many reafons. The firft is, that by the wet lying continually on the ground in winter, the roots of the finer fort of Grafs are rotted and deftroyed; and by letting on of the water, at the feafon when the feeds of Docks, and other bad weeds, which commonly grow by river fides, are falling, thefe feeds are carried upon the Land, where they remain and grow, and fill the ground with bad weeds, which is commonly the cafe with moft of the water meadows in England, the Grafs in general being deftroyed; fo that Rufhes, Docks, and other trumpery, make up the burden of thefe Lands; but if thefe meadows were judicioufly managed, and never floated till March or April, the quantity of fweet good Grafs would be thereby greatly increafed, and the beautiful verdure of the meadows preferved.

Another great improvement of Land is by burning of it, which for four, heathy, and rufhy Land, be it either hot or cold, wet or dry, is a very great improvement; fo that fuch Lands will, in two or three years after burning, yield more, exclufive of the charges, than the inheritance was worth before; but this is not to be practifed on rich fertile Land, for as the fire deftroys the acid juice, which occafions fterility in the poor Land, fo it will in like manner confume the good juices of the richer Land, and thereby impoverifh it, fo that it hath been with great reafon difufed in deep rich countries.

It is alfo a very great improvement where Land is overgrown with Broom, Furz, &c. to ftub them up by the roots; and when they are dry lay them on heaps, and cover them with the parings of the earth, and burn them, and fpread the afhes over the ground. By this method vaft tracts of Land, which at prefent produce little or nothing to their owners, might be made good at a fmall expence, fo as to become good eftates to the proprietors.

LANTANA. Lin. Gen. Plant. 683. American Viburnum.

The

The CHARACTERS are,

The empalement of the flower is cut into four segments. The flower is of an irregular shape, having a cylindrical tube spread open at the bottom, where it is divided into five parts. In the center of the flower is situated the germen, attended by four stamina, two being longer than the other. The germen afterward changes to a roundish fruit, opening into two cells, inclosing a roundish seed.

The SPECIES are,

1. LANTANA (*Aculeata*) foliis oppositis, caule aculeato ramoso, floribus capitato-umbellatis. *Prickly American Viburnum with broad Nettle leaves, and carmine flowers.*

2. LANTANA (*Inermis*) caule inermi, foliis lanceolatis dentatis alternis, floribus corymbosis. *Smooth Lantana with spear-shaped leaves placed alternate, and a smaller flower and fruit.*

3. LANTANA (*Lanuginosa*) caule ramoso lanuginoso, foliis orbiculatis crenatis, oppositis, floribus capitatis. *Lantana with a hairy branching stalk, round crenated leaves placed opposite, and flowers collected in heads.*

4. LANTANA (*Trifolia*) foliis ternis, spicis oblongis imbricatis. Lin. Sp. Plant. 626. *Lantana with leaves placed by threes round the stalks, and oblong imbricated spikes of flowers.*

5. LANTANA (*Urticæfolia*) caule aculeato, foliis oblongo-cordatis serratis oppositis, floribus corymbosis. *Lantana with a prickly stalk, oblong, heart-shaped, sawed leaves placed opposite, and flowers growing in a corymbus.*

6. LANTANA (*Camara*) caule inermi, foliis ovato-lanceolatis, serratis, rugosis, floribus capitatis lanuginosis. *Lantana with a smooth stalk, oval, spear-shaped, rough, sawed leaves, and flowers growing in woolly heads.*

7. LANTANA (*Bullata*) foliis oblongo-ovatis acuminatis serratis rugosis alternis, floribus capitatis. *Lantana with oblong, oval-pointed, sawed leaves, which are tough and placed alternate, and flowers growing in heads.*

8. LANTANA (*Alba*) caule inermi, foliis ovatis serratis, floribus capitatis alaribus sessilibus. *Lantana with a smooth stalk, oval sawed leaves, and flowers growing in heads proceeding from the wings of the leaves, sitting close to the stalks.*

9. LANTANA (*Africana*) foliis alternis sessilibus, floribus solitariis. Hort. Cliff. 320. *Lantana with alternate leaves sitting close to the stalks, and flowers growing singly; commonly called African Jasmine with an Ilex leaf.*

The first sort is pretty common in the English gardens. This grows naturally in Jamaica, and most of the other islands in the West-Indies, where it is called wild Sage. It rises with a woody stalk from four to six feet high, sending out many square branches, armed with sharp crooked spines. The leaves are hairy, spear-shaped, and placed opposite; toward the end of the branches the flowers come out from the wings of the stalks, two foot-stalks arising from the same joint, one on each side, and are terminated by roundish heads of flowers; those which are on the outside, and form the border, are first of a bright red, or scarlet colour; these change to a deep purple before they fall. Those flowers which are in the center are of a bright yellow, but after some time fade to an Orange colour. The flowers are succeeded by roundish berries, which, when ripe, turn black, having a pulpy covering over a single hard seed.

The second sort grows naturally in Jamaica. This rises with a slender, smooth, shrubby stalk about four feet high, dividing into many small branches, which are garnished with spear-shaped leaves indented on their edges, hoary on their under side, and stand alternate upon short foot-stalks. Toward the end of the branches the foot-stalks of the flowers arise alternately from the wings of the leaves, which support small heads of pale purple flowers, succeeded by small purple berries, each having one seed.

The third sort was sent me from La Vera Cruz, by the late Dr. Houstoun. This rises with a shrubby stalk about four feet high, dividing into several woolly branches. The leaves are oblong, and sawed on their edges, standing opposite. The foot-stalks of the flowers come out from the wings of the leaves, sustaining an oblong spike of purple flowers.

The fourth sort rises with a shrubby stalk about three feet high, covered with a gray woolly bark, garnished with oblong leaves, indented on their edges; they are placed opposite at bottom, but by threes on the upper part of the stalk. At the end of the branches arise the foot-stalks of the flowers, which sustain an oblong head of purple flowers, which come out of imbricated cups, and are succeeded by pretty large purple berries, containing one seed. This flowers at the same time with the former sorts, and is an annual plant.

The fifth sort was sent me from La Vera Cruz, by the late Dr. Houstoun. This rises with a prickly branching stalk four or five feet high, garnished with oblong heart-shaped leaves, which are sawed on their edges, ending in acute points. At the end of the branches the flowers come out in round bunches, standing upon slender foot-stalks. The flowers are yellow, and grow in looser bunches or heads than those of the former sorts.

The sixth sort rises with a smooth branching stalk five or six feet high, covered with a dark brown bark. The branches are more divided than those of most other sorts, and are much more ligneous. The leaves are deeply sawed on their edges, their upper surface very rough, and many of them closely set with white prominent spots as if studded; these are placed alternately on the branches. The flowers come out from the wings of the stalk, standing upon pretty long foot-stalks; they are white, and are collected in small woolly heads.

The seventh sort rises with a branching shrubby stalk about four feet high, covered with a dark brown bark, garnished with small, oblong, oval leaves, ending in acute points, standing alternately pretty close to the branches. The flowers come out at the end of the branches, upon short foot-stalks, in close small heads; these are white, and make but little appearance.

The eighth sort was sent me from Campeachy. This hath a slender shrubby stalk, which rises three or four feet high, dividing into many slender branches, garnished with small, oval, sawed leaves, placed opposite; from the wings of the stalk at every joint come out the flowers; they are small, white, and are collected in close heads.

These plants are all of them easily propagated by cuttings, except the fourth, which is an annual plant, so can only be propagated by seeds. They may also be propagated by seeds, which several of the sorts produce in England. These seeds should be sown in pots, and plunged into a hot-bed of tan; the reason for my advising them to be sown in pots, is, because the seeds frequently remain long in the ground before they vegetate; therefore if the plants should not come up the same year, the pots should be placed in a stove in the winter, and the following spring plunged into a new hot-bed, which will bring up the plants. When these are fit to remove, they should be each planted in a small pot, and plunged into another hot-bed, observing to shade them till they have taken new root; then they should have air admitted to them every day, in proportion to the warmth of the season, to prevent their being drawn up with weak stalks; afterward they must be treated in the same manner as other plants from the same country, till they have obtained strength; then they may be removed

moved into an airy glafs-cafe, or a dry ftove, where they may have a large fhare of air in warm weather, but protected from the cold. This is neceffary for the young plants, which fhould not the firft year be expofed to the open air, but afterward they may be placed abroad in the warmeft part of fummer, and in winter placed upon ftands in the dry ftove, where they will continue long in flower, and many of the forts will ripen their feeds; but in winter they fhould be fparingly watered, for much moifture will rot their roots.

If they are propagated by cuttings, the beft time for planting them is in July, after the plants have been expofed to the open air for about a month, by which time the fhoots will be hardened fo as to be out of danger of rotting with a little moifture. Thefe cuttings fhould be planted in fmall pots, and plunged into a moderate hot-bed, and fcreened from the violence of the fun in the middle of the day; in about fix weeks time thefe will be rooted, when they muft be hardened gradually to bear the open air, and afterward treated as the old plants.

The laft fort has been long in the Englifh gardens, and is commonly called the Ilex-leaved Jafmine. This fort rifes with a fhrubby ftalk five or fix feet high, fending out many irregular branches, clofely garnifhed with thin oval leaves ending in points, fawed on their edges, which embrace the branches with their bafe; and from the bofom of each leaf comes out one folitary white flower, which is cut at the top into five parts, and at firft fight has the appearance of a Jafmine flower, but when clofer viewed, the tube will be found curved in the fame manner with thofe which Dr. Linnæus titles ringent flowers. The flowers are not fucceeded by feeds in England, but the plants are eafily propagated by cuttings, which, if planted upon an old hot-bed any time in July, and covered with a bell or hand-glafs, and fhaded from the fun, will put out roots in a month or five weeks; then they may be planted in pots, and placed in the fhade till they have taken frefh root; after which they may be removed to a fheltered fituation, where they may remain till the frofts come on. This plant was brought from the Cape of Good Hope, fo is not very tender, therefore may be preferved in a good green-houfe in winter, but muft have a large fhare of air in mild weather, otherwife it is apt to grow mouldy. In the fummer it may be expofed in the open air, with other green-houfe plants in a fheltered fituation. There is a fucceffion of flowers upon the plants moft part of the year, and the leaves continuing green renders it worthy of a place in every collection of plants.

LAPATHUM. See Rumex.

LAPSANA. Lin. Gen. Plant. 823. Nipplewort.

The Characters are,

The flower is compofed of feveral hermaphrodite florets, included in one common imbricated empalement. The florets have one tubulous petal, ftretched out at the top in fhape of a tongue; thefe have each five fhort hairy ftamina. The germen is fituated at the bottom of the floret, which afterward becomes an oblong three cornered feed, fituated in the fcale of the empalement.

The Species are,

1. Lapsana (*Communis*) calycibus fructûs undulatis, pedunculis tenuibus ramofiffimis. Hort. Cliff. 384. *Common Nipplewort.*

2. Lapsana (*Rhagadiolus*) calycibus fructûs undique patentibus, radiis fubulatis, foliis lanceolatis indivifis. Hort. Upfal. 245. *This is the Rhagadiolus alter,* 511

3. Lapsana (*Stellata*) calycibus fructûs undique patentibus, radiis fubulatis, foliis lyratis. Hort. Upfal. 245. *Rhagadiolus with a Nipplewort leaf.*

4. Lapsana (*Zacintha*) calycibus fructûs torulofis depreffis obtufis feffilibus. Lin. Sp. Plant. 811. *Zacintha, or watered Cichory.*

The firft fort is a common weed, which grows by the fide of foot-paths and hedges in moft parts of England, fo is not permitted to have room in gardens.

The fecond and third forts grow naturally in Portugal. Thefe are annual plants, of no beauty or ufe, but are preferved in botanic gardens for the fake of variety. If the feeds of thefe are permitted to fcatter the plants will come up without trouble, and two or three of them will be enough to leave to keep the forts.

The fourth fort grows naturally in Italy. This is alfo an annual plant of neither ufe or beauty, but is like the others kept for variety. If the feeds of this fort fcatter in the autumn, the plants will come up better than if fown in the fpring. The plants require no culture, but will thrive like weeds.

LARIX. Tourn. Inft. R. H. 586. tab. 353. The Larch-tree.

The Characters are,

It hath male and female flowers, growing feparate on the fame tree. The male flowers are difpofed in a fcaly katkin. The female flowers are difpofed in a conical fhape, having no petals, but a fmall germen, which afterward becomes a nut with a membranaceous wing, inclofed in the fcales of the cones.

The Species are,

1. Larix (*Decidua*) foliis deciduis, conis ovatis obtufis. *Larch-tree with deciduous leaves, and oval obtufe cones.*

2. Larix (*Cedar*) foliis acutis perennantibus, conis obtufis. *Cedar of Libanus.*

The firft fort grows naturally upon the Alps and Apennines, and of late years has been very much propagated in England. This tree is of quick growth, the trunk will rife to the height of fifty feet or more; the branches are flender, their ends generally hang downward, and are garnifhed with long narrow leaves, which arife in clufters from one point, which fpread open above like the hairs of a painter's brufh; they are of a light green, and fall away in autumn like other deciduous trees. In the month of April the male flowers appear, which are difpofed in form of fmall cones; the female flowers are collected into oval obtufe cones, which on fome fpecies have bright purple tops, and in others they are white: thefe differences are accidental; the cones are about one inch long, obtufe at their points; the fcales are fmooth, and lie over each other; under each fcale there is generally lodged two feeds, which have wings.

There are two other varieties of this tree, one of which is a native of America, and the other of Siberia; thefe have different appearances from the common Larch-tree in their fhoots and leaves, but it is doubtful if they are fpecifically different. The American fort thrives pretty well in feveral gardens in England, but the latter requires a colder climate, for they are very apt to die in fummer here, efpecially if they are planted on a dry foil. This fort will often pufh out leaves by the end of February in mild feafons, and if there happens froft later in the year, the fhoots are frequently killed, and their leaves drop off, fo are fometimes naked till June, when the trees put out frefh fhoots. The cones of this fort, which have been brought to England, feem to be in general larger than thofe of the common kind.

The common Larch-tree is now very plenty in moft of the nurferies in England, and of late years there has been great numbers of the trees planted; but thofe which have been planted in the worft foil and in bad fituations have thriven the beft, for where trees of equal fize have been planted in good garden earth at the fame time, the others on the cold ftiff land have in twelve years been twice the height of thefe planted

planted in good ground; which is an encouragement to plant these trees, since they will thrive in the most exposed situations, provided they are planted in clumps near each other, and not single trees; nor should the plants which are planted in very open exposed places be taken from warm nurseries, but rather raised as near to the spot where they are to remain as possible; nor should the plants be more than three or four years growth, when planted where they are designed to grow large; for though trees of greater size will remove very well, and grow several years as well as if they had not been transplanted, yet after twenty or thirty years growth they will frequently fail, where the young planted trees have continued very vigorous.

These trees are raised from seeds, which most years ripen well in England. The cones should be gathered about the end of November, and kept in a dry place till the spring, when they should be spread on a cloth and exposed to the sun, or laid before a fire, which will cause the scales of the cones to open and emit their seeds. These seeds should be sown on a border exposed to the east, where the morning sun only comes on it; or if they are sown on a bed more exposed to the sun, they should be screened with mats in the middle of the day, for when the plants first appear above ground they are very impatient of heat; and when the bed is much exposed to the sun, the surface of the ground will dry so fast as to require to have water very often, which frequently rots the tender roots of the plants. These young plants should be constantly kept clean from weeds, and if they have made good progress they may be transplanted the following autumn, otherwise they may remain in the seed-bed another year, especially if the plants are not too close together. When they are transplanted it should be performed in autumn as soon as their leaves decay; they may be planted in beds at about six inches asunder each way, which will be distance enough for the growth of the plants the two following years, by which time they will be fit to transplant where they are to remain.

When the young trees are planted out for good, they need not be planted at more than eight or ten feet asunder, always planting them closer on exposed situations, than where they are more defended; after the trees are planted, they will require no other care but to keep them clean from weeds for three or four years till they have obtained strength, when they will over-top the weeds, and prevent their growth; the ground between these trees should not be dug, for that I have found has greatly stopped their growth.

The American, or black Larch, thrives pretty well upon moist land, but on dry ground will make but little progress. A few of these trees, by way of variety, may be allowed to have a place in every collection of trees designed for pleasure, but for profit the common Larch is to be preferred.

In Switzerland, where these trees abound, and they have a scarcity of other wood, they build most of their houses with it; and great part of their furniture is also made of the wood, some of which is white, and some red, but the latter is most esteemed. The redness of the wood is supposed to be from the age of the trees, and is not from any difference between them. They frequently cut out the boards into shingles of a foot square, with which they cover their houses, instead of tiles or other covering; these are at first very white, but after they have been two or three years exposed become as black as charcoal; and all the joints are stopped by the resin, which the sun draws out from the pores of the wood, which is hardened by the air, and becomes a smooth shining varnish, which renders the houses so covered impenetrable to either wind or rain; but as this is very combustible, so the magistrates have made an order of police, that the houses so covered should be built at a distance from each other.

In most countries where this wood is in plenty, it is preferred to all the kinds of Fir for every purpose; and in many places there are ships built of this wood, which they say are durable; therefore this may be a very proper tree for planting upon some of the cold barren hills in many parts of England, which at present produce nothing to their proprietors, and in one age may be large estates to their posterity, and a national advantage; which might be effected without a great expence, where the business is properly conducted.

From the Larch-tree is extracted the Venice turpentine, which the inhabitants of the valley of St. Martin, near Lucern, make a considerable merchandize of. They collect this by boring holes in the trunk of the trees, at about two or three feet from the ground, into which they fix narrow troughs about twenty inches long; the end of these are hollowed like a ladle, and in the middle is a small hole bored for the turpentine to run into a receiver, which is placed below it; as the turpentine runs from the trees, it passes along the sloping gutter or trough to the ladle, and from thence runs through the hole into the receiver. The people who gather this visit the trees morning and evening, from the end of May to September, to collect the turpentine from out of the receivers.

The second sort is the Cedar of Libanus, which is a tree of antiquity; and what is remarkable, this tree is not found as a native in any other part of the world, so far as hath come to our knowledge.

The cones of this tree are frequently brought from the Levant, which, if preserved entire, will preserve their seeds good for several years. The time of their ripening is commonly late in the spring, and so consequently are near one year old before we receive them; for which they are not the worse, but rather the better, the cones having discharged a great part of their resin by lying, and the seeds are much easier to get out of them than such as are fresh taken from the tree.

The best way to get the seeds out is, to split the cones, by driving a sharp piece of iron through the center lengthways, which will split the cone, then the seeds may be taken out with ease; these are fastened to a thin leafy substance called wings, like those of the Fir-tree: but before the seeds are taken out, it will be proper to put the cones in water for twenty four hours, which will render them easier to split, so that the seeds may be taken out with greater safety; for there will require care in the doing of it, otherwise many of the seeds will be spoiled, as they are very tender, and will bruise where there is any force employed to get them out.

These seeds should be sown in boxes or pots of light fresh earth, and treated as was directed for the FIRS (to which I refer the reader,) but only shall observe, that these require more shade in summer while young than the Firs.

When the plants come up they must be guarded from the birds, otherwise they will pick off their tops; they must also be constantly kept clean from weeds, and not placed under the drip of trees. The plants may remain in these boxes or pots in which they were sown till the following spring, but it will be proper to place them under a frame in winter, or cover them with mats, for while they are young they are in danger of losing their tops if they are pinched by frost. In the spring, before the plants begin to shoot, they should be carefully taken up and transplanted into beds at about four inches distance, closing the earth gently to their roots; these beds should be arched over with hoops, and covered with mats in the heat of the day, to shade the plants from the sun until they have taken new root; and if the nights prove frosty, it will be proper to keep the mats over them in the night, but in cloudy or moist

moist weather they must be always open. After the plants are well rooted, they will require no other care but to keep them clean from weeds, unless the season should prove very dry, in which case it will be proper to give them some water once a week; but it must be in small quantities, for too much wet is often very injurious to them, so that it will be better to screen them from the sun in hot weather, to prevent the earth from drying too fast, or cover the surface of the ground with Moss to keep it cool, than to water the plants often.

In these beds the plants may stand two years, then they should be either transplanted to the places where they are designed to remain, or to a nursery, where they may grow two years more; but the younger these plants are when they are planted out for good, the better the trees will thrive, and the longer they will continue.

When these plants begin to shoot strong, the leading shoot is very subject to incline to one side; therefore, in order to have them strait, their shoots must be supported with stakes, to keep the leaders strait, until they are grown to the height they are designed, otherwise their branches will extend on every side, and prevent their upright growth.

It is matter of surprize to me, that this tree hath not been more cultivated in England formerly, for till within a few years past there were but few here; for as it grows naturally upon the coldest parts of mount Libanus, where the snow continues most part of the year, so there can be no fear of its being hurt by frost in England. That these trees are of quick growth, is evident from four of them now growing in the physic garden at Chelsea, which (as I have been credibly informed) were planted there in the year 1683, at which time they were not above three feet high; two of which trees are at this time (viz. 1769) more than twelve feet in girt, at two feet above ground, and their branches extend more than thirty feet on every side their trunks.

The soil in which these trees are planted, is a lean hungry sand mixed with gravel, with about two feet surface. They stand at four corners of a pond, which is bricked up within two feet of their trunks, so that their roots have no room to spread on one side, and consequently are cramped in their growth; but whether their standing so near the water may not have been advantageous to them, I cannot say; but sure I am, if their roots had had full scope in the ground, they would have made a greater progress. I have also observed, that lopping or cutting of these trees is very injurious to them (more perhaps than to any other of the resinous trees) in retarding their growth; for two of the four trees above-mentioned, being unadvisedly planted near a green-house, when they began to grow large, had their branches lopped, to let the rays of the sun into the house, whereby they have been so much checked, as at present they are little more than half the size of the other two.

What we find mentioned in Scripture of the lofty Cedars, can be no ways applicable to the common growth of this tree; since, from the experience we have of those now growing in England, as also from the testimony of several travellers, who have visited those few remaining trees on mount Libanus, they are not inclined to grow very lofty, but on the contrary extend their branches very far; to which the allusion made by the Psalmist agrees very well, when he is describing the flourishing state of a people, and says, "They shall spread their branches like the Cedar-tree."

Rauwolf, in his Travels, says, there were not at that time (i. e. anno 1574) upon mount Libanus more than twenty-six trees remaining, twenty-four of which stood in a circle: and the other two, which stood at a small distance, had their branches almost consumed with age; nor could he find any younger trees coming up to succeed them, though he looked about diligently for some. These trees (he says) were growing at the foot of a small hill, on the top of the mountains, and amongst the snow These having very large branches, commonly bend the tree to one side, but are extended to a great length, and in so delicate and pleasant order, as if they were trimmed and made even with great diligence, by which they are easily distinguished at a great distance from Fir-trees. The leaves (continues he) are very like to those of the Larch-tree, growing close together in little bunches upon small brown shoots.

Maundrel, in his Travels, says, there were but sixteen large trees remaining when he visited the mountains, some of which were of a prodigious bulk, but that there were many more young trees of a smaller size; he measured one of the largest, and found it to be twelve yards six inches in girt, and yet found, and thirty-seven yards in the spread of its boughs. At about five or six yards from the ground it was divided into five limbs, each of which was equal to a great tree. What Maundrel hath related, was confirmed to me by a worthy gentleman of my acquaintance, who was there in the year 1720, with this difference only, viz. in the dimensions of the branches of the largest tree, which he assured me he measured, and found to be twenty-two yards diameter. Now, whether Mr. Maundrel meant thirty-seven yards in circumference of the spreading branches, or the diameter of them, cannot be determined by his words, yet either of them well agrees with my friend's account.

The wood of this famous tree is accounted proof against all putrefaction of animal bodies; the saw-dust of it is thought to be one of the secrets used by those mountebanks who pretend to have the embalming mystery. This wood is also said to yield an oil, which is famous for preserving books and writings; and the wood is thought by my Lord Bacon to continue above a thousand years sound. It is also recorded, that in the temple of Apollo at Utica, there was found timber of near two thousand years old. And the statue of the goddess in the famous Ephesian temple, as said to be of this material also, as was most of the timber work of that glorious structure.

This sort of timber is very dry and subject to split, nor does it well endure to be fastened with nails, from which it usually shrinks, therefore pins of the same wood are much preferable to any other.

LARKSPUR. See DELPHINIUM.

LASERPITIUM. Tourn. Inst. R. H. 324. tab. 172. Laserwort.

The CHARACTERS are,

It hath an umbellated flower, composed of many small umbels. The general umbel is uniform; the flowers have five equal heart-shaped petals, and five stamina. The roundish germen is situated under the flower, which afterward becomes an oblong fruit, with eight longitudinal wings or membranes, resembling the fliers of a water-mill.

The SPECIES are,

1. LASERPITIUM (*Commune*) foliolis oblongo-cordatis, inciso-serratis, umbellâ maximâ. *Laserwort with oblong heart-shaped lobes, which are cut like a saw, and a very large umbel.*

2. LASERPITIUM (*Latifolium*) foliolis cordatis inciso-serratis. Hort. Cliff. 96. *Laserwort with heart-shaped lobes cut like a saw.*

3. LASERPITIUM (*Paludapifolium*) foliis ovatis obtosis acutè serratis. *Laserwort with oval obtuse lobes, sharply sawed.*

4. LASERPITIUM (*Gallicum*) foliolis cuneiformibus furcatis. Lin. Sp. Plant. 248. *Laserwort with wedge-shaped forked lobes.*

5. LASERPITIUM (*Angustifolium*) foliolis lanceolatis integerrimis sessilibus. Hort. Cliff. 96. *Laserwort with spear-shaped entire leaves, sitting close to the branches.*

6. LASERPITIUM (*Selinoides*) foliolis trifidis acutis. *Laserwort with acute trifid lobes.*

7. LASERPITIUM (*Trilobum*) foliolis trifidis obtusis, umbellis partialibus contractis. *Laserwort with obtuse trifid leaves, and the small umbels contracted.*

8. LASERPITIUM (*Prutenicum*) foliolis lanceolatis integerrimis extimis coalitis. Hort. Cliff. 96. *Laserwort with entire spear-shaped lobes, whose outer ones coalesce.*

9. LASERPITIUM (*Siler*) foliolis ovato-lanceolatis integerrimis petiolatis. Hort. Cliff. 96. *Laserwort with oval spear-shaped leaves having foot-stalks.*

There are some other varieties, if not distinct species, of this plant; some of which have been put down as distinct species, which differ only in the colour of their flowers, therefore should not be regarded as such; but the number of species has been greatly lessened by some late writers, who have erred as much in lessening, as those before them had done in multiplying of the species.

The plants grow naturally in the south of France, in Italy, and Germany, and are preserved in botanic gardens for the sake of variety; but as they have no great beauty, so are seldom cultivated in other gardens; they require much room, for their roots extend far every way, and the leaves of many sorts will spread three feet when the plants are strong; their flower-stalks rise four or five feet high, and their umbels of flowers are very large; they have all of them perennial roots, but annual stalks. They flower in June, and the seeds ripen in September.

It is generally supposed that the Silphium of the ancients was procured from one species of this genus, but from which of them we are at present ignorant. All the species, if wounded, drop a very acrid juice, which turns to a resinous gummy substance, very acrimonious. This was externally applied by the ancients to take away black and blue spots that came by bruises and blows, as also to take away excresences; it was also by some of the ancients prescribed in internal medicines; but others have cautioned people not to make use of it this way, from the effects which they mention to have seen produced from the violence of its acrimony.

All these plants are extreme hardy, so will thrive in most soils and situations; they are propagated by seed, which if sown in autumn, the plants will come up the following spring, but when they are sown in the spring the seeds commonly remain in the ground a whole year. The plants should be transplanted the following autumn where they are designed to remain, for they send out long deep roots, which are frequently broken by transplanting if they are large; when the plants are removed they should be planted three feet asunder. The roots will continue many years, and require no other culture but to clear them from weeds, and to dig between the roots every spring.

LATHYRUS. Tourn. Inst. R. H. 394. tab. 216, 217. Chichling Vetch.

The CHARACTERS are,

The flower is of the butterfly kind. The standard is heart-shaped, and reflexed at the point. The wings are oblong and blunt; the keel is half round, and the size of the wings. It hath ten stamina, nine of them joined, and one separate. It hath an oblong, narrow, compressed germen, which afterward becomes a long compressed pod, ending in a point, having two valves filled with roundish seeds.

The SPECIES are,

1. LATHYRUS (*Sativus*) pedunculis unifloris, cirrhis diphyllis, leguminibus compressis dorso bimarginatis. Hort. Cliff. 367. *Chichling Vetch with one flower upon a foot-stalk, tendrils having two leaves, and oval compressed pods with two borders on their back part.*

2. LATHYRUS (*Cicera*) pedunculis unifloris, cirrhis diphyllis, leguminibus ovatis compressis, dorso canaliculatis. Lin. Sp. Plant. 730. *Cultivated Chichling Vetch with a purple flower.*

3. LATHYRUS (*Setifolius*) pedunculis unifloris, cirrhis diphyllis, leguminibus teretibus. *Chichling Vetch with one flower upon a foot-stalk, a two-leaved tendril, and a taper pod.*

4. LATHYRUS (*Parisiensis*) pedunculis unifloris, cirrhis tetraphyllis, stipulis dentatis. Flor. Leyd. Prod. 363. *Chichling Vetch with one flower upon a foot-stalk, a four-leaved tendril, and indented stipulæ.*

5. LATHYRUS (*Hispanicus*) pedunculis bifloris, cirrhis polyphyllis, foliolis alternis. Hort. Cliff. 368. *Spanish Chichling Vetch with a variable flower and jointed pod.*

6. LATHYRUS (*Odoratus*) pedunculis bifloris, cirrhis diphyllis, foliis ovato-oblongis, leguminibus hirsutis. Hort. Cliff. 368. *The sweet-scented Pea.*

7. LATHYRUS (*Hirsutus*) pedunculis bifloris, cirrhis diphyllis, foliolis lineari-lanceolatis, leguminibus hirsutis, seminibus scabris. Flor. Leyd. Prod. 363. *Narrow-leaved Chichling Vetch with a hairy pod.*

8. LATHYRUS (*Tingitanus*) pedunculis bifloris, cirrhis diphyllis, foliolis alternis lanceolatis. Flor. Leyd. Prod. 363. *Chichling Vetch of Tangier, with a bitter Vetch pod, and a large red flower.*

9. LATHYRUS (*Annuus*) pedunculis bifloris, cirrhis diphyllis, foliolis lineari-lanceolatis, internodiis membranaceis. *Yellow broad-leaved Chichling Vetch.*

10. LATHYRUS (*Tuberosus*) pedunculis multifloris, cirrhis diphyllis, foliolis ovalibus, internodiis nudis. Hort. Cliff. 367. *Creeping Field Chichling Vetch, with a tuberous root.*

11. LATHYRUS (*Pratensis*) pedunculis multifloris, cirrhis diphyllis, foliolis lanceolatis, cirrhis simplicissimis. Hort. Cliff. 367. *Yellow wild Chichling Vetch of the woods.*

12. LATHYRUS (*Heterophyllus*) pedunculis multifloris, cirrhis diphyllis tetraphyllisque, foliolis lanceolatis. It. W. Goth. 75. *Greater Chichling Vetch of Narbonne with narrow leaves.*

13. LATHYRUS (*Latifolius*) pedunculis multifloris, cirrhis diphyllis foliolis lanceolato-linearibus acuminatis, internodiis membranaceis. *Greater wild Chichling Vetch with a smaller flower, and longer pointed leaves.*

14. LATHYRUS (*Magniflorus*) pedunculis multifloris, cirrhis diphyllis, foliolis lanceolatis, internodiis membranaceis. Hort. Cliff. 367. *Broad-leaved Chichling Vetch; commonly called Everlasting Pea.*

15. LATHYRUS (*Pisiformis*) pedunculis multifloris, cirrhis diphyllis foliolis ovato-lanceolatis, internodiis membranaceis. *Smaller broad-leaved Chichling Vetch with a larger flower; or large, red, flowering Everlasting Pea.*

16. LATHYRUS (*Nissolia*) pedunculis unifloris, foliis simplicibus, stipulis subulatis. Lin. Sp. Pl. 729. *Nissolia, or crimson Grass Vetch.*

17. LATHYRUS (*Amphicarpos*) pedunculis unifloris calyce longioribus, cirrhis diphyllis simplicissimis. Lin. Sp. Pl. 729. *Chichling Vetch with single flowers upon a foot-stalk, which are longer than the empalement, and a two-leaved single tendril.*

18. LATHYRUS (*Aphaca*) pedunculis bifloris, foliis reniformibus simplicissimis subtus venosis. *Chichling Vetch with two flowers upon a foot-stalk, and kidney-shaped single leaves, which are veined on their under side.*

The first sort grows naturally in France, Spain, and Italy; it is an annual plant, with a climbing stalk about two feet high. The leaves come out at each joint alternate: they

they are compoſed of two long narrow lobes, with a tendril or claſper riſing between, which faſtens to any ſupport near. The flowers come ſingly upon foot ſtalks at each joint; they are blue, and ſhaped like thoſe of the Pea; theſe are ſucceeded by oval compreſſed pods, with a double membrane or wing, running longitudinally on the back. It is ſeldom cultivated, unleſs in botanic gardens for the ſake of variety.

The ſecond ſort is cultivated in ſome countries for the ſeeds, which are uſed for feeding of poultry; this grows wild in Italy and Spain. It does not riſe ſo high as the firſt ſort. The leaves are longer, the pods are near twice the length of thoſe, and are channelled on their backſide; this is cultivated in the ſame manner as Vetches or Tares.

The third ſort was ſent me from Verona, where it grows naturally; this is an annual plant, which ſeldom riſes more than ſix or eight inches high. The two lobes of the leaves are ſmall, and end with claſpers. The flowers are of a bright ſcarlet, and are ſucceeded by taper pods filled with roundiſh ſeeds.

The fourth ſort grows naturally in Bavaria; this is alſo an annual plant. The ſtalk riſes from two to three feet high, garniſhed with leaves at each joint, which are compoſed of four oval lobes, ending with claſpers. At the baſe of the foot-ſtalk are two ſmall appendages (called ſtipulæ) which are ſharply indented on their edges. The flowers are ſmall, blue, and ſit cloſe to the ſtalk, ſtanding ſingly; theſe are ſucceeded by compreſſed pods an inch long, containing three or four roundiſh ſeeds.

The fifth ſort grows naturally in Spain and Italy; it is an annual plant, with a climbing ſtalk which riſes three feet high, garniſhed with leaves compoſed of ſeveral ſpear-ſhaped lobes, placed alternate along the midrib, which is terminated by very long claſpers. The foot-ſtalks of the flowers are five or ſix inches long, upon which ſtand two flowers, one above the other, ſhaped like thoſe of the Pea. The ſtandard, which is large, is of a bright red colour, but the keel and wings are white. The flowers are ſucceeded by pretty long jointed pods, filled with roundiſh ſeeds.

The ſixth ſort is commonly known by the title of Sweet Pea; this grows naturally in Ceylon, but is hardy enough to thrive in the open air in England. It is an annual plant with a climbing ſtalk, which riſes from three to four feet high, garniſhed with leaves, compoſed of two large oval lobes, whoſe midrib is terminated by long claſpers. The foot-ſtalks which come out at the joints, ſuſtain the large flowers, with dark purple ſtandards, but the keel and wings of a light blue colour. The flowers have a ſtrong ſweet odour, and are ſucceeded by oblong inflated pods, which are hairy, containing four or five roundiſh ſeeds in each.

There are two other varieties of this, one of which has a Pink-coloured ſtandard with a white keel, and the wings of a pale bluſh colour; this is commonly called Painted Lady Pea. The flowers of the other are all white, which are the only differences between them.

The ſeventh ſort grows naturally in Eſſex. I have found it in places which were ſpread over with Brambles, near Hockerel; this hath a perennial root, ſending out three or four weak ſtalks, which are near two feet long, garniſhed with leaves compoſed of two oblong lobes, whoſe midrib is terminated by claſpers. The foot-ſtalks ſuſtain two purple flowers, which are ſucceeded by rough hairy pods, little more than one inch long, containing three or four roundiſh ſeeds.

The eighth ſort was originally brought from Tangier to England; this is an annual plant, whoſe ſtalk riſes four or five feet high, garniſhed with leaves compoſed of two oval veined lobes, whoſe midrib ends with claſpers. The foot-ſtalks are ſhort, and ſuſtain two large flowers with purple ſtandards, whoſe wings and keel are of a bright red; theſe are ſucceeded by long jointed pods, containing ſeveral roundiſh ſeeds. This is ſometimes titled by the gardeners Scarlet Lupine.

The ninth ſort is an annual plant, which grows naturally about Montpelier; I have alſo received the ſeeds from Siberia; this riſes with a climbing ſtalk five or ſix feet high, which has two membranes or wings running along from joint to joint. The leaves are compoſed of two long narrow lobes, whoſe midrib ends with claſpers. The flowers ſtand upon long foot-ſtalks, each ſuſtaining two pale yellow flowers, which are ſucceeded by long taper pods, containing ſeveral roundiſh ſeeds.

The tenth ſort grows naturally amongſt the Corn in the ſouth of France, and in Italy, but is cultivated in the Dutch gardens for the roots, which are there ſold in the markets, and are commonly eaten; this hath an irregular tuberous root, about as big as theſe of the Pignut, covered with a brown ſkin; theſe ſhoot up ſeveral weak trailing ſtalks garniſhed with leaves compoſed of two oval lobes, ending with claſpers. The foot-ſtalks of the flowers are weak, about three inches long, each ſuſtaining two deep red flowers, which are ſeldom ſucceeded by pods, but the roots increaſe plentifully in the ground. This ſort will grow in moſt ſoils, but will thrive beſt on light ground.

The eleventh ſort grows naturally on the banks and under thickets in moſt parts of England; this hath a perennial creeping root, whereby it propagates ſo faſt as to be a very troubleſome weed, ſo ſhould not be admitted into gardens.

The twelfth ſort grows naturally by the ſide of hedges, and in thickets in ſeveral parts of England; this hath a perennial creeping root, which ſends out many climbing ſtalks, which riſe five or ſix feet high, garniſhed with leaves, which have ſometimes two, and at others four long narrow lobes, terminated by claſpers. The foot ſtalks ſuſtain ſeveral ſmall flowers, with pale ſtandards and blue wings and keels; theſe are ſucceeded by long taper pods, containing ſeveral roundiſh ſeeds.

The thirteenth ſort I found growing naturally in a thicket near Wimbleton in Surry, the ſeeds of which I brought to the Chelſea garden, where the plants have flouriſhed many years, and continued their difference without variation. The ſtalks of this ſort riſe ſix or ſeven feet high, and have a membrane running along on each ſide between the joints. The leaves are compoſed of two narrow ſpear-ſhaped lobes, with long acute points. The foot-ſtalks are very long, and ſuſtain ſeveral ſmall Pea-bloſſom flowers, with pale purple ſtandards, but the keel and wings are of a deep blue colour; theſe are ſucceeded by long taper pods like thoſe of the former ſort.

The fourteenth ſort has been found growing naturally in ſome parts of England, but is frequently cultivated in gardens for ornament. It hath a perennial root, from which ariſe ſeveral thick climbing ſtalks from ſix to eight feet high, which have membranaceous wings on each ſide between the joints. The leaves are compoſed of two ſpear-ſhaped lobes, and the midrib is terminated by claſpers. The foot-ſtalks are eight or nine inches long, and ſuſtain ſeveral large red flowers, which are ſucceeded by long taper pods, containing ſeveral roundiſh ſeeds.

The fifteenth ſort differs from the laſt, in the ſtalks being much ſhorter and ſtronger. The leaves are broader, and of a deeper green. The flowers are larger, and of a brighter red colour, ſo make a much better appearance; theſe differences are laſting from ſeeds, for I have raiſed many plants

plants from feeds within thirty years paft, and have always found them to be the fame as the parent plant.

The fixteenth fort grows naturally in moift meadows in many parts of England; this rifes with an upright ftalk one foot high, garnifhed with long, narrow, fingle leaves, at each joint. The foot-ftalks come out from the joints toward the upper part of the ftalk; they are flender, fome having but one, and others have two bright red flowers on their tops.

The feventeenth fort grows naturally in Syria; this is an annual plant with a trailing ftalk, garnifhed with leaves compofed of two lobes, whofe midrib is terminated by a fingle tendril. The foot-ftalk fupports one flower of a pale purple colour, and when the flowers decay, the germen is thruft into the ground, where the pods are formed, and the feeds ripen.

The eighteenth fort was difcovered by the late Dr. Houftoun, growing naturally at La Vera Cruz; this is an annual plant, with a trailing ftalk a foot long, garnifhed with a fingle kidney-fhaped leaf at each joint. The flowers grow two together upon very fhort foot-ftalks; they are fmall, and of a deep yellow colour; thefe are fucceeded by fhort taper pods, including three or four fmall roundifh feeds.

This fort is tender, fo the feeds fhould be fown upon a hot-bed in the fpring, and when the plants are fit to remove they fhould be each planted into a fmall pot filled with light earth, and plunged into a tan-bed, where they fhould conftantly remain, treating them in the fame manner as other tender plants from warm countries; if they are brought forward in the fpring, they will flower in July, and their feeds will ripen in autumn.

The other forts are preferved in fome curious gardens for the variety of their flowers. Thefe may all of them be propagated by fowing their feeds either in fpring or autumn; but thofe which are fowed in autumn fhould have a light foil and a warm fituation, where the plants will abide the winter, and come to flower early the following fpring, and their feeds will ripen in July; but thofe which are fown in the fpring, fhould have an open expofure, and may be planted upon almoft any foil, for they are not very tender plants, nor do they require much culture: thefe forts fhould all of them be fown where they are defigned to remain, for they feldom fucceed when they are tranfplanted; fo that where they are fown for ornament, many of the forts which are annual, fhould be fix or eight feeds fown in a fmall patch, in different parts of the borders of the flower-garden, and when the plants come up, they fhould be carefully kept clear from weeds; but when they are grown two or three inches high, there fhould be fome fticks put down by them to fupport them, otherwife they will trail on the ground, and become unfightly; befides, they will trail on whatever plants grow near them.

LAVANDULA. Tourn. Inft. R. H. 198. tab. 93. Lavender.

The CHARACTERS are,

The flower is of the lip kind, with one petal, having a cylindrical tube fpreading above; the upper lip is bifid and open, the under lip is cut into three equal fegments. It hath four fhort ftamina, two of which are fhorter than the other. It hath a germen divided in four parts, which afterward turns to four oval feeds, fitting in the empalement.

The SPECIES are,

1. LAVANDULA (*Spicata*) foliis lanceolatis integris, fpicis nudis. Hort. Cliff. 303. *Broad-leaved Lavender.*

2. LAVANDULA (*Anguftifolia*) foliis lanceolato-linearibus, fpicis nudis. *Narrow-leaved Lavender.*

3. LAVANDULA (*Multifida*) foliis duplicato-pinnatifidis. Vir. Cliff. 56. *Cut-leaved Lavender.*

4. LAVANDULA (*Canarienfis*) foliis duplicato-pinnatifidis, hirfutis, fpicis fafciculatis. *Canary Lavender with a longer, narrower, and more elegant cut leaf.*

The firft fort is cultivated in feveral of the Englifh gardens, and has been generally known by the title of Spike, or Lavender Spike; the leaves of this fort are much fhorter, and broader than thofe of the common Lavender; the branches are fhorter, more compact, and fuller of leaves. This fort doth not often produce flowers, but when it does, the flower-ftalks are garnifhed with leaves very different from thofe on the other branches, approaching nearer to thofe of the common fort, but are broader; the ftalks grow taller, the fpikes of flowers are larger; the flowers are fmaller, and are in loofer fpikes.

The fecond fort is the common Lavender, which is fo well known as to require no defcription. Both thefe forts flower in July, at which time the fpikes of the fecond fort are gathered for ufe; there is a variety of this with white flowers.

Thefe are propagated by flips, the beft feafon for which is in March; they fhould be planted in a fhady fituation, or at leaft fhaded with mats until they have taken root; after which they may be expofed to the fun, and when they have obtained ftrength, may be removed to the places where they are defigned to remain. Thefe plants will abide the longeft in a dry, gravelly, or ftony foil, in which they will endure our fevereft winters; though they will grow much fafter in the fummer, if they are planted upon a rich, light, moift foil; but then they are generally deftroyed in winter, nor are the plants half fo ftrong fcented, or fit for medicinal ufes, as thofe which grow upon the moft barren rocky foil.

The third fort grows naturally in Andalufia; this is an annual plant, which rifes with an upright branching ftalk near three feet high; the ftalks are woolly, garnifhed with hoary leaves growing oppofite, which are cut into many divifions to the midrib; thefe fegments are again divided on their borders toward the top into three obtufe fegments, fo that they end in many points. The foot-ftalk of the flower is naked, having four corners or angles, and is terminated by a clofe fpike of flowers about one inch long; the fpike has the rows of flowers twifted fpirally: under this fpike there are commonly two fmall ones proceeding from the fide of the ftalk, at about an inch diftance from the middle fpike. There are two varieties of this, one with blue, and the other with white flowers.

This fort fhould be fown every fpring on borders or beds, and when the plants come up, they may be tranfplanted into other borders of the flower-garden, or into pots, where they are defigned to flower, and will require no farther care but to keep them clean from weeds. Thefe are pretty plants to place in large borders, amongft other plants, for variety, but they are never ufed in England. If the feeds of this fort are permitted to fcatter, the plants will come up the following fpring without care, and may be treated in the manner before directed.

The fourth fort grows naturally in the Canary Iflands. This rifes with an upright branching fquare ftalk four feet high, garnifhed with leaves which are longer, and cut into narrower fegments than thofe of the third fort. They are of a lighter green, and hairy; the naked flower-ftalk is alfo longer, and terminated with a clufter of fpikes of blue flowers. The flowers are fmaller than thofe of the common Lavender, but are of the fame fhape.

This fort being an annual plant, is tenderer than either of the former, fo the feeds muft be fown on a moderate hot-bed in the fpring; and when the plants are fit to remove, they fhould be each planted into a feparate fmall pot, and plunged into another hot-bed, to bring the plants forward;

forward; in the beginning of June they ſhould be inured to the open air, into which they ſhould be placed in a ſheltered ſituation toward the end of that month; in July the plants will flower, and if the autumn proves warm, the ſeeds will ripen in September; but when they do not perfect ſeeds in the open air, the plants may be removed into a glaſs-caſe where the ſeeds will ripen.

LAVATERA. Tourn. Act. Gal. 1706. tab. 3.

The CHARACTERS are,

The flower has a double permanent empalement; the outer is of one leaf, and trifid; the inner is of one leaf, and quinquefid. The flower hath five petals, which are joined at their baſe. It has many ſtamina, which are joined in a column below, but above are looſe. It has an orbicular germen, crowned by many briſtly ſtigmas. The empalement afterward becomes a fruit with ſeveral capſules, covered in front by a hollow ſhield, each capſule having one kidney-ſhaped ſeed.

The SPECIES are,

1. LAVATERA (*Althææfolia*) foliis infimis cordato-orbiculatis, caulinis trilobis acuminatis glabris, pedunculis unifloris, caule herbaceo. *Common Lavatera with the leaf and appearance of Marſhmallow.*

2. LAVATERA (*Africana*) foliis infimis cordato-angulatis, supernè ſagittatis, pedunculis unifloris, caule herbaceo hirſuto. *African Lavatera with a beautiful flower.*

3. LAVATERA (*Trimeſtris*) foliis glabris, caule ſcabro herbaceo, pedunculis unifloris, fructibus orbiculo tectis. Hort. Upſal. 203. *Mallow with a variable leaf.*

4. LAVATERA (*Thuringiaca*) caule herbaceo, fructibus denudatis; calycibus inciſis. Hort. Upſal. 203. *Lavatera with an herbaceous ſtalk, naked fruit, and a cut empalement.*

5. LAVATERA (*Veneta*) caule arboreo, foliis ſeptemangularibus tomentoſis plicatis, pedunculis confertis unifloris axillaribus. Hort. Upſal. 202. *Lavatera, or Tree Mallow, with a ſmall flower.*

6. LAVATERA (*Triloba*) caule fruticoſo, foliis ſubcordatis ſubtrilobis rotundatis crenatis ſtipulis cordatis, pedunculis unifloris. Lin. Sp. Plant. 691. *Shrubby Marſhmallow with a rounder hoary leaf.*

7. LAVATERA (*Olbia*) caule fruticoſo, foliis quinquelobo haſtatis. Hort. Upſal. 202. *Shrubby Marſhmallow with an acute leaf, and a ſmall flower.*

8. LAVATERA (*Hiſpanica*) caule fruticoſo, foliis orbiculatis crenatis tomentoſis, pedunculis confertis unifloris axillaribus. *Lavatera; or Spaniſh ſhrubby Marſhmallow with a rounder leaf.*

9. LAVATERA (*Bryonifolia*) caule fruticoſo, foliis quinquelobatis acutis crenatis tomentoſis, racemis terminalibus. *Shrubby Althæa with a Briony leaf.*

The firſt ſort grows naturally in Syria; it is an annual plant, with an erect, branching, herbaceous ſtalk, riſing two or three feet high; the under leaves are orbicularly heart-ſhaped, ſmooth, and ſtand upon long foot-ſtalks; the upper are divided into three acute lobes; the flowers come out upon long foot-ſtalks from the wings of the leaves; they are very large, and ſpread open like thoſe of the Marſhmallow, and are of a pale red or Roſe colour.

There is a variety of this with white flowers, which has accidentally riſen from ſeeds.

The ſecond ſort grows naturally at the Cape of Good Hope. This differs from the firſt in the ſhape of the leaves, the lower having angles, and the upper being arrow-pointed; the ſtalks are hairy, the flowers larger, and of a brighter red colour.

This ſort is annual, and flowers at the ſame time with the former, and the ſeeds are ripe in the autumn.

The third ſort grows naturally in Spain and Sicily; this is an annual plant, which riſes with ſlender herbaceous ſtalks three or four feet high, covered with a brown bark; the lower leaves are roundiſh, the upper are angular, and ſome are arrow-pointed. The flowers are not half ſo large as thoſe of either of the former, and of a pale red colour.

The fourth ſort hath a perennial root; it riſes four or five feet high, and is woolly, garniſhed with angular heart-ſhaped leaves, ſtanding upon long foot-ſtalks. The flowers come out from the wings of the ſtalk toward the top, ſitting cloſe to the ſtalks at every joint; they are of a purpliſh colour, and ſhaped like thoſe of the Marſhmallow, but are larger. It grows naturally in Auſtria and Bohemia.

The fifth ſort is commonly called Mallow-tree; this riſes with a very ſtrong thick ſtalk to the height of eight or ten feet, dividing into many branches, garniſhed with ſoft woolly leaves that are plaited, and the edges cut into ſeveral angles. The flowers are produced in cluſters at the wings of the leaves, each ſtanding upon a ſeparate foot-ſtalk; they are of a purple colour, and ſhaped like thoſe of the common Mallow, and are ſucceeded by ſeeds of the ſame form.

The ſixth ſort riſes with a ſhrubby ſtalk ſeven or eight feet high, ſending out ſeveral long branches, garniſhed with woolly leaves, differing greatly in ſize and ſhape, the lower being partly heart-ſhaped at their baſe, but divide into five roundiſh lobes; the upper, which are ſmall, have three lobes, which are indented on their edges. The flowers come out from the wings of the ſtalk, three or four at each joint, upon very ſhort foot-ſtalks; they are of a light purple colour, and ſhaped like thoſe of Marſhmallow.

The ſeventh ſort is a ſhrub which grows to the ſame ſize as the ſixth, and differs from it in the ſhape of the leaves, which are divided into three or five acute-pointed lobes; the flowers are ſmaller, but of the ſame ſhape and colour. This grows naturally in the ſouth of France.

The eighth ſort riſes with a ſhrubby ſtalk ſix or eight feet high, ſending out many branches, garniſhed with roundiſh, crenated, woolly leaves, ſtanding upon long foot-ſtalks; the foot-ſtalks of the flowers come out in cluſters from the wings of the leaves, each ſuſtaining one large pale blue flower, of the ſame ſhape with thoſe of the other ſpecies.

The ninth ſort riſes with a ſhrubby ſtalk ſix or ſeven feet high, ſending out ſeveral ſhrubby branches, garniſhed with woolly leaves, divided into five lobes, which end in acute points; the lower part of the branches are adorned with a ſingle flower at each joint, ſitting cloſe to the ſtalk, but the branches are terminated by looſe ſpikes of flowers, which are of a pale blue colour, and ſhaped like thoſe of the former.

The five laſt mentioned ſorts, though they have ſhrubby ſtalks, yet are but of ſhort duration here, ſeldom continuing longer than two years, unleſs when they happen to grow upon dry rubbiſh, where they make but little progreſs; their ſtalks and branches being firmer, are better able to reſiſt the cold; for when they are in good ground, they are very vigorous and full of ſap, ſo are killed by the froſt in common winters.

All the ſhrubby ſorts are eaſily propagated by ſeeds, which ſhould be ſown in the ſpring, upon a bed of light earth; and when the plants are about three or four inches high, they ſhould be tranſplanted to the places where they are deſigned to remain; for as they ſhoot out long fleſhy roots which have but few fibres, ſo they do not ſucceed well if they are tranſplanted after they are grown large. If the ſeeds of theſe plants are permitted to ſcatter on the ground, the plants will come up the following ſpring, and when they happen to fall into dry rubbiſh, and are permitted to grow therein, they will be ſhort, ſtrong, woody, and produce a greater number of flowers than thoſe plants which are more luxuriant. As theſe plants continue a long

time

time in flower, ſo a few plants of each ſort may be allowed a place in all gardens where there is room.

The three firſt ſorts are annual plants; the ſeaſon for ſowing their ſeeds is in March, in the places where they are deſigned to remain, which ſhould be in the middle of the borders in the flower-garden; for, if the ſoil be good, they will grow two or three feet high; when the plants are come up two inches high, they ſhould be thinned if they are too near; after which they will require no other care but to clear them from weeds, and to faſten them to ſtakes, to prevent their being injured by ſtrong winds.

The two firſt ſorts are very ornamental plants in a garden, when placed among other annuals, either in pots or borders.

The fourth ſort hath a perennial root, which abides ſeveral years, but the ſtalks decay in the autumn, and new ones ariſe in the ſpring. This is propagated by ſeeds, which ſhould be ſown upon a bed of light earth in the ſpring, and when the plants are fit to remove, they ſhould be either tranſplanted to the places where they are to remain, or into pots where they may ſtand to get more ſtrength, before they are planted in the full ground. After the plants are well rooted, they will require no other care but to keep them clear from weeds; and if the winter ſhould prove very ſevere, it will be proper to cover the ground about them with old tanners bark, to keep out the froſt; but they will endure the cold of our ordinary winters very well, and will produce their flowers and ripen their ſeeds annually.

LAUREOLA. See DAPHNE.

LAUROCERASUS. See PADUS.

LAURUS. Tourn. Inſt. R. H. 597. tab. 367. The Bay-tree.

It is male and female in different plants; the male flowers have no empalement. They have nine ſtamina which are ſhorter than the petal, ſtanding by threes, terminated by ſlender ſummits. The female flowers have no empalements; they have one petal, which is cut into ſix ſegments at the top. In the bottom is ſituated an oval germen. There are two globular glands, ſtanding upon very ſhort foot-ſtalks, fixed to the baſe of the petal. The germen afterward becomes an oval berry with one cell, incloſing one ſeed of the ſame form.

The SPECIES are,

1. LAURUS (*Nobilis*) foliis lanceolatis venoſis perennantibus, floribus quadrifidis dioeciis. Hort. Cliff. 105. *The common broad-leaved Bay.*

2. LAURUS (*Undulata*) foliis lanceolatis venoſis perennantibus, marginibus undatis. *Common Bay-tree with waved leaves.*

3. LAURUS (*Tenuifolia*) foliis lineari-lanceolatis venoſis perennantibus, floribus quinquefidis dioeciis. *Narrow-leaved Bay-tree.*

4. LAURUS (*Indica*) foliis lanceolatis perennantibus venoſis planis, ramulis tuberculatis cicatricibus, floribus racemoſis. Hort. Cliff. 154. *The Indian Bay.*

5. LAURUS (*Borbonia*) foliis lanceolatis perennantibus marginibus reflexis tranſverſe venoſis, floribus racemoſis. *Carolina Bay-tree with pointed leaves, and blue berries ſitting upon long red foot-ſtalks.*

6. LAURUS (*Benzoin*) foliis ovato-lanceolatis obtuſis integris annuis. *The American Benjamin-tree.*

7. LAURUS (*Saſſafras*) foliis integris trilobiſque. Hort. Cliff. 154. *The Saſſafras.*

8. LAURUS (*Camphora*) foliis trinerviis lanceolato-ovatis, nervis ſupra baſin unitis. Lin. Mat Med. 192. *The Camphire-tree.*

The firſt ſort is the broad-leaved Bay, which grows naturally in Aſia. This is not quite ſo hardy as the common Bay-tree, though it will live in the open air in England through our common winters, if it is planted in a warm ſituation; but ſevere froſts will kill it, therefore many people ſhelter the plants in green-houſes every winter.

The ſecond ſort is the common Bay; of this there are plants with plain leaves, and others which are waved on their edges, but they ſeem to be the ſame ſpecies, for the berries of one have produced a mixture of both ſorts; but this is undoubtedly a different ſpecies from the firſt and third ſorts.

The third ſort hath very long narrow leaves, which are not ſo thick as thoſe of the two former, and are of a lighter green; the branches are covered with a purpliſh bark, and the male flowers come out in ſmall cluſters from the wings of the leaves, ſitting cloſe to the branches. This ſort is too tender to thrive in the open air in England, ſo the plants are generally kept in pots or tubs, and houſed in winter.

The fourth ſort grows naturally at Madeira and the Canary Iſlands, from whence it was formerly brought to Portugal, where it has been propagated in ſo great plenty, as to appear now as if it was a native of that country. In the year 1620, this plant was raiſed in the Farneſian garden, from berries which were brought from India, and was ſuppoſed to be a baſtard ſort of Cinnamon. This grows to the height of thirty or forty feet in temperate countries, but it is too tender to thrive in the open air in England, ſo the plants are kept in pots or tubs, and removed into the green-houſe in winter.

The leaves of this ſort are larger than thoſe of the common Laurel; they are thick, ſmooth, and of a light green, the foot-ſtalks inclining to red: the branches are regularly diſpoſed on every ſide, and the male flowers are diſpoſed in long bunches; they are of a whitiſh green colour; the berries are much larger than thoſe of the other ſorts. It is called by ſome the Royal Bay, and by others the Portugal Bay.

The fifth ſort grows naturally in Carolina, where it is called Red Bay; it alſo is found in other parts of America, but not in ſo great plenty. In ſome ſituations near the ſea, this riſes with a large ſtrait trunk to a conſiderable height, but in the inland parts of the country do not grow ſo large. The wood of this tree is much eſteemed, being of a fine grain, ſo is of excellent uſe for cabinets, &c.

The leaves of this ſort are much longer than thoſe of the common Bay, and are a little woolly on their under ſide, and their edges a little reflexed; the veins run tranſverſly from the midrib to the ſides, and the male flowers come out in long bunches from the wings of the leaves. The female trees produce their flowers in looſe bunches, ſtanding upon pretty long foot-ſtalks which are red; theſe are ſucceeded by blue berries ſitting in red cups.

This ſort is alſo too tender to thrive in the open air in England, for although ſome plants have lived abroad in a mild winter, which were planted in a warm ſituation, yet the firſt ſharp winter has deſtroyed them; ſo that theſe plants muſt be kept in pots or tubs, and houſed in winter like the former.

Theſe five ſorts may be propagated by layers, and the common ſort is generally propagated by ſuckers; but thoſe plants never keep to one ſtem, but generally ſend out a great number of ſuckers from their roots, and form a thicket, but do not advance in height; therefore the beſt way to have good plants, is to raiſe them from the berries, for the plants which come from ſeeds always grow larger than the others, and do not put out ſuckers from their roots, ſo may be trained up with regular ſtems. The beſt way is to ſow the berries in pots, and plunge them into a moderate hot-bed, which will bring up the plants much ſooner than if they were ſown in the full ground, ſo they will have a longer time to get ſtrength before winter; but the plants muſt not be forced with heat, therefore they ſhould be inured to bear the open air the beginning of June, into which they ſhould be removed, where they may remain till autumn; then the pots ſhould be placed under a common frame, that the plants may be protected from hard froſt, but in mild weather may enjoy the free air; for while the plants are ſo young, they are in danger of ſuffering in hard froſt, even the common ſort of Bay. The ſpring following, thoſe plants which will not live in the open air, ſhould be each tranſplanted into ſeparate pots; but the common ſort may be planted in nurſery-beds ſix inches aſunder each way, where they may grow two years, by which time they will be fit to plant where they are deſigned to grow. The other ſorts, which require protection, ſhould be planted in pots, and placed in a ſheltered ſituation till autumn, when they ſhould be placed in the green-houſe.

The common Bay will make a variety in all ever-green plantations, and as it will grow under the ſhade of other trees, where they are not too cloſe, ſo it is very proper to plant in the borders of woods, where it will have a good effect in winter.

The ſixth ſort grows naturally in North America, where it riſes to the height of eight or ten feet, dividing into many branches, garniſhed with oval ſpear-ſhaped leaves, ſmooth on their upper ſurface, but with many tranſverſe veins on their under ſide; theſe leaves fall off in the autumn, like other deciduous trees. The flowers I have but once ſeen, thoſe were all male, and of a white herbaceous colour, but if I remember right they had but ſix ſtamina in each.

The Saſſafras-tree is alſo very common in moſt parts of North America, where it ſpreads greatly by its creeping roots, ſo as to fill the ground with ſuckers wherever they are permitted to grow, but in England this ſhrub is with difficulty propagated. In America it is only a ſhrub, ſeldom riſing more than eight or ten feet high; the branches are garniſhed with leaves of different ſhapes and ſizes, ſome of them are oval and entire, about four inches long and three broad; others are deeply divided into three lobes; theſe are ſix inches long, and as much in breadth from the extremity of the two outſide lobes; they are placed alternately upon pretty long foot-ſtalks, and are of a lucid green; the flowers appear in the ſpring juſt below the leaves, upon ſlender foot-ſtalks, each ſuſtaining three or four ſmall yellow flowers, which have five oval concave petals, and eight ſtamina in the male flowers, which are upon different plants from the female; theſe have an oval germen, that afterward becomes an oval berry, which, when ripe, is blue.

The Camphire-tree grows naturally in Japan, and in ſeveral parts of India, where it riſes to a tree of middling ſtature, dividing into many ſmall branches, garniſhed with oval ſpear-ſhaped leaves, ſmooth on their upper ſide, having three longitudinal veins which unite above the baſe; if theſe are bruiſed, they emit a ſtrong odour of Camphire, as alſo the branches when broken. Theſe have male and female flowers on different trees; I have only ſeen thoſe of the male, which has flowered plentifully in England; thoſe were ſmall, and compoſed of five concave yellow petals, very like thoſe of the Saſſafras-tree, which were produced by threes or fours upon each foot-ſtalk in like manner.

The Saſſafras-tree is commonly propagated by the berries, which are brought from America; but theſe berries generally lie in the ground a whole year, and ſometimes two or three years before they grow, when they are ſown in the ſpring; therefore the ſureſt method of obtaining the plants will be, to get the berries put into a tub of earth ſoon after they are ripe, and ſent over in the earth; and as ſoon as they arrive, to ſow the berries on a bed of light ground, putting them two inches in the earth; and if the ſpring ſhould prove dry, the bed muſt be frequently watered, and ſhaded from the great heat of the ſun in the middle of the day. With this management many of the plants will come up the firſt ſeaſon; but as a great many of the berries will lie in the ground till the next ſpring, ſo the bed ſhould not be diſturbed, but wait until the ſeaſon after, to ſee what will come up: the firſt winter after the plants come up, they ſhould be protected from the froſt, eſpecially in the autumn; for the firſt early froſt at that ſeaſon is apt to pinch the ſhoots of theſe plants, which are tender and full of ſap, and do them more injury than the ſevere froſt of the winter; for when the extreme part of the ſhoots are killed, it generally affects the whole plant.

When the plants have grown a year in the ſeed-bed, they may be tranſplanted into a nurſery, where they may ſtand one or two years to get ſtrength and may then be tranſplanted into the places where they are to remain for good.

There have been ſome of the plants propagated by layers, but theſe are commonly two, and ſometimes three years before they put out roots; and if they are not duly watered in dry weather, they rarely take root; ſo that it is uncertain whether one in three of theſe layers do ſucceed, which makes theſe plants very ſcarce in England at preſent.

The Benjamin-tree, as it is falſly called, may be propagated in the ſame manner as the Saſſafras, by ſowing of the berries: theſe generally lie long in the ground, ſo that unleſs they are brought over in earth, in the ſame way as before directed, they often fail, or at leaſt remain long in the ground; but this ſhrub is now frequently propagated by layers in England, which put out roots pretty freely, when the young ſhoots are choſen for to make layers.

The Camphire-tree is very near akin to the Cinnamon-tree, from which it differs in the leaves, thoſe of the Cinnamon-tree having three ribs running longitudinally from the foot-ſtalk to the point, which are remarkably large; whereas the ribs of the leaves of this tree are ſmall, and extend towards the ſides, and unite before they meet the foot-ſtalks; the leaves have a ſmooth ſhining ſurface: they have male and female flowers in different trees, ſo that there is a neceſſity for both ſexes to ſtand near each other, in order to have good ſeeds.

In Europe this tree is propagated by layers, which are two years, and ſometimes longer, before they take root, ſo that the plants are very ſcarce; and as all thoſe which I have ſeen flower are male trees, ſo there can be no hopes of procuring ſeeds from them here; but if the berries of this, and alſo of the Cinnamon-tree, were procured from the places of their growth, and planted in tubs of earth, as hath been directed for the Saſſafras-tree, there may be a number of theſe plants produced in England: and if they were ſent to the Britiſh colonies in America, they might be there cultivated, ſo as to become a public advantage; eſpecially the Cinnamon-tree, which will grow as well in ſome of our iſlands in the Weſt-Indies, as it doth in the native places of its growth; and in a few years the trees might be had in plenty, for they propagate eaſily by the berries. The Portugueze brought ſome of the Cinnamon-trees from the Eaſt-Indies, and planted them on the iſland of Princes, on the coaſt of Africa, where they now abound, having ſpread over a great part of the iſland; there is alſo one of the trees growing at the Madeiras, and I am credibly informed there are many trees in the Brazils.

The Camphire-tree does not require any artificial heat in winter, ſo that if they are placed in a dry green-houſe, they will thrive very well. During the winter ſeaſon they muſt be ſparingly watered, and in the ſummer they ſhould be placed in a warm ſituation, where they may be defended from ſtrong winds, and not too much expoſed to the direct rays of the ſun; and during this ſeaſon, they muſt be frequently refreſhed with water.

They

They may be propagated by cuttings, which ſhould be planted in pots, and plunged into a moderate hot-bed, covering them cloſe with a hand-glaſs, and ſhading them in the heat of the day.

LAURUS ALEXANDRIA. See RUSCUS.

LAURUS TINUS. See TINUS.

LAWSONIA. Lin. Gen. Pl. 443. Hanna. Ludw. 143.

The CHARACTERS are,

The flower is compoſed of four oval ſpear-ſhaped petals, which ſpread open, and eight ſlender ſtamina. It hath a roundiſh germen, which afterward becomes a globular capſule, ending in a point, having four cells, filled with angular ſeeds.

The SPECIES are,

1. LAWSONIA (*Inermis*) ramis inermibus. Flor. Zeyl. 134. *Broad-leaved Egyptian Privet, called Alhenna, or Henna, by the Arabians.*

2 LAWSONIA (*Spinoſa*) ramis ſpinoſis. Flor. Zeyl. 135. *Lawſonia with prickly branches.*

The firſt ſort grows naturally in India, Egypt, and other warm countries, where it riſes with a ſhrubby ſtalk eight or ten feet high. The branches come out oppoſite; theſe are ſlender, covered with a whitiſh yellow bark, and are garniſhed with oblong ſmall leaves, of a pale green, ending in acute points, placed oppoſite. The flowers are produced in looſe bunches at the end of the branches; they are of a gray or dirty white colour, and are compoſed of four ſmall petals, which turn backward at the top. The flowers are ſucceeded by roundiſh capſules with four cells, filled with angular ſeeds.

The leaves of this ſhrub are much uſed by the Egyptian women to colour their nails yellow, which they eſteem an ornament.

The ſecond ſort grows naturally in both Indies, from whence I have received ſpecimens of it.

This riſes with a woody trunk eighteen feet high or more. The wood is hard and cloſe, covered with a light gray bark. The branches come out alternate, and are garniſhed with oblong oval leaves, which ſtand without order; at the joints where the leaves are placed come out ſingle, ſtrong, ſharp thorns. The flowers are produced in looſe bunches from the ſide of the branches; they are of a pale yellow colour, and of a diſagreeable ſcent; they have four petals, which ſpread open; between each of theſe are ſituated two pretty ſtrong ſtamina, terminated by roundiſh ſummits. After the flowers are paſt, the germen becomes a roundiſh capſule with four cells, including many angular ſeeds.

Theſe plants are both propagated by ſeeds, which ſhould be ſown on a hot-bed early in the ſpring, that the plants may have time to get ſtrength before winter. When the plants are fit to remove, they ſhould be each planted in a ſmall pot, and plunged into a hot-bed of tanners bark, where they muſt be ſcreened from the ſun till they have taken new root, then their treatment ſhould be the ſame as that of the Coffee-tree, with this difference only, not to keep it too warm, nor give them too much water; but eſpecially in the winter, during which ſeaſon it ſhould be given to them very ſparingly, for by over-watering theſe plants I have known many of them deſtroyed; theſe plants are too tender to thrive in the open air in England, ſo they muſt be placed in a moderate ſtove in autumn, but in the warmeſt part of ſummer they may be placed in the open air in a ſheltered ſituation.

LAYERS. Many trees may be propagated by Layers, which do not produce ſeeds here, ſo are not eaſily increaſed by any other method, and a great number of plants are this way increaſed.

Laying of Trees.

The young branches of the former year's ſhoots of trees ſhould be choſen to make Layers, becauſe theſe are generally more inclined to put out roots than the older wood When theſe are produced near the ground, they may with greater facility be layed. Theſe ſhoots ſhould be diveſted of their ſide branches, if they have any; and thoſe ſorts of trees which put out roots with difficulty, ſhould have a ſlit made upward at a joint, in that part which is to lie in the ground; or a piece of wire twiſted cloſe round the branch at the ſame place, which will check the mounting ſap, and cauſe them to put out roots; the ground ſhould be well dug, and the clods broken; then the ſhoots ſhould be layed five or ſix inches under the ſurface, driving down a peg to each to prevent their riſing, leaving the end of the ſhoots five or ſix inches above ground in an erect poſition.

The ſeaſon for laying hardy trees, that ſhed their leaves, is in October, but for ſuch as are tender in March; but for Evergreens, July or Auguſt are good ſeaſons.

However, the ſummer is the beſt time for ſmall plants, becauſe ſuch plants being but ſhort lived, draw root the quicker.

If you would lay young trees from a high ſtandard, the boughs of which cannot be bent down to the ground, then you muſt make uſe of Oſier baſkets, boxes, or pots, filled with fine ſifted mould, mixed with a little rotten Willow duſt, which will keep moiſture to aſſiſt the Layer in taking root; this baſket, box, &c. muſt be ſet upon a poſt, or treſſel, &c. and the bough muſt be laid according to the former way of laying, covering the ſurface with Moſs to prevent the earth from drying.

The harder the wood of the tree is, the younger ſhould be the ſhoots, for ſuch will take root beſt; but if the wood be ſoft, the older boughs will take root the beſt, and are leſs liable to rot.

There are many kinds of trees and plants which will not put out roots from their woody branches, though laid down with the utmoſt care; yet if the young ſhoots of the ſame year are laid in July, they will often put out roots very freely, ſo that when any plants are found difficult to propaga by Layers in the common way, they ſhould be tried at this ſeaſon; but as theſe ſhoots will be ſoft and herbaceous, they muſt not have too much wet, for that will cauſe them to rot; therefore it will be a better method to cover the ſurface of the ground over the Layers with Moſs, which will prevent the ground from drying too faſt, ſo that a little water now and then will be ſufficient.

LEAVES. A Leaf is defined to be a part of a plant extended into length and breadth, in ſuch a manner as to have one ſide diſtinguiſhable from the other; they are properly the moſt extreme part of a branch, and the ornament of the twigs, and conſiſt of a very glutinous matter, being furniſhed every where with veins and nerves; one of their offices is, to ſubtilize and give more ſpirit to the abundance of nouriſhing ſap, and to convey it to the little buds.

If the ſurface of Leaves are altered, by reverſing the branches of trees on which they grow, the plants are ſtopped in their growth, until the foot ſtalks are turned, and the Leaves recover their former poſition. This ſhews how neceſſary it is to ſupport all thoſe weak ſhoots of plants, which are naturally diſpoſed for upright growth, which either twine about the neighbouring trees for ſupport, or that put out claſpers, by which they take hold of whatever trees or plants grow near them, and are thereby ſupported; and, on the contrary, how abſurd is that practice of tying up the ſhoots of thoſe plants which are naturally diſpoſed to trail upon the ground, for in both theſe caſes nature is reverſed, and conſequently the growth of both ſorts of plants is greatly retarded.

This is one of the great functions for which the Leaves of trees and plants are deſigned; but, beſides this, there are others of equal importance to the well-being of plants and fruits

fruits; the first is that of the foot-stalks and Leaves nourishing and preparing the buds of the future shoots, which are always formed at the base of these foot-stalks; and during the continuance of the Leaves in perfect health, these buds increase in their magnitude, and, in the deciduous trees, are brought to maturity before the foot-stalks separate from the buds in autumn; but if by accident the Leaves are blighted, or if the entire surface of the Leaves are cut off, and the foot-stalks are left remaining, the buds will decay for want of that proper nourishment which is conveyed to them from the Leaves; so that whenever trees are divested of their Leaves, or those Leaves are cut, or otherwise impaired, though it may in either case happen when the buds may be nearly formed, yet if it is before the foot-stalks separate naturally from the branches, the future shoots will be weakened in proportion to the time when this is done; therefore from all the experiments which have been made in order to know how serviceable the Leaves of trees and plants are to their well-being, it has been found, that where the plants have been divested of their Leaves, or their Leaves have been eaten or cut, during their growth, the plants have been remarkably weakened thereby. This should teach us not to pull or cut off the Leaves of trees or plants, on any account, while they retain their verdure, and are in health; and this shews how absurd that common practice is, of feeding down Wheat in the winter and spring with sheep; for by so doing, the stalks are rendered very weak, and the ears are in proportion shorter, nor are the grains of Corn so plump and well nourished, as that which is not fed down upon the same ground; this is a fact which I can assert from many years experience. It is very evident, that Grass which is often mowed, the blades will be rendered finer in proportion to the frequency of mowing, yet the species of Grass is the same with that on the richest pastures; so that although this may be a desirable thing for lawns, &c. in gardens, yet where regard is had to the produce, this should be avoided.

Another principal use of the Leaves is to throw off by transpiration, what is unnecessary for the growth of the plants, answering to the discharge made by sweat in animal bodies; for as plants receive and transpire much more, in equal time, than large animals, so it appears how necessary the Leaves are to preserve the plants in perfect health; for it has been found by the most exact calculations, made from repeated experiments, that a plant of the Sun-flower receives and perspires, in twenty-four hours, seventeen times more than a man.

I shall beg leave to mention a few, out of the many experiments which have been made by Monf. Bonnet, of Geneva, to prove that most Leaves imbibe the moisture of the air on their under surface, and not from their upper; they are as follow:

He gathered the Leaves of sixteen sorts of herbaceous plants when fully grown; of each he put several Leaves upon the surface of water in glass vases, some were posited with their upper surface, and others with their under surface upon the water; these were adjusted exactly to the surface of the water, with great care not to let any moisture reach their opposite surfaces, and the same care was taken to prevent their foot-stalks from receiving any moisture. The glasses in which these Leaves were thus placed, were kept in a closet, where the air was very temperate; and as the water in the glasses evaporated, there was from time to time a supply of fresh, which was added with a syringe, so that the Leaves were not disturbed. The Leaves were taken from the following plants; the Plantain, the Mullein, the Wake Robin, the great Mallow, the Nettle, the Marvel of Peru, the Kidney-bean, the Sun-flower, the Cabbage, the Balm, the Cockscomb, the purple leaved Amaranth, Spinach, and the smaller Mallow.

Six of these sorts he found continued green a long time, and these were with different surfaces upon the water; they were of the following sorts, the Wake Robin, the Kidney-bean, the Sun-flower, the Cabbage, the Spinach, and small Mallow; among the others the following sorts were found to draw the moisture better with their upper surface than their under, the Plantain, the Mullein, the great Mallow, the Nettle, the Cockscomb, and the purple Amaranth.

The Leaves of the Nettle, whose under surface was upon the water, were decayed in three weeks, whereas those whose upper surface was next the water continued two months.

The Leaves of Mullein, whose under surface was next the water, did not continue fresh more than five or six days, but those whose upper surface was next the water lasted five weeks.

The Leaves of the purple Amaranth, whose upper surface was next the water, continued fresh three months, whereas those whose under surface was next the water were decayed in a week.

The Leaves of the Marvel of Peru and the Balm, appeared to have the advantage, whose under surfaces were next the water.

The Leaves of Wake Robin and of the Cockscomb, whose foot-stalks only were put into the water, continued fresh a longer time, than those which were placed with either surface next the water.

The Leaves of the great Mallow, the Nettle, the Sun-flower, the Marvel of Peru, and Spinach, whose foot-stalks were plunged into the water, continued fresh a shorter time than those which had either of their surfaces next the water.

The Leaves of the Mullein, of Plantain, and Amaranth, which received the water at their foot-stalk, continued fresh much longer than those whose under surfaces was next the water.

It is not difficult to explain the reason of this fact, for the orifices of the sap-vessels in the foot stalks, are much larger than those of either surface, so that the moisture insinuates in greater quantities and with more ease, the first than by the second way.

After this the same gentleman made experiments on the Leaves of sixteen sorts of trees and shrubs of the following sorts, the Lilac, the Pear-tree, the Vine, the Aspen, the Laurel, the Cherry-tree, the Plumb-tree, the Horse Chestnut, the White Mulberry, the Lime-tree, the Poplar, the Apricot, the Walnut, the Filbert, the Oak, and the Creeper.

Among these species he found that the Lilac and the Aspen imbibed the moisture on their upper surface, equally with the under surface: but in all the other sorts, the under surface imbibed it in much greater quantities than the opposite. The difference was very remarkable in the Leaves of the White Mulberry, for those whose upper surface was laid upon the water faded in five days, whereas the other, whose under surface was next the water, preserved their verdure near six months.

The Vine, the Poplar, and Walnut-tree, are very remarkable instances, how little disposed the upper surface of the Leaves of ligneous plants are to imbibe the moisture; for those of these three sorts, whose upper surfaces were applied to the water, decayed almost as soon as those which had no nourishment.

In all the experiments made by this curious gentleman upon the various Leaves of trees and herbs, it is remarkable, that all those Leaves which imbibed their moisture by their upper surface, were such as had that surface covered either with hairs or down; and on the contrary, where the under surface was garnished with either hair or down, the moisture was imbibed by that surface. He likewise mentions many experiments made by himſelf, and alſo by Monſ. du Hamel de Monceau, of the Royal Academy of Sciences at Paris, in rubbing the Leaves over with varniſh, oil, wax, and honey, to ſee the effect of theſe upon various Leaves, ſome of which were rubbed over on both ſurfaces, others only upon one; ſome only a part of the ſurface, others the edges of the Leaves were rubbed over, and in ſome only the foot-ſtalks of the Leaves were rubbed with theſe. They likewiſe anointed the trunks of ſome trees and ſhrubs, and left the Leaves and branches in their natural ſtate.

The reſult of theſe experiments was, that where the Leaves were anointed on both ſurfaces with varniſh, they decayed preſently; and where they were anointed with the other things, in proportion as thoſe were moſt penetrating, ſo the Leaves continued a ſhorter time than the others; and where one ſurface only was anointed, they continued much longer than thoſe which were anointed on both; and where the pedicle only was anointed, they continued ſtill longer; but the anointing of the trunks made no ſenſible alteration, excepting in very hot weather; when they both imagine, that the anointing them was of ſervice, by hindering the too great tranſpiration which might weaken the trees; for they obſerved, that thoſe trees which were varniſhed, ſuffered leſs from the violent heat, than the trees which were left in their natural ſtate.

Monſ. Bonnet alſo obſerved, that thoſe Leaves which were varniſhed, that the tender parts of the Leaves were deſtroyed by it, and the tough fibres were only left remaining.

We may therefore reaſonably conclude, that one great uſe of Leaves is what has been long ſuſpected by many, viz. to perform, in ſome meaſure, the ſame office for the ſupport of the vegetable life, as the lungs of animals do for the ſupport of animal life; plants, very probably, drawing through their Leaves ſome part of their nouriſhment from the air.

LEDUM. Lin. Gen. Plant. 483. Marſh Ciſtus, or wild Roſemary.

The CHARACTERS are,

The empalement of the flower is indented in five parts. It hath five oval concave petals, and ten ſlender ſtamina, and a roundiſh germen, which afterward becomes a capſule with five cells, opening at the baſe in five parts, filled with ſmall, narrow, acute-pointed ſeeds.

We have but one SPECIES of this genus, viz.

LEDUM (*Paluſtre*) foliis linearibus ſubtus hirſutis, floribus corymboſis. Flor. Suec. 341. *Ledum with very narrow leaves, hairy on their under ſide, and flowers growing in a corymbus.*

This plant grows naturally upon Moſſes and bogs in many parts of Yorkſhire, Cheſhire, and Lancaſhire, where it riſes with a ſlender ſhrubby ſtalk about two feet high, dividing into many ſlender branches, garniſhed with narrow leaves not much unlike thoſe of Heath. The flowers are produced in ſmall cluſters at the end of the branches, which are ſhaped like thoſe of the Strawberry-tree, but ſpread open wider at the top. Theſe are of a reddiſh colour, and in the natural places of their growth are ſucceeded by ſeed-veſſels filled with ſmall ſeeds, which ripen in autumn.

It is with great difficulty this plant is kept in a garden, for as it naturally grows upon bogs, ſo unleſs the plants have ſome ſuch ſoil and a ſhady ſituation, they will not thrive. The plants muſt be procured from the places of their growth, and taken up with good roots, otherwiſe they will not live.

LEEKS. See PORRUM.

LEMON-TREE. See LIMON.

LENS. See ERVUM.

LENS PALUSTRIS. Duck Meat. This is a very common plant, growing upon ſtanding waters in moſt parts of England; where, if it is not diſturbed, it will ſoon cover the whole ſurface.

LENTISCUS. See PISTACIA.

LEONTICE. Lin. Gen. Plant. 381. Lion's Leaf.

The CHARACTERS are,

The flower has ſix acute petals, and ſix nectariums which are fixed by ſmall foot-ſtalks to the baſe of the petals. It has ſix ſtamina. In the center is placed an oval germen, which afterward becomes a globular, ſwollen, ſucculent berry, with one cell, incloſing two or three globular ſeeds.

The SPECIES are,

1. LEONTICE (*Chryſogonum*) foliis pinnatis, petiolo communi ſimplici. Hort. Cliff. 122. *Lion's Leaf with winged leaves, having one common ſingle foot-ſtalk.*

2. LEONTICE (*Leontopetalum*) foliis decompoſitis, petiolo communi trifido. Hort. Cliff. 122. *Lion's Leaf with decompounded leaves, and a common trifid foot-ſtalk.*

Theſe plants both grow naturally in the iſlands of the Archipelago, and alſo in the Corn-fields about Aleppo, where they flower ſoon after Chriſtmas. They have large tuberous roots, about the ſize of thoſe of Cyclamen, covered with a dark brown bark; the leaves ariſe upon ſlender foot-ſtalks immediately from their roots, which grow about ſix inches high; that of the firſt ſort is ſingle, having many ſmall folioli ranged along the midrib, but the foot-ſtalks of the ſecond ſort are branched into three ſmaller; upon each of theſe are ranged ſeveral folioli or ſmall leaves, in the ſame form as the winged leaves. The flowers ſit upon naked foot-ſtalks; thoſe of the firſt ſort ſuſtain many yellow flowers, but the flowers of the ſecond are ſmaller, and of a paler colour. Theſe in their native country appear ſoon after Chriſtmas, but in England they do not flower till the beginning of April, and are never ſucceeded by ſeeds here.

Both theſe plants are propagated by ſeeds, which require to be ſown ſoon after they are ripe, otherwiſe they ſeldom ſucceed; but as they are brought from diſtant countries, they ſhould be preſerved in ſand to be ſent to England. I received a few of theſe ſeeds from the Duke D'Ayen, which were ſent him from Aleppo put up in ſand, and theſe came up better than any of thoſe which came over dry; for of ſeveral parcels of theſe ſeeds which I have ſown of both kinds, I have not had more than two plants ariſe.

The plants are very difficult to preſerve in England, for the roots will not thrive in pots; and when they are planted in the full ground, the froſt frequently deſtroys them in winter, eſpecially while the roots are young. Of late years the winters have proved ſo very unfavourable, as to kill all the young roots which I had raiſed in the Chelſea garden; but before the ſevere froſt in 1740, I had ſome of the roots which were planted in a ſouth-weſt border that flowered ſeveral years, and without any ſhelter ſurvived the winters; but although I covered many of thoſe roots which I had lately raiſed, yet I could not preſerve them.

The leaves of theſe plants decay about Midſummer, and the roots remain in an unactive ſtate till the following ſpring, at which time the flowers and leaves come up nearly at the ſame time.

LEONTODON. Lin. Gen. Plant. 817. Dandelion.

There are four or five ſpecies of this genus which grow naturally in the fields, ſo are not cultivated in gardens; but ſome people in the ſpring gather the roots out of the fields, and blanch them in their gardens for a Sallet herb; however, as they are not cultivated, I ſhall forbear ſaying any thing more of them, than that they are very bad weeds both in gardens and fields, ſo ſhould be rooted out before their ſeeds are ripe, otherwiſe they will ſpread to a great diſtance,

distance, as they have down adhering to them, by which they are wafted about by the wind.

LEONTOPODIUM. See PLANTAGO.

LEONURUS. Tourn. Inst. R. H. 187. tab. 87. Lion's Tail

The CHARACTERS are,

The flowers have one petal of the lip kind; the upper lip is cylindrical, hairy, and entire; the lower lip is reflexed, and cut into three parts. It hath four stamina, two of which are shorter than the other. In the bottom of the tube are situated four germen, which afterward become four oblong angular seeds sitting in the empalement.

The SPECIES are,

1. LEONURUS (*Africana*) foliis lanceolatis, obtuse serratis. Hort. Cliff. 312. *Lion's Tail with spear-shaped leaves, which are bluntly sawed.*

2. LEONURUS (*Nepetæfolia*) foliis ovatis, calycibus decagonis, septem dentatis, inæqualibus. Hort. Cliff. 312. *Lion's Tail with oval leaves, an empalement having ten corners, and seven unequal indentures.*

The first sort is a native of Ethiopia. This rises with a shrubby stalk seven or eight feet high, sending out several four-cornered branches, garnished with oblong narrow leaves, acutely indented on their edges, hairy on their upper side, and veined on their under, standing opposite. The flowers are produced in whorls, each of the branches having two or three of these whorls toward their ends; they are of the lip kind, shaped somewhat like those of the dead Nettle, but are much longer and covered with short hairs: they are of a golden scarlet colour, so make a fine appearance. The flowers commonly appear in October and November, and sometimes continue till the middle of December, but are not succeeded by seeds here.

There is a variety of this with variegated leaves, which is by some admired; but as this seldom produces so large whorls of flowers as the plain sort, is not generally esteemed.

The second sort is a native of the Cape of Good Hope, from whence I have two or three times received the seeds.

This rises with a square shrubby stalk about three feet high, sending out several four-cornered branches, garnished with oval crenated leaves, rough on their under side like the dead Nettle, but veined on their under, placed opposite. The flowers come out in whorls round the branches in like manner as the former, but are not so long nor so deep coloured; they appear at the same season with the first, and continue as long in beauty.

Both these sorts are propagated by cuttings. If the cuttings are planted the beginning of July, after the plants have been exposed to the open air long enough to harden the shoots, they will take root very freely. They should be planted in a loamy border to an east aspect, and if they are covered closely with a bell or hand-glass to exclude the air, and shaded from the sun, it will forward their putting out roots; but when they begin to shoot the glasses should be raised to admit the free air, to prevent their drawing up weak, and by degrees they must be exposed to the open air. As soon as they have taken good root, they must be taken up, and each planted in a separate pot filled with soft loamy earth, and placed in the shade till they have taken new root; then they may be removed to a sheltered situation, where they may remain till October, when they must be removed into the green-house, and afterward treated as the Myrtle, and other hardy green-house plants, observing to water the first sort plentifully.

LEPIDIUM. Tourn. Inst. R. H. 215. tab. 133. Dittander, or Pepperwort.

The CHARACTERS are,

The flower has four oval petals placed in form of a cross, and six awl-shaped stamina, two of which are shorter than the other. In the center is situated a heart shaped germen, which afterward turns to a spear-shaped seed-vessel with two cells, divided by an intermediate partition, containing oblong seeds.

The SPECIES are,

1. LEPIDIUM (*Latifolium*) foliis ovato lanceolatis integris serratis. Hort. Cliff. 330. *Broad-leaved, or common Dittander.*

2. LEPIDIUM (*Arvense*) foliis lanceolatis amplexicaulibus dentatis. Hort. Cliff 331. *Dittander with spear-shaped indented leaves which embrace the stalks.*

3. LEPIDIUM (*Chalepense*) foliis lineari-lanceolatis subdentatis amplexicaulibus radice reptatrice. *Low Dittander of Aleppo with less hoary leaves.*

4. LEPIDIUM (*Iberis*) floribus diandris tetrapetalis, foliis inferioribus lanceolatis serratis, superioribus linearibus integerrimis. Flor. Leyd Prod. 334. *Dittander with flowers having four petals and two stamina, whose under leaves are spear-shaped and sawed, and the upper narrow and entire.*

5. LEPIDIUM (*Perfoliatum*) foliis caulinis pinnato-multifidis, ramiferis cordatis, amplexicaulibus integris. Hort. Cliff. 331. *The true Mithridate Mustard of Dioscorides.*

6. LEPIDIUM (*Virginicum*) floribus subtriandris tetrapetalis, foliis linearibus pinnatis. Lin. Gen Plant. 645. *Dittander with flowers having four petals, and sometimes three stamina, and very narrow winged leaves.*

7. LEPIDIUM (*Lyratum*) foliis lyratis crispis. Lin. Sp. Pl. 644. *Dittander with curled lyre-shaped leaves.*

The first sort grows naturally in moist places in many parts of England, so is seldom cultivated in gardens. It hath small, white, creeping roots, by which it multiplies very fast, so as to render it difficult to eradicate, after it has grown long in any place; the lower leaves are oval, spear-shaped, sawed on their edges, standing upon long foot-stalks. The stalks are smooth, rise two feet high, and send out many side branches; the leaves upon the stalks are longer, narrower, and more acute-pointed than the lower. The flowers grow in close bunches toward the top of the branches, coming out from the side; they are small, and composed of four small white petals. The seeds ripen in the autumn. The whole plant has a hot biting taste like Pepper; and the leaves have been often used by the country people to give a relish to their viands instead of Pepper, from whence it had the appellation of poor Man's Pepper.

This plant is easily propagated, for every piece of the root will grow and multiply wherever it is planted, so will become troublesome to root out after growing for some time in a garden.

The second sort grows naturally in Austria and Italy; this hath a fleshy fibrous root, from which arise several weak stalks about a foot and a half high, garnished with spear-shaped hoary leaves, deeply cut in upon the edges, which embrace the stalks with their base; the flowers are small, white, and grow in loose bunches at the end of the branches.

This is a perennial plant, which propagates very fast by its roots, and is seldom admitted into gardens.

The third sort grows naturally about Aleppo; this hath creeping roots, which extend to a great distance, so will soon spread over a large piece of ground. The leaves of this are longer and narrower than those of the former, and are less hoary; the flowers grow in loose bunches at the end of the branches; they are small, white, and like those of the first. This is a hardy perennial plant, which propagates by its creeping roots in as great plenty as either of the former.

The fourth sort grows naturally in the south of France, Italy, and Sicily. This hath a long fleshy root, which runs

 deep

deep into the ground, sending out many oblong sawed leaves, which spread on the ground; the stalks are slender, stiff, and rise about two feet high, garnished with very narrow entire leaves. The flowers come out in close clusters at the end of the branches; they are white, and appear in June and July, and the seeds ripen in autumn. If these seeds are permitted to scatter the plants will come up early in the spring, and require no other care but to keep them clean from weeds; the roots will abide several years if they are in a dry soil.

The fifth sort grows naturally in Persia and Syria; this is supposed to be the true Mithridate Mustard of Dioscorides. It is an annual plant, whose lower leaves are finely cut into many winged segments; the stalks rise a foot high, dividing at the top into slender branches, garnished with heart-shaped entire leaves, which embrace the stalks with their base. The flowers grow in long loose spikes from the end of the branches; they are small, yellow, and appear in June and July; the seeds ripen in September, soon after which the plant decays.

The seeds of this plant should be sown in the autumn, for those which are sown in the spring seldom flower the same year, and are often killed by the frost in winter; whereas those which are sown in the autumn, or the plants that rise from scattered seeds, will always flower about Midsummer, and their seeds will ripen the September following The plants require no other care but to thin them to a proper distance, and keep them clean from weeds.

The sixth sort is an annual plant, which grows naturally in Virginia, and in most of the islands of the West-Indies, where the inhabitants gather the leaves, and eat them in their sallets as we do the garden Cress.

The lower leaves of this sort are long and sawed on their edges, of a light green, with a biting taste like Cress. The stalk rises a foot and a half high, sending out a great number of small side branches, garnished with narrow leaves regularly sawed on their edges, so as to resemble winged leaves; these sit close to the branches. The flowers are produced at the end of the branches in loose spikes; they are small, white, and are succeeded by roundish, or heart shaped compressed seed vessels, which have a border round them This sort is easily propagated by seeds, which may be sown upon an open bed in April, where the plants are designed to remain, and when they come up they will require no other care but to thin them where they are too close, and keep them constantly clean from weeds; or if the seeds are permitted to scatter in the autumn, the plants will come up very well, and may be treated in the same way as the other.

The seventh sort grows naturally in Asia. This is a biennial plant; the lower leaves which spread on the ground are indented on both sides, and are in shape of a lyre; the stalks rise a foot high, and divide into a great number of slender branches, garnished with small oblong leaves, which are cut and a little curled on their edges; the stalks and leaves are of a gray colour, inclining toward hoariness The flowers are produced in clusters at the end of the branches; they are white, and are succeeded by roundish bordered seed vessels which are compressed, and have two cells each, containing two small oblong seeds, which are ripe in the autumn.

This sort may be propagated by seeds in the same manner as the former; or if the seeds are permitted to scatter in the autumn, the plants will come up without care, and should be treated in the same way as the former sort; but this does not flower till the second year, so the plants should be left farther asunder.

LEPIDOCARPODENDRON. See PROTEA.

LETTUCE. See LACTUCA.

LEUCANTHEMUM. See ANTHEMIS.

LEUCOJUM. Lin. Gen. Plant. 363. Snowdrop.

The CHARACTERS are,

It hath an obtuse compressed spatha. The flower is of the spreading bell-shape, cut into six parts. It hath six short bristly stamina. The roundish germen is situated under the flower, which afterward becomes a turbinated capsule with three cells, opening with three valves, filled with roundish seeds.

The SPECIES are,

1. LEUCOJUM (*Vernum*) spathâ uniflorâ, stylo clavato. Lin. Sp. Pl. 289. *Early great Snowdrop.*

2. LEUCOJUM (*Autumnale*) spathâ multiflorâ, stylo filiformi. Loefl. Lin. Sp. Pl. 289. *Snowdrop with many flowers in a sheath, and a thread-like style.*

The first sort grows naturally in Switzerland and Germany, as also upon the mountains near Turin. This hath an oblong bulbous root, shaped like that of the Daffodil, but smaller; the leaves are flat, of a deep green, four or five in number, considerably broader than those of the small Snowdrop; between these arise an angular stalk near a foot high, which is naked, hollow, and channelled; toward the top comes out a sheath which is whitish, opening on the side, out of which come one, or sometimes two white flowers, hanging upon slender foot stalks; these have but one petal, which is cut into six parts almost to the bottom. They are much larger than those of the small Snowdrop, and the ends of the segments of the petal are tipped with green, where they are of a thicker substance than in any other part. These flowers appear in March, soon after those of the small sort: they have an agreeable scent, not much unlike that of the flowers of Hawthorn; after the flower is past, the germen which is situated below the flower; swells to a Pear-shaped capsule with three cells, inclosing several oblong seeds.

The leaves of this sort decay toward the end of May, after which time the roots may be taken up and transplanted, for they should not be long kept out of the ground. It is propagated here by offsets, which the roots put out pretty plentifully when they are in a situation agreeable for them, and are not too often removed. They should have a soft, gentle, loamy soil, and an exposure to the east; the roots should be planted six inches asunder, and four or five inches deep, and must not be transplanted oftener than every third year.

The second sort is generally known by the title of late, or tall Snowdrop; this grows naturally in the meadows near Pisa in Italy, in Hungary, and also near Montpelier.

The root of this sort is near as large as that of the common Daffodil, and is very like it in shape; the leaves are also not unlike those of the Daffodil, and are more in number than those of the other sort; they are green, and keel-shaped at the bottom, where they fold over each other, and embrace the stalk, which rises a foot and a half high, and at the top is situated a spatha (or sheath) which opens on one side, and lets out three or four flowers, which hang downward, upon pretty long foot-stalks; these are cut into six oval concave segments almost to the bottom, and are of a clear white, with a large green tip to each segment, which is of a thicker consistence than any other part of the petal; within are situated six awl-shaped stamina, with oblong yellow summits, standing erect round a very slender style, crowned by an obtuse stigma. These flowers appear the latter end of April or beginning of May, and are succeeded by large triangular seed-vessels, having three cells, each containing two rows of seeds.

This sort is generally propagated in England by offsets, for the plants raised by seeds will not come to flower in less than four years; and as the roots put out offsets in plenty, so that is the more expeditious method. These roots may be be treated in the same way as the first sort, and should have a soft loamy soil, and be exposed only to the morning sun, where they will flower stronger and continue longer in beauty than when they are in an open situation, though they will thrive almost in any soil.

LEUCOJUM INCANUM. } See CHEIRANTHUS.
LEUCOJUM LUTEUM. }

LEUCOJUM BULBOSUM. See GALANTHUS.

LICHEN. Liverwort.

There being two sorts of this plant, which are sometimes used in medicine, and one of those being accounted a sovereign remedy for the bite of mad dogs, I thought it would not be improper to mention them here, though they are plants which cannot be propagated by any method, except by paring up the turf of Grass whereon they grow, and laying it down in some moist shady place, where, if the turf takes root and thrives, these plants will spread and do well.

The SPECIES are,

1. LICHEN (*Petræus*) petræus latifolius, sive hepatica fontana. C. B. P. *Common broad-leaved Liverwort.*

2. LICHEN (*Officinarum*) terrestris cinereus. Raii Syn. *Ash-coloured ground Liverwort.*

The first sort grows on the sides of wells, and in moist shady places, not only on the ground, but on stones, bricks, or wood. Of this there are several varieties, which are distinguished by the curious in botany; but as most of them are plants of no use, I shall not enumerate them.

The second sort (which is used to cure the bite of mad dogs) grows on commons, and open heaths, where the Grass is short, in most parts of England, especially on declivities, and on the sides of pits. This spreads on the surface of the ground, and, when in perfection, is of an Ash colour, but as it grows old it alters, and becomes a dark colour. This is often carried into gardens with the turf which is laid for walks and slopes, and where the soil is moist and cool it will spread, and be difficult to destroy, so that it renders the Grass unsightly; but this is the only method yet known to have it grow in gardens, where it is desired.

This is esteemed a sovereign remedy for the bite of mad dogs, and hath been for many years used with great success. It was communicated to the Royal Society by Mr. George Dampier, whose uncle had long used this plant to cure the bite of mad dogs, on men and animals, with infallible success. The method of taking it he has delivered as followeth: "Take of the herb, and dry it either in an oven "by the fire, or in the sun; then powder it, and pass it "through a fine sieve; mix this with an equal quantity of "fine powdered Pepper. The common dose of this mix-"ture is four scruples, which may be taken in warm milk, "beer, ale, or broth." He also advises, that the part bitten be well washed, as also the clothes of the person who is bit, lest any of the snivel or drivel of the mad dog should remain. If the person bitten be full grown, he advises, that he be blooded before the medicine is taken, and to use the remedy as soon after the bite as possible, as also to repeat the dose two or three several mornings fasting.

LIGUSTICUM. Tourn. Inst. R. H. 323. tab. 171. Lovage.

The CHARACTERS are,

It hath an umbellated flower. The general umbel is composed of several smaller. The general umbel has an involucrum of seven unequal leaves. The flower hath five equal petals, which are inflexed at their points. It hath five hairy stamina. The germen, which is situated under the flower, afterward turns to an oblong fruit, divided into two parts, which is angular and channelled, containing two oblong smooth seeds.

The SPECIES are,

1. LIGUSTICUM (*Levisticum*) foliis multiplicibus, foliolis supernè incisis. Hort. Cliff. 97. *Common Lovage.*

2. LIGUSTICUM (*Scoticum*) foliis biternatis. Lin. Sp. Pl. 250. *Scotch Lovage with a Smallage leaf.*

3. LIGUSTICUM (*Austriacum*) foliis bipinnatis, foliolis confluentibus incisis integerrimis. *Lovage with double-winged leaves, whose lobes run together, and have entire segments.*

4. LIGUSTICUM (*Lucidum*) foliis pinnatifidis, foliolis linearibus planis. *Lovage of the Pyrenees, with a shining Fennel leaf.*

5. LIGUSTICUM (*Peloponnasiacum*) foliis multiplicato pinnatis, foliolis pinnatim incisis. Lin. Syst. 258. *Broad-leaved, stinking, bastard Hemlock.*

The first sort is the common Lovage of the shops, which was formerly cultivated in the kitchen-gardens as an esculent herb, but has been long disused as such in England. It grows naturally upon the Apennines, and also near the river Liguria, not far from Genoa. It hath a strong perennial root, composed of many strong fleshy fibres, covered with a brown skin, of a strong aromatic smell and taste. The leaves are large, winged, and composed of many large lobes shaped like those of Smallage, but of a deeper green. The lobes toward the top are cut into acute segments. The stalks rise to the height of six or seven feet; they are large, channelled, and divide into several branches, each being terminated by a large umbel of yellow flowers, which are succeeded by oblong striated seeds.

This is easily propagated by seeds, which should be sown in autumn soon after they are ripe, for when they are kept out of the ground till spring they seldom grow the first year; when the plants come up, they may be transplanted into a moist rich border, at about three feet distance from each other, and after they have taken new root they will require no other care but to keep them clean from weeds. The roots will abide many years, and where the seeds are permitted to scatter the plants will come up without care.

The second sort grows naturally near the sea in many parts of Scotland; this hath a perennial root, but of much less size than the former; the leaves are composed of broader and shorter lobes, each leaf having two or three trifoliate leaves, whose lobes are indented on their edges. The stalk rises about a foot high, sustaining a small umbel of yellow flowers on the top, shaped like those of the former; these are succeeded by oblong channelled seeds, which ripen in autumn. This plant may be cultivated in the same manner as the former.

The third sort grows naturally on the Alps; this is a perennial plant. The stalks rise about two feet high, and at every joint are bent alternate, first on one side, then to the opposite; at each joint they are garnished with doubly-winged leaves, composed of small lobes, which run into each other, and just above each leaf comes out a side branch; these, as also the principal stalks, are terminated by umbels of white flowers, which are succeeded by oblong channelled seeds, which are ripe in autumn.

The fourth sort grows naturally on the Pyrenean mountains; this hath a biennial root. The leaves are doubly winged. The lobes are very narrow, and finely divided. The stalks are strong, and rise a foot and a half high, garnished with shining winged leaves, and are terminated by pretty large umbels of yellowish flowers.

The fifth sort grows naturally on the Peloponnesian mountains; this hath a very thick fleshy root like that of Parsnep. The leaves are large, composed of many winged leaves, whose lobes are cut into acute points; these are of a deep green, and, when bruised, emit a fœtid odour. The

stalks

ftalks rife four or five feet high; they are very large and hollow, like thofe of Hemlock, and fuftain at their top large umbels of yellowifh flowers, in fhape of a corymbus; thefe are fucceeded by oblong channelled feeds, which ripen in autumn.

This has by fome perfons been thought to be the Hemlock of the ancients, their conjectures being founded upon the plant, anfwering in many particulars the defcription of Cicuta, and alfo from the poifonous quality of it, together with its fœtid fcent; and as this grows naturally in many parts of Afia, fo they have been induced to believe it might be the fame plant.

LIGUSTRUM. Tourn. Inft. R. H. 596. tab. 367. Privet.

The CHARACTERS are,

The flower hath one funnel-fhaped petal, cut into four oval fegments at the top, which fpread open. It hath two ftamina which ftand oppofite, and a roundifh germen which afterward turns to a fmooth round berry with one cell, inclofing two oblong feeds, flat on one fide, but convex on the other.

The SPECIES are,

1. LIGUSTRUM (*Vulgare*) foliis lanceolato-ovatis obtufis. *The common Privet.*

2. LIGUSTRUM (*Italicum*) foliis lanceolatis acutis. *Privet with fpear-fhaped leaves; commonly called the Italian evergreen Privet.*

The firft fort grows common in the hedges in moft parts of England, where it rifes fifteen or fixteen feet high, with a woody ftem covered with a fmooth gray bark, fending out many lateral branches, garnifhed with fpear-fhaped leaves, ending with obtufe points, and are of a dark green. The flowers are white, and are produced in thick fpikes at the end of the branches, having a tubular petal cut at the top in four parts, which fpread open. Thefe are fucceeded by fmall, round, black berries, which ripen in the autumn. The leaves of this fort frequently remain green till after Chriftmas. There are two varieties of this, one with white, and the other hath yellow variegated leaves; but to preferve thefe varieties they fhould be planted in poor land, for if they are in a rich foil they will grow vigorous, and foon become plain.

The fecond fort grows naturally in Italy; this rifes with a ftronger ftem than the former, the branches are lefs pliable, and grow more erect; their bark is of a lighter colour, the leaves are much larger, and end in acute points, and are alfo of a brighter green; they continue in verdure till they are thruft off by the young leaves in the fpring The flowers of this are rather larger than thofe of the common fort, and are feldom fucceeded by berries in this country.

Both thefe forts are cultivated in the nurferies near London, to furnifh the fmall gardens and balconies in the city the firft being one of the few plants which will thrive in the fmoke of London; but although they will live fome years in the clofe part of the town, yet they feldom produce flowers after the firft year, unlefs it is in fome open places where there is free air.

The Italian Privet is now generally preferred to the common fort for planting in gardens, the leaves being larger, and continuing green all the year, renders it more valuable; and being fo hardy as to refift the greateft cold in this country, it may be planted in any fituation where the common fort will thrive. I have frequently planted it under the dripping of large trees, where I find it will thrive better than moft other fhrubs.

I cannot but think this fort, which is the moft common in Italy, is the Liguftrum mentioned by Virgil in the fecond Eclogue: and my reafon for it is, that as the flowers of this fhrub are of a pure white, but fall off very foon, they are by no means proper to gather for garlands, &c. and the berries being of a fine black colour, and continuing long upon the plants, make a fine appearance. To confirm that thefe berries were gathered for ufe, we find in feveral authors of undoubted credit, that they were ufed in dyeing, as alfo that the beft ink was made of thefe berries.

Befides, is it not much more reafonable to fuppofe, that Virgil would rather draw his comparifon from the flowers and fruit of the fame plant, when he is warning the youth not to truft to his beauty, than to mention two different plants, as has been generally fuppofed? for here are the white flowers of the Privet appearing early in the fpring, which is an allufion to youth; but thefe are of fhort duration, foon falling away; whereas the berries, which may be applied to mature age, are of long continuance, and are gathered for ufe.

Thefe plants are eafily propagated by laying down their tender fhoots in autumn, which in one year's time will be rooted enough to tranfplant; then they may be removed to the places where they are defigned to remain, or planted in a nurfery for two years, where they may be trained for the purpofes defigned.

They may alfo be propagated by cuttings, which if planted in the autumn on a fhady border and in a loamy foil, will take root very freely, and may be afterward treated in the fame way as the layers.

But the ftrongeft and beft plants are thofe which are raifed from feeds; indeed, this is a much more tedious method than the other, fo is feldom practifed, for the feeds generally lie a year in the ground before they vegetate; therefore whoever would propagate the plants in this method, fhould gather the berries and put them in a pot with fand between them, and bury the pots in the ground, as is practifed for Holly-berries and Haws; and after they have laid a year in the ground, take them up and fow them on a border expofed to the eaft, where the plants will come up the following fpring, and thefe will make great progrefs after they have gotten fome ftrength, and will grow upright, and not fend out fuckers like the other.

Formerly thefe plants were greatly in ufe for hedges, but fince fo many other plants of greater beauty have been introduced, they have been almoft rejected.

The two variegated kinds may be propagated by budding, or inarching them upon the plain fort, as alfo by laying down their branches; but as they feldom fhoot fo faft, as to produce many branches proper for layers, the other method is chiefly ufed. The filver ftriped fort is fomewhat tenderer than the plain.

LILAC. See SYRINGA.

LILIASTRUM. See HEMEROCALLIS.

LILIO-ASPHODELUS. See HEMEROCALLIS and CRINUM.

LILIO-FRITILLARIA. See FRITILLARIA.

LILIO-HYACINTHUS. See SCILLA.

LILIO-NARCISSUS. See AMARYLLIS.

LILIUM. Tourn. Inft. R. H. 369. tab 191. The Lily.

The CHARACTERS are,

The flower has no empalement; it hath fix petals reflexed at their points; each petal has a narrow longitudinal nectarium at their bafe. It hath fix ftamina which are erect, with a cylindrical oblong germen, which afterward becomes an oblong capfule, having three cells, which are filled with flat feeds lying above each other in a double order.

The SPECIES are,

1. LILIUM (*Candidum*) foliis fparfis, corollis campanulatis erectis, intus glabris. *Common white Lily with an erect flower.*

2. Lilium (*Peregrinum*) foliis ſparſis, corollis campanulatis cernuis, petalis baſi anguſtioribus. *White foreign Lily with hanging flowers.*

3. Lilium (*Bulbiferum*) foliis ſparſis, corollis campanulatis erectis, intus ſcabris. Hort. Cliff. 120. *Greater Lily with a purple Saffron-coloured flower; commonly called Orange Lily.*

4. Lilium (*Humile*) humile, foliis linearibus ſparſis, corollis campanulatis erectis, caule bulbifero. *Smaller bulb-bearing Lily; by ſome called the fiery Lily.*

5. Lilium (*Pomponium*) foliis ſparſis ſubulatis, floribus reflexis, corollis revolutis. Hort. Cliff. 120. *Narrow-leaved red Lily, or Martagon.*

6. Lilium (*Anguſtifolium*) foliis linearibus ſparſis, pedunculis longiſſimis. *Lily with ſhort graſſy leaves; commonly called Martagon of Pompony.*

7. Lilium (*Chalcedonicum*) foliis ſparſis lanceolatis, floribus reflexis, corollis revolutis. Hort. Cliff. 120. *Lily of Byzantium with a Carmine flower; commonly called the ſcarlet Martagon.*

8. Lilium (*Superbum*) foliis ſparſis lanceolatis, floribus pyramidatis reflexis, corollis revolutis. *The great yellow Martagon.*

9. Lilium (*Martagon*) foliis verticillatis, floribus reflexis, corollis revolutis. Hort. Cliff. 120. *Mountain Lily with reflexed flowers; commonly called purple Martagon.*

10. Lilium (*Hirſutum*) foliis verticillatis hirſutis, floribus reflexis corollis revolutis. *Another Lily with reflexed hairy flowers; commonly called the red Martagon.*

11. Lilium (*Canadenſe*) foliis verticillatis, floribus reflexis, corollis campanulatis. Lin. Sp. Plant. 303. *Martagon of Canada with ſpotted flowers.*

12. Lilium (*Kampſchatenſe*) foliis verticillatis, floribus erectis, corollis campanulatis. Amœn. Acad. 2. p. 348. *Lily with leaves growing in whorls, and an erect bell-ſhaped flower.*

13. Lilium (*Philadelphicum*) foliis verticillatis brevibus, corollis campanulatis, unguibus petalorum anguſtioribus, floribus erectis. Icon. tab. 165. *Lily with very ſhort leaves growing in whorls, and bell-ſhaped flowers whoſe petals are very narrow at their baſe.*

There is a greater variety of Martagons than are here mntioned, but as they are ſuppoſed to be only accidental ariſing from culture, ſo I thought it would be to little purpoſe to inſert them here, therefore I ſhall only give their common titles hereafter.

The common white Lily is ſo well known as to need no deſcription; this grows naturally in Paleſtine and Syria, but has been long cultivated in all the gardens of Europe. It is ſo hardy that no froſt ever injure the roots, and it propagates ſo faſt by offsets from the roots, that it becomes ſo common as to be little regarded, though there is great beauty in the flowers, which have an agreeable odour. Of this ſort there are the following varieties:

The white Lily ſtriped with purple.
The white Lily with variegated leaves.
The white Lily with double flowers.

Theſe are varieties which have accidentally ariſen from culture; the ſort with variegated flowers has been in England more than forty years, but is now very common in moſt of the gardens, and is by ſome perſons eſteemed for the variety of its purple ſtripes; but as the pure white of the flower is ſtained by the purple, ſo as to appear of a dull colour, many prefer the common white Lily to this.

The ſort with variegated leaves is chiefly valued for its appearance in winter and ſpring, for as the leaves come up early in the autumn, which ſpread themſelves on the ground, and being finely edged with broad yellow ſtripes, they make a pretty appearance during the winter and ſpring months. The flowers are the ſame as thoſe of the common ſort, but appear earlier in ſummer, which may be occaſioned by the roots being weaker than thoſe of the plain ſort, for all variegated plants are weaker than thoſe which are plain.

The white Lily with double flowers is leſs valuable than either of the other, becauſe their flowers rarely open well, unleſs they are covered with glaſſes to ſhelter them from the rain and dew, ſo they often rot without expanding. Theſe flowers have none of the agreeable odour which the ſingle ſort is valued for, even when they open the faireſt; for as by the multiplicity of petals in the flowers, the parts of generation are deſtroyed, ſo there is a want of the fecundating powder from whence the odour is ſent out.

The white Lily with dependent flowers was originally brought from Conſtantinople. This is by ſome ſuppoſed to be only a variety of the common ſort, but is undoubtedly a diſtinct ſpecies; the ſtalk is much ſlenderer than the common, the leaves are narrower and fewer in number; the flowers are not quite ſo large, and the petals are more contracted at their baſe; theſe always hang downward, whereas thoſe of the common ſort grow erect The ſtalks of this kind ſometimes are very broad and flat, and appear as if two or three were joined together; when this happens, they ſuſtain from ſixty to one hundred flowers, and ſometimes more; this has occaſioned many to think it a different ſort, who have mentioned this with broad ſtalks and many flowers as a diſtinct ſpecies, though it is accidental, for the ſame root ſcarce ever produces the ſame two years.

Theſe ſorts are eaſily propagated by offsets, which the roots ſend out in ſo great plenty as to make it neceſſary to take them off every other, or at moſt every third year, to prevent their weakening the principal roots. The time for removing of the roots is at the end of Auguſt, ſoon after the ſtalks decay, for if they are left longer in the ground they will ſoon put out new fibres and leaves, when it will be improper to remove them, becauſe that will prevent their flowering the following ſummer. They will thrive in almoſt any ſoil or ſituation, and as they grow tall and ſpread, ſo they muſt be allowed room; therefore in ſmall gardens they take up too much ſpace, but in large borders they are very ornamental.

The common Orange or red Lily is as well known in the Engliſh gardens as the white Lily, and has been as long cultivated here. This grows naturally in Auſtria, and ſome parts of Italy. It multiplies very faſt by offsets from the roots, and is now ſo common as to be almoſt rejected; however, in large gardens theſe ſhould not be wanting, for they make a good appearance when in flower, if they are properly diſpoſed. Of this ſort there are the following varieties:

The Orange Lily with double flowers.
The Orange Lily with variegated leaves.
The ſmaller Orange Lily.

Theſe varieties have been obtained by culture, and are preſerved in the gardens of floriſts. They all flower in June and July, and their ſtalks decay in September, when the roots may be tranſplanted, and their offsets taken off, which ſhould be done once in two or three years, otherwiſe their bunches will be too large, and the flower-ſtalks weak. This doth not put out new roots till toward ſpring, ſo that the roots may be tranſplanted any time after the ſtalks decay till November. It will thrive in any ſoil or ſituation, but will be ſtrongeſt in a ſoft gentle loam not too moiſt.

The bulb-bearing fiery Lily, ſeldom riſes much more than half the height of the former; the leaves are narrower, the flowers are ſmaller, and of a brighter flame colour; they

they are fewer in number, and ftand more erect. Thefe come out a month before the common fort, and the ftalks put out bulbs at every joint, which, if taken off when the ftalks decay, and planted, will produce plants, fo that it may be propagated in plenty. There are feveral varieties of this, which are mentioned as diftinct fpecies, but are fuppofed to have been produced by culture. Thefe are,

The greater, broad-leaved, bulb-bearing Lily.

The many-flowered bulb-bearing Lily.

The fmall bulb-bearing Lily.

The hoary bulb-bearing Lily.

All thefe forts of Lilies will thrive under the fhade of trees, fo may be introduced in plantations, and on the borders of woods, where they will have a good effect during the time they are in flower.

There is alfo a great variety of the Martagon Lily; thefe differ from the common Lilies, in having their petals reflexed backward in form of a Turk's turbant, from whence many give them the title of Turk's Cap. In the gardens of the florifts, particularly thofe in Holland, they make a great variety of thefe flowers, amounting to the number of thirty or upward; but in the Englifh gardens I have not obferved more than a third of that number, and moft of thefe are accidental, for thofe before enumerated are all that I think may be fuppofed fpecifically different. However, for the fake of fuch as are curious in collecting thefe forts of flowers, I fhall here mention thofe varieties which are to be found in the Englifh gardens.

The common Martagon with double flowers.

The white Martagon.

The double white Martagon.

The white fpotted Martagon.

The Imperial Martagon.

The early fcarlet Martagon.

The Conftantinople vermilion Martagon.

The common Martagon with red flowers, which is the fifth fort before enumerated, has very narrow leaves growing without order. The ftalks rife near three feet high, fuftaining at the top eight or ten bright red flowers, which ftand at a diftance from each other.

The fixth fort is called Martagon of Pompony. The ftalks of this rife higher than thofe of the former, the leaves are fhorter and fet clofer upon the ftalks; each of thefe ftalks fuftain from fifteen to thirty flowers, of a very bright red approaching to fcarlet. The foot-ftalks of the flowers are very long, fo that the head of flowers fpread out very wide; thefe hang downward, but their petals are reflexed quite back.

The feventh fort is commonly known by the title of fcarlet Martagon. This rifes with a ftalk from three to four feet high; the leaves are much broader than thofe of the former forts, and appear as if they were edged with white; they are placed very clofe upon the ftalks, but without any order. The flowers are produced at the top of the ftalk; they are of a bright fcarlet, and are feldom more than five or fix in number. This flowers late in July, and in cool feafons will continue in beauty great part of Auguft.

The eighth fort rifes with a ftrong ftalk from four to five feet high, garnifhed with leaves as broad as thofe of the laft mentioned, which ftand without order; the flowers are produced in form of a pyramid, on the upper part of the ftalk. When the roots of this kind are ftrong, they produce forty or fifty flowers upon each ftalk; they are large, of a yellow colour, fpotted with dark fpots, fo make a fine appearance; but the flowers have fuch a difagreeable ftrong fcent, that few perfons can endure to be near them, which has occafioned their being thrown out of moft Englifh gardens.

The ninth fort is frequently called the purple Martagon, though in moft of the old gardens it is known fimply by the title of Turk's Cap. This rifes with a ftrong ftalk from three to four feet high, garnifhed by pretty broad leaves, which ftand in whorls round the ftalk at certain diftances. The flowers are of a dark purple colour, with fome fpots of black; they are produced in loofe fpikes on the top of the ftalks.

The tenth fost is very like the former, but the leaves are narrower; the whorls ftand farther afunder; the leaves and ftalks are fomewhat hairy, and the buds of the flowers are covered with a foft down; the flowers are of a bright colour with few fpots, and come out earlier in the fummer, though the ftalks appear much later above ground.

The eleventh fort is commonly called the Canada Martagon, as it was firft brought to Europe from thence, but it grows naturally in moft parts of North America. The ftalks rife from four to five feet high, garnifhed with oblong pointed leaves, placed in whorls round the ftalk. The flowers are of a yellow colour, fpotted with black, which are fhaped like thofe of the Orange Lily; the petals are not turned backward, like thofe of the other forts of Martagon.

The twelfth fort grows naturally in North America, and is alfo mentioned to grow at Kampfchatfki. The flowers are fhaped like thofe of the Canada Martagon, growing erect, but the petals of this are oval, not narrowed at their bafe as are thofe, and fit clofe to the foot-ftalk; the flowers are of a deeper colour, and not fo much fpotted as thofe of the other fort.

The thirteenth fort grows naturally near Penfylvania. The root of this is fmaller than thofe of the other forts; in the fpring it fends out one upright ftalk a foot and a half high; the leaves come out in whorls round the ftalks; they are fhort, and have obtufe points. The ftalk is terminated by two flowers, which ftand erect upon fhort feparate foot-ftalks; they are fhaped like the flowers of the bulb-bearing fiery Lily, but the petals are narrower at their bafe. The flowers are of a bright purple colour, marked with feveral dark fpots toward their bafe.

All the forts of Lilies may be propagated by offsets from the roots, which fome of the forts produce in plenty; but there are others which fend out very few, which occafions their prefent fcarcity. The roots of all the forts of Martagon may be fafely taken up when their ftalks decay, and if there is a neceffity for keeping the roots out of the ground, if they are wrapped in dry Mofs, they will keep perfectly well for two months; fo that if their roots are to be tranfported to a diftant place, this precaution of wrapping them up is neceffary; but where they are to be planted in the fame garden there will be no occafion for this, efpecially if they are not kept too long out of the ground; for if the place is ready to receive the roots, they fhould be planted the beginning of October; fo if the roots are put in a dry cool place, they will keep very good without any farther care; but if the ground is not ready to receive them till later in the year, then it will be proper to cover the roots with dry fand, or wrap them in Mofs to exclude the air, which, if they are much expofed to, will caufe their fcales to fhrink, which weakens the roots, and is fometimes the occafion of their rotting.

Thefe roots fhould be planted five or fix inches deep in the ground, efpecially if the foil is light and dry; but where the ground is moift, it will be proper to raife the borders in which thefe are to be planted, five or fix inches above the level of the furface of the ground; for if the water rifes fo high in winter as to come near the roots, it will caufe them to rot; and where the foil is naturally ftiff and fubject to bind, there fhould be a good quantity of fea coal afhes or rough fand, well mixed in the border, to feparate the parts, and prevent the ground from binding in the ſpring, otherwiſe the roots will not ſend up very ſtrong ſtalks, nor will they make ſo good increaſe.

As the Canada, Pompony, and the laſt ſort of Martagons, are ſomewhat tenderer than the others, ſo if in very ſevere winters the ſurface of the ground over them is covered with old tanners bark or ſea-coal aſhes, it will be a good way to ſecure them from being injured by the froſt; and in the ſpring the covering may be removed, before the roots ſhoot up their ſtalks.

The roots of all kinds of Martagons muſt never be tranſplanted after they have made ſhoots, for that will ſo much weaken them (if it does not entirely kill them) as not to be recovered in leſs than two or three years, as I have experienced to my coſt; for being obliged to remove a fine collection of theſe roots early in the ſpring, I loſt a great part of them, and the others were long recovering their ſtrength.

All the ſorts of Lilies and Martagons may alſo be propagated by ſowing their ſeeds, by which method ſome new varieties may be obtained, provided the ſeeds are ſaved from the beſt ſorts, eſpecially the Martagons, which are more inclinable to vary than the other Lilies. The manner of ſowing them being the ſame as for TULIPS, the reader is deſired to turn to that article for directions.

LILIUM CONVALLIUM. See CONVALLARIA.

LILIUM PERSICUM. See FRITILLARIA.

LILIUM SUPERBUM. See GLORIOSA.

LIME-TREE. See TILIA.

LIMODORUM. Lin. Gen. Plant. Baſtard Hellebore.

The CHARACTERS are,

The flowers have no empalement, but a ſpatha (or ſheath) ſituated below them. The flower has five oval petals, which are diſſimilar, ſo has much the appearance of a butterfly flower. Within the petals is ſituated a concave nectarium, which is as long as the petals. It hath two ſtamina which are as long as the petals, and a column-ſhaped germen ſituated under the flower, which afterward turns to a capſule of the ſame form, opening with three valves, having one cell, in which are lodged four or five roundiſh ſeeds.

We have but one SPECIES of this plant, viz.

LIMODORUM (*Tuberoſum*) foliis longis anguſtis acuminatis, pedunculis longiſſimis. *Limodorum with long narrow leaves ending in acute points, and a very long foot-ſtalk to the flower.*

This plant grows naturally in Jamaica; it alſo grows in the French iſlands of America, and in the Bahama Iſlands; from ſeveral of theſe places I have received the roots.

The root of this plant is ſhaped like thoſe of the Saffron, but the outer cover is of a darker brown colour; from this comes out three or four leaves, according to the ſize and ſtrength of the root, which are nine or ten inches long, and near three quarters of an inch broad in the middle, being contracted toward both ends, terminating with long acute points; they have five longitudinal furrows, like the firſt leaves of young Palms; theſe leaves come out in the ſpring, and frequently decay the following winter, but when the plants are kept in a warm ſtove, they are ſeldom deſtitute of leaves, and frequently produce flowers. The flower-ſtalk ariſes immediately from the root, on one ſide of the leaves; this is naked, ſmooth, and of a purpliſh colour toward the top. It is near a foot and a half high, terminated by a looſe ſpike of purpliſh red flowers, compoſed of five or ſix petals; the two upper are connected together, forming a ſort of helmet; the two ſide petals expand like the wings of a Butterfly flower, and the lower form a ſort of keel. In the center of the petals is ſituated a column-ſhaped germen, which riſes from their baſe, ſupporting a ſlender ſtyle, to which adhere two ſtamina; after the flowers are faded, the germen becomes a three-cornered column, which becomes a capſule with one cell, opening with three valves, containing ſeveral roundiſh ſeeds.

There are ſeveral other ſpecies of this genus mentioned by Father Plumier, but I have not ſeen more than this here mentioned, which had oval obtuſe leaves, furrowed in the ſame manner as the leaves of this ſort, but were of a thicker conſiſtence; the flowers of this I have not yet ſeen. The root was ſent me from Maryland, where it grew naturally in a thicket.

The ſort firſt deſcribed is too tender to thrive in the open air in England; and although with care it may be preſerved in a warm green-houſe, yet it ſeldom flowers in ſuch a ſituation; ſo that to have it in perfection, it is neceſſary to keep it in the tan-bed in the ſtove in winter; and if in ſummer the pots are plunged in a tan-bed under a deep frame, the plants will thrive and flower as ſtrong as in their native ſoil.

It is propagated by offſets from the root, which are ſent out pretty freely when the plants are in vigour; theſe ſhould be taken off, and the roots tranſplanted, when they are the moſt deſtitute of leaves.

LIMON. Tourn. Inſt. R. H. 621. The Limon-tree.

The CHARACTERS are,

The flower is compoſed of five oblong thick petals, which ſpread open; theſe ſit in a ſmall empalement of one leaf. It hath about ten or twelve ſtamina, which are joined in three or four bodies, and an oval germen, which afterward becomes an oval fruit, with a fleſhy rind, incloſing a thin pulpy fruit with ſeveral cells, each having two hard ſeeds.

The SPECIES are,

1. LIMON (*Vulgaris*) foliis ovato-lanceolatis acuminatis, ſubſerratis. *Limon-tree with oval, ſpear-ſhaped, acute-pointed leaves, which are a little ſawed; or common Limon.*

2. LIMON (*Spinoſus*) foliis ovatis integris, ramis ſubſpinoſis. *The Lime-tree.*

3. LIMON (*Racemoſus*) foliis ovato-lanceolatis ſubſerratis, fructu conglomerato. *Limon with oval ſpear-ſhaped leaves, which are ſomewhat ſawed, and fruit growing in cluſters.*

There are great varieties of this fruit which are preſerved in ſome of the Italian gardens, and in both the Indies there are ſeveral which have not yet been introduced to the European gardens; but theſe, like Apples and Pears, may be multiplied without end from ſeeds; therefore I ſhall only mention the moſt remarkable varieties which are to be found in the Engliſh gardens at preſent, as it would be to little purpoſe to enumerate all thoſe which are mentioned in the foreign catalogues.

The Limon-tree with variegated leaves.

The ſweet Limon.

The Pear-ſhaped Limon.

The imperial Limon.

The Limon called Adam's Apple

The furrowed Limon.

The childing Limon.

The Limon with double flowers.

The common Limon and the ſweet Limon are brought to England from Spain and Portugal in great plenty, but the fruit of the latter are not much eſteemed. The Lime is not often brought to England, nor is that fruit much cultivated in Europe, but in the Weſt Indies it is preferred to the Limon, the juice being reckoned wholeſomer, and the acid is more agreeable to the palate; there are ſeveral varieties of this fruit in the Weſt-Indies, ſome of which have a ſweet juice, but thoſe are not greatly eſteemed; and as the inhabitants of thoſe iſlands do not propagate theſe fruits by grafting or budding, being contented with ſowing their ſeeds, ſo there is no doubt but a great variety of them may be found by any perſon who is curious in diſtinguiſhing them.

The Pear shaped Limon is a small fruit, with very little juice, so is not much propagated any where; the curious who have room and convenience for keeping many of these trees, preserve a plant or two of this sort for the sake of variety.

The fruit of the imperial Limon is sometimes brought to England from Italy, but I do not remember to have seen any of this sort imported from Spain or Portugal, so that I suppose they are not much propagated in either of those countries; for the inhabitants of both those fine countries are so very incurious, especially in horticulture, as to trust almost entirely to nature, therefore the products of their gardens are inferior both in numbers and quality to the gardens in many other parts of Europe, where the climate is much less favourable for these productions. And in the article we are now upon, there are many strong instances of the slothfulness or incuriosity of the Portugueze particularly, for they had many of the most curious sorts of Orange, Limon, and Citron-trees, brought from the Indies to Portugal formerly, which seemed to thrive almost as well there as in their native soil, and yet they have not been propagated; there are a few trees of these sorts still remaining in some neglected gardens near Lisbon almost unnoticed by the inhabitants. As there are also several curious trees and plants, which were formerly introduced from both Indies, some of which thrive and produce fruit amidst the wild bushes and weeds, with which those gardens are spread over.

All the sorts of Limons are propagated by budding or inarching them either on stocks of Limons or Citrons, produced from seeds; but they will not so readily unite on Orange stocks, for which reason the Citrons are preferable to either Oranges or Limons for stocks, as they readily join with either sort, and being of quicker growth, cause the buds of the other sorts to shoot much stronger than if they were on stocks of their own kind. The method of raising these stocks, and the manner of budding them, being already exhibited under the article of AURANTIUM, it would be superfluous to repeat it here.

The culture of the Limon being the same with that of the Orange-tree, it would be also needless to repeat it here; therefore I shall only observe, that the common Limons are somewhat hardier than the Oranges, so will bring their fruit to maturity with us better than the other will do, and require to have a greater share of fresh air in winter.

LIMONIUM. Tourn. Inst. R. H. 341. tab. 177.

The CHARACTERS are,

The flowers have an imbricated perianthium, and are funnel-shaped, composed of five petals. It hath five awl-shaped stamina, crowned by prostrate summits, and a small germen, crowned by pointed stigmas. The empalement of the flower afterward becomes a capsule, shut close at the neck, but expanded above where the seeds are lodged.

The SPECIES are,

1. LIMONIUM (*Vulgare*) foliis ovato-lanceolatis, caule tereti nudo paniculato. *Common great Sea Lavender.*

2. LIMONIUM (*Narbonense*) foliis oblongo-ovatis, caule paniculato patulo, spicis florum brevioribus. *Sea Lavender with oblong oval leaves, a spreading paniculated stalk, and shorter spikes of flowers.*

3. LIMONIUM (*Oleæfolium*) foliis ovatis obtusis, petiolis decurrentibus, caule paniculato, spicis florum erectoribus. *Sea Lavender with oval obtuse leaves, running foot-stalks, a paniculated stalk, and more upright spikes of flowers.*

4. LIMONIUM (*Humile*) foliis lanceolatis, caule humile patulo, spicis florum tenuioribus. *Sea Lavender with spear-shaped leaves, a low spreading stalk, and slender spikes of flowers.*

5. LIMONIUM (*Tartaricum*) foliis lineari lanceolatis, caule ramoso patulo, floribus distantibus uno versu dispositis. *Sea Lavender with narrow spear-shaped leaves, a branching spreading stalk, and flowers ranged thinly on one side the stalk.*

6. LIMONIUM (*Sinuatum*) foliis radicalibus alternatim pinnato-sinuatis, caulinis ternis triquetris subulatis decurrentibus. *Sea Lavender with the lower leaves alternately sinuated like wings, those upon the stalks three-cornered, awl-shaped, and running along the foot-stalk.*

7. LIMONIUM (*Siculum*) caule fruticoso patulo, foliis lineari lanceolatis crassis, floribus solitariis distantibus. *Sea Lavender with a spreading shrubby stalk, narrow, thick, spear-shaped leaves, and flowers growing singly at a distance from each other.*

8. LIMONIUM (*Africanum*) foliis cuneiformibus, caule erecto paniculato, ramis inferioribus sterilibus nudis. *Sea Lavender with wedge-shaped leaves, an upright paniculated stalk, and the under branches sterile and naked.*

9. LIMONIUM (*Cordatum*) caule nudo paniculato, foliis spathulatis retusis. *Sea Lavender with a paniculated naked stalk, and spatula shaped blunt leaves.*

10. LIMONIUM (*Echoiedum*) caule nudo paniculato, tereti, foliis tuberculatis. *Sea Lavender with a naked, taper, paniculated stalk, and leaves set with tubercles.*

11. LIMONIUM (*Fruticosum*) caule erecto fruticoso, foliis lineari-lanceolatis obtusis, floribus alternis. *Sea Lavender with an upright shrubby stalk, narrow spear-shaped leaves, ending in obtuse points, and flowers ranged alternately.*

The first sort grows naturally in the marshes, which are flowed by the sea in several parts of England. The roots of this plant are thick, of a reddish colour, and an astringent taste, sending out many strong fibres, from which comes out several oval, spear-shaped, smooth leaves, of a pretty thick consistence. The stalk is naked, and rises upward of a foot high, divided into many smaller branches at the top, terminated by slender spikes of pale blue flowers, ranged on one side the stalk, coming out of narrow covers like sheaths.

The second sort grows naturally in the south of France on the sea-coast. The leaves of this are of an oblong oval form, smooth, entire, and of a deep green. The stalk rises fifteen or sixteen inches high, dividing into several spreading branches, terminated by short spikes of pale blue flowers, ranged on one side the foot-stalk. This sort seldom flowers till the end of August, so never produces any good seeds in England.

The third sort grows naturally in Narbonne and Provence. This hath small, oval, obtuse leaves, standing on pretty long foot-stalks, which are bordered or winged. The stalk rises a foot and a half high, sending out branches alternately on each side, so as to form a loose kind of pyramid, and are terminated with spikes of pale blue flowers, which are erect.

The fourth sort grows naturally in England. It was first discovered on the sea-banks near Walton in Essex, afterward near Waldon in the same county, and since at the mouth of the river that runs from Chichester in Sussex. The leaves of this sort are spear-shaped, about three inches long, and one broad in the middle, lessening gradually to both ends. The stalk rises four or five inches high, dividing into many spreading branches, which are very thick set with short spikes of pale blue flowers.

The fifth sort was discovered by Dr. Tournefort in th Levant. The leaves of this sort are about four inches long, and three quarters of an inch broad in the middle, diminishing gradually to both ends. The stalks rise about five or six inches high, dividing into several spreading branches, which are terminated by short spikes of pale blue flowers, ranged on one side the foot-stalk. This sort flowers late in August, so never ripens seeds here.

The sixth sort grows naturally in Sicily and Palestine. This is a biennial plant. The lower leaves, which spread on the ground, are indented almost to the middle rib; these indentures are alternate and blunt. The stalks rise a foot and a half high, dividing upward into several branches, garnished at each joint with three narrow leaves sitting close to the stalks, from whose base proceeds a leafy membrane or wing, which runs along on both sides the stalk. The stalks are terminated by panicles of flowers, which sit upon winged foot-stalks, each sustaining three or four flowers of a light blue colour, which continue long without fading.

The seventh sort grows naturally in Sicily. This hath a shrubby stalk, which rises about two feet high, dividing into several ligneous branches, which spread out on every side; the lower part of these are closely garnished with gray leaves, like those of the Sea Purslain, of as thick consistence. The branches are terminated by panicles of blue flowers, which come out singly at a distance from each other, having long tubes, but divide into five segments upward, which spread open.

The eighth sort grows naturally in Sicily, and was found so growing on the border of the sea in Norfolk, by Mr. Henry Scott, a gardener. The lower leaves of this sort are narrow at their base, but enlarge upward, and are rounded at the top, in shape of a wedge. The stalks are slender and stiff, rising from seven to fourteen inches high, sending out many slender side branches; all those which proceed from the lower part of the stalk are barren, having no flowers, but toward the top they have short panicles of whitish flowers, which are small, and sit three or four together upon one foot-stalk.

The ninth sort grows naturally near the sea, about Marseilles and Leghorn. This hath many thick fleshy leaves, which are shaped like a spatula, growing near the root, which are smooth, of a grayish colour, and spread on the ground. The stalks are naked, and rise about six inches high, dividing toward the top into many small branches, which are terminated by short crooked panicles of small flowers, of a pale red colour.

The tenth sort grows naturally about Montpelier and in Italy. This is an annual plant, with long narrow leaves, which are set with rough tubercles like the leaves of Viper Buglofs. The stalks rise about eight inches high, dividing into two or three small branches, which are terminated by reflexed short spikes of pale blue flowers, which come out late in August, so the seeds are seldom perfected in England.

The eleventh sort grows naturally in Egypt. This rises with an upright shrubby stalk to the height of eight or ten feet, divided into many branches, which are garnished with narrow spear-shaped leaves, placed without order, of a thick consistence, and of a gray colour, sitting close to the branches. The flowers are produced at the end of the branches in loose panicles, standing alternate on each side the stalk; they have pretty long tubes, which enlarge upward, where they are cut into five obtuse segments, which spread open; they are of a bright sky blue, but fade to a purple before they fall off.

The first, second, third, fourth, and fifth sorts are abiding plants, which will thrive in the open air in England. These plants may be transplanted at almost any time of the year, provided they are carefully taken up, preserving some earth to their roots. But the plants do not propagate very fast in gardens, and unless they are planted in a moist shady border, do seldom flower well; the best way to have them succeed, is to keep the plants in pots, and in summer to place them in a shady situation, but in winter they may be removed to a place where they may enjoy the sun.

The sixth sort is a biennial plant, which rarely perfect seeds in England; so that unless fresh seeds can be procured from warm countries, where they ripen well, it will be very difficult to continue the sort. If the seeds can be obtained time enough to sow them in the autumn, the plants will come up the following spring; but when they are sown in the spring, they seldom grow the same year. The seeds should be sown on a border of loamy earth, but not stiff or moist, exposed to the south-east. When the plants come up, they must be kept clean from weeds, and if they are too close, some of them should be carefully taken out as soon as they are fit to remove, and planted in small pots, placing them in the shade till they have taken new root; then they may be placed where they may enjoy the morning sun till autumn, when they should be put into a hot bed frame, where they may be screened from hard frost, but enjoy the free air in mild weather: those plants which are left in the border where they were sown, should be covered with mats in hard frost; for though they will often live through the winter in mild seasons, yet hard frost will always destroy them. The following summer the plants will flower, and if the season proves warm and dry they will sometimes ripen seeds.

The seventh and eleventh sorts are shrubby plants, which are too tender to live through the winter in the open air in England, so the plants must be removed into shelter in the autumn, but they only require protection from hard frost: these plants may be placed with Myrtles, Oleanders, and other hardy green-house plants, where they often continue in flower great part of winter, and make a pretty variety These sorts are easily propagated by cuttings, which, if planted in July on a shady border, and duly watered, will take root in six or seven weeks, when they should be taken up and planted in pots filled with light loamy earth, placing them in the shade till they have taken root; then they may be exposed till October, at which time they must be removed into shelter.

LINARIA. Tourn. Inst. R. H. 168. tab. 76. Toad-flax.

The CHARACTERS are,

The flower hath one petal, and is of the ringent kind, having an oblong swelling tube, with two lips above, and the chaps shut. The upper lip is bifid and reflexed on the sides, the lower lip is trifid and obtuse. It hath an oblong nectarium, prominent behind, and four stamina which are included in the upper lip, two of which are shorter than the other, and a roundish germen, which afterward turns to a roundish obtuse capsule with two cells, filled with small seeds.

The SPECIES are,

1. LINARIA (*Vulgaris*) foliis lanceolato-linearibus confertis, caule erecto, spicis terminalibus sessilibus. *Common yellow Toad-flax with a larger flower.*

2. LINARIA (*Triphylla*) foliis ternis ovatis. *Toad-flax with oval leaves placed by threes.*

3. LINARIA (*Lusitanica*) foliis quaternis lanceolatis, caule erecto ramoso, floribus pedunculatis. *Toad-flax with spear-shaped leaves placed by fours, an upright branching stalk, and flowers upon foot-stalks.*

4. LINARIA (*Alpina*) foliis subquaternis linearibus, caule diffuso, floribus racemosis. *Toad-flax with linear leaves placed by fours on the lower part of the stalk, a diffused stalk, and branching flowers.*

5. LINARIA (*Purpurea*) foliis lanceolato-linearibus sparsis, caule florifero erecto spicato. *Toad-flax with spear-shaped linear leaves, and the flower-stalks erect and spiked*

6. LINARIA (*Repens*) foliis linearibus confertis, caule erecto ramoso, floribus spicatis terminalibus. *Toad-flax with linear*

linear leaves in clusters, an erect branching stalk, and flowers in spikes terminating the stalks.

7. Linaria (*Multicaulis*) foliis inferioribus quinis linearibus. *Toad-flax with linear leaves, placed by fives at the lower part of the stalks.*

8. Linaria (*Tristis*) foliis lanceolatis sparsis, inferioribus oppositis, nectariis subulatis, floribus subsessilibus. *Toad-flax with spear-shaped sparsed leaves, which on the lower part of the stalk are opposite, awl-shaped nectariums, and flowers sitting almost close.*

9. Linaria (*Monspesulana*) foliis linearibus confertis, caule nitido paniculato, pedunculis spicatis nudis. *Toad-flax with linear leaves in clusters, a shining paniculated stalk, and flowers in spikes on naked foot-stalks.*

10. Linaria (*Villosa*) foliis lanceolatis hirtis alternis, floribus spicatis, foliolo calycino supremo maximo. *Toad-flax with alternate, hairy, spear-shaped leaves, flowers in spikes, and the upper leaf of the empalement very large.*

11. Linaria (*Pelisseriana*) foliis caulinis linearibus sparsis, radicalibus rotundis. *Toad-flax with linear leaves placed sparsedly on the stalks, and the lower leaves round.*

12. Linaria (*Chalepensis*) foliis lineari-lanceolatis alternis, floribus racemosis, calycibus corollâ longioribus. *Toad-flax with linear spear-shaped leaves placed alternate, branching flowers, and empalements longer than the petals.*

13. Linaria (*Dalmatica*) foliis lanceolatis alternis, caule suffruticoso. *Toad-flax with spear-shaped alternate leaves, and an under-shrub stalk*

14. Linaria (*Genistifolia*) foliis lanceolatis acuminatis, paniculâ virgatâ. *Toad-flax with spear-shaped acute-pointed leaves, and a rod like panicle.*

15. Linaria (*Spuria*) foliis ovatis alternis, caule flaccido procumbente. *Toad-flax with oval leaves placed alternate, and a weak trailing stalk; called Fluellin.*

16. Linaria (*Elatine*) foliis hastatis alternis, caule flaccido procumbente. *Toad-flax with arrow-pointed leaves placed alternate, and a weak trailing stalk.*

17. Linaria (*Cymbalaria*) foliis cordatis quinquelobatis alternis glabris. *Toad-flax with heart-shaped leaves having five lobes, which are alternate and smooth; or common Cymbalaria.*

The first of these plants grow in great plenty upon the sides of dry banks in most parts of England, and is rarely permitted a place in gardens, for it is a very troublesome plant to keep within bounds. This is one of the species mentioned in the catalogue of simples at the end of the College Dispensatory, to be used in medicine.

The second sort grows naturally about Valencia and in Sicily. This is an annual plant, which rises with an upright branching stalk a foot and a half high, garnished with oval, smooth, gray leaves, placed often by threes, and sometimes by pairs opposite; the flowers grow in short spikes at the top of the stalks; they are yellow, and shaped like those of the common sort, but have not so long tubes.

There is a variety of this, whose flowers have a purple standard and spur, which makes a pretty appearance in a garden.

These are propagated by seeds, which should be sown in the spring, on the borders of the flower-garden where they are designed to remain; and when the plants come up, they should be thinned where they are too close, and kept clean from weeds, which is all the culture they require.

The third sort rises with upright stalks two feet high, garnished with spear-shaped smooth leaves, placed sometimes by fours round the stalk, and at others by pairs opposite: the stalks are terminated by large purple flowers with long spurs, standing upon foot-stalks; it seldom ripens seeds in England. This grows naturally in Portugal and Spain.

This is tenderer than the last, so should have a warm situation, otherwise the plants will be destroyed in winter. It is propagated by seeds in the same manner as the former, but it is adviseable always to keep some of these plants in pots, that they may be removed into shelter in winter.

The fourth sort grows naturally about Verona. This is a perennial plant, from whose roots arise several diffused stalks about eight inches long, garnished with narrow, short, gray leaves, placed by fours round the stalks at bottom, but upward they are opposite; the stalks are terminated by short branching tufts of pale yellow flowers, with golden chaps.

The fifth sort grows naturally in the south of France and in Italy. This hath a perennial root, sending out many stalks; those of them which support the flowers are erect, but the other are weaker, and hang loosely on every side the plants; they are garnished with long, narrow, spear-shaped, gray leaves, placed sparsedly. The stalks are terminated by long loose spikes of blue flowers, which appear in June, July, and August, and the seed ripen in the autumn.

The sixth sort grows naturally about Henley in Oxfordshire, and also in some parts of Hertfordshire. This hath a perennial creeping root, from which arise many stalks two feet high, garnished with narrow leaves, growing in clusters toward the bottom, but upward they are sometimes by pairs, and at others single. The flowers are of a pale blue, produced in loose spikes at the end of the stalks.

The seventh sort grows naturally in Sicily. This is an annual plant, from whose root arises many slender stalks about a foot high; their lower part are garnished with five very narrow leaves at each joint, but upward they are sometimes by pairs, and at others they are single: the stalks are garnished with small yellow flowers, coming out single, and are shaped like those of the other species. The flowers appear in July, and the seeds ripen in the autumn. There are two varieties of this, one with a deep yellow, the other a sulphur-coloured flower.

This is propagated by seeds in the same manner as the second sort, or if the seeds are permitted to scatter, the plants will come up without care, and if they are kept clean from weeds, will produce their flowers early in the summer.

The eighth sort grows naturally on the rocks about Gibraltar, from whence the late Sir Charles Wager brought the seeds. This has a perennial root, sending out many slender succulent stalks, which are weak, and hang near the ground, garnished with short, narrow, spear-shaped leaves, of a gray colour and succulent, standing without order. The flowers are produced at the end of the stalks in small bunches; they are yellow, marked with purple stripes, and the chaps of the flower, as also the spur, are of a dark purple colour. They appear in June and July, but do not produce seeds in England.

This plant is easily propagated by planting cuttings in any of the summer months, which, if watered and shaded, will soon take root, and may afterwards be planted in pots filled with fresh, light, undunged earth, in which they will succeed much better than in a richer soil. These must be removed into shelter in winter, where they must have as much free air as possible in mild weather, and be only protected from severe cold; so that if the pots are placed under a hot-bed frame, it will be better than in a green-house, where they are apt to draw too much, which will cause them to decay.

The ninth sort grows naturally in Wales, particularly near Penryn. This hath a perennial root, from which arise many branching stalks two feet high, garnished with very narrow leaves, growing in clusters, of a grayish colour. The flowers are produced in loose spikes at the end of the branches;

branches; they are of a pale blue colour, and smell sweet The seeds ripen in the autumn, which, if permitted to scatter, will furnish a supply of young plants without any further care.

The tenth sort grows naturally in Spain; the seeds of it were sent me by Dr. Hortega from Madrid. This is an annual plant, which rises with a single stalk about a foot and a half high, garnished with hairy spear-shaped leaves, sitting close to the stalk, placed alternate. The flowers grow on the top of the stalks in loose spikes; they are of a pale yellow colour, with a few deep stripes, and the chaps are of a gold colour; the upper segment of the empalement is much larger than the lower.

The seeds of this sort should be sown in the spring, upon a border of light earth, where the plants are designed to remain; and when the plants come up, they must be treated in the same way as those of the second sort.

The eleventh sort grows naturally in France. This is an annual plant, whose bottom leaves are round; the stalks are slender, branching, and rise a foot high, garnished with very narrow leaves at each joint. The flowers are produced in loose spikes at the end of the branches; they are of a bright blue colour; the seeds ripen in the autumn, at which time they should be sown, for those which are sown in the spring, frequently lie in the ground till the spring following before the plants appear. When the plants come up, they must be thinned where they are too close, and kept clean from weeds, which is all the culture they require.

The twelfth sort grows naturally in Sicily. This is an annual plant, which rises with a branching stalk two feet high, garnished with very narrow spear-shaped leaves, placed alternately. The flowers are produced singly all along the branches; they are small, white, and have very long tails or spurs. This flowers in July, and the seeds ripen in the autumn. If the seeds of this sort are permitted to scatter, the plants will come up without care, and require no other culture but to keep them clean from weeds.

The thirteenth sort grows naturally in Crete, and also in Dalmatia. This rises with a strong ligneous stalk three feet high, garnished with smooth spear-shaped leaves placed alternate, sitting close to the stalk. The flowers are produced at the end of the branches in short loose spikes; they are of a deep yellow colour, and much larger than those of the common sort, standing upon short foot-stalks. It is propagated by seeds, which should be sown early in the spring, upon a border of light earth; and when the plants come up, and are fit to remove, some of them should be planted in pots, to be sheltered under a common frame in winter. As these plants only require to be protected from hard frost, so in mild winters they will live abroad without shelter, if they are upon a dry soil; therefore a part of the plants may be planted on a warm border of poor sandy soil, where they will live through our common winters very well; and those plants which grow in rubbish and are stinted, will endure much more cold than the others.

The fourteenth sort grows naturally in Siberia. This is a biennial plant, which rises with an upright branching stalk from three to four feet high, garnished with spear-shaped leaves, ending in acute points, of a grayish colour, placed alternate. The flowers are produced at the end of the branches in loose panicles; they are of a bright yellow colour, shaped like those of the other sorts. This flowers in June and July, and the seeds ripen in autumn, which, if permitted to scatter, the plants will come up the following spring, and require no other care but to thin them where they are too close, and keep them clear from weeds.

The fifteenth sort is frequently called Fluellin, and is sometimes used in medicine; it grows naturally amongst Wheat and Rye in several parts of England. It is an annual plant, with weak, trailing, hairy stalks, which spread on the ground, garnished with oval leaves, placed alternately; at each joint comes out one flower, shaped like those of the other species. The upper lip is yellow, and the under is purple

The sixteenth sort differs from the fifteenth, in nothing but the shape of the leaves, which in this are shaped like the point of an arrow, and those of the other are oval; this is more commonly found in England than the other.

The seventeenth sort was brought from Italy to England, where it now grows in as great plenty in the neighbourhood of London, as if it was in its native country, growing from the joints of walls, wherever the seeds happen to scatter. It is a perennial plant, which will thrive in any soil or situation, so that where it is once established, it will be difficult to root out, for the seeds will get into any joints of walls, or the decayed parts of pales, as also in the hollow of trees, where they grow and propagate plentifully; for the stalks put out roots at their joints, so spread themselves to a great distance.

LINGUA CERVINA. Hart's Tongue.

These plants commonly grow out from the joints of old walls and buildings, where they are moist and shady, and also upon shady moist banks, but are seldom cultivated in gardens. There is a very great variety of these plants, both in the East and West-Indies, but there are very few species of them in Europe; all the hardy sorts may be propagated by parting their roots, and should have a moist soil and shady situation.

LINUM. Tourn. Inst. R. H. tab. 176. Flax.

The CHARACTERS are,

The flower has five large oblong petals, which spread open. It hath five awl-shaped erect stamina. In the center is situated an oval germen, which afterward turns to a globular capsule with ten cells, opening with five valves; in each cell is lodged one oval, plain, smooth seed.

The SPECIES are,

1. LINUM (*Usitatissimum*) calycibus capsulisque mucronatis, petalis crenatis, foliis lanceolatis alternis, caule subsolitario. Lin. Sp. Pl. 277. *Common manured Flax.*

2 LINUM (*Humile*) calycibus capsulisque mucronatis, petalis emarginatis, foliis lanceolatis alternis, caule ramoso. *Low manured Flax with a larger flower.*

3. LINUM (*Narbonense*) calycibus acuminatis, foliis lanceolatis sparsis strictis scabris acuminatis, caule tereti basi ramoso. Lin. Sp. Plant. 278. *Wild blue Flax with an acute leaf.*

4. LINUM (*Tenuifolium*) calycibus acuminatis, foliis sparsis linearibus setaceis retrorsum scabris. Lin. Sp. Pl. 278. *Narrow-leaved wild Flax, with a pale purplish, or flesh-coloured flower.*

5. LINUM (*Anglicum*) calycibus capsulisque acuminatis, caule subnudo scabro, foliis acuminatis. *Greater blue perennial Flax with larger heads.*

6. LINUM (*Perenne*) calycibus capsulisque obtusis, foliis alternis lanceolatis acutis, caulibus ramocissimis. Plaut. 166. *Flax with obtuse empalements and capsules, alternate, spear shaped, acute leaves, and very branching stalks; commonly called Siberian perennial Flax.*

7. LINUM (*Hispanicum*) calycibus acutis, foliis lineari-lanceolatis sparsis, caule paniculato. *Flax with acute empalements, linear spear-shaped leaves placed without order, and a paniculated stalk.*

8. LINUM (*Bienne*) calycibus patulis acuminatis, foliis linearibus alternis, caule ramoso. *Flax with spreading acute-pointed empalements, linear alternate leaves, and a branching stalk.*

9. LINUM

9. LINUM (*Hirsutum*) calycibus hirsutis acuminatis sessilibus alternis, caule ramoso. Lin. Sp. Plant. 277. *Broad-leaved, hairy, wild Flax with a blue flower.*

10. LINUM (*Strictum*) calycibus foliisque lanceolatis strictis mucronatis, margine scabris. Lin. Sp. Plant. 279. *Flax with spear-shaped leaves and empalements, which end in acute points, and have rough edges; or the Passerini Lobellii.* J. B. 3. p. 454.

11. LINUM (*Fruticosum*) calycibus acutis, petalis integris, foliis inferioribus linearibus fasciculatis, superioribus alternis, caule suffruticoso. *Wild Flax with a shrubby stalk and acute leaves.*

12. LINUM (*Nodiflorum*) foliis lanceolatis alternis, floribus alternis sessilibus, caule simplici. *Yellow Flax with single flowers growing from the joints.*

13. LINUM (*Catharticum*) foliis oppositis ovato-lanceolatis, caule dichotomo, corollis acutis. Hort. Cliff. 372. *Meadow Flax with small flowers; commonly called Mountain Flax.*

14. LINUM (*Maritimum*) calycibus ovatis acutis muticis, foliis lanceolatis inferioribus oppositis. Lin. Sp. Pl. 280. *Yellow maritime Flax.*

The first sort is the Flax which is cultivated in most parts of Europe, but particularly in the northern parts. This is an annual plant, which usually rises with a slender unbranched stalk a foot and a half high, garnished with narrow spear-shaped leaves, placed alternate, ending in acute points, of a gray colour. The flowers are produced on the top of the stalks, each stalk sustaining four or five blue flowers, composed of five petals, which are narrow at their base, but broad at the top, where they are slightly crenated. The flowers appear in June and July, and are succeeded by roundish capsules, which have ten cells, opening with five valves, which are terminated by acute points; each cell contains one smooth flattish seed, of a brown colour. The seeds ripen in September, and the plants soon after perish.

When this plant is cultivated in the fields after the usual method, it seldom rises higher than is before mentioned, nor do the stalks branch out; but when they are allowed more room, they will rise between two and three feet high, and put out two or three side branches toward the top, especially if the soil is pretty good where it is sown.

The second sort differs from the first, in having stronger and shorter stalks branching out much more. The leaves are broader, the flowers are larger, and the petals are indented at their extremities. The seed-vessels are also much larger, and the foot-stalks are longer.

The third sort grows naturally in the south of France, in Italy, and Spain. This rises from a foot to eighteen inches high, branching out almost to the bottom into many long slender branches, garnished with narrow spear-shaped leaves. The flowers are produced at the end of the branches, almost in form of an umbel; they are smaller than those of the manured sort, and are of a paler blue colour.

The fourth sort grows naturally about Vienna and in Hungary. This seldom rises more than a foot high, with a slender stalk, which divides into three or four naked foot-stalks at the top, each sustaining two or three flowers, which are of a pale blue colour.

The fifth sort grows naturally in some parts of England. This hath a perennial root, from which arise three or four stalks, garnished with a few short narrow leaves toward their base, but upward have scarce any. The flowers are blue, and are produced at the end of the stalks, which are succeeded by pretty large round seed-vessels, ending in acute points. The roots will continue three or four years.

The sixth sort grows naturally in Siberia. It hath a perennial root, from which arise several strong stalks, in number proportional to the size of the root, and in height according to the goodness of the soil where it grows; for in rich moist ground they will rise near five feet high, but in middling ground about three feet; these divide into several branches upward, garnished with narrow spear-shaped leaves, placed alternate. The flowers are produced at the end of the branches, forming a kind of umbel; they are large, and of a fine blue colour. These appear in June, and are succeeded by obtuse seed-vessels, which ripen in September.

The seventh sort grows naturally in Spain. This hath a biennial root, from which come out several trailing stalks, which never rise much from the ground, but between these come out upright stalks, which rise upward of two feet high, garnished with pretty long, narrow, spear-shaped leaves, placed without order. The flowers grow in a sort of panicle toward the upper part of the branches; they are like those of the common sort, and are of the same colour.

The eighth sort I received from Istria. This hath a biennial root, from which arise two or three stalks, which divide into several branches, garnished with short, narrow, acute-pointed leaves, placed alternately. The flowers come out from the side of the branches, standing upon long foot-stalks. They are of the same size and colour as the common Flax, and appear at the same season. The seeds ripen in the autumn, and the roots abide two years.

The ninth sort grows naturally in Hungary and Austria. This hath a perennial root, from which arise several strong stalks two feet high, dividing into several branches, garnished with broader leaves than the other species, which are hairy. The flowers grow along the stalks alternately; they are large, and of a deep blue colour, appearing at the same time with the common sort, and the seeds ripen in the autumn.

The tenth sort grows naturally in Germany and the south of France, amongst the Corn. This is an annual plant, rising with an upright stalk a foot and a half high, garnished with spear-shaped acute-pointed leaves, which are rough on their edges, placed alternately. The stalks divide into several branches, each sustaining two or three yellow flowers, which appear in July, but unless the autumn proves favourable, the seeds never ripen in England.

The eleventh sort grows naturally in Spain. This hath a shrubby stalk, which rises a foot high, sending out several branches, garnished with very narrow leaves, coming out in clusters. The flowers are produced at the end of the branches, standing erect, upon long slender foot-stalks. The petals of the flower are large, entire, and white, but before the flowers open they are of a pale yellow colour. These flowers appear in July, but unless the autumn proves favourable, the seeds do not ripen in England.

The twelfth sort grows naturally upon the Alps. This hath a perennial root, from which arise two or three slender stiff stalks, which divide into two or three smaller, garnished with spear-shaped leaves, placed alternately. The flowers come out singly at the joints; they are yellow, and appear about the same time with the common sort, and the seeds ripen in the autumn.

The thirteenth sort grows common in many parts of England, upon dry barren hills. It is commonly called Linum catharticum, or purging Flax, and also mountain Flax. This rises with several branching slender stalks about seven or eight inches high, garnished with small, oval, spear-shaped leaves, placed opposite. The flowers are small and white, standing upon pretty long foot-stalks. They appear in July, and are succeeded by small round capsules, containing small flat seeds, which ripen in the autumn.

The fourteenth sort grows naturally about Montpelier. This rises with upright stalks two feet high, the lower part of which are garnished with spear-shaped leaves placed opposite,

posite, but on the upper part they are alternate. The stalks divide into several branches, terminated with yellow flowers, about the size of those of common Flax, which are succeeded by small oval capsules, containing smaller seeds than those of the common Flax.

The first sort is that which is cultivated for use in divers parts of Europe, and is reckoned an excellent commodity; the right tilling and ordering of which, is esteemed a good piece of husbandry.

The ground in which this is to be sown, should be as clean from weeds as possible; in order to have it so, it should be fallowed two winters and one summer, observing to harrow it well between each ploughing, particularly in summer, to destroy the young weeds soon after they appear, that the smallest of them may not stand to ripen their seed; this will also break the clods, and separate their parts so, as that they will fall to pieces on being stirred. If the land should require dung, that should not be laid on till the last ploughing; but this dung should be such as is clear from the seeds of weeds. Just before the season for sowing of the Flax-seed, the land must be well ploughed, laid flat and even, upon which the seeds should be sown about the latter end of March, when the weather is mild and warm.

The common way is to sow the seed in broad-cast, and to allow from two to three bushels of seeds to one acre of land; but from many repeated trials, I have found it a much better method to sow the seeds in drills, at about ten inches distance from each other, by which half the quantity of seed which is usually sown, will produce a greater crop; and when the Flax is thus sown, the ground may be easily hoed between the rows to destroy the weeds, which, if twice repeated in dry weather, will keep the ground clean till the Flax is ripe: this may be performed at half the expence which the hand weeding will cost, and will not tread down the plants, nor harden the ground, which by the other method is always done; for it is undoubtedly necessary to keep the Flax clean from weeds, otherwise they will overbear and spoil the crop.

There are some people who recommend the feeding of sheep with Flax, when it is a good height; and say, they will eat away the weeds and Grass, and do the Flax good and if they should lie in it, and beat it down or flatten it, it will rise again the next rain: but this is a very wrong practice, for if the sheep gnaw or eat the Flax, the plants will shoot up very weak stalks, which never come to half the size they would have done if not cropped; and as to the sheep destroying the weeds, they never are so nice distinguishers, for if they like the crop better than the weeds, they will devour that, and leave the weeds untouched.

Toward the latter end of August the Flax will begin to ripen, when you must be careful that it grow not over ripe; therefore you must pull it up as soon as the heads begin to change brown and hang downwards, otherwise the seeds will soon scatter and be lost, so that the pluckers must be nimble, and tie it up in handfuls, setting them upright till they be perfectly dry, and then house them. If the Flax be pulled when it first begins to flower, it will be whiter than if it stand till the seed is ripe, but then the seed will be lost; but the thread will be stronger when the Flax is left till the seed is ripe, provided it does not stand too long, but the colour of it will not be so good

The Siberian perennial Flax has been made trial of, and answers very well for making of common strong linen, but the thread spun from this is not so fine or white as that which is produced from the common sort; but as the roots of this sort will continue many years, so there will be a great saving in the culture, as it will require no other care but to keep it constantly clean from weeds; which cannot be well done, unless the seeds are sown in rows, that the ground may be constantly kept hoed to destroy the weeds when young, for if they are suffered to grow large, it will be difficult to get the ground clean, and they will weaken the roots. This sort must have the stalks cut off close to the ground when ripe, and tied up in small bundles, managing them afterward in the same way as the common sort.

The eighth sort which I received from Istria produced the finest thread; it grows taller than the common Flax, and has a biennial root; it may be worthy of trial to see how it will thrive in the open fields, for in gardens it lives through the winter without receiving the least injury from the frost. In order to make trial of its goodness, I gave a parcel of the stalks of this, as also of the Spanish and the Siberian perennial Flax, to a person who is well skilled in watering, breaking, and dressing of Flax, who prepared them, and assured me, that the Istrian Flax was by much the finest of the three, and equal in goodness to any he had seen.

The other sorts which are here mentioned, are preserved in gardens for the sake of variety, but none of them are used, except the mountain Flax, which is esteemed a good purger in dropsical disorders, and has of late years been often prescribed.

They are all of them propagated by seeds, which may be sown in the spring, in the places where they are to remain, and will require no other culture but to keep the plants clean from weeds. The annual sorts will flower and perfect their seeds the same year, but the roots of the perennial sorts will continue several years, putting out fresh stalks every spring. The shrubby sorts will live through the winter in the open air, provided it is in a dry soil and a warm situation, but these rarely produce seeds in England.

LINUM UMBILICATUM. See CYNOGLOSSUM.

LIPPIA. Houst. Gen. Nov. Lin. Gen. Plant. 699.

The CHARACTERS are,

The flower hath one petal, which is of the ringent kind. It hath four short stamina, two of which are a little longer than the other, and an oval germen, which afterward turns to a compressed capsule with one cell, opening with two valves, which appear like the scales of the empalement, inclosing two seeds which are joined.

We have but one SPECIES of this genus, viz.

LIPPIA (*Americana*) arborescens, foliis conjugatis oblongis, capitulis squamosis & rotundis. Houst. *Tree Lippia with oblong leaves growing by pairs, having round scaly heads.*

This plant, in the country of its native growth, commonly rises to the height of sixteen or eighteen feet; it has a rough bark; the branches come out opposite, as do also the leaves, which are oblong, pointed, and a little sawed on their edges. From the wings of the leaves come out the foot-stalk, which sustain many round scaly heads, about the size of a large gray Pea, in which are many small yellow flowers appearing between the scales, which are succeeded by seed vessels.

This plant will not thrive in this climate, unless it is preserved in a warm stove, so should be treated in the same manner as other shrubby plants, which are natives of warm countries; which is, to keep them always in the stove plunged in the bark bed, observing to give them a large share of air in warm weather, and frequently refresh them with water; but in winter they must be watered more sparingly, and kept in a moderate degree of warmth, otherwise they will not live through the winter, especially while they are young; but when they have acquired strength they may be preserved with a less share of warmth.

LIQUIDAMBER. Lin. Gen. Pl. 955. Liquidamber, Sweet Gum, or Storax-tree.

The CHARACTERS are,

It hath male and female flowers on the same tree; the male flowers are disposed in long, loose, conical katkins; these have four-leaved empalements, no petals, but a great number of short stamina joined in one body. The female flowers are situated at the base of the male spike, collected in a globe; these have a double empalement. They have no petals, but an oblong germen fastened to the empalement, which afterward turns to a roundish capsule of one cell, with two valves at the top, which are collected in a ligneous globe, containing many oblong acute-pointed seeds.

The SPECIES are,

1. LIQUIDAMBER (*Styraciflua*) foliis quinquelobatis serratis. *Liquidamber with sawed leaves, having five lobes; or Maple-leaved Storax-tree.*

2. LIQUIDAMBER (*Orientalis*) foliis quinquelobatis, sinuatis obtusis. *Liquidamber with leaves having five lobes, which are sinuated and obtuse.*

The first sort has by some writers been ranged with the Maple, but on no other account, except from the similitude of the leaves; for in flower and fruit it is very different from the Maple, and most other genera: nor has it any affinity to the Storax-tree; but the gum which issues from this tree being transparent, and having a great fragrancy, has by some ignorant persons been taken for that.

It grows plentifully in Virginia, and several other parts of North America, where it rises with a strait naked stem to the height of fifteen or sixteen feet, and branches out regularly to the height of forty feet or upward, forming a pyramidal head. The leaves are angular, and shaped somewhat like those of the lesser Maple, having five lobes, but are of a dark green colour; a strong, sweet, glutinous substance exsudes through the pores of the leaves in warm weather, which renders them clammy to the touch.

The flowers are generally produced early in the spring of the year, before the leaves are expanded, which are of a Saffron colour, and grow in spikes from the extremity of the branches; after these are past, the fruit swells to the size of a Walnut, being perfectly round, having many protuberances, each having a small hole and a short tail, which extends half an inch.

This is commonly propagated by layers in England; but those plants which are raised from seeds, grow to be much fairer trees.

The seeds of this tree commonly remain in the ground a whole year before the plants come up, unless they are sown in the autumn; so that the surest way to raise them is, to sow the seeds in boxes or pots of light earth, which may be placed in a shady situation during the first summer, and in autumn they may be removed where they may have more sun; but if the winter should prove severe, it will be proper to cover them with Pease haulm, or other light covering, which should be taken off constantly in mild weather. The following spring, if these boxes or pots are placed upon a moderate hot-bed, it will bring up the plants early, so that they will have time to get strength before winter; the first and second winters it will be proper to screen the plants from severe frost; but afterward they will bear the cold very well.

The seeds of the second sort were sent by Mr. Peyssonel from the Levant. The leaves of this sort differ from those of the first, in having their lobes shorter and deeply sinuated on their borders; they end in blunt points, and are not serrated; but as I have not seen the fruit of this, so I do not know how it differs from the other.

LIRIODENDUM. See TULIPIFERA.

LITHOSPERMUM. Tourn. Inst. R. H. 137. tab. 55. Gromwell, Gromill, or Graymill.

The CHARACTERS are,

The flower hath one petal, with a cylindrical tube divided into five obtuse points at the brim; the chaps are perforated. It hath five short stamina which are shut up in the chaps of the petal; and hath four germen, which afterward turn to so many oval, hard, smooth, acute-pointed seeds, sitting in the spreading empalement.

The SPECIES are,

1. LITHOSPERMUM (*Officinale*) seminibus lævibus, corollis calycem vix superantibus, foliis lanceolatis. Hort. Cliff. 46. *Greater upright Gromwell.*

2. LITHOSPERMUM (*Arvense*) seminibus rugosis, corollis vix calycem superantibus. Hort. Cliff. 46. *Field Gromwell with a red root.*

3. LITHOSPERMUM (*Purpurocæruleum*) seminibus lævibus, corollis calycem multoties superantibus. Hort. Cliff. 46. *Smaller, creeping, broad-leaved Gromwell.*

4 LITHOSPERMUM (*Virginianum*) foliis subovalibus nervosis, corollis acuminatis. Lin. Sp. Plant. 132. *Broad-leaved Gromwell of Virginia, with a longer whitish flower.*

5. LITHOSPERMUM (*Fruticosum*) fruticosum, foliis linearibus hispidis, staminibus corollam subæquantibus. Lin. Sp. 190. *Shrubby Gromwell with rough linear leaves, and the stamina almost equal with the petal.*

The first sort grows naturally upon banks, and in dry fields in many parts of England; so is seldom admitted into gardens. This hath a perennial root, from which arise upright stalks two feet high; garnished with spear-shaped, rough, hairy leaves, placed alternate, sitting close to the stalks. The flowers come out singly at every joint of the small branches; they are white, of one petal, cut into four parts at the top, and stand within the empalement; these are each succeeded by four hard white shining seeds, which ripen in the empalement.

The second sort is an annual plant, which grows among winter Corn in many parts of England. This rises with a slender branching stalk a foot and a half high, garnished with narrow, spear-shaped, rough leaves, placed alternately. The flowers are produced singly on the upper parts of the stalks; they are small, white, and are succeeded by four rough seeds, which ripen in the empalement.

The third sort grows naturally in woods in many parts of England; this has a perennial root, from which come out two or three trailing stalks scarce a foot long; garnished with narrow spear-shaped leaves, placed alternately. The flowers terminate the stalks; they are white, and the petals are much longer than the empalements.

The fourth sort grows naturally in North America; this hath a biennial root, from which arise several very hairy stalks about a foot and a half high, garnished with oval, rough, hairy, veined leaves, sitting close to the stalks alternately. The flowers grow in short reflexed spikes at the end of the branches; these are white, their petals being longer than the empalement, ending in acute points.

The fifth sort grows naturally in the south of France, and also in the Levant. This hath a perennial root, which runs deep in the ground, from which arises in the spring a shrubby erect stalk near three feet high, closely garnished with hairs; the leaves are narrow, and placed alternate; the flowers are produced in short reflexed spikes at the end of the stalk, having hairy empalements; they are of a purplish colour and tubulous, divided into five segments, the two upper are reflexed; the flowers appear in June, but the seeds rarely ripen in England.

This sort stands in the list of medicinal plants, but at present is rarely used: the bark of the root gives a fine purple tincture, for which it is sometimes used, but the colour is not permanent.

These may be cultivated by sowing their seeds soon after they are ripe, in a bed of fresh earth, allowing them at least a foot distance, keeping them clean from weeds, and they will thrive in almost any soil or situation.

LOBELIA. Plum. Nov. Gen. 21. tab. 31.

The CHARACTERS are,

The flower has one petal, which is tubulous, ringent, and cut into five parts at the brim. It hath five awl-shaped stamina the length of the tube, and a pointed germen under the petal, which afterward becomes an oval fleshy berry with two cells, each containing a single seed.

We have but one SPECIES of this genus, viz.

LOBELIA (*Frutescens*) frutescens, foliis ovali oblongis integerrimis. Flor. Zeyl. 313. *Shrubby Lobelia with oblong oval, entire leaves.*

This plant rises with a succulent stalk five or six feet high, garnished with oval, oblong, succulent leaves, placed alternately, which sit close to the stalk. The flowers are produced upon long foot-stalks, which sustain two or three white flowers of one petal, cut into five acute segments at the brim; these are succeeded by oval berries as large as Bullace, containing a stone with two cells, in each of these is lodged a single seed.

This plant grows naturally in many of the islands of the West-Indies. It is propagated by seeds, which must be procured from the countries of its natural growth; these should be sown in pots, and plunged into a hot-bed of tanners bark, where the plants will come up in about five or six weeks, provided the bed is kept warm, and the earth often watered. When the plants are about two inches high, they should be carefully taken out of the pots in which they were sown, and each planted in a separate pot, and then plunged into the hot bed again, observing to shade them in the heat of the day until they have taken new root. In this hot bed the plants may remain until the middle or latter end of September, when they must be removed into the stove, and plunged into the tan-bed, and afterward treated in the same way as other tender exotic plants, which require a stove to preserve them through the winter.

LOBUS ECHINATUS. See GUILANDINA.

LONCHITIS. Rough Spleenwort.

The CHARACTERS are,

The leaves are like those of the Fern, but the pinnulæ are eared at their base; the fruit also is like that of the Fern.

The SPECIES are,

1. LONCHITIS (*Aspera*) aspera. Ger. *Rough Spleenwort.*

2. LONCHITIS (*Major*) aspera major. Ger. Emac *Greater rough Spleenwort.*

The first of these plants is very common in shady woods, by the sides of small rivulets, in many parts of England; but the second sort is not quite so common, and has been brought into several curious botanic gardens from the mountains in Wales. There are also a great variety of these plants in America, which at present are strangers in the European gardens; they are seldom cultivated but in botanic gardens, for the sake of variety, where, these which are hardy, should have a moist soil and a shady situation.

LONICERA. Lin. Gen. Plant. Tourn. Inst. R. H. 609. tab. 379. Upright Honeysuckle.

The CHARACTERS are,

The flower has one petal, with an oblong tube cut into five parts at the brim, and five awl-shaped stamina. Under the petal is situated a roundish germen, which afterward turns to two berries, which join at their base.

The SPECIES are,

1. LONICERA (*Xylosteum*) pedunculis bifloris baccis distinctis, foliis integerrimis pubescentibus. Prod. Leyd. 238. *Dwarf Cherry with twin red fruit; commonly called Fly Honeysuckle.*

2. LONICERA (*Alpigena*) pedunculis bifloris, baccis coadunatis didymis. Lin. Sp. Pl. 174. *Dwarf Alpine Cherry with a red twin fruit, marked with two points; commonly called red-berried upright Honeysuckle.*

3. LONICERA (*Cærulea*) pedunculis bifloris, baccis coadunatis globosis, stylis indivisis. Lin. Sp. Pl. 174. *Mountain dwarf Cherry with a single blue fruit; commonly called single, blue-berried, upright Honeysuckle.*

4. LONICERA (*Nigra*) pedunculis bifloris, baccis distinctis, foliis serratis. Prod. Leyd. 238. *Alpine Dwarf Cherry with a black twin fruit; called black-berried upright Honeysuckle.*

5. LONICERA (*Tartarica*) pedunculis bifloris, baccis distinctis, foliis cordatis obtusis. Hort. Upsal. 42. *Dwarf Cherry with a twin red fruit, and smooth heart-shaped leaves.*

6. LONICERA (*Pyrenaica*) pedunculis bifloris, baccis distinctis, foliis oblongis glabris. Lin. Sp. Pl. 174. *Pyrenean Dwarf Cherry; called Xylosteum.*

7. LONICERA (*Symphoricarpos*) capitulis lateralibus pedunculatis, foliis petiolatis. Lin. Sp. Plant. 175. *Commonly called shrubby St. Peterswort.*

The first sort has been many years cultivated in the English gardens under the title of Fly Honeysuckle. It grows naturally upon the Alps, and in other cold parts of Europe. It rises with a strong woody stalk six or eight feet high, covered with a whitish bark, divided into many branches, garnished with oblong, entire, downy, oval leaves, placed opposite. The flowers come out from the side of the branches opposite, standing upon slender foot-stalks, each sustaining two irregular white flowers, which are succeeded by two red clammy berries, which are joined at their base.

The second sort grows naturally on the Alps; this hath a short, thick, woody stem, which divides near the root into many strong woody branches, which grow erect, garnished with spear-shaped entire leaves placed opposite. The flowers stand upon very long slender foot stalks, which come out opposite on each side the branches, at the base of the leaves; they are red on their outside, but pale within, shaped like those of the former sort, but are larger, and are commonly succeeded by two oval red berries joined at their base, which have two punctures. Sometimes there is but one berry succeeding each other, which is frequently as large as a Kentish Cherry; this I believe has led some to suppose it was a distinct species, as I thought myself when I saw all the fruit upon some of the shrubs were single, but the following years I found they had twin fruit like the others.

The third sort grows naturally upon the Apennines; this is a shrub of humbler growth than either of the former, seldom rising more than four or five feet high. The branches are slender, covered with a smooth purplish bark. The joints are farther asunder, the leaves come out opposite. The foot-stalks of the flowers are very short, each sustaining two white flowers, shaped like those of the former sorts; these are succeeded by blue berries, which are single and distinct.

The fourth sort grows naturally on the Alps and Helvetian mountains; this shrub is very like the former, but the branches are slenderer. The leaves are a little sawed on their edges. The flowers have two berries succeeding them, in which consists their difference.

The fifth sort grows naturally in Tartary; this shrub grows about the same height with the two former, to which this has a great resemblance in its branches, but the leaves of this are heart-shaped, and the berries are red, growing ſometimes ſingle, at others double, and frequently there are three joined together, which are about the ſame ſize with the former.

The ſixth ſort grows naturally on the Pyrenean mountains, and alſo in Canada. This ſeldom riſes more than three or four feet high, dividing into ſeveral irregular branches, garniſhed with oblong ſmooth leaves placed oppoſite. The flowers come out from the ſide of the branches, upon ſlender foot-ſtalks, each ſuſtaining two white flowers, which are cut into five ſegments almoſt to the bottom, and are ſucceeded by berries as the other ſorts.

The ſeventh ſort grows naturally in North America; this hath a ſhrubby ſtalk, which riſes ſix or ſeven feet high, ſending out many ſlender branches, garniſhed with oval hairy leaves placed oppoſite, having very ſhort foot-ſtalks. The flowers are produced in whorls round the ſtalk; they are of an herbaceous colour. The fruit, which is hollow, and ſhaped like a pottage-pot, ripens in the winter.

Theſe ſhrubs are now propagated in the nurſery-gardens near London for ſale, and are commonly intermixed with other flowering ſhrubs for the ſake of variety; but as there is little beauty in their flowers, a few of them only ſhould be admitted to ſet off thoſe which are preferable; they are all of them very hardy plants, ſo will thrive in a cold ſituation better than in a warm one; they love a moiſt light ſoil, in which they will thrive, and produce a greater quantity of fruit than in dry ground.

They may be propagated either by ſeeds or cuttings. The ſeeds commonly lie in the ground a year before they vegetate, but require no particular culture; if they are ſown in autumn, many of them will grow the following ſpring. The cuttings ſhould be planted in autumn in a ſhady border, where they will put out roots the following ſpring, and in the following autumn they may be removed into a nurſery, to grow two years to get ſtrength, after which they ſhould be tranſplanted where they are deſigned to remain.

LORANTHUS. Vaill. Act. R. Sc. 1702. Lin. Gen. Plant. 400.

The CHARACTERS are,

The flower is tubulous, and cut into five narrow ſegments almoſt to the bottom, which are reflexed. It hath four ſtamina. The germen which is ſituated below the empalement, afterward becomes an oval pulpy fruit with one cell, including ſeveral compreſſed ſeeds.

We have but one SPECIES of this genus, viz.

LORANTHUS (*Americanus.*) Lin. Sp. Plant. 331. *Branching Loranthus with a ſcarlet flower, and black berries.*

This plant was diſcovered by Father Plumier, in the French iſlands in America, and was afterward found growing naturally at La Vera Cruz by the late Dr. Houſtoun: it riſes with a ſhrubby ſtalk eight or ten feet high, dividing into ſeveral branches, garniſhed with oblong entire leaves, which have three longitudinal nerves. The flowers are produced at the end of the branches in ſmall cluſters, and are of a ſcarlet colour, cut into five narrow ſegments almoſt to the bottom; theſe are ſucceeded by oval berries, with a pulp covering, a hard ſhell with one cell, incloſing ſeveral compreſſed ſeeds.

This plant is propagated by ſeeds, which ſhould be ſown as ſoon as they are ripe, for if they are kept out of the ground long they often miſcarry; or if they do grow it is not till the year after, ſo that thoſe ſeeds which come from America very ſeldom grow the firſt year; therefore they ſhould be ſown in pots, and kept in a moderate hot-bed the firſt ſummer, and in autumn removed into the ſtove, where, if the pots are plunged in the tan-bed between the plants, and the earth kept moiſt, in the ſpring they may be taken out, and plunged into a moderate freſh hot-bed, which will bring up the plants; theſe muſt be planted in ſeparate ſmall pots, and kept in the bark-ſtove, treating them in the ſame way as other tender plants from the ſame country.

LOTUS. Tourn. Inſt. R. H. 402. Birds-foot Trefoil.

The CHARACTERS are,

The flower is of the butterfly kind. The ſtandard is roundiſh, and reflexed backward. The wings are broad, and ſhorter than the ſtandard, cloſing together at the top. The keel is cloſed on the upper ſide, and convex on the under. It hath ten ſtamina, nine joined and one ſeparate, with an oblong taper germen, which afterward becomes a cloſe cylindrical pod with one cell, opening with two valves, having many tranſverſe partitions, in each of theſe is lodged one roundiſh ſeed.

The SPECIES are,

1. LOTUS (*Corniculatus*) capitulis depreſſis, caulibus decumbentibus, leguminibus cylindricis. Lin. Sp. Pl. 775. *Leſſer, ſmooth, corniculated Birds-foot Trefoil.*

2. LOTUS (*Anguſtiſſimum*) leguminibus ſubbinatis linearibus ſtrictis erectis, caule erecto, pedunculis alternis. Lin. Sp. Plant. 774. *Smaller, five-leaved, hairy Birds-foot Trefoil, with very narrow pods.*

3. LOTUS (*Glaber*) capitulis depreſſis, caulibus decumbentibus, foliis linearibus glabris, leguminibus linearibus. *Birds-foot Trefoil with depreſſed heads, trailing ſtalks, ſmooth linear leaves, and very narrow pods.*

4. LOTUS (*Rectus*) capitulis ſubgloboſis, caule erecto, leguminibus rectis glabris. Hort. Upſal. 221. *Talleſt hairy Birds-foot Trefoil with a conglomerated flower.*

5. LOTUS (*Cretica*) leguminibus ſubternatis, caule fruticoſo, foliis nitidis. Hort. Cliff. 372. *Silvery Birds-foot Trefoil of Crete.*

6. LOTUS (*Hirſutus*) capitulis hirſutis, caule erecto hirſuto, leguminibus ovatis. Hort. Upſal. 220. *Birds-foot Trefoil with hairy heads, an erect hairy ſtalk, and oval pods.*

7. LOTUS (*Candidus*) capitulis ſubgloboſis hirſutis, caule erecto ramoſo, hirſuto, foliis tomentoſis. *Birds-foot Trefoil with globular heads which are hairy, an upright, branching, hairy ſtalk, and woolly leaves.*

8. LOTUS (*Ornithopodoides*) leguminibus ſubquinatis arcuatis compreſſis, caulibus diffuſis. Hort. Cliff. 372. *Birds-foot Trefoil with five arched compreſſed pods, and diffuſed ſtalks.*

9. LOTUS (*Peregrinus*) leguminibus ſubbinatis linearibus compreſſis nutantibus. Hort. Cliff. 372. *Birds-foot Trefoil with two narrow, compreſſed, nodding pods.*

10. LOTUS (*Pratenſis*) leguminibus ſolitariis erectis teretibus terminalibus, caule erecto. Sauv. Monſp. 189. *Yellow meadow Birds-foot Trefoil.*

11. LOTUS (*Edulis*) leguminibus ſubſolitariis gibbis incurvis. Hort. Cliff. 370. *Birds-foot Trefoil with ſingle, convex, incurved pods.*

12. LOTUS (*Maritimus*) leguminibus ſolitariis membranaceo-quadrangulatis, bracteis lanceolatis. Flor. Suec. 610. *Yellow maritime Birds-foot Trefoil with a ſmooth leaf.*

13. LOTUS (*Conjugatus*) leguminibus conjugatis membranaceo-quadrangulatis, bracteis oblongo-ovatis. Lin. Sp. Pl. 774. *Yellow Birds-foot Trefoil with angular pods.*

14. LOTUS (*Tetragonolobus*) leguminibus ſolitariis membranaceo-quadrangulatis, bracteis ovatis. Hort. Upſ. 220. *Red Birds-foot Trefoil with angular pods; commonly called winged Pea.*

15. LOTUS (*Cytiſoides*) capitulis dimidiatis, caule diffuſo ramoſiſſimo, foliis tomentoſis. Prod. Leyd. 387. *Podded, yellow, maritime Birds-foot Trefoil, with the appearance of Cytiſus.*

16. LOTUS (*Jacobæus*) leguminibus ſubternatis, caule herbaceo erecto, foliis linearibus. Hort. Cliff. 372. *Narrow-*

row-leaved Bird's-foot Trefoil of St. James's Island, with a yellow purplish flower.

17. LOTUS (*Dorycnium*) capitulis aphyllis, foliis sessilibus quinatis. Lin. Sp. Pl. 776. *Dorycnium of Montpelier.*

The first, second, and third sorts, grow naturally in many parts of England, so are rarely admitted into gardens. When these grow in moist land and a shady situation, they send out stalks near two feet long; but upon dry, chalky, and gravelly ground, their stalks are not more than three or four inches, and lie flat upon the ground. I have always observed in those pastures where these plants have grown, that the cattle of all sorts have avoided eating them, but the Grass all round them has been eaten very bare. I have cut the plants when young, and given it to various kinds of animals, but could never get them to eat it; and yet the seeds of these have been gathered and sold by some quacks in husbandry, under the title of Lady's Finger Grass, to be sown as an improvement to land for pasture.

The roots of these are perennial, so are difficult to get out when they have had long possession of the land, for they produce great quantities of seeds, which is cast about by the elasticity of the pods when ripe, to a considerable distance.

The fourth sort grows naturally in the south of France, in Italy and Sicily; this has by some been supposed the Cytisus of Virgil, but without foundation; it hath a strong perennial root, from which arise many upright strong stalks from five to six feet high, garnished at every joint by a trifoliate leaf, whose lobes are wedge-shaped; at the base of the foot-stalk are placed two heart-shaped lobes sitting close to the branch; the leaves are hairy on their under side; the flowers are produced at the end of the branches almost in globular heads, sitting close to the foot-stalk: they are of a pale flesh colour, and are succeeded by smooth strait pods almost an inch long, which change to a brown colour when ripe, and contain several roundish seeds. It is rarely cultivated but in botanic gardens for variety, but if any person has an inclination to cultivate this plant for feeding of cattle, it may be done in the same way as the Lucern, for which there is full directions in the article MEDICAGO.

The fifth sort grows naturally in Syria and Crete; this rises with slender stalks which require support, from three to four feet high, sending out a few side branches, garnished at each joint with neat, shining, silvery leaves, which are trifoliate, and have two appendages at the base of their foot-stalks. The foot-stalks of the flowers, which are from two to three inches long, arise from the side of the branches, and sustain heads of yellow flowers which part in the middle, each head containing four or six flowers, which are succeeded by long taper pods filled with roundish seeds, which ripen in the autumn.

This sort has a perennial stalk, but is too tender to live in the open air in England, except the winter proves very mild, so is kept in pots and removed into the green-house in autumn, and treated like other hardy exotic plants, which only require protection from the frost. It is propagated by seeds, which, if sown on a bed of light earth in April, the plants will come up in about a month after, and in another month will be fit to remove, when they should be each put into a separate small pot, placing them in the shade till they have taken new root; then they may be removed to a sheltered situation, where they may remain till autumn.

It may be also propagated by cuttings, which may be planted during any of the summer months, upon a bed of light earth, covering them close with a bell or hand-glass, and screening them from the sun; in about five or six weeks they will have taken root, when they must be inured to bear the open air, and soon after may be planted in pots, and treated in the same way as the seedling plants.

The sixth sort grows naturally in the south of France and Italy; this hath a perennial hairy stalk, which rises three feet high, and divides into several branches, garnished with hoary trifoliate leaves, having two appendages at the base of the stalk; the flowers are collected into heads sitting upon pretty long foot-stalks, which come out of the side of the stalks. They are of a dirty white colour, with a few marks of pale red, and are succeeded by short thick pods of a Chestnut colour, containing several roundish seeds which ripen in the autumn. This is propagated by seeds in the same way as the last sort; the plants will live in the open air in moderate winters, but it will be proper to keep one or two plants in pots to be sheltered, lest those abroad should be destroyed by severe frost.

The seventh sort grows naturally in Sicily; this rises with an upright woody stalk near three feet high, garnished with leaves like the sixth, but much whiter, covered with a short woolly down; the flowers grow in close heads like the last; they are whiter, and are succeeded by short pods, which contain many yellow seeds. This is too tender to live in the open air in England through the winter, so the plants must be kept in pots and housed in autumn. It is propagated in the same way as the fifth sort, and requires the same culture.

The eighth sort grows naturally in Sicily; this is an annual plant, which sends out from the root many stiff stalks from one to two feet high, garnished with trifoliate leaves, having two appendages at their base; the foot-stalks of the flower rise from the wings of the stalks, which are terminated by a cluster of yellow flowers, succeeded by flat pods two inches long, bent like an arch, and have many joints, separating the cells in which the seeds are lodged.

This is propagated by seeds, which should be sown early in April upon an hot-bed or border exposed to the sun, where the plants are to remain; when they come up they must be thinned, leaving them near two feet asunder, and kept clean from weeds, which is all the culture they require.

The ninth sort grows naturally in Spain and Portugal; this is an annual plant like the former, but doth not branch so much; the small leaves are rounder and smoother; the foot-stalks are shorter, and seldom sustain more than two flowers; these are succeeded by two very narrow pods, which hang downward.

The tenth sort grows naturally in the south of France; this hath a perennial root, from which is sent out several hairy stalks near a foot long, garnished with trifoliate hairy leaves, standing upon short foot-stalks, with two appendages at the base of the foot-stalk; the flowers stand upon pretty long foot-stalks singly, which rise from the end of the branches. The flowers are yellow, standing erect, and are succeeded by taper erect pods an inch and a half long. It is propagated by seeds, which should be sown where the plants are to remain, and must be treated as the two former sorts, but the roots of this will continue several years.

The eleventh sort grows naturally in Sicily and Crete, where the pods are eaten by the poorer inhabitants when they are young. It also grows about Nice. This is an annual plant, from whose roots come out several trailing stalks a foot long, garnished at each joint with trifoliate roundish leaves having appendages. The flowers stand singly upon long foot-stalks, which arise from the side of the branches; they are yellow, small, and are succeeded by single pods which are thick, and arched with a deep furrow on the outside. In cold summers the seeds will not ripen here.

This muſt have the ſame culture as the annual ſorts before-mentioned.

The twelfth ſort grows near the borders of the ſea in France, Spain, and Italy; this hath a perennial root, ſending out many ſlender ſtalks about a foot and a half long, which trail upon the ground, garniſhed with trifoliate leaves, which are ſmooth, and have two appendages to the baſe of the foot-ſtalk. The flowers ſtand ſingly upon very long foot-ſtalks, ariſing from the wings of the ſtalk; they are yellow, and are ſucceeded by ſingle pods near two inches long, having four leafy membranes running longitudinally at the four corners. It is propagated by ſeed in the ſame way as the tenth ſort.

The thirteenth ſort grows naturally in the ſouth of France and in Italy; this is an annual plant, from whoſe roots are ſent forth ſeveral branching ſtalks a foot long, garniſhed with trifoliate leaves, whoſe lobes are acute-pointed, and have two oblong oval appendages at the baſe of their foot-ſtalks; the foot-ſtalks of the flowers ariſe from the wings of the branches, each ſuſtaining two yellow flowers, which are ſucceeded by taper pods, having four leafy membranes running longitudinally their length. It is propagated by ſeeds in the ſame way as the annual ſorts before-mentioned.

The fourteenth ſort grows naturally in Sicily, but has been long cultivated in the Engliſh gardens; it was formerly cultivated as an eſculent plant. The green pods of it were dreſſed and eaten as Peaſe, which the inhabitants of ſome of the northern counties ſtill continue, but they are very coarſe, ſo not agreeable to the taſte of thoſe who have been accuſtomed to better fare.

It is an annual plant, which is cultivated in the flower-gardens near London for ornament. This ſends out from the root ſeveral decumbent ſtalks about a foot long, garniſhed with trifoliate oval leaves, with two appendages at the baſe of their foot-ſtalks; from each joint ariſe alternately the foot-ſtalks of the flowers, each ſuſtaining one large red flower at the top, with three leaves juſt under the flower. After the flower fades the germen becomes a ſwelling taper pod, having four leafy membranes or wings running longitudinally.

The ſeeds of this ſort are commonly ſown in patches, five or ſix ſeeds being ſown near each other in the borders of the pleaſure-garden, where they are deſigned to remain. If the ſeeds do all grow, ſome of the plants may be pulled up, leaving only two or three in each patch; afterward they will require no other care but to keep them clean from weeds.

The fifteenth ſort grows near the borders of the ſea, in the ſouth of France and Spain. This is a perennial plant, ſending out from the root many ſtalks, garniſhed with roundiſh trifoliate leaves with two appendages, covered with a woolly down; the flowers ſtand upon ſhort foot-ſtalks, four or ſix growing in a divided head; they are yellow, and are ſucceeded by taper pods filled with roundiſh ſeeds. This is propagated by ſeeds, which ſhould be ſown in the ſpring in the place where the plants are to remain, and muſt be treated in the ſame manner as the hardy perennial ſorts before-mentioned.

The ſixteenth ſort grows naturally in the iſland of St. James. This hath a ſlender ſtalk, which is woody, riſing from two to three feet high, ſending out many ſlender herbaceous branches, garniſhed with narrow gray leaves, which are ſometime trifoliate, and at others there are five narrow lobes to each; theſe ſit cloſe to the branches. The flowers are produced from the ſide of the ſtalks towards their end, upon very ſlender foot-ſtalks, each ſuſtaining four or five flowers collected in a head, of a yellowiſh, deep, purple colour, which are ſucceeded by taper ſlender pods little more than an inch long, containing five or ſix ſmall roundiſh ſeeds. It is too tender to live abroad in England, ſo the plants muſt be kept in pots, and in the winter placed in a warm airy glaſs-caſe, but in the ſummer they ſhould be placed abroad in a ſheltered ſituation. It may be eaſily propagated by cuttings during the ſummer ſeaſon, in the ſame way as the fifth ſort, and alſo by ſeeds; but the plants which have been two or three times propagated by cuttings ſeldom are fruitful.

The ſeventeenth ſort grows naturally about Montpelier. It riſes with weak ſhrubby ſtalks three or four feet high, ſending out many ſlender branches, which are thinly garniſhed with ſmall hoary leaves, compoſed of five lobes in form of a hand, which ſit cloſe to the branches. The flowers are produced at the extremity of the branches in ſmall heads; they are very ſmall, ſo make no great appearance, and are ſucceeded by ſhort pods, containing two or three ſmall round ſeeds. This ſhrub will live in the open air, if it is planted in a dry ſoil and a warm ſituation. It is propagated by ſeeds, which will come up in any common border.

LOTUS ARBOR. See Celtis.

LOVE-APPLE. See Lycopersicon.

LUDVIGIA. Lin. Gen. Plant. 142.

The Characters are,

The flower conſiſts of four ſpear-ſhaped petals, which are equal. In the center of the flower is ſituated four ſtamina. The germen, which ſits under the flower, afterward becomes a four-cornered fruit, crowned with the empalement, and has four cells, which are full of ſmall ſeeds.

We have but one Species of this genus, viz.

Ludvigia (*Alternifolia*) foliis alternis lanceolatis. Lin. Sp. Pl. 118. *Ludvigia with alternate ſpear-ſhaped leaves.*

We have no Engliſh name for this plant, but it is very near akin to the Onagra, or Tree Primroſe.

This plant grows naturally in South America. It is annual, and riſes with an upright branching ſtalk a foot high, garniſhed with ſpear-ſhaped leaves placed alternate. The flowers come out ſingly at the foot-ſtalks of the leaves; they are compoſed of four ſmall yellow petals, ſtanding upon ſhort foot-ſtalks, and are ſucceeded by roundiſh ſeed-veſſels with four leafy membranes; they open in four cells, including many ſmall ſeeds.

The plants muſt be raiſed on a hot-bed in the ſpring, and treated in the ſame manner as hath been directed for the Amaranthus; for if they are not brought forward in the ſpring, they ſeldom produce good ſeeds in England.

LUFFA. See Momordica.

LUNARIA. Tourn. Inſt. R. H. 218. tab. Gen. 105. Sattin-flower, or Honeſty.

The Characters are,

The flower has four petals placed in form of a croſs, which are entire. It hath ſix awl-ſhaped ſtamina; four of theſe are the length of the empalement, the other two are ſhorter; and an oblong oval germen which afterward becomes an erect, plain, compreſſed, eliptical pod, ſitting upon a ſmall foot-ſtalk, terminated by the ſtyle, having two cells opening with two valves, which are parallel, incloſing ſeveral compreſſed kidney-ſhaped ſeeds, which are bordered, ſitting in the middle of the pod.

The Species are,

1. Lunaria (*Redivava*) ſiliculis oblongis. Lin. Sp. Pl. 653. *Sattin-flower with oblong pods.*

2. Lunaria (*Annua*) ſiliculis ſubrotundis. Lin. Sp. Pl. 653. *Sattin-flower with a rounder pod.*

3. Lunaria (*Ægyptiaca*) foliis ſuprà decompoſitis, foliolis trifidis, ſiliculis oblongis pendulis. *Moonwort with leaves decompounded, whoſe lobes are trifid, and oblong hanging pods.*

4. Lunaria

4. LUNARIA (*Perennis*) perennis, siliculis oblongis, foliis lanceolatis incanis. *Perennial Moonwort with oblong pods, and spear-shaped hoary leaves.*

The first sort grows naturally in Hungary, Istria, and Austria. It is a biennial plant, which perishes soon after the seeds are ripe; it rises with a branching stalk from two to three feet high, covered with a reddish hairy bark, sending out branches on every side from the ground upward, garnished with heart-shaped leaves placed alternately, ending in acute points. The flowers terminate the branches in clusters; they are composed of four purplish heart-shaped petals, placed in form of a cross. These are succeeded by large, flat, roundish pods, with two cells, inclosing two rows of flat kidney-shaped seeds, which have a border round them. These pods, when ripe, turn to a clear white or sattin colour, and are transparent.

This is propagated by seeds, which should be sown in the autumn, for those which are sown in the spring often miscarry, or lie a long time in the ground. The plants will grow in almost any soil, but love a shady situation. They require no other culture but to keep them clean from weeds. If the seeds are permitted to scatter, the plants will rise without any furthei care; and if they are left unremoved, they will grow much larger than those which are transplanted.

The second sort grows naturally upon the mountains in Italy. This hath stalks and leaves very like the first, but the flowers are rather larger, and of a lighter purple colour; but the principal difference is in the pods of this being longer and narrower than those of the other; it requires the same culture.

The third sort is an annual plant, which grows naturally in Egypt. This rises with a smooth branching stalk a foot high, garnished with winged leaves, composed of several pair of lobes ranged along the midrib, terminated by an odd one; these lobes are of unequal sizes, and vary in their form; some of them are almost entire, and others are cut at their extremities into three parts; they are smooth, and of a lucid green. The flowers stand upon pretty long slender foot-stalks, which come out from the side and at the end of the branches, in loose small clusters; they are of a purple colour, and are succeeded by oblong compressed pods, which hang downward.

This is propagated by seeds, which should be sown upon an open border where the plants are to remain; if they are sown soon after they are ripe, the plants will come up in the autumn, and live through the winter without shelter, and these will flower early the following summer. When the plants come up, they will require no other care but to keep them clean from weeds, and thin them where they are too close. If the seeds are permitted to scatter in the autumn the plants will rise without care, and may be treated in the same way.

The fourth sort grows naturally in the Archipelago. This is biennial; the stalks rise a foot high, covered with a white hairy bark, garnished with spear-shaped hoary leaves sitting close to the branches. The branches are terminated by loose spikes of yellow flowers, which are succeeded by oblong flat pods, containing flat kidney-shaped seeds, which ripen in the autumn.

This sort is propagated by seeds, which, if sown in the autumn, will succeed better than in the spring; they should be sown on a warm border and a dry poor soil, otherwise they will not live through the winter, but in a rubbishing soil the plants will do best.

LUPINUS. Tourn. Inst. R. H. 392. tab. 213. Lupine.

The CHARACTERS are,

The flower is of the butterfly kind; the standard is roundish, heart-shaped, the sides reflexed and compressed. The wings are nearly oval, and close at their base; the keel is narrow, falcated, and ends in a point. It hath ten stamina joined at their base in two bodies, but are distinct above. In the center is situated a hairy compressed germen, which afterward becomes a large, oblong, thick-shelled pod with one cell, ending with an acute point, including several roundish compressed seeds.

The SPECIES are,

1. LUPINUS (*Varius*) calycibus semiverticillatis appendiculatis, labio superiore bifido, inferiore subtridentato. Hort. Cliff. 499. *Wild Lupine with a purple flower, and round variegated seed; commonly called the lesser blue Lupine.*

2. LUPINUS (*Angustifolius*) calycibus alternis appendiculatis, labio superiore bipartito integro. Lin. Sp. Pl. 721. *Narrow-leaved, taller, blue Lupine.*

3. LUPINUS (*Luteus*) calycibus verticillatis appendiculatis, labio superiore bipartito, inferiore tridentato. Hort. Cliff. 499. *The common yellow Lupine.*

4. LUPINUS (*Hirsutus*) calycibus verticillatis appendiculatis, labio superiore inferioreque integris. Hort. Cliff. 499. *Foreign, greater, hairy Lupine, with a large blue flower; commonly called the great blue Lupine.*

5. LUPINUS (*Albus*) calycibus alternis inappendiculatis, labio superiore integro, inferiore tridentato. Hort. Cliff. 499. *Garden or manured Lupine, with a white flower.*

6. LUPINUS (*Perennis*) calycibus alternis inappendiculatis, labio superiore emarginato, inferiore integro. Lin. Sp. Plant. 721. *Smaller, perennial, creeping, blue Lupine of Virginia.*

The first sort grows naturally amongst the Corn in the south of France and Italy, and in great abundance in Sicily. This is an annual plant, which rises with a firm, strait, channelled stalk, near three feet high, divided toward the top into several branches, garnished with hand-shaped leaves, composed of five, six, or seven oblong lobes, which join in one center at their base. The flowers are produced in spikes at the end of the branches, standing half round the stalk in a sort of whorl; they are of a light blue colour, and are succeeded by strait taper pods with one cell, inclosing a row of roundish seeds.

It is propagated in the borders of the pleasure-garden for ornament, by sowing the seeds in April in the places where they are to remain; and when the plants come up they should be thinned where they are too close, and kept clean from weeds, which is all the culture they require.

The second sort has much the appearance of the first, but the stalks rise higher; the leaves have more lobes, and stand upon longer foot-stalks; the lobes are blunt-pointed, and the seeds are variegated. This requires the same culture as the first, and flowers at the same time.

The third sort is the common yellow Lupine; this grows naturally in Sicily. It rises about a foot high, with a branching stalk, garnished with hand-shaped leaves, composed of nine narrow hairy lobes, which join at their base to the foot-stalks. The flowers are yellow, and are produced in loose spikes at the end of the branches, standing in whorls round the stalks. These are succeeded by flattish hairy pods about two inches long, inclosing four or five roundish seeds, a little compressed on their side. This sort flowers at the same time as the former; but to have a succession of the flowers, the seeds are sown at different times, viz. in April, May, and June, but those only which are first sown will ripen their seeds. It may be cultivated in the same manner as the two former, and is equally hardy.

The fourth sort is supposed to be a native of India. It is an annual plant, which rises with a strong, firm, channelled stalk, from three to four feet high, covered with a soft brownish down, dividing upward into several strong branches, garnished with hand-shaped leaves, composed of nine, ten, or eleven wedge-shaped hairy lobes, which are narrow at their base, where they join the foot-stalk. The flowers are placed in whorls round the stalks above each other, forming a loose spike at the end of the branches; they are large, and of a beautiful blue colour, but have no scent. The pods of this sort are large, almost an inch broad, and three inches long, inclosing three large roundish seeds compressed on their sides, very rough, and of a purplish brown colour. There is a variety of this with flesh coloured flowers, which is commonly called the Rose Lupine; it differs from the blue only in the colour of the flower, but this difference is permanent, for neither of the sorts vary.

This is generally late in the ripening of the seeds, so that unless the autumn proves warm and dry, they do not ripen well in England; therefore the best way to have good seeds is to sow them in September, close to a warm wall on dry ground, where they will live through our ordinary winters; and these plants will flower early the following summer, so there will be time for the seeds to ripen before the rains fall in the autumn, which frequently causes the seeds to rot which are not ripe.

The fifth sort grows naturally in the Levant, but is cultivated in some parts of Italy, as other pulse for food. This hath a thick upright stalk about two feet high, which divides toward the top into smaller branches, garnished with hand-shaped leaves, composed of seven or eight narrow, oblong, hairy lobes, which join at their base, of a dark grayish colour, with a silvery down. The flowers are produced in loose spikes at the end of the branches; they are white, and sit close to the stalk, and are succeeded by hairy strait pods about three inches long, a little compressed on the sides, which contain five or six flattish white seeds, having a little cavity like a navel, in that part which is fixed to the pod. This is an annual plant, which is cultivated for ornament in the pleasure-garden. The seeds must be sown in the places where the plants are to remain, and may be treated in the same way as the first sort.

The sixth sort grows naturally in Virginia, and other of the northern parts of America. It hath a perennial creeping root, from which arise several erect channelled stalks a foot and a half high, garnished with hand-shaped leaves, composed of ten or eleven spear-shaped lobes, which join at their base. The flowers grow in long loose spikes, which terminate the stalks, and are placed without order on every side the stalk; they are of a pale blue colour, and are succeeded by pods, having three or four seeds, which ripen in August, and are soon scattered if they are not gathered; for after a little moisture the sun causes the pods to open with an elasticity, and cast out the seeds to a distance. This sort is propagated by seeds as the former, which should be sown where the plants are to remain; for although the root is perennial, yet it runs so deep into the ground, as that it cannot be taken up entire; and if the root is cut or broken, the plant never thrives well after. I have traced some of the roots of this plant, which have been three feet deep in the ground in one year from seed; they also spread out far on every side, so that they must have room; therefore the young plants should not be left nearer than three feet asunder.

LUPULUS. Tourn. Inst. R. H. 535. tab. 309. The Hop.

The CHARACTERS are,

It has male and female flowers upon different plants. The male flower hath no petal, but has five short hairy stamina. The female flowers have neither petal or stamina, but a small germen situated in the center, which afterward turns to a roundish seed-vessel covered with a thick skin, inclosed in the base of the empalement.

We have but one SPECIES of this plant, viz.

LUPULUS (*Humulus*) mas & femina. C. B. P. 298. *Male and female Hop.*

The male Hop grows wild by the side of hedges and upon banks in many parts of England. The young shoots of these plants are often gathered by the poor people, and boiled as an esculent herb; but these must be taken very young, otherwise they are tough and stringy. This is easily distinguished by the flowers, which are small, and hang in long loose bunches from the side of the stalks, abounding with farina on their summits, and have no Hops succeeding to the flowers.

The female Hop is the sort which is cultivated for use; of this sort, the people who cultivate them reckon three different varieties: as first, the long and square Garlick Hop, the long white Hop, and the oval Hop, all which are indifferently cultivated in England.

There being the greatest plantation of Hops in Kent of any county in England, it is very probable that their method of planting and ordering them should be the best.

As for the choice of their Hop-grounds, they esteem the richest and strongest grounds as the most proper; and if it be rocky within two or three feet of the surface, the Hops will prosper well; but they will by no means thrive on a stiff clay, or spongy wet land.

The Kentish planters account new land best for Hops; they plant their Hop-gardens with Apple-trees at a large distance, and with Cherry-trees between; and when the land hath done its best for Hops, which they reckon it will in about ten years, the trees may begin to bear. The Cherry-trees last about thirty years, and by that time the Apple-trees are large they cut down the Cherry-trees.

As to the situation of a Hop-ground, one that inclines to the south or west is the most eligible; but if it be exposed to the north-east or south-west winds, there should be a shelter of some trees at a distance, because the north-east winds are apt to nip the tender shoots in the spring, and the south-west winds frequently break and blow down the poles at the latter end of the summer, and very much endanger the Hops.

Hops require to be planted in a situation so open, as that the air may freely pass round and between them, to dry up and dissipate the moisture, whereby they will not be so subject to fire-blasts, which often destroy the middles of large plantations, while the outsides remain unhurt.

As for the preparation of the ground for planting, it should, the autumn before, be ploughed and harrowed even; and then lay upon it in heaps a good quantity of fresh rich earth, or well rotted dung and earth mixed together, sufficient to put half a bushel in every hole to plant the Hops in, unless the natural ground be very fresh and good.

The hills where the Hops are to be planted should be ten feet asunder, that the air may freely pass between them, for in close plantations they are very subject to what the Hop-planters call the fire-blast.

If the ground is intended to be ploughed with horses between the hills, it will be best to plant them in squares chequerwise; but if the ground is so small, that it may be done with the breast-plough or spade, the holes should be ranged in a quincunx form. Which way soever you make use of, a stake should be stuck down at all the places where the hills are to be made.

Persons ought to be very curious in the choice of the plants as to the kind of Hop; for if the Hop-garden be planted with a mixture of several sorts of Hops, that ripen at several times, it will cause a great deal of trouble, and be a great detriment to the owner.

The

The two beſt ſorts are the white and the gray bind; the latter is a large ſquare Hop, more hardy, and is the more plentiful bearer, and ripens later than the former.

There is alſo another ſort of the white bind, which ripens a week or ten days before the common; but this is tenderer and a leſs plentiful bearer. but it has this advantage, it comes firſt to market.

But if three grounds, or three diſtant parts of one ground, be planted with theſe three ſorts, there will be this conveniency, that they may be picked ſucceſſively as they become ripe.

If there be a ſort of Hop you value, and would increaſe plants and ſets from, the ſuperfluous binds may be laid down when the Hops are tied, cutting off the tops, and burying them in the hill; or when the Hops are dreſſed, all the cuttings may be ſaved; for almoſt every plant will grow, and become a good ſet the next ſpring.

As to the ſeaſon of planting Hops, the Kentiſh planters beſt approve the months of October and March, both which ſometimes ſucceed very well; but the ſets are not to be had in October, unleſs from ſome ground that is to be deſtroyed; and likewiſe there is ſome danger that the ſets may be rotted, if the winter proves very wet; therefore the moſt uſual time of procuring them is in March, when the Hops are cut and dreſſed.

As to the manner of planting the ſets, there ſhould be five good ſets planted in every hill, one in the middle, and the reſt round about ſloping, the tops meeting at the center; they muſt ſtand even with the ſurface of the ground; let them be preſſed cloſe with the hand, and covered with fine earth, and a ſtick ſhould be placed on each ſide the hill to ſecure it.

The ground being thus planted, all that is to be done more that ſummer is to keep the hills clear from weeds, and to dig up the ground about the month of May, and to raiſe a ſmall hill round about the plants; in June you muſt twiſt the young binds or branches together into a bunch or knot, for if they are tied up to ſmall poles the firſt year, in order to have a few Hops from them, it will not countervail the weakening of the plants.

A mixture of compoſt or dung being prepared for your Hop-ground, the beſt time for laying it on, if the weather prove dry, is about Michaelmas, that the wheels of the dung-cart may not injure the Hops, nor furrow the ground: if this be not done then, you muſt be obliged to wait till the froſt has hardened the ground, ſo as to bear the dung-cart; and this is alſo the time to carry on your new poles, to recruit thoſe that are decayed, and to be caſt out every year.

If you have good ſtore of dung, the beſt way will be to ſpread it in the alleys all over the ground, and to dig it in the winter following. The quantity they will require will be forty loads to an acre, reckoning about thirty buſhels to the load.

If you have not dung enough to cover all the ground in one year, you may lay it on one part one year, and on the reſt in another, or a third; for there is no occaſion to dung the ground after this manner, oftener than once in three years.

Thoſe who have but a ſmall quantity of dung, uſually content themſelves with laying on about twenty loads upon an acre every year; this they lay only on the hills, either about November or in the ſpring; which laſt ſome account the beſt time when the Hops are dreſſed, to cover them after they are cut; but if it be done at this time, the compoſt or dung ought to be very well rotted and fine.

As to the dreſſing of the Hops, when the Hop-ground is dug in February or March, the earth about the hills, and very near them, ought to be taken away with a ſpade, that you may come the more conveniently at the ſtock to cut it.

About the end of February, if the Hops were planted the ſpring before, or if the ground be weak, they ought to be dreſſed in dry weather; but elſe, if the ground be ſtrong and in perfection, the middle of March will be a good time; and the latter end of March, if it be apt to produce over-rank binds, may be ſoon enough.

Then having with an iron picker cleared away all the earth out of the hills, ſo as to clear the ſtock to the principal roots, with a ſharp knife you muſt cut off all the ſhoots which grew up with the binds the laſt year; and alſo all the young ſuckers, that none be left to run in the alley and weaken the hill. It will be proper to cut one part of the ſtock lower than the other, and alſo to cut that part low that was left higheſt the preceding year. By purſuing this method, you may expect to have ſtronger buds, and alſo keep the hill in good order.

In dreſſing thoſe Hops that have been planted the year before, you ought to cut off both the dead tops, and the young ſuckers which have ſprung up from the ſets, and alſo to cover the ſtocks with fine earth a finger's length in thickneſs.

About the middle of April the Hops are to be poled, when the ſhoots begin to ſprout up; the poles muſt be ſet to the hills deep into the ground, with a ſquare iron picker or crow, that they may the better endure the winds; three poles are ſufficient for one hill. Theſe ſhould be placed as near the hills as may be, with their bending tops turned outwards from the hill, to prevent the binds from entangling; and a ſpace between two poles ought to be left open to the ſouth, to admit the ſun beams.

The poles ought to be in length ſixteen or twenty feet, more or leſs, according as the ground is in ſtrength; and great care is to be taken not to overpole a young or weak ground, for that will draw the ſtock too much, and weaken it. If a ground be overpoled, you are not to expect a good crop from it; for the branches which bear the Hops, will grow very little till the binds have over-reached the poles, which they cannot do when the poles are too long. Two ſmall poles are ſufficient for a ground that is young.

If you wait till the ſprouts or young binds are grown to the length of a foot, you will be able to make a better judgment where to place the largeſt poles: but if you ſtay till they are ſo long as to fall into the alleys, it will be injurious to them, becauſe they will entangle one with another, and will not claſp about the pole readily.

Maple and Aſpen poles are accounted the beſt for Hops, on which they are thought to proſper beſt, becauſe of their warmth; or elſe, becauſe the climbing of the Hops is furthered by means of the roughneſs of the bark. But for laſtingneſs Aſhen or Willow poles are preferable; but Cheſtnut poles are the moſt durable of all.

If after the Hops are grown up, you find any of them have been under-poled, taller poles may be placed near thoſe that are too ſhort, to receive the binds from them.

As to the tying of Hops, the buds that do not claſp of themſelves to the neareſt pole when they are grown to three or four feet high, muſt be guided to it by the hand, turning them to the ſun, whoſe courſe they will always follow. They muſt be bound with withered Ruſhes, but not ſo cloſe as to hinder them from climbing up the pole.

This you muſt continue to do till all the poles are furniſhed with binds, of which two or three are enough for a pole; and all the ſprouts and binds that you have no occaſion for are to be plucked up; but if the ground be young, then none of theſe uſeleſs binds ſhould be plucked up, but ſhould be wrapped up together in the middle of the hill.

When

When the binds are grown beyond the reach of your hands, if they forsake the poles, you should make use of a stand-ladder in tying them up.

Toward the latter end of May, when you have made an end of tying them, the ground must have the summer dressing: this is done by casting up with the spade some fine earth into every hill; and a month after this is done, you must hoe the alleys with a Dutch hoe, and make the hills up to a convenient bigness.

When the Hops blow, you should observe if there be any wild barren hills among them, and mark them, by driving a sharpened stick into every such hill, that they may be digged up and replanted.

Hops, as well as other vegetables, are liable to distempers and disasters, and among the rest to the fen.

The Rev. Dr. Hales, in his excellent treatise of Vegetable Staticks, treating of Hops, gives us the following account of the state of Hops in Kent, in the year 1725, that he received from Mr. Austen of Canterbury, which is as follows:

In mid April not half the shoots appeared above ground, so that the planters knew not how to pole them to the best advantage.

This defect of the shoot, upon opening the hills, was found to be owing to the multitude and variety of vermin that lay preying upon the roots; the increase of which, was imputed to the long and almost uninterrupted series of dry weather for three months before. Towards the end of April, many of the Hop vines were infested with flies.

About the 20th of May there was a very unequal appearance, some vines being run seven feet, others not above three or four; some just tied to the poles, and some not visible; and this disproportionate inequality in their size, continued through the whole time of their growth.

The flies now appeared upon the leaves of the forwardest vines, but not in such numbers here as they did in most other places. About the middle of June the flies increased, yet not so as to endanger the crop; but in distant plantations they were exceedingly multiplied, so as to swarm towards the end of the month.

June the 27th some specks of fen appeared. From this day to the 9th of July was very dry weather. At this time, when it was said that the Hops in most parts of the kingdom looked black and sickly, and seemed past recovery, ours held it out pretty well, in the opinion of the most skilful planters.

The great leaves were indeed discoloured, and a little withered, and the fen was somewhat increased. From the 9th of July to the 23d, the fen increased a great deal; but the flies and lice decreased, it raining much daily. In a week more the fen, which seemed to be almost at a stand, was considerably increased, especially in those grounds where it first appeared.

About the middle of August the vines had done growing both in stem and branch, and the forwardest began to be in the Hop, the rest in bloom; the fen continued spreading where it was not before perceived; and not only the leaves, but many of the burs were also tainted with it.

About the 20th of August some of the Hops were infected with the fen, and whole branches corrupted by it. Half the plantations had pretty well escaped hitherto, and from this time the fen increased but little; but several days wind and rain the following week so distorted them, that many of them began to dwindle, and at last came to nothing; and of those that then remained in bloom, some never turned to Hops; and of the rest which did, many of them were so small, that they very little exceeded the bigness of a good thriving bur.

We did not begin to pick till the 8th of September, which is eighteen days later than we began the year before; the crop was little above two hundred on an acre round, and not good. The best Hops sold this year at Way-hill for 16l. the hundred.

About the middle of July Hops begin to blow, and will be ready to gather about Bartholomew-tide. A judgment may be made of their ripeness, by their strong scent, their hardness, and the brownish colour of their seed.

When by these tokens they appear to be ripe, they must be picked with all the expedition possible; for if at this time a storm of wind should come, it would do them great damage, by breaking the branches, and bruising and discolouring the Hops; and it is very well known that Hops, being picked green and bright, will sell for a third part more than those which are discoloured and brown.

The most convenient way of picking them is into a long square frame of wood, called a bin, with a cloth hanging on tenter-hooks within it, to receive the Hops as they are picked.

The frame is composed of four pieces of wood joined together, supported by four legs, with a prop at each end to bear up another long piece of wood, placed at a convenient height over the middle of the bin; this serves to lay the poles upon, which are to be picked.

This bin is commonly eight feet long, and three feet broad; two poles may be laid on it at a time, and six or eight persons may work at it, three or four on each side.

It will be best to begin to pick the Hops on the east or north-side of your ground, if you can do it conveniently; this will prevent the south-west wind from breaking into the garden.

Having made choice of a plot of the ground containing eleven hills square, place the bin upon the hill, which is in the center, having five hills on each side; and when these hills are picked, remove the bin into another piece of ground of the same extent, and so proceed till the whole Hop-ground is finished.

When the poles are drawn up to be picked, you must take great care not to cut the binds too near the hills, especially when the Hops are green, because it will make the sap to flow excessively.

The Hops must be picked very clean, i. e. free from leaves and stalks; and, as there shall be occasion, two or three times in a day the bin must be emptied into a Hop-bag made of coarse linen cloth, and carried immediately to the oast or kiln, in order to be dried; for if they should be long in the bin or bag, they will be apt to heat, and be discoloured.

If the weather be hot, there should no more poles be drawn than can be picked in an hour, and they should be gathered in fair weather, if it can be, and when the Hops are dry; this will save some expence in firing, and preserve their colour better when they are dried.

The best method of drying Hops is with charcoal on an oast or kiln, covered with hair-cloth, of the same form and fashion that is used for drying malt. There is no need to give any particular directions for the making it, since every carpenter or bricklayer, in those countries where Hops grow, or malt is made, knows how to build them.

The kiln ought to be square, and may be of ten, twelve, fourteen, or sixteen feet over at the top, where the Hops are laid, as your plantation requires, and your room will allow. There ought to be a due proportion between the height and the breadth of the kiln, and the beguels of the steddle where the fire is kept, viz. if the kiln be twelve feet square on the top, it ought to be nine feet high from the fire, and the steddle ought to be six feet and a half square, and so proportionable in other dimensions.

The

The Hops must be spread even upon the oast a foot thick or more, if the depth of the curb will allow it, but care is to be taken not to overload the oast, if the Hops be green or wet.

The oast ought to be first warmed with a fire before the Hops are laid on, and then an even steady fire must be kept under them; it must not be too fierce at first, lest it scorch the Hops; nor must it be suffered to sink or slacken, but rather be increased till the Hops be nearer dried, lest the moisture or sweat, which the fire has raised, fall back or discolour them. When they have lain about nine hours, they must be turned, and in two or three hours more they may be taken off the oast. It may be known when they are well dried by the brittleness of the stalks, and the easy falling off of the Hop leaves.

It is found by experience, that the turning of Hops, though it be after the most easy and best manner, is not only an injury or waste to the Hops, but also an expence of fuel and time, because they require as much fuel, and as long a time to dry a small quantity, by turning them, as a large one.

Now this may be prevented, by having a cover (to be let down and raised at pleasure) to the upper bed whereon the Hops lie.

This cover may also be tinned, by nailing single tin plates over the face of it, so that when the Hops begin to dry, and are ready to burn, i. e. when the greatest part of their moisture is evaporated, then the cover may be let down within a foot or less of the Hops (like a reverberatory) which will reflect the heat upon them, so that the top will soon be as dry as the lowermost, and every Hop be equally dried.

As soon as the Hops are taken off the kiln, lay them in a room for three weeks or a month to cool, give, and toughen, for if they are bagged immediately, they will powder, but if they lie awhile (and the longer they lie the better, provided they be covered close with blankets to secure them from the air) they may be bagged with more safety, as not being liable to be broken to powder in treading, and this will make them bear treading the better, and the harder they are trodden the better they will keep.

The common method of bagging is as follows they have a hole made in an upper floor, either round or square, large enough to receive a Hop-bag which consists of four ells and a half of ell-wide cloth, and also contains ordinarily two hundred and a half of Hops, they tie a handful of Hops in each lower corner of the bag, to serve as handles to it, and they fasten the mouth of the hole, so placed that the hoop may rest upon the edges of the hole.

Then he that is to tread the Hops down into the bag, treads the Hops on every side, another person continually putting them in as he treads them, till the bag is full, which being well filled and trodden, they unrip the fastening of the bag to the hoops, and let it down, and close up the mouth of the bag, tying up a handful of Hops in each corner of the mouth, as was done in the lower part.

Hops being thus packed, if they have been well dried, and laid up in a dry place, they will keep good several years; but care must be taken, that they be neither destroyed nor spoiled by the mice making their nests in them.

The crop of Hops being thus bestowed, you are to provide for another, first by taking care of the poles against another year, which are best to be laid up in a shed, having first stripped off the haulm from them; but if you have not that conveniency, set up three poles in the form of a triangle, or six poles (as you please) wide at bottom; and having set them into the ground, with an iron picker, and bound them together at the top, set the rest of your poles about them; and being thus disposed, none but those on the outside will be subject to the injuries of the weather, for all the inner poles will be kept dry, unless at the top; whereas, if they were on the ground, they would receive more damage in a fortnight, than by their standing all the rest of the year.

In the winter time provide your soil and manure for the Hop-ground against the following spring.

If the dung be rotten, mix it with two or three parts of common earth, and let it incorporate together till you have occasion to make use of it in making your Hop-hills; but if it be new dung, then let it be mixed as before, till the spring come twelvemonths, for new dung is very injurious to Hops.

Dung of all sorts was formerly more commonly made use of than it is now, especially when rotted, and turned to mould, and they who have no other manure must use it; which, if they do, cows or hogs-dung, or human ordure mixed with mud, may be a proper compost, because Hops delight most in a manure that is cool and moist.

LUTEOLA. See Reseda.

LYCHNIDEA. See Phlox.

LYCHNIS. Tourn. Inst. R. H. 333. tab. 175. Campion.

The Characters are,

The flower hath five petals, whose tails are the length of the empalement. It hath ten stamina, which are alternately ranged and fastened to the tails of the petals. In the center is situated an almost oval germen. The empalement afterward becomes an oval capsule with one cell, opening with five valves, filled with roundish seeds.

The Species are,

1. Lychnis (*Chalcedonica*) floribus fasciculatis fastigiatis. Hort. Cliff. 174. *Greater hairy Campion with a scarlet flower.*

2. Lychnis (*Viscaria*) petalis integris. Lin. Sp. Plant. 436 *Campion with entire petals, commonly called the Single Catchfly.*

3. Lychnis (*Dioecia*) floribus dioicis. Hort. Cliff. 171. *Campion with male and female flowers on different plants; frequently called Bachelors Button.*

4. Lychnis (*Alba*) floribus dioicis, calycibus inflatis hirsutis. *Wild Campion with a single white flower.*

5. Lychnis (*Flos-cuculi*) petalis quadrifidis, fructu subrotundo. Hort. Cliff. 174. *Campion with quadrifid petals, and a roundish fruit; commonly called Ragged Robin.*

6. Lychnis (*Alpina*) petalis bifidis corymbosis. Lin. Sp. Plant. 436. *Campion with bifid petals, and flowers growing in a corymbus.*

7. Lychnis (*Siberica*) petalis bifidis, caule dichotomo, foliis subhirtis. Lin. Sp. Plant. 437. *Campion with bifid petals, a stalk divided by pairs, and leaves which are somewhat hairy.*

8. Lychnis (*Lusitanica*) caule erecto, calycibus striatis acutis, petalis dissectis. Fig. Plant. Plat. 170. *Campion with an erect stalk, striped acute empalements, and petals cut into many parts.*

The first sort here mentioned, is commonly known by the title of Scarlet Lychnis; of which there is one with double flowers, which is most esteemed for the size of the flowers and multiplicity of the petals; as also for the duration of the flowers, which continue much longer in beauty than the single flowers, so that the latter is not much cultivated at present, though the flowers of this are very beautiful; and as the plants are so easily propagated by seeds, they may soon be had in greater plenty than those with double flowers, which do not produce seeds. Of the single sort there are three varieties, the deep scarlet, the flesh colour, and the white, but the first is the most beautiful.

This is easily propagated by seeds, which should be sown on a border exposed to the east, in the middle of March.

The plants will appear in April, and by the end of June they will be fit to remove, when there should be a bed of common earth prepared to receive them, into which they should be planted at about four inches apart, observing to water and shade them till they have taken root; after which time they will require no other care but to keep them clean from weeds till the following autumn, when they should be transplanted into the borders of the pleasure-garden, where they are to continue. The summer following these plants will flower and produce ripe seeds, but the roots will abide several years, and continue to flower.

The sort with double flowers is a valuable plant; the flowers are very double, and of a beautiful scarlet colour. This hath a perennial root, from which arise two, three, or four stalks, according to the strength of the roots, which in rich moist land grow upwards of four feet high; the stalks are strong, erect, and hairy, garnished the whole length with spear-shaped leaves, embracing the stalks; these are placed opposite. The flowers are produced in close clusters sitting upon the top of the stalk; the flowers are double, and of a bright scarlet colour. They appear the latter end of June, and in moderate seasons continue near a month in beauty. This was originally produced from the seeds of the single sort, and is propagated by slips from the roots in autumn; but as this is a slow method of increasing the plants, so the best way to have them in plenty is to cut off the stalks in June, before the flowers appear, which may be cut into small lengths, each of which should have three joints; these cuttings should be planted on an east border of soft loamy earth, putting two of the joints into the ground, leaving one eye just level with the surface; these must be watered, and then covered close with bell or hand-glasses, so as to exclude the outward air, and shaded with mats when the sun shines hot upon them. The cuttings so managed will put out roots in six weeks or two months, when they must be exposed to the open air. These will make good plants by autumn, when they may be transplanted into the borders of the pleasure-garden, where they will flower the following summer.

I have not seen any double flowers of the two other varieties, but have been informed that there are of both the white and the flesh colour with double flowers in some of the French gardens. These make a variety, but are not so beautiful as the scarlet, so are not much esteemed.

The second sort is commonly called Red German Catchfly. This hath been found growing naturally upon the rocks in Edinburgh park, and in some places in Wales. It was formerly cultivated in flower-gardens for ornament; but since this sort with double flowers hath been produced, the single has been almost banished out of the gardens. This hath long narrow Grass-like leaves, which come out from the root without order, sitting close to the ground; between these come up strait single stalks, which in good ground rise a foot and a half high; at each joint of the stalk come out two leaves opposite, of the same form as the lower, but decrease in their size upward; under each pair of leaves, for an inch in length, there sweats out of the stalk a glutinous liquor, which is almost as clammy as birdlime; so that the flies, which happen to light upon these places, are fastened to the stalk where they die, from whence it had the title of Catchfly. The stalk is terminated by a cluster of purple flowers, and from the two upper joints come out on each side the stalk a cluster of the same flowers, so that the whole form a sort of loose spike. These appear in the beginning of May, and are succeeded by roundish seed-vessels, which are full of small angular seeds, ripening in July.

It may be propagated in plenty by parting of the roots in autumn, at which time every slip will grow; or if the seeds are sown in the same manner as is directed for the first sort, the plants may be raised in plenty. This delights in a light moist soil, and a shady situation.

The double flowering of this sort was accidentally obtained from the seeds of the single. This hath not been known much more than forty years in the English gardens, but it is now so common as to have excluded that with single flowers; it differs only from that in having very double flowers. As this never produces seeds, so it can only be propogated by parting and slipping of the roots; the best time for this is in autumn, at which time every slip will grow. If this is performed in September, the slips will have taken good root before the frost, and will flower well the following summer; but if they are expected to flower strong, the roots must not be divided into small slips, though for multiplying of the plants it matters not how small the slips are. These should be planted on a border exposed to the morning sun, and shaded when the sun is warm, till they have taken root. If the slips are planted in the beginning of September, they will be rooted strong enough to plant in the borders of the flower-garden, by the middle or latter end of October. The roots of this sort multiply so fast, as to make it necessary to transplant and part them every year; for when they are let remain any longer, they are very apt to rot. This requires the same soil and situation as the former.

The third sort grows naturally by the side of ditches, and in moist pastures in many parts of England, so is seldom admitted into gardens. It hath a perennial root, from which arise many branching diffused stalks, from two to three feet high, garnished with oval acute-pointed leaves, placed by pairs at each joint, and are terminated by clusters of purple flowers, which appear in April and May. The male flowers grow upon separate plants from the female. The latter produces seeds, which ripen in July; the stalks decay in autumn, but the roots continue several years.

There is a variety of this with double flowers, which is cultivated in gardens, by the title of red Bachelors Button. This is an ornamental plant, and continues long in flower. It is propagated by slips, which should be planted the beginning of August in a shady border of loamy earth, where they will take root in about six weeks or two months, and may then be transplanted into the borders of the flower-garden. These roots should be annually transplanted, otherwise they frequently rot; and young plants must be propagated by slips, to supply the decay of the old roots, which are not of very long duration. This thrives best in a soft loamy soil, and in a shady situation, where they have only the morning sun.

The fourth sort is very common upon dry banks on the side of roads in most parts of England, so is not admitted into gardens. There is a variety of this with purple flowers, which I find is by some supposed to be the same as the third, but is very different, for the stalks of this are branched out much more; the leaves are longer and more veined, and the flowers of this stand singly upon pretty long footstalks, so are not produced in clusters like those of the third. This is also very hairy, and the empalement of the flowers are swollen like inflated bladders. This flowers near a month after the other, but the male and female flowers grow upon different plants, as in the former.

There is a variety of this with double flowers, which is propagated in gardens by the title of double white Bachelors Button, and is an ornamental plant in the flower-garden, though being white it doth not make so good an appearance as the other; however, it adds to the variety. This is propagated in the same way as the double sort before mentioned, but the plants will thrive in a drier soil, and a more open exposure than that.

The fifth sort grows very common in moist meadows, and by the side of rivers in most parts of England, where it is intermixed with the Grass. This rises with upright unbranched stalks near a foot and a half high, garnished with narrow spear-shaped leaves, placed opposite at each joint. The stalks are slender, channelled, and are terminated by six or seven purple flowers, upon pretty long foot-stalks, which branch out. The empalement of the flower is striped with purple, and the petals of the flowers are deeply jagged in four narrow segments, which appear as if torn; from whence the country people have given it the appellation of Ragged Robin. This sort is never kept in gardens, but there is a variety of it with double flowers, which is propagated by the gardeners for ornament. It only differs from the single in the multiplicity of the petals, and produces no seeds, so is propagated by slips in the same manner as the second sort. It is commonly known by the title of Double Ragged Robin.

The sixth sort grows naturally on the Alps, in Lapland, and the other cold parts of Europe. This is a perennial plant, which delights in a moist soil. The stalks of this are erect, half a foot high, garnished with narrow spear-shaped leaves, placed opposite like the former sort, but are a little shorter and broader. The flowers are produced in a corymbus on the top of the stalk, sitting close together; they are of a purple colour, and the petals are cut in the middle. It is propagated by seeds, and also by parting of the roots; it must have a moist soil and a shady situation, otherwise the plants will not thrive. The time for transplanting the plants and parting the roots, is the same as for the second sort.

The seventh sort grows naturally in Siberia. This hath a perennial root, from which arise many narrow leaves, sitting close to the ground. The stalks rise a foot high, dividing into branches by pairs. The flowers grow out from the division of the branches, as also at the top of the stalks. They are composed of five white petals, which are divided in the middle, and are succeeded by roundish capsules, filled with small angular seeds. This requires the same treatment as the former sort.

The eighth sort was brought from Portugal to England, and is probably a variety of one with single flowers, which grows naturally in that country, but is different from any we have in England. It approaches nearest to the Double Ragged Robin, but is different from that. It hath a perennial root, from which arise many oblong narrow leaves, sitting close to the ground. From these come out upright stalks about nine inches high, which divide upwards by pairs; and from the middle of each division comes out a slender foot-stalk two inches long, sustaining one double purple flower at the top, whose petals are very much jagged at their points; the empalements of the flowers are marked with deep purple stripes. From the side of the stalks there are also foot-stalks come out at the wings, which for the most part sustain but one flower, though sometimes they have two; these flowers being very double, are never succeeded by seeds. It is propagated by slips in the same manner as the third and fourth sorts, but coming from a warm country, it is impatient of much cold, and requires a particular treatment, for it does not thrive well in pots, nor will it live through the winter in open borders: so that the only situation in which I have seen it thrive, was where it was planted as close as possible to a south wall, in dry undunged earth; for in rich or moist ground the roots presently rot, as they also do when they are watered. If they are planted in brick rubbish, they will still do better. I was favoured with this plant by John Browning, Esq; of Lincoln's-Inn, who received it from Portugal.

The other Species of Lychnis are now ranged under the following genera, viz.

Agrostemma, Cucubalus, Saponaria, and Silene, to which articles the reader is desired to turn for those which are not here enumerated.

LYCIUM. Lin. Gen. Plant. 232. Buxthorn.

The Characters are,

The flower is funnel-shaped, of one petal, with an incurved tube, whose brim is cut into five obtuse segments. It has five awl-shaped stamina. In the center is situated a roundish germen, which afterward becomes a roundish berry with two cells, inclosing kidney-shaped seeds fastened to the middle partition.

The Species are,

1. Lycium (*Afrum*) foliis lineari-longioribus, tubo florum longiori, segmentis obtusis. *Boxthorn with longer linear leaves, a longer tube to the flower, and obtuse segments.*

2. Lycium (*Italicum*) foliis lineari-brevioribus, tubo florum breviori, segmentis ovalibus patentissimis. *Boxthorn with shorter linear leaves, a shorter tube to the flower, and oval segments spreading open.*

3. Lycium (*Salicifolium*) foliis cuneiformibus. Vir. Cliff. 14. *Buxthorn with wedge-shaped leaves.*

4. Lycium (*Barbarum*) foliis lanceolatis crassiusculis, calycibus trifidis. Lin. Sp. Plant. 192. *Boxthorn with spear-shaped thick leaves, and trifid empalements.*

5. Lycium (*Chinense*) foliis ovato-lanceolatis, ramis diffusis, floribus solitariis patentibus alaribus, stylo longiori. *Boxthorn with oval spear-shaped leaves, diffused branches, and single spreading flowers proceeding from the sides of the branches, with a longer style.*

6. Lycium (*Halimifolium*) foliis lanceolatis acutis. *Boxthorn with spear-shaped acute leaves.*

7. Lycium (*Capense*) foliis oblongo-ovatis, crassiusculis, confertis, spinis robustioribus. *Boxthorn with oblong, oval, thick leaves, growing in clusters, and stronger spines.*

8. Lycium (*Angustifolium*) foliis lineari-lanceolatis confertis, calycibus brevibus acutis. *Boxthorn with linear spear-shaped leaves growing in clusters, and short acute empalements.*

9. Lycium (*Inerme*) inermis, foliis lanceolatis, alternis, perennantibus. *Smooth Boxthorn with spear-shaped evergreen leaves placed alternate.*

10. Lycium (*Cordatum*) foliis cordato-ovatis, sessilibus, oppositis perennantibus, spinis crassis bigeminis, floribus confertis. *Lycium with oval heart-shaped leaves placed opposite, which are evergreen, and sit close to the stalks, with thick double spines, and flowers growing in clusters.*

The first sort grows naturally in Spain, Portugal, and at the Cape of Good Hope. This rises with irregular shrubby stalks ten or twelve feet high, sending out several crooked knotty branches, covered with a whitish bark, armed with long sharp spines, upon which grow many clusters of narrow leaves; these thorns often put out one or two smaller on their sides, which have some clusters of smaller leaves upon them; the branches are garnished with very narrow leaves an inch and a half long, and at the base of these come out clusters of shorter and narrower leaves. The flowers come out from the side of the branches, standing upon short foot-stalks; they are funnel-shaped, of one petal, with a long incurved tube, cut into five obtuse segments at the brim, of a dull purple colour, and have five stamina almost as long as the tube, with erect summits. In the center is situated a roundish germen, supporting a style which is longer than the stamina, crowned by a bifid stigma. The germen afterward turns to a roundish fleshy berry, of a yellowish colour when ripe, inclosing several hard seeds.

It may be propagated either by seeds, cuttings, or layers. If by seeds, they should be sown in the autumn soon after they are ripe, for if they are kept out of the ground till spring, they seldom come up the first year. If the seeds are sown in pots, the pots should be plunged into some old tan in the winter, and in very severe frost covered with Pease-haulm or straw, but in mild weather should be open to receive the wet; in the spring the pots should be plunged into a moderate hot bed, which will soon bring up the plants; these must be inured to bear the open air as soon as the danger of the frost is over, and when they are three inches high, they may be shaken out of the pots, and each planted in a small separate pot, and placed in the shade till they have taken new root, when they may be removed to a sheltered situation, where they may remain till the autumn; then they should be either removed into the green-house, or placed under a hot-bed frame to shelter them from hard frost; for these plants are too tender to live in the open air in England, so they must be kept in pots, and treated in the same way as Myrtles, and other hardy green-house plants; but when the plants are grown strong, there may be a few of them planted in the full ground in a warm situation, where they will live in moderate winters, but in hard frosts they are commonly destroved. If the cuttings of these plants are planted in a shady border in July, and duly watered, they will take root, and may then be treated in the same way as the seedling plants.

The second sort came from the Cape of Good Hope. This hath an irregular shrubby stalk like the former, but seldom rises more than four or five feet high; the large leaves are shorter and a little broader than those of the first, but the tufts of small leaves are narrower; the tube of the flower is shorter, the brim is deeper, cut into oval segments, which spread open; the empalement is shorter, and cut into acute segments; the flowers and fruit are also smaller.

The third sort grows naturally in the hedges in the south of France, in Spain, and Italy. This hath many irregular shrubby stalks, which rise eight or nine feet high, sending out several irregular branches, covered with a white bark, armed with pretty strong thorns; the leaves are narrow at bottom, growing broader upward, and are of a pale green colour. The flowers come out from the side of the branches; they are of a purplish white colour and small, so make no great appearance.

It may be propagated by cuttings or layers, in the same manner as the first sort. The plants will live abroad in a sheltered warm situation, but in very hard frost they should be covered with straw or litter, otherwise the branches will be killed, and sometimes the roots are destroyed where they have not some cover.

The fourth sort was brought from Africa by the late Dr. Shaw, where it grows naturally. This hath a shrubby stalk, which rises seven or eight feet high, sending out several irregular branches, which are armed with strong spines, and garnished with short, thick, spear-shaped, oval leaves, which stand without order. The flowers come out from the side of the branches; they are small and white, so make little appearance. It may be propagated by cuttings in the same way as the first sort, but is too tender to live in the open air in winter in this country, so the plants must be kept in po s, and removed into the green house in autumn, and treated in the same way as other hardy kinds of green house plants.

The fifth sort grows naturally in China. This has weak, irregular, diffused branches, which rise to a great height, but requir support, otherwise they will trail upon the ground: I have measured some of these branches, which in one year has been upward of twelve feet long; the lower leaves are more than four inches long, and three broad in the middle, of a light green, and a thin consistence; as the shoots advance in length, so the size of the leaves diminish. The flowers come out singly at the joints toward the upper part of the branches, standing upon short slender foot stalks; they are of a dull purple colour with short tubes; the brims are spread open broader than either of the former sorts, and the style is considerably longer than the tube of the flower. The plant is very hardy, and retains its leaves till November before they decay. It propagates fast enough by its creeping roots, which send out suckers at a great distance, and the cuttings thrust into the ground will take root as freely as Willows.

The sixth sort grows naturally in China. This rises with a shrubby stalk to a considerable height, sending out many irregular branches, covered with a very white bark, and armed with a few short spines; the leaves are about three inches long, and one broad in the middle. The flowers of this sort appear in June and July, which are succeeded by small round berries which are as red as coral. This sort is propagated by cuttings, which should be planted in the spring before they begin to shoot, in a border exposed to the morning sun, where they will take root very freely; but these should not be removed till the autumn, when they may be planted to cover high walls, for the branches are too weak to support themselves; and as the leaves continue green as long as most of the deciduous plants, so they are proper plants for such purposes, for they may be trained to a great height.

The seventh sort was brought me from the Cape of Good Hope. This rises with shrubby branching stalks seven or eight feet high, which are armed with long strong thorns, that have several clusters of leaves upon them; the branches are garnished with small, oblong, oval leaves, which are placed without order; sometimes they come out in small clusters from one point, at others they are single; these are of a light green, and a pretty thick consistence. The plants have not as yet flowered here, so I can give no account of them, but by the fruit, which I received entire, I make no doubt of its belonging to this genus.

This sort is pretty bardy, for it has lived abroad four winters, where it was planted against a south-east wall. It may be propagated either by layers or cuttings, in the same manner as the first; and when the plants have obtained strength, they may be planted in a warm situation, where they will live with very little shelter.

The eighth sort has much the appearance of the first, but the branches are not so strongly armed with thorns; the leaves are broader and of a lighter green, standing in clusters at every joint. The flowers are smaller, of a deeper purple colour, and have much shorter empalements, which are cut into acute segments. It flowers at the same time with the first sort, but does not produce any seeds in this country. It is not so hardy as the former sort, so requires protection from very hard frost; therefore the plants should be kept in pots, and housed in the winter, treating them in the same way as other hardy green house plants.

The ninth sort has been long an inhabitant of the Chelsea garden; it was raised from seeds which came from China, and was for many years taken for the Tea-tree, till it produced some flowers, which discovered its true genus. This rises with a strong woody stalk six or seven feet high, sending out many branches, which are covered with a brown bark and are smooth, having no thorns; they are garnished with spear-shaped leaves about three inches long, and near three quarters of an inch broad, placed alternately on the branches, standing upon short foot-stalks; they are of a deep green, and continue all the year. The flowers are white, and of the same shape with the others of this genus, but there has not been any seeds as yet produced in England.

This plant will live in the open air, if it is planted in a warm ſituation and on a dry ſoil; but it is of ſlow growth, ſeldom ſhooting more than three or four inches in a ſeaſon; it is alſo difficult to propagate, for the branches which are laid down, will not take root in leſs than two years, and cuttings are with difficulty made to grow.

The tenth ſort grows naturally at the Cape of Good Hope, from whence the ſeeds were ſent to Holland a few years paſt, where the plants were raiſed. This is a low ſhrubby plant, which ſends out branches from the ground upward, which are covered with a dark green bark, and are armed with ſhort ſtrong thorns, which come out by pairs, and ſometimes there are double pairs upon the ſame foot-ſtalk; theſe are ſituated juſt below the leaves, and where there are four, two of them point upward, and the other two downward. The leaves are heart-ſhaped, not much larger than thoſe of the Box tree, of the ſame conſiſtence and colour, terminating in acute points; they are placed oppoſite upon very ſhort foot-ſtalks, ſtanding pretty cloſe together; theſe continue green all the year. The flowers come out from the ſide of the branches upon ſhort ſlender foot-ſtalks, each ſupporting five or ſix ſmall white flowers, which grow in a cluſter at the top; theſe have very ſhort empalements, and pretty long tubes, divided at the brim into five acute ſegments. Theſe flowers have an agreeable odour, and are ſucceeded by oval ſcarlet berries, each containing two ſeeds

This ſort may be propagated by cuttings in the ſame manner as the firſt ſort, which, if planted in July, and ſhaded from the ſun, will take root very freely; then they ſhould be planted into ſeparate ſmall pots, and placed in the ſhade till they have taken new root, after which they may be treated in the ſame manner as the eighth ſort. This plant has not as yet been planted in the full ground in England, but it lives through the winter under a common frame.

The other ſpecies which were included in this genus, are now removed to CELASTRUS.

LYCOPERSICON. Tourn. Inſt. R. H. 150. tab. 63. Love Apples, or Tomatos.

The CHARACTERS are,

The flower has one wheel-ſhaped petal, with a very ſhort tube, and a large five-cornered brim, which ſpreads open and is plaited. It hath five ſmall ſtamina, which cloſe together. It hath a roundiſh germen, which afterward becomes a roundiſh fleſhy fruit or berry, divided into ſeveral cells, incloſing many flat ſeeds.

The SPECIES are,

1. LYCOPERSICON (*Galeni*) caule inermi herbaceo, foliis pinnatis inciſis, fructu rotundo glabro. *Love Apple with an herbaceous unarmed ſtalk, pinnated cut leaves, and a ſmooth round fruit.*

2. LYCOPERSICON (*Eſculentum*) caule herbaceo, hirſutiſſimo, foliis pinnatis, inciſis, fructu compreſſo ſulcato. *Love Apple with a very hairy herbaceous ſtalk, winged cut leaves, and a compreſſed furrowed fruit; commonly called Tomatos by the Spaniards.*

3. LYCOPERSICON (*Æthiopicum*) caule inermi, herbaceo erecto, foliis ovatis dentato-angulatis, ſubſpinoſis, fructu ſubrotundo ſulcato. *Love Apple with an herbaceous, erect, unarmed ſtalk, oval, angular, indented leaves, with a few ſpines, and a roundiſh furrowed fruit.*

4. LYCOPERSICON (*Pimpinellifolium*) caule inermi herbaceo, foliis inæqualiter pinnatis, foliolis obtuſe-dentatis, racemis reflexis. *Love Apple with an herbaceous unarmed ſtalk, leaves unequally winged, whoſe lobes are bluntly indented, and reflexed ſpikes.*

5. LYCOPERSICON (*Peruvianum*) caule inermi herbaceo, foliis pinnatis, inciſis, undatis, ſtylo longiore perſiſtente. *Love Apple with an unarmed herbaceous ſtalk, winged cut leaves which are waved and a longer permanent ſtyle to the flower.*

6 LYCOPERSICON (*Procumbens*) caule herbaceo, procumbente, foliis pinnatifidis, glabris, floribus ſolitariis alaribus. *Love Apple with an herbaceous trailing ſtalk, wing-pointed ſmooth leaves, and flowers growing ſingly from the wings of the ſtalk.*

7. LYCOPERSICON (*Tuberoſum*) caule inermi herbaceo, foliis pinnatis integerrimis. *Love Apple with an unarmed herbaceous ſtalk, and winged leaves which are entire; commonly called Potatoe, by the Indians Batatas.*

The firſt ſort here mentioned, is ſuppoſed to be the Lycoperſicon of Galen. This is an annual plant, with an herbaceous, branching, hairy ſtalk, which will riſe to the height of five or ſix feet, if ſupported, otherwiſe the branches will fall to the ground, garniſhed with winged leaves, of a very rank diſagreeable odour, compoſed of four or five pair of lobes, terminated by an odd one; theſe are cut on their edges, and end in acute points. The flowers come out from the ſide of the branches, upon pretty long foot-ſtalks, each ſuſtaining ſeveral yellow flowers, ranged in a ſingle long bunch, which are ſucceeded by round, ſmooth, pulpy fruit, about the ſize of a large Cherry. There are two varieties of this, one with yellow, and the other with red fruit. This is the ſort which is uſed in medicine.

The ſecond ſort is very like the firſt, excepting the ſhape and ſize of the fruit, which differ greatly; for thoſe of the ſecond ſort are very large, compreſſed at both ends, and deeply furrowed on the ſides. This never varies to the other, nor that to this; ſo that it is undoubtedly a diſtinct ſpecies. This is the ſort which is commonly cultivated for ſoups; the Portugueze and Spaniards uſe them alſo in many of their ſauces, by whom the fruit are titled Tomatos.

The third ſort is annual; this riſes with an erect herbaceous ſtalk a foot and a half high, dividing into ſeveral branches, garniſhed with oval angular leaves, placed alternately upon pretty long foot-ſtalks, which have one or two ſhort ſpines upon the midrib of the leaves. The flowers are white, and come out ſingly from the ſide of the branches, which are ſucceeded by red ſtriated fruit, firmer than thoſe of the other ſorts, and about the ſize of Cherries.

The fourth ſort is ſomewhat like the firſt, but the leaves are unequally winged, having ſome ſmaller lobes placed between the large ones; the lobes of this are ſhorter, broader, and not cut like thoſe of the firſt, but have ſome obtuſe indentures toward the baſe. Theſe have not that rank diſagreeable odour which the two firſt have; the fruit of it is not ſo large as thoſe of the firſt, but they are round, ſmooth, and are very late before they ripen here; ſo that unleſs the plants are raiſed early in the ſpring, they will not produce ripe fruit in England.

The fifth ſort is annual; this hath a very branching herbaceous ſtalk, ſpreading out into many diviſions, and is not ſo hairy as the two firſt; the leaves are compoſed of a greater number of lobes, which are much ſhorter and more indented on their edges, where they are a little waved The flowers ſtand upon very long foot-ſtalks, which branch out, and ſupport a large number of flowers at the top; theſe have a longer ſtyle than thoſe of the other ſpecies, which is permanent, remaining on the top of the fruit. This is alſo late in ripening the fruit, ſo that unleſs the plants are raiſed early in the ſpring, the fruit will not ripen in England.

The ſixth ſort was raiſed by Mr James Gordon, gardener at Mile End, who gave me ſome of the ſeeds, but from what country it came I could not learn. This hath very weak, trailing, ſmooth ſtalks, not more than a foot long, garniſhed with ſmooth leaves, ſtanding by pairs oppoſite, which are regularly cut on the ſides almoſt to the midrib, in form of a winged leaf; these segments are also indented on their edges, and at their points. The flowers, which are of a pale yellow colour, come out on the side of the stalks singly, and have large spreading empalements, which are deeply cut at the brim into many acute segments, which spread open. The flowers are succeeded by small roundish berries, a little compressed at the top, of an herbaceous yellow colour when ripe.

All these sorts are propagated by sowing their seeds on a moderate hot-bed in March, and when the plants are come up two inches high, they should be transplanted into another moderate hot-bed, at about four inches distance from each other, observing to shade them until they have taken root; after which they must have a large share of fresh air, for if they are too much drawn while young, they seldom do well afterwards.

In May these plants should be transplanted, either into pots or borders near walls, pales, or hedges, to which their branches may be fastened to support them from trailing on the ground, which they otherwise will do, and then the fruit will not ripen; so that where these plants are cultivated for the sake of their fruit, they should be planted to a warm aspect, and the branches regularly fastened as they extend, that the fruit may have the advantage of the sun's warmth to forward them, otherwise it will be late in the season before they are ripe, and they are unfit for use before; but when the plants are brought forward in the spring, and thus regularly trained to the south sun, the fruit will ripen by the latter end of July, and there will be a succession of it till the frost kills the plants.

The third sort is never used either in the kitchen or for medicine, but the plants are preserved for the sake of variety, especially by those persons who are lovers of botany. This sort is propagated by seeds, which should be sown upon a hot-bed in the spring, and the plants afterward treated in the same manner as hath been directed for the Capsicum, with whose culture this plant will thrive, and produce plenty of fruit annually.

The seventh sort is the common Potatoe, which is a plant so well known now, as to need no description. Of this there are two varieties, one with a red and the other with a white root; that whose roots are red, have purplish flowers, but the white root has white flowers; these are supposed to be only accidental variations, and not distinct species.

The common name of Potatoe seems to be only a corruption of the Indian name Batatas. This plant has been much propagated in England within thirty or forty years past, for although it was introduced from America about the year 1623, yet it was but little cultivated in England till of late; the roots being despised by the rich, and deemed only proper food for the meaner sort of persons; however, they are now esteemed by most people, and the quantity of them which are cultivated near London, I believe, exceeds that of any other part of Europe.

This is generally propagated by its roots, which multiply greatly, if planted in a proper soil. The common way is, either to plant the small roots or offsets entire, or to cut the larger roots into pieces, preserving a bud or eye to each; but neither of these methods is what I would recommend, for when the smaller offsets are planted, they generally produce a great number of roots, but these are always small; and the cuttings of the larger roots frequently rot, especially if wet weather happens soon after they are planted; therefore what I would recommend is, to make choice of the fairest roots for this purpose, and to allow them a larger space of ground, both between the rows, as also in the rows, plant from plant; by which method I have observed, the roots have been in general large the following autumn.

The soil in which this plant thrives best, is a light sandy loam, not too dry or over moist; this ground should be well ploughed two or three times, in order to break and divide the parts; and the deeper it is ploughed, the better the roots will thrive. In the spring, just before the last ploughing, there should be a good quantity of rotten dung spread on the ground, which should be ploughed in the beginning of March, if the season proves mild, otherwise it had better be deferred until the middle or latter end of that month; for if it should prove hard frost after the roots are planted, they may be greatly injured, if not destroyed thereby; but the sooner they are planted in the spring, after the danger of frost is over, the better it will be, especially in dry land. In the last ploughing, the ground should be laid even, and then the furrows should be drawn at three feet distance from each other, and about seven or eight inches deep. In the bottom of this furrow the roots should be laid at about one foot and a half asunder; then the furrow should be covered in with the earth, and the same continued through the whole field or parcel of land intended to be planted.

After all is finished, the land may remain in the same state till near the time when the shoots are expected to appear above ground, when the ground should be well harrowed over both ways, which will break the clods, and make the surface very smooth; and by doing of it so late, it will destroy the young weeds, which, by this time, will begin to make their appearance; and this will save the expence of one hoeing, as also stir the upper surface of the ground, which, if much wet has fallen after the planting, is often bound into a hard crust, which retards the appearance of the shoots.

As I have allotted the rows of Potatoes at three feet distance, it was in order to introduce the hoe-plough between them, which will greatly improve these roots; for by twice stirring and breaking of the ground between these plants, it will not only destroy the weeds, but also loosen the ground, whereby every shower of rain will penetrate the ground to the roots, and greatly improve their growth; but these operations should be performed early in the season, before the stems or branches of the plants begin to fall and trail upon the ground, because after that it will be impossible to do it without injuring of the shoots.

If these ploughings are carefully performed, it will prevent the growth of weeds, till the haulm of the plants cover the ground, so that afterward there will be little danger of weeds growing so as to injure the crop; but as the plough can only go between the rows, it will be necessary to make use of a hoe to stir the ground, and destroy the weeds in the rows between the plants; and if this is carefully performed in dry weather, after the two ploughings, it will be sufficient to keep the ground clean until the Potatoes are fit to take up.

In places where dung is scarce, many persons scatter it only in the furrows, where the roots are planted; but this is a very poor method, because when the Potatoes begin to push out their roots, they soon extend beyond the width of the furrows, and the new roots are commonly formed at a distance from the old, so will be out of the reach of this dung, and consequently will receive little benefit from it. And as most of the farmers covet to have a crop of Wheat after the Potatoes are taken off the ground, so the land will not be so thoroughly dressed in every part, nor so proper for this crop, as when the dung is equally spread, and ploughed in all over the land, nor will the crop of Potatoes be so good. I have always observed, where this method of planting the Potatoes has been practised, the land has produced a fine crop of Wheat afterward, and there has scarce one shoot of the Potatoe appeared among the Wheat, which

which I attribute to the farmers planting only the largest roots; for when they have forked them out of the ground the following autumn, there have been six, eight, or ten large roots produced from each, and often many more, and scarce any very small roots; whereas, in such places where the small roots have been planted, there has been a vast number of very small roots produced; many of which were so small, as not to be discovered when the roots were taken up, so have grown the following season, and have greatly injured whatever crop was on the ground.

The haulm of these Potatoes is generally killed by the first frost in the autumn, then the roots should be taken up soon after, and may be laid up in dry sand in any sheltered place, where they may be kept dry, and secure from frost. Indeed the people who cultivate these roots near London, do not wait for the decaying of the haulm, but begin to take up part of them as soon as their roots are grown to a proper size for the market, and so keep taking up from time to time, as they have vent for them. There are others likewise, who do not take them up so soon as the haulm decays, but let them remain much longer in the ground; in which there is no hurt done, provided they are taken up before hard frost sets in, which would destroy them, unless where the ground is wanted for other crops: in which case, the sooner they are taken up the better, after the haulm is decayed. When these roots are laid up, they should have a good quantity of sand or dry earth laid between them, to prevent their heating; nor should they be laid in too large heaps, for the same reason.

LYCOPUS. Water Horehound.

This plant grows in great plenty on moist soils by the sides of ditches, in most parts of England, but is never cultivated in gardens, so that it would be needless to say any thing more of it in this place.

LYSIMACHIA. Tourn. Inst. R. H. 141. tab. 59 Loosstrife.

The CHARACTERS are,

The flower is of one petal, cut into five oblong segments, which spread open. It hath five awl-shaped stamina, and a roundish germen, which afterward turns to a globular capsule with one cell, opening with ten valves, and filled with small angular seeds.

The SPECIES are,

1. LYSIMACHIA (*Vulgaris*) paniculata, racemis terminalibus. Lin. Sp. Pl. 146. *Greater yellow Loosestrife.*

2. LYSIMACHIA (*Thyrsiflora*) racemis lateralibus pedunculatis. Lin. Sp. Pl. 147. *Two-leaved Loosestrife with yellow globular flowers.*

3. LYSIMACHIA (*Atropurpurea*) spicis terminalibus patulis, lanceolatis staminibus corollâ longioribus. Lin. Sp. Plant. 147. *Narrow-leaved Eastern Loosestrife, with a purple flower.*

4. LYSIMACHIA (*Ephemeron*) racemis simplicibus terminalibus, petalis obtusis, staminibus corollâ brevioribus. Lin. Sp. Pl. 146. *Loosestrife with spikes of flowers terminating the stalks, obtuse petals to the flower, and stamina shorter than the petal.*

5. LYSIMACHIA (*Ciliata*) petiolis ciliatis, floribus cernuis. Lin. Sp. Pl. 147. *Loosestrife with hairy foot-stalks and nodding flowers.*

6 LYSIMACHIA (*Salicifolia*) racemis simplicibus terminalibus, petalis obtusis, staminibus corollâ longioribus. *Willow-leaved Loosestrife with a spike of white flowers terminating the stalk.*

7. LYSIMACHIA (*Nummularia*) foliis subcordatis, floribus solitariis, caule repente. Vir. Cliff. 13. *Great Yellow Moneywort.*

8. LYSIMACHIA (*Tenella*) foliis ovatis acutiusculis, pedunculis folio longioribus, caule repente. Lin. Sp. Pl. 148. *Smaller Moneywort with a purplish flower.*

9. LYSIMACHIA (*Nemorum*) foliis ovatis acutis, floribus solitariis, caule procumbente. Hort. Cliff. 52. *Yellow Pimpernel of the woods.*

10. LYSIMACHIA (*Quadrifolia*) foliis subquaternis, pedunculis verticillatis unifloris. Lin. Sp. Plant. 147. *Smaller yellow Loosestrife, with leaves marked with black spots.*

The first sort grows by the side of ditches and rivers in many parts of England, so is not often admitted into gardens, because the roots creep far in the ground, whereby it becomes often a troublesome plant. It rises with upright stalks from two to three feet high, garnished with smooth spear shaped leaves, placed sometimes by pairs, at others there are three, and frequently four of these leaves placed round the stalk at each joint. The upper part of the stalk divides into several foot-stalks, which sustain yellow flowers growing in a panicle; these have one petal which is deeply cut into five segments, spreading open, and are succeeded by roundish seed-vessels, filled with small seeds.

The second sort grows naturally in the northern parts of England. This hath a perennial creeping root, which sends up several erect stalks a foot and a half high, garnished at every joint by two pretty long narrow leaves, placed opposite. The foot-stalks of the flowers come out opposite on each side of the stalks, sustaining at their top a globular or oval thyrse of yellow flowers, whose stamina are much longer than the petals. This is seldom kept in gardens for the same reason as the former.

The third sort is a biennial plant, which was discovered by Dr. Tournefort in the Levant. This rises with an upright stalk about a foot high, garnished with spear-shaped leaves, ending in acute points, placed opposite; they are smooth, and of a lucid green. The flowers are purple, and grow in a loose spike, terminating the stalks.

It is propagated by seeds, which should be sown in autumn soon after they are ripe, and from those the plants will come up the following spring; but if the seeds are kept out of the ground till spring, they will not vegetate till the year after. When the plants come up, they must be kept clean till autumn, then they may be planted into the borders of the pleasure-garden, where they will flower and produce ripe seeds the following summer.

The fourth sort is an annual plant, which grows naturally in the Levant. This hath a shorter stalk than the former. The lower leaves are broader, the spikes of flowers are shorter, and of a pale purple colour. The seeds of this sort should be sown in the autumn, where the plants are to remain; when they come up, they will require no other culture but to keep them clean from weeds, and if they are too close they should be thinned to the distance of four or five inches, which is all the culture they will require.

The fifth sort grows naturally in Canada. This has a perennial creeping root, sending up erect stalks about two feet high, garnished with oblong smooth leaves, placed opposite, which are veined on their under side, and end in acute points. The flowers are produced from the wings of the stalk, each sitting upon a long slender foot-stalk, three or four arising from the short branches, which come out on each side the stalk. The flowers are like those of the first sort, but smaller.

This sort spreads and propagates by roots, in as great plenty as the first, and is equally hardy, so requires no other culture.

The sixth sort grows naturally in Spain. This hath a perennial root, from which arise several upright stalks three feet high, garnished with narrow, smooth, spear-shaped leaves, which stand opposite. The flowers are white, and are produced in a long, close, upright spike at the top of the stalk; they are cut into five oval segments, which spread open; the stamina stand out longer than the petal.

This

This is the fineſt ſpecies of this genus; and as the roots of it do not ſpread like thoſe of the other, ſo deſerves a place in the pleaſure-garden, where it is a very ornamental plant for ſhady borders. It loves a moiſt ſoil and a ſhady ſituation, where it will continue long in beauty. It may be propagated by parting of the roots in autumn, but by this method it increaſes ſlowly, ſo that the only way to have it in plenty is by ſowing of the ſeeds; theſe ſhould be ſown upon an eaſt-aſpected border in autumn, ſoon after they are ripe, then the plants will come up the following ſpring. When the plants come up, they ſhould be kept clean from weeds, and if they are too cloſe, ſome of them may be drawn out and tranſplanted on a ſhady border, which will give the remaining plants room to grow till autumn, when they may be tranſplanted into the borders where they are deſigned to flower.

The ſeventh ſort is commonly called Moneywort, or Herb Two-pence. This is a perennial plant, which grows naturally in moiſt ſhady places in moſt parts of England, ſo is not cultivated in gardens.

The eighth ſort is a ſmall trailing plant, which grows upon bogs in moſt parts of England. The ſtalks ſeldom are more than three or four inches long, and are terminated by three or four ſmall flowers, of a bright purple colour, growing in a bunch.

The ninth ſort is a perennial plant with trailing ſtalks, which grows naturally in moiſt woods in moſt parts of England, ſo is not cultivated in gardens.

The tenth ſort grows naturally among Ruſhes and Reeds, by the river's ſid in Holland. This hath a perennial creeping root like the firſt. The ſtalks riſe a foot high, garniſhed by ſpear-ſhaped leaves, placed ſometimes by pairs, at others by threes, and often four at each joint, ſurrounding the ſtalk. The flowers alſo come out at each joint, four of them ſtand ing round the ſtalk in whorls, which are yellow. It may be treated in the ſame manner as the firſt ſort, and is equally hardy.

LYSIMACHIA GALERICULATA. See Scutellaria.

LYSIMACHIA NON POPPOSA. See Œnothera.

LYSIMACHIA SILIQUOSA. See Epilobium.

LYTHRUM. Lin. Gen. Plant. 532. Willow Herb, or purple Looſtriſe.

The Characters are,

The flower hath ſix oblong blunt petals, which ſpread open, whoſe tails are inſerted in the indentures of the empalement, and ten ſlender ſtamina. In the center is ſituated an oblong germen, which afterward turns to an oblong acute capſule with two cells, filled with ſmall ſeeds.

The Species are,

1. Lythrum (*Salicaria*) foliis oppoſitis cordato-lanceolatis, floribus ſpicatis dodecandriis. Lin. Sp. Plant. 446. *Common purple Willow Herb with oblong leaves.*

2. Lythrum (*Tomentoſum*) foliis cordato ovatis, floribus verticillato-ſpicatis tomentoſis *Purple Willow Herb with roundiſh leaves.*

3. Lythrum (*Hyſſopiſolia*) foliis alternis linearibus, floribus hexandriis. Hort. Upſal. 118. *Willow Herb with a narrow Hyſſop leaf.*

4. Lythrum (*Hiſpanicum*) foliis oblongo ovatis infernè oppoſitis, ſupernè alternis, floribus hexandris. *Spaniſh Willow Herb with a Hyſſop leaf, and oblong, deep, blue flowers.*

The firſt ſort grows naturally by the ſide of rivers and ditches in moſt parts of England. It has a perennial root, from which come forth ſeveral upright angular ſtalks, which riſe from three to four feet high, garniſhed with oblong leaves, placed ſometimes by pairs; at others there are three leaves at each joint, ſtanding round the ſtalk. The flowers are purple, and are produced in a long ſpike at the top of the ſtalk, which make a fine appearance. Although this plant is deſpiſed, becauſe it grows common, yet it merits a place in gardens better than many other which are propagated with care, becauſe they are more rare. It is eaſily cultivated by parting of the roots in autumn, and ſhould be planted in a moiſt ſoil, where it will thrive and flower without any other care than the keeping it clean from weeds.

There is a variety of this with an hexangular ſtalk, and generally with three leaves at each joint; but this is only accidental, for the roots of this, when removed into a garden, come to the common ſort.

The ſecond ſort is like the firſt, but has oval, heart-ſhaped, downy leaves, placed by threes round the ſtalk. The flowers are produced in long ſpikes at the top of the ſtalks, diſpoſed in thick whorls, with ſpaces between each; they are of a fine purple colour, and appear at the ſame time with the former.

The third ſort grows naturally on moiſt bogs in many parts of England, ſo is ſeldom admitted into gardens.

The fourth ſort grows naturally in Spain and Portugal, from both which countries I have received the ſeeds. The root of this is perennial. The ſtalks are ſlender, not more than nine or ten inches long, ſpreading out on every ſide. The lower part of the ſtalks are garniſhed with oblong oval leaves, placed oppoſite. On the upper part of the ſtalks the leaves are narrower, and placed alternate. The flowers come out ſingly from the ſide of the ſtalks at each joint; they are larger than thoſe of the common ſort, and of a deeper purple colour, ſo make a fine appearance in July, when they are in beauty.

This ſort has never produced any ſeeds in England, and the ſevere winter in 1740, killed all the plants here, ſince which time I have not ſeen any of the plants in the Engliſh gardens.

M.

MACALEB. See Cerasus.

MADDER. See Rubia.

MAGNOLIA Plum. Nov. Gen. 38. tab. 7. Lin. Gen. Plant. 610. The Laurel-leaved Tulip-tree.

The Characters are,

The flower is composed of eight or ten oblong, concave, blunt leaves. It hath a great number of short stamina, which are inserted into the germen, and many oblong oval germina, fastened to the receptacle, supporting recurved contorted styles, with hairy stigmas. The germen afterward becomes oval cones, with imbricated capsules, each having one cell, opening with two valves, inclosing one kidney-shaped seed, hanging by a slender thread from the scale of the cone.

The Species are,

1. Magnolia (*Glauca*) foliis ovato-lanceolatis subtus glaucis annuis. *Magnolia with oval spear-shaped leaves which are gray on their under side, and annual; commonly called Small Magnolia.*

2. Magnolia (*Grandiflora*) foliis lanceolatis persistentibus, caule erecto arboreo. Fig. Plant. tab. 172. *Magnolia with spear-shaped leaves, which are evergreen, and an erect tree-like stalk; commonly called Great Magnolia.*

3. Magnolia (*Tripetala*) foliis lanceolatis amplissimis annuis, petalis exterioribus dependentibus. *Magnolia with very large spear-shaped leaves, which are annual, and the outer petals of the flower declining; commonly called Umbrella-tree.*

4. Magnolia (*Acuminata*) foliis ovato-lanceolatis acuminatis annuis, petalis obtusis. *Magnolia with oval, spear-shaped, pointed leaves, which are annual, and obtuse petals to the flower.*

The first sort grows pretty common in Virginia and Carolina, and other parts of North America. In moist places this rises from seven or eight to fifteen or sixteen feet high, with a slender stem. The wood is white and spongy, the bark is smooth and of a greenish white colour; the branches are garnished with thick smooth leaves, resembling those of the Bay, but are of an oval shape, smooth on their edges, and white underneath. The flowers are produced at the extremity of the branches, which are white, and composed of six concave petals, and have an agreeable sweet scent. After these are past the fruit increases in size to be as large as a Walnut with its cover, but of a conical shape, having many cells round the outside; in each of which is lodged a flat seed, about the size of a small Kidney Bean. The fruit is at first green, afterward red, and when ripe of a brown colour. The seeds when ripe, are discharged from their cells, and hang by a slender thread.

When these trees are transplanted from the places of their growth into dry ground they make handsomer trees, and produce a great number of flowers. This is to be understood of America, for in England they do not thrive so well in a dry soil, as in a moist loamy land.

The second sort grows in Florida and South Carolina, where it rises to the height of eighty feet or more, with a strait trunk upward of two feet diameter, having a regular head. The leaves of this tree resemble those of the common Laurel, but are much larger, and of a lucid green on their upper side, and in some trees are of a russet or buff colour on their under side. These leaves continue all the year, so that this is one of the most beautiful evergreen trees yet known. The flowers are produced at the end of the branches, composed of eight or ten petals, which are narrow at their base, but broad at their extremity, where they are rounded, and a little waved; they are of a purple white colour. In the center is situated a great number of stamina and styles, fastened to one common receptaculum; the flowers are succeeded by oblong scaly cones. These trees, in their native places of growth, begin to produce their flowers in May, which are succeeded by others, so that the woods are perfumed with their odour for a long time; but those which have flowered in England, seldom begin till the middle of June, and do not continue long in beauty. There are many large plants of this sort in the gardens of his Grace the Duke of Richmond, at Goodwood in Sussex, which have produced flowers several years; and in the nursery of the late Mr. Christopher Gray, near Fulham, there is one very handsome plant, which has lived in the open air many years, and has abundance of flowers.

As this sort is a native of a warm country, it is a little impatient of cold, especially while young; therefore the plants should be kept in pots, and sheltered in winter for some years, until they have acquired strength, when they may be shaken out of the pots, and planted in the full ground; but they must be planted in a warm situation, where they may be defended from the strong winds, and screened from the north and east, otherwise they will not live abroad.

The third sort grows in Carolina pretty frequent, but in Virginia it is pretty rare. This usually grows from sixteen to twenty feet high, with a slender trunk; the wood is soft and spongy; the leaves of this tree are remarkably large, and are produced in horizontal circles, somewhat resembling an umbrella, from whence the inhabitants of those countries have given it this name. The flowers are composed of ten or eleven white petals, which hang down without any order; the fruit is very like that of the former sort, but longer; the leaves of this sort drop off at the beginning of winter.

This tree is as yet very rare in Europe, but as it is propagated from seeds, we may hope to have it in greater plenty soon, if we can obtain good seeds from Carolina, for it is rarely met with in Virginia.

The fourth sort is also very rare in England. There are but few of the plants at present here, nor is it very common in any of the habitable parts of America; some of these trees have been discovered by Mr. John Bartram, growing on the north branch of Susquehannah river. The leaves of this tree are near eight inches long and five broad, ending in a point. The flowers come out early in the spring, which are composed of twelve white petals, and are shaped like thofe of the fecond fort; the fruit of this tree is longer than thofe of the other fpecies, but in other refpects agrees with them. The wood of this tree is of a fine grain, and an Orange colour.

All thefe forts are propagated by feeds, which muft be procured from the places of their natural growth; thefe fhould be put up in fand, and fent over to England as foon as poffible, for if they are kept long out of the ground, they very rarely grow; therefore the feeds fhould be fown as foon as poffible when they arrive here.

Some years paft I received a good quantity of thefe feeds from Carolina, which I fowed in pots as foon as I received them, and plunged the pots into a moderate hot-bed, and with this management I raifed a great number of plants; but from the feeds which have been lately brought over, there have been but few plants produced; whether the feeds were not perfectly ripe when they were gathered, or from what other caufe this has happened I cannot fay, but it is certain the fault muft be in the feeds, becaufe they have been differently fown and managed by the feveral perfons who received them, and the fuccefs was nearly alike every where.

There have been feveral plants of the firft and fecond fort raifed from layers, and fome from cuttings; but thefe do not thrive fo well as thofe which come from feeds, nor will they grow to near the fize of thofe, fo that it is much the beft way to procure their feeds from America, and propagate them that way.

The firft fort frequently comes up well from feeds, but the young plants are very difficult to keep the two firft years; for if they are expofed much to the fun their leaves change yellow, and the plants decay, fo the beft way is to keep the pots plunged in a moderate hot-bed, and fhade them every day from the fun with mats, giving them air in plenty when the weather is warm, and frequently refrefh them with water; during the winter feafon they muft be fcreened from froft, and in mild weather they muft enjoy the free air to prevent their growing mouldy; they fhould have but little wet after their leaves are fallen. With this management the plants may be trained up, and when they have acquired ftrength they may be planted in the open air, where they will thrive and flower if they have a fheltered fituation.

The fecond fort is not fo difficult to train up; but in order to get them forward it will be proper, when they are removed out of the feed-pots, to plant them each into a feparate fmall pot, and plunge them into a gentle hot-bed of tanners bark, obferving to fhade them from the fun, and admit proper air to them; but at Midfummer they fhould be inured to the open air gradually, and then placed in a fheltered fituation, where they may remain till autumn; but on the firft approach of froft they fhould be removed under fhelter, otherwife the early frofts will pinch their tender fhoots, which often occafions their dying downward after. When the plants have got ftrength, fome of them may be turned out of the pots, and planted in the full ground, in a warm fheltered fituation; but part of them fhould be kept in pots, and fheltered in the winter, to preferve them, left by fevere froft the other fhould be killed.

If the plants make good progrefs, they will be ftrong enough to plant in the full ground in about fix or feven years. The time for removing or fhifting thefe plants is in March, before they begin to fhoot, which may fometimes happen to be too foon to turn them out of the pots into the full ground, efpecially if the feafon proves late; but as there will be no danger in removing them out of the pots, the ball of earth being preferved to their roots, fo it is beft to defer this till the month of April; but it will be neceffary to harden thofe plants which are intended to be planted out, by expofing them to the air as much as poffible, for this will keep the plants backward, and prevent their fhooting; for if they make fhoots in the greenhoufe, thofe will be too tender to bear the fun, until they are by degrees hardened to it, and the leaft froft will greatly pinch them, and fuch often happens very late in the fpring.

The two or three winters after they are planted out, it will be neceffary to lay fome mulch on the furface of the ground about their roots, as alfo to throw fome mats over their heads, efpecially at the beginning of the morning frofts in autumn, for the reafons before given; but they fhould never be too clofely covered up, left the fhoots fhould grow mouldy, for they will certainly kill the leading buds of every fhoot, and prove to the full as injurious to them as the froft. As the plants get ftrength, they will be better able to endure the cold of our climate, though it will be proper to lay fome mulch about their roots every winter, and in very fevere froft to cover their heads and ftems.

It is the fecond fort which requires the moft care, being much tenderer than any of the other forts, for they will endure the cold very well without much care, after they have acquired ftrength; for as thefe lofe their leaves in the winter, the froft will not have fo much force upon them as the fecond fort, whofe leaves are frequently tender toward the end of the fhoots, efpecially when they grow freely, or fhoot late in the autumn.

MAHALEB. See CERASUS.

MAJORANA. See ORIGANUM.

MALABAR NUT. See JUSTICIA.

MALA ÆTHIOPICA. See LYCOPERSICON.

MALA ARMENIACA. See ARMENIACA.

MALACOIDES. See MALOPE.

MALA COTONEA. See CYDONIA.

MALA INSANA. See MELONGENA.

MALLOW. See MALVA.

MALLOW-TREE. See LAVATERA.

MALOPE. Baftard Mallow.

The CHARACTERS are,

The flower is fhaped like that of the Mallow, and hath a double empalement, the outer being compofed of three, and the inner is of one leaf, cut into five fegments; the flower is of one petal, divided into five parts to the bottom. In the center is fituated the column, having a great number of ftamina and ftyles joined clofely to it. The germen afterward becomes a fruit compofed of many cells, which are collected into a head, in each of which is lodged a fingle feed.

We have but one SPECIES of this plant, viz.

MALOPE (*Malacoides*) foliis ovatis crenatis glabris. Lin. Hort. Cliff. 347. *Baftard Mallow with oval fmooth leaves, which are notched.*

The whole plant has greatly the appearance of the Mallow, but differs from it in having the cells collected into a button, fomewhat like a Blackberry; the branches fpread, and lie flat upon the ground, extending themfelves a foot or more each way. The flowers are produced fingly upon long foot-ftalks, from the fetting on of the leaves, which are in fhape and colour like thofe of the Mallow.

This is propagated by feeds, which fhould be fown upon a warm border in Auguft, where the plants will come up before winter, which fhould be planted in fmall pots, and fheltered under a hot-bed frame, for they are too tender to live in the open air in winter; but in fummer they fhould be placed with other hardy foreign plants in a fheltered fituation, where in warm feafons they will produce feeds.

MALPIGHIA. Plum. Nov. Gen. 46. tab. 36. Barbadoes Cherry.

The

The Characters are.

The flower has five kidney-ſhaped petals, which are concave, and ſpread open, and ten awl-ſhaped ſtamina, and two melleous glands adhering to the empalement. It has a ſmall roundiſh germen ſupporting three ſlender ſtyles. The germen afterward turns to a large, ſurrowed, globular berry, with one cell, incloſing three rough, ſtony, angular ſeeds.

The Species are,

1. Malpighia (*Glabra*) foliis ovatis integerrimis glabris, pedunculis umbellatis. Hort. Cliff. 169. *Malpighia with ſmooth, oval, entire leaves, and umbellated foot-ſtalks; commonly called Barbadoes Cherry.*

2. Malpighia (*Punicifolia*) foliis ovato-lanceolatis, acuminatis, glabris, pedunculis umbellatis. *Malpighia with the appearance of the Pomegranate-tree.*

3. Malpighia (*Incana*) foliis lanceolatis ſubtus incanis, pedunculis umbellatis alaribus. *Malpighia with ſpear-ſhaped leaves, hoary on their under ſide, and umbellated foot-ſtalks proceeding from the wings of the ſtalk.*

4. Malpighia (*Urens*) foliis cordato-lanceolatis, ſetis decumbentibus rigidis, pedunculis unifloris aggregatis. *Broad-leaved Malpighia with ſpines growing on the under ſide of the leaf.*

5. Malpighia (*Nitida*) foliis ovatis acutis glabris, pedunculis umbellatis alaribus terminalibuſque. *Malpighia with oval, ſmooth, acute-pointed leaves, and umbellated foot-ſtalks proceeding from the ſides and ends of the branches.*

6. Malpighia (*Anguſtifolia*) foliis lineari-lanceolatis, ſetis decumbentibus rigidis, pedunculis unifloris aggregatis. *Malpighia with linear ſpear-ſhaped leaves, rigid declining briſtles, and foot-ſtalks having umbels proceeding from the ſides of the branches in cluſters.*

7. Malpighia (*Ilicifolia*) foliis lanceolatis dentato-ſpinoſis ſubtus hiſpidis. Lin. Sp. Plant. 426. *Malpighia with ſpear-ſhaped leaves indented and prickly, whoſe under ſide are ſet with ſpiny hairs.*

8. Malpighia (*Coccynia*) foliis ſubovatis dentato-ſpinoſis, pedunculis unifloris. *Malpighia with leaves nearly oval, indented and prickly, and foot-ſtalks with one flower.*

The firſt ſort is commonly cultivated in the Weſt-Indies for the ſake of the fruit; this uſually grows to the height of ſixteen or eighteen feet, having a ſlender ſtem, covered with a light brown bark. The leaves are placed oppoſite; they are oval, ſmooth, ending in acute points, and continue green all the year. The flowers are produced in bunches upon pretty long foot-ſtalks, which come out from the ſide of the branches; they are compoſed of five petals, which are of a Roſe colour, joined at their baſe. The flowers are ſucceeded by red fruit, ſhaped like thoſe of the ſmall wild Cherry, of the ſame ſize, each incloſing four angular furrowed ſtones, ſurrounded by a thin pulp, which has an agreeable acid flavour.

The ſecond ſort grows naturally in Jamaica. This riſes with a ſhrubby ſtalk ten or twelve feet high, dividing into ſeveral ſlender ſpreading branches, covered with a light brown bark, garniſhed with oval, ſpear-ſhaped, ſmooth leaves placed oppoſite, ending in acute points. The flowers are produced in umbels at the end of the branches, ſtanding upon ſhort foot-ſtalks; they are of a pale Roſe colour, compoſed of five obtuſe, concave, indented petals, having long narrow tails. In the center is ſituated the roundiſh germen, ſupporting three ſtyles, attended by ten ſtamina which ſpread aſunder. The germen afterward turns to a roundiſh pulpy berry with many furrows, red when ripe, incloſing three or four hard angular ſeeds.

The third ſort grows naturally at Campeachy. This riſes with a ſtrong woody ſtalk eighteen or twenty feet high, dividing into many branches, covered with a brown ſpotted bark, garniſhed with ſpear-ſhaped leaves placed oppoſite, which are hoary on their under ſide. The flowers come out in umbels from the ſide of the branches; they are of a Roſe colour, and are ſucceeded by oval channelled fruit, like thoſe of the former.

The fourth ſort grows naturally in Jamaica. This riſes with a woody ſtalk from fifteen to eighteen feet high, dividing into many ſtrong branches, which are furrowed, and covered with a brown bark. The leaves are from three to four inches long, and one broad at their baſe, where they are rounded in form of a heart, leſſening gradually to the point; theſe are covered on their under ſides with ſtinging briſtly hairs ſo cloſely, as to render it very troubleſome to handle them, for theſe fallen themſelves into the fleſh, and are difficult to get out again. The flowers are produced in umbels from the ſide of the branches; they are of a light purple colour, and ſhaped like thoſe of the other ſpecies, and are ſucceeded by oval furrowed fruit like thoſe of the former ſort. This is called in the Weſt-Indies Couhage, or Cowitch Cherry.

The fifth ſort grows naturally at Carthagena in New Spain. This riſes with a ſhrubby ſtalk about ten feet high, covered with a light brown ſpotted bark, branching out regularly on every ſide; the leaves are oval, ſmooth, and end in acute points, ſtanding oppoſite, of a light green on their upper ſide, but paler on their under. The flowers come out from the ſide of the ſtalks in ſmall umbels, ſtanding erect; the foot-ſtalks of the umbels are ſcarce an inch long, and come out alternately from the ſide of the branches. The flowers are of a pale bluſh colour, ſhaped like thoſe of the former ſorts, which are ſucceeded by roundiſh furrowed berries with a red ſkin, covering three hard angular ſeeds.

The ſixth ſort grows in the iſland of Barbuda. This riſes with a ſhrubby ſtalk eight or ten feet high, covered with a bright purpliſh bark, which is ſpotted and furrowed, dividing toward the top into ſeveral ſmaller branches, garniſhed with narrow ſpear-ſhaped leaves, of a lucid green on their upper ſide, but of a ruſſet brown on their under, where they are cloſely armed with ſtinging briſtles, which faſten themſelves into the fleſh or clothes of thoſe who touch them. The flowers are produced from the ſide of the branches in cluſters; they are of a pale purple colour, of the ſame form as the other ſpecies, and are ſucceeded by ſmall, oval, furrowed fruit, of a dark purple colour when ripe.

The ſeventh ſort grows naturally in the iſland of Cuba. This riſes with a ſhrubby ſtalk to the height of ſeven or eight feet, ſending out branches the whole length, covered with a gray bark, and garniſhed with narrow prickly leaves like thoſe of the Holly, which have many ſtinging briſtles on their under ſide. The flowers are of a pale bluſh colour, and are produced in ſmall cluſters from the ſide of the branches. The fruit is more pointed than thoſe of the common ſort, and turns to a dark purple colour when ripe.

The eighth ſort grows naturally near the Havannah. This is a very low ſhrub, ſeldom riſing more than two or three feet high; the ſtalk is thick and woody, covered with a rough gray bark, garniſhed with lucid green leaves, which appear as if cut at their ends, where they are hollowed in, and the two corners riſe like horns ending in a ſharp thorn, as do alſo the indentures on the ſides. The flowers come out from the ſide of the branches, upon foot-ſtalks an inch long, each ſuſtaining one ſmall, pale, bluſh flower, of the ſame form with thoſe of the other ſpecies; the fruit is ſmall, conical, and furrowed, changing to a purple red colour when ripe.

The fruit of moſt of the ſpecies here mentioned are promiſcuouſly gathered, and eaten by the inhabitants of the countries

countries where they grow; but the firſt ſort is that which is cultivated in ſome of the iſlands for its fruit, though it is but indifferent; the pulp which ſurrounds the ſtone is very thin, but has a pleaſant acid flavour, which renders it agreeable to the inhabitants of thoſe warm countries, where, to ſupply the want of thoſe Cherries which are cultivated in Europe, they are obliged to eat the fruit of theſe ſhrubs.

Theſe plants are preſerved in the gardens of thoſe perſons who are ſo curious in botanical ſtudies, as to erect hot-houſes for maintaining foreign plants; and where there are ſuch conveniencies, theſe plants deſerve a place, becauſe they retain their leaves all the year, and commonly continue flowering from December till the end of March, when they make a fine appearance at a ſeaſon when there is a ſcarcity of other flowers, and many times they produce ripe fruit here. Thoſe ſorts whoſe leaves are armed with ſtinging briſtles like the Cowitch, are the leaſt worthy of a place in ſtoves, becauſe they are ſo troubleſome to handle, nor do their flowers make ſo good an appearance as many of the other ſorts.

As theſe plants are natives of the warmeſt parts of America, ſo they will not live through the winter in England, unleſs they are preſerved in a warm ſtove; but when the plants have obtained ſtrength they may be expoſed in the open air in a warm ſituation, from the middle or latter end of June till the beginning of October, provided the weather continues ſo long mild; and the plants ſo treated, will flower much better than thoſe which are conſtantly kept in a ſtove.

They are all propagated by ſeeds, which muſt be ſown upon a good hot-bed; and when the plants are fit to tranſplant they muſt be each put into a ſeparate ſmall pot, and plunged into a hot-bed of tanners bark, and muſt be treated in the ſame manner as hath been directed for other tender plants of the ſame country; the two firſt winters it will be proper to keep them in the bark-bed in the ſtove; but afterward they may be placed upon ſtands in the dry ſtove in winter, where they may be kept in a temperate warmth, in which they will thrive much better than in a greater heat; theſe muſt be watered two or three times a week, when they are placed in a dry ſtove, but it muſt not be given to them in large quantities.

MALVA. Tourn. Inſt. R. H. 94. tab. 23. Mallow.

The CHARACTERS are,

The flower has a double empalement; the outer is compoſed of three, the inner is of one leaf, cut into five broad ſegments at the brim. The flower is of one petal. It has a great number of ſtamina which coaleſce at bottom in a cylinder, but ſpread open above. In the center is ſituated an orbicular germen, ſupporting a ſhort cylindrical ſtyle, with many briſtly ſtigmas. The empalement afterward turns to ſeveral capſules, which are joined in an orbicular depreſſed head faſtened to the column, opening on their inſide, each containing one kidney ſhaped ſeed.

The SPECIES are,

1. MALVA (*Sylveſtris*) caule erecto herbaceo, foliis lobatis obtuſis, pedunculis petioliſque piloſis. Lin. Sp. Pl. 689. *Wild Mallow with a ſinuated leaf.*

2. MALVA (*Rotundifolia*) caule repente, foliis cordato-orbiculatis obſoletè quinquelobatis. Hort. Cliff. 347. *Common Mallow with a ſmall flower and a round leaf.*

3. MALVA (*Orientalis*) annua, caule erecto herbaceo, foliis lobatis obtuſis crenatis. *Annual Mallow with an erect herbaceous ſtalk, and obtuſe lobed leaves which are crenated.*

4. MALVA (*Criſpa*) caule erecto, foliis angulatis, floribus axillaribus glomeratis. Vir. Cliff. 356. *Curled or furbelowed Mallow.*

5. MALVA (*Chinenſis*) annua, caule erecto herbaceo ſimplici, foliis ſuborbiculatis obſolete quinquelobatis, floribus confertis alaribus ſeſſilibus. *Upright, annual, China Mallow, with very ſmall white flowers.*

6. MALVA (*Cretica*) caule erecto ramoſo hirſuto, foliis angulatis, floribus alaribus pedunculis brevioribus. *Talleſt annual Mallow of Crete, with ſmall flowers growing in umbels on the ſides of the ſtalk.*

7. MALVA (*Peruviana*) caule erecto herbaceo, foliis lobatis, ſpicis ſecundis axillaribus. Lin. Sp. Pl. 688. *Mallow with an erect herbaceous ſtalk, leaves having lobes, and ſpikes of flowers proceeding from the ſides of the ſtalks.*

8. MALVA (*Alcea*) caule erecto, foliis multipartitis ſcabriuſculis. Hort. Cliff. 347. *Narrow-leaved, curled, Vervain Mallow.*

9. MALVA (*Major*) caule erecto, foliis trilobatis, obtuſis dentatis glabris. *Common, greater, Vervain Mallow.*

10. MALVA (*Moſchata*) foliis radicalibus reniformibus inciſis, caulinis quinquepartitis pinnato-multifidis. Hort. Upſal. 202. *Round, cut-leaved, Vervain Mallow.*

11. MALVA (*Ægyptia*) foliis palmatis dentatis, corollis calyce minoribus. Lin. Sp. Pl. 690. *Egyptian Vervain Mallow with a Crane's-bill leaf.*

12. MALVA (*Tournefortia*) foliis quinquelobatis inciſis, calycibus acutis hiſpidis, pedunculis longiſſimis. *Maritime Vervain Mallow of Provence, with a Crane's-bill leaf.*

13. MALVA (*Capenſis*) foliis ſubcordatis laciniatis hirſutis, caule arboreſcente. *African ſhrubby Mallow with a red flower*

14. MALVA (*Americana*) foliis cordatis crenatis, floribus lateralibus ſolitariis, terminalibus ſpicatis. Prod Leyd. 359. *Low American Marſhmallow, with a yellow ſpiked flower.*

The two firſt ſorts are found wild in moſt parts of England, ſo are rarely cultivated in gardens. The firſt is the ſort commonly uſed in medicine, with which the markets are ſupplied by the herb-folks, who gather it in the fields

The third ſort was diſcovered by Dr. Tournefort in the Levant. This is an annual plant, with an erect ſtalk; the flowers are larger than thoſe of the common ſort, and are of a ſoft red colour.

The fourth ſort is annual. This riſes with an upright ſtalk four or five feet high; the leaves are curled on their edges, for which variety it is preſerved in gardens.

The fifth ſort was formerly ſent from China as a pot-herb, and hath been cultivated in ſome curious gardens in England; though it is not likely to obtain here as an eſculent plant, ſince we have many others which are preferable to it for that purpoſe. This is an annual plant, which will propagate itſelf faſt enough, provided it be permitted to ſcatter its ſeeds.

The ſixth ſort was diſcovered by Dr. Tournefort in the iſland of Candia. This will become a weed, if ſuffered to ſcatter the ſeeds.

The ſeventh ſort grows naturally in Peru. This is an annual plant, riſing with an upright branching ſtalk two feet high, garniſhed with broad hairy leaves, having three lobes. The flowers grow in ſpikes from the wings of the ſtalks, of a pale blue, ſet very cloſely on the ſpikes, and are ſucceeded by ſeeds, which, if permitted to ſcatter, will come up plentifully the following ſpring without care.

The eighth ſort is a biennial plant, which grows naturally in paſtures in many parts of England, ſo is ſeldom admitted into gardens. The ſtalks of this are a foot and a half long, and frequently incline toward the ground. The leaves are finely cut into narrow ſegments almoſt to the midrib, and theſe ſegments are deeply indented. The flowers are ſhaped like thoſe of the common Mallow, and are of a Roſe colour.

The ninth ſort is the common Vervain Mallow, which is mentioned in the catalogue of medicinal plants. This is found

found growing naturally in ſome of the midland counties in England, but not near London. It is a biennial plant; the ſtalks riſe higher than thoſe of the former; the leaves are cut into three obtuſe lobes, which are indented. The flowers are larger than thoſe of the former, but appear at the ſame time, and the ſeeds ripen in autumn.

The tenth ſort is a biennial plant, which grows naturally in France and Italy. The lower leaves of this are rounded and eared ſomewhat like a kidney in ſhape, and are cut on their edges; but thoſe on the ſtalks are divided into five parts, which end with many wing-ſhaped points; the ſtalks of this are ſhorter than thoſe of the other ſorts.

The eleventh ſort grows naturally in Egypt. This is an annual plant, whoſe ſtalks are ſmooth, a foot long, and decline toward the ground. The leaves ſtand upon pretty long foot-ſtalks ſhaped like a hand, having five diviſions, which join at their baſe to the foot-ſtalk. The flowers come out ſingle from the wings of the ſtalk, and at the top in cluſters; they have pretty large acute empalements; they are ſmall, and of a pale blue colour. Theſe appear in June, and the ſeeds ripen in autumn.

The twelfth ſort grows naturally in the ſouth of France. This is an annual plant, which has ſome reſemblance of the former, but the ſtalks are longer and more branched; the leaves are cut into five obtuſe lobes almoſt to the bottom, and are deeply cut on their ſide. The flowers ſtand upon very long foot-ſtalks; the empalement of the flower is large, prickly, and acute-pointed; the flowers are blue, and larger than thoſe of the other ſort. It flowers and ripens its ſeeds about the ſame time with that.

The thirteenth ſort grows naturally at the Cape of Good Hope. This riſes with a woody ſtalk ten or twelve feet high, ſending out branches from the ſide; the ſtalks and branches are cloſely covered with hairs, and are garniſhed with hairy leaves, which are indented on the ſides, ſo as to have the appearance of a trilobate leaf. The flowers come out from the ſide of the branches, upon foot-ſtalks an inch long; they are of a deep red colour, and ſhaped like thoſe of the common Mallow, but are ſmaller. This plant continues flowering great part of the year, which renders it valuable.

There are two other varieties of this plant, which have been mentioned by ſome authors as diſtinct ſpecies. The firſt is, Alcea Africana fruteſcens, groſſulariæ folio ampliore, unguibus fiorum atro-rubentibus. Act. Phil. 1729. Shrubby, African, Vervain Mallow, with a larger Gooſeberry leaf, and the bottoms of the flower of a dark red. The other is, Alcea Africana fruteſcens, folio groſſulariæ flore parvo rubro. Boerh. Ind. alt. 1. 272. Shrubby, African, Vervain Mallow, with a Gooſeberry leaf, and a ſmall red flower. The leaves of the laſt appear very different from either of the other, being deeply divided into three lobes, which are alſo deeply indented, ſo that any perſon, upon ſeeing it, would ſuppoſe it to be a different ſpecies; but I have frequently raiſed all theſe, with ſome other intermediate varieties, from the ſeeds of one plant.

The thirteenth ſort is eaſily propagated by ſeeds, which, if ſown on a common border in the ſpring, the plants will come up; but as it is too tender to live abroad in the winter, ſo when the plants are three or four inches high, they ſhould be each planted into a ſeparate pot, placing them in the ſhade till they have taken freſh root; then they may be removed to a ſheltered ſituation, intermixing them with other hardy exotic plants, where they may remain till the froſt comes on, when they ſhould be removed into the green-houſe, and afterward treated in the ſame way as the hardy plants from the ſame country, always allowing them plenty of free air in mild weather.

The fourteenth ſort grows naturally in moſt of the iſlands in the Weſt-Indies. This is an annual plant, which riſes about a foot high, ſending out a few ſhort woolly branches from the ſide, garniſhed with heart ſhaped woolly leaves, which are notched on their edges, ſtanding alternately upon pretty long foot-ſtalks. The flowers terminate the ſtalks in a cloſe ſpike; they are ſmall, and of a pale yellow colour. The ſeeds ripen in autumn.

This is propagated by ſeeds, which muſt be ſown upon a hot-bed in the ſpring; and when the plants are fit to remove, they ſhould be each planted in a ſeparate ſmall pot, and plunged into a new hot-bed, ſhading them until they have taken freſh root; then they muſt have free air admitted to them in proportion to the warmth of the ſeaſon, and the latter end of June they may be placed in the open air in a ſheltered ſituation, where they will flower and produce ripe ſeeds.

There are ſome other ſorts of Mallows which are natives of this country, but as they are plants of no great beauty or uſe, it is needleſs to mention them in this place.

MALVA ROSEA. See Alcea.

MALUS. The Apple-tree.

The Characters are,

The branches ſpread, and are more depreſſed than thoſe of the Pear-tree. The flower conſiſts of five leaves, which expand in form of a Roſe. The fruit is hollowed about the foot-ſtalk, it is for the moſt part roundiſh and umbilicated at the top; it is fleſhy, and divided into five cells or partitions, in each of which is lodged one oblong ſeed.

The Species are,

1. Malus (*Sylveſtris*) foliis ovatis ſerratis, caule arboreo. *Wild Apple with a very ſour fruit; commonly called Crab.*

2. Malus (*Coronaria*) foliis ſerrato-anguloſis. *Wild Crab of Virginia with a ſweet-ſcented flower.*

3. Malus (*Pumila*) foliis ovatis ſerratis, caule fruticoſo. *Dwarf Apple; commonly called Paradiſe Apple.*

Of the firſt ſort there are two varieties of fruit, one is white, and the other purple toward the ſun, but theſe are accidental variations. There is alſo a variety of this with variegated leaves, which has been propagated in ſome of the nurſeries near London; but when the trees grow vigorous, their leaves ſoon become plain.

The ſecond ſort grows naturally in moſt parts of North America, where the inhabitants plant them for ſtocks to graft other ſorts of Apples upon; the leaves of this are longer and narrower than any of the other ſorts, and are cut into acute angles on their ſides. The flowers of this have a fragrant odour, which perfumes the American woods at the time they appear.

The third ſort is undoubtedly a diſtinct ſpecies from all the others, for it never riſes to any height; the branches are weak, ſcarce able to ſupport themſelves, and this difference is permanent when raiſed from ſeeds.

There is a ſort of Apple called the Fig Apple, which is common to England and North America, but the fruit is not greatly eſteemed; however, as ſome perſons are fond of variety, ſo I have mentioned it. The varieties of French Apples are,

Pomme de Rambour. The Rambour is a very large fruit, of a fine red next the ſun, and ſtriped with a pale red of yellowiſh green. This ripens very early commonly about the end of Auguſt, and ſoon grows meally, therefore is not eſteemed in England.

Pomme de Courpendu, the hanging Body This is a very large Apple, of an oblong figure, having ſome irregular riſings or angles, which run from their baſe to the crown; it is of a red caſt on the ſide toward the ſun, but pale on the other ſide; the foot ſtalk is long and ſlender, ſo that the fruit is always hanging downward

ward, which occasioned the French gardeners to give it this name.

The Renette-blanche, or White Renette, or French Renette. This is a large fine fruit, of a pale green, and a roundish figure, having some small gray spots; the juice is sugary, and it is good for eating or baking; it will keep till after Christmas good.

The Rennette-grise. This is a middle sized fruit, shaped like the Golden Renette, but is of a deep gray colour on the side next the sun, but on the other side intermixed with yellow; it is a very juicy good Apple, of a quick flavour. It ripens in October, and will not keep long.

Pomme d'Api. This is a small hard fruit, of a bright purple colour on the side next the sun, and of a yellowish green on the other side; it is a very firm fruit, but hath not much flavour, so is only preserved by some persons by way of curiosity. It keeps a long time sound, and makes a variety in a dish of fruit.

Le Calville d'Automne, the Autumn Calville. This is a large fruit of an oblong figure, of a fine red colour toward the sun. The juice is vinous, and much esteemed by the French.

Fenouillet ou Pomme d'Anis, the Fennel, or Anise Apple. This is a middle sized fruit, a little longer than a Golden Pippin, of a grayish colour. The pulp is tender, and has a spicy taste like Anise-seed; the wood and the leaves are whitish.

Pomme Violette, the Violet Apple. This is a pretty large fruit, of a pale green, striped with deep red to the sun. The juice is sugary.

The Crab, which is the first sort here mentioned, has been generally esteemed as the best stock for grafting Apples upon, being very hardy, and of long duration; but of late years there have been few persons who have been curious enough to raise these stocks, having commonly sown the kernels of all sorts of Cyder Apples for stocks without distinction, as these are much easier to procure than the other; so the gardeners generally call all those Crabs which are produced from the kernel, and have not been grafted: but were the kernels of the Crabs sown, I should prefer those for stocks, because they are never so luxuriant in their growth as those from Apple kernels, and they will continue longer sound; besides, these will preserve some of the best sorts of Apples in their true size, colour, and flavour, whereas the other free stocks produce larger fruit, which are not so well tasted, nor will they keep so long.

The Paradise Apple some years ago was much esteemed for stocks to graft or bud upon, but these are not of long duration; nor will the trees grafted upon them ever grow to any size, unless they are planted so low as that the cyon may strike its root into the ground, when it will be equal to no stock, for the graft will draw its nourishment from the ground, so that it is only by way of curiosity, or for very small gardens, that these stocks are proper, since there can never be expected any considerable quantity of fruit from such trees.

There is another Apple which is called the Dutch Paradise Apple, much cultivated in the nurseries for grafting Apples upon, in order to have them dwarfs; and these will not decay or canker as the other, nor do they stint the grafts near so much, so are generally preferred for planting espaliers or dwarfs, being easily kept within the compass usually allotted to these trees.

Some persons have also made use of Codlin stocks to graft Apples upon, in order to make them dwarf; but the fruit which are upon these stocks are not so firm, nor do they last so long, therefore the winter fruits should never be grafted upon these stocks.

The Virginian Crab-tree with sweet flowers, is preserved by such persons as are curious in collecting great variety of trees; it may be propagated by budding or grafting it upon the common Crab or Apple-tree; but it is somewhat tender while young, wherefore it should be planted in a warm situation, otherwise it will be subject to suffer by an extreme hard winter. The flowers of this tree are said to be exceeding sweet in Virginia, where it grows in the woods in great plenty; but I could not observe much scent in some of them which have flowered in England, so that I am in doubt whether the sort at present in the gardens is the same with that of Virginia.

The Fig Apple is supposed by many persons to be produced without a previous flower. But this opinion is rejected by more curious observers, who affirm there is a small flower precedes the fruit, which is very fugacious, seldom continuing above a day or two.

The other sorts which are above-mentioned, are what have been introduced from France, but there are not above two or three of them which are much esteemed in England, viz. the French Renette, the Rennette-grise, and the Violet Apple; the other being early fruit which do not keep long, and their flesh is generally meally, so that they do not deserve to be propagated, as we have many better fruits in England, which I shall next mention.

The first Apple which is brought to the London markets is the Codlin. This fruit is so well known in England that it is needless to describe it.

The next is the Margaret Apple: this fruit is not so long as the Codlin, of a middling size; the side next the sun changes to a faint red when ripe, the other side is of a pale green; the fruit is firm, of a quick pleasant taste, but doth not keep long.

The Summer Pearmain is an oblong fruit, striped with red next the sun; the flesh is soft, and in a short time is meally, so that it is not greatly esteemed.

The Kentish Fill Basket is a species of the Codlin, of a large size, longer shaped than the Codlin. This ripens a little later in the season, and is generally used for baking, &c.

The Transparent Apple. This was brought to England about the year 1724, and was esteemed a curiosity; it came from Petersburgh, where it is affirmed to be so transparent, as that the kernels may be perfectly seen when the Apple is held to the light; but in this country it is a meally insipid fruit, so is not worth propagating.

Loan's Pearmain. This is a beautiful fruit to the sight, of a middling size; the side next the sun is of a fine red, striped with the same colour on the other: the flesh is vinous, but as it soon grows meally, it is not greatly esteemed.

The Quince Apple. This is a small fruit, seldom larger than the Golden Pippin, but is longer, and in shape like the Quince, especially toward the stalk; the side next the sun is of a russet colour, on the other side inclining to yellow. This is an excellent Apple for about three weeks in September, but it will not keep much longer.

The Golden Renette is a fruit so well known in England as to need no description. This ripens about Michaelmas, and for about a month is a very good fruit, either for eating raw or baking.

The Aromatic Pippin is also a very good Apple. It is about the size of a Nonparcil, but a little longer; the side next the sun is of a bright russet colour; the flesh is breaking, and hath an aromatic flavour. It ripens in October.

The Hertfordshire Pearmain, by some called the Winter Pearmain. This is a good sized fruit, rather long than round, of a fine red next the sun, and striped with the same colour

colour on the other fide; the flefh is juicy, and ftews well, but is not efteemed for eating by any nice palates. This is fit for ufe in November and December.

The Kentifh Pippin is a large handfome fruit, of an oblong figure; the fkin is of a pale green colour; the flefh is breaking, and full of juice, which is of a thick acid flavour. This is a very good kitchen fruit, and will keep till February.

The Holland Pippin is larger than the former; the fruit is fomewhat longer, the fkin of a darker green, and the flefh firm and juicy. This is a very good kitchen fruit, and will keep late in the feafon.

The Monftrous Renette is a very large Apple, of an oblong fhape, turning red toward the fun, but of a dark green on the other fide; the flefh is apt to be meally, fo it is not much valued by thofe who are curious, and is only prefor the magnitude of the fruit.

The Embroidered Apple is a pretty large fruit, fomewhat fhaped like the Pearmain, but the ftripes of red are very broad, from whence the gardeners have given it this title. It is a middling fruit, and is commonly ufed as a kitchen Apple, though there are many better.

The Royal Ruffet, by fome called the Leather Coat Ruffet, on account of the deep ruffet colour of the fkin. This is a large fair fruit, of an oblong figure, broad toward the bafe; the flefh is inclinable to yellow. This is one of the beft kitchen Apples we have, and is a very great bearer; the trees grow large and handfome, and the fruit is in ufe from October till April, and is alfo a pleafant fruit to eat.

Wheeler's Ruffet is an Apple of a middling fize, flat and round; the ftalk is flender, the fide next the fun of a light ruffet colour, the other fide inclining to a pale yellow when ripe; the flefh is firm, and the juice has a very quick acid flavour; but it is an excellent kitchen fruit, and will keep a long time.

Pile's Ruffet is not quite fo large as the former, but is of an oval figure, of a ruffet colour to the fun, and of a dark green on their other fide. It is a very firm fruit, of a fharp acid flavour, but is much efteemed for baking, and will keep found till April, or later, if they are well preferved.

The Nonpareil is a fruit pretty generally known in England, though there is another Apple which is frequently fold in the markets for it, which is what the French call Haute-bonne. This is a larger fairer fruit than the Nonpareil, more inclining to yellow; the ruffet colour brighter, and it is earlier ripe, and fooner gone: this is not fo flat as the true Nonpareil, nor is the juice fo fharp, though it is a good Apple in its feafon; but the Nonpareil is feldom ripe before Chriftmas, and where they are well preferved, they will keep till May perfectly found. This is juftly efteemed one of the beft Apples that have been yet known.

The Golden Pippin is a fruit peculiar to England. There are few countries abroad where this fucceeds well, nor do they produce fo good fruit in many parts of England, as were to be wifhed. This is in fome meafure owing to their being grafted on free ftocks, which enlarges the fruit, but renders it lefs valuable, becaufe the flefh is not fo firm, nor the flavour fo quick, and it is apt to be dry and meally; therefore this fhould always be grafted upon the Crab-ftock, which will not canker like the others: and though the fruit will not be fo fair to the fight, yet it will be better flavoured.

There are yet a great variety of Apples, which, being inferior to thofe here mentioned, I have omitted, as thofe which are here enumerated will be fufficient to furnifh the table and the kitchen, during the whole feafon of thefe fruits; fo that where thefe forts are to be had, no perfon of tafte will eat the other.

I fhall here mention fome of the Apples, which are chiefly preferred for the making of cyder, though there are in every cyder country new forts frequently obtained from the kernels; but thofe hereafter mentioned, have for fome years been in the greateft efteem.

The Red-ftreak.
Devonfhire Royal Wilding.
The Whitfour.
Herefordfhire Under Leaf.
John Apple, or Deux-annes.
Everlafting Hanger.
Gennet Moyle.

All the forts of Apples are propagated by grafting or budding upon the ftocks of the fame kind, for they will not take upon any other fort of fruit-tree. In the nurferies there are three forts of ftocks generally ufed, to graft Apples upon; the firft are called free ftocks, which are raifed from the kernels of all forts of Apples indifferently, and by fome thefe are alfo termed Crab ftocks; for all thofe trees which are produced from the feeds, before they are grafted, are termed Crabs without any diftinction; but, as I before obferved, I fhould always prefer fuch ftocks as are raifed from the kernels of Crabs, where they are preffed for verjuice; and I find feveral of the old writers on this fubject of the fame mind. Mr. Auften, who wrote a hundred years ago, fays, "The ftock which he accounts "beft for Apple grafts, is the Crab, which is better than "fweeter Apple-trees to graft on, becaufe they are ufu-"ally free from canker, and will become very large trees, "and I conceive will laft longer than ftocks of fweeter "Apples, and will make fruits more ftrong and hardy to "endure frofts." It is very certain, that by frequent grafting fome forts of Apples upon free ftocks, the fruits have been rendered larger, but lefs firm, poignant, and of fhorter duration.

The fecond fort of ftocks is the Dutch Creeper before-mentioned; thefe are defigned to ftint the growth of the trees, and keep them within compafs for dwarfs or efpaliers.

The third fort is the Paradife Apple, which is a very low fhrub, fo only proper for trees which are kept in pots by way of curiofity, for thefe do not continue long.

Some perfons have made ufe of Codlin ftocks for grafting of Apples, in order to ftint their growth; but as thefe are commonly propagated by fuckers, I would by no means advife the ufing of them; nor would I chufe to raife the Codlin-trees from fuckers, but rather graft them upon Crab ftocks, which will caufe the fruit to be firmer, laft longer, and have a fharper flavour. The trees fo propagated will laft much longer found, and never put out fuckers as the Codlins always do, which, if not conftantly taken off, will weaken the trees, and caufe them to canker; and it is not only from the roots, but from the knots of their ftems, there are generally a great number of ftrong fhoots produced, which fill the trees with ufelefs fhoots, and render them unfightly, and the fruit fmall and crumpled.

The method of raifing ftocks from the kernels of Crabs or Apples, is, to procure them where they are preffed for verjuice or cyder, and after they are cleared of the pulp, they may be fown upon a bed of light earth, covering them over about half an inch thick with the fame light earth; thefe may be fown in November or December, where the ground is dry, but in wet ground it will be better to defer it till February; but then the feeds fhould be preferved in dry fand, and kept out of the reach of vermin, for if mice or rats can get to them they will devour the feeds; there fhould alfo be care taken of the feeds when they are fown, to protect them from thefe vermin, by fetting traps to take them, &c. In the spring, when the plants begin to appear, they must be constantly kept clear from weeds, which, if suffered to grow, will soon overtop the plants, and spoil their growth; if these thrive well some of them will be fit to transplant into the nursery the October following, for the sooner the seedling plants are removed from the seed-bed, the less danger there will be of their shooting down tap-roots, which in fruit-trees should always be prevented. The ground where these young stocks are to be planted should be carefully digged, and cleansed from the roots of all bad weeds, and laid level; then the stocks should be planted in rows three feet asunder, and the plants one foot distance in the rows, closing the earth pretty fast to their roots; when the stocks are transplanted out of the seed-bed, the first autumn after sowing, they must not be headed, but such as are inclined to shoot downward, the tap-root must be shortened, in order to force out horizontal roots: if the ground is pretty good in which these stocks are planted, and the weeds constantly cleared away, the stocks will make great progress, so that those which are intended for dwarfs may be grafted the spring twelve months after they are planted out of the seed bed; but those which are designed for standards will require two or three years more growth before they will be fit to graft, by which time they will be upward of six feet high. The other necessary work to be observed in the culture of these trees, while they remain in the nursery, being exhibited under the article of NURSERY, I shall not repeat it in this place.

I shall next treat of the manner of planting such of these trees as are designed for espaliers in the kitchen-garden, where, if there is an extent of ground, it will be proper to plant, not only such sorts as are for the use of the table, but also a quantity of trees to supply the kitchen; but where the kitchen garden is small, the latter must be supplied from standard-trees, either from the orchard, or wherever they are planted; but as many of these kitchen Apples are large, and hang late in the autumn upon the trees, they will be much more exposed to the strong winds, on standard trees than in espaliers, whereby many of the fruit will be blown down before they are ripe, and others bruised, so as to prevent their keeping; therefore where it can be done, I should always prefer the planting them in espaliers.

The distance which I should chuse to allow these trees, should not be less than twenty five or thirty feet, for such sorts as are of moderate growth (if upon Crab or free stocks) but the larger growing sorts should not be allowed less room than thirty-five feet, which will be found full near enough, if the ground is good, and the trees properly trained; for as the branches of these trees should not be shortened, but trained at their full lengths, so in a few years they will be found to meet. Indeed, at the first planting, the distance will appear so great to those persons who have not observed the vigorous growth of these trees, that they will suppose they never can extend their branches so far as to cover the espalier; but if these persons will but observe the growth of standard-trees of the same kinds, and see how wide their branches are extended on every side, they may be soon convinced, that as these espalier trees are allowed to spread but on two sides, they will of course make more progress, as the whole nourishment of the root will be employed in these side branches, than where there is a greater number of branches on every side of the tree, which are to be supplied with the same nourishment.

The next thing to be observed is the making choice of such sorts of fruits as grow nearly alike, to plant in the same espalier. This is of great consequence, because of the distance they are to be placed, otherwise those sorts which make the largest shoots, may be allowed less room to spread than those of smaller growth; besides, when all the trees in one espalier are nearly equal in growth, they will have a better appearance than when some are tall, and others short; but for the better instruction of those persons who are not conversant in these things, I shall divide the sorts of Apples into three classes, according to their different growth.

Largest growing tree.	Middle growing tree.	Smallest growing tree.
All the sorts of Pearmains.	Codlin.	Quince Apple.
Kentish Pippin.	Margaret Apple.	Transparent Apple.
Holland Pippin.	Golden Renette.	Golden Pippin.
Monstrous Renette.	Aromatic Pippin.	Pomme d'Api.
Royal Russet.	Embroidered Apple.	Fenouillet, or Anise Apple.
Wheeler's Russet.	Renette Grise.	
Pile's Russet.		
Nonpareil.		
Violet Apple.		

N. B. These are all supposed to be grafted on the same sort of stocks.

If these Apples are grafted upon Crab stocks, I would willingly place them at the following distance from each other, especially where the soil is good, viz. the largest growing trees at forty feet, the middle growing at thirty feet, and the small growing at twenty feet, which, from constant experience, I find to be full near enough; for in many places where I have planted the trees at twenty-four feet distance, the trees have grown so strong as that in seven years their branches have met; and in some places where every other tree hath been taken up, the branches have almost joined in seven years after; therefore it will be much the better way to plant these trees at a proper distance at first, and between these to plant some Dwarf Cherries, Currants, or other sort of fruit, to bear for a few years, which may be cut away when the Apple-trees have extended their branches to them; for when the Apple-trees are planted nearer together, few persons care to cut down the trees when they are fruitful, so that they are obliged to use the knife, saw, and chizel, more than is proper for the future good of the trees; and many times where the persons are inclinable to take away part of their trees, the distances will be often so irregular (where there was not this consideration in planting) as to render the espalier unsightly.

When the trees are upon the Dutch dwarf stock, the distance should be for the larger growing trees thirty feet; twenty-five for those of middle growth; and the smallest twenty

twenty feet, which will be found full near, where the trees thrive well.

The next is the choice of trees, which should not be more than two years growth from the graft, but those of one year should be preferred; be careful that their stocks are young, sound, and smooth, free from canker, which have not been cut down in the nursery; when they are taken up, all the small fibres should be entirely cut off from their roots, which, if left on, will turn mouldy and decay, so will obstruct the new fibres in their growth; the extreme parts of the roots must be shortened, and all bruised roots cut off; and if there are any misplaced roots, which cross each other, they should also be cut away. As to the pruning of the head of these trees, there need be nothing more done, than to cut off any branches, which are so situated, as that they cannot be trained to the line of the espalier; in the planting there must be care taken not to place their roots too deep in the ground, especially if the soil is moist, but rather raise them on a little hill, which will be necessary to allow for the raising of the borders afterward. The best season for planting these trees (in all soils which are not very moist) is from October to the middle or latter end of November, according as the season continues mild; but so soon as the leaves fall, they may be removed with great safety. After the trees are planted, it will be proper to place down a stake to each tree, to which the branches should be fastened, to prevent the winds from shaking or loosening their roots, which will destroy the young fibres; for when these trees are planted pretty early in the autumn, they will very soon push out a great number of new fibres, which, being very tender, are soon broken, so the trees are greatly injured thereby. If the winter should prove severe, it will be proper to lay some rotten dung, tanners bark, or some sort of mulch, about their roots, to prevent the frost from penetrating of the ground, which might damage these tender fibres; but I would not advise the laying of this mulch before the frost begins, for if it is laid over the roots soon after the trees are planted (as is often practised) it will prevent the moisture entering the ground, and do much harm to the trees.

The following spring, before the trees begin to push out shoots, there should be two or three short stakes put down on one side of the tree, to which the branches should be fastened down as horizontally as possible, never cutting them down, as is by some practised, for there will be no danger of their putting out branches enough to furnish the espalier, if the trees are once well established in their new quarters.

In the pruning of these trees, the chief point is never to shorten any of the branches, unless there is an absolute want of shoots to fill the places of the espalier; for where the knife is much used, it only multiplies useless shoots, and prevents their fruiting; so that the best method to manage these trees is, to go over them three or four times in the growing season, and rub off all such shoots as are irregularly produced, and train the others down to the stakes, in the position they are to remain; if this is carefully performed in summer, there will be little left to be done in the winter, and by bending of their shoots from time to time, as they are produced, there will be no occasion to use force to bring them down, nor any danger of breaking the branches. The distance which these branches should be trained from each other, for the largest sorts of fruit, should be about seven or eight inches, and for the smaller, five or six. If these plain instructions are followed, it will save much unnecessary labour of pruning, and the trees will, at all times, make a handsome appearance; whereas, when they are suffered to grow rude in summer, there will be much difficulty to bring down their shoots without breaking, especially if they are grown stubborn. All the sorts of Apples produce their fruit upon cursons or spurs, so that these should never be cut off, for they will continue fruitful a great number of years.

The method of making the espaliers having been already exhibited under that article, I need not repeat it here, but only observe, that it will be best to defer making the espalier, till the trees have had three or four years growth, for before that time, the branches may be supported by a few upright stakes, so that there will be no necessity to make the espalier, until there are sufficient branches to furnish all the lower part.

I shall now treat of the method to plant orchards, so as to have them produce the greatest profit. And first, in the choice of the soil and situation for an orchard: the best is that on the ascent of gentle hills, facing the south or south-east; but this ascent must not be too steep, lest the earth should be washed down by hasty rains. There are many persons who prefer low situations at the foot of hills, but I am thoroughly convinced from experience, that all bottoms where there are hills on every side, are very improper for this purpose; for the air is drawn down in strong currents, which, being pent in on every side, renders these bottoms much colder than the open situations; and during the winter and spring, these bottoms are very damp, and unhealthy to all vegetables; therefore the gentle rise of a hill, fully exposed to the sun and air, is by much the best situation. As to the soil, a gentle Hazel loam, which is easy to work, and that doth not detain the wet, is the best; if this happens to be three feet deep, it will be the better for the growth of the trees, for although these trees will grow upon very strong land, yet they are seldom so thriving, nor are their fruit so well flavoured, as those which grow on a gentle soil; and on the other hand, these trees will not do well upon a very dry gravel or sand, therefore those soils should never be made choice of for orchards.

The ground intended to be planted should be well prepared the year before, by ploughing it thoroughly, and if some dung is laid upon it the year before, it will be of great service to the trees; if in the precedent spring a crop of Pease or Beans is planted on the ground (provided they are sown or planted in rows, at a proper distance, so as that the ground between them is horse-hoed) it will destroy the weeds, and loosen the ground, so that will be a good preparation for the trees, for the earth cannot be too much wrought or pulverized for this purpose: these crops will be taken off the ground before the season for planting of these trees, which should be performed when the trees begin to shed their leaves.

In chusing of the trees, I would advise the taking such as are but of two years growth from the graft, and never to plant old trees, or such as are grafted upon old stocks; for it is losing of time to plant such, young trees being always more certain to grow, and make a much greater progress than those which are old. As to pruning of the roots, it must be done in the same manner as hath been already directed for the espalier-trees; and in pruning their heads, little more is necessary than to cut such branches as are ill placed, or that cross each other; for I do not approve the heading of them down, as is by some often practised, to the loss of many of their trees.

The distance which these trees should be planted, where the soil is good, must be fifty or sixty feet; and where the soil is not so good, forty feet may be sufficient; but nothing can be of worse consequence, than the crowding trees too close together in orchards: for although there be some persons who may imagine this distance too great, yet I am sure, when they have thoroughly considered the advantages attending this practice, they will agree with me. Nor is it my own authority, for in many of the old writers on this fubject, who have wrote from experience, there is often mention made of the neceffity for allowing a proper diftance to the fruit-trees in orchards; particularly Auften, who fays, "He fhould chufe to prefcribe the planting thefe "trees fourteen or fixteen yards afunder; for both trees "and fruits have many great advantages, if planted a "good diftance from one another." One advantage he mentions is, "The fun refrefhes every tree, the roots, "body, and branches, with the bloffoms and fruits; "whereby trees bring forth more fruit, and thofe fairer "and better." Another advantage he mentions is, "That "when trees are planted at a large diftance, much profit "may be made of the ground under and about thefe trees, "by cultivating garden ftuff, commodious as well for fale "as houfekeeping; as alfo Goofeberries, Rafpberries, "Currants, and Strawberries, may be there planted." Again he fays, "When trees have room to fpread, they "will grow very large and great; and the confequences "of that will be, not only multitudes of fruits, but alfo "long lafting, and thefe two are no fmall advantages." "For (fays he) men are miftaken, when they fay, The "more trees in an orchard, the more fruits; for one or "two larger trees, which have room to fpread, will bear "more fruit than fix or ten (it may be) of thofe that grow "near together, and crowd one another." Again he fays, "Let men but obferve, and take notice of fome Apple-"trees, that grow a great diftance from other trees, and "have room enough to fpread both their roots and branches, "and they fhall fee, that one of thofe trees (being come "to full growth) hath a larger head, and more boughs "and branches, than (it may be) four, or fix, or more, "of thofe which grow near together, although of the "fame age."

And Mr. Lawfon, an ancient planter, advifes to plant Apple-trees twenty yards afunder. As the two authors above quoted have written the beft upon this fubject, and feem to have had more experience than any of the writers I have feen, I have made ufe of them as authorities to confirm what I have advanced; though the fact is fo obvious to every perfon who will make the leaft reflection, that there needs no other proof.

When the trees are planted, they fhould be ftaked, to prevent their being fhaken, or blown out of the ground by ftrong winds; but in doing of this, there fhould be particular care taken to put either ftraw, haybands, or woollen cloth, between the trees and the ftakes, to prevent the trees from being rubbed and bruifed by the fhaking againft the ftakes, for if their bark fhould be rubbed off, it will occafion fuch great wounds, as not to be healed over in feveral years, if they ever recover it.

If the winter fhould prove very fevere, it will be proper to cover the furface of the ground about their roots with fome mulch, to prevent the froft from penetrating the ground, which will injure the young fibres; but this mulch fhould not be laid on too foon, as hath been before-mentioned, left the moifture fhould be prevented from foaking down to the roots of the trees, nor fhould it lie on too long in the fpring, for the fame reafon: therefore where perfons will be at the trouble to lay it on in frofty weather, and remove it again after the froft is over, that the wet in February may have free accefs to the roots of the trees; and if March fhould prove dry, with fharp north or eaft winds, which often happens, it will be proper to cover the ground again with the mulch, to prevent the winds from penetrating and drying the ground, which will be of fingular fervice to the trees. But I am aware, that this will be objected to by many, on account of the trouble, which may appear to be great; but when it is confidered, how much of this bufinefs may be done by a fingle perfon in a fhort time, it can have little force, and the benefit which the trees will receive by this management, will greatly recompenfe the trouble and expence.

As thefe trees muft be conftantly fenced from cattle, it will be the beft way to keep the land in tillage, for by conftant ploughing or digging of the ground, the roots of the trees will be encouraged, and they will make the more progrefs in their growth; but where this is done, whatever crops are fown or planted, fhould not be too near the trees, left the nourifhment fhould be drawn away from them; and if the ground is ploughed, there muft be care taken not to go too near the ftems of the trees, whereby their roots would be injured, or the bark of their ftems rubbed off; but it will be of great fervice to dig the ground about the trees, where the plough doth not come, every fpring or autumn, for five or fix years after planting, by which time their roots will have extended themfelves to a greater diftance.

It is a common practice in many parts of England, to lay the ground down for pafture, after the trees are grown pretty large in their orchards; but this is by no means adviseable, for I have frequently feen trees of above twenty years growth, almoft deftroyed by horfes, in the compafs of one week; and if fheep are put into orchards, they will conftantly rub their bodies againft the ftems of the trees, and their greafe fticking to the bark, will ftint their growth, and in time will fpoil them; therefore wherever orchards are planted, it will be much the better method to keep the ground ploughed or dug annually, and fuch crops put on the ground, as will not draw too much nourifhment from the trees.

In pruning of orchard-trees, nothing more fhould be done but to cut out all thofe branches which crofs each other, which, if left, would rub and tear off the bark from each other, as alfo decayed branches, but never fhorten any of their fhoots. If fuckers, or fhoots from their ftems, fhould come up, they muft be entirely taken off, and when any branches are broken by the wind, they fhould be cut off, either down to the divifion of the branch, or clofe to the ftem from whence it was produced; the beft time for this work is in November, for it fhould not be done in frofty weather, nor in the fpring, when the fap begins to be in motion.

The beft method to keep Apples for winter ufe is, to let them hang upon the trees, until there is danger of froft; to gather them in dry weather, and then lay them in large heaps to fweat for a month or fix weeks; afterward look them over carefully, taking out all fuch as have appearance of decay, wiping all the found fruit dry, and pack them up in large oil jars, which have been thoroughly fcalded and dry, ftopping them down clofe, to exclude the external air; if this is duly obferved, the fruit will keep found a long time, and their flefh will be plump, for when they are expofed to the air, their fkins will fhrink, and their pulp will be foft.

MALUS ARMENIACA. See Armeniaca.

MALUS AURANTIA. See Aurantium.

MALUS LIMONIA. See Limonia.

MALUS MEDICA. See Citreum.

MALUS PERSICA. See Persica.

MALUS PUNICA. See Punica.

MAMMEA. Plum. Nov. Gen. 44. tab. 4. The Mammee-tree.

The Characters are,

The flower has four large concave petals, which fpread open. It hath many awl-fhaped ftamina, and in the center a roundifh germen, which afterward turns to a large flefhy fruit, of a fpherical figure, inclofing one, two, or three large, almoft oval ftones.

We

We have but one SPECIES of this genus, viz.

MAMMEA (*Americana*) staminibus flore brevioribus. Lin. Sp. 512. *Mammee with the stamina shorter than the flower.*

This tree, in the West-Indies, grows to the height of sixty or seventy feet; the leaves are large and stiff, and continue green all the year; the fruit is as large as a man's fist; when ripe, it is of a yellowish green colour, and is very grateful to the taste. It grows in great plenty in the Spanish West-Indies, where the fruit is generally sold in their markets, and is esteemed one of the best fruits of the country. It also grows on the hills of Jamaica, and has been transplanted into most of the Caribbee Islands, where it thrives exceeding well.

In England there are few of these plants, but none of any size. The plants may be propagated by planting the stones, which are often brought from the West-Indies (but these stones should be fresh, otherwise they will not grow;) these should be put into pots, and plunged into a hot-bed of tanners bark. In about two months the plants will appear above ground, after which, in warm weather, the glasses of the hot-bed should be raised to let in the fresh air. In three months after the roots of the plants will have filled the pots, when they should be carefully taken out of the pots, and the outer shell of the nut taken off with all possible care not to injure the roots of the plants; then they must be new potted, and plunged again into the bark-bed, observing to water and shade them until they have taken root, after which they should have air and water in proportion to the warmth of the season. In this bed they may remain till Michaelmas, when they must be removed into the bark-stove, where they must be constantly kept, and may be treated after the manner directed for the Coffee-tree.

If, when the stones of this fruit are brought over, they are put into the tan-bed, under the bottom of any of the pots, they will sprout sooner than those which are planted in the earth.

MANCANILLA. See HIPPOMANE.

MANDRAGORA. Tourn. Inst. R. H. 16. tab. 12. Mandrake.

The CHARACTERS are,

The flower hath one erect bell-shaped petal, which is a little larger than the empalement. It has five awl-shaped stamina, which are arched and hairy at their base. In the center is situated a roundish germen, which afterward turns to a large round berry with two cells, having a fleshy receptacle, convex on each side, filled with kidney-shaped seeds.

We have but one SPECIES of this genus, viz.

MANDRAGORA (*Officinarum.*) Hort. Cliff. 51. *Mandrake with a round fruit.*

This plant grows naturally in Spain, Portugal, Italy, and the Levant, but is preserved here in the gardens of the curious. It hath a long taper root, shaped like a Parsnep, which runs three or four feet deep in the ground; it is sometimes single, and at others divided into two or three branches, almost of the colour of Parsnep, but a little darker; from this arises a circle of leaves, which at first stand erect, but, when grown to their full size, spread open, and lie upon the ground; they are more than a foot in length, and four or five inches broad in the middle, of a dark green colour, and a fœtid scent. These rise immediately from the crown of the root, without any foot-stalk; between them come out the flowers, each standing upon a separate foot-stalk about three inches long, which also arise immediately from the root; the flowers are five-cornered, of an herbaceous white colour, spreading open at the top like a Primrose, having five hairy stamina, with a globular germen in the center, supporting an awl-shaped style. The germen afterward turns to a globular soft berry lying upon the leaves, which, when fully grown, is as large as a Nutmeg, of a yellowish green colour when ripe, full of pulp, in which the kidney shaped seeds are lodged. It flowers in March, and the seeds are ripe in July.

This plant is propagated by seeds, which should be sown upon a bed of light earth soon after they are ripe, for if they are kept until the spring, they seldom succeed well; but those which are sown in autumn will come up in the spring, when they should be carefully cleared from weeds. In this bed they should remain till the autumn, when they should be taken up very carefully, and transplanted into the places where they are to remain, which should be in a light deep soil; for their roots always run downward very deep, so that if the soil be wet, they are often rotted in winter; and if it be too near the gravel or chalk, they seldom thrive well; but if the soil is good, and they are not disturbed, the plants will grow to a large size in a few years, and will produce great quantities of flowers and fruit, which will abide many years.

I have been informed by some persons of credit, that one of these roots will remain sound fifty or sixty years, and be as vigorous as a young plant. I know some plants myself, which are now above forty years old, and in great vigour, which may continue so many years longer, as there are no signs of their decay; but they should never be removed after their roots have arrived to any considerable size, which would break their lower fibres, and so stint the plants, as that if they live they will not recover their former strength in two or three years.

As to the feigned resemblance of a human form, which the roots of this plant are said to have, it is all imposture, owing to the cunning of quacks and mountebanks, who deceive the populace and the ignorant with fictitious images shaped from the fresh roots of Briony and other plants: and what is reported as to the manner of rooting up this plant, by tying a dog thereto, to prevent the certain death of the person who should dare to attempt it, and the groans it emits upon the force offered, &c. is all a ridiculous fable: for I have taken up several large roots of this plant, some of which have been transplanted into other places, but could never observe any accident which attended it, nor was there any difference from that of other deep rooting plants.

MANGIFERA. Lin. Gen. 278. The Mango-tree.

The CHARACTERS are,

The empalement of the flower is cut into five segments; the flower has five petals, and five awl-shaped stamina, crowned by heart-shaped summits, with a round germen, supporting a slender style, terminated by a single stigma. The germen afterward becomes a kidney-shaped Plumb, inclosing an oblong woolly nut of the same form.

We have but one SPECIES of this genus, viz.

MANGIFERA (*Indica.*) Lin. Sp. 290. *The Mango-tree.*

This tree grows naturally in many parts of India, and has by the Portugueze been transplanted to the Brazils and other countries, where it grows to a large size; the wood is brittle, the bark rough when old; the leaves are seven or eight inches long, and more than two broad; the flowers are produced in loose panicles at the end of the branches; these are succeeded by large oblong kidney-shaped Plumbs.

This fruit, when fully ripe, is greatly esteemed by the persons who reside in the countries where it grows, but in Europe we have only the unripe fruit brought over in pickle; however, the account given of the ripe fruit by those who have tasted it, has excited the curiosity of many persons to procure young plants for their gardens in Europe, but hitherto without effect; for I have not heard of any plants now growing here, for their nuts never have grown, nor do I believe they will vegetate, unless they are immediately planted; therefore those who desire to have them, should plant the stones in boxes or tubs of earth, pretty close together, letting them remain in the country until the plants have made some progress, and are of a proper size to bear the voyage, and defend them from salt-water and the spray of the sea in their passage, giving them water sparingly, for too much wet will kill them, especially when they arrive in a cool climate: with carefully observing this, the plants may arrive in a proper state; and when the plants are brought over, they should be placed in a tan-bed in the stove, giving them a moderate degree of warmth, this may be effected.

MANIHOT. See JATROPHA.

MAPLE. See ACER.

MARACOCK. See PASSIFLORA.

MARANTA. Plum. Nov. Gen. 16. tab. 36. Indian Arrow-root.

The CHARACTERS are,

The flower hath one petal, which is of the grinning kind, with an oblong compressed tube, cut into six small segments, representing a lip flower. It has one membranaceous stamina, appearing like a segment of the petal, and a roundish germen, situated under the flower, which afterward turns to a roundish three-cornered capsule with three valves, containing one hard rough seed.

The SPECIES are,

1. MARANTA (*Arundinacea*) segmentis petalorum dentatis. *Indian Arrow-root with the segments of the flower indented.*

2. MARANTA (*Galanga*) segmentis petalorum integerrimis. *Indian Arrow-root with the segments of the petal entire; called Indian Arrow-root.*

The first sort was discovered by Father Plumier in some of the French settlements in America; and the late Dr. William Houston found the same sort growing in plenty near La Vera Cruz in New Spain.

This hath a thick, fleshy, creeping root, which is very full of knots, from which arise many smooth leaves six or seven inches long, and three broad in the middle, terminating in points, which arise immediately from the root; between these come out the stalks, which rise near two feet high, and divide upward into two or three smaller, garnished at each joint with one leaf, of the same shape with the lower, but are smaller. The stalks are terminated by a loose spike of small white flowers, standing upon long footstalks. The flowers are cut into six narrow segments, which are indented on their edges, and sit upon the germen, which afterward turns to a roundish three-cornered capsule, inclosing one hard rough seed.

The other sort was brought from some of the Spanish settlements in America, into the islands of Barbadoes and Jamaica, where it is cultivated in their gardens as a medicinal plant, being a sovereign remedy to cure the bite of wasps, and to extract the poison of the Manchineel-tree. The Indians apply the root to expel the poison of their arrows, which they use with great success. They take up the roots, and after cleansing them from dirt, they mash them, and apply it as a poultice to the wounded part, which draws out the poison, and heals the wound. It will also stop a gangrene, if it be applied before it is gone too far, so that it is a very valuable plant.

This sort is very like the first, but the flowers are smaller, and the segments of the petals are entire, in which their principal difference consists.

These plants being natives of a warm country, are tender, therefore will not live in this climate, unless they are preserved in stoves. They propagate fast by their creeping roots, which should be parted the middle of March, just before they begin to push out new leaves, and planted in pots filled with light earth, then plunged into a moderate hot-bed of tanners bark, observing now and then to refresh them with water; but it must not be given to them in large quantities, for too much moisture will soon rot them when they are in an unactive state. When the green leaves appear above ground, the plants will require frequently to be watered, and should have free air admitted to them every day, in proportion to the warmth of the season and the heat of the bed in which they are placed. Where they are constantly kept in the tan-bed, and have proper air and moisture, they will thrive, so as from a small root to fill the pots, in which they were planted, in one summer. About Michaelmas the first sort will begin to decay, and in a short time after the leaves will die to the ground, but the pots must be continued all the winter in the bark-bed, otherwise the roots will perish; for although they are in an inactive state, yet they will not keep from shrinking very long, when taken out of the ground; and if the pots are taken out of the tan, and placed in any dry part of the stove, the roots often shrivel and decay; but when they are continued in the tan, they should have but little water given to them when their leaves are decayed, lest it rot them.

MARJORAM. See ORIGANUM.

MARLE is a kind of clay, which is become fatter, and of a more enriching quality, by a better fermentation, and by its having lain so deep in the earth as not to have spent or weakened its fertilizing quality by any product.

Marles are of different qualities in different counties of England. There are reckoned to be four sorts of Marles in Sussex, a gray, a blue, a yellow, and a red; of these the blue is accounted the best, the yellow the next, and the gray the next to that; and as for the red, that is the least valuable.

In Cheshire they reckon six sorts of Marle:

1. The cowshut Marle, which is of a brownish colour, with blue veins in it, and little lumps of chalk or lime stone; it is commonly found under clay, or low black land, seven or eight feet deep, and is very hard to dig.

2. Stone, slate, or flat Marle, which is a kind of soft stone, or rather slate of a blue or bluish colour, that will easily dissolve with frost or rain. This is found near rivers, and the sides of hills, and is a very lasting sort of Marle.

3. Peat Marle, or delving Marle, which is close, strong, and very fat, of a brown colour, and is found on the sides of hills, and in wet or boggy grounds, which have a light sand in them about two feet or a yard deep.

4. Clay Marle; this resembles clay, and is pretty near akin to it, but is fatter, and sometimes mixed with chalk stones.

5. Steele Marle, which lies commonly in the bottom of pits that are dug, and is of itself apt to break into cubical bits; this is sometimes under sandy land.

6. Paper Marle, which resembles leaves or pieces of brown paper, but something of a lighter colour; this lies near coals.

The properties of any sorts of Marles, and by which the goodness of them may be best known, are better judged of by their purity and uncompoundedness than their colour: as if it will break in pieces like dice, or into thin flakes, or is smooth like lead ore, and is without a mixture of gravel or sand; if it will shake like slate stones, and shatter after wet, or will turn to dust when it has been exposed to the sun, or will not hang and stick together when it is thoroughly dry, like tough clay, but is fat and tender, and will open the land it is laid on, and not bind, it may be taken for granted, that it will be beneficial to it.

Marles do not make so good an improvement of lands the first year as afterwards.

The quantity of Marle ought to be in proportion to the depth of the earth, for over marling has often proved of worse consequence than under marling, especially where the land is strong; for by laying it in too great quantities, or often

often repeating the marling, the land has become so strong and bound so closely, as to detain the wet like a dish, so that the owners have been obliged to drain the ground at a great expence; but in sandy land there can be no danger in laying on a great quantity, or repeating it often, for it is one of the best dressings for such land.

MARRUBIASTRUM. See SIDERITIS.

MARRUBIUM. Tourn. Inst. R. H. 192. tab. 91. Horehound.

The CHARACTERS are,

The flower is of the lip kind, with a cylindrical tube, divided into two lips. It has four stamina, which are under the upper lip, two of which are a little longer than the other. It hath a four pointed germen, which afterward turns to four oblong seeds, sitting in the empalement.

The SPECIES are,

1. MARRUBIUM (*Vulgare*) dentibus calycinis setaceis uncinatis. Hort. Cliff. 312. *Common white Horehound.*

2. MARRUBIUM (*Peregrinum*) foliis ovato lanceolatis serratis, calycum denticulis setaceis. Hort. Cliff. 311. *Broad-leaved, foreign, white Horehound.*

3. MARRUBIUM (*Creticum*) foliis lanceolatis dentatis, verticillis, minoribus, dentibus calycinis setaceis erectis. *Narrow-leaved, foreign, white Horehound.*

4. MARRUBIUM (*Alyson*) foliis cuneiformibus, quinque verticillis involucro destitutis. Hort. Cliff. 311. *Horehound called Madwort, with leaves which are deeply cut.*

5. MARRUBIUM (*Supinum*) dentibus calycinis setaceis rectis villosis. Hort. Cliff. 312. *Low Spanish Horehound, with silver-coloured sattiny leaves.*

6. MARRUBIUM (*Candidissimum*) foliis subovatis lanatis superne emarginato-crenatis, denticulis calycinis subulatis. Hort. Cliff. 312. *The whitest and most hoary Horehound.*

7. MARRUBIUM (*Hispanicum*) calycum limbis patentibus, denticulis acutis. Hort. Cliff. 312. *Round-leaved Spanish Horehound.*

8. MARRUBIUM (*Crispum*) calycum limbis planis villosis, foliis orbiculatis rugosis, caule herbaceo. *Spanish bastard Dittany, with rough-curled leaves.*

9. MARRUBIUM (*Suffruticosum*) calycum limbis planis villosis, foliis cordatis rugosis incanis, caule suffruticoso. *Spanish bastard Dittany, with a very large hoary leaf.*

10. MARRUBIUM (*Acetabulosum*) calycum limbis planis villosis, foliis cordatis, caule fruticoso. Hort. Cliff. 312. *Whorled, unsavoury, bastard Dittany.*

11. MARRUBIUM (*Acetabulum*) calycum limbis tubo longioribus membranaceis, angulis majoribus rotundatis. Lin. Sp. Plant. 584. *Bastard Dittany, with an empalement like Molucca Baum.*

The first sort is the Prasium, or white Horehound of the shops. This grows naturally in many parts of England, so is seldom propagated in gardens. It hath a ligneous fibrous root, from which come out many square stalks two feet in length, garnished with hoary roundish leaves, indented on the edges, placed opposite. The flowers grow in very thick whorls round the stalks at each joint; they are small, white, and of the lip kind, standing in stiff hoary empalements, cut into ten parts at the top, which end in stiff bristles; these are succeeded by four oblong black seeds, sitting in the empalement.

The second sort grows naturally in Italy and Sicily. This rises with square stalks near three feet high, which branch much more than the first; the leaves are rounder, whiter, and stand farther asunder; the whorls of flowers are not so large, but the flowers have longer tubes.

The third sort grows naturally in Spain and Portugal. This rises with slender hoary stalks three feet high; the leaves are very hoary, much longer and narrower than those of the second; the whorls of flowers are smaller, the bristly indentures of the empalement are longer and erect; the whole plant has an agreeable flavour.

The fourth sort grows naturally in Spain and Italy. This is a biennial plant, whose stalks are about the same length as those of the first; the leaves are wedge-shaped, hoary, and obtusely indented; the whorls of flowers are small, and have no covers. The flowers stand looser in the whorls, and the cuts of the empalement end in very stiff prickles which spread open; the flowers are purple, and larger than those of the first sort.

The fifth sort grows naturally in Spain, and also in the islands of the Archipelago. The stalks of this are seldom above eight or nine inches long, covered with a soft hoary down; the leaves are small, roundish, very soft to the touch, and indented on the edges. The whorls of flowers are small, very downy and white.

The sixth sort grows naturally in Spain. This hath stalks about the same length as the first; the leaves are nearly oval, woolly, and crenated toward the top, and the empalement of the flowers are awl-shaped.

The seventh sort grows naturally in Spain. The stalks of this grow more erect than those of the common sort; the leaves are rounder and more sawed on the edges; the empalement of the flowers spread open, ending in acute segments. The flowers are like those of the common sort; the whole plant is very hoary.

The eighth sort grows naturally in Spain and Sicily. This sends out many stiff roundish stalks, which rise more than two feet high, covered with a white cottony down; the leaves are almost round, rough on their upper side, and woolly on their under; the whorls of flowers are large, the borders of the empalement are flat and hairy; the tube of the flower is scarce so long as the empalement, so the two lips are but just visible.

The ninth sort grows naturally in Spain. The stalks of this are a little shrubby, and rise near three feet high, dividing into small branches; the leaves are heart-shaped, and rough on their upper side, but hoary on their under; the whorls of flowers are large, the borders of the empalements flat and hairy; the tube of the flower is longer, and the flowers are larger than those of the former sort; they are of a pale purple colour, and their upper lips are erect.

The tenth sort grows naturally in Sicily, and the islands of the Archipelago. This rises with a shrubby stalk two feet high, which divides into many branches, garnished with small heart-shaped leaves, sitting pretty close to the stalks; the whorls of flowers are not so large as those of the two former sorts. The rim of the empalements are flat. The flowers are white, and the whole plant is very hoary

The eleventh sort grows naturally in Crete. This hath very woolly stalks, which rise two feet high, garnished with woolly heart-shaped leaves. The whorls of flowers are large, the borders of the empalements flat, and cut into many segments, which are membranaceous, angular, and rounded at the top. The flowers are small, of a pale purple colour, but scarce appear out of their empalements, and their upper lips are erect.

The fourth sort is supposed to be Galen s Madwort. This was by the ancients greatly recommended for its efficacy in curing of madness, and some few of the moderns have prescribed it in the same manner, but at present it is seldom used.

All these plants are preserved in botanic gardens for the sake of variety, but there are not above two of the sorts which are cultivated in other gardens; these are the tenth and eleventh sorts, whose stalks are shrubby, and the plants very hoary, so make a variety when intermixed with other plants; these very rarely produce seeds in England, so are propagated by cuttings, which, if planted in a ſhady border the middle of April, will take root very freely.

They are ſomewhat tender, ſo in very ſevere winters ſhould be ſcreened from the hard froſts, eſpecially thoſe plants which grow in good ground, and are luxuriant, whoſe branches are more replete with juice, ſo are liable to ſuffer by cold; but when they are in a poor, dry, rubbiſhy ſoil, the ſhoots will be ſhort, firm, and dry, ſo are ſeldom injured by the cold, and theſe plants will continue much longer than thoſe in better ground.

The other ſorts are eaſily propagated by ſeeds, which ſhould be ſown on a bed of dry earth in the ſpring; when the plants come up, they muſt be kept clean from weeds, and where they are too cloſe they ſhould be thinned, leaving them a foot and a half aſunder, that their branches may have room to ſpread, after this they require no other culture; they may alſo be propagated by cuttings in the ſame manner as the two other ſorts.

MARRUBIUM NIGRUM. See BALLOTE.

MARTAGON. See LILIUM.

MARTYNIA. Houſt. Gen. Nov. Martyn. Dec. 1. 42.

The CHARACTERS are,

The flower hath one bell-ſhaped petal, and is of the ringent kind, with a ſwelling tube, at the baſe of which is ſituated a gibbous nectarium. It hath four ſlender incurved ſtamina, which are inflexed into each other, and an oblong germen ſituated under the flower. The empalement afterward turns to an oblong gibbous capſule, which divides into two parts, including a hard nut, ſhaped like the body of a ſtag beetle, with two incurved ſtrong horns at the end, having four cells, two of which are generally barren, the other two have one oblong ſeed in each.

The SPECIES are,

1. MARTYNIA (*Annua*) caule ramoſo, foliis angulatis. Lin. Sp. Plant. 618. *Martynia with a branching ſtalk and angular leaves.*

2. MARTYNIA (*Perennis*) caule ſimplici, foliis ſerratis. Lin. Sp. Pl. 618. *Martynia with a ſingle ſtalk, and ſawed leaves.*

3. MARTYNIA (*Louiſianica*) caule ramoſo, foliis cordato-ovatis piloſis. Plat. 286. *Martynia with a branching ſtalk, and oval, heart-ſhaped, hairy leaves.*

The firſt ſort was diſcovered by the late Dr. William Houſtoun near La Vera Cruz in New Spain.

This riſes with a ſtrong herbaceous ſtalk near three feet high, which divides upward into two or three large branches, garniſhed with oblong, oval, hairy, viſcous leaves, cut into angles on their ſides. The flowers are produced in ſhort ſpikes from the end of the branches; they are ſhaped like thoſe of the common Foxglove, but are of a paler purple colour, and are ſucceeded by oblong oval capſules, which are thick, tough, and clammy; when ripe, divide into two parts, and drop off, leaving a large hard nut hanging on the plant, about the ſize and much of the ſame form as the ſtag beetle, with two ſtrong crooked horns at the end. The nut has two deep longitudinal furrows on the ſides, and ſeveral ſmaller croſſing each other in the middle, and is ſo hard, that it is with difficulty cut open, without injuring of the ſeeds; within are four oblong cells, two of which have a ſingle oblong ſeed in each, but the other two are generally abortive. If the plants are brought forward in the ſpring, they will begin to ſhew their flowers in July, which are firſt produced at the diviſion of the branches, and afterward at the extremity of each branch, ſo there will be a ſucceſſion of flowers on the ſame plant till the end of October, when the plants uſually decay.

The ſecond ſort was diſcovered by Mr. Robert Millar, growing naturally about Carthagena in New Spain. This hath a perennial root, and an annual ſtalk. The roots of this plant are thick, fleſhy, and divided into ſcaly knots, ſomewhat like thoſe of Toothwort, from which ariſe ſeveral ſingle, fleſhy, ſucculent ſtalks about a foot high, of a purpliſh colour, garniſhed with oblong thick leaves, whoſe baſe ſits cloſe to the ſtalk; they are ſawed on their edges, rough on their upper ſide, where they are of a dark green, but their under ſide is purpliſh. The ſtalk is terminated by a ſpike of blue flowers, which are bell-ſhaped, and do not ſpread open at the rim ſo much as the former ſort; theſe are not ſucceeded by ſeeds in England.

The third ſort grows naturally in the Miſſiſſippi. This is an annual plant, having a thick fleſhy ſtalk about two feet high, which divides into three or four ſpreading branches, garniſhed with oval heart-ſhaped leaves, of a ſoft green colour, very viſcous and hairy; the flowers terminate the branches in a looſe thyrſe; they are in ſhape like thoſe of the firſt ſort, but of a paler colour, and are ſucceeded by large oval fruit, having very thick covers like the outer ſhell of Walnuts, with two long incurved horns at the end: the outer ſhell drops off when the fruit is ripe, leaving a hard fibrous nut hanging on the plant, which opens in the middle, having four cells, each containing two or three oval ſeeds.

The firſt and third ſorts are annual, ſo are only propagated by ſeed, which ſhould be ſown in pots, and plunged into a hot-bed of tanners bark, where, if the ſeeds are entirely ſeparated from their covers, the plants will appear in about a month, and will grow pretty faſt, if the bed is warm; they ſhould therefore be tranſplanted in a little time after they come up, each into a ſeparate pot, and plunged into the hot-bed again, obſerving to water and ſhade them from the ſun until they have taken new root; then they ſhould have a large ſhare of freſh air admitted to them in warm weather, to prevent their drawing. With this management the plants will make great progreſs, ſo as to fill the pots with their roots in about a month's time, when they ſhould be ſhifted into pots about a foot diameter at the top, and plunged into the hot-bed in the bark-ſtove, where they ſhould be allowed room, becauſe they put out many ſide branches, and will grow three feet high or more, according to the warmth of the bed, and the care which is taken to ſupply them conſtantly with water. When theſe plants thrive well, they will ſend out many ſide branches, which will all of them produce ſmall ſpikes of flowers; but it is only from the firſt ſpike of flowers that good ſeeds can be expected in this country, ſo that particular care ſhould be taken, that none of thoſe are pulled off or deſtroyed, becauſe it is very difficult to obtain good ſeeds here, eſpecially of the firſt ſort: but the third ſort is much more hardy, ſo will perfect its ſeeds here very well, eſpecially if the plants are not too much drawn; but the outer ſkin of the ſeeds ſhould be taken off, otherwiſe they will not grow.

The nuts of the firſt ſort are ſo hard, as to render it very difficult to get out the ſeeds without breaking them; and if the whole nut is put into the ground, the ſeeds do ſeldom grow. I have had ſome of the nuts lying in the ground two years, which after were taken up, and appeared as ſound and good as when they were firſt put in, and ſome of the ſeeds have grown after having been kept four or five years above ground.

The ſecond ſort dies to the root every winter, and riſes again the ſucceeding ſpring. This muſt be conſtantly preſerved in the ſtove, and plunged into the bark-bed, otherwiſe it will not thrive in this country. During the winter ſeaſon, when the ſtalks of the plants are decayed, they ſhould have but little water given them; about the middle of March, juſt before the plants begin to ſhoot, is the proper ſeaſon to tranſplant and part the roots, when they ſhould be planted into pots of a middle ſize, and plunged into the bark-bed, which ſhould at this time be renewed with freſh tan.

tan. When the plants come up, they muſt be treated in the ſame way as other tender exotics, which require the bark-ſtove.

MARVEL OF PERU. See MIRABILIS.

MARUM SYRIACUM. See TEUCRIUM.

MARUM VULGARI. See SATUREJA.

MARYGOLD See CALENDULA.

MARYGOLD (AFRICAN). See TAGETES.

MARYGOLD (FIG). See MESEMBRYANTHEMUM.

MARYGOLD (FRENCH). See TAGETES.

MASTERWORT. See IMPERATORIA.

MASTICHINA. See SATUREJA.

MATRICARIA. Tourn. Inſt. R. H. 493. tab. 281. Feverfew.

The CHARACTERS are,

It hath a compound flower. The ray or border is compoſed of female half florets, and the diſk, which is hemiſpherical, of hermaphrodite florets. The female half florets are tongue-ſhaped, and indented in three parts at the end; theſe have a naked germen. The hermaphrodite florets are tubulous, funnel-ſhaped, and cut into five parts at the brim; they have each five hairy ſhort ſtamina, and an oblong naked germen. The germen of both turn to ſingle, oblong, naked ſeeds.

The SPECIES are,

1. MATRICARIA (*Parthenium*) foliis compoſitis planis, foliolis ovatis inciſis, pedunculis ramoſis. Hort. Cliff. 416. *Common or Garden Feverfew.*

2. MATRICARIA (*Maritima*) receptaculis hemiſphæricis, foliis bipinnatis ſubcarnoſis, ſuprà convexis, ſubtus carinatis. Lin. Sp. Pl. 891. *Feverfew; commonly called dwarf perennial, maritime Chamomile, with ſhort, thick, dark, green leaves.*

3. MATRICARIA (*Indica*) foliis ovatis ſinuatis angulis ſerratis acutis. *Feverfew with oval ſinuated leaves, which are acutely ſawed.*

4. MATRICARIA (*Argentea*) foliis bipinnatis pedunculis ſolitariis. Hort. Cliff. 415. *Feverfew with winged leaves, and ſingle foot-ſtalks.*

5. MATRICARIA (*Americana*) foliis lineari-lanceolatis integerrimis, pedunculis unifloris. *Feverfew with entire ſpear-ſhaped leaves, and foot-ſtalks with one flower.*

The firſt ſort is the common Feverfew, which is directed to be uſed in medicine. It grows naturally in lanes, and upon the ſide of banks in many parts of England, but is frequently cultivated in the phyſic gardens to ſupply the markets. The ſtalks riſe upward of two feet high; they are round, ſtiff, and ſtriated; the leaves are compoſed of ſeven lobes, which are cut into many obtuſe ſegments; the ſtalks and branches are terminated by the flowers, which are diſpoſed almoſt in the form of looſe umbels. The flowers are compoſed of ſeveral ſhort rays, which are white like thoſe of the Chamomile, ſurrounding a yellow diſk, compoſed of hermaphrodite florets, which form a hemiſphere, and are incloſed in one common ſcaly empalement, which are ſucceeded by oblong, angular, naked ſeeds. The whole plant has a ſtrong unpleaſant odour.

The following varieties of this plant are preſerved in botanic gardens, many of which are pretty conſtant, if care is taken of ſaving the ſeeds; but where the ſeeds of theſe plants have been ſuffered to ſcatter, it will be almoſt impoſſible to preſerve the varieties without mixture. I ſhall only juſt inſert them here, for thoſe who are curious in collecting the varieties.

1. Feverfew with very double flowers.
2. Feverfew with double flowers, whoſe borders or rays are plain, and the diſk fiſtular.
3. Feverfew with very ſmall rays.
4. Feverfew with very ſhort fiſtular florets.
5. Feverfew with naked heads, having no rays or border.
6. Feverfew with naked ſulphur-coloured heads.
7. Feverfew with elegant curled leaves.

Theſe plants are propagated by their ſeeds, which ſhould be ſown either in ſpring or autumn; and when the plants are come up, they ſhould be tranſplanted out into nurſery-beds, at about ſix inches aſunder, where they may remain till they have grown large, when they may be taken up, with a ball of earth to their roots, and planted in the middle of large borders, where they will flower, and produce ripe ſeeds.

When the different varieties of theſe plants are intermixed with other plants of the ſame growth in large gardens, they are ornamental; but as their roots ſeldom abide more than two years, freſh plants ſhould be raiſed from ſeeds, to ſupply their places; for although they may be propagated by parting their roots either in ſpring or autumn, yet theſe ſeldom make ſo good plants as thoſe obtained from ſeeds. When theſe plants grow from the joints of walls, or upon dry lime rubbiſh, they will continue much longer than in good ground.

The ſecond ſort grows naturally near the ſea in ſeveral parts of England. I have obſerved it upon the Suſſex coaſt in great plenty, from whence I brought the plants, which were of no longer duration in the garden than two years, though in their native ſoil they may continue longer. The ſtalks of this plant branch out pretty much, and ſpread near the ground; they are garniſhed with dark green leaves, which are compoſed of many double wings or pinnæ, like thoſe of the common Chamomile, but are much thicker in ſubſtance; they have their edges turned backward ſo are convex on their upper ſurface, and concave on their under. The flowers are white, like thoſe of the common Chamomile, and are diſpoſed almoſt in the form of an umbel.

This plant is ſeldom cultivated but in botanic gardens for variety. It may be propagated by ſeeds, which may be ſown either in autumn ſoon after they are ripe, or in the ſpring, upon a bed of common earth, in almoſt any ſituation; and when the plants come up, they will require no other care but to thin them where they are too cloſe, and keep them clean from weeds.

The third ſort I received from Nimpu in India, where it grows naturally, ſo is too tender to live through the winter in the open air in England, therefore the plants ſhould be planted in pots to be ſheltered in winter; the following ſummer they will produce ripe ſeeds, ſoon after which the plants decay.

The fourth and fifth ſorts are preſerved in botanic gardens for variety, but as they have little beauty, ſo are rarely kept in other gardens.

MAUDLIN. See ACHILLEA.

MAUROCENIA. Lin. Gen. Plant. edit. 2 289. The Hottentot Cherry.

The CHARACTERS are,

The flower hath five oval petals, which ſpread open. It hath five ſtamina, which are ſituated between the petals. In the center is ſituated a roundiſh germen, crowned by a trifid ſtigma. The germen afterward turns to an oval berry, with one or two cells, each containing a ſingle oval ſeed.

The SPECIES are,

1. MAUROCENIA (*Frangula*) foliis ſubovatis integerrimis, floribus confertis lateralibus. *Maurocenia with entire leaves, which are almoſt oval, and flowers growing in cluſters on the ſides of the branches; commonly called Hottentot Cherry.*

2. MAUROCENIA (*Phillyrea*) foliis obverſè ovatis ſerratis, floribus corymboſis alaribus & terminalibus. *Maurocenia with obverſe, oval, ſawed leaves, and flowers growing in a corymbus at the ſides and end of the branches; commonly called Cape Phillyrea, and by the Dutch Leplebout.*

3. MAURO-

3. Maurocenia (*Cerasus*) foliis ovatis nervosis integerrimis. *Maurocenia with oval veined leaves, which are entire; commonly called smaller Hottentot Cherry.*

4. Maurocenia (*Americana*) foliis obverse ovatis emarginatis, floribus solitariis alaribus. *Maurocenia with obverse oval leaves, which are indented at the edges, and flowers growing singly from the sides of the branches.*

The first three sorts grow naturally at the Cape of Good Hope; the first rises to a considerable height in that country. The stalk is strong, woody, and covered with a purplish bark, sending out many stiff branches, garnished with very thick leaves, almost oval, standing for the most part opposite, of a dark green colour, and entire. The flowers come out from the side of the old branches in clusters, three, four, or five standing upon one common foot-stalk; they are composed of five plain equal petals, ending in acute points, of a greenish yellow colour, changing to white. In the center is situated the oval germen, crowned by a trifid stigma, and between each petal is situated a stamina, terminated by obtuse summits. The germen afterward turns to an oval pulpy berry, some having but one, and others two cells; in each of these is lodged one oval seed.

The second sort hath a woody stalk, which in this country seldom rises more than eight or ten feet high, dividing into many branches, covered with a dark purplish bark, garnished with pretty stiff leaves, which are sawed on their edges, obversely oval, and stand opposite, of a light green, having short foot-stalks. The flowers are produced in roundish bunches from the side, and at the end of the branches; they are white, and have five small petals, between which are situated the stamina, terminated by obtuse summits. In the center is situated a roundish germen, crowned sometimes by a bifid, and at others by a trifid stigma.

The third sort rises with a woody stalk about the same height as the former, dividing into many branches, garnished with stiff, oval, entire leaves, of a light green colour, having three longitudinal veins; these are sometimes placed opposite, and at others they are alternate, having a strong margin or border surrounding them.

The fourth sort was discovered by the late Dr. Houstoun, growing naturally at the Palisadoes in Jamaica. This rises with a woody stalk from fifteen to twenty feet high, covered with a rough brown bark, which divides into many branches, garnished with stiff leaves, placed alternately, indented at the top with a stiff reflexed border, of a gray colour on their upper side, but of a rusty iron colour on their under, standing upon short foot-stalks. The flowers come out singly along the side of the branches, which have five small white petals, ending in acute points, and five slender stamina, which are terminated by obtuse summits. In the center is situated a roundish germen, supporting a long bifid stigma, which is permanent. The germen afterward turns to a round berry, with one or two cells, each having one oblong seed.

The first sort is too tender to live abroad in England, but as it requires no artificial heat, so may be preserved through the winter in a good green-house, where it deserves a place for the beauty of its leaves, which are very thick, of a deep green, and differing in appearance from every other plant. It may be propagated by laying down those shoots which are produced near the ground, but they are long in putting out roots. The shoots should be twisted in the part which is laid, to facilitate their putting out roots: if these are laid down carefully in autumn, they will put out roots sufficient to remove by the following autumn; it may also be propagated by cuttings, but this is also a tedious method, as they are seldom rooted enough to transplant in less than two years. When this is practised, the young shoots of the former year should be cut off, with a small piece of the old wood at the bottom, and planted in pots filled with loamy earth, which must be plunged into a moderate hot-bed, covering the pots with hand-glasses; which should be close stopped down to exclude the external air; they should be pretty well watered at the time they are planted, but afterward they will require but little wet; the glasses over them should be covered with mats, to screen the cuttings from the sun during the heat of the day, but in the morning before the sun is too warm, and in the afternoon when the sun is low, they should be uncovered, that the oblique rays of the sun may raise a gentle warmth under the glasses. With this care the cuttings will take root, but where this is wanting they seldom succeed. When the cuttings or layers are rooted, they should be each planted in a separate small pot, and placed in the shade till they have taken new root; then they may be removed to a sheltered situation, where they may remain during the summer season; and, before the frosts of the autumn, they must be removed into the green-house, and treated in the same way as the other plants of that country. When the plants have obtained strength, they will produce flowers and fruit, which in warm seasons will ripen perfectly; and if the seeds are sown soon after they are ripe in pots, and plunged into the tan-bed in the stove, the plants will come up the spring following, and may then be treated in the same manner as those which are propagated by cuttings and layers.

The second sort is not altogether so hardy as the first, so must have a warmer place in the green-house in winter, and should not be placed abroad quite so early in the spring, nor suffered to remain abroad so late in the autumn; but if the green-house is warm, the plants will require no additional heat. This may be propagated by layers and cuttings, in the same manner as the first, and requires the same care, for the cuttings are with difficulty made to take root; nor will the branches which are laid put out roots in less than a year, and if these are not young shoots, they will not take root.

The third sort is yet more rare than either of the former, and is with greater difficulty propagated, for the layers and cuttings are commonly two years before they get roots sufficient to remove; and as it never produces seeds here, so can be no other way propagated. This is also tenderer than either of the other sorts, so requires a moderate degree of heat in winter, for without some artificial warmth it will not live through the winters in England. In the middle of summer the plants may be placed abroad in a warm situation, but they must be removed into shelter early in the autumn, before the cold nights come on, otherwise they will receive a check, which they will not recover in winter.

The fourth sort is much more impatient of cold than either of the other, being a native of a warmer country. This is propagated in the same way as the other sorts, but when seeds can be obtained from Jamaica, the plants produced from those will be much the best; but as the seeds seldom come up the first year, so they should be sown in pots, and plunged into a moderate hot-bed of tanners bark, where they may remain all the summer; and in the autumn they should be removed into the bark-stove, and plunged into the tan-bed between the other pots of plants, in any vacant spaces; there they may remain till spring, when they should be taken out of the stove, and plunged into a fresh hot-bed, which will bring up the plants. When these are fit to remove, they should be each transplanted into a separate small pot, and plunged into a hot-bed again, being careful to shade them from the sun till they have taken new root; after which they must be treated in the same manner as other tender plants from the same country.

All the sorts delight in a soft, gentle, loamy soil, not over stiff, so as to detain the wet; nor should the soil be too light, for in such they will not thrive. They retain their leaves all the year, so make a good appearance in the winter season, their leaves being remarkably stiff, and of a fine green, especially the first sort, whose fruit ripens in winter, which, when it is in plenty on the plants, affords an agreeable variety.

MAYS. See ZEA.

MEADIA. Catesb. Carol. 3. p. 1.

The CHARACTERS are,

The flower hath a permanent empalement of one leaf, cut into five long segments, which are reflexed. The flower hath one petal, cut into five parts, whose limb is reflexed backward. It hath five short stamina sitting in the tube, connected at the top with a conical germen, terminated by an obtuse stigma. The empalement afterward becomes an oblong oval capsule with one cell, opening at the top, and filled with small seeds.

This genus of plants was so titled by Mr. Mark Catesby, F. R. S. in honour of the late Dr. Mead, who was a generous encourager of every useful branch of science.

We have but one SPECIES of this genus, viz.

MEADIA (*Dodecatheum.*) Catesb. Hist. Carol. App. 1. tab. 1. *Meadia.*

This plant grows naturally in Virginia, and other parts of North America, from whence many years ago it was sent by Mr. Banister to Dr. Compton, Lord Bishop of London, in whose curious garden I first saw this plant growing in the year 1709; after which it was for several years lost in England till within a few years past, when it was again obtained from America, and has been since propagated in pretty great plenty. It hath a perennial root, from which come out several long smooth leaves, which are six inches long, and two and a half broad; at first standing erect, but afterward they spread on the ground, (especially if the plants are much exposed to the sun.) From between these leaves arise two, three, or four stalks, in proportion to the strength of the roots, eight or nine inches high; they are smooth, naked, and are terminated by an umbel of flowers, under which is situated the many-leaved involucrum. Each flower is sustained by a pretty long slender foot-stalk, which is recurved, so that the flower hangs downward. It has but one petal, which is deeply cut into five spear-shaped segments, which are reflexed backward like the flowers of Cyclamen or Sowbread; the stamina, which are five in number, are short, and sit in the tube of the flower, having five arrow-pointed summits, which are connected together round the style, forming a sort of beak. The flowers are purple, inclining to a Peach blossom colour, and have an oblong germen situated in the bottom of the tube, which afterward becomes an oval capsule inclosed by the empalement, with the permanent style on its apex, which, when ripe, opens at the top to let out the seeds, which are fastened round the style. This plant flowers the beginning of May, and the seeds ripen in July, soon after which the stalks and leaves decay, so that the roots remain inactive till the following spring.

It is propagated by offsets, which the roots put out freely when they are in a loose moist soil and a shady situation; the best time to remove the roots, and take away the offsets, is in August, after the leaves and stalks are decayed, that they may be fixed well in their new situation before the frost comes on. It may also be propagated by seeds, which the plants generally produce in plenty; these should be sown in the autumn soon after they are ripe, either in a shady moist border, or in pots, which should be placed in the shade; in the spring the plants will come up, and must then be kept clean from weeds, and if the season proves dry they must be frequently refreshed with water; nor should they be exposed to the sun, for while the plants are young they are very impatient of heat, so that I have known great numbers of them destroyed in two or three days, which were growing to the full sun. These young plants should not be transplanted till the leaves are decayed, then they may be carefully taken up and planted in a shady border, where the soil is loose and moist, at about eight inches distance from each other, which will be room enough for them to grow one year, by which time they will be strong enough to produce flowers, so may then be transplanted into some shady borders in the flower-garden, where they will appear very ornamental during the continuance of their flowers.

MEADOW SAFFRON. See COLCHICUM.

MEDEOLA. Lin. Gen. Plant. 411.

The CHARACTERS are,

The flower has no empalement; it hath six oblong oval petals, and six awl-shaped stamina terminated by incumbent summits, and three horned germen terminating the style. The germen afterward turns to a roundish trifid berry with three cells, each containing one heart-shaped seed.

The SPECIES are,

1. MEDEOLA (*Asparagoides*) foliis ovato-lanceolatis alternis, caule scandente. *Medeola with oval, spear-shaped, alternate leaves, and a climbing stalk; commonly called Climbing African Asparagus, with a Myrtle leaf.*

2. MEDEOLA (*Angustifolia*) foliis lanceolatis alternis, caule scandente. *Medeola with spear-shaped alternate leaves, and a climbing stalk; or narrow-leaved, climbing, African Asparagus.*

3. MEDEOLA (*Virginiana*) foliis verticillatis, ramis aculeatis. Lin. Sp. Pl. 339. *Medeola with leaves growing in whorls, and prickly branches.*

The first and second sorts grow naturally at the Cape of Good Hope. They have tuberose roots, composed of several dugs or oblong knobs, which unite together at the top, from which arise two or three stiff stalks, which rise four or five feet high, if they meet with any neighbouring support to which they can fasten, otherwise they fall to the ground: the first is garnished with oval spear-shaped leaves, ending in acute points, placed alternately, sitting close to the stalks, of a light green on their under side, and dark on their upper; the leaves of the second are much longer and narrower, in which their difference consists. The flowers come out from the side of the stalks, sometimes singly, at others there are two upon a slender short foot-stalk; they have six oblong equal petals, of a dull white colour; within these sit six stamina, which are as long as the petals, terminated by incumbent summits. In the center is situated a germen with three horns, sitting upon a short style, and crowned by three thick recurved stigmas; the germen afterward turns to a roundish berry with three cells, each containing one heart-shaped seed.

Both these sorts propagate freely by offsets from the roots, so that when they are once obtained there will be no necessity of sowing their seeds, which commonly lie a year in the ground: nor will the plants be strong enough to flower in less than two years more, whereas the offsets will flower the following spring. The time for transplanting and parting of the roots is in July, when their stalks are entirely decayed, for they begin to shoot toward the end of August, and keep growing all the winter. These roots should be planted in pots, and may remain in the open air till there is danger of frost, when they must be removed into shelter, for they are too tender to live through the winter in the open air; for if they are placed in a warm green-house they will thrive and flower very well, but they seldom produce fruit unless they have some heat in winter; therefore where that is desired, the plants should be placed in a stove kept to a moderate degree of warmth.

The third sort is a native of North America. It is by Dr. Linnæus joined to this genus, in which I have followed him; though, if I remember rightly, the characters do not exactly agree with the other, for the flower is either polypetalous, or is cut into many segments, and has but five stamina; being some years since I saw the flowers, I cannot be very certain if I am right. This hath a small scaly root, from which arises a single stalk about eight inches high, garnished with one whorl of leaves at a small distance from the ground, and at the top there are two leaves standing opposite; between these come out three slender foot-stalks, which turn downward, each sustaining one small, pale, herbaceous flower, with a purple pointal.

This plant is hardy enough to live in the open air, but does not propagate fast here, as it produces no seeds, so can only be increased by offsets.

MEDICA. Tourn. Inst. R. H. 410. tab. 231. Medick, or La Lucerne.

The CHARACTERS are,

The flower is of the butterfly kind; the standard is oval and entire; the two wings are oblong, oval, and fixed by an appendix to the keel; the keel is oblong, bifid, and reflexed toward the standard. It has ten stamina, nine of which are joined almost to their tops; the other is single, terminated by small summits. It hath an oblong compressed germen, which is recurved, sitting on a short style; this and the stamina are involved by the keel and standard. The germen afterward turns to a compressed moon-shaped pod, inclosing several kidney-shaped seeds.

The SPECIES are,

1. MEDICA (*Sativa*) caule erecto herbaceo, foliis ternatis, foliolis lanceolatis supernè serratis floribus spicatis alaribus. *Greater upright Medick with purplish flowers; commonly called La Lucerne, and by the French Burgundy Hay.*

2. MEDICA (*Falcata*) caule herbaceo diffuso, foliis ternatis, foliolis lineari lanceolatis, spicis brevioribus alaribus & terminalibus. *Wild Medick with Saffron-coloured flowers.*

3. MEDICA (*Radiata*) caule herbaceo procumbente, foliis ternatis, leguminibus ciliato-dentatis. *Medick with an herbaceous trailing stalk, trifoliate leaves, and pods which have hairy indentures.*

4. MEDICA (*Hispanica*) caule herbaceo procumbente, foliis pinnatis, leguminibus ciliato-dentatis. *Medick with a trailing herbaceous stalk, winged leaves, and pods having hairy indentures.*

5. MEDICA (*Italica*) caule herbaceo prostrato, foliis ternatis, foliolis cuneiformibus supernè serratis, leguminibus margine integerrimis. *Medick with a prostrate herbaceous stalk, trifoliate leaves, whose lobes are wedge-shaped, sawed at the top, and the borders of the pods entire.*

6. MEDICA (*Cretica*) caule herbaceo prostrato, pedunculis racemosis, leguminibus lunatis. *Medick with a prostrate herbaceous stalk, branching foot-stalks, and moon-shaped pods.*

7. MEDICA (*Arborea*) foliis ternatis, foliolis cuneiformibus, caule erecto arboreo. *Medick with trifoliate leaves, whose lobes are wedge-shaped, and an erect tree-like stalk; or Cytisus Virgilii.*

The first sort hath a perennial root, and an annual stalk which rises four feet high in good land, garnished with trifoliate leaves at each joint, whose lobes are spear-shaped, and sawed toward their top. The flowers grow in spikes, which are from two to near three inches in length, standing upon naked foot-stalks, rising from the wings of the stalk; they are of the Pea blossom or butterfly kind, of a fine purple colour; these are succeeded by compressed moon-shaped pods, which contain several kidney-shaped seeds.

There are the following varieties of this plant:

One with Violet-coloured flowers.

Another with yellow flowers.

A third with yellow and Violet flowers mixed.

And a fourth with variegated flowers; but the first is the best for use.

This plant is supposed to have been brought originally from Media, and from thence had its name Medica: it is by the Spaniards called Alfasa; by the French La Lucerne, and Grande Trefle; and by several botanic writers it is called Fœnum Burgundiacum, i. e. Burgundian Hay. But there is little room to doubt of this being the Medica of Virgil, Columella, Palladius, and other ancient writers of husbandry, who have not been wanting to extol the goodness of the fodder, and have given direction for the cultivation of it in those countries where they lived.

But notwithstanding it was so much commended by the ancients, and hath been cultivated to so good purpose by our neighbours in France and Switzerland for many years, it hath not as yet found so good reception in our country as could be wished, nor is it cultivated in any considerable quantity here; though it is evident, it will succeed as well in England as in either of the before-mentioned countries, being extremely hardy, and resisting the severest cold of our climate: as a proof of this, I must beg leave to mention, that the seeds which have happened to be scattered upon the ground in autumn have come up, and the plants have endured the cold of a severe winter, and made very strong plants.

About the year 1650 the seeds of this plant were brought over from France, and sown in England; but whether for want of skill in its culture, whereby it did not succeed, or that the people were so fond of going on in their old beaten road, as not to try the experiment, was the occasion of its being entirely neglected in England, I cannot say, but it is very certain that it was neglected many years, so as to be almost forgotten. However, I hope, before I quit this article, to give such directions for its culture, as will encourage the people of England to make farther trial of this valuable plant, which grows in the greatest heat, and also in very cold countries; with this difference only, that in very hot countries, such as the Spanish West-Indies, &c. where it is the chief fodder for their cattle at this time, they cut it every week, whereas in cold countries it is seldom cut oftener than three or four times a year. It is very certain that this plant will be of great service to the inhabitants of Barbadoes, Jamaica, and the other hot islands in the West-Indies, where one of the greatest things they want is fodder for their cattle; since by the account given of this plant by Pere Feuillé, it thrives exceedingly in the Spanish West-Indies, particularly about Lima, where they cut it every week, and bring it into the market to sell, and is there the only fodder cultivated.

The directions given by all those who have written of this plant, are very imperfect, and generally such as, if practised in this country, will be found entirely wrong; for most of them order the mixing of this seed with Oats or Barley, as is practised for Clover; but in this way it seldom comes up well, and if it does, the plants will draw up so weak by growing amongst the Corn, as not to be recovered under a whole year, if ever it can be brought to its usual strength again.

Others have directed it to be sown upon a low, rich, moist soil, which is found to be the worst next to a clay, of any for this plant; in both which the roots will rot in winter, and in a year or two the whole crop will be destroyed.

But

But the soil in which this plant is found to succeed best in this country is, a light, dry, loose, sandy, deep land, which should be well ploughed and dressed, and the roots of all noxious weeds, such as Couch Grass, &c. destroyed, otherwise they will overgrow the plants while young, and prevent their progress.

The best time to sow the seed is about the middle of April, when the weather is settled and fair; for if you sow it when the ground is very wet, or in a rainy season, the seeds will burst and come to nothing (as is often the case with several of the leguminous plants) therefore you should always observe to sow it in a dry season; and if there happens some rain in about a week or ten days after it is sown, the plants will soon appear above ground.

But the method I should direct for the sowing these seeds is as follows: After having well ploughed and harrowed the land very fine, you should make a drill quite across the ground, almost half an inch deep, into which the seeds should be scattered very thin, covering them a quarter of an inch thick, or somewhat more, with the earth; then proceed to make another drill about a foot and a half from the former, sowing the seeds therein in the same manner as before, and so proceed through the whole spot of ground, allowing the same distance between row and row, and scatter the seeds very thin in the drills. In this manner, an acre of land will require about six pounds of seeds; for when it is sown thicker, if the seed grows well, the plants will be so close as to spoil each other in a year or two, the heads of them growing to a considerable size, as will also the roots, provided they have room. I have measured the crown of one root, which was in my possession, eighteen inches diameter; from which I cut near four hundred shoots at one time, which is an extraordinary increase, and this upon a poor, dry, gravelly soil, which had not been dunged for many years, but the root was at least ten years old; so that if this crop be well cultivated, it will continue many years, and annually improve; for the roots generally run down very deep in the ground, provided the soil be dry; and although they should meet a hard gravel a foot below the surface, yet their roots would penetrate it, and make their way downward, as I have experienced, having taken up some of them which were above a yard in length, and had run above two feet into a rock of gravel, which was so hard as not to be loosened without mattocks and crows of iron, and that with much difficulty.

The reason for directing this seed to be sown in rows is, that the plants may have room to grow; and for the better stirring the ground between them, to destroy the weeds, and encourage the growth of the plants, which may be very easily effected with a Dutch hoe, just after cutting the crop each time, which will cause the plants to shoot again in a very little time, and be much stronger than in such places where the ground cannot be stirred; when the plants first come up, the ground between them should be hoed with a common hoe; and if in doing of this you cut up the plants where they are too thick, it will cause the remaining to be much stronger. This hoeing should be repeated two or three times while the plants are young, according as the weeds are produced, observing always to do it in dry weather, that the weeds may the better be destroyed; for if it be done in moist weather, they will root and grow again.

With this management the plants will grow to the height of two feet or more, by the beginning of August, when the flowers will begin to appear, at which time it should be cut, observing to do it in a dry season; if it is to be made into Hay, it must be often turned that it may soon dry, and be carried off the ground, for if it lie long upon the roots it will prevent their shooting again. After the crop is taken off, you should stir the ground between the rows with a hoe, to kill the weeds and loosen the surface, which will cause the plants to shoots again in a short time, so that by the middle or end of September there will be shoots four or five inches high, when you may turn in sheep upon it to feed it down, for it will not be fit to cut again the same season; nor should the shoots be suffered to remain upon the plants, which would decay when the frosty weather comes on, and fall upon the crown of the roots, and prevent their shooting early the succeeding spring. So that the best way is to feed it until November, when it will have done shooting for that season; but it should not be fed by large cattle the first year, because the roots being young, would be in danger of being destroyed, either by their trampling upon them, or their pulling them out of the ground; but sheep will be of service to the roots by dunging the ground, provided they do not eat it too close, so as to endanger the crown of the roots.

The beginning of February, the ground between the rows should be again stirred with the hoe to make it clean; but in doing of this you should be careful not to injure the crown of the roots, upon which the shoots will be coming out, for it is one of the earliest pabulum for cattle yet known. With this management, if the soil be warm, by the middle of March the shoots will be five or six inches high, when, if you are in want of fodder, you may feed it down till a week in April; after which it should be suffered to grow for a crop, which will be fit to cut the beginning of June, when you should observe to get it off the ground as soon as possible, and stir the ground again with a Dutch hoe, which will forward the plants shooting again; so that by the middle or latter end of July, there will be another crop fit to cut, which must be managed as before; after which it should be fed down again in autumn, and as the roots by this time will have taken deep hold of the ground, there will be little danger of hurting them, if you should turn in large cattle; but you must always observe not to suffer them to remain after the roots have done shooting, lest they should eat down the crown of the roots below the buds, which would considerably damage, if not destroy them.

In this manner you may continue constantly to have two crops to cut, and two feedings upon this plant; and in good seasons there may be three crops cut, and two feedings, which will be a great advantage to grasiers, especially as this plant will grow upon dry barren soils, where Grass will come to little; and this will afford a good feed in dry summers, when Grass is often burnt up; for as it is an early plant in the spring, so it will be of great service when fodder falls short at that season, for it will be fit to feed at least a month before Grass or Clover.

The best places to procure the feed from, are Switzerland, and the northern parts of France, which succeed better with us than that which comes from a more southern climate; but this feed may be saved in England in great plenty, if a small quantity of ground stocked with the plants should be left uncut till the seeds are ripe, when it must be cut, and laid to dry in an open barn where the air may freely pass through, but defended from the wet, for if it be exposed thereto it will shoot while it remains in the pod, whereby it will be spoiled. When it is quite dry it must be threshed out, and cleansed from the husk, and preserved in a dry place till the season for sowing it. The feed saved in England is much preferable to any brought from abroad, as I have several times experienced, the plants produced from it having been much stronger than those produced from France, Switzerland, and Turkey seeds, which were sown at the same time, and on the same soil and situation.

I am inclinable to think that the reaſon of this plant not ſucceeding, when it has been ſown in England, has either been occaſioned by the ſowing it with Corn, with which it will by no means thrive; or by ſowing it at a wrong ſeaſon, or in wet weather, whereby the ſeeds have rotted, and never come up, which hath diſcouraged their attempting it again; but however the ſucceſs has been, I dare aver, that if the method of ſowing and managing of this plant, which is here laid down, be duly followed, it will be found to thrive as well as any other ſort of fodder now cultivated in England, and will continue much longer; for if the ground be duly ſtirred after the cutting each crop, and the laſt crop fed as hath been directed, the plants will continue in vigour forty or fifty years or more without renewing, as there are many large fields now in France much older which are very vigorous. I have had ſome plants in my poſſeſſion more than forty years old in great perfection.

The Hay of this plant ſhould be kept in cloſe barns, it being too tender to be kept in ricks open to the air as other Hay; but it will remain good, if well dried before it be carried in, three years. The people abroad reckon an acre of this fodder ſufficient to keep three horſes all the year round.

And I have been aſſured by perſons of undoubted credit, who have cultivated this plant in England, that three acres of it have fed ten cart horſes from the end of April to the beginning of October, without any other food, though they have been conſtantly worked. Indeed the beſt uſe which can be made of this Graſs is, to cut it, and give it green to the cattle; where this hath been daily practiſed, I have obſerved that by the time the field has been cut over, that part which was the firſt cut, hath been ready to cut again, ſo that there has been a conſtant ſupply in the ſame field from the middle of April to the end of October; when the ſeaſon has continued long mild, and when the ſummers have proved ſhowery, I have known ſix crops cut in one ſeaſon, but in the drieſt ſeaſons there will be always three or four. When the plant begins to flower it ſhould then be cut, for if it ſtands longer the ſtalks will grow large, and the under leaves will decay, ſo that the cattle will not ſo greedily devour it. Where there is a quantity of this cultivated ſome of it ſhould be cut before the flowers appear, otherwiſe there will be too much to cut within a proper time.

When this is made into Hay, it will require a great deal of making; for as the ſtalks are very ſucculent it muſt be often turned, and expoſed a fortnight or more before it will be fit to houſe, ſo this requires a longer time to make than Saint Foin; therefore when it is cut, it ſhould be carried to make upon ſome Graſs ground, becauſe the earth in the intervals of the rows will waſh up, and mix with the Hay in every ſhower of rain, and by carrying it off as ſoon as it is cut, the plants will ſhoot up again ſoon; but it is not ſo profitable for Hay, as to cut green for all ſorts of cattle, eſpecially horſes, which are extremely fond of it, and to them it will anſwer the purpoſe of both Hay and Corn, and they may be worked at the ſame time juſt as much as when they are fed with Corn, or dry food. If milch cows are fed with this plant cut green all the ſummer, they will give a greater quantity of milk than with any other food, and the milk, as alſo the butter, will be better flavoured, as I have frequently experienced: therefore every dairy farm, where there is proper land, ſhould always have a good field ſown with Lucerne, which will be found very advantageous to the poſſeſſor: for I have known twelve cows kept from May-day till October, with leſs than two acres of this plant; whereby the Hay on the other land, has been wholly reſerved for winter.

The ſecond ſort grows naturally in the ſouth of France, in Spain and Italy, and has been ſuppoſed only a variety of the firſt; but I have frequently cultivated this by ſeeds, and have never obſerved it to alter. The ſtalks of this are ſmaller, do not ſtand erect, and never riſe ſo high; the leaves are not half ſo broad; the flowers are produced in ſhort roundiſh ſpikes, and are of a Saffron colour. It hath a perennial root which will continue many years, but is ſeldom cultivated any where.

The third ſort grows naturally in Italy; this is an annual plant, with ſeveral ſlender branching ſtalks a foot and a half long, which ſpread on the ground, garniſhed with trifoliate leaves, whoſe lobes are oval, ſpear-ſhaped, and entire. The flowers are produced ſingly upon ſlender foot-ſtalks, which proceed from the ſide of the branches; they are ſmall, of a yellow colour, and ſhaped like thoſe of the former ſort; theſe are ſucceeded by broad, flat, moon-ſhaped pods, whoſe borders are indented, and terminated by fine hairs; in each of theſe pods is lodged four or five kidney-ſhaped ſeeds.

The fourth ſort grows naturally in Spain; this is alſo an annual plant, whoſe ſtalks grow a foot and a half long, trailing on the ground, garniſhed with winged leaves, compoſed of two pair of ſmall lobes, terminated by one large, oval, ſpear-ſhaped one. The flowers ſtand upon long ſlender foot-ſtalks, each ſuſtaining four or five gold-coloured flowers at the top, which are ſucceeded by compreſſed moon-ſhaped pods, not half ſo large as thoſe of the third ſort, but have hairy indentures like thoſe.

The fifth ſort grows naturally on the borders of the ſea in ſeveral parts of Italy; this is alſo an annual plant, with proſtrate herbaceous ſtalks about a foot long, garniſhed with trifoliate leaves, whoſe lobes are wedge-ſhaped and ſawed toward the top. The flowers are produced upon ſlender foot-ſtalks, each ſuſtaining five or ſix pale yellow flowers, which are ſucceeded by ſmall, thick, moon-ſhaped pods, whoſe borders are entire, containing three or four ſmall kidney-ſhaped ſeeds in each.

The ſixth ſort grows naturally in the Archipelago; this is an annual plant; the ſtalks are ſlender, about a foot long, branching out into ſmaller, garniſhed with winged hoary leaves; thoſe on the lower part of the ſtalk are compoſed of two pair of lobes, terminated by an odd one, but thoſe on the upper part of the ſtalks are trifoliate. The flowers are produced at the end of the ſtalks; they are ſmall, yellow, and ſhaped like thoſe of the other ſorts, and are ſucceeded by compreſſed moon-ſhaped pods, which are acutely indented on their borders, and contain three or four kidney-ſhaped ſeeds.

Theſe annual ſorts are preſerved in the gardens of thoſe who are curious in botany; the ſeeds of theſe ſhould be ſown on an open bed of freſh ground, in the place where the plants are to remain, becauſe they do not bear tranſplanting well, unleſs when they are young. As the plants ſpread their branches on the ground, ſo they ſhould not be ſown nearer than two feet and a half aſunder; when the plants come up, they will require no other care but to keep them clean from weeds.

The ſeventh ſort grows naturally in the iſlands of the Archipelago, in Sicily, and the warmeſt parts of Italy. This riſes with a ſhrubby ſtalk to the height of ſix or ſeven feet, covered with a grayiſh bark, and divides into many branches, which, while young, are covered with a hoary down, garniſhed at each joint with trifoliate leaves; the lobes are ſmall, ſpear-ſhaped, and hoary on their under ſide; theſe remain all the year. The flowers are produced on foot-ſtalks, which ariſe from the ſide of the branches, which are of a bright yellow, each foot-ſtalk ſuſtaining four or five flowers, which are ſucceeded by compreſſed moon-ſhaped pods, each containing three or four kidney-ſhaped ſeeds.

It

It flowers great part of the year, especially when the winters are favourable; or when the plants are sheltered in winter, they are seldom destitute of flowers; but those in the open air begin to flower in April, and continue in succession till December. Those flowers which appear early in summer will have their seeds ripe in August, or the beginning of September, and the others will ripen in succession till the cold stops them.

This plant may be propagated by sowing the seeds, either upon a moderate hot-bed, or on a border of light earth, in the beginning of April; when the plants come up they should be carefully cleaned from weeds, but they should remain undisturbed, if sown in the common ground, till August following; but if on a hot-bed, they should be transplanted about Midsummer into pots, placing them in the shade until they have taken root; after which they may be removed into a situation where they may be screened from strong winds, in which they may abide till the latter end of October, when they must be put under a frame to shelter them from hard frosts, especially while they are young. In April following these plants may be shaken out of the pots, and placed in the full ground, where they are designed to remain, which should be in a light soil and a warm situation, in which they will endure the cold of our ordinary winters extremely well, and continue to produce flowers most part of the year, and retaining their leaves all the winter, renders them the more valuable.

Those also which were sown in an open border may be transplanted in August following, in the same manner; but in doing of this, you must be careful to take them up with a ball of earth to their roots, if possible, as also to water and shade them until they have taken root; after which they will require little more care than to keep them clear from weeds, to prune off the luxuriant branches to keep them within due compass; but you should never prune them early in the spring, nor late in autumn, for if frost should happen soon after they are pruned, it will hurt the tender branches, and many times the whole plant is lost thereby.

These plants have been supposed tender, so were housed every autumn; but I have had large plants of this kind, which have remained in the open air in warm situations many years, without any cover, and have been much stronger, and flowered much better than those which were housed; though, indeed, it will be proper to keep a plant or two in shelter, lest by a very severe winter (which sometimes happens in England) the plants abroad should be destroyed.

They may also be propagated by cuttings, which should be planted in April upon a bed of light earth, and watered and shaded until they have taken root, after which they may be exposed to the open air; but they should remain in the same bed till August following before they are transplanted, by which time they will have made strong roots, so should be removed to the places where they are to remain, observing (as was before directed) to water and shade them until they have taken root; after which you may train them up with strait stems, by fastening them to sticks, otherwise they are apt to grow crooked and irregular; and when you have got their stems to the height you design them, they may then be reduced to regular heads, and with pruning their irregular shoots every year, they may be kept in very good order.

This plant grows in great plenty in the kingdom of Naples, where the goats feed upon it, with whose milk the inhabitants make great quantities of cheese; it also grows in the islands of the Archipelago, where the Turks use the wood of these shrubs to make handles for their sabres, and the Caloyers of Patmos make their beds of this wood.

This is by many people supposed to be the Cytisus of Virgil, Columella, and the old writers on husbandry, which they mention as an extraordinary plant, and worthy of cultivation for fodder, from whence several persons have recommended it as worthy of our care in England. But however useful this plant may be in Crete, Sicily, Naples, or those warmer countries, yet I am persuaded it will never thrive in England, so as to be of any real advantage for that purpose; for in severe frost it is very subject to be destroyed, or at least so much damaged, as not to recover its former verdure before the middle or latter end of May, and the shoots which are produced, will not bear cutting above once in a summer, and then will not be of any considerable length, the stems growing very woody, which renders the cutting of it very troublesome; so that upon the whole, it can never answer the trouble and expence in cultivating it, nor is it worth the trial, since we have so many other plants preferable to it; though in hot, dry, rocky countries, where few other plants will thrive, this may be of great advantage, as it grows from the fissures of rocks, where there is not soil for cultivation, in such situations this plant will live many years, and thrive very well.

But however unfit this may be for such use in England, yet for the beauty of its hoary leaves, which will abide all the year, together with its long continuance in flower, it deserves a place in every good garden, where, being intermixed with shrubs of the same growth, it makes a very agreeable variety.

MEDICAGO Lin. Gen Plant 805. Snail Trefoil.

The CHARACTERS are,

The flower is of the butterfly kind, having an oval erect standard whose borders are reflexed. The wings are oblong, oval, and fixed to the keel by an appendix. The keel is oblong, bifid, obtuse, and reflexed. It hath ten stamina, nine of which are joined, and the other is single; and an oblong germen which sits upon a short style, and is involved with the stamina by the keel, crowned by a very small stigma; it afterward turns to a pod, twisted into the form of a snail, inclosing many kidney-shaped seeds.

The SPECIES are,

1. MEDICAGO (*Maritima*) pedunculis racemosis, leguminibus cochleatis, spinosis, caule procumbente tomentoso. Hort. Cliff. 378. *Medicago with branching foot-stalks, snail-shaped prickly pods, and a trailing woolly stalk; or Sea Medick.*

2. MEDICAGO (*Scutellata*) leguminibus cochleatis inermibus, stipulis dentatis, caule angulose diffuso, foliolis oblongo ovatis acute dentatis. *Snail Trefoil; commonly called Snails.*

3. MEDICAGO (*Tornata*) leguminibus tornatis inermibus, stipulis acutè dentatis, foliolis serratis. *Snail Trefoil with a smaller, turned, smooth fruit.*

4. MEDICAGO (*Intertexta*) leguminibus cochleatis spinosissimis aculeis utrinque tendentibus. *Snail Trefoil with a large fruit, whose spines point upward and downward; commonly called Hedgehog.*

5. MEDICAGO (*Laciniata*) leguminibus cochleatis spinosis, foliolis acutè dentatis tricuspidisque. *Snail-shaped Trefoil with a round prickly capsule, and elegantly cut leaves.*

There are many other species of this genus, which grow naturally in the warm parts of Europe, and are preserved in botanic gardens for the sake of variety; but as these are rarely cultivated in other gardens, so it would be beside my purpose to enumerate them here.

The first sort grows naturally on the borders of the Mediterranean sea; this is a perennial plant, with trailing woolly branches about a foot long, divided into small branches, garnished with trifoliate downy leaves at each joint. The flowers are produced from the side and at the ends

ends of the branches in ſmall cluſters; they are of a bright yellow colour, and are ſucceeded by ſmall, round, ſnail-ſhaped fruit, which are downy, armed with a few ſhort ſpines.

This plant is perennial, and may be propagated by ſeeds, which ſhould be ſown upon a warm border of dry ſoil in the ſpring, where the plants are deſigned to remain; when the plants come up, two or three of them may be tranſplanted into ſmall pots to be ſheltered in winter, becauſe in ſevere froſt, thoſe which are in the open air are ſometimes deſtroyed; though they will endure the cold of our ordinary winters, in a dry ſoil and a ſhady ſituation. Thoſe plants which are left remaining will require no other culture but to thin them where they are too cloſe, and keep them clean from weeds. It may alſo be propagated by cuttings, planted in June and July in a ſhady border, covering them cloſe with a glaſs to exclude the external air; theſe will take root in about ſix weeks time, and may then be either planted in a warm border or in pots, and treated in the ſame way as the ſeedling plants.

The ſecond ſort is an annual plant, which grows naturally in ſome of the warm parts of Europe; but in England it is cultivated in gardens for the oddneſs of its fruit, which is twiſted in the form of a ſnail, and as it ripens turns to a dark brown colour, ſo as to have the appearance of ſnails feeding on the plants at the firſt view. This hath trailing branches; the flowers are of a pale yellow, and come out from the ſide of the branches. Theſe appear in June and July, and the ſeeds ripen in autumn. It is propagated by ſeeds, which ſhould be ſown in the middle of April, where the plants are to remain; the plants ſhould be thinned where they are too cloſe, and kept clean from weeds, which is all the culture they require.

The third ſort is alſo an annual plant, which grows naturally in Spain. This hath trailing branches, and yellow flowers like the ſecond ſort, but the fruit is much longer and cloſer twiſted, ſo as to reſemble the figure of a veſſel called a pipe, being leſs at each end than in the middle. This is frequently kept in gardens for the ſake of variety, and may be propagated and treated in the ſame way as the ſecond ſort.

The fourth ſort is an annual plant, which was formerly more cultivated in the Engliſh gardens than at preſent. The ſtalks, leaves, and flowers are like thoſe of the two former ſorts, but the fruit is much larger, and cloſely armed with long ſpines like a hedgehog, from whence it had the title; theſe ſpines point every way, ſo that it is difficult to handle the fruit without ſmarting for it. It is propagated by ſeeds in the ſame way as the ſecond ſort, and the plants require the ſame treatment.

The fifth ſort grows naturally in Syria; it is an annual plant, with trailing ſtalks like the former; the lobes of the trifoliate leaves are wedge-ſhaped, ſharply indented on the edges, and at the top have three acute points. The flowers are of a pale yellow, and the fruit is ſnail-ſhaped but ſmall, and armed with many weak ſpines. It may be cultivated in the ſame way as the other.

MEDLAR. See Mespilus.

MELAMPYRUM Tourn. Inſt. R. H. 173. tab. 78. Cow Wheat.

The Characters are,

The flower is of the lip kind, having an oblong recurved tube compreſſed at the brim; the upper lip is formed like a helmet; the under lip is plain, erect, and cut into three ſegments. It hath four awl-ſhaped ſtamina which are curved under the upper lip, two of which are ſhorter than the other, terminated by oblong ſummits; in the center is ſituated an acute-pointed germen, with a ſingle ſtyle crowned by an obtuſe ſtigma. The empalement afterward turns to an oblong acute-pointed capſule with two cells, incloſing two pretty large oval ſeeds.

The Species are,

1. Melampyrum (*Pratenſe*) floribus ſecundis lateralibus conjugationibus remotis, corollis clauſis. Flor. Suec. 513. *Broad-leaved yellow Cow Wheat.*

2. Melampyrum (*Criſtatum*) ſpicis quadrangularibus compactis obtuſis imbricatis. Flor. Suec. 510. *Yellow narrow-leaved Cow Wheat.*

3. Melampyrum (*Arvenſe*) ſpicis conicis laxe bracteis dentato-ſetaceis. Prod. Leyd. 298. *Cow Wheat with purpliſh tops.*

4. Melampyrum (*Nemoroſum*) bracteis cordato-lanceolatis, ſummis coloratis ſterilibus, calycibus lanatis. Flor. Suec. 512. *Cow Wheat with blue tops.*

Theſe plants are ſeldom cultivated in gardens. The firſt ſort grows naturally in woods in many parts of England. The ſecond ſort grows plentifully in Bedfordſhire and Cambridgeſhire. The fourth ſort grows in the northern parts of Europe. The third ſort grows naturally in ſome of the ſandy lands in Norfolk, though not in great plenty; but in Weſt Friezland and Flanders it grows very plentifully among the Corn. Cluſius ſays, it ſpoils their bread, making it dark; and that thoſe who eat of it uſed to be troubled with heavineſs of the head, in the ſame manner as if they had eaten Darnel or Cockle: but Tabernæmontanus declares, He has often eaten it without any harm; and ſays, it makes a very pleaſant bread. It is a delicious food for cattle, particularly for fattening of oxen and cows, for which purpoſe it may be cultivated.

The ſeeds of theſe plants ſhould be ſown in the autumn ſoon after they are ripe, otherwiſe they ſeldom grow the firſt year; when the plants come up they muſt be weeded in the ſpring while young, and as ſoon as they begin to ſhew their flowers the cattle may feed upon it; but they ſhould be confined to a certain ſpace, and not permitted to run over the whole field to trample it down, which would deſtroy a great part of it.

The third and fourth ſorts make a pretty appearance, with their purple and blue tops, during the months of July and Auguſt. They are all of them annual plants.

MELASTOMA. Lin. Gen. Plant. 481. The American Gooſeberry-tree.

The Characters are,

The flower has five roundiſh petals, which are inſerted into the border of the empalement, and ten ſhort ſtamina terminated by long recurved ſummits. Under the flower is ſituated a roundiſh germen, which afterward turns to a berry with five cells, covered by the empalement, which crowns it, and contains many ſmall ſeeds.

The Species are,

1. Melastoma (*Plantaginifolia*) foliis denticulatis ovatis acutis. Lin. Sp. Plant. 389. *American Gooſeberry with a large Plantain leaf.*

2. Melastoma (*Acinodendron*) foliis denticulatis ſubtrinerviis ovatis acutis, nervis. Lin. Sp. Plant. 390. *Another Plantain-leaved Gooſeberry of America, bearing a few fruit, which are of a Violet colour.*

3. Melastoma (*Hirta*) foliis denticulatis quinquenervibus, caule hiſpido. Lin. Sp. Plant. 390. *Amerirican Gooſeberry with a hairy, narrow, Plantain leaf, and ſtinging ſtalks.*

4. Melastoma (*Holoſericea*) foliis integerrimis ovato-lanceolatis ſubtus ſericeis, nervis ante baſin coadunatis. Hort. Cliff. 162. *Branching Gooſeberry of the Brazils, with a Malabrathrum leaf.*

5. Melastoma (*Groſſularioides*) foliis lanceolatis utrinque glabris nervis tribus ante baſin coëuntibus. Hort. Cliff.

Cliff. 162. *American Gooſeberry without ſpines, oblong leaves like thoſe of the Malabathrum, and herbaceous flowers growing in long bunches.*

6. Melastoma (*Bicolor*) foliis lanceolatis, nervis tribus longitudinalibus, ſubtus glabris coloratis. Hort. Cliff. 162. *Melaſtoma with ſpear-ſhaped leaves, having three longitudinal veins, which are coloured on their under ſide.*

7. Melastoma (*Malabrica*) foliis lanceolatis-ovatis quinque nervibus ſcabris. Flor. Zeyl. 171. *Melaſtoma with ſpear-ſhaped oval leaves with five veins, which are rough.*

8. Melastoma (*Lævigata*) foliis oblongo-ovatis minutiſſimè dentatis, infernè ſericeis quinquenervibus, floribus racemoſis. *American Gooſeberry without thorns, growing to a large tree, with large ſcarlet leaves like thoſe of Malabathrum, and white flowers growing in long bunches.*

9. Melastoma (*Petiolata*) foliis denticulatis ovatis acuminatis, infernè nitidiſſimis, petiolis longiſſimis. *Melaſtoma with oval acute-pointed leaves, which are indented on their edges, very ſhining on their under ſide, and have very long footſtalks.*

10. Melastoma (*Umbellata*) foliis cordatis acuminatis integerrimis, infernè canis, floribus umbellatis. *American Gooſeberry with entire heart-ſhaped leaves; called Barbadoes Elder.*

11. Melastoma (*Racemoſa*) foliis oblongo-cordatis acuminatis, denticulato-ſerratis, floribus racemoſis ſparſis. *Melaſtoma with oblong, heart-ſhaped, acute-pointed leaves, having ſawed indentures, and flowers growing thinly in long bunches.*

12. Melastoma (*Verticillata*) foliis ovato-lanceolatis, quinquenervibus, ſubtus aureis, floribus verticillatis, caule tomentoſo. *Melaſtoma with oval ſpear-ſhaped leaves, having five veins, of a gold colour on their under ſide, and flowers growing in whorls, with a woolly ſtalk.*

13 Melastoma (*Acuta*) foliis lanceolatis acutis denticulatis infernè incanis trinervibus, floribus racemoſis. *Melaſtoma with acute ſpear-ſhaped leaves, which are indented on their edges, hoary on their under ſide, with three veins, and flowers growing in bunches.*

14. Melastoma (*Glabra*) foliis ovato-lanceolatis acuminatis integerrimis, utrinque glabris trinervibus, floribus racemoſis. *Melaſtoma with entire, oval, ſpear-ſhaped leaves, ending in acute points, with three veins, ſmooth on both ſides, and flowers growing in long bunches.*

15. Melastoma (*Quinquenerva*) foliis ovatis quinquenervibus ſcabris, floribus racemoſis alaribus. *Melaſtoma with oval rough leaves having five veins, and flowers growing in bunches from the ſides of the branches.*

16. Melastoma (*Octandria*) foliis lanceolatis trinervibus glabris, marginibus hiſpidis. *Melaſtoma with ſmooth ſpear-ſhaped leaves having three veins, and hairy prickles on the border.*

The firſt ſort riſes about four or five feet high, the ſtem and branches being covered with long ruſſet hairs; the leaves grow oppoſite: they are five inches long and two broad, having five ribs or veins from end to end, but the three inner join before they reach the baſe; the fruit is produced at the end of the ſhoots, which is a pulpy blue berry, as large as a Nutmeg.

The ſecond ſort grows to be a large tree, having many crooked branches, covered with a brown bark; the leaves are placed oppoſite on the branches; they are ſmooth, entire, five inches long, and two broad in the middle, with three deep veins running through them, and are ſharply indented on their edges, ending in acute points; the fruit grows in looſe ſpikes at the end of the branches, and are of a Violet colour.

The third ſort grows to the height of twenty feet, with a large trunk, covered with a ruſſet bark; the leaves are ſeven inches long, and three and a quarter broad, of a dark ruſſet colour on their upper ſide, but of a yellowiſh ruſſet on their under, ſoft to the touch; the ſtalks are covered with rough hairs, and the leaves are placed by pairs on the branches, which make a beautiful appearance when the trees are viewed at a diſtance.

The fourth ſort ſeldom grows more than eight or ten feet high; the leaves are about four inches long, having three longitudinal veins, which join before they reach the baſe; they are entire, of a ſattin colour on their under ſide, but of a light green on their upper, placed by pairs.

The fifth ſort ſeldom grows more than ſeven or eight feet high; the branches are covered with a ſmooth purple bark, garniſhed with ſpear-ſhaped leaves, five inches long and two broad in the middle; they are ſmooth on both ſides, their edges are entire, and they terminate in acute points. The flowers are produced in pretty long hanging bunches, of an herbaceous colour, with long ſtyles, which are ſtretched out a good length beyond the petals; the fruit is ſmall, and black when ripe.

The ſixth ſort riſes four or five feet high, dividing into many ſmooth ſlender branches, garniſhed with ſpear-ſhaped leaves three inches long, one and a quarter broad, of a lucid green on the upper ſide, but white on the under, having three longitudinal veins, which join before they reach the baſe, and are placed alternately on the branches. The flowers are produced in a looſe panicle at the end of the branches; they are ſmall, white, and have pretty long tubes, which are ſucceeded by ſmall purple fruit.

The ſeventh ſort riſes with an angular ſtalk ſix or ſeven feet high, ſending out branches oppoſite, garniſhed with ſpear-ſhaped, oval, rough leaves placed oppoſite; they are hairy, of a dark green on their upper ſide, but of a pale green on their under. The flowers are produced at the end of the branches, two or three ſtanding together, which are large, of a Roſe colour, inclining to purple, ſitting in large hairy empalements, and are ſucceeded by roundiſh purple fruit, crowned by the empalement, which are filled with a purple pulp ſurrounding the ſeeds.

The eighth ſort grows to the height of twenty feet, with a large ſtrait ſtem, covered with a gray bark; the branches are angular, garniſhed with oblong oval leaves near a foot in length, and ſix inches broad, of a dark green on their upper ſide, but ſilky on their under, with five ſtrong longitudinal veins, placed oppoſite. The flowers are produced in long looſe bunches at the end of the branches, which are white, and are ſucceeded by roundiſh purple fruit, filled with pulp, in which the ſeeds are lodged.

The ninth ſort riſes with a ſtrong erect ſtalk thirty feet high, covered with a gray bark, dividing at the top into ſeveral angular compreſſed branches, garniſhed with oval leaves indented on their edges; they are placed oppoſite on very long foot-ſtalks, of a lucid green on their upper ſide, but of a pale gold colour and ſattiny on their under, with five ſtrong longitudinal veins, and a great number of ſmaller tranſverſe ones. The fruit is produced in looſe panicles at the end of the branches, which are white, and are ſucceeded by purple fruit, about the ſame ſize as thoſe of the former.

The tenth ſort riſes with a ſhrubby ſtalk ten or twelve feet high, covered with a hairy bark, dividing into many branches toward the top, garniſhed with heart-ſhaped leaves, ending in acute points, which are five inches long, and three broad toward their baſe, entire on their borders, of a dark green on their upper ſide, but hoary on their under, with five longitudinal veins, and many ſmaller tranſverſe ones; theſe are placed oppoſite, and ſtand upon long hairy footſtalks. The flowers are produced at the end of the branches, in a ſort of umbel; they are of a pale Roſe colour, and are

are succeeded by small black fruit, a little larger than Elder berries.

The eleventh sort rises with a shrubby stalk eight or nine feet high, covered with a dark brown bark; the branches are straggling, which are garnished with oblong heart-shaped leaves, indented on their edges, ending in acute points; they are smooth on both sides, and of a light green colour. The flowers are produced in very loose bunches at the end of the branches, which are small, of an herbaceous colour, and are succeeded by small fruit, of a dark colour when ripe.

The twelfth sort rises with a shrubby stalk five or six feet high, dividing into small branches, which are covered with a hairy woolly bark, of a rusty iron colour, which are garnished with oval spear-shaped leaves, placed opposite, of a dark green on their upper side, of a rusty iron colour on their under, having five longitudinal veins. The flowers come out in whorls at the joints of the stalks; they are small, of a purplish colour, and are succeeded by small black fruit.

The thirteenth sort is a low shrub, seldom rising more than three feet high, dividing toward the bottom into slender branches, garnished with spear-shaped leaves, ending in acute points, of a dark green on their upper side, but of a hoary white on their under, having three longitudinal veins, and are placed opposite upon short foot-stalks. The flowers are produced in loose bunches at the end of the branches, which are white, and are succeeded by small purple fruit.

The fourteenth sort hath a shrubby stalk eight or nine feet high, divided into many smooth slender branches, garnished with oval spear-shaped leaves; they are entire on their edges, and smooth on both sides, standing opposite, and have three longitudinal veins. The flowers are produced in loose panicles at the end of the branches, which are succeeded by very small purple fruit.

The fifteenth sort rises with several shrubby stalks five or six feet high, dividing into crooked branches, garnished with oval rough leaves, having five longitudinal veins, of a dark green on their upper side, but pale on their under, indented on their edges. The flowers are produced in very loose bunches, which come out from the side of the stalks, of an herbaceous colour, and are succeeded by small purplish fruit, filled with very small seeds.

The sixteenth sort rises with a shrubby stalk seven or eight feet high, dividing into many smooth branches, garnished with smooth spear-shaped leaves, of a dark green colour, with three longitudinal veins; the edges of these leaves are closely set with bristly stinging hairs. The flowers are produced in loose bunches at the end of the branches, of a purplish colour, and are succeeded by very small black fruit.

All the sorts are natives of the warm parts of America, where there are many more species than are here enumerated. Most of these here mentioned, were found by the late Dr. Houstoun, growing naturally in Jamaica, from whence he sent many of their seeds to Europe, some of which succeeded; but most, if not all the plants which were raised from them, were lost in the severe winter in 1740, since which time they have not been recovered in Europe.

There is great beauty in the diversity of the leaves of these plants, many of them being very large, and most of them are of different colours on the two surfaces, their under side being either white, gold colour, or russet, and their upper of different shades of green, so that they make a fine appearance in the hot-house all the year; indeed, their flowers have no great beauty to recommend them, but yet for the singular beauty of their leaves these plants deserve a place in all curious collections, as much as most other sorts.

There are very few of these plants at present in any of the European gardens, which may have been occasioned by the difficulty of bringing over growing plants from the West-Indies; and the seeds being small when they are taken out of the pulp, soon become dry, so rarely succeed. The best way to obtain these plants is, to have the entire fruits put up in dry sand, as soon as they are ripe, and forwarded by the soonest conveyance to England; these should be immediately taken out when they arrive, and the seeds sown in pots of light earth, and plunged into a moderate hot-bed of tanners bark. When the plants come up, and are fit to remove, they must be each planted into a small pot, and plunged into the tan-bed; and afterward should be treated in the manner directed for the ANNONA, to which I shall desire the reader to turn, to avoid repetition.

MELIA. Lin. Gen. Plant. 473. The Bead-tree.

The CHARACTERS are,

The flower hath five spear-shaped petals which spread open, and a cylindrical nectarium of one leaf, indented at the brim in ten parts. It has ten small stamina inserted in the top of the nectarium, with a conical germen, which afterward turns to a soft globular fruit, including a roundish nut, having five rough furrows, and five cells, each containing one oblong seed.

The SPECIES are,

1. MELIA (*Azedarach*) foliis bipinnatis. Flor. Zeyl. 162. *The Bead-tree, or false Sycamore.*

2. MELIA (*Azediracbta*) foliis pinnatis. Hort. Cliff. 161. *Melia with winged leaves, or evergreen Bead-tree.*

The first sort grows naturally in Syria, from whence it was brought to Spain and Portugal, where it is now become as common almost, as if it were a native of those countries. This in warm countries grows to a large tree, spreading out many branches, garnished with winged leaves, composed of three smaller wings, whose lobes are notched and indented on their edges, of a deep green on their upper side, but pale on the under. The flowers come out from the side of the branches in long loose bunches, composed of five long, narrow, spear-shaped petals, of a whitish blue colour, which are succeeded by oblong fruit as large as a small Cherry, green at first, but when ripe changes to a pale yellow, inclosing a nut with five deep furrows, having four or five cells, in each of which is lodged one oblong seed. The pulp which surrounds the nut, is said to have a deadly quality if eaten, and if mixed with grease and given to dogs it will kill them. The nuts are bored through, and strung by the Roman catholicks to serve as beads.

There has been of late years some of these plants introduced to the islands in the West-Indies, where I am informed they continue flowering, and produce their fruit most part of the year. The fruit I have received from thence by the title of Indian Lilac, from which I have raised many of the plants, which flower much stronger than those raised from Portugal seeds.

This sort is propagated by seeds, which should be sown in pots, and plunged into a hot-bed of tanners bark, where (if the seeds are fresh) they will come up in about two months time: in June they should be gradually inured to bear the open air, and soon after placed abroad in a well sheltered situation, that they may be hardened before winter. In October you should remove the pots under a hot-bed frame, where they may enjoy free open air when the weather is mild, and be covered in hard frost.

In April following, you may shake out your plants from the seed-pots and divide them, planting each into a separate small pot, plunging them into a moderate hot-bed, which will greatly promote their rooting, and increase their growth; but you should not draw them too much, but give them a large share of air, when the weather is good; and the beginning

ginning of June, you ſhould remove them out into the open air as before, during the three or four winters, for while the plants are young, they require ſhelter from hard froſt; but when the plants are grown pretty large and woody, they may be planted in the open air. The beſt ſeaſon for this is in April, at which time you ſhould ſhake them out of the pots, being careful not to ſhake the earth from the roots, but only pare off with a knife the outſide of the ball of earth. They ſhould have a dry ſoil and a warm ſituation, otherwiſe they will be liable to miſcarry in ſevere froſty weather.

Some of theſe plants which were planted in an open expoſure, have endured the cold of our ordinary winters very well; but when a ſevere froſt happens, they are generally killed, or at leaſt their ſhoots are deſtroyed to the main ſtem; therefore it is much more ſecure to plant them againſt good aſpected walls, where they will live and produce their flowers annually, and in warm ſeaſons they may have fruit.

The ſecond ſort grows naturally in India, where it becomes a large tree; the ſtem is thick, the wood of a pale yellow, and the bark of a dark purple colour, and very bitter. The branches extend wide on every ſide, which are garniſhed with winged leaves, compoſed of five or ſix pair of oblong acute-pointed lobes, terminated by an odd one, and have a ſtrong diſagreeable odour. The flowers are produced in long branching panicles, which proceed from the ſide of the branches; they are ſmall, white, and cut into five acute ſegments; theſe are ſucceeded by oval fruit the ſize of ſmall Olives, which, when ripe, are yellow; the pulp which ſurrounds the nut is oily, acrid, and bitter; the nut is white, and ſhaped like that of the former. It grows in ſandy land, both in India and the iſland of Ceylon, where it is always green, and produces flowers and fruit twice a year.

This ſort is now very rare in England, and alſo in the Dutch gardens, where ſome years paſt it was more common; it is propagated by ſeeds in the ſame way as the other ſort, but being much tenderer, the plants ſhould be kept conſtantly in the tan-bed.

The firſt ſort is commonly called Zizyphus alba in Portugal and Spain, and in Italy Pſeudocyamorus. It was by moſt of the modern botaniſts titled Azederach, but Dr. Linnæus has altered it to this of Melia, which was by Theophraſtus applied to a ſpecies of Aſh.

MELIANTHUS. Tourn. Inſt. R. H. 430. tab. 245. Honey Flower.

The Characters are,

The flower hath four narrow ſpear-ſhaped petals, divided into two lips, connected on their ſides, and a nectarium of one leaf, ſituated in the lower ſegment of the empalement, and faſtened with it to the receptacle. It hath four erect ſtamina, the two under being ſomewhat ſhorter than the other. In the center is ſituated a four-cornered germen, which afterward becomes a quadrangular capſule with diſtended cells, divided by partitions in the center, each containing one almoſt globular ſeed.

The Species are,

1. Melianthus (*Major*) ſtipulis ſolitariis petiolo adnitis. Hort. Cliff. 492. *African Honey Flower, or greater Melianthus.*

2. Melianthus (*Minor*) ſtipulis geminis diſtinctis. Hort. Cliff. 492. *Smaller, ſtinking, African Honey Flower.*

Theſe plants grow naturally at the Cape of Good Hope, from whence the firſt ſort was brought to Holland in the year 1672; this hath a perennial root, which ſpreads much in light ground, from whence ariſe many hollow ſoft ſtalks ſix or ſeven feet high, garniſhed with large winged leaves, which embrace the ſtalks with their baſe, where they have a large ſingle ſtipulæ faſtened on the upper ſide of the foot-ſtalk, with two ears at the baſe, which alſo embrace the ſtalk. The leaves have four or five pair of very large lobes, terminated by an odd one, which are deeply jagged on their edges into acute ſegments; between the lobes runs a double leafy border, or wing, on the upper ſide of the midrib, ſo as to connect the baſe of the lobes together. The flowers are produced in pretty long ſpikes, which ariſe from the top of the ſtalks; they are of a Chocolate colour, formed like the lip flower, but have four narrow petals, in which it differs from them; theſe are ſucceeded by oblong four-cornered capſules, divided by a central partition into four cells, each containing one roundiſh ſeed.

This plant was formerly preſerved in green-houſes, and treated as a tender exotic, but with that management the plants were ſo much drawn in winter, as to prevent their flowering. But of late years they have been treated in a different way, moſt of them having been planted in the full ground in warm borders, where all thoſe ſtalks which are not killed by froſt ſeldom fail to flower the ſpring following; ſo that the ſureſt method to have them flower, is to cover them with mats or Reeds in froſty weather, to prevent their tops being killed by the cold; and if the plants grow in dry rubbiſh, they will not ſhoot ſo vigorous as in good ground, ſo will be leſs ſucculent, and therefore not ſo liable to ſuffer by froſt.

This plant is eaſily propagated by taking off ſuckers or ſide ſhoots in ſpring, which, if they have good roots, there will be little danger of their growing, ſo ſhould be planted where they are to remain, and will require no other care than is before-mentioned; they may alſo be propagated by cuttings during any of the ſummer months, which, if watered and ſhaded, will take root very well, and may afterwards be tranſplanted where they are deſigned to remain.

The ſecond ſort riſes with round, ſoft, ligneous ſtalks five or ſix feet high, which ſend out two or three branches from their ſide, garniſhed with winged leaves like thoſe of the former ſort, but not half ſo large, and have two diſtinct ſtipulæ adhering to their foot-ſtalks; they are of a deep green on their upper ſide, and whitiſh on their under. The flowers come out from the ſide of the ſtalks in looſe hanging bunches, each ſuſtaining ſix or eight flowers, which are ſhaped like thoſe of the firſt ſort, but ſmaller; the lower part of the petals are green, their upper part are of a Saffron colour, and on their outſide in the ſwelling part of the petals is a bluſh of fine red; theſe have two long and two ſhorter ſtamina, which are terminated by yellow ſummits. The flowers are ſucceeded by four-cornered ſeed-veſſels, which are ſhorter than thoſe of the firſt ſort, in which are lodged four oval ſeeds, in ſeparate apartments.

This ſort does not ſpread its roots ſo much as the firſt, ſo is not propagated with ſo great facility; but cuttings of this ſort planted upon an old hot-bed, whoſe heat is over, and covered cloſe with bell or hand-glaſſes, to exclude the air, will take root pretty freely; theſe may be planted in pots, and ſheltered in the winter under a common frame for a year or two, till they have obtained ſtrength; then they may be planted in a warm border, and treated in the ſame way as the former ſort, with which management I have ſeen them flower much better than any of thoſe which have been treated more tenderly, and theſe plants have perfected their ſeeds in good ſeaſons.

MELILOTUS. See Trigonella.

MELISSA. Tourn. Inſt. R. H. 193. tab. 91. Baum.

The Characters are,

The flower is of the lip kind, having a cylindrical tube; the chaps are gaping; the upper lip is erect, forked, and indented at the end. The under lip is trifid. It hath four awl-ſhaped ſtamina, two of which are as long as the petal, the other are but half ſo long. It hath a quadrifid germen, which afterward turns to four naked ſeeds, ſitting in the empalement.

The SPECIES are,

1. MELISSA (*Officinalis*) racemis axillaribus verticillatis, pedicellis simplicibus. Lin. Sp. Pl. 592. *Garden, or common Baum.*

2. MELISSA (*Romana*) floribus verticillatis sessilibus, foliis hirsutis. *Roman Baum, with soft hairy leaves and a strong smell.*

3. MELISSA (*Grandiflora*) pedunculis axillaribus dichotomis longitudine florum. Lin. Sp. Plant. 592. *Calamint with a large flower.*

4. MELISSA (*Calamintha*) pedunculis axillaribus dichotomis longitudine foliorum. Lin. Sp. Plant. 593. *Common officinal Calamint of the Germans.*

5. MELISSA (*Nepeta*) pedunculis axillaribus dichotomis folio longioribus, caule decumbente. Lin. Sp. Plant. 593. *Calamint with the scent of Pennyroyal.*

6. MELISSA (*Cretica*) racemis terminalibus, pedunculis solitariis breviffimis. Lin. Sp. Plant. 593. *Hoary Calamint with Basil leaves.*

7. MELISSA (*Majoranifolia*) foliis ovatis glabris, floribus verticillatis sessilibus, pedunculis solitariis breviffimis. *Roman Calamint with a Marjoram leaf, and the scent of Pennyroyal.*

8. MELISSA (*Fruticosa*) fruticosa, ramis attenuatis virgatis, foliis subtus tomentosis. Lin. Sp. Pl. 593. *Shrubby Spanish Calamint with a Marum leaf.*

The first sort grows naturally on the mountains near Geneva, and in some parts of Italy, but is cultivated here in gardens as a medicinal and culinary herb. It has a perennial root; the stalks are square, branching, and rises from two to three feet high, garnished with leaves set by pairs, indented about their edges, the lower ones standing upon pretty long foot-stalks. The flowers grow in loose bunches at the wings of the stalks in whorls, standing upon single foot-stalks; they are of the lip kind; the upper lip stands erect, and is forked, the under lip is divided into three parts; the middle one is roundish, and indented at the top. The flowers are white, and the whole plant has a pleasant scent, somewhat like Lemons.

This plant is easily propagated by parting of the root; the best time for this is in October, that the offsets may have time to get root before the frosts come on. They should be planted two feet asunder in beds of common garden earth, in which they will soon spread and meet together; the only culture required is to keep the plants clean from weeds, and cut off the decayed stalks in autumn, stirring the ground between them.

The second sort grows naturally about Rome, and in several parts of Italy; this is very like the former. The stalks are slender, the leaves are much shorter than those of the former sort, and the whole plant is hairy, and of a strong disagreeable odour. It is seldom preserved, except in botanic gardens, but may be cultivated in the same way as the former.

The third sort grows naturally in the mountains of Tuscany and Austria, but is preserved in many English gardens for the sake of variety. It hath a perennial root; the stalks rise about a foot high, garnished at each joint with two leaves standing opposite, sawed on their edges, of a lucid green on their upper side, and whitish on their under: from the wings of the stalks come out single foot-stalks, which divide into two smaller, each of these sustain two flowers upon short separate foot-stalks. The flowers are large, of a purple colour, and shaped like those of the other species. This may be propagated and treated in the same way as the first sort.

The fourth sort is the common Calamint of the shops, which grows naturally in many parts of England, so is seldom kept in gardens.

The fifth sort is found in greater plenty than the fourth in England. The stalks of this are longer, and bend towards the ground. The leaves are larger, and more indented on their edges, and have a very strong scent like Pennyroyal. The whorls of flowers are set closer together than those of the fourth, but in other respects they agree.

The sixth sort grows naturally in the south of France and in Italy; this is not of so long duration as the former sorts, seldom continuing more than two or three years. The stalks are slender, a little ligneous, garnished with small, roundish, hoary leaves, placed opposite. The flowers are produced in whorls toward the upper part of the stalks, which are terminated by a loose spike; they are small and white, shaped like those of the other species, and are succeeded by seeds which ripen in autumn, and if they are permitted to scatter, there will be a sufficient supply of young plants.

The seventh sort grows naturally in Italy; this is a biennial plant, whose stalks are two feet long, declining toward the ground, garnished with roundish leaves, about the size of Marjoram. The flowers come out in close whorls on the upper part of the stalk, each standing on a short separate foot-stalk; they are large, and of a bright purple colour; this is propagated by seeds, which should be sown soon after they are ripe, and then the plants will come up in the spring; but when the seeds are not sown till the spring, they seldom grow till the following year. The plants may also be propagated by cuttings, which, if planted in the summer, and shaded from the sun, will take root very freely. If these plants are planted on a warm border, they will live through the winter, but to preserve the species a plant or two should be kept in pots, and sheltered under a frame in winter.

The eighth sort grows naturally in Spain; this hath slender shrubby stalks about nine inches long, which put out small side branches opposite, garnished with small, hoary, oval-pointed leaves, placed by pairs; these have much the appearance of the Marum Syriacum. The flowers grow in whorled spikes at the end of the stalks; they are small and white, and the seeds ripen in autumn. The whole plant has a strong scent of Pennyroyal; this is of as short duration as the seventh sort, and may be propagated either by seeds or cuttings in the same way, and the plants require the same treatment.

MELISSA TURCICA. See DRACOCEPHALON.

MELO. Tourn. Inst. R. H. 104. tab. 32. Melon.

The CHARACTERS are,

It hath male and female flowers on the same plant. The male flowers have one petal, which is bell-shaped, fastened to the empalement, and cut into five segments at the brim. It hath three or four short stamina inserted in the empalement, joined together, two of which have bifid points. The female flowers have no stamina or summits, but have a large oval germen situated below the flower, which afterward turns to an oval fruit with several cells, filled with oval, acute-pointed, compressed seeds, inclosed in a soft pulp.

There is a great variety of this fruit cultivated in the different parts of the world, and in this country there are too many of them propagated which are of no value, especially by those who supply the markets, where their size is chiefly regarded; so that by endeavouring to augment their bulk, the fruit is rendered of no value: I shall therefore only mention a very few of the varieties, which are the most deserving of care, excluding the common Melons, as being unworthy the trouble and expence of cultivation.

The sort of Melon, which is in the greatest esteem among all the curious in every part of Europe, is the Cantaleupe; which is so called from a place about fourteen miles from Rome, where the Pope has a country-seat, in which place this fruit has been long cultivated; but it was brought thither from that part of Armenia which borders on Persia, where this fruit is in so great plenty, that a horse load is sold for a French crown. The flesh of this Melon, when in perfection, is delicious, and does not offend the most tender stomachs, so may be eaten with safety. The Dutch are so fond of this fruit, as to cultivate very few other sorts, and by way of pre-eminence, call it only by the appellation of Cantaleupe, and never join the title of Melon to it, which they apply indifferently to all the other sorts. The outer coat of this sort is very rough, full of knobs and protuberances like warts; it is of a middling size, rather round than long; the flesh is for the most part of an Orange colour, though there are some with a greenish flesh, but I have never met with any of that colour so good as those of the other.

The Romana is by some much esteemed, and when the fruit is well conditioned, the plants in perfect health, and the season dry, it is a good Melon, and may be brought forwarder in the season than the Cantaleupe, therefore those who are desirous of early Melons may cultivate this sort.

The Succado is also a good sort, and may be cultivated for early fruit, but these must give way to the Cantaleupe when that is in season.

The Zatte is also a very good Melon, but very small. The fruit of this is seldom larger than an Orange; it is a little flatted at the two ends, and the outer coat is warted like the Cantaleupe, but there is so little flesh in one of these fruit, that they are scarce worthy the trouble of propagating.

The small Portugal Melon, which is by some called the Dormer Melon, is a pretty good fruit, and the plants generally produce them in plenty; so by many people this is preferred to most other, especially those who love a plenty, and are not so nice in distinguishing the quality. This may also be cultivated for an early crop.

But the best Melon for this purpose is the Black Galloway, which was brought from Portugal by Lord Galloway many years since, but of late years is rarely to be met with in England, it having been degenerated by growing among other sorts. The fruit of this sort will ripen in a shorter time from its first setting, than any other which I have yet seen; and when suffered to ripen naturally, is not a bad fruit.

The few varieties here mentioned, are sufficient to satisfy the curious, who may be fond of variety, for there are scarce any other which deserve the trouble; and indeed, those who have a true taste for this fruit, seldom cultivate any but the Cantaleupe; but as I before observed, where this fruit is desired early in the season, the Cantaleupe is not so proper as some of the other; therefore a few plants of one of the other sorts should be raised earlier in the spring, but it should be in a different part of the garden from the Cantaleupe Melons; for when two sorts of Melons grow near, they cannot be preserved perfectly right, therefore the Dutch and German gardeners are very careful in this respect; and in order to keep the sort in perfection, do not plant any other sort of Melon, Cucumber, or Gourd, near these, lest by the impregnation of the farina of those other, these fruit should be rendered bad; and in this particular I am convinced, from long experience, they are right; and from the not observing this, many persons who are lovers of this fruit, have gradually diminished their goodness, without knowing the cause, and have imputed it to the long cultivating from the seeds saved in the same garden, believing it absolutely necessary to procure seeds from a distant place frequently to preserve them good; indeed, where a person can securely depend on the care and skill of those he procures the seeds from, it is a very good method to exchange seeds now and then; but there are so few who are exact in making choice of the fruits from which they save the seeds, or careful enough to do it themselves, but often depend on others to clean the seed, that I should advise every one to do it himself, which is the sure way to have it good; for I have frequently been deceived myself, by depending on the fidelity and skill of others; nor could I procure any of these seeds from Cantaleupe, which were good, until my much honoured friend the Chevalier Rathgeb sent me plentifully of it from thence; though I had often been supplied with seeds by persons who I thought could not be deceived in their choice, and who lived near the place of their growth.

Before I quit this head, I beg leave to caution all persons against depending upon seeds which are brought from abroad, either by those persons who import them for sale, or gentlemen who frequently bring or send over these seeds to their friends, for it seldom happens that any of these prove tolerable. I have been so often deceived by these myself, as to determine never to make trial of any of these seeds again, unless I receive them from a person who is skilful, and who eat of the fruit himself, of which he saved the seeds; for in Italy, Spain, Portugal, and many parts of France, the gardeners are very careless in the choice of all their seeds, but of the Melons they are remarkably so; and as for those which come from Constantinople, Aleppo, and other parts of Turkey, I have rarely seen one Melon produced from those seeds which was tolerable.

The seeds of Melons should not be sown until they are three years old, nor would I chuse to sow them when they are more than six or seven; for although they will grow at ten or twelve years old, yet the fruit which are produced from those old seeds, are seldom so thick fleshed as those which come from seeds which are fresher: it is the same of light seeds, which swim upon water when they are taken out of the pulp, for I have made some trials of these, and have had them grow at three years old; but not one of the Melons produced on these plants was near so deep fleshed, as those which grew upon plants raised from heavy seeds, taken out of the same fruit, though they grew in the same bed, and were cultivated exactly in the same manner; nor was their flesh so firm, but rather inclining to be meally; therefore I would not advise the sowing of these light seeds, nor those which are very old.

Having thus largely treated of the choice of the sorts, and of the seeds, I shall next proceed to the method of cultivating them, in order to obtain plenty of good fruit: the method which I am going to prescribe being very different from what has been constantly practised in England, will, I doubt not, be objected to by many; but it is what hath been practised in all the good gardens in Holland and Germany, where the Cantaleupe Melon is produced in great plenty and perfection; and from several years experience, I have found this to be the only method in which these Melons can be cultivated with success; and I am likewise convinced of its being the best way to obtain plenty of any other sort of Melon.

It is common to hear many persons valuing themselves upon having two or three early Melons, which, when brought to the table, are not better than a Pumpkin, and these are produced at a great expence with much trouble; and in order to have them ripe a little earlier than they would naturally come, if suffered to grow to their full size, the stem upon which the fruit grows is commonly twisted, to prevent the nourishment entering the fruit, whereby the growth is checked; then the fruit is closely covered with the mowings of Grass-plats, laid of a sufficient depth to cause a fermentation, by which the fruit becomes coloured; but where this unnatural method is practised, the fruit hath little flesh, and that has neither moisture, firmness, or flavour; so that after four months attendance, with a great expence of dung, &c. there may perhaps be three or four brace of Melons produced, which are fitter for the dung-hill than the table. Therefore my advice is, never to attempt to have these fruit ripe earlier than the middle of June, which is generally soon enough for this climate; and from that time to the end of September, they may be had in plenty, if they are skilfully managed; and when the autumn has continued favourable, I have had them very good in the middle of October.

But in order to continue this fruit so long, the seeds must be sown at two different seasons; or if at three, it will be still better: the first should be sown the second week in March, if the season proves forward; but if it is otherwise, it will be better to defer it till the middle of that month, for the future success greatly depends on the raising the plants in strength; which cannot be so well effected, if the weather should prove so bad after the plants are come up, as that a sufficient quantity of fresh air cannot be admitted to them, therefore it is not adviseable to be too early in sowing the seeds.

These seeds may be sown on the upper side of a Cucumber bed, where there are any; and if there are none, a proper quantity of new horse-dung must be provided, which must be thrown in a heap to ferment, and turned over, that it may acquire an equal heat, in the same manner as hath been directed for CUCUMBERS; and the plants must be raised and managed in the same manner as hath been directed for them, until they are planted where they are to remain for good; to which article the reader is desired to turn, to avoid repetition.

The second season for sowing of these seeds is the end of March; both these sowings must be understood to be planted under frames, for those which are designed for bell or hand-glasses, or to be covered with oil papers, should not be sown till ten or twelve days in April; for when these are sown earlier, if the plants are properly managed, they will extend their shoots to the sides of the glasses, before it will be safe to let them run out; for it often happens in this country, that we have sharp morning frosts in the middle of May; so that if the ends of these vines are then without the glasses, if they are not covered with mats to guard them against the frost, they will be in danger of suffering greatly therefrom; and, on the other hand, if the plants have spread so much as to fill the glasses, and are not permitted to run out, they will be in equal danger of suffering by their confinement from the heat of the sun in the day time; therefore it is, that I should advise the putting of the seed in rather a little later for the glasses, than those which are to be covered with oil papers. Nor will the times here mentioned be found too late, for I have put the seeds of Cantaleupe Melons into a hot-bed the third of May, which were not transplanted, but remained where they were sown, and covered with oiled paper; and from this bed I cut a large crop of good fruit, which ripened about the latter end of August, and continued till the end of October. This I only mention to shew what has and may be done, though it must not be always depended upon.

We next come to the making and preparing of the beds, or, as the gardeners term it, the ridges, into which the plants are to be put out to remain; these should always be placed in a warm situation, where they may be defended from all cold and strong winds, for the east and north winds are generally very troublesome in the spring of the year; so that if the place be exposed to these aspects, it will be difficult to admit a proper share of fresh air to the young plants; and if it is much exposed to the south-west winds, which often are very boisterous in summer and autumn, these will turn up and displace the vines, when they are exposed, whereby they will suffer greatly; therefore the best position for these beds is, where they are open to the south, or a little inclining to the east, and sheltered at a distance by trees from the other points: this place should be inclosed with a good Reed-fence, which is better for this purpose than any other inclosure, because the winds are deadened by the Reeds, and are not reverberated back again, as they are by walls, pales, or other close fences; but in making the inclosure, it should be extended to such distance every way from the beds, as not to obstruct the sun's rays during any part of the day: this should have a door wide enough to admit of wheelbarrows passing, to carry in dung, earth, &c. and should be kept locked, that no persons should be allowed to go in, but those who have business; for ignorant persons, having often curiosity to look into the beds, open the glasses, and let the cold air to the plants, and frequently leave the glasses in part open; or sometimes when they are raised by the gardener to admit the fresh air, the tilts are thrown down, so that the air is excluded; all which are very injurious to the young plants, as is also the handling of the fruit after it is set; therefore none should be admitted, but when the person who is entrusted with the care of them is there.

The next thing is the preparation of the earth for these plants, in which the Dutch and German gardeners are very exact: the mixture which they generally prepare is of the following sorts; of Hazel loam, one third part; of the scouring of ditches or ponds a third part, and of very rotten dung a third part; these are mixed up at least one, and often two years, before they make use of it, frequently turning it over, to incorporate their parts and sweeten it; but the compost in which I find these plants succeed best in England, is two thirds of fresh gentle loam, and one third of rotten neats dung; if these are mixed together one year before it is wanted, so as to have the benefit of a winter's frost and summer's heat, observing to turn it over often, and never suffer weeds to grow upon it, this will be found equal to any other compost whatever.

As these plants succeed best when they are planted young, so before the plants appear there should be a quantity of new dung thrown in a heap, proportionable to the number of lights intended, allowing about fifteen good wheelbarrows full to each light; this must be two or three times turned over to prepare it (as hath been directed for Cucumbers) and in a fortnight it will be fit for use, at which time the trench must be dug to receive the dung, where the bed is intended, which must be made rather wider than the frames, and in length proportional to the number of frames intended. As to the depth, that must be according as the soil is dry or wet; in a dry ground it should not be less than a foot, or a foot and a half deep; for the lower these beds are made, the better they will succeed, where there is no danger of their suffering by wet. In the well laying and mixing of the dung, the same care must be taken as hath been advised already for Cucumbers, which in every respect must be the same for these beds. When the bed is made, the frames should be placed over it to keep out wet; but there should be no earth laid upon it, till after it has been three or four days made, and is found of a proper temperature of heat; for many times these beds will heat so violently when they are first made, as to burn the earth if covered with it; and when this happens, it is much the best way to take off this earth again, for the plants will never thrive in it.

As soon as the bed is found to be of a proper warmth, the earth should be laid upon it, which at first need not be more than two inches thick, except in the middle of each light, where the plants are to be placed, where there must be raised a hill eighteen inches high or more, terminating in a flat cone; in two or three days after the earth is put on the bed,

bed, it will be of a proper temperature to receive the plants; then in the evening you may transplant the plants, but always do it when there is little wind stirring: in taking up of the plants, their roots should be carefully raised with a trowel, so as to preserve all their fibres; for if these are broken off, the plants do not soon recover this; or if they do, they are generally weaker, and never make so good vines as those which are more carefully removed; for these plants are more nice and tender in transplanting than those of Cucumbers, especially the Cantaleupe Melon, which, if it is not planted out, soon after the third (or what the gardeners call the rough) leaf is put out, they are long recovering their vigour; so that when it happens, that the beds cannot be ready for them in time, it will be a good method to plant each plant into a small pot while they are young, and these may be plunged into the hot-bed where they were raised, or in a Cucumber-bed where there is room, so that they may be brought forward; and when the bed is ready they may be turned out of the pots, with the whole ball of earth to their roots, whereby they will receive no check in removing: and this latter method is what I should prefer to any other for the Cantaleupe, because there should never be more than one plant left to grow in each light; therefore by this method there will be no necessity of planting more, as there will be no danger of their succeeding; whereas in the common way most people plant two or more plants in each light, for fear one should miscarry. When the plants are placed on the top of the hills, they should be gently watered, which should be repeated once or twice after till the plants have taken good root, after which they seldom require more; for when they receive too much wet, they often canker at the root, and when that happens they never produce good fruit. When the plants have established themselves well in the new beds, there should be a greater quantity of earth laid on the bed, beginning round the hills where the plants grow, that their roots may have room to strike out; and as the earth is put in from time to time, it must be trodden or pressed down as close as possible, and at last it should be raised at least a foot and a half thick upon the dung all over the bed, observing also to raise the frames, that the glasses may not be too near the plants, lest the sun should scorch them.

When the plants have gotten four leaves, their tops should be pinched off with the finger and thumb, but not bruised or cut with a knife, because in either of these cases the wound will not so soon heal over; this pinching is to cause the plants to put out lateral branches, for these are what will produce the fruit; therefore when there are two or more of these lateral shoots produced, they must also be afterward pinched to force out more, which must be practised often, that there may be a supply of what the gardeners call runners, to cover the bed. The management of these beds must be nearly the same as hath been directed for the Cucumbers, therefore I need not repeat it here; but shall only observe, that the Melons require a greater share of air than Cucumbers, and very little water; and when it is given to them, it should be at a distance from their stems.

If the plants have succeeded well, they will spread over the bed and reach to the frames in about six weeks, at which time the alleys between the beds should be dug out; or where there is but one bed, there should be a trench made on each side of about four feet wide, as low as the bottom of the bed, and hot dung wheeled in to raise a lining to the same height as the dung of the bed, which should be trodden down close, and afterward covered with the same earth as was laid upon the bed, to the thickness of a foot and a half or more, treading it down as close as possible; this will add to the width of the bed, so much as to make it in the whole twelve feet broad, which is absolutely necessary; for the roots of the plants will extend themselves quite through it, so that if the extreme fibres are exposed to the air, it is common to see the vines decay before the fruit is well grown; for where there is no addition made to the width of the bed, the roots will have reached the sides of the bed by the time that the fruit appears, and having no more room to extend themselves, their extremities are dried by the sun and air, which is soon discovered by the plants hanging their leaves in the heat of the day, which is soon after attended with a decay of many of those leaves which are near the stem, and the plants from that time will gradually languish, so that the fruit cannot be supplied with nourishment, but when ripe will be found to have little flesh, that meally and ill flavoured; whereas those plants which have sufficient breadth for their roots to run, and the earth laid of a proper depth and closely trod down, will remain in vigour until the frost destroys them; so that I have had a second crop of fruit on them, which have sometimes ripened well; but all the first were excellent, and of a larger size than these sorts usually grow: the leaves of these plants were very large, of a strong green, so that they were in the utmost vigour; whereas, in most places where the Cantaleupe Melons have been raised in England, the beds have been no wider than they were first made, and perhaps not more than three inches thickness of earth upon them, so that the plants have decayed many times before they have ripened a single fruit; from whence people have imagined, that this sort of Melon was too tender for this climate, when their ill success was entirely owing to their not understanding their culture.

There is also another advantage attending this method of widening the beds, as above directed, which is that of adding a fresh warmth to the beds, by the hot dung which is buried on each side, which will cause the dung in the bed to renew its heat; and as the plants will by this time shew their fruit, this additional heat will be of great service in setting of the fruit, especially if the season should prove cold, as it sometimes happens in this country till the end of May. When the beds are made up in the manner here directed, and the vines have extended so far as to fill the frames, and want more room, the frames should be raised up with bricks about three inches high, to admit the shoots of the vines to run out from under them; for if the plants are strong, they will extend six or seven feet each way from their stems, especially if they are raised from new seeds; for which reason I caution every one to allow them room, and to put but one plant in each light; for when the vines are crowded the fruit seldom will set well, but will drop off when they are as large as an egg; therefore the frames which are designed for Melons should not be made small, for the wider these are, the better will the plants thrive, and produce a greater plenty of fruit.

There is no part of gardening in which the practitioners of this art differ more, than in the pruning and managing of these plants; nor are there any rules laid down in the several books in which the culture of Melons have been treated of, by which any person can be instructed; for there is such inconsistency in all their directions, and what is worse, the greatest part of them are absurd, so that whoever follows them can never hope to succeed; therefore I shall, in as few words as possible, give such plain directions as I hope will be sufficient to instruct any person, who is the least conversant in these things.

I have before advised the pinching off the ends of the plants as soon as they have a joint, in order to get lateral shoots, which are by the gardeners called runners; and when their shoots or runners have two or three joints, to pinch off their tops alſo, to force out more runners, becauſe it is from theſe that the fruit is to be produced; but after a ſufficient number are put out, they ſhould not be ſtopped again, but wait for the appearance of the fruit, which will ſoon come out in plenty; at which time the vines ſhould be carefully looked over three times a week, to obſerve the fruit, and make choice of one upon each runner, which is ſituated neareſt the ſtem, having the largeſt foot-ſtalk, and that appears to be the ſtrongeſt fruit; then pinch off all the other fruit which may appear upon the ſame runner, alſo pinch off the end of the runner at the third joint above the fruit; this will ſtop the ſap and ſet the fruit. There is alſo another method practiſed by ſome gardeners to ſet this fruit, which is the taking off ſome of the male flowers, whoſe farina is juſt ripe and fit for the purpoſe, laying them over the female flowers, which are ſituated on the crown of the young fruit, and with their nails gently ſtrike the male flowers to ſhake the farina into the female flowers, whereby they are impregnated, and the fruit ſoon after will ſwell, and ſhew viſible ſigns of their being perfectly ſet; ſo that where the plants are under frames, and the wind excluded from them, which is neceſſary to convey the farina from the male to the female flowers, this practice may be very neceſſary; but when the fruit appears on the vines, the glaſſes ſhould be conſtantly taken off at all times when the weather is good, for where this is omitted the fruit ſeldom ſets in plenty. The taking off all the other fruit will prevent the nouriſhment being drawn away from the fruit intended to grow, which, if they are all left on, the plant could not ſupply them with ſufficient nouriſhment; ſo that when they come to be as large as the end of a man's thumb they all drop off, and ſcarce one of them ſets, which will be prevented by the method before directed: but there are ſome perſons who are ſo covetous of having a number of fruit, as not to ſuffer any to be taken off, whereby they generally fail in their expectation. My allowing but one fruit to be left upon each runner is, becauſe if half of theſe ſtand there will be full as many fruit as the plant can nouriſh; for if there are more than ſix or eight upon one plant, the fruit will be ſmall, and not ſo well nouriſhed; indeed I have ſometimes ſeen fifteen or twenty Melons upon one plant, but theſe have generally been of the ſmaller kinds, which do not require ſo much nouriſhment as the Cantaleupes, whoſe ſkins are of a thick ſubſtance; ſo that where a greater number are left of them than the plants can well ſupply, their fleſh will be remarkably thin.

As I before adviſed the ſtopping or pinching off the runners three joints above the fruit, ſo by this there will be freſh runners produced a little below the places where the others were pinched; therefore it is that I adviſe the careful looking over the vines ſo often, to ſtop theſe new runners ſoon after they come out, as alſo to pull off the young fruit which will appear; and this muſt be repeated as often as it is found neceſſary, which will be until thoſe intended to ſtand are grown ſo large as to draw all the nouriſhment which the plants can ſupply, for then the plants will begin to abate of their vigour. Theſe few directions, if properly made uſe of, is all the pruning which is neceſſary to be given them; but at the ſame time when this is practiſed, it may be neceſſary to give ſome water to the plants, but at a diſtance from their ſtems, which will be of ſervice to ſet the fruit and cauſe it to ſwell; but this muſt be done with great caution.

When the plants have extended themſelves from under the frames, if the weather ſhould alter to cold, it may be neceſſary to cover the extremities of the vines every night with mats; for if theſe are injured it will retard the growth of the fruit, and often prove very injurious to the plants: after this, what water is given to the plants ſhould be in the alleys between the beds; for as the roots of the vines will by this time have extended themſelves through into the alleys, ſo when the ground there is well moiſtened the plants will receive the benefit of it; and by this method the ſtems of the plants will be preſerved dry, whereby they will continue ſound; but theſe waterings ſhould not be repeated oftener than once a week in very dry warm weather, but be ſure to give as much air as poſſible to the plants when the ſeaſon is warm.

Having given full inſtructions for the management of thoſe Melons which are raiſed under frames, I ſhall next proceed to treat of thoſe which are raiſed under bell or hand-glaſſes. The plants for theſe muſt be raiſed in the ſame manner as hath been already directed, and about the latter end of April, if the ſeaſon proves forward, will be a good time to make the beds; therefore a ſufficient quantity of hot dung ſhould be provided, in proportion to the intended number of glaſſes, allowing eight or nine good wheelbarrows of dung to each glaſs. Where there is but one bed, which is propoſed to be extended in length, the trench ſhould be dug out five feet wide, and the length according to the number of glaſſes intended, which ſhould not be placed nearer than four feet to each other; for when the plants are too near each other, the vines will intermix, and fill the bed ſo cloſely, as to prevent the fruit from ſetting: in digging of the trench, it ſhould be ſo ſituated as to allow for the widening of the bed three or four feet on each ſide; the depth muſt be according as the ſoil is dry or wet; but, as was before obſerved, if the ſoil is ſo dry, as that there is no danger of the beds being hurt by the wet, the lower they are made in the ground the better: in the making of the beds the ſame regard muſt be had to the well mixing and laying of the dung, as was before directed; and after the dung is laid, there ſhould be a hill of earth raiſed, where each plant is to ſtand one foot and a half high; the other part of the bed need not as yet be covered more than four inches thick, which will be ſufficient to keep the warmth of the dung from evaporating; then the glaſſes ſhould be placed over the hills, and ſet down cloſe, in order to warm the earth of the hills to receive the plants: and if the beds work kindly, they will be in a proper temperature to receive the plants, in two or three days after making; then the plants ſhould be removed in the ſame manner as was before directed, and if they are in pots, ſo that there will be no danger of their growing, there ſhould but one plant be put under each glaſs; if they are not in pots, there ſhould be two, one of which may be afterward taken away, if they both grow. Theſe plants muſt be watered at firſt planting, to ſettle the earth to their roots, and ſhaded every day until they have taken new root; and if the nights prove cold, it will be proper to cover the glaſſes with mats, to preſerve the warmth of the bed.

Where there are ſeveral of the beds intended, they ſhould be placed at ten feet diſtance from each other, that there may be a proper ſpace left between them, to be afterward filled up, for the roots of the vines to have room for extending themſelves, for the reaſons before given.

When the plants have taken good root in the beds, their tops muſt be pinched off; and their pruning, &c. muſt, from time to time, be the ſame as for thoſe under the frames: in the day time, when the weather is warm, the glaſſes ſhould be raiſed on the oppoſite ſide to the wind, to admit freſh air to the plants; for where this is not obſerved, they will draw up weak and ſickly; therefore all poſſible care ſhould be taken to prevent this, for if the runners have not proper ſtrength, they can never ſupply the fruit with nouriſhment.

When

When the plants are grown ſo long as to reach the ſides of the glaſſes, if the weather proves favourable, the glaſſes muſt be ſet up on three bricks, ſo as to raiſe them about two or three inches from the ſurface of the beds, to give room for the vines to run out from under them; but when this is done the beds ſhould be covered all over with earth to the depth of one foot and a half, and trod down as cloſe as poſſible; and if the nights ſhould prove cold there ſhould be a covering of mats put over the beds, to prevent the cold from injuring the tender ſhoots of the vines; but as the vines of the Cantaleupe Melons are impatient of wet, it will be neceſſary to arch the beds over with hoops, to ſupport the mats, that they may be ready for covering at all times when they require it; which is the only ſure method to have thoſe Melons ſucceed in England, where the weather is ſo very uncertain and variable; for I have had ſome beds of theſe Melons in as fine order under theſe glaſſes as could be deſired, which were totally deſtroyed by one day's heavy rain in June.

After the thickneſs of earth is laid upon the beds, if the weather ſhould prove cold, it will be adviſeable to dig trenches on each ſide of the beds, into which you ſhould lay a ſufficient quantity of hot dung, to make it of the ſame thickneſs with the bed, after the manner before directed for the frames; or if there is a ſufficient quantity of hot dung ready, the whole ſpace between the beds may be dug out, and filled up with the dung, laying thereon the earth a foot and a half deep, treading it down cloſe; this new dung will add a freſh warmth to the beds, and cauſe the plants to ſhew fruit ſoon after.

The watering of theſe plants muſt be done with great caution, and not given to their ſtems; the pinching of the runners muſt alſo be duly attended to, as alſo the pulling off all ſuperfluous fruit, to encourage thoſe which are deſigned to remain; and in ſhort, every thing before directed for thoſe under frames muſt likewiſe be obſerved for theſe; and the further care is, to cover them in all hard rains and cold nights with mats; which, if performed with care, there will be little danger of their miſcarrying, and theſe vines will remain vigorous until the cold in autumn deſtroys them.

There have been many perſons, who of late years have raiſed their Melons under oiled paper, and in many places they have ſucceeded well; but where this is practiſed there muſt be great care taken not to keep theſe coverings too cloſe over them, for where that is done the vines will draw very weak, and rarely ſet their fruit in any plenty; therefore where theſe coverings are propoſed to be uſed, I ſhould adviſe the bringing up of the plants under hand or bell-glaſſes, in the manner before directed, until they are grown far enough to be let out from under the glaſſes; and then, inſtead of the covering with mats, to put over the oiled paper; and if this covering is prudently managed, it will be the beſt that can be uſed. The beſt ſort of paper for this purpoſe is that which is ſtrong, and not of too dark colour, called Dutch wrapper; it ſhould be done over with Linſeed oil, which will dry ſoon. There ſhould be a proportionable number of the ſheets of this paper paſted together, as will ſpread to the dimenſions of the frame to which it is to be faſtened; and if this is fixed to the frame before the oil is rubbed over it, ſo much the better; but this ſhould be done ſo long before they are uſed, as that the oil may be thoroughly dry, and the ſtench gone off, otherwiſe it will hurt the plants.

There are ſome perſons who make theſe frames of broad hoops, in imitation of the covers of waggons; but as theſe are cumberſome to move, and there are no conveniencies for admitting air to the plants, but by raiſing the whole frame on one ſide, ſo I prefer thoſe made of pantile laths, framed like the ridge of a houſe; each ſlope having hinges, whereby any of the pannels may be raiſed at pleaſure, to admit the air to the plants: but as deſcriptions of theſe things are not well comprehended by perſons not ſo converſant with them, I ſhall exhibit a figure of one of theſe frames, to be added to the article of STOVES.

There are ſome few perſons of quality in England who have made the ſame conveniencies for their hot-beds, to raiſe the Cantaleupe Melons, as is in general practiſed by the Dutch; which is firſt to dig a trench in the ground ten feet wide, and in length proportional to the number of lights which are deſigned to cover the beds: this pit is boarded up with old ſhip planks, ſo as to make the depth full three feet and a half, or more, which pit they fill with tanners bark or dung mixed well together: then they have frames made ſix feet wide in the clear, which are placed in the middle of the hot-bed, whereby there will be two feet left on each ſide the frame within the planking, to allow room for the roots of the Melon plants to extend each way. Upon this hot-bed made with tanners bark and dung, they lay a foot and a half of good loamy earth for the plants to root into. By this method the dung, tan, and earth, is prevented from falling into the walks on the ſides of the beds, ſo the walks will be clean for to paſs between them, which is an advantage. But as theſe ſhip planks will not laſt long ſound, ſo I have made ſeveral of theſe trenches, the ſides of which are bricked up, and the top of the wall covered with an old Oak curb, to prevent the bricks from being diſplaced. Theſe pits I find anſwer the purpoſe better than the other, and when their duration is conſidered, will be found as cheap as the other.

The further management of the Melons, after their fruit is ſet, is to keep pulling off all the ſuperfluous fruit, and to pinch off all weak runners, which may draw away part of the nouriſhment from the fruit; as alſo to turn the fruit gently twice a week, that each ſide may have equal benefit of the ſun and air; for when they are ſuffered to lie with the ſame ſide conſtantly to the ground, that ſide will become of a pale or whitiſh colour, as if it were blanched, for want of the advantages of the ſun and air. The plants will require a little water in very dry weather, but this ſhould be given them in the alleys, at a diſtance from the ſtems of the plants, and not oftener than once in a week or ten days, at which times the ground ſhould be well ſoaked in the alleys. This will encourage the growth of fruit, and cauſe the fleſh to be thick; but the great caution which is neceſſary to be obſerved, is not to over-water the plants, which is certain injury to them; alſo be ſure to give as much free air as poſſible, at all times, when the weather will permit, for this is abſolutely neceſſary to render the fruit good.

When the fruit is fully grown, they muſt be duly watched to cut them at a proper time; for if they are left a few hours too long upon the vines, they will loſe much of their delicacy, therefore they ſhould be looked over at leaſt twice every day; and if thoſe fruit which are intended for the table, are cut early in the morning before the ſun has warmed them, they will be much better flavoured; but if any ſhould require to be cut afterward, they ſhould be put into cold ſpring water or ice, to cool them, before they are brought to the table; and thoſe cut in the morning ſhould be kept in the cooleſt place till they are ſerved up to table. The ſign of this fruit's maturity is, that of its beginning to crack near the foot-ſtalk, and its beginning to ſmell, which never fail; for as theſe Cantaleupe Melons ſeldom change their colour until they are too ripe, that ſhould never be waited for.

The directions here given for the management of the Cantaleupe Melons, will be found equally good for all the

other

other ſorts, as I have fully experienced; for in the common method of managing them, where the earth is laid but three or four inches thick, the plants are very apt to decay before the fruit is ripe; for their roots ſoon reach to the dung, and are extended to the ſides of the bed, where their tender fibres are expoſed to the air and ſun, which cauſes the leaves of the plants to hang down in the heat of the day, ſo it is neceſſary to ſhade the plants with mats, to prevent their ſudden death; for this drooping of the leaves occaſions the watering of the plants often to keep them alive, which is alſo prejudicial to their roots; whereas, when the beds are made of a proper width, and earthed of a ſufficient thickneſs, the plants will bear the ſtrongeſt heat of the ſun in this climate, without ſhewing the leaſt want of moiſture, or their leaves drooping, and they will continue in health till the autumn cold deſtroys them.

In ſaving of the ſeeds I need not repeat here, that only ſuch ſhould be regarded which are taken from the firmeſt fruit, and thoſe which have the higheſt flavour; and if theſe are taken out with the pulp entire, without diſplacing the ſeeds, and ſuffered to remain in the pulp two or three days before it is waſhed out, the better; and then to preſerve only the heavy ſeeds, which ſink in the winter.

MELOCACTUS. } See CACTUS.
MELOCARDUUS. }

MELOCHIA. See CORCHORUS.

MELON See MELO.

MELONGENA. Tourn. Inſt. R. H. 151. tab. 65. Mad Apple, by ſome called Egg Plant.

The CHARACTERS are,

The flower has but one petal, which is cut into five parts, and reflexed. It hath five awl-ſhaped ſtamina. In the center is ſituated an oblong germen, which afterward becomes an oval or oblong fruit with one cell, with a fleſhy pulp, filled with compreſſed roundiſh ſeeds.

The SPECIES are,

1. MELONGENA (*Ovata*) caule inermi herbaceo, foliis oblongo-ovatis tomentoſis integris, fructu ovato. *Mad Apple with an oblong Violet-coloured fruit.*

2. MELONGENA (*Tereti*) caule inermi herbaceo, foliis oblongo-ovatis tomentoſis, fructu tereti *Mad Apple with a taper Violet-coloured fruit.*

3. MELONGENA (*Incurva*) caule inermi herbaceo, foliis oblongis ſinuatis tomentoſis, fructu incurvo. *Mad Apple with an incurved fruit.*

4. MELONGENA (*Spinoſa*) ſpinoſa, foliis ſinuato-laciniatis, fructu tereti, caule herbaceo. *Apple-bearing Nightſhade with prickly leaves and fruit.*

The firſt ſort grows naturally in Aſia, Africa, and America, where the fruit is commonly eaten by the inhabitants; it is cultivated in the gardens in Spain as an eſculent fruit, by the title of Barenkeena; the Turks, who alſo eat the fruit, call it Badinjan, the Italians Melanzana, and the inhabitants of the Britiſh iſlands in America, Brown John, or Brown Jolly. It is an annual plant, with an herbaceous ſtalk, which becomes ligneous, and riſes from two to three feet high, ſending out many ſide branches, garniſhed with oblong, oval, woolly leaves, whoſe borders are very ſlightly ſinuated. The flowers come out ſingly from the ſide of the branches, which have a thick fleſhy empalement, deeply cut into five acute ſegments, armed with ſtrong prickles on the outſide. The flowers have one petal, which is cut at the brim into five ſegments, which expand in form of a ſtar, but are a little reflexed; they are blue, and the ſummits which are connected together in the boſom of the flower are yellow. The flowers are ſucceeded by oval fleſhy fruit, about the ſize and ſhape of a ſwan's egg, of a dark purple on one ſide and white on the other.

There are the following varieties of this ſpecies; one with white fruit, called by ſome the Egg Plant; one with yellow fruit, and another with pale red fruit; all theſe varieties are generally conſtant, the ſeeds producing the ſame fruit as thoſe from which they were taken; but as they only differ in colour, ſo I chuſe not to enumerate them as diſtinct ſpecies.

The ſecond ſort differs from the firſt in the ſhape of the fruit, which is commonly eight or nine inches long, taper and ſtrait, in other reſpects they are the ſame; but as this never varies when propagated in gardens, ſo there can be no doubt of its being a diſtinct ſpecies. There are two varieties of this ſort, one with a purpliſh fruit, and the other white, but the latter is the moſt common in England.

The third ſort differs from the two former in the ſhape of the leaves, which are deeply ſinuated on their borders. The fruit is oblong and incurved, of a yellowiſh colour, and larger at the end than in other part.

The fourth ſort grows naturally in India. This differs greatly from either of the former. The ſtalks and leaves are armed with very ſtrong thorns; the leaves are larger, and deeply jagged on their ſides. The flowers are larger, and of a deeper blue colour. The fruit is long, taper, and white.

Theſe fruit are eaten by moſt of the inhabitants of the warm parts of the globe, and are eſteemed a delicacy, but are ſuppoſed to have a property of provoking luſt.

They are propagated by ſeeds, which ſhould be ſown upon a moderate hot-bed early in March; when the plants come up, they muſt be tranſplanted into another hot-bed about three or four inches aſunder, obſerving to water and ſhade them until they have taken root; after which they muſt have a great ſhare of air when the weather is warm, otherwiſe they will draw up very weak. They muſt alſo be frequently watered, without which they will make but very indifferent progreſs. When the plants are grown ſo ſtrong as to fill the frame (which will be by the middle or end of May) you muſt tranſplant them out into a rich ſpot of ground at two feet diſtance, or into the borders of the pleaſure-garden at the ſame diſtance from other plants, obſerving to preſerve as much earth to the roots as poſſible when you take them up, otherwiſe they are ſubject to miſcarry. They muſt have water plentifully until they have taken root, after which they will require but very little care, more than to keep them clear from weeds, and in very dry weather to give them ſome water.

Theſe plants are only preſerved as curioſities in the Engliſh gardens, the fruit being ſeldom eaten in this country, except by ſome Italians and Spaniards, who have been accuſtomed to eat them in their own countries.

MELOPEPO. See CUCURBITA.

MELOTHRIA. Lin. Gen. Plant. 48.

The CHARACTERS are,

The flower is of one leaf, and wheel-ſhaped. In the center of the flower is ſituated the germen, attended by three conical ſtamina, inſerted in the tube of the flower. The germen afterward becomes an oval ſmall berry, having three diviſions, in which are lodged ſmall flat ſeeds.

We have but one SPECIES of this plant, viz.

MELOTHRIA (*Pendula.*) Lin. Hort. Cliff. 490. *Smalleſt Cucumber with a ſmooth, black, oval fruit.*

This plant grows wild in the woods in Carolina, Virginia, and alſo in many of the iſlands in America; the vine ſpreads upon the ground, having angular leaves, reſembling thoſe of the Melon, but much ſmaller. Theſe vines ſtrike out roots at every joint, which faſten themſelves into the ground, by which means their ſtalks extend to a great diſtance each way. The flowers are very ſmall, in ſhape like thoſe of the Melon, of a pale ſulphur colour

The

The fruit in the West-Indies grows to the size of a Pea, of an oval figure, and changes black when ripe; these are by the inhabitants sometimes pickled when they are green.

In England the fruit are much smaller, and are so hidden by the leaves, as to render it difficult to find them The plants will not grow in the open air here, but must be sown upon a hot bed, and if they are permitted, will soon spread over the surface, when the fruit is ripe; if they scatter their seeds, the plants will come up where the earth happens to be used on a hot bed again. This is preserved in some gardens for the sake of variety, but is of no use.

MENISPERMUM. Tourn. Act. R. Par. 1705. Moonseed.

The CHARACTERS are;

The flower has six oblong, oval, concave petals, and six stamina which are shorter than the petals, with three almost oval germen on the top of the styles, crowned by obtuse indented stigmas. The germen afterward turns to three oval berries with one cell, each inclosing one moon-shaped compressed seed.

The SPECIES are;

1. MENISPERMUM (*Canadense*) foliis peltatis subrotundis angulatis. Hort. Cliff. 140. *Climbing Moonseed of Canada, with a navel-shaped leaf.*

2. MENISPERMUM (*Virginicum*) foliis cordatis peltatis lobatis. Flor. Virg. 40. *Moonseed with an Ivy leaf.*

3. MENISPERMUM (*Carolinianum*) foliis cordatis subtus villosis Lin. Sp. Pl. 340. *Moonseed with heart-shaped leaves, which are hairy on their under side.*

The first sort grows naturally in the woods of Canada, Virginia, and most parts of North America. It hath a thick ligneous root, from which are sent out many climbing stalks, which become ligneous, and rise to the height of twelve or fourteen feet, twisting themselves about the neighbouring plants for support, and are garnished with large, smooth, roundish leaves, whose foot-stalks are placed almost in the middle of them; on the upper side there is a hollow on that part of the leaf, resembling a navel. The flowers come out in loose bunches from the side of the stalks; they are of an herbaceous colour, small, and composed of six oblong oval petals, and six very-short stamina, terminated by single summits; the three germen situated in the center turn to so many channelled berries, each containing one compressed seed.

This sort may be easily propagated by laying down of the branches, which, if performed in autumn, they will have made good roots by that time twelvemonth, when they may be separated from the old plant, and transplanted where they are designed to remain; these plants require support, for their branches are slender and weak. In the country where it grows naturally, they climb up the trees to a considerable height; so that if these are planted near trees in wilderness quarters, where their stalks may have support, they will thrive better than in an open situation.

The second sort differs from the first in the shape of its leaves, which are angular; their foot-stalks join to the base of the leaves, so have no umbilical mark on their surface. The stalks of this become ligneous, and rise as high as those of the first sort; the flowers and leaves do not differ from them. It is propagated after the same manner.

The third sort grows naturally in Carolina. This has by some been supposed the same with the second sort, but it differs from that in its roots not becoming woody as those do. The stalks are also herbaceous; the leaves are entire and hairy, not more than half so large as those of the second, nor is the plant so hardy. This sort does not produce flowers in England, unless the season proves very warm.

This may be propagated by parting of the roots; the best time for doing this is in the spring, a little before the plants begin to shoot; these should be planted in a warm situation and a light soil, for in strong land, where the wet is detained in winter, the roots are apt to rot; therefore if they are planted close to a wall exposed to the south or west, their stalks may be fastened against the wall, to prevent their trailing upon the ground, and in this situation the plants will frequently flower, and by a little shelter in severe frost, their stalks may be preserved from injury.

MENTHA. Tourn. Inst. R. H. 188. tab. 89. Mint.

The CHARACTERS are,

It hath a lip flower of one petal. The mouth is cut into four almost equal segments, the upper being a little larger and indented. It hath four erect stamina, the two nearest being longest. In the bottom of the tube is situated a four-pointed germen, which afterward turns to four naked seeds sitting in the empalement.

The SPECIES are;

1. MENTHA (*Viridis*) floribus spicatis, foliis oblongis serratis. Hort. Upsal. 168. *Mint with spiked flowers; commonly called Spear Mint.*

2. MENTHA (*Glabra*) floribus spicatis, foliis longioribus glabris, supernè minimè serratis. *Narrow-leaved smooth-spiked Mint.*

3. MENTHA (*Candicans*) foliis lanceolatis serratis, subtus incanis, floribus spicatis hirsutissimis. *Mint with spear-shaped sawed leaves, which are hoary on their under side, and very hairy spiked flowers.*

4. MENTHA (*Sylvestris*) spicis confertis, foliis serratis tomentosis sessilibus. Hort. Cliff. 306. *Wild Mint with a longer leaf.*

5. MENTHA (*Aquatica*) spicis crassioribus, foliis ovato-lanceolatis serratis subtus tomentosis petiolatis. *Hairy Water Mint with a thicker spike.*

6. MENTHA (*Piperita*) spicis crassioribus interruptis, foliis lanceolatis acutè serratis. *Blackish hot Mint with a taste like Pepper; commonly called Pepper Mint.*

7. MENTHA (*Crispa*) floribus spicatis, foliis cordatis dentatis undulatis sessilibus. Hort. Cliff. 306. *Danish or German curled Mint.*

8. MENTHA (*Rotundifolia*) spicis confertis, foliis ovatis rugosis sessilibus. *Wild Mint with a rounder rough leaf, and a spiked flower having a strong scent.*

9. MENTHA (*Rubra*) spicis confertis interruptis, foliis oblongo-ovatis acuminatis dentatis sessilibus. *Round-leaved red Mint, smelling like an Orange; commonly called Orange Mint.*

10. MENTHA (*Chalepensis*) foliis oblongis dentatis, utrinque tomentosis sessilibus, spicis tenuioribus. *Narrow-leaved wild Mint of Aleppo, which rarely flowers.*

11. MENTHA (*Palustris*) floribus capitatis, foliis ovatis serratis petiolatis, staminibus corollâ longioribus. Hort. Cliff. 306. *Greater round-leaved Water Mint*

12. MENTHA (*Nigricans*) floribus capitatis, foliis lanceolatis serratis subpetiolatis. Lin. Sp. Plant. 576. *Broad-leaved, blackish, hot Mint, or Pepper Mint.*

13. MENTHA (*Arvensis*) floribus verticillatis, foliis ovatis acutis serratis, staminibus corollâ brevioribus. Lin. Sp. Plant. 577. *Whorled, hairy, Field Mint, or Calamint of the shops.*

14. MENTHA (*Exigua*) floribus verticillatis, foliis ovatis dentatis, staminibus corollâ longioribus. *Smallest Water Mint.*

15. MENTHA (*Gentilis*) floribus verticillatis, foliis ovatis, marginibus ciliatis, staminibus corollam æquantibus. *Whorled Mint with a rounder leaf, smelling like Basil.*

16. MENTHA (*Hirsuta*) floribus verticillatis, foliis ovatis serratis hirsutis, staminibus corollâ longioribus. *Common hairy Water Mint, or Sisymbrium.*

17. MENTHA (*Verticillata*) floribus verticillatis, foliis lanceolatis acutis serratis, rugosis, staminibus corollam æquantibus. *Whorled Mint with a longer acute-pointed leaf, and an aromatic scent.*

The first sort is what the gardeners cultivate to supply the markets, which is used both as a culinary herb, and for medicine. It is generally called Spear Mint, and by some Hart Mint, or Roman Mint. This is a plant so well known as to need no description. There are two varieties of this, one with a curled, and the other has variegated leaves, but both these run from the common sort; these are by some preserved in their gardens for the sake of variety, therefore I have mentioned them here.

The second sort hath smoother and narrower leaves than the first, but in other respects it agrees with that, so that it is frequently cultivated in the gardens for use, without distinction.

The third sort grows naturally in moist places. The leaves of this are shorter, and broader in the middle than either of the former; the serratures on their edges are more acute, and their under sides are woolly. The scent of this sort is very like that of the Garden Mint.

The fourth sort hath longer and broader leaves than either of the former, which are woolly and white. The serratures on their edges are farther asunder; they are hairy, and very sharp pointed. The spikes of flowers are slender, hairy, several of them growing together at the top of the stalk. This is the Mentastrum, or wild Mint of the shops, and is an ingredient in the Trochisci de Myrrha.

The fifth sort grows naturally in several parts of England; it is titled Spiked Horse Mint, or Water Mint. The stalks of this are shorter than those of either of the former; the leaves are oval, spear-shaped, of a pale colour, sawed on their edges. The flowers grow in short thick spikes at the top of the stalks, their stamina being shorter than the petal.

The sixth sort grows naturally by the side of the river between Mitcham and Croydon in Surry. This hath smooth purple stalks; the leaves are smaller than those of common Mint, sawed on their edges, of a darker green colour than either of the former; their midrib and veins are purple, and a little hairy on their under side. The spikes of flowers are shorter and thicker than those of the common Mint, and are broken or interrupted at the bottom. The whole plant has a hot biting taste like Pepper, and a pleasant scent.

The seventh sort was originally brought from Denmark, where it was thought to grow naturally, but Dr. Linnæus fixes it as a native of Siberia. The stalks of this sort are hairy, and rise about the same height with the common. The leaves are heart-shaped, deeply indented on their edges, waved and curled, and sit close to the stalk; they are of a light green. The flowers are purple, growing in thick interrupted spikes at the top of the stalks; their empalements are cut almost to the bottom, and the style of the flower is bifid, standing out beyond the petal.

The eighth sort grows naturally in many parts of England. This rises with a strong hairy stalk about the same height as the common Mint, garnished with oval rough leaves sitting close; they are of a dark green, and crenated on their edges. The spikes of flowers grow in clusters at the top of the stalks, which are short and close; their stamina are stretched out beyond the petal.

The ninth sort is commonly called Orange Mint from its scent, which is somewhat like that of the rind of Orange. This rises with an upright smooth stalk about the same height with the common Mint, but does not branch out like that; the leaves are much broader than those of the common sort, the indentures on their edges are deep, and they end in acute points. The spikes of flowers grow in clusters on the top of the stalks, which are interrupted; their stamina are shorter than the petal. It is commonly cultivated in gardens for its pleasant scent.

The tenth sort grows naturally at Aleppo, but is hardy enough to thrive in the open air in England, in common winters, but in severe cold is often killed. This hath slender stalks, which are purple at bottom, but woolly upward, seldom branching, garnished with oblong indented leaves, which are downy on both sides, sitting close to the stalks. The spikes of flowers are single, and very slender, but do not often appear in England, but when they do, it is late in the summer. It creeps much at the root, so the only way to obtain flowers is to confine their roots in pots.

The eleventh sort grows naturally in ditches in most parts of England, and is commonly known by the name of Water Mint. This hath hairy stalks about a foot high, which branch toward the top, garnished with oval sawed leaves, standing upon pretty long foot-stalks. The flowers grow in roundish spikes at the end of the branches, of a purple colour, and their stamina are longer than the petal. The whole plant has a very strong scent, somewhat like that of Pennyroyal.

The twelfth sort grows naturally in ditches in several parts of England. The stalks of this are purple, smooth, and short; the leaves are small, spear-shaped, of a dark colour, and are slightly sawed on their edges. The flowers grow in roundish heads on the top of the stalks; their stamina are longer than the petal. This sort has a warm biting taste, but not quite so hot as the Pepper Mint before described, but is often used for it.

The thirteenth sort grows naturally in arable land in most parts of England, and is rarely admitted into gardens. This is the Water Calamint of the shops, but is now seldom used in medicine. The stalks of this sort rise about a foot high, are hairy, garnished with oval leaves, ending in acute points, and sawed on their edges. The flowers grow in very thick whorls round the stalks; they are small, of a purple colour, and their stamina are shorter than the petal.

The fourteenth sort grows in watery places in many parts of England. This hath weak trailing stalks a foot and a half long, garnished with small oval leaves, which are indented on their edges, and stand upon pretty long foot-stalks. The flowers grow in thick whorls round the stalks, and their stamina are longer than the petal.

The fifteenth sort grows plentifully on the side of the road between Bocking and Gosfield in Essex. The stalks of this are much smaller, and not so long as those of the former; the leaves are shorter and rounder, and are very little indented on their edges, but have their borders set with hairs. The whorls of flowers are smaller, and the whole plant has the scent of Basil.

The sixteenth sort grows naturally in ditches in many parts of England. This hath hairy stalks a foot or more in height; the leaves are oval, sawed, and very hairy. The flowers grow in large whorls toward the top of the stalks, and their stamina are longer than the petals. This hath a pleasanter scent than the common Water Mint, so is called Sweet Water Mint by way of distinction.

The seventeenth sort grows naturally by the side of the river Medway, between Rochester and Chatham. This hath hairy stalks near two feet high, garnished with spear-shaped leaves, ending in acute points; the stalks are beset with whorls of flowers almost their whole length, so that they have frequently ten or twelve whorls on each. The

flowers

flowers are purplish, and their stamina are equal with the petals: this hath a very pleasant aromatic scent.

All the sorts of Mint are easily propagated by parting the roots in the spring, or by planting cuttings during any of the summer months, but they should have a moist soil; and after the cuttings are planted, if the season should prove dry, they must be often watered, until they have taken root, after which they will require no farther care but to keep them clear from weeds: they should be planted in beds about four feet wide, allowing a path two feet broad, to go between the beds to water, weed, and cut the plants. The distance they should be set is about five or six inches, or more, because they spread very much at their roots; for which reason, the beds should not stand longer than three years, for by that time the roots will be matted so closely, as to rot and decay each other, if permitted to stand longer. There are some people who are very fond of Mint sallet in winter and spring, in order to obtain which, they take up the roots before Christmas, and plant them upon a moderate hot-bed close together, covering them with fine earth about an inch thick, and cover the bed either with mats or frames of glass. In these beds the Mint will come up in a month's time, and be soon fit to cut for that purpose.

When the herb is cut for medicinal use, it should be done in a very dry season, just when it is in flower; for if it stand longer, it will be not near so handsome, nor so well tasted; and if it be cut when it is wet, it will change black and be little worth; this should be hung up to dry in a shady place, where it may remain until it be used.

MENTHA CATARIA. See NEPETA.

MENTZELIA. Plum. Nov. Gen. Plant. 40. tab. 6.

The CHARACTERS are,

The flower hath five petals a little longer than the empalement, and many erect bristly stamina. From the long cylindrical germen, which is situated under the flower, arises a bristly style. The germen afterward turns to a cylindrical long capsule with one cell, containing many small seeds.

We have but one SPECIES of this genus, viz.

MENTZELIA (*Aspera.*) Hort. Cliff. 492. Plumier titles it Mentzelia foliis & fructibus asperis. Nov. Gen. Plant. 41. *Mentzelia with prickly leaves and fruit.*

This plant grows plentifully at La Vera Cruz.

The plant is annual, it rises with a slender, smooth, stiff stalk, a little woody, more than three feet high, branching out alternately at distances; the branches are distorted, and run into one another; they are garnished with leaves shaped like the point of a halbert, standing alternately upon short foot-stalks, covered with short hooked prickles, which fasten themselves into the clothes of those who rub against them. The flowers come out singly from the joints of the stalk, resting upon a cylindrical germen, which is near an inch in length, narrow at the base, but widens upward. Upon the top of it comes out the empalement, after the same manner as those of the Onagra; the flowers are of a pale yellow colour. In the middle arises a great number of stamina which are erect, terminated by single summits; from the germen arises a single style, crowned by a single stigma. The germen afterward turns to a long cylindrical capsule, armed with prickles as the leaves, which also fasten themselves to the clothes of those who rub against them; these have but one cell, which is filled with small seeds.

As this is an annual plant, which perishes soon after the seeds are ripe, so the seeds must be sown on a hot-bed early in the season, that the plants may be brought forward early in the spring, otherwise they will not produce ripe seeds in this country. When the plants are come up about an inch high, they should be each transplanted into a separate pot, and plunged into a hot-bed of tanners bark, being careful to shade them from the sun until they have taken new root; then they must be treated in the same manner as other tender annual plants.

MENYANTHES, is the Trifolium Palustre, or Bog Bean.

This plant is common upon boggy places in divers parts of England; but as it is never cultivated in gardens, so I shall not trouble the reader with any further account of it.

MERCURIALIS. Tourn. Inst. R. H. 534. tab. 308.

The CHARACTERS are,

It is male and female on different plants; the male flowers have a spreading empalement, cut into three concave segments, but have no petals; they have nine or twelve erect hairy stamina. The female flowers have no petals, but have two awl-shaped acute-pointed nectariums; to each of these there is a single broad germen, which afterward turns to a twin capsule shaped like a scrotum, having two cells, each containing one roundish seed.

The SPECIES are,

1. MERCURIALIS (*Annua*) caule brachiato, foliis glabris. Hort. Cliff. 461. *Mercury with spiked and testiculated flowers, which are male and female.*

2. MERCURIALIS (*Perennis*) caule simplicissimo, foliis scabris. Hort. Cliff. 461. *Mountain Mercury, or Dog's Mercury, with spiked and testiculated flowers.*

3. MERCURIALIS (*Tomentosa*) caule subfruticoso, foliis tomentosis. Hort. Cliff. 461. *Shrubby hoary Mercury, having spiked and testiculated flowers.*

The first sort is commonly called French Mercury, from whence it might have been brought into England; for although it is now become a weed in gardens and upon dunghills, yet it is seldom found growing at a distance from habitations. This is an annual plant, with a branching stalk about a foot high, garnished with spear-shaped leaves of a pale or yellowish green colour. The male plants have spikes of herbaceous flowers, growing on the top of the stalk; these fall off soon; but the female plants, which have testiculated flowers proceeding from the side of the stalks, are succeeded by seeds, which, if permitted to scatter, will produce plenty of plants of both sexes.

The second sort grows under hedges and in woods in most parts of England. This hath a perennial root, which creeps in the ground; the stalks are single and without branches, rising ten or twelve inches high, garnished with rough leaves, placed by pairs at each joint, of a dark green colour, indented on their edges; these have their male flowers growing in spikes, upon different plants from those which produce seeds.

The third sort grows naturally in the south of France, in Spain and Italy. This rises with a shrubby branching stalk a foot and a half high, garnished with oval leaves placed by pairs, which are covered with a white down on both sides. The male flowers grow in short spikes from the side of the stalks, upon different plants from the first, which are testiculated and hoary. If the seeds of these are permitted to scatter, the plants will come up the following spring; and if the seeds are sown, it should be performed in the autumn, for those which are sown in the spring never grow the same year. This plant should have a warm situation and a dry rubbishy soil, in which it will live three or four years, but in hard frost these plants are frequently killed.

MESEMBRYANTHEMUM. Dill. Gen. 9. Hort. Elth. 179. Ficoides, or Fig Marygold.

The CHARACTERS are,

The flower hath one petal, which is cut into many linear segments almost to the bottom, and ranged in several series, but are joined together at their base; within these are ranged a great number of hairy stamina. Under the flower is situated an obtuse five-cornered germen, supporting sometimes five, and often ten or more styles. The germen afterward becomes a roundish fleshy fruit, having as many cells as there are styles, filled with small seeds.

The SPECIES are,

1. MESEMBRYANTHEMUM (*Nodiflorum*) foliis alternis teretiusculis obtusis ciliatis. Hort. Upsal. 129. *Fig Marygold of Naples with a white flower; or Egyptian Kali.*

2. MESEMBRYANTHEMUM (*Crystallinum*) foliis alternis ovatis obtusis undulatis. Hort. Cliff. 216. *African Fig Marygold, with a waved Plantain leaf, marked with silvery spots; commonly called the Diamond Ficoides, or Diamond Plant.*

3. MESEMBRYANTHEMUM (*Geniculiflorum*) foliis semiteretibus, floribus sessilibus axillaribus. Lin. Sp. Plant. 481. *Fig Marygold of the Cape, with a taper leaf and a whitish flower.*

4. MESEMBRYANTHEMUM (*Noctiflorum*) foliis semicylindriceis, floribus quadrifidis. Lin. Sp. Pl. 481. *Upright ligneous Fig Marygold of Africa with a radiated flower, which is at first purple, afterward silvery, shut in the day, and opens at night.*

5. MESEMBRYANTHEMUM (*Splendens*) foliis semiteretibus falcatis, caule arborescente. *African Tree Fig Marygold with a taper leaf, and a white flower opening at night, but shut in the day.*

6. MESEMBRYANTHEMUM (*Umbellatum*) foliis subulatis, caule erecto, corymbo trichotoma. Lin. Sp. Plant. 481. *Upright African Fig Marygold with a taper leaf, and white flowers growing in umbels.*

7. MESEMBRYANTHEMUM (*Calimiforme*) acaule, foliis subteretibus connatis, floribus octagynis. Lin. Sp. Plant. 481. *Low Fig Marygold of the Cape, with an Onion leaf and a stamineous flower.*

8. MESEMBRYANTHEMUM (*Tripolium*) foliis lanceolatis planis crenulatis. Hort. Cliff. 217. *Trailing African Fig Marygold, with a Tripolium leaf and a silvery flower.*

9. MESEMBRYANTHEMUM (*Bellidiflorum*) acaule, foliis triquetris linearibus apice trifariàm dentatis. Hort. Cliff. 218. *Dwarf Fig Marygold of the Cape, with a triangular leaf indented at the top, and a smaller purplish flower.*

10. MESEMBRYANTHEMUM (*Subulatum*) acaule, foliis subulatis triquetris dorso supernè serratis. *Dwarf Mesembryanthemum with awl-shaped three-cornered leaves, whose back part is sawed toward the top.*

11. MESEMBRYANTHEMUM (*Deltoides*) caulescens, foliis deltoidibus triquetris dentatis. Hort. Cliff. 218. *African Fig Marygold with a short, thick, gray, triangular leaf, with prickles on the three edges.*

12. MESEMBRYANTHEMUM (*Caulescens*) caulescens, foliis deltoidibus, lateralibus minimè dentatis. *African Fig Marygold with very thick, short, triangular, gray leaves, having small indentures on their edges.*

13. MESEMBRYANTHEMUM (*Barbatum*) acaule, foliis apice barbatis. *African Fig Marygold with a rough spotted leaf, whose point is armed with spines in form of a star*

14. MESEMBRYANTHEMUM (*Stellatum*) caulibus subfruticosis decumbentibus, foliis teretibus apice barbatis. *Shrubby Fig Marygold of the Cape, with a star-pointed tumid leaf, and a purple flower.*

15. MESEMBRYANTHEMUM (*Hispidum*) caule hispido, foliis cylindricis deflexis. Lin. Sp. Pl. 482. *African shrubby Fig Marygold, having stalks adorned with silvery down, and long, small, taper leaves, spotted as it were with silvery drops, and a Violet-coloured flower.*

16. MESEMBRYANTHEMUM (*Villosum*) caule foliisque pubescentibus. Hort. Cliff. 217. *Mesembryanthemum whose stalks and leaves are garnished with downy hairs.*

17. MESEMBRYANTHEMUM (*Scabrum*) foliis subulatis subtus undique scabris. Hort. Cliff. 219. *African Fig Marygold with a long, green, rough, triangular leaf, and a Violet-coloured flower.*

18. MESEMBRYANTHEMUM (*Uncinatum*) articulis caulinis terminatis in folia acuminata subtus dentata. Hort. Cliff. 218. *African Fig Marygold with a short, perfoliated, triangular leaf, whose point is prickly; commonly called Buckshorn Ficoides.*

19. MESEMBRYANTHEMUM (*Perfoliatum*) perfoliatum, foliis majoribus, apicibus triacanthis. Hort. Elth. 251. *Shrubby, perfoliate, African Fig Marygold, with a triangular, gray, spotted leaf, and a thin, white, ligneous bark; commonly called Stagshorn Ficoides.*

20. MESEMBRYANTHEMUM (*Spinosum*) spinis ramosis. Hort. Cliff. 216. *African Fig Marygold with long spines, and smaller leaves arising from the wings of the large leaves.*

21. MESEMBRYANTHEMUM (*Tuberosum*) foliis subulatis papillosis, radice capitatâ. Hort. Cliff. 216. *African Fig Marygold with a triangular recurved leaf, and umbellated flowers of a dark colour, which are purple on their outside.*

22. MESEMBRYANTHEMUM (*Tenuifolium*) foliis subulatis semiteretibus glabris, internodio longioribus. Hort. Cliff. 216. *Low Fig Marygold of the Cape, with a taper leaf and a scarlet flower.*

23. MESEMBRYANTHEMUM (*Stipulaceum*) foliis subulatis subcylindriceis obsoletè papillosis distinctis. Hort. Cliff. 220. *Upright, tree-like, African Fig Marygold, with a jointed stalk and a green leaf.*

24. MESEMBRYANTHEMUM (*Crassifolium*) caule repente semicylindriceo, foliis semicylindricis lævibus connatis, apice triquetris. Hort. Cliff. 217. *Creeping African Fig Marygold with a green triangular leaf, and deep purple-coloured flower.*

25. MESEMBRYANTHEMUM (*Falcatum*) foliis acinaciformibus distinctis lævibus, ramis teretibus. Hort. Cliff. 219. *African Fig Marygold with a triangular, cimiter-shaped, short leaf, and a pale purplish flower.*

26. MESEMBRYANTHEMUM (*Glomeratum*) foliis acinaciformibus connatis lævibus, caule decumbente. *Greater, trailing, African Fig Marygold, with a triangular cimiter-shaped leaf.*

27. MESEMBRYANTHEMUM (*Edule*) falcatum majus, flore amplo luteo. Hort. Elth. 283. *Greater, trailing, African Fig Marygold, with a triangular leaf, and a large eatable fruit.*

28. MESEMBRYANTHEMUM (*Bicolorum*) foliis subtriquetris scabris, corollis bicoloribus. Lin. Sp. Plant. 485. *Shrubby Fig Marygold of the Cape, with a taper leaf having punctures, and flowers with yellow and red petals.*

29. MESEMBRYANTHEMUM (*Acinaciforme*) foliis subulatis triquetris, angulo carinali retrorsum serratis. Hort. Cliff. 218. *African Fig Marygold with a long triangular leaf, which is incurved, and a purple stalk.*

30. MESEMBRYANTHEMUM (*Loreum*) foliis subcylindriceis recurvis basi interiore gibbis connatis caule pendulo, caule scabro. Hort. Cliff. 220. *Fig Marygold of the Cape, with a silvery taper leaf, and flowers having many Orange-coloured petals.*

31. MESEMBRYANTHEMUM (*Serratum*) foliis linearibus obsoletè triquetris distinctis, summis imbricatis, lævibus. Hort. Cliff. 220. *Shrubby Fig Marygold of the Cape, with taper gray leaves growing in clusters, and a white flower.*

32. MESEMBRYANTHEMUM (*Tuberculatum*) acaule, foliis semicylindricis connatis externè tuberculatis. Hort. Cliff. 219. *African Fig Marygold with a long, triangular, succulent leaf, and red stalks.*

33. MESEM-

33. MESEMBRYANTHEMUM (*Veruculatum*) foliis subcylindricis acutis connatis arcuatis lævibus. Hort. Cliff. 220. *African Tree Fig Marygold with a taper gray leaf, having a thick purple top.*

34. MESEMBRYANTHEMUM (*Glaucum*) foliis subulatis triquetris strictis acutis, punctis pellucidis obsoletis sparsis. Hort. Cliff. 220. *African Fig Marygold with an erect ligneous stalk, a triangular, cimiter-shaped, rough leaf, and a large yellow flower.*

35. MESEMBRYANTHEMUM (*Corniculatum*) caulescens, foliis subulatis semicylindricis recurvis connatis longis. Hort. Cliff. 219. *African Fig Marygold with a long triangular leaf, having obtuse borders, and a large flower of a pale yellow within, and marked with a long red streak on the outside.*

36. MESEMBRYANTHEMUM (*Expansum*) caulibus procumbentibus, foliis subtriquetris angulis obtusioribus recurvis connatis, pedunculis brevioribus. *Trailing African Fig Marygold, with a longer, gray, triangular leaf, and a yellowish flower.*

37. MESEMBRYANTHEMUM (*Micans*) foliis planis oppositis ovatis acuminatis connatis integerrimis. *Trailing African Fig Marygold with plain leaves set by pairs, which are lucid, surround the stalk with their base, and a large, whitish, yellow flower.*

38. MESEMBRYANTHEMUM (*Tortuosum*) foliis planis congestis externè punctatis acuminatis, integerrimis. *Trailing Fig Marygold of the Cape with an Olive leaf, and a white flower of a Saffron colour in the middle.*

39. MESEMBRYANTHEMUM (*Ringens*) subacaule, foliis ciliato-dentatis. Lin. Hort. Cliff. 218. *Low Fig Marygold of the Cape, with a triangular leaf indented toward the top, and a yellow flower; commonly called Dogs Chap Ficoides.*

40. MESEMBRYANTHEMUM (*Rostratum*) acaule, foliis crassis triquetris, margines laterales ciliato-dentatis, pedunculis brevibus. *African Fig Marygold with a triangular, cimiter-shaped, short, thick leaf, whose borders have many large spines; commonly called Cats Chap Ficoides.*

41. MESEMBRYANTHEMUM (*Dolabriforme*) foliis dolabriformibus. Hort. Cliff. 219. *Low Fig Marygold of the Cape, with leaves like a stag's horn, and a yellow flower opening at night.*

42. MESEMBRYANTHEMUM (*Difforme*) foliis difformibus. Prod. Leyd. 287. *African Fig Marygold with very broad, thick, shining, deformed leaves.*

43. MESEMBRYANTHEMUM (*Lucidum*) acaule, foliis linguiformibus altero margine crassioribus. Hort. Cliff. 219. *African Fig Marygold without a stalk, broad, thick, shining leaves, growing by pairs, and a very large yellow flower.*

44. MESEMBRYANTHEMUM (*Linguiforme*) acaule, foliis linguiformibus latissimis, pedunculis brevioribus. *African Fig Marygold having no stalk, very broad, thick, shining leaves, placed by pairs, and a large golden flower with a short foot-stalk.*

45. MESEMBRYANTHEMUM (*Pugioneforme*) foliis alternis subulatis triquetris longissimis. Hort. Cliff. 216. *Fig Marygold of the Cape, with a Clove Gilliflower leaf, and a beautiful gold-coloured flower.*

46. MESEMBRYANTHEMUM (*Albidum*) acaule foliis triquetris. *Mesembryanthemum without stalks, and entire gray three-cornered leaves.*

These plants are most of them natives of the Cape of Good Hope, from whence their seeds were first brought to Holland, and the plants raised in many of their curious gardens, and have since been communicated to most parts of Europe.

Most of the plants of this genus have beautiful flowers, which appear at different seasons of the year; some of them flower early in the spring, others in summer, some in the autumn, and there are others which flower in winter; and many of them produce their flowers in such quantity, as that when they are expanded the plants are entirely covered with them; they have all of them thick succulent leaves, but some of the species are much more so than others, and the figures of their leaves vary so much in their several species, that they afford an agreeable variety when they are not in flower.

All the sorts here mentioned are perennial plants, except the two first, which are annual. The perennial sorts are easily propagated by cuttings during any of the summer months; such of them as have shrubby stalks and branches do very readily take root when planted in a bed of light soil, and covered either with mats or glasses, but when they are covered with the latter they must be shaded every day when the sun is warm; the cuttings of the shrubby sorts need not be cut from the plants more than four or five days before they are planted, during which time they should be laid in a dry room, not too much exposed to the sun, that the part which was separated from the old plants may heal over and dry before they are planted, otherwise they are apt to rot. When the cuttings are taken from the old plants, they should be divested of their lower leaves, so far as may be necessary, to allow a naked stalk of sufficient length for planting.

When the cuttings are planted, it will be necessary to give them a little water to settle the ground about them, but it should be done with caution, for too much wet will spoil them; but if there should happen some gentle showers of rain, it will be proper to take off their covers, and let them receive it, but they should be screened from hard rains. The cuttings thus managed will put out good roots in about six weeks, when they should be carefully taken up, and each planted in a separate small pot filled with light sandy earth, and placed in a shady situation, giving them little water to settle the earth to their roots; in this place they may remain about ten days or a fortnight; by which time they will have taken good root, and may be removed to a sheltered place, where they may have more sun, in which they may remain till autumn: during the summer months these may be watered two or three times a week, but it must not be given them in too great plenty; but as the sun declines in autumn, they should not have it so often; for if they are often supplied with it the plants will grow luxuriant, their leaves and branches will be so replete with moisture, that the early frosts in the autumn will destroy them; whereas when they are kept dry, their growth will be stinted, so that they will be hardy enough to resist small frosts; but there must be care taken that they do not shoot their roots through the holes of the pots into the ground, for when they do the plants will grow very luxuriant; and when the pots are removed, and those roots are torn off, their leaves and branches will shrink, so will not recover it in a long time; to prevent which the pots should be removed every month in summer, and where the roots are beginning to come through the pots, they should be cut off. The sorts which grow very freely should be shifted three or four times in the summer, to pare off their roots, and keep them within compass; and these should never be planted in rich earth for the reasons before given, for if the earth is fresh, they will require no dung or other compost, unless it is strong; in which case, sea-sand or lime rubbish will be a good mixture; the quantity of either must be in proportion to the stiffness of the ground, always being careful to render it so light, as that the wet may easily pass off.

We next proceed to treat of those sorts whose stalks and leaves are very succulent. The cuttings of these should be taken from the plants ten days or a fortnight before they are planted, that they may have time for their wounded part to heal over and dry; the lower leaves of these should also be stripped off, that their naked stalks may be of a sufficient length for planting. As these are mostly plants of humble growth, so if their stalks are divested of their leaves an inch and a half, it will be sufficient. The cuttings of these sorts require to be covered with glasses, to keep off the wet; they must also have less water than the other, but in other particulars will require the same treatment. The roots of these do not spread and extend so much as those of the other, so will not require to be shifted oftener than twice a year; they must also be kept in small pots, to confine their roots; the earth in which they are planted should be rather light and not rich. During the summer season they must not have too much wet, and in the winter they must have but little water. If these succulent sorts are placed in an open airy glass-case in winter, where they may have free air admitted to them in plenty in mild weather, and screened from the frost, they will thrive much better than when they are more tenderly treated.

The shrubby kinds may be sheltered in winter under a common frame, where, if they are protected from frost and wet, it is all they require; for the hardier these are treated, the greater quantity of flowers they will produce: some of the sorts are so hardy as to live abroad when planted close to a good aspected wall, in a poor dry soil; so that where there is room to dispose them against a wall, and the border is raised with lime rubbish to prevent their rooting deep and growing luxuriant, they may be preserved through the winter with very little shelter, and these will flower much better than those under cover.

The first sort grows naturally in Egypt, where they cut up the plants, and burn them for pot-ash; and this is esteemed the best sort for making hard soap, and the best sort of glass.

This is an annual plant, which does not perfect seeds in England, for when it is placed in the stove, or kept in the hot-bed, their stalks grow long and slender, so are not productive of flowers; and those which are raised in hot-beds, and afterward exposed in the open air, will flower pretty freely, but do not perfect their seeds. As this plant will thrive in South Carolina as well as in its native soil, so it might turn to the advantage of that colony, and likewise become beneficial to the public, if the inhabitants could be prevailed on to cultivate this plant.

The second sort is annual; it is propagated for the oddness of its leaves and stalks, which are closely covered with pellucid pimples full of moisture, which, when the sun shines on the plants, they reflect the light, and appear like small bubbles of ice; from whence some have called it the Ice Plant, and others have named it the Diamond Plant, or Diamond Ficoides.

This sort is propagated by seeds, which must be sown on a hot-bed early in the spring; and when the plants come up, they must be planted on a fresh hot-bed to bring them forward; after they have taken root in the hot-bed, they should have but little wet. When they are grown large enough to transplant again, they should be each planted into a small pot, filled with light fresh earth, and plunged into a hot-bed of tan, observing to shade them in the heat of the day until they have taken new root; then they should have plenty of fresh air admitted to them every day in warm weather, to prevent their drawing weak. In the latter end of June, some of the plants may be inured to bear the open air, and afterward they may be turned out of the pots, and planted in a warm border, where they will thrive and spread their branches to a great distance upon the ground; but these plants will not be very productive of flowers, therefore some of them must be continued in the small pots, and may at the same time, when the others are planted out, be removed into the stove or glass-case, placing them upon the shelves, that their roots may not get out from the bottom of the pots, so that they may be confined, which will cause them to flower plentifully, and from these good seeds may every year be obtained.

MESPILUS. Tourn. Inst. R. H. 641. tab. 410. The Medlar.

The CHARACTERS are,

The empalement of the flower is permanent. The flower is composed of five roundish concave petals, which are inserted in the empalement. The number of stamina are different in the several species, from ten to twenty or more; these are also inserted in the empalement. The germen is situated under the flower, and supports an uncertain number of styles from three to five; it afterward becomes a roundish or oval berry, carrying the empalement on its top, and inclosing four or five hard seeds.

The SPECIES are,

1. MESPILUS (*Sylvestris*) inermis, foliis lanceolatis dentatis acuminatis, subtus tomentosis, calycibus acuminatis. *Greater Medlar with a Bay-tree leaf, and a smaller less substantial fruit.*

2. MESPILUS (*Germanica*) inermis, foliis lanceolatis integerrimis subtus tomentosis, calycibus acuminatis. Hort. Cliff. 189. *German Medlar with a Bay-tree leaf, which is not sawed.*

3. MESPILUS (*Pyracantha*) spinosa, foliis lanceolato-ovatis crenatis, calycibus fructus obtusis. Hort. Cliff. 189. *Prickly Medlar with spear-shaped, oval, crenated leaves; called Pyracantha.*

4. MESPILUS (*Amelanchier*) inermis, foliis ovalibus serratis, cauliculis hirsutis. Lin. Sp. Pl. 478. *Medlar with thorns, having oval sawed leaves, and hairy stalks; commonly called Amelanchier.*

5. MESPILUS (*Cotoneaster*) foliis ovatis integerrimis. Hort. Cliff. 180. *Medlar with oval entire leaves; commonly called Dwarf Quince.*

6. MESPILUS (*Chamæmespilus*) inermis, foliis ovalibus serratis glabris, floribus capitatis, bracteis deciduis linearibus. Lin. Sp. Plant. 479. *Medlar without thorns, having smooth, oval, sawed leaves, headed flowers, and linear bractea which fall off.*

7. MESPILUS (*Orientalis*) foliis ovatis crassis integerrimis, subtus tomentosis, floribus umbellatis axillaribus. *Medlar with oval, thick, entire leaves, which are woolly on their under side, and flowers growing in umbels from the wings of the stalk; or Dwarf Cherry of Mount Ida.*

8. MESPILUS (*Arbutifolia*) inermis, foliis lanceolatis crenatis subtus tomentosis. Hort. Cliff. 189. *Virginia Medlar with an Arbutus leaf.*

The first sort grows naturally in Sicily, where it becomes a large tree. It rises with a straiter stem, and the branches grow more upright than those of the Dutch Medlar; the leaves are narrower and not sawn on their edges; the flowers are smaller than those of the Dutch Medlar, and the fruit is shaped like a Pear.

The second sort is generally called the Dutch Medlar; this never rises with an upright stalk, but sends out crooked deformed branches at a small height from the ground; the leaves of this are very large, entire, and downy on their under side. The flowers are very large, as are also the fruit, which are rounder, and approach nearer to the shape of an Apple. This being the largest fruit, is now generally cultivated in the gardens; but there is one with smaller fruit,

fruit, which is called the Nottingham Medlar, of a much quicker and more poignant taſte than this, which is ſuppoſed to be only a variety, ſo I have not enumerated it.

The third ſort is moſt commonly known by the title of Pyracantha, or evergreen Hawthorn. This does not riſe to a tree, for the ſtalk and branches are too ſlender and weak to ſtand without ſupport, ſo the plants are generally planted againſt walls or buildings to cover them. It grows naturally in the hedges in the ſouth of France and Italy. The flowers come from the ſide of the branches in large umbels; they are ſmaller than thoſe of the common Hawthorn, and of a dirty white; theſe are ſucceeded by roundiſh umbilicated berries, of a fiery red colour, which ripen in the winter, and being intermixed with the evergreen leaves, make a fine appearance at that ſeaſon of the year.

This is propagated by the ſeeds in the ſame manner as the common Hawthorn, or by laying down of the branches, which, if young, and laid in the autumn, will put out roots fit to remove in one year; but when the old wood is laid down, it ſeldom puts out roots in leſs than two or three years, if they ever do.

The fourth ſort grows naturally in Auſtria, Italy, and France, particularly near Fontainbleau; this riſes with many ſlender ſtalks about three feet high, which put out ſmall ſide branches, covered with a dark purple bark without thorns, which are cloſely garniſhed with oval leaves, ſlightly ſawed on their edges; the ſmall ſide branches which ſuſtain the flowers are very hairy and woolly, as are alſo the foot-ſtalks, and the under ſide of the leaves, but their upper ſides are ſmooth and green. The flowers come out in bunches at the end of the ſhoots, which have five long narrow petals, and about ten ſtamina in each. Theſe are ſucceeded by ſmall fruit, which, when ripe, are black; the gardeners call this New England Quince; there is one of this kind which grows naturally in North America, but the leaves of that are wedge ſhaped, and not ſawed on the edges, ſo I take it to be a different ſpecies.

The fifth ſort grows naturally on the Pyrenean mountains, and in other cold parts of Europe; this riſes with a ſmooth ſhrubby ſtalk about four feet high, dividing into a few ſmall branches, covered with a purple bark, garniſhed with oval entire leaves, with very ſhort foot-ſtalks. The flowers come out from the ſide of the ſtalks two or three together; they are ſmall, of a purpliſh colour, and ſit cloſe to the ſtalks; theſe appear in May, and are ſucceeded by ſmall roundiſh fruit, which are of a bright red colour when ripe.

The ſixth ſort grows naturally in the northern parts of Europe; this hath a ſmooth ſtalk, riſing about four or five feet high, ſending out ſlender branches, which are covered with a purpliſh bark, garniſhed with oval ſmooth leaves, ſawed on their edges, with the teeth pointing upward; they have pretty long ſlender foot-ſtalks, and are of a yellowiſh green on both ſides. The flowers come out from the wings of the ſtalk, four or five joined together in a cloſe head, of a purpliſh colour; between the flowers come out long narrow bractea, which are purpliſh, and fall off as the flowers begin to decay. The fruit is ſmall, and red when ripe.

The ſeventh ſorth grows naturally upon Mount Ida in Crete, where the poor ſhepherds feed upon the fruit when ripe; this hath a ſmooth ſtalk, covered with a brown bark, and riſes eight or ten feet high, dividing into many ſmooth branches, which are garniſhed with oval leaves two inches and a half long, and near two inches broad, of a thick ſubſtance, and a dark green on their upper ſide, but downy on their under, ſtanding upon ſhort foot-ſtalks. The flowers come out from the ſide of the ſtalk; they are of a purple colour, the petals being but little longer than the empalement, which is woolly, and cut into five obtuſe ſegments. The fruit is large, roundiſh, and of a fine red colour when ripe.

The eighth ſort grows naturally in North America, where it rarely riſes more than five feet high, ſending out a few upright branches, garniſhed with ſpear-ſhaped leaves whoſe edges are crenated, and their under ſide downy: the flowers are produced in ſmall bunches on the ſide, and at the extremity of the branches, which are ſucceeded by ſmall roundiſh fruit a little compreſſed, of a purple colour when ripe.

All theſe ſorts are hardy enough to thrive in the open air in England, and ſeveral of them are very ornamental plants for gardens; and all the larger growing kinds are as proper for parks, where, during the ſeaſon of their flowering, they will make a fine appearance; and again in autumn, when their fruit are ripe, they will afford an agreeable variety, and their fruit will be food for the deer and birds; ſo that if clumps of each ſort are planted in different parts of the park, nothing can be more ornamental.

Moſt of the American kinds are uſually propagated in the nurſeries, by grafting or budding them upon the common White Thorns, but the plants ſo propagated will never grow to half the ſize of thoſe which are propagated by ſeeds; ſo that where they are deſigned for parks and large plantations, thoſe plants ſhould always be choſen which have not been grafted or budded, but are upon their own roots, for ſeveral of the American ſorts naturally grow twenty feet high, if they are not ſtinted by grafting; ſo that thoſe ſorts when grafted are only fit to intermix with ſhrubs for gardens, where they are not deſigned to grow large.

But there are many who object to this method of raiſing the plants from ſeeds, on account of their ſeeds not growing the firſt year, as alſo from the tedioufneſs of the plants growth after; but where a perſon can furniſh himſelf with the fruit in autumn, and take out their ſeeds ſoon after they are ripe, putting them into the ground immediately, the plants will come up the following ſpring; and if they are kept clean from weeds, and in very dry weather ſupplied with water, they will make great progreſs; but if theſe are planted in the places where they are to remain, after two years growth from ſeeds, they will ſucceed much better than when the plants are of a greater age; but if they are planted in clumps in parks, the ground ſhould be well trenched, and cleanſed from the roots of all bad weeds; the places muſt alſo be ſecurely fenced, otherwiſe the cattle will ſoon deſtroy them. The beſt time to tranſplant them is in autumn, when their leaves fall off; theſe ſhould be conſtantly kept clean from weeds, and if the ground between the plants is dug every winter for the firſt ſeven years, it will encourage their growth, and by that time the plants will have made ſuch progreſs as to be ſtrong enough to encounter and keep down the weeds; ſo that if the weeds are cut twice in the ſummer, it will be ſufficient; but the incloſure muſt not be taken away, till the ſtems are ſo large, and their bark ſo rough, as that the cattle will not eat it; their ſtems ſhould alſo be trained up ſo high, as that their branches may be out of the reach of cattle, otherwiſe they will crop them, eſpecially thoſe which have no thorns.

If when theſe clumps are planted, the ground be incloſed with an Oak pale, in the ſame manner as that round parks, the fence will laſt as long as the trees will require any protection; but theſe pales ſhould not be quite cloſe, for if they are placed ſo near each other at bottom as to keep out hares, it will be ſufficient; and upward, if they are not ſo cloſe, there will be more air admitted to the plants, which will be of ſervice to them, ſo that ſuch pales may be choſen for theſe

these purposes, which are broader at one end than the other, and the broadest ends turned downward. The plants in these clumps need not be planted at a greater distance from each other than six feet, for by being so near together, they will draw one another up taller than where they have more room to spread.

All the sorts of Mespilus and Cratægus will take by budding or grafting upon each other; they will also take upon the Quince, or Pear stocks, and both these will take upon the Medlars, so that these have great affinity with each other, and might be with more propriety brought together under the same genus than the Pear and Apple, which will not take upon each other; but although the Pear will take upon the White Thorn, yet it is not adviseable to make use of the stocks, because they generally cause the fruit to be small, and often to crack, and renders their flesh stony, so unless it is the very soft melting kinds of Pears which are upon these stocks, the fruit will not be good.

METHONICA. See Gloriosa.

MEUM. See Athamanta.

MEZEREON. See Thymelæa.

MICROPUS. Lin. Gen. Plant. 892. Bastard Cudweed.

The Characters are,

It hath hermaphrodite and female flowers, which are included in the same naked empalement. The female flowers are hid under the scales of the interior empalement, which have each a single oval seed succeeding them, included in the small leaves of the empalement, but have no down about them.

We have but one Species of this genus, viz.

Micropus (*Supinus.*) Hort. Upsal. 275. Prod. Leyd. 145. *Portugal Bastard Cudweed.*

This is an annual plant, which grows naturally in Portugal near the sea. The root sends out several trailing stalks, about six or eight inches long, which are garnished with small, oval, silvery leaves, whose base embrace the stalks. The flowers come out from the wings of the stalks in small clusters; they are very small, white, and sit in a double empalement, the interior being so large as to almost hide the flowers. It flowers in June and July, and the seeds ripen in autumn; this is frequently preserved in gardens, for the beauty of its silvery leaves; if the seeds are sown in autumn, or are permitted to scatter, the plants will come up in the spring, and will require no other care but to keep them clean from weeds, and thin them where they are too close. When the seeds of the plant are sown in the spring, they seldom grow the first year.

MILIUM. Tourn. Inst. R. H. 514. tab. 298. Lin. Gen. Plant. 73. Millet.

The Characters are,

It is of the Corn or Grass tribe, with one flower in each chaff The petal of the flower is bivalve, and smaller than the empalement. It hath three very short hairy stamina, and a roundish germen with two hairy styles. The germen afterward turns to a roundish seed, covered by the petal of the flower.

The Species are,

1. Milium (*Panicum*) paniculâ laxâ flaccidâ, foliorum vaginis pubescentibus. *Millet with a loose hanging panicle, and the sheaths of the leaves hairy*; *Millet with a yellow seed.*

2. Milium (*Sparsum*) panicula sparsâ erectâ, glumis aristatis. *Millet with a loose erect panicle, and bearded chaff.*

The first sort grows naturally in India, but is now cultivated in many parts of Europe as an esculent grain; this rises with a Reed-like stalk three or four feet high, channelled; at every joint there is one Reed-like leaf, which is joined on the top of the sheath, which embraces and covers that joint of the stalk below the leaf; this sheath is closely covered with soft hairs, but the leaf which is expanded has none. The top of the stalk is terminated by a large loose panicle, which hangs on one side, having a chaffy flower, which is succeeded by a small round seed, which is often made into puddings, &c. There are two varieties of this, one with white, and the other hath black seeds, but do not differ in any other particular.

The second sort was discovered growing naturally at La Vera Cruz; this has a slenderer stalk than the former, which rises about three feet high. The sheaths which surround it have no hairs, but are channelled. The leaves are shorter than those of the former. The panicle stands erect, and the chaff has shorter awns or beards.

The common Millet was originally brought from the eastern countries, where it is still greatly cultivated, from whence we are furnished annually with this grain, which is by many persons greatly esteemed for puddings, &c. This is seldom cultivated in England, but by way of curiosity in small gardens, or for feeding of poultry, where the seeds generally ripen very well.

They must be sown the beginning of April upon a warm dry soil, but not too thick, because these plants divide into several branches, and should have much room; and when they come up, they should be cleared from weeds, after which they will, in a short time, get the better of them, and prevent their future growth. In August these seeds will ripen, when it must be cut down and beaten out, as is practised for other grain; but when it begins to ripen, if it be not protected from birds they will soon devour it.

MILLEFOLIUM. See Achillea.

MILLERIA. Houst. Gen. Nov. Martyn. Cent. 4. Lin. Gen. Plant. 881.

The Characters are,

This hath a compound flower, included in a naked empalement, cut into three parts, and is permanent. It hath a large three-cornered germen, without down. The germen afterward turns to an oblong, three-cornered, obtuse seed, inclosed in the empalement.

The Species are,

1. Milleria (*Quinqueflora*) foliis cordatis, pedunculis dichotomis. Hort. Cliff. 426. *Milleria with heart-shaped leaves, and foot-stalks arising from the division of the stalks.*

2. Milleria (*Maculata*) foliis infimis cordato-ovatis acutis rugosis, caulinis lanceolato-ovatis, acuminatis. *Milleria whose lower leaves are oval, heart-shaped, acute-pointed, and rough, and the upper ones oval, spear-shaped, and pointed.*

3. Milleria (*Biflora*) foliis ovatis, pedunculis, simplicissimis. Hort. Cliff. 425. *Milleria with oval leaves, and single foot-stalks.*

4. Milleria (*Triflora*) foliis ovato-lanceolatis acuminatis trinerviis, pedunculis alaribus. *Milleria with oval, spear-shaped, acute-pointed leaves, having three veins, and foot-stalks proceeding from the wings of the leaves.*

The first sort was discovered by the late Dr. William Houstoun at Campeachy, in the year 1731, who sent the seeds to Europe; and as the characters which distinguish this genus are different from all the other genera of the class to which it belongs, so he constituted a new genus with this title.

This rises with an herbaceous branching stalk from three to four or five feet high, garnished with large heart-shaped leaves, slightly sawed on their edges, having two veins on each side the midrib, which join to it near the base, but diverge from it toward the borders of the leaves. The leaves are of a light green, hairy, and stand opposite; their foot-stalks are about an inch long, and have a part of their leaf running on each side like wings. The stalks divide by pairs upward, and the foot-stalks of the flowers come out at the divisions; these branch again by pairs, and terminate in loose spikes of yellow flowers, composed of four or five hermaphrodite florets, which are barren, and one female half floret, which is succeeded by a single, oblong, angular seed, wrapped in the empalement of the flower. It flowers in July and August, and the seeds ripen in autumn.

The second sort was discovered by Mr. Robert Millar at Campeachy, in the year 1734; this approaches near to the first sort, but the stalks rise six or seven feet high, and branch out very wide. The leaves are seven inches long, and four inches and a half broad toward their base, ending in long acute points; they are deeper sawed on their edges, and have several large black spots scattered over them; their surface is rougher, and they are of a darker green than those of the first. The upper leaves are long and spear-shaped; the foot-stalks of the flowers branch out wider, and the spikes of flowers are shorter than those of the first.

The third sort was discovered at Campeachy by the late Dr. Houstoun; this is an annual plant, which rises with an herbaceous stalk about two feet high, branching out at a small distance from the root into three or four slender stalks, which are naked almost to the top, where they have two oval spear-shaped leaves, placed opposite, which are about two inches long, and three quarters of an inch broad near their base, ending in points; they are hairy, and stand upon naked foot-stalks, having three longitudinal veins, and are slightly indented on their edges. The flowers come out at the foot stalks of the leaves in small clusters; the common empalement is composed of three orbicular leaves, which are compressed together; in each of these are situated two or three hermaphrodite florets, which are barren, and one female half floret, which is fruitful, being succeeded by a roundish angular seed, inclosed in the empalement. This flowers and perfects seeds about the same time with the former.

The fourth sort was discovered by the late Mr. Robert Millar at Campeachy. This is an annual plant, which rises with an upright stalk three or four feet high, garnished the whole length with oval spear-shaped leaves near four inches long, and almost two broad near their base; they have three longitudinal veins, and toward the top there are two more which diverge from the midrib, but join again at the point. The upper side of the leaves are of a dark green and smooth, their under are of a pale green, and indented on their edges. The flowers grow from the wings of the leaves in small clusters, standing upon short foot-stalks; these have empalements like the former, but are much smaller, in each of which are situated two hermaphrodite florets which are barren, and one female half floret which is fruitful. This flowers and seeds later in the year than either of the former, so that unless the plants are brought forward in the spring, they will not ripen their seeds in England.

The seeds of these plants should be sown early in the spring on a moderate hot-bed; and when the plants are come up about two inches, they should be each transplanted into a separate pot, filled with light rich earth, and then plunged into a moderate hot-bed of tanners bark, being careful to shade them from the sun until they have taken root, as also to water them frequently. After the plants are rooted, they should have a large share of free air admitted to them; they must also be constantly watered every day in hot weather, for they are very thirsty plants. With this management the plants will, in a month after transplanting, rise to a considerable height; therefore they should be shifted into larger pots, and placed in the stove, plunging them into the bark bed where they may have room to grow, especially the first and second sorts, which usually grow high and branch out, where they are well managed. But the other sorts seldom rise above three or four feet high, and do not spread their branches very far, so these may be allowed less room.

In the middle of July these plants will begin to flower, and the seeds will be ripe about a month after; therefore they must be gathered when they begin to change of a dark brown colour, otherwise they will soon fall off especially those of the two large kinds, which will drop on the least touch when they are ripe. These plants will continue flowering till Michaelmas, or later, if the season proves favourable; but when the cold of the autumn comes on, they will soon decay.

MIMOSA. Tourn. Inst. R. H. 605. tab. 375. Lin. Gen. Plant. 597. The Sensitive Plant.

The Characters are,

The empalement of the flower is small, indented in five parts at the top; the flowers are male and hermaphrodite, included in each head; the male flowers have ten, but the other have seldom more than five stamina, which are long and hairy, and a short slender style, crowned by a truncated stigma. The germen afterward turns to a jointed pod with several transverse partitions, inclosing compressed roundish seeds.

The Species are,

1. Mimosa (*Punctata*) inermis, foliis bipinnatis, spicis cernuis, floribus decandris, inferioribus castratis apetalis. Flor. Zeyl. 505. *Smooth Sensitive Plant with double winged leaves, nodding spikes of flowers, having ten stamina, and the lower without either stamina or petals; or the Sensitive Plant of Jamaica.*

2. Mimosa (*Plena*) inermis, foliis bipinnatis, spicarum floribus pentandris, inferioribus plenis. Hort. Upsal. 145. *Smooth Sensitive Plant with double winged leaves, the flowers of the spikes with five stamina, and the under ones double.*

3. Mimosa (*Pernambucana*) inermis decumbens, foliis bipinnatis, spicis cernuis, floribus pentandris, inferioribus castratis. Hort. Upsal. 145. *Smooth Sensitive Plant with inclining stalks, double winged leaves, nodding spikes of flowers having five stamina, but the under ones without any; spurious Sensitive Plant.*

4. Mimosa (*Aculeata*) aculeata, foliis pinnatis, caule procumbente villoso, siliquis articulatis. *Prickly Sensitive Plant with winged leaves, a hairy trailing stalk, and jointed pods.*

5. Mimosa (*Procumbens*) foliis subdigitatis pinnatis, caule aculeato hispido. Lin. Sp. Plant. 518. *Sensitive Plant with winged handed leaves, and a prickly hairy stalk.*

6. Mimosa (*Pudica*) foliis subdigitatis pinnatis, caule aculeato decumbente, siliculis confertis, involucris hispidis. *Sensitive Plant with winged handed leaves, a prickly declining stalk, and small pods growing in clusters, with prickly coverings.*

7. Mimosa (*Quadrivalvis*) aculeata, foliis bipinnatis, caule quadrangulo, aculeis recurvis, leguminibus quadrivalvibus. Lin. Sp. Plant. 522. *Prickly Sensitive Plant with double winged leaves, a four-cornered stalk, recurved spines, and pods having four valves.*

8. Mimosa (*Sensitiva*) foliis conjugatis pinnatis, partialibus bijugis, intimis minimis, caule aculeato. Lin. Sp. Plant. 518. *Sensitive Plant with conjugated winged leaves, whose wings have two pair of lobes, the inner of which are the least, and a prickly stalk.*

9. Mimosa (*Asperata*) caule fruticoso, foliis bipinnatis, aculeatis, aculeis geminis, siliquis radiatis hirsutis. Fig. Plant. tab. 183. fol. 3. *Sensitive Plant with a shrubby stalk, double winged prickly leaves, whose spines grow in pairs, and hairy radiated pods.*

10. Mimosa (*Viva*) caule inermi herbaceo repente, foliis conjugatis pinnatis, floribus globosis alaribus. *Sensitive Plant with a creeping, herbaceous, unarmed stalk, conjugated winged leaves, and globular flowers proceeding from the wings of the stalks.*

11. Mimosa (*Nilotica*) spinis stipularibus patentibus, foliis bipinnatis, partialibus extimis glandula interstinctis, spicis globosis pedunculatis. Hasselq. It. 475. *Acacia with double winged leaves, and globular spikes of flowers having foot-stalks.*

12. Mimosa (*Farnesiana*) spinis singularibus distinctis, foliis bipinnatis, partialibus octojugis, spicis globosis sessilibus. Hort. Upsal. 146. *Indian Acacia with taper resinous pods.*

13. Mimosa (*Cornigera*) spinis stipularibus geminis connatis, foliis bipinnatis. Hort. Cliff. 208. *Acacia with horned spines, and doubly winged leaves.*

14. Mimosa (*Unguis Cati*) spinosa, foliis bigeminis obtusis. Hort. Cliff. 207. *Prickly Acacia with four obtuse lobes and twisted pods.*

15. Mimosa (*Arborea*) inermis, foliis bipinnatis, pinnis dimidiatis acutis, caule arborea. Lin. Sp. 1503. *Tree Acacia without thorns, and doubly winged leaves, whose pinnæ are acute.*

16. Mimosa (*Purpurea*) inermis, foliis conjugatis pinnatis, foliis intimis minoribus. Lin. Sp. 1500. *Purple flowering Acacia without thorns, whose smallest pinnæ are at bottom.*

17. Mimosa (*Houstoniana*) inermis, foliis bipinnatis glabris, pinnis tenuissimis, siliquis latis villosis. Fig. Pl. 5. *Acacia without thorns, doubly winged, smooth, narrow leaves, and broad hairy pods.*

18. Mimosa (*Lutea*) aculeata, foliis bipinnatis glabris, floribus globosis pedunculatis, aculeis longissimis. *Prickly Acacia with smooth doubly winged leaves, globular flowers, and very long spines.*

19. Mimosa (*Glauca*) inermis, foliis bipinnatis, partialibus sejugis, pinnis plurimis, glandula inter infima. Lin. Sp. 1502. *Acacia without thorns, doubly winged leaves, whose wings are separated, and have small glands between them.*

20. Mimosa (*Angustissima*) inermis, foliis bipinnatis, pinnis angustissimis glabris, leguminibus tumidis. *Narrow-leaved unarmed Acacia, with doubly winged leaves and jointed pods.*

21. Mimosa (*Campechiana*) spinosa, foliis bipinnatis, pinnis angustis, spinis singulis cornu bovinum per longitudinem fissum referentibus. *Acacia with doubly winged leaves, having narrow pinnæ, and single spines like ox's horns, split their length.*

22. Mimosa (*Cinerea*) spinis solitariis, foliis bipinnatis, floribus spicatis. Flor. Zeyl. 215. *Acacia with single spines, doubly winged leaves, and spiked flowers.*

23. Mimosa (*Latifolia*) inermis, foliis conjugatis, pinnis terminalibus oppositis, lateralibus alteris. Lin. Sp. 1499. *Broad-leaved Acacia without thorns, and conjugated leaves, whose upper pinnæ are opposite, but those on the sides are alternate.*

24. Mimosa (*Circinalis*) aculeata, foliis conjugatis pinnatis, pinnis æqualibus, stipulis spinosis. Lin. Sp. 1499. *Prickly Acacia with conjugated winged leaves which are equal, and prickly stipulæ.*

The first sort grows naturally in most of the islands in the West-Indies; it has also been found growing in some warm moist spots, as far north as Virginia. This rises with upright branching stalks six or seven feet high, which become ligneous toward the root, but are not perennial (at least they are not so here in any situation, the plants always decaying in winter;) these are smooth, and garnished with double winged leaves, composed of four or five pair of long winged lobes, which have about twenty pair of small leaves ranged along the midrib; they are smooth, and rounded at their points, of a full green on their upper side, but pale on their under. These small leaves contract themselves together on their being touched. but the foot-stalks do not decline at the same time as those do which are titled Humble Plants; therefore this is called the Sensitive Plant simply, by way of distinction. The flowers are produced upon long foot-stalks, which come out from the wings of the leaves, and are disposed in globular heads, which nod downward; they are yellow, and all those which are hermaphrodite have tubulous petals, with ten stamina in each, but the female flowers, situated round the border, have neither petals or stamina; the hermaphrodite flowers are succeeded by pods an inch and a half long, and a quarter of an inch broad, which change to a dark brown when ripe, inclosing three or four compressed, shining, black seeds.

The second sort was discovered by the late Dr. Houstoun at La Vera Cruz, growing in stagnant waters, where the stalks were very broad and flat, and floated on the surface, in the same way as the pond-weeds do in England; but in those places where the water was dried up, the stalks grew upright and were round, which is always the case when the plants are cultivated in gardens, so that they might easily pass for different plants, to those who never saw them growing in both situations. When this sort is cultivated in gardens, it has great resemblance to the first; but the stalks of this never grow so erect, the wings of the leaves are longer, and stand more horizontal; the heads of flowers are much larger, the stamina are longer, and the flowers on the under side of the spike, which have no stamina, are double: the pods of this sort are shorter, and much broader than those of the first sort. This is also an annual plant in this country. It was since discovered by a friend of mine, growing naturally in a marshy spot of land in the island of Barbuda, from whence he sent me the seeds, with a large branch of the plant in a glass filled with a lixivium, which preserved it in the state it was gathered, with the flowers and pods upon it.

The third sort grows naturally in all the islands of the West-Indies, where it is titled the Slothful Sensitive Plant, because the leaves do not contract on their being touched. The stalks of this sort seldom rise more than two feet and a half high; they are smooth, and garnished with double-winged leaves, which are shorter, and the small leaves are much narrower than those of the two former sorts; the heads of flowers are smaller, and the pods are longer and narrower than those of the other. This sort will live through the winter in a moderate warm air.

The fourth sort was discovered by the late Dr. Houstoun growing naturally at La Vera Cruz. This hath ligneous stalks, which decline to the ground, and send out several side branches, which are armed with short yellowish spines under the foot-stalks of the leaves, and are their whole length closely covered with bristly stinging hairs. It hath single winged leaves, whose base meet in a point, but spread above like the fingers of an open hand, closely garnished with small narrow lobes, set by pairs along the midrib. The flowers come out from the wings of the leaves upon pretty long foot-stalks; they are collected into globular heads, and are of a pale yellowish colour; these are succeeded by small jointed pods, containing two or three shining black seeds.

The fifth sort grows naturally at Campeachy. This hath ligneous declining stalks, which are armed with thorns, and covered with stinging bristly hairs; the leaves are composed of four wings, which join at their base, where they are inserted to the foot-stalk, spreading out like the fingers of an open hand: these wings are much shorter than those of the former, and the small leaves or lobes are broader. The flowers come out from the wings of the stalk upon long foot-stalks, growing in oval heads; they are white, and are succeeded by small prickly pods. This is one of those species whose foot-stalks fall upon being touched.

The

The sixth sort is the most common of any in the islands of the West-Indies, as also in the English gardens; the seeds of this sort are frequently sold in the seed shops, by the title of Humble Plant. The roots of this are composed of a great number of hairy fibres, which mat close together, from which come out several ligneous stalks, which naturally decline toward the ground unless they are supported; they are armed with short recurved spines, and garnished with winged leaves, composed of four, and sometimes five wings, whose base join at a point, where they are inserted to the foot-stalk, spreading upward like the fingers of a hand: these wings are shorter than those of the former sort, and the stalks are not hairy. The flowers come out from the wings of the stalks upon short foot-stalks; they are collected in small globular heads, are yellow, and are succeeded by short, flat, jointed pods, which have two or three orbicular, bordered, compressed seeds in each: these pods are in close clusters, almost covered with stinging hairy covers.

The seventh sort grows naturally at La Vera Cruz. This hath a perennial creeping root, which spreads and multiplies greatly in the sands, where it grows wild; the stalks are slender, and have four acute angles, armed with short recurved spines pretty closely; the leaves stand upon long prickly foot-stalks, which are thinly placed on the branches; they are composed of two pair of wings, standing about an inch asunder; the wings are short, and the small leaves are narrow, and not placed so close together as in many of the other species. The foot-stalks of the flowers come out from the wings of the leaves, sustaining a small globular head of purple flowers; these are succeeded by four-cornered pods about two inches long, which have four cells, opening with four valves, containing several angular seeds in each.

This sort spreads so much at the root, as to render it not so productive of flowers and seeds as most of the others; and the plants which are propagated by parting of the roots, are always weak, so that the best way is to propagate them by seeds, when they can be obtained. This is one of the sorts whose foot-stalks fall on being touched.

The eighth sort grows naturally at La Vera Cruz. This rises with a slender ligneous stalk seven or eight feet high, armed with short recurved thorns. The leaves grow upon long foot-stalks, which are prickly, each sustaining two pair of wings; the exterior pair have two lobes, which join at their base, and are rounded on their outside, but strait on their inner edges, very much shaped like a pair of those shears, used for shearing of sheep; these two outer pair of lobes are much larger than the inner. From the place where these are inserted to the stalk come out small branches, which have three or four globular heads of pale purple flowers upon short foot-stalks, and the principal stalk has many of these heads of flowers on the upper part for more than a foot in length; and this, as also the branches, are terminated by the like heads of flowers, which are succeeded by broad, flat, jointed pods, which open with two valves, some having but one, others two, and some have three orbicular compressed seeds. The leaves of this sort move but slowly when they are touched, but the foot-stalks fall when they are pressed pretty hard.

The ninth sort was also found growing naturally at La Vera Cruz. This hath a shrubby erect stalk about five feet high, which is hairy, and armed with short, broad, strong, white thorns, standing on each side almost opposite, and at others alternately. The leaves are composed of five or six pair of wings, which are ranged opposite along a strong midrib, and between each pair are placed two short strong spines, pointing out each way. The small leaves which compose these wings are extremely narrow, and stand very close to each other. Toward the upper part of the stalk, the flowers are produced from the sides upon short foot-stalks; they are collected into globular heads, and are of a bright purple colour; the stalks are also terminated by smaller heads of the like flowers. These are succeeded by flat jointed pods about two inches long, and a quarter of an inch broad, which spread open like rays, there being commonly five or six of these joined together at their base to the foot-stalk. These pods separate at each articulation, leaving the two side membranes or borders standing, and the seeds which are compressed and square, drop out from the joints of the pods; these pods are hairy at first, but as they ripen become smooth.

This is a perennial plant, which may be preserved through the winter in a warm stove, by which method the seeds may be obtained, for they seldom flower the first year. The foot-stalks of this sort do not fall on being touched, but the small leaves on the wings close up.

The tenth sort grows naturally in Jamaica. This hath trailing herbaceous stalks, which put out roots at every joint, which fasten in the ground, and spread to a great distance; as they will also do here, when placed in a bed of tanners bark. I have had a single plant in one summer, which has spread near three feet square, whose branches were closely joined, so as to cover the surface of the bed; but when they are thus permitted to grow, they seldom produce flowers. These stalks have no thorns, but are garnished with winged leaves, composed of two pair of short wings, whose small leaves or lobes are narrow; these stand upon short foot-stalks, which are smooth. The leaves of this sort contract and fall down upon the least touch, so, that where the plant is extended to a distance, a person may draw any figure with a stick upon the leaves, which will be very visible till the leaves recover again. The flowers come out from the wings of the leaves, upon naked foot-stalks about an inch in length; they are of a pale yellowish colour, and are collected into small globular heads; these are succeeded by short, flat, jointed pods, containing three or four compressed roundish seeds.

These plants are all of them propagated by seeds, which should be sown early in the spring upon a good hot-bed. If the seeds are good, the plants will appear in a fortnight or three weeks, when they will require to be treated with care, for they must not have much wet till they have acquired strength: nor should they be drawn too weak, so that fresh air should be admitted to them, at all times when the weather is temperate. In about a fortnight or three weeks after the plants come up, they will be fit to transplant, especially if the bed, in which they were sown, continues in a proper degree of heat; then there should be a fresh hot-bed prepared to receive them, which should be made a week before the plants are removed into it, that the violent heat may be abated before the earth is laid upon the dung, and the earth should have time to warm before the plants are planted into it. Then the plants must be carefully raised up from the bed to preserve the roots entire, and immediately planted in the new bed, at about three or four inches distance, pressing the earth gently to their roots; then they should be gently sprinkled over with water, to settle the earth to their roots; after this they must be shaded from the sun till they have taken new root, and the glasses of the hot-bed should be covered every night, to keep up the heat of the bed. When the plants are established in their new bed, they must have frequent but gentle waterings; and every day they must have free air admitted to them, in proportion to the warmth of the season, to prevent their being drawn up weak; but they must be constantly kept in a moderate degree of heat, otherwise they will not thrive. In about a month after the plants will be ſtrong enough to remove again, when they ſhould be carefully taken up, preſerving as much earth to their roots as poſſible, and each planted in a ſeparate ſmall pot, filled with good kitchen-garden earth, and plunged into a hot-bed of tan, carefully ſhading them from the ſun till they have taken new root; then they muſt be treated in the ſame manner as other tender exotic plants from very warm countries.

The ſorts which grow upright and tall, will ſoon riſe high enough to reach the glaſſes of the hot bed, eſpecially if they thrive well; therefore they ſhould be ſhifted into larger pots, and removed into the ſtove, and if they are plunged into the tan-bed there, it will greatly forward them. The firſt ſort will often flower here, if the plants are raiſed early in the ſpring, and brought forward by their removal from one hot-bed to another; and two or three times I have had their ſeeds ripen, but this can only be expected in very warm ſeaſons.

The perennial ſorts will live through the winter, if they are preſerved in a warm ſtove, and the following ſummer they will produce flowers and ripen thei ſeeds. Some of theſe may be propagated by laying down their branches, which will put out roots, and then may be ſeparated from the old plants; and I have ſometimes propagated them by cuttings, but the plants which riſe from ſeeds are preferable to either of theſe.

There is no particular management which theſe plants require, different from others of the ſame warm countries; the great care muſt be to keep them in a proper temperature of heat and not to give them too much water, eſpecially in cool weather; nor ſhould they be kept too dry, for many of the ſorts require frequent waterings, as they naturally grow in moiſt places. There ſhould alſo be care taken that they do not root into the tan-bed, for they ſoon put out their roots through the holes at the bottom of the pots, which, when they ſtrike into the tan, will cauſe the plants to grow very luxuriant; but when they are removed, and theſe roots are cut or broken off, the plants ſeldom ſurvive it; therefore the pots ſhould be frequently drawn out of the tan, and if any of the roots are beginning to get through the holes at bottom, they ſhould be cut off cloſe, to prevent their getting into the tan.

Some of thoſe ſorts, whoſe ſtalks ſpread near the ground, may be turned out of the pots in the middle or latter end of June, and planted in a very warm border, where, if they are covered with bell or hand-glaſſes, they will live through the ſummer; but theſe will not grow very large, and upon the approach of cold in autumn they are ſoon deſtroyed; however, thoſe who have not conveniency of ſtoves or tan-beds, may raiſe the plants on common hot beds in the ſpring; and when they have acquired ſtrength, they may be treated in this manner, whereby they will have the pleaſure of theſe plants in ſummer, though not in ſo great perfection as thoſe which have the advantages before-mentioned: but when theſe plants are expoſed to the open air in this country, they will not retain their ſenſibility on being touched.

It is not the light which cauſes them to expand, as ſome have affirmed, who have had no experience of theſe things; for in the longeſt days of ſummer they are generally contracted by five or ſix in the evening, when the ſun remains above the horizon two or three hours longer; and although the glaſſes of the ſtove in which they are placed is covered cloſe with ſhutters to exclude the light in the middle of the day, yet if the air of the ſtove is warm, the leaves of the plants will continue fully expanded, as I have ſeveral times obſerved. Nor do theſe plants continue ſhut till the ſun riſes in the morning, for I have frequently found their leaves fully expanded by the break of day in the morning; ſo that it is plain the light is not the cauſe of their expanſion, nor the want of it that of their contraction.

I have alſo obſerved, that thoſe plants which are placed in the greateſt warmth in winter, continue vigorous, and retain their faculty of contracting on being touched; but thoſe which are in a moderate warmth, have little or no motion.

Some of the ſorts are ſo ſuſceptible of the touch, that the ſmalleſt drop of water falling on their leaves will cauſe them to contract, but others do not move without a much greater preſſure.

The roots of all the ſorts have a very ſtrong diſagreeable odour, almoſt like that of a common ſewer. I have met with ſome accounts of theſe plants, in which it is mentioned, that the leaves and branches have a poiſonous quality; and that the Indians extract a poiſon from them, which kills by ſlow degrees, and that the root of the plant is the only remedy to expel it; but how far this is true I cannot ſay, having never made any experiments on the qualities of theſe plants; but if theſe plants are endued with ſo deadly a quality as related, this ſenſibility in which they are endued, may be deſigned by Providence, to caution perſons from being too free with them; and as many of them are ſtrongly armed with thorns, ſo that is a guard againſt their being eaten by animals; for in all the enquiries which I have made of thoſe perſons who have reſided in the countries where they naturally grow, I could never learn that any animal will browſe upon them.

Theſe plants are all of them natives of America, ſo were unknown to the other parts of the world till that was diſcovered, for I have not heard of any of them being found in any other country; and a few years ago I ſent ſome of the ſeeds of theſe plants to China, which ſucceeded, and occaſioned great admiration in all who ſaw the plants.

The above mentioned ſorts are all of the kind of true Mimoſa, or Senſitive Plants; but as the True Acacia's agrees in their characters with theſe, ſo Dr. Linnæus has joined them together in one genus; therefore I have done the ſame here, as his ſyſtem is now generally adopted.

The eleventh ſort is the tree from whence the True Succus Acaciæ is taken, and the Gum Arabic exſudes from the branches of the ſame: the tree has been generally ſuppoſed to grow only in Egypt, but it has been found growing in ſeveral of the warm parts of America, and abundantly at Senegal in Africa, from whence I have received the ſeeds; and there has been great quantities of the juice imported into England.

This ſort grows to be a pretty large tree in the warm countries, but in England they make little progreſs; it may be propagated by ſeeds, when they are obtained freſh from abroad, which ſhould be ſown in pots, and plunged into a hot-bed of tanners bark; and when the plants are come up, and have obtained ſtrength, they ſhould be each planted into a ſeparate ſmall pot, and plunged into the tan-bed again, obſerving to ſhade them from the ſun, and ſupply them gently with water until they have taken new root; after which they ſhould be treated as other tender plants from warm countries, letting them always remain in the bark-ſtove

The twelfth ſort is much cultivated in the hot countries, for the ſake of their flowers, which have an agreeable odour; ſo that in the warmeſt parts of Italy, Spain, and Portugal, it is cultivated in great plenty, and young plants are annually brought to England by thoſe perſons who import Orange-trees, Jaſmines, &c. in the ſpring; but there are few of them which ſurvive their long paſſage. But as the ſeeds of this ſort are frequently brought from the warm iſlands in America, ſo from thoſe the plants are eaſily raiſed, if managed in the ſame manner as is directed for the eleventh ſort:

ſort; as may all the other ſorts, when freſh ſeeds can be obtained, as they all require to be placed in a ſtove, and not expoſed to the open air, unleſs in very warm weather, for a ſhort time in the middle of ſummer.

MIMULUS. Lin. Gen. Plant. 701. Cynorrhynetrium. Mitch. 3.

The CHARACTERS are,

The flower hath an oblong permanent empalement; it is of the lip or ringent kind, whoſe brim is divided into two lips. The upper lip is erect, divided at the top in two parts; the lower lip is broad and trifid, the middle ſegment is the leaſt; the palate is convex and bifid. It has four ſlender ſtamina, two longer than the other, and a conical germen, which afterward turns to an oval capſule with two cells, filled with ſmall ſeeds.

We have but one SPECIES of this genus, viz.

MIMULUS. Hort. Upſal. 176. tab. 2.

This plant grows naturally in North America, in moiſt ground. It has a perennial root, and an annual ſtalk, which is ſquare, and riſes a foot and a half high, garniſhed at each joint with two oblong ſmooth leaves. The lower part of the ſtalk ſends out two or three ſhort branches, and the upper part is adorned with two flowers at each joint, coming from the boſom of the leaves on each ſide the ſtalk; theſe have an oblong curved empalement with five angles, indented at the top in five parts, out of which ariſes the flower, with a long curved tube, ſpreading open at the top into two lips; the upper lip ſtanding erect, which is ſlightly cut into two parts at the top; the under lip turns downward, and is cut into three ſlight ſegments. The flowers are of a Violet colour, but have no ſcent. Theſe appear in July, and are ſucceeded by oblong capſules with two cells, filled with ſmall ſeeds, which ripen in the autumn.

This plant is very hardy in reſpect to cold, but ſhould have a loamy ſoft ſoil, rather moiſt than dry, and not too much expoſed to the ſun. It may be propagated by parting of the roots in autumn, but they ſhould not be divided too ſmall; it may alſo be propagated by ſeeds, which ſhould be ſown in autumn ſoon after they are ripe, for thoſe which are ſown in the ſpring, ſeldom grow the ſame year; theſe may be ſown on a border expoſed to the morning ſun.

MINT. See MENTHA.

MIRABILIS. Lin. Gen. Plant. 215. Marvel of Peru, or Four o'Clock Flower.

The CHARACTERS are,

The empalement of the flower has five oval ſmall leaves. The flower has one funnel-ſhaped petal, with a long ſlender tube ſitting upon the nectarium. It hath five ſlender unequal ſtamina, which adhere to the petal, with a roundiſh germen within the nectarium, which afterward becomes an oval five-cornered nut, incloſing one ſeed.

The SPECIES are,

1. MIRABILIS (*Jalapa*) floribus congeſtis terminalibus erectis. *Marvel of Peru with an erect ſtalk, and large flowers.*

2. MIRABILIS (*Dichotoma*) floribus ſeſſilibus axillaribus erectis ſolitariis. Amœn. Acad. 4. p. 267. *Marvel of Peru, with an erect ſingle flower ſitting cloſe to the branches.*

3. MIRABILIS (*Longiflora*) floribus congeſtis terminalibus longiſſimis nutantibus, foliis ſubvilloſis. Act. Holmenſ 1756. *Long flowered Marvel of Peru, whoſe flowers terminate the ſtalks in bunches, which nod toward the ground and have hairy leaves.*

The firſt ſort is the Marvel of Peru, which has been many years cultivated in the Engliſh gardens for ornament; of this there are ſeveral varieties, which differ in the colour of their flowers, two of which always retains their difference; one of them has purple and white flowers, which are variable, ſome of them are plain purple, others are plain white, but moſt of them are variegated with the two colours; and all theſe varieties are ſometimes upon the ſame plant, and at others on different plants; the other has red and yellow flowers, which are generally mixed in the ſame flowers, but are often with plain flowers of both colours on the ſame plant, intermixed with thoſe which are variegated; but ſome plants have only plain flowers, and I have never found that the ſeeds of the purple and white ſort ever produced the yellow and red, nor the latter ever vary to the former, and I have conſtantly cultivated both more than fifty years; but although theſe do not change from one to the other, yet as there is no other difference between them than in the colour of their flowers, I have not enumerated them as diſtinct ſpecies.

The ſecond ſort is very common in all the iſlands of the Weſt-Indies, where the inhabitants call it the Four o'Clock Flower, from the flower opening at that time of the day. Of this ſort I have never ſeen any with variable flowers; they are of a purpliſh red colour, and not much more than half the ſize of the other. The ſtalks of this ſort have thick ſwollen joints; the leaves are ſmaller, and the fruit is very rough, ſo there can be no doubt of their being diſtinct ſpecies, for I have never ſeen any alteration in this from ſeed, and I have cultivated it more than forty years. Tournefort was informed by Father Plumier, that the root of this plant was the officinal Jalap, upon which he conſtituted the genus, and gave that title to it; but the late Dr. Houſtoun was fully informed in the Spaniſh Weſt-Indies of the contrary, and brought over a drawing of the plant which was made by a Spaniard at Halapa, and he carried two or three of the plants to Jamaica, where he planted them in a garden, but after he left the iſland they were deſtroyed by hogs; however, he was fully ſatisfied of its being a Convolvulus; indeed the roots of Marvel of Peru are purgative, and when given in a double quantity for a doſe, will anſwer the purpoſe of Jalap.

The third ſort was ſent from Mexico a few years ſince. The ſeeds of this were firſt ſent me from Paris by Dr Monier, of the Royal Academy of Sciences, and afterward I had ſome ſent me from Madrid by Dr. Hortega. The ſtalks of this ſort fall on the ground, if they are not ſupported; theſe grow about three feet long, and divide into ſeveral branches, which are garniſhed with heart-ſhaped leaves, placed oppoſite; theſe, as alſo the ſtalks, are hairy and viſcous, ſticking to the fingers of thoſe that handle them. The flowers come out at the end of the branches; they are white, and have very long ſlender tubes, and a faint muſky odour; theſe are like the other ſorts, cloſely ſhut all the day, but expand every evening when the ſun declines. The ſeeds of this ſort are larger than thoſe of any other ſpecies, and are as rough as thoſe of the ſecond ſort.

The two varieties of the firſt ſort are very ornamental plants in gardens, during the months of July, Auguſt, and September; and if the ſeaſon continues mild, they often laſt till near the end of October. The flowers do not open till toward the evening, while the weather continues warm, but in moderate cool weather, when the ſun is obſcured, they continue open almoſt the whole day. The flowers are ſo plentifully produced at the ends of the branches, as that when they are open the plants ſeem entirely covered with them; and there being ſome plain, and others variegated on the ſame plants, they make a fine appearance.

Theſe plants are propagated by ſeeds, in the choice of which there ſhould be care taken not to ſave any from thoſe plants, whoſe flowers are plain; and thoſe who are deſirous of having only the variegated kinds, are careful to pull off all the plain flowers from thoſe plants which they

they intend for seeds, to prevent their producing any seeds; by this method they rarely have any plants with plain flowers.

The seeds should be sown upon a moderate hot-bed in March, and when the plants come up they should have plenty of air admitted to them when the weather is mild, to prevent their being drawn up weak; and when they are about two inches high, they should be transplanted on another very moderate hot-bed; or if they are each planted in a small pot filled with light earth, and plunged into a moderate hot-bed, it will be a more secure way, for then there will be no danger of shaking them out of the pots, when they are to be planted in the borders, so as to preserve all the earth to their roots; by this method they will not require to be shaded, whereas those which are to be transplanted from the second hot-bed to the borders, often rise with little earth to their roots, so must be carefully shaded, otherwise they often miscarry.

When they are in the second hot-bed, they should be shaded till they have taken fresh root, after which they must have plenty of free air, to prevent their being drawn up weak, and in May they must be gradually inured to bear the open air. The beginning of June, if the season is favourable, they should be transplanted into the borders of the pleasure-garden, giving them proper room; and after they have taken new root they will require no further care. If the seeds are sown in a warm border they will grow very well, but the plants will be late in the season before they flower.

As the seeds of these plants ripen very well every year, so there are not many who are at the trouble of preserving their roots; but if these are taken out of the ground in autumn, and laid in dry sand all the winter, secured from frost, and planted again in the spring, they will grow much larger, and flower earlier than the seedling plants; or if the roots are covered in winter with tanners bark, to keep out the frost, they may remain in the borders, provided the soil be dry. If the roots which are taken out of the ground are planted the following spring in large pots, and plunged into a hot bed under a deep frame, they may be brought forward, and raised to the height of four or five feet, as I have frequently practised; and these plants have come early in the season to flower, so have been intermixed with other ornamental plants to decorate halls and shady courts, where they have appeared very beautiful.

The other two species require the same treatment, but the second sort is not quite so hardy as the other two, so unless the plants are brought forward in the spring, they will not flower till very late, so their seeds will not ripen.

MISLETOE. See Viscum.

MITELLA. Tourn. Inst. R. H. 241. tab. 126. Lin. Gen. Plant. 496. Bastard American Sanicle.

The Characters are,

The flower has a bell-shaped empalement of one leaf. It hath five petals which are inserted in the empalement, as are also the ten awl-shaped stamina, which are shorter than the petals. It hath a roundish bifid germen, with scarce any style. The empalement afterward becomes an oval capsule with one cell, opening with two valves, filled with small seeds.

The Species are,

1. Mitella (*Diphylla*) scapo diphyllo. Lin. Gen. Nov. 79. *Mitella with flower-stalks having two leaves.*

2. Mitella (*Nuda*) scapo nudo. Amœn. Acad. 2. p. 252. *Mitella with a naked stalk.*

The first sort grows naturally in the woods in most parts of North America. It hath a perennial root, from which come out many heart-shaped angular leaves, some of which are obtuse, and others end in acute points; they are indented on their edges, of a lucid green, and stand upon pretty long foot-stalks. The flower-stalks arise immediately from the root, having two or three angular leaves toward the bottom, and about the middle of the stalk come out two small leaves with acute angles, placed opposite. The stalks rise eight or nine inches high, and are terminated by a loose spike of small whitish flowers, whose petals are fringed on their edges. These appear the beginning of June, and are succeeded by roundish capsules, filled with small seeds.

The second sort grows naturally in the northern parts of Asia. This is of humbler growth than the first, seldom rising more than five or six inches high. The leaves are not so angular as those of the first sort, and the flower-stalks are always naked, having no leaves. The spikes of flowers are shorter, and more compact.

Both these are propagated by parting of their roots; the best time for this is in autumn; they should be planted in a shady situation, and a soft loamy soil.

MITELLA MAXIMA. See Bixa.

MOLDAVICA. See Dracocephalum.

MOLLE. See Schinus.

MOLLUGO. Lin. Gen. Plant. 99.

The Characters are,

The empalement of the flower is composed of five oblong small leaves, coloured on their inside, and permanent. The flower has five oval petals shorter than the empalement, and three bristly stamina, which stand near the style, terminated by single summits, with an oval germen having three furrows, supporting three very short styles. The germen afterward becomes an oval capsule with three cells, filled with small kidney-shaped seeds.

The Species are,

1. Mollugo (*Verticillata*) foliis verticillatis cuneiformibus acutis, caule subdiviso decumbente, pedunculis unifloris. Hort. Upsal. 24. *Mollugo with acute wedge-shaped leaves growing in whorls, a trailing divided stalk, and foot-stalks bearing a single flower.*

2. Mollugo (*Quadrifolia*) foliis quaternis obovatis, paniculâ dichotomâ. Hort. Cliff. 28. *Mollugo with four leaves at each joint, which are almost oval, and a panicle arising at the division of the branches.*

There are two or three other species of this genus which are rarely admitted into gardens, so I have not enumerated them here.

Both these sorts are annual; the first is a native of warm countries, so is less hardy than the second; they are both trailing plants, whose stalks lie on the ground; the first spreads out eight or nine inches every way, and at each joint is garnished with six or seven small leaves spread out in form of a star. The flowers are small, like those of Chickweed, one standing upon each foot-stalk; these are succeeded by oval capsules filled with small seeds, which, if permitted to scatter, the plants will come up the following spring without any care; but when the seeds happen to fall upon earth which is thrown upon a hot bed, the plants will be forwarder and stronger than those in the open air. This is preserved in some gardens for the sake of variety, but has no great beauty.

The other sort has been already mentioned under the article Herniaria, but being ranged in this genus by Dr. Linnæus, I have enumeeated it here.

MOLUCCELLA. Lin. Gen. Plant. 643. Molucca Balm.

The Characters are,

The flower hath a large permanent empalement of one leaf, which is deeply indented at the brim, where it spreads open. The flower is of the lip kind, with a short tube The upper lip is erect, concave, and entire. The under lip is trifid, the middle segment being longer than the other. It has four stamina situated under the upper lip, two of which are shorter than the other;

other; and a germen with four parts, which afterward turns to four angular convex seeds, sitting in the empalement.

The SPECIES are,

1. MOLUCCELLA (*Lævis*) calycibus quinquedentatis, denticulis æqualibus. Prod. Leyd. 314. *Molucca Balm with empalements indented in five equal parts; or smooth Molucca Balm.*

2. MOLUCCELLA (*Spinosa*) calycibus septemdentatis. Prod. Leyd. 314. *Molucca Balm whose empalements are indented in seven parts; or prickly Molucca Balm.*

The first sort rises with a square stalk three feet high, spreading out into many smooth branches by pairs, garnished with roundish leaves set opposite, which are deeply notched on their edges, standing upon long foot-stalks; they are smooth, of a light green on both sides, and at the base of their foot-stalks the flowers come out in whorls; these have very large spreading empalements, which are indented in five parts. The flowers are small, and being situated at the bottom of the large empalements, are not visible at a distance; they are white, with a cast of purple, and shaped like those of the other lip flowers, having the the upper lip entire, and hollowed like a spoon, and the under lip is cut into three segments, the middle one being the longest. After the flower is past, the germen turns to four club-shaped angular seeds inclosed in the empalement. It flowers in July, but unless the season proves warm and dry, the seeds do not ripen in England. The smell of this plant is to some persons very disagreeable, and to others very pleasant.

The second sort hath square smooth stalks, of a purplish colour, which do not rise so high as those of the former, but branch out in the same manner. The leaves are smaller, and stand upon shorter foot-stalks; they are deeper, and more acutely indented on their edges. The empalements of the flowers are not so large, and cut into seven segments, each being terminated by an acute spine. The flowers are like those of the former species, as are also the seeds; this is not so hardy as the first sort.

The first grows naturally in several parts of Syria, and the second is a native of the Molucca Islands, from whence this genus received its title. They are both annual plants, which decay soon after their seeds are ripe, and being natives of warm countries they seldom perfect their seeds in England, when they are sown in the spring; therefore the best way is to raise the plants in autumn, and plant them in small pots; these should be placed under a hot-bed frame in winter, where they may have free air in mild weather, by taking off the glasses, but covered in frosty weather, observing to keep them pretty dry, otherwise they are very subject to rot, especially when they are closely covered in frosty weather. In the spring the plants may be turned out of the pots, with all the earth about their roots, and planted in a warm border, defended from strong winds, giving them a little water to settle the earth to their roots; after this they will require no other care, but to keep them clean from weeds, and to support them with stakes, to prevent their being broken by the winds. The plants thus preserved through the winter, will flower the latter end of June, so from these good seeds may be expected.

MOLY. See ALLIUM.

MOMORDICA. Tourn. Inst. R. H. 103. tab. 29, 30. Lin. Gen. Plant. 967. Male Balsam Apple.

The CHARACTERS are,

It hath male and female flowers upon the same plants. The male flowers have an open concave empalement of one leaf. It has three short awl-shaped stamina, which are compressed in a body, and has a reflexed line containing the farina. The female flowers have the same empalement and petal as the male, but sit upon the germen; these have three short filaments without summits. The germen supports one taper trifid style, crowned by three oblong gibbous stigmas, and afterward turns to an oblong fruit, opening with an elasticity, having three membranaceous cells, filled with compressed seeds.

The SPECIES are,

1. MOMORDICA (*Balsamina*) pomis angulatis tuberculatis, foliis glabris patenti-palmatis. Hort. Cliff. 451. *Male Balsam Apple with angular warted fruit, and smooth open-handed leaves.*

2. MOMORDICA (*Charantia*) pomis angulatis tuberculatis, foliis villosis, longitudinaliter palmatis. Hort. Cliff. 451. *Male Balsam Apple with angular warted fruit, and hairy leaves, which are longitudinally hand-shaped.*

3. MOMORDICA (*Zeylanica*) pomis ovatis acuminatis tuberculatis, foliis glabris palmatis serratis. *Male Balsam Apple with an oval, acute-pointed, warted fruit, and smooth hand-shaped leaves which are sawed.*

4. MOMORDICA (*Elaterium*) pomis hispidis, cirrhis nullis. Lin. Sp. Plant. 1010. *Male Balsam Apple with a prickly fruit, and no tendrils to the vines; or wild Cucumber, and Elaterium of Boerhaave.*

The first sort grows naturally in Asia, the second and third in the island of Ceylon; they are all annual plants, which perish soon after they have ripened their fruit; these have trailing stalks like those of the Cucumber and Melon, which extend four or five feet in length, sending out many side branches with tendrils, by which they fasten themselves to any neighbouring plants, to secure themselves from being tost, and blown about by the winds, and are garnished with leaves shaped like those of the Vine. The leaves of the first and third sorts are smooth, and deeply cut into several segments, and are spread open like a hand; but those of the second sort are extended more in length, and are hairy. The fruit of the first species is oval, ending in acute points, having several deep angles, which have sharp tubercles placed on their edges; it changes to a red or purplish colour when ripe, opening with an elasticity, and throwing out its seeds.

The fruit of the second sort is much longer than that of the first, and not so deeply channelled. The tubercles are scattered all over the surface, and are not sharp like those of the other; this fruit is yellow when ripe, and casts out its seeds with an elasticity.

The fruit of the third sort is short and pointed like that of the first, but does not swell so large in the middle. The angles of this are not deep, and the whole surface is closely set with large tubercles; this changes to a deep Orange colour when ripe, and casts out its seeds in the like manner.

The fourth sort is commonly called the Wild or Spurting Cucumber, from its casting out its seeds, together with the viscid juice in which the seeds are lodged, with a violent force, if touched when ripe; and from hence it has sometimes the appellation of Noli me tangere, or Touch me not. This plant grows naturally in some of the warm parts of Europe, but in England it is cultivated in gardens for the fruit, which is used in medicine, or rather the fæcula of the juice of the fruit, which is the Elaterium of the shops.

This plant hath a large fleshy root, somewhat like that of Briony, from which come forth every spring several thick, rough, trailing stalks, which divide into many branches, and extend every way two or three feet; these are garnished with thick, rough, almost heart-shaped leaves, of a gray colour, standing upon long foot-stalks. The flowers come out from the wings of the stalk; these are male and female, growing at different places on the same plant, like those of the common Cucumber, but they are much less, of a pale yellow colour, with a greenish bottom;

tom; the male flowers ſtand upon ſhort thick foot ſtalks, but the female flowers ſit upon the young fruit, which, after the flower is faded, grows of an oval form, an inch and a half long, ſwelling like a Cucumber, of a gray colour like the leaves, and covered over with ſhort prickles. Theſe do not change their colour when ripe, like moſt of the other fruit of this claſs, but if attempted to be gathered, they quit the foot-ſtalk, and caſt out the ſeeds and juice with great violence; ſo that where any plants are growing, and the fruit permitted to ſtand till it is ripe, the ſeeds will be ſcattered all round to a great diſtance, and there will be plenty of the plants produced the following ſpring.

But when the fruit is deſigned for uſe, it ſhould always be gathered before it is ripe, otherwiſe the greateſt part of the juice will be loſt, which is the only valuable part; for the juice which is expreſſed with part of the parenchyma of the fruit, is not to be compared with the other for its virtues; for the Elaterium which is made from the clear juice of the fruit is much whiter, and will retain its virtues much longer than that which is extracted from preſſure.

The three firſt ſorts are annual; their ſeeds muſt be ſown on a hot-bed the beginning of March, and when the plants come up they ſhould be tranſplanted out into a freſh hot-bed, after the manner of Cucumbers or Melons, putting two plants of the ſame kind under each light, and the plants watered and ſhaded until they have taken root; after which they muſt be treated as Cucumbers, permitting their branches to extend upon the ground in the ſame manner, and obſerve to keep them clear from weeds.

With this management (provided you do not let them have too much wet, or expoſe them too much to the open air) they will produce their fruit in July, and their ſeeds will ripen in Auguſt and September, when you muſt obſerve to gather it as ſoon as you ſee the fruit open, otherwiſe it will be caſt abroad, and with difficulty gathered up again.

Theſe plants are preſerved in curious gardens for the oddneſs of their fruit; but as they take up a great deal of room in the hot-beds, requiring frequent attendance, and being of little beauty or uſe, ſo they are not much cultivated in England, except in botanic gardens for variety.

There are ſome perſons who put theſe plants in pots, and faſten them to ſtakes, to ſupport the vines from trailing on the ground, and place the pots in ſtoves; where, when they are ſkilfully managed, they will produce their fruit tolerably well; and in this way they make a better appearance, than when the vines ſpread on the ground like Cucumbers and Melons. But when the plants ſpread on the ground, which is their natural way of growing, they thrive much better, and produce more fruit, than when they are ſupported; for though theſe plants have claſpers, yet theſe are not formed for climbing, but merely to faſten themſelves about any neighbouring ſupport, to ſecure them from being raiſed by the wind and broken, which would often happen where they grow in the open air and are fully expoſed, were it not for this ſecurity.

The fourth ſort is eaſily propagated by ſeeds, which (as was before mentioned) if permitted to ſcatter, there will be a ſupply of plants come up the following ſpring; or if the ſeeds are ſown upon a bed of light earth, the plants will come up in about a month after, and may be tranſplanted to an open ſpot of ground, in rows at three or four feet diſtance, and almoſt as far aſunder in the rows; if theſe are carefully tranſplanted while young, there will be little hazard of their growing; and after they have taken new root, they will require no farther care but to keep them clear from weeds. If the ground is dry in which theſe are planted, the roots will continue three or four years, unleſs the winter ſhould prove very ſevere, which will ſometimes kill them.

MONARDA. Lin. Gen. Plant. 34.

The CHARACTERS are,

The flower has a cylindrical empalement of one leaf, which is cut into five equal parts at the brim. The flower hath one petal, and is of the lip kind, divided at the top into two lips. The upper lip is narrow, entire, and erect; the under lip is broad, trifid, and reflexed. It hath two briſtly ſtamina the length of the upper lip. In the bottom of the tube is ſituated a four-pointed germen, ſupporting a ſlender ſtyle involved with the ſtamina, and crowned by an acute bifid ſtigma. The germen afterward turns to four naked ſeeds, incloſed in the empalement.

The SPECIES are,

1. MONARDA (*Fiſtuloſa*) capitulis terminalibus, caule obtuſangulo. Hort. Upſal. 12. *Monarda with heads of flowers terminating the ſtalks, which have obtuſe angles.*

2. MONARDA (*Didyma*) floribus capitatis ſub-didynamis, caule acutangulo. Lin. Sp. Pl. *Monarda with headed flowers, whoſe ſtamina are almoſt in two bodies, and an acute angular ſtalk.*

3. MONARDA (*Punctata*) floribus verticillatis corollis punctatis. Hort. Upſal. 12. *Monarda with flowers growing in whorls, whoſe petals are ſpotted.*

The firſt ſort grows naturally in Canada, and many other parts of North America. It hath a perennial root. The ſtalks riſe near three feet high, which are hairy, and have obtuſe angles; theſe ſend out two or four ſmall ſide branches toward the top, which, as alſo the principal ſtalk, are garniſhed with oblong leaves, broad at their baſe, but terminate in acute points; they are hairy, a little indented on their edges, and are placed by pairs. The ſtalk and branches are terminated by heads of purple flowers, which have a large involucrum, compoſed of five acute-pointed leaves. The flowers have each two ſtamina which are longer than the petal, with a ſtyle of the ſame length. The flowers appear in July, and are ſucceeded by ſeeds which ripen in autumn.

The ſecond ſort alſo grows naturally in North America, where the inhabitants frequently uſe the leaves for tea, ſo it is commonly called Oſwego Tea, by which title it was brought to England. This hath a perennial root. The ſtalks of this ſort are ſmooth, having four acute angles; they riſe about two feet high, and are garniſhed with ſmooth, oval, ſpear ſhaped leaves, which are indented on their edges, and ſtand oppoſite; theſe, when bruiſed, emit a very grateful refreſhing odour; the ſtalks ſend out toward their top two or four ſmall ſide branches, which are garniſhed with ſmall leaves of the ſame ſhape with the other. The flowers are produced in large heads or whorls at the top of the ſtalk, and there is often a ſmaller whorl of flowers, growing round the ſtalk at a joint below the head; and out of the head ariſes a naked foot-ſtalk, ſuſtaining a ſmall head or whorl of flowers; the flowers are of a bright red colour; they have two lips, the upper lip is long, narrow, and entire, the under lip is cut into three parts; they have each two ſtamina, which are longer than the petal. This plant flowers in July, but in a moiſt ſeaſon, or when the plants are in a moiſt ſoil, they will continue in flower till the middle or latter end of September.

Both theſe ſorts may be propagated by parting of their roots; the firſt does not multiply ſo faſt as the ſecond, but as that produces plenty of ſeeds, ſo it may be eaſily propagated that way: if the ſeeds are ſown in the autumn ſoon after they are ripe, the plants will come up the following ſpring; but if they are not ſown till ſpring, the plants ſeldom riſe till the next year. When the plants are come up, and are fit to remove, they ſhould be tranſplanted into a ſhady border at about nine inches diſtance, and when they have taken new root they will require no other care but to keep them clean from weeds till the autumn, when they should be transplanted into the borders where they are to remain. The following summer they will flower and produce ripe seeds, but the roots will continue several years, and may be parted every other year to increase them. This loves a soft loamy soil, and a situation not too much exposed to the sun.

The second sort seldom ripens seeds in England, but it increases fast enough by its creeping roots, as also by slips or cuttings, which, if planted in a shady border in May, will take root in the same manner as Mint or Balm; but as the roots multiply so fast, there is seldom occasion to use any other method to propagate them.

This sort loves a moist light soil, and in a situation where the plants have only the morning sun, they will continue longer in flower than those which are exposed to the full sun. This is a very ornamental plant in gardens, and the scent of the leaves is very refreshing and agreeable to most people, and some are very fond of the tea made with the young leaves.

The third sort grows naturally in North America; this is seldom more than a biennial root, and probably in its native country may be an annual, for the roots generally perish after the plants have perfected their seeds. It hath square stalks which rise about two feet high, branching out from the bottom to the top, garnished with spear-shaped leaves, which come out in clusters at each joint, where there are two larger leaves placed opposite, and several smaller come out on each side the stalk: the larger leaves are about two inches and a half long, and three quarters of an inch broad, and are slightly indented on their edges. Toward the upper part of the stalk the flowers come out in large whorls, having to each whorl an involucrum, composed of ten or twelve small spear-shaped leaves, of a purplish red colour on their inside; the flowers are pretty large, of the same form with those of the other sorts, of a dirty yellow colour spotted with purple; these have each two long stamina situated under the upper lip, which are terminated by bifid compressed summits, and are succeeded by four naked seeds inclosed in the empalement. It flowers in July, and if the summer proves favourable, the seeds will ripen in the autumn.

This plant is propagated by seeds, which, if sown on a border of light earth exposed to the east, the plants will rise very freely; when they are fit to remove they may be transplanted into a shady border, in the same manner as hath been directed for the first sort; and if they should shoot up stalks to flower, they should be cut down to strengthen the roots, that they may put out lateral buds, for when they are permitted to flower the first year, the roots seldom live through the winter, therefore they should be prevented: in the autumn the plants may be removed, and planted in the open borders of the pleasure-garden, where they will flower the following summer; and if the season should prove dry, they should be duly watered, otherwise they will not be near so beautiful, nor will the plants produce good seeds.

MONBIN. See Spondias.

MONTIA. See Heliocarpus.

MOREA.

The Characters are,

The flower is very like those of the Iris, but it has six spear-shaped petals, which spread open horizontally, three of which are alternately larger than the other, and three erect standards, so these have no falls as the Irises have. The spatha, number of stamina, and the seed-vessel, agrees very nearly to the Iris.

The title of this genus I have added in honour of Robert More, Esq; of Shrewsbury, who is an excellent botanist, and has a garden well stored with plants.

The Species are,

1. Morea (*Juncea*) spathâ biflorâ, foliis subulatis. Fig. Pl. tab. 238. *Morea with two flowers in each sheath, awl-shaped leaves on the stalk, and smaller flowers.*

2. Morea (*Vegeta*) spathâ uniflorâ, foliis gladiolatis. Ibid. *Morea with one flower in a sheath, plain leaves on the stalk, and larger flowers.*

These plants are natives of the Cape of Good Hope, from whence the seeds were brought. They have oblong bulbous roots, which early in the spring send out three or four long, narrow, plain leaves, which end in acute points. Toward the end of March the flower-stalk arises, which is garnished with two or three small leaves, of the same form with those at bottom, and is terminated with one or two spathæ or sheaths, which in those of the first sort is included two flowers, but those of the second have but one. The foot-stalk which sustains the flower is longer than the sheath the six petals of the flower, which spread horizontally, are of a pale blue colour, each having a yellow spot toward the bottom. The three standards which inclose the stamina are white; the flowers appear the latter end of April, but open only in the morning, shutting every day at noon. They continue in flower about a fortnight, then the germen, which is situated immediately under the flower, swells to a turgid three-cornered seed-vessel, having three cells filled with roundish seeds, which ripen the end of June.

There is but little difference in the colour of the flowers, but those of the second are the biggest.

These plants may be propagated by seeds, which should be sown in pots filled with light earth; these should be sown in August soon after they are ripe. The pots should be placed in a hot-bed frame in winter, to screen the seeds from frost. In the spring the plants will appear, when the glasses should be drawn off every day when the weather is mild, to prevent the plants being drawn up weak, and in May the pots should be placed abroad in a sheltered situation, where they may enjoy the morning sun, but screened from the great heat. If the season proves dry the plants must be frequently refreshed with water, but it must not be given them in too great plenty. Toward the latter end of June the leaves will decay, after which, if the roots are too close, they may be transplanted; but as they will be very small, six or seven of them may be planted in one small pot, and then placed on an east border, where they may have only the morning sun, and kept clean from weeds till autumn, when they must be placed into the frame to be screened from frost, but should be always exposed to the open air in mild weather; this should be repeated every winter, and in three years the plants will flower. The time for removing these bulbs is always soon after their stalks and leaves decay. In summer, when they are at rest, they must have very little water, and only require to be kept clean from weeds, and in winter they must be sheltered in a frame.

These may also be propagated by offsets which are produced from the old roots, which will flower the second year.

MORINA. Tourn. Cor 48. tab. 480.

The Characters are,

It hath a double empalement, which is tubulous, bifid, of one leaf, and permanent. The flower hath one petal, with a long tube a little incurved. The top is divided into two lips; the upper lip is small and bifid, the under lip is cut into three equal obtuse segments, the middle one being extended beyond the other. It hath two bristly stamina. The globular germen is situated under the flower, supporting a slender style which is longer than

 the

the ſtamina, crowne by a target-ſhaped ſtigma; the germen afterward becomes a ſingle ſeed, crowned by the empalement of the flower.

We have but one SPECIES of this genus, viz.

MORINA (*Orientalis.*) Hort. Cliff. 14. *Eaſtern Morina.*

This plant was diſcovered by Dr. Tournefort, in his travels in the Levant, who gave it this name in honour of Dr. Morin, a phyſician at Paris.

It grows naturally near Erzeron in Perſia, and was brought into the Engliſh gardens before the ſevere winter in 1740, which killed all the plants. The root of this plant is taper and thick, running deep into the ground, ſending out ſeveral thick ſtrong fibres as large as a finger; the ſtalk riſes near three feet high, it is ſmooth, of a purpliſh colour toward the bottom, but hairy and green at the top, garniſhed at each joint by three or four prickly leaves like thoſe of the Carline Thiſtle, of a lucid green on their upper ſide, armed on their edges with ſpines. The flowers come out from the wings of the leaves on each ſide the ſtalk; theſe have very long tubes, which are ſlender at the bottom, but are enlarged upward, and are a little incurved; the brim opens with two large lips; the upper lip is indented at the top and rounded, the lower lip is cut into three obtuſe ſegments; under the upper lip are ſituated two briſtly ſtamina, which are crooked, and crowned with yellow ſummits. Theſe flowers appear in July, but I never had any ſeeds ſucceed them. Some of the flowers are white, and others of a purpliſh red on the ſame plant.

This plant is propagated by ſeed, which ſhould be ſown ſoon after it is ripe in the autumn, otherwiſe the plants will not come up the following ſummer; for I have ſeveral times obſerved, where the ſeeds have been ſown in the ſpring, they have remained in the ground fourteen or fifteen months before the plants have appeared; this occaſioned its not growing in the Royal Garden at Paris. Theſe ſeeds ſhould be ſown in the places where the plants are to remain, becauſe they ſend forth tap-roots, which run very deep into the ground; and when theſe are broken or injured in tranſplanting, the plants ſeldom thrive after. They may be ſown in open beds or borders of freſh light earth, being careful to mark the places, that the ground may not be diſturbed; for it frequently happens, that the ſeeds do not come up the firſt year when they are ſown in autumn; but when they are ſown in the ſpring, they never come up the ſame year. The ground where the ſeeds are ſown muſt be kept clear from weeds, which is all that is neceſſary to be done until the plants come up; where they are too cloſe together they ſhould be thinned, ſo as to leave them near eighteen inches apart; after which time they will require no other culture but to keep them conſtantly clear from weeds, and in the ſpring, juſt before the plants put out new leaves, to ſtir the ground gently between them, and lay a little freſh earth over the ſurface of the bed to encourage them.

In autumn theſe plants decay to the ground, and ſend forth new leaves the following ſpring; but it will be three years from the time of the plants firſt coming up to their flowering, though after that time they will flower every ſeaſon; and the roots will continue many years, provided they are not diſturbed or killed by very ſevere froſt.

MORUS. Tourn. Inſt. R.H. 589. tab. 363. The Mulberry-tree.

The CHARACTERS are,

It hath male flowers growing at ſeparate diſtances from the female, on the ſame tree. The male flowers are collected in long taper ropes or katkins; theſe have no petals, but have four long, awl ſhaped, erect ſtamina. The female flowers are collected into roundiſh heads; theſe have no petals, but a heart-ſhaped germen ſupporting two long, rough, reflexed ſtyles, crowned by ſingle ſtigmas. The empalement of theſe afterward become large, fleſhy, ſucculent fruit, compoſed of ſeveral protuberances, in each of which is lodged one oval ſeed.

The SPECIES are,

1. MORUS (*Nigra*) foliis cordatis ſcabris. Hort. Cliff. 441. *Mulberry with rough heart-ſhaped leaves; or the common Mulberry.*

2. MORUS (*Laciniatis*) foliis palmatis hirſutis. *Mulberry with hand-ſhaped hairy leaves; ſmaller black Mulberry with elegant cut leaves.*

3. MORUS (*Rubra*) foliis cordatis ſubtus villoſis, amentis cylindricis. Lin. Sp. Pl. 986. *Mulberry with heart-ſhaped leaves which are hoary on their under ſide, and cylindrical katkins.*

MORUS (*Alba*) foliis obliquè cordatis lævibus. Hort. Cliff. 441. *Mulberry with oblique, ſmooth, heart-ſhaped leaves; or Mulberry with a white fruit.*

5. MORUS (*Tinctoria*) foliis obliquè cordatis acuminatis hirſutis. *Mulberry with oblique, heart-ſhaped, acute-pointed, hairy leaves; or Fuſtick wood.*

6. MORUS (*Papyrifera*) foliis palmatis, fructibus hiſpidis. Lin. Sp. Plant. 986. *Mulberry with hand-ſhaped leaves, and prickly fruit.*

7. MORUS (*Tartarica*) foliis ovato-oblongis utrinque æqualibus, inæqualiter. Flor Zeyl. 337. *Mulberry with oval oblong leaves, which are equal on both ſides, but unequally ſawed.*

The firſt ſort is the common black Mulberry-tree, which is cultivated for the delicacy of its fruit. This tree grows naturally in Perſia, from whence it was firſt brought to the ſouthern parts of Europe, but is now become common in every part of Europe, where the winters are not very ſevere, for in the northern parts of Sweden theſe trees will not live in the open air; and in ſeveral parts of Germany they are planted againſt walls, and treated in the ſame way as the Peach, and other tender fruits are here.

Theſe trees are generally of both ſexes, having male flowers or katkins on the ſame tree with the fruit; but it often happens, that ſome of the trees which are raiſed from ſeeds, have only male flowers, and produce no fruit; ſo that thoſe who plant theſe trees for their fruit, ſhould never make choice of ſuch as have been propagated by ſeeds, unleſs they have ſeen them produce fruit in the nurſery. It is alſo the ſureſt way to mark ſuch trees as are fruitful in the nurſery, at the time when their fruit is upon them, becauſe thoſe trees which are propagated by layers, are ſometimes of the male ſort; for I have ſeveral times obſerved, that ſome of the large branches of theſe trees have produced only katkins, when the other parts of the trees have been very fruitful; ſo that unleſs care is taken in the choice of the branches for making the layers, there is the ſame hazard as in ſeedling trees: nor ſhould the ſhoots which come out near the roots of old trees be ever laid down, for theſe rarely produce fruit until they have been planted many years, although the trees from which theſe were produced might be very fruitful. I have obſerved ſome trees which produced only katkins for many years after they were planted, and afterward have become fruitful; the ſame I have obſerved in Walnut-trees, and my honoured friend the Chevalier Rathgeb has informed me, that he has obſerved the ſame in the Lentiſk and Turpentine-trees.

The old Mulberry-trees are not only more fruitful than the young, but their fruit are much larger and better flavoured; ſo that where there are any of theſe old trees, it is the beſt way to propagate from them, and to make choice of thoſe branches which are moſt fruitful. The uſual method of propagating theſe trees is by laying down their branches, which will take root in one year, and are then ſeparated from the old trees; but as the moſt fruitful

6 branches

branches are often so far from the ground as not to be layed, unless by raising of boxes or baskets of earth upon supports for this purpose, so the better way is to propagate them by cuttings, which, if rightly chosen and skilfully managed, will take root very well; and in this method there will be no difficulty in having them from trees at a distance, and from the most fruitful branches. These cuttings should be shoots of the former year, with one joint of the two years wood to their bottom; the cuttings should not be shortened, but planted their full length, leaving two or three buds above ground. The best season for planting them is in March, after the danger of hard frost is over; they should be planted in light rich earth, pressing the ground pretty close about them; and if they are covered with glasses, it will forward their putting out roots; but where there is not such conveniency, the ground about them should be covered with Moss to prevent its drying; and where this is carefully done, the cuttings will require but little water, and will succeed much better than with having much wet. If the cuttings succeed well, and make good shoots, they may be transplanted the following spring into a nursery, where they should be regularly trained to stems, by fixing down stakes to each, to which the principal shoots should be fastened; and most of the lateral branches should be closely pruned off, leaving only two or three of the weakest to detain the sap, for the augmentation of the stem; for when they are quite divested of the side shoots, the sap is mounted to the top, so that the heads of the trees grow too fast for the stems, and become too weighty for their support. In about four years growth in the nursery, they will be fit to transplant where they are to remain; for these trees are transplanted with greater safety while young, than when they are of a large size.

I have two or three times made trial of planting the cuttings of Mulberries on a hot-bed, and have found them succeed extremely well. This I was led to by observing some sticks of Mulberry-trees which were cut for forks, and thrust into the hot-bed, to fasten down the vines of Cucumbers; which, although they had been cut from the tree a considerable time, yet many of them put out roots and shot out branches; so that where any person is in haste to propagate these trees, if the cuttings are planted on a hot-bed, they will take root much sooner than in the common ground.

This tree delights to grow in rich light earth, such as is in most of the old kitchen-gardens about London, where there is also a great depth of earth; for in some of those gardens there are trees of a great age, which are very healthy and fruitful, and their fruit is larger and better flavoured than those of the younger trees. I have never yet seen any of these trees which were planted in a very stiff soil, or on shallow ground either upon clay, chalk, or gravel, which have been healthy or fruitful, but their stems and branches are covered with Moss, so that the little fruit which they sometimes produce are small, ill tasted, and late before they ripen.

If these trees are planted in a situation where they are defended from the strong south and north-west winds, it will preserve their fruit from being blown off; but this shelter, whether it be trees or buildings, should be at such a distance, as not to keep off the sun; for where the fruit has not the benefit of his rays to dissipate the morning dews early, they will turn mouldy and rot upon the trees, especially in bad autumns. There is never any occasion for pruning of these trees, more than to cut off any of the branches which may grow across another, so as to rub and wound their bark, by their motion occasioned by the wind; for their shoots should never be shortened, because their fruit is produced on the young wood.

The second sort grows naturally in Sicily, from whence I received a parcel of the seeds, and raised a good number of the plants; all of these were totally different in their leaves from the common Mulberry, so that I am certain of its being a distinct species. It is also a tree of humbler growth, but the fruit is small and has no flavour, so is not worth propagating; some of the trees produced fruit two or three years in the Chelsea garden.

The white Mulberry is commonly cultivated for its leaves to feed silk worms, in France, Italy, &c. though the Persians always make use of the common black Mulberry for that purpose; and I have been assured by a gentleman of honour, who has made trial of both sorts of leaves, that the worms fed with those of the black produce much better silk than those fed with the white; but he observes, that the leaves of the black sort should never be given to the worms, after they have eaten for some time of the white, lest the worms should burst, which is often the case when they are thus treated.

The trees which are designed to feed silk worms, should never be suffered to grow tall, but rather kept in a sort of hedge; and instead of pulling off the leaves singly, they should be sheared off together with their young branches; which is much sooner done, and not so injurious to the tree.

This white sort may be propagated either from seeds or layers, as the black Mulberry, and is equally hardy; but the most expeditious method of raising these trees in quantity, is from the seeds, which may be procured in plenty from the south of France and Italy. The best way to sow these seeds in England, is to make a moderate hot-bed, which should be arched over with hoops, and covered with mats; upon this bed the seeds should be sown in the middle of March, and covered over with light earth about a quarter of an inch deep: in very dry weather the bed must be frequently watered, and in the heat of the day shaded with mats, and also covered in the nights when they are cold; with this management the plants will come up in five or six weeks, and as they are tender when they first appear, so they must be guarded against frosty mornings, which often happen in May. During the summer they must be kept clean from weeds, which is all the culture they require: but there must be care taken of them the first winter, especially to cover them when the first frosts come in autumn, which will often kill the tender plants to the ground if they are not protected; the following March these plants should be transplanted into the nursery to get strength, where they may remain two or three years, and then should be removed where they are to continue.

There are two or three varieties of this tree, which differ in the shape and size of their leaves, and colour of their fruit; but as they are of no other use than for their leaves, the strongest shooting and that with the largest leaf should be preferred.

The third sort, which is the large-leaved Virginian Mulberry with black fruit, is more uncommon than either of the former. The leaves of this are somewhat like those of the common Mulberry-tree, but are rougher and longer.

This tree is propagated only by seeds, for it will not take by grafting or budding, either on the black or white Mulberries, for it has often been tried on both, but without success. The seeds of this may be procured from North America. This is very hardy, and will endure the cold of our climate in the open air very well, and is coveted as a curiosity by such as delight in the variety of trees and shrub

The fifth sort is the tree whose wood is used by the dyers, and is better known by the title of Fustick, which is given to the wood, than by its fruit, which is of no estimation. This grows naturally in most of the islands in the West-

 Indies,

Indies, but in much greater plenty at Campeachy, where it abounds. This wood is one of the commodities exported from Jamaica, where it grows in greater plenty than in any other of the Britiſh iſlands.

This tree, in the country where it grows naturally, riſes to the height of ſixty feet and upward; it has a light brown bark, which hath ſome ſhallow furrows; the wood is firm, ſolid, and of a bright yellow colour. It ſends out many branches on every ſide, covered with a white bark, garniſhed with leaves about four inches long, which are broad at their baſe, indented at the foot-ſtalks, where they are rounded, but one ſide is broader than the other, ſo that they are oblique to the foot ſtalk; theſe diminiſh gradually, and end in acute points; they are rough like thoſe of the common Mulberry, of a dark green, and ſtand upon ſhort foot-ſtalks. Toward the end of the young branches, come out ſhort katkins of a pale herbaceous colour, and in other parts of the ſame branches the fruit is produced, growing upon ſhort foot-ſtalks; they are as large as Nutmegs, of a roundiſh form, full of protuberances like the common Mulberry, green within, and on the outſide of a luſcious ſweet taſte when ripe.

It is too tender to thrive in this country, unleſs preſerved in a warm ſtove. The ſeeds of this plant come up very freely on a hot-bed, and when the plants are fit to remove, they ſhould be each planted in a ſeparate ſmall pot filled with freſh light earth, and plunged into a hot-bed of tanners bark, and ſhaded from the ſun till they have taken new root; then they ſhould be treated in the ſame way as other plants from thoſe hot countries, always keeping them in the tan-bed in the ſtove, where they will make good progreſs. Theſe plants retain their leaves all the year in the ſtove.

The ſixth ſort grows naturally in China and Japan, where the inhabitants make paper of the bark; they cultivate the trees for that purpoſe on the hills and mountains, much after the ſame manner as Oſiers are cultivated here, cutting down the young ſhoots in autumn for their bark. There were ſeveral of theſe trees raiſed from ſeeds a few years paſt, in the gardens of his Grace the Duke of Northumberland, who was ſo good as to favour me with one of the plants, which throve very well in the open air without any ſhelter, till the year 1768, when the froſt almoſt killed it. This plant makes very ſtrong vigorous ſhoots, but ſeems not to be of tall growth, for it ſends out many lateral branches from the root upward. The leaves are large, ſome of them are entire, others are deeply cut into three, and ſome into five lobes, in form of a hand, eſpecially while the plants are young; they are of a dark green, and rough to the touch, but of a pale green, and ſomewhat hairy on the under ſide, falling off on the firſt approach of froſt in autumn, as do thoſe of the common Mulberry The deſcription which Kœmpfer gives of the fruit is, that they are a little longer than Peaſe, ſurrounded with long purple hairs, are compoſed of acini, or protuberances, and when ripe change to a black purple colour, and are full of ſweet juice. I have alſo raiſed ſeveral plants of this ſort from ſeeds, which were brought from Carolina.

This tree may be propagated by laying down of the branches, in the ſame way as is practiſed for the common Mulberry; or it may be multiplied by planting of the cuttings, in the ſame manner as before directed for the common ſort.

The ſeventh ſort grows naturally in India, where it becomes a large tree. It hath a ſoft, thick, yellowiſh bark, with a milky juice like the Fig-tree, which is aſtringent. The branches come out on every ſide, which are garniſhed with oblong oval leaves, ſtanding up on ſhort foot ſtalks; both ſides of theſe leaves are equal, but their edges are unequally ſawed; they are rough, of a dark green on their upper ſide, but pale on their under, ſtanding alternately on the branches. The flowers come out in round heads at the foot-ſtalks of the leaves on each ſide the branches; they are of an herbaceous white colour; the male flowers have four ſtamina; the female flowers are ſucceeded by roundiſh fruit, which are firſt green, afterwards white, and when ripe turn to a dark red colour.

The plants are too tender to live out of a ſtove in this country, for as I raiſed a good number of the plants, ſo when they had obtained ſtrength, I placed ſome of them in different ſituations, where they were defended from the froſt, but not any of them ſurvived the winter, but thoſe which were in the bark-ſtove, where they are conſtantly kept, and treated in the ſame manner as other tender plants, giving them but little water in winter; with which management the plants thrive, and retain their leaves all the year.

MOSCHATELLINA. See Adoxa.

MOSS. See Muscus.

MOTHERWORT. See Cardiaca.

MOULD, the goodneſs of which may be known by the ſight, ſmell, and touch.

Firſt, by the ſight: thoſe Moulds that are of a bright Cheſtnut, or hazelly colour, are counted the beſt; of this colour are the beſt loams, and alſo the beſt natural earth; and this will be the better yet, if it cuts like butter, and does not ſtick obſtinately, but is ſhort, tolerable light, breaking into ſmall clods, is ſweet, will be tempered without cruſting or chapping in dry weather, or turning to mortar in wet.

The next to that, the dark gray and ruſſet Moulds are accounted the beſt, but the clear tawny is by no means to be approved, and that of a yellowiſh red colour is accounted the worſt of all; this is commonly found in wild and waſte parts of the country, and for the moſt part produce nothing but Furz and Fern, according as their bottoms are more or leſs of a light and ſandy, or of a ſpewy gravel, or clayey nature.

Secondly, by the ſmell: all lands that are good and wholeſome, will, after rain, or breaking up by the ſpade, emit a good ſmell.

Thirdly, by the touch: by this means we may diſcover whether it conſiſts of ſubſtances entirely arenaceous or clammy, or, according as it is expreſſed by Mr. Evelyn, whether it be tender, fatty, deterſive, or ſlippery, or more harſh, gritty, porous, or friable.

That being always the beſt that is between the two extremes, and does not contain the two different qualities of ſoft and hard mixed, of moiſt and dry, of churliſh and mild, that is, neither too unctuous or too lean, but ſuch as will eaſily diſſolve, of a juſt conſiſtence, between ſand and clay, and ſuch as will not ſtick to the ſpade or fingers upon every flaſh of rain.

MULBERRY. See Morus.

MULLEIN. See Verbascum.

MUMMY, is a ſort of grafting wax, made of one pound of common black pitch, and a quarter of a pound of common turpentine, put into an earthen pot, and ſet on the fire in the open air: in doing this, you ought to hold a cover in your hand, ready to cover it, in order to quench it, by putting it thereon, which is to be done ſeveral times, ſetting it on the fire again, that the nitrous and volatile parts may be evaporated. The way to know when it is enough, is by pouring a little of it upon a pewter plate, and if it be ſo it will coagulate preſently; then this melted pitch is to be poured into another pot, and a little common wax is to be added to it, mixing them well together, and then to be kept for uſe.

MUNTINGIA. Plum. Gen. Nov. 41. tab. 6. Lin. Gen. Plant. 575.

The CHARACTERS are,

The empalement of the flower is cut into five segments. The flower hath five heart-shaped petals, inserted in the empalement, and spread open like a Rose. It hath a great number of stamina. In the center is situated a roundish germen, having no style, but is crowned by a stigma divided into many parts. The germen afterward turns to soft fruit with one cell, crowned by the stigma like a navel, and filled with small seeds.

We have but one SPECIES of this genus, viz.

MUNTINGIA (*Calabura.*) Jacq. Amer. tab. 107. *Muntingia with a silky leaf, and large fruit.*

This plant is figured in Sir Hans Sloane's History of Jamaica, by the title of Loti arboris folio angustiore, fructu polyspermo umbilicato. This tree rises to the height of thirty feet or more in its native soil, sending out many branches toward the top, which are garnished with oblong heart-shaped leaves, which end in acute points, are very woolly on their under side, but smooth above, of a lucid green, slightly sawed on their edges, and placed alternately. The flowers come out from the wings of the stalk, standing upon long foot-stalks; they are composed of five heart-shaped petals, which are white, and spread open, resembling those of the Bramble, having many stamina about half the length of the petals, terminated by globular summits, and in the center is situated a roundish germen, crowned by a many-pointed stigma. The germen afterward turns to an umbilicated fruit, as large as the fruit of the Cockspur Hawthorn, and when ripe, of a dark purple colour, inclosing many small, hard, angular seeds. This sort has produced flowers and fruit in England.

This tree is propagated by seed, which should be sown in pots filled with light rich earth, and plunged into a moderate hot-bed of tanners bark. The seeds will often remain in the ground a whole year before the plants will appear, in which case the pots must be kept constantly clear from weeds, and should remain in the hot-bed till after Michaelmas, when they may be removed into the stove, and plunged into the bark-bed, between other pots of tall plants, where there is not room for plants to stand, where they may remain. During the winter season, these pots should be now and then watered when the earth appears dry, and in the beginning of March the pots should be removed out of the stove, and placed into a fresh bark-bed under frames, which will bring up the plants soon after.

When the plants are come up about two inches high, they should be carefully taken out of the pots, and each planted into a separate small pot filled with light rich earth, and then plunged into the hot-bed again, observing to shade them from the sun until they have taken new root; after which time they should be duly watered, and in warm weather they must have a large share of fresh air. In this hot-bed the plants may remain till autumn, when the nights begin to be cold; at which time they should be removed into the stove, and plunged into the bark-bed. During the winter season, these plants must be kept warm, especially while they are young, and frequently refreshed with water; but it must not be given to them in large quantities, lest it rot the tender fibres of their roots. It will be proper to continue these plants in the stove all the year, but in warm weather they should have a large share of air; but as the plants grow in strength they will be more hardy, and may be more exposed in summer, and in winter will live in a dry stove, if kept in a moderate degree of heat.

MURUCUIA. See PASSIFLORA.

MUSA. Plum. Nov. Gen. 24. tab. 34. Lin. Gen. Pl. 1010. The Plantain tree.

The CHARACTERS are,

It hath male and hermaphrodite flowers upon the same stalk; these are produced on a single stalk (or spadix;) the male flowers are situated on the upper part of the spike, and the hermaphrodite below; these are in bunches, each bunch having a sheath or cover, which falls off. The flowers are of the lip kind. The petals constitute the upper lip, and the nectarium the under; they have six awl-shaped stamina, five of which are situated in the petal, and the sixth in the nectarium; this is double the length of the other, terminated by a linear summit, the others have none. The germen is situated under the flower, which is long, having three obtuse angles, supporting an erect cylindrical style, crowned by a roundish stigma. The germen afterward turns to an oblong, three-cornered, fleshy fruit, covered with a thick rind, divided into three parts.

The SPECIES are,

1. MUSA (*Paradisiaca*) spadice nutante, floribus masculis persistentibus. Lin. Sp. Plant. 1477. *The Plantain-tree.*

2. MUSA (*Sapientum*) spadice nutante, floribus masculis deciduis. Lin. Sp. Plant. 1477. *The Banana-tree.*

The first sort is cultivated in all the islands of the West-Indies, where the fruit serves the negroes for bread, and some of the white people also prefer it to most other things, especially to the Yams and Cassada bread.

This plant rises with a soft herbaceous stalk fifteen or twenty feet high; the lower part of the stalk is often as large as a man's thigh, diminishing gradually to the top, where the leaves come out on every side; these are often six feet long, and near two feet broad, with a strong fleshy midrib, and a great number of transverse veins running from the midrib to the borders. The leaves are thin and tender, so that where they are exposed to the open air, they are generally torn by the wind; for as they are large, the wind has great power against them; these leaves come out from the center of the stalk, and are rolled up at their first appearance, but when they are advanced above the stalk, they expand and turn backward; as these leaves come up rolled in the manner before-mentioned, their advance upward is so quick, that their growth may almost be discerned by the naked eye; and if a fine line is drawn across, level with the top of the leaf, in an hour's time the leaf will be near an inch above it. When the plant is grown to its full height, the spikes of flowers will appear in the center, which is often near four feet in length, and nods on one side. The flowers come out in bunches, those on the lower part of the spike being the largest, the others diminish in their size upward; each of these bunches is covered with a spathæ or sheath, of a fine purple colour, which drops off when the flowers open. The upper part of the spike is made up of male or barren flowers, which are not succeeded by fruit, but fall off with their covers. The fruit of this is eight or nine inches long, and above an inch diameter, a little incurved, and has three angles; it is at first green, but when ripe of a pale yellow colour. The skin is tough, and within is a soft pulp of a luscious sweet flavour. The spikes of fruit are often so large, as to weigh upward of forty pounds.

The fruit of this sort is generally cut before it is ripe, and roasted in the embers, then is eaten instead of bread. The leaves are used for napkins and table-cloths, and are food for hogs.

The second sort, which is commonly called Banana, differs from the first, in having its stalks marked with dark purple stripes and spots. The fruit is shorter, straiter, and rounder; the pulp is softer, and of a more luscious taste, so is generally eaten by way of desert, and seldom used in the same way as the Plantain, therefore is not cultivated in such plenty.

Both these plants were carried to the West-Indies from the Canary Islands, to which place it is believed they were carried from Guinea, where they grow naturally; they are also cultivated in Egypt, and in most other hot countries, where they grow to perfection in about ten months, from their first planting to the ripening of their fruit; when their stalks are cut down, there will several suckers come up from the root, which in six or eight months will produce fruit; so that by cutting down the stalks at different times, there is a constant succession of fruit all the year.

In Europe there are some of these plants preserved in the gardens of curious persons, who have hot-houses capacious enough for their reception, in many of which they have ripened their fruit very well; but as they grow very tall, and their leaves are large, they require more room in the stove than most people care to allow them; they are propagated by suckers, which come from the roots of those plants which have fruited; and many times the younger plants, when they are stinted in growth, will put out suckers; these should be carefully taken off, preserving some fibres to their roots, and planted in pots filled with light rich earth, and plunged into the tan-bed in the stove; they may be taken off any time in summer, and it is best to take them off when young, because if their roots are grown large they do not put out new fibres so soon, and when the thick part of the root is cut in taking them off the plants often rot.

During the summer season these plants must be plentifully watered, for the surface of their leaves being large, there is a great consumption of moisture by perspiration in hot weather, but in the winter they must be watered more sparingly, though at that season they must be often refreshed, but it must not be given them in such quantities.

The pots in which these plants are placed should be large, in proportion to the size of the plants, for their roots generally extend pretty far, and the earth should be rich and light. The degree of heat with which these plants thrive best, is much the same with the Anana or Pine Apple, in which I have had many of these plants produce their fruit in perfection, and they were near twenty feet high.

The most sure method to have these plants fruit in England is, after they have grown for some time in pots, so as to have made good roots, to shake them out of the pots with the ball of earth to their roots, and plant them into the tan bed in the stove, observing to lay a little old tan near their roots for their fibres to strike into, and in a few months the roots of these plants will extend themselves many feet each way in the bark, and these plants will thrive a great deal faster than those which are confined in pots or tubs. When the bark-bed wants to be renewed with fresh tan, there should be great care taken of the roots of these plants, not to cut or break them, as also to leave a large quantity of the old tan about them, because if the new tan is laid too near them, it will scorch their roots and injure them. If the plants push out their flower-stems in the spring, there will be hopes of their perfecting their fruit; but when they come out late in the year, the plants will sometimes decay before the fruit is ripe. The stoves in which these plants are placed, should be at least twenty feet in height, otherwise there will not be room for their leaves to expand; for when the plants are in vigour, the leaves are often eight feet in length, and near three feet broad; so that if the stems grow to be fourteen feet to the division of the leaves, and the house is not twenty feet high, the leaves will be cramped, which will retard the growth of the plant; besides, when the leaves are bent against the glass, there will be danger of their breaking them, when they are growing vigorously; for I have had, in one night, the stems of such bent leaves force through the glass, and by the next morning advanced two or three inches above the glass.

I have seen some bunches of fruit of the first sort, which were upward of forty pounds weight, and perfectly ripe in England: but this is not so good a fruit, as to tempt any person to be at the expence of raising them in England: the second sort is preferred to the first, for the flavour of its fruit, in all those hot countries where these plants abound; the bunches of these are not near so large as those of the first sort, nor are the single fruit near so long; these change to a deeper yellow colour as they ripen, but their taste is somewhat like that of mealy Figs. Some persons who have resided in the West-Indies, having eaten some of these fruit which were produced in England, have thought them little inferior to those which grew in America; and I imagine, that the inhabitants of those countries would not esteem these fruits so much, had they variety of other sorts; but for want of better, they eat many kinds of fruit which would not be valued in Europe, could they be obtained in perfection.

MUSCARI. Tourn. Inst. R. H. 347. tab. 180. Musk, or Grape Hyacinth.

The Characters are,

The flower has no empalement. It hath one oval pitcher-shaped petal, which is reflexed at the brim. It hath three nectariums on the top of the germen, and six awl-shaped stamina which are shorter than the petal, whose summits join together. In the center is situated a roundish three-cornered germen, supporting a single style, crowned by an obtuse stigma. The germen afterward turns to a roundish three-cornered capsule, having three cells filled with roundish seeds.

The Species are,

1. Muscari (*Botryoides*) corollis globosis uniformibus, foliis canaliculato cylindricis. *Muscari with uniform globular petals, and cylindrical gutter-shaped leaves; commonly called Grape Hyacinth.*

2. Muscari (*Comosum*) corollis angulato-cylindricis, summis sterilibus long ùs pedicellatis. *Muscari with angular cylindrical petals, which on the top of the spike are barren, and have longer foot-stalks; commonly called Fair-haired Hyacinth.*

3. Muscari (*Racemosum*) corollis ovatis. *Muscari with oval petals; commonly called Musk Hyacinth.*

4. Muscari (*Monstrosum*) floribus paniculatis monstrosis. *Muscari with monstrous flowers growing in panicles; called Feathered Hyacinth.*

5. Muscari (*Orchioides*) paniculâ ramosâ, floribus monstrosis. *Muscari with a branching panicle, and monstrous flowers.*

The first sort grows naturally in the vineyards and arable fields in France, Italy, and Germany; and where it is once planted in a garden, it is not easily rooted out again; for the roots multiply greatly, and if they are permitted to scatter the seeds, the ground will be filled with the roots: there are three varieties of this, one with blue, another with white, and a third with Ash-coloured flowers; these have a small, round, bulbous root, from which come out many narrow gutter-shaped leaves; between these arise the flower-stalk, which is naked below, but toward the top garnished with a close spike of blue flowers, shaped like pitchers, sitting very close to the stalk; these smell like fresh starch, or the stones of Plumbs, which are fresh. They flower in April, and the seeds ripen the latter end of June.

The second sort grows naturally in Spain and Portugal, from whence I have received both roots and seeds. This hath a bulbous root, as large as a middling Onion, from which come out five or six leaves a foot long, and three quarters of an inch broad at their base, diminishing gradually

ally to a point. The flower-ſtalk riſes about a foot high, naked the lower half, but the upper is garniſhed with cylindrical, angular, purple flowers, ſtanding upon foot ſtalks half an inch long; theſe grow horizontally, but the ſtalk is terminated by a tuſt of flowers, whoſe petals are oval, and have neither germen or ſtyle, ſo are barren. This ſort flowers the latter end of April, or the beginning of May; there is a variety of this with white, and another with blue flowers, but the purple is the moſt common.

The third ſort hath pretty large, oval, bulbous roots, from which ariſe ſeveral leaves, which are about eight or nine inches long, and half an inch broad, which end in obtuſe points; theſe embrace each other at their baſe; out of the middle of theſe the naked ſtalk which ſuſtains the flowers ariſes, garniſhed at the top with ſmall flowers growing in a ſpike; theſe have oval pitcher-ſhaped petals, which are reflexed at their brim, and are of an Aſh-coloured purple, or obſolete colour, ſeeming as if faded, but have an agreeable muſky ſcent; the ſtalks do not riſe more than ſix inches high, ſo the flowers make no great appearance; but where they are in ſome quantity, they will perfume the air to a conſiderable diſtance. This ſort flowers in April, and the ſeeds ripen in July.

Of this there are two varieties, one of which has the ſame coloured flowers with this here enumerated, on the lower part of the ſpike, but they are larger, and have more of the purple caſt, but the flowers on the upper part of the ſpike are yellow, and have a very grateful odour. The Dutch gardeners title it Tibcadi Muſcari. As this is ſuppoſed to be only a ſeminal variety of the third, I have not enumerated it as diſtinct. There is another variety of this with very large yellow flowers, that has been lately raiſed from ſeed in Holland, which the floriſts there ſell for a guinea a root.

The fourth ſort hath a large bulbous root, from which come out ſeveral plain leaves. The flower-ſtalks riſe near a foot and a half high, and are terminated by panicles of flowers ſtanding upon long foot-ſtalks, each ſuſtaining three, four, or five flowers, whoſe petals are cut into ſlender filaments like hairs: they are of a purpliſh blue colour, and have neither ſtamina or germen, ſo do never produce ſeeds. It flowers in May, and after the flowers are paſt, the ſtalks and leaves decay at the root, and new ones ariſe the following ſpring.

The fifth ſort has a round, ſolid, bulbous root, covered with a purple ſkin. The leaves are about the ſame length with thoſe of the former ſort, but are narrower, and their borders are incurved, ſo are formed gutter faſhion. The ſtalks riſe about a foot high, but are ſlender, ſo that unleſs they are ſupported, they decline toward the ground, eſpecially when the flowers come out; theſe have two or three long narrow leaves; the ſtalks are naked to the panicle, which is much ſhorter than that of the former ſort, but branches out wide on every ſide. The petals of the flowers are cut into finer filaments, which turn back like the curls of hair; theſe have neither ſtamina or ſtyle, ſo never produce ſeeds; they are of a dark purple colour, and appear in May; in July the ſtalks and leaves decay to the root: this has been an old inhabitant in ſome of the Engliſh gardens, but from whence it originally came is not eaſy to trace.

Theſe five ſorts are very hardy, ſo will thrive in the open air, and require no other culture than any other hardy bulbous-rooted flowers; which is to take up their roots every ſecond or third year, to ſeparate their bulbs, for as ſome of the ſorts multiply pretty faſt, ſo when they are become large bunches, they do not flower ſo ſtrong as when they are ſingle; the beſt time to take them out of the ground, is ſoon after their ſtalks and leaves are decayed; then they ſhould be ſpread on a mat in a dry ſhady room, for a fortnight to dry, after which they may be kept in boxes like other bulbous roots till Michaelmas, when they may be planted again in the borders of the flower garden, and treated in the ſame way as the common hardy kinds of Hyacinths.

The firſt ſort ſhould not be admitted into the flower-garden, becauſe the roots will propagate ſo faſt as to become a troubleſome weed there.

The ſecond ſort has but little beauty, ſo a few of theſe only ſhould be allowed a place merely for the ſake of variety: this is ſo hardy as to thrive in any ſoil or ſituation.

The third ſort merits a place for the extreme ſweetneſs of its flowers, but eſpecially that variety of it with yellow flowers, called Tibcady.

The fourth and fifth ſorts may alſo be allowed to have a place in the common borders of the pleaſure-garden, where they will add to the variety, and are by no means to be deſpiſed.

They are all eaſily propagated by offsets, which moſt of their roots ſend out in pretty great plenty, ſo that there is little occaſion for ſowing of their ſeeds, unleſs it be to gain ſome new varieties.

MUSCIPULA. See SILENE.

MUSCOSE, MUSCOSUS, Moſſy, or abounding with Moſs.

MUSCOSITY. Moſſineſs.

MUSCUS. Moſs.

Theſe, though formerly ſuppoſed to be only excreſcences produced from the earth, trees, &c. yet are no leſs perfect plants than thoſe of greater magnitude, having roots, branches, flowers, and ſeeds, but yet cannot be propagated from the latter by any art.

The botaniſts diſtinguiſh theſe into ſeveral genera, under each of which are ſeveral ſpecies; but as they are plants of no uſe or beauty, it would be to no purpoſe to enumerate them in this place.

MUSHROOMS are, by many perſons, ſuppoſed to be produced from the putrefaction of the dung, earth, &c. in which they are found; but notwithſtanding this notion is pretty generally received amongſt the unthinking part of mankind, yet by the curious naturaliſts they are eſteemed perfect plants, though their flowers and ſeeds have not as yet been perfectly diſcovered. But ſince they may, and are annually propagated by the gardeners near London, and are (the eſculent ſort of them) greatly eſteemed by moſt curious palates, I ſhall briefly ſet down the method practiſed by the gardeners who cultivate them for ſale.

But firſt, it will not be improper to give a ſhort deſcription of the true eatable kind, ſince there are ſeveral unwholeſome ſorts which have been by unſkilful perſons gathered for the table.

The true Champignon, or Muſhroom, appears at firſt of a roundiſh form like a button; the upper part of which, as alſo the ſtalk, is very white, but being opened, the under part is of a livid fleſh colour; but the fleſhy part, when broken, is very white: when theſe are ſuffered to remain undiſturbed, they will grow to a large ſize, and explicate themſelves almoſt to a flatneſs, and the red part underneath will change to a dark colour.

In order to cultivate them, if you have no beds in your own, or in neighbouring gardens, which produce them, you ſhould look abroad in rich paſtures during the months of Auguſt and September, until you find them (that being the ſeaſon when they are naturally produced;) then you ſhould open the ground about the roots of the Muſhrooms, where you will find the earth, very often, full of ſmall white knobs, which are the offsets, or young Muſhrooms; theſe ſhould be carefully gathered, preſerving them in lumps with

with the earth about them: but as this ſpawn cannot be found in the paſture, except at the ſeaſon when the Muſhrooms are naturally produced, you may probably find ſome in old dunghill, eſpecially where there has been much litter amongſt it, and the wet hath not penetrated it to rot it; as likewiſe by ſearching old hot beds it may be often found; for this ſpawn has the appearance of a white mold, ſhooting out in long ſtrings, by which it may be eaſily known, wherever it is met with: or this may be procured by mixing ſome long dung from the ſtable, which has not been thrown on a heap to ferment; which being mixed with ſtrong earth, and put under cover to prevent wet getting to it, the more the air is excluded from it, the ſooner the ſpawn will appear; but this muſt not be laid ſo cloſe together as to heat, for that will deſtroy the ſpawn: in about two months after the ſpawn will appear, eſpecially if the heap is cloſely covered with old thatch, or ſuch litter as hath lain long abroad, ſo as not to ferment, then the beds may be prepared to receive the ſpawn: theſe beds ſhould be made of dung, in which there is good ſtore of litter, but this ſhould not be thrown on a heap to ferment; that dung which hath lain ſpread abroad for a month or longer, is beſt; theſe beds ſhould be made on dry ground, and the dung laid upon the ſurface; the width of theſe beds at bottom ſhould be about two feet and a half, or three feet, the length in proportion to the quantity of Muſhrooms deſired; then lay the dung about a foot thick, covering it about four inches with ſtrong earth. Upon this lay more dung, about ten inches thick; then another layer of earth, ſtill drawing in the ſides of the bed, ſo as to form it like the ridge of a houſe, which may be done by three layers of dung and as many of earth. When the bed is finiſhed, it ſhould be covered with litter or old thatch, to keep out wet, as alſo to prevent its drying; in this ſituation it may remain eight or ten days, by which time the bed will be in a proper temperature of warmth to receive the ſpawn; for there ſhould be only a moderate warmth in it, great heat deſtroying the ſpawn, as will alſo wet; therefore when the ſpawn is found, it ſhould always be kept dry until it is uſed, for the drier it is, the better it will take the bed; for I had a parcel of this ſpawn, which had lain near the oven of a ſtove upward of four months, and was become ſo dry, that I deſpaired of its ſucceſs; but I never have yet ſeen any which produced ſo ſoon, nor in ſo great quantity as this.

The bed being in a proper temperature for the ſpawn, the covering of litter ſhould be taken off, and the ſides of the bed ſmoothed; then a covering of light rich earth, about an inch thick, ſhould be laid all over the bed, but this ſhould not be wet; upon this the ſpawn ſhould be thruſt, laying the lumps two or three inches aſunder; then gently cover this with the ſame light earth, above half an inch thick, and put the covering of litter over the bed, laying it ſo thick as to keep out wet, and prevent the bed from drying: when theſe beds are made in the ſpring or autumn, as the weather is in thoſe ſeaſons temperate, ſo the ſpawn will then take much ſooner, and the Muſhrooms will appear perhaps in a month after making: but thoſe beds which are made in ſummer, when the ſeaſon is hot, or in winter, when the weather is cold, are much longer before they produce.

The great ſkill in managing of theſe beds is, that of keeping them in a proper temperature of moiſture, never ſuffering them to receive too much wet: during the ſummer ſeaſon, the beds may be uncovered to receive gentle ſhowers of rain at proper times; and in long dry ſeaſons the beds ſhould be now and then gently watered, but by no means ſuffer much wet to come to them; during the winter ſeaſon they muſt be kept as dry as poſſible, and ſo cloſely covered as to keep out cold. In froſty or very cold weather, if ſome warm litter ſhaken out of a dung heap is laid on, it will promote the growth of the Muſhrooms; but this muſt not be laid next the bed, but a covering of dry litter between the bed and this warm litter; and as often as the litter is found to decay, it ſhould be renewed with freſh; and as the cold increaſes, the covering ſhould be laid ſo much thicker. If theſe things are obſerved, there may be plenty of Muſhrooms produced all the year; and theſe produced in beds, are much better for the table than any of thoſe which are gathered in the fields.

A bed thus managed, if the ſpawn takes kindly, will continue good for ſeveral months, and produce great quantities of Muſhrooms; from theſe beds, when they are deſtroyed, you ſhould take the ſpawn for a freſh ſupply, which may be laid up in a dry place until the proper ſeaſon of uſing it; which ſhould not be ſooner than five or ſix weeks, that the ſpawn may have time to dry before it is put into the bed, otherwiſe it will not ſucceed well.

Sometimes it happens, that beds thus made do not produce any Muſhrooms till they have lain five or ſix months, ſo that theſe beds ſhould not be deſtroyed, though they ſhould not at firſt anſwer expectation; for I have frequently known theſe to have produced great quantities of Muſhrooms afterward, and have continued a long time in perfection.

MUSTARD. See SINAPIS.

MYAGRUM. Tourn. Inſt. R. H. 211. tab. 99. Gold of Pleaſure.

The CHARACTERS are,

The empalement of the flower is compoſed of four oblong coloured leaves. The flower hath four roundiſh obtuſe petals, placed in form of a croſs. It hath ſix ſtamina the length of the petals, four of which are a little longer than the other. In the center is ſituated an oval germen, which afterward becomes a turbinated, heart-ſhaped, little pod, having two valves, with the rigid ſtyle in the top, incloſing roundiſh ſeeds.

The SPECIES are,

1 MYAGRUM (*Sativum*) ſiliculis ovatis, pedunculatis polyſpermis. Hort. Cliff. 328. *Myagrum with oval pods having foot-ſtalks, incloſing one ſeed; commonly called Gold of Pleaſure.*

2. MYAGRUM (*Alyſſum*) ſiliculis cordatis pedunculatis polyſpermis, foliis denticulatis obtuſis. *Myagrum with heart-ſhaped pods ſtanding upon foot-ſtalks, having many ſeeds and indented leaves.*

3. MYAGRUM (*Rugoſum*) ſiliculis globoſis compreſſis punctato-rugoſis. Hort. Cliff. 328. *Myagrum with globular, compreſſed, ſmall pods, having rough punctures; or Field Charlock with an acute-eared leaf.*

4. MYAGRUM (*Perenne*) ſiliculis biarticulatis diſpermis, foliis extrorſum ſinuatis denticulatis. Hort. Upſal. 182. *Myagrum with ſhort pods, having two joints and two ſeeds, whoſe outer leaves are ſinuated and indented.*

5. MYAGRUM (*Perfoliatum*) ſiliculis cordatis ſubſeſſilibus, foliis amplexicaulibus. Hort. Upſal. 182. *Myagrum with ſmall heart-ſhaped pods ſitting cloſe to the ſtalk, and the leaves embracing it.*

The firſt ſort grows naturally in Corn fields in the ſouth of France and Italy; I have alſo found it growing in the Corn in Eaſthampſted Park, the ſeat of the late William Trumbull, Eſq; but it is not common in this country. It is an annual plant, with an upright ſtalk about a foot and a half high, ſending out two or four ſide branches, which grow erect; they are ſmooth, and have a fungous pith; the lower leaves are of a pale or yellowiſh green, and are eared at their baſe; thoſe upon the ſtalks diminiſh in their ſize all the way up, are entire, and almoſt embrace the ſtalks with their baſe. The flowers grow in looſe ſpikes at the end of the branches, ſtanding upon ſhort foot-ſtalks an

inch

inch long; they are compofed of four fmall yellowifh petals, placed in form of a crofs; thefe are fucceeded by oval capfules, which are bordered and crowned at the top with the ftyle of the flower, having two cells, which are filled with red feeds.

The fecond fort is alfo an annual plant, and differs from the firft in having a taller ftalk; the leaves are much longer, narrower, and are regularly indented on their edges, ending in obtufe points. The flowers are larger, but of the fame form and colour; the capfules are much larger, and are fhaped like a heart. Both thefe plants flower in June and July, and their feeds ripen in September.

The third fort grows naturally on the borders of arable fields, in the fouth of France and Italy. This is an annual plant, whofe lower leaves are five or fix inches long; they are hairy and fucculent; their bafe is eared, and they end in acute points. The ftalks rife a foot and a half high; they are brittle and hairy, branching out toward the top like the two former, and are terminated by fhort loofe fpikes of fmall pale flowers, which are fucceeded by fmall, rough, roundifh capfules, compreffed at the top. It flowers in July, and the feeds ripen in autumn.

The fourth fort grows naturally amongft the Corn, in France and Germany. This is alfo an annual plant; the lower leaves are large, jagged, and hairy; the ftalks branch out from the bottom, and are garnifhed with hairy leaves, unequally jagged. The ftalks are terminated by very long loofe fpikes of yellow flowers, which are fucceeded by fhort pods with two joints, each including one roundifh feed. It flowers about the fame time with the former.

The fifth fort grows naturally in the fouth of France and Italy; this hath a fmooth branching ftalk upward of two feet high; the lower leaves are five or fix inches long, fmooth, fucculent, and a little indented; the upper leaves almoft embrace the ftalks with their bafe. The flowers are produced in long loofe fpikes, which are yellow, and fit clofe to the ftalk; thefe are fucceeded by heart-fhaped compreffed pods, divided into two cells by a longitudinal partition, each containing one roundifh feed. It flowers at the fame time with the former.

If the feeds of all thefe plants are permitted to fcatter in the autumn, the plants will rife without any care, and only require to be thinned and kept clean from weeds. Thefe autumnal plants will always ripen their feeds, whereas thofe which are fown in the fpring fometimes fail.

MYOSURUS. Mouftail.

This plant is very near akin to the Ranunculus, under which genus it is ranged by fome botanifts; the flowers are extremely fmall, and are fucceeded by long flender fpikes of feeds, refembling the tail of a moufe, from whence it had the name It grows wild upon moift grounds in divers parts of England, where it flowers the latter end of April, and the feeds ripen a month after, when the plants decay, being annual.

MYRICA. Lin. Gen. Plant. 981. The Candleberry Myrtle, Gale, or Sweet Willow.

The Characters are,

The male flowers are upon different plants from the female; the male flowers are produced in a loofe, oblong, oval katkin, imbricated on every fide; under each fcale is fituated one moon-fhaped flower, having no petal, but hath four or fix fhort flender ftamina, terminated by large twin fummits, whofe lobes are bifid. The female flowers have neither petal or ftamina, but an oval germen fupporting two flender ftyles, crowned by fingle ftigmas. The germen afterward becomes a berry with one cell, inclofing a fingle feed.

The Species are,

1. Myrica (*Gale*) foliis lanceolatis fubferratis, caule fruticofo. Lin. Sp. Plant. 1024. *Myrica with fpear-fhaped fawed leaves, and a fhrubby ftalk; Sweet Willow, Dutch Myrtle, or Gale.*

2. Myrica (*Cerifera*) foliis lanceolatis fubferratis caule arborefcente. Kalm. *Myrica with entire fpear-fhaped leaves, and a fhrubby ftalk; commonly called Candleberry Myrtle.*

3. Myrica (*Carolinienfis*) foliis lanceolatis ferratis, caule fruticosâ. *Myrica with fpear-fhaped fawed leaves, and a fhrubby ftalk; or Carolina Candleberry-tree, with broader leaves which are more fawed.*

4. Myrica (*Afplenifolia*) foliis oblongis alternatim finuatis. Hort. Cliff. 456. *Myrica with oblong leaves, which are alternately finuated; or Maryland Gale with a Spleenwort leaf.*

5. Myrica (*Quercifolia*) foliis oblongis oppofitè finuatis glabris. *Myrica with oblong fmooth leaves, which are oppofitely finuated.*

6. Myrica (*Hirfuta*) foliis oblongis oppofitè finuatis hirfutis. *Myrica with oblong hairy leaves, which are oppofitely finuated.*

7. Myrica (*Cordifolia*) foliis fubcordatis ferratis feffilibus. Hort. Cliff. 456. *Myrica with fawed leaves, which are almoft heart-fhaped, and fit clofe to the ftalk.*

The firft fort grows naturally upon bogs in many parts of England, particularly in the northern and weftern counties, as alfo in Windfor Park, and near Tunbridge Wells. This rifes with many fhrubby ftalks near four feet high, which divide into feveral flender branches, garnifhed with ftiff fpear-fhaped leaves, of a light or yellowifh green, fmooth, and a little fawed at their points, and emit a fragrant odour when bruifed. The female flowers or katkins are produced from the fide of the branches, growing upon feparate plants from the female, which are fucceeded by clufters of fmall berries, each having a fingle feed. It flowers in July, and the feeds ripen in autumn.

The leaves of this fhrub has been by fome perfons gathered and ufed for tea, but it is generally fuppofed to be hurtful to the brain; but from this ufe of it, a learned phyfician a few years fince, wrote a treatife to prove this to be the true tea, in which he has only fhewn his want of knowledge in thefe things.

It grows naturally in bogs, fo cannot be made to thrive on dry land, for which reafon it is feldom preferved in gardens.

The fecond fort grows naturally in Norh America, where the inhabitants get a fort of green wax from the berries which they make into candles. The method of collecting and preparing of this, is defcribed by Mr. Catefby, in his Hiftory of Carolina.

This grows naturally on bogs and fwampy lands, where it rifes with many ftrong fhrubby ftalks feven or eight feet high, fending out feveral branches, which are garnifhed with ftiff fpear-fhaped leaves, having fcarce any foot-ftalks, of a yellowifh lucid green on their upper fide, but paler on their under; thefe have a very grateful odour when bruifed. The katkins come out upon different plants from the berries; thefe are about an inch long, ftanding erect. The female flowers come out on the fide of the branches in longifh bunches, which are fucceeded by fmall roundifh berries, covered with a fort of meal. This fhrub delights in a moift foft foil, in which it thrives exceeding well, and lives in the open air without any protection.

The third fort grows naturally in Carolina; this doth not rife fo high as the former; the branches are not fo ftrong, and they have a grayifh bark; the leaves are fhorter, broader, and are fawed on their edges, but in other refpects is like the fecond fort; the berries of this alfo is collected for the fame purpofe.

Thefe forts are propagated by feeds, which fhould be fown in the autumn, and then the plants will come up the

 following

following spring; but if the seeds are kept out of the ground till the spring, they seldom grow till the year after. These plants will require water in dry weather, and should be screened from frosts while young, but when they have obtained strength, they will resist the cold of this country very well in the open air

The fourth sort grows naturally in Philadelphia, from whence many of the plants have been brought to England; and those which have been planted on a moist soil, have thriven very well; some of these creep at their roots, and send up suckers plentifully, in the same manner as in their native soil.

This rises with slender shrubby stalks near three feet high, which are hairy, and divide into several slender branches, which are garnished with leaves, alternately indented almost to the midrib, and have a great resemblance to those of Spleenwort; they are of a dark green, hairy on their under side, and sit close to the stalks. The male flowers or katkins come out on the side of the branches between the leaves; these are oval, and stand erect. I have not seen any of these plants in fruit, so I can give no description of it.

This sort will propagate by suckers sent out from the roots, where it is planted in a loose moist soil, and endures the cold full as well as the two former sorts.

The fifth and sixth sorts grow naturally at the Cape of Good Hope; these only differ from each other, in one having very smooth shining leaves, and those of the other are hairy. I do not know if they are really different species, but as I received them from Holland as such, and the plants still retaining their difference, so I have enumerated them both.

These rise with shrubby slender stalks about four feet high, which divide into smaller branches, closely garnished with indented leaves; in one sort they are smooth and shining, and in the other they are hairy, and of a darker green; they sit close to the branches, and end in obtuse points, which are indented: between the leaves come out some oval katkins, which drop off, but the fruitful plants produce berries like the former sorts. These retain their leaves all the year, but are too tender to live through the winter in the open air in England, so must be placed in the greenhouse in winter. They are propagated by layers, but as they do not take root very freely, so the plants are not very common in Europe at present; for I do not find the cuttings of these plants will take root, having made several trials of them in all the different methods; nor have the Dutch gardeners had better success, so that the plants are as scarce there as in England.

When the layers are taken off from the old plants, they should be each put into a separate small pot, filled with soft, rich, loamy earth; and if they are placed under a common frame, shading them from the sun in the middle of the day, it will forward their taking new root; then they may be placed in a sheltered situation during the summer, and in the autumn removed into the green-house, and treated in the same way as other plants from the same country. The best season for laying down the branches, I have observed to be in July, and by the same time the following year they have been fit to remove.

The seventh sort is a native of the Cape of Good Hope; this hath a weak shrubby stalk, which rises five or six feet high, sending out many long slender branches, which are closely garnished their whol length with small heart-shaped leaves, which sit close on the branches, and are slightly indented and waved on their edges. The flowers come out between the leaves in roundish bunches; these are male in all the plants I have yet seen; they have an uncertain number of stamina, and are all included in one common scaly involucrum or cover. These flowers appear in July, but make no great appearance; the leaves of this sort continue all the year.

This is propagated in the same way as the two former sorts, and is difficult to increase, so is not common in the European gardens. It requires the same treatment as the two former sorts.

MYRRHIS. See CHÆROPHYLLUM, SCANDIX, SISON.

MYRTUS. Tourn. Inst. R. H. 640. tab. 409. Myrtle.

The CHARACTERS are,

The empalement of the flower is of one leaf, cut into five points. The flower has five large oval petals, which are inserted in the empalement, and a great number of small stamina, which are also inserted in the empalement, terminated by small summits. The germen is situated under the flower, supporting a slender style, crowned by an obtuse stigma, which afterward turns to an oval berry with three cells, crowned by the empalement, each cell containing one or two kidney-shaped seeds.

The SPECIES are,

1. MYRTUS (*Communis*) foliis ovatis, pedunculis longioribus. *Myrtle with oval leaves, and longer foot-stalks to the flowers; or common broad-leaved Myrtle.*

2. MYRTUS (*Belgica*) foliis lanceolatis acuminatis. *Myrtle with spear-shaped acute-pointed leaves; or broad-leaved Dutch Myrtle.*

3. MYRTUS (*Acuta*) foliis lanceolato-ovatis acutis. *Myrtle with spear-shaped, oval, acute-pointed leaves.*

4. MYRTUS (*Bœtica*) foliis ovato-lanceolatis confertis *Myrtle with oval spear-shaped leaves growing in clusters; commonly called Orange-leaved Myrtle.*

5. MYRTUS (*Italica*) foliis ovato-lanceolatis acutis, ramis erectioribus. *Myrtle with oval, acute-pointed, spear shaped leaves, and erect branches; called upright Myrtle.*

6. MYRTUS (*Tarentina*) foliis ovatis, baccis rotundioribus. *Myrtle with oval leaves and rounder berries; called the Box-leaved Myrtle.*

7. MYRTUS (*Minima*) foliis lineari-lanceolatis acuminatis. *Myrtle with linear, spear-shaped, acute-pointed leaves; commonly called Rosemary-leaved Myrtle.*

8. MYRTUS (*Zeylanica*) pedunculis multifloris, foliis ovatis subpetiolatis. Lin. Sp. Plant. 472. *Myrtle with many flowers on each foot-stalk, and oval leaves having short foot-stalks.*

The first sort is the common broad leaved Myrtle, which is one of the hardiest kinds we have. The leaves of this are an inch and a half long, and one inch broad, of a lucid green, standing upon short foot-stalks. The flowers are larger than those of the other sorts, and come out from the sides of the branches on pretty long foot-stalks; these are succeeded by oval berries, of a dark purple colour, inclosing three or four hard kidney-shaped seeds. It flowers in July and August, and the berries ripen in winter. This sort is by some called the flowering Myrtle, because it generally has a greater quantity of flowers, and those are larger than of any other sort.

The second sort has leaves much less than those of the former, and are more pointed, standing closer together on the branches. The flowers are smaller, and have shorter foot-stalks than those of the first sort; this flowers a little later in the summer, and seldom ripens its berries here.

The double flowering Myrtle I take to be a variety of this, for the leaves and growth of the plant the size of the flowers, and the time of flowering, agree better with this than any of the other sorts.

The third sort grows naturally in the south of France and in Italy; the leaves of this are much smaller than those of the second, ending in acute points, of a dull green, and set pretty close on the branches. The flowers are smaller than either of the former, and come out from the wings of the leaves

leaves toward the end of the branches; the berries are fmall and oval.

The fourth fort hath a ftronger ftalk and branches than either of the former forts, and rifes to a greater height; the leaves are oval, fpear-fhaped, and are placed in clufters round the branches; thefe are of a dark green. The flowers are of a middling fize, and come out fparingly from between the leaves; the berries are oval, and fmaller than thofe of the firft fort, but feldom ripen in England. The gardeners call this the Orange-leaved Myrtle, and by fome it is ftiled the Bay-leaved Myrtle. This fort is not fo hardy as the former.

The fifth fort is the common Italian Myrtle; this hath oval fpear-fhaped leaves, ending in acute points; the branches of this grow more erect than thofe of either of the former forts, as do alfo the leaves, from whence it is called by the gardeners upright Myrtle. The flowers of this fort are not large, and the petals are marked with purple at their points, while they remain clofed; the berries are fmall, oval, and of a purple colour. There is a variety of this with white berries, in which it only differs from it; and I believe the Nutmeg Myrtle is only a variety of this, for I have raifed feveral of the plants from feed, many of which were fo like the Italian Myrtle, as not to be diftinguifhed from it.

The fixth fort is commonly called the Box-leaved Myrtle; the leaves of this are oval, fmall, and fit clofe on the branches; they are of a lucid green, ending in obtufe points; the branches are weak, and frequently hang downward, when they are permitted to grow without fhortening, and have a grayifh bark. The flowers are fmall, and come late in the fummer; the berries are fmall and round.

The feventh fort is called the Rofemary-leaved Myrtle, and by fome it is called the Thyme-leaved Myrtle. The branches of this grow pretty erect; the leaves are placed clofe to the branches; they are fmall, narrow, and end in acute points; they are of a lucid green, and have a fragrant odour when bruifed. The flowers of this are fmall, and come late in the feafon, and are but feldom fucceeded by berries here.

There are fome other varieties of thefe Myrtles, which are propagated in the gardens for fale; but as their difference has been occafioned by culture, fo it would be multiplying their titles to litle purpofe. Thofe which are here enumerated, I believe to be really diftinct; for I have raifed moft of them from feeds, and have not found them change from one to another, though there has been other fmall variations among the plants.

The eighth fort is a native of the ifland of Ceylon; this is much tenderer than either of the former forts, fo cannot be kept through the winter in England without fome artificial heat. This hath a ftrong upright ftalk, covered with a fmooth gray bark, dividing upward into many flender ftiff branches, garnifhed with oval leaves placed oppofite, of a lucid green, and have very fhort foot-ftalks. The flowers come out at the end of the branches, feveral of them being fuftained upon one common foot-ftalk, which branches out, and each flower ftands on a very flender diftinct foot-ftalk; they are very like the flowers of Italian Myrtle, but always appear in December and January, and are never fucceeded by berries here.

As there are feveral varieties of the common forts of Myrtle cultivated in the gardens for fale, I fhall juft mention the titles by which they are known, that the curious may be informed how many there are:

Two forts of Nutmeg Myrtles, one with a broader leaf than the other.

The Bird's Neft Myrtle, the ftriped Nutmeg Myrtle, the ftriped upright Myrtle, the ftriped Rofemary-leaved Myrtle, the ftriped Box-leaved Myrtle, and the ftriped broad-leaved Myrtle.

Thefe plants may all be propagated from cuttings; the beft feafon for which is in the beginning of July, when you fhould make choice of fome of the ftraiteft and moft vigorous young fhoots, which fhould be about fix or eight inches long; the leaves on the lower part muft be ftripped off about two or three inches high, and the part twifted which is to be placed in the ground; then having filled a parcel of pots (in proportion to the quantity of cuttings defigned) with light rich earth, you fhould plant the cuttings therein, at about two inches diftance from each other, obferving to clofe the earth faft about them, and give them fome water to fettle it to the cuttings; then place the pots under a common hot-bed frame, plunging them either into fome cold dung, or tanners bark, which will prevent the earth from drying too faft; but you muft carefully fhade them with mats in the heat of the day, and give them air in proportion to the warmth of the feafon, not forgetting to water them every two or three days, as you fhall find the earth in the pots require it. With this management, in about fix weeks, the cuttings will be rooted, and begin to fhoot, when you muft inure them to the open air by degrees, into which they fhould be removed towards the latter end of Auguft, or the beginning of September, placing them in a fituation where they may be fheltered from cold winds; in which place they may remain till the middle or latter end of October, when the pots fhould be removed into the green-houfe, but fhould be placed in the cooleft part thereof, that they may have air given to them whenever the weather is mild, for they require only to be protected from fevere cold, except the Orange-leaved, and the ftriped Nutmeg Myrtles, which are fomewhat tenderer than the reft, and fhould have a warmer fituation.

If thefe pots are placed under a common hot-bed frame in winter, where they may be fcreened from froft, and have the free air in mild weather, the young plants will fucceed better than in a green-houfe, provided they do not receive too much wet, and are not kept clofely covered, which will occafion their growing mouldy, and dropping their leaves.

The fpring following thefe plants fhould be taken out of the pots very carefully, preferving a ball of earth to the roots of each of them, and every one fhould be placed into a feparate fmall pot, filled with rich light earth, obferving to water them well to fettle the earth to their roots, and place them under a frame until they have taken root; after which they fhould be inured to the open air, and in May they muft be placed abroad for the fummer, in a fheltered fituation, where they may be defended from ftrong winds.

During the fummer feafon, they will require to be plentifully watered, efpecially thofe in fmall pots, which in that feafon foon dry; therefore you fhould obferve to place them where they fhould receive the morning fun; for when they are too much expofed to the fun in the heat of the day, the moifture contained in the earth of thefe fmall pots will foon be exhaled, and the plants greatly retarded in their growth thereby.

In Auguft following you fhould fhift them into pots a fize larger, filling them up with the like rich earth, and obferve to trim the roots which were matted to the fide of the pots, as alfo to loofen the earth from the outfide of the ball with your hands, fome of which fhould be taken off, that the roots may the eafier find paffage into the frefh earth; then you muft water them well, and place the pots in a fituation where they may be defended from ftrong winds; and at this time you may trim the plants, in order to reduce them to a regular figure; fuch of them as are inclinable to make crooked ftems, you fhould thruft down a flender ftrait ftick close by them, to which their stems should be fastened, so as to bring them upright.

If care be taken to train them thus while they are young, the stems afterward, when they have acquired strength, will continue strait without any support, and their branches may be pruned, so as to form either balls or pyramids, which for such plants as are preserved in the green-house, and require to be kept in small compass, is the best method to have them handsome; but then these sheered plants will not produce flowers, for which reason that sort with double flowers should not be clipped, because the chief beauty of that sort consists in its flowers; however it will be necessary to suffer a plant or two of each kind to grow rude, for the use of their branches in nosegays, &c. for it will greatly deface those which have been constantly sheared to cut off their branches.

As these plants advance in stature, they should annually be removed into larger pots, according to the size of their roots; but you must be careful not to put them into pots too large, which will cause them to shoot weak, and many times prove the destruction of them; therefore when they are taken out of the former pots, the earth about their roots should be pared off, and that within side the ball must be gently loosened, that the roots may not be closely confined; and then place them into the same pots again, provided they are not too small, filling up the sides and bottom of them with fresh rich earth, and giving them plenty of water to settle the earth to their roots, which should be frequently repeated, for they require to be often watered both in winter and summer, but in hot weather they must have it in plenty.

The best season for shifting these plants is either in April or August, for if it be done much sooner in the spring, the plants are then in a slow growing state, and so not capable to strike out fresh roots again very soon; and if it be done later in autumn, the cold weather coming on will prevent their taking root; nor is it adviseable to do it in the great heat of summer, because they will require to be very often watered, and also to be placed in the shade, otherwise they will be liable to droop for a considerable time; and that being the season when these plants should be placed amongst other exotics, to adorn the several parts of the garden, these plants, being then removed, should not be exposed until they have taken root again, which, at that time (if the season be hot and dry) will be three weeks or a month.

In October, when the nights begin to be frosty, you should remove the plants into the green-house; but if the weather proves favourable in autumn (as it often happens) they may remain abroad until the beginning of November; for if they are carried into the green-house too soon, and the autumn should prove warm, they will make fresh shoots at that season, which will be weak, and often grow mouldy in winter, if the weather should be so severe as to require the windows to be kept closely shut, whereby they will be greatly defaced; for which reason they should always be kept as long abroad as the season will permit, and removed out again in the spring, before they begin to make shoots; and during the winter season that they are in the green-house, they should have as much free air as possible when the weather is mild.

The three first mentioned sorts I have planted abroad in warm situations, upon a dry soil, where they have endured the cold of our winters for several years very well, with only being covered in very hard frosts with two or three mats, and the surface of the ground about their roots covered with a little mulch to prevent the frost from entering the ground; but in Cornwall and Devonshire, where the winters are more favourable than in most other parts of England, there are large hedges of Myrtle, which have been planted several years, and are very thriving and vigorous, some of which are upward of six feet high; and I believe, if the double-flowering kind were planted abroad, it would endure the cold as well as any of the others, it being a native of the southern parts of France; but in the winter 1768, the Myrtles in Devonshire and Cornwall were killed to the ground. This, and the Orange-leaved kind, are the most difficult to take root from cuttings; but if they are planted toward the latter end of June, making choice of only such shoots as are tender, and the pots are plunged into an old bed of tanners bark which has lost most of its heat, and the glasses shaded every day, they will take root extremely well, as I have more than once experienced. The Orange-leaved sort, and those with variegated leaves, are somewhat tenderer than the ordinary sorts, and should be housed a little sooner in autumn, and placed farther from the windows of the green-house.

The eighth sort is at present rare in Europe, and in very few gardens. This was by Dr. Linnæus separated from the Myrtles in the former editions of his works, and had the title of Myrsine applied to it; but in his Species of Plants, he has joined it to this genus again, to which, according to his system, it properly belongs; for the number of petals, stamina, and style, do agree with those of the Myrtle, but it differs in fructification, this having but one seed in each fruit, and the Myrtle has four or five.

This plant is with difficulty propagated, which occasions its present scarcity; for as it does not produce ripe seeds in Europe, it can only be increased by layers or cuttings. By the former method the layers are commonly two years before they take root, and the cuttings frequently fail, though the latter is preferred, when performed at a proper season, and in a right method; the best time to plant the cuttings is in May; in the choice of them, it should be the shoots of the former year, with a small piece of the two years wood at bottom; these should be planted in a soft loamy earth, and covered with bell or hand-glasses to exclude the air, which will be of great service to promote the cuttings putting out root; and if they are covered with the glasses of the hot-bed above them, it will be yet better: the cuttings should be shaded from the sun in the heat of the day, and gently refreshed with water, but they should by no means have the glasses moved; so the water given to them must be over the whole, which will soak in and moisten the earth under the glasses; in about five or six weeks they will have taken root, when they should be gradually inured to bear the open air, into which it will be proper to remove them about the middle of July, that they may be strengthened before winter. In August they should be carefully taken up, and each planted in a small pot, filled with light earth from a kitchen-garden, and placed in a shady situation till they have taken fresh root; then they should be placed in a sheltered situation, where they may remain till the end of September, and then be removed into the stove.

This plant will not live through the winter in England in a green-house, but if it is placed in a moderate degree of warmth, it will flower well in winter; and in July, August, and September, the plants should be placed abroad in a sheltered situation.

MYRTUS BRABANTICA. See MYRICA.

N.

NAPELLUS. See Aconitum.

NAPUS. See Brassica and Rapa.

NAPÆA. Lin. Gen. Plant. 748.

The Characters are,

It hath male and hermaphrodite flowers in distinct plants. The male flowers hath pitcher shaped empalements of one leaf. The flowers have five oblong petals, which are connected at their base; they have many hairy stamina, which are joined at the bottom into a sort of cylindrical column. The hermaphrodite flowers have the like empalements, petals, and stamina, as the male, and have a conical germen, supporting a cylindrical style, divided at the top into ten parts, crowned by single stigmas. The germen afterward turns to an oval fruit, inclosed in the empalement, divided into ten cells, each containing one kidney-shaped seed.

The Species are,

1. Napæa (*Dioica*) pedunculis involucratis angulatis, foliis scabris floribus dioicis. Lin. Sp. *Napæa with angular foot-stalks, rough leaves, and male flowers upon different plants from the fruitful.*

2. Napæa (*Hermaphrodito*) pedunculis nudis lævibus foliis glabris floribus hermaphroditis. *Napæa with leaves having acute-pointed lobes, and naked foot-stalks with three hermaphrodite flowers.*

Both sorts have perennial roots, which are composed of many thick fleshy fibres, which strike deep into the ground, and are connected at the top into large heads; the leaves of the first sort are rough, hairy, and are deeply cut into six or seven lobes, which are irregularly indented on their edges, each lobe having a strong midrib, which all meet in a center at the foot stalk. The foot-stalks are large and long, arising immediately from the root, and spread out on every side. The stalks rise seven or eight feet high, and divide into smaller branches; they are hairy, and garnished at each joint with one leaf, of the same form as those below, but diminish in their size toward the top, where they seldom have more than three lobes; at the upper part of the stalk come out from the side at each joint a long foot-stalk, sustaining several white flowers, which are tubulous at bottom, where the segments of the petal are connected, but they spread open above, and are divided into five obtuse segments; in the center arises the column, to which their stamina are joined at their base, but spread open above, and in the hermaphrodite flowers the style is connected to the same column. The hermaphrodite flowers are succeeded by compressed orbicular fruit, inclosed in the empalement, and divided into ten cells, each containing a kidney shaped seed, but the male plants are barren. It flowers in July, and the seeds ripen in autumn, soon after which the stalks decay, but the roots will live many years.

The roots of the second sort frequently creep in the ground to some distance from the old plant; the stalks rise about four feet high, and are garnished with smooth leaves, placed alternately, standing upon long slender foot-stalks; they are deeply cut into three lobes, which end in acute points, and are irregularly sawed on their edges. At the base of the leaf comes out the foot-stalk of the flower, which is about three inches long, dividing at the top into three smaller, each sustaining one white flower of the same form with those of the first sort, but are smaller, and the column of stamina is longer, their summits standing out beyond the petal; these plants have some male and hermaphrodite flowers on the same plants.

Both these plants grow naturally in Virginia, and other parts of North America; from the bark of these plants might be procured a sort of hemp, which many of the malvaceous tribe afford; and in some of the sorts which grow naturally in India, the fibres of the bark are so fine, as to spin into very fine threads, of which there might be woven very fine cloth.

These plants are easily propagated by seeds, which, if sown on a bed of common earth in the spring, the plants will rise very freely, and will require no other care but to keep them clear from weeds till autumn, when they may be transplanted into the places where they are to remain; they delight in a rich moist soil, in which they will grow very luxuriantly, so they must be allowed room. The second sort may be propagated by its creeping roots, which may be parted in autumn; but as these plants have no great beauty, so one or two of each sort in a garden, for the sake of variety, will be enough.

NARCISSO LEUCOIUM. See Galanthus.

NARCISSUS. Tourn. Inst. R. H. 353. tab. 185. The Daffodil.

The Characters are,

The flowers are included in an oblong compressed spatha (or sheath) which tears open on the side, and withers. The flowers have a cylindrical funnel-shaped empalement of one leaf, which is spread open at the brim; they have six oval petals on the outside of the nectarium, which are inserted above their base, and six awl-shaped stamina fixed to the tube of the nectarium, terminated by oblong summits; they have a three-cornered, roundish, obtuse germen, situated below the flower, which afterward turns to an obtuse, roundish, three-cornered capsule, with three cells filled with globular seeds.

The Species are,

1. Narcissus (*Pseudonarcissus*) spathâ uniflorâ, nectarii limbo campanulato erecto, petalo æquale. Lin. Sp. Pl. 289. *Daffodil with one flower in each sheath, whose nectarium is erect, bell-shaped, and equal with the petals; or common English Daffodil.*

2. Narcissus (*Poeticus*) spathâ uniflorâ, nectarii limbo rotato brevissimo. Hort. Upsal. 74. *Daffodil with one flower in a sheath, having a very short wheel-shaped nectarium.*

3. Narcissus (*Incomparabilis*) spathâ uniflorâ, nectarii limbo campanulato erecto, petalo dimidio breviore. *Daffodil with one flower in a sheath, having an erect bell-shaped empalement half the length of the petal; or the Incomparable Daffodil.*

4. Narcissus (*Medio-luteus*) spathâ biflorâ, nectarii campanulato, brevissimo, floribus nutantibus. *Daffodil with two flowers in a sheath, a short bell-shaped nectarium, and nodding flowers; called Primrose Peerless.*

5. Nar-

5. NARCISSUS (*Albus*) ſpatha uniflorâ, nectario campanulato breviſſimo, petalis reflexis. *Daffodil with one flower in a ſheath, having a very ſhort bell-ſhaped nectarium, and reflexed petals.*

6. NARCISSUS (*Bulbocodium*) ſpathâ uniflorâ, nectario turbinato maximo, genitalibus declinatis. Lin. Sp. Pl. 289. *Daffodil with one flower in a ſheath, having a very large turbinated nectarium, and declined ſtamina; commonly called the Hoop-Petticoat Narciſſus.*

7. NARCISSUS (*Serotinus*) ſpathâ uniflorâ, nectario breviſſimo ſexpartito. Lœfl. Lin. Sp. Pl. 290. *Daffodil with one flower in a ſheath, having a very ſhort nectarium, which is cu into ſix parts; or ſmall autumnal Daffodil.*

8. NARCISSUS (*Tazetta*) ſpathâ multiflorâ, nectario campanulato, foliis planis. Hort. Upſal. 74. *Daffodil with many flowers in a ſheath, having a bell-ſhaped nectarium, and plain leaves; commonly called Polyanthus Narciſſus.*

9. NARCISSUS (*Jonquilla*) ſpathâ multiflora, nectario companulato brevi, foliis ſubulatis. Hort. Upſal. 75. *Daffodil with many flowers in a ſheath, a ſhort bell-ſhaped nectarium; and awl-ſhaped leaves; called Jonquil.*

The ſorts here enumerated, are all the real ſpecies which I have met with in the Engliſh gardens, though there is a great variety of each ſpecies, which differ ſo much from one another, as to render it very difficult to aſcertain the ſpecies to which they belong; in order to find out as well as I could, from what ſpecies many of thoſe varieties have been raiſed, I endeavoured to degenerate as many of the donble flowering, and others of the beſt kinds, ſo far as I could, by which I have obſerved ſeveral changes, and ſhall here mention, under each ſpecies, the varieties I have obſerved.

The firſt ſort is the common Engliſh Daffodil, which grows naturally by the borders of woods and fields in many parts of England; this hath a large bulbous root, from which comes out five or ſix flat leaves about a foot long, and an inch broad, of a grayiſh colour, a little hollowed in the middle like the keel of a boat. The ſtalk riſes a foot and a half high, having two ſharp longitudinal angles; at the top comes out a ſingle flower, incloſed in a thin ſpatha (or ſheath) which is torn open on one ſide, to make way for the flower to come out, and then withers and remains on the top of the ſtalk. The flower is of one petal or leaf, being connected at the baſe, but is cut into ſix parts almoſt to the bottom, which expand; in the middle of this is ſituated a bell-ſhaped nectarium, called by the gardeners a cup, which is equal in length to the petal, and ſtands erect. The flower nods on one ſide the ſtalk. The petal is of a pale brimſtone colour, and the nectarium yellow. It flowers the beginning of April, and after the flowers are paſt, the germen turns to a roundiſh capſule with three cells, filled with roundiſh black ſeeds, which ripen in July. This ſort propagates very faſt by offsets from the root.

The VARIETIES of this are,

One with white petals, and a pale yellow colour.

One with yellow petals, and a golden cup.

The common, double, yellow Daffodil.

Another double Daffodil with three or four cups within each other.

And, I believe, John Tradeſcant's Daffodil may be referred to this ſpecies.

The ſecond ſort grows naturally in the ſouth of France and in Italy; this has a ſmaller and rounder bulbous root than the former. The leaves are longer, narrower, and flatter than thoſe of that ſort. The ſtalks do not riſe higher than the leaves, which are of a gray colour; at the top of the ſtalk comes out one flower from the ſheath, which nods on one ſide. The petal of this is cut into ſix ſegments, which are rounded at their points; they are of a ſnow white, and ſpread open flat. In the center is ſituated a very ſhort nectarium or cup, which is fringed on the border with a bright purple circle. The flowers have an agreeable odour. This flowers in May, but ſeldom produces ſeeds, however it increaſes faſt enough by offsets.

The double white Narciſſus is the only variety of this which I have obſerved, though there is mentioned in ſome books ſeveral other.

The third ſort grows naturally in Spain and Portugal, from whence I have received the roots. The bulbs of this ſort are very like thoſe of the firſt. The leaves are longer, of a darker green, and the flower-ſtalks riſe higher. The ſegments of the petal are rounder, and ſpread open flatter than thoſe of the firſt ſort. The nectarium or cup, in the middle, is about half the length of the petal, and is edged with a gold-coloured fringe. It flowers in April, but ſeldom produces ſeeds here. This ſort ſports and varies more than any of the other: the following variations I have traced in the ſame roots.

The roots of theſe, the firſt year, produced very double flowers, of the ſort which is commonly called the Incomparable Daffodil. The ſix outer ſegments of the petal were longer than either of the others, and white; the middle was very full of ſhorter petals, ſome of which were white, others yellow, and collected into a globular figure: ſome of theſe roots, the following year, produced flowers leſs double than before, with no white petals in them, but the larger petals were of a ſulphur colour, and the others yellow; from this they afterward degenerated to half double flowers, and at laſt to ſingle flowers, with a cup half the length of the petal, in which manner they have continued to flower many years; ſo that we may conclude, that thoſe varieties were firſt obtained from the ſeeds of this ſingle flower.

The fourth ſort grows naturally in the ſouth of France and in Italy, and has been found growing in the fields in ſome parts of England, but it is likely to have been from ſome roots which have been thrown out of gardens with rubbiſh. The roots of this ſort are not ſo large as thoſe of the firſt, and are rounder; the leaves are long, of a gray colour, and ſmoother than thoſe of the firſt; the flower-ſtalks are of the ſame length with the leaves, and have commonly but one flower in a ſheath, but ſometimes when the roots are ſtrong they have two. The flower nods downward; the ſegments of the petal are a little waved on their edges; the nectarium or cup is ſhort, and bordered with yellow; it flowers in May. The ſcent of theſe flowers is not very agreeable, and as they are not very beautiful, ſo they are ſeldom cultivated in gardens ſince the finer ſorts have been plenty. There is no variety of this, ſo far as I have been able to trace, for I could never obſerve any variation in their flowers.

The fifth ſort has ſome reſemblance of the fourth, but the flowers are whiter; the ſegments of the petal are reflexed, and the border of the nectarium or cup is of a gold colour; this has ſome affinity to the ſecond ſort.

The ſixth ſort grows naturally in Portugal, from whence I have received the roots. The bulbs of this kind are ſmall; the leaves are very narrow, having ſome reſemblance to thoſe of the Ruſh, but are a little compreſſed, and have a longitudinal furrow on one ſide; theſe are ſeldom more than eight or nine inches long. The flower-ſtalk is ſlender, taper, and about ſix inches long, ſuſtaining at the top one fower, which is at firſt incloſed in a ſheath; the petal is ſcarce half an inch long, and is cut into ſix acute ſegments; the nectarium or cup is more than two inches long, very broad at the brim, leſſening gradually to the baſe, being ſomewhat formed like the ladies hoop-petticoats, from whence the flower is ſo called. It flowers in April,

April, but does not produce feeds here. There are no varieties of this fort.

The feventh fort grows naturally in Spain. This hath a fmall bulbous root; the leaves are but few in number, and are narrow; the ftalk is jointed, and rifes about nine inches high, fuftaining at the top one flower, which at firft is inclofed in the fpatha (or fheath;) the flower is cut into fix narrow fegments, which are white; the nectarium or cup is yellow. It flowers late in the autumn, and the roots are tender, fo are often killed by hard frofts in England, which renders it fcarce here.

The eighth fort grows naturally in Portugal, and in the iflands of the Archipelago; of this there are a greater variety than of all the other fpecies, for as the flowers are very ornamental, and come early in the fpring, fo the florifts in Holland, Flanders, and France, have taken great pains in cultivating and improving them; fo that at prefent the catalogues printed by the Dutch florifts, contain more than thirty varieties, the principal of which are thefe hereafter mentioned.

Thefe have yellow petals, with Orange, yellow, or fulphur-coloured cups or nectariums.

The Great Algiers.	The Moft Beautiful.
The Ladies Nofegay.	The Golden Star.
The Greater Bell.	The Mignon.
The Golden Royal.	The Zeylander.
The Golden Scepter.	The Madoufe.
The Triumphant.	The Golden Sun.

The following have white petals, with yellow or fulphur-coloured cups or nectariums.

The Archdutchefs.	The Greater Bozelman.
The Triumphant Nofegay.	The Czarina.
The New Dorothy.	The Grand Monarque.
The Paffe Bozelman.	The Czar of Mufcovy.
The Superb.	The Surpaffante.

There are fome with white petals and white cups, but thefe are not fo much efteemed as the others, though there are two or three varieties with large bunches of fmall white flowers, which have a very agreeable odour, fo are as valuable as any of the other, and comes later to flower than moft of the other forts. There is alfo one with very double flowers, whofe outer petals are white, and thofe in the middle are fome white, and others are of an Orange colour, which have a very agreeable fcent, and is the earlieft in flowering; it is generally called the Cyprus Narciffus, and feems to be a diftinct fpecies from the others. This, like moft other double flowers, never produces any feeds, fo is only propagated by offsets, and is the moft beautiful of all the Narciffufes, when blown upon glaffes of water in a room; but when it is planted in the ground, if the bed in which they are planted is not covered with mats in frofty weather, to prevent their flower-buds from being deftroyed, they feldom flower; for the leaves begin to fhoot early in the autumn, and the flower-buds appear about Chriftmas, which are tender; fo that if hard froft happens when they are coming out of the ground, it generally kills them; but if they are properly fcreened from froft, they will flower in February, and in mild feafons often in January.

The ninth fort is the Jonquil, a flower fo well known as to need no defcription; of this there is the great and fmall Jonquil with fingle flowers, and the common fort with double flowers, which is moft efteemed.

I fhall firft treat of the method of raifing the fine forts of Polyanthus Narciffus from feeds, which is the way to obtain new varieties.

The not practifing this has occafioned our fending abroad annually for great quantities of flower-roots, which were for many years kept up to a high price, on account of the great demand for them in England; whereas, if we were as induftrious to propagate them as our neighbours, we might foon vie with them, if not outdo them, in moft forts of flowers, as may be feen by the vaft variety of Carnations, Auriculas, Ranunculas, &c. which have been produced from feeds in England within a few pears paft, and exceed moft of thofe kinds in any part of Europe.

You muft firft be very careful in faving your feeds, to gather none but from fuch flowers as have good properties, and particularly from fuch only as have many flowers upon a ftalk, that flower tall, and have beautiful cups to their flowers; from fuch you may expect to have good flowers produced, provided the roots are not intermixed with ordinary kinds, which by the mixing of the farina will greatly degenerate the beft fort of flowers; and if you fow ordinary feed, it is only putting yourfelf to trouble and expence to no purpofe, fince from fuch feeds there can be no hopes of procuring any valuable flowers.

Having provided yourfelf with good feeds, you muft procure either fome fhallow cafes or flat pans, made on purpofe for the raifing of feedlings, which fhould have holes in their bottoms, to let the moifture pafs off; thefe muft be filled with frefh light earth about the beginning of Auguft (this being the feafon for fowing the feeds of moft bulbous-rooted flowers;) the earth in thefe muft be levelled very even; then fow the feeds thereon pretty thick, covering them over with fine, fifted, light earth, about half an inch thick, and place the cafes or pans in a fituation where they may have only the morning fun till about ten o'clock, where they fhould remain until the beginning of October, when they muft be removed into a warmer fituation, placing them upon bricks, that the air may freely pafs under the cafes, which will preferve them from being too moift.

They fhould alfo be expofed to the full fun in winter, but fcreened from the north and eaft winds; and if the froft fhould be fevere, they muft be covered either with mats, Peafe-haulm, or fome light covering, otherwife there will be danger of their being deftroyed: in this fituation they may remain until the beginning of April, by which time the plants will be up, when you muft carefully clear them from weeds; and if the feafon fhould prove dry, they muft be frequently watered: the cafes fhould alfo now be removed into their former fhady fituation, or fhaded in the middle of the day, for the heat of the noon-day fun will be too great for the young plants.

The latter end of June, when the leaves of the plants are decayed, you fhould take off the upper furface of the earth in the cafes (which by that time will have contracted a moffinefs, and, if fuffered to remain, will greatly injure the young roots,) obferving not to take it fo deep as to touch the roots; then fift fome frefh light earth over the furface, about half an inch thick, which will greatly ftrengthen the roots; the fame fhould alfo be repeated in October, when the cafes are moved again into the fun.

During the fummer feafon, if the weather fhould prove very wet, and the earth in the cafes appear very moift, you muft remove them into the fun till the earth be dry again; for if the roots receive much wet during the time they are unactive, it very often rots them while young; therefore you muft never give them any water after their leaves are decayed, but only place them in the fhade, as was before directed.

Thus you fhould manage them the two firft feafons, till their leaves are decayed the fecond fummer after fowing; when their leaves are decayed you fhould carefully take up the roots, which may be done by fifting the earth in the cases through a fine sieve, whereby the roots will be easily separated from the earth; then having prepared a bed of good, fresh, light earth, in proportion to the quantity of your roots, you should plant them therein, at about three inches distance every way, and about three inches deep in the ground.

These beds should be raised above the level of the ground, in proportion to the moisture of the soil, which, if dry, three inches will be enough; but if it be wet, they must be raised six or eight inches high, and laid a little rounding, to shoot off the wet.

If these beds are made in August, which is the best time to transplant the roots, the weeds will soon appear very thick; therefore you should clean the surface of the ground to destroy them, being very careful not to disturb any of the roots; and this should be repeated as often as may be found necessary, by the growth of the weeds, observing always to do it in dry weather, that they may be effectually destroyed; and toward the latter end of October, after having entirely cleared the beds from weeds, you should sift a little rich light earth over them, about an inch thick; the goodness of which will be washed down to the roots by the winter's rain, which will greatly encourage their shooting in the spring.

If the cold should be very severe in winter, you should cover the beds, either with old tan or sea-coal ashes, or in want of these with Pease-haulm, or some such light covering, to prevent the frost from penetrating the ground to the roots, which might greatly injure them while they are so young.

In the spring, when the plants begin to appear above ground, you must gently stir the surface of the ground, clearing it from weeds, &c. in doing of which, you should be very careful not to injure the roots; and if the season should prove dry, you should now and then gently refresh them with water, which will strengthen them.

When their leaves are decayed, you should clear the beds from weeds, and sift a little earth over them (as was before directed) which must also be repeated in October in like manner; but the roots should not remain longer in these beds than two years, by which time they will have grown so large as to require more room; therefore they should be taken up as soon as their leaves are decayed, and planted into fresh beds, which should be dug deep, and a little very rotten dung buried in the bottom, for the fibres of the roots to strike into. Then the roots should be planted at six inches distance, and the same depth in the ground. In the autumn, before the frost comes on, if some rotten tan is laid over the beds, it will keep out the frost, and greatly encourage the roots; and if the winter should prove severe, it will be proper to lay a greater thickness of tan over the beds, and also in the alleys, to keep out frost, or to cover them over with straw or Pease-haulm, otherwise they may be all destroyed by the cold. In the spring these coverings should be removed, as soon as the danger of hard frosts is over, and the beds must be kept clean from weeds the following summer: at Michaelmas they should have some fresh earth laid over the beds, and covered again with tan; and so every year continued till the roots flower, which is generally in five years from seed, when you should mark all such as promise well, which should be taken up as soon as their leaves decay, and planted at a greater distance in new prepared beds; but those which do not flower, or those you do not greatly esteem, should be permitted to remain in the same bed; therefore, in taking up those roots which you marked, you must be careful not to disturb the roots of those left, and also to level the earth again, and sift some fresh earth over the beds (as before) to encourage the roots; for it often happens in the seedlings of these flowers, that at their first time of blowing, their flowers seldom appear half so beautiful as they do the second year; for which reason, none of them should be rejected until they have flowered twice, that so you may be assured of their worth.

Thus having laid down directions for the sowing and managing these roots, until they are strong enough to flower, I shall proceed to give some instructions for planting and managing the roots afterwards, so as to cause them to produce large fair flowers.

All the sorts of Narcissus, which produce many flowers upon a stalk, should have a situation defended from cold and strong winds, otherwise they will be subject to be injured by the cold in a severe winter, and their stems broken down when in flower; for notwithstanding their stalks are generally pretty strong, yet the number of flowers upon each renders their heads weighty, especially after rain, which lodges in the flower, and, if succeeded by strong winds, very often destroys their beauty, if they are exposed thereto; so that a border under a hedge, which is open to the south-east, is preferable to any other position for these flowers.

The morning sun rising upon them, will dry off the moisture which had lodged upon them the preceding night, and cause them to expand fairer than when they are planted in a shady situation; and if they are too much exposed to the afternoon sun, they will be hurried out of their beauty very soon; and the strong winds usually coming from the west and south-west points, they will be exposed to the fury of them, which is frequently very injurious to them.

Having made choice of a proper situation, you must then proceed to prepare the earth necessary to plant them in; for if the natural soil of the place be very strong or poor, it will be proper to make the border of new earth, removing the former soil away about three feet deep. The best earth for these flowers is a fresh light Hazel loam, mixed with a little very rotten neats dung: this should be well mixed together, and often turned over in order to sweeten it; then having removed away the old earth to the fore-mentioned depth, you should put a layer of rotten dung or tan in the bottom, about four or five inches thick, upon which you must lay some of the prepared earth about eighteen or twenty inches thick, making it exactly level; then having marked out by line the exact distances at which the roots are to be planted (which should not be less than six or eight inches square,) you must place the roots accordingly, observing to set them upright; then you must cover them over with the before-mentioned earth about eight inches deep, being very careful in doing of it not to displace the roots: when this is done, you must make the surface of the border even, and make up the side strait, which will appear handsome.

The best time for planting these roots is in September, for if they are kept too long out of the ground, it will cause their flowers to be very weak. You should also observe the nature of the soil where they are planted, and whether the situation be wet or dry, according to which you should adapt the fresh earth, and order the beds; for if the soil be very strong, and the situation moist, you should then make choice of a light earth, and raise the beds six or eight inches, or a foot, above the level of the ground, otherwise the roots will be in danger of perishing by too much wet; but if the situation be dry, and the soil naturally light, you should then allow the earth to be a little stronger, and the beds should not be raised above three or four inches high; for if they are made too high, the roots will suffer very much, if the spring should prove dry, nor would the flowers be bear so fair. As also in very severe winters, those beds which are raised much above the level of the ground, will be more exposed to the cold than those which are lower unless the alleys are filled up with rotten tan or litter.

During the summer, the only culture these flowers require is, to keep them free from weeds; and when their leaves are entirely decayed, they should be raked off, and the beds made clean; but by no means cut off their leaves till they are quite decayed, as is by some practised, for that greatly weakens the roots.

Toward the middle of October, if the weeds have grown upon the beds, you should in a dry day gently hoe the surface of the ground to destroy them, observing to rake it over smooth again; and before the frosts come on, the beds should be covered over two inches thick with rotten tan, to keep out the frost; after which they will require no farther care till the spring, when their leaves will appear above ground; at which time you should gently stir the surface of the earth with a small trowel, being very careful not to injure the leaves of the plants, and rake it smooth with your hands, clearing off all weeds, &c. which, if suffered to remain at that season, will soon grow so fast as to appear unsightly, and will exhaust the nourishment from the roots. With this management these roots will flower very strong, some of which will appear in March, and the others in April, which, if suffered to remain, will continue in beauty a full month, and are, at that season, very great ornaments to a flower garden.

After the flowers are past and the leaves decayed, you should stir the surface of the ground, to prevent the weeds from growing; and if at the same time you lay a little very rotten dung over the surface of the beds, the rain will wash down the salts thereof, which will greatly encourage the roots the succeeding year.

During the summer season they will require no farther care but to keep them clear from weeds till October, when the surface of the beds should be again stirred, raking off all weeds, &c. and laying some good fresh earth over the beds about an inch deep, which will make good the loss sustained by weeding, &c. and in the spring you must manage as was directed for the preceding year.

These roots should not be transplanted oftener than every third year, if they are expected to flower strong and make a great increase, because the first year after removing they never flower so strong as they do the second and third; nor will the roots increase so fast, when they are often transplanted; but if you let them remain longer than three years unremoved, the number of offsets, which by that time will be produced, will weaken the large bulbs, and cause them to produce very weak flowers; therefore, at the time of transplanting them, all the small offsets should be taken off, and planted in a nursery-bed by themselves, but the large bulbs may be planted again for flowering. If you plant them in the same bed where they grew before, you must take out all the earth two feet deep, and fill it up again with fresh, in the manner before directed, which will be equal to removing them into another place: this is the constant practice of the gardeners in Holland, who have but little room to change their roots; therefore they every year remove the earth of their beds and put in fresh, so that the same place is constantly occupied by the like flowers. But those people take up their roots every year, for as they cultivate them for sale, the rounder their roots are, the more valuable they will be: the way to have them so is, to take their offsets from them annually, for when the roots are left two or three years unremoved, the offsets will have grown large, and these pressing against each other, will cause their sides to be flatted; so that where the roots are propagated for sale, they should be annually taken up as soon as their leaves decay, and the large bulbs may be kept out of the ground till the middle October; but the offsets should be planted the beginning of September or sooner, that they may get strength, so as to become blowing roots the following year; but where they are designed for ornament, they should not be removed oftener than every third year, for then the roots will be in large bunches, and a number of stalks with flowers coming from each bunch, they will make a much better appearance than where a single stalk rises from each root, which will be the case when the roots are annually removed.

The common sorts of Daffodil are generally planted in large borders of the pleasure-garden, where, being intermixed with other bulbous-rooted flowers, they afford an agreeable variety in their seasons of flowering. These roots are very hardy, and will thrive in almost any soil or situation, which renders them very proper for rural gardens where, being planted under the shade of trees, they will thrive for several years without transplanting, and produce annually in the spring great quantities of flowers, which will make a good appearance before the trees come out in leaf.

The Jonquils should be planted in beds or borders, separate from other roots, because they require to be transplanted at least every year, otherwise their roots are apt to grow long and slender, and seldom flower well after, which is also the case if they are continued many years in the same soil; wherefore the roots should be often removed from one part of the garden to another, or at least the earth should be often renewed, which is the most probable method to preserve these flowers in perfection.

The soil in which these flowers succeed best, is an Hazel loam, neither too light nor over stiff; it must be fresh, and free from roots of trees or noxious weeds, but should not be dunged; for it is very remarkable, that where the ground is made rich, they seldom continue good very long, but are subject to shoot downward, and form long slender roots.

These flowers are greatly esteemed by many people for their strong sweet scent, though there be very few ladies that can bear the smell of them; so powerful is it, that many times it overcomes their spirits, especially if confined in a room; for which reason, they should never be planted too close to a habitation, lest they become offensive, nor should the flowers be placed in such rooms where company is entertained.

NASTURTIUM. See Lepidium.

NASTURTIUM INDICUM. See Tropæolum.

NECTARINE [properly so called of Nectar, the poetical drink of the gods] Nectarine.

This fruit should have been placed under the article of Peaches, to which it properly belongs, differing from them in nothing more than in having a smooth rind, and the flesh being firmer. These the French distinguish by the name of Brugnon, as they do those Peaches which adhere to the stone, by the name of Pavies, retaining the name of Pesche to only such as part from the stone; but since the writers in gardening have distinguished this fruit by the name of Nectarine from the Peaches, so I shall follow their example, lest by endeavouring to rectify their mistakes, I should render myself less intelligible to the reader. I shall therefore mention the several varieties of this fruit which have come to my knowledge:

1. Fairchild's Early Nectarine. This is one of the earliest ripe Nectarines we have; it is a small round fruit, about the size of the Nutmeg Peach, of a beautiful red colour, and well flavoured; it ripens the end of July, or the beginning of August.

2. Elruge Nectarine. The tree has sawed leaves; the flowers are small; it is a middle-sized fruit, of a dark red or purple colour next the sun, but of a pale yellow or greenish colour towards the wall; it parts from the stone, and has a soft melting juice: this ripens in the middle of August.

3. Newington Nectarine. The tree has sawed leaves; the flowers are large and open; it is a fair large fruit (when

planted on a good ſoil,) of a beautiful red colour next the ſun, but of a bright yellow towards the wall; it has an excellent rich juice; the pulp adheres cloſely to the ſtone, where it is of a deep red colour: this ripens the latter end of Auguſt, and is the beſt flavoured of all the ſorts.

4. Scarlet Nectarine is ſomewhat leſs than the laſt, of a fine red or ſcarlet colour next the ſun, but loſes itſelf in paler red towards the wall: this ripens in the end of Auguſt.

5. Brugnon or Italian Nectarine has ſmooth leaves; the flowers are ſmall; it is a fair large fruit, of a deep red colour next the ſun, but of a ſoft yellow towards the wall; the pulp is firm, of a rich flavour, and cloſely adheres to the ſtone, where it is very red: this ripens in the end of Auguſt.

6. Roman Red Nectarine has ſmooth leaves and large flowers; it is a large fair fruit, of a deep red or purple colour towards the ſun, but has a yellowiſh caſt next the wall; the fleſh is firm, of an excellent flavour, cloſely adhering to the ſtone, where it is very red: this ripens in the end of Auguſt.

7. Murry Nectarine is a middle-ſized fruit, of a dirty red colour on the ſide next the ſun, but of a yellowiſh green towards the wall: the pulp is tolerably well flavoured: this ripens the beginning of September.

8. Golden Nectarine is a fair handſome fruit, of a ſoft red colour next the ſun, but of a bright yellow next the wall; the pulp is very yellow, of a rich flavour, and cloſely adheres to the ſtone, where it is of a faint red colour: this ripens the middle of September.

9. Temple's Nectarine is a middle-ſized fruit, of a ſoft red colour next the ſun, but of a yellowiſh green toward the wall; the pulp is melting, of a white colour towards the ſtone, from which it parts, and has a fine poignant flavour: this ripens the end of September

10. Peterborough, or late green Nectarine, is a middle-ſized fruit, of a pale green colour on the outſide next the ſun, but of a whitiſh green towards the wall; the fleſh is firm, and, in good ſeaſons, well flavoured: this ripens the middle of October.

There are ſome perſons who pretend to have more ſorts than I have here ſet down, but I much doubt whether they are different from thoſe here mentioned, there being ſo near a reſemblance between the fruits of this kind, that it requires a very cloſe attention to diſtinguiſh them well, eſpecially if the trees grow in different ſoils and aſpects, which many times alters the ſame fruit ſo much, as hardly to be diſtinguiſhed by perſons who are very converſant with them; therefore, in order to be thoroughly acquainted with their differences, it is neceſſary to conſider the ſhape and ſize of their leaves, the ſize of their flowers, their manner of ſhooting, &c. which is many times very helpful in knowing of theſe fruits.

The culture of this fruit differing in nothing from that of the Peach, I ſhall forbear mentioning any thing on that head in this place, to avoid repetition, but refer the reader to the article of PERSICA, where there is an ample account of their planting, pruning, &c.

NEPETA. Lin. Gen. Plant. 629. Catmint, or Nep.

The CHARACTERS are,

The empalement of the flower is cylindrical, indented into five acute parts at the top. The flower is of the lip kind, with one petal, having an incurved cylindrical tube, gaping at the top. The upper lip is erect, roundiſh, and indented at the point. The under lip is large, concave, entire, and ſawed on the edge. It hath four awl ſhaped ſtamina, ſituated under the upper lip, two of which are ſhorter than the other. In the bottom of the tube is ſituated the quadrifid germen, which afterward turns to four oval ſeeds, ſitting in the empalement.

The SPECIES are,

1. NEPETA (*Cataria*) floribus ſpicatis, verticillis ſubpedicellatis, foliis petiolatis cordatis dentato-ſerratis. Lin. Sp. Pl. 570. *Catmint with ſpiked flowers, whoſe whorls have very ſhort foot-ſtalks, and heart-ſhaped leaves growing on foot-ſtalks, which are indented like the teeth of a ſaw; or common Greater Catmint.*

2. NEPETA (*Minor*) floribus ſpicatis, ſpicis interruptis, verticillis pedicellatis, foliis ſubcordatis ſerratis petiolatis. *Catmint with ſpikes of flowers, which have interrupted whorls ſtanding on foot-ſtalks, and ſawed leaves almoſt heart-ſhaped; or Smaller common Catmint.*

3. NEPETA (*Anguſtifolia*) floribus ſpicatis, verticillis ſubſeſſilibus, foliis cordato-oblongis ſerratis ſeſſilibus. *Catmint with ſpiked flowers, whoſe whorls grow almoſt cloſe to the ſtalks, and oblong, ſawed, heart-ſhaped leaves, ſitting cloſe; or Greater narrow-leaved Catmint.*

4. NEPETA (*Paniculata*) floribus paniculatis, foliis oblongo-cordatis acutis ſerratis ſeſſilibus. *Catmint with panicled flowers, and oblong, heart-ſhaped, acute, ſawed leaves, ſitting cloſe to the ſtalks; or Smaller Catmint, with a Turkey Balm leaf.*

5. NEPETA (*Italica*) floribus ſeſſilibus verticillato-ſpicatis, bracteis lanceolatis longitudine calycis, foliis petiolatis. Lin. Sp. Plant. 571. *Catmint whoſe flowers grow in whorled ſpikes, ſitting cloſe to the ſtalk, with ſpear-ſhaped bracteæ the length of the empalement, and leaves growing upon foot-ſtalks; or Smaller Alpine Catmint.*

6. NEPETA (*Violacea*) verticillis pedunculatis corymboſis, foliis petiolatis cordato-oblongis dentatis. Lin. Sp. Pl. 570. *Catmint with roundiſh whorls ſtanding upon foot-ſtalks, and oblong, heart-ſhaped, indented leaves.*

7. NEPETA (*Tuberoſa*) ſpicatis ſeſſilibus, bracteis ovatis coloratis, verticilla excipientibus, foliis ſeſſilibus. Hort. Cliff. 311. *Catmint with ſpiked flowers ſitting cloſe to the ſtalks, coloured bracteæ receiving the whorls, and leaves ſitting cloſe to the ſtalks.*

8. NEPETA (*Virginica*) foliis lanceolatis, capitulis terminalibus, ſtaminibus flore longioribus. Lin. Sp. Pl. 571. *Catmint with ſpear-ſhaped leaves, whoſe flowers grow in heads, and the ſtamina are longer than the petals.*

9. NEPETA (*Orientalis*) floribus ſpicatis verticillis craſſioribus, foliis cordatis obtuſè dentatis petiolatis. *Catmint with ſpiked flowers, whoſe whorls are very thick, and heart-ſhaped leaves which are obtuſely indented.*

10. NEPETA (*Procumbens*) floribus verticillatis, bracteis ovatis hirſutis, foliis cordato-ovatis crenatis, caule procumbente. *Catmint with whorled flowers, having oval hairy bracteæ, oval heart-ſhaped leaves, which are crenated, and a trailing ſtalk.*

The firſt ſort is the common Nep or Catmint, which grows naturally on the ſide of banks and hedges in many parts of England. This has a perennial root, from which ariſe many ſquare branching ſtalks, about two feet high, garniſhed at each joint by two heart-ſhaped leaves, ſtanding oppoſite, upon pretty long foot-ſtalks; they are ſawed on their edges, and hoary on their under ſide. The flowers grow in ſpikes at the top of the ſtalks, and below the ſpikes are two or three whorls of flowers, which have very ſhort foot-ſtalks. The flowers are white, and have two lips; the upper lip ſtands erect, the lower is a little reflexed, and indented at the point; theſe are each ſucceeded by four oval black ſeeds, which ripen in the empalement.

The whole plant has a ſtrong ſcent between Mint and Pennyroyal. It is called Catmint, becauſe cats are very fond of it, eſpecially when it is withered, for then they will roll themſelves on it, and tear it to pieces, chewing it in in their mouths with great pleaſure. Mr. Ray mentions his having tranſplanted ſome of the plants of this ſort, from the fields into his garden, which were ſoon deſtroyed by the cats, but the plants which came up from ſeeds in his garden eſcaped; which verifies the old proverb, viz. "If "you ſet it the cats will eat it, if you ſow it the cats will "not know it." I have frequently made trial of this, and have always found it true; for I have tranſplanted one of the plants from another part of the garden, within two feet of ſome plants which came up from ſeeds, the latter has remained unhurt, when the former has been torn to pieces and deſtroyed by the cats; but I have always obſerved, where there is a large quantity of the herb growing together, they will not meddle with it. This flowers in June and July, and the ſeeds ripen in autumn. It is uſed in medicine.

The ſecond ſort grows naturally in Italy, and the ſouth of France. The ſtalks of this are ſlenderer, their joints farther aſunder, the leaves are narrower, and the whole plant whiter than the firſt.

The ſtalks of the third ſort do not branch ſo much as either of the former; they are ſlenderer, and their joints farther aſunder; the leaves are ſmall, narrow, and almoſt heart-ſhaped, ſawed on their edges, hoary, and ſtand upon ſhort foot-ſtalks. The ſpikes of flowers are more broken or interrupted, than thoſe of the ſecond, and the whorls ſtand upon foot-ſtalks. It grows naturally in Italy.

The fourth ſort grows naturally in Sicily. This riſes with a ſtrong four-cornered ſtalk near three feet high; the lower joints are four or five inches aſunder. The leaves are long, narrow, and heart-ſhaped, deeply ſawed on their edges, and ſit pretty cloſe to the ſtalk. The flowers grow in panicles along the ſtalks, and are of a pale purpliſh colour. It flowers about the ſame time with the other ſorts.

The fifth ſort grows naturally upon the Alps; the ſtalks of this ſeldom riſe more than a foot and a half high; ſending out very few branches. The whorls of flowers, which form the ſpike, are diſtant from each other, and ſit cloſe to the ſtalk. The leaves are ſhort, oval, heart-ſhaped; and ſtand upon foot-ſtalks; the plant is hoary, and ſtrong ſcented.

The ſixth ſort grows naturally in Spain; the ſtalks of this riſe about two feet high, and have a few ſlender branches coming out from their ſides. The leaves are heart-ſhaped, and indented on their edges. The flowers grow in roundiſh whorls upon foot-ſtalks, and are blue; there is alſo a variety of this with white flowers.

The ſeventh ſort grows naturally in Portugal. This has a thick knobbed root, from which comes out one or two ſtalks, which often decline to the ground. The leaves are oblong, crenated on their edges, ſit cloſe to the ſtalks; and are of a deep green. The upper part of the ſtalk, for more than a foot in length, is garniſhed with whorls of flowers, the lower being two inches aſunder, but are nearer all the way upward; theſe ſit very cloſe to the ſtalks, and are guarded by oval, ſmall, coloured leaves or bracteæ. The flowers are blue, and ſhaped like thoſe of the other ſpecies; there is one of this ſort with an erect ſtalk, which is the only difference between them.

The eighth ſort grows naturally in Sicily. The ſtalks of this grow about two feet high; the branches come out toward the bottom; the leaves are heart-ſhaped, obtuſe, and but little indented, ſtanding upon pretty long footſtalks. The ſtalks are terminated by long ſpikes of whorled flowers, which are ſeparated, and ſit cloſe to them; theſe are wrapped in a hoary down. The flowers are white, and appear in July.

The ninth ſort grows naturally in North America. This hath a perennial root, from which ariſe ſeveral four-cornered ſtalks two feet high, which are garniſhed with hairy leaves, ſomewhat like thoſe of Marjoram, but are larger. The flowers grow in whorls round the ſtalks, and alſo at the extremity of the ſtalk, in a large round whorl or head; they are of a pale fleſh colour, and their ſtamina is longer than the petal. It flowers in July.

The tenth ſort grows naturally in the Levant. The ſtalks of this are ſtrong, and riſe near three feet high. The leaves are heart-ſhaped, and have blunt indentures on their edges, ſtanding upon ſhort foot-ſtalks. The flowers grow in whorled ſpikes at the top of the ſtalks; the whorls are very thick, and ſit cloſe together, terminating in an obtuſe point. The flowers are of a pale fleſh colour; the whole plant is hoary; and has a ſtrong ſcent.

All the ſorts are very hardy, ſo are not injured by froſt; they are eaſily propagated by ſeeds; for if they are permitted to fall, the plants will riſe without trouble; or if the ſeeds are ſown either in the ſpring or autumn, the plants will come up, and require no other culture but to thin them where they are too cloſe, and keep them clean from weeds. If theſe plants are ſown upon a poor dry ſoil, they will not grow too rank, but will continue much longer, and appear handſomer than in rich ground, where they grow too luxuriant, and have not ſo ſtrong a ſcent.

NERIUM. Lin. Gen. Plant. 262. The Oleander, or Roſe Bay.

The CHARACTERS are,

The empalement of the flower is permanent, and cut into five acute ſegments. The flower has one funnel-ſhaped petal cut into five broad obtuſe ſegments, which are oblique. It hath a nectarium terminating the tube, which is torn into hairy ſegments. It hath five ſhort awl-ſhaped ſtamina within the tube. It hath an oblong germen, which is bifid, with ſcarce any ſtyle, crowned by a ſingle ſtigma. The germen afterword turns to two long, taper, acute-pointed pods, filled with oblong ſeeds, lying over each other like the ſcales of fiſh; and crowned with down.

The SPECIES are,

1. NERIUM (*Oleander*) foliis lineari-lanceolatis ternis. *Oleander, or Roſe Bay, with linear ſpear-ſhaped leaves, which are placed by threes round the branches; or Oleander with red flowers.*

2. NERIUM (*Indicum*) foliis linearibus rigidis. *Oleander, or Roſe Bay, with linear rigid leaves; or Indian Roſe Bay, with ſingle ſweet-ſcented flowers.*

3. NERIUM (*Latifolium*) foliis lanceolatis longioribus, flaccidis. *Roſe Bay with longer, ſpear-ſhaped, flaccid leaves; commonly called the Double Oleander.*

The firſt ſort grows naturally in Greece, and in ſeveral parts near the Mediterranean ſea, generally by the ſides of rivers and brooks: there are two varieties of this, one with white, the other with red flowers, but ſeem to have no other difference, ſo may properly be placed together as one ſpecies, though that with white flowers is rarely found growing wild in any place but the iſland of Crete.

This riſes with ſeveral ſtalks to the height of eight or ten feet. The branches come out by threes round the principal ſtalks, and have a ſmooth bark, which in the red flowering is of a purpliſh colour, but the white ſort hath a light green bark. The leaves for the moſt part ſtand by threes round the branches, upon very ſhort foot-ſtalks, and point upward; they are of a dark green, very ſtiff, and end in acute points. The flowers come out at the end of the branches in large looſe bunches, which are in one of a bright purple or crimſon colour, and in the other they are of a dirty white; they have ſhort tubes, which ſpread open at the top, where they are deeply cut into five obtuſe ſegments, which are twiſted at bottom, ſo are oblique to the tube. At the mouth of the tube the torn capillary nectarium is ſituated, and within the tube are the five ſtamina, with the germen at bottom, which afterward turns to a brown, taper, double pod, about four inches long, which opens longitudinally on one ſide, and is filled with oblong ſeeds, crowned with long hairy down, lying over each other like the ſcales of fiſh. This plant flowers in July and Auguſt, and in warm ſeaſons they are ſucceeded by pods, but the ſeeds ſeldom ripen well here.

When the ſummers are warm and dry, theſe plants make a fine appearance, for then they open and flower in great plenty; but in cold moiſt ſeaſons, the flowers often decay without expanding; and the ſort with white flowers, is more tender than the red; ſo that unleſs the weather is warm, and dry at the time the flowers appear, they rot, and make no figure, unleſs they are placed under glaſſes to ſcreen them.

The ſecond ſort grows naturally in India; this riſes with ſhrubby ſtalks ſix or ſeven feet high, which are covered with a brown bark, garniſhed with ſtiff leaves, from three to four inches long, and not more than a quarter of an inch broad; they are of a light green, and their edges are reflexed; theſe are placed ſometimes by pairs oppoſite, at others they are alternate, and ſometimes by threes round the branches. The flowers are produced in looſe bunches at the end of the branches; they are of a pale red, and have an agreeable muſky ſcent. It flowers at the ſame time with the former, but the flowers ſeldom open here in the open air, ſo that unleſs the plants are placed in an airy glaſs-caſe, where they are defended from wet and cold, they will not flower.

The third ſort grows naturally in both Indies; this plant was firſt introduced to the Britiſh iſlands in America from the Spaniſh Main, and is called by the inhabitants of thoſe iſlands South Sea Roſe; the beauty and ſweetneſs of its flowers engaged the inhabitants of the iſlands to cultivate the plants, ſo that in many places they were planted to form hedges; but the cattle browſing upon them, when there was ſcarcity of food, were many of them killed, which has occaſioned their being deſtroyed in all places expoſed to cattle, ſo that now they are only preſerved in gardens, where they make a fine appearance great part of the year, for in thoſe warm countries they are ſeldom deſtitute of flowers: this has been by ſome perſons, who have only a ſuperficial knowledge of plants, thought only a variety of the common ſort, but thoſe who have cultivated both know better; for the firſt will live through the winter in England, if planted in a warm ſituation; but this is too tender to thrive in England, unleſs preſerved in a warm green-houſe; nor will the plants flower without the aſſiſtance of a glaſs caſe in ſummer. The third ſort was not known here till the middle of the laſt century, being a ſtranger in Europe, but the former has been in the Engliſh gardens near two centuries: nor has the ſeeds of the firſt ever produced plants of the third ſort, notwithſtanding it has been poſitively aſſerted by perſons of no ſkill.

The leaves of this ſort are ſix inches long, and one inch broad in the middle, of a much thinner texture than thoſe of the firſt, and their ends are generally reflexed; they are of a light green, and irregularly placed on the branches; ſometimes they are by pairs, at others alternate, and ſometimes by threes round the branches. The flowers are produced in very large bunches at the end of the branches, ſtanding upon very long foot-ſtalks; they have three or four ſeries of petals within each other, ſo are more or leſs double. The flowers are much larger than thoſe of the common ſort, and ſmell like the flowers of Hawthorn. The plain flowers are of a ſoft red or Peach colour; but in moſt they are beautifully variegated with a deeper red, and make a fine appearance. Their uſual time of flowering is in July and Auguſt, but if they are placed in a warm ſtove they will continue to Michaelmas. As the flowers of this are double, they are not ſucceeded by ſeeds, and at preſent we are unacquainted with the ſingle flowering of this kind, for the ſecond is undoubtedly a diſtinct ſpecies.

All the ſpecies of the Roſe Bay are ſuppoſed to have a poiſonous quality; the young branches, when cut or broken, have a milky ſap or juice, and the larger branches, when burnt, emit a very diſagreeable odour, ſo there is great reaſon to believe the plants have ſome noxious quality; but this genus of plants has been confounded by many of the writers on botany, with the Chamærhododendron of Tournefort, and many of the noxious qualities with which the latter abounds, have been applied to the Nerium, but particularly that of the honey, about Trebiſond, which is reckoned very unwholeſome; which has been ſuppoſed to be occaſioned by the bees ſucking it from the flowers of the Nerium, whereas it is from the flowers of the Chamærhododendron, as Tournefort has fully informed us; but the affinity of their names in the Greek language, has occaſioned theſe two plants to be often confounded.

Theſe plants are generally propagated by layers in this country, for although they will take root from cuttings, yet that being an uncertain method, the other is generally purſued; and as the plants are very apt to produce ſuckers, or ſhoots from their roots, theſe are beſt adapted for laying, for the old branches will not put out roots; when theſe are laid down, they ſhould be ſlit at a joint in the ſame manner as is practiſed in laying of Carnations, which will greatly facilitate their taking root; if theſe branches are laid down in autumn, and are properly ſupplied with water, they will have taken root by that time twelvemonth, when they ſhould be carefully raiſed up with a trowel, and if they have taken good root, they ſhould be cut off from the old plant, and each planted in a ſeparate ſmall pot filled with ſoft loamy earth; thoſe of the common ſort will require no other care but to be placed in a ſhady ſituation, and gently watered as the ſeaſon may require, till they have taken new root; but the two other ſpecies ſhould be plunged into a very moderate hot-bed, to forward their taking root, obſerving to ſhade them from the ſun in the heat of the day; after the common ſort has taken new root, the plants may be placed in a ſheltered ſituation with other hardy exotics, where they may remain till the end of October, when they ſhould either be removed into the green-houſe, or placed under a hot-bed frame, where they may be protected from froſt in winter, but enjoy the free air at all times when the weather is mild.

This ſort is ſo hardy as to live abroad in mild winters, if planted in a warm ſituation; but as they are liable to be deſtroyed in ſevere froſt, the beſt way is to keep the plants in pots, or if they are very large, in tubs, that they may be ſheltered in winter, and in the ſummer removed abroad, placing them in a warm ſheltered ſituation. In the winter they may be placed with Myrtles, and other of the hardier kinds of exotic plants, in a place where they may have as much free air as poſſible in mild weather, but ſcreened from ſevere froſt; for if theſe are kept too warm in winter they will not flower ſtrong, and when the air is excluded from them, the ends of their ſhoots will become mouldy, ſo that the hardier they are treated, provided they are not expoſed to hard froſts, the better they will thrive.

The other two ſorts require a different treatment, otherwiſe they will not make any appearance; therefore the young plants, when they have taken new root, ſhould be gradually inured to bear the open air, into which they ſhould be removed in July, where they may remain till October, provided the weather continues mild; but during this

this time, they ſhould be placed in a ſheltered ſituation; and upon the firſt approach of froſt they ſhould be removed into ſhelter, for if their leaves are injured by froſt, they will change to a pale yellow, and will not recover their uſual colour till the following ſummer These ſorts may be preſerved in a good green houſe through the winter, and the plants will be ſtronger than thoſe which are more tenderly treated; but in May, when the flower-buds begin to appear, the plants ſhould be placed in an open glaſs-caſe, where they may be defended from the inclemency of the weather; but when it is warm weather, the air ſhould at all times be admitted to them in plenty. With this management the flowers will expand, and continue long in beauty; and during that time, there are few plants which are equal to them, either to the eye or noſe, for their ſcent is very like that of the flowers of the White Thorn; and the bunches of flowers will be very large, if the plants are ſtrong.

NICOTIANA. Tourn. Inſt. R. H. 117. tab. 41. Tobacco.

The CHARACTERS are,

The empalement of the flower is permanent, of one leaf, cut into five acute ſegments. The flower has one funnel-ſhaped petal, with a long tube ſpread open at the brim, ending in five acute points. It hath five awl-ſhaped ſtamina which are the length of the tube, a little inclined, and terminated by oblong ſummits; and an oval germen ſupporting a ſlender ſtyle, crowned by an indented ſtigma. The germen afterward turns to an oval capſule, having two cells which open at the top, and are filled with rough ſeeds faſtened to the partition.

The SPECIES are,

1. NICOTIANA (*Latiſſima*) foliis ovato lanceolatis rugoſis, ſemiamplexicaulibus. *Tobacco with oval, ſpear-ſhaped, rough leaves, which half embrace the ſtalks.*

2. NICOTIANA (*Tabacum*) foliis lanceolatis ovatis decurrentibus. *Tobacco with ſpear-ſhaped running leaves ſitting cloſe to the ſtalks; or broad-leaved Tobacco.*

3. NICOTIANA (*Anguſtifolia*) foliis lanceolatis acutis, ſeſſilibus, calycibus acutis, tubo floris longiſſimo. Plat. 185. *Tobacco with acute ſpear-ſhaped leaves ſitting cloſe to the ſtalks, ſharp-pointed empalements, and a very long tube to the flower; or narrow-leaved Tobacco.*

4. NICOTIANA (*Fruticoſa*) foliis lineari-lanceolatis acuminatis, ſemiamplexicaulibus, caule fruticoſo. *Tobacco with linear, ſpear-ſhaped, acute-pointed leaves, half embracing the ſtalks, and a ſhrubby ſtalk.*

5. NICOTIANA (*Alba*) foliis ovatis acuminatis ſemiamplexicaulibus, capſulis ovatis obtuſis. *Tobacco with oval acute-pointed leaves half embracing the ſtalk, and oval obtuſe ſeed-veſſels.*

6. NICOTIANA (*Perennis*) foliis petiolatis ovato-integerrimis, caule fruticoſo perenni. *Tobacco with oval ſpear-ſhaped leaves ſitting cloſe to the ſtalks, and a ſhrubby perennial ſtalk.*

7. NICOTIANA (*Ruſtica*) foliis ovatis. Hort. Cliff. 56. *Tobacco with oval leaves; commonly called Engliſh Tobacco.*

8. NICOTIANA (*Rugoſa*) foliis ovatis rugoſis petiolatis. *Tobacco with oval rough leaves having foot-ſtalks.*

9. NICOTIANA (*Paniculata*) foliis cordatis, floribus paniculatis, tubis clavatis. Lin. Sp. Pl. 180. *Tobacco with heart-ſhaped leaves, paniculated flowers, and club-ſhaped tubes.*

10. NICOTIANA (*Glutinoſa*) foliis cordatis, corollis racemoſis ſubringentibus, calycibus inæqualibus. Lin. Sp. Plant. 181. *Tobacco with heart-ſhaped leaves, branching ringent petals, and unequal empalements.*

11. NICOTIANA (*Humilis*) foliis ovato-lanceolatis obtuſis rugoſis calycibus breviſſimis. Plat. 185. *Tobacco with oval, ſpear-ſhaped, obtuſe, rough leaves, and a very ſhort empalement.*

The firſt ſort is the moſt common Tobacco which is ſown in England, and which has been generally taken for the broad-leaved Tobacco of Caſpar Bauhin, and others, but is greatly different from it. The leaves of this ſort are more than a foot and a half long, and a foot broad, the ſurface very rough and glutinous: when theſe plants are in a rich moiſt ſoil, they will grow more than ten feet high; the baſe of the leaves half embrace the ſtalks; the upper part of the ſtalk divides into ſmaller branches, which are terminated by looſe bunches of flowers ſtanding erect; they have pretty long tubes, and are of a pale purpliſh colour. It flowers in July and Auguſt, and the ſeeds ripen in the autumn. This is the ſort of Tobacco which is commonly brought to the markets in pots, to adorn the ſhops and balconies of London, and by ſome it is called Oroonoko Tobacco.

The ſecond ſort is the broad-leaved Tobacco of Caſpar Bauhin; the ſtalks of this ſeldom riſe more than five or ſix feet high, and divide into more branches than the firſt. The leaves are about ten inches long, and three and a half broad, ſmooth, and end in acute points, ſitting cloſe to the ſtalks; the flowers of this are rather larger, and of a brighter purple colour than thoſe of the firſt. It flowers and perfects ſeeds at the ſame time; this is by ſome called ſweet-ſcented Tobacco.

The third ſort riſes with an upright branching ſtalk four or five feet high; the lower leaves are a foot long, and three or four inches broad; thoſe on the ſtalks are much narrower, leſſening to the top, and end in very acute points, ſitting cloſe to the ſtalks; they are very glutinous. The flowers grow in looſe bunches at the top of the ſtalks; they have long tubes, and are of a bright purple or red colour. Theſe appear at the ſame time with the former ſorts, and their ſeeds ripen in the autumn.

The fourth ſort riſes with very branching ſtalks about five feet high; the leaves on the lower part of the ſtalks are a foot and a half long, broad at the baſe, where they half embrace the ſtalks, and are about three inches broad in the middle, terminating in long acute points; the ſtalks divide into many ſmaller branches, which are terminated by looſe bunches of flowers, of a bright purple colour, and are ſucceeded by acute-pointed ſeed-veſſels. This flowers about the ſame time with the former, but if the plants are placed in a warm green-houſe, they will live through the winter. The ſeeds of this ſort were ſent me for Brazil Tobacco.

The fifth ſort grows naturally in the woods in the iſland of Tobago, from whence the ſeeds were ſent me by the late Mr. Robert. Millar. This riſes about five feet high; the ſtalk does not branch ſo much as thoſe of the former; the leaves are large and oval, about fifteen inches long, and two broad in the middle, but diminiſh gradually in their ſize to the top of the ſtalk, and with their baſe half embrace it. The flowers grow in cloſer bunches than thoſe of the former, and are white; theſe are ſucceeded by ſhort, oval, obtuſe ſeed-veſſels. It flowers and perfects ſeeds about the ſame time with the former.

The ſixth ſort grows naturally at Senegal in Africa, from whence the ſeeds were ſent by Mr. Adanſon, to the Royal Garden at Paris. This riſes about four or five feet high; the lower leaves are nine inches long, and four broad in the middle; they are ſmooth, and ſit cloſe to the ſtalks; the upper leaves are of the ſame form, but gradually diminiſh toward the top in their ſize. The flowers are collected in pretty cloſe bunches; they are of a pale purple colour, and are ſucceeded by oblong ſeed-veſſels, incloſed in acute five-pointed empalements. The ſtalks of this ſort are perennial, and put out ſhoots from their joints; if the plants

plants are ſheltered in winter, they will live two or three years.

The ſeventh ſort is commonly called Engliſh Tobacco, from its having been the firſt which was introduced here, and being much more hardy than the other ſorts. The ſeeds ripen very freely, and ſcattering in the autumn, the plants have come up without care, wherever any of the plants have been ſuffered to ſeed, ſo that it has been a weed in many places; but it came originally from America, by the title of Petum. Dodonæus, Tabernemontanus, and others, have titled it Hyoſcyamus luteus, from the affinity there is between this plant and the Henbane; but the flowers of this are tubulous, and not ringent, as are thoſe of the Henbane; nor do the ſeed-veſſels of this open with a lid on the top, as that of Henbane. The ſtalks of this ſeldom riſe more than three feet high; the leaves are placed alternately on the ſtalks, ſtanding upon ſhort footſtalks; they are oval and ſmooth. The flowers grow in ſmall looſe bunches on the top of the ſtalks; they have ſhort tubes, which ſpread open at the top, and are cut into five obtuſe ſegments. They are of an herbaceous yellow colour, appearing in July, and are ſucceeded by roundiſh capſules, filled with ſmall ſeeds, which ripen in the autumn.

The eighth ſort riſes with a ſtrong ſtalk near four feet high; the leaves of this are ſhaped like thoſe of the former, but are greatly furrowed on their ſurface, and near twice the ſize of the former, of a darker green, and have longer foot-ſtalks. The flowers are larger than thoſe of the former, and of the ſame ſhape. This is undoubtedly a diſtinct plant from the former, for I have ſown the ſeeds more than forty years, and have never found any of the plants vary.

The ninth ſort was found growing naturally in the valley of Lima, by Pere Feuille, in the year 1710; and of late years the ſeeds of it were ſent from Peru, by the younger de Juſſieu, to Paris. The ſtalk of this ſort riſes more than three feet high, dividing upward into many ſmaller branches, which are rounder and a little hairy; the leaves are heart-ſhaped, about four inches long, and three broad, ſtanding upon pretty long foot-ſtalks. The flowers are produced in looſe panicles at the end of the branches; theſe have tubes about an inch long, ſhaped like a club; the brim is ſlightly cut into nine obtuſe ſegments, which are reflexed; they are of a yellowiſh green colour, and are ſucceeded by roundiſh capſules, filled with very ſmall ſeeds. It flowers about the ſame time with the other ſorts.

The ſeeds of the tenth ſort were ſent from Peru with thoſe of the former, by the younger de Juſſieu; the ſtalk of this is round, and riſes near four feet high, ſending out two or three branches from the lower part; the leaves are large, heart-ſhaped, and a little waved; they are very clammy, ſtanding upon long foot-ſtalks. The flowers grow in long looſe ſpikes at the top of the ſtalk, having ſhort open tubes, which are curved almoſt like the lip-flowers; they are of a dull purple colour; the empalement is unequally cut, one of the ſegments being twice the ſize of the other.

The eleventh ſort was diſcovered by the late Dr. Houſtoun at La Vera Cruz, who ſent the ſeeds to England. This hath a pretty thick taper root, which ſtrikes deep in the ground; at the top comes out ſix or ſeven oval ſpear-ſhaped leaves which ſpread on the ground; they are about the ſize of thoſe of the common Primroſe, but of a deeper green; the ſtalks riſe about a foot high, branching into three or four diviſions, at each of theſe is placed one ſmall leaf; the branches are terminated by a looſe ſpike of flowers, which are ſmall, tubulous, of a yellowiſh green colour, having very ſhort empalements, which are cut at the brim into five acute ſegments. The ſeed-veſſel is ſmall, oval, and divided into two cells, which are full of ſmall ſeeds.

All the ſorts, except the ſeventh and eighth, require the ſame culture, and are too tender to grow from ſeeds ſown in the full ground, to any degree of perfection in this country, ſo require to be raiſed in a hot-bed, after the following manner:

The ſeeds muſt be ſown upon a moderate hot-bed in March, and when the plants are come up fit to remove, they ſhould be tranſplanted into a new hot-bed of a moderate warmth, about four inches aſunder each way, obſerving to water and ſhade them until they have taken root; after which you muſt let them have air in proportion to the warmth of the ſeaſon, otherwiſe they will draw up very weak, and be thereby leſs capable of enduring the open air: you muſt alſo obſerve to water them frequently, but while they are very young it ſhould not be given them in too great quantities; though when they are pretty ſtrong, they will require to have it often, and in plenty.

In this bed the plants ſhould remain until the beginning of May, by which time (if they have ſucceeded well) they will touch each other, therefore they ſhould be inured to bear the open air gradually; after which they muſt be taken up carefully, preſerving a large ball of earth to each root, and planted into a rich light ſoil, in rows four feet aſunder, and the plants three feet diſtance in the rows, obſerving to water them until they have taken root; after which they will require no farther care, but only to keep them clear from weeds, until the plants begin to ſhew their flower-ſtems; at which time you ſhould cut off the tops of them, that their leaves may be better nouriſhed, whereby they will be rendered larger, and of a thicker ſubſtance. In Auguſt they will be full grown, when they ſhould be cut for uſe; for if they are permitted to ſtand longer, their under leaves will begin to decay. This is to be underſtood for ſuch plants as are propagated for uſe, but thoſe plants which are deſigned for ornament ſhould be planted in the borders of the pleaſure-garden, and permitted to grow their full height, where they will continue flowering from July till the froſt puts a ſtop to them.

The two ſmaller ſorts of Tobacco are preſerved in botanic gardens for variety, but are ſeldom propagated for uſe. The firſt ſort is found growing upon dunghills in divers parts of England. Theſe are both very hardy, and may be propagated by ſowing their ſeeds in March, upon a bed of light earth, where they will come up, and may be tranſplanted into any part of the garden, where they will thrive without farther care.

The laſt ſort, being ſomewhat tenderer than the other, ſhould be ſown early in the ſpring on a hot-bed; and when the plants come up, they ſhould be tranſplanted on another moderate hot-bed, where they muſt be duly watered, and ſhould have a large ſhare of free air in warm weather; and when the plants have obtained a good ſhare of ſtrength, they ſhould be tranſplanted into ſeparate pots, and plunged into a moderate hot-bed to bring them forward. About the middle of June ſome of the plants may be ſhaken out of the pots, and planted into beds of rich earth; but it will be proper to keep ſome plants in pots, which may be placed in the ſtove (in caſe the ſeaſon ſhould prove bad,) that they may ripen their ſeeds, ſo that the ſpecies may be preſerved.

NIGELLA. Tourn. Inſt. R. H. 258. tab. 134. Fennel Flower, or Devil in a Buſh.

The CHARACTERS are,

The flower has no empalement, but a leafy perianthium. It hath five oval, obtuſe, plain petals, which ſpread open, and are contracted at their baſe, and eight very ſhort nectariums ſituated in a circle, each having two lips, the exterior being larger,

larger, the inferior bifid, plain, and convex. It hath a great number of awl-shaped stamina, which are shorter than the petals, terminated by obtuse, compressed, erect summits; and in some five, in others ten, oblong, convex, erect germen. The germen afterward become so many oblong compressed capsules, divided by a furrow, but connected within, filled with rough angular seeds.

The SPECIES are,

1. NIGELLA (*Arvensis*) pistillis quinis, petalis integris, capsulis turbinatis. Lin. Sp. Pl. 534. *Fennel Flower having five pointals, entire petals, and turbinated seed-vessels.*

2. NIGELLA (*Damascena*) floribus involucro folioso cinctis. Hort. Cliff. 215. *Fennel Flower, whose flowers are encompassed with a leafy involucrum.*

3. NIGELLA (*Sativa*) pistillis quinis, capsulis muricatis subrotundis, foliis subpilosis. *Fennel Flower with five prickly pointals, and hairy leaves.*

4. NIGELLA (*Cretica*) pistillis quinis corollâ longioribus, petalis integris. *Fennel Flower with five pointals longer than the petals, which are entire.*

5. NIGELLA (*Latifolia*) pistillis denis corollâ brevioribus. *Fennel Flower with ten pointals which are shorter than the petals.*

6. NIGELLA (*Hispanica*) pistillis denis corollam æquantibus. Hort. Upsal. 154. *Fennel Flower with ten pointals of equal length with the petals.*

7. NIGELLA (*Orientalis*) pistillis denis corollâ longioribus. Hort. Cliff. 215. *Fennel Flower with ten pointals which are longer than the petals.*

The first sort grows naturally among the Corn in France, Italy, and Germany, so is seldom propagated in gardens.

The second sort grows naturally in Spain and Italy among the Corn; this rises with an upright branching stalk a foot and a half high, garnished with leaves much longer and finer than those of the first. The flowers are large, of a pale blue, and have a long leafy involucrum under each; these are succeeded by larger swelling seed-vessels, with horns at the top; of this there is one with single white flowers, and another with double flowers, which is sown in gardens for ornament.

The third sort grows naturally in Crete; this rises about the same height as the former. The leaves are not so finely cut as those of the second, and are a little hairy. At the top of each stalk is one flower, composed of five white petals, which are slightly cut at their end into three points; these are succeeded by oblong swelling seed-vessels, with five horns at the top, filled with small pale-coloured seeds.

The fourth sort also grows naturally in Crete; this rises with branching stalks about a foot high, garnished with shorter and broader leaves than either of the other species. At the top of each branch is one flower, having no involucrum; they are composed of five petals, and have five pointals longer than the petals; the seed-vessel is not much swollen, and has five slender horns at the top; the seeds are of a light yellowish brown colour.

The fifth sort is also a native of Crete; this rises with a branching stalk a foot high, garnished with leaves like those of Larkspur. The flowers have five large oval petals which are entire, and ten pointals which are shorter than the petals, and a great number of green stamina with blue chives; the seed-vessels are like those of the last sort.

The sixth sort rises a foot and a half high; the lower leaves are finely cut, but those on the stalk are cut into broader segments. The flowers are larger than those of the other species, and are of a fine blue colour: the pointals of this are of equal length with the petals; the seed-vessel has five horns, and is of a firmer texture than any of the other. This grows naturally in the south of France and Spain; there is a variety of this with double flowers.

The seventh sort grows naturally in the Corn fields about Aleppo; this rises with a branching stalk a foot and a half high, garnished with pretty long leaves, which are finely divided. The flowers are produced at the end of the branches, they are composed of five yellowish leaves or petals; at the base of these are placed eight nectariums, between which arise a great number of stamina, with an unequal number of germen, some having but five, others having eight or nine; they are oblong and compressed; these afterward become so many oblong compressed seed-vessels, joined together on their inner side, terminating with horns, and open longitudinally, containing many thin compressed seeds, having borders round them.

The varieties of these with double flowers are chiefly propagated in gardens for ornament; but those with single flowers are rarely admitted into any but botanic gardens, where they are preserved for the sake of variety.

All these plants may be propagated by sowing their seeds upon a bed of light earth, where they are to remain (for they seldom succeed well if transplanted;) therefore, in order to have them intermixed amongst other annual flowers in the borders of the flower-garden, the seeds should be sown in patches at proper distances; and when the plants come up they must be thinned where they grow too close, leaving but three or four of them in each patch, observing also to keep them clear from weeds, which is all the culture they require. In July they will produce their flowers, and their seeds will ripen in August, when they should be gathered and dried; then rub out each sort separately, and preserve them in a dry place.

The season for sowing these seeds is in March; but if you sow some of them in August soon after they are ripe, upon a dry soil and in a warm situation, they will abide through the winter, and flower strong the succeeding year; by sowing of the seeds at different times, they may be continued in beauty most part of the summer.

They are all annual plants, which perish soon after they have perfected their seeds; which, if permitted to scatter upon the borders, will come up without any farther care.

NIGELLASTRUM. See AGROSTEMMA.

NIGHTSHADE. See SOLANUM.

NIGHTSHADE, the Deadly. See BELLADONNA.

NIL. See ANIL.

NISSOLIA. See LATHYRUS.

NOLI ME TANGERE. See IMPATIENS.

NONSUCH, or FLOWER of BRISTOL. See LYCHNIS.

NORTHERN ASPECT is the least favourable of any in England, as having very little benefit from the sun, even in the height of summer, therefore can be of little use, whatever may have been advanced to the contrary: for although many sorts of fruit-trees will thrive and produce fruit in such positions, yet such fruit can be of little worth, since they are deprived of the kindly warmth of the sun to correct their crude juices, and render them well tasted and wholesome; therefore it is to little purpose to plant fruit-trees against such walls, except it be for such which are intended for baking, &c. where the fire will ripen, and render those juices wholesome, which, for want of sun, could not be performed while growing.

You may also plant Morello Cherries, for preserving; and white and red Currants to come late, after those which are exposed to the sun are gone; and if the soil be warm and dry, some sorts of summer Pears will do tolerably well on such an exposure, and will continue longer in eating, than if they were more exposed to the sun. But you should by no means plant winter Pears in such an aspect, as hath been practised by many ignorant persons, since we find that the

the beſt ſouth walls, in ſome bad years, are barely warm enough to ripen thoſe fruits.

Duke Cherries planted againſt walls expoſed to the north, will ripen much later in the ſeaſon, and if the ſoil is warm, they will be well flavoured, ſo that hereby this fruit may be continued a month later than is uſual.

NURSERY, or Nurſery-garden, is a piece of land ſet apart for the raiſing and propagating of all ſorts of trees and plants to ſupply the garden, and other plantations. Of this ſort there are a great number in the different parts of this kingdom, but particularly in the neighbourhood of London, which are occupied by the gardeners, whoſe buſineſs it is to raiſe trees, plants, and flowers for ſale; and in many of theſe there is at preſent a much greater variety of trees and plants cultivated, than can be found in any other part of Europe. In France their Nurſeries (which are but few, when compared with thoſe in England) are chiefly confined to the propagation of fruit-trees, from whence they have the appellation of Pepinier. For there is ſcarce any of thoſe gardens, where a perſon can be ſupplied either with evergreens, flowering ſhrubs, or foreſt-trees. And in Holland, their Nurſeries are principally for flowers; ſome few of them, indeed, propagate tender exotic plants. But thoſe Nurſeries in the neighbourhood of London do, ſeveral of them, include all theſe, and from hence moſt of the curious perſons abroad are ſupplied with furniture for their gardens; therefore every planter ſhould begin by making of a Nurſery upon the ground which is intended for planting, where a ſufficient number of the trees may be left ſtanding, after the others have been drawn out, to plant in other places; which, for all large-growing trees, but particularly ſuch as are cultivated for timber, will be found by much the moſt advantageous method; for all thoſe trees which come up from the ſeed, or which are tranſplanted very young into the places where they are deſigned to remain, will make a much greater progreſs, and become larger trees than any of thoſe which are tranſplanted at a greater age, and hereby the expence and trouble of ſtaking, watering, &c. will be ſaved, and the trees will ſucceed much better. Theſe ſhould be thinned gradually, as the trees advance; for by taking away too many at firſt, the cold will check the growth of the remaining trees. But then thoſe trees which are taken out from theſe Nurſeries, after a certain age, ſhould not be depended on for planting; and it will be prudence rather to conſign them for fuel, than by attempting to remove them large, whereby in endeavouring to get them up with good roots, the roots of the ſtanding trees will be often much injured.

What has been here propoſed muſt be underſtood for all large plantations in parks, woods, &c. but thoſe Nurſeries which are only intended for the raiſing of evergreens, flowering ſhrubs, or plants which are deſigned to embelliſh gardens, may be confined to one ſpot, becauſe a ſmall compaſs of ground will be ſufficient for this purpoſe. Two or three acres of land employed this way, will be ſufficient for the moſt extenſive deſigns, and one acre will be full enough for thoſe of moderate extent.

Such a Nurſery as this ſhould be conveniently ſituated for water; for where that is wanting, there muſt be an expence attending the carriage of water in dry weather. It ſhould alſo be as near the houſe as it can with conveniency be admitted, in order to render it eaſy to viſit at all times of the year, becauſe it is abſolutely neceſſary that it ſhould be under the inſpection of the maſter, for unleſs he delights in it, there will be little hopes of ſucceſs.

The many advantages which attend the having ſuch a Nurſery, are ſo obvious to every perſon who has turned his thoughts the leaſt to the ſubject, that it is needleſs for me to mention them here; and therefore I ſhall only beg leave to repeat here what I have ſo frequently recommended, which is the carefully keeping the ground always clean from weeds; for if theſe are permitted to grow, they will rob the young trees of their nouriſhment. Another principal buſineſs is, to dig the ground between the young plants at leaſt once every year, to looſen it for the roots to ſtrike out; but if the ground is ſtiff, it will be better to be repeated twice a year, viz. in October and March, which will greatly promote the growth of the plants, and prepare their roots for tranſplanting.

NUX AVELLANA. See Corylus.

NUX JUGLANS. See Juglans.

NUX VESICARIA. See Staphylodendron.

NYCTANTHES. Lin. Gen. Plant. 16. Arabian Jaſmine.

The Characters are,

The empalement of the flower is cylindrical, of one leaf, cut into eight acute ſegments. The flower is of the ſalver ſhape, of one leaf, with a cylindrical tube longer than the empalement, cut into eight ſegments at the top. It hath two ſmall awl-ſhaped ſtamina ſituated in the bottom of the tube, and one roundiſh depreſſed germen, ſupporting a ſingle ſtyle the length of the tube, crowned by a bifid erect ſtigma. The germen afterward becomes a roundiſh berry with two cells, each containing a large roundiſh ſeed.

The Species are,

1. Nyctanthes (*Sambac*) caule volubili, foliis ſubovatis acutis. Hort. Upſal. 4. *Nyctanthes with a winding ſtalk, and oval acute leaves; or the Arabian Jaſmine.*

2. Nyctanthes (*Hirſuta*) petiolis pedunculiſque villoſis. Lin. Sp. Pl. 6. *Nyctanthes with the foot-ſtalks of the leaves and flowers hairy.*

The firſt ſort grows naturally in India, from whence it has been formerly brought to the iſlands in America, where the plants are cultivated for ornament; this riſes with a winding ſtalk to the height of ten or twelve feet, ſending out many ſmall branches, garniſhed with oval ſmooth leaves, of a light green, ſtanding oppoſite on ſhort footſtalks, which are hairy, ending in acute points. The flowers are produced at the end of the branches upon ſhort footſtalks; each generally ſuſtain three flowers, the two lower being oppoſite, and the middle one is longer; theſe have cylindrical empalements, which are ſhort, and cut almoſt to the bottom into eight narrow ſegments. The tube of the flower is narrow, and is cut at the top into eight obtuſe ſegments, which expand horizontally; they are of a pure white, and have a moſt agreeable odour, ſomewhat like the Orange flower, but ſweeter; theſe flowers, when fully blown, drop out of their cups upon being ſhaken, and frequently fall in the night; ſo that when the plants are in full flower, the place under them is often covered with flowers in the morning, which ſoon change to a purpliſh colour. The plants continue flowering great part of the year, when they are kept in a proper temperature of warmth.

The ſecond ſort grows naturally in India, where it riſes to the height of a tree, dividing into many branches, garniſhed with large, oval, ſmooth leaves, of a lucid green, with hairy foot-ſtalks; theſe come out on every ſide the branches, without order. The flowers are produced on the ſide of the branches, from the wings of the leaves, upon long hairy foot-ſtalks, each ſuſtaining ſeven or eight flowers, which are of a pure white, and very fragrant, but have longer tubes than thoſe of the former ſort. The flowers of this plant open in the evening, and drop off in the morning, which has occaſioned ſome to give it the title of Arbor Triſtis, or the Sorrowful Tree, from its caſting the flowers in the morning; this is very rare in Europe at preſent.

The

The plants of the first sort are frequently brought from Italy, by the Italian gardeners, who bring Orange-trees here for sale; but those plants are always grafted upon stocks of the common Jasmine, which do not keep pace in their growth with the graft, so become very unsightly when the plants are grown to any size; besides, the stocks are very subject to shoot from the bottom, and if these shoots are not constantly rubbed off, they will draw the nourishment from the graft, and starve it; therefore the best method to obtain the plants, is to propagate them by layers or cuttings: the former is the surest method, for unless the cuttings are very carefully managed they will not take root; and as the stalks of this sort are pliable, they may be easily brought down, and laid in pots filled with soft loamy soil, which should be plunged into a hot-bed of tan; if the branches are laid down in the spring and carefully watered, they will put out roots by autumn, when they may be cut from the old plants, and each transplanted into a separate small pot, and then plunged into the tan-bed, where they should be shaded from the sun till they have taken new root.

If these plants are propagated by cuttings, they should be planted in April, into pots filled with the before-mentioned earth, and plunged into a moderate hot-bed of tanners bark. The pots should be pretty large, and there may be ten or twelve cuttings planted in each; if these pots are closely covered with bell or hand-glasses to exclude the air, it will greatly promote their taking root; they must also be shaded from the sun in the heat of the day, and gently refreshed with water when the earth is dry; with this management the cuttings will have taken root by August, when they may be transplanted into separate pots, and treated in the same way as the layers.

These plants may be preserved in a moderate degree of warmth, but if they are plunged in the tan-bed of the bark-stove, they will thrive much better, and produce a greater quantity of flowers; and as the leaves continue all the year, the plants will make a fine appearance in the stove at all seasons, and produce flowers great part of the year.

The second sort requires the same treatment, but is much more difficult to propagate, so is very scarce in the European gardens; there was two or three of these plants brought from Florence a few years since, but they were put into the hands of unskilful persons, so were lost; those now here were brought from Malabar.

NYMPHÆA. Tourn. Inst. R. H. 260. tab. 137, 138. The Water Lily.

The Characters are,

The empalement of the flower is composed of four or five coloured leaves. The flower hath many petals sitting on the side of the germen, for the most part in a single series. It hath a great number of short, plain, incurved stamina, with oblong summits. It hath a large oval germen, but no style, with an orbicular, plain, target-shaped stigma. The germen afterward becomes a hard, oval, fleshy fruit, with a narrow neck, crowned at the top, and divided into ten or fifteen cells full of pulp, with many roundish seeds.

The Species are,

1. Nymphæa (*Lutea*) calyce magno pentaphyllo. Flor. Lap. 218. *Water Lily with a large five-leaved empalement; or greater yellow Water Lily.*

2. Nymphæa (*Alba*) foliis cordatis integerrimis, calyce quadrifido. Lin. Sp. Pl. 510. *Water Lily with heart-shaped entire leaves, and a four-pointed empalement; or greater white Water Lily.*

There are some other species of this genus, which are natives of warm countries, but as they cannot, without great difficulty, be cultivated here, so I shall not enumerate them; for unless there is a contrivance for standing water in the stove, in which the plants may be planted, they will not grow; and such a place would be injurious to most other plants in the stove, by occasioning damps, so that unless a stove was contrived on purpose for some of these aquatic plants, it would be imprudent to attempt their cultivation.

The two sorts here mentioned grow naturally in standing waters in many parts of England; they have large roots, which are fastened in the ground, from which arise the stalks to the surface of the water, where the leaves expand and float; they are large and roundish, those of the second sort are heart-shaped. The flowers arise between the leaves, and swim upon the surface of the water. The white sort has a faint sweet scent; these appear in July, and are succeeded by large roundish seed-vessels, filled with shining black seeds, which ripen toward the end of August, when they sink to the bottom of the water.

The best method to propagate these plants is, to procure some of their seed-vessels just as they are ripe, and ready to open; these should be thrown into canals, or large ditches of standing water, where the seeds will sink to the bottom, and the following spring the plants will appear floating upon the surface of the water, and in June or July will produce their beautiful large flowers. When they are once fixed to the place, they will multiply exceedingly, so as to cover the whole surface of the water in a few years.

In some small gardens I have seen these plants cultivated in large troughs of water, where they have flourished very well, and have annually produced great quantities of flowers; but as the expence of these troughs is pretty great (their insides requiring to be lined with lead, to preserve them,) there are few people who care to be at that charge.

NYSSA. Flor. Virg. 121. The Tupelo-tree.

The Characters are,

It has hermaphrodite and male flowers on the same plant. It has no petal, but has ten awl-shaped stamina, with twin summits as long as the stamina. The oval germen, situated under the flower, supports an awl-shaped incurved style, crowned by an acute stigma. The germen afterward becomes an oval berry of one cell, inclosing an oval acute-pointed nut, with rough, angular, irregular furrows.

We have but one Species of this plant, viz.

1. Nyssa (*Aquatica*) foliis integerrimis. Hort. Cliff. 142. *Tupelo-tree with entire leaves.*

This plant grows naturally in Virginia and several other parts of North America, where it rises with a pretty strong upright stalk near twenty feet high, dividing at the top into several branches, garnished with pretty thick, soft, spear-shaped leaves, placed alternately. The flowers come out from the wings of the stalk, upon long foot-stalks; they have a green empalement without petals. Some of them are male, which have ten stamina, and are barren; others are hermaphrodite, having five stamina, and a longer style arising from the germen, which is situated under the flower; these are succeeded by oval berries about the size of small Olives, inclosing a nut of the same form.

Some of these plants have been introduced of late years into the English gardens, but there are few places where they have made much progress; they may be propagated by seeds, but these must be procured from places where they grow naturally, and should be put into the ground as soon as they arrive, for they always lie a year before the plants come up. The best way is to sow them in pots filled with light loamy earth, placing them where they may have only the morning sun; during the first summer the pots must be kept clean from weeds, and in dry weather duly watered. In autumn the pots should be plunged into the ground, and if the winter should prove severe, they ſhould be covered with old tan, Peaſe-haulm, or other light covering, to prevent the froſt from penetrating of the ground. The following ſpring the pots ſhould be plunged into a moderate hot-bed, which may be hooped over and covered with mats, obſerving conſtantly to keep the earth moiſt. This will bring up the plants by the beginning of May; theſe muſt be gradually hardened to bear the open air, and during the following ſummer the pots ſhould be again plunged into an eaſt border, and in dry weather duly watered. In autumn they ſhould be removed into a frame where they may be ſcreened from the froſt, but in mild weather expoſed to the open air. The ſpring following, before the plants begin to ſhoot, they ſhould be parted carefully, and each planted in a ſeparate ſmall pot filled with loamy earth; and if they are plunged into a moderate hot-bed, it will forward their putting out new roots; then they may be plunged in an eaſt border, and treated in the ſame way as in the former ſummer, and in winter ſheltered again under a frame. The ſpring following, ſuch of the plants as have made the greateſt progreſs, may be turned out of the pots, and planted in a loamy moiſt ſoil, in a ſheltered ſituation, where they will endure the cold of this climate; but unleſs the ground is moiſt they make very little progreſs.

O.

OAK. See Quercus.

OBELISCOTHECA. See Rudbeckia.

OCHRUS. See Pisum.

OCULUS CHRISTI. See Horminum sylvestre.

OCYMUM. Tourn. Inſt. R. H. 203. tab. 96. Baſil.

The Characters are,

The empalement of the flower is ſhort, permanent, of one leaf, divided into two lips. The flower is of the lip kind, of one petal inverted; the riſing lip is broad, and cut into four obtuſe equal parts; the reflexed lip is long, narrow, and ſawed. It hath four ſtamina in the lower lip, which are reflexed, two of which are a little longer than the other, terminated by half-moon-ſhaped ſummits. The germen is divided into four parts, which afterward becomes four naked ſeeds incloſed in the empalement.

The Species are,

1. Ocymum (*Baſilicum*) foliis ovatis glabris, calycibus ciliatis. Hort. Cliff. 315. *Baſil with oval ſmooth leaves, and hairy empalements.*

2. Ocymum (*Minimum*) foliis ovatis integerrimis. Hort. Upſal. 169. *Baſil with oval entire leaves; commonly called Buſh Baſil.*

3. Ocymum (*Medium*) hirſutum, foliis ovato-lanceolatis acuminatis dentatis. *Hairy Baſil with oval ſpear-ſhaped leaves which are indented, and end in acute points.*

4. Ocymum (*Americanum*) foliis ovato-oblongis ſerratis, bracteis cordatis reflexis concavis. Lin. Sp. Pl. 597. *Baſil with oval, oblong, ſawed leaves, and heart-ſhaped, concave, reflexed bractea.*

5. Ocymum (*Campechianum*) foliis lanceolatis ſubtus incanis, petiolis longiſſimis villoſis floribus pedunculatis. *Baſil with ſpear-ſhaped leaves, which are hoary on their under ſide, and very long hairy foot-ſtalks to the flowers.*

6. Ocymum (*Fruteſcens*) racemis ſecundis lateralibus, caule erecto. Lin. Sp. Pl. 597. *Baſil with fruitful ſpikes of flowers on the ſide of the ſtalk, which are erect.*

The three firſt ſorts grow naturally in India and Perſia. Of theſe there is a great variety, which differ in the ſize, ſhape, and colour of their leaves, as alſo in their odour; but as theſe differences are accidental, I have not enumerated them, being convinced from repeated experiments, that the ſeeds of one plant will produce many varieties.

The firſt ſort riſes with a branching ſtalk a foot and a half high; the leaves are large, oval, and ſmooth; the ſtalk is hairy, and four-cornered; the leaves are placed by pairs oppoſite, and the branches alſo come out in the ſame manner; the ſtalk is terminated by a whorled ſpike of flowers, which is five or ſix inches long, and the branches are alſo terminated by ſhort ſpikes of flowers of the ſame ſort; the whole plant has a ſtrong ſcent of Cloves.

Of this there are the following varieties:

1. The fringed-leaved Baſil with purple leaves.
2. The green fringed-leaved Baſil.
3. The green Baſil with ſtudded leaves.
4. The large-leaved Baſil.

The ſecond ſort is a low buſhy plant, which ſeldom riſes more than ſix inches high, ſpreading out into branches from the bottom, forming an orbicular head; the leaves are ſmall, oval, and ſmooth, ſtanding oppoſite on ſhort foot-ſtalks. The flowers are produced in whorls toward the top of the branches: they are ſmaller than thoſe of the former ſort, and are ſeldom ſucceeded by ripe ſeeds in England.

Of this there are ſome varieties, as

1. The ſmalleſt Baſil with black purple leaves.
2. The ſmalleſt Baſil with variable leaves.

The third ſort is the common Baſil which is uſed in medicine, and alſo in the kitchen, particularly by the French cooks, who make great uſe of it in their ſoups and ſauces. This riſes about ten inches high, ſending out branches by pairs oppoſite, from the bottom; the ſtalks and branches are four-cornered; the leaves are oval, ſpear-ſhaped, ending in acute points, and are indented on their edges; the whole plant is hairy, and has a ſtrong ſcent of Cloves, too powerful for moſt perſons, but to ſome it is very agreeable: the whole plant is an ingredient in the compound Briony water.

There are ſome varieties of this Species, viz.

1. Common Baſil with very dark green leaves, and a Violet-coloured flower.
2. Curled-leaved Baſil with ſhort ſpikes of flowers.
3. Narrow-leaved Baſil ſmelling like Fennel.
4. Middle Baſil with a ſcent of Citron.
5. Baſil with ſtudded leaves.

6. Basil with leaves of three colours.

The fourth sort grows naturally in India. This rises with a branching stalk a foot and a half high, which is taper, and of a purplish colour; the leaves are short and hairy; they are of an oval oblong figure, ending in obtuse points; they are sawed on their edges, and stand upon pretty long foot-stalks. The stalks are terminated by three spikes of flowers, that in the middle being longer than the other two; the spikes are long and narrow, and the flowers have short foot-stalks; under each whorl of flowers are two small leaves (or bracteæ) placed opposite, which are heart-shaped, concave, and reflexed. The flowers are small, and in some plants are of a purplish colour, but in general they are white; their empalements are smooth, and cut into five parts at the top; the style of the flower is longer than the petal, and the whole plant has a strong, sweet, aromatic odour.

The fifth sort rises with an upright stalk near two feet high, sending out sometimes two, and at others, four branches toward the top, opposite, garnished with spear-shaped leaves; their foot-stalks are two inches long, and are hairy. The flowers grow in whorled spikes at the top of the stalks, standing upon foot-stalks, each sustaining three flowers; these are about the size of those of the common Basil, and are white; the whole plant has a strong aromatic odour.

The sixth sort grows naturally in the island of Ceylon. This rises with a branching stalk about a foot high, garnished with linear spear-shaped leaves, which are sawed. The flowers grow in whorled spikes at the top of the stalks, which are like those of the common Basil; the whole plant has an dour like Anise-seeds.

These plants, being most of them annual, are propagated from seeds, which should be sown in March, upon a moderate hot-bed; and when the plants are come up they should be transplanted into another moderate hot bed, observing to water and shade them until they have taken root; after which they should have plenty of air in mild weather, otherwise they will draw up very weak. In May they should be taken up with a ball of earth to their roots, and transplanted either into pots or borders, observing to shade them until they have taken root; after which they will require no farther care but to clear them from weeds, and refresh them with water in dry weather. Though these plants are only propagated from seeds, yet if you have any particular sort which may arise from seeds, which you are desirous to increase, you may take off cuttings any time in May or June, and plant them on a moderate hot-bed, observing to water and shade them for about ten days; in which time they will take root, and in three weeks time be fit to remove, either into pots or borders, amongst the seedling plants. In September these plants will perfect their seeds, when those sorts, which appear the most distinct, should have their seeds preserved separate, for sowing the following spring.

The seeds of these plants are usually brought from the south of France or Italy every spring, because they seldom ripen their seeds in this country in the open air. But whoever is curious to preserve the seeds of any of the varieties, should place them in an airy glass case or stove in the autumn, when the weather begins to be cold or wet; and by supplying them with water, and letting them have free air every day in mild weather, they will perfect their seeds very well in this country.

The fifth sort is more tender than any of the other. It was discovered growing wild at Campeachy by the late Dr. William Houstoun. This should be sown on a hot-bed early in the spring, and when the plants are come up, they should be transplanted on another very temperate hot-bed to bring them forward; and when they have obtained strength, they should be each transplanted into a separate pot, and placed either in the stove, or on a moderate hot-bed, where they may have a large share of air in warm weather; but by being sheltered from the cold and wet, the plants will perfect their seeds very well in England.

ŒNANTHE. Tourn. Inst. R. H. 312. tab. 166. Water Dropwort.

The CHARACTERS are,

It is a plant with an umbelliferous flower; the principal umbel has but few rays, but the particular umbels have many short ones; the rays of the principal umbel are difform. Those flowers in the disk are hermaphrodite, and are composed of five heart-shaped inflexed petals, which are almost equal. The germen is situated under the flower, supporting two awl-shaped permanent styles, crowned by obtuse segments. The germen afterward becomes an oval fruit, divided into two parts, containing two almost oval seeds, convex on one side, and plain on the other.

The SPECIES are,

1. ŒNANTHE (*Crocata*) foliis omnibus multifidis obtusis subæqualibus. Hort. Cliff. 99. *Water Dropwort, whose leaves all end in many obtuse points, and are almost equal.*

2. ŒNANTHE (*Fistulosa*) stolonifera, foliis caulinis pinnatis filiformibus fistulosis. Lin. Sp. Plant. 254. *Water Dropwort with slender, fistular, winged leaves, growing on the stalks.*

3. ŒNANTHE (*Pimpinelloides*) foliolis radicalibus ovatis incisis, caulinis integris linearibus longissimis simplicioribus. Hort. Cliff. 99. *Water Dropwort whose lower leaves are oval and cut, but those on the stalks entire, single, narrow, and very long.*

4. ŒNANTHE (*Prolifera*) umbellularum pedunculis marginalibus longioribus ramosis masculis. Hort. Upsal. 63. *Water Dropwort whose foot-stalks on the borders of the umbels are longer, branching, and bear male flowers.*

5. ŒNANTHE (*Globulosa*) fructibus globosis. Hort. Cliff. 99. *Water Dropwort with globular fruit.*

The first of those here mentioned is very common by the sides of the Thames on each side London, as also by the sides of large ditches and rivers in divers parts of England: this plant commonly grows four or five feet high with strong jointed stalks, which, being broken, emit a yellowish fœtid juice; the leaves are somewhat like those of the common Hemlock, but are of a lighter green colour: the roots divide into four or five large taper ones, which, when separated, have some resemblance to Parsneps; for which some ignorant persons have boiled them, whereby themselves and family have been poisoned.

This plant is one of the most poisonous we know; the juice, which is at first like milk, turns afterward to a Saffron colour: if a person should swallow ever so little of this juice, it will so contract every part it touches, that there will immediately follow a terrible inflammation and gangrene; and, which is worse, there has not yet been found any antidote against it; for which reason we ought to be very careful to know this plant, in order to avoid it, for fear we should take it for any other like it, which would certainly prove fatal.

The poisonous quality of this plant hath led some persons to believe it to be the Cicuta of the ancients; but according to Wepfer, the Sium alterum olusatri facie of Lobel, is what the ancients called Cicuta, as may be seen at large in Wepfer's book de Cicuta.

The second sort is very common in moist soils, and by the sides of rivers in divers parts of England: this is not supposed to be near so strong as the first, but is of a poisonous quality.

All the sorts of these plants naturally grow in moist places, so that whoever hath a mind to cultivate them, should sow their seeds soon after they are ripe in autumn, upon a moist soil, where they will come up, and thrive exceedingly the following summer, and require no farther care but to clear them from weeds.

ŒNOTHERA. Lin. Gen. Plant. 424. Tree, or Night Primrose.

The Characters are,

The empalement of the flower is of one leaf, cut into four acute segments at the brim, which turn backward. The flower has four heart-shaped petals, which are lengthways inserted in the divisions of the empalement. It hath eight awl-shaped incurved stamina, which are inserted in the tube of the empalement. The cylindrical germen is situated under the tube of the empalement, supporting a slender style, crowned by a thick, quadrifid, obtuse, reflexed stigma. The germen afterward becomes a four-cornered cylindrical capsule, having four cells, which are filled with small angular seeds.

The Species are,

1. Œnothera (*Biennis*) foliis ovato-lanceolatis planis caule muticato. Vir. Cliff. 33. *Tree Primrose with plain, oval, spear shaped leaves.*

2. Œnothera (*Angustifolia*) foliis lanceolatis dentatis, caule hispido. *Tree Primrose with spear-shaped indented leaves, and a prickly stalk.*

3. Œnothera (*Glabra*) foliis lanceolatis planis, caule glabro. *Tree Primrose with plain spear-shaped. leaves, and a smooth stalk.*

4. Œnothera (*Mollissima*) foliis lanceolatis undulatis. Vir. Cliff. 33. *Tree Primrose with waved spear-shaped leaves.*

5. Œnothera (*Pumila*) foliis radicalibus ovatis, caulinis lanceolatis obtusis, capsulis ovatis sulcatis. Tab. 188. *Tree Primrose with oval leaves at the root, those on the stalks spear-shaped and blunt-pointed, and oval furrowed seed-vessels.*

The other species which have been formerly placed in this genus, are now under Jussiæ and Ludwigia, to which the reader is desired to turn.

The three first sorts grow naturally in Virginia, and in other parts of North America, from whence their seeds were brought to Europe in the beginning of the sixteenth century; but they are now become so common in many parts of Europe, as to be taken for natives. The first hath a long, thick, taper root, which runs deep into the ground, from which arise many oblong leaves, which spread flat on the surface of the ground; between these the stalk comes out, which rises between three and four feet high; the stalk is of a pale green colour, a little hairy, and about the thickness of a finger, full of pith; this is garnished with long narrow leaves set close to the stalk without order. The flowers are produced all along the stalk from the wings of the leaves, the germen sitting close to the stalk, from the top of which arises the tube of the flower, which is narrow; at the top is the empalement, which is cut into four acute segments, reflexed downward. The petal of the flower is cut into four large obtuse segments, which in the evening are expanded quite flat, but are shut in the day; these are of a bright yellow colour. From the flower opening in the evening, many persons call it the Night Primrose. The plants begin to flower about Midsummer, and as the stalks advance in height, so other flowers are produced, whereby there is a succession of flowers on the same plant till autumn.

The second sort hath red stalks, which are set with rough protuberances; it does not rise so high as the first; the leaves are narrower, and the flowers are smaller.

The third sort differs from the first, in having shorter stalks, narrower leaves, and smaller flowers: and from the second, in having smooth stalks, which are of a pale green colour. These differences are permanent, so they are undoubtedly different species.

The fourth sort grows naturally at Buenos Ayres. This hath a shrubby stalk more than two feet high, garnished with narrow, hairy, spear-shaped leaves, ending in acute points, a little waved on their edges. The flowers come out from the wings of the leaves along the stalks, like the other sorts; they are first of a pale yellow; but as they decay change to an Orange colour; they are smaller than those of either of the former sorts, and expand only in the evening; the seed-vessels are slender, taper, and hairy. This flowers at the same time with the former.

The fifth sort grows naturally in Canada, from whence the seeds were brought to Paris a few years past. This is a perennial plant, the root is fibrous; the lower leaves are oval and small, sitting close to the ground; the stalk is slender, near a foot high, and is garnished with small spear-shaped leaves, of a light green, ending in blunt points, sitting close to the stalk. The flowers come out from the wings of the leaves like the other species; these are small, of a bright yellow colour, and appear at the same times as the former, and are succeeded by short, oval, furrowed seed-vessels, filled with small seeds.

The three first sorts are very hardy plants, which if once brought into a garden, and the seeds permitted to scatter there will be a supply of plants without any care. They are biennial, and perish after they have perfected their seeds. The seeds of these plants should be sown in the autumn, for those which are sown in the spring seldom rise the same year; when the plants come up, they should be thinned and kept clean from weeds, which is all the care they require till the autumn, when they should be transplanted to the places where they are designed to flower; but as the roots of these plants strike deep in the ground, so there should be care taken not to cut or break them in removing. The plants will thrive in almost any soil or situation, and will flower in London in small gardens, better than most other plants.

The fourth sort is now become pretty common in the English gardens, for if the seeds of this are permitted to scatter, the plants will come up the following spring, and require no other care but to keep them clean from weeds, and thin them where they grow too close. If these plants are kept in pots, and placed in a green-house in the autumn, they will live through the winter; but as they produce flowers and seeds in the open air, the plants are seldom preserved longer.

The fifth sort is perennial, and may be propagated either by parting of the roots or by seeds; if it is by the former, the best time for doing it is in the spring; but if they are propagated by seeds, these should be sown in the autumn; and the surest way is to sow the seeds in pots, and place them under a hot-bed frame in winter; in the spring the plants will appear, and when they are fit to remove, a few of them may be planted in small pots, to be sheltered under a common frame in the winter; and the others may be planted in a sheltered border, where they will endure the cold of our ordinary winters very well, and the following summer they will produce flowers and seeds in plenty; so there will be little occasion for parting of their roots, because the seedling plants will be much stronger and flower better than those propagated by offsets.

OLDENLANDIA. Plum. Nov. Gen. 42. tab. 36.

The Characters are,

The empalement of the flower is permanent, sitting upon the germen. The flower has four oval petals, which spread open, and four stamina, terminated by small summits. It hath a roundish germen situated under the flower, crowned by an indented stigma.

ſtigma. The germen afterward turns to a globular capſule with two cells, filled with ſmall ſeeds.

We have but one SPECIES of this genus, viz.

OLDENLANDIA (*Corymboſa*) pedunculis multifloris, foliis lineari-lanceolatis. Lin. Sp. Pl. 119 *Oldenlandia with many flowers on a foot-ſtalk, and linear ſpear-ſhaped leaves.*

The ſeeds of this plant were ſent into England by Mr. Robert Millar, who gathered them in Jamaica. It is a low annual plant, which divides into many branches, ſpreading upon the ground. Theſe branches are garniſhed with long narrow leaves, placed oppoſite. From the wings of the leaves ariſes the flower-ſtalk, which grows about an inch, or a little more in length, and divides into three or four ſmaller foot-ſtalks; on the top of each of theſe ſtands one ſmall white flower.

The ſeeds of this plant ſhould be ſown early in the ſpring on a hot-bed, and when the plants are come up, they ſhould be tranſplanted on another hot bed, or into ſmall pots, and plunged into a moderate hot bed of tanners bark, obſerving to water and ſhade them until they have taken root; after which time they muſt have a large ſhare of free air in warm weather, and muſt be frequently refreſhed with water. With this management the plants will flower in June, and their ſeeds will ripen ſoon after, ſo that the ſeeds muſt be gathered from time to time as they ripen; for as the branches grow larger, ſo there will be freſh flowers produced until autumn, when the plants will periſh; but if the ſeeds are permitted to ſcatter in the pots, the plants will ſoon after appear, which will live through the winter, provided they are placed in the ſtove, and will flower early the following ſpring.

OLEA. Tourn. Inſt. R. H. 598. tab. 370. The Olive.

The CHARACTERS are,

It has a ſmall tubulous empalement of one leaf, cut into four ſegments at the top. The flower conſiſts of one petal, which is tubulous, cut at the brim into four ſegments. It has two ſhort ſtamina, terminated by erect ſummits, and a roundiſh germen ſupporting a ſhort ſingle ſtyle, crowned by a thick bifid ſtigma. The germen afterward turns to an oval ſmooth fruit (or berry) with one cell, incloſing an oblong oval nut.

The SPECIES are,

1. OLEA (*Gallica*) foliis lineari-lanceolatis ſubtus incanis. *Olive with linear ſpear-ſhaped leaves, which are hoary on their under ſide; commonly called Provence Olive.*

2. OLEA (*Hiſpanica*) foliis lanceolatis, fructu ovato. *Olive with ſpear-ſhaped leaves, and an egg-ſhaped fruit; called the Spaniſh Olive.*

3. OLEA (*Sylveſtris*) foliis lanceolatis obtuſis rigidis, ſubtus incanis. *Olive with ſpear-ſhaped, obtuſe, rigid leaves, which are hoary on their under ſide; or the Wild Olive.*

4. OLEA (*Africana*) foliis lanceolatis lucidis, ramis teretibus. *Olive with ſpear-ſhaped ſhining leaves, and taper branches; called African Olive.*

5. OLEA (*Buxifolia*) foliis ovatis rigidis ſeſſilibus. *Olive with oval ſtiff leaves, ſitting cloſe to the branches; commonly called Box-leaved Olive.*

The firſt ſort is what the inhabitants of the ſouth of France chiefly cultivate, becauſe from this ſpecies the beſt oil is made, which is a great branch of trade in Provence and Languedoc, and it is the fruit of this ſort which is moſt eſteemed when pickled: of this there are ſome varieties; the firſt is called Olive Picholine; there is another with dark green fruit, one with white fruit, and another with ſmaller and rounder fruit; but as theſe are ſuppoſed to be only accidental varieties, which have riſen from the ſame ſeeds, I have not enumerated them.

The Olive ſeldom riſes to be a large tree, and is rarely ſeen with a ſingle ſtem, but frequently two or three ſtems riſe from the ſame root; theſe grow from twenty to thirty feet high, putting out branches from their ſides almoſt their whole length, which are covered with a gray bark, and garniſhed with ſtiff leaves, of a lively green on their upper ſide, and hoary on their under, ſtanding oppoſite. The flowers are produced in ſmall bunches from the wings of the leaves; they are ſmall, white, and have ſhort tubes, ſpreading open at the top; theſe are ſucceeded by oval fruit, which, in warm countries, ripen in the autumn.

The ſecond ſort is chiefly cultivated in Spain, where the trees grow to a much larger ſize than the former ſort; the leaves are much larger, and not ſo white on their under ſide; and the fruit is near twice the ſize of thoſe of the Provence Olive, but are of a ſtrong rank flavour, and the oil made from theſe is too ſtrong for moſt Engliſh palates.

The third ſort is the wild Olive, which grows naturally in woods in the ſouth of France, Spain, and Italy, ſo is never cultivated; the leaves of this ſort are much ſhorter and ſtiffer than thoſe of the other; the branches are frequently armed with thorns, and the fruit is ſmall and of no value.

The fourth and fifth ſorts grow naturally at the Cape of Good Hope; the fourth riſes to the height of the firſt, to which it bears ſome reſemblance, but the bark is rougher; the leaves are not ſo long, and are of a lucid green on their upper ſide; but as this does not produce fruit in Europe, I can give no account of it.

The fifth ſort is of humbler growth, ſeldom riſing more than four or five feet high, ſending out branches from the root upward, forming a buſhy ſhrub; the branches are taper, and covered with a gray bark; the leaves are oval, very ſtiff, and ſmaller than thoſe of the other ſpecies. This has not produced any fruit in England.

All theſe ſorts are preſerved in the gardens of the curious, but they are rather too tender to thrive in the open air in the neighbourhood of London, where they are ſometimes planted againſt walls, and with a little protection in very ſevere froſt, they are maintained pretty well; but in Devonſhire there are ſome of theſe trees, which have grown in the open air many years, and are ſeldom injured by the froſt, but the ſummers are not warm enough to bring the fruit to maturity. There were ſeveral of theſe trees planted againſt a warm wall at Cambden-houſe, near Kenſington, which ſucceeded very well, till their tops were advanced above the wall; after which they were generally killed in winter, ſo far down as to the top of the wall. Theſe in 1719 produced a good number of fruit, which grew ſo large as to be fit for pickling; but ſince that time, their fruit has ſeldom grown to any ſize.

In Languedoc and Provence, where the Olive-tree is greatly cultivated, they propagate it by truncheons ſplit from the roots of the trees; for as theſe trees are frequently hurt by hard froſts in winter, ſo when the tops are killed, they ſend up ſeveral ſtalks from the root; and when theſe are grown pretty ſtrong, they ſeparate them with an ax from the root; in the doing of which, they are careful to preſerve a few roots to the truncheons; theſe are cut off in the ſpring, after the danger of froſt is over, and planted about two feet deep in the ground, covering the ſurface with litter or mulch, to prevent the ſun and wind from penetrating and drying of the ground; when the plants have taken new root, they are careful to ſtir the ground, and deſtroy the weeds.

This tree will grow in almoſt any ſoil, but when it is planted in rich moiſt ground, they grow larger and make a finer appearance than in poor land; but the fruit is of leſs eſteem, becauſe the oil made from it is not ſo good as that which is produced in a leaner ſoil. The chalky ground is eſteemed the beſt for theſe trees, and the oil which is made from the trees growing in that ſort of land is much finer, and will keep longer than the other.

In

In the countries where the inhabitants are curious in the making of their oil, they are frequently obliged to get truncheons of the ordinary sorts of Olives to plant; but after they have taken good root, they graft them with that sort of Olive which they prefer to the other. In Languedoc they chiefly propagate the Cormeau, the Ampoulan, and Moureau, which are three varieties of the first species; but in Spain the second sort is generally cultivated, where they have more regard to the size of the fruit, and the quantity of oil they will produce, than to their quality.

In England the plants are only preserved by way of curiosity, and are placed in winter in the green-house for variety, so I shall next give an account of the method by which they are here propagated, with their manner of treatment.

These plants may be propagated by laying down their tender branches (in the manner practised for other trees) which should remain undisturbed two years; in which time they will have taken root, and may then be taken off from the old plants, and transplanted either into pots filled with fresh light earth, or into the open ground in a warm situation. The best season for transplanting them is the beginning of April, when you should, if possible, take the opportunity of a moist season; and those which are planted in pots, should be placed in a shady part of the green-house until they have taken root: but those planted in the ground should have mulch laid about their roots, to prevent the earth from drying too fast, and now and then refreshed with water; but you must by no means let them have too much moisture, which will rot the tender fibres of their roots, and destroy the trees. When the plants have taken fresh root, those in the pots may be exposed to the open air, with other hardy exotics, with which they should be housed in winter, and treated as Myrtles, and other less tender trees and shrubs; but those in the open air will require no farther care until the winter following, when you should mulch the ground about their roots, to prevent the frost from penetrating deep into it; and if the frost should prove very severe, you should cover them with mats, which will defend them from being injured thereby; but you must be cautious not to let the mats continue over them after the frost is past, lest by keeping them too close, their leaves and tender branches should turn mouldy for want of free air, which will be of as bad consequence to the trees, as if they had been exposed to the frost, and many times worse; for it seldom happens, if they have taken much of this mould, or have been long covered, so that it has entered the bark, that they are ever recoverable again; whereas it often happens, that the frost only destroys the tender shoots; but the body and larger branches remain unhurt, and put out again the succeeding spring.

These trees are generally brought over from Italy every spring, by the persons who bring over Oranges, Jasmines, &c. from whom they may be procured pretty reasonable; which is a better method than to raise them from layers in this country, that being too tedious; and those which are thus brought over, have many times very large stems, to which size young plants in this country would not arrive in ten or twelve years growth. When you first procure these stems, you should (after having soaked their roots twenty-four hours in water, and cleaned them from the filth they have contracted in their passage) plant them in pots, filled with fresh light earth, and plunge them into a moderate hot-bed, observing to screen them from the violence of the sun in the heat of the day, and also to refresh them with water, as you shall find the earth in the pots dry. In this situation they will begin to shoot in a month or six weeks after, when you should let them have air in proportion to the warmth of the season; and after they have made pretty good shoots, you should inure them to the open air by degrees, into which they should be removed, placing them in a situation where they may be defended from strong winds; in this place they should remain till October following, when they must be removed into the green-house, as was before directed. Having thus managed these plants until they have acquired strong roots, and made tolerable good heads, you may draw them out of the pots, preserving the earth to their roots, and plant them in the open air in a warm situation, where you must manage them as was before directed for the young ones, and these will in two or three years produce flowers, and in very warm seasons some fruit, provided they do well. The Lucca and Box-leaved Olives are the hardiest, for which reason they should be preferred to plant in the open air, but the first sort will grow to be the largest trees.

OMPHALODES. See Cynoglossum.

ONAGRA. See Œnothera.

ONIONS. See Cepa.

ONOBRYCHIS. See Hedysarum.

ONONIS. Lin. Gen. Plant. 772. Rest-harrow, Cammock, Petty-whin.

The Characters are,

The empalement of the flower is cut into five narrow segments, the upper being a little raised and arched, the lower bending under the keel. The flower is of the butterfly kind. The standard is heart-shaped, and larger than the wings. The wings are oval and short; the keel is pointed, and longer than the wings. It hath ten stamina joined together, and an oblong hairy germen, supporting a single style, crowned by an obtuse stigma. The germen afterward becomes a turgid pod with one cell, inclosing kidney-shaped seeds.

The Species are,

1. Ononis (*Spinosa*) floribus subsessilibus solitariis lateralibus, caule spinoso. Hort. Cliff. 359. *Rest-harrow with single flowers sitting close to the sides of the branches, and a prickly stalk; called Cammock, or Petty-whin.*

2. Ononis (*Mitis*) floribus subsessilibus solitariis lateralibus, ramis inermibus. Hort. Cliff. 359. *Rest-harrow with single flowers sitting close to the stalks, and branches without spines.*

3. Ononis (*Repens*) caulibus diffusis, ramis erectis, foliis superioribus solitariis stipulis ovatis. Lin. Sp. 1006. *Trailing maritime Rest-harrow.*

4. Ononis (*Tridentata*) foliis ternatis carnosis sublinearibus tridentatis. Lin. Sp. Plant. 718. *Rest-harrow with trifoliate fleshy leaves, which are narrow, and have three indentures.*

5. Ononis (*Fruticosa*) floribus paniculatis, pedunculis subtrifloris, stipulis vaginalibus, foliis ternatis. Hort. Cliff. 358. *Rest-harrow with paniculated flowers, generally growing three upon a foot-stalk, sheath-like stipulæ, and trifoliate leaves; or purple shrubby Rest-harrow.*

6. Ononis (*Natrix*) pedunculis unifloris aristatis, foliis terminatis stipulis integerrimis. Hort. Cliff. 358. *Rest-harrow with one flower on a foot-stalk, which is terminated by trifoliate leaves.*

7. Ononis (*Viscosa*) pedunculis unifloris, filo terminatis, caule ramoso villoso, foliis ternatis serratis. *Rest-harrow with one flower on each foot-stalk, which are terminated by a thread, a branching hairy stalk, and trifoliate sawed leaves; or broad-leaved erect Rest-harrow of Portugal.*

8. Ononis (*Minutissima*) floribus sessilibus lateralibus, foliis omnibus ternatis petiolatisque, stipulis setaceis. Lin. Sp. Plant. 717. *Rest-harrow with flowers sitting close to the sides of the stalks, all the leaves trifoliate, growing upon foot-stalks, and bristly stipulæ.*

9. Ononis

9. Ononis (*Ornithopodoides*) pedunculis bifloris, ariſtatis, leguminibus linearibus cernuis. *Reſt-harrow with two flowers on a foot-ſtalk, which are terminated by a thread.*

10. Ononis (*Rotundifolia*) pedunculis trifoliis, foliis ternatis ſubrotundis. Hort. Cliff. 358. *Reſt-harrow with naked foot-ſtalks, having three flowers, and trifoliate leaves.*

11. Ononis (*Mitiſſima*) floribus ſeſſilibus ſpicatis, bracteis ſtipularibus, ovatis ventricoſis ſcarioſis imbricatis. Lin. Sp. 1007. *Smooth, annual, purple, Fox-tail Reſt harrow.*

12. Ononis (*Alopecuroides*) ſpicis folioſis ſimplicibus ovatis obtuſis ſtipulis dilatis. Lin. Sp. 1008. *Fox-tail Reſt-harrow of Sicily.*

13. Ononis (*Criſtata*) ſpicis folioſis ſimplicibus obtuſis. Lin. Sp. Pl. 717. *Reſt-harrow with leafy ſpikes, and ſingle obtuſe leaves.*

14. Ononis (*Anil*) foliis ternatis ovatis, petiolis longiſſimis, leguminibus hirſutis. *Reſt-harrow with oval trifoliate leaves, growing on very long foot-ſtalks, and hairy pods.*

The firſt ſort is a common weed in moſt parts of England, ſo is rarely admitted into gardens. It has a ſtrong creeping root, which ſpreads far in the ground, and is with great difficulty eradicated; the ſtalks riſe two feet and a half high; they are ſlender, reddiſh, and hairy, ſending out ſeveral branches on their ſide, which are armed with ſharp prickles. The flowers come out ſingle from the ſide of the branches; they are of the butterfly kind, of a purple colour, which are ſucceeded by ſmall pods, containing one or two kidney-ſhaped ſeeds. It flowers great part of ſummer, and the ſeeds ripen in the autumn. The root of this is one of the five opening roots; the cortical part of it is eſteemed a good medicine for ſtoppage of urine, and to open the obſtructions of the liver and ſpleen; there is a variety of this with white flowers.

The ſecond ſort grows naturally in many parts of England, and has been by ſome ſuppoſed to be only a variety of the firſt; but I have cultivated both by ſeeds, and have always found the plants retain their difference; the ſtalks of this ſort are hairy, and more diffuſed than thoſe of the firſt; the leaves are broader, and ſit cloſer on the branches; the ſtalks grow more upright, aud have no ſpines; the flowers and pods are like thoſe of the firſt. There is alſo a variety of this with white flowers.

The third ſort grows naturally on the borders of the ſea in ſeveral parts of England; this hath a creeping root, from which ariſe many hairy ſtalks, which are near two feet long, ſpreading on every ſide upon the ground, garniſhed with trifoliate hairy leaves, thoſe on the lower part of the ſtalks being pretty large and oval, but the upper are ſmaller and narrower. The flowers are like thoſe of the firſt in ſhape, coming out ſingly from the ſide of the ſtalks, but are of a brighter purple colour; the pods are ſhort, containing two or three ſeeds in each. It flowers in July, and the ſeeds ripen in autumn.

The fourth ſort grows naturally in Spain and Portugal; this riſes with ſhrubby ſtalks two feet and a half high, dividing into ſlender branches, very full of joints, garniſhed with narrow, trifoliate, thick, fleſhy leaves, ſtanding upon ſhort foot-ſtalks. The flowers are produced at the end of the branches in looſe panicles, ſome of the foot-ſtalks ſuſtaining two, and others but one flower; they are of a fine purple colour, and appear in June; the ſeeds ripen in September.

The fifth ſort grows naturally on the Alps; this is a very beautiful low ſhrub; it riſes with ſlender ſhrubby ſtalks about three feet high, dividing into many branches, which are garniſhed with narrow trifoliate leaves, ſawed on their edges, ſitting cloſe to the branches. The flowers come out in panicles at the end of the branches upon long foot-ſtalks, which for the moſt part ſuſtain three large purple flowers; the ſtipula is a kind of ſheath, embracing the foot-ſtalk of the flower. It flowers in May, and the flowers are ſucceeded by turgid pods about an inch long, which are hairy, incloſing three or four kidney-ſhaped ſeeds, which ripen in Auguſt.

The ſixth ſort grows naturally in the ſouth of France and in Spain; this hath a perennial root, and an annual ſtalk which riſes near two feet high, ſending out ſhort branches from the ſide of the lower part; theſe are garniſhed with trifoliate oblong leaves, which are hairy and clammy. The flowers grow in looſe ſpikes at the end of the ſtalks; they are large, and of a bright yellow colour, ſtanding upon pretty long foot-ſtalks, which are extended beyond the flowers, the flowers hanging downward from the middle of the foot-ſtalk. The flowers appear the latter end of June, which are ſucceeded by turgid pods an inch long, containing three or four brown kidney-ſhaped ſeeds, which ripen in September.

The ſeventh ſort grows naturally in Portugal, from whence the ſeeds were ſent to me. This is an annual plant, with a ſtrong, herbaceous, hairy ſtalk, riſing a foot and a half high, ſending out branches the whole length, cloſely garniſhed with trifoliate leaves; the middle lobe being large and oval, the two ſide lobes long and narrow, rounded at their points, and indented on their edges; they are very clammy. The foot-ſtalks of the flowers come out from the wings of the ſtalks ſingly, each ſuſtaining one pale yellow flower, ſtanding erect in the middle of the foot ſtalk, which is extended beyond the flower. This plant flowers in July, and the ſeeds ripen in the autumn.

The eighth ſort grows naturally in the ſouth of France and Italy: this is an annual plant; the ſtalks riſe about nine inches high, ſending out one or two ſide branches toward the bottom; the leaves are ſmall, trifoliate, and oval, ſtanding upon pretty long foot ſtalks, and are indented on their edges. The flowers come out ſingly at the wings of the ſtalk; they are ſmall, yellow, and ſit very cloſe to the ſtalk, having a ſharp briſtly ſtipula under the empalement; the pods are very ſhort and turgid, containing two or three kidney-ſhaped ſeeds. It flowers in July, and the ſeeds ripen in the autumn.

The ninth ſort grows naturally in Sicily, and is an annual plant; the ſtalks riſe about nine inches high, ſending out one or two branches toward the bottom; theſe are garniſhed with ſmall trifoliate leaves, which ſtand on ſhort foot-ſtalks. The flowers come out from the ſide of the branches upon ſhort foot-ſtalks, each ſuſtaining two ſmall yellow flowers which are ſucceeded by jointed compreſſed pods like thoſe of Bird's-foot, having four or five kidney-ſhaped ſeeds in each. This ſort flowers in July, and the ſeeds ripen in the autumn.

The tenth ſort grows naturally on the Alps and Helvetian mountains; this riſes with a ſingle jointed ſtalk a foot and a half high, garniſhed with oval, indented, trifoliate leaves, ſtanding on pretty long foot ſtalks. The foot-ſtalks of the flowers come out from the wings of the leaves; they are long, ſlender, each ſuſtains three pale yellow flowers, which are ſucceeded by ſhort turgid pods, containing two or three ſeeds in each. It flowers in June, and the ſeeds ripen in September.

The eleventh ſort grows naturally in Virginia, from whence I received the ſeeds. This is a biennial plant; from the root comes out many diffuſed ſtalks which trail upon the ground, garniſhed with roundiſh trifoliate leaves, indented on their edges, having ſhort foot-ſtalks; they are of a light green and ſmooth. The flowers come out toward the end of the branches upon very ſlender foot-ſtalks, which ariſe from the wings of the leaves, each ſuſtaining five ſmall yellow flowers; theſe are ſucceeded by compreſſed pods,

 ſhaped

ſhaped like a half-moon, or Medick Trefoil. This flowers in July, and the ſeeds ripen in autumn.

The twelfth ſort came up in earth which was brought from Barbadoes, but it does not ſeem to be a native of that country, for it riſes eaſily from ſeeds in the open air here, and perfects its ſeeds in the autumn, nor will it thrive in greater warmth. This hath an upright ſtalk a foot and a half high, ſending out ſmall ſide branches, which are garniſhed with roundiſh trifoliate leaves, ſawed on their edges, ſtanding upon ſhort foot-ſtalks. The flowers grow in ſhort leafy ſpikes at the end of the branches; they are ſmall, and of a pale purple colour, appearing in July, and are ſucceeded by ſhort turgid pods, containing two or three kidney-ſhaped ſeeds, which ripen in autumn.

The thirteenth ſort grows naturally in Portugal, Spain, and Italy. This is an annual plant, riſing with upright branching ſtalks a foot high, garniſhed with ſingle leaves, ſitting cloſe to the ſtalks; the larger leaves are oval, about one inch long, and three quarters of an inch broad; the upper leaves are narrow, ending in obtuſe points, and are ſlightly indented at their ends. The flowers grow in leafy ſpikes at the end of the ſtalks ſet cloſe together, having hairy empalements; they are pretty large, of a purple colour, and appear in July; theſe are ſucceeded by taper pods about an inch long, incloſing four or five kidney-ſhaped ſeeds. This plant has ſeveral titles, in the different books of botany

The fourteenth ſort grows naturally in the American iſlands. This is an annual plant, riſing with a branching ſtalk two feet high, garniſhed with trifoliate leaves, whoſe lobes are oval, ſtanding upon very long foot-ſtalks, which are hairy. The flowers grow in looſe ſpikes at the end of the branches; they are large, of a purpliſh yellow colour, and are ſucceeded by very turgid hairy pods, each containing five or ſix large kidney-ſhaped ſeeds. This ſort flowers in July and Auguſt, and the ſeeds ripen in the autumn. From this plant Indigo was formerly made, which, I ſuppoſe, was of leſs value than that which is made of the true plants, ſo has not been for many years paſt cultivated in any of the iſlands.

The three firſt ſorts are never cultivated in gardens; theſe are very troubleſome weeds whenever they get into the fields, for the roots ſpread and multiply greatly in the ground, and are ſo tough and ſtrong, that the plough will ſcarcely cut through them, ſo are with great difficulty eradicated when they have once gotten poſſeſſion.

The fourth and fifth ſorts are low ſhrubby plants, which are propagated by ſeeds. The fourth is too tender to thrive in the open air in England, unleſs it is planted in a warm ſituation, and in very ſevere froſt covered to protect it. If the ſeeds of both ſorts are ſown upon a bed of light earth in April, the plants will come up in May, when they muſt be kept clean from weeds; and if they are too cloſe, ſome of them ſhould be carefully drawn up in moiſt weather, and tranſplanted at four or five inches diſtance; thoſe of the fourth ſort upon a warm ſheltered border, but the fifth may be planted in a ſhady border, were they will thrive very well; after theſe have taken root, they will require no other care but to keep them clean from weeds till the following autumn, when they may be tranſplanted to the places where they are to remain; thoſe plants which were left growing in the bed where they were ſown, muſt alſo be treated in the ſame way. Theſe plants will not thrive in pots, therefore ſhould always be planted in the full ground, where the fifth ſort will flouriſh greatly, and frequently ſend up many plants from their roots, but the other is more impatient of cold. Theſe plants will flower the ſecond year, and make a fine appearance during the continuance of their flowers, and the fifth ſort will produce ſeeds in plenty.

The ſixth ſort is propagated by ſeeds, which ſhould be ſown thin in drills upon a bed of light earth, and when the plants come up, they muſt be kept clean from weeds till the autumn, when they ſhould be carefully taken up, and tranſplanted into the borders of the pleaſure-garden, where they are to remain; the ſecond year they will flower and produce ripe ſeeds, but the roots will continue ſome years, and are very hardy.

The ſeventh, ninth, and thirteenth ſorts, are annual hardy plants; theſe are propagated by ſeeds, which ſhould be ſown in the places where the plants are to remain, and will require no other care but to thin them where they are too cloſe, and keep them clean from weeds.

The eleventh ſort is a biennial plant. The ſeeds of this ſhould be ſown on a bed of freſh earth, where the plants are to remain, and when they come up, if they are thinned where they grow too cloſe, and are kept clean from weeds, they will require no other culture.

The twelfth ſort is propagated by ſeeds, which ſhould be ſown either on a moderate hot-bed or a warm border in the ſpring; and when the plants are fit to remove, they ſhould be each tranſplanted into a ſeparate ſmall pot, plunging them into a gentle hot-bed, obſerving to ſhade them till they have taken new root, then they ſhould be gradually inured to the open air; the latter end of May, or the beginning of June, they may be fully expoſed to the open air, but in autumn they ſhould be placed under a common hot-bed frame, to ſcreen them from froſt in winter. The ſpring following they may be ſhaken out of the pots, and planted in the full ground, where they are to remain. As theſe plants have long tap-roots, they will not thrive long in pots; and if they are planted in wet ground, their roots will rot in winter, but in a dry ſoil they are never hurt by cold, and their roots will abide many years.

The thirteenth ſort is an annual plant; the ſeeds of this muſt be ſown upon a moderate hot-bed in the ſpring, and, when the plants are fit to remove, they ſhould be tranſplanted on another hot-bed, to bring the plants forward, treating them in the ſame way as the African and French Marygold. In June they ſhould be taken up with balls of earth to their roots, and tranſplanted into the open borders, where, if they are ſhaded till they have taken root, they will thrive and flower the following month, and perfect their ſeeds in autumn.

ONOPORDUM. Lin. Gen. Plant. 834. Woolly Thiſtle.

The CHARACTERS are,

The common empalement is roundiſh, bellied, and imbricated. The flower is compoſed of many hermaphrodite florets, which are funnel-ſhaped, equal, and uniform, having narrow tubes, ſwelling at the brim, cut into five points; they have five ſhort hairy ſtamina, terminated by cylindrical ſummits, and an oval germen crowned with down, ſupporting a ſlender ſtyle, terminated by a crowned ſtigma. The germen becomes a ſingle ſeed crowned with down, ſitting in the empalement.

The SPECIES are,

1. ONOPORDUM (*Acanthium*) calycibus ſquarroſis, foliis ovato-oblongis ſinuatis. Lin. Sp. Plant. 827. *Woolly Thiſtle with rough empalements, and oblong, oval, ſinuated leaves; or common Woolly Thiſtle.*

2. ONOPORDUM (*Illyricum*) calycibus ſquaroſis, foliis linearibus pinnatifidis. Lin. Sp. Plant. 827. *Woolly Thiſtle with rough empalements, and narrow leaves, ending in many points.*

3. ONOPORDUM (*Arabicum*) calycibus imbricatis. Hort. Upſal. 249. *Woolly Thiſtle with imbricated empalements.*

4. ONOPORDUM (*Orientale*) calycibus ſquarroſis, foliis oblongis, pinnato-ſinuatis decurrentibus, capite magno. *Woolly Thiſtle with rough empalements, oblong, ſinuated,*

nted, wing-pointed leaves running along the stalk, and a large head.

5. ONOPORDUM (*Acaule*) subacaule. Lin. Sp. Pl. 1159. *Woolly Thistle without stalks.*

The first sort grows naturally on uncultivated places in most parts of England. It is a biennial plant; the first year it puts out many large downy leaves, which are sinuated on their edges, and are prickly; these spread on the ground, and continue the following winter, and in the spring arises the stalk in the middle of the leaves, which upon dunghills or good ground, grows five or six feet high, dividing upward into many branches, which have leafy borders running along them; these are indented, and each indenture is terminated by a spine. The stalks are terminated by scaly heads of purple flowers, which appear in June; and to these succeed oblong angular seeds, crowned with a hairy down, which assist their spreading about to a great distance by the wind, so that where the plants are permitted to ripen their seeds, they often become troublesome weeds.

The second sort grows naturally in Spain, Portugal, and the Levant. This rises with a taller stalk than the former, the leaves are much longer and narrower, and the indentures on their sides are regular, ending in sharp spines. The heads of flowers are larger, and the spines of the empalement are longer than those of the first sort.

The third sort grows to the height of nine or ten feet; the stalks divide into many branches; the leaves are longer than of any other species; the heads of flowers are large and of a purple colour; the empalement hath the scales lying over each other like those of fish. This grows naturally in Spain and Portugal.

The fourth sort grows naturally about Aleppo. This rises with an upright branching stalk seven or eight feet high; the leaves are long, and regularly sinuated on their borders, like wing-pointed leaves. The heads of flowers are very large, and the empalement is very rough and prickly.

The fifth sort hath several oblong, oval, woolly leaves, which spread on the ground; between these come out the head of flowers sitting close to the ground; these heads are smaller than any of the other, and the flowers are white. Some of these plants have been formerly cultivated for the table, but it was before the English gardens was well supplied with other esculent plants, for at present they are rarely eaten here. They require no culture, for if the seeds are permitted to fall, the plants will come up fast enough.

OPHIOGLOSSUM. Adders Tongue.

This plant grows naturally in moist meadows, and is not easy to be made to thrive in gardens, so is rarely attempted.

OPHRYS. Tourn. Inst. R. H. 437. tab. 250. Lin. Gen. Plant. 902. Twyblade.

The CHARACTERS are,

It has a single stalk with a vague spatha. The flower hath no empalement; it consists of five oblong petals, which join so as to form a helmet, the under one is bifid. The nectarium is dependent, and keel-shaped behind; it hath two short stamina sitting on the pointal. It hath an oblong contorted germen situated under the flower, with a style adhering to the inner border of the nectarium. The germen afterward turns to an oval, three-cornered, obtuse capsule, with one cell opening with three valves, filled with small seeds like dust.

The SPECIES are,

1. OPHRYS (*Nidus avis*) bulbo fibroso, caule bifolio, foliis ovatis, nectarii labio bifido. Lin. Sp. Plant. 546. *Twyblade with a fibrous root, two oval leaves on the stalk, and a bifid lip to the nectarium; common Twyblade, or Twayblade.*

2. OPHRYS (*Cordata*) bulbo fibroso, caule bifolio, foliis cordatis. Lin. Sp. Plant. 946. *Twyblade with a fibrous root, and two heart-shaped leaves on the stalk; or Smallest Twyblade.*

3. OPHRYS (*Fibroso*) bulbis fibroso fasciculatis, caule vaginato, nectarii labio bifido. Lin. Sp. Plant. 945. *Twyblade with bulbous bunched roots, a sheath-like stalk, and a trifid lip to the nectarium; Birds Nest, or mis-shapen Orchis.*

4. OPHRYS (*Spiralis*) bulbis aggregatis oblongis, caule subfolioso, floribus secundis, nectarii labio indiviso. Act. Upsal. 1740. *Twyblade with oblong clustered bulbs, a leafy stalk, fruitful flowers, and an undivided lip to the nectarium; white, sweet-scented, spiral Orchis, called Triple Ladies Traces.*

5. OPHRYS (*Monorchis*) bulbo globoso, caule nudo, nectarii labio trifido. Act. Upsal. 1740. *Twyblade with a globular bulb, a naked stalk, and a trifid lip to the nectarium; yellow, sweet, or Musk Orchis.*

6. OPHRYS (*Anthropophora*) bulbis subrotundis, caule folioso, nectarii labio lineari tripartito, medio elongato bifido. Lin. Sp. Plant. 948. *Twyblade with roundish bulbs, a leafy stalk, and a narrow three-pointed lip to the nectarium, the middle segment of which is stretched out and bifid; or Man Orchis.*

7. OPHRYS (*Insectifera*) bulbis subrotundis, caule folioso, nectarii labio subquinquelobo. Lin. Sp. Plant. 948. *Twyblade with roundish bulbs, a leafy stalk, and the lip of the nectarium divided almost into five lobes; Greater Fly Orchis.*

8. OPHRYS (*Adrachnitis*) bulbis subrotundis, caule folioso, nectarii labio trifido. *Twyblade with roundish bulbs, a leafy stalk, and a trifid lip to the nectarium; the common Humble Bee Orchis.*

9. OPHRYS (*Sphegodes*) bulbis subrotundis, caule subfolioso, nectarii labio trifido hirsuto. *Twyblade with roundish bulbs, a leafy stalk, and a hairy trifid lip to the nectarium; Humble Bee Satyrion with green wings.*

The first sort grows naturally in woods, and sometimes in moist pastures in several parts of England. The root is composed of many strong fibres, from which arise two oval veined leaves; between these arises a naked stalk about eight inches high, terminated by a loose spike of herbaceous flowers resembling gnats, composed of five petals, with a long bifid lip to the nectarium, with a crest or standard above, and two wings on the side. The flowers sit upon an angular germen, which afterward swells to a capsule, opening when ripe in six parts, and filled with small dusty seeds. This plant refuses culture, but may be transplanted from the places where it grows naturally, into a shady part of the garden, where, if the roots are not disturbed, they will continue several years, and flower in May, but they do not increase in gardens. The best time to remove the roots is in July or August, when the leaves are decaying, for it will be difficult to find the roots after the leaves are quite gone.

The second sort is found in some of the northern counties in England, but is seldom seen growing in the south. This hath a small bulb with many strong fibres to the root, and sends out two small, ribbed, heart-shaped leaves, at bottom. The stalk rises about four inches high, and is terminated by a spike of small herbaceous flowers shaped like those of the first sort.

The third sort grows naturally in shady woods in Kent and Sussex. This has sometimes a single bulbous root, and at others several joined together, from which arises a single stalk near a foot high, embraced the whole length with leaves like sheaths; the top of the stalk is garnished with a loose spike of flowers, shaped like those of the Orchis, and of the colour of decayed leaves. It flowers in June.

The

The fourth ſort grows upon chalky hills in ſeveral parts of England. This hath a globular bulbous root, from which ariſes a ſingle ſtalk ſix inches high, having two oblong leaves at bottom, and rarely any above; the flowers are ſmall, of a yellowiſh green colour, growing in a looſe ſpike on the top of the ſtalk; they have a muſky ſcent. This flowers in Auguſt.

The fifth ſort grows naturally in moiſt paſtures in the northern parts of England; I have alſo found it in great plenty on Endfield Chace, not far from the town. This hath many oblong bulbs joined together at the top, from which ariſe three or four oblong leaves; and between theſe come out a ſlender ſtalk about ſix inches high, having a few narrow leaves, which embrace it like a ſheath. The flowers grow in a cloſe ſpike at the top; they are white, and have an agreeable ſcent. This flowers in Auguſt and September.

The ſixth ſort grows upon the chalk-hills near Northfleet in Kent, and alſo upon Cawſham-hills near Reading. This hath a roundiſh bulbous root, from which come out a few oblong leaves; the ſtalks riſe a foot and a half high, garniſhed with a few narrower leaves; the flowers grow in a looſe ſpike on the top of the ſtalk, they are of a ruſty iron colour, ſometimes inclinable to green. The lip of the nectarium is divided into three parts, the middle ſegment being ſtretched out much longer than the other, and is divided into two; the upper part of the flower being hooded, the whole bears ſome reſemblance to a naked man. This flowers in June.

The ſeventh ſort is commonly called the Fly Orchis. This grows naturally in England, but not in great plenty. Mr. Ray found it growing on the banks of the Devil's-ditch in Cambridgeſhire. I gathered it near Northfleet in Kent. It hath a roundiſh bulbous root, from which ariſe four or five oblong leaves, and a ſtalk about a foot high, having a few narrow acute-pointed leaves, embracing it like a ſheath. The flowers are ranged on the upper part of the ſtalk at a diſtance from each other, they have no ſpur; the creſt and wings are of an herbaceous colour, but the nectarium is very like the body of a fly. It flowers the end of May.

The eighth ſort grows naturally in dry paſtures in ſeveral parts of England, and is commonly called the Humble Bee Orchis; of this there are two or three varieties found wild in England, and ſeveral more in Spain and Portugal. This hath a roundiſh bulbous root, the leaves are like thoſe of the narrow-leaved Plantain. The ſtalk riſes ſix or ſeven inches high, having two or three ſheath-ſhaped leaves embracing it, which are erect; at the top of the ſtalk come out two or three flowers without ſpurs, having purpliſh creſts and wings. The nectarium is large, ſhaped like the body of a humble bee, of a dark ſooty colour, with two or three lines running croſs it, of a darker or lighter colour, which appear brighter or duller according to the poſition of the flower to the ſun. It flowers early in June.

The ninth ſort grows naturally on the chalk-hills near Northfleet in Kent, and in ſeveral other places. This is called the green-winged Humble Bee Orchis. The roots of this are roundiſh, like thoſe of the former ſort; the leaves are narrower and fewer: the ſtalks are ſhorter, the flowers a little ſmaller; the wings are green, and the nectarium of a dark ſooty colour, and hairy. This flowers the end of April.

All theſe ſorts may be preſerved in gardens, though not propagated there. The beſt time to remove the roots from the places where they naturally grow, is juſt before the ſtalks fall, for at that time the roots may be eaſily diſcovered, and then they are beginning to reſt, ſo that the bulb will be fully formed for flowering the following year, and will not ſhrink; but when they are removed at a time of the year when they are in action, the bulb deſigned for flowering the following year, not being fully ripened, will ſhrink, and frequently periſh; or if they ſurvive their removal, do not recover their former ſtrength in leſs time than two years.

When theſe are removed into a garden, the ſoil ſhould be adapted to the ſorts. Such of them as grow naturally in moiſt places, ſhould be planted in ſhady moiſt borders; thoſe which are inhabitants of woods may be planted under trees in wilderneſſes, but ſuch as grow upon chalk hills ſhould have a bed of chalk prepared for them in an open ſituation, and when the plants are fixed in their ſeveral places, they ſhould not be diſturbed after, for if they are kept clean from weeds, the leſs the ground is diſturbed, the better the plants will thrive, and the longer they will continue.

OPUNTIA. Tourn. Inſt. R. H. 239. tab. 122. The Indian Fig, or prickly Pear.

The CHARACTERS are,

The flower is compoſed of ſeveral petals, which are obtuſe, concave, and placed in a circular order, ſitting upon the germen. It hath a great number of awl-ſhaped ſtamina, which are inſerted in the germen. The germen, which is ſituated under the flower, ſupports a cylindrical ſtyle the length of the ſtamina, crowned by a multifid ſtigma. The germen afterward turns to a fleſhy umbilicated fruit with one cell, incloſing many roundiſh ſeeds.

The SPECIES are,

1. OPUNTIA (*Vulgaris*) articulis ovatis compreſſis, ſpinis ſetaceis. *Indian Fig with oval compreſſed joints, and briſtly ſpines; the common Indian Fig.*

2. OPUNTIA (*Ficus Indica*) articulis ovato-oblongis, ſpinis ſetaceis. *Indian Fig with oblong oval joints, and briſtly ſpines.*

3. OPUNTIA (*Tuna*) articulis ovato-oblongis, ſpinis ſubulatis. *Indian Fig with oblong oval joints, and awl-ſhaped ſpines.*

4. OPUNTIA (*Elatior*) articulis ovato-oblongis, ſpinis longiſſimis nigricantibus. *Indian Fig with oblong oval joints, and very long black ſpines.*

5. OPUNTIA (*Maxima*) articulis ovato-oblongis craſſiſſimis, ſpinis inæqualibus. *Indian Fig with oblong, oval, thick joints, and unequal ſpines.*

6. OPUNTIA (*Cochinelifera*) articulis ovato-oblongis ſubinermibus. *Indian Fig with oblong oval joints, almoſt without ſpines; commonly called the Cochineal Fig.*

7. OPUNTIA (*Curaſſavica*) articulis cylindrico-ventricoſis compreſſis, ſpinis ſetaceis. *Indian Fig with compreſſed, cylindrical, bellied joints, and briſtly ſpines; Pinpillow.*

8. OPUNTIA (*Spinoſiſſima*) articulis longiſſimis tenuibus compreſſis, ſpinis longiſſimis confertiſſimis, gracilibus albicantibus armatis. Houſt. MSS. *Stalky Indian Fig with large, narrow, compreſſed leaves, armed with the longeſt, narroweſt, white ſpines, growing in cluſters; this is by the gardeners called Robinſon Cruſoe's Coat.*

9. OPUNTIA (*Polyanthus*) prolifer enſiformi-compreſſus ſerrato-repandus. *Indian Fig with compreſſed ſword-ſhaped joints, whoſe indentures turn backward; Torch Thiſtle with a branching Spleenwort leaf.*

Theſe plants are all of them natives of America, though the firſt ſort is found growing wild on the ſides of the roads about Naples, in Sicily, and Spain, but it is probable that the plants may have been brought from America thither at firſt. This has been long in the Engliſh gardens; the joints or branches are oval or roundiſh, compreſſed on their two ſides flat, and have ſmall leaves coming out in knots on their ſurface, as alſo on their upper edges, which fall off in a ſhort time; and at the ſame knots there are three or four

four short bristly spines, which do not appear, unless they are closely viewed; but on being handled they enter the flesh, and separate from the plant, so are troublesome, and often very difficult to get out. The branches of this sort spread near the ground, and frequently trail upon it, putting out new roots, so are extended to a considerable distance, and never rise in height; these are fleshy and herbaceous while they are young, but as they grow old become drier, of a tough contexture, and have ligneous fibres. The flowers come out on the upper edges of the branches, generally, though sometimes they are produced on their sides; these sit upon the embryo of the fruit, and are composed of several roundish concave petals, which spread open; they are of a pale yellow colour, and within arise a great number of stamina, fastened to the embryo of the fruit, which are terminated by oblong summits; and in the center is situated the style, crowned by a many-pointed stigma; after the flowers are past, the embryo swells to an oblong fruit, whose skin or cover is set with small spines in clusters, and the inside is fleshy, of a purple or red colour, in which are lodged many black seeds. This plant flowers here in July and August, but unless the season is very warm the fruit will not ripen in England.

The second sort hath oblong, oval, compressed branches, which grow more erect than those of the first, armed with longer bristly spines, which come out in clusters from a point on each of the compressed sides, spreading open like the rays of a star. The flowers grow upon the embryo of the fruit, which come out from the upper edges of the leaves like the first, but are larger, and of a brighter yellow colour. The fruit is also larger, and of a deeper purple; the outer skin is also armed with longer spines: this is the most common sort in Jamaica, and upon the fruit of this the wild sort of cochineal feeds, which is called Sylvester.

The third sort hath stronger branches than the second, which are armed with larger thorns, of an awl-shape; they are whitish, and come out in clusters like those of the other sort. The flowers are large, of a bright yellow colour, and the fruit is shaped like the second sort.

The fourth sort grows taller than either of the former; the branches are larger, thicker, and of a deeper green; they are armed with strong black spines, which come out in clusters like those of the other sorts, but the clusters are farther asunder. The flowers are produced from the upper edges of the branches; they are smaller than those of the other sorts, and are of a purplish colour, as are also the stamina; the fruit is of the same form as those of the first, but do not ripen here.

The fifth sort is the largest of all yet known. The joints of these are more than a foot long, and eight inches broad; they are very thick, of a deep green colour, and armed with a few short bristly spines; the older branches of this often become almost taper, and are very strong. The flowers of this sort I have never yet seen, for although I have had many of the plants more than ten feet high, none of them has produced any flowers.

The sixth sort has been always supposed to be the plant, upon which the cochineal insects feed; this hath oblong, smooth, green branches, which grow erect, and rise to the height of eight or ten feet, having scarce any spines on them, and those few which are can scarce be discerned at a distance, and are so soft as not to be troublesome when handled. The flowers of this sort are small, and of a purple colour, standing upon the embryo of the fruit, in the same manner as those of the other sort, but do not expand open like them. The flowers of this appear late in the autumn, and the fruit drops off in winter without coming to any perfection here; this is cultivated in the fields of New Spain, for the increase of the insects, but it grows naturally in Jamaica, where it is probable the true cochineal might be discovered, if persons of skill were to search after the insects.

The seventh sort is said to grow naturally at Curacoa. This hath cylindrical swelling joints, which are closely armed with slender white spines. The branches spread out on every side, and where they have no support fall to the ground, very often separating at the joints of the plants, and, as they lie upon the ground, put out roots, so form new plants; this sort very rarely produces flowers in England. In the West-Indies it is called Pinpillow, from the appearance which the branches have to a pin-cushion stuck full of pins.

The eighth sort was sent me from Jamaica by the late Dr. Houstoun, who found it growing naturally there in great plenty, but could never observe either fruit or flower upon any of the plants, nor have any of them produced either in England. The branches of this sort have much longer joints than most of the other; they are narrower, and more compressed. The spines of this are very long, slender, and of a yellowish brown colour, coming out in clusters all over the surface of the branches, crossing each other, so as to render it dangerous to handle; for upon being touched the spines adhere to the hand, and quit the branches, and penetrate into the flesh, so become very troublesome.

The ninth sort grows naturally in the Brasils. This hath very thin branches, which are indented regularly on their edges like Spleenwort; they are of a light green, and shaped like a broad sword; these are smooth, having no spines. The flowers come out from the side, and at the end of the branches, sitting on the embryos in the same way as the other sorts; they are of a pale yellow colour. The fruit is shaped like those of the first sort, but rarely ripens in England.

All these sorts (except the first) are too tender to thrive in the open air in England, nor can many of them be preserved through the winters here, unless they have artificial heat; for when they are placed in a green-house, they turn to a pale yellow colour, their branches shrink, and frequently rot on the first approach of warm weather in the spring

These plants may all be propagated by cutting off their branches at the joints, during any of the summer months, which should be laid in a warm dry place for a fortnight, that the wounded part may be healed over, otherwise they will rot with the moisture which they imbibe at that part, as is the case with most other succulent plants. The soil in which these plants must be planted, should be composed after the following manner, viz. one third of light fresh earth from a pasture, a third part sea-sand, and the other part should be one half rotten tan, and the other half lime rubbish; these should be well mixed, and laid in a heap three or four months before it is used, observing to turn it over at least once a month, that the several parts may be well united; then you should pass it through a rough screen, in order to separate the largest stones and clods, but by no means sift it too fine, which is a very common fault; then you should reserve some of the smaller stones and rubbish to lay at the bottom of the pots, in order to keep an open passage for the moisture to drain off; which is what must be observed for all succulent plants, for if the moisture be detained in the pots, it will rot their roots, and destroy the plants.

When you plant any of the branches of these plants (except the first sort) you should plunge the pots into a moderate hot-bed, which will greatly facilitate their taking root; you should also refresh them now and then with a little water, but be very careful not to let them have too much, or be too often watered, especially before they are rooted. When the plants begin to shoot, you must give them a large share of air, by raising the glasses, otherwise they will draw up so weak as not to be able to support themselves; and after they have taken strong root, you should inure them to the air by degrees, and then remove them into the stove where they should remain, placing them near the glasses, which should always be opened in warm weather, so that they may have the advantage of a free air, and yet be protected from wet and cold.

During the summer season, these plants will require to be often refreshed with water, but it must not be given to them in large quantities, lest it rot them; and in winter this should be proportioned to the warmth of the stove, for if the air be kept very warm, they will require to be often refreshed, otherwise their branches will shrink; but if the house be kept in a moderate degree of warmth, they should have but little, for moisture at that season will rot them very soon.

The heat in which these plants thrive best, is the temperate point, as marked on botanical thermometers; for if they are kept too warm in winter, it causes their shoots to be very tender, weak, and unsightly. Those sorts which are inclinable to grow upright, should have their branches supported with stakes, otherwise their weight is so great as to break them down.

ORANGE. See Aurantium.

ORCHARD. In planting of an Orchard, great care should be had to the nature of the soil, and such sorts of fruits only should be chosen, as are best adapted to the ground designed for planting, otherwise there can be little hopes of their succeeding; and it is for want of rightly observing this method, that we see in many countries Orchards planted, which never arrive to any tolerable degree of perfection, the trees starving, and their bodies are either covered with Moss, or the bark cracks and divides, both which are evident signs of the weakness of the trees; whereas, if instead of Apples the Orchard had been planted with Pears, Cherries, or any other sort of fruit better adapted to the soil, the trees might have grown very well, and produced great quantities of fruit.

As to the position of the Orchard, (if you are at full liberty to chuse) a rising ground, open to the south-east, is to be preferred; but I would by no means advise planting upon the side of a hill, where the declivity is very great, for in such places the great rains commonly wash down the better part of the ground, whereby the trees will be deprived of proper nourishment; but where the rise is gentle, it is of great advantage to the trees, by admitting the sun and air between them, better than it can upon an entire level; which is an exceeding benefit to the fruit, by dissipating fogs, and drying up the damps; which, when detained amongst the trees, mix with the air, and render it rancid; if it be defended from the west, north, and east winds, it will also render the situation still more advantageous, for it is chiefly from those quarters that fruit-trees receive the greatest injury; therefore, if the place be not naturally defended from these by rising hills (which are always to be preferred,) then you should plant large growing timber-trees at some distance from the Orchard, to answer this purpose.

You should also have a great regard to the distance of planting the trees, which is what few people have rightly considered, for if you plant them too close, they will be liable to blights; the air being hereby pent in amongst them, will also cause the fruit to be ill tasted, having a great quantity of damp vapours from the perspiration of the trees, and the exhalations from the earth mixed with it, which will be imbibed by the fruit, and render their juices crude and unwholesome.

Wherefore I cannot but recommend the method which has been lately practised by some particular gentlemen with very good success, that is, to plant the trees fourscore feet asunder, but not in regular rows. The ground between the trees they plough and sow with Wheat and other crops, in the same manner as if it were clear from trees; and they observe their crops to be full as good as those quite exposed, except just under each tree, when they are grown large, and afford a great shade; and by thus ploughing and tilling the ground, the trees are rendered more vigorous and healthy, scarcely ever having any Moss, or other marks of poverty, and will abide much longer and produce better fruit.

If the ground, in which you intend to plant an Orchard, has been pasture for some years, then you should plough in the green sward the spring before you plant the trees; and if you will permit it to lie a summer fallow, it will greatly mend it, provided you stir it two or three times to rot the sward of Grass, and prevent weeds growing thereon.

At Michaelmas you should plough it pretty deep, in order to make it loose for the roots of the trees, which should be planted thereon in October, provided the soil be dry; but if it be moist, the beginning of March will be a better season. The distance, if designed for a close Orchard, must not be less than forty feet, but the trees planted twice that distance will succeed better.

When you have finished planting the trees, you should provide some stakes to support them, otherwise the wind will blow them out of the ground, which will do them much injury, especially after they have been planted some time; for the ground in the autumn being warm, and for the most part moist, the trees will very soon push out a great number of young fibres, which, if broken off by their being displaced, will greatly retard the growth of the trees.

In the spring following, if the season should prove dry, you should cut a quantity of green sward, which must be laid upon the surface of the ground about their roots, turning the Grass downward, which will prevent the sun and wind from drying the ground, whereby a great expence of watering will be saved; and after the first year they will be out of danger, provided they have taken well.

Whenever you plough the ground betwixt these trees, you must be careful not to go too deep amongst their roots, lest you should cut them off, which would greatly damage the trees: but if you do it cautiously, the stirring the surface of the ground will be of great benefit to them, though you should observe never to sow too near the trees, nor suffer any great rooting weeds to grow about them, which would exhaust the goodness of the soil, and starve them.

If after the turf, which was laid round the trees, be rotted, you dig it in gently about the roots, it will greatly encourage them.

There are some persons who plant many sorts of fruit together in the same Orchard, mixing the trees alternately; but this is a method which should always be avoided, for hereby there will be a great difference in the growth of the trees, which will not only render them unsightly, but also the fruit upon the lower trees ill tasted, by the tall ones overshadowing them; so that if you are determined to plant several sorts of fruit on the same spot, you should observe to place the largest growing trees backward, and so proceed to those of less growth, continuing the same method quite through the whole plantation; whereby it will appear at a distance in a regular slope, and the sun and air will more equally pass throughout the whole Orchard, that every tree may have an equal benefit therefrom; but this can only be practised upon good ground, in which most sorts of fruit-trees will thrive.

The

The ſoil of your Orchard ſhould alſo be mended once in two or three years with dung or other manure, which will alſo be abſolutely neceſſary for the crops ſown between; ſo that where perſons are not inclinable to help their Orchards, where the expence of manure is pretty great, yet, as there is a crop expected from the ground beſides the fruit, they will the more readily be at the charge upon that account.

In making choice of trees for an Orchard, you ſhould always obſerve to procure them from a ſoil nearly akin to that where they are to be planted, or rather poorer; for if you have them from a very rich ſoil, and that wherein you plant them is but indifferent, they will not thrive well, eſpecially for four or five years after planting; ſo that it is a very wrong practice to make the nurſery, where young trees are raiſed very rich, when the trees are deſigned for a middling or poor ſoil. The trees ſhould alſo be young and thriving, for whatever ſome perſons may adviſe to the contrary, yet it has always been obſerved, that though large trees may grow and produce fruit after being removed, they never make ſo good trees, nor are ſo long lived as thoſe which are planted while young.

Theſe trees, after they are planted out, will require no other pruning, but only to cut out dead branches, or ſuch as croſs each other, which render their heads confuſed and unſightly: the pruning them too often, or ſhortening their branches, is very injurious, eſpecially to Cherries and ſtone fruit, which will gum prodigiouſly, and decay in ſuch places where they are cut; and the Apples and Pears, which are not of ſo nice a nature, will produce a greater quantity of lateral branches, which will fill the heads of the trees with weak ſhoots, whenever their branches are thus ſhortened; and many times the fruit is hereby cut off, which, on many ſorts of fruit-trees, is firſt produced at the extremity of their ſhoots.

It may perhaps ſeem ſtrange to ſome perſons, that I ſhould recommend the allowing ſo much diſtance to the trees in an Orchard, becauſe a ſmall piece of ground will admit of very few trees when planted in this method; but they will pleaſe to obſerve, that when the trees are grown up, they will produce a great deal more fruit, than twice the number when planted cloſe, and will be vaſtly better taſted; the trees, when placed at a large diſtance, being never ſo much in danger of blighting as in cloſe plantations; as hath been obſerved in Herefordſhire, the great county for Orchards, where they find, that when Orchards are ſo planted or ſituated, that the air is pent up amongſt the trees, the vapours which ariſe from the damp of the ground, and the perſpiration of the trees, collect the heat of the ſun, and reflect it in ſtreams ſo as to cauſe what they call a fire blaſt, which is the moſt hurtful to their fruits; and this is moſt frequent where the Orchards are open to the ſouth ſun.

But as Orchards ſhould never be planted, unleſs where large quantities of fruit are deſired, ſo it will be the ſame thing to allow twice or three times the quantity of ground, ſince there may be a crop of grain of any ſort upon the ſame place (as was before ſaid,) ſo that there is no loſs of ground; and for a family only, it is hardly worth while to plant an Orchard, ſince a kitchen-garden well planted with eſpaliers will afford more fruit than can be eaten while good, eſpecially if the kitchen-garden be proportioned to the largeneſs of the family; and if cyder be required, there may be a large avenue of Apple-trees extended croſs a neighbouring field, which will render it pleaſant, and produce a great quantity of fruit; or there may be ſome ſingle rows of trees planted to ſurround fields, &c. which will fully anſwer the ſame purpoſe, and be leſs liable to the fire blaſts before-mentioned.

ORCHIS. Tourn. Inſt. R. H. 431. tab. 248, 249. Satyrion, or Fool-ſtones.

The Characters are,

It hath a ſingle ſtalk with a vague ſheath. The flower hath five petals, three without and two within. The nectarium is of one leaf, fixed to the ſide of the receptacle, between the diviſion of the petals. The upper lip is ſhort and erect, the under large, broad, and ſpreading; the tube is pendulous, horn-ſhaped, and prominent behind. It hath two ſhort ſlender ſtamina ſitting upon the pointal, with oval erect ſummits, fixed to the upper lip of the nectarium. It hath an oblong contorted germen under the flower, with a ſhort ſtyle faſtened to the upper lip of the nectarium. The germen afterward turns to an oblong capſule with one cell, having three keel-ſhaped valves, opening on the three ſides, but joined at top and bottom, filled with ſmall ſeeds like duſt.

The Species are,

1. Orchis (*Morio*) bulbis indiviſis, nectarii labio quadrifido crenulato, cornu obtuſo. Act. Upſal. 1740. *Orchis with undivided bulbs, the lip of the nectarium cut into four points, which are ſlightly indented, and an obtuſe horn; or common Female Orchis.*

2. Orchis (*Maſcula*) bulbis indiviſis, nectarii labio æquali, cornu integro, galeæ alis reflexis acutis. *Orchis with undivided bulbs, the lip of the nectarium equal, an entire horn, and the wings of the ſtandard acute and reflexed; the Male Orchis.*

3. Orchis (*Bifolia*) bulbis indiviſis, nectarii labio lanceolato integerrimo, cornu longiſſimo, petalis patentibus. Act. Upſal. 1740. *Orchis with undivided bulbs, the lip of the nectarium entire and ſpear-ſhaped, a very long horn, and petals ſpreading very wide; or Butterfly Orchis.*

4. Orchis (*Militaris*) bulbis indiviſis, nectarii labio quinquefido punctis ſcabro, cornu obtuſo, petalis conſluentibus. Act. Upſal. 1740. *Orchis with undivided bulbs, a five-pointed lip to the nectarium, having rough ſpots, an obtuſe horn, and petals running together; or the Man Orchis.*

5. Orchis (*Pyramidalis*) bulbis indiviſis, nectarii labio trifido antice bidentato, cornu longo, petalis acuminatis. Act. Upſal. 1740. *Orchis with undivided bulbs, a trifid lip to the nectarium, indented with two teeth behind, a long horn, and acute-pointed petals; Mountain military Orchis, with a reddiſh conglomerated ſpike.*

6. Orchis (*Latifolia*) bulbis indiviſis, nectarii labio quadrifido punctis ſcabro, cornu obtuſo, petalis diſtinctis. Act. Upſal. 1740. *Orchis with undivided bulbs, the lip of the nectarium quadrifid, having rough ſpots, an obtuſe horn, and diſtinct petals.*

7. Orchis (*Maculata*) bulbis ſubpalmatis rectis, nectarii cornu conico, labio trilobo, lateralibus reflexo, bracteis flore longioribus. Act. Upſal. 1740. *Orchis with ſtraight, palmated, bulbous roots, a conical horn to the nectarium, the lip cut into three lobes, which are reflexed on the ſides, and bracteæ longer than the flowers.*

8. Orchis (*Conopſea*) bulbis palmatis patentibus, nectarii cornu germinibus breviore, labio plano petalis dorſalibus erectis. Act. Upſal. 1740. *Orchis with handed ſpreading bulbs, the horn of the nectarium ſhorter than the germen, a plain lip, and the hinder part of the petals erect.*

9. Orchis (*Abortiva*) bulbis faſciculatis filiformibus, nectarii labio ovato integerrimo. Act. Upſal. 1740. *Orchis with thread-like bulbs growing in bunches, and the lip of the nectarium oval and entire; or Purple Bird's Neſt.*

The firſt ſort grows naturally in paſtures in moſt parts of England. This hath a double bulbous root, with ſome fibres coming out from the top; it has four or ſix oblong leaves lying on the ground, which are reflexed. The ſtalk riſes nine or ten inches high, having four or ſix leaves which embrace it: this is terminated by a ſhort looſe ſpike of flowers,

flowers, having a four-pointed indented lip to the nectarium, and an obtuse horn. The flowers are of a pale purple colour, marked with deeper purple spots. It flowers in May.

The second sort grows naturally in woods and shady places in many parts of England. This hath a double bulbous root, which is about the size and shape of middling Olives; it hath six or seven long broad leaves, shaped like those of Lilies, which have several black spots on their upper side; the stalk is round, and a foot high, having one or two smaller leaves embracing it. The flowers are disposed in a long spike on the top of the stalk; they are of a purple colour, marked with deep purple spots, and have an agreeable scent. It flowers the latter end of April.

The third sort grows naturally under bushes by the side of pastures in many parts of England. This hath a root composed of two oblong Pear-shaped bulbs, from which come out three or four Lily-shaped leaves, of a pale green, with a few faint spots; the stalk rises near a foot high; it is slender, furrowed, and has a few very small leaves which embrace it: this is terminated by a loose spike of white flowers smelling sweet, which resemble a butterfly with expanded wings. This flowers in June.

The fourth sort is found growing naturally on Cawsham-hills, and in other places where the soil is chalk. The roots of this sort are composed of two bulbs, from which come out four or five oblong leaves; the stalk is about nine inches high, sustaining a loose spike of sweet smelling flowers, each hanging on a pretty long foot-stalk; they have a short obtuse horn, a crest and wings of an Ash colour without, reddish within, and striped with deeper lines; the lip is oblong, divided into five parts, having rough spots. This flowers in June.

The fifth sort grows naturally on chalk-hills in several parts of England. The root of this is composed of two oblong bulbs, from which arise three or four narrow oblong leaves; the stalk rises a foot high, having three or four narrow erect leaves, which embrace it. The flowers are produced in a thick roundish spike at the top; they are of a reddish colour, having long spurs, and the wings are acute-pointed. It flowers in June.

The sixth sort grows naturally on dry pastures in many parts of England. This hath a double bulbous root; the leaves are oblong and narrow; the stalk rises six or seven inches high, having two or three leaves, which embrace it like sheaths. The flowers grow in close short spikes at the top; they are of a purple colour; the lip of the nectarium is divided into four parts, having rough spots; the spur is obtuse, and the petals are distinct. There is a variety of this with a white nectarium. It flowers in June.

The seventh sort grows naturally in moist meadows in many parts of England. The root of this is composed of two fleshy bulbs, which are divided into four or five fingers, so as to resemble an open hand; the stalk rises from nine inches to a foot high, garnished with leaves the whole length, which are three or four inches long, and one broad, embracing the stalk with their base; these are not spotted, and end in acute points. The flowers are disposed in a spike on the top of the stalk, with small narrow leaves (called bracteæ) between them, which are longer than the flowers. The spur is half an inch long, extended backward; the lip of the nectarium is broad, divided into three lobes, two side ones being reflexed; the flowers and bracteæ are of a purplish colour, having deep purple spots.. It flowers in May. There are two varieties of this, differing in the colour of their flowers, and one with a narrow leaf.

The eighth sort grows naturally in meadows in several parts of England. The root of this is composed of two broad fleshy bulbs, both of which are divided into four fingers, which spread asunder. The stalk rises a foot and a half high, and is very strong, inclining to a purple colour; it is garnished with leaves the whole length. The flowers are collected in a close spike at the top of the stalk; they are of a pale purple colour; the spur is about a third part of an inch long; the beard of the nectarium is plain, and divided into three parts, which is marked with deep purple spots; under each foot-stalk is placed a narrow leaf (or bracteæ) of a purplish colour. The leaves and stalks of the plant have many dark spots. It flowers in June. There are two or three varieties of this, which differ in the colour of their flowers.

The ninth sort grows naturally in shady woods in several parts of England, but particularly in Sussex and Hampshire, in both which counties I have several times found it. The root of this plant is composed of many thick, oblique, long fibres, which are fleshy; the stalk rises near two feet high, wrapped round with leaves like sheaths; these are of a purple colour. The flowers are disposed in a loose thyrse at the top of the stalk; they are of a purple colour, having an oval entire lip to the nectarium, the crest terminating in a horn. It flowers in June.

All these sorts of Orchis grow wild in several parts of England, but for the extreme oddness and beauty of their flowers, deserve a place in every good garden; and the reason for their not being cultivated in gardens, proceeds from their difficulty to be transplanted; though this, I believe, may be easily overcome, where a person has an opportunity of marking their roots in their time of flowering, and letting them remain until their leaves are decayed, when they may be transplanted with safety; for it is the same with most sorts of bulbous or fleshy-rooted plants, which, if transplanted before their leaves decay, seldom live, notwithstanding you preserve a large ball of earth about them; for the extreme parts of their fibres extend to a great depth in the ground, from whence they receive their nourishment, which, if broken or damaged by taking up their roots, seldom thrive after; for though they may sometimes remain alive a year or two, yet they grow weaker until they quite decay; which is also the case with Tulips, Fritillarias, and other bulbous roots, when removed after they have made shoots: so that whoever would cultivate them, should search them out in their season of flowering, and mark them; and when their leaves are decayed, or just as they are going off, the roots should be taken up, and planted in a soil and situation as nearly resembling that wherein they naturally grow, as possible, otherwise they will not thrive, so that they cannot be placed all in the same bed; for some are only found upon chalky hills, others in moist meadows, and some in shady woods, or under trees; but if their soil and situation be adapted to their various sorts, they will thrive and continue several years, and during their season of flowering, will afford as great varieties as any flowers which are at present cultivated.

OREOSELINUM. See Athamanta.

ORIGANUM. Lin. Gen. Plant. 645. Origany, or Pot Marjoram.

The Characters are,

The flower is of the lip kind; the upper lip is plain, erect, obtuse, and indented; the under lip is trifid, the segments being nearly equal. The flowers have four slender stamina, two being as long as the petal, the other two are longer, terminated by simple summits; they have a four-cornered germen, supporting a slender style inclining to the upper lip, crowned by a bifid stigma. The germen afterward turns to four seeds shut up in the empalement of the flower.

The Species are,

1. Origanum (*Vulgare*) spicis subrotundis paniculatis conglomeratis, bracteis calyce longioribus ovatis. Lin. Sp. Plant.

Plant. 590. *Pot Marjoram with roundish paniculated spikes gathered in clusters, and oval bracteæ which are longer than the empalement; or common wild Origany.*

2. ORIGANUM (*Heracleoticum*) spicis longis pedunculis aggregatis, bracteis longitudine calycum. Lin. Gen. Pl. 589. *Origany with long spikes growing in bunches, and bracteæ as long as the empalement; or winter sweet Marjoram.*

3. ORIGANUM (*Latifolium*) spicis oblongis paniculatis conglomeratis, foliis ovatis glabris. *Origany with oblong spikes of flowers growing in clustered panicles, and oval smooth leaves; or broad leaved smooth Origany.*

4. ORIGANUM (*Humile*) caule repente, spicis oblongis conglomeratis, bracteis florum longioribus. *Origany with a creeping stalk, and oblong spikes of flowers growing in clusters, with bracteæ longer than the flower; low wild Origany.*

5. ORIGANUM (*Orientale*) caule erecto ramoso, foliis ovatis rugosis, spicis subrotundis conglomeratis, bracteis calycum brevioribus. *Origany with an erect branching stalk, oval rough leaves, roundish spikes of flowers growing in clusters, with bracteæ shorter than the empalement.*

6. ORIGANUM (*Creticum*) spicis aggregatis longis prismaticis rectis bracteis, membranaceis, calyce duplo longioribus. Lin. Sp. Plant. 589. *Origany with long, upright, prismatical spikes growing in clusters, and membranaceous bracteæ twice the length of the empalement; Origany of Crete.*

7. ORIGANUM (*Majorana*) foliis ovalibus obtusis, spicis subrotundis compactis pubescentibus. Hort. Cliff. 304. *Origany with oval obtuse leaves, and roundish, compact, hairy spikes; common, or sweet Marjoram.*

8. ORIGANUM (*Ægyptiacum*) foliis carnosis tomentosis. Lin. Sp. Plant. 588. *Origany with fleshy woolly leaves.*

9. ORIGANUM (*Smyrnæum*) foliis ovatis acutè serratis, spicis congestis umbellatim fastigiatis. Hort. Cliff. 304. *Origany with oval leaves acutely sawed, and spikes of flowers disposed in umbellated bunches.*

10. ORIGANUM (*Dictamnus*) foliis omnibus tomentosis, spicis nutantibus. *Origany with all the leaves woolly, and nodding spikes of flowers; or Dittany of Crete.*

11. ORIGANUM (*Sypileum*) foliis omnibus glabris, spicis nutantibus. Hort. Cliff. 304. *Origany with all the leaves smooth, and nodding spikes of flowers; Dittany of mount Sipylus.*

12. ORIGANUM (*Hybridum*) foliis inferioribus tomentosis, spicis nutantibus. Hort. Cliff. 304. *Origany with the under leaves hoary, and nodding spikes of flowers.*

13. ORIGANUM (*Onites*) spicis oblongis aggregatis hirsutis, foliis cordatis tomentosis. Lin. Sp. Plant. 590. *Origany with oblong hairy spikes growing in bunches, and heart-shaped woolly leaves.*

The first sort grows naturally in thickets, and among bushes in several parts of England; the root is perennial, composed of many small ligneous fibres. The stalks are square, and rise near two feet high; they are ligneous, and garnished with oval leaves, placed by pairs at each joint; from the wings of these come out three or four smaller leaves on each side, which resemble those of Marjoram, sitting close to the stalk; they have an aromatic scent; the flowers are produced in roundish spikes growing in panicles at the top of the stalks, many of the spikes being gathered together; they are of a flesh colour, and peep out of their scaly covering. Their upper lip is cut into two, standing erect, and the lower lip or beard is divided into three; the stamina stand out a little beyond the petals, and are of a purplish colour. It flowers in June and July, and the seeds ripen in the autumn. This sort is sometimes cultivated in gardens, and is by some called Pot Marjoram, as it is generally used in soups.

It will rise plentifully from scattered seeds, or it may be propagated by parting of the roots; the best time for doing this is in autumn, and may be planted in any soil not overmoist, and will thrive in any situation, so require no other care but to keep it clear from weeds. There is a variety of this with white flowers and light green stalks, and another with variegated leaves.

The second sort is now commonly known by the title of winter sweet Marjoram, though this was formerly stiled Pot Marjoram. This hath a perennial root, from which arise many branching four-cornered stalks a foot and a half high, inclining to a purplish colour, garnished with oval, obtuse, hairy leaves, resembling greatly those of sweet Marjoram, standing opposite; the flowers are disposed in spikes about two inches long, several arising together from the divisions of the stalk. They are small, white, and peep out of their scaly covers; these appear in July, and the seeds ripen in autumn. It grows naturally in Greece and the warm parts of Europe, but is hardy enough to thrive in the open air in England, and is chiefly cultivated for nosegays, as it comes sooner to flower than sweet Marjoram, so it is used for the same purposes, till the other comes to maturity. There is a variety of this with variegated leaves. This is generally propagated by parting of the roots in autumn, and should have a dry soil, where it will thrive, requiring no other culture than the first sort.

The third sort grows naturally in France and Italy; this hath a perennial root, from which arise several slender bending stalks near a foot high, garnished with oval smooth leaves, standing on pretty long foot-stalks. The flowers are produced in oblong spikes, which grow in clustered panicles; they are small, of a purplish colour, peeping out of their scaly covering. It flowers in June, and may be propagated by parting of the roots in the same way as the former.

The fourth sort grows plentifully about Orleans in France; this hath a perennial root, from which arise several four-cornered stalks about six inches high, which frequently bend to the ground, and put out roots; they are garnished with oblong hairy leaves, sitting close to the stalk. The flowers grow in oblong clustered spikes at the top of the stalks, having long coloured bracteæ between each; the flowers are some whitish, others purple in the same spikes; they are small, and peep out of their scaly covers. This flowers in June, and may be propagated in the same way as the former.

The fifth sort grows naturally in the Levant; it is a perennial plant. The stalks rise near two feet high, and branch out their whole length; they are purple, garnished with oval rough leaves, somewhat like those of Self-heal, but smaller. The flowers grow in roundish clustered spikes, having short bracteæ; they are purple, and appear in June, but are not succeeded by seeds here. It is propagated by parting of the heads in the same way as the former, and must have a dry soil.

The sixth sort is the Origany of Crete, which is directed to be used in medicine, but there has been great confusion among botanists in distinguishing the species. This rises with four-cornered stalks a foot high, garnished with thick, hoary, oval leaves, of a strong aromatic scent. The flowers grow in long, erect, bunched spikes at the top of the stalks, having membranous bracteæ between, which are twice the length of the empalement; the flowers are small and white like those of the common Origany. It flowers in July, but seldom perfects seeds in England. It is propagated by parting of the roots as the former, but must have a dry soil and a warm situation, otherwise it will not live through the winter here.

The seventh sort is the common sweet Marjoram, which is so well known as to need no description. With us in England it is esteemed an annual plant, though the roots often live through the winter in mild seasons, or if they are sheltered

ſheltered in a green-houſe; but in warm countries, I believe, it is only biennial.

This is propagated by ſeeds, which are generally imported from the ſouth of France or Italy, for they ſeldom ripen in England. Theſe are ſown on a warm border toward the end of March, and when the plants are come up about an inch high, they ſhould be tranſplanted into beds of rich earth, at ſix inches diſtance every way, obſerving to water them duly till they have taken new root, after which they will require no other care but to keep them clean from weeds. The plants will ſpread and cover the ground, in July they will begin to flower, at which time it is cut for uſe, and is then called knotted Marjoram, from the heads of flowers being collected into roundiſh cloſe heads like knots.

The eighth ſort grows naturally in Africa; this is a perennial plant with a low ſhrubby ſtalk, ſeldom riſing more than a foot and a half high, dividing into branches, which are garniſhed with roundiſh, thick, woolly leaves, hollowed like a ladle; they are like thoſe of the common Marjoram, but are of a thicker ſubſtance and woolly, and have much the ſame ſcent. The flowers are produced in roundiſh ſpikes, cloſely joined together at the top of the ſtalks, and, at the end of the ſmall ſide branches; they are of a pale fleſh colour, peeping out of their ſcaly coverings. This ſort flowers in July and Auguſt, but does not ripen ſeeds in England.

It is propagated by ſlips or cuttings, which, if planted in a border of good earth during any of the ſummer months, and ſhaded from the ſun and duly watered, will take root freely, and afterward the plants may be taken up, and planted in ſmall pots, filled with light kitchen-garden earth, and placed in the ſhade till they have taken new root; then they may be removed into an open ſituation, where they may remain till the end of October, when they muſt be placed under ſhelter, for they will not live through the winter in the open air here; but if they are put under a hot-bed frame, where they may be protected from hard froſt, and have as much free air as poſſible in mild weather, they will thrive better than if they are more tenderly treated.

The tenth ſort is the Dittany of Crete, which is uſed in medicine; this grows naturally upon Mount Ida in Candia; it is a perennial plant. The ſtalks are hairy, and riſe about nine inches high, of a purpliſh colour, and ſend out ſmall branches from their ſides by pairs oppoſite; they are garniſhed with round, thick, woolly leaves, which are very white; the whole plant has a piercing aromatic ſcent, and biting taſte; the flowers are collected in looſe leafy heads of a purple colour, which nod downward. They are ſmall, and of a purple colour; the ſtamina ſtands out beyond the petal, two of them being much longer than the other. It flowers in June and July, and in warm ſeaſons the ſeeds ripen in autumn.

This is propagated eaſily by planting cuttings or ſlips, during any of the ſummer months. Theſe ſhould be planted either in pots or a ſhady border, covering them cloſe with a bell or hand-glaſs to exclude the air, and now and then refreſhing them with water, but they muſt not have too much wet. When theſe have taken root, they ſhould be carefully taken up, and each planted into a ſeparate ſmall pot, filled with light earth, and placed in the ſhade till they have taken new root, when they ſhould be removed into an open ſituation, where they may continue till autumn, and then placed under a hot-bed frame, to ſcreen them from the froſt, but they ſhould enjoy the free air at all times in mild weather. The following ſpring ſome of the plants may be ſhaken out of the pots, and planted in a warm border near a good aſpected wall, and in a dry ſoil, where the plants will live through the common winters without any other ſhelter; but as they are liable to be killed by ſevere froſt, ſo it will be proper to keep a few plants in pots, to be ſheltered in winter to preſerve the kind.

The eleventh ſort grows naturally on Mount Sipylus near Magneſia, where it was diſcovered by Sir George Wheeler, who ſent the ſeeds to the Oxford garden, where the plants were raiſed; this hath a perennial root, but an annual ſtalk. The root is compoſed of many ſlender ligneous fibres; the leaves are oval, ſmooth, and of a grayiſh colour; the ſtalks are ſlender, of a purpliſh colour, four-cornered and ſmooth; they riſe near two feet high, ſending out ſlender branches on each ſide oppoſite, which are terminated by ſlender oblong ſpikes of purpliſh flowers, which peep out of their ſcaly covers; the flowers are ſmall, but ſhaped like thoſe of the tenth ſort; their ſtamina are extended out of the petal a conſiderable length. The leaves on the lower part of the ſtalk are almoſt as large as the common Origany, but thoſe on the upper part of the ſtalk and branches are very ſmall, and ſit cloſe to the ſtalk. It flowers in June and July, and in warm ſeaſons the ſeeds ripen here in autumn. It is propagated by cuttings or ſlips, in the ſame way as the Cretan Dittany, and the plants require the ſame treatment.

The twelfth ſort is undoubtedly a variety, which has been produced from the intermixing of the farina of the Cretan Dittany with that of Mount Sipylus; for the plants now in the Chelſea garden, were accidentally produced from the ſeeds of one ſpecies, where both ſorts ſtood near each other, in the garden of John Browning, Eſq; Maſter in Chancery; the ſeeds were dropped from the plant into the border between the two ſorts, ſo that it is uncertain from which ſpecies; but as the ſtalks and heads of flowers bear a greater reſemblance to the Dittany of Mount Sipylus, we may ſuppoſe it aroſe from the ſeeds of that, which had been impregnated by the farina of the Cretan Dittany, which grew near it; for the under leaves of this are round, of a thick texture and woolly, ſo nearly reſembling thoſe of the Cretan Dittany, as not to be diſtinguiſhed from it; but the ſtalks riſe full as high as thoſe of the Dittany of Mount Sipylus, but branch out more their whole length; they are of a purple colour and hairy. The lower leaves on the ſtalks are much larger than thoſe of Mount Sipylus, and are a little hairy, approaching to thoſe of the Cretan Dittany, but are not ſo thick or woolly; the upper leaves are ſmooth, and approach to thoſe of the other ſort, but are larger, as are alſo the ſpikes of flowers, and the ſcaly leaves which cover the flowers are larger and of a deeper purple colour.

I have alſo dried ſamples of another variety, which aroſe from ſeeds in the Leyden garden; the ſeeds were ſent from Paris, by the title which Tournefort gave to that which he found in the Levant, which I have joined to the variety before-mentioned. The leaves of this are as large as thoſe of the Dittany of Crete, but are not ſo thick or woolly; the ſtalks riſe more like thoſe of the Dittany of Mount Sipylus, but branch out wider at the top; the flowers grow in cloſer cluſters, and do not nod downward; they are ſmall, and ſhaped like thoſe of the former ſort, flowering at the ſame time.

By the title which Dr. Linnæus has given to the Cretan Dittany, it may be ſuppoſed he has not ſeen the true ſort, for his title better ſuits the variety to which I have applied it; for all the leaves of the true Dittany are very thick and woolly, even thoſe which are ſituated immediately below the flowers, whereas the lower leaves only are ſo in his title.

The thirteenth ſort grows at Syracuſe; this hath perennial ligneous ſtalks, which riſe a foot and a half high, dividing into many ſmall branches, which are garniſhed with ſmall heart-ſhaped leaves, a little larger than thoſe of Marjoram, which are woolly. The flowers grow in oblong tufted

tufted ſpikes, which are hairy; they are ſmall, white, and peep out of their ſcaly covers; they appear in July, but ſeldom perfect ſeeds in England. This is propagated by cuttings or ſlips, in the ſame way as the tenth ſort, and the plants require the ſame treatment.

The firſt and ſixth ſorts are uſed in medicine, but the firſt being a native of this country, is frequently ſubſtituted for the other, which is pretty rare in England, and is now ſeldom imported here. When the firſt ſort is uſed, thoſe plants which grow upon dry barren ground are to be preferred, as they are much ſtronger and have greater virtue, than thoſe which grow on good land, or are cultivated in gardens.

The Dittany of Crete is alſo uſed in medicine, but the dried herb is generally imported into England, which, by being cloſely packed, and the voyage being long, it loſes much of its virtue; ſo that if the plants of Engliſh growth were uſed, they will be found much better.

ORNITHOGALUM. Tourn. Inſt. R. H. 378. tab. 203. Star of Bethlehem.

The CHARACTERS are,

The flower has no empalement. It is compoſed of ſix petals, whoſe under parts are erect, but ſpread open above. It hath ſix erect ſtamina, about half the length of the petals, crowned by ſingle ſummits, with an angular germen, ſupporting an awl-ſhaped ſtyle, terminated by an obtuſe ſtigma. The germen afterward turns to a roundiſh angular capſule with three cells, filled with roundiſh ſeeds.

The SPECIES are,

1. ORNITHOGALUM (*Pyrenaicum*) racemo longiſſimo, filamentis lanceolatis, pedunculis floriferis patentibus æqualibus, fructiferis ſcapo approximatis. Lin. Sp. Plant. 307. *Star-flower with a very long ſpike of flowers, ſpear-ſhaped filaments, annd foot-ſtalks to the flowers equal and ſpreading, and thoſe of the fruit approaching to the ſtalk*; *or Star-flower with whitiſh green flowers.*

2. ORNITHOGALUM (*Pyramidale*) racemo conico, floribus numeroſis adſcendentibus. Prod. Leyd. 32. *Star-flower with a conical ſpike, having numerous flowers ariſing above each other.*

3. ORNITHOGALUM (*Latifolium*) racemo longiſſimo, foliis lanceolato-enſiformibus. Lin. Sp. Plant. 307. *Star-flower with the longeſt ſpike, and ſpear-ſhaped leaves.*

4. ORNITHOGALUM (*Nutans*) floribus ſecundis pendulis, nectario ſtamineo campaniformi. Lin. Sp. Plant. 308. *Star-flower with fruitful hanging flowers, and a bell-ſhaped nectarium.*

5. ORNITHOGALUM (*Luteum*) ſcapo anguloſo diphyllo, pedunculis umbellatis ſimplicibus. Flor. Suec. 270. *Star-flower with an angular ſtalk having two leaves, and ſingle umbellated foot-ſtalks, with a yellow Star-flower.*

6. ORNITHOGALUM (*Minimum*) ſcapo angulato diphyllo, pedunculis umbellatis ramoſis. Flor. Suec. 271. *Star-flower with an angular ſtalk bearing two leaves, and branching foot-ſtalks, having umbels.*

7. ORNITHOGALUM (*Umbellatum*) floribus corymboſis, pedunculis ſcapo altioribus filamentis emarginatis. Hort. Cliff. 124. *Star-flower with flowers growing in a corymbus, whoſe foot-ſtalks are taller than the ſtalk, and indented filaments.*

8. ORNITHOGALUM (*Arabicum*) floribus corymboſis, pedunculis ſcapo humilioribus, filamentis emarginatis. Prod. Leyd. 32. *Star-flower with flowers growing in a corymbus, foot-ſtalks lower than the ſtalk, and indented filaments.*

9. ORNITHOGALUM (*Capenſe*) foliis cordatis ovatis. Prod. Leyd. 31. *Star-flower with oval heart ſhaped leaves.*

10. ORNITHOGALUM (*Tuberoſum*) racemo, foliis teretibus fiſtuloſis. *Star-flower with a very long ſpike, and taper fiſtular leaves.*

The firſt ſort grows naturally near Briſtol, and alſo near Chicheſter in Suſſex, and ſome other parts of England. This hath a pretty large bulbous root, from which come out ſeveral long keel-ſhaped leaves, which ſpread on the ground; between theſe comes out a ſingle naked ſtalk about two feet long, ſuſtaining a long looſe ſpike of flowers, of a yellowiſh green colour, ſtanding upon pretty long foot-ſtalks, which ſpread wide from the principal ſtalk; the petals of the flowers are narrow, making but little appearance. The flowers have an agreeable ſcent; they appear in May, and when the ſeed-veſſels are formed, the foot-ſtalks which ſuſtain them become erect, and approach near the ſtalk. The ſeeds ripen in Auguſt.

The ſecond ſort grows naturally upon the hills in Portugal and Spain, but has been long cultivated in the Engliſh gardens, by the title of Star of Bethlehem. This hath a very large, oval, bulbous root, from which ariſe ſeveral long keel-ſhaped leaves, of a dark green colour, and in the middle comes out a naked ſtalk, which riſes near three feet high, terminated by a long conical ſpike of white flowers, ſtanding upon pretty long foot-ſtalks, which riſe one above another, inclining to an upright. Theſe appear in June, and are ſucceeded by roundiſh ſeed-veſſels, having three cells, filled with roundiſh ſeeds, which ripen in Auguſt.

The third ſort grows naturally in Arabia; this hath a very large bulbous root, from which come out ſeveral broad keel ſhaped leaves; the ſtalk is thick and ſtrong, riſing between two and three feet high, bearing a long ſpike of large white flowers, ſtanding upon long foot-ſtalks. They are compoſed of ſix petals, which ſpread open in form of a ſtar, and appear in June, but do not ripen their ſeeds in England.

The fourth ſort grows in great abundance naturally in the kingdom of Naples, and is now become almoſt as common in England, for the roots propagate ſo faſt by offſets and ſeeds, as to become troubleſome weeds in gardens; and in many places where the roots have been thrown out of gardens, they have grown upon dunghills and in waſte places as plentifully as weeds. This hath a pretty large compreſſed bulbous root, from which come out many long, narrow, keel-ſhaped leaves, of a dark green colour. The ſtalks are very thick and ſucculent, riſing about a foot high, ſuſtaining ten or twelve flowers in a looſe ſpike, each hanging on a foot-ſtalk an inch long; they are compoſed of ſix petals, which are white within, but of a grayiſh green on their outſide, having no ſcent; within the petals is ſituated the bell-ſhaped nectarium, compoſed of ſix leaves, out of which ariſe the ſix ſtamina, terminated by yellow ſummits. The flowers appear in April, and are ſucceeded by large, roundiſh, three-cornered capſules, which are filled with roundiſh ſeeds; as the capſules grow large, they are ſo heavy as to weigh the ſtalk to the ground.

The fifth ſort grows naturally in Africa, as alſo in the iſland of Zant, from whence I received it; this hath a bulbous root, much ſmaller than either of the former The leaves are long, narrow, keel-ſhaped, and flaccid. The ſtalks riſe almoſt a foot high; they are ſlender, and ſuſtain four or ſix flowers, hanging on long ſlender foot-ſtalks, placed at a diſtance from each other; they are compoſed of ſix petals, of a yellowiſh green colour, the three inner ſtanding erect, and the three outer ſpread open wide. This flowers in July, but does not produce ſeeds here.

The ſixth ſort grows naturally in Yorkſhire, and ſome of the other northern counties in England; this hath a bulbous root, from which come out four or five keel-ſhaped leaves about ſix inches long, and in the middle ariſes an angular ſtalk, having two narrow leaves, which grow about ſix inches high, ſuſtaining at the top ſix or eight yellow flowers in form of an umbel, ſtanding upon ſlender long foot-ſtalks. Theſe appear in April, and are ſucceeded by triangular capsules, having three cells, which are filled with roundish seeds.

The seventh sort grows naturally in most parts of Europe, and is supposed to do so in England, though it is seldom found here, unless in orchards or grounds, where the roots may have been planted, or thrown out of gardens with rubbish. This hath a bulb as large as a small Onion, to which adhere many small offsets; the leaves are long, narrow, and keel-shaped, spreading on the ground, and have a longitudinal white line through the hollow. The stalk rises about six inches high, sustaining an umbel of flowers, which are white within, but have broad green stripes on the outside of the petals; these stand upon long foot-stalks, which rise above the principal stalk. It flowers in April and May, and is succeeded by roundish three-cornered capsules, filled with angular seeds, which ripen in July.

The eighth sort grows naturally in Arabia; this hath a large bulbous root, from which arise many long keel-shaped leaves, which embrace each other with their base; they are of a deep green, and stand erect. The flowers of this kind I have never yet seen, though I have tried many ways to procure them: the roots multiply exceedingly, and are never injured by frost, although the leaves are put out before winter. These roots are frequently brought over from Italy for sale, but I have not heard of any having flowered; and Clusius says, he never saw but one root flower, and that came from Constantinople.

The ninth sort grows naturally at the Cape of Good Hope; this hath a round bulbous root, covered with a white skin, from which come out four or five keel-shaped leaves, embracing each other at their base; they are of a deep green, eight or nine inches long; in the middle of these arises the stalk, which is naked, and about a foot high; just under the flowers come out two or three short leaves, which end in acute points. The flowers stand upon very long foot-stalks; they are formed in a conical spike, and are composed of six oval petals, of a pure white; within these are situated six stamina, which are about half the length of the petals, terminated by roundish summits. The flowers are in beauty in May, and are succeeded by roundish three-cornered capsules with three cells, filled with roundish seeds, which some years ripen here in July.

The tenth sort grows naturally on the dry rocks at the Cape of Good Hope; this hath a large, depressed, bulbous root, as big as a man's fist, covered with an uneven brown skin, putting out several taper hollow leaves, nine or ten inches long, between which comes forth a naked stalk near a foot high, terminated by a loose spike of yellow flowers, of an agreeable sweet scent. It flowers in June, but does not produce seeds in England.

The three sorts first mentioned, are cultivated for ornament in the English gardens. These are propagated by offsets, which their roots commonly produce in great plenty. The best time to transplant their roots is in July or August, when their leaves are decayed; for if they are removed late in autumn, their fibres will be shot out, when they will be very apt to suffer, if disturbed. They should have a light sandy soil, but it must not be over dunged. They may be intermixed with other bulbous-rooted flowers in the borders of the pleasure-gardens, where they will afford an agreeable variety, and continue in flower a long time. Their roots need not be trasplanted oftener than every other year, for if they are taken up every year, they will not increase so fast; but when they are suffered to remain too long unremoved, they will have so many offsets about them as to weaken their blowing roots. These may also be propagated from seeds, which should be sown and managed as most other bulbous rooted flowers, and will produce their flowers in three or four years after sowing.

The fourth sort is scarce worthy of a place in gardens, but as it will thrive in any situation or under trees, so a few plants may be admitted for the sake of variety.

The ninth sort is too tender to thrive in the open air in England, so the roots of this should be planted in pots, filled with light earth, and in the autumn placed under a hot-bed frame, where they may be screened from frost, and in mild weather enjoy the free air. The leaves of this appear in the autumn, and continue growing all the winter, so must not be exposed to frost; nor should be drawn up weak, for then the flowers will be few on a stalk, and not large. If the pots do now and then receive a gentle shower of rain in winter, it will be sufficient, for they should not have much wet during that season. Toward the beginning of July the leaves and stalks decay, and then the roots may be taken up, laying them in a dry cool place till the end of August, when they must be planted again.

The tenth sort was formerly more common in the English gardens than at present. This kind is more tender than either of the former, so should be planted in pots, filled with fresh light earth; and in winter must be placed in an airy glass-case, amongst Sedums, Ficoideses, and such other pretty hardy succulent plants, which require a large share of air in mild weather, and but little wet. In summer they may be removed out of the house, and placed in a warm sheltered situation, observing never to give these plants much water when they are not in a growing state, lest it rot their roots; but when they are growing freely, they must be frequently but gently refreshed with water. These roots should be transplanted every year; the best time to perform this work, is soon after the flower-stems are decayed, when the roots will be in the most inactive state. When this is done, the offsets should be carefully taken off, and each transplanted into a separate small pot, filled with light fresh earth, and may be treated as the old roots.

The other species, which were included in this genus, are now removed to Scilla.

ORNITHOPUS. Lin. Gen. Plant. 790. Birds-foot.

The Characters are,

The empalement of the flower is permanent, of one leaf, tubulous, and indented in five equal segments at the brim. The flower is of the butterfly kind; the standard is heart-shaped and entire; the wings are oval, erect, and almost as large as the standard; the keel is small and compressed. It hath ten stamina, nine of which are joined, and one stands separate, terminated by single summits. The germen is narrow, supporting a bristly ascending style, terminated by a punctured stigma. The germen afterward becomes a taper incurved pod, having many joints connected together, but when ripe separate, each containing one oblong seed.

The Species are,

1. Ornithopus (*Perpucillus*) foliis pinnatis, leguminibus compressis subarcuatis. Hort. Upsal. 234. *Birds-foot with winged leaves, and compressed pods a little arched.*

2. Ornithopus (*Nodosus*) foliis pinnatis, leguminibus confertis pedunculatis. *Birds-foot with winged leaves, and pods growing in clusters upon foot-stalks.*

3. Ornithopus (*Compressus*) foliis pinnatis, pinnis linearibus, leguminibus binis compressis arcuatis. *Birds-foot with linear winged leaves, and compressed arched pods growing in pairs.*

4. Ornithopus (*Scorpioides*) foliis ternatis subsessilibus appendiculatis, impari maximo. Hort. Cliff. 364. *Birds-foot with trifoliate leaves sitting close to the stalk, having appendages, and the middle lobe very large.*

The first sort grows naturally in the south of France, in Spain and Italy. It is an annual plant, having many trailing stalks a foot and a half long, from which come out a few side branches; these are garnished with long winged leaves, composed of about eighteen pair of small oval lobes,

termi-

terminated by an odd one; these lobes stand sometimes opposite, and at others they are alternate and hairy. The flowers are produced in small clusters at the top of foot-stalks, which arise from the wings of the stalks, and are near three inches long, having a small winged leaf, part of which is below, and the other part above the flowers, so that they seem to come from the midrib of the leaf; the flowers are of a deep gold colour, shaped like a butterfly. These appear in July, and are succeeded by flat narrow pods about three inches long, which turn inward at the top like a bird's claw. These are jointed, and a little hairy, containing a single seed in each joint, which ripens in autumn, when the joints separate and fall asunder.

The second sort grows naturally on dry commons and heaths in most parts of England. The root of this sort is composed of two or three strong fibres, to which hang several small tubercles or knobs like grains. There are many slender stalks come out from the root, which spread on the ground, and are from four to eight inches long; these are garnished with small, winged, hairy leaves, composed of six or seven pair of narrow lobes, terminated by an odd one. The flowers stand upon slender foot-stalks, which come out at every joint of the stalk; they are small, of a yellow colour, and are succeeded by clusters of short pods, which are a little incurved at the top. It flowers and seeds about the same time as the former.

The third sort grows plentifully about Messina and Naples. The root of this sort runs deep into the ground, sending out a few small fibres on the side; the stalks are about six inches long, and do not lie flat on the ground like the other; the leaves are hairy, and are composed of ten or twelve pair of narrow lobes placed along the midrib, terminated by an odd one. The flowers grow in small bunches on the top of the branches; they are yellow, and these are generally succeeded by two flat pods, not much more than an inch long, turned inward like a bird's claw. This flowers and seeds about the same time with the former.

The fourth sort grows naturally among the Corn in Spain and Italy. This hath many smooth branching stalks, which rise near two feet high, garnished toward their top with trifoliate oval leaves sitting close, having two small appendages. The lower leaves are often single, and of a grayish colour, the middle lobe being twice the size of the two side ones. The flowers stand upon slender foot-stalks, are yellow, and succeeded by taper pods, which are two inches long, shaped like a bird's claw. This flowers and seeds about the same time with the former.

These plants are propagated by sowing their seeds in the spring upon a bed of light fresh earth, where they are to remain (for they seldom do well when they are transplanted;) and, when the plants come up, they must be carefully cleared from weeds, and where they are too close, some of the plants should be pulled out, so as to leave the remaining ones about ten inches asunder. In June these plants will flower, and the seeds will ripen in August. There is no great beauty in these plants, but, for the variety of their jointed pods, they are preserved by some curious persons in their pleasure-gardens, where, if their seeds are sown in patches in the borders, each sort distinctly by itself, and the plants thinned, leaving only two at each patch, they will require no farther care, and will add to the variety, especially where the snail and caterpillar plants are preserved, which are very proper to intermix with them. They are most of them annual plants, which perish soon after the seeds are ripe.

OROBUS. Tourn. Inst. R. H. 393. tab. 214. Bitter Vetch.

The CHARACTERS are,

The empalement of the flower is tubulous, of one leaf; the brim is oblique, and indented in five parts. The flower is of the butterfly kind; the standard is heart-shaped; the two wings are almost as long as the standard, and join together; the keel is bifid, acute-pointed, rising upwards; the borders are compressed, and the body swollen. It hath ten stamina, nine are joined, and one separate. It hath a cylindrical compressed germen, supporting a crooked rising style, crowned by a narrow downy stigma. The germen afterward becomes a long taper pod, ending in an acute point, having one cell, containing several roundish seeds.

The SPECIES are,

1. OROBUS (*Vernus*) foliis pinnatis ovatis, stipulis semisagittatis integerrimis, caule simplici. Lin. Sp. Pl. 728. *Bitter Vetch with oval winged leaves, entire stipulæ, half arrow-pointed, and a single stalk.*

2. OROBUS (*Tuberosus*) foliis pinnatis lanceolatis, stipulis semisagittatis, caule simplici. Lin. Sp. Pl. 728. *Bitter Vetch with spear-shaped winged leaves, entire half arrow-pointed stipulæ, and a single stalk.*

3. OROBUS (*Sylvaticus*) foliis pinnatis oblongo-ovatis obtusis, stipulis semisagittatis integerrimis, caule hirsuto. *Bitter Vetch with oblong, oval, obtuse, winged leaves, entire stipulæ half arrow-pointed, and a hairy stalk.*

4. OROBUS (*Niger*) caule ramoso, foliis sexjugis ovato-oblongis. Hort. Cliff. 366. *Bitter Vetch with a branching stalk, and leaves composed of six pair of oblong oval lobes.*

5. OROBUS (*Pyrenaicus*) foliis pinnatis lineari-lanceolatis decurrentibus, stipulis semisagittatis, caule simplici. *Bitter Vetch with linear, spear-shaped, winged leaves, running along the stalk, half arrow-shaped stipulæ, and a single stalk.*

6. OROBUS (*Lathyroides*) foliis conjugatis subsessilibus, stipulis dentatis. Hort. Upsal. 220. *Bitter Vetch with leaves placed by couples close to the stalks, and indented stipulæ.*

7. OROBUS (*Luteus*) foliis pinnatis ovato-oblongis, stipulis rotundato-lunatis dentatis, caule simplici. Lin. Sp. Plant. 728. *Bitter Vetch with oval, oblong, winged leaves, roundish, moon-shaped, indented stipulæ, and a single stalk.*

8. OROBUS (*Venetus*) foliis pinnatis ovatis acutis, quatuor-jugatis, caule simplici. Tab. 193. fol. 2. *Bitter Vetch with oval, acute-pointed, winged leaves, having four pair of lobes and a single stalk.*

9. OROBUS (*Americanus*) foliis pinnatis lineari-lanceolatis infernè tomentosis, caule ramosissimo frutescente. *Bitter Vetch with linear, spear-shaped, winged leaves, which are woolly on their under side, and a very branching shrubby stalk.*

10. OROBUS (*Argenteus*) foliis pinnatis oblongo-ovatis infernè sericeis, caule erecto tomentoso, floribus spicatis terminalibus. *Bitter Vetch with oblong, oval, winged leaves, which are silky on their under side, and have an upright woolly stalk, terminated by a spike of flowers.*

11. OROBUS (*Procumbens*) foliis pinnatis, foliis exterioribus majoribus tomentosis, caule procumbente. *Bitter Vetch with winged leaves, whose outer lobes are woolly, and the largest trailing stalk.*

12. OROBUS (*Coccineus*) foliis pinnatis, foliolis linearibus villosis, caule procumbente floribus alaribus & terminalibus. *Bitter Vetch with winged leaves, having hairy linear lobes, a trailing stalk, and flowers growing on the sides at the ends of the branches.*

The first sort grows naturally in the forests of Germany and Switzerland. The root of this is perennial, composed of many strong fibres; the stalks rise a foot high, and are garnished with winged leaves, composed of two pair of oval acute-pointed lobes, and at the base of the foot-stalk is situated a stipula (or small leaf) shaped like the point of an arrow cut through the middle. This embraces the stalk. The flowers stand upon foot-stalks, which arise from the wings of the stalk; they are about three inches long, sustaining

taining fix or feven flowers ranged in a fpike, which are of the butterfly kind. Thefe are at firft of a purple colour, but afterward change blue; they appear early in the fpring, and are fucceeded by flender taper pods an inch and a half long, having one cell, in which are lodged four or five oblong bitter feeds, which ripen in June. There is a variety with pale flowers, which is preferved in fome gardens.

The fecond fort grows naturally in woods and fhady places in moft parts of England. This hath a perennial creeping root, from which arife angular ftalks nine or ten inches long, garnifhed at each joint by one winged leaf, compofed of four pair of fmooth fpear-fhaped lobes, and at the bafe of each is fituated a ftipula like that of the firft fort; from the wings of the ftalks arife the foot-ftalks of the flowers, which are about four inches long, each fuftaining two or three purplifh red flowers, which turn to a deep purple before they fade. Thefe appear in April, and are fucceeded by long taper pods, containing fix or feven roundifh feeds, which ripen the beginning of June

The third fort grows naturally in Cumberland and Wales. The root is perennial and ligneous, from which arife feveral hairy ftalks a foot and a half high, garnifhed at each joint with one winged leaf, compofed of ten or eleven pair of narrow lobes, ranged clofe together along the midrib, at the bafe of which is fituated an acute ftipula embracing the ftalk. The flowers are produced in a clofe fpike ftanding upon foot-ftalks, which arife from the wings of the leaves; they are of a purple colour, and are fucceeded by fhort flat pods, containing two or three feeds. It flowers the beginning of June, and the feeds ripen in July.

The fourth fort grows naturally on the mountains in Germany and Switzerland. This hath a ftrong, ligneous, perennial root, from which arife many branching ftalks two feet high, garnifhed at each joint by one winged leaf, compofed of five or fix fmall, oblong, oval leaves, ranged along the midrib The flowers ftand upen very long foot-ftalks, which arife from the wings of the ftalk; thefe fuftain at their top four, five, or fix purple flowers, which appear in May, and are fucceeded by compreffed pods, containing four or five oblong feeds, which ripen the beginning of July. The ftalks decay in autumn, and new ones arife in the fpring.

The fifth fort grows naturally about Bologna, and in other parts of Italy; this hath a perennial root, compofed of many thick flefhy tubers. The ftalks are cornered, and rife a foot and a half high, garnifhed with winged leaves, compofed of four pair of linear fpear-fhaped lobes, placed along the midrib, which is bordered by the running of the lobes from one to another; at the bafe of each leaf is fituated a ftipula fhaped like that of the firft, and out of this arifes the foot-ftalk of the flower, fuftaining feven or eight flowers ranged in a loofe fpike. They are variegated with purple, blue, and red, appear in May, and are fucceeded by pods, containing two or three feeds, which ripen in July.

The fixth fort grows naturally in Siberia. This hath a perennial root, from which arife three or four branching ftalks about a foot high. The leaves ftand oppofite along the ftalks, to which they fit clofe, having an indented ftipula at their bafe; they are fmooth, ftiff, and of a lucid green. The flowers grow in clofe fpikes upon fhort foot-ftalks, which rife from the wings of the leaves at the top of the ftalks, where are generally three or four of thefe fpikes ftanding together. The flowers are of a fine blue colour, fo make a pretty appearance. Thefe appear in June, and are fucceeded by fhort flattifh pods, containing two or three feeds in each, which ripen in Auguft.

The feventh fort grows naturally in Siberia. This hath a biennial root, from which arife feveral herbaceous ftalks two feet and a half high, garnifhed with winged leaves, compofed of four or five pair of oval oblong lobes, having at their bafe a roundifh moon-fhaped ftipula embracing the ftalk. The flowers come out from the wings of the leaves upon fhort foot-ftalks; they are large, and of a purple colour, appearing in April, and are fucceeded by fwelling pods, containing four or five feeds, which ripen in June.

The eighth fort rows naturally in Italy. This hath a perennial root, from which arife feveral fingle ftalks about a foot high, garnifhed with winged leaves, compofed of four pair of oval lobes, ending in acute points; they are fmooth, and of a pale green colour, placed pretty far diftant on the midrib. The flowers come out upon flender foot-ftalks, which arife from the wings of the leaves, four or five ftanding at the top; they are of a purple colour, and appear in February. Thefe are fucceeded by fwelling pods, each containing three or four roundifh feeds, which ripen in May.

The ninth fort grows naturally in Jamaica, from whence the late Dr. Houftoun fent the feeds. This rifes with a very branching ftalk about three feet high, which is ligneous; the branches are garnifhed with winged leaves, compofed of five or fix pair of narrow fpear fhaped lobes, which are woolly on their under fide. The flowers grow in loofe fpikes at the end of the branches, and are of a pale purple colour; thefe are fucceeded by fmooth compreffed pods, each containing five or fix roundifh feeds.

The tenth fort was difcovered by the late Dr. Houftoun at La Vera Cruz, from whence he fent the feeds to England. This rifes with a fhrubby ftalk five or fix feet high, dividing into many flender branches, which are covered with a brown woolly bark, garnifhed with foft, fatteny, winged leaves; thofe on the young branches are compofed of four pair of oval obtufe lobes, of a brownifh green colour, hairy on their upper fide, but of a filvery filky hue on their under. The leaves on the upper branches are compofed of feven or eight pair of oblong oval lobes, of the fame colour and confiftence as the lower. The flowers are produced in long erect fpikes at the end of the branches; they are of a deep purple colour, and are fucceeded by long, woolly, compreffed pods, each containing four or five feeds.

The eleventh fort was difcovered by Dr. Houftoun at La Vera Cruz. This is a low plant, whofe ftalks bend to the ground, and are feldom more than fix or eight inches long, from which come out a few fhort fide branches; they are garnifhed with winged leaves, compofed of four or five pair of fmall, oblong, oval, woolly lobes, terminated by an odd one, the upper lobes being much larger than the lower. The flowers come out in fmall bunches, ftanding upon fhort foot-ftalks, which arife from the wings of the ftalk; they are fmall, of a bright purple colour; thefe are fucceeded by compreffed pods, each having fix or feven roundifh compreffed feeds.

The twelfth fort was difcovered at the fame time, growing naturally in the fame country as the former, by the fame gentleman. This hath a pretty thick ligneous root, which fends out many flender ftalks a foot and a half long, trailing upon the ground, garnifhed with winged leaves, compofed of three or four pair of narrow hoary lobes, about half an inch long. The flowers come out from the fide and at the end of the ftalks, three or four ftanding upon a fhort foot-ftalk; they are fmall, and of a fcarlet colour, and are fucceeded by fhort taper pods, each containing three or four fmall roundifh feeds.

The eight forts which are firft mentioned have perennial roots, but annual ftalks; feveral of thefe may be propagated by parting their roots; the beft time for doing this is in the autumn, that the plants may be well eftablifhed before the fpring; for as feveral of them begin to put out their ftalks

very

very early in the ſpring, ſo if they are then diſturbed, it will either prevent their flowering, or cauſe their flowers to be very weak. Moſt of theſe plants delight in a ſhady ſituation, and love a loamy ſoil.

They are alſo propagated by ſeeds, but theſe ſhould be ſown in the autumn, for if they are kept out of the ground till ſpring, ſome of the ſorts will never grow, and thoſe which do, ſeldom vegetate the ſame year; and the fourth ſort I could never raiſe from ſeeds ſown in the ſpring, though I have made the trial in different ſituations many times; but the ſeeds which have ſcattered in the ſummer, have come up well the following ſpring, as have alſo thoſe which were ſown in September. When the plants come up, they muſt be kept clean from weeds, and where they are too cloſe together they ſhould be thinned, ſo as they may have room to grow till the autumn, when they ſhould be tranſplanted into the places where they are deſigned to remain. If the roots are ſtrong, they will flower very well the following ſpring, but thoſe which are weak will not flower till the ſecond year; therefore ſuch may be planted in a ſhady border at four or five inches diſtance, where they may grow one year to get ſtrength, and then may be removed to the places where they are to remain. The farther care of them is only to dig the ground between them in ſpring, and in ſummer to keep them clean from weeds.

The firſt, fourth, fifth, ſeventh, eighth, and ninth ſorts, are ornamental plants; and as they are very hardy, requiring little culture, and will thrive in the ſhade, they deſerve a place in every good garden.

The three laſt mentioned ſorts being natives of warm countries, are tender, ſo muſt be preſerved in ſtoves, otherwiſe they will not live in England. Theſe are propagated by ſeeds, which ſhould be ſown early in the ſpring, in ſmall pots filled with light rich earth, and plunged into a hot-bed of tanners bark, obſerving frequently to moiſten the earth, otherwiſe the ſeeds will not grow. When the plants come up, they ſhould be carefully taken out of the pots, and each tranſplanted into ſeparate ſmall pots, filled with rich earth, and then plunged again into the tan-bed, obſerving to ſhade them until they have taken root; after which time they ſhould have freſh air admitted to them every day in warm weather, and muſt be frequently watered. With this management the plants will make great progreſs. When any of the plants are grown too tall to remain in the hot-bed, they ſhould be taken out, and plunged into the bark-bed in the ſtove, where they may have room to grow, eſpecially the tenth and eleventh ſorts; but the other being of humbler growth, may be kept in the hot-bed until Michaelmas, when the nights begin to be cold; at which time they ſhould be removed into the ſtove, and plunged into the bark-bed, where they muſt be treated as other tender exotic plants; by which method they may be preſerved through the winter, and the following ſummer they will produce flowers. Theſe plants are perennial, ſo that if they ſhould not perfect their ſeeds, the plants may be maintained for ſeveral years.

ORYZA. Tourn. Inſt. R. H. 513. tab. 296. Rice.

The Characters are,

The chaff is ſmall, acute-pointed, having two valves nearly equal, incloſing a ſingle flower. The petal has two valves, which are boat-ſhaped, ending in a beard or awn. It has a two-leaved nectarium, and ſix hairy ſtamina the length of the petal, terminated by ſummits, whoſe baſe are bifid, and a turbinated germen, ſupporting two reflexed hairy ſtyles, crowned by feathered ſtigmas. The germen afterward becomes one large, oblong, compreſſed ſeed, having two channels on each ſide, ſitting on the petal of the flower.

We have but one Species of this genus, viz.

Oryza (*Sativa.*) Matth. 403. Rice.

This grain is greatly cultivated in moſt of the eaſtern countries, where it is the chief ſupport of the inhabitants; and great quantities of it are brought into England and other European countries every year, where it is in great eſteem for puddings, &c it being too tender to be produced in theſe northern countries, without the aſſiſtance of artificial heat; but from ſome ſeeds which were formerly ſent to South Carolina, there have been great quantities produced; and it is found to ſucceed as well there as in its native country, which is a very great improvement to our American ſettlements.

This plant grows upon moiſt ſoils, where the ground can be flowed over with water after it is come up; ſo that whoever would cultivate it in England for curioſity, ſhould ſow the ſeeds upon a hot-bed; and when the plants are come up, they ſhould be tranſplanted into pots, filled with rich light earth, and placed into pans of water, which ſhould be plunged into a hot-bed; and as the water waſtes, ſo it muſt, from time to time, be renewed again. In July theſe plants may be ſet abroad in a warm ſituation, ſtill preſerving the water in the pans, otherwiſe they will not thrive; and toward the latter end of Auguſt they will produce their grain, which will ripen tolerably well, provided the autumn proves favourable.

OSIER. See Salix.

OSMUNDA. The Oſmond Royal, or flowering Fern.

This is one of the kinds of Fern which is diſtinguiſhed from the other ſorts, by its producing flowers on the top of the leaves; whereas the others, for the moſt part, produce them on the back of their leaves.

There is but one kind of this plant, which grows wild in England, but there are ſeveral ſorts of them which grow in America; but as they are ſeldom kept in gardens, I ſhall not enumerate their ſpecies.

The common ſort grows on bogs in ſeveral parts of England, therefore whoever hath an inclination to tranſplant it into gardens, ſhould place it in a moiſt ſhady ſituation, otherwiſe it will not thrive.

OSTEOSPERMUM. Lin. Gen. Pl. 887. Hard-ſeeded Chryſanthemum.

The Characters are,

The flower hath an hemiſpherical empalement, and is compoſed of ſeveral hermaphrodite flowers in the diſk, which are tubulous, cut at the brim into five parts. Theſe are ſurrounded by ſeveral female flowers, which are radiated, each having a long narrow tongue, which is cut into three parts at the top. The hermaphrodite flowers have each five ſlender ſhort ſtamina, terminated by cylindrical ſummits, with a ſmall germen ſupporting a ſlender ſtyle, crowned by an obſolete ſtigma; theſe are barren. The female flowers have each a globular germen ſupporting a ſlender ſtyle, crowned by an indented ſtigma; the germen afterward becomes one ſingle hard ſeed.

The Species are,

1. Osteospermum (*Moniliferum*) foliis ovalibus obſoletè ſerratis. Lin. Hort. Cliff. 424. *Hard-ſeeded Chryſanthemum with oval leaves, which are ſlightly ſawed.*

2. Osteospermum (*Piſiferum*) foliis lanceolatis acutè dentatis, caule fruticoſo. Tab. 194. fig. 1. *Hard-ſeeded Chryſanthemum with ſpear-ſhaped leaves, which are acutely indented, and a ſhrubby ſtalk.*

3. Osteospermum (*Spinoſum*) ſpinis ramoſis. Lin Hort. Cliff. 424. *Hard-ſeeded Chryſanthemum with branching ſpines.*

4. Osteospermum (*Polygaloides*) foliis lanceolatis imbricatis ſeſſilibus. Flor. Leyd. Prod. 179. *Hard-ſeeded Chryſanthemum with ſpear-ſhaped leaves ſitting cloſe to the ſtalks, and lying over each other like the ſcales of fiſh.*

The firſt ſort is a native of America, growing in Virginia and New England, in low moiſt ground. This ſort dies to to the root every autumn, and rises again the following spring; and when growing on a moist rich soil, the shoots will rise to the height of five or six feet, and are garnished with very large, angular, divided leaves, placed opposite, which are shaped somewhat like those of the Plane-tree. The flowers are produced at the extremity of the shoots, which are shaped like those of the Sun-flower, but small, so do not make much appearance. This sort never produces any seeds in England, so can only be propagated by parting of the roots; but this should not be done oftener than every third or fourth year. The best season for this, and for transplanting of the roots is in October, soon after the shoots decay. These roots should be planted in light rich earth, and should have a moist situation, where they will thrive extremely well; but in dry ground, if they are not duly watered in dry weather, they will make no progress, and frequently decay in hot weather. It will endure the winter's cold very well in the open air.

If the seeds of this plant are procured from America, they should be sown on a bed of rich earth, and in dry weather they should be watered. These seeds generally remain in the ground a whole year before the plants appear. When the plants come up, they should be treated in the same manner as hath been directed for the old plants.

The second sort grows naturally at the Cape of Good Hope, but has been several years an inhabitant in the English gardens. This rises with a shrubby stalk seven or eight feet high, covered with a smooth gray bark, and divides into several branches, which are garnished with oval leaves, which are unequally indented on their edges; they are placed alternately, and are of a thick consistence, covered with a hoary down, which goes off from the older leaves. The flowers are produced in clusters at the end of the branches, six or eight coming out together; these are yellow, and shaped like those of Ragwort. The border or rays are composed of about ten half florets, which spread open; the disk or middle is composed of tubulous florets, which are cut into five parts at the brim; these are barren, but the half florets round the border have one hard seed, succeeding each of them. This plant flowers but seldom here; the time of its flowering is in July or August.

The third sort grows like the second, but the leaves are more pointed, of a green colour, and acutely sawed on the edges; the foot-stalks of the leaves are bordered, and the leaves are deeply veined. This produces tufts of yellow flowers at the extremity of the shoots from spring to autumn, and frequently ripens seeds.

The fourth sort is a low shrubby plant, which seldom rises above three feet high, and divides into many branches; the ends of the shoots are beset with green branching spines; the leaves are very clammy, especially in warm weather; these are long and narrow, and set on without any order. The flowers are produced singly at the ends of the shoots, which are yellow, and appear in July and August.

These three sorts are too tender to live in the open air in England, so are placed in the green-house in October, and may be treated in the same manner as Myrtles, and other hardy green house plants, which require a large share of air in mild weather; and in the beginning of May the plants may be removed into the open air, and placed in a sheltered situation during the summer season. The second and third sorts must have plenty of water, being very thirsty plants, but the fourth sort must have it given but moderately, especially in winter.

These plants are propagated by cuttings, which may be planted in any of the summer months, upon a bed of light earth, and should be watered and shaded until they have taken root, which they will be in five or six weeks, when they must be taken up, and planted in pots; for if they are suffered to stand long, they will make strong vigorous shoots, and will be difficult to transplant afterward, especially the second and third sorts; but there is not so much danger of the fourth, which is not so vigorous, nor so easy in taking root as the other. During the summer season the pots should be frequently removed, to prevent the plants from rooting through the holes in the bottom of the pots into the ground, which they are very apt to do when they continue long undisturbed, and then they shoot very luxuriantly; and, on their being removed, these shoots, and sometimes the whole plants, will decay.

OSYRIS. Lin. Gen. Plant. 978. Poets Casia.

The Characters are,

It is male and female in different plants; the empalement of the flower is of one leaf, which is divided into three acute segments. The flower hath no petals, but those on the male plants have three short stamina; the female have a germen, which afterward changes to a globular berry, having a single seed.

We have but one Species of this plant, viz.

Osyris (*Alba*) frutescens baccifera. C. B. P. *Shrubby Berry-bearing Poets Casia; and by some Red-berried shrubby Casia.*

This is a very low shrub, seldom rising above two feet high, having ligneous branches, which are garnished with long narrow leaves, of a bright colour. The flowers appear in June, which are of a yellowish colour, and are succeeded by berries, which at first are green, and afterward turn to a bright red colour, somewhat like those of Asparagus.

This plant grows wild in the south of France, in Spain, and some parts of Italy, by the side of roads, as also between the rocks, but is with great difficulty transplanted into gardens, nor does it thrive after being removed; so that the only method to obtain this plant is, to sow the berries where they are to remain. These berries commonly remain a year in the ground before the plants appear, and sometimes they will lie two or three years, so that the ground should not be disturbed under three years, if the plants do not come up sooner. These seeds must be procured from the places where the plants naturally grow, for those which have been brought into gardens never produce any, and it is with great difficulty they are preserved alive.

OTHONNA. Lin. Gen. Plant. 888. Ragwort.

The Characters are,

It hath a radiated flower, composed of hermaphrodite florets which form the disk, and female half florets which form the rays or border; these are included in one common single empalement of one leaf. The hermaphrodite flowers are tubulous, indented at the top in five parts; the female half florets are stretched out like a tongue, and the point has three indentures, which are reflexed. The hermaphrodite florets have short hairy stamina, terminated by cylindrical summits, and an oblong germen supporting a slender style, crowned by a single stigma. The female half florets have oblong germen with a slender style, crowned by a large, bifid, reflexed stigma. The hermaphrodite florets are seldom fruitful, but the female half florets have an oblong seed, which is sometimes naked, and at others crowned with down; these sit in the permanent empalement.

The Species are,

1. Othonna (*Coronopifolium*) foliis pinnatifidis tomentosis, laciniis sinuatis, caule fruticoso. Hort. Upsal. 273. *Othonna with woolly wing-pointed leaves, sinuated jags, and a shrubby stalk; Sea Ragwort.*

2. Othonna (*Calthoides*) foliis cuneiformibus integerrimis sessilibus, caule fruticoso procumbente, pedunculis longissimis. *Othonna with entire wedge-shaped leaves sitting close, a shrubby trailing stalk, and very long foot stalks to the flowers.*

3. Othonna

3. OTHONNA (*Pectinata*) foliis pinnatifidis, laciniis linearibus parallelis. Hort. Cliff. 419. *Othonna with wing-pointed leaves, whose segments are narrow and parallel.*

4. OTHONNA (*Abrotanifolia*) foliis multifido-pinnatis linearibus. Flor. Leyd. Prod. 380. *Othonna with very narrow leaves, ending in many winged points.*

5. OTHONNA (*Bulbosa*) foliis ovato-cuneiformibus dentatis. Lin. Sp. Pl. 926. *Othonna with oval, wedge-shaped, indented leaves.*

There are some other species of this genus, some of which grow naturally in England, and being troublesome weeds, are not admitted into gardens, so I have not enumerated them here.

The first sort is the common Sea Ragwort, which has been supposed to grow naturally near the sea in some parts of England; but I have never yet met with it wild in any parts where I have been, nor have I heard that any other persons has seen it growing naturally here. It grows in great plenty on the sea coast in France and Italy. This sends out many shrubby stalks, which rise from two to three feet high; they are hoary, and are garnished with woolly leaves six or eight inches long, jagged to the midrib in five or six parts; the jags are opposite, narrowest at their base, and at their points are divided into three or four obtuse segments; the stalks have a few leaves toward the bottom of the same shape, and are terminated with yellow flowers growing in a corymbus; these are succeeded by downy seeds, which ripen in the autumn.

The second sort grows naturally at the Cape of Good Hope, from whence the seeds were brought to Holland in 1697, where the plants were raised, and have since been communicated to most of the curious gardens in Europe. This hath a root composed of many small fibres; the stalks are round, branching, and weak; they are herbaceous, hairy, and trail upon the ground, if they are not supported, dividing into a great number of branches, garnished with roundish indented leaves, not unlike those of Ground-ivy, hollowed at their base; the flowers are produced at the end of the stalks in loose umbels; they are radiated; the rays are yellow, like those of common Ragwort; the disk is composed of hermaphrodite flowers, of a dark colour; the seeds have a down on their top.

The third sort grows naturally in Æthiopia. This rises with a shrubby stalk four or five feet high, dividing into several branches, which are garnished with grayish leaves, placed without order; those on their lower part being narrow and entire, but the others are indented on the edges, after the manner of Hartshorn. The flowers are produced in loose umbels at the end of the branches; they are yellow, and are succeeded by downy seeds.

The fourth sort grows naturally at the Cape of Good Hope, from whence the seeds were brought to Holland, and the plants were raised in the Amsterdam garden in 1699. This rises with a shrubby stalk about the thickness of a man's finger, two feet high, which divide into many branches; these are covered with a hoary down, and are garnished with hoary leaves about three inches long and one broad, cut into many narrow segments almost to the midrib; these segments are equal and parallel, and are indented at their ends into two or three points. The flowers are produced on long foot-stalks, which arise from the wings of the stalks; toward the end of the branches they have large yellow rays or borders, with a disk of florets, and are succeeded by oblong purple seeds, crowned with down.

The fifth sort grows naturally on the hills near the Cape of Good Hope, and was raised from seed in the Amsterdam garden. This hath a low, shrubby, branching stalk; the leaves are thick, like those of Samphire, and are cut into many narrow segments. The flowers are produced on short foot-stalks at the end of the branches; they are yellow, and shaped like the other species of this genus, and are succeeded by brown seeds, crowned with soft down.

The first sort was formerly preserved in green-houses, and was supposed too tender to live abroad in the open air; but later experience has taught us, that the cold will not destroy it, provided it is planted on a lean dry soil; but in rich moist ground the plants grow too vigorous in the summer, so their branches being replete with moisture, are sometimes killed in very severe winters. This is easily propagated by slips or cuttings during any of the summer months, which may be planted in a shady border, and now and then refreshed with water; in about six weeks or two months they will have good roots: then they may be transplanted to the places where they are to remain, shading them from the sun, and supplying them with water until they have taken new root; after which they will require no other care but to keep them clean from weeds. This sort flowers most part of summer, but they have little beauty; however, the plants are preserved more for the variety of their hoary divided leaves.

The third, fourth, and fifth sorts, are preserved in green-houses through the winter, but require no artificial warmth; if these are protected from the frost it is sufficient, and in mild weather they must have a large share of free air. In the summer they must be placed abroad in a sheltered situation, among other hardy exotic plants, where they will add to the variety, and flower great part of the summer. These may be all propagated by cuttings during any of the summer months, which may be planted upon an old hot-bed, and covered with glasses, shading them from the sun in the heat of the day. When these have taken root, they should be planted each into a separate pot, filled with soft loamy earth, placing them in the shade till they have taken new root; then they may be removed to a sheltered situation, where they may remain till autumn, treating them in the same way as the old plants.

OXALIS. Lin. Gen. Plant. 515. Wood Sorrel.

The CHARACTERS are,

The empalement of the flower is permanent, and cut into five acute parts. The flower is of one petal, cut into five obtuse indented segments; it hath ten erect hairy stamina, and a germen with five angles, supporting five slender styles. The germen afterward becomes a five-cornered capsule with five cells, which open longitudinally at the angles, containing roundish seeds, which are thrown out with an elasticity on the touch when ripe.

The SPECIES are.

1. OXALIS (*Acetosella*) scapo unifloro, foliis ternatis, radice squamoso-articulatâ. Hort. Cliff. 175. *Common Wood Sorrel with one flower on a foot-stalk, trifoliate leaves, and a scaly-jointed root.*

2. OXALIS (*Corniculata*) caule ramoso diffuso, pedunculis umbelliferis. Hort. Cliff. 175. *Wood Sorrel with a branching diffused stalk, and umbellated foot-stalks.*

3. OXALIS (*Stricta*) caule ramoso erecto, pedunculis umbelliferis. Flor. Virg. 161. *Wood Sorrel with a branching upright stalk, and umbellated foot-stalks.*

4. OXALIS (*Incarnata*) pedunculis unifloris, caule dichotomo. Lin. Sp Pl 443. *Wood Sorrel with one flower on a foot-stalk, and spreading stalks.*

5. OXALIS (*Purpurea*) scapo unifloro, foliis ternatis, radice bulbosâ. Hort. Cliff. 175. *Wood Sorrel with a foot-stalk supporting one flower, trifoliate leaves, and a bulbous root, with large purple flowers.*

6. OXALIS (*Pes-capræ*) scapo umbellifero, foliis ternatis bipartitis. Lin. Sp Plant. 434. *Wood Sorrel with an umbelliferous stalk, and trifoliate leaves, divided in two parts.*

6 7, OXALIS

7 Oxalis (*Frutescens*) caule erecto fruticoso, foliis ternatis, impari maximo. *Wood Sorrel with an upright shrubby stalk, and trifoliate leaves, the middle one being very large.*

The first sort grows naturally in moist shady woods, and close to hedges in many parts of England, so is but seldom admitted into gardens: though whoever is fond of acid herbs in sallads, can scarce find a more grateful acid in any other plant. The roots of this sort are composed of many scaly joints, which propagate in great plenty. The leaves arise immediately from the roots upon single long foot-stalks, are composed of three heart-shaped lobes, which meet in a center, where they join the foot-stalk; they are of a pale green, and hairy; between these come out the flowers upon pretty long foot-stalks, each sustaining one large white flower, of the open bell-shape. They appear in April and May, and are succeeded by five-cornered oblong seed-vessels, having five cells, inclosing small brownish seeds; when these are ripe, the seed-vessels burst open on the least touch, and cast out the seeds to a considerable distance. This is the sort which is directed for medicinal use in the Dispensaries, but those people who supply the market with herbs, generally bring the third sort, which is now become common in the gardens; but this hath no acid, so is unfit for the purpose; but as it rises with an upright branching stalk, so it is soon gathered and tied up in bunches; whereas the leaves of the first grow singly from the root, and require more time in gathering. There is a variety of the first sort with a purplish flower, which grows naturally in the north of England, but as it does not differ from it in any other respect, I have not enumerated it.

The second sort is an annual plant, which grows naturally in woods and shady places in Italy and Sicily. The root of this is long, slender, and fibrous; the stalks trail upon the ground, spreading out eight or nine inches wide on every side, dividing into small branches; the leaves stand upon pretty long foot-stalks, and are composed of three heart-shaped lobes, which have deeper indentures at their points than those of the first sort. The flowers are yellow, growing in form of an umbel upon pretty long slender foot-stalks, arising from the side of the branches. These appear in June and July, and are succeeded by seed-vessels near an inch long, which open with an elasticity, and cast out the seeds.

The third sort grows naturally in Virginia and other parts of North America, from whence the seeds were formerly brought to Europe; but wherever this plant has been once introduced and suffered to ripen seeds, it has become a common weed. This is an annual plant, rising with a branching herbaceous stalk eight or nine inches high; the leaves stand upon very long foot-stalks, and are shaped like those of the second sort. The flowers are yellow, standing in a sort of umbel, upon long, slender, erect foot-stalks; the seed-vessels and seeds are like those of the second sort.

These three sorts require no particular culture; if the roots of the first sort are taken up and transplanted in a shady moist border, they will thrive and multiply exceedingly; and, if they are kept clean from weeds, will require no other care. If the seeds of the other two sorts are sown in an open border, the plants will rise freely, and require no care; for if they are permitted to scatter their seeds, there will be a plentiful supply of the plants.

The fourth sort hath a roundish bulbous root, from which come out slender stalks about six inches high, which divide into branches by pairs, and from the divisions come out the foot-stalks of the leaves; these are long, slender, and sustain a trifoliate leaf, composed of three small, roundish, heart-shaped lobes. The foot-stalks of the flowers are long, slender, and arise from the division of the stalks, each sustaining one purplish flower about the same size and shape as those of the first sort. This flowers in May, June, and July, and sometimes produces ripe seeds in England. It grows naturally at the Cape of Good Hope, so is too tender to live through the winter in the open air in England; but if it is sheltered from hard frost, under a common frame in winter, it will require no other protection. It propagates in plenty by offsets from the root, as also by bulbs, which come out from the side of the stalks.

The fifth sort grows naturally at the Cape of Good Hope in such plenty, that the earth which came from thence, in which some plants were brought to England, was full of it. This hath a roundish bulbous root, covered with a brown skin, sending out strong fibres, which strike deep into the ground; the leaves are trifoliate, composed of three roundish, large, hairy lobes, which are but little indented at the top; these stand upon long slender foot-stalks, which arise from a thick short stalk, which adheres to the root. The foot-stalks of the flowers arise between the leaves, each supporting one large purple flower; these appear in January and February, but are rarely succeeded by seeds here, but the roots put out offsets in great plenty, whereby it is propagated. This will not thrive in winter in the open air here, so the roots should be planted in pots, which may be sheltered under a common frame in winter, where it may have as much free air as possible in mild weather, otherwise the leaves will draw up weak; for the leaves of this plant come out in October, and continue growing till May, when they begin to wither and decay. The roots may be transplanted any time after the leaves decay, till they begin to push out again.

The sixth sort is a native of the same country as the fifth; the roots of this are bulbous; the leaves stand upon long slender foot-stalks, which arise from a head; they are composed of three lobes, which are for the most part divided into two parts almost to their base. The foot-stalks of the flowers are five or six inches long, sustaining several large yellow flowers, ranged in form of an umbel These appear in March, and are sometimes succeeded by seeds here. This requires the same treatment as the fifth.

The seventh sort was discovered by Plumier in some of the French colonies in America, and was since found growing plentifully at La Vera Cruz by the late Dr. Houstoun. It rises with a shrubby stalk a foot and a half high, sending out several slender branches, which are garnished with trifoliate small leaves, composed of three oval lobes, the middle one being twice as large as the side ones. They are placed opposite, and sometimes by threes round the stalk, standing upon short foot-stalks. The foot-stalks of the flowers arise from the wings of the stalks, which are near two inches long, each sustaining four or five yellow flowers, whose petals are not much longer than the empalement; each of these have a smaller foot-stalk which is crooked, so that the flowers hang downward.

This is much tenderer than either of the former, so requires to be placed in a stove kept to a moderate degree of warmth in winter. It is propagated by seeds, which must be sown in pots, and plunged into a moderate hot-bed, and when the plants come up, they should be each planted in a separate pot, filled with light sandy earth, and plunged into a fresh hot-bed, shading them from the sun till they have taken new root; after which they must be treated in the same manner as other tender plants from the same country.

OX-EYE. See Buphthalmum.

OXYACANTHA. See Berberis.

OXYS. See Oxalis.

P.

PADUS. Lin. Gen. Edit. prior. 476. The Bird Cherry, or Cherry Laurel.

The CHARACTERS are,

The empalement of the flower is bell-shaped. The flower hath five large roundish petals, which spread open, and are inserted in the empalement. It hath from twenty to thirty awl-shaped stamina, which are inserted in the empalement, and a roundish germen supporting a slender style. The germen afterward becomes a roundish fruit, inclosing an oval-pointed nut, having rough furrows.

The SPECIES are,

1. PADUS (*Avium*) glandulis duobus, basi foliorum subjectis. Hort. Cliff. 185. *Bird Cherry with two glands at the base of the leaves.*

2. PADUS (*Rubra*) foliis lanceolato-ovatis deciduus, petiolis biglandulosis. Tab. 196. fol. 2. *Bird Cherry with spear-shaped, oval, deciduous leaves, whose foot-stalks have two glands; commonly called by the gardeners Cornish Cherry.*

3. PADUS (*Virginiana*) foliis oblongo-ovatis serratis acuminatis deciduis, basi antice glandulosis. *Bird Cherry with oblong, oval, sawed, acute-pointed, deciduous leaves, and glands on the fore-part of the foot-stalk; American Bird Cherry.*

4. PADUS (*Laurocerasus*) foliis sempervirentibus lanceolato-ovatis. Hort. Cliff. 42. *Bird Cherry with evergreen, spear-shaped, oval leaves; or common Laurel.*

5. PADUS (*Lusitanica*) foliis oblongo-ovatis sempervirentibus eglandulosis. *Bird Cherry with oblong, oval, evergreen leaves, having glands; smaller Portugal Laurel, called Asarero by the Portuguese.*

6. PADUS (*Caroliniana*) foliis lanceolatis acutè denticulatis sempervirentibus. *Evergreen Bird Cherry with spear-shaped leaves, having small acute indentures; called in America Bastard Mahogany.*

The first sort grows naturally in the hedges in Yorkshire, and many of the northern counties in England, as also in some few places near London, but is propagated as a flowering shrub in the nursery-gardens for sale. This rises with several woody stalks to the height of ten or twelve feet. The branches, which grow wide and scattering, are covered with a purplish bark, and garnished with oval spear-shaped leaves, placed alternate, and have two small protuberances or glands at their base. The flowers are produced in long loose bunches from the side of the branches; they have five roundish petals, which are much smaller than those of the Cherry, and are inserted in the border of the empalement; and within these are a great number of stamina, which also are inserted in the empalement; they have a strong scent, which is very disagreeable to most persons. These flowers appear in May, and are succeeded by small roundish fruit, which are first green, afterward turn red, and when ripe, are black, inclosing a roundish furrowed stone or nut, which ripens in August.

The second sort grows naturally in America, from whence I have received the seeds, but has been many years propagated in the nursery gardens about London, where it is generally called Cornish Cherry. This sort has been often confounded with the first, many of the late writers in botany having supposed it was the same species; but I have raised both sorts from seeds, and have always found the young plants to retain their difference. This rises with a strait upright stem more than twenty feet high; the branches are shorter and broader than those of the other, and are not so rough; the flowers grow in closer shorter spikes, which stand more erect; the fruit is larger, and red when ripe. This flowers a little after the first sort.

The third sort grows naturally in Virginia, and other parts of North America. It rises with a thick stem from ten to thirty feet hight, dividing into many branches, which have a dark purple bark, and are garnished with oval leaves, placed alternately on short foot-stalks, of a lucid green, and slightly sawed on their edges, continuing in verdure as late in the autumn as any of the deciduous trees. The flowers come out in bunches like those of the second sort, and are succeeded by large fruit, which is black when ripe, and is soon devoured by the birds. The wood of this tree is beautifully veined with black and white, and will polish very smooth, so is frequently used for cabinet work; as is also the wood of the first sort, which is much used in France, where it is called Bois de Sainte Lucie.

The fourth sort is the common Laurel, which is now so well known as to need no description. This grows naturally about Trebisond, near the Black-sea, and was brought to Europe 1576, and is now become very common, especially in the warmer parts of Europe.

The fifth sort was brought to England from Portugal, but whether it is a native of that country, or was introduced there from some other, is hard to determine. The Portugueze call it Asarero, or Azerero. This was supposed to have been but a low evergreen shrub, but by experience we find, that when it is in a proper soil, it will grow to a large size. There are at present some of these trees whose trunks are more than a foot diameter, and ten or twelve feet high, which are not of many years standing, and are well furnished with branches, which, when young, have a reddish bark; the leaves are shorter than those of the common Laurel, approaching near to an oval form; they are of the same consistence, and of a lucid green, which mixing with the red branches, make a beautiful appearance. The flowers are produced in long loose spikes from the side of the branches; they are white, and shaped like those of the common Laurel, appearing in June, and are succeeded by oval berries smaller than those of the common Laurel; they are first green, afterward red, and when ripe are black, inclosing a stone like the Cherry.

The seeds of the sixth sort were sent from Carolina by the title of Bastard Mahogany, from the colour of the wood, which is somewhat like Mahogany. This seems to be little more than a shrub, if we may judge from the growth here; the stalk does not rise in height, but sends out lateral branches, which spread on every side, covered with a brown bark, and garnished with spear-shaped leaves near two inches long, and three quarters of an inch broad, with small acute indentures on the edges; they ſtand alternately upon very ſhort foot-ſtalks, and are of a lucid green, continuing their verdure all the year. This has not as yet flowered in England, ſo I can give no account of it; but by the ſeeds and deſcription which I received of its flowers, it belongs to this genus.

This plant will live in the open air, if it is planted in a warm ſituation, and ſheltered in ſevere froſt, eſpecially while the plants are young; but when they have acquired ſtrength, there is no doubt of their thriving very well in the open ground. It may be propagated in the ſame manner as the Portugal Laurel from the berries, and if the branches are laid down they will take root.

The three firſt ſorts are eaſily propagated, either by the ſeeds or layers; when they are propagated by the ſeeds, they ſhould be ſown in autumn, for if they are kept out of the ground till ſpring, they ſeldom grow till the ſecond year. They may be ſown upon a bed or border of good ground, in the ſame way as Cherry ſtones which are deſigned for ſtocks; and the young plants may be treated in the ſame manner, planting them out in a nurſery, where they may ſtand two years to get ſtrength, and then they may be tranſplanted to the places where they are to remain. They are uſually intermixed with other flowering ſhrubs in wilderneſs work, where they add to the variety.

If they are propagated by layers, the young ſhoots ſhould be laid down in the autumn, which will have good roots by that time twelvemonth, when they may be ſeparated from the old plants, and tranſplanted into a nurſery for a year or two to get ſtrength, and may then be removed to the places where they are to grow.

The third ſort will grow to be a very large tree, when it is planted in a moiſt ſoil, but in dry ground it rarely riſes more than twenty feet high. There have been ſome plants of late years raiſed from ſeeds which came from Carolina, which have all the appearance of the third ſort, but are of much humbler growth; whether this may proceed from their being brought from a warmer climate, ſo do not agree with the cold of our winters, or whether they are a different ſpecies from that I cannot yet determine, as they have not produced fruit here.

The Laurel may be eaſily propagated by planting of cuttings; the beſt time for doing this is in September, as ſoon as the autumnal rains fall to moiſten the ground; the cuttings muſt be the ſame year's ſhoots, and if they have a ſmall part of the former year's wood to their bottom, they will more certainly ſucceed, and form better roots. Theſe ſhould be planted in a ſoft loamy ſoil about ſix inches deep, preſſing the earth cloſe to them. If they are properly planted, and the ground is good, there will be few of the cuttings fail; and if they are kept clean from weeds the following ſummer, they will make good ſhoots by the following autumn, when they may be tranſplanted into a nurſery, where they may grow two years to get ſtrength, and then ſhould be removed to the places where they are to remain. Theſe plants were formerly kept in pots and tubs, and preſerved in green-houſes in winter; but afterward they were planted againſt warm walls to preſerve them, being frequently injured by ſevere froſt. After this the plants were trained into pyramids and globes, and conſtantly kept ſheered; by which the broad leaves were generally cut in the middle, which rendered the plants very unſightly. Of late years they have been more properly diſpoſed in gardens, by planting them to border woods, and the ſides of wilderneſs quarters, for which purpoſe we have but few plants ſo well adapted; for it will grow under the drip of trees, in ſhade or ſun, and the branches will ſpread to the ground, ſo as to form a thicket; and the leaves being large, and having a fine gloſſy green colour, ſet off the woods and other plantations in winter, when the other trees have caſt their leaves; and in ſummer they make a good contraſt with the green of the other trees. Theſe trees are ſometimes injured in very ſevere winters, eſpecially where they ſtand ſingle, and are much expoſed; but where they grow in thickets, and are ſcreened by other trees, they are ſeldom much hurt; for in thoſe places it is only the young tender ſhoots which are injured, and there will be new ſhoots produced immediately below theſe from the larger branches, to ſupply their place, ſo that in one year the damage will be repaired. But whenever ſuch ſevere winters happen, theſe trees ſhould not be cut or pruned till after the following Midſummer; by which time it will appear what branches are dead, which may then be cut away, to the places where the new ſhoots are produced; for by haſtily cutting theſe trees in the ſpring, the drying winds have free ingreſs to the branches, whereby the ſhoots ſuffer as much, if not more, than they had done by the froſt.

In warmer countries this tree will grow to a large ſize, ſo that in ſome parts of Italy there are large woods of them, but we cannot hope to have them grow to ſo large ſtems in England; for ſhould theſe trees be pruned up, in order to form them into ſtems, the froſt would then become much more hurtful to them, than in the manner they uſually grow, with their branches to the ground: however, if theſe trees are planted pretty cloſe together in large thickets, and permitted to grow rude they will defend each other from the froſt, and they will grow to a conſiderable height: an inſtance of which is now in that noble plantation of evergreen trees, made by his Grace the Duke of Bedford, at Wooburn-Abbey, where there is a conſiderable hill covered entirely with Laurels; and in the other parts of the ſame plantation, there are great numbers of theſe intermixed with the other evergreen trees, where they are already grown to a conſiderable ſize, and make a noble appearance.

The beſt ſeaſon for tranſplanting the plants is in the autumn, as ſoon as the rain has prepared the ground for planting; for although they often grow when removed in the ſpring, yet thoſe do not take ſo well, nor make ſo good progreſs, as thoſe which are removed in the autumn, eſpecially if the plants are taken from a light ſoil, which generally falls away from their roots; but if they are taken up with balls of earth to their roots, and removed but a ſmall diſtance, there will be no danger of tranſplanting them in the ſpring, provided it is done before they begin to ſhoot; for as theſe plants will ſhoot very early in the ſpring, ſo if they are removed after they have ſhot, the ſhoots will decay, and many times the plants entirely fail.

There are ſome perſons, who of late have baniſhed theſe plants from their gardens, as ſuppoſing them poſſeſſed of a poiſonous quality, becauſe the diſtilled water has proved ſo in many inſtances; but, however the diſtilled water may have been found deſtructive to animals, yet from numberleſs experiments, which hath been made both of the leaves and fruit, it hath not appeared that there is the leaſt hurtful quality in either; ſo that the whole muſt be owing to the oil, which may be carried over in diſtillation.

The berries have been long uſed to put into brandy, to make a ſort of ratafia, and the leaves have alſo been put into cuſtards, to give them an agreeable flavour; and although theſe have been for many years much uſed, yet there have been no inſtance of their having done the leaſt injury; and as to the berries, I have known them eaten in great quantities without prejudice.

There are ſome perſons who have grafted the Laurel upon Cherry-ſtocks, with deſign to enlarge the trees; but although they will take very well upon each other, yet they

ſeldom

feldom make much progrefs, when either the Laurel is grafted on the Cherry, or the Cherry upon the Laurel; fo that is only a matter of curiofity, attended with no real ufe: and I would recommend to perfons who have this curiofity, to graft the Laurel upon the Cornifh Cherry, rather than any other fort of ftock, becaufe the graft will unite better with this; and as it is a regular tree and grows large, fo it will better anfwer the purpofe of producing large trees.

The Portugal Laurel may be propagated in the fame way as the common Laurel, either by cuttings, layers, or feeds. If the cuttings are planted at the fame feafon, and in the fame way as hath been directed for the common Laurel, they will take root very freely; or if the young branches are laid in the autumn, they will take root in one year, and then may be removed into a nurfery, where they may grow a year or two to get ftrength, and then tranfplanted where they are to remain.

But although both thefe methods are very expeditious for the propagating thefe plants, yet I would recommend the raifing them from the berries, efpecially where they are defigned for tall ftandards; for the plants which are propagated by cuttings and layers, put out more lateral branches, and become bufhy, but are not fo well inclined to grow upright as thofe which come from feeds; and as there are now many trees in the Englifh gardens, which produce plenty of berries every year, fo that if they are guarded from birds till they are ripe, there may be a fupply of them fufficient to raife plants enough, without propagating them any other way. Thefe berries muft be fown in the autumn, and treated in the fame way as the common Laurel.

This tree is much hardier than the common Laurel, for in the fevere froft of the year 1740, when great numbers of Laurels were entirely killed, and moft of them loft their young fhoots, this remained unhurt in perfect verdure, which renders it more valuable; and as by the appearance of fome trees now growing in the gardens, they feem as if they will grow to a large fize, fo it is likely to be one of the moft ornamental evergreens we have.

PÆONIA. Tourn. Inft. R. H. 275. tab. 146. The Peony.

The CHARACTERS are,

The flower has a permanent empalement, compofed of five concave reflexed leaves. The flower hath five large, roundifh, concave petals, which fpread open, and a great number of fhort hairy ftamina, with two, three, or four oval, erect, hairy germen in the center, having no ftyles. The germen afterward become fo many oval, oblong, reflexed, hairy capfules, having one cell, opening longitudinally, containing feveral oval, fhining, coloured feeds, fixed to the furrow.

The SPECIES are,

1. PÆONIA (*Mafcula*) foliis lobatis ex ovato-lanceolatis. Haller. Helvet. 311. *Peony with lobated leaves, which are oval and fpear-fhaped; or Male Peony.*

2. PÆONIA (*Fœminea*) foliis difformiter lobatis. Haller. Helv. 311. *Peony with difformed lobated leaves; common or Female Peony.*

3. PÆONIA (*Peregrina*) foliis difformiter lobatis, lobis incifis, petalis florum rotundioribus. *Peony with difformed lobated leaves, which are cut, and rounder petals to the flower; Foreign Peony, with a deep red flower.*

4. PÆONIA (*Hirfuta*) foliis lobatis, lobis lanceolatis integerrimis. *Peony with lobated leaves, whofe lobes are fpear-fhaped and entire.*

5. PÆONIA (*Tartarica*) foliis difformiter lobatis pubefcentibus. Tab. 199. *Peony with difformed lobated leaves, which are downy.*

6. PÆONIA (*Lufitanica*) foliis lobatis, lobi ovatis infernè incanis. *Peony with lobated leaves, whofe lobes are oval and hoary on their under fide; or Portugal Peony, with a fingle fweet flower.*

The firft fort here enumerated, is the common Male Peony, which grows naturally in the woods on the Helvetian mountains. The root of this is compofed of feveral oblong knobs, fhaped like the dugs of a cow, which hang by ftrings faftened to the main head; the ftalks rife about two feet and a half high, which are garnifhed with leaves compofed of feveral oval lobes, fome of which are cut into two or three fegments; they are of a lucid green on their upper fide, but are hoary on their under. The ftalks are terminated by large fingle flowers, compofed of five or fix large roundifh red petals, inclofing a great number of ftamina, terminated by oblong yellow fummits. In the center is fituated two, three, or fometimes five germen, which join together at their bafe; they are covered with a whitifh hairy down; thefe afterward fpread afunder, and open longitudinally, expofing the roundifh feeds, which are firft red, then purple, and when perfectly ripe, turn black. The flowers appear in May, and the feeds ripen in the autumn.

There is one variety of this with pale, and another with white flowers, as alfo one whofe leaves have larger lobes; but as thefe are generally fuppofed to be only feminal variations, fo I have not enumerated them here.

The fecond fort is called the Female Peony; the roots of this are compofed of feveral roundifh thick knobs or tubers, which hang below each other, faftened with ftrings; the ftalks are green, and rife about the fame height as the former; thefe are garnifhed with leaves, compofed of feveral unequal lobes, which are varioufly cut into many fegments; they are of a paler green than thofe of the firft, and are hairy on their under fide; the flowers are fmaller, and of a deeper purple colour. It flowers at the fame time as the firft.

There are feveral varieties of this fort with double flowers, which are cultivated in the Englifh gardens; thefe differ in the fize and colour of their flowers, but are fuppofed to have been accidentally obtained from feeds.

The third fort grows naturally in the Levant; the roots of this are compofed of roundifh knobs like thofe of the fecond fort, as are alfo the leaves, but are of a thicker fubftance; the ftalks do not rife fo high, the flowers have a greater number of petals. This flowers a little after the other. The large, double, purple Peony, I fufpect is a variety of this fort.

The fourth fort hath roots like the fecond; the ftalks are taller, and of a purplifh colour; the leaves are much longer, the lobes are fpear-fhaped and entire; the flowers are large, and of a deep red colour. This flowers at the fame time as the two firft forts.

The feeds of the fifth fort were brought from the Levant, and from them there were plants raifed, which produced fingle, and others with double flowers, of the fame fhape, fize, and colour. The roots of thefe are compofed of oblong flefhy tubers or knobs; they are of a pale colour, and hang by ftrings like the other fpecies. The ftalks rife about two feet high, which are of a pale green, and are garnifhed with leaves compofed of feveral lobes, which are irregular in fhape and fize, fome of them having but fix, and others have eight or ten fpear-fhaped lobes; thefe are fome cut into two, fome three fegments, and others are entire; they are of a pale green, and are downy on their under fide. The ftalks are terminated by one flower of a bright red colour, a little lefs than that of the Female Peony, and has fewer petals; they have a great number of ftamina,

mina, and sometimes two, at others three germen, like those of the Female Peony, but shorter and whiter. This flowers a little later than the common Peony.

The seeds of the sixth sort were sent to the Chelsea garden by Dr. de Jussieu, who brought them from Portugal, where the plants grow naturally. The root of this sort is not composed of roundish tubers or knobs, but hath two or three long, taper, forked fangs like fingers. The stalk rises little more than a foot high, and is garnished with leaves composed of three or four oval lobes, of a pale colour on their upper side, and hoary on their under; the stalk is terminated by a single flower, which is of a bright red colour, smaller than either of the former, and of an agreeable sweet scent. This flowers about the same time with the common sort.

The first of these sorts is chiefly propagated for the roots, which are used in medicine; for the flowers being single, do not afford near so much pleasure as those with double flowers, nor will they abide near so long in beauty.

All the sorts with double flowers are preserved in curious gardens for the beauty of their flowers, which, when intermixed with other large growing plants in the borders of large gardens, will add to the variety; and the flowers are very ornamental in basons or flower-pots, when placed in rooms.

They are all extremely hardy, and will grow in almost any soil or situation, which renders them more valuable; for they will thrive under the shade of trees, and in such places they will continue much longer in beauty.

They are propagated by parting their roots, which multiply very fast. The best season for transplanting them is toward the latter end of August, or the beginning of September; for if they are removed after their roots have shot out new fibres, they seldom flower strong the succeeding summer.

In parting of these roots, you should always observe to preserve a bud upon the crown of each offset, otherwise they will come to nothing; nor should you divide the roots too small (especially if you have regard to their blowing the following year;) for when their offsets are weak, they many times do not flower the succeeding summer, or at least produce but one flower upon each root; but where you would multiply them in quantities, you may divide them as small as you please, provided there be a bud to each offset; but then they should be planted in a nursery-bed for a season or two to get strength, before they are placed in the flower-garden.

The single sorts may be propagated from seeds (which they generally produce in large quantities, where the flowers are permitted to remain,) which should be sown soon after they are ripe upon a bed of light fresh earth, covering them over about half an inch thick with the same light earth. The spring following the plants will come up, when they should be carefully cleared from weeds, and in very dry weather refreshed with water, which will greatly forward their growth. In this bed they should remain two years before they are transplanted, observing in autumn, when the leaves are decayed, to spread some fresh rich earth over the beds about an inch thick, and constantly to keep them clear from weeds.

The Portugal Peony may also be propagated either by eeds or parting of the roots, in the same manner as the other sorts, but should have a lighter soil and a warmer situation. The flowers of this kind are single, but smell very sweet, which renders it worthy of a place in every good garden.

PALIURUS. Tourn. Inst. R. H. 616. tab. 387. Christ's Thorn.

The CHARACTERS are,

The flower has no empalement. It hath five petals which are ranged circularly. It hath five stamina, which are inserted in the scales under the petals, terminated by small summits, and a roundish trifid germen, supporting three short styles, crowned by obtuse stigmas. The germen afterward becomes a buckler-shaped nut, divided into three cells, each containing one seed.

We have but one SPECIES of this genus, viz.

PALIURUS (*Spina Christi.*) Dod. Pempt. 848. *Christ's Thorn.*

This plant grows naturally in the hedges near Palestine; it rises with a pliant shrubby stalk to the height of eight or ten feet, sending out many weak slender branches, garnished with oval leaves placed alternately; they have three longitudinal veins, and are of a pale green. The flowers come out at the wings of the stalk in clusters, almost the length of the young branches; they are of a greenish yellow colour, and appear in June, and are succeeded by broad, roundish, buckler-shaped seed-vessels, which have borders like the brims of a hat, the foot-stalks being fastened to the middle, and have three cells, each containing one seed.

This is by many persons supposed to be the plant from which the crown of thorns, which was put upon the head of our Saviour, was composed; the truth of which is supported by many travellers of credit, who affirm that this is one of the most common shrubs in the country of Judea; and from the pliableness of its branches, which may be easily wrought into any figure, it may afford a probability.

This shrub grows wild in most parts of the Levant, as also in Italy, Spain, Portugal, and the south of France, especially near Montpelier, from whence their seeds may be procured, for they do not ripen in England. These seeds should be sown as soon as possible after they arrive, in a bed of light earth, and the plants will come up the following spring; but when the seeds are kept out of the ground till spring, they will not come up till the next year, and very often fail, therefore it is much the best way to sow them in autumn. These seedling plants may be transplanted the following season into a nursery to get strength, before they are planted out for good.

It may also be propagated by laying down its tender branches in the spring of the year, which, if carefully supplied with water in dry weather, will take root in a year's time, and may then be taken off from the old plant, and transplanted where they are to remain.

The best time for transplanting this plant is in the autumn, soon after the leaves decay, or the beginning of April, just before the plants begin to shoot, observing to lay some mulch upon the ground about their roots, to prevent them from drying, as also to refresh them now and then with a little water until they have taken fresh root, after which they will require but very little care. They are very hardy, and will grow to be ten or twelve feet high, if planted in a dry soil and a warm situation. There is little beauty in this plant, but it is kept in gardens as a curiosity.

PALMA. Plum. Gen. 1. The Palm-tree.

The CHARACTERS are,

It hath male and female flowers; in some species on the same plant, and in others on different plants; the empalement of the male flowers are divided into three parts. The flowers have three petals, and six stamina terminated by oblong summits, with an obsolete germen, supporting three short styles, crowned by acute stigmas; these are barren. The female flowers have a common sheath, but no empalement; they have six short petals, and an oval germen sitting upon an awl-shaped style, crowned by a trifid

fid stigma. The germen afterward becomes a fruit of various forms and sizes in different species.

The Species are,

1. Palma (*Dactylifera*) frondibus pinnatis, foliolis angustioribus aculeis terminalibus. *Palm-tree with winged leaves, whose lobes are narrow, and terminated by spines; or Date-tree.*

2. Palma (*Cocos*) frondibus pinnatis, foliolis replicatis, spadicibus alaribus, fructu maximo angulofo. *Palm-tree with winged leaves, whose lobes are folded back, foot-stalks proceeding from the sides of the branches, and a large angular fruit; commonly called Cocoa Nut.*

3. Palma (*Spinosa*) frondibus pinnatis, ubique aculeatis, aculeis nigricantibus fructu majore. *Palm-tree with winged leaves, which are every where armed with black spines, bearing a larger fruit; commonly called Great Macaw-tree.*

4. Palma (*Altissima*) frondibus pinnatis, foliolis replicatis, ramis aculeatis, aculeis sæpius geminatis nigricantibus. *Palm-tree with winged leaves whose lobes are folded back, and prickly branches, whose thorns often come by pairs, and are black.*

5. Palma (*Gracilis*) frondibus pinnatis, caudice æquali, fructu minore. *Palm-tree with winged leaves, an equal trunk, and a smaller fruit; commonly called the Cabbage-tree.*

6. Palma (*Oleosa*) frondibus pinnatis, foliolis linearibus planis, stipitibus spinosis. *Palm-tree with winged leaves, having narrow plain lobes, and prickly midribs; commonly called oily Palm-tree.*

7. Palma (*Prunifera*) frondibus pinnato-palmatis plicatis, caudice squamato. *Palm-tree with hand-shaped winged leaves, which are plaited, and a scaly stalk; called Palmetto, or Thatch.*

8. Palma (*Polypodifolia*) frondibus pinnatis, foliolis lineari-lanceolatis, petiolis spinosis. Hort. Cliff. 482. *Palm-tree with winged leaves, whose lobes are linearly spear-shaped, and prickly foot-stalks.*

9. Palma (*Pumila*) fructu clavato polypyreno. Trew. Dec. tab. 26. *Palm-tree with club-shaped fruit, containing many seeds.*

10. Palma (*Americana*) frondibus pinnatis, foliis lanceolatis plicatis geminatis sparsis. *Palm-tree with winged leaves, whose lobes are spear-shaped, plaited, and come out by pairs from one point, standing thinly along the midrib.*

11. Palma (*Draco*) foliis simplicibus integerrimis flaccidis. *Palm-tree with single, entire, flaccid leaves, commonly called Dragon-tree.*

The first sort here mentioned is the common Date-tree, which grows plentifully in Africa, and some of the eastern countries, from whence the fruit is brought to England. This rises to a great height in the warm countries; the stalks are generally full of rugged knots, which are the vestiges of the decayed leaves, for the trunks of these trees are not solid like other trees, but the center is filled with pith, round which is a tough bark full of strong fibres while young, but as the trees grow old, so this bark hardens and becomes ligneous; to this bark the leaves are closely joined, which in the center rise erect, being closely folded or plaited together; but after they are advanced above the vagina which surrounds them, they expand very wide on every side the stem, and as the older leaves decay, the stalk advances in height. The leaves of these trees, when grown to a size for bearing fruit, are six or eight feet long; these have narrow long leaves (or pinnæ) set on alternately their whole length. The small leaves or lobes are toward the base three feet long, and little more than one inch broad; they are closely folded together when they first appear, and are wrapped round by brown fibres or threads, which fall off as the leaves advance, making way for them to expand; these never open flat, but are hollow like the keel of a boat, with a sharp ridge on their backside; they are very stiff, and, when young, of a bright green, ending with a sharp black spine. These trees have male flowers on different plants from those which produce the fruit, and there is a necessity for some of the male trees to grow near the female, to render them fruitful; or, at least, to impregnate the ovary of the seed, without which the stones, which are taken out of the fruit, will not grow. Most of the old authors, who have mentioned these trees, affirm, that unless the female or fruit-bearing Palm-trees have the assistance of the male, they are barren; therefore in such places, where there are no male trees near the female, the inhabitants cut off the bunches of male flowers when they are just opened, and carry them to the female trees, placing them on the branches near the female flowers to impregnate them; which, they all agree, has the desired effect, rendering the trees fruitful, which would otherwise have been barren. Pere Labat, in his account of America, mentions a single tree of this kind, growing near a convent in the island of Martinico, which produced a great quantity of fruit, which came to maturity enough for eating; but, as there was no other tree of this kind in the island, they were desirous to propagate it, and accordingly planted great numbers of the stones for several years, but not one of them grew; therefore after having made several trials without success, they were obliged to send to Africa, where these plants grew in plenty, for some of the fruit; the stones of which they planted, and raised many of the plants. He then conjectures, that the single tree before-mentioned, might be probably so far impregnated by some neighbouring Palm trees of other species, as to render it capable of ripening the fruit, but not sufficient to make the seeds prolific, as is the case when animals of different kinds copulate.

The flowers of both sexes come out in very long bunches from the trunk between the leaves, and are covered with a spatha (or sheath,) which opens and withers; those of the male have six short stamina, with narrow four-cornered summits filled with farina. The female flowers have no stamina, but have a roundish germen, which afterward becomes an oval berry, with a thick pulp inclosing a hard oblong stone, with a deep furrow running longitudinally. The bunches of fruit are sometimes very large.

This species of Palm is by Dr. Linnæus titled Phœnix, which is the Greek name of it, and he makes it a distinct genus. There are some varieties, if not different species of this tree; in the warm countries, but as we cannot expect to see the trees in perfection in our country, it is not likely we shall come to any certainty how they differ from each other.

These plants may be easily produced from the seeds taken out of the fruit, (provided they are fresh,) which should be sown in pots filled with light rich earth, and plunged into a moderate hot-bed of tanners bark, which should be kept in a moderate temperature of heat, and the earth frequently refreshed with water.

When the plants are come up to a proper size, they should be each planted into a separate small pot, filled with the same light earth, and plunged into a hot-bed again, observing to refresh them with water, as also to let them have air in proportion to the warmth of the season, and the bed in which they are placed. During the summer time they should remain in the same hot-bed, but in the beginning of August you should let them have a great share of air to harden them against the approach of winter; for if they are too much forced, they will be so tender as not to be preserved ferved through the winter without much difficulty, efpecially if you have not the conveniency of a bark-ftove to keep them in.

The beginning of October you muft remove the plants into the ftove, placing them where they may have a moderate fhare of heat (thefe being fomewhat tenderer while young, than after they have acquired fome ftrength,) tho' indeed they may be fometimes preferved alive in a cooler fituation, yet their progrefs would be fo much retarded, as not to recover their vigour the fucceeding fummer. Nor is it worth the trouble of raifing thefe plants from feeds, where a perfon has not the conveniency of a ftove to forward their growth, for where this is wanting, they will not grow to any tolerable fize in twenty years.

Whenever thefe plants are removed, (which fhould be done once a year) you muft be very careful not to cut or injure their large roots, which is very hurtful to them; but you fhould clear off all the fmall fibres which are inclinable to mouldinefs, for if thefe are left on, they will in time decay, and hinder the frefh fibres from coming out, which will greatly retard the growth of the plants.

The foil in which thefe plants fhould be placed, muft be compofed in the following manner, viz. half of light frefh earth taken from a pafture ground, the other half fea-fand, and rotten dung or tanners bark in equal proportion; thefe fhould be carefully mixed, and laid in a heap three or four months at leaft before it is ufed, but fhould be often turned over to prevent the growth of weeds, and to fweeten the earth.

You fhould alfo obferve to allow them pots proportionable to the fizes of the plants, but you muft never let them be too large, which is of worfe confequence than if they are too fmall. During the fummer feafon they fhould be frequently refrefhed with water, but you muft be careful not to give it in too great quantities, but in winter they will require very little.

Thefe plants are very flow growers, even in their native countries, notwithftanding they arrive to a great magnitude; for it has been often obferved by feveral of the old inhabitants of thofe countries, that the plants of fome of thefe kinds have not advanced two feet in height in ten years; fo that when they are brought into thefe countries, it cannot be expected they fhould advance very faft, efpecially where there is not due care taken to preferve them warm in winter; but however flow of growth thefe plants are in their native countries, yet they may be with us greatly forwarded, by placing the pots into a hot-bed of tanners bark, which fhould be renewed as often as is neceffary, and the plants always preferved therein both winter and fummer; obferving to fhift them into larger pots as they advance in growth, as alfo to fupply them with water properly. There are plants now in the Chelfea garden, whofe leaves are nine feet long, which were raifed from feeds more than thirty years ago, and their ftems are not two feet high, one of which has produced fome fmall bunches of male flowers.

The fecond fort here mentioned is the Cocoa-nut, whofe fruit are frequently brought to England, fome of which are of a large fize. The branches of this tree are winged like thofe of the former, but the fmall leaves or lobes are three times as broad; they open flat, their borders fold backward, and are of a lighter green than thofe of the firft fort. The whole leaf (or branch) is often twelve to fourteen feet long; the male flowers grow in different parts of the fame tree with the fruit, proceeding from the trunk between the leaves; they are difpofed in long bunches, as are alfo the females; the nuts growing in very large clufters, which are covered with a thick fibrous coat adhering clofely to them. The nuts are large, oval, and have three holes in the fhell at the top; the kernel is firm, white within, and the fhell contains a quantity of pale juice, which is called the milk.

The Cocoa nut is cultivated in moft of the inhabited parts of the Eaft and Weft-Indies, but is fuppofed a native of the Maldives, and the defert iflands of the Eaft-Indies; from whence it is fuppofed it hath been tranfported to all the warm parts of America, for it is not found in any of the inland parts, or any where far diftant from fettlements. It is one of the moft ufeful trees to the inhabitants of America, who have many of the common neceffaries of life from it. The bark of the nut is made into cordage, the fhell of the nut into drinking bowls, the kernel of the nut affords them a wholefome food, and the milk contained in the fhell, a cooling liquor. The leaves of the trees are ufed for thatching their houfes, and are alfo wrought into bafkets, and moft other things which are made of Ofiers in Europe.

This tree is propagated by planting of the nuts, which in fix weeks or two months after planting will come up, provided they are frefh, and thoroughly ripe, which is what few of them are which are brought to England; for they always gather them before they are ripe, that they may keep during their paffage; fo that the beft way to bring nuts into England for planting, would be to take fuch of them as are fully ripe, and put them up in dry fand in a tub, where the vermin may not come to them; and thefe will often fprout in their paffage, which will be an advantage, becaufe then they may be immediately planted into pots of earth, and plunged into the bark-bed

The third fort is commonly called Macaw-tree by the inhabitants of the Britifh Iflands in America. This rifes to the height of thirty or forty feet. The ftem is generally larger toward the top than the bottom; the branches (or rather the leaves) are winged; the fmall leaves or lobes are long and broad; the ftalk and leaves are ftrongly armed with black fpines of various fizes in every part; the male and female flowers are on the fame tree, coming out in the fame manner as the Cocoa-nut. The fruit is about the fize of a middling Apple, and is inclofed in a hard fhell.

The Macaw-tree is very common in the Caribbee Iflands, where the negroes pierce the tender fruit, from whence flows out a pleafant liquor, of which they are very fond; and the body of the tree affords a folid timber, with which they make javelins, arrows, &c. and is by fome fuppofed to be a fort of ebony. This tree grows very flow, and requires to be kept warm in winter.

The fourth fort grows naturally at La Vera Cruz. This hath winged leaves or branches like the other fort. The fmall leaves or lobes are as narrow as thofe of the firft fort, but are not quite fo ftiff; they fpread open, are flat, and their edges fold backward; their ends are blunt, and have no fpines; the midrib is armed with long black fpines, which frequently come out by pairs from the fame point. The flowers come out from between the leaves, and the fruit grows on the fame plant as the male flowers, which are about the fame fize and fhape as thofe of the former fort; but as the lobes of the leaves are much narrower, and have no fpines on their furface, there can be no doubt of its being a diftinct fpecies.

The fifth fort is commonly called Cabbage-tree in the Weft-Indies. This rifes to a very great height in the countries where it grows naturally. Ligon, in his Hiftory of Barbadoes, fays, there were then fome of the trees growing there, which were more than two hundred feet high; and that he was informed they were a hundred years growing to maturity, fo as to produce feed. The ftalks of thefe trees are feldom larger than a man's thigh; they are fmoother than thofe of moft other forts, for the leaves naturally

turally fall off entire from them, and only leave the vestiga or marks where they have grown. These leaves (or branches) are twelve or fourteen feet long; the small leaves or lobes are near two feet long, and half an inch broad, with several longitudinal plaits or furrows, ending in soft acute points; these are not so stiff as those of the first sort, and are placed alternately. The flowers come out in long loose bunches below the leaves; these branch out into many loose strings, and are near four feet long, upon which the flowers are thinly placed. The female flowers are succeeded by fruit about the size of a Hazel nut, having a yellowish skin, sitting close to the strings of the principal foot-stalk.

As the inner leaves of this encompass the future buds more remarkably than most of the other species, so it is distinguished by this appellation of Cabbage-tree; for the center shoots, before they are exposed to the air, are white and very tender, like most other plants which are blanched; and this is the part which is cut out and eaten by the inhabitants, and is frequently pickled and sent to England by the title of Cabbage; but whenever these shoots are cut out, the plants decay, and never after thrive; so that it destroys the plants, which is the reason that few of the trees are now to be found in any of the islands near settlements, and those are left for ornament.

The sixth sort is called in the West-Indies the Oily Palm, and by some Negroes Oil, for the fruit of this tree was first carried from Africa to America by the negroes. It grows in great plenty on the coast of Guinea, and also in the Cape de Verd Islands, but was not in any of our American colonies till it was carried there; but now the trees are in plenty in most of the islands, where the negroes are careful to propagate them.

The leaves of this tree are winged; the small leaves or lobes are long, narrow, and not so stiff as most of the other sorts; the foot-stalks of the leaves are broad at their base, where they embrace the stem, diminishing gradually upward, and are armed with strong, blunt, yellowish thorns, which are largest at their base. The flowers come out at the top of the stem between the leaves; some bunches have only male flowers, others have female; the latter are succeeded by oval berries, bigger than those of the largest Spanish Olives, but of the same shape; these grow in very large bunches, and when ripe are of a yellowish colour.

From the fruit the inhabitants draw an oil, in the same way as the oil is drawn from Olives; from the body of the tree they extract a liquor, which, when fermented, has a vinous quality, and will inebriate. The leaves of the tree are wrought into mats by the negroes, on which they lie.

The seventh sort is called Palmetto-tree, or Thatch, by the inhabitants of Jamaica, where this tree grows upon all the Honey-comb rocks in great plenty. It rises with a slender stalk ten or twelve feet high, which is naked and smooth, and at top garnished with many fan-shaped leaves placed circularly; these have foot-stalks three or four feet long, which are armed with a few strong, green, crooked spines; the pinnæ, or lobes, do all meet in one center, where they join the foot-stalk, and are joined together a third part of their length from their base; they are at first closely folded into plaits, but afterward spread out like a fan; their ends being pliant, do often hang downward, and between these pinnæ hang down long threads. The flowers and fruit come out from between the leaves; the fruit is of the shape and size of the small Lucca Olives. The leaves of this tree are used for thatch all over the West-Indies.

The eighth sort grows naturally in Japan, and also upon rocky dry mountains at Malabar. This in time rises with a strait trunk about forty feet high, which has many circles round it the whole length, which are occasioned by the vestiga of the leaves, which are placed circularly round the stem; so as these separate entirely and fall off, the circles remain where their base embraced the stalk. The stalks are terminated by an obtuse cone, just below where the leaves are placed: those on the large trees are eight or nine feet long, but those of the small plants are much less; the largest I have seen were not more than two feet long. The base of the foot-stalk, which partly embraces the trunk, is broad and three-cornered, and is armed on each side with short spines, to the place where the lobes, or small leaves, begin. These pinnæ or lobes are long, narrow, and entire, of a lucid green on their upper side, standing by pairs along the midrib very close together. The flowers and fruit are produced in large bunches at the foot-stalks of the leaves; the fruit is oval, about the size of a large Plumb, and nearly of the same shape; the skin or covering changes first yellow, and afterward red when ripe; of a sweet taste, under which is a hard brown shell, inclosing a white nut, which is in taste like the Chestnut.

From the pith of the trunk of this tree is made the sago; this is first pulverized, then it is made into a paste, and afterward granulated.

The ninth sort was discovered by the late Dr. Houstoun, growing naturally in the sands near Old Vera Cruz in America. This hath a thick stem, which seldom rises more than two feet high. The leaves come out round the upper part of the stem, standing upon foot-stalks which are a foot and a half long; they are winged; the lobes, or small leaves, are about eight inches long, and one and a half broad in the middle, drawing to a point at both ends; they are stiff, smooth, and entire, having a few small indentures at their points, and are placed alternate, of a pale green colour; there are fourteen or fifteen of these lobes ranged along the midrib or stalk. The fruit rises up from the side of the stem, upon a short thick foot-stalk, standing upright, and shaped like a club, having many red seeds about the size of large Pease, standing in separate cells round the central foot-stalk, to which they adhere. These plants have their male flowers on separate plants from the fruit, for all those plants which have flowered in England are of the male kind. The plants lose their leaves before the fruit is ripe annually. The first time when Dr. Houstoun saw these plants growing at La Vera Cruz, they were in full leaf, but on his return to the same place three months after, the fruit was then ripe, and all the leaves were fallen off; and this he afterwards observed the following season.

The tenth sort was discovered by the late Dr. Houstoun in the Spanish West-Indies. This rises with a very tall naked trunk, garnished at the top with long winged branches or leaves, whose lobes are spear-shaped and plaited; they are of a softer texture than any of the other sorts, and for the most part come out two from the same point, so stand by pairs on the same side of the midrib; they have two lobes on a side a little above each other, but there is a great space between every four lobes. The flowers come out in long bunches from between the leaves, the male flowers hanging on long slender strings; but the fruit, which is about the size of a middling Plumb, is collected into large bunches.

The eleventh sort grows naturally in the Cape de Verd Islands, from whence I had one of the plants brought me; as also in the Madeiras, from whence I have received the seeds. This is called Dragon-tree, because the inspissated juice of the plants becomes a red powder very like the eastern dragons-blood, and is frequently used instead of it in the shops; but the tree from whence the true dragons-

blood

blood is taken, is of a very different genus from this. Dr. Van Royen, in the Prodromus of the Leyden garden, has ranged this among the Yuccas, I ſuppoſe, from the ſimilitude of the plant to thoſe of that genus; for, as the fruit of this is a berry not unlike thoſe of the Bay-tree, and the ſeeds of the Yucca grow in capſules with three cells, they cannot be of the ſame genus; nor had we any account of the real characters of this plant, ſo as abſolutely to determine the genus, till very lately from ſome trees which have flowered in Portugal. This riſes with a thick trunk nearly equal in ſize the whole length, the inner part of which is pithy; next to this is a circle of ſtrong fibres, and the outſide is ſoft. The ſtalk or trunk riſes twelve or fourteen feet high; there are the circular marks or rings left the whole length, where the leaves are fallen off; for as theſe half embrace the ſtalk with their baſe, ſo when they fall away, the veſtigia where they grew remain. The top of the ſtalk ſuſtains a large head of leaves, which come out ſingly all round it; they are ſhaped like thoſe of the common Iris, but are much longer, being often four or five feet long, and an inch and a half broad at their baſe, where they embrace the ſtalk, and leſſen gradually to the end, where they terminate in a point. Theſe leaves are pliable, and hang down all round the ſtem; they are entire, and of a deep green, ſmooth on both ſurfaces, and greatly reſemble thoſe of the common yellow Iris. As this plant has not flowered in England, I can give no account of its flowers; but ſo far as I can judge from the berries which I have received, it may properly enough be ranged in this genus.

All theſe ſorts of Palms are propagated by ſeeds, which ſhould be ſown in the ſame way as hath been directed for the firſt ſort, and the plants ſhould afterward be treated in the ſame manner, with this difference, that ſuch of them as are natives of very warm countries, will require to be kept in a warmer air. The ſecond, third, fourth, fifth, ſeventh, eighth, ninth, and tenth ſorts, ſhould be conſtantly kept in the bark-bed in the ſtove, otherwiſe they will not make great progreſs in England; and when they do thrive, they grow in about twenty years too tall for moſt of the ſtoves which are at preſent built here, nor can we hope to ſee many of them produce their fruit in England; for the plants are preſerved by the curious for their foliage, which being ſo ſingular and different from that of the European trees, renders them worthy of care.

The other ſorts may be kept in a dry ſtove in winter in a moderate temperature of air, and in the heat of ſummer they may be expoſed to the open air in a warm ſheltered ſituation for about three months; but they ſhould be removed into the ſtove, before the morning froſts come on in the autumn. When theſe plants are kept in a moderate degree of warmth, they ſhould have but little water during the winter ſeaſon; and in the ſummer, when they are expoſed in the open air, they muſt not be often watered, unleſs the ſeaſon is remarkably dry and warm, for too much moiſture will ſoon deſtroy them. The other management of them is nearly the ſame as for the Date Palms, which is not to cut their principal roots when they are ſhifted from one pot or tub to another, nor to confine their roots too much; but as the plants grow in ſize, they ſhould annually be removed into pots or tubs a ſize larger than thoſe they were in the former year. The earth in which they are planted ſhould be light, ſo as to let the moiſture eaſily paſs off; for if it is ſtrong, and detains the moiſture, the tender fibres of the roots will rot.

PANAX. Lin. Gen. Plant. 1031. Ginſeng, or Ninſeng.

The CHARACTERS are,

It hath male and hermaphrodite flowers on diſtinct plants; the male have ſimple globular umbels, compoſed of ſeveral coloured rays, which are equal. The flower has five narrow, oblong, blunt petals, which are reflexed, ſitting on the empalement, and five oblong ſlender ſtamina inſerted in the empalement, terminated by ſingle ſummits. The hermaphrodite umbels are ſimple, equal, and cluſtered; the involucrum is ſmall, permanent, and compoſed of ſeveral awl-ſhaped leaves. The flowers have five oblong equal petals, which are recurved, and five ſhort ſtamina terminated by ſingle ſummits, which fall off, with a roundiſh germen under the empalement, ſupporting two ſmall erect ſtyles, crowned by ſimple ſtigmas. The germen afterward becomes an umbilicated berry with two cells, each containing a ſingle, heart-ſhaped, convex, plain ſeed.

The SPECIES are,

1. PANAX (*Quinquefolium*) foliis ternis quinatis. Flor. Virg. 147. *Panax with trifoliate Cinquefoil leaves; called Ninzin.*

2. PANAX (*Trifolium*) foliis ternis ternatis. Flor. Virg. 35. *Panax with three trifoliate leaves.*

Both theſe plants grow naturally in North America; the firſt is generally believed to be the ſame as the Tartarian Ginſeng, the figures and deſcriptions of that plant, which have been ſent to Europe by the miſſionaries, agreeing perfectly with the American plant.

This hath a fleſhy taper root as large as a man's finger, which is jointed, and frequently divided into two ſmaller fibres downward. The ſtalk riſes above a foot high, naked to the top, where it generally divides into three ſmaller foot-ſtalks, each ſuſtaining a leaf compoſed of five ſpear-ſhaped lobes, which are ſawed on their edges; they are of a pale green, and a little hairy. The flowers ariſe on a ſlender foot-ſtalk, juſt at the diviſion of the foot-ſtalks, which ſuſtain the leaves, and are formed into a ſmall umbel at the top; they are of an herbaceous yellow colour, compoſed of ſmall petals, which are recurved. Theſe appear the beginning of June, and are ſucceeded by compreſſed heart-ſhaped berries, which are firſt green, but afterward turn red, incloſing two hard, compreſſed, heart-ſhaped ſeeds, which ripen the beginning of Auguſt.

The Chineſe hold this plant in great eſteem, according to the accounts which have been tranſmitted to Europe by the miſſionaries. Father Jartoux, in his letters, ſays, that the moſt eminent phyſicians in China have written whole volumes upon the virtues of this plant, and make it an ingredient in almoſt all remedies, which they give to their nobility, for it is of too high price for the common people. They affirm that it is a ſovereign remedy for all weakneſs occaſioned by exceſſive fatigues, either of body or mind; that it cures weakneſs of the lungs and the pleuriſy; that it ſtops vomitings; that it ſtrengthens the ſtomach, and helps the appetite; that it ſtrengthens the vital ſpirits, and increaſes the lymph in the blood; in ſhort, that it is good againſt dizzineſs of the head, and dimneſs of ſight, and that it prolongs life in old age.

This father alſo ſays, he has made trials of the root of this plant himſelf, and has in an hour after taking half one of the roots, found himſelf greatly recovered from wearineſs and fatigue, and much more vigorous, and could bear labour with greater eaſe than before.

This plant has been introduced in the Engliſh gardens from America; where it has been planted in a ſhady ſituation and a light ſoil, the plants have thriven and produced flowers, and ripened their ſeeds annually, but none of theſe ſeeds have grown; for I have ſeveral years ſown them ſoon after they were ripe, without any ſucceſs; I have alſo ſown of the ſeeds which were ſent me from America ſeveral times, in various ſituations, and have not raiſed a ſingle plant from either; and by the accounts which the miſſionaries have ſent from China, it appears they have had no better ſucceſs with the ſeeds of this plant, which they ſay they have

have frequently sown in the gardens in China, but could not raise one plant; so that I believe there is a necessity for the hermaphrodite plants to have some male plants stand near them, to render the seeds prolific; for all those plants which I have seen, or saved the seeds from, were such as had hermaphrodite flowers; and though the seeds seem to ripen perfectly, yet their not growing, though I have waited three years without disturbing the ground, confirms me in this opinion.

The second sort grows naturally in the same countries, but whether it is possessed of the same qualities as the first I cannot say; I have seen but one plant of this sort in England, which was sent me a few years ago from Maryland, and did not live over the first summer, which was remarkably dry, and being planted in a dry soil, was the occasion of its death; the stalk of this was single, and did not rise more than four inches high, dividing into three foot-stalks, each sustaining a trifoliate leaf, whose lobes were longer, narrower, and deeper indented on the edges, than those of the former. The flower-stalk rose from the divisions of the foot-stalk of the leaves, but before the flowers opened, the plant decayed, so I can give no farther account of it.

PANCRATIUM. Dill. Hort. Elth. 221. fol. 289. Sea Daffodil.

The CHARACTERS are,

The flowers are inclosed in an oblong spatha or sheath, and have a funnel-shaped cylindrical nectarium of one leaf, spreading open at the top, with six narrow spear-shaped petals, which are inserted on the outside of the nectarium, and six long stamina inserted in the brim of the nectarium, terminated by oblong prostrate summits. They have a three-cornered obtuse germen situated under the flower, supporting a long slender style, crowned by an obtuse stigma. The germen afterward becomes a roundish three-cornered capsule with three cells, filled with globular seeds.

The SPECIES are,

1. PANCRATIUM (*Maritimum*) spathâ multiflorâ, petalis planis, foliis lingulatis. Lin. Sp. Pl. 291. *Pancratium with a sheath containing many flowers, plain petals, and tongue-shaped leaves; the Sea Daffodil.*

2. PANCRATIUM (*Illyricum*) spathâ multiflorâ, foliis ensiformibus, staminibus nectario longioribus. Flor. Leyd. Prod. 34. *Pancratium with many flowers in a sheath, sword-shaped leaves, and stamina longer than the nectarium; Lily Daffodil of Sclavonia.*

3. PANCRATIUM (*Zeylanicum*) spathâ uniflorâ, petalis reflexis. Flor. Zeyl. 126. *Pancratium with one flower in a sheath, whose petals are reflexed.*

4. PANCRATIUM (*Carribeum*) spathâ biflorâ. Hort. Cliff. 133. *Pancratium with two flowers in a sheath.*

5. PANCRATIUM (*Amboinense*) spathâ multiflorâ, foliis ovatis nervosis. Lin. Sp. Pl. 291. *Pancratium with many flowers in a sheath, and oval veined leaves.*

6. PANCRATIUM (*Carolinianum*) spathâ multiflorâ, foliis linearibus, staminibus nectarii longitudine. Lin. Sp. Pl. 291. *Pancratium with many flowers in a sheath, narrow leaves, and stamina the length of the nectarium.*

7. PANCRATIUM (*Americanum*) spathâ multiflorâ, foliis carinatis angustioribus. *Pancratium with many flowers in a sheath, and narrow keel-shaped leaves.*

8. PANCRATIUM (*Latifolium*) spathâ multiflorâ, foliis carinatis latioribus. *Pancratium with many flowers in a sheath, and broader keel-shaped leaves.*

9. PANCRATIUM (*Ovatum*) foliis ovatis, nervosis, spathâ multiflorâ, staminibus nectario longioribus. *Pancratium with oval veined leaves, and many flowers in a sheath, whose stamina are longer than the nectarium.*

The first sort grows naturally on the sea-coast in Spain, and the south of France. This hath a large, coated, bulbous root, of an oblong form, covered with a dark skin; the leaves are shaped like a tongue; they are more than a foot long, and one inch broad, of a deep green, six or seven of them rising together from the same root, encompassed at bottom with a vagina or sheath; between these arise the stalk, which is a foot and a half long, naked, sustaining at the top six or eight white flowers, inclosed in a sheath, which withers and opens on the side, to make way for the flowers to come out. The germen are situated close to the top of the stalk, from these arise the tube of the flowers, which are three inches long; they are very narrow, swelling at the top, where the cup or nectarium is situated, on the outside of which is fastened the six segments or petals of the flower; these are narrow, and extend a great length beyond the nectarium; from the border of the nectarium arise six long slender stamina, terminated by oblong summits, which are prostrâte, and in the center arises a style the length of the stamina, terminated by an obtuse stigma. The flowers of this sort do not appear in England till the end of August, so are not succeeded by seeds here. The leaves of this sort are green all the winter, and decay in the spring, so the roots should be transplanted in June, after the leaves are decayed. This must be planted in a very warm border, and screened from severe frost, otherwise it will not live through the winter in England.

The second sort grows naturally in Sclavonia, and also in Sicily; this hath a large, coated, bulbous root, covered with a dark skin, sending out many thick strong fibres, which strike deep in the ground; the leaves are sword-shaped, a foot and a half long, and two inches broad, of a grayish colour. The stalks are thick, succulent, and rise near two feet high, sustaining at the top six or seven white flowers, shaped like those of the first sort, but the tube is shorter, and the stamina are much longer. This flowers in June, and frequently produces seeds which ripen in September.

This sort is hardy, and will live through the winter in the full ground, being never injured but in very severe winters; and if, in such seasons, the surface of the ground is covered with tanners bark, sea-coal ashes, straw or Pease-haulm, to keep out the frost, there will be no danger of the roots suffering. It is propagated either by offsets from the roots, or from seeds; the former is the more expeditious method, for the offsets will flower very strong the second year; whereas those which are raised from seeds, seldom flower in less than four or five years.

The roots of this plant should not be removed oftener than every third year, if they are expected to flower strong; the best time to transplant them is in the beginning of October, soon after their leaves decay; they should not be kept long out of the ground, for as they do not lose their fibres every year, so if these are dried by long keeping out of the ground, it greatly weakens the roots. This loves a light sandy soil and a sheltered situation; the roots should be planted nine inches or a foot asunder every way, and five inches deep in the ground.

If the plants are propagated by seeds, they should be sown in pots filled with light earth soon after they are ripe; these pots should be placed under a hot-bed frame in winter, to screen them from frost, but the glasses must be taken off every day in mild weather. The other management being the same as for the Narcissus, I need not repeat it here, so shall only mention, that the young roots will require a little protection in winter, till they have obtained strength.

The third sort grows naturally in Ceylon; this hath a pretty large bulbous root; the leaves are long and narrow, of a grayiſh colour, and pretty thick, ſtanding upright; the ſtalk riſes between them a foot and a half high, naked, ſuſtaining one flower at the top, whoſe petals are reflexed backward; the nectarium is large, and cut at the brim into many acute ſegments; the ſtamina are long, and turn toward each other at their points, in which it differs from the other ſpecies. The flower has a very agreeable ſcent, but is of ſhort duration; this is very rare in the gardens at preſent.

The fourth ſort grows naturally at La Vera Cruz, from whence the late Dr. Houſtoun brought ſome of the roots. The leaves of this ſort are about a foot long, and almoſt two broad, having three longitudinal furrows. The ſtalk riſes about a foot high, then divides like a fork into two ſmall foot-ſtalks, or rather tubes, which are narrow, green, and at firſt are encompaſſed by a thin ſpatha (or ſheath,) which withers and opens to give way to the flowers, which are white, and ſhaped like thoſe of the other ſpecies, but have no ſcent.

The fifth ſort was ſome years paſt in the Engliſh gardens, but I believe is now loſt here; it grows naturally at Amboyna. The root of this ſort is oblong, white, and ſends out ſeveral thick fleſhy fibres, which ſtrike downward; the leaves ſtand upon very long foot-ſtalks, ſome of them are oval, and others heart-ſhaped, about five inches long, and almoſt as many broad, ending in points, having many deep longitudinal furrows; they are of a light green, and their borders turn inward. The ſtalk is thick, round, and ſucculent, riſing near two feet high, ſuſtaining at the top ſeveral white flowers, ſhaped like the other ſpecies, but the petals are broader; the tube is ſhorter, and the ſtamina are not ſo long as the petals. Theſe flowers have a thin ſheath or covering, which ſplits open longitudinally to make way for the flowers.

The ſixth ſort grows naturally on moiſt boggy ſoils in Georgia, where Mr. Cateſby diſcovered it. This hath a roundiſh bulbous root, covered with a light brown ſkin, from which ariſe ſeveral narrow dark green leaves, about a foot long; between theſe come out a thick ſtalk about nine inches high, ſuſtaining ſix or ſeven white flowers, with very narrow petals, having large bell-ſhaped nectariums or cups, which are deeply indented on their brims; the ſtamina do not riſe far above the nectarium, and are terminated by yellow ſummits.

The ſeventh ſort grows naturally in the iſlands of the Weſt-Indies, where it is called White Lily. This hath a pretty large bulbous root, a little flatted at the top, covered with a brown ſkin; the leaves are near a foot and a half long, a little more than one inch broad, of a dark green, and hollowed in the middle like the keel of a boat. The ſtalks riſe near two feet high, they are thick, ſucculent, and naked, ſuſtaining at the top eight or ten white flowers, ſhaped like thoſe of the firſt ſort, but are of a purer white, and have a ſtrong ſweet odour like that of Balſam of Peru. The ſtamina of this are very long, ſpreading out wide each way; the pointal is of the ſame length, ſtanding in the middle of the nectarium. Theſe flowers are of ſhort duration, ſeldom continuing longer in beauty than three or four days, and in very hot weather not ſo long; when theſe fade, the germen, which are ſituated at the bottom of the tubes, turn to ſo many oblong bulbs, which are irregular in form, and when ripe, drop off in the ground, where they put out fibres and become plants.

Theſe foreign ſpecies are moſt, if not all of them, of this kind, bearing bulbs; whereas the two firſt have ſeed-veſſels with three cells, incloſing many roundiſh black ſeeds, ſo that though they agree in the characters of flowers, yet in this particular they differ greatly.

The eighth ſort grows naturally in the Weſt-Indies, where it is not diſtinguiſhed from the former; but as I have frequently propagated both by their bulbs which ſucceed the flowers, and have always found the plants ſo raiſed continue their difference, ſo I make no doubt of their being diſtinct ſpecies. This differs from the former, in the leaves being much longer and broader than that; for theſe are near two feet long, and more than three inches broad, and are hollowed like the keel of a boat. The flowers are larger, the petals longer, and the ſcent is not ſo ſtrong as that of the former, and the roots flower in every ſeaſon of the year. This ſeems to be the ſort figured by Dr. Trew, in the twenty-ſeventh table of his Decades of Rare Plants, but if it is, the leaves in his figure are too flat.

The ninth ſort grows naturally in the Weſt-Indies; this hath a large, roundiſh, bulbous root, from which ariſe ſeveral oval leaves about a foot long, and ſix inches broad in the middle, drawing to a point at both ends; they are of a deep green, and have many longitudinal furrows. The ſtalk is thick, ſucculent, and naked; it riſes a foot and a half high, ſuſtaining at the top ſix or eight white flowers, of an agreeable ſweet ſcent, ſhaped like thoſe of the ſeventh ſort, but are ſmaller; the petals are narrower, the tubes are ſhorter, and ſo are the ſpathæ or ſheaths.

Theſe ſeven ſorts laſt mentioned are tender, ſo will not thrive in England, unleſs they are placed in a warm ſtove. The beſt way to have theſe plants in perfection, is, to plunge the pots into the bark-bed in the ſtove, where they will thrive and flower exceeding well; for though they may be preſerved in a dry ſtove, yet thoſe will not thrive ſo well, nor will their flowers be ſo ſtrong as when they are plunged in the tan-bed, nor will they flower oftener than once a year; whereas when they are in the tan-bed, the ſame roots will often flower two or three times in a year. I have had ſeveral of the ſpecies in flower at all ſeaſons of the year, ſo there has not been a month when ſome of them were not in flower.

They are propagated by offsets from the roots, and alſo by the bulbs which ſucceed the flowers; if the latter are planted in ſmall pots filled with light earth from a kitchen-garden, and plunged into a moderate hot-bed, they will ſoon put out roots and leaves, and with proper management will become blowing roots in one year, ſo that they may be eaſily propagated; and if they are conſtantly kept in the tan-bed in the ſtove, they will put out offsets from their roots, and thrive as well as in their native countries.

PANICUM. Tourn. Inſt. R. H. 515. tab. 298. Panic.

The CHARACTERS are,

There is one flower in each chaff; the chaff opens with three valves, which are oval, ending in acute points. The petals open with two oval acute-pointed valves. The flowers have three ſhort hair-like ſtamina, terminated by oblong ſummits, and a roundiſh germen ſupporting two hair-like ſtyles, crowned by feathered ſtigmas. The germen afterward becomes a roundiſh ſeed, faſtened to the withered petals.

The SPECIES are,

1. PANICUM (*Germanicum*) ſpicâ ſimplici cernuâ, ſetis brevioribus, pedunculo hirſuto. *Panic with a ſingle nodding ſpike, ſhort awns, and a hairy foot-ſtalk; German Panic.*

2. PANICUM (*Italicum*) ſpicâ compoſitâ, ſpiculis glomeratis, ſetis immixtis, pedunculo hirſuto. Lin. Sp. Pl. 56. *Panic with a compounded ſpike, whoſe ſmaller ſpikes grow in cluſters, intermixed with awns, and have a hairy foot-ſtalk; Italian Panic.*

3. PANICUM (*Indicum*) ſpicâ ſimplici longiſſimâ, ſetis hiſpidis, pedunculo hirſuto. *Panic with the longeſt ſingle ſpike, prickly awns, and a hairy foot-ſtalk; Indian Panic with the longeſt ſpike.*

4. PANICUM (*Alopecurodum*) ſpicâ tereti, involucellis bifloris faſciculato-piloſis. Flor. Zeyl. 44. *Panic with a taper ſpike,*

spike, having two flowers in each cover, and hairs growing in clusters.

5. PANICUM (*Cæruleum*) spicâ simplici æquali, pedunculis bifloris. Prod. Leyd. 54. *Panic with an equal single spike, and two flowers growing on each foot-stalk.*

There are several other species of this genus than are here enumerated, some of which grow naturally in England; but as they are not cultivated, so it would be swelling this work too much if they were inserted here.

The first sort grows naturally in Germany and Hungary; of this there are three varieties, one with yellow grain, another with white, and the third has purple grains. This has been formerly cultivated for bread in some of the northern countries. It rises with a jointed Reed-like stalk about three feet high, and the size of the common Reed, garnished at each joint with one Grass-like leaf a foot and a half long, and an inch broad at the base where broadest, ending in acute points. The stalks are terminated by compact spikes, which are about the thickness of a man's finger at their base, growing taper toward their points, closely set with small roundish grain, like that of Millet. This is an annual plant, which perishes soon after the seeds are ripe.

The second sort is frequently cultivated in Italy, and other warm countries. This rises with a Reed-like stalk near four feet high, which is much thicker than that of the former; the leaves are also broader, but of the same shape. The spikes are a foot long, and twice the thickness of those of the former, but not so compact, being composed of several roundish clustered spikes; the grain is also larger, but of the same form.

The third sort grows naturally in both Indies; this hath a Reed like stalk as large as a man's thumb, rising upward of six feet high; the leaves are two inches broad, and more than two feet long, of the same form with those of the former sort; the spikes at the top are a foot and a half long, very compact, and thicker than a man's thumb at the base, growing taper toward the top. The seeds are much larger than those of the other sorts, and are in some white, and others yellow.

The fourth sort grows naturally in both Indies; this hath a strong Reed-like stalk, which rises seven or eight feet high, garnished with leaves more than three feet long; they are near three inches broad at their base, lessening to a point at the end, having a smooth surface; the spikes arise at the wings of the stalk; they are single, but not so compact as those of the former, having soft awns or beards; they are about six inches long, and stand upon very long foot-stalks; the grain of this is pretty large.

The fifth sort grows naturally in Peru; this rises with a Reed like stalk six feet high, which sends out two or three branches from the sides, and is garnished with long leaves two inches broad at their base; the stalks are of a purple colour; the leaves are also inclining to the same colour. The spikes come out from the wings of the stalks, and at the end of the branches; they are about four or five inches long, thicker than a man's thumb, and almost equal at the point with the base. They are of a pale blue colour, having pretty long awns or beards of the same colour, as are also the seeds, which are larger and rounder than those of the other sorts.

The two first sorts are sown in several parts of Europe, in the fields, as Corn, for the sustenance of the inhabitants, but it is reckoned not to afford so good nourishment as Millet; however, it is frequently used in some parts of Germany and Italy, to make cakes and bread, but the German is not so much esteemed as the Italian sort; but as it will ripen better in cold countries than that, it is generally cultivated where a better sort of grain will not succeed.

The seeds of these sorts may be sown in the spring, at the same time as Barley is sown, and may be managed exactly in the same way; but this should not be sown too thick, for these seeds are very small, and the plants grow stronger, therefore require much more room. The German sort doth not grow above three feet high, unless it is sown on very rich land; in which case it will rise to be four feet high, but the leaves and stems of this Corn are very large, so require to stand four or five inches apart, otherwise they will grow up weak, and come to little. These large growing Corns should be sown in drills at about eighteen inches apart, so that the ground may be hoed between the rows of Corn, to keep them clear from weeds, and the stirring of the ground will greatly improve the Corn. In August or September the Corn will ripen, when it may be cut down and dried, and then should be housed.

The Italian Panic grows much larger than the German, and produces much larger spikes; so this should be allowed more room to grow, otherwise it will come to little. This is also later before it ripens, so it is not very proper for cold countries.

The other sorts are natives of very warm countries, where they are used by the inhabitants to make bread. These grow very large, and require a good summer, otherwise they will not ripen in this country. The seeds of these kinds should be sown the latter end of March, or the beginning of April, on a bed of light rich earth, in a warm situation. They should be sown in drills about three feet asunder, and when the plants come up, they must be kept clear from weeds, and thinned where they are too close. When the plants are grown pretty tall, they should be supported by stakes, otherwise the winds will break them down; and when the Corn begins to ripen, the birds must be kept from it, otherwise they will soon destroy it. These sorts are preserved in some curious gardens for the sake of variety, but they are not worth cultivating for use in England. The two last sorts do not ripen here.

PANSIES. See VIOLA TRICOLOR.

PAPAVER. Tourn. Inst. R. H. 237. tab. 119. Poppy.

The CHARACTERS are,

The empalement of the flower is oval, indented, and composed of two almost oval, concave, obtuse leaves, which fall off. The flower has four large roundish petals which spread open, with a great number of hair-like stamina, terminated by oblong, compressed, erect summits. In the center is placed a large roundish germen, having no style, but is crowned by a plain, radiated, target-shaped stigma. The germen afterward becomes a large capsule, crowned by the plain stigma, having one cell, opening in many places at the top under the crown, and is filled with small seeds.

The SPECIES are,

1. PAPAVER (*Rheas*) capsulis glabris globosis, caule piloso multifloro, foliis pinnatifidis incisis. Lin. Sp. Pl. 507. *Poppy with smooth globular heads, a hairy stalk with many flowers, and wing-pointed cut leaves; or common, red, Field Poppy.*

2. PAPAVER (*Hybridum*) capsulis subglobosis torosis hispidis, caule folioso multifloro. Lin. Sp. Plant. 506. *Poppy with globular capsules, which are furrowed and prickly, and a leafy stalk bearing many flowers.*

3. PAPAVER (*Argemone*) capsulis clavatis hispidis, caule folioso multifloro. Lin. Sp. Pl. 506. *Poppy with nail-shaped prickly heads, and a leafy stalk bearing many flowers.*

4. PAPAVER (*Alpinum*) capsulâ hispidâ, scapo unifloro nudo hispido, foliis bipinnatis. Lin. Sp. Plant. 507. *Poppy with prickly heads, a naked prickly stalk bearing one flower, and double winged leaves.*

5. PAPAVER (*Cambricum*) capsulis glabris oblongis, caule multifloro lævi, foliis pinnatis incisis. Lin. Sp. Pl. 508.

Poppy with oblong smooth heads, a smooth stalk bearing many flowers, and cut-winged leaves; yellow Welch Poppy.

6. Papaver (*Nudicaule*) capsulis hispidis, scapo unifloro nudo hispido, foliis simplicibus pinnato-sinuatis. Hort. Upsal 136. *Poppy with prickly heads, a naked rough stalk, having one flower, and single leaves, which are wingedly sinuated.*

7. Papaver (*Orientale*) capsulis glabris, caulibus unifloris, scabris, foliis pinnatis serratis. Hort. Upsal. 136. *Poppy with smooth heads, rough leafy stalks having one flower, and sawed winged leaves.*

8. Papaver (*Somniferum*) calycibus capsulisque glabris, foliis amplexicaulibus incisis. Lin. Sp. Pl. 508. *Poppy with smooth capsules and empalements, and cut leaves embracing the stalks.*

9. Papaver (*Album*) capsulis ovatis glabris, foliis latioribus amplexicaulibus marginibus inciso-serratis. *Poppy with oval smooth heads, and broader leaves embracing the stalks, which are cut on their edges like the teeth of a saw; commonly called white Poppy.*

The first sort is the common red Poppy, which grows naturally on arable land in most parts of England; from the flowers of this sort is drawn a simple water, a tincture, a syrup, and a conserve is also made for medicinal use. It is an annual plant; from the roots rise several rough branching stalks a foot and a half high, garnished with hairy leaves five or six inches long, deeply jagged almost to the midrib, those on the lower part of the leaves being the deepest. At the top of each stalk stand the flowers, which have oval hairy empalements, opening with two valves, and soon fall away. The flowers are composed of four large roundish petals, of a beautiful scarlet colour, which soon fall off. These appear in June, and are succeeded by oblong smooth heads, crowned by the flat target-shaped stigma, perforated in several places at the top, filled with small purplish coloured seeds.

The second sort grows naturally among the Corn in many parts of England; the leaves of this sort are much smaller than those of the first, and cut into much finer segments; the stalks are slender, a little more than a foot high, not so branching as the former. The flowers are not so large, and of a deep red colour, very soon falling away, seldom lasting more than a whole day: these are succeeded by oblong prickly heads, filled with small black seeds. It flowers in June.

The third sort grows naturally among Corn in some parts of England, but not in so great plenty as either of the former. The leaves of this are finer cut and smaller than those of the first sort, but are not so fine as those of the second; the stalks do not rise so high as either of the former, and seldom have many branches. The flowers are not half so large as either of the former, and are of a copper colour, falling away in a few hours. These appear in May, and are succeeded by long, slender, prickly heads, which are channelled, filled with small, black, shrivelled seeds.

The fourth sort grows naturally on the Alps, among the rocks. The leaves of this are smooth and doubly winged, the segments are finely cut; the stalks rise about a foot high, sustaining one small, yellow, or copper colour flower, which is succeeded by roundish prickly heads, filled with small seeds. This flowers about the same time as the former sort.

The fifth sort has a perennial root; it grows naturally in Wales, and also in some of the northern counties in England. I have found it growing plentifully near Kirkby Lonsdale in Westmoreland. Tournefort also found this plant upon the Pyrenean mountains. The leaves of this sort are winged; the lobes are deeply cut on their edges. The stalks rise a foot high; they are smooth, and garnished with a few small leaves, of the same shape as the lower. The upper part of the stalk is naked, sustaining one large yellow flower. These appear in June, and are succeeded by oblong smooth capsules, filled with small purplish seeds.

The sixth sort grows naturally on the confines of Russia near Tartary. The leaves of this sort are single, and sinuated almost to the midrib in form of a winged leaf; they are rough and hairy. The stalk rises near two feet high; it is slender, naked, sustaining one flower at the top, which is composed of four roundish petals, of a pale yellow colour, each having a dark bottom or tail. The flowers have an agreeable scent, but are of a short duration. They come out in June, and are succeeded by long rough capsules, filled with small seeds.

The seventh sort grows naturally in Armenia. The root of this plant is composed of two or three strong fibres as thick as a man's little finger, which are a foot and a half long, of a dark brown on their outside, and full of a milky juice, which is very bitter and acrid. The leaves are winged, and sawed on their edges; they are a foot long, closely covered with bristly white hairs. The stalks rise two feet and a half high; they are very rough and hairy, garnished below with leaves like those at bottom, but smaller; the upper part is naked, sustaining at the top one very large flower, of the same colour with the common red Poppy. These appear in May, and are succeeded by oval smooth capsules, filled by purplish seeds.

The eighth sort is the common black Poppy, the seeds of which are sold in the shops by the title of Maw-seed. The sort with single flowers grows in the warm parts of Europe naturally; this is annual; the stalks rise three feet high; they are smooth, and divide into several branches, garnished with large leaves, which are smooth, and deeply cut or jagged on their edges, embracing the stalks with their base. The flowers grow on the top of the stalks; they are composed of four large roundish petals, of a purplish colour, with dark bottoms, and are succeeded by oval smooth capsules, filled with black seeds. It flowers in June, and the seeds ripen the latter end of August.

There are great varieties in the flowers of this sort, some having very large double flowers, which are variegated of several colours, some are red and white, others purple and white, and some are finely spotted like Carnations; so that during their short continuance in flower, there are few plants whose flowers appear so beautiful, but having an offensive scent, and being of short duration, they are not much regarded.

The ninth sort is the common white Poppy. This is cultivated in gardens for the heads, which are used in medicine. The stalks of this are large, smooth, and rise to the height of five or six feet; they branch out into several smaller stalks, and are garnished with large grayish leaves, whose base embraces the stalks; they are jagged irregularly on their sides. These flowers terminate the stalks; these, when inclosed in the empalement, nod downward, but before the flowers open they are erect. The empalement of the flower is composed of two large oval leaves, of the same grayish colour as the other; these separate and soon drop off. The flower is composed of four large, roundish, white petals, which are of short duration, and are succeeded by large roundish heads as big as Oranges, flatted at both ends, having indented crowns, and are filled with small white seeds. This flowers in June, and the seeds ripen in August.

There are several varieties of this sort, which differ in the colour of their flowers and multiplicity of petals; those with beautiful flowers are preserved in gardens for ornament, but that with the single flowers only is cultivated for use. The seeds of this sort are used in emulsions, being cooling,

cooling, and good in fevers and inflammatory diſtempers, as alſo for the ſtrangury and heat of urine. Of the dry heads infuſed and boiled in wine, is made the Diacodium of the ſhops.

All the ſorts of Poppy are propagated by ſeeds, but the fifth and ſeventh ſorts, which have perennial roots, may alſo be propagated by offſets. The beſt time for ſowing of the ſeeds is in September, when they will more certainly grow than thoſe which are ſown in the ſpring; and thoſe ſorts which are annual, will make larger plants, and flower better than when they are ſown in the ſpring. The beſt way is to ſow the ſeeds of the annual kinds in the places where they are to remain, and to thin the plants where they are too cloſe; thoſe of the large kinds ſhould not be left nearer to each other than a foot and a half, and the ſmaller ſorts may be allowed about half that ſpace. The culture they will require after this, is only to keep them clean from weeds.

Thoſe who are curious to have fine Poppies in their gardens, carefully look over their plants when they begin to flower, and cut up all thoſe plants whoſe flowers are not very double and well marked, before they open their flowers, to prevent their farina mixing with their finer flowers, which would degenerate them; and it is the not being careful of this, that cauſes the flowers to degenerate ſo frequently in many places, which is often ſuppoſed to be occaſioned by the ground.

The yellow Welch Poppy requires a cool ſhady ſituation, where the plants will thrive, and produce plenty of ſeeds annually. If the ſeeds are permitted to ſcatter, the plants will come up better than when ſown by hand; but if they are ſown, it ſhould be always in the autumn, for the ſeeds of this, which are ſown in the ſpring, rarely ſucceed.

The beſt time to tranſplant and part the roots of this ſort, is in the autumn, that the plants may be well eſtabliſhed in their new quarters, before the dry weather comes on in the ſpring.

The eaſtern Poppy will thrive either in ſun or ſhade, for I have ſeveral of theſe plants growing under trees, where they have thriven many years, and flower full as well as thoſe in an open ſituation, but came later in the ſeaſon. This will propagate very faſt by its roots, ſo there is no neceſſity for ſowing of the ſeeds, unleſs to procure new varieties. This ſort ſhould be tranſplanted at the ſame ſeaſon as the former, and if the ſeeds are ſown, it ſhould be at the ſame time, for the reaſons before given.

PAPAVER CORNICULATUM. See GLAUCIUM.

PAPAVER SPINOSUM. See ARGEMONE.

PAPAYA. See CARICA.

PARIETARIA. Tourn. Inſt. R. H. 508. tab. 289. Pellitory.

The CHARACTERS are,

It hath hermaphrodite and female flowers upon the ſame plant. There are two hermaphrodite flowers contained in a ſix-leaved involucrum. They have no petals, but four permanent awl-ſhaped ſtamina, with an oval germen ſupporting a ſlender coloured ſtyle, crowned by a pencil-ſhaped ſtigma. The germen afterward turns to an oval ſeed wrapped up in the empalement. The female flowers have no ſtamina, but in other reſpects are the ſame as the hermaphrodite.

The SPECIES are,

1. PARIETARIA (*Officinalis*) foliis lanceolatis ovatis alternis. Hort. Upſal. 38. *Pellitory with ſpear-ſhaped leaves placed alternately; the officinal Pellitory.*

2. PARIETARIA (*Judiaca*) foliis ovatis caulibus erectiuſculis, corollis hermaphroditis. *Pellitory with oval leaves placed alternately; Pellitory with a Baſil leaf.*

The firſt ſort grows naturally in Germany and Holland, but was not in England till the year 1727, when I brought it here. This is ſuppoſed to be the true ſort, which is recommended by the ancients to be uſed in medicine; it hath a thick perennial root, compoſed of fleſhy reddiſh fibres, from which ariſe many ſtalks a foot and a half high, garniſhed with hairy ſpear-ſhaped leaves. The flowers come out in ſmall cluſters on the ſide of the ſtalks; they are ſmall, of an herbaceous colour, ſo make no figure. Theſe appear in ſucceſſion all the ſummer months, and the ſeeds ripen accordingly, which are caſt out to a diſtance with an elaſticity when ripe.

The ſecond ſort grows plentifully on old walls, and the ſides of dry banks in many parts of England. This differs from the former in having ſhorter ſtalks, and ſmaller oval leaves. The flowers are alſo leſs, and are in ſmaller cluſters, in other reſpects they are the ſame.

They may be propagated in plenty from a ſingle plant, which, if permitted to ſcatter its ſeeds, will fill the ground about it with young plants, for the ſeeds are very difficult to collect, as they are thrown out of their covers as ſoon as they are ripe.

PARIS. Lin. Gen. Pl. 449. True-love, or One-berry.

The CHARACTERS are,

The empalement of the flower is compoſed of four leaves, which expand in form of a croſs. The flower alſo hath four leaves, which ſpread open in the ſame manner. In the center of the flower is ſituated a roundiſh four-cornered germen, ſupporting four ſpreading ſtyles, crowned by ſingle ſummits. This is attended by eight ſtamina, each having an oblong ſummit faſtened by threads on each ſide to the ſtamina. The germen afterward changes to a roundiſh berry, having four cells, which are filled with ſeeds.

We have but one SPECIES of this genus, viz.

PARIS (*Quadrifolia*) foliis quaternis. Flor. Lapp. 155. *Herb Paris, True-love, or One-berry.*

This plant grows wild in moiſt ſhady woods in divers parts of England, but eſpecially in the northern counties, and it is with great difficulty preſerved in gardens. The only method to procure it, is to take up the plants from the places where they grow wild, preſerving good balls of earth to their roots, and plant them in a ſhady moiſt border, where they may remain undiſturbed, in which ſituation they will live ſome years; but as it is a plant of little beauty, it is rarely preſerved in gardens.

PARKINSONIA. Plum. Nov. Gen. 25. tab. 3.

The CHARACTERS are,

The empalement of the flower is of one leaf, indented in five parts at the top. The flower has five equal petals placed circularly; the four upper are oval, the under is kidney-ſhaped. It has ten declining ſtamina, terminated by oblong ſummits, and a long taper germen with ſcarce any ſtyle, crowned by an obtuſe ſtigma. The germen afterward becomes a long taper pod with ſwelling joints, in each of which is lodged one oblong ſeed.

We have but one SPECIES of this plant, viz.

PARKINSONIA (*Aculeata*) foliis minutis, uni coſtæ adnexis. Plum. Nov. Gen. 25. *Prickly Parkinſonia with very ſmall leaves, which are faſtened to one middle rib.*

This plant was diſcovered by Father Plumier in America, who gave it this name in honour of Mr. John Parkinſon, who publiſhed an univerſal hiſtory of plants in Engliſh in the year 1640.

It is very common in the Spaniſh Weſt-Indies, but of late years it has been introduced in the Engliſh ſettlements in America for the beauty and ſweetneſs of its flowers. This, in the countries where it grows naturally, riſes to be a tree of twenty feet high or more, and bears long ſlender bunches of yellow flowers, which hang down after the ſame manner as the Laburnum. Theſe flowers have a moſt agreeable ſweet ſcent, ſo as to perfume the air to a conſiderable diſtance round about the trees; for which reaſon, the inhabi-

tants

tants of the West-Indies plant them near their habitations. And though this plant has not been introduced many years into the English settlements, yet it is now become so common in all the islands, that but few houses are without some of the trees near it; for it produces flowers and seeds in plenty in about two years from seed, so that it may soon be made common in all hot countries; but in Europe it requires a stove, otherwise it will not live through the winter.

This plant is propagated by seeds, which should be sown in small pots filled with light fresh earth early in the spring; the pots must be plunged into a hot-bed of tanners bark, where, in about three weeks or a month's time, the plants will come up, when they should be kept clear from weeds, and frequently refreshed with water. In a little time these plants will be fit to transplant, which should be done very carefully, so as not to injure their roots. They must be each planted into a separate halfpenny pot, filled with light fresh earth, and then plunged into the hot-bed again, observing to stir up the tan; and if it hath lost its heat, there should be some fresh tan added to renew it again. Then shade the plants from the heat of the sun, until they have taken new root; after which time they should have fresh air admitted to them every day, in proportion to the warmth of the season. With this management the plants will grow so fast, as to fill the pots with their roots by the beginning of July, at which time they should be shifted into pots a little larger than the former, and plunged again into the bark-bed to forward their taking new root; after which it will be the best way to inure the plants by degrees to bear the open air, that they may be hardened before winter; for if they are kept too warm in winter, the plants will decay before the next spring. The only method by which I have succeeded in keeping this plant through the winter, was by hardening them in July and August to bear the open air; and in September I placed them on shelves in the dry stove, at the greatest distance from the fire, so that they were in a very temperate warmth; and there they retained their leaves all the winter, and continued in health, when those which were placed in a warmer situation, as also those in the green-house, were entirely destroyed.

PARNASSIA. Tourn. Inst. R. H. 246. tab. 127. Grass of Parnassus.

The CHARACTERS are,

The flower hath a spreading empalement, cut into five parts. The flower has five roundish concave petals, which have five heart-shaped concave nectariums, and five stamina terminated by depressed summits, with a large oval germen having no style, but four obtuse stigmas in their place. The germen afterward turns to an oval four-cornered capsule with one cell, containing several oblong seeds.

The SPECIES are,

1. PARNASSIA (*Palustris*) & vulgaris. Inst. R. H. *Common Marsh Grass of Parnassus.*

2. PARNASSIA (*Vulgaris*) flore pleno. *Common Grass of Parnassus, with a double flower.*

The former of these sorts grows wild in moist meadows in several parts of England, but particularly in the north. It grows on the other side of Watford, in the low meadows by Cassioberry, where it is in pretty great plenty.

The other sort is an accidental variety of the former, which has been discovered wild, and transplanted into gardens. This is but rarely to be found, being in very few gardens at present.

These plants may be taken up from the natural places of their growth, with balls of earth to their roots, and planted into pots filled with pretty strong, fresh, undunged earth, and placed in a shady situation, where, if they are constantly watered, they will thrive very well, and flower every summer; but if the plants are planted in the full ground, it should be in a very moist shady border, otherwise they will not live; and these should be as duly watered as those in the pots in dry weather, to make them produce strong flowers.

They may be propagated by parting of their roots, which should be done in March, before they put out new leaves, but the roots should not be divided too small, for that will prevent their flowering the following summer. These roots should always be planted in pretty strong fresh earth, for they will not thrive in a light rich soil. In the spring they must be constantly watered, if the season should prove dry, otherwise they will not flower; nor should they be parted oftener than every third year, to have them strong. These plants flower in July, and their seeds are ripe the latter end of August.

PARONYCHIA. See ILLECEBRUM.

PARSLEY. See APIUM.

PARSNEP. See PASTINACA.

PARTHENIUM. Lin. Gen. Plant. 939. Bastard Feverfew.

The CHARACTERS are,

It hath a flower composed of hermaphrodite florets and female half florets, which are inclosed in a common five-leaved spreading empalement. The hermaphrodite flowers which form the disk, have one tubulous petal cut into five parts at the brim; they have five hair-like stamina the length of the tube. The germen is situated below the flower, and is scarce visible, supporting a slender style, having no stigma; these flowers are barren. The female flowers, which compose the rays or border, are stretched out on one side like a tongue; these have a large heart-shaped compressed germen, with a slender style, crowned by two long spreading stigmas. These flowers are succeeded by one heart-shaped compressed seed.

The SPECIES are,

1. PARTHENIUM (*Hysterophorus*) foliis composito-multifidis. Lin. Hort. Cliff. 442. *Parthenium with many-pointed compound leaves.*

2. PARTHENIUM (*Integrifolium*) foliis ovatis crenatis. Lin. Hort. Cliff. 442. *Parthenium with oval crenated leaves.*

The first sort grows wild in great plenty in the island of Jamaica, and in some other of the English settlements in the West-Indies, where it is called wild Wormwood, and is used by the inhabitants as a vulnerary herb.

The second sort grows plentifully in several parts of the Spanish West-Indies, from whence the seeds have been brought to Europe.

The first is an annual plant, which may be propagated by sowing the seeds on a hot-bed early in the spring; and when the plants are come up, they should be transplanted on another hot-bed, at about five or six inches distance. When the plants have grown so as to meet each other, they should be carefully taken up, preserving a ball of earth to their roots, and each planted into a separate pot, filled with light rich earth; and if they are plunged into a moderate hot-bed, it will greatly facilitate their taking fresh root; but where this conveniency is wanting, the plants should be shaded from the sun until they have taken new root; after which time they may be exposed, with other hardy annual plants, in a warm situation, where they will flower in July, and their seeds will ripen in August and September.

The second sort is a perennial plant, which dies to the ground every autumn, and shoots up again the following spring. The seeds of this sort were sent me by my good friend Dr. Thomas Dale from South Carolina, where the plants grow wild. This may be propagated by parting of the roots in autumn, and may be planted in the full ground, where it will abide the cold of our ordinary winters very well.

well. This sort flowers in July, but seldom produces good seeds in England.

These plants make no great appearance, so are seldom cultivated but for the sake of variety.

PASQUE-FLOWER. See PULSATILLA.

PASSERINA. Lin. Gen. Plant. 440. Sparrow-wort.

The CHARACTERS are,

The flower has no empalement; it has one withered petal, having a slender cylindrical tube swelling below the middle. It hath eight bristly stamina, sitting on the top of the tube, terminated by erect summits. It has an oval germen under the tube, having a slender style rising on one side of the germen, crowned by a headed stigma, set with prickly hairs on every side. The germen afterward turns to an oval seed pointed at both ends, inclosed in a thick oval capsule of one cell.

The SPECIES are,

1. PASSERINA (*Filiformis*) foliis linearibus convexis quadrifariam imbricatis, ramis tomentosis. Lin. Sp. Pl. 559. *Sparrow-wort with linear convex leaves imbricated four ways, and downy branches.*

2. PASSERINA (*Hirsuta*) foliis carnosis extus glabris, caulibus tomentosis. Lin. Sp. Plant. 559. *Sparrow-wort with fleshy leaves, which are smooth on their outside, and downy stalks.*

3. PASSERINA (*Ciliata*) foliis lanceolatis subciliatis erectis, ramis nudis. Lin. Sp. Plant. 559. *Sparrow-wort with spear-shaped erect leaves, having small hairs and naked branches.*

4. PASSERINA (*Uniflora*) foliis linearibus oppositis, floribus terminalibus solitariis, ramis glabris. Lin. Sp. Plant. 560. *Sparrow-wort with linear leaves placed opposite, single flowers terminating the branches, and smooth stalks.*

The first sort grows naturally at the Cape of Good Hope. This rises with a shrubby stalk five or six feet high, sending out branches the whole height; these, when young, grow erect, but as they advance in length, they incline toward an horizontal position; but more so when the small shoots toward the end are full of flowers and seed-vessels, which weigh down the weak branches from their upright position. The branches are covered with a white down like meal, and are closely garnished with very narrow leaves, which are convex, and lie over each other in four rows like the scales of fish, so as that the young branches seem as if they were four-cornered. The flowers come out at the extremity of the young branches, from between the leaves on every side; they are small and white, so make but little appearance, and are succeeded by small seed-vessels, which seem withered and dry.

This plant may be propagated by cuttings during any of the summer months, which may be planted in a bed of loamy earth, and closely covered with a bell or hand-glass to exclude the air, shading them every day from the sun, and refreshing them now and then with water. With this treatment the cuttings will have taken root in about two months, when they may be taken up, and each planted in a small pot, filled with soft loamy earth, placing them in the shade till they have taken new root; then they may be removed into a sheltered situation, where they may remain till October, when they must be placed in the green-house, for they will not live in the open air through the winter in England; but they require no other treatment than Myrtles and other hardy green-house plants, which is only to screen them from frost. As this plant retains its verdure all the year, so it makes a pretty variety in the green-house in winter.

The second sort grows naturally in Spain and Portugal. This hath shrubby stalks, which rise to a greater height than the former; the branches grow more diffused than those of the former; they are covered with a meally down, and are garnished with short, thick, succulent leaves, lying over each other like the scales of fish; they are smooth, and green on their outside, but downy on their inner. The flowers are small and white, like those of the former. This plant will live abroad in ordinary winters, if it is planted in a dry soil and a warm situation, but in hard frosts they are frequently destroyed; therefore one or two plants should be kept in pots, and sheltered in winter to preserve the species. This may be propagated by cuttings in the same way as the former sort.

The third sort grows naturally in Spain and Portugal, as also at the Cape of Good Hope. This hath a shrubby stalk, rising five or six feet high, sending out many branches, which are naked to their ends, where they are garnished with oblong leaves standing erect, which have hairy points. The flowers are small, white, and come out between the leaves at the end of the branches, but are not succeeded by seeds in England. This may be propagated by cuttings as the two former, and requires the same treatment.

The fourth sort grows naturally at the Cape of Good Hope. It hath a low shrubby stalk, which seldom rises more than a foot high, dividing into many slender branches, which are smooth, and spread out on every side; these are garnished with very narrow leaves placed opposite; they are of a dark green, and have the appearance of those of the Fir-tree, but are narrower. The flowers come out singly at the end of the branches; these are larger than those of the former, and their upper part is spread open almost flat; they are of a purple colour. This may be propagated by cuttings as the other sorts, and the plants must be treated as the first sort.

PASSIFLORA. Lin. Gen. Plant. 910. Passion-flower.

The CHARACTERS are,

The flower has a plain-coloured empalement of five leaves, and five half spear-shaped petals, which are large, plain, and obtuse. The nectarium hath a triple crown; the outer, which is longer, is fastened to the inside of the petal, but is larger and compressed above. It has five awl-shaped stamina, fastened at their base to the column of the style annexed to the germen. The style is an erect cylindrical column, upon whose top sits an oval germen, with three smaller styles which spread out. The germen afterward becomes an oval fleshy fruit with one cell, sitting at the end of the style, filled with oval seeds, fastened longitudinally to the skin or shell.

The SPECIES are,

1. PASSIFLORA (*Incarnata*) foliis trilobis serratis. Amœn. Acad. Vol. I. p. 230. *Passion-flower with leaves having three sawed lobes; commonly called three-leaved Passion-flower.*

2. PASSIFLORA (*Cærulea*) foliis palmatis integerrimis. Amœn. Acad. 1. p. 231. *Passion-flower with hand-shaped entire leaves; or the common Passion-flower.*

3. PASSIFLORA (*Lutea*) foliis trilobis cordatis æqualibus obtusis glabris integerrimis. Amœn. Acad. 1. p. 224. *Passion-flower with heart-shaped leaves, having three equal lobes, which are smooth, obtuse, and entire.*

4. PASSIFLORA (*Glabra*) foliis trilobis integerrimis, lobis sublanceolatis, intermedio productiore. Amœn. Acad. 1. p. 229. *Passion-flower with leaves having three entire lobes, which are somewhat spear-shaped, and the middle one longer than the others.*

5. PASSIFLORA (*Suberosa*) foliis trilobis integerrimis glabris, cortice suberoso. Amœn. Acad. 1. *Passion-flower with leaves having three entire smooth lobes, and a Cork-like bark.*

6. PASSIFLORA (*Olivæformis*) foliis hastatis glabris, petalis florum angustioribus. *Passion-flower with halbert-pointed, smooth leaves, and narrow petals to the flower.*

7. Passiflora (*Fœtida*) foliis trilobis cordatis pilosis, involucris multifido capillaribus. Amœn. Acad. 1. p. 228. *Passion-flower with leaves having three hairy lobes, and the involucrum of the flower composed of many pointed hairs.*

8. Passiflora (*Variegata*) foliis hastatis pilosis amplioribus, involucris multifido-capillaribus. *Passion-flower with the largest halbert-pointed hairy leaves, and empalements composed of many-pointed hairs.*

9. Passiflora (*Holosericea*) foliis trilobis, basi utrinque denticula reflexo. Amœn. Acad. 1. p. 229. *Passion-flower with leaves having three lobes, indented on each side the base, and reflexed.*

10. Passiflora (*Capsularis*) foliis bilobis cordatis oblongis petalis. Lin. Sp. Plant. 957. *Passion-flower with oblong heart-shaped leaves, having two lobes standing upon foot-stalks.*

11. Passiflora (*Vespertilio*) foliis bilobis cuneiformibus, basi biglandulosis, lobis acutis divaricatis. Amœn. Acad. 1. p. 223. *Passion-flower with wedge-shaped leaves having two lobes, and two glands at their base, whose lobes are acute, and spread from each other.*

12. Passiflora (*Normalia*) foliis bilobis obtusis, basi emarginatis petiolatis. *Passion-flower with leaves having two obtuse lobes, which are indented at the base, and have foot-stalks.*

13. Passiflora (*Bicorna*) foliis bilobis glabris rigidis, basi indivisis. *Passion-flower with stiff smooth leaves having two lobes, which are undivided at their base.*

14. Passiflora (*Muricinea*) foliis bilobis transversis, amplexicaulibus. Amœn. Acad. 1. p. 222. *Passion-flower with transverse leaves, having two lobes embracing the stalk.*

15. Passiflora (*Maliformis*) foliis indivisis cordato-oblongis integerrimis, caule triquetro, involucris integerrimis. Amœn. Acad. 1. *Passion-flower with heart-shaped, oblong, entire leaves, a three-cornered stalk, with entire covers to the flowers; commonly called Granadilla in the West-Indies.*

16. Passiflora (*Laurifolia*) foliis oblongis integerrimis, involucris dentatis. *Passion-flower with oblong entire leaves, and the covers of the flowers indented; commonly called Water Lemon in the West-Indies.*

17. Passiflora (*Cuprœa*) foliis indivisis ovatis integerrimis, petiolis æqualibus. Amœn. Acad. Vol. 1. p. 219. *Passion-flower with undivided, oval, entire leaves, and equal foot-stalks.*

18. Passiflora (*Serratifolia*) foliis indivisis serratis. Amœn. Acad. 1. p. 213. *Passion-flower with undivided sawed leaves.*

19. Passiflora (*Multiflora*) foliis indivisis oblongis integerrimis, floribus confertis. Amœn. Acad. 1. p. 221. *Passion-flower with undivided, oblong, entire leaves, and flowers growing in clusters.*

The first sort grows naturally in Virginia and other parts of North America, and was the first known in Europe of all the species, but was not very common in the English gardens till of late years. The root of this plant is perennial, but the stalk is annual in North America, dying to the ground every winter, as it also does in England, unless it is placed in a stove. The stalks of this are slender, rising about four or five feet high, having tendrils or claspers at each joint, which fasten themselves about whatever plants stand near them, whereby the stalk is supported. At each joint comes out one leaf upon a short foot-stalk; these have for the most part three oblong lobes, which join at their base, but the two side lobes are sometimes divided part of their length into two narrow segments, so as to resemble a five-lobed leaf; they are thin, of a light green, and slightly sawed on their edges. The flowers are produced from the joints of the stalk at the foot-stalks of the leaves; these have long slender foot-stalks. The involucrum of the flower is composed of five oblong blunt pointed leaves, of a pale green; these open and disclose five more leaves or petals, which are white, having a fringe or circle of rays in a double order round the style, of a purple colour, the lower row being the longest. In the center of this arises the pillar-like style, with the roundish germen at the top, surrounded at the bottom, where it adheres to the style, with five flattish stamina, which spread out every way, and sustain each of them an oblong summit, which hangs downward, and on their under side are covered with a yellow farina. The flowers have an agreeable scent, but are of short duration, opening in the morning, and fade away in the evening, never opening again, but the plants are succeeded by fresh flowers, which come out at the joints of the stalk above them. When the flowers fade, the roundish germen swells to a fruit as large as a middling Apple, which changes to a pale Orange colour when ripe, inclosing many oblong rough seeds, inclosed in a sweetish pulp.

This sort is usually propagated by seeds which are brought from America, for the seeds do not often ripen in England; though I have had sometimes several fruit perfectly ripe on plants, which were plunged in a tan-bed under a deep frame; but those plants which are exposed to the open air, do not produce fruit here. The seeds should be sown upon a moderate hot-bed, which will bring up the plants much sooner than when they are sown in the open air, so they will have more time to get strength before winter. When the plants are come up two or three inches high, they should be carefully taken up, and each planted in a separate small pot, filled with good kitchen-garden earth, and plunged into a moderate hot-bed to forward their taking new root; after which they should be gradually inured to bear the open air, to which they should be exposed in summer, but in the autumn they must be placed under a garden-frame to screen them from the frost, but they should have the free air at all times in mild weather. The spring following some of these plants may be turned out of the pots, and planted in a warm border, where, if they are covered with tanners bark every winter to keep out the frost, they will live several years, their stalks decaying in the autumn, and new ones arise in the spring, which in warm seasons will flower very well. If those plants, which are continued in pots, are plunged into a tan-bed, some of them may produce fruit; and if the stalks of these are laid down into pots of earth plunged near them, they will take root, so that the plants may be easily propagated this way.

The second sort has not been many years in England, but is now the most common. This grows naturally in the Brasils, yet is hardy enough to thrive in the open air here, and is seldom injured except in very severe winters, which commonly kills the branches to the ground, and sometimes destroys the roots; this rises in a few years to a great height, if they have proper support. I have seen some of these plants, whose branches were trained up more than forty feet high. The stalks will grow almost as large as a man's arm, and are covered with a purplish bark, but do not become very woody. The shoots from these stalks are often twelve or fifteen feet long in one summer; they are very slender, so must be supported, otherwise they will hang to the ground, intermix with each other, and appear very unsightly. These are garnished at each joint with one hand-shaped leaf, composed of five smooth entire lobes. Their foot-stalks are near two inches long, and have two small leaves or ears embracing the stalks at their base; and from the same point comes out a long clasper, which twists round the neighbouring plants, whereby the stalks are supported. The flowers come out at the same joints as the leaves; these have foot-ſtalks almoſt three inches long; they have an outer cover compoſed of three concave oval leaves, of a paler green than the leaves of the plant, which are little more than half the length of the empalement, which is compoſed of five oblong blunt leaves, of a very pale green; within theſe are five petals, nearly of the ſame ſhape and ſize with the empalement, ſtanding alternately between them. In the center of the flower ariſes a thick club-like column about an inch long, on the top of which ſits an oval germen, from whoſe baſe ſpreads out five awl-ſhaped horizontal ſtamina, which are terminated by oblong broad ſummits faſtened in the middle to the ſtamina, hanging downward; theſe may be moved round without ſeparating from the ſtamina; their under ſurface is charged with yellow farina, and, on the ſide of the germen, ariſe three ſlender purpliſh ſtyles near an inch long, ſpreading from each other, terminated by obtuſe ſtigmas. Round the bottom of the column are two orders of rays, the inner, which is the ſhorteſt, inclines upward to the column, the outer, which is near half the length of the petals, ſpreads flat upon them; theſe rays are compoſed of a great number of thread-like filaments, of a purple colour at bottom, but are blue on the outſide. Theſe flowers have a faint ſcent, and continue but one day; after they fade, the germen on the top of the column ſwells to a large oval fruit, about the ſize and ſhape of the Mogul Plumb, and when ripe, is of the ſame pale yellow colour, incloſing a ſweetiſh diſagreeable pulp, in which are lodged oblong ſeeds. This plant begins to flower early in July, and there is a ſucceſſion of flowers daily, till the froſt in autumn puts a ſtop to them.

It may be propagated by ſeeds, which ſhould be ſown in the ſame manner as thoſe of the firſt ſort, and the plants treated in the ſame way till the following ſpring, when they ſhould be turned out of the pots, and planted againſt a good aſpected wall, where they may have height for their ſhoots to extend, otherwiſe they will hang about and entangle with each other, ſo make but an indifferent appearance; but where buildings are to be covered, this plant is very proper for the purpoſe. After they have taken good root in their new quarters, the only care they will requre is to train their ſhoots up againſt the wall, as they extend in length, to prevent their hanging about; and if the winter proves ſevere, the ſurface of the ground about their roots ſhould be covered with mulch to keep the froſt from penetrating of the ground; and if the ſtalks and branches are covered with mats, Peaſe-haulm, ſtraw, or any ſuch light covering, it will protect them in winter; but this covering muſt be taken off in mild weather, otherwiſe it will cauſe the branches to grow mouldy, which will be more injurious to them than the cold. In the ſpring the plants ſhould be trimmed, when all the ſmall weak ſhoots ſhould be entirely cut off, and the ſtrong ones ſhortened to about four or five feet long, which will cauſe them to put out ſtrong ſhoots for flowering the following year.

This plant is alſo propagated by laying down of the branches, which in one year will be well rooted, ſo may be taken off from the old plants and tranſplanted where they are deſigned to remain. The cuttings of this will alſo take root, if they are planted in a loamy ſoil not too ſtiff, in the ſpring, before they begin to ſhoot. If theſe are covered with bell or hand-glaſſes to exclude the air, they will ſucceed much better than when they are otherwiſe treated; but when the cuttings put out ſhoots, the air ſhould be admitted to them, otherwiſe they will draw up weak and ſpoil, they muſt be afterward treated as the layers.

Thoſe plants which are propagated by layers or cuttings, do not produce fruit ſo plentifully as the ſeedling plants, and I have found the plants which have been propagated two or three times, either by layers or cuttings, ſeldom produce fruit, which is common to many other plants.

If in very ſevere winters the ſtalks of theſe plants are killed to the ground, the roots often put out new ſtalks the following ſummer, therefore they ſhould not be diſturbed; and where there is mulch laid on the ground about their roots, there will be little danger of their being killed, although all the ſtalks ſhould be deſtroyed.

There is a variety of this; the lobes of the leaves are much narrower, and are divided almoſt to the bottom. The flowers come later in the ſummer; the petals of the flowers are narrower, and of a purer white, but I believe it is only a ſeminal variation of the other, ſo not worthy of being enumerated.

The third ſort grows naturally in Virginia, and alſo in Jamaica. This hath a perennial creeping root, ſending up many weak ſtalks about three or four feet high, which are garniſhed with leaves ſhaped very like thoſe of Ivy, and are almoſt as large, but of a pale green and very thin conſiſtence. The flowers come out from the wings of the ſtalk, upon ſlender foot-ſtalks an inch and a half long, and at their baſe ariſe very ſlender tendrils, which claſp round any neighbouring ſupport. The flowers are of a dirty yellow colour, and not larger than a ſixpence when expanded, ſo make no great appearance. This may be propagated by its creeping roots, which may be parted in April, and planted where they are to remain. This ſort will live in a warm border, if treated in the ſame way as is directed for the firſt ſort. Some of theſe plants lived many years in the Chelſea garden, in a border to a ſouth-weſt aſpect, but in the year 1740 they were killed by the froſt.

The fourth ſort grows naturally in Jamaica. This hath a perennial root, from which ariſe ſeveral ſlender ſtalks four or five feet high, which have joints four or five inches aſunder; at each of theſe come out one leaf, a tendril, and a flower. The leaves have three lobes. The flowers are ſmaller than thoſe of the laſt mentioned, and are of a greeniſh colour; theſe are ſucceeded by oval fruit, about the ſize of ſmall Olives, which turn purple when they are ripe.

The fifth ſort grows naturally in moſt of the Weſt-India Iſlands; this riſes with a weak ſtalk to the height of twenty feet. As the ſtalks grow old, they have a thick fungous bark, like that of the Cork-tree, which cracks and ſplits. The ſmaller branches are covered with a ſmooth bark, and garniſhed with ſmooth leaves at each joint, ſitting upon very ſhort foot-ſtalks; theſe have three lobes, the middle one being much longer than thoſe on the ſides, ſo that the whole leaf has the form of the point of thoſe halberts uſed by the yeomen of the guards. The flowers are ſmall, of a greeniſh yellow colour, and are ſucceeded by ſmall oval fruit, of a dark purple colour when ripe.

The ſixth ſort grows naturally in the Weſt-Indies. This hath a perennial root, from which ariſe ſeveral ſlender ſtalks, which riſe eight or ten feet high, garniſhed with ſmooth green leaves, ſtanding upon ſlender foot-ſtalks. They are but ſlightly indented into three lobes, which end in acute points, and are ſhaped like the points of halberts, the middle one ſtanding oblique to the foot-ſtalk. The flowers come out from the wings of the leaves on very ſhort foot-ſtalks; they are of a pale yellow. The petals of the flowers are very narrow, and longer than thoſe of the two former ſorts; the fruit is ſmaller, and of an oval form, changing to a dark purple when ripe.

The ſeventh ſort grows naturally in moſt of the iſlands in the Weſt-Indies, where the inhabitants of the Britiſh Iſlands call it Love in a Miſt. The root of this is biennial; the ſtalks riſe eight or ten feet high, when they are ſupported; they are channelled and hairy. The leaves are heart-ſhaped, divided into three lobes, the middle lobe being much the longeſt; they are covered with ſhort brown hairs.

hairs. The tendrils come out at the ſame place as the leaves, as do alſo the flowers, whoſe foot-ſtalks are long, hairy, and pretty ſtrong. The empalement of the flower is compoſed of ſlender hairy filaments, which are wrought like a net; theſe are longer than the petals of the flower, and turn up round them, ſo that the flowers are not very conſpicuous at a diſtance; theſe are white, and of ſhort duration; their ſtructure is the ſame with the other ſorts, and they are ſucceeded by roundiſh oval fruit, about the ſize of an ordinary Golden Pippin, of a yellowiſh green colour, incloſed with the netted empalement. This plant, as alſo the four laſt mentioned ſorts, are propagated by ſeeds, which ſhould be ſown upon a hot-bed early in the ſpring, and when the plants are fit to remove, they ſhould be each tranſplanted into a ſmall pot filled with light kitchen-garden earth, and plunged again into a hot-bed, obſerving to ſhade them from the ſun till they have taken new root; after which time they muſt be treated in the ſame way as other plants from the ſame country, ſhifting them into larger pots as their roots increaſe; and when the plants are too tall to remain under the glaſſes of the hot-bed, they ſhould be removed into an airy glaſs-caſe, where they ſhould have the free air admitted to them in warm weather, but ſcreened from the cold. In this ſituation the plants will flower in July, and their ſeeds will ripen in the autumn. The whole plant has a diſagreeable ſcent when touched.

There is a variety of this, if it is not a diſtinct ſpecies, with hairy leaves not ſo broad as thoſe of the former. The whole leaf is ſhaped more like the point of a halbert, and thoſe leaves which grow toward the upper part of the ſtalks, have very ſmall indentures, ſo approach near to ſimple leaves without lobes. The flowers are alſo ſmaller, but of the ſame form, and the roots are of ſhorter duration, ſo that I am inclined to believe it a diſtinct ſpecies.

The eighth ſort has ſome appearance of the ſeventh, ſo that many perſons have ſuppoſed it was only an accidental variety of it, but there can be no doubt of its being a different ſpecies. The ſtalks of this riſe upward of twenty feet high, and will continue two or three years; the leaves are larger, but of the ſame ſhape, and hairy; the tendrils of this ſort are very long, as are alſo the foot-ſtalks of the flowers, which are ſmooth, not hairy as the former; the empalement of the flowers is netted, but not ſo long as in the former ſort; the flowers are larger, and the rays are of a light blue colour: the fruit is much leſs and rounder than thoſe of the other, and when ripe changes to a deep yellow colour.

The ninth ſort was diſcovered by the late Dr. Houſtoun, growing naturally at La Vera Cruz. This is a perennial plant; the ſtalks riſe twenty feet high, dividing into many ſlender branches, which are covered with a ſoft hairy down. The leaves are ſhaped like the point of a halbert, of a light green; they are ſoft and ſilky to the touch, ſtanding oblique to the foot-ſtalks. The flowers come out at the wings of the leaves like the other ſpecies; theſe are not half ſo large as thoſe of the ſecond ſort, but are of the ſame form. The petals are white, and the rays or filaments are purple, with a mixture of yellow. The fruit of this is ſmall, roundiſh, and yellow when ripe.

The tenth ſort grows naturally in Jamaica. This is a perennial plant; the ſtalks are ſlender, and riſe to twenty feet high when they are ſupported, and divide into many weak branches; the leaves, flowers, and tendrils, come out at each joint. The leaves have three longitudinal veins, which join at the baſe to the foot-ſtalk, but the two outer diverge toward the borders of the leaf in the middle, drawing inward again at the top. The leaves are of a deep green on their upper ſide, but are pale on their under, and ſtand upon ſhort foot-ſtalks; the foot-ſtalks of the flowers are very ſlender, of a purpliſh colour. The flowers are ſhaped like thoſe of the other ſpecies, but when expanded are not more than an inch and a half diameter, of a ſoft red colour, and little ſcent. The fruit is ſmall, oval, and, when ripe, changes to a purple colour.

The eleventh ſort was diſcovered by the late Mr. Robert Millar, growing naturally near Carthagena in New Spain. This hath ſlender ſtriated ſtalks, of a browniſh red colour, dividing into many ſlender branches, which are garniſhed with leaves, ſhaped like the wings of a bat when extended; they are about ſeven inches in length, meaſuring from the two extended points, which may rather be termed the breadth, for from the baſe to the top they are not more than two inches and a half. The foot-ſtalk is ſet half an inch from the baſe of the leaf, from which come out three ribs or veins; two of them extend each way to the two narrow points of the leaf, the other riſes upright to the top, where is the greateſt length of the leaf, if it may be ſo termed. The figure of this leaf is the moſt ſingular of any I have yet ſeen: the flowers come out at the joints of the ſtalk like the others, upon ſhort ſlender foot-ſtalks; they are about three inches diameter when expanded. The petals and rays are white; the rays are twiſted and ſlender, extending beyond the petals. The fruit of this I have not ſeen entire.

The twelfth ſort was diſcovered by the late Dr. Houſtoun, growing naturally at La Vera Cruz in New Spain. This hath ſlender angular ſtalks, which riſe twenty feet high, ſending out many branches, which are garniſhed with moon-ſhaped leaves, and have two blunt lobes ſpreading aſunder each way, ſo as to have the appearance of a half moon. The flowers and tendrils come out from the ſame joints of the ſtalks. The flowers are of a pale colour and ſmall, but ſhaped like thoſe of the other ſorts; theſe are ſucceeded by oval fruit of a purple colour, about the ſize of ſmall oval Grapes.

The thirteenth ſort has ſome reſemblance of the twelfth, but the ſtalks are rounder and become ligneous. The leaves are almoſt as ſtiff as thoſe of the Bay-tree, and are not ſo deeply divided as thoſe of the former. The flowers ſtand upon long foot-ſtalks, which are horizontal; they are ſmall, white, and ſhaped like thoſe of the other ſort. The fruit is oval, ſmall, and of a purple colour, ſitting cloſe to the petals of the flowers, which are permanent.

The fourteenth ſort grows naturally in moſt of the iſlands in the Weſt-Indies. This is by Tournefort ſeparated from this genus, and titled by him Murucuia, which is the Brazilian name for it. This hath ſlender climbing ſtalks, which are channelled, putting out tendrils at the joints, which faſten themſelves about the neighbouring plants for ſupport, and climb to the height of ten or twelve feet, garniſhed with leaves, which are cut into two lobes at their baſe, but at the top are only a little hollowed at a diſtance from each point, riſing again in the middle oppoſite the foot-ſtalk. The baſe of the two lobes ſpread and meet, ſo that they appear as if they embraced the ſtalk, but when they are viewed near, they are found divided to the ſhort crooked foot-ſtalk, which does ſcarcely appear. There are two purpliſh veins ariſing from the foot-ſtalk, which extend each way to the points of the lobes; the tendrils, which come out with the leaves, are long, tough, and of a purple colour. The flowers are produced toward the end of the branches, coming out by pairs on each ſide; they have purple foot-ſtalks, ſuſtaining one flower at the top, whoſe empalement is compoſed of five purple leaves, which form a kind of tube, and within are five very narrow purple petals. The column in the center of the flower is of the ſame length as the petals, but the ſtamina are extended an inch above. When the flowers fade, the germen ſwells to an an oval purple fruit, the size of the small red Goosberry, inclosing a soft pulp, in which are lodged the seeds.

The fifteenth sort grows naturally in the West Indies, where the inhabitants call it Granadilla. The fruit of this sort is commonly eaten there, being served up to their tables in deserts. This hath a thick, climbing, herbaceous, triangular stalk, sending out slender tendrils at each joint, which fasten to the bushes and hedges for support, rising to the height of fifteen or twenty feet, garnished at each joint with one large oval leaf. There are two large stipulæ or ears joined to the stalks, which encompass the foot-stalks of the flowers and leaves, as also the base of the tendril. The leaves are of a lively green, having one strong nerve or midrib running longitudinally, from which arise several small veins, which diverge to the sides, and incurve again toward the top. The flowers stand upon pretty long foot-stalks, which have two small glandules in the middle; the cover of the flower is composed of three soft velvety leaves, of a pale red, with some stripes of a lively red colour; the petals of the flower are white, and the rays are blue. These flowers are large, so make a fine appearance during their continuance, but they are like the other species, of short duration; however, there is a succession of flowers for some time on the plants. After the flowers are past, the germen swells to a roundish fruit, the size of a middling Apple, of a yellow colour when ripe, having a thicker rind than any of the other sorts, inclosing a sweetish pulp, in which are lodged many oblong flat seeds, of a brownish colour, a little rough to the touch.

The sixteenth sort grows naturally in the West-Indies. This hath climbing tough stalks, which put out claspers at every joint like the others, which fasten to the neighbouring trees and hedges for support, and rise upward of twenty feet high, sending out many side branches. The leaves are four or five inches long, and two broad, of a pretty thick consistence, and a bright green on their upper side, but pale on their under. The flowers come out at the joints of the stalks; the buds of the flowers are as large as pigeons eggs, before they begin to expand. The cover of the flower is composed of three large, oval, green leaves, which are indented on their edges, and hollowed like a spoon; within these is the empalement of the flower, which is composed of five oblong leaves, of a pale green on their outside, but whitish within. The petals of the flower are white, and stand alternately with those of the empalement, but are not more than half their breadth, and are marked with several small, brownish, red spots. The rays of the flower are of a Violet colour; the column in the center is yellowish, as is also the round germen at the top, but the three styles are of a purple colour. These flowers have an agreeable odour, and when they fade, the germen swells to the size of a pullet's egg, nearly of the same shape, which turns yellow when ripe. The rind is soft and thick; the pulp has an agreeable acid flavour, which quenches thirst, abates the heat of the stomach, gives an appetite, and recruits the spirits, so is commonly given in fevers. The seeds are heart-shaped and brownish.

The seventeenth sort grows naturally in the Bahama Islands; this hath slender, climbing, three-cornered stalks, which send out tendrils at each joint, fastening themselves to any neighbouring support. The stalks climb to the height of twelve or fourteen feet, and are garnished with oblong oval leaves. Their foot-stalks are slender, from which arise three longitudinal veins, one running through the middle of the leaf, the other two diverge to the sides, drawing toward each other again at the point. The flowers come out from the wings of the stalk upon slender foot-stalks; the empalement of the flower is composed of five oblong, narrow, purplish leaves, and within are five narrower petals of the same colour, which turn backward after they have been some time expanded. The column in the middle of the flower is very long and slender, supporting a round germen, from whose base spread out five slender stamina, terminated by oblong hanging summits; from the top of the germen arise three slender styles, which spread asunder, crowned by roundish summits. When the flowers fade, the germen swells to an oval fruit, about the size of a sparrow's egg, which changes to a purple colour when ripe, filled with oblong seeds, inclosed in a soft sweet pulp.

The nineteenth sort was discovered by the late Dr. Houstoun at La Vera Cruz in New Spain. This hath slender climbing stalks, sending out many small branches, which climb to the height of twenty five or thirty feet, when they meet with neighbouring support, to which they fasten themselves by their tendrils. The stalks by age become ligneous toward the bottom; their joints are not far asunder. The leaves stand upon short slender foot-stalks; they are smooth, entire, and of a lively green colour. The flowers come out from the wings of the leaves, standing upon long foot stalks; the empalement of the flower is composed of five oblong leaves, green on their outside, but whitish within. The flower has five oblong white petals, situated alternately to the leaves of the empalement, which spread open; the rays are of a bluish purple colour, inclining at the bottom to red; the column in the center is short and thick; the germen on the top is oval, and, after the flowers fade, swells to the size of a pullet's egg, and changes to a pale yellow when ripe, having many oblong seeds inclosed in a soft pulp. The flowers of this kind have an agreeable odour, but are of short duration, seldom continuing twenty hours open; but there is a succession of flowers on the plants, from June to September, and sometimes the fruit will ripen here. There are many other sorts now lately discovered, but as I have not had them under my care, so I have not enumerated them.

All the perennial sorts which are natives of the hot parts of America, require a stove to preserve them here, without which they will not thrive; for although some of the sorts will live in the open air during the warm months in summer, yet they make but little progress; nor will the plants produce many flowers, unless the pots in which they are planted, are plunged into the tan-bed of the stove, and their branches are trained against an espalier. The best way to have them in perfection, is to make a border of earth on the back-side of the tan-bed, which may be separated by planks to prevent the earth from mixing with the tan; and when the plants are strong enough, they should be turned out of the pots, and planted in this border; adjoining to which, should be a trellis erected to the top of the stove; against which the stalks of the plants must be trained, and as they advance they will form a hedge to hide the wall of the stove, and most of their leaves continuing green all the year, together with their flowers, which will be very plentifully intermixed in summer, will have a very agreeable effect.

As there will be only a plank partition between the earth and the tan, so the earth will be kept warm by the tan-bed, which will be of great service to the roots of the plants. This border should not be less than two feet broad and three deep, which is the usual depth of the pit for tan; so that where these borders are intended, the pits should not be less than eight feet broad, that the bark-bed, exclusive of the border, may be six feet wide. If the border is fenced off with strong ship planks, they will last some years, especially if they are well painted over with a composition of melted pitch, brick-dust and oil, which will preserve them sound a long time; and the earth should be taken out carefully from between the roots of the plants, at least once a year, putting in fresh; with this management, I have seen these plants in great perfection. But where there has not been this conveniency, I have turned the plants out of the pots, and planted them into the tan, where it was half rotten, into which they have rooted exceedingly, and have thriven for two or three years as well as could be desired; but when their roots extended to a great distance in the tan-bed, they have been injured by renewing of the bark; when it has fermented pretty violently, the roots have been scalded, and the plants have been killed, so that the other method is more eligible.

As these sorts do not often perfect their seeds here, so they may be propagated by laying down their branches, which, if done in April, they will put out roots by the middle of August or September, when they may be separated from the old plants, and either planted in pots to get strength, or into the border of the stove, where they are to remain.

Some of these sorts may also be propagated by cuttings: these should be planted in pots about the middle or latter end of March, and plunged into a moderate hot-bed, observing to screen them from the sun, and refresh them with water gently, as often as the earth may require it; and in about two months or ten weeks, they will put out roots, and may then be treated as the seedling plants.

PASSION-FLOWER. See PASSIFLORA.

PASTINACA. Tourn. Inst. R. H. 319. tab. 170. Parsnep.

The CHARACTERS are,

It hath an umbellated flower; the principal umbel is composed of many smaller, and these are likewise composed of several rays. They have no involucrum, and the empalement is scarce visible; the umbel is uniform. The flowers have five spear-shaped incurved petals, and five hair-like stamina. The germen is situated under the flower, supporting two reflexed styles. The germen afterward becomes an elliptical, plain, compressed fruit, dividing in two parts, having two bordered elliptical seeds.

The SPECIES are,

1. PASTINACA (*Sylvestris*) foliis simpliciter pinnatis hirsutis. *Parsnep with single, winged, hairy leaves; or wild Parsnep.*

2. PASTINACA (*Sativa*) foliis simpliciter pinnatis glabris. *Parsnep with single, winged, smooth leaves; or Garden Parsnep.*

3. PASTINACA (*Opopanax*) foliis decompositis pinnatis. Hort. Cliff. 105. *Parsnep with decompounded winged leaves; or Opopanax.*

The first sort grows naturally on the side of banks, and on dry ground in many parts of England. This is a biennial plant, the first year shooting out hairy leaves, which spread on the surface of the ground, which are singly winged, and the lobes are irregularly cut. The following year the stalks rise four or five feet high; these are channelled, hairy, and garnished with winged leaves like those at the bottom, but smaller; the stalk branches out toward the top, each branch being terminated by a large umbel of yellow flowers, which are succeeded by compressed fruit, having two flat bordered seeds.

The root and seed of this sort is sometimes used in medicine, but it is seldom cultivated in gardens, the markets being supplied from the fields; yet the druggists commonly sell the seeds of the garden kind for it, which they may purchase at an easy price when it is too old to grow, but then the seeds can have no virtue left.

The second sort hath smooth leaves, of a light or yellowish green colour, in which this differs from the former; the stalks also rise higher, and are deeper channelled; the footstalks of the umbels are much longer, and the flowers are of a deeper yellow colour. These two sorts have been thought only varieties, the Garden Parsnep they have supposed to differ from the wild only by culture; but I have cultivated both many years, and have never found that either of the sorts have varied; the seeds of each having constantly produced the same sort as they were taken from, so that I am certain they are distinct species.

This sort is cultivated in kitchen-gardens, the roots of which are large, sweet, and accounted very nourishing. They are propagated by seeds, which should be sown in February or March, in a rich mellow soil, which must be well dug, that their roots may run downward, the greatest excellency being the length and bigness of the roots. These may be sown alone, or with Carrots, as is practised by the kitchen-gardeners near London; some of whom also mix Leeks, Onions, and Lettuce, with their Parsneps; but this I think very wrong, for it is not possible that so many different sorts can thrive well together, except they are allowed a considerable distance; and if so, it will be equally the same to sow the different sorts separate. However, Carrots and Parsneps may be sown together very well, especially where the Carrots are designed to be drawn off very young; because the Parsneps generally spread most towards the latter end of summer, which is after the Carrots are gone, so that there may be a double crop upon the same ground.

When the plants are come up, you should hoe them out, leaving them about ten inches or a foot asunder; observing at the same time to cut up all the weeds, which, if permitted to grow, would soon overbear the plants and choke them. This must be repeated three or four times in the spring, according as you find the weeds grow; but in the latter part of summer, when the plants are so strong as to cover the ground, they will prevent the growth of weeds, so that after that season they will require no farther care.

When the leaves begin to decay, the roots may be dug up for use, before which time they are seldom well tasted; nor are they good for much late in the spring, after they are shot out again; so that those who would preserve these roots for spring use, should dig them up in the beginning of February, and bury them in sand, in a dry place, where they will remain good until the middle of April, or later. These roots are excellent food for hogs.

If you intend to save the seeds of this plant, you should make choice of some of the longest, straitest, and largest roots, which should be planted about two feet asunder, in some place where they may be defended from the strong south and west winds; for the stems of these plants commonly grow to a great height, and are very subject to be broken by strong winds, if exposed thereto; they should be constantly kept clear from weeds, and if the season should prove very dry, you should give them some water twice a week, which will cause them to produce a great quantity of seeds, which will be much stronger than if they were wholly neglected. Toward the latter end of August or the beginning of September, the seeds will be ripe; at which time you should carefully cut off the heads, and spread them upon a coarse cloth for two or three days to dry; after which, the seeds should be beaten off, and put up for use; but you should never trust to these seeds after they are a year old, for they will seldom grow beyond that age.

The leaves of the Garden Parsnep are dangerous to handle, especially in a morning, while the dew remains upon them; at which time, if they are handled by persons who have a soft skin, it will raise it in blisters. I have known some gardeners, when they have been drawing up Carrots from among Parsneps in a morning, when their leaves were wet with dew, draw up the sleeves of their shirts to their shoulders, to prevent their being wet; by doing of which they have had their arms, so far as they were bare, covered over with large blisters; and these were full of a scalding liquor, which has proved very troublesome for several days.

The third sort rises with a green rough stalk seven or eight feet high, garnished with large, decompounded, winged leaves, which are very rough to the touch, and of a dark green colour; the juice is very yellow, which flows out where either the leaf or stalk is broken; the stalks are divided upward into many horizontal branches, each being terminated by a large umbel of yellow flowers. These appear in July, and are succeeded by plain seeds, which are bordered, and a little convex in the middle, which ripen in the autumn. The Opopanax of the shops, is thought to be the concrete juice of this plant.

PASTURE.

Pasture ground is of two sorts: the one is low meadow land, which is often overflowed, and the other is upland, which lies high and dry. The first of these will produce a much greater quantity of Hay than the latter, and will not require manuring or dressing so often; but then the Hay produced on the upland is much preferable to the other, as is also the meat which is fed in the upland more valued than that which is fatted in rich meadows; though the latter will make the fatter and larger cattle, as is seen by those which are brought from the low rich lands in Lincolnshire. But where people are nice in their meat, they will give a much larger price for such as hath been fed on the downs, or in short upland Pasture, than for the other. Besides this, dry Pastures have an advantage over the meadows, that they may be fed all the winter, and are not so subject to poach in wet weather; nor will there be so many bad weeds produced, which are great advantages, and do, in a great measure, recompense for the smallness of the crop.

The first improvement for upland Pasture is, by fencing it, and dividing it into small fields of four, five, six, eight, or ten acres each, planting timber-trees in the hedge-rows, which will screen the Grass from the drying pinching winds of March, which prevent the Grass from growing in large open lands; so that if April proves a cold dry month, the open land produces very little Hay; whereas in the sheltered fields the Grass will begin to grow early in March, and will cover the ground, and prevent the sun from parching the roots of the Grass, whereby it will keep growing, so as to afford a tolerable crop, if the spring should prove dry. But in fencing of land it must be observed (as was before directed) not to make the inclosures too small, especially where the hedge-rows are planted with trees; because when the trees are advanced to a considerable height, they will spread over the land; and, where they are close, will render the Grass sour; so that instead of being an advantage, it will greatly injure the Pasture.

The next improvement of upland Pasture is, to make the turf good, where either from the badness of the soil, or for want of proper care, the Grass hath been destroyed by Rushes, bushes, or mole-hills. Where the surface of the land is clayey and cold, it may be improved by paring it off, and burning it in the manner before directed, under the article of LAND; but if it is a hot sandy land, then chalk, lime, marle, or clay, are very proper manures to lay upon it; but this should be laid in pretty good quantities, otherwise it will be of little service to the land.

If the ground is over-run with bushes or Rushes, it will be a great advantage to the land, to grub them up toward the latter part of the summer; and after they are dried to burn them, and spread the ashes over the ground just before the autumnal rains: at which time the surface of the land should be levelled, and sown with Grass-seed, which will come up in a short time, and make good Grass the following spring. So also where the land is full of mole-hills, these should be pared off, and either burnt for the ashes, or spread immediately on the ground, when they are pared off, observing to sow the bare patches with Grass-seed, just as the autumnal rains begin.

Another improvement of upland Pastures is the feeding of them every other year; for where this is not practised, the land must be manured at least every third year; and where a farmer hath much arable land in his possession, he will not care to part with his manure to the Pasture. Therefore every farmer should endeavour to proportion his Pasture to his arable land, especially where manure is scarce, otherwise he will soon find his error; for the Pasture is the foundation of all the profit, which may arise from the arable land.

These upland Pastures seldom degenerate the Grass which is sown on them, if the land is tolerably good, it will continue so; whereas the low meadows, which are overflowed in winter, in a few years turn to a harsh rushy Grass, but the upland will continue a fine sweet Grass for many years without renewing.

There is no part of husbandry, of which the farmers are in general more ignorant than that of the Pasture; most of them suppose, that when an old Pasture is ploughed up, it can never be brought to have a good sward again; so their common method of managing their land after ploughing, and getting two or three crops of Corn, is to sow with their crop of Barley some Grass-seeds (as they call them;) that is, either the red Clover, which they intend to stand two years after the Corn is taken off the ground, or Rye Grass mixed with Trefoil; but as all these are at most but biennial plants, whose roots decay soon after their seeds are perfected, so the ground having no crop upon it, is again ploughed for Corn; and this is the constant round which the lands are employed in, by the better sort of farmers; for I have never met with one of them, who had the least notion of laying down their land to Grass for any longer continuance; therefore the seeds which they usually sow, are the best adapted for this purpose.

But whatever may have been the practice of these people, I hope to prove, that it is possible to lay down land, which has been in tillage, with Grass, in such manner as that the sward shall be as good, if not better than any natural Grass, and of as long duration. But this is never to be expected, in the common method of sowing a crop of Corn with the Grass-seeds; for wherever this hath been practised, if the Corn has succeeded well, the Grass has been very poor and weak; so that if the land has not been very good, the Grass has scarcely been worth standing; for the following year it will produce but little Hay, and the year after the crop is worth little, either to mow or feed. Nor can it be expected it should be otherwise, for the ground cannot nourish two crops; and if there were no deficiency in the land, yet the Corn being the first, and most vigorous of growth, will keep the Grass from making any considerable progress; so that the plants will be extremely weak, and but very thin, many of them which came up in the spring being destroyed by the Corn; for wherever there are roots of Corn, it cannot be expected there should be any Grass. Therefore the Grass must be thin, and, if the land is not in good heart, to supply the Grass with nourishment, that the roots may branch out after the Corn is gone, there cannot be any considerable crop of Hay.

Therefore, when ground is laid down for Grass, there should be no crop of any kind sown with the seeds; the land should also be well ploughed, and cleaned from weeds; otherwise the weeds will come up the first, and grow so strong as to overbear the Grass, and if they are not pulled up, will entirely spoil it. The best season to sow the Grass-seeds upon dry land is about the beginning of September, if there is an appearance of rain; for the ground being then

warm,

warm, if there happen ſome good ſhowers of rain after the ſeed is ſown, the Graſs will ſoon make its appearance, and get ſufficient rooting in the ground before winter, ſo will not be in danger of having the roots turned out of the ground by the froſt, eſpecially if the ground is well rolled before the froſt comes on, which will preſs it down, and fix the earth cloſe to the roots. Where this hath not been practiſed, the froſt has often looſened the ground ſo much, as to let in the air to the roots of the Graſs, and done it great damage, and this hath been brought as an objection to the autumnal ſowing of Graſs; but it will be found to have no weight, if the above direction is practiſed; nor is there any hazard in ſowing the Graſs at this ſeaſon, but that of dry weather after the ſeeds are ſown; for if the Graſs comes up well, and the ground is well rolled in the end of October or the beginning of November, and repeated again the beginning of March, the ſward will be cloſely joined at bottom; and a good crop of Hay may be expected the ſame ſummer. In very open expoſed cold lands, it is proper to ſow the ſeeds three weeks earlier than is here mentioned, that the Graſs may have time to get good rooting, before the cold ſeaſon comes on to ſtop its growth; for in ſuch ſituations, vegetation is over early in the autumn, ſo the Graſs being weak, may be deſtroyed by froſt: but if the ſeeds are ſown in Auguſt, and a few ſhowers follow ſoon after to bring up the Graſs, it will ſucceed much better than any which is ſown in the ſpring, as I have ſeveral years experienced, on ſome places as much expoſed as moſt in England. But where the ground cannot be prepared for ſowing at that ſeaſon, it may be performed the middle or latter end of March, according to the ſeaſons being early or late; for in backward ſprings, and in cold land, I have often ſowed the Graſs in the middle of April with ſucceſs; but there is danger in ſowing late, of dry weather, and eſpecially if the land is light and dry; for I have ſeen many times the whole ſurface of the ground removed by ſtrong winds at that ſeaſon, ſo that the ſeeds have been driven in heaps to one ſide of the field. Therefore whenever the ſeeds are ſown late in the ſpring, it will be proper to roll the ground well ſoon after the ſeeds are ſown, to ſettle the ſurface, and prevent its being removed.

The ſort of ſeeds which are the beſt for this purpoſe, are the beſt ſort of upland Hay-ſeeds, taken from the cleaneſt Paſtures, where there are no bad weeds; if this ſeed is ſifted to clean it from rubbiſh, three, or at moſt four buſhels, will be ſufficient to ſow an acre of land. The other ſort is the Trifolium pratenſe album, which is commonly known by the names of white Dutch Clover, or white Honeyſuckle Graſs. Eight pounds of this ſeed will be enough for one acre of land. The Graſs-ſeed ſhall be ſown firſt, and then the Dutch Clover-ſeed may be afterward ſown; but they ſhould not be mixed together, becauſe the Clover-ſeeds being the heavieſt, will fall to the bottom, and conſequently the ground will not be equally ſown with them.

After the ſeeds are ſown, the ground ſhould be lightly harrowed to bury the ſeeds; but this ſhould be performed with a ſhort toothed harrow, otherwiſe the ſeeds will be buried too deep. Two or three days after ſowing, if the ſurface of the ground is dry, it ſhould be rolled with a Barley roller, to break the clods and ſmooth the ground, which will ſettle it, and prevent the ſeeds from being removed by the wind.

When the ſeeds are come up, if the land ſhould produce many weeds, theſe ſhould be drawn out before they grow ſo tall as to overbear the Graſs; for where this has been neglected, the weeds have taken ſuch poſſeſſion of the ground, as to keep down the Graſs and ſtarve it; and when theſe weeds have been ſuffered to remain until they have ſhed their ſeeds, the land has been ſo plentifully ſtocked with them, as entirely to deſtroy the Graſs; therefore it is one of the principal parts of huſbandry, never to ſuffer weeds to grow on the land.

As the white Clover is an abiding plant, ſo it is certainly the very beſt ſort to ſow, where Paſtures are laid down to remain; for as the Hay-ſeeds which are taken from the beſt Paſtures, will be compoſed of various ſorts of Graſs, ſome of which may be but annual, and others biennial, ſo when thoſe go off, there will be many and large patches of ground left bare and naked, if there is not a ſufficient quantity of the white Clover, to ſpread over and cover the land. Therefore a good ſward can never be expected, where this is not ſown; for in moſt part of the natural Paſtures, we find this plant makes no ſmall ſhare of the ſward; and is equally good for wet and dry land, growing naturally upon gravel or clay, in moſt parts of England; which is a plain indication how eaſily this plant may be cultivated to great advantage, in moſt ſorts of land throughout this kingdom.

After the ground has been ſown in the manner before directed, and brought to a good ſward, the way to preſerve it good is, by conſtantly rolling the ground with a heavy roller, every ſpring and autumn, and in ſummer after rain, as hath been before directed. This piece of huſbandry is rarely practiſed by farmers, but thoſe who do, find their account in it, for it is of great benefit to the Graſs. Another thing ſhould alſo be carefully performed, which is, to cut up Docks, Dandelion, Knapweed, and all ſuch bad weeds, by their roots, every ſpring and autumn; this will increaſe the quantity of good Graſs, and preſerve the Paſtures in beauty. Dreſſing of theſe Paſtures every third year, is alſo a good piece of huſbandry, for otherwiſe it cannot be expected the ground ſhould continue to produce good crops. Beſides this, it will be neceſſary to change the ſeaſons of mowing, and not to mow the ſame ground every year; but to mow one ſeaſon, and feed the next; for where the ground is every year mown, it muſt be conſtantly dreſſed, as are moſt of the Graſs grounds near London, otherwiſe the ground will be ſoon exhauſted.

PAVIA. Boerh. Ind. alt. 2. p. 260. The ſcarlet flowering Horſe Cheſtnut.

The CHARACTERS are,

The flower has a ſmall empalement of one leaf, indented in five parts at the top. It has five roundiſh petals, waved and plaited on their borders, narrow at their baſe, where they are inſerted in the empalement. It hath eight ſtamina, which are declined, and as long as the petals, terminated by riſing ſummits, and a roundiſh germen ſitting upon an awl ſhaped ſtyle, crowned by an acuminated ſtigma. The germen afterward becomes an oval, Pear-ſhaped, leathery capſule with three cells, in which is ſometimes one, and at others two, almoſt globular ſeeds.

We have but one SPECIES of this genus, viz.

PAVIA (*Octandria.*) Boerh. Ind. alt. 2. p. 260. *The ſcarlet Horſe Cheſtnut.*

This plant grows naturally in Carolina and the Brazils; from the firſt the ſeeds were brought to England, where the plants have been of late years much cultivated in the gardens. In Carolina it is but of humble growth, ſeldom riſing more than eight or ten feet high; the ſtalk is pretty thick and woody, ſending out ſeveral branches, garniſhed with hand-ſhaped leaves, compoſed of five or ſix ſpear-ſhaped lobes, which unite at their baſe, where they join the foot-ſtalk; they are ſawed on their edges, and have long foot-ſtalks. The flowers are produced in looſe ſpikes at the end of the branches, ſtanding upon long naked foot-ſtalks, which ſuſtain five or ſix flowers, which are tubulous at bottom, but ſpread open at the top, where the petals are irregular in ſize and length, having an appearance of a lip flower; they are of a bright red colour, and have ſeven or eight

eight stamina the length of the petals. When the flowers fade, the germen swells to a Pear shaped fruit, with a thick russet cover having three cells, one of which, and sometimes two, are pregnant with globular seeds.

It may be propagated by sowing the seeds in the spring, upon a warm border of light sandy earth; and when the plants come up, they should be carefully cleared from weeds, but they must not be transplanted until the spring following; for as these seedling plants are tender while they are young, so they should be covered with mats the two first winters: this should be carefully performed in autumn, when the early frosts begin; for as the tops of the young plants are very tender, so a small frost at that time will pinch them; and when the tops are killed, they generally decay to the ground; when this happens, they seldom make good plants after. Therefore this should be constantly observed for two years, or three at most, by which time the plants will have gotten strength enough to resist the frost, when they should be removed just before they begin to shoot, and placed either in a nursery to be trained up, or where they are to remain; observing, if the season be dry, to water them until they have taken root, as also to lay some mulch upon the surface of the ground, to prevent the sun and wind from drying it too fast; as the plants advance, so the lateral branches should be pruned off, in order to reduce them to regular stems.

This tree may be propagated by budding or grafting it upon the common Horse Chestnut, which is the common method practised by the nursery-men; but the trees thus raised, seldom make a good appearance long, for the stock of the common Horse Chestnut will be more than twice the size of the other, and frequently put out shoots below the graft, and sometimes the grafts are blown out of the stocks, after ten years growth; but these stocks render the trees hardy, and of a larger growth.

PAULLINIA. Lin. Gen. Plant. 446.

The CHARACTERS are,

The flower has a permanent empalement, composed of four small oval leaves. It hath four oblong oval petals, twice the size of the empalement, and eight short stamina, with a turbinated germen, having three short slender styles, crowned by spreading stigmas. The germen afterward turns to a large three-cornered capsule with three cells, each containing one almost oval seed.

The SPECIES are,

1. PAULLINIA (*Serjana*) foliis ternatis, petiolis teretiusculis, foliolis ovato-oblongis. Lin. Sp. Pl. 365. *Three-leaved Paullinia with taper foot-stalks, and oblong oval leaves.*

2. PAULLINIA (*Mexicana*) foliis biternatis, petiolis marginatis, foliolis ovatis integris. Lin. Sp. Pl. 366. *Paullinia with nine lobes in each leaf, bordered foot-stalks, and oval entire lobes.*

3. PAULLINIA (*Cururu*) foliis ternatis, foliolis cuneiformibus, obtusis subdentatis. Lin. Sp. Plant. 365. *Three-leaved Paullinia with trifoliate leaves, having wedge-shaped lobes, which are indented.*

4. PAULLINIA (*Curassavica*) foliis biternatis, foliolis ovatis subsinuatis. Lin. Sp. Pl. 366. *Paullinia with double trifoliate leaves, having oval sinuated lobes.*

5. PAULLINIA (*Pinnata*) foliis pinnatis, foliolis incisis, petiolis marginatis. Hort. Cliff. 52. *Paullinia with winged leaves, whose lobes are cut, and bordered foot-stalks.*

6. PAULLINIA (*Tomentosa*) foliis pinnatis tomentosis, foliolis ovatis incisis, petiolis marginatis. *Paullinia with winged woolly leaves, whose lobes are oval, cut on their edges, and bordered foot-stalks.*

These plants all grow naturally in the West-Indies. They have climbing stalks with tendrils at each joint, by which they fasten themselves to the neighbouring trees, and rise to the height of thirty or forty feet, garnished at each joint with one leaf, which in some of the species is composed of three lobes like Trefoil, in others of five lobes, and some have nine lobes. These are in some species entire, in others they are indented at the point, and some are cut on their edges; in some species their surface is smooth, in others they are hairy. The flowers come out in long bunches like those of Currants; they are small and white, so make no figure; these are succeeded by three-cornered capsules having three cells, which in the Cururu of Plumier, contain roundish seeds, but those of the Serjana have winged seeds, like those of the Maple reversed, being fastened at the extremity of the wing to the capsule, the seed hanging downward.

As these plants are so tender as not to live through the winter in England, unless they are placed in a warm stove, and requiring a large share of room, they are seldom propagated in Europe, unless in botanic gardens for the sake of variety, for their flowers have very little beauty to recommend them.

PEACH. See PERSICA.

PEAR. See PYRUS.

PEASE. See PISUM.

PEASE EVERLASTING. See LATHYRUS.

PEDICULARIS, Rattle, Cocks Comb, or Lousewort. There are four different kinds of this plant, which grow wild in pastures in several parts of England, and in some low meadows are very troublesome weeds, especially one sort with yellow flowers, which rises to be a foot high or more, and is often in such plenty as to be the most predominant plant; but this is very bad food for cattle, and when it is mowed with the Grass for Hay, renders it of little value. The seeds of this plant are generally ripe by the time the Grass is mowed, so that whenever persons take Grass-seed for sowing, they should be very careful, that none of this seed is mixed with it. As these plants are never cultivated, I shall not trouble the reader with their several varieties.

PEGANUM. Lin. Gen. Plant. 530. Wild Assyrian Rue.

The CHARACTERS are,

The flower has a permanent empalement. It has five oblong oval petals, which spread open, and fifteen awl-shaped stamina about half the length of the petals, whose bases spread into a nectarium under the germen, and are terminated by erect oblong summits. It has a three-cornered roundish germen, elevated at the base of the flower, which afterward becomes a roundish three-cornered capsule, having three cells, filled with oval acute-pointed seeds.

We have but one SPECIES of this genus, viz.

PEGANUM (*Harmala*) foliis multifidis. Hort. Upsal. 144. *Peganum with many-pointed leaves.*

This plant grows naturally in Spain and Syria; it has a root as large as a man's little finger, which by age becomes woody. The stalks decay every autumn, and new ones arise in the spring; these grow about a foot long, and divide into several small branches, which are garnished with oblong thick leaves cut into several narrow segments; they are of a dark green, and of a gummy bitterish taste. The flowers are produced at the end of the branches, sitting close between the leaves, and are composed of five roundish white petals, which open like a Rose, having fifteen awl-shaped stamina. In the center is situated a roundish three-cornered germen, which afterward becomes a roundish three-cornered capsule, having three cells, which contain several oval acute-pointed seeds.

It is propagated by seeds, which should be sown thinly on a bed of light earth, the beginning of April, and when the plants come up, they must be constantly kept clean from weeds, which is all the culture they will require till the end of October, or the beginning of November, when their stalks decay. At which time, if the bed is covered with tanners bark,

bark, ashes, saw-dust, or such like covering, to keep out the frost, it will be a secure way to preserve the roots, which, when young, are somewhat tender. The following March the roots may be taken up, and transplanted into a warm situation and a dry soil, where they will continue several years.

PELECINUS. See Biserrula.

PELLITORY OF SPAIN. See Anthemis.

PELLITORY OF THE WALL. See Parietaria.

PELTARIA. Jac. Vind. 260. Mountain Treacle Mustard.

The Characters are,

The empalement of the flower has five small, concave, coloured leaves; the flower has four petals in form of a cross, having long necks; and six awl-shaped stamina, two of which are shorter than the empalement, terminated by single summits; and a round germen supporting a short style, crowned by an obtuse stigma. The germen afterward becomes a round compressed pod, with one cell, containing one seed.

We have but one Species of this plant, viz.

Peltaria (*Alliacea.*) Jacq. Vind. 260. *Mountain Treacle Mustard with a Woad leaf.*

This plant grows naturally upon the mountains of Austria, and also in Istria; it is a biennial, or at most, a triennial plant: it rises with an upright branching stalk a foot high, garnished with smooth heart-shaped leaves, which embrace the stalks with their base. The stalks are terminated by clusters of white flowers, placed almost in form of umbels; the flowers have four cross-shaped petals, which are succeeded by roundish compressed pods, containing one seed. It flowers in May, and the seeds ripen in July.

It is propagated by seeds, which may be sown in patches in the borders of the flower-garden, in the spring; and when the plants come up, they should be thinned, leaving four or five plants in each, keeping them afterward clean from weeds, which is all the culture they require.

PENNY-ROYAL. See Pulegium.

PENTAPETES. Lin. Gen. Plant. 757.

The Characters are,

The flower has a double empalement, the outer being small and composed of three leaves, the inner is cut into five parts, which are reflexed. It has five oblong petals which spread open, and fifteen narrow stamina joined in a tube at their base. It has a roundish germen, with a cylindrical style the length of the stamina, crowned by a thick stigma. The germen afterward becomes an oval ligneous capsule with five cells, filled with oblong seeds.

We have but one Species of this genus, viz.

Pentapetes (*Phœnicia*) foliis hastato-lanceolatis serratis. Lin. Sp. Pl. 698. *Pentapetes with halbert-pointed, spear-shaped, sawed leaves.*

This plant grows naturally in India, from whence I have several times received the seeds; it is annual, and dies in the autumn, soon after it has ripened seeds. It hath an upright stalk from two to three feet high, garnished with leaves of different forms; the lower leaves are large, and cut on their sides towards the base into two side lobes; the middle is extended two or three inches farther in length, so that the leaves resemble the points of halberts in their shape; they are slightly sawed on their edges, and are of a lucid green on their upper side, standing upon pretty long foot-stalks. The leaves, which are on the upper part of the branches, are much narrower, and some of them have very small indentures on their sides; these sit close to the stalks, and are placed alternately. From the wings of the stalks the flowers come out; they are for the most part single, but sometimes there are two arising at the same place. The foot-stalk of the flower is short and slender. The exterior empalement of the flower is composed of three short leaves; the interior is of one leaf, cut at the top into five acute segments, and are almost as long as the petal. The flower is of one petal, cut into five obtuse segments. In the center of the flower arises a short thick column, to which adhere fifteen short stamina. Between the stamina is situated a roundish germen, supporting a style the length of the stamina, which is crowned by a thick stigma. These are all joined at their base into a sort of column. The flowers are of a fine scarlet colour, and are succeeded by roundish capsules with five cells, each cell inclosing three or four oblong seeds, which ripen in the autumn.

The seeds of this plant must be sown upon a hot-bed in March, and when the plants are fit to transplant, there should be a new hot-bed prepared to receive them, into which should be plunged some small pots filled with good kitchen-garden earth; in each of these should be one plant put, giving them a little water to settle the earth to their roots; they must also be shaded from the sun till they have taken new root, then they should be treated in the same way as other tender exotic plants. When the plants are advanced in their growth, so as to fill the pots with their roots, they should be shifted into larger, and plunged into another hot-bed, where they may remain as long as they can stand under the glasses of the bed, without being injured; afterward they must be removed either into the stove or a glass-case, where they may be screened from the cold, and in warm weather have plenty of fresh air admitted to them. With this management the plants will flower early in July, and there will be a succession of flowers continued till the end of September, during which time they make a good appearance. The seeds ripen gradually after each other in the same succession as the flowers were produced, so they should be gathered as soon as their capsules begin to open at the top.

PENTAPHYLLOIDES. See Potentilla.

PEONY. See Pæonia.

PEPO. See Cucurbita.

PERESKIA. Plum. Nov. Gen. 37. tab. 26. American Goosberry.

The Characters are,

It hath a Rose-shaped flower consisting of several leaves, which are placed orbicularly, whose cup afterward becomes a soft, fleshy, globular fruit, beset with leaves. In the middle of the fruit are many flat roundish seeds inclosed in a mucilage.

We have but one Species of this plant, viz.

Pereskia (*Aculeata*) aculeata, flore albo, fructu flavescente. Plum. Nov. Gen. 37. *Prickly Pereskia with a white flower, and a yellowish fruit.*

This plant grows in some parts of the Spanish West-Indies, from whence it was brought to the English settlements in America, where it is called a Gooseberry, and by the Dutch it is called Blad Apple. It hath many slender branches, which will not support themselves, so must be supported by stakes, otherwise they will trail on the ground. The branches, as also the stem of the plant, are beset with long whitish spines, which are produced in tufts. The leaves are roundish, very thick and succulent, and the fruit is about the size of a Walnut, having tufts of small leaves on it, and hath a whitish mucilaginous pulp.

It may be propagated by planting of the cuttings during any of the summer months, which should be planted in pots filled with fresh light earth, and plunged into a moderate hot-bed of tanners bark, observing to shade them from the sun in the heat of the day. In about two months the cuttings will have made good roots, when they may be carefully taken out of the pots, and each planted in a separate pot, and plunged into the hot-bed again, where they may remain during the summer season; but at Michaelmas, when the nights begin to be cold, they should be removed into

into the ſtove, and plunged into the bark-bed. During the winter ſeaſon, the plants muſt be kept warm; in ſummer they muſt have a large ſhare of air, but they ſhould conſtantly remain in the ſtove, for though they will bear the open air in ſummer, in a warm ſituation, yet they will make no progreſs if they are placed abroad; nor do they thrive near ſo well in the dry ſtove, as when they are plunged in the tan, ſo that the beſt way is to ſet them next a trellis, at the back of the tan-bed, to which their branches may be faſtened, to prevent their trailing on other plants. This plant has not as yet produced either flowers or fruit in England; but as there are ſeveral plants pretty well grown in the gardens of the curious, we may expect ſome of them will flower in a ſhort time.

PERICLYMENUM. Tourn. Inſt. R. H. 608. tab. 578. Honeyſuckle.

The CHARACTERS are,

The empalement of the flower is cut into five parts ſitting upon the germen. The flower is of one petal, having an oblong tube cut at the top into five ſegments, which turn backward. It has five awl-ſhaped ſtamina almoſt the length of the petal, and a roundiſh germen ſituated below the flower, which afterward becomes an umbilicated berry with two cells, each containing one roundiſh ſeed.

The SPECIES are,

1. PERICLYMENUM (*Sempervirens*) floribus capitatis terminalibus, foliis omnibus connatis ſempervirentibus. *Honeyſuckle with flowers growing in heads at the end of the branches, and leaves joined round the ſtalk, which are evergreen; commonly called Trumpet Honeyſuckle.*

2. PERICLYMENUM (*Racemoſum*) racemis lateralibus oppoſitis, floribus pendulis, foliis lanceolatis integerrimis. *Honeyſuckle with flowers in long bunches growing oppoſite, hanging down, and entire ſpear-ſhaped leaves.*

3. PERICLYMENUM (*Verticillatum*) corymbis terminalibus, foliis ovatis verticillatis petiolatis. *Honeyſuckle with round bunches of flowers at the end of the branches, and oval leaves growing in whorls, having foot-ſtalks.*

4. PERICLYMENUM (*Nova Hiſpanicum*) corymbis terminalibus, foliis ovatis acutis. *Honeyſuckle with round bunches of flowers terminating the branches, and oval acute-pointed leaves.*

5. PERICLYMENUM (*Germanicum*) capitulis ovatis imbricatis terminalibus, foliis omnibus diſtinctis. *The German Honeyſuckle.*

6. PERICLYMENUM (*Italicum*) floribus verticillatis terminalibus ſeſſilibus, foliis ſummis connato-perfoliatis. Hort. Cliff. 45. *The Italian Honeyſuckle.*

7. PERICLYMENUM (*Vulgare*) floribus corymboſis terminalibus, foliis hirſutis diſtinctis, viminibus tenuioribus. *Honeyſuckle with a corymbus of flowers terminating the ſtalks, hairy leaves growing diſtinct, and very ſlender branches; commonly called Engliſh Honeyſuckle, or Woodbine.*

8. PERICLYMENUM (*Americanum*) floribus verticillatis terminalibus ſeſſilibus, foliis connato-perfoliatis ſempervirentibus glabris. *Honeyſuckle with whorled flowers ſitting cloſe, terminating the ſtalks, and ſmooth evergreen leaves ſurrounding the ſtalks; or evergreen Honeyſuckle.*

The firſt ſort grows naturally in Virginia, and many other parts of North America, but has been long cultivated in the Engliſh gardens by the title of Virginia Trumpet Honeyſuckle. Of this there are two varieties, if not diſtinct ſpecies, one being much hardier than the other. The old ſort, which came from Virginia, has ſtronger ſhoots; the leaves are of a brighter green; the bunches of flowers are larger, and deeper coloured than thoſe of the other, which came from Carolina. Theſe plants have the appearance of the common Honeyſuckle, but the ſhoots are weaker than any of thoſe, except the wild ſort called Woodbine; they are of a purpliſh red colour, and ſmooth. The leaves are of an oblong oval ſhape inverted, and cloſely ſurround the ſtalk; of a lucid green on their upper ſide, but pale on their under. The flowers are produced in bunches at the end of the branches, having long ſlender tubes, which are enlarged at the top, where they are cut into five almoſt equal ſegments. The outſide of the flower is of a bright ſcarlet, and the inſide yellow.

Theſe plants ſhould be planted againſt walls or pales, to which their branches ſhould be trained for ſupport, otherwiſe they will fall to the ground; for they cannot be reduced to heads like many of the Honeyſuckles, becauſe their branches are too weak and rambling, and are liable to be killed in ſevere winters; therefore they ſhould be planted to a warm aſpect, where they will begin to flower the latter end of June, and there will be a ſucceſſion of flowers till the autumn. They are propagated by laying down their young branches, which will eaſily take root, and may be afterward treated like the common Honeyſuckle.

The ſecond ſort grows naturally in Jamaica. This hath many ſlender branches, which cannot ſupport themſelves, but trail upon any neighbouring buſhes. They grow eight or ten feet long, are covered with a brown bark, and garniſhed with ſpear-ſhaped leaves, of a lucid green on their upper ſide, but pale on their under, ſtanding by pairs oppoſite. The flowers come out from the ſide of the branches at each joint; they are ranged on each ſide the foot-ſtalk in long bunches like Currants. The bunches come out oppoſite; the flowers are ſmall, of a yellowiſh green, and are ſucceeded by ſmall berries of a ſnow white, from whence the plant is called Snowberry Buſh in America.

The third ſort grows naturally in ſome of the iſlands in the Weſt-Indies. This riſes with a ſhrubby ſtalk ten or twelve feet high, ſending out many ſlender branches, covered with a light brown bark, garniſhed with oval leaves, four of them coming out at each joint in whorls round the ſtalk; they ſtand upon ſhort foot-ſtalks, and have one ſtrong midrib, with ſeveral veins running from the midrib to the ſides. The flowers come out in round bunches at the end of the branches; they are of a deep coral colour on their outſide, but of a pale red within. This was found growing in Jamaica by the late Dr. Houſtoun, who brought it to England.

The fourth ſort grows naturally in the kingdom of Chili; Father Feuillée found it near the city of Conception; it was afterward found by the late Dr. Houſtoun, growing at a little diſtance from Carthagena in New Spain. This hath a ſhrubby ſtalk near four inches thick, covered with a gray bark, dividing into many branches, which riſe about twelve feet high, garniſhed with ſtiff leaves, placed oppoſite, of a lucid green on their upper ſide, but pale on their under. The flowers are produced in round bunches at the end of the branches; they are of a deep red colour, cut into four ſegments at the top. Theſe are ſucceeded by oval berries the ſize of ſmall Olives, incloſing a hard ſeed.

The branches of this ſhrub are uſed for dyeing a black in the Spaniſh Weſt-Indies, which is permanent, and cannot be waſhed out. This dye is made with the wood of the ſhrub cut into ſmall pieces, and mixed with a plant called Pangue, and a black earth called Robbo, boiling them in common water till it becomes of a proper conſiſtence.

The three ſorts laſt mentioned are too tender to thrive in this country without artificial heat; they are propagated by ſeeds, which muſt be procured from the countries where they naturally grow. Theſe ſhould be ſown in pots, and plunged into a moderate hot bed, where they may remain till the autumn, for the plants rarely come up the firſt year, ſo the pots ſhould be removed into the ſtove for the winter

 ſeaſon,

ſeaſon, and the following ſpring placed on a freſh hot-bed, which will bring up the plants; when they are fit to remove, they ſhould be each planted in a ſeparate ſmall pot filled with light earth, and plunged into a freſh hot-bed, ſhading them from the ſun till they have taken new root; after which they muſt be treated in the manner as other tender plants from thoſe countries. As the plants obtain ſtrength, they ſhould be more hardily treated, by placing them abroad in a ſheltered ſituation for two months or ten weeks, in the warmeſt part of the ſummer, and in the winter they may be placed in the dry ſtove, kept to a moderate temperature of warmth, where they will thrive, and produce their flowers in the autumn.

The fifth ſort is the common Dutch or German Honeyſuckle, which has been generally ſuppoſed the ſame with the Engliſh wild ſort called Woodbine, but is undoubtedly a very different ſpecies, for the ſhoots of this are much ſtronger. The plants may be trained with ſtems, and formed into heads, which the wild ſort cannot, their branches being too weak and trailing for this purpoſe. The branches of this are ſmooth, of a purpliſh colour, garniſhed with oblong oval leaves, of a lucid green on their upper ſide, but pale on their under, having very ſhort foot-ſtalks; they are placed by pairs, but are not joined at their baſe. The flowers are produced in bunches at the end of the branches, each flower ariſing out of a ſcaly cover, which cover after the flowers fade, forms an oval head, whoſe ſcales lie over each other like thoſe of fiſh. The flowers are of a reddiſh colour on their outſide, and yellowiſh within, of a very agreeable odour. This ſort flowers in June, July, and Auguſt. There are two other varieties of this ſpecies, one is called the long blowing, and the other the late red Honeyſuckle.

The ſixth ſort is commonly called the Italian Honeyſuckle. Of this there are two or three varieties, the early white Honeyſuckle is one; this is the firſt which flowers, always appearing in May. The branches of this are ſlender, covered with a light green bark, garniſhed with oval leaves of a thin texture, placed by pairs ſitting cloſe to the branches, but thoſe which are ſituated toward the end of the branches, join at their baſe, ſo that the ſtalk ſeems as if it came through the leaves. The flowers are produced in whorled bunches at the end of the branches; they are white, and have a very fragrant odour, but are of ſhort duration, ſo that in about a fortnight they are entirely over, and ſoon after the leaves appear as if blighted and ſickly, making an indifferent appearance the whole ſummer, which has rendered them leſs valued than the others. The other variety is the yellow Italian Honeyſuckle, which is the next in ſucceſſion to the white. The ſhoots of this are much like thoſe of the former, but have a darker bark; the leaves are alſo of a deeper green; the flowers are of a yellowiſh red, and appear ſoon after the white; they are not of much longer duration, and are ſucceeded by red berries, containing one hard ſeed incloſed in a ſoft pulp, which ripens in the autumn.

The ſeventh ſort is the common wild Engliſh Honeyſuckle or Woodbine; this grows naturally in the hedges in many parts of England. The branches are very ſlender and hairy, trailing over the neighbouring buſhes, and twining round the boughs of trees; the leaves are oblong, hairy, and diſtinct, not joined at their baſe; they are placed by pairs oppoſite; the flowers are produced in long bunches at the end of the branches. There are two varieties, one with white, and the other yellowiſh red flowers. Theſe appear in July, and there is a ſucceſſion of flowers on the plants till the autumn.

There is alſo a variety of this with variegated leaves, and one with cut leaves ſomewhat like the leaves of Oak, and one of theſe with variegated leaves; but as theſe are accidental varieties, I have not enumerated them.

The eighth ſort is ſuppoſed to grow naturally in North America. This hath ſtrong branches, covered with a purple bark, which are garniſhed with lucid green leaves embracing the ſtalks, which continue their verdure all the year. The flowers are produced in whorled bunches at the end of the branches; they are frequently two, and ſometimes three of theſe bunches riſing one out of another; they are of a bright red on their outſide, and yellow within, of a ſtrong aromatic flavour. This ſort begins to flower in June, and there is a ſucceſſion of flowers till the froſt puts a ſtop to them, ſo that it is the moſt valuable of all the ſorts.

All the ſorts of Honeyſuckles are propagated either by layers or cuttings: when they are propagated by layers, the young ſhoots only ſhould be choſen for that purpoſe. Theſe ſhould be laid in the autumn, and by the following autumn they will have taken root, when they ſhould be cut off from the plants, and either planted where they are to remain, or into a nurſery to be trained up, either for ſtandards, which muſt be done by fixing down ſtakes to the ſtem of each plant, to which their principal ſtalk ſhould be faſtened, and all the other muſt be cut off; the principal ſtalk muſt be trained to the intended height of the ſtem, then it ſhould be ſhortened, to force out lateral branches, and theſe ſhould be again ſtopped to prevent their growing too long; by the conſtant repeating this as the ſhoots are produced, they may be formed into a ſort of ſtandard; but if any regard is had to their flowering, they cannot be formed into regular heads, for by conſtantly ſhortening their branches, the flower-buds will be cut off, ſo that few flowers can be expected; and as it is an unnatural form for theſe trees, ſo there ſhould be but few of them reduced to it, for when they are planted near other buſhes, in whoſe branches the ſhoots of the Honeyſuckles may run and mix, they will flower much better, and have a finer appearance than when they are more regularly trained; therefore when the plants are in the nurſery, if two or three of the principal ſhoots are trained up to the ſtakes, and the others are entirely cut off, they will be fit to tranſplant the following autumn, to the places where they are to remain; for though they may be tranſplanted of a greater age, yet they do not thrive ſo well as when they are removed while they are young.

When theſe plants are propagated by cuttings, they ſhould be planted in September, as ſoon as the ground is moiſtened by rain. The cuttings ſhould have four joints, three of which ſhould be buried in the ground, and the fourth above the ſurface, from which the ſhoots are to be produced. Theſe may be planted in rows, at about a foot diſtance row from row, and four inches aſunder in the rows, treading the earth cloſe to them; and as the evergreen and late red Honeyſuckles are a little more tender than the other ſorts, ſo if the ground between the rows where theſe are planted, is covered with tanners bark or other mulch, to keep out the froſt in winter, and the drying winds of the ſpring, it will be of great advantage to the cuttings; and if the cuttings of theſe ſorts have a ſmall piece of the two years wood at their bottom, there will be no hazard of their taking root. The plants which are raiſed from cuttings, are preferable to thoſe which are propagated by layers, as they have generally better roots.

Theſe plants will grow in almoſt any ſoil or ſituation, (except the ſecond, third, fourth, and the laſt mentioned, which will not thrive where they are too much expoſed to the cold in winter;) they thrive beſt in a ſoft ſandy loam, and will retain their leaves in greater verdure in ſuch ground than if planted in a dry gravelly ſoil, where in warm dry ſeaſons their leaves often ſhrink, and hang in a very diſagreeable manner; nor will thoſe ſorts which naturally flower late in the autumn, continue ſo long in beauty on a dry ground, unleſs the ſeaſon ſhould prove moiſt and cold, as thoſe in a gentle loam, not too ſtiff or wet.

There are few ſorts of ſhrubs which deſerve cultivation better than moſt of theſe, for their flowers are very beautiful, and perfume the air to a great diſtance with their odour, eſpecially in the mornings and evenings, and in cloudy weather, when the ſun does not exhale their odour, and raiſe it too high to be perceptible; ſo that in all retired walks, there cannot be too many of theſe intermixed with the other ſhrubs. I have ſeen theſe plants intermixed in hedges, planted either with Alder or Laurel, where the branches have been artfully trained between thoſe of the hedge; from which the flowers have appeared diſperſed from the bottom of the hedge to the top, and being intermixed with the ſtrong green leaves of the plants, which principally compoſe the hedge, they have made a fine appearance; but the beſt ſorts for this purpoſe are the evergreen and long blowing Honeyſuckles, becauſe their flowers continue in ſucceſſion much longer than the other ſorts.

PERIPLOCA. Tourn. Inſt. R. H. 93. tab. 22. Virginian Silk.

The Characters are,

The flower hath a permanent empalement, cut into five points. The flower has one plain petal, cut into five narrow ſegments, with a ſmall nectarium going round the center of the petal, and the five incurved filaments, which are not ſo long as the petal, with five ſhort ſtamina. It has a ſmall bifid germen with ſcarce any ſtyle, which afterward becomes two oblong bellied capſules with one cell, filled with ſeeds crowned with down, lying over each other like the ſcales of fiſh.

The Species are,

1. Periploca (*Græca*) floribus internè hirſutis. Lin. Sp. Plant. 211. *Virginian Silk, with flowers hairy on their inſide.*

2. Periploca (*Africana*) caule hirſuto. Lin. Sp. Plant. 211. *Virginian Silk with a hairy ſtalk.*

3. Periploca (*Fruticoſa*) foliis oblongo-cordatis pubeſcentibus, floribus alaribus, caule fruticoſo ſcandente. *Virginian Silk, with oblong heart-ſhaped leaves, which are covered with ſoft hairs, and flowers proceeding from the ſides of the ſtalks, which are ſhrubby and climbing.*

The firſt ſort grows naturally in Syria, but is hardy enough to thrive in the open air in England. It hath twining ſhrubby ſtalks, covered with a dark bark, which twiſt round any neighbouring ſupport, and will riſe more than forty feet high, ſending out ſlender branches from the ſide, which twine round each other, and are garniſhed with oval ſpear-ſhaped leaves near four inches long, and two broad in the middle, of a lucid green on their upper ſide, but pale on their under, ſtanding by pairs upon ſhort foot-ſtalks. The flowers come out toward the end of the ſmall branches in bunches; they are of a purple colour, and hairy on their inſide, compoſed of one petal, cut into five ſegments almoſt to the bottom, which ſpread open in form of a ſtar, and within is ſituated a nectarium, which goes round the five ſhort ſtamina and germen, which is hairy. The germen afterward turns to a double long taper pod or capſule, filled with compreſſed ſeeds, lying over each other like the ſcales of fiſh, having a ſoft down fixed to their top. This plant flowers in July and Auguſt, but rarely ripens its ſeeds in England.

It is eaſily propagated by laying down of the branches, which will put out roots in one year, and may then be cut from the old plant, and planted where they are to remain. Theſe may be tranſplanted either in autumn, when their leaves begin to fall, or in the ſpring before they begin to ſhoot, and muſt be planted where they may have ſupport, otherwiſe they will trail on the ground, and faſten themſelves about whatever plants are near them.

The ſecond ſort grows naturally in Africa. This hath many ſlender ſtalks, which twine about each other, or any neighbouring ſupport, and will riſe near three feet high, putting out many ſeveral ſide branches, which are hairy, as are alſo the leaves, ſtanding by pairs upon very ſhort foot-ſtalks. The flowers come out in ſmall bunches from the ſide of the ſtalks; they are ſmall, of a worn-out purpliſh colour, and of a ſweet ſcent, being cut into five narrow ſegments almoſt to the bottom. It flowers in the ſummer months, but does not produce ſeeds here. There is a variety of this with ſmooth leaves and ſtalks, from the ſame country.

The third ſort was diſcovered by the late Dr. Houſtoun, growing naturally at La Vera Cruz in America. This riſes with a ſtrong woody ſtalk to the height of five or ſix feet, covered with a gray bark, putting out many weak branches, which twiſt themſelves about any neighbouring ſupport, and riſe to the height of twenty feet; they are garniſhed with heart-ſhaped leaves of a yellowiſh green, covered with ſilky hairs, which are ſoft to the touch, and ſtand by pairs upon pretty long foot-ſtalks. The flowers come out in ſmall bunches from the wings of the leaves; they are ſmall, white, and of the open bell-ſhape; theſe are ſucceeded by ſwelling taper pods, filled with ſeeds crowned with long feathery down.

The ſecond ſort is hardy enough to thrive in this country, with a little protection from the froſt in winter. If theſe are ſheltered under a common frame, or placed in a green-houſe during the winter ſeaſon, and placed abroad with other hardy exotic plants in ſummer, they will thrive and flower very well; but as all the plants of this genus have a milky juice, ſo they ſhould not have much wet, eſpecially in cold weather, leſt it rot them. They are eaſily propagated by laying down of their branches, which in one year will be rooted enough to tranſplant; theſe ſhould be planted in a light ſandy loam, not rich, and the pots muſt not be too large, for when they are over-potted they will not thrive.

The third ſort is tender, ſo will not thrive in England, unleſs the plants are placed in a warm ſtove. This may be propagated by laying down of the branches, in the ſame manner as the former; or from ſeeds, when they can be procured from the places where they naturally grow. Theſe ſhould be ſown upon a good hot-bed, and when the plants come up, they muſt be treated in the ſame manner as other tender exotic plants.

If theſe plants are conſtantly kept plunged in the tan-bed of the ſtove, they will thrive and flower much better than in any other ſituation, but the ſtove ſhould not be kept too warm in winter; and in the ſummer the plants ſhould have a large ſhare of free air admitted to them; for when they are kept too cloſe, their leaves will be covered with inſects, and the plants will become ſickly in a ſhort time.

PERIWINKLE. See Vinca.

PERSEA. Plum. Nov. Gen. 44. tab. 20. The Avocado, or Avogato Pear.

The Characters are,

The flower hath no empalement, but is compoſed of ſix petals, ending in acute points, which ſpread open. It hath ſix ſtamina about half the length of the petals, terminated by roundiſh ſummits, and a ſhort ſtyle crowned by a pyramidal germen, which afterward becomes a large fleſhy pyramidal fruit, incloſing an oval ſeed, having two lobes.

We have but one Species of this genus, viz.

Persea (*Americana.*) Cluſ. Hiſt. *The Avocado, or Avogato Pear.*

This tree grows in great plenty in the Spanish West-Indies, as also in the island of Jamaica, and hath been transplanted into most of the English settlements in the West-Indies, on account of its fruit; which is not only esteemed by the inhabitants as a fruit to be eaten by way of desert, but is very necessary for the support of life. The fruit of itself is very insipid, for which reason they generally eat it with the juice of Lemons and sugar, to give it a piquancy. It is very nourishing, and is reckoned a great incentive to venery. Some people eat this fruit with vinegar and pepper.

This tree, in the warm countries where it is planted, grows to the height of thirty feet or more, and has a trunk as large as our common Apple-trees; the bark is smooth, and of an Ash colour; the branches are beset with pretty large, oblong, smooth leaves, like those of Laurel, which are of a deep green colour, and continue on the tree throughout the year. The flowers and fruit are, for the most part, produced toward the extremity of the branches. The fruit is as large as one of the largest Pears, inclosing a large seed with two lobes included in a thin shell.

In Europe this plant is preserved as a curiosity, by those persons who delight in collecting exotic plants; and tho' there is little hope of its producing fruit, yet for the beauty of its shining green leaves, which continue through the winter, it deserves a place in every curious collection of plants.

It is propagated by seeds, which should be obtained as fresh as possible, from the countries of its growth; and if they are brought over in sand, they will be more likely to grow, than such as are brought over dry. These nuts or seeds should be planted in pots filled with light earth, and plunged into a hot-bed of tanners bark. In about five or six weeks the plants will come up, when they must, while young, be treated very tender; but when they have grown about four inches high, they should be carefully transplanted; and if there are two or more plants in one pot, they must be parted, being careful to preserve a ball of earth to the root of each, and planted into separate small pots, then plunged into a hot-bed of tanners bark, observing to shade them until they have taken new root; after which time they should have fresh air admitted to them in proportion to the warmth of the season. Towards Michaelmas the plants must be removed into the stove, and plunged into the bark-bed, where, during the winter season, they should be kept warm, and must be gently watered twice a week. In the spring the plants should be shifted into pots a size larger than the former, and the bark-bed should be then renewed with fresh tan, which will set the plants in a growing state early, whereby they will make a fine progress the following summer. These plants should be kept in the stove, for they are too tender to bear the open air in this country, except in the warmest part of summer.

PERSICA. Tourn. Inst. R. H. 624. tab. 402. The Peach-tree.

The Characters are,

The flower has a tubulous empalement of one leaf, cut into five obtuse segments. It hath five oblong obtuse petals, which are inserted in the empalement, and about thirty erect slender stamina, which are shorter than the petals, and are also inserted in the empalement. It hath a roundish hairy germen, which afterwards becomes a roundish, woolly, large, succulent fruit, with a longitudinal furrow, inclosing an oval nut with a netted shell, having many punctures.

There are a great variety of these trees cultivated, in the gardens of those who are curious in collecting the several sorts of fruit: I shall therefore first beg leave to mention only two or three sorts, which are cultivated for the beauty of their flowers; after which, I shall enumerate the several varieties of good fruit which have come to my knowledge.

The Species are,

1. Persica (*Vulgaris*) vulgaris, flore pleno. Tourn. Inst. R. H. 625. *Common Peach-tree with double flowers.*

2. Persica (*Humilis*) Africana nana, flore incarnato simplici. Tourn. Inst. R. H. 625. *Dwarf Almond with single flowers, vulgò.*

3. Persica (*Africana*) Africana nana, flore incarnato pleno. Tourn. Inst. R. H. 625. *Double flowering Dwarf Almond, vulgò.*

The first of these trees is a very great ornament in a garden early in the spring, the flowers being very large, double, and of a beautiful red or purple colour. This may be planted in standards, and if intermixed amongst other flowering-trees of the same growth, makes a very agreeable variety; or it may be planted against the walls of the pleasure-garden, where the beautiful appearance of its flowers early in the spring will be more acceptable in such places than the choicest fruits, which must be exposed to servants, and others, so that they seldom can be preserved in large families until they are ripe. This tree may be propagated by budding it upon the Almond or Plumb stocks, in the same manner as the other sort of Peaches, and should be planted in a good fresh soil that is not over moist.

The other two sorts are of humbler growth, seldom rising above three or four feet high; these may be budded upon Almond stocks, or propagated by layers; they will also take upon Plumb stocks, but they are very apt to canker, after they have stood four or five years upon those stocks, especially that with double flowers, which is tenderer than the other, which sends out suckers from the root, whereby it may be propagated in great plenty.

These shrubs make a very agreeable variety amongst low flowering trees in wilderness-walks. The single sort flowers in the beginning of April, and the double is commonly a fortnight later.

I shall now proceed to mention the several sorts of good Peaches which have come to my knowledge; and though perhaps a greater number of sorts may be found in some catalogues of fruits, yet I doubt whether many of them are not the same kinds called by different names; for, in order to determine the various kinds, it is necessary to observe the shape and size of the flowers, as well as the different parts of the fruit; for this does sometimes determine the kind, when the fruit alone is not sufficient; besides, there is a vast difference in the size and flavour of the same Peach, when planted on different soils and aspects; so that it is almost impossible for a person who is very conversant with these fruits to distinguish them, when brought from various gardens.

1. The white Nutmeg (called by the French, L'Avant Péche Blanche.) This tree has sawed leaves, but generally shoots very weak, unless it is budded upon an Apricot stock; the flowers are large and open, the fruit is small and white, as is also the pulp at the stone, from which it separates; it is a little musky and sugary, but is only esteemed for its being the first sort ripe. It is in eating the end of July, and soon becomes meally.

2. The red Nutmeg (called by the French, L'Avant Péche de Troyes.) This tree has sawed leaves; the flowers are large and open; the fruit is larger and rounder than the white Nutmeg, and is of a bright vermilion colour; the flesh is white, and very red at the stone; it has a rich musky flavour, and parts from the stone. This Peach is well esteemed; it ripens the beginning of August.

3. The early or small Mignon (called by the French, La Double de Troyes, or Mignonette.) This tree has small contracted flowers; the fruit is of a middling size, and round;

round; it is very red on the side next the sun; the flesh is white, and separates from the stone, where it is red; the juice is vinous and rich. It is ripe the beginning of August.

4. The yellow Alberge. This tree has smooth leaves; the flowers are small and contracted; the fruit is of a middling size, somewhat long, the flesh is yellow and dry; it is seldom well flavoured, but should be perfectly ripe before it is gathered, otherwise it is good for little. It is ripe early in August.

5. The white Magdalen. This tree has sawed leaves; the flowers are large and open; the wood is generally black at the pith; the fruit is round, of a middling size; the flesh is white to the stone, from which it separates; the juice is seldom high flavoured; the stone is very small. This ripens early in August.

6. The early Purple (called by the French, La Pourprée hâtive.) This tree has smooth leaves; the flowers are large and open; the fruit is large, round, and of a fine red colour; the flesh is white, but very red at the stone; is very full of juice, which has a rich vinous flavour, and is by all good judges esteemed an excellent Peach. This is ripe the middle of August.

7. The large or French Mignon. The leaves of this tree are smooth; the flowers are large and open; the fruit is a little oblong, and generally swelling on one side; it is of a fine colour; the juice is very sugary, and of a high flavour; the flesh is white, but very red at the stone, which is small. This is ripe in the middle of August, and is justly esteemed one of the best Peaches; it separates from the stone. This sort of Peach is tender, and will not thrive on a common stock, so is generally budded upon some vigorous shooting Peach, or an Apricot, by the nursery-men, which enhances the price of the trees. But the best method is to bud this Peach into some old healthy Apricot, which is planted to a south or south-east aspect, and to cut away the Apricot when the buds have taken, and made shoots; upon some trees which I have seen thus managed, there has been a much greater quantity of fairer, and better flavoured fruit, than I have ever observed in any other management, and the trees have been much more healthy.

8. The Chevreuse, or Belle Chevreuse. This tree has smooth leaves; the flowers are small and contracted; the fruit is of a middling size, a little oblong, of a fine red colour; the flesh is white, but very red at the stone, from which it separates; it is very full of a rich sugary juice, and ripens toward the end of August. This is a very good bearer, and may be ranged with the good Peaches.

9. The red Magdalen (called by the French about Paris, Madeleine de Courson.) The leaves of this tree are deeply sawed; the flowers are large and open; the fruit is large and round, of a fine red colour; the flesh is white, but very red at the stone, from which it separates; the juice is very sugary, and of an exquisite flavour. It is ripe the end of August, and is one of the best sorts of Peaches.

10. The early Newington (or Smith's Newington.) This is very like, if not the same, with what the French call Le Pavie blanc. This tree has sawed leaves; the flowers are large and open; the fruit is of a middling size, of a fine red on the side next the sun; the flesh is firm and white, but very red at the stone, to which it closely adheres. It hath a sugary juice, and is ripe the end of August.

11. The Montauban. This tree has sawed leaves; the flowers are large and open; the fruit is of a middling size, of a deep red, inclining to purple next the sun, but of a pale colour toward the wall; the flesh is melting and white to the stone, from which it separates; the juice is rich, and the tree is a good bearer. It ripens the middle of August, and is well esteemed.

12. The Malta (which is very like, if not the same, with the Italian Peach.) This tree has sawed leaves; the flowers are large and open; the fruit is of a middling size, of a fine red next the sun; the flesh is white and melting, but red at the stone, from which it separates; the stone is flat and pointed; the tree is a good bearer. This ripens the end of August.

13. The Noblest. This tree has sawed leaves; the flowers are large and open; the fruit is large, of a bright red next the sun; the flesh is white and melting, and separates from the stone, where it is of a faint red colour; the juice is very rich in a good season. It ripens the end of August.

14. The Chancellor. The leaves of this tree are smooth; the flowers are small and contracted; the fruit is shaped somewhat like the Belle Chevreuse, but is rounder; the flesh is white and melting, and separates from the stone, where it is of a fine red colour; the skin is very thin, and the juice is very rich. It ripens about the end of August, and is esteemed one of the best sort of Peaches. This tree is very tender, and will not succeed on common stocks, so is budded twice as the Mignon; and if budded on Apricots, as was directed for that sort, will thrive much better than in any other method.

15. The Bellegarde (or as the French call it, the Gallande.) This tree hath smooth leaves; the flowers are small and contracted; the fruit is very large and round, of a deep purple colour on the side next the sun; the flesh is white, melting, and separates from the stone, where it is of a deep red colour; the juice is very rich. This ripens the beginning of September, and is an excellent Peach, but at present not very common.

16. The Lisle (or as the French call it, La petite Violette hâtive.) This tree has smooth leaves; the flowers are small and contracted; the fruit is of a middling size, of a fine Violet colour toward the sun; the flesh is of a pale yellow and melting, but adheres to the stone, where it is very red; the juice is very vinous. This ripens the beginning of September.

17. The Bourdine. This tree hath smooth leaves; the flowers are small and contracted; the fruit is large, round, and of a fine red colour next the sun; the flesh is white, melting, and separates from the stone, where it is of a fine red colour; the juice is vinous and rich. This ripens the beginning of September, and is greatly esteemed by the curious. The tree bears plentifully, and will produce fruit in standards very well.

18. The Rossanna. This tree hath smooth leaves; the flowers are small and contracted; the fruit is large, a little longer than the Alberge; the flesh is yellow, and separates from the stone, where it is red; the juice is rich and vinous. This ripens the beginning of September, and is esteemed a good Peach. This is the same with what some call the purple, and others the red Alberge, it being of a fine purple colour on the side next the sun.

19. The Admirable. This tree hath smooth leaves; the flowers are small and contracted; the fruit is large, round, and red on the side next the sun; the flesh is white, melting, and separates from the stone, where it is of a deep red colour; the juice is sugary and rich. This ripens the beginning of September. This is by some called the early Admirable, but is certainly what the French call L'Admirable; and they have no other of this name which ripens later.

20. The old Newington. This tree has sawed leaves; the flowers are large and open; the fruit is fair and large, of a beautiful red colour next the sun; the flesh is white, hard, and closely adheres to the stone, where it is of a deep red colour; the juice is very rich and vinous. This is esteemed

efteemed one of the beft fort of Pavies. It ripens about the middle of September.

21. The Rambouillet (commonly called the Rambullion.) This tree hath fmooth leaves; the flowers are large and open; the fruit is of a middling fize, rather round than long, deeply divided by a fulcus or furrow in the middle; it is of a fine red colour next the fun, but of a light yellow next the wall; the flefh is melting, of a bright yellow colour, and feparates from the ftone, where it is of a deep red; the juice is rich, and of a vinous flavour. This ripens the middle of September, and is a good bearer.

22. The Bellis (which I believe to be what the French call La Belle de Vitry.) The leaves of this tree are fawed; the flowers are fmall and contracted; the fruit is of a middle fize, round, and of a pale red next the fun; the flefh is white, and adheres to the ftone, where it is red; the juice is vinous and rich. This ripens in the middle of September.

23. The Portugal. This tree has fmooth leaves; the flowers are large and open; the fruit is large, and of a beautiful red colour towards the fun; the fkin is generally fpotted; the flefh is firm, white, and clofely adheres to the ftone, where it is of a faint red colour; the ftone is fmall, but full of deep furrows; the juice is rich and vinous. This ripens the middle of September.

24. La Teton de Venus (or Venus's Breaft) fo called from its having a rifing like a dug, or bubby. This tree has fmooth leaves; the flowers are fmall and contracted; the fruit is of a middling fize, refembling the Admirable, of a pale red colour next the fun; the flefh is melting, white, and feparates from the ftone, where it is red; the juice is fugary and rich. This ripens late in September.

25. La Pourprée (or as the French call it, Pourprée tardive, i. e. the late Purple.) This tree has very large leaves, which are fawed; the fhoots are very ftrong; the flowers are fmall and contracted; the fruit is large, round, and of a fine purple colour; the flefh is white, melting, and feparates from the ftone, where it is red; the juice is fugary and rich. This ripens late in September.

26. The Nivette. This tree has fawed leaves; the flowers are fmall and contracted; the fruit is large, fomewhat longer than round, of a bright red colour next the fun, and of a pale yellow on the other fide; the flefh is melting, full of a rich juice, and is very red at the ftone, from which it feparates. This is efteemed one of the beft Peaches. It ripens in the middle of September.

27. The Royal (La Royale.) This tree has fmooth leaves; the flowers are fmall and contracted; the fruit is large, round, and of a deep red on the fide next the fun, and of a paler colour on the other fide; the flefh is white, melting, and full of a rich juice; it parts from the ftone, where it is of a deep red colour. This ripens the middle of September, and when the autumn is good, is an excellent Peach.

28. The Perfique. This tree has fawed leaves; the flowers are fmall and contracted; the fruit is large, oblong, and of a fine red colour next the fun; the flefh is melting, and full of a rich juice; it feparates from the ftone, where it is of a deep red colour. The ftalk has a fmall knot upon it; this makes a fine tree, and is a good bearer. It ripens the end of September. Many gardeners call this the Nivette.

29. The monftrous Pavy of Pomponne (called by the French, La Pavie rouge de Pomponne.) The leaves of this tree are fmooth; the flowers are large and open; the fruit is very large and round, many times fourteen inches in circumference; the flefh is white, melting, and clofely adheres to the ftone, where it is of a deep red colour; the outfide is a beautiful red next the fun, and of a pale flefh colour on the other fide. This ripens the end of October, and, when the autumn is warm, is an excellent Peach.

30. The Catharine. This tree hath fmooth leaves; the flowers are fmall and contracted; the fruit is large, round, and of a dark red colour next the fun; the flefh is white, hard, and full of a rich juice. It clofely adheres to the ftone, where it is of a deep red colour. It ripens the beginning of October, and in very good feafons is an excellent Peach, but being fo very late ripe, there are not many fituations where it ripens well.

31. The Bloody Peach (called by the French, La Sanguinolle.) This Peach is of a middling fize, of a deep red next the fun; the flefh is of a deep red quite to the ftone, and from thence is, by fome gardeners, called the Mulberry Peach. This fruit rarely ripens in England, fo is not often planted, but it bakes and preferves excellently; for which, as alfo the curiofity, one or two trees may be planted, where there is extent of walling.

There are fome other forts of Peaches which are kept in fome of the nurferies, but thofe which are here enumerated, are the forts moft worth planting, and in this lift the choiceft only fhould be planted; but I fhall juft mention the names of thofe forts omitted, for the fatisfaction of the curious.

The Sion; the Bourdeaux; the Swalch or Dutch; the Carlifle; the Eton; the Péche de Pau; yellow Admirable; the Double Flower. This laft fort is generally planted more for the beauty of the flowers, than for the goodnefs of the fruit, of which fome years the ftandard trees produce great plenty; but they are late ripe, and have a cold, watery, infipid juice. The Dwarf Peach is alfo preferved in fome places as a curiofity. This is a very tender tree, making very weak fhoots, which are very full of flower-buds. The fruit is not fo large as a Nutmeg, and not good, nor will the tree be of long duration, fo it is not worth cultivating.

And indeed, from thefe thirty-one above-named, there are not above ten of them which I would advife to be planted; becaufe, when a perfon can be furnifhed with thofe which are good, or has the beft of the feafon, it is not worth while to plant any which are middling or indifferent, for the fake of variety; therefore the forts which I fhould prefer, are thefe after-mentioned.

The early Purple; the Groffe Mignon; Belle Chevreufe; red Magdalen; Chancellor; Bellegarde; Bourdine; Roffana; Rambouillet, and Nivette. Thefe are the forts beft worth planting, and, as they fucceed each other, they will furnifh the table through the feafon of Peaches; and, where there is room, and the fituation very warm, one or two trees of the Catharine Peach fhould have place, for in very warm feafons it is an excellent fruit.

The French diftinguifh thofe we call Peaches into two forts, viz. Pavies and Peaches; thofe are called Peaches which quit the ftone, and thofe, whofe flefh clofely adheres to the ftone, are called Pavies. Thefe are much more efteemed in France than the Peaches, though in England the latter are preferred to the former by many perfons.

All the different forts of Peaches have been originally obtained from the ftones, which, being planted, produce new varieties, as do the feeds of all other fruits; fo that where perfons have garden enough to allow room for propagating thefe fruits from feeds, there is no doubt but fome good forts may be obtained; though it is true, there will be many of them good for nothing, as is the cafe of moft fruits and flowers which are produced from feeds, amongft which there may be fome valuable kinds, fuperior to thofe from whence the feeds were taken, yet there is always a great number which are little worth; but where perfons are fo curious as to plant the ftones of thefe fruits, great regard fhould be had to the forts; and if the fruits were permitted

permitted to remain upon the trees until they dropped off, the kernels would be fitter for planting, and more likely to grow. The best sorts for sowing are those whose flesh is firm, and cleaves to the stone; and from amongst these you should chuse such as ripen pretty early, and have a rich vinous juice; from which sorts some good fruit may be expected.

These stones should be planted in autumn, on a bed of light dry earth, about three inches deep, and four inches asunder; and in the winter the beds should be covered with mulch, to protect them from the frost, which, if permitted to enter deep into the ground, may destroy them. In the spring, when the plants come up, they should be carefully cleared from weeds, which should also be observed throughout the summer. In this bed they should remain until the following spring, when they should be carefully taken up, so as not to break their tender roots, and transplanted into a nursery, in rows three feet asunder, and one foot distant plant from plant in the rows, observing to lay a little mulch upon the surface of the ground about their roots, to prevent its drying too fast; and if the spring should prove very dry, you should give them a little water once a week, until they have taken root; after which they should be constantly kept clear from weeds, and the ground between the rows carefully dug every spring, to loosen it, so as that the tender fibres may strike out on every side.

In this nursery they may continue one, or at most but two years, according to the progress they make; after which they should be transplanted where they are to remain, to produce fruit.

In removing these trees, you should observe to prune their downright roots (if they have any) pretty short, and to cut off all bruised parts of the roots, as also all the small fibres, which generally dry, and, when left upon the roots after planting again, grow mouldy and decay, so that they are injurious to the new fibres which are to come out from the roots, and very often prevent the growth of the trees; but you should by no means prune their heads, for the plants, which are produced from stones, are generally of a more spongy texture, and so more liable to decay when cut, than those which are budded upon other stocks. Besides, as these trees are designed for standards (for it is not proper to plant them against walls, until you see the produce of the fruit, to shew which of them deserve to be cultivated,) so they will not require any other pruning, but only to cut out decayed branches, or such as shoot out very irregular from the sides, for more than this is generally very injurious to them.

When they have produced fruit, you will soon be a judge of their goodness, therefore such of them as you dislike, may be destroyed, but those which are good, may be propagated by inoculating them upon other stocks, which is the common method now practised to propagate these fruits, therefore I shall now proceed to treat of that more particularly; in the doing of which, I shall set down the method now commonly practised by the nursery-gardeners, and then propose some few things of my own as an improvement thereon, for such persons who are very curious to have good fruit. But first,

You should be provided with stocks of the Muscle and white Pear Plumbs, which are generally esteemed the two best sorts of Plumbs for stocks to inoculate Peaches and Nectarines upon; as also some Almond and Apricot stocks, for some tender sorts of Peaches, which will not grow upon Plumb stocks. These should be all produced from the stone (as hath been already directed in the article Nursery,) and not from suckers, for the reasons there laid down. These stocks should be transplanted, when they have had one year's growth in the seed-bed, for the younger they are transplanted, the better they will succeed, and hereby they will be prevented from sending tap-roots deep in the ground; for by shortening those which seem so disposed, it will cause them to put out horizontal roots. These stocks should be planted at the distance above-mentioned, viz. the rows three feet asunder, and one foot apart in the rows. This is wider than most nursery-men plant them, but I shall give my reasons hereafter for this.

When these stocks have grown in the nursery two years, they will be strong enough to bud; the season for which is commonly any time in July, when the rind will easily separate from the wood; when you should make choice of some good cuttings of the sorts of fruit you intend to propagate, always observing to take from healthy trees, and such as generally produce a good quantity of well tasted fruit; for it is very certain, that any sort of fruit may be so far degenerated, where this care is wanting, as not to be like the same kind. Besides, whenever a tree is unhealthy, the buds taken from that tree will always retain the distemper, in a greater or less degree, according as it hath imbibed a greater or less quantity of the distempered juice. Thus, for instance, where a Peach or Nectarine-tree hath been greatly blighted, so as that the shoots have grown bustled, and the leaves curled up to a great degree, that distemper is seldom recovered again by the greatest art, or at least not without several years good management; for let the seasons prove ever so favourable, these trees will continually shew the same distemper, which many persons are so weak as to suppose a fresh blight; whereas in reality it is no other but the remains of the former sickness, which are spread and intermixed with all the juices of the tree, so that whatever buds are taken from such trees, will always retain a part of the distemper.

Upon the care which is taken in the choice of the buds, the whole success depends; therefore a person, who is curious to have good fruit, cannot be too careful in this particular; for, in general, no more is regarded by those nursery-men who are the most careful in propagating the several sorts of fruit-trees, than the taking their buds or grafts from the true kinds of fruit-trees; but there is still more care required to have sound healthy trees, especially in this of Peach and Nectarines; for if the buds are taken from young plants in the nursery, which have not produced fruit, the shoots of which are generally very strong and vigorous, these buds will have so vicious a habit, as rarely to be corrected and brought into good order; for they will shoot more like the Willow than the Peach, the joints being extended to a great distance from each other, the shoots very gross, and the wood pithy; therefore, where the practice of taking the buds from nursery-trees is long continued, there can be little hopes of the trees so raised. I would therefore recommend it to every curious person, to procure their buds from such trees as have been long growing, whose fruit are well-flavoured, and the trees perfectly sound; as also never to make choice of the strongest or most luxuriant shoots of these trees, but such shoots as are well-conditioned, and whose buds grow pretty close together. For although these do not make so strong shoots the following year, as those which are taken from luxuriant branches, yet they will be better disposed to bear fruit, and will make much better trees.

The cuttings with which you are thus to be provided, should always be taken from the trees either in a morning or evening, or else in a cloudy day; for if they are cut off when the sun is very hot, the shoots will perspire so freely, as to leave the buds destitute of moisture, which is often the cause of their miscarrying; and the sooner they are used, when cut from the trees, the better they will take. The manner of this operation being explained under the article Inocu-

Inoculation, I ſhall not repeat in this place. The management of theſe trees, during their remaining time in the nurſery, is likewiſe fully ſet down under that article. I ſhall therefore proceed to give ſome directions for the choice of theſe trees, when they are to be procured from a nurſery. The firſt care ſhould be to find out a perſon of character to deal with, on whoſe integrity you may depend, for having the trees of thoſe kinds which you propoſe, and either ſee them taken up, or let ſome perſon you can confide in do it for you, becauſe as moſt of the nurſery-men have dealings with each other, if the perſon applied to has not the ſort of fruit deſired in his own nurſery, he procures them from another; and, if the gardener from whom he gets them, is not as honeſt and careful as himſelf, it is a great chance if the trees prove to be of the right kinds.

The trees ſhould alſo be choſen in the autumn, before others have drawn out the beſt; for thoſe who go firſt to the nurſeries, if they have ſkill, will always draw the fineſt plants. In the choice of the trees, you ſhould obſerve the ſtocks upon which they have been budded, that they are of the right ſort, whether Plumb or Apricot; that they are ſound and young; not ſuch as had been budded the preceding year and failed, nor thoſe which have been cut down. If the ſize of the ſtock is near that of a man's finger, it will be better than if they are larger; theſe ſhould be clear of moſs or canker. The buds ſhould be of one year's growth only, and not ſuch as have been cut down in the ſpring, and made a ſecond ſhoot; nor ſhould thoſe trees be choſen, whoſe ſhoots are very ſtrong and luxuriant, but ſuch as have clean ſhoots of a very moderate ſize, whoſe joints are not too far aſunder; and thoſe trees which ſtand in the outſide rows, or near the ends of the rows, where they have moſt air, are generally ſuch; for, where they ſtand cloſe in the nurſery, their ſhoots are drawn up in length; their joints are much farther aſunder, and their buds or eyes flat; for which reaſon, I have before adviſed the planting of the ſtocks at a greater diſtance than the nurſery-men generally allow them; and if a careful diſcreet nurſery-man would be at the trouble and expence in the raiſing of his Peach-trees, according to this method, he would better deſerve three ſhillings per tree, than one in the manner they are uſually raiſed; for every perſon who is at the expence of building walls for fruit, ſhould not think of ſaving a few ſhillings in the purchaſe of their trees; becauſe, if they are bad, or not of the right kinds, there is a great loſs of time and expence to no purpoſe, and the diſappointment will be ſo great, after waiting three or four years, as to diſcourage many from making farther trials, thinking themſelves liable to the ſame ill ſucceſs.

When the trees are choſen in the nurſery, the next care muſt be to have them carefully taken up out of the ground, ſo as not to break or tear the roots, nor injure their bark; for, as theſe trees are very apt to gum in thoſe places where they are wounded, there cannot be too much care taken of this. If the trees are to be tranſported to a diſtant place, their roots ſhould be cloſely wrapped either with Haybands, Straw, or Peaſe-haulm, and mats ſewed over theſe, to prevent the air from drying of their roots and branches. If the leaves of the trees are not fallen when they are taken up, they ſhould be carefully ſtripped off, before the trees are packed up; for, when there are many of theſe left, they are very apt to heat, if they are long in their paſſage, and often occaſion a mouldineſs very hurtful to the branches

We come next to the preparing of the ground to receive the trees. The beſt earth for Peach-trees is ſuch as is taken from a paſture ground, that is neither too ſtiff and moiſt, nor over dry, but of a middling nature, ſuch as is termed Hazel loam. This ſhould be dug from the ſurface of the ground about ten inches deep, taking the turf with it, and ſhould be laid in heaps eight or ten months at leaſt; but that which is prepared one year or more is ſtill better before it is uſed, that it may have the winter's froſt, and ſummer's heat to mellow it, during which time it ſhould be often turned to rot the turf, and break the clods, whereby it will be rendered very light, and eaſy to work, and about the beginning of September you ſhould carry it into the garden, and make the borders, which muſt be raiſed in height, proportionable to the moiſture of the garden; if the ground is very wet, it will be adviſeable to lay ſome rubbiſh in the bottom of the border, to drain off the moiſture, and to prevent the roots of the trees from running downward; and in this caſe it will be proper to make ſome under-drains at the bottom of the border, to convey off the ſuperfluous moiſture, which, if detained about the roots of the trees, will greatly prejudice them; then raiſe the border of earth at leaſt a foot, or in very wet land two feet, above the level of the ground, ſo that the roots of the trees may always remain dry; but if the ground is pretty dry, the borders ſhould not be raiſed above ſix or eight inches higher than the ſurface, which will be ſufficient to allow for their ſinking.

As to the breadth of theſe borders, that cannot be too great; but they ſhould never be leſs than eight feet broad, where fruit-trees are planted; for when the borders are made narrow, the roots of the trees will be ſo confined in four or five years time, that they will ſeldom thrive well after. The depth of theſe borders ſhould not be greater than two feet and a half; for where they are prepared to a great depth, it only entices the roots of the trees downward, which may be the cauſe of their future barrenneſs, for their roots, being got down below the influences of the ſun and ſhowers, imbibe a great quantity of crude juices, which only add to the luxuriant growth of the trees; beſides, whatever fruit are produced from ſuch trees, are not near ſo well taſted as are thoſe which grow upon thoſe trees, whoſe roots lie near the ſurface, and enjoy the kindly benefit of the ſun's heat, to correct and digeſt whatever crudities there may be in the earth.

Where the natural ſoil of the garden is ſhallow, and either chalk, clay, or gravel lies near the ſurface, theſe ſhould not be dug out like pits to receive the earth for the border, as is by ſome practiſed, for this will be no better than planting the trees in tubs or caſes, for their roots will be confined to theſe pits; ſo that when they are extended to the ſides, and can get no farther, the trees will blight and decay; and if it is clay on the ſides, the wet will be detained, as in a baſon, and the earth of the border will be like mud in very wet ſeaſons, ſo unfit for the roots of theſe trees. Therefore, whenever it ſo happens that the ground is of either of the ſorts before-mentioned, it will be the beſt way to raiſe the borders of a proper thickneſs of good earth over theſe, rather than to ſink down into them; for when the roots of the trees lie near the ſurface of the ground, they will extend to a great diſtance in ſearch of nouriſhment; but if they get below the ſtaple of the land, they can find nothing but ſour crude paſture very unfit for vegetation.

Your borders being thus prepared, ſhould lie about three weeks or a month to ſettle, by which time the ſeaſon for planting will be come, which ſhould be performed as ſoon as the leaves begin to decay, that the trees may put out new roots, before the froſt comes on to prevent them. Your ground being ready, and the trees brought carefully to the place, the next work is to prepare them for planting, which is to be formed in the following manner: You muſt ſhorten all the roots, and cut off ſmooth any broken or bruiſed roots, as alſo all the ſmall fibres ſhould be taken off, for the reaſons before given; and where any of the roots croſs each

each other, the worst of them must be cut out, that they may not injure the other.

Having thus prepared your trees, you should measure out their distance, which ought never to be less than twelve feet; but where the ground is very good, they should be planted fourteen feet asunder. This, I doubt not, will be thought too great a distance by many persons, especially since it is contrary to the general practice at this time; but I am satisfied, whoever shall try the experiment, will find it no more than is sufficient for these, where they are rightly managed; for if they take kindly to the soil, their branches may be so trained, as to furnish all the lower part of the wall in a few years, which is what should be principally regarded, and not, as is too often the practice, run up the shoots in height, and leave all the lower part of the tree destitute of bearing wood, so that in a few years there will not be any fruit upon the lower part of the trees; which also must be the case where they are planted too close, because there being no room to extend the branches on either side, they are obliged to lead them upright, which produces the before-mentioned ill effect.

There may be also some persons, who may think this distance too small for these trees, because Plumbs, Cherries, and most other sorts of fruit-trees require much more room; but when it is considered that Peach and Nectarine-trees produce their fruit only upon the former year's wood, and not upon spurs, as Cherries, Plumbs, and Pears do, therefore the shoots of these trees must be annually shortened in every part of them, to obtain bearing wood, whereby the trees may be kept in much less compass than those of other sort of fruit, and every part of the wall may be constantly supplied with bearing branches; for when the trees are planted at a great distance, the branches are often extended to such lengths, as to leave the middle of the trees naked, for there are never any good shoots produced from the old branches of these trees.

In the disposition of the trees, it will not be amiss to plant those sorts of Peaches near each other, which ripen about the same time, for by so doing the fruit may be better guarded from men and insects, and this will save a great deal of trouble in gathering of the fruit; for if a person is obliged to go from one part of the garden to another, or perhaps to look over all the walls of the garden every time the fruit is gathered, it is a great loss of time, which may be avoided by this first care in planting of the trees.

But to return to planting: after you have marked out the places where each tree is to stand, you must with your spade make a hole wide enough to receive the roots of the tree; then you should place it down, observing to turn the bud outwards, that the wounded part of the stock may be hid, and let the stem of the tree be placed about four or five inches from the wall, with its head inclining thereto; then fill in the earth with your hands, observing to break the clods, that the earth may fall in between the roots, so as no void spaces may be left about them. You should also gently shake the tree with your hands, to settle the earth down the better between the roots; then with your foot gently press down the earth about the stem, but do not tread it down too hard, which is many times a very great fault; for when the ground is inclinable to bind, the treading of it close doth often render the ground so hard, as that the tender fibres of the roots cannot strike into it, whereby the tree remains at a stand for some time, and if the earth be not loosened in time, it frequently dies; so that whenever you observe the earth of your borders to be bound, either by great rains, or from any other cause, you should with a fork loosen it again, observing always to do it in dry weather, if in winter or spring; but in summer it should be done in a moist season.

Although I have here given directions for the choice of trees from the nursery, after the usual method of planting these trees, which is that of taking such as have made one year's shoot, yet I would prefer those which were budded the preceding summer, and have made no shoot; for if the bud is sound and plump, and the bark of the stock well closed where the bud is inserted, there will be no danger of its growing; and when the bud has made a shoot the following spring the length of five or six inches, if it is stopped by pinching off the top, it will put out lateral branches, which may be trained to the wall, and this will prevent any cutting off the head, as must be done to those trees which have had one year's growth in the nursery; for these trees do not care for those large amputations, especially some of the more tender sorts; and by this method of planting these trees in bud no time will be lost, when it is considered that the trees which have shot must be cut down, and there is a hazard of their shooting again; therefore I am convinced from experience, that it is the best method.

After you have thus planted your trees which have made their shoots in the nursery, you should fasten their heads to the wall, to prevent their being shaken by the wind, which would disturb their roots, and break off the tender fibres soon after they were produced, to the no small prejudice of the trees; you should also lay some mulch upon the surface of the ground about their roots, before the frost sets in, to prevent it from penetrating the ground, which would injure, if not destroy the whole fibres; but this mulch should not be laid upon the ground too early, lest it prevent the autumnal rains from penetrating to the roots.

These things being duly observed, they will require no farther care till the beginning or middle of March, according as the season is early or late; you must cut off the heads of the new planted trees, leaving only four or five eyes above the bud; in doing of which, you must be very careful not to disturb their roots; to prevent which, you should place your foot down close to the stem of the tree, and take fast hold of that part of the stock below the bud with one hand, to hold it steady, while with the other hand you gently slope off the head of the tree with a sharp knife at the intended place, which should always be just above an eye; this should be done in dry weather, for if there should be much rain soon after, there will be some danger that the wet will enter the wounded part, and damage the tree; nor should it be done in frosty weather, for the same reason, for that would enter the wounded part, and prevent its healing over. After you have headed the trees, you should gently loosen the earth of the borders, for the fibres of the roots to strike out; but you must be very careful in doing of this, not to injure their new roots, which would damage them; and if the mulch which was laid about their roots in autumn be rotten, you may dig it into the border at some distance from the roots of the trees; and when the dry weather comes on, you should pare off some turf from a pasture ground, which should be laid upon the surface of the border about the roots of the trees, turning the Grass downward, which will preserve a gentle moisture in the earth, better than any other sort of mulch; and this will not harbour insects, as most sorts of dung and litter do, to the no small detriment of the trees.

These trees which are planted in bud, and have not made any shoots, should have their stocks cut down at this season just above the bud, for the buds will rarely shoot unless this is performed; and the nearer they are cut to the bud, the sooner will the head of the stock be covered by the buds; for although it may be necessary to leave a part of the stock above the bud, in those trees which are in the nursery, to which the shoots made by the buds may be fastened,

fastened, to prevent their being broken by the wind, yet as these are planted against the wall, to which the shoots may be fastened, there will be no want of any part of the stock.

In watering of these new planted trees, which should not be done unless the spring proves very dry, you should observe to do it with a nossel on the watering-pot, so as to let it out in drops; for when it is hastily poured down, it causes the ground to bind; and if you water over the head of the tree, it will be of great service to it. Your waterings should not be repeated too often, nor should they be given in great quantity, both which are very injurious to new planted trees.

In the middle of May, when these trees will have put out several shoots six or eight inches in length, you should nail them to the wall; observing to train them horizontally, rubbing off all foreright shoots, or such as are weak, whereby those which are preserved will be much stronger; but if there are not more than two shoots produced, and those very strong, you should at the same time nip off their tops, which will cause each of them to push out two or more shoots, whereby the wall will be better supplied with branches; you must also continue to refresh them with water in dry weather during the whole season, otherwise they will be apt to suffer; for their roots having but little hold of the ground the first year after transplanting, if the season should prove very dry it will greatly retard their growth, if due care be not taken to water them.

In the beginning of October, when you observe the trees have done shooting, you should prune them; in doing of which, you must shorten the branches in proportion to the strength of the tree; which, if strong, may be left eight inches long, but if weak, should be shortened to four or five; then you should train them horizontally to the wall (as was before directed,) so that the middle of the trees may be void of branches, for that part of the tree will be easily furnished with wood afterwards; whereas, if the shoots are trained perpendicularly to the wall, those which are the strongest will draw the greatest share of the sap from the roots, and mount upwards; so that the side branches will be deprived of their nourishment and grow weaker, until they many times decay; and this is the reason that we see so many Peach-trees with one or two upright shoots in the middle, and the two sides wholly unfurnished with branches, whereby the middle of each tree cannot produce any fruit, that being filled with large wood, which never produces any bearing shoots. Nor can the two sides of the trees be regularly filled with fruitful branches, when this defect happens to them; therefore this method should be carefully observed in the training up young trees, for when they are permitted to run into disorder at first, it will be impossible to reduce them into a regular healthful state afterwards, the wood of these trees being too soft and pithy to admit of being cut down (as may be practised on many other hardy fruit-trees, which will shoot out vigorously again;) whereas these will gum at the places where they are wounded, and in a few years entirely decay.

The spring following, when the trees begin to shoot, you should carefully look over them, to rub off all foreright shoots, or such as are ill placed, and train those which are designed to remain horizontally to the wall, in their due order as they are produced, for this is the principal season when you can best order the trees as you would have them; whereas, if they are neglected until Midsummer, as is the common practice, a great part of the nourishment will be exhausted by fore-right shoots, and other useless branches, which must afterwards be cut off; and hereby the remaining shoots will be rendered very weak, and perhaps some part of the wall be entirely unfurnished with branches; which might have been easily supplied in the beginning of May, by stopping some of the stronger shoots in such parts of the tree where there is a necessity for more branches; which would cause each of them to shoot out two or more side branches below the ends of the shoots, which may be guided into the vacant parts of the tree as they are produced, so as that every part may be regularly furnished with proper wood, which is the greatest beauty and excellency of wall trees; but you should always forbear stopping the shoots in summer, where there is not a necessity for branches to fill the wall; for there cannot be a greater fault committed, than that of multiplying the number of shoots, so as to cause a confusion, whereby the branches will be too weak to produce good fruit; besides, when they are too close laid in against the wall, the air is excluded from the shoots by the great number of leaves, so that they are never duly ripened; and consequently, what fruit is produced thereon, cannot be so well tasted as those which are produced upon such trees, where the shoots receive all the advantages of sun and air to bring them to maturity.

In the pruning of Peach and Nectarine-trees (which require the same management) the two following rules should be strictly observed, viz. First, That every part of the tree be equally furnished with bearing wood; and secondly, That the branches are not laid in too close to each other, for the reasons before laid down (with some others, which will be hereafter inserted.) As to the first, it must be observed, That all these trees produce their fruit upon the young wood, either of the preceding year, or at most, the two years shoots, after which age they do not bear; therefore the branches should be shortened so, as to cause them to produce new shoots annually in every part of the tree; which cannot be done in the ordinary method of pruning, where persons neglect their trees at the season when they are most capable of management, which is in April, May, and June; at which time the luxuriant growth of branches may be checked by pinching, and new shoots produced where they are wanting, by stopping the neighbouring branches; which shoots, being produced at that season, will have time enough to ripen and gain strength before the autumn comes on; whereas all those shoots which are produced after the middle of June, will be crude and pithy; and though they may sometimes produce a few blossoms, yet those rarely bring fruit. Therefore those persons who only regard their wall-trees at two different seasons, viz. the winter and Midsummer pruning, cannot possibly have them in good order; for when all the branches which were produced in the spring, are permitted to remain until the middle or latter end of June (as is the common practice,) some of the most vigorous will draw the greatest part of the nourishment from the weaker branches, which, when the strong ones are taken off, will be too weak to produce fair fruit; and hereby the strength of the tree is exhausted, to nourish the useless branches which are annually cut off again; and thus are too many trees managed, and at the same time complaints made of their luxuriancy; because two or three shoots, by drawing the greatest share of the nourishment, grow very strong and woody (whereas, if the nourishment had been equally distributed to a regular quantity of branches, there would be no sign of their too great strength,) but by often cutting off these vigorous branches, the trees are entirely destroyed, or at least rendered so weak as not to be able to produce fruit. It is therefore of the greatest consequence to wall-trees, especially of these sorts, to go over them two or three times in the months of April and May, to rub off all irregular shoots, and to train in the branches that are left in due order to the wall, that each shoot may have an equal advantage of sun and air, both of which

which are absolutely necessary to ripen and prepare the wood for the next year's bearing; therefore the oftener the trees are looked over to divest them of these useless branches, from the time they first begin to shoot in the spring till the autumn, the better will the wood be ripened for the succeeding year.

And by duly observing this in summer, there will not be occasion for so much cutting, as is often practised on Peach-trees, to their great injury; for their wood branches are generally soft, tender, and pithy, which, when greatly wounded, are not healed over again so soon as many other sorts of trees.

The distance which the branches of these trees should be allowed against the wall, must be proportioned to the size of the fruit or the length of the leaves; for if we observe how the branches of trees are naturally disposed to grow, we shall always find them placed at a greater or less distance, as their leaves are larger or smaller. And there is no surer guide to a curious artist than nature, from whence a gardener should always be directed in every part of his profession, since his business is to aid and assist nature, where she is not capable of bringing her productions to maturity; or where there is room, to make considerable improvements by art; which cannot be any otherwise effected, than by gently assisting her in her own way.

But to return to pruning of these trees: The branches being carefully trained in, as before directed, in the spring and summer seasons, we now come to treat of the winter pruning, which is commonly performed in February or March. But the best season for this work is in October, when their leaves begin to fall, which will be early enough for their wounds to heal before the frost comes on, so that there will be no danger of their being hurt hereby; and the branches of the trees being proportioned to the strength of the roots at that season, all the ascending sap in the spring will be employed to nourish only those useful parts of the branches which are left; whereas, if they are left unpruned till February, the sap in the branches being then in motion, as may be observed by the swelling of the buds, the greatest part of it will be drawn up to the extreme parts of the branches, to nourish such blossoms as must be afterwards cut off; and this may be easily known, by observing the strongest shoots at that season, when you will find the extreme buds to swell faster than most of the lower ones; for there being no leaves then upon the branches to detain the sap to nourish the lower buds, the upper ones will always draw from those below.

But suppose it were no advantage to the trees to prune them at this season (which I think no one will have reason to doubt after making the trial,) but that it only succeeds as well as the spring pruning; yet there is a great advantage in doing of it in autumn, for that being a much more leisure season with gardeners than the spring, they will have more time to perform it carefully; and then they will not have too many things come together, which may require to be immediately executed; for the spring being the principal season for cropping their kitchen-gardens and attending their hot-beds, if they are disengaged from the business of pruning at that time, it will be of great advantage, especially where there is a great quantity of walling. And here is also another benefit in pruning at this season, which is, the having the borders at liberty to dig and make clean before the spring, so that the garden may not appear in a litter at that season.

In pruning of these trees, you should always observe to cut them behind a wood bud, which may be easily distinguished from the blossom buds, which are shorter, rounder, and more turgid than the wood buds; for if the shoot have not a leading bud where it is cut, it is very apt to die down to the next leading bud; so that what fruit may be produced above that, will come to nothing, there being always a necessity of a leading bud to attract the nourishment; for it is not sufficient that they have a leaf bud, as some have imagined, since that will attract a small quantity of nourishment, the great use of the leaves being to perspire away such crude juices as are unfit to enter the fruit. The length you should leave these branches should be proportioned to the strength of the tree, which, in a healthy strong tree, may be left ten inches or more, but in a weak one, they should not be more than six inches; however, in this you must be guided by the position of a leading bud, for it is better to leave a shoot three or four inches longer, or cut it two or three inches shorter than we would chuse to do, provided there be one of these buds, it being absolutely necessary for the future welfare of the branches; you should also cut out entirely all weak shoots, though they may have many blossom buds upon them; for these have not strength enough to nourish the fruit, so as to give it a kindly flavour, but they will weaken the other parts of the tree.

In nailing the shoots to the wall, you must be careful to place them at as equal distances as possible, that their leaves, when come out, may have room to grow, without shading the branches too much; nor should you ever nail them upright if it can be avoided; for when they are thus trained, they are very subject to shoot from the uppermost eyes, and the lower part of the shoots will thereby become naked.

There is not any thing in the business of gardening which has more exercised the thoughts of the curious, than how to preserve their tender sorts of fruit from being blighted in the spring of the year, and yet there has been little written upon this subject which is worth notice: some have proposed mattresses of straw or Reeds to be placed before the fruit-trees against walls, to prevent their being blasted; others have directed the fixing horizontal shelters in their walls, to prevent the perpendicular dew or rain from falling upon the blossoms of the fruit-trees, which they suppose to be the chief cause of their blighting; but both these contrivances have been far from answering the expectations of those persons who have put them in practice, as I have elsewhere shewn; therefore it may not be improper to repeat some things in this place, which I have before mentioned, in relation to this matter. And,

First, I have already said, That the blights which are so often complained of, do not always proceed from any external cause or inclemency in the season, but from a distemper or weakness in the trees; for if we observe the trees at that season, where they are most subject to what is called a blight, we shall find the branches very small, weak, and not half ripened, as also trained in very close to each other; these branches are, for the most part, full of blossom buds (which is chiefly occasioned by their want of strength.) These buds do indeed open, and to persons not skilled in fruit-trees, shew a great prospect of a plentiful crop of fruit; whereas the whole strength of the branches is spent in nourishing the flowers, and being unable to do more, the blossoms fall off, and the small efforts of the leaf buds are checked, so that many times the greatest part of the branches die away, and this is called a great blight; whereas at the same time it may be often observed, that some trees of a different sort, nay, even some of the same sort, which were stronger, though placed in the same soil, exposed to the same aspect, and subject to the same inclemency of air, have escaped very well, when the weak trees have appeared to be almost dead; which is a plain indication, that it proceeds from some cause within the tree, and not from any external blight. All this will therefore be remedied, by observing the foregoing directi-

 ons

ons in the pruning and management of the trees, so as never to over-burden them with branches, nor to suffer any part of the trees to exhaust the whole nourishment from the root, which will cause the other parts to be very weak; but to distribute the nourishment equally to every shoot, that there may be none too vigorous, at the same time that others are too weak; and by continually rubbing off useless or fore-right shoots as they are produced, the strength of the trees will not be spent, to nourish such branches as must be afterwards cut out, which is too often seen in the management of these trees. And,

Secondly, It sometimes happens, that the roots of these trees are buried too deep in the ground, which, in a cold or moist soil, is one of the greatest disadvantages that can attend these tender fruits; for the sap which is contained in the branches, being by the warmth of the sun put strongly into motion early in the spring, is exhausted in nourishing the blossoms; and a part of it is perspired through the wood branches, so that its strength is lost before the warmth can reach to their roots, to put them into an equal motion in search of fresh nourishment, to supply the expence of the branches; for want of which, the blossoms fall off and decay, and the shoots seem to be at a stand, until the farther advance of the warmth penetrates to the roots, and sets them in motion; when suddenly after, the trees, which before looked weak and decaying, made prodigious progress in their shoots; and before the summer is spent, are furnished with much stronger branches than those trees which have the full advantage of sun and showers, and that are more fruitful and healthy. If therefore this be the case, there is no way of helping this, but by raising up the trees, if they are young; or if they are too old to remove, it is the better way to root them out, and make new borders of fresh earth, and plant down young trees; for it is a great vexation to be at the trouble and expence of pruning and managing these trees, without having the pleasure of reaping any advantage from them, which will always be the case where the trees are thus injudiciously planted. Or,

Thirdly, This may proceed from the trees wanting nourishment, which is many times the case, where they are planted in a hard gravelly soil, in which it is the common practice to dig borders three or four feet wide, and three feet deep into the rock of gravel, which is filled with good fresh earth, into which the trees are planted, where they will thrive pretty well for two years, until their roots reach the gravel, where they are confined, as if planted in a pot; and, for want of proper nourishment, the branches continually decay every year. This cannot be helped, where the trees have been growing some years, without taking them entirely up, or by digging away the gravel from their roots, and adding a large quantity of fresh earth, that may afford them a supply of nourishment a few years longer; but trees so planted, cannot by any art be continued long in health.

But if the unfruitfulness of the trees does not proceed from any of the before-mentioned causes, and is the effect of unkindly seasons, then the best method yet known is, in dry weather, when little dew falls, to sprinkle the branches of the trees gently with water soon after the blossoming season, and while the young-set fruit is tender; which should always be done before noon, that the moisture may evaporate before the night comes on; and if in the night you carefully cover the trees with mats, canvas, or some such light covering, it will be of great service to them: however, where the trees are strong and vigorous, they are not so liable to suffer by a small inclemency, as are those which are weak; so that there will be few seasons in which there may not be hopes of a moderate quantity of fruit from them, though there should be no covering used; for where these coverings are used, if it is not performed with great care and diligence, it is much better to have no covering, but trust to the clemency of the season; for if the coverings are kept too close, or continued too long, the trees will receive more injury hereby, than from being constantly exposed; or if after they have been covered for some time, and then incautiously removed, so as to expose the trees too suddenly to the open air, they will suffer more thereby, than if they had not been covered. However, I must repeat in this place what has been before mentioned under another article, of a management which has been generally attended with success, which is, the putting up two feather-edge deal boards joined together, over the top of the trees, so as to form a penthouse to cast off perpendicular wet. These should be fixed up when the trees begin to blossom, and should remain till the fruit is well set, when they should be taken down, to admit the dew and rain to the leaves and branches of the trees, which must not be longer kept off; and where the wall is long, and exposed to currents of wind, if at the distance of forty feet from each other are fixed some cross Reed hedges, to project about ten feet from the wall, these will break the force of the wind, and prevent its destroying of the blossoms; and these may be removed away, as soon as the danger is over. Where these things have been practised, they were generally attended with success; and as there will be no trouble of covering and uncovering in this method, after they are fixed up, there can be no danger of neglect, as very often is the case when the trouble is great, or to be often repeated.

When your fruit is set, and grown to the bigness of a small nut, you should go over the trees and thin them, leaving them at least five or six inches asunder; for when they are permitted to remain in bunches, as they are often produced, the nourishment, which should be employed wholly to the fruits designed to stand, will be equally spent amongst the whole number, a great part of which must be afterwards pulled off; so that the sooner this is done, the better it will be for the remaining fruit; and if it should sometimes happen, that a part of those left, by any accident should be destroyed, yet the remaining ones will be much the larger and the better tasted for it, and the trees will gain more strength, for a moderate quantity of fruit is always preferable to a great crop; the fruit, when but few, will be much larger, better tasted, and the trees in a condition to bear well the succeeding year; whereas, when they are overcharged with fruit, it is always small, ill tasted, and the trees are generally so much weakened thereby, as not to be in a condition for bearing well for two or three years after; so that upon the whole, it is much better to have a smaller number of fruit than is commonly esteemed a crop, than to have too many, since the fruit and also the trees are benefited thereby. The quantity of fruit to be left on large full grown trees, should never be greater than five dozen upon each; but on middling trees, three or four dozen will be enough.

If the season should prove hot and dry, it will be proper to draw up the earth round the stem of each tree, to form a hollow bason of about six feet diameter, and cover the surface of the ground in this bason with mulch; and once in a week or a fortnight, according to the heat and drought of the season, pour down eight or ten gallons of water to the root of each tree; or where there is an engine, which will disperse the water in gentle easy drops like rain, if the same, or a larger quantity of water, is sprinkled all over the branches of the trees, and this, soaking down to the roots, will keep the fruit constantly growing, which will prevent their falling off the trees, as they generally do where this method is not practised; and the fruit being thus constantly

ſtantly nouriſhed, will be much better taſted, and hereby the trees will be maintained in vigour; ſo that it is what I can from long experience recommend, as one of the moſt neceſſary things to be practiſed by all lovers of good fruit. But this ſhould not be continued longer than while the fruit are growing, for afterward it will be hurtful to the trees and fruit, for a dry autumn ripens both wood and fruit better than in a moiſt latter ſeaſon.

It is a common opinion which has for ſome years prevailed, even among perſons of good underſtanding, that Peach-trees are not long lived, therefore ſhould be renewed every twenty years; which is a great miſtake, for I have eaten ſome of the fineſt Peaches of various kinds, from trees which have been planted above fifty years: and I am convinced by experience, that when the trees are budded upon proper ſtocks, carefully planted and managed, they may be continued fruitful and healthy ſixty years and upward; and the fruit produced on theſe old trees, will be much better flavoured than any of thoſe upon young ones; but I ſuppoſe the foundation of the above opinion was taken from the French, who generally bud their Peaches upon Almond ſtocks, which are of ſhort duration, theſe ſeldom laſting good more than twenty years; but this being ſeldom practiſed in England, the caſe is widely different; nor indeed ſhould we fetch our examples from that nation, where the profeſſors of the art of gardening are at leaſt a century behind the Engliſh; and from their preſent diſpoſition, ſeem unlikely to overtake them; for they depart from nature in almoſt every part of gardening, and are more pleaſed with introducing their little inventions of pruning and managing their fruit-trees, according to their own fancy, than they are careful to draw their inſtructions from nature, from whence the true art is to be obtained; ſo that in a very few inſtances gardeners ſhould deviate from nature, unleſs it be in thoſe particulars where art may be practiſed to the greateſt advantage, which is in the procuring many ſorts of eſculent plants and fruits, earlier and better flavoured than can be obtained without, in which the French are extremely deficient; and herein they truſt too much to nature, and uſe too little art.

I muſt recommend the dunging of the borders for fruit-trees every other year, with this caution, always to uſe ſuch dung for their borders, as is well rotted, and to dig it into the borders in November, that the rain may waſh down the ſalts before the ſpring comes on; and where the ground is very looſe or ſandy, it will be the beſt way to make uſe of neats dung, which is cooler than that of horſes, but for cold ſtrong land the latter is to be preferred.

If the ground is well trenched every year about the roots, it will be of great ſervice to them; and where the ſoil is ſubject to bind very cloſe, if it is forked two or three times in a year to looſen the ſurface, it will greatly help the trees. The borders ſhould not be crouded with any large growing plants, which will draw away the nouriſhment from the trees; therefore when any ſort of kitchen herbs are planted on theſe borders, they ſhould be only ſuch as are of ſmall growth, and which may be taken off early in the ſpring; and if this is carefully obſerved, the cultivating ſmall things on theſe borders can do no harm, becauſe the ground will be ſtirred the oftener, on account of theſe ſmall crops, than perhaps it would have been, when no uſe was to be made of the borders. Theſe rules which are here laid down, if properly obſerved, will direct any curious perſon how to have plenty of good fruit, as alſo to preſerve the trees in vigour a great number of years.

PERVINCA. See VINCA.

PETASITES. See TUSSILAGO.

PETIVERIA. Plum. Nov. Gen. 50. tab. 39. Guinea Henweed.

The CHARACTERS are,

The flower hath a permanent empalement. It hath four ſmall white petals placed in form of a croſs, which ſoon fall off, and ſix awl-ſhaped erect ſtamina, terminated by ſingle ſummits. In the center is ſituated an oblong compreſſed germen, with four awl-ſhaped ſtyles. The germen afterward becomes one oblong ſeed, narrow at the bottom and taper, but broad and indented at the top, reſembling an inverted ſhield, armed with the acute ſtyle, which is reflexed.

The SPECIES are,

1. PETIVERIA (*Alliacea*) floribus hexandris. Hort. Cliff. 141. *Commonly called Guinea Henweed.*

2. PETIVERIA (*Octandra*) floribus octandris. Lin. Sp. Plant. 486. *Guinea Henweed with eight ſtamina.*

The firſt ſort is a very common plant in Jamaica, Barbadoes, and moſt of the other iſlands in the Weſt-Indies, where it grows in ſuch plenty, as to become a very troubleſome weed. The roots are ſtrong, and ſtrike deep in the ground; the ſtalks are jointed, and riſe from three to five feet high, garniſhed with oblong veined leaves, of a deep green, placed alternately upon ſhort foot-ſtalks. The flowers are produced in ſlender ſpikes at the end of the branches; they are very ſmall, ſo make no figure.

The ſecond ſort differs from the firſt, in having a ſhorter and narrower ſtalk, and the flowers having eight ſtamina; but unleſs theſe marks are noticed by a nice obſerver, they may paſs as one ſort.

In Europe, theſe plants are preſerved in the gardens of thoſe perſons who are curious in botany; but there is little beauty in them, and having a ſtrong rank ſcent upon being handled, renders them leſs valuable. They are propagated by ſeeds, which muſt be ſown on a hot-bed early in the ſpring, and when the plants are come up, they ſhould be each tranſplanted into a ſeparate pot, and plunged into a moderate hot-bed to bring them forward. When they have obtained a good ſhare of ſtrength, they ſhould be inured to bear the open air by degrees, into which they may be removed toward the latter end of June, placing them in a warm ſituation, where they may remain till autumn, when they ſhould be removed into the ſtove, and in winter muſt have a moderate degree of warmth, otherwiſe they will not live in this country.

PETREA. Houſt. Gen. Nov. Lin. Gen. Plant. 682.

The CHARACTERS are,

The flower hath an empalement of one leaf, cut into five obtuſe ſegments. It hath one petal, with a ſhort tube, cut above into five almoſt equal ſegments, which are expanded. It hath four ſhort ſtamina, ſituated in the tube, two of which are a little longer than the other, terminated by ſingle ſummits, and four germen ſupporting a ſhort ſtyle, crowned by an obtuſe ſtigma. The germen afterward become four ſeeds wrapped up in a fringed cover.

We have but one SPECIES of this genus, viz.

PETREA (*Volubilis*) fruteſcens foliis lanceolatis rigidis, flore racemoſo pendulo. *Shrubby Petrea with ſtiff ſpear-ſhaped leaves, and flowers growing in long hanging bunches.*

This plant was firſt diſcovered by the late Dr. Houſtoun, growing naturally at La Vera Cruz in New Spain. It riſes with a woody ſtalk to the height of fifteen or ſixteen feet, which is covered with a light gray bark, ſending out ſeveral long branches; theſe have a whiter bark than the ſtem, and are garniſhed with leaves at each joint, which on the lower part of the branches are placed by threes round them, but higher up they ſtand by pairs; they are ſtiff, and their ſurface rough; of a light green, having a ſtrong dark midrib, with ſeveral tranſverſe veins running from the midrib to the borders, which are entire. The flowers are produced at the end of the branches growing in looſe bunches, which are nine or ten inches long, each flower ſtanding upon on a slender foot-stalk about an inch long; the empalement of the flower is composed of five narrow obtuse leaves about an inch long, which are of a fine blue colour, so are much more conspicuous than the petals, which are white, and not more than half the length of the empalement. After the flower is past, the four germen in the center become so many oblong seeds wrapped up in a fringed cover.

This is propagated by seeds, which must be obtained from the places where the trees grow naturally, and these are very few good; for, from the seeds which the Doctor sent to England, there were but two plants raised, though the seeds were distributed to several persons. The seeds must be sown in a good hot-bed, and when the plants come up, they should be each planted in a separate small pot filled with light loamy earth, and plunged into a hot-bed of tanners bark, and afterwards placed in the bark-bed in the stove, where they should constantly remain, and be treated in the same way as the plants of the same country.

PETROSELINUM. See Apium.

PEUCEDANUM. Tourn. Inst. R. H. 318. tab. 169. Hogs-fennel, or Sulphur-wort.

The Characters are,

It hath an umbelliferous flower. The cover of the large umbel is composed of many linear reflexed leaves. The empalement of the flower is small and indented in five parts. The petals of the great umbel are uniform. Each flower is composed of five oblong incurved petals, which are entire; they have each five hair-like stamina, with an oblong germen situated under the flower, supporting two small styles, crowned by obtuse stigmas. The germen afterward turns to an oval fruit channelled on each side, splitting in two parts, containing two seeds, convex on one side, compressed on the other, with three raised furrows, and a broad membranaceous border indented at the top.

The Species are,

1. Peucedanum (*Officinale*) foliis tripartis filiformibus longioribus umbellis circinatis. *Hogs-fennel with leaves which are divided by threes, and these are again divided in three linear parts.*

2. Peucedanum (*Italicum*) foliis tripartitis filiformibus linearibus umbellis difformibus. *Hogs-fennel with leaves cut into three parts, which are long, slender, and have irregular umbels; Greater Italian Hogs-fennel.*

3. Peucedanum (*Alpestre*) foliolis linearibus ramosis. Hort. Cliff. 94. *Hogs-fennel with linear branching leaves.*

4. Peucedanum (*Minus*) foliis pinnatis foliolis pinnatifidis laciniis linearibus oppositis, caule ramosissimo patulo. Flor. Angl. 101. *Smaller Hogs-fennel.*

The first sort is said to grow naturally in England, but I have not been lucky enough to find it, though I have searched the places where it is mentioned to have been found; it grows in several parts of Germany, in marshy meadows. This hath a perennial root, from which arise the foot-stalks of the leaves, which are channelled; these are naked at the bottom, but about four or five inches from the root branches into three smaller foot-stalks, and these again divide into three, and each of these divisions sustain three narrow leaves, which, when bruised, emit a strong scent like sulphur. The stalks rise near two feet high; these are channelled, and divide into two or three branches, each being terminated by a large regular umbel of yellow flowers, composed of several small umbels, which are circular.

The second sort grows naturally on the mountains, and also in the low valleys by the sides of rivers in Italy. The root of this is perennial; the foot-stalks of the leaves are large and furrowed, dividing into three small branches, which are again divided into three, and these end with three long narrow lobes, or small leaves, which are much longer than those of the other sort. The stalks, which sustain the umbels, rise near three feet high, and divide toward the top into several small branches, each sustaining an umbel, composed of several smaller rays or umbels, which stand upon pretty long foot-stalks. The flowers of this are yellow, and shaped like those of the former, but are much larger, as are also the seeds, but have the same form as the other.

The third sort grows naturally in the forest of Fontainebleau, and many other parts of France. It hath a strong perennial root, from which come out leaves, which branch into three divisions, and these divide again into three smaller; each of these smaller divisions are garnished with five short narrow leaves. The stalks are strong, round, and not so deeply channelled as either of the former, sustaining a very large umbel of yellow flowers, shaped like those of the former sorts; the seeds are shorter, but of the same shape as those.

The fourth sort grows naturally on St. Vincent's rock near Bristol. This is a biennial plant, which perishes soon after it has perfected its seeds. The leaves of this sort are short and very narrow, spreading near the surface of the ground; the stalks rise near a foot high, but are branched almost from the bottom; these branches are almost horizontal, and are garnished with a few narrow short leaves, of a lucid green. Each stalk is terminated by a small umbel of flowers, which are of an herbaceous yellow colour, and small. These are succeeded by small channelled seeds.

These sorts are preserved in botanic gardens for the sake of variety; they are all propagated by seeds, which should be sown in the autumn soon after they are ripe; for those which are sown in the spring seldom succeed, or if the plants come up, it is rarely before the following spring: when the plants come up, they must be kept clean from weeds, and the autumn following they may be transplanted where they are to remain; they love a moist soil and a shady situation, but will not thrive under the drip of trees. The roots of the three first sorts will continue several years, and every year produce flowers and seeds. The fourth sort will rarely ripen seeds in a garden, so that I have been obliged to procure them from the place where it grows naturally.

PHACA. Lin. Gen. Plant. 798. Bastard Milk vetch.

The Characters are,

The flower hath a tubulous empalement, cut into five small indentures at the brim. It is of the butterfly kind, having a large, oval, erect standard, with two oblong wings, which are shorter, and a short compressed obtuse keel. It hath ten stamina, nine of which are joined in one body, and the other stands separate. In the center is situated an oblong germen, supporting an awl-shaped style, crowned by a single stigma. The germen afterward becomes an oblong swelling pod, whose upper suture is depressed toward the under, having one cell, containing several kidney-shaped seeds.

The Species are,

1. Phaca (*Bœtica*) caulescens pilosa, leguminibus tereti-cymbiformibus. Lin. Sp. Pl. 755. *Phaca with a hairy stalk, and taper boat-shaped pods.*

2. Phaca (*Alpina*) caulescens erecta glabra, leguminibus semi-ovatis. Lin. Sp. Pl. 755. *Phaca with an upright smooth stalk, and half oval pods.*

The first sort is a native of Portugal and Spain. This has been long preserved in some curious gardens in England, but the other is more rare at present.

The roots of the first sort, which grows naturally in Portugal, will abide many years, and run very deep into the ground; but the branches decay every autumn, and the roots produce fresh stalks every spring, which will rise near four feet high, and grow ligneous. The flowers are produced

produced in ſhort ſpikes from the wings of the leaves; but unleſs the ſeaſon proves very warm, they rarely flower in England, for which reaſon the plants are not much eſteemed.

The ſecond ſort, which is a native of Siberia, hath ſmooth ſtalks, which do not riſe ſo high as the former; the flowers are ſmaller, the pods are much ſhorter, and hang downwards.

Both theſe ſorts are propagated by ſeeds, which ſhould be ſown in the place where the plants are to remain; for as they ſhoot their roots very deep into the earth, ſo it is very difficult to tranſplant them with any ſafety, eſpecially after they have remained any conſiderable time in the ſeed-bed. The plants of the firſt ſort ſhould be left about ſix feet aſunder, that there may be room to dig the ground between them every ſpring, which is all the culture they require, except the keeping them clean from weeds.

PHALANGIUM. See ANTHERICUM.

PHASEOLOIDES. See GLYCINE.

PHASEOLUS. Tourn. Inſt. R. H. 412. tab. 232. Kidney-bean.

The CHARACTERS are,

The empalement of the flower is of one leaf. The flower is of the butterfly kind; it hath a heart-ſhaped, obtuſe, inclined ſtandard, reflexed on the ſides; the wings are oval, the length of the ſtandard, and a narrow ſpiral keel twiſted contrary to the ſun. It hath ten ſtamina, nine joined in one body, and the other ſtanding ſeparate, and an oblong, compreſſed, hairy germen, ſupporting a ſlender, inflexed, ſpiral ſtyle, crowned by an obtuſe hairy ſtigma. The germen afterward becomes a long pod with a thick ſhell, ending in an obtuſe point, incloſing oblong, compreſſed, kidney-ſhaped ſeeds.

It would be to little purpoſe to enumerate all the varieties of this plant, which have come to our knowledge in this place, ſince America does annually furniſh us with ſo many new ſorts, as that there is no knowing what varieties there may be produced in England; beſides, as they are not likely to be much cultivated here, ſince ſome of the old ſorts are preferable to any of the new ones for the uſe of the kitchen-garden, therefore I ſhall only mention a few ſorts, which are cultivated for their flowers, or as curioſities, and then mention thoſe which are moſt eſteemed for the table.

The SPECIES are,

1. PHASEOLUS (*Alatus*) caule volubili, floribus laxè ſpicatis, alis longitudine vexillo. Lin. Sp. Plant. 725. *Kidney-bean with a twining ſtalk, and flowers growing in looſe ſpikes, whoſe wings are as long as the ſtandard.*

2. PHASEOLUS (*Caracalla*) caule volubili, vexillis carinâque ſpiraliter convolutis. Lin. Sp. Plant. 725. *Kidney-bean with a twining ſtalk, whoſe ſtandard and keel are ſpirally twiſted; commonly called Caracalla in Portugal.*

3. PHASEOLUS (*Vexillatus*) caule volubili, vexillis revolutis patulis, leguminibus linearibus ſtrictis. Lin. Sp. Plant. 724. *Kidney-bean with a twining ſtalk, a ſpreading ſtandard which is twiſted backward, and narrow cloſe pods.*

4. PHASEOLUS (*Farinoſus*) caule volubili, pedunculis ſubcapitatis, ſeminibus tetragono-cylindricis pulverulentis. Hort. Upſal. 214. *Kidney-bean with a twining ſtalk, foot-ſtalks ending in flowers, growing in heads, and four-cornered, cylindrical, duſt-coloured ſeeds.*

5. PHASEOLUS (*Vulgaris*) caule volubili, floribus racemoſis geminis, bracteis calyce brevioribus, leguminibus pendulis. Lin. Sp. Plant. 724. *Kidney-bean with a twining ſtalk, branching flowers growing by pairs, bracteæ which are ſhorter than the empalement, and hanging pods; commonly called the Scarlet-bean.*

6. PHASEOLUS (*Coccineus*) caule volubili, floribus racemoſis, ſiliquis brevibus pubeſcentibus. *Kidney-bean with a twining ſtalk, flowers growing in long bunches, and ſhort hairy pods.*

The firſt ſort is an annual plant. The ſeeds of this were brought from Carolina, where it grows naturally. The ſtalks twine about any ſupport, like the common Kidney-bean; they are hairy, and riſe four or five feet high; the leaves are ſhaped like thoſe of the common Kidney-bean, but are narrower. The flowers are produced in looſe ſpikes, ſtanding upon long foot-ſtalks; they are large, and of a purple colour, turning to a blue before they fade.

The ſeeds of this ſort ſhould be ſown on a warm border about the latter end of April, and when the plants begin to run up, they muſt be ſupported either with ſticks, or faſtened to a hedge or wall, to prevent their trailing on the ground, and conſtantly kept clean from weeds. If they are cloſe to a wall or hedge, expoſed to a good aſpect, they will ripen their ſeeds in England, otherwiſe they frequently fail in bad ſeaſons.

The ſecond ſort grows naturally in the Brazils. This has a perennial root, with twining ſtalks which riſe to the height of twelve or fourteen feet; the leaves are ſhaped like thoſe of the common Kidney-bean, but are ſmaller; the flowers are produced in ſlender ſpikes; they are of a purple colour, and have an agreeable odour; theſe are ſucceeded by ſlender pods, which are compreſſed, containing ſeveral compreſſed ſeeds. This is propagated by ſeeds, which ſhould be ſown on a moderate hot-bed in the ſpring; and when the plants come up, they muſt be tranſplanted into pots, filled with light freſh earth, and plunged into a hot-bed to facilitate their taking root; after which they ſhould be inured to bear the open air by degrees, into which they ſhould be removed the end of June or beginning of July, placing them in a ſheltered ſituation; and as they advance in their growth, and fill the pots with their roots, they ſhould be removed into larger pots, which muſt be filled up with freſh light earth.

The third ſort grows naturally in America, and is preſerved in ſome curious gardens for variety, but is a plant of no great beauty. This may be propagated by ſowing the ſeeds in the ſpring upon a hot-bed, and when they come up, they muſt be planted in pots, and treated as the former ſort.

The fourth ſort was brought from America, and is preſerved in ſome curious gardens for the ſake of its long flowering. This is an abiding plant, and ſhould be managed as was directed for the third ſort, but this requires a ſtove to preſerve it through the winter in England.

The fifth ſort has been long cultivated in the Engliſh gardens for the beauty of its ſcarlet flowers. This hath twining ſtalks, which, if properly ſupported, will riſe to the height of twelve or fourteen feet; the leaves are ſmaller than thoſe of the common Kidney-bean. The flowers grow in large ſpikes, and are much larger than thoſe of the common Kidney-bean, and of a deep ſcarlet colour; the pods are large and rough, and the ſeeds are purple, marked with black. This ſort requires no other treatment than the common ſort, but the ſtalks ſhould have tall ſtakes put down by them to twine round, otherwiſe they will fall on the ground, which will ſoon cauſe them to rot.

Although this ſort is chiefly cultivated for the beauty of its flowers at preſent, yet I would recommend it as the beſt ſort for the table; and whoever will make trial of this, I dare ſay muſt prefer it to all the other kinds yet known.

The ſixth ſort grows naturally in the warmeſt part of America, ſo will not thrive in England out of a ſtove; and as the chief beauty of it is in the ſeeds, which are half ſcarlet, and the other half black, ſo theſe may be procured from abroad better than raiſed here.

I ſhall

I ſhall now mention thoſe ſorts of Kidney-beans, which are cultivated in the Engliſh gardens to ſupply the table, which are few in compariſon of the number already known; though theſe are not many of them valuable, and are only cultivated becauſe they require leſs care, or will come a little forwarder in the ſeaſon, for they are inferior in taſte to the others; however, as there are ſome perſons who eſteem them for their qualities before-mentioned, ſo I ſhall put them down in the order of their ripening for uſe.

The three ſorts which are uſually cultivated for early crops, are the ſmall white Dwarf, the Dwarf black, which is called the Negro-bean, and the Liver Colour-bean. The ſtalks of theſe are never very long, ſo may be planted much nearer together than the larger growing kinds, and they require but little ſupport; ſo theſe are planted on hot-beds under frames, or in pots, which are placed in ſtoves, to come early in the ſpring, for which purpoſe they are better adapted than any of the other; but they are not to be compared with ſome of the others for goodneſs, but as they may be had at a time when the others cannot be ſo well obtained, ſo they are generally cultivated in the gardens; and where there are not the convenience of ſtoves or frames for raiſing them very early, they are planted in warm borders near hedges, walls, or pales, where they will be fit for uſe a fortnight earlier than the other ſorts.

The next to theſe are the Batterſea and Canterbury Kidney-beans; theſe do not ramble far, and produce their flowers near the root, ſo bear plentifully for ſome time: the Batterſea Bean is the forwarder of the two, but the other will continue bearing much longer; they are both better flavoured than either of the three former ſorts, but when they begin to be large, are very ſtringy and tough.

There are two or three ſorts of Kidney-beans cultivated with erect ſtalks, which want no ſupport, as they do not put out any twining ſtalk. Theſe are much cultivated by the gardeners for that reaſon, as alſo for their producing a great plenty of pods; but they are inferior in goodneſs to all the other, eſpecially that ſort with black and white ſeeds, whoſe pods have a rank flavour, and, when boiled, become ſoft and meally, ſo this ſhould never be propagated by perſons of taſte.

The beſt ſorts for the table are the ſcarlet bloſſom Bean before-mentioned, and a white Bean of the ſame ſize and ſhape, which appears to be only a variety of the ſcarlet, as it differs in no other reſpect, but the colour of the flowers and ſeeds, being equal in ſize and flavour. And next to theſe is the large Dutch Kidney-bean, which grows as tall as either of theſe, ſo muſt be ſupported by ſtakes, otherwiſe their ſtalks will trail upon the ground and ſpoil. The ſort with ſcarlet flowers is preferable to this in goodneſs, and is alſo hardier; and although it will not come ſo early as ſome of the dwarf kinds, yet as it will continue bearing till the froſt puts a ſtop to it in the autumn, ſo it is much preferable to either of them; for the pods of this ſort, when old, are ſeldom ſtringy, and have a better flavour than the young pods of thoſe ſorts, and will boil greener; and where this is ſown in the ſame ſituation and ſoil as the Batterſea Bean, it will not be a fortnight later.

All the ſorts of Kidney-beans are propagated by ſeeds, which are too tender to be ſown in the open air before the middle of April; for if the weather ſhould be cold and wet after they are in the ground, they will ſoon rot; or if the morning froſt ſhould happen after the plants come up, they will be deſtroyed; therefore the beſt way to have early Kidney-beans, where there is no conveniency of frames for raiſing them, is to ſow the ſeeds in rows pretty cloſe, upon a moderate hot-bed, the latter end of March or the beginning of April. If the heat of the bed is ſufficient to bring up the plants, it will be enough; this bed ſhould be arched over with hoops, that it may be covered with mats every night, or in bad weather. In this bed the plants may ſtand till they have put out their trifoliate leaves, then they ſhould be carefully taken up, and tranſplanted in warm borders near hedges, pales, or walls. If the ſeaſon proves dry at the time of removing them, the plants ſhould be gently watered to forward their taking new root, and afterward they muſt be managed in the ſame way as thoſe which are ſown in the full ground. Theſe tranſplanted Beans will not grow ſo ſtrong as theſe which are not removed, nor will they continue ſo long in bearing, but they will come at leaſt a fortnight earlier than thoſe which are ſown in the full ground.

The firſt crop intended for the full ground, ſhould be put in about the middle of April; but theſe ſhould have a warm ſituation and a dry ſoil, otherwiſe the ſeeds will rot in the ground; or if the weather ſhould prove ſo favourable as to bring up the plants, yet there will be danger of their being killed by morning froſts, which frequently happen the beginning of May.

The ſecond ſort, which ſhould be one of the three large ſorts laſt mentioned, ſhould be ſown about the middle of May. Theſe will come into bearing before the early kinds are over, and if they are of the ſcarlet ſort, will continue fruitful till the froſt deſtroys the plants in the autumn, and theſe will be good as long as they laſt. The manner of planting them is, to draw ſhallow furrows with a hoe, at about four feet diſtance from each other, into which you ſhould drop the ſeeds about two inches aſunder; then with the head of a rake draw the earth over them, ſo as to cover them about an inch deep.

If the ſeaſon is favourable, the plants will begin to appear in about a week's time after ſowing, and ſoon after will raiſe their heads upright; therefore, when the ſtems are advanced above ground, you ſhould gently draw a little earth up to them, obſerving to do it when the ground is dry, which will preſerve them from being injured by ſharp winds; but you ſhould be careful not to draw any of the earth over their ſeed leaves. After this, they will require no farther care but to keep them clear from weeds, until they produce fruit, when they ſhould be carefully gathered two or three times a week; for if they are permitted to remain upon the plants a little too long, the Beans will be too large for eating, and the plants would be greatly weakened thereby.

The large ſorts of Kidney-beans muſt be planted at a greater diſtance row from row; for as theſe grow very tall, ſo if the rows are not at a farther diſtance, the ſun and air will be excluded from the middle rows, therefore theſe ſhould not be leſs than five feet diſtance row from row; and when the plants are about four inches high, the poles ſhould be thruſt into the ground by the ſide of the plants, to which they will faſten themſelves, and climb to the height of eight or ten feet, and bear plenty of fruit from the ground upward. The Dutch and French preſerve great quantities of the large Dutch Beans for winter uſe, which they ſtew and make good with gravy and other ſauces.

There are ſome perſons who raiſe theſe in hot-beds, in order to have them early. The only care to be taken in the management of theſe plants, when thus raiſed, is to allow them room, and give them as much air as can be conveniently, when the weather is mild, as alſo to let them have but a moderate heat; for if the bed be over hot, they will either burn, or be drawn up ſo weak as never to come to good.

The beſt way of ſaving the ſeeds of theſe plants, is to let a few rows of them remain ungathered in the height of the ſeaſon; for if you gather from the plants for ſome time, and afterwards leave the remaining for ſeed, their pods will

not

not be near ſo long and handſome, nor will the ſeed be ſo good. In autumn, when you find they are ripe, you ſhould in a dry ſeaſon pull up the plants, and ſpread them abroad to dry; after which you may threſh out the ſeed, and preſerve it in a dry place for uſe.

PHILADELPHUS. Lin. Gen. Plant. 540. Pipe-tree, or Mock Orange.

The CHARACTERS are,

It hath a permanent empalement, cut into five acute parts, ſitting upon the germen. It hath four or five roundiſh plain petals, and twenty awl-ſhaped ſtamina. The germen is ſituated under the flower, ſupporting a ſlender ſtyle, divided in four parts, which afterward becomes an oval acute-pointed capſule, having four cells, filled with ſmall oblong ſeeds.

The SPECIES are,

1. PHILADELPHUS (*Coronarius*) foliis ſubdentatis. *Philadelphus with oval ſpear-ſhaped leaves, which are acutely indented; the white Syringa, or Mock Orange.*

2. PHILADELPHUS (*Nanus*) foliis ovatis ſubdentatis, flore ſolitario pleno. *Syringa, or Mock Orange, with oval leaves which are ſomewhat indented, and double flowers ſtanding ſingly on the ſides of the branches.*

3. PHILADELPHUS (*Inodorus*) foliis integerrimis. Lin. Sp. Plant. 470. *Philadelphus with entire leaves.*

The firſt ſort has been long cultivated in the Engliſh gardens as a flowering ſhrub, but the place where it naturally grows is uncertain. This ſends up a great number of ſlender ſtalks from the root, which have a gray bark, branching out from their ſide, garniſhed with oval ſpear-ſhaped leaves; they have ſeveral acute indentures on their edges, their ſurface rough, and of a deep green on their upper ſide, but pale on their under, and have the taſte of a freſh Cucumber. The flowers come out from the ſide and at the end of the branches, in looſe bunches, each ſtanding on a ſhort diſtinct foot-ſtalk; they have four oval petals, which ſpread open, with a great number of ſtamina within, ſurrounding the ſtyle. The flowers are white, and have a ſtrong ſcent, which at ſome diſtance reſembles that of Orange flowers, but when near is too powerful for moſt perſons. This ſhrub riſes ſeven or eight feet high.

There is a variety of this with variegated leaves, which ſome people preſerve in their gardens; but as the ſtripes generally diſappear when the plants are in health, ſo it makes little appearance.

The ſecond ſort is of humble growth, ſeldom riſing above three feet high; the leaves are ſhorter than thoſe of the former, and approach near to an oval form; they are but little indented on their edges. The flowers come out ſingly from the ſide of the branches, and have a double or treble row of petals, of the ſame ſize and form as the other, and the flowers have the ſame ſcent; but this ſort flowers very rarely, ſo is not much eſteemed.

Both theſe are extreme hardy, and will thrive in almoſt any ſoil or ſituation, but will grow taller in light good ground than in that which is ſtiff. They are uſually propagated by ſuckers, which are ſent out from their roots in great plenty; they ſhould be taken from the old plants in autumn, and planted in a nurſery to grow one or two years till they have obtained ſtrength, and then they ſhould be tranſplanted to the place where they are deſigned to remain. They are commonly diſpoſed in wilderneſs work, among other ſhrubs of the ſame growth, where they add to the variety.

The third ſort grows naturally in Carolina, and is as yet very rare in Europe. This riſes with a ſhrubby ſtalk about ſixteen feet high, ſending out ſlender branches from the ſides oppoſite, which are garniſhed with ſmooth leaves, ſhaped like thoſe of the Pear-tree, which are entire, ſtanding oppoſite on pretty long foot-ſtalks. The flowers are produced at the end of the branches; they are large, each having four oval petals, which ſpread open, and have large empalements, compoſed of four acute-pointed leaves. The petals are white, and within theſe ſtand a great number of ſhort ſtamina, terminated by yellow ſummits.

This ſhrub is very rare in England, for it will not riſe from ſeeds; I have ſown the ſeeds which were ſent me by the late Dr. Dale from Carolina, two or three times without any ſucceſs, and others have done the ſame, which occaſions its preſent ſcarcity in England; but when the plants are procured from abroad, they may be propagated by laying down their branches. I had one of the ſhrubs which was ſent me by the gentleman before-mentioned, which had thriven in the Chelſea garden near two years; and ſome of the branches which were laid down had put out roots, but they were all deſtroyed by cold in the winter 1740.

PHILLYREA. Tourn. Inſt. R. H. 496. tab. 367. Phillyrea, or Mock Privet.

The CHARACTERS are,

The flowers has a ſmall permanent empalement, cut into five parts at the brim. It has one petal, with a very ſhort tube, cut into five parts, which turn backward, and two ſhort ſtamina, placed oppoſite, terminated by ſingle erect ſummits. It hath a roundiſh germen, ſupporting a ſlender ſtyle, crowned by a thick ſtigma. The germen afterward turns to a globular berry with one cell, incloſing one roundiſh ſeed.

The SPECIES are,

1. PHILLYREA (*Latifolia*) foliis ovato-lanceolatis integerrimis. *Phillyrea with oval, ſpear-ſhaped, entire leaves commonly called the true Phillyrea.*

2. PHILLYREA (*Media*) foliis ovatis ſubintegerrimis. *Phillyrea with oval leaves which are almoſt entire; called broad-leaved Phillyrea.*

3. PHILLYREA (*Spinoſa*) foliis cordato-ovatis ſerratis. Hort. Cliff. 4. *Phillyrea with oval heart-ſhaped leaves which are ſawed; or broad-leaved prickly Phillyrea*

4. PHILLYREA (*Ligustrifolia*) foliis lanceolatis integerrimis. Hort. Cliff. 4. *Phillyrea with ſpear-ſhaped entire leaves; Privet-leaved Phillyrea*

5. PHILLYREA (*Oleæfolia*) foliis lanceolato-ovatis integerrimis, floribus confertis axillaribus. *Phillyrea with ſpear-ſhaped, oval, entire leaves, and flowers growing in cluſters from the ſide of the branches; Olive-leaved Phillyrea.*

6. PHILLYREA (*Anguſtifolia*) foliis lineari-lanceolatis integerrimis, floribus confertis axillaribus. *Phillyrea with narrow, ſpear-ſhaped, entire leaves, and flowers growing in cluſters from the ſides of the branches; narrow-leaved Phillyrea.*

7. PHILLYREA (*Roſmarinifolia*) foliis linearibus. *Phillyrea with very narrow leaves; commonly called Roſemary-leaved Phillyrea.*

The firſt ſort here mentioned, is the moſt common in the Engliſh gardens, where it is known by the title of true Phillyrea; ſo called, to diſtinguiſh it from the Alaternus, which is called ſimply Phillyrea, by the gardeners. This in England riſes with a ſtrong upright ſtem to the height of eighteen or twenty feet, dividing into ſeveral branches, covered with a ſmooth grayiſh bark, garniſhed with oval ſpear-ſhaped leaves, placed oppoſite, which are entire, firm, and of a light green. The flowers come out from the wings of the ſtalk on each ſide; they are of an herbaceous white colour, and grow in ſmall cluſters, and are ſucceeded by globular berries with one cell, incloſing a ſingle ſeed of the ſame form.

The ſecond ſort riſes to an equal height with the firſt, but the branches are more diffuſed, and have a darker bark; the leaves are oval, and of a darker green, a little ſawed on their edges. The flowers come out from the wings of the branches, growing in long bunches; they are of an herbaceous

ceous white colour as the former, and are succeeded by berries of the same form.

The third sort rises with an upright stem as high as the two former, sending out several strong branches which grow erect, covered with a gray bark, garnished with oval heart-shaped leaves, which are firm, of a lucid green, and sawed on their edges, each serrature ending in a spine. The flowers and seeds of this are like those of the two former sorts.

The fourth sort is of humbler growth than either of the former, seldom rising more than ten feet high; the branches are weaker, and spread wider than those, and are covered with a light brown bark; they are garnished with stiff spear-shaped leaves, of a light green, and sit close to the branches. The flowers are produced in small clusters at the wings of the branches on each side; they are small, and whiter than those of the former, and are succeeded by small berries, which ripen in the autumn.

The fifth sort rises about the same height as the fourth; the branches are stronger, and spread out wider; the bark is of a lighter colour; the leaves are stiff, smooth, and entire, standing on very short foot-stalks, of a lucid green, and terminate in a point. The flowers come out in clusters upon pretty long foot-stalks, at the wings of the young branches; they are small, white, and have round berries succeeding them, which ripen in autumn.

The sixth sort rises with a woody stalk ten or twelve feet high, sending out branches by pairs, which are covered with a brown bark, spotted with white, and are garnished with smooth, stiff, narrow, spear-shaped leaves, which are entire, sitting close to the branches, of a light green, and point upward. The flowers come out in large clusters at each joint of the branches, to which they sit close like the whorled flowers, almost surrounding the stalk; they are small, white, and are succeeded by small berries, which ripen in autumn.

The seventh sort is of humbler growth than either of the former, seldom rising more than five or six feet high, send-out out slender branches opposite; the leaves are of a dark green, stiff, very narrow, and entire. The flowers are small, white, and grow in clusters from the side of the branches. The berries of this sort are very small, and rarely ripen in England.

These plants all grow naturally in the south of France, Spain, and Italy, but are hardy enough to thrive in the open air in England, and are never injured except the winters are very severe, which sometimes causes their leaves to fall, and kills a few of the weaker branches, but these are repaired by new shoots the following summer; so that there are but few of the evergreen trees which are hardier than these, or that deserve more to be cultivated for pleasure.

The three first sorts are very proper to intermix with other evergreen trees of the same growth to form clumps in parks, or to plant round the borders of woods, which are filled with deciduous trees, where in the summer time the dark shade of these evergreens will make a fine contrast with the brighter green leaves of the deciduous trees, and in winter, when the latter are destitute of leaves, they will have a fine effect; and these will be a fine harbour for birds. They may be trained up to stems, so as to be out of the reach of cattle, therefore may be planted in open places, where, if they are fenced against cattle till they are grown up, they may be afterwards exposed.

The other sorts, which are of humbler growth, must be confined to gardens or other inclosures, where they may be secured from cattle, hares, rabbits, &c. otherwise they will soon be destroyed.

These plants are propagated either from seeds or layers, but the latter, being the most expeditious method in England, is chiefly preferred. The best time to lay them down is in autumn, when you should dig the ground round the stems of the plants intended to be layed, making it very loose; then making choice of a smooth part of the shoot, you should make a slip upward (in the manner as is practised in laying of Carnations,) and then bend the branch gently down to the ground, making a hollow place with your hand to receive it; and having placed the part which was slit in the ground, so as that the slit may be open, you should fasten it down with a forked stick, that it may remain steady, covering that part of the branch with earth about three inches thick, observing to keep the upper part erect. You must keep them clear from weeds the spring and summer following, which, if suffered to grow up amongst them, will prevent their taking root.

The autumn following most of these plants will be rooted, at which time they may be taken off, and carefully planted in a nursery, where they may be trained up three or four years in the manner you intend them to grow, during which time you should dig the ground between the rows, and cut about the roots of the plants every year, which will cause them to strike out strong fibres, so as to support a good ball of earth when they are removed; you should also support their stems with stakes, in order to make them strait, otherwise they are very apt to grow crooked and unsightly.

When the plants have been thus managed three or four years, you may transplant them into the places where they are designed to remain. The best time for this work is the latter end of September, or the beginning of October; but in removing them, you should dig round their roots, and cut off all downright or strong roots which have shot out to a great distance, that you may the better preserve a ball of earth to each plant, otherwise they are subject to miscarry; and when you have placed them in their new quarters, you should lay some mulch upon the surface of the ground to prevent its drying. You should also support the plants with stakes until they have taken fast hold of the earth, to prevent their being turned out of the ground, or displaced by the winds. These trees delight in a middling soil, which is neither too wet and stiff, nor too dry, though the latter is to be preferred to the former, provided it be fresh.

Those sorts with small leaves are commonly two years before they take root, when laid; therefore they should not be disturbed, for the raising them out of the ground greatly retards their rooting.

PHILLYREA OF THE CAPE. See MAUROCENA.

PHLOMIS. Tourn. Inst. R. H. 177. tab. 82. The Sage-tree, or Jerusalem Sage.

The CHARACTERS are,

The flower hath a permanent empalement with an oblong tube, having five angles. It is of the lip kind. The tube is oblong; the upper lip is oval, forked, and inflexed; the under is cut into three segments, the middle one being large and obtuse. It hath four stamina hid under the upper lip, two being longer than the other, and a germen divided into four parts, supporting a style the length of the stamina. The germen afterward becomes four oblong cornered seeds sitting in the empalement.

The SPECIES are,

1. PHLOMIS (*Fruticosa*) foliis subrotundis tomentosis crenatis, caule fruticoso. *Phlomis with roundish, woolly, crenated leaves, and a shrubby stalk.*

2. PHLOMIS (*Angustifolia*) foliis lanceolatis tomentosis integerrimis, caule fruticoso. *Phlomis with spear-shaped woolly leaves, which are entire, and a shrubby stalk.*

3. PHLOMIS (*Latifolia*) foliis oblongo-ovatis petiolatis tomentosis, floribus capitatis, caule fruticoso. *Phlomis with oblong, oval, woolly leaves, having foot-stalks, flowers growing in large heads, and a shrubby stalk.*

4. PHLOMIS

4. PHLOMIS (*Herba venti*) involucris setaceis hispidis, foliis ovato oblongis scabris, caule herbaceo. Hort. Upsal. 171. *Phlomis with bristly prickly involucrums, oblong, oval, rough leaves, and an herbaceous stalk.*

5. PHLOMIS (*Tuberosa*) involucris hispidis subulatis, foliis cordatis scabris, caule herbaceo. Hort. Upsal. 171. *Phlomis with awl-shaped prickly involucrums, rough heart-shaped leaves, and an herbaceous stalk.*

6. PHLOMIS (*Lychnitis*) foliis lanceolatis tomentosis, floralibus ovatis, involucris setaceis lanatis. Lin. Sp. Pl. 585. *Phlomis with spear-shaped woolly leaves, those under the flower oval, and bristly woolly involucrums.*

7. PHLOMIS (*Purpurea*) foliis ovato-lanceolatis crenatis, subtus tomentosis, involucris setaceis. *Phlomis with oval spear-shaped leaves, which are woolly on their under side, and have a bristly involucrum.*

8. PHLOMIS (*Samia*) foliis cordatis acutis subtus tomentosis, involucris strictis tripartitis. *Phlomis with acute-pointed heart-shaped leaves, which are woolly on their under side, and the covers of the flowers divided into three parts.*

9. PHLOMIS (*Orientalis*) foliis cordatis rugosis subtus tomentosis, involucris lanatis, caule herbaceo. *Phlomis with rough heart-shaped leaves, which are woolly on their under side, woolly covers to their flowers, and an herbaceous stalk.*

10. PHLOMIS (*Flavescenti*) foliis lanceolatis crenatis subtus tomentosis, involucris lanatis, caule fruticoso. *Phlomis with spear-shaped crenated leaves, which are woolly on their under side, woolly covers to their flowers. and a shrubby stalk.*

11. PHLOMIS (*Nissolii*) foliis radicalibus cordatis utrinque tomentosis villosis. Lin. Sp. Plant. 585. *Phlomis whose lower leaves are heart-shaped, woolly, and hairy on every side.*

12. PHLOMIS (*Ferruginea*) involucris lanceolatis, foliis cordatis subtus tomentosis, caule suffruticoso. *Phlomis with spear-shaped involucrums, heart-shaped leaves, which are woolly on their under-side, and a shrubby stalk; whitest, shrubby, Spanish, Jerusalem Sage, with an iron-coloured flower.*

13. PHLOMIS (*Rotundifolia*) involucris subulatis, foliis cordato-ovatis subtus tomentosis, caule fruticoso. *Phlomis with awl-shaped involucrums, oval heart-shaped leaves, which are woolly on their under side, and a shrubby stalk.*

14. PHLOMIS (*Laciniata*) foliis alternatim pinnatis, foliolis laciniatis, calycibus lanatis. Lin. Sp. Plant. 585. *Phlomis with leaves alternately winged, whose lobes are cut, and woolly empalements to the flowers.*

The first sort grows naturally in Spain and Sicily; it hath a pretty thick shrubby stalk, covered with a loose bark, rising five or six feet high, dividing into many irregular, woolly, crenated branches. Their joints are pretty far asunder; at each of these are placed two roundish leaves opposite on short foot stalks, woolly on their under side. The flowers are yellow, and come out in thick whorls round the stalk, having two lips; the upper lip is forked, bending over the under, which is divided into three parts; the middle is broad, and stretched out beyond the two small side segments.

The second sort hath a shrubby stalk like the first, but does not rise so high. The branches are weaker; the leaves are spear-shaped and oval, being longer and narrower than the former; the whorls of flowers are smaller, but the flowers are of the same shape and colour.

These two sorts have been long propagated in the English gardens by the title of Sage-tree, or Jerusalem Sage. The plants were formerly kept in pots, and housed in winter with other exotic plants, but of late years they have been planted in the open air, where they are seldom injured by cold, unless in very severe winters; so they are intermixed with other shrubs of the same growth in quarters of wilderness work, where they add to the variety.

These plants should have a dry soil and a warm sheltered situation, otherwise they will not live in the open air. They may be planted among Cistuses of all the different kind, the shrubby Moon Trefoil, evergreen Cytisus, Wormwood-tree, and some other exotic shrubs of the same countries, which require a warm situation and a dry soil, being too tender for open plantations, which are exposed to strong cold winds; and as they are not of very long duration, they are better when separated from trees and shrubs which continue many years; for these rarely live above twelve or fourteen years in dry ground, and not more than half so long in cold moist land, or where they are not well sheltered.

They are propagated by slips or cuttings, which, if planted in a bed of light earth in April, just before the plants begin to shoot, and covered with mats to screen them from the sun every day, as also to observe when the ground is dry to give them water gently, they will get roots in about two months or ten weeks, when they may be carefully taken up, and transplanted into a nursery, where they may remain one year, and then be transplanted to the places where they are designed to stand, for these plants will not bear transplanting at a greater age.

The third sort hath a shrubby stalk like the former, it seldom rising more than four or five feet high, sending out branches on every side, garnished with broader hoary leaves than either of the former; they are of an oblong oval form, and have pretty long foot-stalks; they are whiter than those of the former. The flowers grow in large whorls or heads, which generally terminate the branches; they are larger than those of the other sorts, and the upper lip is very hairy. The plants are equally hardy with the other, and may be propagated by slips or cuttings in the same way as is before directed for them.

The fourth sort grows naturally in the south of France and Italy; this hath a perennial root and an annual stalk, which rises about two feet high. When the roots are large, they send up a great number of square stalks, which are covered with a hairy down, garnished with oblong, oval, rough leaves, sitting close to the stalks. The flowers grow in whorls round the stalks, having stinging bristly covers; they are of a bright purple colour, so make a pretty appearance.

This may be propagated by parting of the roots; the best time for doing of this is in the autumn, when the stalks begin to decay, that they may get root before the frost comes on, but they should not be parted oftener than every third year, if they are expected to have many flowers. This sort is hardy, so may be planted in exposed places, but not in moist ground.

The fifth sort grows naturally in Tartary; this hath a perennial root. The stalks are purple, have four corners, and rise five or six feet high, garnished with heart-shaped leaves, placed opposite, deeply crenated on their edges. The flowers are purple, and grow in whorls round the stalks; their covers are awl-shaped, and set with stinging hairs. It is propagated by seeds, which should be sown upon an east border in the spring; and when the plants come up, they must be kept clean from weeds the following summer, and in the autumn they should be transplanted where they are to remain.

The sixth sort grows naturally in the south of France, in Spain, and Italy. This root is perennial, the stalk is annual. This sends out long, narrow, woolly leaves from the roots in tufts, which are enveloped at their base by a common covering; they are soft to the touch, and lie upon the ground. The stalks are slender, and near two feet long; their joints are far asunder; at each of these stand two oval leaves opposite, which embrace the stalks with their base.

The whorls of flowers are alſo encompaſſed by theſe leaves, and within them is ſituated a radiated briſtly involucrum, which cover the yellow flowers, ſhaped like thoſe of the other ſorts. The ſtalks decay in the autumn, but the lower leaves continue all the year. It may be propagated by ſlips in the ſpring, for they rarely produce ſeeds here; the plants require a dry ſoil and a warm ſituation.

The ſeventh ſort grows naturally in Portugal and Spain. This hath a ſhrubby ſtalk, which riſes four or five feet high, ſending out ſlender branches, which have four angles covered with a white bark, garniſhed with oval ſpear-ſhaped leaves, crenated on their edges, woolly on their under ſide, ſtanding on very ſhort foot-ſtalks. The flowers come out in whorls at each joint; they have briſtly involucrums, and are of a deep purple colour. It may be propagated by cuttings in the ſame way as the three firſt ſorts, and the plants require the ſame treatment.

The eighth ſort grows naturally in the Levant; this hath a perennial root and an annual ſtalk. The leaves are heart-ſhaped, ending in acute points; they are downy on their under ſide, and have five ſtrong veins. The ſtalks riſe a foot and a half high, garniſhed at each joint with two leaves, placed oppoſite, of the ſame form as the lower, but ſmaller. The flowers grow in whorls round the ſtalks; they are of a worn-out purple colour; their involucrums are cut into ſegments, and are cloſely ſhut.

The ſeeds of the ninth ſort were ſent from Smyrna by the late Conſul Sherard to the Chelſea garden. This has a perennial root, and an annual ſtalk. The lower leaves are very woolly and heart-ſhaped, ſtanding upon long woolly foot-ſtalks. The ſtalks, which are woolly, riſe a foot high; the flowers are large, yellow, and grow in whorls round them; they have very long tubulous empalements, covered with down. This ſort had ſurvived many winters in the open air in the Chelſea garden, but in the year 1740 they were all deſtroyed.

The ſeeds of the tenth ſort were alſo ſent from Smyrna by the ſame gentleman, and ſeveral of the plants were raiſed in the Chelſea garden. This hath ſhrubby ſtalks, which riſe about three feet high, covered with a yellowiſh down, ſending out many ſlender irregular branches, garniſhed with narrow ſpear-ſhaped leaves, which are covered with a yellowiſh down on their under ſide The flowers are produced in heads at the end of the branches; their involucrums are very downy; the flowers are ſmaller than thoſe of either of the three firſt ſorts, and are of a dirty yellow colour. This approaches near to the ſecond ſort, but the leaves are much ſmaller, the branches are ſlenderer, and are covered with a yellow down, eſpecially toward the end of the branches. The whorls of flowers are not near ſo large, and are generally produced at the end of the branches.

This ſort may be propagated by cuttings in the ſame way as the three firſt ſorts, and the plants may be treated in the like manner, with this difference only, of planting them in a warmer ſituation, for it will not bear ſo much cold, though in a warm border the plants have lived ſeveral years abroad in the Chelſea garden.

The eleventh ſort grows naturally in the Archipelago, and alſo in Spain, from both which countries I have received the ſeeds. It hath an annual ſtalk, but the root is perennial, as are alſo the lower leaves, which do not ariſe from the root immediately, but ſtand in cluſters upon ſhort trailing woolly branches; they have very long downy foot-ſtalks; they are heart-ſhaped, and downy on both ſides. The ſtalks are ſlender, and riſe a foot high, garniſhed with oval ſpear-ſhaped leaves, which gradually decreaſe in ſize to the top. The ſtalks generally ſend out two ſide branches oppoſite, near the bottom, and from this diviſion to the top are garniſhed with thin whorls of yellow flowers, which are not cloſely joined together as in the other ſpecies, but each flower ſtands ſeparate. Their empalements are oval, very downy, and cloſely ſhut up. This ſort may be propagated by ſlips in the ſame manner as the ſixth ſort, and the plants ſhould be treated in the like way.

The twelfth ſort grows naturally in Spain and Portugal. This hath a ſhrubby ſtalk, which is a little ligneous, and riſes about two feet and a half high, covered with a thick white down. There are many of the ſtalks which riſe from the ſame root, garniſhed with heart-ſhaped leaves; from the lower part of the ſtalks, at each joint, there are two ſhort ſhoots come out oppoſite, which have four or ſix ſmall leaves, of the ſame ſhape with the others. The flowers, which are of an iron colour, are produced in ſmall whorls toward the upper part of the ſtalk, and have downy ſpear-ſhaped involucrums.

This ſort multiplies by its ſpreading roots, ſo that they may be divided every other year; the beſt time for doing of this is about the middle of September, that the offsets may get root before the froſt comes on; but there ſhould be ſome mulch laid about their roots, to prevent the froſt from penetrating the ground. It may alſo be propagated by cuttings in the ſame way as the three firſt ſorts, during the ſpring and ſummer months. The plants require the ſame treatment as the tenth ſort, for they are not ſo hardy as the three firſt ſorts; therefore if there is ſome tanners bark, or other mulch laid on the ſurface of the ground about their roots every winter, it will be a means of preſerving the roots, ſo that if a ſevere winter ſhould kill the ſtalks, the roots will put out new ones the ſpring following.

The thirteenth ſort grows naturally in Spain and Portugal. This riſes with ſeveral ſhrubby ſtalks from three to four feet high, which divide into ſeveral four-cornered branches, covered with a woolly down, garniſhed with leaves, which on the lower part of the ſtalks are heart-ſhaped, but upward they are of an oval ſpear-ſhape, woolly on their under ſide; they ſtand oppoſite upon ſhort foot-ſtalks. The flowers come out in whorls round the ſtalks; they have awl ſhaped involucrums, ending in acute points, and covered with down; they are of a bright purple colour, but are not ſucceeded by ſeeds in this country. This ſort is propagated by ſlips or cuttings in the ſame way as the three firſt ſorts, and the plants ſhould be treated in the like manner as hath been before directed for the tenth ſort.

The fourteenth ſort grows naturally in the Levant. This hath a perennial root and an annual ſtalk, but the lower leaves continue all the year; theſe are alternately winged, and the ſmall lobes are cut on their edges. The ſtalks riſe a foot and a half high, garniſhed with leaves of the ſame ſhape with the lower, but are ſmaller. The flowers come out in whorls round the ſtalks, like thoſe of the other ſorts, whoſe empalements are downy; they are of a worn-out purple colour, and appear in June, but the ſeeds do not ripen here.

It is propagated by offsets from the root in the ſame way as the eighth ſort, but theſe are ſent out ſparingly alſo, and the plants require the ſame treatment. It is at preſent very rare in England, for the ſevere froſt in the year 1740 deſtroyed all the plants here, which had ſurvived all the winters for twenty years before in the open air.

All the ſpecies of this genus are ornamental plants, when properly diſpoſed in gardens, ſo deſerve a place, for there is generally a ſucceſſion of flowers on them for two or three months, and their hoary down leaves, when intermixed with plants whoſe leaves are green, make a pretty contraſt.

PHLOX. Lin. Gen. Plant. 197. Lychnidea, or Baſtard Lychnis.

The

The CHARACTERS are,

The flower has a cylindrical empalement, which is permanent, with five acute indentures at the top. It has one funnel-shaped petal, with a cylindrical tube, incurved at the base, plain at the top, where it is cut into five equal roundish segments, which spread open. It hath five short stamina situated within the tube, two of which are longer than the tube. It hath a conical germen, supporting a slender style, crowned by an acute trifid stigma. The germen afterward turns to an oval capsule, with three cells sitting in the empalement, each cell containing a single seed.

The SPECIES are,

1. PHLOX (*Glaberrima*) foliis lineari-lanceolatis glabris acuminatis, caule erecto ramoso, corymbo terminali. *Phlox with smooth, narrow, spear-shaped leaves, ending in acute points, and upright branching stalks, terminated by flowers, which grow in a corymbus.*

2. PHLOX (*Carolina*) foliis lanceolatis sessilibus glabris crassis, caule erecto, floribus verticillatis terminalibus. *Phlox with smooth, thick, spear-shaped leaves, sitting close to the stalks, and upright stalks terminated by flowers growing in whorls.*

3. PHLOX (*Maculata*) foliis cordato-lanceolatis lævibus. Lin. Sp. Plant. 152. *Phlox with heart spear-shaped leaves, which are smooth.*

4. PHLOX (*Divaricata*) foliis lato lanceolatis, inferioribus alternis, caule ramoso. Lin. Sp Plant. 152. *Phlox with broad spear-shaped leaves, which are placed alternately at bottom, and a branching stalk.*

5. PHLOX (*Paniculata*) foliis lanceolatis margine scabris, corymbis compositis. Lin. Sp. Pl. 151. *Phlox with spear-shaped leaves having rough borders, and flowers disposed in compound corymbuses.*

6. PHLOX (*Pilosa*) foliis lanceolatis villosis, caule erecto corymbo terminali. Lin. Sp. Plant. 152. *Phlox with hairy spear-shaped leaves, and an upright stalk, terminated by a corymbus of flowers.*

7. PHLOX (*Ovata*) foliis ovatis, floribus solitariis. Lin. Sp. Plant. 152. *Phlox with oval leaves and solitary flowers.*

The first sort grows naturally in Virginia, and in some other parts of North America. This hath a perennial root, which sends up several stalks, in number proportionable to the size of the roots, near a foot and a half high, which divide into three or four small branches toward the top, terminated by a corymbus of flowers. The leaves on the lower part of the stalks are placed opposite; they are smooth, and sit close to the stalks; the leaves on the upper part of the stalks are placed alternate. The flowers grow almost in form of an umbel, standing on short foot-stalks; their empalements are tubulous, and have ten angles or furrows, and are cut at the top into five acute segments; the tube of the flower is twice the length of the empalement, and is divided at the top into five roundish segments, which spread open; these are of a light purple colour.

The second sort grows naturally in Carolina. This hath a perennial root, from which arise several smooth stalks near two feet high, garnished with stiff shining leaves placed opposite; they are spear-shaped, entire, and their edges are reflexed; the upper part of the stalk has generally two slender side branches, and is terminated by a head of flowers, which grow in whorls round the stalks, but the whorls are so nearly placed, as to appear one corymbus at some distance. The empalement of the flower is short, and deeply cut into five acute segments; the tube of the flower is long, and at the top is cut into five roundish segments, which spread open. These flowers are of a deeper purple colour than those of the former.

The third sort grows naturally in Maryland. This hath a perennial root, from which arise several upright stalks, of a purplish colour, closely covered with white spots; these grow about three feet high, garnished with heart, spear-shaped, smooth leaves. Toward the upper part of the stalks are sent out small branches opposite, each being terminated by a small bunch of flowers; but the principal stalk is terminated by a long loose spike of flowers, composed of small bunches, arising from the wings of the stalk at each joint, each cluster having one common foot-stalk; the flowers are of a bright purple colour, but are rarely succeeded by seeds in England.

The fourth sort grows naturally in North America. This hath a perennial root, from which arise several slender stalks, which are apt to incline to the ground if they are not supported; these divide into several small branches, which spread from each other; the lower part of the stalks are garnished with broad spear-shaped leaves, placed alternate, sitting close to the stalks; but on the smaller branches they are narrower, and placed opposite. The flowers grow in loose bunches at the end of the branches; they have sh empalements, which are cut into five narrow acute segments; the tube of the flower is long and slender; the segments at the top are broad and heart-shaped, inverted. They are of a light blue, but are rarely succeeded by seeds in England.

The fifth sort grows naturally in North America. This hath a perennial root and an annual stalk, which is smooth, of a light green, and rises about two feet high, sending out a few side branches, garnished with spear-shaped leaves placed opposite, sitting close to the stalks; they are of a dark green, and their edges are a little rough. The flowers are disposed in a corymbus at the top of the stalks; these are composed of many smaller bunches of flowers, which have each a distinct foot-stalk, and support a great number of flowers, which stand upon short slender foot-stalks; the empalement of the flower is short, cut almost to the bottom into five narrow acute segments; the tube of the flower is long, slender, and is cut at the top into five oval segments, which spread open. The flowers are of a pale purple colour, and are often succeeded by seeds, which ripen in the autumn.

The sixth sort grows naturally in Virginia. This hath a perennial root, from which arise a few single stalks about a foot high, garnished with narrow spear-shaped leaves, ending in acute points, which are a little hairy. The flowers are produced in a loose corymbus at the top of the stalk; their empalements are cut into acute segments almost to the bottom; the tube of the flower is slender, pretty long, and is cut at the top into five oval segments, which spread open. The flowers are of a light purple colour, but are seldom succeeded by seeds in England.

The seventh sort grows naturally in Maryland, and other parts of North America. This hath a perennial root, from which comes out two or three slender stalks about nine inches high, garnished with oval, rough, hairy leaves, placed opposite upon very short foot-stalks The flowers come singly at the top of the stalk; they have very slender tubes, but are cut into five roundish segments, which spread open. They are of a light purple colour, but are not succeeded by seeds in England.

These plants are hardy, so will thrive in the open air in England. They delight in a moist rich soil not too stiff, in which they will grow tall, and produce much larger bunches of flowers than in dry ground; for when the soil is poor and dry, they frequently die in summer, unless they are duly watered.

They are generally propagated by parting their roots, because they do not often produce seeds in England. The best time for this is in autumn, when their stalks begin to decay. These roots should not be divided into small heads,

heads, if they are expected to flower well the following summer; nor should they be parted oftener than every other year, because when they are often removed and parted, it will greatly weaken the roots, so that they will send out but few stalks, and those will be so weak as not to rise their usual height; the bunches of flowers will also be much smaller.

The first, second, and fifth sorts, propagate pretty fast by their spreading roots, but the others increase but slowly this way, therefore the best method to propagate them is by cuttings; if the three first sorts are desired in plenty, they may be easily obtained by this method. The best time to plant the cuttings, is about the latter end of April, or the beginning of May, when the shoots from the roots are about four inches high; these should be cut off close to the ground, and their tops should be shortened; then they must be planted on a border of light loamy earth, and shaded from the sun until they have taken root; or if they are planted close together, and covered with bell or hand-glasses, shading them every day from the sun, they will put out roots in five or six weeks; but when they begin to shoot, the glasses should be gradually raised to admit the free air to them. As soon as they are well rooted, the glasses should be taken off, and the plants inured to the open air; then they should be soon after removed into a bed of good soil, planting them about six inches distance every way, observing to shade them from the sun, and water them duly till they have taken new root; after which, if they are kept clean from weeds, they will require no other care till autumn, when they should be transplanted into the borders of the flower-garden where they are designed to remain.

PHYLICA. Lin. Gen. Plant. 236. Bastard Alaternus.

The CHARACTERS are,

The flowers are collected in a disk, sitting in a common receptacle, each having a permanent empalement, composed of three narrow oblong leaves. They have one perforated petal, with an erect conical tube, cut into five parts at the brim, with an acute scale at each division, which join them together, and five small stamina inserted under the scale, terminated by single summits. The germen is situated at the bottom of the petal, supporting a single style, which afterward becomes a roundish capsule with three lobes, having three cells, each inclosing a single roundish seed, gibbous on one side, and angular on the other.

The SPECIES are,

1. PHYLICA (*Ericoides*) foliis linearibus verticillatis. Lin. Sp. Plant. 195. *Phylica with linear leaves growing in whorls.*

2. PHYLICA (*Plumosa*) foliis lineari-subulatis, summis hirsutis. Prod. Leyd. 199. *Phylica with awl-shaped leaves, which are hairy at the top.*

3. PHYLICA (*Buxifolia*) foliis ovatis sparsis. Lin. Sp. Pl. 195. *Phylica with oval leaves growing scatteringly.*

The first sort grows naturally at the Cape of Good Hope. It also grows wild about Lisbon, where there are large extents of ground covered with it like the heaths in England. This is a low bushy plant, seldom rising more than three feet high; the stalks are shrubby and irregular, dividing into many spreading branches. The young branches are closely garnished with short, narrow, acute-pointed leaves, placed in whorls round the stalks, to which they sit close; they are of a dark green, and continue all the year. At the end of every shoot the flowers are produced in small clusters, sitting close to the leaves; they are of a pure white, and begin to appear in the autumn, continuing in beauty all the winter, and decay in the spring, which renders the plant more valuable.

The second sort grows naturally at the Cape of Good Hope. This hath an erect shrubby stalk, which rises near three feet high, covered with a purplish bark; the leaves are narrow, short, and acute-pointed, sitting close to the branches in alternate order; they are thick, nervous, and of a dark green on their upper side, but hoary on their under. The flowers are collected in small heads at the end of the branches; they are white, woolly, and fringed on their borders, cut into six acute segments at the top. These appear the beginning of winter, and continue long in beauty.

The third sort is a native of the same country as the former. This rises with a shrubby erect stalk five or six feet high; the stalks, when old, are covered with a rough purplish bark, but the younger branches have a woolly down: these are garnished with thick oval leaves, about the size of those of the Box-tree; they are veined, smooth, and of a lucid green on their upper side, but are hoary on their under. The flowers are collected in small heads at the end of the branches; they are of an herbaceous colour, so make no great appearance. These appear at the same time with the former.

As these plants do not produce seeds in England, so they are propagated by cuttings, which, if properly managed, will take root freely. There are two seasons for planting these cuttings, the first is the latter end of March, before the plants begin to shoot; if these are planted in pots, and plunged into a very moderate hot-bed, covering them close with bell or hand-glasses, observing to shade them from the sun in the middle of the day, and to refresh them gently with water, they will put out roots in two months; then they should be inured to the open air, and after they have obtained strength, they should be taken out of these pots, and each planted in a separate small pot, filled with soft loamy earth, and placed in a shady situation until they have taken new root, when they may be removed to a sheltered situation, where they may remain till autumn.

The other season for planting of these cuttings, is about the beginning of August. At this time they may be planted in pots, which may be either plunged into an old hot-bed, or the full ground, covering them close with bell or hand-glasses as before, and treating them in the same way; these will put out roots in about two months, but it will then be too late in the season to transplant them, so they must remain in the same pots till spring.

These plants are generally too tender to thrive in the open air in England, so they must be kept in pots, and housed in winter; for although the first sort will live thro the winter in a warm sheltered situation, when the seasons prove favourable, yet when severe frosts happen, they are always destroyed; but they require no artificial heat to preserve them, if they are sheltered under a hot-bed frame in winter, when they are young, and after they are grown large, kept in a green-house where they may enjoy the free air in mild weather, and are treated in the same way as other hardy exotic plants from the same country; in the summer they must be placed abroad in a sheltered situation; with which management the plants will thrive and continue several years, and as they flower in the winter, they make a good appearance in the green-house during that season.

PHYLLANTHUS. Lin. Gen. Plant. 932. Sea-side Laurel.

The CHARACTERS are,

It hath male and female flowers in the same plant; the empalements of the flowers in both sexes are permanent, bell-shaped, and of one leaf, cut into six parts, which spread open. The flowers have no petals according to some, or no empalements according to others. The male flowers have three short stamina, which join at their base, but spread asunder at their top, and are terminated by twin summits. The female flowers have an angular nectarium surrounding the germen, which afterward

becomes

becomes a roundish capsule with three furrows, having three cells, each containing a single roundish seed.

The Species are,

1. Phyllanthus (*Epiphilanthus*) foliis lanceolatis serratis, crenis floriferis. Hort. Cliff. 439. *Phyllanthus with spear-shaped sawed leaves, having flowers growing on their edges.*

2. Phyllanthus (*Niruri*) foliis pinnatis floriferis, floribus pedunculatis, caule herbaceo erecto. Lin. Sp. *Phyllanthus with winged leaves, bearing flowers upon foot-stalks, and an upright herbaceous stalk.*

3. Phyllanthus (*Frutescens*) foliis pinnatis floriferis, floribus sessilibus, caule herbaceo procumbente. Lin. Sp. *Phyllanthus with winged leaves, bearing flowers sitting close, and a trailing herbaceous stalk.*

4. Phyllanthus (*Emblica*) foliis pinnatis floriferis, caule arboreo, fructu baccato. Lin. Syst. 1265. *Phyllanthus with winged leaves bearing flowers, a tree-like stalk, and the fruit a berry.*

5. Phyllanthus (*Arborescens*) caule arboreo, foliis ovatis obtusis integerrimis. Lin. Syst. 1264. *Tree Phyllanthus with oval, obtuse, entire leaves.*

The first sort grows naturally upon the rocks near the sea, in all the islands of the West-Indies, where the inhabitants title it Sea-side Laurel. This is seldom found growing on the land, which occasions its scarcity in Europe; for the roots strike so deep into the crevices of the rocks, as to render it almost impracticable to transplant the plants, and it is very difficult to propagate by seeds; for unless they are sown soon after they are ripe, they will not grow, and the greatest part of the seed proves abortive, so that it is very rare in Europe. There was formerly a plant of this sort in the gardens at Hampton Court, but this, with many other fine plants, has been destroyed by the ignorance of the gardeners. I also saw a fine plant of this sort in the Amsterdam garden.

This grows with a woody stalk about fifteen or sixteen feet high; the leaves come out without any order; upon the edges of the leaves the flowers are produced, but especially toward the upper part, where they are placed very closely, so as almost to form a sort of border to the leaves; which, together with the shining green colour of the leaves, makes a very beautiful appearance: the leaves continue green all the year, which renders the plant more valuable.

It requires to be placed in a moderate stove in the winter, otherwise it will not live in England; but in summer it may be placed in the open air, in a warm sheltered situation. With this management the plant was in great vigour in the physic-garden at Amsterdam.

The second sort is an annual plant, with an herbaceous stalk about a foot high, which branches out, and has winged leaves, composed of many oval lobes, under which the flowers are produced upon foot-stalks, ranged along the midrib; they are small, of an herbaceous colour, and as they are situated under the leaf, so make no great appearance; however, the plants are cultivated by those who delight in botany.

The third sort is also annual; the stalks trail on the ground, and are garnished with winged leaves, having oblong lobes; under which the flowers are ranged along the midrib, as in the other species, so make little appearance.

The fourth sort is the Nilicamarum of the Hortus Malabaricus, and the Nux Emblica of the shops. This hath a woody tree-like stalk, spreading into many branches, garnished with narrow linear leaves, in shape like those of the deciduous Cypress; but as the plants have not produced flowers in England, so I can give no farther account of the tree.

The fifth sort grows naturally in the West-Indies, where it becomes a tree of middling stature: the leaves are almost oval, of a light green on their upper side, but gray on their under, being very entire. The flowers have not appeared in England, so nothing can be said of them; but the fruit which came over were the size of Walnuts, having three swelling cells, in each of which should be lodged a single seed, but two of them are generally barren; and those which seem to have fair seeds, upon examination, will be found hollow without any germ.

The second and third sorts grow naturally in both Indies: as they are annual, so their seeds may be sown upon a hot-bed in the spring; when the plants come up, and are fit to remove, they should be each put into a separate small pot, and plunged into a hot-bed of tanners bark, shading them from the sun till they have taken new root; after which, their management should be the same as for other plants of the same countries: with this care they will perfect their seeds in autumn, which must be carefully watched, otherwise their husks will open, and scatter the seeds into such pots as are near them, where, if the ground is not disturbed, the plants will come up the following spring.

The fourth sort grows naturally in the East-Indies, where it rises with a ligneous stalk to the height of twelve or fourteen feet; but the plants which have been raised in England, have not exceeded three or four, though there are some of ten years growth; for they frequently lose their leading shoot in winter, and put out lateral branches, which also are apt to lose their tops in winter; so the plants do not advance much, nor have any of them attempted to shew their flowers here, though the plants are in good health.

This and the fifth sort have been raised from seeds which were sown upon a hot-bed, and the plants were put into pots, and plunged into a tan-bed in summer, and in winter removed into the bark-stove, where they have been constantly kept, for they were found to be too tender to live through the winter in a less degree of heat. Their cuttings which have been planted, have failed, and their branches being too strong to make layers, they have not been propagated any other way than from seeds.

PHYLLIS. Lin. Gen. Plant. 286. Simpla Nobla.

The Characters are,

The empalement of the flower is composed of two leaves sitting on the germen. The flower has five obtuse spear-shaped petals, which turn backward. It hath five short hair-like stamina, terminated by oblong summits. The germen, which is situated under the flower, has no style, but is crowned by two awl-shaped reflexed hairy stigmas, and afterward turns to an oblong angular fruit, containing two parallel seeds, convex on their outside, plain on the other, and broad at the top.

We have but one Species of this genus, viz.

Phyllis (*Nobla*) stipulis dentatis. Prod. Leyd. 92. *Phyllis with indented stipulæ; or Simpla Nobla of the Canaries.*

This plant grows naturally in the Canary Islands. It rises with a soft shrubby stalk about two or three feet high, which is seldom thicker than a man's finger, of an herbaceous colour, and full of joints. These send out several small side branches toward the top, garnished with spear-shaped leaves, of a lucid green on their upper side, but pale on their under, having a strong whitish midrib, with several deep veins running from it to the sides. The flowers are produced at the end of the branches in loose panicles; they are small, and of an herbaceous colour at their first appearance, but before they fade, change to a brown or worn-out purple; they are cut into five parts to their base, where they are connected, and fall off without separating, so should be termed a flower of one petal. These segments are reflexed backward so as to cover the germen, which is situated

situated under the flower, and afterward becomes a short turbinated, obtuse, angular fruit, which splits in two parts when ripe, each containing one seed, flat on the inside, convex on the outside, and angular.

It is propagated by seeds, which must be sown on a bed of fresh light earth the beginning of April; the plants will come up by the beginning of May; when they are fit to transplant, they should be put into separate pots, and placed in a shady situation until they have taken root; after which time they should be placed in a sheltered situation, where they may have the morning sun, and in summer will require to be frequently watered. In winter they must be sheltered from the frost, but require to have as much free air as possible in mild weather; the second year the plants will flower, so if in the spring some of the plants are shaken out of the pots, and put into the full ground, they will perfect their seeds much better than those which remain in the pots.

As these plants seldom continue in health above four or five years, it will be proper to raise a supply of young plants to succeed them.

PHYSALIS. Lin. Gen. Plant. 223. Winter Cherry.

The CHARACTERS are,

The flower hath a swelling permanent empalement, which is five-cornered, and cut at the top into five acute points. The flower hath one wheel-shaped petal with a short tube, and a large brim, which is five-cornered and plaited. It hath five small awl-shaped stamina, which join together, and a roundish germen, supporting a slender style, crowned by an obtuse stigma. The germen afterward turns to an almost globular berry with two cells, inclosed in the large inflated empalement, filled with compressed kidney-shaped seeds.

The SPECIES are,

1. PHYSALIS (*Alkekengi*) foliis geminis. Lin. Sp. Plant. 183. *Physalis with two leaves at a joint; or the common Winter Cherry.*

2. PHYSALIS (*Viscosa*) foliis cordatis integerrimis obtusis scabris, corollis glabris. Lin. Sp. Plant. 183. *Physalis with rough, obtuse, entire, heart-shaped leaves, and smooth petals.*

3. PHYSALIS (*Pensylvanica*) radice perenni, caule procumbente, foliis ovatis acutè dentatis, petiolis longissimis. *Physalis with a perennial root, a trailing stalk, and oval leaves which are acutely indented, and have very long foot-stalks.*

4. PHYSALIS (*Virginiana*) caule herbaceo, foliis ovato-lanceolatis acutè dentatis. Tab. 206. fig. 1. *Winter Cherry with an herbaceous stalk, and oval spear-shaped leaves which are acutely indented.*

5. PHYSALIS (*Curassavica*) caule suffruticoso, foliis ovatis tomentosis integerrimis. *Physalis with a shrubby stalk, and oval downy leaves which are entire.*

6. PHYSALIS (*Somnifera*) caule fruticoso, foliis ovatis tomentosis. Lin. Vir. Cliff. 16. *Physalis with a shrubby stalk, and oval woolly leaves.*

7. PHYSALIS (*Flexuosa*) caule fruticoso, ramis rectis, floribus confertis. Lin. Sp. Pl. 180. *Physalis with a shrubby stalk, erect branches, and flowers growing in clusters.*

8. PHYSALIS (*Arborescens*) caule fruticoso, ramis flexuosis, floribus confertis. Lin. Sp. Plant. 182. *Physalis with a shrubby stalk, flexible branches, and flowers growing in clusters.*

9. PHYSALIS (*Ramosa*) foliis ovato-lanceolatis integerrimis oppositis, caule fruticoso. Tab. 206. fig. 2. *Physalis with oval, spear-shaped, entire leaves, which are placed opposite, and a shrubby stalk.*

10. PHYSALIS (*Angulata*) ramosissima, foliis villoso-viscosis pedunculis nutantibus. Lin. Sp. Plant. 183. *The most branching Physalis, with hairy viscous leaves, and nodding foot-stalks.*

11. PHYSALIS (*Minima*) ramosissima, ramis angulatis glabris. Lin. Sp. Pl. 183. *The most branching Physalis, with angular smooth branches.*

12. PHYSALIS (*Ramosissima*) ramosissima, foliis ovatis acuminatis subdentatis petiolis longioribus. *Very branching Physalis, with oval acute-pointed leaves which are somewhat indented, and have longer foot-stalks.*

13. PHYSALIS (*Patula*) ramosissima patula, ramis villosis, foliis ovatis acuminatis subdentatis. *The most branching spreading Physalis with woody branches, and oval acute-pointed leaves which are somewhat indented.*

14. PHYSALIS (*Villosa*) caule erecto ramoso, foliis ovatis serrato-dentatis, petiolis pedunculisque longissimis. *Physalis with an erect branching stalk, oval, indented, sawed leaves, and the foot-stalks of the leaves and flowers very long.*

15. PHYSALIS (*Cordata*) caule erecto ramoso, foliis ovato-lanceolatis viscosis, fructu maximo cordato. *Physalis with an erect branching stalk, oval, spear-shaped, viscous leaves, and a large heart-shaped fruit.*

16. PHYSALIS (*Maxima*) caule erecto ramoso, ramis angulatis, foliis sinuatis, calycibus acutangulis. *Physalis with an erect branching stalk, angular branches, sinuated leaves, and empalements having acute angles.*

The first sort is the common Winter Cherry, which is used in medicine; this grows naturally in Spain and Italy. The roots of this are perennial, and creep in the ground to a great distance, if they are not confined; these shoot up many stalks in the spring, which rise about a foot high or better, garnished with leaves of various shapes; some are angular and obtuse, others are oblong and acute-pointed; they have long foot-stalks. The flowers are produced from the wings of the stalks, standing upon slender foot-stalks; they have one white petal, which has a short tube, and is cut at the brim into five angles spreading open. In the center of the tube is situated a roundish germen, supporting a slender style, crowned by an obtuse stigma; this is accompanied by five stamina of the same length, terminated by oblong, erect, yellow summits, which join together. The flowers are succeeded by round berries about the size of small Cherries, inclosed in an inflated bladder, which turns red in the autumn, when the top opens and discloses the red berry, which is soft, pulpy, and filled with flat kidney-shaped seeds. Soon after the fruit is ripe, the stalks decay to the root.

This plant is easily propagated either by seeds or parting of the roots; the latter being the most expeditious method, is generally practised. These roots may be transplanted and parted, any time after the stalks decay, till the roots begin to shoot in the spring; they love a shady situation, and should be confined, otherwise they will ramble to a great distance in one year, and when the stalks stand at a distance, they make no appearance. Their only beauty is in the autumn, when the fruit is ripe, at which time their red bladders opening and disclosing the Cherry-shaped fruit, make a pretty appearance.

The second sort grows naturally at Buenos Ayres; this hath a creeping root, by which it multiplies very fast, sending up a great number of smooth stalks about two feet high, which divide toward their tops into small spreading branches, garnished with heart-shaped or oval leaves, standing upon pretty long foot-stalks. The flowers come out from the wings of the stalks toward the top, and have long slender foot-stalks; they are of a dirty yellow colour, with purple bottoms, and are succeeded by viscous berries about the size of those of the common sort, of an herbaceous yellow colour, inclosed in a swelling bladder of a light green colour.

This plant is easily propagated by parting of the roots either in the spring or autumn, but is too tender to live abroad

abroad through the winter in England, so should be planted in pots, and sheltered under a hot-bed frame in winter, where they may enjoy the free air at all times in mild weather.

The seeds of the third sort were sent me from Virginia, where the plant grows naturally; this hath a perennial root, but these roots do not creep in the ground like the two former. The stalks of this grow two feet long, and spread on the ground if they are not supported; these are garnished with oval leaves, standing alternately upon very long foot-stalks; they are of a pale green, having several acute indentures on their edges. The flowers come out from the wings of the larger stalk upon very short foot-stalks; they are larger than those of the common sort, and of a pale yellow colour. These are succeeded by very small yellowish berries, which ripen in the autumn, when the season proves warm, but in cool moist summers they come to nothing.

This sort is propagated by seeds, which should be sown upon a warm border about the latter end of March, and when the plants come up, they should be thinned where they are too close, and kept clean from weeds till autumn, when the plants must be transplanted where they are to remain, which should be in a warm situation, where they will live through the winter in mild seasons, but are killed by severe frost if they are not screened.

The seeds of the fourth sort were sent me from Philadelphia by Dr. Bensil, who found the plants growing there naturally. This hath a perennial root, composed of strong fibres, from which arise two or three hairy stalks about nine or ten inches high, dividing into several branches, garnished with oval, spear-shaped, hairy leaves, of a pale green, having several acute indentures on their edges. The flowers come out from the side of the branches at the base of the foot-stalks of the leaves; these have long slender foot-stalks; they have very short tubes, but are larger than most of the species of this genus, of a sulphur colour, with a dark purple bottom. These are succeeded by oval yellowish berries, which ripen in the autumn. This sort may be propagated by seeds in the same way as the third, and the plants require the same treatment.

The fifth sort grows naturally at Curassao in the West-Indies. This hath a perennial creeping root, from which arise several slender stalks about a foot high, which become somewhat ligneous, but do not last above two years; the leaves are small, oval, and hairy: the flowers come out from the wings of the stalk toward the top, standing upon slender foot-stalks; these are of a sulphur colour, and have dark purple bottoms, but are seldom succeeded by berries in England,

This is easily propagated by parting of the roots in the spring, but the plants are too tender to live through the winter in England without artificial warmth, so the pots should be placed in a moderate warmth in winter, but during the months of July, August, and September, they may be placed in the open air in a warm situation.

The sixth sort grows naturally at Curassao. This rises with a shrubby downy stalk about two feet high, dividing into several branches, covered with a thick soft down, garnished with oval woolly leaves. The flowers come out at the end of the branches, at the base of the foot-stalks of the leaves; they are small, of a yellow colour, and sit close to the branches, but are not succeeded by berries in England.

This sort may be propagated by cuttings, which must be planted in small pots filled with light loamy earth, and plunged into a moderate hot-bed the beginning of July; when they are rooted they may be removed into the open air, placing them in a sheltered situation. After they have obtained strength, they should be shaken out of the pots and separated, planting each in a distinct pot; then place them in the shade till they have taken new root, when they may be removed to their former situation, where they may remain till the end of September, and then be removed into the stove, where they should be placed in a moderate temperature of warmth during the winter season.

The seventh sort grows naturally in Crete, Sicily, and Spain. This rises with a shrubby stalk near three feet high, dividing into several branches, which grow erect, and are garnished with oval, spear-shaped, downy leaves. The flowers come out in clusters on the side of the branches; they are small, of an herbaceous white colour, sitting very close to the branches, and are succeeded by small berries, almost as large as those of the first sort, which, when ripe, are red.

This plant is propagated by seeds, which may be sown on a bed of light earth the beginning of April, and when the plants are two or three inches high they should be taken up, and each planted in a separate small pot, and placed in the shade till they have taken new root; then they may be removed to a sheltered situation, where they may remain till the beginning of October; at which time they should be removed into the green-house, for the plants are too tender to live through winter in the open air, so they must be treated like other green-house plants, but should be sparingly watered in winter. These plants will continue several years if they are not too tenderly treated.

The eighth sort grows naturally at Malabar, and also at the Cape of Good Hope. This rises to the height of five or six feet, sending out long flexible branches, covered with a gray bark, garnished with oblong oval leaves, which are often placed opposite. The flowers are produced in clusters at the base of the foot-stalks of the leaves; they are small, of an herbaceous yellow colour, and are succeeded by round purplish berries having ten cells, each including one seed.

This is propagated by seeds, which should be sown upon a moderate hot-bed, and the plants afterward treated in the same way as the last.

The ninth sort grows naturally at Campeachy. This hath a shrubby stalk, which rises ten or twelve feet high, covered with a gray hairy bark, garnished with oval spear-shaped leaves placed alternately, of a pale green, and downy. The flowers come out from the wings of the stalks toward the end of the branches, sometimes one, and at others two, are produced at the same joint opposite; they stand upon short nodding foot-stalks. The flowers are small, of a pale dirty yellow colour, having purple bottoms; these are succeeded by small, spherical, red berries, included in an oval dark purple bladder.

This may be propagated by seeds in the same way as the last mentioned, and the plants require the same treatment, but are not so hardy, therefore they must be kept in a moderate stove in winter, but in the middle of summer they should be placed in the open air, in a sheltered situation, for about three months. It may also be propagated by cuttings, which, if planted in pots during the summer months, and plunged into a gentle warmth, will take root freely, and may be treated in the same way as is before directed for the sixth sort.

The tenth sort is an annual plant, which grows naturally in Virginia. This branches out at bottom on every side; the branches frequently trail upon the ground; they are angular, and full of joints, garnished with hairy, viscous, heart-shaped leaves, standing upon pretty long foot-stalks, acutely indented on their edges. The flowers are produced on the side of the branches upon short, slender, nodding foot-stalks, of an herbaceous yellow colour with dark bottoms,

toms, which are succeeded by large swelling bladders, of a light green, inclosing berries as large as common Cherries, which are yellowish when ripe.

If the seeds of this sort are permitted to scatter, the plants will come up in the spring, and require no other care but to thin them, and keep them clean from weeds; or if the seeds are sown in the spring on a common border, the plants will rise very well, and need no other care.

The eleventh sort is also an annual plant, which grows naturally in the islands of the West-Indies. This rises with an upright branching stalk from two to three feet high. The branches are smooth, angular, and garnished with spear shaped leaves, ending in acute points, sharply indented on their edges. The flowers come out toward the end of the branches upon short slender foot-stalks; they are very small, of a dirty white colour, and are succeeded by berries the size of common Cherries, covered with an angular bladder; they are of a yellowish colour when ripe.

This sort is propagated by seeds, which should be sown on a moderate hot-bed; when the plants come up, and are a little advanced, they should be planted on a fresh hot-bed to bring them forward, and treated in the same way as the Capsicum. When they are grown strong, and are hardened to bear the open air, they may be transplanted with balls of earth to their roots into a warm border, where their seeds will ripen.

The twelfth sort grows naturally in the West-Indies. This is an annual plant with very branching stalks, which seldom rise above a foot high. The leaves are oval, of a deep green, and have long foot-stalks; the flowers are small, white, and stand upon short foot-stalks; the berries are small, and green when ripe.

The seeds of the thirteenth sort were sent me from Barbadoes. This is a low annual plant, seldom rising more than nine or ten inches high. The branches are hairy, and spread out almost horizontally; the leaves are oval, ending in acute points, and are a little indented; the flowers are small, white, and are succeeded by large fruit, inclosed in a long acute-pointed bladder.

The fourteenth sort was discovered at La Vera Cruz. This is an annual plant, with an upright branching stalk near two feet high, garnished with oval leaves, indented on their edges like a saw. They have long foot-stalks, and change to a purplish colour in the autumn. The flowers are small, white, standing upon very long foot-stalks, and are succeeded by large berries almost as large, and of the shape of Heart-cherries, of a yellowish green, with some purple stripes.

The fifteenth sort grows naturally in the same country. This is an annual plant, with a smooth, erect, branching stalk near three feet high, garnished with oval, spear-shaped, viscous leaves, standing on long foot-stalks. The flowers are small, of a pale yellow, and are succeeded by large heart-shaped fruit, of a pale yellow when ripe. The four last mentioned sorts are propagated by seeds in the same manner as the eleventh, and the plants require the same treatment.

The sixteenth sort grows naturally in Peru. This is an annual plant, rising with a strong, herbaceous, angular stalk four or five feet high, of a purplish colour, dividing into several angular branches, garnished with oblong leaves, which are deeply sinuated on their sides, of a deep green. The foot-stalks of the flowers are short; the empalement of the flower is large, bell-shaped, and deeply cut into five segments; the flower is large, of the open bell-shape, of a light blue colour, and is succeeded by berries about the size of common Cherries, inclosed in a large swelling bladder, having five sharp angles. If the seeds are permitted to scatter, the plant will come up the following spring, or if the seeds are sown on a bed of rich earth in the spring, the plants will rise easily, and may be afterward transplanted to the borders of the pleasure-garden, where they must be allowed room, for if the ground is good, the plants will grow very large.

Father Feuillee, who first discovered this plant in Peru, and has given a figure and description of it, recommends it greatly for its virtues, and says the Indians make great use of the berries to bring away gravel, and to relieve persons who have a stoppage of urine; and gives the manner of using them, which is, to bruise four or five of the berries either in common water, or white wine, giving it the patient to drink, and the success is astonishing.

PHYTOLACCA. Tourn. Inst. R. H. 299. tab. 154. American Night-shade.

The CHARACTERS are,

The flower hath no petals, though the cover of the parts of generation being coloured, is by some termed petals; there are five of these which are concave, open, and permanent. It has for the most part ten stamina, which are the same length as the petals, terminated by roundish summits, and ten compressed orbicular germen, joined together on their inside, but are divided on their outside, upon which sit ten very short styles, which are reflexed. The germen afterward turns to an orbicular depressed berry, with ten longitudinal deep furrows, having ten cells, each containing a single smooth seed.

The SPECIES are,

1. PHYTOLACCA (*Vulgaris*) foliis integerrimis, radice perenni. *Phytolacca with entire leaves, and a perennial root; commonly called Virginian Poke, or Porke Physic.*

2. PHYTOLACCA (*Mexicana*) foliis ovato-lanceolatis, floribus sessilibus. *Phytolacca with oval spear-shaped leaves, and flowers sitting close to the stalks.*

3. PHYTOLACCA (*Icosandra*) floribus icosandris decagynis. Lin. Sp. 631. *Phytolacca with the longest spikes of flowers, and an annual root.*

4. PHYTOLACCA (*Dioica*) floribus dioicis caule arborea ramosa. *Phytolacca with a tree-like stem, with male and female flowers on different plants.*

The first sort grows naturally in Virginia, and also in Spain and Portugal. It hath a very thick fleshy root as large as a man's leg, divided into several thick fleshy fibres.

When the roots are become large, they send out several stalks, which are herbaceous, as large as a good walking stick, of a purple colour, and rise to the height of six or seven feet, dividing into many branches at the top, garnished with large entire leaves, rounded at their base, but terminate in a point, and are placed without order, having short foot-stalks; in the autumn the leaves change to a purplish colour. From the joints of the branches, and at their divisions, come out the foot-stalks of the flowers about five inches long; the lower part is naked, but the upper half sustains a number of flowers ranged on each side like common Currants. The flowers have five purplish petals, or covers, within which stand the ten stamina and styles. After the flowers are faded, the germen turns to a depressed berry with ten furrows, having ten cells, filled with smooth seeds.

It may be propagated by sowing the seeds in the spring upon a bed of light earth; and when the plants come up, they should be transplanted into the borders of large gardens, allowing them space to grow, for they must not be planted too near other plants, lest they overbear and destroy them, as they grow to be very large, especially if the soil be very good. When they have taken root, they will require no farther care but only to clear them from weeds, and in the autumn they will produce their flowers and fruit; the stems of these plants constantly decay in the winter, but their roots will

will abide in the ground, and come up again the succeeding spring.

Parkinson says, that the inhabitants of North America make use of the juice of the root as a familiar purge; two spoonfuls of the juice will work strongly. Of late there have been some quacks, who pretend to cure cancers with this herb, but I have not met with one instance of its having been serviceable in that disorder. The inhabitants of North America boil the young shoots of this plant, and eat it like Spinach. The juice of the berries stain paper and linen of a beautiful purple colour, but it will not last long. If there could be a method of fixing the dye, it might be very useful.

The second sort grows naturally in the Spanish West-Indies, where the inhabitants constantly use it for their table. This plant is biennial, seldom continuing longer than two years, and when it flowers and produces plenty of seeds the first year, the plants frequently die before the following spring. This hath an herbaceous stalk about two feet high, dividing at the top into two or three short branches, garnished with oval spear-shaped leaves; they are of a deep green, and have foot-stalks an inch and a half long, placed without order. The foot-stalks of the flowers come out from the side of the branches opposite to the leaves; they are seven or eight inches long; the lower part, about two inches in length, is naked; the remaining part is garnished with white flowers, sitting close to the stalks; these are succeeded by flat berries, having many deep furrows, divided into so many cells, each containing one smooth seed.

Dr. Linnæus has supposed these two species were the same, but whoever sees the two plants growing, or attends to their descriptions, cannot doubt of their being different.

The third sort grows naturally in Malabar. This plant is annual, always perishing soon after it has perfected seeds, so that in this particular it differs greatly from the first: it rises with an herbaceous stalk from three to four feet high, which has several longitudinal furrows, and changes the latter part of summer to a purplish colour. It divides at the top into three or four branches, garnished with spear-shaped, dark, green leaves, standing upon short foot-stalks; sometimes alternately, at others they are placed opposite. The foot-stalks of the flowers come out from the side of the branches opposite to the leaves, the lower part being naked, as in the other sorts; the other part is garnished with larger flowers than those of the other sorts, white on their inside, of an herbaceous colour on their edges, and purplish on their outside, standing upon short foot-stalks; these have not always the same number of stamina, some of them have but eight, and others nine, which are terminated by roundish summits. The flowers are succeeded by orbicular, compressed, soft berries, divided by deep furrows on their outside into ten cells, each containing one smooth shining black seed; the racemus of flowers is very narrow at the top, where it is commonly inclined.

These two sorts are not so hardy as the first, so their seeds should be sown upon a moderate hot-bed in the spring, and when the plants are fit to remove, they should be transplanted to another hot-bed to bring them forward; then they should be treated in the same way as other tender exotic plants; the beginning of July they may be transplanted out upon a warm border, or into pots: they will require to be watered duly in dry weather, and kept clean from weeds. As these plants perfect their seeds every autumn, they may be easily preserved.

The fourth sort grows to the height of fourteen or sixteen feet, having a strong ligneous stem, which divides into many irregular branches, garnished with broad leaves; the flowers grow in a racemus like the others; but as those stopped which were in the Chelsea garden, so I had no opportunity of describing them.

This plant may be propagated by seeds or cuttings, but should be placed in a moderate stove in winter.

PIERCEA. Solanoides. Tourn. Act. Par. 1706.

The CHARACTERS are,

The flower has no petals; the empalement which incloses the parts of generation, is composed of four oblong, oval, coloured leaves. It hath four stamina, which stand erect and close together, terminated by small summits. In the center is situated a large roundish germen, supporting a short style, crowned by an obtuse stigma. The germen afterward turns to a roundish berry, sitting upon the reflexed empalement, having one cell, inclosing a rough seed of the same form.

I have taken the freedom of inscribing this genus of plants to His Grace Hugh Piercy, Duke of Northumberland, who is not only a great encourager of botanical studies, but greatly skilled in the science himself.

The title of Solanoides was applied to this plant, in the Memoirs of the Academy of Sciences for the year 1706. Dr. Linnæus has supposed this to be the same with Plumier's Rivinia, so he has continued that title to this plant, and believed Plumier was mistaken, when he drew eight stamina to the flower; but Plumier's Rivinia is totally different from this plant, and the flowers of it have eight stamina, as he has represented it; so Linnæus is mistaken.

The SPECIES are,

1. PIERCEA (*Glabra*) foliis ovato-lanceolatis glabris. *Piercea with oval, spear-shaped, smooth leaves.*

2. PIERCEA (*Tomentosa*) foliis cordatis pubescentibus. *Piercea with heart-shaped downy leaves.*

These plants grow naturally in most of the islands in the West-Indies, but the first is the most common there. It rises with a slender herbaceous stalk five or six feet high, which by age becomes a little ligneous at the bottom. It divides into many angular herbaceous branches, garnished with oval spear-shaped leaves, of a bright green, with slender foot-stalks. The foot-stalks of the flowers come out from the side of the branches, at the base of the foot-stalks of the leaves, sustaining a great number of small white flowers, ranged along the upper part on both sides. These are succeeded by small red berries full of a red juice, inclosing one hard seed of the same form.

The second sort spreads more than the first; the leaves are smaller, heart-shaped, and covered with short hairy down; the spikes of flowers are not so long, and are not so closely placed together, and have longer foot-stalks.

These plants are propagated by seeds, which should be sown soon after they are ripe, for if they are kept long out of the ground, they seldom grow the same year. They should be sown in pots, filled with light earth, and plunged into a moderate hot-bed. When the plants are come up two inches high, they should be each planted in a small halfpenny pot, and plunged into a moderate hot-bed; then they must be treated in the same way as other exotic plants. When the plants have obtained strength, they should be removed into the stove, and may be placed on shelves, and there they must constantly remain, for they are too tender to thrive in the open air in England in the warmest part of the year.

The juice of the berries of these plants will stain paper and linen of a bright red colour, and I have made many experiments with it to colour flowers, which have succeeded extremely well in the following manner. I pressed out the juice of the berries, and mixed it with common water, putting it into a phial, shaking it well together for some time, till the water was thoroughly tinged; then I cut off the flowers, which were white and just fully blown, and placed their stalks into the phial, and in one night the flowers have been finely variegated with red. The flowers, which I made the experiments on, were the Tuberose, and the double white Narcissus.

PILOSELLA. See HIERACIUM.

PIMPINELLA. Lin. Gen. Pl. 328. Burnet Saxifrage.

The CHARACTERS are,

It hath an umbellated flower; the principal umbel is composed of many rays or smaller umbels, neither of these have any involucrums; the greater umbel is uniform. The flowers have five heart-shaped inflexed petals, nearly equal, and five stamina longer than the petals, terminated by roundish summits. The germen is situated under the flower, supporting two short styles, crowned by obtuse stigmas. The germen afterward becomes an oblong oval fruit, divided into two parts, containing two oblong seeds, plain on the inside, convex on the other, and furrowed.

The SPECIES are,

1. PIMPINELLA (*Major*) foliis pinnatis, foliolis cordatis serratis, summis simplicibus trifidis. *Burnet Saxifrage, whose lower leaves are winged, and single three-pointed leaves at the top; greater Burnet Saxifrage.*

2. PIMPINELLA (*Saxifraga*) foliis pinnatis, foliolis radicalibus subrotundis, summis linearibus. Lin. Sp. Pl. 263. *Burnet Saxifrage with winged leaves, whose bottom lobes are roundish, but those at the top linear.*

3. PIMPINELLA (*Hircina*) foliis pinnatis, foliolis radicalibus pinnatifidis, summis linearibus trifidis. *Burnet Saxifrage with winged leaves, whose lobes of the bottom leaves are wing-pointed, and the upper ones linear and trifid; or lesser Burnet Saxifrage.*

4. PIMPINELLA (*Nigra*) foliis pinnatis hirsutis, foliolis radicalibus cordatis inæqualiter serratis, summis linearibus quinquefidis. *Burnet Saxifrage with hairy winged leaves, whose lobes of the bottom leaves are heart-shaped, unequally sawed, and the upper ones linear and five-pointed; or German Burnet Saxifrage.*

5. PIMPINELLA (*Austriaca*) foliis pinnatis lucidis, foliolis radicalibus lanceolatis, pinnato-serratis, summis linearibus pinnatifidis. *Burnet Saxifrage with shining winged leaves, the lobes of whose bottom leaves are spear-shaped and sawed, and the upper ones linear and wing-pointed; or largest Burnet Saxifrage of Austria.*

6. PIMPINELLA (*Peregrina*) foliis radicalibus pinnatis crenatis, summis cuneiformibus incisis. Lin. Sp. Pl. 164. *Burnet Saxifrage, whose lower leaves are winged, and indented on their edges, and the upper ones wedge-shaped and cut; or foreign Parsley with roundish leaves.*

7. PIMPINELLA (*Anisum*) foliis radicalibus trifidis incisis. Lin. Sp. Plant. 264. *Pimpinel with trifid, cut, lower leaves; or common Anise.*

The first sort grows naturally in chalky woods, and on the side of the banks near hedges, in several parts of England. The lower leaves of this sort are winged; the lobes are sharply sawed on their edges, and sit close to the midrib, of a dark green. The stalks are more than a foot high, dividing into four or five branches; the lower part of the stalk is garnished with winged leaves, shaped like those at the bottom, but are smaller: those upon the branches are short and trifid; the branches are terminated by small umbels of white flowers, which are composed of smaller umbels or rays. The flowers have five heart-shaped petals, which turn inward, and are succeeded by two narrow, oblong, channelled seeds.

The second sort grows naturally in dry pastures in many parts of England. The lower leaves of this are composed of four pair of lobes, terminated by an odd one; they are indented on their edges; the stalks rise near a foot high, sending out three or four slender branches, garnished with very narrow leaves. The umbels of flowers are smaller than those of the first, as are also the flowers and seeds.

The third sort grows naturally in dry gravelly pastures in several parts of England. The lower leaves of this sort have five or six pair of lobes, terminated by an odd one. The stalks are slender, and rise about a foot high, sending out a few small branches, which have a narrow trifid leaf placed at each joint, and are terminated by small umbels of white flowers, composed of several rays, standing upon pretty long foot-stalks.

The seeds of the fourth sort were sent me from Paris by Dr. Bernard de Jussieu. The lower leaves of this sort are composed of six or seven pair of heart-shaped lobes, terminated by an odd one; they are hairy, and of a pale green. The stalk rises near two feet high, dividing into several branches, which have one narrow five-pointed leaf at each joint, and are terminated with umbels of white flowers, like those of the first sort.

The seeds of the fifth sort I gathered in Dr. Boerhaave's private garden near Leyden, who had received the seeds of it from Austria. The lower leaves have five pair of lobes, terminated by an odd one; they are deeply cut in regular jags opposite, in form of a winged leaf, of a lucid green, and have long foot-stalks. The stalks rise more than two feet high, dividing at the top into two or three slender branches, garnished at each joint with one wing-pointed narrow leaf. The umbels of flowers are very like the first.

All these sorts have perennial roots; they are propagated by seeds, which, if sown in the autumn, will more certainly succeed than when they are sown in the spring. When the plants come up, they will require no other culture but to thin them where they are too close, and keep them clean from weeds; the second year they will flower and produce ripe seeds, and the roots will abide some years.

The first sort is directed for medicinal use, but the herb-women either bring the third sort to market for it, or what is worse, substitute Burnet and Meadow Saxifrage in its stead. It enters the pulvis ari compositis, and is esteemed good for the gravel.

The last sort is the common Anise. This is an annual plant, which grows naturally in Egypt, but is cultivated in Malta and Spain, from which countries the seeds are annually brought to England. From these seeds there is a distilled water, and an oil drawn for medicinal use. The pastrycooks also make great use of these seeds in several of their compositions, to give them an aromatic taste and smell. The lower leaves of this plant are divided into three lobes, which are deeply cut on their edges; the stalk rises a foot and a half high, dividing into several slender branches, garnished with narrow leaves, cut into three or four narrow segments, terminated by pretty large loose umbels, composed of many smaller umbels or rays, which stand on pretty long foot-stalks. The flowers are small, and of a yellowish white; the seeds are oblong and swelling.

The seeds of this should be sown the beginning of April upon a warm border, where the plants are to remain; when they come up they should be thinned, and kept clean from weeds, which i all the culture this plant requires, but it is too tender to be cultivated in England for profit.

PINASTER. See PINUS.

PINGUICULA. Butterwort.

This plant is found growing upon bogs in many parts of England, but is never cultivated in gardens, so I shall pass it over with barely mentioning it.

PINUS. Tourn. Inst. R. H. 585. tab. 355. The Pine-tree.

The CHARACTERS are,

The male flowers are collected in a scaly conical bunch; having many stamina, which are connected at their base, terminated by erect summits, included in the scales, which supply the want of petals and empalement. The female flowers are collected in a common

common oval cone, and stand at a distance from the male on the same tree. Under each scale of the cone is produced two flowers, which have no petals, but a small germen, supporting an awl-shaped style, crowned by a single stigma. The germen afterward becomes an oblong oval nut, crowned with a wing, included in the rigid scale of the cone.

The SPECIES are,

1. PINUS (*Sylvestris*) foliis geminis crassiusculis glabris, conis pyramidatis acutis. *Pine-tree with two thick smooth leaves in each sheath, and pyramidal acute cones*; *or Pineaster.*

2. PINUS (*Pinea*) foliis geminis tenuioribus glaucis, conis subrotundis obtusis. *Pine-tree with two narrower gray leaves coming out of each sheath, and roundish blunt cones; the cultivated Pine-tree, commonly called the Stone Pine.*

3. PINUS (*Rubra*) foliis geminis brevioribus glaucis, conis parvis mucronatis. *Pine-tree with two shorter gray leaves proceeding out of each sheath, and small acute-pointed cones; called Scotch Fir or Pine.*

4. PINUS (*Tartarica*) foliis geminis brevioribus latiusculis glaucis, conis minimis. *Pine-tree with two shorter broad leaves in each sheath, which are gray, and the smallest cones; commonly called Tartarian Pine.*

5. PINUS (*Montana*) foliis sæpius ternis tenuioribus viridibus, conis pyramidatis, squamis obtusis. *Pine with three narrow green leaves often in each sheath, and pyramidal cones with blunt scales; called Mugho.*

6. PINUS (*Cembro*) foliis quinis lævibus. H. Scan. 32. Lin. Sp. Plant. 1000. *Pine-tree with five smooth leaves in each sheath; called Cembro.*

7. PINUS (*Maritima*) foliis geminis longioribus glabris, conis longioribus tenuioribusque. *Pine-tree with two longer smooth leaves in each sheath, and longer narrower cones; the maritime Pine.*

8. PINUS (*Halepensis*) foliis geminis tenuissimis, conis obtusis, ramis patulis. Tab. 208. *Pine-tree with two narrow leaves in each sheath, obtuse cones, and spreading branches; Aleppo Pine.*

9. PINUS (*Virginiana*) foliis geminis brevioribus, conis parvis, squamis acutis. *Pine-tree with two shorter leaves in each sheath, and small cones with acute scales; commonly called Jersey Pine.*

10. PINUS (*Rigida*) foliis ternis, conis longioribus squamis rigidioribus. *Pine-tree with three leaves, and longer cones, having rigid scales; commonly called three-leaved Virginian Pine.*

11. PINUS (*Tæda*) foliis longioribus tenuioribus ternis, conis maximis laxis. *Pine-tree with three longer narrower leaves, and the largest loose cones; called the Frankincense-tree.*

12. PINUS (*Echinata*) Virginiana prælongis foliis tenuioribus, cono echinato gracili. Pluk. Alm. 297. *Virginian Pine with longer and narrower leaves, and a slender prickly cone; called three-leaved Bastard Pine.*

13. PINUS (*Strobus*) foliis quinis scabris. Lin. Sp. Pl. 1001. *Pine tree with five rough leaves in each sheath; commonly called Lord Weymouth's Pine.*

14. PINUS (*Palustris*) foliis ternis longissimis. *Pine-tree with the longest leaves, growing by threes out of each sheath; or three-leaved Marsh American Pine.*

There are some other species of this genus in America, which have not been sufficiently examined to ascertain their differences; and it is probable some of the European kinds, which are now supposed to be only varieties of the sorts here enumerated, may be distinct species; but as I have had no opportunity of seeing their cones, so I have omitted them here.

The first sort here enumerated is the Pineaster, or wild Pine, which grows naturally on the mountains in Italy and the south of France, where there are forests of these trees, which, if suffered to stand, grow to a large size; but they are frequently cut for making of pitch; and in the south of France, the young trees are cut for stakes to support their Vines. This grows to a large size; the branches extend to a considerable distance, and while the trees are young, they are fully garnished with leaves, especially where they are not so close as to exclude the air from those within; but as they advance in age, the branches appear naked, and all those which are situated below, become unsightly in a few years, for which reason they have not been much in esteem of late years; for as the wood of the Scotch Fir is much preferable to this, and the branches being generally better garnished with leaves, so the latter has been more generally propagated than the former. The leaves of this sort are of a dark green, and their points are obtuse. The cones are seven or eight inches long, pyramidal, and have pointed scales; the seeds are oblong, a little flatted on their sides, and have narrow wings on their tops.

The second sort, which is generally called the Stone Pine, is very common in Italy, but I much doubt of the country where it grows naturally; for so far as I have been able to learn, there are none of these trees growing in any part of Italy, but where they have been planted, or where the seeds have scattered from planted trees; and I have frequently received the seeds of a Pine from China, which were taken out of the cones so like those of this sort, as not to be distinguished from them; but these have never grown, either by their being too old, or from their having been taken out of the cones; for if the seeds of Pines are kept in the cones, they will grow at ten or twelve years old; but when they are out of their cones, they seldom grow well after two years, and some sorts do not grow after one. The leaves of this are not quite so long as those of the former sort, and are of a grayish or sea-green colour; the cones are not more than five inches long, but are very thick, roundish, and the scales end in an obtuse point; the seeds are more than twice the size of those of the former. The kernels of these are frequently served up in deserts to the table during the winter season in Italy, and formerly they were used in medicine here, but of late years the Pistachia Nuts have been generally substituted in lieu of them. The wood of this tree is white, not so full of resin as many of the other sorts, so is never cultivated for its timber, but chiefly for the beauty of its leaves and for the nuts, which are much esteemed in the south of France and in Italy.

The third sort is generally known here by the title of Scotch Pine, from its growing naturally in the mountains of Scotland; but it is common in most parts of Europe. Monf. du Hamel, of the Royal Academy of Sciences at Paris, mentions his having received cones of this tree from St. Domingo in the West-Indies, so concludes that it grows indifferently in torrid, frozen, and temperate zones. It is by John Bauhin titled, Pinus sylvestris, Genevensis vulgaris; so that it grows commonly in the mountains near that city, and all through Denmark, Norway, and Sweden. The wood of this tree is the red or yellow deal, which is the most durable of any of the kinds yet known. The leaves of this tree are much shorter and broader than those of the former sorts, of a grayish colour, growing two out of each sheath; the cones are small, pyramidal, and end in narrow points; they are of a light colour, and the seeds are small.

This sort grows well upon almost every soil; I have planted numbers of the trees upon peat-bogs, where they have made great progress. I have also planted them in clay, where they have succeeded far beyond expectation; and upon sand, gravel, or chalk, they likewise thrive as well; but as they do not grow near so fast upon gravel and sand, as upon moist ground, so the wood is much preferable; for those trees which have been cut down upon moist soils, foils, where they have made the greatest progress, when they have been sawn out into boards, have not been valuable, the wood has been white, and of a loose texture; whereas those which have grown upon dry gravelly ground, have proved nearly equal to the best foreign deals; and I doubt not but those plantations, which of late years have been made of these trees, will, in the next age, not only turn greatly to the advantage of their possessors, but also become a national benefit.

The fourth sort grows naturally in Tartary, from whence I received the seeds. This hath a great resemblance to the Scotch Pine, but the leaves are broader, shorter, and their points are more obtuse; they emit a strong balsamic odour when bruised; the cones of this are very small, as are also the seeds, some of which were black, and others white; but whether they are from different trees or the same, I could not learn, for the seeds were taken out of the cones, but in the parcel there was one entire cone.

The fifth sort grows naturally upon the mountains in Switzerland. This hath very narrow green leaves, which grow sometimes by pairs, and at others there are three coming out of each sheath; they generally stand erect; the cones are of a middle size, and pyramidal; the scales are flat, having each a small obtuse rising, but are very compact, till they are opened by the warmth of the sun. The seeds of this are much less than those of the Pineaster, but larger than those of the Scotch Pine.

The sixth sort grows naturally in Switzerland, and is by some persons supposed to be the same as the Siberian, but is different; for the cones of this are short and roundish, and the scales are close, whereas those of the Siberian Pine are long, and the scales loose; the leaves have a near resemblance to each other, but when compared, those of this tree are smooth, and the other are rough. The plants, which have been raised from the Switzerland seeds, have made much greater progress than those from the Siberian seeds; the latter are with difficulty kept alive in England. The leaves are long and narrow, smooth to the touch, of a light green, and five of them come out from the same sheath; the branches are closely garnished with them; the cones are about three inches long; the scales are pretty close; the seeds are pretty large, and their shells are easily broken.

The seventh sort grows in the maritime parts of Italy. This hath long smooth leaves, growing by pairs in each sheath; the cones are very long and slender; the seeds are about the size of those of the Pineaster.

The seventh sort grows naturally near Aleppo. This is a tree of middling growth in its native soil, and in England there are none of any large size; for most of the plants, which were growing here before the year 1740, were killed by the frost that severe winter. This tree branches out on every side near the root; the branches at first grow horizontally, but turn their ends upward; their bark is smooth, and of a dark gray colour. The leaves are long and very narrow, of a dark green, and grow by pairs in each sheath; if they are bruised, they emit a strong resinous odour. The cones come out from the side of the branches; they are not much more than half the length of those of the Pineaster, but are full as large at their base; the scales are flatted, and the point of the cone obtuse. The seeds are much less than those of the Pineaster, but of the same shape.

The ninth sort grows naturally in North America. This never rises to any great height, and is the least esteemed in the country of all the sorts. While the plants are young, they make a pretty good appearance, but when they get to the height of seven or eight feet, they become ragged and unsightly, so are not worth cultivating.

The tenth sort grows naturally in Virginia, and other parts of North America, where it rises to a great height; and so far as we can judge by the growth of those trees which are now here, it seems likely to become a large tree in England. The leaves of this are long, three generally standing in each sheath; the cones of this sort come out in clusters round the branches; they are as long as the cones of the Pineaster, and have rigid scales; the seeds are winged, and nearly as large as those of the Pineaster.

The eleventh sort grows naturally in North America. This hath very long narrow leaves, growing by threes out of each sheath; the cones are as large as those of the Stone Pine, but the scales are looser, and the cones more pointed. The scales of this open horizontally, and discharge the seeds. This sort was sent over from America to Mr. Ball of Exeter, and also to Dr. Compton, Bishop of London, by the title of Frankincense Pine.

The twelfth sort grows naturally in Virginia. The cones of this have been brought to England of late years, by the title of Bastard three-leaved Pine. The leaves of this sort are long and narrow; sometimes there are three growing in each sheath, and at others but two; the cones are long, slender, and their scales terminate in sharp points; they are rather longer than those of the Pineaster, and not so thick.

The thirteenth sort grows naturally in North America, where it is called the white Pine. It is one of the tallest trees of all the species, often growing a hundred feet high in those countries, as I have been credibly informed; the bark of this tree is very smooth and delicate, especially when young; the leaves are long and slender, five growing out of each sheath; the branches are pretty closely garnished with them, so make a fine appearance; the cones are long, slender, and very loose, opening with the first warmth of the spring; so that if they are not gathered in winter, the scales open and let out the seeds. The wood of this sort is esteemed for making of masts for ships; it is in England titled Lord Weymouth's, or New England Pine. As the wood of this tree was generally thought of great service to the navy, there was a law made in the ninth year of Queen Anne, for the preservation of the trees, and to encourage their growth in America; and it is within forty years past these trees began to be propagated in England in any plenty, though there were some large trees of this sort growing in two or three places long before, particularly at Lord Weymouth's at Longleet, and Sir Wyndham Knatchbull's in Kent; and it has been chiefly from the seeds of the latter, that the much greater number of these trees now in England have been raised; for although there has annually been some of the seeds brought from America, yet those have been few in comparison to the produce of the trees in Kent; many of the trees, which have been raised from the seeds of those trees, now produce plenty of good seeds, particularly those in the gardens of the late Duke of Argyle at Whitton, which annually produce large quantities of cones, which his grace when living most generously distributed to all the curious.

This sort and the Scotch Pine are the best worth cultivating of all the kinds for the sake of their wood; the others may be planted for variety in parks, &c. where they may make a good appearance in winter, when other trees are destitute of leaves.

All the sorts of Pines are propagated by seeds, which are produced in hard woody cones; the way to get the seeds out of those cones which are close, is to lay them before a gentle fire, which will cause the cells to open, and then the seeds may be easily taken out. If the cones are kept entire, the seeds will remain good some years, so that the surest way to preserve them, is to let them remain in the cones until the time for sowing the seeds; if the cones are kept

kept in a warm place in fummer, they will open and emit the feeds, but if they are not expofed to much heat, many of the forts will remain entire fome years, efpecially thofe which are clofe and compact; and the feeds which have been taken out of cones of feven years old, have grown very well, fo that thefe may be tranfported to any diftance, provided the cones are well ripened and properly put up.

The beft time for fowing the feeds of Pines, is about the end of March; when the feeds are fown, the place fhould be covered with nets to keep off birds, otherwife, when the plants begin to appear with the hufk of the feed on their tops, the birds will pick off the heads of the plants, and deftroy them.

Where the quantity of feeds to be fown is not great, it will be a good way to fow them either in boxes or pots, filled with light loamy earth, which may be removed from one fituation to another, according to the feafon of the year; but if there is a large quantity of the feeds, fo as to require a good fpace to receive them, they fhould be fown on an eaft or north-eaft border, where they may be fcreened from the fun, whofe heat is very injurious to thefe plants at their firft appearance above ground. Thofe feeds, which are fown in pots or boxes, fhould alfo be placed in a fhady fituation, but not under trees; and if they are fcreened from the fun with mats at the time when the plants firft come up, it will be a good method to preferve them.

Moft of the forts will come up in about fix or feven weeks after they are fown, but the feeds of the Stone or cultivated Pine, and two or three of the others, whofe fhells are very hard, frequently lie in the ground a whole year; fo that when the plants do not come up the firft year, the ground fhould not be difturbed, but kept clean from weeds, and the following fpring the plants will rife. This frequently happens in dry feafons, and when they are fown in places a little too much expofed to the fun.

When the plants appear, they muft be conftantly kept clean from weeds, and in very dry feafons, if they are now and then gently refrefhed with water, it will forward their growth; but this muft be done with great care and caution, for if they are haftily watered, it will wafh the tender plants out of the ground, or lay them down flat, which often rots their fhanks; and when this is too often repeated, it will have the fame effect; fo that unlefs it is judicioufly performed, it will be the beft way to give them none, but only fcreen them from the fun.

If the plants come up too clofe, it will be a good method to thin them gently about the beginning of July. The plants which are drawn up, may then be planted on other beds, which fhould be prepared ready to receive them, for they fhould be immediately planted as they are drawn up, becaufe their tender roots are foon dried and fpoiled at this feafon of the year. This work fhould be done (if poffible) in cloudy or rainy weather, and then the plants will draw out with better roots, and will foon put out new fibres again; but if the weather fhould prove clear and dry, the plants fhould be fhaded every day from the fun with mats, and now and then gently refrefhed with water. In drawing up of the plants, there fhould be great care taken not to difturb the roots of the plants left remaining in the feed-beds, &c. fo that if the ground be hard, the beds fhould be well watered fome time before the plants are thinned, to foften and loofen the earth; and if, after the plants are drawn out, the beds are again gently watered to fettle the earth to the roots of the remaining plants, it will be of great fervice to them; but it muft be done with great care, fo as not to wafh out their roots, or lay down the plants. The diftance, which fhould be allowed thefe plants, is four or five inches row from row, and three inches in the rows.

In thefe beds the plants may remain till the fpring twelve months after, by which time they will be fit to tranfplant where they are to remain for good, for the younger plants are, when planted out, the better they will fucceed; for although fome forts will bear tranfplanting at a much greater age, yet young plants, planted at the fame time, will in a few years overtake the large ones, and foon outftrip them in their growth; and there is an advantage in planting young, by faving the expence of ftaking, and much watering, which large plants require. I have feveral times feen plantations of feveral forts of Pines, which were made of plants fix or feven feet high, and at the fame time others of one foot high planted between them, which in ten years were better trees than the old ones, and much more vigorous in their growth; but if the ground, where they are defigned to remain, cannot be prepared by the time before-mentioned, the plants fhould be planted out of the beds into a nurfery, where they may remain two years, but not longer, for it will be very hazardous removing thefe trees at a greater age.

The beft feafon to tranfplant all the forts of Pines, is about the latter end of March, or the beginning of April, juft before they begin to fhoot; for although the Scotch Pine, and fome of the moft hardy forts, may be tranfplanted in winter, efpecially when they are growing in ftrong land, where they may be taken up with balls of earth to their roots, yet this is what I would not advife for common practice, having frequently feen it attended with bad confequences, but thofe which are removed in the fpring rarely fail.

Where thefe trees are planted in expofed fituations, they fhould be fet pretty clofe together, that they may fhelter each other; and when they have grown a few years, part of the plants may be cut down to give room for the others to grow; but this muft be gradually performed, left by too much opening the plantation at once, the air fhould be let in among the remaining trees with too great violence, which will ftop their growth.

Wherever large plantations are defigned to be made, the beft method will be to raife the plants either upon a part of the fame land, or as near to the place as poffible, and alfo upon the fame fort of foil; a fmall piece of ground will be fufficient to raife plants enough for many acres, but, as the plants require fome care in their firft raifing, if the neighbouring cottagers, who have many of them fmall inclofures adjoining to their cottages, or where this is wanting, a fmall inclofure fhould be made them for the purpofe of raifing the plants, and they are furnifhed with the feeds and directions for fowing them, and managing the young plants, till they are fit for tranfplanting, the women and children may be ufefully employed in this work; and the proprietors of land agreeing with them to take their plants, when raifed, at a certain price, it would be a great benefit to the poor, and hereby they would be engaged to have a regard for the plantations when made, and prevent their being deftroyed.

The Scotch Pine, as was before obferved, being the hardieft of all the kinds, and the wood of it the moft ufeful, is the fort which beft deferves care. This will thrive upon the moft barren fands, where fcarce any thing elfe except Heath and Furze will grow; fo that there are many thoufand acres of fuch land lying convenient for water carriage, which at prefent is of little benefit to any body, that might, by plantations of thefe trees, become good eftates to their proprietors, and alfo a national benefit; and as the legiflators have taken this into their confideration, and already paffed fome laws for the encouraging thefe plantations, as alfo for their prefervation and fecurity, fo it may be hoped that this will be undertaken by the gentlemen who are poffeffed of fuch lands in all the different parts of the kingdom,

dom, with proper ſpirit; for although they may not expect to receive much profit from theſe plantations in their own time, yet their ſucceſſors may with large intereſt; and the pleaſure which theſe growing trees will afford them, by beautifying the preſent dreary parts of the country, will in ſome meaſure recompenſe them for their trouble and expence, and by creating employment for the poor, leſſen thoſe rates which are now ſo high in many parts of England as ſcarce to be borne.

The expence of making theſe plantations is what moſt people are afraid of, ſo would not engage in it; but the greateſt of the expence is that of fencing them from the cattle, &c. for the other is trifling, as there will be no neceſſity for preparing the ground to receive the plants; and the charge of planting an acre of land with theſe plants, will not be more than thirty ſhillings where labour is dear, excluſive of the plants, which may be valued at forty ſhillings more. I have planted many acres of land with theſe trees, which was covered with Heath and Furze, and have only dug holes between to put in the plants, and afterward laid the Heath or Furze, which was cut, upon the ſurface of the ground about their roots, to prevent the ground drying, and few of the plants have failed. Theſe plants were moſt of them four years old from ſeed, nor was there any care taken to clean the ground afterward, but the whole left to ſhift, and in five or ſix years the Pines have grown ſo well, as to overpower the Heath and Furze, and deſtroy it.

The diſtance which I have generally planted theſe plants in all large open ſituations was about four feet, but always irregular, avoiding planting in rows as much as poſſible; and in the planting, the great care is not to take up the plants faſter than they can be planted, ſo that ſome men have been employed in digging up of the plants, while others were planting. Theſe who take up the plants, muſt be looked after to ſee they do not tear off their roots, or wound their bark; and as faſt as they are taken up, their roots ſhould be covered to prevent their drying, and put into their new quarters as ſoon as poſſible. In planting them, care ſhould be had to make the holes large enough for their roots, as alſo to looſen and break the clods of earth, and put the fineſt immediately about their roots, then to ſettle the earth gently with the foot to the roots of the plant. If theſe things are duly obſerved, and a proper ſeaſon choſen for performing it, there will be very little hazard of their ſucceeding; but I have ſeen ſome plantations made with plants, which were brought from a great diſtance, and had been ſo cloſely packed up, as to cauſe a heat, whereby moſt of the plants within had their leaves changed yellow, and few of them have grown, which has diſcouraged others from planting, not knowing the true cauſe of their failure.

After the plantations are made, the only care they require for five or ſix years, will be to ſecure the plants from cattle, hares, and rabbits, for if theſe are admitted to them, they will make great deſtruction in a ſhort time; for if the branches are gnawed by hares or rabbits, it will greatly retard the growth of the plants, if not deſtroy them.

In about five or ſix years after planting, the branches of the young trees will have met, and begin to interfere with each other, therefore they will require a little pruning; but this muſt be done with great caution. The lower tier of branches only ſhould be cut off; this ſhould be performed in September, at which time there will be no danger of the wounds bleeding too much, and the turpentine will harden over the wounds as the ſeaſon grows cold, ſo will prevent the wet from penetrating of them. Theſe branches ſhould be cut off cloſe to the ſtem of the plants, and care ſhould be taken in the doing of this, not to break any of the remaining branches of the young trees. This work ſhould be repeated every other year, at each time taking off only the lower tier of branches; for if the plants are much trimmed, it will greatly retard their growth, as it does in general that of all trees; but as theſe trees never put out any new ſhoots where they are pruned, ſo they ſuffer more from amputation than thoſe which do.

In thoſe parts of France, where they have foreſts of theſe trees, the proprietors always give the faggots to thoſe who prune their young trees firſt, for their labour, ſo it coſts them no money. At the ſecond pruning, the proprietor has one-third of the faggots, and the dreſſers have the other two for their work, and afterward the faggots are equally divided between the workmen and proprietors, but there muſt be great care taken that they do not cut off more than ſhould be.

In about twelve or fourteen years theſe will require no more pruning, for their upper branches will kill thoſe below where they have not air; but ſoon after this, if the plants have made good progreſs, it may be neceſſary to thin them; but this ſhould be gradually performed, beginning in the middle of the plantation firſt, leaving the outſide cloſe to ſcreen thoſe within from the cold, ſo by degrees coming to them at laſt, whereby thoſe, which were firſt thinned, will have had time to get ſtrength, ſo will not be in danger of ſuffering from the admiſſion of cold air. When theſe plantations are thinned, the trees ſhould not be dug up, but their ſtems cut off cloſe to the ground, for their roots never ſhoot again, but decay in the earth, ſo there can no harm ariſe by leaving them, and hereby the roots of the remaining plants are not injured. The trees which are now cut, will be fit for many purpoſes; thoſe which are ſtrait, will make good putlocks for the bricklayers, and ſerve for ſcaffolding poles, ſo that there may be as much made by the ſale of theſe, as will defray the whole expence of the planting, and probably intereſt for the money into the bargain.

As the upright growth of theſe trees renders their wood the more valuable, they ſhould be left pretty cloſe together, whereby they will draw each other up, and grow very tall. I have ſeen ſome of theſe trees growing, whoſe naked ſtems have been more than ſeventy feet high, and as ſtrait as a walking cane, and from one of theſe trees there were as many boards ſawed, as laid the floor of a room near twenty feet ſquare. If theſe trees are left eight feet aſunder each way, it will be ſufficient room for their growth; therefore if at the firſt thinning a fourth part of the trees are taken away, the other may ſtand twelve or fourteen years longer, by which time they will be of a ſize for making ladders, and ſtandards for ſcaffolding, and many other purpoſes; ſo that from this ſale, as much may be made, as to not only pay the remaining part of the expence of planting, if any ſhould be wanting in the firſt, but rent for the land with intereſt, and the ſtanding trees for fortunes of younger children. This may be demonſtrated by figures, and there has been ſeveral examples of late years, where the profits have greatly exceeded what is here mentioned.

The fifth ſort is called in Switzerland Torch Pine; the peaſants there make uſe of the wood of this tree inſtead of torches for burning. This tree grows to a great height in its native ſoil, and is well furniſhed with branches. The wood is pretty full of reſin, and when firſt cut is of a reddiſh colour. This is uſed by the inhabitants in their buildings.

The ſixth ſort of Pine makes but ſlow progreſs in England, ſo is not worth cultivating for profit, unleſs upon the ſummits of the northern mountains, where upon the peaty moors, this, and the Siberian Pine, are likely to ſucceed much better than in any other part of Britain, for they naturally grow among ſnow.

The

The eighth ſort is never a large tree in its native country, and in England it grows more like a ſhrub than a tree, and is often greatly injured by cold in winter; and by ſevere froſts ſometimes killed, ſo that this is only kept for the ſake of variety in the Engliſh gardens.

The ninth and tenth ſorts are uſed indifferently by the inhabitants of North America for their buildings, and the ſame purpoſes as the other ſorts of Pine.

There are ſome varieties of theſe in America, if they are not diſtinct ſpecies. Some of them ripen their cones the firſt year, but others are two years, and ſome three before they are ripe; but as theſe have not been well diſtinguiſhed by thoſe who reſide in that country, and there are few of the ſorts ſo large as to produce cones, ſo their differences cannot as yet be aſcertained.

The eleventh and twelfth ſorts I believe are indifferently called red Pine in North America, where their wood is greatly eſteemed; the French at Canada have built a ſixty gun ſhip entirely of this wood, called the Saint Laurent. I have had a little of this wood from America, which was very like that of the Scotch Pine, but had rather more reſin. It may not be amiſs to make trial of ſome of theſe ſorts in plantations, to ſee which of them may deſerve to be propagated; for in ſome places where they are growing, they thrive very well, but theſe will not ſucceed ſo well on dry land as moiſt.

The thirteenth ſort is called the white Pine in moſt parts of North America; of this I believe there are two varieties, if not diſtinct ſpecies; but as they have not been well examined by perſons of ſkill, we cannot take upon us to determine this, for Monſ. Gaultier's deſcription of one ſpecies is very different from that of the Weymouth Pine, and yet he has applied the title of white Pine to both.

This ſort deſerves to be propagated for its beauty, which is ſuperior to all the ſorts of Pines yet known in England. The bark of the young trees, and the branches, are perfectly ſmooth; the branches are well garniſhed with leaves; theſe are long, and of an agreeable green, ſo that in ſummer they have a beauty, and in winter they make a better appearance than any of the ſorts. The wood of this tree is very uſeful, eſpecially for maſts of ſhips, as the trees grow very tall and ſtrait, and are pliable, ſo do not break with the wind, therefore the legiſlature thought proper to paſs a law for the preſervation and increaſe of theſe trees in America; but as theſe trees will thrive in England, they may be propagated in many places where the ſoil is proper for them. This ſort grows beſt upon a moiſt light ſoil, but it ſhould not be too wet; it will alſo thrive on a loamy ſoil, if it is not too much approaching to clay. The ſeeds of this ſort ſhould be ſown with a little more care than thoſe of the Scotch Pine, becauſe their ſtems are not ſo ſtrong, therefore are more apt to go off while young; ſo if theſe are ſown in the full ground, the bed ſhould be ſcreened with mats from the ſun every day, but expoſed to the dews every night. When the plants come up, they ſhould be treated in the ſame way as is before directed for the Scotch Pine; and if all the plants of this kind are tranſplanted into beds in July, it will be a ſecure way to preſerve them; but as theſe plants will grow larger than thoſe of the Scotch Pine, they ſhould be planted farther aſunder; their rows ſhould be ſix inches diſtant, and in the rows they ſhould be four inches apart. This will allow them room to grow till the ſpring twelve month following, when they may be either tranſplanted where they are to remain, or into a nurſery, where they may ſtand two years to get ſtrength; but the ſooner they are planted where they are to ſtand, the leſs danger there will be of their ſucceeding, and the larger they will grow; for although they will bear tranſplanting at a greater age, yet when they are planted young, they will make much greater progreſs, and grow to a greater ſize.

The ſoil in which this ſort of tree thrives beſt, is a ſoft Hazel loam not too wet, in which I have frequently meaſured ſhoots of one year, which were two feet and a half long, and have for ſome years continued growing ſo much: they ſhould have a ſheltered ſituation, for I have obſerved where the trees have been much expoſed to the ſouth-weſt winds, they have not made near ſo great progreſs as thoſe which grew in ſhelter; and where there have been plantations of theſe trees, thoſe on the outſide have not kept pace with the middle, nor have their leaves retained their verdure ſo well.

The fourteenth ſort grows naturally on ſwamps in many parts of North America, where I have been informed they grow to the height of twenty-five or thirty feet. Their leaves are a foot or more in length, growing in tufts at the end of the branches, ſo have a ſingular appearance; but I have not heard the wood was of any uſe but for fuel, and there are few places here where theſe plants do well, for in very ſevere froſts their leading ſhoots are often killed, and in dry ground they will not thrive; ſo that unleſs the ſoil is adapted for them, it is to little purpoſe planting them.

From the wild Pine or Pineaſter is procured the common turpentine, which is chiefly uſed by the farriers, and from it is diſtilled the oil of turpentine. The finer and more valuable part, which comes firſt, is called the ſpirit, what is left at the bottom of the ſtill is the common reſin.

The kernels of the nuts of the manured or ſtone Pine, are of a balſamic nouriſhing nature, good for conſumptions, coughs, and hoarſeneſs, reſtorative, and of ſervice after long illneſs.

PIPER. Lin. Gen. Plant. 42. Pepper, or Lizard's-tail.

The CHARACTERS are,

The flowers are cloſely faſtened to a ſingle ſtalk, and have no complete ſheath; theſe have no petals nor ſtamina, but have two ſummits oppoſite to the root of the germen, which are roundiſh; they have a large oval germen, but no ſtyle, crowned by a prickly triple ſtigma. The germen afterward becomes a roundiſh berry with one cell, containing one globular ſeed.

The SPECIES are,

1. PIPER (*Obtuſifolium*) foliis obverſè ovatis enerviis. Lin. Sp. Pl. 30. *Pepper with obverſe oval leaves, having no veins; or low Lizard's tail with a fleſhy roundiſh leaf.*

2. PIPER (*Pellucidum*) foliis cordatis petiolatis, caule herbaceo. Lin. Sp. Pl. 30. *Pepper with heart-ſhaped leaves, having foot-ſtalks, and an herbaceous ſtalk.*

3. PIPER (*Amalago*) foliis lanceolatis-ovatis quinquenerviis rugoſis. Lin. Sp. Plant. 29. *Pepper with rough, ſpear-ſhaped, oval leaves, having five veins.*

4. PIPER (*Humile*) foliis lanceolatis nervoſis rigidis ſeſſilibus. *Pepper with ſtiff, ſpear-ſhaped, veined leaves ſitting cloſe to the branches.*

5. PIPER (*Peltatum*) foliis peltatis orbiculato-cordatis obtuſis repandis, ſpicis umbellatis. Lin. Sp. Plant. 30. *Pepper with target formed leaves, which are orbicular, heart-ſhaped, obtuſe, recurved, and have ſpikes growing in umbels.*

6. PIPER (*Lauriſolium*) foliis lanceolato-ovatis nervoſis, ſpicis brevibus. *Pepper with ſpear-ſhaped, oval, veined leaves, and ſhort ſpikes.*

7. PIPER (*Tomentoſum*) foliis ovato-lanceolatis tomentoſis, caule arboreſcente. *Pepper with oval, ſpear ſhaped, woolly leaves, and a tree-like ſtalk.*

8. PIPER (*Aduncum*) foliis ovato-lanceolatis, nervis alternis, ſpicis uncinatis. Lin. Sp. Plant. 29. *Pepper with oval ſpear-ſhaped leaves, having alternate veins, and crooked ſpikes.*

9. PIPER (*Decumanum*) foliis ovato-lanceolatis, acuminatis nervis alternis, ſpicis gracilis uncinatis. *Pepper with oval,*

oval, spear-shaped, acute-pointed leaves, having alternate veins, and slender crooked spikes.

10. PIPER (*Siriboa*) foliis cordato-ovatis nervosis acuminatis, spicis reflexis. *Pepper with oval, heart-shaped, nerved, acute-pointed leaves, and reflexed spikes.*

11. PIPER (*Glabrum*) foliis cordatis subseptinervis venosis. Flor. Zeyl. 29. *Pepper with heart-shaped leaves, which are veined, and have almost seven nerves.*

12. PIPER (*Racemosum*) foliis cordatis quinquenerviis reticulatis. Lin. Sp. Plant. 29. *Pepper with heart-shaped netted leaves, having five veins.*

The first sort grows naturally in many of the islands in the West-Indies. This sends out from the root many succulent herbaceous stalks almost as large as a man's little finger; they are jointed, and divide into many branches, never rising above a foot high, but generally spread near the ground, putting out roots at each joint, so propagate very fast, and soon cover a large space of ground. The leaves are very thick, succulent, broad, smooth, and entire. The foot-stalk, which sustains the spike or tail, comes out at the end of the branches; this is also very succulent, the whole length including the spike is about seven inches. The spike is strait, erect, about the size of a goose-quill, and closely covered with small flowers, which require a glass to be distinguished, so have no beauty; but the whole spike much resembles the tail of a lizard, for which Plumier gave that title to the genus.

These spikes appear great part of the year, but they rarely have any seeds in England; the plants increase very fast by their stalks, which put out roots. It requires a warm stove to preserve it in England, and should have but little wet, especially in winter. If the plants are plunged into the tan-bed in the stove, the stalks will put out roots into the tan, so may be cut off to make new plants.

The second sort grows naturally in the West-Indies; this is annual. The stalks are herbaceous and succulent; they rise about seven or eight inches high; the leaves are heart-shaped; the spikes of flowers, which are slender, come out at the end of the stalks; the flowers are very small, and sit close to the foot-stalk. These appear in July, and are succeeded by very small berries, each containing a small seed like dust. If these seeds are permitted to scatter on the pots near it, the plants will come up without trouble; or if the seeds are saved, and sown upon a hot-bed in the spring, the plants will rise easily. These should be transplanted into separate pots, and plunged into a hot-bed of tanners bark, treating them in the same way as other tender plants, but they should not have much wet.

The third sort grows naturally in Jamaica and Barbadoes. This hath several crooked stems, which rise to the height of twelve or fourteen feet, jointed, hollow, and pithy; these divide into many small branches, which are garnished with spear-shaped, oval, rough leaves, with five longitudinal veins. The spikes come out at the end of the branches, having many small flowers sitting close to the foot-stalk, which are succeeded by small berries.

The fourth sort grows naturally in Jamaica. The stalks of this are slender, and frequently trail upon the ground, putting roots out from their joints like the first; they are garnished with stiff spear-shaped leaves, with one strong midrib, and on the backside have several veins running from that to the sides. The spike of flowers is very slender, and shaped like those of the former sorts.

The fifth sort grows naturally in Jamaica. This hath a pretty thick spongy stalk, which rises fifteen feet high, dividing into several branches, which are jointed and pithy. The leaves are almost round; the foot-stalk is fastened to the under side, so that the upper surface has a mark like a navel, where the stalk joins, and from that center run out the veins to the side. The leaves are shaped like a heart, but the other part is round; the stalk being fixed toward the middle, the leaves have the appearance of a target. The spikes are small, and grow in form of an umbel.

The sixth sort grows naturally at La Vera Cruz in America. This hath shrubby jointed stalks, which rise nine or ten feet high, dividing into several branches, which are garnished with spear-shaped oval leaves; they are veined and rough, of the same consistence with Laurel leaves. The spikes of flowers come out from the side of the branch at the joints, opposite to the leaves; they are not more than one inch and a half long, about the thickness of a small quill, and are closely set with flowers like the other sorts.

The seventh sort grows naturally at La Vera Cruz. This hath hollow pithy stalks, which rise twelve or fourteen feet high, dividing into many crooked branches, having swelling joints, garnished with oval spear-shaped leaves, having many veins, and are covered with a woolly down. The spikes of flowers come out from the side of the branches, opposite to the leaves; they are slender, and turn downward.

The eighth sort grows naturally in Jamaica. This hath many hollow stalks, which rise about five feet high; the joints are close and protuberant; they divide into smaller branches, which are garnished with oval, spear-shaped, rough, veined leaves. The spikes of flowers come out from the side of the branches, opposite to the leaves; they are slender, and are closely set with small flowers their whole length. This is called Spanish Elder in the West-Indies.

The ninth sort was sent me by Mr. Robert Millar from Panama, near which place he found it growing naturally. This hath several pithy stalks, which rise about five feet high, divided into many small branches, garnished with oval spear-shaped leaves, ending in acute points. The spikes of flowers are very slender, and incurved.

The tenth sort was sent me from Carthagena. This rises with several shrubby stalks fifteen feet high, dividing into many slender branches, with protuberant joints, garnished with heart-shaped, oval, smooth leaves, ending in acute points, of a dark green on their upper side, but pale on their under. The spikes of flowers come out from the side of the branches; they are extremely slender, and are reflexed at the end like a scorpion's tail.

The eleventh sort grows at Panama. This hath hollow shrubby stalks, which rise about four feet high, divided into many small branches, garnished with heart-shaped leaves, ending in long acute points. The spikes come out from the side of the branches; they are slender, bending in the middle like a bow, and are closely set with small herbaceous flowers, which are succeeded by small berries, inclosing a small single seed.

The twelfth sort grows naturally in Jamaica. This rises with a shrubby pithy stalk about five feet high, sending out several side branches, which have protuberant joints, garnished with heart-shaped leaves, full of small veins, which form a sort of net work. The spikes come out from the side of the branches, opposite to the leaves: they are slender, a little bending in the middle, and are closely set with very small herbaceous flowers.

The ten last mentioned sorts are abiding plants, which require a warm stove to preserve them in England. They may be propagated by seeds, if they can be procured fresh from the countries where the plants grow naturally; these should be sown upon a good hot-bed in the spring, and when the plants come up, and are fit to transplant, they should be each put into a separate small pot, filled with light fresh earth, and plunged into a hot-bed of tanners bark, shading them every day from the sun till they have taken fresh root; then they must be treated in the same way as other tender exotic plants, admitting fresh air to them daily,

daily, in proportion to the warmth of the season, to prevent their drawing up weak; and when the nights are cold, the glasses of the hot-bed should be covered with mats to keep them warm. As the stalks of most of these plants are tender when young, so they should not have much wet, which would rot them; and when water is given to them, it must be with caution, not to beat down the plants, for when that is done they seldom rise again.

In autumn the plants must be plunged into the tan-bed of the bark-stove, and during the winter they must be sparingly watered; they require the same warmth as the Coffee-tree. In summer they require a large share of fresh air in hot weather, but they must be constantly kept in the stove, for they are too tender to bear the inclemency of our weather in summer.

PISONIA. Plum. Nov. Gen. 7. tab. 11. Lin. Gen. Pl. 984. Fingrigo, vulgò.

The CHARACTERS are,

The male flowers grow upon different plants from the fruit. The male flowers are funnel-shaped; the tube is short; the brim is expanded, and cut into five acute parts; they have five awl-shaped stamina, which are longer than the petal, terminated by obtuse summits. The female flowers are of the same form; they sit upon the germen, which is situated under the receptacle, supporting a cylindrical style longer than the petal, crowned by five oblong spreading stigmas. The germen afterward turns to an oval capsule, having five angles and one cell, containing one smooth, oblong, oval seed.

We have but one SPECIES of this genus, viz.

PISONIA (*Aculeata*) spinis axillaribus patentissimis. Lin. Sp. Pl. 1026. *Prickly Pisonia; called Fingrigo in the West-Indies.*

The male plants differ so much in appearance from the female, that those who have not seen them rise from the same seeds, would suppose they were different species, I shall therefore give short descriptions of each.

The male plants have stalks as thick as a man's arm, which rise ten or twelve feet high; the bark is of a dark brown colour, and smooth; these send out many branches opposite, which are much stronger than those of the female, so do not hang about so loose. They are garnished with obverse, oval, stiff leaves, standing opposite on short foot-stalks. From the side of the branches come out short cursons or spurs like those of the Pear-tree, having each two pair of small leaves at bottom, and from the top comes out the foot-stalk of the flowers, which is slender, dividing at the top into three; each of these sustain a small corymbus of herbaceous yellow flowers, having five stamina, standing out beyond the petal, terminated by obtuse summits.

The stalks of the female plants are not so strong as those of the male, so require support. These rise eighteen or twenty feet high, sending out slender weak branches opposite, armed with short strong hooked spines, and garnished with small oval leaves, standing opposite on the larger branches, but on the smaller they are alternate, and have short foot-stalks. The flowers are produced in small bunches at the end of the branches, sitting upon the germen; they are shaped like those of the male, but have no stamina; in the center is situated a cylindrical style, crowned by five spreading stigmas. The germen afterward turns to a channelled, five-cornered, glutinous capsule, armed with small crooked spines, each containing one oblong, oval, smooth seed.

These plants are very common in the savannas, and other low places in the island of Jamaica, as also in several other islands in the West-Indies; where it is very troublesome to whoever passes through the places of their growth, fastening themselves by their strong crooked thorns to the clothes of persons; and their seeds being glutinous and burry, also fasten themselves to whatever touches them; so that the wings of the ground-doves and other birds, are often so loaded with the seeds, as to prevent their flying, by which means they become an easy prey.

In Europe this plant is preserved in the gardens of some curious persons for variety; it is propagated by seeds, which should be sown in pots filled with light rich earth, and plunged into a hot-bed of tanners bark; and when the plants come up, they should be transplanted into separate pots, and plunged into the hot-bed again, where they may remain till Michaelmas, when they should be removed into the stove, and plunged into the bark-bed, and treated in the same manner as hath been directed for several tender plants of the same country.

PISTACIA. Lin. Gen. Plant. 982. Turpentine-tree, Pistachia Nut, and Mastick-tree.

The CHARACTERS are,

The male and female flowers grow upon separate trees; the male flowers are disposed in loose sparsed katkins, having small five-pointed empalements, but no petals; they have five small stamina, terminated by oval, four-cornered, erect summits. The female flowers have small trifid empalements, but no petals; they have each a large oval germen, supporting three reflexed styles, crowned by thick prickly stigmas. The germen afterward turns to a dry berry or nut, inclosing an oval smooth seed.

The SPECIES are,

1. PISTACIA (*Terebinthus*) foliis impari pinnatis, foliolis subovatis recurvis. Lin. Mat. Med. 454. Sp. Pl. 1025. *Pistachia with unequal winged leaves, whose lobes are somewhat oval and recurved; or the Pistachia-tree.*

2. PISTACIA (*Trifolia*) foliis subternatis. Hort. Cliff. 456. *Pistachia with trifoliate leaves; or three-leaved Turpentine-tree.*

3. PISTACIA (*Narbonensis*) foliis pinnatis ternatisque, suborbiculatis. Lin. Sp. Pl. 1025. *Pistachia with winged and trifoliate leaves, which are almost round.*

4. PISTACIA (*Vera*) foliis impari pinnatis, foliolis ovato-lanceolatis. Hort. Cliff. 456. *Pistachia with unequal winged leaves, whose lobes are oval and spear-shaped; or common Turpentine-tree.*

5. PISTACIA (*Lentiscus*) foliis abruptè pinnatis, foliolis lanceolatis. Hort. Cliff. 456. *Pistachia with abrupt winged leaves, and spear-shaped lobes; or common Mastick-tree.*

6. PISTACIA (*Massiliensis*) foliis abruptè pinnatis, foliolis lineari-lanceolatis. *Pistachia with abrupt winged leaves, and narrow spear-shaped lobes; or narrow-leaved Mastick-tree of Marseilles.*

7. PISTACIA (*Americana*) foliis impari pinnatis, foliolis lanceolato-ovatis acuminatis. *Pistachia with unequal winged leaves, whose lobes are spear-shaped, oval, and acute-pointed.*

8. PISTACIA (*Simaruba*) foliis pinnatis deciduis, foliolis oblongo ovatis. *Pistachia with winged deciduous leaves, having oblong oval lobes; commonly called Birch-tree in Jamaica.*

9. PISTACIA (*Orientalis*) foliis impari pinnatis, foliolis lanceolatis, exterioribus majoribus. *Pistachia with unequal winged leaves, whose lobes are spear-shaped, and the outer ones the largest.*

The first sort is the Pistachia Nut-tree, whose fruit is much better known in England than the tree. This grows naturally in Arabia, Persia, and Syria, from whence the nuts are annually brought to Europe. In those countries it grows to the height of twenty-five or thirty feet; the bark of the stem and old branches is of a dark russet colour, but that of the young branches is of a light brown; these are garnished with winged leaves, composed sometimes of two, and at others of three pair of lobes, terminated by an odd one; these lobes approach toward an oval shape, and their edges

edges turn backward; if these are bruised, they emit an odour like the shell of the nut. Some of these trees produce male flowers, others have female, and some, when old, have both on the same tree. The male flowers come out from the side of the branches, in loose bunches or katkins; they are of an herbaceous colour, having no petals, but have each five small stamina, crowned by large four-cornered summits, filled with farina; when that is discharged, the flowers fall off. The female flowers come out in clusters from the side of the branches; these have no petals, but have each a large oval germen, supporting three reflexed styles, and are succeeded by oval nuts. This tree flowers in April, but the fruit seldom ripens in England. It is propagated by the nuts, which should be planted in pots filled with light kitchen-garden earth, and plunged into a moderate hot-bed to bring up the plants; when these appear, they should have a large share of air admitted to them, to prevent their drawing up weak; and by degrees they must be hardened to bear the open air, to which they should be exposed the beginning of June, and may remain abroad till autumn, when they should be placed under a hot-bed frame to screen them from the frost in winter, for while they are young, they are too tender to live through the winter in England without protection, but they should always be exposed to the air in mild weather; the plants shed their leaves in autumn, so should not have much wet in winter; and in the spring, before they begin to shoot, they must be transplanted each into a separate small pot; and if they are plunged into a very moderate hot-bed, it will forward their putting out new roots; but as soon as they begin to shoot, they must be gradually hardened, and placed abroad again: the plants may be kept in pots three or four years till they have got strength, during which time they should be sheltered in winter; afterward they may be turned out of the pots, and planted in the full ground, some against high walls to a warm aspect, and others in a sheltered situation, where they will bear the cold of our ordinary winters very well, but in severe frosts they are often destroyed. The trees flower and produce fruit in England, but the summers are rarely warm enough to ripen the nuts.

The second sort grows naturally in Sicily and the Levant, where it is a tree of middling size, covered with a rough brown bark, dividing into many branches, garnished with leaves, which for the most part have three, but some have four oval lobes; they stand upon long foot-stalks, and are of a dark green colour. The male flowers grow upon different trees from the female, and are like those of the former sort, but of a yellowish green colour. The female flowers of this sort I have not seen, so can give no account of them; these are succeeded by fruit like that of the former, but much smaller. This is propagated by seeds in the same manner as the former, and the plants should be treated in the same way, but require more protection in winter. There were several plants of this kind in the English gardens before the year 1740, which had lived abroad some years against walls, but that severe winter killed most of them.

The third sort grows in Italy and the south of France, but is supposed to have been transplanted there from some other country. This is a tree of a middling size, covered with a light gray bark, sending out many side branches, garnished with leaves, which have sometimes five, and at others but three roundish lobes, standing upon pretty long foot-stalks, of a light green colour. The male flowers grow upon separate trees from the fruit, as in the other sorts; the fruit of this is small, but eatable. This is propagated by the nuts in the same way as the first, and the plants are equally hardy.

The fourth sort grows naturally in Barbary, and also in Spain and Italy. This is a tree of middling size, covered with a brown bark, dividing into many branches, whose bark is very smooth while young; it is garnished with winged leaves, composed of three or four pair of oval spear-shaped lobes, terminated by an odd one. The flowers are male and female on different trees, as the former: the male flowers of this have purplish stamina; they appear in April, but I have not seen any of the female trees in flower. This is propagated by seeds, but unless they are sown in autumn soon after they are ripe, they seldom grow the first year, but remain in the ground a whole year; and unless the seeds are taken from such trees as grow near a male, the seeds will not grow, as I have several times experienced.

The plants of this sort may be treated in the same manner as the first, and are as hardy. There is a tree of this sort now growing in the gardens of the Bishop of London at Fulham, against a wall, which was planted there above sixty years ago, which has endured the winters without cover; and some trees of this kind, which were planted in the open air, in the garden of his Grace the Duke of Richmond, at Goodwood in Sussex, had survived several winters without any protection. From these trees the common turpentine of the shops was formerly taken, but there is little of that now imported, but that from some of the cone-bearing trees is generally substituted for it.

The fifth sort is the common Mastick-tree, which is better known in the gardens by its Latin title of Lentiscus. This grows naturally in Spain, Portugal, and Italy, and being evergreen, the plants have been preserved in the English gardens, to adorn the green-house in winter. This in its native countries rises to the height of eighteen or twenty feet, covered with a gray bark, sending out many branches, which have a reddish brown bark, garnished with winged leaves, composed of three or four pair of small spear-shaped lobes, without an odd one at the end. The midrib which sustains the lobes, has two narrow borders or wings, running from lobe to lobe; these lobes are of a lucid green on their upper side, but pale on their under. The male flowers come out in loose clusters from the side of the branches; they are of an herbaceous colour, appearing in May, and soon fall off. These are generally upon different plants from the fruit, which also grows in clusters, and are small berries of a black colour when ripe.

The plants of this sort are generally propagated by laying down of their young branches, which, if properly managed, will put out roots in one year, and may then be cut off from the old plants, and each transplanted into separate small pots. These must be protected in winter, and in summer placed abroad in a sheltered situation, and treated in the same way as the other hardy kinds of green-house plants. It may also be propagated by seeds, in the same way as the Turpentine-tree, but if the seeds are not taken from trees growing in the neighbourhood of the male, they will not grow; and if they are kept out of the ground till spring, the plants rarely appear till the spring following. When these plants have obtained strength, some of them may be turned out of the pots, and planted against warm walls, where, if the branches are trained against the walls, they will endure the cold of our ordinary winters very well, and with a little shelter in severe winters they may be preserved.

The sixth sort grows naturally about Marseilles, and in some other places in the south of France, where it rises to the same height as the former, from which it differs in having one or two pair of lobes more on each leaf; the lobes are much narrower, and of a paler colour. This difference holds in the plants which are propagated by seeds, so may be be pronounced a diſtinct ſpecies. It is propagated in the ſame way as the former ſort, and is equally hardy.

The ſeventh ſort grows naturally in many of the iſlands in the Weſt-Indies, where it riſes to a middling ſtature, dividing into many branches, covered with a purpliſh bark, and garniſhed with winged leaves, compoſed of two or three pair of ſpear-ſhaped, having oval, acute-pointed lobes, terminated by an odd one; they are very thin and tender, and have long foot-ſtalks. The male flowers come out at the end of the branches; they are diſpoſed in a ſingle racemus (or long bunch;) they are of a purpliſh colour, and have yellow ſummits. The fruit grows upon ſeparate trees from the male flowers; they are ſhaped like the nuts of Piſtachia, but are ſmaller, and not eatable.

The eighth ſort grows naturally in Jamaica, and alſo in moſt of the other iſlands in the Weſt-Indies, where it riſes to the height of thirty or forty feet, covered with a looſe brown bark, which falls off in large pieces; the ſtems are large, and divide into many branches toward the top, which are crooked and unſightly; theſe are garniſhed with winged leaves, compoſed of five or ſix pair of oblong, oval, ſmooth lobes, terminated by an odd one. The flowers come out at the end of the branches, in long looſe bunches of a yellowiſh colour; theſe grow on different trees, or on different parts of the ſame tree from the fruit, which alſo hangs in long bunches, and is about the ſize of a middling Pea, having a dark ſkin, covering a nut about the ſize of a common Cherry-ſtone, and of the ſame colour.

Theſe two trees are tender, ſo will not thrive in this country, unleſs they are kept in a warm ſtove. They are propagated by ſeeds, which muſt be taken from ſuch trees as grow in the neighbourhood of the males, otherwiſe they will not grow, as I have too often found true. Theſe ſhould be ſown in pots filled with light earth, and plunged into a good hot-bed of tanners bark; and when the plants are come up fit to remove, they ſhould be each planted in a ſeparate ſmall pot, and plunged into a freſh hot-bed, treating them in the ſame way as other tender plants from the ſame countries, and in autumn they ſhould be removed into the ſtove, plunging the pots into the tan-bed; during the winter they muſt have but little water, eſpecially if they caſt their leaves, which is generally the caſe after the firſt winter; for the young plants frequently retain their leaves the whole year, but afterward they are deſtitute of leaves for two months, in the latter part of winter. The plants ſhould conſtantly remain in the ſtove, but in warm weather they muſt have a large ſhare of air admitted to them.

The ninth ſort is the true Maſtick-tree of the Levant, from which the Maſtick is gathered. This has been confounded with the common Lentiſcus, by moſt botanic writers; and Tournefort, who was on the ſpot where the Maſtick is collected, has not diſtinguiſhed the ſpecies, though he ſays, the leaves of the trees in the Levant, are larger than thoſe of the common ſort, but takes no notice of their being unequally winged. The ſeeds of this tree were ſent me by Monſ. Richard, gardener to the King of France at Verſailles, who received them from the iſland of Chio in the Levant: the bark of the tree is brown; the leaves are compoſed of two or three pair of ſpear-ſhaped lobes, terminated by an odd one: the outer lobes are the largeſt, the others gradually diminiſh, the innermoſt being the leaſt; theſe turn of a browniſh colour toward autumn, when the plants are expoſed to the open air; but if they are under glaſſes, they keep green. The leaves continue all the year, but are not ſo thick as thoſe of the common ſort, nor are the plants ſo hardy. It is propagated by ſeeds in the ſame way as the common Lentiſcus, but the plants while young ſhould be kept in a gentle temperature of warmth in winter, and require a warm ſheltered ſituation in ſummer. When they have obtained ſtrength, they may be kept in a warm green-houſe in winter, but ſhould have little water during that ſeaſon.

PISUM. Tourn. Inſt. R. H. 394. tab. 215. Pea.

The CHARACTERS are,

The flower hath an empalement cut into five points, the two upper being broadeſt; it hath four petals, and is of the butterfly kind. The ſtandard is broad, heart-ſhaped, reflexed, and indented, ending in a point. The two wings are ſhorter, roundiſh, and cloſe together; the keel is compreſſed, half-moon-ſhaped, and ſhorter than the wings. It hath ten ſtamina in two bodies; the upper ſingle one is plain and awl-ſhaped, the other nine are cylindrical below the middle, awl-ſhaped above, and cut; theſe are joined together. It has an oblong compreſſed germen, with a triangular riſing ſtyle. The germen afterward becomes a large, long, taper pod, terminated by a ſhort riſing point, opening with two valves, having one row of roundiſh ſeeds.

The SPECIES are,

1. PISUM (*Sativum*) ſtipulis infernè rotundatis crenatis, petiolis teretibus, pedunculis multifloris. Hort. Upſal. 215. *Pea whoſe lower ſtipulæ are roundiſh, indented, with taper foot-ſtalks, and many flowers on a foot-ſtalk; greater Garden Pea.*

2. PISUM (*Humile*) caule erecto ramoſo, foliis bijugatis, foliolis rotundioribus. *Pea with an erect branching ſtalk, and leaves having two pair of round lobes; Dwarf Pea.*

3. PISUM (*Umbellatum*) ſtipulis quadrifidis acutis, pedunculis multifloris terminalibus. *Pea with four-pointed acute ſtipulæ, and foot-ſtalks bearing many flowers, which terminate the ſtalks; the Roſe or Crown Pea.*

4. PISUM (*Maritimum*) petiolis ſuprà planiuſculis, caule angulato, ſtipulis ſagittatis, pedunculis multifloris. Flor. Suec. 608. *Pea with foot-ſtalks which are plain on their upper ſide, an angular ſtalk, arrow-pointed ſtipulæ, and foot-ſtalks bearing many flowers.*

5. PISUM (*Americanum*) caule angulato procumbente, foliolis inferioribus lanceolatis acutè dentatis, ſummis ſagittatis. *Pea with an angular trailing ſtalk, whoſe lower leaves are ſpear-ſhaped, ſharply indented, and thoſe at the top arrow-pointed; commonly called Cape Horn Pea.*

6. PISUM (*Ochrus*) petiolis decurrentibus membranaceis diphyllis, pedunculis unifloris. Hort. Cliff. 370. *Pea with membranaceous running foot-ſtalks, having two leaves, and one flower upon a foot-ſtalk.*

There is a great variety of Garden Peaſe now cultivated in England, which are diſtinguiſhed by the gardeners and ſeedſmen, and have their different titles; but as great part of theſe have been ſeminal variations, ſo if they are not very carefully managed, by taking away all thoſe plants which have a tendency to alter before the ſeeds are formed, they will degenerate into their original ſtate; therefore all thoſe perſons who are curious in the choice of their ſeeds, look carefully over thoſe which they deſign for ſeeds at the time when they begin to flower, and draw out all the plants which they diſlike from the other. This is what they call Roguing their Peaſe, meaning hereby, the taking out all the bad plants from the good, that the farina of the former may not impregnate the latter; to prevent which, they always do it before the flowers open; by thus diligently drawing out the bad, reſerving thoſe which come earlieſt to flower, they have greatly improved their Peaſe of late years, and are conſtantly endeavouring to get forwarder varieties; ſo that it would be to little purpoſe in this place, to attempt giving a particular account of all the varieties now cultivated; therefore I ſhall only mention their titles by by which they are commonly known, placing them according to their time of coming to the table, or gathering for use.

The Golden Hotſpur.	Nonpareil.
The Charlton.	Sugar Dwarf.
The Reading Hotſpur.	Sickle Pea.
Maſters's Hotſpur.	Marrowfat.
Eſſex Hotſpur.	Roſe, or Crown Pea.
The Dwarf Pea.	Rouncival Pea.
The Sugar Pea.	Gray Pea.
Spaniſh Morotto.	Pig Pea, with ſome others.

The Engliſh Sea Pea is found wild upon the ſhore in Suſſex, and ſeveral other counties in England, and is undoubtedly a different ſpecies from the common Pea.

The fifth ſort hath a biennial root, which continues two years. This was brought form Cape Horn by Lord Anſon's cook, when he paſſed that Cape, where theſe Peaſe were a great relief to the ſailors. It is kept here as a curioſity, but the Peaſe are not ſo good for eating as the worſt ſort now cultivated in England; it is a low trailing plant; the leaves have two lobes on each foot-ſtalk; thoſe below are ſpear-ſhaped, and ſharply indented on their edges, but the upper leaves are ſmall and arrow-pointed. The flowers are blue, each foot-ſtalk ſuſtaining four or five flowers; the pods are taper, near three inches long, and the ſeeds are round, about the ſize of Tares.

The ſixth ſort is annual. This grows naturally among the Corn in Sicily, and ſome parts of Italy, but is here preſerved in botanic gardens for the ſake of variety. It hath an angular ſtalk riſing near three feet high; the leaves ſtand upon winged foot-ſtalks, each ſuſtaining two oblong lobes. The flowers are of a pale yellow colour, ſhaped like thoſe of the other ſorts of Pea, but are ſmall, each foot-ſtalk ſuſtaining one flower; theſe are ſucceeded by pods about two inches long, containing five or ſix roundiſh ſeeds, which are a little compreſſed on their ſides. Theſe are by ſome perſons eaten green, but unleſs they are gathered very young they are coarſe, and at beſt not ſo good as the common Pea. It may be ſown and managed in the ſame way as the Garden Pea.

I ſhall now proceed to ſet down the method of cultivating the ſeveral ſorts of Garden Peaſe, ſo as to continue them throughout the ſeaſon.

It is a common practice with the gardeners near London, to raiſe Peaſe upon hot-beds, to have them very early in the ſpring; in order to which, they ſow their Peaſe upon warm borders, under walls or hedges, about the middle of October; and when the plants come up, they draw the earth up gently to their ſtems with a hoe, the better to protect them from froſt. In theſe places they let them remain until the latter end of January, or the beginning of February, obſerving to earth them up from time to time as the plants advance in height (for the reaſons before given;) as alſo to cover them in very hard froſt with Peaſe-haulm, ſtraw, or ſome other light covering, to preſerve them from being deſtroyed; then, at the time before-mentioned, they make a hot-bed (in proportion to the quantity of Peaſe intended) which muſt be made of good hot dung, well prepared and properly mixed together, that the heat may not be too great. The dung ſhould be laid from two to three feet thick, according as the beds are made earlier or later in the ſeaſon; when the dung is equally levelled, then the earth (which ſhould be light and freſh, but not over rich) muſt be laid thereon about ſix or eight inches thick, laying it equally all over the bed. This being done, the frames (which ſhould be two feet high on the back ſide, and about fourteen inches in front) muſt be put on, and covered with glaſſes; after which it ſhould remain three or four days, to let the ſteam of the bed paſs off, before you put the plants therein, obſerving every day to raiſe the glaſſes, to give vent for the riſing ſteam to paſs off; then, when you find the bed of a moderate temperature for heat, you ſhould, with a trowel, or ſome other inſtrument, take up the plants as carefully as poſſible, to preſerve the earth to their roots, and plant them into the hot-bed in rows, about two feet aſunder, and the plants about an inch diſtant from each other in the rows, obſerving to water and ſhade them until they have taken root; after which you muſt be careful to give them air at all times when the ſeaſon is favourable, otherwiſe they will draw up very weak, and be ſubject to grow mouldy and decay. You ſhould alſo draw the earth up to the ſhanks of the plants as they advance in height, and keep them always clear from weeds. The water they ſhould have muſt be given them ſparingly, for if they are too much watered, it will cauſe them to grow too rank, and ſometimes rot off the plants at their ſhanks, juſt above ground. When the weather is very hot, you ſhould cover the glaſſes with mats in the heat of the day, to ſcreen them from the violence of the ſun, which is then too great for them: but when the plants begin to fruit, they ſhould be watered oftener, and in greater plenty than before; for by that time the plants will have nearly done growing, and the often refreſhing them will occaſion their producing a greater plenty of fruit.

The ſort of Pea which is generally uſed for this purpoſe, is the Dwarf, for all the other ſorts ramble too much to be kept in frames; the reaſon for ſowing them in the common ground, and afterwards tranſplanting them on a hot-bed, is to check their growth, and cauſe them to bear in leſs compaſs; for if the ſeeds were ſown upon a hot-bed, and the plants continued thereon, they would produce ſuch luxuriant plants as are not to be contained in the frames, and would bear but little fruit.

The next ſort of Pea, which is ſown to ſucceed thoſe on the hot-bed, is the Hotſpur, of which there are reckoned ſeveral varieties; as the Golden Hotſpur, the Charlton Hotſpur, the Maſters's Hotſpur, the Reading Hotſpur, and ſome others; which are very little differing from each other, except in their early bearing, for which the Golden and Charlton Hotſpurs are chiefly preferred; though if either of theſe ſorts are cultivated in the ſame place for three or four years, they are apt to degenerate, and be later in fruiting; for which reaſon, moſt curious perſons procure their ſeeds annually from ſome diſtant place; and in the choice of theſe ſeeds, if they could be obtained from a colder ſituation and a poorer ſoil, than that in which they are to be ſown, it will be much better than on the contrary, and they will come earlier in the ſpring.

Theſe muſt alſo be ſown on warm borders, toward the latter end of October; and when the plants are come up, you ſhould draw the earth up to their ſhanks in the manner before directed, which ſhould be repeated as the plants advance in height (always obſerving to do it when the ground is dry) which will greatly protect the ſtems of the plants againſt froſt; and if the winter ſhould prove very ſevere, it will be of great ſervice to the plants to cover them with Peaſe-haulm, or ſome other light covering, which ſhould be conſtantly taken off in mild weather, and only ſuffered to remain on during the continuance of the froſt; for if they are kept too cloſe, they will draw up very weak and tender, and thereby be liable to be deſtroyed with the leaſt inclemency of the ſeaſon.

In the ſpring you muſt carefully clear them from weeds, and draw ſome freſh earth up to their ſtems, but do not raiſe it too high to the plants, leſt by burying their leaves you ſhould rot their ſtems, as is ſometimes the caſe, efficially in wet ſeaſons. You ſhould alſo obſerve to keep them clean from vermin, which, if permitted to remain amongſt the plants, will increaſe ſo plentifully, as to devour the greateſt part of them. The chief of the vermin which infeſt Peaſe, are the ſlugs, which lie all the day in the ſmall hollows of the earth, near the ſtems of the plants, and in the night-time come out, and make terrible deſtruction of the Peaſe; and theſe chiefly abound in wet ſoils, or where a garden is neglected, and over-run with weeds; therefore you ſhould make the ground clear every way round the Peaſe to deſtroy their harbours; and afterwards in a fine mild morning very early, when theſe vermin are got abroad from their holes, you ſhould ſlack a quantity of lime, which ſhould be ſtrown hot over the ground pretty thick, which will deſtroy the vermin wherever it happens to fall upon them, but will do very little injury to the Peaſe, provided it be not ſcattered too thick upon them. This is the beſt method I could ever find to deſtroy theſe troubleſome vermin.

If this crop of Peaſe ſucceeds, it will immediately follow thoſe on the hot-bed; but for fear this ſhould miſcarry, it will be proper to ſow two more crops at about a fortnight or three weeks diſtance from each other, ſo that there may be the more chances to ſucceed. This will be ſufficient until the ſpring of the year, when you may ſow ſeveral more crops of theſe Peaſe at a fortnight diſtance from each other. The late ſowings will be ſufficient to continue the early ſort of Peaſe through the ſeaſon, but it will be proper to have ſome of the large ſort of Peaſe to ſucceed them for the uſe of the family; in order to which, you ſhould ſow ſome of the Spaniſh Morotto, which is a great bearer, and a hardy ſort of Pea, about the middle of February, upon a clear open ſpot of ground. Theſe muſt be ſown in rows about four feet aſunder, and the Peaſe ſhould be dropped in the drills about an inch diſtance, covering them about two inches deep with earth, being very careful that none of them lie uncovered, which will draw the mice, pigeons, or rooks, to attack the whole ſpot; and it often happens, by this neglect, that a whole plantation is devoured by theſe creatures; whereas, when there are none of the Peaſe left in ſight, they do not ſo eaſily find them out.

About a fortnight after this you ſhould ſow another ſpot, either of this ſort, or any other large ſort of Pea, to ſucceed thoſe, and then continue to repeat ſowing once a fortnight, till the middle or latter end of May, only obſerving to allow the Marrowfats, and other very large ſorts of Peaſe, at leaſt four feet and a half between row and row; and the Roſe Pea ſhould be allowed at leaſt eight or ten inches diſtance plant from plant in the rows, for theſe grow very large, and if they have not room allowed them, they will ſpoil each other by drawing them up very tall, and will produce no fruit.

When theſe plants come up, the earth ſhould be drawn up to their ſhanks (as was before directed,) and the ground kept entirely clear from weeds; and when the plants are grown eight or ten inches high, you ſhould ſtick ſome bruſhwood into the ground cloſe to the Peaſe for them to ramp upon, which will ſupport them from trailing upon the ground, which is very apt to rot the large growing ſorts of Peaſe, eſpecially in wet ſeaſons; beſides, by thus ſupporting them, the air can freely paſs between them, which will preſerve the bloſſoms from falling off before their time, and occaſion them to bear much better, than if permitted to lie upon the ground, and there will be room to paſs between the rows to gather the Peaſe when they are ripe.

The dwarf ſorts of Peaſe may be ſown much cloſer together than thoſe before-mentioned, for theſe ſeldom riſe above a foot high, and rarely ſpread above half a foot in width, ſo that theſe need not have more room than two feet row from row, and not above an inch aſunder in the rows. Theſe will produce a good quantity of Peaſe, provided the ſeaſon be not over-dry; but they ſeldom continue long in bearing, ſo that they are not ſo proper to ſow for the main crop, when a quantity of Peaſe is expected for the table, their chief excellency being for hot-beds, where they will produce a greater quantity of Peaſe (provided they are well managed) than if expoſed to the open air, where the heat of the ſun ſoon dries them up.

The Sickle Pea is much more common in Holland than in England, it being the ſort moſtly cultivated in that country; but in England they are only propagated by curious gentlemen for their own table, and are rarely brought into the markets. This ſort the birds are very fond of, and if they are not prevented, many times deſtroy the whole crop. This ſhould be planted in rows about two feet and a half aſunder, and be managed as hath been directed for the other ſorts.

Although I have directed the ſowing of the large ſorts of Peaſe for the great crop, yet theſe are not ſo ſweet as the early Hotſpur Peaſe; therefore it will alſo be proper to continue a ſucceſſion of thoſe ſorts through the ſeaſon, in ſmall quantities, to ſupply the beſt table, which may be done by ſowing every fortnight; but all thoſe which are ſown late in the ſeaſon, ſhould have a ſtrong moiſt ſoil, for in hot light land they will burn up, and come to nothing.

The large growing ſorts may be cultivated for the common uſe of the family, becauſe theſe will produce in greater quantities than the other, and will endure the drought better, but the early kinds are by far the ſweeter taſted Peaſe.

The beſt of all the large kinds is the Marrowfat, which, if gathered young, is a well taſted Pea; and this will continue good through the month of Auguſt, if planted on a ſtrong ſoil.

The gray and other large winter Peaſe are ſeldom cultivated in gardens, becauſe they require a great deal of room, but are uſually ſown in fields in moſt parts of England. The beſt time for ſowing of theſe is about the beginning of March, when the weather is pretty dry, for if they are put into the ground in a very wet ſeaſon, they are apt to rot, eſpecially if the ground be cold; theſe ſhould be allowed at leaſt three feet diſtance row from row, and muſt be ſown very thin in the rows; for if they are ſown too thick, the haulm will ſpread ſo as to fill the ground, and ramble over each other, which will cauſe the plants to rot, and prevent their bearing.

The common white Pea will do beſt on light ſandy land, or on a rich looſe ſoil. The uſual method of ſowing theſe Peaſe is with a broad-caſt, and ſo harrow them in: but it is a much better way to ſow them in drills about three feet aſunder, for half the quantity of ſeed will do for an acre, and being ſet regularly, the ground may be ſtirred with a hoe to deſtroy the weeds, and earth up the Peaſe, which will greatly improve them, and the Peaſe may be much eaſier cut in autumn when they are ripe. The uſual time for ſowing of theſe Peaſe is about the middle or latter end of March, on warm land, but on cold ground they ſhould be ſown a fortnight or three weeks later. In the common way of ſowing, they allow three buſhels or more to an acre, but if they are drilled, one buſhel will be full enough.

The Green and Maple Rouncivals require a ſtronger ſoil than the white, and ſhould be ſown early in the ſpring; alſo the drills ſhould be made at a greater diſtance from each other, for as theſe are apt to grow rank, eſpecially in a wet ſeaſon, they ſhould be ſet in rows three feet and a half, or four feet aſunder; and the ground between the rows ſhould be ſtirred two or three times with a hoe, which will not only deſtroy the weeds, but, by earthing up the Peaſe, greatly

greatly improve them, and alſo render the ground better to receive whatever crop is put on it the following ſeaſon.

The gray Peaſe thrive beſt on a ſtrong clayey land; theſe are commonly ſown under furrow, but by this method they are always too thick, and do not come up regular; therefore all theſe rank-growing plants ſhould be ſown in drills, where the ſeeds will be more equally ſcattered, and lodged at the ſame depth in the ground; whereas in the common way, ſome of the ſeeds lie twice as deep as others, and are not ſcattered at equal diſtances. Theſe may be ſown toward the end of February, as they are much hardier than either of the former ſorts, but the culture of theſe ſhould be the ſame.

The beſt method to ſow theſe Peaſe is to draw a drill with a hoe by a line about two inches deep, and then ſcatter the ſeeds therein; after which, with a rake, you may draw the earth over them, whereby they will be equally covered: this is a very quick method for gardens, but where they are ſown in fields, they commonly make a ſhallow furrow with the plough, and ſcatter the ſeeds therein, and then with a harrow they cover them over again. After this, the great trouble is to keep them clear from weeds, and draw the earth up to the plants; this, in ſuch countries where labour is dear, is a great expence to do it by the hand with a hoe; but this may be eaſily effected with a ſmall plough, which may be drawn through between the rows, which will entirely eradicate the weeds, and by ſtirring the ſoil, render it mellow, and greatly promote the growth of the plants.

When any of theſe ſorts are intended for ſeed, there ſhould be as many rows of them left ungathered, as may be thought neceſſary to furniſh a ſufficient quantity of ſeed; and when the Peaſe are in flower, they ſhould be carefully looked over to draw out all thoſe plants which are not of the right ſort, for there will always be ſome roguiſh plants (as the gardeners term them) in every ſort, which, if left to mix, will degenerate the kind. Theſe muſt remain until their pods are changed brown, and begin to ſplit, when you ſhould immediately gather them up, together with the haulm; and if you have not room to ſtack them till winter, you may threſh them out as ſoon as they are dry, and put them up in ſacks for uſe; but you muſt be very careful not to let them remain too long abroad after they are ripe, for if wet ſhould happen, it would rot them, and heat, after a ſhower of rain, would cauſe their pods to burſt, and caſt forth their ſeeds, ſo that the greateſt part of them would be loſt; but, as I ſaid before, it is not adviſeable to continue ſowing of the ſame ſeed longer than two years, for the reaſons there laid down, but rather to exchange their ſeeds every year, or every two years at leaſt, whereby you may always expect to have them prove right.

PISUM CORDATUM. See CARDIOSPERMUM.

PITTONIA. See TOURNEFORTIA.

PLANTA, a plant, is defined by the ingenious Mr. John Martyn to be an organical body, deſtitute of ſenſe and ſpontaneous motion, adhering to another body in ſuch a manner, as to draw from it its nouriſhment, and having power of propagating itſelf by ſeed. As to the parts, of which a plant conſiſts, they are the root, ſtalk, leaf, flower, and fruit.

PLANTAGO. Tourn. Inſt. R. H. 126. tab. 48. Plantain.

To this genus Dr. Linnæus has joined the Coronopus and Pſyllium of Tournefort. The firſt of theſe is called Hartſhorn, the latter Fleawort. Of theſe there are ſeveral diſtinct ſpecies, and ſome varieties, but as they are rarely cultivated in gardens, I ſhall not enumerate them here, and ſhall only mention ſuch of them as grow naturally in England. Of the Plantain there are the following ſorts; the common broad-leaved Plantain, called Weybread; the great hoary Plantain, or Lambs-tongue; the narrow-leaved Plantain, or Ribwort, and the following varieties have alſo been found in England, which are accidental; the Beſom Plantain, and Roſe Plantain. The Plantains grow naturally in paſtures in moſt parts of England, and are frequently very troubleſome weeds. The common Plantain, and Ribwort Plantain, are both uſed in medicine, and are ſo well known as to need no deſcription.

Of the Coronopus, or Buckſhorn Plantain, there are two varieties growing in England, viz. the common Buckſhorn, which grows plentifully on heaths every where, and the narrow-leaved Welſh ſort, which is found upon many of the Welſh mountains. The firſt of theſe was formerly cultivated as a ſallad herb in gardens, but has been long baniſhed from thence for its rank diſagreeable flavour; it is ſometimes uſed in medicine. There has been one ſpecies of Pſyllium or Fleawort, found growing naturally in England, which is the ſort uſed in medicine; this was found in the earth, thrown out of the bottom of the canals, which were dug for the Chelſea water-works, where it grew in great plenty. The ſeeds of this muſt have been buried there ſome ages, for no perſon remembers any of the plants growing in that neighbourhood before. The ſeeds of this are ſometimes uſed, which are imported from the ſouth of France.

PLANTAIN-TREE. See MUSA.

PLANTING. Although the method of Planting the various ſorts of trees is fully ſet down under the ſeveral articles where each kind is mentioned, yet it may not be amiſs to ſay ſomething in general upon that head in this place, which ſhall be ſet down as briefly as poſſible. And,

Firſt, The firſt thing in the Planting of trees is to prepare the ground (according to the different ſorts of trees you intend to plant) before the trees are taken out of the earth; for you ſhould ſuffer them to remain as little time out of the ground as poſſible.

In taking up the trees, you ſhould carefully dig away the earth round their roots, ſo as to come at their ſeveral parts to cut them off; for if they are torn out of the ground without care, the roots will be broken and bruiſed very much, to the great injury of the trees. When you have taken them up, the next thing is to prepare them for Planting; in doing of which, there are two things to be principally regarded; the one is to prepare the roots, and the other to prune their heads in ſuch a manner, as may be moſt ſerviceable in promoting the future growth of the trees.

And firſt as to the roots; all the ſmall fibres are to be cut off as near to the place from whence they are produced as may be (excepting evergreens, and ſuch trees as are to be replanted immediately after they are taken up;) otherwiſe the air will turn all the ſmall roots and fibres black, which, if permitted to remain on when the tree is planted, will grow mouldy and decay, and thereby greatly injure the new fibres which are produced, ſo that many times the trees miſcarry for want of duly obſerving this. After the fibres are cut off, you ſhould prune off all the bruiſed or broken roots ſmooth, otherwiſe they are apt to rot, and diſtemper the trees; you ſhould alſo cut out all irregular roots which croſs each other, and all downright roots (eſpecially in fruit-trees) muſt be cut off; ſo that when the roots are regularly pruned, they may in ſome meaſure reſemble the fingers of a hand when ſpread open; then you ſhould ſhorten the larger roots in proportion to the age and ſtrength of the tree, as alſo the particular ſorts of trees are to be conſidered; for the Walnut, Mulberry, and ſome other tender-rooted kinds, ſhould not be pruned ſo cloſe, as the more hardy ſorts of fruit or foreſt-trees, which in young fruit-trees, ſuch as Pears, Apples, Plumbs, Peaches, &c. that are one year old from budding or grafting, may be left about eight or nine inches long, but in older trees they must be left of a much greater length; but this is to be understood of the larger roots only, for the small ones must be cut quite out, or pruned very short. Their extreme parts, which are generally very weak, commonly decay after moving, so that it is the better way entirely to displace them.

The next thing is the pruning of their heads, which must be differently performed in different trees, and the design of the trees must also be considered; for if they are fruit-trees, and intended for walls or espaliers, it is the better way to plant them with the greatest part of their heads, which should remain on until the spring, just before the trees begin to shoot; when they must be cut down to five or six eyes (as is fully set down in the several articles of the various kinds of fruit) being very careful, in doing of this, not to disturb the new roots.

But if the trees are designed for standards, you should prune off all the small branches close to the places where they are produced, as also irregular branches which cross each other; and by their motion, when agitated by the wind, rub and bruise their bark, so as to occasion many times great wounds in those places; besides, it makes a disagreeable appearance to the sight, and adds to the closeness of its head, which should be always avoided in fruit-trees, whose branches should be preserved as far distant from each other, as they are usually produced when in a regular way of growth (which is in all sorts of trees proportionable to the size of their leaves, and magnitude of their fruit.) But to return: After having displaced these branches, you should also cut off all such parts of branches as have by any accident been broken or wounded; for these will remain a disagreeable sight, and often occasion a disease in the tree. But you should by no means cut off the main leading shoots, as is by too many practised, for those are necessary to attract the sap from the root, and thereby promote the growth of the tree.

Having thus prepared the trees for Planting, we must now proceed to the placing them into the ground; but before this, I would advise, if the trees have been long out of the ground, so that the roots are dried, to place them in water eight or ten hours before they are planted, observing to put them in such manner, as that their heads may remain erect, and their roots only immersed therein, which will swell the dried vessels of the roots, and prepare them to imbibe nourishment from the earth. In fixing of them, great regard should be had to the nature of the soil, which, if cold and moist, the trees should be planted very shallow; as also, if it be a hard rock or gravel, it will be much the better way to raise a hill of earth where each tree is to be planted, than to dig into the rock or gravel, and fill it up with earth (as is too often practised,) whereby the trees are planted as it were in a tub, there being but little room for their roots to extend; so that after two or three years growth, when their roots have extended to the sides of the hole, they are stopped by the rock or gravel, so can get no farther, and the trees will decline, and in a few years die. But when they are raised above the surface of the ground, their roots will extend, and find nourishment, though the earth upon the rock or gravel be not three inches thick, as may be frequently observed, where trees are growing upon such soils.

Having thus planted the trees, you should provide a parcel of stakes, which should be driven down by the sides of the trees, and fastened thereto to support them from being blown down, or displaced by the wind, and then lay some mulch upon the surface of the ground about their roots, to prevent the earth from drying.

This is to be understood of standard trees, which cast their leaves, for such as are planted against walls, should have their branches fastened to the wall to prevent the trees from being displaced by the wind, and place their roots about five inches from the wall, inclining their heads thereto; and the spring following, just before they shoot, their heads should be cut down to five or six buds, as is fully directed under the several articles of the different kinds of fruit.

As to the watering of all new-planted trees, I should advise it to be done with great moderation, nothing being more injurious to them than over-watering. Examples enough of this kind may have been seen in many parts of England; and by an experiment made by the late Rev. Dr. Hales, in placing the roots of a dwarf Pear-tree in water, the quantity of moisture imbibed decreased very much daily, because the sap-vessels of the roots, like those of the cut-off boughs in the same experiment, were so saturated and clogged with moisture, by standing in water, that more of it could not be drawn up. And this experiment was tried upon a tree, which was full of leaves, and thereby more capable to discharge a large quantity of moisture than such trees as are entirely destitute of leaves; so that it is impossible such trees can thrive, where the moisture is too great about their roots.

The distance which trees should be planted at, must be proportioned to their several kinds, and the several purposes for which they are intended, all which is explained under their several heads; but fruit-trees, planted either against walls, or for espaliers, should be allowed the following distances: for most sorts of vigorous-shooting Pear-trees, thirty-six or forty-feet; for Apricots, sixteen or eighteen feet; Apples, twenty-five or thirty feet; Peaches and Nectarines, twelve feet; Cherries and Plumbs, twenty five feet, according to the goodness of the soil, or the height of the wall. But as these things are mentioned in their several articles, it will be needless to repeat any more in this place.

It is common to hear persons remarking, that from the present spirit of Planting, great advantages will accrue to the public by the increase of timber; but whoever is the least skilled in the growth of timber must know, that little is to be expected from most of the plantations which have lately been made; for there are few persons who have had this in their view when they commenced planters, and of those few scarce any of them have set out right; for there never was any valuable timber produced from trees which were transplanted of any considerable size, nor is any of the timber of the trees which are transplanted young, equal in goodness to that which has grown from the seeds unremoved. Beside, if we consider the sorts of trees which are usually planted, it will be found, that they are not designed for timber; so that upon the whole, it is much to be doubted, whether the late method of Planting has not rather been prejudicial to the growth and increase of timber than otherwise.

Before I quit this subject of Planting, I must beg leave to observe, that most people are so much in a hurry about Planting, as not to take time to prepare their ground for the reception of trees, but frequently make holes and stick in the trees, amongst all sorts of rubbish which is growing upon the land: and I have frequently observed, that there has not been any care afterward taken to dig the ground, or root out the noxious plants; but the trees have been left to struggle with these bad neighbours, which have had long possession of the ground, and have established themselves so strongly, as not to be easily overcome; therefore what can be expected from such plantations? This is to be understood of deciduous trees, for the Pines and Firs, if once well rooted in the ground, will soon get the better of the plants and destroy them.

Therefore I would advise every person who proposes to plant, to prepare the ground well before-hand, by trenching or deep ploughing it, and clearing of it from the roots of all bad weeds, for by ſo doing there will be a foundation laid for the future ſucceſs of the plantation. Alſo I adviſe no perſon to undertake more of this work than he can afterward keep clean, for all plantations of deciduous trees will require this care, at leaſt for ſeven years after they are made, if they hope to ſee the trees thrive well. Therefore all ſmall plantations ſhould have the ground annually dug between the trees; and as to thoſe which are large, it ſhould be ploughed between them. This will encourage the roots of the trees to extend themſelves, whereby they will find a much greater ſhare of nouriſhment, and by looſening of the ground, the moiſture and air will more eaſily penetrate to the roots, to the no ſmall advantage of the trees. But beſides this operation, it will be abſolutely neceſſary to hoe the ground three or four times in ſummer, either by hand or the hoe-plough. This I am aware will be objected to by many, on account of the expence; but if the firſt hoeing is performed early in the ſpring, before the weeds have gotten ſtrength, a great quantity of ground may be gone over in a ſhort time; and if the ſeaſon is dry when it is performed, the weeds will preſently die after they are cut; and if this is repeated before the weeds come up again to any ſize, it will be found the cheapeſt and very beſt huſbandry; for if the weeds are ſuffered to grow till they are large, it will be a much greater expence to root them out, and make the ground clean; beſide, the weeds will rob the trees of great part of their nouriſhment. I have ſometimes been told, that it is neceſſary to let the weeds grow among trees in ſummer, in order to ſhade their roots, and keep the ground moiſt; but this has come from perſons of no ſkill; but as others may have been deceived by ſuch advice, I imagine it may not be improper to give ſome anſwer to this. And here I muſt obſerve, that if weeds are permitted to grow, they will draw away all moiſture from the roots of the trees for their own nouriſhment, ſo that the trees will be thereby deprived of the kindly dews and gentle ſhowers of rain, which are of great ſervice to young plantations; and theſe will be entirely drawn away by the weeds, which will prevent their penetrating of the ground, ſo that it is only the great rains which can deſcend to the roots of the trees. And whoever has the leaſt doubt of this matter, if they will but try the experiment, by keeping one part of the plantation clean, and ſuffer the weeds to grow on another, they will ſoon be convinced of the truth by the growth of the trees. And though this cleaning is attended with an expence, yet the ſucceſs will overpay this, beſide the additional pleaſure of ſeeing the ground always clean.

PLATANUS. Tourn. Inſt. R. H. 590. tab. 363. The Plane-tree.

The CHARACTERS are,

It hath male and female flowers growing ſeparate on the ſame tree. The male flowers are collected in a round ball; they have no petals, but have oblong coloured ſtamina, which are terminated by four-cornered ſummits. The female flowers have ſmall ſcaly empalements, and ſeveral ſmall concave petals, with ſeveral awl-ſhaped germen ſitting upon the ſtyles, crowned by recurved ſtigmas; theſe are collected in large balls. The germen afterward turns to a roundiſh ſeed ſitting upon the briſtly ſtyle, and ſurrounded with downy hairs.

The SPECIES are,

1. PLATANUS (*Orientalis*) foliis palmatis. Hort. Cliff. 447. *Plane-tree with hand-ſhaped leaves; or Eaſtern Plane-tree.*

2. PLATANUS (*Occidentalis*) foliis lobatis. Hort. Cliff. 447. *Plane-tree with lobated leaves; Occidental, or Virginian Plane-tree.*

Theſe two are undoubtedly diſtinct ſpecies; but there are two others in the Engliſh gardens, which I ſuppoſe to be varieties that have accidentally riſen from ſeed; one is titled the Maple-leaved Plane-tree, and the other is called the Spaniſh Plane-tree.

The firſt ſort, or Eaſtern Plane-tree, grows naturally in Aſia. This riſes to a very great height; the ſtem is tall, erect, and covered with a ſmooth bark, which annually falls off; it ſends out many ſide branches, which are generally a little crooked at their joints; the bark of the young branches is of a dark brown, inclining to a purple colour, which are garniſhed with leaves placed alternate; their foot-ſtalks are long; the leaves are broad, deeply cut into five ſegments, and the two outer are ſlightly cut again into two more; theſe ſegments have many acute indentures on their borders; the upper ſide of the leaves are of a deep green, and the under ſide pale. The flowers come out upon long foot-ſtalks or ropes hanging downward, each ſuſtaining five or ſix round balls of flowers; the upper, which are the largeſt, are more than four inches in circumference; theſe ſit very cloſe to the foot-ſtalks. The flowers are ſo ſmall, as ſcarce to be diſtinguiſhed without glaſſes; they come out at the ſame time as the leaves, which is in June, and in warm ſummers the ſeeds will ripen late in autumn, and if left upon the trees, will remain till ſpring, when the balls fall to pieces, and the briſtly down which ſurrounds the ſeeds, help to tranſport them to a great diſtance with the wind.

The ſecond ſort grows naturally in moſt parts of North America. This tree alſo grows to a large ſize; the ſtem very ſtrait, and of equal girt moſt part of the length; the bark is ſmooth, and annually falls off like that of the other; the foot-ſtalks of the leaves are long; the leaves are broad, and are cut into angles, having ſeveral acute indentures on their borders, with three longitudinal midribs. They are of a light green on their upper ſide, and paler on their under. The flowers grow in round balls like the former, but are ſmaller. The leaves and flowers come out at the ſame time with the former, and the ſeeds ripen in autumn.

That which is called the Maple-leaved Plane, is certainly a ſeminal variety of the Eaſtern Plane, for the ſeeds which ſcattered from a large tree of this kind in the Chelſea garden, have produced plants of that ſort ſeveral times. This differs from the two ſorts before mentioned, in having its leaves not ſo deeply cut as thoſe of the Eaſtern Plane, but they are much deeper cut than thoſe of the Occidental Plane. The foot-ſtalks of the leaves are much longer than thoſe of either of the former, and the upper ſurface of the leaves is rougher, ſo that any perſon might take them for different ſpecies, who had not ſeen them riſe from the ſame ſeeds.

The Spaniſh Plane-tree has larger leaves than either of the other ſorts, which are more divided than thoſe of the Occidental Plane-tree, but not ſo much as the Eaſtern. Some of the leaves are cut into five, and others but three lobes; theſe are ſharply indented on their edges, and are of a light green. This is by ſome called the middle Plane-tree, from its leaves being ſhaped between thoſe of the two other ſorts. It grows rather faſter than either of the other ſorts, but I have not ſeen any very large trees of this kind.

The firſt ſort was brought out of the Levant to Rome, where it was cultivated with much coſt and induſtry: the greateſt orators and ſtateſmen among the Romans took great pleaſure in their villas, which were ſurrounded with Platani; and their fondneſs for this tree became ſo great, that we frequently read of their irrigating them with wine inſtead of water. Pliny affirms, that there is no tree whatſoever which ſo well defends us from the heat of the ſun in ſummer, nor that admits it more kindly in winter, the branches being produced at a proportionable diſtance to the largeneſs of their leaves; ſo that when the leaves are fallen

in

in winter, the branches growing at a great diſtance, eaſily admit the rays of the ſun.

It is generally ſuppoſed, that the introduction of this tree into England, is owing to the great Lord Chancellor Bacon, who planted a noble parcel of them at Verulam, which were there very flouriſhing ſome years ſince, but have lately been deſtroyed.

However, notwithſtanding the Plane-tree is backward in coming out in the ſpring, and the leaves decaying ſoon in autumn, yet for the goodly appearance and great magnitude to which it will grow, it deſerves a place in large plantations, or ſhady receſſes near habitations, eſpecially if the plantation be deſigned on a moiſt ſoil, or near rivulets of water, in which places this tree will arrive to a prodigious magnitude.

The Eaſtern Plane-tree is propagated either from ſeeds, or by layers, the latter of which is generally practiſed in England; though the plants thus raiſed ſeldom make ſo large ſtrait trees, as thoſe which are produced from ſeeds; but it has been generally thought, that the ſeeds of this tree were not productive, becauſe they have not been ſown at a proper ſeaſon, nor managed in a right manner; for I have had thouſands of the young plants ſpring up from the ſeeds of a large tree, which ſcattered upon the ground in a moiſt place; and I ſince find, that if theſe ſeeds are ſown ſoon after they are ripe, in a moiſt ſhady ſituation, they will riſe extremely well; and the plants thus obtained, will make a conſiderable progreſs after the ſecond year, being much hardier and leſs liable to loſe their tops in winter, than thoſe which are propagated by layers. And ſince the ſeeds of this tree ripen well in England, they may be propagated in as great plenty as any other foreſt-tree.

The Virginian Plane-tree will grow extremely well from cuttings, if they are planted the beginning of October upon a moiſt ſoil, and if they are watered in dry weather, will make a prodigious progreſs; ſo that in a few years from the planting, they will afford noble trees for planting of avenues, and other ſhady walks; and their trunks are perfectly ſtrait, growing nearly of the ſame ſize to a conſiderable height, there being the leaſt difference in the girt of this tree, for ſeveral yards upwards, of any other ſort of tree whatſoever.

They are all propagated very eaſily by layers, every twig of them will take root if they are but covered with earth; theſe layers will be well rooted in one year, when they ſhould be cut off from the old trees or ſtools, and planted in a nurſery, where they may remain two or three years to get ſtrength, and then tranſplanted where they are to remain; for the younger theſe trees are planted, the better they will thrive. An experiment of this I made in 1731, when I planted one of theſe trees, whoſe ſtem was eight inches in girt; and near it, in the ſame ſoil and ſituation, I planted another, whoſe girt was not three inches, and the latter is now much larger than the former, and gains more in one year than the other does in three.

PLOUGHING OF LAND.

There is not a greater improvement of arable land than that of well Ploughing it, by which the ſoil is pulverized, and rendered fit to receive the fibres of plants; the oftener this is repeated, and the better it is performed, the greater improvement is made. But there are not many of the practitioners of the art of huſbandry, who attend enough to this part of it, moſt of them contenting themſelves with going on in the old beaten road of their predeceſſors; ſo that the only perſons who have made any conſiderable improvement in this part of agriculture, are the great gardeners near London, who cultivate moſt of their land with the plough, in which they have imitated, as near as poſſible, the uſe of the ſpade in labouring of their ground.

The difference between digging of land with the ſpade, and that of Ploughing, conſiſt in the parts of the earth being much more divided by the former than the latter method; therefore thoſe gardeners, who are curious in the working of their land, oblige their labourers to ſpit the ground as thin as poſſible, that there may remain no large clods unbroken; ſo, when land is ploughed, the ſame regard ſhould be had to break and pulverize the parts as much as poſſible; for when there are great clods left unbroken, the fibres of plants never penetrate farther than the ſurface of them; ſo that all the ſalts, included in theſe lumps of earth, are locked up, that the plants can receive no benefit from them. And theſe clods, in proportion to their ſize, make ſuch interſtices, that the air often penetrates through, and greatly injures the tender fibres of the roots. Therefore the oftener the land is ploughed, and the more the parts are ſeparated and pulverized by the plough and harrow, the better will the plants be nouriſhed and fed; but particularly in all ſtrong land, this part of huſbandry will be the moſt beneficial; but this cannot be effected under four or five Ploughings, and by uſing ſuch ploughs as have either two or four coulters, which will cut and ſeparate the clods much better than it can be performed by the common plough; and in the operation, great care ſhould be had to the breadth of the furrow, for when theſe are made too broad, it will be impoſſible to break and ſeparate the parts ſufficiently. In ſome counties, where the huſbandmen are not very expert in the uſe of the plough, I have ſeen gentlemen oblige them to plough by a line, and they have ſet out the exact width of each furrow. This not only adds a neatneſs to the ground, but likewiſe by keeping the furrows ſtrait, and at equal diſtances, the land will be more equally worked; but many of the good ploughmen, in the counties near London, will direct the plough as ſtrait by their eye, as if they were to uſe a line.

Another thing to be obſerved in Ploughing of land is, that of going to a proper depth; for if the ſurface only be broken up and pulverized, the roots of whatever plants are ſown or planted in it, will in a very ſhort time reach the bottom, and meeting with the hard unbroken ſoil, they are ſtopped from getting farther, and of conſequence the plants will ſtint in their growth; for there are few perſons who have attended enough to the downright growth of the roots of plants; they only have had regard to the roots of thoſe plants, which are of a ſtrong fleſhy ſubſtance, and are called tap-roots, being in form of Carrots. Theſe they ſuppoſe will require to have the land wrought to a greater depth, that the roots may run down, and be the longer, for in that particular their goodneſs conſiſts. But they do not think that the ſmall fibrous-rooted plants ever require ſo much depth to run into the ground, and in this they are greatly miſtaken; for I have traced the ſmall fibres of Graſs and Corn above three feet deep in the ground. And if any perſon is curious to obſerve the length of the fibres of plants, if they will but plant one of each ſort into a ſmall pot of earth, and keep them duly watered till the plants are advanced to flower, and then turn them out of the pots carefully, ſo as not to break any of the fibres of the roots, and after ſeparating the earth from them, meaſure the length of their roots, they will be found much greater than moſt people imagine. I have myſelf frequently traced the roots of plants, which have ſurrounded the pots upwards of twelve times, and the roots of ſome ſtrong growing plants, which have gotten through the holes in the bottom of flower-pots, have in three months time extended themſelves ten or eleven feet from the plant; therefore the deeper the ground is laboured, the greater benefit the plants will receive from it; but it muſt be underſtood only of ſuch land as the ſtaple is deep enough to admit of this, for if the ſoil is ſhallow, and either gravel, chalk, or ſtone lie beneath, it will be very imprudent to turn up either of theſe; therefore the depth of the furrows in ſuch lands, muſt be determined by the ſtaple of the land. By the word ſtaple muſt be underſtood all that depth of ſoil next the ſurface, which is proper for the growth of vegetables. Where clay is next the ſtaple, provided it is not of the blue or ironmould ſort, there will not be the ſame danger of going a little deeper than the ſtaple, as in either of the before-mentioned ſorts of land; for if the clay be of a fat nature, when it hath been well expoſed to air, and often laboured, it will be capable of affording a large ſhare of nouriſhment to the crops.

If between each Ploughing of the land a harrow with long teeth is made uſe of to tear and break the clods, it will be of great ſervice to the land, eſpecially if it is ſtrong, for the more it is ſtirred by different inſtruments, the better will the parts be ſeparated and pulverized; ſo that the common method, as practiſed by the farmers when they fallow their land, is far from anſwering their intention, for they plough up the ground, leaving it in great clods for ſome months; and frequently, during this time, Thiſtles and all bad weeds are ſuffered to grow upon the land, and exhauſt the goodneſs of it, and perhaps, juſt before the ſeeds are ſown, they give it two more Ploughings. This is what the farmers call good huſbandry; but if inſtead of this method they would labour the ground often with the plough, a harrow, and heavy roller, to break and ſeparate the parts, and never ſuffer any weeds to grow upon the land during its lying fallow, I am ſure they would find their account in it; firſt, by the growth and increaſe of their crops, and afterward by a ſaving in the weeding; for if no weeds are ſuffered to grow to ſhed their ſeeds, during the time of fallowing the land, there will but few come up when the ground is ſown, in compariſon with what would otherwiſe, in the common huſbandry.

PLUMBAGO. Tourn. Inſt. R. H. 140. tab. 58. Lin. Gen. Plant. 196. Leadwort.

The CHARACTERS are,

The flower has a tubulous permanent empalement, which is indented at the top in five parts; it hath one funnel ſhaped petal, with a cylindrical tube, narrow at the top. The brim is cut into five oval parts; it has five awl-ſhaped ſtamina, ſituated in the tube, ſitting upon the valves of the nectarium, which includes the germen. The ſmall oval germen ſuſtains a ſingle ſtyle the length of the tube, crowned by a ſlender five-pointed ſtigma. The germen afterward becomes a ſingle oval ſeed, included in the empalement.

The SPECIES are,

1. PLUMBAGO (*Europæa*) foliis amplexicaulibus. Hort. Cliff. 53. *Leadwort with leaves embracing the ſtalks; common Leadwort or Toothwort.*

2. PLUMBAGO (*Zeylanica*) foliis petiolatis. Hort. Cliff. 53. *Leadwort with leaves having foot-ſtalks.*

The firſt ſort grows naturally in the ſouth of France, in Italy, and Spain. This hath a perennial root, which ſtrikes deep into the ground, from which ariſe many ſlender channelled ſtalks three feet high, garniſhed with oval ſpear-ſhaped leaves, whoſe baſe embraces the ſtalks; they are ſmooth, entire, and of a grayiſh colour. The upper part of the ſtalks ſend out many ſhort ſlender branches, garniſhed with ſmall leaves. Theſe, and alſo the principal ſtalks, are terminated by tufts of blue flowers, which are ſmall, funnel ſhaped, and have pretty long tubes; theſe are ſucceeded by oblong, rough, hairy ſeeds. This plant ſeldom flowers till October in England; and for many years, when the autumns have been unfavourable, have not produced flowers, ſo never produces ripe ſeeds here. There is a variety of this with white flowers and pale ſtalks, which is ſuppoſed to have riſen from the ſeeds of the former.

The ſtalks of theſe decay in the winter, and new ones come up the following ſpring; they are propagated here by parting of their roots, which ſend out heads in plenty Theſe may be divided at any time, when the weather is mild, from the time the ſtalks decay, till the roots begin to ſhoot in the ſpring; it ſhould have a light ſoil and a warm ſituation, otherwiſe it will not flower here. The roots ſhould be allowed room to ſpread, and the ſtalks require ſupport; and if the plants are kept clean from weeds, and the ground between them dug every winter, it is all the culture they require.

The ſecond ſort grows naturally in both Indies. This is a perennial plant, with a ſtrong fibrous root, from which ariſe many ſlender ſtalks, which grow near four feet high, garniſhed with ſmooth, oval, ſpear-ſhaped leaves, ending in acute points, placed alternate, ſtanding upon ſhort footſtalks. The upper part of the ſtalk divides into ſmall branches, which are garniſhed with ſmall oval leaves, and terminate in a thyrſe of white flowers, which have long ſlender tubes, cut into five ſegments at the brim; theſe are ſucceeded by oblong ſeeds, covered with the prickly empalement. The upper part of the ſtalk, and the empalements of the flowers, are very glutinous, ſticking to the fingers if touched, and the ſmall flies which ſettle upon them, are faſtened, ſo cannot get off again. This plant is too tender to thrive in the open air in England, ſo requires to be kept in a moderate ſtove, where they will continue flowering great part of the year; and thoſe flowers which appear early in the ſummer, will be ſucceeded by ripe ſeeds in autumn.

This is propagated by ſeeds, which ſhould be ſown on a good hot-bed in the ſpring, where the plants will come up in about five or ſix weeks. When theſe are fit to remove, they ſhould be each planted into a ſeparate ſmall pot, and plunged into a hot-bed of tan, obſerving to ſcreen them from the ſun till they have taken new root; afterward they muſt be treated like other plants from the ſame country. In the ſummer they ſhould have a large ſhare of freſh air admitted to them in warm weather, and require water every other day. In winter they ſhould be kept in a moderate temperature of warmth, and muſt be more ſparingly watered. With this management the roots will abide ſeveral years, and produce plenty of flowers and ſeeds. This ſort is greatly extolled for its medicinal qualities.

PLUMB-TREE. See PRUNUS.

PLUMERIA. Tourn. Inſt. R. H. 659. tab. 439. Red Jaſmine.

The CHARACTERS are,

The flower has a ſmall empalement, divided into five parts; it hath one funnel-ſhaped petal with a long tube, cut into five oblong oval ſegments at the top, which ſpread open; it hath five awl-ſhaped ſtamina, ſituated in the center of the tube, terminated by ſummits which cloſe together, and an oblong bifid germen, with ſcarce any ſtyle, crowned by a double acute ſtigma. The germen afterward becomes a long, ſwelling, acute-pointed capſule with one cell, filled with winged ſeeds, placed over each other like ſcales of fiſh.

The SPECIES are,

1. PLUMERIA (*Rubra*) foliis ovato-oblongis petiolis biglandulofis. Hort. Cliff. 76. *Plumeria with oblong oval leaves; commonly called in the Weſt-Indies Red Jaſmine.*

2. PLUMERIA (*Incarnata*) foliis ovato-oblongis, ramis patulis, floribus corymboſis. *Plumeria with oblong oval leaves, ſpreading branches, and flowers growing in a corymbus; called in the Weſt-Indies the Japan-tree.*

3. PLUMERIA (*Alba*) foliis lanceolatis revolutis, pedunculis ſupernè tuberoſis. Lin. Sp. Plant. 210. *Plumeria with ſpear-ſhaped leaves which turn backward, and the foot-ſtalks having ſwellings on the upper ſide.*

4. Plumeria (*Nivea*) foliis lanceolatis petiolatis obtusis. Lin. Sp. Plant. 210. *Plumeria with spear-shaped obtuse leaves, having foot-stalks.*

5. Plumeria (*Africana*) foliis lineari-lanceolatis longissimis. *Plumeria with very long, narrow, spear-shaped leaves.*

The first sort grows naturally in the Spanish West-Indies, from whence it was transplanted into most of the islands, where it is cultivated in the gardens for ornament. It rises to the height of eighteen or twenty feet; the stalk is covered with a dark green bark, having marks where the leaves have fallen off. The stalks are succulent, and abound with a milky juice, but within they are somewhat ligneous. Toward the top they put out a few thick succulent branches, which are garnished at their ends with oval oblong leaves, of a light green colour, having a large midrib, and many transverse veins; these are full of a milky juice. At the ends of the branches come out the flowers in pretty large clusters; they are shaped like those of the Oleander or Rose Bay, having one petal, which is tubulous, and cut into five oval obtuse segments, which spread open; they are of a pale red colour, and have an agreeable odour. When the flowers are past, the germen becomes a long swelling pod, filled with flat winged seeds, lying over each other like the scales of fish. It usually flowers here in July and August, but is never succeeded by pods in England.

The second sort I received from the island of St. Christopher, by the name of Japan-tree. This sort is very rare in the English settlements at present, having been but lately introduced from the Spanish West-Indies. It is in leaf and stem very like the first, but the stalks do not rise so high; they divide into strong spreading branches, which are filled with a milky juice; the leaves are of a thicker consistence than those of the first, and their veins are larger; the flowers of this are of a deeper colour, and are produced in much larger clusters. It is very common to have upward of twenty of these flowers open in one bunch, and a number to succeed these as they decay, so that the clusters have continued in beauty upward of two months, during which time they make a most beautiful appearance in the stove, and have a very agreeable scent.

The third sort grows plentifully at Campeachy. This is not near so beautiful as the two former sorts, the flowers being smaller, and produced in less bunches, and are moreover of short duration. But for the beauty of their stems and leaves, and for the sake of variety, they deserve room in every curious collection of plants.

The fourth sort was discovered by Dr. Houstoun, growing in great plenty near Carthagena. This sort produces small white flowers, resembling those of the third, so is less valuable than the two first.

The seeds of the fifth sort were sent me by Mr. Richard, gardener to the King of France at Versailles, but I had no account of the country from whence it was sent. This hath a stalk very like the first sort, but the leaves are nine or ten inches long, and not more than one inch broad; they are thick, succulent, and full of a milky juice, a little roundish at their points. The flowers of this sort are said to be yellow, but as the plants have not yet flowered here, I can give no farther account of them.

All these plants may be propagated by seeds; these should be sown in pots filled with light earth, and plunged into a hot-bed of tanners bark; and when the plants are come up about two inches high, they should be transplanted into separate small pots filled with light sandy earth, and plunged into the hot-bed again. They must not have much water, for as all the sorts are very succulent, being full of a milky juice, somewhat like the Euphorbiums, moisture will cause them to rot. In hot weather the plants should have a pretty large share of fresh air admitted to them, by raising the glasses every day, in proportion to the warmth of the season, to prevent their drawing up weak. Toward Michaelmas, when the nights begin to be cold, the plants should be removed into the stove, and plunged into the bark-bed, where they must remain during the winter. As these plants all cast their leaves in the middle of winter, and continue destitute of them till about the beginning of May, so during that time they should be watered very sparingly, because they are in more danger of rotting, while they are in a less active state, by too much moisture, than when they are furnished with leaves, through which the moisture is more freely perspired.

All these sorts are too tender to thrive in the open air of this country in the summer season, therefore should be constantly preserved in the stove, where, in warm weather, they must have a large share of free air, but in cold weather they must be kept very warm. While they are young, it will be proper to continue them in the bark-bed, but when they have obtained strength, they may be placed in a dry stove, where they will thrive well, provided they are kept in a moderate temperature of heat, and have not too much water.

These plants may also be propagated by cuttings, which should be taken from the old plants two months or ten weeks before they are planted, during which time they should be laid on the flues in the stove, that the part which joined to the old plant may be healed over before they are planted, otherwise they will rot. These cuttings should be planted in small pots filled with light sandy earth, and plunged into a moderate hot-bed of tanners bark. If the cuttings succeed, they will have taken root in about two months, when they should have a larger share of air to harden them by degrees to bear the sun and air, and afterward may be treated as the old plants.

PODOPHYLLUM. Lin. Gen. Plant. 571. Ducksfoot, or May Apple.

The Characters are,

The bud of the flower is inclosed in a large three-leaved empalement, in form of a spatha or sheath. It has nine roundish concave petals, plaited on their borders, and smaller than the empalement; it has a roundish germen without a style, crowned by a plaited obtuse stigma. The germen afterward turns to an oval capsule of one cell, crowned by the stigma, filled with roundish seeds.

We have but one Species of this plant, viz.

Podophyllum (*Peltatum*) foliis peltatis lobatis. Lin. Sp. Plant. 505. *Ducksfoot with target-shaped leaves, having lobes.*

This plant grows naturally in many parts of North America. The root is composed of many thick tubers, which are fastened together by fleshy fibres, and propagate greatly under ground, sending out many smaller branches, which strike downward. In the spring arise several foot-stalks about six inches high, which divide into two smaller, each sustaining one leaf, composed of five, six, or seven lobes, the five middle being deeply indented at the top; these join together at their base, where the foot-stalk meets, which is fastened to the under side of the leaf like the handle of a target; the leaves are smooth, and of a light green. At the division of the foot-stalk comes out the flower with a large empalement, covering it like a sheath; the flower hath nine pretty large, concave, white petals, which are roundish at the top, and plaited on their borders. In the center is situated a large, roundish, oval germen, crowned by a plaited obtuse stigma, surrounded by a great number of short stamina, terminated by oblong, erect, yellow summits.

This

This plant propagates so fast by its creeping roots, as that few persons are at the trouble of sowing the seeds. Every part of the root will grow, so they may be annually parted, either in autumn when their leaves decay, or in the spring just before the roots begin to shoot; they require no other culture but to keep them clean from weeds. It loves a light loamy soil, and a shady situation, and is so hardy as seldom to be injured by frost.

POINCIANA. Tourn. Inst. R. H. 619. tab. 391. Barbadoes Flower-fence, or Spanish Carnations.

The CHARACTERS are,

The empalement of the flower is composed of five oblong concave leaves, which fall off. The flower has five unequal petals; four of them are nearly equal and roundish, but the fifth is larger, deformed, and indented. It hath ten long, bristly, rising stigmas, terminated by oblong summits, and an awl-shaped declining germen, which sits upon the style, and is crowned by an acute stigma. The germen afterward becomes an oblong compressed pod, with several transverse partitions; in each of these is lodged a single flattish seed.

We have but one SPECIES of this plant, viz.

POINCIANA (*Pulcherrima*) aculeis geminis. Hort. Upsal. 101. *Flower-fence with double spines.*

There are two varieties of this, which were discovered by the late Dr. Houstoun in the Spanish West-Indies. One of these hath a red, and the other a yellow flower; but as there appears to be no other difference in the plants from the common sort, they must be supposed only accidental variations which have risen from seeds.

This plant grows naturally in both Indies. It is planted in hedges, to divide the lands in Barbadoes, from whence it had the title of Flower-fence; it is also called Spanish Carnations by some of the inhabitants of the British Islands. It rises with a strait stalk from ten to fifteen or twenty feet high, covered with a smooth gray bark, and is sometimes as thick as the small of a man's leg, dividing into several spreading branches at the top, which are armed at each joint with two short, strong, crooked spines; and are garnished with decompound winged leaves, each leaf being composed of six or eight pair of simple winged leaves, the lower pair being composed of four or five pair of lobes, the others gradually increasing in their number toward the top, where they decrease again, and are smaller. The lobes are of a light green colour, and, when bruised, emit a strong odour.

The branches are terminated by loose spikes of flowers, which are sometimes formed into a kind of pyramid, and at others they are disposed more in form of an umbel. The foot-stalk of each flower is near three inches long; the flower is composed of five petals, which are roundish at the top, but are contracted to narrow tails at their base; they spread open, and are beautifully variegated with a deep red, or Orange colour, yellow, and some spots of green; they have a very agreeable odour. In the center of the flower is situated a slender style above three inches long, upon which the germen sits, and is accompanied by ten stamina, nearly of the same length with the style, terminated by oblong summits. After the flower is past, the germen becomes a broad flat pod, about three inches long, divided into three or four cells by transverse partitions, each including one flattish irregular seed. The leaves of this plant are used instead of Sena in the West-Indies to purge, and from thence the plant is by some persons titled Sena.

Ligon says, the seeds of this plant were first carried to Barbadoes from Cape Verd Islands, and the beauty of the flowers was such, that the inhabitants soon spread it over that island, and afterward it was transported into most of the neighbouring islands. This may have been so, but it is very certain that the plant grows naturally in Jamaica, where the late Dr. Houstoun found it in the woods at a great distance from any settlements. He also found it growing naturally at La Vera Cruz, and at Campeachy, where he also found the two varieties with red and yellow flowers.

The seeds of this plant are annually brought over in plenty from the West-Indies, which, if sown upon a hotbed, will rise easily. When the plants are come up, they should be transplanted each into a small pot, and plunged into a hot-bed of tanners bark, observing to shade them from the sun till they have taken root; after which they must have air in proportion to the warmth of the season, and be frequently refreshed with water. When the plants have filled the pots with their roots, they should be shaken out, and placed into larger ones, that they may have room to grow. If care be taken to water and shift them as often as is necessary, they will grow to be three feet high the first season. At Michaelmas the pots should be plunged into a fresh hot-bed of tanners bark in the stove, which should be kept to the Ananas heat marked on the botanical thermometers, and frequently refreshed with water, but they should not have too much water in winter. The earth which these plants should be planted in, must be fresh, light, and sandy, but not over rich, in which they will stand the winter better than if planted in a stronger soil.

These plants must constantly remain in the bark-stove, where in warm weather they should have a large share of air, but they must not be exposed to cold; if damp seizes their top, it very often kills the plants, or at least occasions the loss of their heads. With proper management they will grow much taller here than they usually do in Barbadoes, but their stems will not be larger than a man's finger, which is occasioned by their being drawn up by the glasses of the stove. I have had some of these plants more than eighteen feet high in the Chelsea garden, which have produced their beautiful flowers some years. These flowers have always appeared in December, but in the West-Indies I am informed they flower twice a year, at which times they make a most beautiful appearance.

POKE VIRGINIAN. See PHYTOLACCA.

POLEMONIUM. Tourn. Inst. R. H. 146. tab. 61. Greek Valerian, or Jacob's Ladder.

The CHARACTERS are,

The flower has a permanent empalement, which is cut into five segments; it has one wheel-shaped petal. The tube is very short; the upper part is divided, and spreads open. It hath five slender stamina inserted in the valves of the tube, which are shorter than the petal, and are terminated by roundish summits. In the bottom of the tube is situated an oval acute germen, supporting a slender style, equal with the petal, crowned by a revolving trifid stigma. The germen afterward turns to a three-cornered oval capsule, having three cells, filled with irregular acute-pointed seeds.

The SPECIES are,

1. POLEMONIUM (*Cæruleum*) calycibus corollæ tubo longioribus. Lin. Sp. Plant. 162. *Greek Valerian with an empalement longer than the tube of the flower.*

2. POLEMONIUM (*Reptans*) foliis pinnatis, radicibus reptatricibus. Flor. Virg. 22. *Greek Valerian with winged leaves, and a creeping root.*

The first sort grows naturally in many parts of Europe. It has been discovered growing wild in Carleton Beek, and about Malham Cove, near Craven in Yorkshire. Of this there are three varieties, one with a white, another with a blue, and one with a variegated flower, also another with variegated leaves.

This plant has winged leaves, which are composed of several pair of lobes placed alternately. The stalks rise near a foot and a half high; they are hollow, channelled, and

and are garnifhed with winged leaves, of the fame form with the lower, but decreafe upward in their fize; they are terminated by bunches of flowers, fome are white and others blue, which fit very clofe; they have one petal, which has a fhort tube, cut into five roundifh fegments at the top; they are of a beautiful blue colour, and have each five ftamina, which are terminated by yellow fummits. Thefe flowers appear the latter end of May, and are fucceeded by oval acute-pointed capfules, with three cells, filled with irregular feeds, which ripen in Auguft.

Thefe plants are eafily propagated by fowing their feeds in the fpring upon a bed of light earth, and when they are come up pretty ftrong, they fhould be pricked out into another bed, about four or five inches afunder, obferving to fhade and water them until they have taken root; after which they will require no farther care but to keep them clear from weeds until Michaelmas, at which time they muft be tranfplanted into the borders of the flower-garden, where, being intermixed with different forts of flowers, they will make a beautiful appearance.

This plant is not naturally of long duration, but by taking them up in autumn, and parting of them, they may be continued feveral years; but as the feedling plants always flower much ftronger than the offsets, few perfons ever propagate them by flips.

The fort with white flowers will frequently arife from the feeds of the blue, as will alfo that with variegated flowers; but thefe may be continued by parting of their roots.

The fort with variegated leaves is preferved by parting of their roots, becaufe the plants raifed from feeds would be fubject to degenerate, and become plain. The beft time to part them is about Michaelmas, that they may take good root before the cold weather prevents them. Thefe fhould have a frefh light foil, but if it be too rich, their roots will rot in winter, or the ftripes will go off in the fummer.

The fecond fort grows naturally in Virginia and other parts of North America. This hath creeping roots, by which it multiplies very faft. The leaves have feldom more than three or four pair of lobes, which ftand at a much greater diftance from each other than thofe of the common fort; they are of a darker green. The lobes are narrow, and are placed alternately; the ftalks rife nine or ten inches high, fending out branches their whole length. The flowers are produced in loofe bunches, ftanding upon pretty long foot ftalks; they are fmaller than thofe of the common fort, and are of a lighter blue colour.

This fort may be propagated by feeds in the fame manner as the common fort, or by parting of their roots in autumn, and is equally hardy with the common fort.

POLIANTHES. Lin. Gen. Plant. 384. The Tuberofe.

The CHARACTERS are,

The flower has no empalement; it has one funnel-fhaped petal. The tube is oblong and incurved; the brim is cut into fix oval fegments, which fpread open. It hath fix thick ftamina, fituated in the chaps of the petal, terminated by linear fummits, which are longer than the ftamina. In the bottom of the tube is fituated a roundifh germen, fupporting a flender ftyle, crowned by a thick trifid honey-bearing ftigma. The germen afterward turns to an obtufe, roundifh, three-cornered capfule, having three cells, which are filled with plain half-round feeds, difpofed in a double range.

We have but one SPECIES of this genus, viz.

POLIANTHES (*Tuberofa*) floribus alternis. Hort. Cliff. 127. *Polianthes with flowers placed alternately; commonly called Tuberofe.*

The varieties of this are the Tuberofe with a double flower, the ftriped-leaved Tuberofe, and the Tuberofe with a fmaller flower; the laft is mentioned by feveral authors as a diftinct fpecies, but is certainly a variety.

This fort is frequent in the fouth of France, from whence the roots have been often brought to England early in the fpring, before thofe roots have arrived from Italy which are annually imported; the ftalks of this are weaker, and do not rife fo high, and the flowers are fmaller than thofe of the common Tuberofe, but in other refpects is the fame.

The Tuberofe grows naturally in India, from whence it was firft brought to Europe, where it now thrives in the the warmer parts, as well as in its native foil. The Genoefe are the people who cultivate this plant, to furnifh all the other countries where the roots cannot be propagated without great trouble and care, and from thence the roots are annually fent to England, Holland, and Germany. In moft parts of Italy, Sicily, and Spain, the roots thrive and propagate without care, where they are once planted.

This plant has been long cultivated in the Englifh gardens, for the exceeding beauty and fragrancy of its flowers; the roots of this are annually brought from Genoa, by the perfons who import Orange-trees; for as thefe roots are too tender to thrive in the full ground in England, fo there are few perfons who care to take the trouble of nurfing up their offsets till they become blowing roots, becaufe it will be two or three years before they arrive to a proper fize for producing flowers; and as they muft be protected from the froft in winter, the trouble and expence of covers is greater than the roots are worth, for they are generally fold pretty reafonable by thofe who import them from Italy.

The double flowering is a variety of the firft, which was obtained from feed by Monf. Le Cour, of Leyden in Holland, who for many years was fo tenacious of parting with any of the roots, even after he had propagated them in fuch plenty, as to have more than he could plant, caufed them to be cut in pieces, that he might have the vanity to boaft of being the only perfon in Europe who was poffeffed of this flower; but of late years the roots have been fpread into many parts, and as there is no method to propagate this but by the offsets, moft people who have had of this fort, are careful to multiply and increafe it; which is done by planting the offsets upon a moderate hot-bed early in March, and covering the bed in cold weather with mats or ftraw; in fummer they muft have plenty of water in dry weather. In this bed the roots may remain till the leaves decay in autumn; but if there fhould happen any froft before that time, the bed fhould be covered to guard the roots from the froft, becaufe if the froft enters fo low as to reach the roots it will kill them; and if the leaves are injured by the froft, it will weaken the roots. Where there is due care taken to fcreen them from froft and too much wet, it will be the beft way to let the roots remain in the bed till the end of November, or the beginning of December, provided hard frofts do not fet in fooner; for the lefs time the roots are out of the ground, the ftronger they will be, and the fooner they will flower; when the roots are taken up, they fhould be cleaned from the earth, and laid up in dry fand, where they may be fecure from froft and wet; here they fhould remain until the feafon for planting them again; this fame method fhould be practifed by thofe who are defirous to cultivate the fingle fort in England, and alfo that with ftriped leaves muft be propagated the fame way.

I fhall next give directions for the management of thofe roots which are annually brought from Italy. And firft, in the choice of the roots, thofe which are the largeft and plumpeft, if they are perfectly firm and found, are the beft; and the fewer offsets they have, the ftronger they will flower; but the under part of the roots fhould be particularly examined, becaufe it is there that they firft decay; after the roots

roots are chosen, before they are planted, the offsets should be taken off; for if these are left upon the roots, they will draw away part of the nourishment from the old root, whereby the flower-stems will be greatly weakened.

As these roots commonly arrive in England in the month of February or March, those who are desirous to have these early in flower, should make a moderate hot-bed soon after the roots arrive, which should have good rich earth laid upon the dung, about seven or eight inches deep; this bed should be covered with a frame, and when the bed is in a proper temperature for warmth, the roots should be planted at about six inches distance from each other every way. The upper part of the root should not be buried more than one inch in the ground; when the roots are planted, there should be but little water given them until they shoot above ground, for too much wet will rot them when they are in an inactive state, but afterward they will require plenty of water, especially when the season is warm. When the flower-stems begin to appear, the bed should have a large share of air given to it, otherwise the stalks will draw up weak, and produce but few flowers; for the more air these plants enjoy in good weather, the stronger they will grow, and produce a great number of flowers; therefore, toward the beginning of May, the frame may be quite taken off the bed, and hoops fastened over it, to support a covering of mats, which need not be laid over but in the night, or in very cold weather, so that by enjoying the free open air their stems will be large; and if they are well watered in dry weather, their flowers will be large, and a great number on each stem.

The first planting will require more care than those which are designed to come after them; for in order to have a succession of these flowers, the roots should be planted at three different times, viz. the first the beginning of March, the second the beginning of April, and the third at the end of that month, or the beginning of May; but the latter beds will require a much less quantity of dung than the first, especially that bed which is the last made; for if there is but warmth enough to put the roots in motion, it is as much as will be required; and this last bed will need no covering, for many times those roots which are planted in the full ground at this season, will produce strong flowers in autumn; but in order to secure their flowering, it is always the best way to plant them on a gentle hot-bed. As to the second bed, that should be arched over with hoops, and covered with mats every night, and in bad weather, otherwise the late frosts, which frequently happen in May, will pinch them.

These plants may remain in the beds until the flowers are near expanding, at which time they may be carefully taken up, preserving the earth to their roots, and planted in pots, and then placed in the shade for about a week to recover their removal; after which time the pots may be removed into halls, or other apartments, where they will continue in beauty a long time, and their fragant odour will perfume the air of the rooms where they are placed; and by having a succession of them, they may be continued from Midsummer to the end of October, or middle of November; but as the stems of these plants advance, there should be some sticks put down by each root, to which the stems should be fastened, to prevent their being broken by the wind.

It is a common practice with many people, to plant these roots in pots, and plunge the pots into a hot-bed; but there is much more trouble in raising them in this method, than in that before directed; for if the roots are not planted in very small pots, there will be a necessity of making the beds much larger, in order to contain a quantity of the roots; and if they are first planted in small pots, they should be shaken out of these into pots of a larger size, when they begin to shoot out their flower-stems, otherwise the stalks will be weak, and produce but few flowers; therefore I prefer the other method, as there is no danger in removing the roots, if it is done with care.

When the roots are strong and properly managed, the stems will rise three or four feet high, and each stem will produce twenty flowers or more; and in this the great beauty of these flowers consists, for when there are but a few flowers upon the stalks, they will soon fade away, and must be frequently renewed; for the flowers are produced in spikes, coming out alternately upon the stalk, the lower flowers opening first; and as these decay, those above them open, so that in proportion to the number of flowers upon each stalk, they continue in beauty a longer or shorter time.

The sort with double flowers will require a little more care, in order to have the flowers fair; but this care is chiefly at the time of blowing, for the flowers of this sort will not open, if they are exposed to the open air; therefore when the flowers are fully formed and near opening, the pots should be placed in an airy glass-case, or a shelter of glasses should be prepared for them, that the dews and rains may not fall upon them, for that will cause the flowers to rot away before they open, and the heat of the sun drawn through the glasses, will cause their flowers to expand very fair. With this management, I have had this sort with very double flowers extremely fair, and upward of twenty upon one stem, so that they have made a beautiful appearance; but where this has not been practised, I have rarely seen one of them in any beauty.

POLIUM. Tourn. Inst. R. H. 206. tab. 97. Mountain Poley.

The CHARACTERS are,

The empalement of the flower is cut into five acute segments. The flower is of the lip kind; it hath one petal, with a short tube. The stamina occupy the place of the upper lip, and the lower lip is cut into five segments. It hath four awl-shaped stamina, which are terminated by small summits, and a germen divided into four parts, supporting a slender style, crowned by two narrow stigmas; the germen afterward become four naked seeds, inclosed in the empalement.

The SPECIES are,

1. POLIUM (*Montanum*) foliis lanceolatis integerrimis, caulibus procumbentibus, floribus corymbosis terminalibus. *Mountain Poley with entire spear-shaped leaves, trailing stalks, and flowers growing in a corymbus at the end of the branches.*

2. POLIUM (*Luteum*) spicis oblongis foliis obtusis crenatis tomentosis. *Mountain Poley with oblong spikes of flowers, and obtuse, crenated, woolly leaves; yellow Mountain Poley.*

3. POLIUM (*Angustifolium*) spicis subrotundis, caulibus suffruticosis incanis, foliis linearibus tomentosis. *Mountain Poley with roundish spikes of flowers, hoary shrubby stalks, and very narrow woolly leaves.*

4. POLIUM (*Album*) caule ramoso procumbente, foliis lineari-lanceolatis, floribus corymbosis terminalibus. *Poley with a branching trailing stalk, narrow, spear-shaped, woolly, indented leaves, and flowers growing in a corymbus, terminating the branches.*

5. POLIUM (*Capitatum*) caule erecto diffuso, foliis lineari-lanceolatis crenatis, corymbis terminalibus lateralibusque. *Poley with an erect diffused stalk, narrow, spear-shaped, crenated leaves, and flowers growing in a corymbus, terminating, and proceeding out of the sides of the branches.*

6. POLIUM (*Pyreniacum*) caulibus procumbentibus hirsutissimis, foliis cuneiformi orbiculatis crenatis. *Poley with very hair trailing stalks, and orbicular wedge-shaped leaves, which are crenated.*

7. POLIUM

7. Polium (*Latifolium*) caule erecto ramoso, foliis lanceolatis dentatis subtus tomentosis, floribus confertis terminalibus. *Poley with an upright branching stalk, spear-shaped indented leaves, which are woolly on their under side, and flowers growing in clusters, terminating the branches.*

8. Polium (*Erectum*) caule erecto corymboso, foliis linearibus reflexis, floribus terminalibus. *Poley with an upright stalk, branching out in form of a corymbus, narrow reflexed leaves, and flowers terminating the stalks.*

9. Polium (*Ramosum*) caule ramoso, procumbente, foliis lineari-lanceolatis supernè dentatis, spicis oblongis terminalibus. *Poley with a trailing branching stalk, narrow spear-shaped leaves, which are indented toward the top, and oblong spikes of flowers terminating the stalks.*

10. Polium (*Spicatum*) caule erecto suffruticoso, foliis linearibus confertis, spicis cylindricis fastigiatis terminalibus. *Poley with an upright under-shrub stalk, narrow leaves growing in clusters, and cylindrical spikes of flowers growing in bunches, which terminate the stalks.*

11. Polium (*Fruticosum*) caule erecto fruticoso, foliis lanceolatis tomentosis integerrimis, corymbis terminalibus. *Poley with an upright shrubby stalk, spear-shaped woolly leaves, which are entire, and flowers growing in a corymbus, terminating the stalks.*

12. Polium (*Serratum*) caule procumbente, foliis linearibus serratis, corymbis confertis terminalibus. *Poley with a trailing stalk, narrow sawed leaves, and clustered flowers growing in a corymbus at the end of the stalks.*

13. Polium (*Diffusum*) caule diffuso procumbente, foliis linearibus dentatis tomentosis, spicis subrotundis. *Poley with a trailing diffused stalk, narrow indented woolly leaves, and roundish spikes of flowers.*

14. Polium (*Integerrimum*) caule erecto suffruticoso, foliis lanceolatis integerrimis, corymbis confertis terminalibus. *Poley with an erect shrubby stalk, spear-shaped entire leaves, and clustered flowers growing in a corymbus at the ends of the branches.*

15. Polium (*Smyrneum*) caule diffuso, foliis linearibus pinnato-dentatis, spicis subrotundis lateralibus. *Poley with a diffused stalk, linear, winged, indented leaves, and roundish spikes of flowers proceeding from the sides of the stalks.*

The first sort grows naturally on the mountains about Basil and Geneva, as also in France. The root of this plant is composed of many ligneous fibres, from which arise several weak, trailing, ligneous stalks, eight or nine inches long, sending out many weak branches, garnished with small spear-shaped leaves, of a deep green, and entire, placed by pairs. The flowers are produced in a corymbus at the end of the branches; they are white, and shaped like those of the other species. These appear in June and July, but are seldom succeeded by seeds in England.

The second sort grows naturally in Spain. The stalks of this are rather herbaceous, and trail upon the ground; they are about six inches long, hoary, and garnished with woolly leaves; some of them are wedge-shaped, others are oblong, ending in obtuse points, and are crenated toward their ends. The flowers are collected in oblong thick spikes at the end of the branches; they are of a deep yellow colour, and appear the beginning of June, but are seldom succeeded by seeds in this country.

The third sort grows naturally in Spain and Portugal. The stalks of this are ligneous, erect, and branching, covered with a hoary down; they rise six or eight inches high, garnished with linear woolly leaves about half an inch long, having sometimes two or three slight indentures on their edges. The flowers are collected in roundish spikes at the end of the branches; they are of a bright yellow, and have woolly empalements. These appear in June and July.

The fourth sort grows naturally in the south of France, and in Italy. This hath a trailing branching stalk, which at the bottom is ligneous, but the branches are herbaceous and woolly; they are garnished with linear, spear-shaped, woolly leaves, indented on their edges. The flowers are produced in a corymbus at the end of the branches, which are small, white, and shaped like those of the other species. This flowers in June and July.

The fifth sort grows naturally near the sea, in the south of France and in Italy. This hath an erect branching stalk, which rises a foot high; the lower part becomes ligneous, but the upper is herbaceous; the leaves are linear, spear-shaped, and crenated on their edges, of a pretty thick consistence, and a little woolly. The flowers are collected in a corymbus at the end of the branches; they are white, and like those of the other species. This flowers in July and August.

The sixth sort grows naturally on the Pyrenean mountains. This hath slender shrubby stalks, which trail close upon the ground, and put out roots; the leaves are round at the top, but at their base are contracted in form of a wedge, and are crenated on their edges, so as to resemble at first sight the leaves of Ground-ivy; but they are hairy, and of a thicker consistence. The flowers are collected in round bunches at the end of the branches: one half of their petals are purple, and the other half white; they are larger than those of the other species, but are of the same form. It flowers great part of summer, but seldom produces seeds here.

The seventh sort grows naturally in Spain and Italy. This hath a ligneous, erect, branching stalk, which rises near a foot high; it is very hoary, and branches out toward the top; the leaves are spear-shaped, indented on their edges, and woolly on their under side. The flowers are white, small, and grow in clusters at the end of the branches. It flowers in June and July.

The eighth sort grows naturally in Spain and Italy. This rises with a shrubby stalk nine or ten inches high, branching out toward the top in form of a corymbus; the leaves are linear, and their edges are reflexed. The flowers are collected in roundish woolly heads at the end of the branches; they are white, and smaller than most of the other species This flowers in June and July.

The ninth sort grows naturally in Spain. It hath a trailing branching stalk about six or eight inches long, which is ligneous at bottom, but upward is herbaceous and hoary; the leaves are linear, spear-shaped, and indented toward the ends. The flowers are collected in oblong spikes at the end of the branches; they are of a pale yellow colour, and shaped like those of the other species. This flowers great part of summer.

The tenth sort grows naturally in Sicily. This hath slender shrubby stalks, which rise a foot and a half high; they are smooth and white, sending out a few short branches toward the top, garnished with small linear leaves growing in clusters. The flowers are collected in long cylindrical spikes, which stand in bunches at the top of the stalks, and sometimes come out on the sides; these are small and white. It flowers in July and August.

The eleventh sort grows naturally in Valencia. This hath slender, ligneous, hoary stalks, near two feet high, garnished with small, spear-shaped, entire, woolly leaves at intervals, standing in clusters, and sit close to the stalk; the upper part of the stalk divides into several slender foot-stalks, each sustaining a small corymbus of white flowers. The whole plant has a strong aromatic odour. It flowers late in summer.

The twelfth sort grows naturally in the south of France and in Italy. This hath trailing ligneous stalks about a foot

foot long, garnished with linear, sawed, hoary leaves. The flowers are collected in a corymbus at the end of the branches; they are small and white. This flowers in June and July.

The thirteenth sort grows naturally in Spain and Italy. This hath diffused trailing stalks, which are very woolly, garnished with narrow indented leaves, which are covered with a woolly down, and are terminated by roundish heads of flowers, which are yellow; the whole plant is very hoary. It flowers in July.

The fourteenth sort grows naturally in Spain. This hath erect branching stalks about six or eight inches high; the branches come out opposite the whole length of the stalk; they are garnished with small spear-shaped leaves, of a dark green colour on their upper side, but hoary on their under; the stalks and branches are terminated by clusters of blue flowers, which are collected in roundish heads. This sort flowers in July and August.

The fifteenth sort grows naturally about Smyrna. This hath diffused white stalks, which rise about a foot high, closely garnished with linear leaves, indented regularly on their edges like those of Spleenwort, but the indentures are not deep; they are of a dark green on their upper side, but hoary on their under. The flowers are collected in roundish spikes, which terminate the branches, and also come out from their side; they are white, and shaped like those of the other species. It flowers in July and August.

There are several other species of this genus, which grow naturally in the warmer parts of Europe; but those which are here mentioned, are all that I have yet seen growing in the English gardens, therefore I have omitted the other, as I have had no opportunity to examine them myself.

All the sorts, except the first, are abiding plants; they may be propagated by seeds, when they can be procured fresh. These should be sown upon a bed of fresh light earth in the spring, and when the plants come up, they must be carefully kept clean from weeds; about the middle of July the plants will be fit to remove, when they may be carefully taken up, and part of them planted on a warm border of dry rubbishy soil, observing to shade them from the sun, and water them till they have taken new root; after which they will require no other culture but to keep them clean from weeds. My advising these, and many other aromatic plants, which are natives of the warmer parts of Europe, to be planted in rubbish, is founded upon long experience of their abiding much longer, and resisting the cold of our winters much better than when they are growing in better ground, where they grow much freer, are fuller of moisture, and therefore more liable to be killed by frost.

The other part of the plants may be planted in small pots, filled with fresh, light, undunged earth, and placed in the shade till they have taken new root; then they may be removed into an open situation, where they may remain till the beginning of November, when they should be placed under a common frame, to secure them from the frost in winter, which sometimes destroys these plants in the open air; by this method the species may be preserved.

These plants may be disposed in a garden so as to afford pleasure, by mixing them with Marum, Mastich, and several other aromatic plants, upon the sloping sides of banks, which are exposed to the sun, or upon little hillocks, raised in a sheltered situation, where, by the diversity of their hoary branches, being of various shapes, they will make a pretty appearance; and in such places they will resist the cold much better, than when they are planted in a good soil. I remember to have seen in Dr. Boerhaave's private garden, many small hillocks raised, on which were planted several sorts of aromatic plants.

They may also be propagated by cuttings or slips, which should be planted the beginning of April, just before they shoot, upon a border exposed to the east; and if the season proves dry, they must be watered and shaded until they have taken root; afterward they will require no other care but to keep them clean from weeds, and at Michaelmas the plants should be removed where they are designed to remain; but it will be proper to put a plant of each sort in pots, that they may be sheltered in winter, to preserve the kinds.

The fourth and fifth sorts are sometimes used in medicine.

POLYANTHOS. See PRIMULA.

POLYGALA. Tour. Inst. R.H. 174. tab. 79. Milkwort.

The CHARACTERS are,

The flower has a small permanent empalement. The flower is shaped like those of the butterfly kind; the number of petals is indeterminate. The wings are large, plain, and extend beyond the other petals; the standard is tubulous, short, and reflexed at the brim, where it is bifid. The keel is concave, compressed, and bellied toward the top. It hath eight stamina in two bodies, included in the keel, terminated by single summits, and an oblong germen, supporting an erect style, terminated by a thick bifid stigma. The germen afterward becomes a heart-shaped capsule having two cells, each containing one seed.

The SPECIES are,

1. POLYGALA (*Vulgaris*) floribus cristatis racemosis, caulibus herbaceis simplicibus procumbentibus, foliis lineari-lanceolatis. Amœn. Acad. 2. p. 136. *Milkwort with branching crested flowers, single, trailing, herbaceous stalks, and linear spear-shaped leaves; common Milkwort.*

2. POLYGALA (*Monspeliaca*) floribus cristatis racemosis, caule erecto, foliis lanceolato linearibus acutis. Sauv. Monsp. 53. *Milkwort with branching crested flowers, an erect stalk, and acute, spear-shaped, linear leaves.*

3. POLYGALA (*Myrtifolia*) floribus cristatis, carinâ lunulatâ, caule fruticoso, foliis lævibus oblongis obtusis. Amœn. Acad. 2. p. 138. *Milkwort with crested flowers, a moon-shaped keel, and a shrubby stalk bearing oblong leaves, which end in obtuse points.*

4. POLYGALA (*Chamæbuxus*) floribus imberbibus sparsis, carinæ apice subrotundo, caule fruticoso, foliis lanceolatis. Amœn. Acad. 2. p. 140. *Milkwort with flowers growing thinly and without beards, the point of the keel roundish, a shrubby stalk, and spear-shaped leaves.*

5. POLYGALA (*Senega*) floribus imberbibus spicatis, caule erecto herbaceo simplicissimo, foliis lato-lanceolatis. Amœn. Acad. 2. p. 139. *Milkwort with spiked flowers having no beards, an erect, single, herbaceous stalk, and broad spear-shaped leaves; commonly called Seneka Rattle Snakewort.*

6. POLYGALA (*Mariana*) floribus imberbibus, oblongo-capitatis, caule erecto ramoso, foliis linearibus. *Milkwort with beardless flowers growing in oblong heads, an erect branching stalk, and linear leaves.*

7. POLYGALA (*Americana*) floribus cristatis, racemo terminali, caule erecto ramoso, foliis lanceolatis tomentosis. *Milkwort with crested flowers, an erect branching stalk terminated by a loose spike of flowers, and wooll spear-shaped leaves.*

There are several other species of this genus, some of which grow naturally in Europe, and others in America, but as they are seldom cultivated in gardens, so it would be to little purpose to enumerate them here.

The first sort grows naturally in pastures and upon heaths in many parts of England. Of this there are three varieties, one with a blue, another with a purple, and a third with white flowers, which are frequently found intermixed; and there is another which is larger, and supposed to be a distinct species; but I rather believe this difference is owing to the soil in which they grow, for the large one is generally found growing in moist pastures, and the small one upon dry heaths. This hath a perennial root, from which come

The material originally positioned here is too large for reproduction in this reissue. A PDF can be downloaded from the web address given on page iv of this book, by clicking on 'Resources Available'.

come out three or four ſlender, trailing, herbaceous ſtalks, garniſhed with linear ſpear-ſhaped leaves. The flowers are produced at the top of the ſtalks, branching out; they are ſmall, and of a blue, purple, or white colour, having two wings, a keel and ſtandard like the butterfly flowers. Theſe appear in June, and are ſucceeded by flattiſh heart ſhaped capſules, divided into two cells, each containing one ſeed.

The ſecond ſort grows naturally upon ſterile ground about Montpelier. This ſort is annual; it riſes with an herbaceous ſtalk about ſix inches high, garniſhed with narrow leaves, placed oppoſite, ending in acute points. The flowers are ſmall, of a worn-out purple colour; the keel is bearded like the common ſort. This flowers in July, and has ſeed-veſſels like the firſt ſort, but ſmaller; the ſeeds ripen in autumn.

Theſe ſorts are very rarely admitted into gardens, nor do they thrive ſo well when ſown or tranſplanted there, as in their natural ſituation. If theſe are cultivated, their ſeeds ſhould be ſown ſoon after they are ripe, otherwiſe they rarely grow.

The third ſort grows naturally at the Cape of Good Hope. This hath a ſhrubby ſtalk, covered with a ſmooth brown bark, which riſes five or ſix feet high, ſending out ſeveral ſpreading branches toward the top, cloſely garniſhed with oblong, blunt pointed, ſmooth leaves, of a lucid green, ſitting cloſe to the branches. The flowers are produced at the end of the branches; they are large, white on their outſide, but of a bright purple within; the keel of the flower is hallowed like a half moon, and is bearded; the wings are expanded wide, and the ſtandard is incurved; this plant continueth flowering moſt part of ſummer. The flowers are ſucceeded by compreſſed heart-ſhaped ſeed-veſſels, having two cells, each containing one hard, ſmooth, ſhining ſeed. This plant is propagated by ſeeds, which ſhould be ſown in ſmall pots filled with light loamy earth ſoon after they are ripe. Theſe pots may be placed where they may have the morning ſun only till October, when they ſhould be placed under a hot-bed frame, and plunged into old tanners bark which has loſt its heat, where they may be defended from froſt during the winter, and in the ſpring the pots ſhould be plunged into a moderate hot-bed, which will bring up the plants. When theſe appear, they ſhould not be too tenderly treated, but muſt have a large ſhare of free air admitted to them; and when they are fit to tranſplant, they ſhould be carefully ſhaken out of the pots, and ſeparated, planting each into a ſmall pot, filled with ſoft loamy earth, and plunged into a very moderate hot-bed, to forward their taking new root, obſerving to ſhade them from the ſun, and gently refreſh them with water, but they muſt not have too much wet. When they are rooted, they muſt be gradually inured to bear the open air; in June they may be placed abroad in a ſheltered ſituation, where they may remain till the middle or latter end of October, according as the ſeaſon proves favourable; then they muſt be removed into the green-houſe, and treated in the ſame way as Orange-trees, being careful not to give them too much wet during the winter ſeaſon. In the ſummer they muſt be placed abroad with other green-houſe plants, where, by their long continuance in flower, they will make a fine appearance. The management of this plant is nearly the ſame as for the Orange-tree.

The fourth ſort grows naturally on the Alps, and alſo upon the mountains in Auſtria and Hungary. This riſes with a ſlender, branching, ligneous ſtalk about a foot high, when it grows upon good ground, but on a rocky ſoil, ſeldom more than half that height. The branches are cloſely garniſhed with ſtiff, ſmooth, ſpear-ſhaped leaves, of a lucid green. From between the leaves, toward the top of the branches, the flowers come out upon very ſhort footſtalks; they are white on their outſide, but within are of a purpliſh colour mixed with yellow, and have a grateful odour. Theſe appear in May, and are ſucceeded by ſeed-veſſels, ſhaped like thoſe of the former ſort.

This plant is very difficult to cultivate in gardens, for it commonly grows out of the fiſſures of rocks, ſo cannot be eaſily tranſplanted, and the ſeeds are with difficulty obtained from abroad; nor do theſe vegetate till they have been a whole year in the ground; and when the plants come up, they make very little progreſs here, and are as difficult to tranſplant as almoſt any plant at preſent known, which occaſions its preſent ſcarcity in England.

The beſt method of cultivating this is by ſeeds, which ſhould be procured as freſh as poſſible from the places of its natural growth, and ſown in pots as ſoon as it arrives; theſe pots may be plunged into the ground, where they may have only the morning ſun. If theſe are ſown before Chriſtmas, there will be a chance of the plants coming up the following ſpring, but thoſe which are not ſown till toward ſpring, will remain in the ground a year; therefore the pots ſhould be plunged into the ground, where they may have but little ſun the following ſummer, and in autumn they may be removed, and plunged into an old tan-bed under a hot-bed frame, where they may be protected from ſevere froſt; for although this plant is a native of the Alps and other cold mountains, yet as the ſeeds will not be covered with ſnow here as they are in their native ſituation, they are frequently ſpoiled by the inconſtancy of the weather in England. When the plants come up, they ſhould be placed in ſhade during ſummer, and in autumn they may be turned out of the pots, and planted in a border, where they may have only the morning ſun, for this plant will not thrive long in pots. If the winter proves very ſevere, it will be proper to cover the ſurface of the ground about their roots with mulch to keep out the froſt. If the plants take root in the border, they ſhould remain there undiſturbed, and be only kept clean from weeds, for the ground about their roots ſhould not be dug or dunged.

The fifth ſort grows naturally in moſt parts of North America. This hath a perennial root, compoſed of ſeveral fleſhy fibres, from which ariſe three or four branching ſtalks, which grow erect, garniſhed with ſpear-ſhaped leaves, placed alternately. The flowers are produced in looſe ſpikes at the end of the branches; they are ſmall, white, and ſhaped like thoſe of the common ſort, but their keels have no beards. It flowers here in July, but the plants do not produce ſeeds.

The root of this ſort hath been long uſed by the Seneka Indians, to cure the bite of the Rattle-ſnake, which, if taken in time, is an infallible remedy. And of late years it hath been uſed by the inhabitants of Virginia, in many diſorders which are occaſioned by a thick ſizy blood; ſo that the root of this plant, when its virtues are fully known may become one of the moſt uſeful medicines yet diſcovered. The Seneka Indians uſe this root, which they powder, and generally carry about them when they travel in the woods, leſt they ſhould be bit by the Rattle-ſnake; and whenever this happens, they take a quantity of the powder inwardly, and apply ſome of it to the part bitten, which is a ſure remedy.

The ſixth ſort grows naturally in Maryland. This hath a perennial root, from which ariſe two or three ſtalks about eight inches high, which divide into ſeveral erect branches, garniſhed with ſmall linear leaves, of a dark green colour. The flowers are collected into oblong heads at the end of the ſtalks; they are ſmall, and of a purpliſh blue colour.

Both theſe ſorts are difficult to obtain, for the ſeeds rarely ſucceed, ſo the beſt way is to procure their roots from America;

rica; and when they arrive, plant them in a bed of light earth, in a ſheltered ſituation. In ſummer they muſt be kept clean from weeds, and if the ſurface of the ground about their roots is covered with old tanners bark, or any other kind of mulch in winter, to keep out the froſt, it will be a ſecure method to preſerve them.

The ſeventh ſort was diſcovered by the late Dr. Houſtoun growing naturally at La Vera Cruz. This hath a taper perennial root, which runs deep in the ground, from which ariſe ſeveral ſlender branching ſtalks about ſix or ſeven inches high, garniſhed with downy ſpear-ſhaped leaves. The flowers are produced in looſe ſpikes at the end of the branches; they are larger than thoſe of the common ſort, and are of a bluiſh purple colour. The keel of the flower is bearded, as in the common ſort.

This is too tender to live in the open air in England, and it is one of thoſe plants which will not thrive in pots, ſo is difficult to preſerve here. It is propagated by ſeeds. The ſeeds which I received from Dr. Houſtoun, remained a year in the ground before the plants appeared, and the plants lived one year; but when their roots reached the bottom of the pots, they decayed, and thoſe which were tranſplanted into larger pots, did not ſurvive their removal, though it was performed with great care.

POLYGONATUM. See CONVALLARIA.

POLYPODIUM. Tourn. Inſt. R. H. 540. tab. 316. Polypody.

The CHARACTERS are,

This is one of the Fern tribe, which is diſtinguiſhed from the others, by the fructification being in roundiſh ſpots, diſtributed on the under ſurface of the leaf.

The SPECIES are,

1. POLYPODIUM (*Vulgare*) frondibus pinnatifidis, pinnis oblongis ſubſerratis obtuſis, radice ſquamatâ. Lin. Sp. Plant. 1085. *Polypody with wing-pointed leaves, having oblong obtuſe lobes, which are ſomewhat ſawed, and a ſcaly root; common Polypody.*

2. POLYPODIUM (*Cambricum*) frondibus pinnatifidis, pinnis lanceolatis lacero-pinnatifidis ſerratis. Lin. Sp. Pl. 1086. *Polypody with wing-pointed leaves, whoſe lobes are ſpear-ſhaped, and the jags wing-pointed and ſawed; Welſh Polypody with jagged leaves.*

There are ſeveral other ſpecies of this plant which are natives of America, ſome of which are preſerved in curious botanic gardens for variety; but as they are rarely cultivated in other gardens, it is not worth while to enumerate them in this place.

The firſt ſort is that which is uſed in medicine, and is found growing upon old walls and ſhady banks in divers parts of England.

The ſecond ſort was brought from Wales, where it grows in great plenty, and is the moſt beautiful of all the ſorts.

Theſe plants may be propagated by parting of their roots in the ſpring before they ſhoot, and ſhould be planted in a very poor moiſt ſoil under the ſhade of a wall, for if they are expoſed to the ſun they will not thrive. They chiefly delight to grow out of the joints of walls and old buildings, but are commonly found expoſed to the north.

POMGRANATE. See PSIDIUM.

POMUM ADAMI. See AURANTIUM.

PONTEDERIA. Lin. Gen. Plant. 361.

The CHARACTERS are,

The flowers are included in an oblong ſheath, which opens on one ſide; it hath ſix petals, which are divided; the three upper are erect, and form a kind of lip; the three under are reflexed. It hath ſix ſtamina, which are inſerted to the petals; the three, which are longeſt, are faſtened to the mouth of the tube, the other are inſerted in the baſe; they are terminated by proſtrate ſummits. Under the petals is ſituated an oblong germen ſupporting a ſingle ſtyle, which declines, and is crowned by a ſingle ſtigma. The germen afterward turns to a ſoft fruit, divided into ſix cells, each containing ſeveral ſmall roundiſh ſeeds.

The SPECIES are,

1. PONTEDERIA (*Cordata*) foliis haſtato-cordatis, floribus ſpicatis. *Pontederia with ſpear-pointed heart-ſhaped leaves, and ſpiked flowers.*

2. PONTEDERIA (*Haſtata*) foliis haſtatis, floribus umbellatis. Lin. Sp. Plant. 288. *Pontederia with ſpear-pointed leaves, and flowers growing in umbels.*

The firſt ſort grows naturally in marſhy places in Virginia, and moſt parts of North America, and the late Dr. Houſtoun found it growing plentifully at La Vera Cruz. This hath a perennial root, from which ariſe two or three herbaceous thick ſtalks a foot high, each having one arrow-pointed heart-ſhaped leaf, of a pretty thick conſiſtence. The baſe is deeply indented, and the two ears are rounded; the foot-ſtalk of the leaf cloſely embraces the ſtalk like a ſpatha or ſheath for near three inches in length; above this is another ſheath, which incloſes the ſpike of flowers; this opens on one ſide, and the ſtalk riſes near two inches above it, where the ſpike of flowers begin. The ſpikes are about three inches long; the flowers are blue, ſit very cloſe together, and have the appearance of lip flowers. Theſe appear in June, but are not ſucceeded by ſeeds in England.

As this plant grows naturally on moiſt boggy places, it is very difficult to be preſerved in England; nor does the plant ariſe from ſeeds here, for I have ſowed the ſeeds in various ſituations, and managed them different, but could never get up any of the plants; but I had three or four of the plants ſent me, incloſed in large clods of earth from New England, which I planted in pots, covering them with Moſs, and conſtantly ſupplied them with water. With this management two of them flowered, but the following winter deſtroyed them, as they were not put under ſhelter; ſo that to preſerve them, they ſhould be placed under a hot-bed frame in winter, where they may be expoſed to the open air at all times when the weather is mild.

The ſecond ſort grows naturally about Madraſs in watery places. This riſes with a ſingle ſtalk eight or nine inches high, having one arrow-pointed leaf, whoſe baſe embraces the ſtalk like a ſheath, and from the open ſide of the ſheath comes out the flowers, which are at firſt incloſed in another ſmaller ſheath; theſe grow in a ſmall kind of umbel; they are compoſed of ſix acute-pointed petals, which ſpread open. Each flower ſtands upon a ſlender foot-ſtalk about an inch long; the foot-ſtalk of the leaf riſes a conſiderable height above the flowers, ſo that they appear to come out from the middle of the ſtalk.

This ſort is much more difficult to preſerve in England, becauſe it grows naturally in a hot country, and always in places flowed with water. There was formerly one of theſe plants brought over to Charles Duboiſe, Eſq; at Mitcham, but it was not long lived here.

POPULAGO. See CALTHA.

POPULUS. Tourn. Inſt. R. H. 592. tab. 365. The Poplar-tree.

The CHARACTERS are,

The male and female flowers grow upon ſeparate trees. The male flowers or katkins have one oblong, looſe, cylindrical empalement, which is imbricated. Under each ſcale is ſituated a ſingle flower, without any petal, having a nectarium of one leaf, turbinated at the bottom, and tubulous at the top, and eight ſtamina, terminated by large four-cornered ſummits. The female flowers are in katkins like the male, but have no ſtamina; they have an oval acute-pointed germen, with ſcarce any ſtyle, crowned by a four-pointed ſtigma. The germen afterward becomes an oval capſule with two cells, including many oval ſeeds, having hairy down.

The

The Species are,

1. Populus (*Alba*) foliis lobatis dentatis ſubtus tomentoſis. *Poplar-tree with lobated indented leaves, which are downy on their under ſide; commonly called the Abele-tree.*

2. Populus (*Major*) foliis ſubrotundis, dentato-angulatis ſubtus tomentoſis. Hort. Cliff. 460. *Poplar-tree with roundiſh leaves, which are angularly indented; or white Poplar.*

3. Populus (*Nigra*) foliis ovato-cordatis acuminatis crenatis. *Poplar-tree with oval heart-ſhaped leaves, ending in acute points, which are crenated; the black Poplar.*

4. Populus (*Tremula*) foliis ſubrotundis dentato angulatis utrinque glabris. Hort. Cliff. 460. *Poplar-tree with roundiſh leaves, having angular indentures, and are ſmooth on both ſides; the Aſpen-tree.*

5. Populus (*Balſamifera*) foliis ſubcordatis oblongis crenatis. Hort. Cliff. 460. *Poplar-tree with oblong leaves, which are crenated, and almoſt heart-ſhaped; the Carolina Poplar-tree.*

6. Populus (*Tacamahacca*) foliis ſubcordatis, inferne incanis, ſupernè atroviridis. *Poplar with leaves which are almoſt heart-ſhaped, hoary on their under ſide, and of a dark green above; commonly caled Tacamahacca.*

The firſt ſort grows naturally in the temperate parts of Europe. This and the ſecond ſort are frequently confounded together, but they are certainly different ſpecies; this is commonly called Abele-tree here, and the ſecond white Poplar. The leaves of the firſt are large, and divided into three, four, or five lobes, which are indented on their edges, of a very dark colour on their upper ſide, but very white and downy on their under, ſtanding upon foot-ſtalks, which are about an inch long. The young branches of this tree have a purple bark, and are covered with a white down, but the bark of the ſtem and older branches is gray. In the beginning of April the male flowers or katkins appear, which are cylindrical, ſcaly, and about three inches long; about a week after come out the female flowers on katkins, which have no ſtamina like thoſe of the male. Soon after theſe come out, the male katkins which fall off, and in five or ſix weeks after the female flowers will have ripe ſeeds incloſed in a hairy covering; then the katkins will drop, and the ſeeds will be wafted by the winds to a great diſtance.

The leaves of the ſecond ſort are rounder, and not much above half the ſize of thoſe of the firſt; they are indented on their edges into angles, and are downy on their under ſide, but not ſo white as thoſe of the former, nor are their upper ſurfaces of ſo deep green colour. The ſhoots of this are paler, the katkins are longer, and the down of the ſeeds is whiter and longer.

The leaves of the third ſort are oval, heart-ſhaped, and ſlightly crenated on their edges; they are ſmooth on both ſides, and of a light green colour. The katkins of this are ſhorter than thoſe of the two former.

The leaves of the fourth ſort are roundiſh, angularly indented; they are ſmooth on both ſides, and ſtand upon long ſlender foot-ſtalks, ſo are ſhaken by the leaſt wind, from whence it was titled the trembling Poplar or Aſpen-tree. The katkins of this are much like thoſe of the firſt ſort, but the young ſhoots are of a dark brown colour.

The fifth ſort grows naturally in Carolina, where it becomes a large tree. The ſhoots of this ſort are very ſtrong in England, and are generally angular; they have a light green bark, like ſome ſorts of the Willow. The leaves upon young trees, and alſo thoſe upon the lower ſhoots, are very large, almoſt heart-ſhaped and crenated, but thoſe upon the older trees are ſmaller; as the trees advance, their bark becomes lighter, approaching to a grayiſh colour. The katkins of this ſort are like thoſe of the black Poplar, and the ſummits of the ſtamina are purple.

The ſhoots of this tree, while young, are frequently killed down to a conſiderable length by the froſt in winter; but as the trees grow older, their ſhoots are not ſo vigorous, and become more ligneous, ſo are not liable to the ſame diſaſter; but the trees ſhould be planted in a ſheltered ſituation, for as their leaves are very large, the wind has great power over them, and the branches being tender, they are frequently broken or ſplit down by the winds in the ſummer ſeaſon, where they are much expoſed.

The ſeventh ſort grows naturally in Canada, and in other parts of North America. This ſeems to be a tree of middling growth, and does not ſpire upward, but ſends out many ſhort thick ſhoots on every ſide, which are covered with a light brown bark, garniſhed with leaves, differing from each other in ſhape and ſize, moſt of them almoſt heart-ſhaped, but ſome are oval, and others near to ſpear-ſhaped; they are whitiſh on their under ſide, but of a dark green on their upper. The katkins are like thoſe of the black Poplar, but the number of ſtamina in the male flowers is uncertain, from eighteen to twenty-two. The hermaphrodite flowers come out a month later than the male.

Theſe trees may be propagated either by layers or cuttings, which will readily take root, as alſo from ſuckers, which the white Poplars ſend up from their roots in great plenty. The beſt time for tranſplanting theſe ſuckers is in October, when their leaves begin to decay. Theſe may be placed in a nurſery for two or three years to get ſtrength, before they are planted out where they are deſigned to remain; but if they are propagated from cuttings, it is better to defer the doing of that until February, at which time truncheons of two or three feet long ſhould be thruſt about a foot and a half into the ground. Theſe will readily take root, and if the ſoil be moiſt in which they are planted, will arrive to a conſiderable bulk in a few years.

The black Poplar is not ſo apt to take root from large truncheons, therefore it is the better method to plant cuttings about a foot and a half in length, thruſting them a foot deep into the ground; theſe will take root very freely, and may be afterward tranſplanted where they are to remain. This ſort will grow upon almoſt any ſoil, but will thrive beſt in moiſt places.

The white ſorts, as alſo the Aſpen-tree, likewiſe cauſe a great litter in the ſpring, when their katkins and down fall off; and their roots being very apt to produce a large quantity of ſuckers, but eſpecially thoſe trees that came from ſuckers, which render them unfit to be planted near a houſe or garden; but when they are interſperſed with other trees in large plantations, they afford an agreeable variety, their leaves being very white on their under ſides, which, when blown with the wind, are turned to ſight.

A conſiderable advantage may be made by planting theſe trees upon moiſt boggy ſoils, where few other trees will thrive. Many ſuch places there are in England, which do not at preſent bring in much money to their owners; whereas, if they were planted with theſe trees, they would, in a very few years, over-purchaſe the ground, clear of all expence; but there are many perſons, who think nothing, except Corn, worth cultivating in England; or if they plant timber, it muſt be Oak, Aſh, or Elm; and if their land be not proper for either of theſe, it is deemed little worth; whereas, if the nature of the ſoil was examined, and proper ſorts of plants adapted to it, there might be very great advantage made of ſeveral large tracts of land, which at this time lie neglected.

The wood of theſe trees, eſpecially of the Abele, is very good to lay for floors, where it will laſt many years, and for its exceeding whiteneſs, is by many perſons preferred to Oak; but being of a ſoft contexture, is very ſubject to take the impreſſion of nails, &c. which renders it leſs proper for

for this purpose: it is also very proper for wainscoting of rooms, being less subject to swell or shrink than most other sorts of wood; but for turnery-ware, there is no wood equal to this for its exceeding whiteness, so that trays, bowls, and many other utensils, are made of it; and the bellows-makers prefer it for their use, as do also the shoemakers, not only for heels, but also for the soles of shoes; it is also very good to make light carts, and the poles are very proper to support Vines, Hops, &c. and the lopping will afford good fuel, which in many countries is much wanted.

The Carolina Poplar may also be propagated by cuttings or layers; the latter is generally practised by the nursery-gardeners, being the surest method; and these plants are not so full of moisture as those raised by cuttings, so are less liable to be cut down by the frost when young. There has been no trials made here of the wood of this tree, so I cannot give any account of its worth.

The Tacamahacca sends up a great number of suckers from the roots, by which it multiplies in plenty, and every cutting which is planted will take root; so that when a plant is once obtained, there may soon be plenty of the plants raised. The buds of this tree are covered with a glutinous resin, which smells very strong, which is the Tacamahacca used in the shops.

PORRUM. Tourn. Inst. R. H 382. tab. 204. Leek.

The Characters are,

The flower hath six bell-shaped petals, collected into a spherical head, covered by a common roundish spatha or sheath, which withers. They have six stamina; three of these are alternately broader than the other, and have forked summits in their middle. They have a short, round, three-cornered germen, supporting a single style, crowned by an acute stigma. The germen afterward becomes a short broad capsule with three lobes, having three cells, filled with angular seeds.

The Species are,

1. Porrum (*Sativum*) radice oblongâ tunicatâ, caule planifolio, floribus capitatis, staminibus tricuspidatis. *Leek with an oblong coated root, a plain leaf on the stalk, flowers collected in heads, and three-pointed stamina; commonly called London Leek.*

2. Porrum (*Ampeloprasum*) caule planifolio umbellifero, umbellâ globosâ, staminibus corollâ longioribus. *Leek with a plain leaf on the stalk, which supports a globular umbel of flowers, whose stamina are longer than the petals.*

The first sort is commonly cultivated in the English gardens. Of this there has been generally supposed two sorts, but I have made trial of them both, by sowing their seeds several times, and find they are the same; the difference which has risen between them, has been occasioned by some persons having saved the seeds from old roots, and not from the seedling Leeks, whereby they have degenerated them, and rendered them smaller and narrower leaved, but by care this may be recovered again, as I have experienced.

The other sort grows naturally in Siberia. This hath narrower leaves than the common sort; the stalks are smaller, and do not rise near so high; the heads of flowers are also smaller, and of a purplish colour; the stamina stand out beyond the flower.

Leeks are cultivated by sowing their seeds in the spring, in the same manner as was directed for Onions, with which these are commonly sown, the two sorts of seeds being mixed according to the proportion which is desired of either sort; though the most common method is, to mix an equal quantity of both, for the Onions will greatly out-grow the Leeks in the spring; but these being drawn off in July or August, the Leeks will have time to grow large afterwards, so that there may be a moderate crop of both sorts. The management of Leeks being exactly the same with Onions, I shall not repeat it in this place, but shall only add, that many persons sow their Leeks in beds in the spring; and in June, after some of their early crops are taken off, they dig up the ground, and plant their Leeks out thereon, in rows a foot apart, and six inches asunder in the rows, observing to water them until they have taken root; after which they will require no further culture but to clear the ground from weeds. The Leeks thus planted will grow to a great size, provided the ground be good, and this method is very proper for such persons who have little room.

If you would save the seeds of this plant, you should make choice of some of the largest and best you have, which must remain in the place where they grow until February, when they should be transplanted in a row against a warm hedge, pale, or wall, at about eight inches asunder; and when their stems advance, they should be supported by a string, to prevent their being broken down, to which they are very liable, especially when in head; and the closer they are drawn to the fence in autumn, the better the seeds will ripen; for it sometimes happens in cold summers or autumns, that those which grow in the open garden, do not perfect their seeds in this country, especially if there should be sharp frosts early in autumn, which will entirely spoil the seed.

When it is ripe (which may be known by the heads changing brown) you should cut off their heads with about a foot or more of the stalk to each, and tie them in bundles, three or four heads in each, and hang them up in a dry place, where they may remain till Christmas or after, when you may thresh out the seeds for use. The husk of these seeds is very tough, which renders it very difficult to get out the seeds; therefore some persons, who have but a small quantity, rub it hard against a rough tile, which will break the husks, and get the seeds out better than most other methods I have known used.

PORTULACA. Tour. Inst. R.H. 236. tab. 118. Purslane.

The Characters are,

The empalement of the flower is small, bifid, and permanent, sitting upon the germen. The flower has five plain, erect, obtuse petals, and many hair-like stamina, about half the length of the petals, terminated by single summits, and a roundish germen, supporting a short style, crowned by five oblong stigmas. The germen afterward becomes an oval capsule with one cell, containing many small seeds.

The Species are,

1. Portulaca (*Oleracea*) foliis cuneiformibus, floribus sessilibus. Prod. Leyd. 473. *Purslane with wedge-shaped leaves, and flowers growing close to the stalks; broad-leaved, or Garden Purslane.*

2. Portulaca (*Pilosa*) foliis subulatis alternis, axillis pilosis, floribus sessilibus. Lin. Sp. Plant. 445. *Purslane with awl-shaped leaves placed alternately, hairy joints, and flowers sitting close to the stalks.*

3. Portulaca (*Anacampseros*) foliis ovatis gibbis, pedunculo multifloro, caule fruticoso. Lin. Sp. Plant. 445. *Purslane with oval gibbous leaves, foot-stalks having many flowers, and a shrubby stalk.*

The first sort grows naturally in America, and most of the hot parts of the globe. This is the common Purslane which is cultivated in the gardens, and is so generally known as to need no description. There are two varieties of this, one with deep green leaves, and the other hath yellow leaves, which is called Golden Purslane; but as both these arise from the same seeds, so they are only seminal variations. There is also a third variety, with smaller and less succulent leaves, which is called Wild Purslane, because wherever it is once sown in a garden, and the plants permitted to scatter their seeds, the plants will come up as weeds

weeds the following year; but this I am ſure is a degeneracy from the Garden Purſlane, for I have ſown it ſeveral times, and let the plants ſhed their ſeeds, and it has come up from thoſe ſeeds, and in two years cultivation degenerated to the wild ſort.

Purſlane is propagated from ſeeds, which may be ſown upon beds of light rich earth during any of the ſummer months; but if you intend to have it early in the ſeaſon, it ſhould be ſown upon a hot-bed, for it is too tender to be ſown in the open air before April, and then it muſt be in a warm ſituation. This ſeed is very ſmall, ſo that a little of it will be ſufficient to ſupply a family. There is no other culture which this plant requires, but to keep it clear from weeds, and in dry weather to water it two or three times a week. In warm weather this plant will be fit for uſe in ſix weeks after ſowing; ſo that in order to continue a ſucceſſion of it, you ſhould ſow it at three or four different ſeaſons, allowing a fortnight or three weeks between each ſowing, which will be ſufficient to laſt the whole ſummer, while it is proper to be eaten, for being of a very cold nature, it is unſafe to be eaten; except in the heat of ſummer in England; for which reaſon it is not to any purpoſe to ſow it upon a hot bed, ſince it will come early enough for uſe in the open air.

If the ſeeds are intended to be ſaved, a ſufficient number of the earlieſt plants ſhould be left for this purpoſe, drawing out all thoſe which are weak, or have ſmall leaves, from among them; and when the ſeeds are ripe, the plants ſhould be cut up, and ſpread upon cloths in the ſun to dry, and then the ſeeds may be eaſily beaten out and ſifted, to clear it from the leaves and ſeed-veſſels.

The ſecond ſort grows naturally in moſt of the iſlands of the Weſt-Indies. This is annual; the ſtalks are very ſucculent, of a purple colour, and branches out greatly; the lower branches lie near the ground, but thoſe above them are more erect; the leaves are narrow, awl-ſhaped, and of a lucid green; they are placed alternately on the branches. At the joints there come out tufts of white hairs, and between theſe come out the flowers ſitting cloſe to the branches; they are of a fine Pink colour, but of ſhort duration, ſeldom continuing open longer than five or ſix hours; theſe are ſucceeded by ſhort roundiſh capſules, filled with ſmall black ſeeds. It flowers from the middle of June till autumn.

The third ſort grows naturally at the Cape of Good Hope. This is a perennial plant, with a ſhrubby ſtalk, which riſes four or five inches high, garniſhed with thick, globular, ſucculent leaves; at the top of the ſtalk comes forth a ſlender foot-ſtalk about two inches long, ſuſtaining four or five Roſe-ſhaped flowers, of a reddiſh colour. Theſe appear in July, but are not ſucceeded by ſeeds in England. This plant is too tender to live in the open air in winter, ſo it muſt be kept in pots, and placed in a very moderate ſtove in winter. It is propagated by cuttings in the ſame manner as moſt other ſucculent plants.

POTENTILLA. Lin. Gen. Plant. 558. Cinquefoil.

The CHARACTERS are,

The empalement of the flower is of one leaf, which is ſlightly cut into ten parts; the ſegments are alternately leſs and reflexed. The flower is compoſed of five petals, which are inſerted into the empalement, and ſpread open. It hath twenty awl-ſhaped ſtamina inſerted in the empalement, terminated by moon-ſhaped ſummits. In the center of the flower there are ſeveral germen collected into one head, with very ſlender ſtyles inſerted in the ſide of the germen, crowned by obtuſe ſtigmas. After the flower is paſt, the germen becomes a head of roundiſh ſeeds, included in the empalement.

The SPECIES are,

1. POTENTILLA (*Anſerina*) foliis pinnatis ſerratis, caule repente. Fl. Lap. 210. *Potentilla with winged ſawed leaves, and a creeping ſtalk.*

2. POTENTILLA (*Rupeſtris*) foliis pinnatis alternis, foliolis quinis ovatis crenatis, caule erecto. Hort. Cliff. 193. *Potentilla with alternate winged leaves, having five oval crenated lobes, and an erect ſtalk.*

3. POTENTILLA (*Fruticoſa*) foliis pinnatis, caule fruticoſo. Hort. Cliff. 193. *Potentilla with winged leaves and a ſhrubby ſtalk; commonly called ſhrubby Cinquefoil.*

4. POTENTILLA (*Recta*) foliis ſeptenatis lanceolatis ſerratis utrinque ſubpiloſis, caule erecto. Lin. Sp. Plant. 497. *Potentilla with ſeven lobes to the leaves, which are ſpear-ſhaped, ſawed, hairy on both ſides, with an erect ſtalk.*

5. POTENTILLA (*Argentea*) foliis quinatis cuneiformibus inciſis ſubtus tomentoſis, caule erecto. Lin. Sp. Plant. 497. *Potentilla with five wedge-ſhaped lobes to the leaves, which are woolly on their under ſide, and an erect ſtalk.*

6. POTENTILLA (*Monſpelienſis*) foliis ternatis, caule ramoſo erecto, pedunculis ſupra genicula enatis. Hort. Upſ. 134. *Potentilla with leaves growing by threes, an upright branching ſtalk, and foot-ſtalks riſing above the joints; or Alpine barren Strawberry.*

7. POTENTILLA (*Grandiflora*) foliis ternatis, caule ramoſo, erecto pedunculis ſupra genicula enatis. *Potentilla with leaves growing by threes, the lobes whereof are oval and obtuſely crenated, a branching ſtalk, and longer foot-ſtalks.*

8. POTENTILLA (*Heptaphylla*) foliis ſeptenis quinatiſque, foliolis lanceolatis pinnato-dentatis utrinque piloſis, caule erecto corymboſo, petalis cordatis. *Potentilla with five and ſeven leaves, whoſe lobes are ſpear-ſhaped, wing-indented, and hairy on both ſides, and have an erect branching ſtalk, with heart-ſhaped petals to the flower.*

There are many more ſpecies of this genus, which are preſerved in botanic gardens for the ſake of variety; but as they are not cultivated in other places either for uſe or beauty, I ſhall not enumerate them here.

The firſt ſort here mentioned grows naturally upon cold ſtiff land in moſt parts of England, and is a ſure mark of the ſterility of the ſoil. It ſpreads its ſtalks upon the ground, which ſend out roots from their joints, faſtening into it, and thereby propagates ſo faſt, as in a little time to ſpread over and fill the ground to a great diſtance. It flowers great part of ſummer. The leaves of this plant are uſed in medicine, and are accounted reſtringent and vulnerary. It is never cultivated in gardens, being a very common weed in England.

The ſecond ſort grows naturally on the Alps, and mountains in Germany. This hath a perennial root, which ſends out ſeveral heads joined together; from theſe ariſe the foot-ſtalks of the leaves, which are long, and ſuſtain three pair of roundiſh lobes, terminated by an odd one; theſe are crenated on their edges, and ſit cloſe to the midrib. Out of each head ariſes a hairy ſtalk about nine inches high, divided into ſmall foot-ſtalks, each ſuſtaining two or three white flowers, very like thoſe of the Strawberry. It is eaſily propagated by ſeeds. The beſt time for ſowing them is in the autumn; it loves a moiſt ſoil and a ſhady ſituation.

The third ſort grows naturally in the northern counties of England, and in many of the northern parts of Europe. This hath a ſhrubby ſtalk, which riſes about four feet high, dividing into many branches, garniſhed by winged leaves, compoſed of two or three pair of narrow, acute-pointed, entire, hairy lobes, pale on their under ſide. The flowers are produced at the end of the branches in cluſters; they have five yellow petals, ſpreading open in form of a Roſe, with many germen and ſtamina within. Theſe appear in July, and are ſometimes ſucceeded by ſeeds incloſed in the empalement. This plant is generally cultivated in the nurſery-gardens as a flowering ſhrub, and is commonly propagated by ſuckers, or laying down the tender branches, which will take root in one year, and may then be taken

off

off from the old plants, and planted in a nursery for a year or two to get strength, before they are planted where they are designed to remain. It may also be propagated by cuttings, which may be planted in autumn, in a moist shady border, where they will take root the next spring, and the Michaelmas following may be transplanted into the nursery.

The best season for transplanting of these plants is in October, that they may get new roots before the hard frost sets in; for as this plant grows naturally upon moist boggy land, so when it is removed in the spring, if due care is not taken to water it in dry weather, it is apt to miscarry; nor will this plant live in a hot dry soil, but in a shady situation and on a cool moist soil it will thrive exceedingly.

The fourth sort grows naturally in the south of France and Italy. This hath hand-shaped leaves, composed of five or seven lobes, which join at their base, where they meet the foot stalk; they are deeply crenated on their sides, and are hairy on both. The stalks rise nine or ten inches high, branching toward the top. The flowers are white, shaped like those of the former sort, terminating the stalks, and are succeeded by seeds like the other. This is a biennial plant, which dies soon after the seeds are ripe. It may be propagated as the second sort.

The fifth sort grows naturally on the Alps, and in other rough hilly parts of Europe. This hath a thick fleshy root, from which arise several purple branching stalks about a foot high, garnished with leaves, composed of five wedge-shaped lobes, deeply cut on their edges, which are very hoary on their under side. The flowers grow at the top of the stalk, which branches out into many foot-stalks; they are yellow, and shaped like those of the fourth sort, but smaller. The root is perennial, and the plant may be propagated as the second sort.

The sixth sort grows naturally on the Alps. This is a biennial plant; the stalks grow erect, about a foot high; they are very hairy, garnished with trifoliate oblong leaves, sawed on their edges. The flowers are produced upon foot-stalks, which come out above the joints of the stalk; they are white, and very like those of the Strawberry. This plant flowers in June, and the seeds ripen in autumn, which, if permitted to scatter, will produce plants in plenty the following spring, which will require no other culture but to keep them clean from weeds.

The seventh sort is also a biennial plant, but differs from the other in having taller and stronger stalks, which branch out more; the lobes of the leaves are oval, obtuse, and bluntly indented on their edges; the flowers are larger, and the whole plant is of a deeper green. It flowers in July, and the seeds ripen in autumn. It propagates itself like the former sort.

The eighth sort grows naturally in the south of France and Italy. This is a biennial plant; the stalks are large, and rise near two feet high; they branch very much toward their top; the leaves stand upon very short foot-stalks; they are sometimes composed of five, and at other times of seven lobes, which are regularly indented like winged leaves, and are very hairy on both sides. The flowers are produced at the top of the stalk, each having a foot-stalk an inch and a half long; their empalements are deeply cut into nine segments, which end in acute points. The flowers have sometimes but five, but generally six heart-shaped petals, which are of a pale yellow, and expand like those of the former sorts. It flowers in July, and the seeds ripen in autumn, which, if permitted to scatter, will produce plenty of plants the following spring. This requires no other culture than to keep it clean from weeds.

POTERIUM. Lin. Gen. Plant. 948. Burnet.

The CHARACTERS are,

It hath male and female flowers in the same spike. The male flowers have a three-leaved empalement; they have one petal, which is cut into four parts; these are oval, concave, and permanent; they have a great number of hair-like stamina, terminated by roundish twin summits. The female flowers have one wheel-shaped petal, with a short tube, cut at the brim into four parts; these have no stamina, but two oblong oval germen, with two hairy styles the length of the petal, crowned by coloured pencil-shaped stigmas. The germen afterward becomes two hard seeds, inclosed in the petal of the flower.

The SPECIES are,

1. POTERIUM (*Sanguisorba*) inerme, caulibus subangulosis. Hort. Cliff. 446. *Smooth Poterium with angular stalks; common Burnet.*

2. POTERIUM (*Hybridum*) inerme, caulibus teretibus strictis. Lin. Sp. Plant. 994. *Smooth Poterium with a narrow taper stalk; or large Burnet.*

3. POTERIUM (*Spinosum*) spinis ramosis. Hort. Cliff. 445. *Poterium with branching spines; or prickly evergreen Burnet.*

The first sort is the common Burnet, which grows naturally upon chalky lands in many parts of England. Of this there are two or three varieties, one of them is much smoother than the other, and the third hath larger seeds than either of the former; but these differences are not constant, so they are only seminal variations. This is a perennial plant, from whose root arise a great number of leaves, standing on pretty long foot-stalks, composed of five or six pair of lobes, terminated by an odd one. The lobes are generally ranged a little alternate on the midrib, but sometimes stand by pairs; these are sawed on their edges, and are sometimes smooth, and at others hairy. The stalks rise a foot and a half high, branching out pretty much, and are terminated by long slender foot-stalks, each sustaining an oblong spike of flowers, in which there are some male and others female; they are of a purplish red colour, and appear in June. The female flowers are each succeeded by two hard seeds, which ripen in autumn.

This plant is propagated in gardens; the young tender leaves are put into sallets, and the leaves are used for cool tankards in hot weather. It is used in medicine, and is reckoned to be cordial and alexipharmic. The powder of the root is commended against spitting of blood.

This plant is easily propagated by seeds, which should be sown in autumn, for if it is sown in spring, the seeds frequently lie in the ground till the spring following. If the seeds are permitted to scatter, the plants will come up in plenty; if these are transplanted out in a bed of undunged earth, at about a foot distance every way, and kept clean from weeds, they will continue several years, especially if the soil is dry, and will require no other care. It may also be propagated by parting of the roots in the autumn, but as the plants arise so freely from scattered seeds, the latter method is seldom practised.

The second sort grows naturally in the south of France and Italy. This is a biennial plant, which decays soon after the seeds are ripe. The leaves of this are like those of Agrimony, and are composed of three or four pair of oblong lobes, placed a little alternate on the midrib, and terminated by an odd one; they are deeply sawed on their edges, and have an agreeable scent; the stalks rise two feet high, garnished at each joint with one of those winged leaves, which gradually diminish in their size to the top, and just above the leaf arises a long foot-stalk, which supports two or three small ones, each sustaining a small roundish spike of flowers. These appear in July, and are succeeded by seeds which ripen in autumn. It is propagated by by seeds, which, if sown in autumn, the plants will come up the following spring. These require no other culture than to thin them where they are too close, and keep them clean from weeds; the second year they will flower and ripen their seeds, and soon after decay.

The third sort grows naturally in Crete, and in many of the islands in the Archipelago. This hath a shrubby perennial stalk, which rises about three feet high, dividing into several slender branches, which are armed with branching sharp thorns; the leaves are very small; they are winged, and have six or seven pair of very small lobes, ranged opposite along the midrib, terminated by an odd one; they are of a lucid green, and continue all the year. The flowers are produced in small heads at the end of the branches, and are of an herbaceous colour.

This plant is too tender to live through the winter in the open air, but if it is sheltered under a common hot-bed frame in winter, where it may have the free air at all times when the weather is mild, and sheltered from hard frost, it will thrive better than when it is more tenderly treated. It may be propagated by slips or cuttings during any of the summer months, which, if planted in a bed of light earth, and covered down close with a hand or bell-glass, and shaded from the sun, will sometimes take root, and may then be taken up and planted each into a separate small pot, filled with fresh undunged earth, and placed in the shade till they have taken new root, and then removed to a sheltered situation, where they may stand till the frost comes on, when they should be placed under the hot-bed frame. It requires but little water, especially in cool weather, and wants no particular culture. It may also be propagated by seeds, which should be sown soon after it is ripe.

PRASIUM. Lin. Gen. Plant. 655. Shrubby Hedge-nettle.

The CHARACTERS are,

The flower hath a bell shaped empalement of one leaf, divided into two lips; the upper lip ends in three acute points; the lower lip is cut into two parts. The flower is of the lip kind; it hath one petal; the upper lip is oval, erect, and indented at the end. The lower lip is reflexed, and ends in three points, the middle one being broadest. It has four awl-shaped stamina under the upper lip, two of which are shorter than the other, having oblong summits on their side, and a four-pointed germen, sustaining a slender style the length of the stamina, crowned by a bifid stigma. The germen afterward becomes four berries, each containing a single roundish seed.

The SPECIES are,

1. PRASIUM (*Majus*) foliis ovato oblongis serratis. Lin. Hort. Cliff. 309. *Shrubby stinking Hedge-nettle, with oblong, oval, sawed leaves.*

2. PRASIUM (*Minus*) foliis ovatis, duplici utrinque crena notatis. Lin. Hort. Cliff. 309. *Shrubby stinking Hedge-nettle, with oval leaves, which are indented on every side.*

The first sort grows naturally in Spain and Italy. This rises with a shrubby stalk near three feet high, covered with a whitish bark, divided into many branches, garnished with oblong oval leaves, which are sawed on their edges. The flowers come out from the bosom of the leaves, in whorls round the stalks: they are white, and have large permanent empalements, cut into five points. The flowers are of the lip kind; they appear in June and July, and are succeeded by four small berries sitting in the empalement, which turn black when they are ripe, and have a single roundish seed in each.

The second sort grows naturally in Sicily. This hath a shrubby stalk like the former, but rises a little higher; the bark is whiter, the leaves are shorter and oval, and are doubly crenated on each side; they are of a lucid green. The flowers come out in small whorls from the bosom of the leaves, like the former; they are somewhat larger, and are frequently marked with a few purple spots; these are succeeded by small berries like the other sort, which ripens at the same time.

These plants may be propagated either by cuttings, or from the seeds: if they are propagated by cuttings, they should be planted on a shady border toward the end of April; but the cuttings should not be taken from such plants as have been drawn weak, but rather from those which have been exposed to the open air, whose shoots are short and strong; and if a joint of the former year's wood is cut to each of them, they will more certainly succeed. These cuttings may remain in the same border till they are well rooted, when they may be transplanted into the places where they are to remain, or into pots, that they may be sheltered in winter under a common frame, where they may have as much free air as possible in dry weather, but only require to be screened from hard frost.

If they are propagated by seeds (which the plants produce in plenty every year) they should be sown on a bed of light earth in April; and in May the plants will come up, when they require no other care but that of keeping them clean from weeds; and in the autumn following they may be transplanted in the same manner as before directed for those raised from cuttings, and may be afterward treated more hardily as they acquire strength.

A plant or two of each of these species may by allowed to have a place where there are collections of the different sorts of evergreen shrubs, for the sake of variety, especially where the different sorts of Cistus, Phlomis, Tree-wormwood, and Medicago, are admitted, because these are equally hardy; and when a severe winter happens, which destroys the one, the others are sure of the same fate; but in mild winters they will live abroad, especially if they are planted in a dry rubbishy soil, and have a sheltered situation; but on wet ground the plants will grow very vigorous in summer, so are liable to injury from the early frosts in autumn.

PRENANTHES. Lin. Gen. Plant. 816. Vaill. Mem. Ann. 1721. Wild Lettuce.

The CHARACTERS are,

It hath a smooth cylindrical empalement, having many scales, which are equal, but have three at the base unequal. This common empalement includes from five to eight hermaphrodite florets, disposed in a single round order; they have one petal, which is stretched out like a tongue, and indented in four parts at the end; they have five short hair-like stamina, terminated by cylindrical summits. The germen is situated under the petal, supporting a slender style longer than the stamina, crowned by a bifid reflexed stigma. The germen afterward becomes a single heart-shaped seed, crowned with hairy down.

The SPECIES are,

1. PRENANTHES (*Muralis*) flosculis quinis, foliis lyrato hastatis. Hort. Cliff. 383. *Prenanthes with five florets, and lyre-shaped leaves.*

2. PRENANTHES (*Purpurea*) flosculis quinis, foliis lanceolatis denticulatis. Hort. Cliff. 383. *Prenanthes with five florets, and spear-shaped indented leaves.*

3. PRENANTHES (*Altissima*) erecta, flosculis quinis foliis trilobis. Lin. Sp. Plant. 797. *Upright Prenanthes with five florets, and leaves having three lobes.*

4. PRENANTHES (*Amplexicaulis*) flosculis quinis, caule ramoso, foliis ovato-lanceolatis semiamplexicaulibus. *Prenanthes with five florets, a branching stalk, and oval spear-shaped leaves half embracing the stalk.*

The first sort grows naturally upon walls and dry shady banks in many parts of England, so is seldom cultivated in gardens.

The ſecond ſort grows naturally upon the Helvetian mountains. This hath a creeping root, which ſpreads far in the ground, ſo becomes a troubleſome weed if admitted into gardens. The ſtalks of this riſe four feet high; the leaves are ſpear-ſhaped, and a little indented toward their ends; the flowers are of a purple blue colour, and are produced in panicles from the ſides, and at the top of the ſtalks. Theſe appear in July, and are ſucceeded by ſeeds which ripen in autumn

The third ſort grows naturally in moſt parts of North America, where it is called Dr. Witt's Rattleſnake-root. This ſeldom lives longer than two years. The lower leaves are four or five inches long, and three broad; they are ſometimes divided into five lobes, but generally into three; they are indented a little on their edges, ſmooth, of a dark green on their upper ſide, but pale on their under. The ſtalks riſe three feet high, and are garniſhed with a few ſmall leaves, which are entire; the flowers come out from the ſide of the ſtalk in ſmall bunches; theſe are of a pale yellow colour, and appear in July. They are ſucceeded by ſeeds, crowned with hairy down, which ripen in autumn. There is a variety of this with pale purple flowers, which ariſes from the ſame ſeeds. The roots of theſe plants are ſaid to be an antidote to expel the venom of the Rattleſnake, which induced me to mention theſe plants.

The fourth ſort grows naturally on the mountains in Germany. This hath a perennial root. The ſtalks riſe a foot high, and branch out on each ſide; the leaves are ſpear-ſhaped and oval; their baſe is broad, and half ſurrounds the ſtalk; the flowers grow looſely upon ſlender foot-ſtalks, which come out from the ſide, and at the end of the branches. Theſe appear in June, and the ſeeds ripen in autumn.

Theſe plants are ſeldom admitted into gardens, but if any perſon is deſirous to cultivate them, if they ſow the ſeeds ſoon after they are ripe in a ſhady ſituation, the plants will come up, and require no other care but to keep them clear from weeds.

PRIMULA. Lin. Gen. Plant. 180. The Primroſe.

The CHARACTERS are,

The flower hath a tubulous empalement of one leaf, ending in five acute points; it hath one petal, with a cylindrical tube, but ſpreads open above, where it is cut into five heart-ſhaped ſegments. It has five ſhort ſtamina, ſituated in the neck of the petal, terminated by erect acute-pointed ſummits, and a globular germen ſupporting a ſlender ſtyle, crowned by a globular ſtigma. The germen afterward turns to an oblong capſule, with one cell opening at the top, filled with ſmall angular ſeeds.

The SPECIES are,

1. PRIMULA (*Veris*) foliis dentatis rugoſis, pedunculis unifloris. *Primroſe with rough indented leaves, and foot-ſtalks bearing one flower; or common Primroſe.*

2. PRIMULA (*Elatior*) foliis dentatis rugoſis, floribus faſtigiatis. *Primroſe with rough indented leaves, and flowers growing in bunches; called Cowſlip.*

3. PRIMULA (*Farinoſa*) foliis crenatis glabris, florum limbo plano. *Primroſe with ſmooth crenated leaves, and plain empalements; called Birds-eyen.*

4. PRIMULA (*Polyantha*) foliis petiolatis ſubcordatis crenatis, floribus faſtigiatis pedunculis longiſſimis. *Primroſe, or Cowſlip, with heart ſhaped crenated leaves, having foot-ſtalks, and flowers growing in bunches on very long foot-ſtalks.*

The firſt ſort of Primroſe grows wild in woods, and other ſhady places in moſt parts of England, from whence their roots may be eaſily tranſplanted into the garden, where, if they are placed under hedges, and in ſhady walks, they make a beautiful appearance early in the ſpring, when few other plants are in flower.

This plant is ſo well known as to need no deſcription; the flowers and roots of this are uſed in medicine.

There are ſeveral varieties of this, which have been accidentally obtained; as the common Primroſe with double flowers; the red Primroſe with ſingle and double flowers; theſe have but one flower upon a foot-ſtalk, but the paper-white Primroſe is certainly a diſtinct ſpecies.

The ſecond ſort is the Cowſlip, or Paigle, or Paralyſis of the ſhops. This grows naturally in meadows and moiſt paſtures in many parts of England. The flowers of this ſort grow in bunches on the top of the foot-ſtalk, ſo are eaſily diſtinguiſhed from the former; the flowers are much uſed in medicine, and ſometimes the leaves. As theſe grow wild, their roots may be taken up, and tranſplanted into gardens.

The beſt time to tranſplant them is at Michaelmas, that their roots may have ſtrength to produce their flowers early in the ſpring. Theſe delight in a ſtrong rich ſoil, but will grow in almoſt any ſort of earth, provided they have a ſhady ſituation.

There are a great variety of this at preſent in the gardens, as the Hoſe in Hoſe, the double Cowſlip, and all the ſorts of Polyanthus, which have been ſo much improved within the laſt fifty years, as to almoſt equal the variety of the Auriculas; and in ſome parts of England they are ſo much eſteemed as to ſell for a guinea a root, ſo that there may be ſtill a much greater variety raiſed, as there are ſo many perſons engaged in the culture of this flower.

The ſeveral varieties of Polyanthuſes are produced by ſowing of ſeeds, which ſhould be ſaved from ſuch flowers as have large upright ſtems, producing many flowers upon a ſtalk, which are large, beautifully ſtriped, open flat, and not pin-eyed. From the ſeeds of ſuch flowers there is room to hope for a great variety of good ſorts, but there ſhould be no ordinary flowers ſtand near them, leſt by the mixing of their farina the ſeeds ſhould be degenerated.

Theſe ſeeds ſhould be ſown in boxes filled with ligh rich earth in December, being very careful not to bury the ſeed too deep, for if it be only ſlightly covered with light earth, it will be ſufficient. Theſe boxes ſhould be placed where they may have the benefit of the morning ſun until ten of the clock, but muſt by no means be expoſed to the heat of the day, eſpecially when the plants begin to appear, for at that time one whole day's ſun will entirely deſtroy them. In the ſpring, if the ſeaſon ſhould prove dry, you muſt often refreſh them with water, and as the heat increaſes, you ſhould remove the boxes more in the ſhade, for the heat is very injurious to them.

By the end of May theſe plants will be ſtrong enough to plant out, at which time you ſhould prepare ſome ſhady borders, which ſhould be made rich with neats dung, upon which you muſt ſet the plants about four inches aſunder every way, obſerving to water them until they have taken root; after which they will require no farther care but to keep them clear from weeds, until the latter end of Auguſt following, when you ſhould prepare ſome borders, which are expoſed to the eaſt, with good light rich earth, into which you muſt tranſplant your Polyanthuſes, placing them ſix inches aſunder equally in rows, obſerving, if the ſeaſon proves dry, to water them until they have taken root. In theſe borders your plants will flower the ſucceeding ſpring, at which time you muſt obſerve to mark ſuch of them as are fine to preſerve, and the reſt may be tranſplanted into wilderneſſes, and other ſhady places in the garden, where, although they are not very valuable flowers, they will afford an agreeable variety.

Thoſe which you intend to preſerve, may be removed ſoon after they have done flowering (provided you do not intend to ſave ſeeds from them,) and may be then tranſ-

planted

planted into a fresh border of the like rich earth, allowing them the same distance as before, observing also to water them until they have taken root; after which they will require no farther care, but only to keep them clean from weeds, and the following spring they will produce strong flowers, as their roots will be then in full vigour; so that, if the kinds are good, they will be little inferior to a shew of Auriculas.

These roots should be constantly removed and parted every year, and the earth of the border changed, otherwise they will degenerate, and lose the greatest part of their beauty.

If you intend to save seeds, which is the method to obtain a great variety, you must mark such of them, which, as I said before, have good properties. These should be, if possible, separated from all ordinary flowers, for if they stand surrounded with plain-coloured flowers, they will impregnate each other, whereby the seeds of the valuable flowers will not be near so good, as if the plants had been in a separate border, where no ordinary flowers grew; therefore the best way is to take out the roots of such as you do not esteem as soon as the flowers open, and plant them in another place, that there may be none left in the border but such as you would chuse for seeds.

The flowers of these should not be gathered, except such as are produced singly upon pedicles, leaving all such as grow in large bunches; and if the season should prove dry, you must now and then refresh them with water, which will cause their seeds to be larger, and in greater quantity, than if they were entirely neglected. In June the seed will be ripe, which may be easily known by the pods changing brown, and opening; so that you should at that time look over the plants three times a week, gathering each time such of the seed-vessels as are ripe, which should be laid upon a paper to dry, and may then be put up until the season of sowing.

As the plants which arise from seeds, generally flower much better than offsets, those who would have these flowers in perfection should annually sow their seeds.

PRIMROSE-TREE. See ONAGRA.

PRINOS. Lin. Gen. Plant. 398. Winterberry.

The CHARACTERS are,

The flower hath a permanent empalement of one leaf, which is cut into six small plain segments. It hath one wheel-shaped petal, with no tube, cut into six plain segments; it hath six awl-shaped stamina shorter than the petal, terminated by obtuse summits, and an oval germen sitting upon the style, crowned by an obtuse stigma. The germen afterward turns to a round berry, opening in three parts, including one hard seed.

The SPECIES are,

1. PRINOS (*Verticillatus*) foliis longitudinaliter serratis. Lin. Sp. Plant. 330. *Prinos, or Winterberry, with leaves sawed lengthways.*

2. PRINOS (*Glaber*) foliis apice serratis. Lin. Sp. Plant. 330. *Prinos with leaves sawed at the points.*

The first sort grows naturally in Virginia, and other parts of North America. This rises with a shrubby stalk to the height of eight or ten feet, sending out many branches, garnished with spear-shaped leaves, terminating in acute points, veined on their under side, and sawed on their edges, having slender foot-stalks standing alternately on the branches. The flowers come out from the side of the branches, sometimes single, at others two or three at each joint; they have no tube, but are wheel-shaped, and cut into six parts; they have six awl-shaped erect stamina, terminated by obtuse summits, and an oval germen sitting upon the style, crowned by an obtuse stigma; these are succeeded by berries about the size of those of Holly, which turn purple when ripe. It flowers in July, and the seeds ripen in the winter.

The second sort grows naturally in Canada. This is of lower growth than the former. The leaves are shorter, and sawed at their points, but the flowers of this I have not seen.

It is propagated by seeds, which should be sown soon after they are ripe upon a bed of light earth, covering them about one inch deep. The seeds, which are so soon put into the ground, will many of them come up the following spring, whereas those which are kept longer out of the ground, will remain a whole year before the plants will appear, in the same manner as the Holly, Hawthorn, and some others; therefore the ground should not be disturbed, if the plants do not come up the first year. The young plants may be treated in the same manner as hath been directed for the American Hawthorns, and are full as hardy, but they delight in a moist soil and a shady situation, for in hot land they make but little progress, and rarely produce any fruit.

PRIVET. See LIGUSTRUM.

PROTEA. Lin. Gen. Plant. 104. Silver-tree.

The CHARACTERS are,

The flowers are collected in an oval head; they have one common imbricated scaly periantheum. The flower is of one petal, having a tube the length of the empalement; the brim is cut into four parts, which spread open, and are equal. It has four bristly stamina the length of the petal, terminated by incumbent summits, and a roundish germen, with an erect bristly style, crowned by an obtuse stigma. The germen afterward turns to a roundish naked seed, sitting in a distinct cell of the cone.

The SPECIES are,

1. PROTEA (*Conifera*) foliis lineari-lanceolatis integerrimis, superioribus hirsutis nitidis. Prod. Leyd. 184. *Protea with linear, spear-shaped, entire leaves, the upper of which shine, and are hairy.*

2. PROTEA (*Argentea*) foliis lanceolatis integerrimis acutis hirsutis nitidis. Hort. Cliff. 29. *Protea with entire, spear-shaped, acute, hairy, shining leaves; commonly called Silver-tree.*

3. PROTEA (*Nitida*) foliis oblongo-ovatis hirsutis nitidis integerrimis. *Protea with oblong, oval, hairy, shining leaves, which are entire; called Wageboom.*

These plants are natives of the country near the Cape of Good Hope in Africa, where there is a great number of species. In the catalogue of the Leyden garden there are upward of twenty sorts enumerated; not that they have them growing there, but they have good drawings of them, which were made in the country where they are native The three sorts here mentioned are what I had lately growing in the Chelsea garden.

As these plants are natives of the Cape of Good Hope, they are too tender to live abroad through the winter in England, but the first sort is hardy enough to live in a good green-house. This sort will grow to the height of ten or twelve feet, and may be trained up with a regular strait stem; the branches naturally form a regular large head. The leaves are long and narrow, of a shining silver colour; and as they remain the whole year, the plants make a fine appearance, when they are intermixed with others in the green-house. In the summer these may be placed in the open air in a sheltered situation, for if they are exposed to winds, the plants will be torn and rendered unsightly, nor will they make any progress in their growth. In warm weather they must be frequently but sparingly watered, and in cold weather this must not be too often repeated, lest it should rot their fibres.

The second sort hath a strong upright stalk, covered with a purplish bark, dividing into several branches, which grow erect, garnished with broad, shining, silvery, spear-shaped leaves,

leaves, which make a fine appearance when intermixed with other exotics. This ſhould be placed in an airy dry glaſs-caſe, where it may be protected from cold, and have as much light as poſſible, and in winter ſhould have little water; it riſes eaſily from ſeeds. The ſeeds will ſometimes remain in the ground ſix or eight months, and at other times the plants will appear in ſix weeks; therefore the beſt way is to ſow the ſeeds in ſmall pots filled with ſoft ſandy loam, and plunge them into a moderate hot-bed; and if the plants ſhould not come up ſo ſoon as expected, the pots ſhould remain in ſhelter till the following ſpring, when, if the ſeeds remain ſound, the plants will come up. The pots in which the ſeeds are ſown ſhould have but little wet, for moiſture frequently cauſes them to rot. When the plants appear, they ſhould not be too tenderly treated, for they muſt not be kept too warm, nor ſhould they have much wet; but in warm weather they muſt be expoſed to the open air in a ſheltered ſituation, and in winter protected from froſt.

The third ſort I raiſed from ſeeds which came from the Cape of Good Hope; theſe ſeeds were long and ſlender, very different in ſhape from thoſe of the ſecond ſort, but the plants have great reſemblance to them. The leaves are very ſilky and white; the ſtalks are purple and grow erect, but have not as yet put out any branches.

The firſt ſort may be propagated by cuttings, which ſhould be cut off in April, juſt before the plants begin to ſhoot; they ſhould be planted in ſmall pots, filled with light earth, and plunged into a moderate hot-bed, ſhading them from the ſun, and now and then gently refreſhing them with water; but it muſt be ſparingly given, for much wet will rot them. Theſe cuttings will put out roots by Midſummer, when they may be gently ſhaken out of the pots and parted, planting each in a ſeparate ſmall pot, filled with light earth, and place them in a frame, where they may be ſhaded till they have taken new root; then they ſhould be gradually inured to the open air, into which they ſhould be removed, and treated in the ſame way as the old plants.

PRUNELLA. Lin. Sp. Plant. 600. Self-heal.

The CHARACTERS are,

The flower has a permanent empalement, divided into two lips; it is of the ringent kind, having one petal with a ſhort tube, and oblong chaps; the upper lip is entire and concave; the under lip is trifid; the middle ſegment is broad, and indented at the point; it has four awl-ſhaped ſtamina, two long and two ſhorter, and four germen, ſupporting a ſlender ſtyle, crowned by an indented ſtigma. The germen becomes four ſeeds incloſed in the empalement.

The SPECIES are,

1. PRUNELLA (*Vulgaris*) foliis oblongo-ovatis petiolatis. Lin. Sp. Plant. 600. *Greater Self-heal with oblong oval leaves.*

2. PRUNELLA (*Laciniata*) foliis oblongo-ovatis petiolatis ſupremis quatuor lanceolatis dentatis. Lin. Sp. Plant. 837. *Self-heal with oval, oblong, jagged leaves.*

3. PRUNELLA (*Hyſſopifolia*) foliis lanceolato-linearibus ciliatis ſubſeſſilibus. Sauv. Monſp. 141. *Self-heal with Hyſſop leaves*

4. PRUNELLA (*Canadenſis*) foliis lanceolato-linearibus, internodiis longiſſimis ſpicis interruptis. *Canada Self-heal with narrow leaves, and the joints of the ſtalk very diſtant.*

5. PRUNELLA (*Sulphurea*) foliis oblongis pinnato-ſinuatis villoſis infimis petiolatis ſummis ſeſſilibus. *Self-heal with oblong, wing-pointed, hairy leaves.*

6. PRUNELLA (*Caroliniana*) foliis lanceolatis integerrimis, infimis petiolatis, ſummis ſeſſilibus internodiis prælongis. *Carolina Self-heal.*

7. PRUNELLA (*Novæ Angliæ*) foliis oblongis mucronatis petiolatis ſpicis florum craſſiſſimis. *New England Prunella.*

The firſt ſort which is uſed in medicine, grows naturally in paſtures every where. The leaves are in ſhape a long oval; the ſtalks are ſquare, riſe about eight inches high, and are terminated by a cloſe ſpike of flowers, which are for the moſt part blue, but ſometimes are white. It flowers in June and July, and the ſeeds ripen in Auguſt and September.

The ſecond ſort differs from the firſt, in having jagged leaves.

The third ſort has very narrow leaves, which are covered with ſmall hairs, and ſit cloſe to the ſtalks.

The fourth ſort, which came originally from Canada, has very narrow leaves; the joints of the flower-ſtalk are far aſunder, and the ſpikes of flowers are ſeparated.

The fifth ſort has hairy cut leaves, and ſulphur-coloured flowers.

The ſixth ſort was brought from Carolina. The leaves of this are ſpear-ſhaped and entire; the flower-ſtalk is tall, the joints very diſtant, and is terminated by a thick ſpike of pale blue flowers.

The ſeventh ſort grows naturally in New England, from whence I received the ſeeds at firſt, but have ſince received the ſame from Virginia; and I ſuppoſe it may be found in other parts of North America. The leaves of this are large, ſmooth, ending in ſharp points; the foot-ſtalks of the flowers are brown, and riſe near a foot high, being terminated by thick ſpikes of blue flowers, which appear in July, and are ſucceeded by ſeeds which ripen in September.

All theſe plants are hardy, ſo will thrive in the open air. They are eaſily propagated by ſeeds, which, if ſown in the autumn, will more certainly ſucceed than thoſe ſown in the ſpring; or if their ſeeds are permitted to ſcatter, the plants will riſe without any trouble; but if the ſorts ſtand near each other, there will be ſome difficulty to keep them diſtinct, when their ſeeds fall to the ground. The plants want no other care but to be thinned to proper diſtances, and kept clean from weeds. They are not of long duration, ſo that whoever would keep the ſorts, ſhould ſow their ſeeds annually; but as they are not very ornamental, ſo they are ſeldom preſerved in gardens, unleſs by ſuch as are curious in botany.

PRUNING OF TREES.

There is not any part of gardening which is of more general uſe than that of Pruning, and yet it is very rare to ſee fruit-trees ſkilfully managed. Almoſt every gardener will pretend to be a maſter of this buſineſs, though there are but few who rightly underſtand it; nor is it to be learned by rote, but requires a ſtrict obſervation of the different manners of growth of the ſeveral ſorts of fruit-trees, ſome requiring to be managed one way, and others muſt be treated in a quite different method, which is only to be known from carefully obſerving how each kind is naturally diſpoſed to produce its fruit, for ſome ſorts produce their fruit on the ſame year's wood, as Vines; others produce their fruit, for the moſt part, upon the former year's wood, as Peaches, Nectarines, &c. and others upon curſons or ſpurs, which are produced upon wood of three, four, or five, to fifteen or twenty years old, as Pears, Plumbs, Cherries, &c. therefore, in order to the right management of fruit-trees, there ſhould always be proviſion made to have a ſufficient quantity of bearing wood in every part of the trees, and at the ſame time there ſhould not be a ſuperfluity of uſeleſs branches, which would exhauſt the ſtrength of the trees, and cauſe them to decay in a few years.

The reaſons which have been laid down for Pruning of fruit-trees are as follow: firſt, To preſerve trees longer in a vigorous

vigorous bearing ſtate; the ſecond is, To render the trees more beautiful to the eye; and thirdly, To cauſe the fruit to be larger and better taſted.

1. It preſerves a tree longer in a healthy bearing ſtate; for by Pruning off all ſuperfluous branches, ſo that there are no more left upon the tree than are neceſſary, or than can be properly nouriſhed, the root is not exhauſted in ſupplying uſeleſs branches, which muſt afterwards be cut out, whereby much of the ſap will be uſeleſly expended.

2. By ſkilful Pruning of a tree, it is rendered much more pleaſing to the eye; but here I would not be underſtood to be an advocate for a ſort of Pruning, which I have ſeen too much practiſed of late, viz. the drawing a regular line againſt the wall, according to the ſhape or figure they would reduce the tree to, and cutting all the branches, ſtrong or weak, exactly to the chalked line; the abſurdity of which practice will ſoon appear to every one, who will be at the pains of obſerving the difference of thoſe branches ſhooting the ſucceeding ſpring. All therefore that I mean by rendering a tree beautiful is, that the branches are all pruned according to their ſeveral ſtrengths, and are nailed at equal diſtances, in proportion to the different ſizes of their leaves and fruit, and that no part of the wall (ſo far as the trees are advanced) be left unfurniſhed with bearing wood. A tree well managed, though it does not repreſent any regular figure, yet will appear very beautiful to the ſight when it is thus dreſſed and nailed to the wall.

3. It is of great advantage to the fruit; for the cutting away all uſeleſs branches, and ſhortening all the bearing ſhoots according to the ſtrength of the tree, will render the tree more capable to nouriſh thoſe which are left remaining, ſo that the fruit will be much larger and better taſted. And this is the advantage which thoſe trees againſt walls or eſpaliers have to ſuch as are ſtandards, which are permitted to grow as they are naturally inclined; for it is not their being trained either to a wall or eſpalier, which render their fruit ſo much better than ſtandards, but becauſe the roots have a leſs quantity of branches and fruit to nouriſh, and conſequently their fruit will be larger and better taſted.

The reaſons for Pruning being thus exhibited, the next thing is the method of performing it; but this being fully handled under the ſeveral articles of the different kinds of fruit, I ſhall not repeat it again in this place, and therefore ſhall only add ſome few general inſtructions, which are neceſſary to be underſtood in order to the right management of fruit trees.

There are many perſons who ſuppoſe, that if their fruit-trees are but kept up to the wall or eſpalier during the ſummer ſeaſon, ſo as not to hang in very great diſorder, and in winter to get a gardener to prune them, it is ſufficient, but this is a miſtake; for the greateſt care ought to be employed about them in the ſpring, when the trees are in vigorous growth, which is the only proper ſeaſon to procure a quantity of good wood in the different parts of the tree, and to diſplace all uſeleſs branches as ſoon as they are produced, whereby the vigour of the tree will be entirely diſtributed to ſuch branches only as are deſigned to remain, which will render them ſtrong, and more capable to produce good fruit; whereas, if all the branches are permitted to remain which are produced, ſome of the more vigorous will attract the greateſt ſhare of the ſap from the tree, whereby they will be too luxuriant for producing fruit, and the greateſt part of the other ſhoots will be ſtarved, and rendered ſo weak, as not to be able to produce any thing elſe but bloſſoms and leaves (as hath been before mentioned;) ſo that it is impoſſible for a perſon, let him be ever ſo well ſkilled in fruit-trees, to reduce them into any tolerable order by winter Pruning only, if they are wholly neglected in the ſpring.

There are others who do not entirely neglect their trees during the ſummer ſeaſon, as thoſe before-mentioned, but yet do little more good to them by what they call ſummer Pruning; for theſe perſons neglect their trees at the proper ſeaſon, which is in April and May, when their ſhoots are produced, and only about Midſummer go over them, nailing in all their branches, except ſuch as are produced fore-right from the wall, which they cut out, and at the ſame time often ſhorten moſt of the other branches, which is almoſt as bad as the other; for thoſe branches which are intended for bearing the ſucceeding year, ſhould not be ſhortened during the time of their growth, which will cauſe them to produce one or two lateral ſhoots from the eyes below the place where they were ſtopped, which ſhoots will draw much of the ſtrength from the buds of the firſt ſhoot, whereby they are often flat, and do not produce their bloſſoms; and if thoſe two lateral ſhoots are not entirely cut away at the winter Pruning, they will prove as injurious to the tree; for the ſhoots which theſe produce, will be what the French call water-ſhoots; and in ſuffering thoſe luxuriant ſhoots to remain upon the tree until Midſummer before they are diſplaced, they will exhauſt a great ſhare of the nouriſhment from the other branches (as was before obſerved;) and by ſhading the fruit all the ſpring, when theſe are cut away, and the other branches faſtened to the wall, the fruit, by being ſo ſuddenly expoſed, will receive a very great check, which will cauſe their ſkins to grow tough, and thereby render them leſs delicate. This is to be chiefly underſtood of ſtone-fruit and Grapes, but Pears and Apples being much hardier, ſuffer not ſo much, though it is a great diſadvantage to thoſe alſo to be thus managed.

It muſt alſo be remarked, that Peaches, Nectarines, Apricots, Cherries, and Plumbs, are always in the greateſt vigour when they are the leaſt maimed by the knife, for where theſe trees have large amputations, they are very ſubject to gum and decay; ſo that it is certainly the moſt prudent method carefully to rub off all uſeleſs buds when they are firſt produced, and pinch others where new ſhoots are wanted to ſupply the vacancies of the wall; by which management trees may be ſo ordered, as to want but little of the knife in winter Pruning, which is the ſureſt way to preſerve theſe trees healthful, and is performed with leſs trouble than the common method.

The management of Pears and Apples is much the ſame with theſe trees in ſummer, but in winter they muſt be very differently pruned; for as Peaches and Nectarines for the moſt part produce their fruit upon the former year's wood, therefore they muſt have their branches ſhortened, according to their ſtrength, in order to produce new ſhoots for the ſucceeding year; ſo Pears, Apples, Plumbs, and Cherries, on the contrary, producing their fruit upon curſons or ſpurs, which come out of the wood of five, ſix, or ſeven years old, ſhould not be ſhortened, becauſe thereby thoſe buds which were naturally diſpoſed to form theſe curſons or ſpurs, would produce wood branches, whereby the trees would be filled with wood, but never produce much fruit; and as it often happens that the bloſſom-buds are firſt produced at the extremity of the laſt year's ſhoot, by ſhortening the branches, the bloſſoms are cut away, which ſhould always be carefully avoided.

There are ſeveral authors who have written on the ſubject of Pruning in ſuch a prolix manner, that it is impoſſible for a learner to underſtand their meaning. Theſe have deſcribed the ſeveral ſorts of branches which are produced on fruit-trees, as wood branches, fruit branches, irregular branches, falſe branches, and luxuriant branches, all which they aſſert, every perſon who pretends to Pruning, ſhould diſtinguiſh well; whereas, there is nothing more in all this but a parcel of words to amuſe the reader, without any real meaning,

meaning, for all thefe are comprehended under the defcription already given of luxuriant or ufelefs branches, and fuch as are termed ufeful or fruit-bearing branches; and, where due care is taken in the fpring of the year to difplace thefe ufelefs branches (as was before directed,) there will be no fuch thing as irregular, falfe, or luxuriant branches at the Winter-pruning; therefore it is to no purpofe to amufe people with a cant of words, which, when fully underftood, fignify juft nothing at all.

In Pruning of ftandard-trees you fhould never fhorten their branches, unlefs it be where they are very luxuriant, and grow irregular on one fide of the tree, attracting the greateft part of the fap of the tree, whereby the other parts are unfurnifhed with branches, or rendered very weak, in which cafe the branch fhould be fhortened down as low as is neceffary, to obtain more branches to fill up the hollow of the trees; but this is only to be underftood of Pears and Apples, which will produce fhoots from wood of three, four, or more years old, whereas moft forts of ftone-fruit will gum and decay after fuch amputations.

But from hence I would not have it underftood, that I would direct the reducing of thefe trees into an exact fpherical figure, fince there is nothing more deteftable than to fee a tree (which fhould be permitted to grow as it is naturally difpofed, with its branches produced at proportionable diftances, according to the fize of the fruit,) by endeavouring to make the head exactly regular, fo crouded with fmall weak branches as to prevent the air from paffing between them, which will render it incapable to produce fruit. All that I intend by this ftopping of luxuriant branches, is only when one or two fuch happen on a young tree, where they entirely draw all the fap from the weaker branches, and ftarve them; then it is proper to ufe this method, which fhould be done in time, before they have exhaufted the roots too much.

Whenever this happens to ftone-fruit, which fuffer much more by cutting than the former forts, it fhould be remedied by ftopping or pinching thofe fhoots in the fpring, before they have obtained too much vigour, which will caufe them to pufh out fide branches, whereby the fap will be diverted from afcending too faft to the leading branch (as hath been directed for wall-trees,) but this muft be done with caution.

You muft alfo cut out all dead or decaying branches, which caufe their heads to look very ragged, efpecially at the time when the leaves are upon the tree, thefe being deftitute of them have but a defpicable appearance. In doing of this, you fhould obferve to cut them clofe down to the place where they were produced, otherwife that part of the branch left will decay, and prove equally hurtful to the tree; for it feldom happens, when a branch begins to decay, that it does not die quite down to the place where it was produced, and, if permitted to remain long uncut, does often infect fome of the other parts of the tree. If the branches are large which you cut off, it will be very proper, after having fmoothed the cut part exactly even with a knife, chiffel, or hatchet, to put on a plaifter of grafting clay, which will prevent the wet from foaking into the tree at the wounded part.

All fuch branches as run crofs each other fhould alfo be cut out, for thefe not only occafion a confufion in the head of the tree, but, by laying over each other, rub off their bark by their motion, and very often occafion them to canker, to the great injury of the tree; and on old trees (efpecially Apples) there are often young vigorous fhoots from the old branches near the trunk, which grow upright into the head of the trees. Thefe therefore fhould carefully be cut out every year, left, by being permitted to grow, they will fill the tree too full of wood, which fhould always be guarded againft, fince it is impoffible for fuch trees to produce fo much, or fo good fruit as thofe, whofe branches grow at a farther diftance, whereby the fun and air freely pafs between them in every part of the tree.

Thefe are all the general directions which are proper to be given in this place, fince not only the particular methods, but alfo the proper feafons for Pruning all the different kinds of fruit, are fully exhibited under their feveral articles.

PRUNUS. Tourn. Inft. R. H. 622. tab. 398. The Plumb-tree.

The CHARACTERS are,

The flower hath a bell-fhaped empalement, cut into five points; it hath five large roundifh petals, which are inferted in the empalement, and from twenty to thirty ftamina, which are near as long as the petals, and are alfo inferted in the empalement, terminated by twin fummits. It has a roundifh germen, fupporting a flender ftyle, crowned by an orbicular ftigma. The germen afterward turns to a roundifh fruit, inclofing a nut of the fame form.

The SPECIES are,

1. PRUNUS (*Jaunhâtive*) fructu parvo præcoci. *The white Primordian.* This is a fmall, longifh, white Plumb, of a clear yellow colour, covered over with a white flue, which eafily wipes off. It is a pretty good bearer, and, for its coming very early, one tree may have a place in a large garden of fruit, but it is meally, and has little flavour. This ripens the middle or latter end of July.

2. PRUNUS (*Damas Noir*) fructu magno craffo fubacido. Tourn. *The early Damafk, commonly called the Morocco Plumb.* This is a middle-fized Plumb, of a round fhape, divided with a furrow in the middle (like Peaches.) The outfide is of a dark black colour, covered with a light Violet bloom; the flefh is yellow, and parts from the ftone. It ripens the end of July, and is efteemed for its goodnefs.

3. PRUNUS (*Small Damas*) fructu parvo dulci atro-cœruleo. Tourn. *The little black Damafk Plumb.* This is a fmall black Plumb, covered with a light Violet bloom. The juice is richly fugared; the flefh parts from the ftone, and it is a good bearer. Ripe the beginning of Auguft.

4. PRUNUS (*Grofs Damas*) fructu magno dulci atro-cæruleo. Tourn. *Grofs Damas Violet de Tours, i. e. great Damafk Violet of Tours.* This is a pretty large Plumb, inclining to an oval fhape. The outfide is of a dark blue, covered with a Violet bloom; the juice is richly fugared; the flefh is yellow, and parts from the ftone. Ripe in Auguft.

5. PRUNUS (*Orleans*) fructu rotundo atro-rubente. *The Orleans Plumb.* The fruit is fo well known, that it is needlefs to defcribe it; it is a very plentiful bearer, which has occafioned its being fo generally planted by thofe perfons who fupply the markets with fruit, but it is an indifferent Plumb. It ripens the middle of Auguft.

6. PRUNUS (*Fotheringham*) fructu oblongo atro rubente. *The Fotheringham Plumb.* This fruit is fomewhat long, deepy furrowed in the middle. The flefh is firm, and parts from the ftone; the juice is very rich. This ripens in the middle of Auguft.

7. PRUNUS (*Perdigron*) fructu nigro, carne durâ. Tourn. *The Perdigron Plumb.* This is a middle-fized Plumb, of an oval fhape. The outfide is of a very dark colour, covered over with a Violet bloom; the flefh is firm, and full of an excellent rich juice. This is greatly efteemed by the curious. It ripens the end of Auguft.

8. PRUNUS (*Violet Perdigron*) fructu magno è violaceo rubente fuaviffimo faccharato. Tourn. *The Violet Perdigron Plumb.* This is a large fruit, rather round than long, of a bluifh red colour on the outfide. The flefh is of a yellowifh colour, pretty firm, and clofely adheres to the ftone; the

the juice is of an exquisite rich flavour. This ripens the end of August.

9. PRUNUS (*White Perdigron*) fructu ovato ex albo flavescente. *The white Perdigron Plumb*. This is a middling Plumb, of an oblong figure. The outside is yellow, covered with a white bloom; the flesh is firm, and tolerably well tasted. It is much esteemed for sweetmeats, having an agreeable sweetness, mixed with an acidity. It ripens the end of August.

10. PRUNUS (*Imperial*) fructu ovato magno rubente. Tourn. *The red imperial Plumb, sometimes called the red Bonum Magnum*. This is a large oval-shaped fruit, of a deep red colour, covered with a fine bloom. The flesh is very dry, and very indifferent to be eaten raw, but is excellent for making sweetmeats; this is a great bearer. It ripens the beginning of September.

11. PRUNUS (*Bonum Magnum*) fructu ovato magno flavescente. Tourn. *White imperial, Bonum Magnum, white Holland, or Mogul Plumb*. This is a large oval-shaped fruit, of a yellowish colour, powdered over with a white bloom. The flesh is firm, and adheres closely to the stone; the juice is of an acid taste, which renders it unpleasant to be eaten raw, but it is very good for baking or sweetmeats. It is a great bearer, and is ripe the middle of September.

12. PRUNUS (*Cheston*) fructu ovato cœruleo. *The Cheston Plumb*. This is a middle-sized fruit, of an oval figure. The outside is of a dark blue, powdered over with a Violet bloom; the juice is rich, and it is a great bearer. Ripe the middle of September.

13. PRUNUS (*Apricot*) fructu maximo rotundo flavo & dulci. Tourn. *Prune d'Apricot*, i. e. *the Apricot Plumb*. This is a large round fruit, of a yellow colour on the outside, powdered over with a white bloom. The flesh is firm and dry, of a sweet taste, and comes clean from the stone. This ripens the end of September.

14. PRUNUS (*Maître Claud*) fructu subrotundo, ex rubro & flavo mixto. *The Maître Claude*. Although this name is applied to this fruit, yet it is not what the French so call. This is a middle-sized fruit, rather round than long, of a fine mixed colour, between red and yellow. The flesh is firm, parts from the stone, and has a delicate flavour Ripe the middle of September.

15. PRUNUS (*Diaprée*) fructu rubente dulcissimo. *La Rochecourbon, or Diapree rouge*, i. e. *the red Diaper Plumb* This is a large round fruit, of a reddish colour, powdered over with a Violet bloom; the flesh adheres closely to the stone, and is of a very high flavour. Ripe the middle of September.

16. PRUNUS fructu rotundo flavescente. *La petite Reine Claude*, i. e. *the little Queen Claudia*. This is a small round fruit, of a whitish yellowish colour, powdered over with a pearl-coloured bloom; the flesh is firm and thick, quits the stone, and its juice is richly sugared. Ripe the middle of September.

17. PRUNUS fructu rotundo nigro-purpureo majori dulci. Tourn. *Myrobalan Plumb*. This is a middle-sized fruit, of a round shape; the outside is a dark purple, powdered over with a Violet bloom; the juice is very sweet. It is ripe the beginning of September.

18. PRUNUS fructu rotundo è viridi flavescente, carne durâ suavissima. *La grosse Reine Claude*, i. e. *the large Queen Claudia, by some the Dauphiny*. At Tours it is called the Abricot verd, i. e. green Apricot; at Rouen, La verte bonne, i. e. the good Green; and in other places, Damas verd, i. e. green Damask, or Trompvalet, the Servants Cheat. This is one of the best Plumbs in England; it is of a middle size, round, and of a yellowish green colour on the outside; the flesh is firm, of a deep green colour, and parts from the stone; the juice has an exceeding rich flavour, and it is a great bearer. Ripe the middle of September. This Plumb is confounded by most people in England, by the name of Green Gage; but this is the sort which should be chosen, although there are three or four different sorts of Plumbs generally sold for it, one of which is small, round, and dry; this quits the stone, and is sooner ripe, so not worth planting.

19. PRUNUS fructu amygdalino. Tourn. *Rognon de Coq*, i. e. *Cock's Testicles*. This is an oblong fruit, deeply furrowed in the middle, so as to resemble the testicles; it is of a whitish colour on the outside, streaked with red; the flesh of it adheres firmly to the stone, and it is late ripe.

20. PRUNUS fructu rotundo flavo dulcissimo. *Drap d'Or*, i. e. *the Cloth of Gold Plumb*. This is a middle-sized fruit, of a bright yellow colour, spotted or streaked with red on the outside; the flesh is yellow, and full of an excellent juice. It is a plentiful bearer, and ripens about the middle of September.

21. PRUNUS fructu cerei coloris. Tourn. *Prune de Sainte Catharine*, i. e. *St. Catharine Plumb*. This is an oval-shaped fruit, somewhat flat; the outside is of an Amber colour, powdered over with a whitish bloom, but the flesh is of a bright yellow colour, is dry and firm, adheres closely to the stone, and has a very agreeable sweet taste. This ripens at the end of September, and is very subject to dry upon the tree, when the autumn proves warm and dry. This makes fine sweetmeats, and is a plentiful bearer.

22. PRUNUS fructu ovato rubente dulci. *The Royal Plumb*. This is a large fruit, of an oval shape, drawing to a point next the stalk; the outside is of a light red colour, powdered over with a whitish bloom; the flesh adheres to the stone, and has a fine sugary juice. This ripens the middle of September.

23. PRUNUS fructu parvo ex viridi flavescente. Tourn. *La Mirabelle*. This is a small round fruit, yellow on the outside; the flesh parts from the stone, is of a bright yellow colour, and has a fine sugary juice. This is a great bearer, ripens the end of August, and is excellent for sweetmeats.

24. PRUNUS Brignoniensis, fructu suavissimo. Tourn. *Prune de Brignole*, i. e. *the Brignole Plumb*. This is a large oval-shaped fruit, of a yellowish colour, mixed with red on the outside; the flesh is of a bright yellow colour, is dry, and of an excellent rich flavour. This ripens the middle of September, and is esteemed the best Plumb for sweetmeats yet known.

25. PRUNUS fructu magno è violaceo rubente serotino. Tourn. *Imperatrice*, i. e. *the Empress*. This is a middle-sized oval fruit, of a Violet colour, very much powdered with a whitish bloom; the flesh is yellow, cleaves to the stone, and is of an agreeable flavour. This ripens about the beginning of October.

26. PRUNUS fructu ovato maximo flavo. Tourn. *Prune de Monsieur*, i. e. *Monsieur's Plumb*. This is sometimes called the Wentworth Plumb. It is a large oval shaped fruit, of a yellow colour both within and without, very much resembling the Bonum Magnum, but the flesh of this parts from the stone, which the other doth not. This ripens towards the latter end of September, and is very good to preserve, but the juice is too sharp to be eaten raw. It is a great bearer.

27. PRUNUS fructu majori rotundo rubro. Tourn. *Prune Cerizette*, i. e. *the Cherry Plumb*. This fruit is commonly about the size of the Ox-heart Cherry, is round, and of a red colour; the stalk is long, like that of a Cherry, which this fruit so much resembles, as not to be distinguished therefrom at some distance. The blossoms of this tree come out very early in the spring, and being tender, are very often destroyed by cold, but it affords a very agreeable prospect in

in the ſpring; for as theſe trees are generally covered with flowers, which open about the ſame time as the Almonds, ſo when they are intermixed therewith, they make a beautiful appearance before many other ſorts put out, but by bloſſoming ſo early, there are few years that they have much fruit.

28 PRUNUS fructu albo oblongiuſculo acido. Tourn. *The white Pear Plumb.* This is a good fruit for preſerving, but is very unpleaſant if eaten raw; it is very late ripe, and ſeldom planted in gardens, unleſs for ſtocks to bud ſome tender ſorts of Peaches upon, for which purpoſe it is eſteemed the beſt amongſt all the ſorts of Plumbs.

29. PRUNUS Mytellinum. Park. *The Muſcle Plumb.* This is an oblong flat Plumb, of a dark red colour; the ſtone is large, and the fleſh but very thin, and not well taſted, ſo that its chief uſe is for ſtocks, as the former.

30. PRUNUS fructu parvo violaceo. *The St. Julian Plumb.* This is a ſmall fruit, of a dark Violet colour, powdered over with a meally bloom; the fleſh adheres cloſely to the ſtone, and in a fine autumn will dry upon the tree. The chief uſe of this Plumb is for ſtocks, to bud the more generous kinds of Plumbs and Peaches upon; as alſo for the Bruxelles Apricot, which will not thrive ſo well upon any other ſtock.

31. PRUNUS ſylveſtris major. J. B. *The black Bullace-tree.* This grows wild in the hedges in divers parts of England, and is rarely cultivated in gardens.

32. PRUNUS ſylveſtris, fructu majore albo. Raii Syn. *The white Bullace-tree.* This grows wild as the former, and is ſeldom cultivated in gardens.

33. PRUNUS ſylveſtris. Ger. Emac. *The Black-thorn, or Sloe-tree.* This is very common in the hedges almoſt every where; the chief uſe of this tree is to plant for hedges, as White-thorn, &c. and being of quick growth, is very proper for that purpoſe.

All the varieties of Plumbs are propagated by budding or grafting them upon ſtocks of the Muſcle, white Pear, St. Julian, Bonum Magnum, or any other ſorts of free-ſhooting Plumbs. The manner of raiſing theſe ſtocks hath been already exhibited under the article of NURSERIES, therefore need not be repeated again in this place; but I would obſerve, that budding is much preferable to grafting for theſe ſorts of fruit-trees, becauſe they are very apt to gum, wherever there are large wounds made on them.

The trees ſhould not be more than one year's growth from the bud when they are tranſplanted, for if they are older, they ſeldom ſucceed ſo well, being very ſubject to canker; or if they take well to the ground, commonly produce only two or three luxuriant branches, therefore it is much more adviſeable to chuſe young plants.

The manner of preparing the ground (if for walls) is the ſame as for Peaches; as is alſo the pruning the roots and planting, therefore I ſhall forbear repeating it again. The diſtance which theſe trees ſhould be planted at, muſt not be leſs than twenty-four feet againſt high walls, and if the wall is low they ſhould be placed thirty feet aſunder.

Plumbs ſhould have a middling ſoil, neither too wet and heavy, nor over-light and dry, in either of which extremes they ſeldom do ſo well; thoſe ſorts which are planted againſt walls, ſhould be placed to an eaſt or ſouth-eaſt aſpect, which is more kindly to theſe fruits than a full ſouth aſpect, on which they are ſubject to ſhrivel, and be very dry; and many ſorts will be extreme meally, if expoſed too much to the heat of the ſun, and moſt of the ſorts will ripen extremely well on eſpaliers, if rightly managed.

There are ſome perſons who plant Plumbs for ſtandards, in which method ſeveral of the ordinary ſorts will bear very well; but the fruit will not be near ſo fair as thoſe produced on eſpaliers, and will be more in danger of being bruiſed or blown down by ſtrong winds. The diſtance of placing them for eſpaliers, muſt be the ſame as againſt walls, as muſt alſo their pruning and management; ſo that whatever may be hereafter mentioned for one, ſhould be likewiſe underſtood for both.

Plumbs do not only produce their fruit upon the laſt year's wood, but alſo upon curſons or ſpurs, which come out of wood that is many year's old; ſo that there is not a neceſſity of ſhortening the branches, in order to obtain new ſhoots annually, in every part of the tree (as in Peaches, Nectarines, &c. hath been directed,) ſince the more theſe trees are pruned, the more luxuriant they grow, until the ſtrength of them is exhauſted, and then they gum and ſpoil; therefore the ſafeſt method to manage theſe trees is, to lay in their ſhoots horizontally, annually, as they are produced, at equal diſtances, in proportion to the length of their leaves; and where there is not a ſufficient quantity of branches to fill up the vacancies of the tree, there the ſhoots may be pinched the beginning of May (in the manner as hath been directed for Peaches, &c.) which will cauſe them to produce ſome lateral-branches to ſupply thoſe places; and during the growing ſeaſon, all fore right ſhoots ſhould be diſplaced, and ſuch as are to remain muſt be regularly trained to the wall or eſpalier, which will not only render them beautiful, but alſo give to each an equal advantage of ſun and air; and hereby the fruit will be always kept in a growing ſtate, which they ſeldom are, when overſhaded with ſhoots ſome part of the ſeaſon, and then ſuddenly expoſed to the air, by the taking off or training thoſe branches in their proper poſition.

With thus carefully going over theſe trees in the growing ſeaſon, there will be but little occaſion for cutting them in winter, which (as I before have ſaid) is of ill conſequence to all ſorts of ſtone-fruit; for when the branches are ſhortened, the fruit is cut away, and the number of ſhoots increaſed; becauſe whenever a branch is ſhortened, there are commonly two or more ſhoots produced from the eyes immediately below the cut; ſo that by thus unſkilfully pruning, many perſons crowd their trees with branches, and thereby render the trees unfruitful, and what little fruit the trees produce, are ſmall and ill taſted.

The few rules here laid down will be ſufficient, if due obſervation be joined therewith, to inſtruct any perſon in the right management of theſe ſorts of fruit-trees; therefore I ſhall not ſay more on that ſubject, leſt by multiplying inſtructions, it may render it more obſcure to a learner.

PSEUDOACACIA. See ROBINIA.

PSEUDODICTAMNUS. See MARRUBIUM.

PSIDIUM. Lin. Gen. Plant. 541. The Guava.

The CHARACTERS are,

The flower has a bell-ſhaped empalement, divided into five oval points at the top. It hath five oval, concave, ſpreading petals, indented in the empalement, with a great number of ſtamina which are ſhorter than the petals, and are inſerted in the empalement, terminated by ſmall ſummits. It has a roundiſh germen, ſituated under the flower, ſupporting a long awl-ſhaped ſtyle, crowned by a ſingle ſtigma. The germen afterward becomes a large oval fruit, crowned by the empalement, incloſing a great number of ſmall ſeeds.

The SPECIES are,

1. PSIDIUM (*Pyriferum*) unifloris, foliis ovatis-lanceolatis pedunculis, fructu. *The white Guava with ſingle flowers.*

2. PSIDIUM (*Pomiferum*) foliis ovatis pedunculis trifloris. *The red Guava.*

Both theſe ſorts grow naturally in the Eaſt and Weſt-Indies, and there is alſo a third with a large white fruit, but I do not know whether this is a variety of the common Guava, or of that with the ſmall white fruit; though I am inclined to believe it is the former, becauſe I have raiſed many plants from the feeds of the fmall white Guava, which have produced fruit in the Chelfea garden, and have not varied from their parent plant.

The common red Guava hath a pretty thick trunk, which rifes twenty feet high, covered with a fmooth bark; the branches are angular, garnifhed with oval leaves, having a ftrong midrib, and many veins running toward the fides, of a light green colour, ftanding oppofite upon very fhort foot-ftalks. From the wings of the leaves the flowers come out upon foot-ftalks, about an inch and a half long; they are compofed of five large, roundifh, concave petals, which are inferted in the empalement; and within thefe are a great number of ftamina, which are fhorter than the petals, terminated by fmall fummits; thefe ftamina are alfo inferted in the empalement. Under the flower is fituated a roundifh germen, fupporting a very long awl fhaped ftyle, crowned by a fimple ftigma. After the flower is paft, the germen becomes a large oval fruit, fhaped like a Pomgranate, having one cell, crowned by the empalement of the flower, and filled with fmall feeds; the fruit, when ripe, has an agreeable odour. They are much eaten in the Weft-Indies, both by men and beafts; and the feeds, which pafs whole through the body, and are voided with the excrement in hot countries, grow, whereby the trees are fpread over the ground where they are permitted to ftand. This fruit is very aftringent, and nearly of the fame quality with Pomgranate, fo fhould be avoided by thofe perfons who are fubject to be coftive.

The large white fort grows naturally in the iflands of the Weft-Indies, and is often found intermixed with the former, fo is fuppofed to be only an accidental variety arifing from the fame feeds. This differs from the former, in the colour of the midrib of the leaves, which in this are pale, but thofe of the former are red. The flowers and fruit of of this are larger, and the infide of the fruit is white.

The leaves of the fmall white Guava are like thofe of the larger, but the branches of the tree are not fo angular; the flowers are much fmaller, and the fruit is no larger than a middling Goofeberry, but, when ripe, has a very ftrong aromatic flavour. This flowers in June, and the fruit ripens in autumn.

Thefe plants are propagated by feeds, which, when brought over in the entire fruit, gathered full ripe, they will more certainly fucceed; thefe fhould be fown in pots filled with kitchen garden earth, and plunged into a hot-bed of tanners bark; in about fix weeks the plants will appear (if the feeds are good) when they muft have free air admitted to them, in proportion to the warmth of the feafon; when the plants have obtained ftrength enough to be removed, they fhould be each planted in a fmall pot, filled with the like earth, and plunged into a frefh hot-bed, fhading them from the fun until they have taken new root; then they fhould have a large fhare of free air admitted to them every day in warm weather, to prevent their drawing up weak; they muft alfo be frequently refrefhed with water in fummer. In the autumn they muft be plunged into the tan-bed in the ftove; during the winter they fhould be kept in a moderate warmth, and not have too much water; in fummer they will require plenty of wet, and in hot weather a great fhare of air. With this management the plants will produce flowers and fruit the third year, and may be continued a long time.

PSORALEA. Lin. Gen. Plant. 801. Barba Jovis.

The CHARACTERS are,

The empalement of the flower is cut into five parts, the lower fegments being twice the length of the other. The flower is of the butterfly kind; it hath five petals; the ftandard is roundifh, and indented at the top. The wings are fmall, obtufe, and moon-fhaped; the keel is moon-fhaped, and compofed of two petals. It hath nine ftamina joined together, and one briftly ftamina ftanding feparate, terminated by roundifh fummits, with a linear germen, fupporting an awl-fhaped rifing ftyle, crowned by an obtufe ftigma. The germen afterward turns to a flender compreffed pod, inclofing one kidney-fhaped feed.

The SPECIES are,

1. PSORALEA (*Pinnata*) foliis pinnatis, floribus axillaribus. Hort. Upfal. 225. *Pforalea with winged leaves, and flowers proceeding from the fides of the ftalks.*

2. PSORALEA (*Corylifolia*) foliis fimplicibus ovatis. Hort. Upfal. 225. *Pforalea with fingle oval leaves.*

3. PSORALEA (*Hirta*) foliis ternatis, foliis ovatis, caule fruticofo hirfuto, floribus fpicatis terminalibus. *Pforalea with trifoliate oval leaves, a hairy fhrubby ftalk, and flowers growing in fpikes, terminating the branches.*

4. PSORALEA (*Procumbens*) foliis pinnatis argentis, caulibus procumbentibus, floribus axillaribus. *Pforalea with filvery winged leaves, trailing ftalks, and flowers proceeding from the fides of the ftalks.*

5. PSORALEA (*Scandens*) foliis pinnatis, caule ramofo fcandente, floribus alaribus feffilibus. *Pforalea with winged leaves, a climbing branching ftalk, and flowers fitting clofe at the wings of the ftalk.*

6. PSORALEA (*Capitata*) foliis ternatis, caule fruticofo ramofiffimo, floribus capitatis pedunculatis alaribus. *Pforalea with trifoliate leaves, a very branching fhrubby ftalk, and flowers growing in heads, which have foot-ftalks proceeding from the wings of the leaves.*

7. PSORALEA (*Annua*) foliis pinnatis, fpicis terminalibus. Lin. Sp. Plant. 764. *Pforalea with winged leaves, and flowers growing in fpikes terminating the branches.*

8. PSORALEA (*Humilis*) foliis pinnatis, foliolis rotundioribus villofis, floribus capitatis alaribus terminalibufque, caule fruticofo. *Pforalea with winged leaves, having hairy round lobes, flowers growing in heads from the wings of the leaves, and at the end of the branches, and a fhrubby ftalk.*

9. PSORALEA (*Bitumenofa*) foliis ternatis, foliolis ovatis, floribus capitatis, pedunculis longiffimis. *Pforalea with trifoliate leaves, having oval lobes, and flowers growing in heads on very long foot-ftalks.*

10. PSORALEA (*Anguftifolia*) foliis ternatis, foliolis ovato-lanceolatis, floribus capitatis pedunculis longioribus. *Pforalea with trifoliate leaves, having oval fpear-fhaped lobes, and flowers growing in heads upon long foot-ftalks.*

The firft fort grows naturally at the Cape of Good Hope; it rifes with a foft fhrubby ftalk four or five feet high, garnifhed with deep green winged leaves, compofed of three or four pair of very narrow linear lobes, terminated by an odd one, ftanding upon fhort foot-ftalks. The flowers fit very clofe to the branches, coming out from the wings of the leaves; they are often in clufters. The ftandard, which is erect and reflexed at the top, is of a fine blue; the wings are pale, and the keel white; thefe flowers are fucceeded by fhort pods the length of the empalement, each containing one kidney-fhaped feed. It flowers great part of fummer, and the feeds ripen in autumn. This is eafily propagated by feeds, which fhould be fown upon a moderate hot-bed; and when the plants come up, they muft not be drawn weak, fo they fhould have air and but little heat. When they are fit to remove, they fhould be planted in feparate fmall pots, filled with light earth, and plunged again into the bed, fhading them from the fun till they have taken new root; then they fhould be gradually inured to the open air, into which they fhould be removed about the end of May, and kept abroad till October; then they muft be placed in the green-houfe, and treated in the fame way as other plants from the fame country.

The fecond fort grows naturally in India. This is an annual plant; the ftalks rife a foot and a half high, and are garnished at each joint by one oval leaf. The flowers stand upon long slender foot-stalks, which come out at the wings of the leaves, collected into small round heads, and are of a pale flesh colour. It flowers in July, and the seeds ripen in autumn This is propagated by seeds, which must be sown upon a hot bed in the spring; and when the plants are fit to remove, they should be planted into separate small pots, filled with light earth, and plunged into a moderate hot-bed, shading them from the sun till they have taken new root; after which they must have free air admitted to them in warm weather. When the plants have filled the pots with their roots, they should be removed into larger, and the beginning of July they may be removed into an airy glass-case, where they may be defended from cold, but should have free air in warm weather; with this care the plants will flower, and ripen their seeds.

The third sort was discovered by the late Dr. Houstoun at La Vera Cruz. This rises with a shrubby stalk three or four feet high, sending out a few side branches, garnished with oval, trifoliate, hairy leaves, standing upon slender foot-stalks. The flowers are collected in spikes at the end of the branches; they are of a purple colour, and are succeeded by short pods, each containing one kidney-shaped seed. It is propagated by seeds, which must be sown upon a hot-bed, and the plants afterward treated in the same way as the second sort; but as this is an abiding plant, so they must be removed into the stove in autumn, and kept in a moderate warmth in winter; and in summer they must have a large share of air, but should constantly remain in the stove; the second year they will produce flowers, and sometimes their seeds will ripen in England.

The seeds of the fourth sort came from Malabar. This is an annual plant with trailing stalks, garnished with silvery leaves, composed of three or four pair of narrow lobes, terminated by an odd one. The flowers grow in small clusters at the wings of the leaves; they are small, and of a purple colour; the seed-pods are short, and have one small kidney-shaped seed in each. This is propagated by seeds, in the same manner as the second sort.

The fifth sort was discovered by the late Dr. Houstoun at Campeachy, where it grows naturally. This hath slender, shrubby, climbing stalks, which twine about any neighbouring support, and rise to the height of six or seven feet, garnished with winged leaves, composed of three pair of small, oval, obtuse lobes, terminated by an odd one. The flowers come out in small clusters from the wings of the leaves; they are small, of a bright blue colour, and are succeeded by short pods, including one kidney-shaped seed.

The sixth sort was discovered by the same gentleman, growing naturally at Campeachy. This rises with a shrubby stalk seven or eight feet high, sending out many long slender branches on every side, garnished with trifoliate leaves, whose lobes are small and wedge-shaped. The flowers are produced from the wings of the leaves in close small heads, standing upon pretty long foot-stalks; they are blue, and are succeeded by short pods, each containing a single kidney-shaped seed

These two sorts are propagated by seeds, which must be sown upon a hot-bed, and when the plants come up, they must be treated in the same way as the third sort.

The seventh sort was discovered by the late Dr. Houstoun at La Vera Cruz. This is an annual plant, with a very branching herbaceous stalk, rising a foot and a half high, garnished with winged leaves, composed of five or six pair of narrow wedge-shaped lobes, terminated by an odd one. The flowers are collected in close oblong spikes, terminating the branches; they are small, of a bright blue colour, and are succeeded by short pods, each containing a single kidney-shaped seed. This is propagated by seeds, and requires the same treatment as the second sort.

The eighth sort grows naturally at La Vera Cruz, from whence the late Dr. Houstoun sent the seeds. This hath an upright shrubby stalk which rises five or six feet high, having a few side branches, closely garnished with winged leaves, composed of three or four pair of small roundish hairy lobes, terminated by an odd one. The flowers are collected in small heads, coming out from the wings of the leaves, and at the end of the branches; they are yellow and red intermixed, and are succeeded by short pods, containing one kidney-shaped seed. This sort requires the same treatment as the third.

The ninth sort grows naturally in the south of France and Italy. The root of this is perennial, but the stalk is not of long duration, seldom lasting more than two years; it rises about two feet high, sending out two or three slender branches, garnished with trifoliate leaves, whose lobes are oval, standing upon long foot-stalks; these, if handled, emit a strong scent of bitumen. The flowers are collected in heads, and have foot-stalks seven or eight inches long; they are blue, and are succeeded by short pods, containing one seed.

The tenth sort grows naturally in Sicily, and also in Jamaica, from both which countries I have received the seeds. This hath been supposed to be the same with the former, but I have many years propagated both by seeds, and have never found either of them vary. The leaves of this are narrower than those of the former sort, and are rounded at their base; the stalks are shrubby, and of long duration; the heads of flowers are smaller, and the leaves have not so strong an odour. These are propagated by seeds, which should be sown on a bed of light earth in April, and in May the plants will come up, when they should be kept clean from weeds, and as soon as they are fit to remove they should be transplanted. Those of the ninth sort will live through the winter in the open air, if they are planted in a warm dry border; but the tenth sort requires some shelter in winter, so these should be planted in pots, and put into a common frame in winter, where they may be screened from hard frost. These plants flower from June to autumn, and perfect their seeds annually.

PSYLLIUM. See PLANTAGO.

PTARMICA. See ACHILLEA.

PTELEA. Lin. Gen. Plant. 141. Shrub Trefoil.

The CHARACTERS are,

The empalement of the flower is cut into four acute parts. The flower has four oval spear-shaped petals; it hath four awl-shaped stamina, terminated by roundish summits, and an orbicular compressed germen, supporting a short style, crowned by two obtuse stigmas. The germen afterward becomes a roundish membranaceous capsule with two cells, each containing one obtuse seed.

The SPECIES are,

1. PTELEA (*Trifoliata*) foliis ternatis. Lin. Sp. Pl. 118. *Ptelea with trifoliate leaves; commonly called Carolina Shrub Trefoil.*

2. PTELEA (*Viscosa*) foliis simplicibus. Lin. Sp. Pl. 118. *Ptelea with single leaves.*

The first sort grows naturally in North America. It was first discovered in Virginia by Mr. Banister, who sent the seeds to England, from which some plants were raised at Fulham, and other curious gardens; but being planted in the open air, they were destroyed by a severe winter, so that there were scarce any of the plants left in England; but in 1724, Mr. Catesby sent over a good quantity of the seeds from Carolina, which succeeded so well, as to furnish many gardens with the plants. This rises with an upright woody stem twelve or fourteen feet high, dividing into many branches, covered with a smooth grayish bark, garnished with

with trifoliate leaves; ſtanding upon long foot-ſtalks. The lobes are oval, ſpear ſhaped, ſmooth, and of a bright green on their upper ſide, but pale on their under; theſe come out late in the ſpring, and at the ſame time the bunches of flower-buds appear, which is generally in the beginning of June, the leaves being then but ſmall, and afterward increaſe in their ſize, but are not fully grown till the flowers decay. The flowers are produced in large bunches at the end of the branches; they are of an herbaceous white colour, compoſed of four or five ſhort petals, ending in acute points, faſtened at their baſe to a ſhort empalement, cut into four ſegments almoſt to the bottom. In the center is ſituated an orbicular compreſſed germen, ſupporting a ſhort ſtyle, which is attended by four awl-ſhaped ſtamina; the germen afterward turns to a capſule, ſurrounded by a leafy border, having two cells, each containing one ſeed.

Theſe ſhrubs may be propagated by cuttings, which ſhould be planted in pots, and plunged into a moderate hot-bed. The beſt time for planting them is in the beginning of March; but they muſt be carefully managed, ſo as not to have too much heat, and ſhaded from the ſun in the middle of the day, otherwiſe they will not ſucceed. They may alſo be propagated by layers, but theſe are often two years before they take root; but if good ſeeds can be procured either here or from abroad, the plants raiſed from thoſe will be much ſtronger, than thoſe which are propagated by either of the former methods.

The ſeeds may be ſown in the beginning of April, on a bed of light earth, in a warm ſheltered ſituation, where, if the ground is moiſtened in dry weather, the plants will come up in two months; but if the ſeeds are ſown in pots, and placed on a very moderate hot-bed, the plants will come up ſooner, and make greater progreſs the firſt year, but they muſt not be forced or drawn, for that will make them very tender; therefore in June the plants ſhould be expoſed to the open air, in a ſheltered ſituation, where they may remain till the froſt comes on; when thoſe in the pots ſhould be either placed under a common frame, to ſhelter them from ſevere froſt, or the pots plunged into the ground, near a hedge, that the froſt may be prevented from penetrating through the ſides of the pots to the roots of the plants. The following ſpring the plants may be planted into a nurſery bed, at about one foot diſtance, where they may grow two years, by which time they will be fit to tranſplant where they are deſigned to remain.

Theſe plants are a little tender while they are young, therefore will require ſome protection the firſt and ſecond years, but particularly from the early froſts in autumn, which frequently kill the tops of the tender ſhoots before they are hardened; and the more vigorous the plants have grown the preceding ſummer, the greater danger there is of their being killed; therefore they ſhould be ſcreened either with mats, or ſome other covering; but as they advance in ſtrength, they become more hardy, and are rarely injured by froſt.

The ſecond ſort grows naturally in both Indies. It is very common in moſt of the iſlands in the Weſt-Indies. This ſends up ſeveral ſtalks from the root, about the ſize of a man's arm, ſending out ſeveral upright branches, covered with a light brown bark, which frequently ſeparates from the wood, and hangs looſe; they are garniſhed with ſtiff leaves, which vary greatly in their ſhape and ſize; they are ſpear ſhaped, entire, and of a light green, growing with their points upward, and have very ſhort foot-ſtalks. The flowers are produced at the end of the branches in a ſort of racemus, each ſtanding upon a ſlender foot-ſtalk; they have four ſolid channelled petals, of an herbaceous colour, having four ſtamina, which ſpread open, and in the center is ſituated a roundiſh compreſſed germen, which afterward turns to a compreſſed capſule with three cells, ſurrounded by a broad leafy border, each cell containing one or two roundiſh ſeeds.

This plant is propagated by ſeeds, which, if obtained freſh from abroad, will riſe eaſily upon a hot-bed; when the plants are fit to remove, they ſhould be each planted in a ſeparate ſmall pot, filled with light loamy earth, and plunged into a hot-bed of tanners bark, ſhading them from the ſun till they have taken new root; then they ſhould have free air admitted to them every day in proportion to the warmth of the ſeaſon, for they muſt not be drawn up weak, nor ſhould they have too much water. In the autumn, the plants muſt be removed into the ſtove, where they ſhould have a temperate warmth in winter, but during that ſeaſon little water ſhould be given them; nor ſhould they have too much heat, for either of theſe will ſoon deſtroy them: as the plants obtain ſtrength, ſo they will become more hardy, and may be ſet abroad in the open air for two or three months in the heat of ſummer, but it ſhould be in a ſheltered ſituation; in the winter they muſt be placed in a ſtove, kept to a moderate temperature of warmth, for the plants will not live in a green-houſe here.

This was formerly ſhewn for the Tea-tree in many of the European gardens, where it many years paſſed for it among thoſe who knew no better.

PULEGIUM. Raii Meth. Plant. 61. Pennyroyal, or Pudden-graſs.

The CHARACTERS are,

The empalement of the flower is permanent, cut into five parts. The flower is of the lip kind; it hath one petal, with a ſhort tube, divided at the brim into four parts; the helmet or upper lip of the flower is entire; the lower is cut into three equal ſegments. It hath four ſtamina, two being longer than the other, terminated by roundiſh ſummits, and a four-pointed germen, ſupporting an erect ſtyle, crowned by a bifid ſtigma. The germen afterward becomes four ſmall ſeeds, ſitting in the empalement of the flower.

The SPECIES are,

1. PULEGIUM (*Vulgare*) foliis ovatis obtuſis, ſtaminibus corollam æquantibus, caule repente. *Pennyroyal with oval obtuſe leaves, ſtamina equalling the petal, and a creeping ſtalk; common or broad-leaved Pennyroyal.*

2. PULEGIUM (*Erectum*) foliis lanceolatis, ſtaminibus corollâ longioribus, caule erecto. *Pennyroyal with ſpear-ſhaped leaves, ſtamina longer than the petal, and an upright ſtalk.*

3. PULEGIUM (*Cervinum*) foliis linearibus, floribus verticillatis terminalibuſque. *Pennyroyal with linear leaves, and flowers growing in whorls at the ends of the ſtalks; narrow-leaved Pennyroyal.*

The firſt ſort grows naturally upon moiſt commons, where the water ſtands in winter, in many parts of England. The root is fibrous and perennial; the ſtalks are ſmooth, and trail upon the ground, putting out roots at every joint, whereby it ſpreads and propagates very faſt; the ſtalks are garniſhed at each joint by two oval leaves, which are for the moſt part entire. The flowers grow toward the upper part of the branches, coming out juſt above the leaves at each joint in whorls; they are of a pale purple colour, ſmall and galeated, the helmet being entire; whereas in the Mint, this is indented at the point. The ſtamina of the flowers are of the ſame length with the petal, but the ſtyle is ſomewhat longer; the whole plant has a very ſtrong ſmell, and a hot aromatic taſte. There is a diſtilled water of this plant, and alſo an oil, which is kept in the ſhops for medicinal uſe. There is a variety of this with a white flower, which is ſometimes found growing naturally in England.

The feeds of the fecond fort were fent me from Gibraltar, which fucceeded in the Chelfea garden, but had been before introduced into feveral gardens, where it had been cultivated to fupply the markets. The ftalks of this grow erect, and near a foot high; the leaves are longer and narrower than thofe of the common fort; the whorls of flowers are much larger, and their ftamina are longer than the petals. This fort hath almoft fuperfeded the firft in the markets, for as the ftalks grow erect, fo it is much eafier to cut and tie in bunches than the common fort; it alfo comes earlier to flower, and has a brighter appearance, but whether it is as good for ufe, I fhall leave to be determined by thofe whofe province it belongs to.

The third fort grows naturally in the fouth of France and Italy; it is called Hart's Pennyroyal. This is by fome preferred to the common fort for medicinal ufe; the ftalks of this grow erect, near two feet high, fending out fide branches all their length; the leaves are very narrow, and of a thicker fubftance than thofe of the common fort; the whorls of flowers are rather larger; the fcent is not quite fo ftrong as that of the firft fort, and the ftalks are frequently terminated by whorls of flowers. This is cultivated in gardens here, and flowers about the fame time as the common fort. There is a variety of this with white flowers, which grows taller than that with purple flowers, but I do not believe it is a different fort.

All thefe plants propagate themfelves very faft by their branches trailing upon the ground, which emit roots at every joint, and faften themfelves into the earth, and fend forth new branches; fo that no more is required in their culture, than to cut off any of thofe rooted branches, and plant them out in frefh beds, allowing them at leaft a foot from plant to plant every way, that they may have room to grow; or the young fhoots of thefe planted in the autumn, will take root like Mint.

The beft time for this work is in September, that the plants may be rooted before winter; for if the old roots are permitted to remain fo clofe together, as they generally grow in the compafs of a year, they are fubject to rot in winter; befides, the young plants will be much ftronger, and produce a larger crop the fucceeding fummer, than if they were removed in the fpring. Thefe plants all love a moift ftrong foil, in which they will flourifh exceedingly.

PULMONARIA. Tourn. Inft. R. H. 136. tab. 55. Lungwort.

The CHARACTERS are,

The flower hath a cylindrical permanent empalement, cut into five parts at the top. The flower is of one petal, having a cylindrical tube, cut at the top into five parts, which fpread open, but the chaps are pervious. It hath five fhort ftamina, terminated by erect fummits, which clofe together, and four germen, fupporting a fhort ftyle, crowned by an obtufe indented ftigma. The germen afterward turns to four roundifh feeds, fitting in the bottom of the empalement.

The SPECIES are,

1. PULMONARIA (*Officinalis*) foliis radicalibus ovato-cordatis fcabris. Hort. Cliff. 44. *Lungwort whofe lower leaves are oval, heart-fhaped, and rough; common fpotted Lungwort, or Jerufalem Cowflip.*

2. PULMONARIA (*Alpina*) foliis caulinis ovatis glabris, floribus patulis, fegmentis obtufiufculis. *Lungwort with oval fmooth leaves to the ftalks, fpreading flowers, and obtufe fegments.*

3. PULMONARIA (*Saccharata*) foliis lanceolatis bafi femiamplexicaulibus, calycibus abbreviatis. *Lungwort with fpear-fhaped leaves, whofe bafe half embraces the ftalk, and the empalement fhorter than the tube of the flower.*

4. PULMONARIA (*Anguftifolia*) foliis radicalibus lanceolatis. Hort. Cliff. 44. *Lungwort with the lower leaves fpear fhaped.*

5. PULMONARIA (*Orientalis*) caulibus procumbentibus, floribus fingularibus alaribus, calycibus inflatis corolla longioribus. *Lungwort with trailing ftalks, flowers growing fingly from the fides, and fwollen empalements, which are longer than the petal.*

6. PULMONARIA (*Virginica*) calycibus abbreviatis, foliis lanceolatis obtufiufculis. Lin. Sp. Plant. 135. *Lungwort with fhort empalements to the flowers, and fpear-fhaped obtufe leaves.*

The firft fort grows naturally in woods and fhady places in Italy and Germany, and is cultivated in the Englifh gardens chiefly for medicinal ufe. It hath a perennial fibrous root; the lower leaves are rough, of an oval heart-fhape, and of a dark green on their upper fide, marked with many broad whitifh fpots, but pale and unfpotted on their under; the ftalks rife almoft a foot high, having feveral fmaller leaves on them, ftanding alternately. The flowers are produced in fmall bunches on the top of the ftalks, each having a tubulous hairy empalement as long as the tube of the flower; the brims of the petal are fpread open above them, which are fhaped like a cup; thefe are red, purple, and blue, in the fame bunch, and are fucceeded by four naked feeds, which ripen in the empalement. It is accounted a pectoral balfamic plant, and good for coughs and confumptions, fpitting of blood, and the like diforders of the lungs; it is likewife put into wound-drinks.

The fecond fort grows naturally on the Alps. This hath a perennial fibrous root; the leaves are large, fmooth, and fpotted on their upper fide; the ftalks rife nine inches high, garnifhed with oval leaves, whofe bafe join the ftalks. The flowers grow in fmall bunches on the top of the ftalk; they are purple, and fpread open wider than thofe of the common fort.

The third fort grows naturally upon the Helvetian mountains. It is a perennial plant, whofe leaves are large, fpear-fhaped, and rough. The foot-ftalks of the lower leaves are broad; the ftalks rife a foot high, garnifhed with fpear-fhaped leaves, whofe bafe half embrace the ftalks; they are greatly fpotted with white, appearing as if they were incrufted with fugar-candy; the flowers grow in large bunches on the top of the ftalk; their tubes are longer than the empalement, and their brims are fpread more than thofe of the common fort. They are of a bright blue.

The fourth fort grows naturally in Auftria and Hungary. This hath leaves much narrower than thofe of the common fort, covered with foft hairs. The ftalks rife a foot high, garnifhed with narrow leaves, of the fame fhape with thofe below, but fmaller; thefe almoft embrace the ftalk with their bafe. The flowers are produced in bunches on the top of the ftalks like the others; they are of a red colour before they expand, but, when they are fully blown, of a moft beautiful blue colour.

The fifth fort was difcovered in the Archipelago by Dr. Tournefort. This is an annual plant. The lower leaves are oblong and hairy; the ftalks trail upon the ground, and are garnifhed with oblong hairy leaves, fitting clofe to the ftalks; juft above each leaf comes a fingle flower, of a fullen purple colour, funnel-fhaped, the brims not fpreading; the empalement is fwollen like an inflated bladder, and covers the petal of the flower, fo as not to be feen without a near infpection; after the flowers are paft, the four feeds ripen in the empalement.

The fixth fort grows naturally upon mountains in moft parts of North America. The feeds of this plant were fent many years fince by Mr. Banifter from Virginia, and fome of the plants were raifed in the gardens of the bifhop of London at Fulham, where for feveral years it was growing. This hath a thick, flefhy, perennial root, fending out many fmall fibres. The ftalks rife a foot and a half high; the leaves,

leaves, which are near the root, are long, ſmooth, obtuſe, and of a light green, having ſhort foot-ſtalks; thoſe upon the ſtalk diminiſh in their ſize upward, but are of the ſame ſhape, and ſit cloſe to the ſtalk. Each of the ſmall branches is terminated by a cluſter of flowers, whoſe empalements are very ſhort, and are cut into five ſegments almoſt to the bottom; the tube of the flower is long, and at the top ſpreads open in ſhape of a funnel, the brim being entire, but appears five-cornered from the folding of the petal. The moſt common colour of theſe flowers is blue, but there are ſome purple, others red, and ſome white. The leaves and ſtalks entirely decay in Auguſt, and the roots remain naked till the following ſpring.

There are ſome other ſpecies of this genus, which are preſerved in botanic gardens for the ſake of variety, but, having little beauty, they are ſeldom cultivated in other places.

The firſt, ſecond, third, fifth, and ſixth ſorts have perennial roots, ſo may be cultivated by parting of their roots, which is beſt done in the autumn, that the plants may be well rooted before the dry weather comes on in the ſpring, which will cauſe them to flower much ſtronger.

The ſoil in which they are planted ſhould not be rich, but rather a freſh light ſandy ground, in which they will thrive much better. They ſhould have a ſhady ſituation, and the firſt and third ſorts thrive beſt in a moiſt ſoil, for in a hot dry ſoil they burn and decay in ſummer, unleſs they are duly watered in dry weather. The ſixth ſort ſhould not have a ſoil too moiſt, for as the roots run deep in the ground, they will be in danger of rotting by much wet.

The other ſort is annual, ſo is propagated by ſeeds. The beſt time to ſow theſe is in autumn, ſoon after they are ripe, for the plants will reſiſt the cold of our winters very well, and will flower early the following ſummer, ſo good ſeeds may be obtained; whereas, thoſe which are ſown in the ſpring ſometimes miſcarry, or lie a year in the ground. Theſe ſeeds ſhould be ſown where they are deſigned to remain, for the plants do not ſucceed very well when they are tranſplanted. When the plants come up, they require no other culture but to keep them clear from weeds, and, where they are too cloſe, to thin them. If theſe plants are permitted to ſcatter their ſeeds, they will come up better than when they are ſown.

PULSATILLA Tourn. Inſt. R. H. 284. tab. 148. Paſque-flower.

The CHARACTERS are,

The flower hath a leafy involucrum, ending in many points; it hath two orders of petals, three in each, and a great number of ſlender ſtamina, about half the length of the petals, terminated by erect twin ſummits, and a great number of germen collected in a head. The germen afterward becomes ſo many ſeeds, having long hairy tails ſitting upon the oblong receptacle.

The SPECIES are,

1. PULSATILLA (*Vulgaris*) foliis decompoſitis, flore nutante, limbo erecto. Hort. Cliff. 223. *Paſque-flower with decompounded winged leaves, and a nodding flower, having an erect rim.*

2. PULSATILLA (*Pratenſis*) foliis decompoſitis pinnatis, flore pendulo, limbo reflexo. Hort. Cliff. 223. *Paſque-flower with decompounded winged leaves, and a pendulous flower whoſe border is reflexed.*

3. PULSATILLA (*Vernalis*) foliis ſimpliciter pinnatis, foliolis lobatis, flore erecto. Flor. Suec. 448. *Paſque flower with ſimple winged leaves, whoſe wings have lobes, and an erect flower.*

4. PULSATILLA (*Patens*) foliis digitatis multifidis, flore erecto patente. *Paſque-flower with hand-ſhaped leaves, having many points, and an erect ſpreading flower.*

The firſt of theſe plants is common in divers parts of England; it grows in great plenty on Gogmagog-hills on the left-hand of the highway leading from Cambridge to Haveril, juſt on the top of the hill; alſo about Hilderſham, ſix miles from Cambridge, and on Bernack-heath not far from Stamford, and on Southrop-common adjoining thereto; alſo on mountains and dry paſtures juſt by Leadſtone-hall near Pontefract in Yorkſhire.

This hath a fleſhy taper root, which runs deep in the ground; the leaves are hairy, and finely cut like thoſe of the wild Carrot; the ſtalk riſes a foot high, is pretty thick, hairy, and naked to the top, where there is a leafy involucrum to the flower, which is hairy, ending in many points; it is terminated by one flower, compoſed of ſix petals, ranged in two borders, three without, and three within; they are oblong, thick, and of a purple colour, bell-ſhaped, nodding on one ſide, and their points turn upward. Within the petals are a great number of ſlender yellowiſh ſtamina, terminated by erect ſummits, and in the center a great number of germen are collected in a head, which afterward become ſeeds, each having a long tail, by which they are diſtinguiſhed from Anemone.

There is a variety of this with double flowers, and another with white, but theſe have been obtained from ſeeds of the other.

The ſecond ſort hath ſhorter leaves than the firſt; the ſtalks do not riſe ſo high; the flowers do not expand ſo wide, and hang downward, but their brims are reflexed; they are of a very dark purple colour. This grows naturally in the meadows in Germany.

The third ſort grows naturally on the Alps and Helvetian mountains; this hath a perennial root. The leaves are like thoſe of Smallage, and are ſimply winged; the ſtalk riſes near a foot high, is naked almoſt to the top, where comes out a neat hairy involucrum, and above that one yellow flower, ſhaped like the perennial yellow Adonis, ſtanding erect.

The fourth ſort grows in Siberia; it hath a thick fleſhy root, which ſends out many ſtrong fibres. The leaves are hand-ſhaped, compoſed of ſeveral roundiſh lobes, like ſome of the ſorts of Ranunculi; they are downy, and cut into ſeveral ſegments. The ſtalk riſes nine or ten inches high, having a hairy involucrum a conſiderable diſtance below the flower; it is terminated by one flower, which is large, ſpreading, and of a whitiſh yellow colour, with deep yellow ſtamina.

There are ſome other ſpecies of this plant, but thoſe here mentioned are all the ſorts which I have ſeen growing in England, and therefore I have not enumerated more, as it would be to little purpoſe, ſince it is difficult to procure them from the countries where they naturally grow.

Theſe plants may be propagated by ſeeds, which ſhould be ſown in boxes or pots filled with very light ſandy earth, obſerving not to cover the ſeeds too deep with mould, which will prevent their riſing, for they require no more than juſt to be covered. Theſe boxes ſhould be placed where they may have the morning ſun until ten of the clock, but muſt be ſcreened from it in the heat of the day, and, if the ſeaſon proves dry, the earth ſhould be often refreſhed with water. The beſt time for ſowing of theſe ſeeds is in July, ſoon after they are ripe, for if they are kept till ſpring they ſeldom grow.

The boxes or pots in which the ſeeds are ſown, ſhould remain in this ſhady ſituation until the beginning of October, when they ſhould be moved where they may enjoy the full ſun during the winter ſeaſon. About the beginning of March the plants will begin to appear, at which time the boxes ſhould be again removed, where they may have only the forenoon ſun; for if they are too much expoſed to the heat, the young plants will die away. They ſhould alſo

be

be refreshed with water in dry weather, which will greatly promote their growth; and they must be carefully kept clean from weeds, which, if suffered to grow among them, will in a short time destroy them.

When the leaves of these plants are entirely decayed (which is commonly in July,) you should then take up the roots, which, being nearly the colour of the ground, will be difficult to find while small; therefore you should pass the earth through a fine wire-sieve, which is the best method to separate the roots from the earth (but notwithstanding all possible care taken, yet there will be many small roots left; so that the earth should either be put into the boxes again, or spread upon a bed of light earth, to see what plants will rise out of it the succeeding year.) The roots, being taken up, should be immediately planted again in beds of fresh loamy earth, about three or four inches asunder, covering them about three inches thick with the same earth. The spring following most of these plants will produce flowers, but they will not be so large and fair as in the succeeding years, when the roots are larger.

The roots of these plants generally run down deep in the ground, and are of a fleshy substance, somewhat like Carrots, so will not bear to be kept long out of the ground; therefore, when they are removed, it should be done in autumn, that they may take fresh root before the frost comes on; for if they are transplanted in the spring, they will not produce strong flowers. These plants thrive best in a loamy soil, for in very light dry ground they are apt to decay in summer.

PUMPION. See Pepo.

PUNICA. Tourn. Inst. R. H. 633. tab. 407. The Pomgranate-tree.

The Characters are,

The empalement of the flower is bell-shaped, coloured, and cut into six parts at the top. The flower has five roundish, erect, spreading petals, which are inserted in the empalement, and a great number of slender stamina, which are also inserted in the empalement, terminated by oblong summits. The germen is situated under the flower, supporting a single style, crowned by a headed stigma. It afterward becomes a large almost globular fruit, crowned by the empalement. The fruit is divided into several cells by membranaceous partitions, which are filled with roundish succulent seeds.

The Species are,

1. Punica (*Granatum*) foliis lineari-lanceolatis, caule arborescente, flore majore. *Pomgranate with linear spear-shaped leaves, a tree-like stalk, and a larger flower.*

2. Punica (*Nana*) foliis linearibus, caule frutescente, flore minore. *Pomgranate with linear leaves, a shrubby stalk, and a smaller flower; Dwarf Pomgranate.*

There are the following varieties of the first sort, which are supposed to have been accidentally obtained by culture from the seeds, therefore I have not enumerated them as species; but as many curious persons will expect to find them inserted here, I shall just mention them.

The wild Pomgranate, with single and double flowers.

The sweet Pomgranate.

The small flowering Pomgranate, with single and double flowers.

The Pomgranate with striped flowers.

These plants grow naturally in Spain, Portugal, Italy, and Mauritania. There are also many of them in the West-Indies, but they are supposed to have been transplanted there from Europe. They are so much improved there, as to be much preferable to any in Europe, the fruit being larger and finer flavoured.

This tree rises with a woody stem eighteen or twenty feet high, garnished with narrow spear-shaped leaves, of a light lucid green, and stand opposite. The flowers come out at the end of the branches sometimes singly, and at others three or four together; one of the largest terminating the branch, and immediately under that are two or three smaller buds, which, after the first flower is past, swell larger, and expand, whereby there is a continued succession of flowers for some months. The empalement of the flower is very thick, fleshy, and of one piece, cut at the top into five segments; it is of a fine red colour, and within are included five (in the single flowers, but in the double a great number) of scarlet petals, which are inserted in the empalement. In the center is situated the style, arising from the germen, encompassed by many slender stamina, which are terminated by oblong yellowish summits. After the flower decays, the germen swells to a roundish fruit, crowned by the empalement, having a hard shell, including a pulp, filled with angular seeds.

The Balaustia of the shops, is the empalement of the flower of the double-flowering Pomgranate.

The first of these trees is now pretty common in the English gardens, where formerly it was nursed up in cases, and preserved in green-houses with great care (as was also the double-flowering kind;) but they are both hardy enough to resist the severest cold of our climate in the open air, and, if planted against warm walls in a good situation, the first sort will often produce fruit, which in warm seasons will ripen tolerably well; but as these fruits do not ripen till late in the autumn, they are seldom well tasted in England, for which reason the sort with double flowers is commonly preferred to it. The sort with sweet fruit, as also the wild sort, are less common in the English gardens than the former two.

These plants may be easily propagated by laying down their branches in the spring, which in one year's time will take good root, and may then be transplanted where they are designed to remain. The best season for transplanting of these trees is in spring, just before they begin to shoot; they should have a strong rich soil, in which they flower much better, and produce more fruit, than if planted on dry poor ground; but in order to obtain the flowers in plenty, there should be care taken in the pruning of these trees, for want of which we often see these trees very full of small shoots, but do not find many flowers produced upon them; therefore I shall set down directions for pruning of these trees, so as to obtain a great quantity of flowers and fruit.

The flowers of this tree always proceed from the extremity of the branches, which are produced the same year. This therefore directs, that all weak branches of the former year should be cut out, and that the stronger should be shortened in proportion to their strength, in order to obtain new shoots in every part of the tree. The branches may be laid in against the wall about four or five inches asunder; for, as their leaves are small, there is not a necessity of allowing them a greater distance. The best time for this work is about Michaelmas, or a little later, according to the mildness of the season; for if they are left until spring before they are pruned, they seldom put out their shoots so early, and the earlier they come out, the sooner the flowers will appear, which is of great consequence where fruit is desired. In summer they will require no other dressing, but to cut off very vigorous shoots, which grow from the wall, and never produce flowers (for it is the middling shoots only which are fruitful;) and when the fruit is formed, the branches on which they grow should be fastened to the wall to support them, otherwise the weight of the fruit, when grown large, will be apt to break them down.

Though, as I said before, the fruit of this tree seldom arrives to any perfection in this country, so as to render it valuable,

valuable, yet for the beauty of its ſcarlet-coloured flowers, together with the variety of its fruit, there ſhould be one tree planted in every good garden, ſince the culture is not great which they require; the chief care is to plant them upon a rich ſtrong ſoil, and in a warm ſituation. Upon ſome trees which had theſe advantages, I have obtained a great quantity of fruit, which have arrived to their full magnitude, but I cannot ſay they were well flavoured; however, they made a very handſome appearance upon the trees

The double-flowering kind is much more eſteemed than the other in this country, for the ſake of its large, fine, double flowers, which are of a moſt beautiful ſcarlet colour, and, if the trees are ſupplied with nouriſhment, will continue to produce flowers for near three months ſucceſſively. This tree muſt be pruned and managed in the ſame manner as hath been already directed for the fruit bearing kind, but this ſort may be rendered more productive of its beautiful flowers, by grafting it upon ſtocks of the ſingle kind, which will check the luxuriancy of the trees, and cauſe them to produce flowers upon almoſt every ſhoot; by which method I have had a low tree, which was planted in the open air, extremely full of flowers, which made a very fine appearance.

The ſecond ſort grows naturally in the Weſt Indies, where the inhabitants plant it in their gardens to form hedges. It ſeldom riſes more than five or ſix feet high in thoſe countries, ſo may be kept within compaſs, and in warm countries the plants continue flowering great part of the year. The flowers of this kind are much ſmaller than thoſe of the common ſort; the leaves are ſhorter and narrower, and the fruit is not larger than a Nutmeg, and has little flavour, ſo it is chiefly propagated for the beauty of its flowers. This is undoubtedly a diſtinct ſpecies from the common ſort, and is much tenderer.

This plant may be propagated by layers in the ſame manner as the former ſorts, but they muſt be planted in pots filled with rich earth, and preſerved in a green-houſe, otherwiſe it is too tender to endure the cold of our winters; for though it may live abroad in a warm ſituation, yet it will make little progreſs in the ſummer, when the flowers begin to appear; if the plants are expoſed to the open air, the buds often fall off and never open, ſo that they ſhould not be expoſed to the open air, but placed in an airy glaſs-caſe, giving them a large ſhare of air every day in mild weather. As they will be covered at the top by the glaſſes, the flowers will expand, and the fruit will grow to the full ſize in England with this management, though they are not very deſirable; but hereby the plants may be continued in flower upward of three months, and will make a fine appearance.

PURSLAIN. See Portulaca.

PYRACANTHA. See Mespilus.

PYROLA. Tourn. Inſt. R. H. 256. tab. 132. Lin. Gen. Plant. 490. Winter-green.

The Characters are,

The flower hath a permanent empalement cut into five parts; it hath five roundiſh, concave, ſpreading petals, and ten awl-ſhaped ſtamina, terminated by large nodding ſummits, with two riſing horns, and a roundiſh germen ſupporting a ſlender ſtyle, which is longer than the ſtamina, crowned by a thick ſtigma. The germen afterward becomes a roundiſh, depreſſed, five-cornered capſule, with five cells opening at the angles, filled with ſeeds.

The Species are,

1. Pyrola (*Rotundifolia*) ſtaminibus adſcendentibus, piſtillo declinato. Flor. Suec. 330. *Winter-green with riſing ſtamina, and a declining pointal.*

2. Pyrola (*Secunda*) racemo unilaterali. Flor. Suec. 332. *Winter-green with a bunch of flowers ranged on one ſide the foot-ſtalk.*

3. Pyrola (*Uniflora*) ſcapo unifloro. Flor. Lapp. 167. *Winter-green with one flower on the ſtalk.*

4. Pyrola (*Maculata*) pedunculis bifloris. Lin. Sp. Pl. 396. *Winter-green with two flowers on a foot ſtalk.*

The firſt ſort grows wild in many places in the north of England, particularly near Hallifax in Yorkſhire, on rocky hills and heaths, as alſo in ſhady woods, ſo it is very difficult to preſerve in gardens in the ſouthern parts.

This hath a perennial root, from which ſpring out five or ſix roundiſh leaves about an inch and a half long, and almoſt as broad, of a thick conſiſtence, of a deep lucid green, and entire, ſtanding upon pretty long foot-ſtalks. Between theſe ariſe a ſlender upright ſtalk near a foot high, naked great part of the length, ending in a looſe ſpike of flowers, which are compoſed of five large, concave, white petals, ſpreading open like a Roſe, but the two upper leaves are formed into a kind of helmet. In the center is ſituated a crooked pointal, bending downward, attended by ten ſlender ſtamina, terminated by Saffron-coloured ſummits.

The ſecond ſort grows naturally upon mountains in Italy, particularly near Verona and Genoa, and I have found it growing in Weſtmoreland. This hath a ſlender, creeping, perennial root, from which ariſe two or three very ſlender ligneous ſtalks about five inches high, ſuſtaining at the top four or five oval acute-pointed leaves an inch and a half long, and one broad, of a thinner conſiſtence, and a brighter green than thoſe of the former, each ſtanding upon a ſhort foot-ſtalk; and between theſe, on the ſide of the ſtalk, comes out the foot-ſtalk of the flowers, upon which they are ranged along one ſide; they are ſhaped like the other, but are ſmaller, as are alſo the capſules.

The third ſort grows naturally in ſhady woods in the northern parts of Europe. This hath a perennial creeping root, from which come forth four or five roundiſh leaves, of a pretty thick conſiſtence; and between theſe ariſes a foot-ſtalk about four inches high, ſuſtaining one large white flower at the top, of the ſame ſhape as the others.

The fourth ſort grows naturally in North America. This hath a ligneous perennial root, from which ariſe two or three ligneous ſtalks a foot and a half high, garniſhed with ſtiff leaves two or three inches long, ending in acute points, having ſome ſharp indentures on their borders; the midrib is remarkably broad, and very white, as are alſo the veins which run from it. The flowers are produced at the end of the ſtalk, on ſlender foot-ſtalks, about three inches long, each ſuſtaining two ſmall pale-coloured flowers at the top.

Theſe are all of them very difficult to cultivate in gardens, for as they grow on very cold hills, and in moſſy mooriſh ſoil, when they are removed to a better ſoil and a warmer ſituation, they ſeldom continue long. The beſt time to tranſplant theſe plants into gardens is about Michaelmas, provided the roots can then be found, when they ſhould be taken up with balls of earth to their roots, and planted in a ſhady ſituation, and on a moiſt undunged ſoil, where they ſhould be frequently watered in dry weather, otherwiſe they will not thrive. Some of theſe plants may be planted in pots, which ſhould be filled with earth as nearly reſembling that in which they naturally grow as poſſible, and place them in a ſhady ſituation, where, if they are conſtantly watered in dry weather, they will thrive very well.

The firſt ſort is ordered by the College of Phyſicians to be uſed in medicine, and is generally brought over from Switzerland amongſt other vulnerary plants, in which claſs this plant is ranged, and by ſome hath been greatly commended.

PYRUS. Tourn. Inst. R. H. 628. tab. 404. The Pear-tree.

The CHARACTERS are,

The flower hath a permanent empalement of one leaf, which is divided into five parts at the top; it hath five roundish concave petals, which are inserted in the empalement, and about twenty awl-shaped stamina shorter than the petals inserted in the empalement, and terminated by single summits. The germen is situated under the flower, supporting three or four styles crowned by single summits; it afterward becomes a pyramidal fleshy fruit, indented at the top, but produced at the base, having five membranaceous cells, each containing one smooth oblong seed, pointed at the base.

The Pear and Quince may be joined together with more propriety than the Apple with either, for the fruit of the two former are produced at their base, whereas the Apple is indented; nor will the Apple grow upon either of the other two, or those upon the Apple, when grafted or budded, but the Quince or Pear will grow upon each other, so there is a boundary set by nature between those and the Apple.

The several varieties of Pears which are now cultivated in the curious fruit-gardens, have been accidentally obtained by seeds, so must not be deemed distinct species; but, as they are generally distinguished in the fruit-gardens and nursery by the shape, size, and flavour of their fruit, I shall continue those distinctions, that the work may not appear imperfect to such as delight in the cultivation of these fruits.

The VARIETIES are,

1. PYRUS (*Musk*) sativa, fructu æstivo parvo racemoso odoratissimo. Tourn. *Petit Muscat, i. e. Little Musk Pear; commonly called the Supreme.* This fruit is generally produced in large clusters; it is rather round than long; the stalk short; and when ripe, the skin is of a yellow colour; the juice is somewhat musky, and, if gathered before it is too ripe, is a good Pear. This ripens the beginning of July, and will continue good but for a few days.

2. PYRUS (*Chio*) sativa, fructu æstivo minimo odoratissimo. Tourn. *Poire de Chio, i. e. The Chio Pear; commonly called the little Bastard Musk Pear.* This is smaller than the former, but is in shape pretty much like that. The skin, when ripe, has a few streaks of red on the side next the sun, and the fruit seldom hangs in clusters as the former, but in other respects is nearly like that.

3. PYRUS (*Citron des Carmes*) sativa, fructu æstivo parvo, è virido albido. Tourn. *Poire Hâtiveau, i. e. the Hasting Pear: Poire Madeleine, ou Citron des Carmes.* This is a larger Pear than either of the former, and is produced more toward the pedicle. The skin is thin, and of a whitish green colour when ripe; the flesh is melting, and, if not too ripe, of a sugary flavour, but is apt to be meally. This ripens in the middle of July.

4. PYRUS (*Muscadelle*) sativa, fructu æstivo partim saturatè rubente, partim flavescente. Tourn. *Muscadelles Rouges, i. e. the red Muscadelle. It is also called La Bellissime.* This is a large early Pear, of great beauty; the skin is of a fine yellow colour when ripe, beautifully striped with red; the flesh is half melting, and has a rich flavour if gathered before it be too ripe, but it is apt to be meally.

5. PYRUS (*Muscat*) sativa, fructu æstivo parvo flavescente moschato. Tourn. *Petit Muscat, i. e. the little Muscat.* This is a small Pear, rather round than long; the skin is very thin, and when ripe, of a yellowish colour; the flesh is melting, of a musky flavour, but will not keep long when ripe. This comes the end of July.

6. PYRUS (*Cuisse Madam*) sativa, fructu æstivo oblongo ferrugineo, carne tenerâ moschatâ. Tourn. *Cuisse Madame, Lady's Thigh; in England commonly called Jargonelle.* This is a very long Pear, of a pyramidal shape, having a long foot-stalk; the skin is pretty thick, of a russet green colour from the sun, but towards the sun it is inclined to an iron colour; the flesh is breaking, and has a rich musky flavour. Ripe the beginning of August. This is one of the best early summer Pears yet known, and is certainly what all the French gardeners call the Cuisse Madame, as may be easily observed by their description of this Pear; but I suppose that the titles of this and the Jargonelle, were changed in coming to England, and have been continued by the same names.

7. PYRUS (*Windsor*) sativa, fructu oblongo, è viridi flavescente. *The Windsor Pear.* This is an oblong fruit, which swells toward the crown, but near the stalk is drawn toward a point; the skin is smooth, and when ripe, of a yellowish green colour; the flesh is very soft, and, if permitted to hang out two or three days after it is ripe, grows meally, and is good for nothing.

8. PYRUS (*Jargonelle*) sativa, fructu æstivo oblongo, è viridi albo. *The Jargonelle; now commonly called Cuisse Madame.* This is certainly what the French gardeners call the Jargonelle, which, as I before observed, is now in England given to another fruit much preferable to this, so that the two names are changed; for the Jargonelle is always placed amongst those which the French call bad fruit, and the Cuisse Madame is set down amongst their best fruit, which is certainly the reverse with us, as they are now named. This Pear is somewhat like the Windsor, but is not so swelling toward the crown, and is smaller toward the stalk; the skin is smooth, of a pale green colour; the flesh is apt to be meally, if it stands to be ripe, but, being a plentiful bearer, is much propagated for the London markets.

9. PYRUS (*Orange Musk*) sativa, fructu æstivo globoso sessili moschato, maculis nigris consperso. Tourn. *Orange Mosquée, i. e. the Orange Musk.* This is a middle-sized Pear, of a short globular form; the skin is of a yellowish colour, spotted with black; the flesh is musky, but is very apt to be a little dry and choaky. It ripens the beginning of August.

10. PYRUS (*Blanquet*) sativa, fructu æstivo albido majori. Tourn. *Gros Blanquet, i. e. Great Blanket. This is also called La Mussette d'Anjou, i. e. the Bagpipe of Anjou.* This is a large Pear approaching to a round form; the skin is smooth, of a pale green colour; the flesh is soft and full of juice, which hath a rich flavour; the stalk is short, thick, and spotted; the wood is slender, and the leaf is very much like that of the tree called the Jargonelle. This ripens the beginning of August.

11. PYRUS (*Musk Blanquette*) sativa, fructu æstivo albido saccharato odoratissimo. Tourn. *The Blanquette, or Musk Blanquette; the little Blanket Pear.* This Pear is much less than the former, and more pinched in near the stalk, which is also short, but slenderer than that of the former; the skin is soft, of a pale green colour; the flesh is tender, and full of a rich musky juice; the wood of this tree is much stronger than that of the former, and the shoots are commonly shorter. This ripens the middle of August.

12. PYRUS (*Long-stalked Blanquet*) sativa, fructu æstivo albido, pediculo longo donato. Tourn. *Blanquette à longue queüe, i. e. Long-stalked Blanket Pear.* This Pear is in shape somewhat like the former, but the eye is larger, and more hollowed at the crown; toward the stalk it is somewhat plumper, and a little crooked; the skin is very smooth, white, and sometimes toward the sun is a little coloured; the flesh is between melting and breaking, and is full of a rich sugary juice. This ripens the middle of August.

13. PYRUS (*Skinless*) sativa, fructu æstivo oblongo rufescente saccharato. Tourn. *Poire sans Peau, i. e. the Skin-*

lesi Pear. It is also called Fleur de Guigne, i. e. Flower of Guigne, and by some Rousselet hâtif, i. e. the early Russelet. This is a middle-sized fruit, of a long shape, and a reddish colour, somewhat like the Russelet; the skin is extremely thin; the flesh is melting, and full of a rich sugary juice; the shoots are long and strait. This ripens the middle of August.

14. Pyrus (*Robine*) sativa, fructu æstivo turbinato, carne tenerâ saccharatâ. *Muscat Robine, i. e. the Musk Robine Pear.* This is also called Poire à la Reine, i. e. the Queen's Pear; Poire d'Ambre, i. e. the Amber Pear; and Pucelle de Xaintonge, i. e. the Virgin of Xaintonge. This is a small round Pear, of a yellowish colour when ripe; the flesh is between melting and breaking, It hath a rich musky flavour, and is a middling bearer. It ripens the dle of August.

15. Pyrus (*Le Bourdon Musk*) sativa, fructu æstivo turbinato moschato. *La Bourdon Mosque, i. e. the Musk Drone Pear.* This is a middle-sized round fruit, whose skin is of a yellowish colour when ripe; the flesh is melting, and full of a high musky juice, but it must not hang too long on the tree, for it is subject to grow meally in a short time. This ripens the end of August.

16. Pyrus (*Orange*) sativa, fructu æstivo globoso sessili, è viridi purpurascente saccharato odorato. Tourn. *Orange Rouge, i. e. the red Orange Pear.* This Pear hath been the most common of all the sorts in France, which was occasioned by the general esteem it was in some years since. It is a middle-sized round fruit, of a greenish colour, but the side next the sun changes to a purple colour when ripe; the flesh is melting, and the juice is sugared with a little perfume; the eye is very hollow, and the stalk is short. This ripens the end of August.

17. Pyrus (*Cassolette*) sativa, fructu æstivo oblongo minori cinereo odorato. Tourn. *Cassolette Friolet, Muscat Verd Lechefrion*; this is so called from its being shaped like a perfuming pot. It is a long fruit, in shape like the Jargonelle, of an Ash colour; its flesh is melting, and full of a perfumed juice, but is very apt to rot in the middle as soon as ripe, otherwise it would be esteemed an excellent Pear. It is ripe the end of August.

18. Pyrus (*Orange Musk*) sativa, fructu æstivo turbinato, è viridi albido. Tourn. *Orange Mosquée, i. e. the Musk Orange Pear.* This is a large round Pear, in shape like a Bergamot; the skin is green, and the flesh is melting, but it is very subject to rot upon the tree, which renders it not near so valuable as some others. It ripens the end of August.

19. Pyrus (*Oignonnet*) sativa, fructu æstivo globoso è viridi purpurascente. Tourn. *Gros Oignonnet, i. e. the great Onion Pear.* It is also called Amiré-roux, i. e. Brown Admired; Roy d'Eté, i. e. King of Summer; and Archiduc d'Eté, i. e. the Summer Archduke. This is a middle-sized round Pear, of a brownish colour next the sun; the flesh is melting, and the juice is passably good. This ripens the end of August.

20. Pyrus (*Averat*) sativa, fructu æstivo globoso sessili ex albido flavescente saccharato odorato. Tourn. Robine. *Muscat d'Aoust, i. e. the August Muscat.* It is also called Poire d'Averat, i. e. the Averat Pear; and Poire Royale, i. e. the Royal Pear. This is a roundish flat Pear, in shape very like a Bergamot; the stalk is long, strait, and a little spotted, and the eye is a little hollowed; the skin is smooth, of a whitish yellow colour; the flesh is breaking, but not hard, and its juice is richly sugared and perfumed. It is a great bearer, and is esteemed one of the best summer Pears yet known. It ripens the end of August.

21. Pyrus (*Rose*) sativa, fructu æstivo globoso sessili odorato. Tourn. *Poire-rose, i e. the Rose Pear; and L'Epinerose, i. e. the Thorny Rose.* This is a short round fruit, shaped like the great Onion Pear, but much larger, of a yellowish green colour, but a little inclining to red on the side next the sun; the stalk is very long and slender; the flesh is breaking, and the juice is musky. This ripens the end of August. The shoots and the leaves of this tree are large.

22. Pyrus (*Pouchet*) sativa, fructu æstivo globoso albido saccharato. Tourn. *Poire du Pouchet.* This is a large, round, whitish Pear, shaped somewhat like the Besidéri; the flesh is soft and tender, and the juice is sugary. This ripens the end of August.

23. Pyrus (*Parfume*) sativa, fructu æstivo turbinato sessili saturatiùs rubente punctato. Tourn. *Poire de Parfume, i. e. the perfumed Pear.* This is a middle-sized round fruit, whose skin is somewhat thick and tough, and of a deep red colour spotted with brown; the flesh is melting, but dry, and has a perfumed flavour. This ripens the end of August.

24. Pyrus (*Boncrêtien*) sativa, fructu æstivo oblongo magno, partim rubro, partim albido, odorato. Tourn. *Boncrêtien d'Eté, i. e. the Summer Boncrêtien, or Good Christian.* This is a large oblong fruit, whose skin is smooth and thin; the side next the sun is of a beautiful red colour, but the other side is of a whitish green; the flesh is between breaking and tender, and is very full of juice, which is of a rich perfumed flavour. It ripens the beginning of September.

25. Pyrus (*Salviati*) sativa, fructu æstivo globoso, ex rubro albidoque flavescente saccharato odorato. Tourn. *Salviati.* This Pear is pretty large, round, and flat, very much like the Besideri in shape, but not in colour; the stalk is very long and slender, and the fruit is a little hollowed both at the eye and stalk; the colour is red and yellow next the sun, but on the other side is whitish; the skin is rough; the flesh is tender, but a little soft, and has no core; the juice is sugary and perfumed, somewhat like the Robine, but is not near so moist. This ripens the beginning of September.

26. Pyrus (*Caillot-rosat*) sativa, fructo æstivo globoso sessili rufescente odorato. Tourn. *Caillot-rosat, i. e. Rose-water Pear.* This is a large round Pear, somewhat like the Messire Jean, but rounder; the stalk is very short, and the fruit is hollowed like an Apple, where the stalk is produced; the skin is rough, and of a brown colour; the flesh is breaking, and the juice is very sweet. This ripens the middle of September.

27. Pyrus (*Choak Pear*) sativa, fructu æstivo longo, acerbitate strangulationem minitante. Tourn. *Poire d'Etrangillon, i. e. the Choaky Pear.* The flesh is red. This is seldom preserved in gardens, so there needs no description of it.

28. Pyrus (*Rousselet*) sativa, fructu æstivo oblongo è ferrugineo rubente, nonnunquam maculato. Tourn. *Poire de Rossuelet, i. e. the Russelet Pear.* This is a large oblong Pear; the skin is brown, and of a dark red colour next the sun; the flesh is soft and tender, without much core; the juice is agreeably perfumed, if gathered before it be too ripe. This produces larger fruit on an espalier than on standard-trees. It ripens the middle of September.

29. Pyrus (*Poire de Prince*) sativa, fructu æstivo subrotundo, partim rubro, partim flavescente, odorato. *Poire de Prince, i. e. the Prince's Pear.* This is a small roundish Pear, of a bright red colour next the sun, but of a yellowish colour on the opposite side; the flesh is between breaking and melting: the juice is very high-flavoured. It is a great bearer. It ripens the middle of September, but will keep a fortnight good, which is what few summer fruits will do.

30. Pyrus (*Moüille-bouche*) fativa, fructu æftivo globofo viridi, in ore liquefcente. Tourn. *Gros Moüille-bouche, i. e. the great Mouthwater Pear.* This is a large round Pear, with a fmooth green fkin; the ftalk is fhort and thick; the flefh is melting, and full of juice, if gathered before it be too ripe, otherwife it is apt to grow meally. This ripens the middle of September.

31. Pyrus (*Bergamot d'Eté*) fativa, fructu æftivo rotundo feffili faccharato, è viridi flavefcente. Tourn. *Bergamotte d'Eté, i. e. Summer Bergamot*; this is by fome called the Hamden's Bergamot. It is a pretty large, round, flat Pear, of a greenifh yellow colour, and hollowed a little at both ends like an Apple; the flefh is melting, and the juice is highly perfumed. This ripens the middle of September.

32. Pyrus (*Bergamot Autumn*) fativa, fructu autumnali feffili faccharato odorato, è viridi flavefcente, in ore liquefcente. Tourn. *Bergamotte d'Automne, i. e. the Autumn Bergamot.* This is a fmaller Pear than the former, but is nearly of the fame fhape; the fkin is of a yellowifh green, but changes to a faint red on the fide next the fun; the flefh is melting, and its juice is richly perfumed. It is a great bearer, ripens the end of September, and is one of the beft Pears of the feafon.

33. Pyrus (*Swifs Bergamot*) fativa, fructu autumnali turbinato viridi, ftriis fanguineis diftincta. Tourn. *Bergamotte de Suiffe, i. e. the Swifs Bergamot.* This Pear is fomewhat rounder than either of the former; the fkin is tough, of a greenifh colour, ftriped with red; the flefh is melting, and full of juice, but is not fo richly perfumed as either of the former. This ripens the end of September.

34. Pyrus (*Buerré rouge*) fativa, fructu autumnali fuaviffimo, in ore liquefcente. Tourn. *Buerré rouge, i. e. the red Butter Pear.* It is called l'Amboife, and in Normandy Ifambert; as alfo Buerré gris, i. e. the gray Butter; and Beurré vert, i. e. the green Butter Pear. All thefe different names of Beurrés have been occafioned by the difference of the colours of the fame fort of Pear, which is either owing to the different expofure where they grew, or from the ftock, thofe upon free-ftocks being commonly of a browner colour than thofe which are upon Quince-ftocks; whence fome perfons have fuppofed them to be different fruits, though in reality they are the fame. This is a large long fruit, for the moft part of a brown colour. The flefh is very melting, and full of a rich fugary juice. It ripens the beginning of October, and, when gathered from the tree, is one of the very beft fort of Pears we have.

35. Pyrus (*Doyenné*) fativa, fructu autumnali turbinato feffili flavefcente, & in ore liquefcente. Tourn. *Le Doyenne, i. e. the Dean's Pear.* It is alfo called by all the following names; Saint Michel; Beuerré blanc d'Autonne, i. e. the white autumn Butter Pear; Poire de Neige, i. e. the Snow Pear; Bonne Ente, i. e. a good Graft; the Carlifle and Valencia. It is a large fair fruit, in fhape fomewhat like the gray Beurré, but fhorter and rounder; the fkin is fmooth, and, when ripe, changes to a yellowifh colour; the flefh is melting, and full of juice, which is very cold, but it will not keep good a week after it is gathered, being very fubject to grow meally; it is a very indifferent fruit, but a great bearer, and ripens the beginning of October.

36. Pyrus (*Verte-longue*) fativa, fructu autumnali longo viridique odorato, in ore liquefcente Tourn. *La Verte-longue, i. e. the long green Pear.* It is alfo called Moüille-bouche d'Automne, i. e. the Autumn Mouth-water Pear. This is a long fruit, which is very green when ripe; the flefh is melting, and full of juice, which, if it grows upon a dry warm foil, and a free-ftock, is very fugary, otherwife it is but a very indifferent Pear. It ripens the middle of October, but fome years they will keep till December.

37. Pyrus (*Meffire Jean*) fativa, fructu autumnali tuberofo feffili faccharato, carne durâ. Tourn. *Meffire Jean blanc & gris, i. e. the white and gray Monfieur John.* Thefe, although made two forts of fruit by many perfons, are indubitably the fame, the difference of their colour proceeding from the different foils and fituations where they grow, or the ftocks on which they are grafted. This Pear, when grafted on a free ftock, and planted on a middling foil, neither too wet nor over dry, is one of the beft autumn Pears yet known; but when it is grafted on a Quince-ftock, it is very apt to be ftony, or if planted on a very dry foil, is very apt to be fmall, and good for little, unlefs the trees are watered in dry feafons, which has rendered it lefs efteemed by fome perfons, who have not confidered the caufe of their hardnefs; for when it is rightly managed, there are not many Pears in the fame feafon to be compared with it. This is a large roundifh fruit; the fkin is rough, and commonly of a brown colour; the flefh is breaking, and very full of a rich fugared juice. It ripens the end of October, and will continue good near a month.

38. Pyrus (*Mufcat fleuri*) fativa, fructu autumnali globofo ferrugineo, carne tenerâ fapidiffimâ. Tourn. *Mufcat fleuri, i. e. the flowered Mufcat.* It is alfo called Mufcat à longue queüe d'Automne, i. e. the long-ftalked Mufcat of the Autumn. This is an excellent Pear, of a middling fize, round; the fkin is of a dark red colour; the flefh is very tender, and of a delicate flavour. It ripens the end of October.

39. Pyrus (*Poire de Vigne*) fativa, fructu autumnali globofo ferrugineo, carne vifcidâ. Tourn. *Poire de Vigne, i. e. the Vine Pear.* This is a round fruit, of a middling fize; the fkin is of a dark red colour; the flefh is very melting, and full of a clammy juice; the ftalk is very long and flender. The fruit fhould be gathered before it be full ripe, otherwife it grows meally and foon rots. It ripens the end of October.

40. Pyrus (*Rouffeline*) fativa, fructu autumnali oblongo, dilutè rufefcente, faccharato, odoratiffimo. Tourn. *Poire Rouffeline, i. e. the Rouffeline Pear.* It is alfo called in Touraine, Le Mufcat à longue queüe de la fin d'Automne, i. e. the long-ftalked Mufcat of the end of Autumn. This is by fome Englifh gardeners called the Brute-bonne, but that is a very different fruit from this. It is fhaped fomewhat like the Ruffelet, but the fkin of this is fmooth, and of a greenifh yellow from the fun, but the fide next the fun is of a deep red colour, with fome fpots of gray; the flefh is very tender and delicate; the juice is very fweet, with an agreeable perfume. It ripens the middle of October, but muft not be long kept, left it rot in the middle.

41. Pyrus (*Pendart*) fativa, fructu autumnali oblongo majori cinereo. Tourn. *Poire Pendart, i. e. the Knave's Pear.* This is very like the Caffolette Pear, but is fomewhat larger; the flefh is fine and tender; the juice is very much fugared. It ripens the end of October.

42. Pyrus (*Sucré vert*) fativa, fructu autumnali turbinato tuberofo viridi faccharato, in ore liquefcente. Tourn. *Sucré vert, i. e. the green Sugar Pear.* This Pear is fhaped like the Winter Thorn, but is fmaller; the fkin is very fmooth and green; the flefh is very buttery; the juice is fugared, and of an agreeable flavour; but it is fometimes fubject to be ftony in the middle, efpecially if grafted on a Quince-ftock.

43. Pyrus (*Marquis*) fativa, fructu autumnali tuberofa feffili, è viridi flavefcente, maculis nigris confperfo, carne tenerâ faccharatâ. Tourn. *La Marquife, i. e. the Marquis's Pear.* This is often of two different fhapes, according to the nature of the foil where they are planted, for when the foil is dry, the fruit very much refembles a fine Blanquet; but when the foil is very rich and moift, it grows much larger. It is a well-fhaped Pear, flat at the top; the eye

is small and hallowed; the skin is of a greenish yellow, a little inclining to red on the side next the sun. If this Pear does not change yellow in ripening, it is seldom good; but if it does, the flesh will be tender and delicate, very full of juice, which is sugared. It ripens the beginning of November.

44. Pyrus (*Chat brulé*) sativa, fructu autumnali oblongo, partim albido, partim rufescente. *The Chat brulé, i. e. the burnt Cat.* This is a small oblong Pear, shaped much like the Martin Sec, but differs from it in colour; this being of a pale colour on one side, but of a dark brown on the other; the skin is smooth; the flesh is tender, but dry, and, if kept a short time, is apt to grow meally. It is in eating the beginning of November.

45. Pyrus (*Besideri*) sativa, fructu autumnali globoso sessili, ex albido flavescente. *Le Besideri.* It is so called from Heri, which is a forest in Bretagne, between Rennes and Nantes, where this Pear was found. This is a middle-sized round Pear, of a pale green, inclining to a yellowish colour; the stalk is very long and slender; the flesh is dry, and but very indifferent for eating, but it bakes well. It ripens the middle of November.

46. Pyrus (*Crasan*) sativa, fructu brumali sessili, è viridi flavescente, maculato, utrinque umbilicato, in ore liquescente. Tourn. *The Crasane, or Bergamot Crasane.* It is also called Beurré Plat, i. e. the flat Butter Pear. This is a middle-sized Pear, hollowed at the crown like an Apple; the stalk is very long and crooked; the skin is of a greenish yellow colour when ripe, covered over with a russet coat; the flesh is extremely tender and buttery, and is full of a rich sugared juice, and is the very best Pear of the season. This is in eating the beginning of November.

47. Pyrus (*Dauphine*) sativa, fructu brumali turbinato sessili-flavescente saccharato odorato, in ore liquescente. Tourn. *Lansac ou la Dauphine, i. e. the Lansac or Dauphin Pear.* This Pear is commonly about the size of a Bergamot, of a roundish figure, flat towards the head, but a little produced towards the stalk; the skin is smooth, and of a yellowish green colour; the flesh is yellow, tender, and melting; the juice is sugared, and a little perfumed; the eye is very large, as is also the flower; the stalk is long and strait. When this Pear is upon a free-stock, and planted on a good soil, it is one of the best fruits of the season; but when it is on a Quince-stock, or upon a very dry soil, the fruit will be small, stony, and worth little. It ripens the end of November.

48. Pyrus (*Martinsec*) sativa, fructu brumali oblongo, partim intensè, partim dilutè ferrugineo, saccharato, odorato. Tourn. *Martin Sec, i. e. the Dry Martin.* This is sometimes called the Dry Martin of Champagne, to distinguish it from another Dry Martin of Burgundy. This Pear is almost like the Russelet in shape and colour, which has occasioned some persons to give it the name of Winter Russelet. It is an oblong Pear, whose skin is of a deep russet colour on one side, but the other side is inclining to a red; the flesh is breaking and fine; the juice is sugared, with a little perfume, and if grafted on a free-stock, is an excellent Pear, but if it be on a Quince-stock, it is very apt to be stony. It is in eating the end of November, but if they were permitted to hang their full time on the tree, will keep good two months.

49. Pyrus (*Bigarrade*) sativa, fructu brumali magno sessili, e cinereo flavescente. Tourn. *La Villaine d'Anjou, i. e. the Villain of Anjou.* It is also called Poire Tulipee, i.e. the Tulip Pear, and Bigarrade, i. e. the Great Orange. This is a large round Pear, with a very long slender stalk; the skin is of a pale yellow colour; the flesh is breaking, but not very full of juice. This is in eating the end of November.

50. Pyrus (*Gros queüe*) sativa, fructu brumali flavescente odoratissimo, pediculo crassiori. Tourn. *Poire de gros queüe, i. e. the large-stalked Pear.* This is a large roundish Pear, with a yellow skin; the stalk is very thick, from whence it had the name; the flesh is breaking, dry, and has a very musky flavour; but it is apt to be stony, especially if it be planted in a dry soil, or grafted on a Quince-stock, as are most of the perfumed Pears.

51. Pyrus (*Amadot*) sativa, fructu brumali turbinato rufescente odorato. *L'Amadot, i. e. the Amadot Pear.* This is a middle-sized Pear, somewhat long, but flat at the top, the skin is generally rough, and of a russet colour; the flesh is dry, and high flavoured, if grafted on a free-stock. The wood of this tree is generally thorny, and is esteemed the best sort of Pears for stocks to graft the melting Pears upon, because it gives them some of its fine musky flavour. It is in eating the beginning of December, but will keep good six weeks.

52. Pyrus (*Bouvar*) sativa, fructu brumali, globoso, dilutè virente, tuberoso, punctato, in ore liquescente. Tourn. *Petit Oin, i. e. Little Lard Pear.* It is also called Bouvar and Rousette d'Anjou, i. e. the Russet of Anjou; and Amadont, and Marveille d'Hyver, i. e. the Wonder of the Winter. This Pear is of the size and shape of the Ambret or L'Eschasserie, but the skin is of a clear green colour, and a little spotted; the stalk is pretty long and slender; the eye is large, and deeply hollowed; the flesh is extremely fine, and melting; the juice is much sugared, and has an agreeable musky flavour. It is in eating the middle of December, and is esteemed one of the best fruits in that season. This is better on a free-stock than upon the Quince.

53. Pyrus (*Louisebonne*) sativa, fructu brumali longo è viridi albicante, in ore liquescente. Tourn. *Louisebonne, i. e. the Good Lewis Pear.* This Pear is shaped somewhat like the St. Germain, or the Autumn Verte-longue, but is not quite so much pointed; the stalk is very short, fleshy, and somewhat bent; the eye and the flower are small; the skin is very smooth; the colour is green, inclining to a white when ripe; the flesh is extremely tender and full of juice, which is very sweet, especially when it grows upon a dry soil, otherwise it is apt to be very large and ill-tasted. It is in eating the beginning of December.

54. Pyrus (*Colmar*) sativa, fructu brumali, tuberoso, è viridi flavescente, punctato, saccharato. Tourn. *Poire de Colmar, i. e. the Colmar Pear.* It is also called Poire Manne, the Manna Pear, and Bergamotte tardive, the late Bergamot. This Pear is somewhat like a Boncrétien in shape, but the head is flat; the eye is large, and deeply hollowed: the middle is larger than the head, and is sloped toward the stalk, which is short, large, and a little bent; the skin is green, with a few yellowish spots, but is sometimes a little coloured on the side next the sun; the flesh is very tender, and the juice is greatly sugared. It is in eating the latter end of December, but will often keep good till the end of January, and is esteemed one of the best fruits of that season.

55. Pyrus (*L'Echasserie*) sativa, fructu brumali, globoso, citriformi, flavescente, punctato, in ore liquescente, saccharato, odoratissimo. Tourn. *L'Echasserie.* It is also called Verte-longue d'Hyver, i. e. the Winter long green Pear, and Besidéri, Landri, i. e. the Landry Wilding. This Pear is shaped like a Citron; the skin is smooth, and of a green colour, with some spots while it hangs on the tree, but as it ripens it becomes of a yellowish colour; the stalk is strait and long; the eye is small, and not hollowed: the flesh is melting and buttery; the juice is sugared, with a little perfume. It is in eating the latter end of December.

56. PYRUS (*Virgouleuse*) sativa, fructu brumali longo, è viridi flavescente, in ore liquescente, saccharato. Tourn. *Le Virgoulé, or La Virgouleuse.* It is also called Pujaleuf, and Chambrette; and Poire de Glasse, i. e. the Ice Pear in Gascoigne; but it is called Virgoulé, from a village of that name in the neighbourhood of St. Leonard in Limousin, where it was raised and sent to Paris by the Marquis of Chambert. This Pear is large, long, and of a green colour, inclining to yellow as it ripens; the stalk is short, fleshy, and a little bent; the eye is of a middling size, a little hollowed; the skin is very smooth, and sometimes a little coloured towards the sun; the flesh is melting, and full of a rich juice. It is in eating the latter end of December, and will continue good till the end of January, and is esteemed one of the best fruits of the season; but the tree is very apt to produce vigorous shoots, and the blossoms being generally produced at the extreme part of the shoot, where they are shortened, the fruit will be entirely cut away, which is the reason it is condemned as a bad bearer; but when it is grafted on a free stock, it ought to be allowed at least forty feet to spread; and if upon a Quince-stock, it should be allowed upwards of thirty feet, and the branches trained in against the espalier or wall, at full length, in a horizontal position, as they are produced. Where this tree is thus treated, it will bear very plentifully, and the fruit will be good.

57. PYRUS (*Ambrette*) sativa, spinosa, fructu globoso, sessili, ferrugineo, in ore liquescente, saccharato, odoratissimo. Tourn. *Poire d'Ambrette.* This is so called from its musky flavour, which resembles the smell of the Sweet Sultan-flower, which is called Ambrette in France. This Pear is like the L'Echasserie in shape, but is of a russet colour; the eye is larger, and more hollowed; the flesh is melting; the juice is richly sugared and perfumed; the seeds are large, black, and the cells in which they are lodged are very large; the wood is very thorny, especially when grafted on free-stocks. The fruit is in eating the latter end of December, and continues good till the latter end of January, and is esteemed a very good fruit by most people.

58. PYRUS (*Epine d'Hyver*) sativa, fructu brumali, magno, pyramidato, albido, in ore liquescente, saccharato, odorato. Tourn. *Epine d'Hyver, i. e. Winter-thorn Pear.* This is a large fine Pear, nearly of a pyramidal figure; the skin is smooth, of a pale green colour, inclining to yellow as it ripens; the stalk is short and slender; the flesh is melting and buttery; the juice is very sweet, and in a dry season is highly perfumed; but when it is planted on a moist soil, or the season proves wet, it is very insipid, so that it should never be planted on a strong soil. It ripens the end of December, and will continue good two months.

59. PYRUS (*St. Germain*) sativa, fructu brumali longo, è viridi flavescente, in ore liquescente. Tourn. *La Saint Germain, i. e. the St. Germain Pear.* It is also called L'Inconnüe de la Fare, i. e. the Unknown of La Fare; it being first discovered upon the banks of a river which is called by that name, in the parish of St. Germain. This is a large long Pear, of a yellowish green colour when ripe; the flesh is melting, and very full of juice, which in a dry season, or if planted on a warm dry soil, is very sweet; but when it is planted on a moist soil, the juice is very apt to be harsh and austere, which renders it less esteemed by some persons, though in general it is greatly valued. This is in eating from the end of December till February.

60. PYRUS (*St. Austin*) sativa, fructu brumali tuberoso subacido flavescente punctato. Tourn. *Saint Austin.* This is about the size of a middling Virgoulé Pear, but is somewhat shorter and slenderer near the stalk; the skin is of a fine Citron colour, spotted with red on the side next the sun; the flesh is tender, but not buttery, and is pretty full of juice, which is often a little sharp, which to some persons is disagreeable, but others value it on that account. This is in eating in December, and will continue good two months.

61. PYRUS (*Boncrêtien d'Espagne*) sativa, fructu brumali pyramidato, partim purpureo, punctis nigris consperso, flavescente. Tourn. *Boncrêtien d'Espagne, i. e. the Spanish Boncrêtien.* This is a large Pear, of a pyramidal form, of a fine red or purple colour on the side next the sun, and full of small black spots; the other side is of a pale yellow colour; the flesh is breaking, and when it is on a light rich soil, and grafted on a free-stock, its juice is very sweet. It ripens in the end of December, and will continue good a month or six weeks. If this be grafted on a Quince-stock, it is very apt to be dry and stony. This is a very good fruit for baking.

62. PYRUS (*Poire de Livre*) sativa, fructu brumali, magno, oblongo, turbinato, ferrugineo, utrinque umbilicato. Tourn. *Poire de Livre, i. e. the Pound Pear.* It is also called Gros Ratteau Gris, i. e. the gray raked Pear; and Poire d'Amour, i. e. the Lovely Pear. In England this is called Parkinson's Warden, or the Black Pear of Worcester. This is a very large Pear, each of which commonly weighs a pound or more; the skin is rough, and of an obscure red colour on the side next the sun, but somewhat paler on the other; the stalk is very short, and the eye is greatly hollowed. This is not fit for eating, but bakes or stews exceeding well, and is in season from December to March.

63. PYRUS (*Besi de Cassoy*) sativa, fructu brumali parvo flavescente, maculis rubris consperso. Tourn. *Besi de Cassoy, i. e. the Wilding of Cassoy,* a forest in Bretagne, where it was discovered, and passes under the name of Rousset d'Anjou. It is also called Petit Beurré d'Hyver, i. e. the small Winter Butter Pear. This is a small oblong Pear, of a yellowish colour, spotted with red; the flesh is melting, and the juice is very rich. It is in eating in December and January. This is a prodigious bearer, and commonly produces its fruit in large clusters, provided it be not too much pruned; for it generally produces its blossom-buds at the extremity of its shoots, which, if shortened, the fruit would be cut away. There was a tree of this kind in the gardens of Camden-house near Kensington, which generally produced a great quantity of fruit.

64. PYRUS (*Martin-sire*) sativa, fructu brumali tubinato inæquali, ventre tumido, partim purpureo, partim flavescente. Tourn. *Ronville.* It is also called Hocrenaille and Martin-sire, i. e. the Lord Martin Pear. This Pear is about the size and shape of a large Russelet; the eye is of a middling size, and hollowed a little; the middle of the Pear is generally swelled more on one side than on the other, but is equally extended towards the stalk; the skin is very smooth, soft, and is of a lively red colour next the sun, but on the other side it changes yellow as it ripens. The flesh is breaking, and full of juice, which is very sweet, and a little perfumed; but if grafted on a Quince-stock, is very apt to be small and stony.

65. PYRUS (*Citron d'Hyver*) sativa, fructu brumali citriformi flavescente duro moschato odoratissimo. Tourn. *Citron d'Hyver, i. e. the Winter Citron Pear.* It is also called the Musk Orange Pear, in some places. This is a pretty large Pear, in shape and colour like an Orange or Citron, from whence it had its name. The flesh is hard, dry, and very subject to be stony; for which reasons it is not valued as an eating Pear, but will bake very well. It is in season from December to March.

66. PYRUS (*Rousselet d'Hyver*) sativa, fructu brumali oblongo, è viridi flavescente, saccharato, saporis austeri. Tourn. *Russelet d'Hyver, i. e. the Winter Russelet.* This is

by

by ſome ſuppoſed to be the ſame Pear as is called the Dry Martin, but it is very different from that in ſeveral particulars. The colour of this is a greeniſh yellow, inclining to brown; the ſtalk is long and ſlender; the fleſh is buttery, melting, and generally full of juice, which is very ſweet, but the ſkin is apt to contain an auſtere juice; ſo that if it be not pared, it is apt to be diſagreeable to many perſons palates. It is in eating in January and February.

67. Pyrus (*Portail*) ſativa Pictavienſis, fructu brumali globoſo ſeſſili ſaccharato odorato. Tourn. *Poire Portail, i. e. the Gate Pear.* This Pear was diſcovered in the province of Poictou, where it was ſo much eſteemed, that they preferred it to moſt other fruit, though in the opinion of the moſt curious judges, it does not deſerve the great character which is given to it; for it rarely happens that it proves good for eating, being generally dry, ſtony, and hard, unleſs in extraordinary ſeaſons, and upon a very good ſoil. This muſt always be grafted on a free-ſtock, and ſhould be planted on a light rich-ſoil; and in very dry ſeaſons the trees ſhould be watered, otherwiſe the fruit will be ſtony. It is in ſeaſon from January to March, and bakes well.

68. Pyrus (*Franc-real*) ſativa, fructu brumali magno globoſo flaveſcente, punctis rufis conſperſo. Tourn. *Franc-real.* It is alſo called Fin-or d'Hyver, i. e. the Golden End of Winter. This is a very large Pear, almoſt of a globular figure; the ſkin is yellow, ſpotted with red; the ſtalk is ſhort, and the wood of the tree meally. The fleſh of this Pear is dry, and very apt to be ſtony, but it bakes exceeding well, and continues good from January till March.

69. Pyrus (*Eaſter Bergamot*) ſativa, fructu brumali turbinato ſeſſili ſubacido flaveſcente, punctis aſperioribus conſperſo. Tourn. *Bergamotte Bugi.* It is alſo called Bergamotte de Paſque, i. e. the Eaſter Bergamot. It is a large Pear, almoſt round, but is a little produced in length towards the ſtalk; the eye is flat, and the ſkin is green, having many rough protuberances like ſpots diſperſed all over, but, as it ripens, becomes yellowiſh; the fleſh is breaking, and in a good ſeaſon the juice is ſweet; but it muſt have a free-ſtock, a ſouth-eaſt wall, and a good ſoil, otherwiſe it is apt to be ſtony and auſtere. It is in eating from February till April.

70. Le Muscat d'Alleman, (*Muſcat Germany*) i. e. *the German Muſcat.* This is an excellent Pear, more long than round, of the ſhape of the Winter-royal, but is leſs toward the eye, is more ruſſet, and of a red colour next the ſun; it is buttery, melting, and a little muſky. This is in eating in March and April.

71. Le Bergamotte d'Hollande (*Bergamot d'Holland*) i. e. *the Holland Bergamot*; it is large and round, of the ſhape of the ordinary Bergamot. The colour is greeniſh; the fleſh is half buttery and tender; the juice is highly flavoured. This is a very good Pear, and will keep till March.

72. Le Poire (*Naples*) de Naples, i. e. *the Pear of Naples.* This is a pretty large, long, greeniſh Pear; the fleſh is half breaking; the juice is ſweet, and a little vinous. It is in eating in March. I am in doubt whether this Pear is not in ſome places taken for a Saint Germain, for there is a Pear in ſome gardens very like the Saint Germain, which will keep till April, and this Pear agrees with the characters of that. It is called in England the Eaſter St. Germain.

73. Pyrus (*Boncrêtien d'Hyver*) ſativa, fructu brumali magno pyramidato, è flavo nonnihil rubente. Tourn. *Boncrêcien d'Hyver, i. e. the Winter Boncrêtien Pear.* This Pear is very large and long, of a pyramidal figure; the ſkin is of a yellowiſh colour, but the ſide next the ſun inclines to a ſoft red; the fleſh is breaking, and is very full of rich ſugared juice. This is eſteemed in France one of the beſt winter Pears, but in England it is ſeldom ſo good; though I am fully ſatisfied, if it were grafted on a free-ſtock, and planted in a good ſoil, againſt a wall expoſed to the ſouth-eaſt, and the branches trained at full length, it might be rendered more acceptable than it is at preſent in England.

74. Pyrus (*Cadillac*) ſativa, fructu brumali magno, cydoniæ facie, partim flavo, partim purpureo. Tourn. *Cadillac or Cadillac.* This is a large Pear, ſhaped ſomewhat like a Quince; the ſkin is for the moſt part of a yellow colour, but changes to a deep red on the ſide next the ſun; the fleſh is hard, and the juice auſtere, but it is a very good fruit for baking, and being a plentiful bearer, deſerves a place in every good collection of fruit. It will be good from Chriſtmas to April, or longer.

75. Pyrus (*Paſtorelle*) ſativa, fructu brumali oblongo flaveſcente, punctis rubris conſperſo. *La Paſtorelle.* This Pear is of the ſize and ſhape of a fine Ruſſelet; the ſtalk is ſhort and crooked; the ſkin is ſomewhat rough, of a yellowiſh colour, ſpotted with red; the fleſh is tender, buttery, and when it grows on a dry ſoil, the juice is very ſweet; but on a wet ſoil, or in moiſt years, it is ſubject to have an auſtere taſte. This Pear is in eating in February and March.

76. Pyrus (*Double Flower*) ſativa, fructu brumali ſeſſili, partim flaveſcente, partim purpuraſcente. Tourn. *La Double Fleur, i. e. the double-flowering Pear.* This is ſo called, becauſe the flowers have a double range of petals or leaves. It is a large ſhort Pear; the ſtalk is long and ſtrait; the ſkin is very ſmooth, of a yellowiſh colour, but the ſide next the ſun is commonly of a fine red or purple colour. This is by ſome eſteemed for eating, but it is generally too auſtere in this country for that purpoſe. It is the beſt Pear in the world for baking or compoſts. It is good from February to May.

77. Pyrus (*St. Martial*) ſativa, fructu brumali oblongo, partim flaveſcente, partim purpuraſcente. *St. Martial.* It is alſo called in ſome places Poire Angelique, i. e. the Angelic Pear. This Pear is oblong, in ſhape like the Boncrêtien, but not ſo large, a little flatter at the crown: it has a very long ſtalk; the ſkin is ſmooth and yellowiſh, but on the ſide next the ſun it turns to a purpliſh colour; the fleſh is tender, buttery, and the juice very ſweet. This is in eating in February and March, and will keep very long.

78. Pyrus (*Beſi Chaumontelle*) ſativa, fructu brumali oblongo, partim albido, partim purpureo odorato, ſaccharato. *La Poire de Chaumontelle, or Beſi de Chaumontelle, i. e. the Wilding of Chaumontelle.* This Pear is in ſhape ſomewhat like the Autumn Beurre, but is flatter at the crown; the ſkin is a little rough, of a pale green colour, but turns to a purpliſh colour next the the ſun; the fleſh is melting; the juice is very rich, and a little perfumed. It is in eating from November to January, and is eſteemed by ſome the beſt late Pear yet known.

79. Pyrus (*Carmelite*) ſativa, fructu brumali globoſo ſeſſili cinereo maculis amplis obſcurioribus conſperſo. Tourn. *Carmelite.* This is a middle-ſized Pear, of a roundiſh form; the ſkin is of a gray colour on one ſide, but is inclined to a red on the other, having ſome broad ſpots of a dark colour all over; the fleſh is commonly hard and dry, ſo that it is not very much eſteemed. It is in ſeaſon in March.

80. Pyrus (*Union*) ſativa, fructu brumali maximo pyramidato, dilutè virente. *The Union Pear, otherwiſe called Dr. Uvedale's St. Germain.* This is a very large long Pear, of a deep green colour, but the ſide next the ſun doth ſometimes change to a red as it ripens. This is not fit for eating, but bakes very well; and being a great bearer, and

a very large fruit, deserves a place in every good collection. It is in season from Christmas to April.

The time of each fruit ripening, as here set down, is taken at a medium for seven years, and in the neighbourhood of London, where all sorts of fruit generally ripen a fortnight or three weeks earlier than in almost any part of England; it is very obvious to every person, who will attend to the culture of fruit-trees, that their time of ripening is accelerated by long cultivation; for many of the sorts of Pears, which some years past rarely became ripe in England, unless they grew against the best aspected walls, are now found to ripen extremely well on espaliers and dwarfs; and those Pears, which seldom were in eating till January, are ripe a month earlier. There is also a very great difference in their time of ripening in different seasons, for I have known the fruit of a Pear-tree in one year all ripe and gone by the middle of October, and the very next year the fruit of the same tree has not been fit to eat till the end of December, so that allowance should be made for these accidents. The Besi de Chaumontelle Pear, about forty years past, was seldom fit to eat before February, and has continued good till the middle of April; but now this Pear is commonly ripe in November, and when it is planted on a warm soil, and against a good aspected wall, it is in eating the middle of October. This forwarding of the several kinds of Pears, may be in some measure owing to the stocks upon which they are grafted; for if they are grafted upon early summer Pear-stocks, they will ripen much earlier than when they are upon hard winter Pear-stocks; and if some of the very soft melting Pears were grafted upon such stocks as are raised from the most austere fruit, such as are never fit to eat, and of which the best perry is made, it would improve those fruits, and continue them much longer good; or if the common free-stocks were first grafted with any of these hard Winter Pears, and when they have grown a year, then to graft or bud these soft melting Pears upon them, it would have the same effect; but the Pears so raised, will require a year's more growth in the nursery, and consequently cannot be sold at the same price as those which are raised in the common method, these requiring to be twice budded or grafted, so that there is double labour, beside standing a full year longer; but this difference in the first expence of the trees, is not worth regarding by any person who is desirous to have good fruit; for the setting out in a right way is that which every one should be the most careful of, since by mistaking at first, much time is lost, and an after expence of new trees often attends it.

The ripening of these fruits may also be accelerated by the method of pruning and managing these trees, in which we are greatly improved within the space of a few years past; for if we look into the directions which are given by the best writers on this subject, we shall soon discover how little they knew sixty years ago, of the true method of pruning and managing all sorts of fruit-trees, scarce one of them making any difference in the management of the different kinds of fruit.

Pears are propagated by grafting or budding them upon stocks of their own kind, which are commonly called free-stock, or upon Quince-stocks, or White-thorn, upon all which these fruits will take; but the latter sort of stock is now seldom used, because they never keep pace in their growth with the fruit budded or grafted upon them; as also because the fruit upon such stocks are commonly drier, and more apt to be stony, than when they are upon Pear-stocks. Quince-stocks are greatly used in the nurseries for all sorts of Pears which are designed for dwarfs or walls, in order to check the luxuriancy of their growth, so that they may be kept within compass better than upon free-stocks. But against the general use of these stocks, for all sorts of Pears indifferently, there are very great objections: 1st, Because some sorts of Pears will not thrive upon these stocks, but in two or three years decay, or at most will but just keep alive. 2dly, Most of the sorts of hard-breaking Pears, are rendered stony, and good for little; so that, whenever any of these sorts are thus injudiciously raised, the fruit, although the kind be ever so good, is condemned as good for nothing by such as are not well acquainted with it, when the fault is entirely owing to the stock on which it was grafted. On the contrary, most melting buttery Pears are greatly improved by being upon Quince-stocks, provided they are planted on a strong soil; but, if the ground be very dry and gravelly, no sort of Pear will do well upon Quince-stocks in such places.

These general directions being given, there is no occasion to repeat any part of the method in which these stocks are raised, and the fruits budded or grafted thereon, which has been already mentioned under the article of NURSERIES.

The distance which these trees should be planted, either against walls or espaliers, must not be less than thirty feet, but, if they are planted forty feet, it will be better; for, if they have not room to spread on each side, it will be impossible to preserve them in good order, especially those on free-stocks, for the more these trees are pruned, the more they will shoot; and, as I said before, many sorts of Pears produce their blossom-buds first at the extremity of the former year's shoot; so that when they are shortened, the fruit will be cut away, and this cannot be avoided, where the trees have not room allowed in their first planting.

This distance, I doubt not, will be objected to by many who have not fully attended to the growth of these trees, especially as it hath been the general practice of most gardeners to plant these trees at less than half the distance which is here mentioned; but, whoever will be at the trouble to view any of these trees which have been some years standing, they will always find, if, by accident, one of these trees has been planted against a building, where the branches have had room to spread, that this tree has produced more fruit than twelve trees which have been crouded close, and have not room for their branches to extend. There are some Pear-trees now growing, which spread more than fifty feet in length, and are upward of twenty feet high, which produce a much greater quantity of fruit than six trees in the same room would have done, as there are examples enough to prove, where trees are planted against houses and the ends of buildings at about twelve feet, or much less distance, because there is height of walling for them to grow, which is the reason commonly given by those who plant these trees so close together. But one tree will bear more fruit, when the branches are trained horizontally, than five or six trees, whose branches are led upright: and there never can be any danger of the upper part of the wall being left naked or unfurnished, for I have seen a Pear-tree which has spread more than fifty feet in width, and covered the wall upward of thirty-six feet in height: this was a Summer Boncrêtien Pear, and was extremely fruitful, which rarely happens to this sort when they are not allowed a large share of room. The finest tree of this sort of Pear which I have ever seen, was a large standard-tree in my own possession, whose stem was not more than ten feet high, where the branches came out regularly, and extended near thirty feet on every side from the trunk, many of which were, by the weight of the fruit in summer, brought down to the ground, so were obliged to be supported with props toward the extremity of the branches, to prevent their lying upon the ground; and this tree had its branches so disposed, as to form a natural parabola of forty feet in height, bearing from the lowest to the highest branches; so that in a kindly season, when the blossoms escaped

escaped the frost, it hath produced upward of two thousand Pears, which were much better flavoured than any of the same sort which I have yet tasted. This instance I mention, only to shew how much one of these trees will spread, if proper room be allowed it, and also to observe that, as the branches of this tree had never been shortened, they were fruitful to their extremities. This shews the absurdity of the French gardeners, who do not allow more than ten or twelve feet distance to these trees; and some of their most improved writers on this subject have advised the planting an Apple tree between the Pear-trees, where they are allowed twelve feet, and yet these authors afterward say, that a good Pear-tree will shoot three feet each way in one year; therefore, according to their own observation, the trees so planted will have their branches meet together in two or three years at most; and what must be the case with such trees in five or six years, is not difficult to know But this method of planting has not been peculiar to the French, for most of the gardens in England have been little better planted. Indeed, those persons who were intrusted with the making and planting most of the English gardens, had little skill of their own, so were obliged to follow the directions of the French gardeners, of whom they had so great an opinion, as to get their books translated, and to these have added some trifling notes, which rather betray their weakness; for, where they have objected to the little room which their authors had allowed to these trees, they have, at the most, allowed them but three feet more, from which it is plain, they had not considered the natural growth of the trees, and whoever departs from nature, may be justly pronounced an unskilful gardener.

As most of the English gardens have been made and planted by persons of little judgment, it is very rare to find any of them which produce much fruit; for although many of these gardens have been totally altered and new-planted, yet they have seldom been much altered for the better, and the possessors have been put to the expence of removing the old trees, also the earth of their borders, and to purchase new trees, which have been planted perhaps a foot or two farther asunder than the old trees which were removed; so that, when the young trees have grown a few years, they were in the same condition as the old, and it will be the loss of so many years to the owner; but this will constantly be the case, when it is the interest of the persons employed, who can sell so many young trees, and the planting of three times the number of trees in a garden, more than is proper, may in some measure be ascribed to the same, though in many instances I should be inclinable to think it has proceeded from ignorance, rather than design.

But where fruit-trees have been thus injudiciously planted, if the stocks are healthy and good, the best way to recover this loss is to dig up two or three, and leave every third or fourth tree, according to the distance which they were planted, and spread down the branches of those which are left horizontally, I mean, all such as are capable of being so brought down; but those, which are too stubborn for this, should be cut off near the stem, where there will be new shoots enough produced to furnish the wall or espalier; and, if the sort of fruit is not the same as desired, the young branches may be budded the same summer, or grafted the following spring with any other sort of Pear, and hereby many years may be saved; for one of these old trees will spread to a much greater length, and produce more fruit, when thus managed, in three years, than a new tree will in ten or twelve, especially if the ground is mended. This is a method which I have practised with great success, where I have been employed to amend the blunders of these great gardeners, as they are stiled, and hereby the walls and espaliers have been well furnished in a few years.

But the next thing to be done, after being furnished with proper trees, is the preparing of the ground to receive them; in the doing of which, there should be great regard had to the nature of the soil where the trees are to grow; for if it is a strong stiff land, and subject to wet in the winter, the borders should be raised as much above the level of the ground as you conveniently can. And if under the good soil there is a sufficient quantity of lime rubbish, or stones laid to prevent the roots of the trees from running downward, it will be of great service to them. The borders for these should not be less than eight feet bread, but, if they are twelve, it will be still better. And as these borders may be planted with such sorts of esculent plants, as do not grow large, or whose roots do not run deep, or mat together on the surface, these will do no harm to the Pear-trees, for these are not so nice in their culture as Peach and Nectarine-trees; so the turning of the ground, and mending it for these crops, will rather improve than injure the trees, provided the plants do not shade the trees, or are not suffered to stand too long upon the borders. But all the Cabbage kind, as also Beans, should be excluded from these borders, because they root deep in the ground, and draw much nourishment from the trees.

But if the soil is shallow, and the bottom is either gravel or chalk, there must be a sufficient depth of good earth laid upon the borders, so as to make them two feet and a half deep; for, if the ground is not of this depth, the trees will not thrive well. And, in doing of this, I must caution every person not to dig out the gravel in a trench (as is by some practised,) and fill this trench with good earth, for by so doing, when the roots of the trees are extended to the width of the trench, they will meet with the gravel, which will stop them so that they will be confined as if they were in tubs of earth, whereby the trees will be soon spoiled; therefore, when the gravel or chalk is removed, it should be entirely taken away over the whole garden, otherwise it will be better to raise the whole border above it.

If the garden is to be new-made from a field, then all the good earth on the surface should be carefully preserved; and, if the good ground is taken out where the walks are designed to be made, and laid upon the borders, or in the quarters, it will add to the depth of the soil, and save expence in bringing in of new earth. If the ground can be prepared one year before it is planted, the trees will thrive the better, for by laying the ground in ridges, and turning it over two or three times, it will loosen the soil, and render it much better for planting; but in trenching or ploughing of the ground, there should be great care taken not to go deeper than the ground is good, otherwise all the good soil will be buried below the roots, and the bad ground will be turned on the top, which is what I have known done at a great expence, by persons who have been at the top of their profession, and have thereby entirely ruined the gardens.

Where there is a necessity of bringing in any fresh earth for the borders, it will be proper to do it as soon as possible, and to mix this with the surface of the earth of the borders, that it may be turned over two or three times, that the parts may be well mixed and incorporated before the trees are planted; and, if some very rotten dung is added to this, it will greatly improve it. In chusing of the earth which is to be brought into the garden, there should be this care, viz. That if the natural soil of the garden is light and dry, then the new earth should be loamy and stiff; but where the natural soil is strong or loamy, then the new earth should be light and sandy, which will loosen the parts of the natural soil, and greatly mend it.

There are ſome perſons who recommend the laying the whole depth of the borders with what they call virgin-earth, that is, ſuch as is taken from a paſture where the land has not been ploughed; but if this is not brought into the garden at leaſt one year before the trees are planted, that by turning it over often it may be ſweetened, it will not be ſo good as that which is taken from a kitchen-garden, where the land is good, and has been well wrought, for by often turning and breaking of the ſoil, it will be the better prepared to receive the trees.

Others recommend the mixing a great quantity of rotten dung with the earth of the borders; but this is not ſo proper, for, by making of the ground too rich, it will only encourage the luxuriant growth of the trees; therefore it is always better to mend the borders from time to time as they may require, and not to add ſo much dung in the firſt making them.

Another care is required in the making of the borders on wet ground, which is to contrive ſome covered drains to convey off the water in winter, otherwiſe, by this being detained about the roots of the trees, it will greatly prejudice them; and in the building of the walls round a kitchen-garden, where the ground is inclinable to be wet, there ſhould be ſome arches turned in the foundations of thoſe walls, which are in the loweſt part of the garden, to let off the wet.

The manner of preparing theſe trees for planting, is the ſame as hath been directed for other fruit-trees, viz. to cut off all the ſmall fibres from the roots, and to ſhorten ſome of the longeſt roots, and cut off all the bruiſed ones, or ſuch as ſhoot downright: this being done, you ſhould plant them in the places intended at the before-mentioned diſtance. The beſt time to plant theſe trees (if upon a middling or dry ſoil) is in October or November, leaving their heads on till ſpring, which ſhould be faſtened to the walls or ſtakes, to prevent the wind from diſturbing their roots; and in the beginning of March their heads ſhould be cut off in the manner already directed for Peaches and other fruit-trees, obſerving alſo to lay ſome mulch upon the ſurface of the ground about their roots when they are planted, as hath been ſeveral times already directed for other trees; but in wet ground the trees may be planted in February, or the beginning of March, at any time before the buds are much ſwelled, but theſe may be cut down when they are planted.

The firſt ſummer after planting, the branches ſhould be trained to a wall or eſpalier (againſt which they are planted) in a horizontal poſition, as they are produced, without ſhortening of them, and the Michaelmas following ſome of theſe ſhoots ſhould be ſhortened down to five or ſix eyes, in order to obtain a ſufficient quantity of branches, to furniſh the lower part of the wall or eſpalier; but the ſhoots ought not to be ſhortened, unleſs where there is a want of branches to fill a vacancy, therefore the leſs the knife is uſed to theſe trees, the better they will ſucceed; for, whenever the ſhoots are ſtopped, it occaſions the buds immediately below the cut to ſend forth two or more ſhoots, whereby there will be a confuſion of branches, and rarely any fruit is produced with this management.

The diſtance which the branches of Pears ſhould be trained, muſt be proportioned to the ſize of their fruit. Such ſorts whoſe fruit are ſmall, may be allowed five or ſix inches, but the larger ſorts muſt not be leſs than ſeven or eight inches aſunder. If this be duly obſerved, and the branches carefully trained horizontally as they are produced, there will be no occaſion for much cutting theſe trees, which, inſtead of checking their growth, does, on the contrary, cauſe them to ſhoot the ſtronger.

It is very ſurpriſing to read the tedious methods which moſt of the writers on fruit-trees have directed for pruning of theſe trees, for, by their prolix and perplexed method, one would imagine they had endeavoured to render themſelves as unintelligible as poſſible; and this, I am ſure, may be affirmed, that it is next to impoſſible for a learner ever to arrive at any tolerable ſkill in pruning by the tedious and perplexed directions which are publiſhed by Monſ. Quintiny, and thoſe who have copied from him; for, as theſe have all ſet out wrong in the beginning, by allowing their trees leſs than a third of the diſtance which they ſhould be planted, they have preſcribed rules to keep them within that compaſs, which are the moſt abſurd, and contrary to all reaſon.

I ſhall therefore only lay down a few neceſſary directions for the pruning and managing of theſe trees, which ſhall be done in as few words as poſſible, that a learner may the more eaſily underſtand it, and which (together with proper obſervations) will be ſufficient to any perſon in the right management of them.

Pear-trees generally produce their bloſſom buds firſt at the extremity of the laſt year's ſhoots, ſo that, if theſe are ſhortened, the bloſſoms are cut off; but this is not all the damage, for (as I before ſaid) this occaſions the buds immediately below the cut to put forth two or more ſhoots, whereby the number of branches will be increaſed, and the tree crouded too much with wood; beſides, thoſe buds, which by this management produce ſhoots, would have only produced curſons and ſpurs, upon which the bloſſom-buds are produced, if the leading branch had not been ſhortened; therefore theſe ſhould never be ſtopped, unleſs to furniſh wood to fill a vacancy.

It is not neceſſary to provide a new ſupply of wood in Pear-trees, as muſt be done for Peaches, Nectarines, &c. which only produce their fruit upon young wood, for Pears produce their fruit upon curſons or ſpurs, which continue fruitful many years; ſo that, where theſe trees have been ſkilfully managed, I have ſeen branches which have been trained horizontally upwards of twenty feet from the trunk of the tree, which have been fruitful their whole length. And if we do but carefully obſerve the branches of a healthy ſtandard-tree, which has been permitted to grow without pruning, we ſhall find many of theſe ſpurs that are ten or twelve years old or more, which are very full of bloſſom-buds, and produce a good number of fruit annually.

During the ſummer ſeaſon, theſe trees ſhould be often looked over to train the ſhoots, as they are produced, regularly to the wall or eſpalier, and to diſplace fore-right and luxuriant branches as they ſhoot out, whereby the fruit will be equally expoſed to the air and ſun, which will render the trees more beautiful and the fruit better taſted than when they are ſhaded by the branches; and by thus managing the trees in ſummer, they will always appear beautiful, and in winter they will want but little pruning.

Where Pear-trees are thus regularly trained without ſtopping of their ſhoots, and have full room for their branches to extend on each ſide, there will never be any occaſion for diſbarking of the branches, or cutting off the roots (as hath been directed by ſeveral writers on gardening;) which methods, however they may anſwer the intention for the preſent, yet will certainly greatly injure the trees, as muſt all violent amputations, which ſhould be ever avoided, as much as poſſible, on fruit-trees; and this, I am ſure, can never be wanted, where trees have been rightly planted, and regularly trained, while young.

The ſeaſon for pruning of theſe trees is any time after the fruits are gathered, until the beginning of March, but the ſooner it is done after the fruit is gathered, the better, for reaſons already given for pruning of Peach-trees; though indeed the deferring of theſe until ſpring, where there are large quantities of trees to prune, is not so injurious to them, as to some tender fruits; but if the branches are regularly trained in the summer, and the luxuriant shoots rubbed off, there will be little left to do to them in winter.

All the sorts of summer Pears will ripen very well either on standards, dwarfs, or espaliers, as will all autumn Pears upon dwarfs or espaliers; but, where a person is very curious in his fruit, I would always advise the planting them against espaliers, in which method they take up less room in a garden, and if they are well managed, appear very beautiful, tho fruit larger and better tasted than those produced on dwarfs, as hath been already observed; but some of the winter Pears must be planted against east, south-east, or south-west walls, otherwise they will not ripen well in England in bad seasons.

But although this may be the case with some of the late winter Pears in very bad seasons, yet, in general, most sorts of them will ripen extremely well in all warm situations, when they are planted in espalier, and the fruit will be better flavoured than that which grows against walls, and will keep much longer good; for as the heat against walls which are exposed to the sun, will be very great at some times, and at others there will be little warmth, all fruit, which grow near them, will be hastened unequally, and therefore is never so well flavoured as the same sorts are which ripen well in the open air; and all the fruit which is ripened thus unequally, will decay much sooner than those which ripen gradually in the open air; therefore those winter Pears which grow in espalier, may be kept six weeks longer than those which grow against walls, which is a very desirable thing; for to have plenty of these fruit, at a season when it is very rare to find any other fruit to supply the table but Apples, is what all lovers of fruit must be greatly pleased to enjoy, which is what may be effected by planting many of the late sorts in espalier, where, although the fruit will not be so well coloured as those from the walls, yet they will be found exceeding good. When the Besi de Chaumontelle came first to England, the trees were planted in espalier, and some of them not on a very good soil, or in a warm situation, and yet from these trees I have eaten this Pear in great perfection in April, and sometimes it has kept till May; whereas, all those which have been since planted against walls, ripen their fruit by the beginning of November, and are generally gone by the middle of December, nor are the latter so well tasted as those off the espaliers.

The Virgoleuse and St. Germain, as also the Colmar, are esteemed the most difficult sorts to ripen their fruit, yet these I have eaten in great perfection from espaliers, and often from standard-trees, where they grew upon a warm soil, but the fruit was much smaller on the standard-trees than those of the same sorts which grew against walls or espaliers, but they were full as well flavoured; and some of these sorts I have eaten good in April, which is two months later than they usually keep; but yet I would not advise the planting of these late Pears on standard-trees, because they should hang very late on the trees in autumn, at which season the winds are generally very high; and these standard-trees being much exposed, the fruit is often blown off the trees before they are ripe, and those of them, which may hang on the trees, are frequently bruised by being forced against the branches by the winds, so that they seldom keep well. What I mentioned this for, is to prove that these Pears will ripen very well without the assistance of a wall, so that, if they are planted in espaliers, where the trees are kept low, the fruit will not be so much exposed to the strong winds in autumn as those on the standards, therefore can be in no danger of the fruit coming to perfection; and, as the trees in espaliers will be constantly pruned, and managed in the same manner as those against walls, the fruit will be as large on those trees; therefore, where a person has a warm situation, and a kindly soil, I would not advise the being at an expence to build walls on purpose for Pears, but to plant them against espaliers; and, where there is any one who is very curious to have plenty of these fruit, and will be at the expence to procure them, I should advise the having a sufficient quantity of Reed-mats made to fix up against the back of the espalier in the spring, when the trees are in blossom, which will screen them from cold winds, and preserve the tender fruit until they are past danger, when the Reeds may be taken down, and put under a shed to preserve them from the weather; and if the autumn should prove bad, these Reeds may be fixed up again, which will forward the ripening of the fruit, and also prevent the winds from blowing down and bruising of it. These Reeds may be purchased for one shilling per yard, running measure, at six feet and a half high, and, if they are carefully laid up, and kept from the weather, these Reeds will last seven or eight years, so that the expence will not be very great; and when the advantages which these are of to the fruit are considered, I believe no person will object to the use of them.

But after the fruit is set and growing, there will be farther care necessary in order to have the fruit good, for it is not enough to have preserved a good crop of fruit on the trees, and then leave them entirely to nature during the season of their growth, but there will require some skill and attendance on the trees to help nature, or supply the deficiency of the seasons; for beside the pruning and training of the trees in the manner before directed, there will also be wanting some management of their roots according to the nature of the soil, and the difference of seasons. In all strong land, where the ground is apt to bind very hard in dry weather, the surface of the borders should be now and then forked over to loosen the earth, which will admit the showers and large dews to penetrate and moisten the ground, and be of great service to the trees and fruit, and also prevent the growth of weeds. And if the soil is light and dry, and the season should prove hot and dry, there should be large hollows made round the stems of the trees to hold water; into each of these there should be poured eight or nine pots of water, which should be repeated once in a week or ten days, during the months of June and July, if the season should continue dry. There should also be some mulch laid over the surface of these hollows, to prevent the sun and air from drying the ground. Where this is practised, the fruit will be kept constantly growing, and prove large and plump; whereas, if this is omitted, the fruit will often be small, grow crooked, crack, and fall off from the trees. For if the fruit is once stinted in their growth, and rain should fall plentifully after, it will occasion a great quantity of the fruit to fall off the trees; and those which remain to ripen, will not keep so long as those which never receive any check in their growth; it is from this cause, that some years the fruit in general decays before the usual time. For after it has been for some time stinted in its growth, and then the season proves favourable, whereby it receives a sudden growth, it becomes so replete with juice, as to distend the vessels too suddenly, so that they will not be firm, which occasions their decay; therefore it is always best to keep the fruit constantly in a growing state, whereby it will acquire a proper size, and be rendered better flavoured.

There will also be required some dressing to the ground near the fruit-trees, but this should be laid on in autumn, after the trees are pruned. This dressing should be different according to the nature of the soil; if the land is warm and

4 L dry,

dry, then the dreſſing ſhould be of very rotten dung, mixed with loam; and if this is mixed ſix or eight months before it is laid upon the borders, and three or four times turned over, it will be the better; as will alſo the mixture, if it is made with neats or hogs-dung, both which are colder than horſe-dung, ſo more proper for hot land. But in cold ſtiff land, rotten horſe-dung, mixed with light ſandy earth, or ſea-coal aſhes, will be the moſt proper, as this will looſen the ground, and add a warmth to it.

Theſe dreſſings ſhould be repeated every other year, otherwiſe the trees will not thrive ſo well, nor will the fruit be ſo good. For, notwithſtanding what many perſons have advanced to the contrary, yet experience is againſt them, for the fineſt fruit in England, both as to ſize and flavour, is produced on land which is the moſt dunged and worked. Therefore, I would adviſe the trenching of the ground about the fruit-trees very well every winter, for I am ſure it will be found to anſwer their expectations, who will practiſe this method. And where the ground in the quarters is well dreſſed and trenched, the fruit-trees will partake of the benefit, for as the trees advance in their growth, ſo their roots are extended to a great diſtance from their ſtems; and it is chiefly from the diſtant roots that the trees are ſupplied with their nouriſhment, therefore the dreſſing of the borders only, will not be ſufficient for fruit-trees which are old.

In the gathering of Pears, great regard ſhould be had to the bud which is formed at the bottom of the foot-ſtalk, for the next year's bloſſoms, which, by forcing of the Pear before it be mature, is many times ſpoiled; for while the fruit is growing, there is always a bud formed by the ſide of the foot-ſtalk upon the ſame ſpur, for the next year's fruit; ſo that when the Pears are ripe, if they are gently turned upward, the foot-ſtalk will readily part from the ſpur, without injuring of the bud.

The ſeaſon for gathering all ſummer Pears is juſt as they ripen, for none of theſe will remain good above a day or two after they are taken from the tree; nor will many of the autumn Pears keep good above ten days or a fortnight, after they are gathered. But the winter fruits ſhould hang as long upon the trees as the ſeaſon will permit, for they muſt not receive the froſt, which will cauſe them to rot, and render their juices flat and ill taſted; but if the weather continue mild until the end of October, it will then be a good ſeaſon for gathering them in, which muſt always be done in dry weather, and when the trees are perfectly dry.

In the doing of this you ought carefully to avoid bruiſing them, therefore you ſhould have a broad flat baſket to lay them in as they are gathered; and when they are carried into the ſtore-room, they ſhould be taken out ſingly, and each ſort laid up in a cloſe heap on a dry place, in order to ſweat, where they may remain for ten days or a fortnight, during which time the windows ſhould be open to admit the air, in order to carry off all the moiſture which is perſpired from the fruit; after this, the Pears ſhould be taken ſingly, and wiped dry with a woollen cloth, and then packed up in cloſe baſkets, obſerving to put ſome Wheat-ſtraw in the bottoms and round the ſides of the baſkets, to prevent their bruiſing againſt the baſkets. And if ſome thick ſoft paper is laid double or treble all round the baſket, between the ſtraw and the Pears, this will prevent the Pears from imbibing the muſty taſte which is communicated to them by the ſtraw, when they are contiguous; which taſte often penetrates through the ſkin ſo ſtrongly, that when the fruit is pared the taſte will remain. You ſhould alſo obſerve to put but one ſort of fruit into a baſket, leſt by their different fermentations they ſhould rot each other; but if you have enough of one ſort to fill a baſket which holds two or three buſhels, it will be ſtill better. After you have filled the baſkets, you muſt cover them over with Wheat-ſtraw very cloſe, firſt laying a covering of paper two or three times double over the fruit, and faſten them down; then place theſe baſkets in a cloſe room, where they may be kept dry and from froſt; but the leſs air is let into the room, the better the fruit will keep. It will be very neceſſary to fix a label to each baſket, denoting the ſort of fruit therein contained, which will ſave the trouble of opening them, whenever you want to know the ſorts of fruit; beſides, they ought not to be opened before their ſeaſon to be eaten, for the oftener they are opened and expoſed to the air, the worſe they will keep. I don't doubt but this will be objected to by many, who imagine fruit cannot be laid too thin; for which reaſon they make ſhelves to diſpoſe them ſingly upon, and are very fond of admitting freſh air, whenever the weather is mild, ſuppoſing it very neceſſary to preſerve the fruit; but the contrary of this is found true, by thoſe perſons who have large ſtocks of fruit laid up in their ſtorehouſes in London, which remain cloſely ſhut up for ſeveral months, in the manner before related; and when theſe are opened, the fruit is always found plumper and ſounder than any of thoſe fruits which were preſerved ſingly upon ſhelves, whoſe ſkins are always ſhrivelled and dry. For (as Mr. Boyle obſerves) the air is the cauſe of putrefaction; and, in order to prove this, that honourable gentleman put fruits of ſeveral kinds into glaſſes where the air was exhauſted, in which places they remained ſound for ſeveral months, but, upon being expoſed to the air, rotted in a very ſhort time, which plainly ſhews the abſurdity of the common method now uſed to preſerve fruit.

Q.

QUAMOCLIT. See IPOMOEA.

QUERCUS. Tourn. Inst. R. H. 582. tab. 349. The Oak-tree.

The CHARACTERS are,

It hath male and female flowers on the same tree; the male flowers are disposed in a loose katkin. The female flowers which sit close to the buds, have a hemispherical thick empalement of one leaf, which is rough and entire, almost hiding the flower, which has no petal, but a small oval germen supporting a single five-pointed style, crowned by single permanent stigmas. The germen afterward becomes an oval nut (or acorn,) with a thick cover, having one cell, whose base is fixed into the empalement or cup.

The SPECIES are,

1. QUERCUS (*Robur*) foliis deciduis oblongis, supernè latioribus sinubus acutioribus, angulis obtusis petiolatis glandibus sessilibus. *Oak with oblong deciduous leaves, broader toward the top, having acute indentures, with obtuse angles, which have foot-stalks, and acorns sitting close to the branches; or common Oak.*

2. QUERCUS (*Fœmina*) foliis deciduis oblongis obtusis, pinnato-sinuatis petiolis brevissimis, pedunculis glandorum longissimis. *Oak with oblong, obtuse, deciduous leaves, which are winged-sinuated, very short foot-stalks, with a fruit growing upon long foot-stalks.*

3. QUERCUS (*Sempervirens*) foliis oblongis sinuatis obtusis perennantibus, pedunculis glandorum longissimis. *Oak with oblong, obtuse, indented leaves, which are evergreen, having very long foot-stalks to the acorns; or broad-leaved evergreen Oak.*

4. QUERCUS (*Humilis*) foliis oblongis obtusè-sinuatis, setaceo mucronatis sessilibus, glandibus majoribus. *Oak with oblong obtusely-indented leaves, which have bristly points, and sit close to the stalks, with larger acorns.*

5. QUERCUS (*Cerris*) foliis oblongis lyrato-pinnatifidis laciniis transverse acutis subtus tomentosis. *Oak with oblong wing-indented leaves, which are downy on their under side, and acorns having woolly cups sitting close to the branches.*

6. QUERCUS (*Esculus*) foliis pinnato sinuatis lævibus, fructibus sessilibus. *Dwarf Oak with oblong obtusely-indented leaves, and fruit growing in clusters, sitting close to the branches; or Dwarf Oak.*

7. QUERCUS (*Ægilops*) foliis ovato-oblongis glabris serrato-dentatis, laciniis transversis acutis. Lin. Sp. Pl. 997. *Oak with oblong leaves, which are lyre-shaped, wing-pointed, and have transverse acute jags, which are somewhat downy on their under side.*

8. QUERCUS (*Italica*) foliis obtusè-sinuatis, setaceo mucronatis. Prod. Leyd. 80. *Oak with smooth wing-indented leaves, and fruit sitting close to the branches; commonly called the cut-leaved Italian Oak.*

9. QUERCUS (*Prinus*) foliis obovatis utrinque acuminatis sinuato-serratis, denticulatis rotundatis uniformibus. Hort. Cliff. 448. *Oak with oblong oval leaves which are pointed on both sides, and have sawed sinuses with uniform roundish indentures; the American Chestnut-leaved Oak.*

10. QUERCUS (*Nigra*) foliis cuneiformibus obsoletè trilobis. Flor. Virg. 117. *Oak with wedge-shaped leaves, having three worn-out lobes; the Black Oak.*

11. QUERCUS (*Rubra*) foliorum sinubus obtusis, angulis acutis setâ-terminatis, intermediis vix tridentatis margine integerrimo. Hort. Cliff. 448. *Oak with obtuse sinuses to the leaves, and acute angles terminated by bristles, and an entire border; the Scarlet Oak of Virginia.*

12. QUERCUS (*Alba*) foliis obliquè pinnatifidis, sinubus angulisque obtusis. Lin. Sp. Plant. 996. *Oak with oblique many-pointed leaves, having obtuse sinuses and angles; the White Oak of Virginia.*

13. QUERCUS (*Phellos*) foliis lineari-lanceolatis integerrimis glabris. *Oak with linear, spear-shaped, entire, smooth leaves; the Willow-leaved Oak.*

14. QUERCUS (*Ilex*) foliis oblongo-ovatis subtus tomentosis integerrimis. Prod. Leyd. 81. *Oak with oblong, oval, entire leaves, which are downy on their under side; the narrow-leaved evergreen Oak.*

15. QUERCUS (*Gramuntia*) foliis oblongo-ovatis sinuato-spinosis subtus tomentosis, glandibus pedunculatis. Sauv. Monsp. 96. *Evergreen Oak with oblong, oval, prickly, indented leaves, which are woolly on their under side, and bears acorns with foot-stalks; or the Holly-leaved evergreen Oak.*

16. QUERCUS (*Coccifera*) foliis ovatis indivisis spinoso-dentatis glabris. Prod. Leyd. 80. *Oak with oval, undivided, smooth leaves, which are prickly and indented; or the Kermes Oak.*

17. QUERCUS (*Virginiana*) foliis lanceolato-ovatis integerrimis-petiolatis sempervirentibus. *Oak with spear-shaped, oval, entire-leaves, which are evergreen, and have foot-stalks; commonly called Live Oak in America.*

18. QUERCUS (*Suber*) foliis ovato-oblongis indivisis serratis subtus tomentosis, cortice ramoso fungoso. Hort. Cliff. 448. *Oak with oval, oblong, undivided leaves, which are sawed and woolly on their under side, and have a fungous cleft bark; or Cork-tree.*

The first sort here mentioned, is the most common Oak of this country, which is so well known as to need no description; the leaves of this have pretty long foot-stalks, and the acorns have none, but sit close to the branches.

The second sort is not so common here as the first, but in the wilds of Kent and Sussex I have seen many large trees of this kind. The leaves of this are not so deeply sinuated as those of the first, nor are they so irregular, but the indentures are opposite like the lobes of winged leaves; these have scarce any foot-stalks, but sit close to the branches; the acorns stand upon very long foot-stalks, in which they differ from the common sort. The timber of this sort is accounted better than that of the first, and the trees, when growing, have a better appearance. These have been generally supposed to be seminal varieties, which have accidentally come from acorns of the same tree; I was long of this opinion myself, but having lately seen some young trees with acorns upon them, which were raised from acorns of the

 second

ſecond ſort, and finding they retain their difference, I am inclined to believe they are different.

The third ſort grows upon the Apennines, and alſo in Swabia and Portugal. The leaves of this are broader, and not ſo deeply ſinuated as thoſe of the common Oak; they are of a lighter green on their upper ſide, but pale on their under; they have very ſhort foot-ſtalks, their points are obtuſe, and the acorns have very long foot-ſtalks, which frequently ſuſtain three or four growing in a cluſter.

The fourth ſort grows common in ſome parts of France, where it riſes to be a tall ſtately tree. The leaves are oblong and obtuſely ſinuated, each ſinus being terminated by a briſtly point; the acorns are larger than thoſe of the common Oak.

The fifth ſort grows in the ſouth of France, and in Italy. The leaves of this are ſhorter and broader than thoſe of the common Oak, and are regularly indented on their ſides, the indentures being oppoſite, but not deep; they are of a light green on their upper ſide, and are covered with a ſoft down on their under, ſtanding upon ſhort foot-ſtalks; the acorns grow in cluſters, ſitting cloſe to the branches; their cups are covered with a white down.

The ſixth ſort grows in the ſouth of France, and in Italy. This is a low buſhy Oak, which riſes but ſix or ſeven feet high, ſending out many ſlender branches, garniſhed with oblong leaves, which are obtuſely indented, ſtanding upon ſlender foot-ſtalks; the acorns are ſmall, and grow in cluſters, and the galls grow three or four together.

The ſeventh ſort grows in Burgundy. The leaves of this are oblong and pointed, and are frequently indented in the middle like a lyre; they are jagged and acute-pointed, a little hoary on their under ſide, ſtanding upon ſlender foot-ſtalks. The acorns are ſmall, and have rough prickly cups.

The eighth ſort grows naturally in Virginia, and in other parts of North America. This is a large tree in the countries where it naturally grows; the bark is ſmooth, of a grayiſh colour, but that of the younger branches is darker; the leaves are long and broad; they are obtuſely ſinuated, each ſinus ending with a briſtly point, of a bright green, ſtanding upon ſhort foot-ſtalks. The leaves continue their verdure very late in autumn, ſo that unleſs hard froſt comes on early, they do not fall till near Chriſtmas, nor do they change their colour long before. The acorns of this ſort are a little longer, but not ſo thick as thoſe of the common Oak.

The ninth ſort grows naturally in North America. Of this there ſeems to be two kinds, one of which grows to a much larger ſize than the other, though this may be occaſioned by the ſoil in which they grow; for the largeſt ſort grows in the rich low lands, where it becomes the largeſt tree of any of the Oaks in thoſe countries. The wood is not of a fine grain, but is very ſerviceable; the bark is gray and ſcaly; the leaves are long and broad, indented on the edges, and have many tranſverſe veins running from the midrib to the borders; they are of a bright green, and ſo nearly reſemble thoſe of the Cheſtnut-tree, as ſcarcely to be diſtinguiſhed from it. The acorns of this ſort are very large, and their cups are ſhort. The leaves of the other variety are not ſo large, nor ſo ſtrongly veined, and the acorns are ſmall and a little longer, which may ariſe from the ſoil.

The tenth ſort grows naturally on poor land in moſt parts of North America; this never grows to a large ſize, and the wood is of no value. The bark is of a dark brown colour; the leaves are very broad at the top, where they have two waved indentures, which divide them almoſt into three lobes; they diminiſh gradually to their baſe, where they are narrow; they are ſmooth, of a lucid green, and have ſhort foot-ſtalks. The acorns are ſmaller than thoſe of the common Oak, and have ſhort cups.

The eleventh ſort grows naturally in North America, where it is called the Red Oak, from the leaves changing to a deep red or purple before they fall off. There has been ſuppoſed two ſorts of this Oak, but I believe they are only ſeminal varieties; for from the acorns of the ſame tree I have ſeen plants raiſed, whoſe leaves have been of very different ſhapes and ſizes, and have varied greatly in their colour in autumn, ſome changing to a bright red or ſcarlet, and others to a deep purple colour. The wood is ſoft, ſpongy, and not durable. The acorns of this ſort alſo vary in ſize and ſhape; ſome of them are ſmaller, and others ſhorter and larger than thoſe of the common Oak.

The twelfth ſort grows naturally in North America, where the wood is eſteemed preferable to any of their other ſorts for building, being much more durable than any of them. The bark of this tree is grayiſh; the leaves are long and broad, and are regularly indented almoſt to the midrib; the indentures are obtuſe; they are of a light green, and have ſhort foot-ſtalks. The acorns of this greatly reſemble thoſe of the common Oak.

The thirteenth ſort grows naturally in North America, where they diſtinguiſh two ſorts; one of them is called the Highland Willow Oak, which grows upon poor dry land; the leaves are of a pale green and entire, ſhaped like thoſe of the Willow-tree. The acorns are very ſmall, but have pretty large cups.

The other grows in low moiſt land, and riſes to a much greater height; the leaves are longer and narrower, and the acorns are of the ſame ſize and ſhape, ſo that I ſuſpect their difference is owing to the ſoil in which they grow

The fourteenth ſort is generally known by the title of Ilex or Evergreen Oak. Of this there are ſeveral varieties, which differ greatly in the ſize and ſhape of their leaves; but theſe will all ariſe from acorns of the ſame tree, as I have ſeveral times experienced; nay, the lower and upper branches of the ſame tree are frequently garniſhed with leaves, very different in ſize and ſhape from each other. The leaves are entire, of a lucid green on their upper ſide, but whitiſh and downy on their under, ſtanding upon pretty long foot-ſtalks; theſe remain green all the year, and do not fall till they are thruſt off by young leaves in the ſpring. The acorns are ſmaller than thoſe of the common Oak, but of the ſame ſhape.

The fifteenth ſort is ſuppoſed to be a different ſort. The leaves are ſhorter and broader than the other, approaching in ſhape to thoſe of the Holly-tree, and are ſet with prickles on their edges.

The ſixteenth ſort is the Oak, from which the Kermes, or what is called Scarlet Grain, is collected, which is an inſect that harbours on this tree. It grows naturally in Provence and Languedoc, where it is known by the title d'Avaux. This is of ſmall growth, ſeldom riſing above twelve or fourteen feet high, ſending out branches on every ſide, ſo as to form a buſhy ſhrub; the leaves are oval and undivided; they are ſmooth on their ſurface, but indented on their edges, which are armed with prickles like thoſe of the Holly-tree. The acorns are ſmaller than thoſe of the common Oak.

The ſeventeenth ſort grows naturally in Carolina and Virginia, where it riſes to the height of forty feet. The grain of the wood is hard, tough, and coarſe; the bark is grayiſh; the leaves are entire, ovally ſpear-ſhaped, and of a dark green, of a thick conſiſtence, and continue green all the year. The acorns are ſmall, oblong, and have ſhort cups; they are very ſweet, ſo are eaten by the Indians, who lay them up in ſtore for the winter; they alſo draw a very ſweet oil from them, little inferior to that of sweet Almonds. This is called the Live Oak in America.

The eighteenth sort is the tree whose bark is the Cork. Of this there are two or three varieties, viz. one with a broad, another with a narrow leaf, which are evergreen; and there is one or two which cast their leaves in autumn, but the broad-leaved evergreen is the most common; the other may probably be only a variety, arising by accident. The leaves of this are oblong, oval, undivided, sawed on their edges, and have a little down on their under sides; the foot-stalks are very short; the leaves continue green through the winter till the middle of May, when they generally fall off just before the new leaves come out, so that the trees are very often almost bare for a short time. The acorns are very like those of the common Oak.

The exterior bark of this tree is the Cork; this is taken off from the trees every eight or ten years, but there is an interior bark which nourishes the trees, so that the stripping off the outer is so far from injuring them, that it is necessary to continue the trees, for those whose bark are not taken off, seldom last longer than fifty or sixty years in health; whereas, the trees which are barked every eight or ten years, will live a hundred and fifty years or more. The bark of the young trees is porous and good for little; however it is necessary to take it off when the trees are twelve or fifteen years old, without which the bark will not be good, and after eight or ten years the bark will be fit to take off again; but this second peeling is of little use, but the third peeling the bark will be in perfection, and will continue so many years, for the best Cork is taken from the old trees. The time of year for stripping off this bark is in July, when the second sap flows plentifully. This is performed with an instrument, like that used for disbarking Oaks.

All the sorts of Oaks are propagated by sowing their acorns; the sooner they are put into the ground after they are ripe, the better they will succeed; for they are very apt to sprout where they are spread thin, and if they are laid in heaps, they ferment and rot in a little time; therefore the best season for sowing them is in October, by which time they will be fallen from the trees.

Where Oak-trees are cultivated with a view to profit, the acorns should be sown where the trees are designed to grow, for those which are transplanted will never arrive to the size of those which stand where they are sown, nor will they last near so long sound. For in some places where these trees have been transplanted with the greatest care, and have grown very fast for several years after, yet are now decaying, when those which remain in the place where they came up from the acorns, are still very thriving, and have not the least sign of decay. Therefore, whoever designs to cultivate these trees for timber, should never think of transplanting them, but sow the acorns on the same ground where they are to grow; for the timber of all those trees which are transplanted, is not near so valuable as that of the trees from acorns. I shall give some plain directions for the sowing of acorns, and managing of the young trees during their minority, until they are out of danger, and require no farther care.

The first thing to be done is, that of fencing the ground very well, to keep out cattle, hares, and rabbets; for if either of these can get into the ground, they will soon destroy all the young trees. Indeed, they will in a few years grow to be out of danger from the hares and rabbets, but it will be many years before they will be past injury from cattle, if they are permitted to get into the plantation therefore durable fences should be put round the ground if in the beginning a pale fence is made about the land which may be close at the bottom and open above, and within the pale a quick hedge planted; this will become a good fence by the time the pale decays, against all sorts of cattle, and then the trees will have got above the reach of hares and rabbets, so that they cannot injure them, for the bark of the trees will be too hard for them to gnaw.

After the ground is well fenced, it should be prepared, by ploughing of it three or four times, and after each ploughing to harrow it well, to break the clods, and cleanse the ground from couch, and the roots of all bad weeds. Indeed if the ground is green sward, it will be better to have one crop of Beans, Pease, or Turneps, off the ground, before the acorns are sown, provided these crops are well hoed, to stir the surface and destroy the weeds; for if this is observed, the crop will mend and improve the land for sowing; but in this case the ground should be ploughed as soon as possible, when the crop is taken off, to prepare it for the acorns, which should be sown as soon as may be after they are ripe; for although they may be preserved in sand for some time, yet they will be apt to sprout; and if so, the shoots are in danger of being broken and spoiled; therefore I should advise the sowing early, which is certainly the best method.

In making choice of the acorns, all those should be preferred, which are taken from the largest and most thriving trees; and those of pollard-trees should always be rejected, though the latter are generally the most productive of acorns, but those of the large trees commonly produce the strongest and most thriving plants.

The season for sowing of the acorns being come, and the ground having been ploughed and harrowed smooth, the next work is to sow the acorns, which must be done by drawing of drills across the ground, at about four feet asunder, and two inches deep, into which the acorns should be scattered at two inches distance. These drills may be drawn either with a drill-plough, or by hand with a hoe; but the former is the most expeditious method, therefore in large plantations should be preferred. In the drawing of the drills, if the land has any slope to one side, these should be made the same way as the ground slopes, that there may be no stoppage of the wet by the rows of plants crossing the hanging of the land. This should be particularly observed in all wet ground, or where the wet is subject to lie in winter, but in dry land it is not of much consequence. When the acorns are sown, the drills should be carefully filled in, so as to cover the acorns securely; for if any of them are exposed, they will entice the birds and mice, and if either of these once attack them, they will make great havock with them.

The reason of my directing the drills to be made at this distance, is for the more convenient stirring of the ground between the rows, to keep the young plants clear from weeds; for if this is not carefully done, it cannot be expected that the young plants should make much progress; and yet this is generally neglected by many who pretend to be great planters, and are often at a large expence to plant, but seldom regard them after; so that the young plants have the difficulty to encounter the weeds, which frequently are four or five times the height of the plants, and not only shade and draw them, but also exhaust all the goodness of the ground, and consequently starve the plants. Therefore, whoever hopes to have success in their plantations, should determine to be at the expence of keeping them clean for eight or ten years after sowing, by which time the plants will have obtained strength enough to keep down the weeds; the neglecting of this has occasioned so many young plantations to miscarry, as are frequently to be met with in divers parts of England.

About the middle of April the young plants will appear above ground; but before this, if the ground should produce

duce many young weeds, it will be good husbandry to scuffle the surface over with Dutch hoes, in a dry time, either the latter end of March or the beginning of April, to destroy the weeds, whereby the ground will be kept clean, until all the plants are come up, so as to be plainly discerned, by which time it may be proper to hoe the ground over again; for by doing it early, while the weeds are small, a man will perform more of this work in one day, than he can in three or four when the weeds are grown large; besides, there will be great hazard of cutting off or injuring the young plants, when they are hid by the weeds, and small weeds being cut, are soon dried up by the sun; but large weeds often take fresh root and grow again, especially if rain should fall soon after, and then the weeds will grow the faster by being stirred; therefore it is not only the best method, but also the cheapest husbandry, to begin cleaning early in the spring, and to repeat it as often as the weeds are produced.

The first summer, while the plants are young, it will be the best way to perform these hoeings by hand, but afterward it may be done with the hoe plough; for as the rows are four feet asunder, there will be room enough for this plough to work; and as this will stir and loosen the ground, so it will be of great service to the plants; but there will require a little hand-labour where the plough is used, in order to destroy the weeds, which will come up in the rows between the plants; for these will be out of the reach of the plough, and if they are not destroyed, they will soon overgrow and bear down the young plants.

After the plants have grown two years, it will be proper to draw out some of them where they grow too close; but in the doing of this, great care should be had not to injure the roots of those left, for as the plants which are drawn out, are only fit for plantations designed for pleasure, so these should not be so much regarded in their being removed, as to sacrifice any of those which are designed to remain. In the thinning of these plantations, the plants may at the first time be left about one foot asunder, which will give them room enough to grow two or three years longer, by which time it may be easy to judge which are likely to make the best trees. Therefore these may be then fixed on as standards to remain, though it will be proper to have a greater number at this time marked than can be permitted to grow, because some of them may not answer the expectation; and as it will be improper to thin these trees too much at one time, so the leaving double the number intended at the second thinning will not be amiss. Therefore if they are then left at about four feet distance in the rows, they will have room enough to grow three or four years longer; by which time, if the plants have made good progress, their roots will have spread over the ground, therefore it will be proper to take up every other tree in the rows. But by this I do not mean to be exact in the removing, but to make choice of the best plants to stand, which ever rows they may be in, or if they should not be exactly at the distance here assigned. All that is designed here, is to lay down general rules, which should be as nearly complied with as the plants will permit; therefore every person should be guided by the growth of the trees in the performance of this work.

When the plants have been reduced to the distance of about eight feet, they will not require any more thinning. But in two or three years time, those which are not to remain will be fit to cut down, to make stools for underwood; and those which are to remain, will have made such progress as to become a shelter to each other, for this is what should be principally attended to whenever the trees are thinned; therefore in all such places as are much exposed to the wind, the trees should be thinned with great caution, and by slow degrees, for if the air is let too much at once into the plantation, it will give a sudden check to the trees, and greatly retard their growth; but in sheltered situations there need not be so great caution used as in those places, for the plants will not be in so much danger of suffering.

The distance which I should chuse to allow to those trees which are designed to remain for timber, is from twenty-five to about thirty feet, which will not be too near, where the trees thrive well; in which case their heads will spread, so as to meet in about thirty or thirty-five years; nor will this distance be too great, so as to impede the upright growth of the trees. This distance is intended that the trees should enjoy the whole benefit of the soil; therefore, after one crop of the under-wood, or at the most, two crops are cut, I would advise the stubbing up the stools, that the ground may be entirely clear, for the advantage of the growing timber, which is what should be principally regarded; but in general most people have more regard to the immediate profit of the under-wood than the future good of the timber, and frequently by so doing spoil both; for if the under-wood is left after the trees have spread so far as that their heads meet, it will not be of much worth, and yet, by their stools being left, they will draw away a great share of nourishment from the timber-trees, and retard them in their progress.

The soil in which the Oak makes the greatest progress, is a deep rich loam, in which the trees grow to the largest size; and the timber of those trees which grow upon this land, is generally more pliable than that which grows on a shallow or drier ground, but the wood of the latter is much more compact and hard. Indeed there are few soils in England in which the Oak will not grow, provided there is proper care taken in their cultivation, though this tree will not thrive equally in all soils; but yet it might be cultivated to a national advantage upon many large wastes in several parts of England, as also to the great profit of the estates where these tracts of lands now lie uncultivated, and produce nothing to the owner. And should the present temper of destroying the timber of England continue in practice some years longer, in the same degree which it has for some years past, and as little care is taken to raise a supply, this country, which has been so long esteemed for its naval strength, may be obliged to seek for timber abroad, or be content with such a naval strength as the poor remains of some frugal estates may have left growing; for as to the large forests, from whence the navy has been so long supplied, a few years will put an end to the timber there; and how can it be otherwise, when the persons to whose care these are committed, reap an advantage from the destruction of the timber?

Before I quit this subject, I must beg leave to take notice of another great evil, which is of so much consequence to the public, as to deserve their utmost attention; which is, that of cutting down the Oaks in the spring of the year, at the time when the sap is flowing. This is done for the sake of the bark, which will then easily peel off; and for the sake of this, I think, there is a law, whereby people are obliged to cut down their timber at this season. But by so doing, the timber is not half so durable as that which is fallen in the winter; so that those ships which have been built of this spring-cut timber, have decayed more in seven or eight years, than others which were built with timber cut in winter have done in twenty or thirty. And this our neighbours the French have experienced, and therefore have wisely ordered, that the bark should be taken off the trees standing, at the proper time, but the trees are left till the next, and sometimes until the second winter, before they are cut down; the timber of these are found to be be more durable and better for uſe, than that of any trees which have not been peeled. Therefore, I wiſh we were wiſe enough to copy after them in thoſe things which are for the public good, rather than to imitate them in their follies, which have been too much the faſhion of late years.

The other ſorts of Oak, which are only planted for pleaſure either in parks or gardens, may be raiſed from acorns ſown in beds, and the plants may be trained in a nurſery three or four years; then they ſhould be planted where they are to remain, for although they are not cultivated for their wood, yet the younger they are planted out, the better they will ſucceed, provided they are kept clean from weeds, and ſecured from animals.

QUICK. By the word Quick are generally underſtood all live hedges, of whatever ſort of plants they are compoſed, to diſtinguiſh them from dead hedges; but, in the more ſtrict ſenſe of this word, it is applied to the Hawthorn, or Meſpilus Sylveſtris; under which name, the young plants or ſets are commonly ſold by the nurſery-gardeners who raiſe them for ſale.

In the choice of theſe ſets, thoſe which are raiſed in the nurſery are to be preferred to ſuch as are drawn out of the woods, becauſe the latter have ſeldom good roots, though as they are larger plants than are commonly to be had in the nurſery, many people prefer them on that account; but from long experince I have found, that thoſe hedges which have been planted with young plants from the nurſery, have always made the beſt hedges. Indeed, if perſons would have patience to wait for theſe from ſeed, and to ſow the Haws in the place where the hedge is deſigned, theſe unremoved plants will make a much ſtronger and more durable fence than thoſe which are tranſplanted; but I am aware that moſt people will be for condemning this practice, as being tedious in raiſing; but if the Haws are buried one year in the ground, to prepare them for vegetation before they are ſown, it will not be ſo long before this will become a good fence, as is generally imagined. Nay, from ſome trials of this kind which I have made, I have found that thoſe plants which have remained where they came up from ſeed, have made ſuch progreſs as to overtake, in ſix years, plants of two or three years growth, which were tranſplanted at the time when theſe ſeeds were ſown.

And if the hedges are raiſed from ſeed, it will not be amiſs to mix Holly-berries with the Haws; the berries and Haws ſhould be buried one year, to prepare them, ſo that then both will come up together the following ſpring; and this mixture of Holly with the Quick, will not only have a beautiful appearance in the winter, but will alſo thicken the hedge at the bottom, and make it a better fence.

But where the hedge is to be planted, the ſets ſhould not be more than three years old from the Haws, for when they are older, their roots will be hard and woody; and as they are commonly trimmed off before the ſets are planted, ſo they very often miſcarry, and ſuch of them as do live, will not make ſo good progreſs as younger plants, nor are they ſo durable; for theſe plants will not bear tranſplanting ſo well as many others, eſpecially when they have ſtood long in the ſeed-bed unremoved.

The method of planting, as alſo of plaſhing and pruning of theſe hedges, having been fully explained under the article Hedges, I ſhall not repeat it here.

QUICK-BEAM. See Sorbus Sylvestris.

QUINCE-TREE. See Cydonia.

QUINCUNX ORDER is a plantation of trees, diſpoſed originally in a ſquare, conſiſting of five trees, one at each corner, and a fifth in the middle, which diſpoſition, repeated again and again, forms a regular grove, wood, or wilderneſs, and, when viewed by an angle of the ſquare or parallelogram, preſents equal or parallel alleys.

QUINQUEFOLIUM. See Potentilla.

R.

RADISH. See Raphanus.

RADISH (HORSE.) See Cochlearia.

RAMPIONS. See Campanula radice esculenta.

RAMSONS. See Allium.

RANDIA. Houſt. Gen. Nov. 28. Lin. Gen. Plant. 114.

The Characters are,

The empalement of the flower is of one leaf, cut into five ſhort ſegments at the brim. The flower is funnel-ſhaped, cut into five parts at the top; it has five ſhort ſtamina, terminated by oblong erect ſummits, and an oval germen, ſupporting a cylindrical ſtyle, crowned by two obtuſe unequal ſtigmas. The germen afterward becomes an oval capſule with one cell, having a hard cover, including many compreſſed cartilaginous ſeeds, ſurrounded with pulp.

We have but one Species of this genus, viz.

Randia (*Mitis*) foliis ovatis emarginatis, ſpinis geminatis, caule fruticoſo. *Randia with oval leaves which are indented at the top, ſpines growing by pairs, and a ſhrubby ſtalk.*

This plant grows naturally at La Vera Cruz, where the late Dr. Houſtoun found it in plenty, and ſent the ſeeds to Europe; he gave this title to the genus in honour of Mr. Iſaac Rand, who was a curious botaniſt. This plant was before diſcovered by Sir Hans Sloane growing in the iſland of Barbadoes.

It riſes with a ſhrubby ſtalk to the height of ten or twelve feet, covered with a whitiſh bark. The branches come out oppoſite from the ſide of the ſtalk, each pair croſſing the other; the leaves are of a thick conſiſtence, roundiſh, and a little indented at the top, placed by pairs, ſtanding upon ſhort foot-ſtalks. At the joints immediately under the leaves are two ſhort ſpines ſtanding oppoſite. The flowers are produced from the ſide of the branches; they are ſmall, white, tubulous, and divided at the brim ſlightly into five parts. Theſe are ſucceeded by oval berries about the ſize of a marble, having a brittle ſhell under a thin ſkin, with one cell, inclosing many compressed seeds, surrounded with black pulp. It is propagated by seeds, which should be sown early in the spring in pots, and plunged into a hot-bed of tanners bark. When the plants come up, they must have fresh air admitted to them every day when the weather is warm. In about a month's time after the plants come up, they will be fit to transplant, when they should be carefully shaken out of the pots, and each planted into a separate small pot, and then plunged into the hot-bed again, where they must be screened from the sun until they have taken new root; after which time they must have air and moisture, in proportion to the warmth of the season. The plants may remain in the hot-bed till toward Michaelmas, provided the tan is often turned over, and some fresh earth is added, when the nights begin to be cold; at which time they should be removed into the stove, and, if they are plunged into the bark-bed, it will greatly forward their growth, though they will live in the dry stove, if they are kept in a moderate temperature of heat. During the two first seasons, while the plants are young, it will be proper to keep them constantly in the stove, but then their leaves must be washed, whenever they contract filth; this will bring them forward, but, after the plants have obtained strength, they may be exposed every summer to the open air, in the warmest part of the year, for two or three months, provided they are placed in a warm situation; but in winter they must be constantly placed in a stove, and kept in a moderate warmth, otherwise they will not live in this country.

The leaves of this plant continue green throughout the year, which renders the plant valuable, because it makes an agreeable variety in the winter season, when mixed with other tender plants.

RANUNCULUS. Tourn. Inst. R. H. 285. tab. 149. Crowfoot.

The CHARACTERS are,

The empalement of the flower is composed of five oval concave leaves; the flower has five obtuse petals, which have a narrow base; each of these have an open nectarium upon their tails. It hath many stamina, terminated by oblong, erect, twin summits, and numerous germen collected in a head, having no styles, but are crowned by small reflexed stigmas. The germen afterward become seeds of uncertain irregular figures, fastened to the receptacle by very short foot-stalks.

I shall not here enumerate all the species of this genus, many of which are common weeds in most parts of England, and others are so in several parts of Europe, so are rarely admitted into gardens, therefore I shall only mention those sorts which are cultivated in gardens.

The SPECIES are,

1. RANUNCULUS (*Acris*) calycibus patulis, pedunculis teretibus, foliis tripartito-multifidis, summis linearibus. Lin. Flor. Suec. 466. flore pleno. *Ranunculus with a spreading empalement, a taper foot-stalk, many-pointed leaves, divided by threes, and those at the top linear, bearing a double flower; upright Garden Ranunculus with a double flower.*

2. RANUNCULUS (*Repens*) calycibus patulis, pedunculis sulcatis, stolonibus repentibus, foliis compositis. Flor. Suec. 468. flore pleno. *Ranunculus with a spreading empalement, furrowed foot-stalks, creeping shoots, and compound leaves, with a double flower; or Garden Ranunculus.*

3. RANUNCULUS (*Creticus*) foliis radicalibus reniformibus crenatis sublobatis, caulinis tripartitis lanceolatis integerrimis, caule multifloro. Lin. Sp. Plant. 550. *Ranunculus with kidney-shaped lower leaves, which are crenated and almost divided into lobes, but those upon the stalks divided into three spear-shaped lobes, which are entire, bearing many flowers on a stalk.*

4. RANUNCULUS (*Aconitifolius*) foliis omnibus quinatis lanceolatis inciso-serratis. Hort. Cliff. 229. flore pleno. *Ranunculus with all the leaves divided into five spear-shaped segments, which are sawed, and bear a double flower; commonly called Mountain Ranunculus.*

5. RANUNCULUS (*Gramineus*) foliis lanceolato linearibus sessilibus, caule erecto sulcato, radice globoso. Lin. Sp. 773. *Ranunculus with linear leaves sitting close to the stalk, which is erect, furrowed, having very long foot-stalks to the flowers.*

6. RANUNCULUS (*Rutæfolius*) foliis suprà decompositis, caule simplicissimo unifolio, radice tuberosâ. Hort. Cliff. 230. flore pleno. *Ranunculus with leaves which are decompounded above, a single stalk bearing one leaf, and a tuberous root with a double flower.*

7. RANUNCULUS (*Auricomus*) foliis radicalibus reniformibus crenatis incisis, caulinis digitatis linearibus, caule multiflore. Hort. Cliff. 229. flore pleno. *Ranunculus with kidney-shaped, crenated, lower leaves, those on the stalks hand-shaped, linear, and stalks bearing many flowers.*

8. RANUNCULUS (*Amplexicaulibus*) foliis ovatis acuminatis amplexicaulibus, caule subunifloro, radice fasciculatâ. Hort. Cliff. 229. *Ranunculus with oval acute-pointed leaves which embrace the stalks, one flower upon a stalk, and roots growing in bunches.*

9. RANUNCULUS (*Grandiflorus*) caule erecto bifolio, foliis multifidis, caulinis alternis sessilibus. Flor. Leyd. Prod. 492. *Ranunculus with an erect stalk bearing two leaves, which are many-pointed, those upon the stalks alternate and sitting close.*

10. RANUNCULUS (*Sanguineus*) foliis ternatis biternatisque, foliolis trifidis obtusis, caule simplici. *Ranunculus with leaves placed by threes, which are divided again into twice trifoliate leaves, ending in three obtuse points, and a simple stalk.*

11. RANUNCULUS (*Asiaticus*) foliis ternatis biternatisque, foliolis trifidis incisis, caule infernè ramoso. Lin. Sp. Plant. 552. *Ranunculus with trifoliate and twice trifoliate leaves, whose lobes are trifid and cut, and a stalk branching at the bottom.*

The first sort is a variety of the common upright Meadow Ranunculus, which grows naturally in almost every pasture; but as this hath double flowers, so it is cultivated in gardens. The stalks of this are erect, and rise more than a foot high; the lower leaves have very long foot-stalks; they are divided into several segments, resembling those of the Aconite or Monkshood; the leaves toward the top of the stalks are cut into linear segments to the bottom; the stalk branches at the top into several foot-stalks, which are terminated by double yellow flowers. This is propagated by parting of the roots in autumn, and should be planted in a moist soil and a shady situation.

The second sort is a variety of the common creeping Crowfoot, which grows naturally in cultivated fields in most parts of England. The shoots from the root of this sort trail upon the ground, and put out roots from every joint in the manner of the Strawberry; so that when it is once introduced into a garden, it will multiply fast enough; the leaves and stalks are hairy; the flowers are yellow and double, but small.

The third sort grows naturally in Crete. This hath an Asphodel root; the lower leaves are large, kidney-shaped, and a little hairy, deeply crenated on their borders, and are divided almost into five lobes, having long hairy foot-stalks. The stalks rise about nine or ten inches high, garnished with two or three leaves, which are cut into three entire segments; the top of the stalk divides into several foot-stalks, each sustaining one large pale yellow flower. It is propagated by offsets from the roots, in the same way as the Garden Ranunculus, and should be planted in a warm border, otherwise the frosts will destroy the roots.

The

The fourth ſort grows naturally upon the Alps, with a ſingle flower, but the double has been obtained by ſeeds, and is preſerved in many curious gardens for the beauty of its flowers. This is by ſome gardeners called the Fair Maid of France; it hath a perennial root, compoſed of many ſtrong fibres; the leaves are divided into five ſpear-ſhaped lobes; they are deeply ſawed on their edges, and have ſeveral longitudinal veins. The ſtalks riſe a foot and a half high, and branch out at the top into three or four diviſions; at each of which there is one leaf, of the ſame ſhape with the lower, but ſmaller. The flowers are pure white and very double, each ſtanding upon a ſhort foot-ſtalk. This is propagated by parting the roots in autumn, as ſoon as the leaves decay, and ſhould be planted in an eaſt border, and a loamy ſoil, not too ſtiff.

The fifth ſort grows naturally on the Alps. This has a perennial root; the leaves are long and narrow, like thoſe of Graſs, ſitting cloſe to the ſtalks, which riſe a little more than a foot high; theſe divide at the top into three or four ſlender foot-ſtalks, which are terminated by ſingle yellow flowers, like thoſe of the common Butterflower. There is a double flower of this kind in the Paris garden, but we have not yet got it in England.

The ſixth ſort grows naturally in Auſtria, and alſo in the Levant. This hath a tuberous root; the leaves are decompounded and ſmooth; the ſtalks riſe near a foot high, and have one leaf of the ſame ſhape with the lower, but ſmaller; the ſtalk is terminated by one double flower, about the ſize of the common Butterflower, but of a fine bright yellow colour. It is propagated by offsets from the roots in the ſame way as the Garden Ranunculus, and muſt be planted in a warm border, otherwiſe the froſt will deſtroy the roots in winter.

The ſeventh ſort is a variety of the common ſweet Wood Ranunculus, which hath a double flower. This is a very hardy plant; it may be eaſily propagated by the root, and ſhould have a loamy ſoil and a ſhady ſituation.

The eighth ſort grows naturally upon the Alps and Apennine mountains, where it ſeldom riſes more than ſix inches high; the leaves are narrow, and but one flower upon a ſtalk; but when it is planted in a garden, the ſtalks riſe a foot and a half high, and are garniſhed with oval acute-pointed leaves, ſmooth, of a grayiſh colour, and embrace the ſtalks with their baſe; the ſtalks branch out at the top into ſeveral foot-ſtalks, each ſuſtaining one white flower. It is propagated by parting of the roots in autumn, ſoon after the leaves decay, and may be planted on a ſhady border, where it will thrive exceedingly.

The ninth ſort was diſcovered by Dr. Tournefort in the Levant. This hath a perennial root, from which ariſe ſeveral leaves, cut into many points, like thoſe of Wolfsbane; the ſtalk riſes a foot high, is garniſhed with two leaves, which ſit cloſe, and are alternate. The ſtalk is terminated by one ſingle yellow flower, much larger than that of the Butterflower. It is propagated by parting of the roots in autumn, and ſhould be planted in a light loamy ſoil.

The tenth ſort is common in the Engliſh gardens, and was ſome years paſt more ſo than at preſent; for ſince the Perſian Ranunculus has been introduced here, and ſo many fine varieties have been obtained from ſeeds, they have almoſt baniſhed this ſort out of the gardens. This hath a grumous root, like the Perſian ſort; the leaves are divided by threes, and thoſe are twice again divided by threes; they are obtuſe-pointed; the ſtalk riſes about nine inches high, terminated by one large double red flower.

The eleventh ſort was originally brought from Perſia, but ſince it has been in Europe has been greatly improved by culture, many new flowers have been obtained from ſeeds, amongſt which are many with ſemidouble flowers, which produce ſeeds; and from theſe there are ſuch prodigious varieties of new flowers annually obtained, which are of beautiful colours, ſo as to exceed all other flowers of that ſeaſon, and even vie with the moſt beautiful Carnations; theſe are many of them finely ſcented, and the roots, when ſtrong, generally produce twenty or thirty flowers upon each, which, ſucceeding each other, continue in beauty a full month or longer, according to the heat of the ſeaſon, or the care taken to defend them from the injuries of the weather; all which excellent qualities have rendered them ſo valuable, that the old ſorts are almoſt diſregarded, except in ſome few old gardens.

All the very double flowers of this ſort never produce ſeeds, ſo that they are only multiplied by offsets from their roots, which they generally produce in great plenty, if planted in a good ſoil, and duly attended in winter. The ſeaſon for planting their roots is any time in October, for if they are planted ſooner, they are apt to come up in a ſhort time, and grow pretty rank before winter, whereby they will be in greater danger of ſuffering by froſt; and if they are planted much later, they will be in danger of periſhing under ground; ſo that you ſhould not keep them out of the ground any longer than the beginning or middle of October.

The beds in which the Perſian Ranunculus roots are planted, ſhould be made with freſh light earth, at leaſt three feet deep: the beſt ſoil for them may be compoſed in this manner, viz. Take a quantity of freſh earth from a rich upland paſture, about ſix inches deep, together with the green ſward; this ſhould be laid in a heap to rot for twelve months before it is mixed, obſerving to turn it over very often to ſweeten it, and break the clods; to this you ſhould add a fourth part of very rotten neats dung, and a proportionable quantity of ſea or drift-ſand, according as the earth is lighter or ſtiffer; if it be light, and inclining to a ſand, there ſhould be no ſand added; but if it be a Hazel loam, one load of ſand will be ſufficient for eight loads of earth; but if the earth is ſtrong and heavy, the ſand ſhould be added in a greater proportion; this ſhould be mixed eight months or a year before it is uſed, and ſhould be often turned over, in order to unite their parts well together, before it is put into the beds.

The depth which this ſhould be laid in the beds, muſt be about three feet; this ſhould be ſunk below the ſurface, in proportion to the drineſs or moiſture of the place where they are ſituated, which in dry ground ſhould be two feet eight inches below the ſurface, and the beds raiſed four inches above; but in a moiſt place they ſhould be two feet four inches below, and eight above the ground; and in this caſe, it will be very proper to lay ſome rubbiſh and ſtones in the bottom of each bed, to drain off the moiſture; and if upon this, at the bottom of the beds, ſome very rotten neats dung is laid two or three inches thick, the roots will reach this in the ſpring, and the flowers will be fairer. This earth I would by no means adviſe to be ſcreened very fine, only in turning it over each time, you ſhould be careful to break the clods, and throw out all large ſtones, which will be ſufficient; for if it is made very fine, when the great rains in winter come on, it will cauſe the earth to bind into one ſolid lump, whereby the moiſture will be detained, and the roots, not being able to extend their tender fibres, will rot. Of this I have had many examples, but one particularly to my coſt; when I had procured a fine parcel of theſe roots from abroad, and being deſirous of having them thrive very well, I took great pains to ſcreen the earth of my beds very fine, which I laid above two feet deep, and planted a good part of my roots therein; but the ſeaſon advancing, and having a great deal of other buſineſs upon my hands, I did not ſcreen the earth of all my beds, but planted some of them without doing any thing more than raking them; and the success was, that the roots in those beds which were screened, did, great part of them, entirely rot, and the remaining part were so weak, as not to produce any good flowers; whereas those which were planted in the beds which were not screened, did thrive and flower very well, and scarce any of the roots failed, though the earth of all the beds was the same, and were in the same situation, both with regard to wind and sun; so that the damage which those roots sustained, was owing entirely to the fineness of the earth, and this I have several times since observed in other gardens.

I am aware, that this depth of three feet, which I have here directed to make the beds for these flowers, will be objected to by many persons, on account of the expence and trouble of preparing them, as also supposing it unnecessary to make the beds so deep, for flowers whose roots are small; but if they will give themselves the trouble of making the experiment, by preparing one bed in this manner, and another in the common way, and plant them both with the same flowers, they will soon be convinced of their error, by the success of the flowers. For in the beds which have been prepared of this depth, I have seen one root produce upward of fifty flowers, each of which grew near a foot high, and were extremely large and fair; whereas, in the common method of culture, they are thought to do very well, when they produce eight or ten flowers on each root, and those grow six inches high; but if a person will trace the length of the small fibres of these roots, he will find them to extend three or four feet downwards. And as it is by these distant fibres that the nourishment is taken in, for the increase and strength of the flowers; so if these meet with a poor barren soil below, they shrink, and the flowers are starved for want of proper nourishment in the spring, when it is most required.

The beds being thus prepared, they should lie a fortnight or more to settle, before the roots are planted, that there may be no danger of the earth settling unequally after they are planted, which would prejudice them, by having hollow places in some parts of the bed, to which the water would run and lodge, and so rot the roots. Then having levelled the earth, laying the surface a little rounding, you should mark out the rows by a line, at about six inches distance each way, so that the roots may be planted every way in strait lines; then you should open the earth with your fingers at each cross, where the roots are to be planted about two inches deep, placing the roots exactly in the middle, with their crowns upright; then with the head of a rake you should draw the earth upon the surface of the bed level, whereby the top of the roots will be about an inch covered with earth, which will be sufficient at first. This work should be done in dry weather, because the earth will then work better than if it were wet; but the sooner after planting there happens to be rain, the better it will be for the roots; for if it should prove dry weather long after, and the earth of the beds is very dry, the roots will be subject to mould and decay; therefore in such a case it will be proper to give a little water to the beds, if there should no rain happen in a fortnight's time, which is very rare at that season of the year, so that they will seldom be in danger of suffering that way.

When the roots are thus planted, there will no more be required until toward the end of November, by which time they will begin to heave the ground and their leaves appear, when you should lay a little of the same fresh earth, of which the beds were composed, about half an inch thick all over the beds, which will greatly defend the crown of the root from frost; and when you perceive the leaves to break through this second covering, if it should prove very hard frost, it will be very proper to arch the beds over with hoops, and cover them with mats, especially in the spring, when the flower-buds will begin to appear; for if they are exposed to too much frost, or blighting winds at that season, their flowers seldom open fairly, and many times their roots are destroyed.

In the middle of March the flower-stems will begin to rise, at which time you should carefully clear the beds from weeds, and stir the earth with your fingers between the roots, being very careful not to injure them; this will not only make the beds appear handsome, but also greatly strengthen their flowers. When the flowers are past, and the leaves are withered, you should take up the roots, and carefully clear them from the earth; then spread them upon a mat to dry, in a shady place; after which they may be put up in bags or boxes in a dry room, until the October following, which is the season for planting them again.

These Persian sorts are not only propagated by offsets from the old roots, but are also multiplied by seeds, which the semidouble kinds produce in plenty; therefore, whoever is desirous to have these in perfection, should annually sow their seeds, from which new varieties will be every year produced; but in order thereto, you should be careful in saving your seed, or in procuring it from such persons as understand how to save it; that is, who will be careful not to leave any flowers for seeds, but such as have five or six rows of petals at least, and are well coloured; for since these flowers increase plentifully, it is not worth the trouble to sow any indifferent seeds, because there can be but little hopes of obtaining any good flowers from them.

Being prepared with seeds, about the middle of August, which is the proper season for sowing of them, you should get some large shallow pots, flatted seed-pans, or boxes. These should be filled with light rich earth, levelling the surface very even; then sow the seeds thereon pretty thick, and cover it about a quarter of an inch thick with the same light earth; after which you should remove these pots or boxes into a shady situation, where they may have the morning sun until ten of the clock; and if the season should prove dry, you must often refresh them with water, being very careful in doing of this, not to wash the seeds out of the ground. In this situation the pots should remain until the beginning of October, by which time the plants will begin to come up, though sometimes the seeds will remain in the earth until November; then you should remove the pots into a more open exposure, where they may have full sun, which at that time is necessary to exhale the moisture of the earth; but toward the middle or end of November, when you are apprehensive of frost, the pots should be removed under a common hot-bed frame, where they may be covered with the glasses in the night time, and in bad weather; but in the day, when the weather is mild, they should be entirely opened, otherwise the plants will draw up too weak. The only danger they are in, is from violent rains and frosts, the first often rotting the tender plants, and the frost will often turn them out of the ground, therefore they should be carefully guarded against both these.

In the spring, as the season grows warm, these pots should be exposed to the open air, placing them at first near the shelter of a hedge, to protect them from the cold winds; but towards the beginning or middle of April they should be removed again into a more shady situation, according to the warmth of the season; and if it should prove dry, they must be sometimes refreshed with water; but you should be careful not to give it to them in great quantities, which is very apt to rot these tender roots; the latter end of April or beginning of May, they should be placed where they may have only the morning sun, in which place they may remain till their leaves decay, when they may be be taken out of the earth, and the roots dried in a shady place; after which they may be put up in bags, and preserved in a dry place until the October following, when they must be planted in the manner before directed for the old roots.

The spring following these roots will flower, at which time you should carefully mark such of them as are worthy to be preserved, and the single, or bad coloured flowers, may be pulled up and thrown away, which is the surest method of removing them from the good sorts; for if they are permitted to remain together until their leaves decay, there may be some offsets of the bad sorts mixed with the good flowers. You should not suffer those flowers, which you intend to blow fine the succeeding year, to bear seeds, but cut off the flowers when they begin to decay, for those roots which have produced seeds, seldom flower well afterwards; nor will the principal old root, which has flowered strong, ever blow so fair as will the offsets, which is what should be principally observed, when a person purchases any of these roots, for great part of the complaints made by those who have bought these roots at a dear rate, is principally owing to this. The persons who sold them, being apprized of this matter, have parted with their old roots to their purchasers, and reserved the offsets for their own use; which old roots have often so much degenerated from what they were the preceding year, as to cause a suspicion, whether the persons they were purchased from had not changed the roots: this degeneracy attends these flowers, after having flowered extremely large and fair, or that they have been permitted to seed; so that it is absolutely necessary to sow seeds every year, in order to preserve a succession of good flowers.

The manner of preparing the beds, and the distance and method of planting the roots, having been already directed, I shall not repeat it here; but will only observe, that these flowers being tender, must be protected from hard frosts, and cutting sharp winds, especially after Christmas, when their flower-buds are forming, for if they are neglected at that season, their flowers will rarely prove fair; nor should you suffer them to receive too much wet in winter or spring, which is equally as injurious to them as frost. In planting of these roots, you should observe to place the semidouble kinds, from which you intend to save seeds, in separate beds by themselves, and not intermix them with the double flowers, because they will require to be treated in a different manner; for when the flowers of the semidouble kinds begin to fade, you should carefully guard them from wet, for if they are permitted to receive hard rains, or are watered at that season, the seeds rarely come to maturity, or are so weak, that scarce one in fifty of them will grow.

When the seed begins to ripen (which may be easily known, by their separating from the axis and falling) you should look over the beds every day, gathering it as it ripens, for there will be a considerable distance in the seeds of the same bed coming to maturity, at least a fortnight, three weeks, or a month. When you gather the seed, it should not be exposed to the sun, but spread to dry in a shady place; after which you must put it up where the vermin cannot come to it, until the time of sowing it.

By this method of sowing seeds every year, you will not only increase your stock of roots, but also raise new varieties, which may be greatly mended by changing the seeds into fresh ground; for if a person continually sows his seed in the same garden many years, they will not produce near so fine flowers, as if he procured his seeds at some distance, which is also the case with most other plants.

It will also be necessary to take away all the earth out of the beds in which the roots were blown the preceding year, and put in new, if you intend to plant Ranunculuses there again, otherwise they will not thrive near so well, notwithstanding you may add some new compost to the beds, and this is what all the curious florists continually observe.

RAPA. Tourn. Inst. R. H. 228. tab. 112. Turnep.

The CHARACTERS are,

The empalement of the flower is three leaved. The flower hath four plain spreading petals, which are narrow at their base. It has four oval honey glands, situated between the stamina and style, and six erect awl-shaped stamina; the two which are opposite are the length of the empalement, the other four are longer, terminated by erect acute-pointed summits. It hath a taper germen, supporting a short thick style, crowned by an entire headed stigma. The germen afterward becomes a long taper pod, depressed on the sides, opening in two cells, which are filled with roundish seeds.

The SPECIES are,

1. RAPA (*Rotunda*) radice orbiculatâ depressâ carnosâ. *Turnep with an orbicular, depressed, fleshy root.*

2. RAPA (*Oblonga*) radice oblongâ carnosâ. *Turnep with an oblong fleshy root.*

3. RAPA (*Napus*) radice fusiformi. *Turnep with a spindle-shaped root; commonly called French Turnep.*

The first is the Turnep which is commonly cultivated in the fields, of which there are the following varieties, viz. the round red or purple-topped Turnep, the green-topped Turnep, the yellow Turnep, the black-rooted Turnep, and the early Dutch Turnep. The last sort is commonly sown early in the spring, for to supply the markets in May and June, but is never cultivated for a general crop. The red-rooted Turnep was formerly more cultivated in England than at present, for since the large green-topped Turnep has been introduced, all the skilful farmers prefer it to the other sorts; the roots of the green will grow to a large size, and continue good much longer than the other sorts. The next to this is the red or purple-topped Turnep, which will also grow large, and is extremely good for some time, but the roots of this will become stringy much sooner than those of the green-topped. The long-rooted Turnep, the yellow Turnep, and the blackish-rooted Turnep, are now rarely cultivated in England, neither of them being so good for the table or for feed, as the red and green-topped Turnep, though there are some few persons who sow them for the sake of variety.

The French Turnep is not much cultivated in England, but in France and Holland they are in great esteem, especially for soups, their roots being small, and boiled whole in the soup, and so served up to the table; these must be used while they are young, otherwise they will become rank and stringy.

These are supposed to be only varieties, which have accidentally been obtained from seeds, therefore I have not enumerated them as distinct species; but yet I am certain they are constant, where care is taken in the saving of their seeds, not to suffer any mixture to stand for seeds: I have sown of three or four sorts several years, and have always found them retain their differences; however, it is not easy to determine if some of these were not by culture first obtained from seeds of the common white Turnep. The yellow Turnep seems most unlikely to have been an accidental variety, for I have never known this alter, and the roots are yellow within, whereas all the other have white flesh, notwithstanding their outsides are of very different colours.

The long-rooted Turnep is, I think, a distinct species, the form of the root, and its manner of growth, being totally different from the other sorts. I have seen these roots as long as those of the Parsnep, and nearly of the same shape; these run deep into the ground, so are unfit for feeding of cattle; and unless they are used very young, become

come ſtrong, ſo not proper for the table, which has occaſioned their being rejected of late years.

The green-topped Turnep grows above ground more than any of the other, which renders it preferable for feeding of cattle, and being the ſofteſt and ſweeteſt root when grown large of any of the kinds, is moſt eſteemed for the table; but in very ſevere winters they are in greater danger of ſuffering by froſt than thoſe whoſe roots lie more in the ground, eſpecially if they are not covered by ſnow; for when they are frequently hard frozen and thawed, it cauſes them to rot ſooner than thoſe whoſe fleſh is leſs tender and ſweet. I have ſeen the roots of this ſort, which were more than a foot diameter boiled, and were as ſweet and tender as any of the ſmalleſt roots.

Turneps delight in a light, ſandy, loamy ſoil, which muſt not be rich, for in a rich ſoil they grow rank and are ſticky, but if it be moiſt they will thrive the better in ſummer, eſpecially in freſh land, where they are always ſweeter than upon an old worn-out or a rich ſoil.

The common ſeaſon for ſowing of Turneps, is any time from the beginning of June to the middle of Auguſt, or a little later; though it is not adviſeable to ſow them much after, becauſe, if the autumn ſhould not prove very mild, they will not have time to apple before winter, nor will the roots of thoſe which are ſown after the middle of July, grow very large, unleſs the froſt keeps off long in autumn. But, notwithſtanding this is the general ſeaſon in which the greateſt part of Turneps are ſown in the country, yet about London they are ſown ſucceſſively from March to Auguſt, by thoſe who propagate them to ſupply the markets with their roots; but there is a great hazard of loſing thoſe which are ſown early in the year, if the ſeaſon ſhould prove dry, by the fly, which will devour whole fields of this plant while young; ſo that where a ſmall quantity for the ſupply of a family is wanted, it will be abſolutely neceſſary to water them in dry weather; and where a perſon ſows thoſe ſeeds in April and May, it ſhould always be upon a moiſt ſoil, otherwiſe they ſeldom come to good, the heat of the weather at that ſeaſon being too great for them upon a dry ſoil; but thoſe which are ſown toward the middle or latter end of June, commonly receive ſome refreſhing ſhowers to bring them forward; without which, it is very common to have them all deſtroyed.

Theſe ſeeds ſhould always be ſown upon an open ſpot of ground, for if they are near hedges, walls, buildings, or trees, they will draw up, and be very long topped, but their roots will not grow to any ſize.

They are ſown in great plenty in the fields near London, not only for the uſe of the kitchen, but for food for cattle in winter, when there is a ſcarcity of other food; and this way they are become a great improvement to barren ſandy lands, particularly in Norfolk, where, by the culture of Turneps, many perſons have doubled the yearly value of their ground.

The land upon which this ſeed is ſown, ſhould be ploughed in April, and twy-fallowed in May, that is, once more ploughed and twice well harrowed, and made very fine; then the ſeed ſhould be ſown pretty thin (for it being ſmall, a little will ſow a large piece of ground, one pound is the common allowance for an acre of land.) The ſeed muſt be harrowed in as ſoon as it is ſown, with a ſhort tined harrow, and the ground rolled with a wooden roll, to break the clods, and make the ſurface even. In ten days or a fortnight after ſowing, the plants will come up; at which time, if the ſeaſon ſhould prove dry, they will be in great danger of being deſtroyed by the fly; but if it ſo happen, the ground muſt be ſowed again, for the ſeed being cheap, the chief expence is the labour; but the ground ſhould be firſt harrowed to looſen it, eſpecially if it is ſtiff hard.

When the plants have got four or five leaves, they ſhould be hoed to deſtroy the weeds, and to cut up the plants where they are too thick, leaving the remaining ones about ſix or eight inches aſunder each way, which will be room enough for the plants to ſtand for the firſt hoeing; the ſooner this is performed, when the plants have four leaves, the better they will thrive; but in the ſecond hoeing, which muſt be performed about a month after the firſt, they ſhould be cut up, ſo as that the remaining plants may ſtand fourteen or ſixteen inches diſtance, or more, eſpecially if they are deſigned for feeding of cattle, for where the plants are allowed a good diſtance, the roots will be proportionably large; ſo that what is loſt in number, will be overgained by their bulk, which is what I have often obſerved. But in ſuch places where they are ſown for the uſe of the kitchen, they need not be left at a greater diſtance than ten inches or a foot, becauſe large roots are not ſo generally eſteemed for the table.

It is not many years ſince the practice of ſowing Turneps for feeding of cattle, has been of general uſe; how it happened that this improvement ſhould have been ſo long neglected in every part of Europe, is not eaſy to determine, ſince it is very plain, that this piece of huſbandry was known to the ancients. For Columella, in treating of the ſeveral kinds of vegetables which are proper for the field, recommends the cultivating Rapa in plenty; becauſe (ſays he) thoſe roots which are not wanted for the table, will be eaten by the cattle: yet this plant was not much cultivated in the fields till within the laſt ſeventy or eighty years, nor is the true method of cultivating Turneps yet known, or at leaſt not practiſed, in ſome of the diſtant counties of England, at this time. For in many places the ſeed is ſown with Barley in the ſpring, and thoſe plants which come up, and live till the Barley is cut, produce a little green for the ſheep to pick up, but never have any roots. In other places, where the Turnep-ſeed is ſown by itſelf, the method of hoeing them is not underſtood, ſo that weeds and Turneps are permitted to grow together; and where the Turneps come up thick in patches, they are never thinned, ſo that they draw up to have long leaves, but never can have good roots, which is the principal part of the plant, therefore ſhould be chiefly attended to.

The general method now practiſed in England, for cultivating this plant in the fields, is the ſame as is practiſed by the farming gardeners, who ſupply the London markets with theſe roots, and is the ſame as before directed. But it is only within the compaſs of a few years, that the country-people have been acquainted with the method of hoeing them; ſo that the farmers formerly employed gardeners, who had been bred up in the kitchen-gardens, to perform this work. The uſual price given per acre, for twice hoeing and leaving the crop clean, and the plants ſet out properly, was ſeven ſhillings; at which price the gardeners could get ſo much per week, as to make it worth their while to leave their habitations, and practiſe this in different counties, during the ſeaſon for this work, which always happens, after the greateſt hurry of buſineſs in the kitchen-garden is over; ſo that they uſually formed themſelves in ſmall gangs of ſix or ſeven perſons, and ſet out on their different routs, each gang fixing at a diſtance from the reſt, and undertaking the work of as many farmers in the neighbourhood, as they could manage in the ſeaſon; but as this work is now performed by many country labourers, that practice is loſt to the kitchen-gardeners, the labourers doing it much cheaper.

There has alſo been another method practiſed very lately, by ſome very curious farmers, in cultivating of Turneps, which is, by ſowing the ſeed in rows, with the drill-plough. In ſome places the rows are ſown three feet aſunder, in others

others four, in some five, and some six. The latter has been recommended by some, as the most proper distance; and although the intervals are so large, yet the crop produced on an acre has been much greater than upon the same quantity of land, where the rows have been but half this distance; and upon all the fields which have been tilled, the crops have greatly exceeded those which have been hand-hoed. The late Lord Viscount Townshend was at the expence of making the trial of these two different methods of husbandry, with the greatest care, by equally dividing the same fields into different lands, which were alternately sown in drills, and the intermediate lands in broad-cast. The latter were hoed by hand, in the common method, and the other cultivated by the hoeing-plough; and when the roots were fully grown, his lordship had an equal quantity of land, which had been sowed in the different methods, measured, and the roots drawn up and weighed; those roots which had been cultivated by the plough, were so much larger than the other, that the crop of one acre weighed a ton and a half more than that of an acre in the other husbandry.

But when the Turneps are sown in drills, they will require to be hoed by hand, to separate and cut out the plants, where they are too near together in the rows; as also to cut up the weeds between the plants, where the plough cannot reach them. If this is carefully performed, the ploughing of the intervals will encourage the growth of the roots, by thus stirring of the ground, and make it much better prepared for the crop of Barley, or whatever else is sown the following spring. This method of culture may be supposed to be more expensive than that commonly practised, by those unacquainted with it; but those who have made trials of both, find the horse-hoeing to be much the cheapest, and by far the best. For the country people, who are employed in hand-hoeing of Turneps, are very apt to hurry over their work, so that half the weeds are left growing, and the plants are seldom singled out so well as they should be; nor are they curious enough to distinguish the Charlock (which is one of the most common weeds in arable land) from the Turneps; so that about the middle of September, it is very common to see the fields of Turneps full of the yellow flowers of the Charlock. Now, in the horse-hoeing, all the weeds in the intervals will be entirely destroyed; so that if a few plants in the rows of Turneps should be overlooked, they may be easily drawn when they appear visible, and by this method the land will be sooner and better cleaned from weeds.

The greatest evil which attends a crop of Turneps, is that of their being destroyed by the fly, which usually happens soon after the plants come above ground, or while they are in the seed leaf, for, after they have put out their rough leaves pretty strong, they will be past this danger. This always happens in dry weather, so that, if there should be rain when the Turneps come up, they will grow so fast, as to be in a few days out of danger from the fly; and it hath been found, that those which have been sown in drills, have escaped the fly much better than those sown broad-cast; but, if soot is sown along the surface of each drill, it will be of great service to keep off the fly, and a small quantity of it will be sufficient for a large field, where the drills only are to be covered.

Another danger of the crops being destroyed is from the caterpillars, which very often attack them, when they are grown so large as to have six or eight leaves on a plant. The surest method of destroying these insects, is to turn a large parcel of poultry into the field, which should be kept hungry, and turned early in the morning into the field; these fowls will soon devour the insects, and clear the Turneps. To this evil the Turneps, which are sown in drills, are not so much exposed, for as the ground between the rows will be kept stirred, the plants will be kept growing, so will not be in danger of suffering from these insects, for the parent insects never deposit their eggs upon any plants which are in health; but as soon as they are stinted, they are immediately covered with the eggs of these insects; and this holds in general with vegetables as with animals, who are seldom attacked by vermin when they are in perfect health; whereas, when they become unhealthy, they are soon overspread with them; so that it is the disease which occasions the vermin, and not the vermin the disease, as is commonly imagined.

When the Turneps are sown in drills, it will be the best way to plough between every other row at first, and some time after to plough the alternate intervals, by which method the plants will receive more benefit from the often stirring the ground than they would do, if all the intervals were hoed at one time, and the plants will be in less danger of suffering from the earth being thrown up too high on some rows, while others may be left too bare of earth; but, when the earth has been thrown up on one side of the drill, it may be turned down again soon after the next interval is ploughed. This alternate moving of the earth will prepare the ground very well for the succeeding crop, and greatly improve the Turneps; but, as the plough cannot well be drawn nearer to the drills than two or three inches, the remaining ground should be forked to loosen the parts, and make way for the fibres of the roots to strike out into the intervals; otherwise, if the land is strong, it will become so hard in those places which are not stirred, as to stint the growth of the Turneps. This may be done at a small expence; a good hand will perform a great deal of this work in a day, and, whoever will make the trial, will find their account in practising it, especially on all strong land, where the Turneps are much more liable to suffer from the binding of the ground, than they will be on a loose soil; but yet, in all sorts of ground, it will be of great service to practise this.

When the ground is thus stirred in every part, one ploughing will be sufficient, after the Turneps are eaten, for the sowing of Barley, or any other crop; so that there will be an advantage in this, when the Turneps are kept late on the ground, as will often be the case, especially when they are cultivated for feeding of ewes, because it is often the middle of April before the ground will be cleared; for late feed in the spring, before the natural Grass comes up, is the most wanted, where numbers of sheep or ewes are maintained, and one acre of Turneps will afford more feed than fifty acres of the best pasture at that season.

In Norfolk, and some other counties, they cultivate great quantities of Turneps for feeding of black cattle, which turn to great advantage to their farms, for hereby they procure a good dressing for their land; so that they have extraordinary good crops of Barley upon those lands, which would not have been worth the ploughing, if it had not been thus husbanded.

When the Turneps are fed off the ground, the cattle should not be suffered to run over too much of the ground, for, if they are not confined by hurdles to as much as is sufficient for them one day, the cattle will spoil three times the quantity of Turneps as they can eat, so that it is very bad husbandry to give them too much room; therefore the hurdles should be every day removed forward, and, if the Turneps are drawn out of the ground before the cattle or sheep are turned into the new inclosure, there will be less waste made, for they will then eat up the whole roots; whereas, if they are turned upon the Turneps growing, they will scoop the roots, and leave the rinds, which being hollow, the urine of the sheep will lodge in them; so that, when

when they are forked out of the ground, the sheep will not eat any of those roots which are thus tainted.

I cannot omit taking notice of a common mistake, which has generally prevailed with persons who have not been well informed to the contrary, which is, in relation to the mutton which is fatted with Turneps, most people believing it to be rank and ill-tasted; whereas it is a known fact, that the best mutton this country affords is all fatted on Turneps, and that rank mutton, whose fat is yellow, is what the low marshy lands of Lincolnshire, and other rank pastures, produce.

In order to save good Turnep-seeds, you should transplant some of the fairest roots in February, placing them at least two feet asunder each way, observing to keep the ground clear from weeds, until the Turneps have spread so as to cover the ground, when they will prevent the weeds from growing; when the pods are formed, you should carefully guard them against the birds, otherwise they will devour it, especially when it is near ripe; at which time you should either shoot the birds as they alight upon the seed, or lay some birdlimed twigs upon it; whereby some of them will be caught, and, if they are permitted to remain some time, and afterwards turned loose, they will prevent the birds from coming thither again for some time, as I have experienced. When the seed is ripe, it should be cut up, and spread to dry in the sun; after which it may be threshed out, and preserved for use.

There have been many receipts for preventing the fly taking Turneps, but few of them deserve notice, therefore I shall only mention two or three which I have seen tried with success. The first was steeping the seeds in water with flower of brimstone mixed, so as to make it strong of the brimstone: another was steeping it in water with a quantity of the juice of Horse-aloes mixed, both which have been found of use. The sowing of soot or tobacco-dust, over the young plants, as soon as they appear above ground, has also been found very serviceable: in short, whatever will add vigour to the young plants, will prevent their being destroyed by the fly, for these never attack them till they are stinted in their growth.

RAPHANUS. Tourn. Inst. R. H. 229. tab. 114. Radish.

The CHARACTERS are,

The empalement of the flower is erect. The flower has four heart-shaped petals, placed in form of a cross, which spread open, and are narrow at their base; it hath four honey glands, one on each side the short stamina between them and the style, and one between each of the long stamina and the empalement; it hath six erect stamina, two, which are opposite, are the length of the empalement; the other four are as long as the base of the petals, terminated by single summits, with an oblong swelling germen, crowned by a headed stigma. The germen afterward becomes an oblong, smooth, spongy pod, having an acute point, swelling and almost jointed, having two cells, divided by an intermediate partition, filled with roundish seeds.

The SPECIES are,

1. RAPHANUS (*Sativus*) siliquis teretibus torosis bilocularibus. Lin. Sp. *Radish with an oblong root.*

2. RAPHANUS (*Rotundus*) radice rotundâ. *Round-rooted Radish, or small, round, Naples Radish.*

3. RAPHANUS (*Orbiculatis*) radice orbiculatâ depressâ. *Radish with an orbicular depressed root; commonly called Turnep-rooted, or white Spanish Radish.*

4. RAPHANUS (*Niger*) radice fusiformi. *Radish with a spindle-shaped root; or the black Spanish Radish.*

5. RAPHANUS (*Raphanistrum*) siliquis teretibus articulatis lævibus unilocularibus. Hort Cliff. 340. *Radish with smooth, taper, jointed pods, having one cell; or white flowering Charlock with a jointed pod.*

The last sort grows naturally on arable lands in most parts of Europe, so is seldom admitted into gardens.

The other four sorts are supposed to be only seminal variations, but from forty years experience I have never found either of these to vary from one to the other sort; and I am certain, whoever will make the trial, by saving the seeds of each carefully without mixture, will always find the plants prove the same as the seeds were saved from.

The first sort here mentioned is that which is commonly cultivated in kitchen-gardens for its roots, of which there are several varieties, as the small-topped, the deep red, the pale red or salmon, and the long-topped striped Radish; all which are varieties arising from culture. The small-topped sort is most commonly preferred by the gardeners near London, because they require much less room than those with large tops; for as the forward Radishes are what produce the greatest profit to the gardener, which are commonly sown upon borders near hedges, walls, or pales, if they are of the large-topped sort, they will be apt to grow mostly to a top, and not swell so much in the root as the other, especially if they are left pretty close.

The seasons for sowing this seed are various, according to the time when they are desired for use; but the earliest season is commonly toward the latter end of October, that the gardeners near London sow them to supply the markets; and these, if they do not miscarry, will be fit for use in the beginning of March following, which is full as soon as most people care to eat them. These are commonly sown on warm borders near walls, pales, or hedges, where they may be defended from the cold winds; but there are some who sow Radish-seeds among other crops in the middle of September, and, if these are not destroyed by frost, they will be fit for use soon after Christmas; but these must be eaten while they are young, for they soon grow sticky and strong.

The second sowing is commonly about Christmas, provided the season be mild, and the ground in a fit condition to work; these are also sowed near shelter, but not so near pales and hedges as the first sowing. If these are not destroyed by frost, they will be fit for use the end of March or the beginning of April; but, in order to have a succession of these roots for the table through the season, you should repeat the sowing of their seeds once a fortnight from the middle of January till the beginning of April, always observing to sow the latter crops upon a moist soil and an open situation, otherwise they will run up, and grow sticky, before they are fit for use.

Many of the gardeners near London sow Carrot-seed with the early Radishes, so that when their Radishes are killed, which sometimes happens, the Carrots will remain, for the seeds of Carrots commonly lie in the ground five or six weeks before they come up, and the Radishes seldom lie above a fortnight under ground at that season, so that these are often up and killed, when the Carrot-seed remains safe in the ground; but, when both crops succeed, the Radishes must be drawn off very young, otherwise the Carrots will be drawn up so weak, as not to be able to support themselves when the Radishes are gone.

It is also a constant practice with the kitchen-gardeners to mix Spinach-seed with their latter crops of Radishes, so that when the Radishes are drawn off, and the ground cleaned between the Spinach, it will grow prodigiously, and in a fortnight's time will as completely cover the ground, as though there had been no other crop. And this Spinach, if it be of the broad-leaved kind, will be larger and fairer than it commonly is when sown by itself, because where people have no other crop mixed with them, they commonly sow them too thick, whereby they are drawn

up

up weak; but here the roots standing pretty far apart, so after the Radishes are gone, they have full room to spread, and if the soil be good, it is a prodigious size this Spinach will grow to before it runs up to seed; but this husbandry is chiefly practised by such gardeners as pay very dear for their land, and are obliged to have as many crops in a year as possible, otherwise they could not afford to pay such large rents.

When the Radishes are come up, and have got five or six leaves, they must be pulled up where they are too close, otherwise they will draw up to a top, but the roots will not increase their bulk. In doing of this, some only draw them out by hand, but the best method is to hoe them with a small hoe, which will stir the ground, and destroy the young weeds, and also promote the growth of the plants. The distance which these should be left, if for drawing up small, may be three inches, but if they are to stand until they are pretty large, six inches are full near enough, and a small spot of ground will afford as many Radishes at each sowing as can be spent in a family while they are good.

If you intend to save seeds of your Radishes, you should, at the beginning of May, prepare a spot of ground in proportion to the quantity of seeds intended (but you should always make allowance for bad seasons, because it often happens in a very dry season, that there will not be a fourth part of the quantity of seeds upon the same proportion of ground as there will be in a moist season.) This ground should be well dug and levelled; then you should draw up some of the straitest and best coloured Radishes (throwing away all such as are short, and that branch out in their roots;) the other should be planted in rows three feet distance, and two feet asunder in the rows, observing, if the season is dry, to water them until they have taken root; after which they will require no farther care but only to hoe down the weeds between them, until they are advanced so high, as to spread over the ground, when they will prevent the growth of weeds.

When the seed begins to ripen, you should carefully guard it against the birds, otherwise they will destroy it. When it is ripe (which you may know by the pods changing brown,) you should cut it, and spread it in the sun to dry; after which you should thresh it out, and lay it up for use, where the mice cannot come to it, otherwise they will eat it up.

The small round-rooted Radish is not very common in England, but in many parts of Italy it is the only sort cultivated; the roots of this kind are very white, round, small, and very sweet. This may be propagated in the same manner as the common sort, with this difference, viz. that this must not be sown till the beginning of March, and the plants allowed a greater distance. The seeds of this kind are very subject to degenerate, when saved in England, unless they are at such distance from the common sort, as that the farina of one cannot mix with the other.

The other round-rooted Radishes are rarely cultivated in England, but those who have a mind to have them, may sow them in the same manner as the last.

The black and white Spanish Radishes are commonly cultivated for medicinal use, but there are some persons who are very fond of them for the table. These are commonly sown about the middle of July, or a little earlier, and they are fit for the table by the end of August, or the beginning of September, and will continue good till the frost spoils them. These must be thinned to a greater distance than the common sort, for the roots of these grow as large as Turneps, therefore should not be left nearer together than six inches.

Some persons who are very curious to have these roots in winter, draw them out of the ground before the hard frost comes on, and lay them up in dry sand in the same manner as is practised for Carrots, being careful to guard them from wet and frost, and by this method they preserve them till the spring.

RAPISTRUM. See SINAPIS.

RAPUNCULUS. Tourn. Inst. R. H. 113. tab. 38. Rampion.

The CHARACTERS are,

The empalement of the flower is of one leaf, divided into five acute parts, sitting upon the germen. The flower hath one petal, which is starry, cut into five linear segments, which are recurved; it hath five stamina which are shorter than the petal, terminated by oblong summits. The germen, which is situated under the flower, supports a slender recurved style, crowned by an oblong, twisted, three-pointed stigma, which afterward becomes a roundish capsule with three cells, filled with small roundish seeds.

The SPECIES are,

1. RAPUNCULUS (*Spicatus*) spicâ oblongâ, capsulis bilocularibus, foliis radicalibus cordatis. *Rampion with an oblong spike of flowers, capsules containing two cells, and the lower leaves heart-shaped.*

2. RAPUNCULUS (*Comosus*) fasciculo terminali sessili, foliis dentatis, radicalibus cordatis. *Rampion with flowers growing in bunches terminating the stalks, indented leaves, and those at the bottom heart-shaped.*

3. RAPUNCULUS (*Hemisphericus*) capitulo subrotundo, foliis linearibus integerrimis. *Rampion with roundish heads, and linear entire leaves.*

4. RAPUNCULUS (*Pauciflorus*) capitulo subfolioso, foliis omnibus lanceolatis. *Rampion with heads which are somewhat leafy, and all the leaves spear-shaped.*

5. RAPUNCULUS (*Orbicularis*) capitulo subrotundo, foliis serratis radicalibus cordatis. *Rampion with roundish heads, sawed leaves, the lower ones of which are heart-shaped.*

These are all of them hardy plants, which will thrive in the open air. They are propagated by seed, which should be sown in autumn, for if they are kept out of the ground till the spring, they frequently fail. The seeds should be sown on a bed of fresh undunged earth where they are designed to remain, for they do not thrive so well when they are transplanted; therefore the best method is to make small drills cross the bed about eighteen inches asunder, and sow the seeds therein; then cover them lightly over with earth, for if they are buried too deep they will rot in the ground. In the following spring the plants will come up, when they should be diligently weeded, which is all the care they will require, only they should be thinned where they are too close, so as to leave them six or seven inches apart in the rows, and afterward they require no farther attention but to keep them clear from weeds.

As these plants do not continue above two or three years, they should be sown every other year to continue the sorts, for they are plants which require little trouble to cultivate, and their flowers make a pretty variety in large gardens, therefore they may be allowed a place amongst other hardy flowers.

RAPUNTIUM. Tourn. Inst. R. H. 163. tab. 51. Rampions, or Cardinal-flower.

The CHARACTERS are,

The empalement of the flower is cut into five linear segments, the two upper being larger than the other. The flower is of one petal, with a long cylindrical tube, a little curved, and is divided at the brim into five segments, two of which compose the upper lip, and are smaller than the three lower which compose the under; it hath five awl-shaped stamina, terminated by oblong summits, which coalesce at the top in form of a cylinder, but open in five parts at their base; it has an acute germen, situated below

low the flower, supporting a cylindrical style, crowned by a hairy obtuse stigma. The germen afterward becomes an oval capsule, opening at the top, filled with small seeds.

The SPECIES are,

1. RAPUNTIUM (*Cardinalis*) caule erecto, foliis lanceolatis serratis, spicâ terminali. *Cardinal-flower with an erect stalk, spear-shaped sawed leaves, and a spike of flowers terminating the stalk; commonly called scarlet Cardinal-flower.*

2. RAPUNTIUM (*Americanum*) caule erecto, foliis lineri-lanceolatis integerrimis acuminatis, spicâ terminali. *Cardinal-flower with erect, linear, spear-shaped, entire, acute-pointed leaves, and a spike of flowers terminating the stalks.*

3. RAPUNTIUM (*Siphilitium*) caule erecto, foliis ovato-lanceolatis crenatis, calycum sinubus reflexis. *Cardinal-flower with an erect stalk, oval, spear-shaped, crenated leaves, and the sinuses of the empalements reflexed; commonly called the blue Cardinal-flower.*

4. RAPUNTIUM (*Cliffortianum*) caule erecto, foliis cordatis obsoletè dentatis petiolatis, floribus sparsis thyrso longissimo. *Cardinal-flower with an erect stalk, heart-shaped leaves which are somewhat indented, having foot-stalks, and the longest spike of flowers, which are placed thinly.*

5. RAPUNTIUM (*Urens*) caule erecto, foliis inferioribus subrotundis crenatis, superioribus lanceolatis serratis, spicâ terminali. *Rapuntium with an erect stalk, the lower leaves roundish and crenated, the upper spear-shaped, sawed, and a spike of flowers terminating the stalk.*

6. RAPUNTIUM (*Inflatum*) caule erecto, foliis ovatis subserratis, pedunculo longioribus, capsulis inflatis. *Cardinal-flower with an erect stalk, oval leaves which are somewhat sawed, longer than the foot-stalks, and swelling seed-vessels.*

7. RAPUNTIUM (*Hirtum*) foliis ovalibus crenatis lanatis, floribus lateralibus solitariis. *Cardinal-flower with oval crenated leaves, which are downy, and flowers growing singly from the sides of the stalks.*

8. RAPUNTIUM (*Longiflorum*) foliis lanceolatis dentatis, pedunculis brevissimis lateralibus tubo corollæ longissimo. *Cardinal-flower with spear-shaped indented leaves, very short foot-stalks to the flowers, which proceed from the sides of the stalks, and a very long tube to the petal.*

9. RAPUNTIUM (*Erinum*) caule patulo ramoso foliis lanceolatis subdentatis, pedunculis longissimis. *Rapuntium with a spreading branching stalk, spear-shaped leaves which are somewhat indented, and very long foot-stalks to the flowers.*

10. RAPUNTIUM (*Erinoides*) caulibus procumbentibus, foliis lanceolatis serratis, pedunculis lateralibus. *Cardinal-flower with trailing stalks, spear-shaped sawed leaves, and foot-stalks proceeding from the sides.*

The first sort grows naturally by the side of rivers and ditches in great part of North America, but has been many years cultivated in the European gardens for the great beauty of its scarlet flowers. The root is composed of many white fleshy fibres; the lower leaves are oblong, a little sawed, and of a dark purplish colour on their upper side; the stalks are erect, and rise about a foot and a half high, garnished with spear-shaped leaves sawed on their edges, having very short foot-stalks, and are placed alternately; the stalk is terminated by a spike of flowers of an exceeding beautiful scarlet colour; these have pretty long tubes, which are a little incurved, but at the top are cut longitudinally into five segments; three upper, which are the smallest, are greatly reflexed; the three under, which form the lower lip, are larger, and spread open.

This is propagated by seeds, which, when they ripen in England, should be sown in autumn in pots, and placed under a common hot-bed frame; or if the seeds come from the country where the plants grow naturally, they should be sown in the same way as soon as they arrive; for if they are kept out of the ground till spring, they will lie a year in the ground before they vegetate. The pots in which these seeds are sown, should be exposed to the open air at all times when the weather is mild, but they must be screened from the frost and very hard rain in winter. In the spring the plants will appear, when they should have as much free air as possible in mild weather, and if the spring proves dry, they must be frequently refreshed with water. As soon as they are fit to remove, they should be each planted in a separate small pot, and placed in the shade till they have taken new root; then they may be placed where they may have the morning sun, in which situation they may remain till autumn. During the summer they must be duly watered in dry weather, and when the roots have filled the pots, they should be removed into larger. In autumn they must be placed under a common frame to screen them from hard frost, but they should enjoy the open air at all times when the weather is mild. The spring following they should be new potted, and placed where they may have the morning sun, always observing to water them duly in dry weather, which will cause their stalks to be stronger, and produce larger spikes of flowers, which will continue long in beauty, if they are not too much exposed to the sun; and if the autumn proves warm, the seeds will ripen well in England. The roots of this plant will sometimes last two or three years and produce offsets for increase, but those will not flower so strong as the seedling plants, therefore an annual supply of them should be raised. There are many who propagate this plant by cutting their stalks into proper lengths, and plant them in pots filled with good earth, or into an east border, covering them close with glasses. These frequently take root, so produce young plants, but they are not so good as the seedlings.

The plants of this sort will live in the full ground, if they are protected from hard frost in winter, and they will flower stronger than those in pots.

The second sort grows naturally at Campeachy, from whence the late Mr. Robert Millar sent the seeds; this hath a fibrous root like the first. The stalks are much larger, and rise a foot higher; they are closely garnished with leaves, which are long, smooth, and entire, ending in acute points, and are terminated by short spikes of flowers, which are larger than those of the first sort, but are of the same beautiful scarlet colour.

This is propagated by seeds in the same way as the first, but the plants are not so hardy, therefore require to be placed in a moderate stove in winter, and in summer they should be placed in a deep frame, where they may be covered with glasses in bad weather, but enjoy the free air at all times when the weather is favourable. With this management the plants flowered very well in the Chelsea garden, but they did not perfect seeds.

The third sort grows naturally in Virginia, but has been long an inhabitant of the English gardens; this hath a perennial fibrous root. The leaves are smooth, oval, spear-shaped, and a little indented on their edges; the stalks rise a foot and a half high, garnished with leaves like those at the bottom, which are gradually smaller to the top, sitting close to the stalk. The flowers come out from the wings of the leaves; they are of a pale blue colour, and have large empalements, whose edges are reflexed. The seeds frequently ripen in England.

It is propagated in the same way as the first sort, and the plants require the same culture.

The fourth sort grows naturally in Jamaica, from whence the late Dr. Houstoun sent the seeds; this is an annual plant. The stalk rises a foot high, then divides into four or five smaller, which grow erect. The lower part is garnished

nished with heart-shaped smooth leaves, having small indentures on their borders, and stand upon short foot-stalks. The upper slender stalks are thinly garnished with small purplish flowers to the top, and are succeeded by small seed vessels, which ripen in autumn. When the seeds are permitted to scatter on the pots which stand near them, and those are sheltered from the frost, the plants will come up plentifully the following spring; or if they are sown in pots in autumn, and sheltered in the winter, the plants will come up at the same time; and these should be transplanted into separate small pots, placing them under a frame, where they will flower in June and July, and their seeds will ripen in September, when the plants will decay.

The fifth sort grows naturally in the forests about Blois in France; this is an annual plant. The root is composed of many fleshy fibres; the stalk rises about two feet high, garnished with spear-shaped leaves, which are very thin, and sawed on their edges, sitting close to the stalk; the upper part of the stalk is garnished with very small leaves; from their base arise the flowers, which are of a bright blue colour. These appear in July, and are succeeded by roundish seed-vessels, with holes at the top, which are filled with small red seeds.

The seeds of this plant should be sown in autumn in pots filled with loamy earth, and placed under a hot-bed frame in winter, and when the plants come up in the spring, they should be transplanted either into a border of soft loamy earth, or into separate pots, shading them till they have taken new root; afterward they must be duly watered in dry weather, which will cause them to flower strong, and produce good seeds annually.

The sixth sort grows naturally in North America; this is a biennial plant in England, which rarely flowers the same year as the plants come up, but decays the second year soon after the seeds are ripe. The stalks of this are channelled and hairy; they grow erect to the height of two feet, garnished with thin oval leaves, sitting close to the stalk, of a light green, and a little sawed on their edges. The flowers stand upon long slender foot-stalks, which come out from the wings of the leaves, forming a loose spike, which terminate the stalk; they are small, and of a light blue colour. This is propagated by seeds, which should be sown in autumn, in pots filled with rich earth, and treated in the same way as the first sort.

The seventh sort grows naturally at the Cape of Good Hope. This is a biennial plant; the stalks rise a foot and a half high, covered with a hairy down, and are purplish toward the bottom; the leaves are oval, of a deep green colour, a little hairy on their under side, and sit close to the stalks. The flowers stand upon long slender foot-stalks, which come out from the bosom of the leaves, sometimes one proceeding from a joint, and at others they come out opposite on each side the stalk, each foot-stalk sustaining one pale blue flower, which being small makes but little appearance. This may be propagated in the same way as the first.

The eighth sort grows naturally in moist places on most of the islands of the West-Indies. This is also a biennial plant, whose root is composed of a few strong ligneous fibres; the stalk rises about eight or nine inches high, is closely garnished with leaves on every side, which are hairy, very deeply indented on their edges, of a deep green, and sit close to the stalks. The flowers are white, and come out at every joint from the wings of the leaves, standing upon very short foot-stalks; the tube of the flower is from three to four inches long, very slender, and deeply cut at the top into five segments, which spread open, and are succeeded by turgid seed-vessels, crowned by the five segments of the petal, having three holes at the top, filled with small grayish seeds. The seeds of this sort should be sown soon after it is ripe, in pots filled with rich earth, and plunged into the tan-bed in the stove. In the spring these pots may be removed, and plunged into a hot-bed, which will soon bring up the plants; when these are fit to remove they should be each transplanted into a separate small pot, and plunged into a fresh hot-bed, shading them from the sun till they have taken new root; then they may be treated in the same way as other tender plants from the same country, giving them a large share of air in warm weather. In autumn the plants must be plunged into the tan-bed of the stove, where they will flower the following summer and produce ripe seeds, soon after which the plants will decay. If the seeds of this plant are brought from the West-Indies, they should be sown as soon as they arrive, in pots, and if it happens in the winter, the pots should be plunged into the tan-bed in the stove; but if it is in the spring or summer, they may be plunged into a hot-bed in the common frames. These seeds when sown in the spring, seldom grow the same year, therefore the followiug autumn the pots should be removed into the stove, and managed according to the above directions.

The ninth sort grows naturally at the Cape of Good Hope. This is an annual plant; the stalks are slender, branching, and spread out on every side; they rise about a foot high, garnished with small spear-shaped leaves, which are indented on their edges, and sit close to the branches. The flowers are blue; they stand upon very slender long foot-stalks, and are succeeded by small roundish seed-vessels, filled with small seeds, which ripen in September. If the seeds of this sort are sown in autumn, they will succeed much better than when they are sown in spring; these may be sown in pots, and be sheltered under a common hot-bed frame in winter, exposing them to the open air at all times in mild weather, but screening them from the frost; and in the spring the pots should be plunged into a moderate hot-bed, which will soon bring up the plants; when these are fit to remove, they should be each planted in a separate small pot, and plunged into a moderate hot-bed again, shading them from the sun till they have taken new root; then they must have a large share of free air, at all times when the weather is mild; and as the plants grow strong, they should be gradually hardened to bear the open air, into which they should be removed in June, placing them in a sheltered situation, where they will flower in July; and if the season proves favourable, the seeds will ripen in September, but if the season should prove cold, it will be proper to remove one or two plants into a glass-case, to obtain good seeds.

The tenth sort comes from the Cape of Good Hope. This hath trailing stalks; the leaves are sawed on their edges, and the foot-stalks come out from the side of the branches, in which it differs from the last. It may be propagated by seeds, and treated in the same manner as the last.

RAUVOLFIA. Plum. Nov. Gen. 19. tab. 40.

The Characters are,

The flower has a permanent empalement of one leaf, cut into five parts at the top. The petal is funnel-shaped; the tube is cylindrical, and is cut at the brim into five parts. It has five stamina, which are a little shorter than the tube, terminated by erect summits, and a roundish germen supporting a short style, crowned by a headed stigma. The germen afterward becomes a globular berry with two cells, including one compressed seed in each.

The Species are,

1. Rauvolfia (*Canescens*) subpubescens. Lin. Sp. Pl. 303. *Rauvolfia with hairy leaves.*

2. Rauvolfia (*Nitida*) glaberrima nitidissima. Lin. Sp Plant. *Rauvolfia with smooth leaves.*

Both these sorts grow naturally in the warmest parts of America; Mr. Robert Millar sent the seeds of them from Carthagena in New Spain, where he observed the shrubs growing in great plenty. These rise with several ligneous stalks from the root, which grow ten or twelve feet high, sending out a few small side branches, covered with a smooth green bark when young, but as they are older, their bark changes to a gray. The leaves are placed by fours at each joint round the branches; those of the first sort are two inches and a half long, and an inch and a half broad in the middle; they are of a light green, and have a few slight indentures on their edges; the leaves of the other sort are full as long, but are a third part narrower, and of a thinner substance. These differences continue in the plants which are raised from seeds, for I have several times propagated them both from seeds, and have constantly found the seeds produce the same as the plants from which they were taken. The flowers are produced on slender footstalks, which arise from the wings of the leaves; they are tubulous, globular at their base, and are succeeded by roundish berries about the size of those of the Privet, which turn black when ripe. These plants flower most part of summer, and the fruit ripens in autumn and winter; the leaves and stalks of these plants have a milky juice, which flows out if they are broken.

These plants are propagated by seeds, which should be sown in autumn soon after they are ripe, for if they are kept out of the ground till spring the plants rarely come up the same year; and this is frequently the case with those seeds which are brought to England.

The seeds should be sown in pots, and plunged into a hot-bed of tanners bark, for as they are very hard, they frequently remain a long time in the ground; therefore when they are in pots, they may be shifted from one bed to another, as their heat decays. When the plants come up they should have a large share of fresh air admitted to them in warm weather, and but little water. When they are about two inches high they should be transplanted each into a separate small pot, and plunged into the hot-bed again, observing to shade them from the sun until they have taken new root; after which time they should have free air admitted to them every day, in proportion to the warmth of the season. In this hot-bed the plants may remain till toward Michaelmas, when they should be removed into the stove, and plunged into the tan-bed, where they must be kept warm, and not have too much moisture in cold weather.

As these plants are natives of very hot countries, they will not live in the open air in England, therefore they should constantly remain in the stove; and if they are continued in the bark-bed, they will thrive much faster than when they are placed on stands in a dry stove. But in the summer season they should have a large share of fresh air admitted to them, and the leaves of the plants must be now and then washed with a sponge, to clear them from the filth they are apt to contract, which, if suffered to remain, will retard the growth of the plants. When due care is taken of them they will thrive very fast; the second year they will produce flowers, and continue so to do for many years, and will perfect their seeds in England. They may also be propagated by cuttings, which should be laid to dry for two or three days before they are planted; and then should be plunged into a moderate hot-bed of tanners bark, observing to shade them until they have taken root, after which time they may be treated as the seedling plants.

RESEDA. Tourn. Inst. R. H. 423. tab. 238. Bastard-rocket.

The Characters are,

The empalement of the flower is cut into several parts almost to the bottom, and is permanent. The petals of the flower are unequal, generally trifid, and have a honey gland on their base, the length of the empalement. The honey glands are plain, erect, and produced from the upper side of the receptacle, between the stamina and the place of the upper petal, joining with the base of the petals, dilating from the sides. It hath fifteen or sixteen short stamina, terminated by erect obtuse summits, and a gibbous germen, sitting upon very short styles, crowned by a single stigma. The germen afterward becomes a gibbous angular capsule of one cell, with an aperture between the styles, filled with kidney-shaped seeds, fastened to the angles of the capsule.

The Species are,

1. Reseda (*Vulgaris*) foliis pinnatis, foliolis integris alternis floribus tetragynis. *Bastard-rocket with winged leaves, whose lobes are entire, placed alternate, and have four styles to the flower.*

2. Reseda (*Crispa*) foliis omnibus trifidis, inferioribus pinnatis. Hort. Cliff. 213. *Bastard-rocket with all the leaves trifid, and the lower ones winged.*

3. Reseda (*Phyteuma*) foliis integris trilobisque, calycibus maximis. Hort. Cliff. 412. *Bastard-rocket with entire trifid leaves, having the largest empalement.*

4. Reseda (*Undata*) tetraginisque calycibus quinquepartitis foliis pinnatis undulatis. Lin. Sp. Plant. 644. *Bastard-rocket with difformed indented leaves, and flowers having three styles.*

5. Reseda (*Alba*) foliis pinnatis, floribus tetragynis, calycibus sexpartitis. Hort. Upsal. 149. *Bastard-rocket with winged leaves, and flowers having four styles.*

6. Reseda (*Odorata*) foliis integris trilobisque, floribus tetragynis. Tab. 217. *Bastard-rocket with entire and three-lobed leaves, and flowers having four styles; commonly called Sweet Reseda, or Mignonette d'Egypt.*

7. Reseda (*Canescens*) foliis subulatis sparsis. Sauv. Monsp. 41. *Bastard-rocket with awl-shaped leaves placed thinly.*

8. Reseda (*Sesamoides*) foliis lanceolatis integris, calycibus quadrifidis. Lin. Sp. Plant. 448. *Bastard-rocket with spear-shaped entire leaves, and quadrifid empalements.*

The first sort grows naturally in the south of France, Italy, and Spain. This is a biennial plant, which flowers and seeds the second year, and perishes soon after. The root is long, white, and a little ligneous; the leaves are unequally winged and entire; the stalks are channelled, and garnished with smaller winged leaves; they rise a foot and a half high, terminated by a long loose spike of pale yellow flowers, composed of several unequal petals; the two upper are the largest, the side ones less, and the lower are so small as to be scarce conspicuous; they are all of a singular figure, and appear as if one leaf came out of two others. In the middle are situated many stamina, terminated by yellow summits; at the bottom a three-cornered germen, which afterward turns to a three-cornered seed-vessel, having three or four holes at the top, filled with black seeds.

The second sort grows naturally in chalky land in many parts of England; the lower leaves of this are winged, and every lobe is cut into three smaller; they are curled on their edges, and have some small indentures. The stalks rise about the same height as the former, and are terminated by longer and looser spikes of flowers; the flowers are paler, and approach to a white.

The third sort grows naturally in the south of France and Italy. This is an annual plant, which has generally a single fleshy tap-root, running deep in the ground, sending out several trailing stalks near a foot long, which di-

vide

vide into fmaller branches, garnifhed with fmall leaves, fome of which are wedge-fhaped and entire, others are cut into three obtufe parts. The end of the branches are terminated by loofe fpikes of flowers, ftanding upon pretty long foot-ftalks. The empalement of the flower is large, divided into five fegments almoft to the bottom; the flowers are white, and fhaped like thofe of the other forts.

The fourth fort grows naturally in Italy and Spain. This is a biennial plant; the lower leaves are unequally winged, fome of the intermediate lobes or fegments being much lefs than the other, and of different fhapes. The ftalk rifes a foot and a half high, garnifhed with fmaller difformed winged leaves, indented on their edges. The flowers are produced in flender loofe fpikes at the top of the ftalks; they are fmall and white, and of the fame fhape with the others.

The fifth fort grows naturally in the fouth of France. It is a biennial plant; the lower leaves are large, winged, and compofed of many narrow lobes or fegments, placed alternate, which are of a grayifh colour; the ftalks rife two feet and a half high, garnifhed with the like leaves, which diminifh in their fize to the top; the ftalks are terminated by fhorter and thicker fpikes of flowers than either of the former, which are white, and fhaped like thofe of the other fpecies.

The fixth fort is fuppofed to grow naturally in Egypt. The feeds of this were fent me by Dr. Adrian Van Royen, the late profeffor of botany at Leyden. The root of this plant is compofed of many ftrong fibres, from which come out feveral ftalks about a foot long, which divide into fmall branches, garnifhed with oblong leaves, fome of which are entire, and others are divided into three parts, of a deep green. The flowers are of an herbaceous white colour, produced in loofe fpikes at the end of the branches; they ftand upon pretty long foot-ftalks, and have large empalements, equal with the petals, and fmell very like frefh Rafpberries, which occafions its being much cultivated in the Englifh gardens. This plant is fo like the third fort, as to be taken for the fame by fome, but the flowers of the third have no fcent; fo that thofe who have been impofed on, by having the feeds of the third fort fent them for this, have fuppofed the plant was degenerated.

The feventh fort grows naturally upon the mountains in Spain. This hath a perennial root, from which arife a few flender ligneous ftalks a foot and a half high, which are thinly garnifhed with linear obtufe leaves, of a grayifh colour; the upper part of the ftalk is garnifhed for a good length with fmall, whitifh, purple flowers, ranged in a very loofe fpike, fitting clofe to the ftalk.

The eighth fort grows naturally upon dry banks and old walls in many parts of England, but is cultivated in fome places for the dyers ufe. This is now generally believed to be the plant, with which the ancient inhabitants of this ifland painted themfelves, and not the Woad, as has been by fome fuppofed; for the dyers weed is a native here, whereas the Woad has been fince introduced into this country. This is a biennial plant; the root is compofed of a few ligneous fibres; the leaves are four inches long, and half an inch broad, entire, and ending in obtufe points: thefe the firft year fpread circularly near the ground, and have fome gentle wavings on their edges; the ftalks rife three feet high, garnifhed with leaves of the fame fhape with thofe at bottom. They are terminated by long loofe fpikes of yellowifh flowers, which appear the latter end of June, and the feeds ripen in September.

The five forts firft mentioned, and alfo the feventh, are feldom cultivated in gardens, except for the fake of variety, having very little beauty to recommend them, and being of no ufe; but whoever has a mind to have them, need only fow their feeds in autumn, and when the plants come up, if they are thinned and kept clean from weeds, it is all the culture they require; or if the feeds are permitted to fcatter the plants will come up in plenty, and fometimes become troublefome weeds.

The feeds of the fixth fort fhould be fown on a moderate hot-bed in March, and when the plants are ftrong enough to tranfplant, they fhould be pricked out upon another moderate hot-bed to bring them forward; but the plants fhould have a large fhare of air in warm weather, otherwife they will draw up very weak. About the latter end of May the plants may be planted out, fome into pots, to place near the apartments, and others into warm borders, where they may remain to flower and feed. For the plants which grow in the full ground, often produce more feeds than thofe which are in pots; but at the time when the feed-veffels begin to fwell, the plants are frequently infefted with green caterpillars, which, if they are not deftroyed, will eat off all the feed-veffels.

If the feeds of this plant are fown on a bed of light earth in April, the plants will come up very well, and when they are not tranfplanted will grow larger than thofe which are raifed in the hot-bed, but they will not flower fo early. The plants may be preferved through the winter in a green-houfe, where they will continue flowering moft part of the year, but the fecond year they are not fo vigorous as in the firft.

The eighth fort is the Weld, which is accounted a rich commodity for dyeing; where this is cultivated the feeds are commonly fown with Barley in the fpring, and after the Barley is taken off the ground, the Weld begins to make fome progrefs, and the next feafon is pulled up for ufe. This hath been long practifed, and it will be difficult to prevail on the cultivators of this plant to depart from their old cuftoms; but if any perfon will follow the directions hereafter given, I can from experience promife them much better fuccefs.

The Weld will grow upon very poor foil, but the crop will be in proportion to the goodnefs of the land, for upon very poor ground the plants will not rife a foot high, whereas upon good ground I have meafured them upward of three feet, and the ftalks, leaves, &c. have been in proportion; fo that the better the foil is upon which it is fown, the greater will be the produce.

The beft way to cultivate this plant, is to fow it without any other crop; if the ground is ready by the beginning or middle of Auguft, that will be a good feafon; the land fhould be well ploughed and harrowed fine, but unlefs it is very poor it will not require dung; when the ground is well harrowed and made fine, the feeds fhould be fown; one gallon of the feeds is fufficient to fow an acre of land, for they are fmall. If rain falls in a little time after the feeds are fown it will bring up the plants, and in two months time they will be fo far advanced as to be eafily diftinguifhed from the weeds; then they fhould be hoed in the like manner as Turneps, always obferving to do it in dry weather, for then the weeds will foon die after they are cut up; at this time the plants may be left about fix inches diftance; if this is done in dry weather, and the work well performed, the plants will be clean from weeds till the fpring; but as young weeds will come up in March, fo if in dry weather the ground is hoed again, it may be performed at a fmall expence while the weeds are young, then they will foon decay; and if after this there fhould be many more weeds appear, it will be proper to hoe it a third time, about the beginning of May, which will preferve the ground clean till the Weld is fit to pull. The beft time to pull the Weld for ufe, is as foon as it begins to flower, though moft people ftay till the feeds are ripe, being unwilling to lofe the

the feeds; but it is much better to fow a fmall piece of land with this feed, to remain for a produce of new feeds, than to let the whole ftand for feed, becaufe the plants which are permitted to ftand fo long, will be much lefs worth for ufe than the value of the feeds; befides, by drawing off the crop early, the ground may be fown with Wheat the fame feafon; for the plants may be drawn up the latter end of June, when they will be in the greateft vigour, fo will afford a greater quantity of the dye.

When the plants are pulled, they may be fet up in fmall handfuls to dry in the field, and when it is dry enough it may be tied up in bundles and houfed dry, being careful to ftack it loofely, that the air may pafs between to prevent its fermenting.

That which is left for feeds fhould be pulled as foon as the feeds are ripe and fet up to dry, then beat out for ufe, for if the plants are left too long the feeds will fcatter. The ufual price of the feed is ten fhillings a bufhel.

RHABARBARUM. See RHEUM.

RHABARBARUM MONACHORUM. See RUMEX.

RHAGADIOLUS. See LAPSANA.

RHAMNOIDES. See HIPPOPHÆ.

RHAMNUS. Tourn. Inft. R. H. 593. tab. 366. The Buckthorn.

The CHARACTERS are,

It hath male and female flowers on different plants; thefe have no empalements according to fome, or petals according to others. The cover of the fexes is funnel-fhaped, cut into four parts at the top, which fpread open. The male flowers have four ftamina the length of the tube, terminated by fmall fummits. The female flowers have a roundifh germen, fupporting a fhort ftyle, crowned by a quadrifid ftigma. The germen afterward becomes a roundifh berry, inclofing four hard feeds.

The SPECIES are,

1. RHAMNUS (*Catharticus*) floribus axillaribus, foliis ovato-lanceolatis ferratis nervofis. *Buckthorn with flowers proceeding from the fides of the branches, and oval, fpear-fhaped, fawed, veined leaves; the purging, or common Buckthorn.*

2. RHAMNUS (*Minor*) floribus axillaribus, foliis ovatis acuminatis nervofis integerrimis. *Buckthorn with flowers proceeding from the fides of the branches, and oval, acute-pointed, entire leaves, having veins.*

3. RHAMNUS (*Longifolius*) foliis lanceolatis, floribus axillaribus. *Buckthorn with fpear-fhaped leaves, and flowers growing from the fides of the ftalks.*

4. RHAMNUS (*Africanus*) foliis cuneiformibus confertis perennantibus, floribus corymbofis alaribus. *Buckthorn with wedge-fhaped evergreen leaves, growing in clufters, and flowers growing in roundifh bunches from the fides of the branches.*

The firft fort grows naturally in the hedges in many parts of England; it rifes with a ftrong woody ftalk to the height of twelve or fourteen feet, fending out many irregular branches; the young fhoots have a fmooth, grayifh, brown bark, but the bark of the older branches is darker and rougher, armed with a few fhort thorns. The leaves ftand upon pretty long flender foot-ftalks, of the oval fpear-fhape, fawed on their edges, of a dark green on their upper fide, but of a pale or light green on their under, having a pretty ftrong midrib, and feveral veins proceeding from it. The flowers come out in clufters from the fide of the branches; thofe of the male have as many ftamina as there are divifions in the petal; thofe of the female have a roundifh germen, which afterward turns to a pulpy berry, of a roundifh form, inclofing four hard feeds.

The berries of this are ufed in medicine. From the juice of thefe berries is made a very fine green colour, called by the French Vord-de-veffie, which is much efteemed by the painters in miniature.

The fecond fort grows naturally in the fouth of France. This is an humble fhrub, feldom rifing more than three or four feet high, fending out many irregular branches, covered with a dark brown bark, garnifhed with oval leaves, ending in acute points; they are of a yellowifh green, and a thin confiftence, having feveral veins diverging from the midrib toward the fides. The flowers come out upon fmall curfons or fpurs on the fide of the branches, each ftanding upon a feparate fhort foot-ftalk, of a yellowifh herbaceous colour, having fhort fwelling tubes, cut into five acute fegments at the top, which fpread open; they appear in June, but are not fucceeded by berries here.

Mr. Du Hamel de Monceaux, of the Royal Academy of Sciences at Paris, fays, that the fruit of this fpecies gathered green is the Grained'Avignon, or Avignon Berries, which are ufed in dyeing of yellow; but I have been affured by a gentleman of fkill who refided long in the fouth of France, that the Avignon Berries were the fruit of the narrow-leaved Alaternus; and in order to be fatisfied of the truth, I gathered a quantity of the berries of the narrow-leaved Alaternus before they were full ripe, and carried them to two eminent dealers in this commodity, and afked them if they knew what thofe berries were; they both affured me, after making trial of them, that they were Avignon Berries, and if I had a large quantity of them they would purchafe them all; therefore as the Alaternus before mentioned is one of the moft common fhrubs in the fouth of France, from whence the Avignon Berries are brought, we may fuppofe Mr. Du Hamel has been ill informed.

The third fort grows naturally in Spain and Italy. This grows to a larger fize than the fecond, but not fo high as the firft. The branches are ftronger, and armed with a few long fpines; the leaves are like thofe of the wild Plumb, but a little longer and narrower; the flowers are fmall, of a yellowifh colour, and are produced from the fide of the branches.

The firft fort is fo common in the hedges in many parts of England, that it is feldom cultivated in gardens. This rifes eafily from feeds, if they are fown in autumn foon after the berries are ripe; but if they are kept out of the ground till fpring, the plants will not come up till the year after; thefe will require no particular treatment, but may be managed in the fame way as young Crabs, or any other hardy deciduous tree; it may alfo be propagated by cuttings or layers. If the young fhoots are laid in autumn, they will put out roots by the following autumn, when they may be taken off from the plants, and either planted in a nurfery to remain there to get ftrength for a year or two, or they may be planted where they are defigned to remain. This is not fo proper for hedges as the Hawthorn or Crab, fo thofe fhould be preferred to it.

The two other forts are preferved in botanic gardens for the fake of variety, but as they are not beautiful, few perfons cultivate them here. As thefe do not produce fruit in England, they are propagated either by laying down of the young branches in autumn, or by planting the cuttings in the fpring, before the buds begin to fwell. Thefe will put out roots in the fame manner as the common fort, and may be treated in the fame way, for they are both hardy plants, and will thrive in the open air.

The fourth fort grows naturally at the Cape of Good Hope, fo is too tender to thrive in the open air in England, but if it is placed in a common green-houfe with Myrtles, Olives, and the hardier kinds of exotic plants in winter, and removed to the open air in fummer, it will thrive very well.

well. This rises with a shrubby stalk to the height of four or five feet, sending out many side branches, which, when young, are covered with a green bark; but as they advance, the bark changes to a dark brown, armed with a few long slender thorns, and garnished with wedge-shaped leaves, which come out in clusters at each joint, four, five, or six rising from the same point, which differ in size; they are of a deep green, and continue all the year; their points are rounded, growing narrower to their base, sitting close to the branches. The flowers are produced on the side of the branches at each joint; they are collected into roundish bunches, standing upon foot-stalks an inch long; they are white, and have short tubes; the upper part is cut into five acute segments, which spread open in form of a star. These appear in June, at which time the whole shrub seems covered with flowers, so as to make a fine appearance; and as the leaves continue green all the year, it deserves a place where there is a conveniency to shelter them in winter.

This sort has not as yet produced seeds in England, but it may be easily propagated by cuttings, which should be planted in pots the beginning of April. The pots should be plunged into a moderate hot-bed, and the cuttings should be shaded from the sun in the heat of the day; but they must by no means have too much wet. These cuttings will put out roots in about six weeks; then they must have a large share of air admitted to them, and gradually inured to bear the open air, into which they should be soon after removed; when they are well hardened, they may be shaken out of the pots, and separated, being careful to preserve a ball of earth to each, and plant them into single pots, placing them in the shade till they have taken new root; then they may be removed into a sheltered situation, where they may remain till the frost comes on in autumn, at which time they must be housed, and treated in the same way as the other hardier kinds of green-house plants.

RHEUM. Lin. Gen. Plant. 454. The Rhubarb.

The CHARACTERS are,

The flower has no empalement; it hath one petal, which is narrow at the base, and impervious. The brim is cut into six parts, which are obtuse and alternately smaller; it hath nine hair-like stamina inserted in the petal, and is of the same length, terminated by oblong twin summits, which are obtuse, and a short three-cornered germen with scarce any style, crowned by three feathered stigmas, which are reflexed. The germen afterward becomes a large three-cornered seed, with an acute membranaceous border.

The SPECIES are,

1. RHEUM (*Rhaponticum*) foliis cordatis glabris, spicis obtusis. *Rhubarb with smooth heart-shaped leaves, and obtuse spikes of flowers.*

2. RHEUM (*Undulatum*) foliis subvillosis, undulatis petiolatis potiolis æqualibus. Lin. Diff. 1. tab. 1. Sp. Plant. 372. *Rhubarb with hairy leaves, having equal foot-stalks.*

3. RHEUM (*Compactum*) foliis cordatis glabris, marginibus sinuatis, spicis erectis compactis. Tab. 218. *Rhubarb with heart-shaped smooth leaves, which are sinuated on their borders, and compact spikes of flowers.*

4. RHEUM (*Ribes*) foliis granulatis, petiolis æqualibus. Lin. Sp. Plant. 372. *Rhubarb with granulated leaves, having equal foot-stalks; called by the Arabians Ribes.*

The first sort grows naturally near the Pontic Sea, but has been long an inhabitant of the English gardens. When the seeds were first brought to Europe, they were supposed to be of the true Rhubarb, but upon making trial of the roots, they were found to be greatly inferior to those of the true Rhubarb; and upon examination it was found to be the Rhapontic of Prosper Alpinus, commonly called Pontic Rhubarb. This hath a large thick root, which divides into many less, running deep in the ground; the outside is of a reddish brown colour, and the inside yellow, from which arise several leaves, in number according to the size of the root; these come up folded in the spring, and afterward expand themselves; they are of a roundish heart-shape, smooth, have very thick foot-stalks of a reddish colour, which are a little channelled on their lower side, but flat at the top. When the plant grows in rich land, the foot-stalks of the leaves are near two feet long, and thicker than a man's thumb; the leaves are also often two feet long, and as much in breadth, having several strong longitudinal veins running from the foot-stalks to the borders; they are of a deep green, a little waved on their edges, and have an acid taste, but particularly the foot-stalks, which are now frequently used for making of tarts. From between the leaves arise the flower-stem, which is of a purple colour, garnished with one leaf at each joint, of the same shape with those below, but smaller, and sit close to the stalk. The stalks grow from two to three feet high, according to the strength of the ground, and are terminated by thick close obtuse spikes of white flowers; these are succeeded by large triangular brown seeds, having a border or wing at each angle, which ripen in August.

The seeds of the second sort were sent me from Leyden by the late Dr. Boerhaave, by the title of Rhabarbarum Chinense verum, or true China Rhubarb, which succeeded in the Chelsea garden. The root of this sort divides into a greater number of thick fibres than those of the first, which run deeper into the ground, and are of a deeper yellow within. The leaves appear much earlier in the spring; the foot-stalks are not so much channelled on their under side, and are plain on their upper; they are not so red nor so thick. The leaves are longer, running more to a point, and are waved on their edges, are a little hairy on their upper side, and have many strong veins or ribs on their under. The flower-stem is of a pale brownish colour, rising about four feet high, dividing into several loose panicles or bunches of white flowers, which are succeeded by triangular seeds like those of the first sort, which ripen early in the season.

The seeds of the third sort were sent me from Petersburgh, for the true Tartarian Rhubarb. The roots of this sort are large, and do not divide into so many parts as those of the second, and are yellower within; the leaves appear as early in the spring; the foot-stalks of these are of a pale green, almost as large as those of the first sort; they have scarce any channels, and are flat on their upper side; the leaves are heart-shaped and smooth; they do not run out to so great length in a point as the second, but are longer than those of the first; they are very broad toward their base, and have very large pale green ribs on their under side, a little waved on their edges, and have a sharp acid flavour. The flower-stalk is a pale green; it rises four feet high, as large as a common walking cane, garnished at each joint by one leaf, of the same shape with those below, but smaller, sitting close to the stalk; the upper part of the stalk divides into small branches, which are again divided into less, each sustaining a panicle or spike of white flowers, which are succeeded by large triangular bordered seeds, like those of the first sort.

The roots of this last approach nearer to those of the foreign Rhubarb than either of the other, both in shape and quality; and as the seeds which were sent to Petersburgh, were gathered from the plants growing on the spot where the Rhubarb was taken up, so there is little reason to doubt of its being the true sort, for upon trial of the roots they are found to be as good as the foreign Rhubarb; and as the roots which have grown in England produce great

plenty

plenty of feeds here, fo they may be propagated with great eafe.

It has been learnedly controverted by the botanifts, whether the Rhapontic of the ancients, and the Rhubarb of the moderns, is one and the fame plant; fome affirming, and others denying that there is any agreement; the reafonings on both fides may be feen in the appendix to the fecond volume of John Bauhin's Hiftory of Plants.

There has been lately another fort introduced called the Palmated, and is fuppofed by many to be the true kind, but I believe there are more than one fpecies ufed.

The fourth fort grows naturally on Mount Libanus, and other mountainous parts of Syria. This hath a thick flefhy root, which runs pretty deep in the ground, from which arife feveral leaves in the fpring, which come up folded together, and afterward expand; they have very fhort footftalks, fo fpread near the ground, but during the fpring their borders are erect, and form a fort of hood, having feveral folds, curled and waved on their edges; they are of a purplifh green, and have purple veins and borders; their furface appears ftudded with rough protuberances; when the leaves are fully expanded in fummer, they are a foot long, and about two feet broad; their under fide is paler than the upper, and their borders appear fringed. I have not feen this plant in flower, but the feeds of it were brought from Mount Libanus, by the Right Rev. Dr. Pococke, the late Bifhop of Offory; thefe were large, covered with a fucculent pulp, of a deep red colour, and very aftringent tafte; this fucculent covering may have occafioned its being taken for a berry, by many of the old writers; the fhape of the feed is like that of the other fpecies.

Thefe plants are all propagated by feeds, which fhould be fown in autumn foon after they are ripe, then the plants will come up the following fpring; but if they are kept out of the ground till fpring, the plants feldom come up till the next fpring, fo that a whole year will be loft. The feeds fhould be fown where the plants are defigned to remain, for as the roots are large and flefhy, fo when they are tranfplanted, they do not recover their removal in lefs than two years; nor will the roots of thofe plants which are tranfplanted, ever grow fo large and fair as thofe which remain where they were fown. When the plants appear in the fpring, the ground fhould be hoed over to cut up the weeds; and where the plants are too clofe fome fhould be cut up, to allow room for the others to grow, in the fame manner as is practifed for Carrots and Parfneps, leaving them at the firft time of hoeing fix or eight inches afunder, for fear of accidents, but at the fecond time of hoeing they may be feparated to a foot and a half diftance or more. After this the plants will require no other culture but to keep them clean from weeds, fo that as foon as the weeds appear, if the ground is fcuffled over with a Dutch hoe in dry weather, it may be done for a fmall expence, and thereby the ground will be kept clean. If this is begun early in the fpring before the weeds are large, they will foon die, and by repeating it two or three times at proper intervals during the fpring, the ground will be made clean; and when the plants fpread out their leaves to cover the ground, they will prevent the growth of weeds.

In autumn the leaves of thefe plants decay; then the ground fhould be made clean, and in the fpring before the plants begin to put out their new leaves, the ground fhould be hoed and made clean again; the fecond year after the plants come up, many of the ftrongeft will produce flowers and feeds, but the third year moft of them will flower. The feeds of thefe fhould be carefully gathered when ripe, and not permitted to fcatter, left they fhould grow to injure the old plants. The roots of thefe plants will remain many years without decaying, and I am informed, that the old roots of the true Rhubarb are much preferable to the young ones. They delight in a rich foil, not too dry nor over moift, and where there is a good depth for their roots to run down in fuch land, their leaves will be very large, and their roots will grow to a great fize.

The firft fort is now frequently cultivated in gardens for the foot-ftalks of their leaves, which are peeled and made into tarts in the fpring: it is alfo kept in gardens to fupply the fhops with the roots, which are ufed in medicine.

The true Rhubarb is now fown in many gardens, and may probably fucceed fo well here in time, as that a fufficient quantity of that valuable drug may be raifed to fupply our confumption.

RHEXIA. Gron. Flor. Virg. 41.

The CHARACTERS are,

The empalement of the flower is permanent, oblong, tubulous, and of one leaf, divided into four parts at the brim. The flower has four roundifh petals inferted in the empalement. It hath eight flender ftamina, which are inferted in the empalement, terminated by declining furrowed fummits. It has a roundifh germen, fupporting a declining ftyle the length of the ftamina, crowned by a thick oblong ftigma. The germen afterward becomes a roundifh capfule, with four cells in the fwollen empalement, opening with four valves, filled with roundifh feeds.

The SPECIES are,

1. RHEXIA (*Virginica*) calycibus glabris. Flor. Vir. 41. *Rhexia with fmooth empalements.*

2. RHEXIA (*Mariana*) foliis ciliatis. Lin. Sp. Pl. 346. *Rhexia with fine hairy leaves.*

The firft fort was difcovered by Mr. Banifter in Virginia, from whence he fent the feeds to England, which fucceeded in feveral gardens. This rifes with an erect ftalk near a foot and a half high, is four-cornered and hairy, garnifhed with hairy fpear-fhaped leaves placed oppofite. The ftalk has two foot-ftalks coming out from the fide oppofite, at the upper joint, and is terminated by two other; thefe each fuftain two or three red flowers with heart-fhaped petals, which fpread open in form of a crofs.

The fecond fort grows naturally in Maryland, from whence I received the feeds. This fends up an erect ftalk about ten inches high, garnifhed with fpear-fhaped leaves, fet on by pairs; from every joint of the ftalk comes out two fhort fhoots oppofite, garnifhed with fmall leaves of the fame fhape; the whole plant is covered with ftinging iron-coloured hairs. The ftalk divides at the top into two foot-ftalks, fpreading from each other, having one reddifh flower on each; thefe have four heart-fhaped petals, which fpread open like the other. Thefe plants are propagated by feeds, which muft be procured from the places where they grow naturally. If the feeds arrive before the fpring, and are fown foon after in pots filled with good frefh earth, and placed under a garden frame to guard them from froft, the plants will come up the following fpring; but when the feeds are fown in the fpring, the plants rarely come up the firft year. When the plants come up, and are fit to remove, part of them fhould be planted in an eaft border, where they may have only the morning fun, and the other may be planted into pots, that they may be fheltered under a frame in winter, for they are fometimes deftroyed by fevere froft, though they will live abroad in the common winters very well; the fecond year the plants will flower, and with care they may be continued three or four years.

RHINANTHUS. Lin. Gen. Plant. 658. Rattle, or Loufewort.

There are feveral fpecies of this genus which grow naturally in moift meadows in many parts of Europe, one of which

which is very common in England, where it is one of the most troublesome weeds among the Grass, spreading itself over the whole ground, so that in many of the water meadows, there is more of this plant than Grass. It is an annual plant, which flowers the latter end of May, so that the seeds ripen by the time the Grass of these meadows is mowed, and the seeds scatter and fill the ground with young plants the following spring; therefore in order to destroy it, the Grass should be cut as soon as the flowers of this plant appear.

These plants are with great difficulty kept in gardens; they are biennial, so are only propagated by seeds; these should be sown soon after they are ripe, otherwise they will not succeed, nor will the plants bear removing, therefore should be sown where they are to remain, which should be in a moist rich soil, and a shady situation: when the plants come up, they must be thinned and kept clear from weeds, which is all the culture they require. If the seeds of these plants are permitted to scatter, the plants will come up better than those which are sown by hand.

RHIZOPHORA. Lin. Gen. Plant. 524. This is called Mangrove by the inhabitants of the West-Indies; there are several species of this genus which grow in salt water rivers both in the East and West-Indies, but as they will not grow upon land, it is needless to enumerate them here.

RHODIOLA. Lin. Gen. Plant. 997. Rose-root.

The CHARACTERS are,

It hath male and female flowers in different plants; the male flowers have an empalement of one leaf, cut into four or five segments almost to the bottom; they have four obtuse petals, and four nectariums, which are erect and shorter than the empalement, with eight awl-shaped stamina which are longer than the petals, terminated by obtuse summits. They have four oblong acute germen without style or stigma, so are abortive. The female flowers have the same empalement as the male; they have four obtuse petals equal with the empalement, and have four nectariums like the male, with four oblong acute-pointed germen sitting upon an erect style, crowned by obtuse stigmas. The germen afterward become four horned capsules, compressed on their inner side, filled with roundish seeds.

The SPECIES are,

1. RHODIOLA (*Rosea*) staminibus corollâ duplo longioribus. *Rose-root with stamina twice as long as the petals.*

2. RHODIOLA (*Minor*) staminibus corollâ ferè æquantibus. *Rose-root with stamina scarcely equalling the length of the petals.*

The first sort grows naturally in the clefts of the rocks and rugged parts of the mountains of Wales, Yorkshire, and Westmorland. This has a very thick fleshy root, which, when bruised or cut, sends out an odour like Roses; it has many heads, from whence in the spring come out thick succulent stalks like those of Orpine, about nine inches long, closely garnished with thick succulent leaves, of a gray colour, which are indented on their edges toward the top, and are placed alternately on every side the stalk. The stalk is terminated by a cluster of yellowish herbaceous flowers, which appear early in May; the male flowers have stamina twice the length of the petals. They have a very agreeable scent, but are not of long continuance.

The second sort grows naturally on the Alps; the roots of this are smaller than those of the other sort; the stalks are small, not above nine inches long; the leaves are small, but shaped like those of the other sort; the petals of the flowers are purplish, and the stamina are but little longer than the petals. This flowers later than the other sort.

These plants are preserved in the gardens of the curious, for the sake of variety; they are easily propagated by parting of the roots, which should be performed the beginning of September, at which time their stalks begin to decay; and if the fleshy parts of the roots are cut or broken, they should be laid to dry a few days before they are planted. These plants require a shady situation and a dry undunged soil, in which they will continue many years.

RHODODENDRON. Lin. Gen. Plant. 484. Dwarf Rose-bay.

The CHARACTERS are,

The flower has a permanent empalement cut into five parts; the flower hath one wheel funnel-shaped petal, spreading open at the brim; it has ten slender stamina which decline, and are the length of the petals, terminated by oval summits, and a five-cornered germen supporting a slender style the length of the petal, crowned by an obtuse stigma. The germen afterward becomes an oval capsule with five cells, filled with small seeds.

The SPECIES are,

1. RHODODENDRON (*Hirsutum*) foliis ciliatis nudis, corollis infundibuliformibus. Lin. Sp. Plant. 392. *Rose-bay with naked hairy leaves, and funnel-shaped petals.*

2. RHODODENDRON (*Ferrugineum*) foliis glabris, subtus leprosis, corollis infundibuliformibus. Lin. Sp. Plant. 392. *Rose-bay with smooth leaves, which are hoary on their under side, and have funnel-shaped petals.*

The first sort grows naturally on the Alps, and several mountains in Italy. This is a low shrub, which seldom rises two feet high, sending out many short ligneous branches, covered with a light brown bark, garnished closely with oval spear-shaped leaves, sitting pretty close to the branches; they are entire, having a great number of fine iron-coloured hairs on their edges and under side. The flowers are produced in bunches at the end of the branches, having one funnel-shaped petal; the tube is about half an inch long; the brim is cut into five obtuse segments, which spread half open; they are of a pale red colour, and have ten stamina in each, which are the length of the tube; after the flower is past, the germen in the center turns to an oval capsule with five cells, filled with small seeds.

The second sort grows naturally on the Alps and Apennines. This rises with a shrubby stalk near three feet high, sending out many irregular branches, covered with a purplish bark, closely garnished with smooth, spear-shaped, entire leaves, whose borders are reflexed backward; the upper side is of a light lucid green, their under side of an iron colour; they are placed all round the branches without any order. The flowers are produced in round bunches at the end of the branches; they are funnel-shaped, having short tubes cut into five obtuse segments at the brim, which spread a little open; they are of a pale Rose colour, and make a good appearance.

There are some other species of this genus which grow naturally in the eastern countries, and others are natives of America; but the sorts here mentioned, are all I have seen in the English gardens; these are difficult to propagate and preserve in gardens, for they grow naturally upon barren rocky soils and in cold situations, where they are covered with snow great part of the winter; so that when they are planted in better ground they do not thrive, and for want of their usual covering of snow in winter, they are frequently killed by frost; but could these plants be tamed, and propagated in plenty, they would be great ornaments to the gardens.

They are propagated by seeds, but these are so very small, that if they are covered deep they will not grow. The seeds should be sown as soon as possible after they are ripe, in pots filled with fresh gentle loamy earth, and very lightly covered with a little fine earth; then the pots should be plunged up to their rims in a shady border, and in hard frost they should be covered with bell or hand-glasses, taking them

them off in mild weather. If these seeds are sown early in autumn the plants will come up the following spring; these must be kept shaded from the sun, especially the first summer, and duly refreshed with water; in autumn following they may be transplanted to a shady situation, on a loamy soil, covering the ground about their roots with Moss, which will guard them from frost in winter, and keep the ground moist in summer.

RHUS. Tourn. Inst. R. H. 611. tab. 381. Sumach.

The CHARACTERS are,

The empalement of the flower is permanent, erect, and cut into five parts. The flower has five oval, erect, spreading petals, and five short stamina, terminated by small summits; it has a roundish germen as large as the petals, with scarce any style, crowned by three small stigmas. The germen afterward becomes a roundish hairy berry, inclosing a single hard seed of the same form.

The SPECIES are,

1. RHUS (*Coriaria*) foliis pinnatis obtusiusculè serratis, ovato-lanceolatis subtus villosis. *Sumach with winged leaves, which are obtusely sawed, ovally spear-shaped, and hairy on their under side; Elm-leaved Sumach.*

2. RHUS (*Typhinum*) foliis pinnatis lanceolatis argute serratis subtus tomentosis. *Sumach with spear-shaped winged leaves, which are sharply sawed, and woolly on their under side; Virginian Sumach.*

3. RHUS (*Glabrum*) foliis pinnatis serratis, lanceolatis utrinque glabris. *Sumach with winged leaves, which are spear-shaped, and smooth on both sides.*

4. RHUS (*Carolinianum*) foliis pinnatis serratis lanceolatis, subtus incanis, paniculâ compactâ. *Sumach with sawed, spear-shaped, winged leaves, which are hoary on their under side, with a compact panicle.*

5. RHUS (*Canadense*) foliis pinnati, obsoletè serratis, lanceolatis, utrinque glabris, panicula compositâ. *Sumach with winged spear-shaped leaves, which are slightly sawed, and a compound panicle.*

6. RHUS (*Copallinum*) foliis pinnatis integerrimis, petiolo membranaceo articulato. Flor. Leyd. Prod. 24. *Sumach with entire winged leaves, and a jointed membranaceous foot-stalk; narrow-leaved Sumach.*

7. RHUS (*Chinense*) foliis pinnatis, foliolis ovatis, obtusè serratis, petiolo membranaceo villoso. *Sumach with winged leaves, oval lobes, which are bluntly sawed, and a hairy foot-stalk, having jointed membranes or wings.*

8. RHUS (*Incanum*) foliis ternatis, foliolis ovatis subtus tomentosis. *Three-leaved Sumach with oval leaves, which are downy on their under side.*

9. RHUS (*Tomentosum*) foliis ternatis, foliolis subpetiolatis rhombeis angulatis, subtus tomentosis. Lin. Sp. Pl. 266. *Three-leaved Sumach with angular rhomboid lobes, having foot-stalks, downy on their under side.*

10. RHUS (*Lucidum*) foliis ternatis, foliolis sessilibus cuneiformibus lævibus. Virg. Cliff. 25. *Three-leaved Sumach, whose lobes are smooth, wedge-shaped, and sit close to the stalk.*

11. RHUS (*Africanum*) foliis ternatis, foliolis ovatis nervosis, marginibus sæpius dentatis, utrinque viridibus. *Sumach with trifoliate leaves, having oval veined lobes, which are generally indented on their edges, and green on both sides.*

12. RHUS (*Argenteum*) foliis ternatis, foliolis petiolatis lineari-lanceolatis integerrimis, subtus tomentosis. Hort. Cliff. 111. *Sumach with trifoliate leaves, whose lobes stand upon foot-stalks, are linear, spear-shaped, entire, and downy on their under side.*

13. RHUS (*Radilijawel*) foliis ternatis, lineari-lanceolatis integerrimis sessilibus utrinque viridibus. *Sumach with trifoliate leaves, having linear, spear-shaped, entire lobes, sitting close to the foot-stalk, green on both sides.*

14. RHUS (*Rigidum*) foliis ternatis, foliolis ovatis acuminatis integerrimis, petiolatis, floribus paniculatis terminalibus. *Three-leaved Sumach with oval acute-pointed lobes, which are entire, upon foot-stalks, and flowers growing in panicles, which terminate the branches.*

15. RHUS (*Cotinus*) foliis simplicibus obovatis. Lin. Sp. Plant. 267. *Sumach with single, obverse, oval leaves; Venice Sumach, or Coccygria.*

The first sort of Sumach grows naturally in Italy, Spain, and Turkey. The branches of this tree are used instead of Oak-bark for tanning of leather, and I have been informed that the Turkey leather is all tanned with this shrub. This has a ligneous stalk, which divides at bottom into many irregular branches, which rise to the height of eight or ten feet; the bark is hairy, of an herbaceous brown colour; the leaves are winged, composed of seven or eight pair of lobes, terminated by an odd one, bluntly sawed on their edges, are hairy on their under side, of a yellowish green colour, and placed alternately on the branches; the flowers grow in loose panicles at the end of the branches, which are of a whitish herbaceous colour, each panicle being composed of several spikes of flowers sitting close to the foot-stalks. The leaves and seeds of this sort are used in medicine, and are esteemed very restringent, stiptic, and good for all kinds of fluxes and hæmorrhages; used both inwardly and outwardly they resist putrefaction, and stop gangrenes and mortifications.

The second sort grows naturally in almost every part of North America. This hath a woody stem, with many irregular branches, which are generally crooked and deformed. The young branches are covered with a soft velvet-like down, resembling greatly that of a young stag's horn both in colour and texture, from whence the common people have given it the appellation of stag's horn; the leaves are winged, composed of six or seven pair of oblong heart-shaped lobes, terminated by an odd one, ending in acute points, hairy on their under side, as is also the midrib. The flowers are produced in close tufts at the end of the branches, and are succeeded by seeds, inclosed in purple, woolly, succulent covers, so that the bunches are of a beautiful purple colour in autumn; and the leaves, before they fall in autumn, change to a purplish colour at first, and, before they fall, to a feuillemort. This shrub is used for tanning of leather in America, and the roots are often prescribed in medicine in the countries where the plant grows naturally.

The third sort grows naturally in many parts of North America; this is commonly titled by the gardeners New England Sumach. The stem of this sort is stronger, and rises higher than that of the former; the branches spread more horizontally; they are not quite so downy as those of the last, and the down is of a brownish colour; the leaves are composed of many more pair of lobes, which are smooth on both sides; the flowers are disposed in loose panicles, which are of an herbaceous colour.

The fourth sort grows naturally in Carolina; the seeds of this were brought from thence by the late Mr. Catesby, who has given a figure of the plant in his Natural History of Carolina. This is by the gardeners called the Scarlet Carolina Sumach; it rises commonly to the height of seven or eight feet, dividing into many irregular branches, which are smooth, of a purple colour, and pounced over with a grayish powder, as are also the foot-stalks of the leaves. The leaves are composed of seven or eight pair of lobes, terminated by an odd one; these are not always placed exactly opposite on the midrib, but are sometimes alternate. The upper side of the lobes are of a dark green, and their under heary, but smooth. The flowers are produced at the end

end of the branches in very close panicles, which are large, and of a bright red colour.

The fifth sort grows naturally in Canada, Maryland, and several other parts of North America. This hath smooth branches, of a purple colour, covered with a gray pounce. The leaves are composed of seven or eight pair of lobes, terminated by an odd one; the lobes are spear-shaped, sawed on their edges, of a lucid green on their upper surface, but hoary on their under, and are smooth. The flowers are produced at the end of the branches in large panicles, which are composed of several smaller, each standing upon separate foot stalks; they are of a deep red colour, and the whole panicle is covered with a gray pounce, as if it had been scattered over them.

The sixth sort grows naturally in most parts of North America, where it is known by the title of Beech Sumach, probably from the places where it grows. This is of humbler growth than either of the former, seldom rising more than four or five feet high in England, dividing into many spreading branches, which are smooth, of a light brown colour, closely garnished with winged leaves, composed of four or five pair of narrow lobes, terminated by an odd one; they are of a light green on both sides, and in autumn change purplish. The midrib, which sustains the lobes, has on each side a winged or leafy border, which runs from one pair of lobes to another, ending in joints at each pair, by which it is easily distinguished from the other sorts. The flowers are produced in loose panicles at the end of the branches, of a yellowish herbaceous colour.

These six sorts are hardy plants, and will thrive in the open air here. The first and fourth sorts are not quite so hardy as the others, so must have a better situation, otherwise their branches will be injured by severe frost in the winter; they are easily propagated by seeds, which, if sown in autumn, the plants will come up the following spring, but if they are sown in the spring, they will not come up till the next spring; they may be either sown in pots, or the full ground. If they are sown in pots in autumn, the pots should be placed under a common frame in winter, where the seeds may be protected from hard frost; and in the spring, if the pots are plunged into a very moderate hot-bed, the plants will soon rise, and have thereby more time to get strength before winter. When the plants come up, they must be gradually hardened to bear the open air, into which they should be removed as soon as the weather is favourable, placing them where they may have the morning sun; in the summer, they must be kept clean from weeds, and in dry weather watered; toward autumn it will be proper to stint their growth by keeping them dry, that the extremity of their shoots may harden, for if they are replete with moisture, the early frosts in autumn will pinch them, which will cause their shoots to decay almost to the bottom, if the plants are not screened from them. If the pots are put under a common frame in autumn, it will secure the plants from injury, for while they are young and the shoots soft, they will be in danger of suffering, if the winter proves very severe; but in mild weather they must always enjoy the open air, therefore should never be covered but in frost. The spring following, just before the plants begin to shoot, they should be shaken out of the pots, and carefully separated, so as not to tear the roots, and transplanted into a nursery in rows three feet asunder, and one foot distance in the rows. In this nursery they may stand two years to get strength, and then may be transplanted where they are to remain.

This method of propagating the plants from seeds is seldom practised after a person is once possessed of the plants, for they are very subject to send up a great number of suckers from their roots, whereby they are easily propagated. The suckers of all the sorts may be taken up, and planted in a nursery for a year or two to get strength, and then may be planted where they are to remain.

These shrubs are generally planted in plantations of flowering shrubs in large gardens, where they make a fine variety in autumn, especially the second, fourth, and fifth sorts, with their large purple or red panicles, which have a good effect; but, where these are planted, their suckers should be every year taken off, otherwise they will grow up to a thicket, and destroy the old plants, by robbing them of their nourishment.

The seventh sort grows naturally in the east. The seeds of this were sent to the Royal Garden at Paris, where they succeeded; and from thence I received the plant, which grew very well in the open air at Chelsea three years, but the severe winter in 1740 destroyed it, so that it is not quite so hardy as the other sorts. This rises with a shrubby stalk six or eight feet high, sending out many irregular branches. The young shoots and foot-stalks of the leaves are covered with a soft, brown, hairy down; the leaves are composed of three or four pair of oval lobes, terminated by an odd one; the inner lobes are small, the outer large; the end lobe is heart-shaped, ending in an acute point; they are sawed on their edges, and are hoary on their under side; the midrib, which sustains the lobes, has two leafy membranes running along the sides from joint to joint, which are narrow below, and gradually increase in their breadth to the next joint. When the leaves are broken, they emit a milky juice from the wound. As I have not seen the flowers of this sort, I can give no account of them.

This does not put out suckers from the root so freely as the American kinds, so must either be propagated by layers, or cutting off some of the roots, and planting them upon a gentle hot-bed; by which method it may be propagated, but my plant was too weak for this purpose, when it was destroyed.

The eighth sort grows naturally at the Cape of Good Hope. This hath a strong woody stalk, which rises ten or twelve feet high, covered with a gray bark, sending out many smooth branches, garnished with trifoliate leaves, standing upon pretty long foot-stalks. The lobes of the leaves are oval, entire, hoary on their under side, but smooth, and of a lucid green on their upper; the flowers are produced from the wings of the leaves in small bunches; they are of an herbaceous colour.

The ninth sort also grows naturally at the Cape of Good Hope. This rises with a woody stalk to the height of seven or eight feet, covered with a brown bark, putting out many irregular branches, garnished with trifoliate leaves, standing upon long foot-stalks. The lobes of this sort are angular, shaped like a rhombus, downy on their under side, but of a dark green on their upper. The flowers come out in slender bunches from the side of the branches, of a whitish herbaceous colour, and soon fall away.

The tenth sort grows naturally at the Cape of Good Hope. This rises with a woody stalk like the eighth, dividing into many branches, covered with a brown bark, garnished with trifolitae leaves, whose lobes are wedge or heart-shaped, of a lucid green, and sit close to the branches. This sort does not flower here, so far as I can find, for I have had some of the plants in my care above forty years, but they have not flowered as yet.

The eleventh sort is a native of the Cape of Good Hope. This hath some resemblance of the former, but the lobes of the leaves are twice as large, and are oval, with some indentures on their edges; they have several transverse veins running from the midrib to the edge, and are very stiff, of a bright lucid green on both sides. This sort has not flowered here so far as I can learn.

The

The twelfth fort came from the Cape of Good Hope, where it grows naturally. This rifes with a woody ftalk feven or eight feet high, dividing into feveral irregular branches, covered with a dark brown bark, garnifhed with narrow, fpear-fhaped, trifoliate leaves, ftanding upon pretty long foot-ftalks, downy on their under fide, but of a lucid green on their upper. The flowers are fmall, of an herbaceous colour, and are produced in fmall loofe bunches from the fide of the branches.

The thirteenth fort is a native of the Cape of Good Hope. This rifes with an upright woody ftalk five or fix feet high, fending out many branches, covered with a fmooth brown bark, garnifhed with narrow, fpear-fhaped, trifoliate, entire leaves, ftanding upon fhort foot-ftalks, of a lucid green, and have a deep furrow-lengthways through the middle. This fort has not yet flowered in England.

All thefe African forts are too tender to live through the winter in the open air in England, fo they are planted in pots or tubs, and houfed in autumn; during the winter they muft be treated in the fame way as other hardy green-houfe plants. They all retain their leaves through the year, fo make a good variety when intermixed with other plants in the green-houfe in winter. They may be propagated by cuttings, which fhould be planted in pots the beginning of April, and plunged into a very moderate hot-bed, covering them clofe with hand or bell-glaffes, fcreening them from the fun in the heat of the day. The cuttings fhould be now and then refrefhed with water, but it fhould not be given in too great quantity. If they fucceed, they will put out roots in about two months; when they begin to fhoot, they fhould be gradually hardened to bear the open air, into which they muft be removed, placing them in a fheltered fituation; when the cuttings have filled the pots with their roots, they fhould be fhaken out, and parted carefully, planting each into a feparate fmall pot, placing them in the fhade till they have taken new root, when they may be intermixed with other exotic plants in a fheltered fituation for the fummer, and in autumn be removed into the green-houfe.

The fourteenth fort grows naturally in the ifland of Ceylon. This rifes with a woody ftalk ten or twelve feet high, clothed with trifoliate leaves, ftanding upon pretty long foot-ftalks. The lobes of the leaves are oval; they are thick, fmooth, and of a lucid green. The flowers are fmall, of an herbaceous colour, and are produced in loofe panicles at the end of the branches. Thefe feldom appear in England.

This plant is tender, fo muft be placed in a moderate ftove, otherwife it will not live through the winter in England. It may be propagated by cuttings in the fame way as the former forts, but requires a warmer bed than thofe to promote their putting out roots. When they have good roots, they fhould be each tranfplanted into a feparate fmall pot, and plunged into the tan-bed, and treated in the fame way as other tender exotic plants.

The fifteenth fort grows naturally in Spain, Italy, and the Levant, where the leaves are ufed for tanning of leather; this rifes with an irregular fhrubby ftalk to the height of ten or twelve feet, fending out many fpreading irregular branches, covered with a fmooth brown bark, garnifhed with fingle, obverfe, oval leaves, rounded at their points, which ftand upon long foot-ftalks; they are fmooth, ftiff, and of a lucid green, having a fhort midrib, from whence feveral tranfverfe veins run toward the border. The flowers come out at the end of the branches upon long hair-like foot-ftalks, which divide, and branch into large hair-like bunches, of a purplifh colour: they are fmall, white, and compofed of five fmall oval petals, which fpread open.

This plant is cultivated for fale in the nurfery-gardens near London; it is propagated by layers, which fhould be laid down in autumn, and by the autumn following they will have taken root, when they may be taken off from the old plants, and tranfplanted in a nurfery, where they may grow a year or two to get ftrength, and then be planted where they are to remain. This fhrub is fo hardy, as not to be injured by froft in England.

RIBES. Lin. Gen. Plant. 247. The Currant-tree.

The CHARACTERS are,

The flower has a bellied empalement, cut at the top into five concave obtufe fegments; it hath five fmall, obtufe, erect petals, growing to the border of the empalement, and five awl-fhaped ftamina inferted in the empalement, terminated by incumbent compreffed fummits. The roundifh germen is fituated under the flower, fupporting a bifid ftyle, crowned by obtufe ftigmas, which afterward becomes a globular umbilicated fruit with one cell, containing many roundifh compreffed feeds.

The SPECIES are,

1. RIBES (*Rubrum*) inerme, racemis glabris pendulis, floribus planiufculis. Lin. Sp. Plant. 200. *Currant without thorns, fmooth hanging bunches, and plain flowers; common Currant.*

2. RIBES (*Alpinum*) inerme, racemis erectis, bracteis flore longioribus. Lin. Sp. Pl. 200. *Smooth Currant with erect bunches, and bractea longer than the flower; fweet Alpine Currant.*

3. RIBES (*Nigrum*) inerme, racemis pilofis, floribus oblongis. Lin. Sp. Plant. 201. *Currant without fpines, having hairy branches and oblong flowers; black Currant.*

4. RIBES (*Americanum*) inerme, racemis glabris, floribus campanulatis. *Currant with unarmed fmooth branches, and bell-fhaped flowers.*

The firft fort grows naturally in the northern parts of Europe, but has been long cultivated in the gardens, and greatly improved, fo that at prefent there are the following varieties in the Englifh gardens, viz. the common Currant with fmall red fruit, the fame with white fruit, and another with pale fruit, which is commonly called the Champaign Currant; but, fince the two forts of Dutch Currants have been introduced, and become plenty in the gardens, the old red and white Currants have been almoft banifhed, fo that they are rarely to be found in the Englifh gardens at prefent.

The fecond fort is kept in a few gardens for the fake of variety, but, as the fruit is very fmall, and has little flavour, it is not much cultivated.

The third fort grows naturally in Helvetia, Sweden, and other northern countries, and is fometimes cultivated in gardens for its fruit, of which is made a rob, which is greatly efteemed for fore-throats, from whence the fruit has been called Squinancy Berries for their great ufe in quinfies. As this fruit has a ftrong difagreeable flavour, it is rarely admitted to the table.

The fourth fort grows naturally in Penfylvania, from whence the plants were fent to Mr. Peter Collinfon feveral years paft, and has been difperfed to moft parts of England. This has been by fome thought to be the fame with the common black Currant, but thofe who have long cultivated it, know it is very different. The fhoots of this being much fmaller and more compact, the bark is of a darker colour; the leaves are fmaller, thinner, fmoother, and have not a rank fmell like thofe of the common fort. The flowers are fmaller, bell-fhaped, and grow in thinner bunches; the fruit is fmaller, and not fo round; the plants of this do not produce much fruit, nor is it fo good as to merit cultivation, fo the plant is only kept by way of curiofity. The fruit of the red and white Currants are greatly efteemed for the table, and are alfo very good in fevers; they are cooling and grateful to the ftomach, quench thirft, and are fome-

fomewhat reftringent. The jelly made with the juice of this fruit and fugar is very grateful in fevers, and is ufed as fauce to the table. This fruit may be procured good much longer than moft others upon the plants, by planting them in different fituations; for if they are planted againft pales or walls expofed to the fouth, the fruit will ripen in June, and by planting fome againft north walls, if they are fcreened from birds, and covered in autumn from froft, they may be kept till November; and, as the fruit is greatly ufed for tarts, it is very convenient to have a fucceffion of it for fo long a time.

The Champaign Currant differs from the other only in the colour of the fruit, which is of a pale red or flefh colour. The tafte is fo near to the other, as not to be diftinguifhed; but, this being of a different colour, makes a variety on the table.

There are plants of all thefe forts with variegated leaves, which are kept in fome gardens for the fake of variety, but, as the variegations go off when the plants are vigorous, they fcarce deferve notice.

Thefe forts may be eafily propagated by planting their cuttings any time from the middle of September to the end of October, upon a fpot of frefh earth, either in rows at one foot afunder, or in beds, which in the fpring muft be kept very clear from weeds. Thefe may remain one or two years in the nurfery, during which time they muft be pruned up for the purpofes defigned, i. e. either to clear ftems about one foot high, if for ftandards, or, if for walls, pales, or efpaliers, they may be trained up flat: then they fhould be planted out where they are to remain, for the younger they are planted the better they will fucceed; the beft feafon for which is foon after the leaves begin to decay, that they may take root before winter, fo that they may be in no danger of fuffering from drought in the fpring.

Thefe plants are generally planted in rows at about ten feet afunder, and four diftance in the rows, in thofe gardens where the fruit is cultivated for fale; but the beft method is to train them againft low efpaliers, in which manner they will take up much lefs room in a garden, and their fruit will be much fairer.

The diftance they fhould be placed for an efpalier, ought not to be lefs than ten or twelve feet, that their branches may be trained horizontally, which is of great importance to their bearing.

Thofe that are planted againft pales or walls, fhould alfo be allowed the fame diftance. If they are planted againft a fouth-eaft wall or pale, it will caufe their fruit to ripen at leaft a fortnight or three weeks fooner than thofe in the open air; and thofe which are planted againft a north wall or pale, will be proportionably later, fo that by this method the fruit may be continued a long time in perfection, efpecially if thofe againft the north pales are matted in the heat of the day.

Thefe plants produce their fruit upon the former year's wood, and alfo upon fmall fnags which come out of the old wood; fo that in pruning them, thefe fnags fhould be preferved, and the young fhoots fhortened in proportion to their ftrength. The only method, very neceffary to be obferved in pruning of them, is, not to lay their fhoots too clofe, and never to prune their fnags to make them fmooth This, with a fmall care in obferving the manner of their growth, will be fufficient to inftruct any perfon how to manage this plant, fo as to produce great quantities of fruit.

Thefe plants will thrive and produce fruit in almoft any foil or fituation; and are often planted under the fhade of trees; but the fruit is always beft when they are planted in the open air, and upon a light loamy foil.

RICINOIDES. See IATROPHA.

RICINUS. Tourn. Inft. R. H. 532. tab. 307. Palma Chrifti, vulgo.

The CHARACTERS are,

It hath male and female flowers difpofed in the fame fpike. The male flowers, which are fituated on the lower part of the fpike, have fwelling empalements cut into three parts. The flowers have no petals, but have a great number of flender ftamina, which are connected in feveral bodies, and are terminated by roundifh twin fummits. The female flowers, which are fituated on the upper part of the fpike, have empalements cut into five fegments, and are armed with prickles; they have no petals, but in the center is fituated an oval germen, which is clofely fhut up in the empalement, fupporting three fhort ftyles, which are bifid, crowned by fingle ftigmas. The germen afterward turns to a roundifh fruit, having three furrows, dividing into three cells, opening with three valves, each cell containing one almoft oval feed.

The SPECIES are,

1. RICINUS (*Vulgaris*) foliis peltatis ferratis, fubtus glaucis, petiolis glanduliferis. *Ricinus with target-fhaped fawed leaves, which are gray on their under fide, and footftalks bearing glands.*

2. RICINUS (*Americanus*) foliis peltatis fubferratis, lobis amplioribus utrinque virentibus. *Ricinus with target-fhaped leaves, which are fomewhat fawed, whofe lobes are large, and green on both fides.*

3. RICINUS (*Urens*) foliis peltatis inæqualiter ferratis, capfulis hifpidis. Tab. 219. *Ricinus with target-fhaped leaves, which are unequally fawed, and prickly capfules to the fruit.*

4. RICINUS (*Rugofus*) foliis peltatis ferratis, capfulis rugofis non echinatis. Tab. 220. *Palma Chrifti with target-fhaped fawed leaves, and rough capfules to the fruit, which are not prickly.*

5. RICINUS (*Africanus*) foliis peltatis ferratis, lobis maximis, caule geniculato, capfulis echinatis. *Ricinus with target-fhaped fawed leaves, having the largeft lobes, a jointed ftalk, and prickly covers to the feeds.*

6. RICINUS (*Inermis*) foliis peltatis ferratis, lobis maximis, caule geniculato, capfulis inermis. *Palma Chrifti with fawed target fhaped leaves, having very large lobes, a jointed ftalk, and fmooth covers to the feeds.*

7. RICINUS (*Minor*) foliis palmatis ferratis, profundiùs divifis, capfulis echinatis. *Ricinus with hand-fhaped fawed leaves, which are deeply divided, and prickly covers to the feeds.*

The firft fort grows naturally in Sicily, and other warm parts of Europe. This rifes with a ftrong herbaceous ftalk to the height of ten or twelve feet; the joints are at a great diftance from each other; the ftalk and branches are of a gray colour; the leaves are large, and have long foot-ftalks; they are deeply divided into feven lobes, which are fawed on their edges, and are gray on their under fide; at the divifion of the lobes is a fort of navel, where the foot-ftalk joins the leaves. The flowers are difpofed in long fpikes, which arife at the divifion of the branches; the lower part of the fpikes are garnifhed with male flowers, which have fwollen empalements, divided into three parts, which open, and fhew a great number of flender ftamina, terminated by whitifh fummits, connected at their bafe into feveral fmall bunches. The female flowers, which occupy the upper part of the fpike, have prickly empalements, which inclofe the roundifh germen, upon which fit three fhort ftyles, crowned by oblong ftigmas. The germen afterward becomes an oval capfule with three deep channels, clofely armed with foft fpines, and divided into three cells, each containing one oblong ftriped feed.

The fecond fort grows naturally in the iflands of the Weft-Indies, where it is called Agnus Caftus, or Oil-tree. This is often confounded with the former; moft of the botanifts fuppofe they are the fame plant; but as I have cultivated both more than forty years, in which time I have

 never

never obſerved either of them vary, ſo I think there can be no doubt of their being different plants. This hath brown ſtalks, which riſe ſix or ſeven feet high; the leaves are broader, and not ſo deeply divided as thoſe of the former; they are of a deep green on both ſides, and are unequally ſawed. The ſpikes of flowers are ſhorter, the ſeed-veſſels rounder, and of a browniſh colour; and the ſeeds are much leſs, and of a reddiſh brown colour.

The third ſort grows naturally in the Weſt-Indies, and is often confounded with the former, but is very different. The ſtalk of this ſort is thick, herbaceous, and of a grayiſh green; the joints are cloſer than thoſe of the former ſorts; it riſes about four feet high; the leaves are large, of a deep green on their upper ſide, but grayiſh on their under; they are deeply cut into ſeven lobes or ſegments, which are unequally ſawed on their edges. The ſpikes of flowers are looſe, the covers of the capſules are green, and cloſely armed with ſoft ſpines; the ſeeds are ſmaller and lighter coloured than thoſe of the ſecond ſort.

The fourth ſort grows naturally in both Indies. This riſes with an herbaceous ſtalk about four feet high; the lower part is purpliſh, the upper of a deep green; the joints of this are pretty far aſunder; the leaves are of a deep green on their upper ſide, but are paler on their under; they are not ſo deeply divided as ſome of the other ſorts, and are more regularly ſawed; the ſpikes of flowers are large. The male flowers have more ſtamina, and their ſummits are yellow; the capſules are oval and rough, but have no ſpines; the ſeeds are ſmall, and of a brown colour.

The fifth ſort grows naturally in Africa and both the Indies. This riſes with a large reddiſh ſtalk to the height of ten or twelve feet, with many joints; the leaves are the largeſt of any ſpecies yet known; I have meaſured ſome of them which were more than two feet and a half diameter; they are of a dark green, unequally ſawed on their edges, and not ſo deeply cut as thoſe of ſome other ſorts. The ſpikes of flowers are large; the empalement of the flower is brown; the ſummits on the ſtamina of the male flowers are whitiſh; the capſules are large, oval, and cloſely armed with ſoft ſpines; the ſeeds are very large, and beautifully ſtriped.

The ſixth ſort grows naturally in the Spaniſh Weſt-Indies, from whence the late Mr. Robert Millar ſent me the ſeeds. The plants of this ſort are in every reſpect like thoſe of the fifth, but the capſules which incloſe the ſeeds are ſmooth; and this difference is permanent, therefore may be put down as a diſtinct ſpecies.

The ſeventh ſort grows naturally in Carolina, and ſeveral other parts of America. Of this there are two varieties, if not diſtinct ſpecies; one of them has a red ſtalk, and the other a pale green; they are diſtinguiſhed by the inhabitants of America, by the title of red and white Oil-ſeed. The ſtalks of theſe ſeldom riſe more than three feet high, divided at the top into two or three branches; the leaves are much leſs than thoſe of the other ſorts, and deeper divided; their borders are unequally ſawed; the ſegments of the leaves are frequently cut on their ſides. The ſpikes of flowers are ſmaller and more compact than thoſe of the former ſorts; the capſules are ſmaller, rounder, and of a light green, and are cloſely armed with ſoft ſpines; the ſeeds are ſmall, and re finely ſtriped.

There are ſeveral other ſpecies which grow naturally in both Indies, which have not been examined by any curious botaniſt; for I have received ſeeds of three or four ſorts, which appeared to be very different from any of the known ſorts, but the ſeeds were too old to grow.

The ſorts here enumerated, I have cultivated ſeveral years, and have always found they have kept their difference, ſo that I have no doubt of their being diſtinct ſpecies; and unleſs they are thus tried, there is no poſſibility of determining their ſpecific difference; for when plants are found growing in different ſoils and ſituations, they have ſuch different appearances, as may deceive the moſt ſkilful botaniſt.

The plants are generally annual in theſe countries, though in their native places of growth they continue longer; in England the plants of the firſt ſort are often preſerved thro' the winter, but young plants are much preferable to thoſe which are thus preſerved; therefore few perſons are at the trouble to keep them, unleſs when the ſeaſons prove ſo bad as that their ſeeds do not ripen, whereby the ſpecies might be loſt, if the plants are not preſerved through the winter.

Theſe plants are propagated by ſeeds, which muſt be ſown upon a hot-bed in the ſpring, and when the plants are come up, they ſhould be each planted into a ſeparate pot, and plunged into a freſh hot-bed. As theſe plants grow very faſt, their roots will in a ſhort time fill the pots; therefore they ſhould be ſhifted into larger, toward the latter end of May; when the ſeaſon is warm, they may be hardened to endure the open air by degrees; and then if ſome of the plants are ſhaken out of the pots, and planted out into a very rich border, and in dry weather duly watered, they will grow to a very large ſize, particularly the firſt and fifth ſorts, which I have ſeen upward of ten feet high in one ſeaſon, and theſe plants have produced a great quantity of flowers and ſeeds; but if you intend to preſerve any of the plants through the winter, they muſt be ſhifted into larger pots from time to time as their roots ſhall require, placing them in the open air during the ſummer ſeaſon in ſome warm ſituation, where they may remain until the middle or end of October, when they muſt be removed into the houſe with other exotic plants, obſerving to water them ſparingly in winter, and alſo to admit the free air in mild weather; for they only require to be protected from froſt and cold winds, ſo that they will endure the winter in a warm green-houſe, without any addition of artificial warmth.

Theſe plants deſerve a place in every curious garden for the ſingular beauty of their leaves (notwithſtanding their flowers make no great appearance,) eſpecially thoſe ſorts which may be propagated every year from ſeeds, becauſe thoſe perſons who have no green-houſe to place them into in winter, may cultivate them as other annual plants; amongſt which theſe, being placed either in pots or borders, afford an agreeable variety; but it muſt be obſerved, as theſe are large growing plants, never to place them too near other plants of leſs growth, becauſe theſe will overbear and deſtroy them; and thoſe which are planted in pots, ſhould be allowed room for their roots to expand, and muſt be frequently watered, otherwiſe they will not grow very large.

The inhabitants of the Weſt-Indies draw an oil from the ſeeds of theſe plants, which ſerves for the uſe of their lamps; and as the plants come up as weeds in thoſe warm countries, ſo they are at no trouble to cultivate the plants, but employ their negroes to collect the ſeeds from the plants which grow naturally, whereby they are furniſhed with the oil at a ſmall expence. This oil is good to kill lice in childrens heads.

The ſeeds of the firſt ſort is the Catapucia major of the ſhops; theſe have been formerly given by ſome perſons to purge watery humours, which they do both upward and downward with great violence, ſo that at preſent theſe ſeeds are rarely uſed.

RIPENING OF FRUIT early.

In order to have early fruit, a wall ſhould be erected ten feet high, in length according to the number of trees intended for three years forcing; the method of conſtructing ftructing thefe walls, is fully explained under the article Wall.

This being done, a border may be marked out about four feet wide on the fouth fide of it, and fome fcantlings of wood, about four inches thick, muft be faftened on a low wall, built to prevent the earth of the border from falling into the walk, and alfo to fecure the timber from rotting, to reft the glafs-lights upon; which lights are to flope backward to the wall, to fhelter the fruit as there fhall be occafion.

Bars placed about four inches diftance, cut out of whole deal, muft be placed between thefe glaffes, fo that the lights may reft on them. There muft alfo be a door fhaped to the profile of the frame at each end, that it may be opened at either of the ends, according as the wind blows.

The frame before-mentioned fhould be made fo, that when the firft part has been forced, the frame may be moved the next year forward, and the fucceeding year forward again, fo that the trees will be forced every third year; and having two years to recover themfelves, will continue ftrong for many years.

Thefe trees fhould be well grown before they are forced, otherwife they will foon be deftroyed; and the fruit produced on grown trees, will be much fairer and better tafted than on frefh planted trees.

The fruit which may be planted in thefe frames are,

The Avant, the Albemarle, the Early Newington, and Brown Nutmeg Peaches.

Mr. Fairchild's Early, the Elruge and Newington Nectarines; the Mafculine Apricot; the May Duke and May Cherry.

As for Grapes, the White Chaffelas and Black Sweetwater.

It has been found by experience, that the trees will be injured if the heat be applied before January; and that the time for applying the heat for bringing either Duke or May Cherries, is about the beginning of that month, and applying heat at the fame time would do for Apricots; fo that the Mafculine Apricot will, by the beginning of March, be as large as Duke Cherries, and will be ripe by the beginning of May.

Cherries thus forced, will not hold fo well as Apricots, though the former will laft, perhaps, for feven years in good plight, but Apricots will thrive and profper thus many years.

Fairchild's Early Nectarine commonly ripens about the end of May, if they are forced at the fame time, and the Brugnon Nectarine will follow that.

As for the diftance of thefe trees one from another, it need not be fo great as is directed for thofe planted in the open air, becaufe they will never fhoot fo vigoroufly nor laft fo long, therefore eight or nine feet will be fufficient.

The trees againft that part of the wall, which is defigned for forcing, fhould be pruned as foon as their leaves begin to decay, that the buds on the branches which are left, may be benefited, by receiving all the nourifhment of the branches, whereby they will become turgid and ftrong, by the time the walls are heated.

Apricots, Grapes, Nectarines, Peaches, and Plumbs, if April proves cold, the forcing heat muft be continued till May is fettled, to keep the fruit growing; but fome of the glaffes fhould be opened in the morning in March and April, when the wind is ftill and the fun warm; and they fhould be permitted to receive the fhowers that fall, while the fruit is growing; but while they are in bloffom, no rain fhould come near them, becaufe, if there fhould be any moifture lodged in the bofom of the flowers, and the fun fhould fhine hot through the glaffes, it would be apt to deftroy them.

Another thing which ought to be obferved in planting fruit in thefe frames is, to plant thofe fruits which come forward together, and thofe which come late by themfelves, becaufe it will be prejudicial to the forward fruit, to give them any more heat when they have done bearing, when at the fame time the latter fruits fet amongft them may require more heat, and to be continued longer, fome of them, perhaps, requiring an artificial heat till May.

There may alfo a row or two of fcarlet Strawberries be planted near to the back of this frame, and thefe you may expect will be ripe by the end of March, or beginning of April.

As for the Vines, they may probably be brought to bloffom in April, and have ripe Grapes in June.

RIVINIA. Plum. Nov. Gen. 48. tab. 39.

The Characters are,

The empalement of the flower is permanent, compofed of four oval, concave, coloured leaves. The flower has no petals; it has eight ftamina, which are longer than the empalement, terminated by fmall oval fummits, and a large roundifh germen, fupporting a fhort ftyle, crowned by an obtufe ftigma. The germen afterward turns to a roundifh berry fitting in the empalement, including one hard feed.

The Species are,

1. Rivinia (*Humilis*) foliis lanceolatis petiolatis integerrimis, caule fruticofo ramofo. *Rivinia with fpear-fhaped entire leaves, having foot-ftalks, and a fhrubby branching ftalk.*

2. Rivinia (*Scandens*) fcandens racemofa, amplis folani foliis, baccis violaceis. Plum. Nov. Gen. 48. *Climbing branching Rivinia with Nightfhade leaves, and Violet-coloured berries.*

The firft fort rifes with fhrubby ftalks about four feet high, dividing into feveral fpreading branches, covered with a gray fpotted bark, garnifhed with fpear-fhaped entire leaves, ftanding upon long flender foot-ftalks; they are fmooth, of a lucid green, and pretty thick confiftence, ftanding alternate, at pretty great diftances on the branches. The flowers are produced in long bunches from the fide and at the end of the branches, each ftanding upon a flender foot-ftalk near half an inch long; thefe have no petals, but their empalements are of a fcarlet colour; within thefe are fituated eight ftamina, which are longer than the empalement, terminated by fmall oval fummits; in the center is fituated a roundifh germen, terminating in a point, fupporting a fhort ftyle. The germen turns to a roundifh berry with a thin pulp, furrounding one roundifh hard feed; thefe berries are of a fcarlet colour when ripe, and afterwards change to a purple; they are by the inhabitants called Currants, but are generally efteemed poifonous.

The fecond fort rifes with a climbing woody ftalk to the height of twenty feet, covered with a dark gray bark, garnifhed with oval fpear-fhaped leaves; they are fmooth and entire. The flowers come out in long bunches from the fide of the branches, fhaped like thofe of the other, and are fucceeded by blue berries about the fame fize as thofe of the former. This fort grows naturally in Antigua, from whence I have received the feeds; it was alfo found growing at the Havannah by the late Dr. Houftoun, who found the firft growing in Jamaica.

They are both propagated by feeds, which remain long in the ground before they vegetate; I have had them lie two years before the plants have appeared, but they never rife the fame year the feeds are fown.

Thefe berries fhould be fown in pots, and plunged into a moderate hot-bed. As the plants will not come up the fame year, fo the pots fhould be removed into the ftove before winter, and plunged into the tan-bed; during the winter feafon, the earth muft be fometimes refrefhed, but muft not be be too moist. In the spring the pots may be taken out of the stove, and plunged into a fresh hot bed to bring up the plants; but if they should not then rise, the earth must not be disturbed, because the plants may come up the following season.

When the plants come up and are fit to remove, they should be each transplanted into a separate small pot, and plunged into a hot-bed, and must be treated in the same way as other plants from the same countries.

They retain their leaves all the year, so make a variety in the stove in winter, and when they flower, make a fine appearance, though their flowers are but small; for as they are produced in long bunches, from almost every joint toward the end of the branches, so the whole plant is well adorned during their continuance.

ROBINIA. Lin. Gen. Plant. 775. False Acacia.

The CHARACTERS are,

The empalement of the flower is small, and divided into four parts, the three under segments being narrow, but the upper one is broad. The flower is of the Pea-bloom kind; the standard is large, roundish, obtuse, and spreads open. The two wings are oval, and have short appendixes, which are obtuse. The keel is roundish, compressed, obtuse, and is extended the length of the wings. In the center is situated ten stamina, nine joined together, and the other standing single, terminated by roundish summits. It hath an oblong cylindrical germen, supporting a slender style, crowned by a hairy stigma. The germen afterward becomes an oblong compressed pod, inclosing kidney-shaped seeds.

The SPECIES are,

1. ROBINIA (*Pseudoacia*) racemosa pedicellis unifloris, foliis impari-pinnatis. Hort. Upsal. 212. *Robinia with footstalks supporting long bunches of flowers, and unequal winged leaves; common Bastard Acacia, called in America Locust-tree.*

2. ROBINIA (*Echinata*) pedunculis racemosis, foliis impari pinnatis, leguminibus echinatis. *Robinia with footstalks supporting long bunches of flowers, unequal winged leaves, and prickly pods.*

3. ROBINIA (*Hispida*) foliis impari-pinnatis, foliolis ovatis, racemis pedunculisque hispidis. *Robinia with unequal winged leaves, having oval lobes, whose branches and foot-stalks of the flowers are armed with stinging spines; false Acacia, with a Rose-coloured flower.*

4. ROBINIA (*Rosea*) foliis impari-pinnatis, foliolis ovatis acuminatis, ramis nodosis glabris, pedunculis racemosis. *Robinia with unequal winged leaves, whose lobes are oval and acute-pointed, knobbed smooth branches, and flowers growing in long bunches.*

5. ROBINIA (*Glabra*) foliis impari-pinnatis, foliolis oblongo ovatis, pedunculis racemosis confertis. *Robinia with unequal winged leaves, having oblong oval lobes, and foot-stalks with long bunches of flowers growing in clusters.*

6. ROBINIA (*Alata*) foliis impari-pinnatis, foliolis obversè-ovatis, racemis aggregatis axillaribus, leguminibus membranaceo-tetragonis. *Robinia with unequal winged leaves, whose lobes are obversely oval, long bunches of flowers growing in clusters from the sides of the branches, and pods having four-winged membranes.*

7. ROBINIA (*Pyramidata*) foliis duplicato-pinnatis, foliolis ovatis sessilibus, floribus spicatis terminalibus. *Robinia with doubly-winged leaves, whose lobes are oval, sit close to the midrib, and spikes of flowers terminating the branches.*

8. ROBINIA (*Violacea*) foliis pinnatis, foliolis lanceolatis oppositis, racemis axillaribus pedunculis longioribus. *Robinia with winged leaves, having spear-shaped lobes, placed opposite, and long bunches of flowers on the sides of the branches, upon longer foot-stalks.*

9. ROBINIA (*Latifolia*) foliis impari-pinnatis, foliolis oblongis acuminatis, racemis axillaribus, leguminibus oblongo-ovatis. *Robinia with unequal winged leaves, having oblong acute-pointed lobes, and bunches of flowers proceeding from the sides of the branches.*

10. ROBINIA (*Frutescens*) pedunculis simplicibus, foliis quaternatis petiolatis. Hort. Upsal. 212. *Robinia with single foot-stalks, and leaves growing by fours upon foot-stalks.*

11. ROBINIA (*Caragana*) pedunculis simplicibus, foliis abrupte pinnatis. Hort. Upsal. 212. *Robinia with single foot-stalks, and abrupt winged leaves.*

12. ROBINIA (*Pygmæa*) pedunculis simplicibus, foliis quaternatis sessilibus. Hort. Upsal. 212. *Robinia with simple foot-stalks, and four leaves sitting close to the branches.*

The first sort is the common false Acacia, which is a native of North America. The seeds of this were first brought to Paris from Canada by Monf. Robine, and soon after the seeds were brought from Virginia to England, where the trees were raised in several gardens, which for some years, while young, were in great esteem; but as they grew larger, their branches were frequently broken by strong winds in the summer, which rendered them unsightly, so that for several years they were seldom planted in gardens; but of late years it is become fashionable again, and great numbers of the trees have been raised, so at present there are few gardens in which there are not some of these trees planted.

This sort grows to a very large size in America, where the wood is much valued for its duration; most of the houses which were built at Boston in New England, upon the first settling of the English, was with this timber, which continues very sound at this time.

It grows very fast while young, so that in two or three years from seed, the plants will be eight or ten feet high; and it is not uncommon to see shoots of this tree six or eight feet long in one summer. The branches are armed with strong crooked thorns, garnished with winged leaves, composed of eight or ten pair of oval lobes, terminated by an odd one; they are of a bright green, entire, and sit close to the midrib. The flowers come out from the side of the branches in pretty long bunches, hanging downward like those of Laburnum, each flower standing on a slender foot-stalk. They are of the butterfly or Pea-blossom kind, white, and smell very sweet. They appear in June, and when the trees are well charged with flowers, they make a fine appearance, and their odour perfumes the circumambient air; but they are of short duration, seldom continuing more than one week in beauty; after the flowers fade, the germen becomes oblong compressed pods, which in warm seasons come to perfection in England, but ripen pretty late in the autumn.

The leaves of this tree do not come out till late in the spring, and they fall off pretty early in the autumn, which renders it less valuable than it would otherwise be, were their leaves of longer duration.

The second sort is less common than the first. There was a large tree of this kind some years growing in the garden of the bishop of London at Fulham, which produced plenty of seeds. The pods of this sort are much shorter, and closely beset with short prickles, but in other respects agrees with the first sort.

The third sort grows naturally in Carolina, where it sometimes rises to the height of twenty feet, but in England at present, it seems to be of low growth; the branches spread out near the ground, and produce their flowers very young, which is a sure sign of its not growing tall here. The branches of this tree, and also the foot-stalks of the flowers, are closely armed with small brown spines, like some sorts of Roses; the leaves are like those of the first sort, but their lobes are larger and rounder. The flowers come

come out in bunches like those of the former, but are larger and of a deep Rose colour.

The fourth sort grows naturally at Campeachy, from whence the late Dr. Houstoun sent the seeds. This rises with a strong woody stem to the height of thirty or forty feet, sending out many strong branches on every side, which have large swelling knots, and are closely garnished with single winged leaves, composed of eight or nine pair of oval lobes, ending in points, terminated by an odd one; these are curiously marked with purple spots on their under side, which appear faintly on their upper. The flowers are produced in long close spikes, standing almost erect; they are about half the size of the flowers of the last sort, and are of a fine Rose colour.

The fifth sort was found growing naturally at Campeachy by the late Dr. Houstoun. This rises with a woody branching stalk twelve or fourteen feet high; the old branches are covered with a dark brown bark, but the young shoots and the foot-stalks of the flowers are covered with an iron-coloured down; the leaves are unequally winged; the lobes are oblong, obtuse, and of a pretty thick consistence; they are smooth on their upper side, but have several transverse veins on their under. The flowers are produced at the end of the branches in long close bunches; there are six or seven of them gathered together in clusters. The flowers are but small, and are of a yellowish red colour; the pods of this are like those of the first sort.

The sixth sort grows naturally in Jamaica, where the inhabitants give it the appellation of Dogwood. This hath a strong woody stem, which rises forty feet high, and divides into many branches, covered with a dark brown spotted bark, garnished with unequal winged leaves, composed of three or four pair of obverse oval lobes, terminated by an odd one. The flowers come out in branching bunches from the side of the branches; these generally appear at a time when the trees are destitute of leaves, and as they have large clusters of flowers at every joint, so the trees seem covered with them. The bunches at the extremity of the branches are the largest, and are formed pyramidally. The flowers are but small, and do not open so fully as those of the first sort, but are of a pale Rose colour, so make a fine appearance; these are succeeded by pods, having four broad membranaceous wings running longitudinally at their four corners, and join at their base, covering the pods entirely; each of the pods contain four or five oblong kidney-shaped seeds.

The seventh sort was discovered by Father Plumier, in some of the French settlements of the West-Indies, and was since found by the late Dr. Houstoun growing naturally at Campeachy. This rises with a strong woody stem near thirty feet high, sending out many spreading branches, covered with a light gray bark, spotted with white, garnished with double winged leaves, whose lobes are oval, and sit close to the midrib; they are of a lucid green on their upper side, but of a pale green on their under. The flowers are produced in long loose pyramidal bunches toward the end of the branches, those on the lower part of the bunch having long foot-stalks, which diminish gradually to the top, so as to form a pyramid; these bunches are almost erect. The flowers are of a scarlet colour, so make a fine appearance.

The eighth sort was found growing naturally at Campeachy by the late Dr. Houstoun. This rises with a woody stem to the height of twenty feet, covered with a very light gray bark, garnished with equal winged leaves, composed of ten or eleven pair of oval lobes, placed opposite, of a lucid green on the lower part of the branches, but those toward the end are covered with a soft iron-coloured down. The flowers come out in long bunches from the side of the branches; they are blue, and stand upon long foot-stalks; these are succeeded by pods shaped like those of the first sort, but are downy.

The ninth sort was discovered by the late Dr. Houstoun growing naturally at Campeachy. This rises with a strong woody stem upward of thirty feet high, dividing at the top into many strong branches, covered with a dark grayish bark, spotted with white, garnished with winged leaves, composed of six or seven pair of lobes, terminated by an odd one, of a lucid green on their upper side, but pale on their under. The flowers are produced in long loose bunches from the side of the branches; they are of a pale Rose colour, and have very long foot-stalks; these are succeeded by oval pods swelling in the middle, where is lodged one or two kidney-shaped seeds.

The tenth sort grows naturally in Siberia and Tartary. This hath a shrubby stalk eight or ten feet high, sending out several branches which grow erect, covered with a smooth yellowish bark; the leaves have each two pair of oval pointed lobes, which stand upon short foot-stalks. The flowers are produced upon single foot-stalks, which come out at the joints of the branches; they are yellow, and shaped like those of the Laburnum, but are smaller.

The eleventh sort grows naturally in Siberia. This rises with a shrubby stem ten or twelve feet high, sending out many branches on every side, which are garnished with abrupt winged leaves, composed of four lobes, having short foot-stalks; the flowers are produced singly from the side of the branches which are yellow, and shaped like those of the Laburnum; these appear in May, and are succeeded by pods four inches long, containing four or five seeds in each.

The twelfth sort is also a native of Siberia. This is a low weakly shrub, seldom rising more than three or four feet high in England. This sends out a few slender branches which want support; these are garnished with winged leaves, composed of four oblong lobes, sitting close to the branches; the flowers are produced singly upon foot-stalks, which arise from the wings of the leaves. These are yellow, and appear the beginning of April, but are rarely succeeded by seeds in England.

These two sorts are propagated by seeds, when they can be obtained, and the plants require a moist soil and a shady situation.

The first sort is generally propagated in the English nurseries, by suckers taken from the roots of the old trees, or by cutting off some of the roots, and planting them upon a gentle hot-bed; these will put out shoots, and become plants; but these are not so valuable as those which are raised from seeds, because they do not make near so great progress in their growth, and are very subject to send forth many suckers from their roots, whereby the ground will be filled with them to a great distance.

If this is propagated by seeds, they should be sown on a bed of light earth, about the latter end of March, or the beginning of April; and if the bed is well exposed to the sun, the plants will appear in about five or six weeks, and will require no farther care but to keep them clear from weeds. In this bed the plants may remain till the following spring, when they should be transplanted into a nursery about the latter end of March, placing them in rows at three feet distance row from row, and a foot and a half asunder in the rows. In this nursery they may remain two years, by which time they will be fit to transplant where they are designed to grow; for as these trees send forth long tough roots, so if they stand long unremoved, the roots will extend themselves to a great distance; therefore they must be cut off when the plants are transplanted, which sometimes occasions their miscarrying.

5 These

Theſe trees will grow well upon almoſt every ſoil, but beſt in a light ſandy ground, in which they will ſhoot ſix or eight feet in one year; and while the trees are young they make an agreeable appearance, being well furniſhed with leaves; but when they are old, the branches being frequently broken by winds, render them unſightly, eſpecially if they ſtand in an expoſed place: alſo when the trees grow old, their branches decay and make a bad appearance, which occaſioned their being rooted out of ſeveral gardens ſome years paſt. This is commonly known by the title of Locuſt-tree in America; there are quantities of the ſeeds annually ſent to England with that title.

The ſecond ſort is propagated in the ſame manner as the firſt, and the trees grow to the ſame ſize.

The third ſort is at preſent ſcarce in the gardens about London, but in Devonſhire it is in greater plenty, where the inhabitants give it the title of Raſpberry plant, from the young ſhoots being covered with briſtly hairs like the Raſpperry plants; this does not produce ſeeds in England, ſo it is propagated by layers, and alſo cutting off part of the roots, and planting them upon a gentle hot-bed, where they will put out fibres, ſhoots, and become new plants. This ſort ſhould have a warmer ſituation than the two former, though the ordinary winters in this country never injure it, but in very ſevere winters the young ſhoots are ſometimes killed in expoſed places. It loves a light moiſt ſoil.

The fourth, fifth, ſixth, ſeventh, eighth, and ninth ſorts, are tender, ſo cannot be maintained in England, unleſs they are placed in a ſtove in winter. Theſe are propagated by ſeeds, which ſhould be ſown in ſmall pots, and plunged into a hot-bed of tanners bark; if the ſeeds are good the plants will appear in ſix weeks or two months; when theſe are fit to tranſplant they ſhould be carefully ſhaken out of the pots, and their roots ſeparated, putting each plant into a ſmall pot, then they ſhould be plunged into a hot-bed of tanners bark, and muſt have the ſame treatment as other tender plants from the ſame countries.

The tenth ſort is propagated by ſeeds, which ſhould be ſown in a ſhady ſituation in autumn, then the plants will come up the following ſpring; but if the ſeeds are ſown in the ſpring, the plants ſeldom riſe the ſame ſeaſon. When the plants come up, they will require no other care but to keep them clean from weeds till autumn, when, if the plants have made any progreſs, they ſhould be tranſplanted on a north border at about ſix inches diſtance, where they may grow two years, and then ſhould be planted where they are to remain, which ſhould be in a cool moiſt ſoil.

RONDELETIA. Plum. Nov. Gen. 15. tab. 12.

The CHARACTERS are,

The flower has a permanent empalement ſitting upon the germen, cut into five acute points. It has one funnel-ſhaped petal, with a cylindrical tube longer than the empalement, cut into five roundiſh ſegments at the brim, which are reflexed. It has five awl-ſhaped ſtamina, terminated by ſingle ſummits; the roundiſh germen is ſituated under the flower, ſupporting a ſlender ſtyle the length of the tube, crowned by an obtuſe ſtigma. The germen afterward becomes a roundiſh crowned capſule with two cells, incloſing two or three angular ſeeds in each.

The SPECIES are,

1. RONDELETIA (*Americana*) foliis ſeſſilibus, panicula dichotoma. Lin. Sp. Plant. 172. *Rondeletia with leaves ſitting cloſe to the branches.*

2. RONDELETIA (*Aſiatica*) foliis petiolatis. Flor. Zeyl. 80. *Rondeletia with leaves growing upon foot-ſtalks.*

The ſeeds of this plant were firſt ſent me by Mr. Robert Millar, who collected them on the north ſide of the iſland of Jamaica; he alſo obſerved the trees growing plentifully in the Spaniſh Weſt-Indies; I have alſo ſince received the ſeeds from Barbadoes, which have ſucceeded at Chelſea. This riſes with a woody ſtalk eight or ten feet high, branching out on every ſide, covered with a ſmooth greeniſh bark, garniſhed with oblong leaves, ending in acute points; they are entire, and ſit very cloſe to the branches; their upper ſurface is of a lucid green, the under is paler; they are a little crumpled on their ſurface, and ſtand alternate. The flowers come out in bunches at the end of the branches; they are white, and have little ſcent.

The ſecond ſort grows naturally in Malabar. This riſes with a woody ſtalk ſix or ſeven feet high, dividing into ſeveral branches, covered with a ſmooth bark, garniſhed with ſtiff, oblong, oval leaves, of a lucid green, ſtanding alternate on the lower part of the branches, but by pairs toward the extremity; they have ſhort foot-ſtalks, and are entire. The flowers are produced in large bunches at the end of the branches; they are of a yellowiſh white colour, and have a fragrant odour; theſe are ſucceeded by roundiſh capſules, having two cells, each containing three or four angular ſeeds.

Theſe plants being very tender, cannot be preſerved in England, unleſs they are kept in a warm ſtove. They are propagated by ſeeds, which ſhould be ſown on a hot-bed early in the ſpring; when the plants are come up and fit to remove, they muſt be tranſplanted into ſeparate ſmall pots, and plunged into a moderate hot-bed of tanners bark, where they muſt be treated in the ſame manner as hath been directed for other tender plants from the ſame country.

ROSA. Tourn. Inſt. R. H. 636. tab. 408. The Roſe-tree.

The CHARACTERS are,

The empalement of the flower is divided into five parts at the top, but the baſe is globular and bell-ſhaped. The flower hath five oval heart-ſhaped petals inſerted in the empalement, and a great number of ſhort hair-like ſtamina inſerted in the neck of the empalement, terminated by three-cornered ſummits. It hath many germen ſituated in the bottom of the empalement, each having a ſhort hairy ſtyle, inſerted to the ſide of the germen, crowned by obtuſe ſtigmas. The fleſhy baſe of the empalement afterward becomes a top-ſhaped coloured fruit with one cell, including many hairy oblong ſeeds, faſtened on each ſide to the empalement.

The SPECIES are,

1. ROSA (*Canina*) caule aculeato, petiolis inermibus, calycibus ſemipinnatis. Flor. Suec. 406. *Roſe with a prickly ſtalk, unarmed foot-ſtalks, and empalements which are half-winged; commonly called Wild Briar, Dog Roſe, or Hep-tree.*

2. ROSA (*Spinoſiſſima*) caule petioliſque aculeatis, calycis foliolis indiviſis. Flor. Suec. 407. *Roſe with ſtalks and foot-ſtalks armed with ſpines, and the ſmall leaves of the empalement undivided; commonly called the Burnet-leaved Roſe.*

3. ROSA (*Villoſa*) foliis utrinque villoſis, fructu ſpinoſo. Haller. Helv. 350. *Roſe whoſe leaves are hairy on both ſides, and the fruit prickly: Apple-bearing Roſe*

4. ROSA (*Eglanteria*) aculeata, foliis odoratis, ſubtus rubiginoſis. Haller. Helv. 350. *Roſe with ſpines and ſweet-ſcented leaves, which are ruſty on their under ſide; commonly called Sweet Briar.*

5. ROSA (*Scotica*) caule petioliſque aculeatis, foliis pinnatis, foliolis apice inciſis, fructu globoſo. *Roſe with the ſtalk and foot-ſtalk armed with ſpines, winged leaves, whoſe lobes are cut at their points, and a globular fruit; Scotch Roſe.*

6. ROSA (*Inermis*) caule inermi, pedunculis hiſpidis, calycis foliolis indiviſis, fructibus oblongis. *Roſe with a ſmooth ſtalk, a prickly foot-ſtalk to the flower, the ſmall leaves of the empalement undivided, and oblong fruit.*

7. ROSA (*Hiſpanica*) foliis utrinque villoſis, calycis foliolis acute ſerratis, fructu glabro. *Roſe with leaves which*

re hairy on both ſides, the ſmall leaves of the empalement ſharply ſawed, and a ſmooth fruit.

8. Rosa (*Scandens*) caule aculeato ſcandente, foliolis glabris ſerratis perennantibus. *Roſe with a prickly climbing ſtalk, and ſmooth, ſawed, evergreen leaves.*

9. Rosa (*Sempervirens*) caule aculeato, foliolis quinis glabris perennantibus. Lin. Sp. Plant. 482. *Roſe with a prickly ſtalk, and five ſmooth lobes to the leaves, which are evergreen; evergreen Muſk Roſe.*

10. Rosa (*Virginiana*) inermis, foliis pinnatis ſerratis utrinque glabris, calycis foliolis indiviſis. *Roſe without thorns, having winged leaves, which are ſmooth on both ſides, and the leaves of the empalement undivided; wild Virginian Roſe.*

11. Rosa (*Lutea*) caule aculeato, foliis pinnatis, foliolis ovatis ſerratis utrinque glabris, pedunculis breviſſimis. *Roſe with a prickly ſtalk, winged leaves having oval ſawed lobes, which are ſmooth on both ſides, and ſhort foot-ſtalks to the flower; the ſingle yellow Roſe.*

12. Rosa (*Punicea*) caule aculeato, foliis pinnatis, foliolis rotundioribus ſerratis, petalis emarginatis bicoloribus. *Roſe with a prickly ſtalk, winged leaves having rounder ſawed lobes, the petals of the flower indented at the top, and of two colours; the Auſtrian Roſe.*

13. Rosa (*Moſchata*) caule aculeato ſcandente, foliis ſenis glabris, floribus umbellatis. *Roſe with a prickly climbing ſtalk, leaves having ſeven ſmooth lobes, and flowers growing in umbels; greater Muſk Roſe.*

14. Rosa (*Centifolia*) caule aculeato, pedunculis hiſpidis calycibus ſemipinnatis glabris. Lin. Sp. Pl. 491. *Roſe with a prickly ſtalk, briſtly foot-ſtalks, and ſmooth half-winged empalements; the Dutch hundred-leaved Roſe.*

15. Rosa (*Damaſcena*) caule aculeato, pedunculis hiſpidis, calycibus pinnatifidis hirſutis. *Roſe with a prickly ſtalk, briſtly foot-ſtalks to the flowers, and wing-pointed hairy empalements; Damaſk Roſe.*

16. Rosa (*Alba*) caule aculeato, pedunculis lævibus, calycibus ſemi-pinnatis glabris. Lin. Sp. Plant. 492. *Roſe with a prickly ſtalk, ſmooth foot-ſtalks to the flowers, and ſmooth half-winged empalements; common great white Roſe.*

17. Rosa (*Belgica*) caule aculeato, foliis ſubtus hirſutis, calycibus ſemipinnatis villoſis. *Roſe with a prickly ſtalk, leaves which are hairy on their under ſide, and half-winged hairy empalements to the flowers; the Bluſh Belgic Roſe.*

18. Rosa (*Provincialis*) caule petioliſque aculeatis, foliis ſubtus villoſis, calycibus ſemipinnatis hiſpidis. *Roſe with prickly ſtalks and foot-ſtalks, leaves hairy on their under ſide, and briſtly half-winged empalements; commonly called Provence Roſe.*

19. Rosa (*Incarnata*) caule inermi pedunculis aculeatis calycibus ſemipinnatis. *Roſe with an unarmed ſtalk, prickly foot-ſtalks, and half-winged empalements to the flowers; the Bluſh Roſe.*

20. Rosa (*Gallica*) caule ſubinermi, foliis quinis ſubtus villoſis, calycis foliolis indiviſis. *Roſe with a ſtalk almoſt unarmed, leaves having five lobes, hairy on their under ſide, and the leaves of the empalement undivided; the red Roſe.*

21. Rosa (*Cinnamomea*) foliis pinnatis ſerratis ſubtus villoſis, aculeis oppoſitis, calycis foliolis indiviſis. *Roſe with winged ſawed leaves, which are hairy on their under ſide, ſpines placed oppoſite, and the ſmall leaves of the empalement undivided; the Cinnamon Roſe.*

22. Rosa (*Muſcoſa*) caule petioliſque aculeatis, pedunculis calycibuſque piloſiſſimis. *Roſe with armed ſtalks, the foot-ſtalks and the empalements of the flower very hairy; commonly called Moſs Provence Roſe.*

There are a great variety of double Roſes now cultivated in the Engliſh gardens; moſt of them have been accidentally obtained from ſeeds, ſo that they muſt not be ranged as diſtinct ſpecies, therefore I ſhall only inſert their common names, by which they are known in the gardens, that thoſe who are inclined to collect all the varieties, may be at no loſs for their titles. The ſorts before enumerated, I believe, are diſtinct ſpecies, as their ſpecific characters are different, though it is difficult to determine which of them are really ſo, therefore I do not poſitively aſſert they are diſtinct, though I have great reaſon to believe they are ſo.

The varieties of Garden Roſes which are not before mentioned:

The monthly Roſe,
The ſtriped monthly Roſe,
The York and Lancaſter Roſe,
Mrs. Hart's Roſe,
} are all ſuppoſed to be varieties of the Damaſk Roſe.

The red Belgic Roſe is ſuppoſed a variety of the Bluſh Belgic.

The ſingle Velvet Roſe,
The double Velvet Roſe,
The Royal Velvet Roſe,
} are all varieties; the laſt I raiſed from the ſeeds of the pale Provence Roſe.

The childing Roſe,
The marbled Roſe,
The double Virgin Roſe,
} have great affinity with each other.

The Cabbage Provence is only a variety of the common Provence.

The Bluſh or pale Provence is a variety of the red Provence.

The white monthly,
The white Damaſk,
} are varieties of the Damaſk Roſe.

The Frankfort Roſe may be a diſtinct ſpecies, but is of little value; the flowers rarely open fair, and have no odour.

The double ſweet Briar,
The evergreen ſweet Briar,
The double Bluſh ſweet Briar,
} are varieties of the common ſort.

The Auſtrian Roſe with red and yellow flowers, is only an accidental variety.

The double yellow Roſe is a variety of the ſingle yellow.

The Roſa Mundi is a variety of the red Roſe.

The ſmall, white, and ſemidouble white, are varieties of the common white.

The firſt ſort is very common in hedges in moſt parts of England, ſo is not cultivated in gardens. The heps of this are uſed in medicine for making a conſerve. There are two or three varieties of this Roſe commonly met with in hedges, one with a white, another with a red flower, and one with ſmooth leaves; the two firſt are evidently varieties, but I doubt if the laſt is not a diſtinct ſpecies.

The ſecond ſort grows naturally in many parts of England. This ſeldom riſes above three feet high. The ſtalks are ſlender, cloſely armed with ſmall ſpines; the leaves are ſmall, and are compoſed of three pair of roundiſh lobes, terminated by an odd one; the flowers are white, and have an agreeable muſky ſcent. This propagates faſt by its creeping roots.

The third ſort grows naturally in the northern counties in England; this riſes with ſtrong ſtalks to the height of ſeven or eight feet. The young branches are covered with a ſmooth brown bark; the ſpines are but few, and are very ſtrong; the leaves are large, and hairy on both ſides; they are compoſed of three pair of oblong oval lobes, terminated by an odd one; they are deeply ſawed on their edges; the flowers are large, ſingle, and of a red colour, and are ſucceeded by large roundiſh heps or fruit, which are ſet with ſoft prickles; they have a pleaſant acid pulp ſurrounding the ſeeds, therefore are by ſome perſons preſerved and made into a ſweetmeat, which is ſerved up in deſerts to the table.

The fourth ſort is the common ſweet Briar, which is ſo well known as to need no deſcription. This is found growing naturally in ſome parts of Kent.

The fifth ſort is the Dwarf Scotch Roſe, of which there are two varieties, one with a variegated flower, and the other of a livid red colour. This ſort ſeldom riſes more than a foot high. The ſtalks are covered with a brown bark, and are cloſely armed with ſmall ſpines; the leaves are very ſmall, reſembling thoſe of Burnet; the flowers are ſmall, and ſit cloſe to the branches; the fruit is round, and of a deep purple colour, inclining to black when ripe.

The ſixth ſort riſes to the height of ſix or ſeven feet. The ſtalks and branches have no ſpines, but are covered with a ſmooth reddiſh bark; the leaves are compoſed of three pair of thin oval lobes, terminated by an odd one, of a bright green, and very ſlightly ſawed on their edges, ſtanding pretty far aſunder on the midrib; the foot-ſtalks of the flowers are armed with briſtly hairs; the five leaves of the empalement are long, ſlender in the middle, but terminate in an oval leafy point; the flowers are ſingle, of a bright red colour; theſe are ſucceeded by long ſpear-ſhaped heps, which are ſmooth. The plants produce a ſecond crop of flowers about the end of Auguſt, but theſe fall off, and are not ſucceeded by heps.

The ſeeds of the ſeventh ſort were ſent me by Robert More, Eſq; from Spain, where he found the plants growing naturally; this riſes with ſtrong upright ſtalks about three feet high, armed with ſtrong thorns. The leaves are hairy on both ſides; the lobes are roundiſh, ſawed on their edges; the ſmall leaves of the empalement are acutely ſawed; the flowers are ſingle, of a bright red colour; theſe are followed by large, ſmooth, roundiſh heps, which ripen the end of Auguſt.

The eighth ſort was diſcovered by Sig. Micheli, growing naturally in the woods near Florence, who ſent it to Dr. Boerhaave of Leyden, in whoſe curious gardens I ſaw it growing in the year 1727, who was ſo kind as to communicate to me a plant of it. This hath ſlender ſtalks, which trail upon the ground, unleſs they are ſupported, and if trained up to a pole or the ſtem of a tree, will riſe twelve or fourteen feet high; they are armed with crooked reddiſh ſpines, garniſhed with ſmall leaves, compoſed of three pair of oval acute-pointed lobes, terminated by an odd one, of a lucid green, and ſawed on their edges; they continue green all the year; the flowers are ſmall, ſingle, white, and have a muſky odour; theſe in their natural place of growth continue in ſucceſſion great part of the year, but their time of flowering in England is in June.

The ninth ſort grows naturally in Spain; the ſeeds of this were ſent me by Robert More, Eſq; who found the plants growing there naturally. This riſes with erect ſtalks five or ſix feet high, covered with a green bark, and armed with ſtrong, crooked, white ſpines. The leaves are compoſed of five oval lobes, ending in acute points; they are ſmooth, of a lucid green, and are ſlightly ſawed on their edges; theſe continue all the year, and make a goodly appearance in winter. The flowers grow in large bunches or umbels at the end of the branches; they are ſingle, white, and have a ſtrong muſky odour; they appear in Auguſt, and if the autumn proves favourable, will continue in ſucceſſion till October.

The tenth ſort grows naturally in Virginia and other parts of North America; this riſes with ſeveral ſmooth ſtalks to the height of five or ſix feet. The young branches are covered with a ſmooth purple bark; the leaves are compoſed of four or five pair of ſpear-ſhaped lobes, terminated by an odd one, of a lucid green on their upper ſide, but pale on their under, and are deeply ſawed on their edges; the flowers are ſingle, of a livid red colour; the empalement is divided into five long narrow ſegments, which are entire. This is kept in gardens for the ſake of variety, but the flowers have little ſcent.

The eleventh ſort is the ſingle yellow Roſe; this hath weak ſtalks, which ſend out many ſlender branches, cloſely armed with ſhort, crooked, brown ſpines. The leaves are compoſed of two or three pair of oval thin lobes, terminated by an odd one, of a light green, ſharply ſawed on their edges; the flowers grow upon ſhort foot-ſtalks; they are ſingle, of a bright yellow colour, but have no ſcent.

The twelfth ſort is commonly called the Auſtrian Roſe. The ſtalks, branches and leaves, are like thoſe of the laſt, but the leaves are rounder; the flowers are larger; the petals have deep indentures at their points; they are of a bright yellow within, and of a purpliſh copper colour on the outſide; they are ſingle, have no ſcent, and ſoon fall away. There is frequently a variety of this with yellow flowers upon one branch, and copper colour upon another. This ſort of Roſe loves an open free air and a northern aſpect.

The thirteenth ſort is the Muſk Roſe; this riſes with weak ſtalks to the height of ten or twelve feet, covered with a ſmooth greeniſh bark, armed with ſhort ſtrong ſpines. The leaves are ſmooth, compoſed of three pair of oval ſpear-ſhaped lobes, terminating in points, ending with an odd one; they are of a light green colour, ſawed on their edges; the flowers are produced in large bunches in form of umbels at the end of the branches; they are white, and have a fine muſky odour. There is one with ſingle, and another with double flowers of this ſort. The ſtalks of theſe plants are too weak to ſupport themſelves, ſo the plants ſhould be placed where they may have ſupport.

The fourteenth ſort is the Dutch hundred-leaved Roſe; this riſes with prickly ſtalks about three feet high. The leaves have ſometimes three, and at others five lobes; the lobes are large, oval, ſmooth, of a dark green, with purple edges; the foot-ſtalk of the flower is ſet with brown briſtly hairs; the empalement of the flower is ſmooth, and half-winged; the flowers are very double and of a deep red colour, but have little ſcent.

The fifteenth ſort is the Damaſk Roſe; this riſes with prickly ſtalks eight or ten feet high, covered with a greeniſh bark, armed with ſhort ſpines. The leaves are compoſed of two pair of oval lobes, terminated by an odd one; they are of a dark green on their upper ſide, but pale on their under; the borders frequently turn brown, and are ſlightly ſawed; the foot-ſtalks of the flowers are ſet with prickly hairs; the empalement of the flower is wing-pointed and hairy; the flowers are of a ſoft pale red, and not very double, but have an agreeable odour; the heps are long and ſmooth.

The ſixteenth is the common large white Roſe, ſo well known as to need no deſcription. Of this there are two varieties, one with a half double flower, having but two or three rows of petals, and the other has a ſmaller flower, and the ſhrub is of lower growth.

The ſeventeenth ſort is called the Bluſh Belgic Roſe; this riſes about three feet high, with prickly ſtalks. The leaves are compoſed either of five or ſeven lobes, which are oval, hairy on their under ſide, ſlightly ſawed on their edges; the foot-ſtalks of the flowers, and the empalements are hairy, and without ſpines; the empalements are large and half-winged; the flowers are very double, of a pale fleſh colour, and have but little ſcent. It generally produces great quantities of flowers. The red Belgic Roſe differs from this only in the colour of the flower, which is of a deep red.

The eighteenth ſort is the common Provence Roſe, which is well known in the Engliſh gardens, being cultivated in great

great plenty in the nurseries, and is one of the most beautiful sorts yet known. The flowers of this sort are sometimes very large, and the petals are closely folded over each other like Cabbages, from whence it is called the Cabbage Rose. The flowers of this sort of Rose have the most fragrant odour of all the sorts, therefore is better worth propagating.

The nineteenth sort is the Blush Rose. The stalks of this rise from three to four feet high, and are not armed with spines; the leaves are hairy on their under side; the foot-stalks of the flowers are armed with some small spines; the empalement of the flower is half-winged; the flowers have five or six rows of petals, which are large, and spread open; they are of a pale blush colour, and have a musky scent.

The twentieth sort is the common red Rose, whose flowers are used in medicine. The stalks of this grow erect, and have scarce any spines; they rise from three to four feet high; the leaves are composed of three or five large oval lobes, which are hairy on their under side; the small leaves of the empalement are undivided; the flowers are large, but not very double, spread open wide, and decay soon; they are of a deep red colour, and have an agreeable scent. The Rosa Mundi is a variety of this with striped flowers.

The twenty-first sort is the double Cinnamon Rose; this is one of the smallest flowers, and the earliest of all the kinds. The stalks rise about four feet high, covered with a purplish smooth bark, and have no spines but at the joints immediately under the leaves, where they are placed by pairs; they are short and crooked. The leaves are composed of three pair of oval lobes, terminated by an odd one; they are hairy on their under side, and are sawed on their edges; the leaves of the empalement of the flower are narrow and entire; the flower is small, double, and has a scent like Cinnamon, from whence it has the title of Cinnamon Rose.

The twenty-second sort is called the Moss Provence Rose, from the resemblance which the flowers of this have to those of the common Provence Rose, yet it is undoubtedly a distinct species; for although the stalks and shoots of this are very like those of the common, yet the plants are difficult to propagate, which the common sort is not. This very rarely sends up suckers from the root, and when the branches are laid down, they are long before they put out roots, so that this sort has been frequently propagated by budding it upon stocks of other sorts of Roses, but the plants so raised are not so durable as those which are propagated by layers.

The stalks and branches of this sort are closely armed with brown spines; the foot-stalks of the flowers and the empalements, are covered with long hair-like Moss; the flowers are of an elegant crimson colour, and have a most agreeable odour.

Most of the sorts of Roses are of foreign growth, and have been at various times introduced into the English gardens, but they are generally natives of northern countries, or grow upon the cold mountains in the warmer parts of Europe, so they are very hardy in respect to cold, but love an open free air, especially the yellow Rose, the Austrian Rose, and the monthly Rose. The two former will not flower in a warm soil and situation, nor near the smoke of London; and the monthly Rose will not flower much better, unless it is planted in a free open air, for it seldom flowers in the smoke of London.

The usual time of these shrubs producing their flowers, is from the middle or latter end of May till the middle or end of July.

But in order to continue these beauties longer than they are naturally disposed to last, it is proper to plant some of the monthly Roses near a warm wall, which will occasion their budding at least three weeks or a month before those in the open air; and if you give them the help of a glass before them, it will bring their flowers much forwarder, especially where dung is placed to the backside of the wall (as is practised in raising early fruits; by this method I have seen fair Roses of this kind blown in February, and they may be brought much sooner against hot walls, where people are curious this way.

You should also cut off the tops of such shoots which have been produced the same spring early in May, from some of these sorts of Roses which are planted in the open air, and upon a strong soil; this will cause them to make new shoots, which will flower late in autumn, as will also the late removing the plants in spring, provided they do not suffer by drought.

The next sort of Rose which flowers in the open air is the Cinnamon, which is immediately followed by the Damask Rose, then the Blush, York, and Lancaster come, after which the Provence, Dutch hundred leaved, white, and most other sorts of Roses follow; and the latest sorts are the Musk Roses, which, if planted in a shady situation, seldom flower until September, and, if the autumn proves mild, will continue often till the middle of October.

The plants of these two sorts should be placed against a wall, pale, or other building, that their branches may be supported, otherwise they are so slender and weak as to trail upon the ground. These plants should not be pruned until spring, because their branches are somewhat tender, so that when they are cut in winter, they often die after the knife; these produce their flowers at the extremity of the same year's shoots in large bunches, so that their branches must not be shortened in the summer, lest thereby the flowers should be cut off. These shrubs will grow to be ten or twelve feet high, and must not be checked in their growth, if you intend they should flower well, so that they should be placed where they may be allowed room.

The lowest shrub of all the sorts here mentioned is the Scotch Rose, which rarely grows above one foot high, so that this must be placed among other shrubs of the same growth. The red Rose and the Rosa Mundi, commonly grow from three to four feet high, but seldom exceed that, but the Damask, Provence, and Frankfort Roses, grow to the height of seven or eight feet; so that in planting them great care should be taken to place their several kinds, according to their various growth, among other shrubs, that they may appear beautiful to the eye.

The yellow Rose, as also the Austrian Rose, are both natives of America. These were originally brought from Canada by the French; the other varieties, which are now in the gardens, have been accidentally obtained from seeds, and are preserved by budding them on the other sorts. The shrubs of these Roses seldom shoot so strong as most of the other sorts, especially in the light land near London, where they rarely produce their flowers. These are esteemed for their colour, being very different from all the other sorts of Roses, but as their flowers have no scent, and are of short duration, they do not merit the price they are generally sold at.

The Frankfort Rose is of little value, except for a stock to bud the more tender sorts of Roses upon, for the flowers seldom open fair, and have no scent; but, it being a vigorous shooter, renders it proper for stocks to bud the yellow and Austrian Roses, which will render them stronger than upon their own stocks; but the yellow Roses seldom blow fair within eight or ten miles of London, though in the

 northern

northern parts of Great-Britain they flower extremely well. This sort must have a northern exposure, for if it is planted too warm it will not flower.

The Damask and monthly Rose seldom flower well in small confined gardens, nor in the smoke of London, therefore are not proper to plant in such places, though they frequently grow very vigorously there. These always begin to shoot the first of any of the sorts in the spring, therefore frequently suffer from frosts in April, which often destroys all their flowers.

All the sorts of Roses may be propagated either from suckers, layers, or by budding them upon stocks of other sorts of Roses, which latter method is only practised for some peculiar sorts, which do not grow very vigorous upon their own stocks, and send forth suckers very sparingly, or where a person is willing to have more sorts than one upon the same plant; but where this is designed, it must be observed to bud only such sorts upon the same stock as are nearly equal in their manner of growth, for if there be a bud of a vigorous growing sort, and others of weak growth, budded in the same stock, the strong one will draw all the nourishment from the weaker, and entirely starve them.

If these plants are propagated by suckers, they should be taken off annually in October, and transplanted out either into a nursery in rows (as hath been directed for several other sorts of flowering shrubs,) or into the places where they are to remain; for if they are permitted to stand upon the roots of the old plants more than one year, they grow woody, and do not form so good roots as if planted out the first year, so there is more danger of their not succeeding.

But the best method to obtain good-rooted plants, is to lay down the young branches in autumn, which will take good root by the autumn following, when they may be taken from the old plants, and transplanted where they are to remain. The plants which are propagated by layers, are not so apt to send out suckers from their roots as those which are raised from suckers, therefore should be preferred before them, because they may be much easier kept within compass, and these will also flower much stronger. The plants may be transplanted any time from October to April, but when they are designed to flower strong the first year after planting, they should be planted early; though, as I said before, if they are planted late in the spring, it will cause them to flower in autumn, provided they do not suffer by drought.

Most of these sorts delight in a rich moist soil and an open situation, in which they will produce a greater quantity of flowers, and those much fairer than when they are upon a dry soil, or in a shady situation. The pruning which they require is only to cut out their dead wood, and take off all the suckers, which should be done every autumn; and if there are any very luxuriant branches which draw the nourishment from the other parts of the plant, they should be taken out or shortened, to cause them to produce more branches, if there be occasion for them to supply a vacancy; but you must avoid crowding them with branches, which is as injurious to these plants as to fruit-trees; for if the branches have not equal benefit of the sun and air, they will not produce their flowers so strong, nor in so great plenty, as when they are more open, so that the air may circulate more freely between them.

ROSA SINENSIS. See HIBISCUS.

ROSA THE GELDER. See OPULUS.

ROSMARINUS. Tourn. Inst. R. H. 195. tab. 92. Rosemary.

The CHARACTERS are,

The flower has a tubulous empalement, compressed at the top, the mouth erect, and divided into two lips; the upper lip is entire, and the under bifid. It hath one petal, whose tube is longer than the empalement; the brim is ringent; the upper lip is short, erect, and divided into two parts, whose borders are reflexed; the lower lip is reflexed, and cut into three parts. It hath two awl-shaped stamina, inclining toward the upper lip, terminated by single summits, and a four-pointed germen with a style the shape, length, and in the same situation with the stamina, crowned by an acute stigma. The germen afterward becomes four oval seeds sitting in the bottom of the empalement.

The SPECIES are,

1. ROSMARINUS (*Angustifolia*) foliis linearibus marginibus reflexis, subtus incanis. *Rosemary with linear leaves, which are reflexed on their edges, and hoary on their under side; Garden Rosemary with a narrower leaf.*

2. ROSMARINUS (*Latifolia*) foliis linearibus obtusis, utrinque virentibus. *Rosemary with obtuse linear leaves, which are green on both sides; or broad-leaved wild Rosemary.*

These two sorts have been frequently supposed the same, and their difference accidental, but I have long cultivated both, and have raised them from seeds without finding them vary, so I believe they are distinct species. The leaves of the second sort are broader than those of the first, and their points are obtuse; the flowers are also much larger, and of a deeper colour; the stalks grow larger, and spread out their branches wider; the whole plant has a stronger scent. These differences the gardeners, who cultivate the plants for the market, observe.

There are two other varieties of these plants, one of the first sort with striped leaves, which the gardeners call the Silver Rosemary; the other is of the second sort, which is striped with yellow; this the gardeners call the gold striped Rosemary. The plants of this sort are pretty hardy, so will live in the open air through our common winters, if they are upon a dry soil.

These plants grow plentifully in the southern parts of France, in Spain, and in Italy, where, upon dry rocky soils near the sea, they thrive prodigiously, and perfume the air so as to be smelt at a great distance from the land; but, notwithstanding they are produced in warm countries, yet they are hardy enough to bear the cold of our ordinary winters very well in the open air, provided they are planted upon a poor, dry, gravelly soil, on which they will endure the cold much better than upon richer ground, where the plants will grow more vigorously in summer, and so be more subject to injury from frost, nor will they have so strong an aromatic scent as those upon a dry barren soil.

Those sorts with striped leaves are somewhat tender, especially that with silver stripes, so should either be planted near a warm wall, or in pots, to be sheltered in winter under a frame, otherwise they will be subject to die in frosty weather.

These sorts may be propagated by planting slips or cuttings in the spring of the year, just before the plants begin to shoot, upon a bed of light fresh earth, and when they are rooted they may be transplanted into the places where they are designed to grow; the best time is about the beginning of September, that they may take new root before the frosty weather comes on; for if they are planted too late in autumn, they seldom live through the winter, especially if the weather proves very cold; so that if you do not transplant them early, it will be the better method to let them remain unremoved until March following, when the frost is over, observing never to transplant them at a season when the dry east winds blow, but rather defer the doing of it until the season is more favourable; for if they are planted when there are cold drying winds, their leaves are apt to dry up, which often kills them; but if there happen to be some warm showers soon after they are removed,

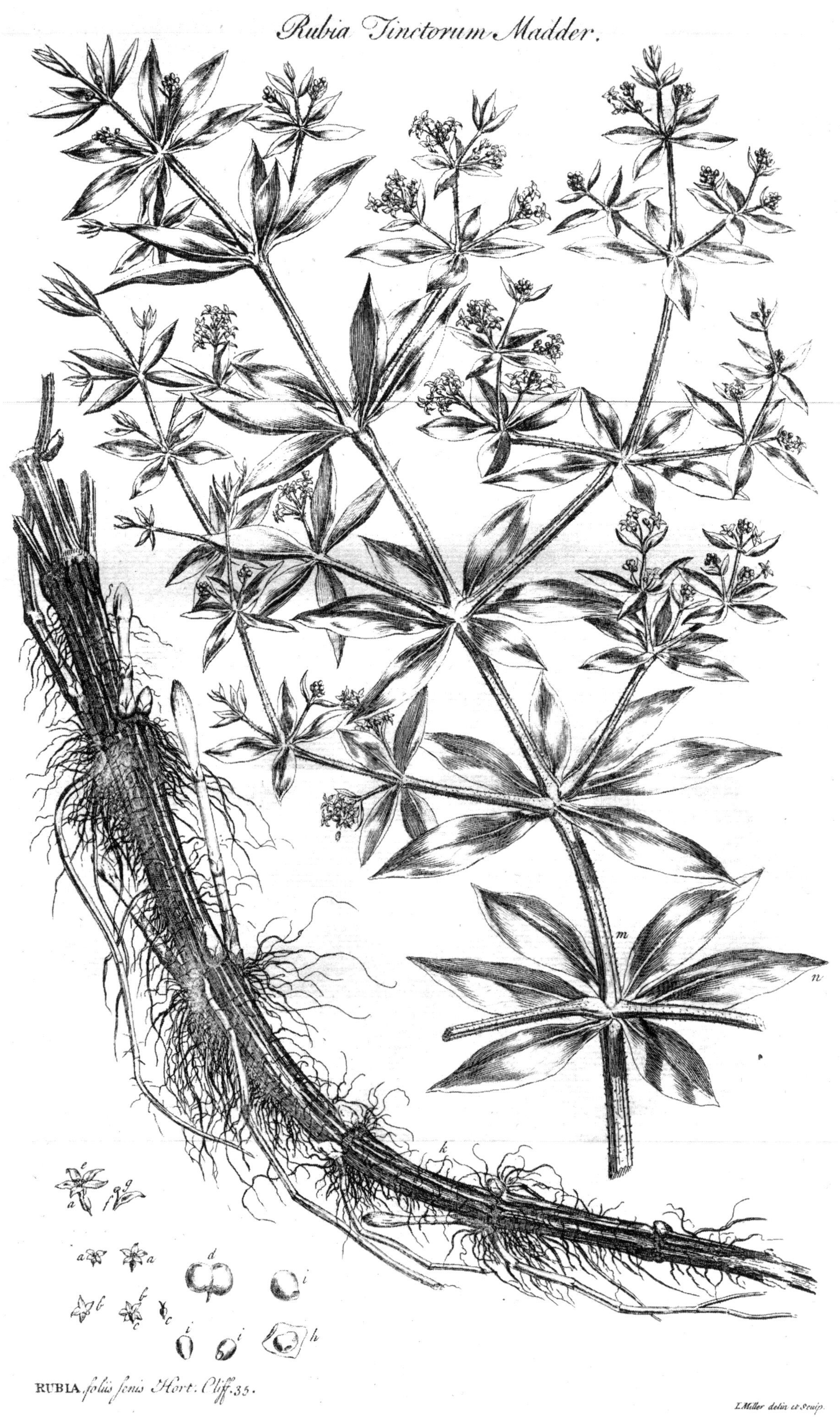

The material originally positioned here is too large for reproduction in this reissue. A PDF can be downloaded from the web address given on page iv of this book, by clicking on 'Resources Available'.

moved, it will cause them to take root immediately, so that they will require no farther care but to keep them clear from weeds.

Although these plants are tender when planted in a garden, yet, when they are by accident rooted in a wall (as I have several times seen them,) they will endure the greatest cold of our winters, though much exposed to the cold winds, which is occasioned by the plants being more stinted and strong, and their roots being drier.

The flowers of the narrow-leaved garden sort are used in medicine, as are also the leaves and seeds.

ROYENA. Lin. Gen. Plant. 491.

The CHARACTERS are,

The flower has a bellied permanent empalement, whose mouth is obtuse and five-pointed. It has one petal, having a tube the length of the empalement, but the brim is divided into five parts, which turn back. It hath ten short stamina growing to the petal, terminated by oblong, erect, twin summits, and an oval hairy germen sitting upon two styles a little longer than the stamina, crowned by single summits. The empalement afterward turns to an oval capsule with four furrows, having one cell with four valves, containing four oblong triangular seeds.

The SPECIES are,

1. ROYENA (*Lucida*) foliis ovatis scabriusculis. Hort. Cliff. 149. *Royena with oval rough leaves.*

2. ROYENA (*Glabra*) foliis lanceolatis glabris. Prod. Leyd. 441. *Royena with smooth spear-shaped leaves.*

3. ROYENA (*Hirsuta*) foliis lanceolatis hirsutis. Prod. Leyd. 441. *Royena with hairy spear-shaped leaves.*

The first sort has been long an inhabitant of some curious gardens in England, but it is not very common here, for it is very difficult to propagate.

This plant grows eight or ten feet high, and puts out its branches on every side, so may be trained up to a regular head: the branches are cloathed with oval shining leaves, which are placed alternately, and continue all the year, so make an agreeable variety among other exotic plants in the green-house during the winter season. The flowers are produced from the wings of the leaves along the branches, but as they have little beauty, few persons regard them. I have not observed any fruit produced by these plants in England.

The second sort grows naturally at the Cape of Good Hope. This rises with a shrubby stalk five or six feet high, sending out many slender branches, covered with a purplish bark, and garnished with small oval leaves, less than those of the Box-tree; they are smooth, entire, and of a lucid green, continuing all the year. The flowers come out from the wings of the leaves round the branches; they are white, and shaped like a pitcher; these are succeeded by roundish purple fruit, which ripen in the winter.

The third sort grows naturally at the Cape of Good Hope. This rises with a strong woody stalk seven or eight feet high, covered with a gray bark, sending out many small branches alternately, which are garnished with spear-shaped hoary leaves, covered with soft hairs. The flowers come out upon short foot-stalks from the side of the branches; they are of a worn-out purple colour, and small.

These plants are too tender to live through the winter in the open air in England, therefore they must be removed into the green-house in autumn, and treated in the same way as Orange-trees, with which culture the plants will thrive.

The first and third sorts are difficult to propagate here, for the branches which are laid down seldom put out roots, and those which do, are two years before they will have made roots sufficient to transplant, and their cuttings very rarely succeed; these are the only methods by which they can be increased in those countries where they do not produce seeds. The best time to plant the cuttings is in September; these should be planted in small pots, and plunged into a very moderate hot-bed. The pots should be closely covered down with hand-glasses to exclude the external air, and the cuttings refreshed with a little water every eighth or tenth day, according as the earth becomes dry, for much moisture will kill them. If the cuttings shoot, they must be gradually inured to bear the open air, and when they are well rooted, they should be each planted in a separate small pot, and afterward treated as the old plants.

If the plants put out any young shoots from the bottom, they should be carefully laid down in the ground while young, because when the shoots are tender they are more apt to put out roots than after they are become woody and hard; these branches should be slit in the same manner as is practised in laying of Carnations; they must be frequently, but gently watered, during the warm weather in summer, but in cold weather it must be sparingly given them; when these are rooted, they may be taken off, and treated in the same way as the cuttings.

The second sort is very apt to send up suckers from the roots, which may be taken off with the roots, and thereby increased; or those which do not put out roots, may be laid down in the same manner as the former; and the cuttings of this more frequently succeed than those of the other, so that this sort is much easier propagated.

RUBIA. Tourn. Inst. R. H. 113. tab. 38. Madder.

The CHARACTERS are,

The empalement of the flower is small, cut into four segments, and sits upon the germen. The flower has one bell-shaped petal, having no tube, but is divided into four parts. It hath four awl-shaped stamina which are shorter than the petals, terminated by single summits, and a twin germen under the flower, supporting a slender style, divided into two parts upward, and crowned by two headed stigmas. The germen afterward becomes two smooth berries joined together, each having one roundish seed with a navel.

The SPECIES are,

1. RUBIA (*Tinctorum*) foliis senis lanceolatis supernè glabris *Madder with six spear-shaped leaves, in whorls, whose upper surfaces are smooth; Dyer's Madder.*

2. RUBIA (*Sylvestris*) foliis inferioribus senis, supernè quaternis binisve, utrinque asperis. *Madder with the lower leaves growing by sixes round the stalks, and the upper ones by fours or pairs, which are rough on both sides; rough wild Madder.*

3. RUBIA (*Peregrina*) foliis quaternis. Prod. Leyd. 2;4. *Madder with four leaves, which are placed round the stalks.*

The first sort, which is cultivated for the root that is used in dyeing and staining of linens, grows naturally in the Levant. This hath a perennial root and an annual stalk; the root is composed of many long, thick, succulent fibres, almost as large as a man's little finger; these are joined at the top in a head like the roots of Asparagus, and run very deep into the ground; I have taken up roots, whose strong fibres have been more than three feet long; from the upper part (or head of the root) come out many side roots, which extend just under the surface of the ground to a great distance, whereby it propagates very fast; for these send up a great number of shoots, which, if carefully taken off in the spring, soon after they are above ground, become so many plants. These roots are of a reddish colour, somewhat transparent, and have a yellowish pith in the middle, which is tough, and of a bitterish taste; from the root arise many large, four-cornered, jointed stalks, which, in good land, will grow five or six feet long, and if supported, sometimes seven or eight; they are armed with short herbaceous prickles

prickles, and at each joint are placed five or six spear-shaped leaves; their upper surfaces are smooth, but their midrib on the under side is armed with rough herbaceous spines; the leaves sit close to the branches in whorls. From the joints of the stalk come out the branches, which sustain the flowers; they are placed by pairs opposite, each pair crossing the other; these have a few small leaves toward the bottom, which are by threes, and upward by pairs opposite; the branches are terminated by loose branching spikes of yellow flowers, which are cut into four parts, resembling stars. These appear in June, and are sometimes succeeded by seeds, which seldom ripen in England.

The second sort grows naturally in Spain. This hath perennial roots like those of the first sort, but are much larger; the stalks of this are smaller than those of the first sort, and are almost smooth; their lower parts are garnished with narrow leaves, placed by sevens in whorls round the stalks, but upward they diminish to four, three, and two toward the top; these are rough on both sides; at each joint of the stalk come out two short foot-stalks opposite, having two small rough leaves, ending with branching foot-stalks, sustaining small yellow flowers.

The third sort grows naturally in Spain and the Balearic Islands; I received the seeds of this sort from Gibraltar, and also from Minorca, where the plants grew out of the crevices of the rocks. The roots of this sort are much smaller than those of the two former, but are less succulent; they strike deep into the ground, and send up several slender four-cornered stalks, which are perennial; these grow a foot and a half long, and divide into many branches, whose joints are very near each other; they are garnished with short, stiff, rough leaves, placed by fours round the stalk, of a lucid green, and continue all the year.

There is a sort which grows naturally in Wales, and also upon St. Vincent's Rock, which has four leaves at each joint, but these are longer and narrower than those of the third sort; the stalks of this are perennial, and the leaves evergreen; so that Mr. Ray has mistaken this plant, having supposed it to be the second, which hath annual stalks rising much higher, therefore I should rather think it might be the third sort, if they were equally hardy; but the third sort is so tender, as to be always killed by severe frosts in England.

The first sort is that which is cultivated for the use of the dyers and callico printers, and is so essential to both manufactories, as that neither of those businesses can be carried on without this commodity; and the consumption of it is so great here, as that upon a moderate computation, there is annually so much of it imported from Holland, as the price of it amounts to more than one hundred and eighty thousand pounds sterling, which might be saved to the public if a sufficient quantity of it were planted in England, where it might be cultivated to greater advantage than in Holland, the lands here being better adapted to grow this plant. But as the growing of this plant in quantity has been for several years discontinued, so the method of culture is not well known to many persons here; and as there is at present an inclination in the public to regain this lost branch of trade (for formerly there was not only enough of this commodity raised in England for our own consumption, but also great quantities of it were sent abroad,) so we shall here give a full account of the culture of the plant, and also of the method of preparing the root for use; and shall begin with the method now practised in Zealand, where the best and greatest quantity of Madder is now raised.

In all the Netherlands, there is no where better Madder cultivated than in Schowen, one of the islands of Zealand, which is performed in the following manner:

The land which is designed for Madder, if it is strong and heavy, is ploughed twice in autumn, that the frost in winter may mellow it and break the clods; then it is ploughed again in the spring, just before the time of planting the Madder; but if the ground is light then it is ploughed twice in the spring; at the last ploughing it is divided into lands of three feet broad, with furrows between each land, four or five inches deep. Madder requires a loamy substantial soil, not too stiff and heavy, nor over light and sandy; for although it may thrive tolerably well in the latter, yet such land cannot have a second crop of Madder planted upon it in less than eight or ten years interval; but in Schowen, where the land is substantial, they need not stay longer than three or four years, in which interval the ground is sown with Corn, or planted with any kinds of pulse. It is granted, that the best land for producing of Madder is in Schowen, where a gemet of land, which is three hundred square rods of twelve feet each, will yield from one thousand pounds to three thousand pounds weight, according to the goodness of the land and the favourableness of the seasons; but in light land the quantity is from five hundred to a thousand pounds weight.

The time for planting of Madder begins toward the end of April, and continues all May, and sometimes in very backward springs there is some Madder planted the beginning of June. The young shoots from the sides of the root are taken off from the mother plant, with as much root as possible; these are called Kiemen, and are planted with an iron dibble in rows at one foot asunder, and commonly four Kiemen in a row each bed.

The quantity of these slips (or Kiemen) as is required to plant one gemet of land, are sold at different prices, according to the price which Madder bears, or to the demand for the plants; they are often sold from sixteen to twenty guilders, and sometimes they have been sold for ten or eleven pounds Flemish, but the lowest price is from fifteen guilders to three pounds Flemish.

The expence of planting out a gemet of land with slips (or Keimen) costs for labour only from sixteen to twenty guilders, according as the land is heavy or light: there are generally employed six men to plant, two to rake the ground, these earn each a guilder a day; and five or six women or boys, called carpers or pluckers of the shoots or Kiemen, these earn twelve Dutch pence a day, or two schillings.

The first year the Madder is planted, it is customary to plant Cabbages or dwarf Kidney-beans in the furrows between the beds, but there is always great care taken to keep the ground clean from weeds; this is generally contracted for at two pounds Flemish at each gemet of land.

In September or October, when the young Madder is cleaned for the last time that season, the green haulm (or stalks) of the plants, is carefully spread down over the beds, without cutting any part off, and in November the haulm of the Madder is covered over with three or four inches of earth.

This covering of the Madder is performed either with the plough or with the spade; if it is done by the first, it costs two guilders and a half, or three guilders in strong land each gemet, and over and above this, one guilder and a half to level the tops of the beds, and make them smooth; but it is better performed with the spade, only it is more chargeable, for that costs from eight to ten guilders each gemet, but at the same time the clods are better broken, and the surface of the beds made smooth and even.

The second year in the beginning of April, which is about the time the Kiemen or young shoots are beginning to come out, the earth on the top of the beds should be scuffled over and raked, to destroy the young weeds, and make the surface smooth and mellow, that the Kiemen may shoot

ſhoot out the eaſier above ground; this labour coſts three ſhillings each gemet.

The ſecond ſummer there muſt be the ſame care taken to keep the Madder clean as in the firſt, and then nothing is planted in the furrows, or ſuffered to grow there; at the laſt time of cleaning the ground in September or October, the green haulm is again ſpread down upon the beds; and in November the Madder is again covered with earth, in the ſame manner as the firſt year.

By this method of culture one can ſee how neceſſary it is to plant the Madder in beds, for thereby it is much eaſier covered with the earth of the furrows; and hereby the earth of the beds is every time heightened, whereby the Madder roots will be greatly lengthened, and the Kiemen or young ſhoots will have longer necks, and by being thus deeply earthed, will put out more fibres and have much better roots, without which they will not grow; and it is of equal uſe to the mother plants, for by this method the roots will be longer; and in this conſiſts the goodneſs and beauty of the Madder, for thoſe which have but few main roots, are not ſo much eſteemed as thoſe which are well furniſhed with ſide roots, called Tengels; a Madder plant that has many of theſe roots, is called a well-bearded Madder plant; therefore one muſt never cut off the ſide roots, for by ſo doing there will be a leſs crop of Madder, and but few Kiemen or young ſhoots can be produced; beſides, by the loſs of moiſture, ſometimes the plants will droop and become weak; and there is great profit in having a large quantity of Kiemen to draw in the ſpring, which are in plenty the ſecond and third years.

The Madder roots are ſeldom dug up the ſecond year, but generally after it has grown three ſummers, therefore the culture of the third year is the ſame as in the ſecond, during the ſpring and ſummer.

Before the firſt day of September, it is forbidden to dig up any Madder in this iſland; but on that day early in the morning a beginning is made, and the perſon who carries the firſt cart load to the ſtove, has a premium of a golden rider, or three ducats.

The digging up the Madder of a gemet of land, coſts from thirty-ſix to one hundred guilders, according to the goodneſs of the crop, and the lightneſs or ſtiffneſs of the ground, but in light land it coſts from nine to ten pounds Flemiſh; the perſons who are adroit in this buſineſs, are generally paid five ſhillings Flemiſh per day.

The Madder produces flowers in the middle of ſummer, and ſometimes a few ſeeds, but they never ripen here; nor would they be of uſe to cultivate the plants, ſince it is ſo eaſily done by the Kiemen.

Some years paſt they began to plant here the great wild Madder, which was called French Madder, but this was not eſteemed ſo good for uſe as the tame Madder, from which it differs much, ſo that was not continued. The more bitter of taſte the roots of the Madder are, when taken out of the ground before it is brought to the ſtove, the leſs it will loſe of its weight in drying, and is the better afterward for uſe.

When the Madder is dug out of the ground, it is carried to the ſtove, and there laid in heaps, in that which is called the cold ſtove, and ſeparated with hurdles made of wicker, and memorandums kept of each parcel, and to what countryman it belongs, that each may be dried in their turns, and prepared or manufactured, for which turn generally lots are caſt beforehand. The Madder thus carried to the ſtove is Relzyn.

This Relzyn is carried about ſix o'clock in the morning, into the tower or ſteeple, hoiſted in baſkets by ropes to the rooms, and divided or ſpread, where it remains till the next day two or three o'clock in the morning, about twenty or twenty-one hours; then thoſe roots which have lain in the hotteſt places are removed to cooler, and thoſe in the cooler are removed to the hotter places nearer the oven. This is continued for four or five days, according as there has been more or leſs carried there; but it is always the goods of one perſon, that every one may have his own, and of as equal quality as poſſible when it is delivered out.

When the Madder is ſufficiently dried in the tower, then it is threſhed on the threſhing floor, which is made clean from dirt or filth, and then it is brought to the kiln, and there ſpread on a hair-cloth for about twenty hours, during which time the kiln is made more or leſs hot, according as the roots are more or leſs thick, or the weather more or leſs cold.

From the kiln the Madder is moved to the pounding-houſe, and is there pounded on an oaken block made hollow, with ſix ſtampers platted at the bottom with iron bands; theſe ſtamps are kept in motion by a mill very much reſembling a griſt-mill, which is turned by three horſes; the preſence of the pounding-maſter is here always required, to ſtir the Madder continually with a ſhovel, to bring it under the ſtampers. When the Madder is thus properly pounded, it is ſifted over a tub till there is enough to fill a caſk: this firſt pounding, which chiefly conſiſts of the thinneſt and ſmalleſt roots, and the outſide huſks with ſome earth, which by drying and threſhing could not be ſeparated, is called Mor Mull.

What remains in the ſieve is put on the block again, and pounded a ſecond time, and when the pounding-maſter gueſſes a third part is pounded, then the Madder is taken out again, and ſifted over another tub, and put into a ſeparate caſk, and this is called Gor gemeens; that which remains in this ſecond operation, not enough pounded in the ſieve, is for the third time put on the block, and pounded till it is all reduced to powder, which is called Kor kraps.

When the Madder is cleanſed from the dirt and Mull, and is entirely pounded at once, then it is called Oor Onberoofde, ſo that this Onberoofde actually conſiſts of the Gemeens and Kraps pounded together, and ſifted without ſeparating them from each other.

When there is two-thirds of Kraps, and one-third Gemeens, which was ſeparately prepared or manufactured, then they are called two and one, or marked $\frac{2}{1}$.

The ſweepings of the ſtove, as alſo of the ground and beams being ſwept together is not loſt, but is put amongſt the Mull, or ſold by itſelf.

The ſweepings of the mill, and every part of the pounding place is alſo gathered together, and put into a caſk; this is called Den Beer.

When the Madder is thus prepared and put into caſks, it is in Zealand examined by ſworn aſſayers and tried, if it is not faulty packed up; that is, whether in the preparing it is properly manufactured, or falſely packed up, and to ſee if every part of the caſk is filled with Madder of equal goodneſs and quality, not burned in the drying, or mixed with dirt; which the aſſayers, by certain trials, and by weighing and waſhing of the Madder, can know, if it is according to the ſtatutes of the country.

There are ſundry ſtatutes made and publiſhed by the ſtates of Zeyland, concerning the preparing of Madder; as one of the 28th of July, 1662, one of the 29th of September, and 31ſt of October, 1671, another on the 23d of September, 1699, and the laſt on the 28th of April, 1735: by which ſtatutes, among other things, it is ſtrictly forbidden, That no perſon ſhall prepare Kraps, in which there ſhall be more than two pounds of dirt in a hundred weight, nor above eight pounds in the like weight of Onberoofde, or in Gemeens more than twelve pounds in a hundred weight.

If the Madder upon trial is found good, the arms of the city or village, and the sign of the stove where the Madder was prepared, is painted on the cask with black paint. The trial of the Madder is in no place more exact, or more religiously observed, than in the city of Zirkzee, therefore the merchants in Germany, who know this, always prefer the Madder of that place to all others, and will not buy any which has not the arms of Zirkzee painted upon the casks, if they are to be had.

We before mentioned the tower, the kiln, &c. where the Madder is dried and prepared for use, the draughts of these are exhibited in the annexed plans, with their explanation; but that a better judgment may be formed of their use, we shall here take notice, that the tower is the place where the Madder is first dried. This tower is heated by fifteen or sixteen pipes or flues of brick-work, which run on each side the tower under the floor, and are covered with low burnt tiles, some of which are loose; so that by taking up these, the heat is moderated, and conducted to any part of the tower, the person who has the care of drying the Madder pleases.

This tower has four or five lofts made of strong laths; they are four or five feet above each other, upon which the Madder is laid; these are heated by an oven, which is placed in the room where the work-people live, and is by them called the Glory.

The kiln is in a room, whose length is equal to the breadth of the stove, and is entirely arched over at the top; the oven, by which the kiln is heated, is called the Hog; this is built upon a stone wall, which rises a foot or two above ground; and the small arch, by which the heat passes through every part, has several square little holes in the brick-work, that the heat may come out; over these holes, on the top of the kiln, are laid wooden laths the whole length, and upon them a hair-cloth, on which the Madder is laid to dry, before it is carried to the pounding place. In the Madder-stove there is no other fuel used but Friezland turf, which gives an equal and moderate heat.

In the Madder-stoves, the people work more by night than day; first, because at the time of year when the Madder is brought into the stoves, the nights are much colder than the days; and secondly, that the master, who must be always attentive to his work, may not be interrupted by visitors; and thirdly, because they see less dust; but principally, because the Madder, which is pounded in the night, is of a much better colour than that which is pounded in the day.

In the Madder-stoves are always constant workmen, one who is the dryer, who has the care of drying the Madder in the tower and the kiln; for the right performance of this, art and experience are required, the goodness of the Madder greatly depending on the right drying. This person is a sort of foreman, and has the direction of all the workmen; his pay is five stivers, for every hundred weight of Madder which is prepared in the stove; he has one person under him for his assistant, to perform part of the laborious work, and to be always at hand; this man is paid eighteen or nineteen shillings per week Flemish, which is the constant wages.

The third person is the pounder, who is always present when the Madder is pounding, who with a particular shovel, which is small, and fitted to the cavity of the pounding block, stirs the Madder from time to time, to bring it under the stampers; he is paid four stivers for every hundred weight of Madder.

The fourth is a driver, who with a team of three horses, causes the mill to turn and pound the Madder; his pay for himself and the three horses, from eight to nine stivers per hundred weight, according as he can bargain.

Besides these four, there are five other assistants, who lay the Madder on and take it off; this is often performed by the wives and boys of the other workmen; these five have fifty stivers for every three thousand pounds of Madder which is prepared, so they have each ten stivers.

There are nineteen or twenty Madder-stoves in the island of Schowen, which, at an average, prepare in one crop, that lasts from September to February, ten thousand weight of Madder each, which in the whole amounts to two million pounds weight; and if we suppose, that the Madder is sold at an average for four pounds Flemish per hundred weight, which is a moderate price, one may soon reckon what advantage the culture of this dyeing commodity produces to this one island.

The countrymen pay to the owners of the Madder-stoves, two guilders for preparing every hundred weight of Mull, and for each hundred weight of hard Madder; that is, of Kraps, Gemeens, or Onberoofde, three guilders, according as they will have them prepared.

In Zealand, Madder is cultivated by every kitchen-gardener, each having a quantity to take up annually, in proportion to the size of the garden, and most of the farmers there do the same; for as the roots require three years growth to have them in perfection, so they every spring plant a certain allotment of land for the purpose. The ground which has produced Madder, will be fresh and well prepared for other crops, in like manner as the land upon which Asparagus-roots have grown to be taken up for forcing upon hot-beds in winter, which is practised by most of the kitchen-gardeners near London, for these roots are commonly allowed three years growth, by those persons who value themselves on having large Asparagus; so there is a great affinity in the culture of both these roots, for the land which will produce one of these crops is always good for the other. It is therefore much to be wished, that the kitchen-gardeners in England would undertake the culture of Madder, who are much more likely to succeed in it than any other persons; but it is feared the trouble of drying and manufacturing the roots for use, is a principal objection to this.

The building of a Madder-stove quite new from the foundation, costs in the whole about twenty-four hundred pounds Flemish, which is twelve hundred pounds sterling.

PLATE I.

An explanation of the plan of the cold stove.

Fig. 1. Is the lower band, whose thickness is fourteen by sixteen inches.
2. The upper band, which is twelve by fourteen inches.
3. The cap and band, which is ten by twelve inches.
4. The upper cap, which is six by seven inches.
5. The two main jaumbs, which are thirteen by fifteen inches of stone.
6. The half bands and posts of nine by seven inches.
7. The uppermost half band, which is small, six by eight inches.

PLATE II.

A plan of the arched room cut through perpendicularly in the middle where the kiln stands, with a representation of the kiln.

AA Is the segment of the arch.
B The oven of the kiln which is called the Hog; this has no chimney; when the fire is first kindled either with turf

A Plan of the Structure of the Cold Stove

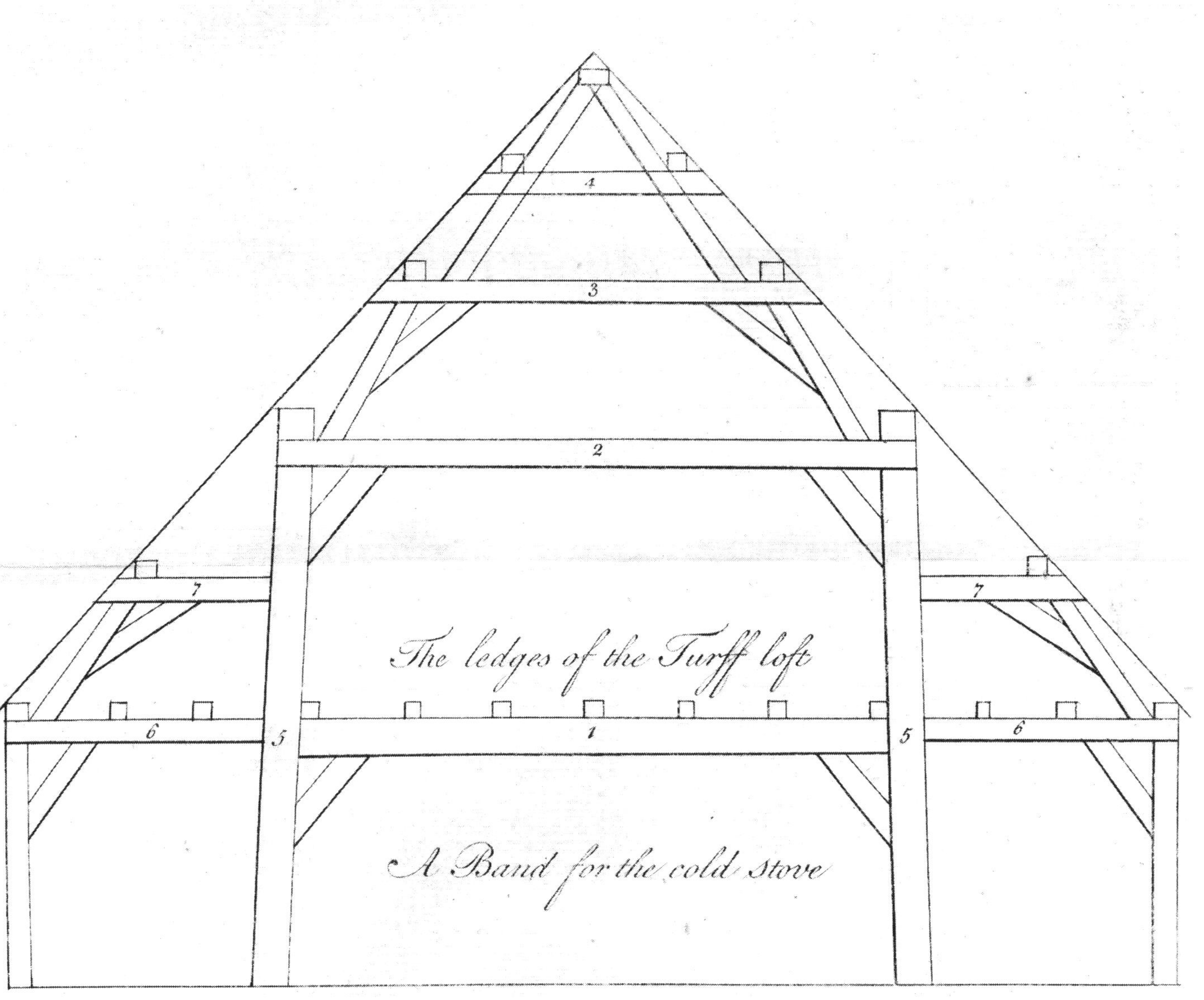

The material originally positioned here is too large for reproduction in this reissue. A PDF can be downloaded from the web address given on page iv of this book, by clicking on 'Resources Available'.

Pl. 2.

The Arched Room cut perpendicularly thro' the middle where the Kiln stands, with a representation of the Kiln.

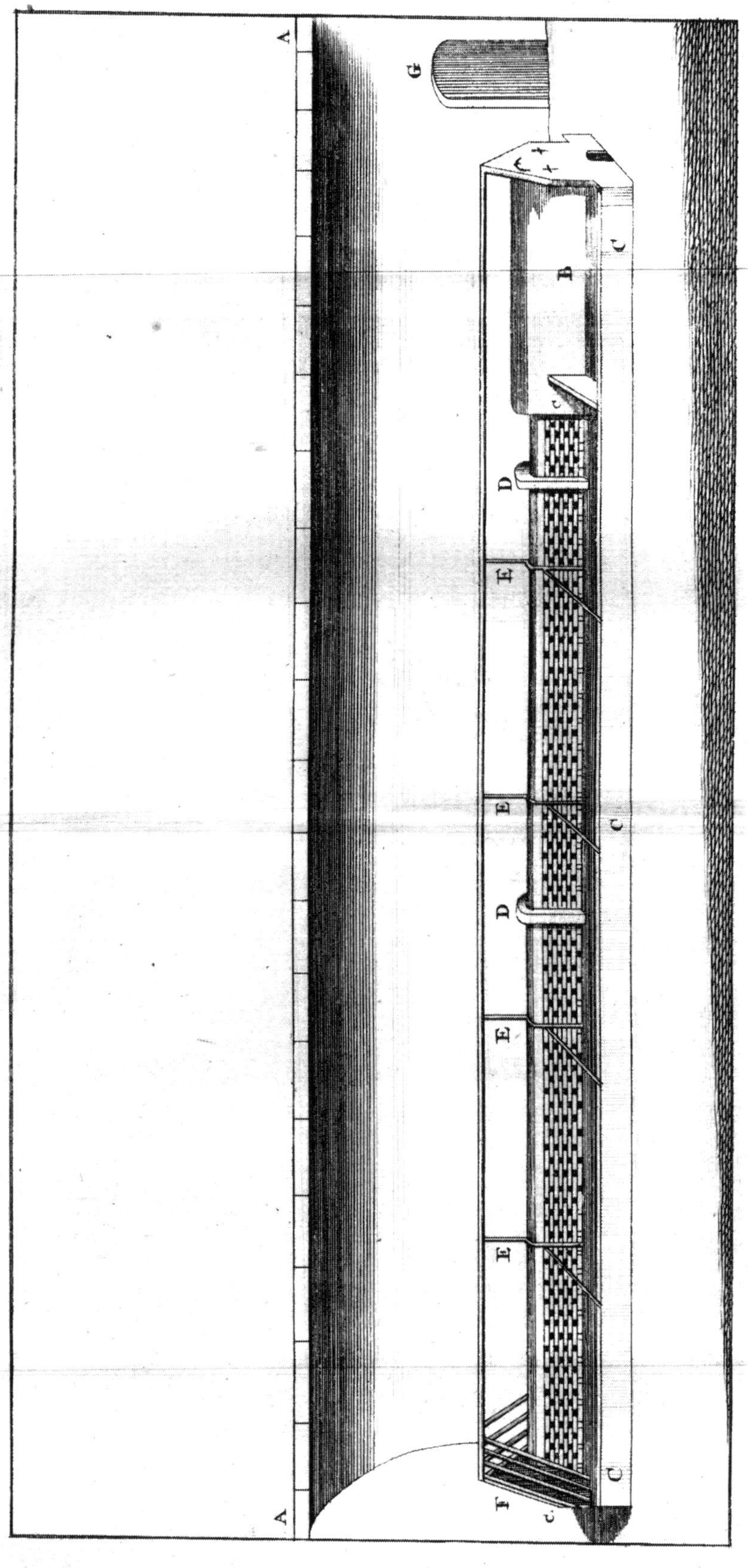

The material originally positioned here is too large for reproduction in this reissue. A PDF can be downloaded from the web address given on page iv of this book, by clicking on 'Resources Available'.

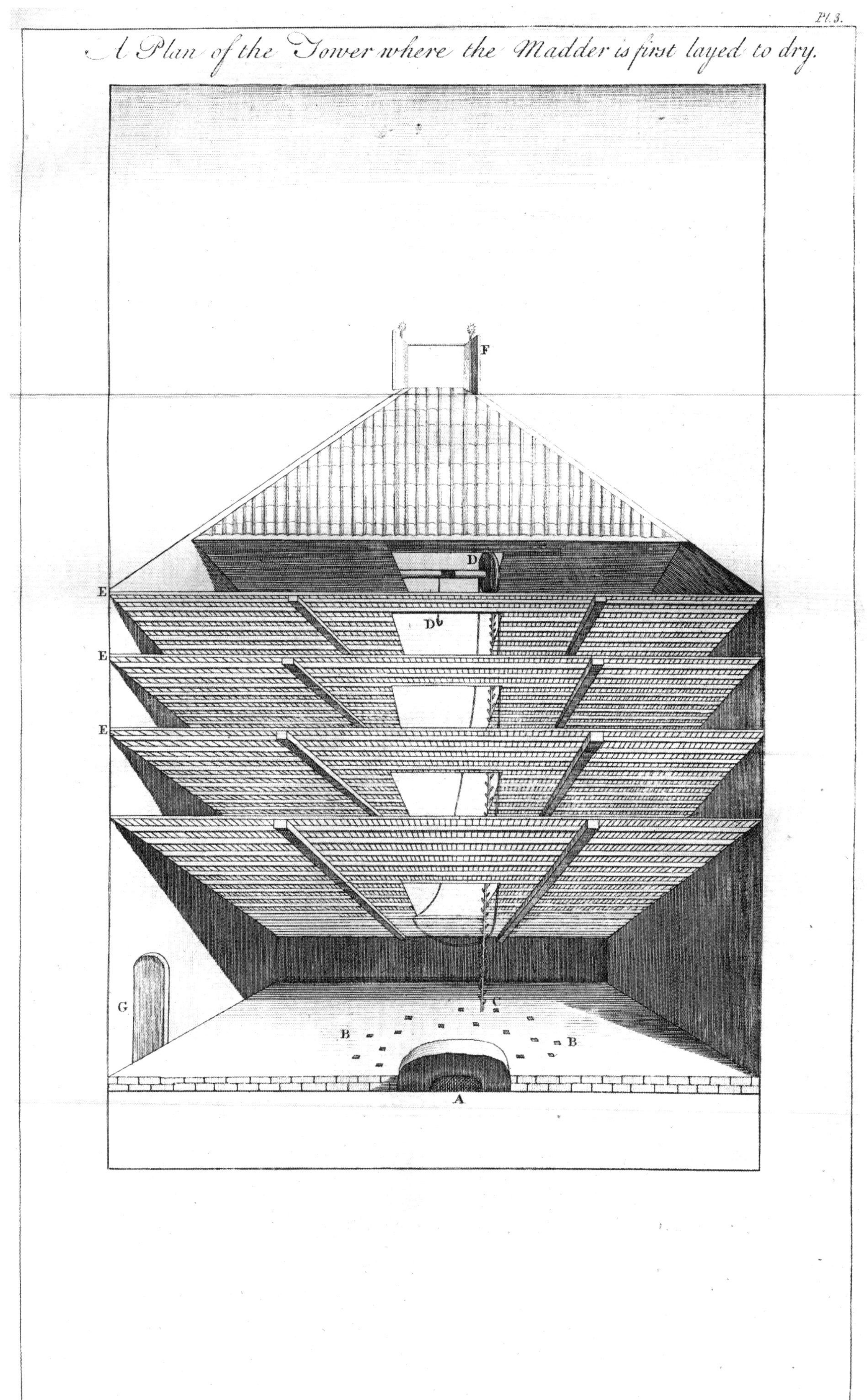

The material originally positioned here is too large for reproduction in this reissue. A PDF can be downloaded from the web address given on page iv of this book, by clicking on 'Resources Available'.

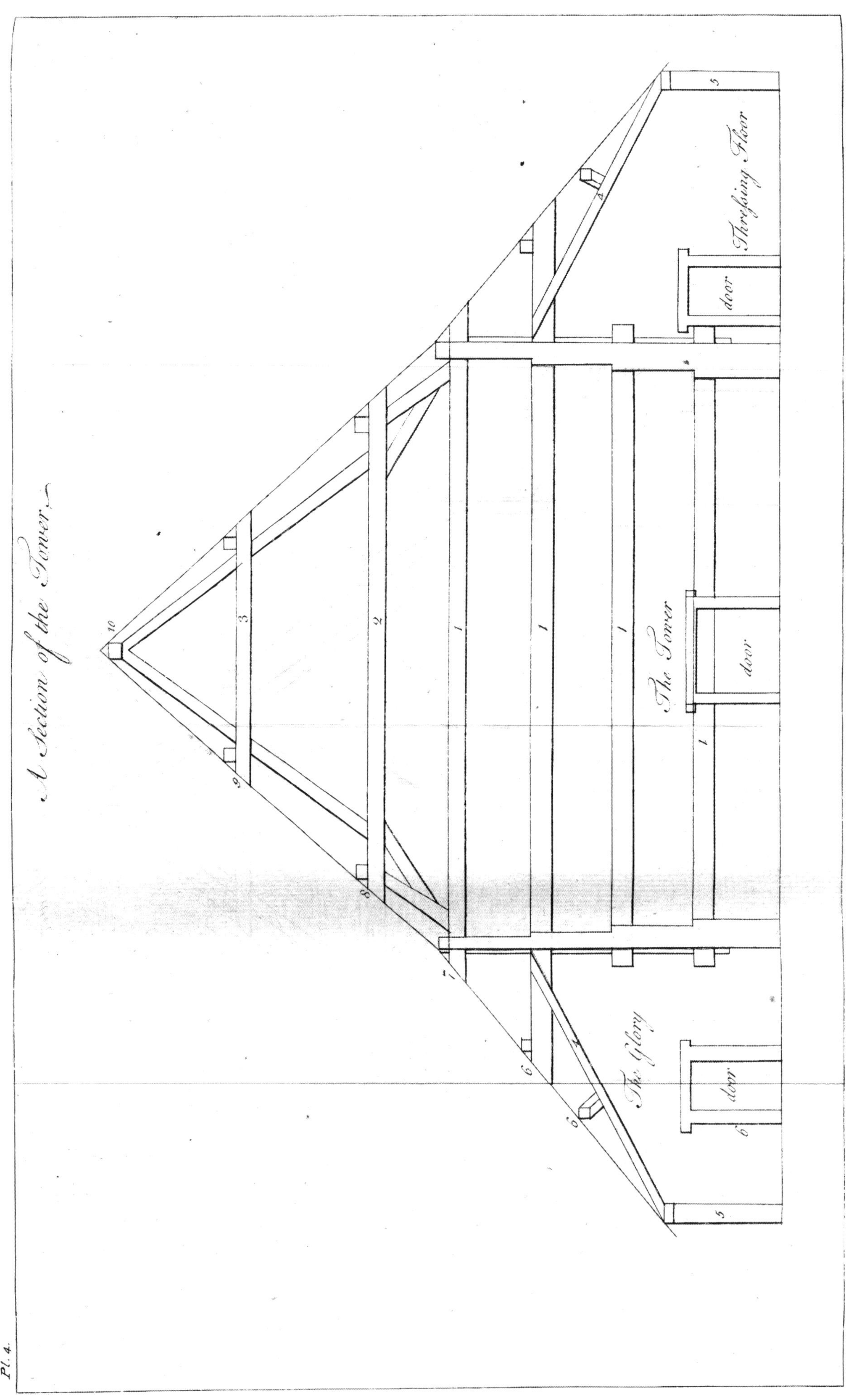

The material originally positioned here is too large for reproduction in this reissue. A PDF can be downloaded from the web address given on page iv of this book, by clicking on 'Resources Available'.

The material originally positioned here is too large for reproduction in this reissue. A PDF can be downloaded from the web address given on page iv of this book, by clicking on 'Resources Available'.

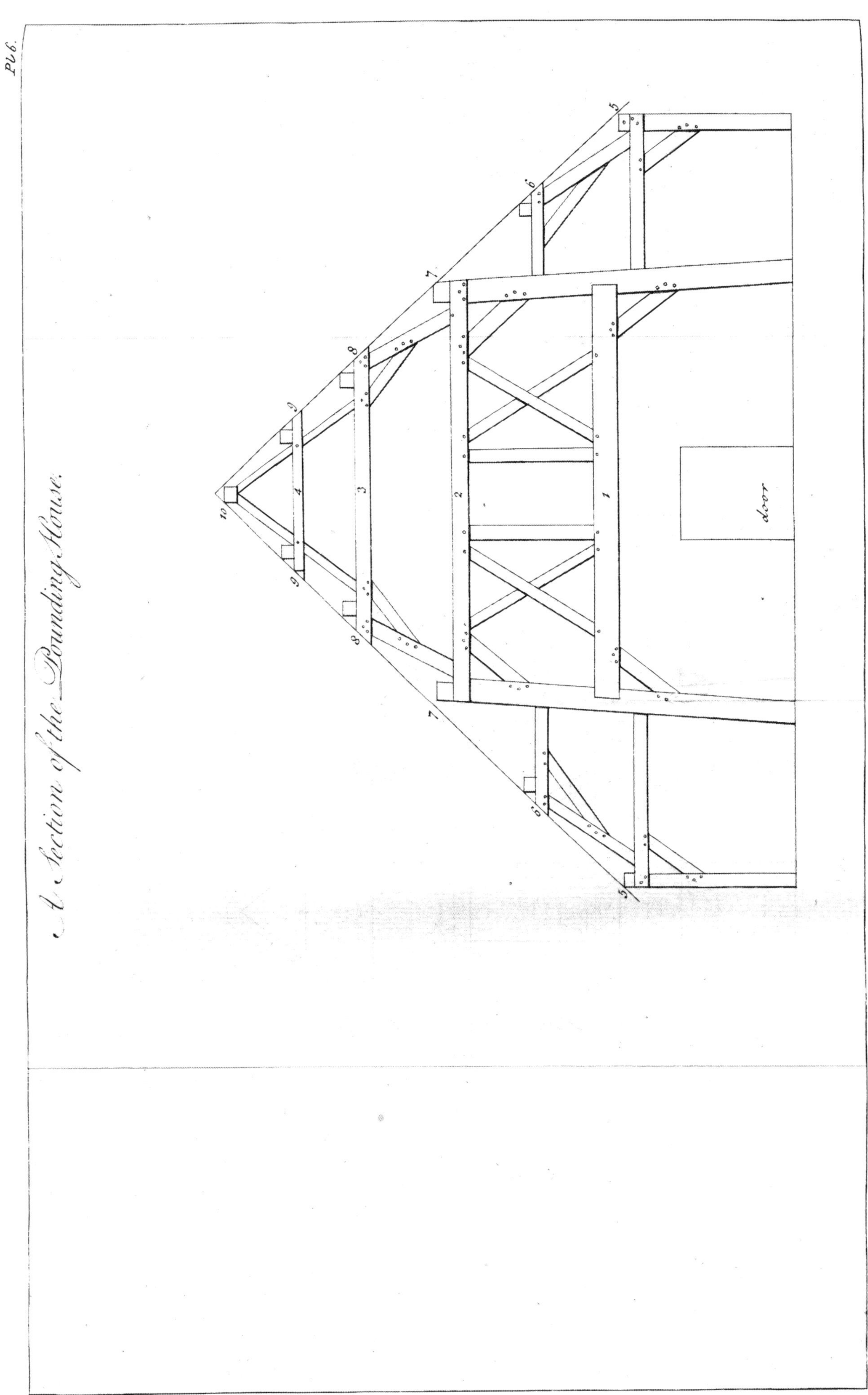

The material originally positioned here is too large for reproduction in this reissue. A PDF can be downloaded from the web address given on page iv of this book, by clicking on 'Resources Available'.

turf or other fuel, the smoke is let out through a small window.

CCC A stone foundation, on which the oven and kiln are built.

CC Is properly the kiln itself, which must be observed in what manner it is built, with little holes to let out the heat.

DD Stone bands made for the greater firmness, about the kiln.

EEEE Iron bars placed to strengthen the kiln, and also to lay the upper long lath upon.

F Small cross laths over the kiln, which lie from one end C to the other end C upon the kiln, but there are few of these represented, that the small holes of the kiln may better appear.

G The door of the entrance.

PLATE III.

A plan of the tower where the Madder is first laid to dry.

A Is the oven of the tower.

BB The pipes whereby the heat spreads itself, is here shewn by the openings where the tiles are taken off.

C A sort of stairs by which they climb.

DD The windlass with its rope and hook, to hoist the Madder to the lofts.

EEEE The four lofts of the lath of the oven.

F The chimney above the roof.

G The door by which they enter.

PLATE IV.

An explanation of the plan of the section of the tower.

Fig. 1. 1. 1. 1. The four bands of the tower, which are sixteen inches square.
2. The cap band ten by twelve inches.
3. The springing band six by eight inches.
4. The interstice to the tower six by seven inches.
5. The spanning plate five by seven inches.
6. 6. The lower and second girder six by seven inches.
7. The third girder seven by nine inches.
8. The fourth girder six by eight inches.
9. The fifth girder six by seven inches.
10. The crown piece of the tower five by six inches.

The ribs in the tower must be laid fourteen inches asunder from middle to middle cornerways, and the laths between an inch and a half distant.

PLATE V.

A plan of the pounding-house, in which is shewn at A the driver, who with his three horses causes the mill to turn, which works the stampers: at B is shewn the pounder, who with his shovel continually brings the Madder under the stampers.

Fig. 1. Is the beam which supports the axletree, which is fourteen by fifteen inches.
2. The hollow oaken block or trough, twenty-seven by twenty-nine inches.
3. The king post eighteen inches square.
4. The upper band six by seven inches.
5. The cross bands five by seven inches.
6. The cross arms six by ten inches.
7. The swaarden six by ten inches.
8. The axis from six to eight inches.
9. The feller six by eight inches of elm wood.
10. The king beam eleven by thirteen inches of fir wood.
11. The drawers under the mill five by six inches.
12. The plate for the running of the truckle three by sixteen inches.
13. The wooden knobs to the wheel of ash.
14. The staves made of box wood.
15. The six stampers six inches square of ash.

PLATE VI.

An explanation of the section of the pounding-house.

Fig. 1. The under band sixteen inches square.
2. The upper band twelve by fourteen inches.
3. The band of the cap post ten by twelve inches.
4. The springing band six by seven inches.
5. The spanning plate five by seven inches.
6. The first girder six by seven inches.
7. The second girder nine by eleven inches.
8. The third girder six by eight inches.
9. The uppermost girder six by seven inches.
10. The top or cap four by five inches.

The above account is the method of cultivating Madder in Zealand, where the best Madder is now produced; to this I shall add, what I have observed of the growing of Madder in other parts of Holland, as also the experience I have had of the growth of Madder in England, with an account of the method of planting it here.

In the year 1727, I observed a great quantity of this plant cultivated in Holland, between Helvoetsluys and the Brill; and it being the first time I had ever seen any considerable parcel of it, I was tempted to make some enquiries about its culture, and take some minutes of it down upon the spot, which I shall here insert, for the use of such as may have curiosity to attempt the culture of it.

In autumn they plough the land, where they intend to plant Madder in the spring, and lay it in high ridges, that the frost may mellow it; in March they plough it again, and at this season they work it very deep, laying it up in ridges eighteen inches asunder, and about a foot high; then about the beginning of April, when the Madder will begin to shoot out of the ground, they open the earth about their old roots, and take off all the side shoots which extend themselves horizontally, just under the surface of the ground, preserving as much root to them as possible; these they transplant immediately upon the tops of the new ridges, at about a foot apart, observing always to do this when there are some showers, because then the plants will take root in a few days, and will require no water.

When the plants are growing, they carefully keep the ground hoed, to prevent the weeds from coming up between them; for if they are smothered by weeds, especially when young, it will either destroy or weaken them so much, that they seldom do well after. In these ridges they let the plants remain two seasons, during which time they keep the ground very clean; and at Michaelmas, when the tops of the plants are decayed, they take up the roots, and dry them for sale. This is what I could learn of their method of cultivating this plant, to which I will subjoin a few observations of my own, which I have since made upon the culture of Madder in England.

The land upon which I have found Madder thrive best, is a soft sandy loam, and if it has been in tillage some years, it will be better than that which is fresh broken up. This

 should

ſhould have at leaſt a depth of two feet and a half, or three feet of good earth, and muſt be quite clear from Couch, or the roots of any bad weeds; for as the roots of Madder ſhould remain three years in the ground, ſo where there are any of thoſe weeds which ſpread and multiply at their roots, they will intermix with the Madder-roots, and in three years will have taken ſuch poſſeſſion of the ground, as to greatly weaken the Madder, and render it very troubleſome to ſeparate when the Madder is taken up.

The ground ſhould be ploughed deep before winter, and laid in ridges to mellow; and if it is not too ſtrong, there will be no neceſſity for ploughing it again, till juſt before the time of planting the Madder, when the land ſhould be ploughed as deep as the beam of the plough will admit: and there ſhould be men following the plough in the furrows, who ſhould dig a full ſpit below the furrow, and turn it up on the top of the baulk; by preparing the ground of this depth, the roots of the Madder will ſtrike down, and be of greater length, in which the goodneſs of the crop chiefly conſiſts. The land being thus prepared and made level, will be fit to receive the plants. The beſt time for planting of the Madder, is about the middle or in the latter end of April, according as the ſeaſon is more or leſs forward, which muſt be determined by the young ſhoots; for when theſe are about two inches above ground, they are in the beſt ſtate for planting.

In the taking up of theſe ſhoots for planting, the ground ſhould be opened with a ſpade, that they may be ſeparated from the mother plants with as much root as poſſible; for if the roots are broken off, they will not ſucceed: theſe plants ſhould be drawn up no faſter than they are planted, for if they lie long above ground they will ſhrink, and their tops will wither, and then they often miſcarry; therefore if they are brought from a diſtant place, there ſhould be great care taken in the packing of them up for carriage; eſpecial regard ſhould be had not to pack them ſo cloſe, or in ſo great quantity, as to cauſe them to heat, for that will ſoon ſpoil them; but if they are a little withered by lying out of the ground, their roots ſhould be ſet upright in water for a few hours, which will ſtiffen and recover them again.

In the planting of Madder, there are ſome who make the rows but one foot aſunder, others one foot and a half, ſome two feet, and others who allow them three feet diſtance. I have made trial of the three laſt diſtances, and have found when the roots have been left three years in the ground, that three feet diſtance row from row is the beſt; but if it is taken up in two years, two feet aſunder may do very well; and the diſtance in the rows plant from plant, ſhould be one foot, or a foot and a half.

If there is no danger of the ground being too wet in winter, the plants may be planted on the level ground; but if on the contrary, the ground ſhould be raiſed in ridges where each row of plants is to be ſet, that their roots may not reach the water in winter, for if they do, it will ſtop their downright growth; and this is the reaſon for the Dutch, who plant Madder in the Low Countries, raiſing their ridges ſo high as two or three feet, and in Zealand, where the ground is drier, they raiſe the beds four or five inches above the intervals, that the wet may drain off from the beds where the Madder is planted.

The method of planting is as follows, viz. The ground being made ſmooth, a line is drawn croſs it to mark out the rows, that they may be ſtrait for the more convenient cleaning, and for the better digging or ploughing of the ground between the rows; then with an iron-ſhod dibble, holes are made, at the diſtance which the plants are to ſtand from each other. The depth of the holes muſt be in proportion to the length of the roots of the plants, which muſt be planted the ſame depth they had ſtood before, while they were upon the mother plants; for if any part of the root is left above ground, the ſun and winds will dry them, which will retard the growth of the plants; and, ſhould any part of the green be buried in the ground, it will not be ſo well, though, of the two, the latter will be leſs prejudicial, eſpecially if there is not too much of the green buried. When the plants are put into the holes, the earth ſhould be preſſed cloſe to them to ſecure them from being drawn out of the ground, for crows and rooks frequently draw the new plants out of the ground, before they get new roots, where there is not this care taken; ſo that in two or three days, I have known half the plants on a large piece of land deſtroyed by theſe birds.

If there happens to be ſome ſhowers of rain fall in a day or two after the plants are planted, it will be of great ſervice to them, for they will preſently put out new roots, and become ſtrong, ſo that, if dry weather ſhould afterward happen, they will not be in ſo much danger of ſuffering thereby, as thoſe which are later planted. There are ſome who, from a covetous temper of making moſt uſe of the ground, plant a row of dwarf Peaſe, or Kidney-beans, between each row of Madder, and pretend that hereby the land is kept cleaner from weeds; but I am very certain the crop of Madder is injured thereby much more than the value of thoſe things which grow between the rows, as I have experienced; therefore I adviſe thoſe perſons who plant Madder, never to ſow or plant any thing between the rows, but to keep the Madder quite clean from weeds, or any other kind of vegetable.

In order to keep the ground thus clean, it ſhould be ſcuffled over with a Dutch hoe, as ſoon as the young weeds appear; at which time a man can perform a great deal of this work in a day, and if it is done in dry weather, the weeds will die as faſt as they are cut down; whereas, when the weeds are left to grow in the ſpring, ſo as to get ſtrength, they are not ſo ſoon deſtroyed, and the expence of hoeing the ground then will be more than double; beſides, there will be danger of cutting down ſome of the weaker plants with the weeds, if the perſons employed to perform this work are not very careful; therefore it is much cheaper, as alſo better for the Madder, to begin this work early in the ſpring, and to repeat it as often as the weeds render it neceſſary; for by keeping the ground thus conſtantly clean, the Madder will thrive the better.

During the firſt ſummer, the only culture which the Madder requires, is that of keeping it clean in the manner before directed, and, when the ſhoots or haulm of the plants decay in autumn, it ſhould be raked off the ground; then the intervals between the rows ſhould be either dug with a ſpade, or ploughed with a hoeing-plough, laying up the earth over the heads of the plants in a roundiſh ridge, which will be of great ſervice to the roots. The Dutch cover the haulm of their Madder with earth, leaving it to rot upon the ground; this perhaps may be neceſſary in their country to keep the froſt out of the ground, but, as I have never found that the ſevereſt winters in England have injured the Madder-roots, ſo there is not the ſame neceſſity for that practice here.

The following ſpring, before the Madder begins to ſhoot, the ground ſhould be raked over ſmooth, that the young ſhoots may have no obſtruction, and, if there ſhould be any young weeds appearing on the ground, it ſhould be firſt ſcuffled over to deſtroy the weeds, and then raked over ſmooth; after this the ſame care muſt be taken in the following ſummer to keep the ground clean, and, if it is performed by the hoe-plough, the earth of the intervals ſhould be thrown up againſt the ſide of the ridges, which will earth up the roots, and greatly increaſe their ſtrength; but, before are the ground of one interval is fo hoed, the haulm of the plants fhould be turned over to the next adjoining interval; and, if they are permitted fo to lie for a fortnight or three weeks, and then turned back again on thofe intervals which were hoed, obferving firft to fcuffle the ground, to deftroy any young weeds which may have appeared fince the ftirring of the ground, then the alternate intervals fhould be ploughed in like manner, turning the earth up againft the oppofite fides of the roots; by this method the intervals will be alternately ploughed, and the plants earthed up, whereby the ground will be kept clean and ftirred, which will greatly promote the growth of the roots; and by this method the fuperficial fhoots will be fubdued, and the principal roots greatly ftrengthened. The following autumn the ground fhould be cleared of the haulm and weeds, and the earth raifed in ridges over the roots, as in the foregoing year.

The third fpring the roots will furnifh a great fupply of young plants, but, before thefe appear, the ground fhould be cleaned and raked fmooth, that the fhoots may have no obftruction to their coming up; and, when the young plants are fit to take off, it fhould be performed with care, always taking off thofe which are produced at the greateft diftance from the crown of the mother plants, becaufe thofe are what rob them moft of their nourifhment; and the wounds made by feparating them from the old roots are not near fo hurtful as thofe near the crown, for by the ftripping off too many of the fhoots it will retard the growth of the plants.

The culture of the Madder in the third fummer muft be the fame as the fecond, but, as the roots will then be much ftronger, the earth fhould be laid up a little higher to them at the times when the ground is cleaned; and, if all the diftant fuperficial fhoots, which come up in the intervals, are hoed or ploughed off, it will be of fervice to ftrengthen the larger downright roots; and, as the haulm will now be very ftrong and thick, the frequent turning it over, from one interval to another, will prevent it rotting, for, if it lies long in the fame pofition, the fhoots which are near the ground, where there will be always more or lefs damp, and being covered with the upper fhoots, the air will be excluded from them, which will caufe them to rot; for the fhoots of Madder are naturally difpofed to climb upon any neighbouring fupport, and in places where they have been fupported, I have feen them more than ten feet high; but the expence of ftaking the plants to fupport their fhoots, would be much too great to be practifed in general, therefore the other method of turning the haulm over from one interval to the other will be found of great ufe, for hereby it is kept from decaying, and by fo doing the fun is alternately admitted to each fide of the roots, which is of more confequence to the growth of the Madder than moft people conceive; from many repeated trials I have found, that where the haulm has decayed or rotted in fummer, it has greatly retarded the growth of the roots. There have been fome ignorant pretenders who have advifed the cutting off the haulm in fummer, in order to ftrengthen the roots; but whoever practifes this, will find to their coft the abfurdity of this method, for I have fully tried this many years ago, and have always found that every other root, upon which this was practifed, was at leaft a third part fmaller than the intermediate roots, whofe haulm was left entire. The occafion of my firft making this experiment was, becaufe the plants had been fet too near each other, and the feafon proving moift, had increafed the number and ftrength of the fhoots, fo that they were fo thick, as that many of them began to rot; to prevent which, I cut off the fhoots of every other plant to give room for the others to fpread; but foon after this was done, the plants produced a greater number of fhoots than before, which were weaker, and the effect it had upon the roots was as before related; fince then I have frequently repeated the experiment on a few roots, and have always found the effect the fame.

As foon as the haulm of the Madder begins to decay in autumn, the roots may be taken up for ufe, becaufe then the roots have done growing for that feafon, and will then be plumper and lefs liable to fhrink than if they are dug up at another feafon; for I have always found that roots of every kind of plant, which are taken out of the ground during the time of their growing, are very apt to fhrink, and lofe more than half their weight in a fhort time.

When the feafon for digging up the Madder-roots is come, it fhould be done in the following manner, viz. A deep trench fhould be dug out at one fide of the ground next to the firft row of Madder, to make a fufficient opening to receive the earth, which muft be laid therein in digging up the row of roots, fo that it fhould be at leaft two feet broad, and two fpits and two fhovelings deep: this fhould be made as clofe as poffible to the roots, being careful not to break or cut them in doing it; then the row of roots muft be carefully dug up, turning the earth into the trench before mentioned. In the doing of this, there fhould be to every perfon who digs, two or three perfons to take out the roots as entire as poffible, that none may be loft, and as much of the earth fhould be fhaken out of the roots as can be done; for after the principal roots are taken up, there will be many of the long fibres remaining below, which break off; therefore, in order to get the roots as clean as poffible, the whole fpot of ground fhould be dug of the fame depth as the firft trench, and the pickers muft follow the diggers to get them all out to the bottom. As the digging of the land to this depth is neceffary, in order to take up the roots with as little lofs as poffible, it is a fine preparation for any fucceeding plants; and I have always found that the ground, where the Madder has grown, produced better crops of all kinds than lands of equal goodnefs, which had not the like culture, efpecially thofe plants with tap-roots, as Carrots and Parfneps.

After the roots are taken up, the fooner they are carried to the place of drying, the finer will be their colour, for if they lie in heaps they are apt to heat, which will difcolour them; or if rain fhould happen to wet them much, it will have the fame effect, therefore no more roots fhould be taken up than can be carried under fhelter the fame day.

The firft place in which the roots fhould be laid to dry, muft be open on the fides to admit the air, but covered on the top to keep out the wet. If a building is to be erected new, fuch as the tanners have for drying their fkins will be as proper as any, for thefe have weather-boards from top to bottom, at equal diftances, to keep out the driving rain, but the fpaces between being open, admit the air freely; and if inftead of plank-floors or ftages above each other, they are laid with hurdles or bafket work, upon which the roots are laid to dry, the air will have freer paffage to the under fide of the roots, which will dry them more equally.

In this place they may remain four or five days, by which time the earth, which adhered to the roots, will be fo dry as to eafily rub off, which fhould be done before the roots are removed to the cold ftove, for the flower the roots are dried, the lefs they will fhrink, and the better will be the colour of the Madder; and the cleaner the roots are from earth, the better the commodity will be for ufe when prepared.

After the roots have laid a fufficient time in this place, they fhould be removed into another building called the cold ftove, in which there fhould be conveniencies of flues paffing through different parts of the floor and the fide walls; in this the roots fhould be laid thin upon the floors, and turned from time to time as they dry, taking those roots away, which are nearest to the flues that convey the greatest heat, placing them in a cooler part of the room, and removing such of them as had been in that situation to the warmer, from whence the other are taken. The constant care in this particular will be of great service to the quality of the Madder, for, when this is properly conducted, the roots will be more equally dried, and the commodity, when manufactured, will be much fairer and better for use.

When the outside of the roots has been sufficiently dried in this cold stove, they should be removed to the threshing floor, which may be the same as in a common barn where Corn is threshed. The floor of this should be swept, and made as clean as possible; then the roots should be threshed to beat off their skins or outside coverings; this is the part which is prepared separately from the inner part of the root, and is called Mull, which is sold at a very low price, being the worst sort of Madder, so cannot be used where the permanency or beauty of the colours are regarded; these husks are separated from the roots, pounded by themselves, and are afterward packed up in separate casks, and sold by the title of Mull. If this is well prepared, and not mixed with dirt, it may be sold for about fifteen shillings per hundred weight, at the price which Madder now bears, and this, as is supposed, will defray the whole expence of drying the crop.

After the Mull is separated from the roots, they must be removed to the warmer stove, where they must be dried with care, for if the heat is too great, the roots will dry too fast, whereby they will lose much in weight, and the colour of the Madder will not be near so bright; to avoid which, the roots should be frequently turned, while they remain in this stove, and the fires must be properly regulated. If some trials are made by fixing a good thermometer in the room, the necessary heat may be better ascertained than can be done any other way; but this will require to be greater at some times than at others, according as the roots are more or less succulent, or the weather more or less cold or damp; but it will always be better to have the heat rather less than over hot, for, though the roots may require a longer time to dry with a slow heat, yet the colour will be the better.

When the roots are properly dried in this stove, they must be carried to the pounding-house, where they must be reduced to powder in the manner before related; but whether it is necessary to separate the Kraps from the Gemeens, as is now practised by the Dutch, the consumers of Madder will be better judges than myself.

There has been some objections of late made to the introducing, or rather retrieving the culture of Madder in England, which it may be proper here to take notice of, lest they should have so much weight as to prevent many persons from engaging in it. The first which has been generally started is, that the land in this country is not so well adapted for growing Madder as that in Holland; to which I can with truth affirm, that there are vast tracts of land here much better adapted for producing Madder than the best land in Zealand, and from the experience which I have had of its growth, will produce a greater crop.

Another objection which I have heard, was the labour in Holland being cheaper than in England. The Dutch will always undersell us, so consequently will maintain this branch of trade; but this is certainly a great mistake, for, though the labourers employed in cultivating Madder, may not earn so great wages as is generally paid in England, sure I am that the difference between an expert English labourer, and that of the best Dutchman, in the ploughing, hoeing, planting, &c. of Madder, is much greater than that of their pay; for I am sure a good English gardener or ploughman will do more business, and perform it better, in four days, than the best workman in Holland can do in six. What I now say is greatly within compass from my own knowledge; so that, supposing we were to proceed in the same manner now practised by the Dutch, this could be no objection to the cultivating of Madder; but we shall soon find ways of performing the most laborious part, at much less expence, by means of the hoeing-plough, which may be used to great advantage in the cultivation of Madder, whereby the expence will be much lessened, and, when once this is well established in England, there can be no doubt but that great improvements will be made both in the culture and method of preparing the commodity for use.

There has been objections made against farther trials of growing Madder, because some who have engaged in it, have not succeeded; but in answer to this, it must be observed, that their ill success was owing to a want of skill. Some of them continued to plant repeated crops of Madder on the same spot of ground, till the roots became so small, as scarce to pay the expence of digging up: and here it is proper to observe, that Madder should not be planted on the same land, till after an interval of seven or eight years, during which interval the ground may be sown with any sort of grain, or kitchen vegetables, which it will produce to great advantage after Madder, because the land will be wrought so deep. The Dutch always sow grain upon their Madder ground in the intervals of four years, and have great crops from it; and they are obliged, from the scarcity of land fit for this purpose, to plant the same ground after an interval of four years; but, as we are not under the same necessity, it will be much better to stay eight years, for the roots of Madder are very similar to those of Asparagus, and draw much the same nourishment from the ground; and it is well known that, when Asparagus-roots are dug up, which have been growing three years, if the same ground is planted with Asparagus again in a few years, it will not thrive equal to that which is planted on ground, upon which Asparagus has not grown for several years; and this is always found to be the case even in kitchen-gardens near London, where, by the well working and frequent dunging of the ground, it may be supposed changed in three or four years, more than the fields can possibly be in eight or ten.

Madder should not be planted in very rich dunged land, for in such there will be very large haulm, but the roots will not be in proportion; and, where there is much dung or sea-coal ashes, the Madder-roots will be of a darker colour, as it also will, where it is cultivated in the smoak of London; which is likewise the case with Liquorice, for that which grows in a sandy loam at a distance from London, is always much brighter and clearer than that which grows in the rich lands in the neighbourhood of London.

If the cultivation of Madder is carried on properly in England, it will employ a great number of hands from the time harvest is over till the spring of the year, which is generally a dead time for labourers, and hereby the parishes may be eased of the poor's rate, which is a consideration worthy of public attention.

RUBECLA. See Asperula, Galium, and Sherardia.

RUBUS. Tourn. Inst. R. H. 614. tab. 315. Bramble, or Raspberry Bush.

The Characters are,

The flower has a permanent empalement, which is cut into five spear-shaped segments; it hath five roundish petals, which are inserted in the empalement, and a great number of stamina, which are also inserted in the empalement, terminated by roundish compressed

preſſed ſummits, with a great number of germen, having ſmall hair-like ſtyles on the ſide of the germen, crowned by permanent ſtigmas. The germen afterward becomes a berry, compoſed of many acini, collected into a head, each having one cell, in which is contained one oblong ſeed.

The SPECIES are,

1. RUBUS (*Fruticoſus*) foliis quinato-digitatis ternatiſque, caule petioliſque aculeatis. Flor. Suec. 409. *Bramble or Blackberry with hand-ſhaped leaves, having five and three lobes, and the foot-ſtalk and branches prickly; the common Blackberry.*

2. RUBUS (*Cæſius*) foliis ternatis nudis, caule aculeato. Hort. Cliff. 192. *Bramble with naked trifoliate leaves, and a prickly ſtalk; the Dewberry.*

3. RUBUS (*Idæus*) foliis quinato-pinnatis ternatiſque, caule aculeato, petiolis canaliculatis. Flor. Suec. 408. *Bramble with winged leaves, having five and three lobes, a prickly ſtalk, and channelled foot-ſtalks; prickly Raſpberry.*

4. RUBUS (*Glaber*) foliis ternatis ſubtus tomentoſis, caule glabro. *Raſpberry with trifoliate leaves, which are woolly on their under ſide, and have a ſmooth ſtalk.*

5. RUBUS (*Occidentalis*) foliis quinato-pinnatis ternatiſque, caule aculeato, petiolis teretibus. Lin. Sp. Pl. 493. *Bramble with winged leaves, having five and three lobes, a prickly ſtalk, and taper foot-ſtalks; Virginia Raſpberry with a black fruit.*

6. RUBUS (*Odoratus*) foliis ſimplicibus palmatis, caule inermi multifolio multifloro. Hort. Cliff. 192. *Raſpberry with ſingle hand-ſhaped leaves, and an unarmed ſtalk, having many flowers; commonly called flowering Raſpberry.*

7. RUBUS (*Hiſpidus*) foliis ternatis nudis, caulibus petioliſque hiſpidis. Lin. Sp. Plant. 493. *Bramble with naked leaves growing by threes, and hairy ſtalks and foot-ſtalks.*

8. RUBUS (*Sexatilis*) foliis ternatis nudis, flagellis repentibus herbaceis. Flor. Suec. 411. *Bramble with naked trifoliate leaves, and creeping herbaceous ſtalks; dwarf Rock Bramble.*

9. RUBUS (*Arcticus*) foliis ternatis, caule inermi unifloro. Flor. Suec. 412. *Bramble with trifoliate leaves, and an unarmed ſtalk, having one flower.*

10. RUBUS (*Chamæmorus*) foliis ſimplicibus lobatis, caule unifloro. Flor. Suec. 413. *Bramble with ſingle leaves, having lobes, and a ſtalk bearing one flower; the Cloudberry.*

The firſt ſort grows naturally on the ſide of banks, and in hedges in moſt parts of England, ſo is not cultivated in gardens; this is ſo well known as to need no deſcription. Of this there are the following varieties:

1. The common Bramble with white fruit, which was found in a hedge near Oxford by Mr. Jacob Bobart. The branches of this ſort are covered with a light green bark; the leaves are of a brighter green than the common ſort, and the fruit is white, but it ſeldom produces fruit in gardens.

2. The Bramble without thorns; this is in every reſpect like the firſt, but the branches and foot-ſtalks have no thorns.

3. The Bramble with elegant cut leaves; this differs from the firſt, by having the leaves more finely cut.

4. The Bramble with double flowers; this differs from the firſt in having very double flowers, ſo is frequently planted in gardens for ornament.

5. The Bramble with variegated leaves; this is by ſome preſerved in gardens, but it is very apt to become plain, if planted in good ground.

Theſe ſorts are eaſily propagated by laying down their branches, which will put out roots at every joint very freely. They may be tranſplanted any time from September to March, and will grow in almoſt any ſoil or ſituation.

The ſecond ſort hath weaker trailing ſtalks than the firſt; the leaves are trifoliate, and the lobes are larger than thoſe of the other; the fruit is ſmaller, the acini larger, and but few in each, which are of a deeper black colour. This grows naturally in England, and is known by the title of Dewberry.

The third ſort is the Raſpberry, which grows naturally in the woods in the northern parts of England, but is cultivated in gardens for its fruit, which ſupplies the table at the ſeaſon when they are ripe. There are two or three varieties of this, one with a red, and the other a white fruit, and the third generally produces two crops of fruit annually; the firſt ripens in July, and the ſecond in October, but thoſe of the latter ſeaſon have ſeldom much flavour. Theſe are accidental varieties, but the fourth ſort I believe to be a diſtinct ſpecies, for the leaves are trifoliate, larger than thoſe of the common ſort, woolly on their under ſide, and the branches and ſtalks have no thorns. This produces but few fruit, and thoſe are ſmall, which has occaſioned its being neglected.

The Raſpberry is generally propagated by ſuckers, though I ſhould prefer ſuch plants as are raiſed by layers, becauſe they will be better rooted, and not ſo liable to ſend out ſuckers as the other, which generally produce ſuch quantities of ſuckers from their roots, as to fill the ground in a year or two; and where they are not carefully taken off or thinned, will cauſe the fruit to be ſmall, and in leſs quantities, eſpecially when the plants are placed near each other; which is too often the caſe, for there are few perſons who allow theſe plants ſufficient room.

In preparing theſe plants, their fibres ſhould be ſhortened; but the buds, which are placed at a ſmall diſtance from the ſtem of the plant, muſt not be cut off, becauſe thoſe produce the new ſhoots the following ſummer. Theſe plants ſhould be planted about two feet aſunder in the rows, and four or five feet diſtance row from row; for if they are planted too cloſe, their fruit is never ſo fair, nor will ripen ſo kindly, as when they have room for the air to paſs between the rows. The ſoil in which they thrive beſt, is a freſh ſtrong loam, for in warm light ground they do not produce ſo great plenty of fruit, for they naturally grow in cold land, and in ſhade; therefore when they are planted in a warm ſituation and a light ſoil, they do not ſucceed.

The ſeaſon for dreſſing of them is in October, at which time all the old wood that produced fruit the preceding ſummer, ſhould be cut down below the ſurface of the ground, and the young ſhoots of the ſame year muſt be ſhortened to above two feet in length; then the ſpaces between the rows ſhould be well dug, to encourage their roots; and if you bury a very little rotten dung therein, it will make them ſhoot vigorouſly the following ſummer, and their fruit will be much fairer. During the ſummer ſeaſon, they ſhould be kept clear from weeds, which, with the before-mentioned culture, is all the management they will require; but it is proper to make new plantations once in three or four years, becauſe when the plants are ſuffered to remain long, they will produce few and ſmall fruit.

The Virginian flowering Raſpberry is commonly propagated in the nurſeries as a flowering ſhrub. The flowers of this ſort are as large as ſmall Roſes, and there is a ſucceſſion of them for two months or more, ſo that they make an agreeable variety during their continuance. This ſort frequently produces fruit in England, which are not ſo large as thoſe of the common ſort, and have little flavour. Theſe ripen in September, or the beginning of October.

The Virginian Raſpberry riſes with purpliſh ſtalks a little higher than the common ſort; the leaves are of a lucid green on their upper ſide, but hoary on their under; their foot-ſtalks are taper; the fruit is ſhaped like thoſe of the common Blackberry, and are of a deep black when ripe; the

the fruit has little flavour, fo the plants are never cultivated for their fruit here. It ripens late in autumn.

The eighth fort grows naturally upon rocky hills in the northern counties of England, and moft of the northern parts of Europe. This hath trailing herbaceous ftalks, which put out roots at their joints, whereby it propagates in plenty; the leaves are trifoliate; the lobes are large, and of a lucid green; the fruit are fmall, fo not worth cultivating.

The ninth fort grows naturally in Norway, Sweden, and Siberia; this hath an upright ftalk about three inches high, garnifhed with fmall trifoliate leaves; the ftalk is terminated by one purple flower, which is fucceeded by a fmall red fruit, having the fcent and flavour of Strawberries. This plant grows naturally upon moffy bogs, fo cannot be cultivated to any purpofe on dry ground, and is preferved in a few gardens for the fake of variety.

The tenth fort grows naturally upon fome of the higheft hills in the north of England and Scotland, alfo upon high boggy places in the northern parts of Europe. This plant cannot be tranfplanted into gardens fo as to thrive; the ftalks rife about fix or eight inches high, and are generally garnifhed with two lobated leaves, ftanding at a diftance from each other. The ftalk is terminated by a fingle flower, which is fucceeded by a fmall black fruit, not much unlike that of the Dewberry.

RUDBECKIA. Lin. Gen. Pl. 878. Dwarf Sun-flower, vulgò.

The Characters are,

It hath female and hermaphrodite florets, inclofed in one common empalement, compofed of two orders of leaves, the fcales of which are plain, broad, and fhort. The rays or border of the flower is compofed of female half florets, which end with two or three indentures; thefe have germen fitting upon proper receptacles, but have neither ftyle or ftamina, and are barren. The hermaphrodite florets are tubulous, funnel-fhaped, and indented in five parts at the brim. They have five fhort hair-like ftamina in each, terminated by cylindrical fummits, and a germen fitting in the common empalement, having a flender ftyle, crowned by a reflexed ftigma, divided in two parts. The germen afterward becomes fingle, oblong, four-cornered feeds, crowned by their proper cup, which has four indentures.

The Species are,

1. Rudbeckia (*Hirta*) foliis indivifis fpatulato-ovatis, radii petalis emarginatis. Lin. Sp. Plant. 907. *Rudbeckia with oval, fpattle-fhaped, undivided leaves, and the petals of the rays indented; commonly called Dwarf American Sun-flower.*

2. Rudbeckia (*Purpurea*) foliis lanceolato-ovatis alternis indivifis, petalis radii bifidis. Flor. Virg. 104. *Rudbeckia with oval, fpear-fhaped, undivided leaves, placed alternate, and the petals of the ray bifid; commonly called Dwarf Carolina Sun-flower.*

3. Rudbeckia (*Triloba*) foliis inferioribus trilobis, fuperioribus indivifis. Hort. Upfal. 269. *Rudbeckia with under leaves, having three lobes, and the upper ones entire.*

4. Rudbeckia (*Laciniata*) foliis inferioribus compofitis acutè dentatis, caulinis fimplicibus dentatifque. *Rudbeckia with compound, indented, lower leaves, thofe upon the ftalks fingle and indented.*

5. Rudbeckia (*Quinata*) foliis omnibus quinatis, acutè dentatis exterioribus trilobatis. *Rudbeckia with all the leaves compofed of five lobes, which are fharply indented, and the outer ones divided into three.*

6. Rudbeckia (*Digitata*) foliis inferioribus compofitis, caulinis quinatis ternatifque, fummis fimplicibus. *Rudbeckia with compound lower leaves, thofe on the ftalks quinquefoliate and trifoliate, and the top ones fingle.*

The firft fort grows naturally in Virginia, and feveral other parts of North America. The root of this will continue four or five years, but unlefs there is care taken to fhelter it in winter, the plants are often deftroyed by cold or too much wet. This fort fends out heads, by which it may be propagated; the leaves are oblong, oval, and hairy; the ftalks rife a foot and a half high, having one or two leaves near the bottom. The foot-ftalk which fupports the flower, is naked near a foot in length, and is terminated by one pretty large yellow flower, fhaped fomewhat like the Sun-flower, from whence it was titled Dwarf Sun-flower. The petals or rays of the flower are very ftiff, and flightly indented at their points; the middle or difk of the flower is very prominent, pyramidal, and of a dark purple colour. Thefe flowers are of long duration, for I have frequently obferved one flower has continued in beauty near fix weeks; and as the plants produce many flowers, fo there is a fucceffion of them on the fame plant, from the middle of July till the froft puts a ftop to them, which renders the plants more valuable. This fort will fometimes produce good feeds in England, when the feafons are favourable; but they are generally propagated here by offsets or flips, unlefs when good feeds can be procured either here or from America. The beft time to feparate the offsets is in the fpring, becaufe the plants continue to flower fo late in autumn, as to render it impracticable to perform it till it is late, fo that the froft will fet in before the flips can have taken root. The plants will live abroad in the open air through the winter, if they are planted in a dry foil and a warm fituation; but it will always be prudent to fhelter two or three plants under a common hot-bed frame in winter to preferve the kind, becaufe in very fevere winters thofe in the open air are fometimes killed.

The fecond fort grows naturally in Carolina, and alfo in Virginia. This is a perennial plant like the former, but very rarely produces feeds in England; nor do the plants put out heads, whereby it may be propagated like the other, fo that it is at prefent not very common here. The leaves of this fort are longer and broader than thofe of the other, and are fmooth, having three veins; the foot-ftalks which fupport the flowers are taller, and have two or three narrow leaves on each, which are placed alternate: on the top is one flower with long narrow Peach-coloured petals, which are reflexed backward; the middle or difk is very prominent, and of a dark purple colour, but the fummits being of a gold colour, adds a luftre to the other. This fort may be treated in the fame manner as the other; it flowers at the fame feafon, but the flowers are not of fo long duration as thofe of the former.

The third fort grows naturally in feveral parts of North America. This is a biennial plant, which in warm fummers perfects its feeds in England; the lower leaves of this fort are divided into three lobes, but thofe upon the ftalks are undivided; they are hairy, and fhaped like thofe of the firft fort; the ftalks branch out on their fides, and are better garnifhed with leaves than either of the other. The flowers are very like thofe of the firft fort, but fmaller; the plants will live through the winter in the open air in mild feafons, and may be propagated by flips or heads; but the beft way is to raife the plants from feeds, becaufe thofe will flower much better than fuch as are procured by flips; the fecond year the feedling plants will flower, and produce ripe feeds.

The fourth fort grows naturally in moft parts of North America, and has been long an inhabitant in European gardens, where it was generally known by the title of Sunflower. The root of this is perennial, but the ftalk is annual; the lower leaves are compofed of five broad lobes, which are deeply cut into acute points, fome of them being jagged almoft to the midrib; the outer lobe is frequently

cut

cut into three deep ſegments. The ſtalks riſe ſeven or eight feet high, and divide upward into ſeveral branches; they are ſmooth, green, garniſhed with ſingle leaves, which are oval, and heart-ſhaped; ſome of theſe are indented on their edges, and others are entire. The foot-ſtalks which ſuſtain the flowers are naked, and terminated by a ſingle flower with yellow petals or rays, ſhaped like thoſe of the Sun-flower, but ſmaller. This does not produce ſeeds here, but is eaſily propagated by parting of the roots in the ſame manner as the perennial Sun-flower. It is very hardy in reſpect to cold, but loves a moiſt ſoil.

The fifth ſort has a perennial root like the former, and is a native of the ſame country. This hath ſmooth green ſtalks, which riſe higher than thoſe of the former; the leaves are all compoſed of five lobes, which are much narrower, and end with ſharper points than thoſe of the former, which are very acutely indented on their ſides. The flowers are ſmaller, and the petals narrower than thoſe of the former ſort, but appear at the ſame ſeaſon. It is equally hardy with the former, and may be propagated in the ſame way.

The ſixth ſort grows naturally in North America, and alſo in Siberia, from both which countries I have received the ſeeds. This hath a perennial root like the two former; the leaves at bottom are compoſed of ſeven or nine lobes, ſome of which are entire, and others are jagged to the midrib; they are ſmooth, and of a dark green; the ſtalks riſe ſix feet high, and divide into many branches. They are of a purple or iron colour, very ſmooth, garniſhed with leaves, which, towards the bottom, are hand-ſhaped, compoſed of five lobes, and the upper have but three; thoſe at the top are ſingle. The flowers are ſmaller than thoſe of the two former ſorts, but are of the ſame ſhape and colour.

The three laſt mentioned ſorts may be propagated in plenty, by parting of their roots; the beſt time for this is in October, when their ſtalks will begin to decay; for if they are removed in the ſpring, they will not produce many flowers the ſame year. They love a moiſt ſoil, and ſhould be allowed room, for if they are too near other plants they will rob them of their nouriſhment. They are proper furniture for large gardens, where they may be allowed room, or in walks round fields, becauſe they require little culture.

RUELLIA. Plum. Nov. Gen. 12. tab. 2.

The CHARACTERS are,

The flower hath a permanent empalement, cut into five narrow acute ſegments at the top, which are erect. It has one petal, with a tube the length of the cup, which inclines at the neck; the brim ſpreads open, and is cut into five ſegments, the two upper being large and reflexed. It hath four ſtamina ſituated in the tube, connected in pairs, terminated by ſhort ſummits, and a roundiſh germen ſupporting a ſlender ſtyle, crowned by a bifid ſtigma. The germen afterward becomes a taper capſule, pointed at each end, having two cells, incloſing roundiſh compreſſed ſeeds.

The SPECIES are,

1. RUELLIA (*Tuberoſa*) foliis ovatis crenatis, pedunculis bifloris. *Ruellia with oval crenated leaves, and foot-ſtalks bearing two flowers.*

2. RUELLIA (*Strepens*) foliis petiolatis, floribus verticillatis ſubſeſſilibus. Hort. Upſal. 178. *Ruellia with leaves having foot-ſtalks, and flowers growing in whorls ſitting cloſe to the ſtalks.*

3. RUELLIA (*Clandeſtina*) foliis petiolatis, pedunculis longis ſubdiviſis nudis. Lin. Hort. Upſal. 179. *Ruellia with leaves having foot-ſtalks, and long naked foot-ſtalks to the flowers, which are divided.*

4. RUELLIA (*Criſpa*) foliis ſubcrenatis lanceolato-ovatis, capitulis ovatis, foliolis hiſpidis. Lin. Sp. Pl. 635. *Ruellia with oval ſpear-ſhaped leaves, which are ſomewhat crenated, oval pods, and prickly, hairy, ſmall leaves.*

The firſt ſort grows naturally in many of the iſlands in the Weſt-Indies; the roots of this are compoſed of many ſwelling fleſhy tubers, which are like thoſe of the Day Lily, but ſmaller. The ſtalk riſes about four or five inches high, ſending out two or three ſhort ſide branches, garniſhed with leaves placed oppoſite; ſome of theſe are ſmall, and ſhaped like a ſpatula, others are much larger; they have ſhort foot-ſtalks, and are a little crenated on their edges. The flowers are produced on the ſide, and at the end of the ſtalk; thoſe on the ſide have two flowers upon each foot-ſtalk, which come out oppoſite, but thoſe at the top ſuſtain three. The flowers have narrow tubes about an inch long, ſpreading out to a ſort of bell-ſhape; at the top they are cut into five obtuſe ſegments, which are large, and ſpread open; they are of a fine blue, but of ſhort duration, each flower ſeldom laſting in beauty one day; after the flower fades, the germen becomes a taper pod one inch and a half long, having two cells, which, when ripe, burſt with a touch, and caſt out the ſeeds to a diſtance.

The ſecond ſort grows naturally in Carolina; the root of this is fibrous and perennial; the ſtalk riſes about a foot high; they are four-cornered, and have two longitudinal furrows; the joints are three or four inches aſunder; at each ſtand two oval leaves upon very ſhort foot-ſtalks. The flowers come out from the wings of the leaves, two or three riſing from the ſame point, ſitting very cloſe to the ſtalks; they are ſmall, and of a pale purple colour, but are very fugacious, as they open early in the morning, but fade by ten or eleven o'clock in the forenoon; theſe are ſucceeded by ſhort taper pods, ſurrounded by the hairy ſegments of the empalement.

The third ſort grows naturally in the Weſt-Indies. This hath a perennial root, compoſed of many fleſhy fibres; the leaves lie cloſe to the ground; the ſtalks grow five or ſix inches high, with leaves placed by pairs at each joint, ſtanding upon foot-ſtalks. The foot-ſtalks which ſuſtain the flowers are naked, and divide into two ſmaller, each ſuſtaining one ſmall purple flower, which is very fugacious; their empalements are cut into very narrow ſegments to the bottom. After the flowers are paſt the germen becomes a taper capſule about an inch long, including roundiſh compreſſed ſeeds.

The fourth ſort grows naturally in both Indies; I received the ſeeds of this from Carthagena in New Spain. This hath a ligneous creeping root; the ſtalks are ſingle, taper, and riſe about five or ſix inches high; the leaves are oval, ſpear-ſhaped, and have very ſhort foot-ſtalks; they are a little waved on their edges, are hairy, and curled. The flowers are produced from the ſide of the ſtalk, which are yellow, coming out between rough, hairy, ſmall leaves. The ſeeds ripen in September.

Theſe plants are propagated by ſeeds, which muſt be ſown in the ſpring in pots, and plunged into a moderate hot-bed; when the plants come up they muſt be tranſplanted each into a ſeparate ſmall pot, and plunged into a freſh hot-bed of tanners bark, and muſt be ſhaded from the ſun until they have taken new root; after which time they muſt have free air admitted to them every day in warm weather. If the plants thrive well, thoſe of the firſt and third ſorts will produce flowers the July following, and will perfect their ſeeds in Auguſt; but the roots will continue, provided they are plunged into the bark-bed in the ſtove, and kept in a moderate temperature of heat.

The

The second sort is not a plant of long continuance, seldom abiding longer than two years; but if it is treated in the same manner as the two other, it will ripen seeds the second year, so may be propagated easily.

The fourth sort does not so constantly produce seeds as the three others, so is not so common in England at present. This requires the same treatment as the other sorts.

If the seeds of these plants are permitted to scatter in the neighbouring pots, the plants will come up without care, so may be transplanted into pots, and plunged into the tan-bed.

RUMEX. Lin. Gen. Plant. 407. Dock.

The CHARACTERS are,

The empalement of the flower is permanent. The flower has three petals, which are larger than the empalement, to which they are very like. It hath six short hair-like stamina, terminated by erect twin summits, and a three-cornered germen supporting three hair-like reflexed styles, thrusting out of the clefts of the petals, crowned by large jagged stigmas. The germen afterward becomes a three-cornered seed, included in the petals of the flower.

The SPECIES are,

1. RUMEX (*Patientia*) floribus hermaphroditis, valvulis integerrimis, foliis oblongo-lanceolatis. Lin. Sp. Pl. 333. *Dock with hermaphrodite flowers having entire valves, and oblong spear-shaped leaves; commonly called Patience Rhubarb.*

2. RUMEX (*Alpinus*) floribus hermaphroditis, valvulis integerrimis graniferis, foliis cordatis obtusis. Lin. Sp. Pl. 333. *Rumex with hermaphrodite flowers, having entire valves bearing grains, and obtuse heart-shaped leaves; called Monks Rhubarb.*

3. RUMEX (*Aquaticus*) floribus hermaphroditis pedicellatis, foliis lanceolatis longissimis. Lin. Sp. Pl. 334. *Rumex with hermaphrodite flowers growing upon small foot-stalks, and the longest spear-shaped leaves; or Water Dock.*

4. RUMEX (*Acutus*) floribus hermaphroditis, valvulis dentatis graniferis, foliis cordato-oblongis. Hort. Cliff. 138. *Rumex with hermaphrodite flowers, indented grain-bearing valves, and oblong heart-shaped leaves; or sharp-pointed Dock.*

5. RUMEX (*Crispus*) floribus hermaphroditis, valvulis integris graniferis, foliis lanceolatis undulatis acutis. Lin. Sp. 335. *Rumex with hermaphrodite flowers, entire grain-bearing valves, and acute, spear-shaped, waved leaves; or curled sharp-pointed Dock.*

6. RUMEX (*Sanguineus*) floribus hermaphroditis, valvulis integerrimis, unica granifera foliis cordato-lanceolatis. Hort. Cliff. 138. *Rumex with hermaphrodite flowers, entire valves, and heart-formed spear-shaped leaves; the bloody Dock.*

7. RUMEX (*Aureus*) floribus hermaphroditis verticillatis, valvulis acutè dentatis, foliis lanceolatis. Hort. Cliff. 138. *Rumex with hermaphrodite flowers growing in whorls, acutely-indented valves, and spear-shaped leaves; sharp-pointed Dock with a golden flower.*

8. RUMEX (*Obtusifolius*) floribus hermaphroditis, valvulis dentatis, foliis cordato-oblongis, obtusiusculis crenulatis. Lin. Sp. 335. *Rumex with hermaphrodite flowers, indented valves, and blunt, oblong, heart-shaped leaves; common broad-leaved, or Butter Dock.*

9. RUMEX (*Pulcher*) floribus hermaphroditis, foliis lyratis. Guet. Stam. 1. p. 7. *Rumex with hermaphrodite flowers, and lyre-shaped leaves; the Fiddle Dock.*

10. RUMEX (*Maritimus*) floribus hermaphroditis, valvulis dentatis graniferis, foliis linearibus. Lech. Scan. 26. *Rumex with hermaphrodite flowers, indented grain-bearing valves, and linear leaves.*

11. RUMEX (*Chalepensis*) floribus hermaphroditis pedunculis longioribus, valvulis profundè dentatis, foliis cordato-oblongis. *Rumex with hermaphrodite flowers growing upon longer foot-stalks, valves which are deeply indented, and oblong heart-shaped leaves; or Aleppo Dock.*

12. RUMEX (*Ægyptiacus*) floribus hermaphroditis, valvulis trifido setaceis, unica granifera. Hort. Upsal. 89. *Rumex with hermaphrodite flowers, and bristly three-pointed valves; or annual Egyptian Dock.*

13. RUMEX (*Lunaria*) floribus hermaphroditis, valvulis lævibus, caule arboreo, foliis subcordatis. Virg. Cliff. 32. *Rumex with hermaphrodite flowers, smooth valves, a tree-like stalk, and leaves which are almost heart-shaped; Tree Sorrel.*

14. RUMEX (*Bucephalophrus*) floribus hermaphroditis, valvulis nudis dentatis planis reflexis. Hort. Upsal. 90. *Rumex with hermaphrodite flowers, and plain, naked, indented, reflexed valves.*

15. RUMEX (*Vesicarius*) floribus hermaphroditis geminatis, valvularum alis maximis membranaceis reflexis, foliis indivisis. Hort. Cliff. 130. *Rumex with hermaphrodite flowers growing by pairs, very large membranaceous wings to the valves, which are reflexed, and undivided leaves.*

16. RUMEX (*Roseus*) floribus hermaphroditis distinctis, valvularum alis maximis membranaceis, foliis erosis. Flor. Leyd. Prod. 230. *Rumex with hermaphrodite flowers growing upon distinct spikes, very large membranaceous wings to the valves, and leaves appearing as if bitten.*

The first sort was formerly much more cultivated in the English gardens than at present; the roots of this has been generally used for the Monks Rhubarb, and has been thought the true, but others suppose the second sort should be used as such. The root is large, and divides into many thick fibres; their outer cover is brown, but they are yellow within, with some reddish veins; the leaves are broad, long, and acute-pointed; their foot-stalks are of a reddish colour; the stalks rise six or seven feet high, and divide toward the top into several erect branches, garnished with a few narrow leaves, terminating with loose spikes of large stamineous flowers. These appear in June, and are succeeded by pretty large three-cornered seeds, whose coverings are entire, which ripen in autumn.

The second sort grows naturally on the Alps, but has been long cultivated in the English gardens. This hath large roots, which spread and multiply by their offsets; they are shorter and thicker than those of the first sort, of a very dark brown on their outside, and yellow within. The leaves are of the round heart-shape, standing upon long foot-stalks. The stalks rise from two to three feet high; they are thick, and have a few small roundish leaves on the lower part, but the upper part is closely garnished with spikes of white flowers, standing erect close to the stalks. These appear the latter end of May, and are succeeded by large triangular seeds, which ripen in August.

The third sort grows naturally in ponds, ditches, and standing waters, in many parts of England; this is supposed to be the Herba Britannica of the ancients. It hath large roots, which strike deep into the loose mud, sending out leaves which are above two feet long, drawing to a point at each end. The stalks rise five or six feet high when the plants grow in water, but in dry land seldom more than three; these are garnished with narrow leaves among the spikes of flowers to the top. The flowers stand upon slender foot-stalks, which are reflexed; they are of an herbaceous colour, appear in June, and the seeds ripen in autumn.

The fourth sort grows naturally in moist places in many parts of England; this is the Oxylapathum of the shops, which is directed by the College to be used in medicine; but the markets are supplied with roots of the common Docks, which are indifferently gathered by those who collect them in the fields, where the eighth fort is much more common than this. The roots of this fort are flender and run downright, fending out a few fmall fibres; the ftalks rife about two feet high, garnifhed at bottom with leaves four inches long, and one and a half broad in the middle; they are rounded at their bafe, where they are flightly indented, but end in acute points; they are plain, and flightly crenated on their edges. From the joints of the ftalk come out alternately flender long foot-ftalks, which fuftain the fpikes of flowers, which grow in fmall whorls round the ftalks, at about an inch diftant; thefe have fcarce any leaves upon the foot-ftalks between the whorls of flowers, fo may be eafily diftinguifhed from the fmall Water Dock, which has many.

The fifth fort is more commonly found growing naturally about London than the fourth; the leaves of this are much longer than thofe of the former, and are indented on their fides, which are alfo waved; the ftalks rife about the fame height as thofe of the former. The fpikes of flowers from their fide are fhorter, and clofer garnifhed with flowers on pretty long foot-ftalks; the covering of the feed is entire.

The fixth fort is very like the fourth in appearance, but the leaves have deep blood-coloured veins, and fome fmall fpots of the fame on their furface; the ftalks are red, and rife about the fame height as the fourth; but the covering of the feed is entire, whereas thofe of the fourth are indented, fo may be readily diftinguifhed. It grows naturally in many parts of England.

The feventh fort grows naturally in feveral parts of England; this is a biennial plant, which perifhes foon after the feeds are ripe; the ftalks rife near two feet high; they are of a deep purple colour, garnifhed with fpear-fhaped leaves toward the bottom, which are four inches long, and almoft one broad in the middle, but thofe on the upper part of the ftalk are very narrow, and not more than two inches long; the fpikes of flowers come out from the fides of the ftalks alternately. The flowers grow in thick whorls, which fit clofe to the ftalks; thefe are of a bright yellow, and the covers of the feeds are fharply indented.

The eighth fort is the moft common Dock by the fides of roads and banks in every part of England; the leaves of this fort are broad and rounded at their points, though fome of them end more acutely than others; they are near a foot long, and five inches broad toward their bafe, having many tranfverfe veins running from the midrib to their borders. The ftalks rife from two to three feet high, branching out on their fides, having a few leaves on their lower part, of the fame fhape with the other, but fmaller. The flowers grow in whorls, fitting very clofe to the ftalks; fome plants have indented coverings to their feed, and others have entire coverings; both thefe are frequently found intermixed, fo that I doubt of their being diftinct fpecies. The leaves of this Dock were formerly much ufed for wrapping up of butter, and from thence the plant was called Butter Dock.

The ninth fort grows naturally in many places near London; this is a biennial plant, which perifhes foon after the feeds are ripe. The ftalks of this rife about a foot high, and branch out from the bottom; the leaves grow near the root; they are about two inches and a half long, and are hollowed on their fides, fo as to refemble the fides of a fiddle; the ftalks are generally bent at their joints. The flowers grow in whorls round the ftalks, to which they fit very clofe; they are hermaphrodite; the covers of the feeds are fharply indented.

The tenth fort is fometimes found growing naturally in England, upon places where the water has ftood in winter. This feldom rifes more than five or fix inches high but divides into two or three branches; the leaves are about three inches long, and a quarter of an inch broad, fmooth, and ftand upon fhort foot-ftalks. The flowers grow in whorls round the branches, to which they fit very clofe; thefe are fucceeded by fmall triangular feeds, having indented covers.

The eleventh fort came originally from Aleppo; this is a biennial plant; the leaves are nine or ten inches long, fmooth, of a light green, and three inches broad at their bafe, where they are indented, and end in acute points. The ftalks rife from two to three feet high, fending out many branches from their fides, garnifhed with large whorls of herbaceous flowers, ftanding upon pretty long foot-ftalks; thefe are fucceeded by three-cornered feeds, whofe coverings are deeply indented.

The twelfth fort grows naturally in Egypt; this is an annual plant; the ftalk rifes about ten inches high, fending out a few horizontal branches toward the bottom; the leaves are about two inches long, and half an inch broad at the broadeft part. The flowers grow in whorls round the ftalks; they are very fmall, and the hair-like beards, which adhere to the covering of the feeds, being long, obfcure the flowers, fo they are fcarce vifible to the naked eye.

The thirteenth fort is commonly known among the gardeners by the title of Sorrel-tree. This came originally from the Fortunate, or Canary Iflands, but has been long an inhabitant in fome Englifh gardens; it rifes with a ligneous ftalk ten or twelve feet high, covered with a fmooth brown bark, fending out many flender branches, garnifhed with fmooth, roundifh, heart-fhaped leaves, ftanding alternately upon pretty long foot-ftalks. The flowers come out in loofe panicles toward the end of the branches; they are of an herbaceous colour, and are fometimes fucceeded by triangular feeds with fmooth covers, but they rarely ripen in England. This plant is eafily propagated by cuttings, which may be planted in any of the fummer months in a bed of loamy earth, and fhaded from the fun until they have taken pretty good root; then they fhould be taken up, and planted in pots, placing them in the fhade till they have all taken new root; after which they may be removed to a fheltered fituation, among other hardy green-houfe plants till autumn, when they muft be removed into the green-houfe, and treated in the fame way as other hardy kind of plants, which only want protection from froft.

The fourteenth fort is a low annual plant, which grows naturally in Italy and Spain. This is generally found on fwampy moift ground; the ftalks are flender, branching at the bottom, and rife about four inches high; the lower part is garnifhed with fmall, oval, fucculent lobes; their upper part is furnifhed with fmall herbaceous flowers growing in whorls, and have no leaves between them: they are fucceeded by fmall feeds, whofe covers are fharply indented and reflexed. Thefe appear in June, and the feeds ripen in Auguft, which, if permitted to fcatter, will furnifh a fupply of young plants the following fpring; or if the feeds are then fown, the plants will come up the following fpring, and require no other care but to thin them and keep them clean from weeds.

The fifteenth fort is an annual plant; this hath pretty thick fucculent ftalks, which rife a foot high, and divide into many branches; the leaves are of the round heart-fhape and undivided, having very long foot-ftalks. The flowers grow in loofe fpikes at the end of the branches; thefe are fucceeded by large covers to the feeds, which are inflated, and have bread membranaceous borders; the feeds are triangular, and ripen in autumn.

The

The ſixteenth ſort grows naturally in Egypt. This is alſo an annual plant, whoſe ſtalks riſe a foot and a half high, dividing upward into ſeveral branches; the ſtalks are garniſhed with arrow-pointed leaves about three inches long, whoſe ſides are irregularly torn, as if they had been gnawed by inſects; they ſtand upon pretty long foot-ſtalks, and have ſmooth ſurfaces; the flowers are diſpoſed in looſe ſpikes; ſome of which have only male flowers, and others have all hermaphrodite flowers. The latter are ſucceeded by triangular ſeeds, incloſed in large inflated covers, of a deep red colour, having membranaceous borders. The ſeeds of this ripen in autumn.

The ſeeds of both theſe ſorts grow very freely, if ſown in a bed of light earth in the ſpring, where the plants are deſigned to remain. When they come up they will require no other care but to keep them clean from weeds, and thin them where they are too cloſe.

All theſe ſorts of Docks riſe eaſily from ſeeds, and if introduced into a garden, will become troubleſome weeds, if their ſeeds are permitted to ſcatter; therefore few perſons care to propagate any of them, except the two firſt ſorts, which are cultivated for their uſe in medicine. The ſeeds of all the Docks ſhould be ſown in autumn ſoon after they are ripe, for thoſe ſeeds which are ſown in the ſpring rarely grow the ſame year: when the plants come up they will require no other care but to thin them where they are too cloſe, and keep them clean from weeds. They all delight in a moiſt rich ſoil.

RUSCUS. Tourn. Inſt. R. H. 79. tab. 15. Knee-holly, or Butchers-broom.

The Characters are,

It hath male and female flowers in diſtinct plants; the male flowers have erect ſpreading empalements, compoſed of ſix oval convex leaves, whoſe borders are reflexed; they have no petals, but an oval nectarium the ſize of the empalement, which is erect and inflated, opening at the mouth; they have no ſtamina, but each has three ſpreading ſummits, ſitting on the top of the nectarium, which are joined at their baſe. The female flowers have empalements, but no petals, and nectariums like the male; they have no ſtamina, but have an oblong oval germen hid within the nectarium, ſupporting a cylindrical ſtyle, crowned by an obtuſe ſtigma, ſtanding above the mouth of the nectarium. The germen afterward becomes a globular berry with two or three cells, incloſing two globular ſeeds.

The Species are,

1. Ruscus (*Aculeatus*) foliis ſuprà floriferis nudis. Hort. Cliff. 465. *Ruſcus with leaves which bear flowers on their upper ſide, and are naked; Knee-holly, or Butchers-broom.*

2. Ruscus (*Hypophyllum*) foliis ſubtus floriferis nudis. Hort. Cliff. 465. *Ruſcus with leaves which bear flowers beneath, and are naked.*

3. Ruscus (*Hypogloſſum*) foliis ſubtus floriferis ſubfoliolo. Hort. Cliff. 465. *Ruſcus with flowers under the leaves.*

4. Ruscus (*Racemoſus*) racemo terminali hermaphroditico. Hort. Cliff. 469. *Ruſcus with hermaphrodite flowers in long bunches, terminating the ſtalks.*

5. Ruscus (*Trifoliatus*) foliis ternis ovatis acuminatis, ſuprà floriferis nudis. *Ruſcus with oval acute-pointed leaves, which are placed by threes, and flowers on their upper ſide.*

6. Ruscus (*Flexuoſus*) foliis ovatis acuminatis, ſuprà floriferis nudis, caulibus flexuoſis. *Ruſcus with acute-pointed leaves, bearing flowers on their upper ſide, and flexible ſtalks.*

7. Ruscus (*Androgynus*) foliis margine floriferis. Hort. Cliff. 464. *Ruſcus with flowers growing on the borders of the leaves.*

8. Ruscus (*Fruteſcens*) caule fruticoſo ramoſo, foliis lanceolatis rigidis, floribus pedunculatis terminalibus. *Ruſcus with a ſhrubby, branching ſtalk, ſpear-ſhaped ſtiff leaves, and flowers growing upon foot-ſtalks, terminating the branches.*

The firſt ſort is very common in the woods in divers parts of England, and is rarely cultivated in gardens. The roots of this kind are ſometimes uſed in medicine, and the green ſhoots are cut, bound into bundles, and ſold to the butchers, who uſe it as beſoms to ſweep their blocks, from whence it had the name of Butchers-broom. It as alſo called by ſome Knee-holly.

This hath roots compoſed of many thick fibres, which twine about each other, from which ariſe ſeveral ſtiff green ſtalks about three feet high, ſending out from their ſide ſeveral ſhort branches, garniſhed with ſtiff, oval, heart-ſhaped leaves, placed alternately on every part of the ſtalk, ending with ſharp prickly points. The flowers are produced in the middle on the upper ſide of the leaves; theſe are male in ſome, and female in other plants; they are ſmall, and cut into ſix parts, of a purple colour, ſitting cloſe to the midrib; they appear in June, and the female flowers are ſucceeded by berries, almoſt as large as Cherries, of a ſweetiſh taſte, which ripen in winter, when they are of a beautiful red colour.

As this plant grows wild in moſt parts of England, it is rarely admitted into gardens; but if ſome of the roots are planted under tall trees in large plantations, they will ſpread into large clumps, and as they retain their leaves in winter, at that ſeaſon they will have a good effect. The ſeeds of this plant generally lie a year in the ground before they vegetate, and the plants ſo raiſed are long before they arrive to a ſize big enough to make any figure, ſo it is not worth while to propagate them that way, eſpecially as the roots may be eaſily tranſplanted from the woods. The roots and ſeeds of this plant have been uſed in medicine; the young ſhoots of this plant in the ſpring are ſometimes gathered and eaten by the poor like thoſe of Aſparagus; the branches of this plant, with their ripe fruit upon them, are frequently cut and put into baſons, mixing them with the ſtalks of ripe ſeeds of male Piony, and thoſe of the wild Iris or Gladwyn, which together make a pretty appearance in rooms, at a ſeaſon of the year when there are few flowers, and theſe will continue a long time in beauty.

The ſecond ſort grows naturally in the mountainous parts of Italy, but is preſerved for the ſake of variety in many Engliſh gardens. The roots of this have large knotty heads with long thick fibres like thoſe of the former ſort, from which ariſe many tough limber ſtalks near two feet high, garniſhed by ſtiff, oblong, oval leaves, ending in points, placed alternately on the ſtalks. The flowers are produced on the under ſurface of the leaves near the middle, ſitting cloſe to the midrib; they are ſmall, of an herbaceous white colour; the female flowers are ſucceeded by ſmall red berries, about the ſize of thoſe of Juniper.

It ſtands in moſt Diſpenſaries among the plants uſed in medicine, and has been commended for opening obſtructions in the kidneys, and to provoke urine.

The third ſort grows naturally upon ſhady mountains in Italy, Hungary, and other parts of Europe. The root of this is compoſed of many thick fibres like thoſe of the former, from which ariſe many tough limber ſtalks, which are about ten inches high, garniſhed with ſpear-ſhaped leaves, having ſeveral longitudinal veins, placed for the moſt part alternate, but ſometimes they are oppoſite. On the middle of the upper ſurface of theſe comes forth a ſmall leaf of the ſame ſhape, and at the ſame point; from the boſom of the ſmall leaves come out flowers, which are of a pale yellow colour. The female flowers are ſometimes ſucceeded by berries almoſt as large as thoſe of the firſt ſort, which are red, and ripen in winter. This is ſometimes called Biſlingua, or double Tongue, from the leaves growing one out of another. It ſtands in Diſpenſaries as a medicinal plant, but is rarely now uſed.

The fourth ſort grows naturally in the Archipelago, but is frequently planted in the Engliſh gardens; it is called Laurus Alexandrina, i. e. Alexandrian Bay, and is ſuppoſed to be the plant with which the ancients crowned their victors and poets. The ſtalks of this being very pliable, may be eaſily wrought into coronets for this purpoſe; and the leaves of this plant having a great reſemblance to thoſe which are repreſented on the ancient buſts, ſeem to confirm this opinion.

The roots of this are like thoſe of the former ſpecies; the ſtalks are ſlender and much more pliable; they riſe about four feet high, and ſend out many ſide branches, garniſhed with oblong acute-pointed leaves, rounded at their baſe, of a lucid green, placed alternately, ſitting cloſe to the branches. The flowers are produced in bunches at the end of the branches, which are hermaphrodite, of an herbaceous yellow colour, and are ſucceeded by berries like thoſe of the firſt ſort, which ripen in autumn.

The fifth ſort grows naturally in Zant and ſome of the other iſlands in the Morea. The roots of this are like thoſe of the former ſorts; the ſtalks riſe about two feet high; they are ſlender, pliable, and garniſhed with oval leaves, placed by threes round the ſtalk, rounded at both ends, terminating in acute points. The flowers grow on the under ſide of the leaves, faſtened to the midrib; they are naked, and have pretty long foot-ſtalks; the ſegments or petals are very narrow; the fruit I have not ſeen, ſo can give no account of it.

The ſixth ſort grows naturally in Italy, where it was diſcovered by Sig. Micheli of Florence. The roots of this are much longer than thoſe of the firſt ſort; the ſtalks riſe near five feet high, are very pliant, and ſend out ſeveral ſide branches, garniſhed with ſtiff oval leaves, ending in acute points. The flowers are produced on the upper ſurface of the leaves, ſitting cloſe to the midrib; they are ſmall, of an herbaceous white colour, and are ſucceeded by berries, which are ſmaller than thoſe of the firſt ſort, of a pale red when ripe.

All theſe ſorts are very hardy, and will thrive in almoſt any ſoil or ſituation, ſo are very proper for planting round the verges of cloſe woods, or under large trees in wilderneſs quarters, for, as they are always green, they make a good appearance in winter, after the deciduous trees have caſt their leaves; they are eaſily propagated by parting of their roots. The beſt time for this is in autumn, but when this is performed, the roots ſhould not be divided into ſmall parts, becauſe that will weaken them ſo much, that they will make but little figure until they have had two or three years growth; they may alſo be propagated by ſowing of their ſeeds; but this is a very tedious method, ſo is ſeldom practiſed.

The ſeventh ſort ſends out pliant ſtalks, which riſe ſeven or eight feet high, and have ſeveral ſhort branches proceeding from the ſides, which are garniſhed with ſtiff leaves, rounded at the foot-ſtalk, ending in acute points. The flowers are produced in cluſters on the edges of the leaves, which are white, and the female flowers are ſucceeded by berries of a yellowiſh red colour, not ſo large as thoſe of the firſt ſort.

This ſort is tender, and muſt therefore be planted in pots, and in winter removed into the green-houſe; but it ſhould be placed where it may have free air in mild weather, for it only requires to be ſcreened from froſt; in the ſummer it muſt be ſet abroad with other hardy green-houſe plants. With this management the plants will ſend forth ſtems ſix or eight feet high, furniſhed with leaves from bottom to top, and will be cloſely ſet with flowers upon their edges, which make a very beautiful and odd appearance. This is alſo propagated by parting the roots as the former, which ſhould not be done very often, becauſe if the roots are not permitted to remain ſome time to get ſtrength, they will produce but weak ſhoots, and very few flowers. This ſort grows plentifully at Madeira, from whence the ſeeds may be procured; but the ſeeds muſt be taken from plants which have the male plants near them, to impregnate the flowers of the female to cauſe their ſeeds to be good. They commonly remain in the ground a year before the plants come up, ſo ſhould be ſown in pots, and placed under a hot-bed frame in winter, to ſcreen the ſeeds from froſt, and the following ſpring the plants will appear.

The eighth ſort was diſcovered by the late Dr. Houſtoun growing naturally at Carthagena in New Spain; this riſes with ſhrubby ſtalks eight or ten feet high, which divide into many branches, garniſhed with ſtiff ſpear-ſhaped leaves, ſometimes ranged in whorls round the ſtalks, and at others they are oppoſite. The flowers are produced in looſe bunches at the end of the branches, ſtanding upon ſlender foot-ſtalks; they are ſmall, of a red colour, and ſhaped like thoſe of the firſt ſort.

This plant is tender, ſo muſt be kept in a ſtove during the winter, otherwiſe it will not live in England.

RUTA. Tourn. Inſt. R. H. 257. tab. 133. Rue.

The CHARACTERS are,

The flower has a permanent empalement cut into five parts; it has four or five oval petals, which are narrow at their baſe, and eight or ten awl-ſhaped ſpreading ſtamina the length of the petals, crowned by erect ſummits, with a gibbous germen having a croſs furrow, ſupporting an erect awl-ſhaped ſtyle, crowned by a ſingle ſtigma. The germen afterward becomes a gibbous capſule with five lobes, and five cells opening in five parts at the top, filled with rough angular ſeeds.

The SPECIES are,

1. RUTA (*Hortenſis*) foliis decompoſitis, floribus octandris, ſtaminibus corollâ longioribus. *Rue with decompounded leaves, and flowers having eight ſtamina, which are longer than the petals; or broad-leaved Garden Rue.*

2. RUTA (*Altera*) foliis decompoſitis, foliolis oblongo-ovatis, ſtaminibus corollâ æquantibus. *Rue with decompounded leaves, the ſmall leaves oblong, oval, and ſtamina equalling the petals.*

3. RUTA (*Sylveſtris*) foliis inferioribus decompoſitis, foliolis linearibus, ſummis quinquefidis trifidiſque. *Rue with decompounded linear leaves below, and the upper ones five or three-pointed; or ſmaller wild Rue.*

4. RUTA (*Chalepenſis*) foliis decompoſitis, floribus decandris, marginibus petalorum ciliatis. *Rue with decompounded leaves, flowers having ten ſtamina, and the borders of the petals of the flowers hairy; or the broad-leaved Aleppo Rue.*

5. RUTA (*Ciliata*) foliis compoſitis, floribus decandris, petalis florum ciliatis. *Rue with compound leaves, flowers having ten ſtamina, and hairy petals to the flower; or narrow-leaved Aleppo Rue.*

6. RUTA (*Linifolia*) foliis ſimplicibus indiviſis. Lin. Sp. Plant. 384. *Rue with ſingle undivided leaves.*

7. RUTA (*Montana*) caule erecto corymboſo, foliis compoſitis, floribus decandris, ſtaminibus corollâ longioribus. *Rue with an erect corymbus ſtalk, compound leaves, and flowers having ten ſtamina, which are longer than the petals.*

8. RUTA (*Patavina*) foliis terminatis ſeſſilibus. Lin. Sp. Plant. 549. *Rue without foot-ſtalks, terminating the branches.*

The firſt ſort is the common Rue, which has been long cultivated in the gardens, and is that which is directed to be be ufed in medicine; but of late years the fecond fort has fo generally prevailed, as almoft to fupplant the firft in the gardens about London, that being hardier than the firft, is not fo liable to be killed by fevere froft.

This rifes with a fhrubby ftalk to the height of five or fix feet, fending out branches on every fide, garnifhed with decompounded leaves, whofe fmall leaves (or lobes) are wedge-fhaped, of a gray colour, and have a ftrong odour. The flowers are produced at the end of the branches, in bunches almoft in form of umbels; they are compofed of four yellow concave petals, which are cut on their edges, and eight yellow ftamina which are longer than the petals, terminated by roundifh fummits. The germen becomes a roundifh capfule, with four lobes punched full of holes, containing rough black feeds.

The fecond fort hath a fhrubby ftalk, which rifes three or four feet high, fending out many branches, garnifhed with decompounded leaves, narrower than thofe of the former fort, of a bluifh gray colour, and have a ftrong odour. The flowers grow in longer and loofer bunches than the former; they have four fhort, concave, yellow petals, and eight fhort ftamina of equal length with the petals. The feed-veffel is like that of the former, but fmaller.

The third fort grows naturally in Spain. The lower leaves of this are compounded of feveral parts, which are joined to the midrib in the fame manner as the branching winged leaves, garnifhed with fmall linear leaves, ftanding without order. The ftalks rife from two to three feet high, branching out from the bottom, and are garnifhed with leaves, divided into five parts, thofe at the top into three, which are as fmall and narrow as thofe at the bottom, of a gray colour, but not fo ftinking as thofe of the other. The flowers grow at the end of the branches in loofe fpikes, which are generally reflexed; the petals of the flower are yellow; thefe are fucceeded by fmall feed-veffels, filled with angular black feeds.

The feeds of the fourth fort came firft from Aleppo, but has been brought from the Cape of Good Hope. This hath ftrong fhrubby ftalks, which rife three feet high, dividing into many branches, garnifhed with decompounded leaves, larger than thofe of the common fort, and have a ftronger odour. The flowers are difpofed almoft in form of an umbel at the end of the branches; they have five concave yellow petals, whofe borders are fet with fine hairs, and ten ftamina, which are of equal length with the petals. The feed-veffels of this are much larger than thofe of the common fort.

The fifth fort grows naturally at Aleppo. This hath fhrubby ftalks, which are fmaller, and do not rife fo high as thofe of the former fort. The leaves are much narrower and grayer than thofe, but have the fame ftrong odour; the flowers are fmaller, having five petals, which are pretty clofe fet with fmall hairs; they have ten thick ftamina, five of which are alternately longer than the petals; the feed-veffels are like thofe of the firft fort.

The fixth fort grows naturally in Spain. This rifes with feveral fingle ftalks from the root near a foot and a half high, garnifhed with fingle narrow leaves, of a yellowifh green colour, placed alternately on the ftalks, to which they fit very clofe; at the bafe of thefe come out one or two very fmall leaves, of the fame fhape and colour. The flowers grow in fmall clufters at the end of the ftalks; they have each five oblong yellow petals, and ten ftamina of equal length, terminated by awl-fhaped fummits.

The eventh fort rifes with an erect ftalk two feet high, garnifhed with compound leaves, whofe fmaller leaves are narrow, obtufe, of a grayifh colour, but have not fo ftrong an odour as the former. The upper part of the ftalk divides in form of a corymbus, fuftaining upon naked footftalks fmall bunches of yellow flowers, which have five concave petals, and ten ftamina which are much longer than the petals, terminated by roundifh fummits.

The eighth fort grows naturally near Padua: this rifes with a branching ftalk three feet high; the branches are garnifhed with very narrow leaves, which are fometimes placed alternately, and at others are oppofite; the flowers are formed almoft in an umbel, terminating the branches; they are fhaped like thofe of the other forts, having five obtufe petals, and ten erect ftamina, but they have little fcent of the Rue.

This may be propagated by feeds, which fhould be fown foon after they are ripe, otherwife they will lie in the ground a whole year. When the plants come up, they fhould be each put into a fmall pot, that they may be fheltered in winter, for the plants will not live abroad in England.

All thefe plants may be propagated either by fowing of their feeds, or by planting flips or cuttings, both of which muft be done in the fpring. The manner of propagating them from cuttings being the fame as for Lavender, Stœchas, and other hardy aromatic plants, need not be here repeated; if they are propagated by feeds, there needs no farther care but to dig a bed of frefh earth in the fpring, making it level, then to fow the feed thereon, and rake the ground fmooth; after which you muft obferve to keep the bed clear from weeds until the plants come up about two inches high, when they fhould be tranfplanted out into frefh beds, where they may remain for ufe. All thefe plants muft have a dry foil, otherwife they are very fubject to be deftroyed in winter. The two Aleppo Rues, and the wild Rue, are fomewhat tenderer than the common fort, but thefe will endure our ordinary winters very well in the open air, efpecially if they are planted on a dry foil.

The fixth and feventh forts are tenderer than either of the other, and are of fhorter duration. The feeds of the feventh fort were fent me from Gibraltar-Hill, where the plant grows naturally. This doth not ripen its feeds here, unlefs the fummers are warm, and in hard winters the plants are generally killed, unlefs they are removed into fhelter.

The fixth fort will live through the winter in the open air, provided it is planted in a poor dry foil, and the fecond year it will perfect feeds; but as it is of fhort duration, young plants fhould be annually raifed to fucceed the others.

All the forts of Rue will live much longer, and are lefs liable to be injured by froft in winter, when they grow in a poor dry rubbifhy foil, than in good ground, for in rich moift land the plants grow very vigoroufly in fummer, and are fo replete with moifture that a fmall froft will kill their tender fhoots; whereas in poor dry ground, their growth will not be great, but their fhoots will be hard and compact, fo more able to refift the cold.

RUTA CANINA. See Scrophularia.

RUTA MURARIA. Wall-rue, or white Maiden-hair.

This plant is found growing out of the joints of old walls in divers parts of England, where it is gathered for medicinal ufe; but as it cannot be cultivated in gardens, fo as to grow to advantage, I fhall not fay any thing more of it in this place.

RUYSCHIANA. Boerh. Ind. alt. 1. p. 172.

The Characters are,

The flower hath a permanent tubulous empalement, cut into five fegments at the top, the upper one being broader and blunter than the other; it is of the lip kind, having a tube longer than

than the empalement. The chaps are large and ſwelling; the upper lip is erect, arched, and gently indented at the top; the lower lip is trifid; the two ſide ſegments are narrow, and ſtand erect; the middle is broad, reflexed, and indented at the point. It hath four ſtamina, two of which are long, ſituated under the upper lip; the other two are ſhorter, and ſituated juſt below them; they are terminated by oblong ſummits, faſtened in the middle; it has four germen, ſituated at the bottom of the empalement, ſupporting a ſlender ſtyle the length of the ſtamina, crowned by a bifid reflexed ſtigma. The germen afterward becomes four oblong ſeeds, which ripen in the empalement.

The Species are,

1. Ruyschiana (*Spicata*) floribus ſpicatis, foliis bracteiſque linearibus glabris indiviſis. *Ruyſchiana with ſpiked flowers, linear leaves, and bracteæ which are ſmooth and undivided.*

2. Ruyschiana (*Laciniata*) floribus ſpicatis, foliis linearibus trifidis hirſutis. *Ruyſchiana with ſpiked flowers, and hairy, linear, three-pointed leaves.*

3. Ruyschiana (*Verticillata*) floribus axillaribus, foliis lanceolatis dentatis glabris. *Ruyſchiana with flowers growing at the wings of the ſtalks, and ſmooth, indented, ſpear-ſhaped leaves.*

The firſt ſort grows naturally in Auſtria and Hungary. This hath a perennial root, and an annual four-cornered ſtalk, which riſes about two feet high; garniſhed with two ſmooth linear leaves. At each joint of the ſtalk come out two or three very narrow ſmall leaves, of the ſame ſhape. The flowers are produced in whorled ſpikes at the top of the ſtalks, having ſmall narrow leaves under each whorl. They have tubulous empalements cut into five ſegments at the top, four of which are narrow, and end in acute points; the other, which is on the upper ſide of the flower, is broader, and rounded at the point. The tube of the flower is longer than the empalement, is ſwelling, and large at the chaps; the upper lip is broad, erect, and arched over the tube; the lower lip is ſhorter, having two ſhort ſide ſegments, which are erect; the middle ſegment is broad, rounded, indented at the point, and is reflexed back to the tube. It has four ſtamina, which lie cloſe under the upper lip, arched in the ſame manner; two of theſe are as long as the ſtyle, which ſtands in the ſame poſition; the other two are ſhorter, and are ſituated juſt below the other; they are terminated by oblong ſummits, faſtened in the middle to the ſtamina. The ſtyle is crowned by a bifid, reflexed, narrow ſtigma; the flowers appear in June, are of a fine blue colour, and are ſucceeded by four oblong ſeeds, which ripen in the empalement.

The ſecond ſort grows naturally in Siberia; this hath a perennial root. The ſtalks are four-cornered, hairy, riſing a foot and a half high, garniſhed with hairy linear leaves, cut into three parts; the flowers grow in ſhort whorled ſpikes at the end of the ſtalk, having ſome very narrow leaves under each whorl; the tube of the flower is longer, more equal in ſize than that of the former, and the middle ſegment of the lower lip is not ſo much reflexed. In other reſpects the flowers are the ſame.

The third ſort grows naturally in Tartary; this hath a perennial root; the ſtalks do not grow erect like the firſt, but ſpread nearer to an horizontal poſition; they have two large leaves oppoſite at each joint, and four ſmaller, two on each ſide between the larger; they are ſmooth, have ſharp indentures on their edges, and ſtand erect. The flowers come out from the ſide of the ſtalks at the baſe of the leaves, two or three ſtanding together on each ſide the ſtalk; the empalements are purple, cut into five acute ſegments at the top, the upper lip having three broad, and the lower two narrower. The flowers are of a paler blue than thoſe of the firſt ſort.

Theſe plants are propagated by ſeed, which ſhould be ſown the latter end of March, in a bed of light earth in an open expoſure; in five or ſix weeks after the plants will appear, when they ſhould be carefully cleared from weeds. When the plants are about two inches high they ſhould be tranſplanted into a bed or border of light undunged earth, obſerving to ſhade them from the ſun until they have taken root; after which time they will require no farther care but to keep them conſtantly clear from weeds till Michaelmas, when they ſhould be removed into the places where they are deſigned to remain for good.

When the plants are firſt tranſplanted into the nurſery-bed, they ſhould be placed about ſix inches aſunder every way, which will be ſufficient room for them to grow the firſt ſeaſon; this will admit of the hoe to come between the plants to deſtroy the weeds, which is by much a better method than pulling out the weeds by hand, and is much ſooner performed. For as the hoe ſtirs the ground between the plants, it not only cuts down the weeds which were up and viſible, but alſo deſtroys all thoſe whoſe ſeeds were ſprouted, and would have ſoon after appeared; ſo that one hoeing, if well performed in dry weather, will more effectually deſtroy the weeds than two hand-weedings would do; beſides, the ſtirring the ground is of great ſervice to the plants.

At Michaelmas, when the plants are tranſpland for good, they ſhould be carefully taken up with balls of earth to their roots, and planted in the middle of the borders, intermixing them with other hardy plants of the ſame growth in the pleaſure-garden, where they will make a pretty appearance when they are in flower.

As theſe plants do not continue many years, it will be proper to raiſe a ſupply of young ones to ſucceed them.

S.

SABINA. See JUNIPERUS.

SACCHARUM. Lin. Gen. Plant. 68. The Sugar-cane.

The CHARACTERS are,

It hath no empalement, but a woolly down longer than the flower incloses it. The flower is bivalve; the valves are acute-pointed, concave, and chaffy. It has three hair-like stamina the length of the valves, terminated by oblong summits, and an awl-shaped germen supporting two rough styles, crowned by single stigmas. The germen afterward becomes an oblong acute-pointed seed, invested by the valves.

We have but one SPECIES of this genus, viz.

SACCHARUM (*Officinarum*) floribus paniculatis. Hort. Cliff. 26. *Sugar-cane with flowers growing in panicles.*

This plant grows naturally in both Indies, where it is cultivated for its juice, which, when boiled, affords that sweet salt which is called Sugar.

The Canes were formerly cultivated in the south of France for the same purpose, but it was in small quantities only, for in sharp winters they were killed, unless they were covered, so they had only the summer for their growth, which was too short for their getting sufficient strength to produce Sugar enough to answer the expence, so the planting of these Canes there has been long discontinued; they were also planted in several parts of Spain before they were introduced to France, and are at present cultivated in plenty in Andalusia, from whence great quantities of Sugar are annually sent to Madrid, but there are few now planted in the other parts of Spain.

The root of this plant is jointed, like those of the other sorts of Cane or Reeds, from which arise four, five, or more shoots in number, proportionable to the age or strength of the root. These rise eight or ten feet high, according to the goodness of the ground in which they grow; for in some moist rich soils there have been Canes measured which were near twenty feet long; but these were not near so good as those of middling growth, as they abounded with juice, which had but a small quantity of the essential salt in it, so that the expence of fuel and trouble of boiling, was more than the Sugar would defray. The Canes are jointed; the joints are more or less distant from each other in proportion to the soil. The leaves are placed at each joint; the base or lower part of the leaf embraces the Cane to the next joint above its insertion, before it expands; they are three or four feet long from the joint where they unfold to their point, according to the vigour of the plant, and have a deep whitish furrow, or hollowed midrib, which is broad, and prominent on the under side; the edges of the leaves are thin, and armed with small sharp teeth, which are scarce to be discerned by the naked eye, but will cut the skin of a tender hand if it be drawn along it. The flowers are produced in panicles at the top of the stalks, composed of many spikes, which are nine or ten inches long, and are again subdivided into smaller spikes; these have long down which inclose the flowers, so as to hide them from sight; afterward the germen becomes an oblong pointed seed, which ripens in the valves of the flower.

This plant is preserved by way of curiosity in several gardens in England, but being too tender to thrive here unless it is preserved in a warm stove, it cannot be brought to any great perfection. I have seen some of the plants growing, which were seven or eight feet high, and at the bottom as large as a common walking-cane, but they have not produced their panicles of flowers here.

It is here propagated by slips taken from the sides of the old plants; those which grow near the root, and have fibres to them, will most certainly grow; so that when the shoots are produced at some distance from the ground, the earth should be raised about them, that they may put out fibres before they are separated from the mother plant. These slips should be planted in pots, and plunged into a moderate hot-bed of tanners bark; then they must be treated in the same way as other tender plants from the same countries, keeping them constantly plunged in the tan-bed in the stove, and as their roots increase in size, the plants should from time to time be shifted into larger pots; but this must be done with caution, for if they are over-potted they will not thrive: they will require to have water frequently in warm weather, but it must not be given them in too great plenty, especially in cold weather. As the leaves of the plants decay, they should be cleared from about the stalks, for if these are left to dry upon them it will greatly retard their growth. The stove in which this plant is placed should be kept in winter to the same temperature of heat as for the Pine-apple, and in hot weather there should be plenty of free air admitted to the plants, otherwise they will not thrive.

I shall here subjoin some account of the method of propagating and cultivating the Sugar-cane in America, with some observations and experiments which have been made by a few curious persons in the British Islands, and shall propose some farther trials to be there made in the culture and management of this useful plant, which are founded upon the experience I have had in the culture of some plants, which are similar in their growth with the Sugar-cane.

The land which is most proper for the growth of Sugar-canes, is such as hath a sufficient depth of soil, and is not too moist and strong, but rather light and easy to work; for although strong moist ground will produce much taller and bigger Canes than the other, yet the quantity of Sugar will be much less, not near so good, and will require a greater quantity of fuel, and a longer time to boil, before the Sugar can be made; which is also the case with all fresh land, where there has not been any Canes growing before; therefore many of the most expert planters burn their land when it is first cleared for planting of Canes, to abate its fertility; but if when land is first cleared of the wood, and the roots of bad weeds, it is sown with Indigo, which such fresh ground will produce much better than the old, or such as has been long cultivated, there may be

two

two or three crops of this taken, which will prepare the land for the Sugar-canes, without being at the trouble of burning it; but the growing of Indigo has been fo little practifed in the Britifh Iflands of America for many years paft, as to be efteemed unworthy the notice of a Sugar-planter; whereas, if they would fometimes change their crops to other fpecies, they would foon find an advantage in the growth of their Canes; but the ufual practice is to continue the Canes always upon the fame land as long as it would produce them, without changing the fpecies, or allowing the ground a fallow to reft and recover itfelf. By this method there are fome plantations fo much exhaufted, as that the crop of Sugar will fcarce defray the expence of culture.

Another thing fhould always be obferved in the planting of frefh land with Canes, which is to allow them much more room than is generally done; for as the ground is ftrong, fo there will a greater number of fhoots come out from each plant, and not having room to fpread at bottom, they will draw each other up to a great height, and be full of watery juice, the fun and external air being excluded from the Canes by the multiplicity of leaves, both which are abfolutely neceffary to ripen and prepare the falts during the growth of the Canes.

If the ground is proper for the Sugar-canes, and they are planted at a good diftance from each other, the fame plantation may be continued above twenty years without replanting, and produce good crops the whole time; whereas in the common method, they are generally re-planted in fix or feven, and in fome of the poor land they are continued but two or three.

The Canes are in thofe warm countries propagated by cuttings or joints of proper lengths; thefe are from fifteen to twenty inches long, in proportion to the nearnefs of their joints or eyes. Thefe cuttings are generally taken from the tops of the Canes, juft below the leaves; but if they were chofen from the lower part, where they are lefs fucculent and better ripened, they would not produce fo luxuriant Canes, but their juice would be lefs crude, and contain a greater quantity of falts, which will be obtained by lefs boiling than thofe which are planted in the common way: this is well known to the judicious to hold true, in moft kinds of vegetables, for it is by thus carefully propagating all kinds of efculent plants, either in the choice of the beft feeds or cuttings, that moft of the kinds have been fo greatly improved of late years.

The diftance which the planters ufually allow to their Canes, is from three to four feet, row from row, and the hills are about two feet afunder in the rows; in each of thefe hills they plant from four to feven or eight cuttings, which is a very great fault, and is the caufe of moft of their blights fo much complained of lately; for if all thefe grow, which is frequently the cafe, they rob each other of their nourifhment; and if a dry feafon happens before they have acquired ftrength, they are very foon ftinted in their growth, fo are attacked by infects, which fpread and multiply fo greatly, as to cover a whole plantation in a little time; when this happens, the Canes are feldom good after, therefore it would be the better way to root them entirely up, when they are fo greatly injured, for they very rarely recover this perfectly; for although the infects are not the caufe of the difeafe, yet they confirm it, and caufe it to fpread.

Therefore, if inftead of planting fo many, there was but one good cutting planted in each hill, or to prevent mifcarriage, two at moft; and if both fucceeded, the weakeft were drawn out foon after they had taken, it will be found of great fervice to prevent thefe blights; and although the number of Canes will not be near fo great from the fame fpace of ground, yet the quantity of Sugar will be full as much, and will require little more than a fourth part of fuel to boil it.

I have been affured by two of the moft fenfible and judicious planters of Sugar in America, that they have made fome experiments of the horfe-hoeing culture for their Canes, which anfwered much beyond their expectations; one of thofe gentlemen told me, he planted one acre in the middle of a large piece of Canes, in rows at five feet afunder, and the hills were two feet and a half diftant, and but one cutting to each hill. The ground between the rows was from time to time ftirred with the horfe-plough, to deftroy the weeds and earth the plants; with this culture the Canes were double the fize of thofe in the fame piece, which were cultivated in the ufual way: and when the Canes were cut, thofe which had been thus planted and managed, were ground and boiled feparately; the produce of Sugar was full as great as the beft acre in the fame piece, and the expence of boiling was little more than a fixth part of the other, and he fold the Sugar for fix fhillings per hundred weight more than he could get for the other.

The time for planting the Canes is always in the rainy feafons, and the fooner they are planted after the rains have began to fall, the more time they will have to get ftrength before the dry weather fits in; for when they have put out good roots, and are well eftablifhed in the ground, they will not be fo liable to fuffer by the drought as thofe which have but newly taken root.

The feafon being come for planting, the ground fhould be marked out by a line, that the rows of Canes may be ftrait, and at equal diftances; but firft it will be proper to divide the piece into lands of fixty or feventy feet broad, leaving intervals between each of about twenty feet; thefe will be found of great ufe when the Canes are cut, for roads in which the carriages may pafs to carry off the Canes to the mill; for where there is not fuch provifion made, the carriages are obliged to pafs over the heads of the Canes, to their no fmall prejudice; befides, by thefe intervals, the fun and air will have freer paffage between the Canes, whereby they will be better ripened, and their juice will be fuller of falts. The middle of thefe intervals may be planted with Yams, Potatoes, or other efculent plants, which may be taken off before the Canes are cut, that the paffage may be clear for the carriages; but a path fhould be left on the fides of each land, for the more convenient riding or walking of the overfeer of the plantation, to view and obferve how the labour is performed.

The common method now practifed in planting of the Canes is, to make a trench with the hoe, which is performed by hand; into this one negro drops the number of cuttings intended for planting, at the diftance the hills are defigned; thefe are by other negroes placed in their proper pofition; then the earth is drawn about the hills with a hoe; all this is performed by hand; but if the right ufe of ploughs was well known, the work might be much better performed, and for lefs than half the expence; therefore inftead of making a trench with a hoe, a deep furrow is made with a plough, and the cuttings properly laid therein, the ground will be deeper ftirred, and there will be more depth for placing the Canes.

If the ground is to be afterward kept clean with the horfe-hoe, the rows of Canes fhould be planted five feet afunder, that there may be room for the horfe and plough to pafs between them; the diftance of the hills from each other fhould be two feet and a half, and but one Cane fhould be permitted to remain in each hill. After the Canes are planted, and have made fome fhoots, the fooner the horfe-plough is ufed the better will the Canes thrive, and the ground will be eafier kept clean from weeds; for

if

if these are torn up when they are young, they will presently die; whereas, when they are suffered to grow large before they are disturbed, they are with great difficulty destroyed.

As the growth of the Canes is promoted according to the cleanness of the ground, so there cannot be too much care taken to keep the Canes perfectly clear of weeds; the beginning of this work soon will render it less troublesome, and it may be performed at a less expence, than when it is neglected for some time. When this is performed with a plough, the earth in the interval should be thrown up to the rows of Canes, first on one side of the row, being careful not to disturb the roots of the Canes, as also not to bury their new shoots; and in the second operation, the earth should be turned over to the other side of the rows, with the same care as before. By this turning and stirring of the land, it will be rendered looser, and the earthing of the plants will greatly strengthen them; so that from each hill there will be as many shoots produced as can be well nourished, and the sun and air will have free ingress among the rows, which will be of the greatest service to the Canes.

When the Canes are from seven to ten feet high, and of a proportionable size; the skin smooth, dry, and brittle, if they are heavy; their pith gray, or inclinable to brown; the juice sweet and glutinous, they are esteemed in perfection.

The time for cutting of the Canes is usually after they have grown six months; but there should not be a fixed period for this, for in some seasons and in different soils, there will be more than a month's difference in their maturity; and those who have made the experiments of cutting their Canes before they were ripe, and letting others stand till after they were ripe, have found the Sugar made from the latter, was much finer than that of the former, though the quantity was not quite so great; however, it will always be best to let them stand till they are in perfection before they are cut, but not longer.

They have also found those Canes which are cut toward the end of the dry seasons, before the rains begin to fall, have produced better Sugar than those which are cut in the rainy seasons, when they are more replete with watery juice; there has also been much less expence of fuel to boil it, which is a material article in large plantations; therefore the better the Canes are nourished in their growth, and the more air and sun is admitted to pass between their rows, the less expence it will be in the boiling and preparing of the Sugar.

In the boiling of Sugar, they use a mixture of wood-ashes and lime, which is called Temper, without which the Sugar will not granulate. The quantity of this mixture, is proportioned to the quality of the ground on which the Canes grew.

SAFFRON. See Crocus.

SAGE. See Salvia.

SAGITTARIA. Lin. Gen. Plant. 946. Arrow-head.

The Characters are,

It hath male and female flowers on the same plant; the male flowers have a permanent empalement of three concave leaves; they have three roundish petals, which spread open, and many awl-shaped stamina collected into a head, terminated by erect summits. The female flowers are situated below the male; these have a three-leaved empalement, and three petals as the male, but no stamina; they have many compressed germen collected in a head, sitting upon very short styles, and have permanent acute stigmas. The germen afterward becomes oblong compressed seeds, having longitudinal borders, and are collected in globular heads.

The Species are,

1. Sagittaria (*Sagitifolia*) foliis omnibus sagittatis acutis petiolis longissimis. *Arrow-head with all the leaves arrow-pointed, and long foot-stalks.*

2. Sagittaria (*Minor*) foliis sagittatis spatulisque, petiolis longioribus. *Arrow-head with arrow-pointed and spattle-shaped leaves, having longer foot-stalks.*

The first sort grows naturally in standing waters in most parts of England; the root is composed of many strong fibres, which strike into the mud; the foot-stalks of the leaves are in length proportionable to the depth of the water in which they grow, so they are sometimes almost a yard long; they are thick and fungous; the leaves which float upon the water are shaped like the point of an arrow, the two ears at their base spreading wide asunder, and are very sharp-pointed. The flowers are produced upon long stalks, which rise above the leaves, standing in whorls round them at the joints; they have each three broad white petals, which spread open, and in the middle is a cluster of stamina with purple summits. The flowers are succeeded by rough heads, containing many small seeds.

The second sort grows plentifully in standing waters near Paris, but has not been found wild in England. This never grows so large as the former; the leaves vary greatly, some of them are oblong, round-pointed, and shaped like a spatula; others are arrow-pointed, but these have their points less acute than those of the former; the flowers are smaller, in which it differs from the former; and as all the plants where this grows retain their difference, so it may be supposed a different species.

SALICARIA. See Lythrum.

SALICORNIA. Tourn. Cor. App. 51. tab. 485. Jointed Glasswort, or Saltwort.

The Characters are,

The flower hath a rugged, swelling, four-cornered empalement, which is permanent. It has no petal, and but one stamina the length of the empalement, crowned by an oblong twin summit, with an oblong oval germen, supporting a single style, crowned by a bifid stigma. The germen afterward becomes a single seed, inclosed in the swelling empalement.

The Species are,

1. Salicornia (*Fruticosa*) articulis apice crassioribus obtusis. Lin. Mat. Med. 8. *Jointed Glasswort with thick obtuse points.*

2. Salicornia (*Perennis*) articulis apice acutioribus, caule fruticoso ramoso. *Glasswort with acute points to the joints, and a shrubby branching stalk.*

The first sort grows plentifully in most of the salt marshes which are overflowed by the tides, in many parts of England. This is an annual plant, with thick succulent jointed stalks, which trail upon the ground. The flowers are produced at the ends of the joints toward the extremity of the branches, which are small, and scarce discernable by the naked eye.

The second sort grows naturally in Sheepey Island; this hath a shrubby branching stalk about six inches long; the points of the articulations are acute, the stalks branch from the bottom, and form a kind of pyramid; they are perennial, and produce their flowers in the same manner as the former.

The inhabitants near the sea-coast where these plants grow, cut them up toward the latter end of summer, when they are fully grown; and after having dried them in the sun, they burn them for their ashes, which are used in making of glass and soap. These herbs are, by the country people, called Kelp, and are promiscuously gathered for use.

From

From the ashes of these plants is extracted the salt, called Sal Kali, or Alkali, which is much used by the chemists.

The manner of gathering and burning of these herbs, is mentioned under the article of SALSOLA, so I shall not repeat it in this place.

SALIX, Tourn. Inst. R. H. 590. tab. 364. The Sallow, or Willow-tree.

The CHARACTERS are,

It hath male and female flowers upon separate plants; the male flowers are disposed in one common imbricated katkin. The scales have each one oblong spreading flower, which has no petal, but a cylindrical nectarious gland in the center. It has two slender erect stamina, terminated by twin summits, having four cells. The female flowers are disposed in katkins as the male; these have neither petals or stamina, but an oval-narrowed germen, scarce distinguishable from the style, crowned by two bifid erect stigmas. The germen afterward becomes an oval awl-shaped capsule with one cell, opening with two valves, containing many small oval seeds, crowned with hairy down.

There are several species of this genus, which grow naturally in the northern parts of Europe, which are of no use, being low creeping shrubs, many of them seldom rising a foot high, so are never cultivated, therefore I shall pass them over, and only enumerate those which are planted for use.

The SPECIES are,

1. SALIX (*Alba*) foliis lanceolatis acuminatis serratis utrinque pubescentibus, serraturis infimis glandulosis. Hort. Cliff. 473. *Willow with spear-shaped, acute-pointed, sawed leaves, which are downy on both sides, and glands below the saws; or common white Tree Willow.*

2. SALIX (*Triandra*) foliis serratis glabris, floribus triandris. Lin. Sp. Plant. 1015. *Willow with smooth sawed leaves, and flowers having three stamina.*

3. SALIX (*Pentandria*) foliis serratis glabris, flosculis pentandris. Hort. Cliff. 454. *Willow with smooth sawed leaves, and flowers having five stamina; or broad-leaved, smooth, sweet Willow.*

4. SALIX (*Vitellina*) foliis serratis ovatis acutis glabris, serraturis cartilagineis, petiolis calloso punctatis. Hort. Upsal. 295. *Willow with smooth, oval, acute, sawed leaves, having cartilaginous indentures, and foot-stalks with callous punctures; or yellow Willow.*

5. SALIX (*Amygdalina*) foliis serratis glabris lanceolatis petiolatis, stipulis trapeziformibus. Flor. Leyd. Prod. 83. *Willow with smooth, spear-shaped, sawed leaves, having foot-stalks, and trapezium-shaped stipulæ; or Almond-leaved Willow.*

6. SALIX (*Fragilis*) foliis serratis glabris ovato-lanceolatis, petiolis dentato-glandulosis. Flor. Lapp. 349. *Willow with oval, spear-shaped, smooth, sawed leaves, and indented glandules to the foot-stalk; the Crack Willow.*

7. SALIX (*Purpurea*) foliis serratis glabris lanceolatis, inferioribus oppositis. H. Scan. 252. *Willow with smooth, spear-shaped, sawed leaves, the lower of which grow opposite.*

8. SALIX (*Viminalis*) foliis subintegerrimis lanceolato-linearibus longissimis acutis subtus sericeis, ramis virgatis. Flor. Suec. 813. *Willow with the longest, linear, spear-shaped, acute leaves, which are almost entire, and silky on their under side, and rod-like branches.*

9. SALIX (*Auriculata*) foliis serratis glabris lanceolatis, omnibus alternis. *Willow with smooth, spear-shaped, sawed leaves, all growing alternate; or Almond-leaved Willow, which casts its bark.*

10. SALIX (*Rubra*) foliis integerrimis lanceolatis longissimis utrinque virentibus. *Willow with the longest, spear-shaped, entire leaves, which are green on both sides; or the least brittle Willow.*

11. SALIX (*Babylonica*) foliis serratis glabris lineari-lanceolatis, ramis pendulis. Hort. Cliff. 454. *Willow with smooth, sawed, linear, spear-shaped leaves, and hanging branches; or the weeping Willow.*

12. SALIX (*Helix*) foliis serratis glabris lanceolato linearibus, superioribus oppositis obliquis. Flor. Leyd. 83. *Willow with linear, spear-shaped, smooth, sawed leaves, the upper of which are placed obliquely opposite; or the yellow dwarf Willow.*

13. SALIX (*Caprea*) foliis ovatis rugosis, subtus tomentosis undatis supernè denticulatis. Flor. Leyd. Prod. 83. *Willow with oval rough leaves, which are waved, woolly on their under side, and indented towards the top; or the broad-leaved Willow, or Sallow.*

14. SALIX (*Acuminata*) foliis oblongo-ovatis acuminatis rugosis, subtus tomentosis. *Willow with oblong, oval, acute-pointed, rough leaves, which are woolly on their under side; or common Sallow.*

The first sort is the common white Willow, which is frequently found growing on the sides of rivers and ditches in many parts of England. It grows to a large size, if the branches are not lopped off; the shoots are covered with a smooth, pale, green bark; the leaves are spear-shaped; they are very white on their under side, and their upper is covered with short, white, woolly hairs, though not so closely as the under; the katkins are short and upper thick. The wood of this is very white, and polishes smooth.

The second sort grows to be a large tree; the young branches are covered with a grayish bark; the leaves are smooth, of a lucid green, eared at their base, ending in acute points, sawed on their edges, and are green on both sides; the branches grow pretty erect, and are flexible, so this is frequently planted in Osier-grounds for the basket-makers. The katkins of this are long, narrow, the scales open, and are acute-pointed.

The third sort hath thick strong shoots, covered with a dark green bark; the leaves are broad, rounded at both ends, very smooth, sawed on their edges, and, when rubbed, have a grateful odour. It is sometimes called the Bay-leaved Willow, and at others the sweet Willow; it grows quick, and is a tree of middling size; the branches are brittle, so are not proper for many purposes.

The fourth sort has slender tough shoots, which are of a yellow colour; the leaves are oval, acute pointed, smooth, and sawed on their edges; the saws are cartilaginous, and the foot-stalks of the leaves have callous punctures. This is very pliable, so is much planted in the Osier-grounds for the basket-makers, but it never grows to a large size.

The fifth sort grows to a pretty large size; the shoots are erect, covered with a light green bark; the leaves are spear shaped, of a lucid green on both sides, sawed on their edges, standing upon short foot-stalks; they have stipulæ in form of a trapezium, at the base of the foot-stalk. The twigs of this sort are flexible, and fit for the use of basket-makers.

The sixth sort grows to a middling size; the shoots of this are very brittle so are unfit for the basket-makers, and are covered with a brownish bark; the leaves are of a lucid green on both sides, and sawed on their edges; the katkins are long and slender; the scales are pretty long, acute-pointed, and stand open. It is commonly called Crack Willow, from the branches being very brittle.

The seventh sort is a tree of middling size; the shoots are very pliable, and fit for the basket-makers, so is much planted in the Osier-grounds; they are of a reddish colour; the leaves are spear-shaped, smooth, and sawed on their edges; those on the lower part of the branches are placed opposite, but on the upper they are alternate, and of a yellowish green.

The eighth ſort makes very long ſhoots, but the tree ſeldom grows to a large ſize; the leaves are very long, entire, and are ſet cloſe upon the branches; they are of a dark green on their upper ſide, but very woolly and white on their under, ending in acute points. The young branches are woolly, and their buds are very turgid. This is pretty much planted in the Oſier grounds, for the uſe of baſket-makers.

The ninth ſort is a tree of middling growth; it caſts its bark annually; the ſhoots are brittle; they have a yellowiſh bark; the leaves are ſpear-ſhaped, ſawed on their edges; they are eared on both ſides at their baſe, and are placed alternate, of a light green on both ſides. This is not very commonly cultivated, the twigs being too brittle.

The tenth ſort hath very pliant branches, ſo is much planted in the Oſier-grounds. The leaves of this are very long, ſpear-ſhaped, entire, and are green on both ſides. It grows to a middling ſize, if planted in moiſt land.

The eleventh ſort grows naturally in the Levant, but has been ſeveral years cultivated in the Engliſh gardens. This will grow to a middling ſize; the branches are long, ſlender, and hang down on every ſide, ſo form natural arches; the leaves are narrow, ſpear-ſhaped, ſmooth, and ſawed on their edg. It is well known in the gardens by the title of weeping Willow.

The twelfth ſort is a tree of lower growth; the branches of this are erect; the leaves are ſmooth, narrow, ſpear-ſhaped, ſawed on their edges, of a dark or bluiſh green, and toward the upper part of the branches are placed oppoſite. It is found by the ſide of ditches, in many parts of England.

The thirteenth ſort grows naturally upon dry land, and on high ſituations, but rarely is of a large ſize; the bark is ſmooth, and of a dark gray colour; the branches are brittle, ſo are unfit for baſket-makers, but it is frequently cultivated in hedges for fuel in many parts of England. It is called Mountain Oſier. The leaves are oval, rough, woolly, and indented toward the top. There is a variety of this in the gardens with variegated leaves.

The fourteenth ſort is the common Sallow; this differs from the laſt, in having longer leaves, which end in acute points; they are woolly on their under ſide, and ſit cloſer to the branches; theſe are not diſtinguiſhed by the farmers, who cultivate them equally.

There are ſome other ſorts of Willows which are planted in the Oſier-grounds, diſtinguiſhed by the baſket-makers and dealers in them, under titles they have applied to them, which are little known to others; theſe are annually cut down, and always kept low, but when they are not cut down, and have room to grow, will become large trees; ſo that they may be planted for the ſame purpoſes as the firſt ſort, and will make a variety when intermixed with it, though they are commonly cultivated for their twigs, which produce good profit to the owners of the land.

All the ſorts of Willows may be eaſily propagated by planting cuttings or ſets, either in the ſpring or autumn, which readily take root. Thoſe ſorts which grow to be large trees, are cultivated for their timber, ſo are generally planted from ſets, which are about ſeven or eight feet long; theſe are ſharpened at their larger end, and thruſt into the ground by the ſides of ditches and banks, where the ground is moiſt; in which places they make a conſiderable progreſs, and are a great improvement to ſuch eſtates, becauſe their tops will be fit to lop every fifth or ſixth year. This is the uſual method now practiſed in moſt parts of England, where the trees are cultivated, as they are generally intended for preſent profit; but if they are deſigned for large trees, or are cultivated for their wood, they ſhould be planted in a different manner, for thoſe which are planted from ſets of ſeven or eight feet long, always ſend out a number of branches toward the top, which ſpread and form large heads fit for lopping; but their principal ſtem never advances in height, therefore where regard is paid to that, they ſhould be propagated by ſhort young branches, which ſhould be put almoſt their whole length in the ground, leaving only two, or at moſt but three buds out of the ground; and, when theſe have made one year's ſhoot, they ſhould be all cut off except one of the ſtrongeſt and beſt ſituated, which muſt be trained up to a ſtem, and treated in the ſame way as timber-trees. If theſe are planted with ſuch deſign, the rows ſhould be ſix feet aſunder, and the ſets four feet diſtance in the rows; by planting them ſo cloſe, they will naturally draw each other upward, and, when they are grown ſo large as to cover the ground and meet, they ſhould be gradually thinned, ſo as at the laſt to leave every other row, and the plants in the rows about eight feet aſunder. If they are ſo treated, the trees will grow to a large ſize, and riſe with upright ſtems to the height of forty feet or more.

When theſe cuttings are planted, it is uſual to ſharpen thoſe ends to a point, which are put into the ground, for the better thruſting of them in; but the beſt way is to cut them horizontally juſt below the bud or eye, and to make holes with an iron inſtrument in the ground where each cutting is to be planted, and, when they are put in, the ground ſhould be preſſed cloſe about the cuttings with the heel to ſettle it, and prevent the air from penetrating to the cuttings.

The after-care muſt be to keep them clear from weeds the two firſt ſeaſons, by which time they will have acquired ſo much ſtrength, as to over power and keep down the weeds; they will alſo require ſome trimming in winter to take off any lateral ſhoots, which, if ſuffered to grow, would retard their upright progreſs.

There are great tracts of land in England fit for this purpoſe, which at preſent produce little to the owners, and might, by the planting of theſe trees, turn to as good account as the beſt Corn land. The larger wood if found, is commonly ſold to the turners for many kinds of light ware, but may be applied to many other purpoſes.

The Sallows are commonly planted in cuttings about three feet long, made from ſtrong ſhoots of the former year; theſe are thruſt down two feet deep into the ground. The cuttings ſhould be placed about three feet row from row, and eighteen inches aſunder in the rows, obſerving always to plant the rows the ſloping way of the ground (eſpecially if the tides overflow the place;) becauſe, if the rows are placed the contrary way, all the filth and weeds will be detained by the ſets, which will choak them up.

The beſt ſeaſon for planting theſe cuttings in the Oſier-grounds is in February, for if they are planted ſooner, they are apt to peel, if it proves hard froſt, which greatly injures them. Theſe plants are always cut every year, and, if the ſoil be good, they will produce a great crop, ſo that the yearly produce of one acre has been often ſold for fifteen pounds, but ten pounds is a common price, which is much better than Corn land; ſo that it is great pity theſe plants are not more cultivated, eſpecially upon moiſt ſoils, upon which few other things will thrive.

SALSOLA. Lin. Gen. Plant. 275. Glaſſwort.

The CHARACTERS are,

The empalement of the flower is permanent, compoſed of five oval obtuſe leaves; the flower has no petals, but hath five ſhort ſtamina, which are inſerted in the diviſions of the empalement; it hath a globular germen, with a ſhort two-pointed ſtyle, crowned by recurved ſtigmas. The germen afterward becomes a globular capſule, with one cell wrapped up in the empalement, incloſing one large ſeed.

The

The SPECIES are,

1. SALSOLA (*Kali*) herbacea, decumbens foliis subulatis spinosis, calycibus ovatis axillaribus. Lin. Sp. Pl. 222. *Herbaceous Salsola with awl-shaped sharp-pointed leaves, and oval empalements proceeding from the sides of the stalks.*

2. SALSOLA (*Tragus*) herbacea, erecta foliis subulatis spinosis lævibus calycibus ovatis. Lin. Sp. *Herbaceous Salsola with linear acute-pointed leaves, and obtuse empalements proceeding from the sides of the stalks.*

3. SALSOLA (*Soda*) herbacea, foliis inermibus. Guett. Stamp. 426. *Herbaceous Salsola with smooth leaves.*

4. SALSOLA (*Vermiculata*) frutescens, foliis ovatis acutis carnosis. Lin. Sp. Plant. 223. *Shrubby Salsola with oval, fleshy, acute-pointed leaves.*

5. SALSOLA (*Rosacea*) herbacea, foliis subulatis mucronatis, calycibus explanatis. Lin. Sp. Plant. 222. *Herbaceous Salsola with pointed awl-shaped leaves, and spreading empalements.*

The first sort grows naturally in the salt marshes in divers parts of England. It is an annual plant, which rises about five or six inches high, sending out many side branches, which spread on every side, garnished with short awl-shaped leaves, which are fleshy, and terminate in acute spines. The flowers are produced from the side of the branches, to which they sit close, and are encompassed by short prickly leaves; they are small, of an herbaceous colour. The seeds are wrapped up in the empalement of the flower, which ripen in autumn, soon after which the plants decay.

The second sort grows naturally on the sandy shores of the south of France, Spain, and Italy. This is also an annual plant, which sends out many diffused stalks, garnished with linear leaves an inch long, ending with sharp spines. The flowers come out from the side of the stalks in the same manner as those of the former; their empalements are blunt, and not so closely encompassed with leaves as those of the other.

The third sort rises with herbaceous stalks near three feet high, spreading wide. The leaves on the principal stalk, and those on the lower part of the branches, are long, slender, and have no spines; those on the upper part of the stalk and branches are slender, short, and crooked. At the base of the leaves are produced the flowers, which are small, and hardly perceptible; the empalement of the flower afterward encompasses the capsule, which contains one cochleated seed.

The fourth sort grows naturally in Spain. This hath shrubby perennial stalks, which rise three or four feet high, sending out many side branches, garnished with fleshy, oval, acute-pointed leaves, coming out in clusters from the side of the branches; they are hoary, and have no stiff prickles. The flowers are produced from between the leaves toward the ends of the branches; they are so small as scarce to be discerned, unless they are closely viewed. The seeds are like those of the other kinds.

The fifth sort grows naturally in Tartary. This is an annual plant, whose stalks are herbaceous, and seldom rise more than five or six inches high. The leaves are awl-shaped, ending in acute points; the empalements of the flowers spread open; the flowers are small, and of a Rose colour, but soon fade: the seeds are like those of the other sorts.

All the sorts of Glasswort are sometimes promiscuously used for making the Sal Alkali, but it is the third sort which is esteemed best for this purpose. The manner of making it is as follows: Having dug a trench near the sea, they lay laths across it, on which they lay the herb in heaps, and, having made a fire below, the liquor, which runs out of the herbs, drops to the bottom, which at length thickening becomes Sal Alkali, which is partly of a black, and partly of an Ash-colour, very sharp and corrosive, and of a saltish taste. This, when thoroughly hardened, becomes like a stone, and is there called Soude or Sode. It is transported from thence to other countries for making of glass.

SALVIA. Tourn. Inst. R. H. 180. tab. 83. Sage.

The CHARACTERS are,

The empalement of the flower is tubulous, of one leaf, large at the mouth, where it is cut into four parts. The flower is of the lip kind, of one petal; the lower part is tubulous, the upper is large and compressed; the lower lip is broad and trifid. It has two short stamina, which stand transverse to the lip, and are fixed in the middle to the tube, to whose tops are fixed glands, upon the upper side of which sit the summits; it has a four-pointed germen, supporting a long slender style, situated between the stamina, crowned by a bifid stigma. The germen afterward becomes four roundish seeds, which ripen in the empalement.

The SPECIES are,

1. SALVIA (*Officinalis*) foliis lanceolatis-ovatis integris crenulatis, floribus verticillato-spicatis. *Sage with spear-shaped, oval, entire leaves, which are slightly crenated on their edges, and flowers growing in whorled spikes.*

2. SALVIA (*Tomentosa*) foliis infimis cordatis, summis oblongo-ovatis serratis tomentosis, floribus verticillato-spicatis. *Sage with heart-shaped lower leaves, the upper of which are oblong, oval, sawed, and woolly, and flowers growing in whorled spikes.*

3. SALVIA (*Auriculata*) foliis lanceolatis sæpius articulatis subtus tomentosis, floribus spicato-verticillatis, calycibus ventricosis. *Sage with spear-shaped leaves, which are frequently eared, and woolly on their under side, flowers growing in whorled spikes, and bellied empalements; commonly called Sage of Virtue.*

4. SALVIA (*Hispanica*) foliis lineari-lanceolatis integerrimis tomentosis, floribus spicatis calycibus brevissimis ventricosis acutis. *Sage with linear, spear shaped, woolly, entire leaves, spiked flowers, and the shortest bellied empalements, ending in acute points.*

5. SALVIA (*Fruticosa*) foliis infimis pinnatis, summis ternatis rugosis, floribus spicatis, caule fruticoso tomentoso. *Sage with winged lower leaves, the upper ones trifoliate and rough, flowers growing in spikes, and a shrubby woolly stalk.*

6. SALVIA (*Pomifera*) foliis lanceolato-ovatis integris crenulatis, floribus spicatis, calycibus obtusis. Hort. Cliff. 12. *Sage with spear-shaped, oval, entire leaves, which are slightly crenated, spiked flowers, and blunt empalements.*

7. SALVIA (*Pinnata*) foliis compositis pinnatis. Hort. Cliff. 13. *Sage with compound winged leaves.*

8. SALVIA (*Orientalis*) foliis infimis pinnatis, summis simplicibus crenatis, floribus verticillatis caulibus procumbentibus hirsutissimis. *Sage with winged lower leaves, the upper ones single and crenated, flowers growing in whorls, and the most hairy trailing stalks.*

9. SALVIA (*Dominica*) foliis cordatis obtusis crenatis subtomentosis, corollis calyce angustioribus. Lin. Sp. Pl. 25. *Sage with heart-shaped, blunt, crenated leaves, which are somewhat woolly, and the petals narrower than the empalement.*

10. SALVIA (*Aurea*) foliis subrotundis integerrimis, basi truncatis dentatis. Hort. Cliff. 13. *Sage with roundish entire leaves, which are torn, and indented at their base.*

11. SALVIA (*Africana*) foliis subrotundis serratis, basi truncatis dentatis. Hort. Cliff. 13. *Sage with roundish sawed leaves, which are torn, and indented at their base.*

12. SALVIA (*Integerrima*) foliis oblongo-ovatis integerrimis, calycibus patulis coloratis. Tab. 225. fig. 2. *Sage with oblong, oval, entire leaves, and spreading coloured empalements.*

The first fort is the common large Sage, which is cultivated in gardens, of which there are the following varieties: 1. The common green Sage. 2. The Wormwood Sage. 3. The green Sage with a variegated leaf. 4. The red Sage. 5. The red Sage with a variegated leaf; thefe are accidental variations, and therefore are not enumerated as fpecies. The common Sage grows naturally in the fouthern parts of Europe, but is here cultivated in gardens for ufe; but that variety, with red or blackifh leaves, is the moft common in the Englifh gardens, and the Wormwood Sage is in greater plenty here than the common green-leaved Sage, which is but in few gardens. The common Sage is fo well known, as to require no defcription.

The fecond fort is generally titled Balfamic Sage by the gardeners. The ftalks of this do not grow fo upright as thofe of the common Sage; they are very hairy, and divide into feveral branches, which are garnifhed with broad, heart-fhaped, woolly leaves, ftanding upon long foot-ftalks; they are fawed on their edges, and their upper furfaces are rough; the leaves, which are upon the flower-ftalks, are oblong and oval, ftanding upon fhorter foot-ftalks, and are very flightly fawed on their edges; the flowers grow in whorled fpikes toward the top of the branches; the whorls are pretty far diftant, and but few flowers in each; they are of a pale blue, about the fize of thofe of the common fort. This Sage is preferred to all the others for making tea.

The third fort is the common Sage of Virtue, which is alfo well known in the gardens and markets. The leaves of this are narrower than thofe of the common fort; they are hoary, and fome of them are indented on their edges toward the bafe, which indentures have the appearance of ears. The fpikes of flowers are longer than thofe of the two former forts, and the whorls are generally naked, having no leaves between them. The flowers are fmaller, and of a deeper blue than thofe of the common red Sage.

The fourth fort grows naturally in Spain. The leaves of this are very narrow and entire, ftanding in clufters on the fide of the ftalks; they are very hoary, and the branches are covered with a hoary down; the leaves on the upper part of the ftalk are narrower than thofe of Rofemary; the flowers grow in clofer fpikes than either of the former, and are of a light blue colour.

The fifth fort grows naturally about Smyrna, from whence the late Dr. William Sherard fent the feeds. This rifes with a fhrubby ftalk four or five feet high, and divides into feveral branches, which grow erect. The leaves on the lower branches are winged, being compofed of two or three pair of fmall lobes, terminated by one large one. Thofe which grow on the flowering branches are trifoliate, the two inner lobes being fmall, and the outer one is large, ending in a point; they have the flavour of Wormwood, and their upper furface is rough. The flowers grow in long fpikes at the end of the branches; the whorls are pretty clofe to each other, and have no leaves between them; the flowers are large, of a flefh colour.

The fixth fort grows naturally in Crete. This hath a fhrubby ftalk, which rifes four or five feet high, dividing into feveral branches, garnifhed with fpear-fhaped, oval, woolly leaves, which are entire, and flightly crenated on their edges. The flowers grow in fpikes at the end of the branches; they are of a pale blue colour, and have obtufe empalements. The branches of this Sage have often punctures made in them by infects, at which places grow large protuberances as big as Apples, in the fame manner as the galls upon the Oak, and the rough balls on the Briar.

The feventh fort grows naturally in the Levant. This is an annual, or at moft a biennial plant with trailing ftalks. The leaves on the lower part of the ftalks are compofed of two or three fmall pair of lobes, terminated by one large one; thofe farther up are trifoliate, the outer lobe being four times the fize of the fide ones. The flowers grow in whorls round the ftalks; they are large, and of a deep blue colour, as are alfo their empalements.

The eighth fort grows naturally about Smyrna, where the late Dr. Sherard gathered the feeds. This is a perennial plant with trailing ftalks, which grow near two feet long, garnifhed toward the bottom with leaves, compofed of two pair of fmall lobes, terminated by a large one, but thofe toward the top are fingle, and ftand oppofite. The flowers are produced in whorls round the ftalks; they are large, and of a flefh colour, but are not fucceeded by feeds here.

The ninth fort grows naturally at Mexico. This is an annual plant, which rifes with an erect, four-cornered, branching ftalk three feet high, garnifhed with large heart-fhaped leaves, of a bright green colour, which are obtufely crenated on their edges, having feveral veins on their lower fide, which diverge from the midrib to the fides. Their foot-ftalks are long and flender; the flowers are produced in clofe fpikes at the end of the branches; they are of a fine blue colour, and their tubes are narrower than the empalement.

The tenth fort grows naturally at the Cape of Good Hope. This rifes with a fhrubby ftalk feven or eight feet high, covered with a light-coloured bark, fending out branches the whole length, which grow almoft horizontally; they are garnifhed with roundifh gray leaves, which are entire, and feem torn at their bafe, where they are alfo indented. The flowers are produced in thick fhort fpikes at the end of the branches; they are very large, and of a dark gold colour.

The eleventh fort grows naturally at the Cape of Good Hope. This rifes with a fhrubby ftalk four or five feet high, dividing into branches, garnifhed with oval fawed leaves, of a gray colour, which have one or two indentures at their bafe that feem torn. The flowers come out in whorls toward the end of the branches; they are of a fine blue colour, and larger than thofe of the common Sage; thefe appear in fucceffion moft of the fummer months, and thofe which come early are often fucceeded by feeds, which ripen in autumn.

The twelfth fort has been lately raifed in the Dutch gardens from feeds, which were brought from the Cape of Good Hope. It has great refemblance to the former, but the branches are ftronger and grow more erect; the leaves are longer, and not fo broad; their edges are not fawed; the flowers grow in long loofe fpikes at the end of the branches; they are larger, and of a paler blue than the other; their empalements are broader, fpread wider, and are of a pale blue colour, in which confifts their difference.

All the forts of Sage may be propagated by feeds, if they can be procured; but, as fome of them do not perfect their feeds in England, and moft of the forts, but efpecially the common kinds for ufe, are eafily propagated by flips, it is not worth while to raife them from feeds. The flips fhould be planted the beginning of April on a fhady border, where, if they are now and then refrefhed with water, if the feafon fhould prove dry, they will foon take root. When the flips have made good roots, they may be taken up with balls of earth to their roots, and tranfplanted where they are to remain, which fhould always be upon a dry foil, and where they may have the benefit of the fun; for if they are planted on a moift foil, or in a fhady fituation, they are very fubject to be deftroyed in winter; nor will thefe plants endure the cold fo well, when planted upon a rich foil, as thofe which have a barren, dry, rocky foil, which is the cafe of moft of the verticillate plants, for thefe will often grow upon walls, where, although they are more

expofed

exposed to the cold than those plants in the ground, they are always found to remain in severe winters when the others are destroyed. The side shoots and tops of these plants may be gathered in the summer, and dried, if designed for tea, otherwise they are best taken green from the plants for most other uses. The roots of the common sorts of Sage will last several years, if they are in a dry warm soil, but, where they are often cropped for use, the plants will become ragged, so there should be a succession of young ones raised every other year.

The fifth, sixth, and eighth sorts, are somewhat tender, so will not live through the winter in the open air in England; therefore these must be planted in pots, and in winter must be removed under a hot-bed frame, that they may have a great share of fresh air whenever the season is mild, for if they are too much drawn, they seldom flower well, and make but an indifferent appearance. In summer they must be exposed amongst other exotic plants in some well-sheltered situation, for they are pretty hardy, and only require to be sheltered from the frost. These plants must be often refreshed with water in warm weather, otherwise they will shrivel and decay; and they should be new-potted at least twice every summer, because their roots will greatly increase, which, if confined in the pots too long, will turn mouldy and decay.

The seventh and ninth sorts are annual plants, so are only propagated by seeds; these may be sown upon a bed of light earth in the places where they are to remain. The seeds of the seventh sort should be sown in autumn, and then the plants will come up the following spring, but, if they are kept out of the ground till spring, the plants will not come up till the next year. Those of the ninth sort may be sown the beginning of April upon a warm border, where the plants will appear in May, and require no other care but to thin them where they grow too close, and keep them clean from weeds, and, if they should grow tall, they must be supported, otherwise the strong winds will break them down; but the seventh sort spreads its branches upon the ground, so will require no support, therefore this only requires to have room, and to be kept clean from weeds.

The tenth, eleventh, and twelfth sorts are natives of a warmer country, so these require protection in winter; they are easily propagated by cuttings in the spring and summer months. If these are planted early in the spring, it will be the better way to plant them in pots, which should be plunged into a very moderate hot-bed; and, if they are shaded from the sun in the heat of the day, and gently refreshed with water, they will put out good roots in about two months, when they should be inured gradually to the open air, into which they should be removed soon after. The cuttings which are raised early in the season, will become strong plants before winter, so will be in a better condition to resist the cold than those which are weak.

If the cuttings are planted in summer, they will require no artificial heat, so that if these are planted on a bed of fresh loamy earth, and covered close down with a bell or hand-glass, and shaded from the sun in the heat of the day, giving them now and then a little water, they will take root freely; and when they begin to shoot, they should have free air admitted to them by raising the glass on one side, and so gradually exposed to the open air. When the cuttings are well rooted, they should be each transplanted into a separate small pot, and placed in a shady situation till they have taken new root; then they may be removed to a sheltered situation, where they may remain till the approach of frost, when they must be carried into shelter, and in winter treated in the same manner as other hardy greenhouse plants, which only require protection from frost, observing not to over-water them during the cold weather, but in summer, when they are in the open air, they will require it often.

SALVIA AGRESTIS. See TEUCRIUM.

SAMBUCUS. Tourn. Inst. R. H. 606. tab. 376. The Elder-tree.

The CHARACTERS are,

The flower has a small permanent empalement of one leaf, cut into five parts; it has one concave wheel-shaped petal cut into five obtuse segments at the brim, which are reflexed, and five awl-shaped stamina the length of the petal, terminated by roundish summits, with an oval germen situated under the flower, having no style, in room of which is a swelling gland, crowned by three obtuse stigmas. The germen afterward becomes a roundish berry with one cell, including three angular seeds.

The SPECIES are,

1. SAMBUCUS (*Nigra*) caule arboreo ramoso, floribus umbellatis. Flor. Leyd. Prod. 243. *Elder with a branching tree-like stalk, and flowers growing in umbels; or common Elder with black berries.*

2. SAMBUCUS (*Laciniata*) foliis pinnatifidis, floribus umbellatis, caule fruticoso ramoso. *Elder with wing pointed leaves, flowers growing in umbels, and a shrubby branching stalk; commonly called Parsley-leaved Elder.*

3. SAMBUCUS (*Racemosa*) racemis compositis ovatis, caule arboreo. Lin. Sp. Plant. 270. *Elder with oval compound bunches of flowers, and a tree-like stalk; or red-berried Mountain Elder.*

4. SAMBUCUS (*Ebulus*) caule herbaceo ramoso, foliolis dentatis. Tab. 226. *Elder with a branching herbaceous stalk, and the small leaves indented; or dwarf Elder.*

5. SAMBUCUS (*Humilis*) caule herbaceo ramoso, foliolis lineari-lanceolatis acutè dentatis. *Elder with an herbaceous branching stalk, and linear spear-shaped lobes, which are sharply indented; dwarf Elder with a cut leaf.*

6. SAMBUCUS (*Canadensis*) cymis quinquepartitis, foliis subpinnatis. Lin. Sp. Plant. 269. *American Elder with leaves almost winged.*

The first sort here mentioned is the common Elder, which is so well known as to need no description. Of this there are the following varieties, viz. The white and green-berried Elder, and the variegated leaved Elder. The latter is undoubtedly a variety, but I much doubt if the white is not a distinct species, for the lobes of the leaves are much less, and very slightly sawed on their edges, whereas those of the common sort are deeply sawed; they are also smoother, and of a lighter green; the plants which have been raised from the berries have not altered, so there is great reason for supposing them different species; but as I have made but one trial of this, I am unwilling to determine upon a single experiment, but shall leave it as a doubt till further trial is made.

The second sort is generally titled Parsley leaved Elder by the gardeners; this is by some supposed to be only a variety of the first, but there can be little reason for doubting of its being a distinct species. The lobes of these leaves are narrower than those of the first, and are cut into several fine segments; these are again deeply indented on their edges regularly, in form of winged leaves. The stalks of this are much smaller than those of the first, and the shoots are short; the leaves have not so strong an odour, and their berries are a little smaller.

The third sort grows naturally upon the mountains in Germany and Italy This sends up many shrubby stalks from the root, which rise ten or twelve feet high, which divide into many branches, covered with a brown bark; the leaves come out opposite; those on the lower part of the branches are composed generally of two pair of lobes, terminated by an odd one; these are shorter and broader than

than thoſe of the common Elder, and are deeply ſawed on their edges; the leaves on the upper part of the branches have frequently but three lobes; they are of a pale green colour, and pretty ſmooth. The flowers come out at the end of the ſhoots in oval bunches, which are compoſed of ſeveral ſmaller; they are of an herbaceous white colour, and are ſucceeded by berries, which are red when ripe.

The fourth ſort grows naturally in many of the midland counties in England, where it is frequently a troubleſome weed in the fields; this is called dwarf Elder, Danewort, and Walwort. It hath creeping roots, which ſpread far in the ground, ſo propagates very faſt wherever the plant once gets poſſeſſion; the ſtalks are herbaceous, and riſe from three to five feet high, in proportion to the goodneſs of the ground, and ſend out a few ſide branches toward the top, garniſhed with winged leaves, compoſed of ſix or ſeven pair of narrow lobes, terminated by an odd one, of a deep green, a little indented on their edges. The flowers grow in umbels at the top of the ſtalks; they are of the ſame form with thoſe of the common Elder, but ſmaller, and are ſpotted with red. Theſe are ſucceeded by black berries like thoſe of the common Elder, but ſmaller.

This plant is frequently uſed in medicine; it purges ſerous watery humours by ſtool, and is therefore much recommended for the dropſy, in which diſorder I have known the juice of this plant perform wonders in a ſhort time; it was adminiſtered three times a week, two ſpoonfuls was the doſe given at each time. It is alſo accounted a good medicine for the gout, and ſcorbutic diſorders. The young ſhoots of the common Elder are frequently ſold for this in the markets, from which it may be eaſily diſtinguiſhed, by the number and ſhape of the lobes on each leaf: the common Elder has ſeldom more than five lobes, which are broader and much ſhorter than thoſe of the dwarf Elder, and are pretty deeply ſawed on their edges; but the leaves of the dwarf Elder have nine, eleven, or thirteen lobes to each leaf, which are long, narrow, and very ſlightly indented on their edges.

The roots of the fifth ſort do not creep ſo much in the ground as thoſe of the fourth; the ſtalks are herbaceous, but do not riſe ſo high, and are cloſer garniſhed with leaves, which have ſeldom more than ſeven lobes to each, and toward the top of the ſtalks but five; theſe are long, and narrower than thoſe of the former, and are deeply cut on their edges, ending with winged acute points. The flowers are produced in umbels at the top of the ſtalks, which are ſhaped like thoſe of the former, and are ſucceeded by the like berries.

The ſixth ſort grows naturally in Canada, and ſeveral other parts of North America, where it grows as large as our common Elder. This ſort, when young, and the ſhoots are full of ſap, is tender, ſo that the froſt often kills them almoſt to the ground; but when the plants become woody, they are rarely hurt. The leaves of this are narrower than thoſe of the common Elder, and are compoſed of many more pinnæ; the berries are alſo ſmaller, in which the difference chiefly conſiſts.

The three firſt and ſixth ſorts may be eaſily propagated from cuttings, or by ſowing their ſeeds; but the former being the moſt expeditious method, is generally practiſed. The ſeaſon for planting of their cuttings is any time from September to March; in the doing of which, there needs no more care than to thruſt the cuttings about ſix or eight inches into the ground, and they will take root faſt enough, and may afterwards be tranſplanted where they are to remain, which may be upon almoſt any ſoil or ſituation; they are extreme hardy, and if their ſeeds are permitted to fall upon the ground, they will produce plenty of plants the ſucceeding ſummer.

The firſt ſort is often planted for making fences becauſe of its quick growth; but as the bottom becomes naked in a few years, it is not ſo proper for that purpoſe; neither would I recommend it to be planted near habitations, becauſe at the ſeaſon when it is in flower, it emits ſuch a ſtrong ſcent, as will occaſion violent pains in the heads of thoſe who abide long near them; beſides, the crude parts which are continually perſpired through the leaves, are accounted unwholeſome, though the leaves, bark, and other parts, are greatly eſteemed for many uſes in medicine.

The fourth ſort propagates itſelf faſt enough wherever it is once planted, by its creeping roots, ſo that it is very difficult to keep it within bounds, therefore is not a proper plant for gardens; but thoſe who are inclined to keep it for medicinal uſe, need only plant one or two of the roots in any abject part of a garden or field, and the place will ſoon be ſpread over with it.

The fifth ſort is preſerved in botanic gardens for the ſake of variety, but is ſeldom admittted into other gardens. This propagates by the root, though not ſo faſt as the other.

The common Elder will grow upon any ſoil, or in any ſituation; the trees are frequently ſeen growing on the top, and out of the ſide of old walls; and they are often ſeen growing cloſe to ditches, and in very moiſt places, ſo that wherever the ſeeds are ſcattered, the plants will come up, as they often do from the hollow of another tree. The leaves and ſtalks of this plant are ſo bitter and nauſeous, that few animals will browſe upon it. I have often ſeen the trees growing in parks, where there has been variety of animals, and have obſerved they were untouched, when almoſt all the other trees within reach have been cropped by the cattle.

The young ſhoots of this tree are ſtrong and very full of pith, but as the trees grow old, their wood becomes very hard, and will poliſh almoſt as well as that of the Box-tree, ſo is often uſed for the ſame purpoſes when Box-wood is ſcarce.

The bark, leaves, flowers, and berries of this tree, are uſed in medicine. The inner bark is eſteemed good for dropſies; the leaves are outwardly uſed for the piles and inflammations. The flowers are inwardly uſed to expel wind, and the berries are eſteemed cordial and uſeful in hyſteric diſorders, and are frequently put into gargariſms for ſore mouths and throats.

SAMOLUS. Tourn. Inſt. R. H. 143. tab. 60. Round-leaved Water Pimpernel.

The CHARACTERS are,

The empalement of the flower is permanent, erect, and cut into five ſegments. It has one petal, with a ſhort ſpreading tube; the brim is plain, obtuſe, and cut into five parts. It has five ſhort ſtamina placed between each ſegment of the petal, terminated by ſummits which join together. The germen is ſituated under the flower, ſupporting a ſlender ſtyle, crowned by a headed ſtigma. The germen afterward becomes an oval capſule with one cell, cut half through into five valves, filled with ſmall oval ſeeds.

We have but one SPECIES of this genus, viz.

SAMOLUS (*Valerandi*) valerandi. J. B. *Round-leaved Water Pimpernel.*

This plant grows wild in ſwampy places, where the water uſually ſtands in winter, and is ſeldom preſerved in gardens. It is an annual plant, which flowers in June, and the ſeeds are ripe in Auguſt; at which time, whoever hath a mind to cultivate this plant, ſhould ſow the ſeeds on a moiſt ſoil, where the plants will come up, and require no farther care but to keep them clear from weeds.

SAMYDA. Lin. Gen. Plant. 525.

The

The CHARACTERS are,

The flower has a rough bell-shaped empalement of one leaf, which is cut at the brim into five points. It has no petal, but has fifteen short awl-shaped stamina inserted in the empalement, terminated by oval summits, and a hairy globular germen, supporting a cylindrical style, crowned by a headed stigma. The germen afterward becomes an oval berry with four furrows, having four cells, including many kidney-shaped seeds, immersed in the oval receptacle.

The SPECIES are,

1. SAMYDA (*Serrulata*) floribus dodecandris foliis ovato-oblongis serrulatis axillaribus. *Samyda with oval sawed leaves, and flowers growing from the wings of the stalks.*

2. SAMYDA (*Parviflora*) floribus decandris foliis ovato-oblongis utrinque glabris. Lin. Sp. 557 *Samyda with compound winged leaves.*

These plants grow naturally in the West-Indies; the first sort rises with a shrubby stalk five or six feet high, sending out several weak branches, garnished with oval leaves drawing to a point, sawed on their edges, of a light green colour. The flowers come out from the wings of the leaves upon short foot-stalks; they have a five-leaved empalement, which is of a bright red within; the stamina, which are about fifteen in number, are inserted in the middle of the empalement, and stand erect; in the center is situated an oval germen, which turns to a berry with four cells, containing small seeds.

The other sort has leaves shaped like those of the Walnut tree, but are smaller, and the inside of the empalement is of a purple colour, in which it differs from the first.

These plants are propagated by seeds, which must be procured from the countries where they naturally grow; these must be sown upon a hot-bed in the spring, and when the plants come up, they must be planted in small pots, and plunged into a hot-bed of tanners bark, and treated in the same way as other tender plants from the same countries. They must be always kept in the bark-bed in the stove, otherwise they will make but little progress in England.

SANGUINARIA. Dill. Hort. Elth. 252. Puccoon.

The CHARACTERS are,

The empalement of the flower is composed of two oval concave leaves, which fall away. It hath eight oblong, obtuse, spreading petals, which are alternately narrow. It has many single stamina, which are shorter than the petals, terminated by single summits, and an oblong compressed germen, having no style, crowned by a permanent thick stigma, with two channels. The germen becomes an oblong bellied capsule with two valves, pointed at both ends, inclosing round acute-pointed seeds.

We have but one SPECIES of this plant, viz.

SANGUINARIA (*Canadensis.*) Hort. Cliff. 202. *Puccoon.*

There are some other varieties of this plant mentioned in the Eltham garden, but they are not distinct species, for they vary annually, therefore it is to no purpose to mention their variations.

It is a native of most of the northern parts of America, where it grows plentifully in the woods; and in the spring, before the leaves of the trees come out, the surface of the ground is, in many places, covered with the flowers, which have some resemblance to our Wood Anemone, but they have short naked pedicles, each supporting one flower at the top. Some of these flowers will have ten or twelve petals, so that they appear to have a double range of leaves, which has occasioned their being termed double flowers; but this is only accidental, the same roots in different years producing different flowers. The roots of this plant are tuberous, and the whole plant has a yellow juice, which the Indians use to paint themselves with.

This plant is hardy enough to live in the open air in England, but it should be planted in a loose soil and a sheltered situation, not too much exposed to the sun. It is propagated by the roots, which may be taken up and parted every other year; the best time for doing of this is in September, that the roots may have time to send out fibres before the hard frost sets in. The flowers of this plant appear in April, and when they decay, the green leaves come out, which will continue till Midsummer; then they decay, and the roots remain unactive till the following autumn; so that unless the roots are marked, it will be pretty difficult to find them, after their leaves decay, for they are of a dirty brown colour on the outside, so are not easily distinguished from the earth.

SANGUIS DRACONIS. See PALMA.

SANGUISORBA. Lin. Gen. Plant. 136. Burnet.

The CHARACTERS are,

The empalement of the flower is composed of two short leaves placed opposite, which fall away. The flower hath one plain petal, cut into four obtuse segments, which join at their base. It has four stamina the length of the petal, terminated by small roundish summits, and a four-cornered germen, situated between the empalement and petal, supporting a short slender style, crowned by an obtuse stigma. The germen afterward turns to a small capsule with two cells, filled with small seeds.

The SPECIES are,

1. SANGUISORBA (*Officinalis*) spicis ovatis Hort. Cliff. 39. *Sanguisorba with oval spikes; or greater Burnet.*

2. SANGUISORBA (*Sabauda*) spicis cylindricis, foliolis cordato-oblongis, rigidis, serratis. *Sanguisorba with cylindrical spikes, the lobes of the leaves oblong, heart-shaped, stiff, and sawed.*

3. SANGUISORBA (*Hispanica*) spicis orbiculatis compactis. *Sanguisorba with round compact spikes.*

4. SANGUISORBA (*Canadensis*) spicis longissimis. Hort. Cliff. 39. *Sanguisorba with the longest spikes; or greatest Canada Burnet.*

The first sort grows naturally in moist meadows in divers parts of England. The stalks of this rise from two to three feet high, branching toward the top, and are terminated by thick oval spikes of flowers, of a grayish brown colour, which are divided into four segments almost to the bottom. These are succeeded by four oblong cornered seeds. The leaves of this sort are composed of five or six pair of lobes placed along a midrib, terminated by an odd one; they are thin, sawed on their edges, and a little downy on their under sides.

The second sort grows naturally in Piedmont. This rises with stiff upright stalks more than three feet high, branching out toward the top, each branch being terminated by a cylindrical spike of brown flowers, shaped like those of the former sort, but smaller. The leaves are long; their foot-stalks are very strong, and much longer than those of the first sort; the leaves have seven or eight pair of stiff lobes, terminated by an odd one; these are heart shaped, deeply sawed on their edges, of a lucid green on their upper side, but pale on their under, having pretty long foot-stalks, at the base of which come out two small roundish leaves or ears, which are deeply indented. This retains its difference when propagated by seeds, so is undoubtedly a distinct species.

The leaves of the third sort are smaller than those of the first, having but four pair of lobes to each, terminated by an odd one; they are bluntly sawed on their edges, and have very short foot-stalks, of a pale green on their upper side, and hoary on their under. The stalks rise about two feet high, branching pretty much toward their top, and are terminated by round heads or spikes of reddish flowers, which are succeeded by seeds in autumn. It grows naturally in Spain.

The

The fourth ſort grows naturally in North America. This hath leaves like thoſe of the firſt ſort, but are a little ſtiffer, compoſed of four or five pair of lobes, terminated by an odd one; thoſe on the lower part of the midrib ſtand alternate, but the two upper pair are oppoſite, of a light green colour, and deeply ſawed on their edges. The ſtalks riſe three feet high, dividing toward the top into ſmall branches, which ſtand erect, and are terminated by long ſpikes of flowers, of an herbaceous white colour, each ſtanding upon a ſhort foot-ſtalk.

There is another with long ſpikes of red flowers, which grows naturally in the ſame countries, whoſe ſtalks riſe higher; the ſpikes of flowers are thicker; the lobes of the leaves are broader, and are whiter on their under ſide; but whether this is a diſtinct ſpecies, or an accidental variety of the fourth, I cannot as yet determine.

All theſe ſorts are very hardy perennial plants, and will thrive in almoſt any ſoil or ſituation. They may be propagated either by ſeeds or parting of the roots; if they are propagated by ſeeds, they ſhould be ſown in the autumn, for when they are ſown in the ſpring, they ſeldom grow the ſame year. When the plants come up, they muſt be kept clean from weeds till they are ſtrong enough to tranſplant, when they may be planted in a ſhady border at about ſix inches diſtance each way, obſerving to water them till they have taken new root; after which they will require no other care but to keep them clear from weeds till autumn, when they may be tranſplanted to the place where they are to remain; the following ſummer they will produce flowers and ſeeds, but their roots will abide many years.

If the roots are parted, it ſhould be done in autumn, that they may get good root before the dry weather comes on in the ſpring.

The other ſorts of Burnet are referred to the article Poterium.

SANICULA. Tourn. Inſt. R. H. 326. tab. 173. Sanicle.

The Characters are,

It is a plant with an umbellated flower. The univerſal umbel hath but few rays; the involucrum is ſituated but half round on the outſide; the partial umbels have many cluſtered rays, and their involucrums ſurround them on every ſide; the flowers have five compreſſed petals, which are bifid, and turn inward; they have five erect ſtamina, which are twice the length of the petals, terminated by roundiſh ſummits, and a briſtly germen, ſituated under the flower, ſupporting two awl-ſhaped ſtyles, which are reflexed. The germen afterward becomes a rough oval-pointed fruit, dividing into two parts, each containing one ſeed.

We have but one Species of this genus, viz.

Sanicula (*Europæa*) officinarum. C. B. P. *Sanicle, or Self-heal.*

This plant is found wild in woods and ſhady places in ſeveral parts of England, but being a medicinal plant, may be propagated in gardens for uſe. It may be increaſed by parting the roots any time from September to March, but it is beſt to do it in autumn, that the plants may be well rooted before the dry weather in ſpring comes on; they ſhould have a moiſt ſoil and a ſhady ſituation, in which they will thrive exceedingly.

SANTOLINA. Tourn. Inſt. R. H. 460. tab. 260. Lavender-cotton.

The Characters are,

It hath a compound flower with a ſcaly hemiſpherical empalement. The flower is uniform, compoſed of many funnel-ſhaped hermaphrodite florets, which are longer than the empalement, cut into five parts at the top, which turn backward; they have five very ſhort hair-like ſtamina, terminated by cylindrical ſummits, and an oblong four-cornered germen, ſupporting a ſlender ſtyle, crowned by two oblong, depreſſed, torn ſtigmas. The germen afterward becomes a ſingle, oblong, four-cornered ſeed, which is either naked or crowned with very ſhort down, ripening in the common empalement.

The Species are,

1. Santolina (*Chamæcypariſus*) pedunculis unifloris, foliis quadrifariàm dentatis. Hort. Cliff. 397. *Lavender-cotton with one flower upon a foot-ſtalk, and leaves indented four ways; or common Lavender-cotton.*

2. Santolina (*Villoſa*) pedunculis unifloris, calycibus globoſis, foliis quadrifariàm dentatis tomentoſis. *Lavender-cotton with one flower upon a foot ſtalk, globular empalements, and woolly leaves, which are indented four ways.*

3. Santolina (*Decumbens*) pedunculis unifloris, caulibus decumbentibus, foliis linearibus quadrifariàm dentatis. *Lavender-cotton with one flower upon a foot-ſtalk, declining ſtalks, and linear leaves, which are four ways indented.*

4. Santolina (*Virens*) pedunculis unifloris, foliis linearibus longiſſimis bifariàm dentatis. *Lavender-cotton with one flower upon a foot-ſtalk, and very long linear leaves, which are two ways indented.*

5. Santolina (*Roſmarinifolia*) pedunculis unifloris, capitulis globoſis, foliis linearibus integerrimis. *Lavender-cotton with one flower upon a foot-ſtalk, globular heads, and linear entire leaves.*

6. Santolina (*Minor*) pedunculis unifloris, foliis linearibus confertis obtuſis. *Lavender-cotton with one flower upon a foot-ſtalk, and linear obtuſe leaves growing in cluſters.*

7. Santolina (*Chamæmelifolia*) pedunculis unifloris, foliis longioribus tomentoſis, duplicato dentatis. *Lavender-cotton with one flower upon a foot-ſtalk, and longer woolly leaves, which are twice indented.*

The firſt ſort is the common Lavender-cotton, which has been long known in the Engliſh gardens; it was formerly titled Abrotanum fœmina, or female Southernwood, and by the corruption of words was called Brotany by the market people; it grows naturally in Spain, Italy, and the warm parts of Europe. This hath a ligneous ſtalk, dividing into many branches, garniſhed with ſlender hoary leaves, that are four ways indented, and have a rank ſtrong odour when handled. The branches are terminated by a ſingle flower, compoſed of many hermaphrodite florets, which are fiſtular, cut into five parts at the top, of a ſulphur colour, and are included in one common ſcaly empalement, having no borders or rays. Theſe are ſucceeded by ſmall, oblong, ſtriated ſeeds, which are ſeparated by ſcaly chaff, and ripen in the empalement; the plants love a dry ſoil and a ſheltered ſituation. The leaves, and ſometimes the flowers, are uſed in medicine, and reputed good to deſtroy worms.

The ſecond ſort has a ſhrubby ſtalk, which branches out like the former, but the plants ſeldom grow ſo tall. The branches are garniſhed very cloſely below with leaves ſhaped like thoſe of the other ſort, but ſhorter, thicker, and whiter; the flowers are much larger, and the brims of the florets are more reflexed; they are of a deeper ſulphur colour than the other. It grows naturally in Spain.

The third ſort is of lower ſtature than either of the former, ſeldom riſing more than fifteen or ſixteen inches high. The branches ſpread horizontally near the ground, and are garniſhed with ſhorter leaves than either of the former, which are hoary, and finely indented; the ſtalks are terminated by ſingle flowers, of a bright yellow colour, which are larger than thoſe of the firſt ſort.

The fourth ſort riſes higher than either of the former. The branches are more diffuſed; they are ſlender, ſmooth, and garniſhed with very narrow long leaves, which are of a deep green colour, but two ways indented; the ſtalks are ſlender, naked toward the top, and terminated by ſingle flowers of a gold colour.

The fifth fort hath fhrubby ftalks which rife about three feet high, fending out long flender branches, garnifhed with fingle linear leaves, of a pale green colour. The ftalks are terminated by large, fingle, globular flowers, of a pale fulphur colour.

The fixth fort is fomewhat like the fifth, but the branches are fhorter, thicker, and clofer garnifhed with leaves, which come out in clufters. The flower-ftalks are fparfedly difpofed, and have leaves to their top; the flowers are fmall, and of a yellow colour.

The feventh fort hath fhrubby ftalks, which rife three feet high, garnifhed with broader leaves than either of the former, whofe indentures are loofer, but double; they are hoary, and when bruifed have an odour like Chamomile. The leaves are placed pretty far afunder, and the ftalks are garnifhed with them to the top. The ftalks are divided likewife at the top into two or three foot-ftalks, each fuftaining one pretty large fulphur-coloured flower

The firft of thefe plants is cultivated in gardens for medicinal ufe, and the fix next are propagated by the gardeners near London for furnifhing balconies, and other little places in and near the city by way of ornament. The feven forts firft mentioned are hardy plants, which will thrive in the open air, provided they are planted in a poor dry foil, for in fuch ground the plants will be ftinted, fo will be hard and better able to refift the cold, and will have a better appearance than thofe which are in rich ground whofe branches will be long and diffufed, fo by hard rains or ftrong winds are difplaced, and fometimes broken down; whereas in poor land they will grow compact, and the plants will continue much longer.

All thefe plants may be cultivated fo as to become ornaments to a garden, particularly in fmall bofquets of evergreen fhrubs, where, if they are artfully intermixed with other plants of the fame growth, and placed in the front line, they will make an agreeable variety, efpecially if care be taken to trim them twice in a fummer to keep them within bounds, otherwife their branches are apt to ftraggle, and in wet weather to be borne down and difplaced, which renders them unfightly; but when they are kept in order, their hoary and different coloured leaves will have a pretty effect in fuch plantations.

They may be propagated by planting flips or cuttings during the fpring, in a border of light frefh earth, but muft be watered and fhaded in hot dry weather until they have taken root; after which they will require no farther care but to keep them clean from weeds till autumn, when they fhould be tranfplanted where they are defigned to remain; but if the ground is not ready by that time to receive them, it will be proper to let them remain in the border until fpring, for if they are tranfplanted late in autumn, they are liable to be deftroyed by cold in winter.

SAPINDUS. Tourn. Inft. R. H. 659. tab. 440. The Sopeberry-tree.

The CHARACTERS are,

The empalement of the flower is compofed of four plain, oval, coloured leaves. The flower has four oval petals, which are lefs than the empalement; it has eight ftamina, which are the length of the petals, terminated by erect fummits, and an oval germen with three or four lobes, fupporting a fhort ftyle, crowned by a fingle ftigma. The germen afterward becomes one, two, or three globular berries, including nuts of the fame form. There is rarely above one of thefe pregnant, the other are abortive.

The SPECIES are,

1. SAPINDUS (*Saponaria*) foliis impari-pinnatis, caule inermi. Lin. Sp. Plant. 526. *Sopeberry-tree with winged leaves.*

2. SAPINDUS (*Rigidus*) foliis quaterno rigidus acutis pinnatis. *Sopeberry-tree with rigid acute-winged leaves.*

The firft fort grows naturally in the iflands of the Weft-Indies, where it rifes with a woody ftalk from twenty to thirty feet high, fending out many branches, garnifhed with winged leaves, compofed of feveral pair of fpear fhaped lobes. The midrib has a membranaceous or leafy border running on each fide from one pair of lobes to the other, which is broadeft in the middle between the lobes; the flowers are produced in loofe fpikes at the end of the branches; they are fmall and white, fo make no great appearance. Thefe are fucceeded by oval berries as large as middling Cherries, fometimes fingle, at others, two, three, or four are joined together; thefe have a faponaceous fkin or cover, which inclofes a very fmooth roundifh nut of the fame form, of a fhining black when ripe. The fkin or pulp which furrounds the nuts is ufed in America to wafh linen, but it is very apt to burn and deftroy it if often ufed, being of a very acrid nature.

The fecond fort grows in India. This hath a ftrong woody ftalk, which rifes about twenty feet high, fending out many ftrong ligneous branches, covered with a fmooth gray bark, and garnifhed with winged leaves, compofed of many fpear-fhaped lobes; they are of a pale green, and fit clofe to the midrib, which has no border or wing like the other. The end of the branches are divided into two or three foot-ftalks, each fuftaining a loofe fpike of flowers like thofe of the other fort; thefe are fucceeded by roundifh berries like thofe of the former.

Thefe plants are propagated by feeds; they muft be put into fmall pots, and plunged into a hot-bed of tanners bark In five or fix weeks the plants will appear, when the glaffes of the hot-bed fhould be raifed every day in warm weather, to admit frefh air to the plants. In three weeks or a month after the plants appear, they will be fit to tranfplant, when they muft be fhaken out of the pots and carefully parted, fo as not to injure their roots, and each planted into a feparate fmall pot, and plunged into the hot bed again, obferving to fhade them from the fun until they have taken new root; after which time they muft have free air admitted to them every day when the weather is warm, and will require to be frequently watered.

After the plants are well rooted they will make great progrefs, efpecially the fecond fort, fo fhould be inured to bear the open air by degrees, for this will live in a greenhoufe in winter, and in fummer may be expofed in the open air; but the firft is not fo hardy, fo muft be treated more tenderly. I have frequently raifed thefe plants from feeds to the height of two feet in one fummer, and the leaves have been a foot and a half in length, fo that they made a fine appearance; but thefe did not furvive the winter, whereas thofe which were expofed to the open air in July, and thereby ftinted in their growth, continued their leaves frefh all the winter. Thefe were placed in a ftove upon fhelves, where the warmth was very moderate, with which thefe plants will thrive better than in a greater heat.

SAPONARIA. Lin. Gen. Plant. 449. Sopewort

The CHARACTERS are,

The flower has a permanent empalement of one leaf, which is cut into five points. It has five petals, whofe tails are narrow, and the length of the empalement; their borders are broad, obtufe, and plain. It has ten awl-fhaped ftamina, which are alternately inferted into the petals, and are terminated by obtufe proftrate fummits, and a taper germen fupporting two erect parallel ftyles, crowned by acute ftigmas. The germen afterward becomes a clofe capfule the length of the empalement; having one cell filled with fmall feeds.

The

The Species are,

1. Saponaria (*Officinalis*) calycibus cylindricis, foliis ovato-lanceolatis. Hort. Cliff. 165. *Sopewort with cylindrical empalements, and oval-spear-shaped leaves; vulgarly called Sopewort.*

2. Saponaria (*Hybrida*) calycibus cylindricis, foliis ovatis nervosis semiamplexicaulibus. *Sopewort with cylindrical empalements, and oval veined leaves half embracing the stalks.*

3. Saponaria (*Vaccaria*) calycibus pyramidatis quinquangularibus, foliis oblongo-ovatis acuminatis. *Sopewort with pyramidal five-cornered empalements, and oblong, oval, acute-pointed leaves.*

4. Saponaria (*Hispanica*) calycibus pyramidatis quinquangularibus, foliis ovato-lanceolatis, semiamplexicaulibus. *Sopewort with pyramidal five-cornered empalements, and oval spear-shaped leaves half embracing the stalks.*

5. Saponaria (*Orientalis*) calycibus cylindricis villosis, caule dichotomo erecto patulo. Hort. Upsal. 106. *Sopewort with cylindrical hairy empalements, and erect spreading stalks, which are divided by pairs.*

The first sort is the common Sopewort of the shops; this grows naturally in many parts of England, and is rarely admitted into gardens; it has a creeping root, so as in a short time to fill a large space of ground, from which arise many purplish stalks about two feet high, which are jointed, and garnished with opposite leaves at each; these are oval, spear-shaped, smooth, and of a pale green. The foot-stalks of the flowers arise from the wings of the leaves opposite; they sustain four, five, or more purple flowers each; these have generally two small leaves placed under them. The stalk is also terminated by a loose bunch of flowers, growing in form of an umbel; they have each a large swelling cylindrical empalement, and five broad obtuse petals, which spread open, of a purple colour. These are succeeded by oval capsules with one cell, filled with small seeds.

The decoction of this plant is used to cleanse and scour woollen cloths: the poor people in some countries use it instead of sope for washing, from whence it had its title.

There is a variety of this with double flowers, which is preserved in gardens, but the roots are very apt to spread, if they are not confined, so these plants should not be placed in borders among better flowers; but as the flowers continue in succession from July to the middle of September, so a few of the plants may be allowed a place in some abject part of the garden, for they will thrive in any situation, and propagate fast enough by their creeping roots.

The second sort was found growing in a wood near Lichbarrow in Northamptonshire, by Mr. Gerard. It has been generally esteemed a Lusus Naturæ, and not a distinct species, but I have never found it alter in forty years; but as it doth not produce seeds, so there is no certainty of its being a distinct species. The roots of this do not spread like those of the first; the stalks are shorter, thicker, and do not grow so erect; they rise a foot high; the joints are very near together and swelling; the leaves are produced singly on the lower part of their stalks, but toward the top they are often placed by pairs; they are oval-shaped, and hollowed like a ladle. The flowers are disposed loosely on the top of the stalk; they have large cylindrical empalements, and are of a purple colour. This plant is preserved for the sake of variety in some gardens; but as there is little beauty in the flowers, it does not merit a place in gardens for pleasure. It is easily propagated by parting of the roots in autumn, and loves a moist shady situation.

The third sort is an annual plant, which grows naturally among Corn in the south of France and Italy. This rises with an upright stalk a foot and a half high, branching out into several divisions by pairs opposite, as are also the leaves; they sit close to the stalks, are smooth, and of a gray colour. The flowers are produced at the end of the branches, each standing upon a long naked foot-stalk; their empalements are large, swelling, and pyramidal, having five acute corners or angles; the petals are but small; they have long necks or tails, which are narrow; the upper part is obtuse, and of a reddish purple colour.

The fourth sort grows naturally in Spain. This is also an annual plant; it rises with a strong smooth stalk about two feet high, garnished with oval spear-shaped leaves, which are fleshy, of a gray colour, and very smooth; these half embrace the stalks with their base; the upper part of the stalk divides into branches, which are again subdivided into long naked foot-stalks, each sustaining a single flower; the empalement of the flower is large, pyramidal, and swelling, having five acute angles. The flowers are composed of five obtuse red petals, which spread open above the empalement.

The fifth sort grows naturally in the Levant. This is a low annual plant, seldom rising more than four inches high, but divides into branches by pairs from the bottom, which spread asunder. The leaves are very small; the flowers come out single from the wings of the leaves; they have hairy cylindrical empalements, out of which the petals of the flower do but just peep, so are not obvious at any distance. The whole plant is very clammy to the touch. As this plant makes no figure, so it is only kept for variety.

These plants are easily propagated by seeds, which should be sown where the plants are to remain, and will require no other care but to keep them clean from weeds, and thin them where they are too close. If the seeds are sown in autumn, or are permitted to scatter, the plants will come up without care.

SAPOTA. Plum. Nov. Gen. 43. tab. 4. The Mammee Sapota.

The Characters are,

The flower hath a permanent empalement, composed of five oval acute-pointed leaves. It has five roundish heart shaped petals, connected at their base, ending in acute points; and five short stamina the length of the tube, terminated by arrow-pointed summits, with an oval germen, supporting a short style, crowned by an obtuse stigma. The germen afterward becomes an oval succulent fruit, inclosing one or two oval hard nuts or stones.

The Species are,

1. Sapota (*Achras*) foliis oblongo-ovatis, fructibus turbinatis glabris. *Sapota with oblong oval leaves, and smooth turbinated fruit.*

2. Sapota (*Mammosa*) foliis lanceolatis, fructu maximo ovato, seminibus utrinque acutis. *Sapota with spear-shaped leaves, a very large oval fruit, and oval seeds which are pointed at both ends.*

The name of Sapota is what these fruit are called by the natives of America, to which some add the appellation of Mammee; but there is no other name given to these fruits by the English, since they have settled in the West Indies, so far as I can learn.

The first of these trees is common about Panama, and some other places in the Spanish West-Indies, but is not to be found in any of the English settlements in America.

The second sort is very common in Jamaica, Barbadoes, and most of the islands in the West-Indies, where the trees are planted in gardens for their fruit, which is by many persons greatly esteemed.

This sort grows in America to the height of thirty-five or forty feet, having a strait trunk, covered with an Ash-coloured

loured bark. The branches are produced on every fide, fo as to form a regular head, and are befet with leaves a foot in length, and near three inches broad in the middle. The flowers which are produced from the branches are of a cream colour; when thefe fall away, they are fucceeded by large oval or top-fhaped fruit, which are covered with a brownifh fkin, under which is a thick pulp of a ruffet colour, very lufcious, called Natural Marmalade, from its likenefs to Marmalade of Quinces.

As thefe trees are natives of very warm countries, they cannot be preferved in England, unlefs they are placed in the warmeft ftoves, and managed with great care. They are propagated by planting the ftones; but as thefe will not keep good long out of the ground, the fureft method to obtain thefe plants is, to have the ftones planted in tubs of earth as foon as they are taken out of the fruit, and the tubs placed in a fituation where they may have the morning fun, and be kept duly watered. When the plants come up, they muft be fecured from the vermin, and kept clear from weeds, but fhould remain in the country till they are about a foot high, when they may be fhipped for England; but they fhould be brought over in the fummer feafon, and, if poffible, time enough for the plants to make good roots after they arrive. During their paffage, they muft have fome water while they continue in a warm climate; but as they come into colder weather they fhould have very little moifture, and muft be fecured from falt water, which will foon deftroy the plants if it gets at them.

When thefe plants arrive in England, they fhould be carefully taken out of the tubs, preferving fome earth to their roots, and planted into pots, and then plunged into a moderate hot-bed of tanners bark, obferving, if the weather is hot, to fhade the glaffes with mats, to fcreen the plants from the fun until they have taken new root; obferving alfo not to water them too much at firft, efpecially if the earth in which they come over is moift, becaufe too much water is very injurious to the plants before they are well rooted; but afterward they muft be frequently refrefhed with water in warm weather, and muft have a large fhare of air admitted to them, otherwife their leaves will be infefted with infects, and become foul; in which cafe they muft be wafhed with a fponge to clean them, without which the plants will not thrive.

In the winter thefe plants muft be placed in the warmeft ftove, and in cold weather they fhould have but little water given to them. As thefe plants grow in magnitude, they fhould be fhifted into pots of a larger fize, but they muft not be over-potted, for that will infallibly deftroy them.

SARRACENA. Tourn. Inft. R. H. 657. tab. 476. The Sidefaddle-flower.

The CHARACTERS are,

The flower has a double empalement; the under is compofed of three leaves, which fall away; the upper has five coloured leaves, which are permanent. It has five oval inflexed petals which inclofe the ftamina, whofe tails are oblong, oval, and erect, and a great number of fmall ftamina terminated by target-fhaped fummits. In the center is fituated a roundifh germen, fupporting a fhort cylindrical ftyle, crowned by a target-fhaped five-cornered ftigma covering the ftamina, and is permanent. The germen afterward becomes a roundifh capfule with five cells, filled with fmall feeds.

The SPECIES are,

1. SARRACENA (*Purpurea*) foliis gibbis. Hort. Cliff. 427. *Sarracena with gibbous leaves.*

2. SARRACENA (*Flava*) foliis ftrictis. Lin. Sp. Pl. 510. *Sarracena with clofed leaves.*

The firft fort grows naturally upon bogs in moft parts of North America. This hath a ftrong fibrous root, which ftrikes deep into the foft earth, from which arife five, fix, or feven leaves, in proportion to the ftrength of the plant; thefe are hollow like a pitcher, narrow at their bafe, but fwell out large at the top; their outer fides are rounded, but on their inner they are a little compreffed, and have a broad leafy border running longitudinally the whole length of the tube; and to the rounded part of the leaf there is on the top a large appendage or ear ftanding erect, of a brownifh colour; this furrounds the outfide of the leaves about two-thirds of the top. From the center of the root, between the leaves, arifes a ftrong, round, naked foot-ftalk about a foot high, fuftaining one nodding flower at the top, which has a double empalement; the outer one is of one leaf, divided into five parts to the bottom, where they are connected to the foot-ftalks; thefe fegments are obtufe, and bend over the flower, fo as to cover the infide of it; they are of a purple colour on the outfide, but green within, having purple edges; the inner empalement, which is compofed of three green leaves, falls off; within thefe are five oval petals of a purple colour, which are hollowed like a fpoon; thefe cover the ftamina and fummits, with part of the ftigma alfo. In the center is fituated a large, roundifh, channelled germen, fupporting a fhort ftyle, crowned by a very broad five-cornered ftigma, faftened in the middle to the ftyle, covering the ftamina like a target; this is green, but the five corners, which are ftretched out beyond the brim, are each cut into two points, and are purplifh. Round the germen are fituated a great number of fhort ftamina, joining the fides of the germen clofely, which are terminated by target-fhaped furrowed fummits, of a pale fulphur colour. When the flower decays, the germen fwells to a large roundifh capfule with five cells, covered by the permanent ftigma, filled with fmall feeds.

The fecond fort grows naturally in Carolina, upon bogs and in ftanding fhallow waters. The leaves of this fort grow near three feet high, fmall at the bottom, but widening gradually to the top. They are hollow, and arched over at the mouth like a friar's cowl. The flowers of this grow on naked pedicles, rifing from the root to the height of three feet; the flowers are green.

Thefe plants are efteemed for the fingle ftructure of their leaves and flowers, which are fo different from all the known plants, as to have little refemblance of any yet difcovered, but there is fome difficulty in getting them to thrive in England, when they are obtained from abroad; for as they grow naturally on bogs, or in fhallow ftanding waters, fo unlefs they are conftantly kept in wet they will not thrive; and although the winters are very fharp in the countries where the firft fort naturally grows, yet being covered with water, and the remains of decayed plants, they are defended from froft.

The beft method to obtain thefe plants is, to procure them from the places of their natural growth, and to have them taken up with large balls of earth to their roots, and planted in tubs of earth, which muft be fometimes watered during their paffage, otherwife they will decay before they arrive; for there is little probability of raifing thefe plants from feeds, fo as to produce flowers in many years, if the feeds do grow, fo that young plants fhould be taken up to bring over, which are more likely to ftand here than thofe which have flowered two or three times. When the plants are brought over they fhould be planted in pots, which fhould be filled with foft fpongy earth, mixed with rotten wood, Mofs, and turf, which is very like the natural foil in which they grow. Thefe pots fhould be put into larger pots which will hold water, with which they muft be conftantly fupplied, and placed in a fhady fituation in fummer; but in the winter they muft be covered with Mofs, or fheltered

tered under a frame, otherwise they will not live in this country; for as the plants must be kept in pots, so if these are exposed to the frost it will soon penetrate through them, and greatly injure, if not destroy the plants; but when they are placed under a common frame, where they may have the open air at all times in mild weather, and be sheltered from hard frost, the plants will thrive and flower very well.

SASSAFRAS. See LAURUS.

SATUREJA. Tourn Inst. R. H. 197. Savory.

The CHARACTERS are,

The flower hath an erect, tubulous, striated empalement of one leaf, indented at the brim in five points; it hath one ringent petal, whose tube is cylindrical and shorter than the empalement; the chaps are single, the upper lip erect and obtuse, having an acute indenture at the point. The under lip is spreading, divided into three equal parts. It has four bristly stamina, two of which are almost the length of the upper lip; the other two are shorter, terminated by summits which touch each other, and a four-pointed germen supporting a bristly style, crowned by two bristly stigmas. The germen afterward becomes four seeds, which ripen in the empalement.

The SPECIES are,

1. SATUREJA (*Hortensis*) pedunculis bifloris. Vir. Cliff. 87. *Savory with two flowers upon each foot-stalk; or Summer Savory.*

2. SATUREJA (*Thymbra*) floribus verticillatis, foliis ovatis acutis. Flor. Leyd. Prod. 324. *Savory with whorled flowers, and oval acute-pointed leaves; or the true Thymbra.*

3. SATUREJA (*Montana*) pedunculis dichotomis lateralibus solitariis, foliis mucronatis. Lin. Sp. Pl. 568. *Savory with single diverging foot-stalks on the sides of the branches, and sharp-pointed leaves; or Winter Savory.*

4. SATUREJA (*Virginiana*) capitulis terminalibus, foliis lanceolatis. Lin. Sp. Pl. 567. *Savory with heads of flowers terminating the stalks, and spear-shaped leaves.*

5. SATUREJA (*Origanoides*) foliis ovatis serratis, corymbis terminalibus dichotomis. Lin. Sp. Plant. 568. *Savory with oval sawed leaves, and flowers growing in a divided corymbus, terminating the stalks.*

6. SATUREJA (*Juliana*) verticillis fastigatis concatenatis foliis lineari-lanceolatis. Lin. Sp. Plant. 567. *Savory with bunched whorls of flowers, and linear spear-shaped leaves; Julian's Thymbra, or the true Savory.*

7. SATUREJA (*Græca*) pedunculis corymbosis lateralibus geminis, bracteis calyce brevioribus. Lin. Sp. Plant. 568. *Savory with corymbuses of flowers growing by pairs from the wings of the leaves, and bracteæ shorter than the empalements.*

8. SATUREJA (*Capitata*) floribus spicatis, foliis carinatis punctatis ciliatis. Lin. Mat. Med. 283. *Savory with spiked flowers, and keel-shaped hairy leaves, having spots.*

The first sort is generally known in the gardens by the title of Summer Savory. It is an annual plant, which grows naturally in the south of France and Italy, but is cultivated in the English gardens for the kitchen, and also for medicinal use. It rises with slender erect stalks a foot high, sending out branches by pairs, garnished with leaves placed opposite; they are stiff, a little hairy, and have an aromatic odour if rubbed. The flowers grow from the wings of the leaves toward the upper part of the branches, each foot-stalk sustaining two flowers of the lip kind, having a short cylindrical tube; the upper lip is erect, and indented at the point; the lower is divided into three almost equal parts; they are of a pale flesh colour, and are succeeded by seeds which ripen in autumn.

The second sort grows naturally in Crete. This rises with a shrubby stalk two feet high, dividing into several slender ligneous branches, garnished with small stiff oval leaves, ending in acute points, of an aromatic odour when bruised. The flowers grow in thick whorls round the stalks toward the top; they have short, hairy, five-pointed empalements; the flower is shaped like that of the former, but is larger, and of a brighter red colour. This plant rarely ripens its seeds in England.

The third sort is well known in the gardens by the title of Winter Savory. It is a perennial plant, which grows naturally in the south of France and Italy, but is here cultivated in gardens both for food and physic. This hath a shrubby, low, branching stalk a foot high, garnished with two very narrow leaves at each joint, which are stiff, and stand opposite; from the base of these come out a few small leaves in clusters. The flowers grow from the wings of the leaves upon short foot-stalks; they are shaped like those of the first sort, but are larger, and of a paler colour, and are succeeded by seeds which ripen in autumn; but the plants will continue several years, especially if they are planted on a poor dry soil.

The fourth sort grows naturally in North America. This hath a perennial root; the stalk rises about a foot and a half high; it is stiff, angular, and branches out toward the top. The leaves are stiff, spear-shaped, pointed, and have a strong scent of Pennyroyal; the stalks are terminated by white flowers, collected into globular heads. These are seldom succeeded by perfect seeds in England.

The sixth sort grows naturally in Spain, and some parts of Italy. It hath very slender ligneous stalks, which grow erect nine inches high, sending out two or three slender side branches toward the bottom, garnished with narrow, spear-shaped, stiff leaves, which are placed opposite. The flowers grow in whorls above each other, for more than half the length of the stalk, and seem as if they were bundled together. They are small, and white, but the seeds seldom ripen here; the whole plant has a pleasant aromatic scent.

The seventh sort grows naturally in Crete. This hath very slender ligneous stalks a foot and a half high, garnished with small, oval, stiff, acute-pointed leaves, whose borders are reflexed. The flowers grow in roundish whorls upon short foot-stalks, which rise by pairs from the wings of the leaves; they are small and white, and if the season proves warm the seeds will ripen in autumn.

The eighth sort grows naturally in Crete. It has a low shrubby stalk, with branches on every side, and are hoary, garnished with stiff, narrow, acute-pointed leaves, which are hollowed like the keel of a boat. The flowers grow in short roundish spikes at the end of the branches; they are small and white; the whole plant is hoary, and very aromatic. This never produces seeds in England.

The first sort is only cultivated by seeds; these should be sown the beginning of April upon a bed of light earth, either where they are to remain, or for transplanting; if the plants are to stand unremoved the seeds should be sown thinly, but if they are to be transplanted they may be sown closer. When the plants appear, they must be kept clean from weeds, and afterward they may be treated in the same way as Marjoram.

The second, sixth, and seventh sorts, are too tender to live through the winter in the open air in England. These are generally propagated by slips or cuttings, which take root very readily during any of the summer months; if these cuttings or slips are planted in a shady border, or are shaded from the sun with mats, they will put out roots in two months fit to be transplanted, when they should be each transplanted into a small pot, and placed in the shade till they have taken new root; then they may be placed in a sheltered situation, where they may remain till the end of October; then they should be placed under a common hot-

bed

bed frame, where they may be exposed to the open air at all times when the weather is mild; but they must be protected from hard frost, which will destroy them.

As these plants seldom live above three or four years, so there should be a supply of young plants raised to preserve the species, otherwise they may be soon lost. In winter they should not have much wet, for they are very subject to grow mouldy by moisture, but especially if the free air is excluded from them, or if their branches are drawn up weak they soon decay.

The third sort is very hardy, so if this is sown or planted upon a dry lean soil, it will endure the greatest cold of our winters. I have seen some of the plants growing upon the top of an old wall, where they were fully exposed to the cold, and these survived severe frost, when most of those which were growing in the ground were destroyed. This may be propagated either by seeds in the same way as the first, or by slips, which, if planted in the spring, will take root very freely. These plants will last several years, but when they are old their shoots will be short and not so well furnished with leaves so will not be so good for use as young plants, therefore it will be proper to raise a supply of young plants every other year.

The fifth sort has a perennial root, but the stalks decay every autumn. There are two varieties of this, one of them has narrow leaves and larger heads than the other, and the leaves have very little scent; whereas those of the common sort smell so like Pennyroyal, as not to be distinguished by those who do not see the plants. This sort sometimes produces good seeds here, from which the plants may be easily propagated; they may also be increased by planting cuttings in the spring, in the same manner as it is practised for Mint; these will take root freely, and if they are afterwards planted in a moist soil, they will thrive exceedingly; but as the plant is never used here, so it is only kept for variety in some curious gardens.

The seventh sort is annual, and so tender as rarely to perfect its seeds here, so that there is great difficulty to preserve it. The cuttings or slips of this will take root, by which the plant may be continued two or three years; but these must be sheltered in winter under a frame and kept dry, for wet at that season will soon destroy them.

SATYRIUM. Lin. Gen. Plant. 901.

The CHARACTERS are,

It hath a single stalk; the flowers have no empalement, but sit upon the germen; they have five oblong oval petals, three outer and two inner, joined in a helmet; they have a one-leaved nectarium, situated on the side between the division of the petals. The upper lip is short and erect, the under is plain and hangs downward; their base represents the hinder part of the scrotum. They have two short slender stamina sitting upon the pointal, having oval summits, with double cells shut up in the upper lip of the nectarium, and an oblong twisted germen situated under the flower, having a short style growing on the upper lip of the nectarium, crowned by an obtuse compressed stigma. The germen afterward becomes an oblong capsule with one cell, having three keels and three cells, opening under the keels three ways, filled with small seeds.

The SPECIES are,

1. SATYRIUM (*Nigrum*) bulbis palmatis, foliis linearibus, nectarii labio resupinato trilobo, intermedia majore. Act. Upsal. 1740. p. 19. *Satyrium with handed bulbs, linear leaves, and the under lip of the nectarium with three lobes, the middle being the largest.*

2. SATYRIUM (*Hircinum*) bulbis indivisis, foliis lanceolatis, nectarii labio trifido, intermedia lineari, obliqua præmorsa. Act. Upsal. 1740. tab. 18. *Satyrium with an undivided bulb, spear-shaped leaves, and the lip of the nectarium trifid, the middle segment being linear and obliquely bitten; the Lizard Flower, or great Goatstones.*

3. SATYRIUM (*Viride*) bulbis palmatis, foliis oblongis obtusis, nectarii labio lineari trifido, intermedia obsoleta. Act. Upsal. 1740. p. 18. *Satyrium with handed bulbs, oblong blunt leaves, and the lip of the nectarium divided into three linear parts, the middle one being obsolete; by some called the Frog Orchis.*

4. SATYRIUM (*Albidum*) bulbis fasciculatis, foliis lanceolatis, nectarii labio trifido acuto, intermedia majore. Act. Upsal. 1740. *Satyrium with clustered bulbs, spear-shaped leaves, and the lip of the nectarium divided into three acute parts, the middle one being the largest.*

The first sort grows naturally upon the Alps. This has a broad-handed bulbous root; the stalk rises nine inches high, garnished with very narrow leaves; those on the lower part are four inches long, but on the upper part they are scarce one inch; their base embraces the stalk. The flowers grow in a thick short spike at the top, of a dark purple colour; the lip of the nectarium has three lobes, the middle one being the largest.

The second sort grows naturally in several parts of England. This has a solid bulbous root; the stalk is strong, fifteen inches high; the lower part is garnished with leaves near five inches long, which embrace the stalk with their base. The spike of flowers which occupy the upper part of the stalk is six inches in length; the flowers are of a dirty white, with some linear stripes and spots of a brown colour; the beard or middle segment of the lip of the nectarium is two inches long, and appears as if it was obliquely bitten off.

The third sort grows naturally on dry pastures, and upon chalk-hills in several parts of England. This has a handed bulbous root; the stalk rises a foot high; the lower part is garnished with leaves three inches long, and half an inch broad, whose base embraces the stalk. The flowers terminate the stalk in a long slender spike; the nectarium of this varies in colour, it is sometimes of a dusky purple, and at others of a yellowish green colour.

The fourth sort grows near Verona, and upon the Alps. This hath several small bulbs, which are joined together; the stalk rises eight inches high; the lower part is garnished with spear-shaped leaves three inches long, which embrace the stalk with their base. The flowers terminate the stalk in a short thick spike, which are of an herbaceous white colour.

All these plants are difficult to propagate, so the best way to obtain them is to take up their roots at a proper season, and transplant them into the gardens, putting the several sorts into different soils, as near to that in which they naturally grow as possible, and to leave the ground undisturbed; for if their roots are injured, the plants seldom thrive after. The management of this plant being the same as for the Orchis, I shall not repeat it here.

SAVINE. See JUNIPERUS.

SAVORY. See SATUREJA.

SAURURUS. Lin. Gen. Plant. 414. Lizard's-tail.

The CHARACTERS are,

The flowers are disposed in a katkin or tail; they have an oblong permanent empalement of one leaf. They have no petal, but have six long hair-like stamina, placed three on each side opposite, terminated by oblong erect summits; an oval germen with three lobes, having no style, but is crowned by three blunt permanent stigmas. The germen afterward becomes an oval berry with one cell, inclosing one oval seed.

We have but one SPECIES of this plant, viz.

SAURURUS (*Cernuus*) foliis cordatis petiolatis, spicis solitariis recurvis. Hort. Upsal. 91. *Lizard's tail with heart-shaped leaves, having foot-stalks, and single recurved spikes of flowers.*

This plant grows naturally in many parts of North America. The root is fibrous and perennial; the stalk rises a

foot

foot and a half high, having some longitudinal furrows; the leaves are heart-shaped, smooth, about three inches long, and two broad at their base, ending in obtuse points, standing upon foot-stalks, which are placed alternately. The spike of flowers comes out from the wings of the leaves toward the top of the stalk; it is taper, about two inches long, but make little appearance, and are not succeeded by seeds in England. The stalk decays in autumn.

This is preserved by botanists for the sake of variety, but as it has no beauty, it is very rarely admitted into other gardens; it is propagated by its creeping root, which may be parted either in autumn soon after the stalks decay, or in the spring before the roots begin to shoot; it loves a moist soil and a shady situation.

SAXIFRAGA. Tourn. Inst. R. H. 252. tab. 129. Saxifrage.

The CHARACTERS are,

The flower hath a short permanent acute empalement, cut into five parts; it has five plain petals, and ten awl-shaped stamina, terminated by roundish summits, with a roundish acute-pointed germen, sitting upon two styles, crowned by obtuse stigmas. The germen afterward becomes an oval capsule with two horns opening between their tops, filled with small seeds.

The SPECIES are,

1. SAXIFRAGA (*Granulata*) foliis caulinis reniformibus lobatis, caule ramoso, radice granulata. Hort. Cliff. 167. *Saxifrage with kidney-shaped leaves upon the stalks, with lobes, a branching stalk, and roots like grains of Corn; or white Saxifrage.*

2. SAXIFRAGA (*Cotyledon*) foliis radicatis aggregatis lingulatis, cartilagineo-serratis, caule ramoso. *Saxifrage with tongue-shaped leaves at the root, which are joined together, having cartilaginous saws, and a branching stalk.*

3. SAXIFRAGA (*Paniculata*) foliis radicatis aggregatis cuneiformibus cartilagineo-serratis, caule paniculato. *Saxifrage with the lower leaves wedge-shaped, joined together, with edges having cartilaginous saws, and a paniculated stalk.*

4. SAXIFRAGA (*Pyramidata*) foliis radicatis aggregatis lingulatis, cartilagineo-serratis, caule pyramidato. *Saxifrage with the lower leaves joined together, which are tongue-shaped, having cartilaginous saws, and a pyramidal stalk.*

5. SAXIFRAGA (*Rotundifolia*) foliis caulinis dentatis reniformibus petiolatis. Lin. Sp. Plant. 403. *Saxifrage with kidney-shaped leaves on the stalks, which stand on foot-stalks, and are indented.*

6. SAXIFRAGA (*Hirsuta*) foliis reniformibus dentatis, caule nudo paniculato. Lin. Sp. Plant. 401. *Saxifrage with indented kidney-shaped leaves, and a naked paniculated stalk.*

7. SAXIFRAGA (*Punctata*) foliis obovatis dentatis petiolatis, caule nudo paniculato. Lin. Sp. Pl. 399. *Saxifrage with oblong, oval, indented leaves, having foot-stalks, and a naked paniculated stalk; commonly called London Pride, or None-so-pretty.*

8. SAXIFRAGA (*Pensylvanica*) foliis lanceolatis denticulatis, caule nudo paniculato, floribus subcapitatis. Lin. Sp. Plant. 401. *Saxifrage with spear-shaped indented leaves, a naked paniculated stalk, and flowers collected in heads.*

9. SAXIFRAGA (*Nivalis*) foliis obovatis crenatis subsessilibus, caule nudo, floribus congestis. Lin. Sp. Plant. 401. *Saxifrage with oblong, oval, crenated leaves, sitting close to the root, a naked stalk, and flowers growing in close bunches.*

10. SAXIFRAGA (*Autumnalis*) foliis caulinis linearibus alternis ciliatis, radicalibus aggregatis. Lin. Sp. Plant. 402. *Saxifrage with linear leaves on the stalk, which are set with fine hairs alternately, and those at the root joined together.*

11. SAXIFRAGA (*Oppositifolia*) foliis caulinis ovatis oppositis imbricatis, summis ciliatis. Flor. Suec. 359. *Saxifrage with oval leaves on the stalks, which are opposite, and lie over each other, and upper leaves having fine hairs.*

12. SAXIFRAGA (*Hypnoides*) foliis caulinis linearibus integris trifidisque, stolonibus procumbentibus, caule erecto nudiusculo. Lin. Sp. Pl. 405. *Saxifrage with linear leaves on the stalks, which are entire or trifid, trailing side shoots, and erect stalks, which are almost naked; commonly called Ladies Cushion.*

There are many more species of this genus than are here enumerated, some of which grow naturally in Great-Britain; but as they are very rarely admitted into gardens, it would be needless to mention them in this work.

The first sort is the common white Saxifrage, which grows naturally in the meadows in most parts of England The roots of this plant are like grains of Corn, of a reddish colour without, from which arise kidney-shaped hairy leaves, standing upon pretty long foot-stalks. The stalks are thick, a foot high, hairy, and furrowed; these branch out from the bottom, and have a few small leaves like those below, which sit close to the stalk; the flowers terminate the stalk, growing in small clusters; they have five small white petals, inclosing ten stamina and the two styles. The roots and leaves of this plant are used in medicine.

There is a variety of this with double flowers, which was found growing wild by Mr. Joseph Blind, gardener at Barnes, who transplanted it into his garden, and afterward distributed it to several curious persons, since which time it hath been multiplied so much, as to become a very common plant in most gardens near London, where it is planted in pots to adorn court-yards, &c. in the spring, and is very ornamental at that season in the borders of the flower-garden.

This plant is propagated by offsets, which are sent forth from the old roots in great plenty. The best season for transplanting them is in July, after their leaves are decayed, when they must be put into undunged earth, and placed in the shade until autumn; but in winter they may be exposed to the sun, which will cause them to flower somewhat earlier in the spring. In April these plants will flower, and if they are in large tufts, will make a very handsome appearance; when they are transplanted they should be put in bunches, that they may produce a greater number of flowers.

The second sort grows naturally on the Alps; this hath a perennial fibrous root. The leaves grow in circular heads, embracing each other at their base after the manner of the common Housleek; they are tongue-shaped, rounded at their points, and have a white cartilaginous sawed border. The stalk rises a foot high, is of a purplish colour, a little hairy, and sends out several horizontal branches. The flowers grow in small clusters at the end of the branches; they are white, and have several small red spots on the inside.

It is easily propagated by offsets, which are sent out in plenty; they may be taken off at almost any season when the weather is mild, and should be planted in a very dry soil and a shady situation.

The third sort grows naturally on the Alps. The leaves of this are gathered into circular heads like the former, but are not more than half an inch long, wedge-shaped, the upper part being broad and rounded, but diminish to their base, where they are narrow; their borders are edged, and indented in the same manner as those of the former. The stalk, in the places where the plant grows naturally, seldom rises more than six inches high, but when transplanted into gardens, is often more than a foot; these have small leaves sitting close to them. The flowers are disposed in loose panicles

panicles on the top of the ſtalks; they are white, ſpotted with red, and may be propagated in the ſame manner as the former.

The fourth ſort grows naturally on the mountains in Italy. The leaves of this are gathered into heads like thoſe of the two former; they are tongue-ſhaped, rounded at their points, and have cartilaginous ſawed borders. The ſtalk riſes two feet and a half high, branching out near the ground, forming a natural pyramid to the top; the flowers have five white wedge-ſhaped petals, and ten ſtamina placed circularly the length of the tube, terminated by roundiſh purple ſummits. When theſe plants are ſtrong they produce very large pyramids of flowers, which make a fine appearance, ſo are very ornamental for halls, or to place in chimnies, where, being kept in the ſhade, and ſcreened from wind and rain, they will continue in beauty much longer than if kept in the open air.

This plant is eaſily propagated by offſets, which are put out from the ſide of the old plants in plenty. They are uſually planted in pots filled with light earth, and in the ſummer ſeaſon placed in the ſhade, but in the winter ſhould be expoſed to the ſun, and all the offsets ſhould be taken off, leaving the plant ſingle, which will cauſe it to produce a much ſtronger ſtem for flowering; for where there are offsets about the old plant, they exhauſt the nouriſhment from it, whereby it is rendered much weaker. Theſe offſets muſt be each planted in a ſeparate ſmall pot, filled with freſh earth, to ſucceed the old plants, which generally periſh after flowering; theſe offsets will produce flowers the ſecond year, ſo there ſhould be annually ſome of them planted to ſucceed the others.

The fith ſort grows naturally on the Helvetian mountains. This hath a perennial root. The ſtalk is erect, a foot high, channelled and hairy; it is garniſhed with kidney-ſhaped leaves, which are ſharply indented; the ſtalks are terminated by ſmall cluſters of flowers, marked with ſeveral red ſpots. It is propagated by parting of the roots; the beſt time for this is in autumn, that the plants may have good roots before the dry weather in the ſpring. It loves a ſhady ſituation and a loamy ſoil.

The ſixth ſort grows naturally on the Alps and Pyrenean mountains. The root is fibrous and perennial; the leaves are thick, kidney-ſhaped, and crenated on their edges, of a deep green on their upper ſide, but pale on their under, ſtanding upon long, thick, hairy foot-ſtalks, which branch into a panicle, ſuſtaining ſeveral ſmall white flowers, marked with red ſpots; the ſtamina of this are longer than the petals. It propagates very faſt by offsets, which ſhould be taken off in autumn, and planted in a ſhady ſituation, where they will thrive faſt enough.

The ſeventh ſort is known by the titles of London Pride, or None-ſo-pretty. It grows naturally on the Alps, and alſo in great plenty upon a mountain called Mangerton, in the county of Kerry in Ireland. The roots of this are perennial; the leaves are oblong, oval, placed circularly at bottom; they have broad, flat, furrowed foot ſtalks, and are deeply crenated on their edges, which are white; the ſtalk riſes a foot high, is of a purple colour, ſtiff, ſlender, and hairy; it ſends out from the ſide on the upper part ſeveral ſhort foot-ſtalks, which are terminated by white flowers ſpotted with red; the ſtamina are longer than the petals of the flower, as are alſo the two ſtyles; theſe have red ſtigmas. It may be propagated in the ſame way as the former, and loves a ſhady ſituation.

The eighth ſort grows naturally in North America. This is a perennial plant with a fibrous root, from which ariſe ſeveral ſpear-ſhaped leaves ſeven or eight inches long, and two broad, having ſmall indentures on their edges; they are a deep green, and of a thick conſiſtence. The ſtalk is naked, and riſes a foot and a half high, branching at the top in form of a panicle, ſuſtaining very ſmall herbaceous flowers, which are collected into ſmall heads. It is propagated by parting the root; the beſt time is in autumn; it loves a moiſt ſoil and a ſhady ſituation, and is never injured by cold.

The ninth ſort grows naturally upon ſome mountains in Wales. This hath a fibrous perennial root, from which come out oblong, roundiſh, leaves, deeply indented, or rather ſawed on their edges, ſitting very cloſe to the root. The ſtalk riſes about five inches high, is naked, and terminated by a cloſe compact cluſter of white flowers; if they are planted in a ſhady ſituation, they will continue almoſt a month. This muſt have a loamy ſoil, otherwiſe it will not thrive.

The tenth ſort grows naturally upon the Auſtrian mountains. It has alſo been found growing in plenty in Knotsford moor in Cheſhire; this is a perennial plant. The leaves are gathered in cluſters at the bottom; they are ſpear-ſhaped; the ſtalks riſe about ſix inches high, garniſhed with narrow leaves the whole length, placed alternately and ſit cloſe to them; the flowers are produced in ſmall cluſters at the top of the ſtalk; they have five yellowiſh petals, having ſeveral red ſpots on their inſide. This plant is difficult to propagate in gardens, for it naturally grows upon bogs, ſo that unleſs it is planted in ſuch looſe rotten earth, and kept conſtantly moiſt, it will not thrive

The eleventh ſort grows naturally upon the Pyrenean and Helvetian mountains, as alſo upon Ingleborough-hill in Yorkſhire, Snowden in Wales, and other high places in the north of England. This is a perennial plant, whoſe ſtalks trail upon the ground, and are ſeldom more than two inches long, garniſhed with ſmall oval leaves ſtanding oppoſite, which lie over each other like the ſcales of fiſh; they are of a brown green colour, and have a reſemblance of Heath. The flowers are produced at the end of the branches, of a deep blue, ſo make a pretty appearance during their continuance, which is great part of March, and the beginning of April. This is propagated by parting of the roots; the beſt time for doing it is in autumn: it muſt have a ſhady ſituation and a moiſt ſoil, otherwiſe it will not thrive in gardens.

The twelfth ſort grows naturally upon the Alps, Pyrenees, and Helvetian mountains; it is alſo found growing plentifully on Ingleborough-hill in Yorkſhire, Snowden in Wales, and ſome other places in the north. This is a perennial plant, whoſe branches ſpread flat upon the ground, and put out roots at their joints, garniſhed with fine ſoft leaves like Moſs, ſome of which are entire, and others cut into three points. The branches join ſo cloſe together as to form a ſoft roundiſh bunch like a pillow or cuſhion, from whence ſome have given it the appellation of Ladies Cuſhion; the flower-ſtalks riſe three or four inches high; they are ſlender, erect, and have two or three ſmall leaves; ſome are entire, and others trifid; they are of a bright green colour, and ſoft to the touch; the flowers grow in ſmall bunches at the top of the ſtalk; they are ſmall, of a dirty colour, ſo make no great appearance.

This ſort propagates faſt enough by its trailing branches, provided it is planted in a moiſt ſoil and a ſhady ſituation, but it will not thrive in dry ground, or where it is much expoſed to the ſun. The beſt time to remove any of theſe plants is in autumn, that they may have the benefit of the winter's rain to eſtabliſh them well before the dry weather of the ſpring comes on; for when they are planted late, they are very ſubject to die, unleſs they are ſupplied with water, and they ſeldom make any figure the firſt year.

SCABIOSA. Tourn. Inſt. R. H. 463. tab. 263, 264. Scabious.

The

The CHARACTERS are,

The common empalement is composed of many leaves, containing many flowers; it has several series of leaves surrounding the receptacle upon which they sit; the inner are gradually smaller. The flowers have a double empalement, and sit upon the germen; the outer is short, membranaceous, folded, and permanent; the inner is divided into five awl-shaped capillary segments. The florets have one erect tubulous petal, cut into four or five parts at the brim; they have four weak, awl-shaped, hair-like stamina, terminated by oblong prostrate summits. The germen is situated under the receptacle of the florets, supporting a slender style, crowned by an obtuse stigma, which is obliquely indented; it afterward becomes an oblong oval seed sitting in the common empalement, and crowned by the cup of the flower.

The SPECIES are,

1. SCABIOSA (*Arvensis*) corollulis quadrifidis radiantibus, caule hispido. Hort. Cliff. 31. *Scabious with quadrifid radiated florets, and a rough hairy stalk; or Meadow Scabious of the shops.*

2. SCABIOSA (*Succissa*) corollulis quadrifidis æqualibus, caule simplici, ramis approximatis, foliis lanceolato-ovatis. Hort. Cliff. 30. *Scabious with quadrifid florets which are equal, a single stalk, and branches growing near, with spear-shaped oval leaves; called Devil's Bit.*

3. SCABIOSA (*Transylvanica*) corollulis quadrifidis æqualibus, squamis calycinis ovatis obtusis. Lin. Sp. Plant. 98. *Scabious with quadrifid florets which are equal, and the scales of the empalement oval and obtuse.*

4. SCABIOSA (*Centauroides*) corollulis quadrifidis fistulosis æqualibus, squamis calycinis acutis, caule paniculato, foliis rigidis pinnatifidis. *Scabious with quadrifid fistulous florets, which are equal, acute scales to the empalement, a paniculated stalk, and stiff wing-pointed leaves.*

5. SCABIOSA (*Montana*) corollulis quadrifidis æqualibus, staminibus longioribus, squamis calycinis acutis, foliis radicalibus lanceolatis integerrimis, caulinis divisis. *Scabious with quadrifid equal florets, longer stamina, acute scales to the empalement, and the lower leaves spear-shaped and entire, but those on the stalks divided.*

6. SCABIOSA (*Altissima*) corollulis quadrifidis radiantibus, caule hispido, foliis lanceolatis pinnatifidis, foliolis imbricatis. Lin. Sp. Plant. 99. *Scabious with radiated quadrifid florets, a rough hairy stalk, and spear-shaped wing-pointed leaves, with lobes set over each other in the manner of tiles.*

7. SCABIOSA (*Alpina*) corollulis quadrifidis æqualibus, calycibus squamosis nitidis obtusis, caule dichotomo, foliis pinnatifidis. *Scabious with equal quadrifid florets, neat scaly empalements, which are obtuse, a stalk divided by pairs, and wing-pointed leaves.*

8. SCABIOSA (*Graminifolia*) corollulis quinquefidis æqualibus, caule erecto hispido, foliis lanceolatis denticulatis hirsutis, semi-amplexicaulibus. *Scabious with equal quinquefid florets, an erect stalk, which is rough, hairy, and spear-shaped hairy leaves, which are somewhat indented, and half embrace the stalks.*

9. SCABIOSA (*Virga pastoris*) corollulis quinquefidis, foliis pinnatis serratis, receptaculis florum globosis. *Scabious with quinquefid florets, winged sawed leaves, and globular receptacles to the flower.*

10. SCABIOSA (*Rigida*) corollulis quinquefidis radiantibus, foliis lineari-lanceolatis integerrimis, caule suffruticoso. *Scabious with radiated quinquefid florets, linear, spear-shaped, entire leaves, and an under-shrub stalk.*

11. SCABIOSA (*Cretica*) corollulis quinquefidis, foliis lanceolatis sub-integerrimis. Hort. Cliff. 31. *Scabious with quinquefid florets, and spear-shaped leaves which are almost entire.*

12. SCABIOSA (*Frutescens*) corollulis quinquefidis radiantibus, foliis bipinnatis linearibus. Lin. Sp. Plant. 101. *Scabious with radiated quinquefid florets, and linear doubly-winged leaves.*

13. SCABIOSA (*Ochroleuco*) corollulis quinquefidis, foliis pinnatis, laciniis lanceolatis, pedunculis nudis lævibus longissimis. Prod. Leyd. 190. *Scabious with quinquefid florets, winged leaves having spear-shaped segments, and long, naked, smooth foot-stalks.*

14. SCABIOSA (*Stellata*) corollulis quinquefidis, foliis dissectis, receptaculis florum subulatis. Hort. Cliff. 31. *Scabious with five-pointed florets, cut leaves, and awl-shaped receptacles to the flowers.*

15. SCABIOSA (*Atropurpurea*) corollulis quinquefidis, foliis dissectis, receptaculis florum subrotundis. Hort. Cliff. 31. *Scabious with five-pointed florets, cut leaves, and roundish receptacles to the flowers.*

16. SCABIOSA (*Africana*) corollulis quinquefidis, foliis inferioribus integris crenatis, caulinis inciso-crenatis, caule fruticoso. *Scabious with five-pointed florets, the lower leaves entire and crenated, those upon the stalks bluntly cut, and a shrubby stalk.*

17. SCABIOSA (*Incisa*) corollulis quinquefidis, foliis inferioribus crenatis, caulinis duplicato-pinnatis, caule fruticoso hirsuto. *Scabious with five-pointed florets, the under leaves crenated, those on the stalks doubly-winged, and a shrubby hairy stalk.*

18. SCABIOSA (*Fimbriata*) corollulis multifidis, calycibus florum longioribus, caule ramoso, foliis dissectis. *Scabious with many-pointed florets, longer empalements to the flowers, a branching stalk, and cut leaves.*

The first sort grows naturally in the fields in divers parts of England; it hath a strong, thick, fibrous root, sending out many branching stalks which rise three feet high; the lower leaves are sometimes almost entire, and at others they are cut into many segments almost to the midrib. The stalks are covered with stiff prickly hairs, garnished with smaller leaves at each joint, which are cut into narrow segments almost to the midrib. The flowers are produced upon naked foot-stalks at the end of the branches; they have a double empalement, which is hairy, and composed of several tubulous florets, cut into four points at the top, each having a particular empalement, resting upon the common placenta. The florets round the border are larger and deeper cut than those which compose the disk or middle; they have four weak stamina, which soon shrink after the flowers open. In the center is situated a style which is longer than the floret, terminated by a roundish stigma. The flowers are of a pale purple colour, and have a strong faint odour. This is the sort intended by the College of Physicians for medicinal use, under the title of Scabiosa.

The second sort grows naturally in moist woods and pastures in most parts of England, and is directed by the College of Physicians to be used, under the title of Morsus Diaboli, or Devil's Bit; this hath a short tap-root, which appears as if the end of it were bitten or cut off, from whence it had the titles of Succisa, and Morsus Diaboli. The leaves are oval, spear-shaped, and smooth; the stalks are single, about two feet high, garnished with two leaves at each joint, shaped like those below, but smaller; they generally send out two short foot-stalks from their upper joint, standing opposite, which are each terminated by one small blue flower, as is also the principal stalk with one larger. These are constructed in the same way as the former. As these plants are to be found plentifully in the fields and woods, they are seldom admitted into gardens.

The third sort grows naturally in Transylvania. It is an annual plant, which is preserved by botanists for variety; but as the flowers have little beauty, so it is rarely allowed a place in other gardens. The stalks rise four or five feet high, dividing into several branches; the leaves are hairy,

cut

cut almoſt to the midrib. The flowers are ſmall, of a pale purpliſh colour; the ſeeds ripen in autumn; if they are permitted to ſcatter, the plants will come up without care.

The fourth ſort grows naturally in Spain and Portugal. It is an annual plant; the ſtalk is ſtiff, and riſes upward of three feet high, dividing toward the top into ſeveral branches, which are again divided into naked foot-ſtalks, each ſuſtaining one ſmall, pale, purpliſh flower, compoſed of many florets; the leaves are ſtiff, and cut into many winged points.

The fifth ſort grows naturally upon the Alps and Apennines. This hath a perennial root, from which come out many entire, ſmooth, ſpear-ſhaped leaves; the ſtalk is ſingle, ſending out two ſhort naked foot-ſtalks from the upper joint; the leaves upon the ſtalks are cut pretty deeply on their edges. The flowers are nearly of the ſame ſize and form with thoſe of the firſt ſort. It may be propagated by ſeeds, and will thrive in a ſhady moiſt border, requiring no other care but to keep the ground clean, and allow them room to ſpread.

The ſixth ſort is a biennial plant, which grows naturally in ſome parts of Italy, and alſo in Tartary. It riſes with a ſtrong branching ſtalk four or five feet high, cloſely armed with ſtiff prickly hairs; the lower leaves are ſpear-ſhaped, cut deeply on the ſides in winged points; thoſe upon the ſtalks are more entire, ſome of them are ſharply ſawed on their edges, and thoſe at the top are linear and entire. The flowers grow from the ſides and at the top of the ſtalks; they are white, and each floret ſits in a briſtly empalement. The ſeeds ripen in autumn; it riſes from ſcattered ſeeds, and requires no care.

The ſeventh ſort grows naturally in Iſtria and upon the Alps. This hath a perennial root; the lower leaves are almoſt entire; the ſtalk is ſtiff, and riſes four feet high, dividing into two upward, which ſpread aſunder; and in the diviſion ariſes a naked foot-ſtalk, which (as alſo the ſide branches) are terminated by flowers, compoſed of many pale florets, incloſed in a ſcaly empalement, whoſe ſcales are obtuſe; the leaves on the ſtalks are wing-pointed and ſtiff.

The eighth ſort grows naturally upon the mountains in Italy. This has a perennial root, from which ariſe three or four ſtalks, whoſe lower parts are garniſhed with linear leaves, of a ſilvery colour, ending in acute points; the upper part of the ſtalk is naked, ſuſtaining at the top one pale blue flower, made up of ſeveral four-pointed florets. It is propagated by ſeeds as the other ſorts, and loves a ſoft loamy ſoil and a ſhady ſituation.

The ninth ſort grows naturally on the Alps. This has a perennial root, from which ariſe ſeveral pretty ſtrong hairy ſtalks three feet high, garniſhed with ſpear-ſhaped leaves, placed oppoſite, which embrace the ſtalks half round with their baſe; they are of a dark green on their upper ſide, but pale on their under, hairy, having a few indentures on their edges, and ending in acute points. The flowers are produced at the top of the ſtalks in the ſame manner as thoſe of the firſt ſort, and are like them. This is hardy, and loves a light loamy ſoil and a ſhady ſituation; it is propagated by ſeeds.

The tenth ſort grows naturally on the Alps. This has a perennial root, compoſed of many ſtrong fibres from which ariſe ſeveral ſtrong channelled ſtalks upward of four feet high, garniſhed with winged leaves, compoſed of four or five pair of lobes, which are unequal in ſize and irregularly placed, ſawed on their edges, and end in acute points. The flowers are produced on naked foot-ſtalks at the end of the branches; the receptacles are globular; the flowers are of a whitiſh yellow, and the ſeeds ripen in autumn. This may be propagated either by ſeeds or parting of the roots; it loves a loamy ſoil.

The eleventh ſort grows naturally in Sicily. This riſes with a ſhrubby ſtalk three feet high, dividing into ſeveral ligneous knotty branches, garniſhed with narrow ſilvery leaves, which are entire. The flowers ſtand upon very long naked foot-ſtalks at the end of the branches; they are made up of many five-pointed tubulous florets, of a fine blue colour. Theſe are not ſucceeded by ſeeds here. It is propagated by ſlips or cuttings, which readily take root if they are planted in any of the ſummer months, if they are ſhaded from the ſun, and duly refreſhed with water. When theſe have made good roots, ſome of them may be planted on a dry border near to a ſouth wall, where they will live in common winters; but as they are frequently deſtroyed by ſevere froſt, ſo ſome of the plants ſhould be planted in pots, and in winter placed under a common frame, where there may be protected from froſt, but in mild weather enjoy the free air.

The twelfth ſort grows naturally in Crete. This hath a ſhrubby ſtalk, which riſes about the ſame height as the former, and divides into many branches; the leaves are ſhorter, much broader, and not ſo white as thoſe of the former ſort; the flowers are not ſo large, and are of a pale purple colour. This flowers from the end of June till autumn, but it does not ripen ſeeds in England. It is propagated by ſlips or cuttings in the ſame way as the former, and requires the ſame treatment.

The thirteenth ſorth grows naturally in Germany. This hath a perennial root, ſending out many leaves near the ground, which are divided into narrow ſegments to the midrib; theſe ſegments are cut on their edges into regular acute points, like winged leaves; the ſtalks riſe two feet high, garniſhed with very narrow cut leaves; they divide into ſeveral long foot-ſtalks, each being terminated by a yellowiſh flower, with radiated borders. This may be propagated by ſeeds, and will thrive any where.

The fourteenth ſort grows naturally in the Levant. This is a low perennial plant with a branching ſtalk; the lower leaves are cut, but the upper leaves are narrow and entire, of a ſilvery colour. The flowers are ſmall, of a pale colour, and have no ſcent, ſo is only kept in botanic gardens for the ſake of variety. It is propagated by ſeeds, and is hardy enough to live in the open air.

The fifteenth ſort grows naturally in India. This is an annual plant, which is commonly cultivated in gardens for ornament. Of this there is a great variety in the colour of their flowers; ſome of them are of a purple, approaching to black, others are of a pale purple, ſome are red, and others have variegated flowers; theſe alſo vary in the ſhape of their leaves, ſome of them having finer cut leaves than others; and ſometimes from the ſide of the flower-cup, there comes out many ſlender foot-ſtalks, ſuſtaining ſmall flowers, in like manner as the Hen and Chicken Daiſies; but as theſe are accidental varieties which come from the ſame ſeeds, they need not be particularly enumerated here.

The flowers of this ſort are very ſweet, and continue a long time. The plants are propagated by ſowing of their ſeeds; the beſt time for which is about the latter end of May or the beginning of June, that the plants may get ſtrength before winter; for if they are ſown too early in the ſpring, they will flower the autumn following, and the winter coming on ſoon, will prevent their ripening ſeeds; beſides, there will be fewer flowers upon thoſe, than if they had remained ſtrong plants through the winter, and had ſent forth their flower-ſtems in ſpring; for theſe will branch out, and produce a prodigious number of flowers, continuing a ſucceſſion of them on the ſame plants from June to September, and produce good ſeeds in plenty.

The feeds of thefe plants fhould be fown upon a fhady border of frefh earth (for if they are fown upon a place too much expofed to the fun, and the feafon fhould prove dry, few of them will grow.) When the plants come up, they may be tranfplanted into other beds or borders, obferving to water and fhade them until they have taken root; after which they will require no farther care but to keep them clear from weeds till Michaelmas, when they may be tranfplanted into the middle of the borders in the pleafure-garden, where the feveral forts being intermixed, will make an agreeable variety.

The fixteenth fort grows naturally in Spain. This is an annual plant; the ftalks rife three feet high; they are hairy, and garnifhed with oblong leaves, which are deeply notched on their edges; thofe on the upper part of the ftalk are cut almoft to the midrib into fine fegments. The flowers ftand upon long foot-ftalks at the top of the ftalks; thefe have globular receptacles; the florets are large, and fpread open like a ftar; they are of a pale purple colour; in favourable feafons the feeds ripen in September, but in cold moift years they do not ripen here. It is propagated by feeds, which fhould be fown in beds of light loamy earth, where the plants are to remain; when the plants come up, they muft be thinned and kept clear from weeds, which is all the culture they require.

The feventeenth fort grows naturally at the Cape of Good Hope. This hath a weak fhrubby ftalk, which divides into feveral branches, rifing about five feet high, garnifhed with oval fpear-fhaped leaves, which are entire, and deeply crenated on their edges, of a light green, and a little hairy. The flower-ftalk is produced at the end of the branches, fuftaining one pale flefh-coloured flower, compofed of many five-pointed florets. This plant continues flowering great part of fummer, and fometimes it produces good feeds in England.

The eighteenth fort is a native of the Archipelago; the ftalks are hairy, and divide into feveral branches, garnifhed toward the bottom with fpear-fhaped, entire, crenated leaves, but thofe on the upper part of the ftalk are doubly winged. The flowers are produced upon long naked ftalks at the end of the branches; they are large, of a pale flefh colour, but have little fcent; the borders of the florets are fringed; they continue in fucceffion all the fummer, and fometimes the early flowers are fucceeded by feeds, which ripen in autumn.

Both thefe forts may be propagated by feeds or cuttings, which may be planted in a fhady border during any of the fummer months. When thefe have put out good roots, they fhould be taken up and planted in pots, and placed in the fhade till they have taken new root; then they may be removed to a fheltered fituation, where they may remain till the froft begins, when they fhould be removed to fhelter, for they are too tender to live in the open air through the winter; but as they only require protection from froft, fo they fhould have as much free air as poffible in mild weather, to prevent their being drawn up weak; and in the middle or latter end of April, they may be placed in the open air in a warm fituation, afterward treating them as other hardy foreign plants.

SCANDIX. Tourn. Inft. R. H. 326. tab. 173. Shepherds Needle, or Venus-comb.

The CHARACTERS are,

It hath an umbelliferous flower; the general umbel is long and has few rays; the particular umbels have many; the general umbel has no involucrum, the particular have a five-leaved one the length of the umbels; the general umbel is deformed, and has hermaphrodite florets in the difk, and female in the rays. The flowers have five inflexed heart-fhaped petals; the inner are fmall, and the outer large; and five flender ftamina, terminated by roundifh fummits with an oblong germen, fupporting two permanent ftyles, crowned by obtufe ftigmas. The germen afterward turns to a long fruit, divided in two parts, each having one furrowed feed, convex on one fide, and plain on the other.

The SPECIES are,

1. SCANDIX (*Pecten*) feminibus lævibus roftro longiffimo Hort. Cliff. 101. *Scandix with fmooth feeds, and the longeft beak; common Shepherds Needle.*

2. SCANDIX (*Auftralis*) feminibus fubulatis hifpidis, floribus radiatis, caulibus lævibus. Lin. Sp. Plant. 257. *Scandix with prickly awl-fhaped feeds, radiated flowers, and fmooth ftalks.*

3. SCANDIX (*Grandiflora*) feminibus pedunculo villofo brevioribus. Flor. Leyd. 111. *Scandix with fhort hairy foot-ftalks.*

4. SCANDIX (*Cretica*) feminibus hifpidis, involucris umbello multifidis, caulibus afperis. *Scandix with briftly feeds, many-pointed involucrums to the umbels, and rough ftalks.*

5. SCANDIX (*Odorata*) feminibus fulcatis angulatis. Hort. Cliff. 101. *Scandix with angular furrowed feeds.*

6. SCANDIX (*Anthrifcus*) feminibus ovatis hifpidis, corollis uniformibus, caule lævi. Lin. Sp. Plant. 257. *Scandix with oval rough feeds, the petals of the flowers uniform, and a fmooth ftalk.*

The firft fort grows naturally in ftiff lands amongft the Corn in many parts of England, fo is not cultivated in gardens. It is an annual plant; the leaves are finely divided into fmall fegments, and have long foot-ftalks; the ftalks rife fix inches high. The flowers are fmall, white, and like thofe of wild Chervil, and fit upon the top of the beak or horns, which are the rudiment of the pod. At the bottom of the fmall umbel five leaves embrace the ftalk, with broad and fhort foot-ftalks, which are afterward cut into fmall fegments like the reft: the feed is long, and runs into a fmall point, refembling a large needle, but the umbels have great refemblance to the umbels of Mufk Crane's-bill. If the feeds are permitted to fcatter, there will be a plentiful fupply of young plants.

The fecond fort grows naturally in the fouth of France, Italy, and Crete. This is an annual plant, with low fpreading ftalks, garnifhed with very narrow fine cut leaves, placed thinly. The flowers are fmall, white, and ftand in fmall umbels at the top of the ftalks; thefe are fucceeded by awl-fhaped rough feeds.

The third fort grows naturally in the Levant. This is an annual plant, with fine cut leaves; the ftalks rife eight inches high, garnifhed at each joint with a fine cut leaf, and terminated by an umbel of white flowers, with large heart-fhaped petals. The horns of this are longer than of any other fort, and their foot-ftalks are very fhort and hairy.

The fourth fort grows naturally in Crete. This hath larger leaves than either of the former, which are finely cut; the ftalks grow a foot high, and divide into many rough channelled branches; the umbels have a many-leaved involucrum, and the feeds are rough.

Thefe four forts will fow themfelves wherever they are once introduced, and require no other care but to thin them, and keep them clean from weeds.

The fifth fort grows naturally in Germany, but has been long kept in the Englifh gardens, and of late years the feeds have been thrown out of gardens, fo that the plants are frequently found growing naturally in the neighbourhood of thofe gardens. It has a very thick perennial root, compofed of many fibres, of a fweet aromatic tafte like Anifeed, from which come forth many large leaves, that branch out fomewhat like thofe of Fern, from whence it was titled Sweet Fern. The ftalks grow four or five feet high; they are hairy and fiftulous. The flowers are difpofed in an umbel at the top of the ftalk; they are white, and have a ſweet aromatic ſcent: the out-petal of the flower is large; the two ſide ones are of a middle ſize, but the two inner are ſmall; theſe are ſucceeded by long, angular, furrowed ſeeds, having the taſte and ſcent of Aniſeed.

This ſort propagates faſt by ſeeds, which, if permitted to ſcatter, there will be plenty of the plants ariſe, and theſe may be tranſplanted to any abject part of the garden, for it will grow in any ſoil or ſituation, and will require no care.

The ſixth ſort grows naturally on the ſide of banks and foot-ways in many parts of England. This is an annual plant, whoſe ſeeds drop early in the ſummer; the plants come up in autumn, and flower early in the ſpring. The leaves of this are finely divided, very like thoſe of the Garden Chervil, but are hairy; the ſtalks riſe a foot and a half, or two feet high, dividing into branches. Theſe ſuſtain umbels of ſmall white flowers, which are ſucceeded by ſhort, hairy, crooked ſeeds.

There have been ſome inſtances of the ill effects of this plant, when taken inwardly; ſome who have eaten this herb in ſoups, by miſtaking it for Garden Chervil, have narrowly eſcaped with their lives.

SCHINUS. Lin. Gen. Plant. 479. Indian Maſtick.

The CHARACTERS are,

The flower hath a ſmall empalement with five indentures at the top; it has five ſmall petals, which ſpread open, and nine or ten ſlender ſtamina, with a roundiſh germen, ſupporting a ſhort thick ſtyle, crowned by a ſingle ſtigma. The germen afterward turns to a globular berry with one cell, incloſing one globular ſeed.

The SPECIES are,

1. SCHINUS (*Molle*) foliis pinnatis, foliolis ſerratis, impari longiſſimo, petiolo æquali. Lin. Sp. Plant. 388. *Schinus with winged leaves, whoſe lobes are ſawed, the end one being very long, and the foot-ſtalks equal.*

2. SCHINUS (*Areira*) foliis pinnatis, petiolo marginato articulato ſubtus aculeato. Lin. Sp. Pl. 389. *Schinus with winged leaves, and jointed bordered foot-ſtalks, having thorns on the under ſide.*

Theſe are all the ſpecies of this genus which I have ſeen growing in the Engliſh gardens, for the Iron Wood of Jamaica, which Dr. Linnæus has ranged in this genus, has male and female flowers on different trees; and the male flowers which have blown here, are Polyandria, ſo cannot be here placed.

The firſt ſort grows naturally in Peru and Mexico, from both theſe countries I have received the ſeeds. This riſes with a woody ſtalk ten or twelve feet high, dividing into many branches, covered with a dark brown bark; the leaves are placed alternate on the branches; they are compoſed of ſeveral pair of lobes, from ten to fifteen, terminated by one lobe, which is longer than the others, of a lucid green, and when bruiſed emit a turpentine odour. The flowers are produced in looſe bunches at the end of the branches; they are very ſmall, white, and compoſed of five ſmall petals, which have ſmall empalements, indented in five parts at the brim.

This plant is propagated by ſeeds, which ſhould be ſown in pots, and plunged into a moderate hot-bed. If the ſeeds are good, the plants will appear in about five or ſix weeks; and if they are properly managed, by admitting freſh air daily to them, according to the warmth of the ſeaſon, they will be fit to tranſplant in about five or ſix weeks after, when they ſhould be each planted in a ſmall pot, and plunged again into a moderate hot-bed, ſhading them from the ſun till they have taken freſh root; then they muſt be gradually inured to the open air, into which they ſhould be removed ſoon after, placing them in a ſheltered ſituation, where they may remain till autumn; but they muſt be removed into ſhelter before the firſt froſts, otherwiſe their tops will be killed, and thereby the plants are frequently deſtroyed.

Theſe plants are tender when young, ſo require a little warmth in winter; but after two or three years growth, they will live in a green-houſe, where, as they retain their leaves all the year, they will make a variety It may alſo be propagated by layers and cuttings; the layers ſhould be put down in the ſpring, and by the following ſpring they will be rooted; the cuttings ſhould be planted in April, which will put out roots in about two months, and may afterward be treated as the ſeedling plants.

The ſecond ſort grows naturally in the Weſt-Indies. This riſes with a ligneous ſtalk eight or ten feet high, ſending out many branches, which have a grayiſh bark, garniſhed with winged leaves, whoſe midrib is bordered and jointed, armed with crooked ſpines under each joint. The lobes are ſmall, oblong, of a lucid green; the ſpines are ſhort and crooked. This ſort has not as yet produced flowers in the Engliſh gardens, ſo I can give no farther deſcription of the plant.

It is propagated by ſeeds, which ſhould be ſown in ſmall pots, and plunged into a hot-bed of tanners bark; theſe ſeeds will often lie three or four months in the ground, and ſometimes a whole year; therefore if the plants ſhould not come up the ſame year, the earth ſhould not be diſturbed in the pots, but placed in the winter in the bark-bed in the ſtove, and the ſpring following plunged again into a freſh hot-bed, which will bring up the plants if the ſeeds are good. When the plants are come up, and are fit to remove, they ſhould be each planted in a ſeparate ſmall pot, and plunged into the tan-bed, where they muſt be ſhaded till they have taken new root, after which they muſt be treated as other tender ſtove-plants.

SCILLA. Lin. Gen. Plant. 378. Squills.

The CHARACTERS are,

The flower has no empalement; it has ſix oval petals, which ſpread open like a ſtar, and ſix awl-ſhaped ſtamina, terminated by oblong proſtrate ſummits. It has a roundiſh germen, ſupporting a ſingle ſtyle, crowned by a ſingle ſtigma. The germen afterward becomes a ſmooth oval capſule with three furrows, divided into three cells, filled with roundiſh ſeeds.

The SPECIES are,

1. SCILLA (*Maritima*) radice tunicatâ. Hort. Cliff. 123. *Squill with a coated root; or common Squill.*

2. SCILLA (*Lilio-hyacinthus*) radice ſquamatâ. Hort. Cliff. 123. *Squill with a ſcaly root.*

3. SCILLA (*Italica*) radice ſolidâ, corymbo conferto hemiſpherico. Lin. Sp. Plant. 308. *Squill with a ſolid root, and an hemiſpherical corymbus of flowers.*

4. SCILLA (*Peruviana*) radice ſolidâ, corymbo conferto conico. Lin. Sp. Plant. 309. *Squill with a ſolid root, and a conical corymbus of flowers.*

5. SCILLA (*Amœna*) radice ſolidâ, floribus lateralibus alternis ſubnutantibus. Hort. Cliff. 123. *Squill with a ſolid root, and flowers growing alternately from the ſides of the ſtalk, which nod.*

6. SCILLA (*Bifolia*) radice ſolidâ, floribus lateralibus erectiuſculis paucioribus. Hort. Cliff. 123. *Squill with a ſolid root, and erect flowers growing thinly.*

7. SCILLA (*Autumnalis*) radice ſolidâ, foliis filiformibus linearibus floribus corymboſis, pedunculis nudis adſcendentibus longitudine floris. Lin. Sp. Plant. 309. *Squill with a ſolid root, ſlender linear leaves, flowers growing in a corymbus, and naked foot-ſtalks.*

8. SCILLA (*Hiſpanica*) radice ſolidâ, floribus paniculatis ſubnutantibus. *Squill with a ſolid root, and flowers growing in panicles, which nod.*

9. SCILLA (*Purpurea*) radice folidâ, racemo conico, floribus numerofis adfcendentibus. *Squill with a folid root, and a conical fpike of many flowers, rifing above each other.*

10. SCILLA (*Eriophora*) radice folidâ, corymbo cónferto hemifpherico, fcapo longiffimo. *Squill with a folid root, an hemifpherical corymbus, and the longeft ftalk.*

The firft is the Squill or Sea Onion, whofe roots are ufed in medicine, of which there are two forts, one with a red, and the other a white root, which are fuppofed to be accidental varieties, but the white are generally preferred for medicinal ufe. The roots are large, fomewhat oval-fhaped, compofed of many coats, lying over each other like Onions; at the bottom come out feveral fibres. From the middle of the root arife feveral fhining leaves, which continue green all the winter, and decay in the fpring; then the flower-ftalk comes out, which rifes two feet high, is naked half way, and terminated by a pyramidal thyrfe of flowers, which are white, compofed of fix petals, which fpread open like the points of a ftar. This grows naturally on the fea-fhores, and in the ditches, where the falt water flows with the tide, in moft of the warm parts of Europe, fo cannot be propagated in gardens, the froft in winter always deftroying the roots, and for want of falt water they do not thrive in fummer. Sometimes the roots, which are brought for ufe, put out their ftems, and produce flowers without being planted in earth, as they lie in the druggifts fhops.

The fecond fort grows naturally in Spain, Portugal, and the Pyrenees. This has a fcaly root like the Lily, for which reafon Tournefort feparated it from the ftarry Hyacinth, and conftituted a genus of it with the title of Lilio-Hyacinthus. The root is oblong and yellow, very like thofe of Martagon; the leaves are fhaped like thofe of the white Lily, but fmaller; the ftalk is flender, and rifes a foot high; it is terminated by blue flowers like thofe of the ftarry Hyacinth.

The third fort grows naturally in Portugal. This hath a roundifh, folid, bulbous root, like the Hyacinth. The leaves come out fparfedly, are very like thofe of the Englifh Hair-bells; the ftalk rifes feven or eight inches high, and is terminated by cluftered flowers of a pale blue colour, which at firft are difpofed in a fort of umbel or depreffed fpike, but afterward draws up to a point, forming a conical corymbus.

The fourth fort grows naturally in Spain and Portugal. This has been long known in the Englifh gardens by the title of Hyacinth of Peru. There are two varieties of this, one with a deep blue, and the other a white flower; the latter is more rare here than the former. The root of this is large, folid, and raifed in the middle a little pyramidal, covered with a brown coat, from which come out five or feven leaves before winter, of a lucid green, keeled, and fpread almoft flat on the ground. From the center of thefe come out one, two, or three ftalks, according to the ftrength of the root; thefe are thick, fucculent, fix or eight inches high, terminated by a conical corymbus of flowers, of a deep blue in fome, and others are white, ftanding upon pretty long foot-ftalks; they are compofed of fix petals, which fpread open like a ftar. In the center of the petals is fituated a large roundifh germen, fupporting a fhort ftyle, crowned by a fingle ftigma, and round the germen come out fix fhort ftamina, which fpread afunder, terminated by oblong proftrate fummits. The germen afterward turns to a roundifh three-cornered capfule, having three cells, which are filled with roundifh feeds.

The fifth fort grows naturally in Byzantium, and was introduced here about the year 1590. The root of this is large, folid, and of a purplifh colour, from which come out five or fix leaves a foot long, which are keeled, channelled, and of a lucid green; between thefe arife two, three, or four purplifh ftalks about eight or nine inches high, fuftaining toward the top five or fix ftar-flowers, which come out fingly from the fide of the ftalk; they are of a Violet blue colour, having a prominent germen in the center, fupporting a flender ftyle, attended by fix flender ftamina, terminated by purple fummits.

The fixth fort is commonly known in the gardens by the title of early ftarry Hyacinth. There are two varieties of this, one with a deep blue, and the other with a white flower; they grow naturally in fome parts of France and Germany. The roots are folid, roundifh, about the fize of a Hazel-nut, from which comes out a flender channelled ftalk four or five inches high, having generally two leaves near the bottom, one fituated above the other, which embrace the ftalk with their bafe. The flowers are thinly placed toward the top of the ftalk; the lower ones have foot-ftalks an inch long, but thofe of the upper fhorten gradually to the top; they are compofed of fix petals, fpreading in form of a ftar, having a turgid germen in the center, fupporting a fhort ftyle, attended by fix ftamina, which in the blue flowers are of the fame colour, and thofe in the white flowers are white.

The feventh fort is the fmall autumnal ftarry Hyacinth, which grows naturally in feveral parts of England, particularly on St. Vincent's Rock near Briftol, at the Lizard-Point in Cornwall, upon Blackheath in Kent, and Richmond-green. This has a round, white, bulbous root, from which come forth a few narrow leaves about fix inches long. In the center of thefe arife one or two flender ftalks five or fix inches high, naked, fuftaining a fmall corymbus of flowers at the top, which are fmall, ftar-pointed, and of a pale blue colour; thefe appear the beginning of September, at which time the leaves come out, and continue growing all the winter, and in the fpring they die away.

The eighth fort grows naturally in Spain and Portugal. It hath an oblong, white, bulbous root, from which come out five or fix leaves a foot long, a little keeled. The flower ftalk rifes nine or ten inches high, is firm, and fuftains many ftarry flowers at the top, difpofed in a loofe panicle, each ftanding upon a pretty long foot-ftalk, which is erect, but the flower nods on one fide; they are of a deep blue Violet colour, having a prominent germen, which afterward turns to a three-cornered capfule, having three cells, filled with roundifh feeds.

The ninth fort grows naturally in Italy. This hath a folid, white, bulbous root, from which arife feveral leaves like thofe of the common fort. The ftalk rifes ten or eleven inches high, terminated by a conical racemus of flowers, of a deep purple colour.

The tenth fort has a very large bulbous root, from which come out feveral leaves, which at firft are upright, but afterward bend toward the earth; they are of a thick fubftance and keeled, of a lucid green, with downy threads on their under fide. Between the leaves arife the flower-ftalk, which is a foot and a half long, round, firm, and naked, fuftaining at the top a large clufter of flowers, formed into an hemifpherical corymbus: thefe have fix petals, which fpread open in form of a ftar, of a purple colour, and have blue bottoms, with a dark blue vein running lengthwife in the middle of each petal.

There is another fort of this, which grows naturally in the Levant, whofe leaves are fhaped like thofe of the Peruvian Hyacinth, but longer, and ftand erect; this propagates very faft by offsets, but never flowers here. I have kept the roots in all fituations almoft forty years, and have not feen one flower.

Thefe plants are all of them hardy, and may be propagated by feeds or offsets, the latter being the more expeditious

ious way, is generally practised. The roots may be transplanted after the leaves are decayed, but, if they are removed after they have put out new fibres, they rarely succeed, at least they will not flower the following spring; they may be treated in every respect like the ordinary kinds of Hyacinths.

If they are propagated by seeds, they should be sown in autumn, soon after they are ripe, either in shallow boxes or pans, in the same manner as has been before directed for Hyacinths, to which the reader is desired to turn, to avoid repetition.

SCLAREA. Tourn. Inst. R. H. 179. tab. 82. Clary.

The CHARACTERS are,

The flower has a tubulous empalement, which widens at the top, and has five acute points at the brim; it is of the lip kind, with one petal, having a crooked tube, which enlarges at the chaps, where it is divided into two lips; the upper lip is erect and arched; the under lip is cut into three segments; it has two stamina, which are situated under the upper lip, terminated by oblong erect summits, and a four-pointed germen, supporting a forked style longer than the upper lip, crowned by a bifid stigma. The germen afterward becomes four roundish seeds, which ripen in the empalement.

The SPECIES are,

1. SCLAREA (*Vulgaris*) foliis rugosis oblongo-cordatis serratis, floribus calyce longioribus concavis acuminatis. *Clary with rough, oblong, heart-shaped, sawed leaves, those among the flowers concave, pointed, and longer than the empalement; or common Clary.*

2. SCLAREA (*Æthiopis*) foliis oblongis erosis lanatis, verticillis lanatis. *Clary with oblong, angular, indented, woolly leaves, and the whorls of the flowers covered with down.*

3. SCLAREA (*Lusitanica*) foliis oblongo-ovatis dentato-serratis tomentosis, verticillis lanatis sessilibus. *Clary with oblong oval leaves, which are woolly, indented like a saw, and woolly whorls of flowers sitting close to the stalk.*

4. SCLAREA (*Pratensis*) foliis cordato-oblongis crenatis summis amplexicaulibus, verticillis subnudis, corollarum galeis glutinosis. *Clary with oblong, heart-shaped, crenated leaves, those on the top embracing the stalk, almost naked whorls, and the helmet of the flower glutinous.*

5. SCLAREA (*Syriaca*) foliis lanceolatis obsolete crenatis subtus tomentosis, verticillis minoribus subnudis. *Clary with spear-shaped leaves, which are slightly crenated, woolly on their under side, and very small whorls of flowers, which are almost naked.*

6. SCLAREA (*Nemerosa*) foliis cordato oblongis crenatis glabris, floribus verticillato-spicatis. *Clary with oblong, heart-shaped, crenated, smooth leaves, and spiked whorled flowers.*

7. SCLAREA (*Sylvestris*) foliis cordato-lanceolatis acutis, bracteis coloratis, flore brevioribus. *Clary with heart-shaped acute-pointed leaves, and coloured bracteæ, which are shorter than the flower.*

8. SCLAREA (*Ceratophylla*) foliis rugosis pinnatifidis lanatis. *Clary with rough, wing-pointed, woolly leaves.*

9. SCLAREA (*Indica*) foliis cordatis acute crenatis, summis sessilibus, verticillis subnudis remotissimis. *Clary with heart-shaped leaves, which are sharply crenated, those on the top sitting close to the stalks, and naked whorls placed far asunder.*

10. SCLAREA (*Orientalis*) foliis lanceolatis acuminatis, serratis, summis sessilibus, floribus verticillato-spicatis. *Clary with spear-shaped, acute-pointed, sawed leaves, the upper ones sitting close to the stalks, and spiked whorled flowers.*

11. SCLAREA (*Glutinosa*) foliis cordato-sagittatis serratis acutis. *Clary with heart-shaped crenated leaves, which are acutely sawed.*

12. SCLAREA (*Tuberosa*) foliis cordato-ovatis rugosis tomentosis, calycibus hispidis, radice tuberosâ. *Clary with oval, heart-shaped, rough, woolly leaves, prickly empalements, and a tuberous root.*

13. SCLAREA (*Tomentosa*) foliis hastato-triangularibus obsolete crenatis, caule tomentoso paniculato. *Clary with triangular halberd-pointed leaves, which are slightly crenated, and a woolly paniculated stalk.*

14. SCLAREA (*Mexicana*) foliis obtusis erosis, staminibus corolla duplo longioribus. *Sclary with obtuse bitten leaves, and stamina twice the length of the petals.*

15. SCLAREA (*Argentea*) foliis dentato-angulatis lanatis, verticillis summis sterilibus, bracteis concavis. *Sclarea with angular indented woolly leaves, concave bractea, and the upper whorls of flowers barren.*

The first sort grows naturally in Syria, but has been long cultivated in the European gardens, both for the kitchen and shops: it is a biennial plant, which perishes after it has borne seeds. The lower leaves of this are large, rough, wrinkled, oblong, and heart-shaped, and are sawed on their edges. The stalks are large, four-cornered, and clammy, about two feet high, garnished with leaves of the same shape as those at bottom, but smaller, sending out small side branches opposite; the flowers are disposed in large loose spikes at the top of the stalks, placed in whorls round them, of a pale blue colour, having two short hollow acute-pointed leaves under each, of a whitish colour. The empalement of the flower is divided into two lips, the upper ending in three, and the under in two spiculæ. The upper lip of the flower stands erect, is arched at the top, under which is the style, which is nearly of the same length, and the two stamina, which are shorter, sit close to the style. After the flowers are past, the germen turns to four roundish seeds, which ripen in the empalement. The whole plant has a very strong scent.

It is propagated by seeds, which should be sown in the spring, and when the plants are fit to remove, they should be either transplanted into beds, or if a large quantity is required, they may be planted in an open spot of ground in rows two feet asunder, and one foot distance in the rows. After the plants have taken root, they will require no farther care but to keep them clean from weeds. The winter and spring following the leaves, which are the only part used, will be in perfection, and in the summer they will run up to flower, and, after they have ripened their seeds decay, so that there should be annually young plants raised for use.

The second sort grows naturally in Istria and Dalmatia. There are two varieties of this, one with very broad leaves, which are slightly indented on the sides; the other has longer leaves, which are deeply jagged. The leaves of both sorts are of a thick substance, very woolly, especially on their under side; their upper sides are rugged, and wrinkled like the first sort. The stalks are square, two feet high, sending out many branches opposite, garnished in the first with entire, oval, acute-pointed leaves, which embrace the stalks with their base, but those of the other are long, narrow, and have several deep indentures on their edges. The stalk and branches are terminated by spikes of flowers in whorls; under each of these whorls are two hollow green leaves, which are shorter than the empalement of the flowers; these empalements are divided into two lips, the upper ending in three, and the under in two spiculæ. The under lip or beard of the flower is white, and the helmet or galea is of a pale blue colour; the plants may be propagated in the same way as the first.

The third sort has some resemblance of the second; but the leaves are larger, very woolly, and glutinous, oblong, oval, deeply indented, and end with very acute points. The stalks are woolly, four-cornered, about two feet and

a

a half high, sending out branches by pairs, terminated in loose spikes of whorled flowers, which are smaller than those of the other sorts. It grows naturally in Portugal and also in Syria; it is propagated by seeds in the same way as the first.

The fourth sort grows naturally in some parts of France and Germany; it is generally found in meadows and rich pastures. It has a perennial root, composed of many strong ligneous fibres, from which come out many oblong heart-shaped leaves, of a deep green colour, whose surfaces are rough, crenated on their edges, and stand upon pretty long foot-stalks. The stalks rise three feet high; they are four-cornered; their lower parts are garnished with leaves, whose base embrace them; the flowers grow in long whorled spikes at the top; they are of a fine blue colour, having scarce any small leaves under the whorls. It is propagated by seeds, but the roots continue long.

The fifth sort grows naturally in Syria; this is an abiding plant, whose roots run deep in the ground. The leaves are spear-shaped, crenated on their edges, and a little woolly on their under side. The stalks are slender, stiff, and rise a foot and a half high, garnished with smaller leaves of the same shape, set by pairs; the flowers grow in small whorls, disposed in loose spikes at the top of the stalks; they are small, blue, and shaped like those of the other sorts. It is propagated by seeds in the same way as the other sorts.

The sixth sort grows naturally on the sides of highways about Vienna and in Hungary. This has an abiding root, sending out smooth leaves about the size and shape of those of broad-leaved Sage, but are indented on their edges. The stalks are slender, four-cornered, and rise a foot and a half high: their lower parts are garnished with leaves like those at the bottom, but smaller, and are terminated by small whorls of blue flowers. It is propagated by seeds in the same way as the first sort, but the roots will continue several years.

The seventh sort grows naturally in Austria and Bohemia. This has an abiding root, from which come out many heart spear-shaped leaves, crenated on their edges, of a bright green colour, and have many white spots dispersed on their surface. The stalks are thick, four-cornered, and rise near three feet high, garnished with leaves by pairs, sitting close to the stalks, which are terminated by loose spikes of flowers in small whorls, whose bracteæ are coloured. It is propagated by seeds as the former sorts.

The eighth sort grows naturally in Syria; this is a biennial plant. The leaves are very thick, woolly, narrow, and wing-pointed, cut into obtuse segments, nearly opposite on their sides, almost to the midrib, somewhat like a stag's-horn in shape; these spread flat on the ground. The stalk rises more than a foot high; it is thick, four-cornered, and very woolly, sending out branches by pairs, garnished with narrow long leaves, placed opposite at each joint, sawed on their edges. The flowers grow in loose whorled spikes at the top of the stalks; they are shaped like those of the fourth sort. It may be propagated by seeds in the same way as the first sort, but should have a dry soil, otherwise the plants are apt to rot in winter.

The ninth sort grows naturally in India, but is hardy enough to live in the open air in England. The root of this will abide several years in a dry soil; the lower leaves are heart-shaped, acutely crenated on their edges. The stalk is four-cornered, and rises four or five feet high, having two or three pair of smaller leaves on the lower part, standing opposite. The upper part of the stalk, for the length of two feet, is garnished with whorls of flowers, which stand at two or three inches distance from each other, having no leaves under them. The empalement of the flower is hairy and blunt; the galea or helmet of the flower is arched, erect, blue, terminating in a blunt point; the two side segments of the under lip are of a Violet colour; the middle segment, which is indented at the point, is white, and curiously spotted with Violet on the inside; the two side indentures turn yellow before the flower drops. When the flower is past, the germen turns to four large roundish seeds, which ripen in the empalement. It is propagated by seeds in the same manner as the other sorts.

The tenth sort grows naturally in the Levant. This hath a perennial root, from which come out many spear-shaped leaves, of a dark green colour, sawed on their edges, ending in acute points. The stalks rise three feet high, sending out branches by pairs, garnished with leaves, which toward the top sit close to the stalk. The flowers grow in whorled spikes at the top; they are of a bright blue colour, and the top of the spike is terminated with very deep blue flowers. It is propagated by seeds in the same manner as the other sorts, and the roots will abide many years.

The eleventh sort grows naturally in moist land both in Germany and Italy; this hath an abiding root, composed of strong ligneous fibres. The leaves are heart-shaped, pointed like a halberd, of a pale or yellowish green colour, and sawed on their edges. The stalks are strong, four-cornered, and rise near four feet high, garnished below with smaller leaves, but the upper part of the stalk is closely set with whorls of large yellow flowers. The whole plant is very clammy and has a strong scent somewhat like the first species. This is propagated by seeds in the same way as the other sorts; it is very hardy, and will continue several years, and may be increased by parting of the roots in autumn.

The twelfth sort grows naturally in Spain. This has a perennial root. The lower leaves are oval and spear-shaped; the stalks rise two feet high; they are four-cornered, and send out branches by pairs; the leaves on the upper part of the stalks are heart-shaped, and embrace the stalks with their base; the flowers are of a brimstone colour, shaped like those of the first sort; the style of this is much longer than the upper lip, and is terminated by a bifid stigma; the empalements are hairy, and end with acute points. It is propagated by seeds, which may be sown in the same way as the other sorts, and the plants may be treated in like manner.

The thirteenth sort grows naturally in the Canary Islands. This hath a perennial shrubby stalk, which rises five or six feet high, dividing into many branches, covered with a flocky down, and garnished with halberd-shaped triangular leaves, placed opposite, standing upon long woolly foot-stalks. The top of the stalk branches out in many foot-stalks, forming a sort of panicle. The flowers are of a light blue colour, ranged in whorled spikes, having two small leaves under each whorl. It is propagated by cuttings, which may be planted any time in summer; after they have put out good roots, they should be each planted in a separate small pot, placing them in the shade till they have taken new root; then they may be placed among other hardy kinds of green-house plants in a sheltered situation till October, when they should be removed into shelter before hard frost comes on; but as they only require protection from hard frost, so they should have as much free air as possible in mild weather.

The fourteenth sort grows naturally in Mexico. This rises with a shrubby stalk eight or ten feet high, sending out slender, four-cornered, hairy branches, of a purplish colour, garnished with oval leaves, pointed at both ends, and sawed on their edges, of a pale green colour, and hairy on their under side. The flowers grow in close thick spikes at the end of the branches, of a fine blue colour, and appear in winter, so make a pretty variety in the green-house at that

 season.

feafon. This plant never produces feeds in England, fo it is only propagated by cuttings, which may be planted during any of the fummer months, in the fame manner as the former fort; and the plants may be treated afterward in the fame way, with this difference, which is to give it a dry fituation in winter, for the young fhoots are very apt to grow mouldy upon being in a damp air.

There are fome other forts of lefs note, which are preferved in botanic gardens for the fake of variety; but thofe here mentioned are worthy of a place in large gardens, where, if they are intermixed among other large growing plants, they will afford a pretty variety, efpecially the fifth, eighth, tenth, and eleventh forts, which produce long fpikes of beautiful flowers, and continue a long time in flower. The flowers of the eleventh fort are ufed in Holland, to give a flavour to the Rhenifh wines, which are brewed at Dort.

SCOLYMUS. Tourn. Inft. 480. tab. 273. The Golden Thiftle.

The CHARACTERS are,

It hath a flower compofed of many hermaphrodite florets, included in an oval imbricated empalement, having many loofe fharp-pointed fcales. The florets are tongue-fhaped, flightly indented in five parts. They have five fhort hair like ftamina, terminated by tubulous fummits. The germen is fituated under the floret, fupporting a flender ftyle crowned with two reflexed ftigmas. The germen afterward becomes a fingle feed, which is oblong, triangular, and ripens in the empalement, the feeds being feparated by plain, roundifh, indented chaff.

The SPECIES are,

1. SCOLYMUS (*Maculatis*) foliis margine attenuatis. Lin. Sp. Plant. 813. *Golden Thiftle with leaves whofe edges are thin; or annual Golden Thiftle.*

2. SCOLYMUS (*Hifpanicus*) foliis margine incraffatis. Lin. Sp. Plant. 813. *Golden Thiftle with leaves which are thicker on the borders.*

The firft fort grows naturally in the fouth of France and Italy. This is an annual plant, which rifes with a branching ftalk five or fix feet high, having two leafy wings running along the fides from joint to joint, which are fcoloped and indented; the borders are thinner than the other parts, and are armed every way with very fharp fpines; they are of a pale green, and fit clofe to the ftalks. The flowers are produced at the top of the ftalks, inclofed in leafy involucrums, which are longer than the flowers, and armed with very ftrong fpines; within thefe are fcaly empalements, which lie over each other like the fcales of fifh, armed with fhort fpines. The flowers are compofed of many golden florets.

The fecond fort grows naturally in Italy and Sicily. This hath a biennial root, from which fpring up thick ftalks about four feet high, branching out on the fides, garnifhed with ftiff jagged leaves, whofe borders are thicker than the other part, armed with fpines like the former fort; the ftalks have leafy borders as the other, which are ftrongly armed with fpines. The flowers are produced at the top of the ftalks, and are fhaped like thofe of the former fort.

They are propagated by feeds, which fhould be fown in March on a bed of frefh undunged earth in an open fituation; when the plants are come up they fhould be kept clear from weeds, and where they grow too clofe, fome of them fhould be pulled out, fo as to leave thofe which are defigned to remain about two feet afunder. This is all the culture which thefe plants require, for as they fend forth taproots, they do not bear tranfplanting well; therefore they muft be fown where they are to remain. When the feafons prove dry the plants will perfect their feeds in autumn, but in wet feafons they rarely ever produce good feeds in England, which renders it difficult to continue the fpecies without procuring frefh feeds from abroad.

Thefe plants are preferved by thofe perfons who are curious in botany for variety's fake, but are rarely admitted in other gardens.

SCOPARIA. Sweet-weed, or Wild Liquorice.

The CHARACTERS are,

The empalement of the flower is cut into four concave fegments; the petal is alfo divided into four equal parts. It has four equal ftamina, crowned by fhort fummits, with a conical ftyle, terminated by an acute ftigma. The flower is fucceeded by a conical capfule, containing many fmall feeds.

We have but one SPECIES of this plant, viz.

SCOPARIA (*Dulcis*) foliis ternis, floribus pedunculatis. Lin. Sp. Plant. 116. *Wild Liquorice, or Sweet-weed.*

This plant grows naturally in moft of the iflands in the Weft-Indies, where it often is very troublefome in the plantations, for there is a great quantity of feeds upon each plant, which, if fuffered to fcatter, the plants appear foon after, and become troublefome weeds.

The ftalks rife about two feet high; they have fix angles, branching out on their upper part, and at each joint have three obtufe fawed leaves, ftanding upon fhort foot-ftalks; the flowers come out from the wings of the ftalk upon foot-ftalks, two, three, or four arifing from the fame place; they are white, a little fringed on their edges, and are fucceeded by conical capfules, opening with two valves, filled with fmall feeds.

There is a great affinity between this plant and the Capraria, fo that they have been often confounded by botanifts, and even thofe who have been more accurate have joined them in the fame genus; but their difference confifts in the form of the flower and the length of their ftamina. The four fegments in the petals of this are equal, as are alfo the four ftamina; whereas the petal of Capraria is ringent, and two of the ftamina are fhorter than the other.

This is an annual plant, preferved in curious gardens for the fake of variety, but has little beauty. The feeds muft be fown upon a hot-bed, and the plants tranfplanted on another hot-bed to bring them forward; toward the end of May, when the weather is good, they may be tranfplanted with balls of earth to their roots to a warm border, where they will flower, and will ripen their feeds in autumn.

SCORDIUM. See TEUCRIUM.

SCORPIURUS. Lin. Gen. Plant. 192. Caterpillars.

The CHARACTERS are,

The empalement of the flower is of one leaf, erect, blown up, lightly compreffed, ending in five acute points. The flower is of the butterfly kind; it has a roundifh ftandard, indented at the point, where it is reflexed and fpreading. The wings are loofe, almoft oval, having obtufe appendages. The keel is half-moon-fhaped; the belly is gibbous, pointed, and erect, cut into two parts below. It hath ten ftamina, nine joined, and one feparate, terminated by fmall fummits; and an oblong taper germen a little reflexed, fupporting a rifing inflexed ftyle, terminated by a point for a ftigma. The germen afterward becomes an oblong, taper, leathery, rough, channelled pod, twifted in many longitudinal cells, divided within, and on the outfide contracted into knotty joints, each cell containing one feed.

The SPECIES are,

1. SCORPIURUS (*Vermiculata*) pedunculis unifloris, leguminibus tectis undique fquamis obtufis. Lin. Sp. Plant. 744. *Caterpillar with one flower upon a foot-ftalk, and a pod covered with obtufe fcales on every fide.*

2. SCORPIURUS (*Muricata*) pedunculis bifloris, leguminibus extrorfum obtufè aculeatis. Lin. Sp. Plant. 745. *Caterpillar with two flowers on each foot-ftalk, and the outfide of the pods armed with blunt fpines.*

3. SCOR-

3. SCORPIURUS (*Falcata*) pedunculis fubtrifloris, leguminibus extrorfum fpinis diftinctis acutis. Lin. Sp. Plant. 745. *Caterpillar with foot-ftalks having three flowers, and the outfide of the pods armed with fharp diftinct fpines.*

4. SCORPIURUS (*Subvillofa*) pedunculis fubquadrifloris, leguminibus extrorfum fpinis confertis acutis. Lin. Sp. Pl. 745. *Caterpillar with four flowers fometimes upon a foot-ftalk, and the outfide of the pods armed with fharp fpines, which grow in clufters.*

5. SCORPIURUS (*Pinnata*) foliis pinnatis. *Caterpillar with a winged leaf.*

The firft fort grows naturally in Italy and Spain. This is an annual plant, with trailing herbaceous ftalks, and at each joint have one fpatule-fhaped leaf with a long foot-ftalk. From the wings of the leaves come out the foot-ftalks of the flowers, which fuftain at the top one yellow butterfly flower, which is fucceeded by a twifted thick pod, in fize and appearance of a large caterpillar, from whence it had this title.

The fecond fort has ftronger ftalks than the firft; the leaves are much broader; the foot-ftalks fupport two fmaller flowers; the pods are flender, longer, and more twifted than thofe of the firft, and are armed with blunt fpines on their outfide.

The third fort hath flenderer ftalks than either of the former: the leaves ftand upon fhorter foot-ftalks, but are fhaped like thofe of the firft fort; the foot-ftalks of the flowers are flender, and frequently fupport three flowers; the pods are flender, not fo much twifted as the former, and are armed on their outfide with fharp diftinct fpines.

The ftalks and leaves of the fourth fort are very like thofe of the firft, but the foot-ftalks of the flowers are longer, and each of them have three or four fmall yellow flowers at the top; the pods are very flender, greatly contorted, and armed with fharp fpines on their outfide.

The fifth fort hath very fhort ftalks; the leaves are winged, compofed of four pair of fmall lobes, terminated by an odd one. The flowers are very fmall, as are alfo their pods, which are lefs twifted than thofe of the three former.

All thefe plants are annual, and grow naturally in moft of the warm countries in Europe, but the firft fort has been long cultivated in the Englifh gardens.

The plants are preferved in feveral curious gardens for their oddnefs more than for any great beauty: they are propagated by fowing their feeds upon a bed of light earth; and when the plants are come up they muft be kept clean from weeds, and fhould be thinned fo as to leave them about ten inches or a foot afunder, becaufe their branches trail upon the ground; and if they have not room they are apt to overbear each other.

Thefe plants feldom thrive well if they are tranfplanted; therefore the beft method is to put in three or four good feeds in each place where you would have the plants remain. When the plants come up there fhould be only one of the moft promifing left in each place, which fhould be conftantly kept clear from weeds; and when their pods are ripe, they fhould be gathered and preferved in a dry place till the following fpring, to be fown.

The firft fort is the beft worth cultivating, the pods being large and more vifible than the other, and are more in form of a caterpillar.

SCORZONERA. Tourn. Inft. R. H. 476. tab. 269. Vipers-grafs.

The CHARACTERS are,

The common empalement is fcaly, cylindrical, and imbricated. The flower is compofed of feveral hermaphrodite florets, which are narrow, tongue-fhaped, and indented in five parts. They have five fhort hair-like ftamina, terminated by cylindrical fummits. The germen is fituated under the floret, fupporting a flender ftyle, crowned by two reflexed ftigmas. The germen afterward turns to a fingle, oblong, channelled feed, crowned with a feathery down.

The SPECIES are,

1. SCORZONERA (*Hifpanica*) caule ramofo, foliis amplexicaulibus integris ferrulatis. Hort. Cliff. 383. *Common Scorzonera.*

2. SCORZONERA (*Humilis*) caule fubnudo unifloro, foliis lato-lanceolatis nervofis planis. Hort. Cliff. 382. *Scorzonera with an almoft naked ftalk, and broad plain leaves having nerves.*

3. SCORZONERA (*Graminifolia*) foliis lineari-enfiformibus integris carinatis. Lin. Sp. Pl. 791. *Portugal Vipers-grafs with a Grafs leaf.*

4. SCORZONERA (*Purpurea*) foliis lineari-fubulatis integris planis, pedunculis cylindricis. Lin. Sp. Plant. 791. *Scorzonera with linear, awl-fhaped, entire, plain leaves, and cylindrical foot-ftalks.*

5. SCORZONERA (*Anguftifolia*) foliis fubulatis integris, pedunculo incraffato, caule bafi villofo. Lin. Sp. Pl. 791. *Scorzonera with awl-fhaped entire leaves, a thick foot ftalk, and the ftalk hairy at its bafe.*

6. SCORZONERA (*Laciniata*) foliis linearibus dentatis acutis, caule erecto, fquamis calycinis patulo-mucronatis. Lin. Sp. Plant. 1114. *Scorzonera with narrow acute-indented leaves, and an erect ftalk.*

7. SCORZONERA (*Refedifolia*) foliis obtufe-dentatis, caule proftrato calycum apicibus tomentofis. Lin. Sp. Pl. 1113. *Scorzonera with obtufe indented leaves, a proftrate ftalk, and the tops of the empalement woolly.*

The firft is the fort which is commonly cultivated in the Englifh gardens for food and phyfic; this grows naturally in Spain. The root is Carrot-fhaped, about the thicknefs of a finger, covered with a dark brown fkin, is white within, and has a milky juice; the lower leaves are long, ending with a long acute point; they are waved and finuated at their edges. The ftalk rifes three feet high, is fmooth, branching at the top, and garnifhed with a few narrow leaves, whofe bafe half embrace the ftalk. The flowers terminate the ftalks in fcaly empalements, compofed of many narrow, tongue-fhaped, hermaphrodite florets, lying imbricatim over each other like the fcales on fifh; they are of a bright yellow colour. After thefe are decayed, the germen which fits in the common empalement, turns to oblong cornered feeds, having a roundifh ball of feathery down at the top.

The fecond fort is like the firft, but the leaves are broader, entire, and are more veined; the ftalk does not rife fo high, and branches more.

The third fort is fhorter than either of the former; the leaves are narrower; the ftalk is almoft naked, and has one yellow flower at the top.

The fourth fort has narrow awl-fhaped leaves, which are fhorter than thofe of the former; the ftalk is taper, and branches at the top; the flowers are of a pale purple colour.

The fifth fort grows a foot and a half high; the leaves are narrow and awl-fhaped; the foot-ftalk immediately under the flower is thicker than below, and the lower part of the ftalk is hairy; the flower is yellow.

The fixth fort rifes with a fmooth branching ftalk two feet high, garnifhed with narrow leaves, having many winged points, refembling thofe of Buckfhorn Plantain, but larger. The flowers are yellow, and ftand upon long naked foot-ftalks at the end of the branches.

The feventh fort is exactly like the former in every refpect, excepting that of the ftalks fpreading on the ground; which is not accidental, for I have cultivated both forts above

above thirty years, and have never found either of them alter.

The firſt ſort only is cultivated for uſe, the others are preſerved in botanic gardens for variety, ſo are ſeldom admitted into other gardens.

Theſe plants may be propagated by ſowing their ſeeds the beginning of April, upon a ſpot of light ground. The beſt method of ſowing them is, to draw ſhallow furrows by a line about a foot aſunder, into which you ſhould ſcatter the ſeeds, thinly covering them over about half an inch thick, with the ſame light earth: when the plants are come up, they ſhould be thinned where they are too cloſe in the rows, leaving them at leaſt ſix inches aſunder; and at the ſame time all the weeds ſhould be cut up. This muſt be repeated as often as is neceſſary, for if the weeds are permitted to grow among the plants, they will draw them up weak.

There are many people who ſow theſe ſeeds promiſcuouſly in a bed, and afterward tranſplant them out at the diſtance they would have them grow; but this is not ſo good a method as the former, becauſe their roots commonly ſhoot downright, which, in being tranſplanted, are often broken, ſo that they never will make ſuch fair roots as thoſe which remain in the ſame place where they are ſown; for when the extreme part of the root is broken, the milky juice flows out ſo faſt, as they ſeldom extend in length afterwards, but only ſhoot out into many forked ſmall roots. Theſe roots may be taken up when the leaves begin to decay, at which time they have done growing, though they may remain in the ground until ſpring, and may be taken up as they are uſed; but thoſe which remain in the ground after February, will ſhoot up their flower ſtems, after which they are not ſo good, being ſticky and ſtrong.

If you intend to ſave ſeeds of theſe plants, you ſhould let a parcel of the beſt remain in the places where they grew; and when their ſtems are grown to their height, they ſhould be ſupported with ſtakes, to prevent their falling to the ground, or breaking. In June they will flower, and about the beginning of Auguſt their ſeeds will ripen, when they ſhould be gathered and preſerved dry till the ſpring following for uſe.

SCROPHULARIA. Tourn. Inſt. R. H. 166. tab. 74. Figwort.

The CHARACTERS are,

The flower has a permanent empalement, cut into five parts at the top. It hath one unequal petal, with a large globular tube. The brim is cut into five ſmall parts; the two upper are large and erect; the two ſide ones ſpread open, and the under is reflexed. It has four ſlender deflexed ſtamina, two of which are the length of the petal, and two are ſhorter, terminated by twin ſummits; and an oval germen ſupporting a ſingle ſtyle, crowned by a ſingle ſtigma. The germen afterward turns to a roundiſh pointed capſule with two cells, which open at the top, and are filled with ſmall ſeeds.

The SPECIES are,

1. SCROPHULARIA (*Nodoſa*) foliis cordatis baſi tranſverſis, caule acutangulo. Lin. Sp. Pl. 863. *Nobbed-rooted ſtinking Figwort.*

2. SCROPHULARIA (*Aquatica*) foliis cordatis petiolatis decurrentibus obtuſis, caule membraneo angulato, racemis terminalibus. Hort. Upſal. 177. *Great Water Figwort, called Water Betony.*

3. SCROPHULARIA (*Sulphurea*) foliis cordato-oblongis, baſi appendiculatis, racemis terminalibus. Lœfl. Lin Sp. Plant. 620. *Yellow Figwort with oblong heart-ſhaped leaves, having appendages at their baſe, and ſtalks terminated by a racemus of flowers.*

4. SCROPHULARIA (*Betonicæfolia*) foliis cordato-ſagittatis, acutè ſerratis, racemis terminalibus. *Scrophularia with heart-ſhaped arrow-pointed leaves, which are acutely ſawed, and ſtalks terminated by a racemus of flowers.*

5. SCROPHULARIA (*Scorodonia*) foliis cordatis duplicato ſerratis, racemo compoſitis. *Figwort with heart-ſhaped ſawed leaves, and flowers proceeding from the wings of the ſtalk.*

6. SCROPHULARIA (*Italica*) foliis cordatis duplicato-ſerratis, racemo compoſito. Flor. Leyd. Prod. 296. *Figwort with heart-ſhaped doubly ſawed leaves, and compound bunches of flowers.*

7. SCROPHULARIA (*Trifoliata*) foliis glabris, inferioribus ternato pinnatis obtuſis, ſuperioribus ſimplicibus, pedunculis ſubtrifloris axillaribus. Lin. Sp. Plant. 865. *Spaniſh Figwort with ſmooth Elder leaves.*

8. SCROPHULARIA (*Sambucifolia*) foliis interrupte pinnatis cordatis inæqualibus, racemo terminali, pedunculis axillaribus geminis dichotomis. Lin. Sp. Pl. 865. *Greateſt Portugal Figwort with woolly leaves.*

9. SCROPHULARIA (*Canina*) foliis inferioribus pinnatis, ſummis integris duplicato-ſerratis, racemis axillaribus. *Figwort with the lower leaves winged, thoſe at the top entire, doubly ſawed, and bunches of flowers at the wings of the ſtalk.*

10. SCROPHULARIA (*Filicifolia*) foliis pinnatis, foliolis acutè dentatis, racemis terminalibus. *Figwort with winged leaves, whoſe lobes are acutely indented, and panicles of flowers terminating the ſtalk.*

11. SCROPHULARIA (*Lucida*) foliis bipinnatis lucidis glaberrimis racemis bipartitis terminalibus. *Figwort with linear winged leaves, which are thick, ſhining, wing-pointed lobes, and ſtalks terminated by panicles of flowers.*

12. SCROPHULARIA (*Orientalis*) foliis lanceolatis ſerratis, racemis oppoſitis. Flor. Leyd. Prod. 294. *Figwort with linear ſpear-ſhaped leaves, which are ſharply ſawed, and a compound racemus of flowers.*

13. SCROPHULARIA (*Verna*) foliis cordatis, pedunculis axillaribus ſolitariis dichotomis. Hort. Cliff. 322. *Figwort with heart-ſhaped leaves, ſingle foot-ſtalks proceeding from the wings, which are divided by pairs.*

14. SCROPHULARIA (*Peregrina*) foliis cordatis, ſuperioribus alternis, pedunculis axillaribus bifloris. Hort. Cliff. 322. *Figwort with heart-ſhaped leaves, the upper of which are alternate, and foot-ſtalks proceeding from the wings of the ſtalks, bearing two flowers.*

15. SCROPHULARIA (*Pinnata*) foliis pinnatis, foliolis inciſis, racemis ſimpliciſſimis terminalibus. *Figwort with winged leaves whoſe lobes are cut, and a ſingle thyrſe of flowers terminating the ſtalks.*

16. SCROPHULARIA (*Marylandica*) foliis cordatis ſerratis acutis baſi rotundatis, caule obtuſangulo. Hort. Upſal. 177. *Figwort with heart-ſhaped, acute, ſawed leaves, which are rounded at their baſe, with obtuſe angles to the ſtalks.*

17. SCROPHULARIA (*Fruteſcens*) foliis lanceolatis obtuſis ſerrato-dentatis, pedunculis bifidis. Lin. Sp. Plant. 866. *Portugal Figwort with ſhrubby ſtalks.*

The firſt ſort grows naturally in woods and under hedges in moſt parts of England, ſo is ſeldom admitted into gardens; but being a medicinal plant, it is here mentioned to introduce the others. This hath a ſpreading root, compoſed of many knobs, from which ariſe ſeveral four-cornered ſtalks near three feet high, garniſhed with heart-ſhaped leaves, ſawed on their edges; they are placed by pairs, of a dark green, or browniſh colour on their upper ſide, but pale on their under, having an odour of Elder. The flowers are produced in ſmall cluſters from the ſides of the ſtalks oppoſite, forming a kind of looſe ſpike to the top; they are of one petal, of a dark purple colour, ſhaped almoſt like a lip flower; the upper lip or creſt being a little arched, the two ſide ſegments ſpread open, and the under

ſegment

ſegment is recurved. Theſe are ſucceeded by roundiſh capſules, ending in acute points, having two cells filled with ſmall ſeeds.

The ſecond ſort grows naturally by the ſide of ditches and watery places in moſt parts of England. This has a fibrous root, ſending out ſtrong four-cornered ſtalks, which grow near four feet high, garniſhed with heart-ſhaped leaves rounded at their points, and crenated on their edges ſomewhat like thoſe of Betony, from whence it has been titled Water Betony. The flowers are larger than thoſe of the former, and a little redder, but of the ſame ſhape. This ſort is ſometimes uſed in medicine, but as it grows wild it is ſeldom admitted into gardens. There is a variety of this with variegated leaves, which is by ſome preſerved in gardens.

The third ſort grows naturally in Italy and Spain, by the ſide of rivers and other moiſt places. The ſtalks of this are ſtronger, taller, and greener than thoſe of the former; the leaves have generally ſmall appendages at their baſe; the flowers are greener, and grow thinner upon the ſtalks than thoſe of the former. In theſe particulars conſiſt their differences.

The fourth ſort grows naturally in Sicily; this has a fibrous root. The ſtalks riſe near four feet high, and have ſharp angles; the leaves are arrow-pointed, heart-ſhaped, and are ſharply ſawed on their edges; the flowers terminate the ſtalks in looſe panicles; they are in ſhape like thoſe of the former, but of a dark red colour.

The fifth ſort grows naturally in Italy; it has a perennial root. The ſtalks riſe four feet high, branch out on their ſide, and are garniſhed with heart-ſhaped ſawed leaves, which on the upper part of the ſtalk are placed alternate. The flowers are produced in looſe panicles at the wings of the ſtalk, each foot-ſtalk ſupporting two flowers; they are ſmall, and of a brown colour.

The ſixth ſort grows naturally in Sicily; this hath a perennial root. The ſtalks riſe four feet high, garniſhed with heart-ſhaped leaves, which are doubly ſawed on their edges; the flowers are diſpoſed in compound ſpikes, which ſit upon long foot-ſtalks, which ariſe from the wings of the ſtalks, and have generally two narrow leaves placed at their baſe, but the flowers terminate the ſtalks like the three firſt ſorts.

The ſeventh ſort grows naturally in Spain; this has a perennial root. The leaves at the bottom are irregularly cut, and have two appendages at their baſe; they are ſmooth, of a lucid green, and ſawed on their edges. The ſtalks riſe four feet high, four-cornered, ſmooth, and garniſhed with oval leaves, ſome of which are entire, and others have ſmall lobes or appendages at their baſe. The flowers grow from the wings of the ſtalks in cluſters, each ſtanding upon a ſeparate foot-ſtalk; they are of a bright red colour with greeniſh bottoms, and much larger than either of the former.

The eighth ſort grows naturally in Portugal; this reſembles the ſeventh, but the ſtalks are larger and riſe higher. The leaves are much longer, have four appendages, and are irregularly ſawed on their edges, running out into longer points; the whole plant is hairy; the flowers grow in compound bunches at the wings of the ſtalks; they are larger than thoſe of the former ſort, and have a greater mixture of green in them.

The ninth ſort grows naturally in Italy; this has a root compoſed of a few thick fleſhy fibres. The ſtalks are ſlender, four-cornered, and riſe about two feet high; the lower leaves are compoſed of ſeveral pinnæ or lobes, which are ſharply ſawed, but thoſe on the ſtalks are entire; on the lower part of the ſtalk they are placed oppoſite, but toward the top they are alternate and ſmall. The flowers come out in bunches from the wings of the ſtalk; they are ſmall, of a dark purple colour, with a mixture of green; the ſeed-veſſels are ſmall and roundiſh.

The tenth ſort grows naturally in Crete; this hath a root compoſed of fleſhy fibres. The lower leaves are broad and jagged, not much unlike thoſe of the Indian Scabious; the ſtalks riſe three feet high, are four-cornered, green, and ſmooth, garniſhed with winged leaves, having very long foot-ſtalks; they are compoſed of two or three-pair of ſmall lobes, terminated by a large one, acutely ſawed on their edges, ending in ſharp points. The ſtalks are terminated by ſlender bunches of ſmall flowers, ſituated ſparſedly, of a purpliſh colour at their rims, edged with white; theſe are ſucceeded by ſmall roundiſh ſeed-veſſels, filled with very ſmall ſeeds.

The eleventh ſort grows naturally in the kingdom of Naples, where it is frequently found upon rocks and old ſtone walls; this is a biennial plant, which periſhes after it has produced ripe ſeeds. The ſtalks riſe fifteen inches high; they are thick, ſmooth, and have ſcarce any corners; the leaves are winged, narrow, of a lucid green, ſucculent, and divided into many ſmall lobes, which are again divided, and wing-pointed; the flowers are produced in looſe panicles on the ſides and at the top of the ſtalk; they are of a dark brown colour, with a mixture of green, and are ſucceeded by pretty large roundiſh capſules, filled with angular dark-coloured ſeeds.

The twelfth ſort grows naturally in the Levant; this has a perennial creeping root. The ſtalks riſe two feet and a half high; their lower parts are cloſely garniſhed with narrow ſpear-ſhaped leaves, ſharply ſawed, and cut at the bottom; the upper part of the ſtalk is garniſhed with compound panicles of ſmall brown flowers, which are ſucceeded by ſmall roundiſh capſules, filled with ſmall ſeeds.

The thirteenth ſort grows naturally in Helvetia; this is a biennial plant, which flowers and produces ſeeds the ſecond year, and then decays. The lower leaves of this ſort are long, heart ſhaped, hairy, and of a pale green colour. The ſtalks riſe three feet high, garniſhed with ſmaller leaves, of the ſame ſhape with thoſe at bottom, placed by threes; the flowers ſtand upon pretty long foot-ſtalks; three of theſe come out at each joint, ſupporting cluſters of pretty large flowers, of a pale yellow colour; theſe appear in April, and are ſucceeded by oval capſules, filled with ſmall ſeeds.

The fourteenth ſort is a biennial plant, which grows naturally in Italy. The leaves of this are heart-ſhaped, ending in acute points, ſawed on their edges, of a lucid green, and on the upper part of the ſtalk are placed alternate; the foot-ſtalks of the flowers come out at the wings of the leaves, each ſuſtaining two or three flowers, of a dark red or purple colour.

The fifteenth ſort grows naturally in the Levant, and alſo upon Gibraltar-hill. The lower leaves of this are doubly winged, variouſly cut and indented; the ſtalk is ſlender; it riſes three feet high, the lower part of which is garniſhed with ſmaller winged leaves, of a lucid green, indented, and ſit cloſe to the ſtalks; the upper part has very ſlender panicles of ſmall flowers coming out of the ſide, and terminate the ſtalks. The flowers are thinly ranged on the foot-ſtalks, are very ſmall, and of a purple colour, with white borders.

The ſixteenth ſort grows naturally in Maryland; this hath a perennial fibrous root. The ſtalks are four-cornered; the leaves are heart-ſhaped, ſharply ſawed on their edges, and rounded at their baſe; the flowers are produced in panicles on the upper part of the ſtalk, and are like thoſe of the firſt ſort, but of an herbaceous colour.

The feventeenth fort was difcovered by the late Dr. Houftoun growing naturally at La Vera Cruz; this is a biennial plant. The ftalk rifes two feet high, garnifhed with oval acute-pointed leaves, fawed on their edges, which fit clofe to the ftalks; thofe at the bottom and top of the ftalk are placed by pairs, but in the middle there are three leaves at each joint, of a pale green; at the top of the ftalk the flowers are produced in roundifh bunches; they are about the fize of thofe of the firft fort, of a fcarlet colour. This fort flowered in the Chelfea garden, but did not perfect its feeds.

Thefe plants are propagated by feeds, which, if fown in the fpring, the plants feldom rife the fame feafon. Some of them may come up in autumn, and others the fpring following, but, if they are fown in autumn foon after they are ripe, the plants will come up the fpring following. They may be moft of them fown in the place where the plants are to remain, for they are hardy enough to bear the cold of our ordinary winters in the open air. When the plants come up, they will require no other care but to thin them where they are too clofe, and keep them clear from weeds. The fecond year the plants will flower and produce ripe feeds; after which thofe forts which are biennial will die, but the others will continue fome years.

The feventh and eighth forts are ornamental plants, fo may be allowed to have a place in the pleafure-garden, where, when the plants are ftrong, they will make a good appearance during their continuance in flower, which generally laft two months, unlefs the feafon proves very hot and dry. The roots of thefe forts will abide fome years, unlefs by a very fevere winter they are deftroyed; but as young plants flower ftronger than the old ones, there fhould be a fucceffion of them annually propagated by feeds.

SCUTELLARIA. Lin. Gen. Plant. 653. Skull-cap.

The CHARACTERS are,

The flower has a very fhort tubulous empalement, whofe brim is entire, having an incumbent fcaly operculum which feems clofed; it is of the lip kind, with a very fhort crooked tube, long compreffed chaps, and a concave trifid upper lip: the under lip is broad and indented; it has four ftamina hid under the upper lip, two of which are longer than the other, terminated by fmall fummits, and a four-pointed germen fupporting a flender ftyle, fituated with the ftamina, crowned by a fingle recurved ftigma. The empalement afterward becomes a helmet-fhaped capfule, including four feeds, which are roundifh.

The SPECIES are,

1. SCUTELLARIA (*Peregrina*) foliis fubcordatis ferratis, fpicis interruptis. Hort. Cliff. 317. *Skull-cap with almoft heart-fhaped fawed leaves, and interrupted fpikes of flowers.*

2. SCUTELLARIA (*Cretica*) foliis cordatis obtufis obtuféque ferratis, fpicis foliofis. Prod. Leyd. 311. *Skull-cap with obtufe heart-fhaped leaves, which are bluntly fawed, and leafy fpikes of flowers.*

3. SCUTELLARIA (*Altiffima*) foliis cordato-oblongis acuminatis ferratis, fpicis fubnudis. Lin. Sp. Pl. 600. *Skull-cap with oblong, acute-pointed, heart-fhaped, fawed leaves, and almoft naked fpikes of flowers.*

4. SCUTELLARIA (*Orientalis*) foliis incifis, fubtus tomentofis, fpicis rotundato tetragonis. Hort. Upfal. 173. *Skull-cap with cut leaves, which are woolly on their under fide.*

5. SCUTELLARIA (*Integrifolia*) foliis ovatis feffilibus, interioribus obfoleté ferratis, fuperioribus integerrimis. Lin. Sp. Plant. 599. *Skull-cap with oval leaves fitting clofe to the ftalks, the under of which are fometimes fawed, and the upper entire.*

6. SCUTELLARIA (*Lupulina*) foliis cordatis incifo-ferratis utrinque glabris, fpicâ rotundato tetragona. Hort. Upfal. 173. *Skull-cap with fawed cut leaves, which are fmooth on both fides, and a roundifh four-cornered fpike of flowers.*

The firft fort grows naturally in Italy. Mr. Ray obferved it about Leghorn and Florence, in the hedges and uncultivated places in plenty; this hath a perennial root; the ftalk is four-cornered, hairy, and rifes two feet high, garnifhed with heart-fhaped leaves placed oppofite, fawed on their edges. The flowers grow in interrupted fpikes at the top of the ftalks, of a purple colour in fome, and in others they are white.

The fecond fort grows naturally in Crete; this hath a ligneous ftalk, which rifes about two feet high, fending out flender fide-branches, garnifhed with obtufe heart-fhaped leaves, bluntly fawed on their edges; they are hoary on their under fide, and of a light green on their upper. The flowers are difpofed in pretty long fpikes at the top of the ftalks; they are white, and have fmall leaves growing between them.

The third fort grows naturally in the Levant; this hath a perennial root. The ftalks rife from three to four feet high, fending out a few flender branches from their fides, garnifhed with oblong heart-fhaped leaves, ending in acute points, fawed on their edges. The flowers are difpofed in naked fpikes at the top of the ftalks; they are purple, and have longer tubes than any of the other forts.

The fourth fort grows naturally in the Levant; this is a perennial plant, with fhrubby ftalks which fpread on the ground, garnifhed with cut leaves placed oppofite, which are almoft triangular, of a light green on their upper fide, and downy on their under, ftanding upon flender foot-ftalks. The flowers are difpofed in fhort fpikes at the end of the branches, of a bright yellow colour, and are fucceeded by gray feeds which ripen in the empalement.

The fifth fort grows naturally in North America; it has a perennial root, from which come forth feveral four-cornered ftalks, two feet high, fending out fide branches. The lower leaves are heart-fhaped, fawed on their edges, ftanding upon pretty long foot-ftalks; the upper leaves are oval and entire. The flowers are difpofed in very long loofe fpikes at the end of the branches, of a blue colour; thefe are fucceeded by feeds, which ripen in the empalement.

The fixth fort grows naturally on the Alps and Apennines; the ftalks of this are fhrubby, trailing on the ground; the leaves are fmooth and cut on their edges; the flowers are difpofed at the top of the ftalks in roundifh four-cornered fpikes; in one they are white, and in another variety they are blue; they are larger than the flowers of any other known fpecies, fo make a pretty appearance in gardens.

Thefe plants are all of them propagated by feeds; if they are fown in autumn foon after they are ripe, they will more certainly fucceed than when they are fown in the fpring, for thefe fometimes mifcarry, and if they fucceed, the plants feldom come up the fame feafon. The feeds may either be fown where the plants are to remain, or in a border to be afterward removed; but, as the fourth fort does not bear tranfplanting well, unlefs they are removed young, the feeds of that had better be fown where the plants are to ftand; this fhould be on a dry warm border of poor earth, where the plants will live much longer, and make a better appearance than on a rich foil. When the plants come up, they will require no other care but to thin them, and keep them clean from weeds.

When the other forts come up, and are fit to remove, they may be tranfplanted into a nurfery bed at five or fix inches diftance, where they may ftand till autumn, but muft be kept clean from weeds during that time: then they may be be tranfplanted into the borders of the flower-garden where they are to remain.

As thefe plants are not of long duration, it will be proper to fow a fucceffion of feeds every other year at leaft to fupply the places of thofe which decay.

SECALE. Tourn. Inft. R. H. 513. tab. 294. Rye.

The CHARACTERS are,

There are two flowers in each involucrum; they have two leaves, which are oppofite, narrow, erect, and fharp-pointed. The petals have two leaves; the outer valve is rigid, bellied, acute-pointed, and compreffed; the lower border is hairy, ending in a long awn, the inner is plain and fpear-fhaped; they have two oval erect nectariums, and three hair-like ftamina, hanging without the flower, terminated by oblong forked fummits, with a top-fhaped germen fupporting two reflexed hairy ftyles, crowned by a fingle ftigma. The germen afterward becomes an oblong almoft cylindrical feed, which ripens in the empalement.

We have but one diftinct SPECIES of this genus which is cultivated in England, though it is often fuppofed the two varieties are effentially different; but, from feveral years cultivating them on the fame land, I could find no real difference between them. Dr. Linnæus titles this Secale glumarum ciliis fcabris. Hort. Upfal. 22. *Rye with rough hairs to the awns.*

The farmers diftinguifh the two varieties by the titles of Winter and Spring Rye; but when thefe are fown three or four years at the fame feafon, and on the fame foil, it will be difficult to know them afunder; but where Rye is fown upon a warm land, it will ripen much earlier than on cold ftiff ground, and by continuing it two or three years, it will be forwarded fo much as to ripen a month earlier than the feeds which have long grown upon a ftrong cold foil; fo thofe who are obliged to fow Rye toward fpring, generally provide themfelves with this early feed.

There are feveral kinds of Grafs which are now ranged under this generical title; but as thefe do not merit cultivation, I fhall not trouble the reader with the mention of them here.

Rye is fo well known to every one who is the leaft acquainted with the different grains, as to need no defcription.

The Winter Rye is what the generality of farmers propagate; it is ufually fown in autumn at the fame feafon with Wheat, and in many of the northern counties, as alfo in Wales, they are often mixed together; though I think it muft be very bad hufbandry, for the Rye will always ripen fooner than Wheat, fo that if the latter is permitted to be fully ripe, the former will fhatter; nor can this be practifed, where the people are not accuftomed to eat Rye bread; for although it is by fome accounted good when mixed, yet being fo very clammy, few people, who have been fed with Wheat, will ever care to eat the bread made of this.

It is generally fown on poor, dry, gravelly, or fandy land, where Wheat will not thrive, and in fuch places may anfwer very well; but on fuch land as will bear Wheat, it is not proper to fow Rye, as the value of it is greatly inferior to that of Wheat.

When Rye is fown, the ground fhould not be too wet; and if it fhould happen that much rain falls before the Rye is come up, it often rots in the ground; but it is not long in coming up, it being much fooner out of the ground than Wheat.

The fmall Rye may be fown in the fpring about the fame time with Oats, and is ufually ripe as foon as the other fort; but if the feafon proves wet, it is apt to run much to ftraw; and the grain is generally lighter than the other, fo the only ufe of this fort is to fow upon fuch lands where the autumnal crop may have mifcarried.

Rye is alfo fown in autumn to afford green feed for ewes and lambs in the fpring, before there is plenty of Grafs. When this is intended, the Rye fhould be fown early in autumn that it may have ftrength to furnifh early feed. The great ufe of this is to fupply the want of Turneps in thofe places where they have failed, as alfo, after the Turneps are over, and before the Grafs is grown enough to fupply green feed for the ewes; fo that in thofe feafons when the Turneps in general fail, it is very good hufbandry to fow the land with Rye, efpecially where there are ftocks of fheep, which cannot be well fupported where green feed is wanting early in the fpring; therefore thofe farmers who have large live ftocks fhould have feveral methods for fupplying themfelves with fufficient feed, left fome fhould fail; for as Turneps are a very precarious crop, fome land fhould be fown with Cole-feed, which will fupply the want of Turneps in winter; and if fome of the ground which was fown late with Turneps which had failed, was fown in autumn with Rye, that would be fit to fupply the want of Cole-feed afterward.

SECURIDACA. Tourn. Inft. R. H. 399. tab. 224. Hatchet-vetch.

The CHARACTERS are,

The empalement of the flower is fhort, compreffed, and cut into two fegments; the flower is of the butterfly kind; the ftandard is heart-fhaped, reflexed on both fides, and fcarce longer than the wings; thefe are oval, joining at the top, but open at the bottom; the keel is compreffed and pointed. It hath ten ftamina, nine joined, and one feparate, terminated by fmall fummits, and an oblong compreffed germen, with a briftly ftyle, crowned by an obtufe ftigma. The germen afterward turns to a long, compreffed, fword-fhaped pod, with a thick border on one fide, plain on the other, opening in two cells, filled with fquare feeds.

We have but one SPECIES of this genus, viz.

SECURIDACA (*Lutea*) herbacea, leguminibus falcato-gladiatis. *Herbaceous Hatchet-vetch with hooked fword-fhaped pods.*

This plant grows naturally in the Corn-fields in Spain and Italy; it is annual, and hath trailing herbaceous ftalks, which grow a foot and a half long, dividing into many branches, which fpread on the ground, garnifhed with winged leaves, compofed of feven or eight pair of oval obtufe lobes terminated by an odd one, of a deep green and fmooth. From the wings of the leaves arife the foot-ftalks of the flowers, by pairs at each joint, which are five or fix inches long, fuftaining at the top a large clufter of yellow flowers of the butterfly kind, fucceeded by compreffed pods near four inches long, ending in acute points, having a future on each fide, one plain and the other rifing, joined at their bafe to the foot-ftalk, but fpread open like the rays of a ftar, and are divided by a longitudinal partition into two cells, each containing a row of fquare flat feeds of a reddifh colour.

It is propagated by fowing the feeds in borders of light earth in the fpring, in the places where the plants are to abide, for they feldom fucceed well if they are tranfplanted; they fhould be allowed at leaft two feet diftance, becaufe their branches trail upon the ground. When the plants come up, they will require no other care but to thin them where they are too clofe, and keep them clean from weeds. A few of thefe plants may be admitted into every good garden for variety, though there is no great beauty in their flowers.

SEDUM. Lin. Gen. Plant. 513. Houfeleek.

The CHARACTERS are,

The empalement of the flower is permanent, and cut into five acute parts. The flower has five plain, spear shaped, acute-pointed petals, and five nectariums, with small single scales indented at the top, each being inserted at their base to the outside of the germen; it has ten awl-shaped stamina the length of the petals, terminated by roundish summits, and five oblong germen ending in slender styles, crowned by obtuse stigmas. The germen afterward becomes five erect spreading capsules, which are compressed, acute-pointed, opening from top to bottom, filled with small seeds.

The SPECIES are,

1. SEDUM (*Album*) foliis oblongis obtusis teretiusculis sessilibus patentibus, cymâ ramosâ. Hort. Cliff. 177. *Houseleek with oblong, obtuse, taper leaves, sitting close to the stalks, spreading open, and a branching stalk.*

2. SEDUM (*Dasyphyllum*) foliis oppositis ovatis obtusis carnosis, caule infirmo, floribus sparsis. Lin. Sp. Pl. 431. *Houseleek with oval, fleshy, blunt leaves, which are placed opposite, a weak stalk, and flowers growing thinly*

3. SEDUM (*Rupestre*) foliis subulatis confertis basi membranaceâ solutis, floribus simosis. Hort. Cliff. 176. *Houseleek with awl-shaped leaves growing in clusters, whose base has a loose membrane, and flowers growing from the top of the branches.*

4. SEDUM (*Hispanicum*) foliis teretibus acutis radicalibus fasciculatis, cymâ pubescente. Hort. Cliff. 176. *Houseleek with fleshy, taper, spreading leaves, a stalk divided by pairs, and erect tops.*

5. SEDUM (*Acre*) foliis subovatis adnato-sessilibus gibbis erectiusculis alternis, cymâ trifidâ. Hort. Cliff. 177. *Stone Crop with oval, gibbous, erect, alternate leaves, sitting close to each other, and a trifid top; or Wall Pepper.*

6. SEDUM (*Rubens*) foliis fuciformibus subdepressis infimis quaternis, cymâ subquadrifidâ, floribus pentandris staminibus reflexis. Lin. Sp. Plant. 619. *Houseleek with cylindrical leaves which grow alternate, and the top always erect.*

7. SEDUM (*Annuum*) caule erecto solitario annuo, foliis ovatis sessilibus gibbis alternis, cymâ recurvâ. Flor. Suec. 319. *Houseleek with an erect, annual, single stalk, oval gibbous leaves which are placed alternate, and a recurved top.*

8. SEDUM (*Reflexum*) foliis subulatis sparsis basi solutis, inferioribus recurvatis. *Stone Crop with trailing stalks, awl-shaped, fleshy, spreading leaves, and flowers growing in reflexed panicles.*

9. SEDUM (*Sexangulari*) foliis subovatis adnato-sessilibus gibbis erectiusculis sexfariàm imbricatis. Flor. Suec. 390. *Stone Crop with almost oval, gibbous, erect leaves, growing close to each other, and imbricated six ways.*

10. SEDUM (*Villosum*) caule erecto, foliis planiusculis, pedunculisque subpilosis. Lin. Sp. Plant. 432. *Houseleek with an erect stalk, plain leaves, and foot-stalks which are somewhat hairy.*

11. SEDUM (*Stellatum*) foliis planiusculis angulatis, floribus lateralibus subsessilibus solitariis. Hort. Cliff. 176. *Houseleek with plain angular leaves, and single flowers sitting close to the sides of the stalk.*

12. SEDUM (*Cepæa*) foliis planis, caule ramoso, floribus paniculatis. Hort. Cliff. 176. *Houseleek with plain leaves, a branching stalk, and flowers growing in panicles.*

13. SEDUM (*Aizoon*) foliis lanceolatis serratis planis, caule erecto, cymâ sessili terminali. Lin. Sp. Plant. 430. *Houseleek with plain, spear-shaped, sawed leaves, and an erect stalk, terminated by a head of flowers sitting close to it.*

14. SEDUM (*Telephium*) foliis planiusculis serratis, corymbo folioso, caule erecto. Lin. Sp. Plant. 430. *Houseleek with plain sawed leaves, a leafy corymbus of flowers, and an erect stalk.*

15. SEDUM (*Hæmatodes*) foliis ovatis integerrimis, summis amplexicaulibus, corymbo terminali. *Houseleek with oval entire leaves, which at the top embrace the stalk, and a corymbus of flowers terminating the branches.*

16. SEDUM (*Anacampseros*) foliis cuneiformibus integerrimis, caulibus decumbentibus, floribus corymbosis. Lin. Sp. Plant. 430. *Houseleek with wedge-shaped entire leaves, trailing stalks, and flowers growing in a corymbus.*

The first sort grows naturally upon old walls in many parts of England, so is seldom planted in gardens, but as it is a medicinal plant I have placed it here. This hath slender trailing branches, garnished with taper succulent leaves about half an inch long, standing alternately round the branches. The flower-stalks rise four or five inches high; their lower part is garnished with leaves, which spread horizontally; the upper part of the stalk divides into small foot-stalks, supporting many white star-pointed flowers, gathered into a sort of umbel.

The second sort also grows upon old walls in many parts of England. The stalks of this are very slender and infirm; the leaves are very short, oval, and of a gray colour, placed opposite. The flowers are set thinly at the top of the stalks; they are small, white, with petals which are obtuse; the summits upon the stamina are pretty large, and of a bright purple colour.

The third sort grows naturally upon St. Vincent's Rock near Bristol, and in several parts of Wales. This has slender purple stalks, which trail upon the ground, closely garnished with short awl-shaped leaves placed round the stalks, having a short loose membrane at their base, which falls off on being touched; the leaves toward the top of the stalk sit close together; they are of a sea-green colour, and not very succulent. The flowers grow at the top of the stalks in roundish bunches, of a bright yellow colour. This plant, when it is once placed upon a wall, will propagate itself in plenty by its trailing branches, which put out roots from their joints.

The fourth sort grows naturally in Spain. This is an annual plant, with upright stalks which rise three or four inches high, garnished with fleshy awl-shaped leaves, of a gray colour. The top of the stalk divides into two slender erect branches, which have small, white, star-pointed flowers ranged above each other, and the top of the stalk at the division of the branches is terminated by two or three flowers sitting close. If the seeds are permitted to scatter, the plants will come up without care.

The fifth sort is the common Stone Crop, or Wall Pepper, so called for the acrid biting quality of the leaves. This grows very common upon old walls and buildings in every part of England, and is so well known as to require no description. There are two varieties of this, one with a large, and the other a small yellow flower.

The sixth sort grows upon moist rocks in several parts of France and Germany, and is seldom seen in gardens. This rises with an erect stalk three inches high, garnished with obtuse, cylindrical, succulent leaves. The stalk divides upward into three or four branches, which sustain small purplish flowers, standing erect.

The seventh sort grows naturally on dry barren rocks in the north of England. This is an annual plant with an erect stalk, garnished with oval leaves, placed alternate. The stalk seldom rises above two or three inches high; the leaves sit close to the stalks, and are of a grayish colour; the flowers grow at the top of the stalk in a reflexed spike; they are small and white.

The eighth sort grows naturally upon old walls and buildings in most parts of England, and is by some called Prick Madam. This has long trailing stalks, garnished with fleshy awl-shaped leaves, spreading out almost horizontally,

zontally, of a gray colour, ending in acute points The flowers grow in reflexed bunches at the top of the ſtalks; they are ſtar-pointed, and of a bright yellow colour.

The ninth ſort is leſs common than either of the former. I have found it growing upon the rocks in Wales. This hath the appearance of common Stone Crop, but the ſtalks and leaves are larger, and have no biting taſte; the leaves are ranged in ſix rows, like the grains of the ſix-rowed Barley; the flowers are yellow, and larger than thoſe of the common Stone Crop.

The tenth ſort grows naturally upon moiſt rocks and boggy ſoils in ſeveral parts of the north of England, and in Wales; this ſeldom riſes more than two or three inches high. The ſtalks are garniſhed with a few plain hairy leaves, terminated by purple flowers growing thinly.

The eleventh ſort grows naturally in Italy and Germany; this is a low annual plant. The leaves are plain and angular; the ſtalks riſe three inches high, dividing at the top into two or three parts; the flowers come out ſingly from the ſide of the ſtalk; they are white, ſtar-pointed, and are ſucceeded by ſtar-pointed rough capſules.

The twelfth ſort is an annual plant, which grows naturally in the ſouth of France and in Italy; this hath plain ſucculent leaves. The ſtalks riſe ſix or ſeven inches high, dividing into ſmaller branches, which ſuſtain ſmall white flowers, growing in large panicles; and if the ſeeds are permitted to ſcatter, the plants will come up without care. This loves a warm dry ſoil.

The thirteenth ſort grows naturally in Siberia; this has a perennial root, from which come out ſeveral ſtalks near a foot high, garniſhed with ſpear-ſhaped, plain, thick leaves, placed alternately, ſlightly ſawed on their edges. The ſtalk is terminated by a flat corymbus of flowers, of a bright yellow colour, ſitting cloſe on the top of the ſtalks.

The fourteenth ſort is not common in the Engliſh gardens; this hath roots like the former. The ſtalks grow more erect, of a purple colour; the leaves are flatter, and more ſawed on their edges, of a dark green colour, and thicker ſubſtance; the flowers are purple.

The fifteenth ſort grows naturally in Portugal. There are two varieties of this, one with white, and the other with purple flowers. The roots of this are compoſed of many thick fleſhy knobs: the ſtalks are thick, ſucculent, and round; they riſe three feet high, garniſhed with oval ſucculent leaves, which are entire, placed by pairs; thoſe on the upper part embrace the ſtalk with their baſe; they are of a pale herbaceous colour. The flowers are collected in large bunches, which terminate the ſtalks.

The ſixteenth ſort grows naturally in Italy. The roots of this are fibrous; the ſtalks trail on the ground, garniſhed with wedge-ſhaped leaves, ſtanding alternately round the ſtalks. The flowers are diſpoſed in a compact corymbus, which ſits cloſe on the top of the ſtalks; they are ſtar ſhaped, of a purple colour, and appear in July.

All the ſorts of Stone Crop are eaſily propagated by planting their trailing ſtalks either in ſpring or ſummer, which ſoon put out roots; but, as theſe thrive much better upon rocks, old walls, or buildings, than in the ground, they may be diſpoſed upon rock-work in ſuch a manner as to have a good effect. If the cuttings or roots of the perennial ſort are planted in ſome ſoft mud laid upon the walls or buildings, they will ſoon take root, and then ſpread into every joint or crevice, and in a ſhort time will cover the place; and, if the ſeeds of thoſe annual ſorts, which grow naturally in dry places, are ſown ſoon after they are ripe on the top of walls, the plants will come up, and maintain themſelves without farther care.

The ſeveral ſorts of Orpine may be eaſily propagated by cuttings during the ſummer months, or by parting of their roots either in ſpring or autumn; theſe thrive beſt in a dry ſoil and a ſhady ſituation, but may alſo be planted for the ſame purpoſes as the other ſorts, eſpecially the ſixteenth ſort, which is evergreen. The ſtalks of this kind hang down, and have a very good effect in rock-work, and the plants require no care, for when they are fixed in the place, they will ſpread and propagate faſt enough.

The ſtalks of the common Orpine are frequently cut in ſummer, and faſtened to laths of the ſize of chimney-boards, which, being framed together, are uſed for ſcreening the ſight of the fire-grates in rooms; theſe ſtalks will ſhoot and ſpread over the frame, and, if the frames are taken out once a week, and the ſtalks watered over to refreſh them, they will continue in verdure for two months.

SEED. The Seed of a plant conſiſts of an embryo, with its coat or cover. The embryo, which contains the whole plant in miniature, and which is called the germ or bud, is rooted in the placenta or cotyledon, which makes the coat or involucrum, and ſerves the ſame purpoſes as the ſecundines, i. e. the chorion and amnis in animals.

The placenta or cotyledon of a plant is always double, and in the middle and common center of the two is a point or ſpeck, which is the embryo or plantule. This plantule, being acted on and moved by the warmth of the ſun and the earth, begins to expand, and protrudes or ſhoots out its radicle or root both upward and downward. By this it abſorbs the nutritious juice from the earth, and ſo grows and increaſes, and, the requiſite heat continuing, the growth continues.

The two placentulæ or cotyledons of a Seed are, as it were, a caſe to this little tender plantule or point, covering it up, ſheltering it from injuries, and feeding it from their own proper ſubſtance; which the plantule receives and draws to itſelf, by an infinite number of little filaments or ramifications, called Funes Umbilicales, or Navel Strings, which it ſends into the body of the placenta.

The cotyledons, for the moſt part, abound with a balſam, diſpoſed in proper cells; and this ſeems to be oil brought to its greateſt perfection, while it remains humid, and then lodged in theſe repoſitories; one part of the compoſition of this balſam is oily and tenacious, and ſerves to defend the embryo from any extraneous moiſture, and, by its viſcidity, to entangle and retain that fine, pure, volatile ſpirit, which is the ultimate production of the plant.

This oil is never obſerved to enter into the veſſels of the embryo, which are too fine to admit ſo thick a fluid. The ſpirit, however, being quickened by an active power, may poſſibly breathe a vital principle into the juices that nouriſh the embryo, and ſtamp upon it the character that diſtinguiſhes the family; after which every thing is changed into the proper nature of that particular plant. That this ſpirit now is truly the efficacious part is evident, for when that is gone off, the oil that remains is quite vapid and inactive. It is this that gives plants their fragrant ſmell and peculiar taſtes, nor do their particular colours a little depend upon it.

Now when the Seed is committed to the earth, the placenta ſtill adheres to the embryo for ſome time, guards it from the acceſs of noxious colds, &c. and even prepares and purifies the cruder juice the plant is to receive from the earth, by ſtraining it, &c. through its own body.

This it continues to do, till the placentula being a little inured to its new element, and its root tolerably fixed in the ground, and fit to abſorb the juice thereof, it then periſhes, and the plant may be ſaid to be delivered; ſo that nature obſerves the ſame method in plants contained in fruits, as in animals in the mother's womb.

It is very ſurpriſing, that many ſorts of Seeds will continue good for ſeveral years, and retain their growing faculty,

culty, whereas many other ſorts will not grow when they are more than one year old; which is, in a great meaſure, owing to their abounding more or leſs with oil, as alſo the nature of the oil, whether it is of a cold or hot quality, and the texture of their outward covering. As for example; the Seeds of Cucumbers, Melons, and Gourds, which have thick horny coverings, and the oil of this Seed being of a cold nature, the Seeds will continue good ten, fifteen, or twenty years; and Radiſh, Turnep, Rape, &c. with other oily Seeds (whoſe coats, though they are not ſo hard and cloſe as the others, yet) abounding with oil, which is of a warmer nature, the Seeds will keep good three or four years; whereas the Seeds of Parſley, Carrots, Parſneps, and moſt other umbelliferous plants, whoſe Seeds are, for the moſt part, of a warm nature, loſe their growing faculty often in one year, but ſeldom remain good longer than two years. Indeed all ſorts of Seeds are preſerved beſt, if kept in the pods or huſks wherein they grow; eſpecially if they are not ſeparated from the placenta, to which they are faſtened by an umbilical cord, through which they receive their nouriſhment in their embryo ſtate; ſo that whoever would ſend Seeds to a diſtant country, ſhould always take care they are full ripe before they are gathered, and that they are preſerved in their pods or huſks; and when they are packed up for exportation, there ſhould be great care taken, that they are not ſhut up too cloſely from the air, which is abſolutely neceſſary to maintain the principle of vegetation in the Seed (though in a leſs degree) as it is to nouriſh the plant when germinated, as I found by trying the following experiment, viz. Having ſaved a parcel of freſh Seeds of ſeveral kinds, I took ſome of each kind, and put into glaſs phials; theſe I ſtopped down cloſe, and ſealed hermetically, then put them up in a trunk; the other parts of the ſame Seeds I put into bags, and hung them up in a dry room, where the air had free admittance, in which place they remained a whole year; and the following ſpring I took out a part of each parcel of Seeds from the phials, as alſo from the bags, and ſowed them at the ſame time, and upon the ſame bed, where they had an equal advantage of the ſun, air, &c. The reſult of this experiment was, that almoſt all the Seeds which I took out of the bags, grew extremely well, but of thoſe which were kept in the phials, not one come up; after which I ſowed the remaining part of the Seeds in the phials, but had not one ſingle plant from the whole, whereas thoſe preſerved in the bags grew very well both the ſecond and third years. And this experiment was afterward tried by one of my particular friends, with whom the effect was the ſame as with me. Some years after this, a gentleman of great eminence for his knowledge of plants, being very deſirous to procure Seeds from every country where the Britiſh nation had any commerce, gave his inſtructions to many of the agents abroad, to ſend him over all the ſorts of Seeds they could collect in their different countries, and to put them up in bottles, ſealing the mouths of the bottles as cloſe as poſſible, to exclude the air; which was done by ſeveral of his correſpondents, who ſent him great quantities of Seeds, but not one of them grew when they were ſown; ſo that thoſe perſons who ſend Seeds to a diſtant country, ſhould never be guilty of the like error.

How the vegetative life is ſo long preſerved in Seeds, when they are deeply immerſed in the ground, is very difficult to explain; but as it is very notorious, that earth taken from the bottom of vaults, houſes, and wells, and from the earth which has been taken at a very great depth in thoſe places, there have been many plants produced, which were not inhabitants of the neighbouring ſoil; and this has been brought as a proof to ſupport the doctrine of ſpontaneous productions, by ſome who have aſſerted, that plants are often produced without Seed.

SELAGO. Lin. Gen. Plant. 687.

The CHARACTERS are,

The flower has a permanent empalement of one leaf, cut into four parts at the top. The flower is of one petal; it has a very ſmall tube; the brim is ſpreading, and cut into five parts; the two upper ſegments are the leaſt. It hath four hair-like ſtamina the length of the petal, to which they are inſerted, two of which are longer than the other, terminated by ſingle ſummits; and a roundiſh germen, ſupporting a ſingle ſtyle, crowned by an acute ſtigma. The germen afterward becomes a ſingle ſeed, wrapped up in the petal of the flower.

We have but one SPECIES of this genus, viz.

SELAGO (*Corymboſa*) corymbo multiplici. Lin. Sp. Pl. 629. *Selago with a multiplied corymbus.*

This plant grows naturally at the Cape of Good Hope. It has ſlender ligneous ſtalks, which riſe ſeven or eight feet high, but are ſo weak as to require ſupport; they ſend out many ſlender branches, garniſhed with ſhort, linear, hairy leaves, that come out in cluſters from the ſame point. The flowers terminate the ſtalks in umbels, the general umbel being compoſed of a multiplicity of ſmall umbels of white flowers, which appear in July and Auguſt, but are not ſucceeded by ſeeds in England.

This plant is preſerved in gardens more for the ſake of variety than for its beauty, for the branches grow very irregular, and hang downward; the leaves are ſmall, ſo make little appearance, and the flowers are ſo ſmall as not to be diſtinguiſhed at any diſtance.

It is propagated by cuttings, which put out roots freely, if they are planted in any of the ſummer months; if theſe are planted in a bed of freſh earth, and covered cloſe down with a bell or hand-glaſs, ſhading them from the ſun, and refreſhing them now and then with water, they will ſoon put out roots; then they muſt be gradually hardened, and afterward tranſplanted into ſmall pots, placing them in the ſhade till they have taken new root; then they may be placed with other hardy green-houſe plants, where they may remain till the end of October, when they muſt be removed into ſhelter, for theſe plants will not live in the open air in England; but they only require protection from hard froſt, ſo they ſhould be treated in the ſame way as other of the hardieſt kinds of green-houſe plants.

SELINUM. Lin. Gen. Plant. 300. Milky Parſley.

The CHARACTERS are,

It has an umbellated flower; the general umbel is plain and ſpreading; the particular umbels are the ſame; the involucrum is compoſed of many linear ſpear-ſhaped leaves, which ſpread open; the umbel is uniform; the flowers have five inflexed heart-ſhaped petals, which are unequal, and five hair-like ſtamina, terminated by roundiſh ſummits. The germen is ſituated under the flower, ſupporting two reflexed ſtyles, crowned by ſingle ſtigmas; it afterward becomes a plain compreſſed fruit, channelled on both ſides, parting in two, containing two oblong elliptical plain ſeeds, channelled in the middle, and have membranes on both ſides.

The SPECIES are,

1. SELINUM (*Sylveſtre*) radice fuſi-formi multiplici. *Milky Parſley with ſpindle-ſhaped roots.*

2. SELINUM (*Paluſtre*) ſublacteſcens radice unicâ. Haller. Helv. 443. *Selinum which is almoſt milky, having a ſingle root.*

The firſt ſort grows by the ſides of lakes and ſtanding waters in ſeveral parts of Germany; this hath many ſpindle-ſhaped roots hanging by fibres, which ſpread and multiply in the ground. The ſtalks riſe five or ſix feet high; they are ſtreaked, and of a purple colour at bottom, ſending out ſeveral

ſeveral branches toward the top; the leaves are finely divided like thoſe of the Carrot, and, when broken, there iſſues out a milky juice; the ſtalks are terminated by umbels of whitiſh flowers, which come out in June, and are ſucceeded by compreſſed bordered ſeeds, which ripen in Auguſt.

The ſecond ſort grows naturally in marſhy places in Germany. The leaves of this are much longer, and cut into narrower ſegments than thoſe of the former; the ſtalks riſe higher; the umbels are larger, as are alſo the ſeeds. The whole plant abounds with a cauſtic milky juice.

Theſe plants are preſerved in botanic gardens for variety, but are rarely cultivated any where elſe; they are eaſily propagated by ſeeds, which ſhould be ſown in autumn, and the plants afterward treated in the ſame way as Angelica.

A SEMINARY is a ſeed-plot, which is adapted or ſet apart for the ſowing of ſeeds. Theſe are of different natures and magnitudes, according to the ſeveral plants intended to be raiſed therein. If it be intended to raiſe foreſt or fruit-trees, it muſt be proportionably large to the quantity of trees deſigned, and the ſoil ſhould be carefully adapted to the various ſorts of trees. Without ſuch a place as this every gentleman is obliged to buy, at every turn, whatever trees he may want to repair the loſſes he may ſuſtain in his orchard, wilderneſs, or larger plantations, ſo that the neceſſity of ſuch a ſpot of ground will eaſily be perceived by every one; but, as I have already given directions for preparing the ſoil, and ſowing the ſeeds in ſuch a Seminary, under the article of NURSERY, I ſhall not repeat it in this place, but refer the reader to that article.

It is alſo as neceſſary for the ſupport of a curious flower-garden, to have a ſpot of ground ſet apart for the ſowing of all ſorts of ſeeds of choice flowers, in order to obtain new varieties, which is the only method to have a fine collection of valuable flowers; as alſo for the ſowing of all ſorts of biennial plants, to ſucceed thoſe which decay in the flower-garden; by which means the borders may be annually repleniſhed, which, without ſuch a Seminary, could not be ſo well done.

This Seminary ſhould be ſituated at ſome diſtance from the houſe, and be entirely cloſed either with a hedge, wall, or pale, and kept under lock and key, that all vermin may be kept out, and that it may not be expoſed to all comers and goers, who many times do miſchief before they are aware of it. As to the ſituation, ſoil, and manner of preparing the ground, it has been already mentioned under the article of NURSERY, and the particular account of raiſing each ſort of plant being directed under their proper heads, it would be needleſs to repeat it here.

SEMPERVIVUM. Lin. Gen. Plant. 538. Houſeleek.

The CHARACTERS are,

The flower has a concave empalement, cut into many acute ſegments; it has ten oblong, ſpear-ſhaped, pointed petals, and twelve or more narrow awl-ſhaped ſtamina, terminated by roundiſh ſummits; it has twelve germen, placed circularly, ſitting upon ſo many ſtyles, which ſpread out, and are crowned by acute ſtigmas. The germen afterward becomes ſo many ſhort compreſſed capſules, pointed on the outſide, and open on the inſide, filled with ſmall ſeeds.

The SPECIES are,

1. SEMPERVIVUM (*Tectorum*) foliis ciliatis, propaginibus patentibus. Lin. Sp. Plant. 464 *Houſeleek with hairy-edged leaves, and ſpreading offsets; or common large Houſeleek.*

2. SEMPERVIVUM (*Globiferum*) foliis ciliatis, propaginibus globoſis. Lin. Sp. Plant. 464. *Houſeleek with hairy-edged leaves, whoſe offsets are globular.*

3. SEMPERVIVUM (*Montanum*) foliis ciliatis, propaginibus patulis. Lin. Sp. Plant. 465. *Houſeleek with entire leaves and ſpreading offsets.*

4. SEMPERVIVUM (*Arachnoideum*) foliis pilis intertextis, propaginibus globoſis. Lin. Sp. Plant. 465. *Houſeleek with threads from leaf to leaf, and globular offsets; commonly called Cobweb Sedum.*

5. SEMPERVIRUM (*Arboreum*) caule arboreſcente lævi ramoſo. Lin. Sp. Plant. 464. *Houſeleek with a ſmooth, tree-like, branching ſtalk; or Tree Houſeleek.*

6. SEMPERVIVUM (*Canarienſe*) caule foliorum ruderibus lacero, foliis retuſis. Lin. Sp. Plant. 464. *Houſeleek with ſtalks torn by the rudiments of the leaves, and blunt-pointed leaves.*

7. SEMPERVIVUM (*Africanum*) foliorum marginibus ſerrato-dentatis, propaginibus patulis. *Houſeleek with leaves whoſe borders are indented like a ſaw, and ſpreading offsets.*

The firſt ſort is our common Houſeleek, which is ſeen in every part of England growing on the tops of houſes and walls, but is not a native of this country; it has many thick ſucculent leaves, ſet together in a round form; they are convex on their outſide and plain within, ſharp-pointed, and their borders are ſet with ſhort fine hairs. The leaves ſpread open, and lie cloſe to the earth, ſending out on every ſide offsets of the ſame form. From the center of theſe heads ariſe the flower-ſtalk, which is about a foot high, ſucculent and round, of a reddiſh colour, garniſhed at bottom with a few narrow leaves; the upper part of the ſtalk divides into two or three parts, each ſuſtaining a ſpike or range of reflexed flowers, compoſed of ſeveral petals, which ſpread open, ending in acute points, of a red colour; in the center is ſituated the germen, which are placed circularly, and, after the petals are fallen off, they ſwell and become ſo many horned capſules, filled with ſmall ſeeds.

This plant is eaſily propagated by offsets, which the plants put out in plenty. If theſe are planted in mud or ſtrong earth placed on a building or old wall, they will thrive without farther care.

The ſecond ſort grows naturally in the northern parts of Europe. The leaves of this ſort are much narrower, and the heads are furniſhed with a greater number of leaves than thoſe of the former, which grow more compact, and are cloſely ſet on their edges with hairs. The offsets of this are globular, their leaves turning inward at the top, and lie cloſe over each other: theſe are thrown off from between the larger heads, and falling on the ground take root, whereby it propagates very faſt. The flower-ſtalks of this are ſmaller than thoſe of the former, and the flowers are of a paler colour.

The third ſort grows naturally upon the Helvetian mountains. This greatly reſembles the firſt, but the leaves are ſmaller, and have no indentures on their edges; the offsets of this ſort ſpread out from the ſide of the older heads, and their leaves are more open and expanded. Out of the center of the heads comes forth the flower-ſtalk, which riſes nine or ten inches high, garniſhed below with ſome narrow leaves; the upper part is divided into three or four branches, which are cloſely furniſhed with deep red flowers, compoſed of twelve ſtar pointed pointals, ſet round the circle or germen, which is attended by twenty-four ſtamina, terminated by purple ſummits.

The fourth ſort grows naturally upon the Alps and Helvetian mountains; this hath much ſhorter and narrower leaves than either of the former. The heads are ſmall and very compact; the leaves are gray, ſharp-pointed, and have ſlender white threads croſſing from one to the other, interſecting each other in various manners, ſo as in ſome meaſure to repreſent a ſpider's web. The flower-ſtalks riſe about ſix inches high, are ſucculent, round, and garniſhed with awl-ſhaped ſucculent leaves, placed alternately; the upper part of of the ſtalk divides into two or three branches, upon each of which is a ſingle row of flowers ranged on one ſide above each other, compoſed of eight ſpear-ſhaped petals, of a bright red colour, with a deep red line running longitudinally in the middle; theſe ſpread open in form of a ſtar, and in the center is ſituated the germen, of an herbaceous colour, ſurrounded by ſixteen purple ſtamina, which are erect, terminated by yellow ſummits.

All the above-mentioned ſorts are extremely hardy, and propagate very faſt by offsets; they love a dry ſoil, ſo are very proper to plant in rock-work, where they will thrive better than in the full ground, as they want no care; for when they are once fixed, they will propagate and ſpread faſt enough, ſo that the larger ſorts may require to be annually reduced to keep them within proper compaſs. When any of theſe heads flower, they die ſoon after, but the offsets ſoon ſupply their place.

The fifth ſort grows naturally at the Cape of Good Hope, and alſo in Portugal; the old walls about Liſbon are covered with this plant. This riſes with a fleſhy ſmooth ſtalk eight or ten feet high, dividing into many branches, terminated by round heads or cluſters of leaves lying over each other like the petals of a double Roſe; they are ſucculent, of a bright green, and have very ſmall indentures on their edges like the teeth of a very fine ſaw. The ſtalks are marked with the veſtiges of the fallen leaves, and have a light brown bark; the flowers riſe from the center of the heads, forming a large pyramidal ſpike; they are of a bright yellow colour, and the petals ſpread open like the points of a ſtar; the other parts are like thoſe of the other ſpecies. This ſort generally flowers in autumn or winter, and the flowers continue long in beauty, during which time they make a fine appearance.

There is a variety of this with variegated leaves, which is much eſteemed by the curious; this was accidentally obtained by a branch which had been accidentally broken from a plant of the plain kind at Badmington, the ſeat of his Grace the Duke of Beaufort, which, after having laid ſome time, was planted, and when the young leaves puſhed out, they were variegated. Theſe plants are eaſily propagated by cutting off the branches, which, when planted, ſoon put out roots; theſe ſhould be laid in a dry place for a week before they are planted, that the bottom may be healed over, otherwiſe they are apt to rot, eſpecially if they have much wet. When the cuttings are planted in pots, they ſhould be placed in a ſhady ſituation, and muſt have but little wet, and, if they are plunged in a ſhady border, they will require no water, for the moiſture of the ground will be ſufficient for them. Some years paſt theſe plants were tenderly treated; their cuttings were put into a hot-bed, to forward their putting out roots, and in winter the plants were kept in ſtoves; but later experience has taught us, that they will thrive better with hardier treatment; for, if they are protected from froſt and wet in winter, and have a good ſhare of air in mild weather, they will thrive better, and flower oftener than when they are tenderly nurſed. I have frequently ſeen the branches of theſe plants, which have been accidentally broken off and fallen on the ground, put out roots as they have laid, and have made good plants. The ſort with ſtriped leaves is tenderer than the other, and more impatient of wet in winter.

The ſixth ſort grows naturally in the Canary Iſlands; this ſeldom riſes high, unleſs the plants are drawn up by tender management. The ſtalk is thick and rugged, chiefly occaſioned by the veſtiges of the decayed leaves; it ſeldom riſes above a foot and a half high, ſupporting at the top one very large crown of leaves, diſpoſed circularly like a full-blown double Roſe. The leaves are large, ending in obtuſe points, are a little incurved, ſucculent, ſoft to the ouch, and pliable. The flower-ſtalk comes out of the center, and riſes two or three feet high, branching out from the bottom, ſo as to form a regular pyramid of flowers, which are of an herbaceous colour, ſhaped like thoſe of the other ſpecies; they are ſucceeded by horned capſules, filled with ſmall ſeeds, which ripen late in autumn or winter, and then the plant dies.

This is propagated by ſeeds, which ſhould be ſown ſoon after it is ripe, in pots filled with light ſandy earth, covering them over very lightly with the ſame. Theſe pots ſhould be placed under a common frame to keep out the froſt, but ſhould be expoſed to the open air at all times in mild weather; here the pots may remain till the ſpring, when the danger of hard froſts is over, when they ſhould be removed to a ſituation where they may have only the morning ſun, and in dry weather the earth ſhould be watered gently. This will ſoon bring up the plants, which muſt be kept clean from weeds, and, when they are fit to remove, they ſhould be planted in pots, filled with light earth, and placed in the ſhade till they have taken new root; then they may be placed with other hardy ſucculent plants in a ſheltered ſituation for the ſummer, and in winter placed in a frame where they may be protected from hard froſt, but enjoy the free air in mild weather, with which they will thrive better than with tender treatment.

The ſeventh ſort grows naturally at the Cape of Good Hope. This is a very low plant, whoſe heads ſpread cloſe on the ground; they are much ſmaller than thoſe of the common Houſeleek. The leaves have white edges, which are indented like the teeth of a ſaw; the flowers are produced in looſe panicles upon naked foot-ſtalks; they are ſmall and white, ſo make but little appearance.

This is propagated by offsets, which are put out in plenty from the ſides of the heads; they muſt be planted in pots, ſheltered from the froſt in winter, and in ſummer placed in the open air with other hardy ſucculent plants.

SENECIO. Tourn. Inſt. R. H. 456. tab. 260. Groundſel.

The CHARACTERS are,

The flower is compoſed of many hermaphrodite florets, which form the diſk, and female half florets, which make the border or rays, included in one common cylindrical empalement, which is rough, ſcaly, and contracted above. The hermaphrodite florets are tubulous, funnel-ſhaped, cut into five parts at the brim, which are reflexed; they have five ſmall hair-like ſtamina, terminated by cylindrical ſummits, and an oval germen, crowned with down, ſituated under the petal, ſupporting a ſlender ſtyle, crowned by two oblong revolving ſtigmas. The germen afterward turns to an oval ſeed, crowned with down, incloſed in the empalement. The female half florets, which form the rays, are ſtretched out like a tongue, and are indented in three parts at the top.

We ſhall not trouble the reader with mentioning thoſe ſpecies of this genus, which are eſteemed common weeds, ſo are not cultivated in gardens, but confine ourſelves to thoſe which are the moſt valuable.

The SPECIES are,

1. SENECIO (*Hieracifolia*) corollis nudis, foliis amplexicaulibus laceris, caule herbaceo erecto. Hort. Upſal. 261 *Groundſel with naked petals, torn leaves embracing the ſtalk, and an erect herbaceous ſtalk.*

2. SENECIO (*Pſeudo-china*) corollis nudis, ſcapo ſubnudo longiſſimo. Flor. Leyd. Prod. 164 *Groundſel with naked florets, and a very long foot ſtalk, which is almoſt naked; called China-root.*

3. SENECIO (*Aureus*) corollis radiantibus, foliis crenatis, infimis, cordatis petiolatis, ſuperioribus pinnatifidis lyratis. Flor. Virg. 98. *Groundſel with radiated flowers and*

 crenated

crenated leaves, the lower ones of which are heart-shaped, and have foot-stalks, but the upper leaves are lyre-shaped and wing-pointed.

4. SENECIO (*Hastulatus*) corollis radiantibus, petiolis amplexicaulibus, pedunculis folio triplo longioribus, foliis pinnato-sinuatis. Flor. Leyd. Prod. 164. *Groundsel with radiated flowers, foot-stalks embracing the stalks, with foot-stalks to the flowers three times the length of the leaves, and winged sinuated leaves.*

5. SENECIO (*Elegans*) corollis radiantibus, foliis pinnatifidis æqualibus patentissimis, rachi inferne angustatâ. Hort. Cliff. 406. *Groundsel with radiated flowers, wing-pointed leaves, which are equal, and the midrib below narrowed.*

6. SENECIO (*Abrotanifolius*) corollis radiantibus, foliis pinnato-multifidis linearibus. Lin. Sp. Plant. 869. *Groundsel with radiated flowers, and wing-pointed, multifid, linear leaves.*

7. SENECIO (*Paludosus*) corollis radiantibus, foliis ensiformibus acutè serratis subtus subvillosis, caule stricto. Lin. Sp. Pl. 870. *Groundsel with radiated flowers, sword-shaped leaves which are acutely sawed, a little hairy on their under side, and a close stalk.*

8. SENECIO (*Sarracenicus*) corollis radiantibus, floribus corymbosis, foliis lanceolatis serratis glabriusculis. Hort. Upsal. 266. *Groundsel with radiated flowers growing in a corymbus, and spear-shaped, sawed, smooth leaves.*

9. SENECIO (*Altissimus*) corollis radiantibus, floribus corymbosis, foliis lanceolatis serratis, semiamplexicaulibus. *Groundsel with radiated flowers growing in a corymbus, and spear-shaped sawed leaves half embracing the stalks.*

10. SENECIO (*Orientalis*) corollis radiantibus, floribus corymbosis, foliis ensiformibus dentatis semiamplexicaulibus. *Groundsel with radiated flowers growing in a corymbus, and sword-shaped indented leaves, which half embrace the stalks.*

11. SENECIO (*Incanus*) corollis radiantibus, foliis utrinque tomentosis semipinnatis laciniis subdentatis, corymbo subrotundo. Haller. Helv. 731. *Groundsel with radiated flowers, half-winged leaves which are downy on both sides, segments which are somewhat indented, and a roundish corymbus of flowers.*

12. SENECIO (*Rigidus*) corollis radiantibus, spatulaceis repandis amplexicaulibus scabris acuminatis serratis, caule fruticoso. Hort. Cliff. 406. *Groundsel with radiated flowers, oblong, heart-shaped, rough-pointed, sawed leaves embracing the stalks, which are shrubby.*

13. SENECIO (*Ilicifolius*) corollis radiantibus, foliis sagittatis amplexicaulibus dentatis, caule fruticoso. Vir. Cliff. 84. *Groundsel with radiated flowers, arrow-pointed indented leaves embracing the stalks, which are shrubby.*

14. SENECIO (*Halimifolius*) corollis radiantibus, foliis obovatis carnosis subdentatis, caule fruticoso. Lin. Sp. Pl. 871. *Groundsel with radiated flowers, oval fleshy leaves, which are somewhat indented, and a shrubby stalk.*

The first sort grows naturally in North America; this is an annual plant. The stalk is round, channelled, and hairy; it rises three feet high, is garnished with torn leaves, which embrace the stalks with their base; the flowers are produced in a sort of umbel on the top of the stalks, composed of florets, having no rays; they are of a dirty white colour, and are succeeded by oblong seeds, crowned with a long down. This plant is preserved in some botanic gardens for the sake of variety, but has little beauty. The seeds of this must be sown upon a hot-bed in the spring, and when the plants are come up fit to remove they should be transplanted to another hot bed to bring them forward, and afterward they may be planted in a warm border, where they will flower in July, and their seeds will ripen in autumn.

The second sort grows naturally at Madrass; this has a perennial root, which has been supposed to be the China-root, but is now generally believed to be a spurious kind. The roots are composed of some thick fleshy tubers, sending out many fibres; the leaves are shaped like those of the Turnep, but are smooth. The flower stalk is slender, almost naked, and rises a foot and a half high, sustaining at the top a few yellow flowers, composed of several hermaphrodite florets, having no rays or borders; these are succeeded by oval seeds, crowned with down, but they rarely ripen in England.

This sort is tender, so will not thrive in this country, unless it is kept in a warm stove; it is propagated by parting of the roots in the spring. The offsets should be planted in pots, filled with light earth, and plunged into the tan-bed in the stove, and treated in the same way as other tender exotics.

The third sort grows naturally in North America; this hath a perennial root, from which come out many roundish leaves upon long, slender, hairy foot stalks, of a purplish colour on their under side, crenated on their edges. The stalks rise near two feet high, garnished with a few leaves, which are indented on each side in form of a lyre. The upper part of the stalk divides into several slender long foot stalks, each sustaining one erect flower, composed of several hermaphrodite florets in the center, and a few female florets form the rays or border. They are yellow, and are succeeded by seeds crowned with down. It is propagated by offsets, which come out in plenty from the root; these may be separated in autumn, and planted in an east border, allowing each plant two feet room to spread. When they have taken new root, they will require no other care but to keep them clean from weeds.

The fourth sort grows naturally in Africa. This has an herbaceous perennial stalk, which rises about two feet and a half high, garnished at bottom with narrow leaves, which are situated on the sides so as to resemble winged leaves. The upper leaves are small, and embrace the stalks; they are very clammy, and stick to the fingers on being handled; the upper part of the stalk divides into several very long foot-stalks, each sustaining one yellow radiated flower. The plants continue in flower most part of the summer, and the seeds sometimes ripen in autumn.

This is propagated by cutting off the side shoots in any of the summer months, and planting them in a shady border, where in five or six weeks they will take root, and may then be taken up, and planted in pots, placing them in the shade till they have taken new root; then they may be removed to an open situation, and in autumn they must be placed under a frame, where they may be screened from hard frost, for they will not live abroad in winter here.

The fifth sort grows naturally at the Cape of Good Hope; it is an annual plant, which hath many herbaceous branching stalks that rise near three feet high, garnished with equal wing pointed leaves. The flowers are produced in bunches on the top of the stalks; they are large and radiated, the border or rays being of a beautiful purple colour, and the middle or disk yellow. These plants flower from July till the frost stops them, and make a fine appearance. The seeds ripen in autumn, which, if permitted to scatter, there will be plenty of plants rise the spring following without care; they may be also sown upon a bed of earth in the spring, and when the plants are fit to remove, they may be transplanted about the borders of the flower-garden. If some of the plants are planted in pots, and housed in winter, they may be preserved till spring.

The

The ſixth ſort grows naturally on the Alps and Pyrenees; this has a perennial root and an annual ſtalk. The root is compoſed of a great number of long ſlender fibres; the ſtalks riſe two feet high, and become a little ligneous in autumn; they are garniſhed with very narrow wing-pointed leaves, reſembling thoſe of Hogs Fennel; the flowers are yellow, and are produceed in bunches on the top of the ſtalks; they have rays or borders, reſembling thoſe of the other ſpecies. It is propagated by ſeeds, which ſhould be ſown upon a bed of loamy earth, where it is expoſed only to the morning ſun, where the plants will riſe better than in a warmer ſituation. When the plants are fit to remove, they may be tranſplanted on a ſhady border, where they may remain till autumn, obſerving to keep them clear from weeds all the ſummer; then they ſhould be tranſplanted to the places where they are to remain. The following ſummer the plants will flower and produce ripe ſeeds, and the roots will continue, if they are in a ſhady ſituation and a loamy ſoil.

The ſeventh ſort grows naturally about Paris, by the ſides of waters and in moiſt meadows. The root is perennial; the ſtalks riſe three or four feet high, are cloſely channelled, and garniſhed with ſword-ſhaped leaves, which are hairy, and ſharply ſawed on their edges. The upper part of the ſtalk divides into ſeveral ſlender foot-ſtalks, ſuſtaining yellow radiated flowers, which are ſucceeded by downy ſeeds in autumn, ſoon after which the ſtalks decay to the root.

The eighth ſort grows naturally on the Helvetian mountains, and is ſometimes found growing in low marſhy places in the Iſle of Ely. This hath a creeping root, by which it propagates and ſpreads wherever it is once eſtabliſhed. The ſtalks of this riſe four feet high, garniſhed with ſmooth ſpear-ſhaped leaves, ſawed on their edges, and placed alternate. The flowers are yellow, radiated, and are produced in a ſort of corymbus on the top of the ſtalk, which are ſucceeded by ſeeds, having down.

The ninth ſort grows naturally in France; this has ſome reſemblance of the eighth, but the root does not creep like that. The leaves are ſhorter; the ſerratures on their edges are very ſmall; they embrace the ſtalks with their baſe, and end in ſharp points. The flowers are produced in larger and looſer bunches on the top of the ſtalk, and are of a paler yellow colour than thoſe of the former.

The tenth ſort grows naturally in the Levant; this has a perennial root. The lower leaves are long, ſmooth, and ſomewhat ſhaped like a ſcymiter, the midrib being curved outward toward the point, and ſlightly indented on their edges. The ſtalk riſes ſix feet high, garniſhed with leaves, growing ſmaller toward the top of the ſtalk. The flowers terminate the ſtalk in a compact corymbus; they are of a deep yellow, and have rays like thoſe of the former ſorts.

Theſe ſorts are eaſily propagated by ſeeds or parting of their roots; the latter is generally practiſed when the plant is once obtained, as that is the moſt expeditious method, eſpecially for the eighth ſort, whoſe roots are apt to ſpread and increaſe too faſt where they are not confined. The beſt time to tranſplant and divide theſe roots is in autumn, when their ſtalks decay, that they may get good rooting before the ſpring. Theſe plants are too large for ſmall gardens, ſo are proper furniture for large borders in extenſive gardens, or to plant on the ſides of woods, where they may be allowed room, for they ſhould have at leaſt four feet allowed to each. When thoſe are intermixed with other tall growing plants in ſuch places, they will add to the variety.

The eleventh ſort grows naturally on the Alps; this is a perennial plant, of low growth. The ſtalks ſeldom riſe a foot high; the whole plant is covered with a very white hoary down; the leaves are winged and indented; the flowers are collected into a cloſe round corymbus on the top of the ſtalk, of a gold colour, and are radiated; theſe are rarely ſucceeded by good ſeeds in England. It is propagated by ſlipping off the heads in the autumn, and planting them in a bed of loamy earth in a ſhady ſituation, where they will put out roots, and may afterward be tranſplanted into an eaſt border, where they may have the morning ſun only, for this plant loves a gentle loamy ſoil, and a ſituation not too much expoſed to the ſun.

The twelfth ſort grows naturally at the Cape of Good Hope. This riſes with a ſhrubby branching ſtalk ſix or ſeven feet high, cloſely garniſhed with rough leaves, whoſe baſe embrace the ſtalks; they are ſtiff, hairy, of a dark green colour, oblong, heart ſhaped, and indented on their edges. The flowers are produced at the end of the branches, which are of a bright yellow colour, and are ſucceeded by ſeeds in autumn.

The thirteenth ſort grows naturally at the Cape of Good Hope. This hath a very branching ſhrubby ſtalk, which riſes four or five feet high, garniſhed with ſtiff leaves, whoſe baſe embraces the ſtalks; they are irregular in their figure, deeply cut on their edges, and of a gray colour on their under ſide. The flowers grow in looſe bunches at the end of the branches, of a pale yellow colour. This ſort flowers great part of ſummer, and the ſeeds ripen in autumn.

The fourteenth ſort grows naturally at the Cape of Good Hope. This has a ſhrubby ſtalk, which riſes ſeven or eight feet high, garniſhed with oblong oval leaves, indented on their edges. The flowers are produced in looſe bunches at the extremity of the branches, almoſt in form of an umbel; they are of a pale yellow colour.

The three ſorts laſt mentioned are too tender to live in the open air through the winter in England, but are ſo hardy as only to require protection from hard froſt, ſo may be kept in pots, and placed either under a frame in winter, or in a common green-houſe with other hardy kinds of plants, which require a large ſhare of air in mild weather. They are all eaſily propagated by ſeeds or cuttings, but the latter being the moſt expeditious method, is generally practiſed here. If the cuttings are planted in a ſhady border during any of the ſummer months, they will readily take root; then they ſhould be taken up, and each planted in a ſeparate pot, and placed in the ſhade till they have taken new root; then they may be removed to a more open ſituation, where they may remain till there is danger of ſharp froſt; when they ſhould be removed into ſhelter, and treated in the ſame way as other hardy kinds of green-houſe plants.

SENNA. Tourn. Inſt. R. H. 618. tab. 319. Senna.

The CHARACTERS are,

The flower has an empalement of five leaves; it has five roundiſh concave petals, and ten declining ſtamina, terminated by oblong arched ſummits. The germen is roundiſh and compreſſed, ſupporting a ſhort ſtyle, crowned by an obtuſe ſtigma. The germen afterward becomes a plain, roundiſh, compreſſed pod, a little incurved, having two cells, divided by an intermediate partition, each containing one or two oblong pointed ſeeds.

The SPECIES are,

1. SENNA (*Alexandrina*) foliolis quadrijugatis lanceolatis acutis. *Senna with four pair of ſpear-ſhaped pointed lobes to the leaves.*

2. SENNA (*Italica*) foliolis quinquejugatis cordatis obtuſis. *Senna with five pair of lobes to the leaves, which are heart-ſhaped and obtuſe.*

The

The firft fort grows naturally in Egypt; this is an annual plant, which rifes with an upright branching ftalk a foot high, garnifhed with winged leaves, compofed of four pair of fmall fpear-fhaped lobes, ending in acute points. The flowers terminate the ftalks in loofe bunches; they are yellow, compofed of five roundifh concave petals, with ten ftamina in the center furrounding the ftyle; after the flower is paft, the germen turns to a roundifh gibbous pod, having two cells, each containing one or two oblong feeds. The leaves of this fort are ufed in medicine, and are commonly known in the fhops by the title of Senna; thefe are annually imported from Alexandria, which occafioned the title of Alexandrina being added to it. This plant is propagated by feeds, which fhould be fown early in the fpring, upon a good hot-bed; and when the plants are come up, and are fit to remove, they fhould be each planted in a fmall pot, and plunged into a frefh hot-bed, fhading them from the fun till they have taken new root, after which they muft be treated in the fame way as the moft tender exotics; for as this is an annual plant, unlefs they are brought forward in the fpring, they will not flower in this country; therefore they muft be conftantly kept in the hot-bed all the fummer, obferving to admit plenty of air in warm weather; by which method I have frequently had thefe plants in flower, but it is very rare that they perfect their feeds in England.

The fecond fort grows naturally in India, from whence I have received the feeds; for though it is called Italian, yet the plant does not grow there naturally. This is alfo an annual plant, rifing with a branching ftalk a foot and a half high; the leaves are winged, each having five pair of heart-fhaped lobes, which are inverted, the point joining the branches, and the obtufe part is upward: they are of a fea-green colour, and of a thick confiftence. The flowers are produced at the end of the branches, fhaped like thofe of the firft fort, but are larger, and of a brighter yellow colour. If the plants are brought forward early in the fpring, they will flower in July, and by fo doing good feeds may be obtained here. This fort is propagated in the fame way as the firft, and the plants require the fame treatment.

SENNA THE BLADDER. See Colutea.

SENNA THE SCORPION. See Emerus.

SENSIBLE PLANT. See Mimosa.

SERAPIAS. Lin. Gen. Plant. 903. Baftard Hellebore.

The Characters are,

It has a fingle ftalk; the fheath of the flower is at a diftance. The germen fuftains the flower, which has no empalement, but has five oblong oval petals. The nectarium is the length of the petal, hollowed at the bafe, oval, and gibbous below, cut into three points. The flower has two fhort ftamina fitting upon the pointal, terminated by erect fummits, placed under the upper lip of the nectarium; and an oblong contorted germen, fituated under the flower, the ftyle growing to the upper lip of the nectarium, crowned by an obfolete ftigma. The germen afterward becomes an oval, obtufe, three-cornered capfule, armed with three keels, opening with a valve under each, having one cell, filled with fmall feeds.

The Species are,

1. Serapias (*Helleborine*) bulbis fibrofis, nectarii labio obtufo crenato petalis breviore. Act. Upfal. 1740. *Serapias with fibrous bulbs, and the lip of the nectarium obtufe, crenated, and fhorter than the petal.*

2. Serapias (*Damafonium*) bulbis fibrofis, petalis nectario longioribus obtufis, foliis lanceolatis nervofis. *Serapias with fibrous bulbs, obtufe petals which are longer than the nectarium, and veined fpear-fhaped leaves.*

3. Serapias (*Paluftre*) bulbis fibrofis, petalis reflexis, nectarii labio obtufo, foliis enfiformibus nervofis. *Serapias with fibrous bulbs, reflexed petals, the lip of the nectarium obtufe, and fword-fhaped veined leaves.*

4. Serapias (*Latifolium*) bulbis fibrofis, nectarii labio quinquefido claufo, foliis lanceolatis nervofis amplexicaulibus. *Serapias with fibrous bulbs, the lip of the nectarium cut into five parts, and fpear-fhaped veined leaves embracing the ftalks.*

There are fome other fpecies of this genus which grow naturally in Great-Britain and Ireland, but as I have not had the good fortune to meet with them, fo I fhall not trouble the reader with an imperfect account of them from books: there are alfo a greater number of them which grow naturally in the Weft-Indies, of which I have famples in my collection; but having never feen any growing plants of them, I fhall not infert them here.

The firft fort grows naturally in woods and fhady places in many parts of England; the roots are compofed of many thick flefhy fibres, from which arifes a fingle jointed ftalk a foot high, garnifhed at each joint with one veined leaf; thofe on the lower part of the ftalk are oval, but thofe above are fpear-fhaped, ending in acute points, embracing the ftalks at their bafe. The ftalk is adorned with flowers toward the top, which have fome refemblance to thofe of Orchis, compofed of two whitifh, and three herbaceous petals, which expand, and in the middle appears the nectarium, which has a refemblance of a difboweled body of a fly, of a purplifh colour. Under the flower is fituated a channelled oblong head, which, after the flower is paft, fwells and becomes a feed-veffel, filled with very fmall feeds.

The fecond fort grows naturally in Stoken-Church woods in Oxfordfhire, and in feveral parts of Weftmoreland and Lancafhire. This has flefhy fibrous roots, not quite fo thick as thofe of the former; the ftalks rife more than a foot high, garnifhed with fpear-fhaped veined leaves, ending in acute points, of a lucid green, and fit clofe to the ftalk. The flowers are white, difpofed alternately on the upper part of the ftalk; the three outer petals are large, and two fmaller within; in the center is fituated the gaping nectarium, which appears to have two wings.

The third fort grows naturally in marfhy woods in many parts of England; this has a flefhy fibrous root, from which arife a fingle ftalk a foot and a half high, garnifhed at bottom with fword-fhaped veined leaves, embracing the ftalk with their bafe, ending in acute points. The upper part of the ftalk is garnifhed with faded purplifh-coloured flowers, difpofed in a loofe fpike; they have five petals, inclofing a large nectarium like the body of a fly, with a yellowifh head, ftriped with purple and a white body; the lip which hangs down is white, and fringed on the edge.

The fourth fort was difcovered firft in Hertfordfhire, but fince it has been found growing in many other places. The root of this is compofed of flefhy fibres; the ftalks rife more than a foot high, garnifhed with fpear-fhaped veined leaves, which embrace the ftalks with their bafe. The ftalk is terminated by a loofe fpike of white flowers, compofed of five petals, and a large five-pointed nectarium, which is fhut; the germen is oblong and channelled: this afterward becomes a capfule of the fame form, filled with fmall feeds.

Thefe plants are rarely kept in gardens, being difficult to propagate, fo there are few who have attempted them. They may be taken up from the places where they naturally grow, when their leaves begin to decay, and planted in a fhady moift place, where they will thrive and flower.

SERJANIA. See Paullinia.

SERPENTARIA. See Aristolochia.

SERPYLLUM. See Thymus.

SERRATULA. Dillen. Nov. Gen. 8. Saw-wort.

The CHARACTERS are,

The flowers are composed of many hermaphrodite florets, contained in one common cylindrical empalement; the scales of which are spear-shaped, ending in acute points. The hermaphrodite florets are equal, funnel-shaped, of one petal. The tube is inflexed, the brim is bellied, and cut into five points; they have each five short hair-like stamina, terminated by cylindrical summits; and an oval germen, supporting a slender style, crowned by two oblong reflexed stigmas. The germen afterward turns to a vertical, oval, single seed, crowned with down, which ripens in the empalement.

The SPECIES are,

1. SERRATULA (*Tinctoria*) foliis pinnatifidis, pinna terminali maxima. Hort. Cliff. 391. *Saw-wort with wing-pointed leaves, whose end lobe is the largest; common Saw-wort.*

2. SERRATULA (*Altissima*) foliis lanceolato-oblongis serratis subtus tomentosis. *Saw-wort with oblong spear-shaped leaves, which are sawed, and downy on their under side.*

3. SERRATULA (*Glauca*) foliis ovato-oblongis acuminatis serratis, floribus corymbosis, calycibus subrotundis. Flor. Virg. 92. *Saw-wort with oblong, oval, acute-pointed, sawed leaves, and flowers in a corymbus, whose empalements are roundish.*

4. SERRATULA (*Squarrosa*) foliis linearibus, calycibus squarrosis sessilibus acuminatis. Hort. Cliff. 392. *Saw-wort with linear leaves and rough empalements, which sit close to the stalks, ending in acute points.*

5. SERRATULA (*Scariosa*) foliis lanceolatis integerrimis, calycibus squarrosis pedunculatis obtusis lateralibus. Lin. Sp. Pl. 818. *Saw-wort with entire spear-shaped leaves, and rough empalements, having obtuse foot-stalks proceeding from the side of the stalks.*

6. SERRATULA (*Spicata*) foliis linearibus, floribus sessilibus lateralibus spicatis, caule simplici. Lin. Sp. Pl. 819. *Saw-wort with linear leaves, flowers in spikes from the side of the stalks sitting close, and a single stalk.*

7. SERRATULA (*Caroliniana*) foliis lanceolatis rigidis, acute serratis, caule corymboso. *Saw-wort with stiff spear-shaped leaves sharply sawed, and stalks forming a corymbus.*

8. SERRATULA (*Præalta*) foliis oblongo-lanceolatis, integerrimis subtus hirsutis. *Saw-wort with oblong, spear-shaped, entire leaves, hairy on their under side.*

9. SERRATULA (*Patula*) foliis oblongo-ovatis obtuse-dentatis, caule ramoso patulo, calycibus subrotundis mollibus. *Saw-wort with oblong oval leaves bluntly indented, a branching spreading stalk, and soft roundish empalements.*

10. SERRATULA (*Alpina*) calycibus subhirsutis ovatis. Lin. Sp Plant. 816. *Saw-wort with oval empalements a little hairy.*

The first sort grows naturally in moist woods and marshes in many parts of England, so is rarely admitted into gardens. There are two varieties of this, one with a white, and the other a purple flower. The root is perennial; the lower leaves are sometimes entire, and at others are cut almost to the midrib into many jags; they are smooth, of a deep green, and neatly sawed on their edges. The stalks rise two feet high, garnished with wing-pointed leaves, whose extreme lobe is much larger than the other; the upper part of the stalk divides into several foot-stalks, sustaining at the top oblong squamous heads or empalements, which include several hermaphrodite florets.

The second sort grows naturally in North America; this has a perennial root, sending out several channelled stalks seven or eight feet high, garnished with spear shaped leaves, slightly sawed on their edges, which are downy on their under side, sitting close to the stalk; the upper part of the stalk divides into foot-stalks, which sustain purple flowers in scaly empalements.

The third sort is a native of North America; the root is perennial; the stalks rise six or seven feet high. The leaves are oblong, oval, stiff, sawed on their edges, and of a light green on both sides. The flowers grow in a loose corymbus at the top of the stalk; they are purple, and have roundish empalements.

The fourth sort grows naturally in Carolina. This has a tuberous root, sending out a single stalk three feet high, garnished with stiff linear leaves, which are entire, and rough to the touch, of a pale green on both sides. The upper part of the stalk is adorned with purple flowers, having oblong, rough, prickly empalements; these sit close to the side of the stalk alternately, and the stalk is terminated by one head, which is larger than the other.

The fifth sort grows naturally in most parts of North America; this has a large tuberous root, from which comes out one strong channelled stalk three or four feet high, closely garnished with narrow spear-shaped leaves, which are entire. The upper part of the stalk is adorned with loose spikes of purple flowers, which come out from the side upon pretty long blunt foot-stalks; they have large rough empalements, composed of wedge-shaped scales. The flowers on the top of the spike blow first, and are succeeded by the other downward, which is contrary to the usual order of the greatest number of plants, whose flowers are ranged in spikes.

The sixth sort is a native of North America; this has a tuberous root, from which comes forth a single stalk from two to three feet high, garnished with very narrow smooth leaves, sitting close to the stalk without any order. The upper part of the stalk is adorned with smaller purple flowers than those of the former, sitting very close to the stalk, forming a long loose spike.

The seventh sort is also a native of North America; this has a perennial fibrous root, from which arise several strong purple stalks upward of six feet high, garnished with stiff spear-shaped leaves, deeply sawed on their edges, of a pale green on their under side. The upper part of the stalk divides into small branches, forming a loose corymbus of purple flowers, which are irregular in height, some of the flowers standing upon shorter foot-stalks than the other; their empalements are round, and the scales terminate in bristly points.

The eighth sort grows naturally in Carolina; this has a fibrous perennial root; the stalk rises four feet high; the leaves are entire, hoary on their under side, sitting close to the stalk. The flowers grow in loose bunches at the end of the branches; they have oval empalements, composed of a few scales, which terminate in bristles. The flowers are of a pale purple colour.

The ninth sort grows naturally in the northern parts of China. This is an annual plant, with an herbaceous stalk a foot and a half high, covered lightly with a white meal; the branches spread out almost horizontally, garnished with smooth, oblong, oval leaves, which have a few blunt indentures on their edges, of a light green colour. The flowers are produced in loose bunches at the extremity of the branches, having roundish soft empalements, whose scales end with hairy points, including hermaphrodite florets of a bluish purple colour.

The tenth sort grows naturally on the tops of mountains in Wales and the north of England, and is but seldom kept in gardens. The root is perennial, from which arise two or three stalks a foot and a half high, of a deep green colour, channelled, and garnished with deep green leaves. From the middle of the stalk upward, there are branches

sent

fent out from the fide, which grow erect, and fuftain at the top fmall bunches of purple flowers, which have oblong flender empalements a little hairy.

The eight forts which are firft mentioned, are hardy perennial plants, fo will thrive in the open air in England. The firft is rarely admitted into gardens, but the other are frequently preferved in the gardens of the curious. The fourth, fifth, and fixth forts, have large knobbed roots, fo are propagated only by feeds which feldom ripen in England. Thefe fhould be fown on an eaft-afpected border, for if they are expofed to the mid-day fun they feldom fucceed well. The feeds will often grow the firft fummer, if they are fown early in the fpring, but fometimes they will remain in the ground a year before the plants appear; fo that if they fhould not come up the firft feafon, the ground fhould not be difturbed, but muft be kept clean from weeds till the following fpring, when, if the feeds were good, the plants will come up; when thefe appear, if they are too clofe, fome of the plants fhould be carefully drawn out while they are young, and planted in another bor er four inches afunder; in this place they may remain till autumn, when thefe, and alfo thofe in the feed-bed, fhould be carefully removed to the places where they are defigned to remain; the following fummer thefe plants will flower, and the roots will abide feveral years, if they are planted in a light loamy foil not over-wet.

The other perennial forts may be propagated by parting of the roots: the beft time for doing this is in autumn, when their ftalks begin to decay; for when they are removed in the fpring, if the feafon fhould prove dry, their roots will not be fufficienly eftablifhed to flower well the fame year. Thefe plants fhould not be removed or parted oftener than every third year, if they are expected to grow ftrong; nor fhould they be parted into fmall heads, for thofe will make no figure the firft year. As thefe plants grow tall, fo they fhould be planted in the middle of large borders, or with other tall plants; they may be planted in fpaces between fhrubs, or on the borders of woods, where they will have a good effect, during their continuance in flower, as they require no other culture than to dig the ground between them every fpring.

The tenth fort is a perennial plant, which may be propagated by its creeping roots, and may be planted in a fhady border, where it will thrive, and annually produce flowers.

The ninth fort is an annual plant; the feeds of this fhould be fown upon a moderate hot-bed the beginning of April, and, when the plants come up, and are fit to remove, they fhould be tranfplanted to a frefh hot-bed to bring them forward; afterward they fhould be treated in the fame way as other hardy annual plants.

SESAMUM. Lin. Gen. Plant. 700. Oily Grain.

The CHARACTERS are,

The flower has an erect permanent empalement, cut at the top into five very fhort equal parts. The flower has one ringent petal, with a roundifh tube the length of the empalement; the chaps are fwollen, and bell-fhaped; the brim is cut into five points, four of which are almoft equal, the other is twice their length. It has four ftamina rifing from the tube, which are fhorter than the petal, the two inner being fhorter than the other, terminated by erect fummits; and an oval hairy germen, fupporting a flender ftyle longer than the ftamina, crowned by a fpear-fhaped ftigma, divided in two parts. The germen afterward becomes an oblong almoft four-cornered capfule, having four cells, filled with oval compreffed feeds.

The SPECIES are,

1. SESAMUM (*Orientale*) foliis ovato-oblongis integris. Hort. Cliff. 318. *Sefamum with oblong, oval, entire leaves.*

2. SESAMUM (*Indicum*) foliis inferioribus trifidis. Prod. Leyd. 292. *Sefamum with trifid lower leaves.*

3. SESAMUM (*Trifoliatum*) foliis omnibus trifidis. *Sefamum with all the leaves trifid.*

The firft fort is cultivated in great plenty in the Levant, but is fuppofed to have been brought there from India. It is an annual plant, rifing with an herbaceous four-cornered ftalk two feet high, fending out a few fhort fide branches; the leaves are oblong, oval, a little hairy, and ftand oppofite. The flowers terminate the ftalks in loofe fpikes; they are fmall, of a dirty white colour, fhaped fomewhat like thofe of the Foxglove. After the flowers are paft, the germen turns to an oval acute-pointed capfule with four cells, filled with oval compreffed feeds, which ripen in autumn.

The fecond fort grows naturally in India; this is alfo an annual plant; the ftalk rifes taller than that of the former; the lower leaves are cut into three parts, which is the only difference between them.

The third fort grows naturally in Africa. This is alfo an annual plant, with a taller and more branched ftalk than either of the former, in which it differs from both the other.

The firft fort is frequently cultivated in all the eaftern countries, and alfo in Africa, as a pulfe; and of late years the feeds have been introduced into Carolina by the African negroes, where they fucceed extremely well. The inhabitants of that country make an oil from the feed, which will keep good many years, without having any rancid fmell or tafte, but in two years becomes quite mild; fo that when the warm tafte of the feed, which is in the oil when firft drawn, is worn off, they ufe it as a fallad oil and for all the purpofes of fweet oil.

The feeds of this plant are alfo ufed by the negroes for food, which feeds they parch over the fire, and then mix them with water, and ftew other ingredients with them, which makes an hearty food. Sometimes a fort of pudding is made of thefe feeds, in the fame manner as with Millet or Rice, and is by fome perfons efteemed, and is rarely ufed for thefe purpofes in Europe. This is called Benny, or Bonny, in Carolina.

In England thefe plants are preferved in botanic gardens as curiofities. Their feeds muft be fown in the fpring upon a hot-bed, and when the plants are come up, they muft be tranfplanted into a frefh hot-bed to bring them forward. After they have acquired a tolerable degree of ftrength, they fhould be planted into pots, and plunged into another hot bed, managing them as hath been directed for Amaranthufes, to which I fhall refer the reader to avoid repetition; for if thefe plants are not thus brought forward in the former part of the fummer, they will not produce good feeds in this country.

The feed of the firft fort is mentioned in the lift of officinal fimples in the College Difpenfatory, but is rarely ufed in medicine in England. From nine pounds of this feed which came from Carolina, there were upwards of two quarts of oil drawn, which is as great a quantity as hath been obtained from any vegetable whatever. This, I fuppofe, might occafion its being called Oily Grain.

SESELI. Boerh. Ind. alt. 1. p. 50. Wild Spignel.

The CHARACTERS are,

It has an umbellated flower; the particular umbels are very fhort, multiplex, and almoft globular. The principal umbel has no involucrum; the particular ones have a many narrow-leaved involucrum, which is as long as the umbel; the principal umbel is uniform. The flowers have five inflexed heart-fhaped petals, which are unequal; they have each five awl-fhaped ftamina, terminated by fingle fummits. The germen is fituated under the flower, fupporting two reflexed ftyles, crowned by obtufe ftigmas.

The

The germen afterward turns to a small, oval, channelled fruit, dividing in two parts, each containing one oval streaked seed, flat on one side, and convex on the other.

The Species are,

1. Seseli (*Montanum*) petiolis ramiferis membranaceis ventricosis emarginatis. Hort. Cliff. 103. *Seseli with bellied, membranaceous, branching foot-stalks, which are indented at the top.*

2. Seseli (*Hippomarathrum*) involucellis conato-monophyllis. Lin. Sp. 374. *Seseli with involucrums of one leaf.*

3. Seseli (*Glaucum*) petiolis ramiferis membranceis oblongis integris, foliolis singularibus binatisque. Guett. 64. *Seseli with branching, oblong, entire, membranaceous foot-stalks, the small leaves single and by pairs.*

4. Seseli (*Pumilum*) petiolis ramiferis membranaceis oblongis integris, foliis caulinis angustissimis., Hort. Cliff. 102. *Seseli with oblong, entire, membranaceous, branching foot-stalks, and very narrow leaves on the stalks.*

5. Seseli (*Tortuosum*) caule alto rigido, foliolis linearibus fasciculatis. Lin. Sp. Pl. 260. *Seseli with a tall stiff stalk, and very narrow leaves in clusters.*

6. Seseli (*Ammoides*) petiolis membrana destitutis. Flor. Leyd. Prod. 112. *Seseli with foot-stalks without membranes.*

The first sort grows naturally in France amongst the Corn; this rises with an erect stalk two feet high, garnished with short leaves, divided into small segments like Hogs Fennel. At the foot-stalk of each branch or leaf is a bellied membrane, which embraces it. The stalk is terminated by an umbel of white flowers, and the seeds ripen in August.

The second sort grows naturally in Germany; this hath a perennial root. The leaves are long, made up of eight or nine pair of winged lobes, which are cut like those of Parsley; the stalk rises two feet and a half high, branching out into several divisions; at each of these there is a membrane embracing the base, and one small leaf, composed of a few linear lobes. The stalks are terminated by compound umbels of yellow flowers, which are succeeded by seeds.

The third sort grows naturally in uncultivated places in the south of France and Italy; this has a perennial root, sending out slender smooth stalks two feet high. The leaves are long and narrow, composed of seven or eight pair of wings, whose lobes are sometimes single, and at others are divided into two parts; they have a membrane embracing their foot-stalks, and are of a gray colour. The stalks are terminated by umbels of flowers, which are purple on their outside, and white within.

The fourth sort grows naturally on the dry hills in many parts of France and Italy; this has a perennial root, from which come out leaves like those of Spignel, but the segments are broader and of a gray colour. The stalks rise a foot high, garnished with a few very narrow leaves, whose foot-stalks are embraced by a long entire membrane, and are terminated by umbels of white flowers.

The fifth sort grows naturally in the south of France, in Italy, and Spain; this has a thick ligneous root, from which arise stiff stalks four feet high, crooked at their joints, and garnished with narrow leaves coming out in bunches. The stalks divide into slender branches, which have small umbels of flowers coming out of their sides, and are terminated by larger. The flowers are small and yellow.

The sixth sort is an annual plant, which grows naturally in Portugal. The leaves of this are like those of Spignel, but much smaller, and have a very acrid biting taste. The stalks rise four inches high, sustaining a small umbel of flowers.

These plants are preserved in the gardens of botanists for the sake of variety, but at present their virtues are unknown, and, as they have little beauty to recommend them, they are rarely admitted into other gardens.

They may be propagated by sowing their seeds in autumn, for when they are sown in the spring, they frequently lie in the ground a year before the plants will appear; whereas those which are sown in autumn, always rise the following spring. They should be sown in drills about eighteen inches asunder, in a bed of fresh earth, where they are designed to remain, and in the spring, when the plants come up, they should be thinned where they are too close, leaving them about six inches distance in the rows; after this the plants will require no farther care but to keep them constantly clear from weeds; the second season they will produce flowers and seeds. The perennial roots, which are permitted to remain after they have seeded, should have the ground gently dug every spring between the rows to loosen the earth, but there should be care taken not to injure their roots with the spade.

SHERARDIA. Dillen. Gen. Nov. 3. Little Field-Madder.

The Characters are,

The flower has a small permanent empalement sitting upon the germen; it has one long tubulous petal, cut into four plain acute parts at the brim; it has four stamina, situated on the top of the tube, terminated by single summits, and an oblong twin germen below the flower, supporting a slender bifid style, crowned by two headed stigmas. The germen afterward becomes an oblong fruit, containing two oblong seeds, which are separated.

We have but one Species of this plant, viz.

1. Sherardia (*Arvensis*) foliis omnibus verticillatis floribus terminalibus. Lin. Sp. Plant. 102. *Sherardia all the leaves placed in whorls, and flowers terminating the stalks.*

This plant grows naturally amongst the Corn in many parts of England; it is an annual plant, with trailing stalks which spread on the ground, garnished with short acute-pointed leaves growing in whorls, some of which have four, others five and six, and some have eight leaves in each whorl. The flowers terminate the stalk; there are generally five or six flowers in each bunch; they are blue, have pretty long tubes, and are cut into four segments at the top, spreading open. Their seeds ripen in autumn.

SHERARDIA. Vaill. See Verbana.

SICYOS. Lin. Gen. Plant. 971. Single-seeded Cucumber.

The Characters are.

It hath male and female flowers on the same plant; the male flowers have a bell-shaped empalement of one leaf, with five indentures. The petal is bell-shaped, growing on the empalement; they have each three stamina, which are united above, terminated by summits, joined in a head. The female flowers are like the male, and sit upon the germen; they have no stamina, but the germen supports a cylindrical style, crowned by a thick three-pointed stigma. The germen afterward becomes an oval fruit set with bristly hairs, having one cell, containing a single seed of the same shape.

The Species are,

1. Sicyos (*Angulata*) foliis angulatis. Hort. Cliff. 452. *Sicyos with angular leaves.*

2. Sicyos (*Laciniata*) foliis laciniatis. Lin. Sp. Pl. 1013. *Sicyos with cut leaves.*

The first sort grows naturally in North America; this is an annual plant, which rises with two large seed leaves like those of the Cucumber. The stalk is trailing, and has tendrils, by which it fastens itself to the neighbouring plants, where, if it has support, will rise fifteen or sixteen feet

feet high, dividing into many branches, garnifhed with angular leaves like thofe of the Cucumber. The flowers come out upon long foot-ftalks from the fide of the branches, ftanding in clufters, fome of which are male or barren flowers, others are female fruitful flowers; they are of a pale fulphur colour; the female flowers are fucceeded by prickly oval fruit, containing one feed. If the feeds are permitted to fcatter, the plants will come up in the fpring better than when fown by hand, and require no other care but to keep them clean from weeds. Thefe plants ramble, and take up too much room for fmall gardens, therefore fhould be allowed a place near a hedge, upon which they may climb: they do not bear tranfplanting well, unlefs when they firft come up.

The fecond fort grows naturally in the Weft-Indies; this is alfo an annual plant, with trailing ftalks like the former, but the leaves of this are cut into feveral fegments. The flowers are larger, and of a deeper colour; the fruit are not quite fo large, nor fo clofely armed with prickly hairs, in which confifts their difference.

This fort is not fo hardy as the firft, therefore whoever has a mind to cultivate it, muft fow the feeds upon a hot-bed in the fpring, and treat the plants in the fame way as Cucumbers and Melons, keeping them under frames, otherwife the feeds will not ripen here; the plants will require more room than either of the former, fo that one or two plants will be enough for curiofity, as they have no great beauty or ufe.

SIDA. Lin. Gen. Plant. 747. Indian Mallow.

The CHARACTERS are,

The empalement of the flower is fingle, permanent, angular, and five-pointed. The flower is of one petal, cut into five broad fegments, which are joined at their bafe, and are indented at their points; it has many ftamina, which are joined in a column at bottom, but fpread open above, terminated by roundifh fummits, and an orbicular germen, fupporting a fhort multifid ftyle, crowned by headed ftigmas. The germen afterward becomes a five-cornered capfule, having five cells, each containing an angular roundifh feed.

The SPECIES are,

1. SIDA (*Ulmifolia*) foliis ovato-lanceolatis ferratis, floribus folitariis axillaribus, femine roftrato bidente. *Indian Mallow with oval, fpear-fhaped, fawed leaves, fingle flowers on the fide of the ftalk, and feeds with two horns.*

2. SIDA (*Carpinifolia*) caule ramofo hirfuto, foliis lanceolatis ferratis floribus confertis axillaribus, femine roftrato fimplici. *Sida with a branching hairy ftalk, fpear-fhaped fawed leaves, flowers in clufters from the wings of the ftalk, and feeds with a fingle horn or tooth.*

3. SIDA (*Anguftifolia*) caule erecto ramofo, foliis lineari-lanceolatis dentatis fubtus villofis, pedunculis axillaribus unifloris. *Sida with an erect branching ftalk, linear fpear-fhaped leaves, hairy on their under fide, and foot-ftalks with one flower at the wings of the ftalks.*

4. SIDA (*Pimpinellifolia*) foliis cordatis ferratis, pedunculis unifloris axillaribus, femine roftrato bidente. *Sida with heart-fhaped fawed leaves, foot-ftalks with one flower from the wings of the ftalk, and feeds with two horns.*

5. SIDA (*Jamaicenfis*) foliis ovato-lanceolatis inæqualiter ferratis, floribus axillaribus feffilibus, femine tridente. *Sida with oval fpear-fhaped leaves which are unequally fawed, flowers fitting clofe at the wings of the ftalks, and feeds with three teeth.*

6. SIDA (*Villofa*) caule erecto hirfuto, foliis fubcordatis feffilibus ferratis fubvillofis, floribus confertis axillaribus feffilibus. *Sida with a hairy ftalk, leaves almoft heart-fhaped, fitting clofe to the ftalk, which are a little woolly, and flowers in clufters fitting clofe at the wings of the ftalk.*

7. SIDA (*Alnifolia*) foliis orbiculatis plicatis ferratis. Hort. Cliff. 346. *Sida with orbicular plaited leaves, which are fawed.*

8. SIDA (*Cordifolia*) foliis cordatis fubangulatis ferratis villofis. Lin. Sp. Pl. 684. *Sida with heart-fhaped leaves almoft angular, which are woolly and fawed.*

9. SIDA (*Hirfuta*) foliis orbiculato-cordatis crenatis, caule petiolifque hirfutis, pedunculis longis axillaribus unifloris. *Sida with orbicular, heart-fhaped, crenated leaves, the ftalks and foot-ftalks of the leaves hairy, and long foot-ftalks from the wings of the ftalk with one flower.*

10. SIDA (*Capitata*) capitulis pedunculatis triphyllis feptemfloris. Lin. Act. Upfal. 1743. p. 137. *Sida with heads on foot-ftalks, which have three leaves and feven flowers.*

11. SIDA (*Hirfutiffima*) foliis lanceolatis ferratis villofis, caule erecto pilofo, pedunculis axillaribus unifloris. *Sida with fpear-fhaped, woolly, fawed leaves, an erect hairy ftalk, and foot-ftalks from the wings of the ftalk with one flower.*

12. SIDA (*Abutifolia*) foliis cordatis crenatis acuminatis villofis, caule petiolifque pilofis, pedunculis axillaribus unifloris. *Sida with heart-fhaped, pointed, crenated, woolly leaves, the ftalks and foot-ftalks hairy, and foot-ftalks with one flower at the wings of the ftalk.*

13. SIDA (*Ciliaris*) caulibus procumbentibus, foliis oblongo-ovatis ferratis hirfutis, floribus feffilibus terminalibus. *Sida with trailing ftalks, oblong, oval, hairy, fawed leaves, and flowers fitting clofe at the end of the branches.*

14. SIDA (*Glabra*) foliis cordatis ferratis acuminatis glabris, caule ramofo, pedunculis axillaribus unifloris. *Sida with heart-fhaped, fawed, acute-pointed, fmooth leaves, a branching ftalk, and foot-ftalks from the wings of the ftalks with one flower.*

15. SIDA (*Sericea*) caulibus procumbentibus, foliis ovatis ferratis tomentofis nitidis, floribus folitariis axillaribus feffilibus. *Sida with trailing ftalks, neat, oval, fawed, woolly leaves, and fingle flowers fitting clofe to the wings of the ftalk.*

16. SIDA (*Pilofa*) foliis fubcordatis nervofis fubtus tomentofis, caule pilofo, pedunculis axillaribus multifloris. *Sida with veined fawed leaves almoft oval, woolly on their under fide, a hairy ftalk, and foot-ftalks with many flowers at the wings of the ftalks.*

17. SIDA (*Fruticofa*) foliis lanceolatis inæqualiter ferratis acuminatis, floribus capitatis terminalibus, caule fruticofo. *Sida with fpear-fhaped acute-pointed leaves unequally fawed, flowers collected in heads at the end of the branches, and a fhrubby ftalk.*

18. SIDA (*Alba*) foliis cordatis acuminatis ferratis nervofis, floribus aggregatis axillaribus feffilibus. *Sida with acute-pointed, heart-fhaped, fawed, veined leaves, and flowers in clufters fitting clofe to the wings of the ftalk.*

19. SIDA (*Frutefcens*) caule erecto fuffruticofo, foliis cordatis crenatis tomentofis, pedunculis axillaribus unifloris. *Sida with an erect under-fhrub ftalk, heart-fhaped, woolly, crenated leaves, and foot-ftalks with one flower from the wings of the ftalk.*

Thefe plants grow naturally in the Weft-Indies, from whence I have received the feeds of three or four fpecies by the title of Broom-weed; and I have been informed that the inhabitants cut thefe plants in the fame manner as we do Heath, and make it up into brooms for fweeping. Sometimes I have received feeds of others by the title of Weft-India Thea, fo that I fuppofe the leaves of fome of thefe plants are fometimes ufed as the Thea. There are certainly more fpecies of this genus than are here mentioned, which have efcaped the notice of thofe who have been in the Weft-Indies in fearch of plants, for we frequently have new

new sorts come up in the earth, which is brought from thence with other plants. Those here enumerated are undoubtedly distinct species, for I have cultivated them several years, and have never observed either of them change, when raised from seeds.

The first sort grows as far north as Virginia, from whence I have several times received the seeds; this has an upright branching stalk three feet high, garnished with oval spear-shaped leaves, sawed on their edges, and sit close to the branches. The flowers come out singly from the wings of the stalks, standing upon very short foot-stalks; they are small, of a pale copper colour, of one petal, which is cut into five parts almost to the bottom, where they are joined. In the center arises a small column, composed of several stamina, and the style which are connected together at bottom, but are separated above. The germen turns to a capsule with five cells, inclosed by the empalement; in each cell is contained one angular seed, gibbous on one side, having two horns or teeth at the point.

The second sort has hairy branching stalks three feet high. The branches of this come out from the bottom, and form a pyramidal bush; the leaves are longer and narrower; the saw on the edges deeper, of a brighter green than those of the former, and stand upon short foot-stalks.

The third sort rises with a slender ligneous stalk two feet high, sending out erect branches, garnished with narrow spear-shaped leaves, indented on their edges, ending in acute points, having pretty long slender foot-stalks. The flowers come out singly from the wings of the stalks; they are small, of a pale yellow colour.

The fourth sort has very slender stalks, which seldom rise more than a foot high, sending out a few slender branches, garnished with small heart shaped leaves, sawed on their edges, a little hoary on their under side, standing upon pretty long foot-stalks. The flowers are small, of a pale yellowish colour, and come out singly from the wings of the stalk.

The fifth sort has a hairy stalk, covered with a dark brown bark three feet high, sending out many branches from the side, garnished with oval spear-shaped leaves, standing upon long foot-stalks, ending in an obtuse point, and are deeply sawed on their edges. The flowers come out by pairs at the foot-stalk of each leaf, sitting close to the stalk; they are larger than those of the former sorts, and of a deeper yellow colour; the seeds of this are larger, and have three teeth.

The sixth sort rises with a ligneous hairy stalk four feet high, sending out a few slender branches toward the top. The leaves are heart-shaped, a little woolly, and sit close to the stalk; they are veined, and sawed on their edges. The flowers come out in clusters on the side of the branches, to which they sit very close; they are small, of a pale yellow colour, and the seeds have two teeth.

The seventh sort has a slender ligneous stalk which rises two feet high, sending out several slender branches, garnished with roundish leaves, having long foot-stalks, hairy on their under side. The flowers come out at the foot-stalks of the leaves, sometimes singly, and at others there are two or three; they are of a pale copper colour.

The eighth sort rises with an herbaceous stalk three feet high, sending out several erect branches, garnished with heart-shaped leaves, sawed on their edges, of a light green colour, soft to the touch, and stand upon very long hairy foot-stalks. The flowers stand upon long foot-stalks, which come out from the wings of the stalk; they are small, and of a sulphur colour.

The ninth sort has very slender stiff stalks, covered with fine hairs, sending out a few side branches, garnished with roundish heart shaped leaves; they are of a light green colour, crenated on their edges, and stand upon long, slender, hairy foot-stalks. The flowers come out upon long foot-stalks from the wings of the stalks singly; they are small and white.

The tenth sort rises with an herbaceous prickly stalk near four feet high, sending out several branches, garnished with rough hairy leaves, standing upon long foot-stalks. These are of different forms, some are divided into five obtuse lobes, others into three, some are hollowed on the sides in shape of a fiddle; they are indented on their edges, and are of a pale green colour. The flowers are collected in heads, which stand upon very long hairy foot-stalks, arising from the wings of the stalks. Under each head are placed three obtuse small leaves, upon which rest seven small pale yellow flowers, which are almost hid by their empalements; these are succeeded by seeds, having acute spines.

The eleventh sort rises with a ligneous stalk three feet high, covered with yellowish hairs, very closely garnished with spear-shaped hairy leaves, sitting close to the stalks, sawed on their edges, and of a pale green on their under side. The flowers come out singly from the wings of the stalk, standing upon short foot-stalks; they are small and white.

The twelfth sort rises with very slender infirm stalks three feet high, covered with long white hairs, garnished with soft, woolly, heart-shaped leaves, sitting upon long, slender, hairy foot-stalks, crenated on their edges. The flowers stand upon long slender foot stalks, which arise from the wings of the stalk, two of them generally coming out at each leaf; they are of a pale yellow colour.

The thirteenth sort has many trailing stalks, which divide into slender branches, covered with a light brown bark, garnished with small, oblong, oval leaves, sawed on their edges, hairy on their under side, standing upon short foot-stalks. The flowers are produced in small clusters, sitting close at the end of the branches; they are of a bright scarlet colour, and are succeeded by seeds, having two stiff bristly teeth.

The fourteenth sort hath smooth round stalks which rise three feet high, sending out long slender branches. The leaves are smooth, heart-shaped, of a light green colour, and stand upon long foot-stalks, sawed on their edges, ending in acute points. The flowers stand upon very long foot-stalks, arising from the wings of the stalks singly; they are small, and of a whitish yellow colour.

The fifteenth sort sends out several stalks from the root, which spread flat on the ground, garnished with oval fatteny leaves, sawed on their edges, having short foot-stalks; the flowers come out singly at the wings of the stalks, sitting very close thereto; they are small, of a yellow colour, and are succeeded by seeds which have no teeth.

The sixteenth sort has a ligneous stalk four feet high, covered with brown hairs, sending out a few long slender branches, the lower parts of which are garnished with oval veined leaves, slightly sawed on their edges, and are downy on their under side. The upper part of the branches are destitute of leaves more than a foot in length; from the side come out foot-stalks two inches long, sustaining several small yellow flowers in clusters, having hairy empalements.

The seventeenth sort grows naturally at La Vera Cruz in New Spain; this rises with a strong shrubby branching stalk six or seven feet high, covered with a rough brown bark, garnished with spear shaped leaves, standing on pretty long foot-stalks, ending in acute points, and are unequally sawed on their edges. The upper surface of the leaves are of a dark green; their under is of a pale or light green colour. The flowers are collected in heads, standing upon long

long naked foot-ſtalks, which terminate the branches; each of theſe heads contain ſeven or eight flowers, whoſe petals extend much beyond their empalements. They are of a pale ſulphur colour when they firſt open, but afterward fade to an almoſt white; the ſeeds have three ſharp teeth, which are burry, and ſtick to the clothes of thoſe who rub againſt them when ripe.

The eighteenth ſort grows naturally in Jamaica; this riſes with a ſhrubby branching ſtalk ſeven or eight feet high, garniſhed at each joint by one large heart-ſhaped leaf, ſtanding upon a pretty long foot-ſtalk, ſawed on their edges, and run out to a long ſharp point, of a light green on their upper ſurface, and pale on their under. The flowers grow in cluſters at the wings of the ſtalks; thoſe on the lower part of the branches are formed in cloſe obtuſe ſpikes; the branches are terminated by one of theſe ſpikes; the flowers are ſmall, and when firſt open are white, but afterward they fade to a browniſh colour.

The nineteenth ſort riſes with ligneous ſtalks, covered with a ſoft woolly bark, garniſhed with heart-ſhaped woolly leaves, ſtanding upon pretty long foot-ſtalks; they are veined, and crenated on their edges. The flowers ſtand upon ſhort foot-ſtalks, which ariſe from the wings of the ſtalk ſingly; their empalements are woolly and obtuſe; the flowers are yellow, and are ſucceeded by ſeeds which have two teeth.

Theſe plants are moſt of them annual in England, but ſome of them are of longer duration in their native countries, and might be ſo here, if they were placed in a warm ſtove in winter; but as moſt of them perfect their ſeeds the ſame year, if the plants are brought forward in the ſpring, ſo few perſons have room in their ſtoves to receive theſe plants, as there are ſo many perennial exotic plants at preſent in the Engliſh gardens, which require a warm ſtove to preſerve them.

Theſe plants are propagated by ſeeds, which ſhould be ſown upon a moderate hot-bed the beginning of April, and when the plants are fit to remove, they ſhould be tranſplanted to another hot-bed four inches diſtance every way; they muſt be ſhaded from the ſun till they have taken new root, and muſt have a large ſhare of free air admitted to them when the weather is mild, to prevent their drawing up weak. If the plants thrive well, they will ſoon have ſtrength enough to be tranſplanted in the open air, for which purpoſe they ſhould be gradually hardened, and taken up with balls of earth to their roots, and planted in a ſheltered part of the garden, obſerving to ſhade and water them until they have taken new root; after which they will require no other care but to keep them clean from weeds.

The ſeventeenth ſpecies will not flower the firſt year, ſo the plants of this muſt be placed in a warm ſtove in autumn, and during the winter they muſt be treated in the ſame way as other tender plants from the ſame country. The following ſummer they will flower and produce ripe ſeeds, but the plants are not of long duration, ſo that there ſhould be a ſucceſſion of young plants raiſed from ſeeds.

SIDERITIS. Tourn. Inſt. R. H. 191. tab. 90. Ironwort.

The CHARACTERS are,

The flower has an oblong tubulous empalement, cut into five ſegments at the top. It is of the lip kind, of one petal, almoſt equal; the tube is cylindrical, the chops oblong and taper. The upper lip is erect, and cut into two acute ſegments, the under lip is cut into three; the two ſide ſegments are acute, the middle is round and crenated. It has four ſtamina within the tube, two of which are as long as the tube, the other are ſhorter; and a four-pointed germen, ſupporting a ſlender ſtyle a little longer than the ſtamina, crowned by two ſtigmas; the upper is cylindrical, concave, and torn; the lower is ſhort, and membranaceous. The germen afterward turns to four ſeeds, which ripen in the empalement.

The SPECIES are,

1. SIDERITIS (*Hirſuta*) caulibus hirſutis procumbentibus, foliis oblongo-ovatis crenatis villoſis, verticillis remotis. *Ironwort with hairy trailing ſtalks, oblong, oval, hairy, crenated leaves, and the whorls of flowers far aſunder.*

2. SIDERITIS (*Lacinia*) herbacea e bractæa caulibus ſpinoſis, labio ſuperiore indiviſo. Lin. Sp. Plant. 575. *Declining herbaceous Ironwort with prickly empalements, and the upper lip of the flower undivided.*

3. SIDERITIS (*Perfoliata*) herbacea hiſpido-piloſa foliis ſuperioribus amplexicaulibus. Lin. Sp. Plant. 575. *Herbaceous, hairy, ſtinging Ironwort, whoſe upper leaves embrace the ſtalk.*

4. SIDERITIS (*Suffruticoſa*) tomentoſa, foliis lineari-lanceolatis ſeſſilibus, calycibus ſpinoſis. *Woolly Ironwort with narrow ſpear-ſhaped leaves ſitting cloſe to the ſtalks, and prickly empalements to the flowers.*

5. SIDERITIS (*Scordioides*) foliis lanceolatis ſubdentatis, bracteis cordatis dentato-ſpinoſis, calycibus æqualibus. Lin. Sp. Plant. 575. *Ironwort with ſpear-ſhaped leaves ſlightly indented, heart-ſhaped, prickly, indented bracteæ, and the empalements of the flower equal.*

6. SIDERITIS (*Syriaca*) fruticoſa tomentoſo-lanata, foliis cuneiformibus ſeſſilibus, calycibus inermibus. *Shrubby, downy, woolly Ironwort, with wedge-ſhaped leaves ſitting cloſe to the ſtalks, and unarmed empalements.*

7. SIDERITIS (*Hiſpanica*) fruticoſa, foliis lanceolatis integerrimis, floribus ſpicatis terminalibus, calycibus ſpinoſis. *Shrubby Ironwort with ſpear-ſhaped entire leaves, and ſpiked flowers terminating the ſtalks, having prickly empalements.*

8. SIDERITIS (*Hyſſopifolia*) foliis lanceolatis glabris integerrimis, bracteis cordatis dentato-ſpinoſis, calycibus æqualibus. Lin. Sp. Plant. 575. *Ironwort with ſmooth, entire, ſpear-ſhaped leaves, prickly, heart-ſhaped, indented bracteæ, and equal empalements.*

9. SIDERITIS (*Canarienſis*) fruticoſa tomentoſa, foliis cordatis pedunculis ante floreſcentiam nutantibus. Lin. Sp. Pl. 574. *Shrubby woolly Ironwort with heart-ſhaped leaves, and the foot-ſtalks nod before the flower opens.*

The firſt ſort grows naturally in France, Spain, and Italy; the root is perennial; the ſtalks are herbaceous, hairy, and trail upon the ground, ſending out branches, garniſhed with oblong, oval, crenated, hairy leaves; the upper part of the ſtalk is furniſhed with whorls of purpliſh flowers, placed pretty far aſunder. It is a plant of no great beauty or uſe, ſo is ſeldom kept in gardens.

The ſecond ſort is an annual plant with trailing ſtalks; the leaves are ſmall, ſpear-ſhaped, and ſit cloſe to the ſtalks. The flowers grow in whorled ſpikes at the end of the branches; they are yellow. It grows in all the ſouthern parts of Europe, and is ſeldom admitted into gardens.

The third ſort grows naturally in the Levant. The roots of this ſort ſeldom continue longer than two years in England; the lower leaves are oblong, entire, and hairy; the ſtalks are ſmooth, hoary, and riſe near four feet high, ſending out ſeveral long ſlender branches, garniſhed with hoary acute pointed leaves, furniſhed with whitiſh flowers in whorls, which are placed far aſunder; the empalements of the flowers are prickly, and the flowers are ſmall.

The fourth ſort grows naturally in Crete; this is a low ſhrubby plant, whoſe ſtalks riſe a foot high, and are ligneous, ſending out branches garniſhed with narrow, ſpear-ſhaped, downy leaves; the upper part of the ſtalk is furniſhed with whorls of whitiſh yellow flowers, having prickly empalements.

The

The fifth sort grows naturally in the south of France and Italy; this hath a perennial root; the stalks rise a foot high, garnished with spear-shaped leaves, which are deeply crenated on their edges, and have short heart-shaped bracteæ, which are prickly. The flowers grow in whorled spikes toward the end of the stalks; they are yellow, and have prickly empalements, which are equal.

The sixth sort grows naturally in Crete; this has a short ligneous stalk, from which is sent out a few branches, garnished with thick, wedge-shaped, downy leaves. The flowers are produced in whorls toward the end of the branches; they are yellow, and have smooth downy empalements.

The seventh sort grows naturally in Spain and Italy; this has a low shrubby stalk, sending out several hairy branches a foot long, garnished with hairy spear shaped leaves. The flowers grow in close whorled spikes at the end of the branches; they are of a sulphur colour, and have very prickly empalements.

The eighth sort grows naturally on the mountains of Valentia; this has a short ligneous stalk, sending out branches a foot and a half long, garnished with narrow smooth leaves, of a strong scent when bruised. The flowers are yellow, and grow in large spiked whorls at the end of the branches.

The ninth sort grows naturally in the Canary Islands; it rises with a soft shrubby stalk five or six feet high, sending out ligneous branches, covered with a soft down, garnished with heart-shaped leaves, having long foot-stalks These differ greatly in size, according to the age and vigour of the plants; they are very woolly, especially on their under side, which is white, but their upper surface is of a dark yellowish green. The flowers grow in thick whorled spikes at the end of the branches; they are of a dirty white colour, and shaped like those of the other sorts.

These plants are preserved in botanic gardens for the sake of variety. The five sorts first mentioned, and also the eighth, are hardy enough to thrive in the open air in England; they are propagated by seeds, which, if sown in autumn, will succeed better than those which are sown in spring. When the plants come up, they should be kept clean from weeds; and when they are fit to remove, part of each sort may be drawn out, and planted in a bed; this will give those which are left in the seed-bed room to grow. The plants which are removed should be shaded and watered until they have taken new root; after which they will require no other care, but to keep them clean from weeds. The fourth sort should have a dry soil and a warm situation, but neither of the sorts should be planted in rich ground, for that will cause them to grow so luxuriant in summer, that the frost or much wet will destroy them in winter.

The annual sort should not be removed, but the plants thinned and left in the place where they are sown, keeping them clean from weeds.

The sixth and seventh sorts will often live through the winter in the open air, especially if their seeds are sown upon dry rubbish; for when either of these happen to grow in the joints of old walls, they will endure the greatest cold of this country, therefore their seeds should be sown in such places. The sixth sort does not produce good seeds in England, so this is propagated by slipping off the heads, planting them in a shady border during the spring or summer months, which will readily take root; some of these may then be taken up and put into pots, that they may be screened under a frame in winter. The other may be removed in autumn, and planted close to warm walls in rubbish, where they will abide some years.

The ninth sort is generally kept in green-houses in England, but in moderate winters I have had the plants live abroad without cover in a warm dry border: however, if they are screened from hard frost under a common frame, where they may be exposed to the open air at all times when the weather is mild, and protected from hard frosts, they will thrive better than with more tender treatment. It is propagated by seeds, which should be sown in autumn, for those which are sown in the spring seldom succeed, or if they do, the plants rarely come up the first year.

SIDEROXYLUM. Iron-wood.

The CHARACTERS are,

The empalement of the flower is permanent, and cut into five segments. The flower is bell-shaped, divided into five parts at the brim. It has five awl-shaped stamina the length of the petal, terminated by single summits, and a round germen supporting an awl-shaped style, crowned by a single stigma. The germen afterward becomes a roundish berry, having one cell, containing four seeds.

The SPECIES are,

1. SIDEROXYLUM (*Inerme*) inerme. Lin. Hort. Cliff. 69. *Smooth Iron-wood.*

2. SIDEROXYLUM (*Oppositifolium*) foliis lanceolatis ex adverso sitis. *Iron-wood with spear-shaped leaves growing opposite.*

These plants grow naturally at the Cape of Good Hope.

The first sort hath large oval leaves, shaped somewhat like those of the Bay-tree, but smoother and blunter at the end. These are placed on the branches without order, as the branches also are produced. The stalks are shrubby, and rise five or six feet high, sending out many branches, covered with a dark brown bark.

The second sort grows more upright and regular; the leaves, which are smaller, and more pointed than those of the first, are placed opposite on the branches, and these continue green through the year.

The wood of these trees being very close and solid, has given occasion for this name being applied to them, it being so heavy as to sink in water; and the title of Iron-wood having been applied to the wood by the inhabitants of the countries where it grows, has occasioned the botanists to constitute a genus by this name. But as the characters of the plants have not been so well examined as could be wished, occasioned by their seldom flowering in Europe, it is very probable, that the plants which have been ranged under this genus, do not properly belong to it; for Dr. Plukenet has figured a plant under the title of Ebenus Jamaicensis, whose characters are very different from those assigned to this genus: and the Jamaica Iron-wood is totally different from both in its characters, for this has male and female flowers on different trees; the male flowers have no petals, as appears by dried samples in my collection.

These plants are natives of warm countries, so cannot be preserved in England, unless they are placed in a warm stove. They are propagated by seeds, when these can be procured from abroad. These must be sown in pots, and plunged into a good hot-bed in the spring, in order to get the plants forward early in the season. When the plants are fit to transplant, they should be each put into a separate small pot, and plunged into a fresh hot-bed. In the winter they must be plunged into the tan-bed in the stove, and treated in the same manner as hath been directed for several tender plants from the same countries. As the plants obtain strength they may be treated more hardily, by placing them in a dry airy glass-case in the winter, giving them free air in mild weather, but in summer they should be placed abroad in a sheltered situation.

I have propagated them by layers, but these were two years before they had made good roots; and sometimes they will take from cuttings, but this is a very uncertain method of propagating them; nor do the plants so raised ever grow so vigorously as those which come from seeds, so that when those can be procured it is the best method.

SIGESBECKIA. Lin. Sp. Plant. 873.

The CHARACTERS are,

The proper involucrum of the flower is composed of five linear, taper, obtuse leaves, which open beyond the petal, and is permanent. The common cover is five-leaved, sitting close; the leaves are oval, concave, equal, and disposed in several series; and between each leaf is contained a floret. The flower is composed of hermaphrodite florets in the disk, and the border or ray is made up of female half florets, which are tongue-shaped. The hermaphrodite florets are funnel-shaped, cut into five parts at the brim; these have five short stamina, with tubulous summits joined together, and an oblong incurved germen as large as the empalement, supporting a slender style crowned by a bifid stigma. The germen afterward turns to an oblong, four-cornered, blunt seed; the female half florets have a short, broad, tongue shaped petal, indented in three parts; these have a germen, style, and stigma, like the hermaphrodite florets, but have no stamina, and are succeeded by single seeds like the other.

We have but one SPECIES of this genus, viz.

SIGESBECKIA (*Orientalis.*) Lin. Hort. Cliff. *Sigesbeckia.*

This plant is annual, perishing at the approach of winter. The seeds of it were brought from the East-Indies, where it is a troublesome weed; in England it is raised on a hot-bed, and brought forward in the spring; then the plants may be planted out in warm borders, and if they are supplied with water in dry weather they will grow four or five feet high, and send out many branches. The flowers are produced at the extremity of the shoots, which are small, and of a yellow colour, so make no great appearance; therefore it is only preserved in the gardens of those persons who are curious in the study of plants.

SILAUM. See PEUCEDANUM.

SILENE. Lin. Gen. Plant. 503. Viscous Campion, or Lychnis.

The CHARACTERS are,

The flower has a permanent empalement, which is indented at the top in five parts. It has five plain obtuse petals, indented at their points, and a nectarium compounded of two small indentures in the neck of each petal, constituting a crown to the chaps; and ten awl shaped stamina inserted alternately to the tail of the petals above each other, terminated by oblong summits. In the center is situated a cylindrical germen, supporting three styles, which are longer than the stamina, crowned by stigmas that are reflexed against the sun. The germen afterward becomes a close cylindrical capsule with three cells, opening at the top five ways, inclosing many kidney-shaped seeds.

The SPECIES are,

1. SILENE (*Quinquevulnera*) petalis integerrimis subrotundis, fructibus erectis alternis. Hort. Cliff. 171. *Silene with entire roundish petals to the flower, and erect fruit alternate; commonly called Dwarf Lychnis.*

2. SILENE (*Noctiflora*) floribus spicatis alternis secundis sessilibus, petalis bifidis. Lin. Sp. Plant. 416. *Silene with fertile spikes of flowers sitting close, and the petals bifid.*

3. SILENE (*Nutans*) petalis bifidis, floribus lateralibus secundis cernuis, caule recurvato. Lin. Sp. Plant. 417. *Silene with bifid petals, nodding flowers growing from the side of the stalks, and a recurved stalk.*

4. SILENE (*Fruticosa*) petalis bifidis, caule fruticoso, foliis lato lanceolatis, panicula trichotoma. Lin. Sp. Plant. 417. *Silene with bifid petals, a shrubby stalk, broad spear-shaped leaves, and panicles divided three ways.*

5. SILENE (*Viridiflora*) petalis semibifidis foliis ovatis scabriusculis acutis, panicula elongata subaphylla. Lin. Sp. Plant. 418. *Silene with bifid petals, bluntest oval leaves, and long panicles without leaves.*

6. SILENE (*Conoidea*) calycibus fructus globosis acuminatis striis triginta, foliis glabris. Hort. Upsal. 110. *Silene with globular acute-pointed capsules to the fruit, and smooth leaves.*

7. SILENE (*Pendula*) calycibus fructus pendulis inflatis, angulis decem scabris. Hort. Upsal 109. *Silene with pendulous swollen empalements to the fruit, with ten rough angles.*

8. SILENE (*Noctiflora*) calycibus decem angularibus, dentibus tubum æquantibus. Lin. Sp. Plant. 419. *Silene with empalements having ten angles, and the indentures as long as the tube.*

9. SILENE (*Vallesia*) caulibus subunifloris decumbentibus foliis laciniatis. *Silene with decumbent stalks, and smooth acute pointed leaves.*

10. SILENE (*Orientalis*) calycibus conicis striis hirsutis, fructibus erectioribus, caule erecto hirsuto, foliis nervosis. *Silene with conical empalements, having hairy stripes, erect fruit, a hairy upright stalk, and veined leaves.*

11. SILENE (*Muscipula*) petalis bifidis, caule dichotomo, floribus axillaribus sessilibus, foliis glabris. Lin. Sp. Plant. 420. *Silene with bifid petals, a stalk divided by pairs, flowers sitting close to the wings of the stalk, and smooth leaves.*

12. SILENE (*Armeria*) floribus fasciculatis fastigiatis, foliis superioribus cordatis glabris. Hort. Upsal. 110. *Silene with flowers gathered into bunches, whose upper leaves are smooth and heart-shaped; commonly called Lobel's Catchfly.*

13. SILENE (*Gigantea*) foliis radicalibus cochleariformibus, simis caule subverticillata. Lin. Sp. Plant. 598. *Silene with obtuse spoon-shaped leaves, and whorled stalks.*

14. SILENE (*Bupleuroides*) petalis bifidis, floribus pedunculatis oppositis bractea brevioribus, foliis lanceolatis acutis glabris. Lin. Sp. Plant. 598. *Silene with bifid petals, flowers placed opposite on foot-stalks, and smooth acute-pointed leaves.*

There are several other species of this genus, whose flowers have no beauty, so are seldom cultivated but in botanic gardens, for the sake of variety, therefore I have not enumerated them, which would swell the work too much.

The first sort grows naturally in Portugal; in England it is well known by the title of Dwarf Lychnis. The seeds of this were formerly sown in drills on the edges of borders, as were several other low annual plants, for edgings of borders; but as they are of short duration, so they soon were rejected for this purpose; after which the seeds were usually sown in patches in the borders, where they made a better appearance than in the former way; but in both these methods the plants were generally left so close as to spoil their growth, for as their stalks were drawn up weak, and had not room to branch out, so their flowers were small, and made little appearance; but to have this plant in beauty, the seeds should be sown thin in autumn, and in the spring the plants should be thinned to the distance of four inches, and afterward kept clean from weeds. When they are so managed, the plants will rise a foot high, with hairy channelled stalks, and divide into many branches, garnished with oval, spear-shaped, hairy leaves, placed opposite. The flowers grow in short spikes at the end of the branches; they are placed alternately, and are of a bright purple colour, edged with white.

The second sort grows naturally in Sicily, and also at the Cape of Good Hope. This is an annual plant, with a low branching stalk, which seldom rises more than eight or nine inches high; the stalks are smooth; the leaves are very narrow and smooth, placed by pairs; the stalks are terminated

nated by ſpikes of dark purple flowers ſtanding alternate, whoſe petals are bifid; they open in the evening, but are cloſely ſhut in the day. If the ſeeds of this plant are ſown in autumn upon a warm border, the plants will live thro' the winter, ſo good ſeeds may be obtained; but when the ſeeds are ſown in the ſpring they often fail.

The third ſort is a perennial plant, which grows naturally on the Alps; the lower leaves of this are ſmooth and ſpear-ſhaped; the ſtalk riſes near two feet high, garniſhed with two narrow leaves, placed oppoſite at each joint, and immediately below them; the ſtalk is very clammy. The flowers come out on ſhort foot ſtalks from the wings of the leaves, each foot-ſtalk for the moſt part ſuſtaining three flowers, with long, white, bifid petals. This plant riſes eaſily from ſeeds if they are ſown in autumn, and the only culture the plants require is to keep them clean from weeds, and allow them room to ſpread.

The fourth ſort grows naturally in Sicily; this has a low ſhrubby ſtalk, which divides into ſeveral ſhort ſhrubby branches, garniſhed with broad, ſmooth, ſpear ſhaped leaves, ending in acute points. The flower-ſtalks riſe about a foot high, and divide into ſpreading panicles by twos and threes. The flowers are of an herbaceous white colour, and are ſucceeded by oval ſmooth capſules, having thick covers, filled with ſmall ſeeds, which ripen in autumn. This ſort riſes eaſily from ſeeds as the former; if the plants are planted in a warm border of dry earth they will live ſeveral years, and require no ſhelter, but in moiſt ground they frequently rot in winter.

The fifth ſort grows naturally in Portugal; this has a perennial root; the lower leaves are roundiſh, and hollowed like a ſpoon; thoſe upon the ſtalks are obtuſe, and ſtand ſometimes by pairs, at others by threes or fours round the ſtalks; they are of a deep green, ſmooth, and ſit cloſe to the ſtalks; the ſtalks are round, ſmooth, and riſe from two to three feet high. The flowers grow in looſe ſpikes at the top, and are of a green colour. This riſes eaſily from ſeeds ſown in autumn, or if the ſeeds are permitted to ſcatter, the plants will come up, and require no other culture but to keep them clean from weeds.

The ſixth ſort grows naturally among Corn in France, Spain, and Italy. It is an annual plant, with an upright branching ſtalk a foot and a half high, having ſwelling viſcous joints, garniſhed with narrow, acute-pointed, ſmooth leaves, ſitting cloſe to the ſtalks. The flowers are ſmall, of a red colour, and produced at the end of the branches: theſe are ſucceeded by globular capſules, ending in acute points, whoſe empalements are ſtriped. The ſeeds of this ſhould be ſown in autumn, and in the ſpring the plants ſhould be thinned and kept clean from weeds, which is all the culture they require.

The ſeventh ſort grows naturally in Sicily and Crete; this is an annual plant, from whoſe root comes out ſeveral branching ſtalks a foot and a half long, which trail upon the ground, garniſhed with oval acute-pointed leaves placed oppoſite. The flowers come out ſingly from the wings of the ſtalk upon ſhort foot-ſtalks; they are large, of a bright red colour, reſembling thoſe of the common wild red Campion. Theſe are ſucceeded by large capſules, included in inflated empalements, having ten rough angles, containing many large roundiſh ſeeds, whoſe weight cauſes the capſules to hang downward. If the ſeeds of this are permitted to ſcatter, the plants will come up without care, and require no more culture but to keep them clean from weeds.

The eighth ſort is an annual plant, which is found naturally in England growing among Corn. It riſes with a thick clammy ſtalk nine inches or a foot high, garniſhed with ſmall oblong leaves by pairs, whoſe baſe embrace the ſtalks; the top of the ſtalk ſuſtains one or two ſmall red flowers, which open only in the night. The ſeeds ripen in Auguſt, which, if permitted to ſcatter, the plants will come up without farther trouble.

The ninth ſort grows naturally in the Archipelago; this riſes with a hairy ſtalk a foot and a half high, garniſhed with narrow ſpear ſhaped leaves, placed by pairs, which are ſmooth, and ſit cloſe to the ſtalks. The flowers are diſpoſed looſely at the top of the ſtalks; they are red, and have long pyramidal ſtriped empalements. If the ſeeds of this ſort are ſown in autumn, and the plants afterward treated in the ſame way as the firſt, they will thrive and flower early in ſummer.

The tenth ſort grows naturally in the Levant; this is an annual plant, with a ſtrong, erect, hairy, branching ſtalk, two feet high. The branches grow erect, as do alſo the flowers, which are red, and have large, conical, ſtriped empalements, whoſe ſtripes are hairy, and of a browniſh colour. This muſt be treated in the ſame way as the firſt ſort.

The eleventh ſort grows naturally in the ſouth of France, Spain, and Italy; this is biennial. The ſtalk is round, clammy, and riſes a foot and a half high, having ſwelling joints; the leaves grow round the ſtalks in cluſters; they are narrow and ſmooth. The upper part of the ſtalk divides into ſpreading branches by pairs, adorned with red flowers coming out ſingly from the wings of the leaves, ſitting cloſe to the ſtalks.

This ſort is eaſily propagated by ſeeds, which, if ſown in autumn, will ſucceed much better than in the ſpring. When the plants come up and are fit to remove, they ſhould be tranſplanted at ſix inches diſtance, ſhading them from the ſun, and watering them until they have taken new root; after which they muſt be kept clean from weeds till autumn, when they ſhould be tranſplanted to the places where they are deſigned to remain for flowering. When the ſeeds of this plant happen to ſcatter upon a wall, and the plants ariſe there, they will continue much longer than in the ground.

The twelfth ſort is an annual plant, which grows naturally in the ſouth of France and Italy. The ſeeds of this have ſpread out upon walls and buildings ſo far, as to induce ſome to believe it a native of England.

There are three varieties of this, which generally retain their differences; one has a bright purple flower, the other a pale red, and the third a white flower; theſe do not differ in any other reſpect, ſo cannot be reckoned as different ſpecies.

Theſe ſeeds ſhould be ſown in autumn, for thoſe which are ſown in the ſpring often fail; and if the plants do come up, they never grow ſo large or make ſo good appearance as the autumnal plants.

The thirteenth ſort is biennial; this grows naturally in Sicily and Crete; the lower leaves of this plant are obtuſe, and are gathered in circular heads like ſome of the Houſeleeks, or thoſe of the Auricula; they are ſmooth, of a pretty thick conſiſtence. The ſtalks riſe five or ſix feet high, and are very viſcous, garniſhed with ſpear-ſhaped leaves, placed oppoſite. The flowers come out upon ſhort foot-ſtalks from the wings of the ſtalks, each foot-ſtalk ſuſtaining three or four greeniſh flowers, which are ſucceeded by oval capſules, ſpreading open at the top, filled with angular ſeeds.

If the ſeeds of this plant are ſown in autumn upon a warm border, they will more certainly ſucceed than thoſe ſown in the ſpring. When the plants come up and are fit to remove, they ſhould be planted on a dry ſoil, and in a warm situation, where they will live through the winter, and the following summer they will flower, ripen their seeds, and then decay.

The fourteenth sort grows naturally in the Levant; this has a perennial root; the lower leaves are narrow, spear-shaped, and smooth; they are gathered in clustered heads, from the middle of which arises an erect clammy stalk a foot and a half high, garnished with very narrow leaves. The flowers come out from the wings of the leaves toward the top of the stalk; their foot stalks are short, each sustains two white flowers, having long tubes, standing erect; the flowers are closed in the day, and expand at night.

As the seeds seldom ripen here, so it is difficult to propagate it: the only way is to slip off the heads in June, and plant them under a glass; these will take root if they are shaded from the sun and duly watered.

SILER. See LASERPITIUM.

SILIQUA. See CERATONIA.

SILIQUASTRUM. See CERCIS.

SILPHIUM. Lin. Gen. Plant. 882. Bastard Chrysanthemum.

The CHARACTERS are,

The common empalement of the flower is oval, imbricated, and permanent; the scales are oval, prominent, and reflexed in the middle. The disk of the flower is composed of hermaphrodite florets, which are tubulous, indented in five parts at the top. These have five short hair-like stamina, terminated by cylindrical summits, and a slender taper germen supporting a long hairy style, crowned by a single stigma; these are barren. The rays of the flower are composed of a few female half florets, which are long, spear-shaped, and for the most part have three indentures at their points; these have a heart-shaped germen with a short single style, having two bristly stigmas of the same length, and are succeeded by single heart-shaped seeds with a membranaceous border, indented at the top, each point ending with a horn or tooth, and are separated by linear chaff, ripening in the empalement.

The SPECIES are,

1. SILPHIUM (*Trifoliatum*) foliis ternis. Prod. Leyd. 181. *Silphium with leaves by threes at a joint.*

2. SILPHIUM (*Asteriscus*) foliis indivisis sessilibus oppositis, inferioribus alternis. Lin. Sp. Pl. 920. *Silphium with undivided leaves set opposite close to the stalks, whose lower leaves are alternate.*

3. SILPHIUM (*Solidaginoides*) foliis oppositis lanceolatis petiolatis acutè serratis. Flor. Virg. 181. *Silphium with sawed leaves having foot-stalks, which grow opposite.*

4. SILPHIUM (*Arborescens*) foliis lanceolatis alternis scabris, obsoletè serratis, caule fruticoso. *Silphium with rough spear-shaped leaves placed alternate, which have slight sawed edges, and a shrubby stalk.*

The first sort grows naturally in many parts of North America; the root is perennial and ligneous; the stalks are annual; these rise six feet high or more in good land; they are of a purplish colour, and branch toward the top. The leaves are oblong, rough, and have some sharp teeth on their edges; toward the bottom of the stalk they stand by fours at each joint; higher up they are by threes, and at the top by pairs, sitting close to the stalks. The foot-stalks are pretty long, each sustaining one flower, whose empalement is composed of three orders of leaves placed imbricatim like the scales of fish, the outer order being the smallest. The ray or border of the flower is composed of thirteen female half florets, which are yellow, tongue shaped, and indented in three points at the end. The disk or middle of the flower is made up of hermaphrodite tubulous flowers, which are slightly cut into five parts at the top; these have five stamina and a style connected together, which are longer than the tube of the floret.

It is propagated by parting of the roots, in the same way as it is practised for the perennial Sunflowers; the best time for this is in autumn when their stalks begin to decay, and the plants may afterward be treated in the same way as the perennial Sunflower.

The second sort grows naturally in Carolina; the root of this is perennial, the stalk is thick, solid, and set with prickly hairs; it rises near three feet high, has many purple spots; the leaves on the lower part of the stalk are placed alternate, but upward they are opposite, and sit close to the stalk; they are rough, having a few slight indentures on their edges. The upper part of the stalk divides into five or six small branches, which are terminated by yellow radiated flowers, like those of the perennial Sunflower, but smaller, having generally nine female half florets which compose the border or ray, the other parts are like those of the former sort. This sort is propagated by parting the roots in the same way as the former, but as this is not quite so hardy, it should be planted in a sheltered situation.

The third sort grows naturally in many parts of North America; this is a perennial plant, whose stalks rise near six feet high, garnished with oblong sawed leaves placed by pairs upon short foot-stalks. The flowers are loosely disposed at the top of the stalks; they are yellow, and have their half florets, which compose the ray, indented in three parts at the end. It may be propagated in the same way as the former, and the plants require the same treatment.

The fourth sort was discovered by the late Dr. William Houstoun, growing naturally at La Vera Cruz in New Spain. This rises with a shrubby stalk to the height of eight or ten feet, sending out ligneous branches, garnished with spear-shaped leaves placed alternately; their surface is rough, and their edges slightly sawed. The flowers are produced at the end of the branches, some of them singly on slender foot-stalks, others have two or three; they are unequal in height, and have short scaly empalements. They are of a deep yellow colour, but are not succeeded by seeds in England.

This sort is with difficulty propagated here, for unless the seeds are procured from the country where the plants grow naturally, they cannot be obtained that way, and the cuttings are not apt to take root. The only method of getting them to grow, is to slip off the young shoots in July, and plant them in a pot, and plunge it into a gentle hot-bed, covering the pot closely with a bell or hand-glass, and shade them from the sun. When the cuttings are rooted, they should be each planted in a separate pot, and during the warm months they may be placed in the open air in a warm situation, but in winter they should be kept in a moderate stove.

SINAPIS. Lin. Gen. Plant. 735. Mustard.

The CHARACTERS are,

The empalement of the flower is composed of four narrow leaves placed in form of a cross, which fall off. The flower has four roundish petals in form of a cross, and four oval nectariums, one on each side of the stamina and the pointal. It has six awl-shaped erect stamina, two of which are opposite and as long as the empalement, the other four are longer. In the center is placed a taper germen, with a style the length of it, crowned by a headed stigma. The germen afterward turns to an oblong pod, which is very rough at bottom, having two cells, opening with two valves, whose intermediate partition is large, compressed, and almost twice the length of the valves; the seeds are globular.

The

The SPECIES are,

1. SINAPIS (*Alba*) siliquis hispidis, rostro obliquo longissimo. Hort. Cliff. 338. *Mustard with prickly pods, and a very long oblique beak; commonly called White Mustard.*

2. SINAPIS (*Nigra*) siliquis glabris apice tetragonis. Hort. Cliff. 338. *Mustard with a smooth four-cornered pod; or common Mustard.*

3. SINAPIS (*Arvensis*) siliquis multangulis toroso-turgidis, rostro longioribus. Hort. Cliff. 338. *Mustard with many angled, rough, swelling pods, having a longer beak.*

4. SINAPIS (*Erucoides*) siliquis lævibus æqualibus, foliis lyratis oblongis glabris, caule scabro. Amœn. Acad. 4. *Mustard with taper, obtuse, smooth pods.*

5. SINAPIS (*Juncea*) ramis fasciculatis, foliis summis lanceolatis integerrimis. Hort. Upsal. 191. *Mustard with bundled branches, and the upper leaves spear-shaped and entire.*

6. SINAPIS (*Hispanica*) foliis duplicato-pinnatis, laciniis linearibus. Hort. Cliff. 338. *Mustard with double-winged leaves, having linear segments.*

The first sort is the common White Mustard, which is generally cultivated as a sallad herb for winter and spring use. This rises with a branched hairy stalk two feet high; the leaves are deeply jagged on their edges and rough. The flowers are disposed in loose spikes at the end of the branches, standing upon horizontal foot-stalks; they have four yellow petals in form of a cross, which are succeeded by hairy pods that end with long, compressed, oblique beaks; the pods generally contain four white seeds.

The second sort is the common Mustard, which is frequently found growing naturally in many parts of England, but is also cultivated in fields for the seed, of which the sauce called Mustard is made. This rises with a branching stalk four or five feet high; the lower leaves are large, rough, and very like those of Turnep; the upper leaves are smaller, and less jagged. The flowers are small, yellow, and grow in spiked clusters at the end of the branches; they have four petals placed in form of a cross, and are succeeded by smooth four-cornered pods.

The third sort grows naturally on arable land in many parts of England. The seed of this is commonly sold under the title of Durham Mustard-seed; of this there are two varieties, if not distinct species; one with cut, the other has entire leaves. The stalks rise two feet high; the leaves are rough, and in one they are jagged like Turnep leaves; the other are long and entire. The flowers are yellow; the pods are turgid, angular, and have long beaks.

The fourth sort grows naturally in Spain; this seldom rises more than eight or nine inches high; the leaves are smooth and much jagged; the stalk branches toward the top, and is terminated by a loose spike of white flowers, which are succeeded by smooth, taper, blunt pods, filled with small brown seeds.

The fifth sort grows naturally in China. The plant is used as a boiled sallad by the Chinese, where it may prove acceptable to those who have not better herbs for that purpose, but in England it is not regarded. The stalks of this rise three feet high, and toward the bottom are garnished with broad, smooth, jagged leaves, but those at the top are entire. The flowers are yellow like those of the first sort, and the pods are smooth and turgid.

The first sort is chiefly cultivated in gardens for a sallad herb in the winter season. The seeds of this are commonly sown very thick in drills, either upon a warm border or in very cold weather upon a moderate hot-bed, with Cresses and other small sallad herbs, which are commonly fit for use in ten days or a fortnight from the time of sowing; for if they are large and have rough leaves, they are too strong to put into sallads. In order to save the seeds of this plant, a spot of ground must be sown with it in the spring, and when the plants have four leaves, the ground should be hoed in the same manner as for Turneps, to cut down the weeds and thin the plants where they are too close; this should be done in dry weather, for then the weeds will soon die after they are cut. If this is well performed, the ground will remain clean for a month, by which time young weeds will spring up again; therefore the ground should be again hoed over, and the plants now left about eight or nine inches asunder, which will be sufficient room for this sort to grow; if this is well performed in dry weather, the ground will remain clean till the seeds are ripe. As soon as the pods change brown the plants should be cut off, and spread upon cloths two or three days to dry, and then threshed out for use.

The second sort is cultivated only for the seeds; these should be sown in the same way as those of the first, and the plants treated in the same manner, with this difference, of allowing the plants twice as much room, because they grow much larger, so these should be hoed out to the distance of eighteen inches; and as the seeds will not ripen so soon as the other so the ground may require to be hoed three times over, but this may be easily seen by the growth of the weeds.

The seeds of these two first species are ordered for medicinal use.

The third sort is a pretty common weed on arable lands in most parts of England; this comes up early in the spring amongst the Corn, so flowers and seeds in May; therefore where it is not weeded out, the seeds will scatter long before the Corn is ripe, and the ground will be stocked with the weed.

The other three sorts are preserved in botanic gardens for variety, but are never cultivated for use; these may be treated in the same way as the two first species.

SINAPISTRUM. See CLEOME.

SISARUM. See SIUM.

SISON. Lin. Gen. Plant. 311. Bastard Stone Parsley.

The CHARACTERS are,

It has an umbellated flower; the general umbel is composed of six thin rays or small umbels, which are unequal, as are also the smaller, which have ten. The involucri of both are four-leaved and unequal; the empalement of the flower is scarce discernible. The outer petals of the general umbel are uniform; the flowers have five equal petals, which are spear-shaped and inflexed. They have five hair-like stamina the length of the petals, terminated by single summits. The oval germen is situated under the flower, supporting two reflexed styles, crowned by obtuse stigmas. The germen afterward becomes an oval streaked fruit, dividing in two parts, each containing one oval streaked seed, convex on one side and plain on the other.

The SPECIES are,

1. SISON (*Amomum*) foliis pinnatis, umbellis erectis. Prod. Leyd. 105. *Sison with winged leaves and erect umbels.*

2. SISON (*Segetum*) foliis pinnatis, umbellis cernuis. Prod. Leyd. 105. *Sison with winged leaves and nodding umbels.*

3. SISON (*Verticillatum*) foliolis verticillatis capillaribus. Lin. Sp. Plant. 253. *Sison with hair-like small leaves in whorls.*

4. SISON (*Canadense*) foliis ternatis. Hort. Cliff. 99. *Canada Myrrh with trilobated leaves.*

The first sort grows on the side of ditche and moist shady banks in many parts of England; it is a biennial plant, which perishes soon after the seeds are ripe. The root is taper; the lower leaves are winged, composed of four pair of lobes, terminated by an odd one, regularly indented on both sides, and the indentures are sawed; they are of a lucid green, and have an aromatic odour. The stalks rise three feet high, and are garnished with leaves of the same form with those below, but smaller; at the end of the branches the white flowers are produced in small umbels, which are succeeded by striated seeds of a hot pleasant aromatic smell and taste, which ripen in August.

This plant is found growing so plentifully wild, as that it is rarely kept in gardens; but whoever is willing to propagate it, should sow the seeds in autumn in a moist shady spot of ground, where the plants will come up, and require no farther care than to keep them clean from weeds; and if the seeds are permitted to scatter the plants will rise without care. The seed of this plant is put into Venice treacle, for a succedanum to the true Amomum.

The second sort grows naturally among Corn on moist land in several parts of England. This is also a biennial plant, which decays soon after the seeds are ripe; it rises with an upright stalk about a foot high, which rarely divides into branches; the leaves stand upon pretty long footstalks; they are winged, but the lobes are small and finer cut than those of the former; the umbels of flowers are more compact, and nod on one side. The plant may be cultivated in the same way as the first.

The third sort grows naturally on the Alps and Apennines; this rises with a swelling jointed stalk two feet high, garnished with very fine slender leaves, standing in whorls like those of the Water Milfoil; it branches out toward the top, each branch being terminated by a pretty large umbel of flowers, which are purplish on their outside, but white within; these are succeeded by seeds, which ripen the end of July. The roots of this plant are composed of thick fleshy knots, somewhat like those of the King's-Spear.

The third sort may be cultivated by seeds, which should be sown in autumn, for those which are sown in the spring seldom grow the first year. The plants require no other culture than to thin them where they are too close, and keep them clean from weeds.

The fourth sort grows naturally in North America; this has perennial roots. The stalks rise three feet and a half high, and are garnished with trifoliate leaves, whose lobes are oval and spear-shaped; the flowers terminate the stalks in umbels, which are irregular in form.

This may be propagated by sowing of the seeds in autumn.

SISYMBRIUM. Tourn. Inst. R. H. 225. tab. 109. Water Cresses.

The CHARACTERS are,

The flower has a spreading empalement, composed of four linear, spear-shaped, coloured leaves, which fall off; it has four oblong spreading petals, placed in form of a cross, and six stamina, four of which are longer than the empalement; the other two, which are opposite, are shorter, terminated by single summits; it has an oblong slender germen, with scarce any style, crowned by an obtuse stigma. The germen afterward becomes a taper, oblong, incurved pod, having two cells, opening with two valves, which are shorter than the intermediate partition, filled with small seeds.

The SPECIES are,

1. SISYMBRIUM (*Nasturtium Aquaticum*) siliquis declinatis, foliis pinnatis, foliolis subcordatis. Hort. Cliff. 336. *Sisymbrium with declining pods, and winged leaves whose lobes are almost heart-shaped; or Water Cress.*

2. SISYMBRIUM (*Sylvestre*) siliquis declinatis, foliis pinnatis, foliolis lanceolatis serratis. Hort. Cliff. 336. *Sisymbrium with declining pods, and winged leaves having spear-shaped sawed lobes.*

3. SISYMBRIUM (*Amphibium*) siliquis declinatis, oblongo-ovatis, foliis pinnatifidis serratis. Lin. Sp. Plant. 657 *Sisymbrium with oblong, oval, declining pods, and wing-pointed sawed leaves.*

4. SISYMBRIUM (*Aquaticum*) foliis simplicibus dentatis serratis. Hort. Cliff. 336. *Sisymbrium with single, indented, sawed leaves.*

5. SISYMBRIUM (*Polyceratium*) siliquis axillaribus sessilibus subulatis aggregatis, foliis repando-dentatis. Hort. Upsal. 193. *Sisymbrium with awl-shaped pods in clusters, sitting close to the stalks, and indented leaves which turn backward.*

6. SISYMBRIUM (*Sophia*) petalis calyce minoribus, foliis decomposito-pinnatis. Flor. Suec. *Sisymbrium with petals smaller than the empalement, and decompounded winged leaves; called Flixweed.*

7. SISYMBRIUM (*Altissimum*) foliis runcinatis flaccidis, foliolis sublinearibus integerrimis, pedunculis laxis. Hort. Upsal. 193. *Sisymbrium with spear, wing-pointed, flaccid leaves, having linear entire lobes with loose foot-stalks.*

8. SISYMBRIUM (*Irio*) foliis runcinatis dentatis, siliquis erectis. Lin. Sp. Plant. 659. *Sisymbrium with spear-shaped, winged, indented leaves, and erect pods.*

9. SISYMBRIUM (*Strictissimum*) foliis lanceolatis dentato-serratis caulinis. Hort. Cliff. 337. *Sisymbrium with spear-shaped, winged, indented leaves on the stalks.*

The first sort is the common Water Cress, which grows naturally in ditches and rills of water in most parts of England. The roots of this plant are composed of a great number of long fibres, which fasten themselves to the mud at the bottom of the ditches, from which arise several stalks, garnished with winged leaves, composed of five or six pair of roundish lobes, terminated by an odd one; these stand almost alternate along the midrib. The stalks rise a foot and a half high; they are channelled, and divide at the top into two or three branches, which are terminated by loose spikes of small white flowers, composed of four petals placed in form of a cross, and are succeeded by taper pods, filled with small brown seeds.

This plant has of late years been generally used as a sallad herb in the spring of the year, and is by many preferred to all other sorts of sallads for the agreeable warm bitter taste, and being accounted an excellent remedy for the scurvy, and to cleanse the blood, as also a good diuretic, it has greatly obtained a preference to most other herbs for winter and spring use with many people. This is generally gathered in the ditches, and in other standing waters near London, to supply the markets; but whoever has a mind to cultivate it may easily do it, by taking some of the plants from the places of their natural growth early in the spring, being careful to preserve their roots as entire as possible, and plant them into mud, and then let the water in upon them by degrees. When they have taken root they will soon flourish, and spread over a large compass of water; they should not be cut the first season, but suffered to run to seed, which will fall into the water, and furnish a sufficient supply of plants afterward.

But where the water is so deep, that it will not be easy to plant them, the best method will be to get a quantity of the plants just as their seeds are ripening, and throw them on the surface of the water where they are designed to grow, and their seeds will ripen, and fall to the bottom, where they will take root, and produce a supply of these plants.

Some of those people who gather this herb for use, either through ignorance or some worse design, have frequently taken the creeping Water Parsnep, and sold it for Water Cress, whereby many persons have suffered who have eaten it; therefore those who make use of Water Cress, should be careful to have the right plant; they may be easily distinguished by the shape of their leaves; those of the Water Cress have roundish almost heart-shaped small leaves or lobes, with a few indentures on their edges, and are of a dark

dark green colour; but thofe of the Water Parfnep have oblong lobes, ending in points, which are of a light green, fawed on their edges.

The fecond fort grows naturally on the borders of the river Thames, and in fome other parts of England. The leaves of this fort are longer than thofe of the firft; the lobes are much narrower, and are fawed on their edges; the flowers ftand upon longer foot-ftalks, and are much fmaller. This fpreads and multiplies in the fame manner as the firft.

The third and fourth forts grows naturally on the banks of the Thames, and in ditches in many parts of England, fo are not admitted into gardens.

The fifth fort grows naturally in the fouth of France and Italy; it is an annual plant, whofe ftalks fpread and decline toward the ground; they grow a foot long, and divide into many branches, garnifhed with fmooth leaves, fhaped like the point of a halberd, deeply finuated on their borders; the indentures turn backward. The flowers come out in clufters at the wings of the ftalk; they are fmall, yellow, and are fucceeded by flender crooked pods, ftanding in clufters. The feeds ripen in autumn.

The fixth fort grows naturally in uncultivated places, and alfo by the fide of footways in many parts of England. The leaves of this are divided into many very narrow fegments; the ftalks rife a foot and a half high, garnifhed with winged leaves whofe lobes are finely cut, refembling thofe of the true Roman Wormwood. The flowers are produced in loofe fpikes at the top of the ftalk; they are fmall, yellow, and compofed of four petals, fet in form of a crofs, and are fucceeded by flender pods, filled with fmall roundifh feeds, which ripen in Auguft, and then the plant dies. The feeds of this plant are ufed in medicine, and are by fome greatly recommended for the gravel and ftoppage of urine.

The feventh fort grows naturally in France and Italy. The lower leaves of this are flaccid, cut in form of winged leaves, ending in arrow-pointed lobes. The ftalk rifes three feet high, garnifhed with linear wing-pointed leaves; it branches out greatly; the flowers grow fparfedly toward the end of the branches, which are fucceeded by very long, flender, fmooth pods, filled with fmall yellowifh feeds. The feeds ripen in autumn, and the plant dies foon after.

The two laft are preferved in botanic gardens for the fake of variety. If their feeds are permitted to fcatter the plants will come up in plenty, and require no other care but to thin them and keep them clean from weeds; or if their feeds are fown in autumn, they will fucceed better than in the fpring.

The eighth fort grows naturally in many parts of England, fo is feldom admitted into gardens; this is an annual plant, which fows itfelf, and comes up without care. It was remarked after the great fire of London, that this plant came up in great plenty on the ruins.

The ninth fort grows naturally on the Helvetian mountains; this hath a perennial root, from which arife feveral branching ftalks three feet high, garnifhed with fpear-fhaped leaves, fawed on their edges, of a deep green, ftanding alternately on the ftalks. The flowers grow in loofe fpikes at the top of the ftalks; they are fmall, yellow, compofed of four petals, placed in form of a crofs, and are fucceeded by taper pods filled with fmall feeds, which ripen in Auguft.

This is preferved in fome gardens for the fake of variety, but has no great beauty; it is propagated by feeds, which fucceed beft when fown in autumn, for thofe which are fown in the fpring feldom come up the fame year. The plants require no care, but to thin them and keep them clean from weeds; they love a cool fhady fituation.

SISYRINCHIUM. Lin. Gen. Plant. 908.

The CHARACTERS are,

The fheath which inclofes the flowers faces both ways, and is compofed of two compreffed keel-fhaped leaves. The flower has fix oblong petals, and three very fhort ftamina, terminated by bifid fummits, which are fixed to the bafe of the ftyle, with an oval germen fituated under the flower, fupporting an awl-fhaped ftyle, crowned by a trifid reflexed ftigma. The germen afterward turns to an oval three-cornered capfule with three cells, filled with roundifh feeds.

The SPECIES are,

1. SISYRINCHIUM (*Bermudiana*) foliis gladiolatis amplexicaulibus, pedunculis brevioribus. *Sifyrinchium with fword-fhaped leaves embracing the ftalks, and fhorter foot-ftalks to the flower.*

2. SISYRINCHIUM (*Anguftifolium*) foliis lineari-gladiolatis, pedunculis longioribus. *Sifyrinchium with linear fword-fhaped leaves, and longer foot-ftalks to the flower.*

3. SISYRINCHIUM (*Bulbofum*) foliis plicatis, fpatha biflora. *Sifyrinchium with a plaited leaf, and two flowers in a fheath.*

The firft fort grows naturally in Bermuda, from whence it had the title of Bermudiana given it by Tournefort; this has a fibrous root, from which arife ftiff, fword-fhaped, entire leaves, of a dark green colour; between thefe come out the foot-ftalks, which rifes fix inches high, compreffed, with two borders or wings running the whole length, having three or four fword-fhaped leaves, which embrace it; thefe grow erect, and are hollowed like the keel of a boat. The ftalk is terminated by a clufter of fix or feven flowers, ftanding upon fhort foot ftalks, inclofed by a two-leaved keel-fhaped fheath before they open; the flowers are of a deep blue colour with yellow bottoms, compofed of fix oval petals, ending in acute points. In the center is fituated an upright ftyle, at the bottom of which are three ftamina, whofe fummits fit clofe to it; the ftigma is cut into three parts, which are reflexed back to the ftyle; thefe are of a gold colour. The germen, which was fituated under the flower, turns to an oval obtufe capfule with three cells, filled with roundifh feeds.

The fecond fort grows naturally in Virginia; this has a perennial fibrous root, from which arife many very narrow, fpear-fhaped, entire leaves, of a light green colour. The ftalks rife about three inches high; they are very flender, compreffed and bordered like thofe of the firft, and have fhort, narrow, fword-fhaped leaves, whofe bafe embrace them; they are terminated by two fmall, pale, blue flowers, inclofed in a two-leaved fheath, ftanding upon longer foot-ftalks than thofe of the other.

Thefe plants are propagated by feeds, and alfo by parting of their roots; if they are raifed from feeds, thefe fhould be fown in autumn foon after they are ripe, upon an eaft-afpected border, where they may have only the morning fun. In the fpring the plants will appear, when their leaves will have much refemblance to Grafs, therefore care fhould be taken that they are not pulled up as weeds by thofe who clean the ground. During the firft fummer they will require no other care but to keep them clean from weeds, unlefs the plants fhould come up fo clofe as not to have room to grow; in which cafe, part of them fhould be drawn out to give room to the others; thefe may be planted in a fhady border at three inches diftance, where they may remain till autumn, when they fhould be tranfplanted to the places where they are to remain; the following fummer they will flower. Thefe plants love a fhady fituation, and a foft, loamy, undunged foil.

The time for tranfplanting and flipping off the fuckers from the old roots is early in autumn, that they may get good root before winter. They are both fo hardy as to thrive in the open air in England, and are very rarely injured but by great cold.

The third ſort grows naturally in the Weſt-Indies; this has a ſmall, oval, bulbous root, covered with a bright red ſkin, from which come out the leaves very like the firſt leaves of Palm-trees, but of a thinner ſubſtance; they have five or ſix longitudinal plaits, and are of a light green colour, ending with points; between theſe ariſe the foot-ſtalk of the flower four inches high, terminated by two or three ſmall blue flowers, incloſed in a ſpatha or ſheath, compoſed of ſix petals, which expand like thoſe of the other ſorts, but do not continue open longer than three or four hours in the morning; when they are expanded, their petals are ſo ſmall as to make but little appearance. This ſort flowers commonly in the middle of ſummer, but does not keep any particular month; they are never ſucceeded by ſeeds in England.

This is propagated by offsets from the roots, which are ſent out in plenty; theſe ſhould be taken off when the roots are tranſplanted: the time for doing of this is ſoon after the leaves decay, or before the roots begin to ſhoot again. They muſt be planted in ſmall pots, and plunged into the tan-bed in the ſtove, where they ſhould conſtantly remain, for they are too tender to thrive in this country, unleſs they are thus treated. Their after-management is the ſame as for other bulbous-rooted plants from the ſame countries.

SIUM. Tourn. Inſt. R. H. 308. tab. 162. Lin. Gen. Plant. 310. Water Parſnep, and Skirret.

The CHARACTERS are,

It hath an umbellated flower; the general umbel is various in different ſpecies; the ſmall ones are plain and ſpreading. The general involucrum is compoſed of ſeveral ſhort, ſpear-ſhaped, reflexed leaves; thoſe of the ſmaller are of very ſmall narrow leaves. The general umbel is uniform; the flowers have five inflexed petals, which are equal; they have five ſtamina, terminated by ſingle ſummits, and a ſmall germen ſituated under the flower, ſupporting two reflexed ſtyles, crowned by obtuſe ſtigmas. The germen afterward becomes a roundiſh, oval, ſtreaked fruit ſplitting in two, each part containing one ſtreaked ſeed, plain on one ſide, and convex on the other.

The SPECIES are,

1. SIUM (*Latifolium*) foliis pinnatis, umbellis terminalibus. Hort. Cliff. 98. *Sium with winged leaves, and the ſtalk terminated by an umbel; or the great Water Parſnep.*

2. SIUM (*Anguſtifolium*) foliis pinnatis ſerratis, umbella terminali. *Sium with winged ſawed leaves, and umbels terminating the ſtalks.*

3. SIUM (*Nodiflorum*) foliis pinnatis, umbellis axillaribus ſeſſilibus. Hort. Cliff. 98. *Sium with winged leaves, and umbels of flowers ſitting cloſe to the wings of the ſtalks.*

4. SIUM (*Siſarum*) foliis pinnatis, floralibus ternatis. Hort. Cliff. 98. *Sium with winged lower leaves, but thoſe under the flowers trifoliate; called Skirrets.*

5. SIUM (*Falcaria*) foliis linearibus decurrentibus connatis. Hort. Cliff. 98. *Sium with linear ſmall leaves, having running membranes, joining at their baſe round the ſtalk.*

6. SIUM (*Siculum*) foliis radicalibus ternatis, caulinis bipinnatis. Prod. Leyd. 105. *Sium with trifoliate bottom leaves, and thoſe on the ſtalks doubly winged.*

The firſt ſort is the great Water Parſnep, which grows naturally in deep ſtanding waters in ſeveral parts of England; it riſes with upright ſtalks five or ſix feet high, garniſhed with large winged leaves, ſhaped like thoſe of the common Parſnep; the ſtalk is terminated by large umbels of pale yellow flowers.

The ſecond ſort is the common upright Parſnep, which grows naturally in ditches in moſt parts of England; this riſes with an upright branching ſtalk three feet high, garniſhed with winged leaves, compoſed of three or four pair of oblong ſawed leaves, terminated by an odd one. The ſtalk is terminated by an umbel of white flowers, which are ſucceeded by ſeeds, which ripen in autumn; this is rarely cultivated, as it is a common weed in ditches and ſtanding waters. Both theſe plants have been recommended by ancient phyſicians for their virtues in medicine, but at preſent they are ſeldom uſed.

The third ſort is very common in ſtanding waters in moſt parts of England. The ſtalks ſpread over the ſurface, and produce umbels of white flowers at their joints. This is the plant which is frequently gathered and ſold for Water Creſs, as is before mentioned under the article SISYMBRIUM.

The fourth ſort is the common Skirret, which was formerly more cultivated in the Engliſh gardens than at preſent. The roots are the only part uſed, and although it is mentioned in moſt Diſpenſaries as a medicinal plant, yet it is rarely uſed as ſuch, being better adapted for the kitchen. It is eſteemed a wholeſome root, affording good nouriſhment, but has a flatulency, and its very ſweet taſte is diſagreeable to many palates.

The root of this plant is compoſed of ſeveral fleſhy fibres, as large as a man's little finger, which join together in one head. The lower leaves are winged, having two or three pair of oblong lobes, terminated by an odd one; the ſtalks riſe a foot high, terminated by an umbel of white flowers, which are ſucceeded by ſtriated ſeeds like thoſe of Parſley, which ripen in autumn.

This plant is cultivated two ways, firſt by ſeeds, and afterward by ſlips from the root: the former method I think the more eligible, becauſe the roots which are raiſed from ſeeds, generally grow larger than thoſe raiſed by ſlips, and are leſs ſubject to be ſticky. The ſeeds ſhould be ſown the latter end of March, or the beginning of April, either in broad-caſt or in drills; the ground ſhould be light and moiſt, for in dry land the roots are generally ſmall, unleſs the ſeaſon proves very moiſt. If the ſeeds are good, the plants will appear in five or ſix weeks after they are ſown, and, when they have put out their leaves ſo as to be well diſtinguiſhed from the weeds, the ground ſhould be hoed over to deſtroy the weeds in the ſame manner as is practiſed for Carrots; and where the ſeeds are ſown in broad-caſt, the plants ſhould be cut up, leaving them two or three inches aſunder. Thoſe ſown in the drills ſhould alſo be thinned to the ſame diſtance, and the ground hoed over to deſtroy the weeds. This ſhould be repeated three times, as is uſually done for Carrots, which, if well performed in dry weather, will keep the ground clean all the firſt part of the ſummer; ſo that, unleſs there ſhould be much rain about Midſummer, there will be ſcarce any neceſſity for farther cleaning of the plants, for their leaves will ſpread, and prevent the growth of weeds afterward. In autumn, when the leaves begin to decay, the roots will be fit for uſe, and may be continued all the winter till they begin to ſhoot in the ſpring, when they will become ſticky, as will alſo any of thoſe which run up to ſeed the firſt ſummer, ſo that all ſuch ſhould be pulled up and thrown away.

The time for propagating this plant by offsets is in the ſpring, before they begin to ſhoot, at which time the old roots ſhould be dug up, and the ſide roots ſhould be ſlipped off, preſerving an eye or bud to each; theſe ſhould be planted in rows one foot aſunder, and four inches diſtant in the rows. If the ground is light, this may be performed with a dibble, but for ſtiff land it will be beſt to make a trench with a ſpade, in the ſame manner as for Aſparagus plants, laying the roots therein at a proper diſtance. The ground muſt be kept clean by hoeing it in the ſame manner as before directed, and at the ſame ſeaſon the roots will be fit for uſe.

The

The fifth sort is a perennial plant, which grows naturally in Germany. The roots of this plant creep and spread very far under ground; the least part of them will grow, so that when it is once brought into a garden, it will soon multiply; they are thick, fleshy, and taste like those of Eryngo. The leaves are divided into linear segments, and their base embrace the stalks, which rise two feet high, and are terminated by large flat umbels of white flowers, but their seeds do not often ripen here.

The sixth sort grows naturally in Sicily, and is preserved in botanic gardens for the sake of variety. The lower leaves are pretty broad, trifoliate, and of a lucid green; the stalks rise two feet high, terminated by an umbel of yellow flowers; the leaves on the stalks are doubly winged, and the seeds ripen in autumn, which should be sown soon after they are ripe.

SMALLAGE. See APIUM.

SMILAX. Tourn. Inst. R. H. 654. tab. 421. Rough Bindweed.

The CHARACTERS are,

It is male and female in different plants. The male flowers have a six-leaved bell-shaped empalement, but no petals; they have six stamina, terminated by oblong summits. The female flowers have the like empalement, but they fall off; they have no petals or stamina, but have an oval germen, supporting three very small styles, crowned by oblong reflexed stigmas. The germen afterward turns to a globular berry with two cells, containing two globular seeds.

The SPECIES are,

1. SMILAX (*Aspera*) caule aculeato angulato, foliis dentato-aculeatis cordatis. Lin. Sp. Plant. 1028. *Smilax with an angular prickly stalk, and heart-shaped, prickly, indented leaves.*

2. SMILAX (*Excelsa*) caule aculeato angulato, foliis cordatis inermibus. *Smilax with an angular prickly stalk, and smooth heart-shaped leaves.*

3. SMILAX (*Sarsaparilla*) caule aculeato angulato, foliis inermibus retuso-cordatis. *Smilax with an angular prickly stalk, and retuse, heart-shaped, unarmed leaves.*

4. SMILAX (*Tamnoides*) caule aculeato tereti, foliis inermibus cordatis oblongis multinerviis. Lin. Sp. Pl. 1030. *Smilax with a taper prickly stalk, and oblong, heart-shaped, unarmed leaves, with many veins.*

5. SMILAX (*China*) caule aculeato teretiusculo, foliis inermibus ovato-cordatis. Lin. Sp. Pl. 1029. *Smilax with a taper prickly stalk, and oval, heart-shaped, unarmed leaves.*

6. SMILAX (*Caduca*) caule subaculeato tereti, foliis inermibus cordatis trinerviis. *Smilax with a taper stalk, having a few small thorns, and unarmed heart-shaped leaves with three veins.*

7. SMILAX (*Aristolochiæfolia*) caule aculeato tereti, foliis inermibus sagittatis obtusiusculis trinerviis. *Smilax with a prickly taper stalk, and very blunt, halberd-pointed, unarmed leaves.*

8. SMILAX (*Spinosa*) caule aculeato tereti, foliis ovato-lanceolatis nervis foliorum infernè aculeatis. *Smilax with a taper prickly stalk, and oval spear-shaped leaves, whose veins on the under side are prickly.*

9. SMILAX (*Virginiana*) caule aculeato angulato, foliis lanceolatis inermibus, acuminatis. *Smilax with an angular prickly stalk, and spear-shaped, acute-pointed, unarmed leaves.*

10. SMILAX (*Canellifolia*) caule inermi tereti, foliis inermibus ovato cordatis quinquenerviis floribus corymbosis. *Rough Bindweed with a taper unarmed stalk, oval, heart-shaped, unarmed leaves, and flowers in a corymbus.*

11. SMILAX (*Humilis*) caule inermi tereti, foliis inermibus ovatis trinerviis. *Smilax with an unarmed taper stalk, and oval unarmed leaves with three veins.*

12. SMILAX (*Hederæfolia*) caule inermi tereti, foliis inermibus oblongo-cordatis trinerviis. *Smilax with a taper unarmed stalk, and oblong, heart-shaped, unarmed leaves with three veins.*

13. SMILAX (*Laurifolia*) caule inermi tereti, foliis inermibus ovato cordatis trinerviis, floribus corymbosis. *Smilax with a taper unarmed stalk, oval, heart-shaped, unarmed leaves, and flowers in a corymbus.*

14. SMILAX (*Cordata*) caule inermi tereti, foliis inermibus cordato-oblongis trinerviis cum acumine. *Smilax with a taper unarmed stalk, and heart-shaped oblong leaves, having three veins, ending with acute points.*

15. SMILAX (*Inermi*) caule inermi tereti, foliis inermibus, caulinis cordatis, racemis ovato-oblongis. Lin. Sp. Plant. 1031. *Smilax with an unarmed taper stalk, unarmed heart-shaped leaves on the stalks, and oval oblong bunches of flowers.*

16. SMILAX (*Caroliniana*) caule inermi tereti, foliis inermibus lanceolatis. *Smilax with a taper unarmed stalk, and spear-shaped unarmed leaves.*

The first sort grows naturally under hedges and in woods in Italy and Spain. The roots are composed of many thick fleshy fibres, from which come out several slender stalks which are angular, armed with short crooked spines, and have claspers on their sides, by which they fasten themselves to any neighbouring plant for support, and rise six or eight feet high. The leaves are stiff, heart-shaped, and acute-pointed, eared at their base; they are of a dark green, and have five longitudinal veins; their edges are set with a f w short reddish spines. The flowers come out from the wings of the stalk in short bunches, which are small and whitish, having no petals. Those on the female plants are succeeded by red berries, which ripen in autumn.

The second sort grows naturally in Syria. The roots of this are like those of the former; the stalks are four-cornered and prickly; these fasten themselves to the trees near them by their claspers, and mount to their tops. The leaves are heart-shaped; they have no spines on their edges, but have five veins running lengthways. The flowers and fruit are like those of the first sort.

The third sort grows naturally in Virginia. The roots of this are like those of the former; the stalks are angular and prickly; the leaves are heart-shaped, turning backward, and unarmed; the flowers are small, in long loose bunches at the wings of the stalks: the berries are small and red.

The fourth sort grows naturally in Carolina. The roots are like the former; the stalks are taper and prickly; the leaves are oblong, heart-shaped, having no spines, but longitudinal veins; the flowers come out in long loose bunches from the side of the stalks, and the berries are black.

The fifth sort grows naturally at Carthagena in New Spain. The stalks are taper, very strong, and armed with short stiff spines; they fasten themselves by their claspers to the neighbouring trees, and rise twenty feet high. The leaves are of a thick substance, and have no spines; they are oval, heart-shaped, ending in an obtuse point, and have three longitudinal veins. The flowers are like those of the other species, but grow in close bunches; the berries are red. This is the same with a plant which I received from China by the title of China-root.

The sixth sort grows naturally at Carthagena in New Spain. This has very strong taper stalks, armed with a very few short spines. The leaves are thick, unarmed, and heart-shaped, their base ending with acute points. This sort climbs on the neighbouring trees, and rises thirty feet high.

The seventh sort grows naturally at La Vera Cruz in New Spain; this hath a thick, taper, prickly stalk, which climbs up the neighbouring trees to the height of thirty or forty feet. The leaves are thick, ſtiff, unarmed, and have two round ears at their baſe; they have three longitudinal veins, and ſtand on ſhort foot-ſtalks.

The eighth ſort grows naturally at La Vera Cruz; this has ſlender, taper, prickly ſtalks, which faſten themſelves to any neighbouring ſupport by their claſpers, and riſe eight or ten feet high. The leaves are oval, ſpear-ſhaped; they have no ſpines on their edges, but their midrib and veins on the under ſide are armed with ſhort reddiſh ſpines.

The ninth ſort grows naturally in Jamaica. The ſtalks of this are ſlender, angular, and prickly; the leaves are ſpear-ſhaped, ending in acute points, having no ſpines; their baſe is a little rounded, but have no ears.

The tenth ſort grows naturally at La Vera Cruz in New Spain. The ſtalks of this are taper and unarmed; the leaves are oval, heart-ſhaped, having five longitudinal veins; they have no ſpines, and ſtand on ſhort foot-ſtalks. The flowers come out from the wings of the ſtalk in round bunches, which are ſucceeded by red berries.

The eleventh ſort grows naturally in Jamaica; this has thick, fleſhy, creeping roots. The ſtalks are taper and unarmed; theſe climb up the neighbouring trees and buſhes to the height of ten or twelve feet. The leaves are oval, ending in acute points, and have three longitudinal veins, but no ſpines.

The twelfth ſort grows naturally in Jamaica. The ſtalks of this are very ſlender and taper, having no thorns; theſe branch out greatly, and riſe over the neighbouring buſhes, having very long claſpers, which twine about their branches. The leaves are oblong, heart-ſhaped, with three longitudinal veins, ending in acute points, of a lucid green, and pretty thick conſiſtence.

The thirteenth ſort grows naturally in Carolina; this has taper unarmed ſtalks, which riſe three or four feet high. The leaves are oval, heart-ſhaped, and have three longitudinal veins. The flowers come out from the wings of the ſtalk at every joint, ſtanding upon ſhort foot-ſtalks, formed in a round bunch; theſe are ſucceeded by roundiſh red berries.

The fourteenth ſort grows naturally in Jamaica. The ſtalks of this are taper, branching, and unarmed; the leaves are heart-ſhaped, oblong, and have three longitudinal veins, ending with acute points, of a lucid green, and ſtand upon ſhort foot-ſtalks.

The fifteenth ſort grows naturally in Jamaica, and alſo in Maryland. The ſtalks of this are ligneous, taper, unarmed, and have very long claſpers, by which they faſten to any neighbouring ſupport, and riſe twenty feet high. The leaves are ſome oval, and others are heart-ſhaped. The flowers come out from the wings of the ſtalk in oblong bunches, which are ſucceeded by red berries.

The ſixteenth ſort grows naturally in Carolina; this has a thick, taper, unarmed ſtalk, which riſes by the help of neighbouring buſhes and trees ten or twelve feet high. The leaves are thick, ſpear-ſhaped, and unarmed. The flowers come out from the wings of the ſtalk in round bunches, which are ſucceeded by black berries.

Theſe plants are many of them preſerved in the gardens of the curious for the ſake of variety, but ſome of them may be diſpoſed as to make them ornamental, eſpecially thoſe ſorts which grow naturally in North America, and alſo the two firſt ſorts, which are ſo hardy as to thrive in the open air in England; and, as they retain their verdure all the year, ſo if the plants are placed on the borders of woods or groves in gardens, and their branches properly ſupported, they will ſkreen the nakedneſs of the ground under the trees from ſight, and in winter, when their leaves are in beauty, they will make a pleaſing variety, when they are properly intermixed with other evergreens; and, as ſome of the ſorts will riſe five or ſix feet high, they will ſhut up from view any diſagreeable objects.

Thoſe ſorts which require a ſtove to protect them in winter, are little eſteemed, becauſe they require much room; and, as their flowers have no beauty to recommend them, few perſons care to be at the trouble of preſerving them for that of their leaves, becauſe there are many other plants whoſe leaves make as good an appearance, and the plants do not require ſo much room, ſo theſe are rather proper furniture of botanic gardens than thoſe of pleaſure.

They are all propagated by ſeeds. Thoſe ſorts which have been brought from North America, frequently produce flowers in England, but the ſummers here are neither warm enough, nor of duration to ripen their ſeeds, ſo that theſe are propagated by parting of their roots; for when the roots have obtained ſtrength, they may be greatly increaſed, by parting of them early in autumn, that the offsets or young plants may have time to get good roots before the froſts come on; and if after they are planted, the cold ſhould come on earlier, or be more ſevere than ordinary, if the ſurface about their roots is covered with ſome old tanners bark or mulch to keep the froſt out of the ground, it will preſerve them; but theſe roots ſhould not be parted oftener than every third or fourth year, for unleſs the roots are large, there will be few ſtalks to each, and therefore will make but little appearance.

The tender ſorts muſt be kept in pots, and plunged into the tan-bed of the bark-ſtove, to have them ſtrong; for although they will live in a moderate warmth in winter, they will make but little progreſs, and their ſtalks will be ſhort, their leaves ſmall, and the plants weak, ſo will make but a poor appearance; therefore, unleſs they can be allowed room in the warm ſtove, and conſtantly kept in the tan-bed, they will not be worth preſerving.

As all the ſorts grow naturally under hedges, and in woods, they ſhould be diſpoſed in ſuch a manner, as to imitate their places of growth, and not place them in the open ſun, where they will not thrive; therefore the hardy kinds ſhould be planted under the ſhade of trees, and the tender ones may be placed between the pots which contain any plants, whoſe branches may ſcreen them from the ſun.

When the ſeeds of theſe plants are obtained from abroad, they ſhould be ſown in pots, and plunged into a moderate hot-bed; theſe generally remain in the ground a whole year before they grow, ſo that when the plants do not come up the firſt ſeaſon, the pots in which the hardy ſorts are ſown, ſhould be in winter ſheltered from froſt under a common frame, and the tender ones plunged into the bark-bed in the ſtove: the following ſpring they muſt be again plunged into the hot-bed, which will bring the plants up very ſoon. When they are come up, they muſt be inured to the open air by degrees, and in June the hardy ſorts may be removed out of the bed, and placed abroad in a ſheltered ſituation, where they ſhould remain till the froſt comes on in autumn, when they muſt be removed into ſhelter. If the pots are plunged into an old tan-bed under a frame where they may be protected from the froſt, and in mild weather be expoſed to the open air, they will thrive much better than with more tender treatment.

The tender ſorts ſhould be plunged between the other pots in the bark-bed of the ſtove, where they ſhould remain all the winter. Theſe plants ſhould remain untranſplanted in the ſeed-pots till the following ſpring, when they ſhould be turned out of the pots, carefully ſeparated, and each planted in a pot; and if the hardy ſorts are plunged into a very temperate hot-bed, it will cauſe them to take new root very ſoon, ſo will greatly ſtrengthen the plants; but the tender

tender forts fhould be plunged into a good hot-bed of tanners bark to bring the plants forward, that they may get ftrength before winter, when they muft be treated in the manner before directed.

The hardy forts fhould be kept in pots for two years, that they may be fheltered in winter, by which time they will have ftrength enough to bear the cold in the open air, fo in the fpring they may be turned out of the pots, and planted where they are defigned to remain, obferving, if the fpring fhould prove dry, to refrefh them now and then with water.

SMYRNIUM. Tourn. Inft. R. H. 315. tab. 168. Alexanders, or Alifanders.

The CHARACTERS are,

It has an umbellated flower; the principal umbel is unequal, the fmall ones are erect; they have no involucrum, and the empalement of the flowers are fcarce difcernible. The flowers have five fpear-fhaped petals, which are a little inflexed, and five ftamina the length of the petals, terminated by fingle fummits. The germen is fituated under the flower, fupporting two ftyles, crowned by headed ftigmas. The germen afterward turns to an almoft globular fruit, which is ftreaked and fplits in two, each containing one moon-fhaped feed, convex on one fide, marked with three ftreaks, and plain on the other.

The SPECIES are,

1. SMYRNIUM (*Olufatrum*) foliis caulinis ternatis petiolatis ferratis. Hort. Cliff. 105. *Smyrnium with trifoliate leaves on the ftalks, which are fawed and have foot-ftalks*; *or common Alexanders, or Alifanders.*

2. SMYRNIUM (*Rotundifolium*) foliis caulinis orbiculatis integerrimis amplexicaulibus. *Smyrnium with orbicular leaves on the ftalks, which embrace them.*

3. SMYRNIUM (*Perfoliatum*) foliis caulinis cordato-ovatis dentatis amplexicaulibus. *Alexanders with heart-fhaped oval leaves, which are indented, embracing the ftalks.*

4. SMYRNIUM (*Creticum*) foliis caulinis ternatis ferratis, fummis oppofitis feffilibus. *Alexanders with leaves by threes, which are fawed, and thofe at the top by pairs, fitting clofe to the ftalks.*

5. SMYRNIUM (*Integerrimum*) foliis caulinis duplicato ternatis integerrimis. Lin. Sp. Plant. 263. *Alexanders with double trifoliate leaves on the ftalks, which are entire.*

The firft fort grows naturally on the rocks by the fea-fhore in Wales, the north of England, and in Scotland. It is alfo found growing wild in many places near London; but here it may be fuppofed to have been thrown out of gardens; for as it was formerly cultivated in gardens for the table, fo the feeds may have been fcattered, which will grow wherever they alight.

The lower leaves of this plant refemble thofe of Smallage, but they are much larger; the lobes are rounder, and are fawed on their edges. The ftalk rifes from three to four feet high; they are furrowed, and branch into many divifions, garnifhed with trifoliate leaves of the fame fhape and form with the lower, but are fmaller. The branches are terminated by large umbels of white flowers, which are fucceeded by large roundifh fruit, containing two moon-fhaped feeds.

The fecond fort grows naturally in Sicily and Crete; the lower leaves of this fort are decompounded of fmall leaves, which divide by threes; their lobes are oval, and indented on their edges; the ftalk is fmooth, hollow, and rifes three feet high, dividing toward the top into two or three branches; at each joint is placed one large orbicular leaf, whofe bafe embrace the ftalk; thefe are of a yellow green colour, and their edges are entire: the branches are terminated by fmall umbels of yellowifh flowers, whofe fmaller umbels or rays are of unequal lengths The feeds are black, and fhaped like thofe of the former, but are fmaller.

The third fort grows naturally in Crete; the lower leaves of this are larger than thofe of the former, but are compofed of feveral winged divifions. The ftalk does not rife fo high as that of the laft mentioned, but is angular, and not fo hollow; the leaves upon the ftalks are much larger; they are of the heart-fhaped oval kind, are indented on their edges, and embrace the ftalks with their bafe; their colour is nearly the fame with the former, but they are of a thinner texture. The umbels of flowers are fmaller, as are alfo the feeds.

Thefe two forts have been frequently blended together by botanifts, who have fuppofed they were but one fpecies; but I have cultivated both many years, and have not found either of them alter.

The fourth fort grows naturally in Crete; the lower leaves of this are fmaller than thofe of the firft fort, and are more like thofe of Smallage; the ftalk rifes higher, and grows more erect than thofe of the firft; the leaves on the lower part of the ftalk are large, and fawed on their edges; they ftand by threes round the ftalk at the joints, their bafe fet clofe, having no foot-ftalks; the upper part of the ftalk and branches are garnifhed with leaves of the fame form, which ftand by pairs. The umbels of flowers are much fmaller, and the feeds are lefs.

The firft of thefe forts is that ordered by the College for medicinal ufe, but is feldom now prefcribed; and at prefent is feldom cultivated in gardens, though formerly it was greatly ufed in the kitchen, before Celery was fo much cultivated, which hath taken place of Alexanders, and entirely fupplanted it. The other forts are preferved in botanic gardens for variety. The fecond fort is much preferable to the firft for blanching, as I have tried, and will be tenderer, and not quite fo ftrong.

All thefe plants may be propagated by fowing their feeds upon an open fpot of ground in Auguft, as foon as they are ripe; for if they are fown in fpring, they often mifcarry, or at leaft do not come up until the fecond year; whereas thofe fown in autumn, rarely fail of coming up foon after Chriftmas, and will make much ftronger plants than the other.

The common fort, when cultivated for the table, fhould be treated in the following manner:

In the fpring the plants fhould be hoed out, fo as to leave them ten inches or a foot apart each way; and, during the following fummer, they muft be conftantly cleared from weeds. In February following the plants will fhoot up again vigoroufly, at which time the earth muft be drawn up to each plant, to blanch them, and in three weeks after they will be fit for ufe; when they may be dug up, and the white part preferved, which may be ftewed and eaten as Celery.

SNAP-DRAGON. See ANTIRRHINUM.

SNEEZ-WORT. See ACHILLEA.

SNOW is defined to be a meteor formed in the middle region of the air, of vapour raifed by the action of the fun, or fubterraneous fire, there congealed, its parts conftipated, its fpecific gravity increafed, and thus returned to the earth in the form of little villi or flakes.

The Snow we receive may properly enough be afcribed to the coldnefs of the atmofphere through which it falls; when the atmofphere is warm enough to diffolve the Snow before it arrives at us, we call it rain; if it preferves itfelf undiffolved, we call it Snow.

Snow is very ufeful; it fructifies the ground; it guards Corn, or other vegetables, from the intenfer cold of the air, efpecially cold piercing winds.

It is fuppofed to abound with falific and fertile particles, as much or more than rain; however, it is accounted more ponderous, and by that means finks deeper in the ground than

than rain does, and therefore is in some cases of more benefit to planting; for which reason, some lay heaps of Snow round the feet of the forest-trees, especially in hot burning lands.

SNOWDROP. See Galanthus.

SOIL. See Earth.

SOLANOIDES. See Piercea.

SOLANUM. Tourn. Inst. R. H. 148. tab. 62. Nightshade.

The Characters are,

The empalement of the flower is permanent, of one leaf. The flower has one wheel-shaped petal, having a very short tube; the brim is large, spreading, and five-pointed. It has five awl-shaped stamina, terminated by oblong summits, which stand together, and a roundish germen, supporting a slender style longer than the stamina, crowned by an obtuse stigma. The germen afterward turns to a roundish berry with two cells, having a convex fleshy receptacle, filled with roundish compressed seeds.

The Species are,

1. Solanum (*Nigrum*) caule inermi herbaceo, foliis ovatis acuminatis glabris, umbellis nutantibus. *Nightshade with an herbaceous unarmed stalk, smooth oval-pointed leaves, and nodding umbels; common Nightshade with black fruit.*

2. Solanum (*Villosum*) caule inermi herbaceo, foliis ovatis dentato, angulatis, umbellis nutantibus. *Nightshade with an herbaceous unarmed stalk, oval, angular, indented leaves, and nodding umbels; Nightshade with red fruit.*

3. Solanum (*Luteum*) caule inermi herbaceo, foliis ovato-lanceolatis acuminatis tomentosis, umbellis nutantibus. *Nightshade with an herbaceous unarmed stalk, oval, spear-shaped, acute-pointed, indented, woolly leaves, and nodding umbels; Nightshade with yellow berries.*

4. Solanum (*Rubrum*) caule inermi herbaceo glabro, foliis oblongo-ovatis acuminatis, dentatis glabris, umbellis nutantibus. *Nightshade with an herbaceous, unarmed, smooth stalk, oblong, oval, acute-pointed, indented, smooth leaves, and nodding umbels.*

5. Solanum (*Americanum*) caule inermi herbaceo, foliis ovatis acuminatis glabris, umbellis erectis. *Nightshade with an herbaceous unarmed stalk, oval, acute-pointed, smooth leaves, and erect umbels.*

6. Solanum (*Scabrum*) caule herbaceo subaculeato, foliis ovatis obtusis integerrimis, petiolis longissimis, umbellis nutantibus. *Nightshade with an herbaceous prickly stalk, oval, obtuse, entire leaves on very long foot-stalks, and nodding umbels.*

7. Solanum (*Guineense*) caule inermi herbaceo, foliis oblongo-ovatis acuminatis glabris subdentatis, umbellis nutantibus. *Nightshade with an herbaceous unarmed stalk, oblong, oval, acute-pointed, smooth leaves, a little indented, and nodding umbels.*

8. Solanum (*Dulcamara*) caule inermi frutescente flexuoso, foliis superioribus hastatis, racemis cymosis. Hort. Cliff. 60. *Nightshade with a shrubby, flexible, unarmed stalk, the upper leaves spear-shaped, and bunches of flowers at the top of the stalk; commonly called Bitter-sweet.*

9. Solanum (*Pseudocapsicum*) caule inermi fruticoso, foliis lanceolatis repandis, umbellis sessilibus. Lin. Sp. Pl. 184. *Nightshade with a shrubby unarmed stalk, spear-shaped leaves turning inward, and the berries sitting close to the stalks; commonly called Amomum Plinii.*

10. Solanum (*Igneum*) caule aculeato fruticoso, foliis lanceolatis anguloso-dentatis. Hort. Cliff. 61. *Nightshade with a shrubby prickly stalk, and spear-shaped leaves which are angularly indented.*

11. Solanum (*Tomentosum*) caule aculeato fruticoso, foliis ovatis dentato-angulatis utrinque tomentosis, pedunculis spinosis. *Nightshade with a shrubby prickly stalk, oval, angular, indented leaves, woolly on every side, and prickly foot-stalks to the flowers.*

12. Solanum (*Sodomæum*) caule aculeato fruticoso, foliis pinnato-laciniatis obtusis utrinque aculeatis. *Nightshade with a shrubby prickly stalk, wing-cut leaves, which are obtuse, and have spines on both sides; commonly called Pomum Amoris.*

13. Solanum (*Indicum*) caule aculeis recurvis, foliis sinuatis subtus tomentosis, utrinque aculeatis, pedunculis aculeatis. Lin. Flor. Zeyl. 95. *Nightshade with recurved thorns on the stalks, and sinuated leaves, downy on their under side, armed with prickles on both sides, and the foot-stalks of the flowers are prickly.*

14. Solanum (*Angustifolium*) caule aculeato herbaceo, foliis sinuatis glabris, utrinque aculeatis, umbellis erectis, calycibus echinatis. *Nightshade with a prickly herbaceous stalk, smooth sinuated leaves armed with spines on both sides, upright umbels, and very prickly empalements.*

15. Solanum (*Quercifolium*) caule aculeato fruticoso, foliis pinnato-laciniatis tomentosis, utrinque aculeatis, pedunculis axillaribus bifloris. *Nightshade with a prickly shrubby stalk, wing-cut leaves which are woolly, prickly on both sides, and foot-stalks with two flowers at the wings of the stalk.*

16. Solanum (*Jamaicense*) caule aculeato fruticoso, foliis oblongis sinuato-pinnatis, aculeatis, umbellis sessilibus. *Nightshade with a prickly shrubby stalk, oblong, wing-sinuated, prickly leaves, and umbels sitting close to the stalks.*

17. Solanum (*Fruticosum*) caule aculeato fruticoso, foliis ovatis tomentosis, anguloso-sinuatis subaculeatis, umbellis sessilibus. *Nightshade with a prickly shrubby stalk, oval, woolly, angular, sinuated leaves a little prickly, and umbels sitting close to the stalks.*

18. Solanum (*Scandens*) caule inermi frutescente flexuoso, foliis ovatis subtus tomentosis, floribus solitariis alaribus. *Nightshade with a shrubby, bending, unarmed stalk, oval leaves which are woolly on their under side, and flowers growing singly from the wings of the stalk.*

19. Solanum (*Laurifolium*) caule inermi fruticoso, foliis ovatis acuminatis integerrimis, subtus tomentosis, umbellis erectis alaribus & terminalibus. *Nightshade with a shrubby unarmed stalk, oval, acute-pointed, entire leaves, which are woolly on their under side, and erect umbels from the wings and the top of the branches.*

20. Solanum (*Caroliniense*) caule aculeato fruticoso, foliis ovatis sinuato-dentatis subtus tomentosis, aculeis utrinque rectis, umbellis sessilibus terminalibus. *Nightshade with a prickly shrubby stalk, oval, sinuated, indented leaves, which are woolly on their under side, the spines every way strait, and umbels sitting close at the end of the branches.*

21. Solanum (*Verbacifolium*) caule inermi fruticoso, foliis ovato lanceolatis integerrimis subtus tomentosis, umbellis erectis pedunculis longissimis. *Nightshade with a shrubby unarmed stalk, oval, spear-shaped, entire leaves, which are woolly on their under side, and erect umbels having very long foot-stalks.*

22. Solanum (*Bonariense*) caule frutescente subinermi, foliis cuneiformibus sinuato-repandis. Lin. Sp. Plant. 185. *Nightshade with a shrubby almost unarmed stalk, and wedge-shaped leaves which are sinuated, and turn backward.*

23. Solanum (*Bahamense*) caule frutescente inermi, foliis lanceolatis sinuato-dentatis glabris, umbellis erectis. *Nightshade with a shrubby unarmed stalk, spear-shaped, sinuated, indented, smooth leaves, and erect umbels.*

24. Solanum (*Sempervirens*) caule inermi fruticoso, foliis ovatis integerrimis, pedunculis lateralibus filiformibus. Lin. Sp. Pl. 185. *Nightshade with a shrubby unarmed stalk, oval entire leaves, and thread like foot-stalks to the flowers, proceeding from the side of the branches.*

25. Solanum

25. SOLANUM (*Africanum*) caule inermi frutescente flexuoso, foliis ovatis subdentatis crassis. *Nightshade with a shrubby, flexible, unarmed stalk, and oval thick leaves somewhat indented.*

26. SOLANUM (*Umbellatum*) caule frutescente inermi, foliis lanceolatis integerrimis subtus pilosis, umbellis erectis terminalibus. *Nightshade with a shrubby unarmed stalk, spear-shaped entire leaves which are hairy on their under side, and erect umbels terminating the branches.*

27. SOLANUM (*Racemosum*) caule inermi fruticoso, foliis ovatis integerrimis, subtus tomentosis, umbellis erectis terminalibus, calycibus obtusis lanuginosis. *Nightshade with a shrubby unarmed stalk, oval entire leaves which are woolly on their under side, erect umbels terminating the branches, and downy obtuse empalements.*

28. SOLANUM (*Trilobatum*) caule aculeato, fruticoso foliis cuneiformibus subtrilobis glabris obtusis inermibus. *Nightshade with a shrubby unarmed stalk, oblong oval leaves, with sinuated indentures, hairy on their under side, and umbels on the side of the branches.*

29. SOLANUM (*Recurvum*) caule aculeato herbaceo, foliis pinnatifidis utrinque aculeatis laciniis sinuatis obtusis. *Nightshade with a prickly herbaceous stalk, leaves with sinuated indentures, bunches of flowers on the side of the branches, and the spines every where recurved.*

30. SOLANUM (*Mammosum*) caule aculeato herbaceo, foliis cordatis quinquelobis, utrinque villosis, aculeatis. *Nightshade with an herbaceous prickly stalk, obtuse sinuated leaves, which are woolly on both sides, and flowers in loose bunches terminating the branches.*

31. SOLANUM (*Schiru schuna*) caule aculeato, foliis pinnato-sinuatis, fructu racemoso. *Nightshade with a prickly stalk, sinuated winged leaves, and fruit growing in long bunches.*

The first sort is now very common upon dunghills, and on rich cultivated soils, in many parts of England, where it often becomes a very troublesome weed. This is the sort which the College of Physicians have directed to be used in medicine, under the title of Solanum hortense: and although it is now become a very troublesome weed in many gardens near London, yet it is not a native of this country, but is supposed to have been brought originally from America, from whence the greater part of the species of this genus have been introduced into Europe.

There are two varieties of this, which are found growing naturally in England. The most common sort is an upright plant with oval, acute-pointed, smooth leaves, and black berries. The other is a low branching plant with indented leaves, and greenish yellow berries; but whether these are only varieties, or distinct species, I cannot say, though I have sown their seeds separately, and have found them keep their difference, but do not know if they will continue it always.

The second sort rises with an erect branching stalk three feet high; the leaves are oval, angular, indented, and smooth; the flowers are white, produced in roundish bunches in form of umbels, having five star-pointed petals, which are reflexed; in the center are five stamina, which are terminated by oblong yellow summits standing close together; after the flowers are past, the germen will swell to round pulpy berries of a deep red colour, standing in nodding umbels on the side of the branches.

The third sort rises with hairy branching stalks two feet high; the leaves are woolly, oval, spear-shaped, acute-pointed, and indented on their edges; the flowers are like those of the former sort; the berries are smaller, and of a dirty yellow colour. The seeds of this came from America.

The seeds of the fourth sort came from the West-Indies; this hath taller and smoother stalks than either of the former; the leaves are of a dark green, smooth, oval, acute-pointed, and indented on their edges in angular indentures; the flowers are produced in umbels on the side of the branches, which are succeeded by smooth red berries.

The fifth sort grows naturally in Virginia; the stalks of this are angular, and rise upward of three feet high, dividing into a few slender spreading branches, garnished with oval, acute-pointed, smooth leaves, of a deep green colour, with a few indentures on their edges; the flowers are very small, and but few in each umbel; they have narrow acute-pointed petals, white on the inside, and purplish without, and are succeeded by small black berries, which ripen late in autumn.

The sixth sort grows naturally in North America. The stalks of this sort rise three feet high, and divide into spreading angular branches, having a few short spines; the leaves are oval and entire, of a dark green colour, and have long foot-stalks; the flowers come out from the side of the branches in small umbels, which nod on one side; they are star-pointed, and are succeeded by small black berries, which ripen late in the autumn.

The seventh sort grows naturally in Guinea. This rises with a strong, thick, angular, herbaceous stalk two feet high, dividing into short thick branches, garnished with oblong, oval, smooth, indented leaves, standing upon pretty long foot-stalks. The flowers are produced in nodding umbels from the side of the stalk; they are like those of the first sort, but are larger. These are succeeded by large black berries the size of the common black Cherry, which ripen in autumn.

These seven sorts are annual, so their seeds should be sown in the spring, on a bed of rich earth, where the plants are designed to remain; and when they come up, they must be thinned, leaving them at least two feet distance, that they may have room to grow; after this they will require no farther care but to keep them clean from weeds; in July and August they will flower, and the seeds will ripen in autumn. Some people plant one or two plants of each sort in pots, whose stalks they train up to sticks to make them strait; and in autumn they remove the pots into the green-house, where they may be preserved till the spring, and during the winter, their fruit being ripe, will make a pretty appearance.

The eighth sort is a climbing woody plant, which grows in the hedges in divers parts of England, and is by some planted in gardens to cover arbours or walls, in London, and other close places, where few other plants will thrive. The cuttings or stalks of this are put into glasses of water, and placed in rooms, where they will put out branches and leaves, and continue a long time green. This plant is also used in medicine, for some particular preparations; but the herb-folks in the markets often sell this instead of the Garden Nightshade, which is a cooling plant, but this a hot acrid one, which renders it contrary to the intention of the ointment, wherein Nightshade is one of the ingredients.

There is a sort of this with white flowers, which is supposed to be a variety of the former, but the leaves are woolly, in which it differs from the other, and this is constant. There is also one with variegated leaves, which is preserved by those who are very curious in collecting the various kinds of striped-leaved plants.

These may be easily propagated by planting their cuttings in the spring upon a moist soil, where they will soon take root, and may afterward be transplanted where they are to remain.

The ninth sort grows naturally at the Madeiras; this rises with a strong woody stalk four or five feet high, dividing into many slender stiff branches, garnished with spear-shaped

 leaves,

leaves, turning backward; the flowers grow in ſmall umbels, or ſingly on the ſide of the branches, to which they ſit cloſe. Theſe are ſucceeded by berries as large as ſmall Cherries, which ripen in winter, when they make a good appearance in the green-houſe. There are two varieties of this, one with a red, and the other has a yellowiſh fruit.

This plant may be propagated by ſowing its ſeeds in a pot of rich earth in the ſpring, placing it upon a moderate hot-bed, which will greatly facilitate their growth. When the plants are come up, you ſhould make a gentle hot-bed, covered with rich earth about ſix inches thick; in this they ſhould be planted about ſix inches diſtance each way, and the bed arched over with hoops, &c. and covered with mats, to ſhade them from the ſun and cold, obſerving frequently to water them.

When the plants have acquired ſtrength, and the ſeaſon becomes favourable, you muſt inure them to bear the open air by degrees, to which they ſhould be fully expoſed in ſummer, when they ſhould be taken up, with a ball of earth to the root, and placed ſeparately in pots, filled with rich earth, and ſet in a ſhady ſituation, until they have taken new root; after which they may be removed into a more open expoſure, and placed among other exotic plants; but they require a great plenty of water in dry weather, without which they ſeldom produce much fruit.

In winter they muſt be removed into the green-houſe, and placed in the coldeſt part of the houſe, where they may have as much free air as poſſible in mild weather; being ſo hardy, as many times to endure the cold of our ordinary winters abroad, when planted in a warm ſituation, ſo that they only require to be ſheltered from ſevere froſt.

The tenth ſort grows naturally in the Weſt-Indies; this riſes with a ſhrubby ſtalk three feet high, dividing at the top into ſeveral branches, cloſely armed with ſtrait gold-coloured ſpines on every ſide. The leaves are angularly indented; their midrib is armed with a row of the like ſpines as thoſe upon the ſtalks, which ſtand erect. The flowers are produced in oblong bunches from the ſide of the ſtalks, which are ſucceeded by red berries almoſt as large as the ſmall black Cherry.

This ſort is much tenderer than either of the former, being brought from the warm parts of America. This is propagated by ſeeds, which muſt be ſown upon a good hot-bed; and when the plants are come up, they ſhould be each tranſplanted into a ſeparate ſmall pot, and plunged into a freſh hot-bed again, obſerving to water and ſhade them until they have taken root; after which they ſhould have air and water in proportion to the heat of the ſeaſon.

In July theſe plants may be inured to bear the open air by degrees, into which they may be removed, if the ſeaſon be warm, otherwiſe they muſt be preſerved either under glaſſes, or in the ſtove; if they are placed in the open air, they ſhould not remain there longer than the middle or latter end of Auguſt, leſt the nights growing cold, ſhould hurt them. During the winter ſeaſon they muſt be preſerved in the ſtove, obſerving to refreſh them frequently with water; but they muſt not have too much each time, eſpecially in cold weather. The ſecond year they will produce flowers and fruit.

The eleventh ſort has a ſhrubby ſtalk, which riſes two feet high, dividing into ſeveral woody branches, armed with ſharp thorns, garniſhed with oval woolly leaves, which have angular indentures on their edges. The flowers are produced in ſmall looſe bunches from the wings of the ſtalks; they are blue, and larger than thoſe of the former ſorts; theſe are ſucceeded by round berries as large as common Cherries, of a gold colour, which turn black when ripe.

The twelfth ſort grows naturally at the Cape of Good Hope. This hath a ſtrong, thick, ſhrubby ſtalk, which riſes from two to three feet high, ſending out many ſhort thick branches, cloſely armed with ſhort, ſtrong, yellow ſpines on every ſide; the leaves are cut almoſt to their midrib in obtuſe ſegments, which are oppoſite, regular, and formed like winged leaves; theſe ſegments have ſeveral obtuſe indentures on their edges; they are of a dark green colour, and armed with the ſame ſort of ſpines as thoſe on the ſtalks, on both ſides. The flowers come out in ſmall bunches on the ſide of the branches; they are blue, and larger than thoſe of the former ſort, and are ſucceeded by round yellow berries as large as Walnuts, which ripen in winter.

The eleventh and twelfth ſorts are not ſo tender as the laſt, but require an open airy glaſs-caſe, or a warm green-houſe in winter, but in ſummer may be expoſed in the open air with other exotic plants. Theſe may be propagated by ſowing their ſeeds on a hot-bed as the former, and ſhould be managed as hath been directed for them, with this difference, that they may be much ſooner expoſed to the air, and ſhould not be treated ſo tenderly.

The thirteenth ſort grows naturally at the Cape of Good Hope; this hath a ſhrubby ſtalk four or five feet high, covered with a white meally down, dividing into a few ſtraggling branches, armed with ſhort, thick, dark, brown, recurved ſpines, with yellowiſh points. The leaves are ſinuated, of a bright green on their upper ſide, but woolly on their under. This ſort has not as yet produced any flowers in England, though there are large plants of it in the Chelſea garden, where they were raiſed from ſeeds, which came from the Cape of Good Hope.

This may be propagated by ſeed, in the ſame way as the two former, and the plants muſt be treated in the ſame way, but they are not ſo hardy as the two former, ſo ſhould be placed in a warm ſtove in winter, and ſhould not have much water in cold weather.

The fourteenth ſort was diſcovered by the late Dr. Houſtoun at La Vera Cruz. This riſes with a prickly herbaceous ſtalk two feet high, dividing into two or three branches, cloſely armed with ſlender yellow ſpines of unequal lengths. The leaves are of a bright green colour, deeply ſinuated; the veins of the leaves are armed with yellow erect ſpines on both ſides. The umbels of flowers ſtand erect at the end of the branches; the flowers are very large, of a fine blue colour, and are ſucceeded by round berries as large as common Cherries, which are marbled with white and green. The empalement of the flower is armed with ſpines like a hedgehog. The fruit ripens late in the autumn, ſo that unleſs the plants are brought forward in the ſpring, they will not produce ripe ſeeds in England.

The fifteenth ſort grows naturally at La Vera Cruz; this hath ſhrubby trailing ſtalks two feet long, armed with long yellow ſpines, covered with a gray bark; the leaves are woolly, very finely cut in form of winged leaves almoſt to their midrib, and armed with long, ſlender, yellowiſh ſpines on their veins on both ſides. The foot-ſtalks of the flowers ariſe from the wings of the ſtalks; they are three inches long, for the moſt part ſuſtaining two large yellow flowers, having very prickly empalements, and are ſucceeded by ſmall round berries the ſize of gray Peaſe, which are marbled with green and white.

The ſixteenth ſort grows at La Vera Cruz; this riſes with a ſhrubby ſtalk five or ſix feet high, armed with ſhort recurved ſpines, covered with a ſmooth browniſh bark, garniſhed with oblong leaves, which are regularly ſinuated on both edges in form of winged leaves, armed with a few ſhort ſpines along their midrib on both ſides. The flowers come out in ſmall looſe bunches from the ſide of the branches, to which they ſit cloſe; they have five white ſtar-pointed petals, and are ſucceeded by ſmall berries about the ſize of thoſe of Juniper, which, when ripe, are red.

The ſeventeenth ſort grows naturally in Jamaica; this riſes with a ſhrubby woolly ſtalk five or ſix feet high, armed with ſhort recurved thorns, garniſhed with oval woolly leaves, angularly ſinuated, and have a very few ſhort crooked ſpines upon the midrib on the under ſide. The flowers are in ſmall umbels ſitting cloſe to the ſide of the branches; they are yellow, and are ſucceeded by ſmall round berries, of a Saffron colour when ripe.

The eighteenth ſort grows naturally at La Vera Cruz; this has a ſtrong ſhrubby ſtalk five feet high, covered with a brown woolly bark, armed with a few ſhort ſpines, ſending out ſeveral ligneous branches, garniſhed with oblong, oval, acute-pointed leaves, covered with a brown woolly down on both ſides. The flowers are large, white, and grow in erect umbels from the ſide of the branches, and have thick woolly empalements; theſe are ſucceeded by yellow berries as large as middling Cherries.

The nineteenth ſort grows naturally at La Vera Cruz; this has a ſhrubby climbing ſtalk ten or twelve feet high, covered with a ſmooth brown bark, divided into ſeveral branches. The leaves are oval, woolly on their under ſide, but of a dark green on their upper. The flowers come out ſingly from the wings of the ſtalk; they are large, of a fine blue colour, and the petal is not divided into ſegments like thoſe of the other ſpecies, but have five angles, each ending in a point; theſe are ſucceeded by round berries about the ſize of gray Peaſe, which are red when ripe.

The twentieth ſort grows naturally at Campeachy; this hath a ſmooth ſhrubby ſtalk ſix or ſeven feet high, ſending out ligneous branches, garniſhed with oval acute-pointed leaves, which are entire, woolly on their under ſide. The flowers are collected into umbels, which ſtand erect; theſe come out from the ſide and at the end of the branches; they are of a light blue colour, and are ſucceeded by round berries the ſize of ſmall black Cherries, which are yellow when ripe.

The twenty-firſt ſort grows naturally at La Vera Cruz; this has a ſhrubby ſtalk four feet high, with a white downy bark, armed on every ſide with ſtrait brown ſpines. The leaves are oval, and have ſinuated indentures, woolly on their under ſide, and have prickly foot-ſtalks. Their midrib is armed with two or three ſmall ſpines, ſometimes on both ſides, and at others but on one. The ſpines are all erect; the flowers are diſpoſed in an umbel ſitting cloſe at the end of the branches; they are large, of a fine blue colour, and have woolly empalements; theſe are ſucceeded by round berries the ſize of large Peaſe, which are red when ripe.

The twenty-ſecond ſort grows naturally at Campeachy; this riſes with a woody ſtalk eight or ten feet high, ſending out ſeveral ligneous furrowed branches, covered with a gray down. The leaves are ſometimes placed alternately on the branches, and at others they are oppoſite, ſtanding upon pretty long foot-ſtalks; their edges are entire, end in acute points, and are woolly on their under ſide. The flowers terminate the branches in large erect umbels, ſtanding upon long foot-ſtalks; they are large, white, and have woolly empalements; theſe are ſucceeded by berries the ſize of Cherries, which turn yellow.

The twelve laſt mentioned ſorts are propagated by ſeeds in the ſame manner as the former, but theſe, being natives of a warm country, muſt be raiſed on a hot-bed early in the ſpring, and when the plants are fit to remove, they muſt be each planted in a ſeparate ſmall pot, and plunged into a moderate hot-bed of tanners bark, obſerving to ſhade them from the ſun until they have taken new root; after which they ſhould have a large ſhare of air admitted to them in warm weather, and muſt be frequently watered. Toward the latter end of June it will be proper to harden the plants gradually, and ſoon after they ſhould be removed into the ſtove, where they may have as much free air as poſſible in warm weather; but as the cold approaches in autumn, they muſt be carefully protected therefrom, and in winter they ſhould be kept in a moderate temperature of warmth, otherwiſe they will not live in this country.

Some of theſe ſorts will bear to be expoſed in the open air in the heat of ſummer, provided they are placed in a warm ſituation, but if the ſeaſon ſhould prove cold, they will not thrive abroad; wherefore it will be better to let them remain in the ſtove, and open the glaſſes in front, and at the top of the ſtove, every day, to admit as much air as poſſible in hot weather, with which management they will thrive much better than in the open air.

The twenty-third ſort grows naturally at Buenos Ayres; this riſes with a woody ſtalk ten or twelve feet high, covered with a purpliſh bark, almoſt ſmooth. At the top it divides into ſeveral erect branches, garniſhed with wedge-ſhaped leaves, which are ſinuated. The flowers are produced in umbels at the end of the branches; they are large, white, and the petal is angular, but not divided at the brim; theſe are often ſucceeded by berries, which change yellow when they are ripe.

The twenty-fourth ſort grows naturally in the Bahama Iſlands; this riſes with a ſmooth ſhrubby ſtalk ſix or eight feet high, covered with a brown bark, dividing into many branches, garniſhed with ſpear-ſhaped leaves, ſinuated on their edges, ending in acute points; they are ſmooth, of a light green colour. The flowers are produced in ſmall umbels from the ſide of the ſtalks, ſtanding erect; they are large, white, and have their petals cut into five ſtar-pointed ſegments; theſe are very rarely ſucceeded by ſeeds in England.

The two laſt mentioned ſorts are not ſo tender as the former, ſo may be treated in the ſame way as the eleventh and twelfth, by houſing them in winter with Oranges and other green-houſe plants, and in ſummer place them abroad in a ſheltered ſituation; they may be propagated by cuttings, which, if planted in a ſhady border during any of the ſummer months, will take root pretty freely, and may then be taken up and potted, placing them in the ſhade till they have taken new root, and then they may be treated in the ſame way as the old plants.

The twenty-fifth ſort grows naturally on the coaſt of Guinea; this has a ſhrubby ſtalk, which riſes ſeven or eight feet high, dividing into many branches. The lower leaves are oblong, oval, ſmooth, of a dark green colour, and ſtand upon ſhort foot-ſtalks; the flowers come out from the ſide of the branches in ſmall bunches, ſtanding upon very ſlender foot-ſtalks; they are of the ſame ſhape and colour with thoſe of the Amomum Plinii, but ſmaller, and are ſometimes ſucceeded by berries about the ſize of ſmall black Cherries, which are yellow when ripe.

This ſort requires a ſtove in winter, and muſt not be expoſed abroad longer than ten or twelve weeks in the warmeſt part of ſummer; it may be propagated by cuttings, which, when planted, muſt be cloſely covered with a bell or hand-glaſs, and ſhaded from the ſun, treating them in the ſame manner as other cuttings of exotic plants.

The twenty-ſixth ſort grows naturally at the Cape of Good Hope; this has ſhrubby flexible plants, requiring ſupport like our common woody Nightſhade, to which the plant has great reſemblance, but the leaves are ſhorter, thicker, and are more indented on their edges. This ſort very rarely flowers in England.

It may be easily propagated by cuttings during any of the summer months, and may be preserved in a green-house in winter, treating it in the same way as the Amomum Plinii.

The twenty-seventh sort grows naturally at Campeachy; this rises with a woody stalk ten or twelve feet high, sending out many branches, having a light gray bark. The leaves are spear-shaped, of a deep green on their upper side, but hoary on their under. The flowers are produced in large umbels at the end of the branches; they are small, star-pointed, and white; their summits, which fill up the mouth of the tube, are purple; these are succeeded by small berries the size of middling Pease, which are yellow when ripe.

The twenty-eighth sort grows naturally at Carthagena; this rises with an herbaceous stalk, which divides into several irregular branches, garnished with large leaves, of a dark green on their upper side, but woolly on their under. The flowers are produced in large erect umbels at the end of the branches, which are white, and are succeeded by round berries the size of small Cherries, sitting in the blunt woolly empalement of the flower, which turn yellow when ripe.

The twenty-ninth sort grows naturally at Carthagena; this rises with a strong shrubby stalk twelve or fourteen feet high. The branches are woody, of a dark brown colour, armed with a few short recurved spines; the leaves are oblong, oval, sinuated on their edges, smooth, of a dark green on their upper side, but their under sides are hairy, of a light green. The flowers come out from the side of the stalk in small umbels; they are white, and the petal is cut into five acute segments almost to the bottom. This has not produced fruit here.

The thirtieth sort grows in the West-Indies, and also at the Cape of Good Hope. The stalk is shrubby, and rises three feet high, dividing into many ligneous branches, closely armed with short, strong, yellow, recurved spines. The leaves are sinuated, and armed with short crooked spines along their midrib. The flowers are produced in long loose bunches from the side of the stalks; they are white, star-pointed, and are succeeded by berries the size of small black Cherries, of a gold colour when ripe. This sort is propagated by seeds, and may be kept in a warm green-house in winter, and in summer placed in the open air.

The thirty-first sort is very common in the islands of the West-Indies, where it is titled Bachelors Pear: this has a prickly herbaceous stalk three or four feet high, dividing into a few branches, closely covered with a hairy down, and armed with short, recurved, brown spines. The leaves are divided into lobes, covered with soft hairs, and armed on both sides with crooked spines. The flowers come out from the side of the branches in small bunches; they are large, of a pale blue colour, and are succeeded by fruit about the size and shape of a Catherine Pear; but the stalk is fixed to the large end, so the fruit becomes inverted. This is of a gold colour when ripe.

These are plants which require to be raised early in the spring upon a hot-bed, and should be treated in the same way as the fifteenth and sixteenth sorts.

This is propagated by seeds, which should be sown upon a hot-bed in the spring, and when the plants come up fit to remove, they must be planted upon a fresh hot-bed to bring them forward, and afterward treated in the way as the tender sorts of Capsicums.

SOLDANELLA. Tourn. Inst. R. H. 82. tab. 16. Soldanel.

The CHARACTERS are,

The flower has an erect permanent empalement, cut into five parts; it has one bell-shaped petal. The brim is cut into acute segments; it has five awl-shaped stamina, terminated by single summits, and a roundish germen, supporting a slender style the length of the petal, which is permanent, crowned by an obtuse stigma. The germen afterward turns to an oblong taper capsule of one cell, obliquely streaked, opening at the top with ten indentures, filled with small acute-pointed seeds.

We have but one SPECIES of this genus, viz.

SOLDANELLA (*Alpina.*) Hort. Cliff. 49. Soldanel.

This plant grows naturally on the Alps, and other mountains in Germany. The root is fibrous and perennial; the leaves are almost kidney-shaped, of a dark green colour, and stand upon long foot-stalks. Between these arise the foot-stalks of the flower, which is naked, about four inches long, sustaining at the top two small, open, bell-shaped flowers, whose brim is cut into many fine segments like a fringe; the most frequent colour of the flower is blue, but it is sometimes found with a snow-white flower. After the flower is past, the germen becomes an oval capsule, with the style coming out at the top, filled with very small acute-pointed seeds.

There is another variety of this, whose leaves are less round.

The best method to propagate these plants is by parting of their roots, because their seeds do not succeed, unless they are perfectly ripe, and well nourished; and this rarely happens in England. Nor do the seeds, which are brought from abroad, succeed, for they seldom grow unless they are sown soon after they are ripe.

The season for transplanting and parting these roots is in September, that they may have time to make good roots before winter; for if they are removed in the spring, they never flower very strong.

The soil in which these plants thrive best, is a strong cool loam; they must have a shady situation, for if they are exposed to the sun, they will not live, nor will they thrive in a warm light soil. In dry weather these plants should be frequently watered, which will cause them to flower strongly, and make a good increase.

SOLIDAGO. Lin. Gen. Plant. 859. Golden-rod, or Woundwort.

The CHARACTERS are,

It has a compound flower made up of hermaphrodite florets, and female half florets inclosed in one oblong imbricated empalement. The hermaphrodite flowers, which compose the disk, are funnel-shaped, cut into five points at the brim; they have five very short hair-like stamina, terminated by cylindrical summits, and a crowned germen, supporting a slender style as long as the stamina, crowned by a bifid open stigma. The germen afterward turns to a single seed, crowned with hairy down. The female half florets are tongue-shaped, indented in three parts; these have a crowned germen with a slender style, crowned by two revolving stigmas, and are succeeded by a single seed like the hermaphrodite florets.

The SPECIES are,

1. SOLIDAGO (*Latifolia*) caule erecto, foliis lanceolatis serratis paniculis corymbosis, lateralibus terminalibusque. *Golden-rod with an erect angular stalk, flowers in clusters upon upright panicles, and the lower leaves spear-shaped.*

2. SOLIDAGO (*Vulgaris*) caule subflexuoso, racemis paniculatis erectis confertis. Lin. Sp. 880. *Golden-rod with an erect panicled stalk, the lower leaves indented like saws, but those on the stalks almost entire.*

3. SOLIDAGO (*Angustifolia*) foliis lineari-lanceolatis subintegerrimis, floribus confertis alaribus sessilibus. *Golden-rod with narrow spear-shaped leaves almost entire, and flowers in clusters sitting close to the stalks.*

4. SOLIDAGO (*Minor*) caule paniculato foliis radicalibus ovatis dentatis, caulinis integerrimis. *Welsh Golden-rod with*

with narrow, spear-shaped, hoary leaves, and the stalk terminated by a corymbus of flowers.

5. SOLIDAGO (*Minuta*) caule erecto, foliis lanceolatis acutis serratis caule corymboso. *Golden-rod with an erect stalk, spear-shaped leaves sharply sawed, and flowers in a corymbus.*

6. SOLIDAGO (*Canadensis*) paniculato-corymbosa racemus recurvatis floribus adscendentibus, foliis trinerviis subserratis scabris. Hort. Upsal. 259. *Golden-rod with a panicled corymbus, a recurved racemus, and rough sawed leaves with three veins.*

7. SOLIDAGO (*Uniflora*) foliis lanceolatis subserratis acuminatis, pedunculis lateralibus unifloris. *Golden-rod with spear-shaped, sharp-pointed, sawed leaves, and lateral foot-stalks with one flower.*

8. SOLIDAGO (*Altissima*) paniculato-corymbosa, racemis recurvatis, floribus adscendentibus, foliis enerviis subintegerrimis. Hort. Upsal. 259. *Golden-rod with a panicled corymbus, a recurved racemus, and leaves almost entire, without veins.*

9. SOLIDAGO (*Pilosa*) caule piloso, foliis lanceolatis scabris sessilibus racemus recurvatis alaribus floribus pedunculatis. *Golden-rod with a panicled corymbus, a recurved racemus, and rough veined leaves.*

10. SOLIDAGO (*Marylandica*) paniculato-corymboso, racemus obtusis patulis, foliis nervosis scabris subintegerrimis. *Woundwort with a corymbus panicle, obtuse spreading spikes, and rough veined leaves.*

11. SOLIDAGO (*Virginiana*) paniculato-corymbosa, racemis longissimis recurvatis, pedunculis foliosis, foliis lanceolatis serratis scabris. *Woundwort with a corymbus panicle, long recurved spikes, leafy foot-stalks, and rough spear-shaped leaves.*

12. SOLIDAGO (*Scrophularifolia*) caule flexuosa, foliis ovatis acuminatis serratis, racemis lateralibus simplicibus. Flor. Leyd. 161. *Woundwort with a flexible stalk, oval acute leaves, and single spikes of flowers.*

13. SOLIDAGO (*Flexicaulis*) caule flexuoso glabro, foliis ovato-lanceolatis glabris dentatis, racemis brevioribus lateralibus simplicibus. *Woundwort with a smooth flexible stalk, oval, spear-shaped, indented leaves, and single spikes of flowers.*

14. SOLIDAGO (*Hirsutissima*) paniculato-corymbosa, racemis recurvatis, caulibus erectis hirsutissimis, foliis lanceolatis serratis trinerviis subtus tomentosis. *Golden-rod with a panicled corymbus, recurved spikes of flowers, a hairy stalk, and spear-shaped sawed leaves, hoary on their under side.*

15. SOLIDAGO (*Humilis*) paniculato-corymbosa, racemis compositis recurvatis, foliis ovato-lanceolatis subdentatis sessilibus. *Golden-rod with a panicled corymbus, recurved spikes of flowers, and spear-shaped indented leaves sitting close to the stalk.*

16. SOLIDAGO (*Rigida*) caule paniculato, foliis inferioribus ovatis dentatis superioribus lanceolatis amplexicaulibus. Lin. Sp. Pl. 878. *Golden-rod with oval rough leaves on the stalk, alternate branches, and bundled spikes of flowers terminating the branches.*

17. SOLIDAGO (*Mexicana*) caule obliquo, pedunculis erectis foliatis ramosis, foliis lanceolatis integerrimis. Hort. Cliff. *Golden-rod with an oblique stalk, erect foot-stalks to the flowers, and entire spear-shaped leaves.*

18. SOLIDAGO (*Fistulosa*) caule piloso ramoso, racemis paniculatis erectis confertis, foliis hirsutis integerrimis. *Golden-rod with a hairy branching stalk, erect paniculated corymbus, a smooth stalk, and very long, entire, smooth leaves.*

19. SOLIDAGO (*Carnosa*) foliis lanceolatis subcarnosis glaberrimis margine scabriusculis, panicula corymbosa. Lin. Sp. Pl. 878. *Golden-rod with spear-shaped, fleshy, very smooth leaves, with rough edges, and a panicled corymbus of flowers.*

20. SOLIDAGO (*Glabra*) foliis lanceolato-linearibus subcarnosis glaberrimis, panicula corymbosa. *Golden-rod with rough, entire, spear-shaped leaves.*

21. SOLIDAGO (*Cæsia*) panicula corymbosa, racemis supra densioribus, caule glabro lævi. Lin. Sp. Pl. 879. *Golden-rod with a panicled corymbus, the upper part of the spikes closer set with flowers, and a smooth stalk.*

22. SOLIDAGO (*Integerrima*) caule paniculato, racemis brevioribus confertis, foliis linearibus glabris integerrimis. *Golden-rod with a panicled stalk, clustered spikes of flowers, the stalks smooth, and sitting close.*

23. SOLIDAGO (*Rugosa*) caule paniculato, racemis lateralibus simplicibus pedunculis foliatis, foliis lanceolatis scabris integerrimis. *Woundwort with a paniculated stalk, single spikes of flowers with leafy foot-stalks.*

24. SOLIDAGO (*Linearia*) caule paniculato, pedunculis erectis, foliis linearibus glabris integerrimis sessilibus. *Golden-rod with a panicled stalk, erect foot-stalks to the flowers, and smooth, narrow, entire leaves.*

25. SOLIDAGO (*Alba*) caule paniculato, racemis erectis, pedunculis foliatis inferioribus ovatis serratis. *Golden-rod with a panicled stalk, erect spikes with flowers in clusters, and spear-shaped, rough, sawed leaves.*

26. SOLIDAGO (*Obtusifolia*) caule paniculato, racemis sparsis, pedunculis erectis, foliis inferioribus lanceolatis serratis caulinis obtusis integerrimis sessilibus. *Golden-rod with a panicled stalk, the spikes of flowers thinly disposed, the foot-stalks erect, the lower leaves spear-shaped and sawed, but those on the stalks obtuse, entire, sitting close.*

The three first sorts grow naturally in England, yet have not been well distinguished by any botanic writer; for in all the books which treat of the English plants, they are put down as one sort, to which they have applied a title of Caspar Bauhin, viz. Virga aurea latifolia serrata, which is a very different plant from either of our English sorts. But the third sort here mentioned, I believe to be what Caspar Bauhin has titled Virga aurea angustifolia minus serrata. As neither of these three English sorts, nor the Welsh sort, are propagated in gardens, so it is needless to trouble the reader with their description.

The sixth sort was first brought from Canada, but has been since found growing naturally in many other parts of North America. The stalks of this rise three feet high, garnished with narrow, acute-pointed, entire leaves, a little rough on their surface, sitting close to the stalk, which is terminated by a close panicle of yellow flowers, making a goodly appearance.

The seventh sort grows naturally on the Alps. The stalks seldom rise more than four or five inches high, garnished with small spear-shaped leaves, sitting close; and at their base the foot-stalk of the flower comes out, which is an inch long, sustaining one yellow flower at the top, so makes but little appearance. This is kept for variety in botanic gardens.

The eighth sort grows naturally in several parts of North America. The stalks of this rise higher than those of the sixth sort; the leaves are broader, without veins, and stand closer upon the stalks; the racemi of the panicles are much longer and more reflexed. The flowers are specious, and come later in the year.

The ninth sort has pretty strong hairy stalks about four feet high, closely garnished with rough, veined, entire leaves, without foot-stalks. The panicle of flowers is very compact, and the racemi are short.

The tenth sort has flexible stalks about three feet high, garnished with oval-pointed leaves, deeply sawed on their edges, standing on short foot-stalks. The racemi of flowers, which are for the most part simple, are produced from the wings of the stalk, which is also terminated by a thick spike.

The eleventh ſort grows naturally in Virginia. The ſtalks are near two feet high, garniſhed with narrow leaves, and moſt part of the ſtalk is adorned with flowers, ſtanding on long foot-ſtalks, which proceed from the wings. The whole plant is very clammy, and ſeldom lives more than two or three years. It is propagated by ſeeds, and will thrive in the full ground, if it has a light dry ſoil and a ſheltered ſituation.

The twelfth ſort has a great reſemblance of the tenth, but the leaves of the twelfth are ſmoother, more unequally ſawed, and their foot-ſtalks are a little longer. The racemi of flowers are longer, and are reflexed.

The thirteenth ſort has ſmooth flexible ſtalks three or four feet high, garniſhed with ſmooth, oval, ſpear-ſhaped leaves, ſawed on their edges. The racemi of flowers are ſhort, and come out on the ſide of the ſtalk.

The fourteenth ſort has hairy ſtalks two feet high, cloſely garniſhed with long ſpear-ſhaped leaves, ſawed on their edges, and very hairy, downy on their under ſide. The ſtalks are terminated by a corymbus of flowers, compoſed of ſeveral reflexed racemi.

The fifteenth ſort has rough hairy ſtalks two feet high, cloſely garniſhed with long ſpear-ſhaped leaves, a little indented on their edges; the ſtalks are terminated by a large corymbus of flowers, compoſed of many long reflexed racemi.

The ſixteenth ſort has rough channelled ſtalks from two to three feet high, garniſhed with large, rough, oval leaves, and are terminated by bunches of yellow flowers, forming almoſt an umbel. The bottom leaves of this ſort are long and ſpear-ſhaped, ſo differ much from thoſe on the ſtalk.

The ſeventeenth ſort was brought from Mexico, where it grows naturally. The ſtalks of this are ſmooth, one foot and a half high, garniſhed with ſmooth, ſpear-ſhaped, entire leaves: the flowers come out on the ſide of the ſtalk, forming a ſmall corymbus at the end of the foot-ſtalk. This is not ſo hardy as the other ſorts, ſo ſhould be planted in a warmer ſituation.

The eighteenth ſort has hairy ſtalks ſix or ſeven feet high, cloſely garniſhed with ſmooth ſpear-ſhaped leaves, ſitting cloſe, and terminated by a compact corymbus of flowers, ranged on ſhort racemi. The lower leaves are very long and ſmooth.

The nineteenth ſort riſes with thick, ſucculent, ſmooth ſtalks five feet high, garniſhed with fleſhy, ſmooth, ſpear-ſhaped leaves, whoſe edges are rough; the flowers terminate the ſtalk in a corymbus; the racemi, which compoſe it, are erect, and below the flowers are cloſely garniſhed with linear ſmooth leaves. The lower leaves of this ſort are very long, ſmooth, and fleſhy.

The twentieth ſort hath ſmooth diffuſed ſtalks three feet high, garniſhed with very narrow ſmooth leaves; the upper part ſends out many long ſide branches, diſpoſed cloſely, which are terminated by ſimple racemi of ſmall yellow flowers, ſtanding erect.

The twenty-firſt ſort ſends out many ſlender hairy ſtalks three feet high, garniſhed with oval, ſpear-ſhaped, rough leaves, which are entire. The upper part of the ſtalk ſends out on each ſide ſingle racemi of bright yellow flowers. This ſort flowers very late in the year.

The twenty-ſecond ſort ſends out at the bottom large oval leaves, ſawed on their edges, whoſe foot-ſtalks are bordered by the appendix of the leaf; the ſtalks are ſlender, ſtiff, and of a purpliſh colour, branching out in looſe racemi of flowers, which are recurved, garniſhed with ſmooth, ſpear-ſhaped, entire leaves.

The twenty-third ſort has ſmooth erect ſtalks two feet high, which are cloſely garniſhed with ſmall, ſpear-ſhaped, ſmooth leaves, ſitting cloſe to the ſtalks, and are entire. The flowers terminate the ſtalk in a cluſtered corymbus, whoſe foot-ſtalks are erect. The lower leaves of this are long, ſpear-ſhaped, ſmooth, and have foot-ſtalks.

The twenty-fourth ſort ſends out ſtrong ſmooth ſtalks two feet high, garniſhed with rough ſpear-ſhaped leaves, indented on their edges; the upper part of the ſtalk divides into many ſlender branches, which are garniſhed with very ſmall leaves, and are terminated by recurved racemi of bright yellow flowers.

The twenty-fifth ſort ſends out ſmooth panicled ſtalks two feet high, garniſhed with ſpear-ſhaped, rough, obtuſe leaves, which are entire, and ſit cloſe to the ſtalk. The flowers terminate the ſtalk in looſe panicles, ſtanding erect.

The twenty-ſixth ſort hath purpliſh ſtalks, which riſe three feet high, and are cloſely garniſhed with rough ſpear-ſhaped leaves, ſlightly ſawed on their edges, ending in acute points. The ſtalks are terminated by erect racemi of flowers, growing in cluſters, of a bright yellow colour.

There are ſeveral other varieties (if not diſtinct ſpecies) of this genus; but it is very difficult to ſettle the ſpecific differences of thoſe now growing in the Engliſh gardens, for of late years there has been a great number of theſe and alſo of Aſters raiſed from ſeeds, which have been ſent from North America, from whence moſt of the ſorts here mentioned originally came. But as the ſeeds have been gathered by perſons little acquainted with the ſcience of botany, ſo they have generally been ſent mixed together, which, when ſown, the plants have riſen promiſcuouſly. So that in order to aſcertain the ſpecies, their ſeeds ſhould be ſaved very carefully and diſtinctly ſown, to ſee if the plants ariſing from each do retain their difference.

Theſe plants are hardy, ſo will thrive in almoſt any ſituation in this country; and as they flower in autumn, when there is a ſcarcity of other flowers, ſo they are proper furniture for large gardens; for they do moſt of them propagate and ſpread ſo much at their roots, as to require more room than can be well ſpared in ſmall gardens. But as they do not require much care in their cultivation, ſo they are fit for wood-walks, and to intermix with ſhrubs, where, when they are properly diſpoſed, they will be very ornamental.

Some of the ſorts begin to flower in the middle of July, which are ſucceeded by others till the end of November; and in favourable ſeaſons there are two or three ſorts, which frequently continue in flower till Chriſtmas, ſo that for near five months theſe plants will, in ſucceſſion, adorn the garden.

They are eaſily propagated by parting of their roots; the beſt time for doing this is ſoon after the flowers decay; for theſe which are parted in the ſpring, will not be well eſtabliſhed in the ground before they begin to put out their ſtalks, ſo will not flower ſtrong, unleſs the ſummer is wet, or the plants are duly watered, which is difficult to perform in large plantations.

When the roots of theſe ſorts are well fixed in the ground, they may remain five or ſix years without tranſplanting; for if the ground about them is dug every winter, and ſuch of the ſorts as ſpread much at their root are reduced ſo as to keep them within proper limits, they will require no other culture. But in five or ſix years it will be proper to take up the roots, becauſe in that time the middle of the bunch of roots will begin to decay; ſo the offsets ſhould be taken off for to plant, and the old decayed roots thrown away.

The plants may alſo be propagated by ſeed; but theſe ſhould be ſown ſoon after they are ripe, for then they will more certainly grow than if ſown in the ſpring, and the plants will come up the following ſpring; whereas thoſe which

which are ſown in the ſpring, generally remain a year in the ground before the plants appear.

When the plants come up, and have ſtrength enough to be removed, they may be planted in a ſhady border at ſix inches diſtance, where they ſhould remain till the following autumn, when they ſhould be tranſplanted to the places where they are deſigned to remain, and the ſummer following they will flower.

SONCHUS. Sowthiſtle.

Theſe are many of them weeds in England, ſo are not planted in gardens; for if their ſeeds are once permitted to ſcatter upon the ground, they will ſoon ſtock it with plants; for which reaſon they ſhould always be extirpated, not only thoſe in the garden, but alſo thoſe in the parts near it; becauſe their ſeeds being furniſhed with down, are wafted in the air to a conſiderable diſtance, where, falling on the ground, they ſoon come up, and prove troubleſome weeds.

SOPHORA. Lin. Gen. Plant. 456.

The CHARACTERS are,

The flower hath a ſhort bell-ſhaped empalement, cut at the brim into five obtuſe ſegments. The flower is of the butterfly kind; the ſtandard is oblong, broad, and reflexed on the ſides. It has two oblong wings with appendages to their baſe; the keel is of two leaves, like thoſe of the wings, whoſe lower borders join like the keel of a boat. It has ten diſtinct ſtamina, which are awl-ſhaped, parallel, and the length of the petals hid in the keel, terminated by ſmall ſummits, and a taper oblong germen, ſupporting a ſtyle the length of the ſtamina, crowned by an obtuſe ſtigma. The germen afterward turns to a long ſlender pod, with ſwellings where each ſeed is poſited.

The SPECIES are,

1. SOPHORA (*Alopecuroides*) foliis pinnatis, foliolis numeroſis villoſis oblongis. Lin. Sp. Pl. 373. *Sophora with winged leaves, having a great number of oblong hairy lobes.*

2. SOPHORA (*Tomentoſa*) foliis pinnatis, foliolis numeroſis ſubrotundis. Lin. Sp. Pl. 373. *Sophora with winged leaves, compoſed of many roundiſh lobes.*

3. SOPHORA (*Tinctoria*) foliis ternatis ſubſeſſilibus, foliolis ſubrotundis glabris. Lin. Sp. Pl. 373. *Sophora with trifoliate leaves ſitting almoſt cloſe to the ſtalks, whoſe lobes are roundiſh and ſmooth.*

The firſt ſort grows naturally in the Levant; this has a perennial creeping root, from which ariſe ſeveral erect ſtalks from three to four feet high, garniſhed with winged leaves, compoſed of a great number of oblong hairy lobes, ranged by pairs along the midrib, terminated by an odd one. The flowers come out from the wings of the ſtalk in long ſpikes, which ſtand erect cloſe to the ſtalk; they are of a pale blue colour, and ſmall. Theſe appear in July, but are rarely ſucceeded by pods in England.

It propagates faſt enough by its creeping root, in the ſame manner as Liquorice, when the plant is once obtained, and is very hardy, ſo ſhould be planted in ſome corner of the garden, at a diſtance from other plants, becauſe the roots of this plant will ſpread, and mix with thoſe of the neighbouring plants, and ſoon over-bear them. It will thrive in almoſt any ſoil or ſituation, for I have frequently ſeen the roots ſpread into the middle of gravel-walks, and ſend up ſtalks.

The ſecond ſort grows naturally in the iſland of Ceylon, and alſo in the Weſt-Indies, but particularly at Jamaica, where the inhabitants call it Sea-ſide Pigeon Pea; this riſes with a downy ſtalk to the height of ſix or ſeven feet, garniſhed with winged leaves, compoſed of five or ſix pair of roundiſh woolly lobes, terminated by an odd one. The flowers come out in ſhort looſe ſpikes from the wings of the ſtalks; they are large and yellow, not much unlike theſe of Spaniſh Broom, but have no ſcent; theſe are ſucceeded by taper woolly pods five or ſix inches long, having five or ſix large ſwellings, in each of which is contained one roundiſh brown ſeed as large as Peaſe.

This plant is tender, ſo will not thrive in England out of a ſtove; it is propagated by ſeeds, which may be eaſily procured from the Weſt-Indies, for the plants do not perfect them in England; theſe ſhould be ſown in pots, and plunged into a good hot-bed, where, if the ſeeds are good, the plants will appear in a month or ſix weeks. When theſe are fit to remove, they ſhould be each tranſplanted into a ſeparate pot, and plunged again into a hot-bed of tanners bark, obſerving to ſhade them from the ſun till they have taken new root; after which they muſt be treated in the ſame way as other tender plants from the ſame countries, always keeping them in the bark-bed in the ſtove, and in the winter they ſhould have but little water.

The third ſort grows naturally in Virginia and Philadelphia; from both theſe places I have received the ſeeds; from this plant there was formerly a coarſe ſort of Indigo made in America, as there was from ſome other plants, before the true Indigo plants were introduced there: this has a perennial root, from which ariſe ſeveral ſtalks about a foot high, ſending out from the bottom a great number of ſmall branches, garniſhed with leaves, compoſed of three oval ſmooth lobes, joined together at the foot-ſtalk like other trifoliate leaves; they ſit cloſe to the branches. The flowers come out toward the end of the branches in ſhort ſpikes; they are of the butterfly kind, yellow, and appear in July; they are often ſucceeded by ſhort ſwelling pods, which in warm ſeaſons come to maturity in England. The ſtalks of this decay to the root in autumn.

This is propagated by ſeeds, which ſhould be ſown on a warm border the beginning of April. The beſt way is to ſow them in ſhallow drills for the more conveniently keeping the plants clean, for they muſt not be removed till the ſtalks decay in autumn, when they ſhould be carefully taken up, and planted in a warm border, where they are deſigned to remain.

SORBUS. Tourn. Inſt. R. H. 633. The Service-tree.

The CHARACTERS are,

The flower has a ſpreading, concave, permanent empalement, indented in five parts; it has five roundiſh concave petals, which are inſerted in the empalement, and about twenty awl-ſhaped ſtamina, which are alſo inſerted in the empalement, terminated by roundiſh ſummits. The germen is ſituated under the flower, ſupporting three ſlender ſtyles, crowned by erect headed ſtigmas; it afterward becomes a ſoft umbilicated fruit, incloſing three or four oblong cartilaginous ſeeds.

The SPECIES are,

1. SORBUS (*Aucuparia*) foliis pinnatis, utrinque glabris. Hall. Helv. 350. *Service-tree with winged leaves, which are ſmooth on both ſides; called Quickbeam, Mountain Aſh, and in the north Roan-tree.*

2. SORBUS (*Domeſtica*) foliis pinnatis, ſubtus tomentoſis. Hall. Helv. 351. *Service-tree with winged leaves, which are woolly on their under ſide; or the cultivated Service.*

The firſt ſort grows naturally in many parts of England, but in the ſouthern counties they are ſeldom ſeen of any great magnitude, for the trees are commonly cut down, and reduced to underwood; but in the north of England and Wales, where they are permitted to grow, there are trees of a very large ſize. The ſtems are covered with a ſmooth gray bark; the branches, while young, have a purpliſh brown bark; the leaves are winged; they are compoſed of eight or nine pair of long narrow lobes, terminated by an odd one, which are ſharply ſawed on their edges; the leaves on the young trees in the ſpring are hoary on their under ſide, which about Midſummer goes off, and thoſe upon the older branches have very little at any ſeaſon. The flowers are produced in large bunches almoſt in form of umbels at the

end

end of the branches; they are composed of five spreading concave petals, shaped like those of the Pear-tree, but smaller; these appear in May, and are succeeded by roundish berries, growing in large bunches, which have a depressed navel on the top, and turn red in autumn when they ripen.

This tree is cultivated in the nursery-gardens, and sold as a flowering shrub; if they were permitted to grow, they would rise to a great height, and have large stems. The leaves of this tree make a pleasant variety when they are mixed with others, during the time of their flowering, and also in autumn. When their fruit is ripe, they make a pretty appearance, but the blackbirds and thrushes are so fond of this fruit, as to devour it as soon as it ripens; so that in those places where there is a plenty of these birds, there will not be any of the fruit left to be perfectly ripe; however, as it is good food for these songsters, where people have a desire of drawing a number of these birds about their habitations, they should plant a quantity of these trees for that purpose.

The second sort grows naturally in the warmer parts of Europe, where it rises to a great height, and becomes a large tree, but in England there are few of any large size. In the south of France and in Italy, the fruit is served up to the table in their deserts, but in England they have not been much esteemed, which has occasioned their being so little cultivated here. There are several varieties of this fruit, which differ from each other in size and shape, as Apples and Pears do; some of these are shaped like Catherine Pears, and are nearly as large; others are depressed at both ends, and shaped like Apples; but both these sorts will arise from seeds of the same tree; so that those, who are desirous of having the largest and best kinds, should propagate them by grafting or budding from those trees whose fruit are the fairest and best flavoured, as is practised for other fruits; these may be grafted or budded upon Pear-stocks, which agree better with this tree than any other except their own, for they will not take upon Apple-stocks, nor do they thrive upon the Hawthorn or Medlar near so well, though the fruit of this tree approaches nearer to those than any other, and are not fit for the table till they are in a state of decay.

The several varieties of this tree differ in the number of their seeds, in the same manner as Pears, Apples, Quinces, and Medlars, some of them having but three seeds in each fruit, and others have four or five; so that, although one of the characters of this genus is, that the fruit has but three seeds, yet that must be understood to be of the wild sort, in which there are seldom more, but those of the cultivated kind are as uncertain as the fruit of Apples and Pears.

In Italy these trees are very common, where they have a great variety of sorts, which have been obtained from seeds; but I have not observed in the English gardens more than three, and those are yet but scarce, for there are at present but few large trees of the true Service in England, one of which was lately growing in the gardens formerly belonging to John Tradescant at South Lambeth, near Vauxhall in Surry, who was a very curious collector of rare plants in King Charles the Second's time; which tree was near forty feet high, and produced a great quantity of fruit annually, which were shaped like Pears; and there are indeed some trees of middling growth in the gardens of Henry Marsh, Esq; at Hammersmith, which produce fruit of the Apple shape (from whence several young plants have been raised of late in the nurseries near London;) but these are small, compared to that in John Tradescant's garden.

There are great numbers of large trees of this Service growing wild about Aubigny in France, from whence his grace the late Duke of Richmond brought a great quantity of the fruit, and from the seeds raised a great number of young plants in his garden at Goodwood in Sussex.

The leaves of this tree differ from those of the first, in their lobes being broader, and not so much sawed; they are also much more downy on their under side, and the young shoots of the tree are covered with a white down. The flowers are produced in larger and more diffused bunches, and are a little larger; but there are seldom more than two or three fruit produced upon each bunch. The stamina of the flowers are also longer than those of the wild sort, which are the only differences I can observe between them.

Both these sorts may be propagated by sowing their seeds in pots soon after the fruit is ripe, sheltering them under a common frame in winter, and plunging the pots into a moderate hot-bed in the spring, which will soon bring up the plants; when they are come up, they should be carefully kept clear from weeds, and in dry weather watered; but they should be soon exposed to the open air, for the only reason of putting them into a hot-bed, is to forward the growth of the seeds; but if, when the plants are come up, the bed is kept covered, it will draw the plants, and spoil them. In this bed the plants should remain until the middle of October, at which time their leaves will decay, when there should be a warm light spot of ground prepared to receive them, into which they should be planted in rows two feet asunder, and a foot distant in the rows, observing to take them up carefully, and to plant them as soon as possible, that their roots may not dry.

During the summer, the ground should be kept constantly clear from weeds, and in winter there should be a little mulch laid upon the surface of the ground about their roots, to protect them from being injured by frost, but in the spring the ground between them should be dug, burying the mulch therein; in doing of which, you must be careful not to cut or injure the roots of the plants.

In this nursery they may continue three or four years, according to their growth, when it will be proper to transplant them out where they are to remain; the best season for which is in October, or in the spring, just before they begin to shoot. The soil should be warm in which they are planted, and the situation defended from cold winds, in which place they will thrive, and produce fruit in a few years.

Those who raise many of these trees from seeds, will procure some varieties of the fruit, from which the best may be selected, and propagated for the table, and the others may be planted for variety in wildernesses or wood-walks, or may be used for stocks to graft the better kinds upon.

The wood of the wild Service-tree is much commended by the wheelwright for being all heart; and it is of great use for husbandmens tools, goads, &c. It is very white and smooth, so will polish pretty well.

There is a sort of this with variegated leaves, which is preserved by such as are curious in collecting the several sorts of striped plants, but there is no great beauty in it; it may be propagated by layers, or by being budded on the plain sort; but they become plain on a very rich soil.

These trees should have a moist strong soil, but will grow in the most exposed places, being extremely hardy, which renders them worthy of care, since they will thrive where few other trees will succeed.

SORREL. See Acetosa.

SOUTHERNWOOD. See Abrotanum.

SOWBREAD. See Cyclamen.

SPARTIUM. Lin. Gen. Plant. 765. The Broom-tree.

The Characters are,

The empalement of the flower is heart-shaped. The flower is of the butterfly kind; the standard is almost heart-shaped, large, and

and wholly reflexed; the wings are oblong, shorter than the standard, and annexed to the stamina; the keel is oblong, longer than the wings; the borders are hairy, and connected together, to which the stamina are inserted. It has ten unequal stamina, nine of which are joined together, and the under stands apart, with an oblong hairy germen, supporting a rising awl-shaped style, to which is fastened an oblong, hairy, inflexed stigma. The germen afterward becomes a long cylindrical obtuse pod, of one cell, opening with two valves, including several globular kidney-shaped seeds.

The Species are,

1. Spartium (*Junceum*) ramis oppositis teretibus apice floriferis, foliis lanceolatis. Hort. Cliff. 356. *Commonly called Spanish Broom.*

2. Spartium (*Radiatum*) sessilibus petiolis persistentibus, ramis oppositis angulans, foliis ternatis linearibus. Lin. Sp. Pl. 998. *Radiated starry Broom.*

3. Spartium (*Monospermum*) ramis angulatis, racemis lateralibus, foliis lanceolatis. Lin. Sp. 995. *The white Spanish Broom.*

4. Spartium (*Scoparium*) foliis ternatis solitariisque, ramis inermibus angulosis. Hort. Cliff. 356. *Common green Broom with a yellow flower.*

5. Spartium (*Lusitanicum*) foliis ternatis, foliolis cuneiformibus, ramis inermibus angulatis. *Portugal Broom with a large flower.*

6. Spartium (*Hirsutum*) foliis ternatis petiolatis foliolis lineari-lanceolatis hirsutis, ramis inermibus angulatis. *Broom with trifoliate leaves upon foot-stalks, linear spear-shaped lobes which are hairy, and angular armed branches.*

7. Spartium (*Glabrum*) foliis ternatis glabris sessilibus, ramis inermibus angulatis, leguminibus glabris. *Broom with trifoliate smooth leaves sitting close to the branches, which are angular and unarmed, and smooth pods.*

8. Spartium (*Angulatum*) foliis solitariis ternatisque, ramis sexangularibus apice floriferis. Lin. Sp. Plant. 709. *Eastern Broom with round, smooth, compressed pods.*

9. Spartium (*Spinosum*) foliis ternatis, ramis angulatis spinosis. Hort. Cliff. 356. *Broom with trifoliate leaves, and angular prickly branches; commonly called prickly Cytisus.*

10. Spartium (*Arborescens*) caule arborescente ramoso aculeato, foliis cuneiformibus confertis, floribus solitariis alaribus. *Prickly Broom with Purslane leaves; or Ebo. of the West-Indies.*

The first sort is the common Spanish Broom, which has been long cultivated in the English gardens for the sweetness of its flowers; of this there are two varieties, if not distinct species, which grow naturally in Spain and Portugal. The first, which is the common sort in England, has larger branches, and broader leaves than the other. The flowers are also larger, of a deeper yellow colour, and appear earlier than those of the other, which has been of late years introduced from Portugal.

Both these sorts have smooth flexible branches, which rise eight or ten feet high. The lower part of the branches are garnished with small, spear-shaped, smooth leaves; the flowers are disposed in a loose spike, terminating the branches; they are large, yellow, of the butterfly kind, have a strong agreeable odour, appear in July, and in cool seasons there is frequently a succession of flowers till September, which are succeeded by compressed pods, containing one row of kidney-shaped seeds, which ripen in autumn.

These plants are easily propagated by seeds, which should be sown in the spring upon a bed of common earth in a shady situation, where the plants will rise very freely; these must be kept clean from weeds the following summer, and in autumn they may be taken up and transplanted in a nursery, which should be chosen in a warm sheltered situation. In the taking up of the plants, there should be care taken not to tear the roots, for these send their roots deep into the ground, and are very apt to be torn if they are not raised out of the ground with a spade; they should be planted in rows three feet asunder, and at one foot distance in the rows. In this nursery they may remain a year or two to get strength, and then may be planted where they are to remain, for they do not succeed if they are removed large.

If the seeds of these sorts are permitted to scatter in autumn, the plants will come up in plenty in the spring without care, which may be transplanted the following autumn, and treated in the same way as those before mentioned. These shrubs are very ornamental to large wood-walks in gardens, but hares and rabbits are very fond of them; so that, unless they are screened from these animals, they will devour them in winter when they have a scarcity of other food.

The second sort grows naturally in Italy; this is a shrub of low growth, seldom rising more than three feet high, but divides into many spreading branches, so as to form a large bush. The branches are small, angular, and come out opposite; the leaves are very narrow, awl-shaped, and are placed round the stalk, spreading out like the points of a star; the flowers are disposed in small clusters at the end of the branches; they are yellow, but not more than half the size of those of the former, and have no scent; they are succeeded by short hairy pods, containing two or three small kidney-shaped seeds in each. This shrub makes a pretty appearance during the time of its continuing in flower, and, as it is hardy, deserves a place in gardens.

It is propagated by seeds, which should be sown in autumn, for those which are sown in the spring seldom grow the same year; these may be sown in a bed of common earth in rows, for the more conveniently keeping the plants clean from weeds. The plants should remain in the seed-bed till the following autumn, when they may be either transplanted to the places where they are to remain, or in a nursery to grow a year or two to get strength, before they are planted out for good; but these plants will not bear transplanting when they are large, so should be removed while they are young.

The third sort has a thick stalk, covered with a rugged bark when old; it rises eight or nine feet high, sending out many slender Rush-like branches of a silvery colour, almost taper, which have a few narrow spear-shaped leaves. The flowers are produced in very short spikes or clusters on the side of the branches; they are small, white, and are succeeded by large oval pods, containing one kidney-shaped seed.

These two sorts grow plentifully in Spain and Portugal, from both which countries the seeds may be easily procured. The seeds should be sown in the middle of April upon a bed of fresh light earth; but the best way will be to sow them in drills about half an inch deep. The drills should not be less than one foot asunder, and the seeds may be laid in the drills at about three inches distance, which will allow room for the plants to grow the first summer, for it will not be safe to remove them till the spring following. Although I have here directed the sowing of these seeds in April, yet it must be understood, if the season proves favourable, otherwise it will be better to defer it longer, for these seeds are as subject to perish in the ground by cold or wet, as are the Kidney-beans: therefore, when the season is favourable for sowing them, the seeds of the Broom may be safely sown.

But at Michaelmas some of the plants of each kind may be taken up and potted, to be sheltered in winter, for while they are young, they are in danger of suffering by frost; so these should be placed under a common hot-bed frame,

frame, to ſcreen them from cold, but in mild weather they muſt be expoſed daily to the open air.

Thoſe plants which are left in the ſeed-bed, may be ſheltered with mats, and ſome mulch laid about their roots to prevent the froſt penetrating the ground; in the ſpring theſe may be tranſplanted in a warm ſituation, where they will do very well; but it is always neceſſary to have a plant or two of each ſort in pots, that they may be ſheltered in winter to preſerve the ſorts.

The fourth ſort is the common Broom, which grows naturally in England, ſo is not often admitted into gardens, though, when it is in flower, it makes a much better appearance than many others which are coſtly; this riſes with a flexible ſtalk four or five feet high, ſending out many Ruſh-like angular branches. The lower part of the branches are garniſhed with trifoliate leaves, but upward they are ſingle. The flowers come out upon ſhort foot-ſtalks ſingly on the ſide of the branches toward the top; theſe are large, of the butterfly kind, and of a bright yellow colour. The flowers and branches of this ſort are uſed in medicine.

The fifth ſort grows naturally in Portugal and Spain; this has ſtronger ſtalks than our common Broom. The branches grow more erect, and have deeper angles; the leaves are all trifoliate, and much larger than thoſe of the fourth, and the lobes are wedge-ſhaped; the flowers are larger, of a deeper yellow colour, and have longer foot-ſtalks. This is not ſo hardy as the laſt.

The ſixth ſort grows naturally in Portugal; this riſes with a ſtrong ſtalk like the former. The branches are angular, and grow erect; they are better furniſhed with leaves than either of the other ſorts, which ſtand upon pretty long foot-ſtalks; the lobes are ſmall, very narrow, and hairy; the flowers grow cloſer together, are larger, and of a deep yellow colour.

The ſeventh ſort was brought from Portugal. The ſtalks and branches of this are ſlender, angular, and ſmooth, and are fully garniſhed with very narrow, trifoliate, ſmooth leaves, ſitting cloſe to the ſtalks. The flowers come out in long looſe ſpikes at the end of the branches; they are large, of a bright yellow colour, and are ſucceeded by ſhort compreſſed pods, which are ſmooth, containing ſmall kidney-ſhaped ſeeds. This is by Tournefort made a Cytiſus.

The eighth ſort grows naturally in the Levant; this has ſlender ſtalks and branches, garniſhed with a few trifoliate, and ſometimes ſingle leaves toward the bottom. The branches have ſix angles or furrows; the flowers are ſmall, of a pale yellow colour, and are produced in looſe ſpikes at the end of the branches, but are rarely ſucceeded by ſeeds in England.

The ninth ſort grows naturally in Italy and Spain near the ſea-coaſt. The ſtalks riſe five or ſix feet high, ſending out many angular flexible branches, armed with long ſpines, upon which grow trifoliate leaves; the flowers are produced at the end of the branches in cluſters, each ſtanding upon a long foot-ſtalk; they are of a bright yellow colour, and are ſucceeded by ſhort ligneous pods, with a thick border on their upper edges, containing three or four kidney-ſhaped ſeeds. This plant will not live abroad in England, unleſs it has a very warm ſituation.

Theſe plants are raiſed from ſeeds in the ſame way as the firſt ſort, and may be treated in the ſame manner.

The tenth ſort is very common in Jamaica, and ſeveral other places in the Weſt-Indies, where the wood is cut, and ſent to England under the title of Ebony, though it is not the true Ebony, which is a native of the eaſtern country, and is a plant of a very different genus. The wood of this American Ebony is of a fine greeniſh brown colour, and poliſhes very well, ſo is much covuted by the inſtrument makers, and is uſed for ſeveral purpoſes, being of a very hard durable nature.

This tree has a pretty thick ſtem, which riſes twelve or fourteen feet high, covered with a rugged brown bark, dividing into ſpreading branches, which grow almoſt horizontal, and are armed with ſhort, brown, crooked ſpines. The leaves are ſmall, ſtiff, and wedge-ſhaped, coming out in cluſters, and ſit cloſe to the branches. The flowers come out upon ſlender foot-ſtalks from the ſide of the branches ſingly; they are of the butterfly kind, of a bright yellow colour, and are ſucceeded by compreſſed moon-ſhaped pods, which incloſe one kidney-ſhaped ſeed.

This plant is propagated by ſeeds, which muſt be procured from the countries of its natural growth. Theſe ſhould be ſown in pots, filled with light freſh earth, early in the ſpring, and plunged into a good hot-bed of tanners bark. In about ſix weeks after the plants will appear, when they muſt be carefully treated (being very tender while young;) they muſt have freſh air admitted to them every day, when the weather is warm, and ſhould be frequently refreſhed with water, when the earth in the pots appears dry. In about five or ſix weeks after the plants appear, they will be fit to tranſplant, when they ſhould be carefully ſhaken out of the pots, and ſeparated, planting each into a ſmall pot, and plunged into the hot-bed again, being careful to ſhade them from the ſun every day until they have taken root; after which time they muſt be treated in the ſame manner as other tender exotic plants, by giving them air every day in warm weather, and watering them once in two or three days gently, and when the nights are cold, to cover the glaſſes. In this hot-bed the plants may remain till autumn, when they muſt be removed into the ſtove, and plunged into the bark-bed. Thoſe of them whoſe roots have filled the pots, ſhould be carefully ſhifted into pots one ſize larger, before they are plunged; but as theſe plants are not of quick growth while young, they do not require to be often ſhifted out of the pots. During the winter ſeaſon theſe plants muſt be kept warm (eſpecially the firſt year) and muſt have but little water; and in cold weather it muſt be given to them in ſmall quantities. As theſe plants are very tender, they will not live in the open air in this country, even in the warmeſt part of the year; therefore they muſt be conſtantly kept in the ſtove, and ſhould be plunged in the bark-bed, at leaſt for the two firſt years, obſerving in the ſummer ſeaſon, when the weather is warm, to admit a large ſhare of freſh air to the plants. With this management the plants will thrive very well, and in a few years will produce their flowers, when they will make a pretty appearance in the ſtove. The ſeeds of this tree are called Organ-berries.

SPERGULA. Dillen. Gen. Nov. 7. Lin. Gen. Pl. 519. Spurrey.

The CHARACTERS are,

The flower has a ſpreading permanent empalement. It has five oval, concave, ſpreading petals, which are larger than the empalement, and ten awl-ſhaped ſtamina ſhorter than the petals, terminated by roundiſh ſummits. It has an oval germen, ſupporting five ſlender, erect, reflexed ſtyles, crowned by thick ſtigmas. The germen afterward turns to an oval cloſe capſule with one cell, opening with five valves, incloſing many depreſſed, globular, bordered ſeeds.

The SPECIES are,

1. SPERGULA (*Arvenſis*) foliis verticillatis, floribus decandris. Hort. Cliff. 173. *Spurrey with leaves in whorls, and flowers with ten ſtamina.*

2. SPERGULA (*Pentandria*) foliis verticillatis, floribus pentandris. Lin. Sp. Pl. 440. *Spurrey with whorled leaves, and flowers with five ſtamina.*

3. Spergula (*Nodosa*) foliis oppositis subulatis lævibus, caulibus simplicibus. Lin. Sp. Pl. 440. *Spurrey with awl shaped smooth leaves placed opposite, and single stalks.*

There are some other species of this genus, which grow naturally as weeds in England, so are not worthy notice here; nor should I have mentioned these, were they not sometimes cultivated.

The first and second sorts are cultivated in Holland and Flanders for feeding their cattle; the usual time for sowing the seeds is in July or August, that the plants may acquire strength before the winter's cold. The use that is made of this is to feed sheep, and other cattle, in winter and spring, when the common Grass fails. This plant seldom rises above six inches high, so will not afford a very great quantity of food; but as it will grow on the poorest land, it may be cultivated in many places to good advantage, where no other Grass will thrive so well, and by feeding it off the ground, the dung of the cattle will improve the land. This pasture, it is affirmed, will make excellent butter, and the mutton fed on it is said to be well tasted, so is by many preferred to that fed on Turneps. Hens will greedily eat this herb, and it makes them lay more eggs.

This plant being annual, must be sown every year; and whoever is willing to save the seeds, should sow it in April, that the plants may flower the beginning of July, and the seeds will ripen in August; when it must be cut before the heads are quite brown, otherwise the seeds will soon scatter.

The seeds being very small, about twelve pounds will be sufficient to sow an acre of land. The ground should be well harrowed before the seeds are sown, for if the larger clods are not broken, there will be an uneven crop of Grass. People in the low country sow this seed after a crop of Corn is taken off the land. The second sort is now much cultivated in Flanders, though it is a much lower plant than the common sort, but they esteem it a much better Grass. The seeds of this kind are smaller and flatter than those of the common sort, and have a white border round each.

SPERMACOCE. Dill. Hort. Elth. 277. Lin. Gen. Plant. 111. Button-weed.

The Characters are,

The flower has a permanent empalement indented in four parts, sitting on the germen. It has one cylindrical petal, whose tube is longer than the empalement, and the brim indented in four parts. It has four awl-shaped stamina shorter than the petal, terminated by single summits, and a roundish compressed germen, situated under the flower, supporting a single style, divided in two parts at the top, crowned by obtuse stigmas. The germen afterward turns to two oblong seeds, which are joined, having two horns, and are convex on one side, and plain on the other.

The Species are,

1. Spermacoce (*Tenuior*) glabro, staminibus inclusis. Lin. Sp. Pl. 102. *Smooth Spermacoce with stamina included in the flower.*

2. Spermacoce (*Verticillata*) glabra, staminibus extantibus. Lin. Sp. Pl. 102. *Smooth Spermacoce with stamina standing out of the flowers.*

The first sort grows to the height of two feet and a half; the stalks are stiff, a little angular, and covered with a brown bark; the branches and leaves come out by pairs; the leaves are smooth, and have one strong vein or midrib. The flowers grow in slender whorls toward the top of the stalks; they are small, white, and sit close to the stalks, having a whorl of small leaves close under them; these are succeeded by two oblong seeds, having small horns, which ripen in the empalement.

The second sort rises with a shrubby stalk three or four feet high, sending out a few slender branches, which are garnished with narrow leaves not so long as those of the former; they are smooth, of a light green, and stand in a kind of whorls round the stalk. The flowers grow in thick globular whorls toward the top of the stalk, one of which terminates the stalk; they are small, very white, and funnel-shaped. The brim is cut into four obtuse segments, which spread open, and the stamina stand out above the tube of the flower. After the flowers are past, the germen turns to two seeds, shaped like those of the former sort.

These plants grow naturally in moist places in Jamaica; the inhabitants call the second sort Button-weed. They are both propagated by seeds, which must be sown upon a hot-bed; and when the plants come up, they must be transplanted on a fresh hot-bed to bring them forward, and afterward treated in the same way as other tender plants, and, if they are placed in a stove, they will live through the winter, and produce good seeds the following year.

SPHÆRANTHUS. Vaill. Act. Par. 1719. Globe-flower.

The Characters are,

The flowers are composed of hermaphrodite florets, and female half florets, which are included in one globular scaly empalement, garnished with them on every side the receptacle. There are several of these florets included in each partial empalement. The hermaphrodite florets are placed in the center; they are funnel-shaped, and cut into five parts at the brim, having five very short hair-like stamina, terminated by cylindrical summits, and a germen supporting a thick style, having a single stigma; these are barren. The female half florets are situated round the border, and have scarce any petals, but an oblong germen supporting a bristly style, crowned by a double stigma; these have one oblong naked seed.

The Species are,

1. Sphæranthus (*Indicus*) pedunculis crispatis. Lin. Sp. 1314. *Globe-flower with curled foot-stalks.*

2. Sphæranthus (*Africanus*) pedunculis lævibus. *Globe-flower with smooth foot-stalks.*

The first sort grows naturally in India; this rises with an herbaceous stalk about a foot high, which rarely branches out, garnished with spear-shaped leaves, whose base sits close to the stalk, and from them is extended a leafy border or wing along the stalk; they are sawed on their edges, and are of a deep green, standing alternate. The foot-stalks of the flowers come out from the side of the stalk, opposite to the leaf; they are about two inches long, and sustain one globular head of flowers at the top, of a purplish red colour; these are succeeded by oblong seeds, situated on the margin, which are naked.

The second sort grows naturally at Madrass, and also at La Vera Cruz in New Spain, where it was discovered by the late Dr. Houstoun. This rises with an herbaceous winged stalk about ten inches high, garnished with oval, spear-shaped, sawed leaves, placed alternately. The upper part of the stalk branches out into small divisions, which are terminated by foot-stalks, sustaining three or four globular flowers, of a pale yellow colour.

These are both annual plants, which require a hot-bed to bring them forward in the spring; and if the summer proves cold, they must be kept in a glass-case, otherwise they will not ripen seeds here.

SPHONDYLIUM. Tourn. Inst. R. H. 219 tab. 170. Cow Parsnep.

The Characters are,

It has an umbellated flower. The principal umbel is composed of many smaller, which are flat; the involucrum of the general umbel has many leaves which decay; the particular ones have from three to seven leaves; the general umbel is difformed. The flowers in the middle or disk have five equal, crooked, inflexed petals,

petals. Those of the rays are unequal. They have five stamina, which are longer than the petals, terminated by small summits, and an oval germen, situated under the flower, supporting two short styles, crowned by single stigmas. The germen afterward turns to an elliptical compressed fruit, furrowed on each side, containing two compressed leafy seeds.

The Species are,

1. Sphondylium (*Vulgare*) foliolis latioribus pinnatifidis, radiis umbellis maximarum. *Common hairy Cow Parsnep.*

2. Sphondylium (*Angustifolium*) foliolis angustioribus pinnatifidis serratis, radiis umbellarum minoribus. *Hairy Cow Parsnep with narrow leaves.*

3. Sphondylium (*Orientale*) foliolis pinnatifidis obtusis, petiolis hispidis, radiis umbellarum inæqualibus. *Greatest eastern Cow Parsnep.*

4. Sphondylium (*Alpinum*) foliis pinnatis utrinque scabris, floribus radiatis. *Small Alpine Cow Parsnep.*

5. Sphondylium (*Glabrum*) foliis simplicibus glabris floribus radiatis. *Smooth Alpine Cow Parsnep.*

The first sort grows naturally by the side of brooks, ditches, and in moist meadows in many parts of England. The root of this plant is taper, fleshy, and shoots deep in the ground. The lower leaves are large and winged, composed of three pair of large lobes, placed along the midrib, terminated by an odd one; the lobes are also cut into two or three pair of wings almost to the midrib, terminated by an odd one. The foot-stalks of the leaves are very hairy; the leaves are of a deep green on their upper side, but are pale on their under, and are rough to the touch; the stalks are garnished at each joint with one leaf, of the same shape with those at bottom, but smaller, whose base embrace the stalks. The flowers terminate the stalks in large umbels, which are composed of about twenty-two partial umbels, every third having longer foot-stalks than the others. The partial umbels have many large flowers, which are barren, and compose the rays; those of the disk or middle are smaller and fruitful, and are each succeeded by two flat bordered seeds.

The second sort grows naturally in moist meadows near Battersea in Surry, and in other parts of England; this has been supposed only a seminal variety of the first, but I have cultivated it in the garden near thirty years, and have always found the plants which were raised from seeds kept their difference. The leaves of this sort are composed of two or three pair of narrow lobes, terminated by an odd one; the wings of the lobes are very narrow, acute-pointed, and are cut almost to the midrib; they are hairy, and of a lighter green than those of the former; the umbels are much smaller, as are also the flowers and seeds, in which it greatly differs from the former.

The third sort grows naturally in the Levant, and also in Siberia. The leaves of this are very broad; their foot-stalks are armed with prickly hairs, and are deeply channelled on their upper side; they are composed of two or three pair of very broad, smooth, obtuse lobes, terminated by an odd one, of a yellowish green colour. The stalks rise eight or ten feet high; they are channelled, and sustain umbels of flowers at the top, which are smaller than those of either of the two former sorts, and the flowers are yellow.

The fourth sort grows naturally on the Alps. The stalks of this do not rise more than one foot and a half high; the leaves are divided to the midrib; the lobes are a little cut on their edges, are of a deep green, and rough on both sides; the umbels of flowers are small and white.

The fifth sort grows naturally on the Alps and Apennines. The stalks of this rise almost three feet high; the leaves of this are smooth, divided into three lobes, but not very deep, and are indented about the edges. The stalks are terminated by small umbels of white flowers, which are succeeded by small, compressed, bordered seeds.

These are all very hardy plants, which may be propagated by seeds; the best time for sowing them is in autumn, soon after they are ripe. They should be sown where the plants are designed to remain, because they send forth tap-roots somewhat like those of the Parsnep, therefore do not thrive so well when transplanted; as if suffered to remain where they are sown. The plants grow very large, therefore the seeds should be sown in drills, at two feet and a half, or three feet distance; and in the spring, when the plants appear, they should be thinned, so as to leave them at least eighteen inches asunder in the rows; after which they will require no farther care but to keep them clear from weeds; and when the plants have obtained strength, they will not easily be injured by weeds, for they will overbear them, and prevent their getting up. The second year these plants will produce flowers and seeds, and their roots soon after die; if their seeds are permitted to scatter, they will fill the neighbouring ground, and become troublesome weeds.

SPIGELIA. Lin. Gen. Plant. 192. Arapabaca. Plum. Nov. Gen. 10. tab. 31. Worm-grass.

The Characters are,

The flower has a permanent empalement of one leaf, which is cut into five acute points; it has one funnel-shaped petal, whose tube is longer than the empalement, cut into five points at the brim, which spread open. It has five stamina, terminated by single summits, and a germen, composed of two globular lobes, supporting one awl-shaped style the length of the tube, crowned by a single stigma. The germen afterward becomes two globular seed-vessels, which are joined, sitting in the empalement.

The Species are,

1. Spigelia (*Anthelmia*) caule erecto, foliis quaternis sessilibus, spicis terminalibus. *Worm-seed with an erect stalk, and leaves growing by fours sitting close to the stalk, which are terminated by spikes of flowers.*

2. Spigelia (*Lonicera*) foliis oppositis ovato-oblongis acuminatis sessilibus, spicis terminalibus. *Worm-grass with oblong acute-pointed leaves growing opposite, sitting close to the stalks, which are terminated by spikes of flowers.*

The first sort grows naturally in moist places in most of the islands of the West-Indies. It is an annual plant with a fibrous root, from which arises a strong, erect, herbaceous stalk near a foot and a half high, which is channelled, and sends out two side branches opposite near the bottom, and a little above the middle is garnished with four oblong, oval, acute-pointed leaves, placed in form of a cross round the stalk, and at the same joint comes out two small side branches opposite; these, and also the principal stalk, have four smaller leaves near the top, sitting round in the same manner as the other; and from these arise short spikes of herbaceous flowers, ranged on one side the foot-stalk. These are succeeded by roundish twin capsules, which contain the seeds.

This plant is esteemed the most efficacious medicine for the worms yet known, and has been long used by the inhabitants of the Brasils as such, and also by the negroes, who taught the inhabitants of the British islands in America the use of it, where it has had great success; and from thence had the appellation of Worm-grass given to it.

It is too tender to thrive in the open air in England, so the seeds should be sown in pots, filled with soft loamy earth in the autumn, and plunged into the bark-bed in the stove, where they should remain till the spring, when they should be plunged into a fresh hot-bed, which will bring up the plants; these must be afterward planted into separate pots, and plunged into another hot-bed, and shaded till they have taken new root; after which they must be treated

in

in the ſame way as other tender annual plants from the ſame countries, keeping them conſtantly in the hot-bed under cover, otherwiſe they will not perfect their ſeeds in England.

The ſecond ſort grows naturally in Carolina, where the inhabitants call it Indian Pink. This has a perennial fibrous root, from which ariſe two or three erect herbaceous ſtalks about ſeven or eight inches high, garniſhed with three or four pair of oval, oblong, acute-pointed, ſmooth leaves, placed oppoſite, ſitting pretty cloſe to the ſtalk. The ſtalk is terminated by a ſhort ſpike of flowers, which are ranged on one ſide; they have ſhort empalements, which are cut into five acute ſegments. The tube of the flower is long, narrow at the bottom, ſwelling upward much larger, and is cut at the brim into five acute ſegments, which ſpread open flat; the outſide of the flower is of a bright red, and the inſide of a deep Orange colour. The ſeeds of this ſort never ripen here.

This plant is uſed in Carolina for the ſame purpoſes as the other in the Weſt-Indies, and is eſteemed the beſt medicine there known for the worms. A particular account of the virtues of this plant is mentioned in the firſt volume of the Philoſophical Eſſays, printed at Edinburgh, communicated by Dr. Garden of Carolina.

This is not eaſily propagated in England, for the roots make but ſlow increaſe, ſo that the plant is not very common in the Engliſh gardens at preſent; for although it is ſo hardy as to endure the cold of our ordinary winters in the open air, yet, as it does not ripen ſeeds, the only way of propagating it is by parting of the roots; and as theſe do not make much increaſe by offſets, ſo the plants are ſcarce. It delights in a moiſt ſoil, and muſt not be often tranſplanted.

SPINA ALBA. See Mespilus.

SPINACIA. Tourn. Inſt. R. H. 533. tab. 308. Spinach, or Spinage.

The Characters are,

It is male and female in different plants; the male flowers have an empalement cut into five oblong, obtuſe, concave ſegments, but no petals, with five hair like ſtamina longer than the empalement, terminated by oblong twin ſummits; theſe are barren. The female flowers have permanent empalements of one leaf, cut into four points, two of which are very ſmall; they have no petals, but a compreſſed roundiſh germen, ſupporting four hair-like ſtyles, crowned by ſingle ſtigmas. The germen afterward turns to a roundiſh ſeed, which is ſhut up in the empalement; in ſome ſpecies they are almoſt ſmooth, and in others they have two or three ſharp thorns.

The Species are,

1. Spinacia (*Oleracea*) foliis ſagittatis ſeminibus aculeatis. *Spinach with arrow-pointed leaves and prickly ſeeds; or common prickly Spinach.*

2. Spinacia (*Glabra*) foliis oblongo-ovatis, ſeminibus glabris. *Spinach with oblong oval leaves, and ſmooth ſeeds.*

The firſt ſort was formerly more cultivated in the Engliſh gardens than at preſent, becauſe it is much hardier, ſo not in much danger from cold, therefore is generally cultivated for uſe in winter. The leaves of this are triangular, and ſhaped like the point of an arrow; the ſtalks are hollow, branching, and herbaceous; they riſe about two feet high. The male flowers are produced in long ſpikes; they are herbaceous, having no petals, but each has five ſlender ſtamina, terminated by oblong twin ſummits, filled with a yellowiſh farina, which, when ripe, flies out on the plants being ſhaken, and ſpreads all round; theſe plants, after their farina is ſhed, ſoon decay. The female flowers, which are upon ſeparate plants, ſit in cluſters cloſe to the ſtalks at every joint; they are ſmall, herbaceous, and have neither ſtamina or petals, but have roundiſh compreſſed germen, which afterward turns to roundiſh ſeeds, armed with ſhort acute ſpines.

There are two or three varieties of this now cultivated in the kitchen-gardens, which differ in the ſize and ſhape of their leaves, and their ſeeds being more or leſs prickly.

The ſeeds of the firſt kind ſhould be ſown upon an open ſpot of ground in Auguſt, obſerving, if poſſible, to do it when there is an appearance of rain; for, if the ſeaſon ſhould prove dry for a long time after the ſeed is ſown, the plants will not come up regularly; part of them may come up ſoon, and a great part of them may remain till rain falls, before they come up, which, if that ſhould not happen in a little time after, many times there will not be half a crop. When the Spinach is come up pretty ſtrong, the ground ſhould be hoed to deſtroy the weeds, and alſo to cut up the plants where they are too cloſe, leaving the remaining plants about three or four inches aſunder; but this ſhould always be done in dry weather, that the weeds may be deſtroyed ſoon after they are cut.

About a month or five weeks after the firſt hoeing, the weeds will begin to grow again; therefore the ground ſhould be then hoed the ſecond time, obſerving, as before, to do it in dry weather. But if the ſeaſon ſhould prove moiſt, it will be proper to gather the weeds up after they are cut, and carry them off the ground; for if the Spinach is not cleaned from weeds before winter, they will grow up, and ſtifle it ſo, that in wet weather the Spinach will rot away.

In the end of October the Spinach will be fit for uſe, when you ſhould only crop off the largeſt outer leaves, leaving thoſe in the center of the plants to grow bigger; and thus you may continue cropping it all the winter and ſpring, until the young Spinach, ſowed in the ſpring, is large enough for uſe, which is commonly in April; at which time the ſpring advancing, the winter Spinach will run up to ſeed; ſo that it ſhould be all cut, leaving only a ſmall parcel to produce ſeeds if wanted.

But the ground in which this winter Spinach is ſown, being commonly planted with early Cabbages, it is not proper to let any of the Spinach remain there for ſeed; therefore it ſhould be cleared off as ſoon as ever the ſpring Spinach is fit for uſe, that the Cabbages may be earthed up, and laid clear, which is of great ſervice to them; wherefore you ſhould ſow a ſmall ſpot of ground with this ſort of Spinach, on purpoſe to ſtand for ſeed, where there ſhould be no other plants among it.

The ſecond ſort differs from the firſt in having oval thick leaves, which are not angular at their baſe; the ſeeds are ſmooth, having no ſpines, and the ſtalks and leaves are much more fleſhy and ſucculent. Of this there are two or three varieties, which differ in the thickneſs and ſize of their leaves, which in one are much rounder and thicker than the other.

Theſe are ſown in the ſpring upon an open ſpot of ground by themſelves, or elſe mixed with Radiſh-ſeed, as is the common practice of the London gardeners, who always endeavour to have as many crops from their land in a ſeaſon as poſſible; but where land is cheap in the country, it will be the better method to ſow it alone without any other ſort of ſeed mixed with it; and when the plants are come up, the ground ſhould be hoed to deſtroy the weeds, and cut out the plants where they are too cloſe, leaving the remaining about three inches aſunder; and when they are grown ſo large as to meet, you may then cut out a part of it for uſe, thinning the plants, that they may have room to ſpread; and this thinning may be twice performed, as there is occaſion for the herb; at the laſt of which, the roots ſhould be left eight or ten inches aſunder; and if then you hoe the ground over again to deſtroy the weeds, it will be of great ſervice to the Spinach, for if the land is good upon which

which it is sown, the sort with broad thick leaves, commonly called Plantain Spinach, will, with this management, many times produce leaves as large as the broad-leaved Dock, and be extremely fine.

But in order to have a succession of Spinach through the season, it will be proper to sow the seed at several different times in the spring; the first in January, which must be on a dry soil; the second the beginning of February, upon a moister soil; the third the beginning of March, which should be on a very moist soil; and the fourth the beginning of April; another in May; but these late sowings should be hoed out thinner at the first time than either of the former, for there will be no necessity to leave it for cutting out thin for use, because the former sowings will be sufficient to supply the table till these are full grown; besides, by leaving it thin at first, it will not be apt to run up to seed so soon as it would, if the plants were close.

These sowings here mentioned, are such as are practised by the kitchen-gardeners near London; but, as this herb is much used in soups, &c. for great tables, there should be some seeds sown every three weeks, during the summer season, to supply the kitchen; but these late sowings should be on moist strong ground, otherwise, if the season proves hot and dry, the Spinach will run to seed before the plants are fit for use, especially if the plants do not stand thin.

In order to save seeds of either of these kinds, you should sow an open rich spot of ground, with the sort you intend, in February, after the danger of frost is over; and when the plants are come up, they should be hoed out to six or eight inches distance, observing to cut down the weeds at the same time; and when the plants have grown about three weeks or a month longer, they should be hoed a second time, when they should be left twelve or fourteen inches asunder at least, for when they have shot out their side branches, they will sufficiently spread over the ground.

You must also observe to keep them clear from weeds, which, if suffered to grow amongst the Spinach, will cause it to run up weak, and greatly injure it. When the plants have run up to flower, you will easily perceive two sorts amongst them, viz. male and female. The male will produce spikes of stamineous flowers, which contain the farina, and are absolutely necessary to impregnate the embryos of the female plants, in order to render the seeds prolific. These male plants are, by the gardeners, commonly called She Spinach, and are often by the ignorant pulled up as soon as they can be distinguished from the female, in order, as they pretend, to give room for the seed-bearing plants to spread; but, from several experiments which I made on these plants, I find, wherever the male plants are entirely removed before the farina is shed over the female plants, the seed will not grow which they produce, so that it is absolutely necessary to leave a few of them in every part of the spot, though there may be a great many drawn out where they are too thick; for a small quantity of male plants (if rightly situated) will be sufficient to impregnate a great number of female, because they greatly abound with the farina, which, when ripe, will spread to a considerable distance, when the plants are shaken by the wind.

When the seeds are ripe (which may be known by their changing their colour, and beginning to shatter,) the plants should be drawn up, and spread abroad for a few days to dry, observing to turn them every other day, that the seeds on both sides may dry equally; you must also guard the seeds from birds, otherwise they will devour them. When it is dry, the seeds should be threshed out, cleaned from the dirt, and laid up for use, where mice cannot come to them, for they are extremely fond of this seed.

SPIRÆA. Tourn. Inst. R. H. 618. tab. 389. Spiræa Frutex.

The Characters are,

The flower has a permanent empalement of one leaf, plain at the base, and cut into five acute segments at the top; it has five roundish oblong petals inserted in the empalement, and twenty or more slender stamina, which are shorter than the petals, and are inserted in the empalement, terminated by roundish summits, and five or more germen, supporting as many slender styles, which are longer than the stamina, crowned by headed stigmas. The germen afterward turns to an oblong, acute-pointed, compressed capsule, opening with two valves, containing a few small acute-pointed seeds.

The Species are,

1. Spiræa (*Salicifolia*) foliis lanceolatis obtusis serratis nudis, floribus duplicato racemosis. Hort. Cliff. 191. *Common Spiræa Frutex.*

2. Spiræa (*Opulifolia*) foliis lobatis serratis, corymbis terminalibus. Lin. Sp. Plant. 489. *Spiræa with lobated sawed leaves, and flowers growing in a corymbus, terminating the stalks; commonly called Virginia Gelder Rose with a Currant leaf.*

3. Spiræa (*Hypericifolia*) foliis obovatis integerrimis, umbellis sessilibus. Hort. Upsal. 131. *Spiræa with entire leaves, and umbels of flowers sitting close to the branches; commonly called Hypericum Frutex.*

4. Spiræa (*Crenata*) foliis oblongiusculis apice serratis, corymbis lateralibus. Lin. Sp. Plant. 489. *Spiræa with oblong leaves, whose points are sawed, and flowers growing in a corymbus on the sides of the branches.*

5. Spiræa (*Tomentosa*) foliis lanceolatis inæqualiter serratis subtus tomentosis, floribus duplicato-racemosis. Lin. Sp. Pl. 480. *Spiræa with spear-shaped leaves, which are unequally sawed, woolly on their under side, and flowers growing in doubly-branching bunches; or Red Spiræa.*

6. Spiræa (*Sorbifolia*) foliis pinnatis, foliolis uniformibus serratis, caule fruticoso, floribus paniculatis. Lin. Sp. Plant. 490. *Spiræa with winged leaves, whose lobes are uniformly sawed, a shrubby stalk, and flowers growing in panicles.*

7. Spiræa (*Frutescens*) foliis lanceolatis supernè serratis nervosis, subtus incanis, floribus racemosis, caule fruticoso. *Spiræa with spear-shaped veined leaves, which are sawed toward their points, and hoary on their under side, flowers growing in long bunches, and a shrubby stalk.*

8. Spiræa (*Paniculatis*) foliis lanceolatis acutè serratis, floribus paniculatis, caule fruticoso. *Spiræa with spear-shaped leaves, which are sharply sawed, flowers growing in panicles, and a shrubby stalk.*

9. Spiræa (*Trifoliata*) foliis ternatis serratis subæqualibus, floribus subpaniculatis. Lin. Sp. Plant. 490. *Spiræa with trifoliate sawed leaves, which are almost equal, and flowers growing in a kind of panicle.*

10. Spiræa (*Filipendula*) foliis pinnatis, foliolis uniformibus serratis, caule herbaceo, floribus cymosis. Lin. Sp. Pl. 480. *Spiræa with winged leaves, having uniform sawed lobes, an herbaceous stalk, and flowers growing on slender foot-stalks at the top; the common Dropwort.*

11. Spiræa (*Ulmaria*) foliis pinnatis, impari majore lobato, floribus cymosis. Flor. Lapp. 201. *Spiræa with winged leaves, whose outer lobe is greater, and divided into lobes, and flowers growing in bunches on weak foot-stalks; Meadow-sweet, or Queen of the Meadows.*

12. Spiræa (*Aruncus*) foliis supra decompositis, spicis paniculatis, floribus divisis. Lin. Sp. Pl. 490. *Spiræa with more than decompounded leaves, paniculated spikes, and male and female flowers.*

The first sort has been long cultivated in the English gardens, but from what country it originally came is not very certain; it is generally sold by the nursery-gardeners with other flowering shrubs; it rises with several shrubby stalks, which

which are very taper, and rough toward the top, covered with a reddiſh bark. The leaves are ſpear-ſhaped, bluntly ſawed on their edges, and of a bright green colour. In rich moiſt ground the ſtalks will riſe ſix or eight feet high, but in moderate land from four to five. The branches are terminated by ſpikes of pale red flowers; the lower part of the ſpikes are branched out into ſmall ſpikes, but the upper parts are cloſe and obtuſe. Each flower is compoſed of five petals, which ſpread open, of a pale red or fleſh colour, and have a great number of ſtamina, ſome of which ſtand out much beyond the petals, but others are not ſo long, terminated by brown headed ſummits; and in the center are ſituated five ſtyles, which are terminated by headed ſtigmas. After the flowers are paſt, the germen turns to pointed capſules, but they rarely come to perfection here.

This is propagated from ſuckers, which are ſent forth in plenty from the ſtems of the old plants, or by laying down the tender branches, which, when rooted, ſhould be tranſplanted out in rows at three feet diſtance, and the plants a foot aſunder in the rows. In this nurſery they may remain two years, obſerving to keep the ground clear from weeds, and in the ſpring to dig up the ground between the rows, ſo that the roots may the more eaſily extend themſelves; but, if they put out ſuckers from their roots, thoſe ſhould be taken off to keep the ſhrubs within bounds, and afterwards they may be tranſplanted where they are to remain, either in ſmall wilderneſs quarters, or in clumps of flowering ſhrubs, obſerving to place them amongſt other ſorts of equal growth. The young ſhoots of this ſhrub, being very tough and pliable, are often uſed for the tops of fiſhing-rods.

The ſecond ſort grows naturally in North America, but is now as common in the Engliſh gardens as the firſt; this riſes with many ſhrubby branching ſtalks, ſometimes eight or ten feet high in good ground, but generally five or ſix, covered with a looſe brown bark, which falls off, and are garniſhed with lobed leaves, about the ſize and ſhape of thoſe of the common Currant Buſh, ending in acute points, ſawed on their edges. The flowers are produced in roundiſh bunches at the end of the branches; they are white, with ſome ſpots of a pale red. This is commonly known in the nurſeries by the title of Virginia Gelder Roſe with a Currant leaf. It may be propagated and managed in the ſame manner as the former, and is equally hardy.

The third ſort came originally from Canada, but is now as common in the nurſery-gardens as either of the former, where it is known by the title of Hypericum Frutex, but has no affinity to St. Johnſwort, and is only ſo called from the reſemblance of their leaves; this riſes with ſeveral ſlender ſhruby ſtalks five or ſix feet high, covered with a dark brown bark, ſending out ſmall ſide branches, garniſhed with ſmall, wedge-ſhaped, entire leaves, which have many punctures on their ſurface like St. Johnſwort. The flowers are diſpoſed in ſmall umbels, which ſit cloſe to the ſtalks, each flower ſtanding upon a long ſlender foot-ſtalk; they are white, compoſed of five roundiſh petals, which ſpread open, and in the center have a great number of ſtamina, almoſt equal in length with the petals. The ſhrubs make a good appearance during the time of their flowering.

This may be propagated by laying down the under branches, which will take root in the compaſs of one year, when they may be taken off, and planted in a nurſery for two or three years (as hath been directed for the former;) after which they may be tranſplanted out where they are deſigned to remain, placing them with the two former, being nearly of the ſame growth, where they will add to the variety.

The fourth ſort grows naturally in Spain; this is not very common at preſent in the Engliſh gardens. The whole appearance of the ſhrub is ſo like the third, as not to be diſtinguiſhed at a ſmall diſtance; the only difference being, that the leaves of this are broader at the point, where they have two or three indentures. The flowers are like thoſe of the former. This may be propagated in the ſame way as the other.

The fifth ſort grows naturally in Philadelphia; this is a ſhrub of lower ſtature than the former. The ſtalks are ſlender, and branch out near the ground; they have a purple bark, covered with a gray meally down. The leaves are ſpear-ſhaped, but ſmaller than thoſe of the firſt ſort, and are unequally ſawed; they are downy, and veined on their under ſide, but are of a bright green above. The branches are terminated by a thick racemus of flowers, which are branched toward the bottom into ſmall ſpikes; the flowers are very ſmall, of a beautiful red colour, and the ſpikes of this are longer than thoſe of the firſt.

The ſixth ſort grows naturally in Siberia upon moiſt land; this is a ſhrub of humble growth in this country, ſeldom riſing more than two feet high, putting out ſome ſide branches, which are covered with a purple bark, and garniſhed with winged leaves, compoſed of three or four pair of oblong, oval, thin lobes, ſawed on their edges. The flowers are produced in panicles at the end of the branches; they are ſmall, white, and of the ſame conſtruction with the former.

The ſeventh ſort grows naturally in North America; this has a ſhrubby ſtalk, which riſes five or ſix feet high, covered with a brown bark, dividing into many ſtrong branches, which are cloſely garniſhed toward their end with ſpear-ſhaped veined leaves, hoary on their under ſide, and are ſawed on their edges toward their points. The flowers are white, and diſpoſed in a racemus.

The eighth ſort grows naturally in North America; this riſes with ſhrubby ſtalks like the firſt, but ſends out horizontal branches which are ſlender, and covered with a brown bark. The leaves are ſpear-ſhaped, of a thin texture, and of a bright green colour on both ſides; they are ſlightly ſawed on their edges, but the ſaws are acute. The flowers are diſpoſed in panicles at the end of the branches; they are ſmall, white, and of the ſame conſtruction of the former.

Theſe ſorts are propagated in the ſame way as the firſt, but, as ſome of them do not put out ſuckers from their roots here in any plenty, their branches ſhould be laid down in autumn, which in one year will take root, and may then be planted where they are deſigned to remain, or into a nurſery, where they may ſtand one or two years to get ſtrength, before they are planted out for good.

The ninth ſort grows naturally in North America; this has a perennial root, but the ſtalks are annual, and riſe about a foot high, ſending out branches from the ſide their whole length; theſe are garniſhed with leaves, which for the moſt part are trifoliate, but are ſometimes ſingle, and at others by pairs, ſharply ſawed on their edges, of a bright green on their upper ſide, and pale on their under. The flowers are diſpoſed in looſe panicles at the top of the ſtalks, ſtanding upon ſlender foot-ſtalks; they have five long ſpear-ſhaped petals, which ſpread open, and a great many ſtamina, which are no longer than the tube of the flower.

It is propagated by ſeeds, which ſhould be ſown on a ſhady border ſoon after they are ripe, for if they are ſown in the ſpring, the plants will not come up till the year after, and many times fail. When the plants appear, they muſt be conſtantly kept clean from weeds; but they ſhould not be removed till autumn, when their leaves begin to decay; then they may be either tranſplanted where they are

deſigned

designed to remain, or into a nursery border, where they may grow a year or two to get strength, before they are planted out for good. This plant loves a shady situation and a moist light soil.

The tenth sort is the common Dropwort, which grows plentifully upon chalky grounds in many parts of England. The roots of this consist of a great number of oval knobs or glandules, which are fastened together by slender fibres, from whence it had the title of Dropwort; the leaves are winged, and composed of many sawed lobes, which are almost placed alternately along the midrib; those near the base are the smallest, the others increase in size to the middle, afterward decrease again to the point. The flower-stalk rises a foot or more in height, and has seldom more than one leaf upon it; the top is garnished with loose bunches of small white flowers, standing upon slender foot-stalks, which are succeeded by several capsules, ranged circularly. The roots of these plants are used in medicine, and are accounted diuretic. It is rarely kept in gardens, but there is a variety of this with double flowers, which was found growing naturally in the north of England, that is kept in gardens for the sake of variety.

The eleventh sort grows naturally on the sides of waters, and in low moist meadows in most parts of England. The stalks are angular, red, and rise three or four feet high, garnished with winged leaves, composed of two or three pair of large indented lobes, terminated by an odd one, which is much larger than the other, and divided into three lobes; they are of a dark green on their upper side, but hoary on their under. The stalks are terminated by large loose bunches of white flowers, which have an agreeable scent, and are succeeded by roundish capsules, twisted like a screw, filled with small seeds.

The leaves and tops of this plant are used in medicine, but the plants are rarely kept in gardens. There is a variety of it with double flowers, which is kept in some gardens, and one with variegated leaves.

The twelfth sort grows naturally upon the mountains in Austria; this has a perennial root and an annual stalk, which rises from three to four feet high, garnished with decompounded winged leaves, composed of several doubly-winged leaves, each having three or four pair of oblong lobes, terminated by an odd one, sawed on their edges, and ending in acute points. The flowers are disposed in long slender spikes, which are formed into loose panicles at the top of the stalks; they are small, white, and of two sexes in the same spike. The seeds rarely ripen here.

This plant is kept in gardens for the sake of variety; it may be propagated by parting of the root in autumn; it loves a moist soil and a shady situation.

SPIRÆA OF AFRICANA. See DIOSMA.

SPONDIAS. The Jamaica Plumb.

The CHARACTERS are,

The flower has five oblong spear-shaped petals, and a small permanent coloured empalement, with ten scaly obtuse nectariums, situated between the petals, and ten bristly stamina shorter than the petals, with oblong summits; the oblong germen, which is immersed in the receptacle, supports five short parallel styles, joined in a five-cornered column, crowned by a simple stigma; and afterward becomes an oblong fleshy berry, inclosing a hard oblong nut, covered with fibres, having five cells.

The SPECIES are,

1. SPONDIAS (*Purpurea*) petiolis communibus compressis. Lin. Sp. 613. *Spondias, whose common foot-stalks are compressed; called Purple Myrabolin Plumb.*

2. SPONDIAS (*Lutea*) foliolis nitidis. Lin. Sp. 613. *Spondias with neat leaves and yellow fruit.*

The first sort is of humble stature, seldom rising more than twelve or fourteen feet high in the West-Indies, but in England is rarely more than half that height; the bark is brown; the leaves are very long, composed of a great number of pinnæ, placed alternate along the midrib, terminated by an odd one; these are sawed on their edges, and end in acute points. The flowers terminate the branches in a racemus, and are of a whitish yellow colour, some of which are succeeded by oblong fleshy berries, of a pale yellow colour, covered with a meally farina; the flesh of which is but thin, of a luscious sweet taste. The nut inclosed appears as if composed of many ligneous fibres.

The fruit of this sort is esteemed by some of the inhabitants of the islands in the West-Indies; but as the flesh is very thin, so a great number of the Plumbs will afford but little meat: however, the wild hogs are very fond of them, and it is their principal food during the season of their ripening.

The second sort is also a native in the warmest parts of America, where it rises to the height of thirty feet, sending out many irregular branches, which are destitute of leaves some months: when the leaves come out, they are unequally winged, having four or six pair of lobes, near two inches long and one broad; the flowers come out before the leaves appear, which are succeeded by yellow Plumbs an inch and a half long; they are disposed in a racemus, and have fibrous stones.

These trees are easily propagated in America, by planting some of the branches, which readily take root in the rainy seasons. The cuttings will also grow in England, if properly managed; but here they are generally raised from the fruit, which are frequently brought from some of the islands in America, which, when fresh, come up very readily. The Leeward Islands generally furnish the best, as their passage is not so long as from Jamaica. The nut should be planted as soon as they arrive in pots, which should be plunged into a hot-bed of tan, where, if the bed is of a good temperature of heat, the plants will appear in a month or five weeks after. When the plants have obtained strength enough to be removed, they should be shaken out of the pots, and carefully parted, planting them in separate small pots, and plunging them into a fresh hot-bed of tan, shading them daily from the sun until they have taken new root; after which they should have air, and be supplied with water in proportion to the warmth of the season; but they are too tender to thrive in the open air in England in the warmest season, therefore should be constantly kept in the bark-stove, and, if carefully managed, will ripen their fruit here. The plants generally drop their leaves in spring, and often remain naked two or three months.

SQUASHES. See PEPO.

SQUILLS. See SCILLA.

STACHYS. Tourn. Inst. R. H. 186. tab. 86. Base Horehound.

The CHARACTERS are,

The flower has a tubulous permanent empalement, cut into five acute parts at the top; it has one lip-shaped petal with a short tube, having oblong chaps. The upper lip is erect, hooked, and a little indented at the point, and cut into three parts. It has four awl-shaped stamina, two of which are longer, and inclined to the upper lip; the other two are shorter, terminated by single summits, and a four-pointed germen, supporting a slender style the length of the stamina, crowned by a bifid acute stigma. The germen afterward turns to four oblong angular seeds, which ripen in the empalement.

The SPECIES are,

1. STACHYS (*Germanica*) verticillis multifloris, foliorum serraturis imbricatis caule lanato. Lin. Sp. 812. *Base Horehound with a woolly stalk, and many flowers in whorls.*

2. STACHYS (*Cretica*) verticillis multifloris, calycibus pungentibus caule hirto. Hort. Upsal. 170. *Base Horehound with*

with many flowers in whorls, prickly empalements, and a hairy ſtalk.

3. STACHYS (*Italica*) foliis lineari-lanceolatis tomentoſis ſubcrenatis, petiolis longiſſimis, caule fruticoſo tomentoſo. *Baſe Horehound with narrow, ſpear-ſhaped, woolly leaves, which are ſomewhat crenated, with very long foot-ſtalks, and have a ſhrubby woolly ſtalk.*

4. STACHYS (*Alba*) foliis oblongo ovatis crenatis piloſis, calycibus pungentibus, labii ſuperiore piloſo. *Baſe Horehound with oblong, oval, crenated, hairy leaves, prickly empalements to the flowers, and the upper lip hairy.*

5. STACHYS (*Alpina*) verticillis multifloris, foliorum ſerratum apice cartilaginis, corollis labio plano. Flor. Suec. 527. *Baſe Horehound with many flowers in whorls, ſawed leaves, and the lip of the flower plain.*

6. STACHYS (*Hiſpanica*) foliis inferioribus ovato-oblongis ſubcrenatis ſubtus tomentoſis, caulinis cordatis acutis ſeſſilibus, calycibus ſpinoſis. *Baſe Horehound with oval, oblong, lower leaves, which are ſlightly crenated, and woolly on their under ſide, thoſe on the ſtalks being heart-ſhaped, acute-pointed, and ſitting cloſe to the ſtalks, and prickly empalements to the flowers.*

7. STACHYS (*Glutinoſa*) ramis ramoſiſſimis, foliis lanceolatis glabris. Hort. Cliff. 310. *Baſe Horehound with very ſpreading branches, and ſmooth ſpear-ſhaped leaves.*

8. STACHYS (*Paluſtre*) verticillis ſexfloris, foliis lineari-lanceolatis ſemi-amplexicaulibus. Flor. Suec. 490. *Baſe Horehound with whorls of ſix flowers, and narrow ſpear-ſhaped leaves, which half embrace the ſtalk.*

The firſt and laſt ſort here mentioned, grow naturally in England; the firſt only in a few particular places, but the latter is common by the ſide of ditches and waters every where, and is here only mentioned, becauſe it is a diſpenſary plant, and has been ſuppoſed a good vulnerary herb. Of this there is another ſpecies, which was found by Mr. Stoneſtreet growing wild, with narrow leaves, ſhorter ſtalks, longer and cloſer ſpikes of flowers, and the leaves ſtand diſtinct upon ſhort foot-ſtalks: this has conſtantly retained its difference in the garden. Both theſe ſorts have creeping roots, ſo will ſoon ſpread over a large ſpot of ground where they have liberty.

The ſeventh ſort grows naturally in Crete; this is a low plant with an herbaceous ſtalk, which is very branchy from the bottom. The ſtalks are ſlender, four-cornered, and ſmooth, garniſhed with a few ſmall ſpear-ſhaped leaves; the whole plant is very clammy, and ſmells like bitumen. The flowers are ſmall, of a dirty white colour, and ſtand in ſmall whorls round the ſtalks; theſe appear in July, and are ſucceeded by roundiſh ſeeds, which ripen in autumn. This is propagated by ſeeds, and requires to be ſheltered under a frame in winter, being too tender to live in the open air here.

The other ſorts are kept in botanic gardens for the ſake of variety, but are not cultivated in other places, ſo it will be needleſs to give a particular deſcription of them here.

They are all propagated by ſeeds, which ſhould be ſown in March upon a bed of light freſh earth; and when the plants are come up, they may be planted out into other beds about ſix inches aſunder, obſerving to water them until they have taken root; after which they will require no farther care but to keep them clear from weeds till Michaelmas, when they ſhould be tranſplanted where they are to remain, which muſt be in an open ſituation, and upon a dry light ſoil, not rich, in which they will endure the winter much better than in good ground. The ſummer following theſe plants will flower, and in Auguſt their ſeeds will ripen, when they may be gathered and preſerved till ſpring for ſowing.

4

STÆHELINA. Lin. Gen. Plant. 844.

The CHARACTERS are,

The common empalement of the flower is oblong, cylindrical, and imbricated; the ſcales are reflexed; the flower is compoſed of ſeveral uniform funnel-ſhaped florets, of one petal. The brim is cut into five equal acute points; they have each five hair-like ſtamina, terminated by cylindrical ſummits, and a ſhort crowned germen, ſupporting a ſlender ſtyle, crowned by a double oblong ſtigma. The germen afterward becomes a ſhort four-cornered ſeed, crowned with a feathery down, which ripens in the empalement.

The SPECIES are,

1. STÆHELINA (*Gnaphalodes*) foliis tomentoſis, ſquamis calycinis lanceolatis. Lin. Sp. Plant. 840. *Stæhelina with woolly leaves, and ſpear-ſhaped ſcales to the empalement.*

2. STÆHELINA (*Dubia*) foliis linearibus denticulatis, ſquamis calycinis lanceolatis, pappo calycibus duplo longioribus. Lin. Sp. 1176. *Stæhelina with linear indented leaves, ſcales to the empalement, and down twice the length.*

The firſt ſort grows naturally at the Cape of Good Hope, from whence it was introduced into the Dutch gardens; this riſes with a ſhrubby ſtalk about three feet high, which divides into ſeveral branches, garniſhed with long taper woolly leaves, ſet thinly. The flowers are produced at the end of the branches in ſingle heads, which are pretty large, and have ſcaly empalements; theſe terminate in ſpines, which are recurved; they are compoſed of ſeveral florets, which are tubulous, hermaphrodite, and of a yellow colour, each of which is ſucceeded by a ſingle four-cornered ſeed, crowned with a feathery down, ripening in the empalement, each being ſeparated by a chaffy ſcale.

The ſecond ſort is a native of the ſame country; this is a low ſhrub, ſeldom riſing more than two feet high, ſending out many ſlender branches, which are garniſhed with leaves placed alternate; there is a ſmall knob or angle juſt under that part where the leaf is inſerted to the branch; the leaves are narrow, and have three blunt angles or corners. The branches are terminated by a ſingle flower, whoſe empalement is oval, and like thoſe of the flowers of Knapweed, being imbricated. The ſcales are oblong, oval, and their points are rounded; ſome of them have a large membranaceous border, whoſe edge is crenated, and ſpread open; the florets are yellow and equal, of the ſame length as the empalements; they are all hermaphrodite, and have a bifid ſtigma, and the ſeeds have a little hairy down on their top.

As theſe plants do not always ripen their ſeeds in England, ſo they are generally propagated by cuttings, which, if planted in any of the ſummer months, and covered cloſe with a bell or hand-glaſs, will take root pretty freely. When theſe have made good roots, they ſhould be taken up carefully, and planted in pots, filled with freſh light earth, not too rich, and placed in the ſhade until they have taken new root; then they ſhould be removed to a ſheltered ſituation, where they may be intermixed with other exotic plants till the autumn, when they muſt be removed into ſhelter, and treated in the ſame way as other plants from the ſame country. Theſe plants do not require any artificial heat in winter, but ſhould have a dry air, for their tender ſhoots are very ſubject to rot with damp; therefore they will thrive better in a glaſs-caſe than a green-houſe in winter.

STAPELIA. Lin. Gen. Plant. 271. Swallow-wort, or Fritillaria craſſa.

The CHARACTERS are,

The flower has a permanent empalement of one leaf, cut into five acute ſegments; it has one large plain petal, cut into five acute ſegments above the middle, and a plain five-pointed ſtarry nectarium, with linear ſegments, whoſe torn points ſurround the parts

parts of generation; it has five plain, broad, erect stamina, with linear summits fastened on each side the stamina, and two oval plain germen, having no style, crowned by a blunt stigma. The germen afterward turns to two oblong taper pods, filled with compressed seeds, crowned with a feathery down, lying over each other like the scales of fish.

The Species are,

1. Stapelia (*Variegata*) denticulis ramorum patentibus. Vir. Cliff. 20. *Stapelia with spreading indentures to the branches; commonly called Fritillaria crassa.*

2. Stapelia (*Hirsuta*) denticulis ramorum erectis. Hort. Cliff. 77. *Stapelia with erect indentures to the branches.*

There is another species of this genus, which has been lately introduced into the English gardens, whose branches are larger, more erect, and the indentures are not so erect as those of the second sort; but, as it has not produced flowers in this country, nor is described in books, so I cannot say more of it, only that it is not so hardy as either of the former, therefore requires a stove in winter to preserve it in England.

There is also a variety of the first sort mentioned in some books, with flat crested branches, and is by some gardeners titled Coxcomb Fritillary; but this is no other than an exuberance of branches joined together, which become flat, so will return back to its original again, therefore is not worthy notice.

The first sort hath many succulent branches arising from the root, which are five or six inches long, having several protuberant indentures on their sides, spreading open horizontally, ending in acute points; the branches, which spread on the ground, emit roots from their joints, so where they have room will extend very wide. The branches abound with a viscous juice of a nauseous taste. From the side of the branches toward their bottom comes out the foot-stalk of the flower at one of the sinuses, and sustains one flower, having a large thick petal cut half way into five points like a star, which spreads open flat, greenish on the outside, but yellow within, having a circle of purple round the nectarii; and the whole petal is finely spotted with purple, resembling the belly of a frog. In the center are the five compressed nectarii, which are prominent, of a livid colour, which include the genital parts. The flower, when blown, has a very fœtid odour, like that of carrion, so like, as that the common flesh-fly deposit their eggs on it, which frequently are hatched, but wanting proper food, die soon after; for I have many years watched the progress of these, to see if the maggots produced from these eggs ever eat any part of the flower, or lived any time; but could never observe either, nor have ever heard that any other person of credit has ever discovered any thing like it; though it has been asserted, by persons of more assurance than knowledge, that they have devoured great part of the petal, and come to maturity, changing afterward into their last state of flies. After the flowers are past, the double germen changes into two taper pods, joined at their base, which are near a span long, and almost as thick as a man's little finger, which are filled with flat seeds, crowned with a feathery down, lying over each other like the scales of fish; but these pods are seldom formed in England, unless the pots in which they grow are plunged in tanners bark; for in upward of forty years which I have cultivated these plants, I never saw them produce their pods but three times, and those were plunged into the tan-bed in the stove.

The branches of the second sort are much larger than those of the first, and stand more erect, but spread and emit roots in the same way; they have four longitudinal furrows, which divide them into four angles, which have protuberant indentures on their edges, whose points are erect; they are nearly of the same colour as those of the first, being of a dark green in summer, but inclining to purple in autumn. The flowers come out upon short foot stalks from the side of the branches; these are of the form with those of the former, but are larger; the petal is of a thicker substance, and on the inside covered with fine purplish soft hairs; the ground of the flower is an herbaceous yellow, streaked and chequered with purplish lines. This sort produces its flowers in much greater plenty than the first sort, so that in summer and autumn they are seldom long destitute of flowers; but I have never seen any of the pods of this sort produced in England.

Both these plants grow naturally upon the rocks near the Cape of Good Hope, where they strike their roots into the crevices of the rocks and spread themselves greatly. They are propagated here very easily, by taking off any of the side branches, during any of the summer months, which, when planted, put out roots very freely. The branches should be slipped off from the plants to the bottom, where they are joined by a small ligature, so will not occasion a great wound, the joints at the place where they are connected being almost closed round; for if they are cut through the branch, the wound will be so great as to occasion their rotting when planted; these should be laid in a dry place under cover for eight or ten days, that the wounded part may dry and heal over before they are planted, otherwise they will rot; then they should be planted in pots, filled with earth, composed of fresh sandy earth, mixed with lime rubbish and sea sand; and if the pots are plunged into a very moderate hot-bed, it will promote their taking root; they should be now and then sprinkled with water, but it must be given them sparingly, and as soon as they have taken root, they must be inured to the open air. If these plants are kept in a very moderate stove in winter, and in summer placed in an airy glass-case, where they may enjoy much free air, but screened from wet and cold, they will thrive and flower very well; for although they will live in the open air in summer, and may be kept through the winter in a good green-house, yet those plants will not flower so well as those managed in the other way. These plants must have little water given them, especially in winter.

STAPHYLÆA. Lin. Gen. Plant. 336. Bladder-nut.

The Characters are,

The empalement is concave, coloured, and so large as to inclose the flower, which has five oblong erect petals, and a pitcher-shaped concave nectarium at the bottom of the flower, with five oblong erect styles, terminated by single summits, and a thick germen, divided in three parts, supporting three styles, to which there are obtuse stigmas contiguous. The germen afterward becomes two hard almost globular seeds, included in three bladders, joined by a longitudinal seam, with an acute point opening within.

The Species are,

1. Staphylæa (*Pinnata*) foliis pinnatis. Hort. Cliff. 112. *Bladder-nut with winged leaves.*

2. Staphylæa (*Trifoliata*) foliis ternatis. Hort. Cliff. 112. *Bladder-nut with trifoliate leaves; or three-leaved Virginia Bladder-nut.*

The first sort grows naturally in woods in several parts of England, but is cultivated as a flowering shrub in the nursery-gardens. This hath several shrubby stalks arising from the same root, which grow ten or twelve feet high, covered with a smooth bark, and divide into several branches, which are pithy, garnished with winged leaves, composed of two pair of oval lobes, terminated by an odd one; these differ greatly in size, according to the strength and vigour of the shrubs. They are smooth, entire, and of a light green colour, standing upon pretty long foot-stalks. The flowers come out upon long slender foot-stalks, which hang downward; these spring from the wings of the stalks near their extremity,

extremity, and are diſpoſed in oblong bunches; they have each five oblong white petals, which expand in form of a Roſe; theſe are ſucceeded by inflated capſules or bladders, compoſed of three cells, one or two of which have a roundiſh, ſmooth, hard ſeed, and the other are barren.

This ſhrub makes a variety when intermixed with others, though their flowers are not very beautiful. The nuts of this tree being hard and ſmooth, are ſtrung for beads by the Roman Catholicks in ſome countries; and the children of the poor inhabitants eat the nuts, though they have a diſagreeable taſte.

The ſecond ſort grows naturally in North America, but is now become as common in the nurſery-gardens about London as the other. This hath a more ſubſtantial ſtalk than the firſt; the bark of the older branches and ſtalks are ſmooth, and of a gray colour; that of the young is of a light green and very ſmooth; the leaves are by threes on each foot-ſtalk; the lobes are oval, ending in a point, and their edges are ſawed; they are of different ſizes, according to the age and ſtrength of the plants; they are ſmooth, and of a light green colour. The flowers are produced from the ſide of the branches, in longer bunches than thoſe of the former ſorts, but their foot-ſtalks are much ſhorter; the flowers are of a cleaner white, and their petals are ſomewhat larger than thoſe of the firſt, as are alſo the bladder capſules; the ſeeds are larger, and ripen better than thoſe of the common ſort.

Both theſe ſorts are uſually propagated by ſuckers from the root, which the firſt ſort ſends out in plenty; theſe ſhould be taken from the old plants in autumn, and their roots trimmed, then planted in a nurſery, in rows at three feet diſtance, and one foot aſunder in the rows. In this nurſery the plants ſhould ſtand one or two years, according to their ſtrength, and then be tranſplanted to the places where they are to remain.

The plants which are propagated in this manner from ſuckers, are very ſubject to put out ſuckers in greater plenty from their roots than thoſe which are raiſed from ſeeds, or propagated by layers or cuttings, ſo are not to be choſen when the other can be had; therefore thoſe who propagate them for their own uſe, ſhould prefer the other methods. If they are propagated by layers, the young branches ſhould be laid down in autumn, in the ſame manner as is practiſed for other trees and ſhrubs; theſe will have put out roots by the following autumn, when they may be taken from the old plants, and planted in a nurſery, where they may grow one or two years to get ſtrength, and then may be removed to the places where they are to ſtand.

When theſe are propagated by cuttings, it ſhould be the ſhoots of the former year, and if they have a ſmall piece of the two years wood at the bottom, they will more certainly ſucceed; for as the young ſhoots are ſoft and pithy, ſo they are very ſubject to rot when they have no part of the old wood to them. They ſhould be planted in autumn on a ſhady border, but muſt not have too much wet.

They may alſo be propagated by ſowing their ſeeds early in autumn, in beds of light freſh earth; and when the plants are come up, they muſt be carefully kept clear from weeds, and, in very dry weather, if they are now and then refreſhed with water, it will greatly promote their growth; in theſe beds they may remain until October following, at which time they ſhould be carefully dug up, and planted in a nurſery, placing them in rows three feet aſunder, and the plants one foot diſtance in the rows; and, if the following ſpring ſhould prove very dry, it will be convenient to give them a little water, to encourage their taking root; after which they will require no farther care but to keep the ground clear from weeds in ſummer, and every ſpring to prune off irregular branches, and dig the ground between the rows to looſen the earth, that their roots may the more eaſily extend. In this nurſery they may remain two years, by which time it will be proper to tranſplant them out where they are to remain, either in wilderneſs quarters, or in clumps of various trees, where they will add to the diverſity. The beſt ſeaſon for tranſplanting theſe trees is in autumn, with other deciduous trees. When theſe ſeeds are ſown in the ſpring, the plants ſeldom come up till the following year.

AFRICAN BLADDER NUT. See ROYENA.

LAUREL-LEAVED AMERICAN BLADDER NUT. See PTELEA.

STAR-FLOWER. See ORNITHOGALUM.

STARWORT. See ASTER.

STATICE. Tourn. Inſt. R. H. 341. tab. 177. Thrift, or Sea Pink.

The CHARACTERS are,

The flowers are collected in a roundiſh head, ſurrounded by a common ſcaly empalement; each flower has alſo a funnel-ſhaped empalement of one leaf; the flowers are funnel-ſhaped; the baſe of the petals are narrow, their points broad, and ſpread open; they have five ſtamina which are ſhorter than the petals, terminated by proſtrate ſummits; and a ſmall germen, ſupporting five ſtyles, which ſtand apart, crowned by acute ſtigmas. The germen afterward turns to one ſmall roundiſh ſeed, incloſed in the empalement.

The SPECIES are,

1. STATICE (*Armeria*) ſcapo ſimplici capitulo, foliis linearibus. Lin. Sp. 394. *Thrift with ſingle ſtalks, linear leaves, and flowers terminating the ſtalks.*

2. STATICE (*Montana*) foliis linearibus ſubulatis, ſquamis calycinis obtuſis. *Thrift with linear awl-ſhaped leaves, and obtuſe ſcales to the empalement.*

3. STATICE (*Maritima*) foliis linearibus planis, ſquamis calycinis obtuſis. *Thrift with plain linear leaves, and obtuſe ſcales to the empalement; or Sea Pink.*

The firſt ſort grows naturally on the Alps, and other cold mountains in ſeveral parts of Europe. This has a perennial fibrous root, from which come out many narrow, ſmooth, ſpear-ſhaped leaves, of a dark green colour, ſitting cloſe over each other at their baſe. The foot-ſtalks of the flowers are naked, and riſe about a foot high, terminated by one globular head, containing ſeveral ſmall, pale, red flowers, which are included in one common ſcaly empalement; immediately under the flower is placed five narrow leaves, which afterward fall off. The flowers are ſucceeded by oblong ſeeds, which are cloſely wrapped up in the particular empalement of the flower. There are two varieties of this, one with white flowers, and the other a bright red.

The ſecond ſort is alſo a native of the Alps, and other cold mountains, where it ſeldom riſes more than two inches high, but when it is planted in gardens it becomes much larger. The roots of this are fibrous and perennial; they divide into heads, which have a great number of narrow Graſs-like leaves, ſitting cloſe round the heads, whoſe baſe embrace the ſtems, and lie over each other. The ſtalks are naked, and riſe about ſix inches high, ſuſtaining on their tops heads of pale purpliſh flowers, incloſed in one common ſcaly empalement, whoſe ſcales are broad, and rounded at their points.

The third ſort grows naturally in ſalt marſhes, where the ſea flows over them frequently, in many parts of England, ſo is very rarely admitted into gardens. The leaves of this ſort are very narrow, ſhort, and plain; the ſtalks ſeldom riſe more than three or four inches high; the heads of flowers are ſmall, and the flowers are of a very pale fleſh colour, ſo make but little appearance.

There was fome years paft another fpecies of this genus in the Englifh gardens which came from Portugal, with a thick perennial ftalk, which by age became fhrubby, and rofe a foot and a half in height; the leaves like thofe of the firft fort, but much larger; the foot-ftalks of the flowers were a foot and a half long, naked, and terminated by one large globular head of flowers, of a pale red colour; but all the plants of this kind which were in England, the fevere froft in the beginning of the year 1740 deftroyed, fince which time I have not feen one plant.

The fecond fort was formerly planted in gardens, to make edgings on the fides of borders in the flower-gardens; for which purpofe they were then in great efteem, but of late they have been very juftly rejected, becaufe there was a neceffity of tranfplanting thefe edgings every year, otherwife they could not be kept within due bounds; befides, wherever a plant failed, which was no extraordinary thing, there always appeared a large unfightly gap; however, though they are not in ufe at prefent for that purpofe, yet a few plants of the firft and fecond fhould have a place in fome part of the flower-garden for variety, efpecially the variety with red flowers; they will grow in almoft any foil or fituation, and their flowers will continue a long time in beauty, efpecially in a fhady fituation.

All thefe forts may be propagated by parting their roots; the beft time for which is in autumn, that they may take root before the froft, for thefe will flower much ftronger than thofe tranfplanted in the fpring; and the plants will not be in fo much danger of mifcarrying, efpecially when the fpring happens to prove dry. After thefe plants have taken root, they will require no farther care but to keep them clear from weeds, and to tranfplant and part their roots annually, for if they are permitted to ftand longer unremoved, they are very fubject to rot and decay, efpecially when they are planted in good ground.

STEWARTIA. Lin. Gen. Plant. 758.

The CHARACTERS are,

The flower has a permanent empalement, cut into five oval concave fegments; it has five large oval petals, which fpread open, and a great number of flender ftamina, joined in a cylinder at bottom, and are fhorter than the petals, to which they are connected at their bafe, terminated by roundifh proftrate fummits, with a roundifh hairy germen, fupporting five ftyles the length of the ftamina, crowned by obtufe ftigmas. The germen afterward turns to a five-cornered capfule with five cells, opening with five valves, whofe cells are clofed, each containing one oval compreffed feed.

We have but one SPECIES of this plant, viz.

STEWARTIA (*Malacodendron.*) Act. Upfal. 1741. *Stewartia.*

This fhrub grows naturally in Virginia, where it rifes with ftrong ligneous ftalks to the height of ten or twelve feet, covered with a brown bark, and garnifhed with oval fpear-fhaped leaves, fawed on their edges, and pretty much veined, ftanding alternately. The flowers are produced from the wings of the ftalks; their empalements are of one leaf, cut into five obtufe fegments almoft to the bottom. The flower is of one petal (according to Ray and Tournefort) which is cut into five parts almoft to the bottom, but their bafe are connected together, and fall off united; the fegments are narrow at their bafe, but fpread open, are broad, and obtufe at their points, and hollowed like a fpoon in the middle; they are white, but one of the fegments in each flower is ftained with an herbaceous yellow colour. In the center of the flower arife five ftyles, which are furrounded by a circle of purple ftamina, terminated by roundifh blue fummits. The ftamina are inferted to the bafe of the petals, fo form at their bafe one body, being there connected together. The fruit of this is a conical, dry, ligneous capfule, having five fharp angles, and five cells, which open at the top with five valves, each cell containing one oblong fmooth feed.

This fhrub is at prefent very rare in the Englifh gardens. The feeds are feldom brought to England, and thofe frequently fail, either by their not having been properly impregnated, or duly ripened, for I have examined feveral which have been hollow, having only a fhell; and the plants which do come up, are very difficult to maintain while young, for if they are expofed to too much fun, they will foon be deftroyed, nor do they thrive when expofed to the open air. The only way in which I have feen the young plants fucceed was, when they were fown under glaffes, and the furface of the ground between the plants was covered with Mofs to keep the ground moift, and the glaffes were conftantly fhaded every day when the fun was bright. With this management the plants feemed in good health, but made little progrefs in their growth.

STOCK GILLIFLOWER. See CHEIRANTHUS.

STOEBE. Lin. Gen. Plant. 839.

The CHARACTERS are,

The flower is compofed of many hermaphrodite florets, included in one common empalement, whofe fcales are awl-fhaped and permanent; between each fcale is fituated one floret, whofe empalement is compofed of five narrow acute leaves, which are equal and erect. The florets are funnel-fhaped, of one petal, cut into five points at the brim; they have five fhort hair-like ftamina, terminated by cylindrical fummits, and an oblong germen, fupporting a flender ftyle, crowned by a bifid acute ftigma. The germen afterward becomes a fingle feed, crowned with a long feathery down, fitting in the common empalement.

We have but one SPECIES of this plant, viz.

STOEBE (*Æthiopica.*) Hort. Cliff. 390. *Stoebe.*

This plant grows naturally at the Cape of Good Hope; it is perennial, having a ligneous ftalk which rifes two or three feet high, fending out flender branches from the fides, garnifhed with fhort linear leaves, that are for the moft part hooked, of a grayifh colour, and placed irregularly round the branches. The flowers are produced in fingle heads at the end of the branches; they are of a pale yellow colour, and are compofed of feveral hermaphrodite florets, included in one common empalement, whofe fcales lie over each other like thofe of fifh. The florets are fingle, and peep out between the fcales of the empalement.

It is propagated by cuttings or flips, which fhould be planted in July upon a bed of foft loam, and covered clofe down either with a bell or hand-glafs, fhading them every day from the fun till they have taken root; then they muft be gradually inured to the open air, and afterward taken up, and planted in pots, placing them in the fhade till they have taken new root; then they may be placed in a fheltered fituation with other tender exotic plants, and in autumn they muft be removed into fhelter, for they are too tender to live through the winter in the open air in England.

STOECHAS. Tourn. Inft. R. H. 201. tab. 95. Caffidony, French Lavender, or Stickadore.

The CHARACTERS are,

The flower has an oval permanent empalement, whofe brim has fome obfcure indentures; it is of the lip kind, having a cylindrical tube longer than the empalement, whofe brim fpreads open. The upper lip is large, bifid, and open; the under lip is cut into three roundifh almoft equal fegments. It has four ftamina within the tube, which are turned afide, two of which are fhorter than the other, terminated by fmall fummits, and a quadrifid germen, fupporting a flender ftyle the length of the tube, crowned by an obtufe indented ftigma. The germen afterward turns to four almoft oval feeds, which ripen in the empalement. To which the following notes muft be added: the flowers are

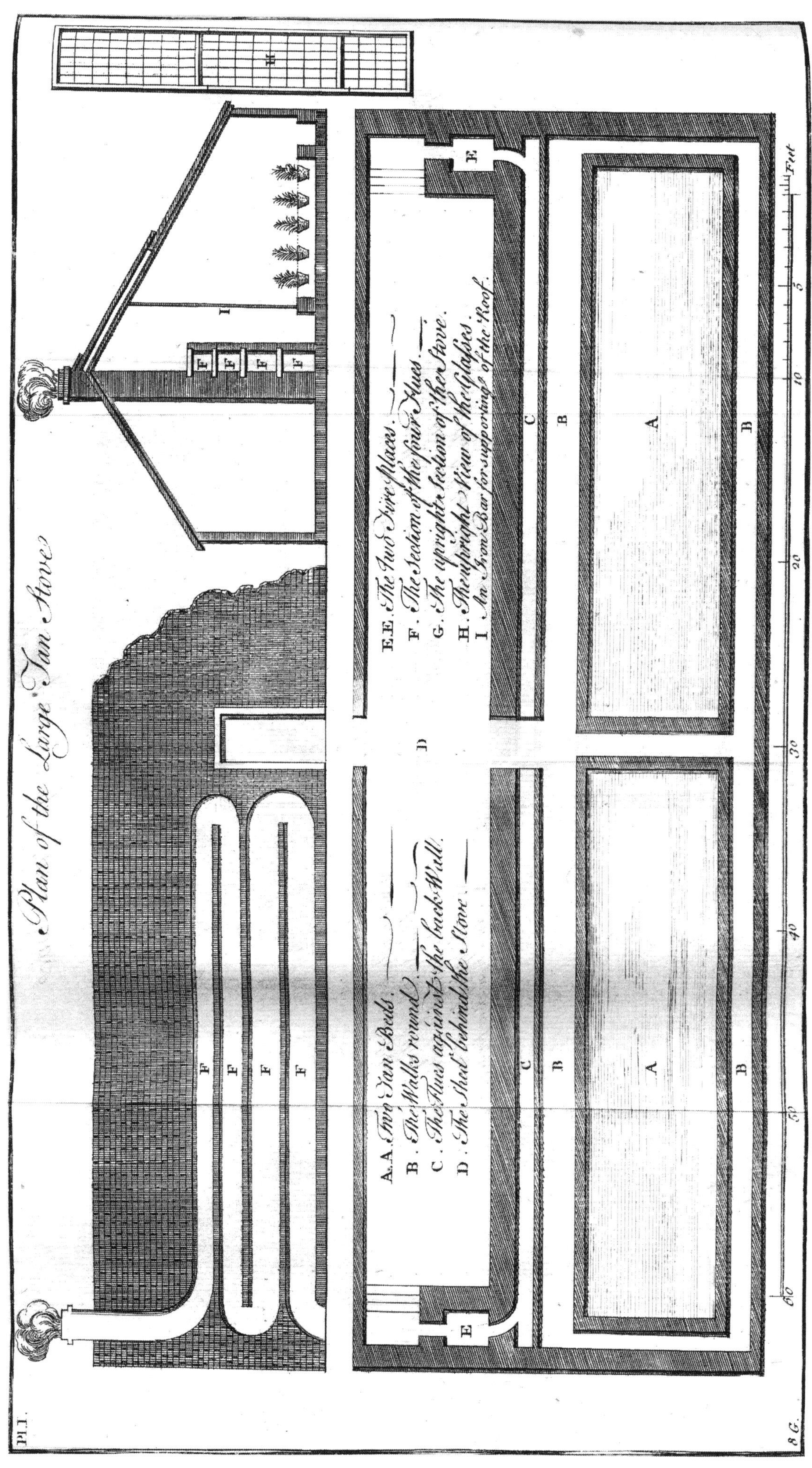

The material originally positioned here is too large for reproduction in this reissue. A PDF can be downloaded from the web address given on page iv of this book, by clicking on 'Resources Available'.

ranged in several series, and the spikes are terminated by tufts of leaves.

The Species are,

1. Stoechas (*Officinarum*) foliis lanceolato-linearibus, pedunculis brevioribus. *Stoechas with spear-shaped linear leaves, and shorter foot-stalks to the flowers.*

2. Stoechas (*Pedunculata*) foliis lanceolato-linearibus, pedunculis longissimis. *Stoechas with spear-shaped linear leaves, and the longest foot-stalks to the flowers.*

3. Stoechas (*Dentata*) foliis pinnato-dentatis. *Stoechas with wing-indented leaves.*

The first sort grows naturally in the south of France and Spain, from whence the tops or heads of flowers are imported to England for medicinal use; this has a low, thick, shrubby stalk, which rises about two feet high, sending out ligneous branches the whole length, which are garnished with spear-shaped linear leaves about an inch long, which are hoary and pointed, of a strong aromatic scent, and stand opposite on the branches at each joint, with smaller leaves of the same shape, coming out at the same places. The branches are terminated with scaly spikes of purple flowers about an inch in length; the spikes are four-cornered; the scales lie over each other like those of fish; out of each scale peeps one lip flower, whose tube is the length of the scale, so the two lips only appear; the under is spread open, and the upper stands erect. The spike of flowers is terminated by a small tuft of purple leaves like the Clary of Matthiolus; the flowers are succeeded by oval seeds, which ripen in autumn. The whole plant has a very strong, aromatic, agreeable odour.

The heads of flowers of this kind are used in some of the capital medicines directed by the College of Physicians, which are commonly brought from the south of France, where the plants are in great plenty; but, as these are seldom imported, and very little care taken in the drying and packing them, they are very apt to take a mouldiness in their passage, and are not near so good for use as those which are gathered fresh in England, where the plants may be cultivated to great advantage.

The second sort grows naturally in Spain. The difference between this and the first consists in the foot-stalks, which sustains the spikes of flowers, being three times the length of those of the first sort, and naked, having no leaves. The spikes of flowers are longer, and not so thick, and have more coloured leaves on their tops, which are longer, and of a brighter purple colour. These differences are not accidental, for I have many years propagated this plant by seeds, and have always found the plants produced were the same. Of both these there are some plants which vary in the colour of their flowers, some producing white, and others purplish flowers, but the most common colour is blue.

These plants may be cultivated by sowing their seeds upon a bed of light dry soil in March, and when they come up, they should be carefully cleared from weeds until they are two inches high, at which time they should be removed; therefore there must be a spot of light dry ground prepared, and laid level, which must be trodden out in beds, into which the plants should be planted at about five or six inches distance each way, observing to water and shade them until they have taken root; after which they will require no farther care but to keep them clear from weeds the following summer; but, if the winter should prove severe, it will be proper to cover them with mats, Pease-haulm, or some other light covering, to guard them against the frost, which otherwise would be apt to injure them while they are so young; but in March, or the beginning of April, the following spring, they must be removed into the places where they are to remain, observing, if possible, to transplant them in a warm moist season, and not to let them remain long above ground, for if their roots are dried, they seldom grow well after. The soil in which these are planted, should be a dry warm sand or gravel; and the poorer the soil is in which they are planted, the better they will endure the cold of the winter, provided the ground be dry, though indeed the plants will thrive better in summer upon a rich moist ground; but then they will not produce so many flowers, nor will the heads or spikes have near so strong an aromatic scent, as is the case with most sorts of aromatic plants.

These plants may also be propagated by planting slips or cuttings of any of the kinds in the spring, observing to refresh them with water until they have taken root, after which they may be managed as hath been directed for the seedling plants; but, as those plants raised from seeds are much better than these, it is hardly worth while to propagate them this way, especially since their seeds ripen so well in this country.

The heads of the first sorts may be gathered for use when the flowers are in full perfection, and spread to dry in a shady place, after which they may be put up for use.

The third sort grows naturally in Andalusia in Spain, and also about Murcia; this has a ligneous stalk, which rises two or three feet high, furnished with branches on every side, which are four-cornered, and garnished with leaves, placed opposite by pairs, indented regularly on both sides, almost to the midrib, in form of winged leaves; they are of a grayish colour, have a pleasant aromatic odour, and biting warm taste. The flowers are produced in scaly spikes at the end of the branches, standing upon long naked foot-stalks; they are four-cornered, hairy, and about an inch long, terminated by a few purplish leaves in the like manner as the other sorts, which inclined me to keep it joined to them. It flowers great part of summer, but the seeds very rarely ripen in England.

As this plant seldom produces seeds in England, it is propagated by slips or cuttings, which, if planted in April, and treated in the same way as those of the two other sorts, will take root very freely; but these plants, when rooted, must be planted in pots, that they may be sheltered from severe frost in winter, because they are too tender to live in the open air through the winter in England, especially while they are young; but, when they have obtained strength, some of them may be turned out of the pots, and planted in a warm situation, upon a dry rubbishy soil, where they will be stinted from growing too vigorously, so will endure the cold much better than if they were growing in better ground.

STONECROP. See Sedum.

STONECROP-TREE. See Chenopodium.

STOVES are contrivances for the preserving such tender exotic plants, which will not live in these northern countries without artificial warmth in winter. These are built in different methods, according to the ingenuity of the artist, or the different purposes for which they are intended, but in England they are at present reducible to two or three.

The first is called a dry Stove, being so contrived, that the flues through which the smoke passes, are either carried under the pavement of the floor, or else are erected in the back part of the house, over each other, and are returned six or eight times the whole length of the building, according to the height. In these Stoves the plants are placed on shelves of boards laid on a scaffold, rising above each other like the seats in a theatre, for the greater advantage of their standing in sight, and enjoying an equal share of light and air. In these Stoves are commonly placed the tender sorts of Aloes, Cereuses, Euphorbiums, Tithymals, and other

other ſucculent plants, which are impatient of moiſture in winter, and therefore require for the moſt part to be kept in a ſeparate Stove, and not placed among trees, or herbaceous plants, which perſpire freely, and thereby often cauſe a damp air in the houſe, which is imbibed by the ſucculent plants to their no ſmall prejudice. Theſe Stoves ſhould be regulated by a thermometer, ſo as not to overheat them, nor to let the plants ſuffer by cold; in order to which, all ſuch plants as require nearly the ſame degree of heat, ſhould be placed by themſelves in a ſeparate houſe; for if in the ſame Stove there are plants placed of many different countries, which require as many different heats, by making the houſe warm enough for ſome plants, others, by having too much heat, are drawn and ſpoiled.

The other ſort of Stoves are commonly called Bark-ſtoves, to diſtinguiſh them from the dry Stoves already mentioned. Theſe have a large pit, nearly the length of the houſe, three feet deep, and ſix or ſeven feet wide, according to the breadth of the houſe; which pit is filled with freſh tanners bark to make a hot-bed, and in this bed the pots of the moſt tender exotic plants, and herbaceous plants, are plunged. The heat of this bed being moderate, the roots of the plants are always kept in action, and the moiſture detained by the bark, keeps the fibres of their roots in a ductile ſtate, which in the dry Stove, where they are placed on ſhelves, are ſubject to dry too faſt, to the great injury of the plants. In theſe Stoves, if they are rightly contrived, may be preſerved the moſt tender exotic-trees and plants, which, before the uſe of the bark was introduced, were thought impoſſible to be kept in England; but, as there is ſome ſkill required in the ſtructure of both theſe Stoves, I ſhall not only deſcribe them as intelligibly as poſſible, but alſo annex plans of both Stoves hereto, by which it is hoped every curious perſon will be capable of directing his workmen in their ſtructure.

The dimenſion of theſe Stoves ſhould be proportioned to the number of plants intended to be preſerved, or the particular fancy of the owner; but their length ſhould not exceed forty feet, unleſs there are two fire-places, and in that caſe it will be proper to make a partition of glaſs in the middle, and to have two tan-pits, that there may be two different degrees of heat for plants from different countries (for the reaſons before given in the account of dry Stoves;) and were I to erect a range of Stoves, they ſhould be all built in one, and only divided with glaſs partitions, at leaſt the half way toward the front, which will be of great advantage to the plants, becauſe they may have the air in each diviſion ſhifted by ſliding the glaſſes of the partitions, or by opening the glaſs-door, which ſhould be made between each diviſion, for the more eaſy paſſage from one to the other.

Theſe Stoves ſhould be raiſed above the level of the ground, in proportion to the dryneſs of the place; for if they are built on a moiſt ſituation, the whole ſhould be placed upon the top of the ground, ſo that the brick-work in front muſt be raiſed three feet above the ſurface, which is the depth of the bark-bed, ereby none of the bark will be in danger of lying in water; but, if the ſoil be dry, the brick-work in front need not be more than one foot above ground, and the pit may be ſunk two feet below the ſurface. Upon the top of this brick-work in front muſt be laid the plate of timber, into which the wood-work of the frame is to be mortiſed; this ſhould be of ſound Oak without ſap, the dimenſion ten inches wide, and ſix deep, and the upright timbers in front muſt be placed four feet aſunder, or ſomewhat more, which is the proportion of the width of the glaſs-doors or ſaſhes; theſe ſhould be about eight or nine feet long; their dimenſion ſhould be ten inches by ſix, of yellow Fir; from the top of theſe ſhould be ſloping glaſſes, which ſhould reach within three feet of the back of the Stove, where there ſhould be a ſtrong crown piece of timber placed, in which there ſhould be a groove made for the glaſſes to ſlide into; the dimenſion of the ſloping timbers ſhould be one foot by nine inches, of yellow Fir, and the crown plate one foot by nine or ten inches of the ſame timber. The wall in the back part of the Stove ſhould be at leaſt eighteen or twenty-two inches, which is two bricks, or two bricks and a half: for the greater thickneſs there is in the brick-wall, the more heat will be thrown to the front, whereby the air of the Stove will be better warmed, and the building will be ſo much ſtronger; for to this back wall the flues, through which the ſmoke is to paſs, muſt be joined. This back-wall ſhould be carried up about ſixteen feet high or more for tall Stoves, that they may be of a proper height to ſupport the timbers of the back roof, which covers the ſhed behind the Stove. This roof is faſtened into the crown-piece before mentioned, which in tall Stoves ſhould be about thirty feet above the ſurface of the tan-bed, which will give a ſufficient declivity to the ſloping glaſſes to carry off the wet, and be of a reaſonable height for containing many tall plants. The back roof may be ſlated, covered with lead, or tiled, according to the fancy of the owner; but the manner of the outſide building is better expreſſed by the annexed plan, than is poſſible to be deſcribed in words.

In the front of the houſe, before the tan-bed, there ſhould be a walk about two feet wide, for the conveniency of walking; next to which the bark-pit muſt be placed, which ſhould be in width proportionable to the breadth of the houſe. If the houſe is fifteen feet wide, which is a due proportion, the pit may be eight feet wide, and behind the pit ſhould be a walk two feet wide to paſs, in order to water the plants, &c. then there will be two feet left next the back wall to erect the flues, which muſt be all raiſed above the level of the bark-bed. Theſe flues ought to be one foot wide in the clear, that they may not be too ſoon ſtopped with the ſoot, as alſo for the more conveniently cleaning them; the lower flue, into which the ſmoke firſt enters from the fire, ſhould be two feet and a half deep in the clear; this ſhould be covered with broad tiles, which ſhould be a foot and a half ſquare, that they may be wide enough to extend over the wall in form of the flues, and to take ſufficient hold of the back-wall; over this the ſecond flue muſt be returned back again, which may be twenty inches deep, and covered on the top as before, and ſo in like manner the flues may be returned over each other ſix or eight times, that the heat may be ſpent before the ſmoke paſſes off. The thickneſs of the wall in front of theſe flues need not be more than four inches, but it muſt be well-jointed with mortar, and pargitered within ſide to prevent the ſmoke from getting into the houſe, and the outſide ſhould be faced with mortar, and covered with a coarſe cloth, to keep the mortar from cracking, as is practiſed in ſetting up coppers. If this be carefully done, there will be no danger of the ſmoke entering the houſe, which cannot be too carefully guarded againſt; for there is nothing more injurious to plants than ſmoke, which will cauſe them to drop their leaves, and, if it continue long in the houſe, will entirely deſtroy them.

The fire-place muſt be made at one end, where there is but one; but, if the Stove is ſo long as to require two, they ſhould be placed at each end of the ſhed, which muſt be made the length of the Stove, that the fires and the back of the flues may not ſuffer from the outer air, for it will be impoſſible to make the fires burn equally, where the wind has full ingreſs to them; and it will be troubleſome to attend the fires in wet weather, where they are expoſed to the rain.

The

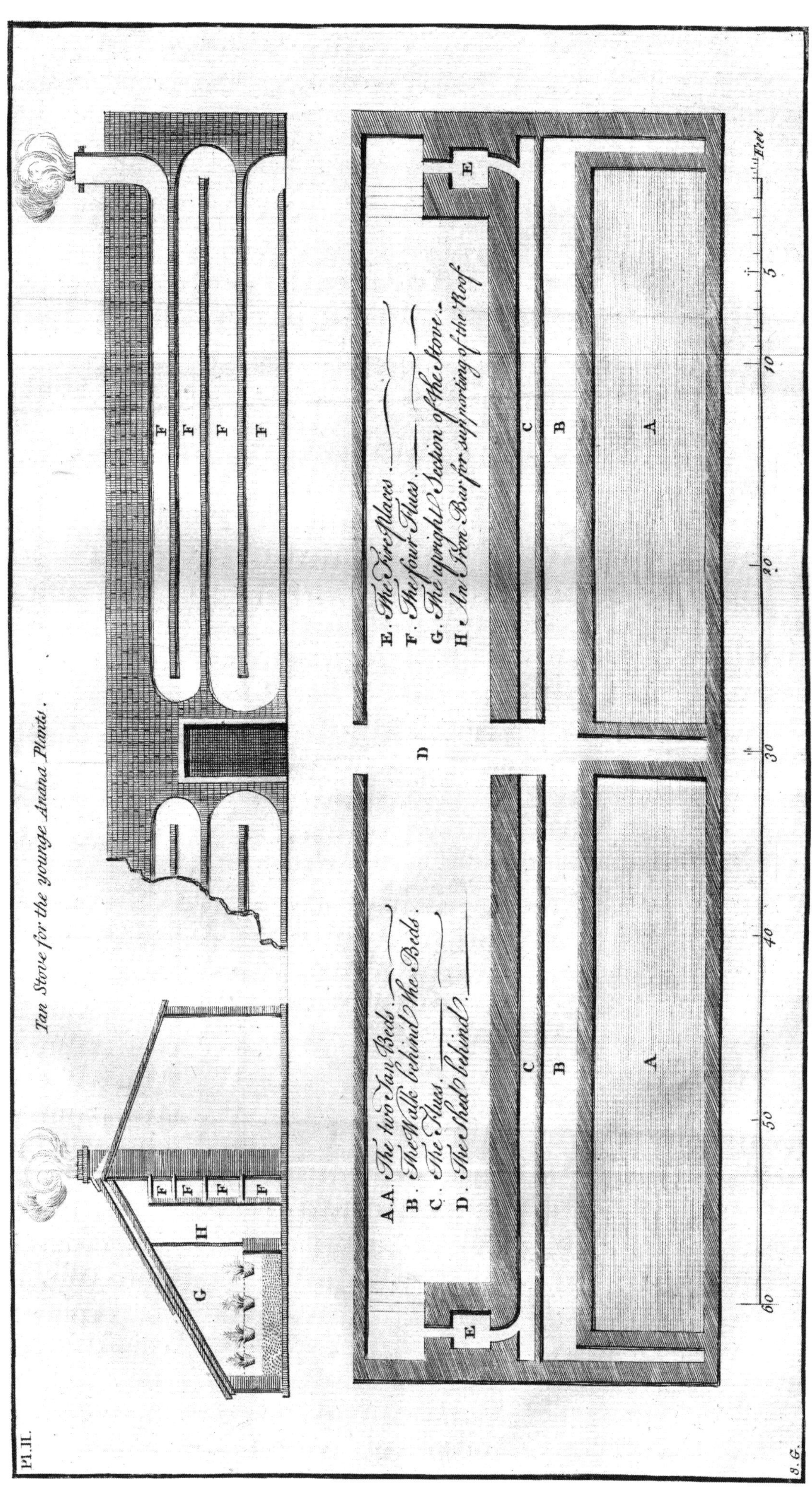

The material originally positioned here is too large for reproduction in this reissue. A PDF can be downloaded from the web address given on page iv of this book, by clicking on 'Resources Available'.

The contrivance of the furnace muſt be according to the fuel which is deſigned to burn; but as turf is the cheapeſt firing for Stoves, where it can be had, many prefer it, becauſe it laſts longer than any other ſort of fuel, ſo requires leſs attendance. I ſhall deſcribe a proper ſort of furnace for that purpoſe.

The whole of this furnace ſhould be erected within the houſe, which will be a great addition to the heat, and the front wall on the outſide of the fire-place, next the ſhed, ſhould be three bricks thick, the better to prevent the heat from coming out that way. The door of the furnace, at which the fuel is put in, muſt be as ſmall as conveniently may be to admit of the fuel; and this door ſhould be placed near the upper part of the furnace, and made to ſhut as cloſe as poſſible, ſo that there may be but little of the heat paſs off through it. This furnace ſhould be about twenty inches deep, and ſixteen inches ſquare at bottom, but may be ſloped off on every ſide, ſo as to be two feet ſquare at the top; and under this furnace ſhould be a place for the aſhes to fall into, which ſhould be about a foot deep, and as wide as the bottom of the ſurface; this ſhould alſo have an iron door to ſhut as cloſe as poſſible, but juſt over the aſh-hole, above the bars which ſupport the fuel, ſhould be a ſquare hole about four or ſix inches wide to let in air to make the fire burn; this muſt alſo have an iron frame, and a door to ſhut cloſe when the fire is perfectly lighted, which will make the fuel laſt longer, and the heat will be more moderate.

The top of this furnace ſhould be nearly equal to the top of the bark-bed, that the loweſt flue may be above the fire, ſo that there may be a greater draught for the ſmoke, and the furnace ſhould be arched over with bricks. The beſt materials for this purpoſe are what the Bricklayers call Windſor Bricks, which ſhould be laid in loam of the ſame kind as that the bricks are made with; and this, when burnt by fire, will cement the whole together, and become like one brick; but you ſhould be very careful, wherever the fire is placed, that it be not too near the bark-bed, for the heat of the fire will, by its long continuance, dry the bark, ſo that it will loſe its virtue, and be in danger of taking fire; to prevent which, it will be the beſt method to continue a hollow, between the brick-work of the fire and that of the pit, about eight inches wide, which will effectually prevent any damage ariſing from the heat of the fire; nor ſhould there be any wood-work placed near the flues, or the fire-place, becauſe the continual heat of the Stove may in time dry it ſo much, as to cauſe it to take fire, which ought to be very carefully guarded againſt.

The entrance into this Stove ſhould be either from a green-houſe, the dry Stove, or elſe through the ſhed where the fire is made, becauſe in cold weather the front glaſſes muſt not be opened. The inſide of the houſe ſhould be clean white-waſhed, becauſe the whiter the back part of the houſe is, the better it will reflect the light, which is of great conſequence to plants, eſpecially in winter, when the Stove is obliged to be ſhut up cloſe.

Over the top ſliding glaſſes there ſhould be either wooden ſhutters, or tarpawlins fixed in frames, to cover them in bad weather, to prevent the wet from getting through the glaſſes, and to ſecure them from being broken by ſtorms and hail; and theſe outer coverings will be very ſerviceable to keep out the froſt; and if in very ſevere cold there is a tarpawlin hung before the upright glaſſes in the front, it will be of great ſervice to the Stove, and much leſs fire will preſerve a heat in the houſe.

As in this Stove will be placed the plants of the hotteſt parts of the Eaſt and Weſt-Indies, the heat ſhould be kept up equal to that marked Anana upon the botanical thermometers, and ſhould never be ſuffered to be above eight or ten degrees cooler at moſt; nor ſhould the ſpirit be raiſed above ten degrees higher in the thermometer during the winter ſeaſon, both which extremes will be equally injurious to the plants.

But in order to judge more exactly of the temper of the air in the Stove, the thermometer ſhould be hung at a good diſtance from the fire, nor ſhould the tube be expoſed to the ſun; but, on the contrary, as much in ſhade as poſſible, becauſe, whenever the ſun ſhines upon the ball of the thermometer but one ſingle hour, it will raiſe the liquor in the tube conſiderably, when perhaps the air of the houſe is not near ſo warm, which many times deceives thoſe who are not aware of this.

In the management of the plants placed in the bark-bed, there muſt be a particular regard had to the temper of the bark, and the air of the houſe, that neither be too violent; as alſo to water them frequently, but ſparingly, in cold weather, becauſe when they are in continual warmth, which will cauſe them to perſpire freely, if they have not a proper ſupply to anſwer their diſcharge, their leaves will decay, and ſoon fall off. As to the farther directions concerning the culture of the particular plants, the reader is deſired to turn to their ſeveral articles, where they are diſtinctly treated of.

The other ſort of Stove, which is commonly called the dry Stove, as was before ſaid, may be either built with upright and ſloping glaſſes at the top, in the ſame manner, and after the ſame model of the Bark-ſtove, which is the moſt convenient; or elſe the front glaſſes, which ſhould run from the floor of the cieling, may be laid ſloping, to an angle of 45 degrees, the better to admit the rays of the ſun in ſpring and autumn. The latter method has been chiefly followed, by moſt perſons who have built theſe ſorts of Stoves; but were I to have the contrivance of a Stove of this kind, I would have it built after the model of the Bark-ſtove, with upright glaſſes in front, and ſloping glaſſes over them, becauſe this will more eaſily admit the ſun at all the different ſeaſons; for in ſummer, when the ſun is high, the top glaſſes will admit the rays to ſhine almoſt all over the houſe; and in winter, when the ſun is low, the front glaſſes will admit its rays; whereas, when the glaſſes are laid to any declivity in one direction, the rays of the ſun will not fall directly thereon above a fortnight in autumn; and about the ſame time in ſpring, and during the other parts of the year, they will fall obliquely thereon; and in ſummer, when the ſun is high, the rays will not reach above five or ſix feet from the glaſſes. Beſides, the plants placed toward the back part of the houſe, will not thrive in the ſummer ſeaſon for want of air; whereas, when there are ſloping glaſſes at the top, which run within four feet of the back of the houſe, theſe, by being drawn down in hot weather, will let in perpendicular air to all the plants; and of how much ſervice this is to all ſorts of plants, every one who has had opportunity of obſerving the growth of plants in a Stove will eaſily judge; for when plants are placed under cover of a cieling, they always turn themſelves toward the air and light, and thereby grow crooked; and if, in order to preſerve them ſtrait, they are turned every week, they will nevertheleſs grow weak, and look pale and ſickly; for which reaſons, I am ſure, whoever has made trial of both ſorts of Stoves, will readily join with me to recommend the model of the Bark-ſtove for every purpoſe.

As to the farther contrivance of this Stove, it will be neceſſary to obſerve the temper of the place, whether the ſituation be dry or wet; if it be dry, then the floor need not be raiſed above two feet above the level of the ground; but if it be wet, it will be proper to raiſe it three feet, eſpecially if the flues are to be carried under the floor; for when they are erected cloſe upon the ſurface of the ground,

ground, they will raiſe a damp, which will prevent the flues drawing ſo well as when they are more elevated. The furnace of this Stove muſt be placed at one end of the houſe, according to the directions before given. This muſt be made according to the fuel intended to burn, which, if for coals or wood, may be made according to the common method for coppers, but only much larger; becauſe, as the fire is to be continued in the night chiefly, if there is not room to contain a proper quantity of fuel, it will occaſion a great deal of trouble in tending upon the fire in the night, which ſhould be avoided as much as poſſible; becauſe, whenever the trouble is made very great or difficult, and the perſon who is intruſted with the care of it, has not a very great affection for the thing, and is withal not very careful, there will be great hazard of the fire being neglected, which in a little time may be of dangerous conſequence to the plants; but, if the fuel intended be turf, then the contrivance of the furnace may be the ſame as for the Bark-ſtove already mentioned. The flues of this Stove, if they are carried under the pavement, may be turned after the following manner,

which will cauſe them to draw better than if ſtrait; and by this method of diſpoſing them, they may be ſo much turned as to reach them from the back to the front of the houſe.

The depth of them ſhould not be leſs than eighteen inches, and the width nearly equal, which will prevent their being choked up with ſoot, as is often the caſe when the flues are made too ſmall. The ſpaces between the flues ſhould be filled up either with dry brick rubbiſh, lime, or ſand, from which there will little moiſture ariſe, and the flues ſhould be cloſely plaſtered with loam both within and without, and the upper part of them covered with a coarſe cloth under the floor, to prevent the ſmoke from getting into the houſe.

When the flue is carried from the furnace to the end of the houſe, it may be returned in the back above the floor three or four times in ſtrait lines, which may be contrived to appear like ſo many ſteps; by which means the ſmoke will be continued in the houſe until all its heat is ſpent, which will conſequently warm the air of the houſe the better; and the chimneys, through which the ſmoke is to paſs off, may be either at both ends, or in the middle, carried up in the thickneſs of the brick-work of the flues, ſo as not to appear in ſight in the houſe. The flues ſhould be firſt covered with broad tiles, and then a bed of ſand laid over them about two inches thick, upon which the plain tiles ſhould be laid to correſpond with the reſt of the floor. This thickneſs of cover will be full enough to prevent the too ſudden riſe of the heat from the flues.

But if the furnace is placed under the floor, the thickneſs of ſand between the brick arch which covers it and the floor, ſhould not be leſs than four or ſix inches, ſo that the bottom of the furnace ſhould be ſunk the lower; and if from the fire-place to the end of the houſe, the flues are laid a little riſing, it will cauſe them to draw the better; but this riſe muſt be allowed in the placing them lower under the floor next the fire, becauſe the floor muſt be laid perfectly level, otherwiſe it will appear unſightly.

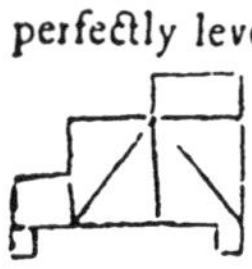

In this Stove there ſhould be a ſtand or ſcaffold erected for placing ſhelves above each other, in the manner annexed, that the plants may be diſpoſed above each other, ſo as to make a handſome appearance in the houſe; but theſe ſhelves ſhould be made moveable, ſo as to be raiſed or ſunk, according to the various heights of the plants, otherwiſe it will be very troubleſome to raiſe or ſink every particular plant according to their heights, or every year as they advance in their growth.

In placing the feet of this ſtand, you muſt be careful not to ſet them too near the fire, nor directly upon the top of the flue, eſpecially that end next the fire, leſt by the conſtant heat of the tiles the wood ſhould take fire, which cannot be too much guarded againſt, ſince ſuch an accident would go near to deſtroy all the plants, if the houſe eſcaped being burnt. This ſtand or ſcaffold ſhould be placed in the middle of the houſe, leaving a paſſage about two feet and a half in the front, and another of the ſame width in the back, for the more conveniently paſſing round the plants to water them, and that the air may freely circulate about them. In diſpoſing the plants, the talleſt ſhould be placed backward, and the ſmalleſt in front, ſo that there will not be occaſion for more than four ſhelves in height at moſt; but the ſcaffold ſhould be ſo contrived, that there may be two or three ſhelves in breadth laid upon every riſe whenever there may be occaſion for it, which will ſave a deal of trouble in diſpoſing of the plants.

In the erection of theſe Stoves, it will be of great ſervice to join them all together with only glaſs partitions between them, as was before obſerved; and where ſeveral of theſe Stoves and green-houſes are required in one garden, then it will be very proper to have the green-houſe in the middle, and the Stoves at each end, either in the manner directed in the plan of the green-houſe exhibited in that article, or carried on in one ſtrait front.

By this contrivance in the ſtructure of theſe houſes, a perſon may paſs from one to the other of them, without going to the open air, which, beſides the pleaſure to the owner, is alſo of great uſe; becauſe there will be no occaſion of making a back-way into each of them, which otherwiſe muſt be, ſince the front glaſſes of the Stove ſhould not be opened in cold weather, if it can poſſibly be avoided on any account, otherwiſe the cold air ruſhing in, will greatly prejudice the very tender plants.

But beſides the Stoves here deſcribed, and the green-houſe, it will be very neceſſary to have a glaſs-caſe or two, wherever there are great collections of plants. Theſe may be built exactly in the manner already deſcribed for the Stoves, with upright glaſſes in front, and ſloping glaſſes over the top of them; which ſhould run within three feet of the back of the houſe. The height, depth, and other dimenſions, ſhould be conformable to that of the Stoves, which will make a regularity in the building. Theſe may be placed at the end of the range on each hand beyond the Stoves; and if there be a flue or two carried along round each of theſe, with an oven to make a fire in very cold weather, it will ſave a great deal of labour, and prevent the froſt from ever entering the houſe, be the winter ever ſo ſevere; but the upper glaſſes of theſe houſes ſhould have either ſhutters or tarpawlins in frames, to cover them in froſty weather; and if there is a contrivance to cover the upright glaſſes in froſt, either with mats or tarpawlins, it will be of great uſe in winter, otherwiſe the flue muſt be uſed when the froſt comes on, which ſhould not be done but upon extraordinary occaſions; becauſe the deſign of theſe houſes is, to keep ſuch plants as require only to be preſerved from froſt, and need no additional warmth, but at the ſame time require more air than can conveniently be given them in a green-houſe. In one of theſe houſes may be placed all the ſorts of African Sedums, Cotyledons, and other ſucculent plants from the Cape of Good Hope. In the other may be placed the ſeveral kinds of Arctotis, Oſteoſpermum, Royena, Lotus, and other woody or herbaceous plants from the ſame country, or any other in the ſame latitude.

Thus by contriving the green-houſe in the middle, and one Stove and a glaſs-caſe at each end, there will be a conveniency to keep plants from all the different parts of the world, which can be no otherwiſe maintained but by placing them in different degrees of heat, according to the places of their native growth.

The Stoves before deſcribed, are ſuch as are uſually built to maintain exotic plants, which will not live in England, unleſs they enjoy a temperature of air, approaching to that of the ſeveral countries from whence they are brought; therefore, whoever is inclinable to preſerve a large collection of plants from different countries, muſt contrive to have two or three of theſe Stoves, each of which ſhould be kept in a different temperature of warmth; and the plants ſhould be alſo adapted to the ſeveral degrees of heat, as they ſhall require, to preſerve them. But as the far greateſt number of Stoves, which have been erected in England, are deſigned for the culture of the Ananas only, ſo I ſhall add a deſcription and plans of two ſorts of Stoves, of the leaſt expence in building for this purpoſe; ſo that whoever is inclinable to erect a Stove for ripening of the Ananas, may, by attending to the plans and deſcriptions, direct the building and contriving ſuch Stoves as they are deſirous to have, or according to the number of fruit propoſed to be ripened annually.

The firſt ſort of Stove is that which is deſigned for the plants which produce the fruit the ſame year; for as the plants do not generally fruit, until the ſecond year from their being taken from the old plants, whether they are ſuckers from the ſide of the plants, or crowns taken from the fruit, if they fruit the ſucceeding year, the fruit will be ſmall; therefore, when they are properly managed, they will not produce their fruit until the ſecond year; by which time they will have obtained ſtrength to produce large fruit, in which their greateſt value conſiſts; for although there are ſeveral varieties of this fruit, which differ in degrees of goodneſs, as in moſt other fruits, yet they may all of them be improved in their ſize without diminiſhing of their excellence in taſte; though I know there are ſome perſons of a contrary opinion, and who believe, that the ſmall fruit are always better flavoured than the large; but from long experience I can aſſert, that the larger and better nouriſhed this fruit is, the higher will be its flavour, ſuppoſing the ſorts are the ſame; therefore every perſon who cultivates this fruit, ſhould endeavour to have it improved to the greateſt perfection; in order to which, it will be proper to have a ſmall Stove, in which the young plants may be placed to bring them forward for fruiting, and the following autumn they ſhould be removed into the larger Stove for ripening: but I ſhall return to the deſcription of the larger Stove. The length of this muſt be proportionable to the quantity of fruit deſired in one ſeaſon; for as to their width, that ſhould not be much varied; the tan-bed ſhould never be narrower than ſix, nor ſhould it be more than eight feet wide; for when it is more, there will be difficulty in reaching thoſe plants which are in the middle of the bed, to water or clean them; and if there is room enough on each ſide of the bed for a walk a foot and a half broad, it will be ſufficient for perſons to water and do every thing which is neceſſary to the plants; and as theſe places are not deſigned for walking in, ſo it is to no purpoſe to have broad walks, which will take up too much ſpace; for the fires muſt be larger, in proportion to the dimenſion of the houſe, otherwiſe the air cannot be kept in a proper temperature of warmth. If the Stove is made thirty-ſix feet long in the clear, then the tan-bed may be thirty-three feet long, and a walk left at each end a foot and a half wide, which will be ſufficient to walk round the bed to water and attend the plants, and ſuch a tan-bed will contain eighty fruiting plants very well; and this Stove may be very well warmed with one fire; but if the Stove is built much larger, there muſt be two fire-places contrived, one at each end, otherwiſe the air of the houſe cannot be kept in a proper temperature of heat. The quantity of fuel which will be wanting for a Stove of thirty-ſix feet long in the clear, is about three chaldron and a half of coals, or in ſuch proportion for any other ſort of fuel; where coals can be had reaſonable it is the beſt kind of fuel; and the pit or Scotch coal is preferable to the Newcaſtle coal, becauſe the latter is very ſubject to melt or run into clinkers, when the oven is very hot, which the pit-coal never does, but always burns away with a white aſh, making but little ſoot; ſo that the flues will not require to be ſo often cleaned, as when the other coal is uſed. The next beſt fuel for Stoves is Peat, where it can be procured good; but the ſcent of this fuel is diſagreeable to many people. There are ſome perſons who burn wood in their Stoves; but this fuel requires much greater attendance than any other, therefore is not very proper for this purpoſe; but in the building of the Stoves, the ovens muſt be contrived for the ſort of fuel which is to be uſed in them; but theſe will be afterward deſcribed, and the places where they ſhould be ſituated, are delineated in the plan.

The Stoves deſigned for ripening the fruit of the Ananas, ſhould have upright glaſſes in their front, which ſhould be high enough to admit a perſon to walk upright under them on the walk in the front of the houſe; or where this cannot be admitted, the front walk may be ſunk one foot lower than that on the back of the tan-bed, ſo that the ſurface of the bed will be a foot above the walk, which will be rather an advantage, as the plants will be ſo much nearer the glaſs; and a perſon may with great eaſe water and attend the plants, when they are thus raiſed above the walk; therefore, when a Stove is ſo ſituated, as that the raiſing of it high above ground might be attended with inconvenience, the walks quite round the tan-bed may be ſunk a foot or eighteen inches below the top of the bed, which will admit of the Stove being built ſo much lower; for if there is height for a perſon to walk under the glaſſes, it will be as much as is required; but as the flues, when returned four times againſt the back wall, will riſe near ſeven feet, ſo the bottom of the lower flue ſhould be on the ſame level with the walk, to admit room enough for the whole under the roof. Over the upright glaſſes there muſt be a range of ſloping glaſſes, which muſt run to join the roof, which ſhould come ſo far from the back wall as to cover the flues, and the walk behind the tan-pit; for if the ſloping-glaſſes are of length ſufficient to reach nearly over the bed, the plants will require no more light, therefore theſe glaſſes ſhould not be longer than is abſolutely neceſſary, which will render them more manageable; but the annexed plan will render this more intelligible than any written deſcription can do.

The other ſort of Stove, which is deſigned for raiſing of young plants until they are of a proper ſize to produce fruit, need not be built ſo high as the former; therefore there will not be wanting any upright glaſſes in the front, but the frames may be made in one ſlope, as in the annexed plan; indeed of late years, many perſons have made tan-beds, with two flues running through the back wall, to warm the air in winter; and theſe beds have been covered with glaſſes, made in the ſame manner as thoſe for common hot-beds, but larger; theſe were contrived to ſave expence, and have in many places anſwered the intention; but to theſe there are ſeveral objections. 1. That of having no paſſage into them, ſo that the glaſſes muſt be taken off, when the plants want water, &c. 2. The damps very often riſe in the winter ſeaſon, when the glaſſes are cloſely ſhut,

shut, which often proves very injurious to the plants. 3. There is danger of the tan taking fire, where there is not great care taken that it doth not lie near the flues; so that although the small Stoves here proposed require more expence in their building, yet, being greatly preferable to those pits, and the after-expence being the same, they will be found so much more convenient as to render them more general where this fruit is cultivated.

Where there is no danger of the wet settling about the tan in winter, the bark-pit may be sunk two feet deep in the ground, and raised one foot above the surface; the only walk which is necessary in these Stoves, is that on the back of the tan-bed, which may be on the level with the surface of the ground; so that the tan-bed will be more than one foot above the walk, and the flues beginning from the level of the walk, there will be room to return them three times, which will warm the air much more with the same fire, than when they are carried but twice the length of the Stove.

But in wet land the tan-bed should be wholly raised above the level of the ground, in order to preserve the tan from being chilled by moisture; and in such places the walk on the back should be raised near two feet above the level of the ground, because the tan-bed should not rise much more than one foot above the walk; for if it is higher, it will be more difficult to reach the plants when they require water; the brick wall of the pit, on the side next the walk, need not be more than four inches thick, so far as rises above the walk, but below that it should be nine inches thick. The reason for reducing the wall above, is to gain room for the walk, which would otherwise be too much contracted; and if there is a kirb of Oak laid on the top of the four inch wall, it will secure the bricks from being displaced, and sufficiently strengthen the wall, which being but one foot above the walk, will not be in any danger of falling; and on this kirb there may be two or three upright iron bars fixed with claws, to support the crown-piece of timber, which will secure it from camphering in the middle, which in a great length is very often the case, where there are no supports placed under it; there may be more or less of these bars according to the length of the Stove, but if they are about ten feet asunder it will be near enough. If these iron bars are one inch and a half square, they will be strong enough to answer the design.

But as it is hoped that the annexed plan of this small Stove will convey a clear idea of the whole contrivance, this will render it unnecessary to add any farther description here.

An explanation of the plate which represents the two sorts of frames with oiled paper for covering of Melons.

The first of these frames is contrived like the covers of waggons; it has a frame of wood at the base, to which are fastened broad hoops, which are bent over circularly, as is represented at Fig. 1. The width of this frame should be six feet, for less will not be sufficient to cover the bed; and if they are more than six feet broad, they will be too heavy and troublesome to move. *a* shews the section of the width, *b* the frame of wood at the base, *Cc* the arch of hoops, and *d* a small slip of wood, which is fastened to the under side of the hoops to keep them in their proper position.

The distance between each hoop should not be more than one foot, and there should be two rows of strong packthread or rope-yarn on each side of the arch running from hoop to hoop at the places marked, *e. e. e. e.* to keep the oiled paper from sinking down with wet. The length of each frame should not be much more than ten feet, which will be sufficient length for covering three plants, that being about the size of a three light frame, for if they are longer, they will be heavy and troublesome to move; therefore there should be as many of these frames made, as may be necessary for covering the quantity of plants designed. Fig. 2. represents two lengths of these frames joined, *G* shews the profile of the frame, and *H* represents the paper turned back, that it may be seen how it is laid over the frame.

Fig. 3. represents the other sort of frame which is contrived like the roof of a house, *a* shews a section of the base; *b b* the two slopes, *c* one of the sides which is contrived to be raised at any time to admit air to the plants, *d* shews the place where this shuts down, and *e* the prop which supports it. If in the making of these frames every other light is made with hinges so as to be raised, and on the opposite side they are contrived to rise alternately, it will be a very good method, for then air may be given at the side contrary to the wind; and in very warm weather, when the plants require a large share of air, they may all be raised on both sides, which will make a thorough air to the whole bed. Fig. 4. shews the plan of these frames; and Fig. 5. the same erected; *g* represents the profile of it, and *f* the covering of paper. This sort of frame may be made of pantile laths, or of slips of deal of like dimensions, because they should not be too heavy; but the base of the frame to which these are fastened should be more substantial. Some persons who have made trial of both, recommend the latter for the convenience of giving air to the plants, for there is no other contrivance in the first sort for admitting the air, but by raising the whole frame on one side in proportion to the quantity of air intended to be admitted; and when the season is warm, they generally raise those frames on both sides, and permit the plants to run out from under them.

When these frames are made, if they are well painted over with the following composition, it will greatly preserve them, viz. To every six pounds of melted pitch, add half a pint of Linseed oil, and a pound of brick-dust; these should be well mixed together, and used warm; when this dries it becomes a hard cement, so that no moisture can penetrate through it, and is the best sort of pigment for all timber exposed to the weather, I have ever seen used; so that where the colour is not offensive to the sight, it should be preferred to every other.

When the frames are thoroughly dry, the paper should be pasted on to the frames. The best sort of paper for this purpose, is what they call Dutch Wrapper; this is strong, and when oiled over becomes pellucid, so admits the rays of light through it extremely well. After the paste is well dried, the paper should be oiled over on the outside, which, if well done with Linseed oil, will be sufficient, for the oil will soak quite through the paper, so there will be no necessity for oiling both sides, nor for doing it over more than once. The oil should be dry before the frames are exposed to the wet, otherwise the paper will tear. In the pasting of the paper on the frames, there should be care taken to stretch it very smooth, and also to paste it to all the ribs of the frames, and also to the packthreads, to prevent the wind from raising the paper, which would soon tear it when it became loose.

The above description, together with the annexed plan, it is hoped, will be sufficient instructions for any one who is desirous of making these covers; and what has been before mentioned under the article MELON, will be directions enough for the use of them; so that I shall only add one caution which may be necessary to repeat here, which is, not to keep these covers too close down over the plants, lest it draw them too weak, so that air should always be admitted to the plants at all times in proportion to the warmth of the season.

These

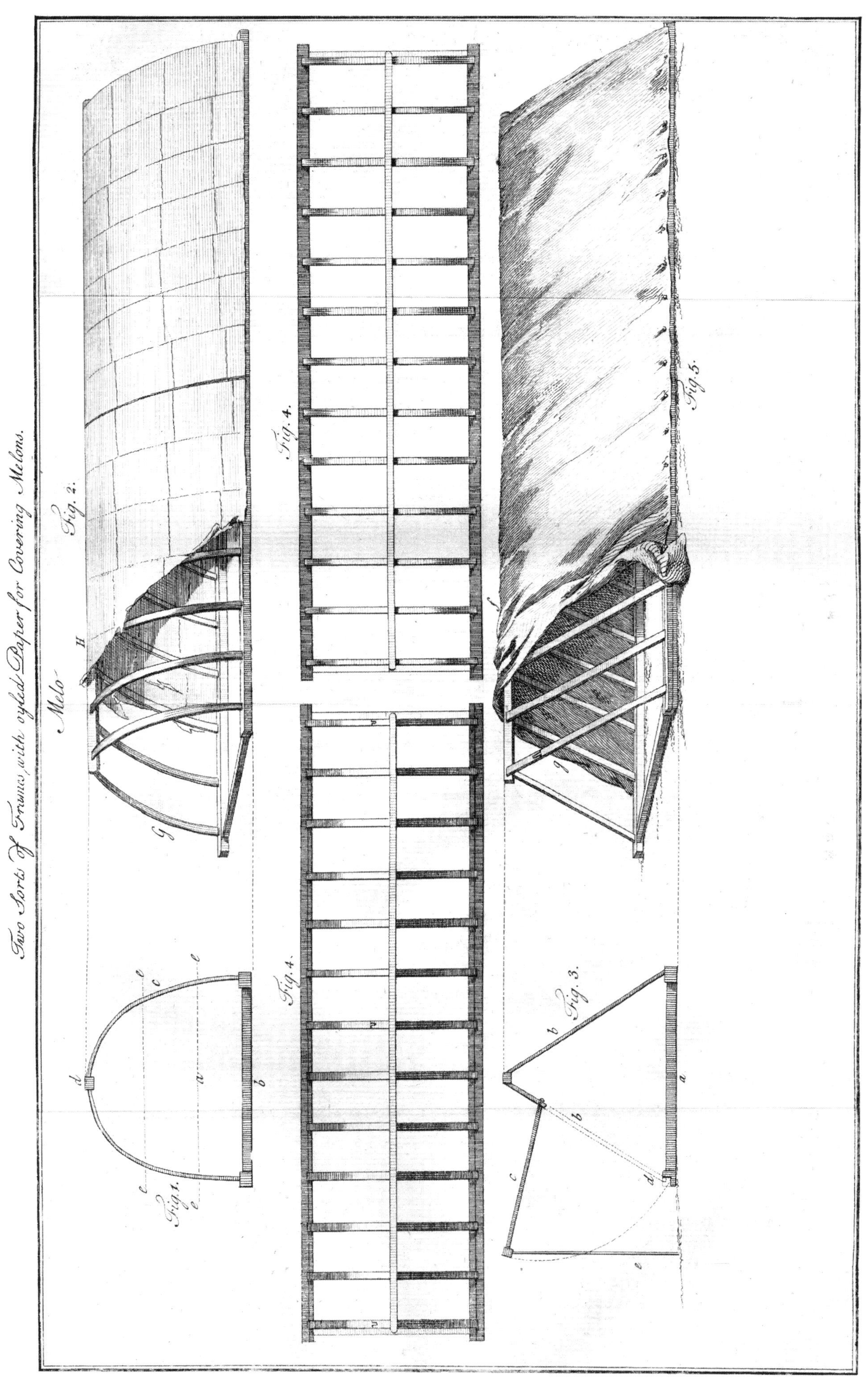

The material originally positioned here is too large for reproduction in this reissue. A PDF can be downloaded from the web address given on page iv of this book, by clicking on 'Resources Available'.

These covers of oiled paper are not only useful for covering of Melons, but are the best things to cover cuttings of exotic plants, when planted, that can be contrived; and are also capable of being used for many other purposes.

The paper will seldom last longer than one season, so it will require a new covering every spring; but if the frames are well made, and when they are out of use, laid up in shelter from the wet, they will last several years, especially if there is a band of Straw laid round the Melon-bed, upon which the frames may stand, so they will not rest upon the ground, and the Straw-bands will prevent the damp from rising so as to rot them. These Straw-bands are such as are recommended for the hot-beds of Asparagus in winter.

STRAMONIUM. See Datura.

STRATIOTES. Lin. Gen. Plant. 607. Water Soldier.

The Characters are,

It has one flower, inclosed in a compressed obtuse sheath, composed of two leaves, which are keel-shaped and permanent. The empalement of the flower is of one leaf, trifid, and erect. It has three almost heart-shaped petals, which are twice the size of the empalement, and about twenty stamina inserted in the receptacle of the flower, terminated by single summits. The germen is situated under the empalement, supporting six styles, divided in two parts, crowned by single stigmas. The germen afterward becomes an oval capsule, narrowed on every side, having six angles, and as many cells, filled with oblong incurved seeds.

We have but one Species of this genus, viz.

Stratiotes (*Aloides.*) Lin. Flor. Lapp. 222. Water Soldier, Water Aloe, or Fresh Water Soldier.

This plant is in shape like the Aloe, but the leaves are thinner, and serrated on the edges very sharply; they are of a grayish colour, and about a foot long; between the leaves, from the center of the plant, arise one, two, and sometimes three stalks, almost the length of the leaves, each being terminated by a three-forked sheath, out of which bursts one white flower, composed of three roundish heart-shaped petals, with many yellow stamina in the middle. Below the flower is situated a conical germen, which is reversed, the broad end standing upward, and the narrow downward. This becomes a six-angled capsule, having six cells, filled with seeds. It grows plentifully in standing waters in the Isle of Ely, and many places in the north of England, from whence young plants may be procured in spring, when they first rise on the surface of the water; and these being placed in large ponds or canals, will strike down their roots, and propagate without any farther care. In autumn the plants sink down to the bottom of the water, and rise again in the spring.

STRAWBERRY. See Fragaria.

STRAWBERRY-TREE. See Arbutus.

STYRAX. Tourn. Inst. R. H. 598. tab. 369. Storax-tree.

The Characters are,

The flower has a short cylindrical empalement, indented in five parts; it has one funnel-shaped petal, with a short cylindrical tube, whose brim is cut into five large obtuse segments, which spread open; it has ten or twelve awl-shaped stamina, disposed circularly, inserted in the petals, and terminated by oblong summits; and a roundish germen, supporting a single style the length of the stamina, crowned by a ragged stigma. The germen afterward turns to a roundish fruit with one cell, including two nuts, which are plain on one side, and convex on the other.

We have but one Species of this genus, viz.

Styrax (*Officinale.*) Hort. Cliff. 187. *The Storax-tree.*

This plant grows plentifully in the neighbourhood of Rome, and also in Palestine, and several of the islands in the Archipelago, from whence the fruit has been brought to England, where there have been many plants raised of late years in some curious gardens.

It has a woody stalk, which rises twelve or fourteen feet high, covered with a smooth grayish bark, sending out many slender ligneous branches, garnished with oval leaves, shaped like those of the Quince-tree, of a bright green on their upper side, but hoary on their under, placed alternately on short foot-stalks. The flowers come out from the side of the branches upon foot-stalks, which sustain five or six flowers in a bunch; these have one very white petal, which is funnel-shaped, the lower part being tubulous and cylindrical, and the upper divided into five obtuse segments, which spread open, but not flat, rather inclining to an angle. These are often succeeded by berries in England.

It may be propagated by sowing the seeds in pots, filled with fresh light earth, and plunged into a moderate hot-bed. If they are sown the latter end of summer, and the pots kept in a moderate hot-bed of tanners bark all the winter, the plants will come up the succeeding spring; whereas those sown in the spring, often remain in the ground a whole year before the plants come up.

When the plants are come up, they should be hardened gradually to the open air, into which they should be removed in June, placing them in a sheltered situation, observing to keep them clean from weeds, as also to supply them with water duly in dry weather. In this place they may remain till autumn, when they should be placed under a common hot-bed frame, where they may be screened from hard frost in winter, but in mild weather enjoy the free air as much as possible; for if they are kept too close, their tops are very subject to grow mouldy. The leaves of these plants fall off in autumn, and in the spring, before they begin to shoot, they should be shaken out of the pots, and their roots carefully parted, and each transplanted into a separate small pot, and plunged into a very moderate hot-bed, observing to water and shade them until they have taken root; after which they should be inured to the open air by degrees, into which they must be removed in June, placing them in a warm situation; in which place they may remain till the end of October, at which time they should be removed into shelter for the winter season. These plants are tolerably hardy, and only require to be sheltered from severe frost while they are young, for in Italy they grow extremely well in the open air, and produce fruit in great plenty. When the plants have grown three or four years in pots, and are become strong, some of them may be turned out of the pots, and planted in the full ground, against a wall to the south aspect, to which their branches should be trained, in the same manner as is practised with fruit-trees; in which situation they will bear the cold of our winters very well, but in very severe frost it will be proper to cover the branches either with mats, Straw, or other light covering to protect them. With this care the plants will flower annually, and in warm seasons ripen their seeds.

The gum of this tree is used in medicine, which is obtained by making incisions in the tree. It is brought from Turkey, but it is so adulterated by mixing saw-dust or other stuff with it, that it is very difficult to meet with any that is pure. It has a most pleasant fragrant odour; it is called Styrax Calamita, because it was transported in hollow canes.

Of late years there has been another species of Storax imported here from North America, which is collected from the liquid Amber-tree. This has been titled Liquid Storax by some, but is very different from that which is brought from Turkey, and is clear, inclining to yellow; it is brought sometimes liquid, and at others it is dried in the sun to a concrete resin before it is transported.

SUBER. See QUERCUS.

SUCCORY. See CICHORIUM.

SUMACH. See RHUS.

SURIANA. Plum. Nov. Gen. 37. tab. 40.

The CHARACTERS are,

The empalement of the flower is permanent, composed of five spear-shaped small leaves. The flower has five oval petals the length of the empalement, which spread open; it has five slender stamina, which are shorter than the petals, terminated by single summits, and five roundish germen, supporting a slender style the length of the stamina, which is inserted in the middle to the side of the germen, crowned with an obtuse stigma. The germen afterward becomes five roundish seeds joined together.

We have but one SPECIES of this genus, viz.

SURIANA (*Maritima.*) Hort. Cliff. 492. *Suriana.*

This plant grows naturally by the sea-side in most of the islands of the West-Indies, where it rises with a thick shrubby stalk eight or nine feet high, covered with a dark brown bark, dividing into branches; the upper part of which are closely garnished with leaves on every side, standing without order; they are rounded at their points, and sit close to the branches, having no foot-stalks, and of a dirty green colour. From between the leaves come out the foot-stalks of the flowers, which are about an inch long; these do each sustain two, three, or four yellow flowers, which have some four, and others five petals, rounded at their points, and almost heart-shaped; these are succeeded by roundish seeds, which are joined together, sitting in the empalement. Some flowers have two, others three, four, or five seeds to each.

It is propagated by seeds, which must be sown on a hot-bed early in the spring; and when the plants are come up, they must be carefully cleared from weeds, and frequently refreshed with water. In warm weather the glasses of the hot-bed should be raised every day, to admit fresh air to the plants, to prevent their drawing up too weak. When the plants are fit to remove, they should be taken up carefully, and each planted in a separate small pot, and plunged into a hot-bed of tanners bark, observing to shade them until they have taken new root; after which they must have fresh air admitted to them every day, in proportion to the warmth of the season. In this hot-bed the plants may remain till autumn, when the nights begin to be cold; at which time they should be removed into the stove, and plunged into the bark-bed. During the winter season these plants must be kept very warm, especially while they are young, otherwise they will not live through the winter in this country. They must also be frequently refreshed with water; but it must not be given to them in large quantities in cold weather, for too much moisture in winter will soon destroy them. These plants make but slow progress the first year; afterwards they will grow pretty freely, if they are not stinted. In summer they must have a large share of air, by opening the glasses of the stoves; and if their leaves are covered with filth (which the plants in stoves often contract,) they should be carefully washed with a sponge, otherwise the plants will not only appear unsightly, but it will retard their growth.

SYCAMORE. See ACER MAJUS.

SYMPHYTUM. Tourn. Inst. R. H. 138. tab. 56. Comfrey.

The CHARACTERS are,

The flower has a five-cornered, erect, permanent empalement, cut into five acute parts; it has one petal with a short tube, above which the limb has a swelling belly and thicker tube; the brim is indented in five obtuse parts, which are reflexed; the chaps are armed with five awl-shaped rays, which are connected in a cone; it has five awl-shaped stamina, which are alternate with the rays of the chaps, terminated by erect acute summits, and four germen, supporting a slender style the length of the petal, crowned by a single stigma. The germen afterward turns to four gibbous acute-pointed seeds, which ripen in the empalement.

The SPECIES are,

1. SYMPHYTUM (*Officinale*) foliis ovato-lanceolatis decurrentibus. Hort. Cliff. 47. *Comfrey with oval, spear-shaped, running leaves; Comfrey with a purple flower.*

2. SYMPHYTUM (*Tuberosum*) foliis summis oppositis. Lin. Sp. Plant. 136. *Comfrey with the upper leaves placed opposite.*

3. SYMPHYTUM (*Orientale*) foliis ovatis subpetiolatis. Lin. Sp. Plant. 136. *Comfrey with oval leaves and short foot-stalks.*

There are several other species of this genus, but those which are here enumerated, are all the sorts at present to be found in the English gardens.

The first sort grows naturally in England, but the most common here is that with a whitish yellow flower, which is found growing by the side of ditches and other moist places in great plenty, but that with purple flowers is the most common in Holland and Germany; these are supposed to be accidental varieties, which differ in the colour of their flowers; however, this difference is permanent in the plants raised from seeds, as I have many times found; nor are the two kinds ever found mixed where they grow wild, for in those places where the blue is found, the white is never seen, and vice versa: but as there are no specific differences between them, I shall not separate them.

The common Comfrey has thick roots, composed of many fleshy fibres or fangs, which run deep in the ground; they are black on the outside, but white within, full of a slimy tenacious juice. The lower leaves are large, long, sharp-pointed, hairy, and rough. The stalks rise two feet high, which are garnished with oval spear-shaped leaves, ending in acute points; they are hairy, rough, and from their base runs a leafy border along the stalk. From the upper part of the stalk are sent out some side branches, which are commonly garnished with two smaller leaves, and are terminated by loose bunches of flowers, which are reflexed; each flower has one tubulous petal, whose upper petal is bellied and thicker than the lower, and the chaps are closed by the stamina and rays, which cross it, and shuts up the tube. These in the common English sort are of a yellowish white, and the foreign one is of a purple colour.

The second sort grows naturally in Germany. The roots of this are composed of many thick fleshy knobs or tubers, which are joined by fleshy fibres: the stalks incline on one side; they rise a foot and a half high; the leaves on the lower part are six inches long, and two and a half broad in the middle, ending in acute points, and are not so rough and hairy as those of the other species; they are placed alternate, and sit close to the stalks. The two upper leaves on every branch stand opposite, and just above them are loose spikes or bunches of pale yellow flowers, whose petals are stretched out farther beyond the empalement than those of the other.

The third sort grows naturally on the side of rivers near Constantinople; this has a perennial root like the first; the stalks grow two feet high; the leaves are rounder, and are armed with rough prickly hairs. The flowers are blue, and grow in bunches like those of the first sort.

These plants may be cultivated either by sowing their seeds in the spring, or by parting of their roots; the latter way being the more expeditious, is chiefly practised where they are planted for use. The best season for parting the roots is in autumn, at which time almost every piece of a root will grow. They should be planted about two feet and a half asunder, that they may have room to spread, and

and will require no farther care but to keep them clear from weeds, for they are extremely hardy, and will grow upon almoſt any ſoil or in any ſituation.

SYRINGA. Lin. Gen. Plant. 22. Lilac.

The CHARACTERS are,

The flower has a ſmall, tubulous, permanent empalement of one leaf, indented in four parts; it has one petal, with a long cylindrical tube, cut into four obtuſe ſegments at the brim, which ſpread open, and two very ſhort ſtamina, terminated by ſmall ſummits, ſtanding within the tube; it has an oblong germen, ſupporting a ſhort ſlender ſtyle, crowned by a thick bifid ſtigma. The germen afterward turns to an oblong, compreſſed, acute-pointed capſule with two cells, opening with two valves contrary to the partition, including in each cell one oblong acute-pointed ſeed, with a membranaceous border.

The SPECIES are,

1. SYRINGA (*Vulgaris*) foliis ovato-cordatis. Hort. Cliff. 6. *Syringa with oval heart-ſhaped leaves; or blue Lilac.*

2. SYRINGA (*Perſica*) foliis lanceolatis. Lin. Sp. Pl. 9. *Syringa with ſpear-ſhaped leaves; commonly called Perſian Jaſmine.*

3. SYRINGA (*Laciniata*) foliis lanceolatis integris diſſectiſque laciniata. Hort. Cliff. 6. *Syringa with entire ſpear-ſhaped leaves, and others which are cut and jagged; commonly called cut-leaved Perſian Jaſmine.*

The firſt ſort is very common in the Engliſh gardens, where it has been long cultivated as a flowering ſhrub. It is ſuppoſed to grow naturally in ſome parts of Perſia, but is ſo hardy as to reſiſt the greateſt cold of this country. There are three varieties of this ſhrub, which are commonly cultivated in the Engliſh gardens, which differ in the colour of their flowers, and alſo in that of their ſhoots and leaves; one of theſe has white flowers, one blue, and the third has purple flowers; the latter is commonly known by the title of Scotch Lilac, to diſtinguiſh it from the other. This is the moſt beautiful of the three, and is probably called the Scotch Lilac, becauſe it was firſt mentioned in the Catalogue of the Edinburgh garden. Whether this was raiſed from ſeeds, or which other way it was obtained I could never learn, but I take it to be a diſtinct ſpecies from the others, though there is not marks enough upon them to diſtinguiſh their ſpecific differences, becauſe I have raiſed many of the plants from ſeeds, which have always retained their difference; as have alſo the white, when they were propagated by ſeeds, ſo that they may be rather eſteemed as diſtinct ſorts, although by the rules now admitted for determining ſpecific differences, they may not have ſufficient marks whereby to diſtinguiſh them; and as they have been by moſt of the modern botaniſts joined together, I ſhall not ſeparate them again, but ſhall mention the particulars in which they differ.

Theſe ſhrubs grow to the height of eighteen or twenty feet in good ground, and divide into many branches; thoſe of the white ſort grow more erect than the other, and the purple or Scotch Lilac has its branches more diffuſed than either. The branches of the white are covered with a ſmooth bark, of a gray colour; thoſe of the other two are darker. The leaves of the white are of a very bright green, but thoſe of the other are of a dark green; their ſhape and ſize are ſo near as not to be diſtinguiſhed thereby. They are heart-ſhaped, and are placed oppoſite. The buds of the future ſhoots, which are very turgid before the leaves fall, are of a very bright green in the white ſort, but thoſe of the other two are of a dark green. The flowers are always produced at the ends of the ſhoots of the former year, and below the flowers come out ſhoots to ſucceed them; for that part upon which the flowers ſtand, decays down to the ſhoots below every winter. There are generally two bunches or panicles of flowers joined at the end of each ſhoot; thoſe of the blue are the ſmalleſt, and are placed thinner than either of the other. The bunches on the white are larger; the flowers are cloſer placed, and larger than the blue; but thoſe of the Scotch are larger, and the flowers are fairer than thoſe of either of the other, ſo make a much finer appearance. The panicles of flowers grow erect, and being intermixed with the fine green leaves, have a fine effect; and if we add to this the fragrancy of their flowers, it may be ranged among the moſt beautiful ſhrubs which now decorate the Engliſh gardens. They flower in May, and when the ſeaſon is cool, theſe ſhrubs will continue three weeks in beauty, but in hot ſeaſons the flowers ſoon fade. Their ſeeds are ripe in September, which, if ſown ſoon after, the plants will come up the following ſpring; but as their roots ſend out great plenty of ſuckers annually, ſo few perſons ever take the trouble to propagate theſe plants by ſeeds. I have raiſed ſeveral plants of the three ſorts from ſeeds, and conſtantly found them prove the ſame as the ſhrubs from which the ſeeds were taken. Theſe plants do generally flower the third year from ſeed; and I have always found theſe plants not ſo apt to ſend out ſuckers, as thoſe which were produced by ſuckers, ſo are much more valuable, for the others put out ſuch plenty of ſuckers, as that if they are not annually taken from the plants, they will ſtarve them.

Theſe plants thrive beſt upon a light rich ſoil, ſuch as the gardens near London are for the moſt part compoſed of; and there they grow to a much larger ſize, where they are permitted to ſtand unremoved than in any other part of England; for in ſtrong loam, or upon chalky land, they make little progreſs. If the ſuckers are ſmall, when they are taken from the old plants, they ſhould be planted in a nurſery, in rows three feet aſunder, and one foot diſtance in the rows, where they may ſtand a year or two to get ſtrength, and then they ſhould be removed to the places where they are to remain. The beſt time to tranſplant theſe ſhrubs is in autumn.

There is a variety or two of theſe ſhrubs with blotched leaves, which ſome perſons are fond of; but as theſe variegations are the effect of weakneſs, ſo whenever the ſhrubs become healthy, their verdure returns again.

The ſecond ſort grows naturally in Perſia, but has been long cultivated in the Engliſh gardens, where it is beſt known among the gardeners by the title of Perſian Jaſmine. This is a ſhrub of much lower growth than the former, ſeldom riſing more than ſix or eight feet high. The ſtalks of this ſhrub are woody, covered with a ſmooth brown bark; the branches are ſlender, pliable, and extend wide on every ſide; theſe frequently bend downward where they are not ſupported; they are garniſhed with narrow ſpear-ſhaped leaves, placed oppoſite, of a deep green colour, ending in acute points. The flowers are produced in large panicles at the end of the former year's ſhoots, in like manner as the former; they are of a pale purple colour, and have a very agreeable odour. Theſe appear the latter end of May, ſoon after thoſe of the common ſort, and continue longer in beauty; but theſe do not perfect their ſeeds in England.

There is a variety of this with almoſt white flowers, which has of late years been obtained; but whether it came from ſeeds, or was accidentally produced by ſuckers from the purple kind, I cannot ſay.

The third ſort differs from the ſecond in having two ſorts of leaves; thoſe on the lower part of the branches are for the moſt part entire; theſe are broader and ſhorter than thoſe of the ſecond, and do not end in ſo ſharp points. The leaves on the younger branches are cut into three or five ſegments, like winged leaves almoſt to the midrib. The branches of this ſort are ſlenderer and weaker than thofe of the fecond; their bark is of a darker brown, and the flowers of a brighter purple colour.

This was brought into Europe before the other, and came by the Perfian title Agem. Both thefe forts are ufually propagated by fuckers, which their roots fend out in great plenty; thefe fhould be carefully taken off from the old plants in the autumn, and planted in a nurfery in the fame manner as is before directed for the firft, where they may grow two years to get ftrength, and may then be tranfplanted to the places where they are defigned to remain. The plants which are fo propagated, are always very prolific in fuckers; for which reafon it will be a better way to raife them by laying down their young branches, which in one year will be fufficiently rooted to tranfplant, and may then be treated in the fame way as the fuckers.

T.

TABERNÆMONTANA. Plum. Gen. Nov. 18. tab. 30.

The CHARACTERS are,

The flower has a fmall empalement, cut into five acute parts; it has one funnel-fhaped petal, with a long cylindrical tube, which is bellied at both ends, and at the brim is cut into five oblique fegments; it has five fmall ftamina in the middle of the tube, terminated by fummits which join together, and two germen, fupporting an awl-fhaped ftyle, crowned by decayed ftigmas. The germen afterward turns to two bellied capfules, which are horizontally reflexed, opening with one valve, having one cell, filled with oblong oval feeds, lying imbricatim, and furrounded with pulp.

The SPECIES are,

1. TABERNÆMONTANA (*Citrifolia*) foliis glomerato-umbellatis oppofitis ovatis lateralibus corymbofis lateralibus. Lin. Sp. 303. *Tabernæmontana with oval leaves, which are placed oppofite, and flowers growing in a corymbus on the fides of the branches.*

2. TABERNÆMONTANA (*Alba*) foliis oblongo-ovatis acuminatis oppofitis, floribus corymbofis terminalibus. *Tabernæmontana with oblong, oval, acute-pointed leaves, which are placed oppofite, and flowers growing in a corymbus, terminating the branches.*

3. TABERNÆMONTANA (*Laurifolia*) foliis oppofitis ovalibus obtufiufculis. Lin. Sp. 308. *Tabernæmontana with oval obtufe leaves placed oppofite.*

4. TABERNÆMONTANA (*Amfonia*) foliis alternis, caulibus fubherbaceis. Lin. Sp. 308. *Tabernæmonta with herbaceous ftalks and alternate leaves.*

The firft fort grows naturally in Jamaica, and fome of the other iflands in the Weft-Indies. Sir Hans Sloane has figured it in his Hiftory of Jamaica, under the title of Nerium arboreum folio latiore obtufo, flore luteo minore. Tab. 186. fig. 2. *Tree-like Nerium with a broader obtufe leaf, and a fmaller yellow flower.*

This rifes with an upright woody ftalk to the height of fifteen or fixteen feet, covered with a fmooth gray bark, which abounds with a milky juice, and fends out feveral branches from the fide, which grow erect, garnifhed with thick leaves, which have a milky juice, of a lucid green, and have many tranfverfe veins from the midrib to the border, ftanding oppofite on foot-ftalks an inch long. The flowers come out in roundifh bunches from the wings of the ftalk; they are fmall, of a bright yellow colour, and have an agreeable odour. The tube of the flower is half an inch long; the brim is cut into five acute points, which fpread open like thofe of the common Jafmine. Thefe flowers, in their native foil, are fucceeded by two fwelling capfules, joined at their bafe, but fpread from each other horizontally, and are filled with oblong feeds, lying over each other like the fcales of fifh, and are included in a foft pulp.

The fecond fort was difcovered by the late Dr. William Houftoun in the year 1730, growing naturally at La Vera Cruz. This rifes with a woody ftalk ten or twelve feet high, covered with a wrinkled gray bark, fending out many branches toward the top, which are garnifhed with oblong oval leaves, of a lucid green, and of a thick confiftence; they are five inches long, and two and a half broad, rounded at both ends, but terminate with an acute point. Thefe are placed oppofite, and have fhort foot-ftalks. The flowers come out in pretty large roundifh bunches at the end of the branches; they are white, and fmaller than thofe of the firft fort, having an agreeable fcent. Thefe are fucceeded by fhorter and rounder pods, which fpread from each other horizontally like the former.

Both thefe plants are very impatient of cold, fo will not live in this country, unlefs they are placed in a warm ftove. They may be propagated by feeds or cuttings; if by feeds, they muft be procured from the countries where the plants grow naturally, for the feeds are never ripened in England; thefe fhould be fown early in the fpring on a hot-bed, and when the plants are come up, they muft be carefully tranfplanted into fmall pots, filled with light rich earth, and plunged into a hot-bed of tanners bark, being careful to fhade them in the heat of the day until they have taken new root; after which time, they muft have free air admitted to them every day when the weather is warm; but if the nights fhould prove cold, the glaffes of the hot-bed fhould be covered with mats every evening, foon after the fun goes off from the bed. Thefe plants muft be often refrefhed with water, but it muft not be given to them in large quantities, efpecially while they are young, for as they are full of a milky juice, they are very fubject to rot with much moifture.

The plants may remain during the fummer feafon in the hot-bed, provided the tan is ftirred up to renew the heat when it wants, and a little new tan added; but at Michaelmas, when the nights begin to be cold, the plants fhould be removed, and plunged into the bark-bed in the ftove; where, during the winter feafon, they muft be kept in a moderate degree of warmth, and in cold weather they

fhould

should have but little water given them. As these plants are too tender to live in the open air in this country, they should constantly remain in the stove, where, in warm weather, they may have free air admitted to them, by opening the glasses, but in cold weather they must be kept warm. With this management the plants will thrive and produce their flowers; and as their leaves are always green, they will make a pleasant diversity amongst other tender exotic plants in the stove.

The third sort grows naturally in Jamaica, and also in some of the other warm islands of America. This rises with a shrubby stalk fourteen feet high, sending out a few branches toward the top, garnished with oval obtuse leaves, placed opposite. The flowers are produced in a racemus; they are yellow, and have an agreeable odour. This plant requires the same treatment as the two others.

The fourth sort is a perennial plant, whose roots will live a few years, but the stalks decay every autumn; it grows naturally in Virginia, where it was discovered by Mr. Clayton, who gave it the title of Amsonia. This sends up in the spring two or three stalks a foot high, which are white, having no odour, but terminate the stalks; it produces no seeds in this country, so is only propagated by offsets, which renders this plant rare in England.

The plants of the three first sorts may also be propagated by cuttings during the summer season, which should be cut off from the old plants, and laid to dry in the stove five or six days before they are planted, that the wounded parts may heal, otherwise they will rot. These cuttings should be planted in pots, filled with fresh light earth, and plunged into the hot-bed of tanners bark, and closely covered with a hand-glass, observing to shade them from the sun in the middle of the day in hot weather, as also to refresh them now and then with a little water. When the cuttings have taken root, they may be transplanted into separate pots, and treated in the same manner as those which are raised from seeds.

TACAMAHACA. See Populus.

TAGETES. Tourn. Inst. R. H. 478. tab. 278. African, or French Marigold.

The Characters are,

The common empalement of the flower is single, oblong, erect, and five-cornered; the flower is compound and radiated; the ray or border is composed of five female half florets, which are tongue-shaped. The disk or middle is made up of hermaphrodite florets, which are tubulous, cut into five obtuse segments; these have five short hair-like stamina, terminated by cylindrical summits, and an oblong germen, supporting a slender style, crowned by a bifid reflexed stigma. The germen afterward becomes a single, linear, compressed seed, almost the length of the empalement, crowned by five acute-pointed unequal scales.

The Species are,

1. Tagetes (*Erecta*) caule simplici erecto, pedunculis nudis unifloris. Hort. Cliff. 418. *Tagetes with a single erect stalk, and naked foot-stalks bearing one flower; or upright African Marigold.*

2. Tagetes (*Patula*) caule subdiviso diffuso. Hort. Cliff. 418. *Tagetes with a diffused subdivided stalk; commonly called French Marigold.*

3. Tagetes (*Minuta*) caule simplici erecto, pedunculis squamosis multifloris. Hort. Cliff. 419. *Tagetes with a single erect stalk, and scaly foot-stalks bearing many flowers.*

4. Tagetes (*Rotundifolia*) caule simplici erecto, foliis cordatis simplicibus, pedunculis nudis unifloris. *Tagetes with a single stalk, simple heart-shaped leaves, and naked foot-stalks, having one flower.*

The first sort grows naturally in Mexico, but has been long cultivated in the English gardens, where it is commonly titled African, or African Marigold; of this there are the following varieties:

1. Pale yellow, or brimstone colour, with single, double, and fistulous flowers.

2. Deep yellow with single, double, and fistulous flowers.

3. Orange-coloured, with single, double, and fistulous flowers.

4. Middling African, with Orange-coloured flowers.

5. Sweet-scented African.

These are all very subject to vary, so that unless the seeds are very carefully saved from the finest flowers, they are very apt to degenerate: nor should their seeds be too long sown in the same garden without changing it, for the same reason; therefore, those who are desirous to have these flowers in perfection, should exchange their seeds with some person of integrity at a distance where the soil is of a different nature, at least every other year. If this is done, the varieties may be continued in perfection.

This plant is so well known as to need no description It flowers from the beginning of July, till the frost puts a stop to it.

The second sort grows naturally in Mexico, but has been long in the English gardens, where it is distinguished from the first by the title of French Marigold.

Of this there are several varieties, some of which have much larger flowers than others, and their colour varies greatly; there are some which are beautifully variegated, and others quite plain; but as these are accidents arising from culture, so they do not merit farther distinction; for I have always found that seeds saved from the most beautiful flowers will degenerate, especially if they are sown in the same garden for two or three years together, without changing the seed.

These plants are annual, so must be propagated from seeds every spring, which may be sown upon a moderate hot-bed the beginning of April; and when the plants are come up, they should have plenty of fresh air, for if they are drawn too much, they will not afterward become handsome, notwithstanding they have all possible care taken of them. When they are about three inches high, they should be transplanted on a very moderate hot-bed, which may be arched over with hoops and covered with mats, for these plants are hardy enough to be brought up without glasses; in this bed they should be planted about six inches asunder each way, observing to water and shade them until they have taken root; but as the plants acquire strength, they should be inured to bear the open air by degrees; and about the end of May they should be taken up with a ball of earth to the root of each plant, and planted into the borders of the parterre-garden, or into pots, for furnishing the courts, &c. shading them carefully from the sun till they have taken new root, and also supplying them duly with water. When their flowers appear, if any should prove single, the plants should be destroyed, and then those in pots may be removed to the court where the several varieties, being intermixed with other annual plants, afford an agreeable variety.

These plants have a strong disagreeable scent, especially when handled, for which reason they are not so greatly esteemed for planting near habitations; but the flowers of the sweet-scented sort being more agreeable, are generally preferred, especially for planting in small gardens.

The third sort grows naturally in Chili in the Spanish West-Indies. This is a plant of taller growth than either of the former. The stalk is single, erect, and branches a little toward the top; it rises about ten feet high; the branches grow erect. The leaves are narrower than either of the other. The foot-stalks of the flowers are scaly, and

stand

ſtand erect cloſe to the ſtalk; theſe ſuſtain three or four ſmall white flowers, which appear very late in autumn; ſo that unleſs it is kept in a glaſs-caſe, the ſeeds will not ripen here. This plant has very little beauty, ſo is only preſerved for the ſake of variety.

The fourth ſort riſes with an upright ſtalk about two feet high, ſending out a few branches toward the top, garniſhed with heart-ſhaped leaves, ſtanding upon long ſlender foot-ſtalks, ending in very acute points, being in ſhape like thoſe of the black Poplar, rough to the touch, and are ſlightly crenated on their edges; the branches and ſtalks are each terminated by one large yellow flower, ſtanding upon a long naked foot-ſtalk. The empalement of the flower is ſhort; the leaves of which it is compoſed are oblong and oval, drawing to a point. The female florets, which compoſe the rays or border, are much longer than the empalement. The hermaphrodite florets, in the diſk or middle, are equal; they are of a deep yellow colour, and make a good appearance, for the flowers are double. This plants was diſcovered by the late Dr. Houſtoun growing naturally at La Vera Cruz in New Spain, from whence he ſent the ſeeds to England.

The two laſt ſorts are not ſo hardy as the former, ſo the ſeeds of theſe ſhould be ſown earlier in the ſpring upon a good hot-bed; and when the plants are fit to remove, they ſhould be tranſplanted on a freſh hot bed, at about three inches diſtance each way, obſerving to ſhade them from the ſun till they have taken new root; then they ſhould be treated in the ſame way as the Amaranthus, and other tender annual plants, being careful not to draw them up weak; when they have ſpread ſo as to meet each other, they ſhould be taken up with balls of earth to their roots, and planted in pots filled with light rich earth, and plunged into a hot-bed under a deep frame, where the plants may have room to grow, being careful to ſhade them from the ſun till they have taken new root; after which they muſt have air and water in proportion to the warmth of the ſeaſon; and when the plants have grown too tall to remain longer in the frame, they ſhould be removed to an airy glaſs-caſe, where they may ſtand to flower and ripen their ſeeds.

TAMARINDUS. Tourn. Inſt. R. H. 660. tab. 445. The Tamarind-tree.

The CHARACTERS are,

The empalement of the flower is compoſed of five oval plain leaves, which are equal; the flower has five petals, which are almoſt like thoſe of the butterfly kind, one of them ſtanding erect; two are placed like wings on each ſide, and two reflect downward; it has three awl-ſhaped ſtamina, ſituated in the ſinuſes of the empalement, and are arched toward the upper petal, terminated by ſingle ſummits, and an oblong oval germen, ſupporting an awl-ſhaped aſcending ſtyle, crowned by a ſingle ſtigma. The germen afterward becomes a long, ſwelling, compreſſed pod, having a double cover, and one cell containing three, four, or five angular compreſſed ſeeds, ſurrounded with pulp.

We have but one SPECIES of this genus, viz.

TAMARINDUS (*Indica.*) Hort. Cliff. 18. *The Tamarind-tree.*

This tree grows naturally in both Indies, and alſo in Egypt; but it has been ſuppoſed by ſome eminent botaniſts, that the Tamarind which grew in the Eaſt-Indies, was different from that of the Weſt, becauſe the pods of the firſt are almoſt double the length of thoſe of the latter. The pods which have been brought me from the Eaſt-Indies, have generally been ſo long as to contain five, ſix, and ſometimes ſeven ſeeds, whereas thoſe of the Weſt-Indies have very rarely more than four; but the plants which I have raiſed from the ſeeds of both ſorts, are ſo like as not to be diſtinguiſhed, therefore I ſuppoſe it may be owing to the ſoil or culture, that one is ſo much larger than the other.

This tree grows to a very large ſize in thoſe countries where it is a native, but in England it will not thrive out of a ſtove, eſpecially in winter. The ſtem is very large, covered with a brown bark, and divides into many branches at the top, which ſpread wide every way, and are cloſely garniſhed with winged leaves, compoſed of ſixteen or eighteen pair of lobes, without a ſingle one at the end. The lobes are about half an inch long, and a ſixth part of an inch broad, of a bright green, a little hairy, and ſit cloſe to the midrib. The flowers come out from the ſide of the branches five, ſix, or more together upon the ſame foot-ſtalk in looſe bunches; theſe are compoſed of five reddiſh petals, one of which is reflexed upward like the ſtandard in ſome of the butterfly flowers, two others ſtand on each ſide like the wings, and the other two are turned downwards; theſe (in the countries where the plants grow naturally) are ſucceeded by thick compreſſed pods, two, three, four, or five inches long, having a double ſkin or cover, and ſwell in every place where the ſeeds are lodged, full of an acid ſtringy pulp, which ſurrounds ſmooth, compreſſed, angular ſeeds.

The Tamarinds which are brought from the Eaſt-Indies, are darker and drier, but contain more pulp, being preſerved without ſugar, and are fitter to be put into medicines than thoſe from the Weſt-Indies, which are redder, have leſs pulp, and are preſerved with ſugar, ſo are pleaſanter to the palate.

The plants are preſerved in the gardens of thoſe who have conveniency to maintain rare exotic-trees and ſhrubs.

They are eaſily propagated by ſowing their ſeeds on a hot-bed in the ſpring; and when the plants are come up, they ſhould be planted each into a ſeparate ſmall pot, filled with light rich earth, and plunged into a hot-bed of tanners bark to bring them forward, obſerving to water and ſhade them until they have taken root; and as the earth in the pots appear dry, they muſt be watered from time to time, and ſhould have air given to them in proportion to the warmth of the ſeaſon, and the bed in which they are placed; when the pots in which they are planted are filled with their roots, the plants ſhould be ſhifted into pots of a larger ſize, which muſt be filled up with rich light earth, and again plunged into the hot-bed, giving them air, as before, according to the warmth of the ſeaſon; but in very hot weather the glaſſes ſhould be ſhaded with mats in the heat of the day, otherwiſe the ſun will be too violent for them through the glaſſes, nor will the plants thrive, if they are expoſed to the open air, even in the warmeſt ſeaſon; ſo that they muſt be conſtantly kept in the bark-ſtove both ſummer and winter, treating them as hath been directed for the Coffee-tree, with whoſe culture they will thrive exceeding well.

Theſe plants, if rightly managed, will grow very faſt; for I have had them upwards of three feet high in one ſummer from ſeed, and have had two plants which produced flowers the ſame ſeaſon they were ſown; but this was accidental, for none of the older plants have produced any flowers, although I have ſeveral plants of different ages, ſome of which are ſixteen or eighteen years old, and about twelve feet high, with large ſpreading heads.

TAMARIX. Lin. Gen. Plant. 337. The Tamariſk.

The CHARACTERS are,

The empalement of the flower is obtuſe, erect, and permanent; it is cut into five parts; the flower has five oval concave petals, which ſpread open, and five hair-like ſtamina, terminated by roundiſh ſummits; it has an acute-pointed germen without a ſtyle, crowned by three oblong, feathery, twiſted ſtigmas. The germen afterward turns to an oblong acute-pointed capſule with three corners, having one cell, opening with three valves, containing many ſmall downy ſeeds.

The

The SPECIES are,

1. TAMARIX (*Gallica*) floribus pentandris. Hort. Cliff. 111. *Tamarisk with flowers having five stamina; or French Tamarisk.*

2. TAMARIX (*Germanica*) floribus decandris. Hort. Cliff. 111. *Tamarisk whose flowers have ten stamina; or German Tamarisk.*

The first sort grows naturally in the south of France, in Spain, and Italy, where it rises to a middling size, but in England is seldom more than fourteen or sixteen feet high. The bark is rough, and of a dark brown colour; it sends out many slender branches, most of which spread out flat, and hang downward at their ends; these are covered with a Chestnut-coloured bark, and garnished with very narrow finely divided leaves, which are smooth, of a bright green colour, and have small leaves or indentures which lie over each other like scales of fish. The flowers are produced in taper spikes at the end of the branches, several of them growing on the same branch. The spikes are about an inch long, and as thick as a large earth-worm. The flowers are set very close all round the spike; they are very small, and have five concave petals, of a pale flesh colour, with five slender stamina, terminated by roundish red summits, and are succeeded by oblong, acute-pointed, three-cornered capsules, filled with small downy seeds, which seldom ripen in England.

The wood, bark, and leaves of this tree, are used in medicine, and are accounted specific for all disorders of the spleen, as being believed to lessen it much. The bark is sometimes used for rickets in children.

The second sort grows naturally in Germany, in moist land; this is rather a shrub than tree, having several ligneous stalks arising from the same root, which grow erect, sending out many side branches, which are also erect, having a pale green bark when young, which afterward changes to a yellowish colour. The leaves are shorter, and set closer together than those of the other sort, and are of a lighter green, approaching to a gray colour; the flowers are produced in long loose spikes at the end of the branches, standing erect; they are larger than those of the former, and have ten stamina standing alternately.

These both cast their leaves in autumn, and it is pretty late in the spring before the young ones push out, which renders them less valuable; they are now frequently planted in gardens for ornament, and, when they are mixed with other shrubs, make a pretty variety.

They may be easily propagated by laying down their tender shoots in autumn, or by planting cuttings in an east border in the spring before they begin to shoot, which, if supplied with water in dry weather, will take root in a short time; but they should not be removed until the following autumn, at which time they may be either placed in a nursery to be trained up two or three years, or else into the quarters where they are designed to remain, observing to mulch their roots, and water them according as the season may require, until they have taken root; after which, the only culture they will require is to prune off the straggling shoots, and keep the ground clean about them.

TAMUS. Lin. Gen. Plant. 991. The black Briony.

The CHARACTERS are,

It has male and female flowers on different plants. Those of the male plants have empalements, composed of six oval, spear-shaped, spreading leaves; they have no petals, but have six short stamina, terminated by erect summits; the flowers of the female have bell-shaped empalements of one leaf, cut into six spear-shaped segments, which sit upon the germen; these have no petals, but have oblong punctured nectarii sitting on the inside of each segment of the empalement, and a large, oblong, oval, smooth germen under the empalement, with a cylindrical style, crowned by three reflexed indented stigmas. The germen afterward becomes an oval berry with three cells, including two globular seeds.

The SPECIES are,

1. TAMUS (*Communis*) foliis cordatis indivisis. Hort. Cliff. 458. *Black Briony with heart-shaped undivided leaves; common black Briony.*

2. TAMUS (*Cretica*) foliis trilobis. Lin. Sp. Pl. 1028. *Black Briony with leaves which are divided into three lobes.*

The first sort is rarely cultivated in gardens, but grows wild under the sides of hedges in divers parts of England, and is there gathered for medicinal use. The root is very large, fleshy, and has a dark brown skin or cover; the stalks are smooth, and twine round any neighbouring support, whereby they rise to the height of ten or twelve feet, garnished with smooth heart-shaped leaves, of a lucid green, which are alternate. The flowers are produced in long bunches from the side of the stalks; those of the male plants fall off soon after their farina is cast abroad, but the female flowers are succeeded by oval smooth berries, which are red when ripe.

It may be easily propagated by sowing the seeds soon after they are ripe, under the shelter of bushes, where, in the spring, the plants will come up, and spread their branches over the bushes, and support themselves, requiring no farther care, and their roots will abide many years in the ground without decaying.

The second sort was discovered in the island of Crete by Dr. Tournefort, who sent the seeds to the Royal Garden at Paris; this has a rounder root than the other. The stalks twine round any neighbouring support in like manner; the leaves of this are divided into three lobes, in which the principal difference consists. This is an abiding plant, which is hardy enough to live in the full ground in England, and may be propagated as the other.

TANACETUM. Tourn. Inst. R. H. 461. tab. 261. Tansey.

The CHARACTERS are,

It has a flower composed of hermaphrodite and female florets, contained in one common, hemispherical, imbricated empalement, whose scales are compact and acute-pointed. The hermaphrodite florets, which compose the disk of the flower, are funnel-shaped, cut into five segments, which are reflexed; these have five short hair-like stamina, terminated by cylindrical tubulous summits, and a small oblong germen, supporting a slender style, crowned by a bifid revolved stigma. The germen afterward becomes an oblong naked seed. The female florets are trifid, which compose the rays or border, and are deeply divided within; these have an oblong germen with a slender style, crowned with two reflexed stigmas, but no stamina.

The SPECIES are,

1. TANACETUM (*Vulgare*) foliis bipinnatis incisis serratis. Hort. Cliff. 398. *Tansey with doubly-winged cut leaves, which are sawed; or common yellow Tansey.*

2. TANACETUM (*Sibericum*) foliis pinnatis, laciniis lineari-filiformibus, corymbis glabris, caule herbaceo. Lin. Sp. Plant. 844. *Tansey with winged leaves, which are cut into linear thread-like segments, a smooth corymbus, and an herbaceous stalk.*

3. TANACETUM (*Balsamita*) foliis ovatis integris serratis. Hort. Cliff. 398. *Tansey with oval, entire, sawed leaves; Costmary, or Alecoast.*

4. TANACETUM (*Fruticosum*) foliis pinnatifidis, laciniis lanceolatis obtusiusculis integerrimis. Lin. Sp. Plant. 844. *Tansey with wing-pointed leaves, having spear-shaped, entire, obtuse segments; or African Tree Tansey.*

5. TANACETUM (*Suffruticosum*) foliis pinnato-multifidis, laciniis linearibus subdivisis. Hort. Cliff. 398. *Tansey with many-*

many-pointed winged leaves, having linear segments, which are divided and acute; or shrubby African Tansey, with leaves like the cut-leaved Lavender.

6. TANACETUM (*Crithmifolium*) foliis pinnatis, pinnis linearibus remotis integerrimis. Lin. Sp. Plant. 843. *Tansey with winged leaves, whose lobes are linear, grow at a distance from each other, and are entire.*

The first sort is the common Tansey, which is used in medicine and the kitchen; this grows naturally by the sides of roads, and the borders of fields in many parts of England. It has a fibrous creeping root, which will spread to a great distance, where they are not confined, from which arise many channelled stalks from two to almost four feet high, according to the goodness of the soil, garnished with doubly-winged leaves, whose lobes are cut and sharply sawed, of a deep green colour, and have a pleasant grateful odour. The stalks divide near the top into three or four branches, which stand erect, and are terminated by umbels of naked yellow flowers, composed of many florets, which are included in hemispherical scaly empalements.

There are three varieties of this, one with a curled leaf, which is titled Double Tansey by gardeners; another with variegated leaves; and a third with larger leaves, which have little scent; but, as these have accidentally been produced from seeds of the common Tansey, they are not enumerated as distinct species.

This sort is easily propagated by the creeping roots, which, if permitted to remain undisturbed, will, in a short time, overspread the ground where they are permitted to grow; so that wherever this is planted in a garden, the slips should be placed two feet asunder, and in particular beds, where the paths round them may be often dug, to keep their roots within bounds. They may be transplanted either in spring or autumn, and will thrive in almost any soil or situation.

The common Tansey is greatly used in the kitchen early in the spring, at which season, that which is in the open ground, or especially in a cold situation, is hardly forward enough to cut, so that where this is much wanted at that season, it is the best way to make a gentle hot-bed in December, and plant the old roots thereon without parting them, and arch the bed over with hoops, to cover it with mats in cold weather, by which method the Tansey will come up in January, and be fit to cut in a short time after.

The second sort grows naturally in Siberia; this has a perennial fibrous root. The stalks rise three feet high; the leaves are narrow and winged; the lobes are very narrow, and end in two or three points, which are entire; the flowers are produced in small thin umbels from the side, and at the top of the stalk; they are yellow and but small, the umbels having few florets in each. It may be propagated in the same way as the first.

The third sort grows naturally in the south of France and Italy, but is here planted in gardens, and was formerly pretty much used in the kitchen, and also in medicine. The roots of this are hard, fleshy, and creep in the ground; the lower leaves are oval, sawed on their edges, of a grayish colour, and have long foot-stalks. The stalks rise from two to three feet high, and send out branches from the side, garnished with oval sawed leaves like those at the bottom, but smaller, and sit close to the stalk. The flowers are produced at the top of the stalks in a loose corymbus; they are naked, and of a deep yellow colour. The whole plant has a soft pleasant odour.

It is propagated easily by parting of their roots; the best time for this is autumn, that they may be well established in the ground before the spring. Where the plant is cultivated for use, the plants should be planted in beds at two feet distance every way, that they may have room to grow, for in two years the roots will meet, so every other year they should be transplanted and parted to keep them within compass; they will thrive in almost any soil or situation, but will continue longest in dry land.

The fourth sort grows naturally at the Cape of Good Hope; this rises with a shrubby stalk eight or ten feet high, sending out branches on every side, garnished with wing-pointed leaves, whose segments are spear-shaped, entire, and blunt-pointed. The flowers are produced in small roundish bunches at the end of the branches; they are of a sulphur colour. The seeds rarely ripen in England.

The fifth sort was brought from the Cape of Good Hope, where it grows naturally; this rises with a branching under shrub stalk two feet high, garnished with wing-pointed leaves, whose lobes are very narrow, and frequently cut into acute segments. The flowers are produced in small roundish bunches at the end of the branches; they are larger than those of the former sort, and are of a bright yellow colour.

The sixth sort grows naturally at the Cape of Good Hope; this has a thick shrubby stalk, covered with a gray bark, which rises seven or eight feet high, sending out many branches on every side, which are closely garnished with linear-winged leaves, whose lobes or pinnæ are very narrow, and spread from each other. The leaves sit close to the stalks on every side; the branches are terminated by close, large, roundish bunches of bright yellow flowers. Some of the foot-stalks sustain but one, others two, three, or four flowers upon each; there is a succession of them on the same plants till late in autumn, and those which come early in the season will be succeeded by seeds.

The fourth and sixth sorts are too tender to live through the winter in the open air, so must be kept in pots, and removed into shelter before hard frosts come on; they are all of them easily propagated by cuttings, which may be planted in a bed of loamy earth during any of the summer months; these should be shaded from the sun until they have taken root, and must be frequently refreshed with water. When they have good roots, they should be taken up with balls of earth, and planted in pots, placing them in a shady situation till they have taken new root; then they may be removed to a sheltered situation, placing them among other hardy exotic plants, where they may remain till late in October, when they must be put into shelter. These plants are so hardy, as only to require protection from hard frost, so must not be tenderly treated, and in mild weather should always be as much exposed to the air as possible, to prevent their drawing weak.

TAN, or TANNERS BARK, is the bark of the Oak-tree, chopped or ground into coarse powder, to be used in tanning or dressing of skins, after which it is of great use in gardening: first, by its fermentation (when laid in a proper quantity,) the heat of which is always moderate, and of long duration, which renders it of great service for hot-beds; and secondly, after it is well rotted, it becomes excellent manure for all sorts of cold stiff land, upon which one load of Tan is better than two of rotten dung, and will continue longer in the ground.

The use of Tan for hot-beds has not been many years known in England. The first hot-beds of this sort which were made in England were at Blackheath in Kent, above fourscore years ago; these were designed for the raising of Orange-trees; but the use of these hot-beds being but little known at that time, they were made but by two or three persons, who had learned the use of them in Holland and Flanders, where the gardeners seldom make any other hot-beds; but in England there were very few hot-beds made of Tanners Bark, before the Ananas plants were introduced into this country, which was in 1719, since which time the

the uſe of theſe hot-beds have been more general, and are now made in all thoſe gardens where the Ananas plants are cultivated, or where there are collections of tender exotic plants preſerved; and the gardeners here are now better ſkilled in the making and managing of theſe hot-beds, than in moſt other countries, which might render it leſs neceſſary to give a full deſcription of them here; but yet, as there may be ſome perſons in the remote parts of England, who have not had an opportunity of informing themſelves of the uſe of Tanners Bark for this purpoſe, I ſhall inſert the ſhorteſt and plaineſt method of making and managing them, as they are practiſed by the moſt knowing perſons, who have long made uſe of theſe hot-beds; and firſt I ſhall begin with the choice of the Tan.

The tanners, in ſome parts of England, do not grind the bark to reduce it into ſmall pieces, as is commonly practiſed by the tanners near London, where there is great difference in the ſize of the Bark, ſome being ground much ſmaller than the other, according to the different purpoſes for which it is intended; but in many places the Bark is only chopped into large pieces, which renders it very different for the uſe of hot-beds; for if the Tan is very coarſe, it will require a longer time to ferment than the ſmall Tan; but when it begins to heat, it will acquire a much greater degree, and will retain the heat a much longer time than the ſmall; therefore where there is choice, the middling-ſized Tan ſhould be preferred, for it is very difficult to manage a hot-bed when made of the largeſt Tan; the heat of which is often ſo great, as to ſcald the roots of plants, if the pots are fully plunged into the bed; and I have known this violent heat continue upward of two months, ſo that it has been unſafe to plunge the pots more than half their depth into the Tan, till near three months after the beds have been made; therefore where the perſons, who have the care of theſe beds, do not diligently obſerve their working, they may in a ſhort time deſtroy the plants which are placed in the beds: on the other hand, if the Tan is very ſmall, it will not retain the heat above a month or ſix weeks, and will be rotten and unfit for a hot-bed in a ſhort time; ſo that where the middle-ſized Tan can be procured, it ſhould always be preferred to any other, otherwiſe it will be proper to mix the ſmall with the large Tan.

The Tan ſhould be always ſuch as has been newly taken out of the pits, for if it lies long in the tanners yard before it is uſed, the beds ſeldom acquire a proper degree of heat, nor do they continue their heat long; ſo that when it has been more than three weeks or a month out of the pit, it is not ſo good for uſe as that which is new. If the Tan is very wet, it will be proper to ſpread it abroad for two or three days, to drain out the moiſture, eſpecially if it is in autumn or winter ſeaſon, becauſe then, as there will be little ſun to draw a warmth into the Tan, the moiſture will prevent the fermentation, and the beds will remain cold, but in the ſummer ſeaſon there is no great danger from the moiſture of the Tan. The heat of the ſun through the glaſſes will be then ſo great, as ſoon to cauſe a fermentation in the Tan.

Theſe Tan-beds ſhould always be made in pits, having brick-walls round them, and a brick pavement at the bottom, to prevent the earth from mixing with the Tan, which will prevent the Tan from heating. Theſe pits muſt not be leſs than three feet deep, and ſix feet in width; the length muſt be in proportion to the number of plants they are to contain; but if they are not ten feet in length, they will not retain their heat long, for where there is not a good body of Tan, the outſide of the bed will ſoon loſe its heat; ſo that the plants, which are there plunged, will have no benefit of the warmth, nor will the middle of theſe beds retain their heat long, ſo that they will not anſwer the purpoſe for which they are intended.

When the Tan is put into the bed, it muſt not be beaten or trodden down too cloſe, for that will cauſe it to adhere, and form one ſolid lump, ſo that it will not acquire a proper heat; nor ſhould it be trodden down at the time when the pots are plunged into the beds; to avoid which, there ſhould be a board laid croſs the bed, which ſhould be ſupported at each end, to prevent its reſting upon the Tan, upon which the perſon ſhould ſtand who plunges the pots, ſo that the Tan will not be preſſed down too cloſe. When the Tan is quite freſh, and has not been out of the pits long enough to acquire a heat, the beds will require a fortnight or three weeks time, or ſometimes a month, before they will be of a proper temperature of warmth to receive the plants; but in order to judge of this, there ſhould be three or four ſticks thruſt down in the Tan about eighteen inches deep, in different parts of the bed, ſo that by drawing out the ſticks, and feeling them at different depths, it will be eaſy to judge of the temper of the bed; and it will be proper to let a few of theſe ſticks remain in the bed, after the plants are plunged, in order to know the warmth of the Tan, which may be better judged of by feeling theſe ſticks, than by drawing out the pots, or plunging the hand into the Tan.

When the Tan is good, one of theſe beds will retain a proper degree of heat for near three months; and when the heat declines, if the Tan is forked up and turned over, and ſome new Tan added to it, the heat will renew again, and will continue two months longer; ſo that by turning over the Tan, and adding ſome new Tan every two months, or thereabouts, as the bed is found to decline of its heat, they may be continued one year; but every autumn it will be proper to take out a good quantity of the old Tan, and to add as much new to the bed, that the heat of the bed may be kept up in winter; for if the heat is ſuffered to decline too much during the cold ſeaſon, the plants will ſuffer greatly; to prevent this, there ſhould always be ſome new Tan added to the bed in winter, when the heat is found to decline; but the Tan ſhould be laid in a dry place a week or ten days to dry before it is put into the bed, otherwiſe the moiſture will chill the old Tan in the bed, and prevent the fermentation; ſo that unleſs the Tan is turned over again, there will be little or no heat in the beds, which often proves fatal to the plants which are plunged in them; therefore whoever has the management of theſe beds, ſhould be very careful to obſerve conſtantly the warmth of the Tan, ſince, upon keeping the beds in a due temperature of warmth, their whole ſucceſs depends; and where this caution is not taken, it frequently happens that the Ananas plants run into fruit very ſmall, or the plants are infeſted by inſects; both which are occaſioned by the growth of the plants being ſtopped, either by the decline of the heat of the Tan, or the heat being too great; therefore great regard muſt be had to that, eſpecially in winter.

The great advantages which theſe Tan-beds have of thoſe which are made of horſe-dung, are the moderate degree of heat which they acquire, for their heat is never ſo violent as that of horſe-dung, and they continue this heat much longer; and when the heat declines, it may be renewed, by turning the beds over, and mixing ſome new Tan with the old, which cannot be ſo well done with horſe-dung, and likewiſe the beds will not produce ſo great ſteams, which are often injurious to tender plants, ſo that theſe Tan-beds are much preferable to thoſe of horſe-dung for moſt purpoſes.

Tan, when it is well rotted, is alſo an excellent manure for all cold and ſtiff lands; and if it is laid upon Graſs ground in autumn, that the rains in winter may waſh it into the ground, it will greatly improve the Graſs; but

 when

when it is used new, or in the spring of the year, when dry weather comes, it is apt to cause the Grass to burn, which has occasioned the disuse of Tan in many places, but if properly used, it will be found an excellent dressing for all stiff lands.

TAPIA. See CRATEVA.

TARCHONANTHUS. Lin. Gen. Plant. 846.

The CHARACTERS are,

It has a flower composed of several hermaphrodite florets, included in one common top-shaped empalement, which is short, permanent, and hairy. The florets are uniform, funnel-shaped, and of one petal, indented in five parts at the top; they have each five very short hair-like stamina, terminated by cylindrical tubulous summits longer than the petal, and an oblong germen, supporting a style the length of the stamina, crowned by two awl-shaped stigmas, which open lengthways. The germen afterward turns to a single oblong seed, crowned with down, which ripens in the empalement.

We have but one SPECIES of this genus, viz.

TARCHONANTHUS (*Camphoratus*.) Hort. Cliff. 398. *Shrubby African Fleabane with Sage leaves, smelling like Camphire.*

This plant grows naturally at the Cape of Good Hope, and also in India; it has a strong woody stalk, which rises to the height of twenty feet, sending out many ligneous branches at the top, garnished with leaves, which are in shape like those of the broad-leaved Sallow, having a downy surface like those of Sage, and their under sides are white; these resemble in smell the Rosemary leaves when bruised. The flowers are produced in spikes at the extremity of the shoots, which are of a dull purple colour, so do not make any great appearance. The usual time of its flowering is in autumn; but they continue great part of winter, and are not succeeded by seeds here. These plants are preserved to make a variety in the green-house, during the winter season, by those who are curious in collecting of foreign plants; they retain their leaves all the year.

It is too tender to live through the winter in the open air in England, but requires no artificial heat, therefore may be placed in a common green-house with Myrtles, Oleanders, and other hardy exotic plants in winter, and in summer may be exposed with them in the open air, and treated in the same manner as they are.

It may be propagated by cuttings, which should be planted in May, in pots filled with light earth, and if they are plunged into a moderate hot-bed, it will promote their putting out roots. These should be shaded with mats, or covered with oiled paper, to screen them from the sun until they are rooted. By the middle of July these cuttings will have taken root, when they should be each transplanted into a separate pot, and placed in the shade until they have taken new root; after which time they may be placed with other hardy exotic plants in a sheltered situation, where they may remain till the middle or end of October, when they should be removed into the green-house, placing them where they may have a large share of air in mild weather. This plant is very thirsty, so must be often watered, and every year the plants must be shifted, and, as they increase in size, should be put into larger pots.

TARRAGON. See ABROTANUM.

TAXUS. Tourn. Inst. R. H. 589. tab. 362. The Yew-tree.

The CHARACTERS are,

The male flowers are for the most part produced on separate trees from those with fruit; they have neither empalement or petals, but the germ is like a four-leaved cover; they have a great number of stamina, which are joined at the bottom in a column longer than the germ, terminated by depressed summits, having eight points, opening on each side their base, casting their farina. The female flowers are like the male, having no empalement or petals, but have an oval acute-pointed germen, but no style, crowned by an obtuse stigma. The germen afterward becomes a berry lengthened from the receptacle, globular at the top, covered by a proper coat at bottom, open at the top, full of juice, and of a red colour, but, as it dries, wastes away, including one oblong oval seed, whose top without the berry is prominent.

We have but one SPECIES of this genus, viz.

TAXUS (*Baccata*) foliis approximatis. Lin. Sp. Plant. 1040. *Yew-tree with leaves growing near each other; or the common Yew.*

This tree grows naturally in England, and also in most of the northern countries of Europe, and in North America, where, if it is suffered to stand long, wiill rise to a good height, and have very large stems; it naturally sends out branches on every side, which spread out, and are almost horizontal; these are closely garnished with narrow, stiff, blunt-pointed leaves, of a very dark green. The flowers come out from the side of the branches in clusters; the male flowers having many stamina, are more conspicuous than the female; these, for the most part, are upon different trees, but sometimes are upon the same tree; they appear the latter end of May, and the berries ripen in autumn.

There is hardly any sort of ever-green tree, which has been so generally cultivated in the English gardens as the Yew, upon the account of its being so tonsile, as to be with ease reduced into any shape the owner pleased; and it may be too often seen, especially in old gardens, what a wretched taste of gardening prevailed formerly in England, from the monstrous figures of beasts, &c. we find these trees reduced into; but of late this taste has been justly exploded by persons of superior judgment, for what could be more absurd than the former methods of planting gardens? where, those parts next the habitation, were crouded by a large quantity of these and other sorts of evergreen trees, all of which were clipped into some trite figure or other, which, besides the obstructing the prospect from the house, and filling up the ground, so that little room was left for other shrubs and flowers. Beside, it occasioned an annual expence to render the trees disagreeable, for there never was a person, who had considered the beauty of a tree in its natural growth, with all its branches diffused on every side, but must acknowledge such a tree infinitely more beautiful than any of those shorn figures, so much studied by persons of a grovelling imagination.

The only use this tree is fit for in gardens, is to form hedges for the defence of exotic plants; for which purpose, when it is necessary to have hedges, it is the most proper of any tree in being; the leaves being small, the branches are produced very closely together; and if carefully shorn, they may be rendered so close, as to break the winds better than any other sort of fence whatever, because they will not be reverberated, as against walls, pales, and other close fences; therefore consequently, are much to be preferred for such purposes.

These trees may be easily propagated by sowing their berries in autumn as soon as they are ripe (without clearing them from the pulp which surrounds them, as hath been frequently directed,) upon a shady bed of fresh undunged soil, covering them over about half an inch thick with the same earth.

In the spring the bed must be carefully cleared from weeds; and if the season prove dry, it will be proper to refresh the bed with water now and then, which will promote the growth of the seeds; many of which will come up the same spring, but others will remain in the ground until autumn or spring following; but where the seeds are preserved above ground till spring before they are sown, the

the plants never come up till the year after, ſo that by ſowing the ſeeds as ſoon as they are ripe, there is often a whole year ſaved.

Theſe plants, when they come up, ſhould be conſtantly cleared from weeds, which, if permitted to grow amongſt them, would cauſe their bottoms to be naked, and frequently deſtroy the plants when they continue long undiſturbed.

In this bed the plants may remain two years; after which, in autumn, there ſhould be a ſpot of freſh undunged ſoil prepared, into which they ſhould be removed the beginning of October, planting them in beds about four or five feet wide, in rows about a foot aſunder, and ſix inches diſtance from each other in the rows, obſerving to lay a little mulch upon the ſurface of the ground about their roots, as alſo to water them in dry weather until they have taken root; after which they will require no farther care but to keep them clear from weeds in ſummer, and to train them according to the purpoſe for which they are deſigned.

In theſe beds they may remain two or three years, according as they have grown, when they ſhould again be removed into a nurſery, placing them in rows at three feet diſtance, and the plants eighteen inches aſunder in the rows, obſerving to do it in autumn, as was before directed, and continue to trim them in the ſummer, for what they are intended; after they have continued three or four years in this nurſery, they may be tranſplanted where they are to remain, always obſerving to remove them in autumn where the ground is very dry; but on cold moiſt land it is better in the ſpring.

Theſe trees are very ſlow in growing, but yet there are many very large trees upon ſome barren cold ſoils in divers parts of England. The timber of theſe trees is greatly eſteemed for many uſes.

TELEPHIOIDES. See ANDRACHNE.

TELEPHIUM. Tourn. Inſt. R. H. 248. tab. 128. Orpine.

The CHARACTERS are,

The empalement of the flower is permanent, compoſed of five leaves, which are obtuſe, and the length of the petals. The flower has five oblong obtuſe petals, and five awl-ſhaped ſtamina which are ſhorter than the petals, terminated by proſtrate ſummits, with a three-cornered acute germen, having no ſtyle, crowned by three acute ſpreading ſtigmas. The germen afterward turns to a ſhort three-cornered capſule with one cell, opening with three valves, containing many round ſeeds.

We have but one SPECIES of this genus, viz.

TELEPHIUM (*Imperati.*) Hort. Upſal. 70. *Orpine, Livelong, or the true Orpine of Imperatus.*

This plant grows naturally in the ſouth of France and Italy. The root is compoſed of ligneous fibres, of a yellowiſh colour. The branches or ſtalks are ſlender, and trail upon the ground; they are garniſhed with ſmall, oval, ſmooth leaves, of a grayiſh colour, which are ranged alternately along the ſtalk, having one longitudinal nerve running through the middle. The flowers are produced at the end of the branches in ſhort thick ſpikes, which are reflexed like thoſe of the Heliotropium. They are compoſed of five white petals, which ſpread open, and are the length of the empalement, having five very ſlender ſtamina, terminated by yellow ſummits.

This is propagated by ſeeds, which ſhould be ſown in autumn on a bed of freſh light earth, in an open ſituation, for if they are ſown in the ſpring, the plants will not come up till the following ſpring. When the plants are come up, they ſhould be thinned, ſo as to leave them ſix or eight inches aſunder, and ſhould be conſtantly kept clear from weeds; for if theſe are permitted to grow, they will ſoon overbear the plants, and deſtroy them. Theſe plants do not bear removing well, ſo ſhould ſtand in the place where they were ſown. In the ſummer they will flower, and the ſeeds will ripen in autumn, which will ſcatter ſoon if it is not gathered when ripe; and, if the ground is not diſturbed, the plants will come up in plenty, and require no other care than to keep them clear from weeds.

TEREBINTHUS. See PISTACHIA.

TERNATEA. See CLITORIA.

TERRACES. A Terrace is a bank of earth, raiſed on a proper elevation, ſo that any perſon who walks round a garden, may have a better proſpect of all that lies round him; and theſe elevations are ſo neceſſary, that thoſe gardens which are flat, and that have them not, are deficient.

When the Terraces are rightly ſituated, they are great ornaments, eſpecially when they are well made, and their aſcent not too ſteep.

There are ſeveral kinds of Terrace-walks:

1. The great Terrace, which generally lies next to the houſe.

2. The ſide Terrace, which is commonly raiſed above the level of the parterre, lawn, &c.

3. Thoſe Terraces which encompaſs a garden.

As to the breadth of ſide Terraces, this is uſually decided by its correſpondence with ſome pavilion, or ſome little jette or building; but moſt of all by the quantity of ſtuff that is to ſpare for thoſe purpoſes.

The ſide Terrace of a garden ought not to be leſs than twenty feet, and never wider than forty.

As for the height of a Terrace, ſome allow it to be but five feet high, but others more or leſs, according to their fancies; but more exact perſons never allow above five or ſix feet, but in a ſmall garden, and a narrow Terrace-walk, three feet; and ſometimes three feet and a half high are ſufficient for a Terrace twelve feet wide, and four feet are ſufficient for a Terrace of twenty feet wide; but when the garden is proportionably large, and the Terrace is thirty or forty feet wide, then it muſt be at leaſt five or ſix feet high.

The nobleſt Terrace is very deficient without ſhade, for which Elm-trees are very proper; for no ſeat can be ſaid to be complete, where there is not an immediate ſhade almoſt as ſoon as out of the houſe, and therefore theſe ſhady trees ſhould be detached from the body and wings of the edifice.

TETRACERA. Lin. Gen. Plant. 604.

The CHARACTERS are,

The flower has a permanent empalement of ſix roundiſh ſpreading leaves; the three outer are alternate, and ſmaller than the other; it has ſix ſmall petals, which ſoon fall off, and a great number of ſtamina, which are permanent, the length of the empalement, terminated by ſingle ſummits; it has four oval germen, ſupporting a ſhort awl-ſhaped ſtyle, crowned by an obtuſe ſtigma. The germen afterward becomes four oval reflexed capſules, each having one cell, opening at the ſeam on the upper ſide, incloſing one roundiſh ſeed.

We have but one SPECIES of this genus, viz.

TETRACERA (*Volubilis.*) Hort. Cliff. 214.

This plant grows naturally at La Vera Cruz, where it was diſcovered by the late Dr. Houſtoun, who ſent it to England. It has a woody ſtalk, which riſes to the height of twelve or fourteen feet, covered with a gray bark, ſending out ſeveral ſlender ligneous branches, which twine about any neighbouring ſupport, garniſhed with oblong oval leaves, whoſe ſurface are very rough, ſlightly indented on their edges toward their points, having many tranſverſe veins running from the midrib to the edges, placed alternate on the branches, ſtanding upon ſhort foot-ſtalks, of a grayiſh colour on their upper ſurface, and brown on their under. The flowers are produced in panicles at the end of the branches; theſe panicles are compoſed of three or four ſhort thick ſpikes, which branch out from the lower part of the principal ſpike, which is much longer and thicker than the other. The flowers have ſix thin purple petals, of the ſame length as the empalement, which are very fugacious, ſo that they ſoon fall off; theſe ſit upon the germen. After the flowers are paſt, the four germen become ſo many oval capſules, which are reflexed backward; theſe open lengthways on the upper ſide, and have each one oblong ſeed incloſed.

This ſhrub is very different from that which Dr. Plukenet titles Fagus Americanus ulmi ampliſſimis foliis, capſulis bigemellis. Amalth. 87. though Dr. Linnæus has added this ſynonime to it.

This is propagated by ſeeds, which muſt be procured from the countries where the plant naturally grows, which may probably be found in ſome of the Britiſh iſlands in the Weſt-Indies. I have received it from the iſland of Barbuda, where it was found by the late Dr. Creſſy, who ſent me ſpecimens and ſeeds. Theſe ſeeds are frequently abortive, for, upon examining them, there was ſcarce more than a twentieth part which had any kernels; the others appeared fair, but were hollow. The ſeeds ſhould be ſown in pots filled with light earth, and plunged into a moderate hot-bed of tanners bark, where they muſt be treated in the ſame way as other exotic ſeeds from the ſame countries; and as the plants ſeldom come up the ſame year, the pots ſhould be removed into the ſtove before winter, and plunged into the tan-bed, between the other pots of plants, where they ſhould remain till ſpring, when they ſhould be taken out and plunged into a freſh hot-bed of tanners bark, which will bring up the plants if the ſeeds were good. When the plants are fit to remove, they ſhould be each planted in a ſeparate ſmall pot, filled with light earth, and plunged into a good bed of tan, ſhading them from the ſun till they have taken new root; after which, their treatment muſt be the ſame as for the Annona, and the like tender exotic plants, which require to be kept always in the tan-bed.

TETRAGONIA. Lin. Gen. Plant. 551.

The CHARACTERS are,

The flower has a permanent coloured empalement, compoſed of four oval plain leaves, ſitting upon the germen. It has no petals, but about twenty hair-like ſtamina, which are ſhorter than the empalement, terminated by oblong proſtrate ſummits, and a roundiſh four-cornered germen under the flower, ſupporting four awl-ſhaped ſtyles, which are recurved and as long as the ſtamina, with hairy ſtigmas the length of the ſtyles. The germen afterward becomes a thick capſule with four cells, having four angles, which have narrow wings or borders, containing one hard oblong ſeed in each.

The SPECIES are,

1. TETRAGONIA (*Fruticoſa*) foliis linearibus. Flor. Leyd. Prod. 250. *Tetragonia with linear leaves.*

2. TETRAGONIA (*Decumbens*) foliis ovatis integerrimis, caule fruticoſo decumbente. *Tetragonia with oval entire leaves, and a ſhrubby trailing ſtalk.*

3. TETRAGONIA (*Herbacea*) foliis ovatis. Flor. Leyd. Prod. 250. *Tetragonia with oval leaves.*

Theſe plants grow naturally at the Cape of Good Hope. The firſt ſort has ſlender ligneous ſtalks, which riſe three or four feet high, if they are ſupported, otherwiſe they trail upon the ground, covered with a light gray bark, and divide into a great number of trailing branches, which, when young, are ſucculent, of an herbaceous colour, covered with ſmall pellucid drops, ſomewhat like the Diamond Ficoides, which reflect the light. As the branches are older, they become more ligneous; they are garniſhed with thick, ſucculent, narrow, concave, blunt leaves, placed alternate, and at their baſe come out a cluſter of ſmaller leaves. The flowers are produced from the wings of the ſtalks at every joint toward the end of the branches; ſometimes they come ſingly, at others there are two, and ſometimes three flowers at each joint; theſe have empalements of five leaves, which ſpread open, and are a little reflexed; they are green without, and yellow within, each having about forty ſtamina, which are terminated by oblong proſtrate ſummits, which fill up the middle of the flower. They appear in July and Auguſt, and are ſucceeded by large four-cornered capſules, having four wings or borders, and four cells, each containing one oblong ſeed, which ripens in winter.

The ſecond ſort has larger ſtalks than the former, which branch out in like manner; the branches trail upon the ground where they are not ſupported; the young branches are very ſucculent, and almoſt as thick as a man's little finger; the leaves are two inches long, and one broad; their ſurface are covered with very ſmall pellucid drops, as are the ſtalks. The flowers are larger, and ſtand upon pretty long foot-ſtalks, three or four ariſing from the ſame points; the empalement, and alſo the ſummits, are of a pale ſulphur colour.

Theſe may be propagated by cuttings, which ſhould be cut from the plants a few days before they are planted, that the part where they are cut may be healed, otherwiſe they will rot, for the leaves and ſtalks of this plant are very full of moiſture. The beſt time to plant theſe cuttings is in July, that they may have time to make good roots before winter. They may be planted on a bed of freſh earth; and if they are ſhaded from the ſun in the heat of the day, it will be of ſervice to them. They ſhould be frequently refreſhed with water, but they muſt not have it in too great plenty, for that will rot them. In about ſix weeks after planting, the cuttings will be ſufficiently rooted to tranſplant, therefore they ſhould be taken up, and planted in pots filled with light, freſh, undunged earth, and placed in a ſhady ſituation until they have taken new root; after which time they may be placed with other hardy exotic plants, in a ſheltered ſituation, where they may remain till the middle or latter end of October; at which time they ſhould be removed into the green-houſe, and placed where they may enjoy as much free air as poſſible in mild weather, for they only require to be protected from the froſt, being pretty hardy with reſpect to cold; but they ſhould not have too much moiſture in winter. If theſe plants are planted in the full ground in the ſummer ſeaſon, they will grow prodigiouſly rank and large, as they alſo will if they are permitted to root into the ground through the holes at the bottom of the pots; therefore the pots ſhould be frequently removed to prevent it, for when they grow too freely, their leaves will be very full of moiſture; which, together with the weight of the fruit, which are always produced at the extremity of the branches, will weigh the branches upon the ground, and render the plants very unſightly. The plants of this kind commonly grow very ſtraggling; therefore the more their roots are confined in the pots, the more cloſe and ſtinted will be the heads of the plants, which is what they ſhould always be kept to, in order to render them ſightly. The flowers of this plant have no great beauty, but as the whole face of the plant is peculiar, it may be allowed a place in every collection of plants for the ſake of variety, ſince it requires no great trouble to cultivate it.

Theſe plants may alſo be propagated by ſeeds, which ſhould be ſown on a warm border of light freſh earth, where ſometimes they will remain a whole year before the plants come up; therefore when they do not come up the firſt ſeaſon, the borders ſhould not be diſturbed, but kept conſtantly clear from weeds; and when the plants are come up about four inches high, they ſhould be taken up and planted in pots (and treated in the ſame manner as hath been directed

rected for the cuttings;) for if they are suffered to grow in the border till they are large, they will not transplant so well, nor will they make so handsome plants.

The third sort hath large fleshy roots; the branches are weak, and trail upon the ground; these generally decay about Midsummer, and new shoots are produced late in autumn. The leaves of this come out in bunches; they are oval, plain, and not so thick and succulent as those of the other sorts. The flowers are produced from the wings of the leaves in February; these are like those of the second sort, and have pretty long slender foot-stalks. The cuttings of this sort will grow, if they are planted early in the spring, so that it may be propagated with the same facility as either of the other kinds.

All these sorts require protection in winter; but if they are placed in an airy glass-case, or under a frame in winter, with Ficoides, and other hardy plants, where they may have a large share of free air in mild weather, and protected from the frost, they will thrive much better than when they are more tenderly treated.

TETRAGONOTHECA. Hort. Elth. 283. Bastard Sunflower.

The CHARACTERS are,

The flower is composed of hermaphrodite and female florets, which are included in one large common empalement, cut into four plain, triangular, heart-shaped segments, which spread open. The disk or middle of the flower is made up of hermaphrodite florets, which are funnel-shaped, cut into five parts at the brim, which are reflexed; they have five short hair-like stamina, terminated by cylindrical summits, and a naked germen, supporting a slender style, crowned by two reflexed stigmas. The germen afterward becomes one naked roundish seed. The female half florets, which compose the ray or border of the flower, have their petals stretched out like a tongue on one side, and are cut at their points into three equal acute parts. These have no stamina, but a naked germen, supporting a slender style with two twisted stigmas, and are succeeded by single naked seeds.

We have but one SPECIES of this genus, viz

TETRAGONOTHECA (*Helianthoides.*) Lin. Sp. Plant. 903. *Dwarf or Bastard Sun-flower.*

This plant is a native of Carolina; the roots of this plant are perennial, but the stalks are annual, and perish in autumn on the approach of cold. The roots will abide through the winter in the full ground, if they are planted in a warm situation, so do not require any shelter, except in very severe winters; when, if they are covered over with rotten tan, or Pease-haulm, to keep out the frost, there will be no danger of their being killed.

About the latter end of April, or the beginning of May, the roots will send forth new shoots, which are garnished with large, oblong, rough leaves, placed by pairs, closely embracing the stalks; these are a little sinuated on their edges, and are covered with small hairs. The stalks usually grow about two feet high in England, and branch out toward the top into several smaller stalks, each having one large yellow flower at their top, shaped like a Sun-flower, which, before it expands, is covered with the inflated empalement, which is four-cornered. The seeds of this plant rarely ripen in England, but when they are obtained from abroad, they should be sown in the full ground in the spring of the year, where sometimes they will remain a year before the plants come up; so that if they do not come up the same year, the ground should not be disturbed, but kept clean from weeds, and wait till the second year to see what plants will come up. When the plants appear, they must be kept clean from weeds, and if the season should prove dry, they will require to be frequently watered. In autumn the plants should be transplanted into the places where they are to remain.

These plants will live three years in a proper soil and situation; but as it does not ripen seeds here, the best method is to procure good seeds from abroad annually.

TEUCRIUM. Lin. Gen. Plant. 625. Tree Germander.

The CHARACTERS are,

The empalement of the flower is of one leaf, cut into five acute equal segments at the top, and is permanent. The flower is of the lip kind with one petal, having a short cylindrical tube a little incurved at the chaps. The upper lip is erect, and deeply cut into two acute segments. The lower lip spreads, and is cut into three; the middle one is large and roundish, the two side ones are acute and erect. It has four awl-shaped stamina, which are longer than the upper lip, and are prominent between the segments, terminated by small summits. It has a germen divided in four parts, supporting a slender style, crowned by two slender stigmas. The germen afterward turns to four roundish naked seeds, which ripen in the empalement.

The SPECIES are,

1. TEUCRIUM (*Flavum*) foliis cordatis obtuse serratis, floralibus integerrimis concavis, caule fruticoso. Lin. Sp. Plant. 565. *Tree Germander with heart-shaped leaves, which are bluntly sawed, those between the flowers concave and entire, and a shrubby stalk; common Tree Germander.*

2. TEUCRIUM (*Lucidum*) foliis ovatis acute inciso-serratis glabris, floribus axillaribus geminis caule erecto. Lin. Sp. 790. *Germander with oval smooth leaves, which are acutely sawed, and two flowers proceeding from the side of the stalk.*

3. TEUCRIUM (*Fruticans*) foliis integerrimis oblongo-ovatis petiolatis, suprà glabris, subtus tomentosis pedunculis unifloris. Lin. Sp. Plant. 790. *Tree Germander with entire, oblong, oval leaves, having foot-stalks, smooth above, and hoary underneath; or Spanish Tree Germander.*

4. TEUCRIUM (*Latifolium*) foliis integerrimis, rhombeis, acutis, villosis, subtus tomentosis. Hort. Upsal. 195. *Tree Germander with entire leaves, which are hairy, shaped like an acute rhombus, and woolly on their under side.*

5. TEUCRIUM (*Campanulatum*) foliis multifidis, floribus solitariis. Lin. Sp. Plant. 562. *Germander with many-pointed leaves, and flowers growing singly.*

6. TEUCRIUM (*Botrys*) foliis multifidis, pedunculis axillaribus ternis. Lin. Sp. Plant. 562. *Germander with many-pointed leaves, and flowers growing in whorls by threes.*

7. TEUCRIUM (*Chamædrys*) foliis ovatis inciso-crenatis petiolatis, floribus subverticillatis. Hort. Cliff. 302. *Germander with oval leaves on foot-stalks, with crenated cuts, and flowers growing almost in whorls; or smaller creeping Germander.*

8. TEUCRIUM (*Nisolianum*) foliis trifidis quinquefidisque filiformibus, floribus pedunculatis solitariis oppositis, caule decumbente. Lin. Sp. 782. *Germander with trifid and quinquefid leaves, and flowers upon solitary foot-stalks.*

9. TEUCRIUM (*Massiliense*) foliis ovatis rugosis incisocrenatis incanis, caulibus erectis, racemis erectis. Lin. Sp. 789. *Germander with oval rough leaves, which are hoary, crenated, and erect stalks and spikes of flowers.*

10. TEUCRIUM (*Scorodonia*) foliis cordatis serratis petiolatis, racemis lateralibus secundis caule erecto. Lin. Sp. 789. *Germander with heart-shaped sawed leaves, and many long bunches of flowers growing from the wings of the stalk; or wild Sage.*

11. TEUCRIUM (*Scordium*) foliis oblongis sessilibus dentato-serratis floribus geminis lateralibus pedunculatis caule diffuso. Lin. Sp. 790. *Germander with oblong, crenated, sawed leaves, and twin flowers.*

12. TEUCRIUM (*Marum*) foliis integerrimis ovatis subtus tomentosis, utrinque acutis, racemis secundis villosis. Lin. Sp. 789. *Germander with oval, entire, woolly leaves; called Marum Syriacum.*

13. Teucrium (*Chamæpytis*) foliis trifidis linearibus integerrimis, floribus sessilibus lateralibus solitariis caule diffuso. Mat. Med. 287. *Germander with linear entire leaves; or common Ground Pine.*

14. Teucrium (*Iva*) foliis tricuspidatis linearibus, floribus sessilibus. Lin. Sp. 787. *Germander with linear tricuspid leaves, and flowers sitting close to the stalks.*

15. Teucrium (*Moschatum*) foliis linearibus tomentosis integerrimis, floribus sessilibus. *Germander with linear, woolly, entire leaves, and flowers sitting close to the branches; or Musk Ground Pine with entire leaves.*

16. Teucrium (*Chamædrifolium*) foliis oblongo-ovatis obtusè dentatis, floribus solitariis alaribus pedunculatis, calycibus acutis. *Germander with oblong oval leaves, which are bluntly indented, and flowers placed singly at the wings of the stalks, having acute empalements.*

17. Teucrium (*Canadense*) foliis ovato-lanceolatis, inæqualiter serratis, racemis alaribus terminalibusque calycibus inflatis. *Germander with oval spear-shaped leaves which are unequally sawed, and long bunches of flowers springing from the wings, and terminating the stalks, and inflated empalements.*

18. Teucrium (*Virginicum*) foliis ovatis inæqualiter serratis racemis terminalibus. Flor. Virg. 64. *Germander with oval leaves unequally sawed, and a racemus of flowers terminating the stalks.*

The first sort grows naturally in the south of France, in Spain, and in Italy; it rises with a shrubby stalk two or three feet high, sending out many ligneous branches, garnished with heart-shaped leaves, a little waved, bluntly sawed on their edges, of a lucid green on their upper side, but a little hoary on their under, standing upon short foot-stalks. The upper part of the branches, for six or eight inches in length, are adorned with flowers, which come out from the wings of the stalk, two or three standing on each side at every joint; they are of a dirty white colour, and stand upon slender foot-stalks; under each of these whorls stand two smaller leaves, which are entire and concave.

This sort was formerly preserved in green-houses with great care, but of late years it hath been planted in the full ground, and is found hardy enough to endure the cold of our severest winters without shelter, provided it is planted on a dry soil.

This may be propagated by planting cuttings in the spring, on a bed of fresh light earth, observing to shade and water them until they have taken root; after which they will require no farther care but to keep them clear from weeds until the following autumn, when they may be transplanted where they are to remain, being very careful in removing them not to shake off all the earth from their roots, as also to water them, if the season should prove dry, until they have taken fresh root; after which, the only care they require is to keep the ground clean about them, and to prune off such shoots as are ill situated, and the flowering branches when they decay, whereby their heads will appear more regular.

It may also be propagated by seeds, which generally are produced in plenty. If these are sown upon a bed of light earth in April, the plants will come up in six weeks after, and may be transplanted in autumn, where they are designed to remain.

The second sort grows naturally on the Alps, but in the lower parts, where the cold is not very severe, and generally on moist ground; this hath a shrubby stalk like the former, and rises about the same height, but branches out more than that. The stalks are covered with a short hairy down; the lower leaves are oval, crenated, and of a lucid green on their upper side, but a little hoary on their under; the leaves between the flowers are spear-shaped and entire; the spikes of flowers are much longer; the flowers are larger, and their colour more inclining to a yellow than those of the former, and may be propagated in the same way.

The third sort grows naturally in Spain and Sicily, near the borders of the sea; this has a shrubby branching stalk, which rises six or eight feet high, covered with a hoary bark. The branches are garnished with small oval leaves, placed opposite, sitting close to them; they are smooth on their upper side, of a lucid green, but their under sides are hoary. The flowers come out singly from the wings of the stalk at the upper part of the branches, one on each side, standing upon short foot-stalks; their empalements are short and hoary. The middle segment of the lower lip is large, and indented at the point; the stamina are long-hooked, and supply the place of the upper lip; the flowers are blue, and come in succession great part of summer, but the plants seldom produce good seeds in England.

There is a variety of this with variegated leaves, which is preserved in some gardens.

The third sort is tenderer than the former, though this will endure the cold of our ordinary winters, if planted on a dry soil and in a warm situation, but in severe frost it is generally destroyed; for which reason the plants are often preserved in pots, and removed into the green-house in winter. This is propagated by cuttings in the same manner as the former.

The fourth sort grows naturally in Spain; this has a great resemblance of the third, but the branches spread more horizontally. The leaves are sometimes heart-shaped, and at others in form of a rhombus; the lower leaves, which are the largest, are an inch and a half long, and three quarters of an inch broad; the upper are smaller, and of a different shape; these are downy on both sides, but the lower leaves are only so on the under. The flowers come out at the upper parts of the branches in like manner as the former, but are larger, and of a paler blue colour.

This is propagated in the same way as the other, and the plants require the same treatment.

The fifth sort grows naturally in Spain and Italy upon moist ground. The stalks of this are herbaceous, and trail upon the ground; they grow about a foot in length, and are garnished with deep green leaves cut in many points almost to the midrib; they are smooth, and stand opposite. The flowers are white, and come out on each side the stalks singly; these are succeeded by four seeds, which ripen in autumn.

This plant is preserved in botanic gardens for variety; it is propagated by seeds, which may be sown in the spring in the place where the plants are to remain, and, when they come up, will require no other culture but to thin them where they are too close, and keep them clean from weeds. These plants ripen their seeds the first year, but, if they are in a warm situation, they will live through the winter.

The sixth sort grows naturally in the south of France, in Italy and Germany, in the Corn-fields; this is an annual plant, which perishes soon after the seeds are ripe. The stalks are four-cornered and hairy, about a foot long, garnished at every joint by leaves placed opposite, which are hairy, and almost cut to the midrib; the segments are cut into three points. The flowers come out at the wings of the stalks in whorls, three standing together on each side upon short foot-stalks; they are white, and shaped like those of the other species; the seeds ripen in August and September.

This is propagated by seeds in the same way as the last; but if the seeds of this are sown in autumn, or permitted

to

to fcatter when ripe, they will fucceed better than if fown in the fpring, and the plants will come earlier to flower.

The feventh fort grows naturally in the fouth of France, and in Germany; this has a creeping fibrous root, which fpreads in the ground, and multiplies greatly, fending out many four-cornered hairy ftalks, which are eight or nine inches long, having a few fhort branches, garnifhed with oval leaves, which are deeply crenated on their borders, and upon foot-ftalks; they are of a light green above, but hoary on their under fide. The flowers grow from the wings of the ftalks towards the upper part almoft in whorls, ftanding chiefly to one fide of the ftalk; they are of a reddifh colour, the lower lip turning inward. The feeds ripen in autumn.

It is a perennial plant, and propagates very faft by its creeping roots, and will thrive in almoft any foil or fituation: the beft time to tranfplant it is in autumn. This was a few years fince in great requeft as a fpecific for the gout, but is at prefent in little efteem.

The eighth fort grows naturally in Spain; this is a perennial plant, having fome refemblance of the former, but the roots do not creep. The ftalks are taller and more erect; the leaves are narrower, pointed at both ends, and not fo deeply indented; the indentures are fharper; the ftalks are garnifhed with flowers great part of their length, which come out in bunches at the wings; they are longer than thofe of the former, and of a brighter red colour.

It may be propagated by parting of the roots in autumn, or by fowing of the feeds at the fame feafon, which will more certainly fucceed than thofe which are fown in the fpring. It loves an open fituation expofed to the fun, but will thrive in almoft any foil which is not too moift.

The ninth fort grows naturally in Italy and near Marfeilles; this is like the feventh fort, but the ftalks grow almoft twice the length of thofe, and fend out a greater number of branches. The leaves of this are more acutely indented on their edges; they are hairy, of a light green on their upper fide, and hoary on their under. The flowers grow almoft in whorls from the wings of the ftalks, to which they fit very clofe; they are fometimes red, and at others white, and both colours are often on the fame plant. This fort may be propagated in the fame way as the former.

The tenth fort is the common wild or Wood Sage, which grows naturally in woods and thickets in many parts of England, fo is rarely admitted into gardens; this has a creeping perennial root, from which arife ftiff, ligneous, four-cornered ftalks a foot and a half high, garnifhed at each joint by two heart-fhaped leaves, placed oppofite, flightly fawed on their edges, and ftand upon foot-ftalks. The upper part of the ftalks have three or four long fpikes of flowers, which incline to one fide of the ftalk; they are of an herbaceous white colour, and the ftamina are terminated by purple fummits. It flowers in July, and the feeds ripen in autumn. This plant will grow in any foil or fituation, and was formerly ufed in medicine.

The eleventh fort is the common Water Germander, which grows naturally in the ifle of Ely, and fome other fenny parts of England; it has a fmall, ftringy, fibrous, creeping root, which is perennial, from which arife many four-cornered, trailing, diffufed ftalks, garnifhed with oblong, hairy, indented leaves, fitting clofe to the ftalks. The flowers are produced at the wings of the ftalks, two arifing on each fide, at every joint; they are of a purple colour, and fit very clofe to the bottom of the leaves; thefe appear in July, but are feldom fucceeded by feeds. The whole plant has an odour like that of Garlick. The herb is ufed in medicine.

This plant may be propagated by its creeping roots, or planting the young fhoots in the fpring, in the fame manner as Mint, Penny-Royal, &c. and fhould have a moift foil, otherwife it will not thrive in gardens.

The twelfth fort is the common or Syrian Marum, which grows naturally in Syria, and alfo in the kingdom of Valencia; this has a low fhrubby ftalk, fending out many ligneous branches, which in warm countries will rife three or four feet high, but in England it is rarely feen half that height. The ftalks are very hoary, garnifhed with fmall oval leaves oppofite at each joint, about the fize of thofe of Thyme, and are pointed at both ends; they are hoary, and have a piercing grateful fcent, fo quick as to caufe fneezing. The flowers grow in loofe whorled fpikes at the end of the branches, of a bright red colour; they appear in July and Auguft, but are not fucceeded by feeds in England.

This plant is eafily propagated by flips or cuttings, which, if planted during the fummer months on a bed of light loamy earth, covering them down clofe either with bell or hand-glaffes, and fhading them from the fun, will put out roots very freely. When thefe have made good roots, they may be tranfplanted either into feparate fmall pots, or on a warm border at about fix inches diftance every way, obferving to fhade them from the fun, and fupply them with water till they have taken new root; after which they will require no other care but to keep them clean from weeds. Thefe plants will live through the winter in the open air, if they are planted in a dry foil and a warm fituation, when the frofts are not very fevere; but in very hard winters they are frequently killed, if they are not protected by mats or fome other covering. There was, about forty years ago, a great number of thefe plants growing in the warm borders of the royal gardens at Kenfington, which were clipped into conical forms, and were near three feet high; but now there are few plants of a large fize to be found in the Englifh gardens, becaufe their branches are annually cut to keep them fhort.

The cats are very fond of this plant, and where there are but few of thefe plants will deftroy them, unlefs they are protected from them; but where there is a great number of the plants together, the cats feldom touch them.

The thirteenth fort is the common Ground Pine, which is ufed in medicine; it grows naturally on chalky arable land in feveral parts of England; it is an annual plant, with a fingle ligneous root, fending out a few flender fibres from the fide, from which arife many weak, trailing, hairy ftalks, garnifhed with narrow leaves, ending with three points, fet by pairs, and crofs each other at every joint: they are hairy, and, when bruifed, emit a ftrong refinous odour. The flowers fit clofe to the ftalks at the wings of the leaves; there are two or three of them at each joint, of a bright yellow colour, and fhaped like the other fpecies. If the feeds are permitted to fcatter, the plants will come up better than if fown, and require no other care but to thin them, and keep them clean from weeds.

This plant is greatly recommended for its virtues; there is fcarce a better herb than this for opening obftructions; it is a ftrong diuretic, and an excellent remedy for the rheumatifm.

The fourteenth fort grows naturally in the fouth of France, in Italy, and Spain; it is an annual plant, with a fingle ligneous root, fending out a few fibres. The ftalks are about fix inches high, clofely garnifhed with very hairy narrow leaves, which are indented towards their points. The flowers come out from the wings of the ftalks, to which they fit very clofe; they are large, of a bright purple colour, and appear in July; but unlefs the feafon proves favourable, they are not fucceeded by feeds in England.

The fifteenth fort grows naturally about Nice in Italy; this is alfo an annual plant, much like the former, but the leaves are narrower and entire. The whole plant is covered with white woolly hairs, and the flowers are smaller than thofe of the former.

Both thefe plants fucceed beft, if, when they perfect their feeds, they are permitted to fcatter in the fame manner as the thirteenth fort; or, if the feeds are fown, it fhould be in autumn, for they rarely fucceed when they are fown in the fpring.

The fixteenth fort was difcovered by the late Dr. Houftoun, growing naturally at La Vera Cruz; this is an annual plant, with an erect four-cornered ftalk a foot and a half high, garnifhed with fmooth, oblong, oval leaves, which are bluntly indented. The flowers come out from the wings of the ftalk, two of them arifing at each joint, upon fhort flender foot-ftalks; they are fmall and white, having fhort empalements, which are cut at the brim into five very acute points. The flowers appear in July, and are fucceeded by feeds which ripen in autumn.

The feventeenth fort grows naturally in North America; this is a perennial plant, very like our Scorodonia or Wood Sage, but does not creep at the root as that does; the ftalks are erect, and garnifhed with oval fpear-fhaped leaves, which are white on their under fide, and deeply fawed on their edges; the ftalks are terminated by racemi of yellow flowers, and the whorls have fix leaves.

This is a very hardy plant, fo will thrive in the open air; it may be propagated by parting of the roots, or by fowing of the feeds, which is beft if done in autumn.

The eighteenth fort grows naturally in Virginia; this is alfo a perennial plant, having oval leaves, which are unequally fawed; the ftalk is annual, and rifes near a foot high, which is terminated by a long fpike of red flowers, which appear in July and Auguft, when the plants make a pretty appearance.

This is eafily propagated by feeds, which are produced in plenty; if thefe are fown in the autumn on a bed of light earth, they will fucceed better than if fown in the fpring.

THALICTRUM. Tourn. Inft. R. H. 270. tab. 143. Meadow Rue.

The CHARACTERS are,

The flower has no empalement, but has four or five roundifh concave petals, which fall off foon, and a great number of broad ftamina, which are compreffed toward their tops, terminated by twin fummits, with feveral very fhort ftyles fitting fingly upon roundifh germen, crowned by thick ftigmas. The germen afterward turn to fo many keel-fhaped capfules collected in a head, each containing one oblong feed.

The SPECIES are,

1. THALICTRUM (*Flavum*) caule foliofo fulcato, paniculâ multiplici erectâ. Hort. Cliff. 226. *Meadow Rue with a furrowed leafy ftalk, and many erect panicles of flowers.*

2. THALICTRUM (*Speciofum*) caule angulofo, foliis linearibus bifidis trifidifque, paniculâ multiplici erectâ. *Meadow Rue with an angular ftalk, narrow leaves ending in two or three points, and many erect panicles of flowers.*

3. THALICTRUM (*Aquilegifolium*) fructibus pendulis triangularibus rectis, caule tereti. Lin. Sp. Plant. 547. *Meadow Rue with a pendulous triangular fruit, and a taper ftalk; commonly called Feathered Columbine.*

4. THALICTRUM (*Lucidum*) caule foliofo fulcato, foliis linearibus carnofis. Dalib. Prif. 162. *Meadow Rue with a furrowed leafy ftalk, and linear flefhy leaves.*

5. THALICTRUM (*Canadenfe*) floribus pentapetalis, radice fibrosâ. Flor. Leyd. Prod. 486. *Meadow Rue with flowers having five petals, and a fibrous root; Canada Meadow Rue.*

6. THALICTRUM (*Tuberofum*) floribus pentapetalis, radice tuberosâ. Hort. Cliff. 227. *Meadow Rue with flowers having five petals, and a tuberous root.*

7. THALICTRUM (*Minus*) foliis fexpartitis, floribus cernuis. Lin. Sp. Plant. 546. *Meadow Rue with leaves cut into fix fegments, and pendulous flowers.*

8. THALICTRUM (*Fœtidum*) caule paniculato ramofiffimo foliofo. Lin. Sp. Plant. 545. *Meadow Rue with a very branching paniculated leafy ftalk; the leaft ftinking Meadow Rue.*

9. THALICTRUM (*Dioicum*) floribus dioicis. Lin. Sp. Plant. 545. *Meadow Rue with male and female flowers upon different plants.*

10. THALICTRUM (*Anguftifolium*) foliolis lanceolato-linearibus integerrimis. Hort. Cliff. 226. *Meadow Rue with fpear-fhaped linear leaves, which are entire.*

11. THALICTRUM (*Alpinum*) caule fimpliciffimo fubnudo, racemo fimplici terminali. Hort. Cliff. 227. *Meadow Rue with a fingle ftalk, which is almoft naked, and terminated by a fingle bunch of flowers.*

The firft fort grows naturally by the fide of rivers and in moift meadows in many parts of England. This has a yellow creeping root, from which arife feveral furrowed ftalks four or five feet high, garnifhed at each joint with leaves compofed of many lobes, which differ in their form and fize; fome are fpear-fhaped and entire, others are obtufe, and cut into three points; they are of a deep green colour on their upper fide, but pale on their under. The flowers are of an herbaceous white colour, and formed into many panicles, ftanding erect on the top of the ftalks. Thefe appear in July, and are fucceeded by fhort triangular capfules, containing one oblong feed.

The fecond fort grows naturally in the meadows about Montpelier. The root of this is like the former; the ftalks are angular, and rife five feet high; they are better furnifhed with leaves, whofe lobes are very narrow, fome of them ending with two, and others with three points, of a bright green colour. The flowers are yellow, and are formed into many panicles which terminate the ftalks. This fort flowers about the fame time with the former.

The third fort grows naturally upon the Alps; of this there are two varieties; one with a green ftalk and white ftamina, the other has purple ftalks and ftamina. Thefe two are propagated in gardens, by the title of Feathered Columbine; this has a thick fibrous root; the ftalks are taper, and rife three feet high; the leaves are like thofe of the Columbine. The flowers grow in large panicles at the top of the ftalk. It flowers in June, and the feeds, which are in triangular capfules, ripen in Auguft.

The fourth fort grows naturally in the meadows about Paris; this hath upright channelled ftalks, which rife three feet high, garnifhed at each joint with winged leaves, compofed of many linear flefhy lobes, which are for the moft part entire, ending in acute points. The flowers are of a yellowifh white colour; they appear in July, and are fucceeded by fmall angular capfules, with one fmall oblong feed in each, which ripens in Auguft.

The fifth fort grows naturally in North America; this has a fibrous root, of a dark colour. The ftalks are fmooth, of a purple colour, and rife three or four feet high, branching toward the top. The leaves are like thofe of Columbine, of a grayifh colour, and fmooth. The flowers are produced in large panicles at the top of the ftalks; they are larger than thofe of the former forts, and have five white petals which foon fall off, and a great number of white ftamina with yellow fummits. This flowers in June, and the feeds ripen in Auguft.

The fixth fort grows naturally in Spain; this has knobbed roots; the leaves are fmall, obtufe, and indented in three parts at their points; they are of a grayifh colour, and fmooth. The ftalks rife a foot and a half high, naked almoft to the top, where they divide into two or three ſmall ones, under which is ſituated one leaf. Each diviſion of the ſtalk is terminated by a ſmall bunch of pretty large flowers, having five white petals. The flowers are almoſt diſpoſed in form of an umbel. They appear in June, and are ſucceeded by ſmall angular capſules, containing one oblong ſeed in each, which ripen in Auguſt.

The ſeventh ſort grows naturally in ſome parts of Cambridgeſhire; this has a creeping fibrous root. The ſtalks riſe about a foot high, and are garniſhed with winged leaves, compoſed of many obtuſe ſhort lobes, which are cut into ſix ſegments. The ſtalks branch out wide; the flowers grow in looſe panicles; they are ſmall and nodding. The ſtamina are of an herbaceous white, and the ſummits are yellowiſh. It flowers in June.

The eighth ſort grows naturally in the ſouth of France; this hath a very branching ſtalk, which riſes about ſix or ſeven inches high, garniſhed with winged leaves which are downy, compoſed of a great number of ſmall lobes which are bluntly indented, and have a fœtid ſcent. The flowers grow in looſe panicles; they are ſmall, of an herbaceous white colour, with yellowiſh ſtamina. This flowers in June.

The ninth ſort grows naturally in North America. The root of this is fibrous; the ſtalks riſe near a foot high, and are almoſt naked to the top, where they have one leaf, compoſed of many ſmall lobes, of a grayiſh colour, indented at their points. The flowers are produced in ſmall bunches at the top of the ſtalks; they are male and female in different plants. Theſe appear in June.

The tenth ſort grows naturally in Italy and ſome parts of Germany; this hath a perennial root. The ſtalks riſe from two to three feet high; the leaves are winged like thoſe of the other ſorts; their lobes are narrow and entire. The flowers are ſmall, and are collected in panicles at the top of the ſtalks, and are of an herbaceous white colour.

The eleventh ſort grows naturally on the Alps; this hath a fibrous creeping root; the leaves are ſmall, blunt, and of a grayiſh colour. The ſtalks riſe about ſix inches high, and are almoſt naked; they are terminated by a looſe ſingle ſpike of flowers, each having four petals. This flowers the latter end of April or the beginning of May.

Theſe plants are generally propagated by parting their roots. The beſt time for this work is in September, when their leaves and ſtalks begin to decay, that they may take freſh root before the froſt comes on to prevent them; they ſhould alſo be planted in a freſh light ſoil, and have a ſhady ſituation, in which they will thrive exceedingly, though they may be planted in almoſt any ſoil or ſituation, provided it be not too hot and dry; but moſt of them creep ſo much under ground, as to become very troubleſome in a garden, for which reaſon there are but few of the ſorts admitted into gardens. The third, fifth, and ſixth ſorts are frequently cultivated in gardens. The roots of theſe do not creep like the others, and their flowers have ſome beauty to recommend them, but the others are only kept in botanic gardens for the ſake of variety: therefore when they are admitted, their roots ſhould be confined in pots, otherwiſe they cannot be kept within bounds.

THAPSIA. Tourn. Inſt. R. H. 321. tab. 171. The deadly Carrot, or ſcorching Fennel.

The CHARACTERS are,

It has an umbellated flower; the general umbel is large, compoſed of about twenty rays, which are nearly equal, theſe have no involucri; the general umbel is uniform. The flowers have five ſpear-ſhaped incurved petals, and five hair-like ſtamina the length of the petals, terminated by ſingle ſummits. It has an oblong germen ſituated under the flower, ſupporting two ſhort ſtyles, crowned by obtuſe ſtigmas. The germen afterward becomes an oblong fruit, girt with a longitudinal membrane, dividing into two parts, each containing one oblong ſeed, pointed at both ends, having plain borders on both ſides.

The SPECIES are,

1. THAPSIA (*Villoſa*) foliolis dentatis baſi coadunatis. Hort. Cliff. 105. *Scorching Carrot with indented lobes, which are joined at their baſe.*

2. THAPSIA (*Maxima*) foliis pinnatis, foliolis latiſſimis pinnatifidis ſubtus villoſis petiolis decurrentibus. *Scorching Carrot with winged leaves, having very broad wing-pointed lobes, which are hairy on their under ſide, and running footſtalks.*

3. THAPSIA (*Fœtida*) foliolis multifidis baſi anguſtatis. Hort. Cliff. 105. *Scorching Carrot with many-pointed lobes, which are narrowed at their baſe.*

4. THAPSIA (*Apulia*) foliis digitatis foliolis bipinnatis multifidis ſetaceis. Hort. Cliff. 106. *Scorching Carrot with many-pointed briſtly lobes.*

5. THAPSIA (*Trifoliata*) foliis ternatis ovatis. Lin. Sp. Plant. 262. *Scorching Carrot with oval trifoliate leaves.*

6. THAPSIA (*Altiſſima*) foliis decompoſitis lobis maximis lucidis, umbella maxima. *The talleſt ſcorching Carrot, with decompounded lucid leaves and the largeſt umbel.*

The firſt ſort grows naturally in Spain, Portugal, and the ſouth of France; this hath a thick fleſhy root in ſhape of a Carrot, which has an outward blackiſh ſkin; the inſide is white, bitter, and very acrid, with a little aromatic taſte. The leaves are winged; the lobes are thick, hairy, and indented; they are regularly cut into oppoſite ſegments like other winged leaves. The ſtalk is ſpungy, and riſes about two feet high, dividing upward into two or three ſmall branches, each being terminated by a large umbel of yellow flowers. Theſe appear in June, and are ſucceeded by large, flat, bordered ſeeds, which ripen in Auguſt.

The ſecond ſort grows naturally in Spain, particularly all over Old Caſtile, quite to the Pyrenean mountains. The root of this ſort is large, thick, and of a dark colour without. The leaves are very thick, and hairy on their under ſide; they ſpread circularly on the ground, and are divided into broad lobes, like moſt of the other umbelliferous plants. The ſtalks riſe three or four feet high; they are large, jointed, and full of pith, having one leaf at each joint, ſhaped like thoſe at the bottom, but are ſmaller as they are nearer the top. The ſtalk is terminated by a large umbel of yellow flowers, which appear the latter end of June, and the ſeeds ripen two months after.

The third ſort grows naturally in Italy and Spain. The leaves of this ſort are cut into many narrow ſegments, almoſt as ſmall as thoſe of the Garden Carrot, but are rough and hairy; their ſegments are always oppoſite, and are narrower at their baſe than their points. The ſtalks riſe about two feet high, and are terminated by umbels of ſmall yellow flowers, which appear in July; theſe are ſucceeded by flat bordered ſeeds, which ripen the beginning of September.

The fourth ſort grows naturally in Apulia. The root of this is about the thickneſs of a man's thumb; the bark is yellow and wrinkled, the inſide white, and abounds with an acrid milky juice; the leaves are finely divided like thoſe of Fennel; they are hairy, and ſit cloſe to the root. The ſtalk riſes from two to three feet high; it is naked, and branches into two or three ſmaller ſtalks, each being terminated by a ſmall umbel of flowers, which are large, yellow, and appear in July; theſe are ſucceeded by flat ſeeds, having cartilaginous borders, which ripen in September.

The fifth ſort grows naturally in North America. The ſeeds were ſent me by Dr. Benſel from Philadelphia. This has a ſlender tap-root, which is ſhaped like thoſe of Parſley; the leaves at the bottom are heart-ſhaped. The ſtalk is fingle, and does not branch; it rifes near two feet high, is of a purple colour, and flender, garnifhed at each joint with one trifoliate leaf, whofe lobes are oval and crenated. The ftalk is terminated by a fmall umbel of purple flowers, which appear in July, and are fucceeded by compreffed channelled feeds which ripen in September. Dr. Gronovius thinks this plant very like that which is figured by Kempfer, by the title of Nindzi.

The fixth fort grows naturally in Auftria; this has a taper root as large as a man's thumb. The leaves fpread circularly on the ground, and are divided into feveral parts; the lobes are very fmall, cut into many acute fegments or points, which are oppofite, like winged leaves; they are rough and hairy. The foot-ftalks of the leaves are broad, and are clofely fet with prickly hairs; the ftalk rifes near two feet high, and is terminated by an umbel of yellow flowers, which appear in July, and are fucceeded by bordered compreffed feeds, which ripen in September.

Thefe plants are only propagated by feeds, which fhould be fown in autumn; for if they are kept out of the ground till fpring, they often mifcarry; or if they grow, they commonly lie a whole year in the ground before the plants come up; whereas thofe feeds which are fown in autumn, generally grow the following fpring. Thefe fhould be fown in drills, in the place where they are defigned to remain. The drills fhould be at leaft two feet and a half afunder, becaufe the plans fpread their leaves very wide. When the plants come up in the fpring, they muft be carefully cleared from weeds; and where they are too clofe together, fome of them fhould be drawn out, to give room for the others to grow; but at this time they need not be left more than two or three inches apart; for the firft year that the plants arife from feeds, they make but flow progrefs, fo the autumn following the remaining part of the plants may be taken up, leaving thofe which are defigned to remain about eighteen inches afunder; and thofe plants which are taken up, may be tranfplanted into another bed, if they are wanted. After the firft year thefe plants will require no farther care but to keep them clean from weeds; and every fpring, juft before the plants begin to pufh out new leaves, the ground fhould be carefully dug between them, to loofen it, but the roots muft not be injured, left it fhould caufe them to decay. The plants being thus managed, will continue feveral years, and produce flowers and feeds annually, from which new plants may be raifed. They delight in a foft loamy foil, and if they are expofed only to the morning fun, they will thrive better than if they have a warmer fituation, for they endure the cold of our winters very well.

The roots of the third fort were formerly ufed in medicine, but are now never ordered, being fuppofed to have a poifonous quality. Boerhaave fays it has much the fame qualities as Euphorbium; it burns the bowels, and produces a diarrhœa.

THELIGONUM. Lin. Gen. Plant. 947. Dogs Cabbage.

The Characters are,

It has male and female flowers on the fame plant. The male flowers have a turbinated empalement of one leaf, cut into two fegments, which turn backward. It has no petal, but feveral erect ftamina the length of the empalement, terminated by fingle fummits. The female flowers have a fmall bifid empalement of one leaf, which is permanent. It has no petals, but has a globular germen, fupporting a fhort ftyle, crowned by an obtufe ftigma. The germen afterward becomes a thick globular capfule with one cell, inclofing one globular feed.

We have but one Species of this genus, viz.

Theligonum (*Cynocrambe.*) Sauv. Monf. 129. *This is the Cynocrambe, or Dogs Cabbage of Diofcorides.*

This plant grows naturally in the fouth of France, in Italy, and Tartary. It is annual. The ftalks trail on the ground like thofe of Chickweed; they grow about a foot long; their joints are pretty clofe, garnifhed with oval acute-pointed leaves, ftanding on pretty long foot ftalks, which are bordered. At each joint is placed one of thefe leaves, and from the fame point come out feveral fmaller, of the fame fhape, on fhorter foot-ftalks. The flowers are produced from the wings of the ftalk in clufters, fitting very clofe; they are fmall, of an herbaceous white colour, fo make no great appearance. The male and female flowers grow from the fame joint. The female flowers are fucceeded by a fingle roundifh feed, which ripens in autumn.

It is preferved in botanic gardens for the fake of variety. The feeds of this muft be fown in autumn, in the place where the plants are to remain; for when they are fown in the fpring, the plants rarely come up the fame year. They require no other culture but to keep them clean from weeds, and thin them where they are too clofe.

THEOBROMA. Lin. Gen. Plant. 806. Baftard Cedar.

The Characters are,

The empalement of the flower is compofed of three oval concave leaves, which are reflexed. The flower has five oval petals, which fpread open, and are hollowed like a fpoon; from the top of each petal come out a bifid briftly ligula, divided like two horns. It has a great number of fhort ftamina joined in five bodies, which are terminated by roundifh fummits, and a roundifh germen, fupporting a fingle ftyle the length of the petals, crowned by a fingle ftigma. The germen afterward turns to a roundifh fruit with five angles, opening in five cells, each containing feveral feeds.

We have but one Species of this genus, viz.

Theobroma (*Guazuma*) foliis ferratis. Hort. Cliff. 379. *Theobroma with fawed leaves.*

This grows naturally in moft of the iflands in the Weft-Indies, where it rifes to the height of forty or fifty feet, having a trunk as large as a middle-fized man's body, covered with a dark brown furrowed bark, fending out many branches toward the top, which fpread wide on every hand, garnifhed with oblong heart-fhaped leaves, placed alternate, of a bright green on their upper fide, and pale on their under, fawed on their edges, with a ftrong midrib, and feveral tranfverfe veins, ftanding upon fhort foot-ftalks. The flowers come out in bunches from the wings of the leaves; they are fmall, of a yellow colour, having five concave petals, which fpread open circularly, and a great number of ftamina, which at their bafe are joined in five bodies, terminated by roundifh fummits. In the center is fituated a roundifh germen, fupporting a flender ftyle the length of the ftamina, crowned by a fingle ftigma. The germen afterward turns to a roundifh warted fruit, having five obtufe angles, and five cells, which contain feveral irregular feeds.

The wood of this tree is white and ductile, fo is frequently cut into ftaves for cafks. The fruit and leaves are good fodder for cattle, therefore when the planters clear the land from wood, they leave the trees of this fort ftanding for the feed, which is of great ufe in dry feafons, when the common fodder is fcarce.

There are fome plants of this fort in England preferved in the gardens of curious perfons; it is propagated by feeds, which muft be procured as frefh as poffible, from the countries where the plants grow naturally. Thefe fhould be fown upon a good hot-bed in the fpring; and when the plants are fit to remove, they fhould be each planted in a feparate fmall pot, and plunged into a hot-bed of tanners bark, obferving to fhade them from the fun till they have taken new root; then they fhould be treated in the fame way as the Coffee-tree, keeping them always in the tan-bed in the ftove.

THERMOMETERS, or THERMOSCOPES, are inſtruments of very great uſe to gardeners in the management of ſtoves. They ſhew, by inſpection, the preſent condition of the air, whether it be hot or cold; which day in ſummer is the hotteſt, and in the winter which is the coldeſt, or any part of the day, and from thence many uſeful experiments have and may be made, viz. how much one ſpring exceeds another in coldneſs; which baths are the hotteſt or coldeſt; and, if being held in the hand of a perſon in a fever, or otherwiſe applied, will nicely ſhew the abatement or increaſe of a fever.

The common Thermometer, which is uſed for hot-houſes, has a tube of about two feet in length, and about the eighth part of an inch diameter; and in this it is remarked, that the air is cold for the plants when the ſpirit riſes to fifteen inches above froſt; that it is temperate at ſixteen inches and a half; that it is warm when it riſes to eighteen inches, which is the ſtandard for Pine-apple heat. It is marked for hot air at twenty inches, and ſultry hot at twenty-one and a half; but in the common Thermometers theſe degrees are differently marked; this temperate air is about our warm, this warm air our hot, and our hot air is about the ſame as the ſultry.

Theſe Thermometers are marked with the names of ſome of the remarkable plants which are preſerved in the hot-houſes; but as the number of theſe plants has been greatly increaſed in England of late years, I have directed ſome Thermometers to be made with a ſcale, divided into degrees, and with three different points of heat marked in claſſes, which correſpond with theſe Thermometers; and under each claſs, I have drawn up liſts of the ſeveral plants, ranged according to the degrees of heat in which they are found to ſucceed, whereby the culture of them is made eaſy to perſons of ſmall ſkill.

By this means every gardener may know when it is proper to apply his heat in its full force, and what degree of heat ought to be uſed for the welfare of any plant from any part of the world.

Mr. Boyle, by placing a Thermometer in a cave, which was cut ſtrait into the bottom of a cliff, fronting the ſea, to the depth of 130 feet, found the ſpirit ſtood, both in winter and ſummer, at a ſmall diviſion above temperate; the cave had 80 feet depth of earth above it.

I, ſays Dr. Hales, marked ſix Thermometers numerically, 1, 2, 3, 4, 5, 6. The Thermometer, number 1, which was the ſhorteſt, I placed to a ſouth aſpect in the open air; the ball of number 2, I ſet two inches under ground; that of number 3, four inches; number 4, eight inches; number 5, ſixteen inches; and number 6, twenty-four inches: and that the heat of the earth at thoſe ſeveral depths may the more accurately be known, it is proper to place near each Thermometer a glaſs tube, ſealed at both ends, of the ſame length with the ſtems of the ſeveral Thermometers, and with tinged ſpirit of wine in them to the ſame height as in each correſponding Thermometer; the ſcale of degrees of each Thermometer being marked on a ſliding ruler, with an index to the back of it, pointing to the correſponding tube.

When at any time an obſervation is to be made, by moving the index to point to the top of the ſpirit in that tube, an accurate allowance is hereby made for very different degrees of heat and cold in the ſtems of the Thermometers at all depths; by which means the ſcale of degrees will ſhew truly the degrees of heat in the balls of the Thermometers, and conſequently the reſpective heats of the earth at the ſeveral depths where they are placed.

The ſtems of theſe Thermometers, which were above the ground, were fenced from weather and injuries, by ſquare wooden tubes. The ground they were placed in, was a brick earth in the middle of my garden.

July the 30th he began to keep a regiſter of their riſe and fall: during the following month of Auguſt he obſerved, that when the ſpirit in the Thermometer, number 1, (which was expoſed to the ſun) was about noon riſen to 48 degrees, then the ſecond Thermometer was 45, the fifth 33, and the ſixth 31; the third and fourth at intermediate degrees: the fifth and ſixth Thermometers kept nearly the ſame degree of heat, both night and day, till towards the latter end of the month, when, as the days grew ſhorter and cooler, and the nights longer and cooler, they then fell to 25 and 27 degrees.

Now ſo conſiderable a heat of the ſun, at two feet depth under the earth's ſurface, muſt needs have a ſtrong influence in raiſing the moiſture at that and greater depths, whereby a very great and continual reek muſt always be aſcending during the warm ſummer ſeaſon, by night as well as by day; for the heat at two feet deep is nearly the ſame night and day, the impulſe of the ſun-beams giving the moiſture of the earth a briſk undulating motion; which watery particles, when ſeparated and rarefied by heat, aſcend in the form of a vapour; and the vigour of the warm and confined vapour (ſuch as is that which is one, two, or three feet deep in the earth) muſt be very conſiderable, ſo as to penetrate the roots with ſome vigour, as we may reaſonably ſuppoſe from the vaſt force of confined vapour in æolipiles, in the digeſter of bones, and the engine to raiſe water by fire.

If plants were not in this manner ſupplied with moiſture, it were impoſſible for them to ſubſiſt under the ſcorching heats within the tropics, where they have no rain for many months together; for though the dews are much greater there than in theſe more northern climates, yet, doubtleſs, where the heat ſo much exceeds ours, the whole quantity, evaporated in a day there, does as far exceed the quantity that falls by night in dew, as the quantity evaporated here in a ſummer's day is found to exceed the quantity of dew which falls in the night.

But the dew which falls in the hot ſummer ſeaſon, cannot poſſibly be of any benefit to the roots of trees, becauſe it is remanded back from the earth by the following day's heat, before ſo ſmall a quantity of moiſture can have ſoaked to any conſiderable depth.

The great benefit therefore of dew in hot weather muſt be, by being plentifully imbibed into vegetables, thereby not only refreſhing them for the preſent, but alſo furniſhing them with a freſh ſupply of moiſture towards the great expences of the ſucceeding day.

It is therefore probable, that the roots of trees and plants are thus, by means of the ſun's warmth, conſtantly irrigated with freſh ſupplies of moiſture, which, by the ſame means, inſinuates itſelf with ſome vigour into the roots; for if the moiſture of the earth were not thus actuated, the roots muſt then receive all their nouriſhment merely by imbibing the next adjoining moiſture from the earth; and conſequently the ſhell of the earth, next the ſurface of the roots, would always be conſiderably drier the nearer it is to the root, which I have not obſerved to be ſo.

But when, towards the latter end of October, the vigour of the ſun's influence is ſo much abated, that the firſt Thermometer was fallen to three degrees above the freezing point, the ſecond to ten degrees, the fifth to fourteen degrees, and the ſixth to ſixteen degrees; then the briſk undulations of the moiſture of the earth, and alſo of the aſcending ſap, much abating, the leaves faded and fell off.

The greateſt degree of cold, in the following winter, was in the firſt twelve days of November; during which time, the ſpirit in the firſt Thermometer was fallen four degrees below the freezing point, the deepest Thermometer ten degrees; the ice on ponds was an inch thick; the sun's greatest warmth, at the winter solstice, in a very serene, calm, frosty day, was, against a south aspect of a wall, 19 degrees, and, in a free open air, but 11 degrees above the freezing point.

From the 10th of January to the 29th of March was a very dry season, when the green Wheat was generally the finest that was ever remembered: but from the 29th of March 1725, to the 29th of September following, it rained more or less every day, except ten or twelve days about the beginning of July; and that whole season continued so very cool, that the spirit in the first Thermometer rose but to 24 degrees, except now and then a short interval of sun-shine; the second only to 20 degrees, the fifth and sixth to 24 and 23 degrees, with very little variation; so that, during this whole summer, those parts of roots which were two feet under ground, had three or four degrees more warmth than those which were but two inches under ground; and, at a medium, the general degree of heat through this whole summer, both above and under ground, was not greater than the middle of the preceding September.

THLASPI. Tourn. Inst. R. H. 212. tab. 101. Mithridate, or Treacle Mustard.

The CHARACTERS are,

The empalement of the flower is composed of four oval concave leaves, which fall off. The flower has four oval petals, double the size of the empalements, placed in form of a cross; it has six stamina half the length of the petals, two of which are shorter than the others, terminated by acute summits, and a roundish compressed germen, supporting a single style the length of the stamina, crowned by an obtuse stigma. The germen afterward becomes an oval, heart-shaped, compressed little pod, with an acute border, divided into two cells by an intermediate partition, containing two or three seeds in each.

The SPECIES are,

1. THLASPI (*Campestre*) siliculis subrotundis, foliis sagittatis dentatis incanis. Hort. Cliff. 330. *Mithridate Mustard with roundish pods, and arrow-pointed, hairy, and indented leaves.*

2. THLASPI (*Arvense*) siliculis orbiculatis, foliis oblongis dentatis glabris. Flor. Lapp. 251. *Treacle Mustard with orbicular pods, and oblong, indented, smooth leaves; Treacle Mustard, or Penny Cress.*

3. THLASPI (*Perfoliatum*) siliculis obcordatis, foliis caulinis cordatis glabris subdentatis, petalis longitudine calycis, caule ramoso. Lin. Sp. 902. *Treacle Mustard with heart-shaped indented leaves, and a branching stalk.*

4. THLASPI (*Alpestre*) siliculis obcordatis, foliis subdentatis, caulinis amplexicaulibus, petalis longitudine calycis, caule simplici. Lin. Sp. 903. *Treacle Mustard with heart-shaped indented leaves embracing the stalks, which are single.*

5. THLASPI (*Peregrinum*) siliculis suborbiculatis, foliis lanceolatis integerrimis. Lin. Sp. 903. *Treacle Mustard with orbicular pods, and spear-shaped leaves.*

6. THLASPI (*Alliaceum*) siliculis subovatis ventricosis, foliis oblongis obtusis dentatis glabris. Prod. Leyd. 333. *Treacle Mustard with oval swelling pods, and oblong, obtuse, smooth, indented leaves.*

7. THLASPI (*Hirtum*) siliculis subrotundis pilosis, foliis caulinis sagittatis hirsutis. Prod. Leyd. 333. *Treacle Mustard with roundish hairy pods, and hairy arrow-pointed leaves.*

8. THLASPI (*Montanum*) siliculis obcordatis, foliis glabris radicalibus carnosis obovatis integerrimis, caulinis amplexicaulibus, corollis calyce majoribus. Lin. Sp. 902. *Treacle Mustard with heart-shaped pods, smooth, entire, fleshy leaves at bottom, the upper embracing the stalks.*

The first sort grows naturally amongst the Corn in divers parts of England, as also on the side of dry banks; it is a biennial plant, which perishes soon after it has ripened its seeds. The root is composed of ligneous fibres, which spread in the ground; the bottom leaves are long, narrow at their base, and broader toward their points, where they have several indentures, and are hoary on both sides. The stalk rises about a foot high, branching toward the top, and is pretty closely garnished with leaves, placed alternately, whose ears embrace the stalk. The flowers are produced in short spikes at the end of the stalks; they are small, white, and composed of four petals, placed in form of a cross; these are succeeded by roundish capsules, having two cells, containing two or three seeds in each. The whole plant has a warm biting taste. The seeds of this are frequently used instead of those of the next, which is the sort directed to enter the composition of Venice Treacle.

The second sort is an annual plant, which grows naturally in several parts of England: I have found it growing in plenty in the meadows on the right hand side of Godalming in Surry. The root of this is composed of slender fibres; the stalk rises a foot high, is angular, channelled, and smooth; the leaves are smooth and indented, of a deep green colour, and sit close to the stalks; the flowers are produced in loose spikes toward the upper part of the stalks, which are small, white, and composed of four petals, placed crosswise like the former; these are succeeded by broad, flat, roundish, compressed pods, having leafy borders, which have two cells, each containing two or three dark brown seeds, tasting like Garlick. The seeds are an ingredient in Theriaca.

The third sort is an annual plant, which grows naturally in the northern counties of England. Of this there are two sorts mentioned in books, which differ only in size, so that I believe it is owing to the different soils in which they grow; for I have frequently sown the seeds of both in the garden, where, when the plants came up, they have proved to be the same. The stalks of this rise about nine inches high, divided at the top into several branches, which are cloathed with smooth, oblong, heart-shaped, entire leaves, whose base embrace the stalks. The flowers are small, white, and are produced in loose short spikes at the end of the branches.

The fourth sort grows naturally in Sicily; this is a biennial plant, whose stalks rise eight or nine inches high, branching out toward the top, garnished with blunt thick leaves, of a grayish colour, which are spear-shaped and entire, placed opposite, sitting close to the stalk; they have a bitter warm taste. The flowers are produced in loose spikes at the top of the stalks; they are small, and of a purple colour, having four heart-shaped petals, placed in form of a cross; these are succeeded by heart-shaped pods, of a fine green colour, which are divided into two cells, each containing three or four small, oblong, yellowish seeds, which have an acrid taste.

The fifth sort is an annual plant, which grows naturally in the northern parts of Europe; this rises about six or eight inches high. The stalk branches toward the top, and is garnished with oblong, smooth, blunt leaves, which are a little indented; these sit close to the stalk, which, if bruised, have a strong scent of Garlick. The branches are terminated by loose spikes of small white flowers, composed of four roundish petals, placed in form of a cross, and are succeeded by swelling roundish pods, containing a few dark brown seeds.

The sixth sort grows naturally in Wales, and in a few places in England; this has a perennial creeping root. The lower leaves are oblong and hoary; they are very slightly sinuated on the edges. The stalks are about five or six inches long, and incline toward the ground; the flowers are rather larger than those of the first sort, but are of the same form; the pods are hoary, but not hairy. This grows

natu-

naturally on the ſide of a bank beyond Wandſworth in the road to Putney.

The ſeventh ſort grows naturally upon the Alps, and in ſome parts of Yorkſhire, in dry ſtony paſtures. The root of this is perennial and creeping; the ſtalks riſe four or five inches high; the lower leaves are wedge-ſhaped, being broad, and rounded at their points, but narrow at their baſe, of a deep green colour, and entire; thoſe upon the ſtalks are rounder, and ſit very cloſe. The flowers are produced in looſe ſpikes at the end of the branches; they are ſmall and white, ſhaped like thoſe of the other ſorts, which are ſucceeded by roundiſh heart-ſhaped pods, divided into two cells, each containing two or three brown ſeeds.

The eighth ſort grows naturally in rocky places in the ſouth of France, in Spain, and Italy; it is a biennial plant with us. The root is compoſed of ligneous tough fibres, which penetrate the crevices of the rocks; the lower leaves are roundiſh, fleſhy, and entire; the ſtalks riſe about five inches high, and divide into ſmall branches, garniſhed with fleſhy, linear, ſpear-ſhaped, entire leaves, of a deep green colour, having ſmooth ſurfaces. The flowers grow in looſe ſpikes at the end of the branches; they are of a beautiful red colour, with ſome dark bloody ſtripes; theſe are ſucceeded by oblong elliptical pods, which contain ſeveral ſmall red ſeeds.

Theſe plants are propagated by ſeeds, which ſhould be ſown where the plants are to remain; this may be performed either in the ſpring or autumn; but the latter is to be preferred, becauſe the ſeeds at that ſeaſon never fail, and the plants, which come up before winter, will grow much ſtronger, and produce a greater quantity of ſeeds than thoſe which are ſown in the ſpring, eſpecially if the ſeaſon proves dry; and there is very little danger of the plants being injured by froſt in winter, if they are upon dry ground. When the plants come up, they will require no other care but to thin them where they are too cloſe, and keep them clean from weeds.

The two ſorts which are firſt mentioned, may be cultivated for their ſeeds to be uſed in medicine, ſo may be ſown thin upon beds of light ground, in the ſame way as for other garden plants; and when they come up, the ground ſhould be hoed to deſtroy the weeds; and where the plants are too thick, they ſhould be cut up in the ſame manner as is practiſed for Onions, Carrots, &c. leaving them three or four inches apart; and by twice hoeing the ground, if it is well performed, and in dry weather, it will keep the ground clean till the ſeeds are ripe.

The other ſorts are ſeldom cultivated but in botanic gardens for variety, ſo a few plants of each will be ſufficient; therefore theſe may be ſown in drills, and when the plants come up, they muſt be thinned, and kept clean from weeds. If the ſeeds of theſe plants are permitted to ſcatter, the plants will come up without care.

THISTLE. See CARDUUS.

THORN APPLE. See DATURA.

THORN, the Glaſtenbury. See MESPILUS.

THUYA. Tourn. Inſt. R. H. 586. tab. 358. The Arbor Vitæ.

The CHARACTERS are,

It has male and female flowers in the ſame plant; the male flowers are produced in an oval katkin, placed oppoſite upon the common foot-ſtalk, each flower embracing it with its baſe; theſe come out of an oval concave ſcale; they have no petals, but have four ſtamina, which are ſcarce diſcernible; their ſummits adhere to the baſe of the ſcale of the empalement. The female flowers are collected in a common almoſt oval cone, two flowers ſtanding oppoſite on each ſcale; they have no petals, but have a ſmall germen, ſupporting a ſlender ſtyle, crowned by a ſingle ſtigma; theſe are ſucceeded by an oblong oval cone, opening longitudinally, whoſe ſcales are almoſt equal, convex on the outſide, and obtuſe, each containing an oblong ſeed.

The SPECIES are,

1. THUYA (*Occidentalis*) ſtrobilis lævibus, ſquamis obtuſis. Hort. Cliff. 449. *Thuya with ſmooth cones, and obtuſe ſcales; the common Arbor Vitæ.*

2. THUYA (*Orientalis*) ſtrobilis ſquarroſis, ſquamis acuminatis reflexis. Hort. Upſal. 289. *Thuya with rugged cones, and acute-pointed reflexed ſcales; or China Arbor Vitæ.*

The firſt ſort grows naturally in Canada, Siberia, and other northern countries. In ſome of the Engliſh gardens which have not been altered, there are ſome of theſe trees which are of a large ſize: it has a ſtrong woody trunk, which riſes to the height of forty feet or more. The bark, while young, is ſmooth, and of a dark brown colour, but, as the trees advance, the bark becomes cracked, and leſs ſmooth. The branches are produced irregularly on every ſide, ſtanding almoſt horizontal, and the young ſlender ſhoots frequently hang downward, thinly garniſhed with leaves; ſo that when the trees are grown large, they make but an indifferent appearance. The young branches are flat, and the ſmall leaves lie imbricatim over each other like the ſcales of fiſh; the flowers are produced from the ſide of the young branches pretty near to the foot-ſtalk; the male flowers grow in oblong katkins, and between theſe the female flowers are collected in form of cones. When the former have ſhed their farina, they ſoon after drop off, but the female flowers are ſucceeded by oblong cones, having obtuſe ſmooth ſcales, containing one or two oblong ſeeds. The leaves of this tree have a rank oily ſcent when bruiſed.

The ſecond ſort grows naturally in the northern parts of China, where it riſes to a conſiderable height; but this has not been long enough in Europe to have any trees of large ſize. The ſeeds of this ſort were firſt ſent to Paris by ſome of the miſſionaries, and there are ſome of the trees growing in the gardens of ſome curious perſons there, which are more than twenty feet high. The branches of this ſort grow cloſer together, and are much better adorned with leaves, which are of a brighter green colour, ſo make a much better appearance than the other, and, being very hardy, is eſteemed preferable to moſt of the evergreen trees with ſmall leaves, for ornament in gardens. The branches of this tree croſs each other at right angles; the leaves are flat, but the ſingle diviſions of the leaves are ſlender, and the ſcales are ſmaller, and lie cloſer over each other than thoſe of the firſt ſort. The cones are alſo much larger, and of a beautiful gray colour; their ſcales end in acute reflexed points.

Theſe trees may be propagated by ſeeds, layers, or cuttings. The firſt ſort is commonly propagated by cuttings; theſe ſhould be planted in September, upon a ſhady border, and in a loamy ſoil; the cuttings ſhould be choſen from the ſhoots of the ſame year, with a ſmall joint of the former year's wood at the bottom of each. Theſe ſhould be planted three or four inches deep, in proportion to their length, treading the ground cloſe to them to prevent the admiſſion of air. If the following ſpring ſhould prove dry, there ſhould be a little mulch laid over the ſurface of the ground to prevent its drying; where this is performed in time, it will ſave the trouble of watering the cuttings; and it will be much better for them, becauſe when theſe are putting out their young fibres, if they are much watered, it will rot them while they are tender. Theſe cuttings will be rooted enough to tranſplant by the next autumn, when they may be either planted in beds, or in nurſery-rows to be trained up.

When they are propagated by layers, the young branches only ſhould be laid down in autumn, which will alſo put out

out roots by the next autumn, when they may be taken up, and tranfplanted in the fame manner as thofe raifed from cuttings; but although thefe are very expeditious methods of propagating this tree, yet thofe who are defirous to have large trees, fhould always propagate them by feeds, for the plants fo raifed will be much preferable to the other.

There is a variety of the firft fort with variegated leaves, which fome people keep in their gardens; but as this proceeds from a weaknefs in the plants, fo whenever the plants become ftrong and vigorous, they always return to their plain colour again; to prevent which, they generally plant them in very poor ground. This variety can only be preferved by propagating the plants, either by cuttings or layers.

The China fort is generally propagated by layers in the fame way as the former, but the cuttings of this, if rightly managed, will take root very freely; but moft people have over nurfed them. If thefe are planted in September in a border of foft loam, expofed to the eaft, and if before the hard froft fets in, the furface of the ground is covered with old tanners bark about two inches thick, it will prevent the froft from penetrating the ground; and if this remains in the fpring, it will alfo keep the ground moift, for if the cuttings or layers of this fort are watered too much in the fpring, when they are beginning to put out young fibres, it will certainly rot them, as I have frequently experienced; therefore I advife every one not to water thefe cuttings or layers, nor fhould the plants be much watered when they are tranfplanted, for the fame reafon; but as there are many plants now in England which ripen their feeds, fo thofe who can be fupplied with them, fhould prefer them to the other; for, after the firft two years, the feedling plants will greatly outftrip the other in growth, and the plants will be much handfomer.

Thefe feeds fhould be fown foon after they are ripe, which is in the fpring. They fhould be fown in pots, filled with foft loamy earth, and plunged into the ground in an eaft border, where they may have only the morning fun, obferving always to keep the pots clean from weeds. Sometimes thefe feeds will come up the fame year, but they often lie in the ground till the next fpring; therefore the pots fhould be put in a common hot-bed frame in winter, and in the fpring the plants will come up; thefe muft not be too much expofed to the fun the firft year, and if in the next winter they are fheltered under a frame, it will be a good way to preferve them; and the fpring following they may be tranfplanted into beds, and treated in the fame way as thofe propagated by cuttings.

THYMBRA. Lin. Gen. Plant. 627.

The Characters are,

It has an empalement of one leaf, whofe brim is cut into two lips. The flower is of one petal, of the lip kind. The upper lip is concave, cut into two obtufe fegments. The lower lip ends with three almoft equal points; it has four flender ftamina, the two under being fhorter than the other, terminated by twin fummits, and a four-pointed germen, fupporting a flender half bifid ftyle, crowned by acute ftigmas. The germen afterward becomes four feeds, which ripen in the empalement.

The Species are,

1. Thymbra (*Spicata*) floribus fpicatis. Lin. Sp. Pl. 569. *Thymbra with fpiked flowers; or Mountain Macedonian Hyffop.*

2. Thymbra (*Verticillata*) floribus verticillatis. Lin. Sp. Plant. 569. *Thymbra with whorled flowers; or rough, narrow-leaved, Mountain Hyffop.*

The firft fort grows naturally on Mount Libanus, in Macedonia, and alfo in Spain; it is a low fhrubby plant like Heath, branching out into flender ligneous ftalks, which are fix or eight inches long covered with a brown bark, garnifhed with narrow acute-pointed leaves, fitting clofe to the ftalks oppofite; they have an aromatic odour when bruifed. The ftalks are terminated by thick clofe fpikes of purple flowers, near two inches long. The empalements are ftiff and hairy, cut half their length into acute fegments, out of thefe the flowers peep with their two lips; the upper is concave and arched, the under is cut into three equal portions, a little reflexed.

The fecond fort grows naturally in Spain and Italy; this has a fhrubby ftalk, which feldom rifes much more than a foot high, putting out many fmall ligneous branches, garnifhed with narrow fpear-fhaped leaves, which have many punctures; they ftand oppofite, and are of an aromatic flavour. The flowers grow in whorled fpikes at the end of the branches. The leaves which ftand under each whorl, are broader than thofe below, and are covered with fine hairs. The flowers are purple, and fit clofe to the ftalks; the upper lip is concave, ending with two obtufe points; the lower ends with three equal points.

Thefe plants are propagated by feeds, which fhould be fown in the fpring on a bed of light earth, where, if the feeds are good, the plants will appear in about fix or eight weeks. When they come up, they muft be kept clean from weeds, and in July they will be fit to remove; at which time part of them fhould be planted in fmall pots, and the other may be planted in a warm border of dry ground, being careful to fhade them from the fun, and fupply them with water until they have taken new root; after which thofe in the full ground will require no other care but to keep them clean from weeds, and, if the winter fhould prove very fevere, they fhould be covered with mats, or fome other light covering, to protect them; for the young plants are in greater danger of being deftroyed than thofe which are older. Thofe plants in the pots fhould be fheltered under a common frame in winter, where they may enjoy the free air in mild weather, and be protected from hard froft.

Thefe plants will live in the open air in England, unlefs the winters prove very fevere, efpecially if they are planted in a poor, dry, ftony foil.

THYMELÆA. See Daphne and Passerina.

THYMUS. Tourn. Inft. R. H. 196. tab. 93. Thyme.

The Characters are,

The flower has a permanent empalement, divided into two lips, whofe chaps are hairy and fhut. The upper lip is broad, plain, erect, and indented in three parts; the under lip ends in two equal briftles. The flower is of the lip kind; it has one petal, with a tube the length of the empalement. The chaps are fmall; the upper lip is fhort, erect, obtufe, and indented at the point; the lower lip is long, broad, and divided into three parts. It has four incurved ftamina, two being longer than the other, terminated by fmall fummits, and a four-pointed germen, fupporting a flender ftyle, crowned by a bifid acute ftigma. The germen afterward turns to four fmall roundifh feeds, ripening in the empalement.

The Species are,

1. Thymus (*Vulgaris*) erectus, foliis revolutis ovatis, floribus verticillato-fpicatis. Hort. Cliff. 305. *Upright Thyme with oval leaves, which turn backward, and flowers growing in whorled fpikes; or common broad-leaved Thyme.*

2. Thymus (*Tenuifolius*) foliis lineari-lanceolatis incanis, floribus verticillato-fpicatis. *Thyme with linear, fpear-fhaped, hoary leaves, and flowers growing in whorled fpikes; or common Thyme with narrow leaves.*

3. Thymus (*Cephalotos*) capitulis imbricatis magnis, bracteis ovatis, foliis lanceolatis. Lin. Sp. Pl. 592. *Thyme with large imbricated heads, oval bracteæ, and fpear-fhaped leaves.*

4. Thymus

4. Thymus (*Villosus*) capitulis imbricatis magnis, bracteis dentatis, foliis setaceis pilosis. Lin. Sp. Plant. 592. *Thyme with large imbricated heads, indented bracteæ, and bristly hairy leaves.*

5. Thymus (*Serpyllum*) floribus capitatis, caulibus decumbentibus, foliis planis obtusis basi ciliatis. Lin. Sp. *Thyme with flowers growing in heads, creeping stalks, and oval hairy leaves; or broad-leaved, hairy, Mother of Thyme.*

6. Thymus (*Glaber*) floribus capitatis, caulibus decumbentibus, foliis lanceolatis glabris. *Thyme with flowers growing in heads, spreading stalks, and smooth spear-shaped leaves; or common greater Mother of Thyme, with a purple flower.*

7. Thymus (*Ovatus*) caulibus decumbentibus, foliis ovatis glabris, floribus verticillato-spicatis. *Thyme with strong creeping stalks, oval smooth leaves, and flowers growing in whorled spikes; or common greater Mother of Thyme, with a smaller flower.*

8. Thymus (*Lanuginosus*) caulibus decumbentibus, foliis ovato lanceolatis rigidis lanuginosis, floribus capitatis. *Thyme with creeping stalks, oval, spear-shaped, stiff leaves, which are downy, and flowers growing in heads; or hairy Rock Mother of Thyme.*

9. Thymus (*Odoratissimus*) caulibus decumbentibus, foliis lineari-lanceolatis glabris, floribus alaribus terminalibusque. *Thyme with trailing stalks, linear, spear-shaped, smooth leaves, and flowers growing at the wings and tops of the stalks.*

The first sort is the common Thyme, which is cultivated in the gardens for the kitchen, and also for medicine. This grows naturally on stony rocky places in the south of France, in Spain and Italy, and is so well known here as to need no description.

It is propagated either by seeds, or parting the roots; the season for the latter is in March or October. If it is propagated by seeds, they should be sown upon a bed of light earth in the spring, observing not to bury the seeds too deep, nor to sow them too thick, for the seeds are very small. When the plants are come up, they should be carefully cleared from weeds; and if the spring should prove dry, if they are watered twice a week, it will greatly promote their growth. In June the plants should be thinned, leaving them about six inches asunder each way, that they may have room to spread; and those plants which are drawn out, may be transplanted into fresh beds at the same distance, observing to water them until they have taken root; after which they will require no farther care but to keep them clear from weeds; and when the plants are big enough, they may be drawn up for use.

But if the plants are propagated by parting their roots, the old plants should be taken up at the times before-mentioned, and slipt into as many parts as can be taken off with roots; these should be transplanted into beds of fresh light earth at six or eight inches distance, observing, if the season is dry, to water them until they have taken root; after which they must be duly weeded, which will cause them to thrive, and soon be fit for use.

In order to save the seeds of these plants, some of the old roots should remain unremoved in the place where they were sown the preceding year; these will flower in June, and in August the seed will ripen, which must be taken as soon as it is ripe, and beat out, otherwise the first rain will wash it all out of the husks.

These plants root greatly in the ground, and thereby draw out the goodness of the soil sooner than most other plants; so that whatever is sown or planted upon a spot of ground, whereon Thyme grew the preceding year, will seldom thrive, unless the ground be trenched deeper than the Thyme rooted, and well dunged.

If this plant grows upon walls, or on dry, poor, stony land, it will endure the greatest cold of this country; but in rich ground, where the plants grow vigorously, they are sometimes destroyed by very severe frost.

There is a variety of this with variegated leaves, which is by some preserved in their gardens.

The second sort has shorter stalks; the leaves are longer, narrower, and end in sharper points than the first, and the whole plant is hoary. The flowers grow in long whorled spikes, and are larger than those of the common Thyme. This may be propagated and treated in the same way as the first sort.

The third sort grows naturally in Spain and Portugal; this has a low woody stalk, from which come out many stiff branches about five or six inches long, garnished with small, narrow, spear-shaped leaves, placed opposite, terminated by pretty large heads of flowers, which come out from oval scaly leaves, lying over each other like the scales of fish; they are white and small, so make no great appearance. The whole plant is of a hoary colour, and has an aromatic weak scent.

The fourth sort grows naturally in Portugal; this has slender, ligneous, hairy stalks, which grow erect about six inches high, garnished with very narrow, bristly, hairy leaves, which, at the lower part of the stalk, come out in clusters, but upward they are placed by pairs. The stalks are terminated by large scaly heads. The leafy scales are indented in acute points; these lie over each other in the same order as the other, and between them the flowers peep out, which are of a purple colour, shaped like those of the common Thyme.

These two sorts may be propagated by slips, if they are planted in April on an east border, and closely covered with a bell or hand-glass, they will soon put out roots, when some of them may be transplanted into pots; to be sheltered under a frame in winter; but the others should be planted on a warm border of dry ground, observing to shade and water them till they have taken new root. These plants will live through the winter in the open air in a warm dry situation, but in severe frost they are generally destroyed; therefore to preserve the kinds, a few plants of each should be sheltered under a frame in winter; they may be propagated by seeds, when they can be procured. If these are sown on a bed of light earth, in the same way as common Marjoram, the plants will come up, and may be treated as those raised from slips.

The fifth sort is the common Mother of Thyme, which is frequently titled wild Thyme; it grows naturally upon dry commons and pastures in most parts of England, so is very rarely admitted into gardens. This is so well known as to need no description. There is a very common mistake which has prevailed in regard to this plant, which is, that the sheep and deer which feed upon it, have much finer flavoured flesh than others, whereas no cattle will meddle with it; for in the places where it grows, when the Grass is as closely eaten down as possible, the wild Thyme will be found in flower with all its stalks entire.

Of this there are the following varieties. The small creeping Mother of Thyme without scent. Narrow-leaved Mother of Thyme, smelling like the leaves of the Walnut-tree. Shrubby Mother of Thyme with pale red flowers, and the Lemon Thyme. The last is frequently kept in gardens for the agreeable odour of its leaves. But when this is propagated by seeds, the plants have not the same scent; so it is an accidental variety, which is maintained by propagating it by slips and cuttings.

The sixth sort has broader and smoother leaves than the common sort; the stalks grow much longer; the joints are farther distant; the heads of flowers are larger, and the flowers

flowers are of a brighter purple colour. There is a variety of this with variegated leaves, which is propagated in gardens, and was formerly planted for edgings to borders; but it is now frequently brought in pots to the markets to supply the London gardens.

The feventh fort has creeping ftalks like the common kind, but they grow longer, and their joints are farther afunder; their leaves are oval, fmooth, and of a lucid green. The flowers grow in clofe thick whorls, which are diftant from each other, forming a loofe fpike five or fix inches long. The flowers of this fort are much fmaller than thofe of the common fort, appearing but little beyond their empalements. This is pretty common in the neighbourhood of Paris, but is rarely found growing naturally in England.

The eighth fort grows naturally in the foreft of Fontainbleau in France; this has creeping flender ftalks like the firft, which are garnifhed with fmall, oval, fpear-fhaped, hoary leaves; the young fhoots of the fame year are alfo very white and hoary. The leaves are ftiffer than thofe of the other forts. The flowers are produced in round heads at the end of the branches; they are of a bright purple colour, and appear at the fame time as thofe of the other forts.

The ninth fort grows naturally in Tartary; this is a biennial plant. The ftalks are long, flender, and trail upon the ground, but do not emit roots from their joints as moft of the others do. The ftalks are fmooth, of a light brown colour, garnifhed with narrow fpear-fhaped leaves, which are fmooth. The fmall whorls of flowers come out at the wings of the leaves, and the ftalks are terminated by oblong heads of flowers, whofe empalements are hoary. The flowers are of a bright purple colour. The whole plant has an agreeable aromatic fcent.

All thefe forts may be eafily propagated by thofe who are defirous to have them in their gardens, either by flips, or parting of their roots, in the fame manner as Thyme, or their feeds may be fown in the fpring. They delight in dry undunged ground, where they will propagate themfelves by their trailing ftalks, and require no other care but to keep them clean from weeds.

THYME, THE MARUM. See TEUCRIUM.

THYME, THE MASTICH. See SATUREJA.

TIARELLA. Lin. Gen. Plant. 495. Sanicle.

The CHARACTERS are,

The flower has a permanent empalement, divided into five oval acute parts; it has five oval petals the length of the empalement, and ten awl-fhaped ftamina, which are much longer than the petals, terminated by roundifh fummits, and a bifid germen, ending with two ftyles, crowned by fingle ftigmas. The germen afterward becomes an oblong capfule with one cell, opening with two valves, containing feveral fmall oval feeds.

The SPECIES are,

1. TIARELLA (*Cordifolia*) foliis cordatis. Lin. Gen. Nov. 188. Sp. Pl. 405. *Tiarella with heart-fhaped leaves.*

2. TIARELLA (*Trifoliata*) foliis ternatis. Lin. Gen. Nov. 188. Sp. Pl. 405. *Tiarella with trifoliate leaves.*

The firft fort grows naturally in North America; this has a perennial fibrous root, which creeps and multiplies, from which come out many heart-fhaped leaves upon flender foot-ftalks. The leaves are unequally indented on their edges, and are of a light green colour. The flowers ftand upon flender naked foot-ftalks, which arife immediately from the root between the leaves, which is about four inches long, and is terminated by a loofe fpike of fmall herbaceous white flowers, but are feldom fucceeded by feeds in England.

This plant is propagated by its creeping roots which fpread in the ground, and fhoot up heads; thefe may be taken off and tranfplanted in the autumn. It loves a moift foil and fhady fituation, and requires no other care but to keep it clean from weeds.

The fecond fort grows naturally in the northern parts of Afia; this has a perennial fibrous root, from which fpring up a few trifoliate leaves upon foot-ftalks, like thofe of the Bilberry, but much fmaller. The ftalk is flender, and rifes five or fix inches high; it is rough and hairy, garnifhed with two leaves at the bottom, and another toward the top, a little below the fpikes of flowers; they are angular, and fawed on their edges. The ftalk is terminated by a loofe fpike of flowers, which are compofed of five white fmall petals, inferted in the empalement, and ten awl-fhaped ftamina which are longer than the petals, terminated by roundifh fummits.

This fort is propagated by parting of the root, in the fame manner as the former, and delights in a moift foil and a fhady fituation.

TILIA. Tourn. Inft. R. H. 611. tab. 381. The Lime, or Linden-tree.

The CHARACTERS are,

The flower has a concave coloured empalement, which is cut into five parts; it has five oblong blunt petals, which are crenated at their points, and many awl-fhaped ftamina, terminated by fingle fummits, with a roundifh germen, fupporting a flender ftyle the length of the ftamina, crowned by an obtufe five-cornered ftigma. The germen afterward becomes a thick globular capfule with five cells, opening at the bafe with five valves, each containing one roundifh feed.

The SPECIES are,

1. TILIA (*Cordata*) foliis cordatis acuminatis, inæqualiter ferratis, fructibus quinque locularibus tomentofis. *Lime-tree with heart-fhaped acute-pointed leaves, which are unequally fawed, and a woolly fruit having five cells; Lime-tree with a fmaller leaf.*

2. TILIA (*Æuropæa*) foliis acuminatis, ferratis, fubhirfutis, fructibus quadri-locularibus fubpilofis. *Lime-tree with acute-pointed leaves, which are fawed, fomewhat hairy, and a hairy fruit having four cells; the red-twiged Lime-tree.*

3. TILIA (*Americana*) foliis cordatis acuminatis ferratis, fubtus pilofis floribus nectario inftructis. *Lime-tree with heart-fhaped, acute-pointed, fawed leaves, which are hairy on their under fide, and flowers furnifhed with nectarii. This is called the American black Lime.*

4. TILIA (*Caroliniana*) foliis cordatis obliquis glabris fubferratis cum acumine, floribus nectario inftructis. *Lime-tree with heart-fhaped fmooth leaves, which are oblique to the foot-ftalk, fomewhat fawed on their edges, ending in acute points, and flowers having nectarii; or Carolina Lime-tree.*

The firft fort grows naturally in the woods in many parts of England; of this there are two or three varieties, which differ in fize and fmoothnefs of their leaves, fome of them having much larger and rougher leaves than the others: I raifed plants of three of thefe varieties from feeds, but have conftantly found them vary from one to the other; and I much doubt if the fecond is more than a feminal variety, but as I have not had an opportunity of raifing any of the plants from feeds, I cannot pofitively determine this.

The large-leaved Dutch Lime was generally preferred to our common fort for the fize of its leaves; but of late years all thefe trees are little efteemed, becaufe it is late in fpring before their leaves come out, and they begin to decay the firft in autumn; and when the trees are planted in a dry foil, their leaves frequently decay in July, fo are continually falling off, making a litter all the remaining part of fummer.

The third fort was brought from New England, by the title of Black Lime. The branches of this fort are covered with a dark brown bark. The leaves are large, heart-ſhaped, and end in acute points; they are deeply ſawed on their edges, of a deep green on their upper ſide, but of a pale green, and a little hairy on their under ſide, ſtanding upon long ſlender foot-ſtalks. The flowers are produced in bunches, in the ſame manner as thoſe of the common Lime-tree; but the petals of the flowers are narrower, and have nectarii growing to their baſe. The flowers of this ſort do not appear till late in July, ſo are a full month after the common ſort. The capſules are ſmaller, rounder, and leſs hairy than thoſe of the common ſort.

The ſeeds of the fourth ſort were brought from Carolina by the late Mr. Cateſby. This tree ſeems to be of much ſmaller growth than either of the other ſorts; the branches ſpread more horizontally. The leaves are ſmaller, and have a ſmoother ſurface than either of the other; they are heart-ſhaped, but the midrib runs oblique to the foot-ſtalk, ſo that one ſide is much larger than the other. Their edges are ſlightly ſawed, and their tops run out into long acute points. The bunches of flowers ſtand upon long ſlender foot-ſtalks; the petals of the flowers are narrow, and end in acute points; theſe have a narrow nectarium faſtened to their baſe on the inſide, which ſtands erect, cloſe to the petals. The flowers emit a very fragrant odour, and are continually haunted by bees during their continuance.

All theſe trees are eaſily propagated by layers, which in one year will take good root, and may then be taken off, and planted in a nurſery, at four feet diſtance row from row, and two feet aſunder in the rows. The beſt time to lay them down, and to remove them, is when their leaves begin to fall, that they may take root before the froſt comes on, though they may be tranſplanted any time from September to March, in open weather; but if the ſoil is dry, it is much the better way to remove them in autumn, becauſe it will ſave a great expence in watering, eſpecially if the ſpring ſhould prove dry. In this nurſery they may remain four or five years, during which time the ground ſhould be dug every ſpring, and conſtantly kept clear from weeds, and the large ſide ſhoots pruned off, to cauſe them to advance in height; but the ſmall twigs muſt not be pruned off from the ſtems, becauſe theſe are abſolutely neceſſary to detain the ſap, for the augmentation of their trunks, which are apt to ſhoot up too ſlender, when they are entirely diveſted of all their lateral twigs. If the ſoil, in which they are planted, be a fat loam, they will make a prodigious progreſs in their growth, ſo that in three years time they will be fit to tranſplant out where they are to remain.

They may alſo be propagated by cuttings; but, as this method is not ſo certain as by layers, the other is generally practiſed. In order to obtain proper ſhoots for laying down a Lime-tree, it is cut down cloſe to the ground, from the roots of which a great number of ſtrong ſhoots are produced the following year; theſe will be large enough to lay down the following autumn, eſpecially if the ſmalleſt of them are cut off cloſe early in the ſummer; for when too many ſhoots are ſuffered to grow all the ſummer, they will be much weaker, than if only a ſufficient quantity is left. The manner of laying down theſe ſhoots having been already directed under the article of LAYERS, I need not repeat it here.

There are ſome perſons who raiſe theſe trees from ſeeds, which, although it is a ſlower way, yet when the trees are deſigned to grow large, is the beſt method; and if they are only once tranſplanted, and this performed while they are young, it will be ſtill the better way; for all trees that are tranſplanted, are ſhorter lived than thoſe which remain in the places where they aroſe from the ſeeds, and their timber will be ſounder, and grow to a much larger ſize.

When this method is practiſed, the ſeeds ſhould be ſown in autumn, ſoon after they are ripe, upon a ſhady border of moiſt light ſoil, where the plants will come up the following ſpring; but, when the ſeeds are kept out of the ground till ſpring, the plants will not come up till the year after. When the plants appear, they ſhould be conſtantly kept clean from weeds till the following autumn; then they ſhould be carefully taken up, and tranſplanted into a nurſery, where they may grow two or three years to get ſtrength, and then may be planted where they are deſigned to remain; for the younger they are planted out to remain, the more they will thrive.

The timber of the Lime-tree is uſed by the carvers, it being a ſoft light wood, as alſo by architects for framing the models of their buildings; the turners likewiſe uſe it for making light bowls, diſhes, &c. but it is too ſoft for any ſtrong purpoſes.

Theſe trees will continue growing, and remain ſound a great number of years, and, if planted in a good loamy ſoil, will grow to a conſiderable bulk. I have meaſured one of theſe trees, which was near ten yards in girth two feet above the ground, and was then in a thriving condition; and Sir Thomas Brown mentions one of theſe trees which grew in Norfolk, that was ſixteen yards in circuit, a foot and a half above ground, in height thirty yards, and in the leaſt part of the trunk it was eight yards and a half.

TINUS. See VIBURNUM.

TITHYMALUS. Tithymaloides. Tourn. Inſt. App. 654. Spurge.

The CHARACTERS are,

The flower has an empalement of one leaf, indented in three parts; it has one petal which is ſhaped like a ſlipper, of a thick fleſhy conſiſtence. Under the upper part of the flower are ſituated the ten ſtamina, which are inſerted in the receptacle of the flower; they are ſlender, and terminated by globular ſummits; in the center is ſituated a roundiſh three-cornered germen, ſupporting three bifid ſtyles, crowned by oblong ſtigmas. The germen afterward becomes a roundiſh capſule having three cells, each containing one oval ſeed.

The SPECIES are,

1. TITHYMALUS (*Myrtifolius*) foliis ovatis acuminatis. *Spurge with oval acute-pointed leaves.*

2. TITHYMALUS (*Lauro-ceraſifolius*) foliis oblongo ovatis obtuſis ſucculentibus. *Spurge with oblong, oval, obtuſe leaves, which are very ſucculent.*

The firſt ſort grows naturally near Carthagena in America, from whence Mr. Robert Millar, ſurgeon, ſent the branches, which were planted here, and ſucceeded; this riſes with ſhrubby ſucculent ſtalks to the height of twelve or fourteen feet, which are too weak to ſtand without ſupport, though they are frequently as large as a man's little finger; but their leaves being ſucculent, are ſo heavy as to weigh down the branches, if they are not ſupported. The leaves are oval, and terminate in acute points; and are ranged alternately on two ſides of the branches, to which they ſit cloſe. The flowers are produced at the end of the branches three or four together; they are of a ſcarlet colour, of one petal, in ſhape of a ſlipper; theſe are ſucceeded by roundiſh capſules with three furrows, dividing them into three cells, each containing one oblong ſeed. The whole plant abounds with an acrid milky juice.

The ſecond ſort grows naturally in Barbadoes, and moſt of the other iſlands in the Weſt-Indies, where the Engliſh inhabitants know it by the title of Poiſon Buſh; this hath thick, ſhrubby, ſucculent ſtalks, which will grow to the height of ten or twelve feet, larger than thoſe of the firſt ſort, and are garniſhed with oblong oval leaves, ending in blunt points, of a very thick conſiſtence, and of a dark green colour, ranged alternately on two ſides of the ſtalk. The flowers grow at the end of the branches; they are ſhaped like thoſe of the firſt ſort, and are of a deep red colour,

lour; these are succeeded by roundish capsules divided into three cells, each containing one oblong seed.

This whole plant abounds with an acrid milky juice, which will draw blisters on the flesh wherever it is applied; and if it mixes with the blood, I have been credibly informed it becomes a deadly poison; for that if the points of arrows or the edges of swords are rubbed over with this juice, it becomes deadly to any animal wounded with those weapons.

These plants are both propagated by cuttings, which may be taken from the plants during any of the summer months; and after having laid in a dry place for a fortnight or three weeks, until the wounded part be healed over, they should be planted into small pots, filled with light sandy earth, mixed with lime rubbish, and then plunged into a hot-bed of tanners bark, observing now and then to refresh them gently with moisture, but they should never receive much wet, which will rot them.

After they have taken root, they may have a greater share of air by raising the glasses, but they must never be wholly exposed to the open air. In this bed they may remain until the beginning of October, when they must be removed, and placed with the Melon and Torch Thistle in a warm dry stove, and during the winter season they should have very little water, which, if given in plenty, seldom fails to rot them.

These plants are too tender to thrive in the open air in England, therefore should constantly remain in the stove, observing in the summer season, when the air is warm, to admit a large share of fresh air to them, and in the winter to place them in a warm part of the stove, otherwise they cannot be preserved.

They must be shifted every summer, and fresh earth given to them. If the earth is light or sandy, it will require no mixture, for rich or strong ground is very improper for them; therefore where the soil is inclinable to either of these, there should be a good mixture of sand and lime-rubbish to prevent its binding, or detaining moisture.

These plants are preserved for their odd appearance amongst other succulent plants, their leaves being very large, thick, and full of a milky acrid juice.

TITHYMALUS. See Euphorbia.

TOAD FLAX. See Linaria.

TOBACCO. See Nicotiana.

TOLUIFERA. Lin. Gen. Plant. 470. Balsam of Tolu-tree.

The Characters are,

The flower has a bell-shaped empalement of one leaf, which is slightly indented in five parts at the brim; it has five petals inserted in the receptacle of the flower, four of which are narrow and equal, being a little longer than the empalement, and the fifth is much larger, and almost heart-shaped, having a tail the length of the empalement; it has ten short stamina, terminated by oblong erect summits, and a roundish germen, supporting a very short style, crowned by an acute stigma. The germen afterward turns to a roundish fruit with four cells, each containing one oval seed.

We have but one Species of this genus, viz.

Toluifera (*Balsamum*.) Lin. Mat. Med. *The Balsam-tree of Tolu.*

This tree grows naturally near Carthagena in America, from whence the late Dr. Houstoun sent the seeds to England: in its native place this grows to a tree of a large size. The bark is very thick, rough, and of a brown colour; the branches spread out wide on every side, and are garnished with winged leaves, composed of several oblong oval lobes, placed alternately along the foot-stalk, terminated by an odd one, rounded at both ends, but run out to an acute point at the top; they are smooth, of a light green colour, and sit close to the foot-stalk. The flowers are produced in small bunches at the wings of the branches, each standing upon a slender foot-stalk almost an inch long; their empalements are of the round bell-shape, being of one leaf, which is slightly scollopped at the brim into five obtuse parts. The flower has four narrow petals of a yellow colour, which are a little longer than the empalement, and one more whose tail is of the same length of the other petals; the top is of an oval heart-shape, stretched out beyond the other parts; it has ten short stamina within the tube of the flower, which are terminated by oblong erect summits, of a sulphur colour; and at the bottom of the tube is situated a roundish germen, having a very short style, crowned by an acute-pointed stigma. After the flower is past, the germen turns to a roundish fruit the size of a large Pea, divided into four cells, each containing one oblong oval seed.

This tree may be propagated by seeds, which must be procured from the country where it grows naturally, and should be fresh, otherwise they will not grow. When they are gathered from the tree, they should be put up in sand to preserve them; for when they are sent over in papers, the insects generally devour them. These seeds must be sown in pots filled with light earth as soon as they arrive, and plunged into the tan. If it should happen in autumn or winter they must be plunged into the stove, but in spring or summer they may be plunged into the tan-bed under a frame; they should be taken out of their covers, otherwise they will be long in the ground before they vegetate. When the plants come up, and are fit to remove, they should be carefully transplanted, each into a separate pot, and plunged into a good hot-bed of tanners bark, shading them from the sun till they have taken new root; after which they should be treated in the same way as the Coffee-tree, with which management the plants will succeed.

TORDYLIUM. Tourn. Inst. R. H. 320. tab. 170. Lin. Gen. Plant. 293. Hartwort.

The Characters are,

It has an umbellated flower; the principal umbel is composed of many small ones, compounded of many rays; the involucrum of the greater umbel is composed of narrow leaves, as long as the rays of the umbel; those of the rays are half the length; the umbels are difformed; the flowers have five heart-shaped inflexed petals, which are equal; they have each five hair-like stamina, terminated by single summits, and a roundish germen, situated under the flower, supporting two small styles, crowned by obtuse stigmas. The germen afterward turns to a roundish compressed fruit longitudinally indented, dividing in two parts, each containing one roundish compressed seed with an indented border.

The Species are,

1. Tordylium (*Maximum*) umbellâ confertâ radiatâ, foliolis lanceolatis inciso-serratis. Hort. Cliff. 90. *Hartwort with the rays of the umbel closed together, the lobes of the leaves spear-shaped, and cut like saws.*

2. Tordylium (*Officinale*) involucris partialibus longitudine petalorum, foliolis ovatis laciniatis. Hort. Cliff. 90. *Hartwort with the involucri of the rays as long as the petals of the flower, and oval jagged leaves.*

3. Tordylium (*Syriacum*) involucris umbellâ longioribus. Hort. Cliff. 90. *Hartwort with longer involucri to the umbels.*

4. Tordylium (*Apulum*) umbellulis remotis, foliis pinnatis, pinnis subrotundis laciniatis. Hort. Cliff. 90. *Hartwort with umbels placed distant, and winged leaves with roundish lobes.*

5. Tordylium (*Secacul*) umbellulis remotis, foliis duplicato pinnatis, pinnis incisis tomentosis. Hort. Cliff. 90. *Hartwort with the umbels growing at a distance, winged leaves having roundish lobes, which are cut on their edges.*

6. Tory-

6. TORDYLIUM (*Nodosum*) umbellis simplicibus sessilibus seminibus exterioribus hispidis. Lin. Gen. Pl. 240. *Hartwort with single umbels sitting close to the stalks, and the outer side of the seeds prickly*; or *Knotted Parsley*.

7. TORDYLIUM (*Anthiscus*) umbellâ confertâ, foliolis ovato-lanceolatis pinnatifidis. Hort. Cliff. 90. *Hartwort with closed umbels, and oval, spear-shaped, wing-pointed lobes*; *called Hedge Parsley*.

8. TORDYLIUM (*Latifolium*) umbellis confertis nudiusculis, foliis pinnatis, foliolis lanceolatis inciso serratis. Lin. Sp. 345. *Hartwort with naked umbels, and winged leaves, whose lobes are spear-shaped and sawed*.

The first sort grows in Italy and Spain; this is a biennial plant, which dies soon after it has perfected its seeds. The lower leaves of this sort are large and winged, each having three or four pair of lobes, terminated by an odd one. They are rough and hairy, having many deep indentures on their edges like the teeth of a saw; the stalk rises three feet high, sending out two or three branches from the side, garnished at each joint by one winged leaf; those on the lower part of the stalk have two pair of small lobes, terminated by an odd one, but those toward the top have one pair, and the middle lobe is long and narrow. The stalk and branches are terminated by umbels of white flowers, whose rays are closed together; these are succeeded by oval compressed seeds, having a thick white border.

The second sort grows plentifully about Rome, and also in the south of France; this is mentioned in the last edition of Ray's Synopsis as an English plant, growing naturally in Oxfordshire, where I have found it growing on the side of banks, but the seeds were sown there by Mr. Jacob Bobart, gardener at Oxford. The leaves of this sort are composed of three or four pair of oval lobes, terminated by an odd one; they are soft and hairy, bluntly indented on their edges. The stalks rise a foot and a half high, and divide into three or four branches, having one small leaf at each joint, and are terminated by umbels of white flowers, composed of several small umbels or rays, which stand upon long foot-stalks, spreading out wide from each other. The flowers are succeeded by smaller compressed seeds, which are bordered.

The third sort grows naturally in Syria; this is a low plant, whose stalks seldom rise a foot high. The lower leaves are composed of two pair of oval lobes, terminated by one large one; these are hairy, and slightly crenated on their edges; they branch out into two or three divisions, and are terminated by umbels of white flowers, which have large involucrums, for the most part trifid. The points are spear-shaped, and at their base is situated a small umbel, composed of a few flowers, sitting very close to the tails of the involucri. The flowers are succeeded by large, oval, compressed, bordered seeds.

The fourth sort grows naturally in Italy. The stalks of this sort branch out from the bottom; the stalk seldom rises more than a foot high; it is hairy and rough; the lower leaves are composed of three pair of roundish lobes, terminated by an odd one; the general umbel is composed of eight smaller, which have long foot-stalks. The flowers are white, and the exterior petal of each is much larger than those of the other sorts; the seeds are roundish and bordered.

The fifth sort grows naturally about Aleppo, and in other parts of Syria. The bottom leaves are doubly winged, each leaf being composed of four pair of wings, terminated by an odd one. The wings are composed of seven oval lobes, standing alternately, which are deeply jagged; they are of a yellowish green colour, and a little hairy. The stalks are taper, and not channelled; they rise two feet and a half high, and have a few small hairs scattered over them, and at each joint are garnished with one smaller winged leaf; they send out one or two short branches toward the top, terminated by large umbels of yellow flowers, composed of ten small umbels, whose foot-stalks are alternately longer, spreading open wide from each other. The flowers are succeeded by compressed oval seeds, shaped like those of Parsneps, of a yellowish colour.

The sixth sort grows naturally in arable land, in several of the maritime counties in England, so is rarely admitted into gardens; this has trailing stalks which spread flat on the ground. The leaves are like those of Parsley, but are cut into finer segments; the umbels of flowers are small, and sit close to the joints of the stalks; the flowers are small, white, and are succeeded by short seeds a little compressed, set with sharp burry prickles on their outside.

The seventh sort grows naturally on the side of banks and foot-paths in many parts of England; this rises with a slender stalk three feet high. The leaves are like those of Parsley; their lobes are spear-shaped, and have winged points; they are hairy, and stand thinly on the stalks. The flowers are produced in small umbels at the top of the stalks, which are composed of several smaller umbels or rays, which close together, of a pale red colour; these are succeeded by small prickly seeds.

The eighth sort grows naturally among the Corn in Cambridgeshire, and in some other parts of England. This rises with a channelled stalk three feet high, garnished with one winged leaf at each joint, composed of two pair of lobes, terminated by a long one; they are broad, spear-shaped, and deeply sawed on their edges. The umbels of flowers which terminate the stalks are clustered together; the seeds are broad, rough, and have borders round them.

All these plants may be termed annual, because they do not live more than one year; but some of them are called biennial, from the young plants which come up in autumn living through the winter, and producing their flowers and fruits the following summer; but, as the seeds which are sown, or permitted to scatter, perfect their seeds in the compass of one year, they should be termed annual, for this is the property of many of the plants with umbellated flowers, whose seeds should be sown in autumn, otherwise, if they come up (which frequently does not happen the same year when they are sown in the spring,) the plants generally decay before their seeds ripen; but, as their whole growth is performed within the year, they should be esteemed as annual plants.

They are propagated by seeds, which should be sown in autumn soon after they are ripe, when the plants will soon appear, and are very hardy, so that they require no farther care but to keep them clear from weeds, and where they come up too close together, they should be thinned so as to leave them six inches asunder. In June following the plants will flower, and their seeds will ripen in autumn, which, if permitted to scatter on the ground, will produce a supply of plants without any trouble. These plants will grow on any soil or situation, so may be put into any obscure part of the garden.

TORMENTILLA. Tourn. Inst. R. H. 298. tab. 153. Tormentil.

The CHARACTERS are,

The flower has a plain empalement of one leaf, divided into eight segments at the top; it has four oval heart-shaped petals, whose tails are inserted in the empalement, and many awl-shaped stamina, which are inserted in the empalement, terminated by single summits; it has eight small germen collected in a head, which have slender styles the length of the germen, inserted to their sides, crowned by obtuse stigmas. The germen afterward turns to a fruit, containing many small seeds included in the empalement.

The

The SPECIES are,

1. TORMENTILLA (*Erecta*) caule erecto. Lin. Sp. Pl. 500. *Tormentil with an erect stalk; common Tormentil.*

2. TORMENTILLA (*Reptans*) caule repente. Lin. Sp. Pl. 500. *Tormentil with a creeping stalk; or creeping winged Cinquefoil.*

The first sort grows wild on dry pastures and commons in most parts of England, so is never cultivated in gardens; this is so commonly known as to need no description. The roots of this plant have been frequently used for tanning of leather, in places where Oak-bark is scarce. This root is also much used in medicine, and is accounted the best astringent in the whole vegetable kingdom.

The second sort is found in some particular places of England growing wild, but particularly in Oxfordshire. The stalk of this sort spreads on the ground, and emit roots from their joints, whereby they propagate very fast: this is rarely preserved, unless in some botanic gardens for the sake of variety. It requires no care to propagate these plants, since, if their roots are once planted in almost any soil or situation, the plants will flourish without any other care but to prevent their being over-run with great weeds.

TOURNEFORTIA. Lin. Gen. Plant. 176.

The CHARACTERS are,

The empalement of the flower is permanent, of one leaf, cut into five small segments at the top. The flower is of one petal, of the globular bell-shape, cut at the brim into five acute points, which spread open horizontally; it has five awl-shaped stamina the length of the tube, terminated by single summits, and a globular germen, supporting a single style the length of the stamina, crowned by a single stigma. The germen afterward becomes a spherical succulent berry, inclosing four oblong oval seeds, resting upon the empalement.

The SPECIES are,

1. TOURNEFORTIA (*Fœtidissima*) foliis ovato-lanceolatis, hirtis pedunculis ramosis spicis. *Tournefortia with oval spear-shaped leaves, and pendulous spikes of flowers.*

2. TOURNEFORTIA (*Hirsutissima*) foliis ovatis petiolatis, caule hirsuto, spicis ramosissimis terminalibus. Lin. Sp. Pl. 140. *Tournefortia with oval leaves growing upon foot-stalks, and a hairy stalk, terminated by very branching spikes of flowers.*

3. TOURNEFORTIA (*Volubilis*) foliis ovatis acuminatis, glabris petiolis reflexis, caule volubili. Lin. Sp. Pl. 143. *Tournefortia with oval acute-pointed leaves, having reflexed foot-stalks, and a twining stalk.*

4. TOURNEFORTIA (*Scandens*) foliis cordatis hirsutis, spicis racemosis reflexis, caule volubili. *Tournefortia with hairy heart-shaped leaves, branching reflexed spikes of flowers, and a twining stalk.*

5. TOURNEFORTIA (*Tomentosa*) foliis cordatis subtus tomentosis, spicis racemosis brevibus, caule volubili. *Tournefortia with heart-shaped leaves, which are woolly on their under side, very short branching spikes of flowers, and a twining stalk.*

6. TOURNEFORTIA (*Carnosa*) foliis ovatis rugosis petiolatis, spicis racemosis axillaribus, caule fruticoso. *Tournefortia with oval rough leaves growing upon foot-stalks, branching spikes of flowers proceeding from the wings of the stalks, and a shrubby stalk.*

7. TOURNEFORRIA (*Suffruticosa*) foliis sub-lanceolatis incanis suffruticosa. *Tournefortia with spear-shaped hoary leaves, and a branching stalk.*

8. TOURNEFORTIA (*Humilis*) foliis lanceolatis sessilibus, spicis simplicibus recurvis lateralibus. Lin. Sp. Pl. 141. *Tournefortia with spear-shaped leaves sitting close to the stalk, and single recurved spikes of flowers growing at the wings.*

The first sort grows naturally in Jamaica, and in some of the other islands in the West-Indies, where it rises with shrubby stalks ten or twelve feet high, sending out many branches, garnished with oval spear-shaped leaves, placed alternately round the stalks, hairy on their under side, and stand upon short foot-stalks. The branches are terminated by long branching spikes of flowers, which are ranged on one side the foot-stalks in the same manner as those of the Heliotrope or Turnsol. Some of the foot-stalks sustain two, others three, and some four spikes of flowers, reflexed like a scorpion's tail at the top. The flowers are of a dirty white colour, small, and closely ranged on one side the spike; these are succeeded by small succulent fruit, inclosing four oblong seeds in each.

The second sort is also a native of the islands in the West-Indies. The stalks of this are shrubby, taper, and rough; they rise to the height of eight or ten feet, dividing into many branches, covered with a light, brown, hairy, rough bark, garnished with oval leaves, placed alternately, having many transverse veins running from the midrib to the sides; they are of a deep green on their upper side. The branches are terminated by branching spikes of flowers, succeeded by small, roundish, succulent fruit, each inclosing four oblong seeds.

The third sort grows naturally in Jamaica, and some of the islands in America; this hath a twining ligneous stalk, which twists about the neighbouring trees for support, and rises to the height of ten or twelve feet, sending out several slender ligneous branches, garnished with oval acute-pointed leaves, whose foot-stalks are reflexed. The flowers are produced in branching spikes from the side and the top of the branches; they are small, white, and are succeeded by small, white, succulent berries, having one or two black spots on each.

The fourth sort was discovered by the late Dr. Houstoun, growing naturally in Jamaica; this hath shrubby branching stalks, which twine about the neighbouring trees, and rise to the height of ten or twelve feet. The branches are garnished with heart-shaped hairy leaves, ending in acute points, of a thinner contexture than those of the former species, and stand upon short foot-stalks. The flowers come out at the end of the branches in very slender spikes; they are small, and of a dirty brown colour, ranged along the upper side of the foot-stalk; these are succeeded by small pulpy berries, each containing four seeds.

The fifth sort grows naturally near Carthagena in New Spain; this has climbing stalks, which twine about any neighbouring support, and rise to the height of ten or twelve feet. The branches are garnished with heart-shaped leaves, downy on their under side, and stand upon very short foot-stalks. The flowers are produced in short branching spikes, which come out from the wings of the branches; they are of a dirty white colour, small, and are succeeded by small succulent berries, inclosing two, three, and sometimes four seeds.

The sixth sort grows naturally near Carthagena in New Spain; this has strong ligneous stalks, which rise near twenty feet high, sending out several, strong, ligneous branches, covered with a light, brown, rough bark, garnished with thick oval leaves, rough on their upper surface, and of a dark green colour, but pale and smoother on their under side, standing upon pretty-long foot-stalks. The flowers are produced in spikes from the wings of the branches; they are small, white, and shaped like those of the other species, and are succeeded by small succulent berries, each including two or three oblong seeds.

The seventh sort grows in the same country; this has woody stalks which rise five or six feet high, from which spring out many slender ligneous branches, garnished with oval spear-shaped leaves, which are rounded at each end, but have acute points, of a dark green on their upper surface,

face, but have a white down on their under side, sitting close to the branches. The flowers are produced from the wings of the stalks, and also at the top, formed in slender branching spikes, which are recurved; they are white, and are succeeded by small succulent berries, which contain two or three seeds.

The eighth sort grows naturally at Campeachy; this plant has low shrubby stalks, which seldom rise more than three feet high, sending out a few, slender, ligneous branches, garnished with rough spear-shaped leaves, sitting close to the branches, of a dark green on their upper side, but pale on their under. The flowers come out in single spikes from the wings of the stalk; they are white, and are succeeded by small succulent berries like the former sort.

These plants are propagated by seeds or cuttings; if by seeds, they must be procured from the countries where they grow naturally; these should be sown in small pots filled with light earth, and plunged into a hot-bed of tanners bark. These seeds sometimes grow the first year, but they often remain in the ground a whole year; therefore, if the plants should not come up the same season, the pots should be plunged in autumn into the tan-bed in the stove, where they should remain all the winter; and in the spring they should be removed out, and plunged into a fresh tan-bed, which will soon bring up the plants if the seeds were good. When these are fit to remove, they should be each planted in a small pot, and plunged into a tan-bed, where they must be shaded from the sun till they have taken new root; afterward they must be treated in the same way as other tender plants from the same countries, which require to be kept constantly in the bark-stove. The plants raised from cuttings, must be treated in the same way.

TOXICODENDRON. Tourn. Inst. R. H. 610. tab. 381. Poison-tree.

The CHARACTERS are,

The male flowers are upon different plants from the female; they have a small empalement, cut into five points at the brim, but have no petals; they have five short stamina, terminated by roundish summits. The female flowers have empalements like the male; they have no stamina, but in the center is situated a roundish germen, supporting three small styles, crowned with globular stigmas. The germen afterward turns to a berry with one or two cells, inclosing one seed in each.

The SPECIES are,

1. TOXICODENDRON (*Vulgare*) foliis ternatis, foliolis obcordatis, glabris, integerrimis, caule radicante. *Poison-tree with trifoliate leaves, having roundish, heart-shaped, smooth, entire leaves, and a stalk putting out roots; called Poison Oak.*

2. TOXICODENDRON (*Pubescens*) foliis ternatis, foliolis ovatis inciso-angulatis pubescentibus. *Poison-tree with trifoliate leaves, whose lobes are oval, angularly cut, and covered with soft short hairs.*

3. TOXICODENDRON (*Glabrum*) foliis ternatis, foliolis ovato-lanceolatis glabris, caule erecto fruticoso. *Poison-tree with trifoliate leaves, whose lobes are oval, spear-shaped, smooth, and an erect shrubby stalk.*

4. TOXICODENDRON (*Pinnatifolium*) foliis pinnatis, foliolis ovato-lanceolatis integerrimis. *Poison-tree with winged leaves, whose lobes are oval, spear-shaped, and entire; called Poison Ash.*

5. TOXICODENDRON (*Crenatum*) foliis ternatis, foliolis ovatis crenato-dentatis glabris. *Poison-tree with trifoliate leaves, whose lobes are oval, smooth, and bluntly indented.*

6. TOXICODENDRON (*Volubilis*) foliis ternatis, foliolis ovatis inciso-sinuatis glabris, caule volubili radicante. *Poison-tree with trifoliate leaves, whose lobes are oval, smooth, and cut into sinuses, and a twining rooting stalk.*

7. TOXICODENDRON (*Serratum*) foliis sæpiùs ternatis, foliolis oblongo-ovatis rugosis serratis, caule radicante. *Poison-tree with leaves which are generally trifoliate, oblong, oval, rough, sawed lobes, and a rooting stalk.*

8. TOXICODENDRON (*Arboreum*) foliis ternatis, foliolis lanceolatis supernè inæqualiter serratis, subtus tomentosis, caule arborescente. *Poison tree with trifoliate leaves, spear-shaped lobes, unequally sawed toward their points, downy on their under side, and a tree-like stalk.*

9. TOXICODENDRON (*Arborescens*) foliis ternatis, foliolis ovato-lanceolatis acuminatis glabris, caule fruticoso ramoso. *Poison-tree with trifoliate leaves, having oval, spear-shaped, acute-pointed, smooth lobes, and a shrubby branching stalk.*

The first sort grows naturally in most parts of North America; it has a low shrubby stalk, which seldom rises more than three or four feet high, sending out shoots near the bottom, which trail upon the ground, putting out roots from their joints, whereby it multiplies and spreads greatly; so that when it is not confined, or trained up to a support, the stalks seldom rise upward. If the stalks happen to be close to a wall, they emit roots which fasten to the joints in the wall, and support themselves when they are severed from the root; and the stalks of such plants will become more ligneous, and rise much higher, than those which grow in the ground. The foot-stalks of the leaves are near a foot long; the leaves are composed of three smooth, oval, heart-shaped lobes, which are entire. The flowers come out from the side of the stalk in loose panicles, of an herbaceous colour, and small, so make no great appearance. Some plants have only male flowers with five stamina in each; these decay without producing fruit; but upon other plants there are only female flowers, which have a germen and three very short styles; these are succeeded by roundish, channelled, smooth berries, of a yellowish gray colour, which inclose one or two seeds.

This plant, when once planted in a garden, will propagate fast enough by its trailing branches, which put out roots at every part. It will thrive in almost any soil or situation.

The second sort grows naturally in most parts of North America. The stalks of this rise higher than those of the former; the branches are slender, but ligneous; they have a brown bark, and are garnished with downy leaves, standing upon pretty long foot-stalks, composed of three oval lobes, indented angularly, and hoary on their under side. The male flowers, which are produced on separate plants from the fruit, come out from the side of the stalks in close short spikes; these are of an herbaceous colour, and have five short stamina in each. The female flowers are produced in loose panicles; these are in shape and colour like the male, but larger, and have a roundish germen, supporting three very short styles, and are succeeded by roundish berries, which ripen in autumn.

The third sort grows naturally in North America; this has a shrubby, erect, branching stalk, which rises six or seven feet high, covered with a brown bark, garnished with smooth trifoliate leaves, whose lobes are oval, spear-shaped, and have a few sinuated indentures on their borders. The male and female flowers grow upon separate plants; their shape and colour like those of the former, and the fruit is also like that.

The fourth sort grows naturally in Virginia, Pensylvania, New England, and Carolina; from all these countries I have received seeds: it also grows in Japan. This, in the countries where it grows naturally, rises with a strong woody stalk to the height of twenty feet or upward, but in England we seldom see any of them more than ten or twelve; the reason of this is from the plants being tender, so are destroyed in severe winters; but I have seen some plants, which

which were kept in pots and sheltered in winter, upward of ten feet high, in the garden of Samuel Reynardson, Esq; at Hillendon, which, after his death, were purchased with all his other exotic plants, by Sir Robert Walpole. This has a strong woody stalk, covered with a light brown bark, inclining to gray. The branches are garnished with winged leaves, composed of three or four pair of lobes, terminated by an odd one. The lobes vary greatly in their shape, but for the most part they are oval and spear-shaped; they are sometimes rounded at their base, but generally end in acute points; their upper surface is smooth, of a lucid green, but their under side is pale, and a little hairy. The foot-stalks of the leaves change to a bright purple colour toward the latter part of summer, and in autumn all the leaves are of a beautiful purple colour before they fall off. The male flowers are produced upon loose panicles from the wings of the branches; they are small, of an herbaceous white colour, composed of five small roundish petals, and have five short stamina within, terminated by roundish summits. The female flowers are upon separate plants from the male, and are disposed on loose panicles; these are shaped like the male, but are somewhat larger, and have in their center a roundish germen, supporting three very short styles, crowned with globular stigmas. The germen afterward turns to a berry, variable in shape, sometimes almost oval, at others shaped like a small spear; but the most general form is roundish, with a protuberance almost like the Cicer; these include one seed. It flowers in July, and in warm seasons the female plants produce fruit, but they seldom ripen here.

This is undoubtedly the same plant which is mentioned by Dr. Kempfer in his Amœnitates Exoticarum, by the title of Sitz, vel Sitz Adju, or Arbor vernicifera legitima, folio pinnato juglandis, fructu racemosa Ciceris facie. Fasc. v. p. 791, 792. *The true Varnish-tree, with a Walnut-tree leaf, and a branching fruit-like Cicers.* But the figure he has exhibited of it, is the most inaccurate of any perhaps to be found in any of the modern books of botany; it is drawn from a side shoot of a branch which has been cut off, so has neither flower or fruit to it, and being a vigorous shoot, the leaves are very different in size and shape from those on plants which have not been headed; and his description of the leaves seems to have been taken from this branch, otherwise he could not have compared them to those of the Walnut-tree. He seems to have been conscious of this fault, by his adding another figure of the plant in small under his own, taken from a Japan herbal, in which there is a much better representation of it than his own conveys. How a person who was employing himself in making drawings of plants in a country, where the natural history of it was so well known, should make choice of such an imperfect sample for his figure, is amazing; for there can be no doubt of his meeting with perfect plants in flower or fruit, in a place where the shrubs are cultivated so plentifully as he mentions; and in his description of it, he sets out by comparing the height of the shrubs to those of Willow, than which he could not have chosen any plant by way of comparison, which would have conveyed a more indetermined idea; for it is well known there are different species of Willow, whose growth is from four to forty feet high; therefore there can be no other way of reconciling his description with what he afterward mentions, unless when he is giving an account of the method used by the natives in collecting the varnish; where he says the shrubs are cut down every third year, but by comparing their growth with that of the Willows, which are cut down for fuel, &c. every four or five years.

However, as the dried samples of the plant which he brought over, agrees with the American Toxicodendron, and the milky juice of both have the same qualities of staining, so there can be no doubt of the plants being the same; therefore if it is thought that varnish may be of public utility, it may be collected in plenty in most of the English settlements in North America.

Kempfer has also given a figure and description of a spurious Varnish tree, which is called Fasi-no-Ki by the natives, and is by him titled Arbor vernicifera spuria, sylvestris angustifolia. *Spurious wild Varnish-tree with a narrow leaf*; which he says agrees with the other in every part, excepting the lobes of the leaves, which are narrower. This led me into a mistake in the former editions of the Gardeners Dictionary, by supposing their difference might arise from culture only; but having since raised from seeds a shrub which has all the appearance of his spurious Varnish-tree, and is evidently a distinct species, if not of a different genus from the true sort, I am certain Kempfer has been guilty of a great mistake in this particular. The seeds of this were sent from China, for those of the Varnish-tree; but when I sowed them, I remarked they were pretty much shaped like those of the Beech-tree, but smaller; being thick on one side and slender on the other, in shape of a wedge, from whence I supposed there were three of the seeds included in one capsule. There is a shrub of this kind now growing in the Chelsea garden, which is more than fifteen feet high; but, as it has not yet produced flowers, I was at a loss where to range it, till one plant produced flowers in the garden of William Sharp, Esq; on Endfield Chase.

The fifth sort grows naturally in North America, from whence the seeds were a few years since brought to England. This has a shrubby stalk, which sends out many ligneous branches, covered with a smooth purple bark, garnished with smooth trifoliate leaves, standing upon foot-stalks an inch long; the lobes are oval, of a deep lucid green on their upper side, but of a pale green on their under, deeply crenated or indented on their edges, their base joining close to the foot-stalks. The leaves, when bruised, emit an odour like that of Orange-peal, from whence the gardeners have titled it the sweet-scented Toxicodendron. The male flowers are produced in short close panicles; they are small, of an herbaceous white colour; they are upon separate plants from the fruit, which grow in sparsed panicles, and are of an oval shape.

The sixth sort grows naturally in North America. The stalks of this sort emit roots their whole length, whereby they fasten to trees or any neighbouring support, and climb to the height of six or eight feet, garnished with trifoliate oval leaves, which are smooth, and cut into sinuses on their edges. The flowers are produced in short panicles from the side of the branches; they are male and female on different plants like the other species.

The seventh sort was sent me by Mr. John Bartram from Philadelphia, by the title of Great Toxicodendron; this hath trailing roots, which run near the surface of the ground, sending up stalks in different places; the leaves stand upon long foot-stalks, composed of three or four obtuse rough lobes, sawed on their edges.

The first, sixth, and seventh sorts propagate in plenty by their creeping stalks and roots; the others are propagated by laying down their branches, which will put out roots in one year, and may then be taken off and transplanted, either in the places where they are to remain, or in a nursery to grow two or three years, to get strength before they are planted out for good; they may also be propagated by seeds, which should be sown on a bed of light earth, and when the plants come up, they must be kept clean from weeds the following summer; and before the frost comes on in autumn, the bed should be hooped over, that the plants may be covered with mats, for otherwise the early frosts will

will kill their tops, which frequently causes their stalks to decay to the ground; for as the young plants are tender, and generally shoot late the first year, they are in much greater danger than when they get more strength. In spring the plants may be transplanted into nursery-beds to grow a year or two, and may then be transplanted for good.

These plants are preserved by the curious in botany, for the sake of variety; but as there is little beauty in them, there are not many of the sorts cultivated in England. The wood of these trees, when burnt, emits a noxious fume, which will suffocate animals when they are shut up in a room where it is burnt: an instance of this is mentioned in the Philosophical Transactions by Dr. William Sherard, which was communicated to him in a letter from New England by Mr. Moore, in which he mentions some people who had cut some of this wood for fuel, which they were burning, and in a short time they lost the use of their limbs, and became stupid; so that if a neighbour had not accidentally opened the door, and seen them in that condition, it is generally believed they would soon have perished. This should caution people from making use of this wood for such purpose.

When a person is poisoned by handling this wood, in a few hours he feels an itching pain, which provokes a scratching, which is followed by an inflammation and swelling. Sometimes a person has had his legs poisoned, which have run with water. Some of the inhabitants of America affirm, they can distinguish this wood by the touch in the dark, from its extreme coldness, which is like ice; but what is mentioned of this poisonous quality, is most applicable to the fourth sort here mentioned, which, by the description, agrees with this species.

The juice of the tree is milky, when it first issues out of the wounded part; but soon after it is exposed to the air, it turns black, and has a very strong fœtid scent, and is corroding; for I have observed, on cutting off a small branch from one of these shrubs, that the blade of the knife has been changed black in a moment's time, so far as the juice had spread over it, which I could not get off without grinding the knife.

The eighth sort grows naturally in Jamaica on the red hills, and at Campeachy in great plenty. It has a thick woody stem, which rises near thirty feet high, with a smooth Ash-coloured bark, sending out ligneous branches on every side, which have a hairy rusty-coloured bark, garnished with trifoliate leaves, which have hairy foot-stalks. The lobes are spear shaped, unequally sawed toward the top, and have many transverse veins running from the midrib to the borders, and have a brown woolly down on their under side. The flowers are ranged in a single racemus, which springs from the wings of the branches; they are small, of a yellowish colour, and the female flowers are succeeded by small oval berries, of an Orange colour when ripe.

The ninth sort grows naturally about Carthagena in New Spain; this rises with a shrubby stalk twelve or fourteen feet high, covered with a gray bark, sending out a great number of branches on every side, which are garnished with trifoliate smooth leaves, whose lobes are oval, spear-shaped, and oblique to their foot-stalks. The male and female flowers are upon different plants; they are formed in loose panicles, are small, and of a dirty white colour. The female flowers are succeeded by small, oval, smooth berries, each including one seed.

The two last sorts are tender plants, so will not thrive in this country, without the assistance of artificial heat; they are propagated by seeds, when these can be procured from the countries where the plants grow naturally. These should be sown as soon as they arrive here, in pots filled with light earth, and plunged into a tan-bed. Sometimes the plants will come up the same year, but the seeds often lie long in the ground when they are sown in the spring; and when they do not grow the first year, the pots should be plunged in the bark-bed in the stove in autumn, where they may remain all the winter; and in the spring they should be plunged into a fresh hot-bed under a frame, which will soon bring up the plants. When these are fit to remove, they should be each planted in a small pot, filled with light earth, and plunged into a new tan-bed, observing to shade them from the sun till they have taken new root; then they should be treated in the same way as other tender exotic plants, which are constantly kept in the bark-stove.

TRACHELIUM. Tourn. Inst. R. H. 130. tab. 50. Lin. Gen. Plant. 204. Throatwort.

The CHARACTERS are,

The flower has a small empalement, cut at the top in five parts, sitting upon the germen. The flower has one petal, which is funnel-shaped, having a long, slender, cylindrical tube, cut at the top into five small oval segments, which spread open; it has five hair-like stamina the length of the petal, terminated by single summits, and a roundish three-cornered germen, situated under the flower, supporting a long slender style, crowned by a globular stigma. The germen afterward turns to a roundish obtuse capsule with three lobes, having three cells, which are filled with small seeds.

We have but one SPECIES of this genus, viz.

TRACHELIUM (*Cæruleum.*) Hort. Upsal. 41. *Blue Mountain Throatwort.*

This plant grows naturally in shady woods in many parts of Italy. It has a perennial root, which is fleshy and tuberous. The leaves are oval, spear-shaped, sawed on their edges, ending in acute points. The stalks rise a foot and a half high, garnished with leaves, shaped like those at the bottom, but come out irregularly. Sometimes there are two pretty large leaves, and one or two smaller rising from the same joint; at others, one large and three smaller; these come out alternate, and the upper part of the stalk, immediately under the umbel, is naked of leaves, except two or three narrow ones, which are close to the foot-stalks of the flowers; these are disposed in form of an umbel, composed of many small umbels. The flowers are small, funnel-shaped, and of an azure blue colour; these are succeeded by roundish capsules, with three cells, filled with small seeds, which ripen in autumn.

This plant is propagated by seeds, which should be sown in autumn soon after they are ripe; for when they are kept out of the ground till spring, they frequently fail; or if they do grow, it is not before the following spring. When the plants come up, they should be kept clean from weeds, and as soon as they are big enough to remove, should be transplanted on an east-aspected border of light undunged earth, placing them in rows six inches apart, and four inches distant in the rows, shading them from the sun till they have taken new root; after which they require no other care but to keep them clean from weeds till autumn, when they may be transplanted into the borders of the flower-garden, where they will flower the following summer.

But as these plants thrive better on old walls, when by accident they have arisen there from seeds, so their seeds, when ripe, may be scattered on such walls as are old, or where there is earth lodged sufficient to receive the seeds, where the plants will come up and resist the cold much better, and continue longer than when sown in the full ground; and when a few of the plants are established on the walls, they will shed their seeds, so that they will maintain themselves without any farther care. I have observed some plants

plants of this kind, which have grown from the joints of a wall, where there has not been the least earth to support them, which have resisted the cold, though they have been greatly exposed to the winds, when most of those in the full ground were killed; so that these plants are very proper to cover the walls of ruins, where they will have a very good effect.

TRADESCANTIA. Lin. Gen. Plant. 360. Spiderwort.

The CHARACTERS are,

The flower has a difformed sheath. The proper empalement is permanent, composed of three oval concave leaves. The flower has three roundish equal petals, which spread open, and six hairy slender styles, terminated by kidney-shaped summits, and an oval, obtuse, three-cornered germen, supporting a slender style, crowned by a triangular blunt stigma. The germen afterward turns to an oval capsule, shut up in the empalement, having three cells, containing a few angular seeds.

We have but one SPECIES of this genus, viz.

TRADESCANTIA (*Virginiana.*) Hort. Cliff. 127. *Spiderwort.*

The root of this plant is composed of several fleshy fibres, which spread wide, from which arise many long, narrow, keeled leaves, which embrace each other at their base; they are veined, rough on their edges, of a grayish colour, and succulent; between the leaves arise a thick jointed stalk about a foot long, garnished at each joint with one leaf, whose base embraces it. At the top of the stalk are two leaves, which spread asunder; above these come out many flowers almost in a sort of umbel; these have a three-leaved empalement, and three large roundish petals, of a deep blue colour, which in the morning spread open flat, but in the middle of the day they shrink up, and do not open again; but there is a succession of flowers from the same bunch daily, for a considerable time. The germen afterward swells to a roundish capsule with three angles, having three cells, including a few angular seeds.

We have two other varieties of this plant, one with a white, the other has a purple flower; but these are supposed changeable from seeds.

It is easily propagated by seeds, which, if they are permitted to scatter, will produce plenty of young plants the following spring; or if the seeds are sown soon after they are ripe, the plants will come up the spring after; and when they are fit to remove, they should be planted in a nurse.y-bed at about nine inches distance, and the ground kept clean from weeds. In autumn they should be removed into the borders of the flower-garden, where they will flower and produce seeds, and the roots will continue several years.

TRAGACANTHA. Tourn. Inst. R. H. 417. tab. 234. Goats-thorn.

The CHARACTERS are,

The empalement of the flower is indented in five parts, the lower segments being the shortest. The flower is of the butterfly kind; the standard is long, erect, indented at the point; the borders are reflexed. The wings are shorter than the standard. The keel is of the same length with the wings, and is indented; it has ten stamina, nine are joined and one is separated, terminated by roundish summits, and a sharp taper germen, supporting an awl-shaped style, crowned by an obtuse stigma. The germen afterward becomes a short swelling pod, having two longitudinal cells, inclosing kidney-shaped seeds.

The SPECIES are,

1. TRAGACANTHA (*Massiliensis*) petiolis longioribus spinescentibus, foliolis ovatis obtusis. *Goats-thorn with longer foot-stalks ending in spines, having oval obtuse lobes to the leaves.*

2. TRAGACANTHA (*Hispanica*) foliolis lanceolatis, floribus solitariis axillaribus, filiculis ovatis inflatis. *Goats-thorn with spear-shaped lobes, flowers proceeding singly from the sides of the branches, and oval, inflated, bladder pods.*

3. TRAGACANTHA (*Argentea*) foliolis lanceolatis acuminatis tomentosis, floribus alaribus terminalibusque. *Goats-thorn with spear-shaped, acute-pointed, woolly leaves, flowers growing on the sides and at the ends of the branches.*

4. TRAGACANTHA (*Glabra*) foliolis linearibus glabris, floribus congestis axillaribus. *Goats thorn with very narrow smooth leaves, and flowers growing in clusters on the sides of the branches.*

The first sort grows naturally on the sea shore about Marseilles, and in Italy; this has a thick, short, ligneous stalk, which branches out greatly on every side. The young branches are woolly, closely garnished with winged leaves, whose foot-stalks end in acute thorns. The lobes are small, oval, obtuse, and of a silvery colour. The flowers are large, white, and shaped like a butterfly; they are produced in clusters at the end of the branches, and are succeeded by short pods, having two longitudinal cells, containing two or three kidney-shaped seeds, which seldom ripen in England.

The second sort grows naturally in the islands of Majorca and Minorca; this hath a thick woody stalk, rising about two feet high, sending out many ligneous branches, closely garnished with small, spear-shaped, hoary leaves, ranged by pairs along a very strong foot-stalk, ending with a sharp thorn. The flowers are produced singly from the sides of the branches; they are large, white, and are succeeded by oval bladder pods, containing four kidney-shaped seeds, which do not ripen in England.

The third sort grows naturally in the islands of the Archipelago; this has a very low shrubby stalk, divided into many downy branches, garnished with winged leaves, composed of nine or ten pair of spear-shaped woolly lobes, ending in acute thorns. The flowers are produced from the side and at the top of the branches; they are white, and shaped like those of the other species, but smaller.

The fourth sort grows naturally in Spain; this is a very low plant. The stalks are pretty thick and woody, but seldom rise more than five or six inches high, dividing into several branches, closely garnished with small winged leaves, composed of several pair of small, linear, smooth lobes, of a bright green colour. The foot-stalks of these end in very sharp thorns, which stand out beyond the lobes: the flowers grow in clusters from the side of the stalks; they are smaller than those of the other species, and of a dirty white colour.

These sorts may be propagated by seeds, when they can be procured from the countries where the plants grow naturally, which should be sown on a bed of fresh earth in April; and when the plants come up, they should be carefully kept clean from weeds. If the season should prove very dry, it will be of great service to water the plants now and then; when they are large enough to transplant, they should be carefully taken up, and some of them planted in small pots, filled with fresh earth, placing them in the shade until they have taken new root; then they may be removed into an open situation, where they may remain till the latter end of October, when they should be placed under a common frame, where they may be sheltered from severe frost, but may have free air in mild weather.

The remainder of the plants may be planted on a warm dry bank, where they must be shaded until they take root; and if the season should continue dry, they must be refreshed with water, otherwise they will be in danger, because, while they are so young, their roots will not have established themselves in the ground sufficiently to nourish them in great droughts.

Those

Those plants which were planted in pots, may be preserved under frames in winter, until they have obtained strength, when they may be shaken out of the pots, and planted in a lean dry soil and a warm situation, where they will endure the cold of our ordinary winters very well; but, as they are sometimes destroyed by hard frost, it will be proper to keep a plant of each kind in pots, which may be sheltered in winter to preserve the species.

These plants may also be propagated by slips, for as they rarely produce seeds in this country, the latter method is generally used here. The best time for this work is in April, just as the plants begin to shoot, at which time the tender branches of the plants should be slipped off, and their lower parts divested of the decayed leaves; then they should be planted on a very moderate hot-bed, which should be covered with mats, to screen them from the great heat of the sun by day and the cold by night. These cuttings should be gently watered until they have taken root; then they may be exposed to the open air, observing always to keep them clear from weeds.

On this bed they may remain until the following spring, where, if the winter should be very severe, they may be covered with mats as before; and in April they may be transplanted out either into pots, filled with sandy light earth, or into warm borders, where, if the soil be dry, gravelly, and poor, they will endure the severest cold of our climate; but if they are planted in a very rich soil, they often decay in winter.

From one species of this genus, Monsieur Tournefort says, the Gum Adragant, or Dragon, is produced in Crete.

TRAGIA. Plum. Gen. Nov. 14. tab. 12.

The CHARACTERS are,

It hath male and female flowers in the same plant. The empalement of the male flower is cut into three oval acute-pointed segments; it has no petals, but there are three stamina in each the length of the empalement, terminated by roundish summits. The empalement of the female flowers are permanent, cut into five oval concave segments. The flowers have no petals or stamina, but a roundish germen, having three furrows, supporting an erect style, crowned by a bifid spreading stigma. The germen afterward turns to a roundish three-lobed capsule, having three cells, each containing one globular seed.

The SPECIES are,

1. TRAGIA (*Volubilis*) foliis cordato-oblongis, caule volubili. Lin. Sp. Pl. 980. *Tragia with oblong heart-shaped leaves, and a twining stalk.*

2. TRAGIA, (*Involucrata*) involucris fœmineis pentaphyllis pinnatifidis. Lin. Sp. Pl. 980. *Tragia with five-leaved involucri to the female flowers, which are wing-pointed.*

The first sort grows plentifully in the Savannahs in Jamaica, and other warm parts of America, where it twines round whatever plants or trees it grows near, and rises seven or eight feet high, having tough stems. The leaves are oblong, heart-shaped, ending in acute points, deeply sawed on their edges, standing alternately upon pretty long foot-stalks. The male flowers come out from the wings of the stalk in bunches of about two inches in length; the female flowers are produced on separate foot stalks, arising from the same point as the male; these are succeeded by roundish capsules with three cells, each inclosing one roundish seed. The whole plant is covered with burning spines, like those of the Nettle, which renders it very unpleasant to handle.

The second sort grows naturally in India; this rises with an erect ligneous stalk about three feet high, which rarely sends out any side branches, garnished with oblong spear-shaped leaves, which end in very long acute points, sharply sawed on their edges, ranged alternately on the stalk, and are closely covered with yellowish stinging hairs. The flowers are produced in small clusters from the wings of the stalk, standing several together upon the same foot-stalk; the upper are all male, and the under female; the latter are succeeded by roundish capsules with three cells, each inclosing one seed.

As these are plants of no great beauty, they are seldom preserved in this country, except in some botanic gardens for variety; they are propagated by seeds, which must be sown on a hot-bed early in the spring, and afterward transplanted into pots, and plunged into a hot bed of tanners bark, and treated in the same manner as other tender plants, which require to be kept in the bark-stove.

TRAGOPOGON. Tourn. Inst. R. H. 477. tab. 270. Goats-beard.

The CHARACTERS are,

The common empalement of the flower is single, composed of eight acute-pointed leaves, which are alternately large, joined at their base. The flower is composed of many hermaphrodite florets, which are uniform, of one petal, stretched out like a tongue, indented at their points in five parts, and lie over each other like the scales of fish; these have each five short hair-like stamina, terminated by cylindrical summits, and an oblong germen, situated under the floret, supporting a slender style the length of the stamina, crowned by two revolving stigmas. The empalement of the flower afterward swells to a belly, inclosing many oblong, angular, rough seeds, slender at both ends, crowned by a feathery down.

The SPECIES are,

1. TRAGOPOGON (*Pratense*) calycibus corollæ radium æquantibus, foliis integris strictis. Lin. Sp. Pl. 789. *Goats-beard with an empalement equal to the rays of the flower, and entire leaves; or common Goats-beard.*

2. TRAGOPOGON (*Minus*) calycibus corollæ radiis longioribus, foliis linearibus strictis. *Goats-beard with the empalement longer than the rays of the flower, and linear closed leaves; or small, yellow, Meadow Goats-beard.*

3. TRAGOPOGON (*Porrifolium*) calycibus corollæ radiis longioribus, foliis integris strictis, pedunculis supernè incrassatis. Hort. Upsal. 243. *Goats-beard with an empalement longer than the rays of the flower, entire closed leaves, and the foot-stalk thicker at the upper part; commonly called Salsafy.*

4. TRAGOPOGON (*Picroides*) calycibus corollæ radio brevioribus, foliis linearibus strictis, caule hirsuto. *Goats-beard with the empalement shorter than the rays of the flower, narrow closed leaves, and a hairy stalk; hairy Goats-beard.*

5. TRAGOPOGON (*Dalechampii*) calycibus corolla brevioribus aculeatis, foliis pinnato-hastatis. Hort. Cliff. 382. *Goats-beard with prickly empalements, which are shorter than the petals, and arrow wing-pointed leaves; or rough Sowthistle of Crete.*

The first sort grows naturally in the meadows of Austria and Germany; this is very different from the second, which grows naturally in England, for I have sown the seeds of both sorts several years in the same bed of earth, and have always found the plants have retained their difference. The lower leaves of this are three quarters of an inch broad at their base, where they embrace the stalk, and more than a foot long, closed together, ending in acute points. The stalk rises near three feet high, garnished at each joint with one leaf, of the same shape with those below, but smaller; it is terminated by one large yellow flower, composed of hermaphrodite florets, which lie over each other like the scales of fish; these are included in one common simple empalement, which is equal in length to the rays of the flower. Each floret is succeeded by an oblong seed, which is larger at the base than at the point, where it is crowned with a large feathery down. The seeds of the border or ray are crooked and rough, but those of the disk are strait and smooth.

The second sort grows naturally in moist pastures in many parts of England; it is by the common people titled Sleep-at-noon, or Go-to-bed-at-noon, because the flowers are generally closed up before that time every day. The lower leaves of this sort are almost as long as those of the first sort, but are not more than a third part so broad; they are of a deep green colour, and end in acute points. The stalks rise about a foot high, and sustain one yellow flower at the top, not more than half so large as those of the first; the empalements of these flowers are longer than the rays, and the seeds are much smaller than those of the other.

When this sort shoots up in stalks four inches high, the common people gather it out of the fields, and boil it in the same way as Asparagus, and some give it the preference.

The third sort is cultivated in gardens by the title of Salsafy. The roots of this are dressed in different ways, and served up to the table; but of late years some persons cultivate it for the stalks, which are cut in the spring when they are four or five inches high, which are dressed like Asparagus, in the like manner as the second sort. The stalks of this are much longer, and are tenderer than the other, so are better for the purpose than those of the second sort; the leaves are broad; the flowers are large and blue; the foot-stalk immediately under the flower is much thicker than below, and the empalement is longer than the rays of the flower.

The fourth sort grows naturally in Istria; this has narrow hairy leaves. The stalks rise about a foot and a half high, are naked most part of their length, very hairy, and sustain one pretty large yellow flower, whose empalement is much shorter than the rays of the flower, which is also very hairy.

The fifth sort grows naturally in Crete, and also in Italy; this is an annual plant, very like the Sowthistle in stalk and leaf, but the empalement of the flower is prickly. It is seldom admitted into gardens, because the seeds are wafted by the winds to a great distance, and thereby fill the garden with the plants.

These plants are propagated from seeds, which should be sown in April upon an open spot of ground, in rows about nine or ten inches distance; and when the plants are come up, they should be hoed out, leaving them about six inches asunder in the rows. The weeds should also be carefully hoed down as they are produced, otherwise they will soon overbear the plants and spoil them. This is all the culture required, and if the soil be light and not too dry, the plants will have large roots before winter; at which time the Salsafy, whose roots are eaten at that season, will be fit for use, and may be taken up any time after their leaves begin to decay; but, when they begin to shoot again, they will be sticky, and not fit for use; but many persons cultivate this sort for the shoots, as was before mentioned.

The common yellow sort, whose shoots are sold in the market, will be fit for use in April or May, according to the forwardness of the season. The best time to cut them is, when their stems are about four inches long, for if they stand too long, they are never so tender as those which are cut while young.

Some people in cultivating these plants, sow their seeds in beds pretty close; and when they come up, they transplant them out in rows at the before mentioned distance; but, as they form tap-roots, which abound with a milky juice, when the extreme part of their roots is broken by transplanting, they seldom thrive well afterward; therefore, it is by far the better way to make shallow drills in the ground, and scatter the seeds therein, as before directed, whereby the rows will be at a due distance, and there will be nothing more to do than to hoe out the plants when they are too thick in the rows, which will be much less trouble than the other method of transplanting, and the plants will be much larger and fairer.

TRAGOSELINUM. See Pimpinella.

TRIBULUS. Tourn. Inst. R. H. 265. tab. 141. Caltrops.

The Characters are,

The empalement of the flower is cut into five acute parts; there are five oblong blunt petals to the flower, which spread open, and ten small awl-shaped stamina, terminated by single summits, and an oblong germen the length of the stamina, having no style, but crowned by a headed stigma. The germen afterward turns to a roundish prickly fruit, divided into five capsules, armed with three or four angular thorns on one side, joining together. The cells are transverse, and contain two or three Pear-shaped seeds.

The Species are,

1. Tribulus (*Terrestris*) foliolis sexjugatis subæqualibus. Hort. Cliff. 160. *Caltrops with six pair of lobes to each leaf, which are almost equal.*

2. Tribulus (*Maximus*) foliolis quadrijugis exterioribus majoribus. Lin. Sp. Pl. 386. *Caltrops with four pair of lobes to each leaf, of which the outer are the largest.*

3. Tribulus (*Cistoides*) foliolis octojugatis subæqualibus. Lin. Sp. Pl. 387. *Caltrops with eight pair of lobes to each leaf, which are almost equal.*

The first sort is a very common weed in the south of France, in Spain, and Italy, where it grows among Corn, and on most of the arable land, and is very troublesome to the feet of cattle; for the fruit being armed with strong prickles, run into the feet of the cattle, which walk over the land. This is certainly the plant which is mentioned in Virgil's Georgicks, under the name of Tribulus, though most of his commentators have applied it to other plants.

It is called in English Caltrops, from the form of the fruit, which resembles those instruments of war that were cast in the enemies way to annoy their horses.

This hath a slender fibrous root, from which spring out four or five slender hairy stalks, which spread flat on the ground, garnished at each joint with winged leaves, composed of six pair of narrow hairy lobes, almost of equal size; those on the lower part of the stalk stand alternately, but toward the top they are placed opposite. The flowers come out from the wings of the stalk, standing upon short foot-stalks; they are composed of five broad, obtuse, yellow petals, which spread open. In the center is situated an oblong germen, crowned by a headed stigma, attended by ten short stamina, terminated by single summits, and are succeeded by roundish, five-cornered, prickly fruit, which, when ripe, divides into five parts, each having a transverse cell, containing one or two seeds.

This plant is preserved in botanic gardens for variety. It is propagated by seeds, which should be sown in autumn, for those which are kept out of the ground till spring, commonly remain in the ground a whole year before the plants come up. These seeds should be sown on an open bed of light earth, where they are designed to remain; for, as it is an annual plant, it doth not bear transplanting very well, unless it be done when the plants are very young. In the spring, when the plants come up, they should be carefully cleared from weeds; and where they come up too close, some of the plants should be pulled out to give room for the remaining plants to grow; after this they will require no other culture but to keep them clear from weeds. If the seeds are permitted to scatter, the plants will come up the following spring, and maintain their place, if they are not overborne with weeds.

The second sort grows naturally in Jamaica, and some of the other islands in the West-Indies; this is an annual plant, with pretty thick, compressed, channelled stalks, which

which trail upon the ground, garnished with smooth winged leaves, placed by pairs opposite; they are sometimes composed of three, but most commonly of four pair of lobes, the outer being the largest. The flowers come out from the wings of the stalk; they are composed of five large yellow petals, which spread open, and have an agreeable odour; these are succeeded by roundish prickly fruit, ending in a long point, but seldom ripen in England.

The third sort grows naturally in the West-Indies; it was found by the late Dr. Houstoun at the Havannah; this has a ligneous root, from which spring out many stalks, which are hairy, jointed, and trail upon the ground, garnished at each joint by winged leaves, which differ greatly in size, some being composed of eight pair of oblong lobes which are nearly equal, but opposite to these come out small leaves composed of but four pair of lobes. The large leaves stand alternately upon the stalks, and the small ones on the opposite side; at the wings of the stalks come out the foot-stalks of the flowers, which are hairy, and near two inches long, each sustaining one pale yellow flower, composed of five large petals, which have narrow tails, but are very broad and rounded at their points. The flowers are succeeded by roundish fruit armed with very acute spines, but these rarely ripen in England.

The two last sorts, being natives of hot countries, are very tender, so must be sown in pots in autumn, and plunged into the tan-bed in the stove; when the plants are come up, they must each be transplanted into a separate pot, and then plunged into a hot-bed of tanners bark, where they must be treated in the same manner as other tender exotic plants, being careful to bring them forward as early as possible in the summer, otherwise they will not perfect their seeds in this country.

The third sort will live through the winter, if it is plunged in the bark-stove, and treated in the same manner as other tender plants, and the following summer they will flower earlier, so there will be more time for the seeds to ripen.

TRICHOMANES. Maiden-hair.

There are three or four varieties of this plant, which grow naturally in Europe; but in America there is a great number of species, which are remarkably different from each other, as also from the European kinds.

These being of the tribe of Ferns, or capillary plants, are seldom preserved in gardens. Their roots should be planted in moist shady places, especially the European sorts, which commonly grow from between the joints of old walls, and in other very moist shady situations; but those sorts which are brought from hot countries, must be planted in pots filled with rubbish, and strong earth mixed, and in winter they must be screened from hard frost, to which, if they are exposed, it will destroy them.

The common sort in England is generally sold in the markets for the true Maiden-hair, which is a very different plant, and not to be found in England, it being a native of the south of France, and other warm countries, so is rarely brought to England.

TRICOSANTHES. Lin. Gen. Plant. 966.

The CHARACTERS are,

It has male and female flowers, at separate distances, on the same plant. The male flowers have a long smooth empalement of one leaf, cut into five small segments which are reflexed; the petal is plain, spreading, and cut into five parts, ending in long branching hairs; they have three short stamina arising from the point of the empalement, terminated by cylindrical erect summits joined in a body, and three small styles fastened to the empalement. The female flowers sit upon the germen, and have empalements and petals like the male flowers, but have no stamina; they have a long slender germen situated under the flower, supporting a style the length of the empalement, crowned by three oblong stigmas. The germen afterward turns to a long succulent fruit, having three cells, inclosing many compressed seeds.

We have but one SPECIES of this genus, viz.

TRICOSANTHES (*Anguina*) pomis teretibus oblongis. Hort. Cliff. 450. *Tricosanthes with a taper, oblong, incurved fruit.*

This plant grows naturally in China, it is an annual, and of the Cucumber tribe. The stalks run to a great length, and, if they are not supported, trail upon the ground, in the same manner as Cucumbers and Melons. The leaves are angular and rough; the flowers come out from the side of the stalks; they are white, and cut into many small filaments or threads. The fruit is taper, near a foot long, incurved, and divided into three cells, which include many compressed seeds like those of Cucumber.

It is propagated by seeds, which must be sown on a hot-bed early in the spring, and afterwards treated in the same way as Cucumbers and Melons, keeping them covered with glasses, otherwise they will not ripen their fruit here.

TRICHOSTEMMA. Gron. Flor. Virg. 64. Lin. Gen. Plant. 652.

The CHARACTERS are,

The flower has a liped empalement; the upper lip is twice as large as the under, cut into three equal acute segments, the under lip in two. The flower is of the lip kind, and has a very short tube; the upper lip is compressed and hooked, the under is cut into three parts, the middle one being the least; it has four hair-like stamina, which are long and incurved, two of them being a little shorter than the other, terminated by single summits, and a four-pointed germen, supporting a slender style, crowned by a bifid stigma. The germen afterward turns to four roundish seeds, inclosed in the swollen empalement of the flower.

The SPECIES are,

1. TRICHOSTEMMA (*Dicotoma*) staminibus longissimis exsertis. Lin. Sp. Plant. 598. *Trichostemma with the longest stretched-out stamina.*

2. TRICHOSTEMMA (*Brachiata*) staminibus brevibus inclusis. Lin. Sp. Pl. 598. *Trichostemma with shorter stamina included in the petal.*

The first sort grows naturally in many parts of North America; it is an annual plant, which rises about six or eight inches high, dividing into small branches, garnished with small roundish leaves, not unlike those of sweet Marjoram, placed opposite, covered with fine, small, downy hairs. The flowers are produced at the wings of the branches; they are small, of a purple colour, gaping with two lips; the upper lip is arched, and is much larger than the lower; it is cut into three acute points; the lower lip is small, and cut into two points. These appear late in August, so that unless the season proves warm, the seeds will not ripen in England.

The second sort grows naturally in Virginia; this hath an herbaceous, angular, branching stalk, which rises from nine inches to a foot high; the leaves stand by pairs on the branches, shaped like those of the wild Marjoram, are a little hairy, and sit close to the branches; the flowers are produced at the top of the branches; they are small, of a purple colour. The four stamina stand within the tube of the flower; these flowers do not appear till the end of summer, so the seeds seldom ripen here.

They are propagated by seeds, which should be sown in pots in autumn; and in winter the pots should be placed under a frame to shelter them from severe frost, but should be exposed to the open air at all times when the weather is mild. In the spring the plants will appear; and when they are fit to remove, they should be planted on a bed of light earth, ſhading them from the ſun till they have taken freſh root, then they will require no other culture but to keep them clean from weeds.

TRIDAX. Lin. Gen. Pl. 872. American Starwort.

The CHARACTERS are,

The flower has a common, cylindrical, imbricated empalement. The ſcales are acute-pointed, and erect. The flowers are compoſed of hermaphrodite florets in the diſk, and the rays of female half florets. The hermaphrodite florets are funnel-ſhaped, cut at the brim into five points; theſe have five ſhort hair-like ſtamina, terminated by cylindrical ſummits joined together, and an oblong crowned germen, ſupporting a briſtly ſtyle, crowned by an obtuſe ſtigma. The germen afterward becomes an oblong ſingle ſeed, crowned with a ſimple down. The female half florets are plain, of one petal, cut into three ſegments at the top; theſe have an oval germen like the hermaphrodite florets, but no ſtamina, and are ſucceeded by ſingle ſeeds of the ſame ſhape.

We have but one SPECIES of this genus, viz.

TRIDAX (*Procumbens.*) Hort. Cliff. 418. *Trailing Starwort with a whitiſh copper-coloured flower, and hairy jagged leaves.*

This plant was diſcovered by the late Dr. Houſtoun, growing naturally by the road-ſide leading to old La Vera Cruz in America. The ſtalks are herbaceous, hairy, and trail upon the ground, emitting roots at their joints, whereby it ſpreads and propagates, garniſhed with rough hairy leaves, placed by pairs, ending in acute points, and are acutely jagged on their edges. The flowers are produced upon long naked foot ſtalks, which terminate their branches. They have one common empalement, compoſed of oval ſcales, ending in acute points, which lie over each other like the ſcales of fiſh; within which are ranged many female half florets, which compoſe the border or rays, and a good number of hermaphrodite florets, which form the diſk or middle; theſe are of a pale copper colour, inclining to white, and are each ſucceeded by a ſingle oblong ſeed, crowned with down.

This plant is propagated by ſeeds, which ſhould be ſown in pots, and plunged into a hot-bed; when the plants come up, and are fit to remove, they ſhould be each planted in a ſmall pot, and plunged into a hot-bed of tanners bark, obſerving to ſhade them from the ſun till they have taken new root; then they muſt be treated in the ſame way as other tender plants from the Weſt-Indies, placing them in the bark-ſtove in autumn, where they ſhould conſtantly remain.

It may alſo be propagated by its trailing ſtalks, which frequently put out roots at their joints; if theſe are cut off and planted, they will make new plants. This plant does not produce flowers in plenty here, and but rarely perfects its ſeeds in England.

TRIFOLIUM. Tourn. Inſt. R. H. 404. tab. 228. Trefoil, or Clover.

The CHARACTERS are,

This flower has a tubulous permanent empalement of one leaf, It is of the butterfly kind, drying in the empalement. The ſtandard is reflexed, the wings are ſhorter than the ſtandard, and the keel is ſhorter than the wings; it has ten ſtamina, nine are joined, and one is ſeparate, terminated by ſingle ſummits, and an almoſt oval germen ſupporting an awl-ſhaped ſtyle, crowned by a ſingle ſtigma. The germen afterward becomes a ſhort pod with one valve, containing a few roundiſh ſeeds.

There are a great number of ſpecies of this genus, ſeveral of which grow naturally in England, and others in many parts of Europe; but as great part of them are plants of ſmall eſtimation, they are rarely cultivated either in the field or garden; therefore it would be ſwelling this work too much to enumerate them all here, ſo I ſhall ſelect only ſuch of them as are cultivated either for uſe or beauty.

The SPECIES are,

1. TRIFOLIUM (*Pratenſe*) ſpicis ſubvilloſis, cinctis ſtipulis oppoſitis membranaceis, corollis monopetalis. Lin. Sp. 1082. *The red or Dutch Clover.*

2. TRIFOLIUM (*Repens*) capitulis umbellaribus leguminibus tetraſpermis, caule repente. Lin. Sp. Pl. 767. *Trefoil with umbellated heads, pods having four ſeeds, and a creeping ſtalk; White Meadow Trefoil, Honeyſuckle-graſs, or White Dutch Clover.*

3. TRIFOLIUM (*Agrarium*) ſpicis ovalibus imbricatis, vexillis deflexis perſiſtentibus, calycibus nudis, caule erecto. Flor. Suec. 617. *Trefoil with oval imbricated ſpikes of flowers, having deflexed permanent ſtandards, naked empalements, and an erect ſtalk; Yellow Meadow Trefoil, or Hop Clover.*

4. TRIFOLIUM (*Filiforme*) ſpicis imbricatis, vexillis deflexis perſiſtentibus, calycibus pedicillatis, caulibus procumbentibus. Lin. Sp. Pl. 773. *Trefoil with imbricated ſpikes of flowers, having deflexed permanent ſtandards, empalements ſtanding upon foot-ſtalks, and trailing ſtalks; the leaſt Yellow Hop Trefoil, called None-ſuch, or Black-ſeed.*

5. TRIFOLIUM (*Ochroleucrum*) ſpicis ovatis, calycibus foliatis, caule erecto villoſo, foliolis lanceolatis. *Trefoil with oval ſpikes of flowers, having leafy empalements, an erect hairy ſtalk, and ſpear-ſhaped leaves; greater hairy Meadow Trefoil, with a whitiſh ſulphur or copper-coloured flower.*

6. TRIFOLIUM (*Rubens*) ſpicis villoſis oblongis obtuſis aphyllis, foliolis ſubrotundis. Flor. Leyd. 380. *Trefoil with oblong, blunt, hairy ſpikes of flowers without leaves, and roundiſh lobes.*

7. TRIFOLIUM (*Squarroſum*) ſpicis ſubpiloſis, calycum infimo dente longiſſimo reflexo, caule herbaceo. Lin. Sp. 1087. *Narrow-leaved Spaniſh Trefoil.*

8. TRIFOLIUM (*Anguſtifolium*) ſpicis villoſis conico-oblongis, dentibus calycinis ſetaceis, ſubæqualibus, foliolis linearibus. Hort. Cliff. 375. *Trefoil with oblong, conical, hairy ſpikes, having briſtly indentures to the empalements, which are almoſt equal, and linear lobes to the leaves.*

9. TRIFOLIUM (*Arvenſe*) ſpicis villoſis ovalibus, dentatis calycinis ſetaceis æqualibus. Hort. Cliff. 375. *Trefoil with oval hoary ſpikes, and briſtly indentures to the empalements, which are equal; or Hare's-foot Trefoil.*

10. TRIFOLIUM (*Fragiferum*) capitulis ſubrotundis, calycibus inflatis bidentatis reflexis, caulibus repentibus. Hort. Cliff. 373. *Trefoil with roundiſh heads, reflexed bladder empalements with two teeth, and a creeping ſtalk; Strawberry Trefoil.*

11. TRIFOLIUM (*Officinalis*) leguminibus racemoſis nudis diſpermis, caule erecto. Hort. Cliff. 376. *Trefoil with long naked bunches of pods containing two ſeeds, and an erect ſtalk; or common Melilot.*

12. TRIFOLIUM (*Cæruleum*) ſpicis oblongis, leguminibus ſeminudis mucronatis, caule erecto. Hort. Cliff. 375. *Trefoil with oblong ſpikes, half naked acute-pointed pods, and an upright ſtalk; Sweet Melilot Trefoil.*

The firſt ſort, which is well known in England by the title of Red Clover, needs no deſcription; this has been frequently confounded with the red Meadow Trefoil by the botaniſts, who have ſuppoſed they were the ſame ſpecies; but I have often ſown the ſeeds of both in the ſame bed, which have conſtantly produced the two ſpecies without varying. The ſtalks of the Meadow Trefoil are weak and hairy; the ſtipulæ, which embrace the foot-ſtalks of the leaves, are narrow and very hairy; the heads of flowers are rounder, and not ſo hairy as thoſe of the Clover, whoſe ſtalks are ſtrong, almoſt ſmooth, furrowed, and riſe twice the height of the other; the heads of flowers are large, oval, and hairy; the petal of the flowers open much wider, and their tubes are ſhorter than thoſe of the other; but the Clover has been ſo much cultivated in England for near a hundred

hundred years paſt, that the ſeeds have been ſcattered over many of the Engliſh paſtures, ſo that there are few of them which have not Clover mixed with the other Graſſes; and this has often deceived the botaniſts, who have ſuppoſed that the Meadow Trefoil has been improved to this by dreſſing of the land.

Since the red Clover has been cultivated in England, there has been great improvement made of the clay lands, which before produced little but Rye-graſs, and other coarſe bents, which, by being ſown with red Clover, have produced more than ſix times the quantity of fodder they formerly had on the ſame land; whereby the farmers have been enabled to feed a much greater ſtock of cattle than they could do before, with the ſame extent of ground, which has enriched the ground, and prepared it for Corn; ſo where the land is kept in tillage, it is the uſual method now amongſt the farmers to lay down their ground with Clover, after having had two crops of Corn, whereby there is a conſtant rotation of Wheat, Barley, Clover, or Turneps, on the ſame land. The Clover-ſeed is generally ſown with the Barley in the ſpring, and when the Barley is taken off, the Clover ſpreads and covers the ground, and this remains two years; after which the land is ploughed again for Corn.

The Clover is a biennial plant, whoſe roots decay after they have produced ſeeds; but by eating it down, or mowing it when it begins to flower, it cauſes the roots to ſend out new ſhoots, whereby the plant is continued longer than it would naturally do. The common allowance of ſeed for an acre of ground is ten pounds. In the choice of the ſeeds, that which is of a bright yellow colour, inclining to brown, ſhould be preferred, and the pale-coloured thin ſeed ſhould be rejected. The Clover-ſeed ſhould be ſown after the Barley is harrowed in, otherwiſe it will be buried too deep; and after the ſeeds are ſown, the ground ſhould be rolled, which will preſs the ſeeds into the ground; but this ſhould be done in dry weather, for moiſture will often cauſe the ſeeds to burſt, and when the ground is wet the ſeeds will ſtick to the roll. This is the method which is generally practiſed by moſt people in ſowing of this ſeed with Corn, but it will be much better if ſown alone; for the Corn prevents the growth of the plants until it is reaped and taken off the ground, ſo that one whole ſeaſon is loſt; and many times, if there be a great crop of Corn upon the ground, it ſpoils the Clover, ſo that it is hardly worth ſtanding; whereas, when it is ſown without any other ſeed, the plants will come up more equal, and come on much faſter than that which was ſown the ſpring before under Corn.

Therefore, from many years trial, I would adviſe the ſeeds to be ſown in Auguſt, when there is a proſpect of rain ſoon after; for as the ground is at that ſeaſon warm, ſo the firſt ſhower of rain will bring up the plants, and theſe will have time enough to get ſtrength before the winter: and if ſome time in October, when the ground is not too wet, the Clover is well rolled, it will preſs the ground cloſe to the roots, and cauſe the plants to ſend out more ſhoots; the ſame ſhould be repeated in March, which will be found very ſerviceable to the Clover. The reaſon of my preferring this ſeaſon for the ſowing of the ſeeds rather than the ſpring is, becauſe the ground is cold and wet in ſpring, and if much rain fall after the ſeeds are ſown, they will rot in the ground; and many times when the ſeed is ſown late in the ſpring, if the ſeaſon ſhould prove dry, the ſeeds will not grow; ſo that I have always found the other ſeaſon has been the ſureſt.

About the middle of May this Graſs will be fit to cut, when there ſhould be great care taken in making it; for it will require a great deal more labour and time to dry than common Graſs, and will ſhrink into leſs compaſs; but if it be not too rank, it will make extraordinary rich food for cattle. The time for cutting it is when it begins to flower; for if it ſtands much longer, the lower part of the ſtems, and the under leaves will begin to dry, whereby it will make a leſs quantity of Hay, and that not ſo well flavoured.

Some people cut three crops in one year of this Graſs; but the beſt way is to cut but one in the ſpring, and feed it the remaining part of the year, whereby the land will be enriched, and the plants will grow much ſtronger.

One acre of this plant will feed as many cattle as four or five acres of common Graſs; but great care ſhould be taken of the cattle, when they are firſt put into it, leſt it burſt them: to prevent which, ſome turn them in for a few hours only at firſt, and ſo ſtint them as to quantity; and this by degrees, letting them at firſt be only one hour in the middle of the day, when there is no moiſture upon the Graſs, and ſo every day ſuffer them to remain a longer time, until they are fully ſeaſoned to it; but great care ſhould be had never to turn them into this food in wet weather; or if they have been for ſome time accuſtomed to this food, it will be proper to turn them out at night in wet weather, and let them have Hay, which will prevent the ill conſequences of this food; but there are ſome who give ſtraw to their cattle while they are feeding upon this Graſs, to prevent the ill effects of it; which muſt not be given them in the field, becauſe they will not eat it where there is plenty of better food. There are others who ſow Rye-graſs amongſt their Clover, which they let grow together, in order to prevent the ill conſequences of the cattle feeding wholly on Clover; but this is not a commendable way, becauſe the Rye-graſs will greatly injure the Clover in its growth, and the ſeeds will ſcatter and fill the ground with bents.

Where the ſeeds are deſigned to be ſaved, the firſt crop in the ſpring ſhould be permitted to ſtand until the ſeeds are ripe, which may be known by the ſtalks and heads changing to a brown colour; then it ſhould be cut in a dry time; and when it is well dried, it may be houſed until winter, when the ſeeds ſhould be threſhed out; but if the ſeeds are wanted for immediate ſowing, it may be threſhed before it be houſed or ſtacked; but then it muſt be well dried, otherwiſe the ſeeds will not quit their huſks.

It has been a great complaint amongſt the farmers, that they could not threſh out theſe ſeeds without great labour and difficulty; which I take to be chiefly owing to their cutting the ſpring crop when it begins to flower, and ſo leave the ſecond crop for ſeed, which ripens ſo late in autumn, that there is not heat enough to dry the huſks ſufficiently; whereby they are tough, and the ſeeds rendered difficult to get out; which may be entirely remedied by leaving the firſt crop for ſeed, as hath been directed; and then the ground will be ready to plough, and prepare for Wheat the ſame year, which is another advantage.

When cattle are fed with this Hay, the beſt way is to put it in racks, otherwiſe they will tread a great quantity of it down with their feet. This feed is much better for moſt other cattle than milch cows, ſo that theſe ſhould rarely have any of it, leſt it prove hurtful to them; though when it is dry, it is not near ſo injurious to any ſort of cattle as when green.

The ſecond ſort grows naturally in moſt of the paſtures in England, and is generally known among the country people, by the title of white Honeyſuckle.

This is an abiding plant, whoſe branches trail upon the ground, and ſend out roots from every joint, ſo that it thickens and makes the cloſeſt ſward of any of the ſown Graſſes; and it is the ſweeteſt feed for all ſorts of cattle yet known; therefore when land is deſigned to be laid down for paſture, with intent to continue ſo, it ſhould be ſown with the ſeeds of this plant. The uſual allowance of this feed is eight pounds to one acre of land; but this fhould never be fown with Corn, for if there is a crop of Corn, the Grafs will be fo weak under it, as to be fcarce worth ftanding; but fuch is the covetoufnefs of moft farmers, that they will not be prevailed on to alter their old cuftom of laying down their grounds with a crop of Corn, though they lofe twice the value of their Corn by the poornefs of the Grafs, which will never come to a good fward, and one whole feafon is alfo loft; for if this feed is fown in the fpring without Corn, there will be a crop of Hay to mow by the middle or latter end of July, and a much better after-feed for cattle the following autumn and winter, than the Grafs which is fown with Corn will produce the fecond year. The feed of this fort may alfo be fown in autumn, in the manner before directed for the common red Clover; and this autumnal fowing, if the feeds grow kindly, will afford a good early crop of Hay the following fpring; and if, after the Hay is taken off the land, the ground is well rolled, it will caufe the Clover to mat clofe upon the ground, and become a thick fward.

The feeds of this white Dutch Clover is annually imported from Flanders, by the way of Holland, from whence it received the name of Dutch Clover; not that it is more a native of that country than of this, for it is very common in moift paftures in every county in England; but the feeds were never collected for fowing in England till of late years; nor are there many perfons at prefent here who fave this feed, although it may be done, if the fame care as is practifed for the red Clover is taken of this fort; therefore it fhould be recommended to every farmer, who is defirous to improve his land, carefully to fow an acre or two of this white Clover by itfelf for feeds, which will fave him the expence of buying for fome years when the price is great, and there will be no want of fale for any quantity they may have to fpare.

The farther account of this Grafs may be feen under the article of PASTURE.

The third fort grows naturally among the Grafs in moft of the upland paftures in this country; but the feeds are frequently fold in the fhops by the title of Hop Clover, and are by many people mixed with the other forts of Clover and Grafs feeds, for laying down ground to pafture; this grows with upright branching ftalks about a foot high, garnifhed with trifoliate leaves, whofe lobes are oblong and heart-fhaped, but reverfed, the narrow point joining the foot-ftalks. The flowers, which are yellow, grow from the wings of the ftalk upon long foot ftalks, collected into oval imbricated heads, having naked empalements, lying over each other like fcales, fomewhat like the flowers of Hops, from whence this plant had the title of Hop Clover. But there are two forts of this which grow naturally in England. The other, which is the fourth fort, is a much fmaller plant than this, and has trailing ftalks. The heads of flowers are fmaller, and the flowers are of a deeper yellow colour; thefe are not abiding plants, fo are by no means proper to be fown, where the ground is defigned to continue in pafture; but in fuch places where one or two crops only are taken, and the land is ploughed again for Corn, it may do well enough when it is mixed with other feeds, though the cattle are not very fond of it green, unlefs when it is very young. The large fort is the moft profitable, but this is rarely to be had without a mixture of the fmall kind, and alfo of the fmaller Melilot, which is commonly called None-fuch, or fometimes Black-feeds; for thofe who fave the feeds for fale, are feldom curious enough to diftinguifh the forts; but where the beauty of the verdure is confidered, there muft not be any of the feeds fown, becaufe their yellow heads of flowers are very unfightly among the Grafs; and if it is in gardens, where the Grafs is conftantly mowed, the flowers of thefe plants will come out near the root in fuch clufters, as to occafion large, unfightly, yellow patches; and as the heads decay, they turn brown, and have a very difagreeable appearance.

The fifth fort grows naturally on chalky lands in many parts of England; and in fome countries the feed is fown after the fame manner as the common red Clover, efpecially on chalky ground, where it will thrive, and produce a better crop than Clover The ftalks of this are hairy, and grow erect to the height of two feet or more, garnifhed with trifoliate leaves, ftanding upon long foot-ftalks, whofe lobes are longer than thefe of the red Clover, and have no marks of white; they are of a yellowifh green colour, and are covered with foft hairs. The flowers grow in oval fpikes at the end of the branches; they are of a pale copper colour; their petals are long and tubulous, but the brim is divided into two lips as the other forts.

This is known by the title of Trefoil, in the places where it is cultivated; but the feedfmen fell the Hop Clover by that name, fo they make no diftinction between this, the Hop Clover, and None-fuch; therefore, by which of thefe three titles the feeds are bought, they often prove the fame. This fort of Trefoil is much cultivated in that part of Effex which borders on Cambridgefhire.

The fixth fort grows naturally in Spain and Italy; this has upright ftalks near two feet high, which are hairy, garnifhed with trifoliate leaves, having roundifh lobes, which are fawed at their points. The flowers are produced at the top of the ftalk in long, obtufe, hairy fpikes, of a bright red colour, fo make a pretty appearance during their continuance. It is an annual plant, fo is not proper for fowing as fodder.

The feventh fort is an annual plant, which grows naturally in the fouth of France and Italy; it rifes with a ftrong fmooth ftalk near three feet high, garnifhed with trifoliate leaves, whofe lobes are two inches and a half long, and near a quarter broad, ftanding upon long foot-ftalks, which are embraced by ftipulæ or fheaths their whole length. The flowers are produced at the top of the ftalks in very long fpikes; they are of a beautiful red colour, fo make a fine appearance. It flowers in July, and the feeds ripen in autumn.

The eighth fort grows naturally in Spain and Italy; this rifes with a flender ftiff ftalk near two feet high, garnifhed with trifoliate leaves, whofe lobes are very narrow and hairy. The flowers are produced at the top of the ftalks in oblong conical fpikes; the indentures of their empalements end in long briftly hairs, which are almoft equal in length; the fpikes are hairy, and the flowers of a pale red colour.

The ninth fort is the common Haresfoot Trefoil, which grows naturally upon dry gravelly land in moft parts of England, and is a fure indication of the fterility of the foil, for it is rarely feen upon good ground This plant is feldom eaten by cattle, fo is unfit for pafture, and is only mentioned here becaufe it is fometimes ufed in medicine; it is an annual plant, whofe root decays foon after it has perfected feeds.

The tenth fort grows naturally on arable land in many parts of England; this has trailing ftalks, which put out roots at their joints. The leaves ftand upon long flender foot-ftalks; the lobes are roundifh, and fawed on their edges; the flowers are collected in roundifh heads, ftanding upon flender foot-ftalks, which rife from the wings of the ftalks; thefe have bladdery empalements, which terminate in two teeth. When thefe lie on the ground, their globular heads, having a little blufh of red on their upper fide toward

toward the sun, and the other part being white, have a great resemblance of Strawberries, and from thence it was titled Strawberry Trefoil.

These sorts are preserved in botanic gardens for variety; they are easily propagated by seeds, which may be sown on an open bed of ground, either in autumn or spring. The plants which come up in autumn, will grow much larger, and flower early in the summer than those which are sown in the spring, so from those good seeds may be always obtained, whereas the other sometimes miscarry. When the plants come up, they require no other care than to keep them clean from weeds, and thin them where they are too close.

The eleventh sort is the common Melilot, which is used in medicine; it grows naturally among the Corn in many parts of England, particularly in Cambridgeshire in great plenty, where it is a most troublesome weed; for in reaping it is scarce possible to separate it from the Melilot, so that it is carried in with the Corn; and the seeds of the Melilot being ripe about the same time with the Corn, they are threshed out with it, and being heavy are difficult to separate from it; and when a few of the seeds are ground with the Corn, it spoils the flour, for the bread, or whatever else is made with it, will have a strong taste like Melilot plaster.

The roots of this plant are strong and ligneous, from which spring out several stalks, which rise from two to four feet high, according to the goodness of the land. The stalks branch out, and are garnished with trifoliate leaves, having oval sawed lobes, of a deep green colour. The flowers are produced in long slender spikes, which spring from the wings of the stalks; they are of a bright yellow, and shaped like the other butterfly flowers; these are succeeded by naked seeds.

The twelfth sort grows naturally in Bohemia and Austria, but has been long cultivated in England as a medicinal plant, though at present it is rarely used; it is annual. The stalks are large, hollow, and channelled; they rise about a foot high, garnished with trifoliate leaves, whose lobes are oval, and slightly sawed on their edges, standing upon pretty long foot stalks. The flowers are collected in oblong spikes, which stand upon very long foot-stalks, springing from the wings of the stalk at every joint; they are of a pale blue colour, shaped like those of the common Melilot; these appear in June and July, and are succeeded by small yellow seeds, of a kidney shape, two or three being included in each short pod. The whole plant has a very strong scent like that of Fenugreek, and perishes soon after the seeds are ripe.

If the seeds of these two sorts are permitted to scatter, the plants will rise without care, and require no other culture but to keep them clean from weeds, and thin them where they grow too close.

TRIGONELLA. Lin. Gen. Plant. 804. Fenugreek.

The CHARACTERS are,

The empalement of the flower is bell-shaped, of one leaf, cut at the top into five almost equal segments. The flower is of the butterfly kind; the standard is oval, obtuse, and reflexed: the two wings are oblong, reflexed, and spreading flat like the standard, so as outwardly to appear like a regular flower of three petals; the keel is very short, obtuse, and occupies the navel of the flower. It has ten short rising stamina, nine of which are joined, and one stands separate, terminated by single summits, and an oval oblong germen, supporting a single style, crowned by a rising stigma. The germen afterward turns to an oblong oval pod compressed, filled with kidney-shaped seeds.

The SPECIES are,

1. TRIGONELLA (*Fœnum Græcum*) leguminibus sessilibus strictis erectiusculis subfalcatis acuminatis, caule erecto. Hort. Cliff. 229. *Trigonella with sithe-shaped acute-pointed pods, which are close, erect, and sit close to the stalks; or common Fenugreek.*

2. TRIGONELLA (*Spinosa*) leguminibus subpedunculatis congestis declinatis erectis subfalcatis compressis pedunculis communibus spinosis brevissimis. Lin. Sp. Pl. 1094. *Trigonella with linear, erect, parallel pods growing in clusters, having foot-stalks; or wild Fenugreek.*

3. TRIGONELLA (*Polycerates*) leguminibus sessilibus arcuatis consertis, caulibus procumbentibus. *Trigonella with arched pods growing in clusters, sitting close to the stalks, which trail on the ground.*

4. TRIGONELLA (*Platycarpos*) leguminibus pedunculatis congestis pendulis ovalibus compressis, caule diffuso, foliolis subrotundis. Hort. Cliff. 229. *Trigonella with clustered, oval, compressed, hanging pods, having foot-stalks, diffused stalks, and roundish lobes.*

5. TRIGONELLA (*Ruthenica*) leguminibus pedunculatis congestis pendulis linearibus rectis, foliolis sublanceolatis. Lin. Sp. Pl. 776. *Trigonella with linear strait pods, which hang down in clusters upon foot-stalks, and spear-shaped lobes to the leaves.*

The first sort is the common Fenugreek, whose seeds are used in medicine. Where this plant grows naturally is uncertain, but it is cultivated in the fields, in the south of France, and in Germany, from whence great quantities of the seeds are annually imported here for use. It is also much cultivated in India.

It is an annual plant, which rises with a hollow, branching, herbaceous stalk a foot and a half high, garnished with trifoliate leaves, placed alternately, whose lobes are oblong, oval, indented on their edges, and have broad furrowed foot-stalks. The flowers come out singly at each joint from the wings of the stalk; they are white, of the butterfly kind, and sit very close to the stalk; these are succeeded by long compressed pods, shaped somewhat like a broad sword, ending in long points, having a broad membrane on one edge, filled with square yellow seeds, indented on one side like a kidney. The whole plant has a very strong odour.

This plant has not as yet been cultivated in any quantity for use in England, as it has generally proved a very uncertain crop, occasioned by the inconstancy of the weather here, for in cold wet seasons the plants are frequently killed before the seeds ripen; and if any of them live long enough to perfect their seeds, the pods change of a dirty colour, and the seeds turn black and unsightly, especially when much rain falls about the time of their ripening; therefore the seeds, which are imported from the continent, are always preferred to those of our own growth.

But as the consumption of these seeds is very great in England, there are some persons who are inclinable to make fresh trials to cultivate the plants here. As I have many years cultivated this in small quantities, and have made trials by sowing the seeds at different seasons, and after various manners, by which I have acquired a knowledge of its culture, so I shall here give such directions for the management of this plant, as from experience has been found to succeed best.

The ground in which this plant thrives best, is a light hazel loam, not enriched with dung; this should be cleaned from the roots of weeds, and well ploughed twice, and harrowed fine before the seeds are sown. The best time to sow the seeds is in the beginning of September, in shallow drills like Pease. The rows should be two feet asunder, and the seeds must be scattered one inch distant from each other in the drills; for if the plants are too close together in the spring, they may be easily thinned with the hoe, when the ground is cleaned. When the seeds are sown at the before mentioned

mentioned time, the plants will appear in three weeks or a month after; and if the weeds appear at the fame time, the ground fhould be hoed over as foon as poffible in dry weather, to deftroy the weeds; and when the plants are grown an inch high, the earth fhould be drawn up to their ftems, in the fame manner as is practifed for Peafe. This will fecure their ftems from being injured by fharp cutting winds; and if a ridge of earth is drawn up on the north or eaft fide of each row, it will protect the plants from the pinching winds which blow from both thofe quarters; for although this plant will not be in any danger from the froft in the ordinary winters, yet in very fevere frofts they are fometimes killed; but, as this plant will live in any fituation, where Peafe will ftand through the winter, there will be no greater hazard of the one crop than the other.

In the fpring of the year the ground muft be hoed again in dry weather to kill the weeds, and the plants fhould be again earthed up in the like manner as Peafe, with whofe culture this plant will thrive; but there muft be great care taken to keep the ground as clean from weeds as poffible, for if they are permitted to grow, they will foon advance above the plants, and greatly weaken them; and when their pods begin to form, they cannot be too much expofed to the fun and air, whereby they will be lefs liable to fuffer from moifture.

When the feeds are fown in autumn, the plants will grow much ftronger, and have many more fide branches than thofe which come up in the fpring, fo will produce a much greater crop of feeds; and thefe will produce their flowers five or fix weeks earlier, fo will have a better feafon to ripen; but in order to have them better ripened, the top of the plants fhould be cut off with garden-fhears about the middle of June, by which time the pods will be formed on the lower part of the ftalks, which will be greatly forwarded by topping of the ftalks in the fame way as is commonly practifed for Garden-beans; for where the plants are fuffered to extend in length, the lower pods often mifcarry, or are lefs nourifhed, and thofe on the top of the ftalks are late before they ripen; fo where the topping of the plants is omitted, the pods at bottom will open and caft out their feeds, before thofe above will be ripe; therefore to preferve the firft and cut off the other, will be found the beft method, for by fo doing the pods will ripen equally, and much earlier in the feafon.

If the fummer proves warm, the feeds will ripen in Auguft, and the plants fhould then be cut off, and laid to dry for five or fix days, in which time they fhould be turned two or three times, that the pods may dry equally; then the feeds may be either threfhed out in the field, or the haulm may be houfed in a barn, to be threfhed in a more convenient time.

The fecond fort grows naturally in Spain and Sicily. The ftalks of this are flender, and rife near a foot high, fending out two or three flender branches, garnifhed with trifoliate leaves, whofe lobes are wedge-fhaped, fawed at their ends where they are indented; thefe ftand upon long flender foot-ftalks. The flowers are produced in clufters at the end of the branches, upon fhort foot-ftalks, which ftand erect; they are fmall, of a pale colour, and are fucceeded by narrow pods, ftanding parallel and erect. This is an annual plant, which flowers in July. The feeds ripen the end of Auguft, and the plants decay foon after.

The third fort grows naturally in Spain and Italy; this is alfo an annual plant, whofe root decays foon after the feeds are ripe. The ftalks trail upon the ground, and extend a foot and a half in length, fending out feveral fide branches, garnifhed with fmall trifoliate leaves, whofe lobes are wedge-fhaped, and fawed at their points. The flowers are produced in clufters at the wings of the ftalk; they are fmall, of a pale yellow colour, and fit very clofe to the ftalks; thefe are fucceeded by fhort hooked pods, which fit clofe to the ftalks in clufters. It flowers in July, and the feeds ripen in autumn.

The fourth fort grows naturally in Siberia. The root of this is biennial; the ftalks trail upon the ground, extend a foot in length, and fend out many fide branches, garnifhed with trifoliate leaves, having roundifh lobes, which are fawed on their edges. The flowers come out from the wings of the ftalks upon foot-ftalks, growing in clufters; they are fmall, of a yellowifh white colour, and are fucceeded by oval compreffed pods, containing two feeds in each.

The fifth fort grows naturally in Ruffia; this is alfo a biennial plant, whofe roots decay foon after the feeds are ripe. The ftalks of this are very flender, and trail upon the ground; they extend a foot and a half in length, and divide into feveral branches. The leaves are trifoliate; the lobes are wedge-fhaped, indented at the point, fawed, and are narrower than either of the former. The flowers are fmall, yellow, and are produced in clufters upon flender foot-ftalks, which fpring from the wings of the ftalk, and are fucceeded by narrow erect pods, which contain three or four fmall feeds.

Thefe plants are frequently cultivated in botanic gardens for variety, but I do not know any ufe is made of either of the forts except the firft. The feeds of thefe fhould be fown in the places where the plants are defigned to ftand, for they will not bear tranfplanting If they are fown in autumn, in the fame way as is before directed for the firft fort, the plants will come earlier to flower, and good feeds may be obtained with more certainty than from the fpring plants. All the culture thefe require is to thin them where they ftand too clofe, and keep them clean from weeds. A few plants of each fort in a garden will be fufficient, as they have no great beauty.

TRILLIUM. Lin. Gen. Plant. 412. American Herb Paris.

The CHARACTERS are,

The flower has a three-leaved permanent empalement; it has three oval petals, which are a little larger than the empalement, and fix awl fhaped ftamina, which are erect, fhorter than the petals, and terminated by oblong fummits, with a roundifh germen, having three flender recurved ftyles, crowned by fingle ftigmas. The germen afterward becomes a roundifh berry with three cells, filled with roundifh feeds.

The SPECIES are,

1. TRILLIUM (*Cernuum*) flore pedunculato cernuo. Lin. Sp. Pl. 339. *Trillium with a nodding flower growing upon a foot-ftalk.*

2. TRILLIUM (*Erectum*) flore pedunculato erecto. Lin. Sp. Pl. 340. *Trillium with a flower growing erect upon a foot-ftalk.*

3. TRILLIUM (*Seffile*) flore feffili erecto. Lin. Sp. Pl. 340. *Trillium with an erect flower having no foot-ftalk.*

Thefe plants grow naturally in the woods in many parts of North America; the firft was fent me from Philadelphia by Dr. Benfel, who found it growing in plenty there. The root of this plant is tuberous, fending out many fibres; the ftalk is fingle, naked, and rifes five or fix inches high, with three oval fmooth leaves, placed at the top upon fhort foot-ftalks, which fpread out in a triangle, of a deep green colour. From the center of the foot ftalks of the three leaves comes out one flower upon a fhort foot-ftalk, which nods downward; this has a three-leaved green empalement fpreading open, and within are three petals about the fize of the empalement, of a whitifh green on their outfide, and purple within, having fix ftamina in the center, furrounding the ftyle, which have oblong fummits. The flowers of this appear in April, and are fucceeded by roundifh fucculent berries,

berries, having three cells, filled with roundish seeds, which ripen in June.

The second sort has a taller stalk than the first. The three leaves are placed at a distance from the flower, which stands upon a long foot-stalk, and is erect; the petals of the flower are larger, and end with sharper points than those of the first.

The third sort grows in shady thickets in Carolina. The stalk of this is purple; the three leaves grow at the top like the first, but they are much longer, and end in acute points; the petals of the flowers are long, narrow, and stand erect.

These plants are propagated by seeds, which should be sown upon a shady border soon after they are ripe; and when the young plants come up the following spring, they must be kept clean from weeds; and in autumn, after their leaves decay, the roots may be transplanted to a moist shady place, where they are to remain; if the seeds are sown in the spring, they will not vegetate till the next year.

TRIOSTEUM. Lin. Gen. Plant. 211. Dr. Tinkar's Weed, or false Ipecacuana.

The Characters are,

The flower has a permanent empalement of one leaf, cut into five segments; it has a tubulous flower of one petal, with a short brim, cut into five parts which stand erect, and five slender stamina the length of the tube, terminated by oblong summits, with a roundish germen, supporting a cylindrical style, crowned by a thick stigma. The germen afterward becomes an oval berry with three cells, each including one hard, three-cornered, obtuse seed.

The Species are,

1. Triosteum (*Perfoliatum*) floribus verticillatis sessilibus. Lin. Sp. Pl. 176. *Triosteum with flowers growing in whorls, sitting close to the stalks; commonly called Dr. Tinkar's Weed, or false Ipecacuana.*

2. Triosteum (*Angustifolium*) floribus oppositis pedunculatis. Lin. Sp. Pl. 175. *Triosteum with flowers growing opposite upon foot-stalks.*

The first sort grows naturally in the woods in several parts of North America; the root is composed of thick fleshy fibres, which are contorted and rough, from which spring several strong herbaceous stalks a foot and a half high, garnished at each joint by two oblong broad leaves, sitting close to the stalk. From the bosoms of these come out the flowers in whorls, sitting very close to the stalks; the empalements are cut into five segments. The flowers are small, tubulous, of a dark red colour, and cut slightly at the brim into five obtuse segments, and are succeeded by roundish berries, which turn yellow when ripe, having three cells, in each of which is contained one hard seed. The root is perennial, but the stalks decay every autumn.

The second sort differs from the first in its leaves being longer and narrower. The flowers stand single upon short foot-stalks, having two at each joint, whereas the other has many growing in whorls round the stalks, but the roots of both are indifferently used in America by the title of Dr. Tinkar's Weed.

Both these plants are natives of Virginia, and some other northern parts of America, where their roots have been frequently used as an emetic, and are commonly called Ipecacuana. One of the first persons who brought the roots into use was Dr. Tinkar, from whence many of the inhabitants have called them by the name of Dr. Tinkar's Weed. The leaves of the first sort greatly resemble those of the true Ipecacuana, but the roots are of a different form; but so far as I can judge by the imperfect fruit of a specimen in my collection of the true Ipecacuana, as also by the figure and description given by Piso in his History of Brasil, it seems to belong to this genus.

The first grows on low marshy grounds near Boston in New England, very plentifully, where the roots are taken up every year, and are continued in use amongst the inhabitants of Boston.

This plant is preserved in several curious gardens in England, and is hardy enough to thrive in the open air; but it should be planted on a moist light soil, for if it is on a dry ground, there must be care taken to water the plants constantly in dry weather, otherwise they will not thrive. It may be propagated by seeds, which should be sown in autumn on a border of light earth, exposed to the morning sun, for if the seeds are sown in the spring, they will remain in the ground a whole year before the plants will come up, so that during this time the border must be constantly kept clear from weeds; the following spring, when the plants appear, they should be duly watered in dry weather, which will greatly promote their growth; but if the seeds are sown in autumn, the plants will come up the following spring, and must be constantly kept clean from weeds, which, if permitted to grow amongst them, will soon overbear the plants while they are young, and either quite destroy them, or so much weaken them, that they will not recover in a long time.

The plants may remain in this seed border until the Michaelmas following, when they should be carefully taken up, and transplanted where they are designed to remain. Some of them should be planted in pots, that they may be sheltered in winter, lest those which are in the full ground should be destroyed by severe frost.

They may be also propagated by parting of the roots. The best season for this work is in the spring, just before the plants begin to shoot, which is commonly about the middle or latter end of March; but in doing of this, the roots must not be parted too small, for that will prevent their flowering strong.

These plants perfect their seeds in this country every year, which, if sown in autumn as soon as they are ripe, is the best way to propagate them. The seedling plants will not flower until the third year, and then they are seldom so strong as the older plants.

TRIPOLIUM. See Aster.

TRITICUM. Tourn. Inst. R. H. 512. tab. 292, 293. Wheat.

The Characters are,

It has an oval chaffy empalement with two valves, which inclose two or three flowers. The petals have a double valve as large as the empalement; the outer valve is bellied and acute-pointed; the inner is plain. The flowers have three hair-like stamina, terminated by oblong forked summits, and a top-shaped germen, supporting two hairy reflexed styles, crowned by feathered stigmas. The germen afterward becomes an oblong oval seed, obtuse at both ends, convex on one side, and channelled on the other, wrapped up in the petal of the flower.

The Species are,

1. Triticum (*Hybernum*) calycibus quadrifloris ventricosis lævibus, imbricatis submuticis. Hort. Upsal. 21. *Winter Wheat without awns.*

2. Triticum (*Æstivum*) calycibus quadrifloris ventricosis glabris imbricatis aristis. Hort. Upsal. 21. *Summer or Spring Wheat.*

3. Triticum (*Turgidum*) calycibus quadrifloris ventricosis villosis imbricatis subaristatis. Hort. Upsal. 21. *Gray Pollard, or Duckbill Wheat.*

4. Triticum (*Quadratum*) glumis ventricosis villosis imbricatis, spicis oblongis pyramidatis. *Wheat with hairy, bellied, imbricated husks, and oblong pyramidal spikes; commonly called Cone Wheat.*

5. TRITICUM (*Polonicum*) calycibus bifloris nudis, flosculis longissime aristatis, racheos dentibus barbatis. Lin. Sp. 127. *Polonian Wheat.*

There are some other varieties of Wheat, which the farmers in different parts of England distinguish by different titles; but they are only seminal variations, which have risen from culture. Some of these differ in the colour of their chaff, and others in the form of their spikes; but as they are subject to vary, we shall not enumerate them as different species. The varieties are, the red Wheat without awns, the red-eared bearded Wheat, many-eared Wheat, and naked Barley. But the five sorts above enumerated, I have sown several years, and have always found them constant without variation.

Where Wheat grows naturally, is very hard to determine at present; but it is generally supposed that Africa is the country, because in the earliest accounts we have of it, there is mention of its being transported from thence to other countries, and Sicily was the first country in Europe where this grain was cultivated; but although the country of its natural growth is in a very warm climate, yet it is found to bear the inclemency of rough climates very well; and in countries more north than England, where the summers are long enough to ripen the grain, it is found to succeed.

The first sort is the common Wheat, which is sown in most parts of England, and is so well known as to need no description. The spikes or ears of this are long; the grains are ranged in four rows, and lie over each other like the scales of fish; the chaff is smooth, bellied, and is not terminated by awns or beards.

The second sort is called Summer or Spring Wheat; this will ripen much earlier than the other, so has often been sown in the spring of the year, at the same time with Oats; but if the season proves wet, it is very subject to grow tall, and have very thin grains, which has discouraged people from sowing it at that season; so that unless from the severity of the winter, or some other accident, the winter Corn is injured, the sowing Wheat in the spring is rarely practised.

The third sort is called in some places gray Wheat, in others Duckbill Wheat and gray Pollard; but in Sussex it is generally known by the title of Fullers Wheat. This sort grows very tall, and if it is sown too thick, is very apt to lodge with rain and wind, for the ears are large and heavy; they nod on one side as the grain increases in weight. The awns are long; the chaff hairy, which detains the moisture, all which help to lodge it; for which reason, many people do not chuse to cultivate this sort; but where the roots are at a proper distance from each other, they will put out many stalks from each; the stalks will be stronger, support themselves better, and the grain produces more flour in proportion than any of the other sorts. The awns of this sort frequently drop off when the grain is full grown.

The fourth sort is more cultivated in Oxfordshire and Berkshire than in any other part of England. The ears of this sort are formed like a cone, ending with a slender point, from whence it had the title of Cone Wheat. Of this there are the white and red, which I believe are only varieties, for I have generally seen them mixed in the field. The awns of this are long and rough, so the farmers say it guards the grain from birds, which has been a recommendation to sow it, especially near inclosures, where there is shelter for birds. Mr. Tull prefers this sort for sowing in drills, but I have seen the third sort answer much better in the horse hoeing husbandry.

The Polonian Wheat grows tall, the ears are long and heavy, so that where it is sown too thick, it is very subject to be lodged, therefore the farmers little regard it; but it produces much flour, and therefore worthy of cultivation.

The season for sowing of Wheat is autumn, and always when the ground is moist. In the downs of Hampshire, Wiltshire, and Dorsetshire, the farmers begin sowing of their Wheat in August, if there happens rain; so that when they are in their harvest, if the weather stops them, they employ their people in sowing; for if the Corn is not forward in autumn, so as to cover the ground before winter, it seldom succeeds well on those dry lands, especially if the spring should prove dry; but in the low strong lands, if they get their Wheat into the ground by the middle of November, the farmers think they are in good season; but sometimes it so happens, from the badness of the season, that in many places the Wheat is not sown till Christmas or after; but this late-sown Wheat is subject to run too much to straw, especially if the spring should prove moist.

The usual allowance of Seed-wheat to one acre of land, is three bushels; but from repeated experiments it has been found, that half that quantity is sufficient; therefore, if the farmers have regard to their own interest, they should save this expence of seed, which amounts to a considerable article in large farms, especially when it is to be purchased, which most of the skilful farmers do, at least every other year, by way of change; for they find that the seeds continued long upon the same land, will not succeed so well as when they procure a change of seeds from a distant country. And the same is practised by the husbandmen of the Low-Countries, who commonly procure fresh seeds from Sicily every second or third year, which they find succeed better with them than the seeds of their own country. In the choice of the seeds, particular regard should be had to the land upon which it grew, for if it is light land, the Wheat which grew upon strong land is the best, and so vice versa.

There have been some persons in England curious enough to procure their Seed-wheat from Sicily, which has succeeded very well; but the grain of this has proved too hard for our English mills to grind, which has occasioned their neglecting to procure their seeds from thence; nor do I think there can be much advantage in procuring the seeds from abroad, since the lands of England are so various as to afford as much change of seeds as will be necessary. And the less we purchase from abroad, the greater will be the saving to the public; so that it should be the business of skilful farmers to want as few seeds as possible, since, by exchange with each other, they may so contrive as not to part with ready money for any seeds.

The land which is usually allotted for Wheat, is laid fallow the summer before the Corn is sown; during which time it is ploughed two or three times, to bring it into a tilth; and the oftener and better the ground is ploughed, and the more it is laboured with harrows between each ploughing to break and divide the clods, the better will be the crop, and the fewer weeds will be produced. But in this article most of the farmers are deficient; for after they have given their lands one ploughing, they frequently leave it to produce weeds, which sometimes are permitted to stand until they shed their seeds, whereby the ground will be plentifully stocked with weeds; and as an excuse for this, they say that these weeds will supply their sheep with some feed, and the dung of the sheep will mend their land; but this is a very bad piece of husbandry, for the weeds will draw from the land more than the dung of the sheep will supply; so that it is undoubtedly the best method to keep the ground as clean from weeds as possible, and to stir it often to separate and break the clods, and render the land fine; and where the land can enjoy a winter's fallow, it will be of much greater service to it than the summer; by

thus

thus labouring of the land, it will be of equal ſervice to it as a moderate dreſſing of dung. Therefore if the farmers could be prevailed on to alter their method of huſbandry, they would find their advantage in it; for the expence of dreſſing in ſome countries is ſo great, as to take away the whole profit of the crop.

There is alſo a very abſurd method in common practice with the farmers, which is the carrying of their dreſſing, and ſpreading it on the land in the ſummer, where it lies expoſed till the ſun has dried out all the goodneſs of it, before it is ploughed into the ground, ſo that the dreſſing is of little value; therefore the dung ſhould never be laid on the land faſter than it can be ploughed in, for one load of dung ſo managed, is better than three in their uſual method.

As Wheat remains a longer time upon the ground than moſt other ſorts of Corn, it requires a greater ſtock of nouriſhment to lengthen and fill the ears: therefore, if the dreſſing is exhauſted in winter, the Corn will have but ſhort ears, and thoſe but lean, nor will the grain afford much flour; ſo that it frequently happens, that a light dreſſing of ſoot in the ſpring, at the time the Wheat is beginning to ſtalk, proves of greater ſervice to the crop than a dreſſing of dung laid on the land before it is ploughed, eſpecially if the dung is not very good. Deep ploughing (where the ſtaple of the ground will admit of it) is alſo of great ſervice to the Corn; for the ſmall fibres of the roots, which are the mouths that ſupply the nouriſhment, extend themſelves very deep into the ground. I have traced many of them upward of three feet, and believe they ſpread much farther where the ground is light; therefore it is of great advantage to the crop, to have the ground ſtirred and looſened to a proper depth; for by ſo doing the roots will find a ſupply of paſture for the nouriſhment and augmentation of the ears, at the time they are forming, when it is moſt required; for if the ground is ploughed ſhallow, the roots will have extended themſelves to that depth by the ſpring; ſo that when the nouriſhment is wanted to ſupply the ſtalks, the roots are ſtinted by the hardneſs of the ſoil, which they cannot penetrate; when this is the caſe, the colour of the blade is frequently ſeen to change in April, and ſeldom recovers its verdure again; and when this happens, the ſtalks are always weakened in proportion to the decay of the blade; for it is well known from long experience, that the leaves or blade of Corn are neceſſary to draw in nouriſhment from the air and dews, for the increaſe of the ſtalk and ear; but in order to aſcertain this, I have made trial of it, by cutting off the leaves of ſome roots of Wheat alternately, early in the ſpring, and have conſtantly found the ſtalks upon thoſe roots much ſmaller, the ears ſhorter, and the grain thinner than thoſe of the intermediate roots, whoſe blades were not cut. This ſhews the abſurdity of that practice of feeding ſheep upon Corn in the winter and ſpring. I have frequently ſeen in ſome gardens, plants diveſted of their lower leaves, which ignorant perſons have ſuppoſed to draw away the nouriſhment from the head; but wherever this has been practiſed, I have always obſerved, that in proportion to the number of leaves cut off, the plants have been weakened by it, ſo that until thoſe leaves decay naturally, they ſhould never be taken off.

Of late years, many compoſts have been advertiſed for the ſteeping of the ſeeds of Corn, in order to improve their growth; ſome of theſe have been ſold at a dear rate; but as ſo great ſucceſs was aſſured by the inventors, to thoſe who ſhould make uſe of them, there were numbers of perſons who made the trial; but ſo far as I have been able to get information of their experiments, they did not ſucceed ſo well as to encourage the uſe of theſe compoſitions; and from ſeveral trials which I made myſelf with great care, I always found, that the Wheat which had been ſteeped in theſe compoſitions came up ſooner, and grew much ranker in the winter than that which had not been ſteeped; but in the ſpring the unſteeped Wheat had a greater number of ſtalks to each plant, and the ears were better fed than thoſe which had been ſteeped; therefore theſe ſorts of compoſts have been found of no real uſe to the crop.

My experiments were made in the following manner. The Wheat was ſown in drills, on the ſame ſpot of ground; the ſeeds which had been ſteeped were ſown in alternate rows, and the intermediate rows were ſown with unſteeped Corn. The rows were a foot and a half aſunder, and the grains were all taken out of one meaſure, and ſown as equally as poſſible: the ſteeped Corn appeared above ground three days before the other, and continued to grow faſter than the unſteeped Corn, during the winter; but in the ſpring, the blade of the ſteeped Corn changed its colour, and their points became brown; then I gave a light dreſſing to one of the rows, which ſoon recovered its verdure, and cauſed it to be the ſtrongeſt row of the whole; but the other which had not this dreſſing, produced weaker ſtalks and ears than that which was not ſteeped.

I have before obſerved, that in general the farmers ſow more than double the quantity of Corn on their lands than is neceſſary; therefore there is a great waſte of grain, which in ſcarce years amounts to a conſiderable ſum in large farms, and to a whole country it is an object worthy the attention of the public; but I fear whatever may be ſaid to prevent this, will have but little weight with the practitioners of agriculture, who are ſo fond of old cuſtoms, as rarely to be prevailed upon to alter them, though they are extremely abſurd. But if theſe people could be prevailed on to make the trial with care, they muſt be ſoon convinced of their error; for if they will but examine a field of Corn ſown in the common way, they will find but few roots which have more than two or three ſtalks, unleſs by chance, where there may be ſome few roots which have room to ſpread, upon which there may be ſix, eight, or ten ſtalks, and frequently many more; but I have ſeen a field of Wheat which had not a greater allowance than one buſhel of Corn to an acre, ſo that the roots had room to ſpread, produce from ſix to twelve or fourteen ſtalks, which were ſtrong, and had long well-nouriſhed ears; and the produce of flour was much greater than in any of thoſe fields in the neighbourhood, which were ſown with the common allowance. Where the land is good, and the roots ſtand at a proper diſtance from each other, there will be few roots which will not produce as many ſtalks as I have here mentioned, and the ears will be better nouriſhed.

But if the land is not covered with the blades of Corn by the ſpring, the farmers think they ſhall have no crop; whereas, if they would have patience to wait till the roots put out their ſtems, they will ſoon be convinced of the contrary, eſpecially if they could be prevailed on to draw a weighty roller over the Wheat in March, which will cauſe it to ſpread; and by ſettling of the looſe ground to the roots, the drying winds in the ſpring would be prevented from penetrating to their fibres, ſo will produce the more ſtalks; but before this operation, it will be proper to have the Corn cleaned from weeds, for if theſe are permitted to grow, they will draw away much nouriſhment from it; and if, at this ſeaſon, the land is made clean from weeds, the Corn will ſoon after ſpread, and cover the ground, whereby the growth of weeds will be greatly leſſened.

There is not any part of huſbandry which requires the farmer's attention more than that of keeping his land clean from weeds; and yet there are few who trouble themſelves about it, or who underſtand the proper method of doing it: few of them know thoſe weeds which are annual, ſo as to diſtinguiſh them from thoſe which are perennial; and without this knowledge, it will be much more difficult for a perſon to clean his land, let his induſtry be ever ſo great, for annual weeds may be ſoon deſtroyed, if taken in time; whereas, if they are neglected, their ſeeds will ſoon ripen and ſcatter; after which it will require three times the labour and expence to get rid of them, as would have been ſufficient at the beginning; and then the crop would have had no bad neighbours to rob it of its nouriſhment. The common method now practiſed is a very abſurd one, for the weeds are left to grow till the Wheat is beginning to ear, by which time many of the weeds are in flower, and ſome will have ripened ſeeds. Beſide the ground being covered by the Corn, all the low weeds are hid, and theſe are left to ripen and ſcatter their ſeeds; the tall weeds only are taken out, and if the people employed are not careful, many of theſe will eſcape them, as they will be ſo intermixed with the ſtalks of Wheat, as not to appear, unleſs diligently ſought after. By this method the weeds are permitted to ſtand, and rob the Corn of its nouriſhment, during the principal time of its growth, and the humble weeds are never deſtroyed; and by going amongſt the ſtalks when they are tall, great numbers of them are broken and trod under the work-peoples feet; yet however obvious this is to every farmer, none of them have thought of altering this practice. I would therefore recommend a method which is now in common practice among the kitchen-gardeners, which has been found of great benefit to their crops, and has alſo been a great ſaving to them in the expence of weeding, which is making uſe of the ſmall kind of hoes for cleaning the Wheat early in the ſpring, before the ground is covered with the blades of Corn. With this inſtrument, all the low as well as the tall weeds will be cut up, and if it is performed in dry weather, the weeds being then ſmall will ſoon die. Where the ground happens to be very full of weeds, it may be neceſſary to go over it a ſecond time, at about three weeks after the firſt, to cut up any weeds which may have before eſcaped. By laying the ground clean at this time, the Corn will not be robbed of its nouriſhment; and there will not be time for the weeds to grow ſo as to prejudice it much after, for the ground will be ſo much ſhaded by the Corn, as to keep down the weeds, ſo that they cannot have time to ripen their ſeeds before harveſt.

If, at the time of this operation, the roots of Corn are cut up where they are too cloſe, it will be found of great ſervice to the other; but this, I fear, few of the old farmers will ever agree with me in, though what I mention is not from theory but experiments, which have been repeated with great care; and where it was practiſed, the produce of twenty rods of ground was much greater both in weight and meaſure than the ſame quantity of ground in the beſt part of the field, where this was not practiſed, and the ſtalks ſtood upright, when a great part of the Corn in the ſame field was lodged.

I have often obſerved in thoſe fields where foot-paths are made through the Corn, that by the ſide of thoſe paths where the Corn is thin, and has been trodden down in the winter and ſpring, the ſtalks have ſtood erect, when moſt of the Corn in the ſame field has been laid flat on the ground, which was owing to the ſtalks being ſo much ſtronger from their having more room, the other having been drawn up tall and ſlender by being ſo cloſe. There is alſo another great advantage in keeping Corn clean from weeds, and giving it room to ſpread, which is, that the Corn is not ſo liable to take the ſmut as when it is full of weeds, and the roots too much crowded, this I have frequently obſerved; ſo that cleanneſs and free air is as eſſential to the growth of vegetables as animals, and the changing of the ſeed annually is alſo as neceſſary as the change of air is to all ſorts of animals; for where this has been carefully practiſed, there has rarely happened any ſmutty Corn in the field.

Brining of the Seed-wheat is what the farmers generally practiſe to prevent the ſmut, which in moſt years anſwers very well; but there is nothing which contributes more to this than keeping the plants in good health, which is better effected by the method before propoſed, for by ſtirring of the ground with the hoe between the roots of Corn in the ſpring, they will be better ſupplied with nouriſhment; for in ſtrong lands, where the water may have lain in the winter, the ſurface of the ground will bind ſo hard on the firſt dry weather, as to ſtint the Corn, and frequently cauſe it to change colour. When this happens, the roots ſeldom put out many ſtalks, and thoſe which are put out are weak; but where the ſurface of the ground can be ſtirred to looſen the parts, the Corn will ſoon recover its colour and ſtrength, and cover the land with ſhoots.

What has been here directed, muſt be underſtood to relate to Wheat ſown in broad-caſt, which is the uſual method practiſed by farmers in every part of England; for the horſe-hoeing huſbandry, which was practiſed by Mr. Tull, has been almoſt univerſally rejected by the farmers in every country, it being ſo oppoſite to their accuſtomed practice, that few of them can be prevailed upon to make trial of it; and indeed, by the abſurdity of the author in a few particulars, he has diſcouraged many from engaging in it, who would otherwiſe have practiſed it; but upon finding Mr. Tull poſitively aſſerting, that the ſame land would nouriſh the ſame ſpecies of plants without changing the crops for ever, and this without manure, which being contrary to all experience, led them to believe his other principles had no better foundation. And he practiſed this method of ſowing the ſame ſpecies upon the ſame ground till his crops failed, and were much worſe than thoſe of his neighbours, who continued their old method of huſbandry; and hereby his horſe hoeing huſbandry was ridiculed by them, and laid aſide by gentlemen who were engaging in it. But notwithſtanding thoſe and ſome other particulars which have been advanced by Mr. Tull, yet it is much to be wiſhed that this new huſbandry might be univerſally practiſed; for ſome few perſons who have made ſufficient trial of it, have found their crops anſwer much better than in the common or old method of huſbandry; and the French, who have learned it from Mr. Tull's book, are engaging in the practice of it with greater ardour than thoſe of our own country; and although they had not the proper inſtruments of agriculture for the performance, and met with as ſtrong oppoſition from the perſons employed to execute the buſineſs, as in England, yet the gentlemen ſeem determined to perſiſt in the practice of it; though, as yet, few of their experiments have had the ſucceſs they hoped for, partly from the aukwardneſs of their labourers, and partly from their averſeneſs to practiſe this huſbandry, and alſo from their being made in lands not well conditioned; but yet their produce has been equal to that of the old huſbandry, and they ſay, that if the produce of the land in the new method of huſbandry, does not exceed that in the old way, yet by ſaving ſeven parts from eight of the Seed-corn, it is a great affair to a whole country, eſpecially in times of ſcarcity.

As Mr. Tull has given full directions for the practice of this huſbandry, I ſhall refer the reader to his book for inſtruction, and ſhall only mention two or three late experiments which have been made in his method, whereby the utility of it will more fully appear.

The firſt was in a field of Wheat, which was ſown partly in broad-caſt in the common method, and partly according to Tull's method; the ſpots thus ſown were not regular in lands, but interſperſed indifferently in many directions.

Thoſe

Thofe parts of the field in Tull's method, were in rows at two feet diftance, and ftood thin in the rows. The roots of the Wheat in thefe fpots had from ten to thirty ftalks upon a root, and continued upright till it was reaped; whereas few of the roots in the common method had more than two or three ftalks, and thefe were moft of them lodged before harveft; fo that upon trial of the grain when threfhed, there was near a third part more in weight and meafure, than from the fame extent of ground, taken in the beft part of the field fown in the common way.

Another trial was made in fowing of the Corn in rows at different diftances, with fome fown in two parts of the ground broad-caft. The event was, that all which was fown broad-caft in the ufual way was lodged, as was alfo moft of that where the rows were fix or nine inches afunder; thofe which ftood a foot diftance efcaped better, but the rows two feet afunder were the beft, and the produce much greater than any of the other; which plainly fhews the abfurdity of that practice, in fowing a great quantity of feeds, to have a better produce, which is the opinion of moft of the old farmers; and it was formerly the prevailing opinion among gardeners, who allowed near eight times the quantity of feeds for the fame fpace of ground, as is now ufually fown, and thefe crops are greatly fuperior to any of thofe.

The produce of an acre of Wheat is various, according to the goodnefs of the foil. In fome of the fhallow, chalky, down lands, where there have been near four bufhels of Corn fown, I have known the produce not more than double of the feed; but when this is the cafe, the farmer had much better let his land lie wafte, fince the produce will not defray the expence, fo that more than the rent of the land is loft: and although thefe fort of crops are frequently feen on fuch land, yet fuch is the paffion for ploughing among the hufbandmen at prefent, that if they were not reftrained by their landlords, they would introduce the plough into every field, notwithftanding they are fure to lofe by it.

But although the produce of thefe poor downs is fo fmall, as before related, yet upon good land, where the Corn has ftood thin upon the ground, I have known eight and ten quarters reaped from an acre over the whole field, and fometimes much more. And I have been informed by perfons of great credit, that on good land, which was drilled and managed with the horfe-hoe, they have had twelve quarters from an acre of land, which is a great produce; and this is with greater certainty, if the feafon proves bad, than can be expected by the common hufbandry.

The price of Corn varies continually, and this variation is often very great in the fpace of one or two years; fo that from being fo cheap, as that the farmers could not pay their rents in the compafs of a year or two, the price has been more than doubled; for one or two plentiful harvefts have lowered the price of Wheat fo much, as to make it difficult for the needy farmer to go on with his bufinefs, who wants ready money for his crops as foon as he can prepare them for the market. This has eftablifhed a fet of people called dealers in Corn, who have taken the advantage of the farmer's neceffity, and engroffed their Corn to keep it for better markets; and thefe dealers have of late years increafed greatly in their numbers, to the great prejudice of the raifers and alfo the confumers of Corn, which may in time prove fatal to the country, by monopolizing the greateft part of the produce, and then fet their own price upon it; fo that between thefe Corn-factors, as they are called, and the diftillers, the price of bread may be too great for the labouring poor, which is an affair that requires more public attention than has yet been given to it.

The French have been, and are building public granaries for the confervation of their Corn, in moft of their provinces; for as in fome years they have great plenty of Corn, and at other times as great fcarcity, they are contriving to prevent any great want of it.

When the Wheat is fold much under four fhillings the bufhel, the farmer cannot pay his rent, and live; nor can the poorer fort of people afford to purchafe good bread, when the Wheat is fold at a price much higher than fix fhillings the bufhel; therefore when it is at a medium between thefe, there can be no great caufe of complaint on either fide.

TRIUMFETTA. Plum. Gen. Nov. 40. tab. 8.

The Characters are,

The flower has no empalement, but it has five linear, erect, obtufe petals, which turn inward; it has ten awl-fhaped rifing ftamina the length of the petals, terminated by fingle fummits, and a roundifh germen, fupporting a ftyle the length of the ftamina, crowned by an acute bifid ftigma. The germen afterward becomes a globular capfule, fet with long prickles on every fide, having four cells, each containing one feed, which is convex on one fide, and angular on the other.

We have but one Species of this genus, viz.

Triumfetta (*Lappula.*) Hort. Cliff. 210. *Triumfetta with prickly branching fruit.*

This plant grows naturally in Jamaica and moft of the other iflands in the Weft-Indies; it rifes with an upright ftem to the height of fix or feven feet, which becomes ligneous toward the bottom, dividing upward into four or five branches, garnifhed with leaves, placed alternately, divided almoft into three lobes toward the top, ending in acute points; they are covered with a foft brown down on their under fide, but their upper is of a yellowifh green; their borders are acutely but unequally fawed, ftanding upon foot-ftalks an inch long. The branches are terminated by long fpikes of flowers, and from the fide of the ftalk come out feveral fmall clufters. The flowers are fmall, the petals narrow, of a yellow colour; thefe are fucceeded by burry capfules, fomething like thofe of the Agrimony, but rounder; the prickles are longer than thofe, and are placed on every fide. This plant generally flowers here in July and Auguft, and in warm feafons the feeds do fometimes ripen.

This fort is propagated by feeds, which muft be fown on a hot-bed early in the fpring; and when the plants come up, and have four or five leaves, they fhould be each tranfplanted into a feparate pot, and plunged into a moderate hot-bed of tanners bark, fhading them from the fun until they have taken new root; then they muft be treated in the fame manner as hath been directed for other tender exotic plants. During the fummer feafon the plants may remain in this hot-bed, but in autumn they muft be removed into the ftove, and plunged into the bark-bed. If the plants live through the winter, they will flower the following fummer, fo will ripen their feeds in autumn; but they may be continued two or three years, provided they are carefully managed.

TROLLIUS. Lin. Gen. Plant. 620. Globe Ranunculus, or Locker Gowlans.

The Characters are,

The flower has no empalement; it has about fourteen oval petals, whofe points meet together, with nine nectariums, which are narrow, plain, incurved, and umbilicated, perforated at their bafe, and a great number of briftly ftamina, terminated by erect fummits, with numerous germina fitting clofe like a column, having no ftyles, but are crowned by pointed ftigmas. The germen afterward become fo many capfules, collected into an oval head, each containing one feed.

The Species are,

1. Trollius (*Europæus*) corollis conniventibus, nectariis longitudine ftaminum. Lin. Sp. Plant. 556. *Trollius*

with

with the petals of the flower meeting, and nectarii the length of the ſtamina; commonly called Globe Flower, or Locker Gowlans.

2. Trollius (*Aſiaticus*) corollis patentibus nectariis longitudine petalorum. Lin. Sp. Pl. 557. *Trollius with an open ſpreading flower, and nectariums the length of the petals.*

The firſt ſort grows naturally in the northern counties in England, and in many parts of Wales. I found it in great plenty growing in the park of Burrow-hall in Lancaſhire; it has a perennial, fibrous, black root, from which ſpring up many leaves, which reſemble thoſe of Wolfſbane, cut into five ſegments almoſt to the bottom; the ſtalk riſes near two feet high; it is ſmooth, hollow, and branches toward the top; each branch is terminated by one large yellow flower, ſhaped like thoſe of Crow-foot, which has no empalement, compoſed of ſeveral concave petals, whoſe points turn inward toward each other, covering the parts of generation, ſo are of a globular form; whence it had the title of Globe Ranunculus. It flowers the latter end of May and the beginning of June, and the ſeeds ripen in Auguſt. This plant is frequently kept in gardens about London, and is eaſily propagated by parting of the roots; the beſt time for doing this is the latter end of September, when the leaves are beginning to decay. The roots ſhould not be divided into ſmall parts, if they are expected to flower ſtrong the following year, and ſhould be planted at a foot diſtance from each other; they require a ſhady ſituation and a moiſt ſoil. The roots need not be removed or parted oftener than once in three years, unleſs there is a deſire of increaſing them.

The ſecond ſort grows naturally in Siberia, from whence it was brought to the imperial garden at Peterſburgh, and has been communicated ſince to ſeveral parts of Europe; this differs from the firſt in having larger leaves, which are of a lighter green colour; their ſegments are fewer and larger, reſembling thoſe of the yellow Monks-hood. The petals of the flower ſpread open, and do not converge at their points like thoſe of the firſt ſort. The flowers, ſtamina, and nectariums are of an elegant Saffron colour. It flowers in May.

This ſort may be propagated and treated in the ſame way as the firſt, but it requires a moiſter ſoil, and ſhould have a ſhady ſituation, but not under the drip of trees; it thrives beſt on a north border where the ſoil is loamy, but not too ſtiff. In ſuch ſituations the plants will produce ſeeds in England, but if they are in a dry ſoil, or much expoſed to the ſun, they frequently die in ſummer. I have ſeen this ſort in the moſt flouriſhing ſtate, where the ſurface of the ground was covered with Moſs to keep it moiſt.

As the flowers of both theſe plants make a pretty appearance during their continuance, they deſerve a place in every good garden for the ſake of variety, eſpecially as they will thrive in moiſt ſhady places where few better plants will live; and by thus ſuiting the plants to the different ſoils and ſituations of a garden, every part may be furniſhed with beauties, and a great variety may be preſerved.

TROPÆOLUM. Lin. Gen. Plant. 421. Indian Creſs.

The Characters are,

The empalement of the flower is of one leaf, ending in five points; it is erect, ſpreading, and falls off. The two under ſegments are narrow; their tail ends in a nectarious horn, which is longer than the empalement. The flower has five roundiſh petals inſerted in the ſegments of the empalement; the two upper ſit cloſe to the foot-ſtalk, but the lower have oblong hairy tails. It has eight ſhort awl-ſhaped ſtamina, which decline, and are unequal, terminated by oblong riſing ſummits, having four cells, and a roundiſh germen with three lobes, which are ſtreaked, ſupporting a ſingle erect ſtyle, crowned by an acute trifid ſtigma. The germen afterward becomes a ſolid fruit in three parts, convex on the outſide, angular within, having many furrows, each part or cell including one furrowed ſeed, convex on one ſide, and angular on the other.

The Species are,

1. Tropæolum (*Minus*) foliis ſubquinquelobis, petalis obtuſis. Hort. Upſal. 93. *Tropæolum with leaves which are almoſt divided into five lobes, and obtuſe petals to the flower; the common, or ſmall Indian Creſs.*

2. Tropæolum (*Majus*) foliis ſubquinque lobis petalis obtuſis. Hort. Upſal. 93. *Tropæolum with five lobed leaves, and acute-pointed briſtly petals to the flower; commonly called greater Indian Creſs.*

The firſt ſort grows naturally in Peru; this was firſt brought to Europe in 1684, and was raiſed in the gardens of Count Bevening in Holland.

It has a trailing herbaceous ſtalk, garniſhed with leaves almoſt circular. The foot-ſtalk is inſerted in the center of the leaf, like a buckler, as in the Navelwort; they are ſmooth, of a grayiſh colour; the flowers come out from the wings of the ſtalks, ſtanding upon very long ſlender foot-ſtalks, of an admirable ſtructure, compoſed of five acute-pointed petals; the two upper are large and rounded, the three under are narrow; their tails join together, and are lengthened into a tail two inches long. After the flower is paſt, the germen turns to a roundiſh fruit, which is furrowed, and divided into three lobes, each including one ſtreaked ſeed. It flowers from Midſummer till the froſt ſtops it in autumn.

There are two varieties of this, one with a deep Orange-coloured flower, inclining to red, and the other with a pale yellow flower.

The ſecond ſort grows naturally about Lima; this has larger ſtalks than the former. The leaves are alſo larger, and their borders are indented almoſt into lobes; the flowers are larger, and their petals are rounded at their points. There are two colours of this ſort as in the former, and one with double flowers, which is propagated by cuttings, for it does not produce ſeeds.

The firſt ſort is leſs common at preſent in the Engliſh gardens than the ſecond, though it was formerly more ſo; the flowers of the latter being larger make a finer appearance, for which it is preferred; they are both eſteemed annual plants, though they may be continued through the winter if they are kept in pots, and ſheltered in a good green-houſe, in like manner as that with double flowers is preſerved, ſo may be propagated by cuttings as that is; but, as theſe ripen their ſeeds conſtantly every year, the plants are generally raiſed from them; theſe may be ſown in April in the places where they are to remain, which ſhould be where their ſtalks may have ſupport, for they will climb ſix or eight feet high, when they are trained up, and then their flowers will make a good appearance; but when they trail upon the ground, they will ſpread over the neighbouring plants, and become unſightly.

The flowers of theſe plants are frequently eaten in ſallads; they have a warm taſte like the Garden Creſs, and are eſteemed very wholeſome; they are likewiſe uſed for garniſhing diſhes. The ſeeds are pickled, and by ſome are preferred to moſt kinds of pickles for ſauce.

TUBEROSE. See Polyanthes.

TULIPA. Tourn. Inſt. R. H. 373. tab. 199, 200. Lin. Gen. Plant. 376. Tulip.

The Characters are,

The flower has no empalement; it is of the bell-ſhape, compoſed of ſix oblong, oval, concave, erect petals; it has ſix awl-ſhaped ſtamina, which are ſhorter than the petals, terminated by oblong four-cornered ſummits, and a large, oblong, taper, three-cornered germen, having no ſtyle, crowned by a triangular, three-lobed, permanent ſtigma. The germen afterward turns to a three-cornered capſule, having three cells, which are filled

filled with compressed seeds, lying over each other in a double order.

The Species are,

1. Tulipa (*Sylvestris*) flore subnutante, foliis lanceolatis. Lin. Sp. Pl. 305. *Tulip with a nodding flower, and spear-shaped leaves; or the smaller, yellow, Italian Tulip.*

2. Tulipa (*Gesneriana*) flore erecto, foliis ovato-lanceolatis. Lin. Sp. Pl. 306. *Tulip with an erect flower, and oval spear-shaped leaves. This is the common Tulip with all its varieties.*

The first sort was formerly preserved in the English gardens, but since there has been so many varieties of the second sort propagated in England, the first has been rejected, and is now only to be found in old neglected gardens. The petals of this flower end in acute points; the flower is yellow, and nods on one side, and the leaves are narrower than those of the common sort.

The common Tulip is so well known as to need no description; and it would be to little purpose to enumerate the several varieties of these flowers, which may be seen in one good garden, since there is no end of their numbers; and what some people may value at a considerable rate, others reject; beside there are annually a great variety of new flowers obtained from breeders, so those which are old, if they have not very good properties to recommend them, are thrown out and despised. I shall therefore point out the properties of a good Tulip, according to the characteristics of the best florists of the present age. 1. It should have a tall strong stem. 2. The flower should consist of six leaves, three within, and three without; the former ought to be larger than the latter. 3. Their bottom should be proportioned to their top, and their upper part should be rounded off, and not terminate in a point. 4. These leaves, when opened, should neither turn inward, nor bend outward, but rather stand erect, and the flower should be of a middling size, neither over large, nor too small. 5. The stripes should be small and regular, arising from the bottom of the flower, for if there are any remains of the former self-coloured bottom, the flower is in danger of losing its stripes again. The chives should not be yellow, but of a brown colour. When a flower has all these properties, it is esteemed a good one.

Tulips are generally divided into three classes, according to their seasons of flowering; as Præcoces, or early blowers, Medias, or middling blowers, and Serotines, or late blowers; but there is no occasion for making any more distinctions than two, viz. early and late flowers.

The early-blowing Tulips are not near so fair, nor rise half so high, as the late ones, but are chiefly valued for appearing so early in the spring; some of which will flower the middle of March in mild seasons, if planted in a warm border near a wall, pale, hedge, or other shelter, and a month after the others will succeed them; so that they keep flowering until the general season for the late flowers to blow, which is toward the end of April.

The roots of the early-blowing Tulips should be planted the beginning of September in a warm border, near a wall, pale, or hedge, because if they are put into an open spot of ground, their buds are in danger of suffering by morning frosts in the spring. The soil for these should be renewed every year, where people intend to have them fair. The best soil for this purpose is that which is taken from a light loamy pasture, with the turf rotted amongst it; and to this should be added a fourth part of sea sand. This mixture may be laid about eighteen inches deep, which will be sufficient, for these need not be planted more than four or five inches deep at most. The offsets should not be planted amongst the blowing roots, but in a border by themselves, where they may be planted pretty close together, especially if they are small; but these should be taken up when their leaves decay, in the same manner as the blowing roots, otherwise they would rot; for these are not so hardy as the late blowers, nor do they increase half so fast as those, so that a greater care is required to preserve the offsets of them.

When these Tulips come up in the spring, the earth upon the surface of the borders should be gently stirred and cleared from weeds; and as the buds appear, if the season should prove severe, it will be of great service to cover them with mats, for want of which many times they are blighted, and their flower-buds decay before they blow, which is often injurious to the roots, as is also the cropping of the flowers, so soon as they are blown, because their roots, which are formed new every year, are not at that time arrived to their full magnitude, and are hereby deprived of proper nourishment.

If, when these flowers are blown, the season should prove very warm, it will be proper to shade them with mats, &c. in the heat of the day; as also if the nights are frosty, they should be in like manner covered, whereby they may be preserved a long time in beauty; but, when their flowers are decayed, and their seed-vessels begin to swell, they should be broken off just at the top of the stalks, because if they are permitted to seed, it will injure the roots.

When the leaves of these flowers are decayed (which will be before the late blowers are out of flower,) their roots should be taken up, and spread upon mats in a shady place to dry; after which they should be cleared from their filth, and put up in a dry place, where vermin cannot come to them, until the season for planting them again, being very careful to preserve every sort separate, that you may know how to dispose of them at the time for planting them again, because it is the better way to plant all the roots of each sort together (and not to intermix them, as is commonly practised in most other kinds of flowers;) for as there are few of them which blow at the same time, so, when the several roots of one sort are scattered through a whole border, they make but an indifferent appearance; whereas, when twenty or thirty roots of the same sort are placed together, they will all flower at the same time, and have a better effect.

There are many curious persons, who, in order to preserve their several kinds of Tulips, and other bulbous-rooted flowers separate, have large flat boxes made, which are divided into several small partitions, each of which is numbered in the same manner as the divisions of their beds; so that when a catalogue of their roots is made, and the numbers fixed to each sort in the beds, there is nothing more to do when the roots are taken up, but to put every kind into the division marked with the same number, which was placed to each sort in the bed, which saves a great deal of trouble, and effectually answers the purpose of preserving the kinds separate.

The late blowing Tulips are so numerous, that, as I before observed, it would be to no purpose to attempt to give a catalogue of them. These are generally obtained from breeders, which is a term applied to all such flowers as are produced from seeds, which are of one self-colour, and have good bottoms and chives; these in time break into various beautiful stripes, according to the ground of their former self-colour; but this must be entirely thrown off, otherwise they do not esteem a flower well broken.

Of these breeders there hath been a great variety brought into England from Flanders of late years, which is the grand nursery for most sorts of bulbous-rooted flowers; but there are some curious persons, who have lately obtained many valuable breeders from seed in England; and doubtless, were we as industrious to sow the seeds of these flowers as the people of Holland and Flanders, we might in a few years

years have as great variety as is to be found in any part of Europe; for although it is six or seven years from the sowing before these flowers blow, yet, if after the first sowing there is every year a fresh parcel sown, when the seven years are expired, there will be constantly a succession of roots to flower every year, which will reward the expectation, and keep up the spirit of raising; but it is the length of time at first, which deters most people from this work.

The manner of propagating these flowers from seeds is as follows: You should be careful in the choice of the seed, without which there can be little success expected. The best seed is that which is saved from breeders which have all the good properties before related, for the seeds of striped flowers seldom produce any thing that is valuable.

The best method to obtain good seeds is to make choice of a parcel of such breeding Tulip-roots as you would save seeds from, and plant them in a separate bed from the other breeders, in a part of the garden where they may be fully exposed to the sun, observing to plant them at least eight or nine inches deep; for if they are planted too shallow, their stems are apt to decay before their seed is perfected.

These flowers should always be exposed to the weather, for if they are shaded with mats, or any other covering, it will prevent their perfecting the seed. About the middle of July (a little sooner or later, as the summer is hotter or colder) the seeds will be fit to gather, which may be known by the driness of their stalks, and the opening of the seed-vessels; at which time it may be cut off, and preserved in the pods till the season for sowing it, being careful to put it up in a dry place, otherwise it will be subject to mould, which will render it good for little.

Having saved a parcel of good seed, about the beginning of September is the best season for sowing it, when there should be provided a parcel of shallow seed-pans or boxes, six or eight inches deep, which should have holes in their bottoms to let the moisture pass off; these must be filled with fresh light earth, laying the surface very even, upon which the seeds should be sown as regularly as possible, that they may not lie upon each other; then there should be some of the same light earth sifted over them, about half an inch thick. These boxes or pans should be placed where they may have the morning sun till eleven of the clock, in which situation they may remain until the middle of October, at which time they should be removed into a more open situation, where they may enjoy the benefit of the sun all the day, and be sheltered from the north winds, where they should remain until winter, when they must be placed on a south border, to screen them from frost; but in the spring, when the plants are up, they should be again removed to their first situation; and if the season should be dry, they must be refreshed with water, while the plants remain green, but as soon as their tops begin to decay, there must be no more given them, lest it rot their tender bulbs: therefore the boxes should be placed in a shady situation during the summer season, but not under the drip of trees.

These plants, at their first appearance, have very narrow grassy leaves very like those of Onions, and come up with bending heads, in the same manner as they do; so that persons, who are unacquainted with them, may pull them up instead of Grass, whilst they are very young, before their leaves are a little more expanded; which is not performed the first year, for they seldom appear before the middle of March, and they commonly decay about the latter end of May, or the beginning of June, according as the season is hotter or colder.

The weeds and Moss should also be cleared off from the surface of the earth in the boxes, and a little fresh earth sifted over them soon after their leaves decay, which will be of great service to the roots. These boxes should be constantly kept clear from weeds, which, if permitted to grow therein, when they are pulled up, their roots will be apt to draw the bulbs out of the ground. At Michaelmas they should be fresh earthed again, and as the winter comes on, they must be again removed into the sun as before, and treated in the same manner, until the leaves decay, when the bulbs should be carefully taken up, and put in a cool shady room till the end of August, when they should be planted in beds of fresh sandy earth, which should have tiles laid under them, to prevent the roots from shooting downward, which they often do when there is nothing to stop them, and thereby they are destroyed. The earth of these beds should be about five inches thick upon the tiles, which will be sufficient for nourishing these roots while they are young.

The distance which these young bulbs should be allowed, need not be more than two inches, nor should they be planted above two inches deep; but toward the end of October, it will be proper to cover the bed over with a little tanners bark about two inches deep, which will preserve the roots from the frost, and prevent Moss or weeds from growing over them; but, if the winter should be very severe, it will be proper to cover the bed either with mats or Pease-haulm, to prevent the frost from entering the ground, because these roots are much tenderer while young, than they are after they have acquired strength.

In the spring the surface of the ground should be gently stirred to make it clean, before the plants come up; and if the spring should prove dry, they must be frequently refreshed with water, during the time of their growth; but this must not be given to them in great quantities, lest it rot their tender bulbs; and when their leaves are decayed, the roots should be taken up, and treated in the same way as before.

When the bulbs are large enough to blow, they should be planted in fresh beds at the distance, and in the same manner as old roots, where, when they flower, such of them as are worthy to be preserved should be marked with sticks; and at the season for taking up the bulbs, they must be separated from the others, in order to be planted as breeders in different beds; but you should by no means throw out the rest until they have flowered two or three years, because it is impossible to judge exactly of their value in less time; for many, which at first flowering appear beautiful, will afterwards degenerate so as to be of little value; and others, which did not please at first, will many times improve, so that they should be preserved until their worth can be well judged of.

Having thus given an account of the method of raising these flowers from seeds, I shall now proceed to the management of the roots which are termed breeders, so as to have some of them every year break out into fine stripes.

There are some who pretend to have a secret how to make any sort of breeders break into stripes whenever they please; but this, I dare say, is without foundation; for from many experiments which I and others have made of this kind, I never could find any certainty in this. All that can be done by art is, to shift the roots every year into fresh earth of different mixtures and to different situations, by which method I have had very good success.

The earth of these beds should be every year different, for although it is generally agreed that lean, hungry, fresh earth does hasten their breaking, and cause their stripes to be the finer and more beautiful, yet, if they are every year planted in the like soil, it will not have so much effect upon them, as if they were one year planted in one sort of earth, and the next year in a very different one, as I have several times experienced; and if some fine striped Tulips are planted in the same beds with the breeders intermixing them

them together, it will alſo cauſe the breeders to break the ſooner.

The beſt compoſt for theſe roots is a third part of freſh earth from a good paſture, which ſhould have the ſward rotted with it, a third part of ſea ſand, and the other part ſifted lime rubbiſh; theſe ſhould be all mixed together ſix or eight months at leaſt before it is uſed, and ſhould be frequently turned to mix the parts well together. With this mixture the beds ſhould be made about two feet deep, after the following manner: After the old earth is taken from out of the bed to the depth intended, then ſome of the freſh earth ſhould be put in about eighteen inches thick; this ſhould be levelled exactly, and then lines drawn each way of the bed chequerwiſe, at ſix inches diſtance; upon the center of each croſs ſhould be placed the Tulip-roots, in an upright poſition; and after having finiſhed the bed in this manner, the earth muſt be filled in, ſo as to raiſe the bed ſix or eight inches higher, obſerving, in doing this, not to diſplace any of the roots, and alſo to lay the top of the beds a little rounding, to throw off the wet.

There are many perſons who are ſo careleſs in planting their Tulip-roots, as only to dig and level the beds well, and then with a blunt dibble to make holes, into which they put the roots, and then fill up the holes with a rake; but this is by no means a good method; for the dibble, in making the holes, preſſes the earth cloſely on each ſide, and at the bottom, whereby the moiſture is often detained ſo long about the roots as to rot them, eſpecially if the ſoil is inclinable to bind; beſides, the earth being hard at the bottom of the bulbs, they cannot ſo eaſily emit their fibres, which muſt certainly prejudice the roots.

Theſe beds ſhould be ſunk, more or leſs, below the ſurface, according to the moiſture or drineſs of the ground; for the roots ſhould be ſo elevated as never to have the water ſtand near the reach of their fibres in winter, for moiſture is very apt to rot them; ſo that where the ſoil is very wet, it will be proper to lay ſome lime rubbiſh under the earth, in order to drain off the wet, and the beds ſhould be entirely raiſed above the level of the ground; but to prevent their falling down into the walks, after froſt, or hard rains, it will be proper to raiſe the paths between them, either with ſea-coal aſhes or rubbiſh, eight or ten inches, which will ſupport the earth of the beds; and theſe paths may ſlope at each end from the middle, which will make paſſage for the water to run off as it falls. But where the ſoil is dry, the beds may be ſunk eighteen or twenty inches below the ſurface, for in ſuch places the beds need not be more than four or ſix inches above the ſurface, which will be allowance enough for their ſettling.

During the winter ſeaſon there will be no farther care required. The roots being planted thus deep, will be in no danger of ſuffering by ordinary froſts; but if the winter ſhould prove very ſevere, ſome rotten tan or Peaſe-haulm may be laid over the beds to keep out the froſt during its continuance, but this muſt be removed when the froſt is over; and in the ſpring, when their leaves begin to appear above ground, the earth upon the ſurface of the beds ſhould be ſtirred to clear it from weeds, Moſs, &c. and when the flower-buds begin to come up, they ſhould be guarded from froſt, otherwiſe they are very ſubject to blight and decay ſoon after they appear, if the froſt pinches their tops; but they need only be covered in ſuch nights when there is a proſpect of froſt, for at all other times they ſhould have as much open air as poſſible, without which they will draw up weak, and produce very ſmall flowers.

When theſe breeders are in flower, you ſhould carefully examine them to ſee if any of them have broken into beautiful ſtripes, which, if you obſerve, there ſhould be a ſtick put into the ground by every ſuch root, to mark them, that they may be ſeparated from the breeders, to plant amongſt the ſtriped flowers the following year; but you ſhould carefully obſerve, whether they have thrown off their former colour entirely, as alſo when they decay, to ſee if they continue beautiful to the laſt, and not appear ſmeared over with the original colour; in both which caſes they are very ſubject to go back to their old colour the next year; but if their ſtripes are diſtinct and clear to the bottom, and continue ſo to the laſt (which is what the floriſts call dyeing well,) there is no great danger of their returning back again, as hath been by ſome confidently reported; for if one of theſe flowers is quite broken (as it is termed,) it will never loſe its ſtripes, though ſometimes they will blow much fairer than at others, and the flowers of the offsets will be often more beautiful than thoſe of the old roots.

This alteration in the colour of theſe flowers may be ſeen long before they are blown, for the green leaves will appear of a fainter colour, and ſeem to be ſtriped with white, or of a browniſh colour, which is a plain proof, that the juices of the whole plant are altered, or, at leaſt, the veſſels through which the juice is ſtrained: ſo that hereby particles of a different figure are capable of paſſing through them, which, when entered into the petals of the flower, reflect the rays of light in a different manner, which occaſions the variety we ſee in the colours of flowers. This breaking of the colours in flowers proceeds from weakneſs, or at leaſt is the cauſe of weakneſs in plants; for it is obſervable, that after Tulips are broken into fine ſtripes, they never grow ſo tall as before, nor are the ſtems, leaves, or flowers ſo large; and it is the ſame in all other variegated plants and flowers whatever, which are alſo much tenderer than they were before they were ſtriped; ſo that many ſorts of exotic plants, which by accident have become variegated in their leaves, are often rendered ſo tender, as not to be preſerved without much more care, though indeed the ſtriping of Tulips doth never occaſion ſo great weakneſs in them as to render them very tender. The greateſt effect it hath on them, is in leſſening their growth; the more beautifully their ſtripes appear, the ſhorter will be their ſtems, and the weaker their flowers.

There is nothing more to be obſerved in the culture of ſtriped flowers than what has been directed for breeders, excepting that theſe ſhould be arched over with tall hoops and rails, that they may be ſhaded from the ſun in the day time, and protected from ſtrong winds, hard rains, and froſty mornings, otherwiſe the flowers will continue but a ſhort time in beauty; but where theſe inſtructions are duly followed, they may be preſerved in flower a full month, which is as long as moſt other flowers continue.

There are ſome perſons who are ſo extremely fond of theſe flowers, as to be at a great expence in erecting large frames of iron work to cover their beds of Tulips, in ſuch a manner, that they may walk between two beds under the frames, over which are ſpread tarpawlins, ſo as to keep off ſun, rain, and froſt, whereby they can view the flowers without being at the trouble of taking off or turning up the tarpawlins, or being incommoded by the ſun or rain, which cannot be avoided where the covering is low; beſides, by thus raiſing the covers, the flowers have a greater ſhare of air, ſo that they are not drawn ſo weak as they are when the covering is low and cloſe to them; but theſe frames being expenſive, can only be made by perſons of fortune; however, there may be ſome of wood contrived at a ſmaller expence, which, being arched over with hoops, may anſwer the purpoſe as well as the iron frames, though they are not ſo ſightly or laſting.

When the flowers are faded, the heads of all the fine ſorts ſhould be broken off, to prevent their ſeeding; for if this is not obſerved, they will not flower near ſo well the following

following year, nor will their ftripes continue fo perfect: this will alfo caufe their ftems to decay fooner than otherwife they would do, fo that their roots may be taken up in June; for they fhould not remain in the ground after their leaves are decayed. In taking the roots out of the ground, you muft be very careful not to bruife or cut them, which will endanger their rotting, and, if poffible, it fhould be done a day or two after rain. When thefe roots are taken out of the ground, they muft be cleared from their old covers, and all forts of filth, and fpread upon mats in a fhady place to dry; after which they fhould be put up in a dry place, where vermin cannot get to them, obferving to keep every fort feparated; but they fhould not be kept too clofe from the air, nor fuffered to lie in heaps together, left they fhould grow mouldy, for if any of the roots once take the mould, they commonly rot when they are planted again.

The offsets of thefe roots, which are not large enough to produce flowers the fucceeding year, fhould be alfo put by themfelves, keeping each fort diftinct; thefe fhould be planted about a month earlier in autumn than the blowing roots, in particular beds by themfelves in the flower-nurfery, where they may not be expofed to public view; but the earth of the beds fhould be prepared for them in the fame manner as for larger roots; thefe fhould not be planted above five inches deep, becaufe they are not ftrong enough to pufh through fo great covering of the earth as the old roots; they may alfo be placed much nearer together than thofe which are to flower, and in one year moft of them will become ftrong enough to flower, when they may be removed into the flower-garden, and placed in the beds amongft thofe of the fame kinds.

TULIPIFERA. Herm. Hort. Leyd. The Tulip-tree.

The CHARACTERS are,

The proper involucrum of the flower is compofed of two angular leaves, which fall off; the empalement is compofed of three oblong plain leaves like petals, which fall away. The flower is nearly of the bell-fhape, and has fix petals, which are obtufe and channelled at their bafe; the three outer fall off; it has a great number of narrow ftamina, which are inferted to the receptacle of the flower, having long narrow fummits faftened to their fide, and many germen difpofed in a cone, having no ftyle, crowned by a fingle globular ftigma. The germen afterward becomes fcaly feeds, lying over each other like the fcales of fifh, and form the refemblance of a cone.

We have but one SPECIES of this genus, viz.

TULIPIFERA (*Liriodendron.*) *The Virginian Tulip-tree.*

This is a native of North America, where it grows to be a tree of the firft magnitude, and is generally known through all the Englifh fettlements by the title of Poplar. Of late years there has been great numbers of thefe trees raifed from feeds in the Englifh gardens, fo that now they are become common in the nurferies about London, and there are many of the trees in feveral parts of England which do annually produce flowers. The firft tree of this kind which flowered here, was in the gardens of the late Earl of Peterborough at Parfons-Green near Fulham, which was planted in a wildernefs among other trees; before this was planted in the open air, the few plants which were then in the Englifh gardens, were kept in pots or tubs, and houfed in winter, fuppofing they were too tender to live in the open air; but this tree, foon after it was planted in the full ground, convinced the gardeners of their miftake, by the great progrefs it made, while thofe which were kept in pots and tubs increafed flowly in their growth; fo that afterward there were many others planted in the full ground, which are now arrived to a large fize, efpecially thofe which were planted in a moift foil. One of the handfomeft trees of this kind near London, is in the garden of Waltham-Abbey; and at Wilton, the feat of the Earl of Pembroke, there are fome trees of great bulk; but the old tree at Parfons-Green is quite deftroyed by the other trees which were fuffered to overhang it, and rob it of its nourifhment, from a fear of taking down the neighbouring trees, and admitting the cold air to the Tulip-tree it fhould injure it.

The young fhoots of this tree are covered with a fmooth purplifh bark, garnifhed with large leaves, whofe foot-ftalks are long; they are ranged alternate; the leaves are of a fingular form, being divided into three lobes; the middle lobe is blunt and hollowed at the point, appearing as if it had been cut with fciffars The two fide lobes are rounded, and end in blunt points. The upper furface is fmooth, and of a lucid green; the under is of a pale green. The flowers are produced at the end of the branches; they are compofed of fix petals, three without and three within, which form a fort of bell-fhaped flower, from whence the inhabitants of North America gave it the title of Tulip. Thefe petals are marked with green, yellow, and red fpots, fo make a fine appearance when the trees are well charged with flowers. The time of this tree flowering is in July, and when the flowers drop, the germen fwells and forms a kind of cone; but thefe do not ripen in England.

This tree is propagated by feeds, which are now annually imported in great plenty from America. Thefe fhould be fown as foon as they arrive, in pots or tubs, filled with light earth from the kitchen-garden, or in a bed in the full ground. Thofe which are fown in the firft way, may be placed on a very gentle hot-bed, which will forward their growth; fo that if they come up the fame feafon, the plants will acquire more ftrength before winter. When the plants appear, they muft be fhaded in the heat of the day from the fun, but frefh air muft be admitted daily to prevent their drawing up weak; and as the feafon advances, they muft be gradually hardened to bear the open air. While the plants are young, they do not care for much fun, fo they fhould be either fhaded, or placed where the morning fun only fhines upon them; they muft alfo be conftantly fupplied with water, but not have it in too great plenty. As the young plants commonly continue growing late in the fummer, fo when there happens early frofts in autumn, it often kills their tender tops, which occafions their dying down a confiderable length in winter; therefore they fhould be carefully guarded againft thefe firft frofts, which are always more hurtful to them than harder frofts afterward, when their fhoots are better hardened; however, the firft winter after the plants come up, it will be the better way to fhelter them in a common hot-bed frame, or to arch them over with hoops, and cover them with mats, expofing them always to the open air in mild weather.

The following fpring, juft before the plants begin to fhoot, they fhould be tranfplanted into nurfery-beds, in a fheltered fituation, where they are not too much expofed to the fun. The foil of thefe beds fhould be a foft gentle loam, not too ftiff, nor over light; this fhould be well wrought, and the clods well broken and made fine. There muft be great care taken not to break the roots of the plants in taking them up, for they are very tender; then they fhould be planted again as foon as poffible, for if their roots are long out of the ground, they will be much injured thereby. Thefe may be planted in rows at about a foot diftance, and at fix inches diftance in the rows; for as they fhould not remain long in thefe nurfery-beds, fo this will be room enough for them to grow; and by having them fo clofe, they may be fhaded in the fummer, or fheltered in the winter, with more eafe than when they are farther apart.

When the plants are thus planted, if the furface of the beds is covered with rotten tanners bark, or with Mofs, it will prevent the earth from drying too faft; fo that the plants will not require to be fo often watered, as they muft be where they are exposed to the sun and air; after this, the farther care will be to keep them clean from weeds, and if the latter part of summer should prove moist, it will occasion the plants growing late in autumn, so their tops will be tender and liable to be killed by the first frosts. In this case they should be covered with mats to protect them.

If the plants make great progress the first summer, they may be transplanted again the following spring; part of them may be planted in the places where they are to remain, and the other should be planted in a nursery where they may grow two years, to acquire strength before they are planted out for good; though the younger they are planted in the places where they are to stand, the larger they will grow, for the roots run out into length; and when they are cut, it greatly retards their growth, so that these trees should never be removed large, for they rarely succeed, if transplanted, when they are grown to a large size. Some trees I have seen removed pretty large, which have survived their removal; but young plants of two years old, which were planted near them, were much larger in fifteen years than the old ones.

When the seeds are sown upon a bed in the full ground, the bed should be arched over with hoops, and shaded in the heat of the day from the sun, and frequently refreshed with water; as should also the plants when they appear, for when they are exposed much to the sun while young, they make but small progress. The care of these in summer must be to keep them clean from weeds, supplying them duly with water, and shading them from the sun in hot weather; but as these seeds will not come up so soon as those which were placed on a hot-bed, they generally continue growing later in autumn, therefore will require shelter from the early frosts in autumn; for as the shoots of these will be much softer than those of the plants which had longer time to grow, so if the autumnal frosts should prove severe, they will be in danger of being killed down to the surface of the ground, by which the whole summer's growth will be lost, and sometimes the plants are entirely killed the first winter, if they are not protected.

As these plants will not have advanced so much in their growth as the other, they should remain in the seed-bed to have another year's growth before they are removed; therefore all that will be necessary to observe the second year, is to keep them clean from weeds; and now they will not be in so much danger of suffering from the warmth of the sun as before, therefore will not require such constant care to shade them, nor should the watering of them be continued longer than the spring; for if the autumn should prove dry, it will prevent the plants from shooting late, and harden those shoots which were made early in the year, whereby the plants will be in less danger from the early frosts.

After the plants have grown two years in the seed-bed, they will be strong enough to remove; therefore, in the spring, just at the time when their buds begin to swell, they should be carefully taken up, and transplanted into nursery-beds, and treated in the same way as has been before directed for the plants which were raised in pots.

There are some people who propagate this tree by layers, but the layers are commonly two or three years before they take root; and the plants so raised, seldom make such strait trees as those raised from seeds, though indeed they will produce flowers sooner, as is always the case with stinted plants.

This tree should be planted on a light loamy soil, not too dry, on which it will thrive much better than upon a strong clay, or a dry gravelly ground; for in America they are chiefly found upon a moist light soil, where they will grow to a prodigious size, though it will not be proper to plant these trees in a soil which is too moist, in England, because it might endanger the rotting of the fibres of the roots, by the moisture continuing too long about them, especially if the bottom be a clay, or a strong loam, which will detain the wet.

TURKS CAP. See LILIUM and CACTOS.

TURKY WHEAT. See ZEA.

TURNEP. See RAPA.

TURNERA. Plum. Gen. Nov. 15. tab. 12.

The CHARACTERS are,

The empalement of the flower is funnel-shaped, of one leaf; having an oblong, cylindrical, angular tube, cut into five segments. The flower has five heart-shaped plain petals, with narrow tails, which are inserted in the tube of the empalement; it has five awl-shaped stamina, which are shorter than the petals, inserted in the empalement, terminated by acute-pointed erect summits, and a conical germen, supporting three slender styles, crowned by hairy many pointed stigmas. The germen afterward turns to an oval capsule with one cell, which opens at the top with three valves, and contains several oblong obtuse seeds.

The SPECIES are,

1. TURNERA (*Ulmifolia*) floribus sessilibus petiolaribus foliis basi biglandulosis. Lin. Sp. 337. *Turnera with spear-shaped hairy leaves, which are obtusely indented and acute-pointed.*

2. TURNERA (*Angustifolia*) floribus sessilibus petiolaribus foliis lanceolatis rugosis acuminatis. *Turnera with oval spear-shaped leaves, which are sawed and rough.*

These plants are both of them natives of the warm parts of America. The second species was found by Father Plumier in Martinico, who gave it the name of Turnera, from Dr. Turner, a famous English physician, who lived in queen Elizabeth's reign.

The first sort was discovered by Sir Hans Sloane, who has figured it in his Natural History of Jamaica, under the following title; Cistus urticæ folio, flore luteo, vasculis trigonis. Vol. 1. p. 202. but both these sorts were observed by my late friend Dr. William Houstoun, in several parts of America.

The first sort rises with a shrubby stalk to the height of eight or ten feet, sending out branches on every side the whole length, garnished with narrow, spear-shaped, hairy leaves, terminating in acute points, sawed on their edges; these, when rubbed, emit a disagreeable odour. The flowers grow from the foot-stalks of the leaves, to which they sit very close, having two pretty large leafy appendages to their empalements. The flowers are of a pale yellow colour, composed of five large oval petals, whose tails are twisted and join; these are succeeded by short tubular capsules, having one cell, which opens at the top with three valves, which turn back, and let out the seeds.

The second sort has a shrubby stalk like the first, and rises to near the same height. The branches of this are slender, and stiffer than those of the former. The leaves are oval, spear-shaped, rough on their upper side, and of a lucid green; their under side has many strong veins, and is of a lighter colour; they are sawed on their edges, and have longer foot-stalks than those of the first species. The flowers sit close upon the foot-stalks of the leaves, in like manner as the former, but the flowers are larger, and of a brighter yellow than those. These differences remain constant, and never alter when raised from seeds; so that from near thirty years experience in sowing the seeds, I may pronounce them different species.

The plants are easily propagated, by sowing their seeds on a hot-bed early in the spring; and when the plants are come up two inches high, they should be transplanted into small pots, and plunged into a hot-bed of tanners bark, observing to water and shade them until they have taken root; after which they must be treated, as hath been directed for the Guavas, and other tender plants from the same countries, to which the reader is desired to turn, to avoid repetition. The seeds of these plants will often fall into the pots which are placed near them in the stove, which will grow, and soon furnish plants enough, after a person is once possessed of them. As they are too tender to live in the open air in England, they must be placed in the bark-bed in the stove, where, during the winter season, they must be kept warm, and frequently watered; but in the summer season they must have a great share of air, otherwise they will draw up tender, and not produce many flowers.

When the plants are grown pretty large, they may be treated more hardily, by placing them in the dry stove, where, if they are kept in a moderate degree of heat, they will thrive and flower very well. Those who would save the seeds of these plants, must watch them carefully, because, when they are ripe, they soon scatter if they are not gathered.

These plants produce their flowers great part of the year, if they are kept in a proper degree of warmth, so that there are some of the flowers in beauty for at least nine or ten months, which renders the plants more valuable.

TURNSOLE. See HELIOTROPIUM and CROTON.

TURRITIS. Tourn. Inst. R. H. 223. Tower Mustard.

The CHARACTERS are,

The empalement of the flower is composed of four oblong oval leaves, which close together. The flower has four oblong, oval, entire petals, placed in form of a cross, and six erect awl-shaped stamina the length of the tube, two of which are shorter than the other, terminated by single summits, and a taper germen a little compressed, having no style, but is crowned by an obtuse stigma. The germen afterward becomes a long four-cornered pod with two cells, which are divided by an intermediate partition, opening with two valves, and filled with small, roundish, indented seeds.

The SPECIES are,

1. TURRITIS (*Glabra*) foliis radicalibus dentatis hispidis, caulinis integerrimis amplexicaulibus glabris. Hort. Cliff. 339. *Tower Mustard with prickly lower leaves, which are indented, and the upper ones smooth, entire, and embracing the stalk.*

2. TURRITIS (*Hirsuta*) foliis omnibus hispidis, caulinis amplexicaulibus. Hort. Cliff. 339. *Tower Mustard with all the leaves prickly, and the upper ones embracing the stalk.*

The first sort grows naturally in several parts of England, upon walls and dry banks; this hath its lower leaves much jagged on their edges, and rough. The stalks rise two feet high, garnished with smooth grayish leaves, ending in points, which embrace the stalks with their base. The upper part of the stalk has slender branches, proceeding from the wings of the leaves, which sustain spikes of small white flowers, having four petals, placed in form of a cross. These are succeeded by long, slender, compressed, four-cornered pods, which grow erect close to the stalk, filled with small seeds.

The second sort grows naturally upon old walls and buildings in the northern counties of England; the lower leaves are shaped like those of the Daisy, but rough. The stalks rise eight or ten inches high, garnished with oval leaves, whose base embrace the stalks; they are as rough as the lower leaves. The upper part of the stalks branch into slender shoots, which sustain short spikes of white flowers like those of the former sort, which are succeededed by slender pods, having four corners shorter than those of the first sort.

These plants are kept in botanic gardens for variety but if their seeds are scattered upon an old wall or building in autumn, soon after they are ripe, the plants will come up, and thrive without farther care, and their seeds will scatter on the walls and spread, so there will be no danger of the plants maintaining their situation, if they are not purposely destroyed.

The other species inserted in the former editions of this work, are referred to ARABIAS, BRASSICA, and HESPERIS, under which articles they will be found.

TUSSILAGO. Tourn. Inst. R. H. 487. tab. 276. Lin. Gen. Plant. 856. Colt's-foot.

The CHARACTERS are,

The flower has one common cylindrical empalement, whose scales are linear, spear-shaped, and equal. The flower is made up of hermaphrodite florets which compose the disk, and female half florets which form the rays or border. The hermaphrodite florets are funnel-shaped, cut at the brim into five segments; these have five short hair-like stamina, terminated by cylindrical summits, and a short crowned germen, supporting a slender style, crowned by a thick stigma. The germen afterward becomes an oblong compressed seed, crowned with a hairy down. The female half florets are stretched out on one side with a narrow tongue-shaped segment; these have no stamina, but have a short crowned germen, which turns to a seed like those of the hermaphrodite florets, which ripen in the empalement.

The SPECIES are,

1. TUSSILAGO (*Farfara*) scapo imbricato uniflora, foliis subcordatis, angulatis denticulatis. Lin. Hort. Cliff. 411. *Colt's-foot with an imbricated stalk bearing one flower, and angular indented leaves, which are nearly heart-shaped; or common Colt's-foot.*

2. TUSSILAGO (*Anandria*) scapo unifloro, foliis lyrato ovatis. Lin. Sp. Plant. 865. *Colt's-foot with one flower on each stalk, and oval lyre-shaped leaves.*

3. TUSSILAGO (*Alpina*) scapo subnudo unifloro, foliis cordato-orbiculatis crenatis. Hort. Cliff. 411. *Colt's-foot with an almost naked stalk bearing one flower, and orbicular, heart-shaped, crenated leaves.*

The first of these sorts is very common in watery places in almost every part of England, and is rarely kept in gardens; for the roots creep under ground, and increase so fast, that in a short time they will spread over a large spot of ground. This plant is so well known as to need no description.

The second sort grows naturally in Siberia; this is a very low plant, whose leaves grow close to the ground, of an oval form, indented on the sides like a lute. The flowers stand upon short foot-stalks, which rise between the leaves, each sustaining one flower at the top, of a dirty purplish colour. These are succeeded by downy seeds.

The third sort grows naturally on the Alps; this is a low perennial plant, whose leaves are round, indented at the foot-stalk in form of a heart; their edges are crenated; their upper surface is smooth, of a bright green colour; their under sides are a little downy and whitish. The foot-stalks of the flowers, which arise from the root, sustain one purplish flower at the top, which is made up of hermaphrodite and female florets like those of the other sorts.

The two last are frequently kept in gardens for the sake of variety; they are easily propagated by parting their roots in autumn, and must be planted in a moist shady border, where they will thrive, and require no farther care but to keep them clean from weeds.

V.

VACCARIA. See SAPONARIA.

VACCINIUM. Lin. Gen. Plant. 434. The Bill-berry, Whortle-berry, or Cran-berry.

The CHARACTERS are,

The flower has a small permanent empalement sitting upon the germen; it is bell-shaped, of one petal, slightly cut into four segments at the brim, which turn backward; it has eight stamina, terminated by horned summits, having two awns on their backside which spread asunder. The germen is situated below the flower, supporting a single style longer than the stamina, crowned by an obtuse stigma; it afterward turns to an umbilicated globulary berry with four cells, containing a few small seeds.

The SPECIES are,

1. VACCINIUM (*Myrtillus*) pedunculis unifloris, foliis ovatis serratis deciduis, caule angulato. Flor. Lapp. 143. *Whortle-berry with one flower upon each foot-stalk, oval sawed leaves, which fall off in winter, and an angular stalk.*

2. VACCINIUM (*Vitis Idæa*) racemis terminalibus nutantibus, foliis obovatis revolutis integerrimis subtus punctatis. Lin. Sp. Pl. 351. *Whortle-berry with nodding bunches of flowers terminating the branches, and oval leaves which are entire, turned back, and punctured on their under side; Red-whorts, or Whortle-berries.*

3. VACCINIUM (*Pensylvanicum*) foliis ovatis mucronatis, floribus alaribus nutantibus. *Whortle-berries with oval-pointed leaves, and nodding flowers proceeding from the wings of the stalks.*

4. VACCINIUM (*Hispidulum*) foliis integerrimis revolutis ovatis, caulibus repentibus, filiformibus, hispidis. Lin. Sp. Plant. 352. *Whortle-berries with oval entire leaves turning back, and a slender, creeping, bristly stalk.*

5. VACCINIUM (*Oxycoccos*) foliis integerrimis revolutis ovatis, caulibus repentibus filiformibus nudis. Lin. Sp. Plant. 351. *Whortle-berries with oval, entire, reflexed leaves, and naked, slender, creeping stalks; Moss-berries, or Moor-berries, by some called Cran-berries.*

The first sort grows very common upon large wild heaths in many parts of England, but is never cultivated in gardens, it being with great difficulty transplanted, nor will it thrive long when moved thither; for from many trials which I have made, by taking up the plants at different seasons with balls of earth to their roots, and planting them in gardens, I could never succeed so as to preserve the plants above two years, and those never produced any fruit, so that it is not worth the trouble of cultivating.

The fruit of this sort is gathered by the poor inhabitants of those villages which are situated in the neighbourhood of their growth, and carried to the market-towns. These are by some eaten with cream or milk; they are also put into tarts, and much esteemed by the people in the north, but they are seldom brought to London. The shrub on which these grow, rises about two feet high, having many stems, which are garnished with oblong leaves, shaped like those of the Box-tree, but somewhat longer, and are a little sawed on their edges. The flowers are shaped like those of the Arbutus, or Strawberry-tree, of a greenish white colour, changing to a dark red toward the top. The fruit are about the size of large Juniper-berries, and of a deep purple colour, having a flue upon them when they are untouched, like the blue Plumbs, which rub off with handling.

The second sort is of a much humbler growth, seldom rising above six or eight inches high. The leaves of this are so like that of the dwarf Box, as that, at a distance, the plants are often taken for it even by persons of skill. This is an evergreen, which grows upon moory ground in several parts of the north, but it is full as difficult to transplant into gardens as the other sort, though I have been assured by persons of credit, that they have seen this sort planted to make edgings to the borders of the gardens in Norway and Sweden, where the plants may grow much better from the cold of those climates, than they will do in England, for this is a native of very cold countries. I have several times received plants of this sort from Greenland, by the whale-ships. The berries of this sort are red, and have a more agreeable acid flavour than those of the first sort. The fruit is frequently used for tarts in several of the northern counties, where the plants grow wild upon the moors.

The third sort grows naturally in Virginia, and other parts of North America; this has a low shrubby stalk like the second. The leaves are small, oval, pointed, and not unlike some sorts of Myrtle; they continue green all the year. The flowers come out from the wings of the leaves at every joint; their foot-stalks are pretty long, and nod downward; they sustain but one flower; they are small, white, and are succeeded by small red berries which seldom ripen here.

The fourth sort grows naturally in marshy grounds in most parts of North America. The stalks of this are slender, imbricated, and trail upon the ground; the scales are bristly; the leaves are oval, entire, and their edges turn backward; the flowers come out from the wings of the stalk, of an herbaceous white colour, and in their native soil are succeeded by large red berries, but in England the fruit never comes to perfection.

The plants of this sort are difficult to preserve in England, for they require a moorish, boggy soil, which should be covered with Moss, and constantly kept wet, otherwise they will not thrive.

The fifth sort produces long slender branches, not bigger than thread, which trail upon the mossy bogs, so are often hid by the Moss. These branches are thinly garnished with small leaves, about the size and shape of those of Thyme, having their upper surface of a shining green colour, but are white underneath. The flowers are generally produced toward the extremity of the shoots, which are in shape like those of the former sorts, but are smaller, and of a red colour; these grow upon long slender foot-stalks, and are succeeded by round, red, spotted berries, of a sharp acid flavour, which are much esteemed by the inhabitants of the places near the bogs where they grow. Some use them for tarts, and others eat them with milk or cream.

This

This sort is a native of bogs, therefore cannot by any art be propagated upon dry land; but where there are natural bogs, the plants may be taken up carefully, preserving some of the soil to their roots, and transplanted into the bogs in the autumn; and if they are once fixed in the place, they will spread and propagate themselves in great plenty, and require no farther care.

The two sorts first mentioned also propagate very fast by their creeping roots, so that when they are fixed in a proper soil, they will soon overspread the ground, for the heaths, upon which they naturally grow, are generally covered with the plants. The first sort grows with the Heath, their roots intermixing together, and frequently is found upon sandy heaths in divers parts of England; but the second sort grows only upon moorish land, where, by its creeping roots, the ground is soon covered with the plants.

There are several other species of this genus, some of which are natives of Spain and Portugal, others of Germany and Hungary, and several of the northern parts of America, from whence those large fruit are brought to England, which are used by the pastrycooks of London, during the winter season, for tarts; but as all these sorts naturally grow upon swamps and bogs, they are not easy to transplant into gardens in their native country, so as to thrive, or produce fruit, therefore there can be little hopes of cultivating them to advantage in England.

VALANTIA.

The CHARACTERS are,

It hath solitary hermaphrodite flowers in the place of the germen, which are divided into four segments; it hath four stamina the length of the petals, terminated by small summits, and a large germen with a slender style, crowned by headed stigmas; the empalement becomes a thick compressed capsule, containing one seed.

The SPECIES are,

1. VALANTIA (*Hispida*) floribus masculis trifidis, hermaphroditici germini hispido insidentibus. Lin. Sp. 1490. *Valantia with three male flowers sitting on the hispid germen of the hermaphrodite flowers.*

2. VALANTIA (*Muralis*) floribus masculis trifidis hermaphroditici germine glabro insidentibus. Sauv. Monsp. 162. *Valantia with three male flowers sitting on the smooth germen of the hermaphrodite flowers.*

3. VALANTIA (*Aparina*) floribus masculis trifidis pedicillatis hermaphroditici pedunculo insidentibus. Hort. Upsal. 302. *Valantia with trifid male flowers sitting on the footstalks of the hermaphrodite flowers.*

4. VALANTIA (*Articulata*) floribus masculis quadrifidis, pedunculis dichotomis nudis, foliis cordatis. Hort. Upsal. 302. *Valantia with quadrifid male flowers, the pedicles of the flowers forked, and heart-shaped leaves.*

5. VALANTIA (*Cruciata*) floribus masculis quadrifidis, pedunculis diphyllis. Hort. Upsal. 302. *Valantia with quadrifid male flowers, whose foot-stalks have two leaves; or hairy Crosswort.*

These plants are rarely cultivated in gardens, unless those of botany for the sake of variety. The four first sorts are trailing annual plants, and if their seeds are sown in the autumn, they will succeed better than in the spring; and if their seeds are permitted to scatter, the plants will come up, and require no other culture than to thin them, and keep them clean from weeds.

The fifth sort is the common Crosswort, which grows plentifuly on the side of banks in many parts of England, so is rarely admitted into gardens; but as it stands in the list of medicinal plants, so I have thought proper to mention it.

VALERIANA. Tourn. Inst. R. H. 131. tab. 52. Valerian.

The CHARACTERS are,

The flower has a small empalement; it has one tubulous petal cut into five obtuse segments at the brim, with a gibbous honey gland on the inside; it has three small, erect, awl-shaped stamina the length of the petals, terminated by roundish summits. The germen is situated under the flower, supporting a slender style, crowned by a thick stigma; it afterward turns to a crowned capsule which falls off, in which is lodged a single seed.

The SPECIES are,

1. VALERIANA (*Phu*) floribus triandris, foliis caulinis pinnatis, radicalibus indivisis. Hort. Upsal. 13. *Valerian with flowers having three stamina, winged leaves to the stalks, but those at the root undivided; Garden Valerian.*

2. VALERIANA (*Officinalis*) floribus triandris, foliis omnibus pinnatis. Hort. Cliff. 15. *Valerian with three stamina to the flowers, and all the leaves winged; Greater wild Valerian.*

3. VALERIANA (*Rubra*) floribus monandris caudatis, foliis lanceolatis integerrimis. Hort. Cliff. 15. *Valerian with flowers having tails, but one stamina, and spear-shaped entire leaves; Red Valerian.*

4. VALERIANA (*Angustifolia*) floribus monandris caudatis, foliis linearibus integerrimis. *Valerian with tailed flowers having one stamina, and linear entire leaves; Narrow-leaved red Valerian.*

5. VALERIANA (*Calcitrapa*) floribus monandris, foliis pinnatifidis. Hort. Upsal. 14. *Valerian with flowers having one stamina, and wing-pointed leaves.*

6. VALERIANA (*Pyrenaica*) floribus triandris, foliis cordatis serratis petiolatis, summis ternatis. Hort. Cliff. 15. *Valerian with three stamina to the flowers, and heart-shaped sawed leaves growing on foot-stalks, placed by threes at the top.*

7. VALERIANA (*Celtica*) floribus triandris, foliis ovato-oblongis obtusis integerrimis. Lin. Mat. Med. 23. *Valerian with three stamina to the flowers, and oblong, oval, blunt, entire leaves; Celtic Nard.*

8. VALERIANA (*Siberica*) floribus tetandris æqualibus, foliis pinnatifidis, seminibus paleâ ovali adnatis. Hort. Upsal. 13. *Valerian with four equal stamina to the flowers, wing-pointed leaves, and seeds fastened to an oval husk.*

9. VALERIANA (*Locusta*) floribus triandris, caule dichotomo, foliis linearibus. Flor. Suec. 32. *Valerian with a forked stalk, and spear-shaped entire leaves; Corn-sallad, or Lambs-lettuce.*

10. VALERIANA (*Vesicaria*) caule dichotomo, foliis lanceolatis serratis, calycibus inflatis. Hort. Cliff. 16. *Valerian with a forked stalk, spear-shaped sawed leaves, and swollen empalements.*

11. VALERIANA (*Coronata*) caule dichotomo, foliis lanceolatis dentatis, fructu sexdentato. Hort. Cliff. 16. *Valerian with a forked stalk, spear-shaped indented leaves, and a fruit having six indentures.*

12. VALERIANA (*Cornucopia*) floribus diandris ringentibus, foliis ovatis sessilibus. Hort. Cliff. 15. *Valerian with a ringent flower having two stamina, and oval leaves set close to the stalk.*

There are some other species of this genus which grow naturally in England, and others in different parts of Europe; but, as they are seldom cultivated in gardens, they are omitted, lest the work should swell beyond its intended bulk.

The first of these sorts is propagated in England for medicinal use, and is called in the shops by the name of Phu, to distinguish it from the Mountain Valerian, which is also used in medicine, and is preferred to all the other sorts by the modern physicians, though the roots of the first are still continued in some of the capital medicines, and are by some

ſome eſteemed equal in virtue, if not ſuperior, to the wild ſort.

This hath thick, fleſhy, jointed roots, which ſpread near the ſurface of the ground in a very irregular manner, croſſing each other, and matting together by their ſmall fibres; they have a very ſtrong ſcent, eſpecially when dry. The lower leaves, which riſe immediately from the root, are many of them entire; others are divided into three, five, or ſeven obtuſe ſmooth lobes, of a pale green colour. The ſtalks riſe three or four feet high; they are hollow, and ſend out branches from their ſide by pairs, garniſhed with winged leaves, placed oppoſite at each joint, which are compoſed of four or five pair of long narrow lobes, terminated by an odd one. The branches are terminated by flowers, diſpoſed in form of an umbel, which are ſmall, tubulous, white, and cut ſlightly at the brim into five parts; theſe are ſucceeded by oblong flat ſeeds, having a downy crown. It grows naturally in Alſatia, but has been long cultivated in our gardens.

It is propagated by parting of its roots, either in the ſpring or autumn, but the latter is much preferable; theſe ſhould be planted in beds of freſh earth about two feet aſunder, for they commonly ſpread and multiply very faſt. If the ſeaſon is dry, you muſt water the plants until they have taken root; after which they will require no farther care but to keep them clean from weeds, and in autumn, when their leaves are decayed, the roots ſhould be taken up and dried for uſe.

The ſecond ſort is generally found upon dry chalky ſoils, in ſhady places, in divers parts of England. The roots of this, which grow wild upon ſuch ſoils, are much preferable to thoſe of the ſame kind which are cultivated in gardens, when gathered from their native places of growth, where they are ſmaller, but have a ſtronger flavour.

The roots of this plant are compoſed of long fleſhy fibres, which unite in heads. All the leaves of this ſort are winged; thoſe at the bottom are compoſed of broader lobes than thoſe on the ſtalks; they are notched on their edges, and are hairy. The ſtalks, in their natural ſituation, ſeldom grow more than a foot high, but, when the roots are cultivated in a garden, they grow twice that height; theſe are channelled, hollow, hairy, and garniſhed at each joint with two winged leaves placed oppoſite, whoſe lobes are very narrow and almoſt entire. At the upper part of the ſtalk comes out two ſmall ſide branches oppoſite; theſe, and alſo the principal ſtalk, are terminated by cluſters of flowers, formed into a kind of umbel, which are ſhaped like thoſe of the firſt ſort but ſmaller, and have a tinge of purple on their outſide.

This plant may alſo be propagated by parting the roots either in ſpring or autumn, as was directed for the firſt ſort, but ſhould always be planted upon a dry, freſh, undunged ſoil, in which, though the roots will not make near ſo great progreſs as in a rich moiſt ſoil, yet they will be much preferable for uſe. Theſe roots ſhould alſo be taken up, when the leaves decay in autumn, and preſerved dry until uſed.

The third ſort grows naturally in rough ſtony places in the ſouth of France, and in Italy, but has been long cultivated in the Engliſh gardens for ornament.

The roots of this ſort are ligneous, and as thick as a man's finger, ſpreading very wide. The ſtalks riſe three feet high; they are ſmooth, of a grayiſh colour, and hollow, garniſhed at each joint with ſmooth ſpear-ſhaped leaves. The upper part of the ſtalk ſends out branches by pairs, which, with the principal ſtalk, are terminated with red flowers growing in cluſters, which have long tubes, cut into five parts at the top, and from the tube is ſent out a ſpur or heel, like the flowers of Larkſpur. It flowers moſt part of ſummer, and the ſeeds ripen accordingly in ſucceſſion.

There is a variety of this with white flowers, and one with pale fleſh-coloured flowers, but they do not differ in any other reſpect.

It is eaſily propagated by parting of the roots in autumn, or by ſowing of the ſeeds ſoon after they are ripe, in a ſhady border, where the plants will ſometimes come up the ſame autumn, eſpecially if the ſeaſon proves moiſt, otherwiſe they will not appear till the following ſpring. When theſe are fit to remove, they ſhould be tranſplanted into beds at about nine inches or a foot aſunder, obſerving to water them till they have taken new root; after which they will require no farther care but to keep them clear from weeds, and in autumn they muſt be tranſplanted where they are to remain.

Theſe plants grow large, therefore ſhould have room, ſo are not proper furniture for ſmall gardens. When the ſeeds of theſe plants light on joints of old walls or buildings, the plants will come up, and thrive as well as in the ground, and will continue much longer, ſo the ſeeds May be ſcattered between the ſtones of grottos and ſuch like buildings, where the plants will flower from May till the froſt ſtops them, and will make a good appearance.

The fourth ſort grows about Montpelier, and upon mount Baldus in Italy. The root of this is ligneous, but not ſo large as that of the former ſort; the ſtalks riſe about two feet high or better, and branch out on each ſide from the root to within ſix inches of the top, garniſhed with leaves which are as narrow as thoſe of Flax. The upper part of the ſtalk is naked, and terminated by a compact cluſter of bright red flowers, ſhaped like thoſe of the former ſort, but ſmaller.

The fifth ſort grows naturally in Spain and Portugal; it is an annual plant, which periſhes ſoon after the ſeeds are ripe. The lower leaves, which ſpread on the ground, are cut into many obtuſe ſegments; the ſtalks, when the plants are in good ground, will riſe near a foot and a half high, but on dry ſtony ſoils not half ſo high, and when they grow out of the joints of old walls, not more than three inches; they are hollow, ſmooth, and ſend out branches by pairs from the upper joints, garniſhed with wing-pointed leaves, whoſe lobes or ſegments are very narrow. The ſtalk and branches are terminated by tufts of flowers ſhaped like thoſe of the Garden Valerian, but ſmaller, and have a fleſh-coloured tinge at the top. The ſeeds have a down, which helps to ſpread them, ſo it propagates without care.

The ſixth ſort grows naturally on the Pyrenean mountains; this has a fibrous perennial root, from which come out many heart-ſhaped leaves, ſtanding upon foot-ſtalks more than a foot in length. The leaves are bluntly ſawed on their edges, ſmooth, and of a bright green on their upper ſide, but their under ſide is pale and a little hairy. The ſtalks riſe three feet high; they are hollow, channelled, and ſend out branches oppoſite toward the top, garniſhed with leaves placed oppoſite, which are ſhaped like thoſe below, but pointed, and frequently at the top there are three leaves placed round the ſtalk, ſtanding upon ſhort foot-ſtalks. The ſtalk and branches are terminated by pale fleſh-coloured flowers, diſpoſed in form of umbels, which have very ſhort ſpurs or heels. The ſeeds ripen in July, which are crowned with down, whereby they are tranſported to a diſtance.

This plant delights in ſhade and a moiſt ſoil; it may be propagated by ſowing of the ſeeds on a ſhady border ſoon after they are ripe, and when the plants come up, they ſhould be treated in the ſame way as is before directed for the third ſort.

The

The ſeventh ſort grows naturally upon the Alps and Styrian mountains; this was ſent me by Dr. Allione from Turin, who gathered it on the Alps near that place; it is a very humble plant. The ſtalks trail upon the ground among the Moſs, and put out roots at their joints, which ſwell into knobs or tubers. The leaves are oblong, oval, and entire; the flower-ſtalks riſe three or four inches high, garniſhed with two or three pair of ſmall oval leaves; the flowers are ſmall, of a pale incarnate colour, and are formed in a looſe ſpike ſitting very cloſe to the ſtalk.

This plant is difficult to preſerve in gardens, for it naturally grows upon rocky mountains which are covered with Moſs, where the ſnow continues ſix or ſeven months, ſo it requires a very cold ſituation and a ſtony ſoil.

The eighth ſort grows naturally in Siberia; this is a biennial plant, which flowers and produces ſeeds the ſecond year, and then decays. The leaves of this are winged; the lobes of the lower leaves are oblong, oval, ending in roundiſh points; the ſtalks riſe a foot high, garniſhed with leaves, compoſed of four or five pair of lobes, terminated by a broad one, which is cut into three or five points. The lobes are acute-pointed; the leaves are ſmooth, placed by pairs, and ſit cloſe to the ſtalks. The upper part of the ſtalk has two pair of branches; the lower pair are near three inches long, but the upper are not half that length; theſe, and alſo the principal ſtalk, are terminated by bright yellow flowers, collected in a ſort of umbel, ſhaped like thoſe of the firſt ſort. It is propagated by ſeeds, which ſhould be ſown where the plants are to remain; this may be performed either in autumn, ſoon after they are ripe, or in the ſpring; they have ſucceeded with me equally at both ſeaſons. When the plants come up, they muſt be thinned where they are too cloſe, and kept clean from weeds, which is all the culture they require.

The ninth ſort is the common Corn-ſallad which is cultivated in gardens, but is found growing naturally upon arable land amongſt the Corn in many parts of England; this is an annual plant, which dies when it has perfected its ſeeds. The lower leaves of this are oblong, and broad at their points, which are rounded, and narrowed at their baſe, where they embrace each other. From between the leaves ariſes an angular ſtalk, from three to eight or nine inches high, which divides into two branches which ſpread from each other, and theſe both divide again into two other in like manner, garniſhed with leaves ſhaped like thoſe at the bottom, but ſmaller, placed by pairs at each joint. The branches are terminated by cluſters of white flowers, ſhaped like thoſe of the other ſpecies, which are ſucceeded by pretty large roundiſh ſeeds a little compreſſed on one ſide. The ſeeds are very apt to drop before they have changed colour.

It is cultivated as a ſallad-herb for the ſpring, but, having a ſtrong taſte which is not agreeable to many palates, it is not ſo much in uſe as it was formerly; it is propagated by ſeeds, which ſhould be ſown the latter end of Auguſt, then the firſt rains will bring up the plants, which ſhould be hoed to thin them where they are too cloſe, and to deſtroy the weeds. Early in the ſpring the plants will be fit for uſe. The younger the plants are when uſed, the leſs ſtrong will be their taſte, ſo they may ſupply the table in a ſcarcity of other herbs. When the ſeeds of this ſort are ſown in the ſpring, if the ſeaſon proves dry, the plants will not appear till autumn or the ſpring following; beſides, in ſummer the herb is not ſo fit for uſe. I have known the ſeeds of this plant lie in the ground many years, when they have happened to be buried deep, and upon being turned up to the air, the plants have come up as thick as if the ſeeds had been newly ſown.

There are two other ſpecies of this which grow naturally in England, but, as they are ſeldom admitted into gardens, I have not enumerated them; theſe are by ſome ſuppoſed to be only accidental varieties, but I have ſown them all ſeveral years, and have never found either of them alter.

The tenth ſort grows naturally in Candia; this is an annual plant, whoſe ſtalks riſe ſix or eight inches high, and divide by pairs like the former. The leaves are much narrower than thoſe of the former, end in acute points, and are ſawed on their edges; the flowers are like thoſe of the former ſort, but have a ſwollen bladder empalement, which incloſes the ſeeds.

The eleventh ſort grows naturally in Italy. The leaves at bottom are long, round-pointed, and deeply notched on their edges; the ſtalk riſes near a foot high, ſending out branches oppoſite; the upper part divides by pairs in the ſame manner as the former. The flowers are collected in globular heads, are of an herbaceous white colour, and are ſucceeded by ſtarry fruit having ſix indentures. This and the former ſort are ſuppoſed to be only varieties ariſing from the ſame ſeeds, but I have ſown them more than thirty years, and have not obſerved either of them vary.

The twelfth ſort grows naturally in the arable fields in Sicily and Spain; this is an annual plant. The ſtalks are channelled, of a purpliſh colour, eight or nine inches high, garniſhed with oval ſmooth leaves, placed by pairs at each joint, ſitting cloſe to the ſtalks, of a lucid green. From each ſide of the ſtalk ſprings out ſlender branches, but the upper part divides into two ſpreading branches like the other ſorts. The joints are ſwelling, and the branches divide again by pairs, which are terminated by cluſters of red flowers, ſhaped like thoſe of the red Valerian, but larger; they have two leaves cloſe under the bunches, embracing the ſtalks with their baſe. When the flowers are paſt, the fruit ſtretches out in ſhape of a cornucopia, or horn of plenty.

Theſe three ſorts are propagated by ſeeds, which ſhould be ſown in autumn where the plants are to remain. When theſe come up, they will require no other culture but to thin them where they are too cloſe, and keep them clean from weeds. The plants, which riſe in autumn, will live through the winter, and come early to flower the following ſummer, ſo will produce good ſeeds; whereas thoſe which are ſown in the ſpring, do not ripen their ſeeds unleſs the ſeaſon proves warm.

VALERIANA GRÆCA. See POLEMONIUM.

VALERIANELLA. See VALERIANA.

VANILLA. Plum. Gen. Nov. 25. tab. 28.

The CHARACTERS are,

It has a ſingle ſtalk. The flowers are included in ſheaths ſitting upon the germen, and have no empalement; they have five oblong petals which ſpread open very wide, with turbinated nectariums, whoſe baſe are tubulous, ſituated on the back ſide of the petals; their brims are oblique and bifid; the upper lip is ſhort and trifid; the under one runs out in a long point; they have two very ſhort ſtamina ſitting upon the pointal, and the ſummits are faſtened to the upper lip of the nectarium; they have long, ſlender, contorted germen, ſituated under the flower, ſupporting a ſhort ſtyle faſtened to the upper lip of the nectarium, crowned by an obſolete ſtigma. The germen afterward becomes a long, taper, fleſhy pod, including many ſmall ſeeds.

The SPECIES are,

1. VANILLA (*Mexicana*) foliis oblongo-ovatis mucronatis, nervoſis, floribus alternis. *Vanilla with oblong, oval, acute-pointed, veined leaves, and flowers growing alternately.*

2. VANILLA (*Axillaris*) foliis oblongis obtuſis, compreſſis articulatis, floribus alaribus. *Vanilla with oblong, blunt, compreſſed, jointed leaves, and flowers proceeding from the ſides of the ſtalks.*

The first sort is that which the Spaniards cultivate in the West-Indies, which we shall describe hereafter.

The second sort was sent me from Carthagena in New Spain, where it grows naturally; this has a climbing stalk, sending out roots from the joints, which fasten to the stems of trees or any neighbouring support, and climb to a great height. The leaves, which come out singly at each joint, are oblong, smooth, and jointed. The flowers come out from the side of the branches, shaped like those of the great Bee Orchis, but longer: the galea or helmet of the flower is of a pale Pink colour, and the labia is purple. This plant flowered in the Chelsea garden, but wanting its proper support, it lived but one year.

There are two or three varieties of the first sort, which differ in the colour of their flowers and the length of their pods; and there are many other species which grow naturally in both the Indies, which have been brought to this genus, but those above-mentioned are all I have seen growing.

The plant, which produces the fruit called Vanilla or Bahilla by the Spaniards, hath a trailing stem, somewhat like common Ivy, but not so woody, which fastens itself to whatever tree grows near it, by small fibres or roots produced at every joint, which fasten to the bark of the tree, and by which the plants are often nourished, when they are cut or broken off from the root a considerable height from the ground, in like manner as the Ivy is often seen in England. The leaves are as large as those of the common Laurel, but are not of so thick a substance; they are produced altern tely at every joint, of a lively green colour on the upper side, but paler underneath. The stems of these plants shoot into many branches, which fasten themselves also to the branches of the trees, by which means they rise to the height of eighteen or twenty feet, and spread quite over some of the smaller trees, to which they are joined. The flowers are of a greenish yellow colour, mixed with white, which, when fallen, are succeeded by the fruit, which are six or seven inches long.

The sort which is manufactured, grows not only in the Bay of Campeachy, but also at Carthagena, at the Caraccas, Honduras, Darien, and Cayen, at all which places the fruit is gathered and preserved, but is rarely found in any of the English settlements in America, though it might be easily carried thither and propagated; for the shoots of these plants are full of juice, so may be easily transported, because they will continue fresh out of the ground for several months. I had some branches of this plant, which were gathered by Mr. Robert Millar at Campeachy, and sent over between papers by way of sample, and had been at least six months gathered when I received them; but upon opening the papers I found the leaves rotten, with the moisture contained in them, and the paper was also perished with it, but the stems appeared fresh; upon which I planted some of them in small pots, and plunged them into a hot-bed of tanners bark, where they soon put out leaves, and sent forth roots from their joints; but, as these plants naturally fasten themselves to the stems of the trees, in the woods where they grow naturally, so it is with great difficulty that they are kept alive, when they have not the same support; therefore, whoever would preserve any of these plants in Europe, should plant them in tubs of earth, near the stem of some vigorous American tree, which requires a stove, and can bear a great deal of water, because the Vanillas must be plentifully watered in the summer season, otherwise they will not thrive. They require also to be shaded from the sun by trees, so that if these are planted at the foot of the Hernandia, or Jack-in-a-Box, whose leaves are very large, and afford a good shade, they will succeed better than when they are exposed in single pots alone; and as these plants require the same degree of heat in winter, they will agree well together.

When these plants are designed for propagation in the warm parts of America, there is nothing more required than to make cuttings of about three or four joints in length, which should be planted close to the stems of trees, in low marshy places, and to keep down other troublesome plants, which, if permitted to grow about the cuttings before they are well rooted, would overbear and destroy them; but, after they are established, and have fastened their shoots to the stems of the trees, they are not in much danger of being injured by neighbouring plants, though, when the ground is kept clear from weeds, the plants will be much better nourished.

These plants do not produce flowers until they are grown strong; so that the inhabitants affirm, that it is six or seven years from the planting to the time of their bearing fruit; but when they begin to flower and fruit, they continue for several years bearing, and this without any culture; and as it is a commodity which bears a good price, it is well worth cultivating in several of the English settlements, especially as they will grow on moist woody places, where the land is not cleared from timber.

The method used to preserve the fruit is, when it turns of a yellow colour, and begins to open, to gather it, and lay it in small heaps to ferment two or three days, in the same manner as is practised for the Cocoa or Chocolate pods; then they spread them in the sun to dry, and when they are about half dried, they flat them with their hands, and afterwards rub them over with the oil of Palma Christi, or of the Cocoa; then they expose them to the sun again to dry, and afterward they rub them over with oil a second time; then they put them in small bundles, covering them with the leaves of the Indian Reed to screen them from air.

These plants produce but one crop of fruit in a year, which is commonly ripe in May, fit for gathering, for they do not let them remain on the plants to be perfectly mature, because then they are not fit for use; but when they are about half changed yellow, they esteem them better for keeping than when they are changed to a dark brown colour; at which time the fruit splits, and shews a great quantity of small seeds, which are inclosed within it. While the fruit is green, it affords no remarkable scent, but as it ripens, it emits a most grateful aromatic scent. When the fruit begins to open, the birds attack them and devour all the seeds very greedily, but do not eat any other part of the fruit.

The fruit which are brought to Europe, are of a dark brown colour, about six inches long, and scarce an inch broad; they are wrinkled on the outside, and full of a vast number of black seeds, like grains of sand, of a pleasant smell, like Balsam of Peru.

The fruit is only used in England as an ingredient in Chocolate, to which it gives a pleasant flavour to some palates, but to others it is very disagreeable; but the Spanish physicians in America use it in medicine, and esteem it grateful to the stomach and brain, for expelling of wind, to provoke urine, to resist poison, and cure the bite of venomous animals.

As this plant is so easily propagated by cuttings, it is very strange that the inhabitants of America should neglect to cultivate it, especially as it is an ingredient in their Chocolate, which is so much drank all over America; but, as the English have in a manner quite neglected the culture of the Cocoa, it is no wonder they should neglect this, since the former was cultivated in great plenty by the Spaniards in Jamaica, while that island remained in their possession; so that the English had an example before them, if they would have followed it; whereas the Vanilla was not found growing there; and therefore it is not to be ſuppoſed, that the perſons, who were ſo indolent as to quit the culture of many valuable plants then growing on the ſpot, ſhould be at the trouble of introducing any new ones.

VEGETABLE, a term applied to all plants, conſidered as capable of growth, i. e. to all natural bodies, which have parts organically formed for generation and accretion, but not for ſenſation.

VEGETATION is the act whereby plants receive nouriſhment and grow, and ſignifies the way of growth, or increaſe of bulk, parts, and dimenſions proper to all trees, ſhrubs, plants, and herbs, &c.

In ſome trees it is chiefly the roots which vegetate; ſo that if they are cut into as many pieces as reaſonably may be, if theſe pieces are but planted in the ground, they quickly grow, as is ſ.en in the Elm, and many other trees.

In ſome it is ſeated both in the roots, and all over the trunk and branches, as in the Willow and other kinds, which, if cut into a thouſand pieces, it is ſcarce poſſible to deſtroy or kill them, unleſs they are ſtripped of both their barks; for if they are in the earth but the length of three or four inches, they will put out roots and branches, ſo will certainly grow again.

The uſe of this principle of life is accounted to be for the concoction of the indigeſted ſalts, which aſcend through the roots, where they are ſuppoſed to aſſimulate the nature of the tree they are helping to form, though perhaps the root may likewiſe aſſiſt in the work.

Theſe things being preſuppoſed in the ſpring of the year, as ſoon as the ſun begins to warm the earth, and the rains melt the latent ſalts, the whole work of Vegetation is ſet on foot; then the emulgent fibres ſeek for food, which has been prepared as aforeſaid.

There are ſome who ſuppoſe that ſubterraneous fires are concerned in the work of Vegetation, or the growth of plants; yet as, upon the beſt obſervation that can be made, none can pretend to have diſcovered any heat or fumigation to iſſue from the bowels of the earth, adequate to the meaneſt artificial fire, it is plain that the ſun is the principle, and ſo may be called the Father of Vegetation, and the earth the Mother, the rain and air being neceſſary co-efficients in this ſurpriſing work.

The curious Malpighius has very accurately delivered the proceſs of nature in the Vegetation of plants to the effect following:

The ovum or ſeed of the plant being excluded out of the ovary (which is called the pod or huſk,) and requiring farther foſtering and brooding, is committed to the earth. The earth, like a kind mother, having received it into her boſom, does not only perform the office of incubation, by her own warm vapours and exhalations, in conjunction with the heat of the ſun, but gradually ſupplies what the ſeed requires to its farther growth, as abounding every-where with canals and ſinuſe , in which the dew and rain water, impregnated with fertile ſalts, glide like the chyle and blood in the arteries, &c. of animals.

This moiſture, meeting with the new-depoſited ſeed, is percolated or ſtrained through the pores or pipes of the outer rind or huſk, anſwering to the ſecundines of fœtuſes, on the inſide whereof lie one or more, commonly two, thick ſeminal leaves, correſponding to the placenta in women, and the cotyledons in brutes.

The ſeed-leaves conſiſt of a great number of little veſiculæ or bladders, with a tube correſponding to the navel-ſtrings in animals.

The moiſture of the earth, ſtrained through the rind of the ſeed, is received into theſe veſiculæ, which cauſes a ſlight fermentation with the proper juice before contained therein.

This fermented liquor is conveyed by the umbilical veſſel to the trunk of the little plant, and to the germ or bud which is contiguous to it, upon which a Vegetation and increaſe of the plant ſucceed.

As to the vegetable matter, or the food where the plants grow, there is ſome doubt. It hath been a general opinion amongſt almoſt all the modern naturaliſts, that the Vegetation of plants, and even of minerals too, is principally owing to water, which not only ſerves as a vehicle to convey to them the fine rich earth, &c. proper for their nouriſhment, but being tranſmuted into the body of the plant, affords the greateſt part, if not all the matter with which they are nouriſhed, and by which they grow and increaſe in bigneſs. This opinion is countenanced by very great names, particularly by the ingenious Dr. Woodward, who, in order to aſcertain this point, made many curious experiments; an account of which may be ſeen in the Tranſactions of the Royal Society.

The vegetable matter being very fine and light, is ſurpriſingly apt and diſpoſed to attend water in all its motions, and follow it into each of its receſſes, as appears from many inſtances, percolate it with all the care imaginable, filter it with ever ſo many filtrations, yet ſome terreſtrial matter will remain.

Dr. Woodward filtred water through ſeveral ſheets of thick paper, and after that through very cloſe fine cloth, twelve times double, and this over and over; and yet a conſiderable quantity of this matter diſcovered itſelf in the water after all.

Now if it thus paſſes interſtices that are ſo very ſmall and fine along with the water, it is leſs ſtrange it ſhould attend it in its paſſage through the ducts and paſſages of plants. It is true, filtring and diſtilling of water interrupts, and makes it quit ſome of the earthy matter it was before impregnated withal; but then that which continues with the water after this, is fine and light, and ſuch conſequently as is in a peculiar manner fit for the growth and nouriſhment of vegetables.

And this is the caſe of rain water. The quantity of terreſtrial matter it bears up into the atmoſphere is not great; but what it doth bear up is chiefly of that light kind, or vegetable matter, and that too perfectly diſſolved, and reduced to ſingle corpuſcles, all fit to enter the tubes and veſſels of plants; on which account it is that this water is ſo very fertile and prolific.

Hence it is, that in agriculture, be the earth never ſo rich, good, and fit for the production of Corn, or other vegetables, little will come of it, unleſs the particles be ſeparated and looſe; and it is on this account ſuch pains are beſtowed in the digging, tilling, ploughing, fallowing, harrowing, and breaking the clodded lumps of earth; and it is the ſame way that ſea ſalt, nitre, and other ſalts promote Vegetation.

It is evident to obſervation, how apt all ſorts of ſalts are to be wrought upon by moiſture, how eaſily they run with it; and when theſe are drawn off, and have deſerted the lumps with which they are incorporated, they muſt moulder immediately, and fall aſunder in courſe.

Lime likewiſe is in the ſame way ſerviceable in this affair. The huſbandmen ſay, it does not fatten, but only mellows the ground; by which they mean, it doth not contain any thing in itſelf, that is of the ſame nature with the vegetable mould, or afford any matter fit for the formation of plants, but merely ſoftens and relaxes the earth; by that means rendering it more capable of entering the ſeeds and vegetables ſet in it, in order to their nouriſhment, than otherwiſe it would have been.

If therefore the ſoil, wherein any vegetable or ſeed is planted, contains all or moſt of the ingredients, and thoſe

in

in due quantity, it will grow and thrive, otherwise it will not. If there be not as many sorts of corpuscles as are requisite for the constitution of the main and more essential parts of the plant, it will not prosper at all. If there are these, and not in sufficient plenty, it will never arrive to its natural stature; or if any of the less necessary and essential corpuscles are wanting, there will be some failure in the plant. It will be defective in smell, taste, colour, and some other way.

Indeed it is inconceivable, how one uniform homogeneous matter, having its principles, or original parts, of the same substance, constitution, magnitude, figure, and gravity, should constitute bodies so unlike in all those respects, as vegetables of different kinds are, nay even as the different parts of the same vegetable; that one should carry a resinous, another a milky, a third a yellow, and a fourth a red juice in its veins; that one affords a fragrant, another an offensive smell; one sweet to the taste, another acid, bitter, acerb, austere, &c. that one should be nourishing, another poisonous; one purging, another astringent; and these all receive their nourishment from the same soil.

But a proof of this matter is, that the soil once proper for the protection of some sort of vegetables, does not ever continue so, but in tract of time loses its property; sooner in some lands, and later in others.

As for example: If Wheat be sown upon land proper for that grain, the first crop will succeed very well, and perhaps the second and third, as long as the ground is in heart, as the farmers call it; but in a few years it will produce no more, if sowed with that Corn; some other grain it may, as Barley; and after this has been sown so oft, that the land can bring no more of it, it may afterward yield some good Oats, and perhaps Pease after them.

At length it becomes barren; the vegetative matter that at first it abounded with, being reduced by the successive crops, and most of it borne off, each sort of grain takes out that peculiar matter that is proper for its own nourishment.

It may be brought to bear another series of the same vegetables, but not till it is supplied with another fund of matter of the like sort with what it first contained, either by the ground's lying fallow for some time, till the rain hath poured a fresh stock upon it, or by the manuring it.

That this supply is of the like sort, is evident by the several manures found best to promote the Vegetation, which are chiefly either of parts of vegetables, or of animals; of animals, which either derive their own nourishment immediately from vegetable bodies, or from other animals that do so; in particular, the blood, excrements, and urine of animals that do so; shaving of horns and hoofs, hair, feathers, calcined shells, lees of wine and beer, ashes of all sorts of vegetable bodies, leaves, straw, roots, and stubble, turned into the earth by ploughing, or otherwise, to rot and dissolve there.

These are our best manures, and, being vegetable substances, when refunded back again into the earth, serve for the formation of other bodies.

But to apply this to gardens, where the trees, shrubs, and herbs, after their having continued in one station till they have derived thence the greatest part of the matter fit for their increase, will decay and degenerate, unless either fresh earth, or some fit manure, be applied to them.

It is true they may maintain themselves there for some time, by sending forth roots farther and farther, to an extent all round, to fetch in more provision; but at last they must have a fresh supply brought to them, or they will decay.

All these instances argue a particular terrestrial matter, and not water, for the subject to which plants owe their increase; were it water only, there would be no need of manures, or changing the species; the rain falls in all places, in this field and in that, indifferently, on one side of an orchard or garden, as well as the other; nor could there be any reason, why a tract of land should yield Wheat one year and not the next, since the rain showers down all alike upon the earth.

That the concourse of heat is really necessary in Vegetation, appears from all the experiments, and also from nature, from the fields and forests, gardens and orchards. We see in autumn, as the sun's power is gradually less and less, so its effect on plants is emitted, and Vegetation slackens by little and little.

Its failure is first discernible in trees, which, being raised highest above the earth, require a more intense heat to elevate the water charged with nourishment to their tops; so that, for want of fresh support and nutriment, they shed their leaves, unless supported by a very firm and hard constitution, as our evergreens are. Next, the shrubs part with theirs; then the herbs and lower tribes, the heat at length not being sufficient to supply even to these, though so near the earth, the fund of their nourishment.

As the heat returns the succeeding spring, they all recruit again, and are furnished with fresh supplies and verdure; but first, those which are lowest and nearest the earth, and that require a less degree of heat to raise the water with its earthy charge into them, then the shrubs and higher vegetables in their turn, and lastly the trees.

As the heat increases, it grows too powerful, and hurries the matter with too great rapidity through the finer and more tender plants; these therefore go off and decay, and others, that are more hardy and vigorous, and require a greater degree of heat, succeed in their order. By which mechanism, provident nature furnishes us with a very various and different entertainment, and what is best suited to each season all the year round.

As the heat of the several seasons affords us a different face of things, the several distant climates shew the different scenes of nature, and productions of the earth.

The hotter countries ordinarily yield the largest and tallest trees, and those too in a much greater variety than the colder; even those plants common to both, attain to a much greater bulk in the southern than in the northern climates.

Nay, there are some regions so cold, that they raise no vegetables at all to a considerable size; this we learn from Greenland, Iceland, and other parts of like cold situation and condition: in these there are no trees, and the shrubs are poor, little, and low.

Again, in the warmer climates, and such as furnish trees and the large vegetables, if there happen a remission or diminution of the usual heat, their productions are impeded in proportion. Our own summers give us proof enough of this, for though at such times there is heat sufficient to raise the vegetative matter into the lower plants, as Wheat, Barley, Pease, and the like, and we have plenty of Strawberries, Raspberries, Gooseberries, Currants, and the fruits of such vegetables as are low, and near the earth, and a moderate store of Cherries, Plumbs, &c. and some others, that grow at something of a greater height, yet our Apples, Pears, Peaches, Nectarines, and Grapes, and the production of warmer countries, have been fewer, and those not so thoroughly ripened and brought to perfection, as they are in more benign seasons.

Nor is it that heat only which promotes Vegetation, but any other indifferently, according to its power and degree, as we find from our stoves, hot-beds, &c.

And by the rightly adapting of these artificial heats, the English gardeners have of late years so much improved their

art,

art, as in a great meafure to fupply the want of natural heat, and to vie with the people who inhabit countries feveral degrees fouth of England, in the early produćts of efculent plants, and the accelerating and ripening the fruits of the warmeft climates. And as the knowledge of Vegetation is improved, and the praćtitioners of the art are better acquainted with the theory, it may be hoped the art may be farther extended and improved; therefore it is highly neceffary, that the theory of Vegetation fhould be ftudied by every perfon who propofes to make any proficiency in gardening and agriculture.

VELLA. Lin. Gen. Plant. 714. Spanifh Crefs.

The Characters are,

The empalement of the flower is cylindrical, and compofed of four linear obtufe leaves, which drop off. The flower has four petals, placed in form of a crofs, whofe tails are the length of the empalement, and fix ftamina of the fame length, two of which are a little fhorter, terminated by fingle fummits, and an oval germen, fupporting a conical ftyle, crowned by a fingle ftigma. The germen afterward turns to a globular capfule with two cells, divided by an intermediate partition twice as large as the pod, oval, erect, ftretching beyond the capfule, each cell containing one feed.

The Species are,

1. Vella (*Annua*) foliis pinnatifidis, filiculis pendulis. Lin. Sp. Plant. 641. *Vella with wing-pointed leaves, and hanging pods.*

2. Vella (*Pfeudo Cytifus*) foliis integris filiculis erećtis. Lin. Sp. Plant. 641. *Vella with entire leaves, and erect pods.*

The firft fort grows naturally in Valencia; it is an annual plant, which feldom rifes more than half a foot high. The ftalk divides toward the top into feveral branches, each ending in loofe fpikes of flowers, which are followed by round fwelling pods, having a leafy border or creft on the top, which is hollowed like a helmet. The pod opens with two valves, and has two cells, which contain roundifh feeds like thofe of Muftard. The leaves are jagged, and end in many points.

This plant is preferved in botanic gardens for variety, but as it is not very beautiful, nor of ufe, it is feldom cultivated in other gardens. If the feeds are permitted to fcatter, the plants will come up and thrive very well; or if they are fown in autumn, they will fucceed much better than thofe which are fown in the fpring; for when the feafon proves dry, thofe feeds which are fown in the fpring frequently lie in the ground till the following autumn before the plants appear; whereas thofe which are fown in autumn, always come up foon after, or early in the fpring, fo will more certainly produce ripe feeds. The feeds fhould be fown where the plants are to remain, and if they are kept clean from weeds, and thinned where they are too clofe, they will require no other culture.

The fecond fort grows naturally in Spain. The leaves of this are entire, hairy, and fit clofe to the ftalk; they are of a grayifh colour. The ftalks become ligneous, and rife about two feet high, terminated by roundifh bunches of pale yellow flowers, which ftretch out in length; the flowers have four crofs-fhaped petals, and are fucceeded by pods like the former. This plant will continue two or three years; it is propagated by feeds in the fame manner as the former.

VERATRUM. Tourn. Inft. R. H. 272. tab. 145. White Hellebore.

The Characters are,

It has hermaphrodite and male flowers intermixed in the fame fpike. The flowers have no empalement; they have fix oblong fpear-fhaped petals, which are permanent, and fix awl-fhaped ftamina fitting on the point of the germen, fpreading afunder, terminated by quadrangular fummits; they have three oblong erect germen fitting upon the ftyle, which are fcarce vifible, crowned by a fingle fpreading ftigma. The germen afterward becomes three oblong, erect, compreffed capfules with one cell, opening on the infide, including many oblong, compreffed, membranaceous feeds. The male flowers have the fame characters of the hermaphrodite, but are barren.

The Species are,

1. Veratrum (*Album*) racemo fupradecompofito, corollis erećtis. Lin. Sp. Plant. 1044. *White Hellebore with a decompounded fpike, and erect petals; or White Hellebore with a greenifh flower.*

2. Veratrum (*Nigrum*) racemo compofito, corollis patentiffimis. Lin. Sp. Plant. 1044. *White Hellebore with a compound fpike, and very fpreading petals; or White Hellebore with a dark red flower.*

3. Veratrum (*Luteum*) racemo fimpliciffimo, foliis feffilibus. Lin. Sp. Plant. 1044. *White Hellebore with a fingle fpike, and leaves fitting clofe to the ftalk.*

4. Veratrum (*Americanum*) racemo fimpliciffimo, corollis patentibus, ftaminibus longioribus. *White Hellebore with a fingle fpike of flowers, fpreading petals, and longer ftamina.*

The firft fort grows naturally on the mountains in Auftria, Helvetia, and Greece. The root is perennial, compofed of many thick fibres gathered into a head; the leaves are ten inches long, and five broad in the middle, rounded at the points, having many longitudinal plaits like thofe of Gentian; the ftalks rife three or four feet high, and branch out on every fide almoft their whole length; under each of thefe branches is placed a narrow plaited leaf, which diminifhes in its fize as it is nearer the top of the ftalk. The branches and principal ftalk are terminated by fpikes of flowers fet very clofe together, which are compofed of fix green erećt petals; in their center is fituated three obtufe germen. From the point of thefe arife fix ftamina, which fpread afunder, terminated by four-cornered fummits. Thefe are fucceeded by oblong compreffed capfules with one cell, filled with membranaceous feeds.

The fecond fort grows naturally in Hungary and Siberia; it has a perennial root like the former. The leaves are longer and thinner than thofe of the firft fort; they are plaited in the like manner, but are of a yellowifh green colour, and appear fooner in the fpring; the ftalks rife higher than thofe of the former. It has fewer leaves upon it, and does not branch into fo many fpikes. The flowers of this are of a dark red colour, and the petals fpread open flat, in which it differs from the former.

The third fort grows naturally in Virginia, and other parts of North America, where it is fometimes called Rattle-fnake-root. The root of this is tuberous; the leaves are oblong, and fhaped like thofe of Plantain, having feveral longitudinal furrows or plaits, fpreading themfelves on the ground. Between thefe come out a fingle ftalk which rifes near a foot high, having a few very fmall leaves or fheaths, placed alternately; and at the top the flowers are produced in a fingle, thick, clofe fpike; they are fmall, and of a yellowifh white colour, but are rarely fucceeded by feeds here.

The fourth fort was fent me from Philadelphia by Mr. John Bartram, who found it growing naturally in that country. The root of this is compofed of thick flefhy fibres; the leaves are oblong, oval, of a light green colour, having fix longitudinal veins or plaits, fpreading on the ground, rounded at their points, and continue all the year. In the center of the leaves fprings up a fingle erećt ftalk a foot high, having a few veftiges, or fmall leaves, ftanding alternately clofe to the ftalk, which end in acute points. The ftalk is terminated by a thick obtufe fpike of dark red flowers, whofe petals fpread open flat. In the center of the petals

petals is situated three obtuse germen joined together, from whose point arises six stamina, which spread asunder, and are longer than the petals; these are terminated by four-cornered summits, of a purple colour. This plant flowers the latter end of June, and in warm seasons the seeds will ripen here.

The first of these plants is that which is ordered for medicinal use, and is by much the stronger and more acrid plant of the two; for when both sorts are placed near each other, the snails will entirely devour the leaves of the second sort, when at the same time they scarcely touch those of the first.

The plants are also very pretty ornaments, when planted in the middle of open borders in the pleasure-garden; for if they are placed near hedges or walls, where snails generally harbour, they will greatly deface the leaves, especially of the second sort, by eating them full of holes before they are unfolded; and as a great part of the beauty of these plants is in their broad folded leaves, so, when they are thus defaced, the plants make but an indifferent appearance.

Both these sorts may be propagated by parting their roots in autumn, when their leaves decay; but they should not be parted too small, for that will prevent their flowering the following summer; these heads should be planted in a light fresh soil, in which they will thrive exceedingly, and produce strong spikes of flowers. The roots should not be removed oftener than once in four or five years, by which time (if they like the soil) they will be very strong, and produce many heads to be taken off; but if they are frequently transplanted, it will prevent their increasing, and cause them to flower very weak.

They may also be propagated by seeds, which should be sown as soon as ripe, either in a bed or box, filled with fresh light earth. In the spring the plants will appear, at which time, if the season proves dry, you should now and then refresh them with water, which will greatly promote their growth; you must carefully clear them from weeds, which, if permitted to grow, will soon overspread and destroy these plants while young. The autumn following, when their leaves decay, you should prepare a bed of fresh light earth; then carefully take up the young plants (observing not to break their roots,) and plant them therein about six inches square, where they may remain until they are strong enough to flower, when they should be transplanted into the borders of the pleasure-garden; but, as these plants seldom flower in less than four years from seeds, this method of propagating them is not much practised.

The two American sorts are at present scarce in the English gardens, but, as they are hardy enough to thrive in the open air, in a few years they may be more plenty; these may be propagated by offsets or seeds, in the same manner as the former.

VERBASCUM. Tourn. Inst. R. H. 146. tab. 161. Mullein.

The CHARACTERS are,

The flower has a small permanent empalement of one leaf, cut into five parts; it hath one wheel-shaped petal, with a very short cylindrical tube, the brim cut into five oval obtuse segments, and five awl-shaped stamina, which are shorter than the petal, terminated by roundish, compressed, erect summits, with a roundish germen, supporting a slender style inclining to the stamina, crowned by a thick obtuse stigma. The germen afterward becomes a roundish capsule with two cells, opening at the top, having an half oval receptacle fixed to the partition, filled with angular seeds.

The SPECIES are,

1. VERBASCUM (*Thapsus*) foliis decurrentibus utrinque tomentosis. Vir. Cliff. 13. *Mullein with running leaves, which are woolly on both sides; or white Mullein, Hig-taper, or Cows-lungwort.*

2. VERBASCUM (*Lychnitis*) foliis cuneiformi-oblongis. Hort. Upsal. 45. *Mullein with oblong wedge-shaped leaves.*

3. VERBASCUM (*Album*) foliis cordato-oblongis, subtus incanis, spicis racemosis. *Mullein with oblong heart-shaped leaves, which are hoary on their under side, and branching spikes of flowers; or Female Mullein.*

4. VERBASCUM (*Luteum*) foliis radicalibus ovatis petiolatis, caulinis oblongis sessilibus subtus tomentosis serratis. *Mullein with oval lower leaves growing on foot-stalks, but those on the stalks oblong, sawed, woolly on their under side, and sitting close.*

5. VERBASCUM (*Grandiflorum*) foliis ovato-acutis utrinque tomentosis, floribus in spicâ densissimâ sessilibus. Haller. Helvet. 507. *Mullein with oval acute-pointed leaves, which are woolly on both sides, and flowers disposed in thick spikes sitting close to the stalk; or Female Mullein with a large yellow flower.*

6. VERBASCUM (*Nigrum*) foliis serratis supernè rugosis, infernè subhirsutis, petiolis ramosis, staminum barbâ purpurascente. Haller. Helvet. 511. *Mullein with sawed leaves, whose upper sides are rough, those on the under side hairy, branching foot stalks, and purplish beards to the stamina; commonly called Sage-leaved black Mullein.*

7. VERBASCUM (*Sinuatum*) foliis radicalibus pinnatifido-repandis tomentosis, caulinis amplexicaulibus nudiusculis ramis primus oppositis. Lin. Sp. Plant 25. *Mullein with the lower leaves oblong, sinuated, woolly, waved, and those on the stalks heart-shaped, embracing the stalks with their base, and almost naked; or black Mullein with a horned Poppy leaf.*

8. VERBASCUM (*Glabrum*) foliis amplexicaulibus oblongis glabris, pedunculis solitariis. Hort. Upsal. 46. *Mullein with oblong smooth leaves embracing the stalks, and single foot-stalks to the flowers; or white Moth Mullein.*

9. VERBASCUM (*Blattaria*) foliis radicalibus pinnato sinuatis, caulinis dentatis acuminatis semi-amplexicaulibus, pedunculis solitariis. *Mullein with the lower leaves jagged like wings, those on the stalks acute pointed, indented, and half embracing the stalks, and single foot-stalks to the flowers; yellow Moth Mullein.*

10. VERBASCUM (*Ferrugineum*) foliis ovato-oblongis obsoletè crenatis, utrinque virentibus petiolatis, caule ramoso. *Mullein with oblong oval leaves having obsolete crenatures, and both sides green, with a branching stalk; or Moth Mullein with an iron-coloured flower.*

11. VERBASCUM (*Annuum*) foliis radicalibus oblongis integerrimis, utrinque viridibus, caulinis acutis sessilibus, pedunculis aggregatis. *Mullein with oblong, entire, lower leaves, which are green on both sides, those on the stalks acute-pointed, sitting close, and clustered foot-stalks.*

12. VERBASCUM (*Phœniceum*) foliis ovatis crenatis radicalibus, caule subnudo racemoso. Lin. Sp. Plant. 178. *Mullein with naked, oval, crenated, lower leaves, and an almost naked branching stalk; or purple Moth Mullein.*

13. VERBASCUM (*Myconi*) foliis lanatis radicalibus, scapo nudo. Lin. Sp. Plant. 179. *Mullein with woolly lower leaves, and a naked stalk; commonly called Borage-leaved Auricula.*

The first is the common Mullein or Hig-taper which is used in medicine, which grows naturally by the side of highways and on banks in many parts of England; it is a biennial plant, which perishes soon after it has perfected seeds. The lower leaves, which spread on the ground, are long and broad, very woolly, and of a yellowish white colour, having scarce any foot-stalks. The stalk rises four or five feet high; the lower part is garnished with leaves, shaped like those below, but smaller, whose base half embraces

braces the ſtalk, and have wings running along the ſtalk. The upper part is cloſely garniſhed with yellow flowers, ſitting very cloſe, formed into a long thick ſpike, compoſed of five obtuſe roundiſh petals, having five ſtamina in the center, of an agreeable odour.

The ſecond ſort grows naturally in ſome parts of England; I have obſerved it in plenty in ſome parts of Nottinghamſhire: this is alſo a biennial plant. The lower leaves are oblong, indented on their edges, ending in acute points. The ſtalk riſes three or four feet high, ſending out from every joint ſhort ſpikes of ſmall yellow flowers, which are paler than thoſe of the firſt, and have a pleaſanter odour. At the baſe of each ſpike is ſituated a ſmall, oblong, acute-pointed leaf, covered with a white powder which waſhes off. When the flowers decay, they are ſucceeded by oval capſules, filled with ſmall ſeeds, which ripen in autumn.

The third ſort grows naturally in Italy and Spain. The lower leaves of this are broad, rough on their upper ſide, and a little hoary; their under ſide is pale and woolly. The ſtalk riſes ſix or ſeven feet high, ſending out ſome erect ſide branches; the flowers are diſpoſed in long branching ſpikes; they are white, having the moſt agreeable ſcent of all the ſpecies.

The fourth ſort has oval leaves, ſtanding upon thick foot-ſtalks; they are of a ſoft texture, of a pale green on their upper ſide, but hoary on their under. The ſtalk riſes three our four feet high; the upper part is garniſhed with ſmaller leaves, of the ſame ſhape with thoſe below; the upper part of the ſtalk is garniſhed with pale yellow flowers, diſpoſed in a looſe ſpike, having ſmall leaves intermixed with the flowers.

The fifth ſort has oval leaves which terminate in a point; they are of a yellowiſh green colour, and woolly on both ſides. The ſtalks riſe about four feet high; they are of a purpliſh colour, covered with a hoary down. The flowers ſit very cloſe to the ſtalk, forming a very thick ſpike, having no leaves between them; they are much larger than thoſe of the firſt ſort, and are of a deeper yellow colour.

The ſixth ſort grows naturally in ſeveral parts of England The lower leaves of this are ſpear-ſhaped, and rounded at the foot-ſtalk, where they are indented like a heart; they are of a pale green on their upper ſide, and hoary on their under, indented on their edges; thoſe upon the ſtalk are oblong, acute-pointed, and ſawed. The ſtalks riſe three or four feet high, the upper part ending in a long ſpike of yellow flowers, which ſtand in ſhort ſpikes or cluſters; theſe have purpliſh ſtamina which are bearded; they have an agreeable odour at a ſmall diſtance, but, when ſmelt too near, become leſs agreeable.

The ſeventh ſort grows naturally in Italy, Greece, and alſo upon the rocks at Gibraltar. The lower leaves are oblong, ſinuated on their borders, waved and hoary. The ſtalk riſes four or five feet high, ſending out many ſlender branches, garniſhed with heart-ſhaped leaves, whoſe baſe embrace the ſtalk; the upper part of the ſtalk and branches have no leaves, but the flowers are diſpoſed in ſmall cluſters at diſtances; they are ſmall, yellow, and have little odour.

The eighth ſort grows naturally in the ſouth of France and Italy. The leaves of this are oblong, ſmooth, and of a dark green colour; the ſtalk riſes three or four feet high, ſending out two or three ſide branches, garniſhed with oblong, ſmooth, green leaves, whoſe baſe embrace the ſtalk. The flowers come out ſingly from the ſide of the ſtalk, upon foot-ſtalks an inch long; they have one petal, cut into five obtuſe ſegments almoſt to the bottom; they are white within, and have a little bluſh of red on the outſide: the ſeed veſſels of this ſort are round, and filled with ſmall ſeeds.

The ninth ſort grows naturally in ſome parts of England; this differs from the former, in the lower leaves being much longer, and deeply ſinuated on their edges in a regular manner, in imitation of the rangement of the lobes of winged leaves; they are of a brighter green colour than thoſe of the former. The ſtalks riſe much taller; the flowers are of a bright yellow colour, and the ſtamina, which are hairy, are of a purple colour.

The tenth ſort is commonly cultivated in gardens, and is known by the title of iron coloured Moth Mullein; this has a perennial root, in which it differs from all the former ſorts, though there are ſome who ſuppoſe it to be only a variety of the laſt mentioned, but it differs greatly from that in other reſpects The bottom leaves are oblong, oval, crenated on their edges, but entire; they are of a dark green on their upper ſide, of a pale green on their under, ſtanding upon pretty long foot-ſtalks. The ſtalk riſes three or four feet high, branching out on each ſide, and has a few ſharp-pointed ſmall leaves on the lower part ſitting cloſe to the ſtalk. The flowers are diſpoſed in a long looſe ſpike on the upper part of the ſtalk, having ſhort ſlender foot-ſtalks; they are of one petal cut almoſt to the bottom into five obtuſe ſegments, of a ruſty iron colour, and are larger than thoſe of the common ſort. This plant does not produce ſeeds here.

The eleventh ſort grows naturally in Sicily; this is a biennial plant, which periſhes ſoon after the ſeeds are ripe. The lower leaves are long, rounded at their points, are entire, and of a deep green on both ſides. The ſtalk is ſtrong, and riſes five or ſix feet high, garniſhed with ſmall acute-pointed green leaves, whoſe baſe ſits cloſe to it. The flowers form a very long looſe ſpike at the top, coming out in cluſters from the ſide of the ſtalk; they are large, of a deep yellow colour, and are ſucceeded by large round capſules, which are brown, opening in two parts, filled with ſmall dark-coloured ſeeds.

The twelfth ſort grows naturally in Spain and Portugal. The root of this is perennial; the leaves are oval, entire, of a light green colour, and a little hairy; the ſtalk riſes three feet high, and is almoſt naked of leaves, but the flowers, which come out ſingly, are ranged along it almoſt the whole length, ſtanding upon ſhort foot-ſtalks. They are of a dark blue, inclining to purple; theſe appear in June and July, but are ſeldom ſucceeded by ſeeds here.

The thirteenth ſort grows naturally upon the Alps and Pyrenean mountains; it is a very humble plant, whoſe leaves ſpread on the ground. The roots are compoſed of ſlender fibres; the leaves are oval, thick, fleſhy, and hairy, crenated on their edges, and have compreſſed hairy foot-ſtalks. Between them ariſe ſlender naked foot ſtalks about four inches long, which divide into three or four ſmall ones at the top, each ſuſtaining one large blue flower, compoſed of five oval petals, which ſpread open flat, and five thick erect ſtamina which ſtand erect. After the flowers are paſt, the germen turns to an oblong-pointed capſule which opens in two parts, and is filled with ſmall ſeeds.

The root of this is perennial, and the plant is uſually propagated by offſets which come out from the ſide of the old plant; theſe ſhould be taken off in autumn, and planted in ſmall pots, filled with light ſandy earth; they muſt always have a ſhady ſituation, for they will not thrive when they are expoſed to the ſun.

The firſt nine and the eleventh ſorts are biennial plants, theſe may be all cultivated by ſowing their ſeeds in Auguſt, on a bed of light earth, in an open ſituation, where the plants will come up the ſucceeding ſpring. In ſpring the plants ſhould be tranſplanted where they are to remain, allowing them a great diſtance, for as they grow large, they muſt not be planted nearer than two feet from other plants.

The

The following year they will flower, and their feeds will be ripe in August or September. Notwithstanding some of these plants grow wild in England, yet two or three of each kind may be admitted into large gardens, for the variety of their hoary leaves, together with the extreme sweetness of their flowers, which have a scent something like Violets; and, as they require little care, they may be allowed a place in the borders of large gardens, where, during their continuance in flower, they will add to the variety; and, if their feeds are permitted to scatter, will come up without care, but the seventh sort seldom produces good seeds in England.

The tenth and twelfth sorts have perennial roots, and as they do not produce good seeds here, they are propagated by offsets, which should be taken off in autumn, time enough to get good root before winter, otherwise they will not flower the following summer. These plants thrive best in a sandy loam, and should be planted on an east border, where they may have only the morning sun, for they do not thrive well when they are too much exposed to the sun.

VERBENA. Tourn. Inst. R. H. 200. tab. 94. Vervain.

The CHARACTERS are,

The flower has an angular permanent empalement, indented in five parts at the brim; it has one petal, with a cylindrical tube the length of the empalement, cut into five points at the brim, which spread open, and are nearly equal; it has sometimes two, at others four very short bristly stamina within the tube, two of which are shorter than the other, with as many incurved summits as stamina, with a four-cornered germen, supporting a slender style the length of the tube, crowned with an obtuse stigma. The germen afterward becomes two or four oblong seeds, closely shut up in the empalement.

The SPECIES are,

1. VERBENA (*Officinalis*) tetrandra, spicis filiformibus paniculatis foliis multifido-laciniatis, caule solitario. Lin. Sp. Plant. 20. *Vervain with four stamina, slender spikes disposed in panicles, leaves having many-pointed jags, and a single stalk; this is the common Vervain.*

2. VERBENA (*Hastata*) tetranda, spicis longis acuminatis, foliis hastatis. Hort. Upsal. 8. *Vervain with four stamina to the flowers, long acute-pointed spikes, and halbert-shaped leaves.*

3. VERBENA (*Supina*) tetrandra, spicis filiformibus solitariis, foliis bipinnatifidis. Lin. Sp. Plant. 21. *Vervain with four stamina to the flowers, single slender spikes, and double wing-pointed leaves; or narrow-leaved Vervain.*

4. VERBENA (*Urticæfolia*) tetrandra, spicis filiformibus paniculatis, foliis indivisis serratis petiolatis. Hort. Ups. 9. *Vervain with four stamina to the flowers, slender spikes growing in panicles, and undivided sawed leaves having foot-stalks; Canada Vervain.*

5. VERBENA (*Spuria*) tetrandra, spicis filiformibus, foliis multifido laciniatis, caulibus numerosis. Hort. Ups. 8. *Vervain with four stamina to the flowers, slender spikes, leaves with many-jagged points, and numerous stalks; Nettle-leaved Vervain of Canada.*

6. VERBENA (*Bonariensis*) tetrandra, spicis fasciculatis, foliis lanceolatis amplexicaulibus. Hort. Upsal. 8. *Vervain with four stamina to the flower, spikes disposed in bunches, and spear-shaped leaves embracing the stalks; tallest Vervain of Buenos Ayres.*

7. VERBENA (*Carolina*) tetrandra, spicis filiformibus paniculatis, foliis infernè cordato oblongis caulinis lanceolatis serratis petiolatis. *Vervain with four stamina to the flowers, slender spikes growing in panicles, the under leaves oblong and heart-shaped, and those on the stalks spear-shaped, sawed, having foot-stalks.*

8. VERBENA (*Nodiflora*) tetrandra, spicis capitato-conicis, foliis serratis, caule repente. Flor. Zeyl. 399. *Vervain with four stamina to the flowers, spikes growing in conical heads, sawed leaves, and a creeping stalk.*

9. VERBENA (*Indica*) diandra, spicis longissimis carnosis subnudis. Lin. Sp Plant. 19. *Vervain with two stamina to the flowers, and very long fleshy spikes, which are almost naked.*

10. VERBENA (*Americana*) diandra, spicis carnosis subnudis, foliis ovatis obtusis, obsolete crenatis petiolatis. *Vervain with two stamina to the flowers, fleshy spikes which are almost naked, and oval obtuse leaves growing upon foot-stalks, having slight indentures.*

11. VERBENA (*Orubica*) diandra, spicis longissimis foliolis. Lin. Sp. Plant. 18. *Vervain with two stamina to the flowers, and the longest leafy spikes.*

12. VERBENA (*Jamaicensis*) diandra, spicis brevioribus, foliis ovatis serratis, subtus incanis. *Vervain with two stamina to the flowers, shorter spikes, and oval sawed leaves, which are hoary on their under side.*

13. VERBENA (*Stæchadifolia*) diandra, spicis ovatis, foliis lanceolatis serrato-plicatis, caule fruticoso. Prod. Leyd. 377. *Vervain with two stamina to the flowers, oval spikes, and spear-shaped leaves which are plaited, and a shrubby stalk.*

14. VERBENA (*Fruticosa*) diandra, spicis rotundis, foliis ovatis serratis, caule fruticoso ramoso. *Vervain with two stamina to the flowers, round spikes, oval sawed leaves, and a shrubby branching stalk.*

15. VERBENA (*Angustifolia*) diandra, spicis carnosis subnudis, foliis lineari-lanceolatis obsoletè serratis. *Vervain with two stamina to the flowers, naked fleshy spikes, and narrow spear-shaped leaves, slightly sawed on their edges.*

16. VERBENA (*Mexicana*) diandra, spicis laxis, calycibus fructûs reflexo pendulis subglobosis hispidis. Lin. Sp. Plant. 19. *Vervain with two stamina to the flowers, loose spikes, the empalement of the fruit almost globular, prickly, and reflexed downward.*

17. VERBENA (*Curassavica*) diandra, spicis laxis, calycibus aristatis, foliis ovatis argutè serratis. Lin. Sp. Pl. 19. *Vervain with two stamina to the flowers, loose spikes, bearded empalements, and oval leaves which are sharply sawed.*

18. VERBENA (*Rugosa*) diandra, spicis ovatis, foliis subrotundis serratis & rugosis, caule fruticoso ramoso. *Vervain with two stamina to the flowers, oval spikes, roundish, sawed, rough leaves, and a shrubby branching stalk.*

The first is very common on the side of roads, foot-paths, and farm-yards near habitations; for although there is scarce any part of England, in which this is not found in plenty, yet it is never found above a quarter of a mile from a house; which occasioned its being called Simpler's Joy, because wherever this plant is found growing, it is a sure token of a house being near; this is a certain fact, but not easy to be accounted for. It is rarely cultivated in gardens, but is the sort directed by the College of Physicians for medicinal use, and is brought to the markets by those who gather it in the fields.

There is another species which approaches near to this, but is taller; the leaves are broader, and the flowers larger. It came from Portugal, and is by Tournefort titled Verbena Lusitanica, latifolia procerior. Inst. R. H. 200. *Taller, broad-leaved, Portugal Vervain.* But I am in some doubt of its being specifically different from the common sort, though the plants in the garden grow much taller, branch more, and the flowers are larger than the first, yet as there is so near an affinity, I cannot be sure it is a different species.

The second sort grows naturally in most parts of North America; this sends up many four-cornered furrowed stalks from the root, which rise five or six feet high, garnished with oblong leaves, ending in acute points, deeply sawed on their edges, and stand upon slender foot-stalks; from the joints

joints come out short branches, set with smaller leaves of the same form. The stalks are terminated by spikes of blue flowers in clusters, which appear in August, and if the autumn proves favourable, the seeds will ripen the middle of October.

The third sort grows naturally in Spain and Portugal; this is a biennial plant, which perishes soon after the seeds are ripe. The stalks branch much, and rise near two feet high. The leaves are double wing pointed, and sit close to the stalks. The flowers are disposed in loose spikes, of a light blue colour, and larger than those of the common sort.

The fourth sort grows naturally in most parts of North America; this is a biennial plant. The stalks are four-cornered, about three feet high. The leaves are long, ending in acute points, and sawed on their edges. The stalks are terminated by panicles of spikes, which are long, slender, and sustain small white flowers, which are ranged loosely, and are succeeded by seeds which ripen in autumn.

The fifth sort also grows naturally in North America; this is a biennial plant, whose lower leaves are long, deeply jagged, and sawed on their edges, of a deep green colour. The stalks rise two feet high, garnished with small leaves of the same shape. The upper part of the stalk branches out into numerous foot-stalks, which sustain panicles of spiked blue flowers, and if the season proves favourable, the seeds will ripen in autumn.

The sixth sort grows naturally at Buenos Ayres; this has four-cornered stalks, which rise six or seven feet high, branching from the side, garnished with long spear-shaped leaves, whose base embrace the stalks, of a pale green colour. The stalks are terminated by spikes of blue flowers, which are clustered together. These appear late in summer, so do not always produce good seeds in England in the open air.

The seventh sort grows naturally at Philadelphia. The seeds were sent me by Dr. Bensel; this is a perennial plant. The lower leaves are heart-shaped, rough, of a dark green colour, are sawed on their edges, ending in acute points. The stalks rise six feet high, branch toward the top, and are terminated by slender spikes of white flowers, formed into panicles; these appear late in autumn, so that unless the season proves favourable, the seeds do not ripen here.

The eighth sort grows naturally in Virginia, and also in Jamaica. The stalks of this trail upon the ground, and emit roots from their joints, whereby they spread, and propagate greatly; and from these arise other branches about eight or nine inches high, garnished with oval spear-shaped leaves, sawed on their edges, and sit close to the stalks. The flowers are collected in conical heads, standing upon naked foot-stalks, which spring from the wings of the branches; they are of a yellowish white colour, and come late in autumn, so are rarely succeeded by good seeds here.

The ninth sort grows naturally in most of the islands in the West-Indies; it is an annual plant. The stalk rises a foot and a half high, garnished with oblong oval leaves, placed opposite, of a light green, and sawed on their edges. The stalk is terminated by a long fleshy spike of blue flowers, which are succeeded by two oblong seeds ripening late in autumn. The spikes of flowers are from a foot to a foot and a half in length.

The seeds of the tenth sort were sent me from Panama, where it grows naturally in moist places; this is an annual plant, whose stalks rise a foot high, garnished with oval, blunt-pointed, fleshy leaves, standing upon long foot-stalks; they are notched slightly on their edges, and are of a light green. The stalks are terminated by thick spikes of blue flowers, which appear late in autumn; so that unless the season proves warm, the seeds do not ripen in England.

The seeds of the eleventh sort were also sent me from Panama; this rises with a shrubby stalk three feet high, which divides into three or four branches, garnished with oblong oval leaves placed opposite, of a deep green on their upper side, but hoary on their under, and are deeply sawed; their foot stalks are short, and have leafy borders running from the base of the leaves. The flowers grow in thick spikes a foot long, which terminate the branches. They are large, and of a fine blue colour, so make a good appearance, and have small acute-pointed leaves intermixed with them on the spikes. This plant, when the season is warm, will perfect seeds in autumn.

The seeds of the twelfth sort were sent me from Paris, and were said to come from Senegal in Africa; this is a perennial plant, with a branching stalk two feet high, garnished with oval sawed leaves placed opposite, of a deep green on their upper side, but hoary on their under. The flowers are disposed in fleshy spikes at the end of the branches, which are shorter, and not so thick as those of the former sorts. The flowers are small, white, so make but little appearance; the seeds ripen in autumn, but the plants may be preserved two years in a warm stove.

The thirteenth sort grows naturally in Jamaica, and in several other places in the West-Indies. This rises with a shrubby branching stalk five or six feet high, adorned with spear-shaped leaves, sawed on their edges, standing upon short foot-stalks. The flowers have long naked foot-stalks, which arise from the wings of the stalk; they are blue, and collected in oval heads; these appear late in autumn, so unless the season proves warm, the seeds do rarely ripen in England, but the plants may be kept two or three years in a warm stove.

The fourteenth sort grows naturally at Campeachy; this has a shrubby branching stalk four feet high, garnished with oval sawed leaves, of a light green colour. The flowers are of a pale blue, collected into oval heads, which stand upon long naked foot-stalks, springing from the wings of the branches; this flowers late in autumn, so is not succeeded by seeds in England.

The fifteenth sort grows naturally at La Vera Cruz; this is an annual plant, with a branching stalk a foot and a half high, garnished with pale green leaves, ending in acute points, slightly sawed on their edges. The branches are terminated by fleshy spikes of blue flowers which are naked, and in warm seasons are succeeded by seeds which ripen in autumn.

The sixteenth sort grows naturally in Mexico; this has a shrubby stalk which rises five or six feet high, dividing into several branches, garnished with oblong sawed leaves which end in acute points, sitting close to the branches, of a light green on both sides. The branches are terminated by slender loose spikes of pale flowers which are very small, whose empalements afterward become swelled, and almost globular; they are reflexed downward, and set with stinging hairs. It flowers late in the summer, and in good years the seeds ripen in England.

The seventeenth sort grows naturally at La Vera Cruz; this has a slender ligneous stalk, which branches and rises near three feet high, adorned with small oval leaves, of a light green, which are sharply indented on their edges. The flowers stand sparsedly upon slender foot-stalks, arising from the wings of the branches; toward the top, the flowers are ranged at a distance from each other in a loose spike; they are small, and of a bright blue colour, sitting very close; these are succeeded by two seeds inclosed in the empalement, which is terminated by short awns or beards.

The eighteenth sort grows naturally at Campeachy; this has a strong woody stalk ten or twelve feet high, covered with

with a light brown bark, fending out many ligneous branches, garnifhed with roundifh, fawed, rough leaves, of a light green colour, ftanding upon fhort foot-ftalks. The flowers are fmall, of a pale blue colour, collected into oval heads, ftanding upon naked foot-ftalks; thefe feldom appear in this country, and are not fucceeded by feeds here; but the plants are eafily propagated by cuttings, if planted during the fummer months, fo may be preferved many years in a moderate ftove.

The firft fort, as was before obferved, being a common weed in England, is not kept in gardens.

The third fort may be eafily propagated by feeds which fhould be fown in autumn, and requires no other culture than to keep it clean from weeds, and thin the plants where they are too clofe.

The fourth and fifth forts may alfo be propagated in the fame manner, and are equally hardy. If the feeds are permitted to fcatter, the plants will come up the following fpring.

The fecond and feventh forts have perennial roots, and are hardy enough to thrive in the open air; thefe may be propagated by feeds, which fhould be fown in autumn, for when they are fown in the fpring, they rarely grow the fame year; the plants require no other culture but to keep them clean from weeds, and allow them proper room to fpread; they may alfo be propagated by parting their roots in autumn. They love a foft loamy foil not too dry.

The other forts being natives of warmer climates require more care. The feeds of thefe fhould be fown upon a hot-bed early in the fpring, and when the plants are fit to remove, they fhould be each tranfplanted into a feparate fmall pot, and plunged into a frefh hot-bed to bring them forward; they muft be fhaded in the day time with mats until they have taken new root, then they muft be treated in the fame way as other tender plants from the fame countries.

Thofe forts which are annual muft be removed into the ftove, or a good glafs-cafe, when they are become too tall to remain longer under the frames; for if they are placed abroad in the open air, they will not ripen their feeds here unlefs the fummer is very warm; therefore where there is a conveniency of having a bark-bed in a glafs-cafe, for plunging fome of thefe tender annual plants, they will thrive much better, and come to greater perfection, than thefe which are placed on fhelves.

The feventeenth fort is by much the tendereft plant of all the fpecies, and is very difficult to preferve when young. The feeds of this fhould be fown in a fmall pot, and plunged into a good hot-bed of tanners bark. When the plants appear, they fhould be fhaded from the fun in the heat of the day. They muft alfo be frequently refrefhed with water, but it muft be given to them fparingly, for much wet will kill them. When they are tranfplanted into fmall pots, they muft be carefully fhaded till they have taken new root, and they muft be conftantly kept in the bark-bed.

VERBESINA. Lin. Gen. Plant. 873. Indian Hemp Agrimony.

The CHARACTERS are,

The common empalement of the flower is compofed of a double order of leaves, which are channelled. The flower is made up of hermaphrodite florets in the difk, and female half florets in the border. The hermaphrodite florets are funnel-fhaped; they have five very fhort hair-like ftamina, terminated by cylindrical fummits, and a germen the fame figure as the feed, fupporting a flender ftyle, crowned by two reflexed ftigmas. The germen afterward becomes a thick angular feed, crowned by a few three-pointed chaff.

The SPECIES are,

1. VERBESINA (*Alata*) foliis alternis decurrentibus undulatis obtufis. Hort. Cliff. 411. *Verbefina with alternate running leaves, which are obtufe and waved.*

2. VERBESINA (*Nodiflora*) foliis lanceolatis ferratis feffilibus. Hort. Cliff. 500. *Verbefina with fpear-fhaped fawed leaves, which are obtufe.*

3. VERBESINA (*Lavenia*) foliis ovatis trinerviis glabris petiolatis, feminibus tricornibus. Flor. Zeyl. 310. *Verbefina with oval leaves having three veins, placed oppofite on foot-ftalks, and feeds with three horns.*

4. VERBESINA (*Proftrata*) foliis oppofitis lanceolatis integerrimis, caulibus procumbentibus, floribus feffilibus. *Verbefina with fpear-fhaped entire leaves, which are placed oppofite, trailing ftalks, and flowers fitting clofe to the branches.*

5. VERBESINA (*Pfeudo-acmella*) foliis oppofitis lanceolatis argutè dentatis, caule ramofo pilofo. *Verbefina with fpear-fhaped, acutely indented leaves placed oppofite, and an erect, branching, hairy ftalk.*

6. VERBESINA (*Acmella*) foliis oblongo-ovatis trinerviis fubdentatis, pedunculis unifloris dichotomiæ caulis. Flor. Zeyl. *Verbefina with oval, fpear-fhaped, fawed leaves, placed oppofite, and fingle flowers upon each foot ftalk, produced from the divifions of the ftalk.*

The firft fort grows naturally in moft of the iflands of the Weft-Indies; it is an annual plant, with an upright winged ftalk about two feet high, from the fides of which fpring out toward the top a few fhort branches. The leaves are oval, blunt, and waved on their edges; they are placed alternate, and from the bafe of each leaf is extended a leafy border, running along two fides of the ftalk. The flowers ftand upon long naked foot-ftalks, arifing from the top and the wings of the ftalk; they are of a deep Orange colour, and are compofed of hermaphrodite and female florets, included in one common fpherical empalement, which are both fruitful: thefe are fucceeded by broad, compreffed, bordered feeds with two teeth, which ripen in the empalement.

The fecond fort grows naturally in the Weft-Indies; this has an upright branching ftalk a foot and a half high. The leaves are fpear-fhaped, a little fawed on their edges, fitting clofe to the ftalk oppofite. The flowers arife from the wings of the ftalk upon flender foot-ftalks, three, four, or more fpringing from the fame joint; each of thefe fuftain one white radiated flower, compofed of many florets, which are fucceeded by oblong black feeds.

The third fort grows naturally in both Indies; this rifes with an upright branching ftalk two or three feet high. The leaves are oval, acute-pointed, and fmooth, having three longitudinal veins; they ftand oppofite upon pretty long foot-ftalks. The flowers fpring from the wings and ends of the branches; they are yellow, and ftand upon fhort foot-ftalks.

The fourth fort grows naturally in India; this has trailing ftalks, which fpread on the ground; they extend two feet or more in length, and put out roots from their joints, fending out many fide branches. The leaves are long, broad, fmooth, and entire. The flowers are very fmall and white; thefe fit clofe to the ftalks at the bafe of the leaves.

The fifth fort grows naturally in the Weft-Indies; this has a purplifh, hairy, branching ftalk, which rifes a foot and a half high. The leaves are long and broad, ending in acute points; they have a few fharp indentures on their edges, and ftand oppofite. The flowers are white, ftanding upon flender foot-ftalks, which fpring from the wings of the ftalk, fometimes fingle, and at others two or three at the fame joint.

The fixth fort grows naturally in both Indies; the ftalks of this branch out their whole length, and decline downward. The leaves are fmooth, heart-fhaped, and have three veins; they are indented on their edges, and ftand oppofite. The flowers ftand upon long naked foot-ftalks, which fpring from the wings of the branches; they are of a yellow colour, and have oblong prominent difks, with a few very fmall rays.

Thefe plants are propagated by feeds, which fhould be fown upon a moderate hot-bed in the fpring; and when the plants are fit to remove, they fhould be tranfplanted on a frefh hot-bed to bring them forward, and muft be afterward treated in the fame way as other tender annual plants, being careful not to draw them up too weak. In June they may be taken up with balls of earth, and planted in a warm border, where they muft be fhaded and watered till they have taken new root; after which they will require but little care, and will produce good feeds in autumn.

VERONICA. Tourn. Inft. R. H. 143. tab. 60. Male Speedwell, or Fluellin.

The CHARACTERS are,

The flower has a permanent empalement, cut into five acute fegments; it has one tubulous petal the length of the empalement; the brim is cut into four oval plain fegments, which fpread open, and two ftamina, which are terminated by oblong fummits, with a compreffed germen, fupporting a flender declining ftyle, crowned by a fingle ftigma. The germen afterward becomes a compreffed heart-fhaped capfule with two cells, filled with roundifh feeds.

The SPECIES are,

1. VERONICA (*Officinalis*) fpicis lateralibus pedunculatis, foliis oppofitis, caule procumbente. Lin. Mat. Med. 11. *Speedwell with fpikes of flowers growing upon foot-ftalks, fpringing from the fides of the ftalks, leaves placed oppofite, and a trailing ftalk; or common Male Speedwell, or Fluellin.*

2. VERONICA (*Spuria*) fpicis terminalibus, foliis ternis æqualiter ferratis. *Speedwell with fpikes of flowers terminating the ftalks, and narrow, fpear-fhaped, fawed leaves, placed oppofite; or narrow-leaved fpiked Speedwell.*

3. VERONICA (*Longifolia*) fpicis terminalibus, foliis oppofitis lanceolatis ferratis acuminatis. Hort. Upfal. 7. *Speedwell with fpikes of flowers terminating the ftalks, and acute-pointed fawed leaves, which are lance-fhaped, placed oppofite; or greater broad-leaved upright Speedwell.*

4. VERONICA (*Spicata*) fpicâ terminali, foliis oppofitis crenatis obtufis, caule adfcendente fimpliciffimo. Lin. Sp. Pl. 10 *Speedwell with a fpike of flowers terminating the ftalk, obtufe crenated leaves placed oppofite, and a fingle afcending ftalk; or fmaller fpiked Speedwell.*

5. VERONICA (*Pannonica*) fpicis lateralibus paniculatis, foliis ovatis inæqualiter crenatis feffilibus. *Speedwell with fpikes of flowers in panicles from the wings of the ftalk, and oval leaves, which are unequally notched, and fit clofe; or Hungarian Speedwell.*

6. VERONICA (*Hybrida*) fpicis terminalibus, foliis oppofitis obtusè ferratis fcabris, caule erecto. Lin. Sp. Pl. 11. *Speedwell with fpikes of flowers terminating the ftalk, rough, obtufe, fawed leaves, which are placed oppofite, and an erect ftalk; or Welfh fpiked Speedwell.*

7. VERONICA (*Virginica*) fpicis terminalibus, foliis quaternis quinifve. Lin. Sp. Plant. 9. *Speedwell with fpikes of flowers terminating the ftalks, and four or five leaves at each joint; or tall Virginian Speedwell with many fpikes of white flowers.*

8. VERONICA (*Maritima*) fpicis terminalibus, foliis ternis æqualiter ferratis. Hort. Upfal. 7. *Speedwell with fpikes of flowers terminating the ftalks, and leaves growing by threes, which are equally fawed; or long-leaved fpiked Speedwell.*

9. VERONICA (*Auftriaca*) fpicis lateralibus pedunculatis laxis argute dentatis, foliis oppofitis ferratis. Lin. Sp. Pl. 10. *Speedwell with loofe fpikes of flowers growing on foot-ftalks, and leaves growing by threes, which are unequally fawed.*

10. VERONICA (*Orientalis*) fpicis terminalibus, foliis pinnato incifis acuminatis. *Speedwell with fpikes of flowers terminating the ftalks, pinnated leaves growing oppofite, and an erect ftalk.*

11. VERONICA (*Maxima*) racemis lateralibus, foliis cordatis rugofis dentatis, caule ftricto. *Speedwell with loofe fpikes of flowers growing upon foot-ftalks, fpringing from the wings of the ftalk, and very narrow fharply-fawed leaves placed oppofite; or Auftrian Speedwell.*

12. VERONICA (*Incana*) fpicis terminalibus, foliis oppofitis crenatis obtufis, caule erecto tomentofo. *Speedwell with fpikes of flowers terminating the ftalks, crenated leaves, and hoary ftalks, which are erect; or eaftern Speedwell.*

13. VERONICA (*Fruticulofa*) fpicis longiffimis lateralibus pedunculatis, foliis oppofitis inæqualiter ferratis. *Speedwell with long fpikes of flowers proceeding from the wings of the ftalk, and a ftrait ftalk; or greateft Speedwell.*

15. VERONICA (*Hybrida*) fpicis terminalibus, foliis oppofitis crenatis obtufis, caule erecto tomentofo. Hort. Upfal. 7. *Speedwell with fpikes of flowers terminating the ftalks, crenated obtufe leaves placed oppofite, and an erect woolly ftalk; or hoary, woolly, fpiked Speedwell.*

15. VERONICA (*Becabunga*) racemis lateralibus, foliis ovatis planis, caule repente. Flor. Suec. 11. *Speedwell with lateral fpikes of flowers, oval plain leaves, and a creeping ftalk; or greater Water Speedwell, commonly called Brooklime.*

There are feveral other fpecies of this genus, fome of which grow naturally in England; but, as they are rarely admitted into gardens, it is befide the intention of this work to mention them.

The firft fort grows wild in woods, and other fhady places, in divers parts of England, and is a plant of little beauty; but, as it is the fort which is ufed in medicine, under the title of Paul's Betony, I thought it neceffary to infert it here. This is a low plant, whofe ftalks trail upon the ground, and put out roots from their joints, whereby it fpreads and propagates. The leaves are oval, about an inch long, fawed on their edges, and placed oppofite. The flowers are difpofed in fpikes, which arife from the wings of the ftalk; they are fmall, of a pale blue colour, and have one petal, which is cut at the brim into four fegments; when they decay, the germen turns to a capfule, not unlike that of Shepherd's Pouch in fhape, filled with fmall feeds.

This is generally brought to market by fuch perfons as make it their bufinefs to gather herbs in the fields, fo that it is not often cultivated in gardens; but thofe who have a mind to propagate it, may do it with much eafe, for as the branches trail upon the ground, they pufh out roots from their joints, which branches being cut off and planted, will take root, and grow in almoft any foil or fituation. The whole herb is ufed in medicine, and is one of the Wound-herbs which are brought from Switzerland. A tea of this herb is much recommended for the gout and rheumatifm.

The fecond fort grows naturally in Italy and Spain; this has a perennial root, which fends out many offsets, by which it is eafily propagated. The lower leaves are long and hairy; the ftalks rife a foot high, and are garnifhed with very narrow fpear-fhaped leaves, placed oppofite, which have a few flight ferratures on their edges. The ftalks are terminated by long fpikes of blue flowers, which are fucceeded by feeds in capfules like the former. It has been doubted if this was fpecifically different from the common upright Speedwell; but I have many times propagated this by feeds, and have always found the plants fo raifed, maintain their difference. There is a variety of this with a flesh-coloured flower.

The third sort grows naturally in Austria and Hungary. The lower leaves of this are long, broad in the middle, drawing to a point at each end; they are sawed on their edges, and are of a lucid green colour. The stalks rise a foot and a half high, and are garnished with leaves of the same shape with the lower, but smaller, placed opposite; they are terminated by long spikes of blue flowers, which appear in June, and are succeeded by flat seed-vessels, filled with compressed seeds, which ripen in autumn.

The fourth sort grows naturally in the northern parts of Europe, and also in several closes near Newmarket-heath. The lower leaves of this are about an inch and a half long, and three quarters of an inch broad, of a pale green colour, and notched on their edges. The stalks rise a foot and a half high; they do not branch; the leaves on the lower part stand opposite, but on the upper they are alternate; the stalks are terminated by short spikes of blue flowers, which appear about the same time as the former.

The fifth sort grows naturally in Hungary. The lower leaves of this are unequally notched; the stalks rise a foot high, garnished with leaves, placed opposite, of a lucid green, which sit close to the stalks. The flowers are disposed in panicled spikes, which stand upon long naked foot-stalks that spring from the upper wings of the stalk; they are larger than those of the other species, and are of a beautiful blue colour, so make a fine appearance, but are of short duration.

The sixth sort grows naturally on the Alps and Pyrenean mountains, and also upon the mountains in Wales. The lower leaves of this are rough and hairy, blunt-pointed, and obtusely sawed on their edges, standing upon pretty long foot-stalks; the stalks grow erect about six or eight inches high, garnished with oval notched leaves, placed opposite. From the side of the stalk spring out two or three branches, which toward the bottom are garnished with small leaves, placed opposite, but terminate in long spikes of pale blue flowers. The spikes on the side branches are four or five inches long, but those of the principal stalk are eight or nine.

The seventh sort grows naturally in Virginia. The stalks of this are erect, and rise four or five feet high, garnished at each joint by four or five spear-shaped sawed leaves, which stand round the stalk in whorls, ending in acute points. The stalks are terminated by long slender spikes of white flowers, which appear late in July; these are succeeded by compressed capsules, filled with seeds, which ripen in autumn.

The eighth sort grows naturally in Italy and the south of France. The stalks of this rise three feet high garnished with leaves, placed by fours toward the bottom, but at the top by threes at each joint; they are deeply sawed on their edges, ending in acute points, of a bright green colour; the stalks are terminated by spikes of blue flowers.

The ninth sort grows near the sea in several parts of Europe. The stalks of this do not rise so high as those of the former; the leaves are placed by fours and threes round the stalk, and have longer foot-stalks; they are broader at the base, and run out into long acute points; they are unequally sawed on their edges, and are of a bright green colour. The flowers are disposed in spikes, which terminate the stalks, of a bright blue colour.

The tenth sort grows naturally in many parts of France and Germany. The stalks of this are single, and do not branch; they are round, hairy, and rise a foot and a half high, garnished with spear-shaped hairy leaves. The stalk is terminated by a long spike of blue flowers.

The eleventh sort grows naturally in Austria. The lower leaves of this are narrow, and cut into fine segments; the stalks are slender, and incline downward, garnished with linear leaves, which are acutely notched on their edges; the flowers are disposed in long loose spikes, which spring from the wings of the stalk, of a bright blue colour, and stand upon foot-stalks.

The twelfth sort grows naturally in the Levant; this has slender branching stalks which decline, and are garnished with narrow leaves, which are acutely cut on their edges, of a pale green colour, and smooth. The flowers are disposed in loose spikes on the top and from the side of the stalk, of a pale blue colour.

The thirteenth sort grows naturally upon Mount Baldus in Italy. The stalks of this are slender, stiff, and upright, garnished by rough heart-shaped leaves, which are indented, and placed opposite; those on the lower part of the stalk are small; in the middle they are much larger, and diminish again in their size toward the top. The flowers come out in spikes from the wings of the stalk toward the top, of a bright blue colour.

The fourteenth sort grows naturally in the Ukrain Tartary. The stalks of this are very white and woolly; they rise about a foot and a half high, garnished with oblong hoary leaves, placed opposite, notched on their edges, and sit close to the stalks, which are terminated by spikes of deep blue flowers, which stand erect.

The fifteenth sort is the common Brooklime, which grows naturally in brooks and streams of water in most parts of England, so is not cultivated in gardens, but as it is much used in medicine, I have given it a place here. The stalks of this are thick, succulent, and smooth, emitting roots from their joints, whereby they spread and propagate. The leaves are oval, flat, succulent, and smooth; they stand opposite; the flowers come out in long bunches from the wings of the stalk; they are of a fine blue colour, and stand upon short foot-stalks; these appear great part of summer, and are succeeded by heart-shaped seed-vessels, filled with roundish seeds. The whole herb is used, and is esteemed an excellent antiscorbutic.

These plants may all be propagated by parting their roots, which may be done every other year, for if they are not often parted or divided, they will many of them grow too large for the borders of small gardens; but yet they should not be parted into very small heads, because when they have not a number of stems, so as to form a good bunch, they are soon past their beauty, and have but a mean appearance. The best time to part these roots is at Michaelmas, that they may be well rooted again before winter; for when they are removed in the spring, they seldom flower strong the same year, especially if the season should prove dry. Those sorts which grow pretty tall, are very proper to plant on the sides of open wilderness quarters, but those with trailing branches are fit for the sides of banks or irregular shady slopes, where they will make an agreeable variety; they are all of them very hardy, so are in no danger of suffering by cold, and require no other care but to keep them clean from weeds, and to be transplanted every second or third year.

They may also be propagated by seeds, which should be sown in autumn, for when they are sown in the spring, the plants rarely come up the same year; but, as most of the sorts propagate very fast by their offsets, their seeds are seldom sown

If these plants are placed in a shady border, they will thrive much better than when they are more exposed to the sun, and their flowers will continue much longer in beauty.

VIBURNUM. Lin. Gen. Plant. 332. The Wayfaring, or pliant Meally-tree.

The CHARACTERS are,

The flower has a small permanent empalement, which is cut into five parts; it has one bell-shaped petal, cut at the brim into five obtuse segments, which are reflexed; it has five awl-shaped stamina the length of the petal, terminated by roundish summits, and a roundish germen, situated under the flower, having no style, but the place is occupied by a roundish gland, and crowned by three obtuse stigmas. The germen afterward turns to a compressed fruit with one cell, inclosed in one hard seed.

The SPECIES are,

1. VIBURNUM (*Lantana*) foliis cordatis serratis venosis subtus tomentosis. Vir. Cliff. 25. *Wayfaring-tree with heart-shaped, sawed, veined leaves, which are woolly on their under side.*

2. VIBURNUM (*Prunifolium*) foliis subrotundis crenato-serratis glabris. Flor. Virg. 33. *Wayfaring-tree with roundish, crenated, sawed leaves, which are smooth; commonly called Black Haw.*

3. VIBURNUM (*Dentatum*) foliis ovato-orbiculatis profundè serratis venosis. *Wayfaring-tree with oval round leaves, which are deeply sawed and veined.*

4. VIBURNUM (*Tinus*) foliis ovatis integerrimis, ramificationibus subtus villoso-glandulosis. Lin. Sp. Pl. 267. *Wayfaring-tree with oval entire leaves, whose branches are hairy, and glandulous on the under side; or hairy-leaved Laurustinus.*

5. VIBURNUM (*Lucidum*) foliis ovato-lanceolatis integerrimis, utrinque virentibus lucidis. *The shining-leaved Laurustinus.*

6. VIBURNUM (*Nudum*) foliis ovato-lanceolatis integerrimis, subtus venosis. *American Tinus with oval leaves, which are entire.*

7. VIBURNUM (*Opulus*) foliis lobatis petiolis glandulosis. Lin. Sp. Pl. 261. *Common Guelder Rose.*

8. VIBURNUM (*Americanum*) foliis cordato-ovatis acuminatis serratis, petiolis longissimis lævibus. *American Guelder Rose with acute-pointed sawed leaves, and white flowers.*

The first sort grows naturally in many parts of Europe, and is the common Viburnum or Lantana of the old botanists. The leaves are heart-shaped, much veined, irregularly sawed on their edges, and are very woolly on their under side. The stalks are woody, and rise twelve or fourteen feet high, sending out strong ligneous branches, which are covered with a light coloured bark; these are terminated by umbels of white flowers, whose summits are red, and are succeeded by roundish compressed berries, which turn first to a bright red colour, and are black when ripe, inclosing one seed of the same shape.

There is a variety of this with variegated leaves, which is preserved in some of the gardens near London; but when she plants are removed into good ground, and are vigorous, their leaves become plain.

The second sort grows naturally in most parts of North America, where it is commonly called Black Haw. This rises with a woody stalk ten or twelve feet high, covered with a brown bark, and sends out branches from the side; these, when young, are covered with a purple smooth bark, and are garnished with oval smooth leaves, which are slightly sawed on their edges, and stand upon short slender foot-stalks, sometimes opposite, and at others without order. The flowers are disposed in small umbels, which come out from the side and at the end of the branches; they are white, and smaller than those of the common Viburnum; these are succeeded by berries, which ripen in autumn.

The third sort grows naturally in North America. The stalks of this are soft and pithy, and branch out greatly from the bottom upward. The bark is of a gray colour; the leaves are roundish, oval, strongly veined, and sawed on their edges, of a light green colour, and placed opposite upon pretty long foot-stalks. The flowers are disposed in a corymbus at the end of the branches, which are white, and almost as large as those of the common sort; these appear the latter end of June, but are seldom succeeded by seeds in England.

The fourth sort is the Laurustinus with small leaves, which are hairy on their under side; this plant is so well known as to need no description, but as it is frequently confounded with the next, it may be necessary to point out its difference. The leaves of this are seldom more than two inches and a half long, and one and a quarter broad; they are rounded at their base, but end in acute points; they are veined and hairy on their under side, and are not of so lucid a green colour on their upper side. The umbels of flowers are smaller, and appear in autumn, continuing all the winter, and the plants are much hardier.

The fifth sort is commonly known in the nursery-gardens by the title of shining-leaved Laurustinus. The stalks of this rise higher; the branches are much stronger than those of the former sort. The bark is smoother, and turns of a purplish colour; the leaves are larger, of a thicker consistence, and of a lucid green colour; the umbels are much larger, and so are the flowers; these seldom appear till the spring, and when the winters are sharp, the flowers are killed, so never open unless they are sheltered. The plants of this sort were formerly kept in tubs, and housed in winter, and, when they were so treated, made a fine appearance early in the spring, and in very mild seasons the plants in the open air do the same.

There is a variety of this with variegated leaves, which makes as good a figure as any of the striped plants which are preserved in gardens.

The sixth sort is a native of North America, where it rises to the height of ten or twelve feet, sending out branches on every side their whole length; these have a smooth purplish bark, garnished with oval entire leaves, of a thick consistence, and of a lucid green; they stand opposite. The flowers are produced in umbels at the end of the branches; they are white, and not unlike the flowers of Laurustinus; these appear in July, and are succeeded by berries, which seldom ripen in England.

There seems to be two sorts of this in the gardens, one of which comes from the more northern parts of America, and sheds its leaves in winter; the other, which grows in Carolina and Virginia, is an evergreen, but both are so much alike in summer, as scarce to be distinguished.

The seventh sort is the common Marsh Elder, which grows naturally in marshy grounds, and on the sides of rivers in many parts of England, so is not often kept in gardens; it is called by some of the nursery-gardeners Guelder Rose with flat flowers, to distinguish it from the other, whose flowers are globular. The Marsh Elder is the original species, and the Guelder Rose is a variety which accidentally arose from it. The former has a border of male flowers, which are large, and the middle of the umbel is composed of hermaphrodite flowers, which are succeeded by oval red berries; the latter has all male flowers, of the same size and shape with those of the border of the first, so that they swell out into a round figure, which has occasioned some country people giving it the title of Snowball-tree. This sort is cultivated in gardens for the beauty of its flowers, which make a fine appearance during their continuance.

It will rise to the height of eighteen or twenty feet, if it is permitted to stand. The stem becomes large, woody, and grows irregular; they have a gray bark. The leaves are placed opposite; they are divided into three or four lobes; the branches come out opposite, and are apt to be divided somewhat like those of the Maple, jagged on their edges, and of a light green colour. The flowers come out at the

those

end of the branches; those of the first in large umbels, and those of the second in a corymbus; they are very white, and appear the beginning of June; those of the first have oval berries succeeding the hermaphrodite flowers, which turn of a scarlet colour when ripe, but the other, having only male flowers, are barren.

The eighth sort grows naturally in Carolina and some other parts of North America; this rises with a shrubby stalk eight or ten feet high, sending out many branches, which are covered with a smooth purple bark, garnished with heart-shaped oval leaves, ending in acute points; they are deeply sawed on their edges, have many strong veins, and stand upon very long slender foot-stalks opposite. The flowers are collected into large umbels at the end of the branches; those ranged on the border are male and barren, but the middle is composed of hermaphrodite flowers, which are succeeded by oval berries. The flowers are white, and the berries are red when ripe.

The first sort may be propagated either from seeds, or by laying down the tender branches; but the former method being tedious, is seldom practised, because the seeds seldom grow the first year, unless they are sown in autumn, and as the branches easily put out roots, that is the more expeditious method.

The best time for laying these branches is in autumn, just as the leaves begin to fall (the manner of laying them being the same as for other hardy trees, need not be here repeated.) By the succeeding autumn the layers will be rooted, when you may take them off from the old plants, and transplant them into a nursery for two or three years, in which they may be trained up to regular stems and heads, and may afterward be planted where they are to remain.

The second sort is generally propagated by layers here, because the seeds do not often ripen in England. The young shoots of this take root very freely; the cuttings will also take root, if they are planted in autumn; the seeds, when they are brought to England, always remain in the ground a year like those of the other sorts, so that the propagating the plants by seeds is a tedious method.

The Laurustinuses are propagated by laying down their young branches, which put out roots very freely; so that when they are layed in autumn, they will be well rooted by that time twelve months, when they should be taken off from the old plants, and may be either planted where they are to remain, or into a nursery to grow two years to get strength. The best season to transplant these is at Michaelmas, that they may get new root before winter; for as these plants begin to flower early in winter, it is a plain indication of their growing at that season, so they will more surely succeed than at any other time of the year, though they may be removed in the spring with balls of earth to their roots, provided it is done before they begin to shoot; they may also be removed the latter end of July or the beginning of August, if rain happens at that time, for after they have done shooting, which is soon after Midsummer, they will be in no danger, provided they are not kept out of the ground any time.

These plants may also be propagated by seeds, which should be mixed with earth in autumn, soon after they are ripe; these should be exposed to the open air, and receive the rain in winter, and in the spring they may be sown upon a gentle hot-bed, which will bring up the plants; these should remain in the bed till autumn, and then may be transplanted and treated in the same way as the layers. I have raised many of these plants from seeds, which I find hardier than those raised by layers.

Some people train up the Laurustinus with naked stems to have round heads, but if these are planted in the open air, they will be in more danger of suffering by severe frost, than those whose branches grow rude from the bottom; for if the frost kills the outer part of the shoots, the stems will be protected, so will soon put out new branches: but where the stems are naked, the frost frequently kills them to the root.

The sixth sort may be propagated in the same way as the Laurustinus, and requires the same treatment; it loves a soft loamy soil, and should have a sheltered situation.

The seventh and eighth sorts are easily propagated by layers or cuttings. The common Guelder Rose sends out plenty of suckers from the roots, by which it is frequently propagated; but as the plants so raised are very subject to put out suckers, they are not so good as those which come from layers or cuttings. Both these sorts love a moist soil, in which they will make much greater progress, and produce their flowers in greater plenty than on a dry soil.

They are both very hardy, so will thrive in the coldest situations, but not within the spray of the sea. The common Guelder Rose is seldom suffered to stand very long in gardens, but I have seen one in an old garden, whose stem was more than two feet and a half round.

VICIA. Tourn. Inst. R. H. 396. tab. 221. Vetch, or Tare.

The CHARACTERS are,

The flower is of the butterfly kind; the standard is oval, broad at the tail, indented at the point, and the borders are reflexed; the two wings are shorter than the standard; the keel is shorter than the wings; the tail is oblong. It has ten stamina, nine joined and one separated, terminated by erect summits, and a linear compressed germen, supporting a slender style, crowned by an obtuse stigma, which is bearded on the under side. The germen afterward turns to a long pod with one cell, opening with two valves, ending with an acute point, containing several roundish seeds.

The SPECIES are,

1. VICIA (*Cracca*) pedunculis multifloris, floribus imbricatis, foliolis lanceolatis pubescentibus, stipulis integris. Lin. Sp. Plant. 735. *Vetch with many imbricated flowers on a foot-stalk, spear-shaped hairy lobes to the leaves, and entire stipula; or many-flowered Vetch.*

2. VICIA (*Sylvatica*) pedunculis multifloris, foliolis ovalibus, stipulis denticulatis. Lin. Sp. Pl. 734. *Vetch with foot-stalks supporting many flowers, oval lobes to the leaves, and indented stipulæ; or the largest many-flowered Wood Vetch.*

3. VICIA (*Cassubica*) pedunculis subsexfloris, foliolis denis ovatis acutis, stipulis integris. Lin. Sp. Plant. 735. *Vetch with foot-stalks, having about six flowers, leaves with ten oval acute lobes, and entire stipulæ.*

4. VICIA (*Biennis*) pedunculis multifloris, petiolis sulcatis, subdodecaphyllis, foliolis lanceolatis glabris. Lin. Sp. Pl. 736. *Many-flowered Vetch with furrowed foot-stalks, and for the most part twelve spear-shaped smooth lobes to each leaf.*

5. VICIA (*Sativa*) leguminibus sessilibus subbinatis erectis, foliis retusis, stipulis notatis. Lin. Sp. Plant. 736. *Vetch with erect pods growing by pairs, and sitting close to the stalks, blunt lobes to the leaves, and spotted stipulæ; or common cultivated Vetch, with a black seed, frequently called Tares.*

There are many more species of this genus, some of which grow naturally in England; but as they are rarely cultivated except in botanic gardens for the sake of variety, they are omitted, as they are plants of little use or beauty.

The first sort here mentioned grows naturally among bushes, and by the sides of woods in most parts of England. The root is perennial, but the stalks are annual; these are weak, requiring support; they rise five or six feet high, fastening their tendrils, which grow at the end of their leaves, to the bushes or hedges, whereby they climb; they are hairy, as are also the leaves, which are composed of about ten pair of spear-shaped lobes, terminated by a tendril. The flowers stand upon long foot-stalks, which spring from

from the wings of the ſtalk; the ſpikes are long; the flowers lie one over the other; they are of a fine blue colour, ſo make a pretty appearance, when they come out from between the buſhes or ſhrubs which ſupport them; they appear in July, and are ſucceeded by compreſſed pods, filled with round ſeeds, which ripen in autumn.

The ſecond ſort grows naturally in the woods near Bath and Briſtol; this has a perennial root. The ſtalks are weak, and climb by the help of their tendrils over the neighbouring buſhes and hedges, riſing to the height of ſeven or eight feet. The leaves are compoſed of ſeven or eight pair of oval ſmooth lobes, terminated by tendrils. The flowers are produced in long ſpikes from the wings of the ſtalks; they are of a pale blue colour, and are larger than thoſe of the former ſort; they appear in July, and are ſucceeded by ſhort ſmooth pods, filled with round ſeeds, which ripen in autumn.

The third ſort grows naturally in Caſſubia; this has a ligneous creeping root; the ſtalks trail upon the ground; they grow three feet long, and their lower part become ligneous toward autumn, but they die to the root in winter. The leaves are compoſed of ten pair of oval acute-pointed lobes. The flowers come out from the wings of the ſtalk; they are diſpoſed in ſhort ſpikes, each containing, for the moſt part, ſix pale blue flowers, which appear in July, and are ſucceeded by ſhort ſmooth pods like thoſe of Lentils, including three or four round ſeeds, which ripen in autumn.

Theſe ſorts have been recommended to be ſown in the fields for fodder for cattle; but as their ſtalks are ſlender and leſs ſucculent than thoſe of the common Vetch, ſo it is doubtful if theſe will anſwer the purpoſe of farmers to cultivate them; for as their ſtalks trail to a great length, ſo if they have not ſupport, they will be ſubject to rot by lying upon the ground; and although their roots are perennial, yet as it is late in the ſpring before they ſhoot to a height ſufficient to cut for uſe, ſo there is little want of green feed for cattle at that time.

However, a few of theſe plants may be allowed a place in large gardens for the ſake of variety, where, if they are properly placed, they may be ornamental, particularly on the borders of wood-walks, or in thickets of ſhrubs. If ſome of the firſt ſort are allowed to climb up upon their branches, they will have a good effect during their continuance in flower.

Theſe ſorts are propagated by ſeeds, which ſhould be ſown in autumn ſoon after they are ripe, for if they are kept out of the ground till ſpring, the ſeeds often fail, or at leaſt remain in the ground a year before they vegetate; they ſhould be ſown in the places where the plants are deſigned to remain, for they do not bear tranſplanting well. Theſe plants grow naturally in woods and thickets of buſhes, where their roots are ſcreened from the ſun, and their ſtalks furniſhed with ſupport by the buſhes, point out the places where the ſeeds ſhould be ſown, which ſhould be where they are ſheltered by ſhrubs. If three or four ſeeds are ſown in each patch, it will be ſufficient, for if one or two plants come up in each place, it will be enough. When the plants come up, they will require no other culture but to keep them clean from weeds, and their ſtalks muſt be permitted to climb upon the neighbouring ſhrubs; for if they trail upon the ground, they will produce few flowers, and in wet ſeaſons the ſtalks will rot, ſo the plants will be rather unſightly.

The fourth ſort grows naturally in Siberia; this is a biennial plant, which promiſes fairly to become a uſeful one for fodder; for the ſtalks of this grow to a great length, and are well furniſhed with leaves: theſe do not decay in autumn, but continue green through the winter in defiance of the moſt ſevere froſt; ſo that in February and March, when there is often a ſcarcity of green feed for ewes and lambs, this may be of great ſervice.

The ſtalks of this riſe five or ſix feet high, if ſupported. The leaves are compoſed of five or ſix pair of ſmooth ſpear-ſhaped lobes, terminated by tendrils. The foot-ſtalks are deeply furrowed. The flowers are produced in ſpikes upon long foot-ſtalks, which ſpring from the wings of the ſtalks; they are of a light blue colour, and appear in July; theſe are ſucceeded by ſhort compreſſed pods, containing three or four round ſeeds, which ripen in autumn.

This ſort is propagated by ſeeds, which may be ſown in the ſpring or autumn; and when the plants come up, they will require no other culture but to keep them clean from weeds; and if they are ſupported from trailing upon the ground, they will continue in verdure all the winter, and the following ſummer they will flower and produce ripe ſeeds.

If this plant is deſigned for feed, the ſeeds ſhould be ſown in rows at four feet diſtance, and ſhould be dropped thin in the rows; for as the ſtalks ſend out many branches, and extend to a great length, ſo when the plants are too cloſe, the branches will intermix, and mat ſo cloſely together, as to rot each other by excluding the air. When the plants come up, they muſt be kept clean from weeds, which, while they are young, ſhould be performed with Dutch hoes, but afterward it may be done by the hoeing-plough, which will ſave expence; and with this inſtrument the plants may be earthed up in the ſame manner as Peaſe and Beans, which will greatly ſtrengthen their ſtalks, and make them and the leaves larger and more ſucculent, ſo increaſe the quantity of feed. If this is practiſed as often as may be found neceſſary to deſtroy the weeds in ſummer, it will prepare the ground for any crop which may afterward be put upon the land; and as this will be in no danger of ſuffering from froſt, ſo it ſhould be preſerved till the ſpring, when there is a want of green feed for ewes, at which time it may be cut as it is wanted; but a part of the plants ſhould be permitted to ſtand for ſeeds, for thoſe which are cut, if they do ſhoot again, will flower ſo late in ſummer, that unleſs the autumn proves very warm, the ſeeds will not ripen; therefore it will be a better way to ſow a ſufficient quantity of ſeeds for this purpoſe, in a ſeparate ſpot of ground, becauſe, when the other is cut, the ground may be ploughed for other crops; and if in mild ſeaſons there may be ſo great plenty of other green feed as not to want this, if the plants are ploughed into the ground, it will be a good dreſſing for other crops.

This is what I am now beginning to try in the field, where I have not as yet had experience of its culture; but what I have here adviſed, is founded upon experiments which I have for ſeveral years made, on ſmall patches ſown in gardens in different ſituations. In all theſe patches I have found the plants continue in great verdure, when moſt of the perennial plants in the ſame ſituation have ſuffered greatly by the froſt; and from eight of theſe plants, I could have cut as much feed as would have been equivalent to half a truſs of green Clover.

The fifth ſort is the common Vetch or Tare, which is much cultivated in the fields for fodder; of this there are two varieties, if not diſtinct ſpecies. The firſt, which is the moſt common, has a black ſeed; the other has ſeeds as white, if not whiter than the whiteſt Peaſe; and this difference is permanent, for I have ſown both ſorts many years, and have never found either of them vary. Theſe plants are annual, and periſh ſoon after they have perfected their ſeeds. The ſtalks are angular, ſtreaked, and hairy; they are weak and want ſupport, ſo generally decline where they have nothing near to faſten themſelves to. The leaves are compoſed of ſeveral pair of blunt lobes, and are terminated

nated by tendrils. The flowers come out from the wings of the ſtalk, ſitting very cloſe to the baſe of the foot-ſtalks of the leaves; two of theſe generally ſpring from the ſame joint; they are pretty large, and of the butterfly ſhape; they are purple, and appear in June and July, which are ſucceeded by erect pods, containing three or four round ſeeds in each.

There is another kind of Vetch which is cultivated in the fields, with a ſmaller black ſeed; this is called in ſome counties Rath-ripe Vetch, and in others Pebble, or Summer Vetch; but this being much tenderer than the common Vetch, is not much cultivated, as it muſt always be ſown in the ſpring, and will ripen its ſeeds the ſame ſummer, but it will not afford near ſo much fodder as the other.

Vetches are generally ſown at two ſeaſons, one is in autumn, and the other early in the ſpring; but the beſt time is in Auguſt, for the ſeeds which are ſown then will come up ſoon, and the plants will have time to get ſtrength before winter, ſo will be in leſs danger of ſuffering by froſt than thoſe which are ſown later, and will be fit to cut for feed much earlier in the ſpring, for that is the time when green feed is moſt wanted; and if they are deſigned for ſeed, and not to be cut for fodder, thoſe early-ſown Vetches will come ſoon into flower, and the ſeeds will be ripe early, ſo they may be cut and ſtacked in good weather, which is a great advantage, for thoſe which ripen late are often ſtacked or houſed wet, and then the ſeeds frequently ſprout in the mow and are ſpoiled.

The uſual method of ſowing Vetches is in broad-caſt, ploughing them lightly in; in this way, the common allowance of ſeeds for one acre of land is two buſhels, but there are ſome who ſow two buſhels and a half; this practice may do well enough for thoſe Vetches which are deſigned to be cut for fodder in the ſpring, but thoſe which are ſown with an intent to ſtand for ſeeds, will do much better if they are ſown in drills, in the ſame way as is practiſed for Peaſe, and then leſs than half the quantity of ſeeds will be ſufficient, for the drills ſhould not be nearer to each other than three feet, that the hoe-plough may have room to go between them, to deſtroy the weeds, and earth up the plants, for by this management they will produce a much greater crop, and ripen earlier in the ſeaſon. Theſe drills ſhould be about the ſame depth as thoſe uſually made for Peaſe, and the ſeeds ſhould be ſcattered about the ſame diſtance in the drills. Theſe ſeeds ſhould be carefully covered as ſoon as they are ſown; for if they are left open, the rooks will diſcover them; and when they once find the rows, if they are not carefully watched, they will entirely devour them. Indeed theſe being ſown early in autumn, will be in leſs danger than thoſe which are ſown late, or in the ſpring, becauſe there is more food for rooks and pigeons in the open fields at this ſeaſon, and the plants will appear much ſooner above ground. The beſt time to ſow them is about the beginning of Auguſt, for the rains which uſually fall about that ſeaſon, will bring them up in a ſhort time. Toward the latter end of October the plants will have obtained conſiderable ſtrength, therefore they ſhould be earthed up with the hoeing-plough. This work ſhould be performed in dry weather, and in doing it care muſt be had to lay the earth up as high to the ſtems of the plants as poſſible, ſo as not to cover their ſtalks, becauſe this will ſecure them againſt froſt. The whole ſpace of ground between the rows ſhould alſo be ſtirred, in order to deſtroy the weeds, which, if carefully performed in dry weather, will lay the land clean till March; at which time the crop ſhould be earthed a ſecond time, and the ground cleaned again between the rows, which will cauſe the plants to grow vigorous, and in a little time they will ſpread ſo as to meet, and cover the ſpaces; whereas thoſe ſown in the ſpring, will not grow to half this ſize, and will be much later in flowering.

Some people ſow theſe Vetches, and when they are fully grown, plough them into the ground to manure it. Where this is deſigned, there will be no occaſion to ſow them in drills at this diſtance, nor to huſband them in the manner before directed; but in this caſe it will be the beſt method to ſow them in autumn, becauſe they will be fit to plough in much ſooner the following year, ſo that the land may be better prepared to receive the crops for which it is intended. In ſome parts of France, and in Italy, theſe Vetches are ſown for feeding of cattle while green, and are accounted very profitable, and in many parts of England they are cultivated to feed cart-horſes, &c. though upon ſuch land where Lucern will thrive. it will be much better huſbandry to cultivate that for this purpoſe.

Where theſe plants are cultivated for their ſeeds, they ſhould be cut ſoon after the pods change brown; and when they are dry, they muſt be immediately ſtacked, for if they are ſuffered to lie out in the field to receive wet, and there come one hot day after it, the pods will moſt of them burſt, and caſt out the ſeeds. When the ſeeds are threſhed out, the haulm is eſteemed very good for cattle; and ſome have recommended the ſeeds for horſes, and affirm they are as proper for thoſe animals as Beans; which, if true, will render them more valuable, becauſe theſe will grow on the lighteſt ſandy land, where Beans will not thrive, ſo may be a good improvement to ſome counties in England, where they do not attempt to cultivate Beans.

VINCA. Lin. Gen. Plant. 261. Periwincle.

The CHARACTERS are,

The empalement of the flower is permanent. The flower has one ſalver-ſhaped petal, whoſe tube is longer than the empalement. The brim is broad, ſpreading open, and ſlightly cut into five obtuſe ſegments; it has five very ſhort inflexed ſtamina, terminated by erect obtuſe ſummits, and two roundiſh germen, which have two roundiſh corpuſcles on their ſide, ſupporting one common ſtyle the length of the ſtamina, crowned by two ſtigmas. The germen afterward turns to a fruit, compoſed of two taper acute-pointed huſks, opening lengthways with one valve, and filled with oblong cylindrical ſeeds.

The SPECIES are,

1. VINCA (*Minor*) caulibus procumbentibus, foliis lanceolato-ovatis. Lin. Sp. Pl. 209. *Periwincle with trailing ſtalks, and oval ſpear-ſhaped leaves; or common narrow-leaved Periwincle.*

2. VINCA (*Major*) caulibus erectis, foliis ovatis floribus pedunculatis. Lin. Sp. Pl. 209. *Periwincle with erect ſtalks, and oval leaves; or broad-leaved Periwincle.*

3. VINCA (*Roſea*) foliis oblongo-ovatis integerrimis, tubo floris longiſſimo, caule ramoſo fruticoſo. Tab. 86. *Periwincle with oblong, oval, entire leaves, a very long tube to the flower, and a ſhrubby branching ſtalk.*

The firſt ſort grows naturally under hedges and buſhes in many parts of England The ſtalks are ſlender, and trail upon the ground, emitting fibres from their joints, which take root, whereby the plant multiplies and ſpreads greatly. The leaves are placed oppoſite on their ſtalks; they are oval, ſpear-ſhaped, of a thick conſiſtence, very ſmooth, and entire; their upper ſide is of a lucid green, and their under of a paler colour. The flowers ſtand ſingly upon foot-ſtalks, which ſpring from the wings of the ſtalks; they are nearly of a funnel-ſhape, but ſpread more at the brim, which is almoſt flat like a ſalver; their brim is divided into five broad obtuſe ſegments. The moſt common colour of the flower is blue; but it is often found with a white flower, and ſometimes the flowers are variegated with both colours. Theſe flowers begin to appear in April, and there is often a succession of them continued great part of summer. The flowers are very rarely succeeded by seeds. Tournefort says, he was at a loss for the fruit of this plant, to engrave the figure of it in his Elements of Botany, which he obtained by planting some plants in small pots to confine their roots and prevent their stalks from trailing upon the ground. This experiment I tried several years without success; but I afterward planted three or four plants in the full ground, and constantly cut off their lateral shoots, leaving only the upper stalks, and these plants the second year produced plenty of the pods.

There are two varieties of this plant with variegated leaves; one has white, and the other yellow stripes; these are by some preserved in their gardens for the sake of variety. There is also one with double purple flowers, which I believe to be only an accidental variation, therefore have not enumerated it here.

The second sort is also found growing naturally in several parts of England. The stalks of this are larger than those of the former, and do not trail so close to the ground; they rise two feet high, but their tops decline again to the ground, and often put out roots when they are suffered to lie on the ground. The leaves of this sort are oval, heart-shaped, and and opposite upon thick foot-stalks; their upper surface is of a lucid green, their under is of a lighter green colour; they are of a thick consistence, and entire. The flowers come out from the wings of the stalk in like manner as the former, and are of the same shape, but much larger. The usual colour is blue, but they are sometimes seen with white flowers. This sort flowers earlier in the spring than the former, and there is a succession of them great part of summer.

As these plants delight to grow under the cover of trees and bushes, so they may be made ornamental in large gardens, if they are planted on the verges of wildernesses, where they will spread and cover the ground; and as their leaves continue green all the year, they will have a good effect in winter, and their flowers appearing great part of summer, will add to the variety.

They are easily propagated by their trailing stalks, which put out roots very freely, especially those of the first sort; and if the stalks of the large sort are laid in the ground, they will root very soon, and may be cut off and transplanted where they are to remain; when they are once rooted, they will spread and multiply very fast without farther care. The first sort is used in medicine, and is esteemed a good vulnerary plant.

The third sort grows naturally in the island of Madagascar, from whence the seeds were brought to the Royal Garden at Paris, where the plants were first raised, and produced their flowers the following summer; from these plants good seeds were obtained, which were sent me by Mr. Richard, gardener to the king at Versailles and Trianon. It rises with an upright branching stalk to the height of three or four feet, which, when young, are succulent, jointed, and of a purple colour; but as the plants advance, their lower parts become ligneous. The branches which come out from the side, have their joints very close; they have a smooth purple bark, and are garnished with oblong, oval, entire leaves, which are smooth and succulent, sitting pretty close to the branches. The flowers come out from the wings of the branches singly, standing upon very short foot-stalks; their tube is long and slender; their brim spreads open flat, which is divided into five broad obtuse segments, which are reflexed at their points. The upper surface of the petal is of a bright crimson or Peach colour, and their under side is of a pale flesh colour. There is a succession of these flowers upon the same plant, from February to the end of October. Those flowers which appear early in the summer, are succeeded by taper seed-vessels, filled with roundish black seeds, which ripen in autumn.

This sort is propagated by seeds or cuttings; those plants which arise from seeds grow more upright, and do not branch so much as the plants which are propagated by cuttings. The seeds of this should be sown upon a moderate hot bed in the spring, and when the plants come up, and are fit to remove, they should be transplanted on a fresh hot-bed at about four inches distance, shading them from the sun till they have taken new root; then they must be treated in the same way as other tender plants which are natives of warm countries; but there must be great care had to prevent their drawing up weak, nor should they have water in too great plenty. When the plants have obtained strength, they should be carefully taken up with balls of earth to their roots, and planted in pots filled with good earth, and plunged into a moderate hot-bed to facilitate their taking new root, observing to screen them from the sun; and when they are well rooted in the pots, they must be gradually hardened to bear the open air; but unless the summer proves warm, these plants should not be placed abroad, for they will not thrive if they are exposed to cold or wet; therefore during the summer they should be placed in an airy glass-case, and in winter they must be removed into the stove, where the air is kept to a temperate heat, without which they will not live through the winter in England.

If these plants are propagated by cuttings, they should be planted in pots during any of the summer months. The pots should be plunged into a moderate hot-bed, and if they are closely covered with bell or hand-glasses, it will cause them to put out roots sooner than they otherwise would do; when these have put out roots, they must be gradually hardened, and afterward planted in pots, and treated in the same way as the seedling plants.

This plant deserves a place in the stove, as much as any of the exotic plants we have in England, because the flowers are very beautiful, and there is a constant succession of them all the summer.

VINCITOXICUM See ASCLEPIAS.

VINE. See VITIS.

VIOLA. Tourn. Inst. R H. 419. tab. 236. Violet.

The CHARACTERS are,

The flower has a short permanent empalement of five leaves, which are differently ranged in the different species. The flower is of the ringent kind, and is composed of five unequal petals; the upper is broad, obtuse, and indented at the point, having a horned nectarium at the base; two side petals are opposite; the two lower are larger, and reflexed; it has five small stamina, which are annexed as appendages to the entrance of the nectarium, terminated by obtuse summits, which are sometimes connected, and a roundish germen, supporting a slender style, which stands out beyond the summits, and is crowned by an oblique stigma. The germen afterward turns to an oval three-cornered capsule with one cell, opening with three valves, including many oval seeds.

The SPECIES are,

1. VIOLA (*Odorata*) acaulis, foliis cordatis, stolonibus reptantibus. Lin. Sp. Pl. 934. *Violet having stalks, heart-shaped leaves, and creeping shoots; or Purple March Violet.*

2. VIOLA (*Hirta*) acaulis, foliis cordatis piloso-hispidis. Flor. Suec. 718. *Violet without stalk, having heart-shaped leaves with stinging hairs; or hairy, scentless, March Violet.*

3. VIOLA (*Palustris*) acaulis, foliis reniformibus. Haller. Helvet. 5501. *Violet without stalk, and kidney-shaped leaves; or Violet with round smooth leaves.*

4. VIOLA (*Mirabilis*) caule triquetro, foliis reniformis cordatis floribus cauline petalis. Lin. Sp. Pl. 934. *Violet without stalks, and spear-shaped notched leaves; or Acadian Violet.*

5. VIOLA

5. VIOLA (*Multifida*) acaulis, foliis pedatis feptempartitis. Lin. Sp. Plant. 933. *Violet without ftalks, and leaves growing like feet, divided into feven parts; or three-coloured Virginia Violet.*

6. VIOLA (*Pinnata*) acaulis, foliis pinnatifidis palmatis quinque lobis dentatis indivifìfque. Lin. Sp. Plant. 933. *Violet without ftalks, and hand-fhaped leaves, with five indented undivided lobes; or Virginia Violet, with leaves like thofe of the Plane-tree.*

7. VIOLA (*Pinnata*) acaulis, foliis pinnatifidis. Lin. Sp. Pl. 734. *Violet without ftalks, and leaves having many points; or Alpine Violet.*

8. VIOLA (*Cenifia*) acaulis, grandiflora, foliis ovalibus uniformibus integerrimis. Allion. *Violet without a ftalk, having a large flower, and oval entire leaves, which are uniform.*

9. VIOLA (*Montana*) caulibus erectis, foliis cordatis oblongis. Lin. Sp. Pl. 935. *Violet with erect ftalks, and oblong heart-fhaped leaves; or tree-like purple Violet.*

10. VIOLA (*Tricolor*) caule triquetro diffufo, foliis oblongis dentatis, ftipulis multifidis. *Violet with a four-cornered diffufed ftalk, oblong indented leaves, and many-pointed ftipulæ; commonly called Hearts-eafe or Panfies.*

11. VIOLA (*Calcarata*) caule diffufo decumbente, foliis oblongis incifis, ftolonibus reptatricibus. *Violet with a diffufed trailing ftalk, oblong cut leaves, and creeping fhoots; or yellow Mountain Violet with a large flower.*

The firft fort, which is the common fweet Violet, grows naturally under hedges in the neighbourhood of London; but in feveral of the diftant counties, the Violet without fcent is the fort moft frequent. Of the common Violet there are the following varieties. The fingle blue and white; the double blue and white; and the pale purple. Thefe are all of them commonly preferved in gardens for the odour of their flowers, and are fo well known as to need no defcription.

The fecond fort is found growing naturally in many parts of England. The leaves of this are larger, and are covered with rough ftinging hairs. The flowers are larger, and have no fcent, which are the only differences.

The third fort grows naturally in marfhes and on bogs in feveral parts of England. The leaves of this are fmall, kidney-fhaped, and fmooth. The flowers are fmall, and of a pale blue colour; they appear in June, and are fucceeded by fmall oblong capfules, filled with roundifh feed.

The fourth fort grows naturally in North America. The leaves of this are fpear-fhaped, and deeply notched on their edges, ftanding upon fhort foot-ftalks. The flowers are larger than thofe of the common fort, but have no fcent.

The fifth fort is alfo a native of North America. The leaves of this are divided into feven parts or lobes, which are united at the foot-ftalk. The flowers ftand upon naked foot-ftalks; they are of the Panfy kind, and have no fcent; they appear in June, but are not fucceeded by feeds here.

The fixth fort grows naturally in Virginia. The leaves of this are moft of them divided into five lobes like the fingers of a hand, but fome of the lower leaves are entire. The flowers are fmall, white, and have no fcent.

The feventh fort grows naturally on the Alps; this was fent me by Dr. Allione from Turin; it is a very low plant, feldom rifing two inches high. The leaves are fmall, and cut into winged points. The flowers are of a pale blue colour, and appear in June.

The eighth fort was fent me by the fame gentleman, who found it growing on the Alps; this is alfo an humble plant, with oval, entire, uniform leaves, not more than half an inch long, and a quarter broad, ftanding upon fhort foot-ftalks. The flowers are large, of a light blue colour, and appear in June. Thefe have no fcent.

The ninth fort grows naturally on the Alps, and the mountains in Auftria. The root of this is perennial, but the ftalks and leaves decay in autumn; it has erect ftalks, which rife more than a foot high, garnifhed with oblong heart-fhaped leaves. The flowers ftand upon long foot-ftalks, which fpring from the wings of the ftalks; they are fhaped like thofe of the Dog Violet, and are of a pale blue colour; thefe appear the end of May, and are fucceeded by roundifh capfules, filled with fmall feeds, which ripen in Auguft.

The tenth fort is the Hearts-eafe or Panfies, which grows naturally in fome of the northern counties of England, but is generally cultivated in gardens near London. Of this there are many varieties, which differ greatly in the fize and colour of their flowers. Some of thefe varieties have very large beautiful flowers, which have an agreeable odour; others have fmall flowers without fcent; whether thefe are diftinct fpecies or accidental varieties, I have not been able to determine; for I have faved the feeds of moft of the varieties as carefully as poffible, and have fown them feparate, but have always had a mixture arife, which may have come from feeds lying in the ground; for in gardens, where thefe plants have been permitted to fcatter their feeds, it is impoffible to know how long the feeds may lie in the ground, and when they are turned up to the furface, they will grow, which renders it difficult to determine the fpecific differences of thefe plants in fuch places.

This is an annual plant, whofe roots decay after they have flowered and perfected their feeds. The lower leaves are roundifh or oblong, and are indented on their edges; the ftalks rife feven or eight inches high, fending out many diffufed branches, garnifhed with leaves which are longer and narrower than thofe below, notched on their edges, and fit clofe to the branches. The flowers ftand upon long naked foot-ftalks, which fpring from the wings of the ftalk, fhaped like thofe of the common Violet. Some of the varieties have flowers much larger, and others are of the fize of March Violets; fome of them have the two upper petals of a deep yellow colour, with a purple fpot in each, the two middle of a paler yellow with a deep yellow fpot, and the lower petal of a velvet colour; in others the petals are white, with yellow and purple fpots; in fome the yellow is the moft prevailing colour, and in others the purple.

The eleventh fort grows naturally upon mountains in the north of England and in Wales; this has a perennial root, fending out fhoots from the fide, which fpread and propagate, in which it differs from all the Panfies. The lower leaves are oblong and jagged; the ftalks feldom rife more than four or five inches high; they decline, and are garnifhed with narrower leaves than thofe below, which are deeper cut on their fides. The flowers ftand upon naked foot-ftalks two inches long; they are much larger than thofe of the common fort, and are of a deep yellow colour, with a few purple ftreaks in the center. This plant continues flowering great part of fummer, but the flowers have no fcent.

The common Violets are eafily propagated by parting of their roots; this may be done at two feafons. The firft or moft common feafon for removing and parting of thefe roots is at Michaelmas, that the young plants may be well rooted before winter; this is generally practifed where the plants are put on the borders of wood-walks in large plantations; but in the gardens where they are cultivated for their flowers, the gardeners tranfplant and part their plants, foon after their flowering feafon is over. Thefe will have all the remaining fummer to grow and get ftrength, fo will produce a greater quantity of flower the following spring than those which are removed in autumn ; but this is not to be practised where they cannot be supplied with water till they have taken new root, unless in moist seasons.

Violets may also be propagated by seeds, which should be sown soon after they are ripe, which is about the end of August. The plants will come up the following spring, and when they are fit to remove, they should be transplanted in shady borders to grow till autumn, and then they may be planted where they are to remain; but the double-flowering Violets do not produce seeds. Although the white, blue, and purple Violets are generally supposed to be varieties which have accidentally sprung from seeds, yet I have several years sowed the seeds of all the three sorts, and have not found either of them vary.

The other sorts of Spring Violets are sometimes preserved in botanic gardens for the sake of variety; these may be propagated in the same way as the common sort, but require a moist soil and a shady situation.

The upright sort does not send out shoots like the common Violet, so increases but slowly by offsets; this may be propagated by seeds in plenty, and is as hardy as the common sort.

The several varieties of Pansies will scatter their seeds in a short time after the flowers are past; and from these self-sown seeds the plants, which come up in autumn, will flower very early in the spring, and these will be succeeded by the spring plants; so that where they are indulged in a garden, and their seeds are permitted to scatter, there will be a constant succession of their flowers the greatest part of the year, for they will flower all the winter in mild seasons, and most part of the summer in shady situations, which renders them worthy of a place in every good garden; but then they must not be allowed to spread too far, lest they become troublesome weeds, for their seeds, when ripe, are cast out of their covers with great elasticity to a considerable distance, and the plants will soon spread over a large space of ground, if they are permitted to stand.

The common Pansy stands in the College Dispensatory as a medicinal plant, but is rarely used in England.

The great yellow Violet propagates by offsets in pretty great plenty, if it has a moist soil and a shady situation; this may be transplanted in autumn, and the offsets may then be taken off; but the roots should not be divided into small heads, nor should they be too often transplanted, because they will not produce many flowers unless the plants are strong, and have good root in the ground. This sort will not live in a dry soil, nor in a situation much exposed to the sun.

VIORNA. See Clematis.

VIRGA AUREA. See Solidago.

VISCUM. Tourn. Inst. R. H. 609. tab. 380. Misleto.

The Characters are,

It has male and female flowers upon separate plants. The male flowers have no petals, but have four summits, which are oblong and acute-pointed, fastened to the leaves of the empalement. The female flowers have an empalement of four small oval leaves sitting upon the germen, but have no petals or stamina, with an oblong three cornered germen, situated under the flower, having no style, but is crowned by an obtuse stigma. The germen afterward turns to a globular smooth berry with one cell, including a fleshy heart-shaped seed.

We have but one Species of this genus, viz.

Viscum (*Album*) foliis lanceolatis obtusis, caule dichotomo, spicis axillaribus. Lin. Sp. Pl. 1033. *Misleto with blunt spear-shaped leaves, stalks dividing by pairs, and spikes of flowers rising from the wings of the stalk.*

This plant, instead of rooting and growing in the earth like other plants, fixes itself, and takes root on the branches of trees; it spreads out with many branches, and forms a large bush. The branches are ligneous; they have a yellow green bark; the largest is about the thickness of a man's finger; the other are gradually smaller, full of joints, which easily part asunder; at each of which grow two thick fleshy leaves, which are broad and rounded at their points, and narrow at their base. The flowers come out from the wings of the stalk in short spikes; they have four yellow leaves, which are by some called petals, and by others the empalement. The female flowers are succeeded by round white berries, which are almost pellucid, about the size of large white Currants, full of a tough viscid juice, in the middle of which lies one h art shaped flat seed.

It grows upon the white Thorn, the Apple, the Crab, the Hazel, the Ash, and Maple, but is rarely found upon the Oak, though the Misleto of the last has been always accounted the best of all; which opinion, as Mr. Ray well observes, may be owing to the superstitious honour the ancient Druids of this island gave to this Misleto, to whom nothing was more sacred.

This plant is always produced from seed, and is not to be cultivated in the earth as most other plants, but will always grow upon trees, from whence the ancients accounted it a super-plant, most of whom thought it was an excrescence on the tree, without the seed being previously lodged there; which opinion is now generally confuted from a repeated number of experiments.

The manner of its being propagated is this, viz. the Misleto thrush, which feeds upon the berries in winter, when they are ripe, often carry the seeds from tree to tree; for the viscous part of the berry, which immediately surrounds the seed, doth sometimes fasten it to the outward part of the bird's beak, which, to get disengaged of, he strikes his beak against the branches of a neighbouring tree, and thereby leaves the seed sticking by this viscous matter to the bark, which, if it lights upon a smooth part of the tree, will fasten itself thereto, and the following winter will put out and grow; and in the same manner it may be propagated by art, for if the berries, when full ripe, are rubbed upon the smooth part of the bark of a tree, they will adhere closely thereto, and, if not destroyed, will produce plants the following winter.

The trees which this plant doth most readily take upon, are the Apple, the Ash, and other smooth-rinded trees before-mentioned; but I have several times tried it upon the Oak without success, for the bark of that tree is of too close a texture to admit the seeds sticking therein, which is also the reason it is so rarely found upon that tree; and notwithstanding the great encomiums which have been given to the Misleto of the Oak for its medicinal virtues, yet I cannot help thinking that it is equally good from whatever tree it be taken; nor is it possible to find this plant growing in any quantity upon the Oak; so that those persons, who pretend to furnish the town with it for physical use, do but impose upon the world, for it is so rarely met with, that whenever a branch of an Oak-tree hath any of these plants growing upon it, it is cut off, and preserved by the curious in their collections of natural curiosities, and of these there are but few to be seen in England.

As to what some persons have asserted of the manner how it is propagated from tree to tree, by the Misleto thrushes, which eat the berries and void the seed in their dung upon the branches of trees, whereby the seeds are stuck thereon, and take root into the bark and produce fresh plants, I can by no means agree to, since, if it were only this way propagated, it would always be found on the upper part of the sides of such branches, upon which the dung can only be supposed to lodge, whereas it is generally found upon the under side of branches, where it is almost impossible for these birds to cast their dung; besides, I believe the stomachs of these birds are too powerful digesters, to suffer any seeds to pass so entire through the intestines as to afterwards grow; but I shall leave this to such as have leisure to make observations in those places where this plant abounds.

Of the berries of this plant birdlime was formerly made in England. This was done by boiling the berries in water till they burst, when they were well beaten in a mortar, and afterward washed till all the branny husks were cleared away.

VISNAGA. See DAUCUS.

VITEX. Tourn. Inst. R. H. 603. tab. 373. Agnus Castus, or the Chaste-tree.

The CHARACTERS are,

The empalement of the flower is cylindrical, and indented in five parts. The flower has one ringent petal; the brim is plain, and divided into two lips, which are trifid; the middle segment is the broadest in both. It has four hair-like stamina, two being shorter than the other, terminated by moveable summits, and a roundish germen, supporting a slender style, crowned by two awl-shaped spreading stigmas. The germen afterward turns to a globular berry with four cells, each containing one oval seed.

The SPECIES are,

1. VITEX (*Agnus Castus*) foliis digitatis, spicis verticillatis. Lin. Sp. Pl. 938. *Chaste-tree with fingered leaves, and whorled spikes of flowers; or common Chaste-tree.*

2. VITEX (*Latifolia*) foliis digitatis serratis, spicis paniculatis. *Chaste-tree with fingered sawed leaves, and spikes in panicles; Chaste-tree with a broader sawed leaf.*

3. VITEX (*Indica*) foliis ternatis quinatisve, paniculis dichotomis. Lin. Sp. Pl. 938. *Chaste-tree with trifoliate and quinate leaves, and panicles of flowers rising from the division of the branches; smaller Indian Chaste-tree.*

4. VITEX (*Chinensis*) foliis ternatis quinatisque pinnato-incisis, spicis verticillatis terminalibus. *Chaste-taste with ternate and quinate leaves which are cut like wings, and whorled spikes of flowers terminating the branches.*

The first sort grows naturally in Sicily near Naples, by the sides of rivers, and in the Archipelago in moist places; it has a shrubby stalk ten or twelve feet high, sending out branches opposite the whole length, which are angular, pliable, and have a grayish bark, garnished with leaves for the most part placed opposite, composed of five, six, or seven lobes, which unite at the foot stalk, and spread out like the fingers of a hand, ending in blunt points, of a dark green on their upper side, but hoary on their under. The flowers are produced in spikes at the extremity of the branches, from seven to fifteen inches long, disposed in whorls round the stalks, with intervals between each whorl; they are of the lip kind; the two lips are each cut into three segments, the middle being larger than the two sides; in some plants white, and in others blue; these are generally late before they appear, so that in bad seasons they do not open fair. The flowers have an agreeable odour, and make a good appearance in autumn, when the flowers of most other shrubs are gone, for in warm mild seasons I have seen these shrubs in full flower the middle of October.

The second sort grows naturally in the south of France, and in Italy; this is a lower shrub than the first; it seldom rises more than four or five feet high, coming up with several stalks from the root, which do not branch so much as the former; their bark is also whiter. The leaves are fingered, and composed of five or seven lobes which unite at the foot-stalk; these are not so disproportionate in their length, are sawed on their edges, and are not so stiff as those of the former. The flowers come out in panicled spikes toward the end of the branches; the spikes are shorter, and the flowers smaller than those of the first sort, and appear sooner; they are all of them blue which I have seen.

The third sort grows naturally in both Indies; this has a shrubby stalk, which rises nine or ten feet high, sending out many side branches, which have a brown bark, garnished with leaves which have sometimes three, and at others five, oval acute-pointed lobes which are entire, and a little downy on their under side. The flowers are disposed in panicles which arise at the division of the branches; these are small and white, but are not succeeded by any seeds in England.

The fourth sort grows naturally in the northern parts of China, where it rises with woody stalks eight or ten feet high, having a gray bark. The branches come out opposite, garnished with leaves placed opposite upon long foot-stalks; these are composed of three or five spear-shaped lobes, which are deeply sawed on their edges, and end in very acute points, of a dark green on their upper side, but gray on their under. The flowers are disposed in whorled spikes, which come out opposite from the wings of the stalk; these are blue, and about the size of those of the first.

The first sort sort is pretty common in many English gardens, where it has been long an inhabitant, but was not much propagated till of late years. The second sort is less common, and only in some curious gardens at present. These plants are very hardy, and may be propagated by planting their cuttings early in the spring, before they shoot; they require a fresh light soil, and must be frequently refreshed with water until they have taken root; after which they must be carefully cleared from weeds during the summer season, and if the following winter prove severe, you must lay a little mulch upon the surface of the ground between the plants, to prevent the frost from penetrating to their roots, which would injure them while they are young; and as these cuttings are apt to shoot late in the year, their tops will be very tender, and the early frosts in autumn often kill them down a considerable length, if they are not protected, therefore they should then be covered with mats, which will be of great service to them. Toward the middle of March, if the season is favourable, you should transplant them either into the places where they are designed to remain, or into a nursery to grow two or three years to get strength, where they must be pruned up, in order to form them into regular stalks, otherwise they are very subject to shoot out their branches in a straggling manner.

They may also be propagated by laying down their branches in the spring of the year; in doing of which, you must be very careful not to break them, for their shoots are very apt to split if they are much forced; these will take root in one year, provided they are watered in very dry weather, and may then be transplanted out, and managed, as was directed for those plants raised from cuttings.

The third sort is too tender to live in the open air in England, so must be planted in pots, and in winter kept in the stove; it is propagated both by cuttings and layers, but the cuttings of this must be planted in pots, and plunged into a moderate hot-bed, covering them close with a bell or hand glass, to exclude the air; they should be refreshed with water now and then, but it must not be given them too freely. The best time to plant the cuttings is about the middle or latter end of April, for if they succeed they will put out roots in six or seven weeks, and will then begin to shoot, so they should have the free air gradually admitted to them, to prevent their shooting weak; then they may be carefully taken up, and each planted in a separate small pot filled with light earth, and plunged into the hot-bed again, shading them from the sun till they have taken new root, after which they should have plenty of free air at all times when the weather is good, treating them in the same manner as other tender plants. In winter they must be kept in a moderate temperature of heat, but in the fummer they fhould be removed into the open air.

The fourth fort has been lately introduced into the Englifh gardens from Paris, where the plants were raifed from feeds, which were fent from China by the miffionaries. I was favoured with fome young plants by Monfieur Richard, gardener to the king at Verfailles. The two forts with white and blue flowers have fucceeded in the Chelfea garden, but that with red flowers was injured in the way and mifcarried.

This is propagated by cuttings, which muft be planted in the fpring in pots, plunging them into a moderate hot-bed, and, when the cuttings are well rooted, they fhould be carefully taken up, and each planted in a feparate fmall pot, filled with light earth, and placed in the fhade until they have taken new root; then they may be removed to a fheltered fituation, placing them with other green-houfe plants, where they may remain all the fummer; but in autumn they muft be put into fhelter, for they will not live in the open air in this country; but as they caft their leaves early in autumn, fo they muft not have much wet in winter. The plants are late in putting out new leaves in the fpring, and, before thefe appear, they have fo much the appearance of dead plants, that they have been turned out of the pots by fome, fuppofing they were fo.

VITIS. Tourn. Inft. R. H. 613. tab. 384. The Vine.

The CHARACTERS are,

The flower has a fmall empalement indented in five parts; it has five fmall petals which drop off, and five awl-fhaped ftamina, which fpread open and fall away, terminated by fingle fummits, with an oval germen having no ftyle, crowned by a headed obtufe ftigma. The germen afterward turns to an oval or roundifh berry with one cell, including five hard feeds or ftones.

I fhall not trouble the reader with an enumeration of all the forts of Grapes which are at prefent known in England, which would fwell this work much beyond its intended bulk, and be of little ufe, fince many of them are not worth the trouble of cultivating; fo I fhall only felect thofe which ripen well in this country, or that merit a little affiftance to bring them to perfection by artificial heat.

The July Grape; this is called by the French, Morillon noir hatif, is a fmall, round, black berry, growing loofe on the bunches. The juice is fugary, but has little flavour, and has no merit but that of ripening early. It ripens the beginning of Auguft.

The black Sweet Water is a fmall roundifh berry, growing clofe in the bunches, which are fhort. The fkin is thin, the juice very fweet, and the birds and flies are very apt to devour them if they are not guarded. It ripens foon after the other.

The white Sweet Water is a large round berry when in perfection, but thefe are very different in fize on the fame bunch; fome of them will be of a large fize, and others extremely fmall, for which reafon it is not much efteemed. The juice is fugary, but not vinous. This ripens about the fame time with the former.

The Chaffelas Blanc, or Royal Mufcadine, as it is called by fome, is an excellent Grape; the bunches are generally large, and at the upper part divide with two fmaller fide bunches or fhoulders. The berries are round, and, when perfectly ripe, turn of an amber colour. The juice is rich and vinous; it ripens in September, but, if carefully preferved, they will hang very late and become excellent.

The Chaffelas Mufque, or Le Cour Grape, as it is here called, by fome called the Frankindal, is an excellent Grape, and generally ripens well in England, if it has a good afpected wall. The berries are very like thofe of the former in fhape, fize, and colour, but are flefhy, and have a little mufky flavour. It ripens at the fame time with the former.

The black Clufter, or Munier Grape, as it is called by the French, from the hoary down of the leaves in fummer, is a good fruit, and ripens well here. The bunches are fhort, the berries are oval, and are very clofe to each other, fo that many of thofe which grow on the infide continue green when the outer are perfectly ripe. It ripens in September, and is by fome called the Burgundy Grape.

The Auverna, or true Burgundy Grape, fometimes called black Morillon, is an indifferent fruit for the table, but is efteemed one of the beft forts for making wine. The berries of this are oval, and hang loofer on the bunches than thofe of the Clufter Grape, fo ripen equally, which gives it the preference.

The Corinth, or as it is vulgarly called the Currant Grape, is a fmall roundifh berry, generally without ftone, of a deep black colour, and much cluftered on the bunches, which are fhort; it has a fugary juice, and ripens in September, but will not laft long.

The red Chaffelas is very like the white in fize and fhape, but is of a dark red colour; it is a very good grape, but ripens later than the white, and is pretty rare in England.

The white Mufcadine is fomewhat like the Chaffelas, but the berries are fmaller, and hang loofer on the bunches, which are longer, but not fo thick as thofe of the Chaffelas. The juice is fweet, but not fo rich as the Chaffelas.

The black Frontinac, or Mufcat noir, is a round berry of good fize; they grow loofe on the bunches, yet do not ripen equally. The bunches are fhort, the berries, when fully ripe, are very black, and are covered with a meal or flue, like the black Plumbs. The juice of this is very rich and vinous. It ripens the end of September, or the beginning of October.

The red Frontinac, or Mufcat rouge, is an excellent Grape when fully ripe, but unlefs the feafon proves very warm, they rarely ripen without artificial heat in England. The bunches of this fort are longer than thofe of the former; the berries are large and round; when they are fully ripe, they are of a brick colour, but before they are gray with a few dark ftripes, and this is frequently taken for a different kind, and is commonly called grifley Frontinac; but I am convinced it is the fame Grape. The juice of this has the moft vinous flavour of all the forts, and is greatly efteemed in France.

The white Frontinac has larger bunches than either of the former; the berries are round, and are fo clofely cluftered on the bunches, as that unlefs they are carefully thinned early in the feafon, when the berries are very fmall, the fun and air will be excluded from many of them, fo that they will not ripen, and the moifture will be detained in the autumn, which will caufe them to rot. The juice of this is excellent, and if the fruit is perfectly ripe is inferior to none. This the French call Mufcat blanc.

The Alexandrian Frontinac, or Mufcat d'Alexandrie, is by fome called Mufcat of Jerufalem. The berries of this are oval, and hang loofe on the bunches; thefe are long, and are not fhouldered. There are two forts, one with white and the other has red berries; their juice is very rich and vinous, but they feldom ripen in England without artificial heat.

The red and black Hamburgh, by fome called the Warner Grape, from the perfon who brought it to England. Thefe have middle-fized berries inclining to an oval fhape. The bunches are large, and their juice, when ripe, is fugary, with a vinous flavour. This ripens in October.

The St. Peter's Grape has a large oval berry, of a deep black colour when ripe. The bunches are very large, and make a fine appearance at the table, but the juice is not rich,

rich, and it ripens late in the year. The leaves of this ſort are much more divided than thoſe of the other ſorts, approaching to thoſe of the Parſley-leaved Grape, ſo it may be diſtinguiſhed before the fruit is ripe.

The Claret Grape, Bourdelais, or Verjuice Grape, the Raiſin Grape, the ſtriped Grape, and many other ſorts which never come to perfection here, are not worthy of a place in gardens, unleſs for the ſake of variety; for when they have the aſſiſtance of heat to bring them to maturity, their juice is harſh, and without flavour, ſo they ſhould not occupy the room of better fruit.

All the ſorts of Grapes are propagated either from layers or cuttings, the former of which is greatly practiſed in England, but the latter is what I would recommend, as being much preferable to the other; for the roots of Vines do not grow ſtrong and woody, as in moſt ſorts of trees, but are long, ſlender, and pliable; therefore when they are taken out of the ground they ſeldom ſtrike out any fibres from the weak roots, which generally ſhrivel and dry; ſo that they rather retard than help the plants in their growth, by preventing the new fibres from puſhing out; for which reaſon I had rather plant a good cutting than a rooted plant, provided it be well choſen, for there is little danger of its growing.

But as there are few perſons who make choice of proper cuttings, or at leaſt that form their cuttings rightly in England, ſo it will be proper to give directions for this in the firſt place, before I proceed. You ſhould always make choice of ſuch ſhoots as are ſtrong and well ripened of the laſt year's growth; theſe ſhould be cut from the old Vine, juſt below the place where they were produced, taking a knot, or piece of the two-years wood to each, which ſhould be pruned ſmooth; then you ſhould cut off the upper part of the ſhoots, ſo as to leave the cutting about ſixteen inches long. When the piece or knot of old wood is cut at both ends, near the young ſhoot, the cuttings will reſemble a little mallet; from whence Columella gives the title of Malleolus to the Vine-cuttings. In making the cuttings after this manner, there can be but one taken from each ſhoot; whereas moſt perſons cut them into lengths of about a foot, and plant them all, which is very wrong, for the upper part of the ſhoots are never ſo well ripened as the lower, which was produced early in the ſpring, and has had the whole ſummer to harden; ſo that if they take root, they never make ſo good plants; for the wood of thoſe cuttings being ſpongy and ſoft, admits the moiſture too freely, whereby the plants will be luxuriant in growth, but never ſo fruitful as ſuch whoſe wood is cloſer and more compact.

When the cuttings are thus prepared, if they are not then planted, they ſhould be placed with their lower part in the ground in a dry ſoil, laying ſome litter upon their upper parts to prevent them from drying: in this ſituation they may remain till the beginning of April, (which is the beſt time for planting them) when you ſhould take them out, and waſh them from the filth they have contracted; and if you find them very dry, you ſhould let them ſtand with their lower parts in water ſix or eight hours, which will diſtend their veſſels, and diſpoſe them for taking root. Then the ground being before prepared where the plants are deſigned to remain (whether againſt walls or for ſtandards, for they ſhould not be removed again) the cuttings ſhould be planted; but in preparing the ground you ſhould conſider the nature of the ſoil, which, if ſtrong, and inclinable to wet, is by no means proper for Grapes; therefore where it ſo happens, you ſhould open a trench where the cuttings are to be planted, which ſhould be filled with lime rubbiſh, the better to drain off the moiſture; then raiſe the border with freſh light earth about two feet thick, ſo that it may be at leaſt a foot above the level of the ground; then you ſhould open the holes at about ſix feet diſtance from each other, putting one good ſtrong cutting into each hole, which ſhould be laid a little ſloping, that their tops may incline to the wall; but it muſt be put in ſo deep, as that the uppermoſt eye may be level with the ſurface of the ground, for when any part of the cutting is left above ground, as is the common method uſed by the Engliſh gardeners, moſt of the buds attempt to ſhoot, ſo that the ſtrength of the cuttings is divided to nouriſh ſo many ſhoots, which muſt conſequently be weaker than if only one of them grew; whereas, on the contrary, by burying the whole cutting in the ground, the ſap is all employed on one ſingle ſhoot, which conſequently will be much ſtronger; beſides, the ſun and air are apt to dry that part of the cutting which remains above ground, and ſo often prevents their buds from ſhooting.

Then having placed the cutting into the ground, you ſhould fill up the hole gently, preſſing down the earth with your foot cloſe about it, and raiſe a little hill juſt upon the top of the cutting, to cover the upper eye quite over; which will prevent it from drying; this being done, there is nothing more neceſſary but to keep the ground clear from weeds until the cuttings begin to ſhoot; at which time you ſhould look over them carefully, to rub off any ſmall ſhoots, if ſuch are produced, faſtening the firſt main ſhoot to the wall, which ſhould be conſtantly trained up, as it is extended in length, to prevent its breaking or hanging down; you muſt continue to look over theſe once in about three weeks during the ſummer ſeaſon, conſtantly rubbing off all lateral ſhoots which are produced; and be ſure to keep the ground conſtantly clear from weeds, which, if ſuffered to grow, will exhauſt the goodneſs of the ſoil and ſtarve the cuttings.

The Michaelmas following, if your cuttings have produced ſtrong ſhoots, you ſhould prune them down to two eyes, which, though by ſome people may be thought too ſhort, yet I am ſatisfied, from ſeveral experiments, to be the beſt method. The reaſon for adviſing the pruning vines at this ſeaſon, rather than deferring it till ſpring, is, becauſe the tender parts of thoſe young ſhoots, if left on, are ſubject to decay in winter, for they are apt to grow late in the year, ſo the tops of their ſhoots are tender, and the early froſts will pinch them, and then they are frequently killed down a conſiderable length, which weakens their roots; but if they are cut off early in autumn, the wounds will heal over before the bad weather, and thereby the roots will be greatly ſtrengthened.

In the ſpring, after the cold weather is paſt, you muſt gently dig up the borders to looſen the earth; but you muſt be very careful in doing this, not to injure the roots of your Vines; you ſhould alſo raiſe the earth up to the ſtems of the plants, ſo as to cover the old wood, but not ſo deep as to cover either of the eyes of the laſt year's wood. After this they will require no farther care until they begin to ſhoot, when you ſhould look over them carefully, to rub off all weak dangling ſhoots, leaving no more than the two ſhoots, which are produced from the two eyes of the laſt year's wood, which ſhould be faſtened to the wall; and ſo from this, until the Vines have done ſhooting, you ſhould look them over once in three weeks or a month, to rub off all lateral ſhoots as they are produced, and to faſten the main ſhoots to the wall as they are extended in length, which muſt not be ſhortened before the middle or latter end of July, when it will be proper to nip off their tops, which will ſtrengthen the lower eyes, and during the ſummer ſeaſon you muſt conſtantly keep the ground clear from weeds; nor ſhould you permit any ſort of plants to grow near the Vines, which would not only rob them of nouriſhment, but

but ſhade the lower parts of the ſhoots, and thereby prevent their ripening; which will not only cauſe their wood to be ſpongy and luxuriant, but render it leſs fruitful.

As ſoon as the leaves begin to drop in autumn, you ſhould prune theſe young Vines again, leaving three buds to each of the ſhoots, provided they are ſtrong; otherwiſe it is better to ſhorten them down to two eyes if they are good, for it is a very wrong practice to leave much wood upon young Vines, or to leave their ſhoots too long, which greatly weakens the roots; then you ſhould faſten them to the wall, ſpreading them out horizontally each way, that there may be room to train the new ſhoots the following ſummer, and in the ſpring the borders muſt be digged as before.

The third ſeaſon you muſt go over the Vines again, as ſoon as they begin to ſhoot, to rub off all danglers as before, and train the ſtrong ſhoots in their proper places, which this year may be ſuppoſed to be two from each ſhoot of laſt year's wood; but if they attempt to produce two ſhoots from one eye, the weakeſt of them muſt be rubbed off, for there ſhould never be more than one allowed to come out of each eye. If any of them produce fruit, as many times they will the third year, you ſhould not ſtop them ſo ſoon as is generally practiſed upon the bearing ſhoots of old Vines, but permit them to ſhoot forward till a month after Midſummer, at which time you may pinch off the tops of the ſhoots; for if this were done too ſoon, it would ſpoil the buds for the next year's wood, which in young Vines muſt be more carefully preſerved than on older plants, becauſe there are no other to be laid in for a ſupply of wood, as is commonly practiſed on old Vines.

During the ſummer you muſt conſtantly go over your Vines, and diſplace all weak lateral ſhoots as they are produced, and carefully keep the ground clear from weeds, as was before directed, that the ſhoots may ripen well, which is a material thing to be obſerved in moſt ſorts of fruit-trees, but eſpecially in Vines, which ſeldom produce any fruit from immature branches. Theſe things being duly obſerved, are all that is neceſſary in the management of young Vines; I ſhall therefore proceed to lay down rules for the government of grown Vines, which I ſhall do as briefly as poſſible. And,

Firſt, Vines rarely produce any bearing ſhoots from wood that is more than one year old, therefore great care ſhould be taken to have ſuch wood in every part of the trees; for the fruit are always produced upon the ſhoots of the ſame year, which come out from buds of the laſt year's wood. The method commonly practiſed by the gardeners in England is, to ſhorten the branches of the former year's growth down to three or four eyes, at the time of pruning; though there are ſome perſons who leave theſe ſhoots much longer, and affirm, that by this practice they obtain a greater quantity of fruit; but however this may be, it is a very wrong practice, ſince it is impoſſible that one ſhoot can nouriſh forty or fifty bunches of Grapes, ſo well as it can ten or twelve, ſo that what is gotten in number is loſt in their magnitude; beſides, the greater quantity of fruit there is left on Vines, the later they are ripened, and their juice is not ſo rich; and this is well known in the wine countries, where there are laws enacted to direct the number and length of ſhoots that are to be left upon each Vine, leſt by overbearing them, they not only exhauſt and weaken the roots, but thereby render the juice weak, and ſo deſtroy the reputation of their wine.

Wherefore the beſt method is to ſhorten the bearing ſhoots to about four eyes in length, becauſe the lowermoſt ſeldom is good, and three buds are ſufficient, for each of theſe will produce a ſhoot, which generally has two or three bunches of Grapes; ſo that from each of thoſe ſhoots, there may be expected ſix or eight bunches, which is a ſufficient quantity. Theſe ſhoots muſt be laid about eighteen inches aſunder, for if they are cloſer, when the ſide-ſhoots are produced, there will not be room enough to train them againſt the wall, which ſhould always be provided for; and as their leaves are very large, the branches ſhould be left at a proportionable diſtance from each other, that they may not croud or ſhade the fruit.

At the winter-pruning of your Vines, you ſhould always obſerve to make the cut juſt above the eye, ſloping it backward from it, that if it ſhould bleed the ſap might not flow upon the bud; and where there is an opportunity of cutting down ſome young ſhoots to two eyes, in order to produce vigorous ſhoots for the next year's bearing, it ſhould always be done; becauſe in ſtopping thoſe ſhoots which have fruit upon them as ſoon as the Grapes are formed, which is frequently practiſed, it often ſpoils the eyes for producing bearing branches the following year, and this reſerving of new wood is what the Vignerons abroad always practiſe in their vineyards. The beſt ſeaſon for pruning of Vines is about the middle or end of October, for the reaſons before laid down.

The latter end of April, or the beginning of May, when the Vines begin to ſhoot, you muſt carefully look them over, rubbing off all ſmall buds which may come from the old wood, which only produce weak dangling branches; as alſo when two ſhoots are produced from the ſame bud, the weakeſt of them ſhould be diſplaced, which will cauſe the others to be ſtronger; and the ſooner this is done, the better it is for the Vines.

In the middle of May you muſt go over them again, rubbing off all the dangling ſhoots as before; and at the ſame time you muſt faſten up all the ſtrong branches, ſo that they may not hang from the wall; for if their ſhoots hang down, their leaves will be turned with their upper ſurfaces the wrong way, and when the ſhoots are afterwards trained upright, they will have their under ſurface upward, and until the leaves are turned again, and have taken their right poſition, the fruit will not thrive; ſo that the not obſerving this management, will cauſe the Grapes to be a fortnight or three weeks later before they ripen; beſides, by ſuffering the fruit to hang from the wall, and be ſhaded with the cloſeneſs of the branches, it is greatly retarded in its growth; therefore, during the growing ſeaſon, you ſhould conſtantly look over the Vines, diſplacing all dangling branches and wild wood which come from the ſide of the buds, and faſten up the other ſhoots regularly to the wall, as they are extended in length, and towards the middle of June you ſhould ſtop the bearing branches, which will ſtrengthen the fruit, provided you always leave three eyes above the bunches; for if you ſtop them too ſoon it will injure the fruit, by taking away that part of the branch which is neceſſary to attract the nouriſhment to the fruit, as alſo to perſpire off the crudities of the ſap, which is not proper for the fruit to receive.

But although I recommend the ſtopping thoſe ſhoots which have fruit at this ſeaſon, yet this is not to be practiſed upon thoſe ſhoots which are intended for bearing the next year, for theſe muſt not be ſtopt before the middle of July, leſt, by ſtopping them too ſoon, you cauſe the eyes to ſhoot out ſtrong lateral branches, whereby they will be greatly injured.

During the ſummer ſeaſon you ſhould be very careful to rub off all dangling branches, and train up the ſhoots regularly to the wall as before, which will greatly accelerate the growth of the fruit, and alſo admit the ſun and air to them, which is abſolutely neceſſary to ripen, and give the fruit a rich flavour; but you muſt never diveſt the branches of of their leaves, as is the practice of some persons; for although the admitting of the sun is necessary to ripen them, yet if they are too much exposed thereto, their skins will be tough, and they will rarely ripen; besides, the leaves being absolutely necessary to nourish the fruit, by taking them off the fruit is starved, and seldom comes to any size, as I have several times observed; therefore a great regard should be had to the summer management of the Vines, where persons are desirous to have their fruit excellent, and duly ripened.

When the fruit are all gathered you should prune the Vines, whereby the litter of their leaves will be entirely removed at once, and their fruit will be the forwarder the succeeding year, as has been before observed.

As many of the richest and best sorts of Grapes will not ripen in England, unless the season proves very warm, or the soil and situation are very favourable, there have been many hot-walls built to accelerate the ripening of this fruit, and bring it to full perfection by artificial heat; and as these succeed very well, when they are properly contrived, and the Vines rightly managed, I shall here give proper directions, which, if duly attended to, will be sufficient to instruct persons in both.

The method of building hot-walls will be treated under the article Wall, so I shall pass it over in this place, and proceed to the preparing of the ground for planting. The borders against these hot-walls should have the earth taken out two feet deep (provided the ground is dry,) otherwise one foot will be sufficient, because in wet land the borders should be raised at least two feet above the level of the ground, that the roots of the Vines may not be injured by the wet. When the earth is taken out, the bottom of the trench should be filled with stones, lime-rubbish, &c. a foot and a half or two feet thick, which should be levelled and beaten down pretty hard, to prevent the roots of the Vines from running downward. The trenches should be made five feet wide at least, otherwise the roots of the Vines will in a few years extend themselves beyond the rubbish, and, finding an easy passage downwards, will run into the moist ground, and thereby imbibe so much wet, as to lessen the vinous flavour of the Grapes; but before the rubbish is filled into the trench, it is a better method to raise a nine inch wall, at five feet distance from the back wall, which will keep the rubbish from intermixing with the neighbouring earth, and also confine the roots of the Vines to the border in which they are planted, so that they cannot reach to the moist ground. This nine inch wall should be raised to the height of this intended border, so will be of great use to lay the plate of timber of the frames upon, which will be necessary to cover the Vines when they are forced, whereby the timbers will be better preserved from rotting; and where the borders are raised to any considerable height above the level of the ground, they should be a brick and a half thick; these walls will preserve the borders from falling down into the walks; but in carrying up these walls it will be proper to leave little openings, about eight or ten feet distance, to let the water pass off, because when the rubbish at the bottom of the trench unites and binds very hard, the water cannot easily find a passage through it; therefore it will be the better method to leave these small passages in the wall, lest the moisture being confined at the bottom, should be pent up as in a ditch, which will be of ill consequence to the Vines.

When the walls are finished and thoroughly dry, the rubbish should be filled in, as before directed; then there should be fresh earth laid upon it two feet thick, which will be a sufficient depth of soil for the Vines to root in. These borders should be thus prepared at least a month or six weeks before the Vines are planted, that they may have time to settle. The best time to plant them is about the end of September, or the beginning of October, if planted with rooted plants, for when these are removed in the spring, their roots are very subject to bleed, which will greatly weaken them. The distance these Vines should be allowed to remain is the same as for common walls, i. e. about six feet; afterward lay a little mulch on the surface of the ground about their roots, to prevent the sun and air from drying the earth, and if the following spring should prove very dry, they should have some water once a week, which will be as often as they require it, for nothing will destroy them sooner than too much water.

The management of these Vines, for the three first years after planting, being the same as is practised for those against common walls, I shall not repeat it in this place, having fully treated of that already, only will observe that, during these three years, the Vines should be encouraged as much as possible, and the shoots not left too long, nor too many in number on each root, that they may be duly ripened and prepared for bearing the fourth year, which is the soonest they should be forced; for when any sort of fruit-trees are forced by fire too young, they seldom continue long in health, so that what fruit they produce is small, and not well flavoured; therefore, in being over hasty to save a year or two, very often the whole design miscarries; for unless the trees are in a proper condition to bear much fruit, it is not worth while to make fires for a small quantity of starved ill-tasted fruit, the expence and trouble being the same for ten or twelve bunches of Grapes, as it will be for a hundred or more.

These Vines should not be forced every year, but with good management they may be forced every other year, though it would be better, if it were done only every third year; therefore, in order to have a supply of fruit annually, there should be a sufficient quantity of walling built, to contain as many Vines as will be necessary for two or three years, and by making the frames in front moveable, they may be shifted from one part of the wall to another, as the Vines are alternately forced; therefore I would obvise about forty feet length of walling to be each year forced, which is as much as one fire will heat, and when the Vines are in full bearing, will supply a reasonable quantity of Grapes for a middling family, but for great families twice this length will not be too much.

In most places where these hot-walls have been built, they are commonly planted with early kinds of Grapes, in order to have them early in the season; but this, I think, is hardly worth the trouble, for it is but of little consequence to have a few Grapes earlier by a month or six weeks than those against common walls; therefore I should advise, whenever a person is willing to be at the expence of these walls, that they may be planted with some of the best kinds of Grapes, which rarely come to any perfection in this country without the assistance of some artificial heat, of which the following sorts are the most valuable

The red Muscat of Alexandria.
The white Muscat of Alexandria.
The red Frontinac.
The white Frontinac.
The black Frontinac.

When the Vines which are planted against the hot-walls are grown to full bearing, they must be pruned and managed after the same manner as hath been directed for those against common walls, with this difference only, viz. that those seasons when they are not forced, the Vines should be carefully managed in the summer, for a supply of good wood against the time of their being forced; so that it will be the better method to divest the Vines of their fruit, in order to encourage the wood, for as these sorts will not

ripen

ripen without heat, it is not worth while to leave them on the Vines during the ſeaſon of reſting, except it be the common Frontinacs, which in a good ſeaſon will ripen without artificial heat; but, even theſe, I would not adviſe many Grapes to be left on them during the years of their reſting, becauſe as the deſign of this is to encourage and ſtrengthen them, therefore all poſſible care ſhould be had, that the young wood is not robbed by overbearing; for thoſe years when the Vines are forced, the joints of the young wood are generally drawn farther aſunder than they ordinarily grow in the open air; ſo that when they are forced two or three years ſucceſſively, the Vines are ſo much exhauſted, as not to be recovered into a good bearing ſtate for ſome years, eſpecially if they are forced early in the ſeaſon, or where great care is not taken in the ſummer to let them have a proper ſhare of free air, to prevent their being drawn too much, and alſo to ripen their ſhoots. Thoſe years when the Vines are forced, the only care ſhould be to encourage the fruit, without having much regard to the wood, ſo that every ſhoot ſhould be pruned for fruit, and none of them ſhortened for a ſupply of young wood, becauſe they may be ſo managed by pruning in the years of their reſting, as to repleniſh the Vines with new wood. Thoſe Vines which are deſigned for forcing in the ſpring, ſhould be pruned early the autumn before, that the buds which are left on the ſhoots, may receive all poſſible nouriſhment from the root, and at the ſame time the ſhoots ſhould be faſtened to the trellis in the order they are to lie, but the glaſſes ſhould not be placed before the Vines till about the middle or end of January, at which time alſo the fires muſt be lighted; for if they are forced too early in the year, they will begin to ſhoot before the weather will be warm enough to admit air to the Vines, which will cauſe the young ſhoots to draw out weak, and thereby their joints will be too far aſunder, ſo conſequently there will be fewer Grapes on them; and thoſe bunches which are produced will be ſmaller, than when they have a ſufficient quantity of air admitted to them every day.

If the fires are made at the time before directed, the Vines will begin to ſhoot the latter end of February, which will be ſix weeks earlier than they uſually come out againſt the common walls, ſo that by the time that other Vines are ſhooting theſe will be in flower, which will be early enough to ripen any of theſe ſorts of Grapes perfectly well. The fires ſhould not be made very ſtrong in theſe walls, for if the air is heated to about ten degrees above the temperate point, on the botanical thermometers, it will be ſufficiently warm to force out the ſhoots leiſurely, which is much better than to force them violently. Theſe fires ſhould not be continued all the day time, unleſs the weather ſhould prove very cold, and the ſun does not ſhine to warm the air, at which times it will be proper to have ſmall fires continued all the day; for where the walls are rightly contrived, a moderate fire made every evening, and continued till ten or eleven of the clock at night, will heat the wall, and warm the incloſed air to a proper temperature; and as theſe fires need not be continued longer than about the end of April (unleſs the ſpring ſhould prove very cold,) ſo the expence of fuel will not be very great, becauſe they may be contrived to burn either coal, wood, turf, or almoſt any other ſort of fuel; though where coal is to be had reaſonable, it makes the eveneſt and beſt fires, and will not require much attendance.

When the Vines begin to ſhoot they muſt be frequently looked over to faſten the new ſhoots to the trellis, and to rub off all dangling ſhoots; in doing of which great care muſt be taken, for the ſhoots of theſe forced Vines are very tender, and very ſubject to break when any violence is offered. They ſhould alſo be trained very regular, ſo as to lie as near as poſſible at equal diſtances, that they may equally enjoy the benefit of the air and ſun, which is abſolutely neceſſary for the improvement of the fruit. When the Grapes are formed, the ſhoots ſhould be ſtopped at the ſecond joint beyond the fruit, that the nouriſhment may not be drawn away from the fruit, which muſt be avoided as much as poſſible in theſe forced Vines, upon which no uſeleſs wood ſhould be left, which will ſhade the fruit, and exclude the air from it by their leaves.

As the ſeaſon advances and the weather becomes warm, there ſhould be a proportionable ſhare of free air admitted to the Vines every day, which is abſolutely neceſſary to promote the growth of the fruit, but the glaſſes ſhould be ſhut cloſe every night, unleſs in very hot weather, otherwiſe the cold dews in the night will retard the growth of the fruit. The bunches of the white Frontinac ſhould alſo be carefully looked over, and the ſmall Grapes cut out with very narrow-pointed ſciſſars, in order to thin them, for theſe berries grow ſo cloſe together on the bunches, that the moiſture is detained between them, which often occaſions their rotting, and the air being excluded from the middle of the bunches, the Grapes never ripen equally, which by this method may be remedied, if done in time; and as theſe Grapes are protected by the glaſſes from the blights which frequently take thoſe which are expoſed, there will be no hazard in thinning theſe Grapes ſoon after they are ſet; at which time it will be much eaſier performed than when the Grapes are grown larger, and conſequently will be cloſer together; but in doing of this the bunches muſt not be roughly handled, for if the Grapes are the leaſt bruiſed, or the farina, which there naturally is upon them, be rubbed off, their ſkins will harden and turn of a brown colour, ſo the fruit will never thrive after; therefore the ſciſſars which are uſed for this purpoſe, ſhould have very narrow points, that they may be more eaſily put between the Grapes, without injuring the remaining ones. The other ſorts of Grapes, which I have recommended for theſe hot-walls, do not produce their fruit ſo cloſe together on the bunches, ſo they will not require this operation, unleſs by any accident they ſhould receive a blight, which often occaſions a great inequality in the ſize of the Grapes, which, whenever it thus happens, will require to be remedied by cutting off the ſmall Grapes, that the bunches may ripen equally, and appear more ſightly.

By the middle of June theſe Grapes will be almoſt full-grown, therefore the glaſſes may be kept off continually in the day time, unleſs the ſeaſon ſhould prove very cold and wet; in which caſe they muſt be kept on, and only opened when the weather is favourable; for as the racy vinous flavour of theſe fruits is increaſed by a free air, ſo, during the time of their ripening, they ſhould have as large a ſhare as the ſeaſon will admit to be given them.

Before the Grapes begin to ripen, they muſt be carefully guarded againſt birds, waſps, and other inſects, otherwiſe they will be deſtroyed in a ſhort time; to prevent which, the Vines ſhould be carefully covered with nets, ſo as to exclude the birds, who make great havock with the Grapes, by breaking their ſkins; and if there are a few twigs covered with birdlime, placed here and there on the outſide of the nets, it will be of ſervice, becauſe the birds are often ſo bold as to attempt to break the nets to get to the Grapes, which, if they attempt, they may be ſo entangled on theſe twigs as not to get looſe; and whenever that happens, they ſhould not be diſengaged, but ſuffered to remain to keep off their companions; and if they get off themſelves, it will have the deſired effect, for there will few other birds come to the ſame place that ſeaſon, as I have more than once experienced.

As

As to the wafps, the beft method is to hang up fome phials about half filled with fugared water, and rub the necks of the phials with a little honey, which will draw all the wafps and flies to them, which, by attempting to get at the liquor, will fall into the phials and be drowned; thefe phials fhould be carefully looked over once in three or four days to take out the wafps, and deftroy them, and to replenifh the phials with liquor. If this be duly obferved, and the phials placed in time, before the Grapes are attacked, it will effectually prevent their being injured; but where thefe precautions are not taken, the Grapes will be in danger of being abfolutely deftroyed, for as thefe early Grapes will ripen long before any others againft common walls, they will be in much more danger, there being no other fruit for them at that feafon in the neighbourhood; whereas, when Grapes in general begin to ripen, there is a quantity in almoft every garden; fo that if they deftroy a part in each garden, yet there will be a greater chance to have fome efcape, than where there is only one wall for them to attack.

Thefe forts of Grapes, being forced in the manner before directed, will begin to ripen early in Auguft, efpecially the Black and Red Frontinacs, which will be fit for the table a fortnight earlier than the other forts; but, as the defign of forcing them is to have them in as great perfection as poffible in this climate, they fhould not be gathered until they are thorough ripe, for which reafon fome of the later forts fhould be left on the Vines till September; but then the glaffes fhould be kept over them in wet and cold weather, to protect the fruit from it; but whenever the weather is fair, the glaffes muft be opened to let in the free air, otherwife the damps arifing from the earth at that feafon, will caufe a mouldinefs upon the Grapes, which will rot them; fo that if the feafon fhould prove very cold and wet, while the fruit are upon the Vines, it will be proper to make a fmall fire every night to dry off the damps, and prevent this injury. Moft people in England gather their Grapes too foon, never fuffering them to remain on the Vines to ripen perfectly, even in the warmeft feafons, when, if they are left on till after Michaelmas, they will be good.

Of late years many perfons have planted Grapes againft efpaliers, which in fome places have fucceeded very well in good feafons; but if they are not planted in a good foil and to a proper afpect, and the forts rightly chofen, they feldom produce any fruit which are fit to be eaten. The foil proper to plant Vines in efpaliers, fhould be the fame as is hereafter directed for vineyards, viz. either a chalky, or gravelly bottom, with about a foot and a half, or two feet of light hazel earth on the top, a little floping to the fouth or fouth-eaft, that the wet may eafily find a paffage, fo as not to remain on the ground. In fuch a foil fituated to the fun, and fcreened from cold winds, there are feveral forts of Grapes, which in warm feafons will ripen very well in England.

But there are fome curious perfons who line the back-fide of their efpaliers with low Reed-hedges, and others who do it with thin flit deals; both of which are a good defence to the Vines againft blights in the fpring, and accelerate the ripening of the Grapes, fo that in tolerable feafons they will come to good maturity. Neither of thefe methods are very expenfive, for thefe clofe fences need not be more than four feet high; becaufe the Vines being to be managed after the fame manner as thofe in vineyards, the branches which carry the fruit will never rife above that height; for the bearing fhoots muft always be trained about two feet above the furface of the ground, fo that the fruit will be always below the top of the clofe fences; and as for the upright fhoots which are defigned for the next year's bearing, it matters not how much they rife above the fence; fo thefe may have a loofe trellis to which they may be faftened, to prevent their overhanging the fruit.

In the making of thefe kinds of clofe efpaliers for Grapes, it will be proper to lay one ftrong oaken plank, (fuch as are procured in breaking up old fhips or barges) next the furface of the ground, which will laft many years found, and be very ufeful in fupporting the fences. If thefe plank are fifteen inches broad, as they may always be readily procured, and the upper part of the fence be Reeds, there may be two lengths cut out of them (provided the Reeds are of a due length,) without including their tops. In the front of thefe hedges fhould be a flight trellis to faften the Vines to, which may be made of Afh-poles. The upright poles of thefe trelliffes need not be nearer together than eighteen inches; and if there are three crofs poles, at about a foot afunder, they will be fufficient to faften the bearing fhoots of the Vines at proper diftances, in the manner they are defigned to be trained, which fhould be in fuch pofitions, that the fruit may not be overfhadowed by the branches; and if the upright poles are cut fo long, as to be a foot and a half above the Reeds, they will be tall enough to fupport the upright fhoots for the next year's bearing, which, being trained fingly at proper diftances, will have the advantage of the fun and air to ripen the wood much better than where four or five fhoots are faftened to the fame pole.

To this trellis the Reeds may be faftened with hoops on the back-fide, after the manner ufually practifed in making common Reed-fences; and if on the top of the Reeds there is faftened a thin flip of deal, to fecure their tops from being broken, it will preferve them a long time. In making of thefe fences, the Reeds fhould not be laid too thick, for that will not only be more expence, but will be troublefome to faften, and not laft fo long as when they are made of a moderate thicknefs: therefore as the Reeds will be cut into two lengths, each bundle will fpread about fix feet in length, obferving firft to fpread the bottom parts of the bundles, which contain the largeft ends of the Reeds the whole length; and then the upper parts of the other Reeds fhould be reverfed, and fpread in front of them, which will make the upper part of the fence almoft as thick as the bottom. But neither of thefe, nor the boarded fences, need be made till the Vines are in full bearing, which will be the fourth or fifth year after planting, according to the progrefs they make; during which time the fhoots may be fupported by any common ftakes, for if the fences are made before the Vines are planted, as is frequently practifed, they will be half decayed by the time the Vines are fit to bear, and before this time the fences are of no ufe to them.

The forts of Grapes which are proper to plant againft thefe fences are,

The Miller Grape.
The Chaffelas White.
The White Mufcadine.
The Sweet Water, and
Le Cour Grape.

Thefe, if well managed, will ripen very well, provided the feafon is tolerably good, and will come foon after thofe of the walls; fo that if they are taken care of, by hanging of mats before them, when the nights prove cold in autumn, and are permitted to hang till October, the fruit will prove very good. But where the Sweet Water Grape is planted againft thefe fences, they will require to be covered in the fpring, at the time when they are in flower, if there fhould be cold nights, otherwife the bunches will receive a blaft, which will deftroy the greateft part of the Grapes; fo that many times there will not be more than fix or eight good Grapes on each bunch, and the others will be fmall ftarved fruit, hardly fo large as the fmalleft Peafe.

In planting of these Vines, either for open espaliers or the close fences, it should be performed in the same manner as for walls: the cuttings should be planted six feet asunder; and as these are only designed for the table, a single row of Vines of a moderate length will be sufficient to supply a family, where there are others against walls to come before them. But where a person is inclinable to have more rows than one, they should be placed at least twelve feet asunder, that they may equally enjoy the sun and air.

As to the pruning and other management of these Vines, that being the same as for those against walls, I shall not repeat it in this place, it being fully treated of before; and to which I have nothing here to add.

In the folio edition of the Gardeners Dictionary, we inserted the several methods of planting and managing vineyards in the principal parts of Europe, where the best wines are produced; but as this volume is an abridgment of that work, we have omitted such articles as we supposed might be of least utility to the public; and have frequently shortened others, so as not to render this imperfect: therefore as there may be some purchasers who may be inclined to make trials of vineyards in England, so we have here given the best directions we can, for planting and managing them in this country to the best advantage; which we have extracted from the practice of those persons who reside in countries where there is good wine made, and where the climate approaches nearest to that of England; and also from many repeated trials which have been made with success in different parts of this country, from which any diligent person may readily engage in the practice.

The first and great thing to be considered in planting vineyards is the choice of soils and situations, without which there will be little hopes of success, for upon this the whole affair greatly depends. The best soil for a vineyard in England is such, whose surface is a light sandy loam, and not above a foot and a half, or two feet, with a gravelly or chalky bottom, either of which are equally good for Vines; but if the soil is deep, upon either clay or a strong loam, it is by no means proper for this purpose; for although the Vines may shoot vigorously, and produce a great quantity of Grapes, yet these will be later ripe, fuller of moisture, and so consequently their juice not mature nor well digested, but will abound with crudity, which in fermenting will render the wine sour and ill-tasted, which is the common complaint of those who have made wine in England.

Nor is a very rich, light, deep soil, such as is commonly found near London, proper for this purpose, because the roots of these Vines will be enticed down too deep to receive the influences of sun and air, and hereby will take in much crude nourishment, whereby the fruit will be later ripe, and replete with moisture, which must necessarily contribute greatly to render the juices less perfect, therefore great attention should be had to the nature of the soil upon which they are planted.

The next thing necessary to be considered is the situation of the place, which, if possible, should be on the north-side of a river, upon an elevation inclining to the south, with a small gradual descent, that the moisture may the better drain off; but if the ground slopes too much, it is by no means proper for this purpose; but if at a distance from this place, there are larger hills which defend it from the north and north-west wind, it will be of great service, because hereby the sun's rays will be reflected with a greater force, and the cold winds being kept off, will render the situation very warm. Add to this a chalky surface, which, if those hills do abound with (as there are many such situations in England,) it will still add to the heat of the place, by reflecting a greater quantity of the sun's rays.

The country about this should be open and hilly, for if it be much planted, or low and boggy, the air will constantly be filled with moist particles, occasioned by the plentiful perspiration of the trees, or the exhalations from the adjoining marshes, whereby the fruit will be greatly prejudiced (as was before observed.) These vineyards should always be open to the east, that the morning sun may come on them to dry off the moisture of the night early, which, by lying too long upon the Vines, greatly retards the ripening of their fruit, and renders it crude and ill-tasted. And since the fruit of Vines are rarely injured by easterly winds, there will be no reason to apprehend any danger from such a situation, the south-west, north-west, and north winds being the most injurious to vineyards in England (as indeed they are to most other fruit) so that, if possible, they should be sheltered therefrom.

Having made choice of a soil and situation proper for this purpose, the next thing to be done is to prepare it for planting. In doing of which the following method should be observed: in the spring, if the ground is green sward, it should be ploughed as deep as the surface will admit, turning the sward into the bottom of each furrow; then it should be well harrowed to break the clods, and cleanse it from the roots of noxious weeds; and after this, it must be kept constantly ploughed and harrowed for at least one year, to render the surface light; and hereby it will be rendered fertile, by imbibing the nitrous particles of the air (especially if it be long exposed thereto before it is planted;) in the next March the ground should be well ploughed again, and after having made the surface pretty even, the rows should be marked out from south-east to north-west, at the distance of ten feet from each other; and these rows should be crossed again at five or six feet distance, which will mark out the exact places where each plant should be placed; so that the Vines will be ten feet row from row, and five or six feet asunder in the rows, nearer than which they ought never to be planted. For herein most people, who have planted vineyards, have greatly erred, some having allowed no more than five feet row from row, and the plants but three feet asunder in the rows, and others, who think they have been full liberal in this article, have only planted their Vines at six feet distance every way; but neither of these have allowed a proper distance to them, as I shall shew; for, in the first place, where the rows are placed too close, there will not be room for the sun and air to pass in between them, when the Vines are fully grown, to dry up the moisture, which, being detained amongst the Vines, must produce very ill effects: and, secondly, where the Vines are placed in exact squares so near together as six feet, the effect will be much the same; for the autumns in England are often attended with rains, cold dews, or fogs, so proper care should be taken to remove every thing which may obstruct the drying up the damps which arise from the ground.

The skilful Vignerons abroad are also sensible how much it contributes to the goodness of their Vines, to allow a large space between the rows; and therefore where the quality of the wine is more regarded than the quantity, they never plant their Vines at less than ten feet row from row, and some allow twelve. It was an observation of Bellonius, almost two hundred years since, that in those islands of the Archipelago, where the rows of Vines were placed at a great distance, the wine was much preferable to those which were close planted; and this he positively affirms to be the case in most countries where he had travelled. Indeed we need not have recourse to antiquity for the certainty of such facts, when we are daily convinced of this truth in all close plantations of any kind of fruit, where it is constantly observed, that the fruits in such places are never so well coloured, so early ripe, nor near so well flavoured, as those produced

on

on trees, where the air can freely circulate about them, and the rays of the fun have free accefs to the branches, whereby the juices are better prepared before they enter the fruit.

Having thus confidered the diftance which is neceffary to be allowed to thefe plants, we come next to the planting; but in order to this, the proper forts of Grapes fhould be judicioufly chofen, and in this particular we have egregioufly erred in England. Moft of the vineyards at prefent planted here, are of the fweeteft and beft fort of Grapes for eating, which is contrary to the general practice of the Vignerons abroad, who always obferve, that fuch Grapes never make good wine; and therefore from experience, make choice of thofe forts of Grapes, whofe juice, after fermenting, affords a noble rich liquor: thefe Grapes are always auftere, and not fo palatable. This is alfo agreeable to the conftant practice of our cyder-makers in England, who obferve, that the beft eating Apples feldom make good cyder; whereas the more rough and auftere forts, after being preffed and fermented, afford a ftrong vinous liquor. And I believe it will be found true in all fruits, that where the natural heat of the fun ripens and prepares their juices, fo as to render them palatable, whatever degree of heat thefe juices have more, either by fermentation, or from any other caufe, will render them weaker and lefs fpirituous. Of this we have many inftances in fruits; for if we tranfplant any of our fummer or autumn fruits, which ripen perfectly in England without the affiftance of art, into a climate a few degrees warmer, thefe fruits will be meally and infipid; fo likewife if we bake or ftew any of thefe fruits, they will be good for little, lofing all their fpirit and flavour by the additional heat of the fire; and fuch fruits as are by no means eatable raw, are hereby rendered exquifite, which, if tranfplanted into a warmer climate, have, by the additional heat of the fun, been alfo altered fo as to exceed the moft delicious of our fruits in this country.

From whence it is plain, that thofe Grapes which are agreeable to the palate for eating, are not proper for wine; in making of which, their juices muft undergo a ftrong fermentation; therefore fince we have in England been only propagating the moft palatable Grapes for eating, and neglected the other forts, before we plant vineyards, we fhould take care to be provided with the proper forts from abroad, which fhould be chofen according to the fort of wines intended to be imitated; though I believe the moft probable fort to fucceed in England is the Auvernat, or true Burgundy Grape, (which is rarely found in the Englifh vineyards, though it is a common Grape in the gardens againft walls.) This fort of Grape is moft preferred in Burgundy, Champaign, Orleans, and moft of the other wine countries in France; and I am informed, that it fucceeds very well in feveral places to the north of Paris, where proper care is taken of their management; fo that I fhould advife fuch perfons as would try the fuccefs of vineyards in England, to procure cuttings of this Grape; but herein fome perfon of integrity and judgment fhould be employed to get them from fuch vineyards where no other forts of Grapes are cultivated, which is very rare to find, unlefs in fome particular vineyards of thofe perfons who are very exact to keep up the reputation of their wines, nothing being more common than for the Vignerons to plant three or four forts of Grapes in the fame vineyard, and at the time of vintage to mix them all together; which renders their wines lefs delicate, than in fuch places where they have only this one true fort of Grape. And here I would caution every one againft mixing the juice of feveral Grapes together, which will caufe the wine to ferment at different times, and in different manners.

The cuttings being thus provided (for I would always prefer thefe to layers, or rooted plants, for the reafons given at the beginning of this article) about the beginning of April is the beft feafon for planting, when it will be proper to put the lower ends of the cuttings in water about three inches, fetting them upright for fix or eight hours before they are ufed; then at the center of every crofs mark already made by a line, to the diftance the Vines are defigned, fhould be a hole made with a fpade, or other inftrument, about a foot deep; into each of which fhould be put one ftrong cutting, placing it a little floping; then the hole fhould be filled up with earth, preffing it gently with the feet to the cutting, and raifing a little hill to each about three inches, fo as juft to cover the uppermoft eye or bud, which will prevent the wind and fun from drying any part of the cuttings, and this upper eye only will fhoot; the under ones, moft of them, will pufh out roots, fo that this fhoot will generally be very ftrong and vigorous.

After they are thus planted, they will require no other care until they fhoot, except to keep the ground clear from weeds, which fhould be conftantly obferved; but as the diftance between the rows of Vines is very great, fo the ground between them may be fown or planted with any kind of efculent plants, which do not grow tall, provided there is proper diftance left from the Vines, and care taken that the Vines are not injured by the crops, or in the gathering, and carrying them off the ground; and this hufbandry may be continued two or three years, till the Vines come to bearing; after which time there fhould be no fort of crop put between them in fummer, becaufe the cleaner the ground is kept between the Vines from weeds or other plants, the more heat will be reflected to the Grapes; but after the Grapes are gathered, there may be a crop of Coleworts for fpring ufe planted between the rows of Vines, and the cultivating of thefe will be of ufe to the Vines, by ftirring of the ground; but as to watering, or any other trouble, there will be no occafion for it, notwithftanding what fome people have directed, for in England there is no danger of their mifcarrying by drought. When the cuttings begin to fhoot, there fhould be a ftick of about three or four feet long ftuck down by each, to which the fhoot fhould be faftened, to prevent their breaking or lying on the ground; fo that as the fhoots advance, the faftening fhould be renewed, and all fmall lateral fhoots (if there are any fuch produced) fhould be conftantly difplaced, and the ground between the Vines always kept clean. This is the whole management which is required the firft fummer.

But at Michaelmas, when the Vines have done fhooting, they fhould be pruned; for if they are left unpruned till fpring, their fhoots being tender (efpecially toward their upper parts) will be in danger of fuffering if the winter fhould prove fevere.

This pruning is only to cut down the fhoots to two or three eyes; and if, after this is done, the earth be drawn up in a hill about each plant, it will ftill be a greater defence againft froft.

At the beginning of March the ground between the Vines fhould be well dug to loofen it, and render it clean; but you fhould be careful not to dig deep clofe to the Vines, left thereby their roots fhould be cut or bruifed; and at the fame time the earth fhould be again laid up in a hill about each plant; but there muft be care taken not to bury the young eyes of the former year's fhoot, which were left to produce new wood.

At the beginning of May, when the Vines are fhooting, there fhould be two ftakes fixed down to the fide of each plant, which muft be fomewhat taller and ftronger than thofe of the former year; to thefe the two fhoots (if fo many are produced) fhould be faftened, and all the fmall trailing or lateral fhoots fhould be conftantly difplaced, to ſtrengthen the ſhoots; the ground ſhould alſo be kept very clear from weeds as before.

The autumn following theſe Vines ſhould be pruned again in the following manner; thoſe of them which have produced two ſtrong ſhoots of equal vigour, muſt be cut down to three eyes each; but in ſuch as have one ſtrong ſhoot and a weak one, the ſtrong one muſt be ſhortened to three eyes, and the weak one to two; and ſuch Vines as have produced but one ſtrong ſhoot, ſhould be ſhortened down to two eyes alſo, in order to obtain more wood againſt the ſucceeding year.

In the ſpring, about the middle of March, the ground between the Vines ſhould be again dug, as before, and two ſtakes ſhould be placed down by the ſide of all ſuch Vines as have two ſhoots, at ſuch diſtance on each ſide of the plant as the ſhoots will admit to be faſtened thereto, and the ſhoots ſhould be drawn out on each ſide to the ſtakes, ſo as to make an angle of about forty-five degrees with the ſtem: but by no means ſhould they be bent down horizontally, as is by ſome practiſed, for the branches lying too near the earth, are generally injured by the damps which ariſe from thence, but eſpecially when they have fruit, which is never ſo well taſted, nor ſo early ripe upon thoſe branches, as when they are a little more elevated.

In May, when the Vines begin to ſhoot, they muſt be carefully looked over, and all the weak dangling ſhoots ſhould be rubbed off as they are produced; and thoſe ſhoots which are produced from ſtrong eyes, ſhould be faſtened to the ſtakes to prevent their being broken off by the wind.

This management ſhould be repeated at leaſt every three weeks or a month, from the beginning of May to the end of July; by which means the ſhoots which are trained up for the ſucceeding year, will not only be ſtronger, but alſo better ripened and prepared for bearing, becauſe they will have the advantage of ſun and air, which is abſolutely neceſſary to prepare their juices; whereas, if they are crouded by a number of ſmall dangling weak branches, they will ſhade and exclude the rays of the ſun from the other ſhoots, and ſo by detaining the moiſture a longer time amongſt the branches, occaſion the veſſels of the young wood to be of a larger dimenſion; and hereby the crude juice finds an eaſy paſſage through them, ſo that the ſhoots in autumn ſeem to be moſtly pith, and are of a greeniſh immature nature; and wherever this is obſerved, it is a ſure ſign of a bad quality in the Vines.

The ſoil alſo ſhould be conſtantly kept clean, becauſe, if there are any vegetables (either weeds or plants of other kinds) growing between the Vines, it will detain the dews longer, and by their perſpiration occaſion a greater moiſture than would be, if the ground were entirely clear; ſo that thoſe who plant other things between their rows of Vines, are guilty of a great error.

In autumn the Vines ſhould be pruned, which ſeaſon I approve of rather than the ſpring (for reaſons before given) and this being the third year from planting, the Vines will now be ſtrong enough to produce fruit, therefore they muſt be pruned accordingly. Now ſuppoſe the two ſhoots of the former year, which were ſhortened to three eyes, have each of them produced two ſtrong branches the ſummer paſt, then the uppermoſt of theſe ſhoots upon each branch ſhould be ſhortened down to three good eyes (never including the lower eye, which is ſituate juſt above the former year's wood, which ſeldom produces any thing, except a weak dangling ſhoot;) and the lower ſhoots ſhould be ſhortened down to two good eyes each, theſe being deſigned to produce vigorous ſhoots for the ſucceeding year, and the former are deſigned to bear fruit; but where the Vines are weak, and have not produced more than two or three ſhoots the laſt ſeaſon, there ſhould be but one of them left with three eyes for bearing, the other muſt be ſhortened down to two or if weak to one good eye, in order to obtain ſtrong ſhoots the following ſummer; for there is nothing more injurious to Vines than the leaving too much wood upon them, eſpecially while they are young; or the overbearing them, which will weaken them ſo much, as not to be recovered again to a good ſtate in ſeveral years, though they ſhould be managed with all poſſible ſkill.

In March the ground between the Vines ſhould be well dug as before, obſerving not to injure their roots by digging too deep near them; but where there are ſmall horizontal roots produced on or near the ſurface of the ground, they ſhould be pruned off cloſe to the places where they were produced; theſe being what the Vignerons call day-roots, and are by no means neceſſary to be left on: after having dug the ground, the ſtakes ſhould be placed down in the following manner: On each ſide of the Vine ſhould be a ſtake put in at about ſixteen inches from the root, to which the two branches, which were pruned to three eyes, each for bearing, ſhould be faſtened, (obſerving, as was before directed, not to draw them down too horizontally;) then another taller ſtake ſhould be placed down near the foot of the Vine, to which the two ſhoots which were pruned down to two eyes ſhould be faſtened, provided they are long enough for that purpoſe; but if not, when their eyes begin to ſhoot, theſe muſt be trained upright to the ſtakes, to prevent their trailing on the ground, hanging over the fruit-branches, or being broke by the wind.

In May the Vines ſhould be carefully looked over again, at which time all weak lateral branches ſhould be rubbed off as they are produced; and thoſe ſhoots which ſhow fruit muſt be faſtened with baſs to the ſtakes, to prevent their being broken, until they are extended to three joints beyond the fruit, when they ſhould be ſtopped; but the ſhoots which are deſigned for bearing the following ſeaſon ſhould be trained upright to the middle ſtake, by which method the fruit-branches will not ſhade theſe middle ſhoots, nor will the middle ſhoots ſhade the fruit, ſo that each will enjoy the benefit of the ſun and air.

This method ſhould be repeated every three weeks or a month, from the beginning of May to the middle of July, which will always keep the ſhoots in their right poſition, whereby the leaves will not be inverted, which greatly retards the growth of the fruit; and by keeping the Vines conſtantly clear from horizontal ſhoots, the fruit will not be crouded with leaves and ſhaded, but will have conſtantly the advantage of the ſun and air equally, which is of great conſequence; for where the fruit is covered with theſe dangling ſhoots in the ſpring, and are afterwards expoſed to the air, either by diveſting them of their leaves, or elſe diſplacing their branches entirely, as is often practiſed, the fruit will become hard, and remain at a perfect ſtand for three weeks, and ſometimes will never advance afterward, as I have ſeveral times obſerved; therefore there cannot be too much care taken to keep them conſtantly in a kindly ſtate of growth, as the Vignerons abroad well know; though in England it is little regarded by the generality of gardeners, who, when their Grapes ſuffer by this neglect, immediately complain of the climate, or the untowardneſs of the ſeaſon, which is too often a cover for neglects of this nature. And here I cannot help taking notice of the abſurd practice of thoſe, who pull off their leaves from their Vines, which are placed near the fruit, in order to let in the rays of the ſun to ripen them, not conſidering how much they expoſe their fruit to the cold dews, which fall plentifully in autumn, which, being imbibed by the fruit, greatly retard them; beſides, no fruit will ripen ſo well when entirely expoſed to the ſun, as when they are gently ſcreened with leaves, which are abſolutely neceſſary to prepare the juices before

before they enter the fruit, the groſs parts of which are perſpired away by the leaves; the fruit muſt either be deprived of nouriſhment, or elſe ſome of the groſs particles will enter with the more refined parts of the juice, and thereby render the fruit worſe than it would otherwiſe be, were the leaves permitted to remain upon the branches; for if the weak dangling ſhoots are conſtantly diſplaced as they are produced, the fruit will not be too much ſhaded by the leaves that are upon the bearing branches.

When the fruit is ripe, if the ſtalks of the bunches are cut half through a fortnight before they are gathered, it will cauſe the juice to be much better, becauſe there will not be near ſo great a quantity of nouriſhment enter the fruit, whereby the watery particles will have time to evaporate, and the juice will be better digeſted. This is practiſed by ſome of the moſt curious Vignerons in the ſouth of France, where they make excellent wine. But if after the fruit be cut, it is hung up in a dry room upon ſtrings, ſo as not to touch each other, for a month before they are preſſed, it will alſo add greatly to the ſtrength of the wine, becauſe in that time a great quantity of the watery parts of the juices will evaporate. This is a conſtant practice with ſome perſons who inhabit the Tiroleſe, on the borders of Italy, where is made a moſt delicious rich wine, as hath been atteſted by Dr. Burnet in his travels; and I have heard the ſame from ſeveral gentlemen who have travelled that road ſince.

But with all the care that can poſſibly be taken, either in the culture of the Vines, or in making the wine, it will not be near ſo good while the vineyard is young, as it will be after it has been planted ten or twelve years; and it will be conſtantly mending until it is fifty years old, as is atteſted by ſeveral curious perſons abroad, as alſo by the moſt ſkilful wine-coopers at home, who can tell the produce of a young vineyard from that of an old one, after it is brought to England, by the colour of the wine. This difference is very eaſily accounted for, from the different ſtructure of the veſſels of the plants; thoſe of the young Vines being larger, and of a looſer texture, eaſily admit a larger quantity of groſs nouriſhment to paſs through them; whereas thoſe of old Vines, which are more woody, are more cloſely conſtricted, and thereby the juice is better ſtrained in paſſing through them, which muſt conſequently render it much better, though the Grapes from a young vineyard will be larger, and afford a greater quantity of juice, ſo that people ſhould not be diſcouraged if their Wines at firſt are not ſo good as they would wiſh; ſince afterward, when the vineyard is a few years older, the wine may anſwer their expectation. As to the fermenting and managing the wine, that is treated of particularly under the article Wines, to which the reader is deſired to turn.

The vineyard being now arrived to a bearing ſtate, ſhould be treated after the following manner: Firſt, in the pruning there ſhould never be too many branches left upon a root, nor thoſe too long, for although by doing of this there may be a greater quantity of fruit produced, yet the juice of theſe will never be ſo good as when there is a moderate quantity, which will be better nouriſhed, and the roots of the plants not ſo much weakened; which is found to be of ſo bad conſequence to vineyards, that when gentlemen abroad let out vineyards to Vignerons, there is always a clauſe inſerted in their leaſes, to direct how many ſhoots ſhall be left upon each Vine, and the number of eyes to which the branches muſt be ſhortened; becauſe were not the Vignerons thus tied down, they would overbear the Vines, ſo that in a few years they would exhauſt their roots, and render them ſo weak as not to be recovered again in ſeveral years; and their wine would be ſo bad, as to bring a diſreputation on the vineyard, to the great loſs of the proprietor.

The number of branches which the Italians generally agree to leave upon a ſtrong Vine are four; two of the ſtrongeſt have four eyes, and the two weaker are ſhortened down to two eyes each; which is very different from the common practice in England, where it is uſual to ſee ſix or eight branches left upon each root, and thoſe perhaps left with ſix or eight eyes to each; ſo that if theſe are fruitful, one root muſt produce near four times the number of bunches which the Italians do ever permit, and ſo conſequently the fruit will not be ſo well nouriſhed, and the roots will alſo be greatly weakened; as is the caſe of all ſorts of fruit-trees, when a greater number of fruit is left on than the trees can nouriſh.

The next thing is, conſtantly to keep the ground perfectly clean between the Vines, never permitting any ſort of plants or weeds to grow there. The ground ſhould alſo be carefully dug every ſpring, and every third year have ſome manure, which ſhould be of different ſorts, according to the nature of the ground, or which can be moſt conveniently procured.

If the land is ſtiff, and inclinable to bind on the ſurface, then ſea-ſand, or ſea-coal aſhes, are either of them very good manure for it; but if the ground be looſe and dry, then a little lime mixed with dung is the beſt manure for it. This muſt be ſpread thin upon the ſurface of the ground before it is dug, and in digging ſhould be buried equally in every part of the vineyard. Theſe are much preferable to that of all dung for Vines, ſo that it will be worth the expence to procure either of them; and as they require manuring but every third year, where the vineyard is large it may be divided into three equal parts, each of which may be manured in its turn, whereby the expence will be but little every year; when the whole is manured together it will add to the expence, and in many places there cannot be a ſufficient quantity procured to manure a large vineyard in one year.

This digging and manuring ſhould always be performed about the middle of March, at which time all the ſuperficial or day-roots, as they are called, muſt be cut off, but the larger roots muſt not be injured by the ſpade, &c. therefore the ground cloſe to the ſtem of the Vines muſt not be dug very deep. After this is done, the ſtakes ſhould be placed down, one on each ſide the Vines, at about ſixteen inches from their ſtems, to which the longeſt bearing branches ſhould be faſtened, and one ſtake cloſe to the ſtem, to which the two ſhorter branches ſhould be trained upright, to furniſh wood for the ſucceeding year.

In the ſummer they muſt be carefully looked over as before, rubbing off all weak dangling ſhoots, and training the good ones to the ſtakes regularly as they are produced; and thoſe of them which have fruit ſhould be ſtopped in June, about three joints beyond the bunches, but the upright ſhoots, which are deſigned for bearing the following year, muſt not be ſtopped till the middle of July, when they may be left about five feet long; for if they are ſtopped ſooner in the year, it will cauſe them to ſhoot out many dangling branches from the ſides of the eyes, which will not only occaſion more trouble to diſplace them, but alſo will be injurious to the eyes or buds.

N. B. "All this ſummer dreſſing ſhould be performed "with the thumb and finger, and not with knives, be-"cauſe the wounds made by inſtruments in ſummer do not "heal ſo ſoon as when ſtopped by gently nipping the lead-"ing bud, which, if done before the ſhoot is become "woody, may be effected with great eaſe, being very ten-"der while young."

When a vineyard is thus carefully dreſſed, it will afford as much pleaſure in viewing it as any plantation of trees and ſhrubs whatever, the rows being regular; and if the ſtakes

ſtakes are exactly placed, and the upright ſhoots ſtopped to an equal height, there is nothing in nature, which will make a more beautiful appearance; and during the ſeaſon that the Vines are in flower they emit a moſt grateful ſcent, eſpecially in the morning and evening; and when the Grapes begin to ripen, there will be freſh pleaſure ariſing in the viewing of them.

But as the beauty of vineyards ariſes from the regular diſpoſition of the branches of the Vines, great care ſhould be taken in their management to train them regularly, and to provide every year for new wood to bear the ſucceeding year; becauſe the wood which has produced fruit is commonly cut quite away after the fruit is gathered, or at leaſt is ſhortened down to two eyes to force out ſhoots for the next year, where there is not a ſufficient number of branches upon the Vine of thoſe trained upright; ſo that in ſummer, when the Vines are in perfection, there ſhould be ſix upright ſhoots trained for the next year's wood, and three or four bearing branches with fruit on them; more than theſe ought never to be left upon one Vine, for the reaſons before given.

N. B. The Auvernat, or true Burgundy Grape, is valued in France before any other ſort, becauſe the fruit never grows very cloſe upon the bunches, therefore are more equally ripened, for which reaſon it ſhould alſo be preferred in England; though in general thoſe ſorts are moſt eſteemed with us that have always cloſe bunches, which is certainly wrong; for it may be obſerved, that the Grapes on ſuch bunches are commonly ripe on one ſide and green on the other, which is a bad quality for ſuch as are preſſed to make wine.

I ſhall now ſubjoin a few ſorts of Vines, which are preſerved in ſome curious gardens, more for the ſake of variety than the value of their fruit: theſe are,

1. Vitis foliis cordatis dentatis ſubtus villoſis, cirrhis racemiferis. Flor. Zeyl. 99. *Wild Indian Vine, with round berries.*

2. Vitis foliis cordatis ſubtrilobis dentatis ſubtus tomentoſis. Lin. Sp. Pl. 203. *Wild Virginia Grape.*

3. Vitis foliis cordatis dentato-ſerratis utrinque nudis. Lin. Sp. Pl. 203. *The Virginia Fox-grape.*

4. Vitis foliis quinatis, foliolis multifidis. Hort. Cliff. 74. *Vine with jagged leaves; commonly called the Parſley-leaved Grape.*

5. Vitis foliis ſupradecompoſitis, foliis lateralibus pinnatis. Lin. Sp. Pl. 203. *Climbing Virginia ſhrub with Parſley leaves, ſending out tendrils; this is the Reynardſonia.* Rand. Ind. Hort. Chelſ. *Falſly called the Pepper-tree.*

The firſt ſort grows naturally in both Indies. The ſtalks of this are woody, and ſend out many ſlender branches, which are furniſhed with tendrils, by which they faſten themſelves to the neighbouring trees, and are thereby ſupported. The leaves are heart-ſhaped, indented on their edges, and hairy on their under ſide. The flowers are diſpoſed in bunches like thoſe of the other ſpecies, and are ſucceeded by round berries or Grapes, of an auſtere taſte.

The ſecond ſort hath ligneous ſtalks which ſend out many branches, that faſten themſelves by tendrils to any neighbouring ſupport. The leaves of this are large, and for the moſt part divided into three lobes, which are indented on their edges. The under ſide of the leaves is covered with a white down. The fruit is diſpoſed in bunches like the other Grapes. The berries are round and black; the juice has a rough flavour.

The third ſort has heart ſhaped leaves, which are indented on their edges, and are ſmooth on both ſides. The plants climb on trees by the help of their tendrils, like thoſe of the other ſorts. The fruit is diſpoſed in bunches. The berries are black, and their juice has a flavour reſembling the ſcent of a fox, from whence the inhabitants have given it the title of Fox-grape.

The fourth ſort is ſuppoſed to grow naturally in Canada, but it has been long cultivated in the European gardens for its fruit; but as it has little flavour, and ripens late in autumn, ſo it has been almoſt baniſhed the Engliſh gardens, where at preſent there are only a few plants preſerved for the ſake of variety. The ſtalks and branches of this are like thoſe of the common Grape, but the leaves are cut into many ſlender ſegments. The Grapes are round and white, and are diſpoſed in looſe bunches.

The fifth ſort is by Dr. Linnæus ranged under this genus of Vitis, but the characters of this plant are not ſufficiently known in Europe, to determine the proper genus to which it belongs, for the plant ſeldom produces flowers here, and has never produced any fruit in England, for which reaſon I have ranged it under the ſame genus, upon Dr. Linnæus's authority.

The ſtalk of this plant is ligneous, and ſends out many ſlender branches furniſhed with tendrils, which faſten themſelves to any neighbouring plants for ſupport, and are garniſhed with leaves compoſed of many ſmaller winged leaves, ſo that they are divided ſomewhat like thoſe of common Parſley; they are of a lucid green on their upper ſide, but are much paler on their under. The flowers ſpring from the wings of the ſtalks in looſe bunches; they are very ſmall, white, and are compoſed of five ſmall petals, which expand and ſoon fall off; theſe are not ſucceeded by any fruit in England, but the berries which I have received from America had generally three ſeeds in each.

The firſt ſort being a native of warm countries, will not live in England without artificial heat; it is eaſily propagated by ſeeds, when they are brought from the countries where the plants grow naturally, for they do not produce any here; theſe muſt be ſown in ſmall pots, which ſhould be plunged into a hot-bed of tanners bark. When the plants come up and are fit to remove, they ſhould be each tranſplanted into a ſeparate ſmall pot filled with light earth, and plunged into a freſh hot-bed of tanners bark, ſhading them from the ſun till they have taken new root; then they muſt be treated in the ſame way as other tender exotic plants from the ſame countries, always continuing them in the ſtove, otherwiſe they will not thrive. Theſe plants caſt off their leaves every winter.

The ſecond and third ſorts grow in great plenty in the woods of America, where, I have been informed, are many other ſorts, which produce fruit very little inferior to ſome of the fine ſorts which are cultivated in Europe; notwithſtanding which, it is generally thought impoſſible to make wine in America; but this, I dare ſay, muſt proceed from a want of ſkill, rather than any bad quality in the ſoil or climate; ſo that inſtead of planting vineyards on their looſe rich lands (as hath been generally practiſed by the inhabitants of thoſe countries,) if they would plant them upon riſing ground, where the bottom was rocky or hard near the ſurface, I dare ſay they would have very good ſucceſs; for the great fault complained of in thoſe countries is, that the Grapes generally burſt before they are fully ripe, which muſt certainly be occaſioned by their having too much nouriſhment; therefore, when they are planted on a poorer ſoil, this will be in part remedied. Another cauſe of this may proceed from the moiſture of the air (occaſioned by the perſpiration of trees, &c.) which being imbibed by the fruit, may break their ſkins. This cannot indeed be prevented, until the country is better cleared of the timber; but, however, this ſhould caution people not to plant Vines

in

in fuch places where there are great quantities of woods, becaufe of this effect which it hath on the Grapes. But to return:

Thefe two Vines are preferved in the gardens of thofe who are curious in botany, but I have not feen either of them produce fruit in this country. They may be propagated by layers in the fame manner as the common Grapes, which will take root in one year, and may be taken off, and tranfplanted where they are to remain, which fhould be againft a warm wall; becaufe if they are expofed to much cold in winter, they are often deftroyed, efpecially while they are young.

The pruning and management is the fame with any other forts of Grapes, but only they fhould have fewer fhoots, and thofe fhortened down very low, otherwife they will make very weak fhoots the following year, and never arrive to any confiderable ftrength, fo will not be capable of producing any fruit.

The fourth fort is planted againft walls, and treated in the fame way as the common Vines, and may be propagated by cuttings or layers in like manner.

The fifth fort is preferved in fome gardens for the fake of variety, but as it rarely produces flowers in England, fo it has not much beauty; it is a native in Virginia and Carolina. From both of thefe countries I have received the feeds. As this fort does not produce feeds here, it is generally propagated by laying down the young branches, which will put out roots in one year fit to remove, when they may be taken off, and tranfplanted where they are to remain. Thefe require fupport; and as their young branches are tender and liable to be killed by froft, fo if they are planted againft a wall or pale expofed to the fouth, they will fucceed much better than when they are fully expofed to the open air, and fupported by props. The young fhoots of thefe plants fhould be fhortened down to two or three buds in the fpring, which will caufe the fhoots of the following fummer to be much ftronger; and when they are regularly trained againft the wall or pale, they will produce flowers in warm feafons.

This plant is very apt to pufh out fuckers from the root, by which it is often propagated, but the plants fo raifed are very fubject to fend out fuckers again, whereby they are robbed of their nourifhment, and do not thrive fo well as thofe which come from layers.

VITIS IDÆA. See VACCINIUM.

VITIS SYLVESTRIS. See CLEMATIS.

ULEX. Lin. Gen. Plant. 86. Furze, Gorfe, or Whins.

The CHARACTERS are,

The flower has a two-leaved empalement; it has five petals, and is of the butterfly kind. The ftandard is large, erect, oval, heart-fhaped, and indented at the point. The wings are fhorter and obtufe. The keel is compofed of two obtufe petals, whofe borders are joined at bottom; it has ten ftamina, nine joined, and one feparate, terminated by fingle fummits, with an oblong cylindrical germen, fupporting a rifing ftyle, crowned by a fmall obtufe ftigma. The germen afterward turns to an oblong turgid pod with one cell, opening with two valves, inclofing a row of kidney-fhaped feeds.

The SPECIES are,

1. ULEX (*Europæus*) foliis villofis acutis fpinis fparfis. Lin. Sp. Plant. 741. *Ulex with acute-pointed hairy leaves, and fparfed fpines; the common Furze, Whins, or Gorfe.*

2. ULEX (*Capenfis*) foliis obtufis folitariis, fpinis fimplicibus terminalibus. Flor. Leyd. Prod. 372. *African Furze, or Whins, with fingle blunt leaves ending with fpines.*

The common Furze, Gorfe, or Whins, as it is called in the different counties in England, is fo well known as to need no defcription.

There are two or three varieties of this, which are frequently met with on the commons and heaths in moft parts of England; but as they are not fpecifically different they are not worth enumerating here, efpecially as they are plants which are feldom cultivated.

Thefe plants propagate themfelves very plentifully by feeds, fo that when they are eftablifhed in a fpot of ground they foon fpread over the place; for as the feeds ripen, the pods open with the warmth of the fun, and the feeds are caft out with an elafticity to a great diftance all round, and thefe foon vegetate; whereby the ground is filled with young plants, which are not eafily deftroyed when they are well rooted in the ground.

Some years ago the feeds of this plant were fown to form hedges about fields, where, if the foil was light, the plants foon became ftrong enough for a fence againft cattle; but as thefe hedges in a few years became naked at the bottom, and fome of the plants frequently failed, there became gaps in the hedges, therefore the raifing of them for that purpofe has been of late years little practifed. But there are fome perfons who have fown the feeds of this plant upon very poor, hungry, gravel or fandy land, which has produced more profit than they could make of the ground by any other crop, efpecially in fuch places where fuel of all forts is dear; for this Furze is frequently ufed for heating ovens, burning lime and bricks, and alfo for drying malt. And in fome places where there has been a fcarcity of fuel, I have known poor land, which would have let for two fhillings per acre, which has been fown with Furze, produce one pound per acre per ann. fo that there has been a confiderable improvement made by this plant. But this is not worth practifing in fuch countries where fuel of any kind is cheap, or upon fuch land as will produce good Grafs or Corn; therefore it is only mentioned here to fhew, that poor lands may be fo managed, as to bring an annual profit to their proprietors.

The fecond fort is a native of the country near the Cape of Good Hope, where it ufually grows to the height of five or fix feet; but in Europe, where it is preferved as a curiofity in fome gardens, it feldom rifes fo high. The ftalk is ligneous and hard, covered with a greenifh bark when young, but it afterward becomes grayifh. The branches are flender and ligneous; the leaves are fingle, obtufe, and the fhoots terminate with fpines. This has been feveral years in the Englifh gardens, but has not produced any flowers.

This plant is too tender to live in the open air through the winter in England, therefore it is preferved in greenhoufes, with the hardier forts of exotic plants, which do not require any artificial heat to preferve them.

It is very difficult to propagate either by layers or cuttings, for the layers are generally two or three years before they have fufficient root to tranfplant, and the cuttings do very rarely take root; and as the plant does not produce feeds in Europe, it is very rare in the European gardens. It is a plant of no great beauty, but, as it is an evergreen, it is admitted into the gardens of thofe who are curious in botany for the fake of variety.

ULMARIA. See SPIRÆA.

ULMUS. Tourn. Inft. R. H. 601. tab. 372. The Elmtree.

The CHARACTERS are,

The flower has a rough permanent empalement of one leaf, cut into five points, and coloured within; it has no petals, but has five awl-fhaped ftamina twice the length of the empalement, terminated by fhort erect fummits, having four furrows, and an orbicular erect germen fupporting two ftyles which are reflexed, and crowned by hairy ftigmas. The germen afterward turns to a roundifh, compreffed, bordered capfule, including one roundifh compreffed feed.

The

The SPECIES are,

1 ULMUS (*Campestris*) foliis oblongis acuminatis, duplicato serratis, basi inæqualibus. *Elm with oblong acute-pointed leaves, which are doubly sawed on their edges, and unequal at their base; the common rough, or broad-leaved Witch Elm.*

2. ULMUS (*Scaber*) foliis oblongo-ovatis inæqualiter serratis, calycibus foliaceis. *Elm with oblong oval leaves which are unequally sawed, and have leafy empalements to the flowers; the Witch Hazel, or very broad-leaved Elm by some unskilful persons called the British Elm.*

3. ULMUS (*Sativa*) foliis ovatis acuminatis duplicato-serratis, basi inæqualibus. *Elm with oval acute-pointed leaves which are doubly sawed, and unequal at their base; the small-leaved or English Elm.*

4. ULMUS (*Glaber*) foliis ovatis glabris, acutè serratis. *Elm with oval smooth leaves, which are sharply sawed on their edges; the smooth-leaved Witch Elm.*

5. ULMUS (*Hollandica*) foliis ovatis acuminatis rugosis, inæqualiter serratis, cortice fungoso. *Elm with oval, acute-pointed, rough leaves, which are unequally sawed, and a fungous bark; the Dutch Elm.*

6. ULMUS (*Minor*) foliis oblongo-ovatis glabris acuminatis duplicato-serratis. *Elm with oblong, smooth, acute-pointed leaves, which are doubly sawed; the smooth narrow-leaved Elm, by some called the upright Elm.*

The first sort is very common in the north-west counties of England, where it is generally believed to grow naturally in the woods; this grows to a very large size. The bark of the young branches is smooth and very tough, but that of the old trees cracks and is rough. The branches spread, and do not grow so erect as those of the third sort. The leaves are rough, and are doubly sawed on their edges. Their base is unequal, standing on short foot-stalks. The flowers come out in March upon the slender twigs in clusters, of a deep red colour, and are succeeded by oval bordered capsules, containing one roundish compressed seed, which ripens in May. The wood of this tree is very good for all the purposes of any kind of Elm, and the trees grow to a very large size, but the leaves do not come out till late in the spring, so there are few persons who plant these trees for ornament.

The second sort grows naturally in some of the northern counties in England, where it is frequently called Witch Hazel, from the resemblance of the young shoots and leaves to those of Hazel. This grows to a tree of great magnitude. The bark of the young shoots is very smooth and tough; it is of a yellowish brown colour, with spots of white. The leaves are oval, unequally sawed on their edges. The flowers grow in clusters toward the end of the twigs; they have long leafy empalements, of a green colour, appearing in the spring before their leaves, and the seeds ripen the latter end of May. The wood of this tree is not so good for use as that of the first sort. Formerly, when long bows were in use, many of them were made of the boughs of this tree.

The third sort is commonly known in the nursery-gardens by the title of English Elm, which is far from being a right appellation; for it is not a native of England, and is only found in plantations where the young trees were procured from the neighbourhood of London. Where this tree grows naturally is not easy to determine; some persons have supposed it was brought from Germany. As this tree is well known it requires no description. The flowers of this are of a purplish red colour, and generally appear the beginning of March, but I could never observe any seeds upon this sort.

The fourth sort is very common in several parts of Hertfordshire, Essex, and other north-east counties of England; this grows to a large tree, and is much esteemed. The branches spread out like those of the first sort. The leaves are sharply sawed on their edges; they are smoother than most of the other sorts, and do not appear till the middle or latter end of May, so the trees are seldom planted for ornament.

The fifth sort is well known by the title of Dutch Elm; this was brought from Holland the beginning of King William's reign, and was for some time a fashionable tree, and has been recommended for its quick growth; it was some years ago in request for forming hedges in gardens, for which purpose it was one of the most improper trees that could be chosen; for they made very strong irregular shoots, which are distant from each other. The leaves were very large and rough, and the branches covered with a fungous rough bark, which was disagreeable, so that when the hedges were sheared, they appeared naked and disagreeable the whole summer after. The wood of this tree is good for nothing, so it is almost banished this country.

The sixth sort is found growing in hedge-rows in several parts of England. The branches of this sort have a smooth grayish bark, and grow erect. The leaves are narrower, and more pointed than those of the English Elm, and are smoother; they are later in coming out in the spring than those, but continue longer in autumn; this has been by some called the Irish Elm.

There are some other varieties of this tree which are preserved in the nursery-gardens, but their difference is not remarkable enough to deserve notice; therefore they are omitted, as are also those with variegated leaves, of which there are several varieties propagated in the nurseries about London; these are by some persons esteemed.

All the sorts of Elm may be either propagated by layers or suckers taken from the roots of the old trees, the latter of which is generally practised by the nursery-gardeners; but as these are often cut up with indifferent roots, they often miscarry, and render the success doubtful; whereas those which are propagated by layers are in no hazard, and always make better roots, and come on faster than the other, and do not send out suckers from their roots in such plenty, for which reason this method should be more universally practised. And since a small compass of ground filled with stools of these plants will be sufficient to furnish a nursery of a considerable extent annually, with layers to be transplanted, it is richly worth every person's while, who would cultivate these trees, to allot a spot of ground for this purpose.

The best soil for such a nursery is a fresh Hazel loam, neither too light and dry, nor over moist and heavy; this ground should be well trenched, and if a little rotten dung is buried therein it will be of service; in doing of this, great care should be taken to pick out all roots of pernicious weeds, which, if left in the ground, would be very injurious to the layers, and cannot afterwards be so easily rooted out; then having laid the ground level, the plants must be planted at about three feet asunder each way. The best season for this work is in autumn, as soon as the leaves begin to decay, that they may take root before the dry weather in the spring comes on, whereby a great expence of watering them will be saved; for if they are well settled in the ground before the dry weather, they will require little more than to mulch their roots to keep the earth from drying.

These plants should be permitted to grow rude two years, during which time the ground between should be carefully cleaned and dug every spring, by which time they will be well rooted, and have made pretty strong shoots, so that they may be laid in the ground. The manner of perform-

ing

ing this being already defcribed in the article LAYERS, I fhall forbear repeating it in this place.

When thefe layers are well rooted, which will be in one year, they fhould be taken off, and tranfplanted out into a nurfery, which fhould be upon a good foil, and well prepared (as before for the ftools.) The plants fhould be planted in rows four feet afunder, and two feet diftance plant from plant in rows. This fhould be done in autumn, as foon as the leaves begin to decay and; if there is fome mulch laid upon the furface of the ground about their roots, it will preferve them from being hurt by froft in winter, and from drying winds in fpring, and thereby fecure them from all hazard.

In this nurfery they may remain four or five years, obferving conftantly to dig the ground between them every fpring, and to trim them up to ftems, which will promote their growth, and render them ftrong enough to tranfplant out where they are to remain in the time beforementioned.

Thefe trees are very proper to plant in hedge-rows, upon the borders of fields, where they will thrive much better than when planted in a wood or clofe plantation, and their fhade will not be very injurious to whatever grows under them; but when thefe trees are tranfplanted out upon banks after this manner, the banks fhould be well wrought and cleared from all other roots, otherwife the plants, being taken from a better foil, will not make much progrefs in thefe places. About Michaelmas will be a good time for this work for the reafons before affigned; but when they are planted there fhould be fome ftakes fixed in by them, to which they fhould be faftened to prevent their being difplaced by the winds, and part of their heads fhould be taken off before they are planted, which will alfo be of ufe in preventing their being eafily overturned by winds; but by no means fhould their leading fhoot be ftopped, nor the branches too clofely cut off; for if there are not fome fhoots left on to draw and attract the fap, they will be in danger of mifcarrying.

Thefe trees are alfo proper to plant at a diftance from a garden or building, to break the violence of winds; for which purpofe there is not any tree more ufeful, for they may be trained up in form of a hedge, keeping them cut every year, which will caufe them to grow very clofe and handfome, to the height of forty or fifty feet, and be a great protection againft the fury of winds; but they fhould not be planted too near a garden, where fruit-trees, or other plants are placed, becaufe the roots of the Elms run fuperficially near the furface of the ground to a great diftance, and will intermix with the roots of the other trees, and deprive them of nourifhment; nor fhould they be planted near gravel or Grafs-walks, which are defigned to be well kept, becaufe the roots will run into them, and fend forth fuckers in great plenty, which will deface the walks, and render them unfightly.

But for large gardens, where fhade is required, there is fcarce any tree fo proper for that purpofe, being eafy to remove when grown to a confiderable fize; fo that a perfon who is willing to have his plantations for fhade in a fhort time, may procure trees of near one foot circumference in their trunk, which will be in little danger of fucceeding, provided they are removed with care. And thefe will take root and grow very well, though not fo well as young plants, which is what few other forts of trees will do; but then they fhould be fuch trees as have been thus regularly trained up in a nurfery, and have good roots, and not fuch as are taken out of hedge-rows (as is by fome practifed,) which feldom rife with any tolerable roots, and confequently often mifcarry; and this has been the occafion of fo many plantations of thefe trees failing; for though fome of them may live a few years, yet few of them are of long duration, and they rarely increafe much in their ftems, but frequently grow hollow, their heart decaying firft, fo that they are fupported only by their bark or fhell for a few years, and the firft fevere winter, or very dry fummer, they are generally deftroyed.

But although I have faid, that Elms which are trained up in a nurfery may be removed with fafety, at a larger fize than moft other trees, yet I would not have it underftood, that by this I would recommend the planting of them when large; for if people would have a little patience when they plant, and never plant any of thefe trees which are more than five or fix inches in girt of their ftems, they will in a few years become better trees than any of thofe which are tranfplanted of a much larger growth, and will grow to a much larger fize; befides, they are much more eafily removed, and do not require to be fo ftrongly fupported, nor is there much danger of the young trees mifcarrying; therefore it is much more eligible to make choice of young thriving trees (but not out of a better foil than that where they are to be planted,) and never to plant any large trees, unlefs where a fmall number may be wanted for an immediate fhade; and in fuch cafes it is always proper to plant fome young trees amongft the large ones, to fucceed them when they fail.

In planting of thefe trees, great care fhould be taken not to bury their roots too deep, which is very injurious to them, efpecially if they are planted on a moift loam or clay; in which cafe, if the clay is near the furface, it will be the beft way to raife the ground in a hill, where each tree is to be planted, which will advance their roots above the furface of the ground, fo that they will not be in danger of rotting in winter with moifture.

When thefe trees are propagated by fuckers taken from old trees, they are commonly laid into the ground in rows pretty clofe together in beds, where, in dry weather, they may be frequently watered to encourage their putting out roots. In thefe beds they are left commonly two years, by which time thofe that live will be rooted (though a great many of them generally die;) then they are tranfplanted into the nurfery, and managed as hath been directed for the layers.

There are fome who raife the Witch Elm from feeds, which it generally produces in great plenty, and are ripe in May; thefe fhould be fown upon a bed of frefh loamy earth, and gently covered. In dry weather they fhould be watered, and if the bed is fhaded from the violent heat of the fun, it will be of great fervice to the feeds (for I always obferve the plants to come up better in the fhade than when expofed to the fun.) When the plants come up they fhould be carefully cleared from weeds, and after they have ftood two years in the feed-bed, they will be fit to plant out into the nurfery, where they muft be managed as the former.

VOLKAMERIA. Lin. Gen. Plant. 706.

The CHARACTERS are,

It hath a ringent flower of one petal, whofe tube is much longer than the empalement; the brim is cut into five parts. It has four longer flender ftamina, two being fhorter than the other, with a four-cornered germen, having a long ftyle, crowned by a bifid ftigma. The germen turns to a roundifh berry with two cells, including a nut with two cells.

The SPECIES are,

1. VOLKAMERIA (*Aculeata*) fpinis petiolorum rudimentis. Lin. Sp. Plant. 637. *Prickly Volkameria.*

2. VOLKAMERIA (*Inermis*) ramis inermibus. Lin. Sp. Plant. 637. *Smooth Volkameria.*

The firft fort grows naturally in the Weft-India iflands, where it rifes to the height of twenty feet, having many pliable branches which are much diffused, covered with a light smooth bark, garnished with oval, spear-shaped, lucid leaves, placed opposite. The flowers come out from the side of the stalk, five or six standing on the same foot-stalk, almost in form of an umbel; they are in shape somewhat like those of the common Jasmine, but the tube is curved, and two of the stamina are longer than the other, so it comes under the class of ringent flowers. They have no scent, and are not succeeded by seeds in England, nor are the plants very free to flower here.

The second sort is also a native of both Indies; this rises higher than the former, the stem and branches are stronger, and grow more erect; the bark is very white, but have short crooked spines immediately under the foot-stalk of the leaves. The leaves frequently grow round the branches in clusters or whorls. The flowers are set upon long foot-stalks arising from the wings of the stalk, each supporting several flowers, which generally stand erect. They are shaped like those of the former, have no scent, nor are succeeded by seeds here.

As these do not produce seeds in England, so the plants are propagated by cuttings, which readily put out roots, when they are planted in pots and plunged into a moderate hot-bed, covering them close with hand-glasses to screen them from the external air. The cuttings may be planted any time from the middle of May to the end of July; if they are planted later in the season, there will not be summer enough for them to get strong roots before the cold of autumn; nor sooner than May, because their shoots will not be hardened enough for planting.

When they have put out roots, the plants should be carefully separated, and each planted into a separate small pot, and plunged into a gentle hot-bed, to get fresh roots in the pots; then they may be inured to the open air, provided the weather is warm, and may remain abroad in a sheltered situation until the nights begin to be cold, when they must be removed into the house, for cold will soon destroy them.

In winter these plants will require some warmth, so should be placed in a stove where the air is never greatly warmed, because in heat they are very subject to shoot and grow weak; but in a common green-house they will not live through the winter.

URENA. Hort. Elth. 319. Lin. Gen. Pl. 754. Indian Mallow.

The CHARACTERS are,

It hath a malvaceous flower with a double empalement, the outer being of one leaf slightly cut at the brim into five parts, but the inner is five-leaved, permanent, and cut to the bottom. The flower is composed of five leaves which are oblong, and blunt at their extremity, but narrow at their base, where they coalesce. In the center there are many stamina which are joined, and form a column at their base, but spread open above. It has a roundish five-cornered germen with a single style, and ten hairy reflexed stigmas. The germen change to a pentagonal fruit which is burry, and divides into five cells, each having one angular seed.

The SPECIES are,

1. URENA (*Lobata*) foliis angulatis. Hort. Cliff. 348. *Indian Mallow with angular leaves.*

2. URENA (*Aculeata*) foliis inferioribus angulatis, superioribus trilobis quinquelobisque acutè serratis. *Urena with angular lower leaves, the upper ones divided into three or five lobes, which are sharply sawed; or Indian shrubby Vervain Mallow from Bengal.*

3. URENA (*Sinuata*) foliis sinuato-multifidis villosis. Flor. Zeyl. 257. *Urena with sinuated hairy leaves, having many points.*

The first sort grows naturally in China; this rises with an upright stalk two feet high, which becomes ligneous toward the autumn. It sends out a few side branches, which are taper, stiff, and have a dark green bark, garnished with roundish angular leaves, standing upon pretty long foot-stalks, of a dark green on their upper side, and pale on their lower. The flowers come out single from the wings of the stalk, sitting close to it; they are shaped like those of the Mallow, but are small, and of a deep blush colour; these are succeeded by roundish capsules, armed with prickly hairs, divided into five cells, each containing one kidney-shaped seed.

The second sort grows naturally on the coast of Malabar, from whence I received the seeds; this rises with a ligneous stalk three feet high, dividing into four or five branches, which have a grayish bark, garnished with leaves of different forms; those on the lower part are angular, those above are cut some into three, and others have five angular obtuse lobes, of a dark green on their upper side, but pale on their under, sharply sawed on their edges, and stand upon long foot stalks. The flowers come out singly from the wings of the stalk; they are shaped like those of the other, but are larger. The petals are narrower at their base, and have deep red bottoms.

The seeds of the third sort came from Malabar; the stalks of this are hairy, and divide into many branches: it rises about two feet high, and is garnished with oblong leaves, divided into three obtuse lobes to the midrib. The lobes are indented in several parts; they are of a light green on both sides, and hairy. The flowers sit close to the stalks singly at the wings; they are shaped like those of the former, but are of a pale blush colour, with a deep red bottom.

These plants are propagated by seeds, which should be sown on a hot-bed early in the spring; and when the plants are fit to remove, they should be transplanted into pots, and plunged into a fresh hot-bed to bring them forward; and afterward they must be treated in the same manner as hath been directed for the tender sorts of HIBISCUS, to which the reader is desired to turn. If the plants are brought forward in the spring, and afterward placed in the stove, or under a deep frame, they will ripen seeds the first season; but if they should not, they may be preserved through the winter in the stove, and will ripen their seeds the following season, after which the plants seldom continue.

URTICA. Tourn. Inst. R. H. 534, tab. 308. The Nettle.

The CHARACTERS are,

It has male and female flowers at remote distances, sometimes on the same, and at others on separate plants. The male flowers have empalements composed of four roundish concave leaves, and have a pitcher-shaped nectarium in the center of the flower and four awl shaped spreading stamina, terminated by summits with two cells. The female flowers have an oval permanent empalement with two valves; they have neither petals nor stamina, but an oval germen without any style, crowned by a hairy stigma. The germen afterward turns to an oval compressed seed, which ripens in the empalement.

The SPECIES are,

1. URTICA (*Dioica*) foliis oppositis cordatis, racemis geminis. Lin. Sp. Plant. 984. *Nettle with heart-shaped leaves which are placed opposite, and double spikes of flowers; the great stinging Nettle.*

2. URTICA (*Urens*) foliis oppositis ovalibus. Lin. Sp. Plant. 984. *Nettle with oval leaves which are placed opposite; smaller stinging Nettle.*

3. URTICA (*Pilulifera*) foliis oppositis cordatis, amentis fructiferis globosis. *Nettle with heart-shaped leaves placed opposite,*

ppoſite, and ſeed-bearing globular katkins; commonly called Roman Nettle.

4. Urtica (*Dodartia*) foliis oppoſitis ovato-lanceolatis, ſubintegerrimis, amentis fructiferis globoſis. *Nettle with oval ſpear-ſhaped leaves, which are almoſt entire, and placed oppoſite, and globular ſeed-bearing katkins; commonly called Spaniſh Marjoram.*

5. Urtica (*Cannabina*) foliis oppoſitis tripartitis inciſis. Hort. Upſal. 282. *Nettle with leaves placed oppoſite, which are cut into three parts.*

6. Urtica (*Cylindrica*) foliis oppoſitis oblongis, amentis cylindricis ſolitariis indiviſis. Lin. Sp. Pl. 984. *Nettle with oblong leaves which are placed oppoſite, and ſingle, cylindrical, undivided katkins.*

7. Urtica (*Mariana*) foliis oppoſitis ovato-lanceolatis acuminatis crenatis, amentis cylindricis indiviſis. *Nettle with oval, ſpear-ſhaped, acute-pointed, crenated leaves, which are placed oppſite, and cylindrical undivided katkins.*

8. Urtica (*Canadenſis*) foliis alternis cordato-ovatis, amentis racemoſis diſtichis erectis. Hort. Cliff. 441. *Nettle with oval heart-ſhaped leaves, which are placed alternate, and erect, branching, double katkins.*

9. Urtica (*Nivea*) foliis alternis orbiculato utrinque acutis ſubtus tomentoſis. Hort. Cliff. 441. *Nettle with orbicular leaves, pointed at both ends, placed oppoſite, and woolly on their under ſide.*

10. Urtica (*Balearica*) foliis oppoſitis cordatis ſerratis, amentis fructiferis globoſis. Lin. Sp. 1395. *Balearick Nettle with heart-ſhaped ſawed leaves, placed oppoſite, and globular fruit.*

The firſt of theſe ſorts is a very common weed upon the ſides of banks, ditches, and other uncultivated places, where its roots will ſpread, and over-run the grounds, ſo that it ſhould always be carefully extirpated from gardens; it is ſometimes uſed in medicine, but may eaſily be pro cured from the fields at almoſt any ſeaſon.

The ſecond ſort is alſo a very common weed in gardens and cultivated fields; but it being an annual plant, is not ſo difficult to eradicate as the former.

Theſe plants are ſo well known as to need no deſcription.

The third ſort grows naturally in Romney Marſh, and near Yarmouth; this is an annual plant, which riſes three feet high. The ſtalk is herbaceous, thick, of a purple colour, armed in every part with ſtinging hairs. The branches come out oppoſite. The leaves are heart-ſhaped, ending in acute points, deeply ſawed on their edges, and ſtand oppoſite upon long foot-ſtalks; theſe are alſo armed with ſtinging hairs on both ſides. The male and female flowers come out from the wings of the leaves, at the ſame joint on each ſide the ſtalk; the male ſtanding above the female, upon long ſlender foot-ſtalks or katkins, placed very looſely. The female flowers have ſhorter foot-ſtalks, and are in globular heads; theſe are ſucceeded by ſmooth ſhining ſeeds like thoſe of the Flax.

There is a variety of this growing naturally in the Balearick iſlands, which was diſcovered by Mr. Salvadore, an apothecary in Barcelona, who ſent the ſeeds to England, which were ſown in the Chelſea garden; but the plants, when cultivated, approached ſo near to the laſt mentioned ſort, in every part except the colour of the ſtalk, as to make it doubtful of its being a diſtinct ſpecies. This is the tenth ſort mentioned.

The fourth ſort grows naturally in Spain and Italy; this is alſo an annual plant, whoſe ſtalks are much ſlenderer than thoſe of the former, and ſeldom branch. The leaves are placed by pairs, upon very ſlender foot-ſtalks; they are oval, ſpear-ſhaped, and for the moſt part entire, and have male and female flowers ſpringing from the wings of the leaves, which are ſhaped like the former, the whole plant being armed with ſtinging hairs.

Theſe plants may be eaſily propagated, by ſowing their ſeeds in March upon a bed of light earth; and when the plants are come up, they ſhould be tranſplanted into beds, or the borders of the pleaſure-garden, interſperſing them amongſt other plants, that they may not be eaſily diſcovered by perſons whom there is a deſign to deceive, by gathering a ſprig for them to ſmell to. After the plants have taken root, they will require no farther care but only to keep them clear from weeds.

The ſeeds of the third ſort are ſometimes uſed in medicine.

The fifth ſort grows naturally in Tartary; this has a perennial root, from which ſpring up many ſquare ſtalks, which riſe five or ſix feet high, garniſhed with oblong leaves, deeply cut into three lobes, which are acutely indented on their edges; theſe ſtand oppoſite upon long foot-ſtalks. The flowers are produced from the wings of the leaves in long cylindrical katkins; the male are produced on the lower part of the ſtalk, and the female on the upper; the latter are ſucceeded by ſeeds like thoſe of Flax, incloſed in the three-cornered empalement of the flower.

This plant is eaſily propagated either by ſeeds or parting of the roots, and will thrive in moiſt ſoils or ſituations.

The ſixth ſort grows naturally in Canada and other parts of North America; it is an annual plant, with a lucid herbaceous ſtalk, which divides into ſeveral branches, garniſhed with oblong ſawed leaves, having three longitudinal veins; they are placed oppoſite upon pretty long foot-ſtalks. The flowers are produced from the wings of the ſtalks in ſingle katkins, which are not divided; they appear late in the year, and unleſs the autumn is very favourable, the ſeeds will not ripen in England.

The ſeventh ſort grows naturally in North America; this has a perennial root, from which ſpring out many ſtalks from two to three feet high, garniſhed with oval ſpear-ſhaped leaves, placed oppoſite, ſtanding upon long foot-ſtalks; they are crenated on their edges, and end in acute points. The flowers come out from the wings of the leaves in long, cylindrical, undivided katkins; the ſeeds do not ripen in England.

The eighth ſort grows naturally in Canada and Virginia. The root is perennial; the ſtalks riſe two feet high; the leaves are oval, heart-ſhaped, and ſtand alternately upon the ſtalks; the flowers come out in branching katkins from the wings of the ſtalks, but are not ſucceeded by ſeeds in this country.

The two laſt ſorts are common in many Engliſh gardens, where they are preſerved more for the ſake of variety than for any beauty. They may be propagated by parting their roots in the ſpring, and planted in almoſt any ſoil or ſituation, and will endure the ſevereſt cold of this climate in the open air.

The ninth ſort grows naturally in China, where it is titled Peama; this is a perennial plant, ſending up many ſtalks from the root, which riſe three or four feet high, garniſhed with oval leaves, drawing to points at both ends, ſawed on their edges, of a deep green on their upper ſide, but very white on their under, and have five longitudinal veins; they are placed alternately, and ſtand upon very long ſlender foot-ſtalks. The flowers ſpring from the wings of the ſtalk in looſe katkins; theſe are not ſucceeded by ſeeds in England.

This may alſo be propagated by parting of the roots, which ſhould be done in the ſpring, for at that ſeaſon this plant is in its leaſt vigour, the winter being the time when it is moſt flouriſhing.

The plants muſt be planted in pots filled with light earth, and as they are too tender to thrive without artificial heat in England, they ſhould be kept in a temperate ſtove, and only expoſed to the open air for three months in the heat of ſummer.

UVA URSI. See ARBUTUS.

VULNERARIA. See ANTHYLLIS.

UVULARIA. Lin. Gen. Plant. 373.

The CHARACTERS are,

The flower has no empalement; it has ſix oblong, erect, ſpear-ſhaped petals, and ſix awl-ſhaped ſtamina, terminated by oblong, erect, four-cornered ſummits; it has an oblong, obtuſe, three-cornered germen, ſupporting a ſtyle longer than the ſtamina, crowned by a triple, obtuſe, ſpreading ſtigma. The germen afterward turns to an oblong obtuſe capſule, with three lobes and as many cells, filled with flat orbicular ſeeds, ranged in a double order.

The SPECIES are,

1. UVULARIA (*Amplexicaulis*) foliis amplexicaulibus. Lin. Sp. Plant. 304. *Uvularia with leaves embracing the ſtalk.*

2. UVULARIA (*Perfoliata*) foliis perfoliatis. Amœn. Acad. 2. p. 3. *Uvularia with perfoliate leaves.*

The firſt ſort grows naturally in Bohemia and Saxony. The root is perennial, but the ſtalk is annual; it riſes about two feet high, ſending out one or two branches from the lower part, garniſhed with oblong ſmooth leaves, ending in acute points, whoſe baſe embrace the ſtalks. The flowers come out ſingly from the boſom of the leaves upon long ſlender foot-ſtalks; they are compoſed of ſix oblong naked petals, of a yellowiſh colour; theſe hang downward, but are rarely ſucceeded by ſeeds here.

The ſecond ſort grows naturally in North America; this has a perennial root and an annual ſtalk. The root is compoſed of many thick fleſhy fibres, from which ſpring up ſeveral ſtalks, which for the moſt part divide into two at a ſmall height from the ground, and are garniſhed with oblong, ſmooth, pointed leaves, which are broad at their baſe, ſurrounding the ſtalk in ſuch a manner, as if the ſtalk run through them. The flowers are compoſed of ſix oblong yellow petals, ending in acute points; theſe ſtand upon ſlender foot-ſtalks, which ariſe from the boſom of the leaves, and hang downward, but are not ſucceeded by ſeeds in England.

They are both very hardy plants, and will live in the full ground; but as the flowers have not much beauty, ſo they are only cultivated for the ſake of variety; they are propagated by parting of their roots. The beſt ſeaſon for removing them is about Michaelmas, when their roots may be ſeparated, and planted in the borders of the flower-garden; but this ſhould only be done every third year, for if they are often removed, the plants will not thrive ſo well or flower ſo ſtrong, as when they ſtand two or three years unremoved; they delight in a ſoil not too wet or ſtiff, but a gentle Hazel loam.

W.

WACHENDORFIA. Burman.

The CHARACTERS are,

The flowers are ranged alternately in cluſters on the ſide of the ſtalk, each cluſter having acute-pointed ſpathæ; each flower has ſix oblong petals, the three upper are erect, ſpread open, and are joined at bottom; two on the ſide ſpread open like wings, the lower forms a kind of keel; it has a protuberant nectarium on each ſide the upper petal, with three awl-ſhaped ſtamina, which decline, terminated by horizontal ſummits, and an oval three-cornered germen, ſupporting one awl-ſhaped ſtyle, crowned by a ſingle ſtigma. The germen afterward turns to an oval, three-cornered, hairy capſule, with three cells, containing three oval ſeeds.

The SPECIES are,

1. WACHENDORFIA (*Thyrſiflora*) ſcapo ſimplici. Lin. Sp. Plant. *Wachendorfia with a ſimple ſtalk.*

2. WACHENDORFIA (*Paniculata*) ſcapo polyſtachio. Lin. Sp. Plant. *Wachendorfia with ſtalks ſending out many flower-ſtems.*

Theſe plants grow naturally at the Cape of Good Hope; the roots of both have many ſtrong fibres ſpringing from a fleſhy head; out of theſe heads ariſe ſeveral plaited leaves; thoſe of the firſt are much ſmaller than the ſecond ſort; from the center ariſes the flower-ſtalk, which in the firſt ſort is ſtrait, ſimple, and about a foot and a half high, garniſhed with white flowers, diſpoſed in looſe ſpikes; but the ſtalk of the ſecond ſort riſes more than three feet high, ſending out alternately cluſters of flower, each cluſter being covered with a ſpatha, which withers and remains on the ſtalk till it decays. The flowers of this are larger than thoſe of the firſt ſort, and are of an herbaceous white, inclining to yellow. The petals are divided into ſix parts almoſt to the bottom; they have each three ſtamina and one ſtyle, which ſits on the germen, crowned by horizontal ſummits. The germen afterward becomes an oval three-cornered capſule, having three cells, containing three oval ſeeds.

Theſe plants are uſually propagated by offsets, which come out from the ſide of the roots, becauſe their ſeeds do not often come to perfection in England; but when they do, if they are ſown in pots ſoon after they are ripe, and the pots placed in a garden-frame in the autumn to ſcreen the ſeeds from froſt, the plants will come up the following ſpring; and when they are ſtrong enough to remove, they ſhould be each planted in a ſeparate ſmall pot, and may be expoſed abroad till the autumn; then they ſhould be placed under a frame to ſcreen them from froſt, for they will not live in the open air through the winter in England. The ſecond year the plants will flower, and if the ſeaſon is warm, the ſeeds will ripen in autumn.

The offsets which are taken from the old roots, muſt be planted in ſeparate pots, and after they have taken root, ſhould be treated in the ſame way.

WALKERIA.

The CHARACTERS are,

The empalement of the flower is of one piece, cut half way into five ſegments, which are reflexed; the corolla of the flower is of one petal, deeply divided into five ſegments, which are concave; it has five incurved ſtamina, which are ſhorter than the petal, crowned by oval ſummits, and a conical germen without any ſtyle, crowned by a ſmall blunt ſtigma. The germen afterward turns to a conical capſule, divided into five cells, each containing one angular ſeed.

The title of this genus is given in honour of Doctor Richard Walker, Vice-Maſter of Trinity-College in Cambridge, who was a great lover of botany, and has eſtabliſhed a botanic garden in Cambridge, for the public uſe of the univerſity.

We have but one ſpecies of this genus, the ſeeds of which were brought from the Eaſt Indies, but from what particular part we are not acquainted.

It is an annual plant, whoſe branches are diffuſed and trail upon the ground, garniſhed with roundiſh leaves about the ſize of thoſe of Chickweed, but of a thicker conſiſtence, and of a bluiſh colour, ſtanding upon ſhort foot-ſtalks: the flowers come out from the wings of the ſtalk at each joint, having very ſhort foot-ſtalks; the flower is of one petal, ſhaped like thoſe of the Winter Cherry, but of a fine blue colour; theſe are each ſucceeded by conical capſules, divided into five cells, each containing one angular ſeed.

It is propagated by ſeeds, which muſt be ſown upon a hot-bed early in the ſpring; but as the ſeeds do frequently lie long in the ground, ſo it is the ſureſt way to ſow them in ſmall pots, and plunge them into a hot-bed, becauſe if the plants do not riſe in due time, the pots may be removed to another hot-bed, which will cauſe the ſeeds to vegetate. When the plants come up, and are ſtrong enough to remove, they ſhould be planted into ſeparate ſmall pots, and plunged into a hot-bed of tanners bark, ſhading them until they have taken root again; after which time they ſhould have a large ſhare of freſh air admitted to them in warm weather, and duly watered. The beginning of July they will flower, and the ſeeds will ripen in autumn; but if the plants ſhould come up late in ſummer, and not perfect their ſeeds, if the pots are removed in autumn, and plunged into the tan bed in the ſtove, the plants may be preſerved through the winter, ſo will flower early the next ſummer, and thereby good ſeeds may be obtained; but when the plants come up early, and produce good ſeeds the ſame year, they ſeldom continue longer.

WALKS are made either of gravel, ſand, or Graſs; theſe three ſorts of Walks are the moſt common in England; but where gravel or ſand cannot be procured, they are ſometimes laid with powdered coal, ſea coal aſhes, and ſometimes of powdered brick; but theſe are rarely uſed, when either gravel or ſand can be procured; however, where ſea-coal aſhes can be had, it is preferable to the powdered coal or bricks, becauſe they bind very hard, and never ſtick to the feet in froſty weather, which is a good quality, but the darkneſs of its colour has been an objection to the uſe of it in gardens; however, for Wilderneſs walks I think it is preferable to moſt other materials; but I ſhall proceed to give directions for the making of the ſeveral ſorts of Walks, and firſt of the Gravel-walks.

In order to the laying of Walks in gardens, when they are marked out, the earth ſhould be taken away to a certain depth, that the bottom of them may be filled with ſome lime-rubbiſh, or coarſe gravel, flint ſtones, or other rocky materials, which will be ſerviceable to prevent weeds from growing through the gravel, and alſo to keep away worm-caſts. This bottom ſhould be laid ten inches or a foot thick, over which the coat of gravel ſhould be ſix or eight inches; which gravel ſhould be fine, but yet not ſcreened, becauſe that ſpoils it. This ſhould be laid on a heap, rounding, that the larger rough ſtones may run down on the ſides, which being every now and then raked off, the gravel by that means will be ſufficiently fine.

After the gravel has been laid to the thickneſs above-mentioned, then the Walks muſt be exactly levelled, and raked true from all great drips, as well as little holes. By this means moſt of the ſtones of the Walks will be raked under your feet, which ſhould rather be gently ſprinkled back again, over the laſt length that is raked, than buried (as is the practice of many gardeners;) for by this means the Walk will lie much harder, and the coarſeſt ſtones will very much contribute to its firmneſs.

There is alſo a great fault committed frequently, in laying Walks too round, and ſome to that degree, that they cannot be walked on with that eaſe and pleaſure that ought to be.

The common allowance for a Gravel-walk of five feet breadth, is an inch riſe in the crown; ſo that if a Walk be twenty feet wide, according to this proportion, it will be four inches higher in the middle than on each ſide, and a Walk of twenty-five feet will be five inches, one of thirty feet ſix inches, and ſo on.

When a Walk has been thus carefully laid, trodden down, and raked, or rather, after every length or part of it (which commonly is about fifteen feet each,) then it ſhould be rolled well, both in length and alſo croſs-ways. The perſon who rolls it, ſhould wear ſhoes with flat heels, that he may not make holes in the Walks, for when theſe are once made in a new Walk, it will not be eaſy to roll them out again.

In order to lay Gravel-walks firm, it will be neceſſary to give them three or four water-rollings, that is, they muſt be rolled when it rains ſo very faſt, that the Walks ſwim with water; this will cauſe the gravel to bind, ſo that when the Walks come to be dry, they will be as hard as terrace.

Iron-mould gravel is accounted the beſt for binding, or gravel with a little binding loam amongſt it; which latter, though it be apt to ſtick to the heels of ſhoes in wet weather, yet nothing binds better in dry weather.

When the gravel is over ſandy, ſome ſharp loam is frequently mixed with it, which, if they be caſt together in heaps, and well mixed, will bind like a rock; whereas looſe gravel is as uncomfortable and uneaſy to walk on, as any other fault in a Walk can render it.

The beſt gravel for Walks is ſuch as abounds with ſmooth pebbles (as is that dug at Blackheath,) which, being mixed with a due proportion of loam, will bind like a rock, and is never injured by wet or dry weather, and the pebbles being ſmooth, are not ſo liable to be turned up, and looſened by the feet in walking, as are thoſe which are angular and rough; for where Walks are laid with ſuch gravel as is full of irregular ſtones, they appear unſightly in a day's time after rolling, becauſe the ſtones will riſe upon the ſurface wherever they are walked upon, but the ſmooth pebbles will remain handſome two or three days without rolling.

Gravel-walks are not only very neceſſary near the houſe, but there ſhould always be one carried quite round the garden, becauſe, being ſoon dry after rain, they are proper for walking on in all ſeaſons; but then theſe ſhould be narrow, and thoſe adjoining to the houſe ought to be large and magnificent, proportionable to the grandeur of the houſe and garden. The principal of theſe Walks ſhould be elevated, and carried parallel with the houſe, ſo as to form a terrace; this ſhould extend itſelf each way, in proportion to the width of the garden; ſo that from this there may be a communication with the Side-walks, without going on the Graſs, that there may be a dry Walk continued quite through the gardens; but there is not a more ridiculous ſight, than that of a ſtrait Gravel-walk leading to the front of the houſe, interſecting the Graſs, ſo as to make it appear like the ſtiff formal Graſs-plats frequently made in little court yards by perſons of low taſte.

Graſs-walks in gardens were formerly in great eſteem, and looked upon as neceſſary ornaments to a garden; but of late years they have juſtly been baniſhed by every perſon of true taſte, for thoſe narrow ſlips of Graſs were very unſightly, and far from being ornamental, and for the moſt part uſeleſs, being generally too damp for perſons of tender conſtitutions to walk upon; and whenever they were conſtantly uſed, they became bare in the places frequently trodden, ſo were rendered more unſightly; and as the intention of Walks in gardens is to have at all ſeaſons a dry communication throughout the garden, for exerciſe and recreation, Graſs-walks were very improper, becauſe every ſhower of rain made them ſo wet, as not to be fit for uſe a conſiderable time, and the dews render them too damp for uſe either in the morning or evening; and if the Graſs of Walks is not very fine and ſhort, like that of the downs, it will be very troubleſome to walk upon; beſides, whenever the ground is ſo dry, as that perſons may with ſafety walk upon Graſs, the lawns and other parts of verdure in gardens are better adapted for uſe than any of thoſe formal ſtiff Walks, which were ſo much eſteemed in the laſt age.

Having given directions for the making of Gravel-walks, I ſhall come next to treat of Sand-walks, which are now very frequently made in gardens, as being leſs expenſive in the making, and alſo in keeping, than the former; and in very large irregular gardens, which are ſuch as moſt perſons eſteem, this is a very great article; for as the greateſt part of the Walks which are made in gardens, are carried about in an irregular manner, it would be very difficult to keep them handſome, if they were laid with gravel, eſpecially where they are ſhaded by trees; for the dripping of the water from their branches, in hard rains, is apt to waſh the gravel in holes, and render the walks very unſightly; and when theſe Wood-walks are of Graſs, they do not appear ſightly, nor are they very proper for walking on; for after rain they continue ſo long damp as to render them unfit for uſe, and the Graſs generally grows ſpiry and weak for want of air, and by the continual dropping of the trees, will by degrees be deſtroyed; therefore it is much better to lay theſe Walks with ſand, which will be dry and wholeſome; and whenever they appear moſſy, or any weeds begin to grow on them, if they are ſcuffled over with a Dutch hoe in dry weather, and then raked ſmooth, it will deſtroy the weeds and Moſs, and make the Walks appear as freſh and handſome as if they had been new laid.

In the modern way of laying out gardens, the Walks are carried through woods and plantations, ſo that theſe are ſhady and convenient for walking in the middle of the day. Theſe are uſually carried about, winding as much as the ground will admit of, ſo as to leave a ſufficient thickneſs of wood to make the Walks private; and that the perſons who are walking in one part of them, may not be ſeen by thoſe who are in any of the other parts. Where theſe Walks are contrived with judgment, a ſmall extent of ground will admit of a great many turns, ſo that a perſon may walk ſome miles in a ſmall garden. But theſe turns ſhould be made as natural as poſſible, ſo as not to appear too much like a work of art, which will never pleaſe ſo long as the former.

The breadth of theſe Walks muſt be proportioned to the ſize of the ground, which in a large extent may be eight or ten feet wide; but in ſmall gardens five or ſix feet will be ſufficient. As the Walks are deſigned to wind as much as the ground will allow, ſo this width will be ſufficient, becauſe the wider they are, the greater muſt be the turns, otherwiſe the Walks will not be private for any diſtance. Beſides, as it will be proper to line the ſides of theſe Walks with Honeyſuckles, Sweet-briar, Roſes, and many other ſweet-flowering ſhrubs, ſo the tall trees ſhould be placed at leaſt five or ſix feet from the Walk, to allow room for theſe. But as I ſhall particularly treat of the method of laying out wilderneſſes, and planting of them, in ſuch a manner as to render them as nearly reſembling a natural wood as poſſible, under its proper head, I ſhall add nothing more in this place, except a few common directions for making of theſe Sand walks.

When the ground is traced out in the manner as the Walks are deſigned, the earth ſhould be taken out of the Walks, and laid in the quarters. The depth of this muſt be proportioned to the nature of the ſoil, for where the ground is dry, the Walks need not be elevated much above the quarters, ſo the earth ſhould be taken out four or five inches deep in ſuch places; but where the ground is wet, the bottom of the Walks need not be more than two inches below the ſurface, that the Walks may be raiſed ſo high, as to throw off the wet into the quarters, which will render them more dry and healthy to walk on.

After the earth is taken out to the intended depth, the bottom of the Walks ſhould be laid with rubbiſh, coarſe gravel, or whatever of the like nature can be moſt readily procured. This ſhould be four, five, or ſix inches thick, and beaten down as cloſe as poſſible, to prevent the worms from working through it; then the ſand ſhould be laid upon this about three inches thick, and after treading it down as cloſe as poſſible, it ſhould be raked over to level and ſmooth the ſurface. In doing of this, the whole ſhould be laid a little rounding to throw off the wet, but there will be no neceſſity of obſerving any exactneſs therein; for as the whole ground is to have as little appearance of art as poſſible, the rounding of theſe Walks ſhould be as natural, and only ſo contrived as that the water may have free paſſage from them.

The ſand with which theſe Walks are laid ſhould be ſuch as will bind, otherwiſe it will be very troubleſome to walk on them in dry weather; for if the ſand be of a looſe nature, it will be moved with ſtrong gales of wind, and in dry weather will ſlide from under the feet. If, after theſe Walks are laid, they are well rolled two or three times, it well ſettle them, and cauſe them to be firm. If the ſand is too much inclinable to loam, it will alſo be attended with as ill conſequence as that which is too looſe, for this will ſtick to the feet after every rain; ſo that where ſand can be obtained of a middle nature, it ſhould always be preferred.

In ſome countries where ſand cannot be eaſily procured, theſe Walks may be laid with ſea-ſhells well pounded, ſo as to reduce them to a powder, which will bind extremely well, provided they are rolled now and then; but where none of theſe can be eaſily procured, ſea coal aſhes, or whatever elſe can be gotten, which will bind and be dry to the feet, may be uſed for this purpoſe; and where any of theſe can only be had in ſmall quantities, the Walks ſhould have a greater ſhare of rubbiſh laid in their bottom, and theſe ſpread thinly over them; and in moſt places rubbiſh, rough ſtones, or coarſe gravel, may be eaſily procured.

WALLS are abſolutely neceſſary in gardens, for the ripening of all ſuch fruits as are too delicate to be perfected in this country without ſuch aſſiſtance. Theſe are built with different materials; in ſome countries they are built of ſtone,

ſtone, in others with brick, according as the materials can be procured beſt and cheapeſt.

Of all materials proper for building Walls for fruit-trees, brick is the beſt, in that it is not only the handſomeſt, but the warmeſt and kindeſt for the ripening of fruit; beſides that, it affords the beſt conveniency of nailing, for ſmaller nails will ſerve in them than in ſtone Walls, eſpecially if the joints are not too large; and brick Walls, with copings of free-ſtone, and ſtone pilaſters or columns, at proper diſtances, to ſeparate the trees, and break off the force of the winds, make not only the moſt beautiful, but the moſt durable Walls.

In ſome parts of England there are Walls built both of brick and ſtone, which have been very commodious. The bricks of ſome places are not of themſelves ſubſtantial enough for Walls, nor are they any where ſo durable as ſtone; and therefore ſome perſons, that they might have Walls both ſubſtantial and wholeſome, have built double ones, the outſide being of ſtone and the inſide brick, or a ſtone Wall lined with brick; but when theſe are built, there muſt be great care taken to bind the bricks well into the ſtone, otherwiſe they are very apt to ſeparate one from the other, eſpecially when hard froſt comes after much wet; which ſwells the mortar, and frequently throws down the bricks, when the Walls are only faced with them, and not well tied into the ſtone.

Where the Walls are built entirely of ſtone, there ſhould be trelliſſes fixed up againſt them, for the more convenient faſtening the branches of the trees; the timbers of theſe eſpaliers need not be more than an inch and a half thick, and about two inches and a half broad; theſe ſhould be fixed croſs each other, at about four inches diſtance; for if they are at a much greater diſtance, it will be difficult to faſten the ſhoots of the trees properly. As this trellis will be laid cloſe to the Wall, the branches of the trees will lie about two inches from the Wall; in which poſition the fruit will ripen better than when it lies cloſe to the Wall; ſo that there ſhould always be theſe eſpaliers framed againſt them, which will render theſe Walls very good for fruit-trees, which, without the eſpaliers, ſeldom are found to anſwer the purpoſe of ripening the fruits well, beſides the inconvenience of having no good faſtening for the branches of the trees.

There have been ſeveral trials made of Walls built in different forms; ſome of them having been built ſemicircular, others in angles of various forms, and projecting more towards the north, to ſcreen off the cold winds; but there has not been any method as yet, which has ſucceeded near ſo well, as that of making the Walls ſtrait and building them upright.

The faireſt trial which I have ſeen made of circular Walls was at Goodwood in Suſſex, the ſeat of the Duke of Richmond, where, in the middle of two ſouth Walls, there were two large ſegments of circles, in which there were the ſame ſorts of fruit-trees planted, as againſt the ſtrait parts of the Walls; but there never was any fruit upon the trees in the circular part of the Walls which came to maturity, nor were the trees of long continuance, being blighted every ſpring, and in a few years were totally deſtroyed; and when the branches of thoſe trees, which grew upon the ſtrait part of the Walls, had extended themſelves ſo far as to admit of their being led into the circular parts of the Walls, they were conſtantly blighted and killed.

When the trees which had been planted in the circular parts were deſtroyed, the Walls were filled with Vines; but the Grapes of the ſame ſort were a full month later, than thoſe growing againſt the ſtrait part of the Walls; ſo that they rarely ripened, which occaſioned their being rooted out, and Figs were afterwards planted, but the fruit of theſe ſucceeded little better; nor can it be ſuppoſed, that any trees or plants will thrive ſo well in theſe circles, where there is a conſtant draught of air round them, which renders the ſituation much colder than the open free air.

I have alſo ſeen, at Mr. Le Cour's garden in Holland, ſome Walls built in angles of different forms, but theſe ſucceeded no better than the circles before-mentioned; for I did not find one tree in health againſt the Walls, nor did they produce fruit.

There are ſeveral other ſchemes which have been propoſed by different perſons, for the building of Walls to accelerate the ripening of fruits; among which there was a very ingenious book written ſome years ago, intitled, "Fruit-walls improved, by inclining them to the horizon;" in which the author has ſhewn, by calculation, that there will be a much greater number of the rays of the ſun fall upon ſuch Walls, than upon thoſe which are built perpendicular; and from thence has drawn calculations, that Walls ſo built will be of great ſervice in the accelerating of fruit; and he has taken the trouble of calculating the different inclinations, which ſuch Walls ſhould have in the different climates, in order to receive the greateſt number of the ſun's rays. This theory ſeems to have all the demonſtration neceſſary for its ſupport, but upon trial has not ſucceeded in the leaſt; for as theſe Walls muſt be built againſt banks of earth, the damps which ariſe from the ground overbalance the advantage of the ſun's rays; beſides, theſe ſloping Walls being much more expoſed to the cold dews in the night, the fruit will be more chilled thereby; and in the ſpring the morning froſts will prove much more deſtructive to the tender bloſſoms of the fruit-trees, as they will be more expoſed to them than againſt an upright Wall; add to this, their being much more expoſed to the winds and the rain; and it will be found, by comparing the advantages propoſed from theſe Walls, with the diſadvantages to which the fruit-trees will be expoſed, that upright Walls will have the preference; for it is not the ſtrongeſt rays of the ſun, in the heat of ſummer, which are ſo much wanting for ripening of fruit, as the continuance of a moderate ſhare of warmth; and, above all, the having of the ſun in a morning, to dry off the cold dews of the night early, is of the greateſt uſe; and in this reſpect the upright Walls are much preferable to the ſloping, as they will have the direct rays of the ſun in the morning, which will be oblique on the other.

There are ſome perſons who recommend the painting of Walls black, or of a dark colour, as they ſuppoſe the dark colour will imbibe more of the ſun's rays, ſo will retain the warmth longer; this alſo anſwers better in theory than in practice; for although it muſt be allowed, that a black Wall is warmer to the touch than a common brick Wall, yet, as the fruit generally is ſituated at a ſmall diſtance from the Wall, it receives no benefit from the warmth of the Wall, but it is the reflected heat which accelerates the ripening of fruit; therefore I would adviſe every one to make fair trials of theſe things before they put them in practice, and not take upon truſt what they may be told by perſons, who are too ſanguine in recommending to others ſchemes, which they have adopted upon very ſlight principles, or perhaps upon a ſingle trial; this painting of the Walls is recommended by the ſame perſon who wrote upon inclining Walls, and he has propoſed this upon the ſame principles; but the introducing of theſe ſchemes ſhould be avoided, until there have been ſufficient trials made to warrant their uſe.

Where perſons are willing to be at the expence in the building of their Walls ſubſtantial, they will find it anſwer much better than thoſe which are ſlightly built, not only in their duration, but alſo in their warmth; therefore a Wall two bricks thick, will be found to anſwer better than one brick and a half; and if in the building of garden Walls they are grouted with ſoft mortar, to fill and cloſe all the joints, the Walls will be much ſtronger, and the air will not ſo eaſily penetrate through them, as it does through thoſe which are built in the common way.

According to the modern taſte in gardening, there are very few Walls built round gardens, which is certainly very right; not only with regard to the pleaſure of viewing the neighbouring country from the garden, but alſo in regard to the expence, therefore the quantity of walling ſhould be proportioned to the fruit conſumed in the family; but as it will be neceſſary to incloſe the kitchen-garden, for the ſecurity of the garden ſtuff, ſo, if that be walled round, it will contain as much fruit as will be uſually wanted in the family, becauſe the kitchen-garden is always proportioned to the number of perſons maintained; but if the quantity of walling which ſurrounds the kitchen-garden, ſhould be judged too little for the ſupply of fruit, there may be a croſs Wall built through the middle of the kitchen-garden; or, where the ſize of the garden will admit, there may be two croſs Walls built; but this muſt not be done, where there is not room to place the Walls at leaſt eighty or one hundred feet aſunder, and if they are allowed a much greater diſtance it will be better; and as the kitchen garden ſhould always be placed out of ſight from the houſe, the Walls may be hid by plantations of ſhrubs at ſome little diſtance.

The beſt aſpect for Walls in England is, to have one point to the eaſtward of the ſouth; for theſe will enjoy the benefit of the morning ſun, and will be leſs expoſed to the weſt and ſouth-weſt winds (which are very injurious to fruits in England) than thoſe Walls which are built due ſouth. I know there are many perſons who object to the turning of Walls the leaſt point to the eaſt, on account of the blights which they ſay come from that quarter in the ſpring; but from many years experience and obſervation I can affirm, that blights as often attack thoſe Walls which are open to the ſouth-weſt, as thoſe which are built to any other aſpect; and I believe, whoever will be at the trouble to obſerve for ſeven years, which aſpected Walls ſuffer moſt from blights, will find thoſe which are built with a point to the eaſtward of the ſouth, as ſeldom blighted as thoſe which are turned to any other aſpect; therefore, in the contrivance of a kitchen-garden, there ſhould be as great length of theſe Walls built as the ſituation of the ground will admit.

The next beſt aſpect is due ſouth, and the next to that ſouth-eaſt, which is preferable to the ſouth-weſt for the reaſons before aſſigned; but as there will, for the moſt part, be ſouth-weſt and weſt walls in every garden, theſe may be planted with ſome ſorts of fruit, which do not require ſo much heat to ripen them as thoſe deſigned for the beſt Walls; but wherever there are north Walls, thoſe will only be proper for baking Pears, Plumbs, and Morello Cherries for preſerving; or ſome Duke Cherries may be planted againſt theſe Walls, to continue them longer in the ſeaſon, which will be found uſeful in ſupplying the table till Peaches, Nectarines, and Plumbs are ripe.

Where perſons are very curious to have good fruit they erect a trellis againſt their Walls, which projects out about two inches from them, to which they faſten their trees; which is an excellent method, becauſe the fruit will be at a proper diſtance from the Walls, ſo as not to be injured by them, and will have all the advantage of their heat; and by this method the Walls will not be injured by driving nails into their joints, which, by every year being drawn out, draws out the mortar from between the bricks, and thereby makes holes, in which ſnails and other vermin will harbour and deſtroy the fruit, and alſo impair the Wall.

Theſe trelliſſes may be contrived according to the ſorts of fruit which are planted againſt them. Thoſe which are deſigned for Peaches, Nectarines, and Apricots (which, for the moſt part, produce their fruit on the young wood,) ſhould have their rails three, or, at moſt, four inches aſunder every way; but for the other ſorts of fruit, which continue bearing on the old wood, they may be five or ſix inches apart; and thoſe for Vines may be eight or nine inches diſtance. For as the ſhoots of Vines are always trained at a much greater diſtance than thoſe of any other ſort of fruit, the trelliſſes for theſe need not be near ſo cloſe, eſpecially as thoſe for Peaches and Nectarines, whoſe ſhoots are generally ſhortened to about five or ſix inches or leſs; ſo that if the rails are not pretty cloſe, many of the ſhort branches cannot be faſtened to them.

Theſe trelliſſes may be made of any ſort of timber, according to the expence which the owner is willing to beſtow; but Fir is moſt commonly uſed for this purpoſe, which, if made of yellow deal, well dried and painted, will laſt many years; but if any perſon will go to the expence of Oak, it will laſt ſound much longer; but thoſe who are unwilling to be at the expence of either, a trellis may be made of Aſh-poles, in the ſame manner as is practiſed in making eſpaliers for counter borders, with this difference only, that every fourth upright rail or poſt ſhould be ſtronger, and faſtened with iron hooks to the Wall, which will ſupport the whole; and as theſe rails muſt be laid much cloſer together than is generally practiſed for eſpaliers, theſe ſtrong upright rails or poſts will not be farther diſtant than four or five feet from each other. To theſe the croſs rails which are laid horizontally ſhould be well nailed, which will ſecure them from being diſplaced, and alſo ſtrengthen the trelliſſes; but to the other ſmaller upright poles they need only be faſtened with wire. To theſe trelliſſes the ſhoots of the trees ſhould be faſtened with Oſier-twigs, rope-yarn, or any other ſoft bandage, for they muſt not be nailed to it, becauſe that will decay the wood work.

Theſe trelliſſes need not be erected until the trees are well ſpread, and begin to bear fruit plentifully; before which time the young trees may be trained up againſt any ordinary low eſpaliers, made of a few ſlender Aſh-poles, or any other ſlender ſticks; by which contrivance the trelliſſes will be new when the trees come to bearing, and will laſt many years after the trees have overſpread them; whereas, when they are made before the trees are planted, they will be half rotten before the trees attain half their growth.

When theſe trelliſſes are intended to be made againſt new Walls, it will be proper to faſten ſome ſtrong iron hooks into the Wall as it is built, at the diſtance which the upright poſts are intended to be placed; becauſe when theſe are afterwards driven into the Wall, they diſplace the mortar in the joints, and injure the Wall.

In the building of the Walls round a kitchen-garden, the inſides, which are deſigned to be planted with fruit-trees, ſhould be made as plain as poſſible, ſo that the piers ſhould not project on thoſe ſides above four inches at moſt, and theſe ſhould be placed about fourteen feet aſunder, in ſuch Walls as are deſigned for Peach and Nectarine-trees; ſo that each tree may be planted exactly in the middle between the piers, which will render them more ſightly, and be better for the trees; but where Apricots, Plumbs, or Cherries, are to be planted, the piers may be only ten feet aſunder; and againſt every other pier the trees ſhould be planted, which will allow them ſufficient room to ſpread; as the trellis will

project

project as forward as the piers, the branches of the trees may be fpread as on a plain; but when the piers project no more on the infide of the garden, they fhould be built ftronger on the outfide, for the better fupporting of the Walls.

The ufual thicknefs which Garden-walls are allowed, if built with bricks, is thirteen inches, which is one brick and a half, but this fhould be proportionable to the height; for if they are built twelve or fourteen feet high or more, as is fometimes practifed, then the foundations of the Walls fhould be carried up at leaft two bricks and a half thick, a foot or more above the level of the furface of the ground; then may be diminifhed on each fide, to reduce them to the thicknefs of two bricks, which muft be continued to the top of the Walls; and the piers in thefe high Walls, fhould alfo be proportionably ftronger than is commonly allowed to lower Walls; for as thefe will be much more expofed to ftrong gales of wind, if they are not well built they will be in danger of being blown down; therefore the piers of thefe Walls fhould be projected the length of a brick on their backfide, and the thicknefs of a brick on their front; and if thefe are built about ten or twelve feet afunder, they will greatly ftrengthen the Walls.

Bot there is no neceffity for building Walls higher than nine or ten feet, unlefs it be for Pears, which, if properly managed, will fpread over a great compafs of walling; but as only fome of the lateft winter Pears require the affiftance of a Wall, there need no more but that part of the Wall, where thefe are defigned, to be built higher; for Peaches and Nectarines never require a Wall higher than nine or ten feet, provided they are rightly managed; becaufe whenever they are carried to a greater height, the lower part of the Wall is unfurnifhed with bearing branches; and although Apricots, Plumbs, and Cherries, will frequently grow higher, yet, if they are planted at a proper diftance, and the branches trained horizontally from the bottom, they will not foon cover a Wall of this height; and Vines may be kept as low as any fort of fruit, for when they are planted againft low Walls, they muft be treated fomewhat after the fame manner as thofe in vineyards, which is, to cut off the greateft part of the wood which produced fruit the preceding year, and train in new fhoots for the next year's bearing, which are rarely left a yard in length, therefore will not require very high Walls.

If the Pears which are defigned to be planted, are allowed a fouth-weft afpect, on which they will ripen very well, then the Wall to this afpect fhould be built fourteen feet high or more; for as thefe trees fpread very far, when on free-ftocks, they fhould not be fhortened and ftopped in their growth, which will prevent their bearing. But I fhall now proceed to give fome directions for the building of Hot-walls, to accelerate the ripening of fruits, which is now pretty much practifed in England.

In fome places thefe Walls are built at a very great expence, and fo contrived as to confume a great quantity of fuel; but where they are judicioufly built, the firft expence will not be near fo great, nor will the charge of fuel be very confiderable, becaufe there will be no neceffity of making fires more than three months, beginning about the middle or latter end of January, and ending by the middle of May, when there will be no want of fires, if the glaffes are clofe fhut every night, or in bad weather; for half an hour's fun-fhine on the glaffes at that feafon, will fufficiently warm the air inclofed in the glaffes, for the growth of any of our European fruits.

There are fome perfons who plant Vines, and other fruit-trees, by the fides of ftoves, and draw fome of their branches into the ftove, in order to obtain early fruit, and very often train the Vines over the whole tan-bed; which is very wrong, where the ftove is defigned for the Ananas, becaufe the air muft be kept much warmer for them, than is required for Grapes or other fruits, fo they can never fucceed well together; for when there is only a fufficient quantity of air admitted for the growth of Grapes, the Ananas fuffer for want of proper heat; and fo, on the contrary, when the ftove is kept up to the proper heat for the Ananas, it will be too hot for Grapes; therefore it will be proper to have the Vines on a particular Wall by themfelves, becaufe thefe require to have a greater fhare of air admitted to them, when they begin to fhoot, than moft other forts of fruits, fo that it is by much the better method to have them feparate.

The ordinary height of thefe Hot-walls is eight or nine feet, which will be fufficient for any of thofe forts of fruits which are generally forced; for by forcing of the trees, they are commonly weakened in their growth, fo that they will not grow fo vigoroufly as thofe which are always expofed to the open air; and where there is not a quantity of walling planted fufficient to let one part reft every other year, the trees will never be very healthy, and will laft but a few years. The quantity of walling to produce early fruit for a middling family, cannot be lefs than eighty or one hundred feet in length; therefore where a perfon is defirous to have the fruit in perfection, and the trees to continue in a good condition many years, there fhould be three times this quantity of walling built; fo that by dividing it into three parts, there will be two years for the trees to recover their vigour between the times of their being forced; whereby a greater quantity of bearing wood may be obtained, and the fruit will be fairer and in larger quantities than when they are forced every year, or every other year; and as the glaffes may be contrived fo as to move from one to the other, the expence of building the Walls fo much longer will not be very great, becaufe the frames and glaffes need not be more than for one year's fruit.

The foundations of thefe Walls fhould be made four bricks and a half thick, in order to fupport the flues; otherwife, if part of them reft on brick-work, and the other part on the ground, they will fettle unequally, and foon be out of order; for wherever there happen any cracks in the flues, through which the fmoke can make its efcape, it will prevent their drawing; and if the fmoke gets within the glaffes, it will greatly injure the fruit. This thicknefs of Wall need not be continued more than fix inches above the ground, where fhould be the foundation or bottom of the firft flue, which will raife it above the damps of the earth; then the Walls may be fet off four inches on each fide, which will reduce it to the thicknefs of three bricks and a half, fo that the back wall may be two bricks thick, which is abfolutely neceffary to throw the heat out more in front; for when the back Walls are built too thin, the heat will efcape through them. The Wall in front next to the fruit will be only four inches thick, whereby there will be allowance of nine inches for the flues, which may be covered with twelve inch tiles; for if they have an inch and a half bearing on each fide, it will be fufficient.

The ovens in which the fires are made, muft be contrived on the backfide of the Walls, which fhould be in number proportionable to the length of the Walls. The length ufually allowed for each fire to warm is forty feet, though they will do very well for fifty; but I would not advife the flues to be longer than this to each fire, becaufe when the ovens are made at a great diftance, there is a neceffity of making the fires fo much ftronger to warm the Walls, which will occafion the heat to be too violent near the fires. Thefe ovens fhould be fhedded over, to keep out the wind and rain, otherwife the fires will not burn equally. Some people make thefe fheds of timber, but it is much better to build them of brick, and tile them over; becaufe the wooden fheds will in a few years decay, and afterwards will be a conftant charge to keep in repair; and befides they may be in danger of firing, if great care is not conftantly taken of the fires. As it is abfolutely neceffary to have ovens below the foundation of the firft flues, there muft be fteps down into the fheds, to come to the mouth of the ovens to fupply the fuel; therefore the fheds fhould not be narrower than eight feet in the clear, for as the fteps will require four feet fpace, there fhould be at leaft four feet more for the perfon who attends the fire, to have room to turn himfelf to clear out the afhes and to put in the fuel. Where the length of walling requires two ovens, it will be proper to have them in the middle included in one fhed, which will fave expence, and allow more room to attend the fires; for in this cafe the fheds muft be at leaft ten feet long, and then they need not be more than fix in breadth. The fteps down into thefe fhould be at one end, fo that the door opening into the fheds will not be oppofite to the mouths of the ovens, therefore the fires will burn more regular: for whenever the doors are contrived to front the mouth of the ovens, if the winds fet directly againft them it will caufe the fire to burn too fiercely, and the fuel will be foon confumed.

Thefe ovens may be contrived in the fame manner as thofe which are already defcribed for ftoves, wherefore I fhall not repeat it again in this place; but muft obferve, that when the two ovens are joined together, there fhould be a partition Wall at leaft three bricks thick between them, otherwife the fires will foon deftroy it; and if there fhould be the leaft hole in the Wall, through which the fmoke of the two fires can communicate, it will prevent their drawing.

The lower flue, through which the fmoak firft paffes from the fire, may be two feet and a half deep; therefore the back Wall fhould be at leaft two and a half or three bricks thick, as high as to the top of this flue; then it may be fet off to two bricks or two and a half thicknefs, which muft be continued to the top of the Wall. The fecond flue, which fhould return over the firft, may be made two feet, the third a foot and a half, and the fourth one foot deep; which four flues, with their coverings, will rife near eight feet in height, fo that there will be juft room left for the fixing of the frames at the top to fupport the glaffes under the coping of the Wall: and thefe four returns will be fufficient to warm the air in the frames, for the fmoke will have left its heat by the time it has paffed thus far.

In the carrying up of thefe Walls there fhould be fome ftrong iron hooks faftened at convenient diftances, which fhould project one inch from the Wall to which the trellis muft be faftened, which is to fupport the trees. Thefe hooks fhould be long enough to faften into the back Wall, for the Wall in front, being but four inches thick, will not be ftrong enough to fupport the trellis; but in placing of them, care fhould be taken not to lay them crofs the middle of the flues, becaufe they would obftruct the clearing them of foot whenever there fhould be occafion; fo that the beft way is to lay them juft under the tiles which cover each flue at about three feet afunder, which will be near enough, provided the bars of the hooks are made fufficiently ftrong; but thefe fhould be flat, left they obftruct the fmoke. As the flues muft be well plaftered with loam on their infide, fo likewife fhould the loam be fpread under the tiles which cover them, to the thicknefs of the hooks, that the flues may be very fmooth. It will alfo be very proper to cover thefe flues on the fide next the trellis with hop-bags, or fome fuch coarfe cloth, in the manner as hath been directed for the ftoves, which will make them fo tight, that no fmoke will find its way, which, without this covering, it is very apt to do through the joints of the Walls, efpecially when they are fo thin as thefe muft be built; and this covering will alfo ftrengthen the Wall of the flues, and join the whole work together. If at each end of thefe flues there are fmall arches turned in the back Walls, in fuch a manner that there may be holes opened to clean the flues of foot whenever there is a neceffity for it, the trouble will be much lefs than to open the flues in front; and there will be no damage done to the trees, nor will the flues be in the leaft injured by this, which they muft be when they are opened in front.

The borders in front of thefe Hot-walls fhould be about four feet wide, which will make a fufficient declivity for the floping glaffes; and in thefe borders there may be a row of dwarf Peafe planted to come early, or a row of dwarf Kidney-beans, either of which will fucceed very well; and if they are not planted too near the trees, will not do them much injury. On the outfide of thefe borders fhould be low Walls erected, which fhould rife three or four inches above the level of the borders, upon which the plate of timber fhould be laid on which the floping glaffes are to reft; and this Wall will keep up the earth of the border, and alfo preferve the wood from rotting.

The glaffes which are defigned to cover thefe Walls muft be divided into two ranges; for as they muft reach from the ground-plate (juft above the level of the border) to almoft the top of the Wall, they will be more than twelve feet long, which will be too great a length for fingle frames; which, when they are much more than fix feet long, are too heavy to move, efpecially if the frames are made of a proper ftrength to fuftain the glafs. Thefe frames fhould be contrived in fuch a manner, as that the upper row may flide down; and by making on one fide three fmall holes in the wood work which fupports the frames, at about a foot diftance, and having a fmall iron pin to fix into them, the top glaffes may be let down one, two, or three feet, according as there may be occafion to admit air. The lower row of glaffes may be contrived fo as to eafily take out, but as they muft lie floping, and the upper row muft bear on them, they cannot be contrived to flide upwards; nor indeed will there be any occafion for their moving, becaufe it is much better to let the air in at the top than in the front of the trees.

The floping timbers, which are to fupport the glafs-frames, muft be faftened at bottom into the ground-plate in the front of the border, and at the top into ftrong iron cramps fixed in the upper part of the Wall for that purpofe. Thefe timbers fhould be made of Fir, which will not twift as Oak and fome other wood will, where it is laid in fuch pofition. They muft be made fubftantial, otherwife they will not laft many years, efpecially as they are defigned to be moveable. On the top of thefe fhould be fixed a ftrong board, under which the upper row of glaffes fhould flide. The ufe of this board is, to fecure the upper part of the glaffes from being raifed by the winds, and alfo to keep the wet from getting to the trees; therefore it fhould be joined as clofe as poffible to the Wall, and fhould project about two inches over the glafs-frames, which will be enough to throw the wet on the glaffes, and likewife to fecure them faft down.

The breadth of thefe frames for the glaffes may be about three feet or a little more, according as the divifions of the length of the Wall will admit; for a fmall matter in their width is of no confequence, provided they are not too wide to be eafily moved; for when they are wider than a man can eafily reach with his arms to manage, they will be very troublefome to carry from one place to another. The bars of thefe frames, which are to fupport the glafs, fhould be placed lengthwife of the frames; for when they are placed acrofs, they ftop the moifture which is lodged on the infide inside of the glasses, and cause it to fall in drops on the borders at every bar, which will be very injurious to any thing under them; and if it falls on the trees, will greatly damage them, especially when they are in blossom. The lead into which the glasses of these frames are fixed, should be very broad, and the joints well cemented; otherwise the wet will find an easy passage through, and do great damage to the fruit.

At each end of the range of glasses, there will be an angular space between the glasses and the Wall, which must be closely stopped to prevent the air from getting in, which might greatly injure the fruit. These are by some persons closely boarded up; but if they are closed with glasses, so contrived as to open to let in air at proper times, it will be of great advantage; because when the wind may be too strong against the front glasses, one or both of these end glasses may be opened, according to the warmth of the air inclosed, which will be often very useful to cool the air, and to admit a small quantity of fresh air to the fruit.

The sorts of fruit which are usually planted for forcing, are Cherries, Plumbs, Peaches, Apricots, and Nectarines; but the last-mentioned seldom succeed well, nor will the trees continue long, so is scarce worth planting against Hot-walls. As for the Vines, I would propose they should be planted by themselves against a particular Wall; for as they will require more air to be admitted to them when they begin to shoot, than any of the above-mentioned fruits, they will not succeed if they are included in the same frame; but the others will do very well in the same border, and require the same temperature of warmth. The best of these sorts to plant against Hot-walls, are those here mentioned:

Cherries.

The Early May, and May Duke.

Plumbs.

The Early Black Damask, or Morocco.
The Great Damask Violet of Tours.
The Drap d'Or.

Peaches.

The Red Nutmeg.
The Red Magdelain.
The Montauban.

Nectarines.

Fairchild's Early Nutmeg.
The Elruge.

Apricot.

The Masculine.

These being the sorts which ripen early, are the most proper to plant against these Walls, although they are not so valuable as some other sorts of these fruits; yet, as they naturally ripen three weeks or a month earlier in the season, they will be very early ripe when they are brought forward by artificial warmth.

In the preparing of the borders for planting these fruit-trees, there should be the same care taken as for those against open borders; which, being fully treated of in another part of this work, I shall not repeat here. There must also be the same care in training up the trees when they shoot; but the trellisses need not be made against these Walls, until the trees are grown large enough to spread, and produce a great quantity of fruit; till which time they may be supported by any low ordinary trellis, which will do very well till the time that the trees will have strength enough to force, which will not be until the fourth or fifth year after planting, according to the progress they have made; for if they are forced too young, it will weaken them so much, as that they seldom make vigorous trees afterwards; besides, the quantity of fruit which such young trees produce, is not worth the expence and trouble of forcing; for the quantity of fuel used, and the trouble, will be the same for small trees, which are not capable of producing more than six or eight fruit each, as for those trees which may produce three or four dozen; so that the greater time the trees have to grow before they are forced, the better they will pay for the trouble and expence.

When the trees have acquired strength enough to produce a quantity of fruit, that part which is designed to be forced the following spring, should be carefully pruned early in autumn; when the very weak shoots must be either entirely cut out, or pruned very short, because these, by being forced, will for the most part decay; and though some of them may be full of flower-buds, yet these shoots being weak, cannot nourish them; so that the flowers having exhausted all the sap, the shoots die soon after, and rarely produce any fruit, or at least do not bring them to perfection. The other more vigorous shoots should also be shortened to a proper length, after the same manner as is directed for those trees in the open air; with this difference only, viz. that these which are designed for forcing, should not have their shoots left so long, because the forcing of them will weaken them; and consequently, should there be as great a length of branches, there will probably be a greater number of fruit on them; because, as these will be screened from the open air, they will not be liable to blasts, or the injuries of frost; and the having too many fruit on the trees, will render them small, and also too much weaken the trees; then the shoots should be all regularly fastened to the trellis, at a proper distance from each other; so that when the branches shoot the following spring, they may not over-hang each other. The reason for my advising these trees to be pruned so early in the season is, that those branches which are left on, may enjoy the whole nourishment of the sap; so that the buds will become very turgid during the winter season, and will be prepared to open when the fires are set to work.

The time for beginning to make the fires is about the middle or latter end of January, according as the season is more or less favourable; for if the trees are forced too early into flower, they will be in some danger of miscarrying, if the weather should prove severe; so that it is by much the surest method to begin about the time here directed, because there will be a necessity of admitting fresh air to the trees when they are in flower, which cannot be done with safety when they flower in very bad weather. And those trees which are forced into flower by the middle of February, will ripen their fruit as early as most people will desire to eat them; for the Cherries will ripen early in April, and the Apricots by the beginning of May; and soon after the Plumbs, Peaches, and Nectarines will be ripe.

There are some persons who plant Strawberries in their borders before the fruit-trees, in order to have early fruit, which often succeed very well; but wherever this is practised, great care should be taken to keep them from spreading over the border, because these plants will exhaust the principal goodness of the earth, and thereby injure the trees; so that when it is designed to have Strawberries in these borders, I would advise, that the roots should be either planted in pots, or singly at a good distance on a shady border of loamy earth, one year before they are designed to be forced; during which time the runners should be diligently pulled off, to encourage the main roots for fruiting; and at Michaelmas these plants may be transplanted with large balls of earth to their roots, into the borders, before the fruit-trees which are to be forced the following spring, so that they may have time to get new root before that season; and if these plants are carefully watered when they begin to shew their flower-buds, they will produce a good quantity of fruit, which will ripen the latter end of April, or the beginning of May; but then I would alſo adviſe, that theſe plants be taken away as ſoon as they have done bearing, that they may not rob the trees of their nouriſhment.

Since I have mentioned this method of having early Strawberries, I ſhall take the liberty to inſert another method, which is often practiſed to obtain this fruit early in the ſpring, though it doth not ſo properly come under this article; which is, to train up the plants either in pots or borders, after the manner before directed, for at leaſt one year or more; then, about the beginning of February, there ſhould be a moderate hot-bed prepared, in length proportionable to the number of plants deſigned to be forced, and the breadth ſhould be proportionable to the width of the frames which are deſigned to cover them. Theſe frames may be ſuch as are uſed for common hot-beds, to raiſe early Cucumbers, &c. This hot-bed muſt be covered with freſh loamy earth about eight inches thick, into which the Strawberry plants ſhould be placed, with large balls of earth to the roots, as cloſe as they can conveniently be planted (for as they muſt be kept clear from runners, they will not ſpread much during the time they remain in the bed, which will be no longer than until their fruit is gone.) Then they ſhould be gently watered to ſettle the earth to their roots, which muſt be frequently repeated as the earth becomes dry, otherwiſe they will produce but few fruit. While the nights continue cold, the glaſſes of the hot-bed ſhould be covered with mats, to preſerve a kindly warmth in the beds; but in the day-time, when the weather is favourable, the glaſſes ſhould be raiſed to admit freſh air to the plants; for if they are too much drawn (eſpecially when they begin to flower,) they will not produce much fruit. If the ſeaſon ſhould continue long cold, and the heat of the beds ſhould decline, it will be proper to lay ſome freſh hot dung round the ſides of the beds to renew their heat, being always careful not to make them too hot, for that will ſcorch their roots, and prevent their fruiting. If the plants which are planted in theſe beds are ſtrong, and in a good condition for bearing, and care is taken in tranſplanting of them to preſerve good balls of earth to their roots, as alſo to keep a due temperature of warmth in the beds, they will produce ripe fruit by the end of April, or the beginning of May, in plenty, and will continue bearing until ſome of thoſe in the open air come in to ſucceed them.

The beſt kind of Strawberries to plant for forcing, is the Scarlet, for the Hautboys grow too rampant for this purpoſe.

But to return to the ſubject of Hot-walls; what I have here inſerted concerning the forcing of fruits, has been only to obtain theſe fruits earlier in the ſeaſon, than they would naturally ripen againſt common Walls. But in ſome parts of England, where moſt of our good kinds of fruit ſeldom ripen, it might be very well worth while to build ſome of theſe Walls, to obtain good fruit from the beſt kinds of Peaches, Plumbs, &c. eſpecially in ſuch places where fuel is plenty, becauſe there the expence will not be great after the firſt building of the Walls. For I would not propoſe to have coverings of glaſs, excepting for a ſmall proportion of the Walls; the reſt may have frames of canvas, to ſhut over them, in the ſame manner as the glaſſes are contrived, which will ſucceed very well, where proper care is taken; for as there will not be occaſion to cover theſe trees until the beginning of March, at which time alſo the fires muſt be made, ſo, before the trees are in flower, the weather may be frequently warm enough to open the covers to admit ſun and air to the trees in the middle of the day; for if theſe covers are kept too cloſely ſhut, the ſhoots of the trees will draw very weak, and their leaves will turn pale, for want of light and air. And as the deſign of theſe contrivances is only to bring the trees into flower, three, or at moſt, four weeks earlier than they would naturally come againſt common Walls, there will be no neceſſity of making very large fires, or keeping the covers too cloſely over the trees.

Inſtead of canvas for theſe covers, oiled papers may be uſed, which ſhould be done in the manner directed for raiſing of Melons, by paſting as many ſheets of paper together, as will fit the frames on which they are to be fixed; and when the paſte is dry, the paper ſhould be faſtened into the frames, and then the oil rubbed over on the outſide with a bruſh, which will ſoak through the paper, and when the paper is dry, the cover may be uſed. This paper ſhould be of Dutch wrapper, which will laſt very well one ſeaſon, and the expence of repairing it will not be very great; wherefore theſe are to be preferred to the canvas, becauſe all ſorts of plants will thrive much better under them than they will under canvas, or any other cloſe covering, which will not admit the rays of light ſo well through to the plants

The frames deſigned for either canvas or paper may be made much ſlighter than thoſe for glaſs, becauſe theſe being very light, will not require ſo much ſtrength to ſupport them; and if theſe are well painted, and every year, when their uſe is over, carried into ſhelter, they will laſt a long time, for they will not be wanted abroad longer than three or four months, viz. from the beginning of March to the middle of June; for after this time the fruit will not require any covering, the trees being then full of leaves, and the young ſhoots will by that time have made ſuch progreſs, as to become a good defence for the fruit; but theſe covers ſhould not be too ſuddenly taken away, but by degrees the trees ſhould be inured to the open air, otherwiſe the change will be too great, and may occaſion moſt of the fruit to fall off, eſpecially if cold nights ſhould follow.

By this method gentlemen may be ſupplied with moſt of the beſt kinds of fruit, in the northern parts of England, where, without ſome ſuch care, they cannot expect much good fruit in their gardens. And as coal is in great plenty in thoſe places, the expence will be very little; therefore I am ſurpriſed that moſt of the gentlemen, who live in the north, do not put this method in practice. That there are ſome of theſe Walls built in the north is well known, but then they are chiefly deſigned to produce a little early fruit, more for curioſity than any real uſe; and theſe Walls are, for the moſt part, ſo ill contrived, that four times the fuel is expended, as will be requiſite when the Walls are built after the manner here directed; and where the heat is not pretty equally diſtributed through every part of the Wall, ſome of the trees will have too much heat, while others will have little benefit from the fires.

Where the Walls are planted with the beſt kinds of fruit, which are deſigned to ripen in perfection, if the autumns ſhould prove cold, or very wet, before the fruit be ripe, it will be proper to put the covers over the trees; and if there are ſome ſlow fires made to dry off the damps, it will be of great uſe to prevent the fruit from growing mouldy, and to haſten their ripening; but when this is practiſed, the covers ſhould be taken off whenever the weather will admit of it, that the fruit may enjoy the benefit of the free air, without which they will be inſipid or ill-taſted.

Although in the former directions for forcing trees, in order to have early fruit, I have adviſed, that ſuch trees ſhould have one or two years reſt, in order to recover vigour; yet that is not to be underſtood of thoſe trees, which are only deſigned to be brought forward enough to produce their fruit in perfection; for as the fires are not deſigned to be made till the beginning of March, the trees will not be weakened

weakened thereby, becaufe they will be inured to the open air long before their fruit is ripe, and will have time to ripen their fhoots, and form their buds, for the next year's bearing; therefore thefe trees may be thus forced every year, without doing them much injury, provided they are carefully managed.

There are fome perfons near London who make it their bufinefs to raife early fruit to fupply the markets, which they perform by the heat of dung only, having no Fire-walls in their gardens. The method which thefe people follow is to have a good quantity of new dung laid in a heap to warm (after the fame manner as is practifed for making of hot-beds.) When this dung is in a proper temperature of heat, they lay it clofe on the backfide of their Fruit-wall, about four feet thick at the bottom, and floping to about ten inches or a foot thick at the top. This dung fhould be gently beat down with a fork to prevent the heat going off too foon; but it fhould not be trodden down too hard, left that fhould prevent its heating. The outfide of the dung fhould be laid as fmooth as poffible, that the wet may run off more eafily; and if there is a covering of thatch, as is fometimes practifed, it preferves the dung from rotting too foon, whereby the heat is continued the longer. The time for laying this dung to the back of the Wall, is the fame as for making the fires, i. e. about the middle or end of February. The firft parcel of dung will continue warm about a month or five weeks, when there fhould be a fupply of new dung prepared, and the old taken quite away, or mixed up with this new dung, to renew the heat, which, if it works kindly, will be fufficient to laft the feafon. Thefe Walls are covered with glaffes or oiled paper, in the fame manner as the Fire-walls, and the trees muft be treated the fame way: but there muft be more care taken to open the glaffes againft thefe Walls, whenever the weather will permit, otherwife the fteam of the dung will occafion a great dampnefs through the Wall, which, if pent in about the trees, will be very pernicious to them, efpecially at the time they are in flower.

By this method fome gardeners have forced long Walls, filled with old well-grown fruit-trees, which have produced great quantities of fruit annually, which has well anfwered their expence; but, as in many parts of England it will be very difficult to procure a fufficient quantity of new dung for this purpofe, the Fire-walls are moft ufeful, and leaft expenfive.

WALL FLOWER. See CHEIRANTHUS.

WALNUT. See JUGLANS.

WALTHERIA. Lin. Gen. Plant. 741.

The CHARACTERS are,

The flower has a cup-fhaped permanent empalement of one leaf, cut into five points at the brim; it has five heart-fhaped petals and five ftamina, joined in a cylinder, terminated by loofe fummits, and an oval germen, fupporting a fingle ftyle, crowned by a bifid ftigma. The germen turns to an oval capfule with one cell, inclofing one obtufe feed.

The SPECIES are,

1. WALTHERIA (*Americana*) foliis ovalibus ferratis plicatis dentatis floribus pedunculatis capitulis. Lin. Sp. 941. *Waltheria with oblong, oval, fawed leaves, and flowers growing in clufters upon very long foot-ftalks, at the wings of the branches.*

2. WALTHERIA (*Indica*) foliis ovatis ferratis nervofis, floribus confertis alaribus feffilibus. *Waltheria with oval, fawed, veined leaves, and cluftered flowers fitting clofe at the wings of the ftalk.*

The firft fort grows naturally in both Indies; this rifes with a fhrubby branching ftalk to the height of eight or ten feet, covered with foft hairs. The leaves are placed alternately upon long foot-ftalks; they are hairy and foft, having feveral longitudinal veins. From the wings of the branches arife the foot-ftalks, terminated by clufters of very fmall yellow flowers, which juft peep out of their hairy foft empalements; under each clufter is placed a fmall leaf of the fame fhape with thofe below. The flowers are fucceeded by a fingle feed, wrapped in the empalement of the flower.

The fecond fort grows naturally at Campeachy, from whence the feeds were fent me. The ftalks of this are ligneous; they rife fix or feven feet high, dividing into feveral branches, which are lefs hairy than thofe of the former fort. The leaves are oval, of a yellowifh green colour, fawed on their edges, and hairy, but are not fo foft as thofe of the former, having many veins running from the midrib, ftanding upon long foot-ftalks. The flowers are very fmall, yellow, and are collected into round clufters, having very fhort foot-ftalks clofe to the wings of the leaves.

Thefe plants are propagated by feeds, which muft be fown on a hot-bed; and when the plants are fit to tranfplant, they muft be each planted into a feparate fmall pot, and plunged into a frefh hot-bed, and afterward treated in the fame manner as other tender plants from the fame country, fo muft be kept in the bark-ftove, otherwife they will not thrive in England. The fecond year the plants will flower, and produce good feeds; but they may be continued three or four years, if the plants are often fhifted, and their roots pared, to keep them within compafs; for if they are permitted to remain long undifturbed in the tan-bed, their roots will run out through the holes in the bottom of the pots, and extend to a great diftance in the tan-bed; and when this happens, if their roots are torn, or cut off, the plants feldom furvive it. When the plants root into the tan, they grow very luxuriant, and cannot be kept within reafonable compafs; but on their roots being difturbed, their branches will hang, and their leaves fhrivel up, and drop off; therefore, to keep thefe plants within bounds, they fhould be drawn up out of the tan, at leaft once in fix weeks, during the fummer feafon, and the plants fhifted out of the pots once in three months; with this management the laft fort may be continued feveral years, but the firft feldom lives longer than two years.

WARNERIA. Yellow-root.

The CHARACTERS are,

The flower has no empalement; it has three roundifh petals, which fall off very foon, and a great number of awl-fhaped ftamina, terminated by oval fummits, with feveral roundifh germen, fupporting a fhort ftyle, crowned by a permanent bifid ftigma. The germen afterward turns to a roundifh fruit, compofed of many acini like the Strawberry, each having one cell, including a fingle feed.

We have but one SPECIES of this genus, viz.

WARNERIA (*Canadenfis.*) *Warneria of Canada; Yellow-root, or Hydraftis.* Lin.

I have given the title of Warneria to this genus in honour of Richard Warner, Efq; of Woodford in Effex, who is a very curious botanift, and is poffeffed of a large collection of curious plants, of which he is very communicative to all lovers of plants. This title of the plant was given by me to it long before Dr. Linnæus's fecond volume of his Syftema Naturæ was publifhed, in which he has given to this plant the title of Hydraftis; the characters of which were fent him from England, but he had not feen the plant.

It grows naturally in Penfylvania, from whence the roots were fent me by Dr. Benfel. The root is flefhy, of an irregular form, and a deep yellow colour, fending out one or two ftalks about ten inches high; toward the bottom of thefe is one large, roundifh, indented leaf, ftanding upon a foot-ftalk; the upper part of each ftalk is garnifhed by a fmaller leaf, of the fame form as the lower, which embraces the ftalk, which is terminated by fingle white flowers, compofed posed of three petals, which drop off in a few hours after they expand, leaving a great number of stamina with the styles naked. The germen soon after swell, and compose a fruit very like that of the Dewberry, having many acini of a bright red colour when ripe, each acini including one seed.

This plant may be propagated by seeds, which should be sown in a pot of loamy earth soon after they are ripe, placing the pot in shade till autumn, when it may be put under a frame for the winter; the plants will appear in the spring, then should be placed in shade till their leaves decay, when the roots may be transplanted in a shady border, where they may remain to flower.

WATER is one of the most considerable requisites belonging to a garden; if a garden be without it, it brings a certain mortality upon whatsoever is planted. By waterings the great droughts in summer are allayed, which would infallibly burn up most plants, had we not the help of Water to qualify the excessive heats; besides, as to noble seats, the beauty that Water will add, in making jet d'eau, and cascades, which are some of the noblest ornaments of a garden.

The qualities of Water.

Sir Isaac Newton defines Water (when pure) to be a very fluid salt, volatile, and void of all savour and taste, and it seems to consist of small, hard, porous, spherical particles of equal diameters, and equal specific gravities; and also that there are between them spaces so large, and ranged in such a manner, as to be pervious on all sides.

Their smoothness accounts for their sliding easily over the surfaces of one another.

Their sphericity keeps them from touching one another in more points than one, and by both these their frictions, in sliding over one another, are rendered the least possible.

The hardness of them accounts for the incompressibility of Water, when it is free from the intermixture of air.

The porosity of Water is so very great, that there is at least forty times as much space as matter in it, for Water is nineteen times specifically lighter than gold, and of consequence rarer in the same proportion; but gold will, by pressure, let Water pass through its pores, and therefore may be supposed to have (at least) more pores than solid parts.

Dr. Boerhaave is of opinion, that if Water could be had alone and pure, it would have all the requisites of an element, and be as simple as fire; but there has been no expedient hitherto found out for making it such.

Rain-water, which seems to be the purest of all those we know of, is replete with infinite exhalations of all kinds, which it imbibes from the air; so that though it be filtred and distilled ever so often, yet there still remain fæces.

The purest of all Waters we can any way arrive at, is that distilled from snow, gathered in a clear, still, pinching night, in some very high place, taking none but the outer, or superficial part thereof. By a number of repeated distillations thereof, the greatest part of the earth, and other fæces, may be separated from it; and this is what we must be content to call pure Water.

Of the fluidity of Water.

Water, says Dr. Boerhaave, is fluid, but the fluidity is not natural thereto; for naturally it is of the chrystalline kind, and accordingly, wherever a certain degree of fire is wanting, there we see Water become ice. That this ice is the proper effect of the want of heat, and not of any additional spicula introduced into the Water, as Mariotte and others contend, is evident enough, were it only hence, that on this supposition it could not penetrate the substance of all bodies, as we find it does, and even that of metals.

This Water, in its state of solution, never remains at rest; its parts are in perpetual motion, as was first discovered by the French with the help of microscopes; and is farther confirmed by this, that if a little saffron be suspended in the middle of a vessel full of Water, the saffron colour will in a little time form, as it were, a kind of atmosphere around, and at length be diffused through the whole Water. Now this could no way be effected without a motion of the watery particles among each other. Add, that if you cast a quantity of the driest salt, in the coldest weather into Water, it will soon be dissolved, which argues the continual motion of the particles of that element.

He adds, that he had more than once filled a large wide vessel with Water, and narrowly watched with a good microscope, but could never perceive it without some sort of undulatory motion.

Water scarce ever continues two moments exactly of the same weight, but is always varying more or less, by reason of the air and fire contained in it. Thus, if you lay a piece of pure limpid ice in a nice balance, you will find it continue in equilibrio. The expansion of Water in boiling, shews what effect the different degree of fire has on the gravity of Water.

This uncertainty makes it difficult to fix the specific gravity of Water, in order to settle its degree of purity; but this we may say in the general, that the purest Water we can procure is, that which weighs 880 times as much as air.

However, neither have we any tolerable standard for air, for Water being so much heavier than air, the more Water is contained in air, the heavier of course must it be; as in effect, the principal part of the weight of the atmosphere seems to arise from the Water.

Of all Waters, the purest is that which falls in rain in a cold season, and a still day; and this we must be content to take for elementary Water. The Rain-water in summer, or when the atmosphere is in commotion, it is certain, must contain infinite kinds of heterogeneous matter. Thus, if you gather the Water that falls after a thunder-clap in a sultry summer's day, and let it stand and settle, you will find a real salt sticking at the bottom; but in winter, especially when it freezes, the exhalations are but few, so that the rain falls without much adulteration; and hence, what is thus gathered in the morning, is found of good use for taking away spots in the face, and that gathered from snow, against inflammations in the eye. Yet this Rain-water, with all its purity, may be filtred and distilled a thousand times, and it will still leave some fæces behind it; so that to procure the purest Water possible, a man must look for it in a spacious plain in the winter time, when the earth is covered with snow, and its pores locked up with frost.

The next in point of purity is Spring-water. This, according to Dr. Halley, is collected from the air itself, which, being saturated with Water, and coming to be condensed by the evening's cold, is driven against the cold tops of mountains, where, being farther condensed and collected, it gleets down or distils, as much as in an alembic.

Spring-water becomes the better by running, for during all its course, it is depositing what heterogeneous matters it contained; but while the river drives on its Waters in an uninterrupted stream, all its salts, with all the vegetable and animal matters drained into it, either from exhalations, or from the ground it washes gradually, either sink to the bottom, or are driven to the shore.

But what Water descends from springs on the tops of mountains, is generally pretty free from heterogeneous bodies.

Having treated of the properties of Water philosophically, I shall next consider it as essentially necessary in gardens for use, as also of the beauty which Water adds to gardens, where it can be obtained in plenty, if it is properly disposed; and first of its use.

In the kitchen-garden Water is absolutely necessary, for without it there can be little expected; therefore in such places where there cannot be a supply of Water obtained for basons or ponds, wells must be dug; and where the depth to the Water is too great to be raised by pumps, there must be either machines for raising it contrived, or it must be drawn by hand; but in such places which are so unhappily situated as to require machines for the raising of Water from a great depth, there is but small encouragement to make kitchen-gardens, because the constant supplying of Water in those dry situations, will be attended with great expence; and generally the produce of such land is of little worth, especially in dry seasons.

Where kitchen-gardens are supplied with Water from wells, there should be a contrivance of large cisterns, into which the Water should be raised, to be exposed to the sun and air some time before it is used; for the rawness of this Water, when fresh drawn from wells, is not agreeable to the growth of vegetables; so that where large ponds are in the neighbourhood of these gardens, from whence the Water can be led into them, that is by much the best for the growth of vegetables; next to this, River-water is to be preferred, especially from those rivers which run through or near large towns, where the Water is fattened by the soil thrown into the rivers; but the Water of some very clear rivers is as hard as that from the deepest springs, rising through gravel or sand; but the springs issuing through chalk, are generally much softer.

If good Water can be obtained in plenty from the neighbourhood of the kitchen-garden, then there should be two or three basons made in different parts of the garden; so that no part of the garden should be too far distant from the Water, for where the Water is to be carried to a considerable distance, the expence of labour will be great, and there will be great danger of the plants suffering from their being but sparingly watered, labourers being very apt to slight their work, when attended with trouble, if they are not well looked after. The size of these basons should be in proportion to the quantity of Water which will be required, or that they can be supplied with; but their depth should not be more than four feet, for when they are deeper, there is danger of persons being drowned, if by accident they should fall into them; besides Water, when very deep, is not so well warmed and tempered by the sun and air as when it is shallow; therefore the Water of shallow basons is best for the use of gardens.

In making of these basons, there must be particular regard had to the natural soil of the garden, for in loose sandy land there will require much care in making of the clay-walls so as to hold Water; but where the ground is loamy, or inclining to clay, there will be little difficulty in making basons, and the clay-walls need not be so thick. Where the ground is loose, the clay-walls at the bottom should not be less than two feet thick, and those on the sides one foot and a half. The clay should be well wrought over and trod after it is taken from the pit, before it is used in building the Wall. The true sign of good clay is, that it be close and firm, without any mixture of sand, and that it be tenacious and fat in handling; as for the colour, it is no matter whether it be green, yellow, blue, or red; but before the clay is brought to the place, the bason should be dug out and formed, for if the clay is too long exposed to the sun and air, it will not be so fit for use, especially if it be laid in small parcels.

The best time of the year for making basons is in autumn, when the sun is declining, and the weather temperate; for in the spring of the year the east and north-east winds generally blow, which are drying; so that the clay-walls, which are not very carefully covered as fast as they are made, very often crack in many places, and these small cracks often grow wider, and the Water will find a passage through them. The same inconveniency happens from the violent heat of the sun in summer; for when the clay dries fast, it will be very difficult (not to say impossible) to prevent its cracking, and these will let off the Water; and if the clay-wall should not be well made at first, it will be very difficult to mend it after; besides the uncertainty there is in finding out the places through which the Water finds a passage, which is seldom done without strictly examining every part of the clay.

When the ground is dug out level, where the bason is designed, the clay must be brought in, and laid very carefully in the bottom, being very careful that no dirt, or small stones, be mixed with the clay; and there must be some Water thrown from time to time upon it, as it is closely trod by mens naked feet, and then it must be rammed very close: in the performance of this, there must be great care taken that every part of the clay is equally kneaded and rammed, without which there will be great danger of the Water making its way through those parts of the clay which are not well wrought. After the bottom is finished with clay, there should be a stratum of coarse gravel laid over it about four or five inches thick, which will greatly secure the clay-wall, and render the Water clear; but where the basons are large, so that the clay-walls are long in making, the clay should be covered with moist litter, to prevent its drying, which may be taken off when the whole is finished, to lay on the gravel; but if part of the side-walls are finished before this is done, it will be the better, because there may be some Water let into the bason as soon as the gravel is laid, which will prevent the clay from cracking; then the walls round the side of the bason must be carried up with the same care as hath been directed for the bottom, observing also to cover the clay first with litter while the work is carrying on, and afterward lay it with coarse gravel; and as the walls are finished round, the Water may be let in, to secure the clay from drying or cracking.

When the whole is finished, the upper part of the walls must be laid with turf, which will secure them from being broken, and prevent the sun from penetrating the clay; but before this is done, there must be a stratum of sand laid upon the clay four or five inches thick, and upon this a thin stratum of good earth laid, for the Grass to take root in. The bed of sand will prevent the Grass from rooting into the clay; and this will also keep out the frost, which will penetrate the clay, where there is not a covering of sand to secure it, and by being frozen and swelled, and afterward drying, the clay is very apt to crack in many places. The turf on the side of the bason should be laid as far down as the Water is apt to shrink, that no part of the clay may be wholly exposed to the weather, for the reasons before given.

Where these basons are made, there should be no trees growing near, for the roots of trees or shrubs will extend themselves to the clay-walls, and by penetrating them, will occasion fissures, through which the Water will find an easy passage; and where tall trees are growing near basons or ponds, the shaking of the trees with violent winds is apt to loosen the clay-walls, and occasion cracks in them, therefore these cautions are necessary to be observed.

In some countries, where clay cannot be easily procured, the walls of these basons are frequently made of chalk, which is beaten into fine powder, and made into a sort of mortar; and with this the walls are made, by ramming and working it very hard and firm. These basons hold Water very well where they can be well supplied with Water, so as not to be too long dry, for when it so happens, the sun and wind dry the chalk, and cause it to crack, and these cracks commonly extend through the thickness of the walls, so as to let off the Water.

There are others who build their walls with brick laid in terrace, which is a good method for such places where the ground is very loose and sandy, because the walls, when well built, will support the loose earth from falling or settling away from the sides; but where terrace is used, the walls should not be long dry and exposed, for the heat is apt to crack the terrace.

Some persons make a cement of powdered tile and lime, two-thirds of the former to one-third of the latter, being very careful in the mixing of it not to add too much Water, but to labour it well in the beating, which is a principal thing to be observed. With this cement they cover the surface of the walls of basons about two inches thick, laying the plaster very smooth, and being very careful that no sticks, straws, or stones, are mixed with it; this plastering is commonly performed in dry weather, and as soon as it is finished, it is rubbed over with oil or bullocks blood, and the Water let into the bason as soon as possible. This cement has the property of hardening under Water, so as to be equal to stone, and will continue as long found.

Whatever the materials are with which the walls are made, there must be great care taken that they are built so strong, as that they may resist the weight of the Water; so that where the ground about the bason is not very solid, the walls should be thicker, and supported on the backside by buttresses of the same materials, placed at proper distances; or if the walls are made of clay, there should be planks supported by strong timbers, placed at proper distances to support the clay, otherwise there will be great danger of their being broken down, especially where the basons are large, so as that the winds have room to act upon the surface of the Water, and drive it in large waves against the banks.

The directions here given are only for basons or reservoirs of Water for use, so must not be supposed for large pieces of Water for beauty; for where the ground is of a loose sandy nature, so as not to hold Water, the expence of claying the bottom and sides will be too great, if the Water is of a large extent; therefore it would be imprudent to attempt it in such places, but where there is a supply of Water, and the ground is well adapted to hold it. There can be no greater beauty than that which Water affords to a seat, provided it is properly disposed; therefore I shall give some general hints, by which persons may be directed in the forming of large pieces of Water, so as to render them beautiful.

In those places where there is a command of running Water, it will be a great additional beauty, because the Water will always be much clearer, so more beautiful than still Water; besides, if it moves with any degree of velocity, there may be one or more falls of Water contrived, which will still add to the beauty. In the conducting of this Water, the level of the ground must be carefully taken, for the great skill in the contriving of rivers, or other pieces of Water, is in the saving of expence in the digging; therefore where the ground is naturally low, the Water should be conducted through these low parts, and never endeavour to carry it through higher ground; for in such places the banks will be so high as to shut out the sight of the Water, to persons who stand at a little distance from it on either side, unless the Water is very broad; and where it is so, the eye is thrown to a considerable distance over the surface of the Water, by the steepness of the banks, therefore the slopes on the side of Water should always be made as easy as possible; nor should they be made flat, with sharp edges on the top (as is too generally practised;) for these stiff regular slopes are not near so pleasing as those which are made gently convex, for the eye will slide over these to the Water, having no ridge to cut the sight, and at a small distance there will be no appearance of a cut, as will always be seen where the upper part of the slope is finished in a sharp angle; and the great skill is to contrive, that as much of the surface of the Water may appear to the sight as possible.

In most of the old gardens, where there are pieces of Water, there is nothing more common than to see them brought into regular figures, such as long strait canals, or basons, either round or polygonal, so that all the boundaries of the Water are seen at one view; but these, however large may be their extent, are not near so pleasing as where the Water is so conducted, as that the termination may be seen as little as possible; for when the Water is lost from the sight by some gentle easy turns, the imagination may be led to suppose the surface of the Water extended to a considerable distance; so that sometimes small pieces of Water are so artfully contrived, as to make them appear very considerable.

As in the old stile of laying out gardens the Water was generally wrought into regular strait canals, which corresponded with the strait walks, hedges, and regular lines of trees, which were then chiefly studied, so, as the taste altered from this stiff method of disposing gardens, to that which approached nearer to nature in the forming of rivers, or other large pieces of Water, those who have succeeded best have always had great regard to the natural situation of the ground, so as to lead the Water through the natural hollows, whereby the great expence of digging is saved; and by contriving to make the head in some narrow part of the ground, it may be done at a much less expence, and will be better secured than where the head is of great extent; therefore it is better either to shorten the extent of the Water, or to carry it farther, according to the natural situation of the ground, than to terminate it where it may occasion great expence; and it is always observed, that where there is the greatest expence in the making of these large heads, the whole will appear less beautiful than where nature is chiefly consulted; for nothing can be more unsightly than those extensive heads which are sometimes made to pieces of Water which rise six or eight feet, and sometimes much more, above the surface of the ground, whereby the Water is hid from the sight, to those persons who are situated on that side of the head, and a large bank of earth shuts up the view; and sometimes these heads are so situated, as to appear in sight of the house, or from a principal part of the gardens, which is a very great absurdity.

Since the taste has been altered in the disposition of gardens, and a more natural method has been pursued by persons of judgment, there have been great improvements made in the distribution of Water, so as to render it truly ornamental to the seats where they are placed; but there are some, who, by pretending to imitate or copy from these works, have erred as much in making so many short unnatural turns in their Water, as those before-mentioned have done by their regular strait sides; for in what is usually termed serpentine rivers, nothing is more common than to see a small surface of Water twisted in so many short turns, as that many of them appear at one view; and these windings are often made like parts of circles, with such an air of stiffness, as to render them equally disagreeable with any the most studied figures, to persons of good taste. Another thing is also common to these unnatural pieces of Water, which is, their being made of the same width in every part, which

which ſhould always be avoided, for nothing is more beautiful than to ſee the Water extend to a large ſurface in ſome places, and to have it in others more contracted; and this may be generally done at a much leſs expence than the other, where the natural ſite of the ground is well conſidered, which ſhould be done with the utmoſt care, before any work of this ſort is begun, for want of which many perſons have repented after having been at great expence.

There is alſo another material thing to be obſerved, in the ſituation of large pieces of Water, which is, never to extend them ſo near to the houſe, as that they may annoy it, by the damp, which the vapours exhaling from the Water may occaſion, eſpecially when expoſed to the wind, which will at times drive the vapours toward the houſe, and thereby render the habitation unhealthy, and deſtroy the furniture; therefore it is much better to walk out to ſee the Water, than to ſacrifice the habitation for the pleaſure of ſeeing it from the houſe; nor ſhould the Water be ſo ſituated, as that the ſurface may be level with the floor of the houſe, for there is generally ſome moiſture, which will percolate through the veins of the earth, enough to occaſion ſo much damp, as to render the lower part of the houſe unwholeſome; and where there is a conſiderable damp in the foundation of a houſe, part of it will aſcend upward, and render the apartments ſo, therefore great care ſhould be had as to this.

Where perſons are not ſo happily ſituated, as to have the command of a conſtant running Water, but yet from ſome neighbouring reſervoirs or ponds can be ſupplied with it, there may be ſome agreeable pieces of Water contrived, both for uſe and beauty, eſpecially where there is a large ſupply, for otherwiſe it will be better to contract the deſign; for nothing can be more ridiculous than that of having either ponds or rivers deſigned, where they cannot be ſupplied with Water in the dry ſeaſons, when there is the greateſt want of it, both for uſe and pleaſure.

In thoſe places where there is a great ſcarcity of Water, there ſhould be large reſervoirs contrived, into which the Water which deſcends from the hills and riſing grounds may be led; ſo that a large body of Water may be collected during the rainy ſeaſon, for a ſupply in time of drought; theſe reſervoirs, when large, may contain as much Water as may be neceſſary for the uſe of the houſe and gardens; but theſe can rarely ſupply Water enough for beauty, therefore in ſuch ſituations it ſhould not be attempted.

As Water never appears ſo well, as when it is ſituated near woods, ſo in the contrivance of rivers, or pieces of Water, they ſhould be ſo placed as to have planting near, that the contraſt between the wood and Water may appear as perfect as poſſible; and in ſome places, where the Water can be ſeen through the open groves, between the ſtems of large trees, it will add greatly to the beauty of the place; but where the Water is deſigned to terminate, the head ſhould be as much concealed as poſſible, by cloſe plantations of evergreen trees, which may be faced with Alders and weeping Willows, planted cloſe on the ſides of the Water, ſo as that their branches may hang over; and if the Water is contracted, and led through theſe trees with a gentle winding, it may ſeem to run much farther, and to communicate with a larger body of Water at a diſtance; in the contriving of which, the greateſt art is to make it appear as natural as poſſible; for the leſs art there appears in theſe things, the longer they will pleaſe, and the more they will be eſteemed by perſons of good judgment.

WATSONIA.

The title of this genus is given to it in honour of my learned friend Dr. William Watſon, F. R. S. whoſe knowledge in the ſcience of botany juſtly demands this tribute.

The CHARACTERS are,

The flower has a permanent ſpatha (or ſheath) which divides into two parts almoſt to the bottom; it is of one petal. The tube is long, a little curved, and ſwells at the upper part; the rim is cut into ſix obtuſe ſegments, which ſpread open. It has three long ſlender ſtamina, which are terminated by proſtrate oblong ſummits, and a roundiſh three-cornered germen, ſupporting a ſlender ſtyle a little longer than the ſtamina, crowned by three bifid ſtigmas. The germen afterward turns to a roundiſh three-cornered capſule, having three cells, opening with three valves, each containing three or four roundiſh ſeeds.

This has been titled by Dr. Trew, Mariana flore rubello, before he had been acquainted with the name which I had applied to it; but he has ſince informed me by a letter, that as I had raiſed the plants from ſeeds, he would ſuppreſs his title, and adopt mine, who he thought had the moſt right to give it; and that he rather chuſes to do ſo, becauſe the figure he has publiſhed of it was drawn from the plant in the Chelſea garden.

The SPECIES are,

1. WATSONIA (*Meriana*) floribus infundibuliformibus ſubæqualibus. *Watſonia with funnel-ſhaped leaves, and equal flowers.*

2. WATSONIA (*Humilis*) foliis gladiolatis majoribus. *Dwarf Watſonia with ſword-ſhaped leaves, and larger leaves.*

Theſe plants grow naturally at the Cape of Good Hope, and were raiſed from ſeeds, which were brought from thence in the Chelſea garden. The deſcription of the firſt ſort:

The root is bulbous, compreſſed, and ſhaped like a kidney, covered with a fibrous brown ſkin. The leaves are ſword-ſhaped, about a foot long, and an inch broad, ending in points; the two ſides have ſharp edges, but the middle is thicker, and has a prominent midrib; they are of a dark green colour, and riſe immediately from the root. The ſtalk comes out from the root between the leaves, and riſes a foot and a half high. The flowers are produced from the ſide, ſtanding alternately at about an inch and a half diſtance from each other; they have each a ſpatha or ſheath, compoſed of two leaves, which are joined at their baſe, where they are broad, but gradually leſſen to their points. Before the flowers appear, they are of the ſame green colour with the ſtalk, and are divided, a ſmall part of their length incloſing the flower, but afterward they are ſplit almoſt to the bottom, and wither before the flowers decay, becoming dry, and wrap round the ſeed-veſſel. The tube of the flower is an inch and a half long, narrow at the baſe, a little curved, ſwelling much larger above. The rim is divided into ſix obtuſe ſegments, which ſpread open; the flower is of a red copper colour on the outſide, but of a deeper red within; it has three ſtamina, which are incurved, terminated by oblong ſummits, of a dark brown colour, faſtened in the middle to the apex of the ſtamina, lying proſtrate. At the bottom of the tube of the petal is ſituated an oval three-cornered germen, ſupporting a ſlender ſtyle a little longer than the ſtamina, crowned by three bifid reflexed ſtigmas. The flowers generally appear in May, and the ſeeds ripen in July.

The ſecond ſort is much leſs than the firſt; the leaves are ſhorter, and not quite ſo broad; the flower-ſtalk ſeldom riſes above a foot high, and the flowers are ranged cloſer upon the ſtalk; they are alſo of a deeper red colour.

Theſe plants are propagated by offsets from the root, in the ſame manner as the Crocus or Gladiolus. The time for tranſplanting of the roots is in Auguſt, ſoon after the ſtalks decay; the larger roots muſt be each put into a ſeparate pot filled with light freſh earth, and may be placed in the open air till toward the end of October, when the leaves will begin to appear above ground, at which time it will be proper to remove them into shelter; for as this plant is a native of a warm country, it will require to be screened from frost.

The best way of treating these roots is to plunge the pots into an old bed of tanners bark, which has lost its heat, in October; this bed should be covered with a frame, the glasses of which should be drawn off every day in mild weather, that they may enjoy as much free air as possible, to prevent their drawing up weak; but they must be covered in bad weather, and screened from frost. The latter end of April, when they begin to put out their flower-stalks, the pots should be removed to an airy glass-case, where they may stand to flower; and when the flowers are decayed, they should be placed in the open air to perfect their seeds.

The offsets and small roots may be planted three or four in a pot, according to their size, and should have the same treatment as the larger roots the first year, and by that time twelvemonth they will be strong enough to flower, so should have separate pots.

WILDERNESSES, if rightly situated, artfully contrived, and judiciously planted, are very great ornaments to a fine garden, but it is rare to see these so well executed in gardens as could be wished, nor are they often judiciously situated; for they are frequently so situated as to hinder a distant prospect, or else are not judiciously planted; the latter of which is scarce ever to be found in any of our most magnificent gardens, very few of their designers ever studying the natural growth of trees, so as to place them in such manner as not to obstruct the sight to the view; I shall therefore briefly set down what has occurred to me from time to time, when I have considered these parts of gardens; whereby a person will be capable to form an idea of the true beauties, which ought always to be studied in the contrivance of Wildernesses.

1. Wildernesses should always be proportioned to the extent of the gardens in which they are made, that they may correspond in magnitude with the other parts of the garden, for it is very ridiculous to see a large Wilderness planted with tall trees in a small spot of ground; and, on the other hand, nothing can be more absurd than to see little paltry squares, or quarters of Wilderness work, in a magnificent large garden.

Wildernesses should never be placed too near the habitation, because the great quantity of moisture, which is perspired from the trees, will cause a damp unwholesome air about the house, which is often of ill consequence. Nor should they be situated so as to obstruct any distant prospect of the country, which should always be preserved wherever it can be obtained, there being nothing so agreeable to the mind as an unconfined prospect of the adjacent country; but where the sight is confined within the limits of the garden from its situation, then there is nothing so agreeable to terminate the prospect, as a beautiful scene of the various kinds of trees judiciously planted; and if it is so contrived, that the termination is planted circularly with the concave toward the sight, it will have a much better effect than if it end in strait lines or angles, which are never so agreeable to the mind, therefore those lines should be broken.

The trees should also be adapted to the size of the plantation, for it is very absurd to see tall trees planted in small squares of a little garden; and so likewise, if in large designs the plantation has only small shrubs, it will have a mean appearance.

The walks must also be proportioned to the size of the ground, and not make large walks in a small Wilderness, nor too many walks, though smaller, whereby the greatest part of the ground is employed in walks; nor should the grand walks of a large Wilderness be too small, both of which are equally faulty. These walks should not be entered immediately from those of the pleasure-garden, but rather be led into by a small private walk, which will render it more entertaining.

The old formal method of contriving Wildernesses was to divide the whole compass of ground, either into squares, angles, circles, or other figures, making the walks correspondent to them, planting the sides of the walks with hedges of Lime, Elm, Hornbeam, &c. and the quarters within were planted with various kinds of trees promiscuously without order; but this can by no means be esteemed a judicious method, because first hereby there will be a great expence in keeping the hedges of a large Wilderness in good order by shearing them, which, instead of being beautiful are rather the reverse; for as these parts of a garden should, in a great measure, be designed from nature, whatever has the stiff appearance of art, does by no means correspond therewith; besides, these hedges are generally trained up so high, as to obstruct the sight from the stems of the tall trees in the quarters, which ought never to be done.

In the next place the walks are commonly made to intersect each other in angles, which also shew too formal and trite for such plantations, and are by no means comparable to such walks as have the appearance of meanders or labyrinths, where the eye cannot discover more than twenty or thirty yards in length. These should now and then lead into an open piece of Grass; and if in the middle part of the Wilderness there is contrived a large opening, in the center of which may be erected a dome or banquetting-house, surrounded with a green plat of Grass, it will be a considerable addition to the beauty of the place.

From the sides of the walks and openings the trees should rise gradually above each other, to the middle of the quarters, where should always be planted the largest-growing trees which should appear to view, it will have a very different effect from the common method, where the trees are planted large and small without order.

In these plantations there may be planted next the walks and openings Roses, Honeysuckles, Spiræa frutex, and other kinds of low flowering shrubs, which may be always kept within compass; and at the foot of them, near the sides of the walks, may be planted Primroses, Violets, Daffodils, and many other sorts of wood flowers, to appear as in a natural wood. Behind these should be planted Syringas, Cytisuses, Althæa frutex, Mezereons, and other flowering shrubs of a middle growth, which may be backed with other flowering shrubs of a large growth.

In small gardens, where there is not room for these magnificent Wildernesses, there may be some rising clumps of evergreens, so designed as to make the ground appear much larger than it is in reality; and if in these there are some serpentine walks well contrived, it will greatly improve the places, and deceive those who are unacquainted with the ground as to its size.

In Wildernesses there is but little trouble or expence after their first planting, which is an addition to their value; the only labour required is to keep the walks free from weeds. And in the quarters, if the weeds are hoed down three or four times in a summer, it will still add to their neatness. The trees should also be pruned to cut out all dead wood, or irregular branches, where they cross each other, and just to preserve them within due bounds; and, as was before observed, if the ground be slightly dug between the trees, it will greatly promote their vigour. This being the whole labour of a Wilderness, it is no wonder they are so generally esteemed, especially when we consider the pleasure they afford.

SWEET WILLIAMS. See Dianthus.

WILLOW. See Salix.

WILLOW, the French. See EPILOBIUM.

WIND is defined to be the ſtream or current of the air, together with ſuch vapours as the air carries along with it, or it is a ſenſible agitation of the air, whereby a large quantity thereof flows out of one place or region to another.

The ancients made but four Winds, according to the four cardinal points, but this was quickly looked upon as too groſs a diviſion. The following age added eight more to this number, which was thought too nice a ſubdividing, and therefore they reduced the laſt number to four, taking every other or middle Wind, and adding them to the old account; but our ſailors, who are far beyond the ancients for their ſkill in navigation, have divided the horizon into thirty-two equal parts, adding twenty-eight to the four cardinal Winds; a thing uſeful in navigation, but of no great concern in natural philoſophy, unleſs it be to give a hint, that the Wind blows from all parts of the heavens.

Of the qualities of Winds.

1. "A Wind that blows from the ſea is always moiſt; "in ſummer it is cold, in winter warm, unleſs the ſea be "frozen up." This is well demonſtrated thus: There is vapour continually riſing out of all water (as appears even hence, that a quantity of water, being left a little while in an open veſſel, is found ſenſibly diminiſhed,) but eſpecially if it be expoſed to the ſun's rays, in which caſe the evaporation is beyond all expectation. By this means the air incumbent on the ſea becomes impregnated with a deal of vapour, but the Winds blowing from off the ſea, ſweep theſe vapours along with them, and conſequently are always moiſt.

Again, water in ſummer, &c. conceives leſs heat than terreſtrial bodies, expoſed to the ſame rays of the ſun; but in winter ſea-water is warmer than the earth, covered with froſt, ſnow, &c. Wherefore, as the air, contiguous to any body, is found to partake of its heat and cold, the air, contiguous to ſea-water, will be warmer in winter and colder in ſummer, than that contiguous to the earth: or thus; vapours raiſed from water by the ſun's warmth in winter, are warmer than the air they riſe in, as appears from the vapours condenſing, and becoming viſible, almoſt as ſoon as they are got out into the air. Freſh quantities of vapours, therefore, continually warming the atmoſphere over the ſea, will raiſe its heat beyond that over the land.

Again, the ſun's rays reflected from the earth into the air in ſummer are much more than thoſe from the water into the air. The air therefore over the earth, warmed by the reflection of more rays than that over water, is warmer. Hence ſea Winds make cloudy hazy weather.

2. "Winds which blow from the Continent are always "dry, in ſummer warm, and cold in winter;" for there is much leſs vapour ariſing from the earth than from water, and therefore the air over the Continent will be impregnated with much fewer vapours: add, that the vapours or exhalations raiſed by a great degree of heat out of the earth, are much finer and leſs ſenſible than thoſe from water. The Wind therefore, blowing over the Continent, carries but little vapour with it, and is therefore dry.

Our northern and ſoutherly Winds, however, which are commonly eſteemed the cauſes of cold and warm weather, Dr. Derham obſerves, are really the effects of the cold or warmth of the atmoſphere: hence it is, that we frequently ſee a warm ſoutherly Wind on a ſudden changed to the north, by the fall of ſnow or hail; and that in a cold froſty morning we ſee the Wind north, which afterward wheels about toward the ſoutherly quarters, when the ſun has well warmed the air, and again in the cold evening turns northerly or eaſterly.

Some Winds are drying, others are moiſt; ſome gather clouds, others diſperſe them; ſome are warm, others cold; but their influence is not one and the ſame in all places, for ſuch Winds as are warm in one country are cold in another; thoſe that are wet with us are dry with other nations, and on the contrary.

The dry Winds are ſuch as carry but a few vapours along with them, and therefore lick off the moiſt particles from the bodies over which they paſs; and thus in Holland the north and eaſt Winds, with the intermediate points, are drying, becauſe the cold northern ſea yields but few vapours in compariſon of thoſe that come from warmer parts of the ocean; but the weſterly Winds and others are moiſt, becauſe they iſſue from warm and vaporous parts, the weſtern Wind ſeldom failing to ſend rain.

Such Winds gather clouds which blow from the quarters where the vapours ariſe, which, in conjunction with the vapours of our own region, fill the air; and, on the contrary, thoſe that bring little vapour along with them, and bear away that which hangs over us, bring fair weather.

Winds are either warm or cold, as the countries are from whence they blow; and therefore when a briſk Wind blows from a cold quarter, it allays the heat of ſummer, which is very troubleſome in ſtill weather. Thus a quick blaſt of a pair of bellows will put out a flame, which a gentle blowing increaſes; for the quick blaſt drives all the flame to one ſide, where it is ſtifled by the force of the incumbent air for want of aliment, but a gentle Wind augments the motion of the flame every way, and makes it ſeize on more parts of fuel.

Now, becauſe all the heat or cold of Wind proceeds from the heat or cold of the country where it blows, therefore the ſame Winds are cold or hot every where. Beyond the line they are juſt the reverſe of what they are with us; their cold Winds are from the ſouth, ours from the north; and as our ſouth Winds are warm, from no other reaſon but becauſe they bring us an air heated by the ſun; for the very ſame reaſon the north Winds are warm to our antipodes.

From what has been ſaid, it is evident that the ſun is the cauſe of the Wind, and motion the cauſe of the vapours.

Of WINES, and vinous liquors.

Wine is a briſk, agreeable, and ſpirituous juice, drawn from vegetable bodies, and fermented.

Dr. Boerhaave characterizes Wine, that the firſt thing that it affords by diſtillation be a thin, fatty, inflammable, &c. fluid, called a ſpirit; and in this it is diſtinguiſhed from another claſs of fermented vegetable juices, viz. vinegars; which, inſtead of ſuch ſpirit, yield for the firſt thing an acid, uninflammable matter.

In order to the making Wines, it will be of great advantage to be well acquainted with the buſineſs of fermentation. This Dr. Boerhave defines and explains as follows:

Fermentation is a change produced in vegetable bodies, by means of an inteſtine motion excited therein; the effect whereof is this, that the part which firſt riſes from them in diſtillation is either a thin, fat, acrid, hot, tranſparent, volatile, and inflammable fluid, that will mix with water; or elſe a thin, acid, pellucid, leſs volatile, uninflammable liquor, capable of extinguiſhing fire.

The liquor, obtained by means of fermentation, is called thin, becauſe none appears to be thinner than the ſpirit of fermented vegetables; acid, becauſe it acts almoſt like fire, when applied to the tongue or other parts of the body; volatile, becauſe there appears to be no liquor that is raiſed with greater eaſe; but it is this liquor being totally inflammable, and at the ſame time capable of mixing with water, that ultimately diſtinguiſhes fermentation from all other operations

rations in nature; for neither putrefaction, digestion, effervescence, nor any thing of that kind, will ever afford a liquor at once possessed of those qualities.

Putrefaction, indeed, as well as fermentation, is performed by means of an intestine motion; but the former will never produce either of the liquors above described, as the effect of fermentation; that is, neither a vinous nor acetous liquor.

We see then, that there are two different effects of fermentation, the production of an inflammable spirit, and an uninflammable acid; and whatever operation will afford neither of these liquors, is improperly called fermentation; which therefore can only take place in the vegetable kingdom; for all the art in the world, so far as hitherto appears, will never gain such spirits from animals or fossils, and consequently never excite an actual and real fermentation in them; for fermentation is the single operation in nature, by which such spirits can be obtained.

2. Any vegetable liquor so fermented, as to afford the inflammable spirit above-mentioned for the first thing in distillation, we call Wine; but if the liquor be so fermented, as first to afford the acid uninflammable one, it is called vinegar; by which we mean every thin, acid, volatile, vegetable liquor, capable of extinguishing fire. So likewise, under the name of Wine, we include beer or ale, mead and metheglin, cyder, perry, all sorts of artificial Wines, and whatever liquors afford spirits possessed of the properties before set down.

The like is to be understood of vinegar, which is obtainable from all the same bodies that afford Wine; so that we have either the Wine or vinegar of all sorts of fruits, as of Grapes, Currants, Mulberries, Cherries, &c. all sorts of grain, as Barley, Wheat, Oats, &c. all sorts of pulse, as Beans, Pease, Tares, &c. all sorts of roots, as Turneps, Carrots, Radishes, &c. and in short, of all sorts of vegetable substances, even Grass itself.

3. All the bodies capable of being changed by fermentation, either into Wine or vinegar, are said to be fermentable bodies; and because such a change can only be wrought, so far as we know at present, upon vegetables, these alone are accounted fermentable.

4. Any matter, which, being mixed with a fermentable body, increases its intestine motion, or excites or forwards the fermentation, is called the ferment; and, according to the doctrine before delivered, nothing can properly be called so, but what will produce either Wine or vinegar.

These fermentable bodies may be reduced to the following classes.

The first class will consist of the meally seeds, i. e. all the grain, which, being fully ripe and well dried, may be reduced, by grinding, to a light meal or flour, that is neither clammy nor unctuous.

The second class consists of all the pulpy summer-fruits, which, when ripe, affect the tongue with the sense of acidity and sharpness, as Apples, Pears, Grapes, Goosberries, &c. Under this class may be ranged all manner of bulbous pulpy roots growing in the ground, if they are first deprived of their volatile, alkaline salt, which is apt to determine them to putrefaction.

The third class takes in all the juicy parts of plants, as the leaves, flowers, stalks and roots, provided they are not too oily, or too alkaline; in which cases vegetables will rather putrify than ferment.

The fourth class contains the fresh, expressed, and native juices of all kinds of vegetables; to which may be added all the native, saline liquors that distil from wounded plants, as the tears of the Vine, the Walnut, the Birch-tree, &c.

Under the fifth class come the most perfect of all the vegetable juices, viz. those that are unctuous, condensed, and elaborated by nature herself, such as honey, manna, sugar, and all other kinds of concocted juices capable of dissolving in water.

In order to fit any of the fermentable bodies of fermentation, there are several particulars requisite:

1. Maturity; the juice of unripe berries, as of Currants or Goosberries, for instance, will scarce be brought to ferment at all; while it is very difficult to hinder their juice, when fully ripe, from falling spontaneously into fermentation.

Thus the juice of unripe Grapes, being incapable of fermenting, is a rough, acid liquor, called verjuice, that will for several years remain in the same unactive state; but after they come to maturity, it can no sooner be pressed into the vessel than it becomes a fermentable spirituous fluid.

2. Another requisite to prepare a body for fermentation is, that it should contain only a moderate proportion of oil; for if it either exceeds in the quantity, or be entirely destitute of oil, it will never be brought to ferment at all. Thus Almonds, Fennel-seeds, &c. are always deprived of their oil before they are attempted to be fermented.

3. The bodies intended for fermentation must not be too acid or austere; as is plain from the acid juices of unripe fruit, which are greatly indisposed to ferment.

4. The last thing required to fit and prepare a body to undergo fermentation, is the property of dissolving in water; for want of which all acid bodies, and such woods, roots, and herbs as are dry and hard, become unfit for this operation; for unless the parts of these bodies are dissolved, the requisite intestine motion thereof will not ensue; but without such motion fermentation cannot subsist.

Hence honey itself can never be made to ferment, whilst it retains its native, thick consistence; but, being dissolved by heat, or let down with water, it immediately enters the state of fermentation. On the other hand, so violently as the juice of Grapes affects this state, yet if, immediately after it is expressed, it be reduced, by boiling, to the consistence of a jelly, it will lie quiet and never ferment at all, unless it be again diluted and let down with water.

Ferments are of two kinds; the natural or spontaneous, and those produced by fermentation.

The spontaneous or natural ferments are,

1. All the fresh expressed juices of fully ripened fruits, which easily run into fermentation.

2. Honey, manna, sugar, and the like thick and inspissated vegetable juices, which cause a strong fermentation.

3. The ferments produced by fermentation are, the fresh flowers or yeast of any fermenting vegetable juice or liquor, as of Wine, beer, &c. By flowers or yeast is to be understood that light frothy matter, which covers the surface of the fermenting liquor in the nature of a tender crust; and which, being added to any other fermentable juices, will excite a fermentation in them.

4. The fresh fæces or lees of any fermenting liquor, as of Wine, ale, beer, &c. For all fermentation divides the liquor, which is the subject of it, into three parts, viz. the flowers or yeast, which possess the uppermost place; the operating or fermenting fluid, which lies in the middle; and the gross and seemingly exhausted matter, which, falling to the bottom of the vessel, is known by the name of lees, sediments, feculence, or mother, that will, if raised again out of the liquor into which it was precipitated, cause it to work afresh.

Thus, when a hogshead of Wine has done fermenting, and is fined down, if the vessel be any way shaken or disturbed, it will grow turbid again, and ferment anew, as vintners very well know. For as such as were the flowers in the act of fermention, such is the mother after the action is over.

5. Acid

5. Acid paste, or bakers leaven, which is no more than any kind of meal brought into a close lump by means of water, after the same manner as common bread is made; for this being set in a warm place, during the space of four or five days it will first swell, then turn very acid, and at length become a ferment.

6. Those ferments which reside in, or stick to the sides of the casks that have contained fermenting liquors; for such casks will of themselves raise a fermentation in the liquors committed to them; and Helmont was of opinion, they might be capable of doing this for ever.

Upon account of this inherent ferment it is, that old-seasoned vessels, or such as have been long employed by vintners or brewers, bear so great a price among them.

It is very remarkable, though a thing well known to brewers and vintners, that a new cask checks the fermentation of vinous liquors, and renders them weak and spiritless; for which reason they never chuse to make use of such a cask before it is seasoned, as they call it, by having first contained some spirituous or fermented liquor or other; which being plentifully drank in by the wood, the original liquor comes to be deprived of a large proportion of its spirit, and more fermentable part, whence the remainder must needs taste flat and vapid.

This is certain, that even must itself will not easily ferment in a new pure vessel, but with the greatest facility, if put into one that has before contained fermenting juices; for the parts of the fermenting liquors, with which such a vessel must have been impregnated, presently rouze and determine it to action.

7. There are some ferments that appear to be heterogeneous, or which are improperly called ferments; as the white of an egg beat into a frost, which is used when the liquor to be fermented proves too dilute or thin to sustain the operation. For in this case the fermentable parts of the fluid easily extricate themselves, and so fly off for want of something to detain and keep them in the body of the liquor; which therefore requires some viscid substance to be mixed with it, in order to prevent this avolation of its subtile parts. And this cannot be more commodiously effected, than by the white of an egg.

8. Of the like heterogeneous kind of ferments are all fixed and acid salts. Thus, if the liquor designed for fermentation be too acid to work kindly, the addition of an alkaline salt, as that of Vine branches, or any saponaceous substance, will, by taking off from the acidity, fit it for and so promote the operation; but if the liquor be of itself too alkaline, then tartar, or the like, ought to be added to it, to promote the fermentation.

But this does not happen, because either the acid or alkaline salt is an actual ferment, as some chymists have vehemently contended for the alkaline, because the salts employed respectively temper and take down the predominant acid or alkali, which before hindered the fermentation of the liquor.

And if such salts should in due quantities be mixed with any proper subject of fermentation, possessed of all the qualities before set down, as requisite to it, the operation would be entirely checked and prevented; so that alkaline bodies may as well be said to hinder as promote fermentation.

9. And lastly: Of the same sort are certain austere or rough tasted substances, as all harsh and green fruit, Pomegranate bark and flowers, the Tamarisk bark, Crab Apples, unripe Medlars, &c. which, when the liquor designed for the fermentation is too much broken in its parts, or dissolved in its texture, bind it together again by its astringent quality; so that though it was before too thin and aqueous, it is now reduced to a proper consistence for fermentation.

Thus, when must proves thin and watery, it will not ferment kindly, unless some austere or astringent ingredient, as red Rose leaves, or the like, be added to it, to thicken and improve its consistence, and at the same time prevent the air it contains from making too easy an escape.

But when a liquor is too austere, or its roughness proves so great, that it cannot ferment, the addition of a fixed alkali, in a proper quantity, will remove the obstruction, and leave it at liberty to work.

So likewise when the operation is prevented by too large a proportion of acid in the liquor, the method is to throw chalk, crabs-eyes, bole armoniack, or the like, into it; but if it be too unctuous or oily, as is the case of some Spanish Wines, salt of tartar is made choice of; and thus, as circumstances alter, different bodies are employed to stop or promote fermentation in liquors.

In order for fitting the subjects of the second class for fermentation, and making vinous liquors, viz. pulpy summer fruits, and the roots of bulbous plants, in case they prove crude or hard, they are to be first boiled in water, and afterwards bruised, which will dispose them for fermentation; but if such subjects are juicy, they may be directly ground to a pulp, or have their juice pressed from them; or if they are very succulent, there may be no occasion to bruise them, only directly to commit them to the press, and squeeze out all their juice.

But if the flesh or substance be strong and tough, it may be proper to rasp, shave, or cut them into small pieces, which will be of service in some bulbous roots, and make them yield their juice with the greater ease, and in greater plenty

Prepared fruits seldom stand in need of any thing to make them ferment, for they generally begin to work of their own accord; but if the weather should prove exceeding cold, or the operation proceed but languidly, it may not be amiss to quicken it by adding a small proportion of a ferment, as a little yeast, the lees or mother of Wine, or even a little new Wine may serve the turn.

The subjects of the third class, viz. the succulent parts of plants, need only, in order to their fermentation, be beat to a thick kind of pulp, while they are fresh, and mixed with a proper porportion of rain-water, that is just enough to dilute them, for if much water be employed, the spirit will be the weaker for it.

These require but very little ferment, or none at all, to make them work in the summer season, and no large proportion in the winter; but in case any at all be required, nothing will prove more serviceable than honey or sugar.

The subjects of the fourth and fifth classes, viz. the fresh native juices, and weeping liquors of vegetables, with the condensed and unctuous juices of the same, are to be diluted and let down with rain-water, to a due consistence, which is then thought to be obtained, when the compound liquor will just keep a new laid egg afloat; but some vegetable juices may naturally be of this very density or consistence, and in that case they will require no water at all. If any be thicker or denser, they ferment not so kindly, and if thinner or rarer, they afford but a weak spirit. Thus, in order to ferment sugar, treacle, or any common syrup, we first let down the matter with water, to the consistence above-mentioned; and then, if there be occasion, put yeast to it, to quicken the fermentation, and make it proceed kindly.

The subjects of the fourth class, viz. the prepared recent juices, and spontaneous tears of vegetables, are so far from requiring any ferment, that it often proves very difficult to restrain or check the fermentation they naturally fall into, especially if the season be warm and the juices rich; at most, if the weather should prove cold, they need only be set in a warm place to make them work.

The

The ſubjects of the fifth claſs, viz. the prepared or inſpiſſated juices of vegetables, require no ferment at all in the ſummer, and but a ſmall proportion in winter, to ſet them on working; leſs than an ounce of yeaſt to twenty pints of prepared liquor, will uſually do for that purpoſe in the coldeſt ſeaſon; but in hot countries, or ſultry ſeaſons, theſe prepared juices, and eſpecially ſugar, are of themſelves apt to fall into a too violent fermentation; which therefore ought to be abated by the contrary means.

All the vegetable bodies of the ſeveral claſſes deſigned for fermentation, and prepared for it in the foregoing manner, ought, together with their ferments, to be committed to caſks of Oak already ſeaſoned with the ſame kind of fermented liquor, or ſome other conſiſting of ſubtil and penetrating parts. Then thoſe caſks or veſſels having their bung-holes lightly covered with a thin or ſingle cloth, and being ſet in a warm place the liquor will ferment.

The mouths of the veſſels are thus ſlightly covered over, that the air may have a free paſſage in and out of them, for they are here deſigned to ſerve as vent-holes; and theſe veſſels are ordered of wood, becauſe fermentation is never obſerved to be ſo well carried on in thoſe of glazed earth or glaſs, though, on account of their tranſparency, it is ſometimes performed in the latter, that the phænomena may be better obſerved.

The preparatory buſineſs of fermentation hitherto deſcribed, has been carried on by art, but nature muſt now perform the reſt of the work; ſo that we are here only concerned to obſerve the phænomena which ariſe in the operation.

When therefore any fermentable body is prepared after the manner above delivered, and with its due proportion of a ferment committed to a large ſtrong glaſs veſſel ſtanding in a warm place;

1. The whole body of the liquor ſoon begins to ſwell, heave, rarefy, and ſend up little bubbles to the top of the veſſel, where they burſt with an audible noiſe, and form into froth. Now the liquor which was before tranſparent grows opaque, and a violent uninterrupted inteſtine motion manifeſts itſelf therein.

2. The parts of the fermenting fluid appear incredibly elaſtic, and the motion of them exceeding violent. Indeed, by means of this property of fermentation, very terrifying and ſurpriſing actions may be performed. Thus, if a hundred pints of muſt were, on ſome warm day in autumn, to be confined cloſe in a veſſel of Oak above an inch thick in the ſides, and made ever ſo tight and ſtrong with iron hoops, yet could not this prevent the working of the liquor; but in ſpite of ſo great a reſiſtance, it would burſt the veſſel with a report as loud as that of a cannon.

And therefore the way to preſerve new Wine in the ſtate of muſt, is to put it up in very ſtrong but ſmall caſks, firmly cloſed on all ſides, by which means it will be kept from fermenting, and then it goes by the name of ſtum; but if it ſhould happen to fall into fermentation, the readieſt and only way to ſtop it is by the fume of ſulphur, or ſomething of the like nature.

Were it not for the knowledge of this property of burning ſulphur, the Wine-merchants and vintners might frequently ſuſtain great damages from the burſting of the veſſels when the liquor is upon the fret, or, by ſome alteration in the air, or other accident, begins to ferment again; but the ſmoke of a little common brimſtone, or a lighted match dipped in it, and held under a caſk of Wine that is juſt ready to burſt its hoops, will calm its fury, and make it ſubſide as ſuddenly as a ſpoonful of oil thrown into a large foaming copper of boiling ſugar, takes down its heat, and prevents the miſchief it might otherwiſe occaſion.

3. A thick ſkin or cruſty ſcurf forms itſelf on the ſurface, through which the elaſtic or fermenting matter is continually breaking. This cruſt appears to be the principal cauſe of fermentation, for it keeps in, or prevents the ſpirituous part of the liquor from flying off; and if it be frequently broken, it puts a check to the fermentation, and will often entirely ſtop it if wholly taken away.

4. This ſkin or cruſt, which we now call flowers or yeaſt, gradually conſumes and precipitates to the bottom of the liquor, in which caſe it is called by the name of fæces or mother; and after this, the fluid above it immediately becomes tranſparent again, ceaſes to hiſs and bubble, has a very penetrating, pungent, ſpirituous, or vinous taſte and ſcent, with a mixture of acidity and ſweetneſs. And now the liquor, having undergone the operation of fermentation, is become Wine.

The vapour ariſing from the liquor during its fermentation, ought not to be approached too near, or breathed in too great a quantity, becauſe it is highly poiſonous; and if it prove not mortal, may at leaſt render the perſon apoplectic and paralytic. We have accounts in the French and German Tranſactions, of people who were immediately ſtruck dead, by receiving at the noſe the fumes that iſſued from large veſſels of Wine in the ſtate of fermentation.

And now if the liquor thus fermented be ſtopped down cloſe, it will begin to feed upon and digeſt its own lees and mother, and at length conſume them; in which caſe we commonly ſay the Wine begins to ripen, and afterwards this mother ſhoots to the ſides of the containing veſſel, and there appears in the form of an eſſential ſalt, which is then called tartar.

The ſpace of time required for finiſhing the fermentation differs with the ſubject matter, the ſeaſon of the year, the nature of the place, and other circumſtances; but it is known to be perfectly performed by the ſeveral phænomena juſt now mentioned.

As ſoon as the flowers fall to the bottom, the veſſel ſhould be bunged down, otherwiſe the volatile part would fly off, and the fermented liquor become vapid and flat.

In this ſtate it ought to ſtand for ſome weeks in a cool place, by which means it will grow ſtronger and more liquid; for during this time it imbibes and conſumes its own fæces, which abound in ſubtile ſpirituous parts, and grows ſoft and loſes of its acidity by throwing off its tartar.

And the longer it is thus ſuffered to ſtand, the more ſtrength it gains, or the more ſpirit it will yield in diſtillation.

Thus, for inſtance, malt liquors newly brewed afford but a ſmall quantity of inflammable ſpirit; but if ſuffered to remain for ſome weeks in the veſſel, till they become fine and clean, they will yield much greater proportion; though, to avoid ſo great an apparatus of veſſels as would then be required, malt liquors brewed, in order to make ſpirits, are ſeldom kept, but immediately after fermentation committed to the ſtill. And hence we are furniſhed with a reaſon why all ſtale, vinous liquors are ſtronger, and inebriate ſooner, than ſuch as are new.

Some ſhort general directions as to the making of Wines.

Wine is made of Grapes, by ſtamping them in a vat, or cruſhing and expreſſing the juice out of them in a preſs, and then fermenting, &c.

In the ſouthern parts of France their method is, for red Wines, to tread the Grapes, or ſqueeze them between their hands, and let the whole ſtand, juice and huſks, till the tincture be in colour as they would have it, and then they preſs it; but for white Wines they preſs the Grapes immediately.

When

When they have been pressed they tun the must and stop up the vessel, leaving the cask empty about the depth of half a foot, or better, to give room for its working.

At the end of ten days they fill this space with some other proper Wine, that will not provoke it to work again, repeating this every ten days for some time: new Wine spending itself a little before it is perfect.

About Paris, and in the northern parts of France, they let the marc and must stand two days and nights for white Wines, and at least a week for claret Wines, before they tun it, and while it continues working they keep it as warm as possible.

Some, upon stopping it up for good and all, roll the cask abou the cellar to mix it with the lees, and after it has been settled a few days rack it off with great improvement.

To fine it down they put shavings of green Beech into the cask; but they first take off all the rind, and boil them an hour in water to extract their rankness, and afterward dry them in the sun or an oven. A peck of these will serve for a hogshead of Wine; they put it in a gentle working, and purify it in twenty-four hours; they also give it an agreeable flavour.

Some sweeteen their Wines with Raisins of the sun, trod in the vat with the Grapes, they having been first plumped by boiling; others by boiling half the must, scumming it, and tunning it up hot with the other.

Wine is distinguished, from the several degrees and steps of its preparation, into

1. Mere-goute, mother-drop, which is the virgin Wine, or that which runs of itself out of the tap of the vat, before the Grapes are trodden.

2. The must, surmoust, or scum, which is the Wine or liquor in the vat, after the Grapes have been trodden in the vat.

3. The pressed Wine, or Vin de Pressurage, which is that squeezed with a press out of the Grapes half-bruised by treading.

4. Boisson, or draught Wine. This is made of the husks left of the Grapes, which are called rape or marc, which, by throwing water upon, and pressing afresh, they make a liquor for servants.

Wines are also distinguished into

Vin doux, or sweet Wine, which is that which has not yet worked nor boiled.

Bourou; that which has been prevented working by casting in cold water.

Wine of the cuve, or worked Wine, i. e. that which has been let to work in the vat to give it a colour.

Vin cuit, i. e. boiled Wine; that which has had a boiling before it worked, and which, by that means, still retains its native sweetness.

Vin passé, i. e. strained Wine; that which is made by steeping dry Grapes in water, and letting it ferment of itself.

The goodness of Wine consists in its being neat, dry, clear, fine, brisk, without any taste of the soil, of a clean steady colour; in its having a strength, without being heady; a body, without being sour; and its keeping, without growing hard.

After Wines have been made, they require to be managed according to their different state and circumstances. We shall therefore consider them under these four general heads following:

1. The natural purification or clarification of Wines, whereby, of themselves, they pass from the state of crudity and turbulency to that of maturit, by degrees growing clear, fine, and potable.

2. The unseasonable workings, frettings, and other sicknesses, to which, from either internal or external accidents, they are afterward subject.

3. Their state of declination or decay, wherein they degenerate from their goodness and pleasantness, becoming palled, or turning into vinegar.

4. The several artifices used to them in each of these states and conditions. As to the first, viz. the natural clarification of new Wines, two things occur which deserve consideration; the manner how, and the cause by which, the same is effected.

As for the manner, it is to be observed that Wine, while yet in the must, is usually put into open vessels, the abundance and force of the spirits, i. e. the more subtile and active parts therein contained being then so great, as not to endure being imprisoned in close ones; at which time it appears troubled, thick, and feculent, all parts of it being violently moved and agitated, so that the whole mass of the liquor seems to boil like water in a caldron over the fire.

This tumult being in some degree composed, and the gas sylvestre (as Van Helmont calls it,) or wilder spirit, sufficiently evaporated, they then pour the must into close vessels, there to be farther defecated by continuance of the same motion of fermentation, reserving the froth or flower of it, and putting the same into small casks hooped with iron, lest otherwise the force of it might break them.

This flower, thus separated, is what they call stum, either by transposition of the letters in the word must, or from the word stum, which in High Dutch signifies mute; because this liquor (as one may say) is hindered from that maturity, by which it should speak its goodness and wholesomeness.

This being done, they leave the rest of the Wine to finish its own fermentation; during which it is probable, that the spirituous parts impel and diffuse the grosser and feculent parts up and down in a confused and tumultuous manner, until, all being disposed in their proper regions, the liquor becomes more pure in substance, more transparent to the eye, more piquant and gustful to the palate, more agreeable to the stomach, and more nutritive to the body.

The impurities, being thus separated from the liquor, are, upon chemical examinations, found to consist of salt, sulphur (each of which is impregnated with some spirits,) and much earth, which, being now dissociated from the purest spirits, either mutually cohere, coagulate, and affix themselves to the sides of the vessels, in form of a stony crust, which is called tartar and argol, or sink to the bottom in a muddy substance, like the grounds of ale or beer, which is called the lees of Wine. And this is the process of nature in the clarification of all Wines, by an orderly fermentation.

As for the principal agent or efficient cause of this operation, it seems to be no other but the spirit of the Wine itself; which, moving every way in the mass of the liquor, thereby dissolves that common tie of mixture, whereby all the heterogeneous parts thereof were combined and blended together; and having gotten itself free, at length abandons them to the tendency of their gravity and other properties, which, they soon obeying, each kind consorts with its like, and betaking themselves to their several places or regions, leave the liquor to the possession and government of its noblest principle, the spirit. For this spirit, as it is the life of the Wine, doubtless is also the cause of its purity and vigour, in which the perfection of that life seems to consist.

From the natural fermentation of Wines we pass to the accidental; from their state of soundness to that of their sickness, which is the second general head.

We have the teftimony of experience, that frequently even thofe Wines that are good and generous are invaded by unnatural and fickly commotions, or (as the Wine coopers call them) workings; during which they are turbulent in motion, thick of confiftence, unfavoury in tafte, unwholefome in ufe, after which they undergo fundry alterations for the worfe.

The caufes of this may be either internal or external.

Among the internal, the chief place may be affigned to the exceffive quantity of tartar or of lees, which contain much falt and fulphur, and continually fend forth into the liquor abundance of quick and active particles, that, like ftum, or other adventitious ferment, put it into a frefh tumult or confufion, which, if not in time allayed, the Wine either grows rank or pricking, or elfe turns four, by reafon that the fulphur, being too much exalted above the reft of the elements or ingredients, predominates over the pure fpirits, and infects the whole mafs of liquor with fharpnefs or acidity; or elfe it comes to pafs, that the fpirits being fpent and flown away in the commotion, the falt, diffolved and fet afloat, obtains the maftery over the other fimilar parts, and introduceth ranknefs or ropinefs.

Nay, if thofe commotions chance to be fuppreffed before, the Wine is thereby much depraved, yet do they always leave fuch ill impreffions, as, more or lefs, alienate Wine from the goodnefs of its former ftate, in colour, confiftence, and tafte.

For hereby all Wines acquire a deeper tincture, i. e. a thicker body or confiftence, facks and white Wines changing from a clear white to a cloudy yellow; and claret lofing its bright red for a dufkifh Orange colour, and fometimes for a tawney. In like manner they degenerate alfo in tafte, and affect the palate with foulnefs, roughnefs, and rancidity, very unpleafant.

Among the external are commonly reckoned the too frequent or violent motion of Wines after their fettlement in their veffels, immoderate heat, thunder, or the report of cannon, and the admixture of any exotic body, which will not fymbolize, or agree and incorporate with them; efpecially the flefh of vipers, which has been frequently obferved to induce a very great acidity upon even the fweeteft and fulleft-bodied Malaga and Canary Wines; or by putting new Wines in a ftate of fermentation into vaults with old Wines, in them more or lefs according to their different ages, but in all enough to make it turbid.

This brings us in the next place to the third previous thing confiderable, viz. the palling or flatting of Wines, and their declining towards vinegar, before they have attained to their ftate of maturity and perfection.

Of this the greateft and neareft caufe feems to be their jejunenefs and poverty of fpirits, either native or adventitious:

Native, when the Grapes themfelves are of a poor and hungry kind, or gathered unripe, or nipt by early frofts, or half ftarved in their growth by a dry and unkindly feafon, or too full of watery parts:

Adventitious, when the liquor, rich perhaps and generous enough at firft, comes afterwards to be impoverifhed by lofs of fpirits, either by oppreffion, or by exhauftion.

The fpirits of Wine may be oppreffed, when the quantity of impurities or dregs, with which they are combined, is fo great, and their crudity, vifcofity, and tenacity fo ftubborn, that they can neither overcome them, nor deliver them from their adhefion; but are forced to yield to the obftinacy of the matter on which they fhould operate, and fo to remain unactive and clogged, as may be exemplified in the coarfe Wines of Moravia, which, by reafon of their great aufterity and roughnefs, feldom attain to a due exaltation of their fpirits, but ftill remain turbulent, thick, and in a ftate of crudity, and therefore eafily pall; in which refpect they are condemned by fome German phyficians as bad for generating the fcurvy, and adminiftring matter for the ftone and gout, they yielding more of tartar than other Wines.

The fpirits of Wine may be exhaufted or confumed, either fuddenly or gradually; fuddenly, by lightening, which fpoils Wine, not by congelation or fixation of its fpirits; for then fuch Wines might be capable of being reftored by fuch means as are apt to reinforce and volatilize the fpirits again, contrary to what has been found by experience; but perhaps by difgregation, and putting them to flight, fo as to leave the liquor dead, palled, and never to be revived by any fupply.

Gradually, two ways, viz. by unnatural fermentation, of the ill effects of which fomething has already been faid; or by heat from without, of which we have an inftance in the making of vinegar; which commonly is done by fetting the veffels of Wine againft the hot fun, which, beating upon the mafs of liquor, and rarifying the finer parts thereof, gives wings to the fugitive fpirits to fly away together with the purer and more volatile fulphur, leaving the remainder to the dominion of the falt, which foon debafeth and infecteth it with fournefs.

This being the common manner of turning Wine int vinegar, in all ages and all countries, it may be doubted whether fpirit of Wine may be drawn out of vinegar, notwithftanding it hath been delivered as practicable by Senuertus himfelf.

The times of the year when Wines are obferved to be moft prone to ferment and fret, and then to grow qually (as it is called,) that is, turbulent and foul, are Midfummer and Allhallowtide, when our vintners are wont to rack them from their grofs lees, efpecially Rhenifh, which commonly grows fick in June if not racked; and they chufe to do it in the wane of the moon, and fair weather, the wind being northerly.

Having thus fuccinctly recounted the moft remarkable diftempers of Wines, gueffed at their refpective caufes, and touched upon the times, it is proper to proceed to their ufual remedies, fuch, at leaft, as may be collected from Wine coopers and vintners; which is the fourth and laft part propofed to be treated of.

To begin, therefore, with fome of the artifices ufed to Wines when yet in muft: it is obfervable, that though to raifing a fermentation in them at that time, there is not fo much need of any additional ferment, as there is in the wort of ale, beer, hydromel, metheglin, and other forts of drinks, familiar to us in England; becaufe the juice of the Grape is replenifhed with generous fpirits, fufficient of themfelves to begin that work; yet it is ufual in fome countries to put quick lime either upon the Grapes, when they are preffing, or into the muft; to the end that, by the force and quicknefs of its faline and fiery particles, the liquor may be both accelerated and affifted in working.

For the fame reafon, perhaps, it is that the Spaniards mix with their Wines, while they are yet flowing from the prefs, a certain thing they call gieffo, which probably is a kind of gypfum or plafter, whereby the Wines are made more durable, of a paler colour, and pleafanter tafte; others put into the cafk fhavings of Fir, Oak, or Beech, for the fame purpofe.

Again; though the firft fermentation fucceeds generally well, fo that the whole mafs of liquor is thereby delivered from the grofs lee; yet fometimes it happens either through fcarcity of fpirits at firft, or through immoderate cold, that fome part of thofe impurities remain confufed and floating therein.

Now,

Now, in this case, Wine-coopers put into the Wine certain things to hasten and help its clarification; such as, being of gross and viscous parts, may adhere to the floating lee, and, sinking, carry it with them to the bottom; of which sort are isinglass and the whites of eggs, or such as, meeting with the grosser and earthly particles of the lee, dissociate and sink them by their gravity; of which kind are the powders of alabaster, calcined flints, white marble, roach allum, &c.

The Grecians, at this day, have a peculiar way of spurring nature, in fining and ripening their strongest and most generous Wines; and this is done by adding to them, when they begin to work, a proportionate quantity of sulphur and allum; not (as is very probable) to prevent their fuming up to the head, and inebriating, according to the conjecture of that great man the Lord of St. Alban's, for notwithstanding this mixture, they cause drunkenness as soon, if not sooner, than other Wines; nor are men intoxicated by the vapours of Wine flying up immediately from the stomach into the brain, but only to excite and promote fermentation, and hasten their clarification that ensues thereupon; the sulphur perhaps, helping to attenuate and divide those gross and viscid parts, wherewith Greek Wine abounds, and the allum conducing to the speedier precipitation of them afterwards. And a learned traveller relates, that some merchants put into every pipe of their Greek Wine a gill, or thereabouts, of the chemical oil of sulphur, in order to preserve it the longer clear and sound; which, though it is very probable, because the sulphur is known to resist putrefaction in liquors, yet one would decline the use of Wines so preserved, unless in time of pestilential infection.

But of all ways of the hastening the clarification and ripening of Wine, none seems to be more easy, or less noxious, than that borrowed from one of the ancients by the Lord Chancellor Bacon; which is, by putting the Wine into vessels well stopped, and letting it down into the sea.

That this practice was very ancient is manifest from that discourse of Plutarch, Quæst. Natur. 27. about the efficacy of cold upon must; whereof he gives this reason, that cold, not suffering the must to ferment, by suppressing the activity of the spirits therein contained, conserveth the sweetness thereof a long time; which is not improbable, because experience teaches, that such as make their vintage in a rainy season cannot get their must to ferment well in a vault, unless they cause great fires to be made near the casks; the rain mixed with the must, together with the ambient cold, hindering the motion of fermentation, which arises chiefly from heat.

That the same is frequent at this day also, may be collected from what Mr. Boyle has observed in his History of Cold, on the relation of a Frenchman, viz. that the way to keep Wine long in the must (in which the sweetness makes many to desire it) is to tun it up immediately from the press, and, before it begins to work, to let down the vessels, closely and firmly stopped, into a well or deep river, there to remain for six or eight weeks; during which time the liquor will be so confirmed in its state of crudity as to retain the same, together with its sweetness, for many months after, without any sensible fermentation.

But it may be objected, how can these two so different effects, the clarification of new Wine, and the conservation of Wine in the must, be derived from one and the same cause, the cold of the water?

But this may be conceived without much difficulty; for it seems not unreasonable, that the same cold which hinders must from fermenting, should yet accelerate and promote the clarification of Wine after fermentation; in the first, by giving a check to the spirit, before it begins to move and act upon the crude mass of liquor, so that it cannot in a long time after recover strength enough to work: in the latter, by keeping in the pure and genuine spirit, otherwise apt to exhale, and rendering the flying lee more prone to subside, and so making the Wine much sooner clear, fine, and potable. Thus much concerning the helps of new Wine.

The general and principal remedy for the preternatural or sickly commotions incident to Wines after their first clarification, and tending to their impoverishment or decay, is racking, i. e. drawing them from their lees into fresh vessels.

Which yet being sometimes insufficient to preserve them, vintners find it necessary to pour into them a large quantity of new milk as well to blunt the sharpness of the sulphureous parts now set afloat and exalted, as to precipitate them and other impurities to the bottom by adhesion.

But, taught by experience, that by this means the genuine spirits of the Wine also are much flatted and impaired (for the lee, though it makes the liquor turbid, doth yet keep the Wine in heart and conduce to its duration;) therefore lest such Wines should pall and die upon their hands, as of necessity they must, they draw them for sale as fast as they can vend them.

For the same disease they have divers other remedies, particularly accommodated to the nature of the Wine that needs them: to instance a few;

For Spanish Wines disturbed by a flying lee they have this receipt: Make a parell (as they call it) of the whites of eggs, bay salt, milk, and conduit water; beat them well together in a convenient vessel, then pour them into a pipe of Wine, (having first drawn out a gallon or two to make room,) and blow off the froth very clean; hereby the tumult will in two or three days be composed, the liquor refined and drink pleasantly, but will not continue to do so long; and therefore they advise to rack it from the milky bottom after a week's settlement, lest otherwise it should drink foul and change colour.

If sacks or Canary Wines chance to boil over, draw off four or five gallons; then putting into the Wine two gallons of milk, from which the cream hath been skimmed, beat them till they are thoroughly mixed together, and add a pennyworth of roach allum, dried in a fire-shovel and powdered, and as much of white starch; after this take the whites of eight or ten eggs, a handful of bay salt, and having beaten them together in a tray put them also into the Wine, filling up the pipe again, and letting the Wine stand two or three days, in which time the Wine will recover to be fine and bright to the eye, and quick to the taste; but you must be sure to draw it off that bottom very soon, and spend it as fast as you can.

For claret, in like manner distempered with a flying lee, they make use of this artifice:

They take two pounds of the powder of pebble-stones baked in an oven, the whites of ten or twelve eggs, a handful of bay salt, and having beaten them well together, in two gallons of the Wine, they mix them with that in the cask, and after two or three days draw off the Wine from the bottom.

The same parell serves also for white Wines upon the fret, by the turbulency and rising of their lee.

To cure Rhenish of its fretting (to which it is most prone a little after Midsummer, as was before observed,) they seldom use any other art but giving it vent, and covering the oaken bung with a tile or slate, from which they carefully wipe off the filth purged from the Wine by exhalations; and after the commotion is by this means composed, and much of the fretting matter cast forth, they let it remain quiet for a fortnight, or thereabouts, and then rack it into a fresh cask, newly fumed with a sulphurated match.

As for the various accidents that frequently ensue and vitiate Wine (after those before-mentioned reboilings, notwithstanding their suppression before they were incurable;) you may remember they have all been referred to such as alter and deprave Wines, either in colour or consistence, or taste or smell. Now for each of these maladies our vintners are provided of a cure.

To restore Spanish and Austrian Wines grown yellow or brownish, they add to them sometimes milk alone, and sometimes milk, and isinglass well dissolved therein; sometimes milk and white starch; by which they force the exalted sulphur to separate from the liquor, and sink to the bottom, so reducing the Wine to its former clearness and whiteness.

The same effect they produce with a composition of Iris-roots and salt-petre, of each four or five ounces, the whites of eight or ten eggs, and a competent quantity of common salt, mixed and beaten in the Wine.

To amend claret decayed in colour, first they rack it upon a fresh lee, either of Alicant or red Bourdeaux Wine; then they take three pounds of Turnsole, and steep it all night in two or three gallons of the same Wine, and having strained the infusion through a bag, they pour the tincture into a hogshead (sometimes they suffer it first to fine itself in a rundlet) and then cover the bung-hole with a tile, and so let it stand for two or three days, in which time the Wine usually becomes well-coloured and bright.

Some fuse only the tincture of Turnsole.

Others take half a bushel of full-ripe Elder-berries, pick them from their stalks, bruise them, and put the strained juice into a hogshead of discoloured claret, and so make it drink brisk, and appear bright.

Others, if the claret be otherwise sound, and the lee good, overdraw three or four gallons; then replenish the vessel with as much good red Wine, and roll it upon its bed, leaving it reversed all night; and then next morning they turn it again, so as the bung-hole may be uppermost; which stopped, they leave the Wine to fine.

But in all these cases they observe to set such newly recovered Wines abroach the very next day after they are fined, and to draw them for sale speedily.

To correct Wines faulty in consistence, i. e. such as are lumpish, foul, or ropy;

They generally make use of the powders of burnt allum, lime-chalk plaster, Spanish white, calcined marble, bay salt, and other the like bodies, which cause a precipitation of the gross and viscid parts of the Wine then afloat: as for example;

For attenuation of Spanish Wines that are foul and lumpish, having first racked them into a newly-scented cask, they make a parell of burnt allum, bay salt, and conduit water; then they add to these a quart of Bean-flower, or powder of Rice; and if the Wine be brown and dusky, milk, otherwise not; and beating all these well together with the Wine, blow off the froth, and cover the bung with a clean tile or stone. Lastly, they rack the Wine again after a few days, and put it into a cask well scented.

The manner of scenting casks is as follows:

They take four ounces of brimstone, one ounce of burnt allum, and two ounces of aqua vitæ; these may be put together in an earthen pan or pipkin, and hold them over a chafing-dish of glowing coals, till the brimstone is melted and runs; then they dip therein a little piece of new canvas, and instantly sprinkle thereon the powders of Nutmegs, Cloves, Corianders, and Anise-seeds. This canvas they fire, and let it burn out in the bung-hole, so as the fume may be received into the vessel; and this is said to be the best scent for all Wines.

To prevent the foulness and ropiness of Wines, the old Romans used to mix sea-water with their must.

To cure the ropiness of claret, the vintners, as wel French as English, have many remedies; of which these tha follow are the most usual:

First they give the Wine the parell, then draw it from the lee, after the clarification by that parell; this done, they infuse two pounds of Turnsole in good sack all night; and the next day, putting the strained infusion into a hogshead of Wine with a spring funnel, leave it to fine, and after draw it for excellent Wine.

Another is this: they make a lee of the ashes of Vine branches, or of Oaken leaves, and pour it into the Wine hot, and after stirring leave it to settle; the quantity of a quart of lee to a pipe of Wine.

A third is only a spirit of Wine, which, put into a muddy claret, serves to the refining it effectually and speedily; the proportion being a pint of spirit to a hogshead; but this is not to be used in sharp and eager Wines.

When white Wines grow foul and tawny, they only rack them on a fresh lee, and give them time to fine.

For the mending of Wines that offend in taste, vintners have few other correctives, but what conduce to clarification; nor do they indeed much need variety in the case, seeing all unsavouriness of Wines whatever proceeds from their impurities set afloat, and the dominion of others, their sulphureous or saline parts, over the finer and sweeter; which causes are removed chiefly by precipitation.

For all clarification of liquors may be referred to one of these three causes:

1. Separation from the grosser parts of the liquor from the finer.

2. The equal distribution of the spirits of the liquor, which always render bodies clear and untroubled.

3. The refining of the spirit itself.

And the two latter are consequents of the first, which is effected chiefly by precipitation, the instruments whereof are weight and viscosity of the body mixed with it; the one causing it to cleave to the gross parts of the liquor flying up and down in it, the other sinking them to the bottom.

But this being more than vintners commonly understand, they rest not in clarification alone, having found out certain specifics, as it were, to palliate the several vices of Wines of all sorts, which make them disgustful. Of these I shall recite two or three of the greatest use and esteem amongst them.

To correct rankness, eagerness, and pricking of sacks, and other sweet Wines, they take twenty or thirty of the whitest lime-stones, and slack them in a gallon of the Wine; then they add some more Wine, and stir them together in a half tub, with a parelling staff; next they pour this mixture into the hogshead, and having again used the parelling instrument, leave the Wine to settle and then rack it.

This Wine may probably be no ill drink for gross bodies and rheumatic pains, but injurious to good fellows of a hot and dry constitution and meagre habits.

Against the pricking of French Wines they prescribe this easy and cheap composition: Take of the powder of Flanders tile one pound, of roach allum half a pound, mix them and beat them well, with a convenient quantity of Wine; then put them into the hogshead, as the former.

When their Rhenish Wines prick, they first rack them off into a clean and strongly-scented cask or vat, then they add to the Wine eight or ten gallons of clarified honey, with a gallon or two of skim-milk, and, beating all together, leave them to settle.

Sometimes it happens, that claret loses much of its briskness and piquantness; and in such case they rack it upon a good lee of red Wine, and put into it a gallon of Sloes or Bullace, which, after a little fermentation and rest, makes the Wine drink brisk and rough.

To

To meliorate the taſte of hungry and too eager white Wines, they draw off three or four gallons of it, and infuſing therein as many pounds of Malaga Raiſins ſtoned, and bruiſed in a ſtone-mortar, till the Wine has ſufficiently imbibed their ſweetneſs and tincture (which it will do in a day's time,) they run it through an Hippocras bag; then put it into a freſh caſk well ſcented, together with the whole remainder of the Wine in the hogſhead, and ſo leave it to fine.

To help ſtinking Wines, the general remedy is racking them from their old and corrupt lee; beſides which, ſome give them a fragrant ſmell or flavour, by hanging in them little bags of ſpices, ſuch as Ginger, Zedoary, Cloves, Cinnamon, Orris-roots, Cubebs, Grains of Paradiſe, Spikenard, and other aromatics.

Others boil ſome of theſe ſpices in a pottle of good ſound Wine of the ſame ſort, and tun up the decoction hot.

Others correct the ill ſavour of rank-leed French Wine with only a few Cinnamon canes hung in them.

Others again, for the ſame purpoſe, uſe Elder-flowers and tops of Lavender.

Having thus run over the Vintners Diſpenſatory, and deſcribed many of their principal receipts or ſecrets, for the cure of the acute diſeaſes of Wine, we ſhall come to the fourth head, which contains medicaments proper for their chronic diſtempers, viz. loſs of ſpirits, and decay of ſtrength.

Concerning theſe, therefore, it is obſervable, that as when Wines are in preternatural commotions, from an exceſs and predomination of their ſulphureous parts, the grand medicine is to rack them from the lees; ſo on the contrary, when they decline and tend towards palling, by reaſon of the ſcarcity of their ſpirits and ſulphur, the moſt effectual preſervative is to rack them upon other lees, richer and ſtronger than their own; that being from thence ſupplied with the new ſpirits, they may acquire ſomewhat more of vigour and quickneſs.

I ſay preſervative; becauſe there is, in truth, no reſtoring of Wines after they are perfectly palled and dead, for nothing that is paſt perfection, and hath run its natural race once, can receive much amendment.

But beſides reinforcing of impoveriſhed Wines, by new and more generous lees, there are ſundry confections, by which alſo, as by cordials, the languiſhing ſpirits of many of them may be ſuſtained, and, to ſome degree, recruited, of which the following are examples:

When ſacks begin to languiſh (which doth not often happen, eſpecially in this city, where it is drank in plenty,) they refreſh them with a cordial ſyrup, made of moſt generous Wine, ſugar, and ſpices.

For Rheniſh and white Wines, a ſimple decoction of Raiſins of the ſun, and a ſtrong-ſcented caſk, uſually ſerve the turn.

For claret inclining to a conſumption, they preſcribe a new and richer lee, and the ſhavings of Fir-wood, that the ſpirit, being recruited by the additional lee, may be kept from the exhaling by the unctuous ſpirit of the turpentine.

This artifice is uſed in Paris in the moſt delicate and thin-bodied Wines of France, and is very probably the cauſe of that exceeding dulneſs and pain of the head, which always attends debauches with ſuch Wines.

Nor is it a modern invention, but well known to, and frequently uſed by the Romans, in the time of their greateſt wealth and luxury; for Pliny (Hiſt. Nat. lib. 14. cap. 2.) takes ſingular notice of the cuſtom of the Italian vintners, in mixing with their Wines turpentine of ſeveral ſorts.

The Grecians long before had their vina picata and reſinata, as is evident by the commendation of ſuch Wines by Plutarch, and the preſcription of them to women, in ſome caſes, by Hippocrates; and they were ſo much delighted with their vinum piſſites, that they conſecrated the Pitch-tree to Bacchus; but I ſhall next take ſome notice of the more diſingenuous practices of vintners in the tranſmutation or ſophiſtication, which they call trickings or compaſſings.

They transform poor Rochelle and Coniac white Wines into Rheniſh; Rheniſh into ſack; the lags of ſack and malmſeys into muſcadels.

They counterfeit Raſpie Wine with Fleur-de-lys roots; Verdea with decoctions of Raiſins; they ſell decayed Xeres, vulgarly ſherry, for Luſenna Wine; in all theſe impoſtures deluding the palate ſo nearly, that few are able to diſcern the fraud, and keeping theſe arcana ſo cloſe, that few can come to the knowledge of them.

As for their metamorphoſis of white into claret, by daſhing it with red, nothing is more commonly either done or known.

For their converſion of white into Rheniſh, they have ſeveral artifices to effect it, among which this is the moſt uſual:

They take a hogſhead of Rochelle or Coniac, or Nantz white Wine; rack it into a freſh caſk ſtrongly ſcented; then give the white parell; put into it eight or ten gallons of clarified honey, or forty pounds of coarſe ſugar, and, beating it well, leave it to clarify.

To give this mixture the delicate flavour, they ſometimes add the decoction of the yellow Clary-flowers, or Galitricum, of which drugs there is an incredible quantity uſed yearly at Dort, where the ſtaple of Rheniſh Wines was; and this is that drink with which the Engliſh ladies were wont to be ſo delighted, under the ſpecious name of Rheniſh in the muſt.

The manner of making adulterate baſtard is thus:

Take four gallons of white Wine, three gallons of old Canary, five pounds of baſtard ſyrup; beat them well together, put them into a clean rundlet well ſcented, and give them time to fine.

Sack is made of Rheniſh, either by a ſtrong decoction of Malaga Raiſins, or by a ſyrup of ſack, ſugar, and ſpices.

Muſcadel is ſophiſticated with the lags of ſack or Malmſey thus:

They diſſolve it in a convenient quantity of Roſe-water, of muſk two ounces, of Calamus Aromaticus powdered one ounce, of Coriander beaten half an ounce, and while this infuſion is yet warm, they put it into a rundlet of old ſack or Malmſey, and this they call a flavour for Muſcadel.

There are many other ways of adulterating Wines in this city; but becauſe they all tend to the above-mentioned alterations, and are not ſo general, I ſhall paſs them over, and mention the obſervations of a certain curious author on this ſubject.

The myſtery of Wines conſiſts in the making and meliorating of natural Wines.

Melioration is either of ſound or vicious Wines. Sound Wines are bettered,

1. By preſerving.
2. By timely fining.
3. By mending colour, ſmell, and taſte.

1. To preſerve Wines, care muſt be taken that after the preſſing they may ferment well, for without good fermentation they become qually, i. e. cloudy, thick, and duſky, and will never fine themſelves, as other Wines do; and when they are fined by art, they muſt be ſpeedily ſpent, or elſe they will become qually again, and then will not be recoverable by any art.

To preſerve Spaniſh Wines, and chiefly Canary, and therefore principally that which is razie, which will not keep long, they make a layer of Grapes and Gieſſe, whereby it acquires a better durance and taſte, and a white colour, moſt pleaſing to the Engliſh.

Razie Wine is ſo called, becauſe it comes from Rheniſh Vine cuttings, ſometimes renewed. The Grapes of this Vine are fleſhy, yielding but a little juice.

The French and Rheniſh Wines are chiefly and commonly preſerved by the match, thus uſed at Dort in Holland:

They take twenty or thirty pounds of brimſtone, rack into it melted, as Cloves, Cinnamon, Mace, Ginger, and Coriander-ſeeds; and ſome, to ſave charges, uſe the reliques of the Hippocras bag, and, having mixed theſe well with the brimſtone, they draw through this mixture, long, ſquare, narrow pieces of canvas, which pieces they light, and put into the veſſel at the bung-hole, and preſently ſtop it cloſe: great care is to be had in proportioning the brimſtone to the quantity and quality of the Wine, for too much makes it rough. This ſmoking keeps the Wine long white and good, and gives it a pleaſant taſte.

There is another way for French and Rheniſh Wines, viz. firing it. It is done in a ſtove, or elſe a good fire made round about the veſſel, which will gape wide, yet the Wine never runs out. It will boil, and afterwards may ſoon be racked.

Secondly, For timely fining of Wines. All Wines in the muſt are more opacous and cloudy. Good Wine ſoon fines, and the groſs lees ſettle quickly, and alſo the flying lee in time. When the groſſer lees are ſettled, they draw off the Wine; this is called racking. The uſual times for racking are Midſummer and Allhallowtide.

The practice of the Dutch and Engliſh to rid the Wine of the flying lees ſpeedily, and which ſerves moſt for French and Spaniſh Wine, is thus performed:

Take of iſinglaſs half a pound; ſteep it in half a pint of the hardeſt French Wine that can be got, ſo that the Wine may fully cover it; let them ſtand twenty-four hours; then pull and beat the iſinglaſs to pieces, and add more Wine; four times a day ſqueeze it to jelly, and as it thickens add more Wine. When it is full and perfectly jellied, take a pint or quart to a hogſhead, and ſo proportionably; then overdraw three or four gallons of that Wine you intend to fine, which mix well with the ſaid quantity of jelly; then put this mixture to the piece of Wine, and beat it with a ſtaff, and fill it top full.

Note, That French Wines muſt be bunged up very cloſe, but not the Spaniſh; and that iſinglaſs raiſes the lees to the top of ſtrong Wines, but in weaker precipitates them to the bottom.

They mend the colour of ſound clarets by adding thereto red Wine, tent, or Alicant; or by an infuſion of Turnſole, made in two or three gallons of Wine, and then putting it into the veſſel, to be then (being well ſtopped) rolled for a quarter of an hour.

This infuſion is ſometimes twice or three times repeated, according as more colour is to be added to the Wine; about three infuſions of the Turnſole are ſufficient; but then it muſt be rubbed and wringed.

Claret over-red is amended with the addition of white Wines.

White Wines coming over ſound, but brown, are thus remedied:

Take of alabaſter powder, overdraw the hogſhead three or four gallons; then put this powder into the bung, and ſtir and beat it with a ſtaff, and fill it top full. The more the Wine is ſtirred, the finer it will come upon the lee, that is, the finer it will be.

To colour ſack-white: Take of white ſtarch two pounds, of milk two gallons, boil them together two hours; when cold, beat them well, with a handful of white ſalt, and then put them into a clean but ſweet butt, beating them with a ſtaff, and the Wine will be pure and white.

One pound of the before-mentioned jelly of iſinglaſs takes away the brownneſs of French and Spaniſh Wines, mixed with two or three gallons of Wine; according as it is brown and ſtrong, more or leſs to be uſed. Then overdraw the piece of Wine about eight gallons, and uſe the rod; then fill the veſſel full, and in a day or two it will be fine, and be white, and mend if qually.

The firſt buds of Ribes nigra, i. e. black Currants, infuſed in Wines, eſpecially Rheniſh, make it diuretic, and more fragrant in ſmell and taſte, and ſo doth Clary.

The inconvenience is that the Wine becomes more heady: a remedy for which is Elder-flowers added to the Clary, which alſo betters the fragrancy thereof, as it is manifeſt in Elder-vinegar; but theſe flowers are apt to make the Wine ropy.

To help brown Malagas and Spaniſh Wine: Take powder of Oris-roots and ſalt-petre, of each four ounces, the whites of eight eggs, to which add as much ſalt as will make a brine; put this mixture into Wine, and mix them with a ſtaff.

To meliorate muddy and tawny clarets: Take of rain-water two pints, the yolks of eight eggs, ſalt a handful; beat them well, let them ſtand ſix hours before you put them into the caſk, then uſe the rod, and in three days it will come to itſelf.

To amend the taſte and ſmell of Malaga Wines: Take of the beſt Almonds four pounds, make an emulſion of them with a ſufficient quantity of the Wine to be cured; then take the whites and yolks of twelve eggs, beat them together with a handful of ſalt, put them into the pipe, uſing the rod.

To amend the ſmell and taſte of French and Rheniſh Wines, which are foul: Take one pound of honey, a handful of Elder-flowers, an ounce of Orris-powder, one Nutmeg, a few Cloves to an auln of the Wine, boil them in a ſufficient quantity of the Wine to be cured, to the conſumption of half, and when it is cold ſtrain it, and uſe it with the rod; ſome add a little ſalt. If the Wine be ſweet enough, add one pound of the ſpirits of Wine to a hogſhead, and give the caſk a ſtrong ſcent. Spirit of Wine makes any Wine briſk, and fines it, without the former mixture.

A lee of the aſhes of Vine branches, viz. a quart to a pipe, being beaten into Wine, cures the ropineſs of it, and ſo infallibly doth a lee of Oaken aſhes.

For Spaniſh ropy Wine: Rack it from the lees into a new-ſcented caſk, then take of allum one pound, of Orris-roots powdered half a pound, beat them well into the Wine with a ſtaff; ſome add fine and well-dried ſand, put warm to the Wine. If the Wine beſides prove brown, add three pottles of milk to a pipe: this cures ropy Wine, before it begins to fret.

To amend and preſerve the colour of clarets: Take red Beet-roots, q. ſ. ſcrape them clean, and cut them into ſmall pieces; then boil them in q. ſ. of the ſame Wine, to the conſumption of the third part; ſcum it well, and when cool decant off what is clear, and uſe the rod.

Firing of Wines in Germany is thus performed: They have in ſome vaults three or four ſtoves, which they heat very hot; others make fires almoſt before every vat; by this means the muſt fermenteth with that vehemency, that the Wine appears between the ſtaves; when this ebullition, fermentation, and working ceaſe, they let the Wine ſtand ſome days, and then rack it. This firing is only uſed in cold years, when the Wine falls out green.

To ſet old Wine a fretting, being deadiſh and dull of taſte: Take of ſtum two gallons to a hogſhead, put it hot upon the Wine, then ſet a pan of fire before the hogſhead, which will then ferment till all the ſweetneſs of the ſtum is communicated to the Wine, which thereby becomes briſk and pleaſant.

Some use this ftumming at any time; fome in Auguft only, when the Wine hath a difpofition to fret of itfelf, more or lefs ftum to be added, as the Wine requires.

The beft time to rack Wine is in the decreafe of the moon, and when the Wine is free from fretting, the wind being at north-eaft or north-weft, and not at fouth, the fky ferene, free from thunder and lightning.

Having thus given an account of the different practices of the vignerons, vintners, and Wine-coopers, in the management of their feveral Wines, I fhall next offer a few things which have occurred to me from fome obfervations and experiments, relating to the making of Wines in England.

The Grapes being ripe, fhould be cut when they are perfectly dry, and carried into a large dry room, where they muft be fpread upon Wheat-ftraw, in fuch a manner as not to lie upon each other; in this place they may remain a fortnight, three weeks, or a month, according as there is conveniency, obferving to let them have air every day, that the moifture perfpired from the Grapes may be carried off. Then, having the preffes and other things in order, you fhould proceed in the following manner: Firft, all the Grapes fhould be pulled off the bunches, and put into tubs, being careful to throw away fuch as are mouldy, rotten, or not ripe, which, if mixed with the others, will render the Wine lefs delicate; and if the ftalks of the bunches are preffed with the Grapes, there will be an auftere juice come from them, which will render the Wine acid and fharp; this, I fear, has fpoiled a great quantity of Wine which was made in England, which, if otherwife managed, might have proved very good; for we find in France, and other Wine countries, where perfons are defirous of having good Wine, they always pick the Grapes from off the ftalks before they are preffed, though indeed the common vignerons, who have more regard to the quantity than the quality of their Wines, do not practife this. But as in England we labour under the inclemency of climate, we fhould omit nothing of art which may be neceffary to help the want of fun.

The Grapes being thus carefully picked off, fhould be well preffed, and if it is defigned for red Wine, the hufks and ftones fhould be put into the liquor; and if the feeds or ftones of the Grapes are broken in the prefs, the Wine will have more ftrength, which muft be put into a large vat, where the whole fhould ferment together five or fix days; after which the Wine fhould be drawn off, and put into large cafks, leaving the bung-hole open to give vent to the air which is generated by the fermentation. But it muft be remarked, that after the Wine is preffed out, and put into the vat with the hufks, if it does not ferment in a day or two at moft, it will be proper to add a little warmth to the room by fires, which will foon put it into motion; and for want of this it often happens, where people prefs their Wine, and leave it to ferment in open cold places, that the nights being cold, check the fermentation, and fo caufe the Wine to be foul, and almoft ever after upon the fret. This hufbandry is much practifed upon the Rhine, where they always have ftoves placed in the houfes where the Wine is fermented, wherein they keep fires every night, as the feafon is cold, while the Wines are fermenting.

If white Wine is defired, then the hufks of the Grapes fhould not remain in the liquor above twelve hours, which will be long enough to fet it a fermenting; and when it is drawn off, and put into other veffels, it fhould not remain there above two days before it is drawn off again; and this muft be repeated three or four times, which will prevent its taking any tincture from the hufks in fermenting.

When the greateft fermentation is over, the Wine fhould be drawn off into frefh cafks, which muft be filled to the top, but the bung-hole fhould be left open three weeks or a month, to give vent to the generated air, and that the fcum may run over; and as the Wine fubfides in the cafks, they fhould be carefully refilled with Wine of the fame fort from a ftore-cafk, which fhould be provided for that purpofe; but this muft be done with much care, left, by haftily refilling the cafks, the fcum, which is naturally produced upon all new Wines, fhould be broken thereby, which will mix with the Wine, and foul it, caufing it to take an ill tafte; therefore it would be proper to have a funnel, which fhould have a plate at the fmall end, bored full of little holes, that the Wine may pafs through in fmall drops, which will prevent its breaking the fcum.

After the Wine has remained in this ftate a month or fix weeks, it will be neceffary to ftop up the bung-hole, left, by expofing it too much to the air, the Wine fhould grow flat, and lofe much of its fpirit and ftrength; but it muft not be quite ftopped up, but rather fhould have a pewter or glafs tube, of about half an inch bore, and two feet long, placed in the middle of the bung-hole. The ufe of this tube is to let the air, which is generated by the fermentation of the Wine, pafs off, becaufe this being of a rancid nature, would fpoil the Wine, if it were pent up in the cafk; and in this tube there may always remain fome Wine, to keep the cafk full, as the Wine fubfides; and, as it fhall be neceffary, the Wine in the tube may be eafily replenifhed. For want of rightly underftanding this affair, a great quantity of the choiceft Wines of Italy, and other countries, have been loft. A great complaint of this misfortune I received from a very curious gentleman in Italy, who fays, "Such "is the nature of this country Wines in general (nor are "the choiceft Chianti's excepted,) that at two feafons of "the year, viz. the beginning of June and September, the "firft when the Grapes are in flower, and in the other "when they begin to ripen, fome of the beft Wines are "apt to change, efpecially at the latter feafon; not that "they turn eager, but take a moft unpleafant tafte, like "that of a rotten Vine leaf, which renders them not only "unfit for drinking, but alfo to make vinegar of, and is "called the Settembrine. And what is moft ftrange is, "that one cafk, drawn out of the fame vat, fhall be in-"fected, and another remain perfectly good, and yet both "have been kept in the fame cellar.

"As this change happens not to Wines in flafks (though "that will turn eager,) I am apt to attribute it to fome "fault in refilling the cafk, which muft always be kept "full, which, either by letting alone too long, till the "decreafe be too great, and the fcum there naturally is on "all Wines, thereby being too much dilated, is fubject to "break, or elfe, being broken by refilling the cafk gives it "that vile tafte. But againft this there is a very ftrong "objection, i. e. that this defect feizes the Wine only at a "particular feafon, viz. September; over which if it gets, "it will keep a good many years, fo the cafe is worthy "the enquiry of naturalifts, fince it is evident that moft "Wines are more or lefs affected with this diftemper, "during the firft year after making."

Upon receiving this information from Italy, I confulted the Rev. Dr. Hales of Teddington, who was then making many experiments on fermenting liquors, and received from him the following curious folution of the caufe of this change in Wine, which I fent over to my friend in Italy, who has tried the experiment, and it has accordingly anfwered his expectation, in preferving the Wine, which was thus managed, perfectly good. He has alfo communicated the experiment to feveral vignerons in feveral parts of Italy, who are repeating the fame, which take in Dr. Hales's words:

"From many experiments which I made the laft fummer, "I find that all fermented liquors generate air in large "quantities, during the time of their fermentation; for, "from

“ from an experiment made on twelve cubic inches of Malaga Raiſins, put into eighteen cubic inches of water the beginning of March, there were 411 cubic inches of air generated by the middle of April; but afterwards, when the fermentation was over, it reſorbed a great quantity of this air; and from forty-two cubic inches of ale from the tun (which had fermented thirty-four hours before it was put into the bolt-head) had generated 639 cubic inches of air from the beginning of March to the middle of June; after which it reſorbed thirty-two cubic inches of air; from whence it is plain, that fermented liquors generate air, during the time of their fermentation, but afterwards they are in an imbibing ſtate, which may perhaps account for the alteration of the nice Italian Wines; for Wine, during the firſt year after making, continues fermenting more or leſs, during which time a great quantity of air is generated, until the cold in September put a ſtop to it, after which it is in an imbibing ſtate. Now the air thus generated is of a rancid nature (as the Grotto del Cano,) and will kill a living animal, if put into it. So that if, during the fermentation of the Wine, there are two quarts of this rancid air generated, which is cloſely pent up in the upper part of the veſſel, when the cold ſhall ſtop the fermentation, the Wine, by abſorbing this air, becomes foul, and acquires this rancid taſte; to prevent which, I would propoſe the following experiment:

X b A

“ Suppoſe the veſſel A filled with Wine, in the bung-hole of this veſſel b, I would have a glaſs tube of two feet long, and about two inches bore, fixed with a pewter ſocket cloſely cemented, ſo as that there may be no vacuities on the ſides; and into this tube ſhould be another, of about half an inch bore, cloſely fixed; the lower tube ſhould always be kept about half full of Wine, up to X, which will ſupply the veſſel, as the Wine therein ſhall ſubſide; ſo that there will be no room left in the upper part of the veſſel to contain any generated air, which will paſs off through the upper ſmall tube, which muſt be always left open for this purpoſe; and the tube being ſmall, there will be no danger of letting in too much air to the Wine.

“ As the Wine in the lower tube ſhall ſubſide, it may be refilled by introducing a ſlender funnel through the ſmall tube, down to the ſcum upon the ſurface of the Wine in the larger tube, ſo as to prevent its being broken, by the Wine falling too violently upon it. This experiment being tried with glaſs tubes, will give an opportunity to obſerve what impreſſion the different ſtates of the air have upon the Wine, by its riſing or falling in the tubes; and if it ſucceeds, it may be afterwards done by wooden or metal tubes, which will not be in danger of breaking.”

This curious experiment, having ſucceeded wherever it has yet been tried, will be of great ſervice in the management of the Wines, there being many uſeful hints to be taken from it, particularly with regard to fermenting Wines; for, ſince we find that Wines too long fermented (eſpecially thoſe which are produced in cool countries) ſeldom keep well, ſo, by letting them ſtand in a cool place, the fermentation will be checked, which is agreeable to the practice of the Champagnois, who keep the Wines in winter in cellars above ground; but when the weather grows warmer in ſpring, they then carry them down into their vaults, where they are cooler than in the cellars; and this method of removing their Wines from the cellars to the vaults, and back again into the cellars, as the ſeaſons of the year ſhall require, is found of great ſervice in preſerving the Wines in perfection; for theſe Wines being weak (when compared with thoſe produced in more ſouthern countries,) have not body enough to maintain them, if they are permitted to ferment all the ſucceeding ſummer, which the heat of the ſeaſon will promote where the Wine is expoſed to its influence; and this ſurely muſt be worth the trial by thoſe who make Wine in this country, ſince it is the practice of the northern countries, which is the moſt proper for our imitation, and not that of the moſt ſouthern.

But after the Wine has paſſed its fermentation in the vat, and is drawn off in the caſks, it will require ſomething to feed upon; ſo that you ſhould always preſerve a few bunches of the beſt Grapes, which may be hung up in a room for that purpoſe, until there be occaſion for them; when they ſhould be picked off the ſtalks, and two or three good handfuls put into each caſk, according to their ſeveral ſizes; for want of this, many times people make uſe of other things, which are by no means ſo proper for this purpoſe.

The vignerons of different countries do alſo put various ſorts of herbs into the vat, when the Wine is fermenting, to give it different flavours. Thoſe of Provence make uſe of Sweet-marjoram, Balm, and other ſorts of aromatic herbs; and upon the Rhine they always put ſome handfuls of a peculiar kind of Clary into the vats, from whence ariſe the different flavours we obſerve in Wines, which, it is poſſible, were made in the ſame manner, and from the ſame ſorts of Grapes. How far this might be thought worth practiſing in England, a few experiments would inform us; though it is to be queſtioned, whether theſe herbs mend the Wine, becauſe it ſeems to obtain amongſt the vignerons, purely to alter the flavour of their Wines, in order to render them agreeable to the palate of their particular cuſtomers; but, however this be, it is yet certain, that there is ſome art uſed to alter the flavour of the Wine, in moſt of the different Wine countries in France, for it is the ſame ſort of Grape which the curious always plant in Orleans, Champagne, and Burgundy; and how different theſe Wines are in their flavour and quality, every one who is acquainted with them well knows; and this difference can never be effected by the ſituation of the places, ſince there is no very great difference in the heat of thoſe countries; nor do I believe their different ways of making the Wine can alter their flavour ſo much, eſpecially thoſe of Orleans and Burgundy, where there is little difference in their management; but in Champagne there is this difference from the reſt, that they always cut their Grapes in a morning, before the dew is gone off, or in cloudy weather; whereas the vignerons of all the other places never cut any till they are perfectly dry, which may occaſion a great alteration in the Wine.

The method commonly practiſed to give the red colour to Wine, is to let it ferment a few days upon the ſkins, which they always obſerve to preſs two or three times, in order to make them diſcharge their contents; but where a deep-coloured rough Wine is deſired, there they put a quantity of a certain ſort of Grape, whoſe juice is red, into each vat; this is well known in England by the name of Claret Grape; the leaves of this Vine always change to a deep purple colour as the fruit ripens, and the Grapes are of a fine blue colour, with a blue over them like fine Plumbs; but the juice of them is very auſtere, eſpecially if they are not very ripe.

This red Wine will not require to be drawn off into caſks more than at firſt from the vat, for it may remain in the ſame veſſel until it is fit to bottle off, which, I think, ſhould not be done till the Wine is two years old; the greater quantity of Wine there is in each veſſel, the more force it will have, and ſo conſequently be in leſs danger of ſuffering from the injuries of weather, eſpecially if the before-mentioned method be practiſed; but where there are large quantities of Wine preſerved in cloſe vaults, people ſhould be very cautious how they at firſt enter them, after they have

have been shut up for some time; because the air of this vault will become rancid from the mixture of the generated air proceeding from the Wines, which has often killed people who have incautiously entered them.

WINTER. [Prognostics of a hard Winter.] The Lord Bacon gives these as signs or forerunners of a hard Winter.

If stone or wainscot, that has been used to sweat (as it is called,) be more dry in the beginning of Winter, or the drops of eaves of houses come down more slowly than they used to do, it portends a hard and frosty Winter. The reason is, that it shews an inclination in the air to dry weather, which, in the winter time, is always joined with frost.

Generally a moist and cool summer betokens a hard Winter likely to ensue. The reason is, that the vapours of the earth, not being dissipated by the sun in the summer, do rebound upon the Winter.

A hot and dry summer, especially if the heat and drought extend far in September, betokens an open beginning of Winter, and cold to succeed towards the latter part of the Winter, and in the beginning of the spring; for all that time the former heat and drought bear the sway, and the vapours are not sufficiently multiplied.

An open and warm Winter portends a hot and dry summer; for the vapours disperse into the Winter showers, whereas cold and frost keep them in, and transport them into the late spring and summer following.

When birds lay up Haws and Sloes, and other stores, in old nests and hollow trees, it is a sign of a hard Winter approaching.

If fowls or birds, which used at certain seasons to change countries, come earlier than the usual time, they shew the temperature of the weather, according to that country from whence they came; as the Winter birds, fieldfares, snipes, woodcocks, &c.

If they come earlier, and out of the northern countries, they intimate cold Winters likely to ensue with us. And if it be in the same country, they shew a temperature of season, like that of the season in which they come; as bats, cuckoos, nightingales, and swallows, which come towards summer if they come early, it is a sign of a hot summer to follow. Cold dews and morning rains, about Bartholomew tide, and hoar frost in the morning about Michaelmas, fortel a hard Winter.

WOODS and groves are the greatest ornaments to a country-seat, therefore every seat is greatly defective without them; wood and water being absolutely necessary to render a place agreeable and pleasant. Where there are Woods already grown to a large size, so situated as to be taken into the garden or park, or so nearly adjoining, as that an easy communication may be made from the garden to the Wood; they may be so contrived by cutting of winding walks through them, as to render them the most delightful and pleasant parts of a seat (especially in the heat of summer) when those walks afford a goodly shade from the scorching heat of the sun.

As I have already treated of the use and beauty of wildernesses, and have given directions for the making and planting of them, I shall not enlarge much upon that head in this place; therefore I shall only give some short instructions for the cutting and making of these Wood-walks, in those places where persons are so happy as to have any grown Woods so situated as to be near the habitation, and are either taken into the garden, or walks made from the house or garden leading to them; as also how to plant or decorate the sides of these walks with shrubs and flowers, so as to render them agreeable and pleasant; and then I shall more fully treat of the method to raise and improve Woods, so as to be of the greatest advantage to the possessor, and a public benefit to the nation.

Where persons have the convenience of grown Woods near their habitation, so as that there may be an easy communication from one to the other, there will be little occasion for wildernesses in the garden; since the natural Woods may be so contrived, as to render them much pleasanter than any new plantation can possibly arrive to within the compass of twenty years, where the trees make the greatest progress in their growth; and in such places where their growth is slow, there cannot be expected shade equal to the grown Woods in double that number of years; but there is not only the pleasure of enjoying a present shade from these Woods, but also a great expence saved in the planting of wildernesses, which, if they are large, and the trees to be purchased, will amount to no small sum.

If the Wood is so situated, as that the garden may be contrived between the house and that, then the walk into the Wood should be made as near to the house as possible, that there may not be too much open space to walk through in order to get into the shade; if the Wood is of small extent, then there will be a necessity for twisting the walks pretty much, so as to make as much walking as the compass of ground will admit; but there should be care taken not to bring the turns so near each other as that any two walks may be exposed to each other, for want of a sufficient thickness of Wood between; but where the Wood is large, the twists of the walks should not approach nearer to each other than fifty or sixty feet; or in very large Woods they should be at a greater distance; because, when the under Wood is cut down, which will be absolutely necessary every tenth or twelfth year, according to its growth, then the walks will be quite open until the under Wood grows up again, unless a border of shrubs, intermixed with some evergreens, is planted by the sides of the walks; which is what I would recommend, as this will greatly add to the pleasure of these walks.

These Wood-walks should not be less than eight or nine feet broad in small Woods; but in large ones fifteen feet will not be too much, and on each side of the walks. The border of shrubs and evergreens may be nine or ten feet broad; which may be so managed, as to shut out the view from one part of the walk to the other at those times when the under Wood is cut down; at which times there will be an absolute occasion for such plantations, and at all times they will afford great pleasure by adding to their variety, as also by their fragrant odour.

The former method, which was practised in cutting these walks through Woods, was to have them as strait as possible; so that there was much trouble to make sights through the Woods, for direction how to cut them; but where this was practised, every tree which stood in the line, good and bad, was cut down, and many times boggy or bad ground was taken into the walks, so that an expence of draining and levelling was necessary to render them proper for walking on; besides this, there were many other inconveniencies attending these strait cuts through Woods, as, first, by letting in a great draught of air, which in windy weather renders the walks unpleasant; and these cuts will also appear at a great distance from the Woods, which will have a very bad effect; therefore the modern practice of twisting the walks through Woods is to be preferred. In the cutting of these walks there should be particular care taken to lead them over the smoothest and soundest part of the ground, as also to avoid cutting down the trees; so that whenever these stand in the way, it will be better to lead the walk on one side than to have the tree stand in the middle; for although some persons may contend for the beauty of such trees which are left standing in walks, yet it must be allowed, that unless the walk is made much broader in those places than in the other, the trees will occasion obstructions,

ſtructions to the walkers or riders, eſpecially when ſeveral perſons are walking together, ſo that it will be much better to have the walks entirely clear from trees; and where any large ſpreading tree ſtands near the walk, to cut away the ſmall wood, ſo as to make an opening round the trees, where there may be ſome ſeats placed for perſons to reſt under the ſhade. The turns made in theſe walks ſhould be as eaſy and natural as poſſible; nor ſhould there be too many of them, for that will render the walking through them diſagreeable; therefore the great ſkill in making of theſe walks is, to make the turns ſo eaſy as not to appear like a work of art, nor to extend them ſtrait to ſo great length, as that perſons who may be walking at a great diſtance may be expoſed to the ſight of each other; both theſe extremes ſhould be avoided as much as poſſible, ſince they are equally diſagreeable to perſons of true taſte. When a Wood is properly managed in this way, and a few places properly left like an open grove, where there are ſome large trees ſo ſituated as to form them, there can be no greater ornament to a fine ſeat than ſuch a Wood.

We ſhall now treat of the culture of Woods for profit to the poſſeſſor, and public benefit of the nation.

The great deſtruction of the Woods and foreſts which has been of late years made in this country, ſhould alarm every perſon who wiſhes well to it; ſince there is nothing which ſeems more fatally to threaten a weakening, if not a diſſolution, of the ſtrength of this once famous and flouriſhing nation, than the notorious decay of its timber: and as this devaſtation has ſpread through every part of the country, ſo unleſs ſome expedient be ſeriouſly and ſpeedily reſolved on, to put a ſtop to this deſtruction of the timber, and alſo for the future increaſe of it, one of the moſt glorious bulwarks of this nation will, in a few years, be wanting to it.

And as there are ſmall hopes of this being remedied by thoſe entruſted with the care of public Woods, ſince their private intereſt is ſo much advanced by the deſtroying the timber, which they were appointed to preſerve; therefore, unleſs private perſons can be prevailed on to improve their eſtates, by encouraging the growth of timber, it is greatly to be feared, that in an age there will be a want of it for the ſupply of the navy; which, whenever it happens, muſt put a period to the trade of this country.

It has been often urged, by perſons whoſe judgment in other affairs might be depended on, that the great plantations, which for ſeveral years paſt have been carried on in ſeveral parts of this kingdom, will be of public benefit by the propagation of timber; but in this they are greatly miſtaken, for in moſt of the plantations which have been made for years paſt, there has been little regard had to the propagation of timber, preſent ſhade and ſhelter have been principally conſidered; and in order to obtain theſe ſoon, great numbers of trees have been taken out of Woods, hedge rows, &c. which, if they had remained in their firſt ſituation, might have afforded good timber, but by being tranſplanted large, are abſolutely rendered unfit for any uſe but fuel; ſo that the great quantity of plantations which have been made, I fear, will rather prejudice than be of uſe to the improvement of timber; nor is there any other method of increaſing the uſeful timber of this country, than by ſowing the ſeeds in the places where they are to remain, or in ſuch ſituations where there are plenty of Oaks in the neighbourhood; if the ground is properly fenced, to keep out cattle and vermin, the Acorns which drop from thoſe trees will ſoon produce plenty of young trees, which, if properly taken care of, will ſoon grow to be large.

The two moſt ſubſtantial timbers of this country are the Oak and Cheſtnut; though the latter has been of late years almoſt entirely deſtroyed in England, ſo that there are ſcarce any remains of trees of ſize in the Woods at preſent; but there can be no doubt of this tree having been one of the moſt common trees of this country, as may be proved by the old buildings in many parts of England, in which the greateſt part of the timber is Cheſtnut. But as I have already treated largely of the method of propagating both theſe trees for profit, under their reſpective titles, I ſhall not repeat it here. Next to theſe, the Elm is eſteemed as a profitable timber; but of theſe there are few cultivated in Woods, eſpecially in the ſouth part of England, where they chiefly grow in hedge rows, or plantations near houſes; but in the north-weſt part of England, there are numbers of very large trees of the Witch Elm growing in parks, and ſome in Woods, as if that tree was a native of this country, which has been much doubted; though as this tree propagates itſelf by ſeeds, it may be deemed an indigenous plant in England.

The Beech is another tree common in the Woods, eſpecially upon the chalky hills of Buckinghamſhire, Kent, Suſſex, and Hampſhire, where there are ſome very large Woods entirely of this ſort; ſome of which have been of long ſtanding, as appears by the age of the trees; but whether this tree is a native of this country, has been a point often diſputed.

The Aſh is a very profitable tree, and of quick growth; ſo that in leſs than an age, the trees will arrive to a large ſize from the ſeeds, therefore a perſon may hope to reap the profits of his labour, who ſow the ſeeds; but this is not a beautiful tree to ſtand near a habitation, being late in the ſpring in putting out its leaves, and the firſt that ſheds them in autumn; nor is a friendly tree to whatever grows near it, the roots drawing away all the nouriſhment of the ground, whereby the trees or plants which grow near are deprived of it; ſo that where the Aſh-tree grows in hedge rows, the hedges in a few years are entirely deſtroyed; and if they are in paſture-grounds, and the cows browze on them, the butter made with their milk will be bad; for which reaſon the Aſh ſhould be ſown ſeparate in lands which are incloſed, where cattle are not permitted to come, and at a diſtance from the habitation.

Upon ſandy or rocky ſoils, the Scotch Pine will thrive exceedingly, and turn to great advantage to the planter, provided the plants are planted young, and treated in the manner directed in the article Pinus, to which the reader is deſired to turn to avoid repetition.

There are alſo ſeveral aquatic trees, which are very profitable to thoſe who have low marſhy lands, where the harder kinds of timber will not thrive; theſe are the Poplars of ſeveral ſorts, the Willow, Alder, &c. but as theſe, and all the other kinds of trees, have been fully treated of, both as to their propagation and uſes, and alſo an account of the different ſoils in which each will thrive beſt, under their reſpective titles, the reader is referred to them for farther information; and I ſhall next treat of the general management of Woods, of whatever kinds of trees they are compoſed.

Where there are young Woods, great care muſt be taken of the fences; for if cattle ſhould get in among the trees, they will, in a ſhort time, do infinite damage to them, by browzing on the branches, or barking the trees; and during the firſt ten years of their growth, they ſhould be ſecured from hares and rabbets, otherwiſe, in ſevere froſt, when the ground is covered with ſnow, whereby they are deprived of other food, they will get into the Woods, and eat off the bark from the young trees, and gnaw all the branches within their reach; ſo that in a few days, where there are plenty of theſe animals, there may be ſuch deſtruction made among the young trees as cannot be retrieved, but by cutting them down to the ground, which will be a loſs of ſeveral years; therefore thoſe perſons who have the care of young Woods, ſhould be very diligent in froſty weather in looking over the trees, and ftopping the holes in the fences, to keep out all vermin.

Another care to be taken of young Woods, is the thinning the trees from time to time, as they increafe in their growth; but in doing of this, there muft be great caution ufed, for it fhould be gradually performed, fo as not to lay open the trees too much, to let the cold air among them, which will greatly retard their growth; nor fhould the trees be left fo clofe, as to draw each other up like may-poles, but rather obferve a medium in this work, cutting down a few each year, according as there may be neceffity for it, being careful not to permit thofe to ftand, which may fpoil the growth of the neighbouring trees, always obferving to leave thofe trees which are the moft promifing.

The young trees in thefe Woods fhould not be lopped or pruned, for the more they are cut the lefs they will increafe in bulk; every branch which is cut off will rob the tree of its nourifhment, in proportion to the fize of the branch; therefore the hatchet fhould not be fuffered to come into young Woods, unlefs in the hands of fkilful perfons.

Where perfons have more regard to the future welfare of their timber than their immediate profit, the under Wood fhould be grubbed up as the trees advance, that the roots may have the whole benefit of the foil, and their ftems enjoy the free air, without which their iftems are generally covered with Mofs, and their growth greatly ftinted; as may be obferved in all fuch Woods where there is any quantity of under Wood remaining, in which places it is rarely found that the trees do ever grow to a large fize; therefore where large timber is defired, the trees muft have room to extend their roots and branches, without which it cannot be expected; but from a covetous temper, many people let their under Wood remain as long as it will live; for as the timber increafes, the under Wood will be gradually decaying by the fhade and drip of the large trees, fo that by this method the timber fuffers more in a few years than the value of the under Wood; therefore by endeavouring to have both, neither of them can be fo good as where they are feparately preferved.

If perfons who have eftates would be careful to nurfe up trees in their hedge rows, it would in time become a fortune to their fucceffors; as hereby the timber growing in the hedges may be worth more than the freehold of the eftate, which has often been the cafe with eftates from which their poffeffors have cut down timber for fortunes for their younger children; the frequency of this fhould encourage perfons to be a little more attentive to the growth and prefervation of young Woods, fince the expence and trouble is not great, and the future profit very certain; befides, the pleafure of feeing trees of a man's own fowing make yearly advances, muft be very great to thofe who have any relifh for country amufements.

There are feveral perfons who plant copfes for cutting every ten or twelve years, according to their growth. Thefe are ufually planted in autumn, either with ftools or young plants, which are drawn out of the Woods; the latter fhould always be preferred to the former. Thefe copfes are commonly planted with feveral forts of trees, as Oak, Beech, Cheftnut, Afh, Birch, Willow, &c. but the Afh and Cheftnut are the moft profitable, where they grow kindly, becaufe the poles of Afh are very valuable; thefe alfo are good for hoops, fo that there is no danger of having fale for thefe Copfe-woods when they are fit for cutting; but where the copfes are intended to remain, there fhould be no ftandard-trees left for timber, becaufe as the heads of the trees fpread, and over-top the under Wood, it will caufe that to decay; and where the ftandards are left upon the ftumps of the Copfe-wood they will never grow to a large fize, nor will the timber be fo valuable as that produced immediately from a young root; therefore whoever will make the experiment, will be convinced that it is more for the advantage of both to keep them in diftinct Woods.

But where perfons plant copfes upon land free from trees, it will be the better method to fow the feeds, efpecially if Cheftnut, Oak, or Beech are the trees intended; for although it is a prevailing opinion with the generality of perfons, that by planting they fave time, yet I am fure of the contrary; for if the feedling plants are kept clear from weeds, they will, in eight or ten years, outgrow thofe which are planted, and thefe unremoved copfes will continue much longer in vigour than the other; fo that for either timber or Copfe-wood, the beft method is to prepare the ground well, to fecure the fences and fow the feeds, which is fo far from lofing, that in twenty years it will be found to gain time, which is what every planter wifhes to do.

The ufual time of felling timber is from November to February, at which time the fap in the trees is hardened; for when the fap is flowing in the trees, if they are cut down the worm will take the timber, and caufe it to decay very foon; therefore if the durablenefs of the timber is confidered, the trees fhould always be cut in the winter months; but as the bark of the Oak is fo valuable for tanning leather, there has been a law paffed to oblige perfons to cut thefe trees during the fpring feafon, when the bark will readily peel off; by which the timber is rendered unfit for building either fhips or houfes, as it will be very fubject to caft, rift, or twine, and the worm will foon take it; therefore it would be more for the public benefit, if a law was enacted to oblige every perfon to ftrip off the bark of fuch trees as are defigned to be cut down in the fpring, when the bark will run, leaving the trees with their branches ftanding till the following winter, which will be found to anfwer both purpofes well.

X.

XANTHIUM. Tourn. Inst. R. H. 438. tab. 252. Lesser Burdock.

The Characters are,

It hath male and female flowers on the same plant. The male flowers have a common scaly empalement; they are composed of several tubulous funnel-shaped florets, which are equal, and disposed in a hemisphere, cut into five segments at the top, and have each five very small stamina, terminated by erect parallel summits. The female flowers are situated under the male by pairs; they have no petals or stamina, but they are succeeded by oblong, oval, prickly fruit, having two cells, each including one oblong seed, convex on one side, and plain on the other.

The Species are,

1. Xanthium (*Strumarium*) caule inermi, aculeis fructibus erectis. *Lesser Burdock with an unarmed stalk, and the spines of the fruit erect; or lesser Burdock.*

2. Xanthium (*Canadense*) caule inermi, foliis cuneiformi ovatis sublobis. *Lesser Burdock with an unarmed stalk, and incurved spines to the fruit; or Canada Burdock.*

3. Xanthium (*Spinosum*) spinis ternatis. Hort. Upsal. 283. *Lesser Burdock, having triple spines.*

4. Xanthium (*Chinense*) caule inermi, aculeis fructibus, longissimis erectis simplicibus. *Lesser Burdock with an unarmed stalk, and very long erect spines to the fruit which are single.*

The first sort grows naturally in Europe, and also in India, from whence I have received the seeds; it has been found growing wild in a few places in England, but of late years it has not been seen in those places. It grew some years past in the road near Dulwich college. The stalk of this plant is round, spotted with black; it rises in good ground two feet high, sending out a few side branches. The leaves stand upon long slender foot-stalks. From the wings of the stalks arise the foot-stalks of the flowers, which are produced in loose spikes, the male flowers growing at the top, and the female flowers under them; they are of an herbaceous colour, collected into roundish heads. The female flowers are succeeded by oblong oval fruit, closely armed with short erect spines.

This plant has been much esteemed by some physicians for the cure of scrophulous tumours, and also in leprosies, but is rarely now used.

The second sort grows narurally in North America. The stalks of this are much thicker, and rise higher than those of the first; the leaves are not hollowed at their base, nor are they divided so deeply on their sides as those of the first; they are unequally indented on their edges, and have three longitudinal veins. The flowers are produced in shorter and looser spikes. The fruit are much larger, and are armed with stronger spines, which are incurved.

The third sort grows naturally in Portugal and Spain. The stalks of this rise three feet high, sending out many branches; these are garnished with oblong leaves which are indented on their edges, ending in acute prickles, of a dark green on their upper side, but hoary on their under, having very short foot-stalks. The flowers come out from the sides of the branches, two or three at each place, one of which is female, and is succeeded by oblong oval fruit, armed with slender sharp spines, which are erect. The stalks and branches are armed with long, stiff, triple thorns on every side, which renders it dangerous to handle them.

The fourth sort was discovered by the late Dr. Houstoun in the year 1730, growing naturally at La Vera Cruz; this plant rarely grows more than six or seven inches high. The leaves are small, and shaped like those of the second sort; the flowers are produced in loose spikes at the top of the stalks; the fruit is as large as those of the second sort, but the spines are slender, single, and strait.

All these plants are annual. The first will come up from the seeds which fall in autumn, and requires no other care but to thin the plants where they are too close, and keep them clear from weeds. The second sort formerly was as easily cultivated, and came up from the self-sown seeds as readily, and rarely failed to ripen its seeds; but of late years the autumns have proved so bad, that the seeds have not come to maturity.

The third sort will perfect its seeds some years on self-sown plants; but as they sometimes fail, the sure way is to raise the plants on a gentle hot-bed, and, after they have obtained strength, plant them on a warm border on a lean soil, which will stint the plants in their growth, and cause them to be more fruitful; for when they are planted in rich ground the plants will grow to a large size, and will not produce flowers till late in autumn, so the seeds will not ripen.

The fourth sort must be raised on a hot-bed in the spring, and the plants should be transplanted each into a small pot, and plunged into a fresh hot-bed to bring them forward. After they have obtained strength they should be inured to the free air gradually, and in June some of the plants may be turned out of the pots, preserving the ball of earth to their roots, and planted in a south border, where, if the season proves favourable, they will perfect their seeds.

XANTHOXYLUM. Lin. Gen. Plant. 335. The Tooth-ach-tree.

The Characters are,

The flower has no empalement, but has five oval petals, ard five slender stamina, which are longer than the petals, terminated by furrowed summits; it has three germen, which are united at their base, having each a lateral style, crowned by obtuse stigmas. The germen afterward become so many capsules, each containing one roundish, hard, shining seed.

The Species are,

1. Xanthoxylum (*Clava Herculus*) foliis pinnatis. *Tooth-ach-tree with winged leaves, having spear-shaped sawed lobes growing upon foot-stalks.*

2. Xanthoxylum (*Americanum*) foliis pinnatis, foliolis oblongo-ovatis, integerrimis sessilibus. *Tooth-ach-tree with winged leaves, having oblong, oval, entire lobes without foot-stalks, commonly called broad-leaved Tooth-ach-tree.*

The first sort grows naturally in South Carolina, where it rises to the height of fifteen or sixteen feet. The stem is woody,

woody, covered with a whitiſh rough bark, armed with ſhort thick ſpines; theſe grow to a large ſize as the trunk increaſes in bulk, ſo as to become protuberances terminating in ſpines. The leaves are ſometimes placed by pairs, and at others they ſtand without order; they are compoſed of three, four, or five pair of ſpear-ſhaped lobes placed oppoſite, terminated by an odd one, of a deep green on their upper ſide, and of a yellowiſh green below, a little ſawed on their edges, and ſtand upon ſhort foot-ſtalks. At the end of the branches come forth the foot-ſtalks, which ſuſtain the flowers; theſe branch out, and form a looſe panicle. The flowers are compoſed of five white petals, which are ſmall, and having no cover, they are by ſome called the empalement, but being of a different colour from the leaves, I ſhall take the liberty to ſtyle them petals. Within theſe are ſituated five ſtamina, which are terminated by reddiſh ſummits, and in the hermaphrodite flowers there are three ſtyles faſtened to the ſide of the germen. After the flower is paſt, the germen turns to a roundiſh four-cornered capſule, each containing one roundiſh, hard, ſhining ſeed. It is ſometimes called Pellitory-tree.

This has been generally confounded with the prickly Yellow Wood, or Yellow Hercules of Sir Hans Sloane, but is very different from that; for in the Weſt-Indies it is one of their largeſt timber trees, and the ſpecimens which I have received from Jamaica, are very different from thoſe of Carolina. The leaves of the former are twice as large as thoſe of the latter; the lobes of the leaves are almoſt three inches long, and an inch and a half broad; they ſit cloſe to the foot-ſtalk, and the leaves are equally winged, having no ſingle lobe at the end. The flowers of this I have not ſeen, but the capſules have five cells, each containing one black, ſhining, hard ſeed.

The ſecond ſort grows naturally in Penſylvania and Maryland; this hath a woody ſtem, which riſes ten or twelve feet high, ſending out many branches toward the top; theſe have a purpliſh bark, and are armed with ſhort thick ſpines ſtanding by pairs. The leaves are unequally winged, and are compoſed of four or five pair of oblong oval lobes, terminated by an odd one; theſe ſtand cloſe to the midrib, having no foot-ſtalks. The midrib is armed on the under ſide with ſome ſmall ſpines. The upper ſide of the leaves are of a deep green, and their under of a pale green; they have a warm biting taſte. The bark of the tree is uſed for curing the tooth-ach, from whence it has the name. The flowers grow in looſe panicles like thoſe of the former ſort, and theſe are ſucceeded by fruit with five cells, each including one hard ſhining ſeed.

Theſe plants are generally propagated by ſeeds, but as they never ripen in this country, they muſt be procured from thoſe places where they naturally grow. When the ſeeds arrive in England, they ſhould be ſown in pots as ſoon as poſſible, for they do not grow the firſt year; and when they are kept out of the ground till ſpring, they frequently lie two years in the ground before the plants appear; therefore the pots ſhould be plunged into the ground up to their rims, in an eaſt-aſpected border, where they may remain during the ſummer; this will prevent the earth in the pots from drying too faſt, which it is very apt to do when the pots are ſet upon the ground in the ſun. The only care to be taken of the ſeeds, is to keep the pots conſtantly clean from weeds, and in very dry weather refreſh them now and then with water. In autumn the pots ſhould be placed under a common hot-bed frame, where they may be ſcreened from froſt, or elſe plunged into the ground in a warm border, and covered with tan to keep out the froſt, and the following ſpring they ſhould be plunged into a hot-bed, which will bring up the plants. When theſe appear, they muſt be frequently, but ſparingly watered, and kept clean from weeds; and, as the ſummer advances, they ſhould be gradually inured to bear the open air, into which they ſhould be removed in June, placing them in a ſheltered ſituation, where they may remain till autumn, when they muſt be placed in a hot-bed frame to ſhelter them in winter. The ſpring following, before the plants begin to ſhoot, the pots ſhould be carefully taken up, and each plunged into a gentle hot-bed, which will forward them greatly in putting out new roots. The after care muſt be to ſhelter them for a year or two in winter, until the plants have gotten ſtrength; then in the ſpring, after the danger of froſt is over, ſome of them may be turned out of the pots, and planted in the full ground in a ſheltered ſituation, where the ſecond ſort will thrive very well, and reſiſt the cold; but the firſt is not quite ſo hardy, ſo theſe may be planted againſt a ſouth wall, where they will thrive very well. Some of the plants of this ſort had been planted in the open air in the Chelſea garden ſome years, where they had thriven and endured the cold without any covering; but the ſevere winter in 1740, deſtroyed them all. Theſe plants may be increaſed by cutting off ſome of their ſtrong roots, preſerving their fibres to them, and planted in pots filled with light earth, plunging them into a moderate hot-bed, which will cauſe them to puſh out, and become plants; but theſe will not thrive ſo well, nor grow near ſo large as thoſe which are raiſed from ſeeds; the roots will alſo put out ſuckers, whereby the plants may be increaſed.

XERANTHEMUM. Tourn. Inſt. R. H. 499. tab. 284. Eternal-flower.

The CHARACTERS are,

The flower is compoſed of hermaphrodite and female florets, which have one common ſcaly empalement. The hermaphrodite florets, which form the diſk, are funnel-ſhaped, and cut into five points; the female florets, which compoſe the border or rays, are tubulous, and cut into five leſs equal points; the hermaphrodite florets have five ſhort ſtamina, terminated by cylindrical ſummits, and a ſhort germen, ſupporting a ſlender ſtyle, crowned by a bifid ſtigma. The germen afterward becomes an oblong ſeed, crowned with hairs, which ripens in the empalement.

The SPECIES are,

1. XERANTHEMUM (*Annuum*) foliis lanceolatis patentibus, caule herbaceo. Lin. Sp. Plant. 1201. *Eternal-flower with ſpreading ſpear-ſhaped leaves; commonly called Ptarmica.*

2. XERANTHEMUM (*Inapertum*) foliis lineari-lanceolatis utrinque tomentoſis. *Eternal-flower with linear ſpear-ſhaped leaves, which are downy on their under ſide.*

3. XERANTHEMUM (*Orientale*) foliis lineari lanceolatis, capitulis cylindraceis, ſemine maximo. *Eternal-flower with linear ſpear-ſhaped leaves, cylindrical heads, and a very large ſeed.*

4. XERANTHEMUM (*Speciosiſſimum*) fruticoſum erectum, foliis lanceolatis, ramis unifloris ſubnudis. Lin. Sp. Plant. 1202. *Shrubby erect Eternal-flower with ſpear-ſhaped leaves, and almoſt naked branches, bearing one flower.*

5. XERANTHEMUM (*Retortum*) caulibus fruteſcentibus provolutis, foliis tomentoſis recurvatis. Lin. Sp. Pl. 1202. *Eternal-flower with ſhrubby trailing ſtalks, and downy recurved leaves.*

6. XERANTHEMUM (*Seſamoides*) ramis unifloris imbricatis, foliis obſoletis. Lin. Sp. Pl. 859. *Eternal-flower with branches terminated by one imbricated flower, and obſolete leaves.*

The firſt ſort grows naturally in Auſtria, and ſome parts of Italy, but has been long cultivated in the Engliſh gardens for ornament. Of this there are the following varieties; one with a large ſingle white flower; the purple and white with double flowers; though theſe only differ in the colour and multiplicity of petals in their flowers, ſo are not mentioned as diſtinct ſpecies, yet where their ſeeds are carefully ſaved ſeparate, they are generally conſtant.

 Theſe

These plants are annual; they have a slender, furrowed, angular, branching stalk, covered with a white down, rising two feet high, garnished with spear-shaped hoary leaves, sitting close to the stalk, which divides into four or five branches, garnished with a few leaves at their lower parts, of the same shape with the other, but less. The upper part of the branches is naked, and sustains one flower at the top, composed of several female and hermaphrodite florets, included in one common scaly empalement, of a silvery colour. The florets are succeeded by oblong seeds, crowned with hairs. The petals of these flowers are dry, so if they are gathered when perfectly dry, and kept from the air, they will retain their beauty a long time.

The second sort grows naturally in Italy. The stalks of this do not rise much more than a foot high, and do not branch so much as the former. The leaves are narrower, and the whole plant very hoary. The flowers are not half so large as those of the former, and the scales of their empalements are very neat and silvery.

The third sort grows naturally in the Levant; this rises about the same height as the first sort. The leaves are narrower, and are placed closer on the stalks to the top. In other respects the plants are very like, but the flowers are much less, of a paler purple colour, and have a cylindrical empalement. The seeds are very large, and seldom more than three or four in each head.

These flowers were formerly much more cultivated in the English gardens than at present, especially the two sorts with double flowers, which the gardeners near London propagated in great plenty for their flowers, which they brought to market in the winter season, to adorn rooms, to supply the place of other flowers, which are not easy to be procured at that season; for these being gathered when they are fully blown, and carefully dried, will continue fresh and beautiful many months; but as there are no other colours in these flowers but white and purple, the gardeners had a method of dipping them into various tinctures, so as to have some of a fine blue, others scarlet, and some red, which made a pretty variety; and, if they were rightly stained, and afterwards hung up till they were thorough dry, they would continue their colours as long as the flowers lasted. The stalks of the flowers were not set in water, but the pots or glasses were half filled with dry sand, into which the stalks were placed, and in these they would continue in beauty the whole winter.

These plants are propagated by seeds, which may be sown either in the spring or autumn on a border of light earth; but the latter season is preferable, for those plants which come up in autumn will flower sooner. The flowers will be doubler and much larger than those which are sown in the spring, and from these good seeds may be always obtained; whereas the spring plants many times fail in cold years, and in hot dry seasons the plants do not grow to any size.

When the plants come up, and are about two inches high, they should be pricked out into another border under a warm wall, pale, or hedge, at about four or five inches distance from each other, where the plants will endure the cold of our ordinary winters very well, and in the spring will require no farther care but to keep them clear from weeds, for they may remain in the same place for good. In June they will begin to flower, and the middle or latter end of July they will be fit to gather for drying; but a few of the best and most double flowers of each kind should be suffered to remain for seed, which, in about six weeks or two months time, will be ripe, and the plants will perish soon after; so that the seeds must be annually sown, in order to preserve the kinds.

The fourth sort grows naturally at the Cape of Good Hope. This rises with a shrubby stalk three or four feet high, dividing into four or five branches, whose lower parts are garnished with thick-pointed leaves on their under side, and are ranged without order. The upper part of the branches are naked, and are terminated by one large yellow flower, composed of many oblong acute-pointed rays in the border, and the middle or disk, which is prominent, is made up of hermaphrodite florets, which are of a splendid yellow colour.

The fifth sort grows naturally at the Cape of Good Hope. The stalks of this sort are very slender, ligneous, and trail upon the ground, garnished with small silvery leaves placed without order, which sit close to the stalks, and are reflexed. The flowers are produced from the wings of the branches, sometimes one, and at others two or three flowers arise at the same point; these have scaly empalements; their border or rays are composed of many female florets, of a white colour. and their middle of hermaphrodite florets; these are succeeded by oblong seeds, crowned with hairs.

The sixth sort is a native of the country near the Cape of Good Hope; this has a shrubby branching stalk, which rises three or four feet high. The branches are slender, and like those of the Spanish Broom, but hoary; these have very small leaves resembling scales, which sit close to the branches; they are hoary, ending in acute points. The stalks are each terminated by one large silvery flower, having a stiff, dry, scaly empalement. The rays of the flower are composed of many dry female florets, and the disk or middle is made up of hermaphrodite florets.

As the plants of this sort do not ripen their seeds in England, they are propagated by cuttings, which, if planted on a bed of light earth, during any of the summer months, and shaded from the sun, will put out roots. When these have gotten sufficient roots, they should be carefully taken up, and planted into separate pots, and placed in the shade till they have taken fresh root; then they may be removed to a sheltered situation, where they may have more sun; and here they may remain till autumn, when they must be removed into shelter, for they are too tender to live abroad through the winter in England, though they do not require any artificial warmth. I have kept these plants in a common hot-bed frame all winter. exposing them always to the open air in mild weather, but covering them in frost; and these plants have been stronger, and have flowered better than those which were placed in the green-house; so that I would recommend this method of treatment as the best, for the plants are apt to draw up weak in a green-house, and that prevents their flowering; nor are the plants near so handsome as those which are more exposed to the open air.

In the summer time they should be placed abroad in a sheltered situation with other hardy exotic plants, and in dry weather they will require to be often watered, for they are pretty thirsty plants, but in winter it should be sparingly given to them. As these plants are not of long duration, there should be young plants propagated to succeed them, for if they live four or five years, it is full long enough, because after that age they become unsightly.

XIMENIA.. Plum. Gen. Nov. 6. tab. 21.

The Characters are.

The flower has a small empalement of three leaves, which falls off; it has one bell-shaped petal, cut into three segments at the top, which turn backward; it has three short awl-shaped stamina, terminated by single summits, and a small oval germen, situated under the flower, supporting a very short style, crowned by a headed stigma. The germen afterward turns to an oval fleshy berry, including an oval nut with one cell, containing one seed of the same form.

The

The Species are,

1. Ximenia (*Americana*) foliis oblongis pedunculis multifloris. Lin. Sp. 1193. *Ximenia with oblong leaves and many flowers.*

2. Ximenia (*Agehalid*) foliis geminis lanceolatis. Lin. Sp. Plant. 1194. *Ximenia with twin leaves.*

The firft fort grows naturally in the iflands of the Weft-Indies; it rifes with a woody ftalk twenty feet high, fending out feveral branches, which are armed with thorns, garnifhed with fpear-fhaped leaves, ftanding round the branches without order. The flowers are produced at the end of the branches; they have one bell-fhaped petal, cut almoft to the bottom into three fegments, which are rolled backward, and are hairy; within they are of a yellow colour, and are fucceeded by oblong, oval, flefhy fruit, fhaped like a Plumb, including a hard nut of the fame form.

The fecond fort grows naturally in Egypt, where it becomes a tree of middling fize. The ftem is large and woody; the branches are flender and ftiff; they have a green bark, and are armed with ftrong fpines; the leaves come out by pairs; they are larger than thofe of the Box-tree, and end in points, but are of the like confiftence and colour. The flowers come out on the fide of the branches; they are fhaped like thofe of Hyacinth, but are fmall, and of a white colour; thefe are fucceeded by oblong black berries, including an oval nut, having one kernel or feed.

Both thefe forts are propagated by feeds, which muft be procured from the countries where they grow naturally; thefe fhould be fown in pots filled with light earth, and plunged into a hot-bed of tanners bark. If the feeds are frefh, the plants will appear in fix weeks or two months. When thefe are about three inches high, they muft be each carefully tranfplanted into a feparate fmall pot filled with light earth, and plunged into a good hot-bed of tanners bark, where they muft be fhaded from the fun till they have taken new root; then they muft be treated in the fame manner as other tender plants from the warm countries. During the firft fummer they may be kept in the tan-bed under frames, where they will thrive better than in the ftove; but in autumn, when the nights grow cool, they fhould be removed into the ftove, and plunged into the tan-bed; and in this they fhould always be kept, obferving to fhift them into larger pots when they require it; and in fummer, when the feafon is warm, they fhould have a large fhare of free air admitted to them. With this management the plants will thrive well, but they cannot be expected to flower very foon in this country.

XIPHION, or XIPHIUM. Tourn. Inft. R. H. 362. tab. 189. Bulbous Iris, or Flower-de-luce.

The Characters are,

The flowers have each a permanent fpatha or fheath; they have fix petals, the three outer broad, obtufe, and reflexed, and the inner erect, pointed, and joined to the other at their bafe; they have three awl-fhaped ftamina, which lie upon the reflexed petal, and are terminated by oblong depreffed fummits, and an oblong germen under the flower, fupporting a fhort ftyle, crowned by a tripartite ftigma. The germen afterward becomes an oblong angular capfule with three cells, filled with roundifh feeds.

The Species are,

1. Xiphium (*Perficum*) foliis carinatis caule longioribus. *Bulbous Iris, with keel-fhaped leaves which are longer than the ftalk; Perfian Iris.*

2. Xiphium (*Vulgare*) foliis fubulato-canaliculatis, caule brevioribus. *Bulbous Iris, with channelled awl-fhaped leaves which are fhorter than the ftalk; Bulbous Iris with a blue Violet flower.*

3. Xiphium (*Latifolium*) foliis fubulato-canaliculatis, floribus majoribus. *Bulbous Iris with channelled awl-fhaped leaves, and larger flowers; or broad-leaved Bulbous Iris.*

4. Xiphium (*Planifolium*) foliis planis caule longioribus. *Bulbous Iris, with plain leaves which are longer than the ftalk; or broad-leaved Bulbous Iris with a blue flower.*

The firft fort grows naturally in Perfia, but has been many years cultivated in the Englifh gardens for the beauty of its flowers; it has an oval bulbous root, from which come out five or fix pale green leaves, which are hollowed like the keel of a boat, ending in points. Between thefe the flower-ftalk arifes, which is feldom above three inches high, fupporting one or two flowers, which are included in fpathæ (or fheaths;) thefe have three erect petals called ftandards, which are of a pale fky-blue colour, and three reflexed petals called falls, which on their outfide are of the fame colour, but the lip has a yellow ftreak running through the middle, and on each fide are many dark fpots, with one large deep purple fpot at the bottom. Thefe flowers have a very fragrant fcent, and generally appear in February, which renders them more valuable.

The fecond fort grows naturally in the warm parts of Europe. There are feveral varieties of this fpecies; the moft common fort is blue, but there is one with a yellow, and another with a white flower; one with a blue flower, having white falls, another with yellow falls, one with a Violet-coloured flower, having blue falls, with fome others; but thefe are all fuppofed to be varieties, which have been produced by culture.

The root of this is bulbous; the leaves are hollow or channelled, ending in points, where their two fides meet; thefe are not fo long as the flower-ftalk, which rifes between them, and is embraced by the bafe of the leaves. This fupports two or three flowers, which are each inclofed in a feparate fheath at the top of the ftalk. The flowers are fhaped like thofe of the firft fort, but differ in their colour.

The third fort has much larger bulbous roots than either of the former. The leaves are fhaped like thofe of the fecond fort, but are much larger; the flower ftalk is near twice the height of the fecond fort, and the flowers are more than double their fize. This is by fome fuppofed to be only a variety of the fecond, but I think it a diftinct fpecies, for I have many years raifed a great number of the plants from feed, and have never found a fingle one degenerated to the fecond fort, and have raifed many of the fecond fort from feeds, without one inftance of a plant improving to the third.

There is a great variety of this fpecies; which differ in the colours of their flowers. Some are of a deep blue, others of a light or fky-blue, fome of a deep purple, and others with fine variegated flowers, which make a fine appearance during their continuance, which is not long, unlefs the feafon proves cold, or the flowers are fhaded from the fun.

The fourth fort grows naturally in Spain and Portugal. The root of this has a dark-coloured coat, but is white within, and of a fweet tafte. The leaves are eight or nine inches long, and more than an inch broad at their bafe; they are almoft plain, but toward their bafe are hollowed like the keel of a boat, and end in points, being of a pale green on their upper fide, and a little hoary on their under. The flowers ftand upon naked foot-ftalks, which arife from the root, and grow five or fix inches high, fuftaining two or three flowers at the top, which are each wrapped up in a feparate fheath; thefe are fhaped like thofe of the other forts, and have a very agreeable odour.

There are four or five varieties of this fpecies, which differ in the colour of their flowers; but the moft common colour is blue.

They are all propagated by offsets from their roots, but to obtain new varieties, they muft be propagated by feeds in the following manner:

Having

Having procured a parcel of feeds from good flowers, you fhould provide fome flat pans or boxes, which muft have holes in their bottoms to let the moifture pafs off; thefe fhould have pieces of tile or oyfter-fhells laid over each hole to prevent their being ftopped; then they muft be filled with frefh, light, fandy earth, and the beginning of September the feeds fhould be fown thereon pretty thick, obferving to fcatter them as equally as poffible; then cover them over about half an inch thick with the fame light frefh earth, and place the boxes or pans where they may have the morning fun till eleven o'clock; and if the feafon fhould prove very dry, they muft be now and then refrefhed with water.

In this fituation they may remain until the middle of October, when they fhould be removed into a warmer fituation, where they may have the full fun moft part of the day, and fcreened from fevere frofts; in which place they muft abide all the winter, obferving to keep them clear from weeds and Mofs, which, at this feafon, are very apt to fpread over the furface of the earth, in tubs, pans, or pots, when they are expofed to the open air.

In the fpring the plants will appear above ground, when, if the feafon is dry, they muft be now and then refrefhed with water, and conftantly kept clear from weeds; and as the feafon advances, and the weather becomes warm, they fhould be again removed into their former fhady fituation, where they may enjoy the morning fun only. When the leaves begin to decay (which will be in June,) they muft be cleared from weeds and dead leaves, and fome frefh earth fifted over them about half an inch thick, ftill fuffering them to abide in the fame fituation all the fummer feafon; during which time they will require no farther care but to keep them clear from weeds, until the beginning of October, when they muft be again removed into a warm fituation, and the furface of the earth lightly taken off, and fome frefh earth fifted over them.

In this place they muft remain all the winter, as before; and in the fpring they muft be treated as was directed for the former years.

When the leaves are decayed, the bulbs fhould be carefully taken up (which may be beft done by fifting the earth through a fine fieve,) and a bed or two of good light frefh earth fhould be prepared, into which the bulbs muft be planted, at about three inches afunder each way, and three inches deep. Thefe beds muft be conftantly kept clean from weeds and Mofs; and if the winter fhould prove fevere, the bed fhould be covered with rotten tanners bark, or Peafe-haulm, to keep out the froft; and in the fpring, juft before the plants come up, the furface of the beds fhould be ftirred, and fome frefh earth fifted over them about half an inch thick, which will greatly ftrengthen the roots.

During the fpring and fummer, they muft be conftantly weeded; and at Michaelmas the earth fhould be again ftirred, and fome frefh fifted over the beds again, as before, obferving in winter and fpring ftill to keep the beds clean, which is the whole management they will require, and in June following the greateft part of the roots will flower; at which time you fhould carefully look over them, and put down a ftick by all thofe whofe flowers are beautiful, to mark them; and as foon as their leaves are decayed, thefe roots may be taken up to plant in the flower-garden amongft other choice forts.

But the nurfery-beds fhould ftill remain, obferving to keep them clear from weeds, as alfo to fift frefh earth over them, as was before directed; and the following feafon the remaining part of the roots, which did not flower the foregoing feafon, will now fhew their bloffoms; fo that you may know which of them are worth preferving in the flower-garden, which fhould now be marked; and when their leaves are decayed, they muft be taken up and planted with the other fine forts, in an eaft border of light frefh earth; but the ordinary forts may be intermixed with other bulbous-rooted flowers in the larger borders of the pleafure-garden, where, during their continuance in flower, they will afford an agreeable variety.

But after thefe choice flowers are obtained from feeds, they may be increafed by offsets, as other bulbous flowers are. The offsets fhould be planted in a feparate border from the blowing roots, for one year, until they have ftrength enough to produce flowers, when they may be placed in the flower-garden with the old roots.

Thefe bulbs need not be taken up oftener than every other year, which fhould always be done foon after their leaves decay, otherwife they will fend forth frefh fibres, when it will be too late to remove them; nor fhould they be kept long out of the ground, two months is full enough; for when they are kept longer, their bulbs are fubject to fhrink, which caufes their flowers to be weak the following year.

The Perfian Iris is greatly efteemed for the beauty and extreme fweetnefs of its flowers, as alfo for its early appearance in the fpring, it generally being in perfection in February, or the beginning of March, according to the forwardnefs of the feafon, at which time there are few other plants in beauty.

This may be propagated by feeds in the fame manner as the other forts; but the boxes in which they are fown, fhould be put under a garden-frame in winter, to fhelter them from hard froft, becaufe, while the plants are young, they are fomewhat tender. From the feeds of this kind I could never obtain any varieties, their flowers being always the fame.

Thefe plants are alfo propagated by offsets, in the fame manner as the other forts, but their roots fhould not be tranfplanted oftener than every third year; nor fhould they be ever kept out of the ground long, becaufe their roots will fhrink and entirely decay when they are long above ground, fo as not to be recovered again. This fort was formerly more common in the gardens near London than at prefent, which, I fuppofe, has been occafioned by the keeping the roots above ground too long, which deftroyed them.

XYLON. See BOMBAX.

Y.

YEW-TREE. See TAXUS.

YUCCA. Dillen. Gen. Nov. 5. The Indian Yucca, or Adam's Needle.

The CHARACTERS are,

The flower has no empalement; it is bell shaped, composed of six large petals, whose tails are joined, and six short reflexed stamina, terminated by small summits, and an oblong three-cornered germen which is longer than the stamina, having no style, crowned by an obtuse stigma with three furrows. The germen afterward turns to an oblong three-cornered capsule, divided into three cells, filled with compressed seeds, lying over each other in a double arrangement.

The SPECIES are,

1. YUCCA (*Gloriosa*) foliis integerrimis. Vir. Cliff. 29. *Yucca with entire leaves; commonly called Adam's Needle.*

2. YUCCA (*Aloifolia*) foliis crenulatis strictis. Lin. Sp. Pl. 319. *Yucca with narrow leaves, which are slightly crenated.*

3. YUCCA (*Draconis*) foliis crenatis nutantibus. Lin. Sp. Plant. 319. *Yucca with nodding crenated leaves.*

4. YUCCA (*Filamentosa*) foliis serrato-filamentosis. Lin. Sp. Plant. 319. *Yucca with sawed thready leaves.*

The first of these plants is a native of Virginia and other parts of North America.

This sort seldom rises with a stem above two feet high, which is garnished with leaves almost to the ground. The leaves are broad, stiff, and have the appearance of those of the Aloe, but are narrower, of a dark green colour, ending in a sharp black spine. The plants frequently produce spikes of flowers, which rise from the center of the leaves. The stalks grow three feet high, branch out to a considerable distance, but the flowers are placed very sparsedly on the branches, which renders it less beautiful than the flowers of the other kinds; they are white within, but each petal is marked with a purple stripe on the outside, bell-shaped, and hang downward.

The second sort rises with a thick, tough, fleshy stalk to the height of ten or twelve feet, having a head or tuft of leaves at the top; these are narrower and stiffer than those of the former sort, and are of a lighter green colour; their edges are slightly sawed, and their points end with sharp thorns. The flower-stalk rises in the center of the leaves, and is from two to three feet long, branching into a pyramidal form. The flowers grow close on the branches, and form a regular spike; they are of a bright purple colour on the outside, and white within, making a fine appearance. The plants of this do not flower so often as the other sort; and when they flower, the head decays, but one or two young heads come out from the side of the stalk, below the old one.

The third sort grows naturally in South Carolina. The stalks of this sort rise about three or four feet high; the leaves are narrow, of a dark green colour, and hang downward; they are sawed on their edges, and end in acute spines. I never saw the flowers of this sort, but have been informed they are white.

The threaded sort is not so common as the others in the English gardens; but as it is a native of Virginia, it might easily be procured in plenty from thence. The stalk and leaves are like those of the first sort, but the leaves are obtuse, and have no spine at their ends. From the side of the leaves come out long threads, which curl and hang down.

All these plants are either propagated by seed, when obtained from abroad, or else from offsets or heads taken from the old plants, after the manner of Aloes in colder countries.

When they are raised from seeds, they should be sown in pots, and plunged into a moderate hot-bed, where the plants will come up in five or six weeks after: and when they are two or three inches high, they should be transplanted each into a separate small pot, and plunged into the hot-bed, where the plants should have air and water in proportion to the warmth of the season, and the bed wherein they are placed.

In June they should be inured, by degrees, to bear the open air; into which they must be removed, to harden them before winter, placing them in a well-sheltered situation, where they may remain until the beginning of October, when they must be removed into the green-house, where they may be arranged amongst the hardier sorts of Aloes, and should be treated in the same manner as hath been already directed for them; to which the reader is desired to turn for further instructions.

When these plants have acquired strength, those of the common sort, and also the threaded, may be afterwards turned out into a warm border, where they will endure the cold of our ordinary winters very well; but the other sorts must be kept in pots, that they may be sheltered in winter; and if they are treated in the same way as the large American Aloe, they will thrive very well.

The offsets taken from the old plants should be laid in a dry place, for a week or ten days before they are planted, that their wounds may heal, otherwise they will be subject to rot with moisture.

As the second and third sorts do not put out offsets so plentifully as the first, in order to propagate them, therefore the heads of the plants may be cut off in June; and after the wounded part is dry, the heads may be planted, which will soon take root, provided the pots are plunged into a moderate hot-bed; and this cutting off the heads will occasion the stems to put out suckers, which they seldom do without, until they flower; so that by this method the plants may be obtained in plenty.

Z.

ZANTHOXYLUM. See XANTHOXYLUM.

ZEA. Lin. Gen. Plant. 926. Turkey Corn.

The CHARACTERS are,

It has male and female flowers situated at remote distances on the same plant. The male flowers are disposed in a loose spike, having oval, oblong, chaffy empalements, opening with two valves, each inclosing two flowers; these have two short compressed nectariums, and three hair-like stamina, terminated by quadrangular summits, which open in four cells at the top. The female flowers, which are situated below the male, are disposed in a thick spike, inclosed with leaves; these have thick chaffy empalements with two valves. The flowers are composed of two short membranaceous broad valves, which are permanent, and a small germen with a slender style, crowned by a single stigma, which is hairy toward the point. The germen afterward turns to a roundish compressed seed, angular at the base, and half inclosed in its proper receptacle.

The SPECIES are,

1. ZEA (*Americana*) caule altissimâ, foliis latioribus pendulis, spicâ longissimâ. *Indian Corn with the tallest stalk, broader hanging leaves, and the longest spike.*

2. ZEA (*Alba*) caule graciliore, foliis carinatis pendulis, spicâ longâ gracili. *Indian Corn with slender stalks, keel-shaped hanging leaves, and a long slender spike.*

3. ZEA (*Vulgare*) caule humiliori, foliis carinatis pendulis spicâ breviore. *Indian Corn with a lower stalk, hanging keel-shaped leaves, and a shorter spike.*

These three species have been generally supposed but one, and only accidental variations; but from long experience I can affirm they are different, and do not alter by culture.

The first sort grows naturally in the islands of the West-Indies; this hath a very large strong stalk, which rises to the height of ten or twelve feet. The leaves are long, broad, and hang downward; they have a broad white midrib. The male flowers come out in branching spikes at the upper part of the stalks; these are eight or ten inches long. The female flowers come out from the bottom of the leaves on the side of the stalk; they are disposed in a close, long, thick spike, and are covered closely with thin leaves; out of the end of the covers hang a small long bunch of filaments or threads, which are supposed to receive and convey the farina of the male flowers to the germen of the female. When the seeds of this sort are ripe, the spikes or ears are nine or ten inches long, and sometimes a foot; but these rarely ripen in England.

I have not seen any variety of colours in this species, though it is very probable there are the same varieties in the colour of the grain, as in the other species; but as this is less common in Europe, we are not so well acquainted with it.

The second sort is cultivated in Italy, Spain, and Portugal. The stalks of this sort are slenderer than those of the former, and seldom rise more than six or seven feet high. The leaves are narrower than those of the first sort, and are hollowed like the keel of a boat, and their tops hang downwards. The spikes of male flowers are shorter than those of the first, and the ears or spikes of grain are slenderer, and not more than six or seven inches long. The grains of this sort do not come to maturity in England, unless the season proves very warm, and the grains are planted early and in a warm soil and situation.

The third sort is cultivated in the northern parts of America, and also in Germany. The stalks of this are slender, and seldom rise more than four feet high. The leaves are shorter and narrower than those of the two former; they are hollowed like the keel of a boat, and their tops hang down. The spikes of male flowers are short, and the ears or spikes of grain are seldom more than four or five inches long. This sort ripens its grain perfectly well in England, in as little time as Barley, so may be cultivated here to advantage.

There are several varieties of the two last species, which differ in the colour of their grain. The most common colour is that of a yellowish white; but there are some with deep yellow, others with purple, and some with blue grains; and when the different colours are planted near each other, the farina will mix, and the ears will have grains of several colours intermixed; but when the grains of the different varieties are planted at a proper distance from each other, the produce will be the same with the grains which were sown.

These plants are seldom cultivated in England for use, but in Italy and Germany it is the food of the poor inhabitants; as it is also in many parts of North America, where it is treated in the following manner:

They first dig the ground well in the spring, and after having made it level, they draw a line across the whole piece intended to be planted; then they raise little hills at about three or four feet distance, into each of which they put two or three good seeds, covering them about an inch thick with earth; then they move the line four feet farther, continuing to do the same through the whole spot of ground, so that the rows may be four feet asunder, and the hills three or four feet distance. Six quarts of this seed is generally allowed to an acre of land, which, if the soil be good, will commonly produce fifty bushels of Corn.

In the planting of this Corn, where they observe to plant the grain of any one colour in a field by itself, and no other coloured grain stand near it, it will produce all of the same colour again, as hath been affirmed by many curious persons who have tried the experiment; but, if the rows are alternately planted with the grain of different colours, they will interchange, and produce a mixture of all the sorts in the same row, and frequently on one and the same spike; and some do affirm they will mix with each other, at the distance of four or five rods, provided there is no tall fence or building between to intercept them.

There is nothing more observed in the culture of this grain, but only to keep it clear from weeds, by frequent hoeing of the ground; and when the stems are advanced, to draw the earth up in a hill about each plant, which, if done, will greatly strengthen them, and preserve the ground about their roots moist for a considerable time.

When

When the Corn is ripe, they cut off the ftalks clofe to the ground, and after having gathered off the fpikes of grain, they fpread the ftalks in the fun, to harden and dry, which they afterward ufe in the fame manner as Reeds in England for making fences, covering fheds, &c. for which purpofe they are very ufeful to the inhabitants of warm countries; and when there is a fcarcity of forage, they feed their cattle with them green, as faft as the Corn is gathered off.

The Corn is ground to flour, and the pooreft fort of people in America, and alfo in Italy and Germany, make their bread of this flour; and in many of the warmer countries, the inhabitants roaft the whole fpikes, and drefs them many different ways, making feveral difhes of it; but this grain feldom agrees with thofe who have not been accuftomed to eat it; however, in times of fcarcity of other grain, this would be a better fubftitute for the poor than Bean-flour, or other forts, which have been ufed in England, and at all times will be found a hearty food for cattle, hogs, and poultry; fo that in light fandy lands, where Beans and Peas fucceed not well, this grain may be cultivated to anfwer both purpofes to advantage.

If this grain is cultivated by the horfe-hoeing hufbandry, it may be done at lefs expence than in any other method; for this is one of the plants which is more particularly adapted to this hufbandry.

The time for fowing this Corn, is about the fame as for Barley; in light warm land it may be fown the latter end of March or the beginning of April, but in cold ground the middle or end of April will be early enough, for the grain is fubject to rot in cold land, efpecially if the feafon proves wet. When the large forts are planted in a garden for curiofity, their feeds fhould be fown upon a moderate hot-bed the beginning of March; and when the plants are fit to remove, they fhould be tranfplanted on another moderate hot-bed to bring them forward; but they muft not be kept too clofely covered, for that will draw them up weak; therefore, when the weather is mild, they fhould be inured to bear the open air; and the middle of May, they fhould be taken up with balls of earth to their roots, and tranfplanted into a warm border at three or four feet diftance, carefully watering them if the weather proves dry, until they have taken new root; after which they will require no other care but to keep them clean from weeds. If the feafon fhould prove warm, thefe plants will ripen the Corn late in autumn.

ZINNIA. Lin. Gen. Plant. 1161.

The CHARACTERS are,

The common empalement of the flower is imbricated with fcales, and permanent, of a cylindrical oval form. The flower is compofed of feveral funnel-fhaped hermaphrodite florets, having each five fhort ftamina, whofe fummits coalefce; the ray is compofed of feveral large, fpreading, female, half florets, which are permanent. The hermaphrodite florets have an oblong bearded germen with a flender ftyle, crowned by two fhort reflexed ftigmas; thefe florets are fucceeded by oblong fingle feeds with two horns, which ripen in the empalement.

The SPECIES are,

1. ZINNIA (*Pauciflora*) floribus feffilibus verbefina foliis, oppofitis ovatis acutis integerrimis fcabris. Act. Petrop. 1763. *Zinnia with flowers growing without foot-ftalks, and oval entire leaves.*

2. ZINNIA (*Multiflora*) floribus pedunculatis. Jacq. Obfer. tab. 40. *Zinnia with flowers having foot-ftalks.*

The firft fort grows naturally in Peru; it is an annual plant, which rifes from two to four feet high; the lower part of the ftalk becomes ligneous in autumn. The branches come out oppofite, and are garnifhed with oblong leaves, which vary both in form and fize: fome are broad at their bafe, and terminate in acute points; others are of an oblong oval; they are fmooth without foot-ftalks, and placed oppofite. The flowers are produced fingly upon pretty long foot-ftalks, moft of them terminating the branches; but fometimes they rife from the divifion of the branches. The empalement of the flower is of a cylindrical form, compofed of fcales, lying over each other imbricatim, clofely inclofing feveral hermaphrodite florets, which form the difk, furrounded by half florets or rays, which are large, fpread open, and of unequal number: in fome flowers there are but five, in others ten or more, of a yellow colour, but afterward change to a brown, remaining till the feeds are ripe.

The fecond fort grows naturally in Louifiana; this is alfo an annual plant. The ftalk is more erect than that of the former, as are alfo the branches; they are covered with foft hairs, and are channelled; the leaves are placed oppofite; they are oval, fpear-fhaped, having three longitudinal veins, and are hairy. The flowers terminate the branches; the ray or border, which is compofed of female half florets, fpread open, and are of a deep gold colour on the upper fide when firft open, but afterward change to a dark copper, but on the back fide of a pale ftraw colour. The florets which compofe the difk are tubulous, and have five ftamina ftretched out beyond the corolla, whofe fummits, which are yellow, are connected; the half florets which compofe the border or ray are permanent, remaining till the feeds are ripe, which are of the fame ftructure with thofe of the former fort.

Both thefe plants are propagated by feeds, which muft be fown upon a moderate hot-bed in March. When the plants come up, they muft have air admitted to them by tilting of the lights of the bed every day when the weather is not too cold, otherwife the plants will draw up weak: when the plants are about an inch high, they fhould be planted on another hot-bed to bring them forward; but they fhould not be treared too tenderly, for they are very fubject to grow too luxuriant in branches; and the firft fort will not produce many flowers, if the plants are not ftinted in their growth while young, which may be effected by planting them in fmall pots to confine their roots, otherwife the feeds will not ripen in England.

The fecond fort is much more prolific of flowers than the firft, fo may be treated with lefs care: thefe, when they have been brought forward on the hot-bed, may be inured gradually to bear the open air in May, and in June fhould be tranfplanted into the borders of the flower-garden, where they will continue flowering all the fummer, and will perfect their feeds very well; but the plants of the firft fort fhould be placed either in the ftove or a glafs-cafe to obtain good feeds.

ZIZIPHORA. Lin. Gen. Plant. 33. Field Bafil.

The CHARACTERS are,

The flower hath a long, rough, cylindrical empalement, which is flightly cut into five parts at the brim. The flower is of the labiated kind, having a long cylindrical tube. The upper lip is oval, reflexed, and entire; the under lip (or beard) is divided into three equal fegments; it has two fpreading ftamina, terminated by oblong fummits, and a quadrifid germen, fupporting a briftly ftyle, crowned by a fharp-pointed inflexed ftigma. The germen afterward turns to four oblong feeds, which ripen in the empalement.

The SPECIES are,

1. ZIZIPHORA (*Capitata*) capitulis terminalibus, foliis ovatis. Lin. Sp. Plant. 21. *Field Bafil with heads terminating the ftalks, and oval leaves.*

2. ZIZIPHORA (*Tenuior*) floribus lateralibus, foliis lanceolatis. Lin. Sp. Plant. 21. *Field Bafil with flowers growing on the fides of the ftalk, and fpear-fhaped leaves.*

3. Ziziphora (*Hispanica*) foliis lanceolatis, floribus terminalibus. Hort. Cliff. 305. *Field Basil with spear-shaped leaves, and flowers terminating the stalks.*

The first sort grows naturally in Virginia; this is an annual plant, which has a four-cornered stalk, sending out side branches, which stand opposite; these are terminated by a cluster of small flowers, surrounded by oval leaves, ending in acute points. The flowers have a slender cylindrical empalement, out of which they just peep; they are purple, of the lip kind, and have but two stamina.

The second sort grows naturally in the Levant; this sends up many slender ligneous branches, which rise near a foot high, garnished with spear-shaped leaves, about the size of those of Summer Savory, and have a scent like them. The flowers are produced in whorls round the stalks, which are like those of the former sort.

The third sort grows naturally on the Alps and Apennine mountains. The stalks of this rise about six inches high, garnished with small spear-shaped leaves placed opposite. The flowers are produced in a cluster at the top of the stalks, which are of the same shape and colour as those of the first sort, and are surrounded with spear-shaped leaves.

These plants are all of them annual, so are propagated only by seeds.

The seeds may be sown in a border of light earth, either in spring or autumn. Those plants which come up in autumn, will abide through the winter, and will grow much larger than those which come up in the spring, though neither of them rise very high. The seeds should be sown where the plants are to remain, for they do not thrive well when they are transplanted, unless the earth remains to their roots. These have a pretty strong aromatic scent, somewhat resembling Summer Savory; but as they are plants of little beauty, they are seldom cultivated but in botanic gardens for sake of variety.

The seeds of those plants which come up in autumn, will be ripe in July or August; but those of the spring plants will not ripen till the latter end of August, or the beginning of September, when, if the seeds are permitted to scatter, the plants will come up, and require no farther care but to clear them from weeds, and thin them where they are too close.

ZIZIPHUS. Tourn. Inst. R. H. 627. tab. 433. Rhamnus. Lin. Gen. Plant. 235. The Jujube.

The Characters are,

The flower has no empalement; it has one funnel-shaped petal, which spreads open at the top, and is cut into four or five segments; it has five awl-shaped stamina, whose base are inserted to the petal, and are terminated by small summits, and an oval germen, supporting two slender styles, crowned by obtuse stigmas. The germen afterward becomes an oblong oval berry, inclosing a single nut of the same form, which has two cells, each containing an oblong seed.

The Species are,

1. Ziziphus (*Jujuba*) aculeis geminatis rectis, foliis oblongo-ovatis serratis. *Jujube with strait thorns growing by pairs, and oblong, oval, sawed leaves; or the common Jujube.*

2. Ziziphus (*Sylvestris*) aculeis geminatis, altero recurvo, foliis ovatis nervosis. *Jujube with twin spines, one of which is recurved, and oval veined leaves; or the wild Jujube.*

3. Ziziphus (*Oenoplia*) aculeis solitariis recurvis, pedunculis, aggregatis, foliis cordato-rotundis nervosis, subtus tomentosis. *Jujube with single recurved spines, foot-stalks in clusters, and round, heart-shaped, veined leaves, which are downy on their under side.*

4. Ziziphus (*Africana*) aculeis geminatis rectis, foliis ovatis nervosis. *Jujube with double strait thorns, and oval veined leaves.*

The first sort grows naturally in the warm parts of Europe; it has a woody stalk, which divides into many crooked irregular branches, which are armed with strong strait thorns set by pairs at each joint. The leaves are two inches long and one broad, slightly sawed on their edges, and stand upon short foot-stalks. The flowers are produced on the side of the branches, two or three arising from the same place, which sit close; they are small, and of a yellow colour; these are succeeded by an oval fruit, about the size of a middling Plumb, of a sweetish taste, and are clammy, including a hard oblong stone, pointed at both ends.

The fruit of this tree was formerly used in medicine; it is reckoned pectoral, and good for coughs, pleurisies, and hot sharp humours, but is now seldom to be found in the shops. In Italy and Spain, this fruit is served up at the table in deserts during the winter season as a dry sweetmeat.

The second sort grows naturally about Tunis in Africa, this has slender woody stalks, which send out many weak branches, covered with a grayish bark, and armed with spines, which come out by pairs at each joint, one of which is longer than the other, and is strait; the other is short and recurved. The leaves are small, oval, and veined; they are half an inch long, and as much in breadth, sitting close to the branches. The flowers of this sort I have not seen, so can give no farther description of this plant.

The third sort grows naturally in India; this rises with shrubby stalks ten or twelve feet high, sending out many slender branches, which have a yellowish bark, and are armed with single recurved thorns at each joint. The leaves are round, heart-shaped, about two inches long, and as much in breadth, and are indented at the foot-stalk; they have three longitudinal veins, and are covered with a yellowish down on their under side. The flowers come out in clusters from the wings of the branches; they are small, and of a yellowish colour; these are succeeded by oval fruit about the size of small Olives, inclosing a stone of the same shape.

The fourth sort grows naturally in Syria, from whence I have received the seeds; this sends up several shrubby stalks from the root, which divide into slender branches; these are armed with strait spines, which are set by pairs at each joint. The leaves are small, oval, and veined, and are placed alternate, standing upon very short foot-stalks. The flowers are small, of a yellow colour, arising at the wings of the branches. The fruit is round, and about the size of Sloes.

These plants are preserved in the gardens of some curious persons only for the sake of variety, for they do not produce fruit in England. The first and fourth sorts, which are the most hardy, will scarce live through the winters in England, even when they are planted against south walls; in which situation I have kept the plants two or three years, when the winters have proved mild, but they were afterwards killed by a sharp frost. They may be propagated by putting their stones into pots of fresh light earth, soon after their fruits are ripe; and in winter they should be placed under a common hot-bed frame, where they may be sheltered from severe frost. In the spring these pots should be plunged into a moderate hot-bed, which will greatly forward the growth of the seeds; and when the plants are come up, they should be inured to the open air by degrees, into which they must be removed in June, placing them near the shelter of a hedge; and in very dry weather they must be frequently refreshed with water.

In this situation they may remain till the beginning of October, when they must be removed either into the greenhouse, or placed under a hot bed frame, where they may be defended from frost, but should have as much free air as possible in mild weather.

During

During the winter feafon they fhould be now and then refrefhed with water; but after their leaves are fallen (as they always fhed them in winter,) they muft not be over-watered, which would rot the tender fibres of their roots, and caufe the plants to decay.

In March, juft before the plants begin to fhoot, they fhould be tranfplanted each into a feparate fmall pot, filled with light frefh earth; and if they are plunged into a moderate hot-bed, it will greatly promote their taking root; but in May they muft be inured to the open air by degrees, into which they fhould be foon after removed.

Thus thefe plants fhould be managed while young, at which time they are tender; but when they are three or four years old, fome of them may be planted in the full ground, againft a warm wall or pale, where, if they have a dry foil, they will endure the cold of our ordinary winters pretty well, but in hard frofts they will require to be fheltered, fo it will be proper to keep a plant or two in pos, which may be houfed in winter.

Thefe plants may be alfo propagated by fuckers, which the old ones many times fend forth from their roots; but thefe are feldom fo well rooted as thofe produced from feeds, nor do they make fo good plants, for which reafon they are but rarely propagated that way.

The fecond fort is not fo hardy as the firft, fo thefe plants muft be kept in pots, and in the winter placed into the green-houfe, and treated in the fame way as other hardy exotic plants, being careful not to over-water them at that feafon, but efpecially when they have fhed their leaves.

This fort is propagated by feeds, which muft be procured from the country where it naturally grows; thefe fhould be fown in pots filled with light earth, and plunged into a hot-bed of tanners bark, which will bring up the plants in about fix weeks, if the feeds are good. When the plants begin to advance in height, they fhould be gradually hardened, and in June they may be placed in the open air in a fheltered fituation; but in autumn they muft be removed into fhelter, where they muft remain all the winter; and in the fpring, before the plants begin to pufh out their leaves, they fhould be carefully tranfplanted each into a feparate fmall pot, and plunged into a gentle hot-bed to forward their putting out new roots. In fummer they muft be expofed abroad, but in winter they muft be houfed.

The third and fourth forts are tenderer than the former, fo will not thrive in this country unlefs the plants are kept in a warm ftove. Thefe are propagated in the fame way as the former, but the plants muft be more tenderly treated, for they fhould not be wholly expofed abroad at any time of the year; in fummer they muft have a large fhare of air in warm weather, and in winter they muft be kept in a warm ftove.

ZYGOPHYLLUM. Lin. Gen. Plant. 474. Fabago. Tourn. Inft. R. H. 258. tab. 135. Bean Caper.

The Characters are,

The empalement of the flower is compofed of five oval obtufe leaves. The flower has five obtufe petals, which are longer than the empalement, and are indented at their points; it has a clofed nectarium, which includes the germen, compofed of feveral fcales or little leaves, to which the bafes of the ftamina are faftened; it has ten awl-fhaped ftamina, terminated by oblong fummits, and an oblong germen, fupporting an awl-fhaped ftyle, crowned by a fingle ftigma. The germen afterward becomes an oval five-cornered capfule with five cells, containing feveral roundifh feeds.

The Species are,

1. Zygophyllum (*Fabago*) capfulis prifmatico-pentandris. Hort. Upfal. 103. *Bean Caper with a prifmatical capfule and five ftamina.*

2. Zygophyllum (*Seffilifolium*) capfulis globofo-depreffis. Lin. Sp. Pl. 385. *Bean Caper with globular depreffed capfules.*

3. Zygophyllum (*Morgfana*) capfularum angulis compreffo-membranaceis. Lin. Sp. Plant. 385. *Bean Caper with compreffed membranaceous angles to the capfules.*

4. Zygophyllum (*Fulvum*) capfulis ovatis acutis. Lin. Sp. Pl. 386. *Bean Caper with oval acute pointed capfules.*

The firft fort grows naturally in Syria; this has been long an inhabitant of fome curious gardens in England. The root is thick, flefhy, and ftrikes deep into the ground, and will grow as thick as a man's arm when old. The ftalks decay every autumn to the root, from which fpring new fhoots every year, in number proportional to the fize of the root; they rife three or four feet high, fending out a few fide branches, which are fmooth, green, and jointed; they are garnifhed with fmooth flefhy leaves like thofe of Purflane, two ftanding together upon the fame foot-ftalk, which is an inch long; they are of a bluifh green colour. The flowers are produced from the wings of the ftalk, two or three arifing at the fame joint, upon fhort foot-ftalks; they are compofed of five roundifh concave petals, of a reddifh colour on their outfide, and ten ftamina, which are twice the length of the petals. The flowers are fucceeded by oblong prifmatical capfules with five fides, which have cells filled with roundifh feeds. This fort flowers in June and July, and the feeds ripen in autumn.

The fecond fort grows naturally at the Cape of Good Hope. This rifes with a thick woody ftalk three or four feet high, fending out many branches, which are garnifhed with fucculent leaves, placed by fours, fitting clofe to them. From the wings of the ftalks the flowers are produced upon pretty long flender foot-ftalks; they are compofed of five fulphur-coloured petals, which have a brown fpot on each of their tails; thefe are fucceeded by roundifh depreffed fruit, having five cells, each containing two roundifh feeds. This plant continues flowering all fummer and autumn, and the feeds ripen in winter.

The third fort grows naturally at the Cape of Good Hope; this has a fhrubby ftalk, which divides into many irregular jointed branches, which rife four or five feet high, and are garnifhed with thick fucculent leaves, which are larger and more obtufe than thofe of the fecond fort; they are placed by fours at each joint, two on each fide the ftalk oppofite. The flowers come out from the wings of the ftalk upon flender foot-ftalks; thefe have but four petals, which are broader than thofe of the fecond fort, but of the fame colour, each having a brown fpot at their tails. The fruit has four broad membranaceous wings to it, refembling the fails of a mill. This plant flowers moft part of fummer, but the fruit feldom ripens well in England.

The fourth fort is a native of the Cape of Good Hope. The ftalks of this branch out greatly from the bottom; they are fhrubby, jointed, and irregular. The leaves are of the confiftence of thofe of Purflane; they are narrow at their tails, but oval toward their points, and are placed by fours at each joint like the former. The flowers come out from the wings of the ftalk upon flender foot-ftalks; they are of a pale yellow colour, each petal having a pretty large red fpot at their tails. The fruit is oval, about three quarters of an inch long, having five deep furrows, and is divided into five cells, which are filled with roundifh feeds. This plant flowers great part of the year, and the fruit ripens in autumn and winter.

The firft fort is propagated only by feeds, which ripen very well in England in warm feafons; thefe may be either fown upon a moderate hot-bed in the fpring, or on a warm border of light ground; thofe which are fown upon the hot-bed will come up in three weeks or a month; and about a month after, the plants will be fit to remove, when they ſhould be each planted in a ſeparate ſmall pot, filled with freſh light earth, and plunged into a gentle hot-bed to promote their taking root, and ſhaded from the ſun in the day time; afterward they muſt be gradually hardened to bear the open air, to which they ſhould be expoſed all the ſummer; but in autumn, when their ſtalks begin to decay, they ſhould be placed in a hot-bed frame to ſhelter them from the froſt in winter, for while they are young they are a little tender. The ſpring following they may be turned out of the pots, and planted in a ſouth-border cloſe to the wall, in a dry rubbiſhy ſoil, where they will endure the cold without covering. There is a plant of this kind in the Chelſea garden, which is near fifty years old, and has reſiſted the ſevereſt cold without any covering, and produces great plenty of flowers and fruit annually.

Thoſe plants which come up in the full ground, will require no other care but to keep them clean from weeds, and thin them where they come up too cloſe, giving them room to grow the firſt year; and when their ſtalks decay in autumn, the ſurface of the ground ſhould be covered with tan to prevent the froſt from penetrating to the roots; or in froſty weather, they may be covered with ſtraw or Peaſe-haulm, which will anſwer the ſame purpoſe; and in the ſpring the roots ſhould be carefully taken up, planting them cloſe to a warm wall, as was before directed.

The other three ſorts are too tender to live through the winter in the open air in this country, ſo they muſt be kept in pots, and houſed in autumn. Theſe plants may be propagated either by ſeeds or cuttings.

The ſecond and fourth ſorts ripen their ſeeds pretty well in England, ſo theſe may be propagated by ſowing them on a moderate hot-bed in the ſpring; and when the plants are about an inch high, they ſhould be each tranſplanted into a ſmall pot filled with light earth, and plunged into a moderate hot-bed, ſhading them from the ſun till they have taken new root; then as the ſeaſon advances, they ſhould be gradually hardened to bear the open air, into which they ſhould be removed the latter end of May, placing them in a warm ſheltered ſituation, where they may remain till autumn, when they ſhould be placed in an airy dry glaſs-caſe, where they will ſucceed better than in the green-houſe; for they require a large ſhare of air in mild weather, otherwiſe their ſhoots are apt to be weak and tender, ſo are often injured by damp air in winter; but they do not require any artificial heat. If they are ſcreened from the froſt, and have plenty of air, they will thrive very well.

The third ſort ſeldom produces good ſeeds in England, ſo is propagated by cuttings, and the two others are generally increaſed in the gardens the ſame way, that method being very expeditious, though the ſeedling plants grow ſtronger, and riſe to a greater height. Theſe cuttings may be planted in a bed of light earth during any of the ſummer months; if theſe are covered cloſe down with bell or hand glaſſes, and ſhaded from the ſun, they will put out roots in five or ſix weeks, and then they may be taken up carefully and potted, placing them in the ſhade till they have taken new root; after which they may be removed to a warm ſheltered ſituation, and treated in the ſame way as thoſe plants raiſed from ſeeds.

ZYLOSTEUM. See LONICERA.

INDEX LATINUS.

Consolida

Mala

N.

O.

P.

Vinca

THE

ENGLISH NAMES

OF THE

PLANTS mentioned in this WORK,

Referring to their LATIN NAMES.

Beech-

I N D E X

C.

Cockſcomb,

D.

E.

F.

Horfe

I N D E X.

N.

O.

P.

T.

INDEX.

THE END.